Moments of Inertia of Common Geometric Shapes

Rectangle $\bar{I}_{x'} = \frac{1}{12}bh^3$ $\bar{I}_{y'} = \frac{1}{12}b^3h$ $I_x = \frac{1}{3}bh^3$ $I_y = \frac{1}{3}b^3h$ $J_C = \frac{1}{12}bh(b^2 + h^2)$	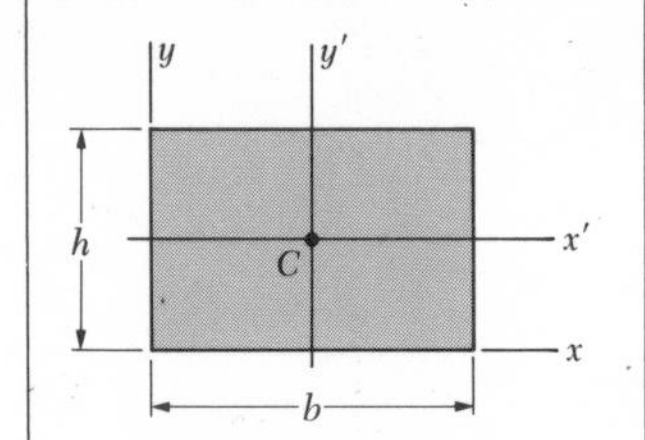
Triangle $\bar{I}_{x'} = \frac{1}{36}bh^3$ $I_x = \frac{1}{12}bh^3$	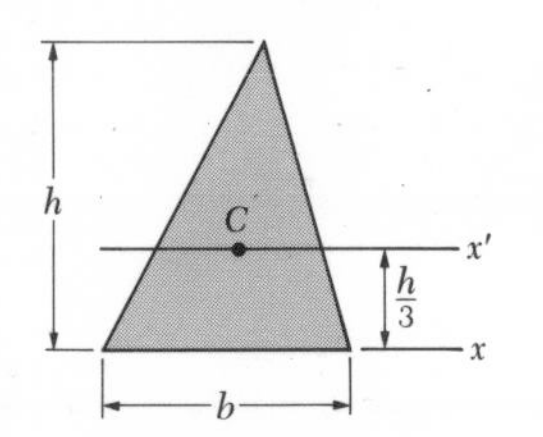
Circle $\bar{I}_x = \bar{I}_y = \frac{1}{4}\pi r^4$ $J_O = \frac{1}{2}\pi r^4$	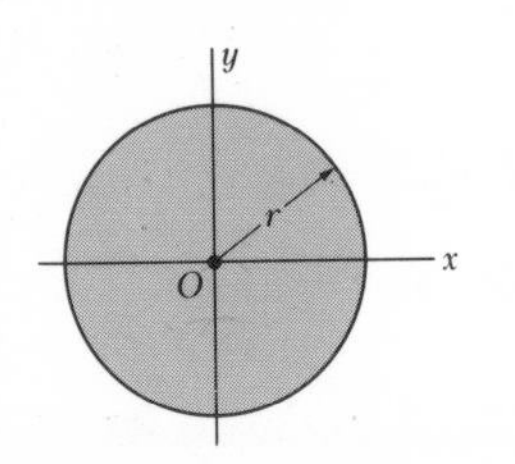
Semicircle $I_x = I_y = \frac{1}{8}\pi r^4$ $J_O = \frac{1}{4}\pi r^4$	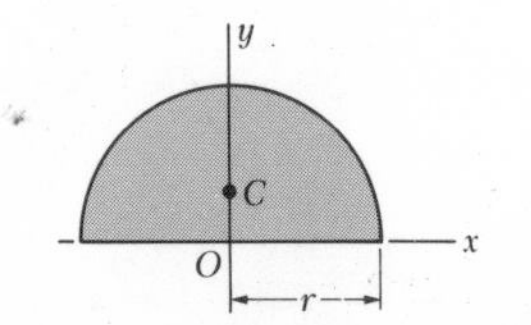
Quarter circle $I_x = I_y = \frac{1}{16}\pi r^4$ $J_O = \frac{1}{8}\pi r^4$	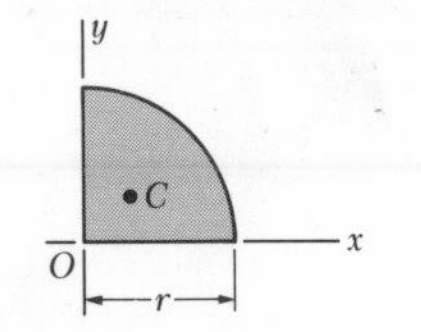
Ellipse $\bar{I}_x = \frac{1}{4}\pi ab^3$ $\bar{I}_y = \frac{1}{4}\pi a^3b$ $J_O = \frac{1}{4}\pi ab(a^2 + b^2)$	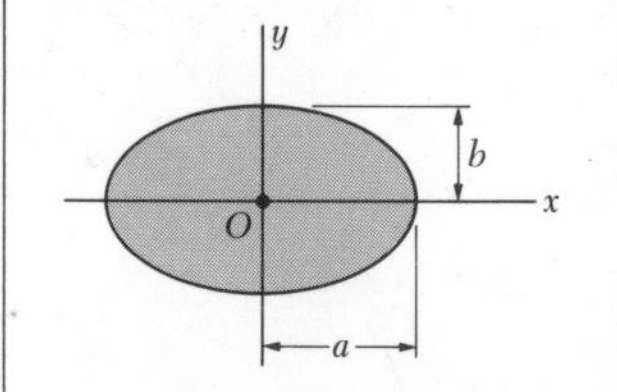

Mass Moments of Inertia of Common Geometric Shapes

Slender rod $I_y = I_z = \frac{1}{12}mL^2$	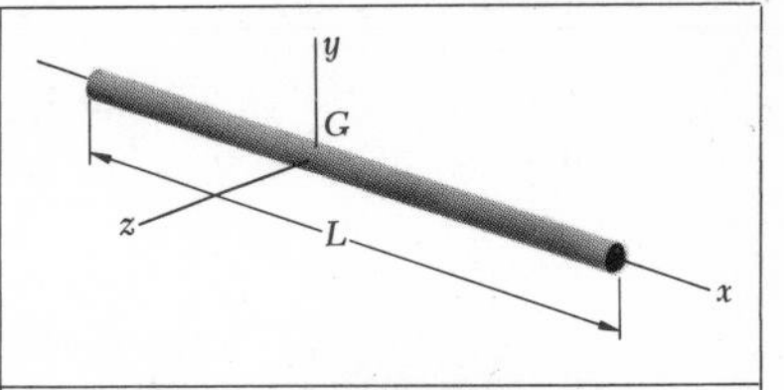
Thin rectangular plate $I_x = \frac{1}{12}m(b^2 + c^2)$ $I_y = \frac{1}{12}mc^2$ $I_z = \frac{1}{12}mb^2$	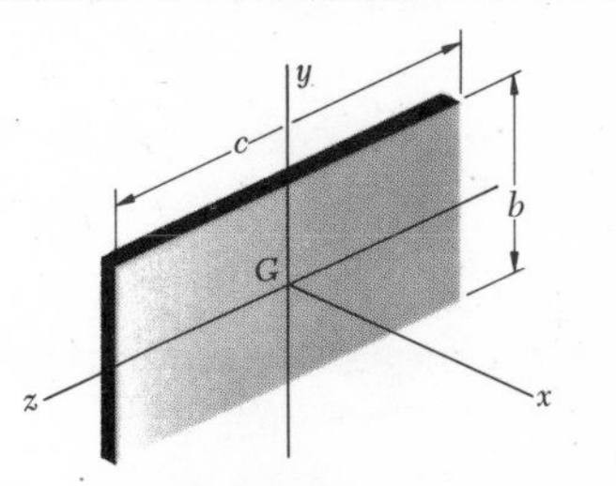
Rectangular prism $I_x = \frac{1}{12}m(b^2 + c^2)$ $I_y = \frac{1}{12}m(c^2 + a^2)$ $I_z = \frac{1}{12}m(a^2 + b^2)$	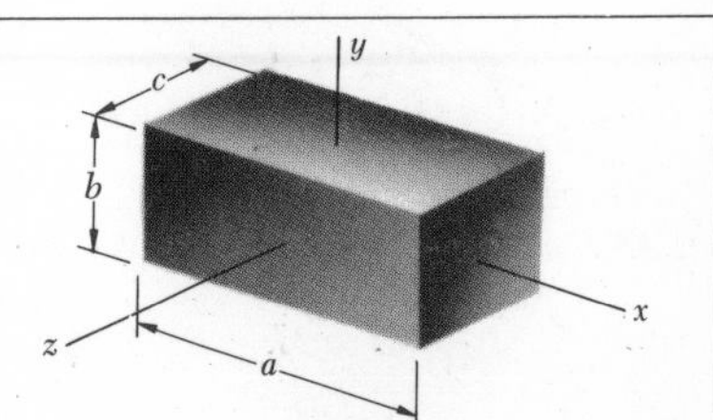
Thin disk $I_x = \frac{1}{2}mr^2$ $I_y = I_z = \frac{1}{4}mr^2$	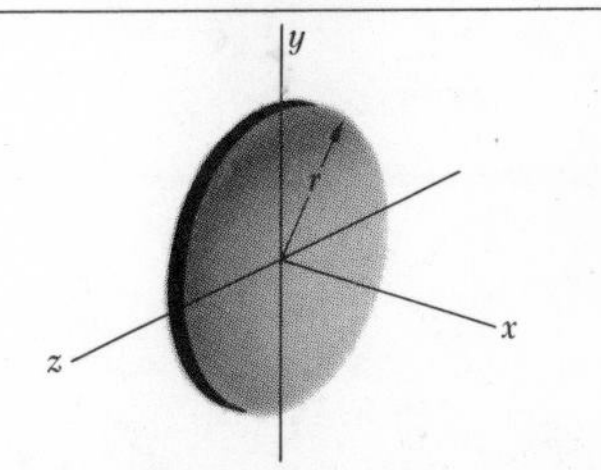
Circular cylinder $I_x = \frac{1}{2}ma^2$ $I_y = I_z = \frac{1}{12}m(3a^2 + L^2)$	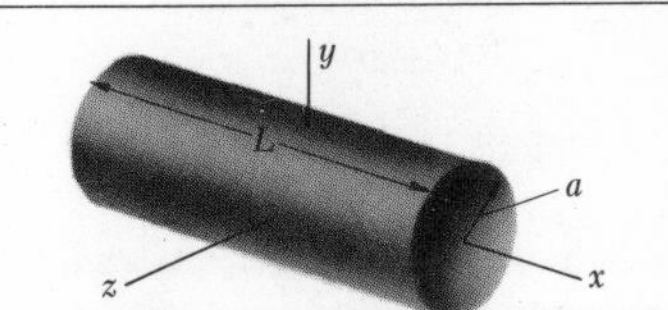
Circular cone $I_x = \frac{3}{10}ma^2$ $I_y = I_z = \frac{3}{5}m(\frac{1}{4}a^2 + h^2)$	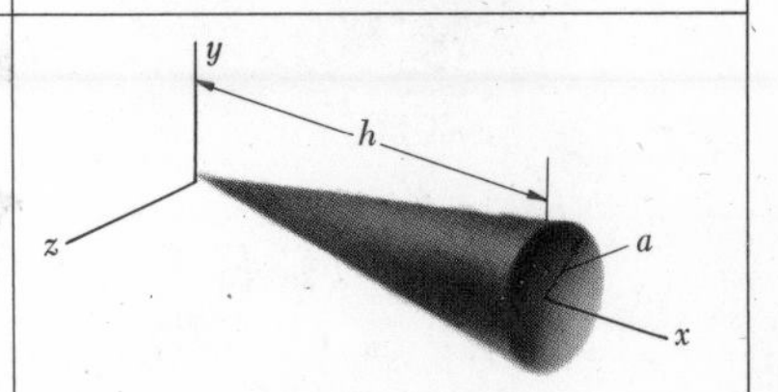
Sphere $I_x = I_y = I_z = \frac{2}{5}ma^2$	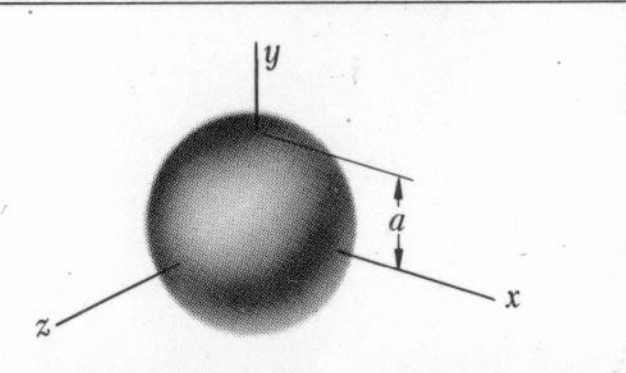

Vector Mechanics for Engineers

Statics

Fifth Edition

Ferdinand P. Beer
Lehigh University

E. Russell Johnston, Jr.
University of Connecticut

McGraw-Hill Book Company
New York St. Louis San Francisco Auckland Bogotá Hamburg
London Madrid Mexico Milan Montreal New Delhi
Panama Paris São Paulo Singapore Sydney Tokyo Toronto

The cover photograph is of Florida's Sunshine Skyway Bridge as it was nearing completion. This bridge, which rises 175 feet above Tampa Bay, is Florida's first cable-stayed bridge, and its 1200-foot main span is the longest concrete span in the Americas. The Sunshine Skyway Bridge was designed by Figg and Muller Engineers, Inc., for the Florida Department of Transportation. *(Photograph by Fred Fox.)*

Back cover photographs: (top) Men on girders *(© Don Klumpp/The Image Bank);* (bottom) Brooklyn Bridge (© Robert Semeniuk/The Stock Market).

VECTOR MECHANICS FOR ENGINEERS: Statics

4567890 KGPKGP 892109

P/N 004507-0
Part of
ISBN 0-07-079946-6

This book was set in Laurel by York Graphic Services, Inc.
The editors were Anne Duffy and David A. Damstra;
the designer was Merrill Haber;
the production supervisor was Joe Campanella.
The drawings were done by Felix Cooper.
Arcata Graphics/Kingsport was printer and binder.

Library of Congress Cataloging-in-Publication Data

Beer, Ferdinand Pierre
Vector mechanics for engineers/Ferdinand P. Beer, E. Russell Johnston, Jr.—5th ed.
p. cm.
Bibliography: v. [1], p.
Includes index.
Contents: [1] Statics
ISBN: 0-07-079946-6
1. Mechanics, Applied. 2. Vector analysis. 3. Mechanics, Applied—Problems, exercises, etc. I. Johnston, E. Russell (Elwood Russell) (date). II. Title.
TA350.B3552 1988
620. 1—dc19 87-18633

ABOUT THE AUTHORS

"How did you happen to write your books together, with one of you at Lehigh and the other at UConn, and how do you manage to keep collaborating on their successive revisions?" These are the two questions most often asked of our two authors.

The answer to the first question is simple. Russ Johnston's first teaching appointment was in the department of civil engineering and mechanics at Lehigh University. There he met Ferd Beer, who had joined that department two years earlier and was in charge of the courses in statics and dynamics. Born in France and educated in France and Switzerland (he holds an M.S. degree from the Sorbonne and a Sc.D. degree in the field of theoretical mechanics from the University of Geneva), Ferd had come to the United States after serving in the French army during the early part of World War II and had taught for four years at Williams College in the Williams-MIT joint arts and engineering program. Born in Philadelphia, Russ had obtained a B.S. degree in civil engineering from the University of Delaware and a Sc.D. degree in the field of structural engineering from MIT.

Ferd was delighted to discover that the young man who had been hired chiefly to teach graduate structural engineering courses was not only willing but eager to help him reorganize the courses in statics and dynamics. Both believed that these courses should be taught from a few basic principles and that the various concepts involved would be best understood and remembered by the students if they were presented in a graphic way. Together they wrote lecture notes, to which they later added problems they felt would appeal to future engineers, and soon they had produced the manuscript of the first edition of *Mechanics for Engineers.*

The second edition of their text found Russ Johnston at Worcester Polytechnic Institute and the third at the University of Connecticut. In the meantime, both Ferd and Russ had assumed administrative responsibilities in their departments, and both were involved in research, consulting, and the supervision of graduate students, Ferd in the area of stochastic processes and random vibrations, and Russ in the area of elastic stability and structural analysis and design. However, their interest in improving the teaching of the basic mechanics courses had not subsided and they both taught sections of these courses as they kept revising their texts.

This brings us to the second question: How did the authors manage to work together so effectively after Russ Johnston had left Lehigh? Part of the answer may be provided by their phone bills and the money they spend on postage. As the publication date of a new edition approaches, they call each other daily and rush to the post office with express-mail packages in order to double-check their work. There are also frequent visits between the two families. At one time there were even joint camping trips, with both families pitching their tents next to each other. The Beers were the first to graduate to a trailer, which was used to illustrate a problem in one of the early editions of their text, but was replaced by the Johnstons' trailer in the next one. Now this trailer has also been replaced, both authors preferring the comforts of a motel and its dining room to those of a camping ground and its fireplaces.

Contents

Preface

The main objective of a first course in mechanics should be to develop in the engineering student the ability to analyze any problem in a simple and logical manner and to apply to its solution a few, well-understood basic principles. It is hoped that this text, designed for the first course in statics offered in the sophomore year, and the volume that follows, *Vector Mechanics for Engineers: Dynamics*, will help the instructor achieve this goal.†

Vector algebra is introduced early in the text and used in the presentation and the discussion of the fundamental principles of mechanics. Vector methods are also used to solve many problems, particularly three-dimensional problems where their application results in a simpler and more concise solution. The emphasis in this text, however, remains on the correct understanding of the principles of mechanics and on their application to the solution of engineering problems, and vector algebra is presented chiefly as a convenient tool.‡

One of the characteristics of the approach used in these volumes is that the mechanics of *particles* has been clearly separated from the mechanics of *rigid bodies*. This approach makes it possible to consider simple practical applications at an early stage and to postpone the introduction of more difficult concepts. In this volume, for example, the statics of particles is treated first (Chap. 2); after the rules of addition and subtraction of vectors have been introduced, the principle of equilibrium of a particle is immediately applied to practical situations involving only concurrent forces. The statics of rigid bodies is considered in Chaps. 3 and 4. In Chap. 3, the vector and scalar products of two vectors are introduced and used to define the moment of a force about a point and about an axis. The presentation of these new concepts is followed by a thorough and rigorous discussion of equivalent systems of forces leading, in Chap. 4, to many practical applications involving the equilibrium of rigid bodies under general force systems. In the volume on dynamics, the same division is observed. The basic concepts of force, mass, and acceleration, of work and energy, and of impulse and momentum are

† Both texts are also available in a single volume, *Vector Mechanics for Engineers: Statics and Dynamics*, fifth edition.

‡ In a parallel text, *Mechanics for Engineers: Statics*, fourth edition, the use of vector algebra is limited to the addition and subtraction of vectors.

introduced and first applied to problems involving only particles. Thus students may familiarize themselves with the three basic methods used in dynamics and learn their respective advantages before facing the difficulties associated with the motion of rigid bodies.

Since this text is designed for a first course in statics, new concepts have been presented in simple terms and every step explained in detail. On the other hand, by discussing the broader aspects of the problems considered, a definite maturity of approach has been achieved. For example, the concepts of partial constraints and of static indeterminacy are introduced early in the text and used throughout.

The fact that mechanics is essentially a *deductive* science based on a few fundamental principles has been stressed. Derivations have been presented in their logical sequence and with all the rigor warranted at this level. However, the learning process being largely *inductive*, simple applications have been considered first. Thus the statics of particles precedes the statics of rigid bodies, and problems involving internal forces are postponed until Chap. 6. Also, in Chap. 4, equilibrium problems involving only coplanar forces are considered first and solved by ordinary algebra, while problems involving three-dimensional forces and requiring the full use of vector algebra are discussed in the second part of the chapter.

Free-body diagrams are introduced early, and their importance is emphasized throughout the text. Color has been used to distinguish forces from other elements of the free-body diagrams. This makes it easier for the students to identify the forces acting on a given particle or rigid body and to follow the discussion of sample problems and other examples given in the text. Free-body diagrams are used not only to solve equilibrium problems but also to express the equivalence of two systems of forces or, more generally, of two systems of vectors. This approach is particularly useful as a preparation for the study of the dynamics of rigid bodies. As will be shown in the volume on dynamics, by placing the emphasis on "free-body-diagram equations" rather than on the standard algebraic equations of motion, a more intuitive and more complete understanding of the fundamental principles of dynamics may be achieved.

Because of the current trend among American engineers to adopt the international system of units (SI metric units), the SI units most frequently used in mechanics have been introduced in Chap. 1 and are used throughout the text. Approximately half the sample problems and 60 percent of the problems to be assigned have been stated in these units, while the remainder retain U.S. customary units. The authors believe that this approach will best serve the needs of the students, who will be entering the engineering profession during the period of transition from one system of units to the other. It also should be recognized that the passage from one system to the other entails more than the use of conversion factors. Since the SI system of units is an absolute system based on the units of time, length, and mass, whereas the U.S. customary system is a gravitational system based on the units of time, length, and force, different approaches are required for the solution of many problems. For example, when SI units are used, a body is generally specified by its mass expressed in kilograms; in most problems of statics it will be necessary to determine the weight of the body in newtons, and an additional calculation will be required for this purpose. On the other hand, when U.S. customary units are used, a body is specified by its weight in pounds and, in dynamics problems, an additional calculation will be required to determine its mass in slugs (or $\text{lb} \cdot \text{sec}^2/\text{ft}$). The authors, therefore, believe that problem assignments should include both systems of units. The actual distribution of assigned problems between the two systems of units, however, has been left to the instructor, and a sufficient number of problems of each type have been provided so that four complete lists of assignments may be selected with the proportion of problems stated in SI units set anywhere between 50 and 75 percent. If so desired, two complete lists of assignments may also be selected from problems stated in SI units only and two others from problems stated in U.S. customary units.

A large number of optional sections have been included. These sections are indicated by asterisks and may thus easily be distinguished from those which form the core of the basic statics course. They may be omitted without prejudice to the understanding of the rest of the text. Among the topics covered in these additional sections are the reduction of a system of forces to a wrench, applications to hydrostatics, shear and bending-moment diagrams for beams, equilibrium of cables, products of inertia and Mohr's circle, mass products of inertia and principal axes of inertia for three-dimensional bodies, and the method of virtual work. The sections on beams are especially useful when the course in statics is immediately followed by a course in mechanics of materials, while the sections on the inertia properties of three-dimensional bodies are primarily intended for the students who will later study in dynamics the motion of rigid bodies in three dimensions.

The material presented in the text and most of the problems require no previous mathematical knowledge beyond algebra, trigonometry, and elementary calculus, and all the elements of vector algebra necessary to the understanding of the text have been carefully presented in Chaps. 2 and 3. In general, a greater emphasis has been placed on the correct understanding of the basic mathematical concepts involved than on the nimble manipulation of mathematical formulas. In this connection, it should be mentioned that the determination of the centroids of composite areas precedes the calculation of centroids by integration, thus making it possible to establish the concept of moment of area firmly before introducing the use of integration. The presentation of numerical solutions takes into account the universal use of calculators by engineering students and instructions on the proper use of calculators for the solution of typical statics problems have been included in Chap. 2.

Each chapter begins with an introductory section setting the purpose and goals of the chapter and describing in simple terms the material to be covered and its application to the solution of engineering problems. The body of the text has been divided into units, each consisting of one or several theory sections, one or several sample problems, and a large number of problems to be assigned. Each unit corresponds to a well-defined topic and generally may be covered in one lesson. In a number of cases, however, the instructor will find it desirable to devote more than one lesson to a given topic. Each chapter ends with a review and summary of the material covered in that chapter. Marginal notes have been added to help students organize their review work, and cross-references have been included to help them find the portions of material requiring their special attention.

The sample problems have been set up in much the same form that students will use in solving the assigned problems. They thus serve the double purpose of amplifying the text and demonstrating the type of neat and orderly work that students should cultivate in their own solutions. Most of the problems to be assigned are of a practical nature and should appeal to engineering students. They are primarily designed, however, to illustrate the material presented in the text and to help students understand the basic principles of mechanics. The problems have been grouped according to the portions of material they illustrate and have been arranged in order of increasing difficulty. Problems requiring special attention have been indicated by asterisks. Answers to all even-numbered problems are given at the end of the book.

The introduction in the engineering curriculum of instruction in computer programming and the increasing availability of personal computers or mainframe terminals on most campuses make it now possible for engineering students to solve a number of challenging statics and dynamics problems. Only a few years ago these problems would have been considered inappropriate for an undergraduate course because of the large number of computations their solutions require. In this new edition of *Vector Mechanics for Engineers*, a group of four problems designed to be solved with a computer has been added to the review problems at the end of each chapter. In *Statics*, these problems may involve the analysis of a structure for various configurations or loadings of the structure, or the determination of the equilibrium positions of a given mechanism. In *Dynamics*, they may involve the determination of the motion of a particle under various initial conditions, the kinematic or kinetic analysis of mechanisms in successive positions, or the numerical integration of various equations of motion. Developing the algorithm required to solve a given mechanics problem will benefit the students in two different ways: (1) it will help them gain a better understanding of the mechanics principles involved; (2) it will provide them with an opportunity to apply the skills acquired in their computer programming course to the solution of a meaningful engineering problem.

A diskette containing interactive software designed for the IBM PC or any compatible computer has been placed in the pocket inside the rear cover of this text. This software includes twelve "Tutorials," each consisting of a problem in statics for which students are more likely to require help; thus, the software plays a role similar to that of the Sample Problems. However, being interactive, the Tutorials allow the students to vary the data, let them participate in the solution of the problem, and check the results obtained. By allowing the students to vary the data, the Tutorials provide them with an inexhaustible source of problems of the same type. They also allow the more inquisitive students to test the "sensitivity" of a given problem to a change in data or to find the conditions which lead to special or critical situations. The authors wish to thank Professor Raymond P. Canale of the University of Michigan for helping them prepare the Tutorials and Mr. Tad Slawecki of EnginComp Software, Inc., for programming them.

The authors gratefully acknowledge the many helpful comments and suggestions offered by the users of the previous editions of *Mechanics for Engineers* and of *Vector Mechanics for Engineers.*

Ferdinand P. Beer
E. Russell Johnston, Jr.

List of Symbols

a Constant; radius; distance
A, B, C, . . . Reactions at supports and connections
$A, B, C, \ldots$ Points
A Area
b Width; distance
c Constant
C Centroid
d Distance
e Base of natural logarithms
F Force; friction force
g Acceleration of gravity
G Center of gravity; constant of gravitation
h Height; sag of cable
i, j, k Unit vectors along coordinate axes
$I, I_x, \ldots$ Moment of inertia
$\bar{I}$ Centroidal moment of inertia
$I_{xy}, \ldots$ Product of inertia
J Polar moment of inertia
k Spring constant
k_x, k_y, k_O Radius of gyration
k Centroidal radius of gyration
l Length
L Length; span
m Mass
M Couple; moment
$\mathbf{M}_O$ Moment about point O
$\mathbf{M}_O^R$ Moment resultant about point O
M Magnitude of couple or moment; mass of earth
M_{OL} Moment about axis OL
N Normal component of reaction
O Origin of coordinates
p Pressure
P Force; vector
Q Force; vector
r Position vector

r	Radius; distance; polar coordinate
$\mathbf{R}$	Resultant force; resultant vector; reaction
R	Radius of earth
$\mathbf{s}$	Position vector
s	Length of arc; length of cable
$\mathbf{S}$	Force; vector
t	Thickness
$\mathbf{T}$	Force
T	Tension
U	Work
$\mathbf{V}$	Vector product; shearing force
V	Volume; potential energy; shear
w	Load per unit length
$\mathbf{W}, W$	Weight; load
x, y, z	Rectangular coordinates; distances
$\bar{x}, \bar{y}, \bar{z}$	Rectangular coordinates of centroid or center of gravity
α, β, γ	Angles
γ	Specific weight
δ	Elongation
$\delta \mathbf{r}$	Virtual displacement
δU	Virtual work
$\boldsymbol{\lambda}$	Unit vector along a line
η	Efficiency
θ	Angular coordinate; angle; polar coordinate
μ	Coefficient of friction
ρ	Density
ϕ	Angle of friction; angle

Vector Mechanics for Engineers

Statics

Introduction

CHAPTER

1

1.1. What Is Mechanics? Mechanics may be defined as that science which describes and predicts the conditions of rest or motion of bodies under the action of forces. It is divided into three parts: mechanics of *rigid bodies*, mechanics of *deformable bodies*, and mechanics of *fluids*.

The mechanics of rigid bodies is subdivided into *statics* and *dynamics*, the former dealing with bodies at rest, the latter with bodies in motion. In this part of the study of mechanics, bodies are assumed to be perfectly rigid. Actual structures and machines, however, are never absolutely rigid and deform under the loads to which they are subjected. But these deformations are usually small and do not appreciably affect the conditions of equilibrium or motion of the structure under consideration. They are important, though, as far as the resistance of the structure to failure is concerned and are studied in mechanics of materials, which is a part of the mechanics of deformable bodies. The third division of mechanics, the mechanics of fluids, is subdivided into the study of *incompressible fluids* and of *compressible fluids*. An important subdivision of the study of incompressible fluids is *hydraulics*, which deals with problems involving liquids.

Mechanics is a physical science, since it deals with the study of physical phenomena. However, some associate mechanics with mathematics, while many consider it as an engineering subject. Both these views are justified in part. Mechanics is the foundation of most engineering sciences and is an indispensable prerequisite to their study. However, it does not have the *empiricism* found in some engineering sciences, i.e., it does not rely on experience or observation alone; by its rigor and the emphasis it places on deductive reasoning it resembles mathematics. But, again, it is not an *abstract* or even a *pure* science; mechanics is an *applied* science. The purpose of mechanics is to explain and predict physical phenomena and thus to lay the foundations for engineering applications.

1.2. Fundamental Concepts and Principles. Although the study of mechanics goes back to the time of Aristotle (384–322 B.C.) and Archimedes (287–212 B.C.), one has to wait until Newton (1642–1727) to find a satisfactory formulation of its fundamental principles. These principles were later expressed in a modified form by d'Alembert, Lagrange, and Hamilton. Their validity remained unchallenged, however, until Einstein formulated his *theory of relativity* (1905). While its limitations have now been recognized, *newtonian mechanics* still remains the basis of today's engineering sciences.

The basic concepts used in mechanics are *space, time, mass,* and *force.* These concepts cannot be truly defined; they should be accepted on the basis of our intuition and experience and used as a mental frame of reference for our study of mechanics.

The concept of *space* is associated with the notion of the position of a point P. The position of P may be defined by three lengths measured from a certain reference point, or *origin,* in three given directions. These lengths are known as the *coordinates* of P.

To define an event, it is not sufficient to indicate its position in space. The *time* of the event should also be given.

The concept of *mass* is used to characterize and compare bodies on the basis of certain fundamental mechanical experiments. Two bodies of the same mass, for example, will be attracted by the earth in the same manner; they will also offer the same resistance to a change in translational motion.

A *force* represents the action of one body on another. It may be exerted by actual contact or at a distance, as in the case of gravitational forces and magnetic forces. A force is characterized by its *point of application,* its *magnitude,* and its *direction;* a force is represented by a *vector* (Sec. 2.3).

In newtonian mechanics, space, time, and mass are absolute concepts, independent of each other. (This is not true in *relativistic mechanics,* where the time of an event depends upon its position, and where the mass of a body varies with its velocity.) On the other hand, the concept of force is not independent of the other three. Indeed, one of the fundamental principles of newtonian mechanics listed below indicates that the resultant force acting on a body is related to the mass of the body and to the manner in which its velocity varies with time.

We shall study the conditions of rest or motion of particles and rigid bodies in terms of the four basic concepts we have introduced. By *particle* we mean a very small amount of matter which may be assumed to occupy a single point in space. A *rigid body* is a combination of a large number of particles occupying fixed positions with respect to each other. The study of the mechanics of particles is obviously a prerequisite to that of rigid bodies. Besides, the results obtained for a particle may be used directly in a large number of problems dealing with the conditions of rest or motion of actual bodies.

The study of elementary mechanics rests on six fundamental principles based on experimental evidence.

The Parallelogram Law for the Addition of Forces. This states that two forces acting on a particle may be replaced by a single force, called their *resultant,* obtained by drawing the diagonal of the parallelogram which has sides equal to the given forces (Sec. 2.2).

The Principle of Transmissibility. This states that the conditions of equilibrium or of motion of a rigid body will remain unchanged if a force acting at a given point of the rigid body is replaced by a force of the same magnitude and same direction, but acting at a different point, provided that the two forces have the same line of action (Sec. 3.3).

Newton's Three Fundamental Laws. Formulated by Sir Isaac Newton in the latter part of the seventeenth century, these laws may be stated as follows:

FIRST LAW. If the resultant force acting on a particle is zero, the particle will remain at rest (if originally at rest) or will move with constant speed in a straight line (if originally in motion) (Sec. 2.10).

SECOND LAW. If the resultant force acting on a particle is not zero, the particle will have an acceleration proportional to the magnitude of the resultant and in the direction of this resultant force.

As we shall see in Sec. 12.2, this law may be stated as

$$\mathbf{F} = m\mathbf{a} \tag{1.1}$$

where **F**, m, and **a** represent, respectively, the resultant force acting on the particle, the mass of the particle, and the acceleration of the particle, expressed in a consistent system of units.

THIRD LAW. The forces of action and reaction between bodies in contact have the same magnitude, same line of action, and opposite sense (Sec. 6.1).

Newton's Law of Gravitation. This states that two particles of mass M and m are mutually attracted with equal and opposite forces **F** and $-$**F** (Fig. 1.1) of magnitude F given by the formula

$$F = G\frac{Mm}{r^2} \tag{1.2}$$

where r = distance between the two particles
G = universal constant called the *constant of gravitation*

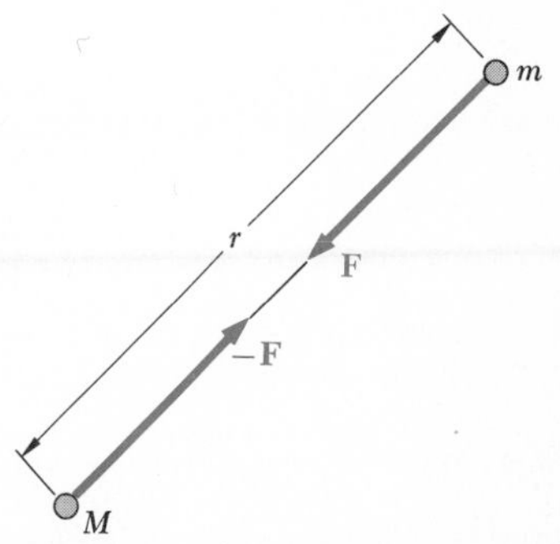

Fig. 1.1

Newton's law of gravitation introduces the idea of an action exerted at a distance and extends the range of application of Newton's third law: the action **F** and the reaction $-$**F** in Fig. 1.1 are equal and opposite, and they have the same line of action.

A particular case of great importance is that of the attraction of the earth on a particle located on its surface. The force **F** exerted by the earth on the particle is then defined as the *weight* **W** of the particle. Taking M equal to the mass of the earth, m equal to the mass of the particle, and r equal to the radius R of the earth, and introducing the constant

$$g = \frac{GM}{R^2} \tag{1.3}$$

the magnitude W of the weight of a particle of mass m may be expressed as†

$$W = mg \tag{1.4}$$

The value of R in formula (1.3) depends upon the elevation of the point considered; it also depends upon its latitude, since the earth is not truly spherical. The value of g therefore varies with the position of the point considered. As long as the point actually remains on the surface of the earth, it is sufficiently accurate in most engineering computations to assume that g equals 9.81 m/s^2 or 32.2 ft/s^2.

The principles we have just listed will be introduced in the course of our study of mechanics as they are needed. The study of the statics of particles, carried out in Chap. 2, will be based on the parallelogram law of addition and on Newton's first law alone. The principle of transmissibility will be introduced in Chap. 3 as we begin the study of the statics of rigid bodies, and Newton's third law in Chap. 6 as we analyze the forces exerted on each other by the various members forming a structure. It should be noted that the four aforementioned principles underlie the entire study of the statics of particles, rigid bodies, and systems of rigid bodies. In the study of dynamics, Newton's second law and Newton's law of gravitation will be introduced. It will then be shown that Newton's first law is a particular case of Newton's second law (Sec. 12.2) and that the principle of transmissibility could be derived from the other principles and thus eliminated (Sec. 16.5). In the meantime, however, Newton's first and third laws, the parallelogram law of addition, and the principle of transmissibility will provide us with the necessary and sufficient foundation for the study of the entire field of statics.

As noted earlier, the six fundamental principles listed above are based on experimental evidence. Except for Newton's first law and the principle of transmissibility, they are independent principles which cannot be derived mathematically from each other or from any other elementary physical principle. On these principles rests most of the intricate structure of newtonian mechanics. For more than two centuries a tremendous number of problems dealing with the conditions of rest and motion of rigid bodies, deformable bodies, and fluids have been solved by applying these fundamental principles. Many of the solutions obtained could be checked exper-

† A more accurate definition of the weight **W** should take into account the rotation of the earth.

imentally, thus providing a further verification of the principles from which they were derived. It is only recently that Newton's mechanics was found at fault, in the study of the motion of atoms and in the study of the motion of certain planets, where it must be supplemented by the theory of relativity. But on the human or engineering scale, where velocities are small compared with the velocity of light, Newton's mechanics has yet to be disproved.

1.3. Systems of Units. With the four fundamental concepts introduced in the preceding section are associated the so-called *kinetic units,* i.e., the units of *length, time, mass,* and *force*. These units cannot be chosen independently if Eq. (1.1) is to be satisfied. Three of the units may be defined arbitrarily; they are then referred to as *base units*. The fourth unit, however, must be chosen in accordance with Eq. (1.1) and is referred to as a *derived unit*. Kinetic units selected in that way are said to form a *consistent system of units*.

International System of Units (SI Units†). In this system, which will be in universal use after the United States has completed its current conversion, the base units are the units of length, mass, and time, and are called, respectively, the *meter* (m), the *kilogram* (kg), and the *second* (s). All three are arbitrarily defined. The second, which is supposed to represent the 1/86 400 part of the mean solar day, is actually defined as the duration of 9 192 631 770 cycles of the radiation associated with a specified transition of the cesium atom. The meter, originally intended to represent one-ten-millionth of the distance from the equator to the pole, is now defined as 1 650 763.73 wavelengths of the orange-red line of krypton 86. The kilogram, which is approximately equal to the mass of 0.001 m^3 of water, is actually defined as the mass of a platinum standard kept at the International Bureau of Weights and Measures at Sèvres, near Paris, France. The unit of force is a derived unit. It is called the *newton* (N) and is defined as the force which gives an acceleration of 1 m/s^2 to a mass of 1 kg (Fig. 1.2). From Eq. (1.1) we write

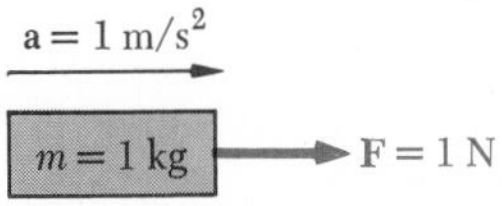

$$1\text{ N} = (1\text{ kg})(1\text{ m/s}^2) = 1\text{ kg}\cdot\text{m/s}^2 \tag{1.5}$$

$a = 1\text{ m/s}^2$
$m = 1\text{ kg}$ $F = 1\text{ N}$

Fig. 1.2

The SI units are said to form an *absolute* system of units. This means that the three base units chosen are independent of the location where measurements are made. The meter, the kilogram, and the second may be used anywhere on the earth; they may even be used on another planet. They will always have the same significance.

The *weight* of a body, or *force of gravity* exerted on that body, should, like any other force, be expressed in newtons. From Eq. (1.4) it follows that the weight of a body of mass 1 kg (Fig. 1.3) is

$$\begin{aligned} W &= mg \\ &= (1\text{ kg})(9.81\text{ m/s}^2) \\ &= 9.81\text{ N} \end{aligned}$$

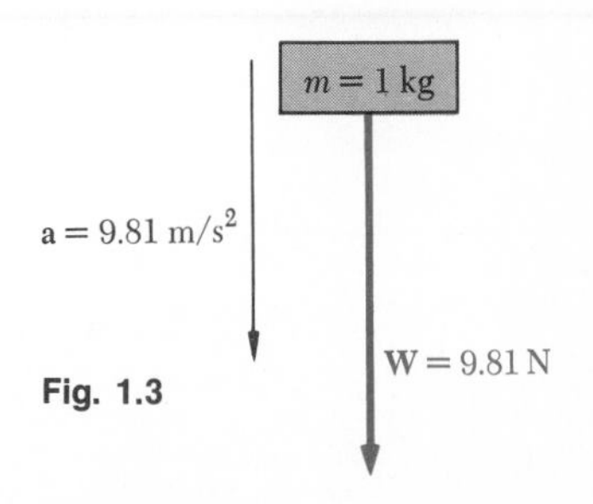

Fig. 1.3

†SI stands for *Système International d'Unités* (French).

Table 1.1 SI Prefixes

Multiplication Factor	Prefix†	Symbol
$1\,000\,000\,000\,000 = 10^{12}$	tera	T
$1\,000\,000\,000 = 10^{9}$	giga	G
$1\,000\,000 = 10^{6}$	mega	M
$1\,000 = 10^{3}$	kilo	k
$100 = 10^{2}$	hecto‡	h
$10 = 10^{1}$	deka‡	da
$0.1 = 10^{-1}$	deci‡	d
$0.01 = 10^{-2}$	centi‡	c
$0.001 = 10^{-3}$	milli	m
$0.000\,001 = 10^{-6}$	micro	μ
$0.000\,000\,001 = 10^{-9}$	nano	n
$0.000\,000\,000\,001 = 10^{-12}$	pico	p
$0.000\,000\,000\,000\,001 = 10^{-15}$	femto	f
$0.000\,000\,000\,000\,000\,001 = 10^{-18}$	atto	a

† The first syllable of every prefix is accented so that the prefix will retain its identity. Thus, the preferred pronunciation of kilometer places the accent on the first syllable, not the second.

‡ The use of these prefixes should be avoided, except for the measurement of areas and volumes and for the nontechnical use of centimeter, as for body and clothing measurements.

Multiples and submultiples of the fundamental SI units may be obtained through the use of the prefixes defined in Table 1.1. The multiples and submultiples of the units of length, mass, and force most frequently used in engineering are, respectively, the *kilometer* (km) and the *millimeter* (mm); the *megagram*† (Mg) and the *gram* (g); and the *kilonewton* (kN). According to Table 1.1, we have

$$1 \text{ km} = 1000 \text{ m} \qquad 1 \text{ mm} = 0.001 \text{ m}$$
$$1 \text{ Mg} = 1000 \text{ kg} \qquad 1 \text{ g} = 0.001 \text{ kg}$$
$$1 \text{ kN} = 1000 \text{ N}$$

The conversion of these units into meters, kilograms, and newtons, respectively, can be effected by simply moving the decimal point three places to the right or to the left. For example, to convert 3.82 km into meters, one moves the decimal point three places to the right:

$$3.82 \text{ km} = 3820 \text{ m}$$

Similarly, 47.2 mm is converted into meters by moving the decimal point three places to the left:

$$47.2 \text{ mm} = 0.0472 \text{ m}$$

Using the scientific notation, one may also write

$$3.82 \text{ km} = 3.82 \times 10^{3} \text{ m}$$
$$47.2 \text{ mm} = 47.2 \times 10^{-3} \text{ m}$$

† Also known as a *metric ton.*

The multiples of the unit of time are the *minute* (min) and the *hour* (h). Since

$$1 \text{ min} = 60 \text{ s} \qquad \text{and} \qquad 1 \text{ h} = 60 \text{ min} = 3600 \text{ s}$$

these multiples cannot be converted as readily as the others.

By using the appropriate multiple or submultiple of a given unit, one may avoid writing very large or very small numbers. For example, one usually writes 427.2 km rather than 427 200 m, and 2.16 mm rather than 0.002 16 m.†

Units of Area and Volume. The unit of area is the *square meter* (m^2), which represents the area of a square of side 1 m; the unit of volume is the *cubic meter* (m^3), equal to the volume of a cube of side 1 m. In order to avoid exceedingly small or large numerical values in the computation of areas and volumes, one uses systems of subunits obtained by respectively squaring and cubing not only the millimeter but also two intermediate submultiples of the meter, namely, the *decimeter* (dm) and the *centimeter* (cm). Since, by definition,

$$\begin{aligned} 1 \text{ dm} &= 0.1 \text{ m} = 10^{-1} \text{ m} \\ 1 \text{ cm} &= 0.01 \text{ m} = 10^{-2} \text{ m} \\ 1 \text{ mm} &= 0.001 \text{ m} = 10^{-3} \text{ m} \end{aligned}$$

the submultiples of the unit of area are

$$\begin{aligned} 1 \text{ dm}^2 &= (1 \text{ dm})^2 = (10^{-1} \text{ m})^2 = 10^{-2} \text{ m}^2 \\ 1 \text{ cm}^2 &= (1 \text{ cm})^2 = (10^{-2} \text{ m})^2 = 10^{-4} \text{ m}^2 \\ 1 \text{ mm}^2 &= (1 \text{ mm})^2 = (10^{-3} \text{ m})^2 = 10^{-6} \text{ m}^2 \end{aligned}$$

and the submultiples of the unit of volume are

$$\begin{aligned} 1 \text{ dm}^3 &= (1 \text{ dm})^3 = (10^{-1} \text{ m})^3 = 10^{-3} \text{ m}^3 \\ 1 \text{ cm}^3 &= (1 \text{ cm})^3 = (10^{-2} \text{ m})^3 = 10^{-6} \text{ m}^3 \\ 1 \text{ mm}^3 &= (1 \text{ mm})^3 = (10^{-3} \text{ m})^3 = 10^{-9} \text{ m}^3 \end{aligned}$$

It should be noted that when the volume of a liquid is being measured, the cubic decimeter (dm^3) is usually referred to as a *liter* (L).

Other derived SI units used to measure the moment of a force, the work of a force, etc., are shown in Table 1.2. While these units will be introduced in later chapters as they are needed, we should note an important rule at this time: When a derived unit is obtained by dividing a base unit by another base unit, a prefix may be used in the numerator of the derived unit but not in its denominator. For example, the constant k of a spring which stretches 20 mm under a load of 100 N will be expressed as

$$k = \frac{100 \text{ N}}{20 \text{ mm}} = \frac{100 \text{ N}}{0.020 \text{ m}} = 5000 \text{ N/m} \qquad \text{or} \qquad k = 5 \text{ kN/m}$$

but never as $k = 5$ N/mm.

†It should be noted that when more than four digits are used on either side of the decimal point to express a quantity in SI units—as in 427 200 m or 0.002 16 m—spaces, never commas, should be used to separate the digits into groups of three. This is to avoid confusion with the comma which is used in many countries in place of a decimal point.

Table 1.2 Principal SI Units Used in Mechanics

Quantity	Unit	Symbol	Formula
Acceleration	Meter per second squared	. . .	m/s^2
Angle	Radian	rad	†
Angular acceleration	Radian per second squared	. . .	rad/s^2
Angular velocity	Radian per second	. . .	rad/s
Area	Square meter	. . .	m^2
Density	Kilogram per cubic meter	. . .	kg/m^3
Energy	Joule	J	N·m
Force	Newton	N	$kg \cdot m/s^2$
Frequency	Hertz	Hz	s^{-1}
Impulse	Newton-second	. . .	kg·m/s
Length	Meter	m	‡
Mass	Kilogram	kg	‡
Moment of a force	Newton-meter	. . .	N·m
Power	Watt	W	J/s
Pressure	Pascal	Pa	N/m^2
Stress	Pascal	Pa	N/m^2
Time	Second	s	‡
Velocity	Meter per second	. . .	m/s
Volume, solids	Cubic meter	. . .	m^3
Liquids	Liter	L	10^{-3} m^3
Work	Joule	J	N·m

† Supplementary unit (1 revolution = 2π rad = 360°).

‡ Base unit.

U.S. Customary Units. Most practicing American engineers still commonly use a system in which the base units are the units of length, force, and time. These units are, respectively, the *foot* (ft), the *pound* (lb), and the *second* (s). The second is the same as the corresponding SI unit. The foot is defined as 0.3048 m. The pound is defined as the *weight* of a platinum standard, called the *standard pound* and kept at the National Bureau of Standards in Washington, the mass of which is 0.453 592 43 kg. Since the weight of a body depends upon the gravitational attraction of the earth, which varies with location, it is specified that the standard pound should be placed at sea level and at the latitude of 45° to properly define a force of 1 lb. Clearly the U.S. customary units do not form an absolute system of units. Because of their dependence upon the gravitational attraction of the earth, they form a *gravitational* system of units.

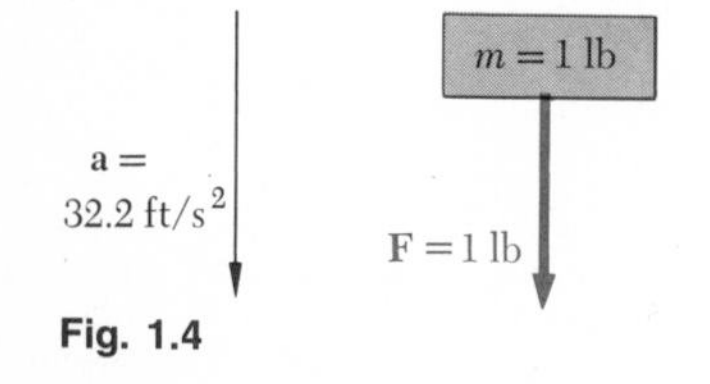

Fig. 1.4

While the standard pound also serves as the unit of mass in commercial transactions in the United States, it cannot be so used in engineering computations since such a unit would not be consistent with the base units defined in the preceding paragraph. Indeed, when acted upon by a force of 1 lb, that is, when subjected to the force of gravity, the standard pound receives the acceleration of gravity, $g = 32.2\ ft/s^2$ (Fig. 1.4), not the unit acceleration required by Eq. (1.1). The unit of mass consistent with the foot, the pound, and the second is the mass which receives an acceleration of 1 ft/s^2 when a force of 1 lb is applied to it (Fig. 1.5). This unit, some-

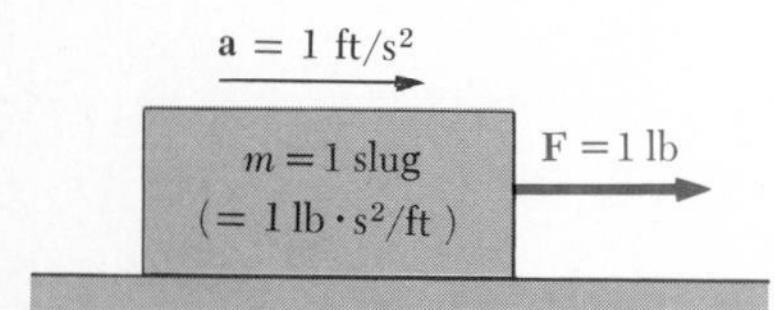

Fig. 1.5

times called a *slug*, can be derived from the equation $F = ma$ after substituting 1 lb and 1 ft/s^2 for F and a, respectively. We write

$$F = ma \qquad 1\text{ lb} = (1\text{ slug})(1\text{ ft/s}^2)$$

and obtain

$$1\text{ slug} = \frac{1\text{ lb}}{1\text{ ft/s}^2} = 1\text{ lb}\cdot\text{s}^2/\text{ft} \tag{1.6}$$

Comparing Figs. 1.4 and 1.5, we conclude that the slug is a mass 32.2 times larger than the mass of the standard pound.

The fact that in the U.S. customary system of units bodies are characterized by their weight in pounds rather than by their mass in slugs will be a convenience in the study of statics, where we shall constantly deal with weights and other forces and only seldom with masses. However, in the study of dynamics, where forces, masses, and accelerations are involved, we shall have to express the mass m in slugs of a body of which the weight W has been given in pounds. Recalling Eq. (1.4), we shall write

$$m = \frac{W}{g} \tag{1.7}$$

where g is the acceleration of gravity ($g = 32.2\text{ ft/s}^2$).

Other U.S. customary units frequently encountered in engineering problems are the *mile* (mi), equal to 5280 ft; the *inch* (in.), equal to $\frac{1}{12}$ ft; and the *kilopound* (kip), equal to a force of 1000 lb. The *ton* is often used to represent a mass of 2000 lb but, like the pound, must be converted into slugs in engineering computations.

The conversion into feet, pounds, and seconds of quantities expressed in other U.S. customary units is generally more involved and requires greater attention than the corresponding operation in SI units. If, for example, the magnitude of a velocity is given as $v = 30$ mi/h, we shall proceed as follows to convert it to ft/s. First we write

$$v = 30\,\frac{\text{mi}}{\text{h}}$$

Since we want to get rid of the unit miles and introduce instead the unit feet, we should multiply the right-hand member of the equation by an expression containing miles in the denominator and feet in the numerator. But, since we do not want to change the value of the right-hand member, the expression used should have a value equal to unity. The quotient (5280 ft)/(1 mi) is such an expression. Operating in a similar way to transform the unit hour into seconds, we write

$$v = \left(30\,\frac{\text{mi}}{\text{h}}\right)\left(\frac{5280\text{ ft}}{1\text{ mi}}\right)\left(\frac{1\text{ h}}{3600\text{ s}}\right)$$

Carrying out the numerical computations and canceling out units which appear both in the numerator and the denominator, we obtain

$$v = 44\,\frac{\text{ft}}{\text{s}} = 44\text{ ft/s}$$

1.4. Conversion from One System of Units to Another. There are many instances when an engineer wishes to convert into SI units a numerical result obtained in U.S. customary units or vice versa. Because the unit of time is the same in both systems, only two kinetic base units need be converted. Thus, since all other kinetic units can be derived from these base units, only two conversion factors need be remembered.

Units of Length. By definition the U.S. customary unit of length is

$$1 \text{ ft} = 0.3048 \text{ m} \tag{1.8}$$

It follows that

$$1 \text{ mi} = 5280 \text{ ft} = 5280(0.3048 \text{ m}) = 1609 \text{ m}$$

or

$$1 \text{ mi} = 1.609 \text{ km} \tag{1.9}$$

Also

$$1 \text{ in.} = \tfrac{1}{12} \text{ ft} = \tfrac{1}{12}(0.3048 \text{ m}) = 0.0254 \text{ m}$$

or

$$1 \text{ in.} = 25.4 \text{ mm} \tag{1.10}$$

Units of Force. Recalling that the U.S. customary unit of force (pound) is defined as the weight of the standard pound (of mass 0.4536 kg) at sea level and at the latitude of 45° (where $g = 9.807 \text{ m/s}^2$), and using Eq. (1.4), we write

$$\begin{aligned} W &= mg \\ 1 \text{ lb} &= (0.4536 \text{ kg})(9.807 \text{ m/s}^2) = 4.448 \text{ kg}\cdot\text{m/s}^2 \end{aligned}$$

or, recalling (1.5),

$$1 \text{ lb} = 4.448 \text{ N} \tag{1.11}$$

Units of Mass. The U.S. customary unit of mass (slug) is a derived unit. Thus, using (1.6), (1.8), and (1.11), we write

$$1 \text{ slug} = 1 \text{ lb}\cdot\text{s}^2/\text{ft} = \frac{1 \text{ lb}}{1 \text{ ft/s}^2} = \frac{4.448 \text{ N}}{0.3048 \text{ m/s}^2} = 14.59 \text{ N}\cdot\text{s}^2/\text{m}$$

and, recalling (1.5),

$$1 \text{ slug} = 1 \text{ lb}\cdot\text{s}^2/\text{ft} = 14.59 \text{ kg} \tag{1.12}$$

Although it cannot be used as a consistent unit of mass, we recall that the mass of the standard pound is, by definition,

$$1 \text{ pound mass} = 0.4536 \text{ kg} \tag{1.13}$$

This constant may be used to determine the *mass* in SI units (kilograms) of a body which has been characterized by its *weight* in U.S. customary units (pounds).

To convert into SI units a derived U.S. customary unit, one simply multiplies or divides the appropriate conversion factors. For example, to convert into SI units the moment of a force which was found to be $M = 47 \text{ lb}\cdot\text{in.}$, we use formulas (1.10) and (1.11) and write

$$\begin{aligned} M = 47 \text{ lb}\cdot\text{in.} &= 47(4.448 \text{ N})(25.4 \text{ mm}) \\ &= 5310 \text{ N}\cdot\text{mm} = 5.31 \text{ N}\cdot\text{m} \end{aligned}$$

The conversion factors given in this section may also be used to convert into U.S. customary units a numerical result obtained in SI units. For example, if the moment of a force was found to be $M = 40 \text{ N}\cdot\text{m}$, we write, following the procedure used in the last paragraph of Sec. 1.3,

$$M = 40 \text{ N}\cdot\text{m} = (40 \text{ N}\cdot\text{m})\left(\frac{1 \text{ lb}}{4.448 \text{ N}}\right)\left(\frac{1 \text{ ft}}{0.3048 \text{ m}}\right)$$

Carrying out the numerical computations and canceling out units which appear both in the numerator and the denominator, we obtain

$$M = 29.5 \text{ lb}\cdot\text{ft}$$

The U.S. customary units most frequently used in mechanics are listed in Table 1.3 with their SI equivalents.

1.5. Method of Problem Solution. The student should approach a problem in mechanics as he would approach an actual engineering situation. By drawing on his own experience and on his intuition, he will find it easier to understand and formulate the problem. Once the problem has been clearly stated, however, there is no place in its solution for the student's particular fancy. *The solution must be based on the six fundamental principles stated in Sec. 1.2 or on theorems derived from them.* Every step taken must be justified on that basis. Strict rules must be followed, which lead to the solution in an almost automatic fashion, leaving no room for the student's intuition or "feeling." After an answer has been obtained, it should be checked. Here again, the student may call upon his common sense and personal experience. If not completely satisfied with the result obtained, he should carefully check his formulation of the problem, the validity of the methods used for its solution, and the accuracy of his computations.

The *statement* of a problem should be clear and precise. It should contain the given data and indicate what information is required. A neat drawing showing all quantities involved should be included. Separate diagrams should be drawn for all bodies involved, indicating clearly the forces acting on each body. These diagrams are known as *free-body diagrams* and are described in detail in Secs. 2.11 and 4.2.

Table 1.3 U.S. Customary Units and Their SI Equivalents

Quantity	U.S. Customary Unit	SI Equivalent
Acceleration	ft/s^2	0.3048 m/s^2
	in./s^2	0.0254 m/s^2
Area	ft^2	0.0929 m^2
	in^2	645.2 mm^2
Energy	ft·lb	1.356 J
Force	kip	4.448 kN
	lb	4.448 N
	oz	0.2780 N
Impulse	lb·s	4.448 N·s
Length	ft	0.3048 m
	in.	25.40 mm
	mi	1.609 km
Mass	oz mass	28.35 g
	lb mass	0.4536 kg
	slug	14.59 kg
	ton	907.2 kg
Moment of a force	lb·ft	1.356 N·m
	lb·in.	0.1130 N·m
Moment of inertia		
Of an area	in^4	0.4162×10^6 mm^4
Of a mass	lb·ft·s^2	1.356 kg·m^2
Momentum	lb·s	4.448 kg·m/s
Power	ft·lb/s	1.356 W
	hp	745.7 W
Pressure or stress	lb/ft^2	47.88 Pa
	lb/in^2 (psi)	6.895 kPa
Velocity	ft/s	0.3048 m/s
	in./s	0.0254 m/s
	mi/h (mph)	0.4470 m/s
	mi/h (mph)	1.609 km/h
Volume	ft^3	0.02832 m^3
	in^3	16.39 cm^3
Liquids	gal	3.785 L
	qt	0.9464 L
Work	ft·lb	1.356 J

The *fundamental principles* of mechanics listed in Sec. 1.2 *will be used to write equations* expressing the conditions of rest or motion of the bodies considered. Each equation should be clearly related to one of the free-body diagrams. The student will then proceed to solve the problem, observing strictly the usual rules of algebra and recording neatly the various steps taken.

After the answer has been obtained, it should be *carefully checked.* Mistakes in *reasoning* may often be detected by checking the units. For example, to determine the moment of a force of 50 N about a point 0.60 m from its line of action, we would have written (Sec. 3.12)

$$M = Fd = (50\ \text{N})(0.60\ \text{m}) = 30\ \text{N}\cdot\text{m}$$

The unit N·m obtained by multiplying newtons by meters is the correct unit for the moment of a force; if another unit had been obtained, we would have known that some mistake had been made.

Errors in *computation* will usually be found by substituting the numerical values obtained into an equation which has not yet been used and verifying that the equation is satisfied. The importance of correct computations in engineering cannot be overemphasized.

1.6. Numerical Accuracy. The accuracy of the solution of a problem depends upon two items: (1) the accuracy of the given data; (2) the accuracy of the computations performed.

The solution cannot be more accurate than the less accurate of these two items. For example, if the loading of a bridge is known to be 75,000 lb with a possible error of 100 lb either way, the relative error which measures the degree of accuracy of the data is

$$\frac{100\ \text{lb}}{75{,}000\ \text{lb}} = 0.0013 = 0.13\ \text{percent}$$

In computing the reaction at one of the bridge supports, it would then be meaningless to record it as 14,322 lb. The accuracy of the solution cannot be greater than 0.13 percent, no matter how accurate the computations are, and the possible error in the answer may be as large as $(0.13/100)(14{,}322\ \text{lb}) \approx 20\ \text{lb}$. The answer should be properly recorded as $14{,}320 \pm 20$ lb.

In engineering problems, the data are seldom known with an accuracy greater than 0.2 percent. It is therefore seldom justified to write the answers to such problems with an accuracy greater than 0.2 percent. A practical rule is to use 4 figures to record numbers beginning with a "1" and 3 figures in all other cases. Unless otherwise indicated, the data given in a problem should be assumed known with a comparable degree of accuracy. A force of 40 lb, for example, should be read 40.0 lb, and a force of 15 lb should be read 15.00 lb.

Pocket electronic calculators are widely used by practicing engineers and engineering students. The speed and accuracy of these calculators facilitates the numerical computations in the solution of many problems. However, the student should not record more significant figures than can be justified, merely because they are easily obtained. As noted above, an accuracy greater than 0.2 percent is seldom necessary or meaningful in the solution of practical engineering problems.

Statics of Particles

CHAPTER 2

2.1. Introduction. In this chapter we shall study the effect of forces acting on particles. First we shall learn how to replace two or more forces acting on a given particle by a single force having the same effect as the original forces. This single equivalent force is the *resultant* of the original forces acting on the particle. Later we shall derive the relations which exist among the various forces acting on a particle in a state of *equilibrium* and use these relations to determine some of the forces acting on the particle.

The use of the word "particle" does not imply that we shall restrict our study to that of small corpuscles. What it means is that the size and shape of the bodies under consideration will not significantly affect the solution of the problems treated in this chapter, and that all the forces acting on a given body will be assumed applied at the same point. Since such an assumption is verified in many practical applications, we shall be able to solve a number of engineering problems in this chapter.

The first part of the chapter is devoted to the study of forces contained in a single plane, and the second part to the analysis of forces in three-dimensional space.

FORCES IN A PLANE

2.2. Force on a Particle. Resultant of Two Forces. A force represents the action of one body on another and is generally characterized by its *point of application*, its *magnitude*, and its *direction*. Forces acting on a given particle, however, have the same point of application. Each force considered in this chapter will thus be completely defined by its magnitude and direction.

The magnitude of a force is characterized by a certain number of units. As indicated in Chap. 1, the SI units used by engineers to measure the magnitude of a force are the newton (N) and its multiple the kilonewton (kN), equal to 1000 N, while the U.S. customary units used for the same purpose are the pound (lb) and its multiple the kilopound (kip, or k), equal to 1000 lb. The direction of a force is defined by the *line of action* and the *sense* of the force. The line of action is the infinite straight line along which the force acts; it is characterized by the angle it forms with some fixed axis (Fig. 2.1). The force itself is represented by a segment of that line; through the use of an appropriate scale, the length of this segment may be chosen to represent the magnitude of the force. Finally, the sense of the force should be indicated by an arrowhead. It is important in defining a force to indicate its sense. Two forces, such as those shown in Fig. 2.1*a* and *b*, having the same magnitude and the same line of action but different sense, will have directly opposite effects on a particle.

Experimental evidence shows that two forces **P** and **Q** acting on a particle *A* (Fig. 2.2*a*) may be replaced by a single force **R** which has the same effect on the particle (Fig. 2.2*c*). This force is called the *resultant* of the forces **P** and **Q** and may be obtained, as shown in Fig. 2.2*b*, by constructing a parallelogram, using **P** and **Q** as two sides of the parallelogram. *The diagonal that passes through A represents the resultant.* This is known as the *parallelogram law* for the addition of two forces. This law is based on experimental evidence; it cannot be proved or derived mathematically.

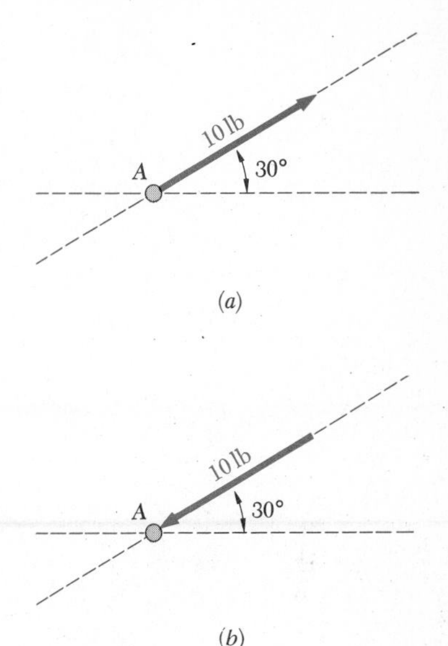

Fig. 2.1

2.3. Vectors.

It appears from the above that forces do not obey the rules of addition defined in ordinary arithmetic or algebra. For example, two forces acting at a right angle to each other, one of 4 lb and the other of 3 lb, add up to a force of 5 lb, *not* to a force of 7 lb. Forces are not the only expressions which follow the parallelogram law of addition. As we shall see later, *displacements, velocities, accelerations, momenta* are other examples of physical quantities possessing magnitude and direction and which are added according to the parallelogram law. All these quantities may be represented mathematically by *vectors,* while those physical quantities which do not have direction, such as *volume, mass,* or *energy,* are represented by ordinary numbers or *scalars.*

Vectors are defined as *mathematical expressions possessing magnitude and direction, which add according to the parallelogram law.* Vectors are represented by arrows in the illustrations and will be distinguished from scalar quantities in this text through the use of boldface type (**P**). In longhand writing, a vector may be characterized by drawing a short arrow above the letter used to represent it ($\vec{P}$), or by underlining the letter ($\underline{P}$). The last method may be preferred since it can also be used on a typewriter. The magnitude of a vector defines the length of the arrow used to represent the vector. In this text, italic type will be used to denote the magnitude of a vector. Thus, the magnitude of the vector **P** will be referred to as *P*.

A vector used to represent a force acting on a given particle has a well-defined point of application, namely, the particle itself. Such a vector is said to be a *fixed,* or *bound,* vector and cannot be moved without modifying the conditions of the problem. Other physical quantities, however, such as couples (see Chap. 3), are represented by vectors which may be freely moved in space; these vectors are called *free* vectors. Still other

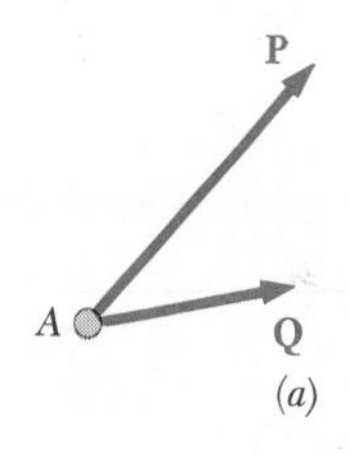

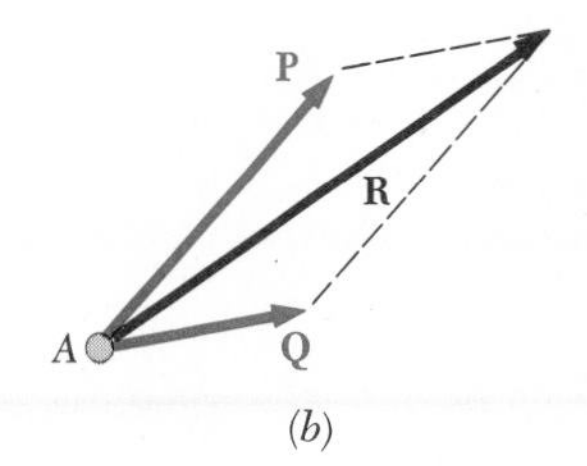

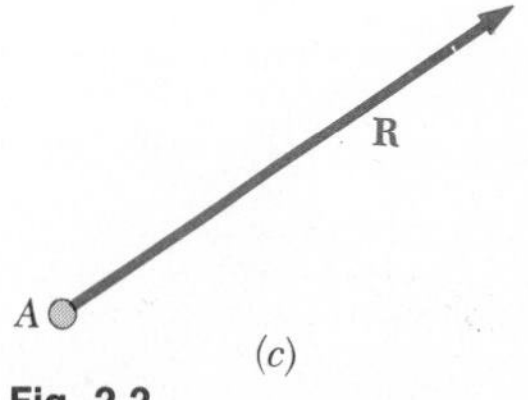

Fig. 2.2

physical quantities, such as forces acting on a rigid body (see Chap. 3), are represented by vectors which may be moved, or slid, along their line of action; they are known as *sliding* vectors.†

Two vectors which have the same magnitude and the same direction are said to be *equal,* whether or not they also have the same point of application (Fig. 2.4); equal vectors may be denoted by the same letter.

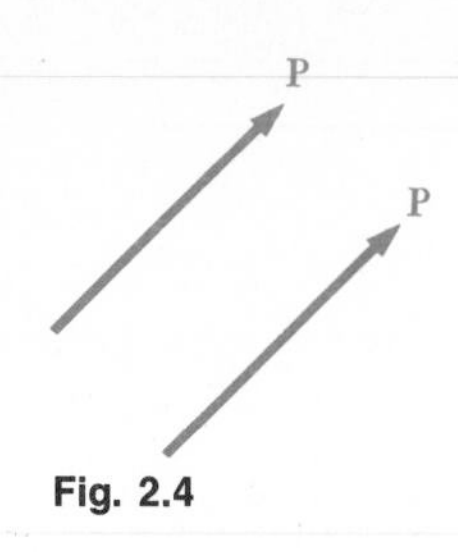

Fig. 2.4

The *negative vector* of a given vector **P** is defined as a vector having the same magnitude as **P** and a direction opposite to that of **P** (Fig. 2.5); the negative of the vector **P** is denoted by $-\mathbf{P}$. The vectors **P** and $-\mathbf{P}$ are commonly referred to as *equal and opposite* vectors. Clearly, we have

$$\mathbf{P} + (-\mathbf{P}) = 0$$

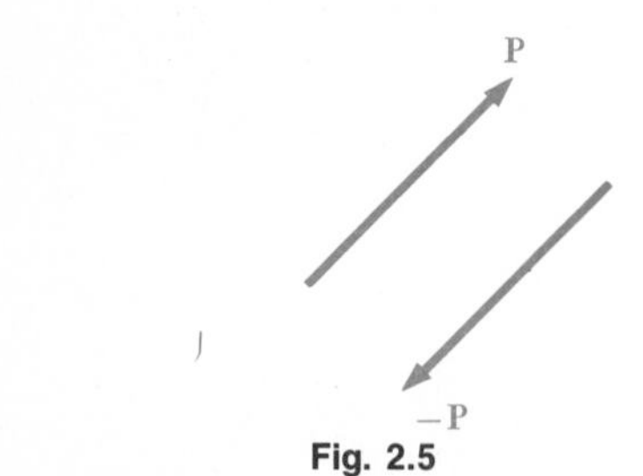

Fig. 2.5

2.4. Addition of Vectors. We saw in the preceding section that, by definition, vectors add according to the parallelogram law. Thus the sum of two vectors **P** and **Q** is obtained by attaching the two vectors to the same point A and constructing a parallelogram, using **P** and **Q** as two sides of the parallelogram (Fig. 2.6). The diagonal that passes through A represents the sum of the vectors **P** and **Q**, and this sum is denoted by $\mathbf{P} + \mathbf{Q}$. The fact that the sign + is used to denote both vector and scalar addition should not cause any confusion if vector and scalar quantities are always carefully distinguished. Thus, we should note that the magnitude of the vector $\mathbf{P} + \mathbf{Q}$ is *not,* in general, equal to the sum $P + Q$ of the magnitudes of the vectors **P** and **Q**.

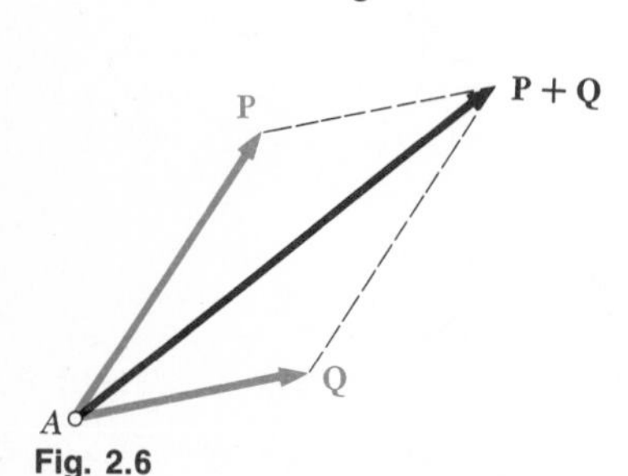

Fig. 2.6

Since the parallelogram constructed on the vectors **P** and **Q** does not depend upon the order in which **P** and **Q** are selected, we conclude that the addition of two vectors is *commutative,* and we write

$$\mathbf{P} + \mathbf{Q} = \mathbf{Q} + \mathbf{P} \tag{2.1}$$

† Some expressions have magnitude and direction, but do not add according to the parallelogram law. While these expressions may be represented by arrows, they *cannot* be considered as vectors.

A group of such expressions are the finite rotations of a rigid body. Place a closed book on a table in front of you, so that it lies in the usual fashion, with its front cover up and its binding to the left. Now rotate it through 180° about an axis parallel to the binding (Fig. 2.3*a*); this rotation may be represented by an arrow of length equal to 180 units and oriented as shown. Picking up the book as it lies in its new position, rotate it now through 180° about a horizontal

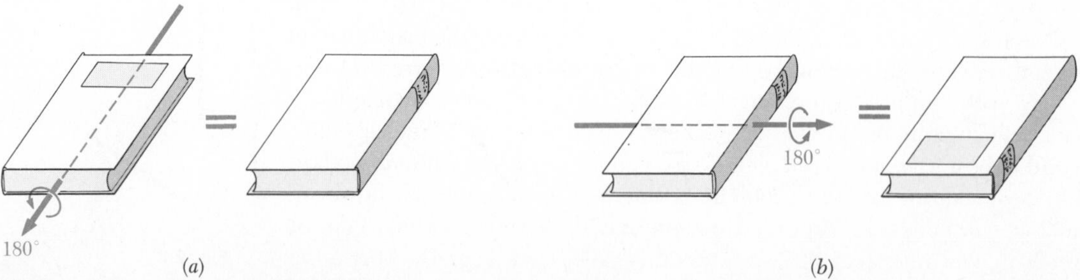

Fig. 2.3 Finite rotations of a rigid body.

From the parallelogram law, we can derive an alternate method for determining the sum of two vectors. This method, known as the *triangle rule,* is derived as follows: Consider Fig. 2.6, where the sum of the vectors **P** and **Q** has been determined by the parallelogram law. Since the side of the parallelogram opposite **Q** is equal to **Q** in magnitude and direction, we could draw only half of the parallelogram (Fig. 2.7*a*). The sum of the two vectors may thus be found by *arranging* **P** *and* **Q** *in tip-to-tail fashion and then connecting the tail of* **P** *with the tip of* **Q**. In Fig. 2.7*b*, the other half of the parallelogram is considered, and the same result is obtained. This confirms the fact that vector addition is commutative.

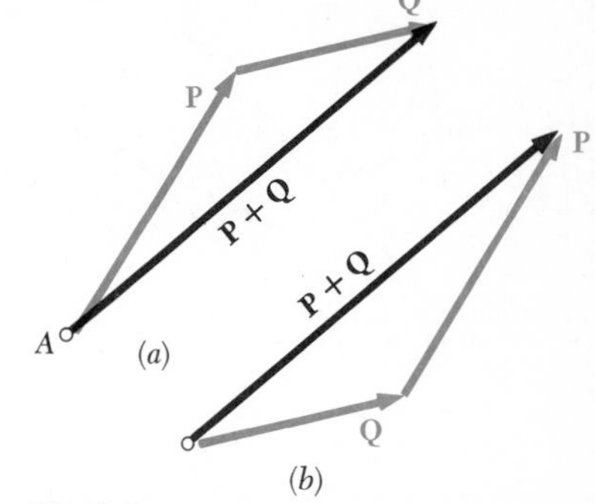

Fig. 2.7

The *subtraction* of a vector is defined as the addition of the corresponding negative vector. Thus, the vector **P** − **Q** representing the difference between the vectors **P** and **Q** is obtained by adding to **P** the negative vector −**Q** (Fig. 2.8). We write

$$\mathbf{P} - \mathbf{Q} = \mathbf{P} + (-\mathbf{Q}) \tag{2.2}$$

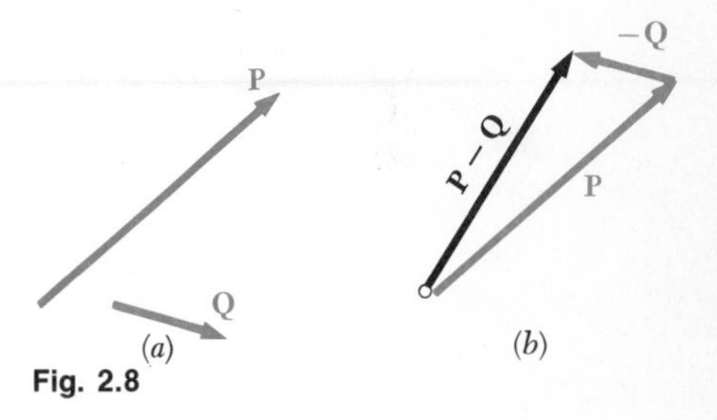

Fig. 2.8

Here again we should observe that, while the same sign is used to denote both vector and scalar subtraction, confusion will be avoided if care is taken to distinguish between vector and scalar quantities.

We shall now consider the *sum of three or more vectors.* The sum of three vectors **P**, **Q**, and **S** will, *by definition,* be obtained by first adding the vectors **P** and **Q**, and then adding the vector **S** to the vector **P** + **Q**. We thus write

$$\mathbf{P} + \mathbf{Q} + \mathbf{S} = (\mathbf{P} + \mathbf{Q}) + \mathbf{S} \tag{2.3}$$

Similarly, the sum of four vectors will be obtained by adding the fourth vector to the sum of the first three. It follows that the sum of any number of vectors may be obtained by applying repeatedly the parallelogram law to successive pairs of vectors until all the given vectors are replaced by a single vector.

---

axis perpendicular to the binding (Fig. 2.3*b*); this second rotation may be represented by an arrow 180 units long and oriented as shown. But the book could have been placed in this final position through a single 180° rotation about a vertical axis (Fig. 2.3*c*). We conclude that the sum of the two 180° rotations represented by arrows directed respectively along the z and x axes is a 180° rotation represented by an arrow directed along the y axis (Fig. 2.3*d*). Clearly, the finite rotations of a rigid body *do not* obey the parallelogram law of addition; therefore they *cannot* be represented by vectors.

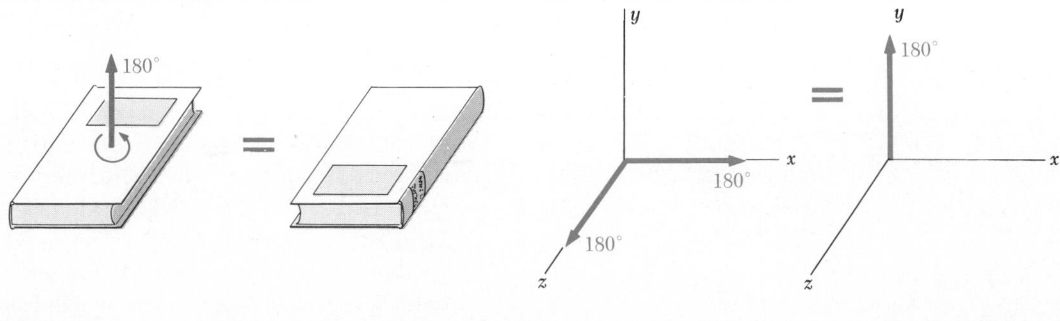

If the given vectors are *coplanar,* i.e., if they are contained in the same plane, their sum may be easily obtained graphically. In that case, the repeated application of the triangle rule will be preferred to the application of the parallelogram law. In Fig. 2.9 the sum of three vectors **P**, **Q**, and **S** was obtained in that manner. The triangle rule was first applied to obtain the sum **P** + **Q** of the vectors **P** and **Q**; it was applied again to obtain the sum of the vectors **P** + **Q** and **S**. The determination of the vector **P** + **Q**, however, could have been omitted and the sum of the three vectors could have been obtained directly, as shown in Fig. 2.10, by *arranging the given vectors in tip-to-tail fashion and connecting the tail of the first vector with the tip of the last one.* This is known as the *polygon rule* for the addition of vectors.

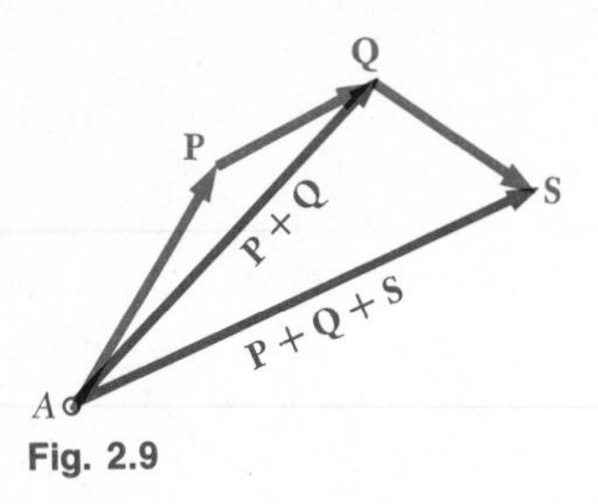

Fig. 2.9

We observe that the result obtained would have been unchanged if, as shown in Fig. 2.11, the vectors **Q** and **S** had been replaced by their sum **Q** + **S**. We may thus write

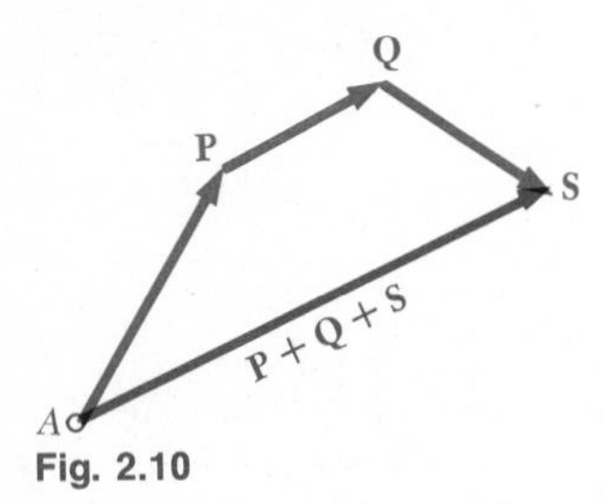

Fig. 2.10

$$\mathbf{P} + \mathbf{Q} + \mathbf{S} = (\mathbf{P} + \mathbf{Q}) + \mathbf{S} = \mathbf{P} + (\mathbf{Q} + \mathbf{S}) \tag{2.4}$$

which expresses the fact that vector addition is *associative.* Recalling that vector addition has also been shown, in the case of two vectors, to be commutative, we write

$$\begin{aligned} \mathbf{P} + \mathbf{Q} + \mathbf{S} &= (\mathbf{P} + \mathbf{Q}) + \mathbf{S} = \mathbf{S} + (\mathbf{P} + \mathbf{Q}) \\ &= \mathbf{S} + (\mathbf{Q} + \mathbf{P}) = \mathbf{S} + \mathbf{Q} + \mathbf{P} \end{aligned} \tag{2.5}$$

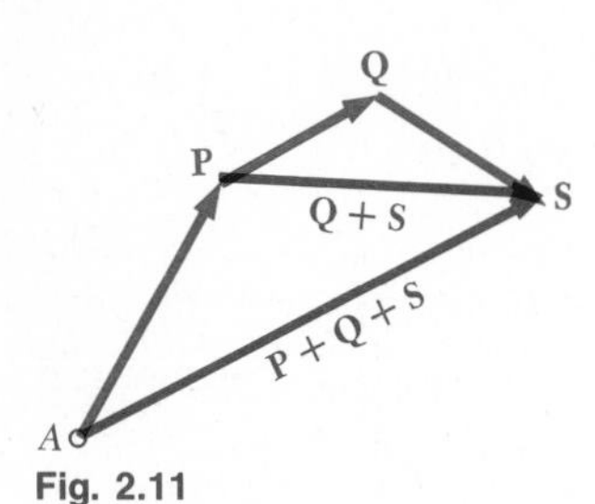

Fig. 2.11

This expression, as well as others which may be obtained in the same way, shows that the order in which several vectors are added together is immaterial (Fig. 2.12).

Product of a Scalar and a Vector. Since it is convenient to denote the sum **P** + **P** by 2**P**, the sum **P** + **P** + **P** by 3**P**, and, in general, the sum of n equal vectors **P** by the product n**P**, we shall define the product n**P** of a positive integer n and a vector **P** as a vector having the same direction as **P** and the magnitude nP. Extending this definition to include all scalars, and recalling the definition of a negative vector given in Sec. 2.3, we define the product k**P** of a scalar k and a vector **P** as a vector having the same direction as **P** (if k is positive), or a direction opposite to that of **P** (if k is negative), and a magnitude equal to the product of P and of the absolute value of k (Fig. 2.13).

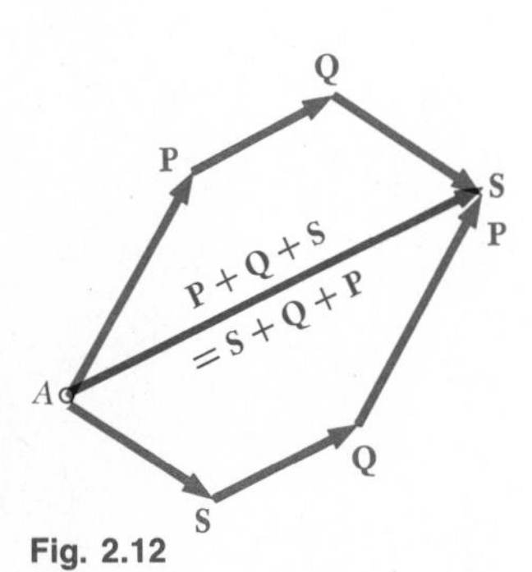

Fig. 2.12

2.5. Resultant of Several Concurrent Forces. Consider a particle A acted upon by several coplanar forces, i.e., by several forces contained in the same plane (Fig. 2.14*a*). Since the forces considered here all pass through A, they are also said to be *concurrent.* The vectors representing the forces acting on A may be added by the polygon rule (Fig. 2.14*b*). Since the use of the polygon rule is equivalent to the repeated application of the parallelogram law, the vector **R** thus obtained represents the resultant of the given concurrent forces, i.e., the single force which has

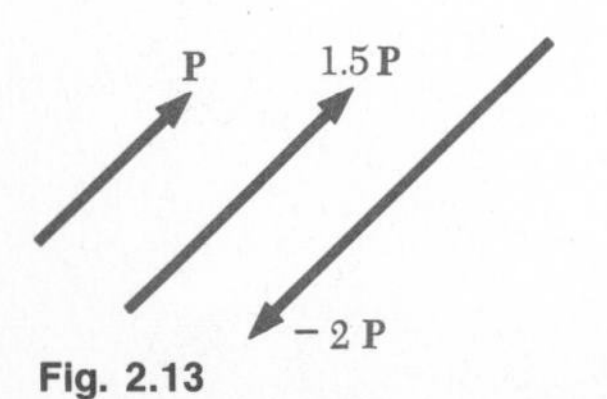

Fig. 2.13

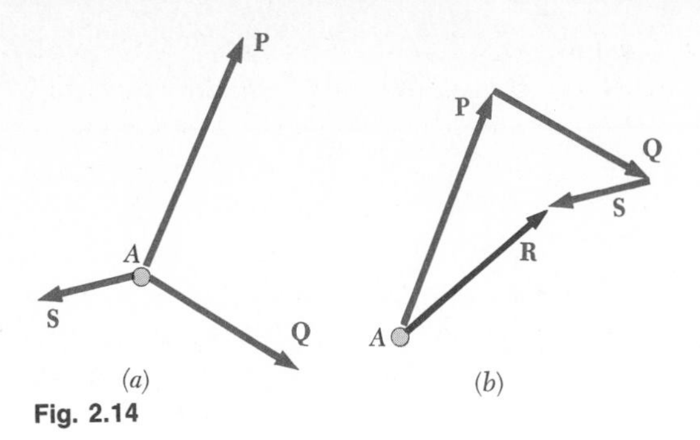

Fig. 2.14

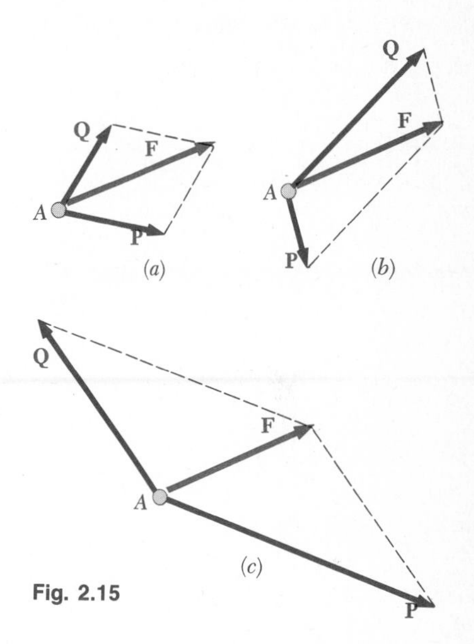

Fig. 2.15

the same effect on the particle A as the given forces. As indicated above, the order in which the vectors **P**, **Q**, and **S** representing the given forces are added together is immaterial.

2.6. Resolution of a Force into Components. We have seen that two or more forces acting on a particle may be replaced by a single force which has the same effect on the particle. Conversely, a single force **F** acting on a particle may be replaced by two or more forces which, together, have the same effect on the particle. These forces are called the *components* of the original force **F**, and the process of substituting them for **F** is called *resolving the force* **F** *into components*.

Clearly, for each force **F** there exist an infinite number of possible sets of components. Sets of *two components* **P** *and* **Q** are the most important as far as practical applications are concerned. But, even then, the number of ways in which a given force **F** may be resolved into two components is unlimited (Fig. 2.15). Two cases are of particular interest:

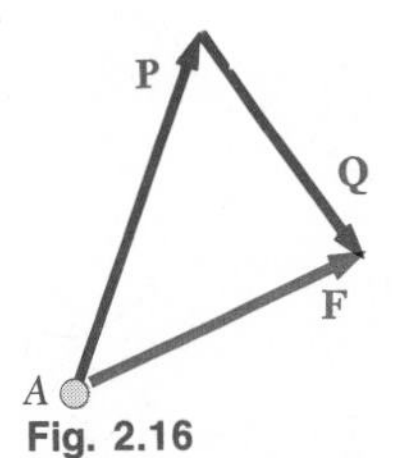

Fig. 2.16

1. *One of the Two Components,* **P**, *Is Known.* The second component, **Q**, is obtained by applying the triangle rule and joining the tip of **P** to the tip of **F** (Fig. 2.16); the magnitude and direction of **Q** are determined graphically or by trigonometry. Once **Q** has been determined, both components **P** and **Q** should be applied at A.
2. *The Line of Action of Each Component Is Known.* The magnitude and sense of the components are obtained by applying the parallelogram law and drawing lines, through the tip of **F**, parallel to the given lines of action (Fig. 2.17). This process leads to two well-defined components, **P** and **Q**, which may be determined graphically, or trigonometrically, by applying the law of sines.

Many other cases may be encountered; for example, the direction of one component may be known while the magnitude of the other component is to be as small as possible (see Sample Prob. 2.2). In all cases the appropriate triangle or parallelogram is drawn, which satisfies the given conditions.

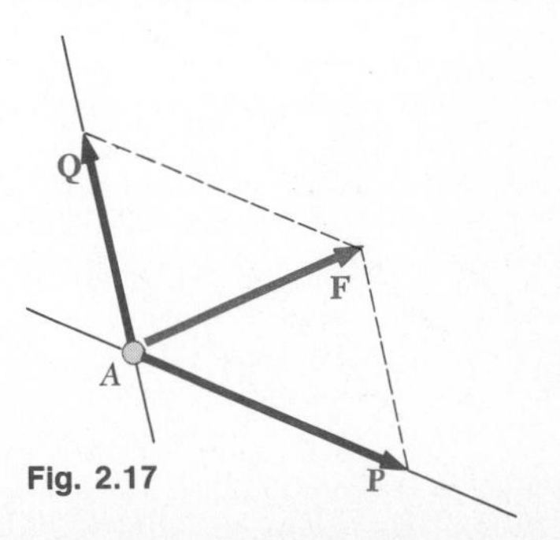

Fig. 2.17

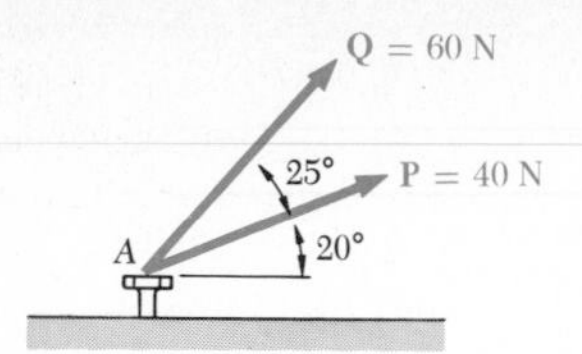

SAMPLE PROBLEM 2.1

The two forces **P** and **Q** act on a bolt A. Determine their resultant.

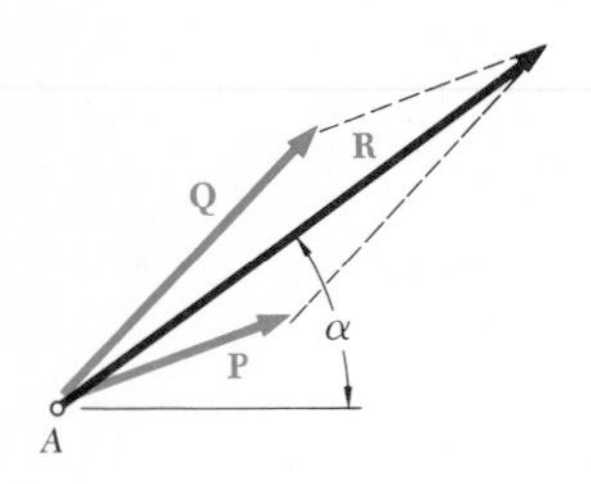

Graphical Solution. A parallelogram with sides equal to **P** and **Q** is drawn to scale. The magnitude and direction of the resultant are measured and found to be

$$R = 98 \text{ N} \qquad \alpha = 35° \qquad \mathbf{R} = 98 \text{ N} \measuredangle 35° \quad ◀$$

The triangle rule may also be used. Forces **P** and **Q** are drawn in tip-to-tail fashion. Again the magnitude and direction of the resultant are measured.

$$R = 98 \text{ N} \qquad \alpha = 35° \qquad \mathbf{R} = 98 \text{ N} \measuredangle 35° \quad ◀$$

Trigonometric Solution. The triangle rule is again used; two sides and the included angle are known. We apply the law of cosines.

$$R^2 = P^2 + Q^2 - 2PQ \cos B$$
$$R^2 = (40 \text{ N})^2 + (60 \text{ N})^2 - 2(40 \text{ N})(60 \text{ N}) \cos 155°$$
$$R = 97.73 \text{ N}$$

Now, applying the law of sines, we write

$$\frac{\sin A}{Q} = \frac{\sin B}{R} \qquad \frac{\sin A}{60 \text{ N}} = \frac{\sin 155°}{97.73 \text{ N}} \tag{1}$$

Solving Eq. (1) for $\sin A$, we have

$$\sin A = \frac{(60 \text{ N}) \sin 155°}{97.73 \text{ N}}$$

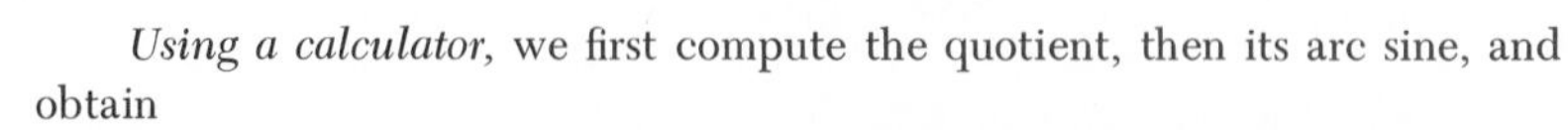

Using a calculator, we first compute the quotient, then its arc sine, and obtain

$$A = 15.04° \qquad \alpha = 20° + A = 35.04°$$

We use 3 significant figures to record the answer (cf. Sec. 1.6):

$$\mathbf{R} = 97.7 \text{ N} \measuredangle 35.0° \quad ◀$$

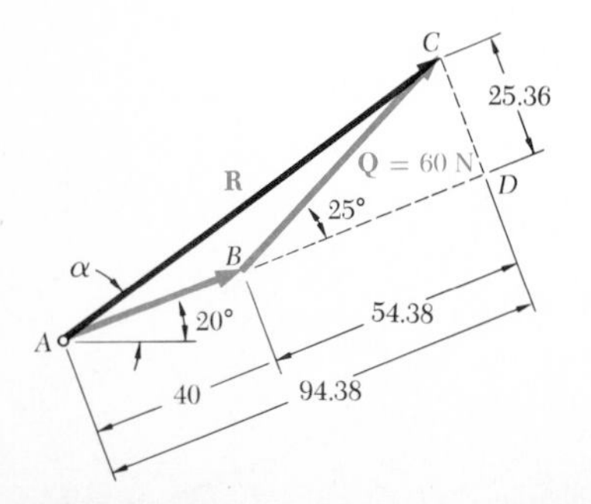

Alternate Trigonometric Solution. We construct the right triangle BCD and compute

$$CD = (60 \text{ N}) \sin 25° = 25.36 \text{ N}$$
$$BD = (60 \text{ N}) \cos 25° = 54.38 \text{ N}$$

Then, using triangle ACD, we obtain

$$\tan A = \frac{25.36 \text{ N}}{94.38 \text{ N}} \qquad A = 15.04°$$
$$R = \frac{25.36}{\sin A} \qquad R = 97.73 \text{ N}$$

Again, $\alpha = 20° + A = 35.04°$ $\quad \mathbf{R} = 97.7 \text{ N} \measuredangle 35.0° \quad ◀$

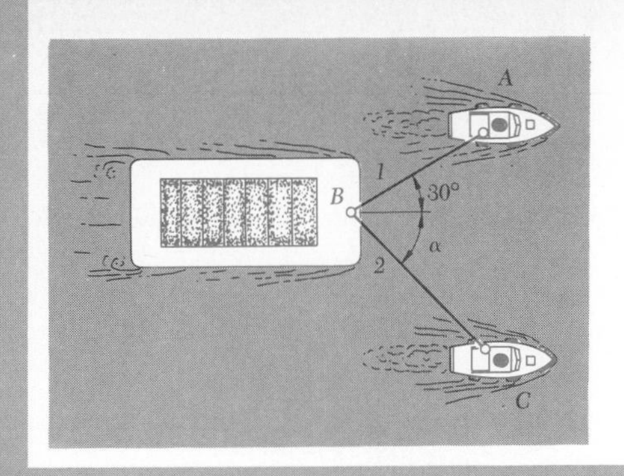

SAMPLE PROBLEM 2.2

A barge is pulled by two tugboats. If the resultant of the forces exerted by the tugboats is a 5000-lb force directed along the axis of the barge, determine (*a*) the tension in each of the ropes, knowing that $\alpha = 45°$, (*b*) the value of α such that the tension in rope *2* is minimum.

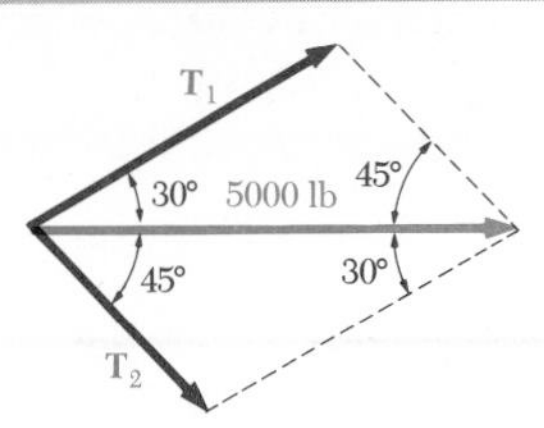

***a.* Tension for $\alpha = 45°$.** *Graphical Solution.* The parallelogram law is used; the diagonal (resultant) is known to be equal to 5000 lb and to be directed to the right. The sides are drawn parallel to the ropes. If the drawing is done to scale, we measure

$$T_1 = 3700 \text{ lb} \qquad T_2 = 2600 \text{ lb} \blacktriangleleft$$

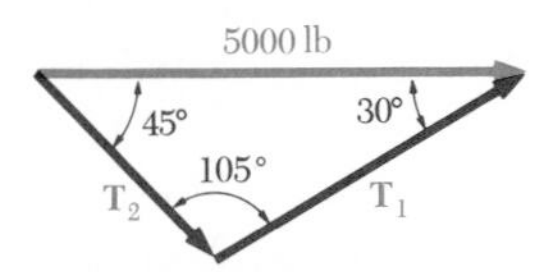

Trigonometric Solution. The triangle rule may be used. We note that the triangle shown represents half of the parallelogram shown above. Using the law of sines, we write

$$\frac{T_1}{\sin 45°} = \frac{T_2}{\sin 30°} = \frac{5000 \text{ lb}}{\sin 105°}$$

With a calculator, we first compute and store the value of the last quotient. Multiplying this value successively by sin 45° and sin 30°, we obtain

$$T_1 = 3660 \text{ lb} \qquad T_2 = 2590 \text{ lb} \blacktriangleleft$$

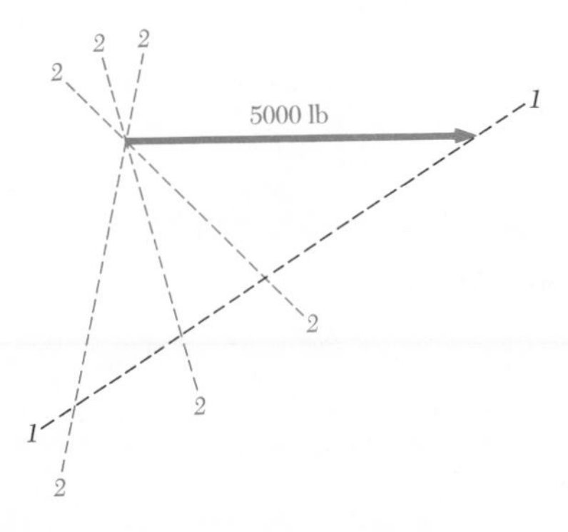

***b.* Value of α for Minimum T_2.** To determine the value of α such that the tension in rope *2* is minimum, the triangle rule is again used. In the sketch shown, line *1-1* is the known direction of $\mathbf{T}_1$. Several possible directions of $\mathbf{T}_2$ are shown by the lines *2-2*. We note that the minimum value of T_2 occurs when $\mathbf{T}_1$ and $\mathbf{T}_2$ are perpendicular. The minimum value of T_2 is

$$T_2 = (5000 \text{ lb}) \sin 30° = 2500 \text{ lb}$$

Corresponding values of T_1 and α are

$$T_1 = (5000 \text{ lb}) \cos 30° = 4330 \text{ lb}$$
$$\alpha = 90° - 30° \qquad \alpha = 60° \blacktriangleleft$$

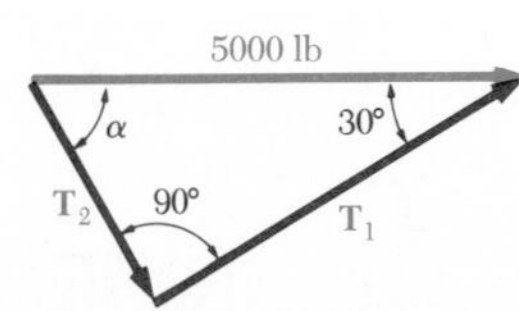

Problems†

2.1 and 2.2 Determine graphically the magnitude and direction of the resultant of the two forces shown, using (*a*) the parallelogram law, (*b*) the triangle rule.

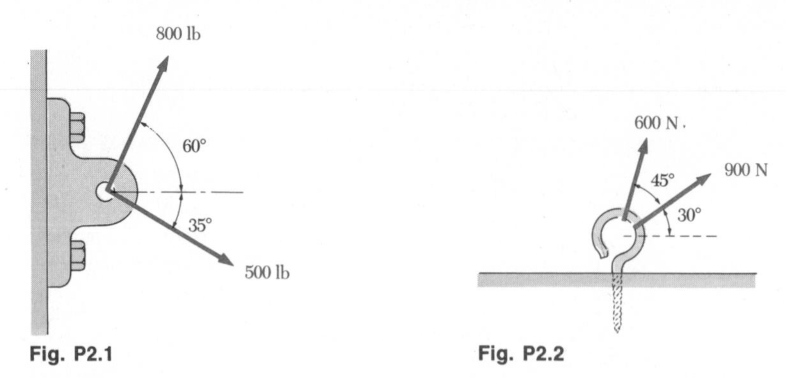

Fig. P2.1 **Fig. P2.2**

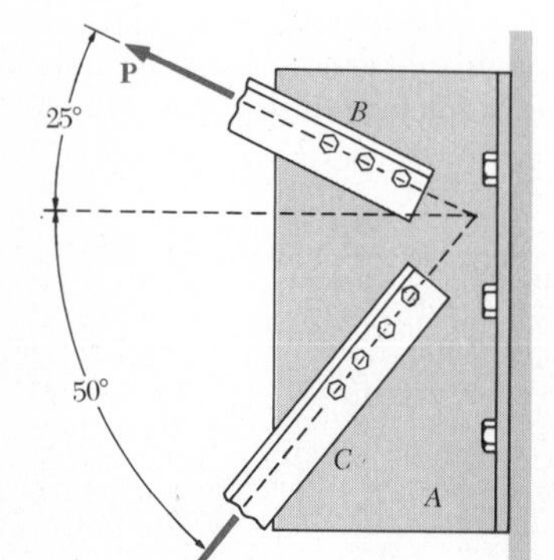

Fig. P2.3 and P2.4

2.3 Two structural members B and C are bolted to bracket A. Knowing that both members are in tension and that $P = 10$ kN and $Q = 15$ kN, determine graphically the magnitude and direction of the resultant force exerted on the bracket.

2.4 Two structural members B and C are bolted to bracket A. Knowing that both members are in tension and that $P = 6$ kips and $Q = 4$ kips, determine graphically the magnitude and direction of the resultant force exerted on the bracket.

2.5 Solve Prob. 2.3, assuming that $P = 15$ kN and $Q = 12$ kN.

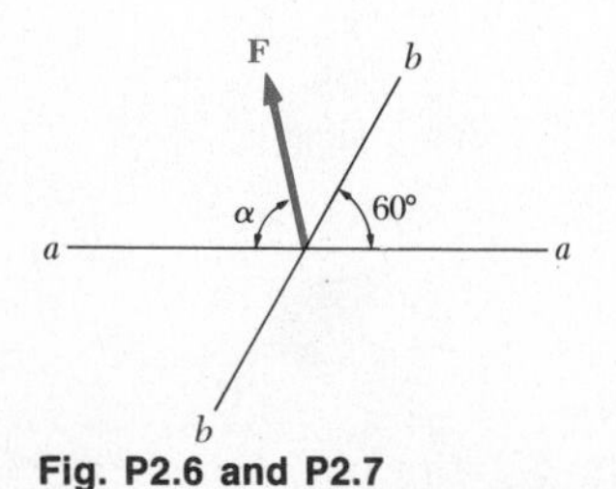

Fig. P2.6 and P2.7

2.6 The force **F** of magnitude 300 lb is to be resolved into two components along lines *a-a* and *b-b*. Determine by trigonometry the angle α, knowing that the component of **F** along line *a-a* is to be 240 lb.

2.7 The force **F** of magnitude 240 N is to be resolved into two components along lines *a-a* and *b-b*. Determine by trigonometry the angle α, knowing that the component of **F** along line *b-b* is to be 90 N.

†Answers to all even-numbered problems are given at the end of the book.

2.8 A trolley which moves along a horizontal beam is acted upon by two forces as shown. (*a*) Knowing that $\alpha = 25°$, determine by trigonometry the magnitude of the force **P** so that the resultant force exerted on the trolley is vertical. (*b*) What is the corresponding magnitude of the resultant?

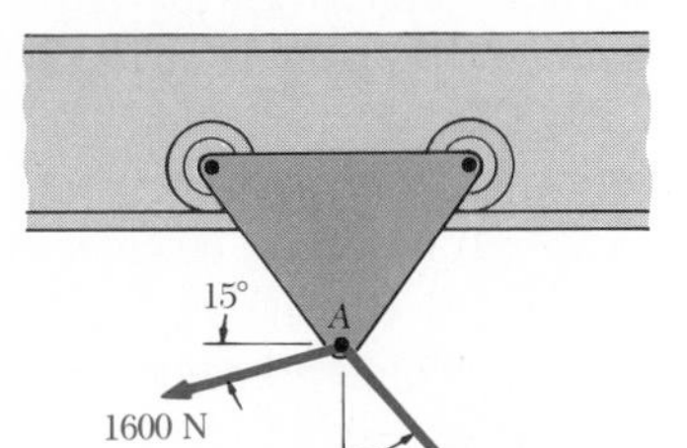

Fig. P2.8 and P2.13

2.9 A disabled automobile is pulled by means of two ropes as shown. The tension in AB is 500 lb and the angle α is 25°. Knowing that the resultant of the two forces applied at A is directed along the axis of the automobile, determine by trigonometry (*a*) the tension in rope AC, (*b*) the magnitude of the resultant of the two forces applied at A.

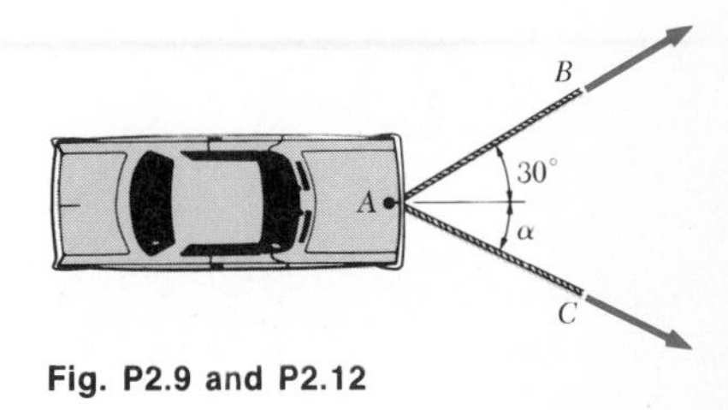

Fig. P2.9 and P2.12

2.10 Solve Prob. 2.9, assuming that the tension in rope AB is 4 kN and that $\alpha = 20°$.

2.11 Solve Prob. 2.8, assuming that $\alpha = 35°$.

2.12 A disabled automobile is pulled by means of two ropes as shown. Knowing that the tension in rope AB is 750 lb, determine by trigonometry the tension in rope AC and the value of α so that the resultant force exerted at A is a 1200-lb force directed along the axis of the automobile.

2.13 A trolley which moves along a horizontal beam is acted upon by two forces as shown. Determine by trigonometry the magnitude and direction of the force **P** so that the resultant is a vertical force of 2500 N.

2.14 Solve Prob. 2.2 by trigonometry.

2.15 Determine by trigonometry the magnitude and direction of the resultant of the two forces shown.

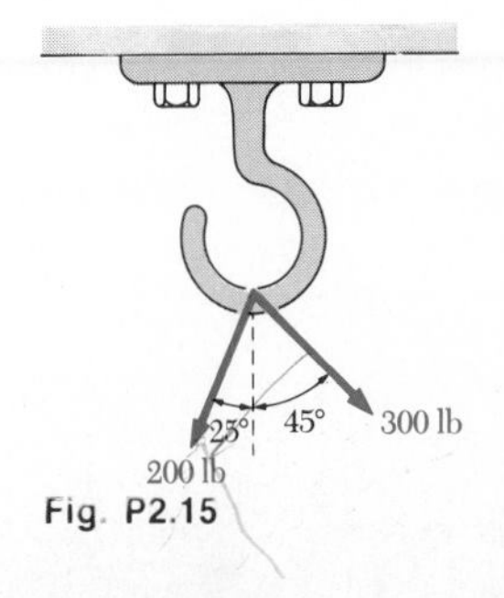

Fig. P2.15

2.16 If the resultant of the two forces exerted on the trolley of Prob. 2.8 is to be vertical, determine (*a*) the value of α for which the magnitude of **P** is minimum, (*b*) the corresponding magnitude of P.

2.7. Rectangular Components of a Force. Unit Vectors.†

In many problems it will be found desirable to resolve a force into two components which are perpendicular to each other. In Fig. 2.18, the force **F** has been resolved into a component $\mathbf{F}_x$ along the x axis and a component $\mathbf{F}_y$ along the y axis. The parallelogram drawn to obtain the two components is a *rectangle,* and $\mathbf{F}_x$ and $\mathbf{F}_y$ are called *rectangular components.*

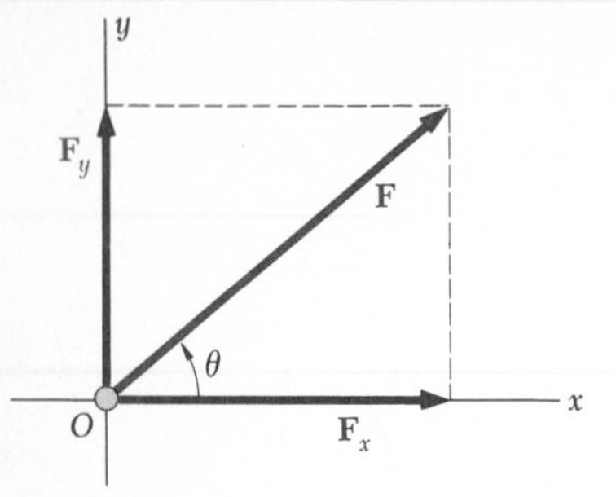

Fig. 2.18

The x and y axes are usually chosen horizontal and vertical, respectively, as in Fig. 2.18; they may, however, be chosen in any two perpendicular directions, as shown in Fig. 2.19. In determining the rectangular components of a force, the student should think of the construction lines shown in Figs. 2.18 and 2.19 as being *parallel* to the x and y axes, rather than *perpendicular* to these axes. This practice will help avoid mistakes in determining *oblique* components as in Sec. 2.6.

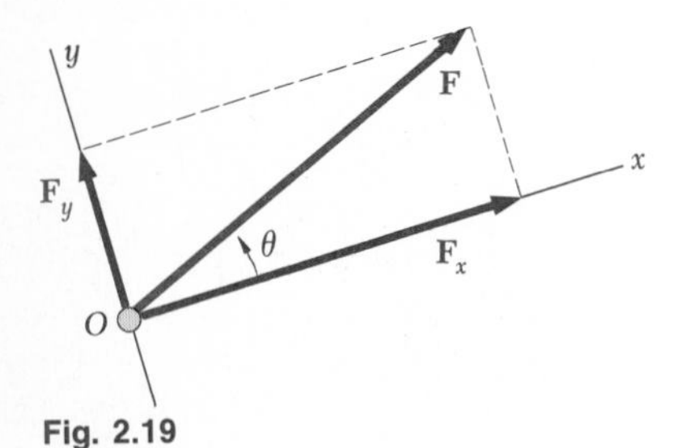

Fig. 2.19

We shall, at this point, introduce two vectors of magnitude 1, directed respectively along the positive x and y axes. These vectors are called *unit vectors* and are denoted by **i** and **j**, respectively (Fig. 2.20). Recalling the definition of the product of a scalar and a vector given in Sec. 2.4, we note that the rectangular components $\mathbf{F}_x$ and $\mathbf{F}_y$ of a force **F** may be obtained by multiplying respectively the unit vectors **i** and **j** by appropriate scalars (Fig. 2.21). We write

$$\mathbf{F}_x = F_x\mathbf{i} \qquad \mathbf{F}_y = F_y\mathbf{j} \tag{2.6}$$

and

$$\mathbf{F} = F_x\mathbf{i} + F_y\mathbf{j} \tag{2.7}$$

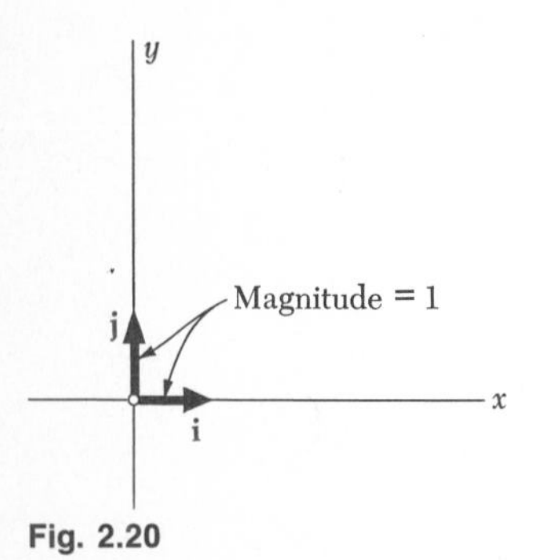

Fig. 2.20

While the scalars F_x and F_y may be positive or negative, depending upon the sense of $\mathbf{F}_x$ and of $\mathbf{F}_y$, their absolute values are respectively equal to the magnitudes of the component forces $\mathbf{F}_x$ and $\mathbf{F}_y$. The scalars F_x and F_y are called the *scalar components* of the force **F**, while the actual component forces $\mathbf{F}_x$ and $\mathbf{F}_y$ should be referred to as the *vector components* of **F**. However, when there exists no possibility of confusion, the vector as well as the scalar components of **F** may be referred to simply as the *components* of **F**. We note that the scalar component F_x is positive when the vector component $\mathbf{F}_x$ has the same sense as the unit vector **i** (i.e., the same sense as the positive x axis) and negative when $\mathbf{F}_x$ has the opposite sense. A similar conclusion may be drawn regarding the sign of the scalar component F_y.

Denoting by F the magnitude of the force **F** and by θ the angle between **F** and the x axis, measured counterclockwise from the positive x axis (Fig. 2.21), we may express the scalar components of **F** as follows:

$$F_x = F\cos\theta \qquad F_y = F\sin\theta \tag{2.8}$$

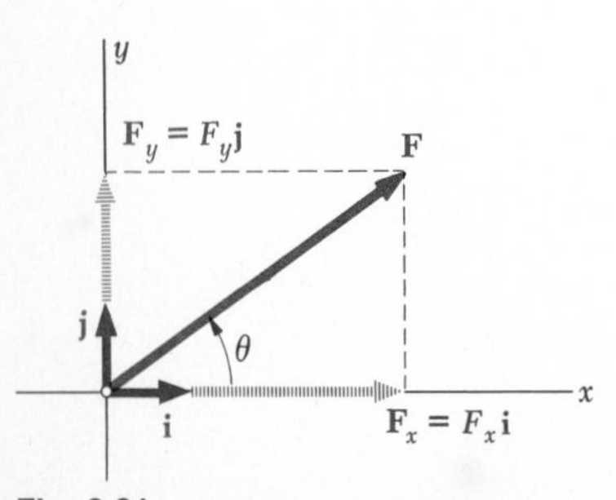

Fig. 2.21

† The properties established in Secs. 2.7 and 2.8 may be readily extended to the rectangular components of any vector quantity.

We note that the relations obtained hold for any value of the angle θ from $0°$ to $360°$, and that they define the signs as well as the absolute values of the scalar components F_x and F_y.

Example 1. A force of 800 N is exerted on a bolt A as shown in Fig. 2.22*a*. Determine the horizontal and vertical components of the force.

In order to obtain the correct sign for the scalar components F_x and F_y, the value $\theta = 180° - 35° = 145°$ should be substituted for θ in the relations (2.8). However, it will be found more practical to determine by inspection the signs of F_x and of F_y (Fig. 2.22*b*) and to use the trigonometric functions of the angle $\alpha = 35°$. We write therefore

$$F_x = -F \cos \alpha = -(800 \text{ N}) \cos 35° = -655 \text{ N}$$
$$F_y = +F \sin \alpha = +(800 \text{ N}) \sin 35° = +459 \text{ N}$$

The vector components of **F** are thus

$$\mathbf{F}_x = -(655 \text{ N})\mathbf{i} \qquad \mathbf{F}_y = +(459 \text{ N})\mathbf{j}$$

and we may write **F** in the form

$$\mathbf{F} = -(655 \text{ N})\mathbf{i} + (459 \text{ N})\mathbf{j}$$

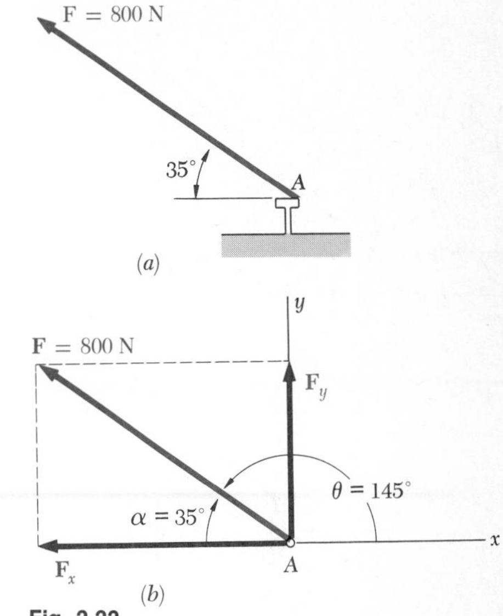

Fig. 2.22

Example 2. A man pulls with a force of 300 N on a rope attached to a building, as shown in Fig. 2.23*a*. What are the horizontal and vertical components of the force exerted by the rope at point A?

It is seen from Fig. 2.23*b* that

$$F_x = +(300 \text{ N}) \cos \alpha \qquad F_y = -(300 \text{ N}) \sin \alpha$$

Observing that $AB = 10$ m, we find from Fig. 2.23*a*

$$\cos \alpha = \frac{8 \text{ m}}{AB} = \frac{8 \text{ m}}{10 \text{ m}} = \frac{4}{5} \qquad \sin \alpha = \frac{6 \text{ m}}{AB} = \frac{6 \text{ m}}{10 \text{ m}} = \frac{3}{5}$$

We thus obtain

$$F_x = +(300 \text{ N})\tfrac{4}{5} = +240 \text{ N} \qquad F_y = -(300 \text{ N})\tfrac{3}{5} = -180 \text{ N}$$

and write

$$\mathbf{F} = (240 \text{ N})\mathbf{i} - (180 \text{ N})\mathbf{j}$$

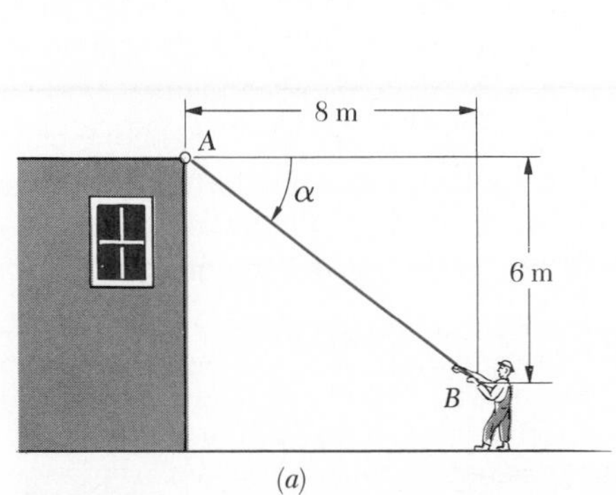

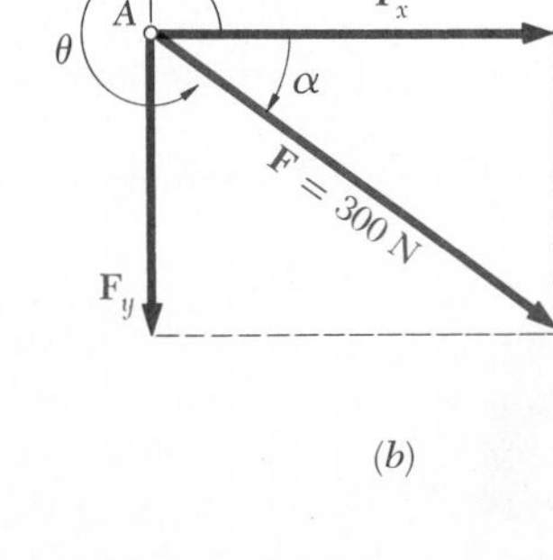

Fig. 2.23

When a force **F** is defined by its rectangular components F_x and F_y (see Fig. 2.21), the angle θ defining its direction can be obtained by writing

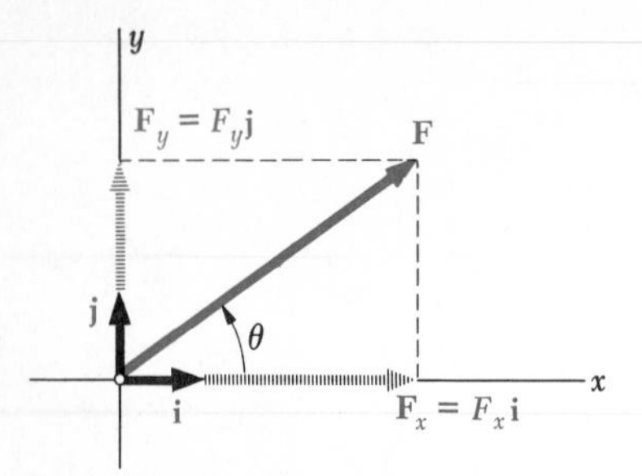

Fig. 2.21 ***(repeated)***

$$\tan \theta = \frac{F_y}{F_x} \tag{2.9}$$

The magnitude F of the force may be obtained by applying the Pythagorean theorem and writing

$$F = \sqrt{F_x^2 + F_y^2} \tag{2.10}$$

or by solving one of the formulas (2.8) for F.

Example 3. A force $\mathbf{F} = (700 \text{ lb})\mathbf{i} + (1500 \text{ lb})\mathbf{j}$ is applied to a bolt A. Determine the magnitude of the force and the angle θ it forms with the horizontal.

First we draw a diagram showing the two rectangular components of the force and the angle θ (Fig. 2.24). From Eq. (2.9), we write

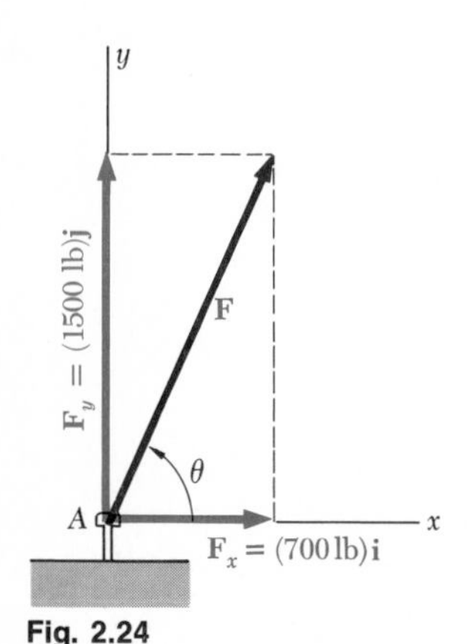

Fig. 2.24

$$\tan \theta = \frac{F_y}{F_x} = \frac{1500 \text{ lb}}{700 \text{ lb}}$$

Using a calculator,† we enter 1500 lb and divide by 700 lb; computing the arc tangent of the quotient, we obtain $\theta = 65.0°$. Solving the second of Eqs. (2.8) for F, we have

$$F = \frac{F_y}{\sin \theta} = \frac{1500 \text{ lb}}{\sin 65.0°} = 1655 \text{ lb}$$

The last calculation is facilitated if the value of F_y is stored when originally entered; it may then be recalled to be divided by $\sin \theta$.

2.8. Addition of Forces by Summing *x* and *y* Components. It was seen in Sec. 2.2 that forces should be added according to the parallelogram law. From this law, two other methods, more readily applicable to the *graphical* solution of problems, were derived in Secs. 2.4 and 2.5: the triangle rule for the addition of two forces and the polygon rule for the addition of three or more forces. It was also seen that the force triangle used to define the resultant of two forces could be used to obtain a *trigonometric* solution.

When three or more forces are to be added, no practical trigonometric solution may be obtained from the force polygon which defines the resultant of the forces. In this case, an *analytic* solution of the problem may be obtained by resolving each force into two rectangular components. Con-

†It is assumed that the calculator used has keys for the computation of trigonometric and inverse trigonometric functions. Some calculators also have keys for the direct conversion of rectangular coordinates into polar coordinates, and vice versa. Such calculators eliminate the need for the computation of trigonometric functions in Examples 1, 2, and 3 and in problems of the same type.

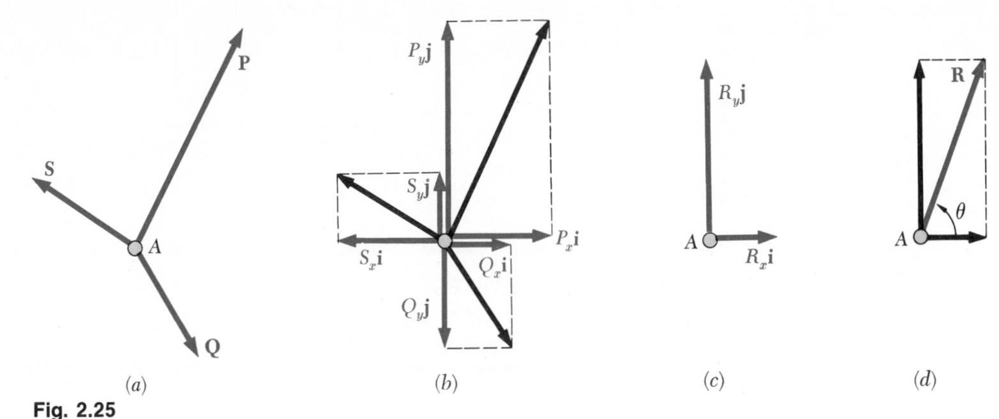

Fig. 2.25

sider, for instance, three forces **P**, **Q**, and **S** acting on a particle A (Fig. 2.25a). Their resultant **R** is defined by the relation

$$\mathbf{R} = \mathbf{P} + \mathbf{Q} + \mathbf{S} \tag{2.11}$$

Resolving each force into its rectangular components, we write

$$\begin{aligned} R_x\mathbf{i} + R_y\mathbf{j} &= P_x\mathbf{i} + P_y\mathbf{j} + Q_x\mathbf{i} + Q_y\mathbf{j} + S_x\mathbf{i} + S_y\mathbf{j} \\ &= (P_x + Q_x + S_x)\mathbf{i} + (P_y + Q_y + S_y)\mathbf{j} \end{aligned}$$

from which it follows that

$$R_x = P_x + Q_x + S_x \qquad R_y = P_y + Q_y + S_y \tag{2.12}$$

or, for short,

$$R_x = \Sigma F_x \qquad R_y = \Sigma F_y \tag{2.13}$$

We thus conclude that *the scalar components* R_x *and* R_y *of the resultant* **R** *of several forces acting on a particle are obtained by adding algebraically the corresponding scalar components of the given forces.*†

In practice, the determination of the resultant **R** is carried out in three steps as illustrated in Fig. 2.25. First, the given forces shown in Fig. 2.25a are resolved into their x and y components (Fig. 2.25b). Adding these components, we obtain the x and y components of **R** (Fig. 2.25c). Finally, the resultant $\mathbf{R} = R_x\mathbf{i} + R_y\mathbf{j}$ is determined by applying the parallelogram law (Fig. 2.25d). The procedure just described will be carried out most efficiently if the computations are arranged in a table. While it is the only practical analytic method for adding three or more forces, it is also often preferred to the trigonometric solution in the case of the addition of two forces.

†Clearly, this result also applies to the addition of other vector quantities, such as velocities, accelerations, or momenta.

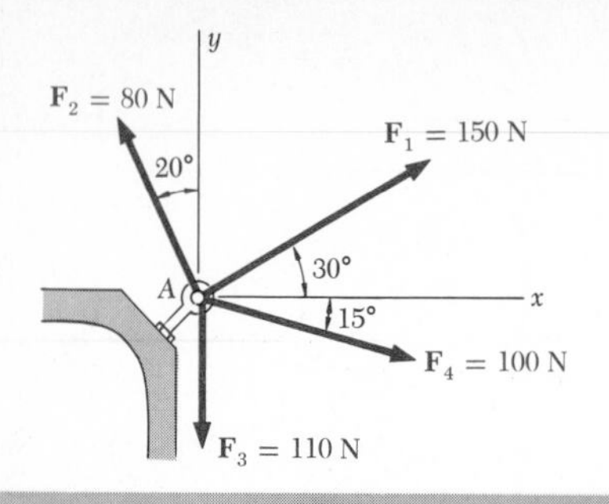

SAMPLE PROBLEM 2.3

Four forces act on bolt A as shown. Determine the resultant of the forces on the bolt.

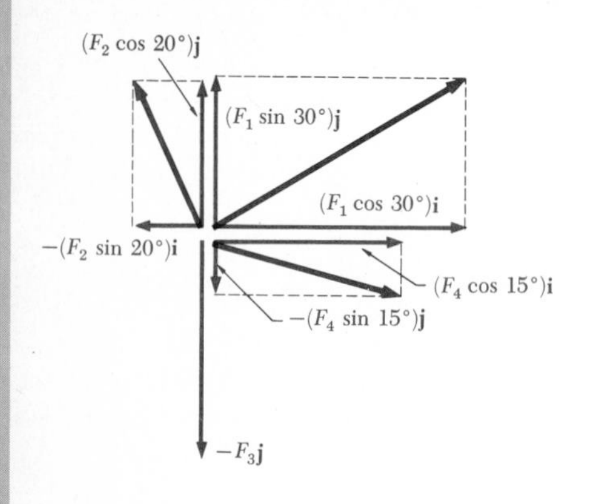

Solution. The x and y components of each force are determined by trigonometry as shown and are entered in the table below. According to the convention adopted in Sec. 2.7, the scalar number representing a force component is positive if the force component has the same sense as the corresponding coordinate axis. Thus, x components acting to the right and y components acting upward are represented by positive numbers.

Force	Magnitude, N	x Component, N	y Component, N
$\mathbf{F}_1$	150	+129.9	+75.0
$\mathbf{F}_2$	80	−27.4	+75.2
$\mathbf{F}_3$	110	0	−110.0
$\mathbf{F}_4$	100	+96.6	−25.9
		$R_x = +199.1$	$R_y = +14.3$

Thus, the resultant $\mathbf{R}$ of the four forces is

$$\mathbf{R} = R_x\mathbf{i} + R_y\mathbf{j} \qquad \mathbf{R} = (199.1 \text{ N})\mathbf{i} + (14.3 \text{ N})\mathbf{j} \blacktriangleleft$$

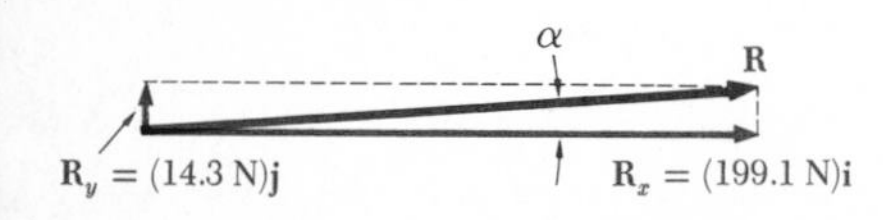

The magnitude and direction of the resultant may now be determined. From the triangle shown, we have

$$\tan \alpha = \frac{R_y}{R_x} = \frac{14.3 \text{ N}}{199.1 \text{ N}} \qquad \alpha = 4.1°$$

$$R = \frac{14.3 \text{ N}}{\sin \alpha} = 199.6 \text{ N} \qquad \mathbf{R} = 199.6 \text{ N} \measuredangle 4.1° \blacktriangleleft$$

With a calculator, the last computation may be facilitated if the value of R_y is stored when originally entered; it may then be recalled to be divided by $\sin \alpha$. (Also see the footnote on p. 26.)

Problems

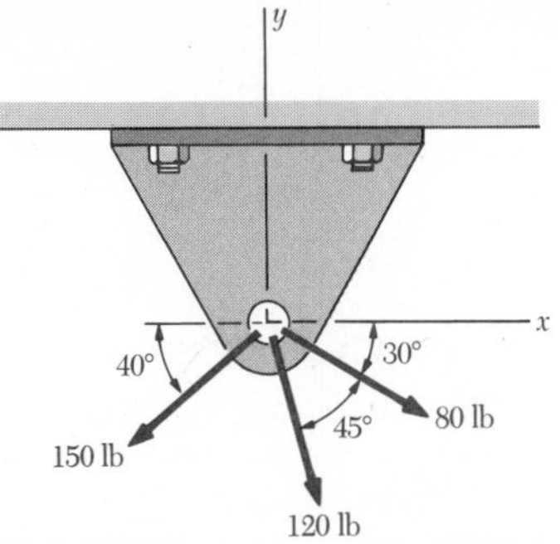

Fig. P2.17

Fig. P2.18

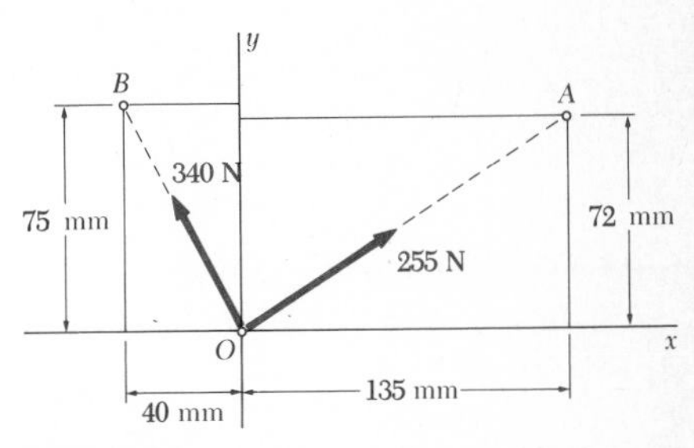

Fig. P2.19

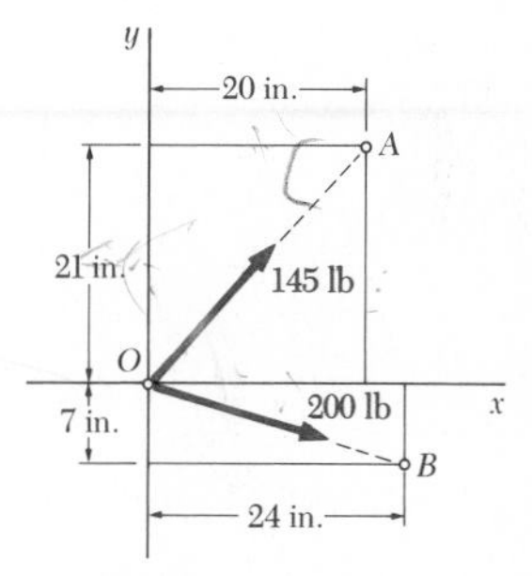

Fig. P2.20

2.17 through 2.20 Determine the x and y components of each of the forces shown.

2.21 Cable AC exerts on beam AB a force **P** directed along line AC. Knowing that **P** must have a 350-lb vertical component, determine (*a*) the magnitude of the force **P**, (*b*) its horizontal component.

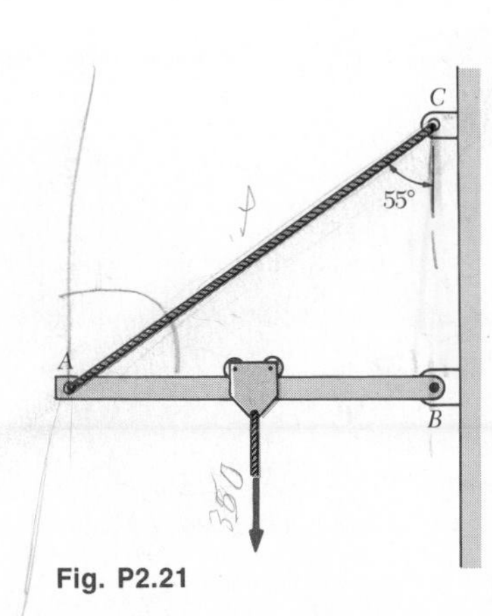

Fig. P2.21

2.22 The hydraulic cylinder BD exerts on member ABC a force **P** directed along line BD. Knowing that **P** must have a 750-N component perpendicular to member ABC, determine (*a*) the magnitude of the force **P**, (*b*) its component parallel to ABC.

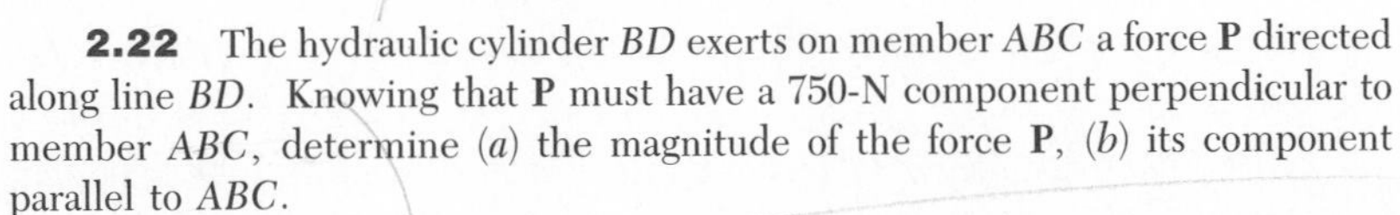

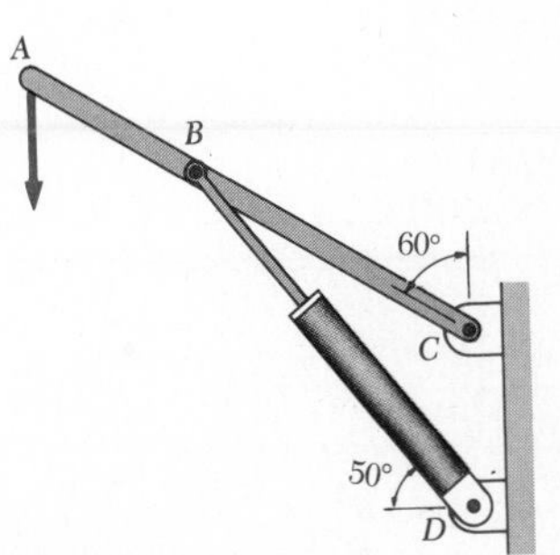

Fig. P2.22

2.23 The compression member BC exerts on the pin at C a 485-N force directed along line BC. Determine the horizontal and vertical components of that force.

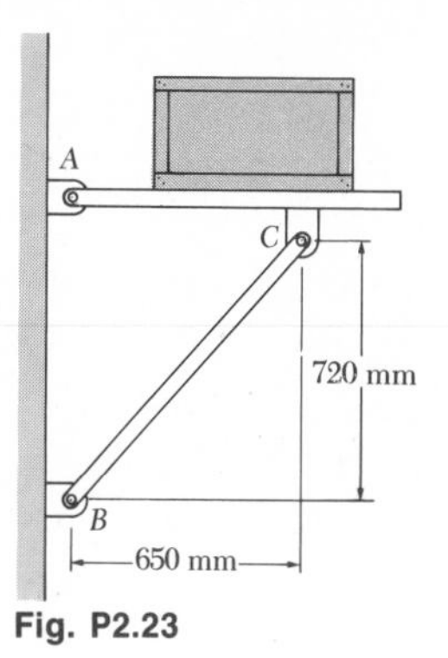

Fig. P2.23

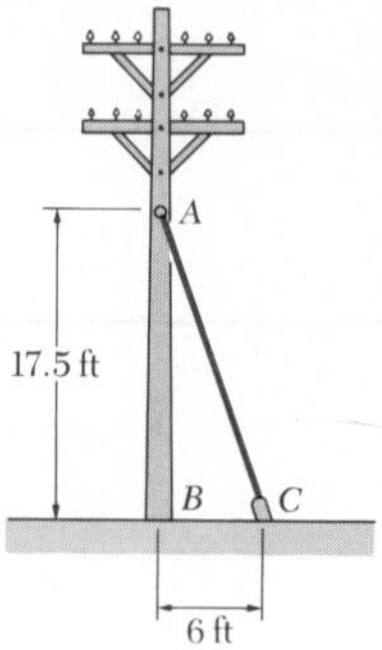

Fig. P2.24

2.24 The tension in the telephone-pole guy wire is 555 lb. Determine the horizontal and vertical components of the force exerted on the anchor at C.

2.25 Using x and y components, solve Prob. 2.1.

2.26 Using x and y components, solve Prob. 2.2.

2.27 Determine the resultant of the three forces of Prob. 2.18.

2.28 Determine the resultant of the three forces of Prob. 2.17.

2.29 Determine the resultant of the two forces of Prob. 2.19.

2.30 Determine the resultant of the two forces of Prob. 2.20.

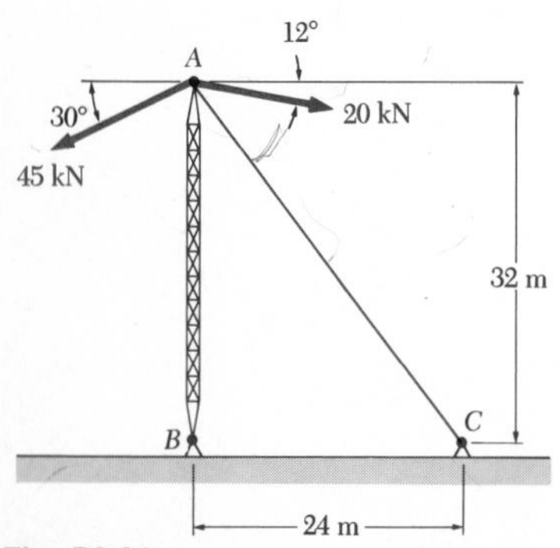

Fig. P2.31

2.31 Two cables which have known tensions are attached to the top of pylon AB. A third cable AC is used as a guy wire. Determine the tension in AC, knowing that the resultant of the forces exerted at A by the three cables must be vertical.

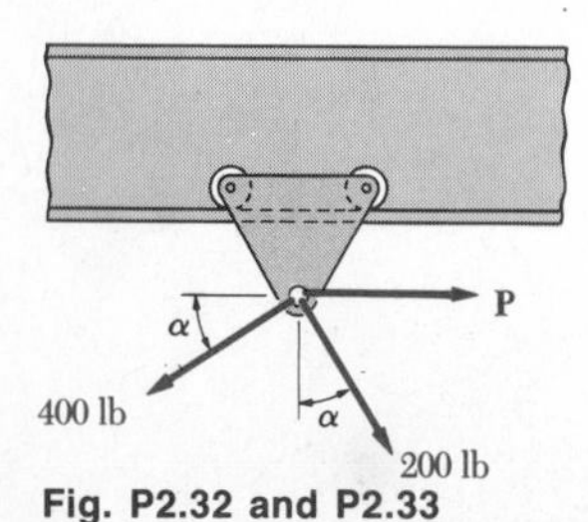

Fig. P2.32 and P2.33

2.32 A hoist trolley is subjected to the three forces shown. Knowing that $\alpha = 40°$, determine (*a*) the magnitude of the force **P** for which the resultant of the three forces is vertical, (*b*) the corresponding magnitude of the resultant.

2.33 A hoist trolley is subjected to the three forces shown. Knowing that $P = 250$ lb, determine (*a*) the value of the angle α for which the resultant of the three forces is vertical, (*b*) the corresponding magnitude of the resultant.

2.34 A collar which may slide on a vertical rod is subjected to the three forces shown. Determine (*a*) the value of the angle α for which the resultant of the three forces is horizontal, (*b*) the corresponding magnitude of the resultant.

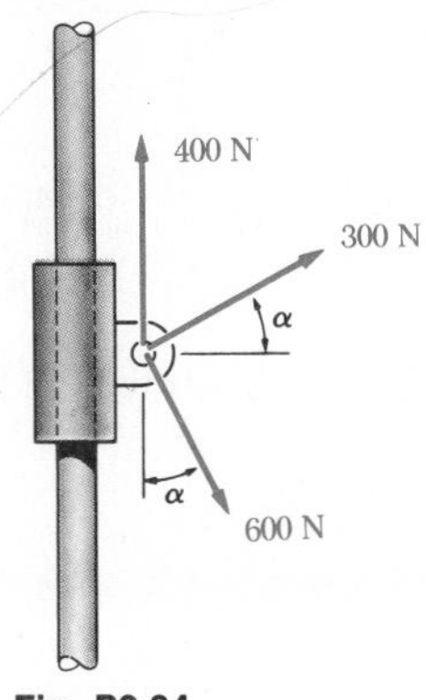

Fig. P2.34

2.9. Equilibrium of a Particle.

In the preceding sections, we discussed the methods for determining the resultant of several forces acting on a particle. Although it has not occurred in any of the problems considered so far, it is quite possible for the resultant to be zero. In such a case, the net effect of the given forces is zero, and the particle is said to be in equilibrium. We thus have the following definition: *When the resultant of all the forces acting on a particle is zero, the particle is in equilibrium.*

A particle which is acted upon by two forces will be in equilibrium if the two forces have the same magnitude, same line of action, and opposite sense. The resultant of the two forces is then zero. Such a case is shown in Fig. 2.26.

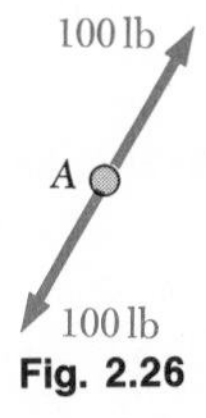

Fig. 2.26

Another case of equilibrium of a particle is represented in Fig. 2.27, where four forces are shown acting on *A*. In Fig. 2.28, the resultant of the given forces is determined by the polygon rule. Starting from point *O* with $\mathbf{F}_1$ and arranging the forces in tip-to-tail fashion, we find that the tip of $\mathbf{F}_4$ coincides with the starting point *O*. Thus the resultant **R** of the given system of forces is zero, and the particle is in equilibrium.

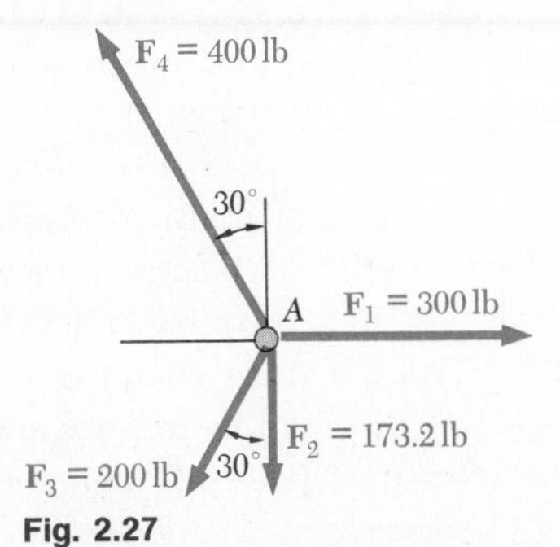

Fig. 2.27

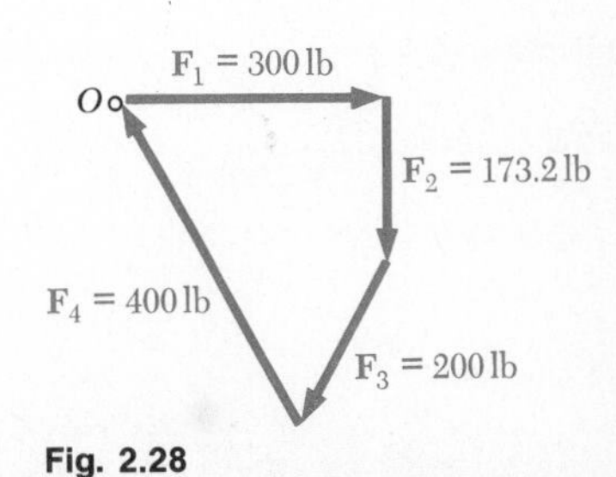

Fig. 2.28

The closed polygon drawn in Fig. 2.28 provides a *graphical* expression of the equilibrium of A. To express *algebraically* the conditions for the equilibrium of a particle, we write

$$\mathbf{R} = \Sigma \mathbf{F} = 0 \tag{2.14}$$

Resolving each force $\mathbf{F}$ into rectangular components, we have

$$\Sigma(F_x \mathbf{i} + F_y \mathbf{j}) = 0 \qquad \text{or} \qquad (\Sigma F_x)\mathbf{i} + (\Sigma F_y)\mathbf{j} = 0$$

We conclude that the necessary and sufficient conditions for the equilibrium of a particle are

$$\Sigma F_x = 0 \qquad \Sigma F_y = 0 \tag{2.15}$$

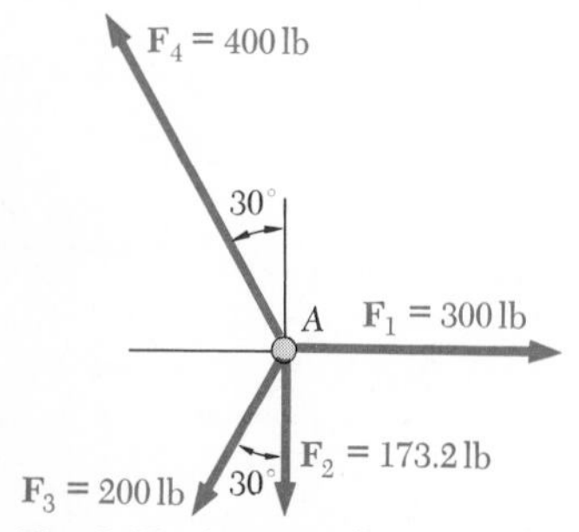

Fig. 2.27 ***(repeated)***

Returning to the particle shown in Fig. 2.27 we check that the equilibrium conditions are satisfied. We write

$$\begin{aligned}\Sigma F_x &= 300 \text{ lb} - (200 \text{ lb}) \sin 30° - (400 \text{ lb}) \sin 30° \\ &= 300 \text{ lb} - 100 \text{ lb} - 200 \text{ lb} = 0\end{aligned}$$

$$\begin{aligned}\Sigma F_y &= -173.2 \text{ lb} - (200 \text{ lb}) \cos 30° + (400 \text{ lb}) \cos 30° \\ &= -173.2 \text{ lb} - 173.2 \text{ lb} + 346.4 \text{ lb} = 0\end{aligned}$$

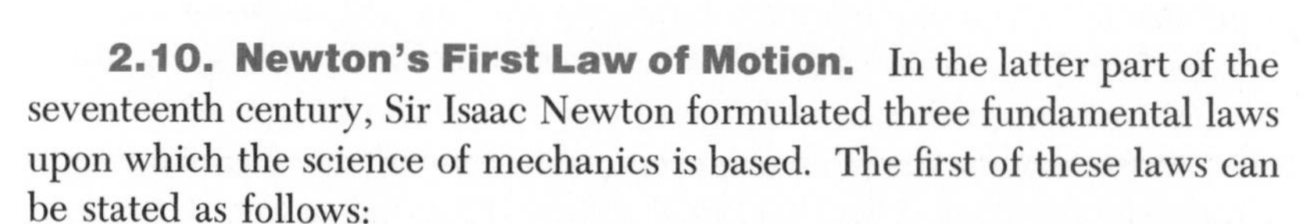

2.10. Newton's First Law of Motion. In the latter part of the seventeenth century, Sir Isaac Newton formulated three fundamental laws upon which the science of mechanics is based. The first of these laws can be stated as follows:

If the resultant force acting on a particle is zero, the particle will remain at rest (if originally at rest) or will move with constant speed in a straight line (if originally in motion).

From this law and from the definition of equilibrium given in Sec. 2.9, it is seen that a particle in equilibrium either is at rest or is moving in a straight line with constant speed. In the following section, various problems concerning the equilibrium of a particle will be considered.

2.11. Problems Involving the Equilibrium of a Particle. Free-Body Diagram. In practice, a problem in engineering mechanics is derived from an actual physical situation. A sketch showing the physical conditions of the problem is known as a *space diagram.*

The methods of analysis discussed in the preceding sections apply to a system of forces acting on a particle. A large number of problems involving actual structures, however, may be reduced to problems concerning the equilibrium of a particle. This is done by choosing a significant particle and drawing a separate diagram showing this particle and all the forces acting on it. Such a diagram is called a *free-body diagram.*

As an example, consider the 75-kg crate shown in the space diagram of Fig. 2.29*a*. This crate was lying between two buildings, and it is now being lifted onto a truck, which will remove it. The crate is supported by a vertical cable, which is joined at A to two ropes which pass over pulleys attached to the buildings at B and C. It is desired to determine the tension in each of the ropes AB and AC.

In order to solve this problem, a free-body diagram must be drawn, showing a particle in equilibrium. Since we are interested in the rope tensions, the free-body diagram should include at least one of these tensions and, if possible, both tensions. Point A is seen to be a good free body for this problem. The free-body diagram of point A is shown in Fig. 2.29*b*. It shows point A and the forces exerted on A by the vertical cable and the two ropes. The force exerted by the cable is directed downward and is equal to the weight W of the crate. Recalling Eq. (1.4), we write

$$W = mg = (75\text{ kg})(9.81\text{ m/s}^2) = 736\text{ N}$$

and indicate this value in the free-body diagram. The forces exerted by the two ropes are not known. Since they are respectively equal in magnitude to the tension in rope AB and rope AC, we denote them by $\mathbf{T}_{AB}$ and $\mathbf{T}_{AC}$ and draw them away from A in the directions shown in the space diagram. No other detail is included in the free-body diagram.

Since point A is in equilibrium, the three forces acting on it must form a closed triangle when drawn in tip-to-tail fashion. This *force triangle* has been drawn in Fig. 2.29*c*. The values T_{AB} and T_{AC} of the tension in the ropes may be found graphically if the triangle is drawn to scale, or they may be found by trigonometry. If the latter method of solution is chosen, we use the law of sines and write

$$\frac{T_{AB}}{\sin 60°} = \frac{T_{AC}}{\sin 40°} = \frac{736\text{ N}}{\sin 80°}$$

$$T_{AB} = 647\text{ N} \qquad T_{AC} = 480\text{ N}$$

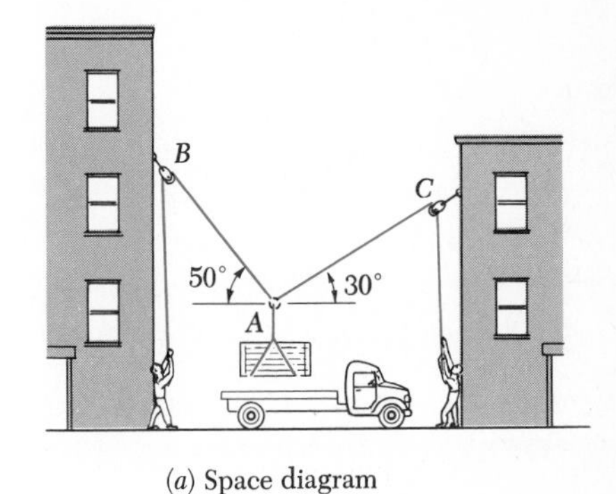

(*a*) Space diagram

(*b*) Free-body diagram

(*c*) Force triangle

Fig. 2.29

When a particle is in *equilibrium under three forces*, the problem may always be solved by drawing a force triangle. When a particle is in *equilibrium under more than three forces*, the problem may be solved graphically by drawing a force polygon. If an analytic solution is desired, the *equations of equilibrium* given in Sec. 2.9 should be solved:

$$\Sigma F_x = 0 \qquad \Sigma F_y = 0 \tag{2.15}$$

These equations may be solved for no more than *two unknowns;* similarly, the force triangle used in the case of equilibrium under three forces may be solved for two unknowns.

The more common types of problems are those where the two unknowns represent (1) the two components (or the magnitude and direction) of a single force, (2) the magnitude of two forces each of known direction. Problems involving the determination of the maximum or minimum value of the magnitude of a force are also encountered (see Probs. 2.44 and 2.45).

SAMPLE PROBLEM 2.4

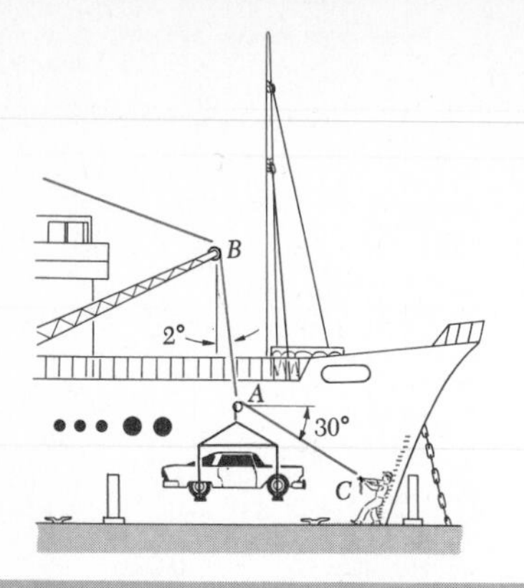

In a ship-unloading operation, a 3500-lb automobile is supported by a cable. A rope is tied to the cable at A and pulled in order to center the automobile over its intended position. The angle between the cable and the vertical is 2°, while the angle between the rope and the horizontal is 30°. What is the tension in the rope?

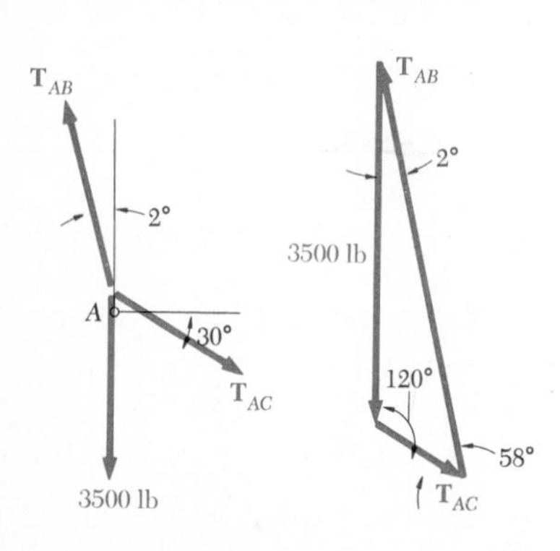

Solution. Point A is chosen as a free body, and the complete free-body diagram is drawn. T_{AB} is the tension in the cable AB, and T_{AC} is the tension in the rope.

Equilibrium Condition. Since only three forces act on the free body, we draw a force triangle to express that it is in equilibrium. Using the law of sines, we write

$$\frac{T_{AB}}{\sin 120°} = \frac{T_{AC}}{\sin 2°} = \frac{3500 \text{ lb}}{\sin 58°}$$

With a calculator, we first compute and store the value of the last quotient. Multiplying this value successively by sin 120° and sin 2°, we obtain

$$T_{AB} = 3570 \text{ lb} \qquad T_{AC} = 144 \text{ lb} \blacktriangleleft$$

SAMPLE PROBLEM 2.5

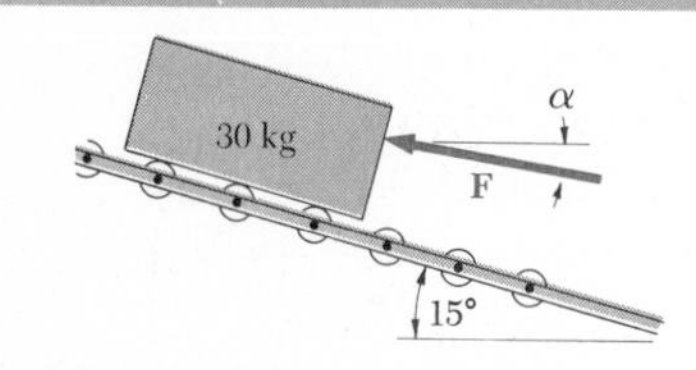

Determine the magnitude and direction of the smallest force **F** which will maintain the package shown in equilibrium. Note that the force exerted by the rollers on the package is perpendicular to the incline.

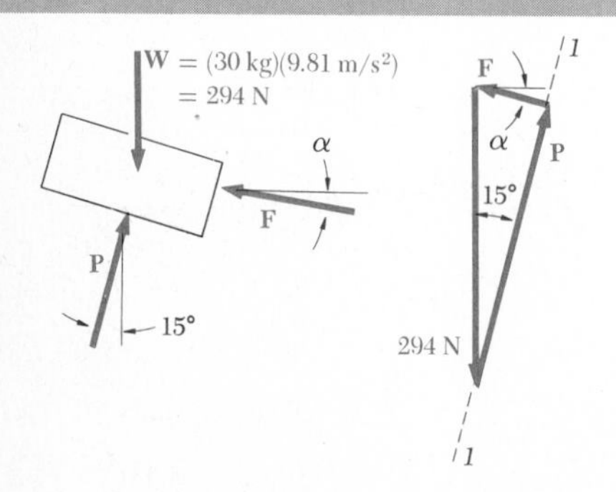

Solution. We choose the package as a free body, assuming that it may be treated as a particle. We draw the corresponding free-body diagram.

Equilibrium Condition. Since only three forces act on the free body, we draw a force triangle to express that it is in equilibrium. Line *1-1* represents the known direction of **P**. In order to obtain the minimum value of the force **F**, we choose the direction of **F** perpendicular to that of **P**. From the geometry of the triangle obtained, we find

$$F = (294 \text{ N}) \sin 15° = 76.1 \text{ N} \qquad \alpha = 15°$$

$$\mathbf{F} = 76.1 \text{ N} \measuredangle 15° \blacktriangleleft$$

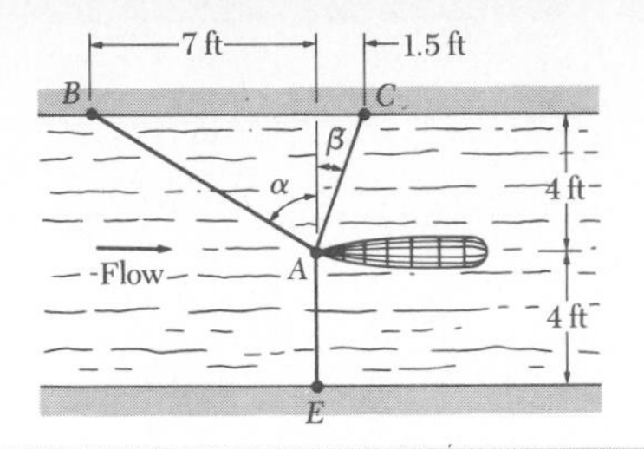

SAMPLE PROBLEM 2.6

As part of the design of a new sailboat, it is desired to determine the drag force which may be expected at a given speed. To do so, a model of the proposed hull is placed in a test channel and three cables are used to keep its bow on the centerline of the channel. Dynamometer readings indicate that for a given speed, the tension is 40 lb in cable AB and 60 lb in cable AE. Determine the drag force exerted on the hull and the tension in cable AC.

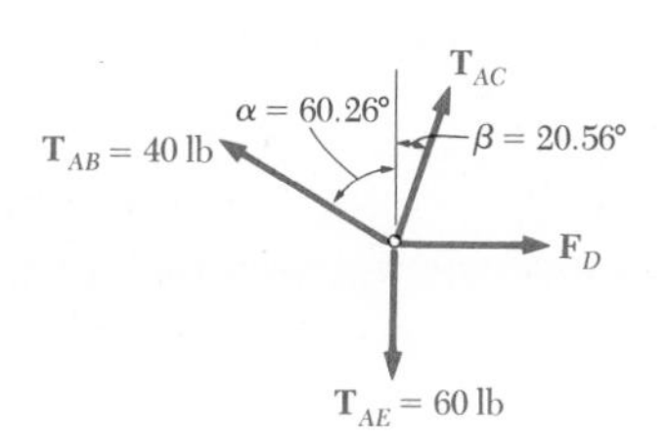

Solution. First, the angles α and β defining the direction of cables AB and AC are determined. We write

$$\tan \alpha = \frac{7 \text{ ft}}{4 \text{ ft}} = 1.75 \qquad \tan \beta = \frac{1.5 \text{ ft}}{4 \text{ ft}} = 0.375$$

$$\alpha = 60.26° \qquad \beta = 20.56°$$

Choosing the hull as a free body, we draw the free-body diagram shown. It includes the forces exerted by the three cables on the hull, as well as the drag force $\mathbf{F}_D$ exerted by the flow.

Equilibrium Condition. We express that the hull is in equilibrium by writing that the resultant of all forces is zero:

$$\mathbf{R} = \mathbf{T}_{AB} + \mathbf{T}_{AC} + \mathbf{T}_{AE} + \mathbf{F}_D = 0 \tag{1}$$

Since more than three forces are involved, we shall resolve the forces into x and y components:

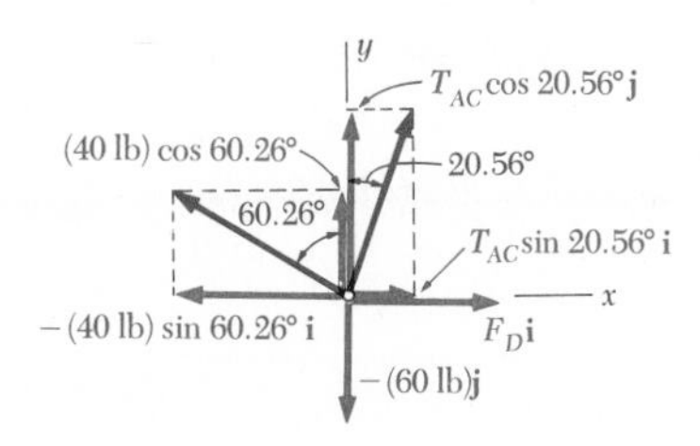

$$\begin{aligned}
\mathbf{T}_{AB} &= -(40 \text{ lb}) \sin 60.26°\mathbf{i} + (40 \text{ lb}) \cos 60.26°\mathbf{j} \\
&= -(34.73 \text{ lb})\mathbf{i} + (19.84 \text{ lb})\mathbf{j} \\
\mathbf{T}_{AC} &= T_{AC} \sin 20.56°\mathbf{i} + T_{AC} \cos 20.56°\mathbf{j} \\
&= 0.3512 T_{AC}\mathbf{i} + 0.9363 T_{AC}\mathbf{j} \\
\mathbf{T}_{AE} &= -(60 \text{ lb})\mathbf{j} \\
\mathbf{F}_D &= F_D\mathbf{i}
\end{aligned}$$

Substituting the expressions obtained into Eq. (1) and factoring the unit vectors **i** and **j**, we have

$$(-34.73 \text{ lb} + 0.3512 T_{AC} + F_D)\mathbf{i} + (19.84 \text{ lb} + 0.9363 T_{AC} - 60 \text{ lb})\mathbf{j} = 0$$

This equation will be satisfied if, and only if, the coefficients of **i** and **j** are equal to zero. We thus obtain the following two equilibrium equations, which express, respectively, that the sum of the x components and the sum of the y components of the given forces must be zero.

$$(\Sigma F_x = 0:) \qquad -34.73 \text{ lb} + 0.3512 T_{AC} + F_D = 0 \tag{2}$$

$$(\Sigma F_y = 0:) \qquad 19.84 \text{ lb} + 0.9363 T_{AC} - 60 \text{ lb} = 0 \tag{3}$$

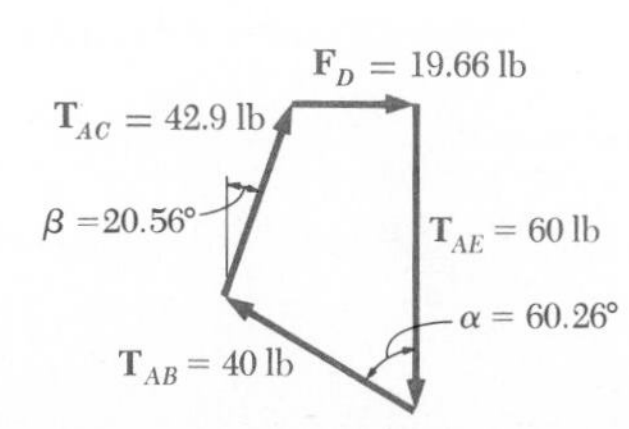

From Eq. (3) we find $\qquad T_{AC} = +42.9 \text{ lb}$ ◄

and, substituting this value into Eq. (2), $\qquad F_D = +19.66 \text{ lb}$ ◄

In drawing the free-body diagram, we assumed a sense for each unknown force. A positive sign in the answer indicates that the assumed sense is correct. The complete force polygon may be drawn to check the results.

Problems

2.35 through 2.38 Two cables are tied together at C and loaded as shown. Determine the tension in AC and BC.

Fig. P2.35

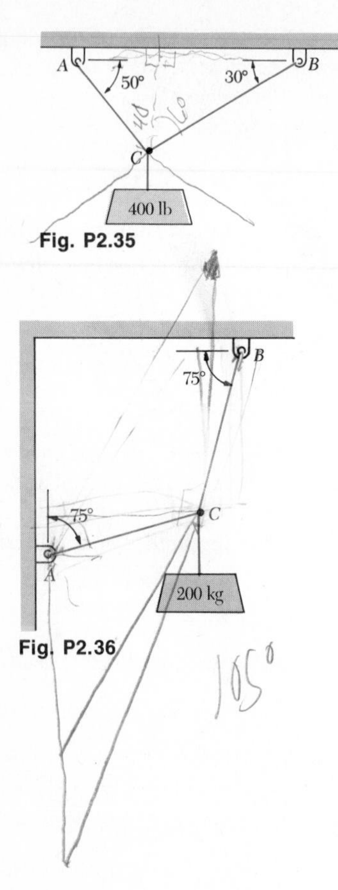

Fig. P2.36

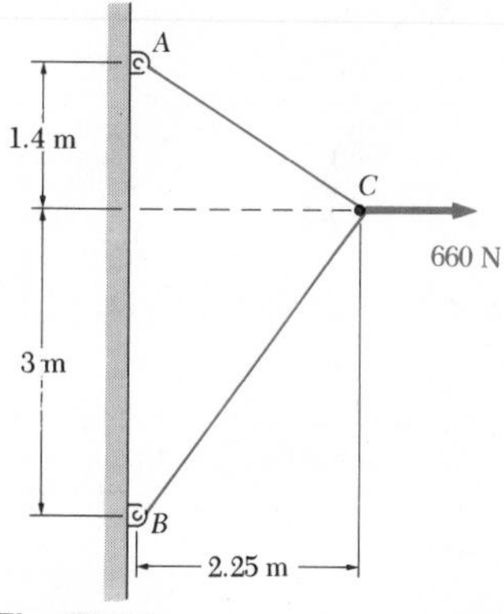

Fig. P2.37

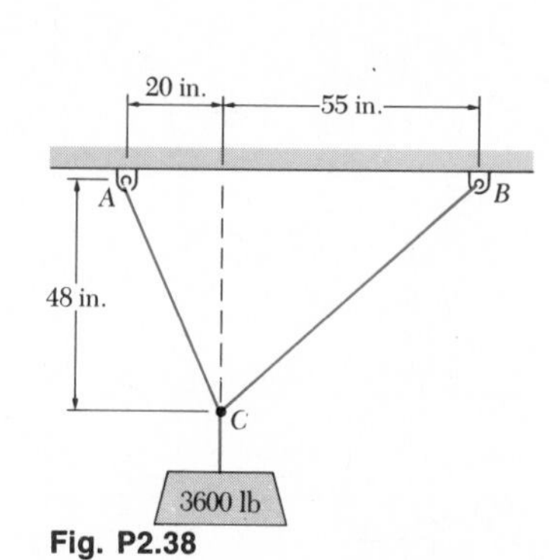

Fig. P2.38

2.39 Two cables are tied together at C and loaded as shown. Knowing that $P = 500$ N and $\alpha = 60°$, determine the tension in AC and BC.

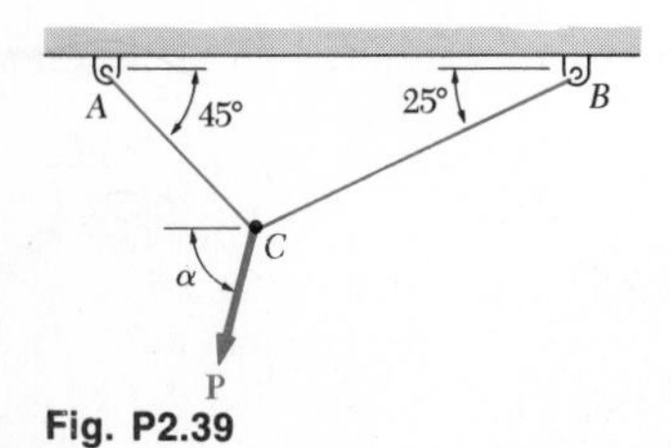

Fig. P2.39

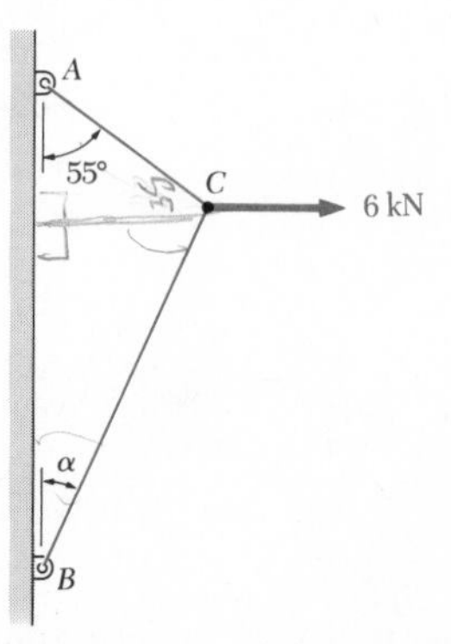

Fig. P2.40

2.40 Two cables are tied together at C and loaded as shown. Knowing that $\alpha = 30°$, determine the tension in AC and BC.

2.41 Two forces of magnitude $T_A = 8$ kips and $T_B = 15$ kips are applied as shown to a welded connection. Knowing that the connection is in equilibrium, determine the magnitudes of the forces T_C and T_D.

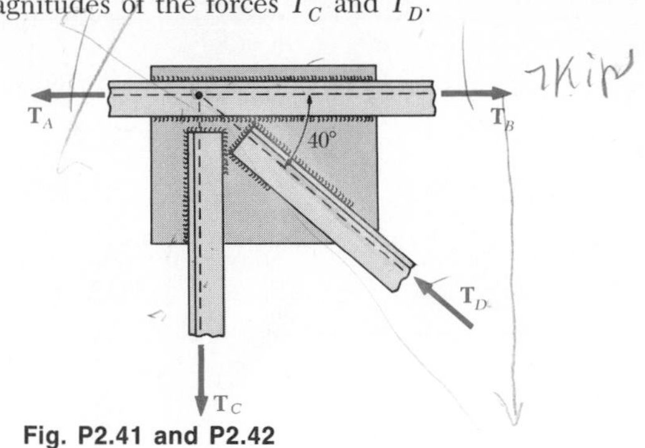

Fig. P2.41 and P2.42

2.42 Two forces of magnitude $T_A = 6$ kips and $T_C = 9$ kips are applied as shown to a welded connection. Knowing that the connection is in equilibrium, determine the magnitudes of the forces T_B and T_D.

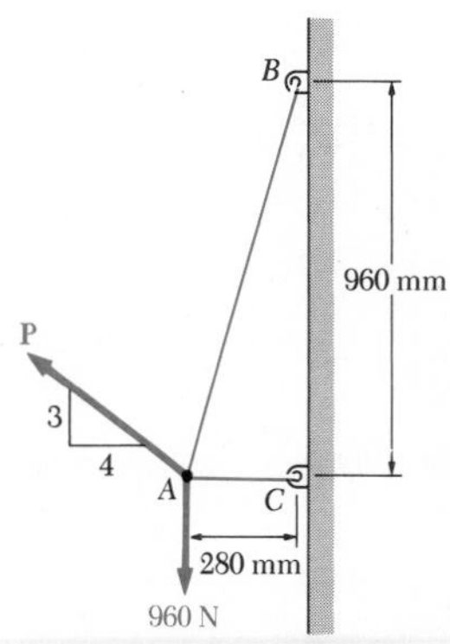

Fig. P2.43 and P2.44

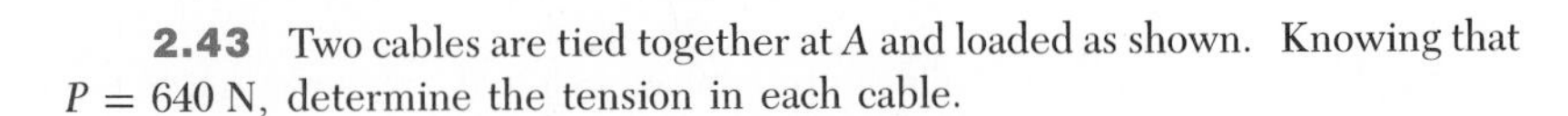

2.43 Two cables are tied together at A and loaded as shown. Knowing that $P = 640$ N, determine the tension in each cable.

2.44 Two cables are tied together at A and loaded as shown. Determine the range of values of P for which both cables remain taut.

2.45 For the cables of Prob. 2.40, find the value of α for which the tension is as small as possible (*a*) in cable BC, (*b*) in both cables simultaneously. In each case determine the tension in both cables.

2.46 For the cables of Prob. 2.39, it is known that the maximum allowable tension is 600 N in cable AC and 750 N in cable BC. Determine (*a*) the maximum force **P** that may be applied at C, (*b*) the corresponding value of α.

2.47 The 60-lb collar A may slide on a frictionless vertical rod and is connected as shown to a 65-lb counterweight C. Determine the value of h for which the system is in equilibrium.

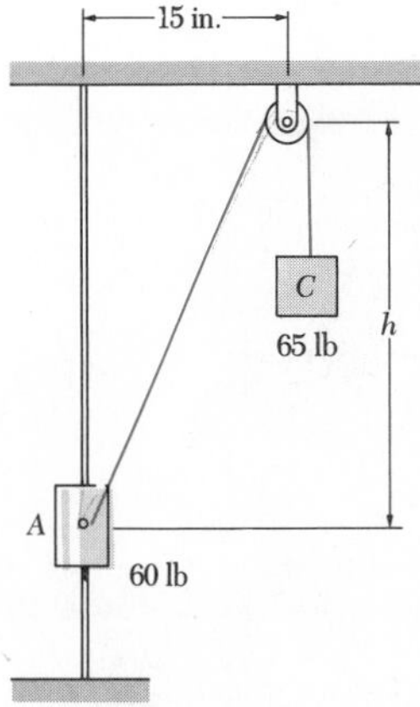

Fig. P2.47

2.48 A movable bin and its contents weigh 700 lb. Determine the shortest chain sling ACB which may be used to lift the loaded bin if the tension in the chain is not to exceed 1250 lb.

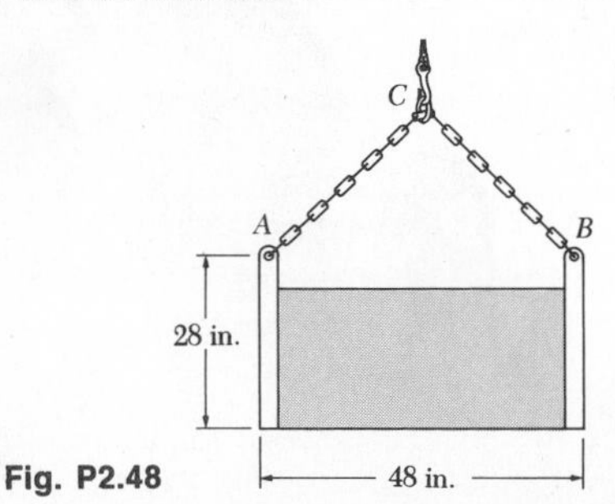

Fig. P2.48

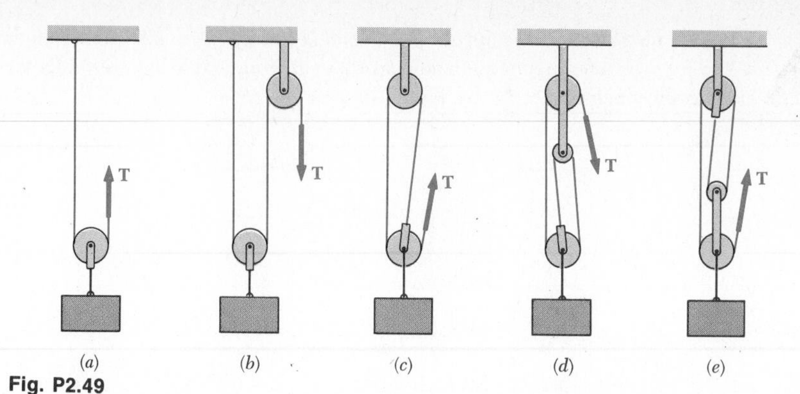

Fig. P2.49

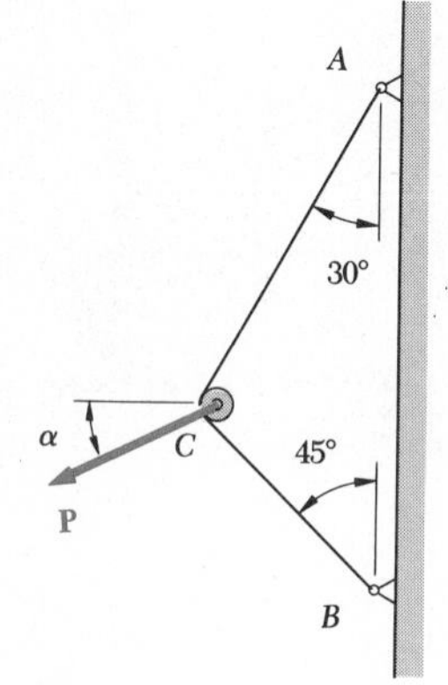

Fig. P2.51

2.49 A 250-kg crate is supported by several rope-and-pulley arrangements as shown. Determine for each arrangement the tension in the rope. (The tension in the rope is the same on each side of a simple pulley. This can be proved by the methods of Chap. 4.)

2.50 Solve parts *b* and *d* of Prob. 2.49 assuming that the free end of the rope is attached to the crate.

2.51 The force **P** is applied to a small wheel which rolls on the cable *ACB*. Knowing that the tension in both parts of the cable is 600 N, determine the magnitude and direction of **P**.

2.52 A 450-lb crate is to be supported by the rope-and-pulley arrangement shown. Determine the magnitude and direction of the force **F** which should be exerted on the free end of the rope.

***2.53** The collar *A* may slide freely on the horizontal frictionless rod. The spring attached to the collar has a constant of 10 lb/in. and is undeformed when the collar is directly below support *B*. Determine the magnitude of the force **P** required to maintain equilibrium when (*a*) $c = 18$ in., (*b*) $c = 32$ in., (*c*) $c = 45$ in.

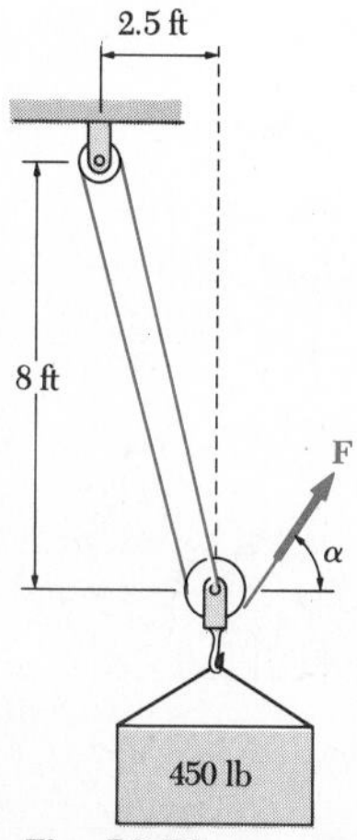

Fig. P2.52

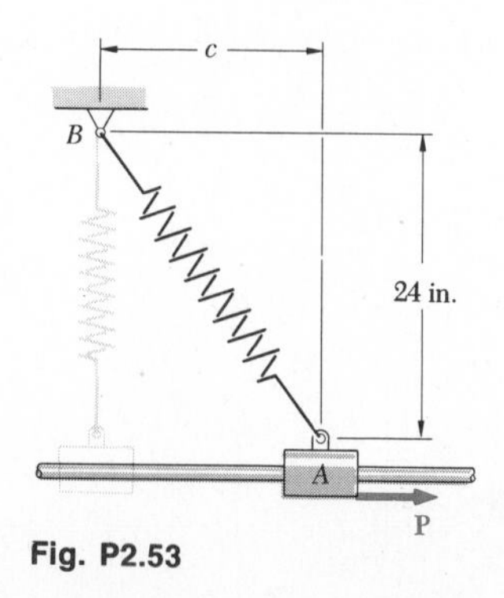

Fig. P2.53

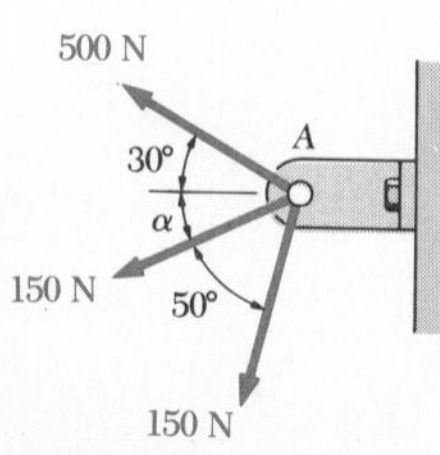

Fig. P2.54

2.54 Three forces are applied to a bracket. The directions of the two 150-N forces may vary, but the angle between these forces is always 50°. Determine the range of values of α for which the magnitude of the resultant of the forces applied to the bracket is less than 600 N.

FORCES IN SPACE

2.12. Rectangular Components of a Force in Space. The problems considered in the first part of this chapter involved only two dimensions; they could be formulated and solved in a single plane. In this section and in the remaining sections of the chapter, we shall discuss problems involving the three dimensions of space.

Consider a force **F** acting at the origin O of the system of rectangular coordinates x, y, z. To define the direction of **F**, we may draw the vertical plane $OBAC$ containing **F** and shown in Fig. 2.30*a*. This plane passes through the vertical y axis; its orientation is defined by the angle ϕ it forms with the xy plane, while the direction of **F** within the plane is defined by the angle θ_y that **F** forms with the y axis. The force **F** may be resolved into a vertical component $\mathbf{F}_y$ and a horizontal component $\mathbf{F}_h$; this operation, shown in Fig. 2.30*b*, is carried out in plane $OBAC$ according to the rules developed in the first part of the chapter. The corresponding scalar components are

$$F_y = F\cos\theta_y \qquad F_h = F\sin\theta_y \tag{2.16}$$

But $\mathbf{F}_h$ may be resolved into two rectangular components $\mathbf{F}_x$ and $\mathbf{F}_z$ along the x and z axes, respectively. This operation, shown in Fig. 2.30*c*, is carried out in the xz plane. We obtain the following expressions for the corresponding scalar components:

$$\begin{aligned} F_x &= F_h\cos\phi = F\sin\theta_y\cos\phi \\ F_z &= F_h\sin\phi = F\sin\theta_y\sin\phi \end{aligned} \tag{2.17}$$

The given force **F** has thus been resolved into three rectangular vector components $\mathbf{F}_x$, $\mathbf{F}_y$, $\mathbf{F}_z$, directed along the three coordinate axes.

Applying the Pythagorean theorem to the triangles OAB and OCD of Fig. 2.30, we write

$$\begin{aligned} F^2 &= (OA)^2 = (OB)^2 + (BA)^2 = F_y^2 + F_h^2 \\ F_h^2 &= (OC)^2 = (OD)^2 + (DC)^2 = F_x^2 + F_z^2 \end{aligned}$$

Eliminating F_h^2 from these two equations and solving for F, we obtain the following relation between the magnitude of **F** and its rectangular scalar components:

$$F = \sqrt{F_x^2 + F_y^2 + F_z^2} \tag{2.18}$$

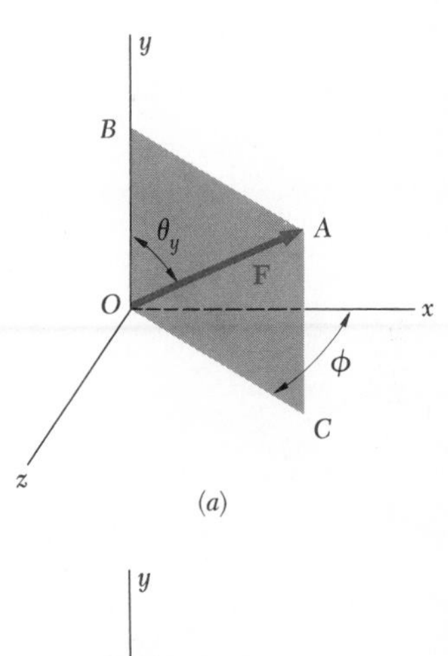

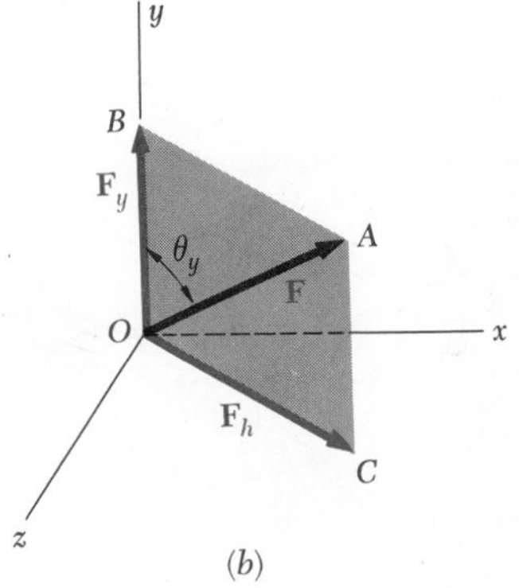

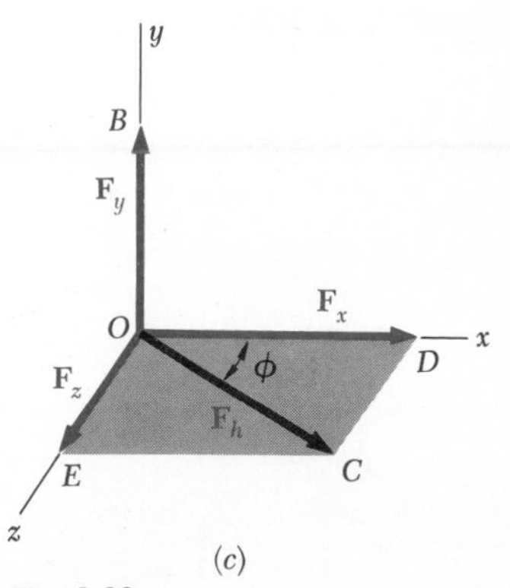

Fig. 2.30

The relationship existing between the force **F** and its three components $\mathbf{F}_x$, $\mathbf{F}_y$, $\mathbf{F}_z$ is more easily visualized if a "box" having $\mathbf{F}_x$, $\mathbf{F}_y$, $\mathbf{F}_z$ for edges is drawn as shown in Fig. 2.31. The force **F** is then represented by the diagonal *OA* of this box. Figure 2.31*b* shows the right triangle *OAB* used to derive the first of the formulas (2.16): $F_y = F \cos \theta_y$. In Fig. 2.31*a* and *c*, two other right triangles have also been drawn: *OAD* and *OAE*. These triangles are seen to occupy in the box positions comparable with that of triangle *OAB*. Denoting by θ_x and θ_z, respectively, the angles that **F** forms with the *x* and *z* axes, we may derive two formulas similar to $F_y = F \cos \theta_y$. We thus write

$$F_x = F \cos \theta_x \qquad F_y = F \cos \theta_y \qquad F_z = F \cos \theta_z \tag{2.19}$$

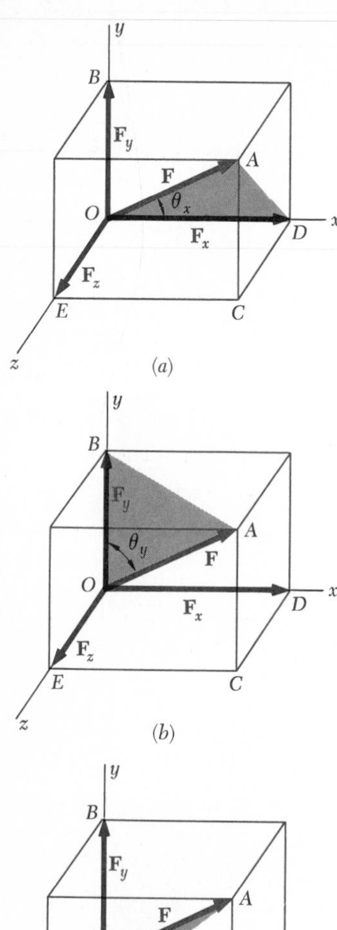

Fig. 2.31

The three angles θ_x, θ_y, θ_z define the direction of the force **F**; they are more commonly used for this purpose than the angles θ_y and ϕ introduced at the beginning of this section. The cosines of θ_x, θ_y, θ_z are known as the direction cosines of the force **F**.

Introducing the unit vectors **i**, **j**, and **k**, directed respectively along the *x*, *y*, and *z* axes (Fig. 2.32), we may express **F** in the form

$$\mathbf{F} = F_x\mathbf{i} + F_y\mathbf{j} + F_z\mathbf{k} \tag{2.20}$$

where the scalar components F_x, F_y, F_z are defined by the relations (2.19).

Example 1. A force of 500 N forms angles of 60, 45, and 120°, respectively, with the *x*, *y*, and *z* axes. Find the components F_x, F_y, and F_z of the force.

Substituting $F = 500$ N, $\theta_x = 60°$, $\theta_y = 45°$, $\theta_z = 120°$ into formulas (2.19), we write

$$\begin{aligned} F_x &= (500 \text{ N}) \cos 60° = +250 \text{ N} \\ F_y &= (500 \text{ N}) \cos 45° = +354 \text{ N} \\ F_z &= (500 \text{ N}) \cos 120° = -250 \text{ N} \end{aligned}$$

Carrying into Eq. (2.20) the values obtained for the scalar components of **F**, we have

$$\mathbf{F} = (250 \text{ N})\mathbf{i} + (354 \text{ N})\mathbf{j} - (250 \text{ N})\mathbf{k}$$

As in the case of two-dimensional problems, the plus sign indicates that the component has the same sense as the corresponding axis, and the minus sign that it has the opposite sense.

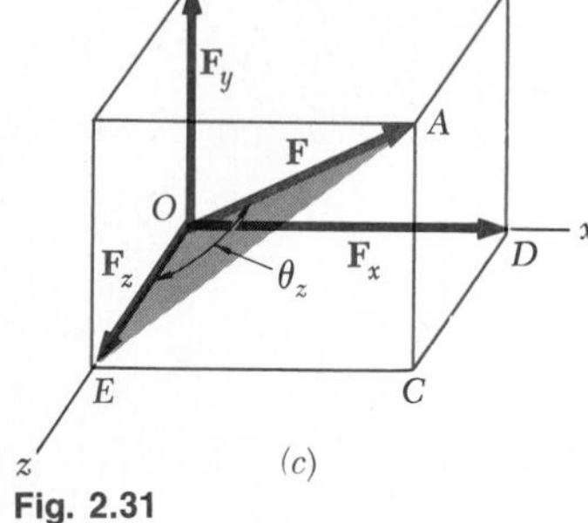

Fig. 2.32

The angle a force **F** forms with an axis should be measured from the positive side of the axis and will always be comprised between 0 and 180°. An angle θ_x smaller than 90° (acute) indicates that **F** (assumed attached to *O*) is on the same side of the *yz* plane as the positive *x* axis; $\cos \theta_x$ and F_x will then be positive. An angle θ_x larger than 90° (obtuse) would indicate that **F** is on the other side of the *yz* plane; $\cos \theta_x$ and F_x would then be negative. In Example 1 the angles θ_x and θ_y are acute, while θ_z is obtuse: consequently, F_x and F_y are positive, while F_z is negative.

Substituting into (2.20) the expressions obtained for F_x, F_y, F_z in (2.19), we write

$$\mathbf{F} = F(\cos\theta_x\mathbf{i} + \cos\theta_y\mathbf{j} + \cos\theta_z\mathbf{k}) \tag{2.21}$$

which shows that the force **F** may be expressed as the product of the scalar F and of the vector

$$\boldsymbol{\lambda} = \cos\theta_x\mathbf{i} + \cos\theta_y\mathbf{j} + \cos\theta_z\mathbf{k} \tag{2.22}$$

Clearly, the vector $\boldsymbol{\lambda}$ is a vector of magnitude equal to 1 and of the same direction as **F** (Fig. 2.33). We shall refer to $\boldsymbol{\lambda}$ as the *unit vector* along the line of action of **F**. It follows from (2.22) that the components of the unit vector $\boldsymbol{\lambda}$ are respectively equal to the direction cosines of the line of action of **F**:

$$\lambda_x = \cos\theta_x \qquad \lambda_y = \cos\theta_y \qquad \lambda_z = \cos\theta_z \tag{2.23}$$

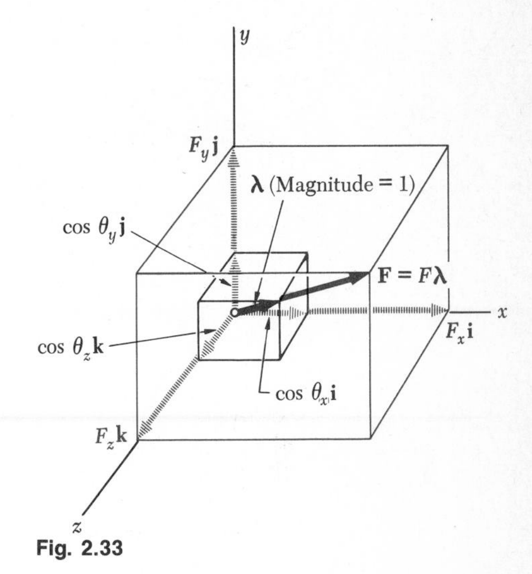

Fig. 2.33

We should observe that the values of the three angles θ_x, θ_y, θ_z are not independent. Expressing that the sum of the squares of the components of $\boldsymbol{\lambda}$ is equal to the square of its magnitude, we write

$$\lambda_x^2 + \lambda_y^2 + \lambda_z^2 = 1$$

or, substituting for λ_x, λ_y, λ_z from (2.23),

$$\cos^2\theta_x + \cos^2\theta_y + \cos^2\theta_z = 1 \tag{2.24}$$

In Example 1, for instance, once the values $\theta_x = 60°$ and $\theta_y = 45°$ have been selected, the value of θ_z *must* be equal to 60° or 120° in order to satisfy identity (2.24).

When the components F_x, F_y, F_z of a force **F** are given, the magnitude F of the force is obtained from (2.18).† The relations (2.19) may then be solved for the direction cosines,

$$\cos\theta_x = \frac{F_x}{F} \qquad \cos\theta_y = \frac{F_y}{F} \qquad \cos\theta_z = \frac{F_z}{F} \tag{2.25}$$

and the angles θ_x, θ_y, θ_z characterizing the direction of **F** may be found.

Example 2. A force **F** has the components $F_x = 20$ lb, $F_y = -30$ lb, $F_z = 60$ lb. Determine its magnitude F and the angles θ_x, θ_y, θ_z it forms with the axes of coordinates.

From formula (2.18) we obtain†

$$\begin{aligned} F &= \sqrt{F_x^2 + F_y^2 + F_z^2} \\ &= \sqrt{(20\text{ lb})^2 + (-30\text{ lb})^2 + (60\text{ lb})^2} \\ &= \sqrt{4900}\text{ lb} = 70\text{ lb} \end{aligned}$$

† With a calculator programmed to convert rectangular coordinates into polar coordinates, the following procedure will be found more expeditious for computing F: First determine F_h from its two rectangular components F_x and F_z (Fig. 2.30*c*), then determine F from its two rectangular components F_h and F_y (Fig. 2.30*b*). The actual order in which the three components F_x, F_y, F_z are entered is immaterial.

Substituting the values of the components and magnitude of **F** into Eqs. (2.25), we write

$$\cos\theta_x = \frac{F_x}{F} = \frac{20 \text{ lb}}{70 \text{ lb}} \qquad \cos\theta_y = \frac{F_y}{F} = \frac{-30 \text{ lb}}{70 \text{ lb}} \qquad \cos\theta_z = \frac{F_z}{F} = \frac{60 \text{ lb}}{70 \text{ lb}}$$

Calculating successively each quotient and its arc cosine, we obtain

$$\theta_x = 73.4° \qquad \theta_y = 115.4° \qquad \theta_z = 31.0°$$

The computations indicated may easily be carried out with a calculator.

2.13. Force Defined by Its Magnitude and Two Points on Its Line of Action. In many applications, the direction of a force **F** is defined by the coordinates of two points, $M(x_1, y_1, z_1)$ and $N(x_2, y_2, z_2)$, located on its line of action (Fig. 2.34). Consider the vector $\overrightarrow{MN}$ joining M

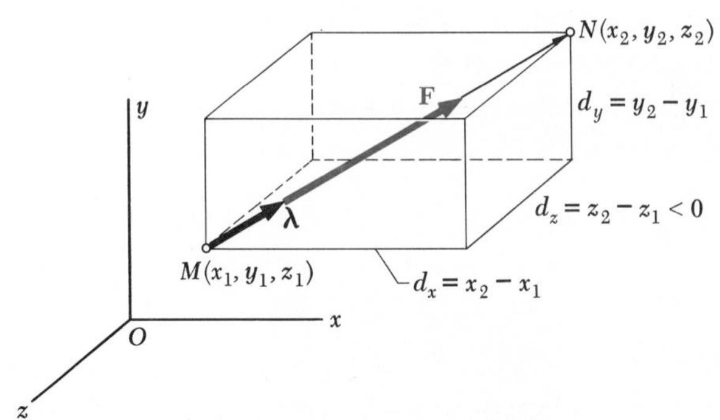

Fig. 2.34

and N and of the same sense as **F**. Denoting its scalar components by d_x, d_y, d_z, respectively, we write

$$\overrightarrow{MN} = d_x\mathbf{i} + d_y\mathbf{j} + d_z\mathbf{k} \tag{2.26}$$

The unit vector $\boldsymbol{\lambda}$ along the line of action of **F** (i.e., along the line MN) may be obtained by dividing the vector $\overrightarrow{MN}$ by its magnitude MN. Substituting for $\overrightarrow{MN}$ from (2.26) and observing that MN is equal to the distance d from M to N, we write

$$\boldsymbol{\lambda} = \frac{\overrightarrow{MN}}{MN} = \frac{1}{d}(d_x\mathbf{i} + d_y\mathbf{j} + d_z\mathbf{k}) \tag{2.27}$$

Recalling that **F** is equal to the product of F and $\boldsymbol{\lambda}$, we have

$$\mathbf{F} = F\boldsymbol{\lambda} = \frac{F}{d}(d_x\mathbf{i} + d_y\mathbf{j} + d_z\mathbf{k}) \tag{2.28}$$

from which it follows that the scalar components of **F** are, respectively,

$$F_x = \frac{Fd_x}{d} \qquad F_y = \frac{Fd_y}{d} \qquad F_z = \frac{Fd_z}{d} \tag{2.29}$$

The relations (2.29) considerably simplify the determination of the components of a force **F** of given magnitude F when the line of action of **F** is defined by two points M and N. Subtracting the coordinates of M from those of N, we first determine the components of the vector $\overrightarrow{MN}$ and the distance d from M to N:

$$d_x = x_2 - x_1 \qquad d_y = y_2 - y_1 \qquad d_z = z_2 - z_1$$
$$d = \sqrt{d_x^2 + d_y^2 + d_z^2}$$

Substituting for F and for d_x, d_y, d_z, and d into the relations (2.29), we obtain the components F_x, F_y, F_z of the force.

The angles $\theta_x, \theta_y, \theta_z$ that **F** forms with the coordinate axes may then be obtained from Eqs. (2.25). Comparing Eqs. (2.22) and (2.27), we may also write

$$\cos\theta_x = \frac{d_x}{d} \qquad \cos\theta_y = \frac{d_y}{d} \qquad \cos\theta_z = \frac{d_z}{d} \tag{2.30}$$

and determine the angles $\theta_x, \theta_y, \theta_z$ directly from the components and magnitude of the vector MN.

2.14. Addition of Concurrent Forces in Space. We shall determine the resultant **R** of two or more forces in space by summing their rectangular components. Graphical or trigonometric methods are generally not practical in the case of forces in space.

The method followed here is similar to that used in Sec. 2.8 with coplanar forces. Setting

$$\mathbf{R} = \Sigma\mathbf{F}$$

we resolve each force into its rectangular components and write

$$\begin{aligned} R_x\mathbf{i} + R_y\mathbf{j} + R_z\mathbf{k} &= \Sigma(F_x\mathbf{i} + F_y\mathbf{j} + F_z\mathbf{k}) \\ &= (\Sigma F_x)\mathbf{i} + (\Sigma F_y)\mathbf{j} + (\Sigma F_z)\mathbf{k} \end{aligned}$$

from which it follows that

$$R_x = \Sigma F_x \qquad R_y = \Sigma F_y \qquad R_z = \Sigma F_z \tag{2.31}$$

The magnitude of the resultant and the angles $\theta_x, \theta_y, \theta_z$ it forms with the axes of coordinates are obained by the method of Sec. 2.12. We write

$$R = \sqrt{R_x^2 + R_y^2 + R_z^2} \tag{2.32}$$

$$\cos\theta_x = \frac{R_x}{R} \qquad \cos\theta_y = \frac{R_y}{R} \qquad \cos\theta_z = \frac{R_z}{R} \tag{2.33}$$

SAMPLE PROBLEM 2.7

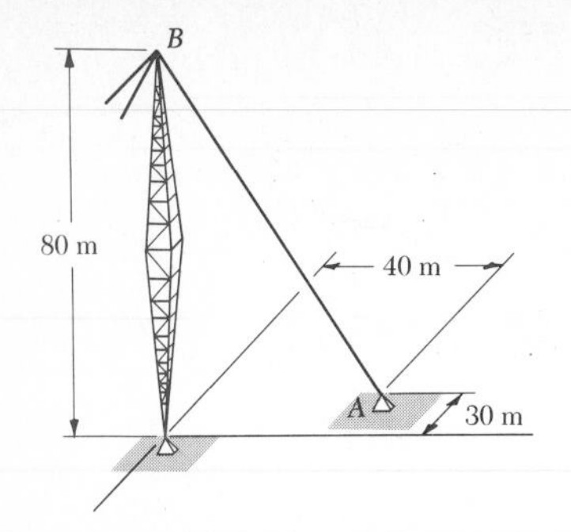

A tower guy wire is anchored by means of a bolt at A. The tension in the wire is 2500 N. Determine (*a*) the components F_x, F_y, F_z of the force acting on the bolt, (*b*) the angles θ_x, θ_y, θ_z defining the direction of the force.

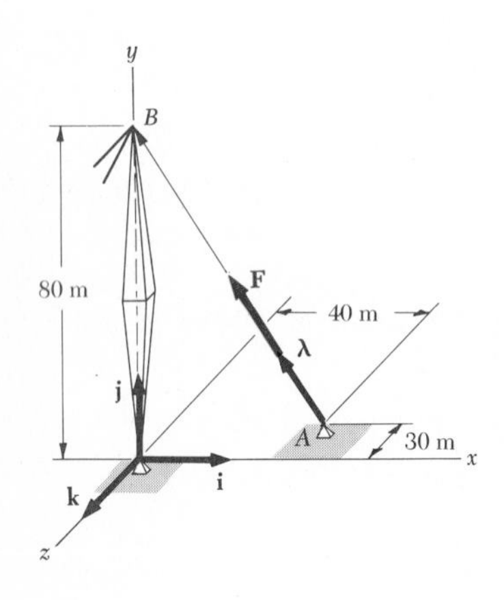

***a.* Components of the Force.** The line of action of the force acting on the bolt passes through A and B, and the force is directed from A to B. The components of the vector $\overrightarrow{AB}$, which has the same direction as the force, are

$$d_x = -40 \text{ m} \qquad d_y = +80 \text{ m} \qquad d_z = +30 \text{ m}$$

The total distance from A to B is

$$AB = d = \sqrt{d_x^2 + d_y^2 + d_z^2} = 94.3 \text{ m}$$

Denoting by **i**, **j**, **k** the unit vectors along the coordinate axes, we have

$$\overrightarrow{AB} = -(40 \text{ m})\mathbf{i} + (80 \text{ m})\mathbf{j} + (30 \text{ m})\mathbf{k}$$

Introducing the unit vector $\boldsymbol{\lambda} = \overrightarrow{AB}/AB$, we write

$$\mathbf{F} = F\boldsymbol{\lambda} = F\frac{\overrightarrow{AB}}{AB} = \frac{2500 \text{ N}}{94.3 \text{ m}}\overrightarrow{AB}$$

Substituting the expression found for $\overrightarrow{AB}$, we obtain

$$\mathbf{F} = \frac{2500 \text{ N}}{94.3 \text{ m}}[-(40 \text{ m})\mathbf{i} + (80 \text{ m})\mathbf{j} + (30 \text{ m})\mathbf{k}]$$

$$\mathbf{F} = -(1060 \text{ N})\mathbf{i} + (2120 \text{ N})\mathbf{j} + (795 \text{ N})\mathbf{k}$$

The components of **F**, therefore, are

$$F_x = -1060 \text{ N} \qquad F_y = +2120 \text{ N} \qquad F_z = +795 \text{ N} \quad ◄$$

***b.* Direction of the Force.** Using Eqs. (2.25), we write

$$\cos\theta_x = \frac{F_x}{F} = \frac{-1060 \text{ N}}{2500 \text{ N}} \qquad \cos\theta_y = \frac{F_y}{F} = \frac{+2120 \text{ N}}{2500 \text{ N}}$$

$$\cos\theta_z = \frac{F_z}{F} = \frac{+795 \text{ N}}{2500 \text{ N}}$$

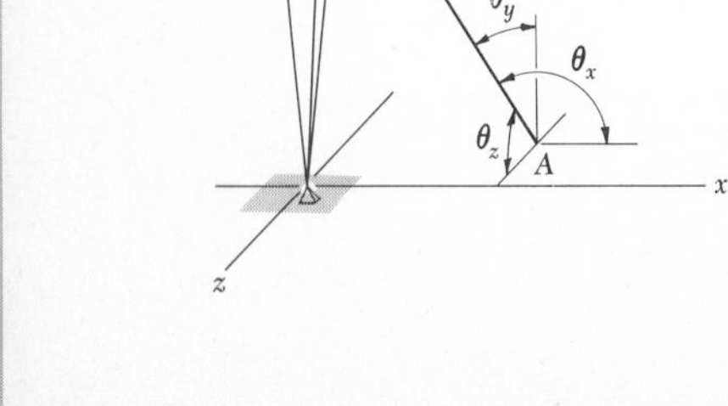

Calculating successively each quotient and its arc cosine, we obtain

$$\theta_x = 115.1° \qquad \theta_y = 32.0° \qquad \theta_z = 71.5° \quad ◄$$

(*Note.* This result could have been obtained by using the components and magnitude of the vector $\overrightarrow{AB}$ rather than those of the force **F**.)

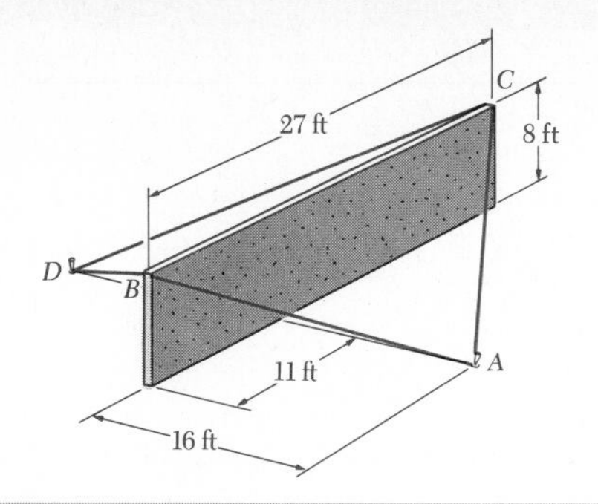

SAMPLE PROBLEM 2.8

A precast-concrete wall section is temporarily held by the cables shown. Knowing that the tension is 840 lb in cable AB and 1200 lb in cable AC, determine the magnitude and direction of the resultant of the forces exerted by cables AB and AC on stake A.

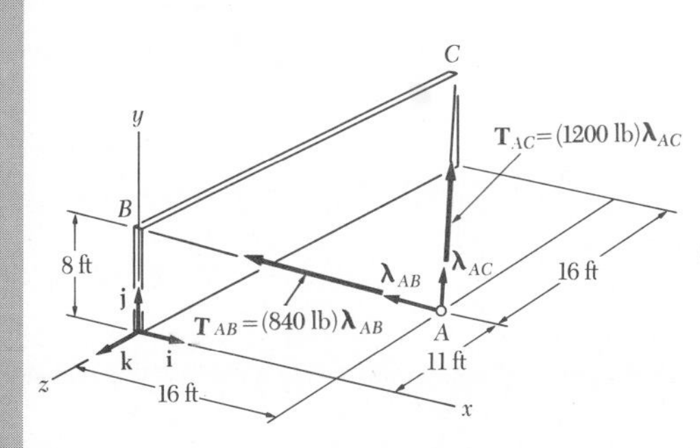

Solution. The force exerted by each cable on stake A will be resolved into x, y, and z components. We first determine the components and magnitude of the vectors $\overrightarrow{AB}$ and $\overrightarrow{AC}$, measuring them from A toward the wall section. Denoting by **i**, **j**, **k** the unit vectors along the coordinate axes, we write

$$\overrightarrow{AB} = -(16 \text{ ft})\mathbf{i} + (8 \text{ ft})\mathbf{j} + (11 \text{ ft})\mathbf{k} \qquad AB = 21 \text{ ft}$$

$$\overrightarrow{AC} = -(16 \text{ ft})\mathbf{i} + (8 \text{ ft})\mathbf{j} - (16 \text{ ft})\mathbf{k} \qquad AC = 24 \text{ ft}$$

Denoting by $\boldsymbol{\lambda}_{AB}$ the unit vector along AB, we have

$$\mathbf{T}_{AB} = T_{AB}\boldsymbol{\lambda}_{AB} = T_{AB}\frac{\overrightarrow{AB}}{AB} = \frac{840 \text{ lb}}{21 \text{ ft}}\overrightarrow{AB}$$

Substituting the expression found for $\overrightarrow{AB}$, we obtain

$$\mathbf{T}_{AB} = \frac{840 \text{ lb}}{21 \text{ ft}}[-(16 \text{ ft})\mathbf{i} + (8 \text{ ft})\mathbf{j} + (11 \text{ ft})\mathbf{k}]$$

$$\mathbf{T}_{AB} = -(640 \text{ lb})\mathbf{i} + (320 \text{ lb})\mathbf{j} + (440 \text{ lb})\mathbf{k}$$

Denoting by $\boldsymbol{\lambda}_{AC}$ the unit vector along AC, we obtain in a similar way

$$\mathbf{T}_{AC} = T_{AC}\boldsymbol{\lambda}_{AC} = T_{AC}\frac{\overrightarrow{AC}}{AC} = \frac{1200 \text{ lb}}{24 \text{ ft}}\overrightarrow{AC}$$

$$\mathbf{T}_{AC} = -(800 \text{ lb})\mathbf{i} + (400 \text{ lb})\mathbf{j} - (800 \text{ lb})\mathbf{k}$$

The resultant **R** of the forces exerted by the two cables is

$$\mathbf{R} = \mathbf{T}_{AB} + \mathbf{T}_{AC} = -(1440 \text{ lb})\mathbf{i} + (720 \text{ lb})\mathbf{j} - (360 \text{ lb})\mathbf{k}$$

The magnitude and direction of the resultant are now determined:

$$R = \sqrt{R_x^2 + R_y^2 + R_z^2} = \sqrt{(-1440)^2 + (720)^2 + (-360)^2}$$

$$R = 1650 \text{ lb} \blacktriangleleft$$

From Eqs. (2.33) we obtain

$$\cos\theta_x = \frac{R_x}{R} = \frac{-1440 \text{ lb}}{1650 \text{ lb}} \qquad \cos\theta_y = \frac{R_y}{R} = \frac{+720 \text{ lb}}{1650 \text{ lb}}$$

$$\cos\theta_z = \frac{R_z}{R} = \frac{-360 \text{ lb}}{1650 \text{ lb}}$$

Calculating successively each quotient and its arc cosine, we have

$$\theta_x = 150.8^\circ \qquad \theta_y = 64.1^\circ \qquad \theta_z = 102.6^\circ \blacktriangleleft$$

Problems

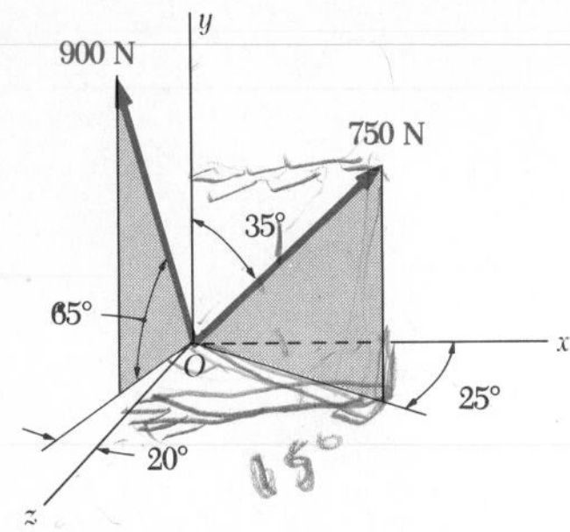

Fig. P2.55 and P2.56

2.55 Determine (*a*) the *x*, *y*, and *z* components of the 750-N force, (*b*) the angles θ_x, θ_y, and θ_z that the force forms with the coordinate axes.

2.56 Determine (*a*) the *x*, *y*, and *z* components of the 900-N force, (*b*) the angles θ_x, θ_y, and θ_z that the force forms with the coordinate axes.

2.57 A gun is aimed at a point *A* located 20° west of north. Knowing that the barrel of the gun forms an angle of 35° with the horizontal and that the maximum recoil force is 800 N, determine (*a*) the *x*, *y*, and *z* components of that force, (*b*) the values of the angles θ_x, θ_y, and θ_z defining the direction of the recoil force. (Assume that the *x*, *y*, and *z* axes are directed, respectively, east, up, and south.)

2.58 Solve Prob. 2.57, assuming that point *A* is located 25° north of west and that the barrel of the gun forms an angle of 30° with the horizontal.

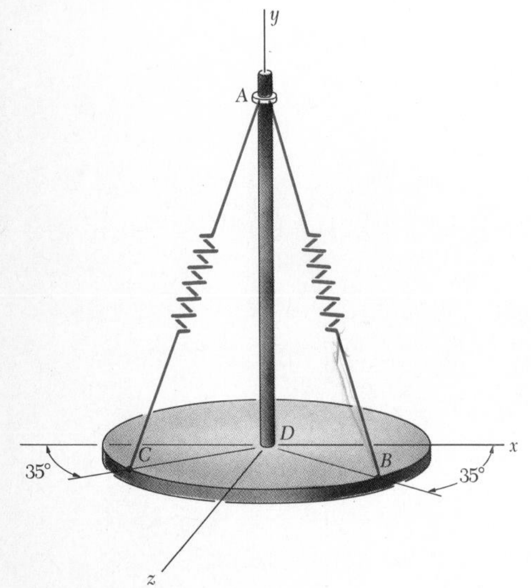

Fig. P2.59 and P2.60

2.59 The angle between the spring *AB* and the post *DA* is 30°. Knowing that the tension in the spring is 50 lb, determine (*a*) the *x*, *y*, and *z* components of the force exerted on the circular plate at *B*, (*b*) the angles θ_x, θ_y, and θ_z defining the direction of the force at *B*.

2.60 The angle between the spring *AC* and the post *DA* is 30°. Knowing that the tension in the spring is 40 lb, determine (*a*) the *x*, *y*, and *z* components of the force exerted on the circular plate at *C*, (*b*) the angles θ_x, θ_y, and θ_z defining the direction of the force at *C*.

2.61 Determine the magnitude and direction of the force $\mathbf{F} = (240 \text{ N})\mathbf{i} - (270 \text{ N})\mathbf{j} + (680 \text{ N})\mathbf{k}$.

2.62 Determine the magnitude and direction of the force $\mathbf{F} = (690 \text{ lb})\mathbf{i} + (300 \text{ lb})\mathbf{j} - (580 \text{ lb})\mathbf{k}$.

2.63 A force acts at the origin in a direction defined by the angles $\theta_x = 70°$ and $\theta_z = 130°$. Knowing that the *y* component of the force is +400 lb, determine (*a*) the other components and the magnitude of the force, (*b*) the value of θ_y.

2.64 A force acts at the origin in a direction defined by the angles $\theta_y = 65°$ and $\theta_z = 40°$. Knowing that the *x* component of the force is −750 N, determine (*a*) the other components and the magnitude of the force, (*b*) the value of θ_x.

2.65 Two cables BG and BH are attached to frame ACD as shown. Knowing that the tension in cable BG is 540 N, determine the components of the force exerted by cable BG on the frame at B.

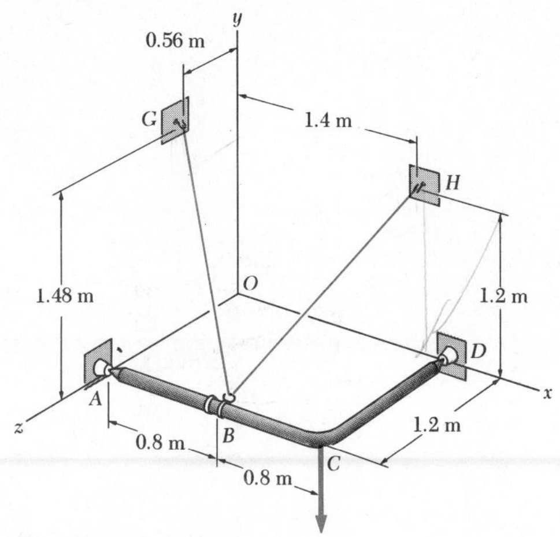

Fig. P2.65 and P2.66

2.66 Two cables BG and BH are attached to frame ACD as shown. Knowing that the tension in cable BH is 750 N, determine the components of the force exerted by cable BH on the frame at B.

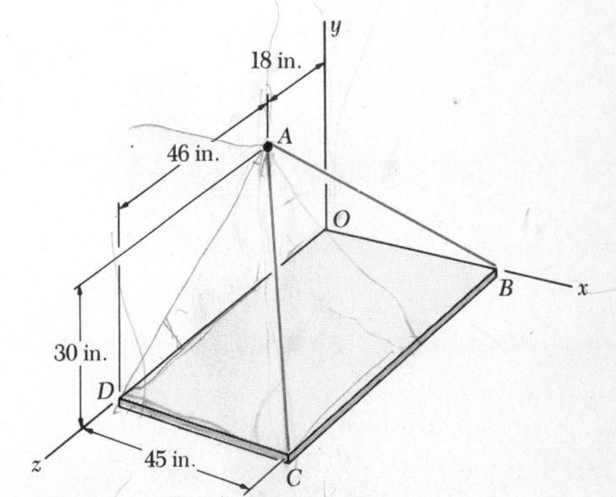

Fig. P2.67, P2.68, and P2.69

2.67 Knowing that the tension in cable AB is 285 lb, determine the components of the force exerted on the plate at B.

2.68 Knowing that the tension in cable AC is 426 lb, determine the components of the force exerted on the plate at C.

2.69 Knowing that the tension is 285 lb in cable AB and 426 lb in cable AC, determine the magnitude and direction of the resultant of the forces exerted at A by the two cables.

2.70 Referring to the frame of Prob. 2.65 and knowing that the tension is 540 N in cable BG and 750 N in cable BH, determine the magnitude and direction of the resultant of the forces exerted by the cables on the frame at B.

2.71 Determine the resultant of the two forces shown.

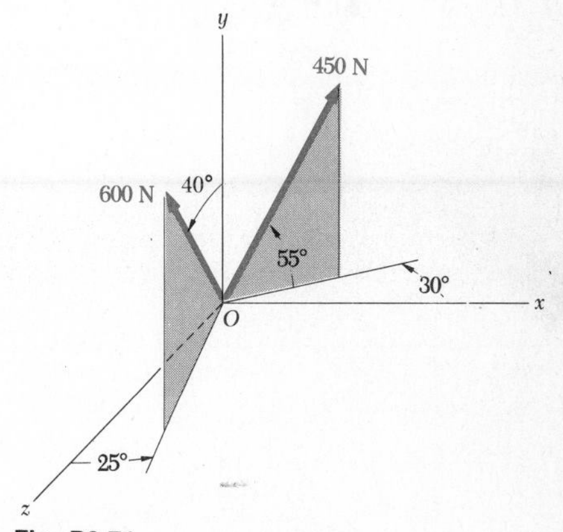

Fig. P2.71

2.72 The angle between each of the springs AB and AC and the post DA is 30°. Knowing that the tension is 50 lb in spring AB and 40 lb in spring AC, determine (*a*) the magnitude and direction of the resultant of the forces exerted by the springs on the post at A, (*b*) the point where the resultant intersects the plate.

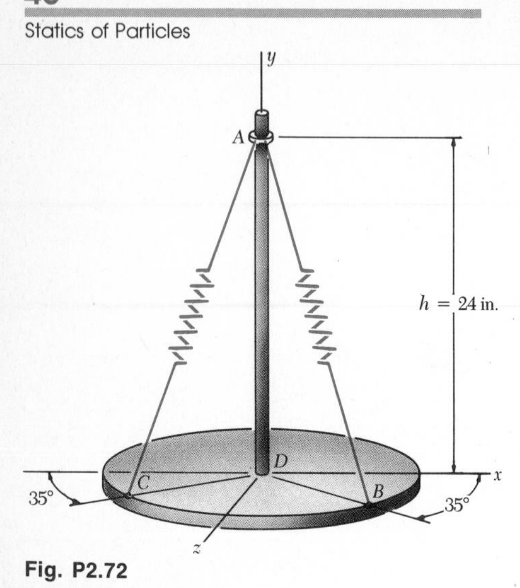

Fig. P2.72

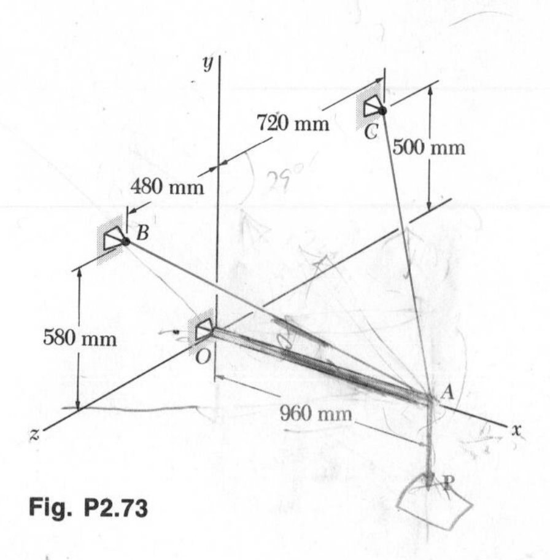

Fig. P2.73

2.73 The boom OA carries a load **P** and is supported by two cables as shown. Knowing that the tension in cable AB is 732 N and that the resultant of the load **P** and of the forces exerted at A by the two cables must be directed along OA, determine the tension in cable AC.

2.74 For the boom and loading of Prob. 2.73, determine the magnitude of the load **P**.

2.15. Equilibrium of a Particle in Space. According to the definition given in Sec. 2.9, a particle A is in equilibrium if the resultant of all the forces acting on A is zero. The components R_x, R_y, R_z of the resultant are given by the relations (2.31); expressing that the components of the resultant are zero, we write

$$\Sigma F_x = 0 \qquad \Sigma F_y = 0 \qquad \Sigma F_z = 0 \tag{2.34}$$

Equations (2.34) represent the necessary and sufficient conditions for the equilibrium of a particle in space. They may be used to solve problems dealing with the equilibrium of a particle involving no more than three unknowns.

To solve such problems, we first should draw a free-body diagram showing the particle in equilibrium and *all* the forces acting on it. We may then write the equations of equilibrium (2.34) and solve them for three unknowns. In the more common types of problems, these unknowns will represent (1) the three components of a single force or (2) the magnitude of three forces each of known direction.

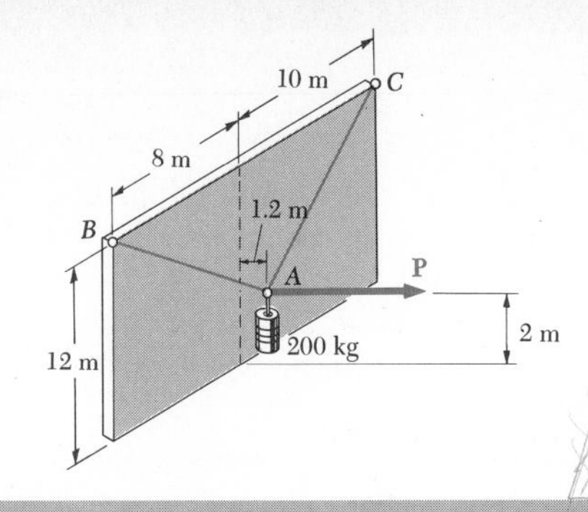

SAMPLE PROBLEM 2.9

A 200-kg cylinder is hung by means of two cables AB and AC, which are attached to the top of a vertical wall. A horizontal force $\mathbf{P}$ perpendicular to the wall holds the cylinder in the position shown. Determine the magnitude of $\mathbf{P}$ and the tension in each cable.

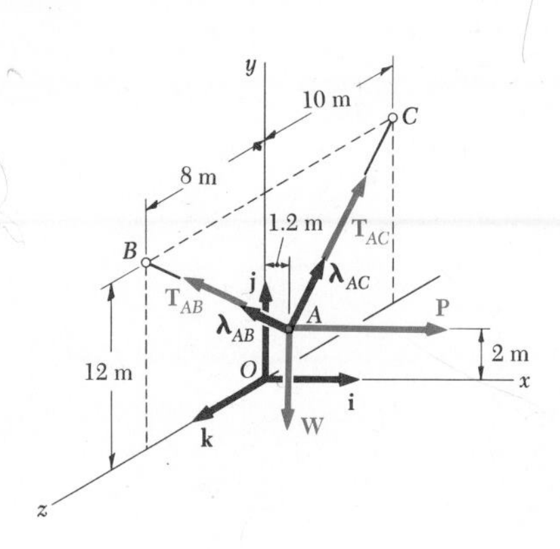

Solution. Point A is chosen as a free body; this point is subjected to four forces, three of which are of unknown magnitude.

Introducing the unit vectors $\mathbf{i}, \mathbf{j}, \mathbf{k}$, we resolve each force into rectangular components.

$$\mathbf{P} = P\mathbf{i}$$
$$\mathbf{W} = -mg\mathbf{j} = -(200 \text{ kg})(9.81 \text{ m/s}^2)\mathbf{j} = -(1962 \text{ N})\mathbf{j} \tag{1}$$

In the case of $\mathbf{T}_{AB}$ and $\mathbf{T}_{AC}$, it is necessary first to determine the components and magnitudes of the vectors $\overrightarrow{AB}$ and $\overrightarrow{AC}$. Denoting by $\boldsymbol{\lambda}_{AB}$ the unit vector along AB, we write

$$\overrightarrow{AB} = -(1.2 \text{ m})\mathbf{i} + (10 \text{ m})\mathbf{j} + (8 \text{ m})\mathbf{k} \qquad AB = 12.86 \text{ m}$$
$$\boldsymbol{\lambda}_{AB} = \frac{\overrightarrow{AB}}{12.86 \text{ m}} = -0.0933\mathbf{i} + 0.778\mathbf{j} + 0.622\mathbf{k}$$
$$\mathbf{T}_{AB} = T_{AB}\boldsymbol{\lambda}_{AB} = -0.0933T_{AB}\mathbf{i} + 0.778T_{AB}\mathbf{j} + 0.622T_{AB}\mathbf{k} \tag{2}$$

Denoting by $\boldsymbol{\lambda}_{AC}$ the unit vector along AC, we write in a similar way

$$\overrightarrow{AC} = -(1.2 \text{ m})\mathbf{i} + (10 \text{ m})\mathbf{j} - (10 \text{ m})\mathbf{k} \qquad AC = 14.19 \text{ m}$$
$$\boldsymbol{\lambda}_{AC} = \frac{\overrightarrow{AC}}{14.19 \text{ m}} = -0.0846\mathbf{i} + 0.705\mathbf{j} - 0.705\mathbf{k}$$
$$\mathbf{T}_{AC} = T_{AC}\boldsymbol{\lambda}_{AC} = -0.0846T_{AC}\mathbf{i} + 0.705T_{AC}\mathbf{j} - 0.705T_{AC}\mathbf{k} \tag{3}$$

Equilibrium Condition. Since A is in equilibrium, we must have

$$\Sigma\mathbf{F} = 0: \qquad \mathbf{T}_{AB} + \mathbf{T}_{AC} + \mathbf{P} + \mathbf{W} = 0$$

or, substituting from (1), (2), (3) for the forces and factoring $\mathbf{i}, \mathbf{j}, \mathbf{k}$,

$$(-0.0933T_{AB} - 0.0846T_{AC} + P)\mathbf{i} + (0.778T_{AB} + 0.705T_{AC} - 1962 \text{ N})\mathbf{j} + (0.622T_{AB} - 0.705T_{AC})\mathbf{k} = 0$$

Setting the coefficients of $\mathbf{i}, \mathbf{j}, \mathbf{k}$ equal to zero, we write three scalar equations, which express that the sums of the x, y, and z components of the forces are respectively equal to zero.

$$(\Sigma F_x = 0:) \qquad -0.0933T_{AB} - 0.0846T_{AC} + P = 0$$
$$(\Sigma F_y = 0:) \qquad +0.778T_{AB} + 0.705T_{AC} - 1962 \text{ N} = 0$$
$$(\Sigma F_z = 0:) \qquad +0.622T_{AB} - 0.705T_{AC} = 0$$

Solving these equations, we obtain

$$P = 235 \text{ N} \qquad T_{AB} = 1401 \text{ N} \qquad T_{AC} = 1236 \text{ N} \blacktriangleleft$$

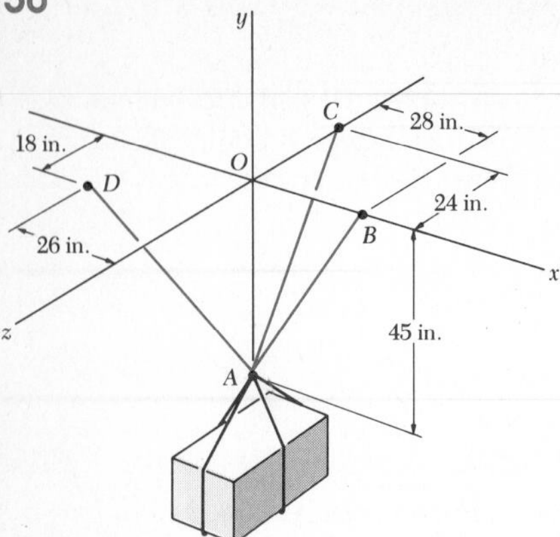

Fig. P2.75 and P2.76

Problems

2.75 A crate is supported by three cables as shown. Determine the weight W of the crate, knowing that the tension in cable AD is 924 lb.

2.76 A crate is supported by three cables as shown. Determine the weight W of the crate, knowing that the tension in cable AB is 1378 lb.

2.77 A container is supported by three cables which are attached to a ceiling as shown. Determine the weight W of the container, knowing that the tension in cable AB is 6 kN.

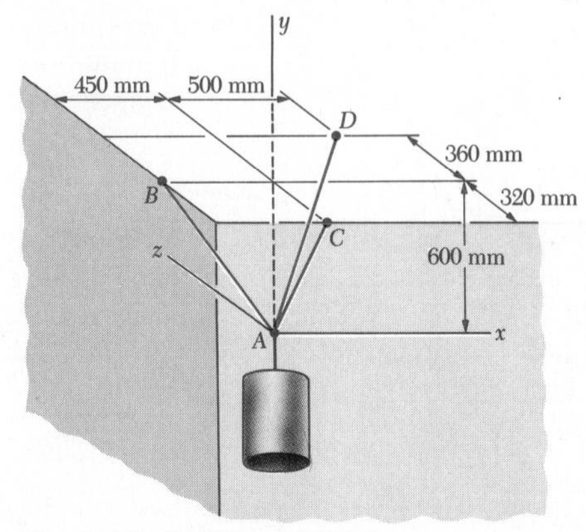

Fig. P2.77, P2.78, and P2.79

2.78 A container is supported by three cables which are attached to a ceiling as shown. Determine the weight W of the container, knowing that the tension in cable AD is 4.3 kN.

2.79 A container of weight $W = 9.32$ kN is supported by three cables which are attached to a ceiling as shown. Determine the tension in each cable.

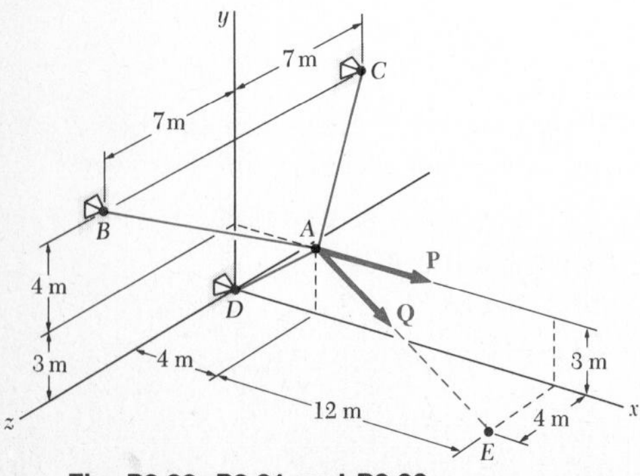

Fig. P2.80, P2.81 and P2.82

2.80 Three cables are connected at A, where the forces **P** and **Q** are applied as shown. Determine the tension in each of the cables when $P = 28$ kN and $Q = 0$.

2.81 Three cables are connected at A, where the forces **P** and **Q** are applied as shown. Determine the tension in each of the cables when $P = 0$ and $Q = 36.4$ kN.

2.82 Three cables are connected at A, where the forces **P** and **Q** are applied as shown. Knowing that $Q = 36.4$ kN and that the tension in cable AD is zero, determine (*a*) the magnitude and sense of **P**, (*b*) the tension in cables AB and AC.

2.83 A 36-lb triangular plate is supported by three wires as shown. Determine the tension in each wire, knowing that $a = 6$ in.

2.84 Solve Prob. 2.83, assuming that $a = 8$ in.

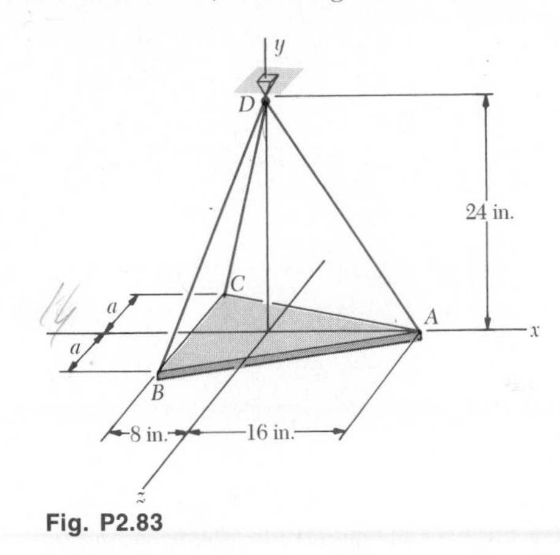

Fig. P2.83

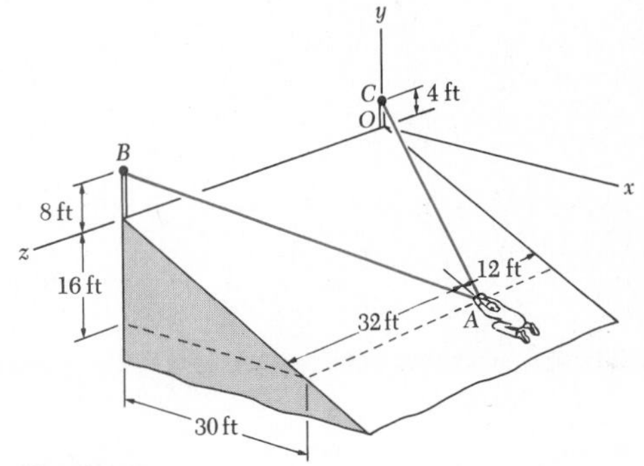

Fig. P2.85

2.85 In trying to move across a slippery icy surface, a 175-lb man uses two ropes *AB* and *AC*. Knowing that the force exerted on the man by the icy surface is perpendicular to that surface, determine the tension in each rope.

2.86 Solve Prob. 2.85, assuming that a friend is helping the man at *A* by pulling on him with a force $\mathbf{P} = -(45 \text{ lb})\mathbf{k}$.

2.87 A container of weight $W = 360$ N is supported by cables *AB* and *AC*, which are tied to ring *A*. Knowing that $\mathbf{Q} = 0$, determine (*a*) the magnitude of the force **P** which must be applied to the ring to maintain the container in the position shown, (*b*) the corresponding values of the tension in cables *AB* and *AC*.

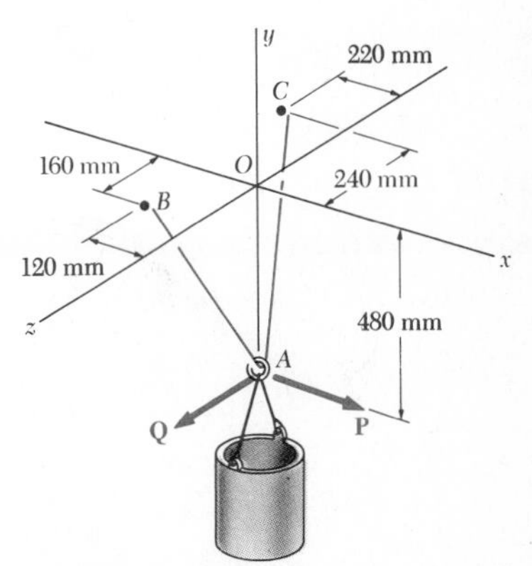

Fig. P2.87 and P2.89

2.88 Solve Prob. 2.87, knowing that $\mathbf{Q} = (60 \text{ N})\mathbf{k}$.

2.89 A container is supported by a single cable which passes through a frictionless ring *A* and is attached to fixed points *B* and *C*. Two forces $\mathbf{P} = P\mathbf{i}$ and $\mathbf{Q} = Q\mathbf{k}$ are applied to the ring to maintain the container in the position shown. Knowing that the weight of the container is $W = 660$ N, determine the magnitudes of **P** and **Q**. (*Hint.* The tension must be the same in portions *AB* and *AC* of the cable.)

2.90 Determine the weight *W* of the container of Prob. 2.89, knowing that $P = 478$ N.

2.91 Cable *BAC* passes through a frictionless ring *A* and is attached to fixed supports at *B* and *C*, while cables *AD* and *AE* are both tied to the ring and are attached, respectively, to supports at *D* and *E*. Knowing that a 200-lb vertical load **P** is applied to ring *A*, determine the tension in each of the three cables.

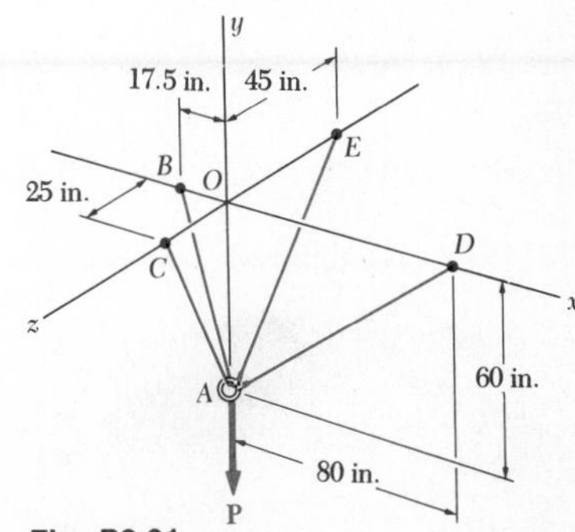

Fig. P2.91

2.92 Knowing that the tension in cable *AE* of Prob. 2.91 is 75 lb, determine (*a*) the magnitude of the load **P**, (*b*) the tension in cables *BAC* and *AD*.

2.93 A 6-kg circular plate of 200-mm radius is supported as shown by three wires of equal length L. Knowing that $\alpha = 30°$, determine the smallest permissible value of the length L if the tension is not to exceed 35 N in any of the wires.

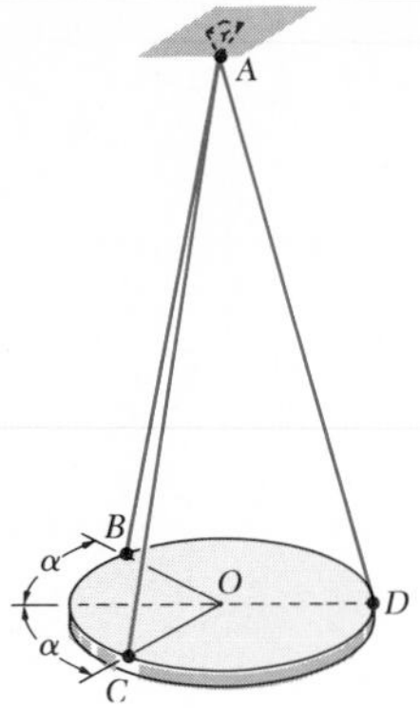

Fig. P2.93

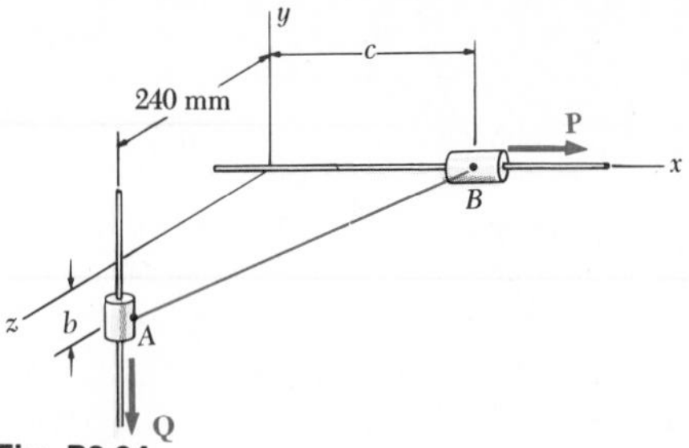

Fig. P2.94

2.94 Collars A and B are connected by a 440-mm wire and may slide freely on frictionless rods. If a force **Q** of magnitude 450 N is applied to collar A as shown, determine (*a*) the tension in the wire when $c = 80$ mm, (*b*) the corresponding magnitude of the force **P** required to maintain the equilibrium of the system.

2.95 Solve Prob. 2.94, assuming that $c = 280$ mm.

Review and Summary

In this chapter we have studied the effect of forces on particles, i.e., on bodies of such shape and size that all forces acting on them may be assumed applied at the same point.

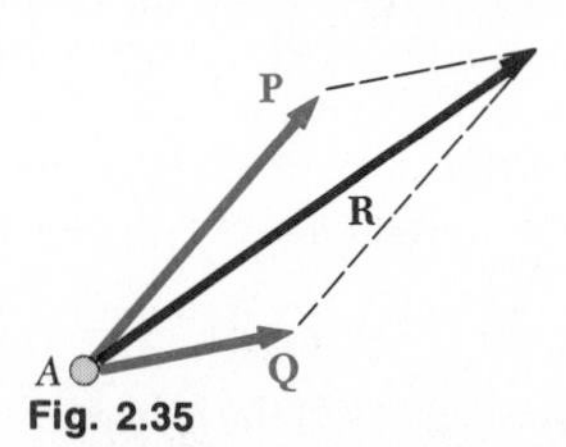

Fig. 2.35

Resultant of two forces

Forces are *vector quantities;* they are characterized by a *point of application*, a *magnitude*, and a *direction*, and add according to the *parallelogram law* (Fig. 2.35). The magnitude and direction of the resultant **R** of two forces **P** and **Q** may be determined either graphically or by trigonometry, using successively the law of cosines and the law of sines [Sample Prob. 2.1].

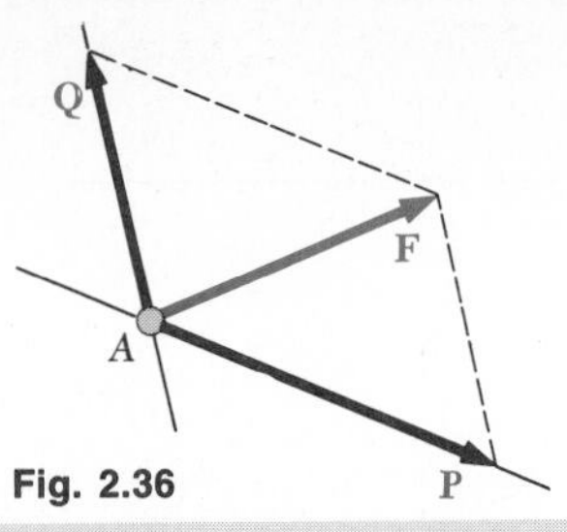

Fig. 2.36

Components of a force

Any given force acting on a particle may be resolved into two or more *components*, i.e., it may be replaced by two or more forces which have the same effect on the particle. A force **F** may be resolved into two components **P** and **Q** by drawing a parallelogram which has **F** for its diagonal; the components **P** and **Q** are then represented by the two adjacent sides of the parallelogram (Fig. 2.36) and may be determined either graphically or by trigonometry [Sec. 2.6].

Rectangular components

Unit vectors

A force **F** is said to have been resolved into two *rectangular components* if its components $\mathbf{F}_x$ and $\mathbf{F}_y$ are perpendicular to each other and directed along the coordinates axes (Fig. 2.37). Introducing the *unit vectors* **i** and **j** along the x and y axes, respectively, we write [Sec. 2.7]

$$\mathbf{F}_x = F_x\mathbf{i} \qquad \mathbf{F}_y = F_y\mathbf{j} \tag{2.6}$$

and

$$\mathbf{F} = F_x\mathbf{i} + F_y\mathbf{j} \tag{2.7}$$

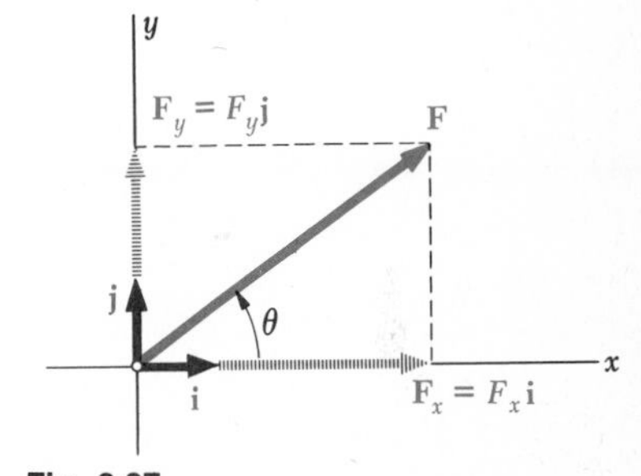

Fig. 2.37

where F_x and F_y are the *scalar components* of **F**. These components, which may be positive or negative, are defined by the relations

$$F_x = F\cos\theta \qquad F_y = F\sin\theta \tag{2.8}$$

When the rectangular components F_x and F_y of a force **F** are given, the angle θ defining the direction of the force may be obtained by writing

$$\tan\theta = \frac{F_y}{F_x} \tag{2.9}$$

The magnitude F of the force may then be obtained by solving one of the equations (2.8) for F or by applying the Pythagorean theorem and writing

$$F = \sqrt{F_x^2 + F_y^2} \tag{2.10}$$

Resultant of several coplanar forces

When *three or more coplanar forces* act on a particle, the rectangular components of their resultant **R** may be obtained by adding algebraically the corresponding components of the given forces [Sec. 2.8]. We have

$$R_x = \Sigma F_x \qquad R_y = \Sigma F_y \tag{2.13}$$

The magnitude and direction of **R** may then be determined from relations similar to Eqs. (2.9) and (2.10) [Sample Prob. 2.3].

Forces in space

A force **F** in *three-dimensional space* may be resolved into rectangular components $\mathbf{F}_x$, $\mathbf{F}_y$, and $\mathbf{F}_z$ [Sec. 2.12]. Denoting by θ_x, θ_y, and θ_z, respectively, the angles that **F** forms with the x, y, and z axes (Fig. 2.38), we have

$$F_x = F \cos \theta_x \qquad F_y = F \cos \theta_y \qquad F_z = F \cos \theta_z \qquad (2.19)$$

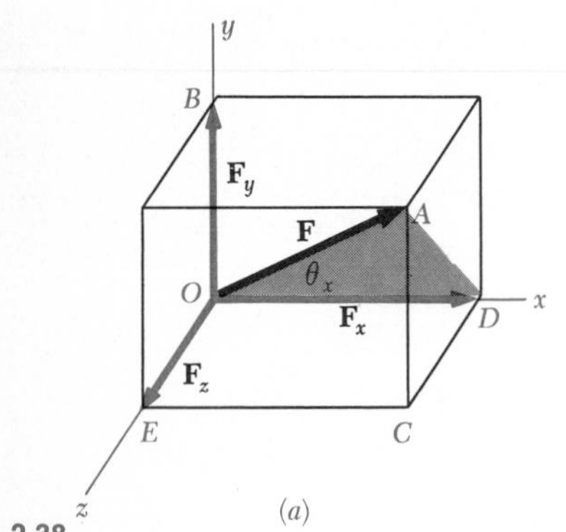

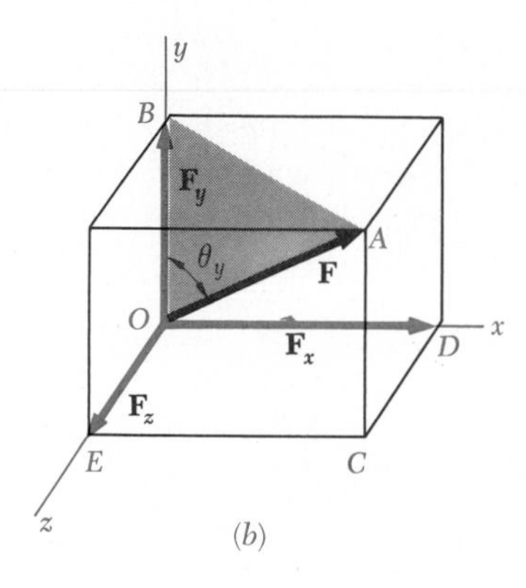

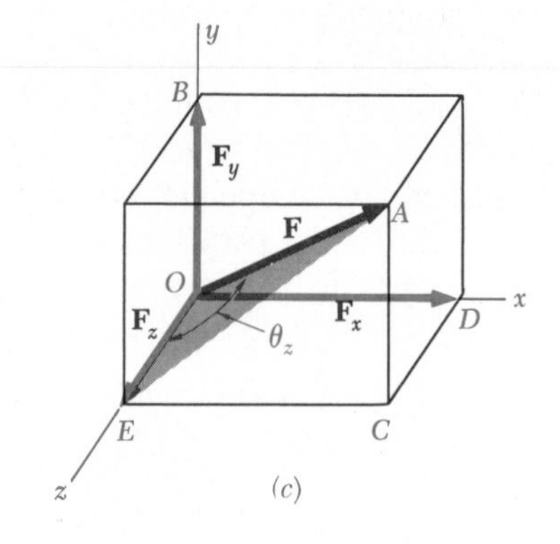

Fig. 2.38

Direction cosines

The cosines of θ_x, θ_y, θ_z are known as the *direction cosines* of the force **F**. Introducing the unit vectors **i**, **j**, **k** along the coordinate axes, we write

$$\mathbf{F} = F_x\mathbf{i} + F_y\mathbf{j} + F_z\mathbf{k} \qquad (2.20)$$

or

$$\mathbf{F} = F(\cos \theta_x\mathbf{i} + \cos \theta_y\mathbf{j} + \cos \theta_z\mathbf{k}) \qquad (2.21)$$

which shows (Fig. 2.39) that **F** is the product of its magnitude F and of the unit vector

$$\boldsymbol{\lambda} = \cos \theta_x\mathbf{i} + \cos \theta_y\mathbf{j} + \cos \theta_z\mathbf{k} \qquad (2.22)$$

Since the magnitude of $\boldsymbol{\lambda}$ is equal to unity, we must have

$$\cos^2 \theta_x + \cos^2 \theta_y + \cos^2 \theta_z = 1 \qquad (2.24)$$

When the rectangular components F_x, F_y, F_z of a force **F** are given, the magnitude F of the force is found by writing

$$F = \sqrt{F_x^2 + F_y^2 + F_z^2} \qquad (2.18)$$

and the direction cosines of **F** are obtained from Eqs. (2.19). We have

$$\cos \theta_x = \frac{F_x}{F} \qquad \cos \theta_y = \frac{F_y}{F} \qquad \cos \theta_z = \frac{F_z}{F} \qquad (2.25)$$

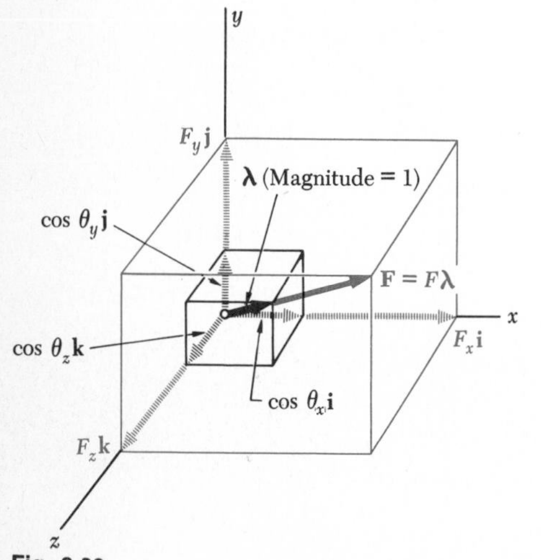

Fig. 2.39

When a force **F** is defined in three-dimensional space by its magnitude F and two points M and N on its line of action [Sec. 2.13], its rectangular components may by obtained as follows: We first express the vector $\overrightarrow{MN}$ joining points M and N in terms of its components d_x, d_y, and d_z (Fig. 2.40); we write

$$\overrightarrow{MN} = d_x\mathbf{i} + d_y\mathbf{j} + d_z\mathbf{k} \tag{2.26}$$

We next determine the unit vector $\boldsymbol{\lambda}$ along the line of action of **F** by dividing $\overrightarrow{MN}$ by its magnitude $MN = d$:

$$\boldsymbol{\lambda} = \frac{\overrightarrow{MN}}{MN} = \frac{1}{d}(d_x\mathbf{i} + d_y\mathbf{j} + d_z\mathbf{k}) \tag{2.27}$$

Recalling that **F** is equal to the product of F and $\boldsymbol{\lambda}$, we have

$$\mathbf{F} = F\boldsymbol{\lambda} = \frac{F}{d}(d_x\mathbf{i} + d_y\mathbf{j} + d_z\mathbf{k}) \tag{2.28}$$

from which it follows [Sample Probs. 2.7 and 2.8] that the scalar components of **F** are, respectively,

$$F_x = \frac{Fd_x}{d} \qquad F_y = \frac{Fd_y}{d} \qquad F_z = \frac{Fd_z}{d} \tag{2.29}$$

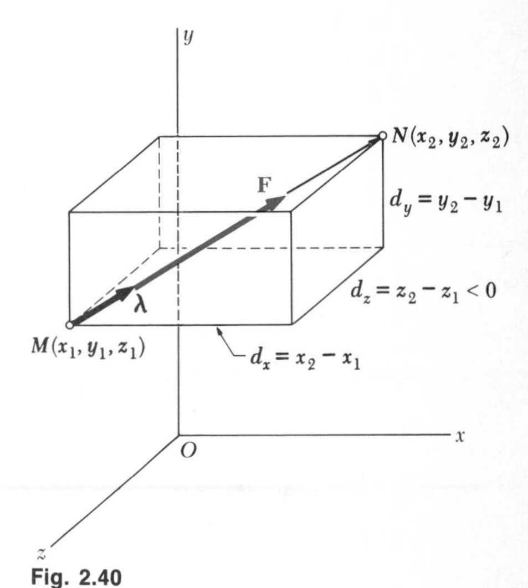

Fig. 2.40

Resultant of forces in space

When *two or more forces* act on a particle in *three-dimensional space*, the rectangular components of their resultant **R** may be obtained by adding algebraically the corresponding components of the given forces [Sec. 2.14]. We have

$$R_x = \Sigma F_x \qquad R_y = \Sigma F_y \qquad R_z = \Sigma F_z \tag{2.31}$$

The magnitude and direction of **R** may then be determined from relations similar to Eqs. (2.18) and (2.25) [Sample Prob. 2.8].

Equilibrium of a particle

A particle is said to be in *equilibrium* when the resultant of all the forces acting on it is zero [Sec. 2.9]. The particle will then remain at rest (if originally at rest) or move with constant speed in a straight line (if originally in motion) [Sec. 2.10].

Free-body diagram

To solve a problem involving a particle in equilibrium, one first should draw a *free-body diagram* of the particle showing all the forces acting on it [Sec. 2.11]. If *only three coplanar forces* act on the particle, a *force triangle* may be drawn to express that the particle is in equilibrium. This triangle may be solved graphically or by trigonometry for no more than two unknowns [Sample Prob. 2.4]. If *more than three coplanar forces* are involved, the equations of equilibrium

$$\Sigma F_x = 0 \qquad \Sigma F_y = 0 \tag{2.15}$$

should be used and solved for no more than two unknowns [Sample Prob. 2.6].

When a particle is in *equilibrium in three-dimensional space* [Sec. 2.15), the three equations of equilibrium

Equilibrium in space

$$\Sigma F_x = 0 \qquad \Sigma F_y = 0 \qquad \Sigma F_z = 0 \tag{2.34}$$

should be used and solved for no more than three unknowns [Sample Prob. 2.9].

Review Problems

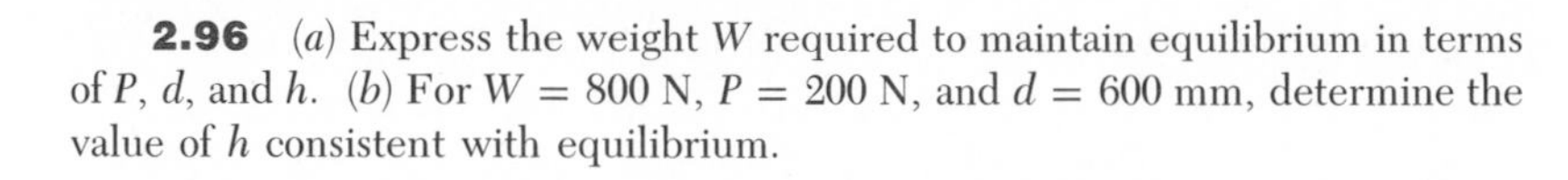

2.96 (*a*) Express the weight W required to maintain equilibrium in terms of P, d, and h. (*b*) For $W = 800$ N, $P = 200$ N, and $d = 600$ mm, determine the value of h consistent with equilibrium.

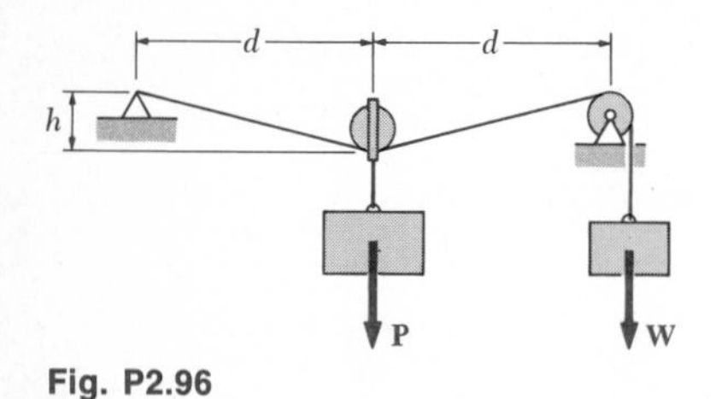

Fig. P2.96

2.97 A cable loop of length 1.5 m is placed around a crate. Knowing that the mass of the crate is 300 kg, determine the tension in the cable for each of the arrangements shown.

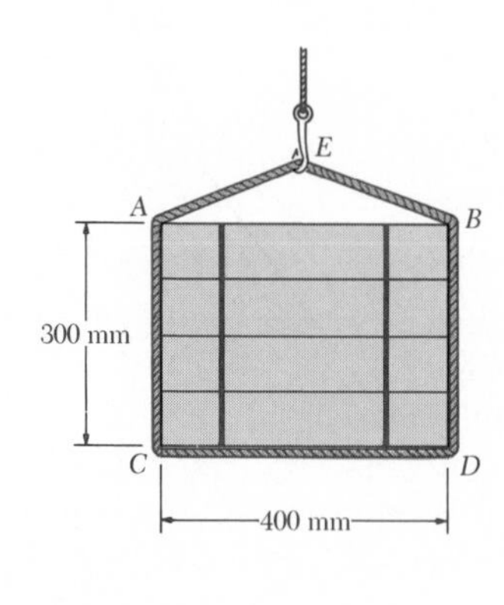

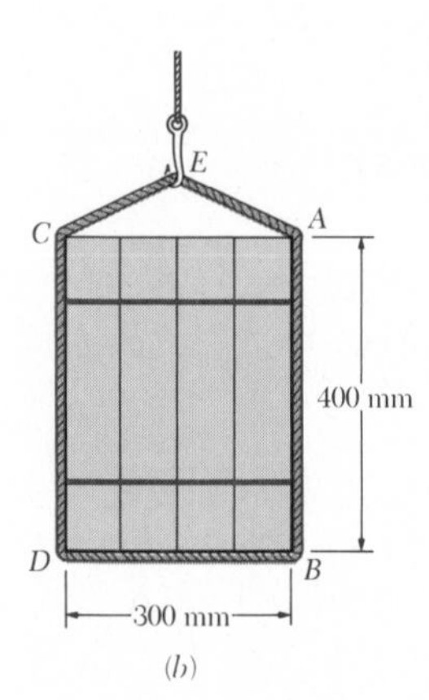

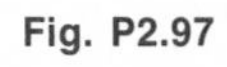
Fig. P2.97

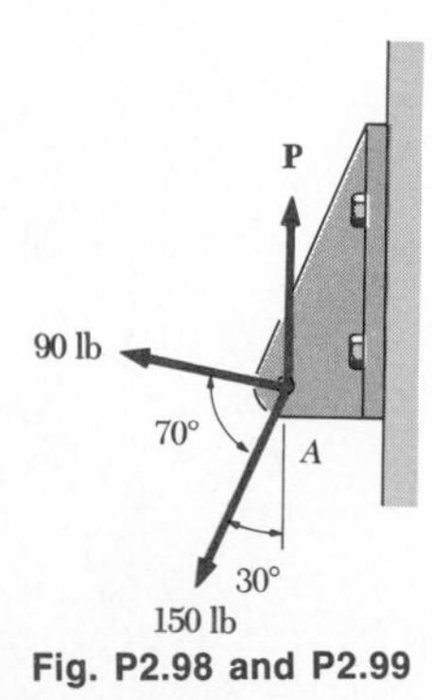

Fig. P2.98 and P2.99

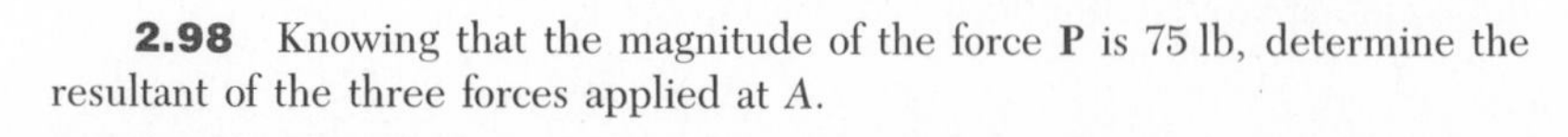

2.98 Knowing that the magnitude of the force **P** is 75 lb, determine the resultant of the three forces applied at A.

2.99 Determine the range of values of P for which the resultant of the three forces applied at A does not exceed 175 lb.

2.100 A container is supported by three cables as shown. Determine the weight W of the container, knowing that the tension in cable AB is 500 N.

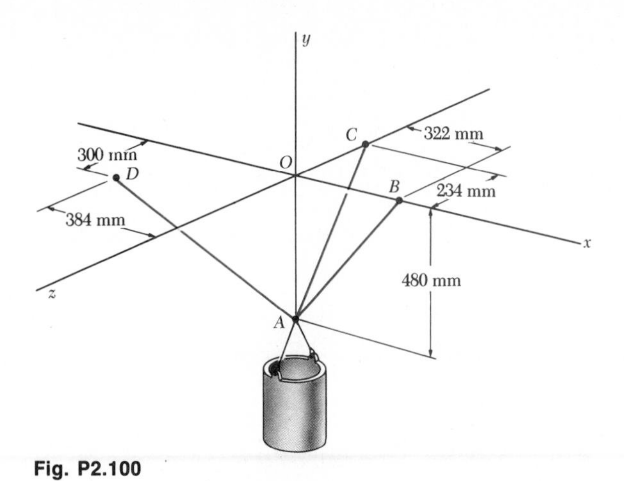

Fig. P2.100

Fig. P2.102 and P2.103

2.101 In Prob. 2.100, determine the angles θ_x, θ_y, and θ_z for the force exerted at D by cable AD.

2.102 Knowing that $P = 300$ N, determine the tension in cables AC and BC.

2.103 Determine the range of values of **P** for which both cables remain taut.

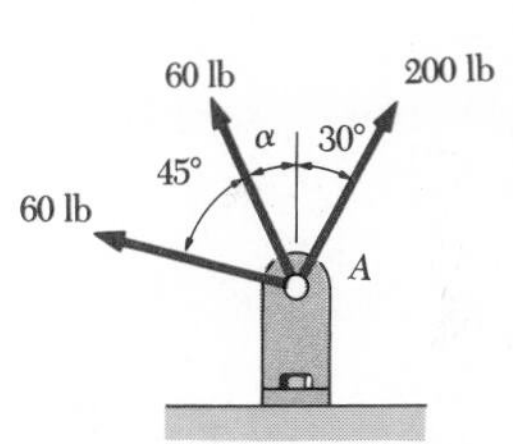

Fig. P2.104 and P2.105

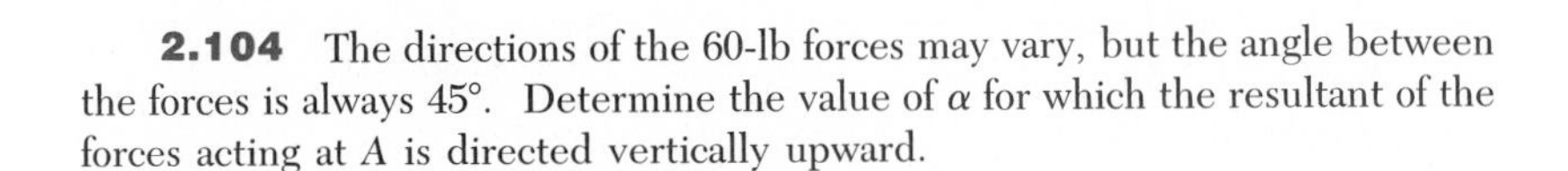

2.104 The directions of the 60-lb forces may vary, but the angle between the forces is always 45°. Determine the value of α for which the resultant of the forces acting at A is directed vertically upward.

2.105 Determine the resultant of the three forces acting at A when (*a*) $\alpha = 0$, (*b*) $\alpha = 25°$.

2.106 A 1200-N force acts at the origin in a direction defined by the angles $\theta_x = 65°$ and $\theta_y = 40°$. It is also known that the z component of the force is positive. Determine the value of θ_z and the components of the force.

2.107 Three wires are connected at point D, which is located 18 in. below the T-shaped pipe support ABC. Determine the tension in each wire when a 180-lb cylinder is suspended from point D as shown.

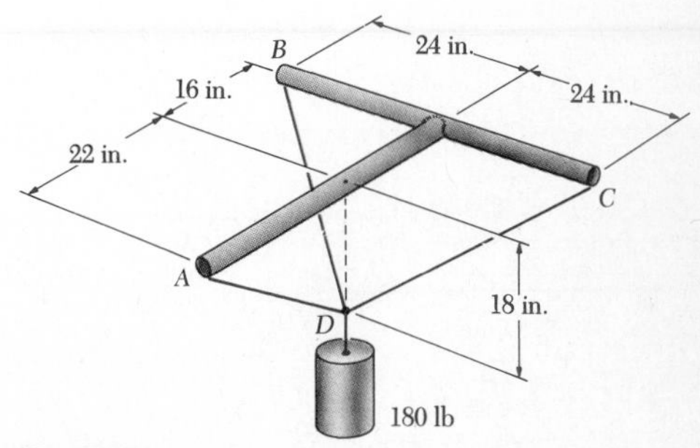

Fig. P2.107

The following problems are designed to be solved with a computer.

2.C1 Write a computer program which can be used to determine the magnitude and direction of the resultant of n forces which act at a fixed point A. Use this program to solve (*a*) Sample Prob. 2.3, (*b*) Prob. 2.28.

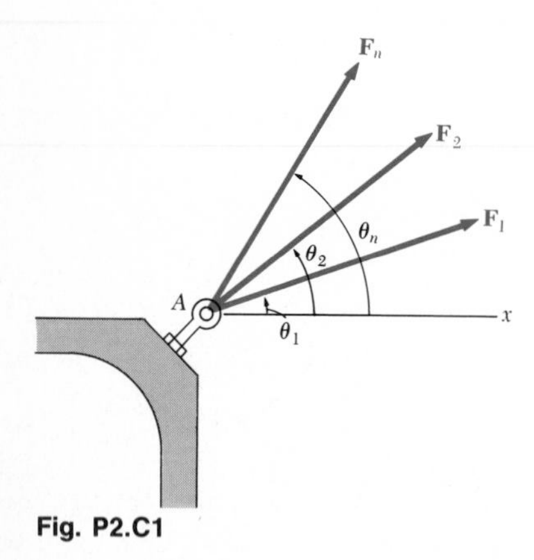

Fig. P2.C1

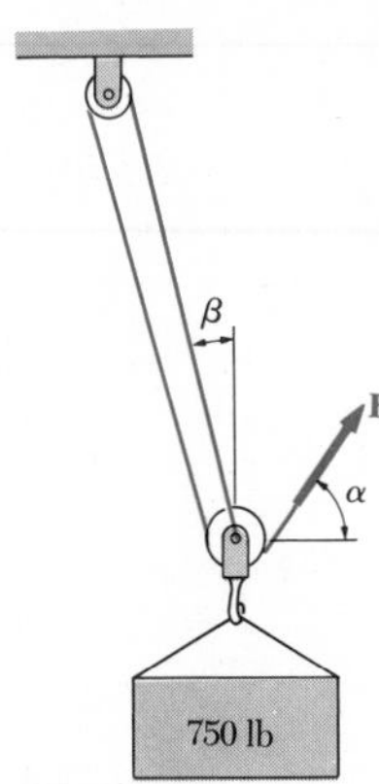

Fig. P2.C2

2.C2 A 750-lb crate is supported by the rope-and-pulley arrangement shown. Write a computer program which can be used to determine, for a given value of β, the magnitude and direction of the force **F** which should be exerted on the free end of the rope. Use this program to calculate F and α for values of β from 0 to 30° at 5° intervals.

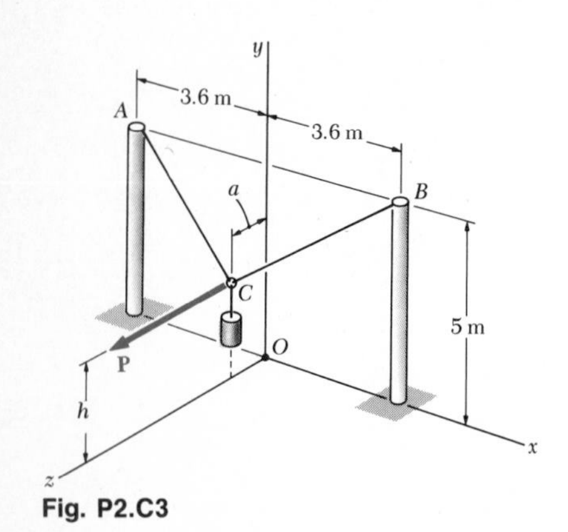

Fig. P2.C3

2.C3 A cylinder of weight $W = 650$ N is supported by two cables AC and BC, each 4.6 m long, which are attached to the top of vertical posts. A horizontal force **P**, perpendicular to the plane containing the posts, holds the cylinder in the position shown. Write a computer program and use it to calculate the elevation h, the magnitude of **P**, and the tension in each cable for values of a from 0 to 2.8 m at 0.4-m intervals.

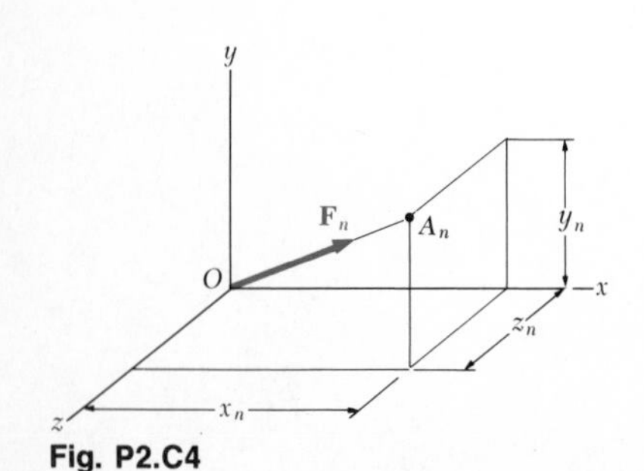

Fig. P2.C4

2.C4 Write a computer program which can be used to calculate the magnitude and direction angles of the resultant of n forces which are applied at the origin and act along lines OA_n. For the data given below, use this program to determine the magnitude and direction of (*a*) $\mathbf{F}_1 + \mathbf{F}_2$, (*b*) $\mathbf{F}_1 + \mathbf{F}_2 + \mathbf{F}_3$, (*c*) $\mathbf{F}_1 + \mathbf{F}_2 + \mathbf{F}_3 + \mathbf{F}_4$.

	F_n, lb	x_n, in.	y_n, in.	z_n, in.
$\mathbf{F}_1$	100	20	10	9
$\mathbf{F}_2$	150	25	15	−10
$\mathbf{F}_3$	200	−9	−3	15
$\mathbf{F}_4$	50	6	14	−20

Rigid Bodies: Equivalent Systems of Forces

CHAPTER

3

3.1. Introduction. In the preceding chapter it was assumed that each of the bodies considered could be treated as a single particle. Such a view, however, is not always possible, and a body, in general, should be treated as a combination of a large number of particles. The size of the body will have to be taken into consideration, as well as the fact that forces will act on different particles and thus will have different points of application.

Most of the bodies considered in elementary mechanics are assumed to be *rigid*, a *rigid body* being defined as one which does not deform. Actual structures and machines, however, are never absolutely rigid and deform under the loads to which they are subjected. But these deformations are usually small and do not appreciably affect the conditions of equilibrium or motion of the structure under consideration. They are important, though, as far as the resistance of the structure to failure is concerned, and are considered in the study of mechanics of materials.

In this chapter we shall study the effect of forces exerted on a rigid body and we shall learn how to replace a given system of forces by a simpler equivalent system. Our analysis will rest on the fundamental assumption that the effect of a given force on a rigid body remains unchanged if that force is moved along its line of action (*principle of transmissibility*). It follows that forces acting on a rigid body may be represented by *sliding vectors*, as indicated earlier in Sec. 2.3.

Two important concepts associated with the effect of a force on a rigid body are the *moment of a force about a point* (Sec. 3.6) and the *moment of a force about an axis* (Sec. 3.11). Since the determination of these quantities involves the computation of vector products and scalar products of two vectors, the fundamentals of vector algebra will be introduced in this chapter and applied to the solution of problems involving forces acting on rigid bodies.

Another concept introduced in this chapter is that of a *couple*, i.e., the combination of two forces which have the same magnitude, parallel lines of action, and opposite sense (Sec. 3.12). As we shall see, any system of forces

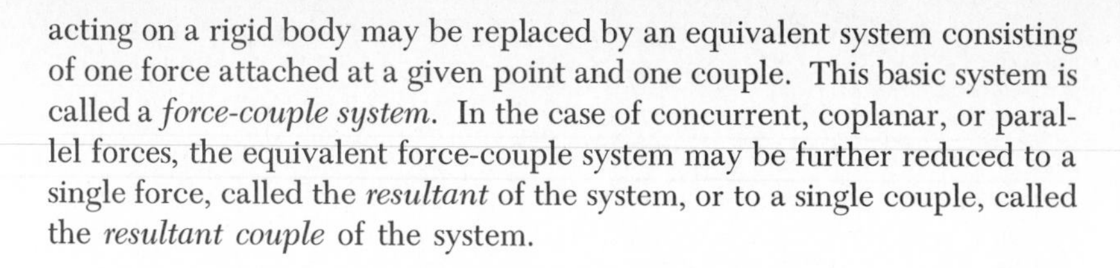

acting on a rigid body may be replaced by an equivalent system consisting of one force attached at a given point and one couple. This basic system is called a *force-couple system*. In the case of concurrent, coplanar, or parallel forces, the equivalent force-couple system may be further reduced to a single force, called the *resultant* of the system, or to a single couple, called the *resultant couple* of the system.

3.2. External and Internal Forces. Forces acting on rigid bodies may be separated into two groups (1) *external forces*, (2) *internal forces*.

1. The *external forces* represent the action of other bodies on the rigid body under consideration. They are entirely responsible for the external behavior of the rigid body. They will either cause it to move or assure that it remains at rest. We shall be concerned only with external forces in this chapter and in Chaps. 4 and 5.
2. The *internal forces* are the forces which hold together the particles forming the rigid body. If the rigid body is structurally composed of several parts, the forces holding the component parts together are also defined as internal forces. Internal forces will be considered in Chaps. 6 and 7.

Fig. 3.1

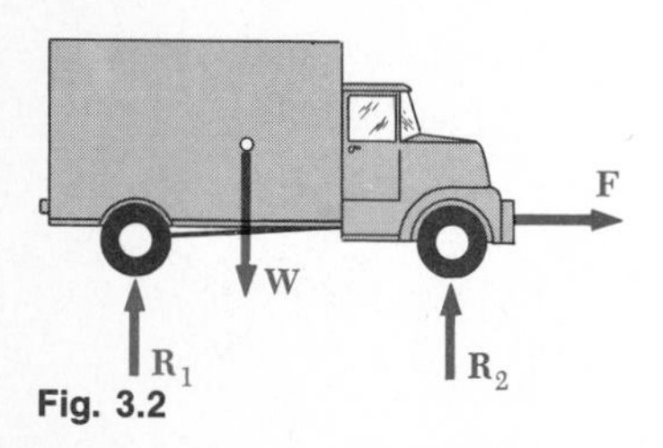

Fig. 3.2

As an example of external forces, we shall consider the forces acting on a disabled truck that men are pulling forward by means of a rope attached to the front bumper (Fig. 3.1). The external forces acting on the truck are shown in a *free-body diagram* (Fig. 3.2). Let us first consider the *weight* of the truck. Although it embodies the effect of the earth's pull on each of the particles forming the truck, the weight may be represented by the single force **W**. The *point of application* of this force, i.e., the point at which the force acts, is defined as the *center of gravity* of the truck. It will be seen in Chap. 5 how centers of gravity may be determined. The weight **W** tends to make the truck move vertically downward. In fact, it would actually cause the truck to move downward, i.e., to fall, if it were not for the presence of the ground. The ground opposes the downward motion of the truck by means of the reactions $\mathbf{R}_1$ and $\mathbf{R}_2$. These forces are exerted *by* the ground *on* the truck and must therefore be included among the external forces acting on the truck.

The men pulling on the rope exert the force **F**. The point of application of **F** is on the front bumper. The force **F** tends to make the truck move forward in a straight line and does actually make it move, since no external force opposes this motion. (Rolling resistance has been neglected here for simplicity.) This forward motion of the truck, during which all straight lines remain parallel to themselves (the floor of the truck remains horizontal, and its walls remain vertical), is known as a *translation*. Other forces might cause the truck to move differently. For example, the force exerted by a jack placed under the front axle would cause the truck to pivot about its rear axle. Such a motion is a *rotation*. It may be concluded, therefore, that each of the *external forces* acting on a *rigid body* can, if unopposed, impart to the rigid body a motion of translation or rotation, or both.

3.3. Principle of Transmissibility. Equivalent Forces. The *principle of transmissibility* states that the conditions of equilibrium or of motion of a rigid body will remain unchanged if a force **F** acting at a given point of the rigid body is replaced by a force **F**′ of the same magnitude and same direction, but acting at a different point, *provided that the two forces have the same line of action* (Fig. 3.3). The two forces **F** and **F**′ have the same effect on the rigid body and are said to be *equivalent*. This principle, which states in fact that the action of a force may be *transmitted* along its line of action, is based on experimental evidence. It *cannot* be derived from the properties established so far in this text and must therefore be accepted as an experimental law. However, as we shall see in Sec. 16.5, the principle of transmissibility may be derived from the study of the dynamics of rigid bodies, but this study requires the introduction of all Newton's three laws and of a number of other concepts as well. We shall therefore base our study of the statics of rigid bodies on the three principles introduced so far, i.e., the parallelogram law of addition, Newton's first law, and the principle of transmissibility.

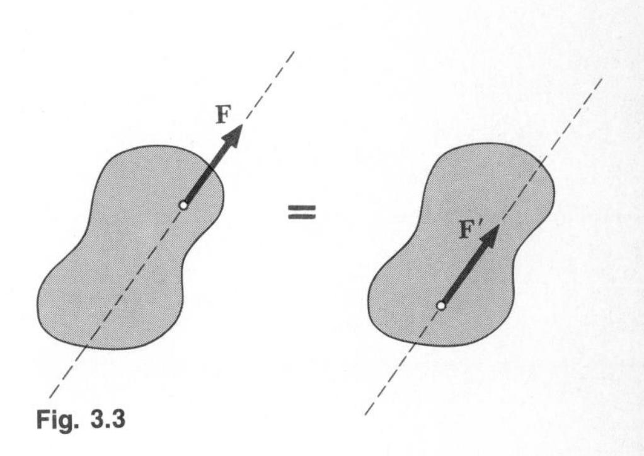

Fig. 3.3

It was indicated in Chap. 2 that the forces acting on a particle could be represented by vectors. These vectors had a well-defined point of application, namely, the particle itself, and were therefore fixed, or bound, vectors. In the case of forces acting on a rigid body, however, the point of application of the force does not matter, as long as the line of action remains unchanged. Thus forces acting on a rigid body must be represented by different kinds of vectors, known as *sliding vectors,* since these vectors may be allowed to slide along their line of action. We should note that all the properties which will be derived in the following sections for the forces acting on a rigid body will be valid more generally for any system of sliding vectors. In order to keep our presentation more intuitive, however, we shall carry it out in terms of physical forces rather than in terms of mathematical sliding vectors.

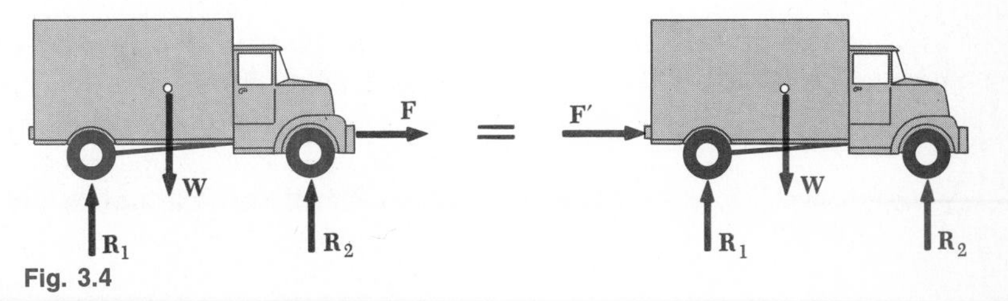

Fig. 3.4

Returning to the example of the truck, we first observe that the line of action of the force **F** is a horizontal line passing through both the front and the rear bumpers of the truck (Fig. 3.4). Using the principle of transmissibility, we may therefore replace **F** by an *equivalent force* **F**′ acting on the rear bumper. In other words, the conditions of motion are unaffected, and all the other external forces acting on the truck ($\mathbf{W}, \mathbf{R}_1, \mathbf{R}_2$) remain unchanged if the men push on the rear bumper instead of pulling on the front bumper.

The principle of transmissibility and the concept of equivalent forces have limitations, however. Consider, for example, a short bar *AB* acted upon by equal and opposite axial forces $\mathbf{P}_1$ and $\mathbf{P}_2$, as shown in Fig. 3.5*a*. According to the principle of transmissibility, the force $\mathbf{P}_2$ may be replaced by a force $\mathbf{P}'_2$ having the same magnitude, same direction, and same line of action, but acting at A instead of B (Fig. 3.5*b*). The forces $\mathbf{P}_1$ and $\mathbf{P}'_2$ acting on the same particle may be added according to the rules of Chap. 2, and, being equal and opposite, their sum is found equal to zero. The original system of forces shown in Fig. 3.5*a* is thus equivalent to no force at all (Fig. 3.5*c*) from the point of view of the external behavior of the bar.

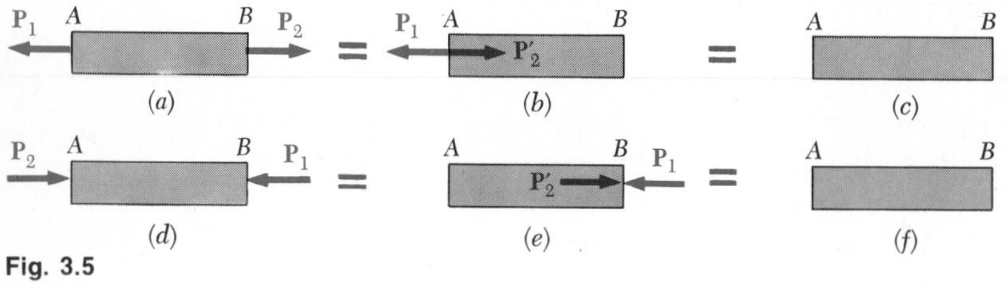

Fig. 3.5

Consider now the two equal and opposite forces $\mathbf{P}_1$ and $\mathbf{P}_2$ acting on the bar AB as shown in Fig. 3.5*d*. The force $\mathbf{P}_2$ may be replaced by a force $\mathbf{P}'_2$ having the same magnitude, same direction, and same line of action, but acting at B instead of A (Fig. 3.5*e*). The forces $\mathbf{P}_1$ and $\mathbf{P}'_2$ may then be added, and their sum is found again to be zero (Fig. 3.5*f*). From the point of view of the mechanics of rigid bodies, the systems shown in Fig. 3.5*a* and *d* are thus equivalent. But the *internal forces* and *deformations* produced by the two systems are clearly different. The bar of Fig. 3.5*a* is in *tension* and, if not absolutely rigid, will increase in length slightly; the bar of Fig. 3.5*d* is in *compression* and, if not absolutely rigid, will decrease in length slightly. Thus, while the principle of transmissibility may be used freely to determine the conditions of motion or equilibrium of rigid bodies and to compute the external forces acting on these bodies, it should be avoided, or at least used with care, in determining internal forces and deformations.

3.4. Vector Product of Two Vectors. In order to gain a better understanding of the effect of a force on a rigid body, we shall introduce a new concept, the concept of *moment of a force about a point*. This concept will be more clearly understood, and we shall be able to apply it more effectively, if we first add to the mathematical tools at our disposal by defining the *vector product* of two vectors.

The vector product of two vectors $\mathbf{P}$ and $\mathbf{Q}$ is defined as the vector $\mathbf{V}$ which satisfies the following conditions:

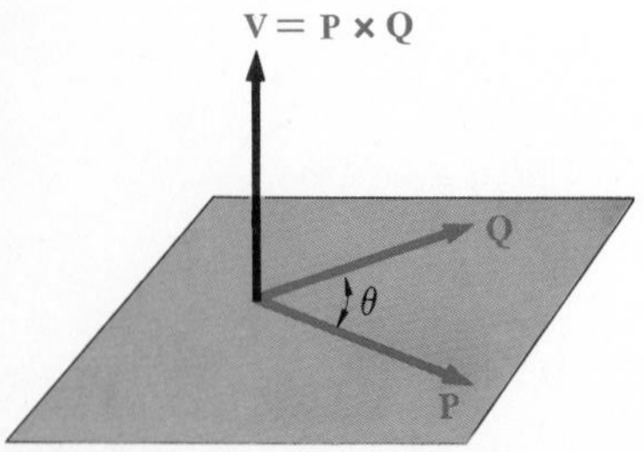

Fig. 3.6

1. The line of action of $\mathbf{V}$ is perpendicular to the plane containing $\mathbf{P}$ and $\mathbf{Q}$ (Fig. 3.6).
2. The magnitude of $\mathbf{V}$ is the product of the magnitudes of $\mathbf{P}$ and $\mathbf{Q}$ and of the sine of the angle θ formed by $\mathbf{P}$ and $\mathbf{Q}$ (the measure of which will always be 180° or less); we thus have

$$V = PQ \sin \theta \tag{3.1}$$

3. The sense of $\mathbf{V}$ is such that a person located at the tip of $\mathbf{V}$ will observe as counterclockwise the rotation through θ which brings the vector $\mathbf{P}$ in line with the vector $\mathbf{Q}$; note that if $\mathbf{P}$ and $\mathbf{Q}$ do not have a common point of application, they should first be redrawn from the same point. The three vectors $\mathbf{P}$, $\mathbf{Q}$, and $\mathbf{V}$—taken in that order—are said to form a *right-handed triad.*†

As stated above, the vector $\mathbf{V}$ satisfying these three conditions (which define it uniquely) is referred to as the vector product of $\mathbf{P}$ and $\mathbf{Q}$; it is represented by the mathematical expression

$$\mathbf{V} = \mathbf{P} \times \mathbf{Q} \tag{3.2}$$

Because of the notation used, the vector product of two vectors $\mathbf{P}$ and $\mathbf{Q}$ is also referred to as the *cross product* of $\mathbf{P}$ and $\mathbf{Q}$.

It follows from Eq. (3.1) that, when two vectors $\mathbf{P}$ and $\mathbf{Q}$ have either the same direction or opposite directions, their vector product is zero. In the general case when the angle θ formed by the two vectors is neither $0°$ nor $180°$, Eq. (3.1) may be given a simple geometric interpretation: The magnitude V of the vector product of $\mathbf{P}$ and $\mathbf{Q}$ measures the area of the parallelogram which has $\mathbf{P}$ and $\mathbf{Q}$ for sides (Fig. 3.7). The vector product $\mathbf{P} \times \mathbf{Q}$ will therefore remain unchanged if we replace $\mathbf{Q}$ by a vector $\mathbf{Q}'$ coplanar with $\mathbf{P}$ and $\mathbf{Q}$ and such that the line joining the tips of $\mathbf{Q}$ and $\mathbf{Q}'$ is parallel to $\mathbf{P}$. We write

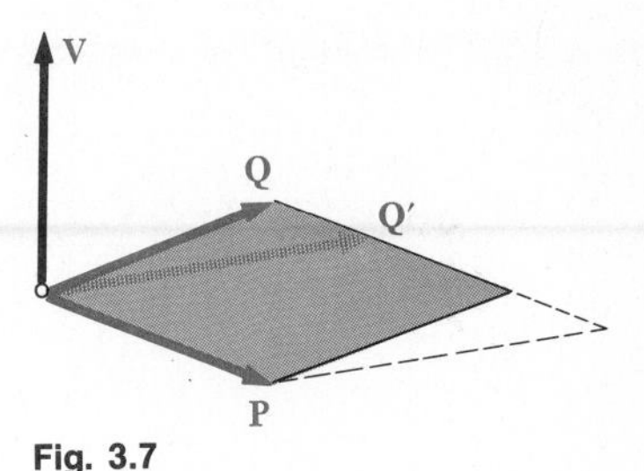

Fig. 3.7

$$\mathbf{V} = \mathbf{P} \times \mathbf{Q} = \mathbf{P} \times \mathbf{Q}' \tag{3.3}$$

From the third condition used to define the vector product $\mathbf{V}$ of $\mathbf{P}$ and $\mathbf{Q}$, namely, the condition stating that $\mathbf{P}$, $\mathbf{Q}$, and $\mathbf{V}$ must form a right-handed triad, it follows that vector products *are not commutative*, i.e., $\mathbf{Q} \times \mathbf{P}$ is not equal to $\mathbf{P} \times \mathbf{Q}$. Indeed, we may easily check that $\mathbf{Q} \times \mathbf{P}$ is represented by the vector $-\mathbf{V}$ equal and opposite to $\mathbf{V}$. We thus write

$$\mathbf{Q} \times \mathbf{P} = -(\mathbf{P} \times \mathbf{Q}) \tag{3.4}$$

Example. Let us compute the vector product $\mathbf{V} = \mathbf{P} \times \mathbf{Q}$ of the vector $\mathbf{P}$ of magnitude 6 lying in the zx plane at an angle of $30°$ with the x axis, and of the vector $\mathbf{Q}$ of magnitude 4 lying along the x axis (Fig. 3.8).

It follows immediately from the definition of the vector product that the vector $\mathbf{V}$ must lie along the y axis, have the magnitude

$$V = PQ \sin\theta = (6)(4) \sin 30° = 12$$

and be directed upward.

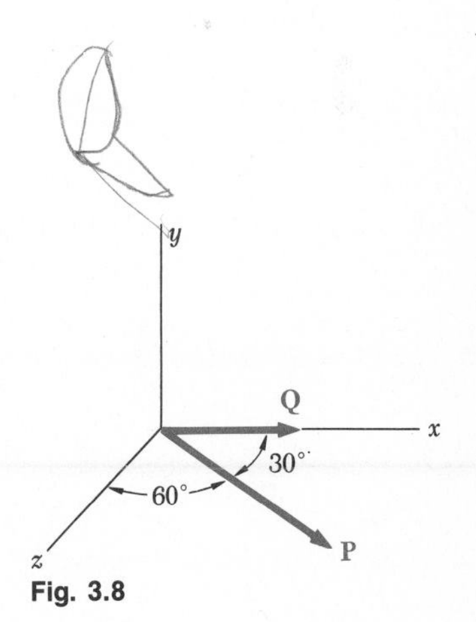

Fig. 3.8

† We should note that the x, y, and z axes used in Chap. 2 form a right-handed system of orthogonal axes and that the unit vectors $\mathbf{i}$, $\mathbf{j}$, $\mathbf{k}$ defined in Sec. 2.12 form a right-handed orthogonal triad.

We saw that the commutative property does not apply to vector products. We may wonder whether the *distributive* property holds, i.e., whether the relation

$$\mathbf{P} \times (\mathbf{Q}_1 + \mathbf{Q}_2) = \mathbf{P} \times \mathbf{Q}_1 + \mathbf{P} \times \mathbf{Q}_2 \tag{3.5}$$

is valid. The answer is *yes*. Many readers are probably willing to accept without formal proof an answer which they intuitively feel is correct. However, since the entire structure of vector algebra and of statics depends upon the relation (3.5), we shall take time out to derive it.

We may, without any loss of generality, assume that **P** is directed along the y axis (Fig. 3.9a). Denoting by **Q** the sum of $\mathbf{Q}_1$ and $\mathbf{Q}_2$, we drop perpendiculars from the tips of **Q**, $\mathbf{Q}_1$, and $\mathbf{Q}_2$ onto the zx plane, defining in this way the vectors $\mathbf{Q}'$, $\mathbf{Q}'_1$, and $\mathbf{Q}'_2$. These vectors will be referred to, respectively, as the *projections* of **Q**, $\mathbf{Q}_1$, and $\mathbf{Q}_2$ on the zx plane. Recalling the property expressed by Eq. (3.3), we note that the left-hand member of Eq. (3.5) may be replaced by $\mathbf{P} \times \mathbf{Q}'$ and that, similarly, the vector products $\mathbf{P} \times \mathbf{Q}_1$ and $\mathbf{P} \times \mathbf{Q}_2$ may respectively be replaced by $\mathbf{P} \times \mathbf{Q}'_1$ and $\mathbf{P} \times \mathbf{Q}'_2$. Thus, the relation to be proved may be written in the form

$$\mathbf{P} \times \mathbf{Q}' = \mathbf{P} \times \mathbf{Q}'_1 + \mathbf{P} \times \mathbf{Q}'_2 \tag{3.5'}$$

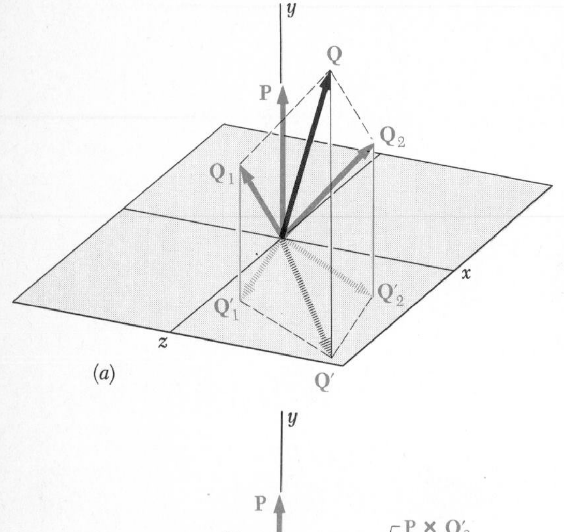

Fig. 3.9

We now observe that $\mathbf{P} \times \mathbf{Q}'$ may be obtained from $\mathbf{Q}'$ by multiplying this vector by the scalar P and rotating it counterclockwise through 90° in the zx plane (Fig. 3.9b); the other two vector products in (3.5′) may be obtained in the same manner from $\mathbf{Q}'_1$ and $\mathbf{Q}'_2$, respectively. Now, since the projection of a parallelogram onto an arbitrary plane is a parallelogram, the projection $\mathbf{Q}'$ of the sum **Q** of $\mathbf{Q}_1$ and $\mathbf{Q}_2$ must be the sum of the projections $\mathbf{Q}'_1$ and $\mathbf{Q}'_2$ of $\mathbf{Q}_1$ and $\mathbf{Q}_2$ on the same plane (Fig. 3.9a). This relation between the vectors $\mathbf{Q}'$, $\mathbf{Q}'_1$, and $\mathbf{Q}'_2$ will still hold after the three vectors have been multiplied by the scalar P and rotated through 90° (Fig. 3.9b). Thus the relation (3.5′) has been proved and we can now be sure that the distributive property holds for vector products.

The third property, the associative property, does not apply to vector products; we have in general

$$(\mathbf{P} \times \mathbf{Q}) \times \mathbf{S} \neq \mathbf{P} \times (\mathbf{Q} \times \mathbf{S}) \tag{3.6}$$

3.5. Vector Products Expressed in Terms of Rectangular Components. We shall now determine the vector product of any two of the unit vectors, **i**, **j**, **k**, defined in Chap. 2. Consider first the product $\mathbf{i} \times \mathbf{j}$ (Fig. 3.10a). Since both vectors have a magnitude equal to 1 and since they are at a right angle to each other, their vector product will also be a unit vector. This unit vector must be **k**, since the vectors **i**, **j**, **k** are mutually perpendicular and form a right-handed triad. On the other hand, the product $\mathbf{j} \times \mathbf{i}$ will be equal to $-\mathbf{k}$ since the 90° rotation which brings **j** into **i** is observed as counterclockwise by a person located at the tip of $-\mathbf{k}$ (Fig. 3.10b). Finally it should be observed that the vector product of a unit vector by itself, such as $\mathbf{i} \times \mathbf{i}$, is equal to zero, since both vectors have the same direction. The vector products of the various possible pairs of unit vectors are

$$\begin{array}{lll} \mathbf{i} \times \mathbf{i} = 0 & \mathbf{j} \times \mathbf{i} = -\mathbf{k} & \mathbf{k} \times \mathbf{i} = \mathbf{j} \\ \mathbf{i} \times \mathbf{j} = \mathbf{k} & \mathbf{j} \times \mathbf{j} = 0 & \mathbf{k} \times \mathbf{j} = -\mathbf{i} \\ \mathbf{i} \times \mathbf{k} = -\mathbf{j} & \mathbf{j} \times \mathbf{k} = \mathbf{i} & \mathbf{k} \times \mathbf{k} = 0 \end{array} \tag{3.7}$$

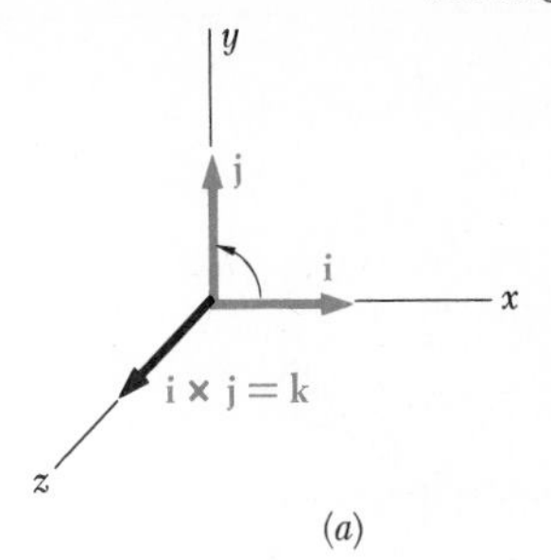

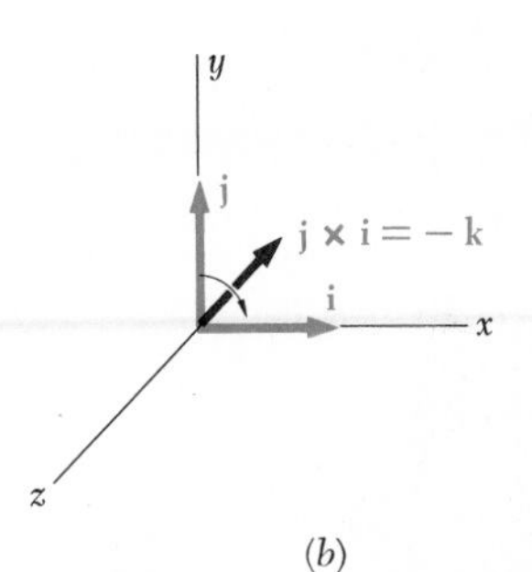

Fig. 3.10

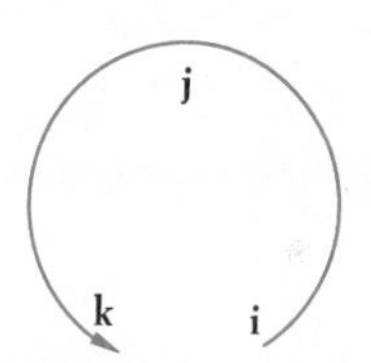

Fig. 3.11

By arranging in a circle and in counterclockwise order the three letters representing the unit vectors (Fig. 3.11), we may simplify the determination of the sign of the vector product of two unit vectors: The product of two unit vectors will be positive if they follow each other in counterclockwise order, and negative otherwise.

We may now easily express the vector product **V** of two given vectors **P** and **Q** in terms of the rectangular components of these vectors. Resolving **P** and **Q** into components, we first write

$$\mathbf{V} = \mathbf{P} \times \mathbf{Q} = (P_x\mathbf{i} + P_y\mathbf{j} + P_z\mathbf{k}) \times (Q_x\mathbf{i} + Q_y\mathbf{j} + Q_z\mathbf{k})$$

Making use of the distributive property, we express **V** as the sum of vector products such as $P_x\mathbf{i} \times Q_y\mathbf{j}$. Observing that each of the expressions obtained is equal to the vector product of two unit vectors, such as $\mathbf{i} \times \mathbf{j}$, multiplied by the product of two scalars, such as P_xQ_y, and recalling the identities (3.7), we obtain, after factoring **i**, **j**, and **k**,

$$\mathbf{V} = (P_yQ_z - P_zQ_y)\mathbf{i} + (P_zQ_x - P_xQ_z)\mathbf{j} + (P_xQ_y - P_yQ_x)\mathbf{k} \tag{3.8}$$

The rectangular components of the vector product **V** are thus found to be

$$\begin{aligned} V_x &= P_yQ_z - P_zQ_y \\ V_y &= P_zQ_x - P_xQ_z \\ V_z &= P_xQ_y - P_yQ_x \end{aligned} \tag{3.9}$$

Returning to Eq. (3.8), we observe that its right-hand member represents the expansion of a determinant. The vector product **V** may thus be expressed in the following form, more easily memorized:†

$$\mathbf{V} = \begin{vmatrix} \mathbf{i} & \mathbf{j} & \mathbf{k} \\ P_x & P_y & P_z \\ Q_x & Q_y & Q_z \end{vmatrix} \tag{3.10}$$

† Any determinant consisting of three rows and three columns may be evaluated by repeating the first and second columns and forming products along each diagonal line. The sum of the products obtained along the colored lines is then subtracted from the sum of the products obtained along the black lines.

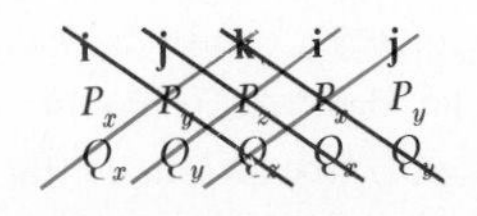

3.6. Moment of a Force about a Point.

Let us now consider a force **F** acting on a rigid body (Fig. 3.12). As we know, the force **F** is represented by a vector which defines its magnitude and direction. However, the effect of the force on the rigid body depends also upon its point of application A. The position of A may be conveniently defined by the vector **r** which joins the fixed reference point O with A; this vector is known as the *position vector* of A.† The position vector **r** and the force **F** define the plane shown in Fig. 3.12*a*.

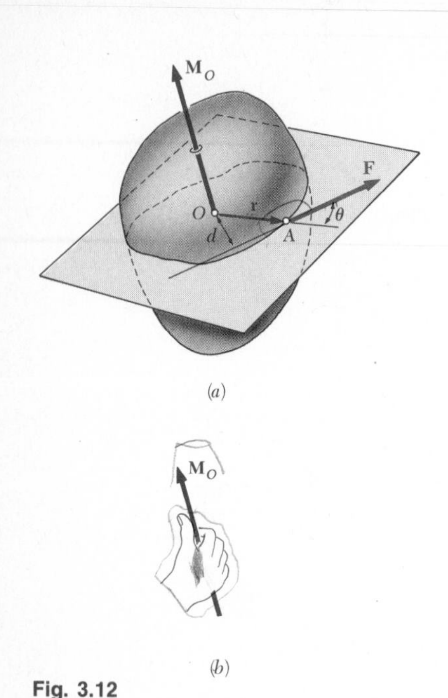

Fig. 3.12

We shall define the *moment of* **F** *about* O as the vector product of **r** and **F**:

$$\mathbf{M}_O = \mathbf{r} \times \mathbf{F} \tag{3.11}$$

According to the definition of the vector product given in Sec. 3.4, the moment $\mathbf{M}_O$ must be perpendicular to the plane containing O and the force **F**. The sense of $\mathbf{M}_O$ is defined by the sense of the rotation which would bring the vector **r** in line with the vector **F**; this rotation should be observed as *counterclockwise* by an observer located at the tip of $\mathbf{M}_O$. Another way of defining the sense of $\mathbf{M}_O$ is furnished by the *right-hand rule:* Close your right hand and hold it so that your fingers are curled in the sense of the rotation that **F** would impart to the rigid body about a fixed axis directed along the line of action of $\mathbf{M}_O$; your thumb will indicate the sense of the moment $\mathbf{M}_O$ (Fig. 3.12*b*).

Finally, denoting by θ the angle between the lines of action of the position vector **r** and the force **F**, we find that the magnitude of the moment of **F** about O is

$$M_O = rF \sin\theta = Fd \tag{3.12}$$

where d represents the perpendicular distance from O to the line of action of **F**. Since the tendency of a force **F** to make a rigid body rotate about a fixed axis perpendicular to the force depends upon the distance of **F** from that axis, as well as upon the magnitude of **F**, we note that *the magnitude of* $\mathbf{M}_O$ *measures the tendency of the force* **F** *to make the rigid body rotate about a fixed axis directed along* $\mathbf{M}_O$.

In the SI system of units, where a force is expressed in newtons (N) and a distance in meters (m), the moment of a force will be expressed in newton-meters (N·m). In the U.S. customary system of units, where a force is expressed in pounds and a distance in feet or inches, the moment of a force will be expressed in lb·ft or lb·in.

†We may easily verify that position vectors obey the law of vector addition and, thus, are truly vectors. Consider, for example, the position vectors **r** and **r**′ of A with respect to two reference points O and O', and the position vector **s** of O with respect to O' (Fig. 3.40*a*, Sec. 3.16). We check that the position vector $\mathbf{r}' = \overrightarrow{O'A}$ may be obtained from the position vectors $\mathbf{s} = \overrightarrow{O'O}$ and $\mathbf{r} = \overrightarrow{OA}$ by applying the triangle rule for the addition of vectors.

We may observe that the moment $\mathbf{M}_O$ of a force about a point, while it depends upon the magnitude, the line of action, and the sense of the force, does *not* depend upon the actual position of the point of application of the force along its line of action. Conversely, the moment $\mathbf{M}_O$ of a force $\mathbf{F}$ does not characterize the position of the point of application of $\mathbf{F}$.

However, as we shall see presently, the moment $\mathbf{M}_O$ of a force $\mathbf{F}$ of given magnitude and direction *completely defines the line of action of* $\mathbf{F}$. Indeed, the line of action of $\mathbf{F}$ must lie in a plane through O perpendicular to the moment $\mathbf{M}_O$; its distance d from O must be equal to the quotient M_O/F of the magnitudes of $\mathbf{M}_O$ and $\mathbf{F}$; and the sense of $\mathbf{M}_O$ determines whether the line of action of $\mathbf{F}$ is to be drawn on one side or the other of the point O.

We recall from Sec. 3.3 that the principle of transmissibility states that two forces $\mathbf{F}$ and $\mathbf{F}'$ are equivalent (i.e., have the same effect on a rigid body) if they have the same magnitude, same direction, and same line of action. This principle may now be restated as follows: *Two forces* $\mathbf{F}$ *and* $\mathbf{F}'$ *are equivalent if, and only if, they are equal* (i.e., have the same magnitude and same direction) *and have equal moments about a given point O.* The necessary and sufficient condition for two forces $\mathbf{F}$ and $\mathbf{F}'$ to be equivalent is thus

$$\mathbf{F} = \mathbf{F}' \qquad \text{and} \qquad \mathbf{M}_O = \mathbf{M}'_O \tag{3.13}$$

We should observe that it follows from this statement that if the relations (3.13) hold for a given point O, they will hold for any other point.

Problems Involving Only Two Dimensions. Many applications deal with two-dimensional structures, i.e., structures which have length and breadth but only negligible depth, and which are subjected to forces contained in the plane of the structure. Two-dimensional structures and the forces acting on them may be readily represented on a sheet of paper or on a blackboard. Their analysis is therefore considerably simpler than that of three-dimensional structures and forces.

Consider, for example, a rigid slab acted upon by a force $\mathbf{F}$ (Fig. 3.13). The moment of $\mathbf{F}$ about a point O chosen in the plane of the figure is represented by a vector $\mathbf{M}_O$ perpendicular to that plane and of magnitude Fd. In the case of Fig. 3.13*a* the vector $\mathbf{M}_O$ points *out* of the paper, while in the case of Fig. 3.13*b* it points *into* the paper. As we look at the figure, we observe the action of $\mathbf{F}$ in the first case as counterclockwise, and in the second case as clockwise. Therefore, it is natural to refer to the sense of the moment of $\mathbf{F}$ about O in Fig. 3.13*a* as counterclockwise ↺, and in Fig. 3.13*b* as clockwise ↻.

Since the moment of a force $\mathbf{F}$ acting in the plane of the figure must be perpendicular to that plane, we need only specify the *magnitude* and the *sense* of the moment of $\mathbf{F}$ about O. This may be done by assigning to the magnitude M_O of the moment a positive or negative sign, according to whether the vector $\mathbf{M}_O$ points out of or into the paper.

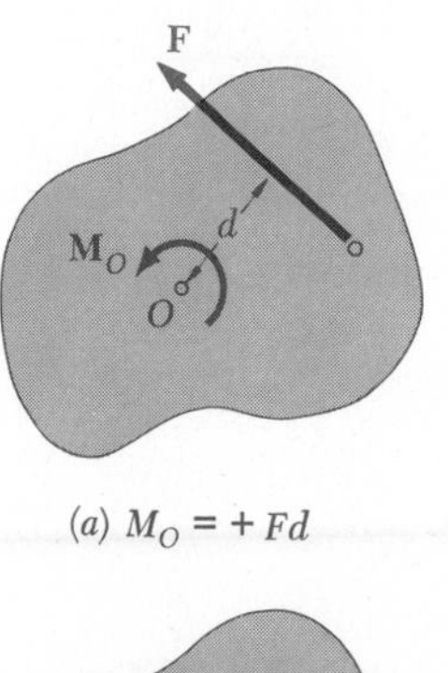

(*a*) $M_O = +Fd$

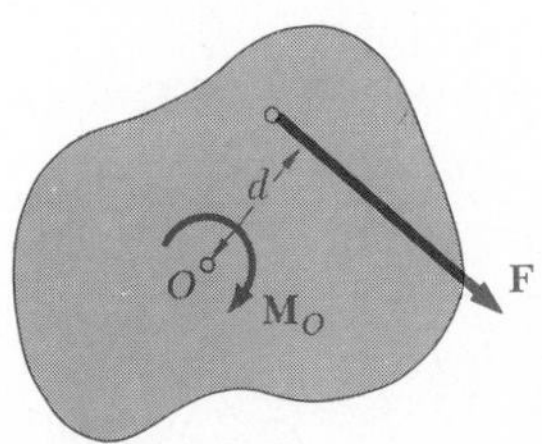

(*b*) $M_O = -Fd$

Fig. 3.13

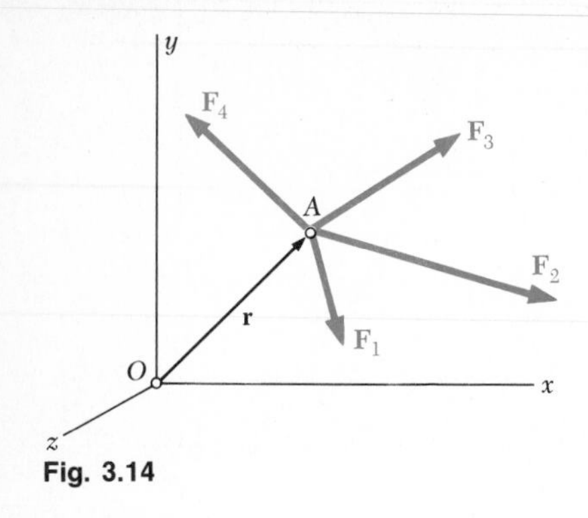

Fig. 3.14

3.7. Varignon's Theorem. The distributive property of vector products may be used to determine the moment of the resultant of several *concurrent forces*. If several forces $\mathbf{F}_1$, $\mathbf{F}_2$, . . . are applied at the same point A (Fig. 3.14), and if we denote by $\mathbf{r}$ the position vector of A, it follows immediately from formula (3.5) of Sec. 3.4 that

$$\mathbf{r} \times (\mathbf{F}_1 + \mathbf{F}_2 + \cdots) = \mathbf{r} \times \mathbf{F}_1 + \mathbf{r} \times \mathbf{F}_2 + \cdots \tag{3.14}$$

In words, *the moment about a given point O of the resultant of several concurrent forces is equal to the sum of the moments of the various forces about the same point O.* This property was originally established by the French mathematician Varignon (1654–1722), long before the introduction of vector algebra, and is known as *Varignon's theorem.*

The relation (3.14) makes it possible to replace the direct determination of the moment of a force $\mathbf{F}$ by the determination of the moments of two or more component forces. As we shall see in the next section, $\mathbf{F}$ will generally be resolved into components parallel to the coordinate axes. However, it may be found more expeditious in some instances to resolve $\mathbf{F}$ into components which are not parallel to the coordinate axes (see Sample Prob. 3.3).

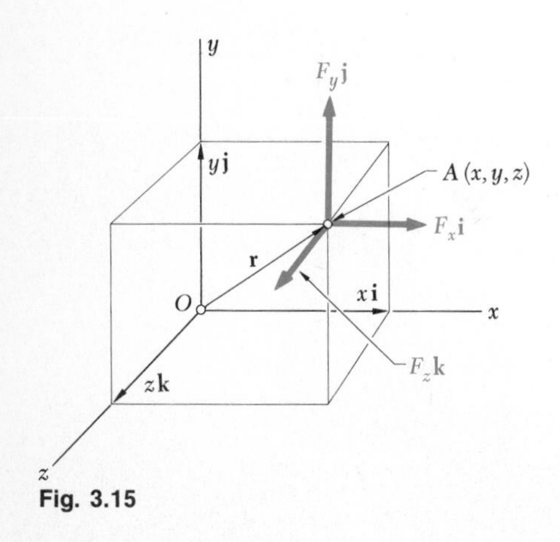

Fig. 3.15

3.8. Rectangular Components of the Moment of a Force. In general, the determination of the moment of a force in space will be considerably simplified if the force and the position vector of its point of application are resolved into rectangular x, y, and z components. Consider, for example, the moment $\mathbf{M}_O$ about O of a force $\mathbf{F}$ of components F_x, F_y, and F_z, applied at a point A of coordinates x, y, and z (Fig. 3.15). Observing that the components of the position vector $\mathbf{r}$ are respectively equal to the coordinates x, y, and z of the point A, we write

$$\mathbf{r} = x\mathbf{i} + y\mathbf{j} + z\mathbf{k} \tag{3.15}$$
$$\mathbf{F} = F_x\mathbf{i} + F_y\mathbf{j} + F_z\mathbf{k} \tag{3.16}$$

Substituting for $\mathbf{r}$ and $\mathbf{F}$ from (3.15) and (3.16) into

$$\mathbf{M}_O = \mathbf{r} \times \mathbf{F} \tag{3.11}$$

and recalling the results obtained in Sec. 3.5, we write the moment $\mathbf{M}_O$ of $\mathbf{F}$ about O in the form

$$\mathbf{M}_O = M_x\mathbf{i} + M_y\mathbf{j} + M_z\mathbf{k} \tag{3.17}$$

where the components M_x, M_y, and M_z are defined by the relations

$$\begin{aligned} M_x &= yF_z - zF_y \\ M_y &= zF_x - xF_z \\ M_z &= xF_y - yF_x \end{aligned} \tag{3.18}$$

As we shall see in Sec. 3.11, the scalar components M_x, M_y, and M_z of the moment $\mathbf{M}_O$ measure the tendency of the force $\mathbf{F}$ to impart to a rigid body a motion of rotation about the x, y, and z axes, respectively. Substituting from (3.18) into (3.17), we may also write $\mathbf{M}_O$ in the form of the determinant

$$\mathbf{M}_O = \begin{vmatrix} \mathbf{i} & \mathbf{j} & \mathbf{k} \\ x & y & z \\ F_x & F_y & F_z \end{vmatrix} \tag{3.19}$$

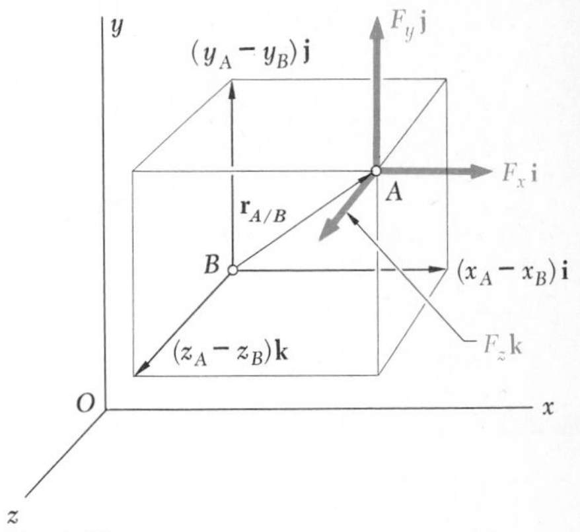

Fig. 3.16

To compute the moment $\mathbf{M}_B$ about an arbitrary point B of a force $\mathbf{F}$ applied at A (Fig. 3.16), we must replace the position vector $\mathbf{r}$ in Eq. (3.11) by a vector drawn from B to A. This vector is the *position vector of* A *relative to* B and will be denoted by $\mathbf{r}_{A/B}$. Observing that $\mathbf{r}_{A/B}$ may be obtained by subtracting $\mathbf{r}_B$ from $\mathbf{r}_A$, we write

$$\mathbf{M}_B = \mathbf{r}_{A/B} \times \mathbf{F} = (\mathbf{r}_A - \mathbf{r}_B) \times \mathbf{F} \tag{3.20}$$

or, using the determinant form,

$$\mathbf{M}_B = \begin{vmatrix} \mathbf{i} & \mathbf{j} & \mathbf{k} \\ x_{A/B} & y_{A/B} & z_{A/B} \\ F_x & F_y & F_z \end{vmatrix} \tag{3.21}$$

where $x_{A/B}$, $y_{A/B}$, and $z_{A/B}$ denote the components of the vector $\mathbf{r}_{A/B}$:

$$x_{A/B} = x_A - x_B \qquad y_{A/B} = y_A - y_B \qquad z_{A/B} = z_A - z_B$$

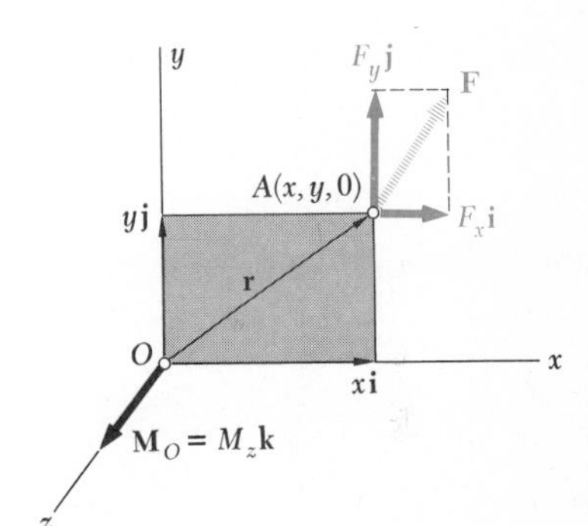

Fig. 3.17

In the case of *problems involving only two dimensions*, the force $\mathbf{F}$ may be assumed to lie in the xy plane (Fig. 3.17). Carrying $z = 0$ and $F_z = 0$ into the relations (3.19), we obtain

$$\mathbf{M}_O = (xF_y - yF_x)\mathbf{k}$$

We check that the moment of $\mathbf{F}$ about O is perpendicular to the plane of the figure and that it is completely defined by the scalar

$$M_O = M_z = xF_y - yF_x \tag{3.22}$$

As noted earlier, a positive value for M_O indicates that the vector $\mathbf{M}_O$ points out of the paper (the force $\mathbf{F}$ tends to rotate the body counterclockwise about O), and a negative value that the vector $\mathbf{M}_O$ points into the paper (the force $\mathbf{F}$ tends to rotate the body clockwise about O).

To compute the moment about $B(x_B, y_B)$ of a force lying in the xy plane and applied at $A(x_A, y_A)$ (Fig. 3.18), we carry $z_{A/B} = 0$ and $F_z = 0$ in the relations (3.21) and check that the vector $\mathbf{M}_B$ is perpendicular to the xy plane and is defined in magnitude and sense by the scalar

$$M_B = (x_A - x_B)F_y - (y_A - y_B)F_x \tag{3.23}$$

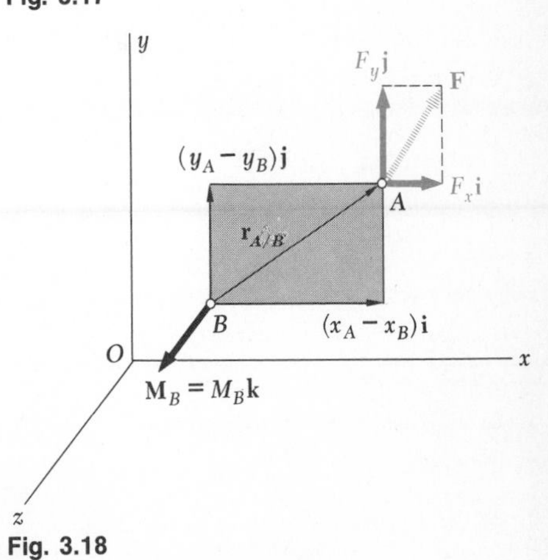

Fig. 3.18

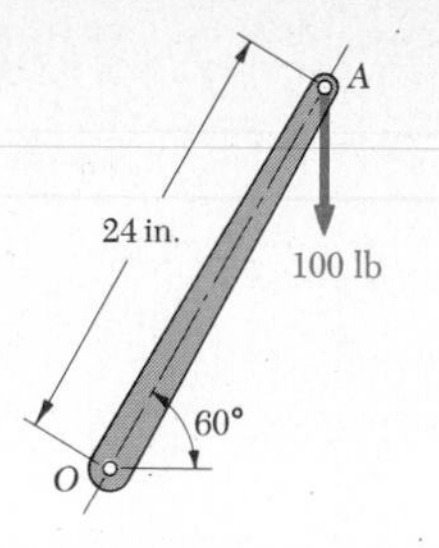

SAMPLE PROBLEM 3.1

A 100-lb vertical force is applied to the end of a lever which is attached to a shaft at O. Determine (*a*) the moment of the 100-lb force about O; (*b*) the magnitude of the horizontal force applied at A which creates the same moment about O; (*c*) the smallest force applied at A which creates the same moment about O; (*d*) how far from the shaft a 240-lb vertical force must act to create the same moment about O; (*e*) whether any one of the forces obtained in parts *b*, *c*, and *d* is equivalent to the original force.

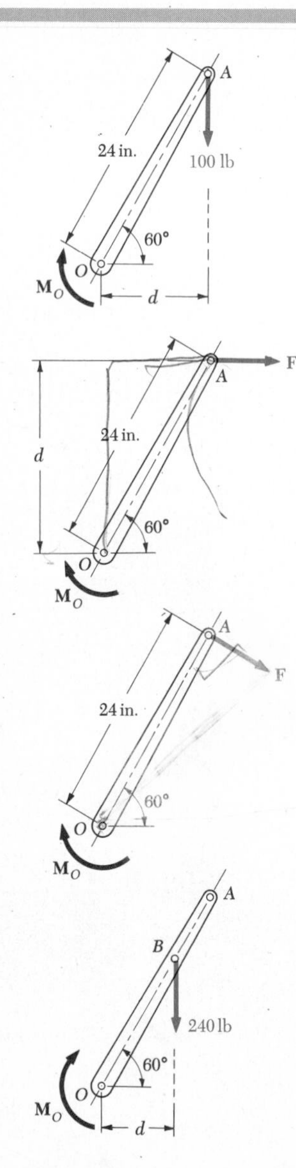

***a.* Moment about *O*.** The perpendicular distance from O to the line of action of the 100-lb force is

$$d = (24 \text{ in.}) \cos 60° = 12 \text{ in.}$$

The magnitude of the moment about O of the 100-lb force is

$$M_O = Fd = (100 \text{ lb})(12 \text{ in.}) = 1200 \text{ lb}\cdot\text{in.}$$

Since the force tends to rotate the lever clockwise about O, the moment will be represented by a vector $\mathbf{M}_O$ perpendicular to the plane of the figure and pointing *into* the paper. We express this fact by writing

$$\mathbf{M}_O = 1200 \text{ lb}\cdot\text{in.} \circlearrowright \quad ◀$$

***b.* Horizontal Force.** In this case, we have

$$d = (24 \text{ in.}) \sin 60° = 20.8 \text{ in.}$$

Since the moment about O must be 1200 lb·in., we write

$$M_O = Fd$$
$$1200 \text{ lb}\cdot\text{in.} = F(20.8 \text{ in.})$$
$$F = 57.7 \text{ lb} \qquad \mathbf{F} = 57.7 \text{ lb} \rightarrow \quad ◀$$

***c.* Smallest Force.** Since $M_O = Fd$, the smallest value of F occurs when d is maximum. We choose the force perpendicular to OA and find $d = 24$ in.; thus

$$M_O = Fd$$
$$1200 \text{ lb}\cdot\text{in.} = F(24 \text{ in.})$$
$$F = 50 \text{ lb} \qquad \mathbf{F} = 50 \text{ lb} \measuredangle 30° \quad ◀$$

***d.* 240-lb Vertical Force.** In this case $M_O = Fd$ yields

$$1200 \text{ lb}\cdot\text{in.} = (240 \text{ lb})d \qquad d = 5 \text{ in.}$$

but

$$OB \cos 60° = d \qquad OB = 10 \text{ in.} \quad ◀$$

e. None of the forces considered in parts *b*, *c*, and *d* is equivalent to the original 100-lb force. Although they have the same moment about O, they have different x and y components. In other words, although each force tends to rotate the shaft in the same manner, each causes the lever to pull on the shaft in a different way.

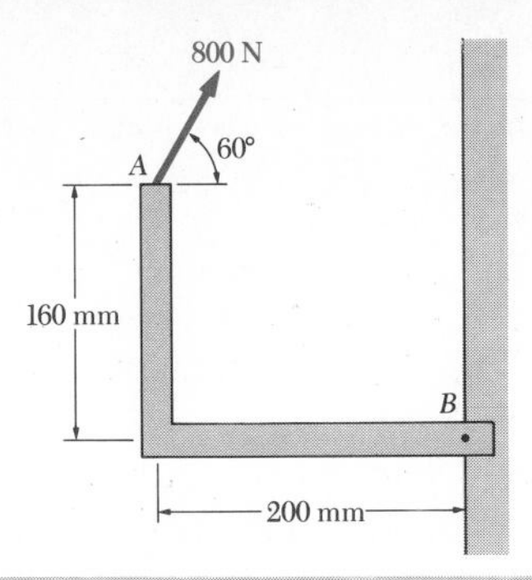

SAMPLE PROBLEM 3.2

A force of 800 N acts on a bracket as shown. Determine the moment of the force about B.

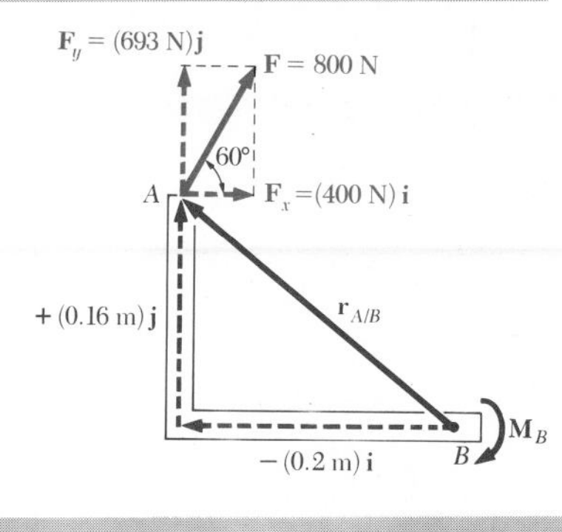

Solution. The moment $\mathbf{M}_B$ of the force $\mathbf{F}$ about B is obtained by forming the vector product

$$\mathbf{M}_B = \mathbf{r}_{A/B} \times \mathbf{F}$$

where $\mathbf{r}_{A/B}$ is the vector drawn from B to A. Resolving $\mathbf{r}_{A/B}$ and $\mathbf{F}$ into rectangular components, we have

$$\begin{aligned} \mathbf{r}_{A/B} &= -(0.2 \text{ m})\mathbf{i} + (0.16 \text{ m})\mathbf{j} \\ \mathbf{F} &= (800 \text{ N}) \cos 60°\mathbf{i} + (800 \text{ N}) \sin 60°\mathbf{j} \\ &= (400 \text{ N})\mathbf{i} + (693 \text{ N})\mathbf{j} \end{aligned}$$

Recalling the relations (3.7) for the cross products of unit vectors (Sec. 3.5), we obtain

$$\begin{aligned} \mathbf{M}_B = \mathbf{r}_{A/B} \times \mathbf{F} &= [-(0.2 \text{ m})\mathbf{i} + (0.16 \text{ m})\mathbf{j}] \times [(400 \text{ N})\mathbf{i} + (693 \text{ N})\mathbf{j}] \\ &= -(138.6 \text{ N}\cdot\text{m})\mathbf{k} - (64.0 \text{ N}\cdot\text{m})\mathbf{k} \\ &= -(202.6 \text{ N}\cdot\text{m})\mathbf{k} \end{aligned}$$

$$M_B = 203 \text{ N}\cdot\text{m} \circlearrowright \blacktriangleleft$$

The moment $\mathbf{M}_B$ is a vector perpendicular to the plane of the figure and pointing *into* the paper.

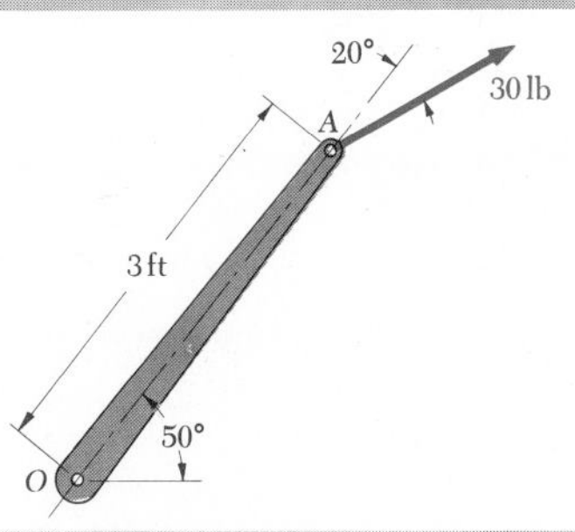

SAMPLE PROBLEM 3.3

A 30-lb force acts on the end of the 3-ft lever as shown. Determine the moment of the force about O.

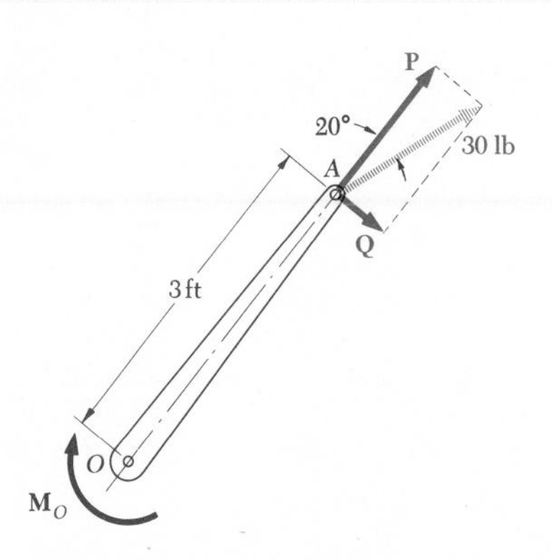

Solution. The force is replaced by two components, one component $\mathbf{P}$ in the direction of OA and one component $\mathbf{Q}$ perpendicular to OA. Since O is on the line of action of $\mathbf{P}$, the moment of $\mathbf{P}$ about O is zero and the moment of the 30-lb force reduces to the moment of $\mathbf{Q}$, which is clockwise and, thus, represented by a negative scalar.

$$\begin{aligned} Q &= (30 \text{ lb}) \sin 20° = 10.26 \text{ lb} \\ M_O &= -Q(3 \text{ ft}) = -(10.26 \text{ lb})(3 \text{ ft}) = -30.8 \text{ lb}\cdot\text{ft} \end{aligned}$$

Since the value obtained for the scalar M_O is negative, the moment $\mathbf{M}_O$ points *into* the paper. We write

$$\mathbf{M}_O = 30.8 \text{ lb}\cdot\text{ft} \circlearrowright \blacktriangleleft$$

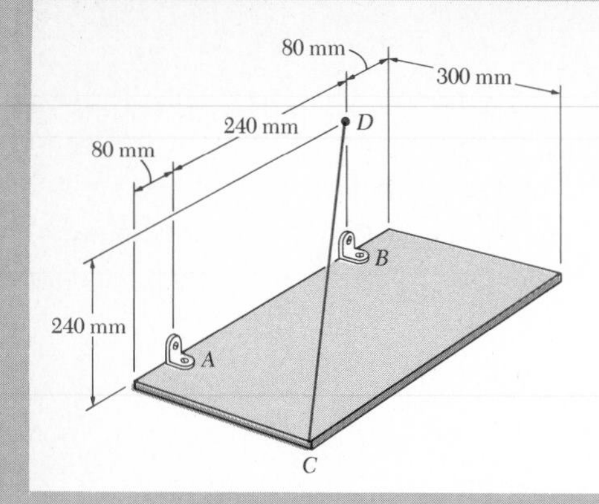

SAMPLE PROBLEM 3.4

A rectangular plate is supported by brackets at A and B and by a wire CD. Knowing that the tension in the wire is 200 N, determine the moment about A of the force exerted by the wire on point C.

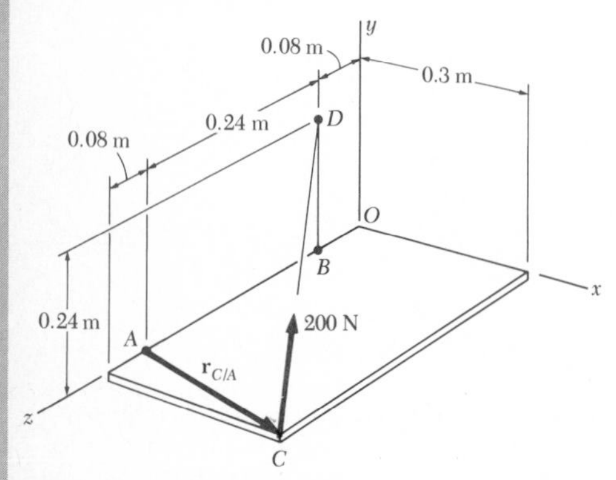

Solution. The moment $\mathbf{M}_A$ about A of the force $\mathbf{F}$ exerted by the wire on point C is obtained by forming the vector product

$$\mathbf{M}_A = \mathbf{r}_{C/A} \times \mathbf{F} \tag{1}$$

where $\mathbf{r}_{C/A}$ is the vector drawn from A to C,

$$\mathbf{r}_{C/A} = \overrightarrow{AC} = (0.3 \text{ m})\mathbf{i} + (0.08 \text{ m})\mathbf{k} \tag{2}$$

and where $\mathbf{F}$ is the 200-N force directed along CD. Introducing the unit vector $\boldsymbol{\lambda} = \overrightarrow{CD}/CD$, we write

$$\mathbf{F} = F\boldsymbol{\lambda} = (200 \text{ N})\frac{\overrightarrow{CD}}{CD} \tag{3}$$

Resolving the vector $\overrightarrow{CD}$ into rectangular components, we have

$$\overrightarrow{CD} = -(0.3 \text{ m})\mathbf{i} + (0.24 \text{ m})\mathbf{j} - (0.32 \text{ m})\mathbf{k} \qquad CD = 0.50 \text{ m}$$

Substituting into (3), we obtain

$$\mathbf{F} = \frac{200 \text{ N}}{0.50 \text{ m}}[-(0.3 \text{ m})\mathbf{i} + (0.24 \text{ m})\mathbf{j} - (0.32 \text{ m})\mathbf{k}]$$
$$= -(120 \text{ N})\mathbf{i} + (96 \text{ N})\mathbf{j} - (128 \text{ N})\mathbf{k} \tag{4}$$

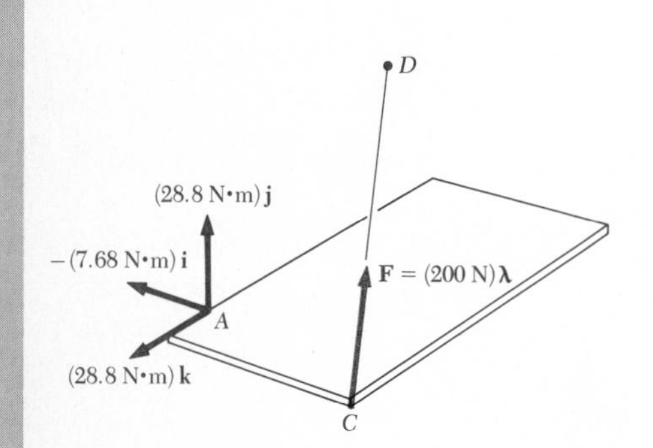

Substituting for $\mathbf{r}_{C/A}$ and $\mathbf{F}$ from (2) and (4) into (1), and recalling the relations (3.7) of Sec. 3.5, we obtain

$$\mathbf{M}_A = \mathbf{r}_{C/A} \times \mathbf{F} = (0.3\mathbf{i} + 0.08\mathbf{k}) \times (-120\mathbf{i} + 96\mathbf{j} - 128\mathbf{k})$$
$$= (0.3)(96)\mathbf{k} + (0.3)(-128)(-\mathbf{j}) + (0.08)(-120)\mathbf{j} + (0.08)(96)(-\mathbf{i})$$
$$\mathbf{M}_A = -(7.68 \text{ N}\cdot\text{m})\mathbf{i} + (28.8 \text{ N}\cdot\text{m})\mathbf{j} + (28.8 \text{ N}\cdot\text{m})\mathbf{k} \blacktriangleleft$$

Alternate Solution. As indicated in Sec. 3.8, the moment $\mathbf{M}_A$ may be expressed in the form of a determinant:

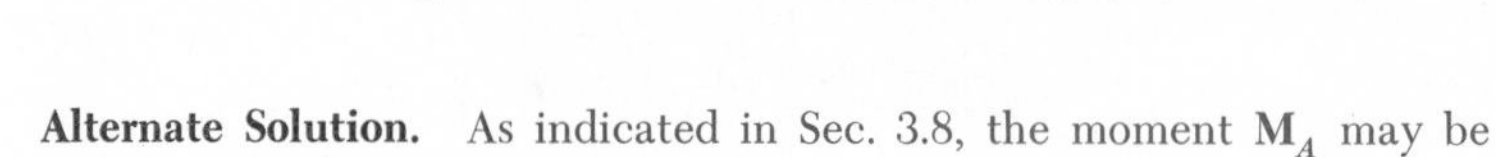

$$\mathbf{M}_A = \begin{vmatrix} \mathbf{i} & \mathbf{j} & \mathbf{k} \\ x_C - x_A & y_C - y_A & z_C - z_A \\ F_x & F_y & F_z \end{vmatrix} = \begin{vmatrix} \mathbf{i} & \mathbf{j} & \mathbf{k} \\ 0.3 & 0 & 0.08 \\ -120 & 96 & -128 \end{vmatrix}$$

$$\mathbf{M}_A = -(7.68 \text{ N}\cdot\text{m})\mathbf{i} + (28.8 \text{ N}\cdot\text{m})\mathbf{j} + (28.8 \text{ N}\cdot\text{m})\mathbf{k} \blacktriangleleft$$

Problems

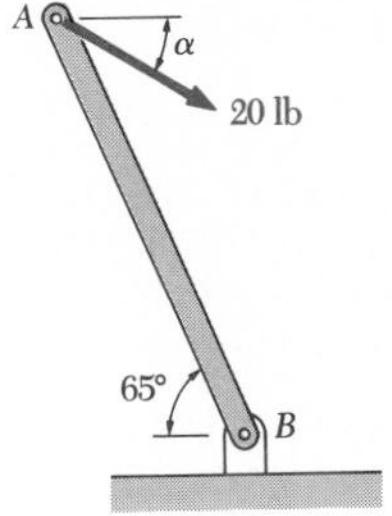

Fig. P3.1 and P3.3

3.1 A 20-lb force is applied to the control rod AB as shown. Knowing that the length of the rod is 9 in. and that $\alpha = 25°$, determine the moment of the force about point B by resolving the force into horizontal and vertical components.

3.2 Solve Prob. 3.1 by resolving the force into components along AB and in a direction perpendicular to AB.

3.3 A 20-lb force is applied to the control rod AB as shown. Knowing that the length of the rod is 9 in. and that the moment of the force about B is 120 lb·in. clockwise, determine the value of α.

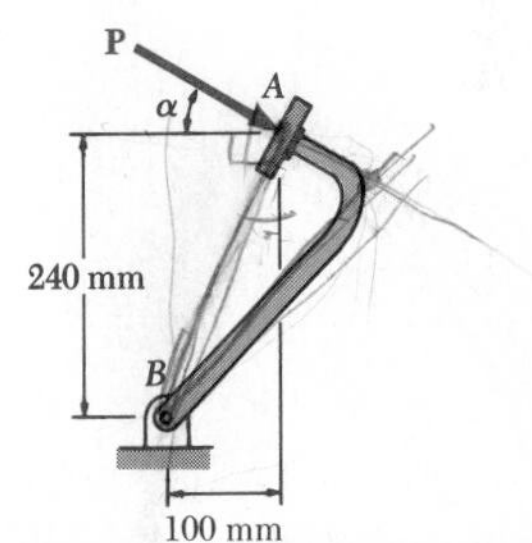

Fig. P3.4 and P3.5

3.4 For the brake pedal shown, determine the magnitude and direction of the smallest force **P** which has a 104-N·m clockwise moment about B.

3.5 A force **P** is applied to the brake pedal at A. Knowing that $P = 450$ N and $\alpha = 30°$, determine the moment of **P** about B.

3.6 A 300-N force is applied at A as shown. Determine (*a*) the moment of the 300-N force about D, (*b*) the smallest force applied at B which creates the same moment about D.

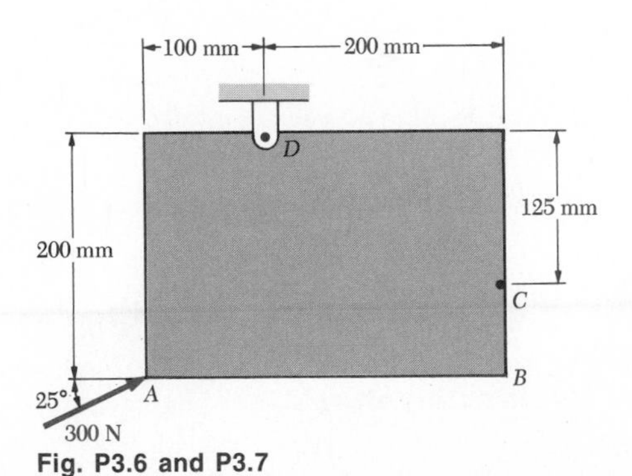

Fig. P3.6 and P3.7

3.7 A 300-N force is applied at A as shown. Determine (*a*) the moment of the 300-N force about D, (*b*) the magnitude and sense of the horizontal force applied at C which creates the same moment about D, (*c*) the smallest force applied at C which creates the same moment about D.

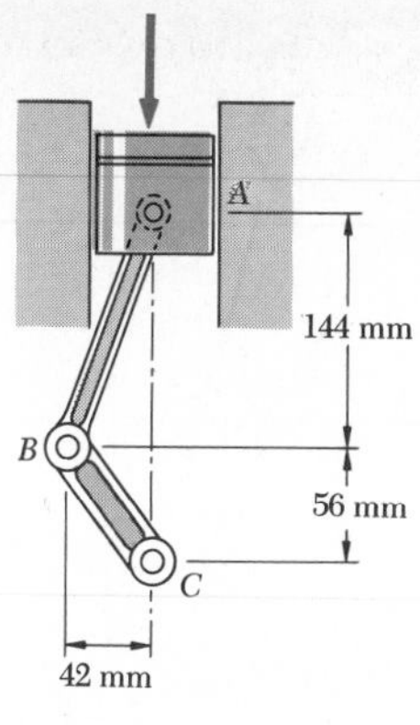

Fig. P3.8

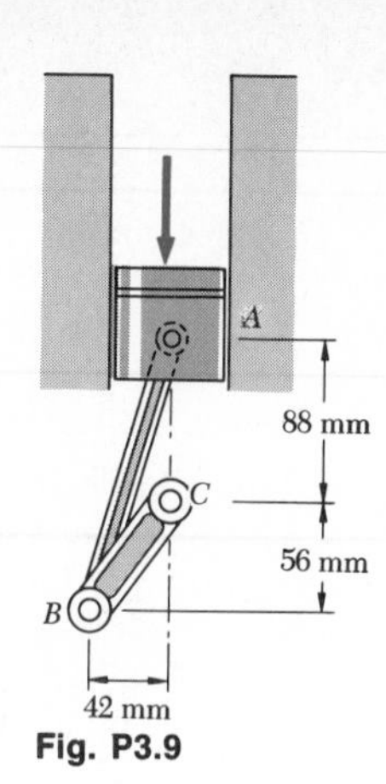

Fig. P3.9

3.8 and 3.9 It is known that the connecting rod AB exerts on the crank BC a 2.5-kN force directed down and to the left along the centerline of AB. Determine the moment of that force about C.

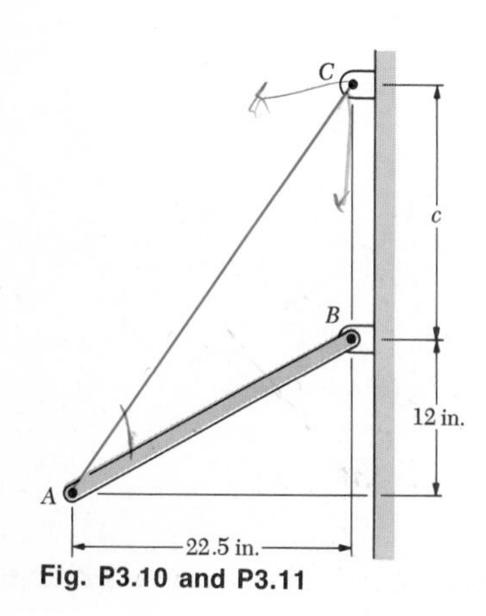

Fig. P3.10 and P3.11

3.10 Rod AB is held in place by the cord AC. Knowing that the tension in the cord is 300 lb and that $c = 18$ in., determine the moment about B of the force exerted by the cord at point A by resolving that force into horizontal and vertical components applied (*a*) at point A, (*b*) at point C.

3.11 Rod AB is held in place by the cord AC. Knowing that $c = 42$ in. and that the moment about B of the force exerted by the cord at point A is 700 lb·ft, determine the tension in the cord.

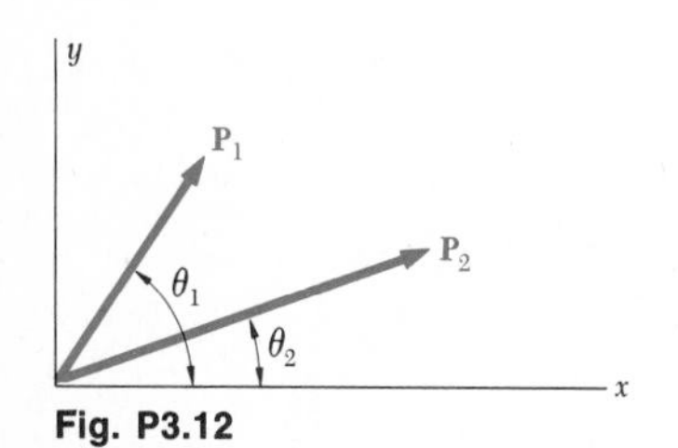

Fig. P3.12

3.12 Form the vector product $\mathbf{P}_1 \times \mathbf{P}_2$ and use the result obtained to prove the identity $\sin(\theta_1 - \theta_2) = \sin\theta_1 \cos\theta_2 - \cos\theta_1 \sin\theta_2$.

3.13 A force $\mathbf{F} = F_x\mathbf{i} + F_y\mathbf{j}$ acts at a point of coordinates x and y. Derive an expression for the perpendicular distance d from the line of action of $\mathbf{F}$ to the origin O of the system of coordinates.

3.14 The line of action of a force $\mathbf{P}$ passes through two points $A(x_1, y_1)$ and $B(x_2, y_2)$. If the force is directed from A to B, determine the moment of the force about the origin.

3.15 Determine the moment about the origin O of the force $\mathbf{F} = 4\mathbf{i} + 5\mathbf{j} - 3\mathbf{k}$ which acts at a point A. Assume that the position vector of A is (*a*) $\mathbf{r} = 2\mathbf{i} - 3\mathbf{j} + 4\mathbf{k}$, (*b*) $\mathbf{r} = 2\mathbf{i} + 2.5\mathbf{j} - 1.5\mathbf{k}$, (*c*) $\mathbf{r} = 2\mathbf{i} + 5\mathbf{j} + 6\mathbf{k}$.

3.16 Determine the moment about the origin O of the force $\mathbf{F} = -2\mathbf{i} + 3\mathbf{j} + 5\mathbf{k}$ which acts at a point A. Assume that the position vector of A is (*a*) $\mathbf{r} = \mathbf{i} + \mathbf{j} + \mathbf{k}$, (*b*) $\mathbf{r} = 2\mathbf{i} + 3\mathbf{j} - 5\mathbf{k}$, (*c*) $\mathbf{r} = -4\mathbf{i} + 6\mathbf{j} + 10\mathbf{k}$.

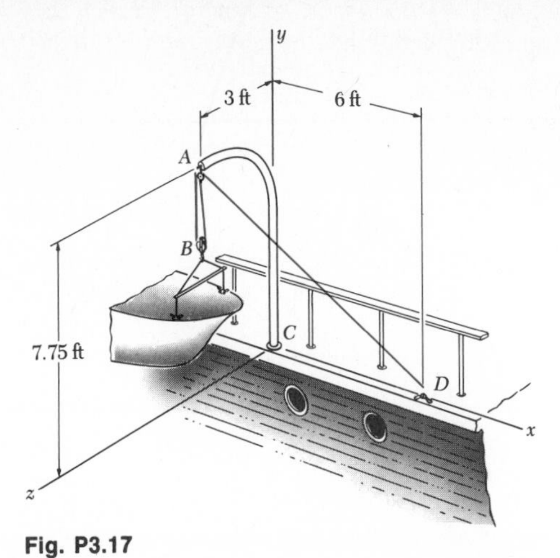

Fig. P3.17

3.17 A small boat hangs from two davits, one of which is shown in the figure. The tension in line *ABAD* is 82 lb. Determine the moment about *C* of the resultant force $\mathbf{R}_A$ exerted on the davit at *A*.

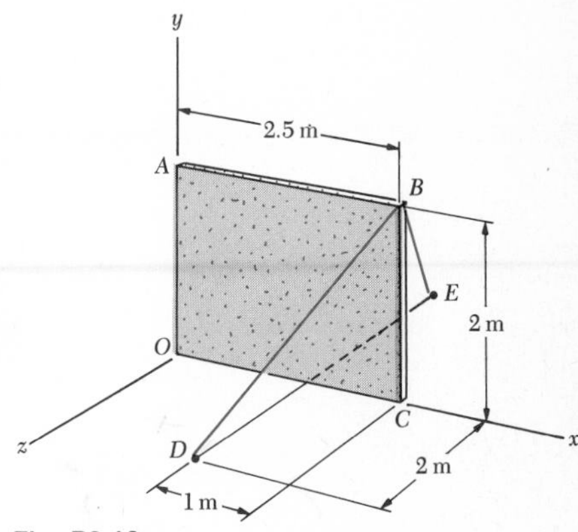

Fig. P3.18

3.18 A precast concrete wall section is temporarily held by two cables as shown. Knowing that the tension in cable *BD* is 900 N, determine the moment about point *O* of the force exerted by the cable at *B*.

3.19 A 200-N force is applied as shown to the bracket *ABC*. Determine the moment of the force about *A*.

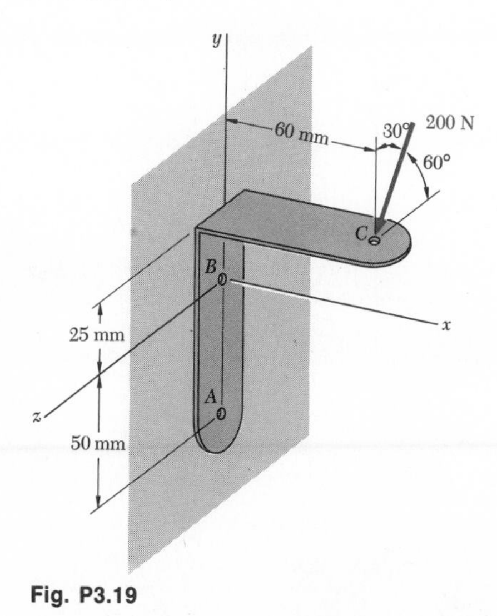

Fig. P3.19

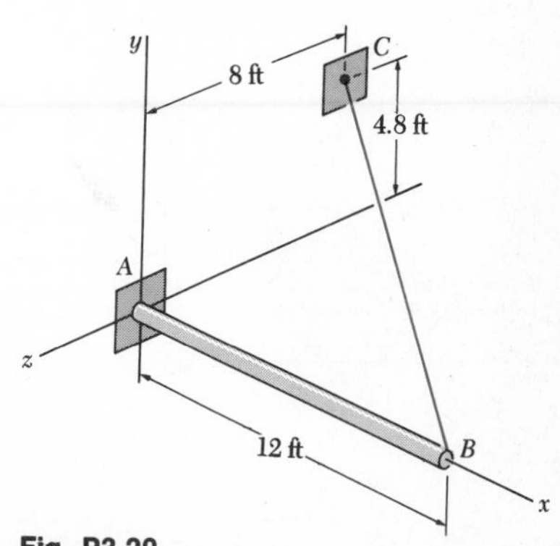

Fig. P3.20

3.20 The 12-ft boom *AB* has a fixed end *A*. A steel cable is stretched from the free end *B* of the boom to a point *C* located on the vertical wall. If the tension in the cable is 380 lb, determine the moment about *A* of the force exerted by the cable at *B*.

3.21 In Prob. 3.18, determine the perpendicular distance from point O to cable BD.

3.22 In Prob. 3.18, determine the perpendicular distance from point C to cable BD.

3.23 In Prob. 3.20, determine the perpendicular distance from point A to cable BC.

3.24 In Prob. 3.17, determine the perpendicular distance from point C to portion AD of the line $ABAD$.

3.25 In Sample Prob. 3.4, determine the perpendicular distance from point A to wire CD.

3.26 A single force $\mathbf{F}$ acts at a point A of coordinates $x = y = z = a$. Show that $M_x + M_y + M_z = 0$; that is, show that the *algebraic* sum of the rectangular components of the moment of $\mathbf{F}$ about O is zero.

3.9. Scalar Product of Two Vectors.

We shall now expand our knowledge of vector algebra and introduce the *scalar product* of two vectors.

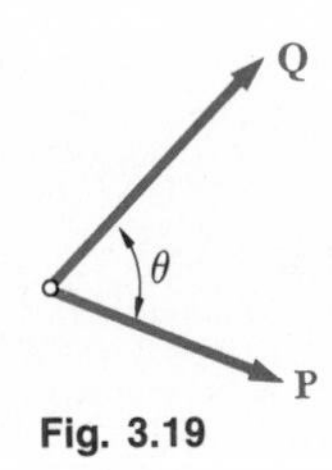

Fig. 3.19

The scalar product of two vectors $\mathbf{P}$ and $\mathbf{Q}$ is defined as the product of the magnitudes of $\mathbf{P}$ and $\mathbf{Q}$ and of the cosine of the angle θ formed by $\mathbf{P}$ and $\mathbf{Q}$ (Fig. 3.19). The scalar product of $\mathbf{P}$ and $\mathbf{Q}$ is denoted by $\mathbf{P} \cdot \mathbf{Q}$. We write therefore

$$\mathbf{P} \cdot \mathbf{Q} = PQ \cos \theta \tag{3.24}$$

Note that the expression just defined is not a vector but a *scalar*, which explains the name *scalar product;* because of the notation used, $\mathbf{P} \cdot \mathbf{Q}$ is also referred to as the *dot product* of the vectors $\mathbf{P}$ and $\mathbf{Q}$.

It follows from its very definition that the scalar product of two vectors is *commutative*, i.e., that

$$\mathbf{P} \cdot \mathbf{Q} = \mathbf{Q} \cdot \mathbf{P} \tag{3.25}$$

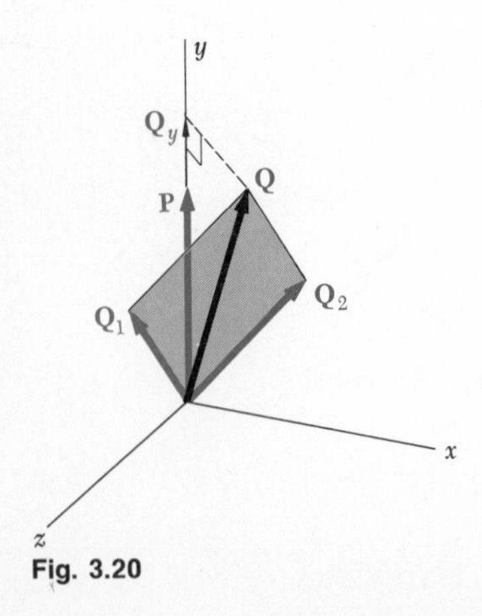

Fig. 3.20

To prove that the scalar product is also *distributive*, we must prove the relation

$$\mathbf{P} \cdot (\mathbf{Q}_1 + \mathbf{Q}_2) = \mathbf{P} \cdot \mathbf{Q}_1 + \mathbf{P} \cdot \mathbf{Q}_2 \tag{3.26}$$

We may, without any loss of generality, assume that $\mathbf{P}$ is directed along the y axis (Fig. 3.20). Denoting by $\mathbf{Q}$ the sum of $\mathbf{Q}_1$ and $\mathbf{Q}_2$, and by θ_y the angle $\mathbf{Q}$ forms with the y axis, we express the left-hand member of (3.26) as follows:

$$\mathbf{P} \cdot (\mathbf{Q}_1 + \mathbf{Q}_2) = \mathbf{P} \cdot \mathbf{Q} = PQ \cos \theta_y = PQ_y \tag{3.27}$$

where Q_y is the y component of $\mathbf{Q}$. We may, in a similar way, express the right-hand member of (3.26) as

$$\mathbf{P} \cdot \mathbf{Q}_1 + \mathbf{P} \cdot \mathbf{Q}_2 = P(Q_1)_y + P(Q_2)_y \tag{3.28}$$

Since $\mathbf{Q}$ is the sum of $\mathbf{Q}_1$ and $\mathbf{Q}_2$, its y component must be equal to the sum of the y components of $\mathbf{Q}_1$ and $\mathbf{Q}_2$. Thus, the expressions obtained in (3.27) and (3.28) are equal and the relation (3.26) has been proved.

As far as the third property—the associative property—is concerned, we note that this property cannot apply to scalar products. Indeed, $(\mathbf{P} \cdot \mathbf{Q}) \cdot \mathbf{S}$ has no meaning, since $\mathbf{P} \cdot \mathbf{Q}$ is not a vector but a scalar.

We shall now express the scalar product of two vectors $\mathbf{P}$ and $\mathbf{Q}$ in terms of their rectangular components. Resolving $\mathbf{P}$ and $\mathbf{Q}$ into components, we first write

$$\mathbf{P} \cdot \mathbf{Q} = (P_x\mathbf{i} + P_y\mathbf{j} + P_z\mathbf{k}) \cdot (Q_x\mathbf{i} + Q_y\mathbf{j} + Q_z\mathbf{k})$$

Making use of the distributive property, we express $\mathbf{P} \cdot \mathbf{Q}$ as the sum of scalar products such as $P_x\mathbf{i} \cdot Q_x\mathbf{i}$ and $P_x\mathbf{i} \cdot Q_y\mathbf{j}$. But we may easily check from the definition of the scalar product that the scalar products of the unit vectors are either zero or one.

$$\begin{array}{lll} \mathbf{i} \cdot \mathbf{i} = 1 & \mathbf{j} \cdot \mathbf{j} = 1 & \mathbf{k} \cdot \mathbf{k} = 1 \\ \mathbf{i} \cdot \mathbf{j} = 0 & \mathbf{j} \cdot \mathbf{k} = 0 & \mathbf{k} \cdot \mathbf{i} = 0 \end{array} \tag{3.29}$$

Thus, the expression obtained for $\mathbf{P} \cdot \mathbf{Q}$ reduces to

$$\mathbf{P} \cdot \mathbf{Q} = P_xQ_x + P_yQ_y + P_zQ_z \tag{3.30}$$

In the particular case when $\mathbf{P}$ and $\mathbf{Q}$ are equal, we check that

$$\mathbf{P} \cdot \mathbf{P} = P_x^2 + P_y^2 + P_z^2 = P^2 \tag{3.31}$$

Applications

1. ***Angle formed by two given vectors.*** Let two vectors be given in terms of their components:

$$\mathbf{P} = P_x\mathbf{i} + P_y\mathbf{j} + P_z\mathbf{k}$$
$$\mathbf{Q} = Q_x\mathbf{i} + Q_y\mathbf{j} + Q_z\mathbf{k}$$

To determine the angle formed by the two vectors we shall equate the expressions obtained in (3.24) and (3.30) for their scalar product and write:

$$PQ \cos \theta = P_xQ_x + P_yQ_y + P_zQ_z$$

Solving for $\cos \theta$, we have

$$\cos \theta = \frac{P_xQ_x + P_yQ_y + P_zQ_z}{PQ} \tag{3.32}$$

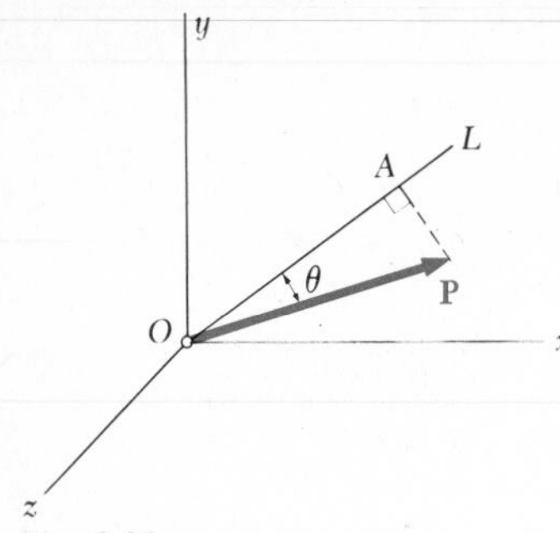

Fig. 3.21

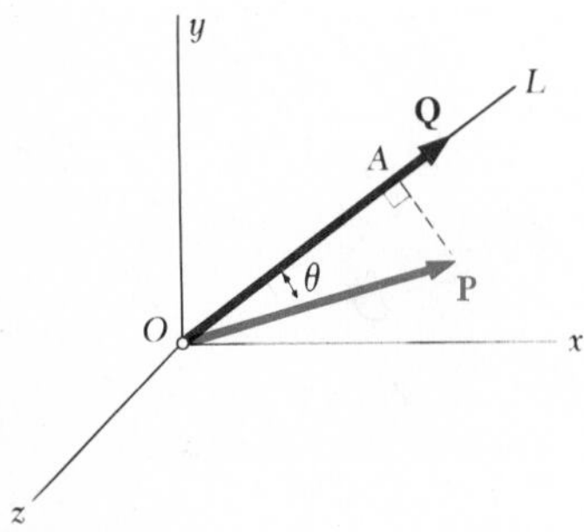

Fig. 3.22

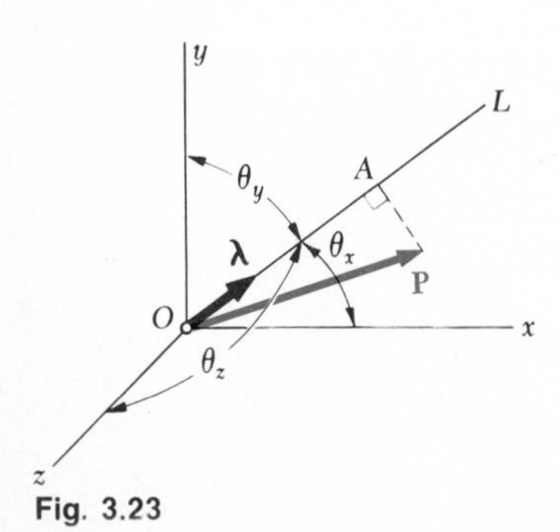

Fig. 3.23

2. *Projection of a vector on a given axis.* Consider a vector **P** forming an angle θ with an axis, or directed line, OL (Fig. 3.21). The *projection of* **P** *on the axis OL* is defined as the scalar

$$P_{OL} = P \cos \theta \tag{3.33}$$

We note that the projection P_{OL} is equal in absolute value to the length of the segment OA; it will be positive if OA has the same sense as the axis OL, that is, if θ is acute, and negative otherwise. If **P** and OL are at a right angle, the projection of **P** on OL is zero.

Consider now a vector **Q** directed along OL and of the same sense as OL (Fig. 3.22). The scalar product of **P** and **Q** may be expressed as

$$\mathbf{P} \cdot \mathbf{Q} = PQ \cos \theta = P_{OL} Q \tag{3.34}$$

from which it follows that

$$P_{OL} = \frac{\mathbf{P} \cdot \mathbf{Q}}{Q} = \frac{P_x Q_x + P_y Q_y + P_z Q_z}{Q} \tag{3.35}$$

In the particular case when the vector selected along OL is the unit vector $\boldsymbol{\lambda}$ (Fig. 3.23), we write

$$P_{OL} = \mathbf{P} \cdot \boldsymbol{\lambda} \tag{3.36}$$

Resolving **P** and $\boldsymbol{\lambda}$ into rectangular components, and recalling from Sec. 2.12 that the components of $\boldsymbol{\lambda}$ along the coordinate axes are respectively equal to the direction cosines of OL, we express the projection of **P** on OL as

$$P_{OL} = P_x \cos \theta_x + P_y \cos \theta_y + P_z \cos \theta_z \tag{3.37}$$

where θ_x, θ_y, and θ_z denote the angles that the axis OL forms with the coordinate axes.

3.10. Mixed Triple Product of Three Vectors. We define the *mixed triple product* of the three vectors **S**, **P**, and **Q** as the scalar expression

$$\mathbf{S} \cdot (\mathbf{P} \times \mathbf{Q}) \tag{3.38}$$

obtained by forming the scalar product of **S** with the vector product of **P** and **Q**.†

A simple geometrical interpretation may be given for the mixed triple product of **S**, **P**, and **Q** (Fig. 3.24). We first recall from Sec. 3.4 that the vector $\mathbf{P} \times \mathbf{Q}$ is perpendicular to the plane containing **P** and **Q**, and that its

†Another kind of triple product will be introduced later (Chap. 15): the *vector triple product* $\mathbf{S} \times (\mathbf{P} \times \mathbf{Q})$.

magnitude is equal to the area of the parallelogram constructed on **P** and **Q**. On the other hand, Eq. (3.34) indicates that the scalar product of **S** and **P** × **Q** may be obtained by multiplying the magnitude of **P** × **Q** (i.e., the area of the parallelogram built on **P** and **Q**) by the projection of **S** on the vector **P** × **Q** (i.e., by the projection of **S** on the normal to the plane containing the parallelogram). The mixed triple product is thus equal, in absolute value, to the volume of the parallelepiped having the vectors **S**, **P**, and **Q** for sides (Fig. 3.25). We may check that the sign of the mixed triple product will be positive if **S**, **P**, and **Q** form a right-handed triad, and negative if they form a left-handed triad [that is, **S** · (**P** × **Q**) will be negative if the rotation which brings **P** into line with **Q** is observed as clockwise from the tip of **S**]. The mixed triple product will be zero if **S**, **P**, and **Q** are coplanar.

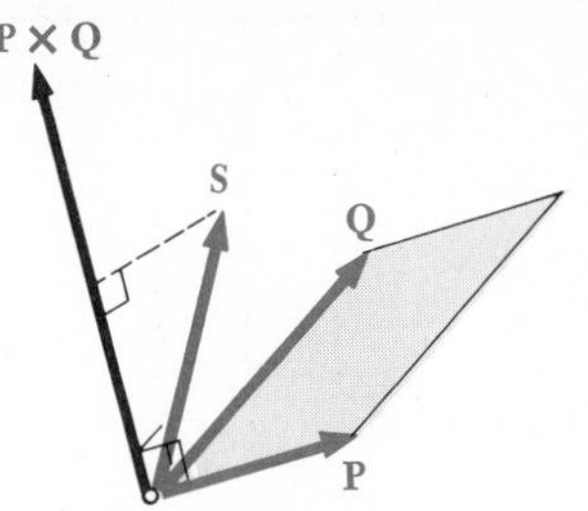

Fig. 3.24

Since the parallelepiped defined in the preceding paragraph is independent of the order in which the three vectors are taken, the six mixed triple products which may be formed with **S**, **P**, and **Q** will all have the same absolute value, although not the same sign. We check that

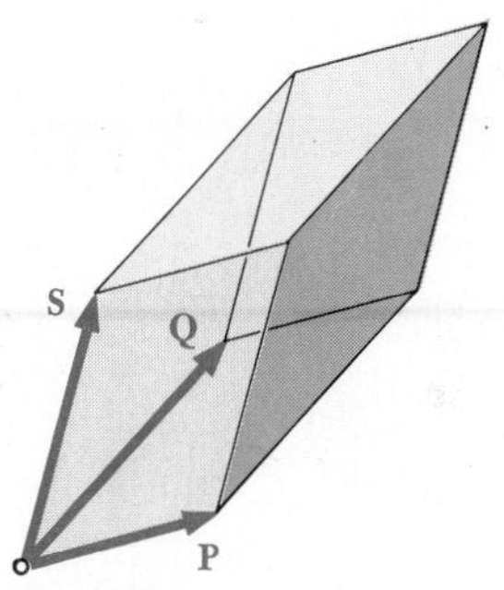

Fig. 3.25

$$\mathbf{S}\cdot(\mathbf{P}\times\mathbf{Q}) = \mathbf{P}\cdot(\mathbf{Q}\times\mathbf{S}) = \mathbf{Q}\cdot(\mathbf{S}\times\mathbf{P}) = -\mathbf{S}\cdot(\mathbf{Q}\times\mathbf{P}) = -\mathbf{P}\cdot(\mathbf{S}\times\mathbf{Q}) = -\mathbf{Q}\cdot(\mathbf{P}\times\mathbf{S}) \tag{3.39}$$

Arranging in a circle and in counterclockwise order the letters representing the three vectors (Fig. 3.26), we note that the sign of the mixed triple product is conserved if the vectors are permuted in such a way that they are still read in counterclockwise order. Such a permutation is said to be a *circular permutation*. It also follows from (3.39) that the mixed triple product of **S**, **P**, and **Q** may be defined equally well as **S** · (**P** × **Q**) or (**S** × **P**) · **Q**.

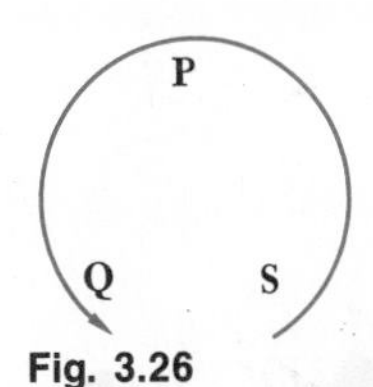

Fig. 3.26

We shall now express the mixed triple product of the vectors **S**, **P**, and **Q** in terms of the rectangular components of these vectors. Denoting **P** × **Q** by **V**, and using formula (3.30) to express the scalar product of **S** and **V**, we write

$$\mathbf{S}\cdot(\mathbf{P}\times\mathbf{Q}) = \mathbf{S}\cdot\mathbf{V} = S_xV_x + S_yV_y + S_zV_z$$

Substituting from the relations (3.9) for the components of **V**, we obtain

$$\mathbf{S}\cdot(\mathbf{P}\times\mathbf{Q}) = S_x(P_yQ_z - P_zQ_y) + S_y(P_zQ_x - P_xQ_z) + S_z(P_xQ_y - P_yQ_x) \tag{3.40}$$

This expression may be written in a more compact form if we observe that it represents the expansion of a determinant:

$$\mathbf{S}\cdot(\mathbf{P}\times\mathbf{Q}) = \begin{vmatrix} S_x & S_y & S_z \\ P_x & P_y & P_z \\ Q_x & Q_y & Q_z \end{vmatrix} \tag{3.41}$$

By applying the rules governing the permutation of rows in a determinant, we could easily verify the relations (3.39) which were derived earlier from geometrical considerations.

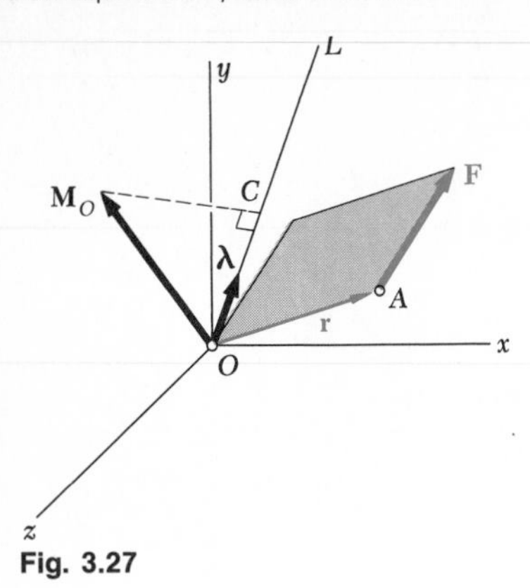

Fig. 3.27

3.11. Moment of a Force about a Given Axis. Now that we have further increased our knowledge of vector algebra, we shall introduce a new concept, the concept of *moment of a force about an axis*. Consider again a force **F** acting on a rigid body and the moment $\mathbf{M}_O$ of that force about O (Fig. 3.27). Let OL be an axis through O; *we define the moment M_{OL} of* **F** *about OL as the projection OC of the moment* $\mathbf{M}_O$ *on the axis OL.* Denoting by $\boldsymbol{\lambda}$ the unit vector along OL, and recalling from Secs. 3.9 and 3.6, respectively, the expressions (3.36) and (3.11) obtained for the projection of a vector on a given axis and for the moment $\mathbf{M}_O$ of a force **F**, we write

$$M_{OL} = \boldsymbol{\lambda} \cdot \mathbf{M}_O = \boldsymbol{\lambda} \cdot (\mathbf{r} \times \mathbf{F}) \tag{3.42}$$

which shows that the moment M_{OL} of **F** about the axis OL is the scalar obtained by forming the mixed triple product of $\boldsymbol{\lambda}$, **r**, and **F**. Expressing M_{OL} in the form of a determinant, we write

$$M_{OL} = \begin{vmatrix} \lambda_x & \lambda_y & \lambda_z \\ x & y & z \\ F_x & F_y & F_z \end{vmatrix} \tag{3.43}$$

where $\lambda_x, \lambda_y, \lambda_z$ = direction cosines of axis OL
x, y, z = coordinates of point of application of **F**
F_x, F_y, F_z = components of force **F**

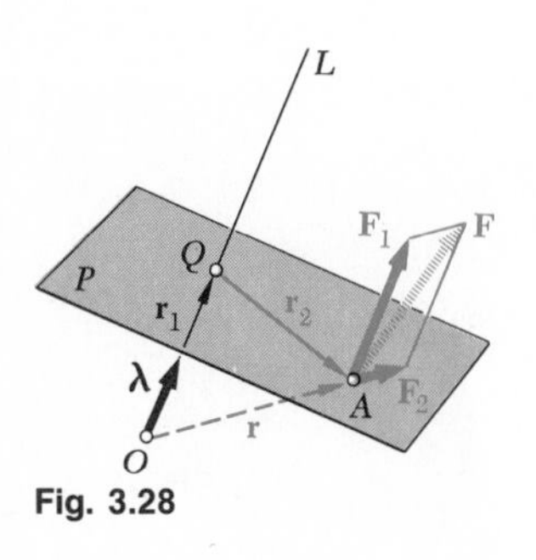

Fig. 3.28

The physical significance of the moment M_{OL} of a force **F** about a fixed axis OL becomes more apparent if we resolve **F** into two rectangular components $\mathbf{F}_1$ and $\mathbf{F}_2$, with $\mathbf{F}_1$ parallel to OL and $\mathbf{F}_2$ lying in a plane P perpendicular to OL (Fig. 3.28). Resolving **r** similarly into two components $\mathbf{r}_1$ and $\mathbf{r}_2$, and substituting for **F** and **r** into (3.42), we write

$$\begin{aligned} M_{OL} &= \boldsymbol{\lambda} \cdot [(\mathbf{r}_1 + \mathbf{r}_2) \times (\mathbf{F}_1 + \mathbf{F}_2)] \\ &= \boldsymbol{\lambda} \cdot (\mathbf{r}_1 \times \mathbf{F}_1) + \boldsymbol{\lambda} \cdot (\mathbf{r}_1 \times \mathbf{F}_2) + \boldsymbol{\lambda} \cdot (\mathbf{r}_2 \times \mathbf{F}_1) + \boldsymbol{\lambda} \cdot (\mathbf{r}_2 \times \mathbf{F}_2) \end{aligned}$$

Noting that all mixed triple products, except the last one, are equal to zero, since they involve vectors which are coplanar when drawn from a common origin (Sec. 3.10), we have

$$M_{OL} = \boldsymbol{\lambda} \cdot (\mathbf{r}_2 \times \mathbf{F}_2) \tag{3.44}$$

The vector product $\mathbf{r}_2 \times \mathbf{F}_2$ is perpendicular to the plane P and represents the moment of the component $\mathbf{F}_2$ of **F** about the point Q where OL intersects P. Therefore, the scalar M_{OL}, which will be positive if $\mathbf{r}_2 \times \mathbf{F}_2$ and OL have the same sense, and negative otherwise, measures the tendency of $\mathbf{F}_2$ to make the rigid body rotate about the fixed axis OL. Since the other component $\mathbf{F}_1$ of **F** does not tend to make the body rotate about OL, we conclude that *the moment M_{OL} of* **F** *about OL measures the tendency of the force* **F** *to impart to the rigid body a motion of rotation about the fixed axis OL.*

It follows from the definition of the moment of a force about an axis that the moment of **F** about a coordinate axis is equal to the component of $\mathbf{M}_O$ along that axis. Substituting successively each of the unit vectors **i**, **j**, and **k** for $\boldsymbol{\lambda}$ in (3.42), we check that the expressions thus obtained for the *moments of* **F** *about the coordinate axes* are respectively equal to the expressions obtained in Sec. 3.8 for the components of the moment $\mathbf{M}_O$ of **F** about O:

$$\begin{aligned} M_x &= yF_z - zF_y \\ M_y &= zF_x - xF_z \\ M_z &= xF_y - yF_x \end{aligned} \qquad (3.18)$$

We observe that just as the components F_x, F_y, and F_z of a force **F** acting on a rigid body measure, respectively, the tendency of **F** to move the rigid body in the x, y, and z directions, the moments M_x, M_y, and M_z of **F** about the coordinate axes measure the tendency of **F** to impart to the rigid body a motion of rotation about the x, y, and z axes, respectively.

More generally, the moment of a force **F** applied at A about an axis which does not pass through the origin is obtained by choosing an arbitrary point B on the axis (Fig. 3.29) and determining the projection on the axis BL of the moment $\mathbf{M}_B$ of **F** about B. We write

$$M_{BL} = \boldsymbol{\lambda} \cdot \mathbf{M}_B = \boldsymbol{\lambda} \cdot (\mathbf{r}_{A/B} \times \mathbf{F}) \qquad (3.45)$$

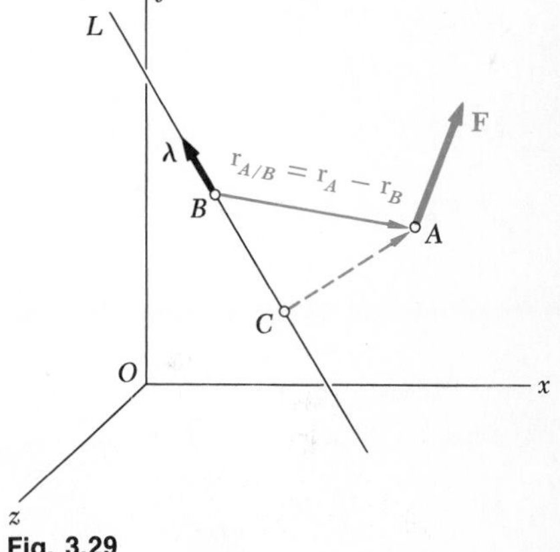

Fig. 3.29

where $\mathbf{r}_{A/B} = \mathbf{r}_A - \mathbf{r}_B$ represents the vector drawn from B to A. Expressing M_{BL} in the form of a determinant, we have

$$M_{BL} = \begin{vmatrix} \lambda_x & \lambda_y & \lambda_z \\ x_{A/B} & y_{A/B} & z_{A/B} \\ F_x & F_y & F_z \end{vmatrix} \qquad (3.46)$$

where $\lambda_x, \lambda_y, \lambda_z$ = direction cosines of axis BL

$x_{A/B} = x_A - x_B \qquad y_{A/B} = y_A - y_B \qquad z_{A/B} = z_A - z_B$

F_x, F_y, F_z = components of force **F**

It should be noted that the result obtained is independent of the choice of the point B on the given axis. Indeed, denoting by M_{CL} the result obtained with a different point C, we have

$$\begin{aligned} M_{CL} &= \boldsymbol{\lambda} \cdot [(\mathbf{r}_A - \mathbf{r}_C) \times \mathbf{F}] \\ &= \boldsymbol{\lambda} \cdot [(\mathbf{r}_A - \mathbf{r}_B) \times \mathbf{F}] + \boldsymbol{\lambda} \cdot [(\mathbf{r}_B - \mathbf{r}_C) \times \mathbf{F}] \end{aligned}$$

But, since the vectors $\boldsymbol{\lambda}$ and $\mathbf{r}_B - \mathbf{r}_C$ lie in the same line, the volume of the parallelepiped having the vectors $\boldsymbol{\lambda}$, $\mathbf{r}_B - \mathbf{r}_C$, and **F** for sides is zero, as is the mixed triple product of these three vectors (Sec. 3.10). The expression obtained for M_{CL} thus reduces to its first term, which is the expression used earlier to define M_{BL}.

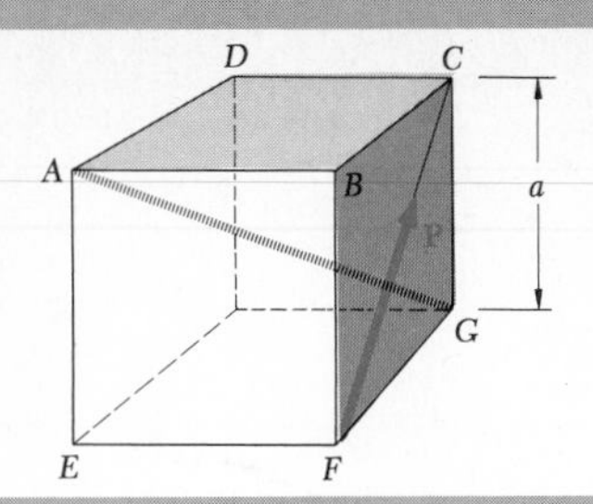

SAMPLE PROBLEM 3.5

A cube of side a is acted upon by a force **P** as shown. Determine the moment of **P** (*a*) about *A*, (*b*) about the edge *AB*, (*c*) about the diagonal *AG* of the cube. (*d*) Using the result of part *c*, determine the perpendicular distance from *AG* to *FC*.

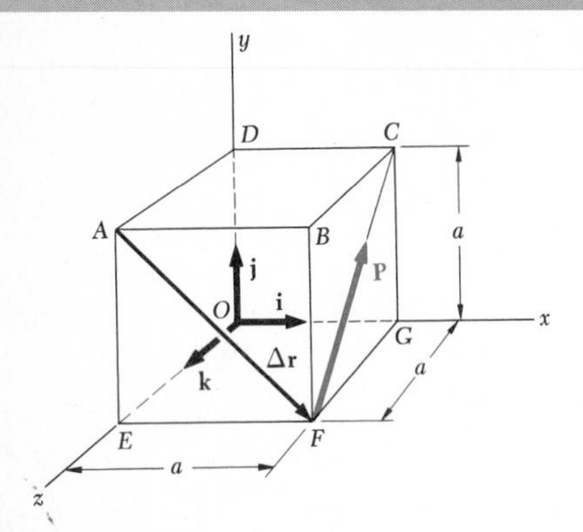

a. **Moment about *A*.** Choosing x, y, and z axes as shown, we resolve into rectangular components the force **P** and the vector $\mathbf{r}_{F/A} = \overrightarrow{AF}$ drawn from *A* to the point of application *F* of **P**.

$$\mathbf{r}_{F/A} = a\mathbf{i} - a\mathbf{j} = a(\mathbf{i} - \mathbf{j})$$
$$\mathbf{P} = (P/\sqrt{2})\mathbf{j} - (P/\sqrt{2})\mathbf{k} = (P/\sqrt{2})(\mathbf{j} - \mathbf{k})$$

The moment of **P** about *A* is

$$\mathbf{M}_A = \mathbf{r}_{F/A} \times \mathbf{P} = a(\mathbf{i} - \mathbf{j}) \times (P/\sqrt{2})(\mathbf{j} - \mathbf{k})$$
$$\mathbf{M}_A = (aP/\sqrt{2})(\mathbf{i} + \mathbf{j} + \mathbf{k}) \quad ◄$$

b. **Moment about *AB*.** Projecting $\mathbf{M}_A$ on *AB*, we write

$$M_{AB} = \mathbf{i} \cdot \mathbf{M}_A = \mathbf{i} \cdot (aP/\sqrt{2})(\mathbf{i} + \mathbf{j} + \mathbf{k}) \qquad M_{AB} = aP/\sqrt{2} \quad ◄$$

We verify that, since *AB* is parallel to the x axis, M_{AB} is also the x component of the moment $\mathbf{M}_A$.

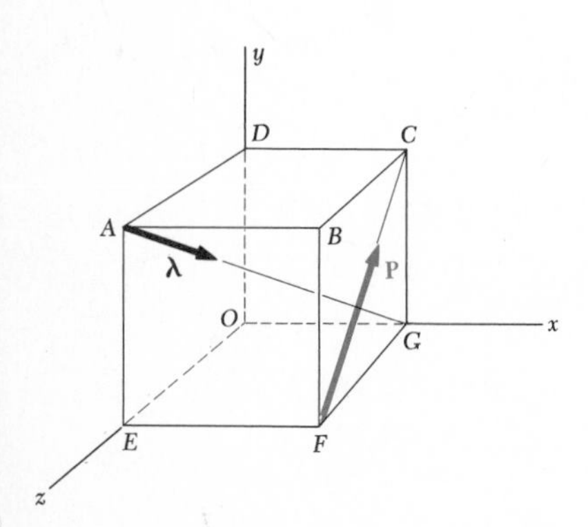

c. **Moment about Diagonal *AG*.** The moment of **P** about *AG* is obtained by projecting $\mathbf{M}_A$ on *AG*. Denoting by $\boldsymbol{\lambda}$ the unit vector along *AG*, we have

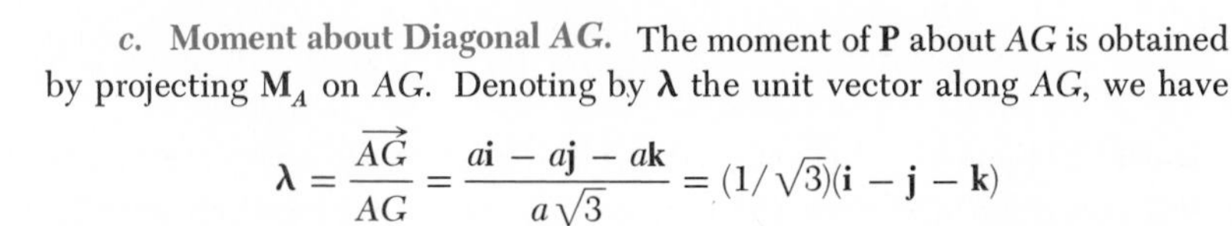

$$\boldsymbol{\lambda} = \frac{\overrightarrow{AG}}{AG} = \frac{a\mathbf{i} - a\mathbf{j} - a\mathbf{k}}{a\sqrt{3}} = (1/\sqrt{3})(\mathbf{i} - \mathbf{j} - \mathbf{k})$$

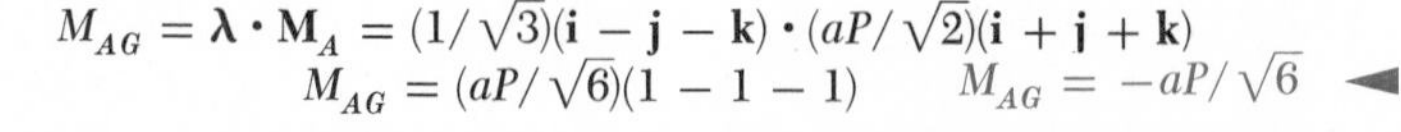

$$M_{AG} = \boldsymbol{\lambda} \cdot \mathbf{M}_A = (1/\sqrt{3})(\mathbf{i} - \mathbf{j} - \mathbf{k}) \cdot (aP/\sqrt{2})(\mathbf{i} + \mathbf{j} + \mathbf{k})$$
$$M_{AG} = (aP/\sqrt{6})(1 - 1 - 1) \qquad M_{AG} = -aP/\sqrt{6} \quad ◄$$

Alternate Method. The moment of **P** about *AG* may also be expressed in the form of a determinant:

$$M_{AG} = \begin{vmatrix} \lambda_x & \lambda_y & \lambda_z \\ x_{F/A} & y_{F/A} & z_{F/A} \\ F_x & F_y & F_z \end{vmatrix} = \begin{vmatrix} 1/\sqrt{3} & -1/\sqrt{3} & -1/\sqrt{3} \\ a & -a & 0 \\ 0 & P/\sqrt{2} & -P/\sqrt{2} \end{vmatrix} = -aP/\sqrt{6}$$

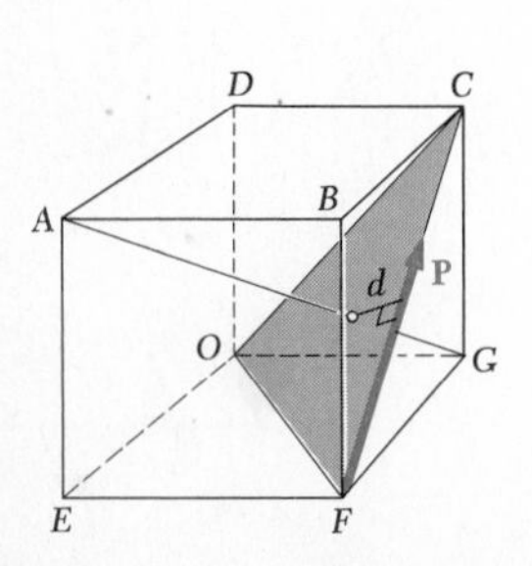

d. **Perpendicular Distance from *AG* to *FC*.** We first observe that **P** is perpendicular to the diagonal *AG*. This may be checked by forming the scalar product $\mathbf{P} \cdot \boldsymbol{\lambda}$ and verifying that it is zero:

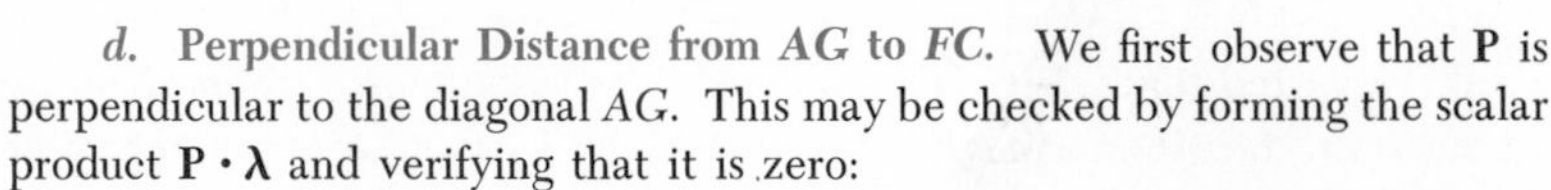

$$\mathbf{P} \cdot \boldsymbol{\lambda} = (P/\sqrt{2})(\mathbf{j} - \mathbf{k}) \cdot (1/\sqrt{3})(\mathbf{i} - \mathbf{j} - \mathbf{k}) = (P/\sqrt{6})(0 - 1 + 1) = 0$$

The moment M_{AG} may then be expressed as $-Pd$, where d is the perpendicular distance from *AG* to *FC*. (The negative sign is used since the rotation imparted to the cube by **P** appears as clockwise to an observer at *G*.) Recalling the value found for M_{AG} in part *c*,

$$M_{AG} = -Pd = -aP/\sqrt{6} \qquad d = a/\sqrt{6} \quad ◄$$

Problems

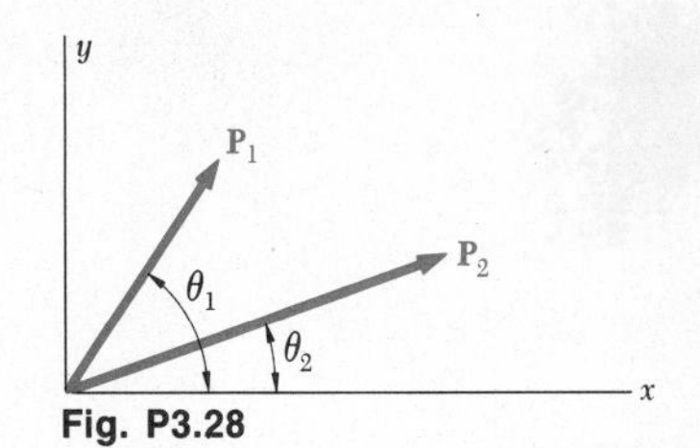

Fig. P3.28

3.27 Given the vectors **P** = 2**i** + 3**j** − **k**, **Q** = 5**i** − 4**j** + 3**k**, and **S** = −3**i** + 2**j** − 5**k**, compute the scalar products **P**·**Q**, **P**·**S**, and **Q**·**S**.

3.28 Form the scalar product $\mathbf{P}_1 \cdot \mathbf{P}_2$ and use the result obtained to prove the identity $\cos(\theta_1 - \theta_2) = \cos\theta_1 \cos\theta_2 + \sin\theta_1 \sin\theta_2$.

3.29 Three cables are used to support a container as shown. Determine the angle formed by cables *AB* and *AD*.

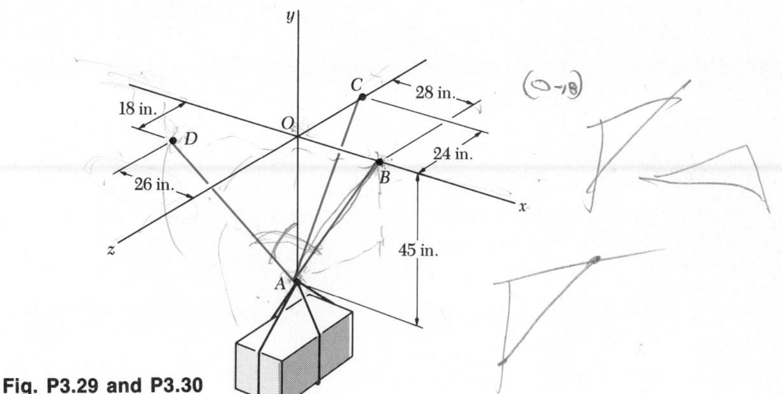

Fig. P3.29 and P3.30

3.30 Three cables are used to support a container as shown. Determine the angle formed by cables *AC* and *AD*.

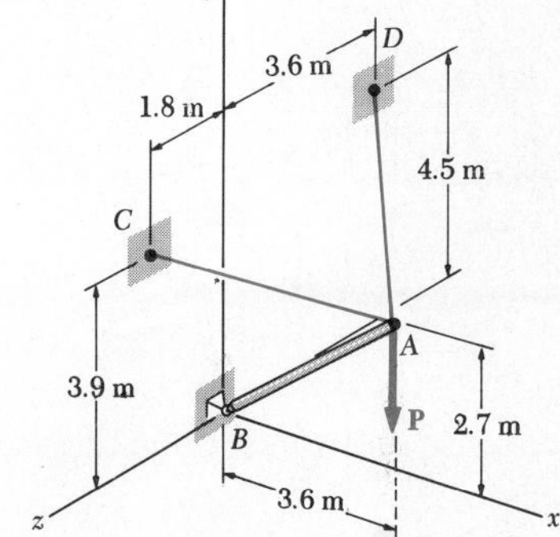

Fig. P3.31 and P3.32

3.31 Knowing that the tension in cable *AC* is 945 N, determine (*a*) the angle between cable *AC* and the boom *AB*, (*b*) the projection on *AB* of the force exerted by cable *AC* at point *A*.

3.32 Knowing that the tension in cable *AD* is 540 N, determine (*a*) the angle between cable *AD* and the boom *AB*, (*b*) the projection on *AB* of the force exerted by cable *AD* at point *A*.

3.33 The 500-mm tube *AB* may slide along a horizontal rod. The ends *A* and *B* of the tube are connected by elastic cords to the fixed point *C*. For the position corresponding to x = 275 mm, determine the angle formed by the two cords, (*a*) using Eq. (3.32), (*b*) applying the law of cosines to triangle *ABC*.

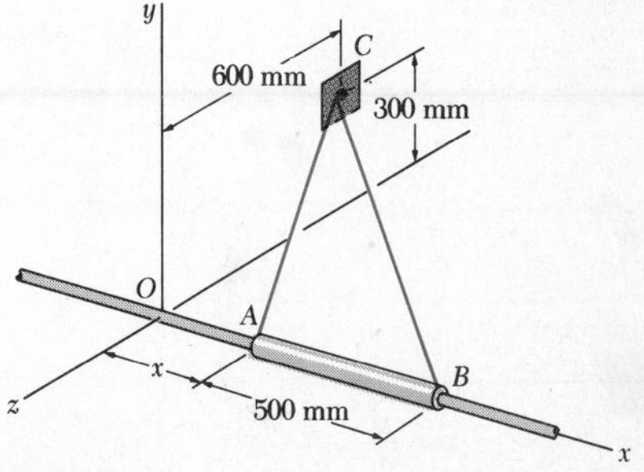

Fig. P3.33

3.34 Solve Prob. 3.33 for the position corresponding to x = 100 mm.

3.35 Given the vectors **P** = 2**i** + 3**j** − **k**, **Q** = 5**i** − 4**j** + 3**k**, and **S** = −3**i** + 2**j** − 5**k**, compute **P**·(**Q** × **S**), (**P** × **Q**)·**S**, and (**S** × **Q**)·**P**.

3.36 Given the vectors **P** = 4**i** − 2**j** + 3**k**, **Q** = 2**i** + 4**j** − 5**k**, and **S** = 7**i** − **j** + S_z**k**, determine the value of S_z for which the three vectors are coplanar.

3.37 The rectangular platform is hinged at A and B and supported by a cable which passes over a frictionless hook at E. Knowing that the tension in the cable is 1349 N, determine the moment about each of the coordinate axes of the force exerted by the cable at C.

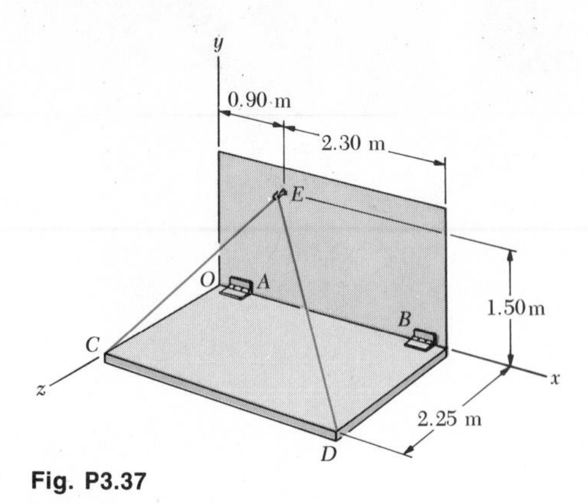

Fig. P3.37

3.38 For the platform of Prob. 3.37, determine the moment about each of the coordinate axes of the force exerted by the cable at D.

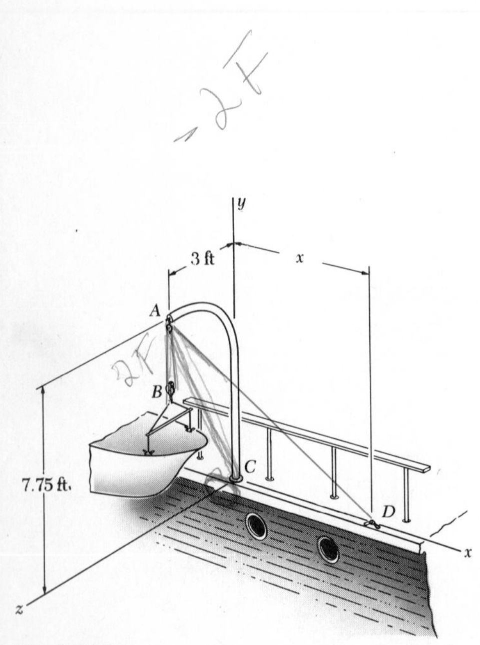

Fig. P3.39

3.39 A small boat hangs from two davits, one of which is shown in the figure. It is known that the moment about the z axis of the resultant force $\mathbf{R}_A$ exerted on the davit at A must not exceed 279 lb · ft in absolute value. Determine the largest allowable tension in line $ABAD$ when $x = 6$ ft.

3.40 For the davit of Prob. 3.39, determine the largest allowable distance x when the tension in line $ABAD$ is 60 lb.

3.41 A single force $\mathbf{P}$ acts at C in a direction perpendicular to the handle BC of the crank shown. Knowing that $M_x = +20$ N · m, $M_y = -8.75$ N · m, and $M_z = -30$ N · m, determine the magnitude of $\mathbf{P}$ and the values of ϕ and θ.

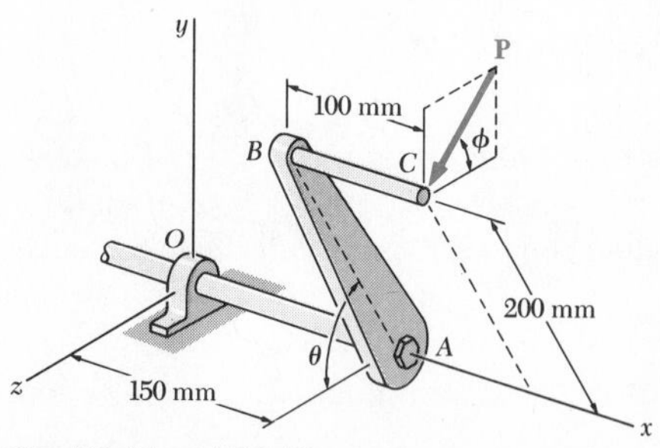

Fig. P3.41 and P3.42

3.42 A single force $\mathbf{P}$ acts at C in a direction perpendicular to the handle BC of the crank shown. Determine the moment M_x of $\mathbf{P}$ about the x axis when $\theta = 65°$, knowing that $M_y = -15$ N · m and $M_z = -36$ N · m.

3.43 The 23-in. vertical rod CD is welded to the midpoint C of the 50-in. rod AB. Determine the moment about AB of the 235-lb force **P**.

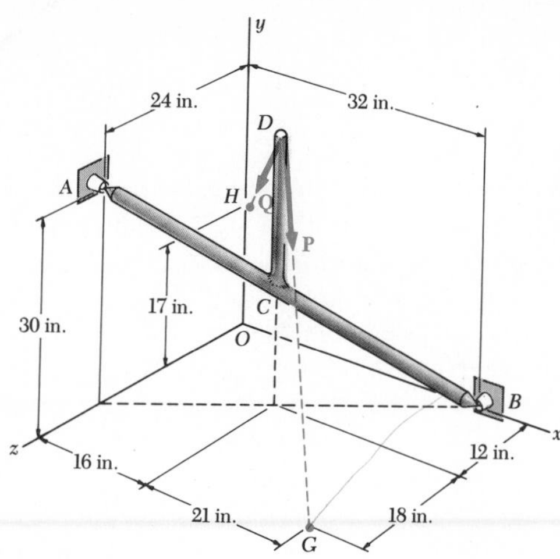

Fig. P3.43 and P3.44

3.44 The 23-in. vertical rod CD is welded to the midpoint C of the 50-in. rod AB. Determine the moment about AB of the 174-lb force **Q**.

3.45 The frame ACD is hinged at A and D and supported by a cable which passes through a ring at B and is attached to hooks at G and H. Knowing that the tension in the cable is 1125 N, determine the moment about the diagonal AD of the force exerted on the frame by portion BH of the cable.

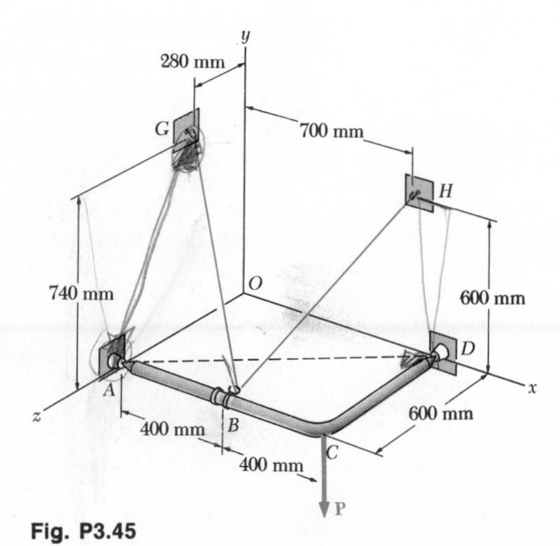

Fig. P3.45

3.46 In Prob. 3.45, determine the moment about the diagonal AD of the force exerted on the frame by portion BG of the cable.

3.47 Two forces $\mathbf{F}_1$ and $\mathbf{F}_2$ in space have the same magnitude F. Prove that the moment of $\mathbf{F}_1$ about the line of action of $\mathbf{F}_2$ is equal to the moment of $\mathbf{F}_2$ about the line of action of $\mathbf{F}_1$.

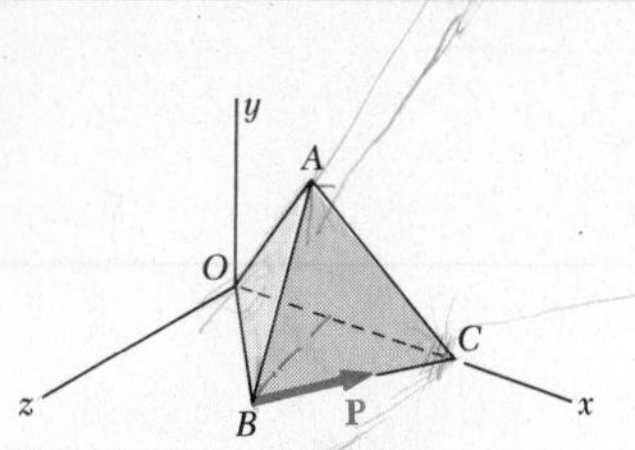

Fig. P3.48 and P3.49

3.48 A regular tetrahedron has six edges of length a. A force $\mathbf{P}$ is directed as shown along edge BC. Determine the moment of $\mathbf{P}$ about edge OA.

3.49 A regular tetrahedron has six edges of length a. (*a*) Show that two opposite edges, such as OA and BC, are perpendicular to each other. (*b*) Use this property and the result obtained in Prob. 3.48 to determine the perpendicular distance between the edges OA and BC.

***3.50** In Prob. 3.43, determine the perpendicular distance between rod AB and the line of action of $\mathbf{P}$.

***3.51** In Prob. 3.45, determine the perpendicular distance between portion BH of the cable and the diagonal AD.

***3.52** Use the result obtained in Prob. 3.46 to determine the perpendicular distance between portion BG of the cable of Fig. P3.45 and the diagonal AD.

***3.53** Use the result obtained in Prob. 3.44 to determine the perpendicular distance between rod AB and the line of action of $\mathbf{Q}$.

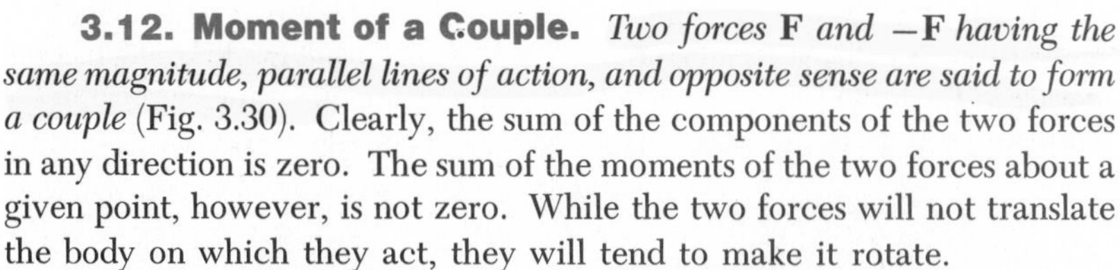

Fig. 3.30

3.12. Moment of a Couple. *Two forces* $\mathbf{F}$ *and* $-\mathbf{F}$ *having the same magnitude, parallel lines of action, and opposite sense are said to form a couple* (Fig. 3.30). Clearly, the sum of the components of the two forces in any direction is zero. The sum of the moments of the two forces about a given point, however, is not zero. While the two forces will not translate the body on which they act, they will tend to make it rotate.

Denoting by $\mathbf{r}_A$ and $\mathbf{r}_B$, respectively, the position vectors of the points of application of $\mathbf{F}$ and $-\mathbf{F}$ (Fig. 3.31), we find that the sum of the moments of the two forces about O is

$$\mathbf{r}_A \times \mathbf{F} + \mathbf{r}_B \times (-\mathbf{F}) = (\mathbf{r}_A - \mathbf{r}_B) \times \mathbf{F}$$

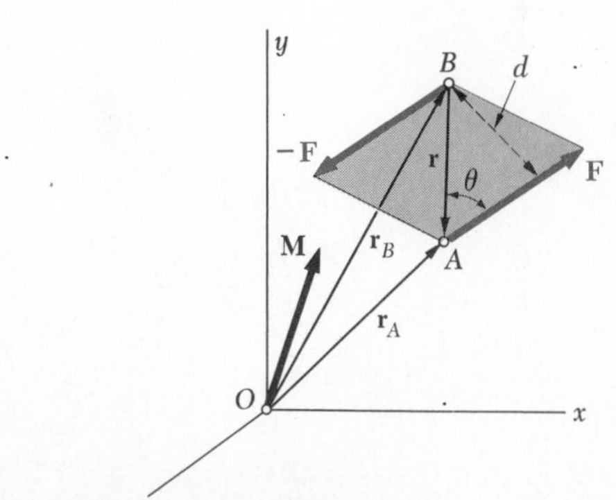

Fig. 3.31

Setting $\mathbf{r}_A - \mathbf{r}_B = \mathbf{r}$, where $\mathbf{r}$ is the vector joining the points of application of the two forces, we conclude that the sum of the moments of $\mathbf{F}$ and $-\mathbf{F}$ about O is represented by the vector

$$\mathbf{M} = \mathbf{r} \times \mathbf{F} \tag{3.47}$$

The vector $\mathbf{M}$ is called the *moment of the couple;* it is a vector perpendicular to the plane containing the two forces and its magnitude is

$$M = rF \sin \theta = Fd \tag{3.48}$$

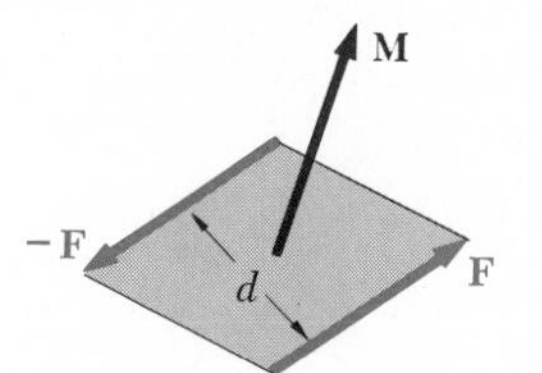

Fig. 3.32

where d is the perpendicular distance between the lines of action of $\mathbf{F}$ and $-\mathbf{F}$. The sense of $\mathbf{M}$ is defined by the right-hand rule.

Since the vector $\mathbf{r}$ in (3.47) is independent of the choice of the origin O of the coordinate axes, we note that the same result would have been obtained if the moments of $\mathbf{F}$ and $-\mathbf{F}$ had been computed about a different point O'. Thus, the moment $\mathbf{M}$ of a couple is a *free vector* (Sec. 2.3) which may be applied at any point (Fig. 3.32).

From the definition of the moment of a couple, it also follows that two couples, one consisting of the forces $\mathbf{F}_1$ and $-\mathbf{F}_1$, the other of the forces $\mathbf{F}_2$ and $-\mathbf{F}_2$ (Fig. 3.33), will have equal moments if

$$F_1 d_1 = F_2 d_2 \tag{3.49}$$

and if the two couples lie in parallel planes (or in the same plane) and have the same sense.

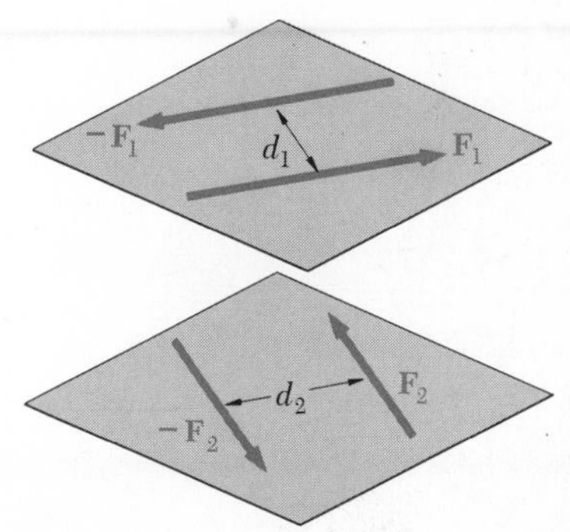

Fig. 3.33

3.13. Equivalent Couples. Consider the three couples shown in Fig. 3.34, which are made to act successively on the same rectangular box. As seen in the preceding section, the only motion a couple may impart to a rigid body is a rotation. Since each of the three couples shown has the same moment $\mathbf{M}$ (same direction and same magnitude $M = 120$ lb·in.), we may expect the three couples to have the same effect on the box.

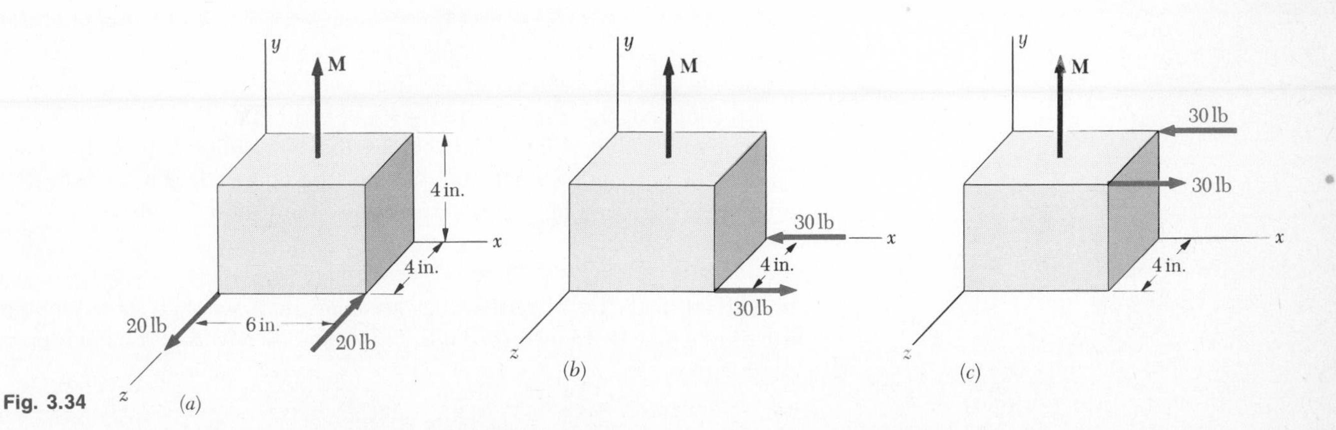

Fig. 3.34

As reasonable as this conclusion may appear, we should not accept it hastily. While intuitive feeling is of great help in the study of mechanics, it should not be accepted as a substitute for logical reasoning. Before stating that two systems (or groups) of forces have the same effect on a rigid body, we should prove that fact on the basis of the experimental evidence introduced so far. This evidence consists of the parallelogram law for the addition of two forces (Sec. 2.2) and of the principle of transmissibility (Sec. 3.3). Therefore, we shall state that *two systems of forces are equivalent* (i.e., they have the same effect on a rigid body) *if we can transform one of them into the other by means of one or several of the following operations:* (1) replacing two forces acting on the same particle by their resultant; (2) resolving a force into two components; (3) canceling two equal and opposite forces acting on the same particle; (4) attaching to the same particle two equal and opposite forces; (5) moving a force along its line of action. Each of these operations is easily justified on the basis of the parallelogram law or the principle of transmissibility.

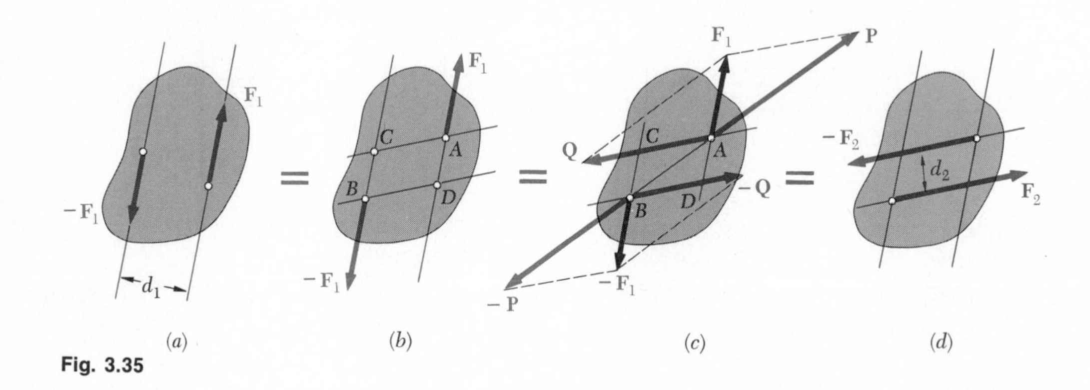

Fig. 3.35

Let us now prove that *two couples having the same moment* **M** *are equivalent*. First, we shall consider two couples contained in the same plane, and we shall assume that this plane coincides with the plane of the figure (Fig. 3.35). The first couple consists of the forces $\mathbf{F}_1$ and $-\mathbf{F}_1$, of magnitude F_1 and at a distance d_1 from each other (Fig. 3.35*a*), and the second couple of the forces $\mathbf{F}_2$ and $-\mathbf{F}_2$, of magnitude F_2 and at a distance d_2 from each other (Fig. 3.35*d*). Since the two couples have the same moment **M** perpendicular to the plane of the figure, they must have the same sense (assumed here counterclockwise) and the relation

$$F_1 d_1 = F_2 d_2 \qquad (3.49)$$

must be satisfied. To prove that they are equivalent, we shall show that the first couple may be transformed into the second by means of the operations listed above.

Denoting by A, B, C, D the points of intersection of the lines of action of the two couples, we first slide the forces $\mathbf{F}_1$ and $-\mathbf{F}_1$ until they are attached, respectively, at A and B, as shown in Fig. 3.35*b*. The force $\mathbf{F}_1$ is then resolved into a component $\mathbf{P}$ along line AB and a component $\mathbf{Q}$ along AC (Fig. 3.35*c*); similarly, the force $-\mathbf{F}_1$ is resolved into $-\mathbf{P}$ along AB and $-\mathbf{Q}$ along BD. The forces $\mathbf{P}$ and $-\mathbf{P}$ have the same magnitude, same line of action, and opposite sense; they may be moved along their common line of action until they are applied at the same point and then canceled. Thus the couple formed by $\mathbf{F}_1$ and $-\mathbf{F}_1$ reduces to a couple consisting of $\mathbf{Q}$ and $-\mathbf{Q}$.

We shall now show that the forces $\mathbf{Q}$ and $-\mathbf{Q}$ are respectively equal to the forces $-\mathbf{F}_2$ and $\mathbf{F}_2$. The moment of the couple formed by $\mathbf{Q}$ and $-\mathbf{Q}$ may be obtained by computing the moment of $\mathbf{Q}$ about B; similarly, the moment of the couple formed by $\mathbf{F}_1$ and $-\mathbf{F}_1$ is the moment of $\mathbf{F}_1$ about B. But, by Varignon's theorem, the moment of $\mathbf{F}_1$ is equal to the sum of the moments of its components $\mathbf{P}$ and $\mathbf{Q}$. Since the moment of $\mathbf{P}$ about B is zero, the moment of the couple formed by $\mathbf{Q}$ and $-\mathbf{Q}$ must be equal to the moment of the couple formed by $\mathbf{F}_1$ and $-\mathbf{F}_1$. Recalling (3.49), we write

$$Qd_2 = F_1 d_1 = F_2 d_2 \qquad \text{and} \qquad Q = F_2$$

Thus the forces $\mathbf{Q}$ and $-\mathbf{Q}$ are respectively equal to the forces $-\mathbf{F}_2$ and $\mathbf{F}_2$, and the couple of Fig. 3.35*a* is equivalent to the couple of Fig. 3.35*d*.

Next we shall consider two couples contained in parallel planes P_1 and P_2 and prove that they are equivalent if they have the same moment. In view of the foregoing we may assume that the couples consist of forces of the same magnitude F acting along parallel lines (Fig. 3.36*a* and *d*). We propose to show that the couple contained in plane P_1 may be transformed into the couple contained in plane P_2 by means of the standard operations listed above.

Let us consider the two planes defined respectively by the lines of action of $\mathbf{F}_1$ and $-\mathbf{F}_2$, and of $-\mathbf{F}_1$ and $\mathbf{F}_2$ (Fig. 3.36*b*). At a point on their line of intersection we attach two forces $\mathbf{F}_3$ and $-\mathbf{F}_3$, respectively equal to $\mathbf{F}_1$ and $-\mathbf{F}_1$. The couple formed by $\mathbf{F}_1$ and $-\mathbf{F}_3$ may be replaced by a couple consisting of $\mathbf{F}_3$ and $-\mathbf{F}_2$ (Fig. 3.36*c*), since both couples have clearly the same moment and are contained in the same plane. Similarly, the couple formed by $-\mathbf{F}_1$ and $\mathbf{F}_3$ may be replaced by a couple consisting of $-\mathbf{F}_3$ and $\mathbf{F}_2$. Canceling the two equal and opposite forces $\mathbf{F}_3$ and $-\mathbf{F}_3$, we obtain the desired couple in plane P_2 (Fig. 3.36*d*). Thus, we conclude that two couples having the same moment $\mathbf{M}$ are equivalent, whether they are contained in the same plane or in parallel planes.

The property we have just established is very important for the correct understanding of the mechanics of rigid bodies. It indicates that when a couple acts on a rigid body, it does not matter where the two forces forming the couple act, or what magnitude and direction they have. The only thing which counts is the *moment* of the couple (magnitude and direction). Couples with the same moment will have the same effect on the rigid body.

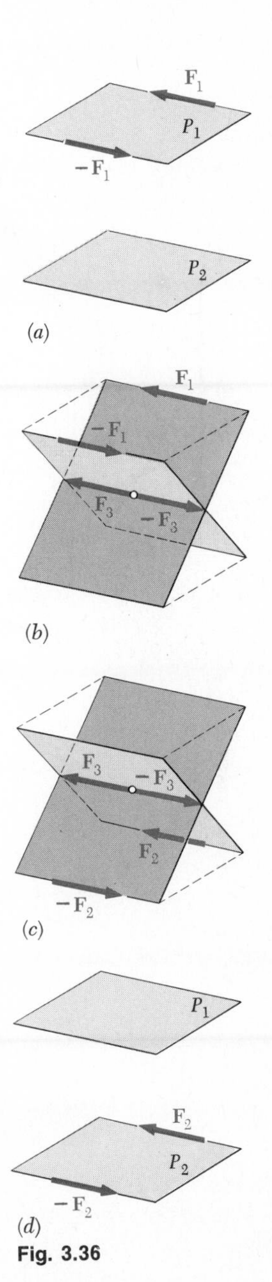

Fig. 3.36

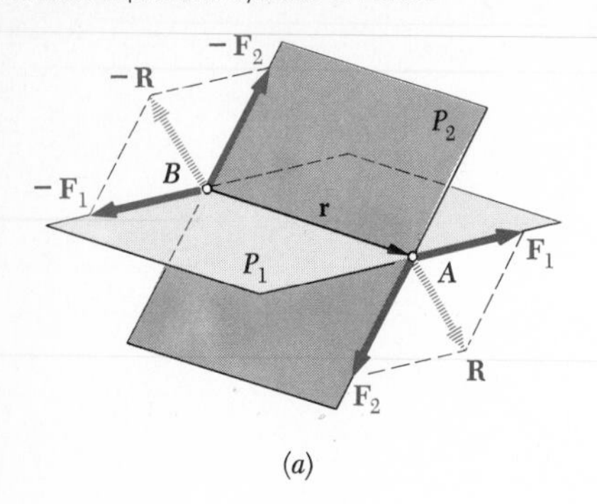

(a)

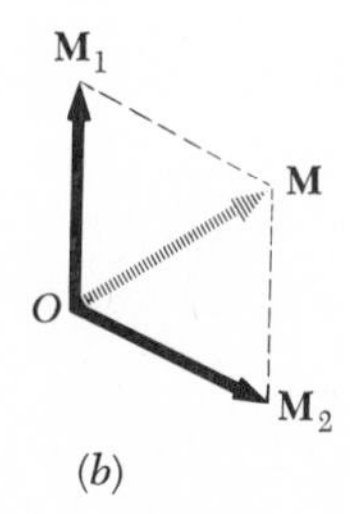

(b)

Fig. 3.37

3.14. Addition of Couples. Consider two intersecting planes P_1 and P_2 and two couples acting respectively in P_1 and P_2. We may, without any loss of generality, assume that the couple in P_1 consists of two forces $\mathbf{F}_1$ and $-\mathbf{F}_1$ perpendicular to the line of intersection of the two planes and acting respectively at A and B (Fig. 3.37*a*). Similarly, we assume that the couple in P_2 consists of two forces $\mathbf{F}_2$ and $-\mathbf{F}_2$ perpendicular to AB and acting respectively at A and B. It is clear that the resultant $\mathbf{R}$ of $\mathbf{F}_1$ and $\mathbf{F}_2$ and the resultant $-\mathbf{R}$ of $-\mathbf{F}_1$ and $-\mathbf{F}_2$ form a couple. Denoting by $\mathbf{r}$ the vector joining B to A, and recalling the definition of the moment of a couple (Sec. 3.12), we express the moment $\mathbf{M}$ of the resulting couple as follows:

$$\mathbf{M} = \mathbf{r} \times \mathbf{R} = \mathbf{r} \times (\mathbf{F}_1 + \mathbf{F}_2)$$

and, by Varignon's theorem,

$$\mathbf{M} = \mathbf{r} \times \mathbf{F}_1 + \mathbf{r} \times \mathbf{F}_2$$

But the first term in the expression obtained represents the moment $\mathbf{M}_1$ of the couple in P_1, and the second term the moment $\mathbf{M}_2$ of the couple in P_2. We have

$$\mathbf{M} = \mathbf{M}_1 + \mathbf{M}_2 \tag{3.50}$$

and we conclude that the sum of two couples of moments $\mathbf{M}_1$ and $\mathbf{M}_2$ is a couple of moment $\mathbf{M}$ equal to the vector sum of $\mathbf{M}_1$ and $\mathbf{M}_2$ (Fig. 3.37*b*).

3.15. Couples May Be Represented by Vectors. As we saw in Sec. 3.13, couples which have the same moment, whether they act in the same plane or in parallel planes, are equivalent. There is therefore no need to draw the actual forces forming a given couple in order to define its effect on a rigid body (Fig. 3.38*a*). It is sufficient to draw an arrow equal in magnitude and direction to the moment $\mathbf{M}$ of the couple (Fig. 3.38*b*). On the other hand, we saw in Sec. 3.14 that the sum of two couples is itself a couple and that the moment $\mathbf{M}$ of the resultant couple may be obtained by forming the vector sum of the moments $\mathbf{M}_1$ and $\mathbf{M}_2$ of the given couples.

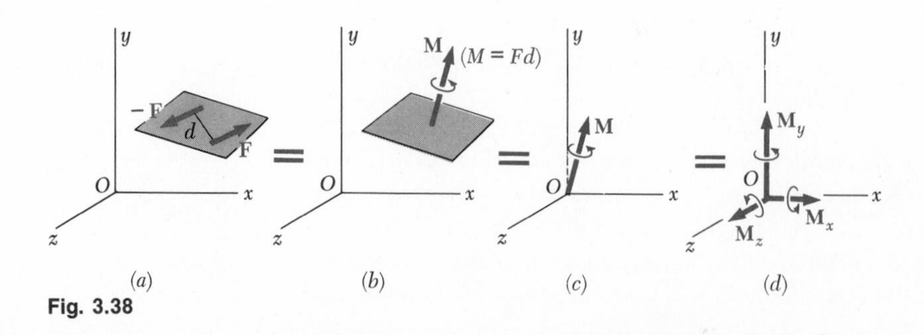

Fig. 3.38

Thus, couples obey the law of addition of vectors, and the arrow used in Fig. 3.38*b* to represent the couple defined in Fig. 3.38*a* may truly be considered a vector.

The vector representing a couple is called a *couple vector*. Note that, in Fig. 3.38, a colored arrow is used to distinguish the couple vector, *which represents the couple itself*, from the vector representing the *moment* of the couple, and that the symbol ↰ is added to avoid any confusion with vectors representing forces. A couple vector, like the moment of a couple, is a free vector. Its point of application, therefore, may be chosen at the origin of the system of coordinates, if so desired (Fig. 3.38*c*). Furthermore, the couple vector $\mathbf{M}$ may be resolved into component vectors $\mathbf{M}_x$, $\mathbf{M}_y$, and $\mathbf{M}_z$, directed along the axes of coordinates (Fig. 3.38*d*) and representing couples acting, respectively, in the yz, zx, and xy planes.

3.16. Resolution of a Given Force into a Force at *O* and a Couple. Consider a force $\mathbf{F}$ acting on a rigid body at a point A defined by the position vector $\mathbf{r}$ (Fig. 3.39*a*). Suppose that for some reason we would rather have the force act at point O. We know that we can move $\mathbf{F}$ along its line of action (principle of transmissibility); but we cannot move it to a point O away from the original line of action without modifying the action of $\mathbf{F}$ on the rigid body.

We may, however, attach two forces at point O, one equal to $\mathbf{F}$ and the other equal to $-\mathbf{F}$, without modifying the action of the original force on the rigid body (Fig. 3.39*b*). As a result of this tranformation, a force $\mathbf{F}$ is now applied at O; the other two forces form a couple of moment $\mathbf{M}_O = \mathbf{r} \times \mathbf{F}$. Thus, *any force* $\mathbf{F}$ *acting on a rigid body may be moved to an arbitrary point O, provided that a couple is added, of moment equal to the moment of* $\mathbf{F}$ *about O.* The couple tends to impart to the rigid body the same motion of rotation about O that the force $\mathbf{F}$ tended to produce before it was transferred to O. The couple is represented by a couple vector $\mathbf{M}_O$ perpendicular to the plane containing $\mathbf{r}$ and $\mathbf{F}$. Since $\mathbf{M}_O$ is a free vector, it may be applied anywhere; for convenience, however, the couple vector is usually attached at O, together with $\mathbf{F}$, and the combination obtained is referred to as a *force-couple system* (Fig. 3.39*c*).

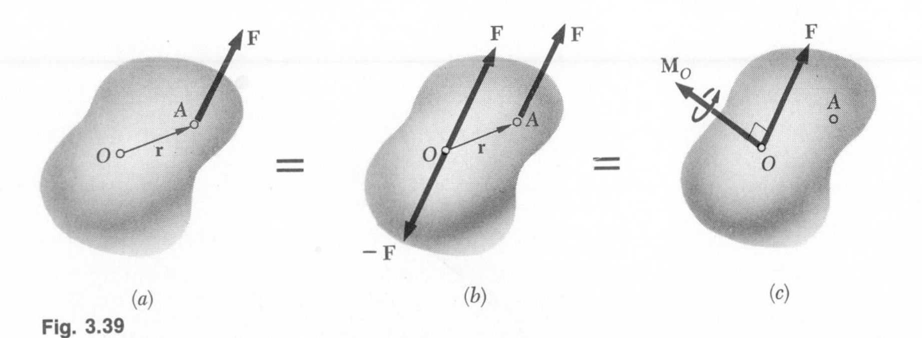

Fig. 3.39

If the force **F** had been moved from A to a different point O' (Fig. 3.40a and c), the moment $\mathbf{M}_{O'} = \mathbf{r}' \times \mathbf{F}$ of **F** about O' should have been computed, and a new force-couple system, consisting of **F** and of the couple vector $\mathbf{M}_{O'}$, would have been attached at O'. The relation existing between the moments of **F** about O and O' is obtained by writing

$$\mathbf{M}_{O'} = \mathbf{r}' \times \mathbf{F} = (\mathbf{r} + \mathbf{s}) \times \mathbf{F} = \mathbf{r} \times \mathbf{F} + \mathbf{s} \times \mathbf{F}$$

$$\mathbf{M}_{O'} = \mathbf{M}_O + \mathbf{s} \times \mathbf{F} \tag{3.51}$$

where **s** is the vector joining O' to O. Thus, the moment $\mathbf{M}_{O'}$ of **F** about O' is obtained by adding to the moment $\mathbf{M}_O$ of **F** about O the vector product $\mathbf{s} \times \mathbf{F}$ representing the moment about O' of the force **F** applied at O.

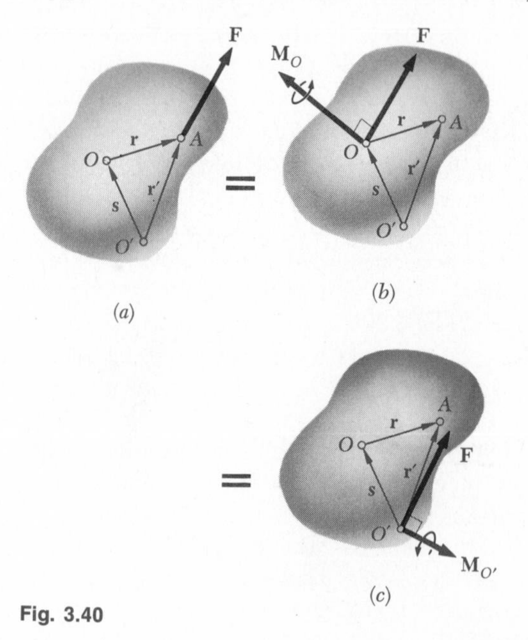

Fig. 3.40

This result could also have been established by observing that, in order to transfer to O' the force-couple system attached at O (Fig. 3.40b and c), the couple vector $\mathbf{M}_O$ may be freely moved to O'; to move the force **F** from O to O', however, it is necessary to add to **F** a couple vector of moment equal to the moment about O' of the force **F** applied at O. Thus, the couple vector $\mathbf{M}_{O'}$ must be the sum of $\mathbf{M}_O$ and of the vector $\mathbf{s} \times \mathbf{F}$.

As noted above, the force-couple system obtained by transferring a force **F** from a point A to a point O consists of **F** and of a couple vector $\mathbf{M}_O$ perpendicular to **F**. Conversely, any force-couple system consisting of a force **F** and of a couple vector $\mathbf{M}_O$ which are *mutually perpendicular* may be replaced by a single equivalent force. This is done by moving the force **F** in the plane perpendicular to $\mathbf{M}_O$ until its moment about O becomes equal to the moment of the couple to be eliminated.

SAMPLE PROBLEM 3.6

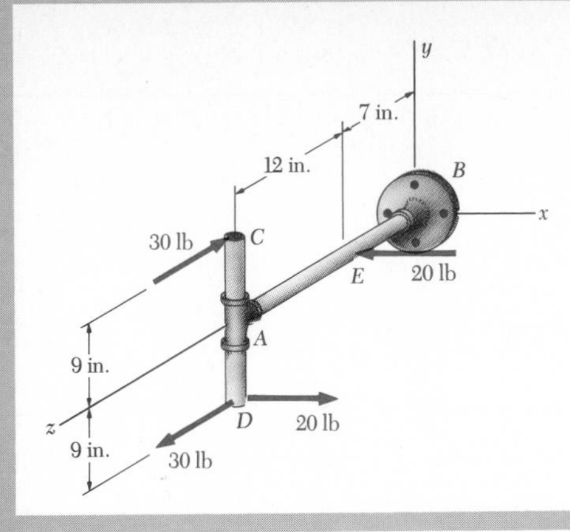

Determine the components of the single couple equivalent to the two couples shown.

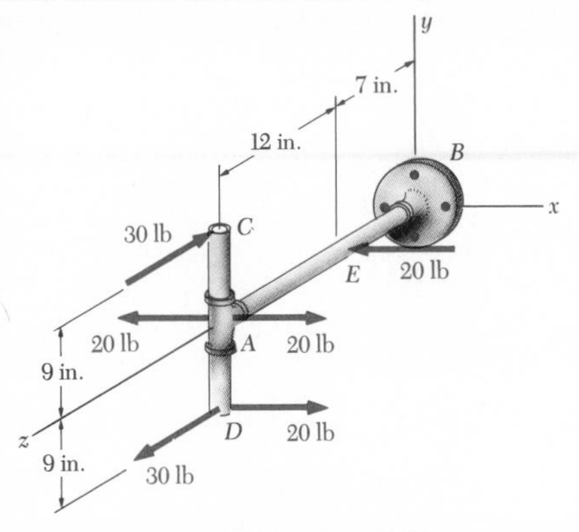

Solution. Our computations will be simplified if we attach two equal and opposite 20-lb forces at A. This enables us to replace the original 20-lb-force couple by two new 20-lb-force couples, one of which lies in the xz plane and the other in a plane parallel to the xy plane. The three couples shown in the adjoining sketch may be represented by three couple vectors $\mathbf{M}_x$, $\mathbf{M}_y$, and $\mathbf{M}_z$ directed along the coordinate axes. The corresponding moments are

$$M_x = -(30 \text{ lb})(18 \text{ in.}) = -540 \text{ lb} \cdot \text{in.}$$
$$M_y = +(20 \text{ lb})(12 \text{ in.}) = +240 \text{ lb} \cdot \text{in.}$$
$$M_z = +(20 \text{ lb})(9 \text{ in.}) = +180 \text{ lb} \cdot \text{in.}$$

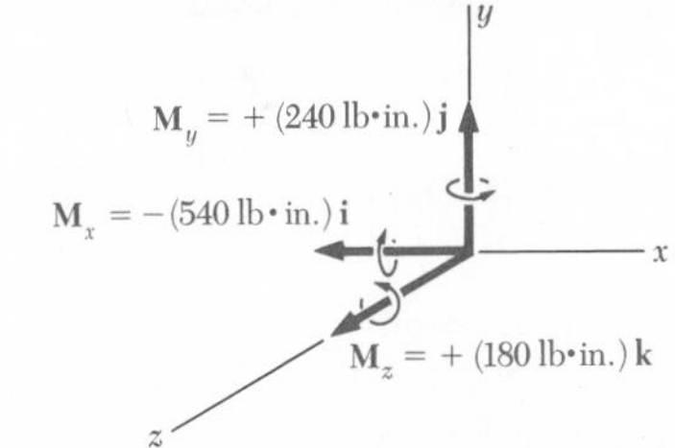

These three moments represent the components of the single couple $\mathbf{M}$ equivalent to the two given couples. We write

$$\mathbf{M} = -(540 \text{ lb} \cdot \text{in.})\mathbf{i} + (240 \text{ lb} \cdot \text{in.})\mathbf{j} + (180 \text{ lb} \cdot \text{in.})\mathbf{k} \quad ◀$$

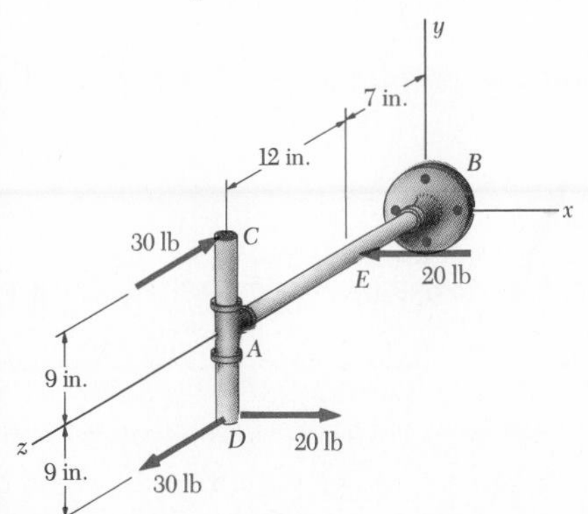

Alternate Solution. The components of the equivalent single couple $\mathbf{M}$ may also be obtained by computing the sum of the moments of the four given forces about an arbitrary point. Selecting point D, we write

$$\mathbf{M} = \mathbf{M}_D = (18 \text{ in.})\mathbf{j} \times (-30 \text{ lb})\mathbf{k} + [(9 \text{ in.})\mathbf{j} - (12 \text{ in.})\mathbf{k}] \times (-20 \text{ lb})\mathbf{i}$$

and, after computing the various cross products,

$$\mathbf{M} = -(540 \text{ lb} \cdot \text{in.})\mathbf{i} + (240 \text{ lb} \cdot \text{in.})\mathbf{j} + (180 \text{ lb} \cdot \text{in.})\mathbf{k} \quad ◀$$

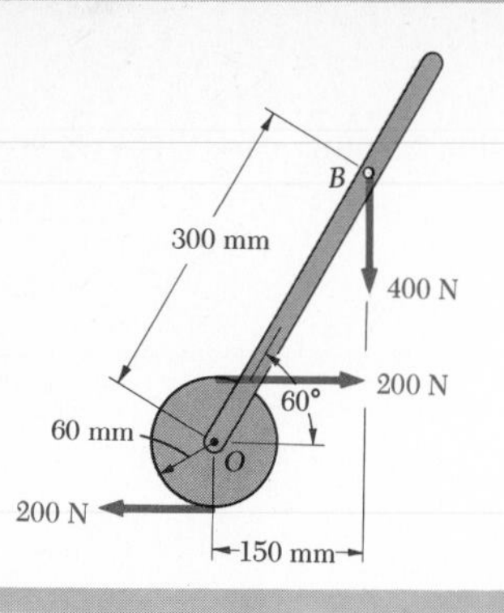

SAMPLE PROBLEM 3.7

Replace the couple and force shown by an equivalent single force applied to the lever. Determine the distance from the shaft to the point of application of this equivalent force.

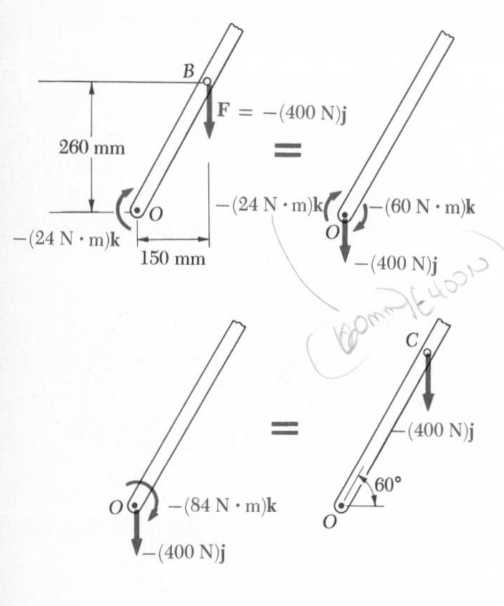

Solution. First the given force and couple are replaced by an equivalent force-couple system at O. We move the force $\mathbf{F} = -(400\text{ N})\mathbf{j}$ to O and at the same time add a couple of moment $\mathbf{M}_O$ equal to the moment about O of the force in its original position.

$$\mathbf{M}_O = \overrightarrow{OB} \times \mathbf{F} = [(0.150\text{ m})\mathbf{i} + (0.260\text{ m})\mathbf{j}] \times (-400\text{ N})\mathbf{j}$$
$$= -(60\text{ N}\cdot\text{m})\mathbf{k}$$

This couple is added to the couple of moment $-(24\text{ N}\cdot\text{m})\mathbf{k}$ formed by the two 200-N forces, and a couple of moment $-(84\text{ N}\cdot\text{m})\mathbf{k}$ is obtained. This last couple may be eliminated by applying $\mathbf{F}$ at a point C chosen in such a way that

$$-(84\text{ N}\cdot\text{m})\mathbf{k} = \overrightarrow{OC} \times \mathbf{F}$$
$$= [(OC)\cos 60°\mathbf{i} + (OC)\sin 60°\mathbf{j}] \times (-400\text{ N})\mathbf{j}$$
$$= -(OC)\cos 60°(400\text{ N})\mathbf{k}$$

We conclude that

$$(OC)\cos 60° = 0.210\text{ m} = 210\text{ mm} \qquad OC = 420\text{ mm} \blacktriangleleft$$

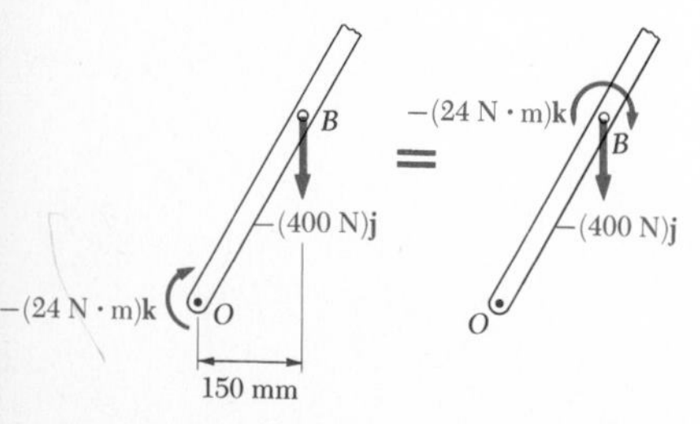

Alternate Solution. Since the effect of a couple does not depend on its location, the couple of moment $-(24\text{ N}\cdot\text{m})\mathbf{k}$ may be moved to B; we thus obtain a force-couple system at B. The couple may now be eliminated by applying $\mathbf{F}$ at a point C chosen in such a way that

$$-(24\text{ N}\cdot\text{m})\mathbf{k} = \overrightarrow{BC} \times \mathbf{F}$$
$$= -(BC)\cos 60°(400\text{ N})\mathbf{k}$$

We conclude that

$$(BC)\cos 60° = 0.060\text{ m} = 60\text{ mm} \qquad BC = 120\text{ mm}$$
$$OC = OB + BC = 300\text{ mm} + 120\text{ mm} \qquad OC = 420\text{ mm} \blacktriangleleft$$

Problems

3.54 Two 80-N forces are applied as shown to the corners B and D of a rectangular plate. (*a*) Determine the moment of the couple formed by the two forces by resolving each force into horizontal and vertical components and adding the moments of the two resulting couples. (*b*) Use the result obtained to determine the perpendicular distance between lines BE and DF.

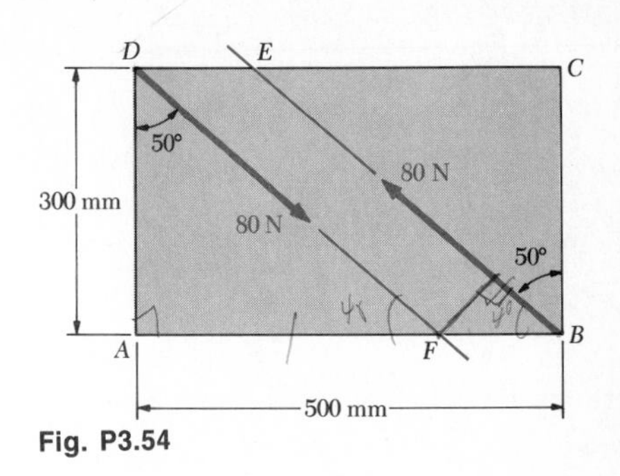

Fig. P3.54

3.55 A piece of plywood in which several holes are being drilled successively has been secured to a workbench by means of two nails. Knowing that the drill exerts a 12-N · m couple on the piece of plywood, determine the magnitude of the resulting forces applied to the nails if they are located (*a*) at A and B, (*b*) at B and C, (*c*) at A and C.

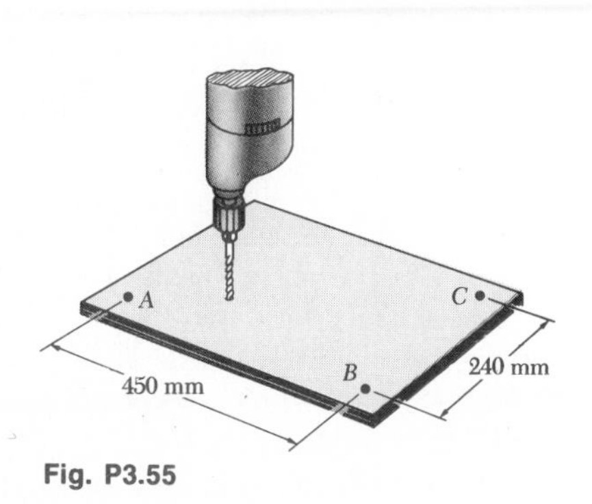

Fig. P3.55

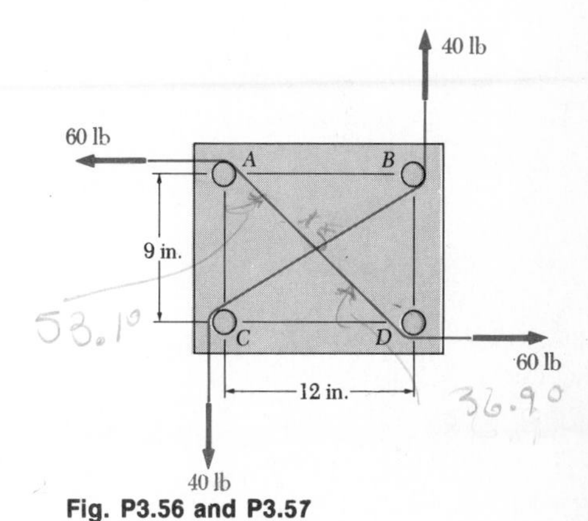

Fig. P3.56 and P3.57

3.56 Four $1\frac{1}{2}$-in.-diameter pegs are attached to a board as shown. Two strings are passed around the pegs and pulled with the forces indicated. (*a*) Determine the resultant couple acting on the board. (*b*) If only one string is used, around which pegs should it pass and in what directions should it be pulled to create the same couple with the minimum tension in the string? (*c*) What is the value of that minimum tension?

3.57 Four pegs of the same diameter are attached to a board as shown. Two strings are passed around the pegs and pulled with the forces indicated. Determine the diameter of the pegs, knowing that the resultant couple applied to the board is 1132.5 lb · in. counterclockwise.

3.58 The two shafts of a speed-reducer unit are subjected to couples of magnitude $M_1 = 12$ lb · ft and $M_2 = 5$ lb · ft, respectively. Replace the two couples by a single equivalent couple, specifying its magnitude and the direction of its axis.

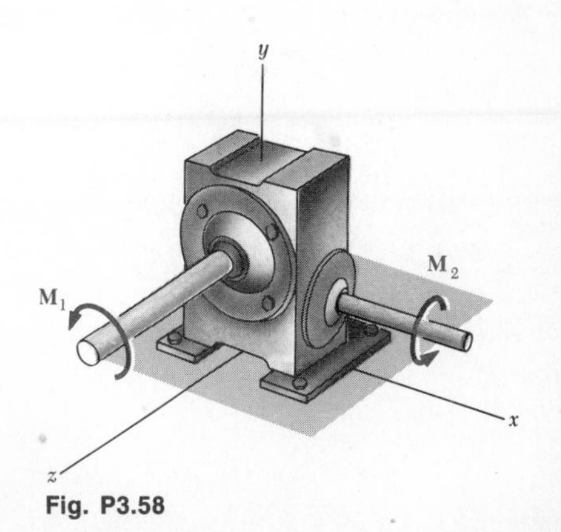

Fig. P3.58

3.59 Replace the two couples shown by a single equivalent couple, specifying its magnitude and the direction of its axis.

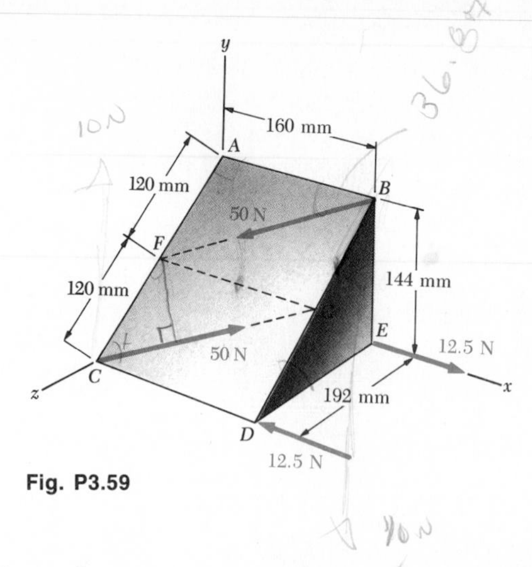

Fig. P3.59

3.60 Solve Prob. 3.59, assuming that two 10-N vertical forces have been added, one acting upward at C and the other downward at B.

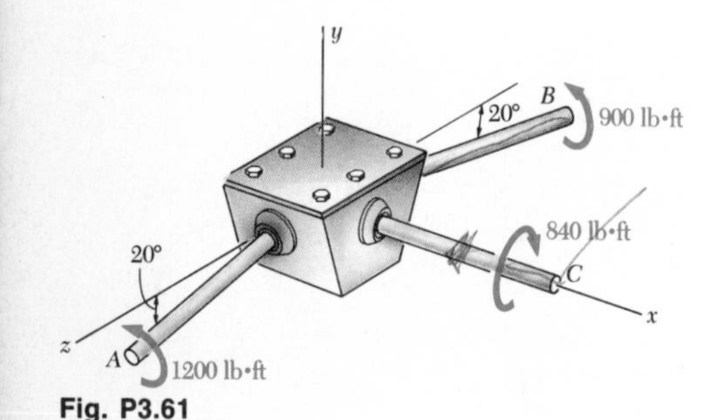

Fig. P3.61

3.61 Shafts A and B connect the gear box to the wheel assemblies of a tractor, and shaft C connects it to the engine. Shafts A and B lie in the vertical yz plane, while shaft C is directed along the x axis. Replace the couples applied to the shafts by a single equivalent couple, specifying its magnitude and the direction of its axis.

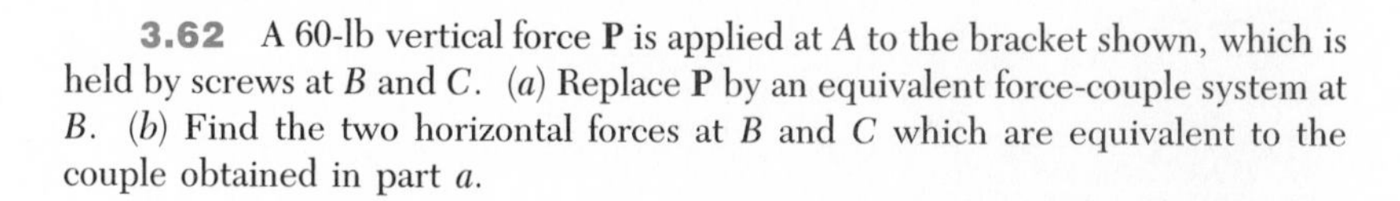

3.62 A 60-lb vertical force **P** is applied at A to the bracket shown, which is held by screws at B and C. (*a*) Replace **P** by an equivalent force-couple system at B. (*b*) Find the two horizontal forces at B and C which are equivalent to the couple obtained in part *a*.

P
7.5 in.
A
3 in.
B
4.5 in.
C

Fig. P3.62

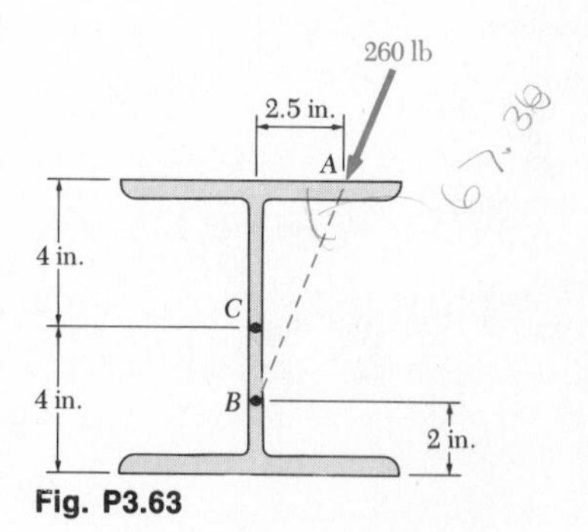

Fig. P3.63

3.63 A 260-lb force is applied at A to the rolled-steel section shown. Replace that force by an equivalent force-couple system at the center C of the section.

3.64 Force **P** has a magnitude of 300 N and is applied at A in a direction perpendicular to the handle ($\alpha = 0$). Assuming $\beta = 30°$, replace force **P** by (*a*) an equivalent force-couple system at B, (*b*) an equivalent system formed by two parallel forces applied at B and C.

3.65 Solve Prob. 3.64, assuming that force **P** is horizontal ($\alpha = \beta = 30°$).

3.66 Force **P** is applied to the handle at A. Replace **P** by an equivalent system formed by two parallel forces applied at B and C and show that (*a*) these forces are parallel to **P**, (*b*) their magnitudes are independent of both α and β.

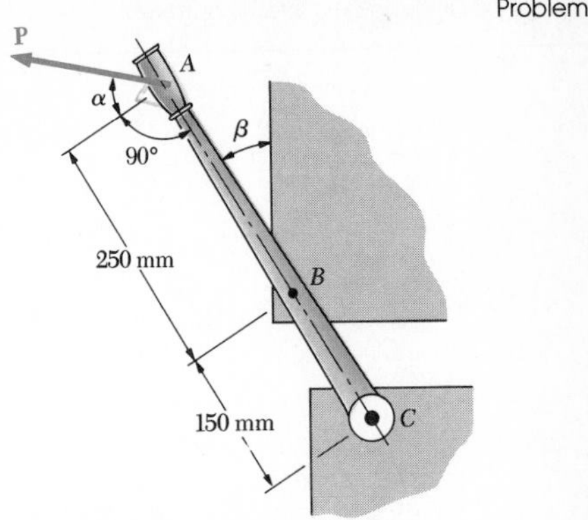

Fig. P3.64 and P3.66

3.67 The shearing forces exerted on the cross section of a steel channel may be represented by a 900-N vertical force and two 250-N horizontal forces as shown. Replace this force and couple by a single force **F** applied at point C and determine the distance x from C to line BD. (Point C is defined as the *shear center* of the section.)

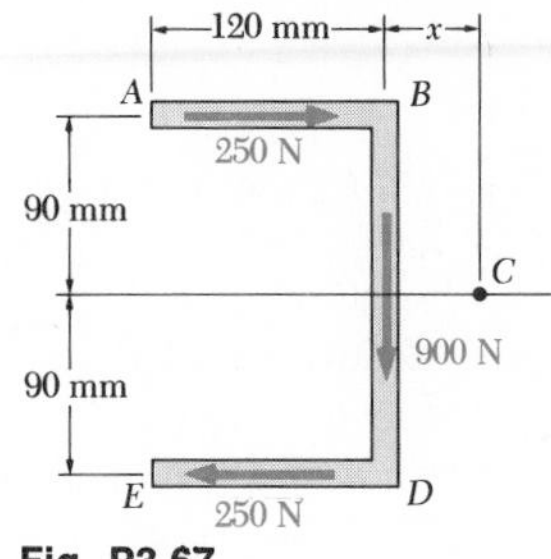

Fig. P3.67

3.68 A force and couple act as shown on a square plate of side $a = 25$ in. Knowing that $P = 60$ lb, $Q = 40$ lb, and $\alpha = 50°$, replace the given force and couple by a single force applied at a point located (*a*) on line AB, (*b*) on line AC. In each case determine the distance from A to the point of application of the force.

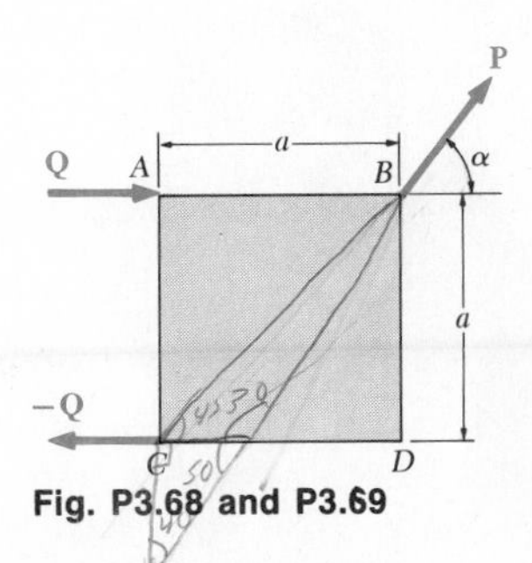

Fig. P3.68 and P3.69

3.69 The force and couple shown are to be replaced by an equivalent single force. Knowing that $P = 2Q$, determine the required value of α if the line of action of the single equivalent force is to pass through (*a*) point A, (*b*) point C.

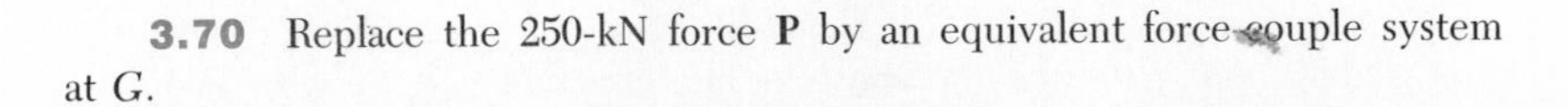

3.70 Replace the 250-kN force **P** by an equivalent force-couple system at G.

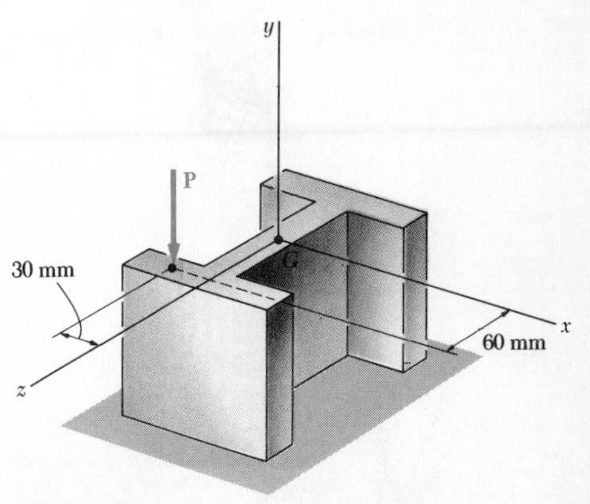

Fig. P3.70

3.71 The 12-ft boom AB has a fixed end A and the tension in cable BC is 570 lb. Replace the force that the cable exerts at B by an equivalent force-couple system at A.

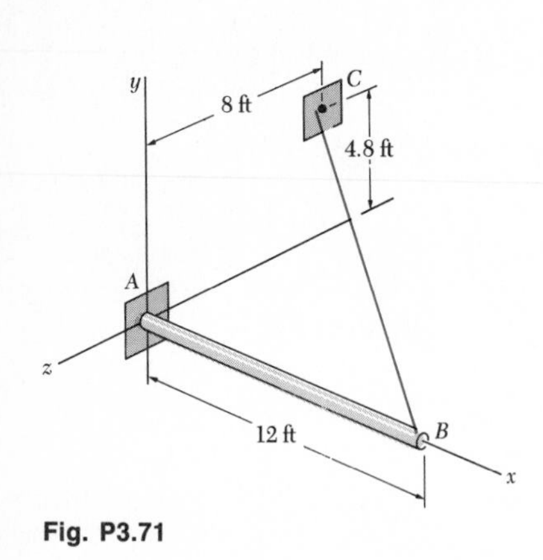

Fig. P3.71

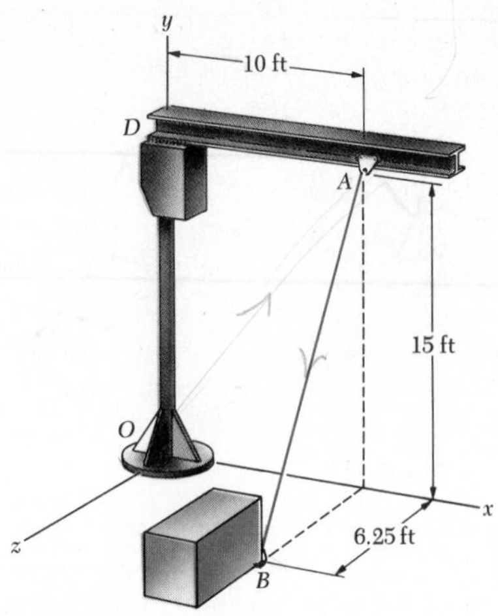

Fig. P3.72

3.72 The jib crane shown is oriented so that its boom AD is parallel to the x axis and is used to move a heavy crate. Knowing that the tension in cable AB is 2.6 kips, replace the force exerted by the cable at A by an equivalent force-couple system at the center O of the base of the crane.

3.73 Replace the 150-N force by an equivalent force-couple system at A.

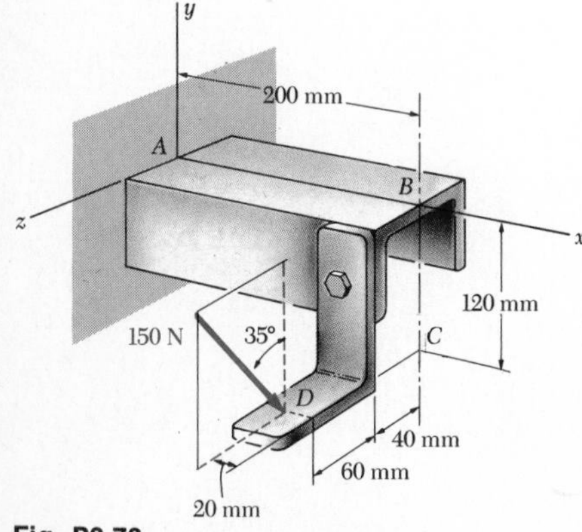

Fig. P3.73

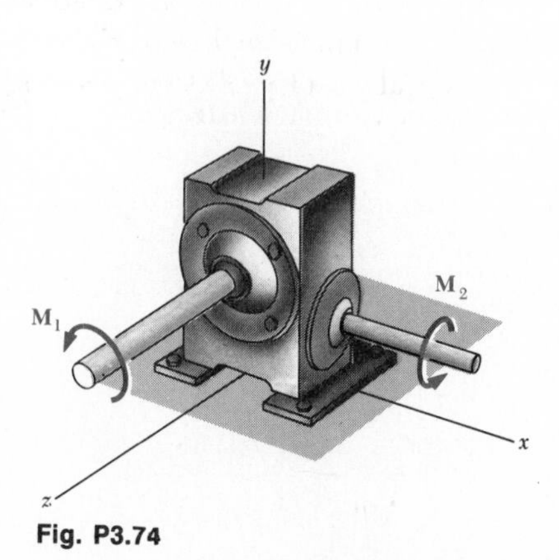

Fig. P3.74

3.74 The speed-reducer unit shown weighs 75 lb and its center of gravity is located on the y axis. Show that the weight of the unit and the two couples acting on it, of magnitude $M_1 = 20$ lb·ft and $M_2 = 4$ lb·ft, respectively, may be replaced by a single equivalent force and determine (*a*) the magnitude and direction of that force, (*b*) the point where its line of action intersects the floor.

3.75 Five separate force-couple systems act at the corners of a bent plate as shown. Find two force-couple systems which are equivalent.

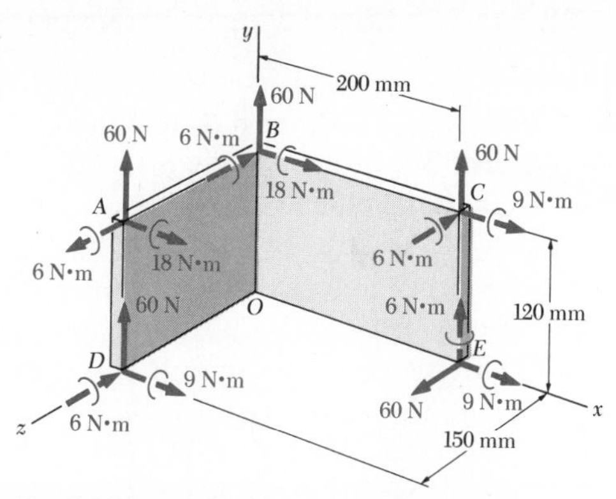

Fig. P3.75 and P3.76

3.76 Five separate force-couple systems act at the corners of a bent plate as shown. Determine which of these systems is equivalent to a force $\mathbf{F} = (60 \text{ N})\mathbf{k}$ and a couple $\mathbf{M} = (9 \text{ N}\cdot\text{m})\mathbf{i} - (6 \text{ N}\cdot\text{m})\mathbf{j}$ located at the origin O.

3.17. Reduction of a System of Forces to One Force and One Couple. Consider a system of forces $\mathbf{F}_1$, $\mathbf{F}_2$, $\mathbf{F}_3$, etc., acting on a rigid body at the points A_1, A_2, A_3, etc., defined by the position vectors $\mathbf{r}_1$, $\mathbf{r}_2$, $\mathbf{r}_3$, etc. (Fig. 3.41*a*). As seen in the preceding section, $\mathbf{F}_1$ may be moved from A_1 to a given point O if a couple of moment $\mathbf{M}_1$ equal to the moment $\mathbf{r}_1 \times \mathbf{F}_1$ of $\mathbf{F}_1$ about O is added to the original system of forces. Repeating this procedure with $\mathbf{F}_2$, $\mathbf{F}_3$, etc., we obtain the system shown in Fig. 3.41*b*, consisting of forces acting at O and of couples. Since the forces are now concurrent, they may be added vectorially and replaced by their resultant $\mathbf{R}$. Similarly, the couple vectors $\mathbf{M}_1$, $\mathbf{M}_2$, $\mathbf{M}_3$, etc., may be added vectorially and replaced by a single couple vector $\mathbf{M}_O^R$. Any system of forces, however complex, may thus be reduced to an *equivalent force-couple system acting at a given point O* (Fig. 3.41*c*). We should note that while each of the couple vectors $\mathbf{M}_1$, $\mathbf{M}_2$, $\mathbf{M}_3$, etc., in Fig. 3.41*b* is perpendicular to the corresponding force, the resultant force $\mathbf{R}$ and the resultant couple vector $\mathbf{M}_O^R$ in Fig. 3.41*c* will not, in general, be perpendicular to each other.

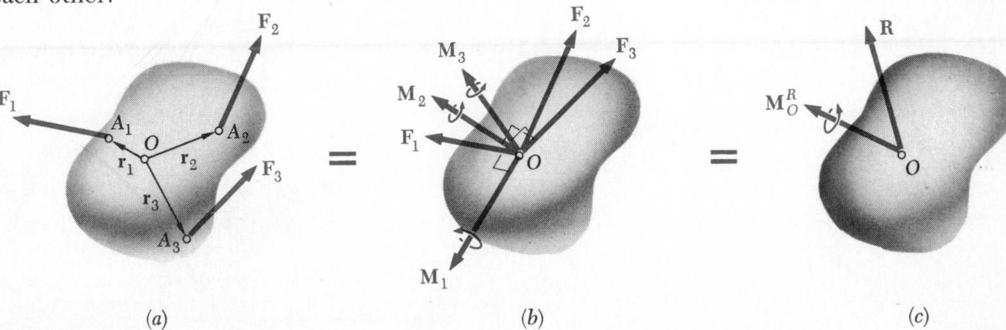

Fig. 3.41

The equivalent force-couple system is defined by the equations

$$\mathbf{R} = \Sigma\mathbf{F} \qquad \mathbf{M}_O^R = \Sigma\mathbf{M}_O = \Sigma(\mathbf{r} \times \mathbf{F}) \tag{3.52}$$

which express that the force **R** is obtained by adding all the forces of the system, while the moment $\mathbf{M}_O^R$ of the couple, called *moment resultant* of the system, is obtained by adding the moments about O of all the forces of the system.

Once a given system of forces has been reduced to a force and a couple at a point O, it may easily be reduced to a force and a couple at another point O'. While the resultant force **R** will remain unchanged, the new moment resultant $\mathbf{M}_{O'}^R$ will be equal to the sum of $\mathbf{M}_O^R$ and of the moment about O' of the force **R** attached at O (Fig. 3.42). We have

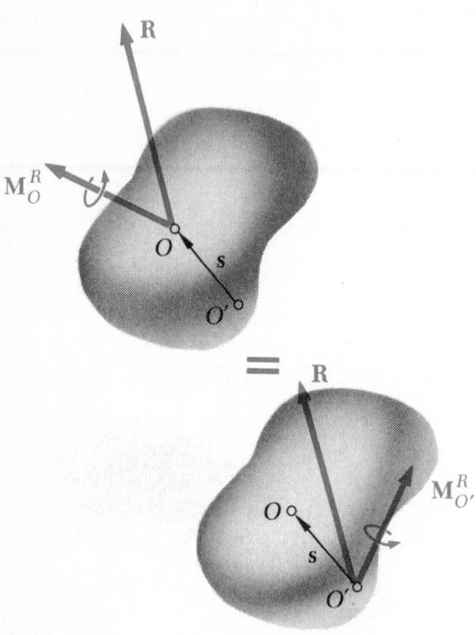

Fig. 3.42

$$\mathbf{M}_{O'}^R = \mathbf{M}_O^R + \mathbf{s} \times \mathbf{R} \tag{3.53}$$

In practice, the reduction of a given system of forces to a single force **R** at O and a couple vector $\mathbf{M}_O^R$ will be carried out in terms of components. Resolving each position vector **r** and each force **F** of the system into rectangular components, we write

$$\mathbf{r} = x\mathbf{i} + y\mathbf{j} + z\mathbf{k} \tag{3.54}$$
$$\mathbf{F} = F_x\mathbf{i} + F_y\mathbf{j} + F_z\mathbf{k} \tag{3.55}$$

Substituting for **r** and **F** into (3.52) and factoring the unit vectors **i**, **j**, **k**, we obtain **R** and $\mathbf{M}_O^R$ in the form

$$\mathbf{R} = R_x\mathbf{i} + R_y\mathbf{j} + R_z\mathbf{k} \qquad \mathbf{M}_O^R = M_x^R\mathbf{i} + M_y^R\mathbf{j} + M_z^R\mathbf{k} \tag{3.56}$$

The components R_x, R_y, R_z represent, respectively, the sums of the x, y, and z components of the given forces and measure the tendency of the system to impart to the rigid body a motion of translation in the x, y, or z direction. Similarly, the components M_x^R, M_y^R, M_z^R represent, respectively, the sums of the moments of the given forces about the x, y, and z axes and measure the tendency of the system to impart to the rigid body a motion of rotation about the x, y, or z axis.

If the magnitude and direction of the force **R** are desired, they may be obtained from the components R_x, R_y, R_z by means of the relations (2.18) and (2.19) of Sec. 2.12; similar computations will yield the magnitude and direction of the couple vector $\mathbf{M}_O^R$.

3.18. Equivalent Systems of Forces. We have seen in the preceding section that any system of forces acting on a rigid body may be reduced to a force-couple system at a given point O. This equivalent force-couple system characterizes completely the effect of the given system on the rigid body. *Two systems of forces are equivalent, therefore, if they may be reduced to the same force-couple system at a given point O.* Recalling that the force-couple system at O is defined by the relations (3.52), we state: *Two systems of forces* $\mathbf{F}_1$, $\mathbf{F}_2$, $\mathbf{F}_3$, *etc., and* $\mathbf{F}'_1$, $\mathbf{F}'_2$, $\mathbf{F}'_3$, *etc., are equivalent if, and only if, the sums of the forces and the sums of the moments about a given point O of the forces of the two systems are, respectively, equal.* Expressed mathematically, the necessary and sufficient conditions for the two systems of forces to be equivalent are

$$\Sigma\mathbf{F} = \Sigma\mathbf{F}' \qquad \text{and} \qquad \Sigma\mathbf{M}_O = \Sigma\mathbf{M}'_O \tag{3.57}$$

Note that to prove that two systems of forces are equivalent, the second of the relations (3.57) must be established with respect to *only one point O*. It will hold, however, with respect to *any point* if the two systems are equivalent.

Resolving the forces and moments in (3.57) into their rectangular components, we may express the necessary and sufficient conditions for the equivalence of two systems of forces acting on a rigid body as follows:

$$\begin{array}{lll} \Sigma F_x = \Sigma F'_x & \Sigma F_y = \Sigma F'_y & \Sigma F_z = \Sigma F'_z \\ \Sigma M_x = \Sigma M'_x & \Sigma M_y = \Sigma M'_y & \Sigma M_z = \Sigma M'_z \end{array} \tag{3.58}$$

These equations have a simple physical significance. They express that two systems of forces are equivalent if they tend to impart to the rigid body (1) the same translation in the x, y, and z directions, respectively, and (2) the same rotation about the x, y, and z axes, respectively.

3.19. Equipollent Systems of Vectors. When two systems of vectors satisfy Eqs. (3.57) or (3.58), i.e., when their resultants and their moment resultants about an arbitrary point O are respectively equal, the two systems are said to be *equipollent*. The result established in the preceding section may thus be restated as follows: *If two systems of forces acting on a rigid body are equipollent, they are also equivalent.*

It is important to note that this statement does not apply to *any* system of vectors. Consider for example a system of forces acting on a set of independent particles which do *not* form a rigid body. A different system of forces acting on the same particles may happen to be equipollent to the first one, i.e., it may have the same resultant and the same moment resultant. Yet, since different forces will now act on the various particles, their effects on these particles will be different; the two systems of forces, while equipollent, are *not equivalent*.

3.20. Further Reduction of a System of Forces. We saw in Sec. 3.17 that any given system of forces acting on a rigid body may be reduced to an equivalent force-couple system at O, consisting of a force $\mathbf{R}$ equal to the sum of the forces of the system, and of a couple vector $\mathbf{M}_O^R$ of moment equal to the moment resultant of the system.

When $\mathbf{R} = 0$, the force-couple system reduces to the couple vector $\mathbf{M}_O^R$. The given system of forces may then be reduced to a single couple, called the *resultant couple* of the system.

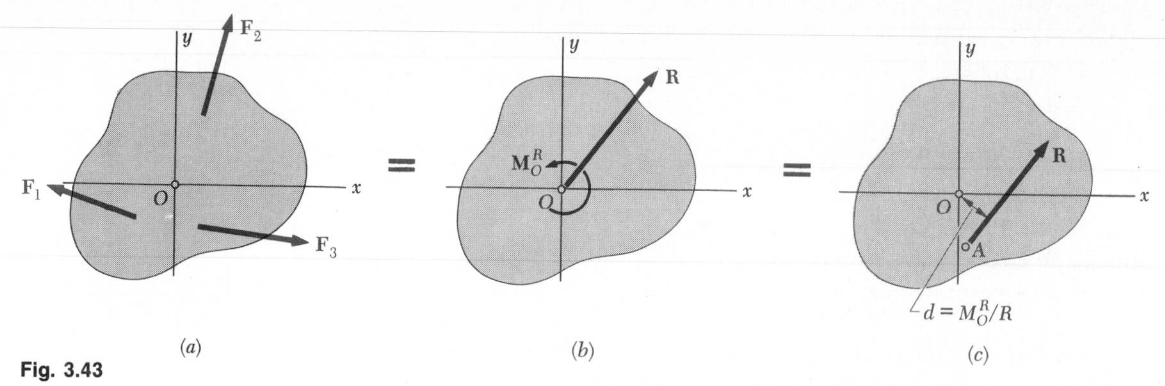

Fig. 3.43

We shall now investigate the conditions under which a given system of forces may be reduced to a single force. It follows from Sec. 3.16 that the force-couple system at O may be replaced by a single force $\mathbf{R}$ acting along a new line of action if $\mathbf{R}$ and $\mathbf{M}_O^R$ are mutually perpendicular. The systems of forces which may be reduced to a single force, or *resultant,* are therefore the systems for which the force $\mathbf{R}$ and the couple vector $\mathbf{M}_O^R$ are mutually perpendicular. While this condition *is generally not satisfied* by systems of forces in space, it *will be satisfied* by systems consisting of (1) concurrent forces, (2) coplanar forces, or (3) parallel forces. We shall discuss these cases separately.

1. *Concurrent forces* are applied at the same point and may therefore be added directly into their resultant $\mathbf{R}$. Thus, they always reduce to a single force. Concurrent forces have been discussed in detail in Chap. 2.
2. *Coplanar forces* act in the same plane, which we shall assume here to be the plane of the figure (Fig. 3.43*a*). The sum $\mathbf{R}$ of the forces of the system will also lie in the plane of the figure, while the moment of each force about O, and thus the moment resultant $\mathbf{M}_O^R$, will be perpendicular to that plane. The force-couple system at O consists therefore of a force $\mathbf{R}$ and a couple vector $\mathbf{M}_O^R$ which are mutually perpendicular (Fig. 3.43*b*).† They may be reduced to a single force $\mathbf{R}$ by moving $\mathbf{R}$ in the plane of the figure until its moment about O becomes equal to $\mathbf{M}_O^R$. The distance from O to the line of action of $\mathbf{R}$ is $d = M_O^R/R$ (Fig. 3.43*c*).

   As noted in Sec. 3.17, the reduction of a system of forces is considerably simplified if the forces are resolved into rectangular components. The force-couple system at O is then characterized by the components (Fig. 3.44*a*)

$$R_x = \Sigma F_x \qquad R_y = \Sigma F_y \qquad M_z^R = M_O^R = \Sigma M_O \tag{3.59}$$

†Since the couple vector $\mathbf{M}_O^R$ is perpendicular to the plane of the figure, it has been represented by the symbol ↺. A counterclockwise couple ↺ corresponds to a vector pointing out of the paper, and a clockwise couple ↻ to a vector pointing into the paper.

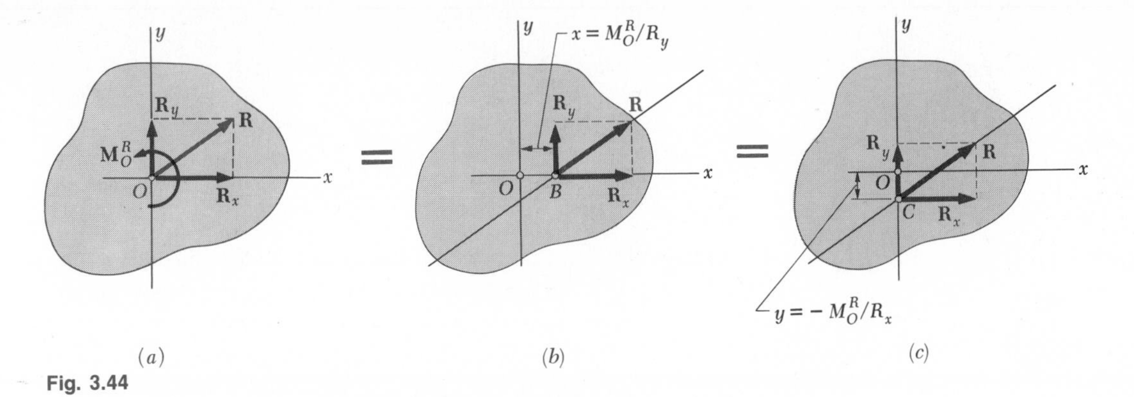

Fig. 3.44

To reduce the system to a single force **R** we shall express that the moment of **R** about O must be equal to $\mathbf{M}_O^R$. Denoting by x and y the coordinates of the point of application A of the resultant, and recalling formula (3.22) of Sec. 3.8, we write

$$xR_y - yR_x = M_O^R$$

which represents the equation of the line of action of **R**. We may also determine directly the x and y intercepts of the line of action of the resultant by noting that $\mathbf{M}_O^R$ must be equal to the moment about O of the y component of **R** when **R** is attached at B (Fig. 3.44*b*), and to the moment of its x component when **R** is attached at C (Fig. 3.44*c*).

3. *Parallel forces* have parallel lines of action and may or may not have the same sense. Assuming here that the forces are parallel to the y axis (Fig. 3.45*a*), we note that their sum **R** will also be parallel to the y axis. On the other hand, since the moment of a given force must be perpendicular to that force, the moment about O of each force of the system, and thus the moment resultant $\mathbf{M}_O^R$, will lie in the zx plane. The force-couple system at O consists therefore of a force **R** and a couple vector $\mathbf{M}_O^R$ which are mutually perpendicular (Fig. 3.45*b*). They may be reduced to a single force **R** (Fig. 3.45*c*) or, if $\mathbf{R} = 0$, to a single couple of moment $\mathbf{M}_O^R$.

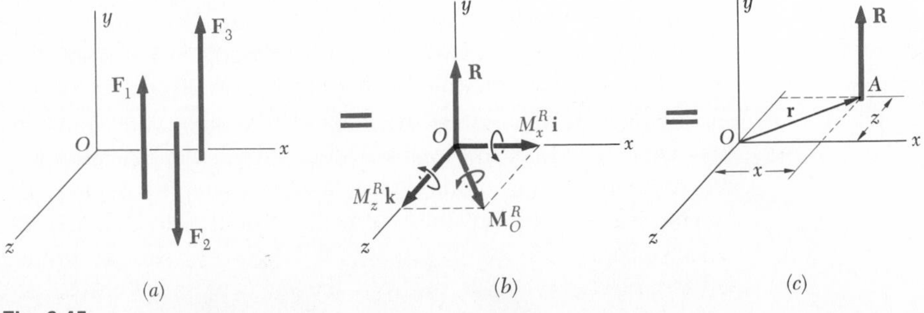

Fig. 3.45

In practice, the force-couple system at O will be characterized by the components

$$R_y = \Sigma F_y \qquad M_x^R = \Sigma M_x \qquad M_z^R = \Sigma M_z \tag{3.60}$$

The reduction of the system to a single force may be carried out by moving $\mathbf{R}$ to a new point of application $A(x, 0, z)$ chosen so that the moment of $\mathbf{R}$ about O is equal to $\mathbf{M}_O^R$. We write

$$\mathbf{r} \times \mathbf{R} = \mathbf{M}_O^R$$
$$(x\mathbf{i} + z\mathbf{k}) \times R_y\mathbf{j} = M_x^R\mathbf{i} + M_z^R\mathbf{k}$$

Computing the vector products and equating the coefficients of the corresponding unit vectors in both members of the equation, we obtain two scalar equations which define the coordinates of A:

$$-zR_y = M_x^R \qquad xR_y = M_z^R$$

These equations express that the moments of $\mathbf{R}$ about the x and z axes must, respectively, be equal to M_x^R and M_z^R.

***3.21. Reduction of a System of Forces to a Wrench.** In the general case of a system of forces in space, the equivalent force-couple system at O consists of a force $\mathbf{R}$ and a couple vector $\mathbf{M}_O^R$ which are not perpendicular, and neither of which is zero (Fig. 3.46*a*). Thus, the system of forces *cannot* be reduced to a single force or a single couple. The couple vector, however, may be replaced by two other couple vectors obtained by resolving $\mathbf{M}_O^R$ into a component $\mathbf{M}_1$ along $\mathbf{R}$ and a component $\mathbf{M}_2$ in a plane perpendicular to $\mathbf{R}$ (Fig. 3.46*b*). The couple vector $\mathbf{M}_2$ and the force $\mathbf{R}$ may then be replaced by a single force $\mathbf{R}$ acting along a new line of action.

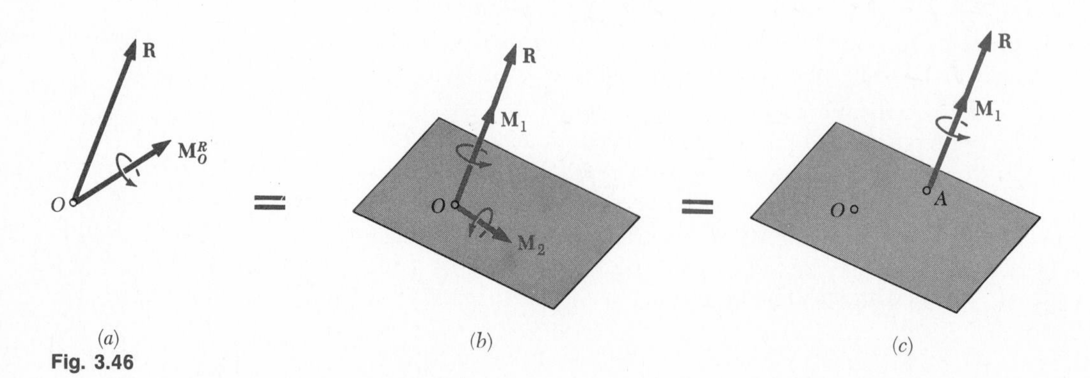

Fig. 3.46

The original system of forces thus reduces to $\mathbf{R}$ and to the couple vector $\mathbf{M}_1$ (Fig. 3.46*c*), i.e., to $\mathbf{R}$ and a couple acting in the plane perpendicular to $\mathbf{R}$. This particular force-couple system is called a *wrench* because the resulting combination of push and twist is the same that would be caused by an actual wrench. The line of action of $\mathbf{R}$ is known as the *axis of the wrench*, and the ratio $p = M_1/R$ is called the *pitch* of the wrench. A wrench, therefore, consists of two collinear vectors, namely, a force **R** and a couple vector

$$\mathbf{M}_1 = p\mathbf{R} \tag{3.61}$$

Recalling the expression (3.35) obtained in Sec. 3.9 for the projection of a vector on the line of action of another vector, we note that the projection of $\mathbf{M}_O^R$ on the line of action of **R** is

$$M_1 = \frac{\mathbf{R} \cdot \mathbf{M}_O^R}{R}$$

Thus, the pitch of the wrench may be expressed as†

$$p = \frac{M_1}{R} = \frac{\mathbf{R} \cdot \mathbf{M}_O^R}{R^2} \tag{3.62}$$

To define the axis of the wrench we may write a relation involving the position vector **r** of an arbitrary point P located on that axis. Attaching the resultant force **R** and couple vector $\mathbf{M}_1$ at P (Fig. 3.47), and expressing that the moment about O of this force-couple system is equal to the moment resultant $\mathbf{M}_O^R$ of the original force system, we write

$$\mathbf{M}_1 + \mathbf{r} \times \mathbf{R} = \mathbf{M}_O^R \tag{3.63}$$

or, recalling Eq. (3.61),

$$p\mathbf{R} + \mathbf{r} \times \mathbf{R} = \mathbf{M}_O^R \tag{3.64}$$

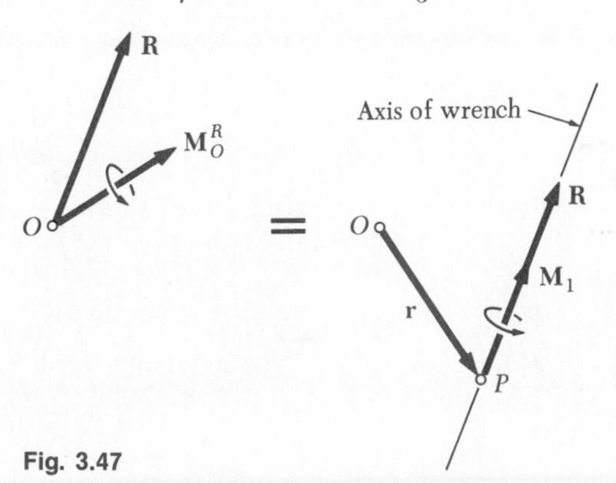

Fig. 3.47

† The expressions obtained for the projection of the couple vector on the line of action of **R** and for the pitch of the wrench are independent of the choice of point O. Using the relation (3.53) of Sec. 3.17, we check that if a different point O' had been used, the numerator in (3.62) would be

$$\mathbf{R} \cdot \mathbf{M}_{O'}^R = \mathbf{R} \cdot (\mathbf{M}_O^R + \mathbf{s} \times \mathbf{R}) = \mathbf{R} \cdot \mathbf{M}_O^R + \mathbf{R} \cdot (\mathbf{s} \times \mathbf{R})$$

Since the mixed triple product $\mathbf{R} \cdot (\mathbf{s} \times \mathbf{R})$ is identically equal to zero, we have

$$\mathbf{R} \cdot \mathbf{M}_{O'}^R = \mathbf{R} \cdot \mathbf{M}_O^R$$

Thus, the scalar product $\mathbf{R} \cdot \mathbf{M}_O^R$ is independent of the choice of point O.

SAMPLE PROBLEM 3.8

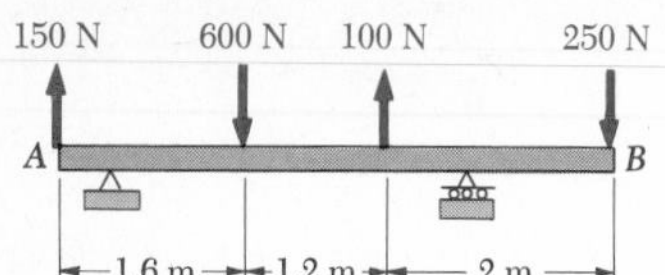

A 4.80-m beam is subjected to the forces shown. Reduce the given system of forces to (*a*) an equivalent force-couple system at *A*, (*b*) an equivalent force-couple system at *B*, (*c*) a single force or resultant.

Note. Since the reactions at the supports are not included in the given system of forces, the given system will not maintain the beam in equilibrium.

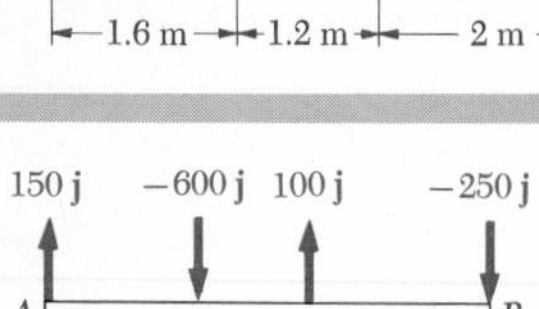

a. **Force-Couple System at *A*.** The force-couple system at *A* equivalent to the given system of forces consists of a force **R** and a couple $\mathbf{M}_A^R$ defined as follows:

$$\mathbf{R} = \Sigma \mathbf{F}$$
$$= (150\text{ N})\mathbf{j} - (600\text{ N})\mathbf{j} + (100\text{ N})\mathbf{j} - (250\text{ N})\mathbf{j} = -(600\text{ N})\mathbf{j}$$
$$\mathbf{M}_A^R = \Sigma(\mathbf{r} \times \mathbf{F})$$
$$= (1.6\mathbf{i}) \times (-600\mathbf{j}) + (2.8\mathbf{i}) \times (100\mathbf{j}) + (4.8\mathbf{i}) \times (-250\mathbf{j})$$
$$= -(1880\text{ N}\cdot\text{m})\mathbf{k}$$

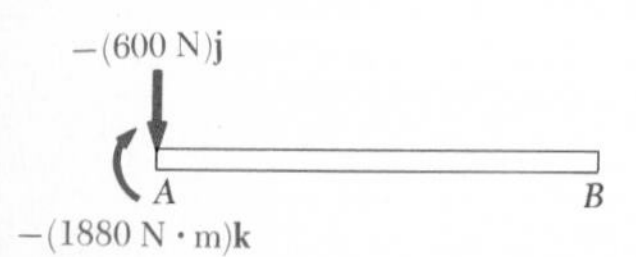

The equivalent force-couple system at *A* is thus

$$\mathbf{R} = 600\text{ N}\downarrow \qquad \mathbf{M}_A^R = 1880\text{ N}\cdot\text{m} \circlearrowright \quad ◀$$

b. **Force-Couple System at *B*.** We shall find a force-couple system at *B* equivalent to the force-couple system at *A* determined in part *a*. The force **R** is unchanged, but a new couple $\mathbf{M}_B^R$ must be determined, the moment of which is equal to the moment about *B* of the force-couple system determined in part *a*. Thus, we have

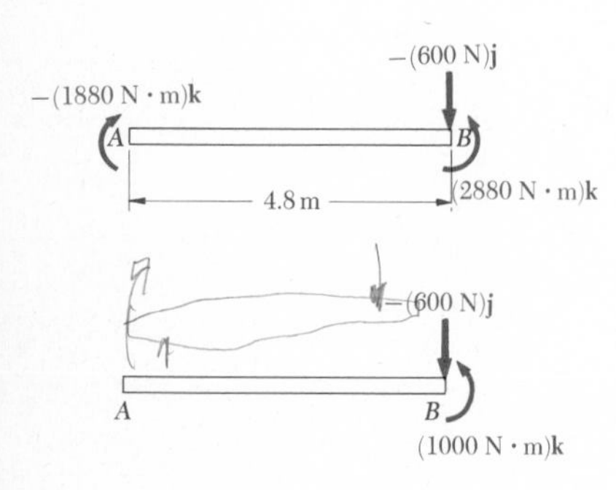

$$\mathbf{M}_B^R = \mathbf{M}_A^R + \overrightarrow{BA} \times \mathbf{R}$$
$$= -(1880\text{ N}\cdot\text{m})\mathbf{k} + (-4.8\text{ m})\mathbf{i} \times (-600\text{ N})\mathbf{j}$$
$$= -(1880\text{ N}\cdot\text{m})\mathbf{k} + (2880\text{ N}\cdot\text{m})\mathbf{k} = +(1000\text{ N}\cdot\text{m})\mathbf{k}$$

The equivalent force-couple system at *B* is thus

$$\mathbf{R} = 600\text{ N}\downarrow \qquad \mathbf{M}_B^R = 1000\text{ N}\cdot\text{m} \circlearrowleft \quad ◀$$

c. **Single Force or Resultant.** The resultant of the given system of forces is equal to **R** and its point of application must be such that the moment of **R** about *A* is equal to $\mathbf{M}_A^R$. We write

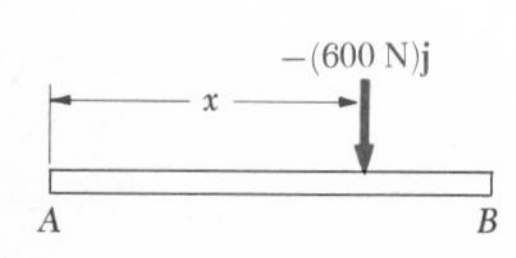

$$\mathbf{r} \times \mathbf{R} = \mathbf{M}_A^R$$
$$x\mathbf{i} \times (-600\text{ N})\mathbf{j} = -(1880\text{ N}\cdot\text{m})\mathbf{k}$$
$$-x(600\text{ N})\mathbf{k} = -(1880\text{ N}\cdot\text{m})\mathbf{k}$$

and conclude that $x = 3.13$ m. Thus, the single force equivalent to the given system is defined as

$$\mathbf{R} = 600\text{ N}\downarrow \qquad x = 3.13\text{ m} \quad ◀$$

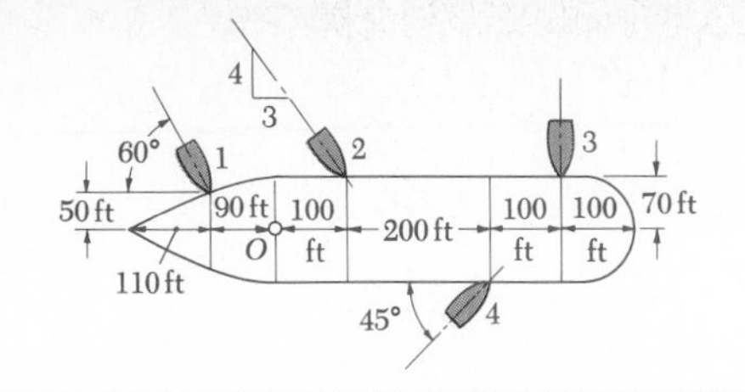

SAMPLE PROBLEM 3.9

Four tugboats are used to bring an ocean liner to its pier. Each tugboat exerts a 5000-lb force in the direction shown. Determine (*a*) the equivalent force-couple system at the foremast *O*, (*b*) the point on the hull where a single, more powerful tugboat should push to produce the same effect as the original four tugboats.

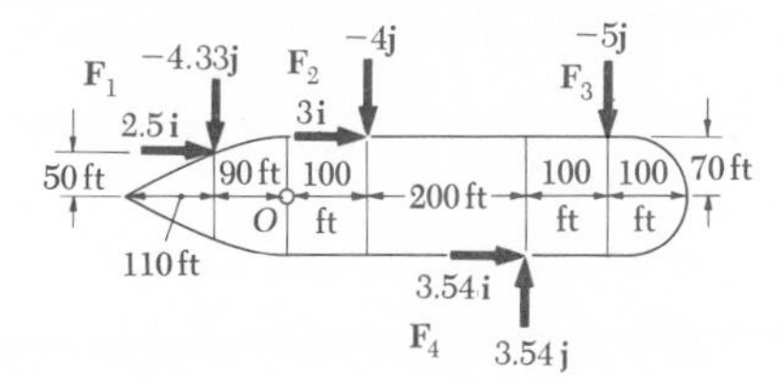

a. **Force-Couple System at *O*.** Each of the given forces is resolved into components in the diagram shown (kip units are used). The force-couple system at *O* equivalent to the given system of forces consists of a force **R** and a couple $\mathbf{M}_O^R$ defined as follows:

$$
\begin{aligned}
\mathbf{R} &= \Sigma\mathbf{F} \\
&= (2.50\mathbf{i} - 4.33\mathbf{j}) + (3.00\mathbf{i} - 4.00\mathbf{j}) + (-5.00\mathbf{j}) + (3.54\mathbf{i} + 3.54\mathbf{j}) \\
&= 9.04\mathbf{i} - 9.79\mathbf{j} \\
\mathbf{M}_O^R &= \Sigma(\mathbf{r} \times \mathbf{F}) \\
&= (-90\mathbf{i} + 50\mathbf{j}) \times (2.50\mathbf{i} - 4.33\mathbf{j}) \\
&\quad + (100\mathbf{i} + 70\mathbf{j}) \times (3.00\mathbf{i} - 4.00\mathbf{j}) \\
&\quad + (400\mathbf{i} + 70\mathbf{j}) \times (-5.00\mathbf{j}) \\
&\quad + (300\mathbf{i} - 70\mathbf{j}) \times (3.54\mathbf{i} + 3.54\mathbf{j}) \\
&= (390 - 125 - 400 - 210 - 2000 + 1062 + 248)\mathbf{k} \\
&= -1035\mathbf{k}
\end{aligned}
$$

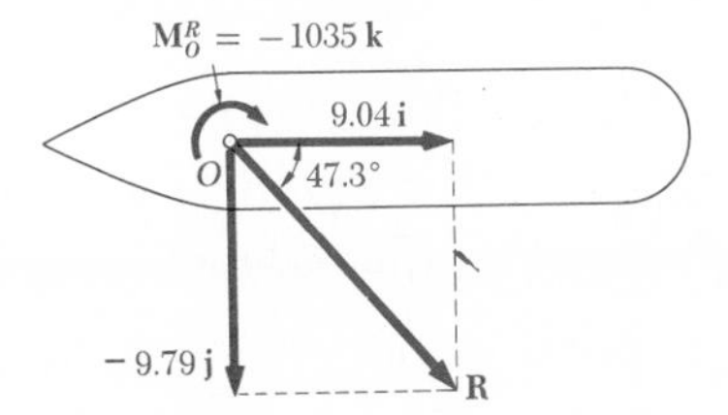

The equivalent force-couple system at *O* is thus

$$\mathbf{R} = (9.04 \text{ kips})\mathbf{i} - (9.79 \text{ kips})\mathbf{j} \qquad \mathbf{M}_O^R = -(1035 \text{ kip}\cdot\text{ft})\mathbf{k}$$

or

$$R = 13.33 \text{ kips} \measuredangle 47.3° \qquad M_O^R = 1035 \text{ kip}\cdot\text{ft} \circlearrowright \quad \blacktriangleleft$$

Remark. Since all the forces are contained in the plane of the figure, we could have expected the sum of their moments to be perpendicular to that plane. Note that the moment of each force component could have been obtained directly from the diagram by forming the product of its magnitude and its perpendicular distance to *O*, and assigning to this product a positive or a negative sign, depending upon the sense of the moment.

b. **Single Tugboat.** The force exerted by a single tugboat must be equal to **R** and its point of application *A* must be such that the moment of **R** about *O* is equal to $\mathbf{M}_O^R$. Observing that the position vector of *A* is

$$\mathbf{r} = x\mathbf{i} + 70\mathbf{j}$$

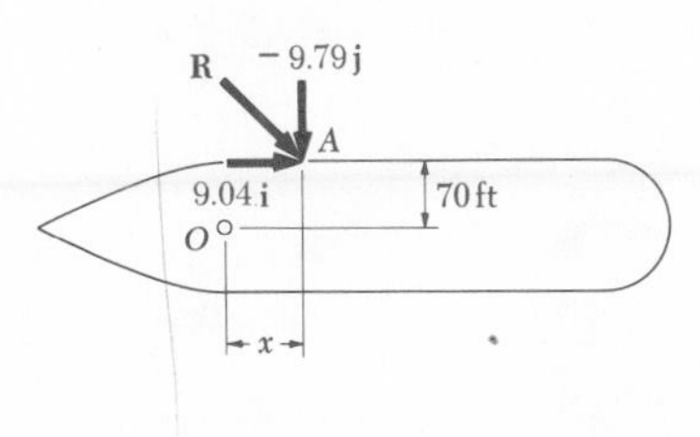

we write

$$
\begin{aligned}
\mathbf{r} \times \mathbf{R} &= \mathbf{M}_O^R \\
(x\mathbf{i} + 70\mathbf{j}) \times (9.04\mathbf{i} - 9.79\mathbf{j}) &= -1035\mathbf{k} \\
-x(9.79)\mathbf{k} - 633\mathbf{k} &= -1035\mathbf{k}
\end{aligned}
$$

$$x = 41.1 \text{ ft} \quad \blacktriangleleft$$

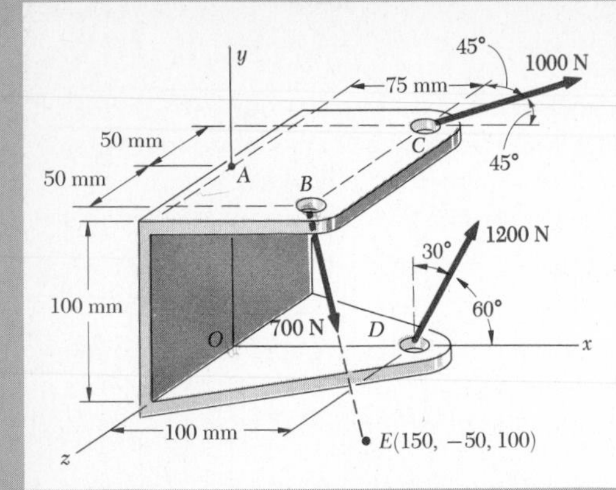

SAMPLE PROBLEM 3.10

Three cables are attached to a bracket as shown. Replace the forces exerted by the cables by an equivalent force-couple system at A.

Solution. We first determine the relative position vectors drawn from point A to the points of application of the various forces and resolve the forces into rectangular components. Observing that $\mathbf{F}_B = (700\ \text{N})\boldsymbol{\lambda}_{BE}$ where

$$\boldsymbol{\lambda}_{BE} = \frac{\overrightarrow{BE}}{BE} = \frac{75\mathbf{i} - 150\mathbf{j} + 50\mathbf{k}}{175}$$

we have, using meters and newtons,

$$\begin{aligned} \mathbf{r}_{B/A} &= \overrightarrow{AB} = 0.075\mathbf{i} + 0.050\mathbf{k} & \mathbf{F}_B &= 300\mathbf{i} - 600\mathbf{j} + 200\mathbf{k} \\ \mathbf{r}_{C/A} &= \overrightarrow{AC} = 0.075\mathbf{i} - 0.050\mathbf{k} & \mathbf{F}_C &= 707\mathbf{i} - 707\mathbf{k} \\ \mathbf{r}_{D/A} &= \overrightarrow{AD} = 0.100\mathbf{i} - 0.100\mathbf{j} & \mathbf{F}_D &= 600\mathbf{i} + 1039\mathbf{j} \end{aligned}$$

The force-couple system at A equivalent to the given forces consists of a force $\mathbf{R} = \Sigma\mathbf{F}$ and a couple $\mathbf{M}_A^R = \Sigma(\mathbf{r} \times \mathbf{F})$. The force $\mathbf{R}$ is readily obtained by adding respectively the x, y, and z components of the forces:

$$\mathbf{R} = \Sigma\mathbf{F} = (1607\ \text{N})\mathbf{i} + (439\ \text{N})\mathbf{j} - (507\ \text{N})\mathbf{k} \quad ◄$$

The computation of $\mathbf{M}_A^R$ will be facilitated if we express the moments of the forces in the form of determinants (Sec. 3.8):

$$\mathbf{r}_{B/A} \times \mathbf{F}_B = \begin{vmatrix} \mathbf{i} & \mathbf{j} & \mathbf{k} \\ 0.075 & 0 & 0.050 \\ 300 & -600 & 200 \end{vmatrix} = 30\mathbf{i} \qquad -45\mathbf{k}$$

$$\mathbf{r}_{C/A} \times \mathbf{F}_C = \begin{vmatrix} \mathbf{i} & \mathbf{j} & \mathbf{k} \\ 0.075 & 0 & -0.050 \\ 707 & 0 & -707 \end{vmatrix} = \qquad 17.68\mathbf{j}$$

$$\mathbf{r}_{D/A} \times \mathbf{F}_D = \begin{vmatrix} \mathbf{i} & \mathbf{j} & \mathbf{k} \\ 0.100 & -0.100 & 0 \\ 600 & 1039 & 0 \end{vmatrix} = \qquad 163.9\mathbf{k}$$

Adding the expressions obtained, we have

$$\mathbf{M}_A^R = \Sigma(\mathbf{r} \times \mathbf{F}) = (30\ \text{N}\cdot\text{m})\mathbf{i} + (17.68\ \text{N}\cdot\text{m})\mathbf{j} + (118.9\ \text{N}\cdot\text{m})\mathbf{k} \quad ◄$$

The rectangular components of the force $\mathbf{R}$ and the couple $\mathbf{M}_A^R$ are shown in the adjoining sketch.

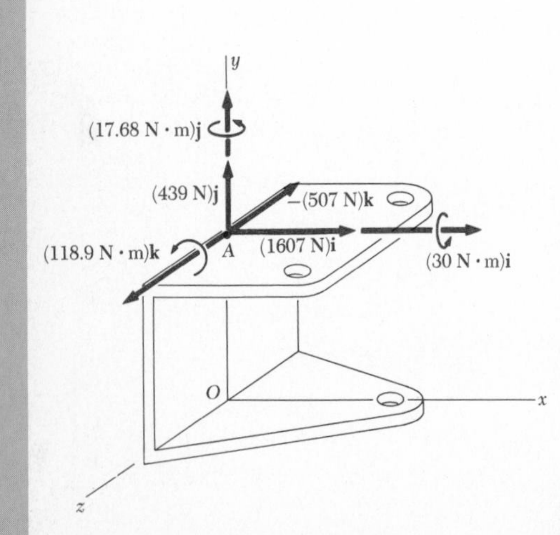

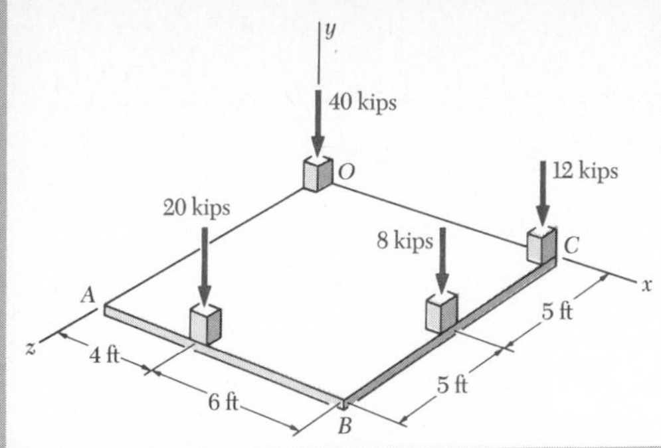

SAMPLE PROBLEM 3.11

A square foundation mat supports the four columns shown. Determine the magnitude and point of application of the resultant of the four loads.

Solution. We shall first reduce the given system of forces to a force-couple system at the origin O of the coordinates. This force-couple system consists of a force $\mathbf{R}$ and a couple $\mathbf{M}_O^R$ defined as follows:

$$\mathbf{R} = \Sigma\mathbf{F} \qquad \mathbf{M}_O^R = \Sigma(\mathbf{r} \times \mathbf{F})$$

The position vectors of the points of application of the various forces are determined and the computations are arranged in tabular form.

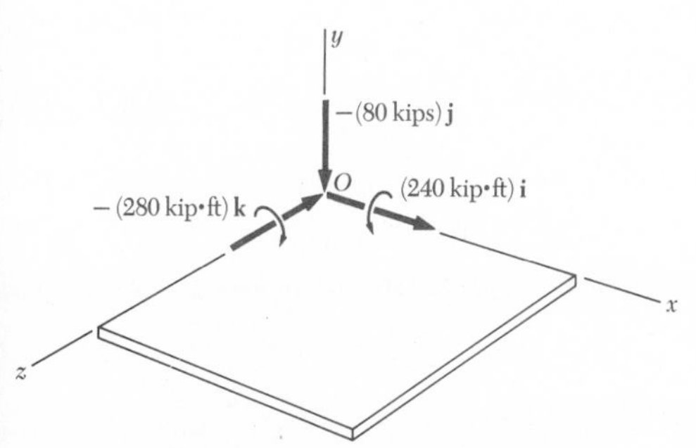

r, ft	F, kips	r × F, kip · ft
0	$-40\mathbf{j}$	0
$10\mathbf{i}$	$-12\mathbf{j}$	$-120\mathbf{k}$
$10\mathbf{i} + 5\mathbf{k}$	$-8\mathbf{j}$	$40\mathbf{i} - 80\mathbf{k}$
$4\mathbf{i} + 10\mathbf{k}$	$-20\mathbf{j}$	$200\mathbf{i} - 80\mathbf{k}$
	$\mathbf{R} = -80\mathbf{j}$	$\mathbf{M}_O^R = 240\mathbf{i} - 280\mathbf{k}$

Since the force $\mathbf{R}$ and the couple vector $\mathbf{M}_O^R$ are mutually perpendicular, the force-couple system obtained may be reduced further to a single force $\mathbf{R}$. The new point of application of $\mathbf{R}$ will be selected in the plane of the mat and in such a way that the moment of $\mathbf{R}$ about O will be equal to $\mathbf{M}_O^R$. Denoting by $\mathbf{r}$ the position vector of the desired point of application, and by x and z its coordinates, we write

$$\mathbf{r} \times \mathbf{R} = \mathbf{M}_O^R$$
$$(x\mathbf{i} + z\mathbf{k}) \times (-80\mathbf{j}) = 240\mathbf{i} - 280\mathbf{k}$$
$$-80x\mathbf{k} + 80z\mathbf{i} = 240\mathbf{i} - 280\mathbf{k}$$

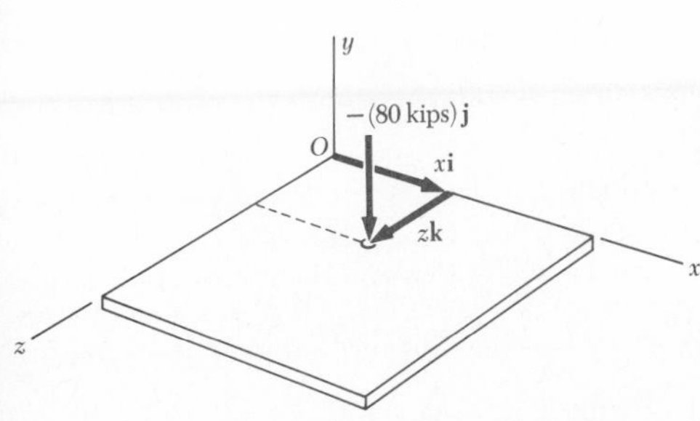

from which it follows that

$$-80x = -280 \qquad 80z = 240$$
$$x = 3.50 \text{ ft} \qquad z = 3.00 \text{ ft}$$

We conclude that the resultant of the given system of forces is

$$\mathbf{R} = 80 \text{ kips} \downarrow \qquad \text{at } x = 3.50 \text{ ft}, z = 3.00 \text{ ft} \blacktriangleleft$$

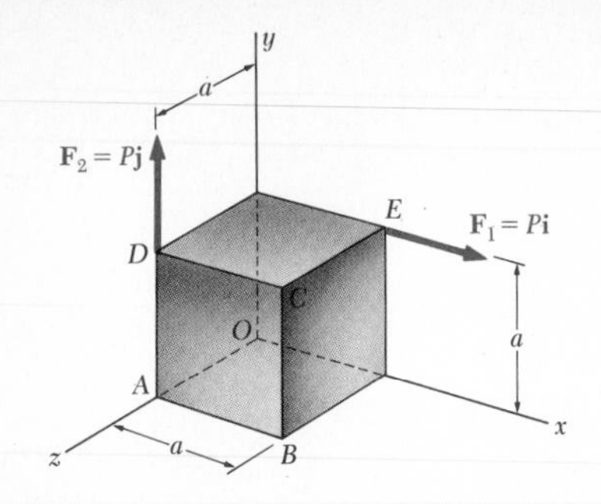

SAMPLE PROBLEM 3.12

Two forces of the same magnitude P act on a cube of side a as shown. Replace the two forces by an equivalent wrench and determine (*a*) the magnitude and direction of the resultant force $\mathbf{R}$, (*b*) the pitch of the wrench, (*c*) the point where the axis of the wrench intersects the yz plane.

Solution. We first determine the equivalent force-couple system at the origin O. We observe that the position vectors of the points of application E and D of the two given forces are $\mathbf{r}_E = a\mathbf{i} + a\mathbf{j}$ and $\mathbf{r}_D = a\mathbf{j} + a\mathbf{k}$. The resultant $\mathbf{R}$ of the two forces and their moment resultant $\mathbf{M}_O^R$ about O are

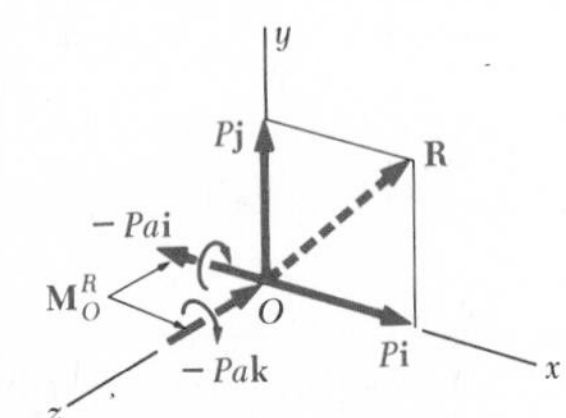

$$\mathbf{R} = \mathbf{F}_1 + \mathbf{F}_2 = P\mathbf{i} + P\mathbf{j} = P(\mathbf{i} + \mathbf{j}) \qquad (1)$$

$$\mathbf{M}_O^R = \mathbf{r}_E \times \mathbf{F}_1 + \mathbf{r}_D \times \mathbf{F}_2 = (a\mathbf{i} + a\mathbf{j}) \times P\mathbf{i} + (a\mathbf{j} + a\mathbf{k}) \times P\mathbf{j}$$

$$= -Pa\mathbf{k} - Pa\mathbf{i} = -Pa(\mathbf{i} + \mathbf{k}) \qquad (2)$$

(*a*) ***Resultant Force* R.** It follows from Eq. (1) and the adjoining sketch that the resultant force $\mathbf{R}$ has the magnitude $R = P\sqrt{2}$, lies in the xy plane, and forms angles of 45° with the x and y axes. Thus

$$R = P\sqrt{2} \qquad \theta_x = \theta_y = 45° \qquad \theta_z = 90° \quad ◀$$

=

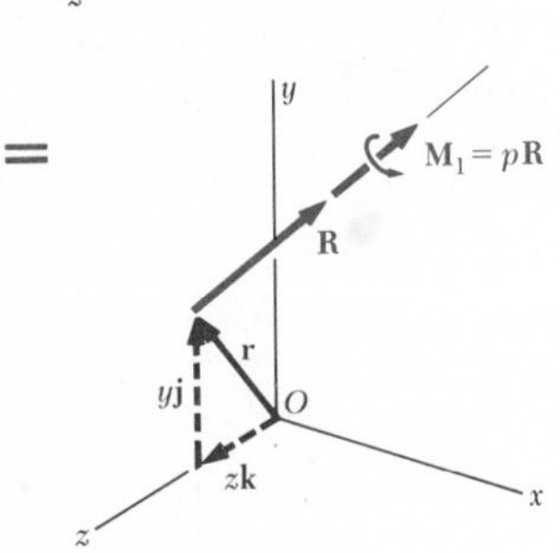

(*b*) ***Pitch of Wrench.*** Recalling formula (3.62) of Sec. 3.21 and Eqs. (1) and (2) above, we write

$$p = \frac{\mathbf{R} \cdot \mathbf{M}_O^R}{R^2} = \frac{P(\mathbf{i} + \mathbf{j}) \cdot (-Pa)(\mathbf{i} + \mathbf{k})}{(P\sqrt{2})^2} = \frac{-P^2a(1 + 0 + 0)}{2P^2} \qquad p = -\frac{a}{2} \quad ◀$$

(*c*) ***Axis of Wrench.*** It follows from the above and from Eq. (3.61) of Sec. 3.21 that the wrench consists of the force $\mathbf{R}$ found in (1) and of the couple vector

$$\mathbf{M}_1 = p\mathbf{R} = -\frac{a}{2}P(\mathbf{i} + \mathbf{j}) = -\frac{Pa}{2}(\mathbf{i} + \mathbf{j}) \qquad (3)$$

To find the point where the axis of the wrench intersects the yz plane, we shall express that the moment of the wrench about O is equal to the moment resultant $\mathbf{M}_O^R$ of the original system:

$$\mathbf{M}_1 + \mathbf{r} \times \mathbf{R} = \mathbf{M}_O^R$$

or, noting that $\mathbf{r} = y\mathbf{j} + z\mathbf{k}$ and substituting for $\mathbf{R}$, $\mathbf{M}_O^R$, and $\mathbf{M}_1$ from Eqs. (1), (2), and (3),

$$-\frac{Pa}{2}(\mathbf{i} + \mathbf{j}) + (y\mathbf{j} + z\mathbf{k}) \times P(\mathbf{i} + \mathbf{j}) = -Pa(\mathbf{i} + \mathbf{k})$$

$$-\frac{Pa}{2}\mathbf{i} - \frac{Pa}{2}\mathbf{j} - Py\mathbf{k} + Pz\mathbf{j} - Pz\mathbf{i} = -Pa\mathbf{i} - Pa\mathbf{k}$$

Equating the coefficients of $\mathbf{k}$, and then the coefficients of $\mathbf{j}$, we find

$$y = a \qquad z = a/2 \quad ◀$$

Problems

3.77 A 3-m beam is loaded in the various ways represented in the figure. Find two loadings which are equivalent.

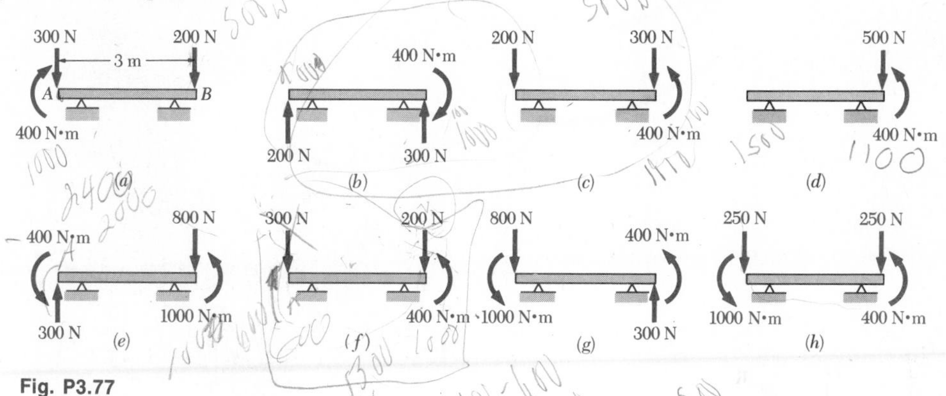

Fig. P3.77

3.78 A 3-m beam is loaded as shown. Determine the loading of Prob. 3.77 which is equivalent to this loading.

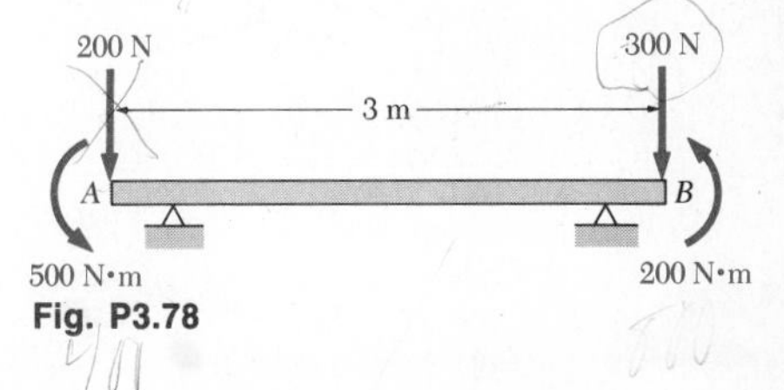

Fig. P3.78

3.79 Determine the resultant of the loads and the distance from point A to its line of action for the beam and loading of (*a*) Prob. 3.77*a*, (*b*) Prob. 3.77*b*, (*c*) Prob. 3.78.

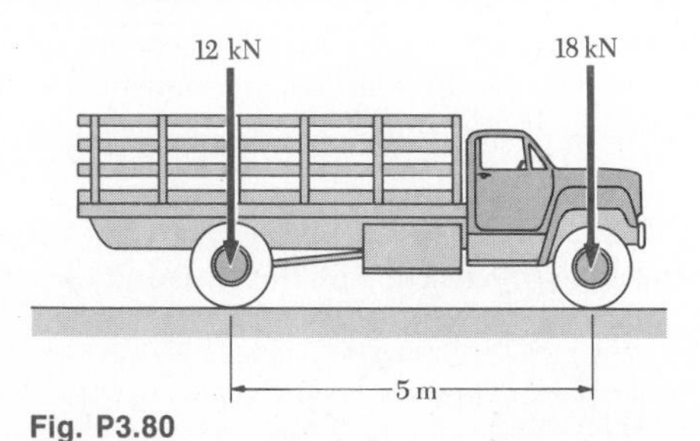

Fig. P3.80

3.80 By driving the truck shown over a scale, it was determined that the loads on the front and rear axles are, respectively, 18 kN and 12 kN when the truck is empty. Determine (*a*) the location of the center of gravity of the truck, (*b*) the weight and location of the center of gravity of the heaviest load which may be carried by the truck if the load on each axle is not to exceed 40 kN.

3.81 Determine the distance from point A to the line of action of the resultant of the three loads shown when (*a*) $x = 1.25$ ft, (*b*) $x = 4$ ft, (*c*) $x = 8$ ft.

3.82 Express as a function of x the distance d from point A to the line of action of the resultant of the three loads shown.

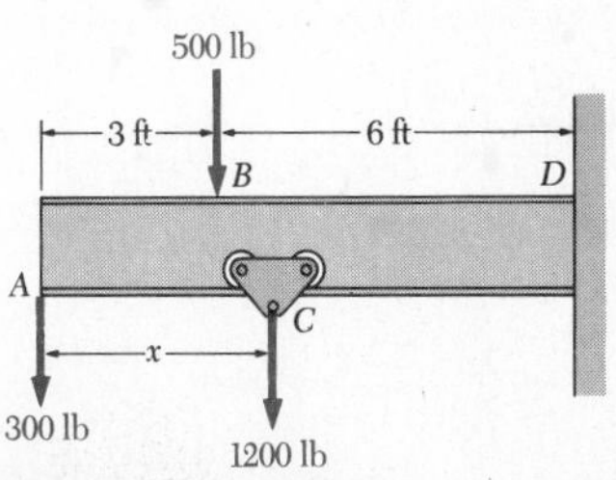

Fig. P3.81 and P3.82

3.83 Four forces act on a 700 × 375 mm plate as shown. (*a*) Find the resultant of these forces. (*b*) Locate the two points where the line of action of the resultant intersects the edge of the plate.

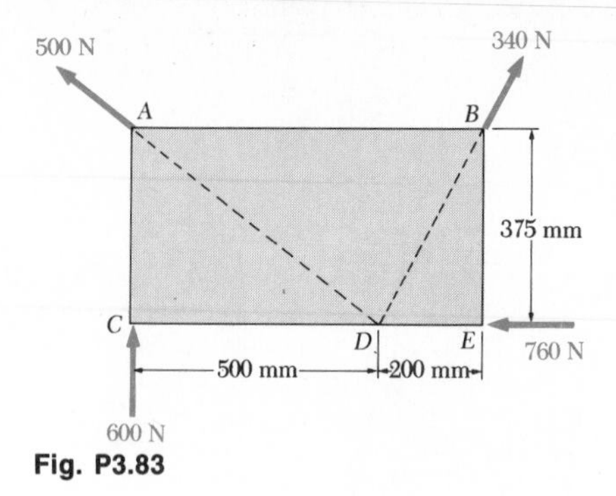

Fig. P3.83

3.84 Solve Prob. 3.83, assuming that the 760-N force is directed to the right.

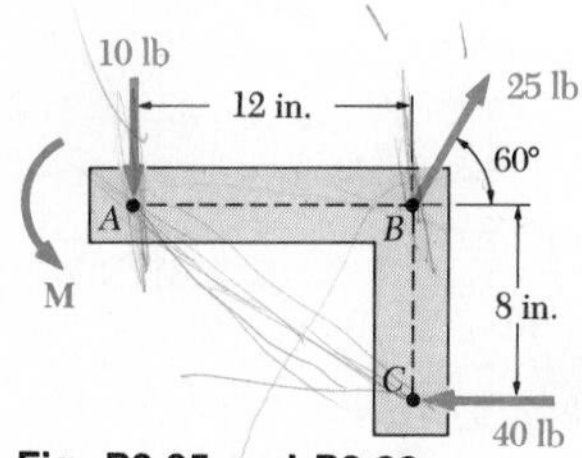

Fig. P3.85 and P3.86

3.85 The three forces shown and a couple of magnitude $M = 80$ lb · in. are applied to an angle bracket. (*a*) Find the resultant of this system of forces. (*b*) Locate the points where the line of action of the resultant intersects line *AB* and line *BC*.

3.86 The three forces shown and a couple **M** are applied to an angle bracket. Find the moment of the couple if the line of action of the resultant of the force system is to pass through (*a*) point *A*, (*b*) point *B*, (*c*) point *C*.

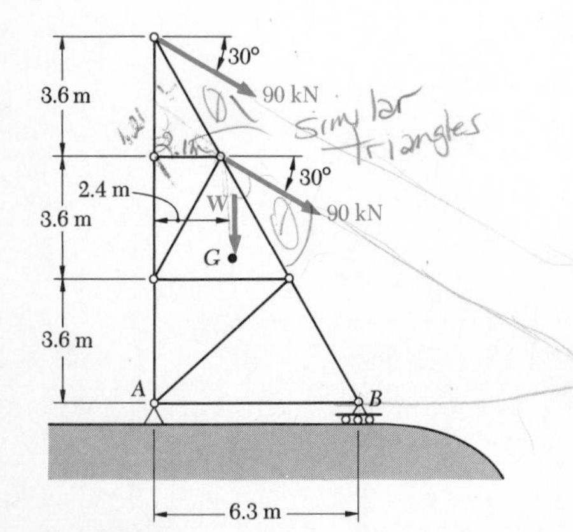

Fig. P3.87

3.87 Two cables exert forces of 90 kN each on a truss of weight $W = 200$ kN. Find the resultant force acting on the truss and the point of intersection of its line of action with line *AB*.

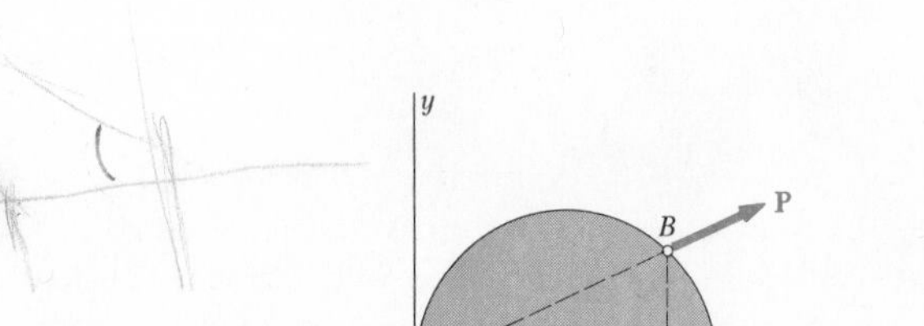

Fig. P3.88

3.88 A force **P** of given magnitude *P* is applied to the edge of a semicircular plate of radius *a* as shown. (*a*) Replace **P** by an equivalent force-couple system at the point *D* obtained by drawing the perpendicular from *B* to the *x* axis. (*b*) Determine the value of *θ* for which the moment of the equivalent force-couple system at *D* is maximum.

3.89 A blade held in a brace is used to tighten a screw at A. (*a*) Determine the forces exerted at B and C, knowing that these forces are equivalent to a force-couple system at A consisting of $\mathbf{R} = -(25\text{ N})\mathbf{i} + R_y\mathbf{j} + R_z\mathbf{k}$ and $\mathbf{M}_A^R = -(13.5\text{ N}\cdot\text{m})\mathbf{i}$. (*b*) Find the corresponding values of R_y and R_z. (*c*) What is the orientation of the slot in the head of the screw for which the blade is least likely to slip when the brace is in the position shown?

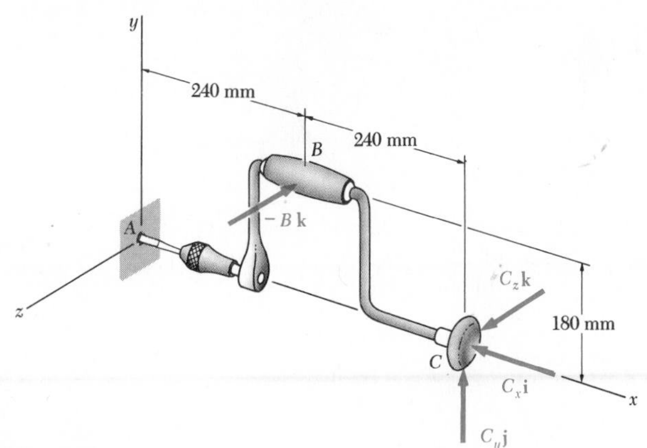

Fig. P3.89

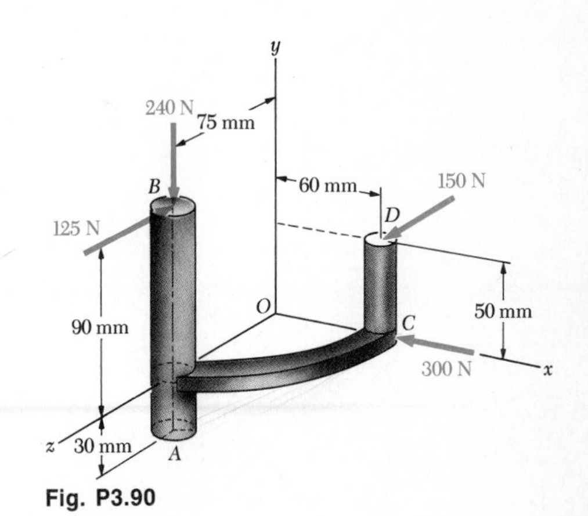

Fig. P3.90

3.90 A machine component is subjected to the forces shown, each of which is parallel to one of the coordinate axes. Replace these forces by an equivalent force-couple system at A.

3.91 In order to unscrew the tapped faucet A, a plumber uses two pipe wrenches as shown. By exerting a 40-lb force on each wrench, at a distance of 10 in. from the axis of the pipe and in a direction perpendicular to the pipe and to the wrench, he prevents the pipe from rotating, and thus avoids loosening or further tightening the joint between the pipe and the tapped elbow C. Determine (*a*) the angle θ that the wrench at A should form with the vertical if elbow C is not to rotate about the vertical, (*b*) the force-couple system at C equivalent to the two 40-lb forces when this condition is satisfied.

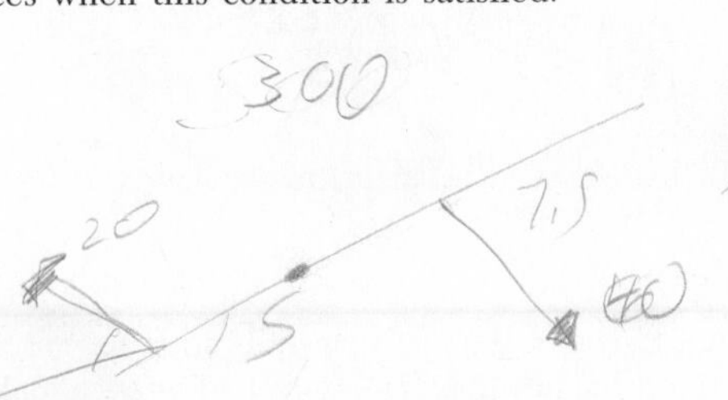

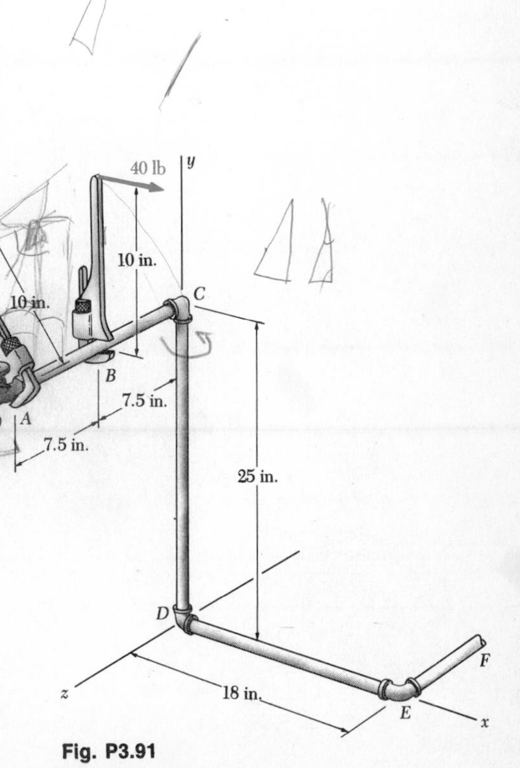

Fig. P3.91

3.92 Assuming $\theta = 60°$ in Prob. 3.91, replace the two 40-lb forces by an equivalent force-couple system at D and determine whether the plumber's action tends to tighten or loosen the joint between (*a*) pipe CD and elbow D, (*b*) elbow D and pipe DE. Assume all threads to be right-handed.

3.93 Assuming $\theta = 60°$ in Prob. 3.91, replace the two 40-lb forces by an equivalent force-couple system at E and determine whether the plumber's action tends to tighten or loosen the joint between (*a*) pipe DE and elbow E, (*b*) elbow E and pipe EF. Assume all threads to be right-handed.

3.94 A rectangular concrete foundation mat supports four column loads as shown. Determine the magnitude and point of application of the resultant of the four loads.

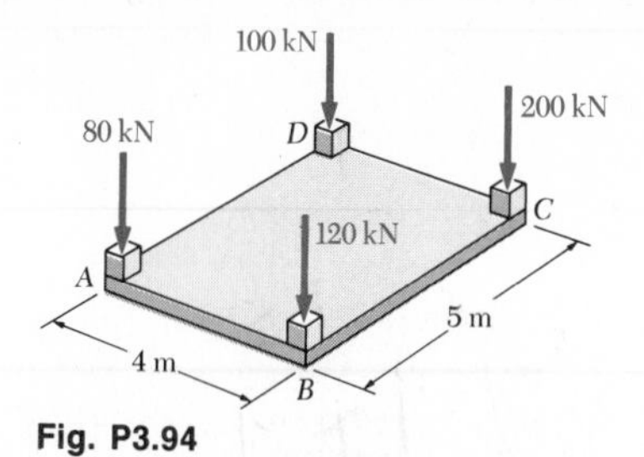

Fig. P3.94

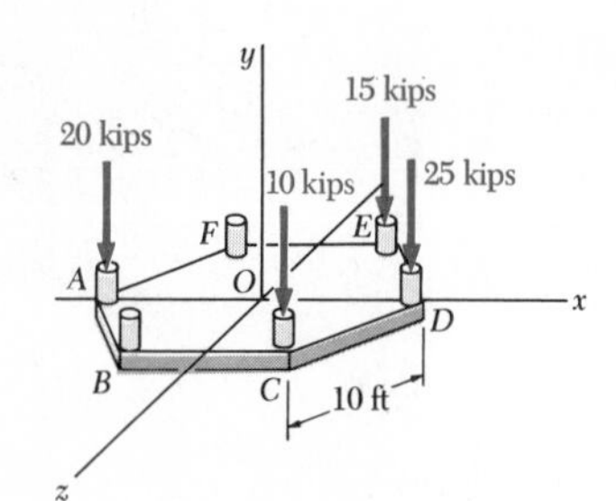

Fig. P3.95 and P3.96

3.95 A concrete foundation mat in the shape of a regular hexagon of side 10 ft supports four column loads as shown. Determine the magnitude and point of application of the resultant of the four loads.

3.96 Determine the magnitudes of the additional loads which must be applied at B and F if the resultant of all six loads is to pass through the center of the mat.

3.97 In Prob. 3.94, determine the magnitude and point of application of the smallest additional load which must be applied to the foundation mat if the resultant of the five loads is to pass through the center of the mat.

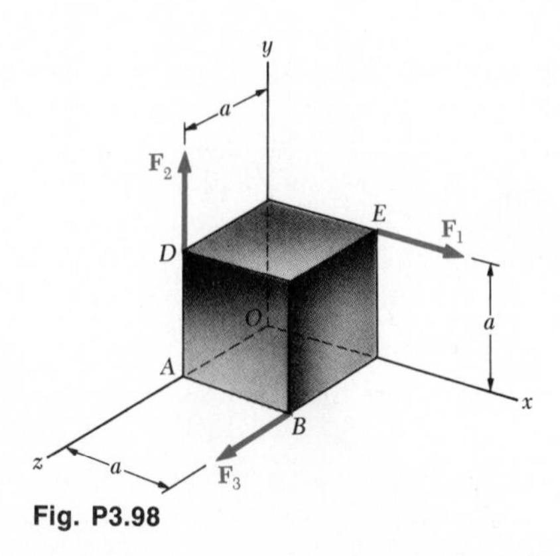

Fig. P3.98

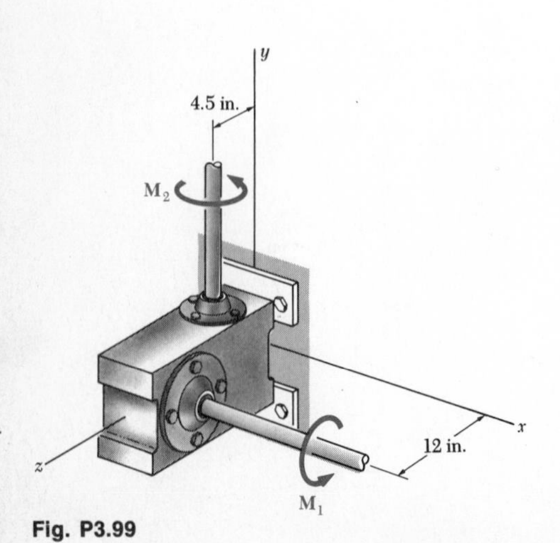

Fig. P3.99

***3.98** Three forces of the same magnitude P act on a cube of side a as shown. Replace the three forces by an equivalent wrench and determine (a) the magnitude and direction of the resultant force $\mathbf{R}$, (b) the pitch of the wrench, (c) the axis of the wrench.

***3.99** The two shafts of a speed-reducer unit are subjected to couples of magnitude $M_1 = 12$ lb·ft and $M_2 = 5$ lb·ft, respectively. The unit weighs 80 lb and its center of gravity is located on the z axis at $z = 9$ in. Replace the weight and the two couples by an equivalent wrench and determine (a) the resultant force $\mathbf{R}$, (b) the pitch of the wrench, (c) the point where the axis of the wrench intersects the xz plane.

***3.100** Two forces of magnitude P act as shown along the diagonals GC and AE of the horizontal faces of a rectangular parallelepiped. Replace the two forces by an equivalent system consisting of (*a*) a force and couple at O, (*b*) a wrench. (Specify the pitch and axis of the wrench.)

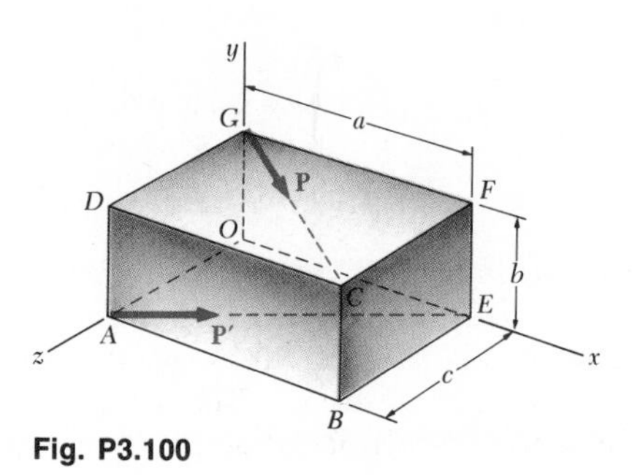

Fig. P3.100

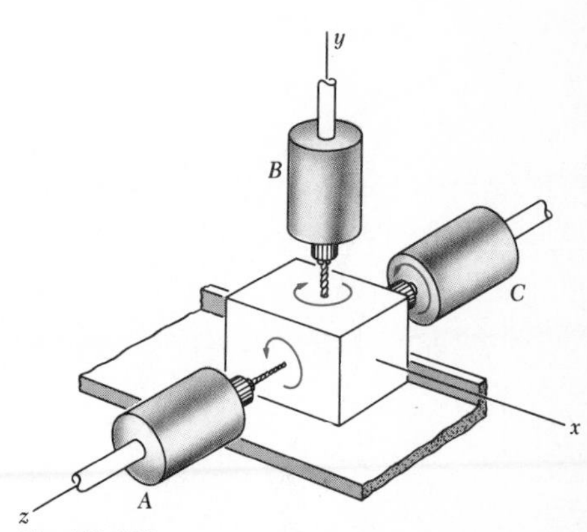

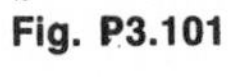
Fig. P3.101

***3.101** In an automated manufacturing process three holes are drilled simultaneously as shown in an aluminum block. Each drill exerts a 50-N force and a 0.100-N · m couple on the block. Knowing that drill A rotates counterclockwise, and drills B and C clockwise (as observed from each drill), reduce the forces and couples exerted by the drills on the block to an equivalent wrench and determine (*a*) the resultant force **R**, (*b*) the pitch of the wrench, (*c*) the point where the axis of the wrench intersects the xz plane.

***3.102** If only drills A and B are used in the manufacturing process of Prob. 3.101, show that the forces and couples exerted by the two drills on the aluminum block may be reduced to a single force **R** and determine (*a*) the magnitude and direction of **R**, (*b*) the line of action of **R**.

***3.103** In Sample Prob. 3.12 an additional force $\mathbf{Q} = Q\mathbf{k}$ is applied at the corner B of the cube. Knowing that the three forces acting on the cube may be reduced to a single force **R**, determine (*a*) the force **Q**, (*b*) the magnitude and direction of **R**, (*c*) the point where the line of action of **R** intersects the yz plane.

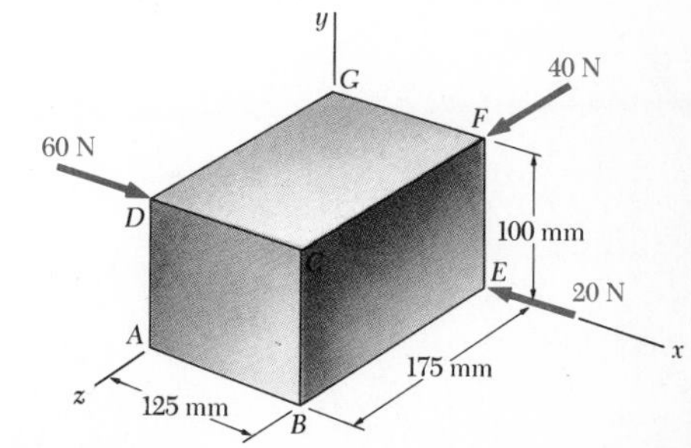

Fig. P3.104

***3.104** A rectangular block is acted upon by the three forces shown, which are directed along its edges. Replace these forces by an equivalent wrench and determine (*a*) the magnitude and direction of the resultant force **R**, (*b*) the pitch of the wrench, (*c*) the point where the axis of the wrench intersects the yz plane.

***3.105** Replace the two wrenches shown by a single equivalent wrench and determine (*a*) the resultant force **R**, (*b*) the pitch of the single equivalent wrench, (*c*) the point where its axis intersects the xz plane.

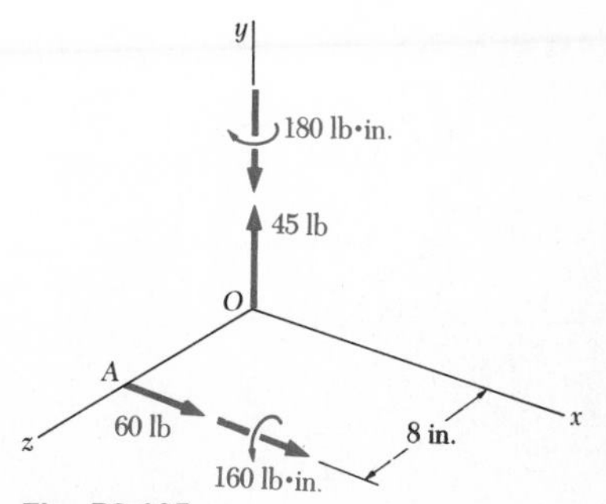

Fig. P3.105

***3.106** Assuming $\theta = 60°$ in Prob. 3.91, replace the two 40-lb forces by an equivalent wrench. Determine (*a*) the magnitude and direction of the resultant force **R**, (*b*) the pitch of the wrench, (*c*) the point where the axis of the wrench intersects the yz plane.

***3.107** In Prob. 3.90, replace the forces applied to the machine component by an equivalent wrench. Determine (*a*) the magnitude and direction of the resultant force **R**, (*b*) the pitch of the wrench, (*c*) the point where the axis of the wrench intersects the *xz* plane.

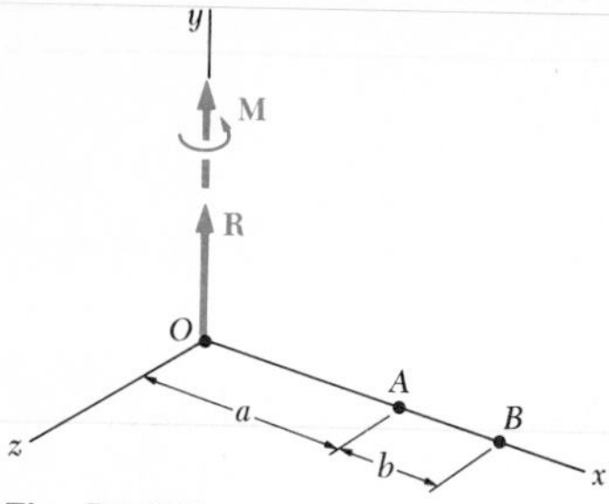

Fig. P3.108

***3.108** (*a*) Replace the wrench shown by an equivalent system consisting of two forces perpendicular to the *x* axis and applied respectively at *A* and *B*. (*b*) Solve part *a* assuming $R = 150$ N, $M = 20$ N·m, $a = 125$ mm, and $b = 50$ mm.

***3.109** Show that, in general, a wrench may be replaced by two forces chosen in such a way that one force passes through a given point while the other force lies in a given plane.

***3.110** Show that a wrench may be replaced by two perpendicular forces, one of which is applied at a given point.

***3.111** Show that a wrench may be replaced by two forces, one of which has a prescribed line of action.

Review and Summary

Principle of transmissibility

In this chapter we studied the effect of forces exerted on a rigid body. We first learned to distinguish between *external* and *internal* forces [Sec. 3.2] and saw that, according to the *principle of transmissibility*, the effect of an external force on a rigid body remains unchanged if that force is moved along its line of action [Sec. 3.3]. In other words, two forces **F** and **F′** acting on a rigid body at two different points have the same effect on that body if they have the same magnitude, same direction, and same line of action (Fig. 3.48). Two such forces are said to be *equivalent*.

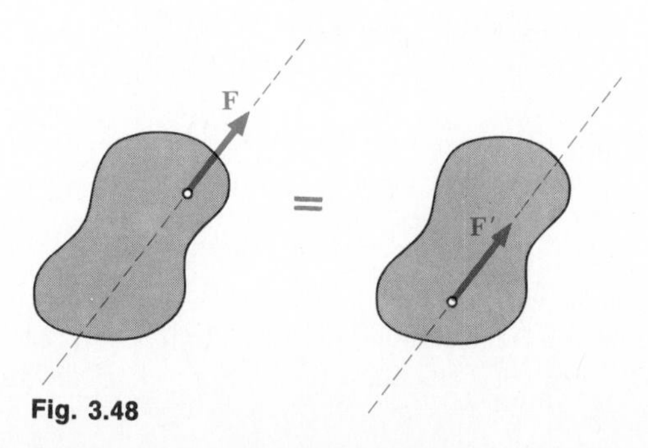

Fig. 3.48

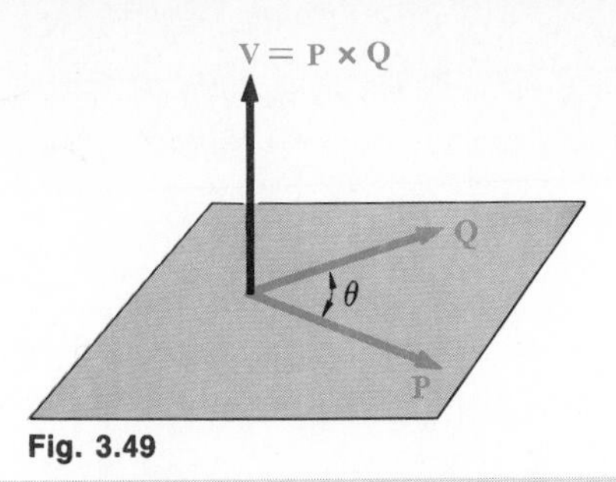

Fig. 3.49

Vector product of two vectors

Before proceeding with the discussion of *equivalent systems of forces,* we introduced the concept of *vector product of two vectors* [Sec. 3.4]. The vector product

$$\mathbf{V} = \mathbf{P} \times \mathbf{Q}$$

of the vectors **P** and **Q** was defined as a vector perpendicular to the plane containing **P** and **Q** (Fig. 3.49), of magnitude

$$V = PQ \sin \theta \qquad (3.1)$$

and directed in such a way that a person located at the tip of **V** will observe as counterclockwise the rotation through θ which brings the vector **P** in line with the vector **Q**. The three vectors **P**, **Q**, and **V**—taken in that order—are said to form a *right-handed triad.* It follows that the vector products **Q × P** and **P × Q** are represented by equal and opposite vectors. We have

$$\mathbf{Q} \times \mathbf{P} = -(\mathbf{P} \times \mathbf{Q}) \qquad (3.4)$$

It also follows from the definition of the vector product of two vectors that the vector products of the unit vectors **i**, **j**, and **k** are

$$\mathbf{i} \times \mathbf{i} = 0 \qquad \mathbf{i} \times \mathbf{j} = \mathbf{k} \qquad \mathbf{j} \times \mathbf{i} = -\mathbf{k}$$

and so on. The sign of the vector product of two unit vectors may be obtained by arranging in a circle and in counterclockwise order the three letters representing the unit vectors (Fig. 3.50): The product of two unit vectors will be positive if they follow each other in counterclockwise order, and negative otherwise.

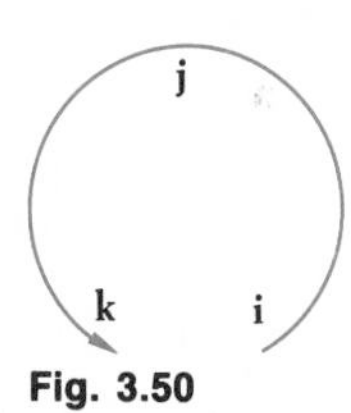

Fig. 3.50

Rectangular components of vector product

The *rectangular components of the vector product* **V** of two vectors **P** and **Q** were expressed [Sec. 3.5] as

$$\begin{aligned} V_x &= P_y Q_z - P_z Q_y \\ V_y &= P_z Q_x - P_x Q_z \\ V_z &= P_x Q_y - P_y Q_x \end{aligned} \qquad (3.9)$$

Using a determinant form, we also wrote

$$\mathbf{V} = \begin{vmatrix} \mathbf{i} & \mathbf{j} & \mathbf{k} \\ P_x & P_y & P_z \\ Q_x & Q_y & Q_z \end{vmatrix} \qquad (3.10)$$

Moment of a force about a point

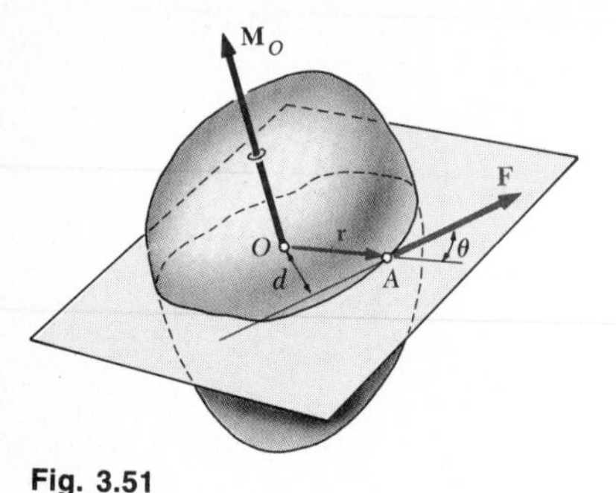

Fig. 3.51

The *moment of a force* **F** *about a point O* was defined [Sec. 3.6] as the vector product

$$\mathbf{M}_O = \mathbf{r} \times \mathbf{F} \tag{3.11}$$

where **r** is the *position vector* drawn from O to the point of application A of the force **F** (Fig. 3.51). Denoting by θ the angle between the lines of action of **r** and **F**, we found that the magnitude of the moment of **F** about O may be expressed as

$$M_O = rF \sin\theta = Fd \tag{3.12}$$

where d represents the perpendicular distance from O to the line of action of **F**.

Rectangular components of moment

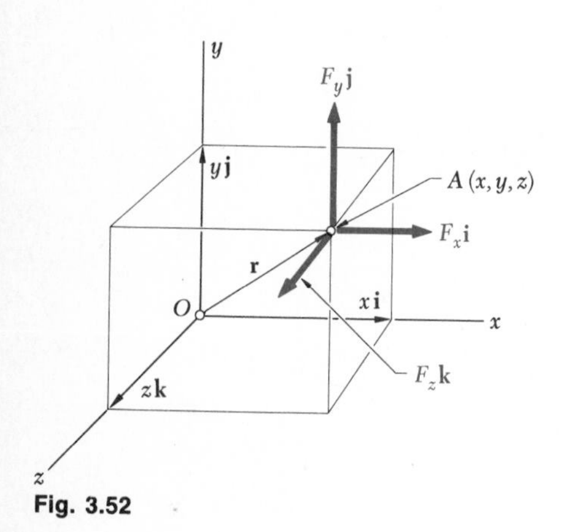

Fig. 3.52

The *rectangular components of the moment* $\mathbf{M}_O$ *of a force* **F** were expressed [Sec. 3.8] as

$$\begin{aligned} M_x &= yF_z - zF_y \\ M_y &= zF_x - xF_z \\ M_z &= xF_y - yF_x \end{aligned} \tag{3.18}$$

where x, y, z are the components of the position vector **r** (Fig. 3.52). Using a determinant form, we also wrote

$$\mathbf{M}_O = \begin{vmatrix} \mathbf{i} & \mathbf{j} & \mathbf{k} \\ x & y & z \\ F_x & F_y & F_z \end{vmatrix} \tag{3.19}$$

In the more general case of the moment about an arbitrary point B of a force **F** applied at A, we had

$$\mathbf{M}_B = \begin{vmatrix} \mathbf{i} & \mathbf{j} & \mathbf{k} \\ x_{A/B} & y_{A/B} & z_{A/B} \\ F_x & F_y & F_z \end{vmatrix} \tag{3.21}$$

where $x_{A/B}$, $y_{A/B}$, and $z_{A/B}$ denote the components of the vector $\mathbf{r}_{A/B}$:

$$x_{A/B} = x_A - x_B \qquad y_{A/B} = y_A - y_B \qquad z_{A/B} = z_A - z_B$$

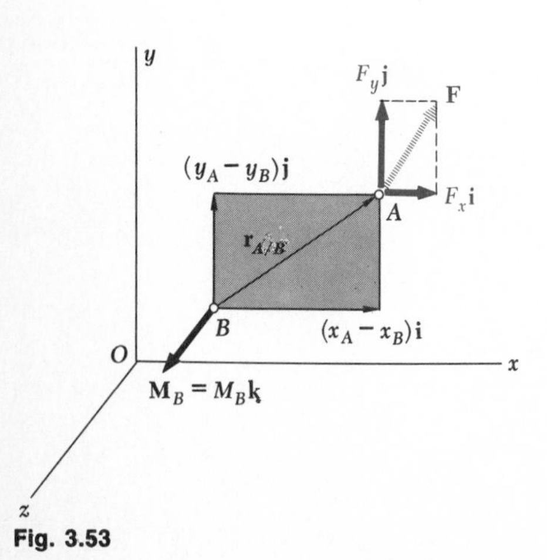

Fig. 3.53

In the case of *problems involving only two dimensions*, the force **F** may be assumed to lie in the *xy* plane. Its moment $\mathbf{M}_B$ about a point B in the same plane is perpendicular to that plane (Fig. 3.53) and is completely defined by the scalar

$$M_B = (x_A - x_B)F_y - (y_A - y_B)F_x \tag{3.23}$$

Various methods for the computation of the moment of a force about a point were illustrated in Sample Probs. 3.1 through 3.4.

Scalar product of two vectors

The *scalar product* of two vectors **P** and **Q** [Sec. 3.9] was denoted by $\mathbf{P} \cdot \mathbf{Q}$ and defined as the scalar quantity

$$\mathbf{P} \cdot \mathbf{Q} = PQ \cos \theta \qquad (3.24)$$

where θ is the angle between P and Q (Fig. 3.54). Expressing the scalar product of **P** and **Q** in terms of the rectangular components of the two vectors, we had

$$\mathbf{P} \cdot \mathbf{Q} = P_x Q_x + P_y Q_y + P_z Q_z \qquad (3.30)$$

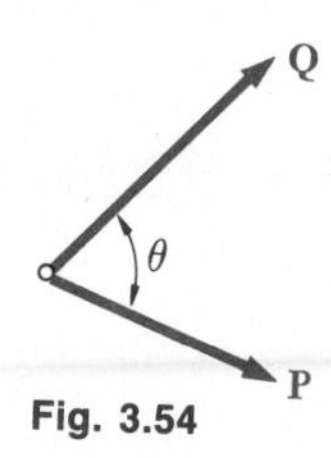

Fig. 3.54

Projection of a vector on an axis

The *projection of a vector* **P** *on an axis OL* (Fig. 3.55) may be obtained by forming the scalar product of **P** and the unit vector $\boldsymbol{\lambda}$ along OL. We have

$$P_{OL} = \mathbf{P} \cdot \boldsymbol{\lambda} \qquad (3.36)$$

or, using rectangular components,

$$P_{OL} = P_x \cos \theta_x + P_y \cos \theta_y + P_z \cos \theta_z \qquad (3.37)$$

where θ_x, θ_y, and θ_z denote the angles that the axis OL forms with the coordinate axes.

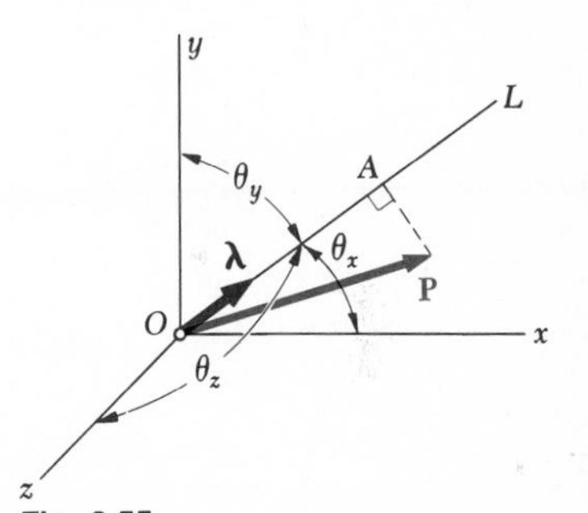

Fig. 3.55

Mixed triple product of three vectors

The *mixed triple product* of the three vectors **S**, **P**, and **Q** was defined as the scalar expression

$$\mathbf{S} \cdot (\mathbf{P} \times \mathbf{Q}) \qquad (3.38)$$

obtained by forming the scalar product of **S** with the vector product of **P** and **Q** [Sec. 3.10]. It was shown that

$$\mathbf{S} \cdot (\mathbf{P} \times \mathbf{Q}) = \begin{vmatrix} S_x & S_y & S_z \\ P_x & P_y & P_z \\ Q_x & Q_y & Q_z \end{vmatrix}$$

where the elements of the determinant are the rectangular components of the three vectors.

Moment of a force about an axis

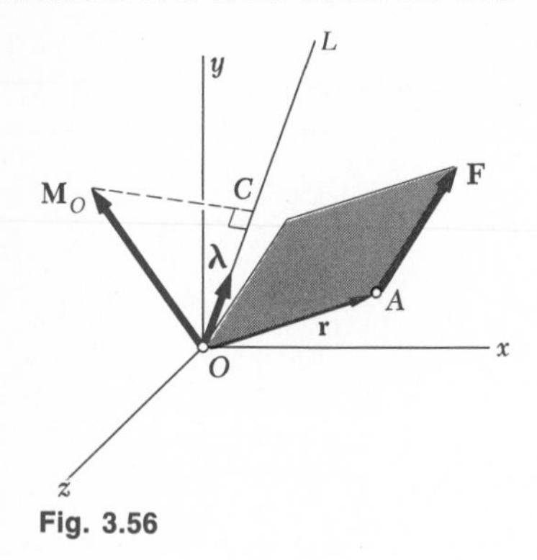

Fig. 3.56

The *moment of a force* **F** *about an axis OL* [Sec. 3.11] was defined as the projection *OC* on *OL* of the moment $\mathbf{M}_O$ of the force **F** (Fig. 3.56), i.e., as the mixed triple product of the unit vector **λ**, the position vector **r**, and the force **F**:

$$M_{OL} = \boldsymbol{\lambda} \cdot \mathbf{M}_O = \boldsymbol{\lambda} \cdot (\mathbf{r} \times \mathbf{F}) \tag{3.42}$$

Using the determinant form for the mixed triple product, we have

$$M_{OL} = \begin{vmatrix} \lambda_x & \lambda_y & \lambda_z \\ x & y & z \\ F_x & F_y & F_z \end{vmatrix} \tag{3.43}$$

where $\lambda_x, \lambda_y, \lambda_z$ = direction cosines of axis *OL*
x, y, z = components of **r**
F_x, F_y, F_z = components of **F**

An example of the determination of the moment of a force about a skew axis was given in Sample Prob. 3.5.

Couples

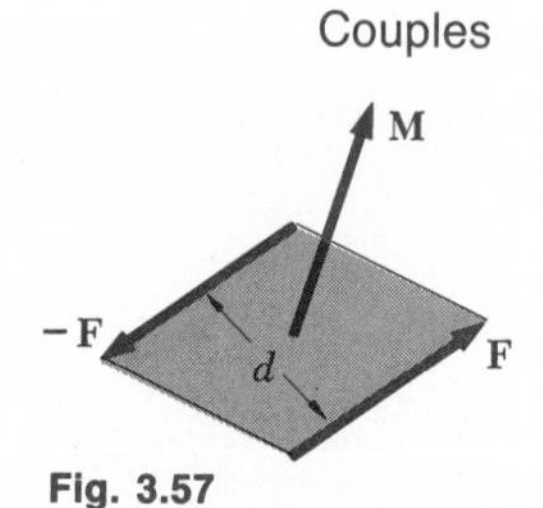

Fig. 3.57

Two forces **F** *and* −**F** *having the same magnitude, parallel lines of action, and opposite sense are said to form a couple* [Sec. 3.12]. It was shown that the moment of a couple is independent of the point about which it is computed; it is a vector **M** perpendicular to the plane of the couple and equal in magnitude to the product of the common magnitude *F* of the forces and the perpendicular distance *d* between their lines of action (Fig. 3.57).

Two couples having the same moment **M** are *equivalent*, i.e., they have the same effect on a given rigid body [Sec. 3.13]. The sum of two couples is itself a couple [Sec. 3.14] and the moment **M** of the resultant couple may be obtained by adding vectorially the moments $\mathbf{M}_1$ and $\mathbf{M}_2$ of the original couples [Sample Prob. 3.6]. It follows that a couple may be represented by a vector, called a *couple vector*, equal in magnitude and direction to the moment **M** of the couple [Sec. 3.15]. A couple vector is a *free vector* which may be attached to the origin *O* if so desired and resolved into components (Fig. 3.58).

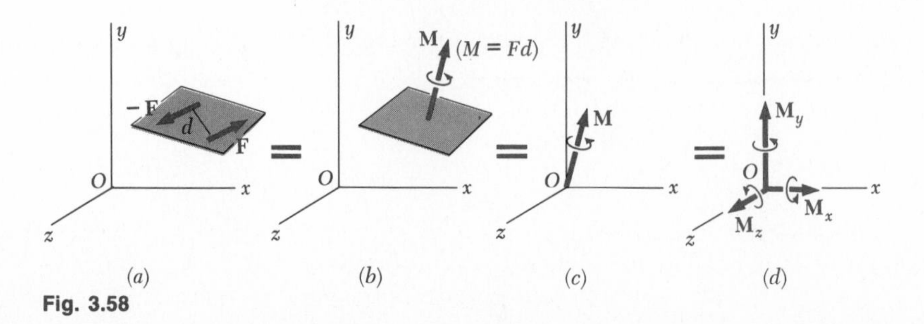

Fig. 3.58

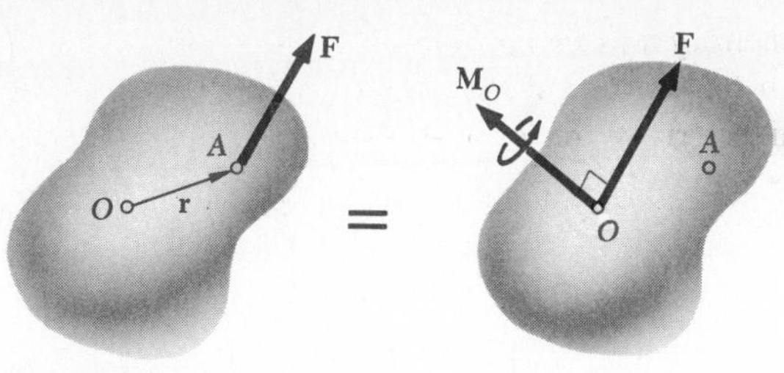

Fig. 3.59

Force-couple system

Any force $\mathbf{F}$ acting at a point A of a rigid body may be replaced by a *force-couple system* at an arbitrary point O, consisting of the force $\mathbf{F}$ applied at O and a couple of moment $\mathbf{M}_O$ equal to the moment about O of the force $\mathbf{F}$ in its original position [Sec. 3.16]; it should be noted that the force $\mathbf{F}$ and the couple vector $\mathbf{M}_O$ are always perpendicular to each other (Fig. 3.59).

Reduction of a system of forces to a force-couple system

It follows [Sec. 3.17] that *any system of forces may be reduced to a force-couple system at a given point O* by replacing each of the forces of the system by an equivalent force-couple system at O (Fig. 3.60) and adding all the forces and all the couples obtained in this manner into a resultant force $\mathbf{R}$ and a resultant couple $\mathbf{M}_O^R$ [Sample Probs. 3.8 through 3.11]. Note that, in general, the resultant $\mathbf{R}$ and the couple vector $\mathbf{M}_O^R$ will not be perpendicular to each other.

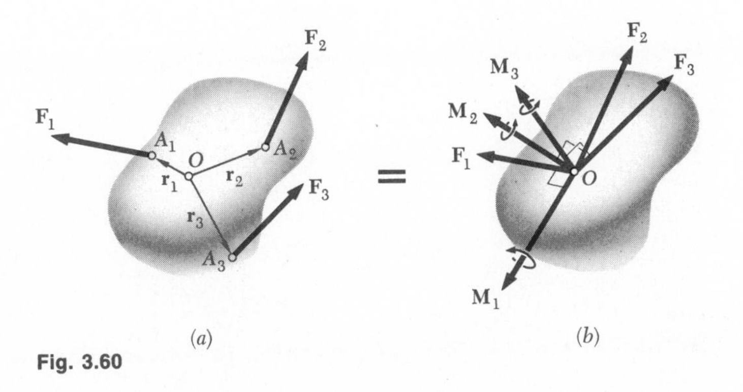

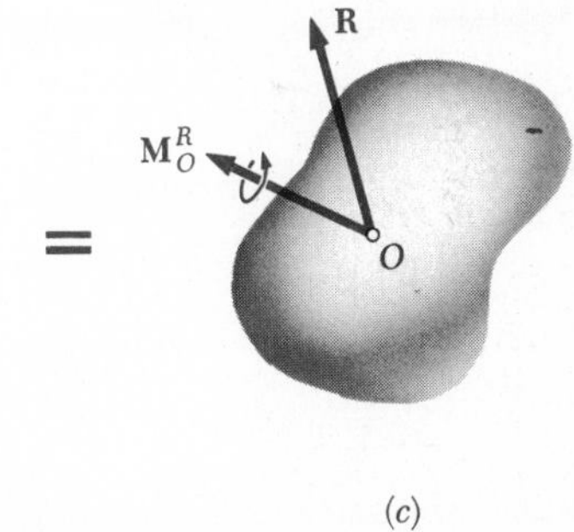

Fig. 3.60

Equivalent systems of forces

We concluded from the above [Sec. 3.18] that, as far as rigid bodies are concerned, *two systems of forces* $\mathbf{F}_1$, $\mathbf{F}_2$, $\mathbf{F}_3$, *etc., and* $\mathbf{F}'_1$, $\mathbf{F}'_2$, $\mathbf{F}'_3$, *etc., are equivalent if, and only if,*

$$\Sigma\mathbf{F} = \Sigma\mathbf{F}' \qquad \text{and} \qquad \Sigma\mathbf{M}_O = \Sigma\mathbf{M}'_O \tag{3.57}$$

Further reduction of a system of forces

If the resultant force $\mathbf{R}$ and the resultant couple vector $\mathbf{M}_O^R$ are perpendicular to each other, the force-couple system at O may be further reduced to a single resultant force [Sec. 3.20]. This will be the case for systems consisting either of (*a*) concurrent forces (cf. Chap. 2), (*b*) coplanar forces [Sample Probs. 3.8 and 3.9], or (*c*) parallel forces [Sample Prob. 3.11]. If the resultant **R** and the couple vector $\mathbf{M}_O^R$ are *not* perpendicular to each other, the system *cannot* be reduced to a single force. It may, however, be reduced to a special type of force-couple system called a *wrench,* consisting of the resultant **R** and of a couple vector $\mathbf{M}_1$ directed along **R** [Sec. 3.21 and Sample Prob. 3.12].

Review Problems

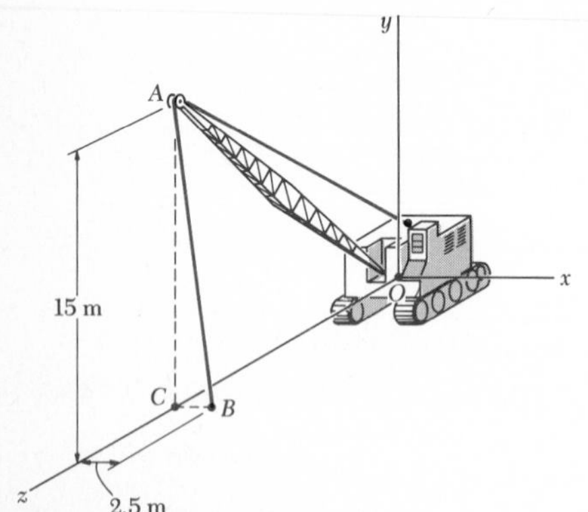

Fig. P3.112 and P3.113

3.112 A crane is oriented so that the end of the 25-m boom *AO* lies in the *yz* plane. At the instant shown the tension in cable *AB* is 4 kN. Determine the moment about each of the coordinate axes of the force exerted on *A* by cable *AB*.

3.113 The 25-m crane boom *AO* lies in the *yz* plane. Determine the maximum permissible tension in cable *AB* if the absolute value of the moments about the coordinate axes of the force exerted on *A* by cable *AB* must be as follows: $|M_x| \leq 60$ kN · m, $|M_y| \leq 12$ kN · m, $|M_z| \leq 8$ kN · m.

3.114 A force **P** of magnitude 520 lb acts on the frame shown at point *E*. Determine the moment of **P** (*a*) about point *D*, (*b*) about a line joining points *O* and *D*.

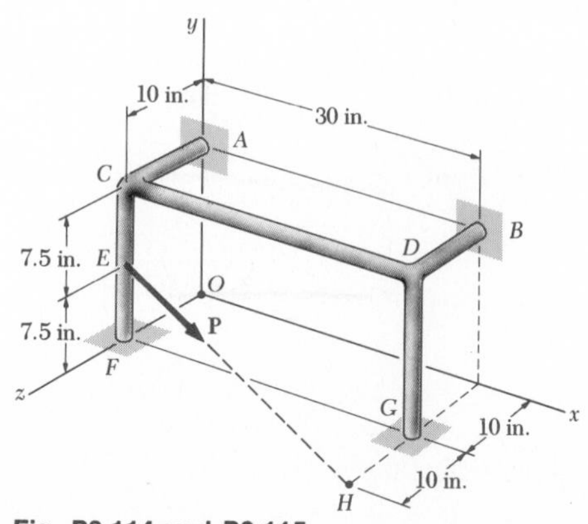

Fig. P3.114 and P3.115

3.115 A force **P** acts on the frame shown at point *E*. Knowing that the absolute value of the moment of **P** about a line joining points *F* and *B* is 300 lb · ft, determine the magnitude of the force **P**.

3.116 and 3.117 It is known that the slider B exerts on rod AB a 162.5-lb force perpendicular to rod BC and directed down and to the right. Determine the moment of that force about A.

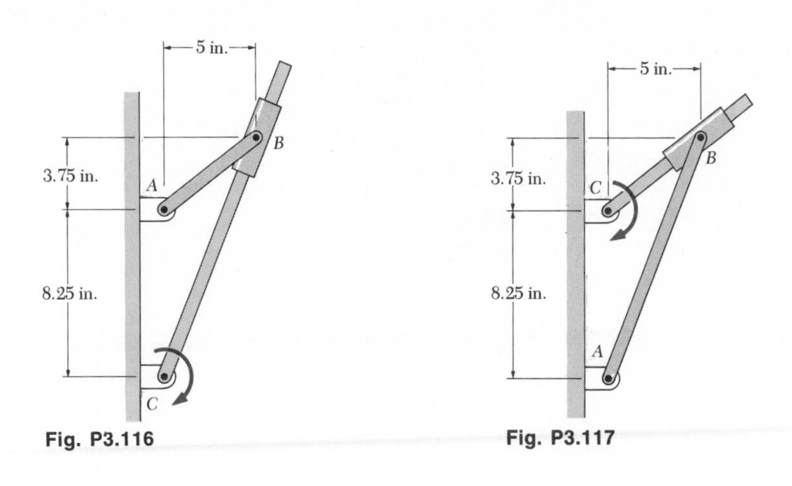

Fig. P3.116

Fig. P3.117

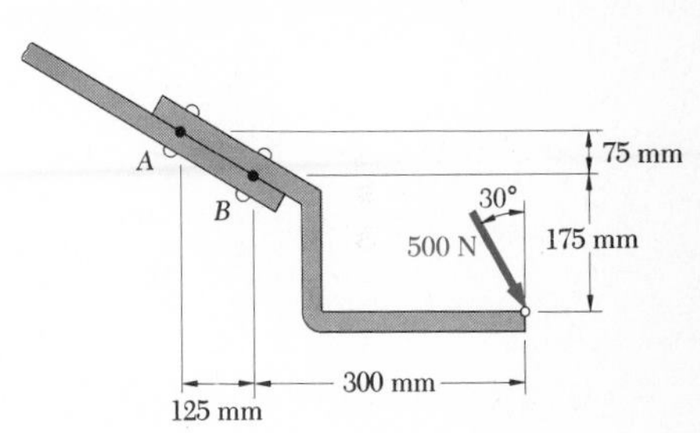

Fig. P3.118

3.118 A 500-N force is applied to a bent plate as shown. Determine (*a*) an equivalent force-couple system at B, (*b*) an equivalent system formed by a vertical force at A and a force at B.

3.119 A 6×12 in. plate is subjected to four loads. Find the resultant of the four loads and the two points at which the line of action of the resultant intersects the edge of the plate.

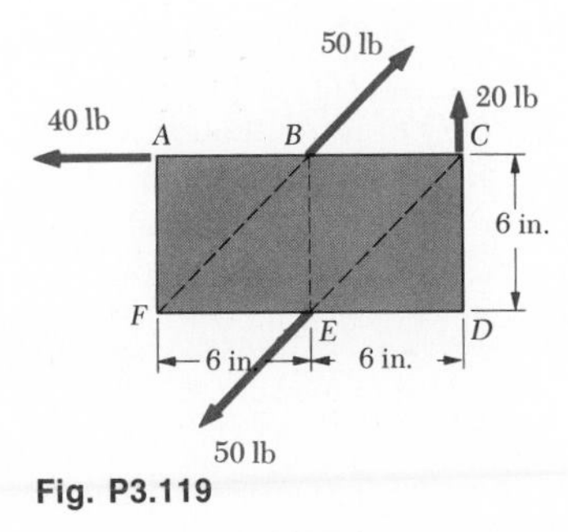

Fig. P3.119

3.120 A concrete foundation mat of 5-m radius supports four equally spaced columns, each of which is located 4 m from the center of the mat. Determine the magnitude and the point of application of the resultant of the four loads.

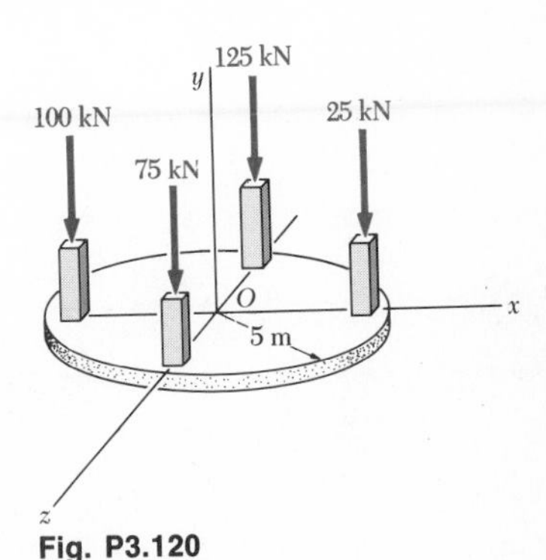

Fig. P3.120

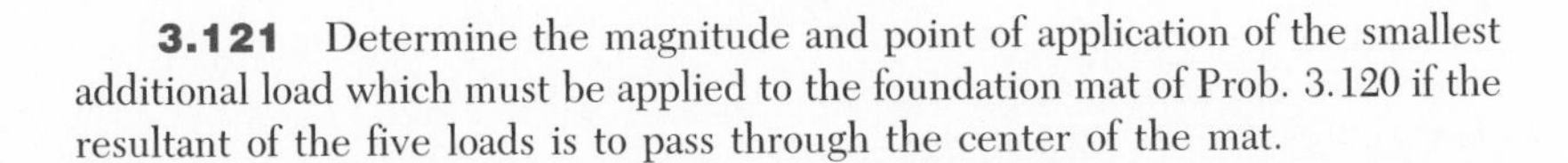

3.121 Determine the magnitude and point of application of the smallest additional load which must be applied to the foundation mat of Prob. 3.120 if the resultant of the five loads is to pass through the center of the mat.

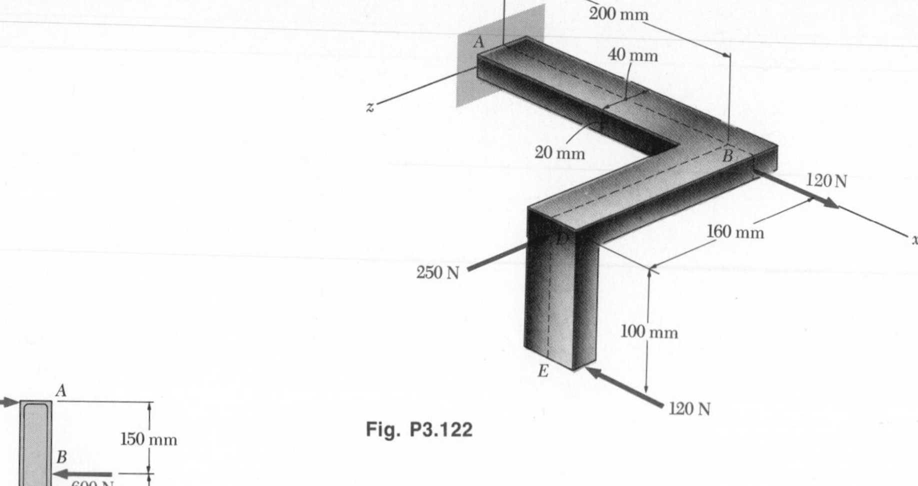

Fig. P3.122

Fig. P3.123

3.122 Replace the three forces shown by (*a*) a force-couple system at A, (*b*) a wrench. (Specify the pitch and axis of the wrench.)

3.123 Three horizontal forces are applied as shown to a vertical cast-iron arm. Determine the resultant of the forces and the distance from the ground to its line of action when (*a*) $P = 160$ N, (*b*) $P = 1000$ N, (*c*) $P = 5200$ N.

The following problems are designed to be solved with a computer.

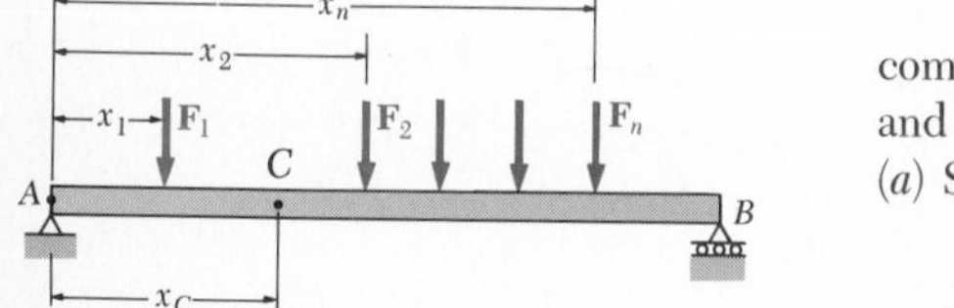

Fig. P3.C1

3.C1 A beam AB is subjected to several vertical forces as shown. Write a computer program which can be used to determine the magnitude of the resultant and point C where its line of action intersects AB. Use this program to solve (*a*) Sample Prob. 3.8*c*, (*b*) Prob. 3.81.

3.C2 Two forces **P** and **Q** of magnitude $P = 50$ N and $Q = 120$ N are applied as shown to corners A and B of a square plate of side $a = 500$ mm. Write a computer program which can be used to calculate the equivalent force-couple system at D. Use this program to determine the magnitude and the direction of **R** and the couple $\mathbf{M}_D$ for values of β from 0 to 360° at 30° intervals.

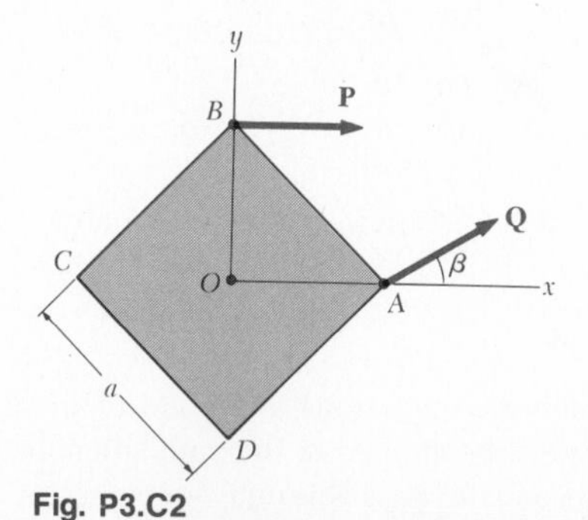

Fig. P3.C2

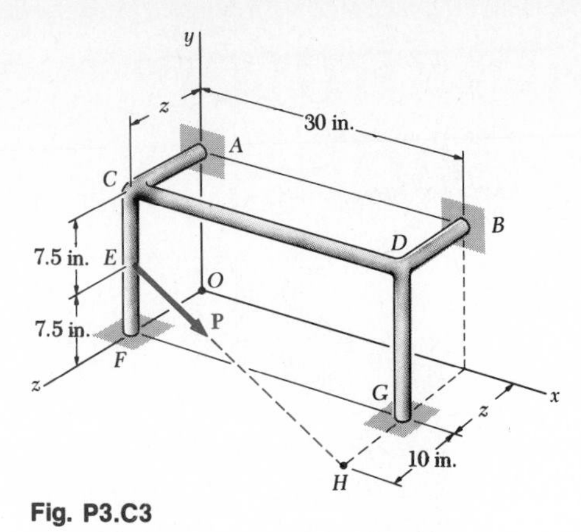

Fig. P3.C3

3.C3 A force **P** of magnitude 520 lb acts on the frame shown at point E. Several comparative designs are to be made, using various lengths z for portions AC and BD of the frame. Write a computer program and use it to calculate, for values of z from 0 to 20 in. at 1-in. intervals, (*a*) the rectangular components and magnitude of the moment of **P** about point O, (*b*) the moment of **P** about a line joining points O and D.

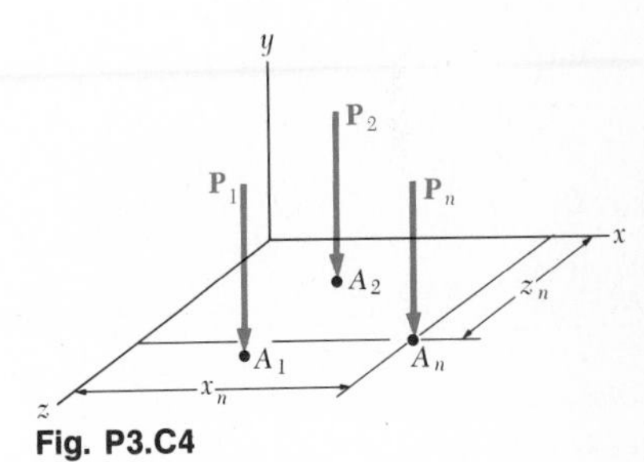

Fig. P3.C4

3.C4 Write a computer program which can be used to determine the magnitude and point of application of the resultant of the vertical forces $\mathbf{P}_1$, $\mathbf{P}_2$, . . . , $\mathbf{P}_n$ which act at points A_1, A_2, . . . , A_n located in the xz plane. Use this program to solve (*a*) Sample Prob. 3.11, (*b*) Prob. 3.94, (*c*) Prob. 3.95.

Equilibrium of Rigid Bodies

CHAPTER

4

4.1. Introduction. We saw in the preceding chapter that the external forces acting on a rigid body may be reduced to a force-couple system at some arbitrary point O. When the force and the couple are both equal to zero, the external forces form a system equivalent to zero and the rigid body is said to be in *equilibrium*.

The necessary and sufficient conditions for the equilibrium of a rigid body, therefore, may be obtained by setting $\mathbf{R}$ and $\mathbf{M}_O^R$ equal to zero in the relations (3.52) of Sec. 3.17:

$$\Sigma\mathbf{F} = 0 \qquad \Sigma\mathbf{M}_O = \Sigma(\mathbf{r} \times \mathbf{F}) = 0 \tag{4.1}$$

Resolving each force and each moment into its rectangular components, we may express the necessary and sufficient conditions for the equilibrium of a rigid body by the following six scalar equations:

$$\Sigma F_x = 0 \qquad \Sigma F_y = 0 \qquad \Sigma F_z = 0 \tag{4.2}$$
$$\Sigma M_x = 0 \qquad \Sigma M_y = 0 \qquad \Sigma M_z = 0 \tag{4.3}$$

The equations obtained may be used to determine unknown forces applied to the rigid body or unknown reactions exerted on it by its supports. We note that Eqs. (4.2) express the fact that the components of the external forces in the x, y, and z directions are balanced; Eqs. (4.3) express the fact that the moments of the external forces about the x, y, and z axes are balanced. The system of the external forces, therefore, will impart no motion of translation or rotation to the rigid body considered.

In order to write the equations of equilibrium for a rigid body, it is essential to first identify correctly all the forces acting on that body and draw the corresponding *free-body diagram*. In this chapter we shall first consider the equilibrium of *two-dimensional structures* subjected to forces contained in their plane and learn to draw their free-body diagram. In addition to the forces *applied* to the structure, we shall consider the *reactions* exerted on the structure by its supports. We shall learn to associate a specific reaction with each type of support and determine whether the structure is properly supported, so that we may know in advance whether the equations of equilibrium may actually be solved for the unknown forces and reactions. Later in the chapter, we shall consider the equilibrium of three-dimensional structures and subject them and their supports to the same kind of analysis.

4.2. Free-Body Diagram. In solving a problem concerning the equilibrium of a rigid body, it is essential to consider *all* the forces acting on the body; it is equally important to exclude any force which is not directly applied on the body. Omitting a force or adding an extraneous one would destroy the conditions of equilibrium. Therefore, the first step in the solution of the problem should consist in drawing a *free-body diagram* of the rigid body under consideration. Free-body diagrams have already been used on many occasions in Chap. 2. However, in view of their importance to the solution of equilibrium problems, we shall summarize here the various steps which must be followed in drawing a free-body diagram.

First, a clear decision is made regarding the choice of the free body to be used. This body is then detached from the ground and separated from any other body. The contour of the body thus isolated is sketched.

All external forces are then indicated. These forces represent the action exerted *on* the free body *by* the ground and the bodies which have been detached; they should be applied at the various points where the free body was supported by the ground or connected to the other bodies. The *weight* of the free body should also be included among the external forces, since it represents the attraction exerted by the earth on the various particles forming the free body. As will be seen in Chap. 5, the weight should be applied at the center of gravity of the body. When the free body is made of several parts, the forces the various parts exert on each other should *not* be included among the external forces. These forces are internal forces as far as the free body is concerned.

The magnitude and direction of the *known external forces* should be clearly marked on the free-body diagram. Care should be taken to indicate the sense of the force exerted *on* the free body, not that of the force exerted *by* the free body. Known external forces generally include the *weight* of the free body and *forces applied* for a given purpose.

Unknown external forces usually consist of the *reactions*—sometimes called *constraining forces*—through which the ground and other bodies oppose a possible motion of the free body and thus constrain it to remain in the same position. Reactions are exerted at the points where the free body is *supported* or *connected* to other bodies. They will be discussed in detail in Secs. 4.3 and 4.8.

The free-body diagram should also include dimensions, since these may be needed in the computation of moments of forces. Any other detail, however, should be omitted.

4.3. Reactions at Supports and Connections for a Two-Dimensional Structure. In the first part of this chapter we shall consider the equilibrium of a two-dimensional structure; i.e., we shall assume that the structure considered and the forces applied to it are contained in the plane of the figure. Clearly, the reactions needed to maintain the structure in the same position will also be contained in the plane of the figure.

The reactions exerted on a two-dimensional structure may be divided into three groups, corresponding to three types of *supports*, or *connections:*

1. *Reactions Equivalent to a Force with Known Line of Action.* Supports and connections causing reactions of this group include *rollers, rockers, frictionless surfaces, short links and cables, collars on frictionless rods,* and *frictionless pins in slots.* Each of these supports and connections can prevent motion in one direction only. They are shown in Fig. 4.1, together with the reaction they produce. Reactions of this group involve *one unknown,* namely, the magnitude of the reaction; this magnitude should be denoted by an appropriate letter. The line of action of the reaction is known and should be indicated clearly in the free-body diagram. The sense of the reaction must be as shown in Fig. 4.1 in the case of a frictionless surface (away from the surface) or of a cable (tension in the direction of the cable). The reaction may be directed either way in the case of double-track rollers, links, collars on rods, and pins in slots. Single-track rollers and rockers are generally assumed to be reversible, and thus the corresponding reactions may also be directed either way.

2. *Reactions Equivalent to a Force of Unknown Direction.* Supports and connections causing reactions of this group include *frictionless pins in fitted holes, hinges,* and *rough surfaces.* They can prevent translation of the free body in all directions, but they cannot prevent the body from rotating about the connection. Reactions of this group involve *two unknowns* and are usually represented by their x and y components. In the case of a rough surface, the component normal to the surface must be directed away from the surface.

3. *Reactions Equivalent to a Force and a Couple.* These reactions are caused by *fixed supports* which oppose any motion of the free body and thus constrain it completely. Fixed supports actually produce forces over the entire surface of contact; these forces, however, form a system which may be reduced to a force and a couple. Reactions of this group involve *three unknowns,* consisting usually of the two components of the force and the moment of the couple.

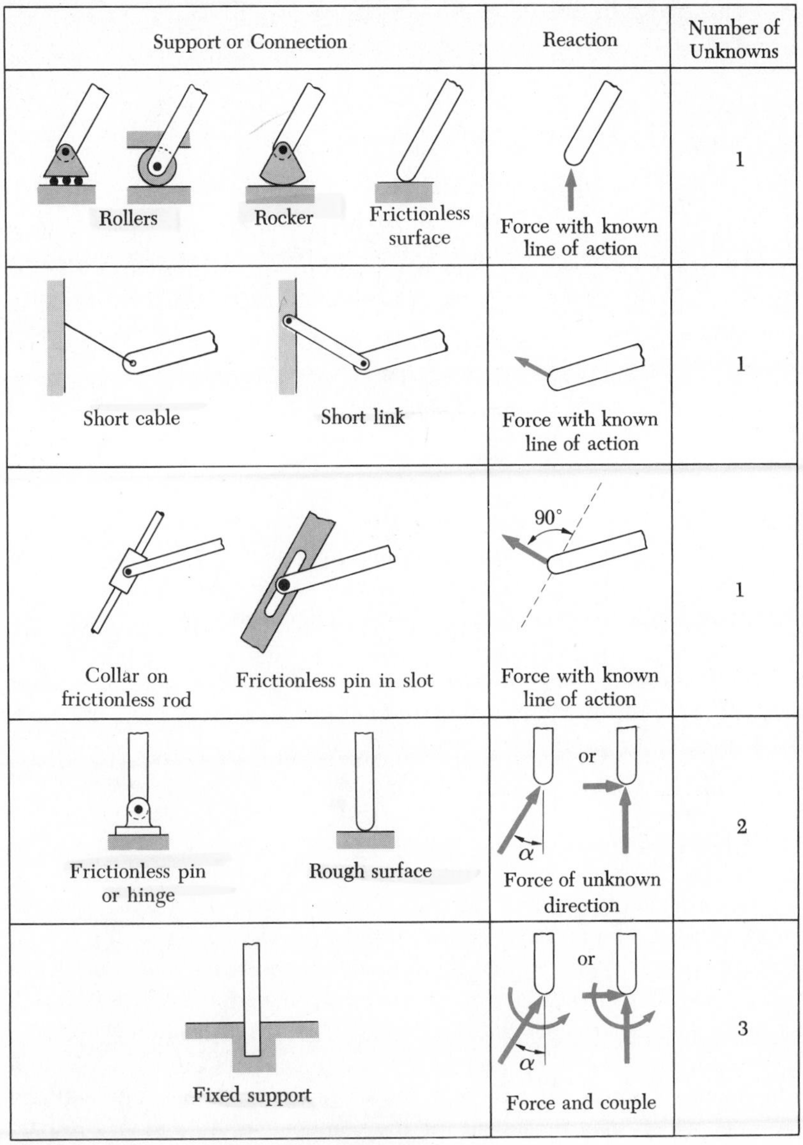

Support or Connection	Reaction	Number of Unknowns
Rollers; Rocker; Frictionless surface	Force with known line of action	1
Short cable; Short link	Force with known line of action	1
Collar on frictionless rod; Frictionless pin in slot	90° Force with known line of action	1
Frictionless pin or hinge; Rough surface	α or Force of unknown direction	2
Fixed support	α or Force and couple	3

Fig. 4.1 Reactions at supports and connections.

When the sense of an unknown force or couple is not clearly apparent, no attempt should be made to determine it. Instead, the sense of the force or couple should be arbitrarily assumed; the sign of the answer obtained will indicate whether the assumption is correct or not.

read

4.4. Equilibrium of a Rigid Body in Two Dimensions. The conditions stated in Sec. 4.1 for the equilibrium of a rigid body become considerably simpler in the case of a two-dimensional structure. Choosing the x and y axes in the plane of the structure, we have

$$F_z = 0 \qquad M_x = M_y = 0 \qquad M_z = M_O$$

for each of the forces applied to the structure. Thus, the six equations of equilibrium derived in Sec. 4.1 reduce to

$$\Sigma F_x = 0 \qquad \Sigma F_y = 0 \qquad \Sigma M_O = 0 \tag{4.4}$$

and to three trivial identities $0 = 0$. Since the third of the equations (4.4) must be satisfied regardless of the choice of the origin O, we may write the equations of equilibrium for a two-dimensional structure in the more general form

$$\Sigma F_x = 0 \qquad \Sigma F_y = 0 \qquad \Sigma M_A = 0 \tag{4.5}$$

where A is any point in the plane of the structure. The three equations obtained may be solved for no more than *three unknowns.*

We saw in the preceding section that unknown forces usually consist of reactions, and that the number of unknowns corresponding to a given reaction depends upon the type of support or connection causing that reaction. Referring to Sec. 4.3, we check that the equilibrium equations (4.5) may be used to determine the reactions of two rollers and one cable, or of one fixed support, or of one roller and one pin in a fitted hole, etc.

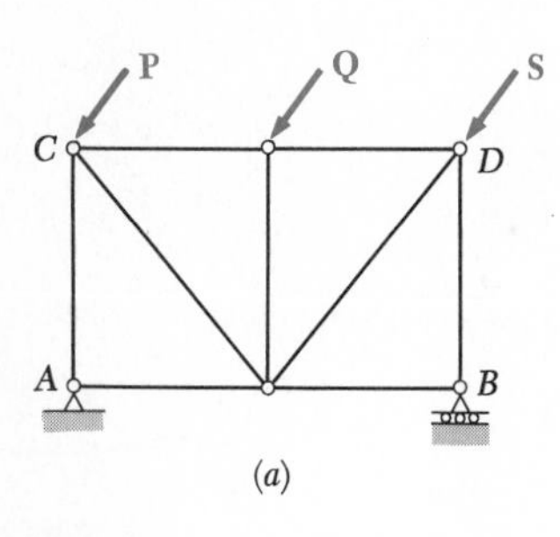

(*a*)

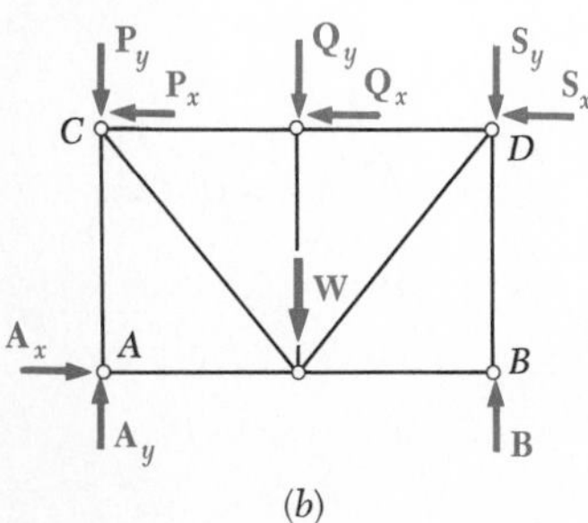

(*b*)

Fig. 4.2

Consider, for instance, the truss shown in Fig. 4.2*a*, which is subjected to the given forces **P**, **Q**, and **S**. The truss is held in place by a pin at A and a roller at B. The pin prevents point A from moving by exerting on the truss a force which may be resolved into the components $\mathbf{A}_x$ and $\mathbf{A}_y$; the roller keeps the truss from rotating about A by exerting the vertical force **B**. The free-body diagram of the truss is shown in Fig. 4.2*b*; it includes the reactions $\mathbf{A}_x$, $\mathbf{A}_y$, and **B** as well as the applied forces **P**, **Q**, **S**, and the weight **W** of the truss. Expressing that the sum of the moments about A of all the forces shown in Fig. 4.2*b* is zero, we write the equation $\Sigma M_A = 0$, which may be solved for the magnitude B since it does not contain A_x or A_y. Expressing, then, that the sum of the x components and the sum of the y components of the forces are zero, we write the equations $\Sigma F_x = 0$ and $\Sigma F_y = 0$, which may be solved for the components A_x and A_y, respectively.

Additional equations could be obtained by expressing that the sum of the moments of the external forces about points other than A is zero. We could write, for instance, $\Sigma M_B = 0$. Such a statement, however, does not contain any new information, since it has already been established that the system of the forces shown in Fig. 4.2*b* is equivalent to zero. The additional equation *is not independent* and cannot be used to determine a fourth unknown. It will be useful, however, for checking the solution obtained from the original three equations of equilibrium.

While the three equations of equilibrium cannot be *augmented* by additional equations, any of them may be *replaced* by another equation. Therefore, an alternate system of equations of equilibrium is

$$\Sigma F_x = 0 \qquad \Sigma M_A = 0 \qquad \Sigma M_B = 0 \tag{4.6}$$

where the line AB is chosen in a direction different from the y direction (Fig. 4.2*b*). These equations are sufficient conditions for the equilibrium of the truss. The first two equations indicate that the external forces must reduce to a single vertical force at A. Since the third equation requires that the moment of this force be zero about a point B which is not on its line of action, the force must be zero and the rigid body is in equilibrium.

A third possible set of equations of equilibrium is

$$\Sigma M_A = 0 \qquad \Sigma M_B = 0 \qquad \Sigma M_C = 0 \tag{4.7}$$

where the points A, B, and C are not in a straight line (Fig. 4.2*b*). The first equation requires that the external forces reduce to a single force at A; the second equation requires that this force pass through B; the third, that it pass through C. Since the points A, B, C are not in a straight line, the force must be zero, and the rigid body is in equilibrium.

The equation $\Sigma M_A = 0$, which expresses that the sum of the moments of the forces about pin A is zero, possesses a more definite physical meaning than either of the other two equations (4.7). These two equations express a similar idea of balance, but with respect to points about which the rigid body is not actually hinged. They are, however, as useful as the first equation, and our choice of equilibrium equations should not be unduly influenced by the physical meaning of these equations. Indeed, it will be desirable in practice to choose equations of equilibrium containing only one unknown, since this eliminates the necessity of solving simultaneous equations. Equations containing only one unknown may be obtained by summing moments about the point of intersection of the lines of action of two unknown forces or, if these forces are parallel, by summing components in a direction perpendicular to their common direction. In the case of the truss of Fig. 4.3, for example, which is held by rollers at A and B and a short link at D, the reactions at A and B may be eliminated by summing x components. The reactions at A and D will be eliminated by summing moments about C and the reactions at B and D by summing moments about D. The equations obtained are

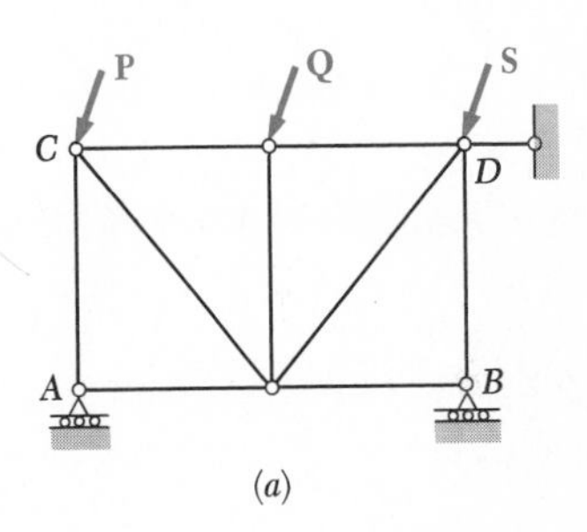

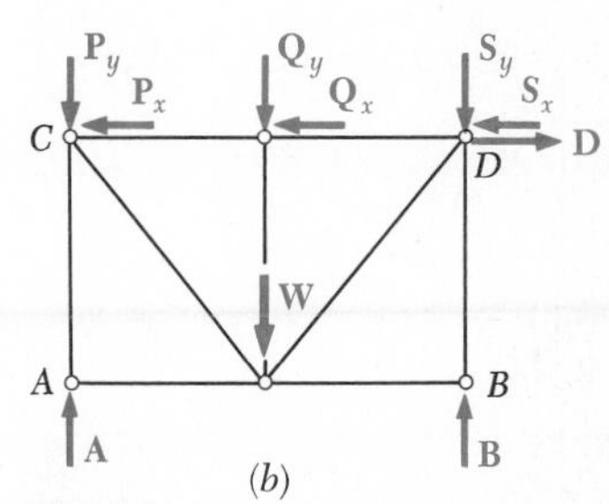

Fig. 4.3

$$\Sigma F_x = 0 \qquad \Sigma M_C = 0 \qquad \Sigma M_D = 0$$

Each of these equations contains only one unknown.

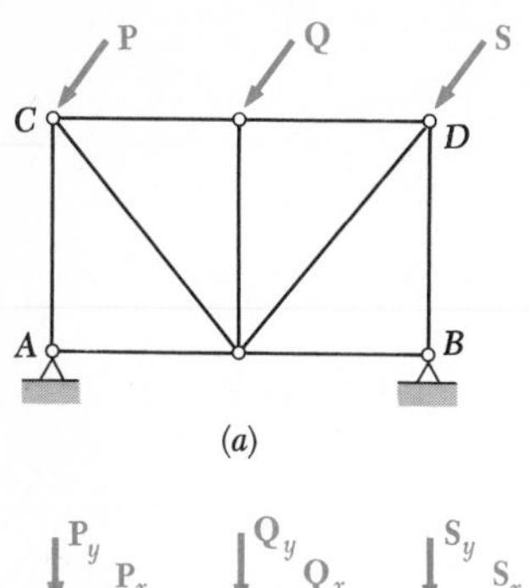

(a)

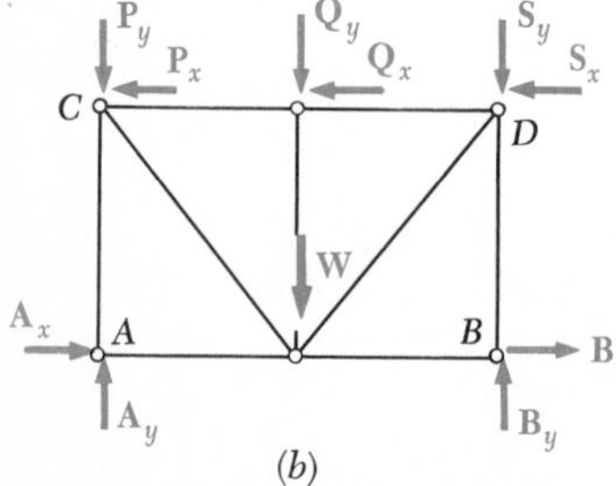

(b)

Fig. 4.4 Statically indeterminate reactions.

4.5. Statically Indeterminate Reactions. Partial Constraints. In each of the two examples considered in the preceding section (Figs. 4.2 and 4.3), the types of supports used were such that the rigid body could not possibly move under the given loads or under any other loading conditions. In such cases, the rigid body is said to be *completely constrained.* We also recall that the reactions corresponding to these supports involved *three unknowns* and could be determined by solving the three equations of equilibrium. When such a situation exists, the reactions are said to be *statically determinate.*

Consider now the truss shown in Fig. 4.4*a*, which is held by pins at A and B. These supports provide more constraints than are necessary to keep the truss from moving under the given loads or under any other loading conditions. We also note from the free-body diagram of Fig. 4.4*b* that the corresponding reactions involve *four unknowns*. Since, as was pointed out in Sec. 4.4, only three independent equilibrium equations are available, there are *more unknowns than equations*, and all the unknowns cannot be determined. While the equations $\Sigma M_A = 0$ and $\Sigma M_B = 0$ yield the vertical components B_y and A_y, respectively, the equation $\Sigma F_x = 0$ gives only the sum $A_x + B_x$ of the horizontal components of the reactions at A and B. The components A_x and B_x are said to be *statically indeterminate.* They could be determined by considering the deformations produced in the truss by the given loading, but this method is beyond the scope of statics and belongs to the study of mechanics of materials.

The supports used to hold the truss shown in Fig. 4.5*a* consist of rollers at A and B. Clearly, the constraints provided by these supports are not sufficient to keep the truss from moving. While any vertical motion is prevented, the truss is free to move horizontally. The truss is said to be *partially constrained.*† Turning our attention to Fig. 4.5*b*, we note that the reactions at A and B involve only *two unknowns*. Since three equations of equilibrium must still be satisfied, there are *fewer unknowns than equations,* and one of the equilibrium equations will not be satisfied. While the equations $\Sigma M_A = 0$ and $\Sigma M_B = 0$ can be satisfied by a proper choice of reactions at A and B, the equation $\Sigma F_x = 0$ will not be satisfied unless the sum of the horizontal components of the applied forces happens to be zero. We thus check that the equilibrium of the truss of Fig. 4.5 cannot be maintained under general loading conditions.

It appears from the above that if a rigid body is to be completely constrained and if the reactions at its supports are to be statically determinate, *there must be as many unknowns as there are equations of equilibrium.* When this condition is *not* satisfied, we may be sure that the rigid body is not completely constrained, or that the reactions at its supports are not statically determinate, or both.

We should note, however, that, while *necessary*, the above condition is *not sufficient*. In other words, the fact that the number of unknowns is equal to the number of equations is no guarantee that the body is completely constrained or that the reactions at its supports are statically deter-

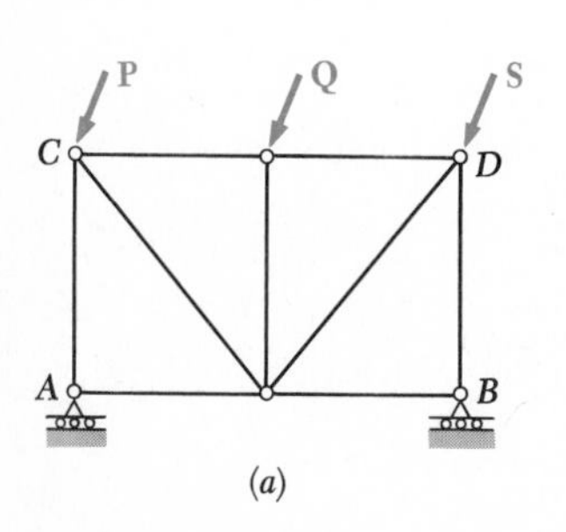

(a)

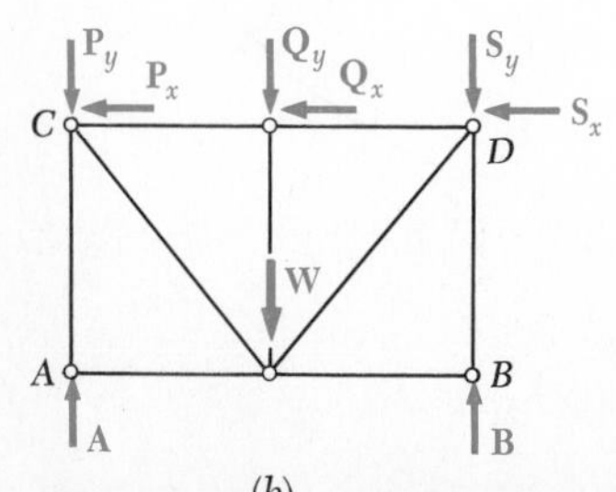

(b)

Fig. 4.5 Partial constraints.

† Partially constrained bodies are often referred to as *unstable*. However, to avoid confusion between this type of instability, due to insufficient constraints, and the type of instability considered in Chap. 10, which relates to the behavior of a rigid body when its equilibrium is disturbed, we shall restrict the use of the words *stable* and *unstable* to the latter case.

minate. Consider the truss shown in Fig. 4.6*a*, which is held by rollers at *A*, *B*, and *E*. While there are three unknown reactions, **A**, **B**, and **E** (Fig. 4.6*b*), the equation $\Sigma F_x = 0$ will not be satisfied unless the sum of the horizontal components of the applied forces happens to be zero. There is a sufficient number of constraints, but these constraints are not properly arranged, and the truss is free to move horizontally. We say that the truss is *improperly constrained*. Since only two equilibrium equations are left for determining three unknowns, the reactions will be statically indeterminate. Thus, improper constraints also produce static indeterminacy.

Another example of improper constraints—and of static indeterminacy—is provided by the truss shown in Fig. 4.7. This truss is held by a pin at *A* and by rollers at *B* and *C*, which altogether involve four unknowns. Since only three independent equilibrium equations are available, the reactions at the supports are statically indeterminate. On the other hand, we note that the equation $\Sigma M_A = 0$ cannot be satisfied under general loading conditions, since the lines of action of the reactions **B** and **C** are required to pass through *A*. We conclude that the truss can rotate about *A* and that it is improperly constrained.†

The examples of Figs. 4.6 and 4.7 lead us to conclude that *a rigid body is improperly constrained whenever the supports*, even though they may provide a sufficient number of reactions, *are arranged in such a way that the reactions must be either concurrent or parallel.*‡

In summary, to be sure that a two-dimensional rigid body is completely constrained and that the reactions at its supports are statically determinate, we should check that the reactions involve three—and only three—unknowns, and that the supports are arranged in such a way that they do not require the reactions to be either concurrent or parallel.

Supports involving statically indeterminate reactions should be used with care in the *design* of structures, and only with a full knowledge of the problems they may cause. On the other hand, the *analysis* of structures possessing statically indeterminate reactions often may be partially carried out by the methods of statics. In the case of the truss of Fig. 4.4, for example, the vertical components of the reactions at *A* and *B* were obtained from the equilibrium equations.

For obvious reasons, supports producing partial or improper constraints should be avoided in the design of stationary structures. However, a partially or improperly constrained structure will not necessarily collapse; under particular loading conditions, equilibrium may be maintained. For example, the trusses of Figs. 4.5 and 4.6 will be in equilibrium if the applied forces **P**, **Q**, and **S** are vertical. Besides, structures which are designed to move *should* be only partially constrained. A railroad car, for instance, would be of little use if it were completely constrained by having its brakes applied permanently.

† Rotation of the truss about *A* requires some "play" in the supports at *B* and *C*. In practice such play will always exist. Besides, we may note that if the play is kept small, the displacements of the rollers *B* and *C* and, thus, the distances from *A* to the lines of action of the reactions **B** and **C**, will also be small. The equation $\Sigma M_A = 0$ will then require that **B** and **C** be very large, a situation which may well result in the failure of the supports at *B* and *C*.

‡ Because this situation arises from an inadequate arrangement or *geometry* of the supports, it is often referred to as *geometric instability*.

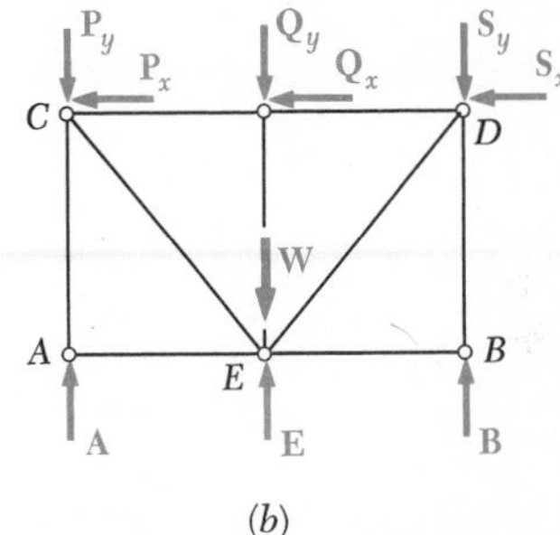

(*a*)

(*b*)

Fig. 4.6 Improper constraints.

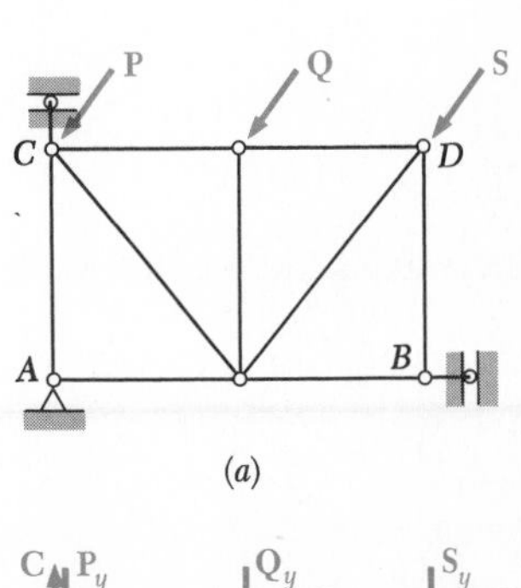

(*a*)

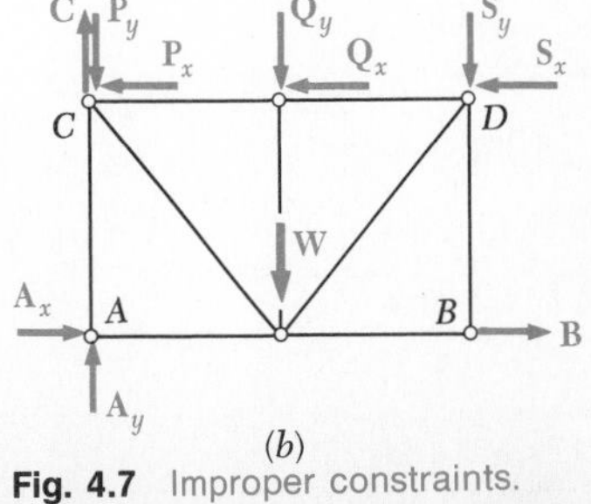

(*b*)

Fig. 4.7 Improper constraints.

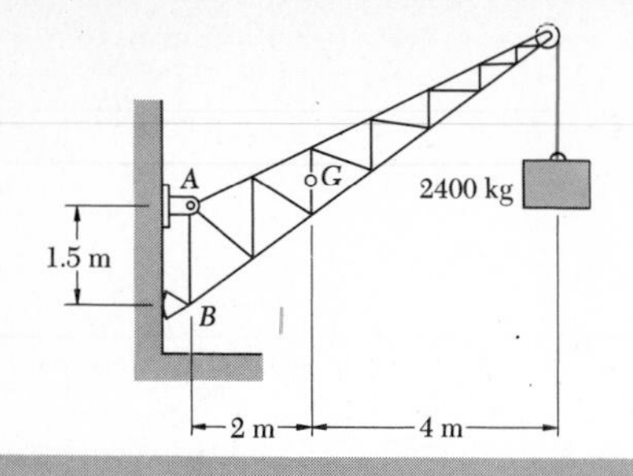

SAMPLE PROBLEM 4.1

A fixed crane has a mass of 1000 kg and is used to lift a 2400-kg crate. It is held in place by a pin at A and a rocker at B. The center of gravity of the crane is located at G. Determine the components of the reactions at A and B.

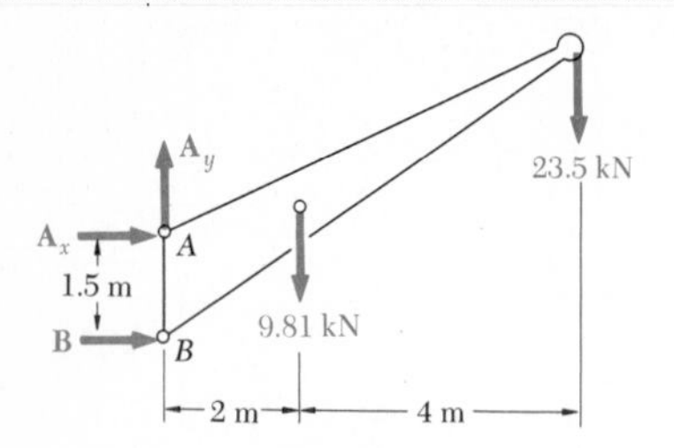

Solution. A free-body diagram of the crane is drawn. Multiplying the masses of the crane and of the crate by $g = 9.81 \text{ m/s}^2$, we obtain the corresponding weights, that is, 9810 N or 9.81 kN, and 23 500 N or 23.5 kN. The reaction at pin A is a force of unknown direction, represented by its components $\mathbf{A}_x$ and $\mathbf{A}_y$. The reaction at the rocker B is perpendicular to the rocker surface; thus it is horizontal. We assume that $\mathbf{A}_x$, $\mathbf{A}_y$, and $\mathbf{B}$ act in the directions shown.

Determination of **B**. We express that the sum of the moments of all external forces about point A is zero. The equation obtained will contain neither A_x nor A_y since the moments of $\mathbf{A}_x$ and $\mathbf{A}_y$ about A are zero. Multiplying the magnitude of each force by its perpendicular distance from B, we write

$$+\circlearrowleft \Sigma M_A = 0: \qquad +B(1.5 \text{ m}) - (9.81 \text{ kN})(2 \text{ m}) - (23.5 \text{ kN})(6 \text{ m}) = 0$$
$$B = +107.1 \text{ kN} \qquad \mathbf{B} = 107.1 \text{ kN} \rightarrow \blacktriangleleft$$

Since the result is positive, the reaction is directed as assumed.

Determination of $\mathbf{A}_x$. The magnitude A_x is determined by expressing that the sum of the horizontal components of all external forces is zero.

$$\xrightarrow{+} \Sigma F_x = 0: \qquad A_x + B = 0$$
$$A_x + 107.1 \text{ kN} = 0$$
$$A_x = -107.1 \text{ kN} \qquad \mathbf{A}_x = 107.1 \text{ kN} \leftarrow \blacktriangleleft$$

Since the result is negative, the sense of $\mathbf{A}_x$ is opposite to that assumed originally.

Determination of $\mathbf{A}_y$. The sum of the vertical components must also equal zero.

$$+\uparrow \Sigma F_y = 0: \qquad A_y - 9.81 \text{ kN} - 23.5 \text{ kN} = 0$$
$$A_y = +33.3 \text{ kN} \qquad \mathbf{A}_y = 33.3 \text{ kN} \uparrow \blacktriangleleft$$

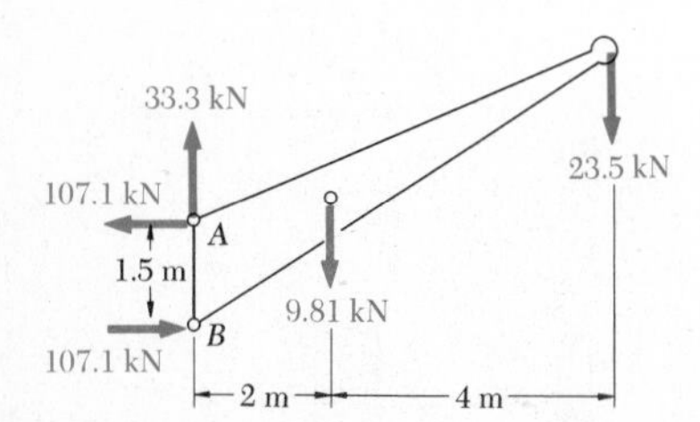

Adding vectorially the components $\mathbf{A}_x$ and $\mathbf{A}_y$, we find that the reaction at A is 112.2 kN $\nwarrow$17.3°.

Check. The values obtained for the reactions may be checked by recalling that the sum of the moments of all external forces about any point must be zero. For example, considering point B, we write

$$+\circlearrowleft \Sigma M_B = -(9.81 \text{ kN})(2 \text{ m}) - (23.5 \text{ kN})(6 \text{ m}) + (107.1 \text{ kN})(1.5 \text{ m}) = 0$$

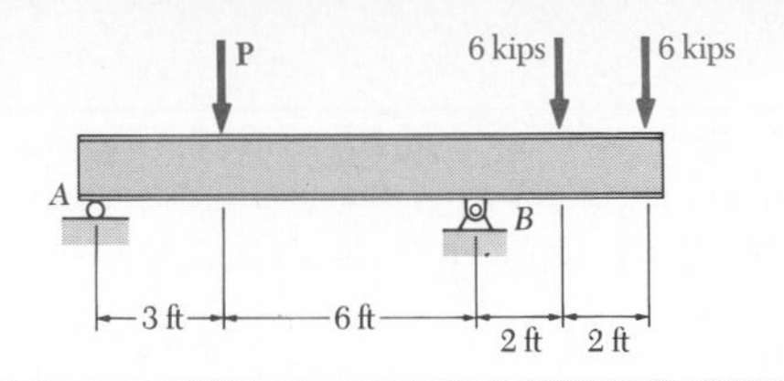

SAMPLE PROBLEM 4.2

Three loads are applied to a beam as shown. The beam is supported by a roller at A and by a pin at B. Neglecting the weight of the beam, determine the reactions at A and B when $P = 15$ kips.

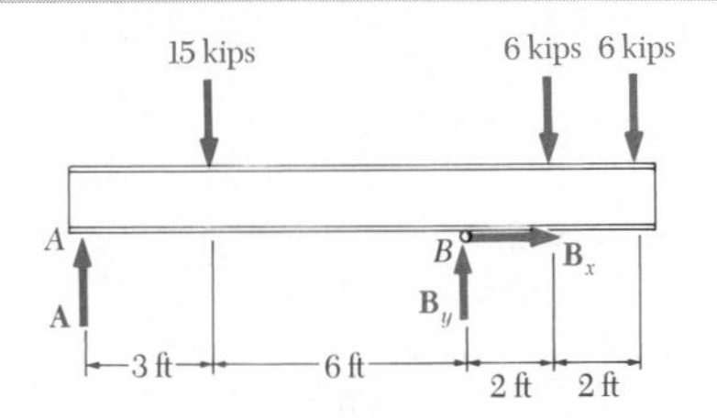

Solution. A free-body diagram of the beam is drawn. The reaction at A is vertical and is denoted by **A**. The reaction at B is represented by components $\mathbf{B}_x$ and $\mathbf{B}_y$. Each component is assumed to act in the direction shown.

Equilibrium Equations. We write the following three equilibrium equations and solve for the reactions indicated:

$$\xrightarrow{+} \Sigma F_x = 0: \qquad B_x = 0 \qquad \mathbf{B}_x = 0 \quad ◀$$

$$+\circlearrowleft \Sigma M_A = 0:$$
$$-(15 \text{ kips})(3 \text{ ft}) + B_y(9 \text{ ft}) - (6 \text{ kips})(11 \text{ ft}) - (6 \text{ kips})(13 \text{ ft}) = 0$$
$$B_y = +21.0 \text{ kips} \qquad \mathbf{B}_y = 21.0 \text{ kips} \uparrow \quad ◀$$

$$+\circlearrowleft \Sigma M_B = 0:$$
$$-A(9 \text{ ft}) + (15 \text{ kips})(6 \text{ ft}) - (6 \text{ kips})(2 \text{ ft}) - (6 \text{ kips})(4 \text{ ft}) = 0$$
$$A = +6.00 \text{ kips} \qquad \mathbf{A} = 6.00 \text{ kips} \uparrow \quad ◀$$

Check. The results are checked by adding the vertical components of all the external forces:

$$+\uparrow \Sigma F_y = +6.00 \text{ kips} - 15 \text{ kips} + 21.0 \text{ kips} - 6 \text{ kips} - 6 \text{ kips} = 0$$

Remark. In this problem the reactions at both A and B are vertical; however, these reactions are vertical for different reasons. At A, the beam is supported by a roller; hence the reaction cannot have any horizontal component. At B, the horizontal component of the reaction is zero because it must satisfy the equilibrium equation $\Sigma F_x = 0$ and because none of the other forces acting on the beam has a horizontal component.

We could have noticed at first glance that the reaction at B was vertical and dispensed with the horizontal component $\mathbf{B}_x$. This, however, is a bad practice. In following it, we would run the risk of forgetting the component $\mathbf{B}_x$ when the loading conditions require such a component (i.e., when a horizontal load is included). Also, the component $\mathbf{B}_x$ was found to be zero by using and solving an equilibrium equation, $\Sigma F_x = 0$. By setting $\mathbf{B}_x$ equal to zero immediately, we might not realize that we actually make use of this equation and thus might lose track of the number of equations available for solving the problem.

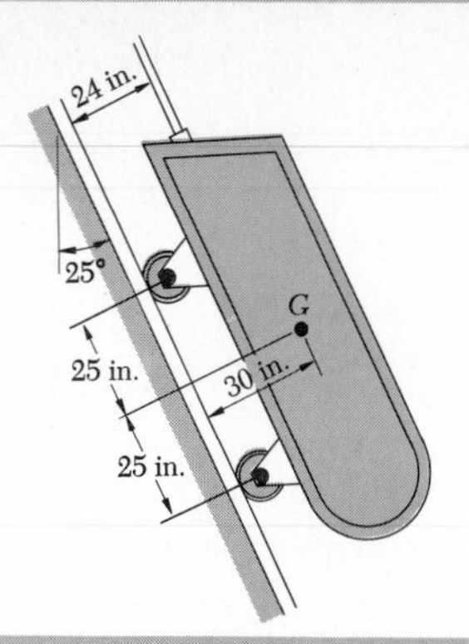

SAMPLE PROBLEM 4.3

A loading car is at rest on a track forming an angle of 25° with the vertical. The gross weight of the car and its load is 5500 lb, and it is applied at a point 30 in. from the track, halfway between the two axles. The car is held by a cable attached 24 in. from the track. Determine the tension in the cable and the reaction at each pair of wheels.

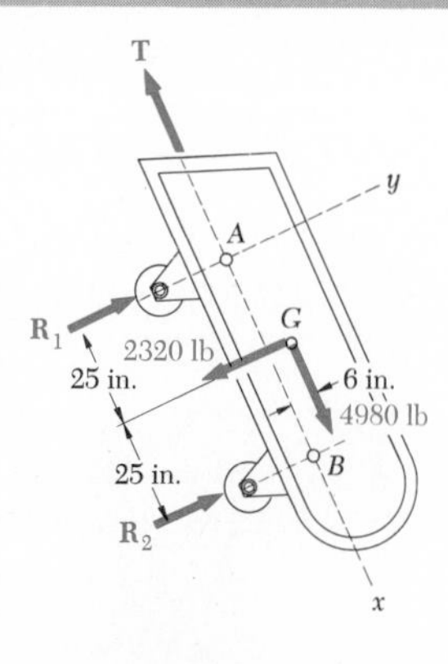

Solution. A free-body diagram of the car is drawn. The reaction at each wheel is perpendicular to the track, and the tension force **T** is parallel to the track. For convenience, we choose the x axis parallel to the track and the y axis perpendicular to the track. The 5500-lb weight is then resolved into x and y components.

$$W_x = +(5500 \text{ lb}) \cos 25° = +4980 \text{ lb}$$
$$W_y = -(5500 \text{ lb}) \sin 25° = -2320 \text{ lb}$$

Equilibrium Equations. We take moments about A to eliminate **T** and $\mathbf{R}_1$ from the computation.

$$+\circlearrowleft \Sigma M_A = 0: \quad -(2320 \text{ lb})(25 \text{ in.}) - (4980 \text{ lb})(6 \text{ in.}) + R_2(50 \text{ in.}) = 0$$
$$R_2 = +1758 \text{ lb} \qquad \mathbf{R}_2 = 1758 \text{ lb} \nearrow \blacktriangleleft$$

Now, taking moments about B to eliminate **T** and $\mathbf{R}_2$ from the computation, we write

$$+\circlearrowleft \Sigma M_B = 0: \quad (2320 \text{ lb})(25 \text{ in.}) - (4980 \text{ lb})(6 \text{ in.}) - R_1(50 \text{ in.}) = 0$$
$$R_1 = +562 \text{ lb} \qquad \mathbf{R}_1 = 562 \text{ lb} \nearrow \blacktriangleleft$$

The value of T is found by writing

$$\searrow +\Sigma F_x = 0: \quad +4980 \text{ lb} - T = 0$$
$$T = +4980 \text{ lb} \qquad \mathbf{T} = 4980 \text{ lb} \nwarrow \blacktriangleleft$$

The computed values of the reactions are shown in the adjacent sketch.

Check. The computations are verified by writing

$$\nearrow +\Sigma F_y = +562 \text{ lb} + 1758 \text{ lb} - 2320 \text{ lb} = 0$$

A check could also have been obtained by computing moments about any point except A or B.

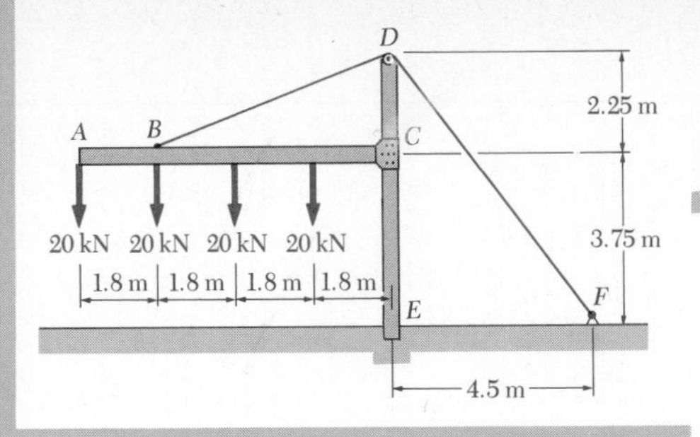

SAMPLE PROBLEM 4.4

The frame shown supports part of the roof of a small building. Knowing that the tension in the cable is 150 kN, determine the reaction at the fixed end E.

Solution. A free-body diagram of the frame and of the cable BDF is drawn. The reaction at the fixed end E is represented by the force components $\mathbf{E}_x$ and $\mathbf{E}_y$ and the couple $\mathbf{M}_E$. The other forces acting on the free body are the four 20-kN loads and the 150-kN force exerted at end F of the cable.

Equilibrium Equations. Noting that $DF = \sqrt{(4.5\text{ m})^2 + (6\text{ m})^2} = 7.5\text{ m}$, we write

$$\overset{+}{\rightarrow}\Sigma F_x = 0: \qquad E_x + \frac{4.5}{7.5}(150\text{ kN}) = 0$$

$$E_x = -90.0\text{ kN} \qquad \mathbf{E}_x = 90.0\text{ kN} \leftarrow \quad ◄$$

$$+\uparrow\Sigma F_y = 0: \qquad E_y - 4(20\text{ kN}) - \frac{6}{7.5}(150\text{ kN}) = 0$$

$$E_y = +200\text{ kN} \qquad \mathbf{E}_y = 200\text{ kN} \uparrow \quad ◄$$

$$+\circlearrowleft\Sigma M_E = 0: \qquad (20\text{ kN})(7.2\text{ m}) + (20\text{ kN})(5.4\text{ m}) + (20\text{ kN})(3.6\text{ m})$$

$$+ (20\text{ kN})(1.8\text{ m}) - \frac{6}{7.5}(150\text{ kN})(4.5\text{ m}) + M_E = 0$$

$$M_E = +180.0\text{ kN}\cdot\text{m} \qquad \mathbf{M}_E = 180.0\text{ kN}\cdot\text{m} \circlearrowleft \quad ◄$$

SAMPLE PROBLEM 4.5

A 400-lb weight is attached to the lever AO as shown. The constant of the spring BC is $k = 250$ lb/in., and the spring is unstretched when $\theta = 0$. Determine the position of equilibrium.

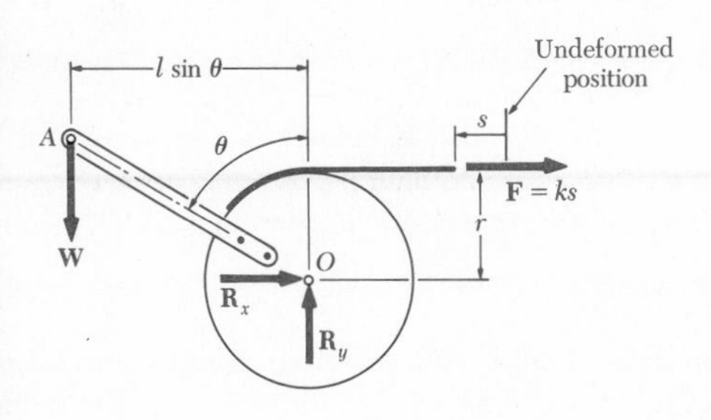

Solution. ***Force Exerted by Spring.*** Denoting by s the deflection of the spring from its undeformed position, and noting that $s = r\theta$, we write

$$F = ks = kr\theta$$

Equilibrium Equation. Summing the moments of $\mathbf{W}$ and $\mathbf{F}$ about O,

$$+\circlearrowleft\Sigma M_O = 0: \qquad Wl\sin\theta - r(kr\theta) = 0 \qquad \sin\theta = \frac{kr^2}{Wl}\theta$$

Substituting the given data, we obtain

$$\sin\theta = \frac{(250\text{ lb/in.})(3\text{ in.})^2}{(400\text{ lb})(8\text{ in.})}\theta \qquad \sin\theta = 0.703\theta$$

Solving by trial and error, we find $\qquad \theta = 0 \qquad \theta = 80.3° \quad ◄$

Problems

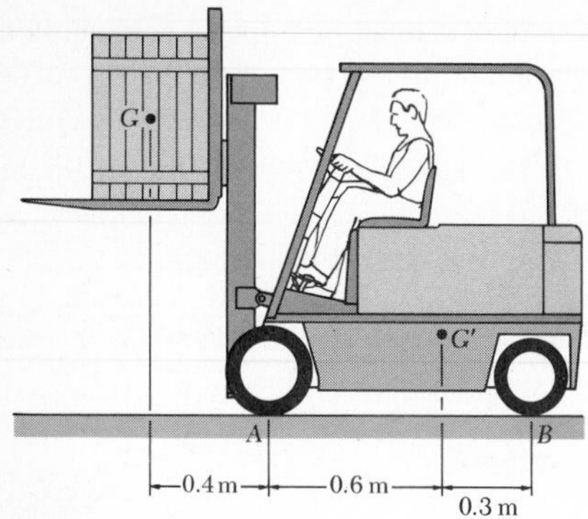

Fig. P4.1

4.1 A 2800-kg forklift truck is used to lift a 1500-kg crate. Determine the reaction at each of the two (*a*) front wheels *A*, (*b*) rear wheels *B*.

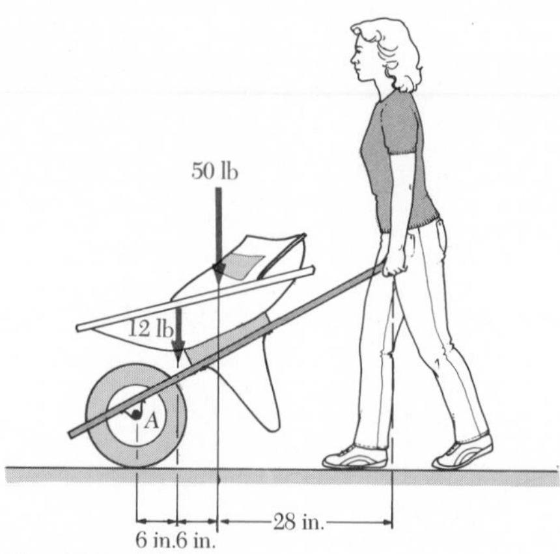

Fig. P4.2

4.2 A gardener uses a 12-lb wheelbarrow to transport a 50-lb bag of fertilizer. What force must she exert on each handle?

4.3 The gardener of Prob. 4.2 wishes to transport a second 50-lb bag of fertilizer at the same time as the first one. Determine the maximum allowable horizontal distance from the axle *A* of the wheelbarrow to the center of gravity of the second bag if she can hold only 15 lb with each arm.

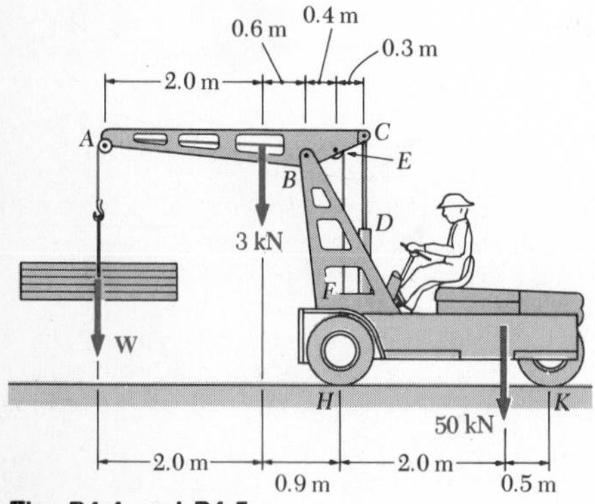

Fig. P4.4 and P4.5

4.4 A load of lumber of weight $W = 25$ kN is being raised by a mobile crane. The weight of the boom *ABC* and the combined weight of the truck and driver are as shown. Determine the reaction at each of the two (*a*) front wheels *H*, (*b*) rear wheels *K*.

4.5 A load of lumber of weight $W = 25$ kN is being raised as shown by a mobile crane. Knowing that the tension is 25 kN in all portions of cable *AEF* and that the weight of boom *ABC* is 3 kN, determine (*a*) the tension in rod *CD*, (*b*) the reaction at pin *B*.

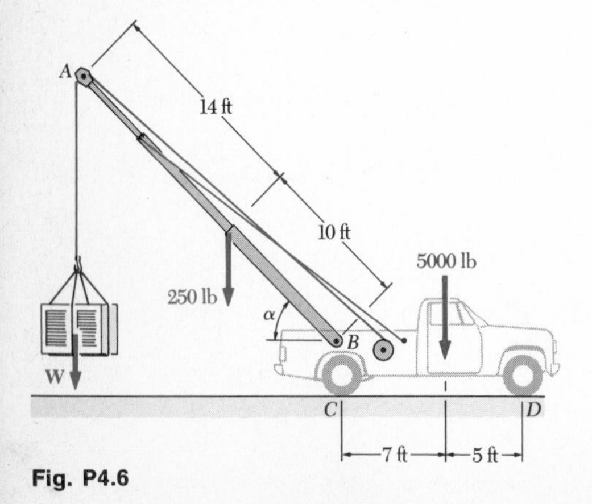

Fig. P4.6

4.6 A truck-mounted crane is used to lift a 750-lb compressor. The weights of the boom *AB* and of the truck are as shown, and the angle the boom forms with the horizontal is $\alpha = 40°$. Determine the reaction at each of the two (*a*) rear wheels *C*, (*b*) front wheels *D*.

4.7 For the truck-mounted crane of Prob. 4.6, determine the smallest allowable value of α if the truck is not to tip over when a 3000-lb load is lifted.

4.8 For the beam of Sample Prob. 4.2, determine the range of values of **P** for which the beam will be safe, knowing that the maximum allowable value for each of the reactions is 25 kips and that the reaction at A must be directed upward.

4.9 Three loads are applied as shown to a light beam supported by cables attached at B and D. Neglecting the weight of the beam, determine the range of values of Q for which neither cable becomes slack when $P = 0$.

4.10 Three loads are applied as shown to a light beam supported by cables attached at B and D. Knowing that the maximum allowable tension in each cable is 12 kN and neglecting the weight of the beam, determine the range of values of Q for which the loading is safe when $P = 0$.

4.11 For the beam of Prob. 4.10, determine the range of values of Q for which the loading is safe when $P = 5$ kN.

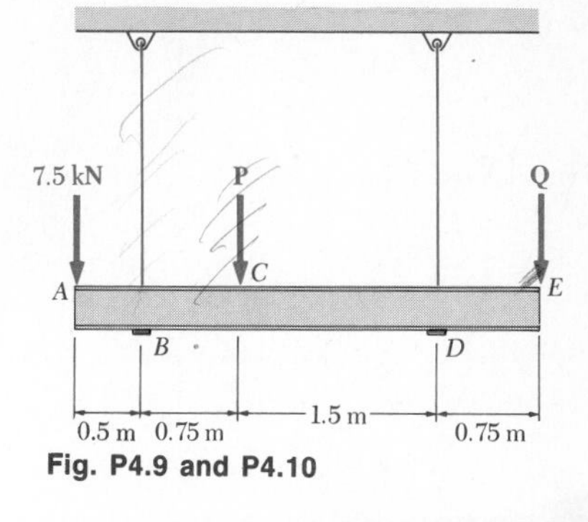

Fig. P4.9 and P4.10

4.12 Determine the maximum load which may be raised by the mobile crane of Prob. 4.4 without tipping over, knowing that the largest force which can be exerted by the hydraulic cylinder D is 100 kN, and that the maximum allowable tension in cable AEF is 35 kN.

4.13 Determine the tension in cable ABD and the reaction at the support C.

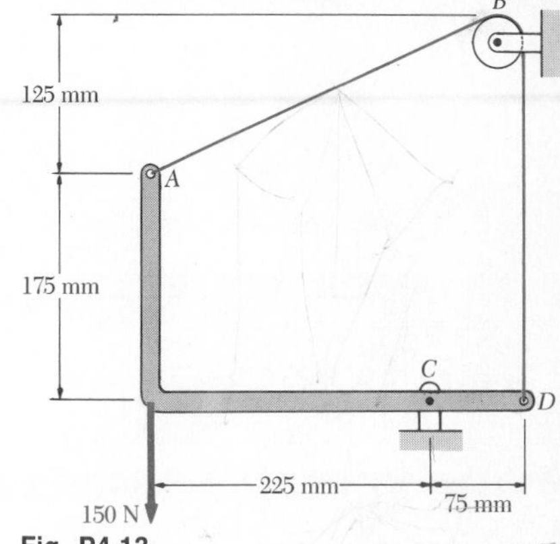

Fig. P4.13

4.14 Two links AB and DE are connected by a bell crank as shown. Knowing that the tension in link AB is 180 lb, determine (*a*) the tension in link DE, (*b*) the reaction at C.

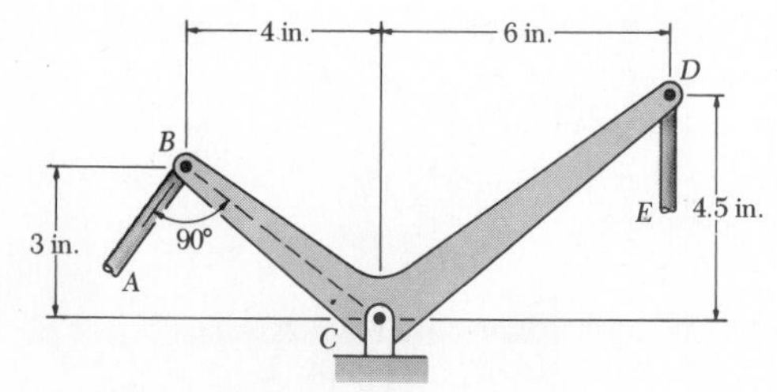

Fig. P4.14 and P4.15

4.15 Two links AB and DE are connected by a bell crank as shown. Determine the maximum force which may be safely exerted by link AB on the bell crank if the maximum allowable value for the reaction at C is 400 lb.

4.16 The lever AB is hinged at C and attached to a control cable at A. If the lever is subjected at B to a 500-N horizontal force, determine (*a*) the tension in the cable, (*b*) the reaction at C.

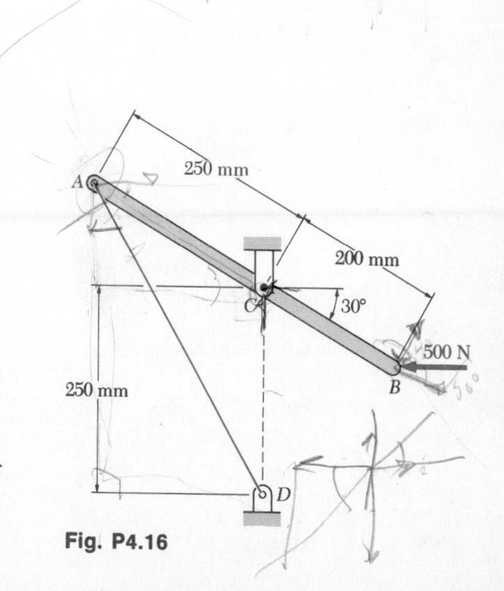

Fig. P4.16

4.17 A truss may be supported in three different ways as shown. In each case, determine the reactions at the supports.

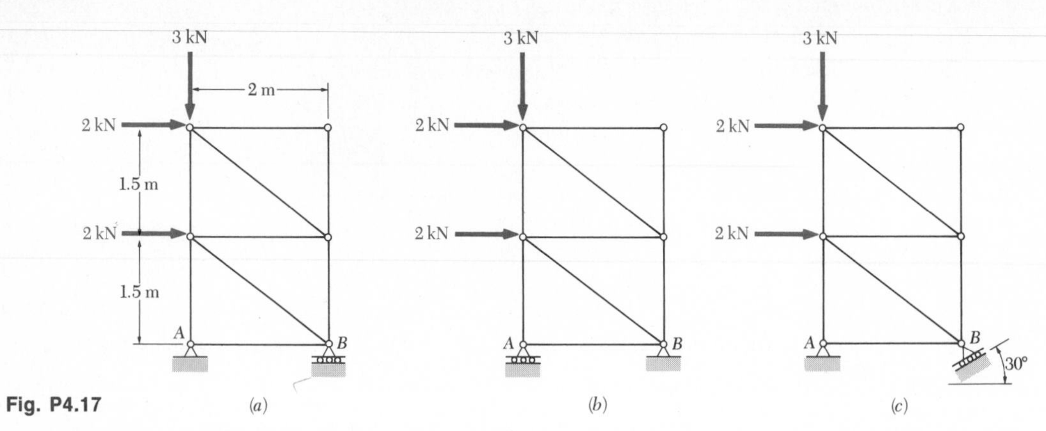

Fig. P4.17

4.18 Determine the reactions at A and B when (*a*) $\alpha = 0$, (*b*) $\alpha = 90°$, (*c*) $\alpha = 30°$.

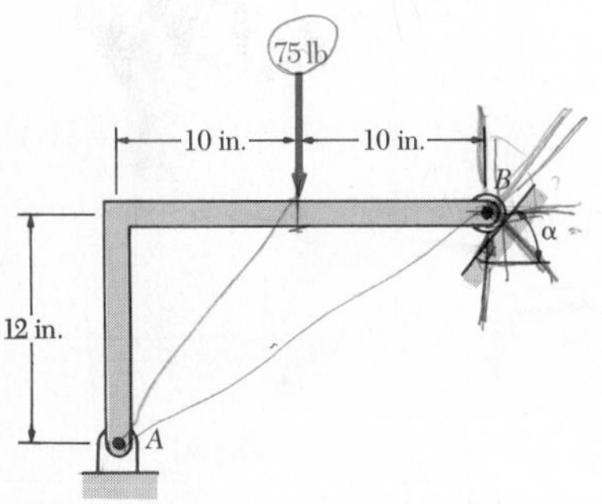

4.19 Solve Prob. 4.18, assuming that the 75-lb force has been replaced by a 750-lb · in. clockwise couple.

4.20 The mechanism shown is designed to measure the tension in a heavy paper tape used in a manufacturing process. Determine (*a*) the force exerted at E by the spring when the tension in the tape is 150 N, (*b*) the corresponding reaction at D.

75 lb
10 in.
10 in.
B
α
12 in.
A

Fig. P4.18

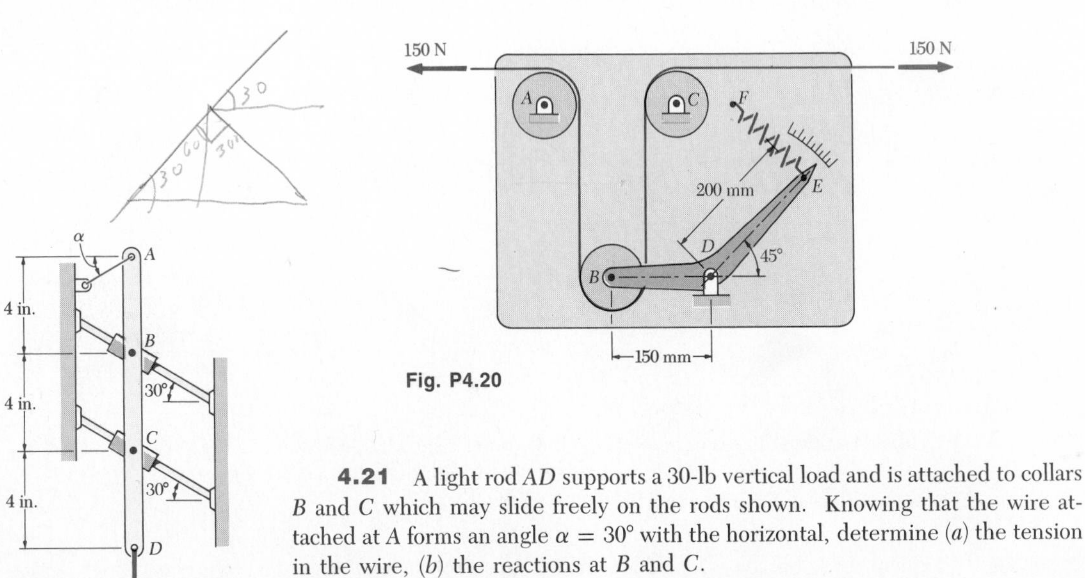

Fig. P4.21

Fig. P4.20

4.21 A light rod AD supports a 30-lb vertical load and is attached to collars B and C which may slide freely on the rods shown. Knowing that the wire attached at A forms an angle $\alpha = 30°$ with the horizontal, determine (*a*) the tension in the wire, (*b*) the reactions at B and C.

4.22 Solve Prob. 4.21, assuming that $\alpha = 0$.

4.23 A light bar AD is suspended from a cable BE and supports a 20-kg block at C. The extremities A and D of the bar are in contact with frictionless, vertical walls. Determine the tension in cable BE and the reactions at A and D.

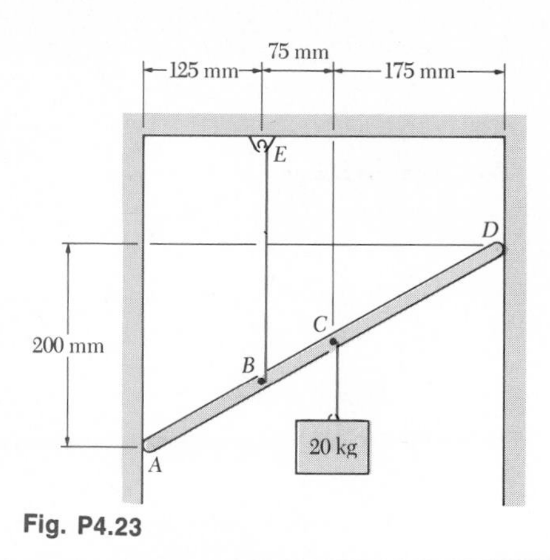

Fig. P4.23

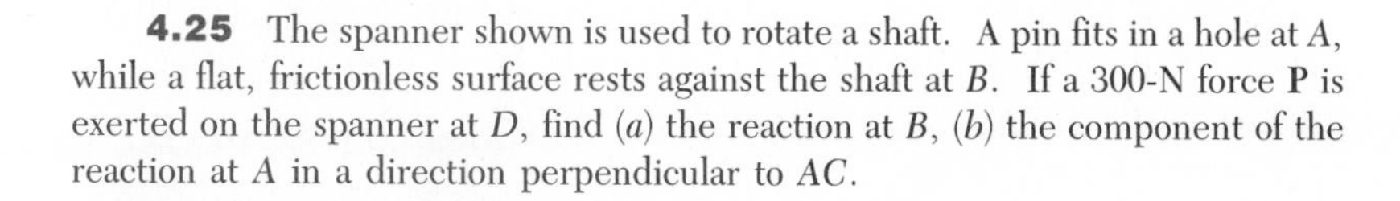

4.24 A movable bracket is held at rest by a cable attached at C and by frictionless rollers at A and B. For the loading shown, determine (*a*) the tension in the cable, (*b*) the reactions at A and B.

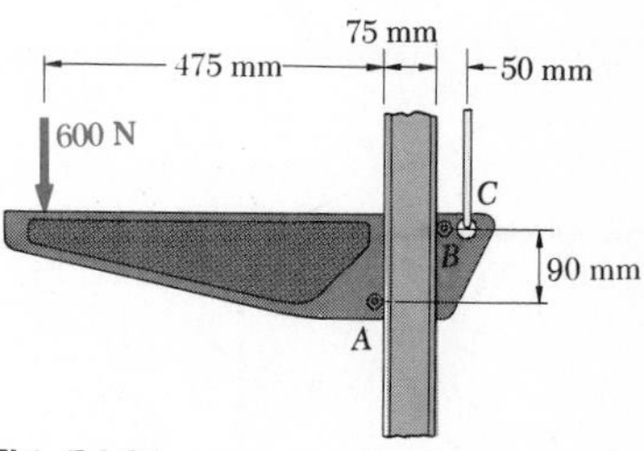

Fig. P4.24

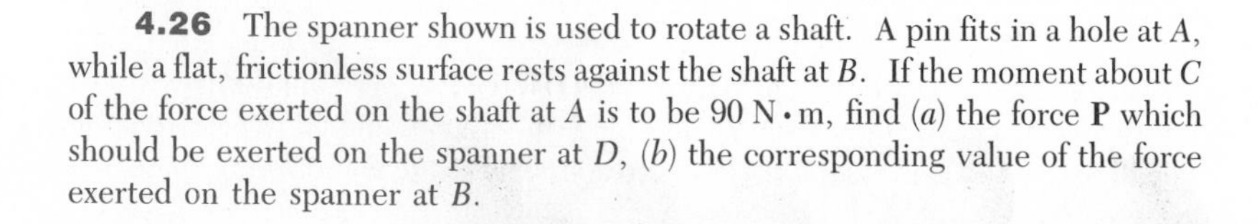

4.25 The spanner shown is used to rotate a shaft. A pin fits in a hole at A, while a flat, frictionless surface rests against the shaft at B. If a 300-N force **P** is exerted on the spanner at D, find (*a*) the reaction at B, (*b*) the component of the reaction at A in a direction perpendicular to AC.

4.26 The spanner shown is used to rotate a shaft. A pin fits in a hole at A, while a flat, frictionless surface rests against the shaft at B. If the moment about C of the force exerted on the shaft at A is to be 90 N · m, find (*a*) the force **P** which should be exerted on the spanner at D, (*b*) the corresponding value of the force exerted on the spanner at B.

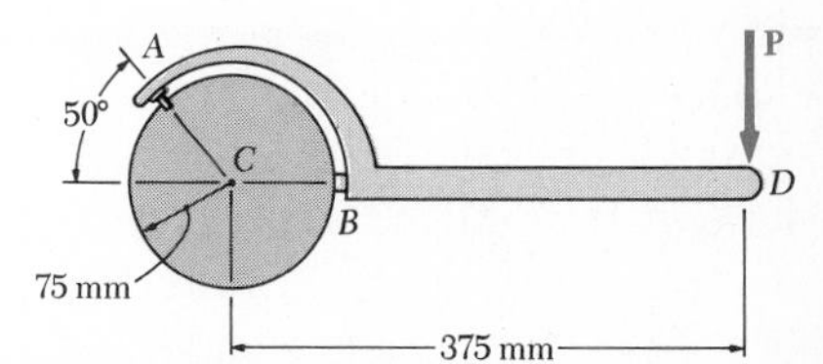

Fig. P4.25 and P4.26

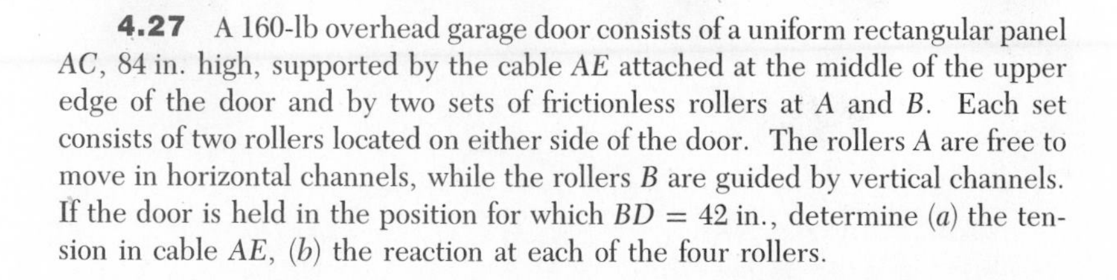

4.27 A 160-lb overhead garage door consists of a uniform rectangular panel AC, 84 in. high, supported by the cable AE attached at the middle of the upper edge of the door and by two sets of frictionless rollers at A and B. Each set consists of two rollers located on either side of the door. The rollers A are free to move in horizontal channels, while the rollers B are guided by vertical channels. If the door is held in the position for which $BD = 42$ in., determine (*a*) the tension in cable AE, (*b*) the reaction at each of the four rollers.

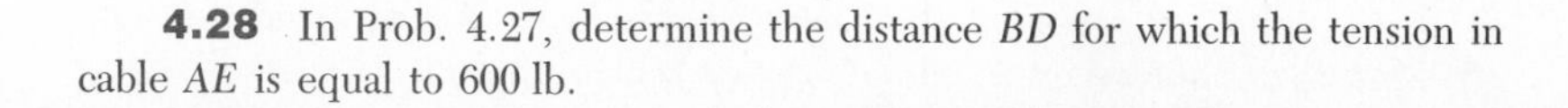

4.28 In Prob. 4.27, determine the distance BD for which the tension in cable AE is equal to 600 lb.

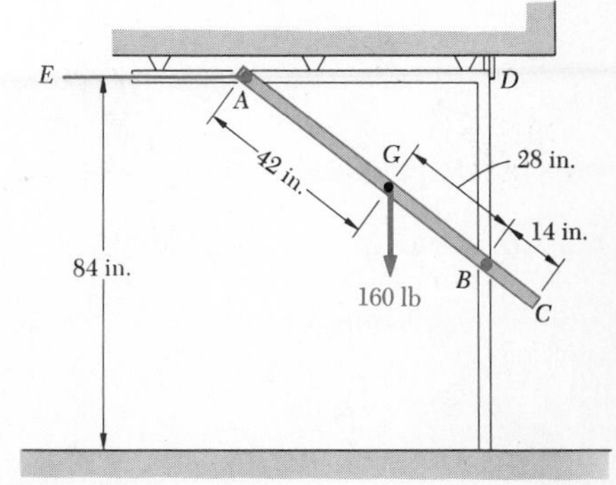

Fig. P4.27

4.29 The tensioner shown is used to maintain a given tension in the transmission belt CDE. The triangular receptacle contains a helical spring which exerts a couple $\mathbf{M}$ on arm AB, causing the 50-mm-radius idler pulley at B to press against the belt. Knowing that, in the position shown, the tension in the belt is $T = 200$ N, determine (a) the couple $\mathbf{M}$ exerted on arm AB, (b) the horizontal and vertical components of the force exerted at A on arm AB.

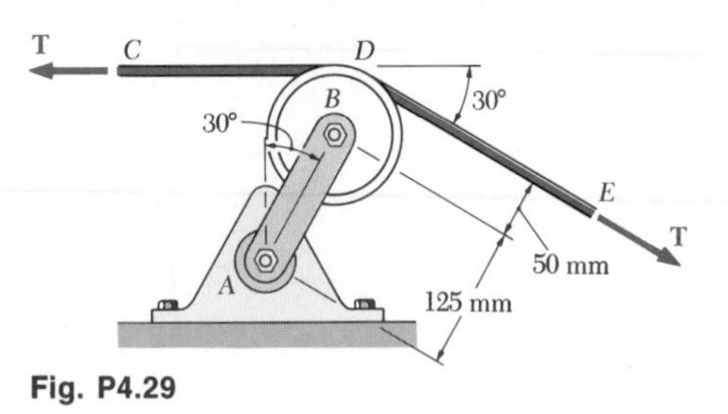

Fig. P4.29

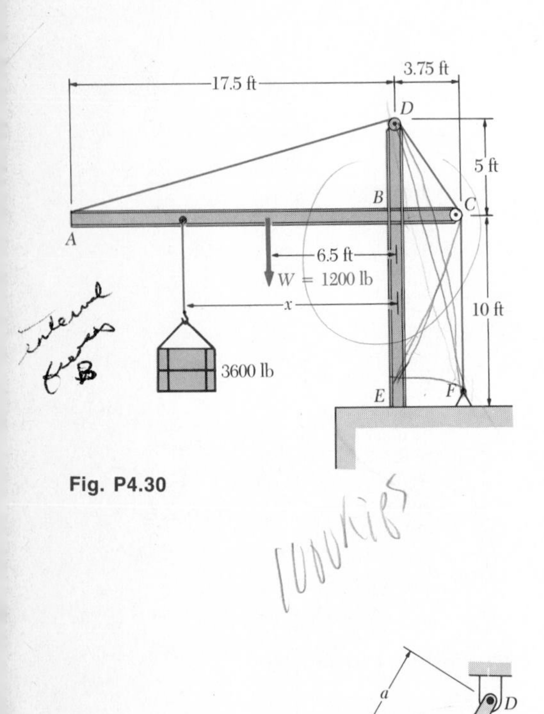

Fig. P4.30

4.30 The rig shown consists of a 1200-lb horizontal member ABC and a vertical member DBE welded together at B. The rig is being used to raise a 3600-lb crate at a distance $x = 12$ ft from the vertical member DBE. If the tension in the cable is 4 kips, determine the reaction at E, assuming that the cable is (a) anchored at F as shown in the figure, (b) attached to the vertical member at a point located 1 ft above E.

4.31 For the rig and crate of Prob. 4.30, and assuming that the cable is anchored at F as shown, determine (a) the required tension in cable $ADCF$ if the maximum value of the couple at E as x varies from 1.5 to 17.5 ft is to be as small as possible, (b) the corresponding maximum value of the couple.

4.32 For the frame of Sample Prob. 4.4, determine the range of allowable values of the tension in cable BDF if the magnitude of the couple at the fixed end E is not to exceed 150 kN · m.

4.33 Rod AD is acted upon by a vertical force $\mathbf{P}$ at end A and by two equal and opposite horizontal forces of magnitude Q at points B and C. Neglecting the weight of the rod, express the angle θ corresponding to the equilibrium position in terms of P and Q.

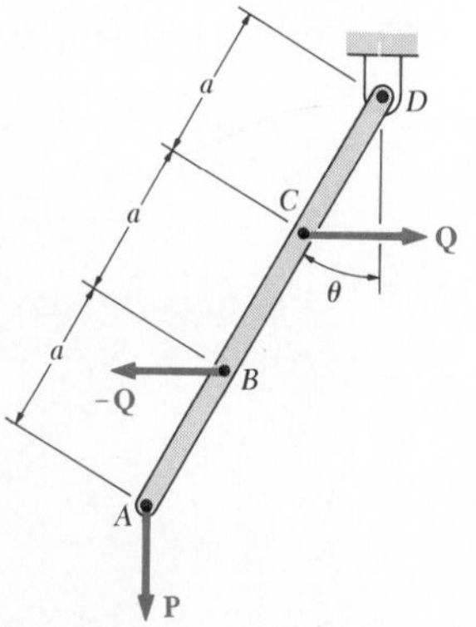

Fig. P4.33

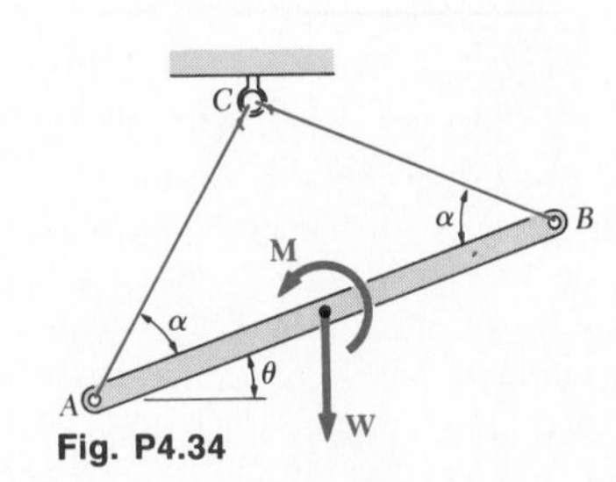

Fig. P4.34

4.34 A uniform rod AB of length l and weight W is suspended from two cords AC and BC of equal length. Determine the angle θ corresponding to the equilibrium position when a couple $\mathbf{M}$ is applied to the rod.

4.35 and 4.36 A vertical load **P** is applied at end B of rod BC. (*a*) Neglecting the weight of the rod, express the angle θ corresponding to the equilibrium position in terms of P, l, and the counterweight W. (*b*) Determine the value of θ corresponding to equilibrium if $P = 2W$.

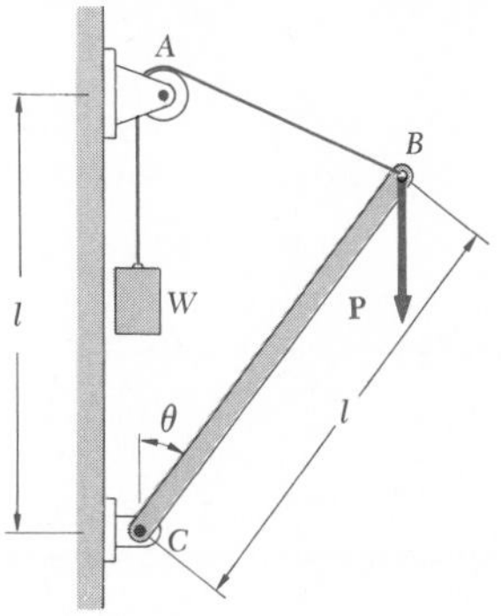

Fig. P4.35

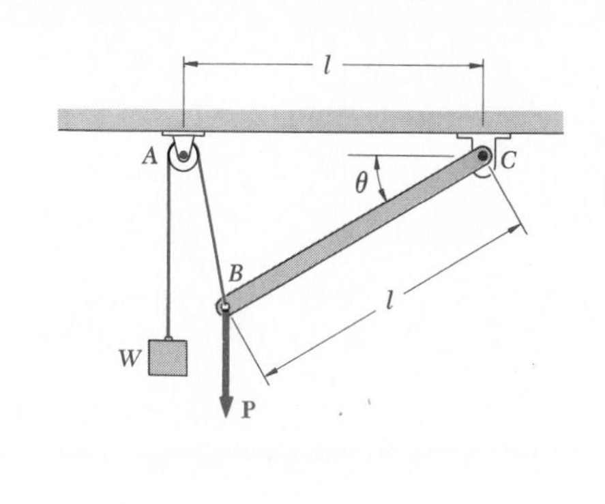

Fig. P4.36

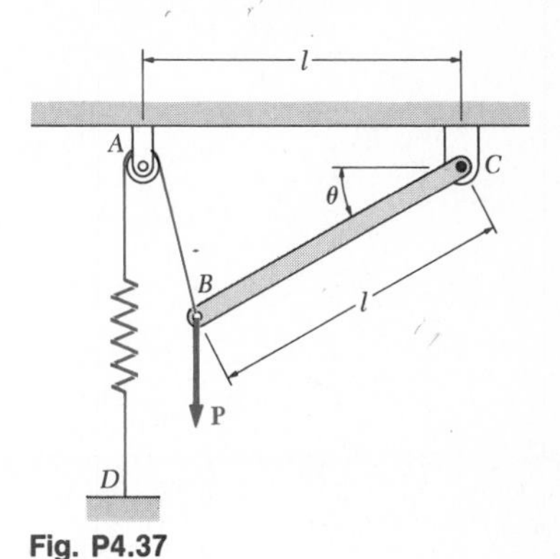

Fig. P4.37

4.37 A vertical load **P** is applied at end B of rod BC. The constant of the spring is k and the spring is unstretched when $\theta = 0$. (*a*) Neglecting the weight of the rod, express the angle θ corresponding to the equilibrium position in terms of P, k, and l. (*b*) Determine the value of θ corresponding to equilibrium if $P = 2kl$.

4.38 A force **P** of magnitude 80 lb is applied to end E of cable CDE which passes under pulley D and is attached to the mechanism at C. Neglecting the weight of the mechanism and the radius of the pulley, determine the value of θ corresponding to equilibrium. The constant of the spring is $k = 20$ lb/in., and the spring is unstretched when $\theta = 90°$. (*Hint.* Solve by trial and error.)

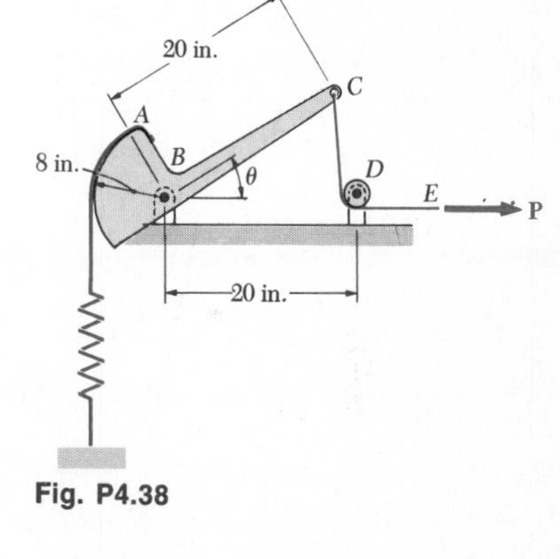

Fig. P4.38

4.39 A slender rod AB, of weight W, is attached to blocks A and B which may move freely in the guides shown. The constant of the spring is k and the spring is unstretched when AB is horizontal. (*a*) Neglecting the weight of the blocks, derive an equation in θ, W, l, and k which must be satisfied when the rod is in equilibrium. (*b*) Find the value of k, knowing that $W = 40$ N, $l = 0.8$ m, and that the rod is in equilibrium when $\theta = 36°$.

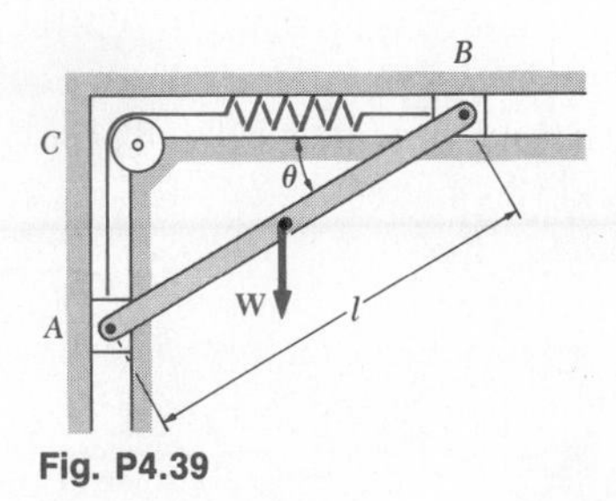

Fig. P4.39

4.40 The internal spring AC is of constant k and is undeformed when $\theta = 60°$. (*a*) Derive an equation in θ, W, a, and k which must be satisfied when the system is in equilibrium. (*b*) Find the value of k, knowing that $W = 80$ N, $a = 300$ mm, and that the system is in equilibrium when $\theta = 25°$.

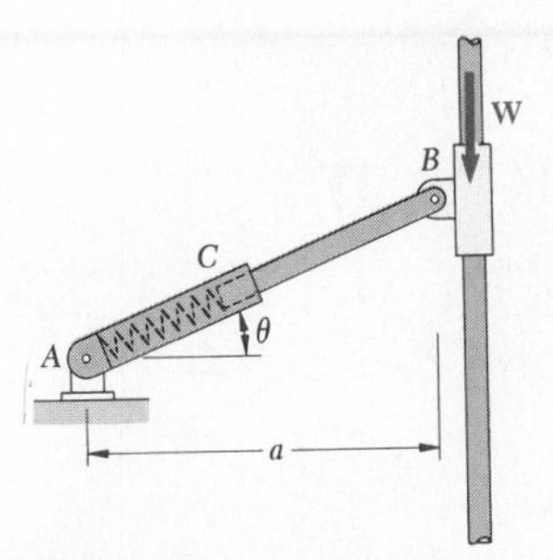

Fig. P4.40

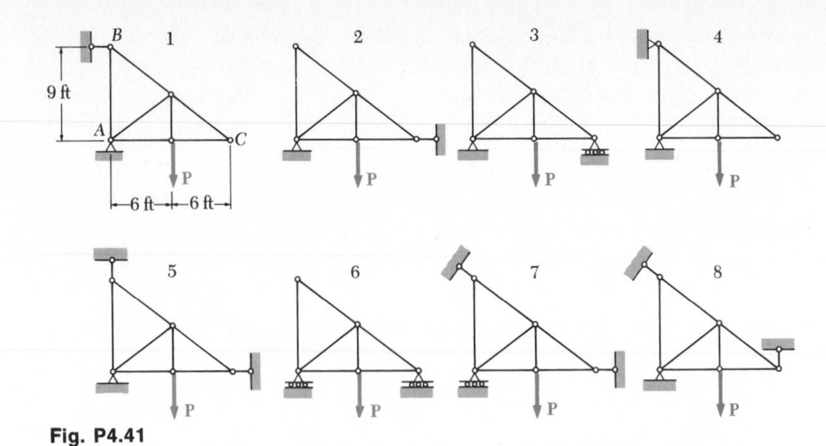

Fig. P4.41

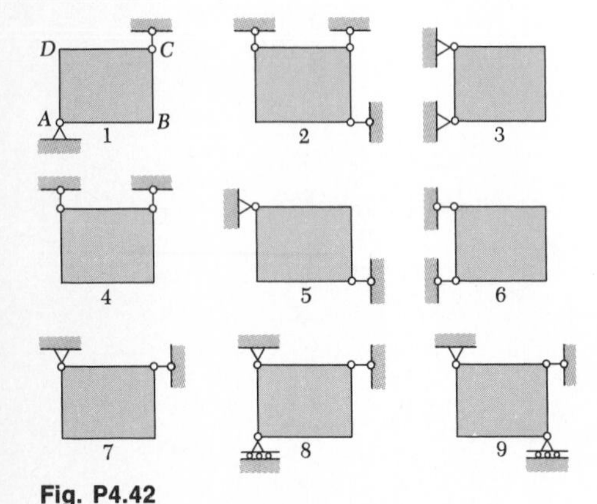

Fig. P4.42

4.41 A truss may be supported in eight different ways as shown. All connections consist of frictionless pins, rollers, and short links. In each case, determine whether (*a*) the truss is completely, partially, or improperly constrained, (*b*) the reactions are statically determinate or indeterminate, (*c*) the equilibrium of the truss is maintained in the position shown. Also, wherever possible, compute the reactions, assuming that the magnitude of the force **P** is 12 kips.

4.42 Nine identical rectangular plates, 500 × 750 mm, and each of mass $m = 40$ kg, are held in a vertical plane as shown. All connections consist of frictionless pins, rollers, or short links. For each case, answer the questions listed in Prob. 4.41, and, wherever possible, compute the reactions.

4.6. Equilibrium of a Two-Force Body. A particular case of equilibrium which is of considerable interest is that of a rigid body subjected to two forces. Such a body is commonly called a *two-force body.* We shall show that *if a two-force body is in equilibrium, the two forces must have the same magnitude, same line of action, and opposite sense.*

Consider a corner plate subjected to two forces $\mathbf{F}_1$ and $\mathbf{F}_2$ acting at A and B, respectively (Fig. 4.8*a*). If the plate is to be in equilibrium, the sum of the moments of $\mathbf{F}_1$ and $\mathbf{F}_2$ about any axis must be zero. First, we sum moments about A: Since the moment of $\mathbf{F}_1$ is obviously zero, the moment of $\mathbf{F}_2$ must also be zero and the line of action of $\mathbf{F}_2$ must pass through A (Fig. 4.8*b*). Summing moments about B, we prove similarly that the line of

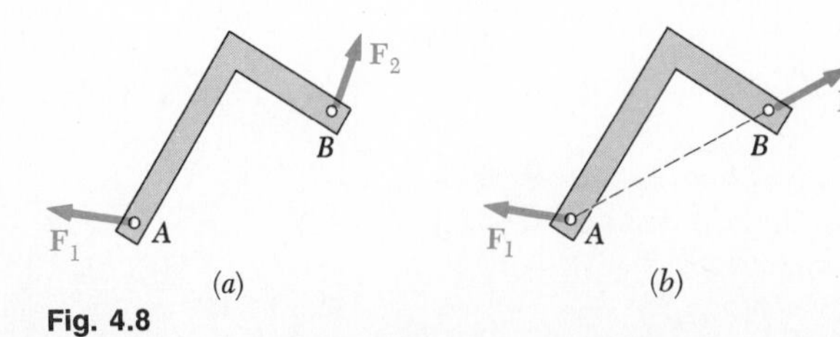

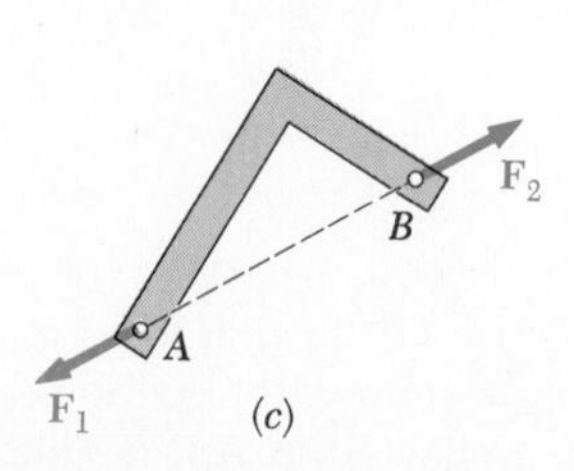

Fig. 4.8

action of $\mathbf{F}_1$ must pass through B (Fig. 4.8*c*). Both forces have the same line of action (line AB). From the equation $\Sigma F_x = 0$ or $\Sigma F_y = 0$, it is seen that they must have also the same magnitude but opposite sense.

If several forces act at two points A and B, the forces acting at A may be replaced by their resultant $\mathbf{F}_1$ and those acting at B by their resultant $\mathbf{F}_2$. Thus a two-force body may be more generally defined as *a rigid body subjected to forces acting at only two points*. The resultants $\mathbf{F}_1$ and $\mathbf{F}_2$ then must have the same line of action, same magnitude, and opposite sense.

In our study of structures, frames, and machines, we shall see that the recognition of two-force bodies will enable us to simplify the solution of certain problems.

4.7. Equilibrium of a Three-Force Body. Another case of equilibrium that is of great interest is that of a *three-force body*, i.e., a rigid body subjected to three forces or, more generally, *a rigid body subjected to forces acting at only three points*. Consider a rigid body subjected to a system of forces which may be reduced to three forces $\mathbf{F}_1$, $\mathbf{F}_2$, and $\mathbf{F}_3$ acting at A, B, and C, respectively (Fig. 4.9*a*). We shall show that if the body is in equilibrium, *the lines of action of the three forces must be either concurrent or parallel.*

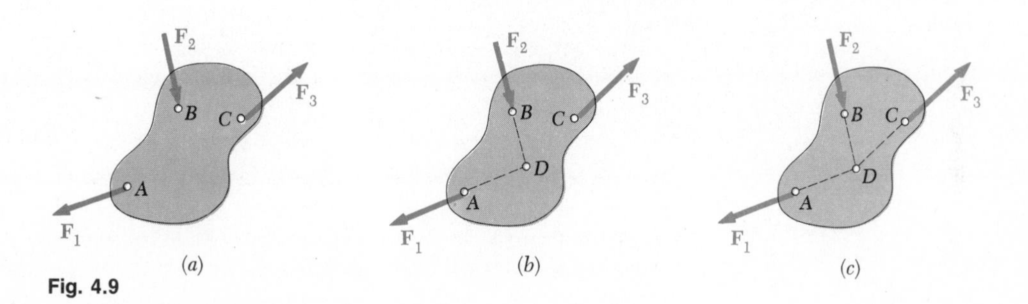

Fig. 4.9

Since the rigid body is in equilibrium, the sum of the moments of $\mathbf{F}_1$, $\mathbf{F}_2$, and $\mathbf{F}_3$ about any axis must be zero. Assuming that the lines of action of $\mathbf{F}_1$ and $\mathbf{F}_2$ intersect, and denoting their point of intersection by D, we sum moments about D (Fig. 4.9*b*); since the moments of $\mathbf{F}_1$ and $\mathbf{F}_2$ about D are zero, the moment of $\mathbf{F}_3$ about D must also be zero and the line of action of $\mathbf{F}_3$ must pass through D (Fig. 4.9*c*). The three lines of action are concurrent. The only exception occurs when none of the lines intersect; the lines of action are then parallel.

Although problems concerning three-force bodies may be solved by the general methods of Secs. 4.3 to 4.5, the property just established may be used to solve them either graphically, or mathematically from simple trigonometric or geometric relations.

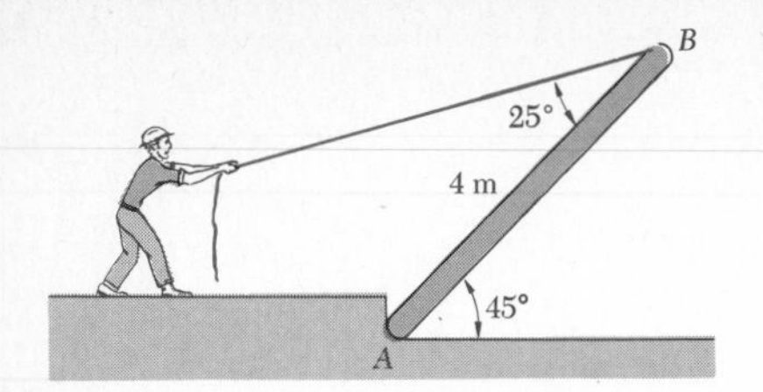

SAMPLE PROBLEM 4.6

A man raises a 10-kg joist, of length 4 m, by pulling on a rope. Find the tension T in the rope and the reaction at A.

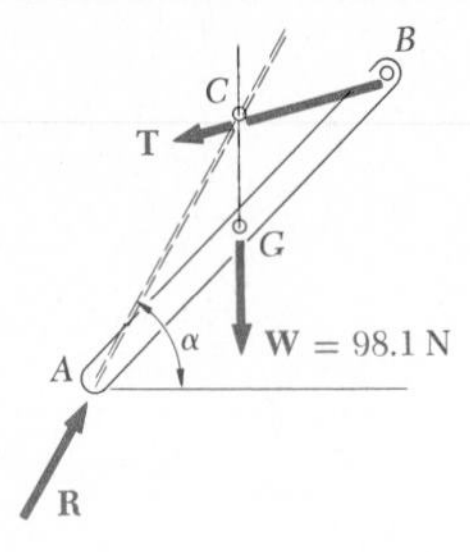

Solution. The joist is a three-force body since it is acted upon by three forces: its weight **W**, the force **T** exerted by the rope, and the reaction **R** of the ground at A. We note that

$$W = mg = (10 \text{ kg})(9.81 \text{ m/s}^2) = 98.1 \text{ N}$$

Three-Force Body. Since the joist is a three-force body, the forces acting on it must be concurrent. The reaction **R**, therefore, will pass through the point of intersection C of the lines of action of the weight **W** and of the tension force **T**. This fact will be used to determine the angle α that **R** forms with the horizontal.

Drawing the vertical BF through B and the horizontal CD through C, we note that

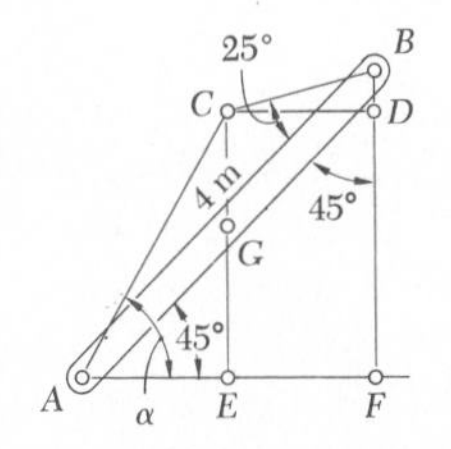

$$\begin{aligned}
AF &= BF = (AB)\cos 45° = (4 \text{ m})\cos 45° = 2.828 \text{ m} \\
CD &= EF = AE = \tfrac{1}{2}(AF) = 1.414 \text{ m} \\
BD &= (CD)\cot(45° + 25°) = (1.414 \text{ m})\tan 20° = 0.515 \text{ m} \\
CE &= DF = BF - BD = 2.828 \text{ m} - 0.515 \text{ m} = 2.313 \text{ m}
\end{aligned}$$

We write

$$\tan \alpha = \frac{CE}{AE} = \frac{2.313 \text{ m}}{1.414 \text{ m}} = 1.636$$

$$\alpha = 58.6° \quad ◄$$

We now know the direction of all the forces acting on the joist.

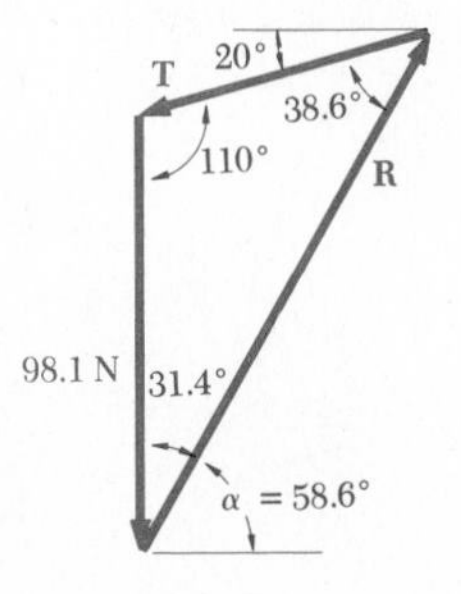

Force Triangle. A force triangle is drawn as shown, and its interior angles are computed from the known directions of the forces. Using the law of sines, we write

$$\frac{T}{\sin 31.4°} = \frac{R}{\sin 110°} = \frac{98.1 \text{ N}}{\sin 38.6°}$$

$$T = 81.9 \text{ N} \quad ◄$$
$$\mathbf{R} = 147.8 \text{ N} \measuredangle 58.6° \quad ◄$$

Problems

4.43 The spanner shown is used to rotate a shaft. A pin fits in a hole at A, while a flat, frictionless surface rests against the shaft at B. If a 300-N force **P** is exerted on the spanner at D, find the reactions at A and B.

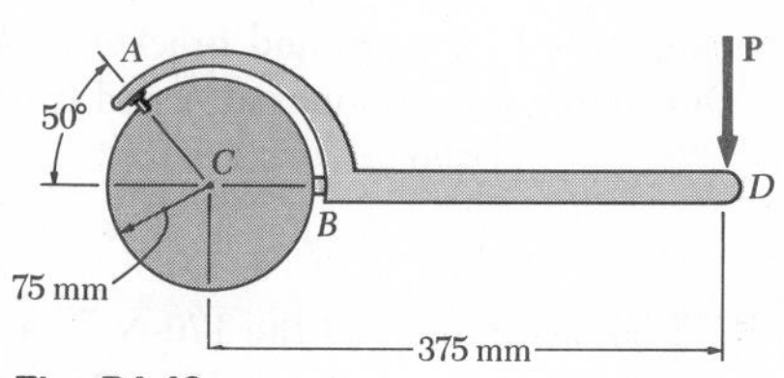

Fig. P4.43

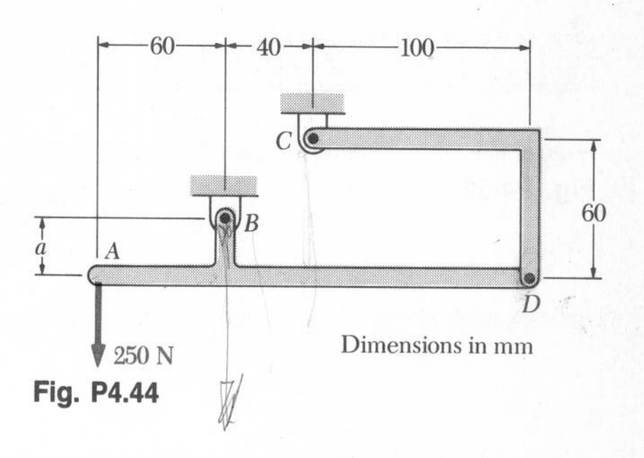

Fig. P4.44

4.44 Determine the reactions at B and C when $a = 30$ mm.

4.45 Using the method of Sec. 4.7, solve Prob. 4.18*b*.

4.46 Using the method of Sec. 4.7, solve Prob. 4.18*c*.

4.47 A T-shaped bracket supports a 300-N load as shown. Determine the reactions at A and C when (*a*) $\alpha = 90°$, (*b*) $\alpha = 45°$.

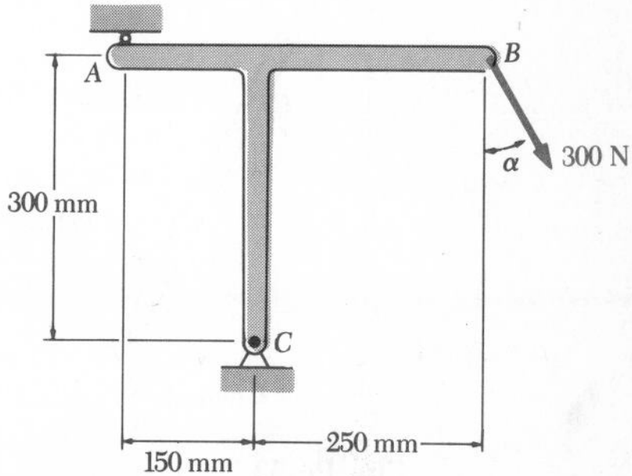

Fig. P4.47

4.48 One end of rod AB rests in the corner A and the other is attached to cord BD. If the rod supports at 200-N load at its midpoint C, find the reaction at A and the tension in the cord.

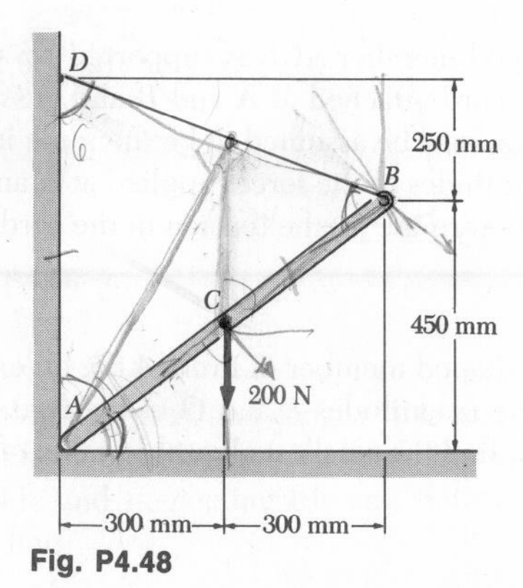

Fig. P4.48

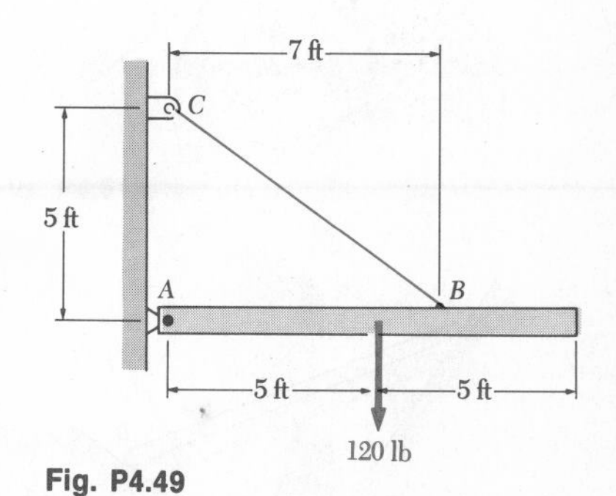

Fig. P4.49

4.49 A 10-ft wooden beam weighing 120 lb is supported by a pin and bracket at A and by cable BC. Find the reaction at A and the tension in the cable.

4.8. Reactions at Supports and Connections for a Three-Dimensional Structure. The reactions on a three-dimensional structure range from the single force of known direction exerted by a frictionless surface to the force-couple system exerted by a fixed support. Consequently, the number of unknowns associated with the reaction at a support or connection may vary from one to six in problems involving the equilibrium of a three-dimensional structure. Various types of supports and connections are shown in Fig. 4.10 with the corresponding reactions. A simple way of determining the type of reaction corresponding to a given support or connection and the number of unknowns involved is to find which of the six fundamental motions (translation in x, y, and z directions, rotation about the x, y, and z axes) are allowed and which motions are prevented.

Ball supports, frictionless surfaces, and cables, for example, prevent translation in one direction only and thus exert a single force of known line of action; they each involve one unknown, namely, the magnitude of the reaction. Rollers on rough surfaces and wheels on rails prevent translation in two directions; the corresponding reactions consist of two unknown force components. Rough surfaces in direct contact and ball-and-socket supports prevent translation in three directions; these supports involve three unknown force components.

Some supports and connections may prevent rotation as well as translation; the corresponding reactions include, then, couples as well as forces. The reaction at a fixed support, for example, which prevents any motion (rotation as well as translation), consists of three unknown forces and three unknown couples. A universal joint, which is designed to allow rotation about two axes, will exert a reaction consisting of three unknown force components and one unknown couple.

Other supports and connections are primarily intended to prevent translation; their design, however, is such that they also prevent some rotations. The corresponding reactions consist essentially of force components but *may* also include couples. One group of supports of this type includes hinges and bearings designed to support radial loads only (for example, journal bearings, roller bearings). The corresponding reactions consist of two force components but may also include two couples. Another group includes pin-and-bracket supports, hinges, and bearings designed to support an axial thrust as well as a radial load (for example, ball bearings). The corresponding reactions consist of three force components but may include two couples. However, these supports will not exert any appreciable couples under normal conditions of use. Therefore, *only* force components should be included in their analysis, *unless* it is found that couples are necessary to maintain the equilibrium of the rigid body, or unless the support is known to have been specifically designed to exert a couple (see Probs. 4.87 through 4.90).

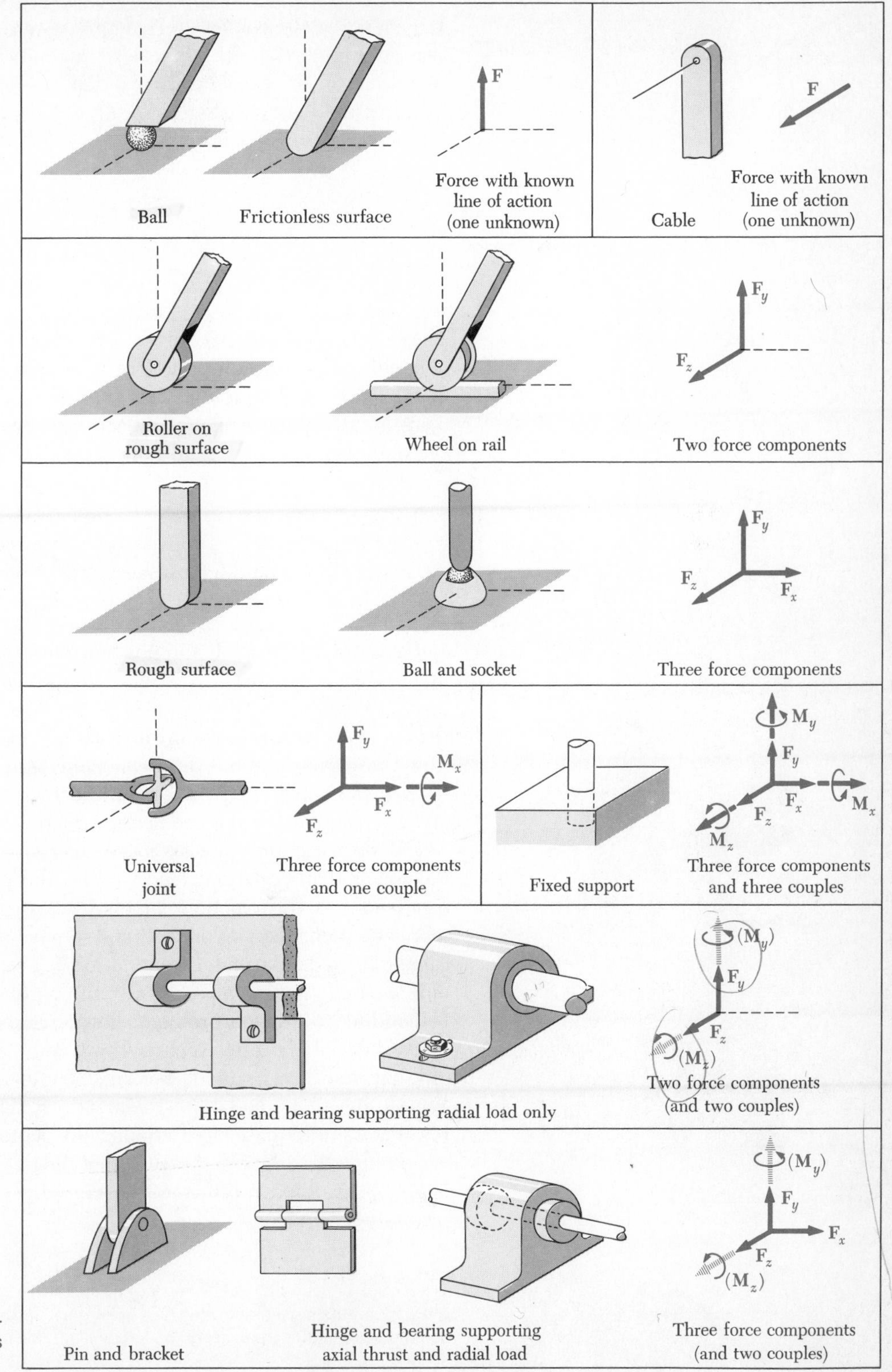

Fig. 4.10 Reactions at supports and connections.

4.9. Equilibrium of a Rigid Body in Three Dimensions.

We saw in Sec. 4.1 that six scalar equations are required to express the conditions for the equilibrium of a rigid body in the general three-dimensional case:

$$\Sigma F_x = 0 \qquad \Sigma F_y = 0 \qquad \Sigma F_z = 0 \tag{4.2}$$
$$\Sigma M_x = 0 \qquad \Sigma M_y = 0 \qquad \Sigma M_z = 0 \tag{4.3}$$

These equations may be solved for no more than *six unknowns*, which generally will represent reactions at supports or connections.

In most problems the scalar equations (4.2) and (4.3) will be more conveniently obtained if we first express in vector form the conditions for the equilibrium of the rigid body considered. We write

$$\Sigma \mathbf{F} = 0 \qquad \Sigma \mathbf{M}_O = \Sigma(\mathbf{r} \times \mathbf{F}) = 0 \tag{4.1}$$

and express the forces $\mathbf{F}$ and position vectors $\mathbf{r}$ in terms of scalar components and unit vectors. Next we compute all vector products, either by direct calculation or by means of determinants (see Sec. 3.8), observing that as many as three unknown reaction components may be eliminated from these computations through a judicious choice of point O. Equating to zero the coefficients of the unit vectors in each of the two relations (4.1), we obtain the desired scalar equations.†

If the reactions involve more than six unknowns, there are more unknowns than equations, and some of the reactions are *statically indeterminate.* If the reactions involve less than six unknowns, there are more equations than unknowns, and some of the equations of equilibrium cannot be satisfied under general loading conditions; the rigid body is only *partially constrained.* Under the particular loading conditions corresponding to a given problem, however, the extra equations often reduce to trivial identities such as $0 = 0$ and may be disregarded; although only partially constrained, the rigid body remains in equilibrium (see Sample Probs. 4.7 and 4.8). Even with six or more unknowns, it is possible that some equations of equilibrium will not be satisfied. This may occur when the supports are such that the reactions are forces which must either be parallel or intersect the same line; the rigid body is then *improperly constrained.*

† In some problems, it will be found convenient to eliminate the reactions at two points A and B from the solution by writing the equilibrium equation $\Sigma M_{AB} = 0$, which involves the determination of the moments of the forces about the axis AB joining points A and B (see Sample Prob. 4.10).

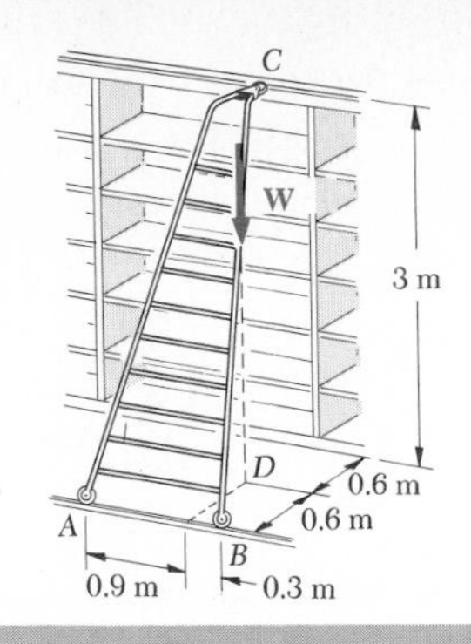

SAMPLE PROBLEM 4.7

A 20-kg ladder used to reach high shelves in a storeroom is supported by two flanged wheels A and B mounted on a rail and by an unflanged wheel C resting against a rail fixed to the wall. An 80-kg man stands on the ladder and leans to the right. The line of action of the combined weight **W** of the man and ladder intersects the floor at point D. Determine the reactions at A, B, and C.

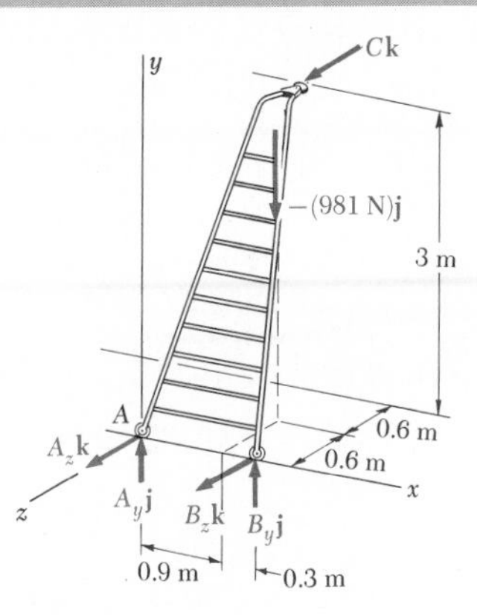

Solution. A free-body diagram of the ladder is drawn. The forces involved are the combined weight of the man and ladder,

$$\mathbf{W} = -mg\mathbf{j} = -(80\text{ kg} + 20\text{ kg})(9.81\text{ m/s}^2)\mathbf{j} = -(981\text{ N})\mathbf{j}$$

and five unknown reaction components, two at each flanged wheel and one at the unflanged wheel. The ladder is thus only partially constrained; it is free to roll along the rails. It is, however, in equilibrium under the given load since the equation $\Sigma F_x = 0$ is satisfied.

Equilibrium Equations. We express that the forces acting on the ladder form a system equivalent to zero:

$$\Sigma \mathbf{F} = 0: \qquad A_y\mathbf{j} + A_z\mathbf{k} + B_y\mathbf{j} + B_z\mathbf{k} - (981\text{ N})\mathbf{j} + C\mathbf{k} = 0$$
$$(A_y + B_y - 981\text{ N})\mathbf{j} + (A_z + B_z + C)\mathbf{k} = 0 \qquad (1)$$

$$\Sigma \mathbf{M}_A = \Sigma(\mathbf{r} \times \mathbf{F}) = 0: \qquad 1.2\mathbf{i} \times (B_y\mathbf{j} + B_z\mathbf{k}) + (0.9\mathbf{i} - 0.6\mathbf{k}) \times (-981\mathbf{j})$$
$$+ (0.6\mathbf{i} + 3\mathbf{j} - 1.2\mathbf{k}) \times C\mathbf{k} = 0$$

Computing the vector products, we have†

$$1.2B_y\mathbf{k} - 1.2B_z\mathbf{j} - 882.9\mathbf{k} - 588.6\mathbf{i} - 0.6C\mathbf{j} + 3C\mathbf{i} = 0$$
$$(3C - 588.6)\mathbf{i} - (1.2B_z + 0.6C)\mathbf{j} + (1.2B_y - 882.9)\mathbf{k} = 0 \qquad (2)$$

Setting the coefficients of **i**, **j**, **k** equal to zero in Eq. (2), we obtain the following three scalar equations, which express that the sum of the moments about each coordinate axis must be zero:

$$3C - 588.6 = 0 \qquad C = +196.2\text{ N}$$
$$1.2B_z + 0.6C = 0 \qquad B_z = -98.1\text{ N}$$
$$1.2B_y - 882.9 = 0 \qquad B_y = +736\text{ N}$$

The reactions at B and C are therefore

$$\mathbf{B} = +(736\text{ N})\mathbf{j} - (98.1\text{ N})\mathbf{k} \qquad \mathbf{C} = +(196.2\text{ N})\mathbf{k} \blacktriangleleft$$

Setting the coefficients of **j** and **k** equal to zero in Eq. (1), we obtain two scalar equations expressing that the sums of the components in the y and z directions are zero. Substituting for B_y, B_z, and C the values obtained above, we write

$$A_y + B_y - 981 = 0 \qquad A_y + 736 - 981 = 0 \qquad A_y = +245\text{ N}$$
$$A_z + B_z + C = 0 \qquad A_z - 98.1 + 196.2 = 0 \qquad A_z = -98.1\text{ N}$$

We conclude that the reaction at A is $\qquad \mathbf{A} = +(245\text{ N})\mathbf{j} - (98.1\text{ N})\mathbf{k} \blacktriangleleft$

†The moments in this Sample Problem and in Sample Probs. 4.8 and 4.9 could also be expressed in the form of determinants (see Sample Prob. 3.10).

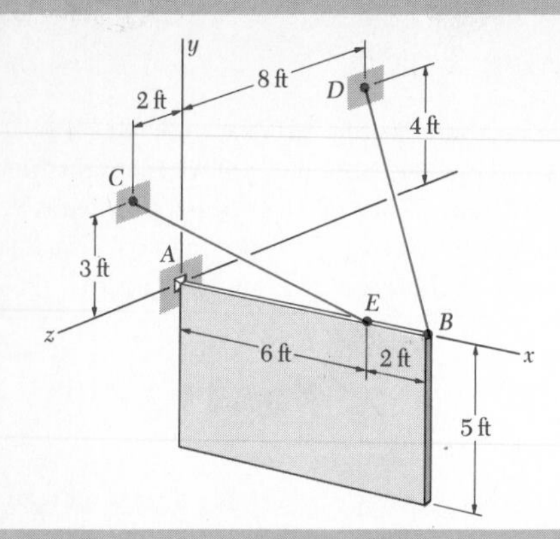

SAMPLE PROBLEM 4.8

A 5×8 ft sign of uniform density weighs 270 lb and is supported by a ball and socket at A and by two cables. Determine the tension in each cable and the reaction at A.

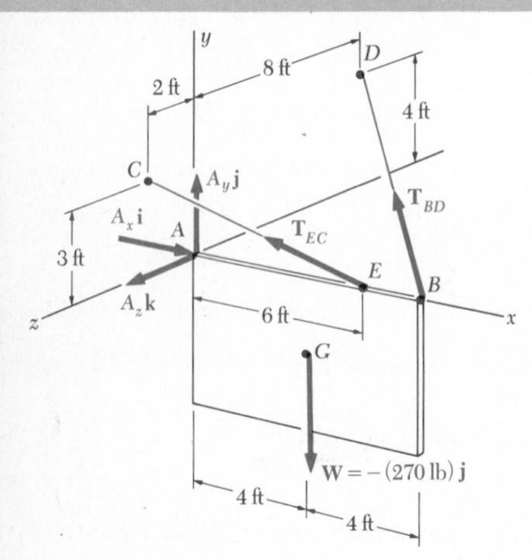

Solution. A free-body diagram of the sign is drawn. The forces acting on the free body are the weight $\mathbf{W} = -(270 \text{ lb})\mathbf{j}$ and the reactions at A, B, and E. The reaction at A is a force of unknown direction and is represented by three unknown components. Since the directions of the forces exerted by the cables are known, these forces involve only one unknown each, namely the magnitudes T_{BD} and T_{EC}. Since there are only five unknowns, the sign is partially constrained. It may rotate freely about the x axis; it is, however, in equilibrium under the given loading since the equation $\Sigma M_x = 0$ is satisfied.

The components of the forces $\mathbf{T}_{BD}$ and $\mathbf{T}_{EC}$ may be expressed in terms of the unknown magnitudes T_{BD} and T_{EC} by writing

$$\overrightarrow{BD} = -(8 \text{ ft})\mathbf{i} + (4 \text{ ft})\mathbf{j} - (8 \text{ ft})\mathbf{k} \qquad BD = 12 \text{ ft}$$
$$\overrightarrow{EC} = -(6 \text{ ft})\mathbf{i} + (3 \text{ ft})\mathbf{j} + (2 \text{ ft})\mathbf{k} \qquad EC = 7 \text{ ft}$$
$$\mathbf{T}_{BD} = T_{BD}\left(\frac{\overrightarrow{BD}}{BD}\right) = T_{BD}(-\tfrac{2}{3}\mathbf{i} + \tfrac{1}{3}\mathbf{j} - \tfrac{2}{3}\mathbf{k})$$
$$\mathbf{T}_{EC} = T_{EC}\left(\frac{\overrightarrow{EC}}{EC}\right) = T_{EC}(-\tfrac{6}{7}\mathbf{i} + \tfrac{3}{7}\mathbf{j} + \tfrac{2}{7}\mathbf{k})$$

Equilibrium Equations. We express that the forces acting on the sign form a system equivalent to zero:

$$\Sigma \mathbf{F} = 0: \qquad A_x\mathbf{i} + A_y\mathbf{j} + A_z\mathbf{k} + \mathbf{T}_{BD} + \mathbf{T}_{EC} - (270 \text{ lb})\mathbf{j} = 0$$
$$(A_x - \tfrac{2}{3}T_{BD} - \tfrac{6}{7}T_{EC})\mathbf{i} + (A_y + \tfrac{1}{3}T_{BD} + \tfrac{3}{7}T_{EC} - 270 \text{ lb})\mathbf{j}$$
$$+ (A_z - \tfrac{2}{3}T_{BD} + \tfrac{2}{7}T_{EC})\mathbf{k} = 0 \qquad (1)$$

$$\Sigma \mathbf{M}_A = \Sigma(\mathbf{r} \times \mathbf{F}) = 0:$$
$$(8 \text{ ft})\mathbf{i} \times T_{BD}(-\tfrac{2}{3}\mathbf{i} + \tfrac{1}{3}\mathbf{j} - \tfrac{2}{3}\mathbf{k})$$
$$+ (6 \text{ ft})\mathbf{i} \times T_{EC}(-\tfrac{6}{7}\mathbf{i} + \tfrac{3}{7}\mathbf{j} + \tfrac{2}{7}\mathbf{k})$$
$$+ (4 \text{ ft})\mathbf{i} \times (-270 \text{ lb})\mathbf{j} = 0$$
$$(2.667T_{BD} + 2.571T_{EC} - 1080)\mathbf{k} + (5.333T_{BD} - 1.714T_{EC})\mathbf{j} = 0 \qquad (2)$$

Setting the coefficients of $\mathbf{j}$ and $\mathbf{k}$ equal to zero in Eq. (2), we obtain two scalar equations which may be solved for T_{BD} and T_{EC}:

$$T_{BD} = 101.3 \text{ lb} \qquad T_{EC} = 315 \text{ lb} \quad ◄$$

Setting the coefficients of $\mathbf{i}$, $\mathbf{j}$, and $\mathbf{k}$ equal to zero in Eq. (1), we obtain three more equations which yield the components of $\mathbf{A}$. We have

$$\mathbf{A} = +(338 \text{ lb})\mathbf{i} + (101.2 \text{ lb})\mathbf{j} - (22.5 \text{ lb})\mathbf{k} \quad ◄$$

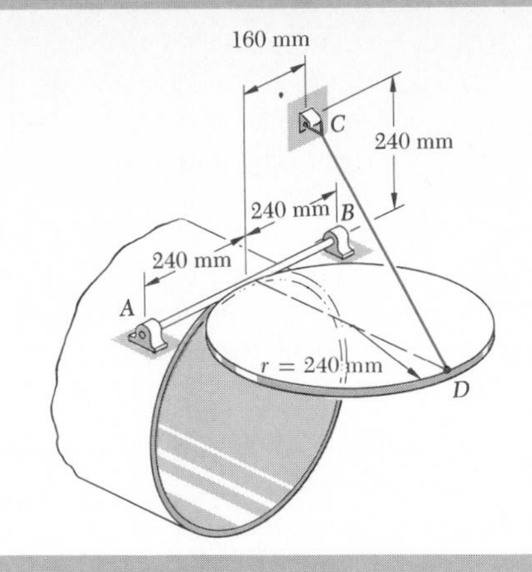

SAMPLE PROBLEM 4.9

A uniform pipe cover of radius $r = 240$ mm and mass 30 kg is held in a horizontal position by the cable CD. Assuming that the bearing at B does not exert any axial thrust, determine the tension in the cable and the reactions at A and B.

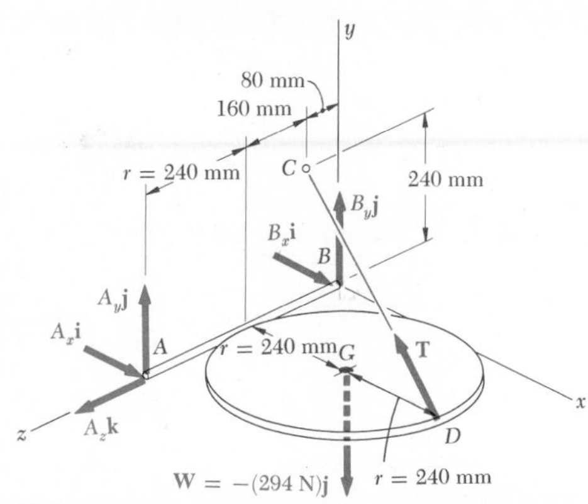

Solution. A free-body diagram is drawn with the coordinate axes shown. The forces acting on the free body are the weight of the cover,

$$\mathbf{W} = -mg\mathbf{j} = -(30\text{ kg})(9.81\text{ m/s}^2)\mathbf{j} = -(294\text{ N})\mathbf{j}$$

and reactions involving six unknowns, namely, the magnitude of the force $\mathbf{T}$ exerted by the cable, three force components at hinge A, and two at hinge B. The components of $\mathbf{T}$ are expressed in terms of the unknown magnitude T by resolving the vector $\overrightarrow{DC}$ into rectangular components and writing

$$\overrightarrow{DC} = -(480\text{ mm})\mathbf{i} + (240\text{ mm})\mathbf{j} - (160\text{ mm})\mathbf{k} \qquad DC = 560\text{ mm}$$

$$\mathbf{T} = T\frac{\overrightarrow{DC}}{DC} = -\tfrac{6}{7}T\mathbf{i} + \tfrac{3}{7}T\mathbf{j} - \tfrac{2}{7}T\mathbf{k}$$

Equilibrium Equations. We express that the forces acting on the pipe cover form a system equivalent to zero:

$$\Sigma\mathbf{F} = 0: \qquad A_x\mathbf{i} + A_y\mathbf{j} + A_z\mathbf{k} + B_x\mathbf{i} + B_y\mathbf{j} + \mathbf{T} - (294\text{ N})\mathbf{j} = 0$$

$$(A_x + B_x - \tfrac{6}{7}T)\mathbf{i} + (A_y + B_y + \tfrac{3}{7}T - 294\text{ N})\mathbf{j} + (A_z - \tfrac{2}{7}T)\mathbf{k} = 0 \qquad (1)$$

$$\Sigma\mathbf{M}_B = \Sigma(\mathbf{r} \times \mathbf{F}) = 0:$$

$$2r\mathbf{k} \times (A_x\mathbf{i} + A_y\mathbf{j} + A_z\mathbf{k}) + (2r\mathbf{i} + r\mathbf{k}) \times (-\tfrac{6}{7}T\mathbf{i} + \tfrac{3}{7}T\mathbf{j} - \tfrac{2}{7}T\mathbf{k}) + (r\mathbf{i} + r\mathbf{k}) \times (-294\text{ N})\mathbf{j} = 0$$

$$(-2A_y - \tfrac{3}{7}T + 294\text{ N})r\mathbf{i} + (2A_x - \tfrac{2}{7}T)r\mathbf{j} + (\tfrac{6}{7}T - 294\text{ N})r\mathbf{k} = 0 \qquad (2)$$

Setting the coefficients of the unit vectors equal to zero in Eq. (2), we write three scalar equations, which yield

$$A_x = +49.0\text{ N} \qquad A_y = +73.5\text{ N} \qquad T = 343\text{ N} \blacktriangleleft$$

Setting the coefficients of the unit vectors equal to zero in Eq. (1), we obtain three more scalar equations. After substituting the values of T, A_x, and A_y into these equations, we obtain

$$A_z = +98.0\text{ N} \qquad B_x = +245\text{ N} \qquad B_y = +73.5\text{ N}$$

The reactions at A and B are therefore

$$\mathbf{A} = +(49.0\text{ N})\mathbf{i} + (73.5\text{ N})\mathbf{j} + (98.0\text{ N})\mathbf{k} \qquad \mathbf{B} = +(245\text{ N})\mathbf{i} + (73.5\text{ N})\mathbf{j} \blacktriangleleft$$

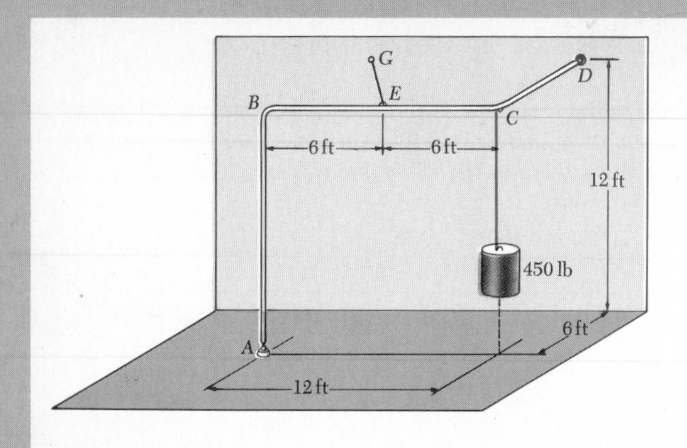

SAMPLE PROBLEM 4.10

A 450-lb load hangs from the corner C of a rigid piece of pipe $ABCD$ which has been bent as shown. The pipe is supported by the ball-and-socket joints A and D fastened, respectively, to the floor and to a vertical wall, and by a cable attached at the midpoint E of the portion BC of the pipe and at a point G on the wall. Determine (*a*) where G should be located if the tension in the cable is to be minimum, (*b*) the corresponding minimum value of the tension.

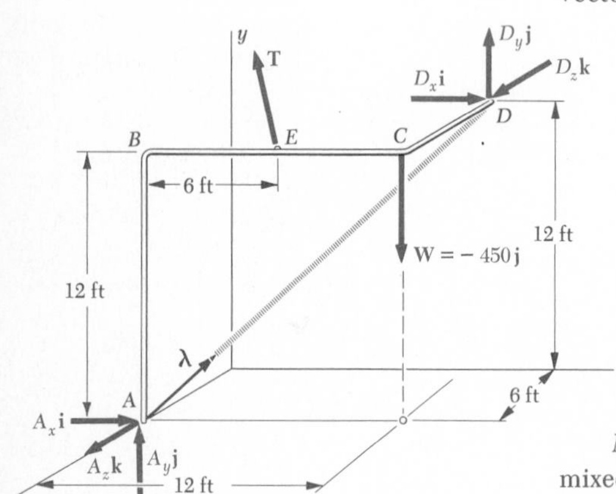

Solution. The free-body diagram of the pipe includes the load $\mathbf{W} = -450\mathbf{j}$, the reactions at A and D, and the force $\mathbf{T}$ exerted by the cable. To eliminate the reactions at A and D from the computations, we express that the sum of the moments of the forces about AD is zero. Denoting by $\boldsymbol{\lambda}$ the unit vector along AD, we write

$$\Sigma M_{AD} = 0: \qquad \boldsymbol{\lambda} \cdot (\overrightarrow{AE} \times \mathbf{T}) + \boldsymbol{\lambda} \cdot (\overrightarrow{AC} \times \mathbf{W}) = 0 \qquad (1)$$

The second term in Eq. (1) may be computed as follows:

$$\overrightarrow{AC} \times \mathbf{W} = (12\mathbf{i} + 12\mathbf{j}) \times (-450\mathbf{j}) = -5400\mathbf{k}$$

$$\boldsymbol{\lambda} = \frac{\overrightarrow{AD}}{AD} = \frac{12\mathbf{i} + 12\mathbf{j} - 6\mathbf{k}}{18} = \tfrac{2}{3}\mathbf{i} + \tfrac{2}{3}\mathbf{j} - \tfrac{1}{3}\mathbf{k}$$

$$\boldsymbol{\lambda} \cdot (\overrightarrow{AC} \times \mathbf{W}) = (\tfrac{2}{3}\mathbf{i} + \tfrac{2}{3}\mathbf{j} - \tfrac{1}{3}\mathbf{k}) \cdot (-5400\mathbf{k}) = +1800$$

Substituting the value obtained into Eq. (1), we write

$$\boldsymbol{\lambda} \cdot (\overrightarrow{AE} \times \mathbf{T}) = -1800 \text{ lb} \cdot \text{ft} \qquad (2)$$

Minimum Value of Tension. Recalling the commutative property for mixed triple products, we rewrite Eq. (2) in the form

$$\mathbf{T} \cdot (\boldsymbol{\lambda} \times \overrightarrow{AE}) = -1800 \text{ lb} \cdot \text{ft} \qquad (3)$$

which shows that the projection of $\mathbf{T}$ on the vector $\boldsymbol{\lambda} \times \overrightarrow{AE}$ is a constant. It follows that $\mathbf{T}$ is minimum when parallel to the vector

$$\boldsymbol{\lambda} \times \overrightarrow{AE} = (\tfrac{2}{3}\mathbf{i} + \tfrac{2}{3}\mathbf{j} - \tfrac{1}{3}\mathbf{k}) \times (6\mathbf{i} + 12\mathbf{j}) = 4\mathbf{i} - 2\mathbf{j} + 4\mathbf{k}$$

The corresponding unit vector being $\tfrac{2}{3}\mathbf{i} - \tfrac{1}{3}\mathbf{j} + \tfrac{2}{3}\mathbf{k}$, we write

$$\mathbf{T}_{\min} = T(\tfrac{2}{3}\mathbf{i} - \tfrac{1}{3}\mathbf{j} + \tfrac{2}{3}\mathbf{k}) \qquad (4)$$

Substituting for $\mathbf{T}$ and $\boldsymbol{\lambda} \times \overrightarrow{AE}$ in Eq. (3), we find $T = -300$. Carrying this value into (4), we obtain

$$\mathbf{T}_{\min} = -200\mathbf{i} + 100\mathbf{j} - 200\mathbf{k} \qquad T_{\min} = 300 \text{ lb} \blacktriangleleft$$

Location of G. Since the vector $\overrightarrow{EG}$ and the force $\mathbf{T}_{\min}$ have the same direction, their components must be proportional. Denoting by x, y, 0 the coordinates of G, we write

$$\frac{x - 6}{-200} = \frac{y - 12}{+100} = \frac{0 - 6}{-200} \qquad x = 0 \quad y = 15 \text{ ft} \blacktriangleleft$$

Problems

4.63 A 4 × 8 ft sheet of plywood weighing 40 lb has been temporarily propped against column CD. It rests at A and B on small wooden blocks and against protruding nails. Neglecting friction at all surfaces of contact, determine the reactions at A, B, and C.

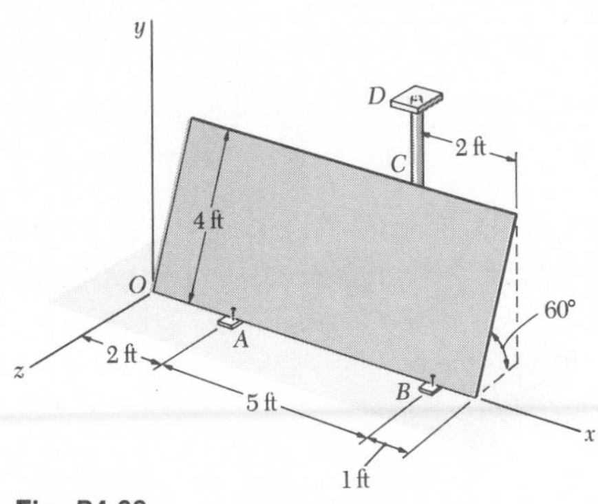

Fig. P4.63

4.64 Two transmission belts pass over a double-sheaved pulley which is attached to an axle supported by bearings at A and D. The radius of the inner sheave is 125 mm and the radius of the outer sheave is 250 mm. Knowing that when the system is at rest, the tension is 90 N in both portions of belt B and 150 N in both portions of belt C, determine the reactions at A and D. Assume that the bearing at D does not exert any axial thrust.

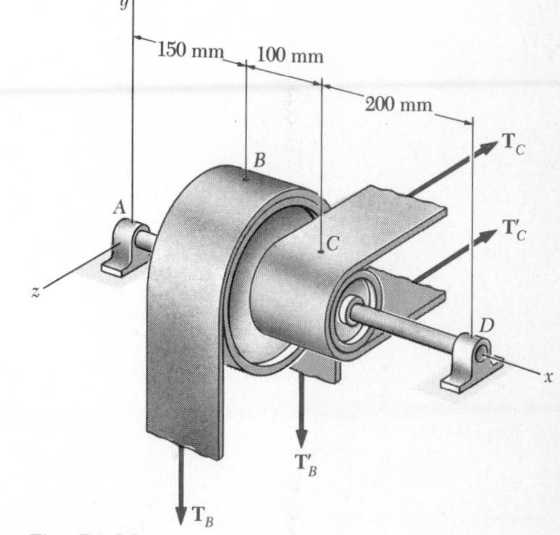

Fig. P4.64

4.65 Solve Prob. 4.64, assuming that the pulley rotates at a constant rate and that $T_B = 104$ N, $T'_B = 84$ N, and $T_C = 175$ N.

4.66 A small winch is used to raise a 120-lb load. Find (*a*) the magnitude of the vertical force **P** which should be applied at C to maintain equilibrium in the position shown, (*b*) the reactions at A and B, assuming that the bearing at B does not exert any axial thrust.

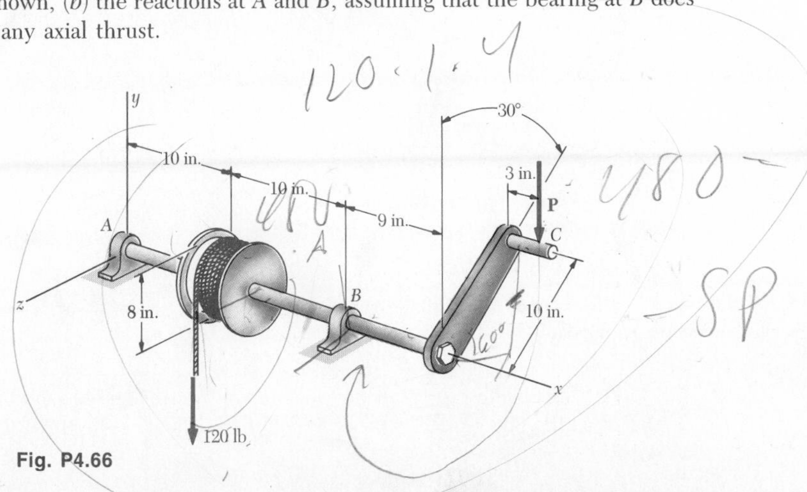

Fig. P4.66

4.67 A 200-mm lever and a 240-mm-diameter pulley are welded to the axle *BE* which is supported by bearings at *C* and *D*. If a 720-N vertical load is applied at *A* when the lever is horizontal, determine (*a*) the tension in the cord, (*b*) the reactions at *C* and *D*. Assume that the bearing at *D* does not exert any axial thrust.

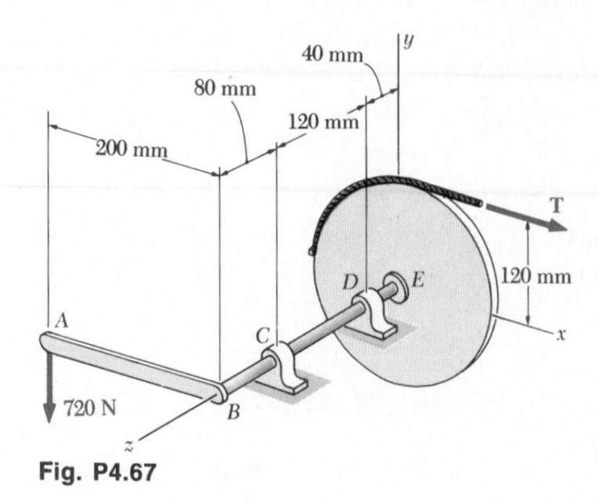

Fig. P4.67

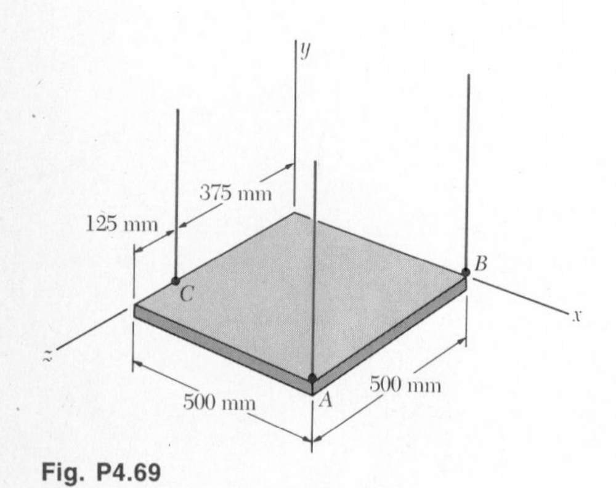

Fig. P4.69

4.68 Solve Prob. 4.67, assuming that the axle has been rotated clockwise in its bearings by 30° and that the 720-N load remains vertical.

4.69 The 20-kg square plate shown is supported by three wires. Determine the tension in each wire.

4.70 The rectangular plate shown weighs 80 lb and is supported by three wires. Determine the tension in each wire.

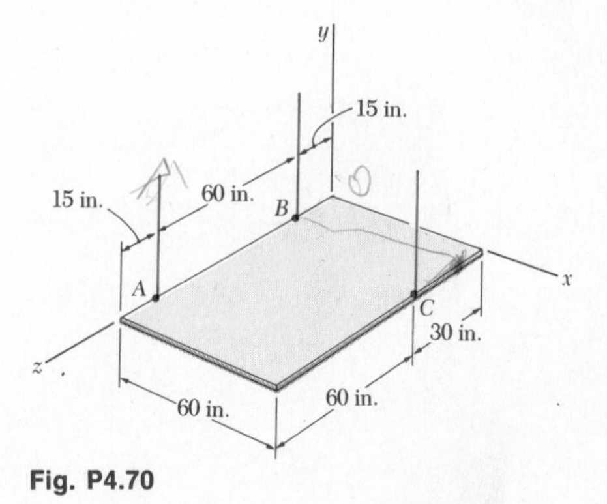

Fig. P4.70

4.71 A load *W* is to be placed on the 80-lb plate of Prob. 4.70. Determine the magnitude of *W* and the point where it should be placed if the tension is to be 60 lb in each of the three wires.

4.72 Determine the mass and location of the smallest block which should be placed on the 20-kg plate of Prob. 4.69 if the tensions in the three wires are to be equal.

4.73 A 12-m pole supports a horizontal cable CD and is held by a ball and socket at A and two cables BE and BF. Knowing that the tension in cable CD is 14 kN and assuming that CD is parallel to the x axis ($\phi = 0$), determine the tension in cables BE and BF and the reaction at A.

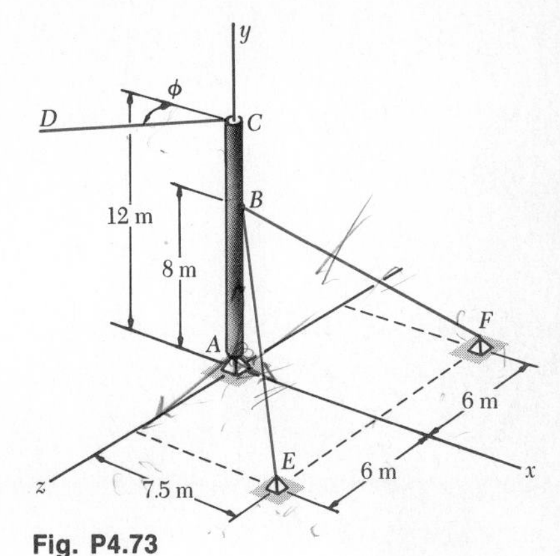

Fig. P4.73

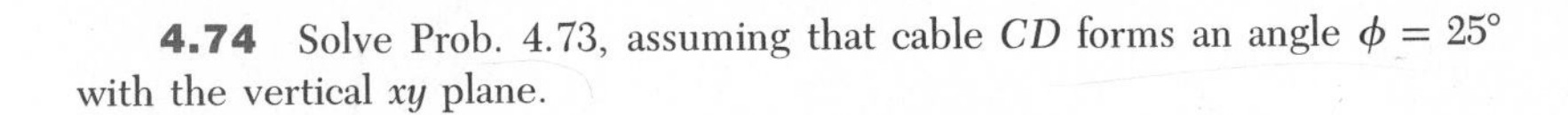
4.74 Solve Prob. 4.73, assuming that cable CD forms an angle $\phi = 25°$ with the vertical xy plane.

4.75 The 12-ft boom AB is acted upon by the 850-lb force shown. Determine (*a*) the tension in each cable, (*b*) the reaction at the ball and socket at A.

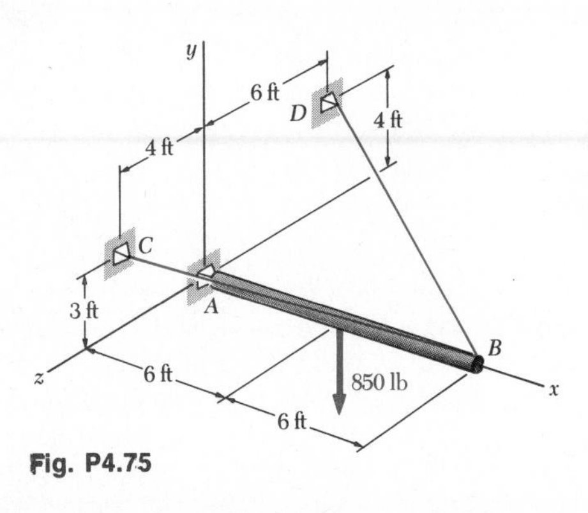

Fig. P4.75

4.76 Solve Prob. 4.75, assuming that the 850-lb load is applied at point B.

4.77 A 300-kg crate hangs from a cable which passes over a pulley B and is attached to a support at H. The 100-kg boom AB is supported by a ball and socket at A and by two cables DE and DF. The center of gravity of the boom is located at G. Determine (*a*) the tension in cables DE and DF, (*b*) the reaction at A.

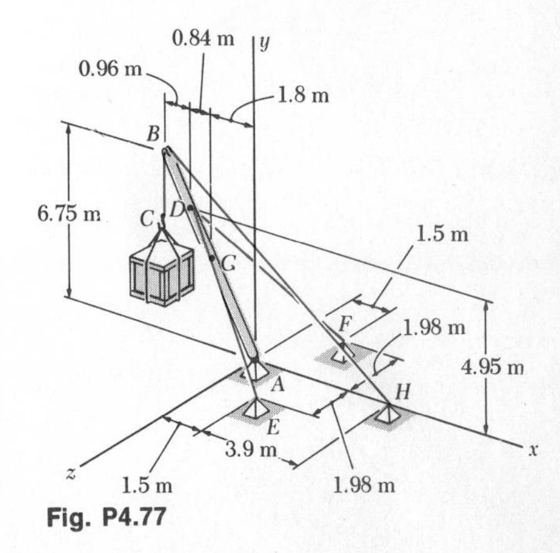

Fig. P4.77

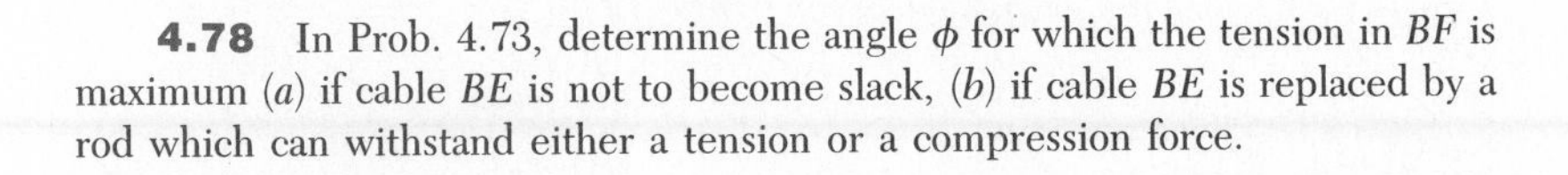
4.78 In Prob. 4.73, determine the angle ϕ for which the tension in BF is maximum (*a*) if cable BE is not to become slack, (*b*) if cable BE is replaced by a rod which can withstand either a tension or a compression force.

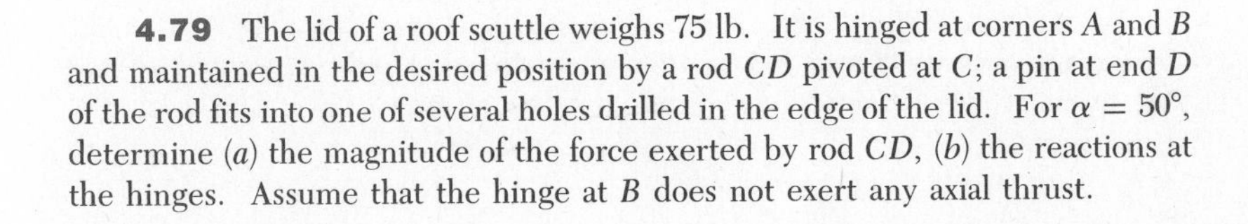
4.79 The lid of a roof scuttle weighs 75 lb. It is hinged at corners A and B and maintained in the desired position by a rod CD pivoted at C; a pin at end D of the rod fits into one of several holes drilled in the edge of the lid. For $\alpha = 50°$, determine (*a*) the magnitude of the force exerted by rod CD, (*b*) the reactions at the hinges. Assume that the hinge at B does not exert any axial thrust.

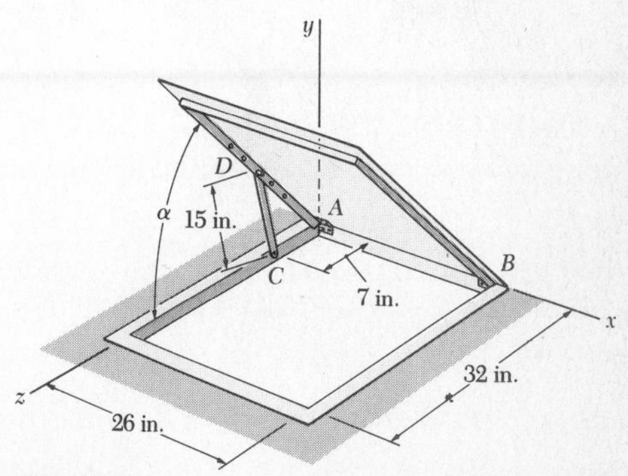

Fig. P4.79

4.80 Solve Prob. 4.79, assuming $\alpha = 30°$.

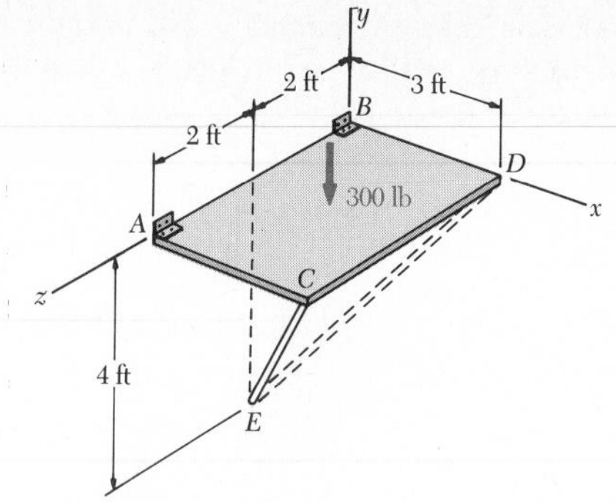

Fig. P4.81

✓**4.81** The horizontal platform $ABCD$ weighs 60 lb and supports a 240-lb load at its center. The platform is normally held in position by hinges at A and B and by braces CE and DE. If brace DE is removed, determine the reactions at the hinges and the force exerted by the remaining brace CE. The hinge at A does not exert any axial thrust.

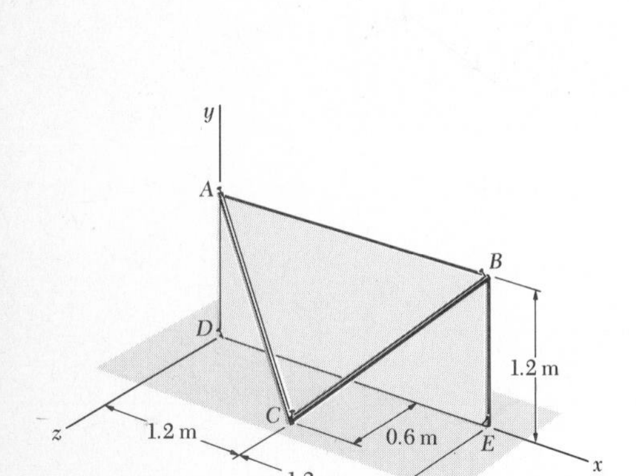

Fig. P4.82

4.82 A 1.2 × 2.4 m sheet of plywood is temporarily held by nails at D and E and by two wooden braces nailed at A, B, and C. Wind is blowing on the hidden face of the plywood sheet and it is assumed that its effect may be represented by a force $P\mathbf{k}$ applied at the center of the sheet. Knowing that each brace becomes unsafe with respect to buckling when subjected to a 1.8-kN axial force, determine (*a*) the maximum allowable value of the magnitude P of the wind force, (*b*) the corresponding value of the z component of the reaction at E. Assume that the nails are loose and do not exert any couple.

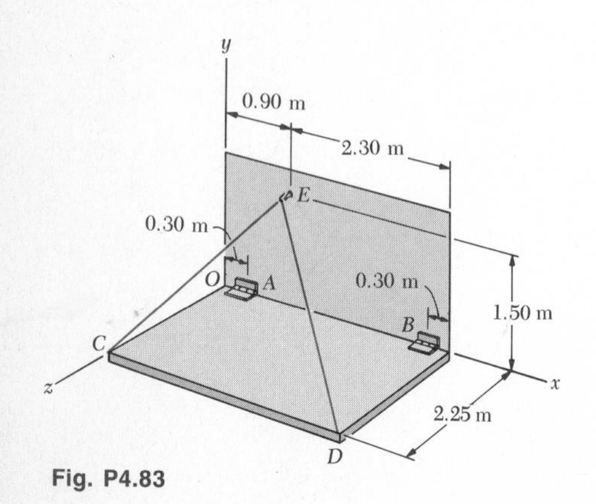

Fig. P4.83

4.83 A 250-kg uniform rectangular platform, 3.20 m long and 2.25 m wide, is supported by hinges at A and B and by a cable attached to corners C and D of the platform and passing over a frictionless hook E. Assuming that the tension is the same in both parts of the cable and that the hinge at A does not exert any axial thrust, determine (*a*) the tension in the cable, (*b*) the reactions at A and B.

4.84 Solve Prob. 4.83, assuming that cable CED is replaced by a cable attached to corner C and to hook E.

4.85 Solve Prob. 4.83, assuming that cable CED is replaced by a cable attached to corner D and to hook E.

4.86 Solve Prob. 4.83, assuming that cable CED is replaced by two cables, CE and DE, attached, respectively, to corners C and D and tied to hook E, and that the hinge at A is replaced by a roller exerting on the platform a force parallel to the z axis.

4.87 Solve Prob. 4.81, assuming that the hinge at A has been removed and that the hinge at B can exert couples about axes parallel to the x and y axes, respectively.

4.88 Solve Prob. 4.81, assuming that the hinge at B has been removed and that the hinge at A can exert an axial thrust, as well as couples about axes parallel to the x and y axes, respectively.

4.89 Solve Prob. 4.83, assuming that the hinge at A has been removed and that the hinge at B can exert couples about axes parallel to the y and z axes, respectively.

4.90 Solve Prob. 4.83, assuming that the hinge at B has been removed and that the hinge at A can exert an axial thrust, as well as couples about axes parallel to the y and z axes, respectively.

4.91 The uniform rod AB weighs 18 lb. It is supported by a ball and socket at A and by cord CD which is attached to the midpoint C of the rod. Knowing that the rod leans against a frictionless vertical wall at B, determine (*a*) the tension in the cord, (*b*) the reactions at A and B.

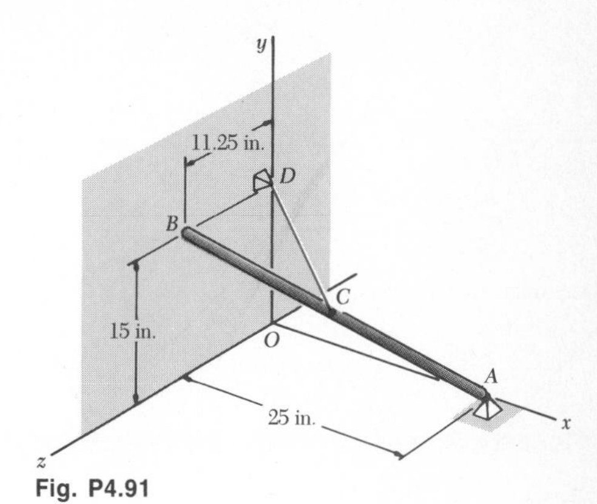

Fig. P4.91

4.92 The rigid L-shaped member ABC is supported by a ball and socket at A and by three cables. Determine the tension in each cable and the reaction at A caused by the 500-lb load applied at G.

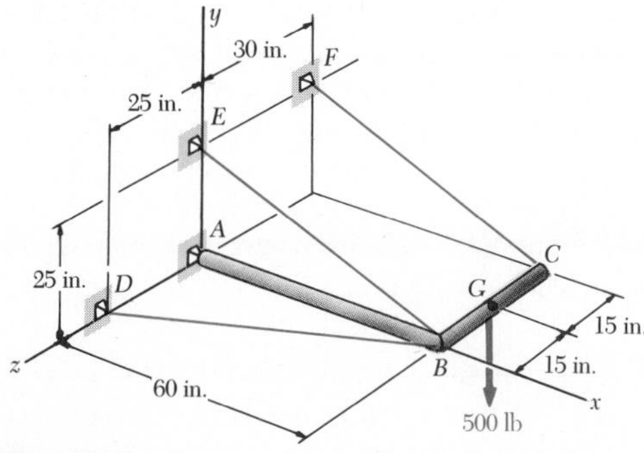

Fig. P4.92

4.93 A 180-kg catamaran is sailing in a straight course and at a constant speed. The effect of the wind on the sail may be represented by the force $\mathbf{F} = (420\text{ N})\mathbf{i} - (560\text{ N})\mathbf{k}$ applied at the point of coordinates $x = -0.48$ m, $y = 3$ m, $z = -0.36$ m. Due to the efforts of the crew, the mast is maintained in a vertical position and the twin hulls are subjected to equal forces, consisting of buoyant forces $\mathbf{B} = B\mathbf{j}$ along the lines of action $x = -0.6$ m, $z = \pm 1.2$ m, drag forces $\mathbf{D} = -D\mathbf{i}$ along the lines of action $y = -0.6$ m, $z = \pm 1.2$ m, and side forces $\mathbf{S} = S\mathbf{k}$ along the common line of action $x = -d$, $y = -0.6$ m. Determine (*a*) the magnitude D of the drag on each of the hulls, (*b*) the magnitude S of each of the side forces, and the distance d defining their line of action.

Fig. P4.93

4.94 The catamaran of Prob. 4.93 is manned by two sailors, one of mass 75 kg and the other 60 kg. The center of gravity of the boat is located at $x = -0.9$ m, $z = 0$, and the 75-kg sailor is sitting at $x = -1.5$ m, $z = +1.2$ m, while the 60-kg sailor leans overboard to maintain the mast in a vertical position. Determine the x and z coordinates of the center of gravity of the second sailor.

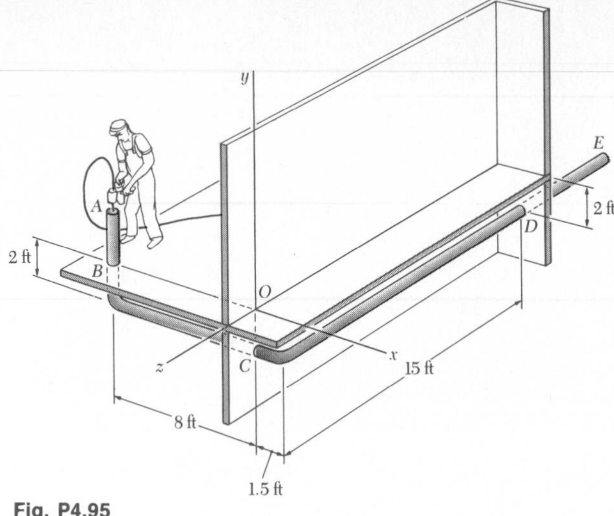

Fig. P4.95

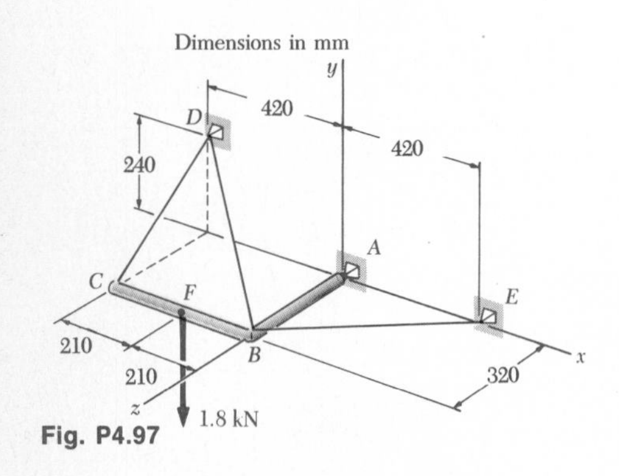

Fig. P4.97

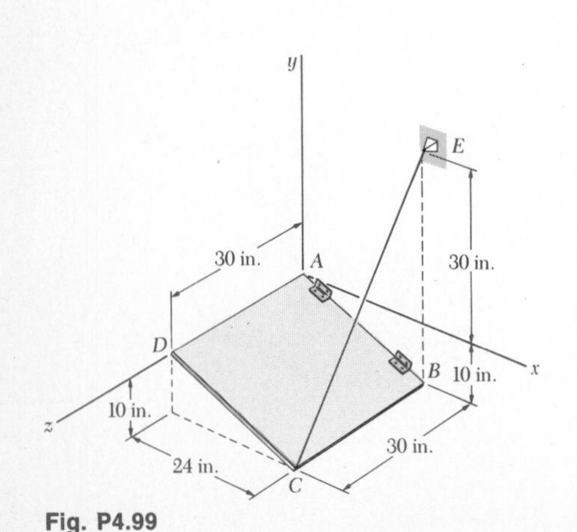

Fig. P4.99

4.95 In order to clean the clogged drainpipe AE, a plumber has disconnected both ends of the pipe and introduced a power snake through the opening at A. The cutting head of the snake is connected by a heavy cable to an electric motor which keeps it rotating at a constant speed while the plumber forces the cable into the pipe. The forces exerted by the plumber and the motor on the end of the cable may be represented by the wrench $\mathbf{F} = -(6.5 \text{ lb})\mathbf{j}$, $\mathbf{M} = -(97.5 \text{ lb} \cdot \text{ft})\mathbf{j}$. Determine the additional reactions at B, C, and D caused by the cleaning operation. Assume that the reaction at each support consists of two force components perpendicular to the pipe.

4.96 Solve Prob. 4.95, assuming that the plumber exerts no force on the end of the cable and that the motor exerts a couple $\mathbf{M} = -(97.5 \text{ lb} \cdot \text{ft})\mathbf{j}$.

4.97 The rigid L-shaped member ABC is supported by a ball and a socket at A and by three cables. If a 1.8-kN load is applied at F, determine the tension in each cable.

4.98 Solve Prob. 4.97, assuming that the 1.8-kN load is applied at C.

4.99 The 26 × 30 in. plate $ABCD$ is supported by hinges along edge AB and by wire CE. Knowing that the plate is uniform and weighs 100 lb, determine the tension in the wire.

4.100 Solve Prob. 4.99, assuming that the wire CE is replaced by a wire connecting E and D.

4.101 The bent rod ABC is hinged to a vertical wall by means of two brackets and bears at C against another vertical wall. The upper bracket fits in a groove in the rod to prevent the rod from sliding down. Neglecting friction, determine the reaction at C when a 150-N load is applied at D as shown.

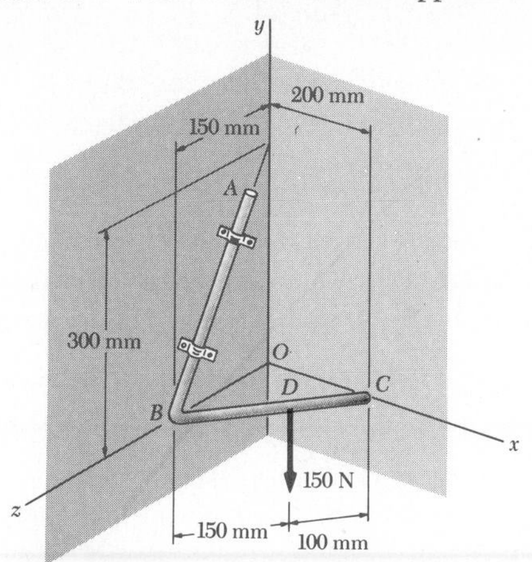

Fig. P4.101

4.102 The frame ACD is supported by ball-and-socket joints at A and D and by a cable which passes through a ring at B and is attached to hooks at G and H. Knowing that the frame supports at point C a load of magnitude $P = 335$ N, determine the tension in the cable.

4.103 Solve Prob. 4.102, assuming that the cable GBH is replaced by a cable GB attached at G and B.

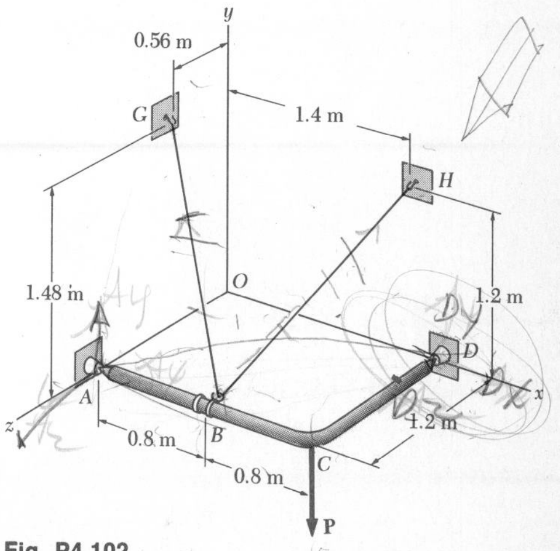

Fig. P4.102

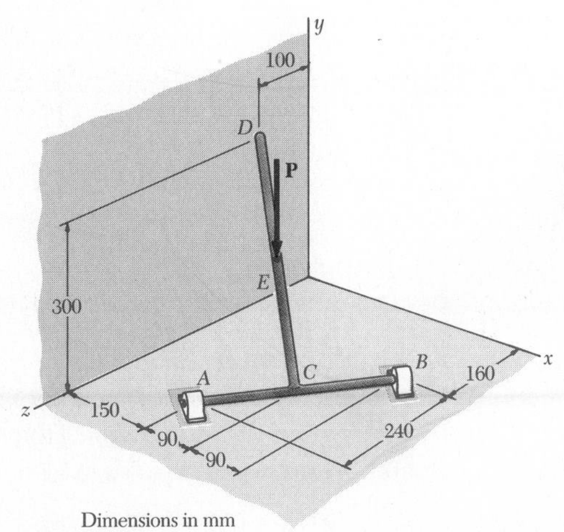

Fig. P4.104

4.104 Two rods are welded together to form a T-shaped lever which leans against a frictionless vertical wall at D and is supported by bearings at A and B. A vertical force $\mathbf{P}$ of magnitude 600 N is applied at the midpoint E of rod DC. Determine the reaction at D.

Review and Summary

Equilibrium equations

This chapter was devoted to the study of the *equilibrium of rigid bodies*, i.e., to the situation when the external forces acting on a rigid body *form a system equivalent to zero* [Sec. 4.1]. We then have

$$\Sigma \mathbf{F} = 0 \qquad \Sigma \mathbf{M}_O = \Sigma(\mathbf{r} \times \mathbf{F}) = 0 \tag{4.1}$$

Resolving each force and each moment into its rectangular components, we may express the necessary and sufficient conditions for the equilibrium of a rigid body by the following six scalar equations:

$$\Sigma F_x = 0 \qquad \Sigma F_y = 0 \qquad \Sigma F_z = 0 \tag{4.2}$$
$$\Sigma M_x = 0 \qquad \Sigma M_y = 0 \qquad \Sigma M_z = 0 \tag{4.3}$$

These equations may be used to determine unknown forces applied to the rigid body or unknown reactions exerted by its supports.

Free-body diagram

It is essential, when solving a problem involving the equilibrium of a rigid body, to consider *all* the forces acting on the body. Therefore, the first step in the solution of the problem should be to draw a *free-body diagram* showing the body under consideration and all the forces acting on it, unknown as well as known [Sec. 4.2].

Equilibrium of a two-dimensional structure

In the first part of the chapter, we considered the *equilibrium of a two-dimensional structure*, i.e., we assumed that the structure considered and the forces applied to it were contained in the plane of the figure. We saw that each of the reactions exerted on the structure by its supports could involve one, two, or three unknowns, depending upon the type of support [Sec. 4.3].

In the case of a two-dimensional structure, Eqs. (4.1), or Eqs. (4.2) and (4.3), reduce to *three equilibrium equations*, namely

$$\Sigma F_x = 0 \qquad \Sigma F_y = 0 \qquad \Sigma M_A = 0 \tag{4.5}$$

where A is an arbitrary point in the plane of the structure [Sec. 4.4]. These equations may be solved for three unknowns. While the three equilibrium equations (4.5) cannot be *augmented* by additional equations, any of them may be *replaced* by another equation. Therefore, we may write alternative sets of equilibrium equations, such as

$$\Sigma F_x = 0 \qquad \Sigma M_A = 0 \qquad \Sigma M_B = 0 \tag{4.6}$$

where the line AB is chosen in a direction different from the y direction, or

$$\Sigma M_A = 0 \qquad \Sigma M_B = 0 \qquad \Sigma M_C = 0 \tag{4.7}$$

where the points A, B, and C are not in a straight line.

Statical indeterminacy

Partial constraints

Improper constraints

Since any set of equilibrium equations may be solved for only three unknowns, the reactions at the supports of a rigid two-dimensional structure may not be completely determined if they involve *more than three unknowns;* they are said to be *statically indeterminate* [Sec. 4.5]. On the other hand, if the reactions involve *fewer than three unknowns*, equilibrium will not be maintained under general loading conditions; the structure is said to be *partially constrained.* The fact that the reactions involve exactly three unknowns is no guarantee that the equilibrium equations can be solved for all three unknowns. If the supports are arranged in such a way that the reactions *must be either concurrent or parallel,* the reactions are statically indeterminate and the structure is said to be *improperly constrained.*

Two-force body

Three-force body

Two particular cases of equilibrium of a rigid body were given special attention. In Sec. 4.6, a *two-force body* was defined as a rigid body subjected to forces at only two points, and it was shown that the resultants $\mathbf{F}_1$ and $\mathbf{F}_2$ of these forces must have the *same magnitude, same line of action, and opposite sense* (Fig. 4.11), a property which will simplify the solution of certain problems in later chapters. In Sec. 4.7, a *three-force body* was defined as a rigid body subjected to forces at only three points, and it was shown that the resultants $\mathbf{F}_1$, $\mathbf{F}_2$, and $\mathbf{F}_3$ of these forces must be *either concurrent* (Fig. 4.12) *or parallel.* This property provides us with an alternative approach to the solution of problems involving a three-force body [Sample Prob. 4.6].

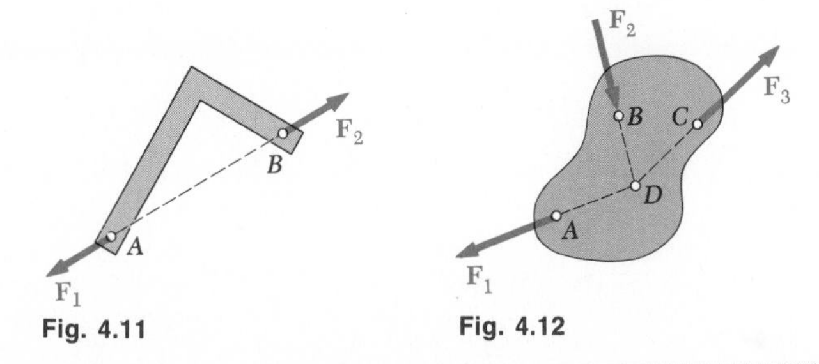

Fig. 4.11

Fig. 4.12

Equilibrium of a three-dimensional body

In the second part of the chapter, we considered the *equilibrium of a three-dimensional body* and saw that each of the reactions exerted on the body by its supports could involve from one to six unknowns, depending upon the type of support [Sec. 4.8].

In the general case of the equilibrium of a three-dimensional body, all of the six scalar equilibrium equations (4.2) and (4.3) listed at the beginning of this review should be used and solved for *six unknowns* [Sec. 4.9]. In most problems, however, these equations will be more conveniently obtained if we first write

$$\Sigma \mathbf{F} = 0 \qquad \Sigma \mathbf{M}_O = \Sigma(\mathbf{r} \times \mathbf{F}) = 0 \tag{4.1}$$

and express the forces $\mathbf{F}$ and position vectors $\mathbf{r}$ in terms of scalar components and unit vectors. The vector products may then be computed either directly or by means of determinants, and the desired scalar equations obtained by equating to zero the coefficients of the unit vectors [Sample Probs. 4.7 through 4.9].

We noted that as many as three unknown reaction components may be eliminated from the computation of $\Sigma\mathbf{M}_O$ in the second of the relations (4.1) through a judicious choice of point O. Also, the reactions at two points A and B may be eliminated from the solution of some problems by writing the equation $\Sigma M_{AB} = 0$, which involves the computation of the moments of the forces about an axis AB joining points A and B [Sample Prob. 4.10].

If the reactions involve more than six unknowns, some of the reactions are *statically indeterminate;* if they involve less than six unknowns, the rigid body is only *partially constrained.* Even with six or more unknowns, the rigid body will be *improperly constrained* if the supports are such that the reactions must either be parallel or intersect the same line.

Review Problems

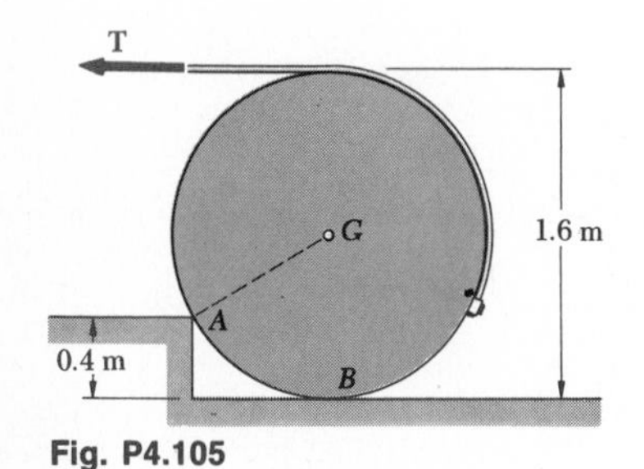

Fig. P4.105

4.105 A 150-kg cylindrical tank, 1.6 m in diameter, is to be raised over a 0.4-m obstruction. A cable is wrapped around the tank and pulled horizontally as shown. Knowing that the corner of the obstruction at A is rough, find the required tension in the cable and the reaction at A.

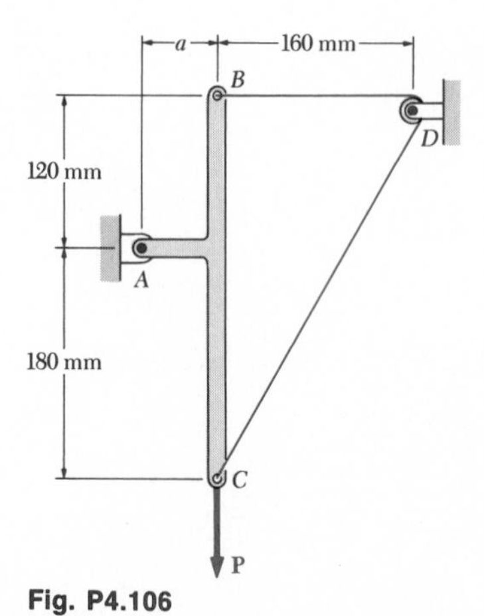

Fig. P4.106

4.106 A force **P** of magnitude 450 N is applied to member ABC which is supported by a frictionless pin at A and by cable BDC. Since the cable passes over a pulley at D, the tension may be assumed to be the same in portions BD and CD of the cable. Determine (*a*) the tension in the cable, (*b*) the reaction at A when $a = 60$ mm.

4.107 Knowing that the tension in all portions of the belt is 600 N, determine the reactions at supports A and B of the plate when (*a*) $\alpha = 0$, (*b*) $\alpha = 90°$, (*c*) $\alpha = 30°$.

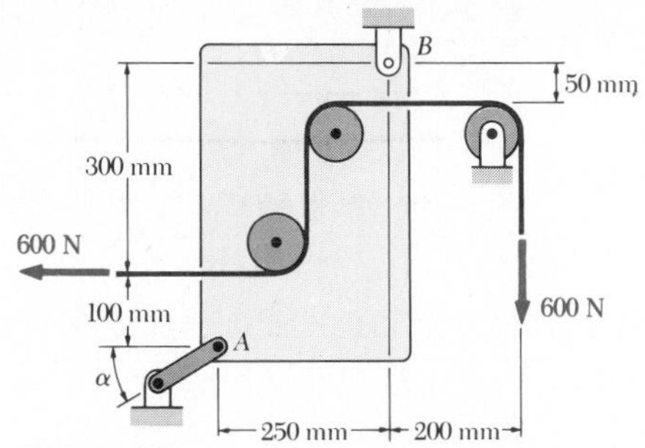

Fig. P4.107

4.108 A 3 × 5 ft storm window weighing 15 lb is supported by hinges at A and B. In the position shown, it is held away from the side of the house by a 2-ft stick CD. Assuming that the hinge at A does not exert any axial thrust, determine (*a*) the magnitude of the force exerted by the stick, (*b*) the reactions at A and B.

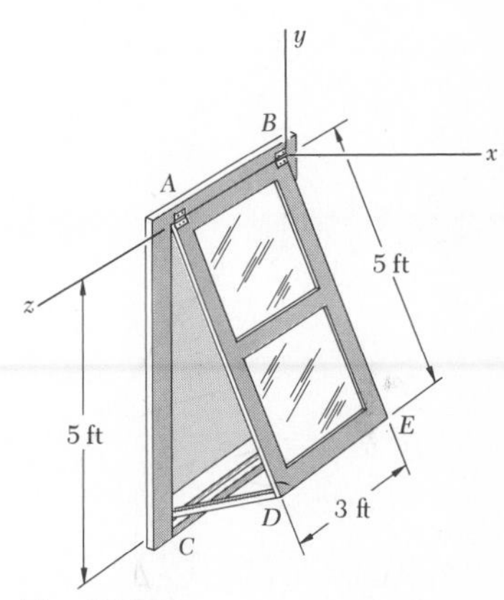

Fig. P4.108

4.109 Solve Prob. 4.108, assuming that the hinge at B has been removed and that the hinge at A can exert an axial thrust, as well as couples about axes parallel to the x and y axes, respectively.

4.110 A worker is pushing a 120-lb wheelbarrow up an 18° incline when he encounters a 1.4-in. obstruction at D. Determine the magnitude and direction of (*a*) the force he must apply to each handle to start the wheel over the obstruction, (*b*) the corresponding reaction at D. (*Hint*. As it starts rotating about D, the wheel is a two-force body.)

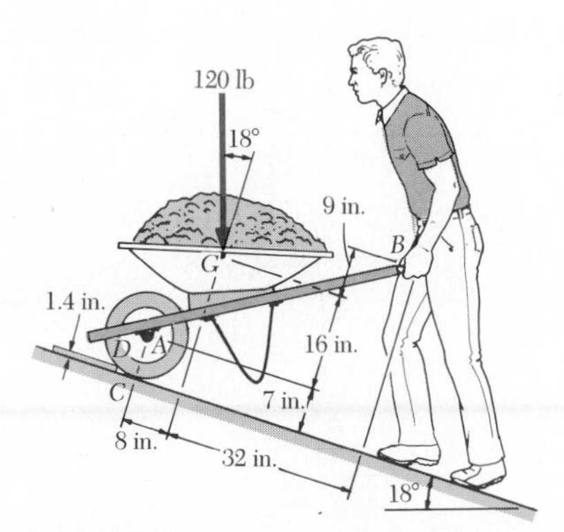

Fig. P4.110

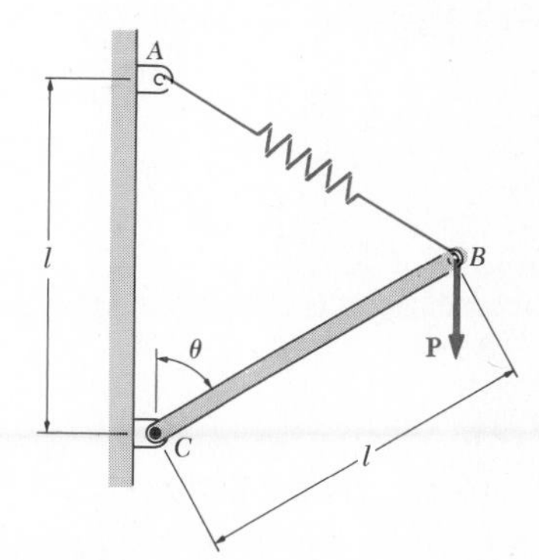

Fig. P4.111

4.111 A vertical load **P** is applied at end B of rod BC. The constant of the spring is k and the spring is unstretched when $\theta = 60°$. (*a*) Neglecting the weight of the rod, express the angle θ corresponding to the equilibrium position in terms of P, k, and l. (*b*) Determine the values of θ corresponding to equilibrium if $P = \frac{1}{4}kl$.

4.112 A slender rod BC of length L and weight W is held by two cables as shown. Knowing that cable AB is horizontal and that the rod forms an angle of 40° with the horizontal, determine (*a*) the angle θ that cable CD forms with the horizontal, (*b*) the tension in each cable.

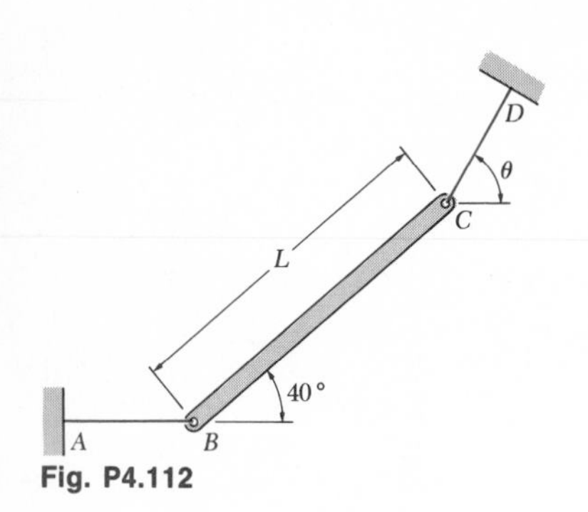

Fig. P4.112

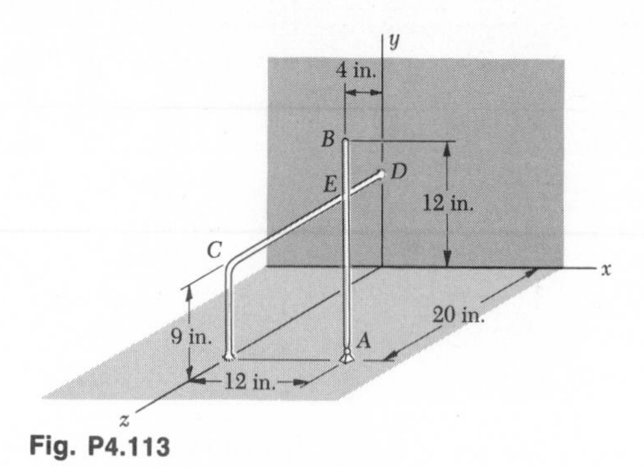

Fig. P4.113

4.113 The uniform rod AB weighs 22.5 lb. It is supported by a ball and socket at A and leans against both rod CD and the vertical wall. Assuming the wall and the rods to be frictionless, determine (*a*) the force which rod CD exerts on AB, (*b*) the reactions at A and B. (*Hint.* The force exerted by CD on AB must be perpendicular to both rods.)

4.114 A force **P** is applied to a bent rod AD which may be supported in four different ways as shown. In each case determine the reactions at the supports.

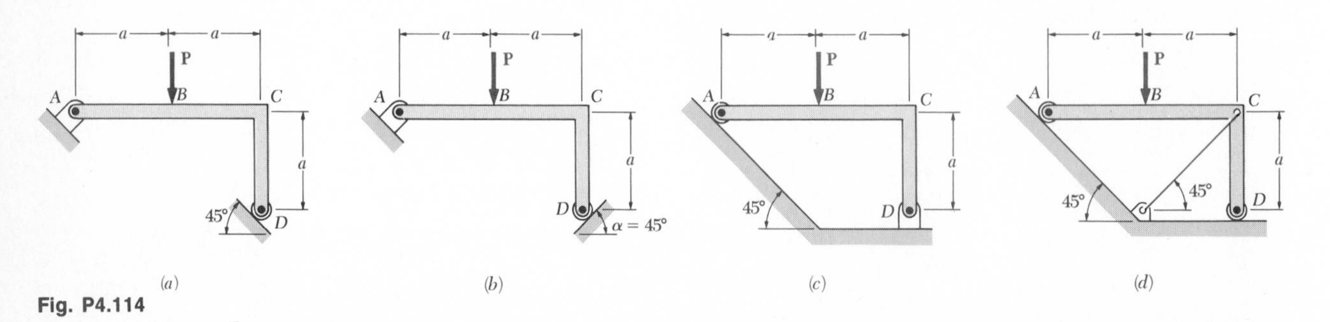

Fig. P4.114

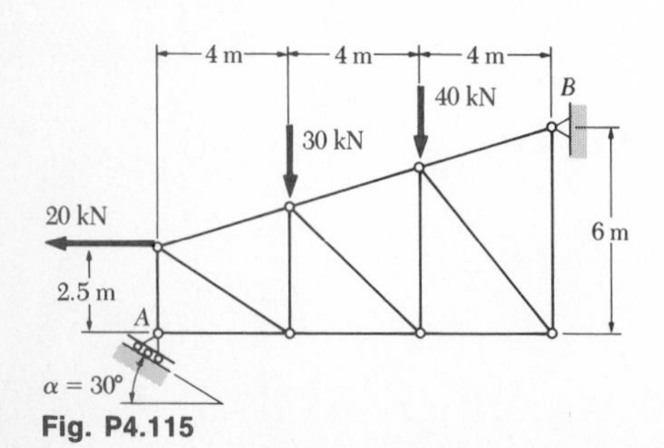

Fig. P4.115

4.115 Determine the reactions at A and B for the truss and loading shown.

***4.116** In the problems listed below, the rigid bodies considered were completely constrained, and the reactions were statically determinate. For each of these rigid bodies it is possible to create an improper set of constraints by changing either a dimension of the body or the direction of a reaction. In each problem determine the value of α or of a which results in improper constraints. (*a*) Prob. 4.21, (*b*) Prob. 4.44, (*c*) Prob. 4.106, (*d*) Prob. 4.114*b*, (*e*) Prob. 4.115.

The following problems are designed to be solved with a computer.

4.C1 Two wheels A and B, of weight W_A and W_B, respectively, are connected by a rod of negligible weight and are free to roll on the surface shown. Write a computer program which can be used to determine the angle α that the rod forms with the horizontal in its equilibrium position. Use this program to calculate α for values of θ from 0 to 90° at 5° intervals. Consider the cases when (*a*) $W_A = W_B$, (*b*) $W_A = 2W_B$.

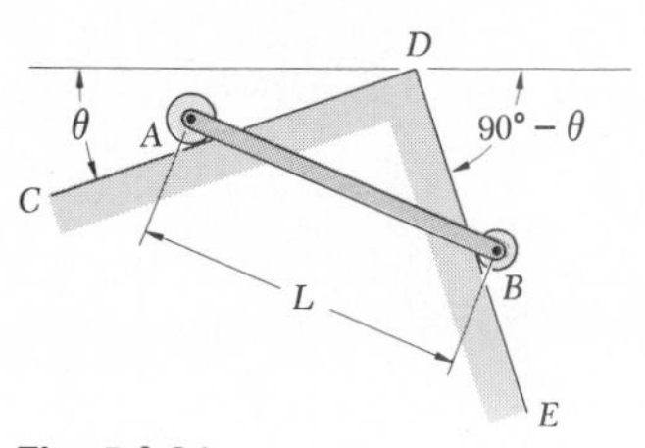

Fig. P4.C1

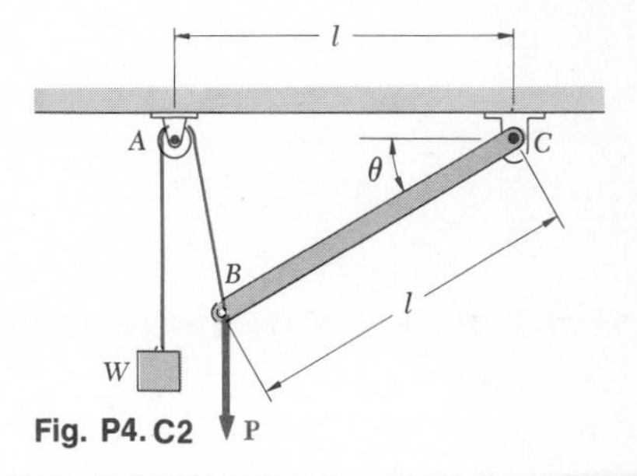

Fig. P4.C2

4.C2 A vertical load **P** is applied to the end of rod BC. Write a computer program which can be used to calculate the magnitude P of the load required for the equilibrium of the system. For $W = 10$ lb and $l = 25$ in., use this program to calculate P for values of θ from 0 to 90° at 5° intervals.

4.C3 The derrick shown supports a 2000-kg crate. It is held by a ball and socket at A and by two cables attached at D and E. Knowing that the derrick stands in a vertical plane forming the angle ϕ with the xy plane, write a computer program and use it to calculate the tension in each cable for values of ϕ from 0 to 40° at 4° intervals.

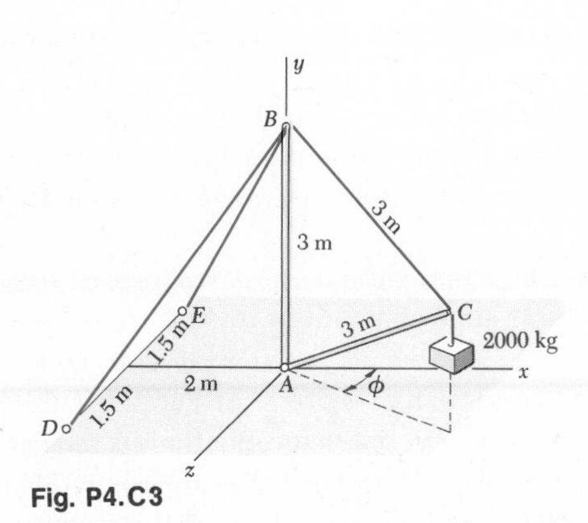

Fig. P4.C3

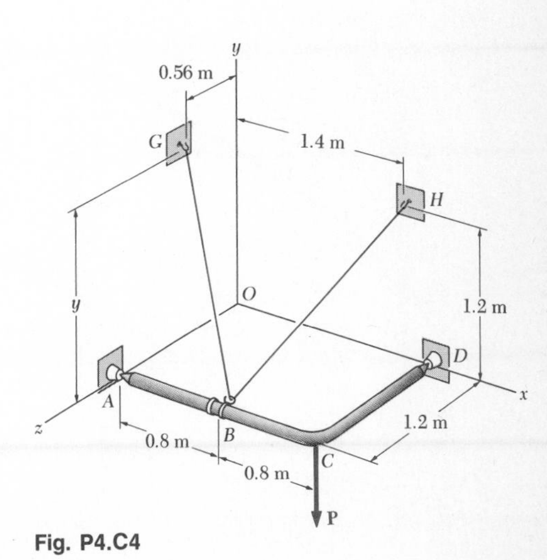

Fig. P4.C4

4.C4 The frame ACD is supported by ball-and-socket joints at A and D and by a cable which passes through a ring at B and is attached to hooks at G and H. Knowing that the frame supports at point C a load of magnitude $P = 335$ N, write a computer program and use it to calculate the tension in the cable for values of y from 0 to 1.5 m at 100-mm intervals.

Distributed Forces: Centroids and Centers of Gravity

CHAPTER

5

5.1. Introduction. We have assumed so far that the attraction exerted by the earth on a rigid body could be represented by a single force **W**. This force, called the force of gravity, or the weight of the body, was to be applied at the *center of gravity* of the body (Sec. 3.2). Actually, the earth exerts a force on each of the particles forming the body. The action of the earth on a rigid body should thus be represented by a large number of small forces distributed over the entire body. We shall see in this chapter, however, that all these small forces may be replaced by a single equivalent force **W**. We shall also learn to determine the center of gravity, i.e., the point of application of the resultant **W**, for various shapes of bodies.

In the first part of the chapter, we shall consider two-dimensional bodies, such as flat plates and wires contained in a given plane. We shall introduce two concepts closely associated with the determination of the center of gravity of a plate or wire, namely, the concept of *centroid* of an area or line, and the concept of *first moment* of an area or line with respect to a given axis.

We shall also find that the computation of the area of a surface of revolution or of the volume of a body of revolution is directly related to the determination of the centroid of the line or area used to generate that surface or body of revolution (Theorems of Pappus-Guldinus). And, as we shall see in Secs. 5.8 and 5.9, the determination of the centroid of an area simplifies the analysis of beams subjected to distributed loads and the computation of the forces exerted on submerged rectangular surfaces such as hydraulic gates and portions of dams.

In the last part of the chapter, we shall learn to determine the center of gravity of a three-dimensional body, as well as the centroid of a volume and the first moments of that volume with respect to the coordinate planes.

5.2. Center of Gravity of a Two-Dimensional Body. Let us first consider a flat horizontal plate (Fig. 5.1). We may divide the plate into n small elements. The coordinates of the first element are denoted by x_1 and y_1, those of the second element by x_2 and y_2, etc. The forces exerted by the earth on the elements of plate will be denoted, respectively, by $\Delta\mathbf{W}_1, \Delta\mathbf{W}_2, \ldots, \Delta\mathbf{W}_n$. These forces or weights are directed toward the center of the earth; however, for all practical purposes they may be assumed parallel. Their resultant is therefore a single force in the same direction. The magnitude W of this force is obtained by adding the magnitudes of the elementary weights.

$$\Sigma F_z: \quad W = \Delta W_1 + \Delta W_2 + \cdots + \Delta W_n$$

To obtain the coordinates $\bar{x}$ and $\bar{y}$ of the point G where the resultant $\mathbf{W}$ should be applied, we write that the moments of $\mathbf{W}$ about the y and x axes are equal to the sum of the corresponding moments of the elementary weights,

$$\begin{aligned} \Sigma M_y: &\quad \bar{x}W = x_1\,\Delta W_1 + x_2\,\Delta W_2 + \cdots + x_n\,\Delta W_n \\ \Sigma M_x: &\quad \bar{y}W = y_1\,\Delta W_1 + y_2\,\Delta W_2 + \cdots + y_n\,\Delta W_n \end{aligned} \tag{5.1}$$

If we now increase the number of elements into which the plate is divided and simultaneously decrease the size of each element, we obtain at the limit the following expressions:

$$W = \int dW \qquad \bar{x}W = \int x\,dW \qquad \bar{y}W = \int y\,dW \tag{5.2}$$

These equations define the weight $\mathbf{W}$ and the coordinates $\bar{x}$ and $\bar{y}$ of the center of gravity G of a flat plate. The same equations may be derived for a wire lying in the xy plane (Fig. 5.2). We shall observe, in the latter case, that the center of gravity G will generally not be located on the wire.

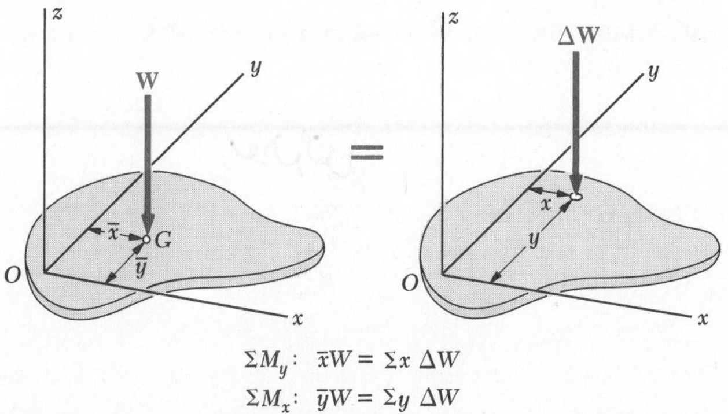

Fig. 5.1 Center of gravity of a plate.

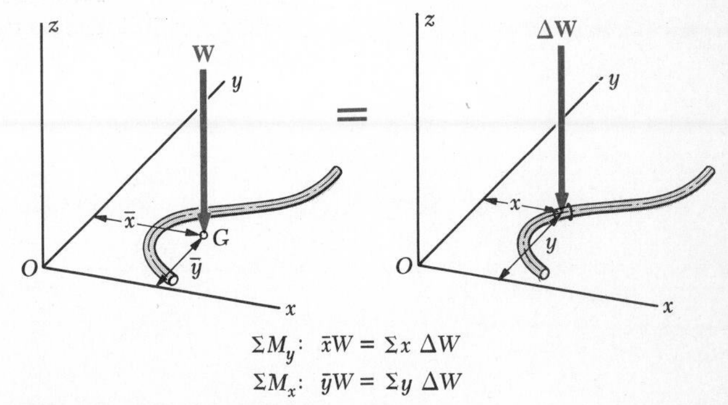

Fig. 5.2 Center of gravity of a wire.

5.3. Centroids of Areas and Lines. In the case of a homogeneous plate of uniform thickness, the magnitude ΔW of the weight of an element of plate may be expressed as

$$\Delta W = \gamma t \, \Delta A$$ Plate

where γ = specific weight (weight per unit volume) of material
t = thickness of plate
ΔA = area of element

Similarly, we may express the magnitude W of the weight of the entire plate in the form

$$W = \gamma t A$$

where A is the total area of the plate.

If U.S. customary units are used, the specific weight γ should be expressed in lb/ft^3, the thickness t in feet, and the areas ΔA and A in square feet. We check that ΔW and W will then be expressed in pounds. If SI units are used, γ should be expressed in N/m^3, t in meters, and the areas ΔA and A in square meters; the weights ΔW and W will then be expressed in newtons.†

Substituting for ΔW and W in the moment equations (5.1) and dividing throughout by γt, we write

$$\Sigma M_y: \quad \bar{x}A = x_1 \, \Delta A_1 + x_2 \, \Delta A_2 + \cdots + x_n \, \Delta A_n$$
$$\Sigma M_x: \quad \bar{y}A = y_1 \, \Delta A_1 + y_2 \, \Delta A_2 + \cdots + y_n \, \Delta A_n$$

If we increase the number of elements into which the area A is divided and simultaneously decrease the size of each element, we obtain at the limit

$$\bar{x}A = \int x \, dA \qquad \bar{y}A = \int y \, dA \tag{5.3}$$

These equations define the coordinates $\bar{x}$ and $\bar{y}$ of the center of gravity of a homogeneous plate. The point of coordinates $\bar{x}$ and $\bar{y}$ is also known as the *centroid C of the area A* of the plate (Fig. 5.3). If the plate is not homogeneous, the equations cannot be used to determine the center of gravity of the plate; they still define, however, the centroid of the area.

In the case of a homogeneous wire of uniform cross section, the magnitude ΔW of the weight of an element of wire may be expressed as

$$\Delta W = \gamma a \, \Delta L$$ Wire

where γ = specific weight of material
a = cross-sectional area of wire
ΔL = length of element

† It should be noted that in the SI system of units a given material is generally characterized by its density ρ (mass per unit volume) rather than by its specific weight γ. The specific weight of the material can then be obtained by writing

$$\gamma = \rho g$$

where $g = 9.81 \text{ m/s}^2$. Since ρ is expressed in kg/m^3, we check that γ will be expressed in $(kg/m^3)(m/s^2)$, that is, in N/m^3.

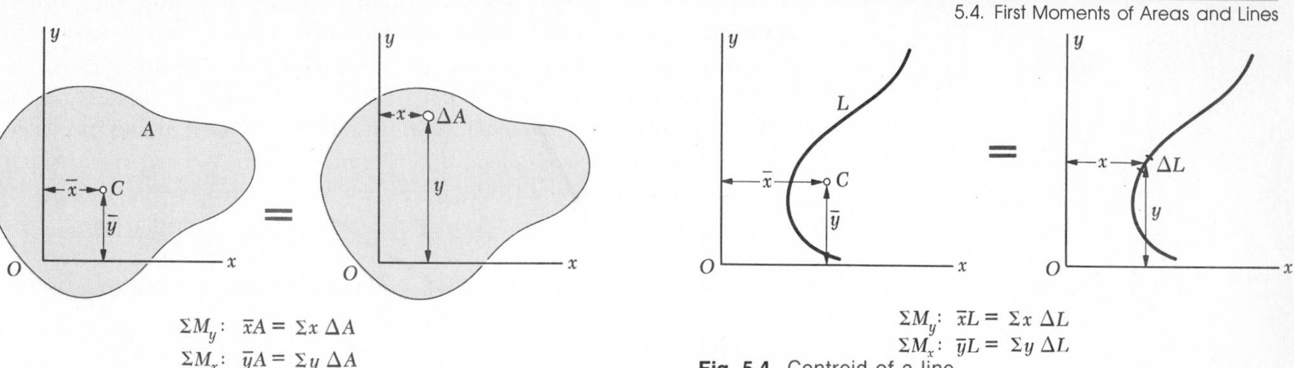

Fig. 5.3 Centroid of an area.

Fig. 5.4 Centroid of a line.

The center of gravity of the wire then coincides with the *centroid C of the line L* defining the shape of the wire (Fig. 5.4). The coordinates $\bar{x}$ and $\bar{y}$ of the centroid of the line L are obtained from the equations

$$\bar{x}L = \int x\,dL \qquad \bar{y}L = \int y\,dL \tag{5.4}$$

5.4. First Moments of Areas and Lines. The integral $\int x\,dA$ in Eqs. (5.3) of the preceding section is known as the *first moment of the area A with respect to the y axis* and is denoted by Q_y. Similarly, the integral $\int y\,dA$ defines the *first moment of A with respect to the x axis* and is denoted by Q_x. We write

$$Q_y = \int x\,dA \qquad Q_x = \int y\,dA \tag{5.5}$$

Comparing Eqs. (5.3) with Eqs. (5.5), we note that the first moments of the area A may be expressed as the products of the area and of the coordinates of its centroid:

$$Q_y = \bar{x}A \qquad Q_x = \bar{y}A \tag{5.6}$$

It follows from Eqs. (5.6) that the coordinates of the centroid of an area may be obtained by dividing the first moments of that area by the area itself. The computation of the first moments of an area is also useful in the determination of shearing stresses under transverse loadings in mechanics of materials. Finally, we observe from Eqs. (5.6) that if the centroid of an area is located on a coordinate axis, the first moment of the area with respect to that axis is zero, and conversely.

Relations similar to Eqs. (5.5) and (5.6) may be used to define the first moments of a line with respect to the coordinate axes and to express these moments as the products of the length L of the line and of the coordinates $\bar{x}$ and $\bar{y}$ of its centroid.

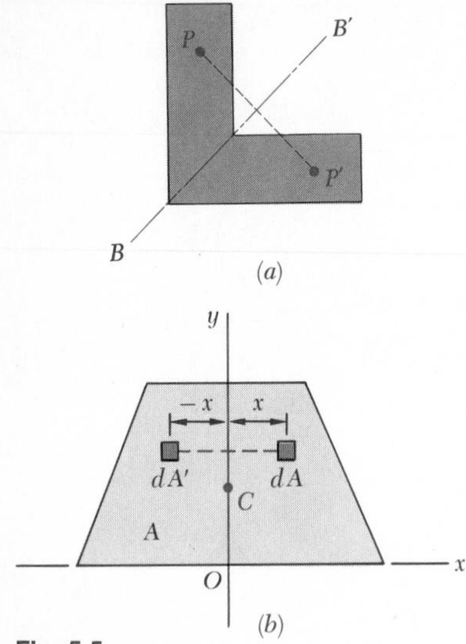

Fig. 5.5

An area A is said to be *symmetric with respect to an axis BB′* if to every point P of the area corresponds a point P' of the same area such that the line PP' is perpendicular to BB' and is divided into two equal parts by that axis (Fig. 5.5*a*). A line L is said to be symmetric with respect to BB' if it satisfies similar conditions. When an area A or a line L possesses an axis of symmetry BB', its first moment with respect to BB' is zero and its centroid is located on that axis. Indeed, considering the area A of Fig. 5.5*b*, which is symmetric with respect to the y axis, we observe that to every element of area dA of abscissa x corresponds an element of area dA' of abscissa $-x$. It follows that the integral in the first of Eqs. (5.5) is zero and, thus, that $Q_y = 0$. It also follows from the first of the relations (5.3) that $\bar{x} = 0$. Thus, if an area A or a line L possesses an axis of symmetry, its centroid C is located on that axis.

We further note that if an area or line possesses two axes of symmetry, its centroid C must be located at the intersection of the two axes of symmetry (Fig. 5.6). This property enables us to determine immediately the centroid of areas such as circles, ellipses, squares, rectangles, equilateral triangles, or other symmetric figures, as well as the centroid of lines in the shape of the circumference of a circle, the perimeter of a square, etc.

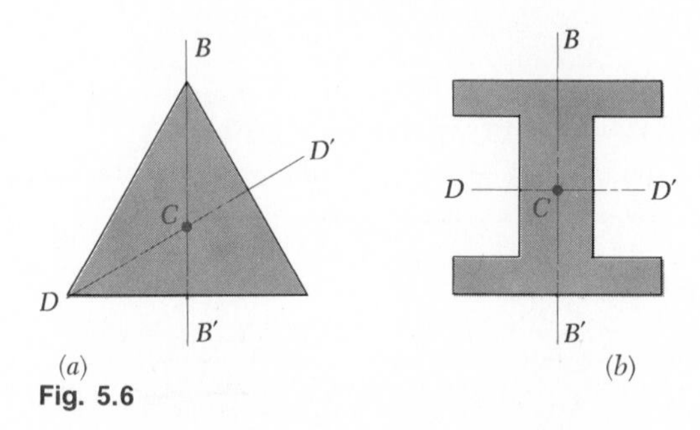

Fig. 5.6

An area A is said to be *symmetric with respect to a center O* if to every element of area dA of coordinates x and y corresponds an element of area dA' of coordinates $-x$ and $-y$ (Fig. 5.7). It then follows that the integrals in Eqs. (5.5) are both zero, and that $Q_x = Q_y = 0$. It also follows from Eqs. (5.3) that $\bar{x} = \bar{y} = 0$, that is, the centroid of the area coincides with its center of symmetry O. It may be shown in a similar way that the centroid of a line symmetric with respect to a center O coincides with O.

It should be noted that a figure possessing a center of symmetry does not necessarily possess an axis of symmetry (Fig. 5.7), while a figure possessing two axes of symmetry does not necessarily possess a center of symmetry (Fig. 5.6*a*). However, if a figure possesses two axes of symmetry at a right angle to each other, the point of intersection of these axes will be a center of symmetry (Fig. 5.6*b*).

Centroids of unsymmetrical areas and lines and of areas and lines possessing only one axis of symmetry will be determined by the methods of Secs. 5.6 and 5.7. Centroids of common shapes of areas and lines are shown in Fig. 5.8*A* and *B*. The formulas defining these centroids will be derived in the Sample Problems and Problems following Secs. 5.6 and 5.7.

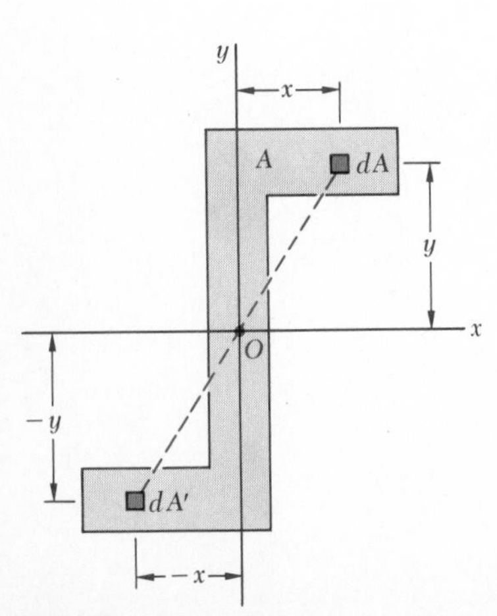

Fig. 5.7

Shape		$\bar{x}$	$\bar{y}$	Area
Triangular area	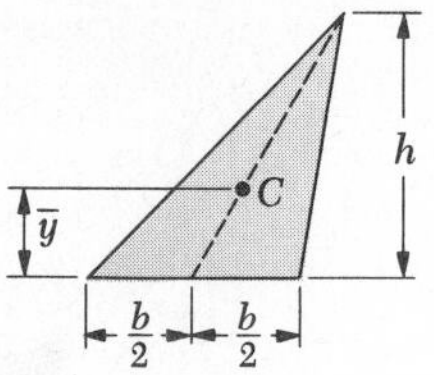		$\frac{h}{3}$	$\frac{bh}{2}$
Quarter-circular area	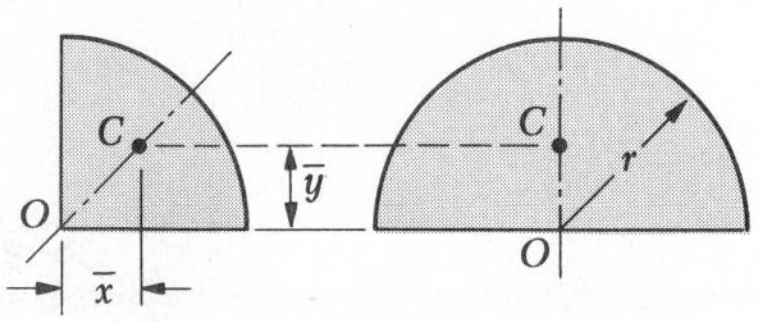	$\frac{4r}{3\pi}$	$\frac{4r}{3\pi}$	$\frac{\pi r^2}{4}$
Semicircular area		0	$\frac{4r}{3\pi}$	$\frac{\pi r^2}{2}$
Quarter-elliptical area	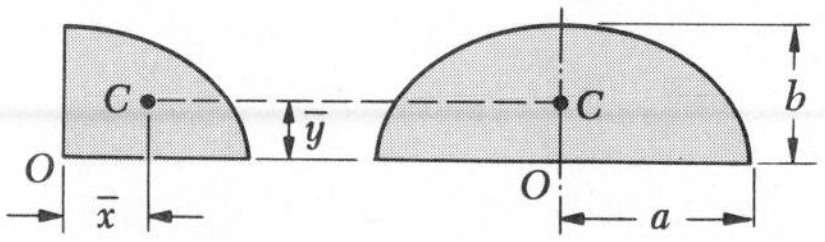	$\frac{4a}{3\pi}$	$\frac{4b}{3\pi}$	$\frac{\pi ab}{4}$
Semielliptical area		0	$\frac{4b}{3\pi}$	$\frac{\pi ab}{2}$
Semiparabolic area	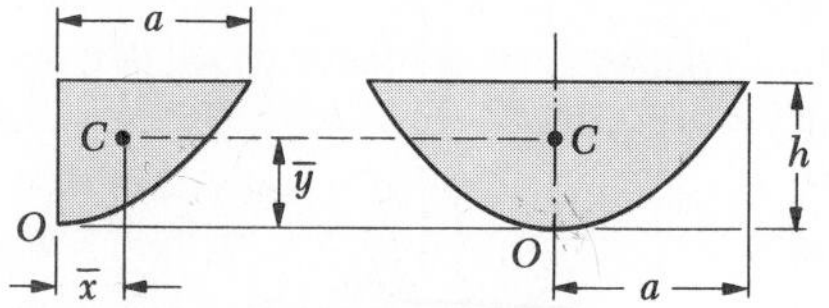	$\frac{3a}{8}$	$\frac{3h}{5}$	$\frac{2ah}{3}$
Parabolic area		0	$\frac{3h}{5}$	$\frac{4ah}{3}$
Parabolic spandrel	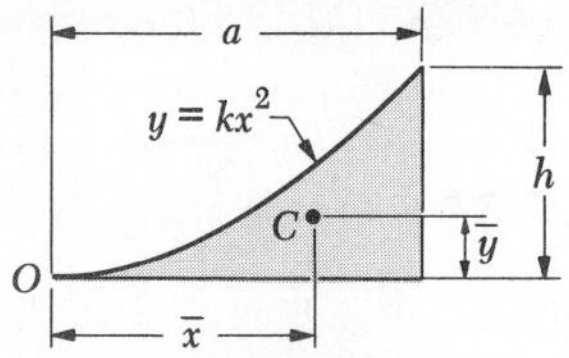 $y = kx^2$	$\frac{3a}{4}$	$\frac{3h}{10}$	$\frac{ah}{3}$
General spandrel	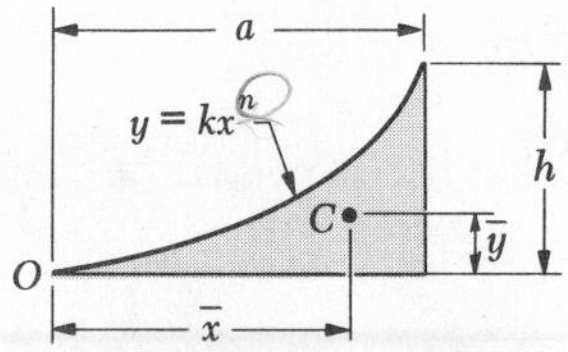 $y = kx^n$	$\frac{n+1}{n+2}a$	$\frac{n+1}{4n+2}h$	$\frac{ah}{n+1}$
Circular sector	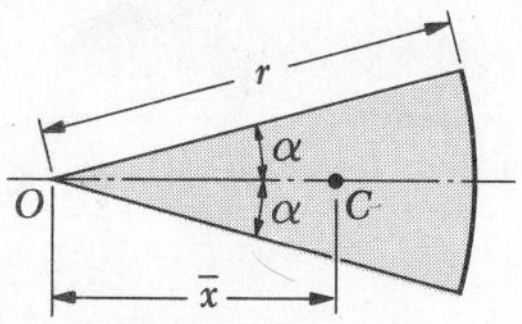	$\frac{2r \sin \alpha}{3\alpha}$	0	αr^2

Fig. 5.8A Centroids of common shapes of areas.

Shape		$\bar{x}$	$\bar{y}$	Length
Quarter-circular arc	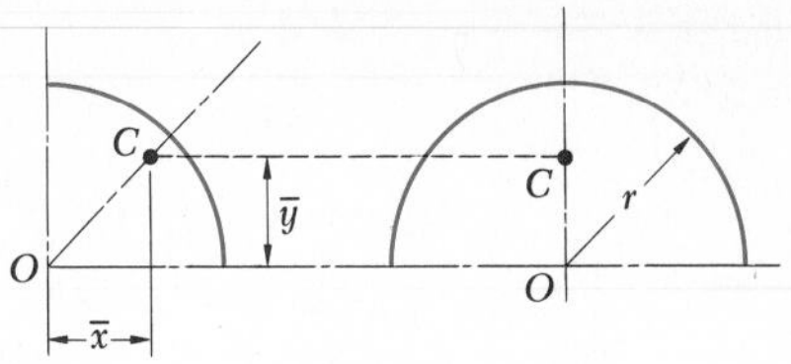	$\frac{2r}{\pi}$	$\frac{2r}{\pi}$	$\frac{\pi r}{2}$
Semicircular arc		0	$\frac{2r}{\pi}$	πr
Arc of circle	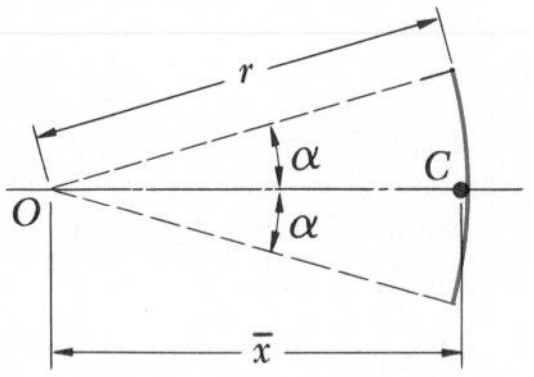	$\frac{r \sin \alpha}{\alpha}$	0	$2\alpha r$

Fig. 5.8B Centroids of common shapes of lines.

5.5. Composite Plates and Wires. In many instances, a flat plate may be divided into rectangles, triangles, or other common shapes shown in Fig. 5.8A. The abscissa $\bar{X}$ of its center of gravity G may be determined from the abscissas $\bar{x}_1, \bar{x}_2, \ldots$ of the centers of gravity of the various parts by expressing that the moment of the weight of the whole plate about the y axis is equal to the sum of the moments of the weights of the various parts about the same axis (Fig. 5.9). The ordinate $\bar{Y}$ of the center of gravity of the plate is found in a similar way by equating moments about the x axis. We write

$$\Sigma M_y: \quad \bar{X}(W_1 + W_2 + \cdots + W_n) = \bar{x}_1 W_1 + \bar{x}_2 W_2 + \cdots + \bar{x}_n W_n$$
$$\Sigma M_x: \quad \bar{Y}(W_1 + W_2 + \cdots + W_n) = \bar{y}_1 W_1 + \bar{y}_2 W_2 + \cdots + \bar{y}_n W_n$$

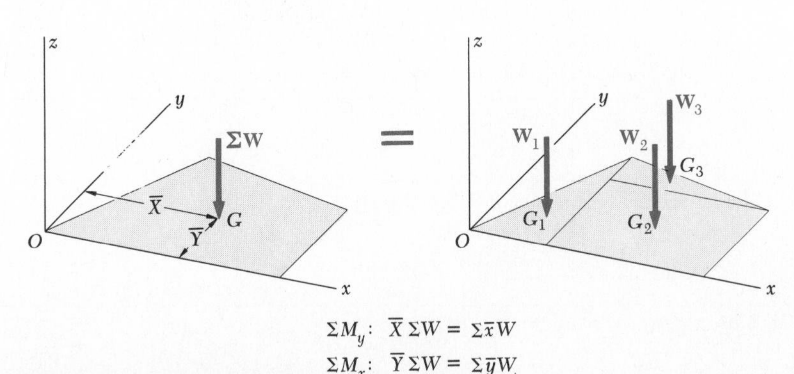

Fig. 5.9 Center of gravity of a composite plate.

or, for short,

$$\bar{X}\Sigma W = \Sigma\bar{x}W \qquad \bar{Y}\Sigma W = \Sigma\bar{y}W \tag{5.7}$$

These equations may be solved for the coordinates $\bar{X}$ and $\bar{Y}$ of the center of gravity of the plate.

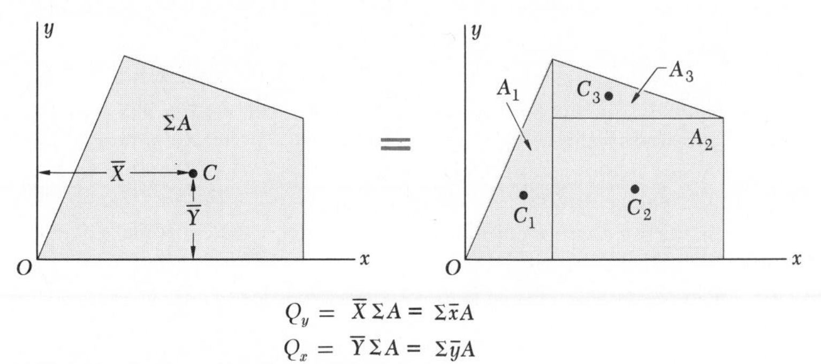

Fig. 5.10 Centroid of a composite area.

If the plate is homogeneous and of uniform thickness, the center of gravity coincides with the centroid C of its area. The abscissa $\bar{X}$ of the centroid of the area may be determined by noting that the first moment Q_y of the composite area with respect to the y axis may be expressed both as the product of $\bar{X}$ and the total area and as the sum of the first moments of the elementary areas with respect to the y axis (Fig. 5.10). The ordinate $\bar{Y}$ of the centroid is found in a similar way by considering the first moment Q_x of the composite area. We have

$$Q_y = \bar{X}(A_1 + A_2 + \cdots + A_n) = \bar{x}_1A_1 + \bar{x}_2A_2 + \cdots + \bar{x}_nA_n$$
$$Q_x = \bar{Y}(A_1 + A_2 + \cdots + A_n) = \bar{y}_1A_1 + \bar{y}_2A_2 + \cdots + \bar{y}_nA_n$$

or, for short,

$$Q_y = \bar{X}\Sigma A = \Sigma\bar{x}A \qquad Q_x = \bar{Y}\Sigma A = \Sigma\bar{y}A \tag{5.8}$$

These equations yield the first moments of the composite area, or they may be solved for the coordinates $\bar{X}$ and $\bar{Y}$ of its centroid.

Care should be taken to record the moment of each area with the appropriate sign. First moments of areas, just like moments of forces, may be positive or negative. For example, an area whose centroid is located to the left of the y axis will have a negative first moment with respect to that axis. Also, the area of a hole should be recorded with a negative sign (Fig. 5.11).

Similarly, it is possible in many cases to determine the center of gravity of a composite wire or the centroid of a composite line by dividing the wire or line into simpler elements (see Sample Prob. 5.2).

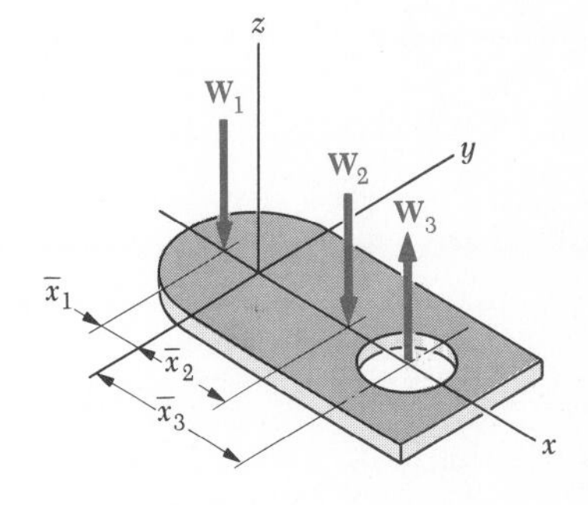

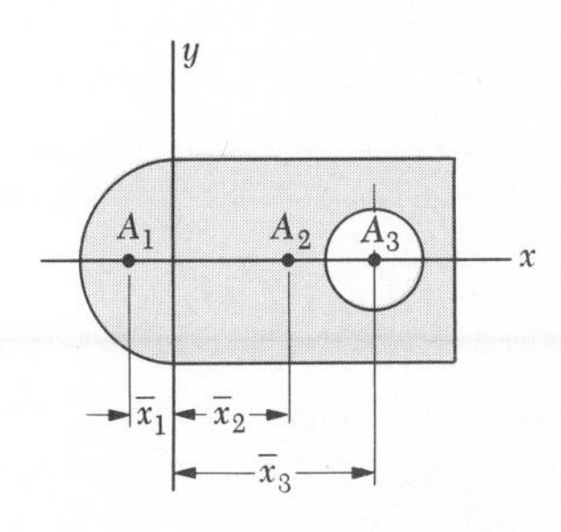

	$\bar{x}$	A	$\bar{x}A$
A_1 Semicircle	−	+	−
A_2 Full rectangle	+	+	+
A_3 Circular hole	+	−	−

Fig. 5.11

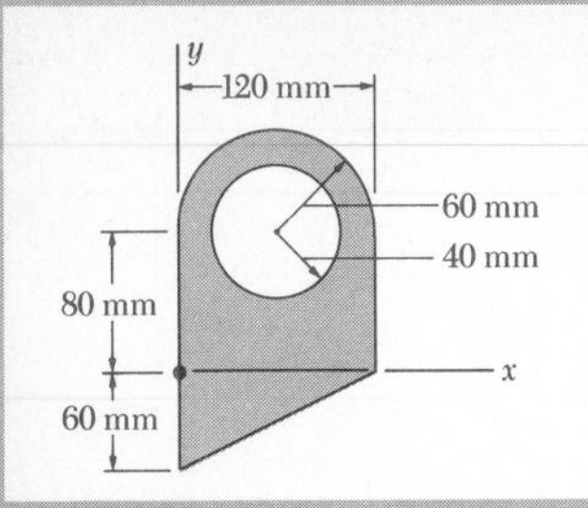

SAMPLE PROBLEM 5.1

For the plane area shown, determine (*a*) the first moments with respect to the x and y axes, (*b*) the location of the centroid.

Solution. The area is obtained by adding a rectangle, a triangle, and a semicircle and subtracting a circle. Using the coordinate axes shown, the area and the coordinates of the centroid of each of the component areas are determined and entered in the table below. The area of the circle is indicated as negative, since it is to be subtracted from the other areas. We note that the coordinate $\bar{y}$ of the centroid of the triangle is negative for the axes shown. The first moments of the component areas with respect to the coordinate axes are computed and entered in the table.

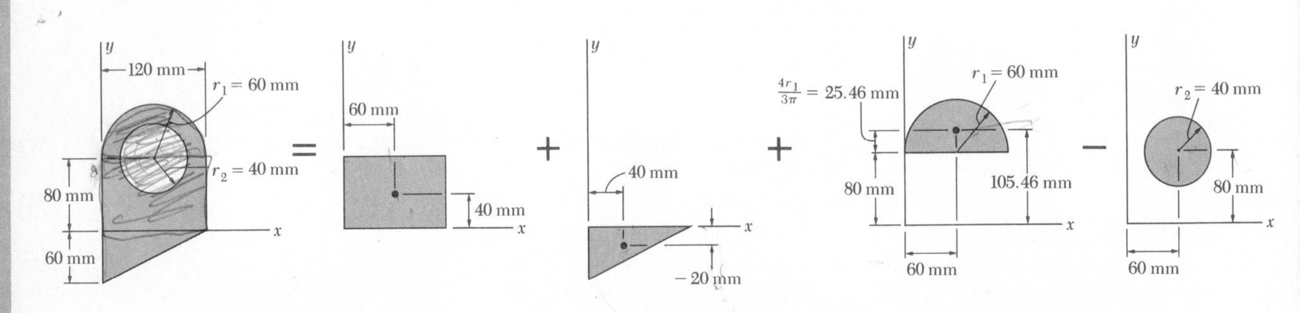

Component	A, mm^2	$\bar{x}$, mm	$\bar{y}$, mm	$\bar{x}A$, mm^3	$\bar{y}A$, mm^3
Rectangle	$(120)(80) = 9.6 \times 10^3$	60	40	$+576 \times 10^3$	$+384 \times 10^3$
Triangle	$\frac{1}{2}(120)(60) = 3.6 \times 10^3$	40	−20	$+144 \times 10^3$	-72×10^3
Semicircle	$\frac{1}{2}\pi(60)^2 = 5.655 \times 10^3$	60	105.46	$+339.3 \times 10^3$	$+596.4 \times 10^3$
Circle	$-\pi(40)^2 = -5.027 \times 10^3$	60	80	-301.6×10^3	-402.2×10^3
	$\Sigma A = 13.828 \times 10^3$			$\Sigma\bar{x}A = +757.7 \times 10^3$	$\Sigma\bar{y}A = +506.2 \times 10^3$

a. First Moments of Area. Using Eqs. (5.8), we write

$$Q_x = \Sigma\bar{y}A = 506.2 \times 10^3 \text{ mm}^3$$

$$Q_x = 506 \times 10^3 \text{ mm}^3 \quad ◀$$

$$Q_y = \Sigma\bar{x}A = 757.7 \times 10^3 \text{ mm}^3$$

$$Q_y = 758 \times 10^3 \text{ mm}^3 \quad ◀$$

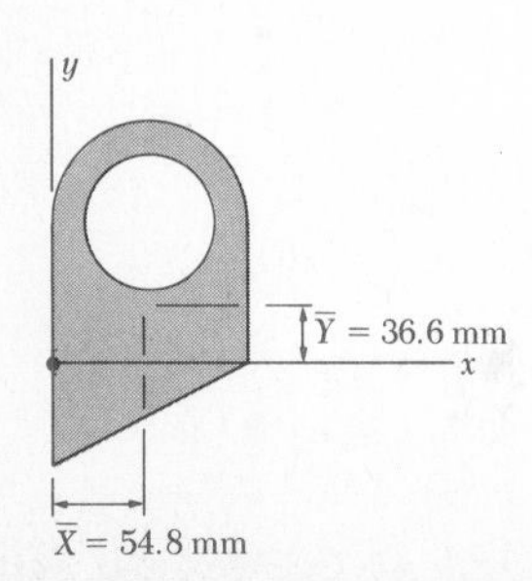

b. Location of Centroid. Substituting the values obtained from the table into the equations defining the centroid of a composite area, we obtain

$$\bar{X}\Sigma A = \Sigma\bar{x}A: \qquad \bar{X}(13.828 \times 10^3 \text{ mm}^2) = 757.7 \times 10^3 \text{ mm}^3$$

$$\bar{X} = 54.8 \text{ mm} \quad ◀$$

$$\bar{Y}\Sigma A = \Sigma\bar{y}A: \qquad \bar{Y}(13.828 \times 10^3 \text{ mm}^2) = 506.2 \times 10^3 \text{ mm}^3$$

$$\bar{Y} = 36.6 \text{ mm} \quad ◀$$

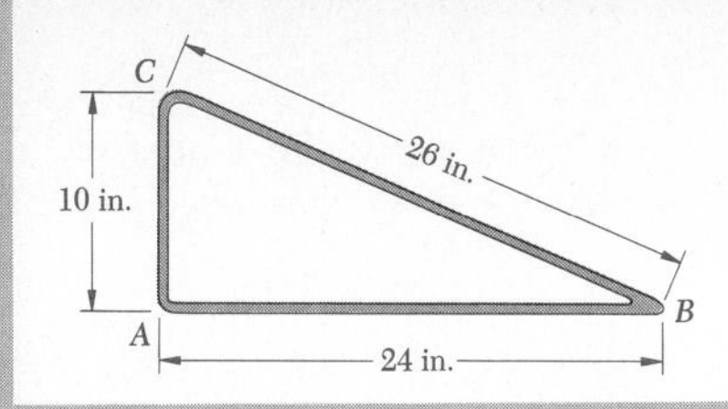

SAMPLE PROBLEM 5.2

The figure shown is made of a thin homogeneous wire. Determine its center of gravity.

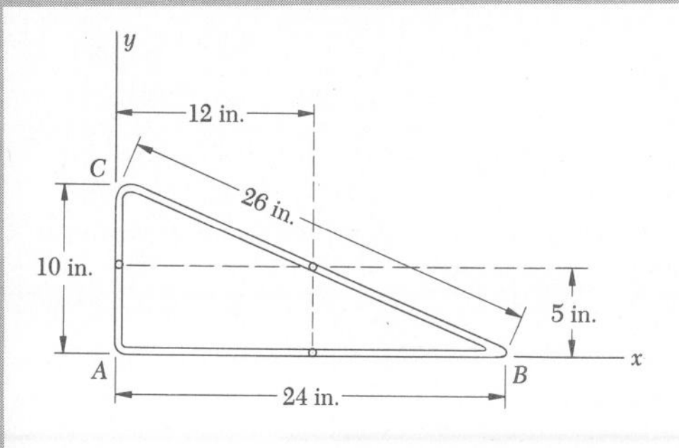

Solution. Since the figure is formed of homogeneous wire, its center of gravity may be located by determining the centroid of the corresponding line. Choosing the coordinate axes shown, with origin at A, we determine the coordinates of the centroid of each segment of line and compute its first moments with respect to the coordinate axes.

Segment	L, in.	$\bar{x}$, in.	$\bar{y}$, in.	$\bar{x}L$, in^2	$\bar{y}L$, in^2
AB	24	12	0	288	0
BC	26	12	5	312	130
CA	10	0	5	0	50
	$\Sigma L = 60$	. . .	. . .	$\Sigma \bar{x}L = 600$	$\Sigma \bar{y}L = 180$

Substituting the values obtained from the table into the equations defining the centroid of a composite line, we obtain

$\bar{X}\Sigma L = \Sigma \bar{x}L$: $\quad \bar{X}(60 \text{ in.}) = 600 \text{ in}^2 \qquad \bar{X} = 10 \text{ in.}$ ◄

$\bar{Y}\Sigma L = \Sigma \bar{y}L$: $\quad \bar{Y}(60 \text{ in.}) = 180 \text{ in}^2 \qquad \bar{Y} = 3 \text{ in.}$ ◄

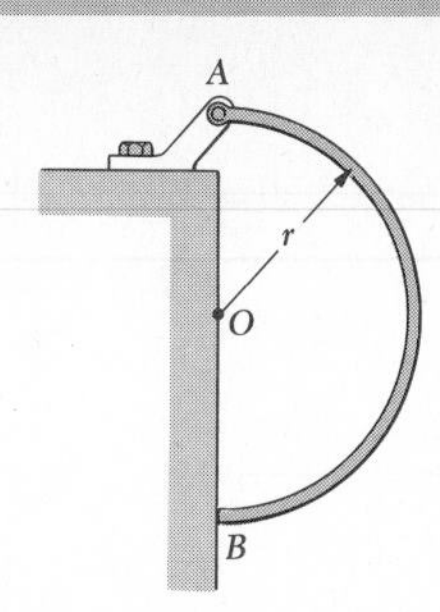

SAMPLE PROBLEM 5.3

A uniform semicircular rod of weight W and radius r is attached to a pin at A and bears against a frictionless surface at B. Determine the reactions at A and B.

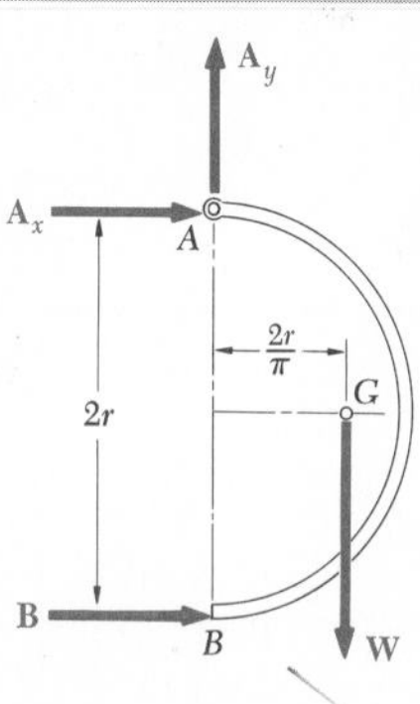

Solution. A free-body diagram of the rod is drawn. The forces acting on the rod consist of its weight **W** applied at the center of gravity G, whose position is obtained from Fig. 5.8B, of a reaction at A represented by its components $\mathbf{A}_x$ and $\mathbf{A}_y$, and of a horizontal reaction at B.

$$+\circlearrowleft \Sigma M_A = 0: \qquad B(2r) - W\left(\frac{2r}{\pi}\right) = 0$$

$$B = +\frac{W}{\pi} \qquad \qquad \mathbf{B} = \frac{W}{\pi} \rightarrow \blacktriangleleft$$

$$\xrightarrow{+} \Sigma F_x = 0: \qquad A_x + B = 0$$

$$A_x = -B = -\frac{W}{\pi} \qquad \qquad \mathbf{A}_x = \frac{W}{\pi} \leftarrow$$

$$+\uparrow \Sigma F_y = 0: \qquad A_y - W = 0 \qquad \qquad \mathbf{A}_y = W \uparrow$$

Adding the two components of the reaction at A:

$$A = \left[W^2 + \left(\frac{W}{\pi}\right)^2\right]^{1/2} \qquad \qquad A = W\left(1 + \frac{1}{\pi^2}\right)^{1/2} \blacktriangleleft$$

$$\tan \alpha = \frac{W}{W/\pi} = \pi \qquad \qquad \alpha = \tan^{-1} \pi \blacktriangleleft$$

The answers may also be expressed as follows:

$$\mathbf{A} = 1.049W \angle 72.3^\circ \qquad \mathbf{B} = 0.318W \rightarrow \blacktriangleleft$$

Problems

5.1 through 5.11 Locate the centroid of the plane area shown.

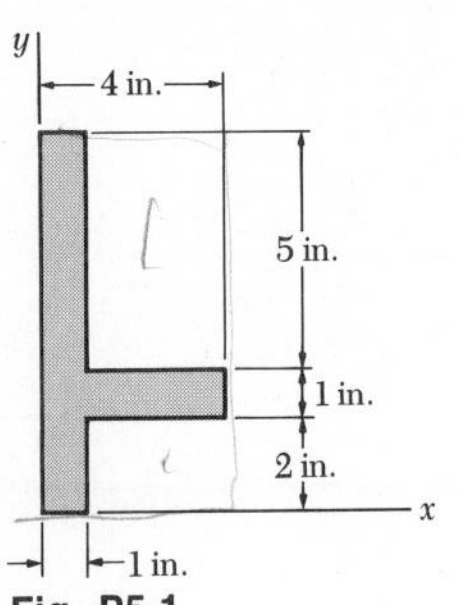

Fig. P5.1

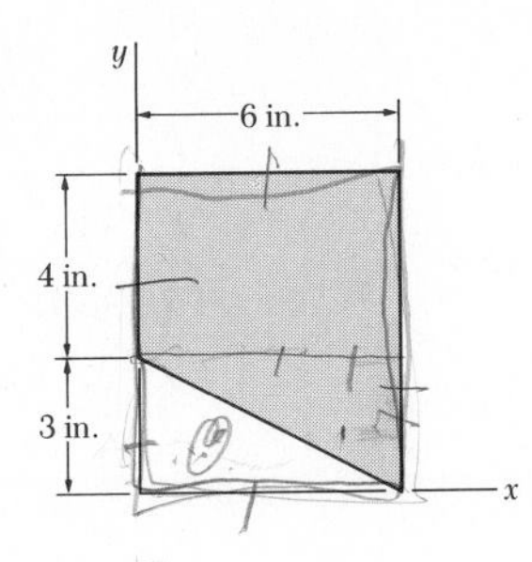

Fig. P5.2

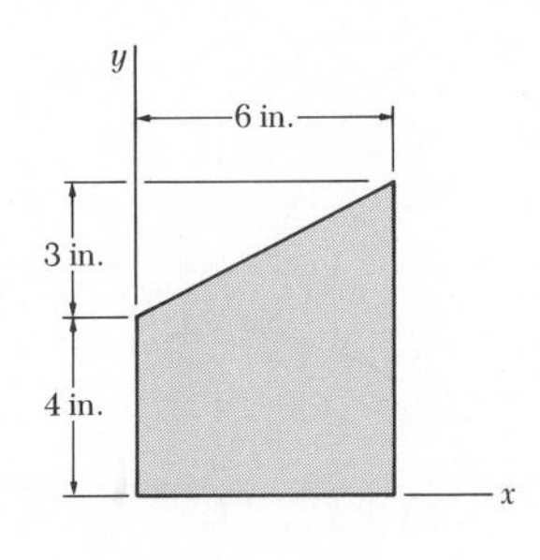

Fig. P5.3

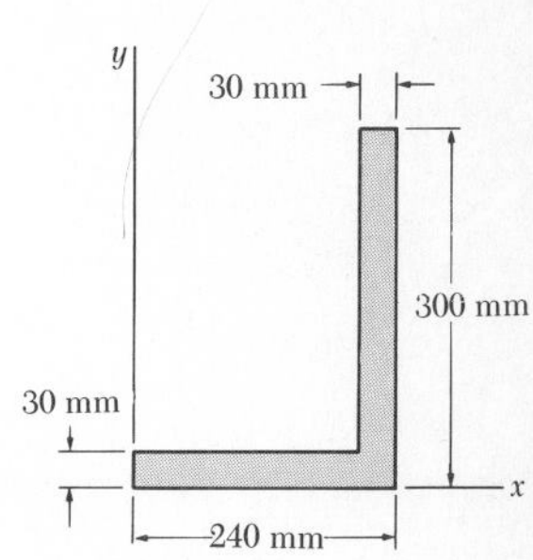

Fig. P5.4

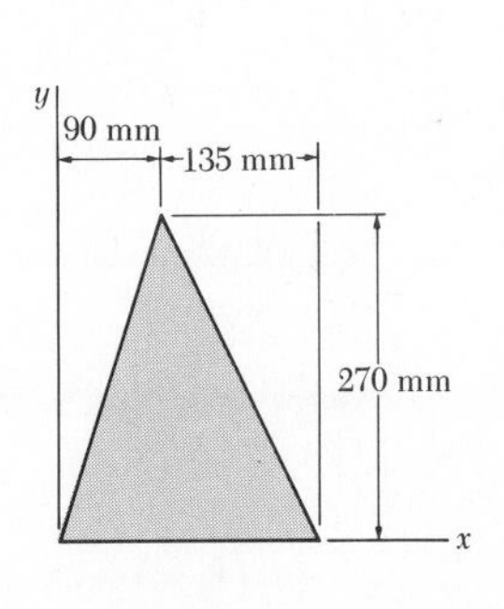

Fig. P5.5

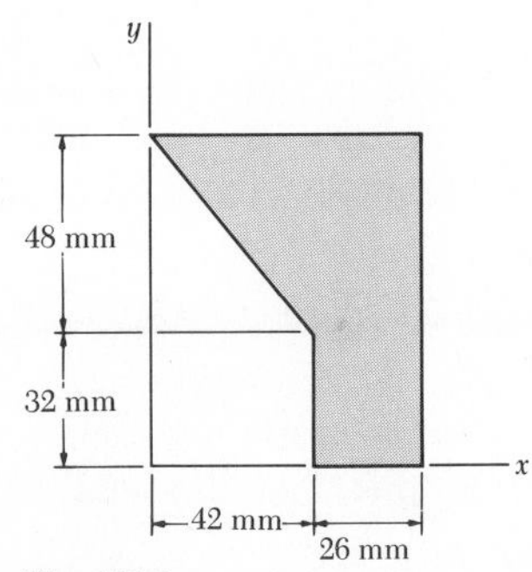

Fig. P5.6

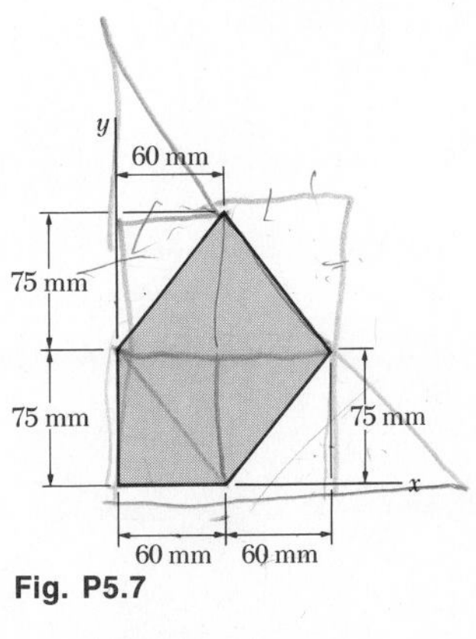

Fig. P5.7

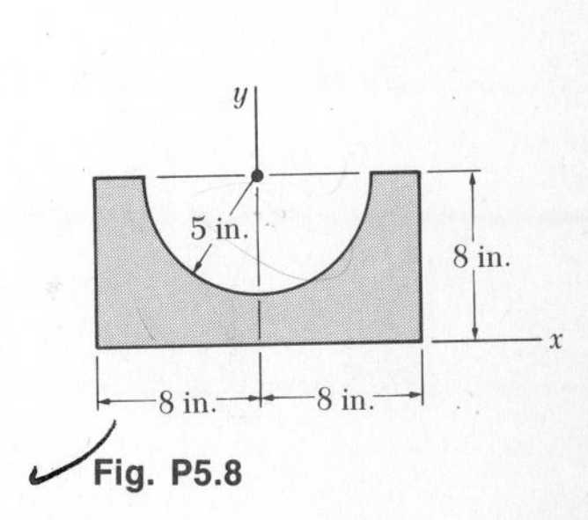

Fig. P5.8

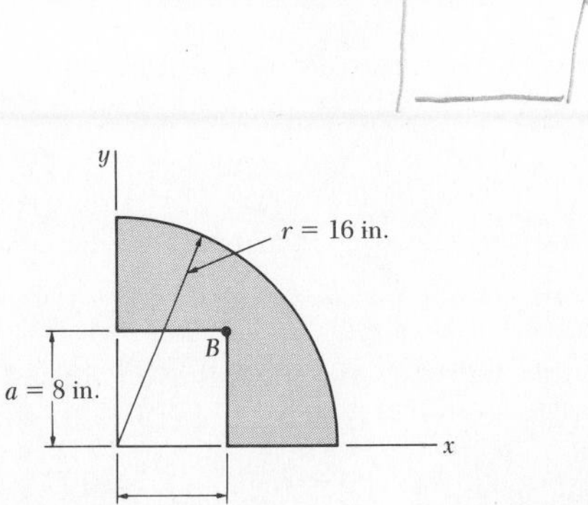

Fig. P5.9

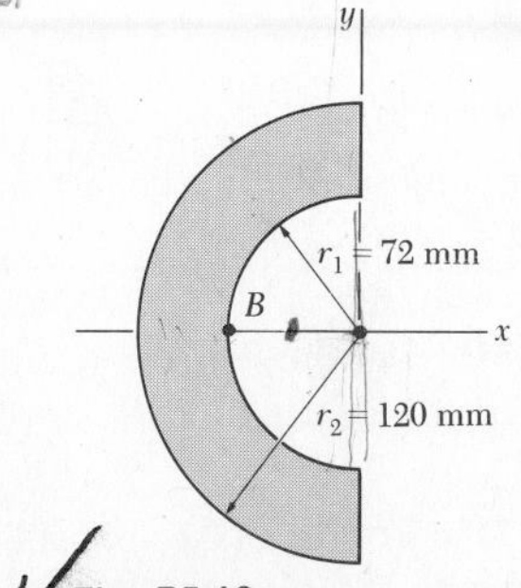

Fig. P5.10

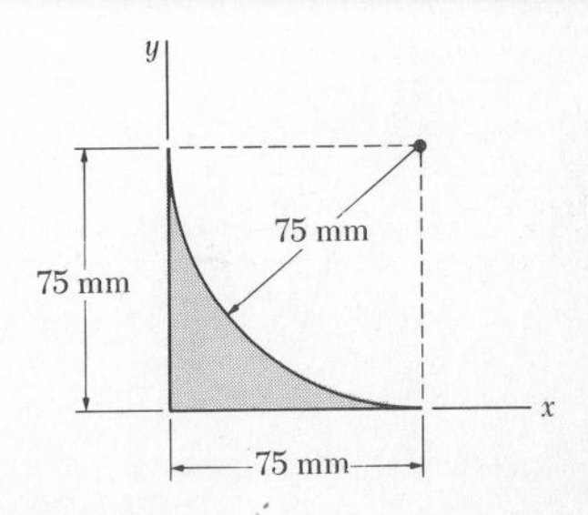

Fig. P5.11

5.12 through 5.15 Locate the centroid of the plane area shown.

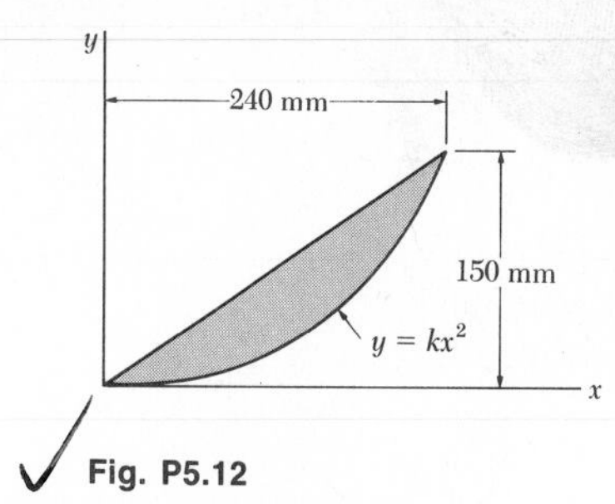

Fig. P5.12

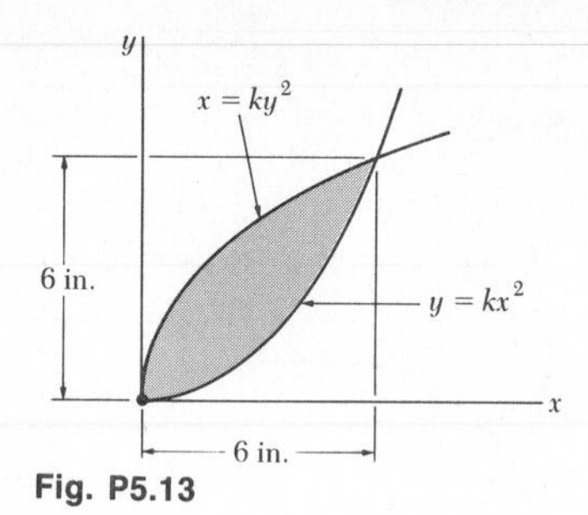

Fig. P5.13

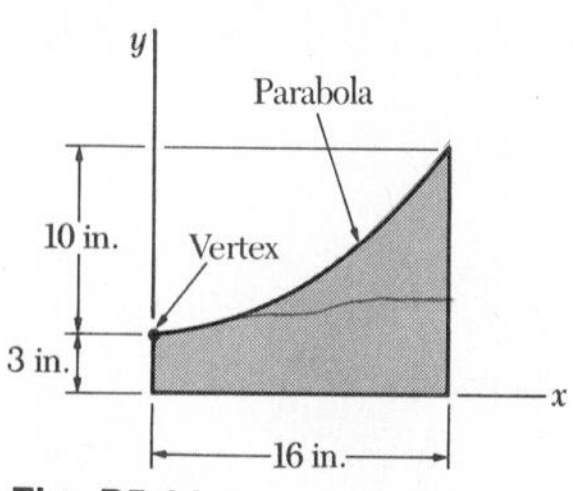

Fig. P5.14

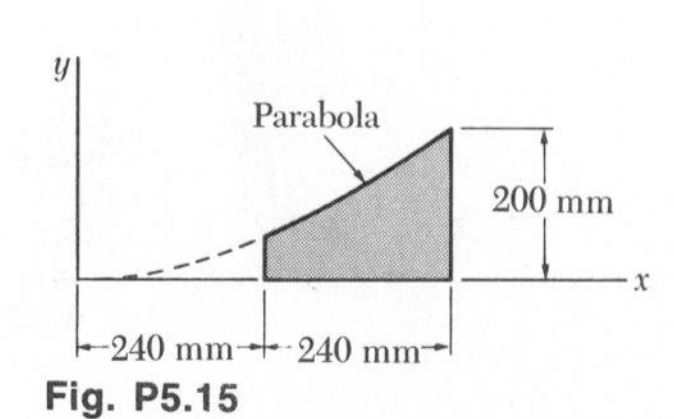

Fig. P5.15

5.16 Determine the abscissa of the centroid of the circular segment in terms of r and α.

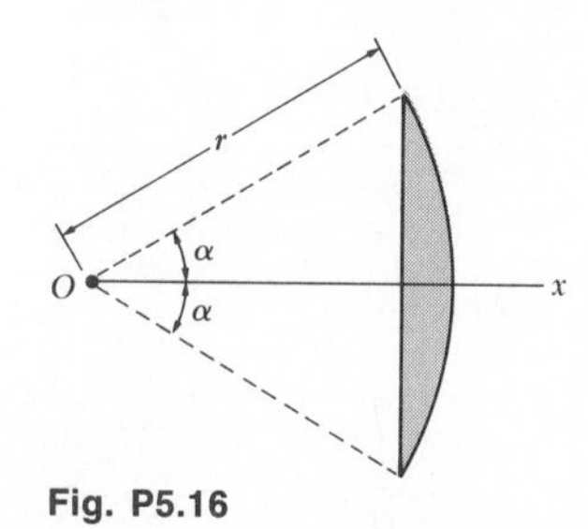

Fig. P5.16

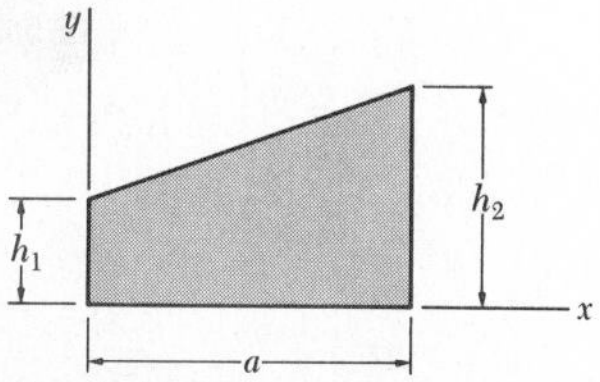

Fig. P5.17

5.17 Determine the abscissa of the centroid of the trapezoid shown in terms of h_1, h_2, and a.

5.18 Locate the centroid C in Prob. 5.10 in terms of r_1 and r_2 and show that as r_1 approaches r_2, the location of C approaches that for a semicircular arc of radius $\frac{1}{2}(r_1 + r_2)$.

5.19 For the semiannular area of Prob. 5.10, determine the ratio of r_1 to r_2 for which the centroid of the area is located at $x = -\frac{1}{2}r_2$ and $y = 0$.

5.20 For the plane area of Prob. 5.9, determine the ratio of a to r for which the centroid of the area is located at point B.

5.21 For the semiannular area of Prob. 5.10, determine the ratio of r_1 to r_2 for which the centroid of the area is located at point B.

5.22 A built-up beam has been constructed by nailing together seven planks as shown. The nails are equally spaced along the beam and the beam supports a vertical load. As proved in mechanics of materials, the shearing forces exerted on the nails at A and B are proportional to the first moments with respect to the centroidal x axis of the shaded areas shown, respectively, in parts a and b of the figure. Knowing that the force exerted on the nail at A is 120 N, determine the force exerted on the nail at B.

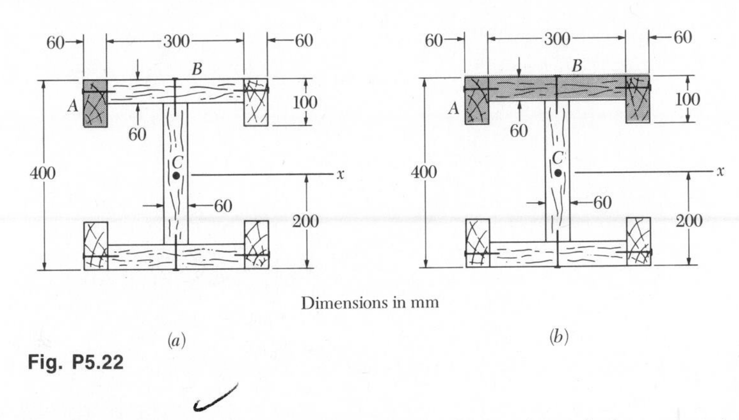

Fig. P5.22

5.23 and 5.24 The horizontal x axis is drawn through the centroid C of the area shown and divides it into two component areas A_1 and A_2. Determine the first moment of each component area with respect to the x axis and explain the result obtained.

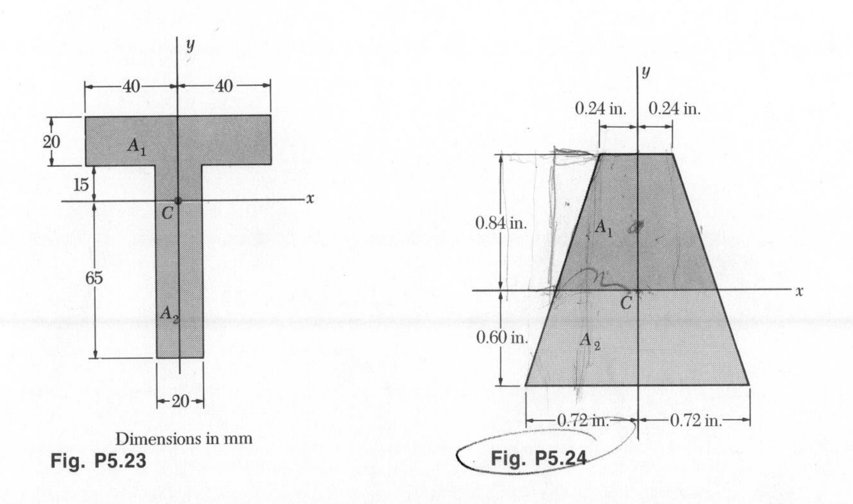

Fig. P5.23

Fig. P5.24

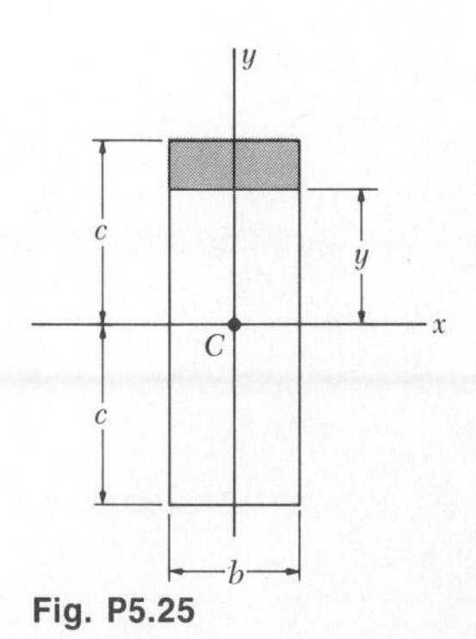

Fig. P5.25

5.25 The first moment of the shaded area with respect to the x axis is denoted by Q_x. (*a*) Express Q_x in terms of b, c, and the distance y from the base of the shaded area to the x axis. (*b*) For what value of y is Q_x maximum and what is that maximum value?

5.26 through 5.29 A thin homogeneous wire is bent to form the perimeter of the figure indicated. Locate the center of gravity of the wire figure thus formed.

5.26 Fig. P5.2
5.27 Fig. P5.3
5.28 Fig. P5.4
5.29 Fig. P5.5

5.30 A uniform circular rod of weight 8 lb and radius 10 in. is attached to a pin at *C* and to the cable *AB*. Determine (*a*) the tension in the cable, (*b*) the reaction at *C*.

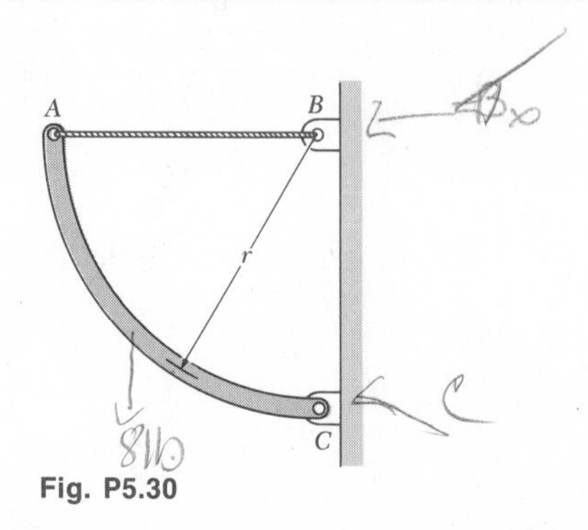

Fig. P5.30

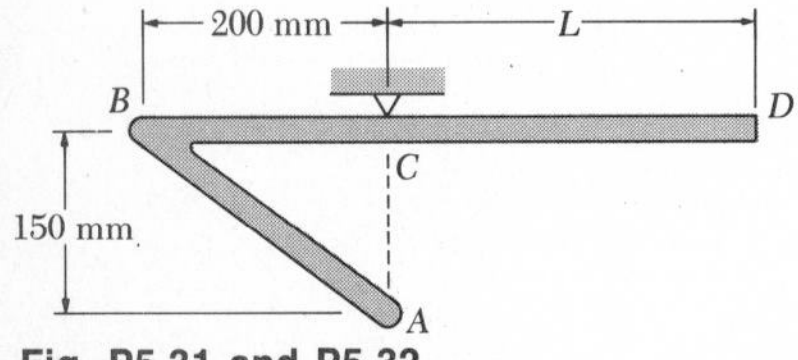

Fig. P5.31 and P5.32

5.31 The homogeneous wire *ABCD* is bent as shown and is attached to a hinge at *C*. Determine the length *L* for which portion *BCD* of the wire is horizontal.

5.32 The homogeneous wire *ABCD* is bent as shown and is attached to a hinge at *C*. Determine the length *L* for which portion *AB* of the wire is horizontal.

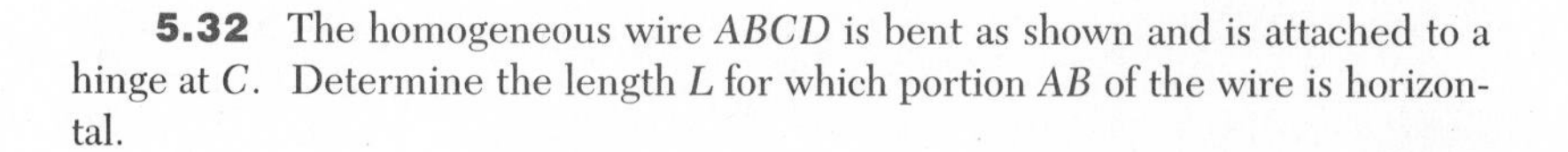

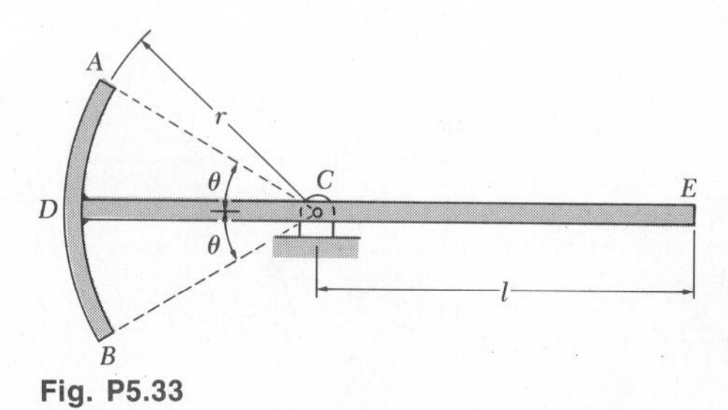

Fig. P5.33

5.33 Knowing that the figure shown is formed of a thin homogeneous wire, determine the length l of portion *CE* of the wire for which the center of gravity of the figure is located at point *C*, when (*a*) $\theta = 15°$, (*b*) $\theta = 60°$.

5.34 Determine the distance h for which the centroid of the shaded area is as high above line BB' as possible when (*a*) $k = 0.10$, (*b*) $k = 0.80$.

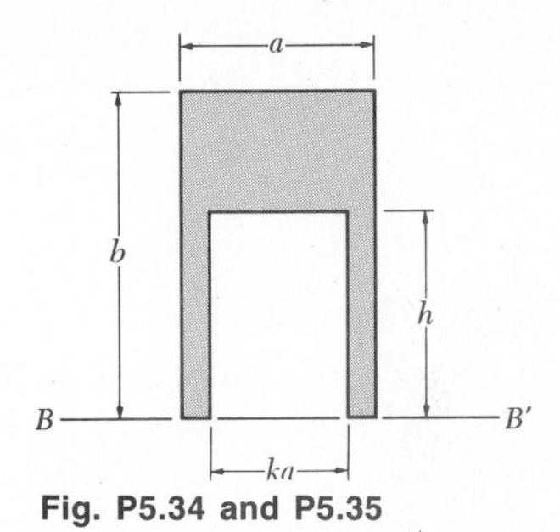

Fig. P5.34 and P5.35

5.35 Show that when the distance h is selected to maximize the distance $\bar{y}$ from line BB' to the centroid of the shaded area, we also have $\bar{y} = h$.

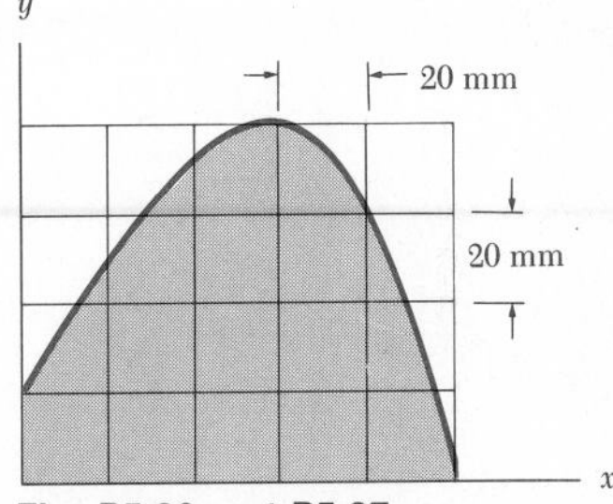

Fig. P5.36 and P5.37

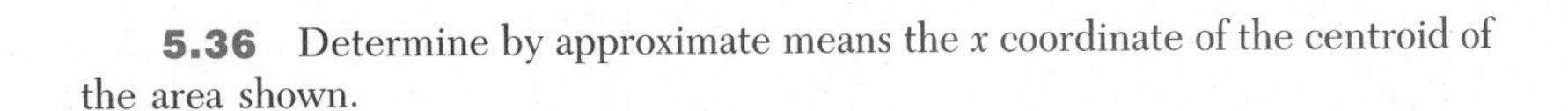

5.36 Determine by approximate means the x coordinate of the centroid of the area shown.

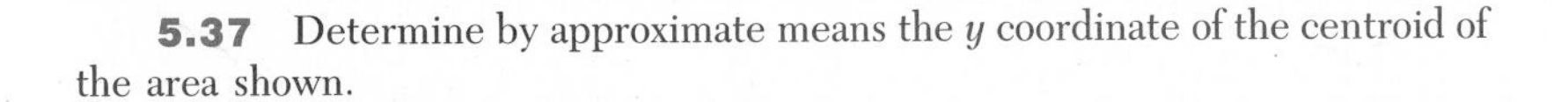

5.37 Determine by approximate means the y coordinate of the centroid of the area shown.

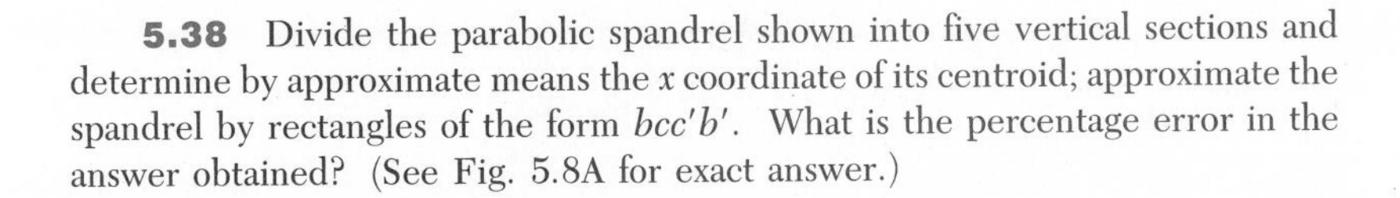

5.38 Divide the parabolic spandrel shown into five vertical sections and determine by approximate means the x coordinate of its centroid; approximate the spandrel by rectangles of the form $bcc'b'$. What is the percentage error in the answer obtained? (See Fig. 5.8A for exact answer.)

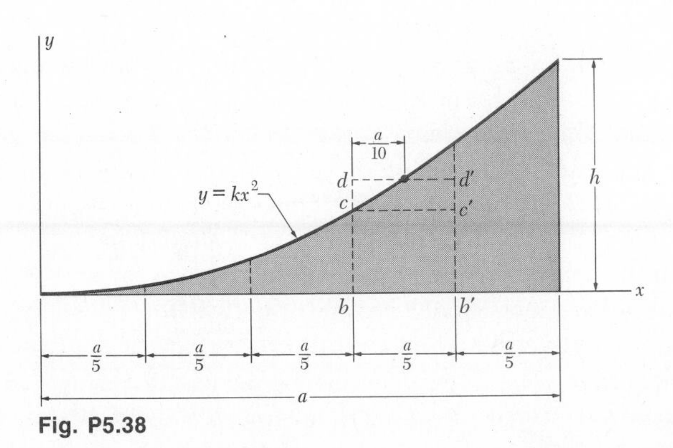

Fig. P5.38

5.39 Solve Prob. 5.38, using rectangles of the form $bdd'b'$.

5.6. Determination of Centroids by Integration. The centroid of an area bounded by analytical curves (i.e., curves defined by algebraic equations) is usually determined by computing the integrals in Eqs. (5.3) of Sec. 5.3:

$$\bar{x}A = \int x\,dA \qquad \bar{y}A = \int y\,dA \tag{5.3}$$

If the element of area dA is chosen equal to a small square of sides dx and dy, the determination of each of these integrals requires a *double integration* in x and y. A double integration is also necessary if polar coordinates are used and if dA is chosen equal to a small square of sides dr and $r\,d\theta$.

In most cases, however, it is possible to determine the coordinates of the centroid of an area by performing a single integration. This is achieved by choosing for dA a thin rectangle or strip, or a thin sector or pie-shaped element (Fig. 5.12); the centroid of the thin rectangle is located at its center, and the centroid of the thin sector at a distance $\frac{2}{3}r$ from its vertex (as for a triangle). The coordinates of the centroid of the area under consideration are then obtained by expressing that the first moment of the entire area with respect to each of the coordinate axes is equal to the sum (or integral) of the corresponding moments of the elements of area. Denoting by $\bar{x}_{el}$ and $\bar{y}_{el}$ the coordinates of the centroid of the element dA, we write

$$\begin{aligned} Q_y &= \bar{x}A = \int \bar{x}_{el}\,dA \\ Q_x &= \bar{y}A = \int \bar{y}_{el}\,dA \end{aligned} \tag{5.9}$$

If the area itself is not already known, it may also be computed from these elements.

The coordinates $\bar{x}_{el}$ and $\bar{y}_{el}$ of the centroid of the element of area should be expressed in terms of the coordinates of a point located on the curve bounding the area under consideration. Also, the element of area dA should be expressed in terms of the coordinates of the point and their differentials. This has been done in Fig. 5.12 for three common types of elements; the pie-shaped element of part c should be used when the equation of the curve bounding the area is given in polar coordinates. The appropriate expressions should be substituted in formulas (5.9), and the equation of the curve should be used to express one of the coordinates in terms of the other. The integration is thus reduced to a single integration which may be performed according to the usual rules of calculus. Once the area and the integrals in Eqs. (5.9) have been determined, these equations may be solved for the coordinates $\bar{x}$ and $\bar{y}$ of the centroid of the area.

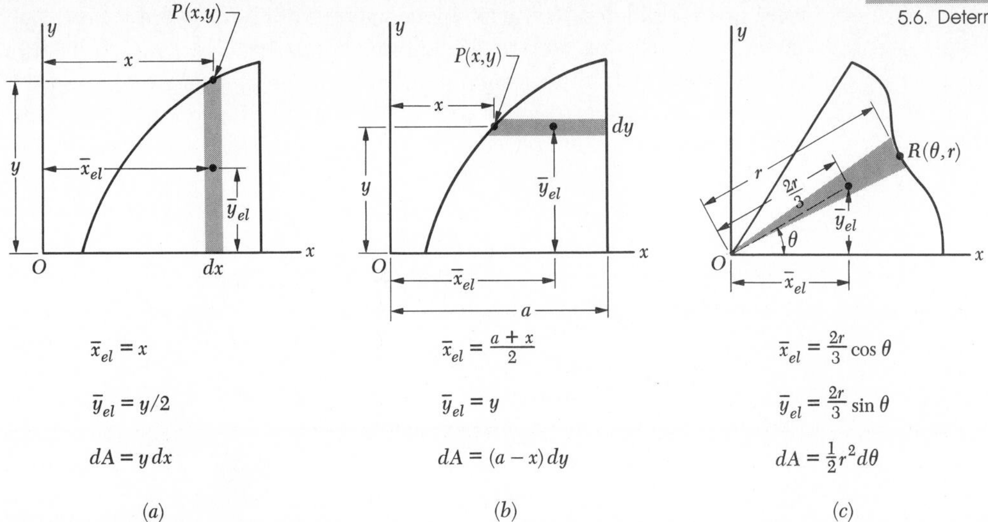

Fig. 5.12 Centroids and areas of differential elements.

The centroid of a line defined by an algebraic equation may be determined by computing the integrals in Eqs. (5.4) of Sec. 5.3:

$$\bar{x}L = \int x\,dL \qquad \bar{y}L = \int y\,dL \tag{5.4}$$

The element dL should be replaced by one of the following expressions, depending upon the type of equation used to define the line (these expressions may be derived by using the Pythagorean theorem):

$$dL = \sqrt{1 + \left(\frac{dy}{dx}\right)^2}\,dx$$

$$dL = \sqrt{1 + \left(\frac{dx}{dy}\right)^2}\,dy$$

$$dL = \sqrt{r^2 + \left(\frac{dr}{d\theta}\right)^2}\,d\theta$$

The equation of the line is then used to express one of the coordinates in terms of the other, the integration may be performed by the methods of calculus, and Eqs. (5.4) may be solved for the coordinates $\bar{x}$ and $\bar{y}$ of the centroid of the line.

5.7. Theorems of Pappus-Guldinus. These theorems, which were first formulated by the Greek geometer Pappus during the third century A.D. and later restated by the Swiss mathematician Guldinus, or Guldin (1577–1643), deal with surfaces and bodies of revolution.

A *surface of revolution* is a surface which may be generated by rotating a plane curve about a fixed axis. For example (Fig. 5.13), the surface of

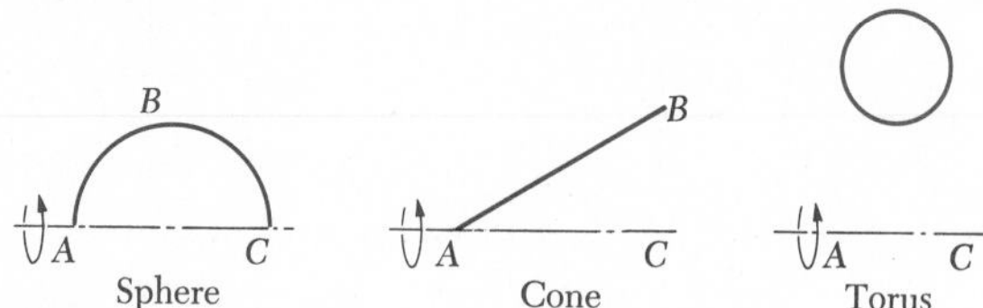

Fig. 5.13 Generating a surface of revolution.

a sphere may be obtained by rotating a semicircular arc ABC about the diameter AC; the surface of a cone by rotating a straight line AB about an axis AC; the surface of a torus or ring by rotating the circumference of a circle about a nonintersecting axis. A *body of revolution* is a body which may be generated by rotating a plane area about a fixed axis. A solid sphere may be obtained by rotating a semicircular area, a cone by rotating a triangular area, and a solid torus by rotating a full circular area (Fig. 5.14).

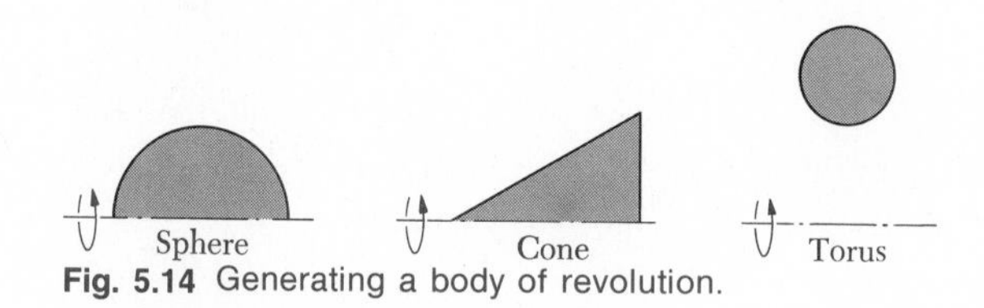

Fig. 5.14 Generating a body of revolution.

THEOREM I. *The area of a surface of revolution is equal to the length of the generating curve times the distance traveled by the centroid of the curve while the surface is being generated.*

Proof. Consider an element dL of the line L (Fig. 5.15) which is revolved about the x axis. The area dA generated by the element dL is equal to $2\pi y\, dL$. Thus, the entire area generated by L is $A = \int 2\pi y\, dL$. But we saw in Sec. 5.3 that the integral $\int y\, dL$ is equal to $\bar{y}L$. We have therefore

$$A = 2\pi\bar{y}L \tag{5.10}$$

where $2\pi\bar{y}$ is the distance traveled by the centroid of L. It should be noted

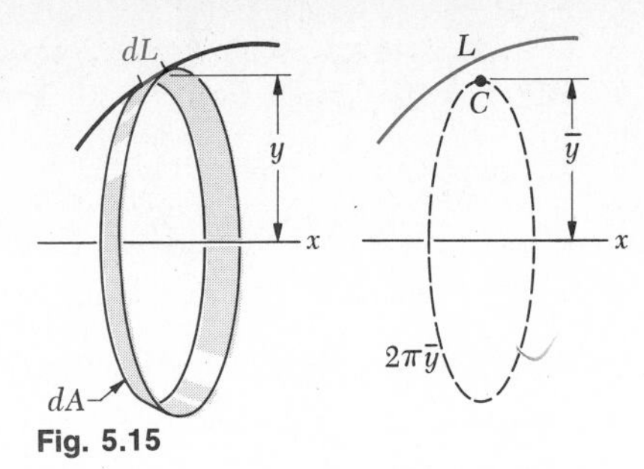

Fig. 5.15

that the generating curve should not cross the axis about which it is rotated; if it did, the two sections on either side of the axis would generate areas of opposite signs and the theorem would not apply.

THEOREM II. *The volume of a body of revolution is equal to the generating area times the distance traveled by the centroid of the area while the body is being generated.*

Proof. Consider an element dA of the area A which is revolved about the x axis (Fig. 5.16). The volume dV generated by the element dA is equal

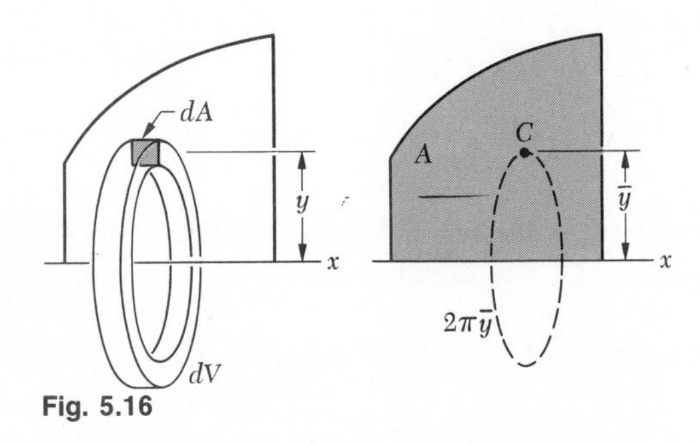

Fig. 5.16

to $2\pi y\, dA$. Thus, the entire volume generated by A is $V = \int 2\pi y\, dA$. But since the integral $\int y\, dA$ is equal to $\bar{y}A$ (Sec. 5.3), we have

$$V = 2\pi\bar{y}A \tag{5.11}$$

where $2\pi\bar{y}$ is the distance traveled by the centroid of A. Again, it should be noted that the theorem does not apply if the axis of rotation intersects the generating area.

The theorems of Pappus-Guldinus offer a simple way for computing the area of surfaces of revolution and the volume of bodies of revolution. They may also be used conversely to determine the centroid of a plane curve when the area of the surface generated by the curve is known or to determine the centroid of a plane area when the volume of the body generated by the area is known (see Sample Prob. 5.8).

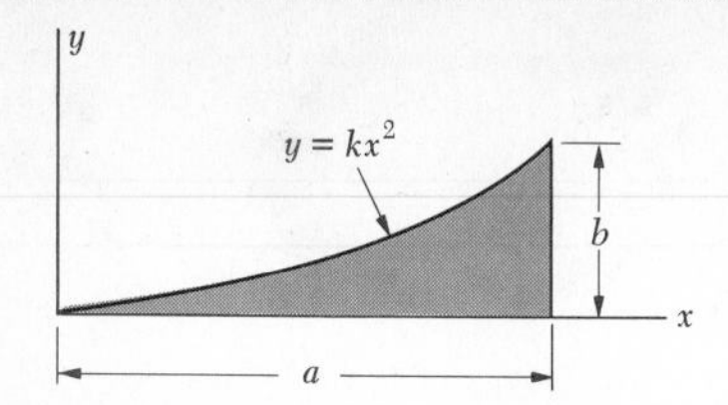

SAMPLE PROBLEM 5.4

Determine by direct integration the centroid of a parabolic spandrel.

Solution. The value of k is determined by substituting $x = a$ and $y = b$ in the given equation. We have $b = ka^2$ and $k = b/a^2$. The equation of the curve is thus

$$y = \frac{b}{a^2}x^2 \qquad \text{or} \qquad x = \frac{a}{b^{1/2}}y^{1/2}$$

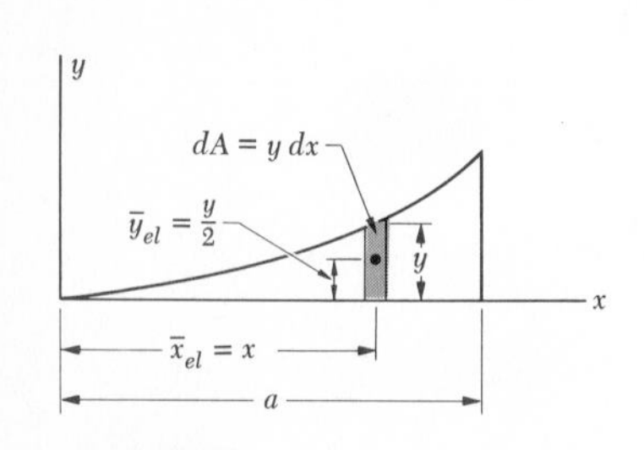

Vertical Differential Element. We choose the differential element shown and find the total area of the figure.

$$A = \int dA = \int y\,dx = \int_0^a \frac{b}{a^2}x^2\,dx = \left[\frac{b}{a^2}\frac{x^3}{3}\right]_0^a = \frac{ab}{3}$$

The first moment of the differential element with respect to the y axis is $\bar{x}_{el}\,dA$; hence, the first moment of the entire area with respect to this axis is

$$Q_y = \int \bar{x}_{el}\,dA = \int xy\,dx = \int_0^a x\left(\frac{b}{a^2}x^2\right)dx = \left[\frac{b}{a^2}\frac{x^4}{4}\right]_0^a = \frac{a^2b}{4}$$

Since $Q_y = \bar{x}A$, we have

$$\bar{x}A = \int \bar{x}_{el}\,dA \qquad \bar{x}\frac{ab}{3} = \frac{a^2b}{4} \qquad \bar{x} = \tfrac{3}{4}a \quad ◀$$

Likewise, the first moment of the differential element with respect to the x axis is $\bar{y}_{el}\,dA$, and the first moment of the entire area is

$$Q_x = \int \bar{y}_{el}\,dA = \int \frac{y}{2}y\,dx = \int_0^a \frac{1}{2}\left(\frac{b}{a^2}x^2\right)^2 dx = \left[\frac{b^2}{2a^4}\frac{x^5}{5}\right]_0^a = \frac{ab^2}{10}$$

Since $Q_x = \bar{y}A$, we have

$$\bar{y}A = \int \bar{y}_{el}\,dA \qquad \bar{y}\frac{ab}{3} = \frac{ab^2}{10} \qquad \bar{y} = \tfrac{3}{10}b \quad ◀$$

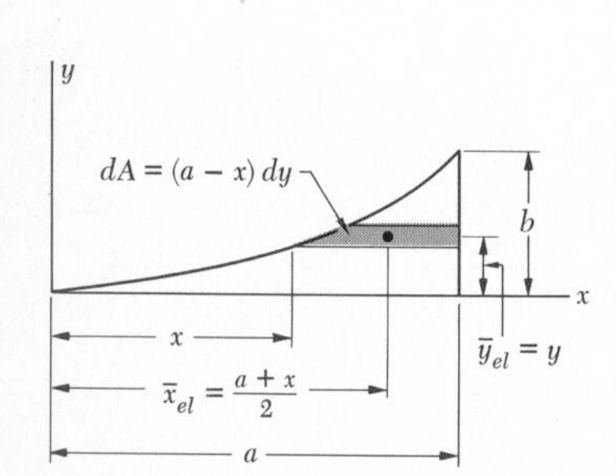

Horizontal Differential Element. The same result may be obtained by considering a horizontal element. The first moments of the area are

$$Q_y = \int \bar{x}_{el}\,dA = \int \frac{a+x}{2}(a-x)\,dy = \int_0^b \frac{a^2 - x^2}{2}dy$$

$$= \frac{1}{2}\int_0^b \left(a^2 - \frac{a^2}{b}y\right)dy = \frac{a^2b}{4}$$

$$Q_x = \int \bar{y}_{el}\,dA = \int y(a-x)\,dy = \int y\left(a - \frac{a}{b^{1/2}}y^{1/2}\right)dy$$

$$= \int_0^b \left(ay - \frac{a}{b^{1/2}}y^{3/2}\right)dy = \frac{ab^2}{10}$$

The expressions obtained are again substituted in the equations defining the centroid of the area to obtain $\bar{x}$ and $\bar{y}$.

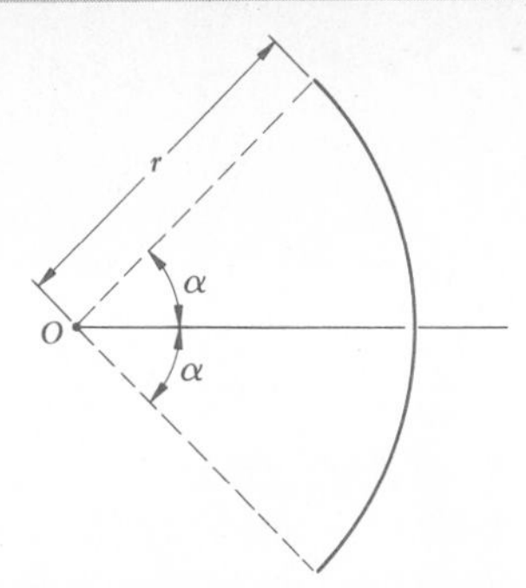

SAMPLE PROBLEM 5.5

Determine the centroid of the arc of circle shown.

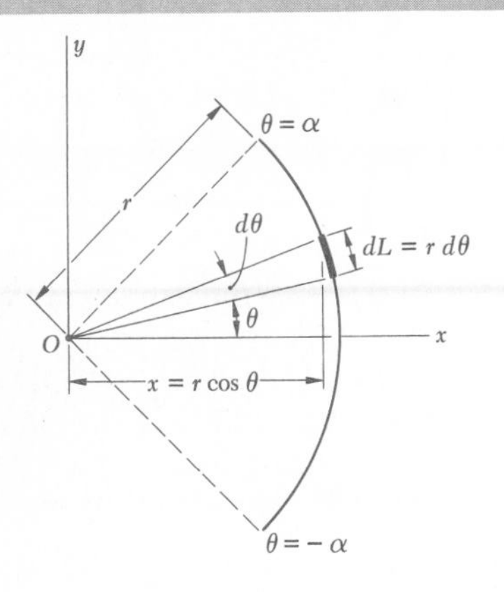

Solution. Since the arc is symmetrical with respect to the x axis, we have $\bar{y} = 0$. A differential element is chosen as shown, and the length of the arc is determined by integration.

$$L = \int dL = \int_{-\alpha}^{\alpha} r\, d\theta = r \int_{-\alpha}^{\alpha} d\theta = 2r\alpha$$

The first moment of the arc with respect to the y axis is

$$Q_y = \int x\, dL = \int_{-\alpha}^{\alpha} (r \cos\theta)(r\, d\theta) = r^2 \int_{-\alpha}^{\alpha} \cos\theta\, d\theta$$

$$= r^2[\sin\theta]_{-\alpha}^{\alpha} = 2r^2 \sin\alpha$$

Since $Q_y = \bar{x}L$, we write

$$\bar{x}(2r\alpha) = 2r^2 \sin\alpha \qquad \bar{x} = \frac{r \sin\alpha}{\alpha} \quad ◀$$

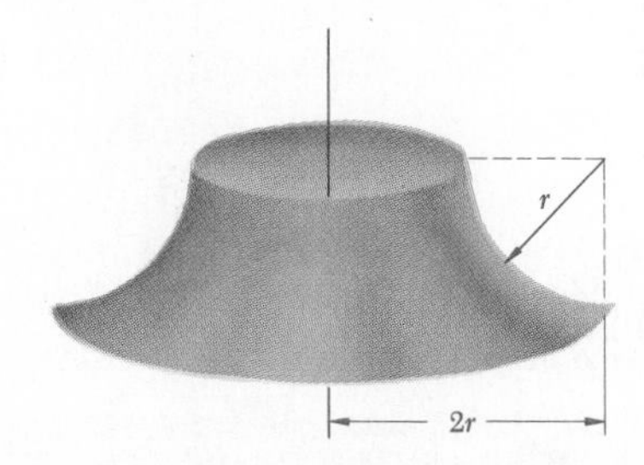

SAMPLE PROBLEM 5.6

Determine the area of the surface of revolution shown, obtained by rotating a quarter-circular arc about a vertical axis.

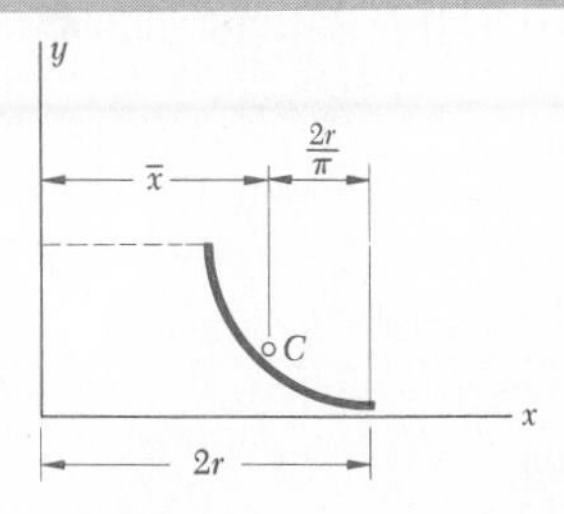

Solution. According to Theorem I of Pappus-Guldinus, the area generated is equal to the product of the length of the arc and the distance traveled by its centroid. Referring to Fig. 5.8*B*, we have

$$\bar{x} = 2r - \frac{2r}{\pi} = 2r\left(1 - \frac{1}{\pi}\right)$$

$$A = 2\pi\bar{x}L = 2\pi 2r\left(1 - \frac{1}{\pi}\right)\frac{\pi r}{2}$$

$$A = 2\pi r^2(\pi - 1) \quad ◀$$

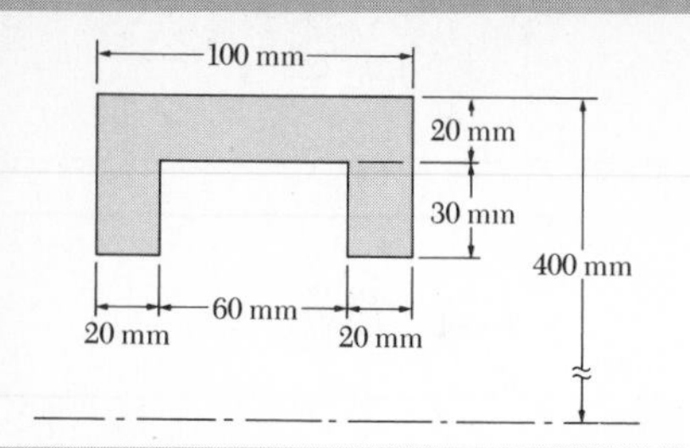

SAMPLE PROBLEM 5.7

The outside diameter of a pulley is 0.8 m, and the cross section of its rim is as shown. Determine the mass and weight of the rim, knowing that the pulley is made of steel and that the density of steel is $\rho = 7.85 \times 10^3$ kg/m^3.

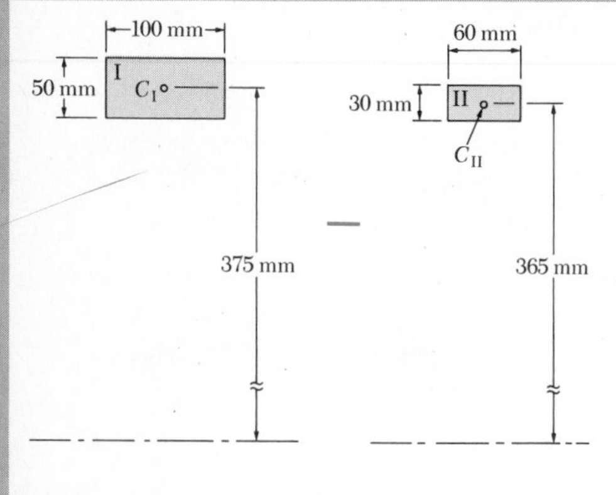

Solution. The volume of the rim may be found by applying Theorem II of Pappus-Guldinus, which states that the volume equals the product of the given cross-sectional area and of the distance traveled by its centroid in one complete revolution. However, the volume will be more easily obtained if we observe that the cross section consists of rectangle (I) with a positive area and rectangle (II) with a negative area.

	Area, mm^2	$\bar{y}$, mm	Distance Traveled by C, mm	Volume, mm^3
I	+5000	375	$2\pi(375) = 2356$	$(5000)(2356) = 11.78 \times 10^6$
II	−1800	365	$2\pi(365) = 2293$	$(-1800)(2293) = -4.13 \times 10^6$
				Volume of rim $= 7.65 \times 10^6$

Since 1 mm $= 10^{-3}$ m, we have 1 mm$^3 = (10^{-3}\text{ m})^3 = 10^{-9}$ m^3, and we obtain $V = 7.65 \times 10^6\text{ mm}^3 = (7.65 \times 10^6)(10^{-9}\text{ m}^3) = 7.65 \times 10^{-3}\text{ m}^3$.

$$m = \rho V = (7.85 \times 10^3\text{ kg/m}^3)(7.65 \times 10^{-3}\text{ m}^3) \qquad m = 60.0\text{ kg}$$

$$W = mg = (60.0\text{ kg})(9.81\text{ m/s}^2) = 589\text{ kg}\cdot\text{m/s}^2 \qquad W = 589\text{ N}$$

SAMPLE PROBLEM 5.8

Using the theorems of Pappus-Guldinus, determine (*a*) the centroid of a semicircular area, (*b*) the centroid of a semicircular arc. We recall that the volume of a sphere is $\frac{4}{3}\pi r^3$ and that its surface area is $4\pi r^2$.

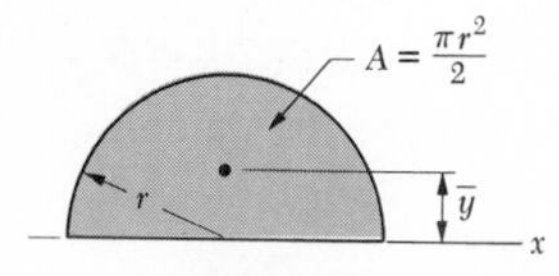

Solution. The volume of a sphere is equal to the product of the area of a semicircle and of the distance traveled by the centroid of the semicircle in one revolution about the x axis.

$$V = 2\pi\bar{y}A \qquad \tfrac{4}{3}\pi r^3 = 2\pi\bar{y}(\tfrac{1}{2}\pi r^2) \qquad \bar{y} = \frac{4r}{3\pi}$$

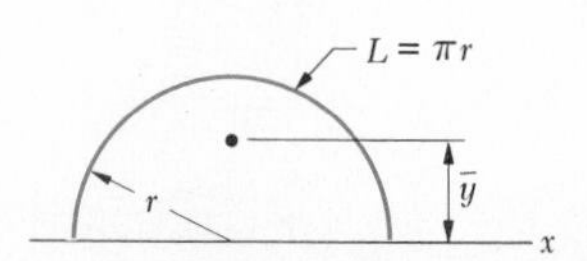

Likewise, the area of a sphere is equal to the product of the length of the generating semicircle and of the distance traveled by its centroid in one revolution.

$$A = 2\pi\bar{y}L \qquad 4\pi r^2 = 2\pi\bar{y}(\pi r) \qquad \bar{y} = \frac{2r}{\pi}$$

Problems

5.40 through 5.43 Determine by direct integration the centroid of the area shown.

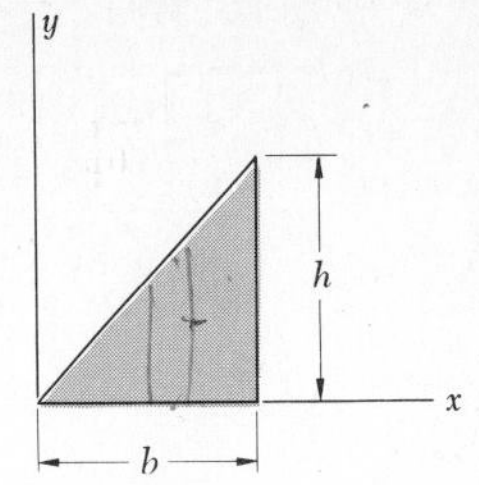

Fig. P5.40

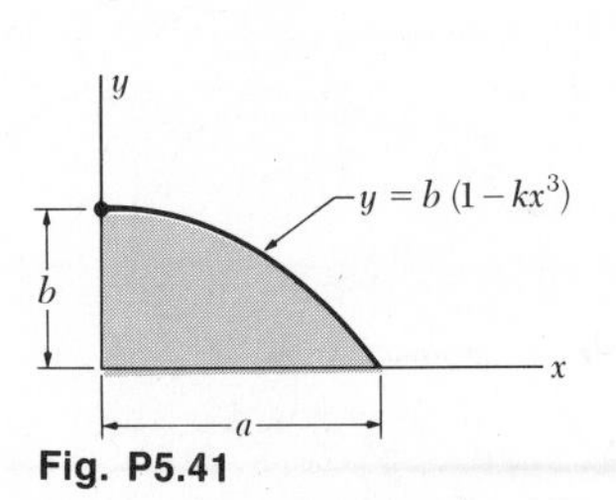

Fig. P5.41

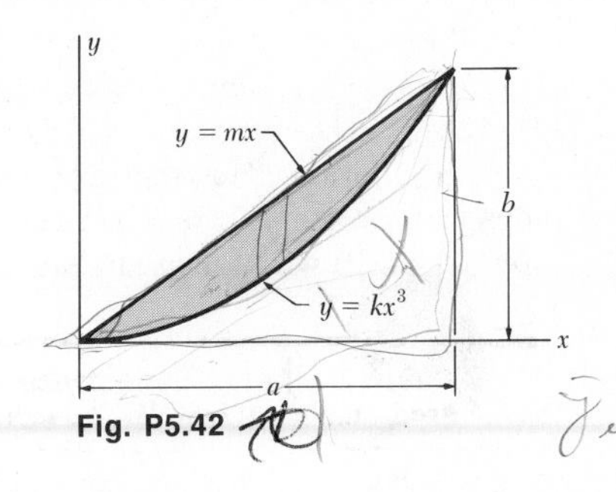

Fig. P5.42

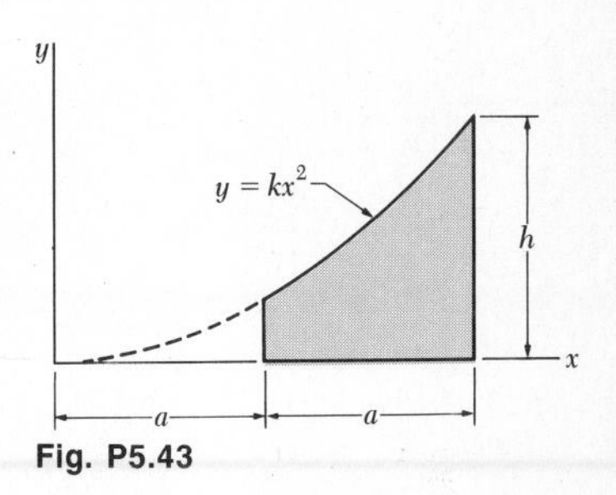

Fig. P5.43

5.44 through 5.49 Derive by direct integration the expressions for $\overline{x}$ and $\overline{y}$ given in Fig. 5.8 for:

5.44 A general spandrel ($y = kx^n$)
5.45 A quarter-elliptical area
5.46 A semicircular area
5.47 A semiparabolic area
5.48 A circular sector
5.49 A quarter-circular arc

5.50 Determine by direct integration the x coordinate of the centroid of the area shown.

5.51 Determine by direct integration the y coordinate of the centroid of the area shown.

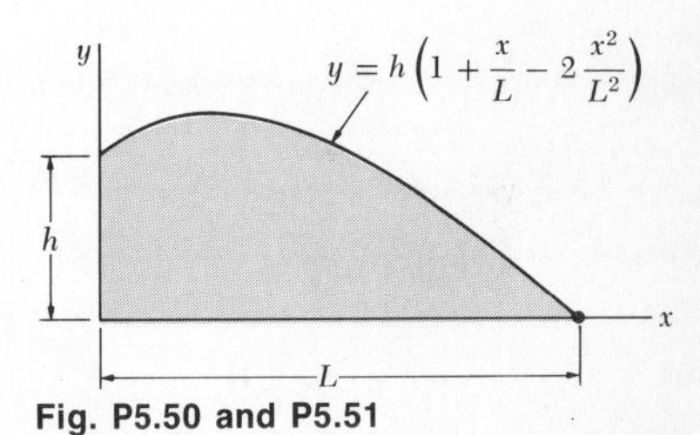

Fig. P5.50 and P5.51

***5.52 and *5.53** Determine by direct integration the centroid of the area shown.

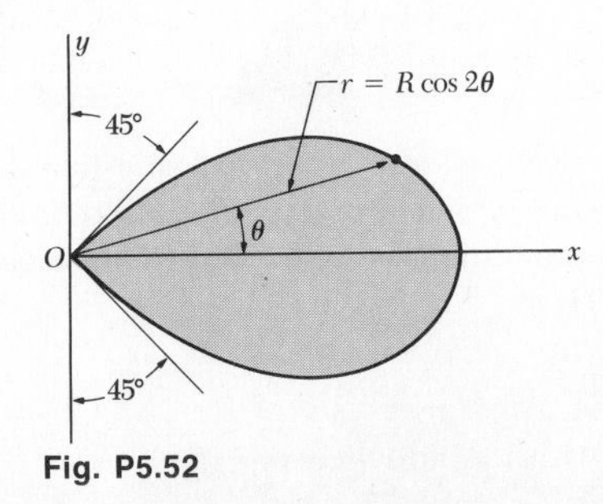

Fig. P5.52

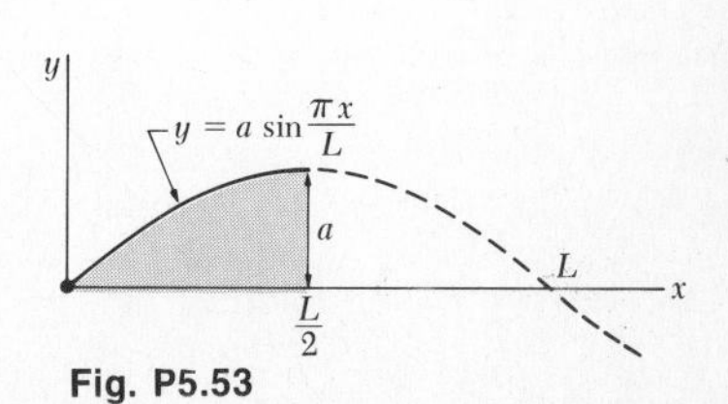

Fig. P5.53

5.54 Determine the centroid of the area shown when $a = 4$ in.

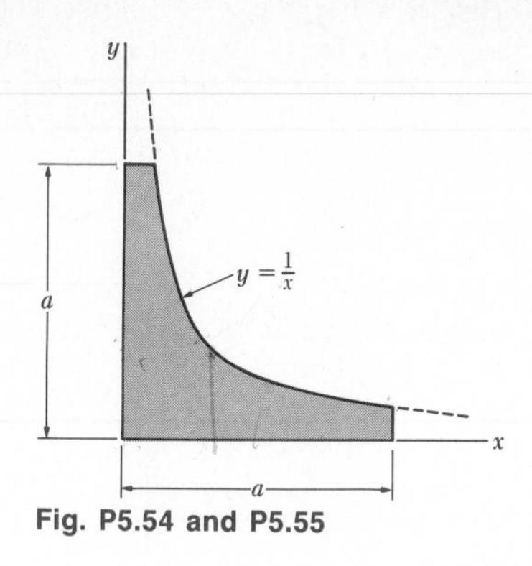

Fig. P5.54 and P5.55

5.55 Determine the centroid of the area shown in terms of a.

5.56 Determine the volume of the solid obtained by rotating the area of Prob. 5.7 about (*a*) the x axis, (*b*) the y axis.

5.57 Determine the volume of the solid obtained by rotating the trapezoid of Prob. 5.3 about (*a*) the x axis, (*b*) the y axis.

5.58 Determine the volume of the solid generated by rotating the parabolic spandrel shown about (*a*) the x axis, (*b*) the axis AA'.

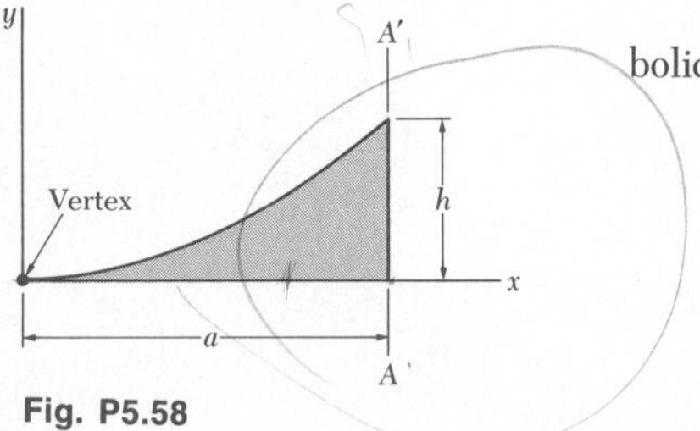

Fig. P5.58

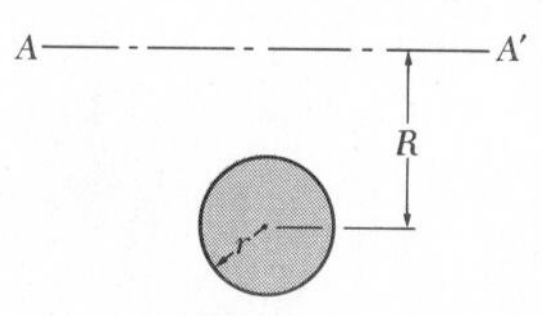

Fig. P5.59

5.59 Determine the surface area and the volume of the torus obtained by rotating the circular area shown about the axis AA'.

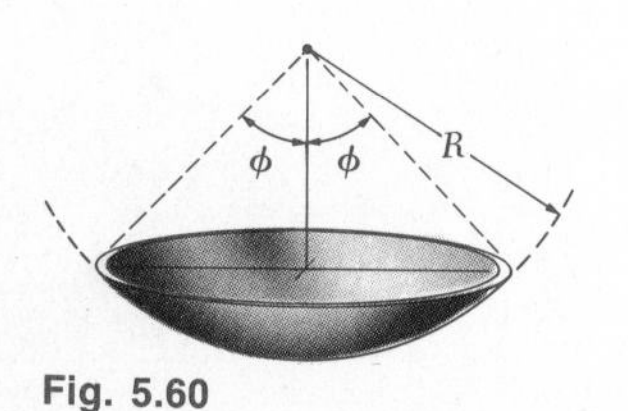

Fig. 5.60

5.60 A spherical dish is formed by passing a horizontal plane through a spherical shell of radius R. Knowing that $R = 10$ in. and $\phi = 60°$, determine the area of the inside surface of the dish.

5.61 Determine the volume and weight of water required to completely fill the spherical dish of Prob. 5.60. (Specific weight of water $= 62.4\ \text{lb/ft}^3$.)

5.62 A spherical pressure vessel has an inside diameter of 0.8 m. Determine (*a*) the volume of liquefied propane required to fill the vessel to a depth of 0.6 m, (*b*) the corresponding mass of the liquefied propane. (Density of liquefied propane $= 580\ \text{kg/m}^3$.)

5.63 For the pressure vessel of Prob. 5.62, determine the area of the surface in contact with the liquefied propane.

5.64 Determine the volume and total surface area of the body shown.

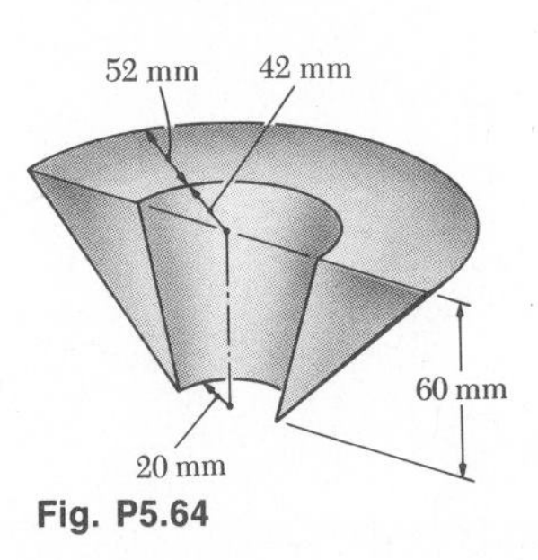

Fig. P5.64

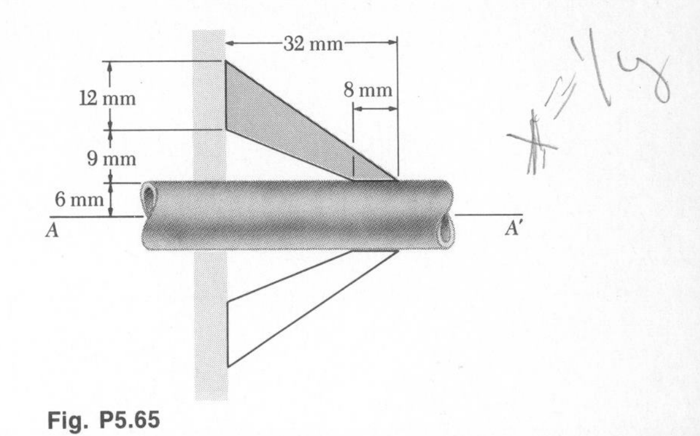

Fig. P5.65

5.65 Determine the volume and mass of the cast-iron pipe collar obtained by rotating the shaded area about the centerline of the 12-mm-diameter pipe. (Density of cast iron = 7200 kg/m^3.)

5.66 Determine the volume and weight of the solid brass knob shown. (Specific weight of brass = 0.306 lb/in^3.)

5.67 Determine the total surface area of the solid brass knob shown.

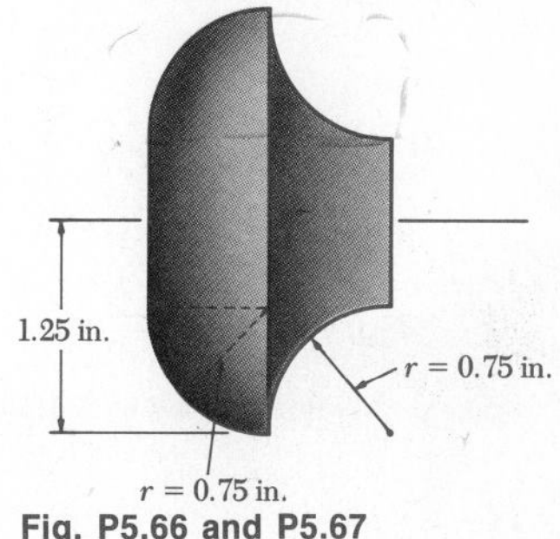

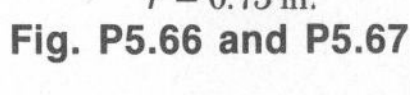

Fig. P5.66 and P5.67

5.68 The rim of a double-groove V-belt sheave of 10-in. outside diameter is made of cast iron and weighs 4.50 lb. Knowing that the cross-sectional area of the rim is 0.570 in^2, determine the distance $\bar{y}$ from the axis of rotation AA' to the centroid of the cross section of the rim. (Specific weight of cast iron = 0.264 lb/in^3.)

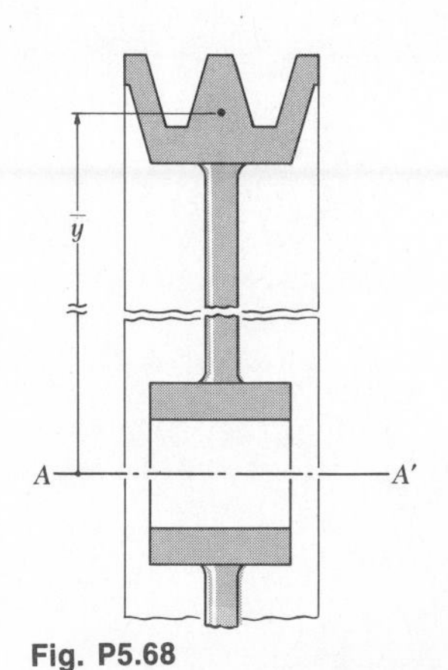

Fig. P5.68

5.69 An oil-storage tank in the form of a shell of constant strength has the cross section shown. (This is also the cross section of a drop of water on an unwetted surface.) Determine by approximate means (*a*) the volume of the tank, (*b*) the surface area of the tank.

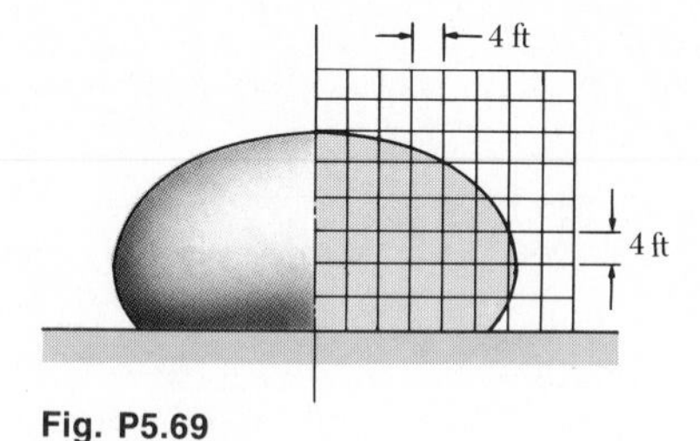

Fig. P5.69

Fig. P5.70

5.70 A 60-W standard light bulb has the cross section shown. For the portion of the bulb above the base and neglecting the thickness of the glass, determine (*a*) the volume of inert gas inside the bulb, (*b*) the outside surface area of the bulb.

5.71 Determine the volume of the solid of revolution formed by revolving each of the plane areas shown about its vertical edge AB. Show that the volumes of the solids formed are in the ratio 6:4:3:2:1.

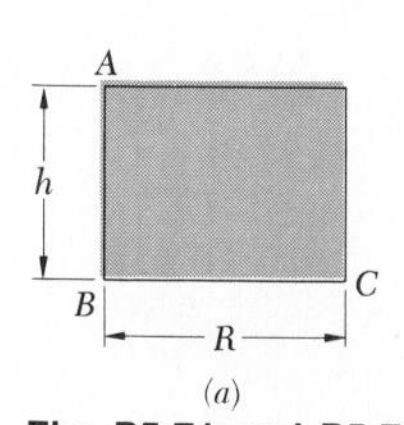

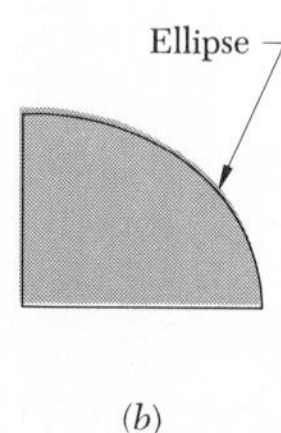

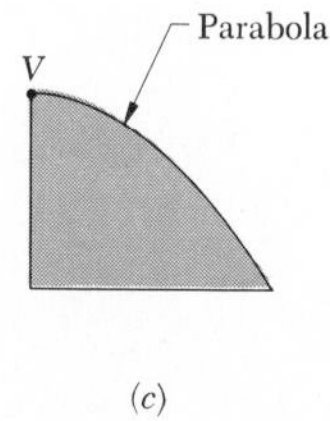

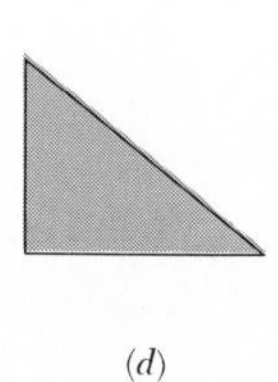

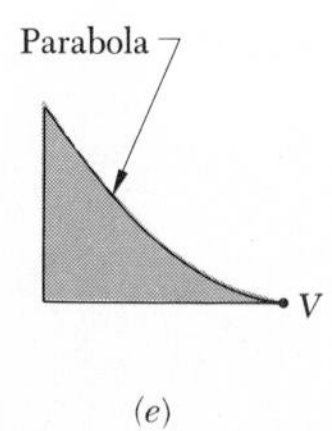

Fig. P5.71 and P5.72

5.72 Determine the volume of the solid formed by revolving each of the plane areas shown about its base BC. Show that the volumes of the solids formed are in the ratio 15:10:8:5:3.

***5.8. Distributed Loads on Beams.** The concept of centroid of an area may be used to solve other problems besides those dealing with the weight of flat plates. Consider, for example, a beam supporting a *distributed load;* this load may consist of the weight of materials supported directly or indirectly by the beam, or it may be caused by wind or hydrostatic pressure. The distributed load may be represented by plotting the load w supported per unit length (Fig. 5.17); this load will be expressed in N/m or in lb/ft. The magnitude of the force exerted on an element of beam of length dx is $dW = w\,dx$, and the total load supported by the beam is

$$W = \int_0^L w\,dx$$

But the product $w\,dx$ is equal in magnitude to the element of area dA shown in Fig. 5.17*a*, and W is thus equal in magnitude to the total area A under the load curve,

$$W = \int dA = A$$

We shall now determine where a *single concentrated load* **W**, of the same magnitude W as the total distributed load, should be applied on the beam if it is to produce the same reactions at the supports (Fig. 5.17*b*). This concentrated load **W**, which represents the resultant of the given distributed loading, should be equivalent to this loading as far as the free-body diagram of the entire beam is concerned. The point of application P of the equivalent concentrated load **W** will therefore be obtained by expressing that the moment of **W** about point O is equal to the sum of the moments of the elementary loads $d\mathbf{W}$ about O:

$$(OP)W = \int x\,dW$$

or, since $dW = w\,dx = dA$ and $W = A$,

$$(OP)A = \int_0^L x\,dA \tag{5.12}$$

Since the integral represents the first moment with respect to the w axis of the area under the load curve, it may be replaced by the product $\bar{x}A$. We have therefore $OP = \bar{x}$, where $\bar{x}$ is the distance from the w axis to the centroid C of the area A (this is *not* the centroid of the beam).

A distributed load on a beam may thus be replaced by a concentrated load; the magnitude of this single load is equal to the area under the load curve, and its line of action passes through the centroid of that area. It should be noted, however, that the concentrated load is equivalent to the given loading only as far as external forces are concerned. It may be used to determine reactions but should not be used to compute internal forces and deflections.

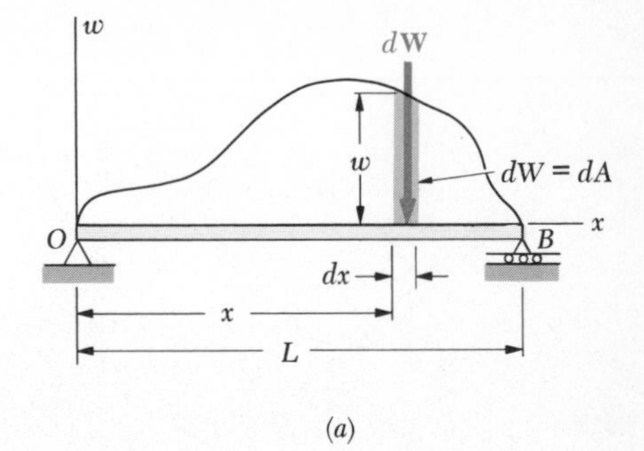

(*a*)

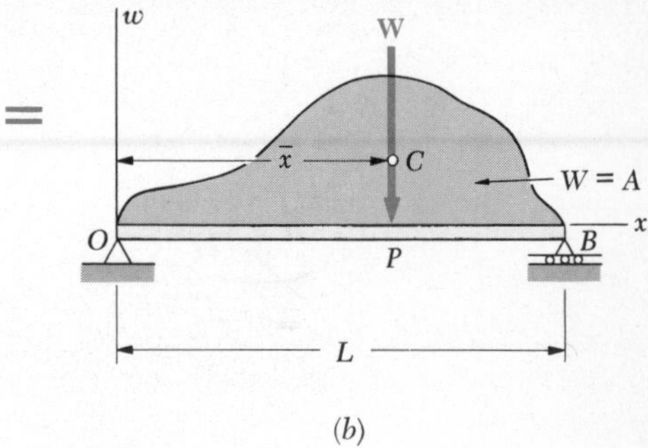

(*b*)

Fig. 5.17

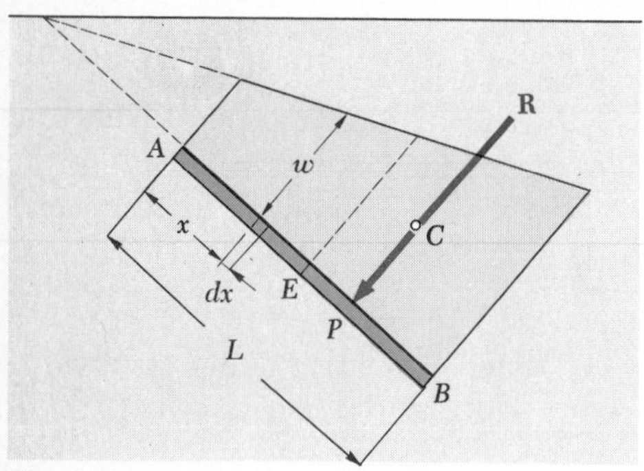

Fig. 5.18

***5.9 Forces on Submerged Surfaces.** The approach used in the preceding section may be used to determine the resultant of the hydrostatic pressure forces exerted on a *rectangular surface* submerged in a liquid. Consider the rectangular plate shown in Fig. 5.18, of length L and width b perpendicular to the plane of the figure. As noted in Sec. 5.8, the load exerted on an element of plate of length dx is $w\,dx$, where w is the load per unit length. But this load may also be expressed as $p\,dA = pb\,dx$, where p is the gage pressure in the liquid† and b is the width of the plate; thus, $w = bp$. Since, on the other hand, the gage pressure in a liquid is $p = \gamma h$, where γ is the specific weight of the liquid and h is the vertical distance from the free surface, it follows that

$$w = bp = b\gamma h \tag{5.13}$$

which shows that the load per unit length w is proportional to h and, thus, varies linearly with x.

Recalling the results of Sec. 5.8, we observe that the resultant $\mathbf{R}$ of the hydrostatic forces exerted on one side of the plate is equal in magnitude to the trapezoidal area under the load curve and that its line of action passes through the centroid C of that area. The point P of the plate where $\mathbf{R}$ is applied is known as the *center of pressure.*‡

We shall consider next the forces exerted by a liquid on a curved surface of constant width (Fig. 5.19*a*). Since the determination of the resultant $\mathbf{R}$ of these forces by direct integration would not be easy, we shall consider the free body obtained by detaching the volume of liquid ABD bounded by the curved surface AB and by the two plane surfaces AD and DB shown in Fig. 5.19*b*. The forces acting on the free body ABD consist of the weight $\mathbf{W}$ of the volume of liquid detached, the resultant $\mathbf{R}_1$ of the forces exerted on AD, the resultant $\mathbf{R}_2$ of the forces exerted on BD, and the resultant $-\mathbf{R}$ of the forces exerted *by the curved surface on the liquid.* The resultant $-\mathbf{R}$ is equal and opposite to, and has the same line of action as the resultant $\mathbf{R}$ of the forces exerted *by the liquid on the curved surface.* The forces $\mathbf{W}$, $\mathbf{R}_1$, and $\mathbf{R}_2$ may be determined by standard methods; after their values have been found, the force $-\mathbf{R}$ will be obtained by solving the equations of equilibrium for the free body of Fig. 5.19*b*. The resultant $\mathbf{R}$ of the hydrostatic forces exerted on the curved surface will then be obtained by reversing the sense of $-\mathbf{R}$.

The methods outlined in this section may be used to determine the resultant of the hydrostatic forces exerted on the surface of dams or on rectangular gates and vanes. Resultants of forces on submerged surfaces of variable width should be determined by the methods of Chap. 9.

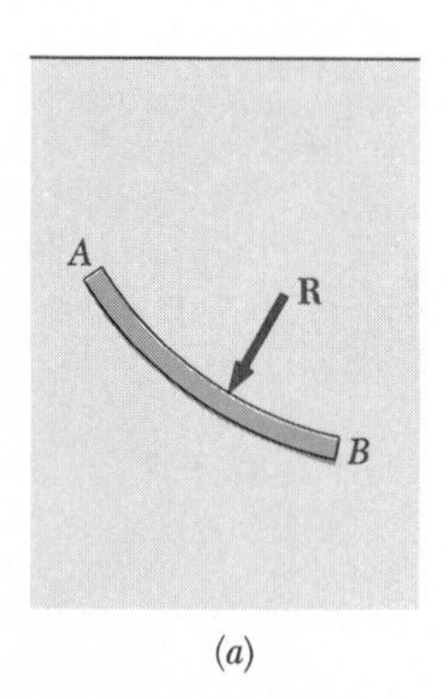

(*a*)

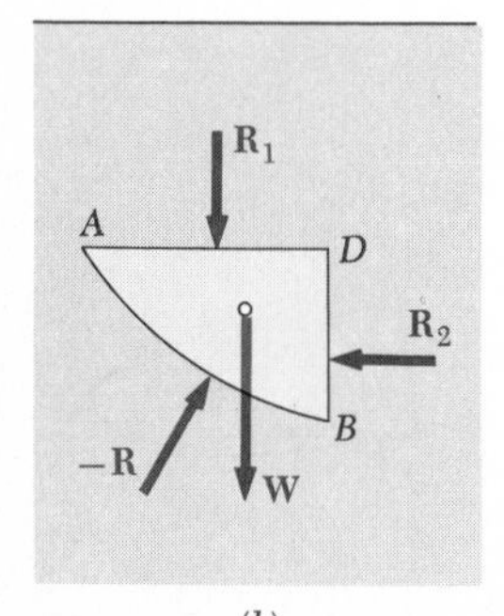

(*b*)

Fig. 5.19

† The pressure p, which represents a load per unit area, is expressed in N/m^2 or in lb/ft^2. The derived SI unit N/m^2 is called a *pascal* (Pa).

‡ Noting that the area under the load curve is equal to $w_E L$ and recalling Eq. (5.13), we may write

$$R = w_E L = (bp_E)L = p_E(bL) = p_E A$$

where A denotes the area of the *plate.* Thus, the magnitude of $\mathbf{R}$ may be obtained by multiplying the area of the plate by the pressure at its center E. The resultant $\mathbf{R}$, however, *should be applied at P, not at E.*

SAMPLE PROBLEM 5.9

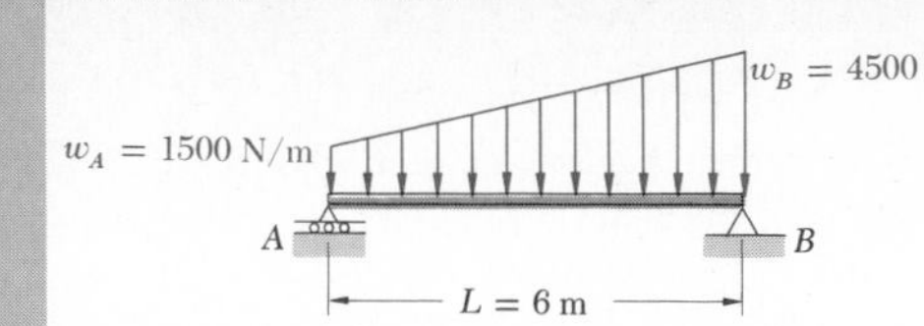

A beam supports a distributed load as shown. (*a*) Determine the equivalent concentrated load. (*b*) Determine the reactions at the supports.

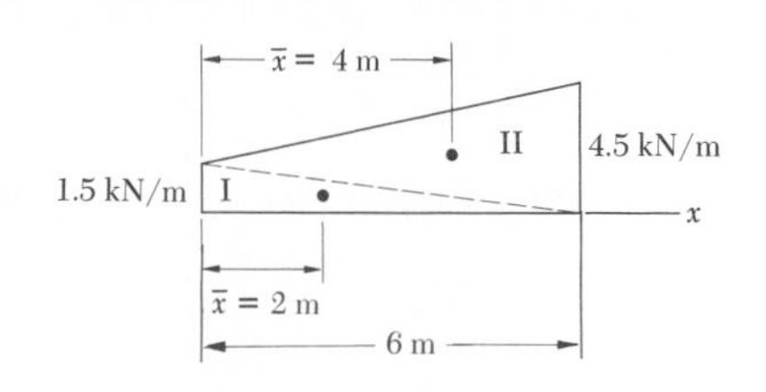

***a*. Equivalent Concentrated Load.** The magnitude of the resultant of the load is equal to the area under the load curve, and the line of action of the resultant passes through the centroid of the same area. We divide the area under the load curve into two triangles and construct the table below. To simplify the computations and tabulation, the given loads per unit length have been converted into kN/m.

Component	A, kN	$\bar{x}$, m	$\bar{x}A$, kN·m
Triangle I	4.5	2	9
Triangle II	13.5	4	54
	$\Sigma A = 18.0$	. . .	$\Sigma \bar{x}A = 63$

Thus, $\bar{X}\Sigma A = \Sigma \bar{x}A$: $\quad \bar{X}(18\text{ kN}) = 63\text{ kN}\cdot\text{m} \qquad \bar{X} = 3.5\text{ m}$

The equivalent concentrated load is

$$\mathbf{W} = 18\text{ kN}\downarrow \blacktriangleleft$$

and its line of action is located at a distance

$$\bar{X} = 3.5\text{ m to the right of } A \blacktriangleleft$$

***b*. Reactions.** The reaction at A is vertical and is denoted by $\mathbf{A}$; the reaction at B is represented by its components $\mathbf{B}_x$ and $\mathbf{B}_y$. The given load may be considered as the sum of two triangular loads as shown. The resultant of each triangular load is equal to the area of the triangle and acts at its centroid. We write the following equilibrium equations for the free body shown:

$$\xrightarrow{+}\Sigma F_x = 0: \qquad \mathbf{B}_x = 0 \blacktriangleleft$$

$$+\circlearrowleft \Sigma M_A = 0: \qquad -(4.5\text{ kN})(2\text{ m}) - (13.5\text{ kN})(4\text{ m}) + B_y(6\text{ m}) = 0$$

$$\mathbf{B}_y = 10.5\text{ kN}\uparrow \blacktriangleleft$$

$$+\circlearrowleft \Sigma M_B = 0: \qquad +(4.5\text{ kN})(4\text{ m}) + (13.5\text{ kN})(2\text{ m}) - A(6\text{ m}) = 0$$

$$\mathbf{A} = 7.5\text{ kN}\uparrow \blacktriangleleft$$

Alternate Solution. The given distributed load may be replaced by its resultant, which was found in part *a*. The reactions may be determined by writing the equilibrium equations $\Sigma F_x = 0$, $\Sigma M_A = 0$, and $\Sigma M_B = 0$. We again obtain

$$\mathbf{B}_x = 0 \qquad \mathbf{B}_y = 10.5\text{ kN}\uparrow \qquad \mathbf{A} = 7.5\text{ kN}\uparrow \blacktriangleleft$$

SAMPLE PROBLEM 5.10

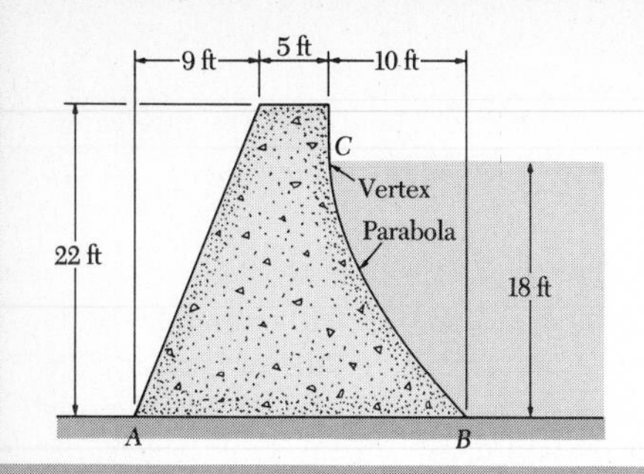

The cross section of a concrete dam is as shown. Consider a section of the dam 1 ft thick, and determine (*a*) the resultant of the reaction forces exerted by the ground on the base of the dam *AB*, (*b*) the resultant of the pressure forces exerted by the water on the face *BC* of the dam. Specific weight of concrete = 150 lb/ft^3; of water = 62.4 lb/ft^3.

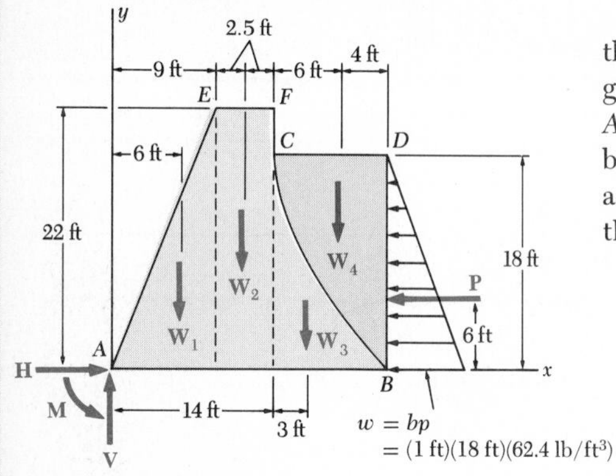

***a*. Ground Reaction.** As a free body we choose a section *AEFCDB*, 1 ft thick, of the dam and water as shown. The reaction forces exerted by the ground on the base *AB* are represented by an equivalent force-couple system at *A*. Other forces acting on the free body are the weight of the dam, represented by the weights of its components $\mathbf{W}_1$, $\mathbf{W}_2$, and $\mathbf{W}_3$, the weight of the water $\mathbf{W}_4$, and the resultant **P** of the pressure forces exerted on section *BD* by the water to the right of section *BD*. We have

$$
\begin{aligned}
W_1 &= \tfrac{1}{2}(9\text{ ft})(22\text{ ft})(1\text{ ft})(150\text{ lb/ft}^3) = 14{,}850\text{ lb}\\
W_2 &= (5\text{ ft})(22\text{ ft})(1\text{ ft})(150\text{ lb/ft}^3) = 16{,}500\text{ lb}\\
W_3 &= \tfrac{1}{3}(10\text{ ft})(18\text{ ft})(1\text{ ft})(150\text{ lb/ft}^3) = 9000\text{ lb}\\
W_4 &= \tfrac{2}{3}(10\text{ ft})(18\text{ ft})(1\text{ ft})(62.4\text{ lb/ft}^3) = 7488\text{ lb}\\
P &= \tfrac{1}{2}(18\text{ ft})(1\text{ ft})(18\text{ ft})(62.4\text{ lb/ft}^3) = 10{,}109\text{ lb}
\end{aligned}
$$

Equilibrium Equations

$\xrightarrow{+}\Sigma F_x = 0$: $\quad H - 10{,}109\text{ lb} = 0 \qquad \mathbf{H} = 10{,}110\text{ lb} \rightarrow$ ◀

$+\uparrow\Sigma F_y = 0$: $\quad V - 14{,}850\text{ lb} - 16{,}500\text{ lb} - 9000\text{ lb} - 7488\text{ lb} = 0$

$\mathbf{V} = 47{,}840\text{ lb}\uparrow$ ◀

$+\circlearrowleft \Sigma M_A = 0$: $\quad -(14{,}850\text{ lb})(6\text{ ft}) - (16{,}500\text{ lb})(11.5\text{ ft})$
$- (9000\text{ lb})(17\text{ ft}) - (7488\text{ lb})(20\text{ ft}) + (10{,}109\text{ lb})(6\text{ ft}) + M = 0$

$\mathbf{M} = 520{,}960\text{ lb}\cdot\text{ft}\circlearrowleft$ ◀

We may replace the force-couple system obtained by a single force acting at a distance *d* to the right of *A*, where

$$d = \frac{520{,}960\text{ lb}\cdot\text{ft}}{47{,}840\text{ lb}} = 10.89\text{ ft}$$

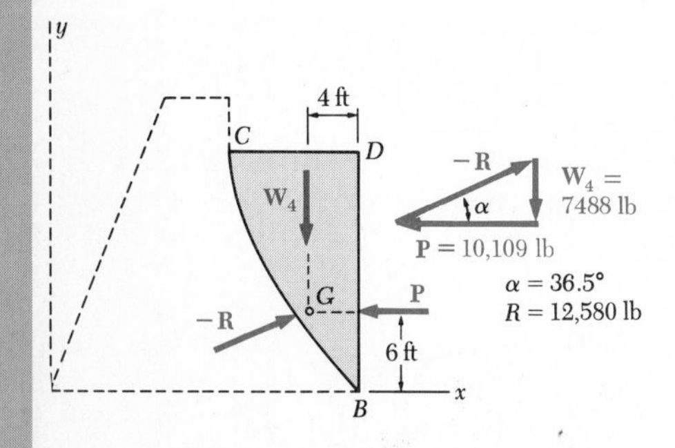

***b*. Resultant R of Water Forces.** The parabolic section of water *BCD* is chosen as a free body. The forces involved are the resultant $-\mathbf{R}$ of the forces exerted by the dam on the water, the weight $\mathbf{W}_4$, and the force **P**. Since these forces must be concurrent, $-\mathbf{R}$ passes through the point of intersection *G* of $\mathbf{W}_4$ and **P**. A force triangle is drawn from which the magnitude and direction of $-\mathbf{R}$ are determined. The resultant **R** of the forces exerted by the water on the face *BC* is equal and opposite:

$\mathbf{R} = 12{,}580\text{ lb} \measuredangle 36.5°$ ◀

Problems

5.73 and 5.74 Determine the magnitude and location of the resultant of the distributed load shown. Also calculate the reactions at A and B.

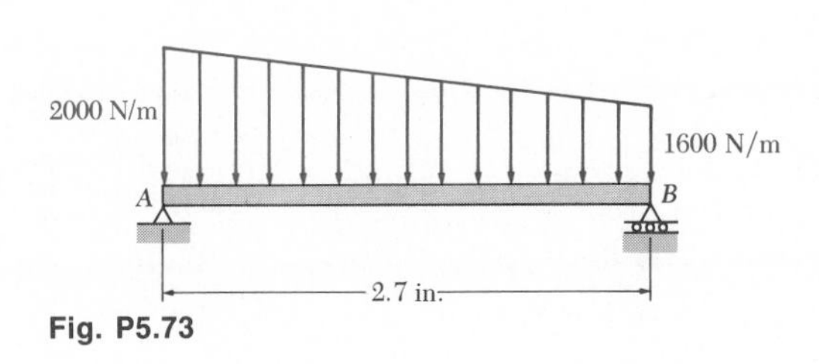

Fig. P5.73

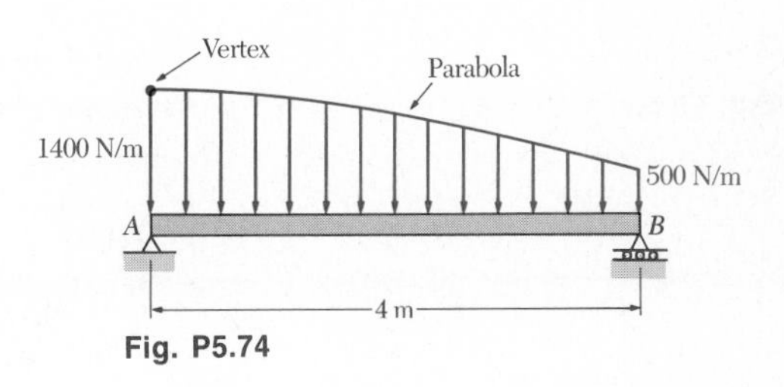

Fig. P5.74

5.75 through 5.80 Determine the reactions at the beam supports for the given loading condition.

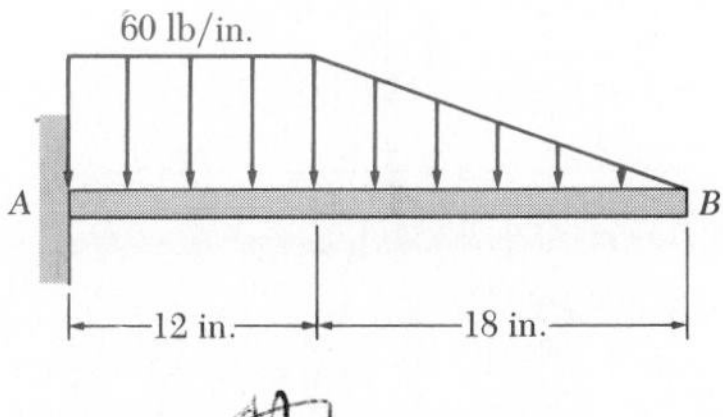

Fig. P5.75

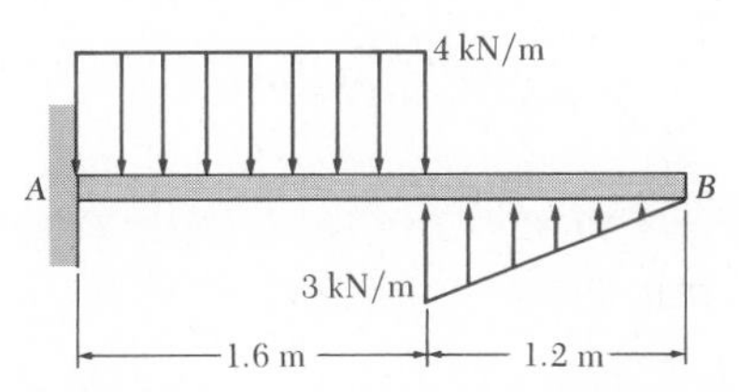

Fig. P5.76

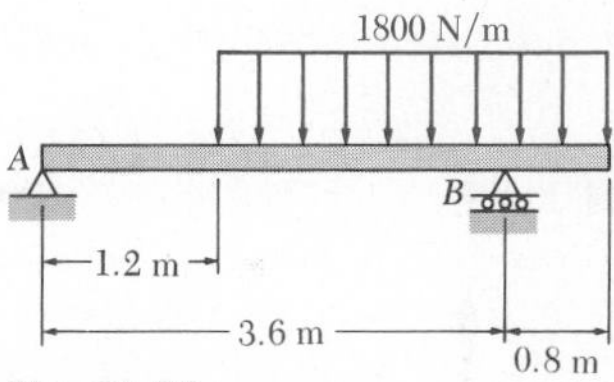

Fig. P5.77

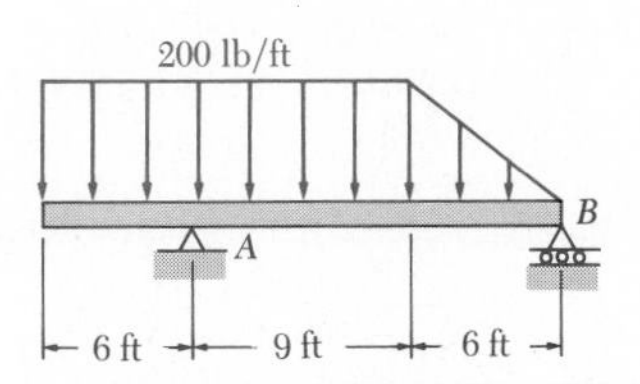

Fig. P5.78

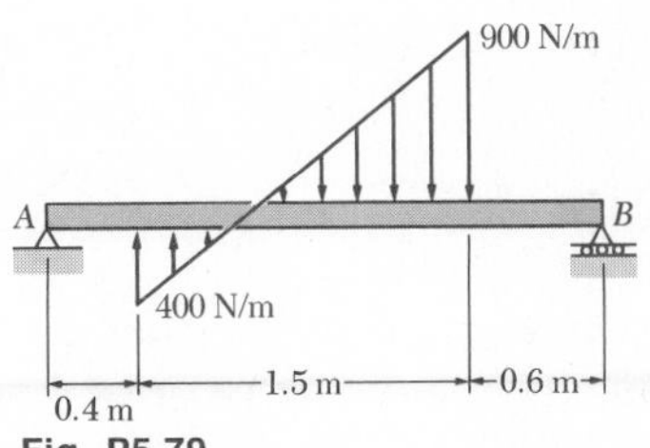

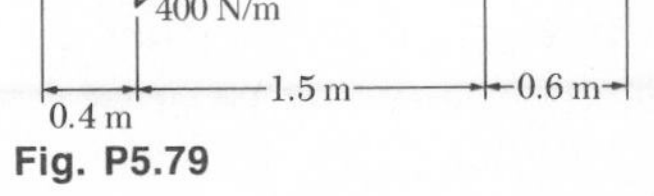

Fig. P5.79

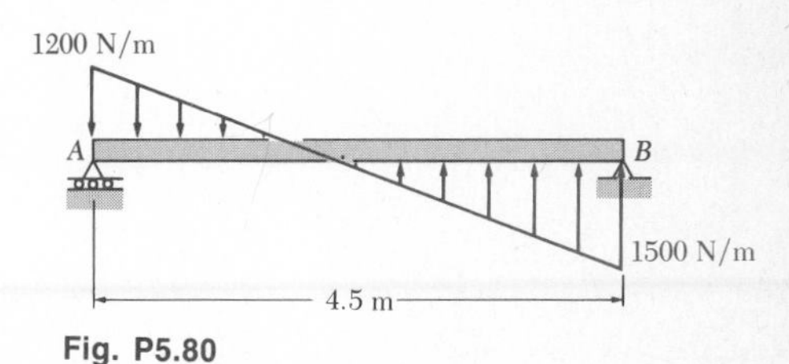

Fig. P5.80

5.81 Determine the reactions at the beam supports for the given loading condition when $w_0 = 450$ lb/ft.

5.82 Determine (*a*) the distributed load w_0 at the end C of the beam ABC for which the reaction at C is zero, (*b*) the corresponding reaction at B.

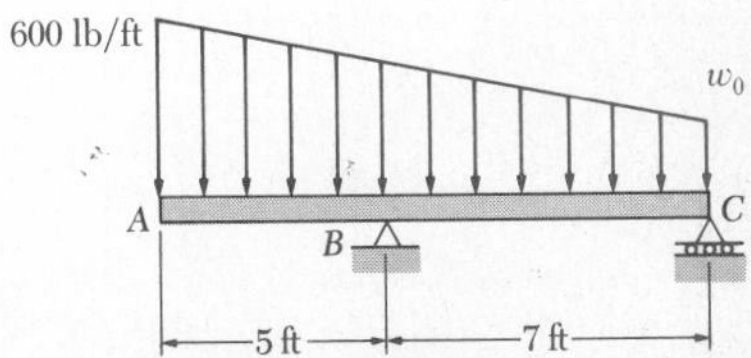

Fig. P5.81 and P5.82

5.83 The beam AB supports two concentrated loads and rests on soil which exerts a linearly distributed upward load as shown. Determine the values of w_A and w_B corresponding to equilibrium.

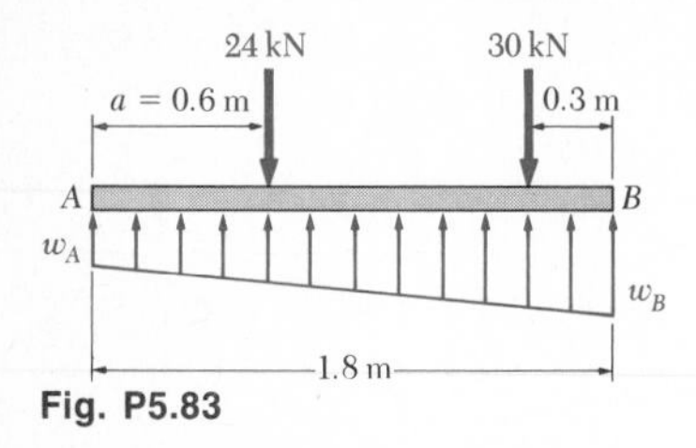

Fig. P5.83

5.84 For the beam and loading of Prob. 5.83, determine (*a*) the distance a for which $w_A = 18$ kN/m, (*b*) the corresponding value of w_B.

In the following problems, use $\gamma = 62.4$ lb/ft^3 for the specific weight of fresh water and $\gamma_c = 150$ lb/ft^3 for the specific weight of concrete if U.S. customary units are used. With SI units, use $\rho = 10^3$ kg/m^3 for the density of fresh water and $\rho_c = 2.40 \times 10^3$ kg/m^3 for the density of concrete. (See footnote page 172 for the determination of the specific weight of a material from its density.)

5.85 and 5.86 The cross section of a concrete dam is as shown. Consider a section of the dam 1 m thick, and determine (*a*) the resultant of the reaction forces exerted by the ground on the base AB of the dam, (*b*) the resultant of the pressure forces exerted by the water on the face BC of the dam.

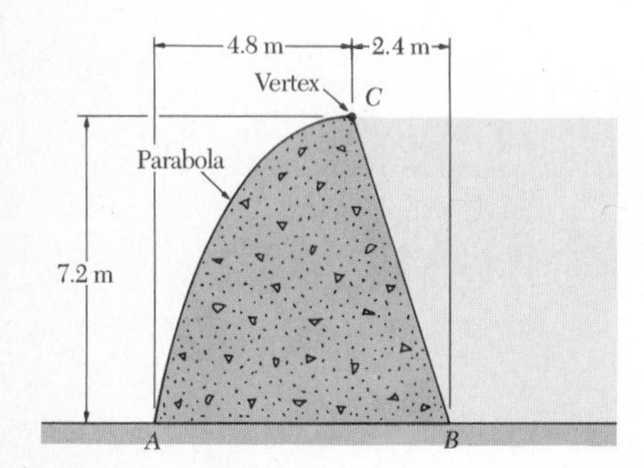

Fig. P5.85

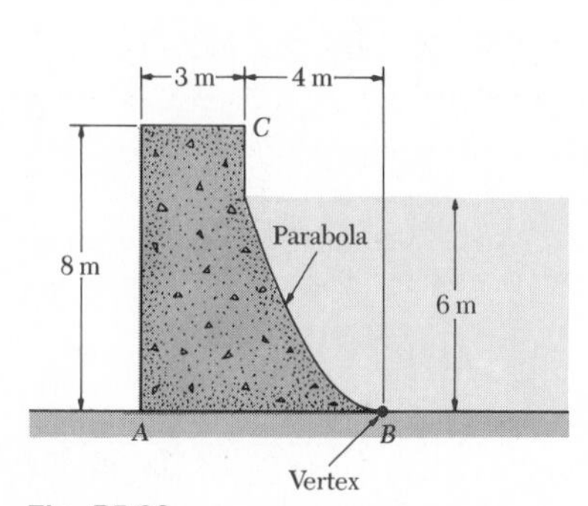

Fig. P5.86

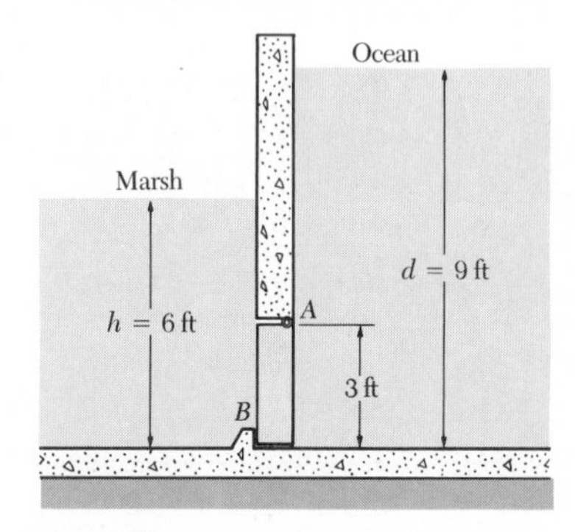

Fig. P5.87

5.87 A freshwater marsh is drained to the ocean through an automatic tide gate which is 4 ft wide and 3 ft high. The gate is held by hinges located along its top edge at A and bears on a sill at B. At a given time, the water level in the marsh is $h = 6$ ft and in the ocean $d = 9$ ft. Determine the force exerted by the sill on the gate at B and the hinge reaction at A. (Specific weight of salt water = 64 lb/ft^3.)

5.88 The automatic tide gate described in Prob. 5.87 is used to drain a freshwater marsh into the ocean. If the water level in the marsh is $h = 6$ ft, determine the ocean level d for which the gate will open.

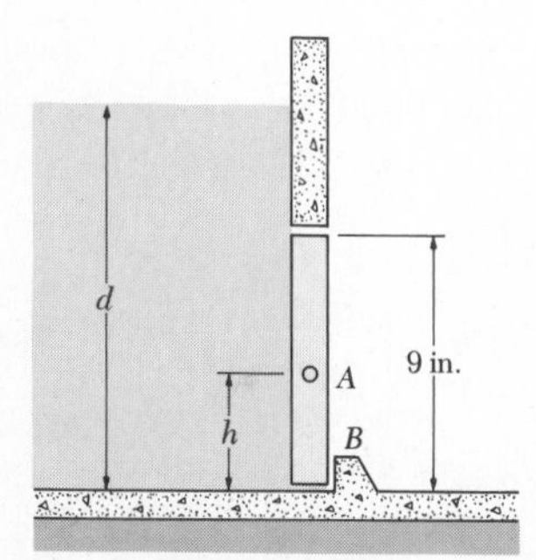

Fig. P5.89 and P5.90

5.89 An automatic valve consists of a 9×9 in. square plate which is pivoted about a horizontal axis through A located at a distance $h = 3.6$ in. above the lower edge. Determine the depth of water d for which the valve will open.

5.90 An automatic valve consists of a 9×9 in. square plate which is pivoted about a horizontal axis through A. If the valve is to open when the depth of water is $d = 18$ in., determine the distance h from the bottom of the valve to the pivot A.

5.91 The gate AB is located at the end of a 2-m-wide water channel and is supported by hinges along its top edge A. Knowing that the floor of the channel is frictionless, determine the reactions at A and B when $d = 3$ m.

5.92 Solve Prob. 5.91 when $d = 2.1$ m.

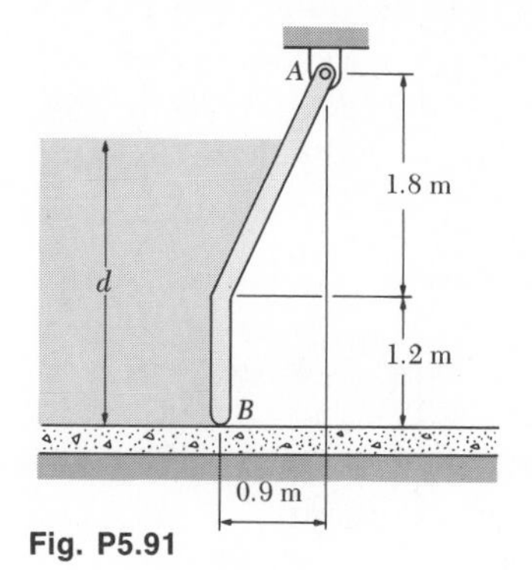

Fig. P5.91

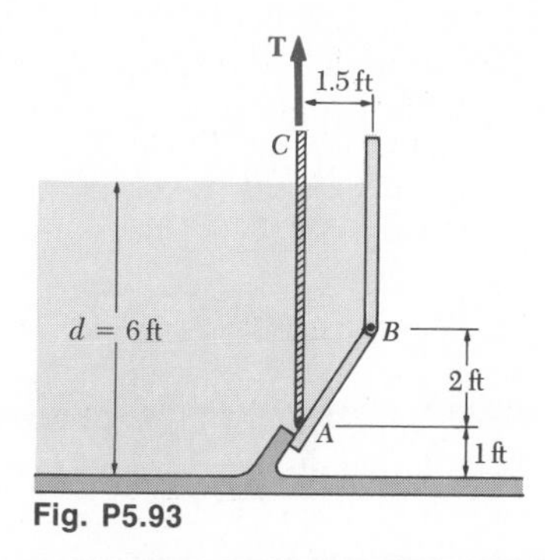

Fig. P5.93

5.93 The quick-acting gate AB is 1.75 ft wide and is held in its closed position by a vertical cable and by hinges located along its top edge B. For a depth of water $d = 6$ ft, determine the minimum tension required in cable AC to prevent the gate from opening.

5.94 For the gate of Prob. 5.93 determine the depth d for which the gate will open when $T = 800$ lb.

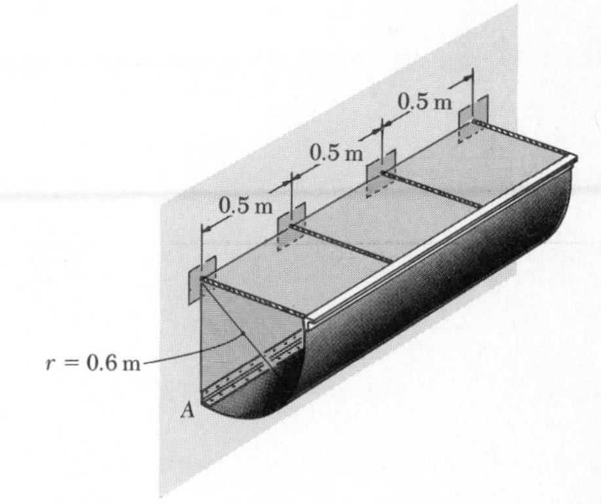

Fig. P5.95

5.95 A long trough is supported by a continuous hinge along its lower edge and by a series of horizontal cables attached to its upper edge. Determine the tension in each of the cables, at a time when the trough is completely full of water.

5.96 Determine the minimum allowable value of the width a of the rectangular concrete dam if the dam is not to overturn about point A when $d = 12$ ft. Assume that a seal exists at B and that no water pressure is present under the dam.

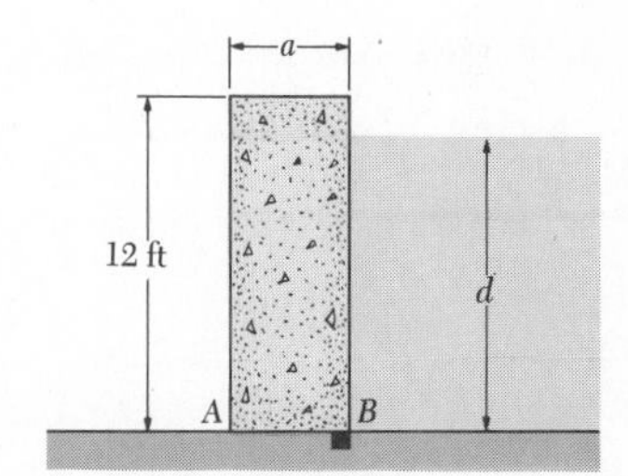

Fig. P5.96 and P5.98

5.97 Solve Prob. 5.96, assuming that a seal exists only at A and that the full hydrostatic pressure at B is present under the dam from A to B.

5.98 Knowing that the width of the rectangular concrete dam is $a = 3.5$ ft, determine the maximum allowable value of the depth d of water if the dam is not to overturn about A. Assume that a seal exists at B and that no water pressure is present under the dam.

5.99 The end of a freshwater channel consists of a plate $ABCD$ which is hinged at B and is 0.5 m wide. Knowing that $a = 0.6$ m, determine the reactions at A and B.

5.100 The end of a freshwater channel consists of a plate $ABCD$ which is hinged at B and is 0.5 m wide. Determine the length a for which the reaction at A is zero.

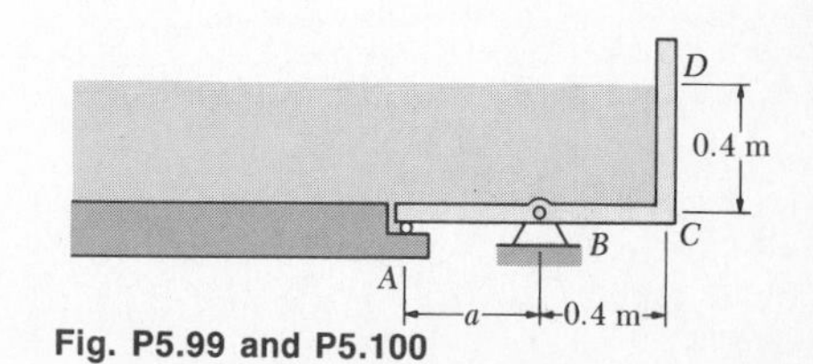

Fig. P5.99 and P5.100

5.10. Center of Gravity of a Three-Dimensional Body. Centroid of a Volume. The *center of gravity* G of a three-dimensional body is obtained by dividing the body into small elements and expressing that the weight $\mathbf{W}$ of the body attached at G is equivalent to the system of distributed forces $\Delta\mathbf{W}$ representing the weights of the small elements. Choosing the y axis vertical with positive sense upward (Fig. 5.20), and denoting by $\bar{\mathbf{r}}$ the position vector of G, we write that $\mathbf{W}$ is equal

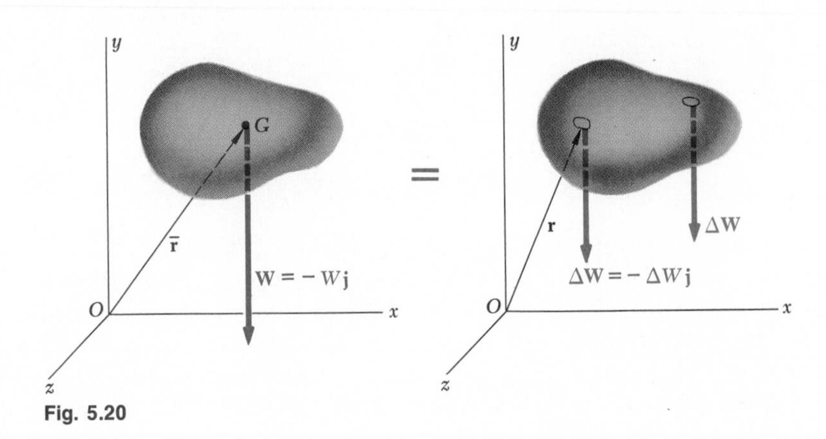

Fig. 5.20

to the sum of the elementary weights $\Delta\mathbf{W}$ and that its moment about O is equal to the sum of the moments about O of the elementary weights:

$$\Sigma\mathbf{F}: \qquad -W\mathbf{j} = \Sigma(-\Delta W\mathbf{j}) \tag{5.13}$$
$$\Delta\mathbf{M}_O: \qquad \bar{\mathbf{r}} \times (-W\mathbf{j}) = \Sigma[\mathbf{r} \times (-\Delta W\mathbf{j})]$$

Rewriting the last equation in the form

$$\bar{\mathbf{r}}W \times (-\mathbf{j}) = (\Sigma\mathbf{r}\,\Delta W) \times (-\mathbf{j}) \tag{5.14}$$

we observe that the weight $\mathbf{W}$ of the body will be equivalent to the system of the elementary weights $\Delta\mathbf{W}$ if the following conditions are satisfied:

$$W = \Sigma\,\Delta W \qquad \bar{\mathbf{r}}W = \Sigma\mathbf{r}\,\Delta W$$

Increasing the number of elements and simultaneously decreasing the size of each element, we obtain at the limit

$$W = \int dW \qquad \bar{\mathbf{r}}W = \int \mathbf{r}\,dW \tag{5.15}$$

We note that the relations obtained are independent of the orientation of the body. For example, if the body and the axes of coordinates were rotated so that the z axis pointed upward, the unit vector $-\mathbf{j}$ would be replaced by $-\mathbf{k}$ in Eqs. (5.13) and (5.14), but the relations (5.15) would remain unchanged. Resolving the vectors $\bar{\mathbf{r}}$ and $\mathbf{r}$ into rectangular components, we verify that the second of the relations (5.15) is equivalent to the three scalar equations

$$\bar{x}W = \int x\,dW \qquad \bar{y}W = \int y\,dW \qquad \bar{z}W = \int z\,dW \tag{5.16}$$

If the body is made of a homogeneous material of specific weight γ, the magnitude dW of the weight of an infinitesimal element may be expressed in terms of the volume dV of the element, and the magnitude W of the total weight in terms of the total volume V. We write

$$dW = \gamma\,dV \qquad W = \gamma V$$

Substituting for dW and W in the second of the relations (5.15), we write

$$\bar{\mathbf{r}}V = \int \mathbf{r}\,dV \tag{5.17}$$

or, in scalar form,

$$\bar{x}V = \int x\,dV \qquad \bar{y}V = \int y\,dV \qquad \bar{z}V = \int z\,dV \tag{5.18}$$

The point of coordinates $\bar{x}$, $\bar{y}$, $\bar{z}$ is also known as the *centroid C of the volume V* of the body. If the body is not homogeneous, Eqs. (5.18) cannot be used to determine the center of gravity of the body; they still define, however, the centroid of the volume.

The integral $\int x\,dV$ is known as the *first moment of the volume with respect to the yz plane*. Similarly, the integrals $\int y\,dV$ and $\int z\,dV$ define the first moments of the volume with respect to the zx plane and the xy plane, respectively. It is seen from Eqs. (5.18) that if the centroid of a volume is located in a coordinate plane, the first moment of the volume with respect to that plane is zero.

A volume is said to be symmetrical with respect to a given plane if to every point P of the volume corresponds a point P' of the same volume, such that the line PP' is perpendicular to the given plane and divided into two equal parts by that plane. The plane is said to be a *plane of symmetry* for the given volume. When a volume V possesses a plane of symmetry, the first moment of V with respect to that plane is zero and the centroid of the volume must be located in the plane of symmetry. When a volume possesses two planes of symmetry, the centroid of the volume must be located on the line of intersection of the two planes. Finally, when a volume possesses three planes of symmetry which intersect in a well-defined point (i.e., not along a common line), the point of intersection of the three planes must coincide with the centroid of the volume. This property enables us to determine immediately the centroid of the volume of spheres, ellipsoids, cubes, rectangular parallelepipeds, etc.

Centroids of unsymmetrical volumes or of volumes possessing only one or two planes of symmetry should be determined by integration (Sec. 5.12). Centroids of common shapes of volumes are shown in Fig. 5.21. It should be observed that the centroid of a volume of revolution in general *does not coincide* with the centroid of its cross section. Thus, the centroid of a hemisphere is different from that of a semicircular area, and the centroid of a cone is different from that of a triangle.

Shape		$\bar{x}$	Volume
Hemisphere	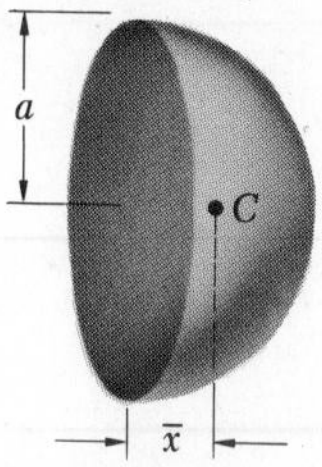	$\frac{3a}{8}$	$\frac{2}{3}\pi a^3$
Semiellipsoid of revolution	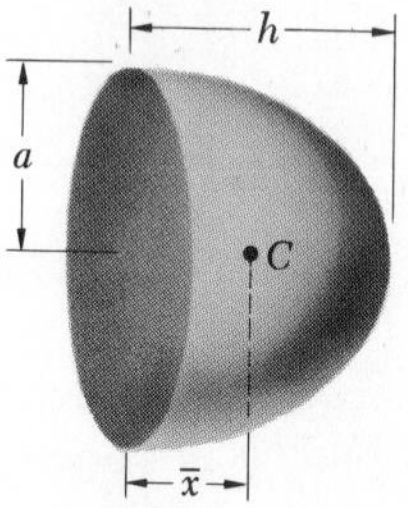	$\frac{3h}{8}$	$\frac{2}{3}\pi a^2 h$
Paraboloid of revolution	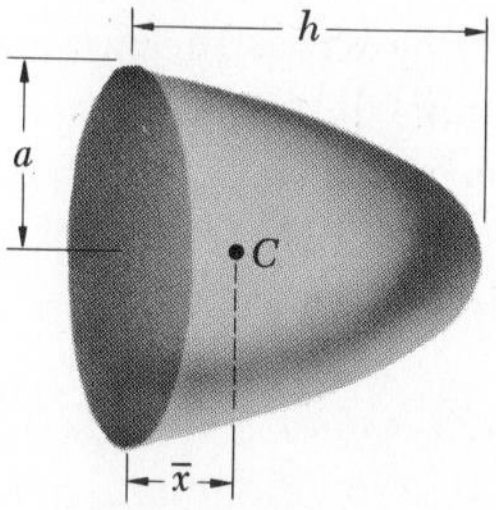	$\frac{h}{3}$	$\frac{1}{2}\pi a^2 h$
Cone	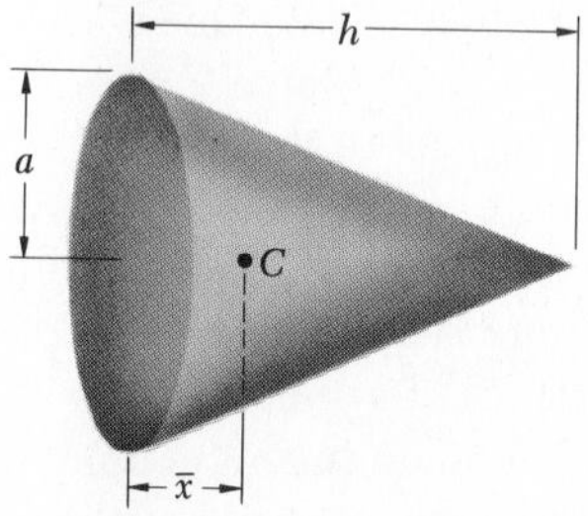	$\frac{h}{4}$	$\frac{1}{3}\pi a^2 h$
Pyramid	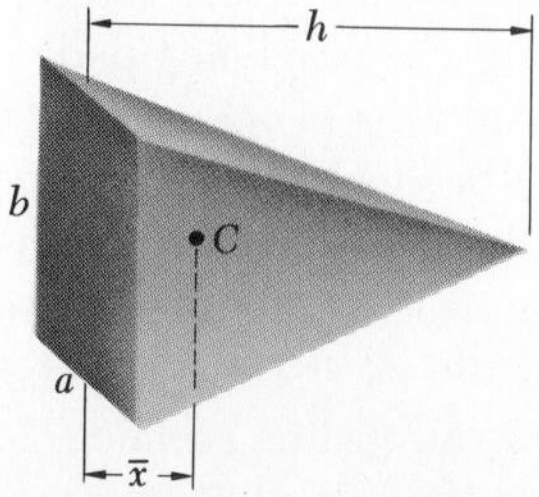	$\frac{h}{4}$	$\frac{1}{3}abh$

Fig. 5.21 Centroids of common shapes of volumes.

5.11. Composite Bodies. If a body can be divided into several of the common shapes shown in Fig. 5.21, its center of gravity G may be determined by expressing that the moment about O of its total weight is equal to the sum of the moments about O of the weights of the various component parts. Proceeding as in Sec. 5.10, we obtain the following equations defining the coordinates, $\bar{X}$, $\bar{Y}$, $\bar{Z}$ of the center of gravity G.

$$\bar{X}\Sigma W = \Sigma \bar{x} W \qquad \bar{Y}\Sigma W = \Sigma \bar{y} W \qquad \bar{Z}\Sigma W = \Sigma \bar{z} W \tag{5.19}$$

If the body is made of a homogeneous material, its center of gravity coincides with the centroid of its volume and the following equations may be used:

$$\bar{X}\Sigma V = \Sigma \bar{x} V \qquad \bar{Y}\Sigma V = \Sigma \bar{y} V \qquad \bar{Z}\Sigma V = \Sigma \bar{z} V \tag{5.20}$$

5.12. Determination of Centroids of Volumes by Integration. The centroid of a volume bounded by analytical surfaces may be determined by computing the integrals given in Sec. 5.10:

$$\bar{x}V = \int x\,dV \qquad \bar{y}V = \int y\,dV \qquad \bar{z}V = \int z\,dV \tag{5.21}$$

If the element of volume dV is chosen equal to a small cube of sides dx, dy, and dz, the determination of each of these integrals requires a *triple integration* in x, y, and z. However, it is possible to determine the coordinates of the centroid of most volumes by *double integration* if dV is chosen equal to the volume of a thin filament as shown in Fig. 5.22. The coordinates of the centroid of the volume are then obtained by writing

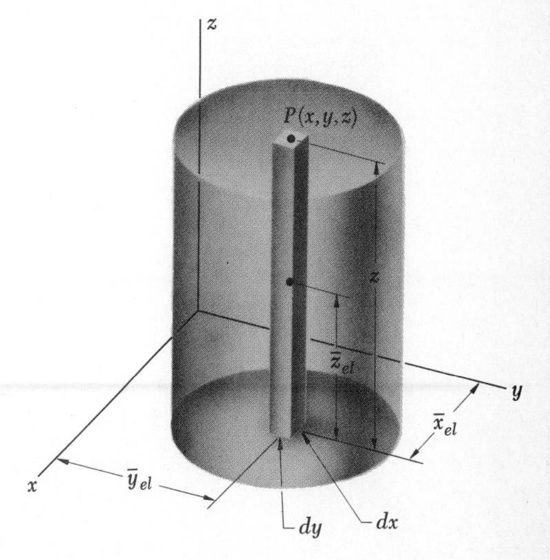

$\bar{x}_{el} = x$, $\bar{y}_{el} = y$, $\bar{z}_{el} = \frac{z}{2}$

$dV = z\,dx\,dy$

Fig. 5.22 Determination of the centroid of a volume by double integration.

$$\bar{x}V = \int \bar{x}_{el}\,dV \qquad \bar{y}V = \int \bar{y}_{el}\,dV \qquad \bar{z}V = \int \bar{z}_{el}\,dV \tag{5.22}$$

and substituting for the volume dV and the coordinates $\bar{x}_{el}$, $\bar{y}_{el}$, $\bar{z}_{el}$ the expressions given in Fig. 5.22. Using the equation of the surface to express z in terms of x and y, the integration is reduced to a double integration in x and y.

If the volume under consideration possesses *two planes of symmetry*, its centroid must be located on their line of intersection. Choosing the x axis along this line, we have

$$\bar{y} = \bar{z} = 0$$

and the only coordinate to determine is $\bar{x}$. This will be done most conveniently by dividing the given volume into thin slabs parallel to the yz plane. In the particular case of a body of revolution, these slabs are circular; their volume dV is given in Fig. 5.23. Substituting for $\bar{x}_{el}$ and dV into the equation

$$\bar{x}V = \int \bar{x}_{el}\,dV \tag{5.23}$$

and expressing the radius r of the slab in terms of x, we may determine $\bar{x}$ by a single integration.

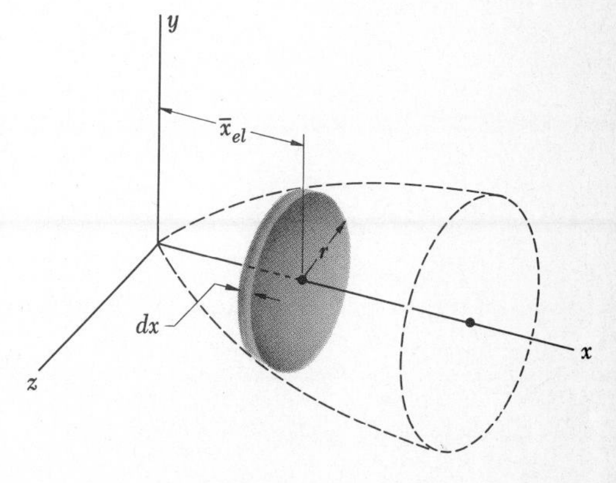

$\bar{x}_{el} = x$

$dV = \pi r^2\,dx$

Fig. 5.23 Determination of the centroid of a body of revolution.

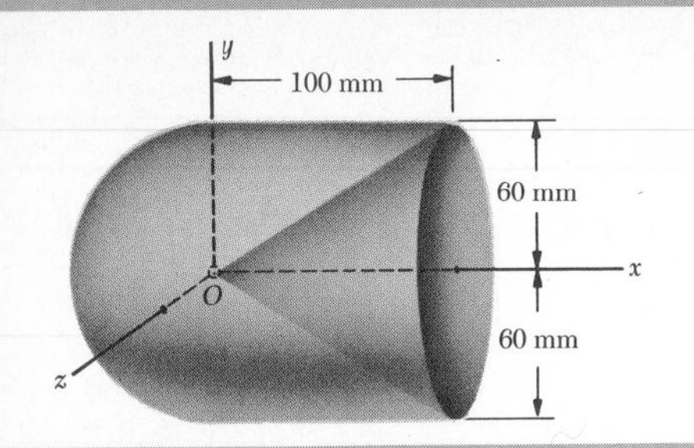

SAMPLE PROBLEM 5.11

Determine the center of gravity of the homogeneous body of revolution shown.

Solution. Because of symmetry, the center of gravity lies on the x axis. The body is seen to consist of a hemisphere, plus a cylinder, minus a cone, as shown. The volume and the abscissa of the centroid of each of these components are obtained from Fig. 5.21 and entered in the table below. The total volume of the body and the first moment of its volume with respect to the yz plane are then determined.

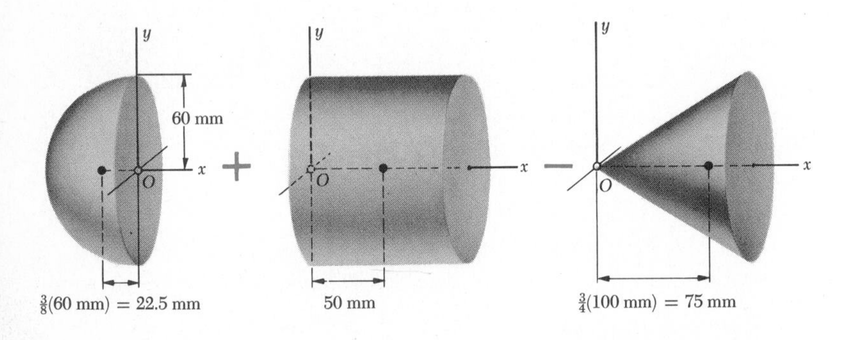

Component	Volume, mm^3	$\bar{x}$, mm	$\bar{x}V$, mm^4
Hemisphere	$\frac{1}{2}\frac{4\pi}{3}(60)^3 = 0.452 \times 10^6$	-22.5	-10.17×10^6
Cylinder	$\pi(60)^2(100) = 1.131 \times 10^6$	$+50$	$+56.55 \times 10^6$
Cone	$-\frac{\pi}{3}(60)^2(100) = -0.377 \times 10^6$	$+75$	-28.28×10^6
	$\Sigma V = 1.206 \times 10^6$	. . .	$\Sigma\bar{x}V = +18.10 \times 10^6$

Thus,

$\bar{X}\Sigma V = \Sigma\bar{x}V$: $\quad \bar{X}(1.206 \times 10^6 \text{ mm}^3) = 18.10 \times 10^6 \text{ mm}^4$

$\bar{X} = 15 \text{ mm}$ ◄

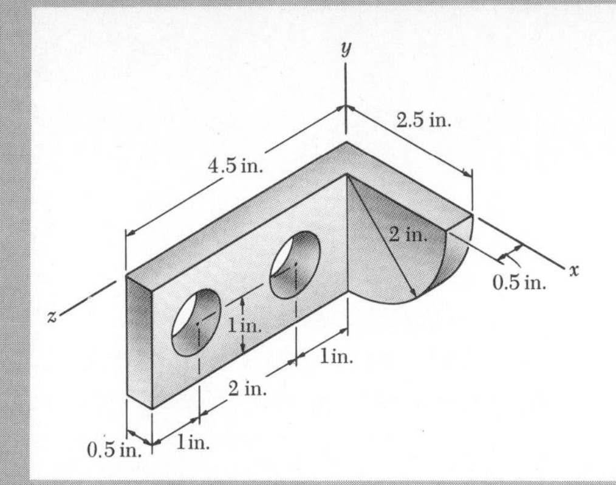

SAMPLE PROBLEM 5.12

Locate the center of gravity of the steel machine element shown. Both holes are of 1-in. diameter.

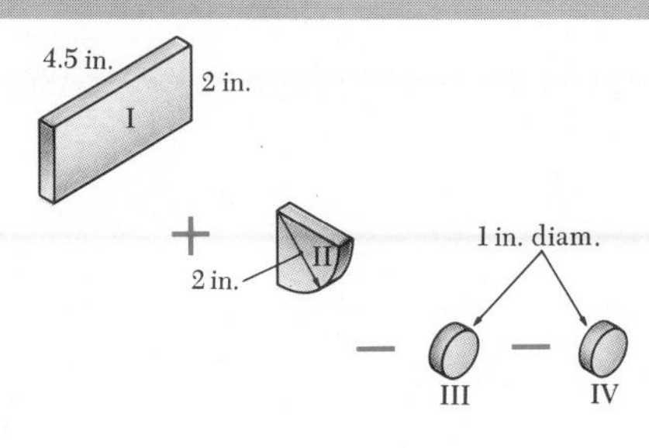

Solution. The machine element is seen to consist of a rectangular parallelepiped (I) plus a quarter cylinder (II) minus two 1-in.-diameter cylinders (III and IV). The volume and the coordinates of the centroid of each component are determined and entered in the table below. Using the data accumulated in the table, we then determine the total volume and the moments of the volume about each of the coordinate planes.

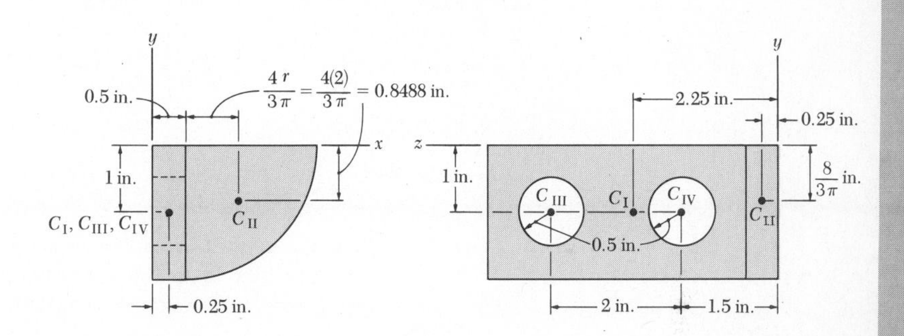

	V, in^3	$\bar{x}$, in.	$\bar{y}$, in.	$\bar{z}$, in.	$\bar{x}V$, in^4	$\bar{y}V$, in^4	$\bar{z}V$, in^4
I	$(4.5)(2)(0.5) = 4.5$	0.25	-1	2.25	1.125	-4.5	10.125
II	$\frac{1}{4}\pi(2)^2(0.5) = 1.571$	1.3488	-0.8488	0.25	2.119	-1.333	0.393
III	$-\pi(0.5)^2(0.5) = -0.3927$	0.25	-1	3.5	-0.098	0.393	-1.374
IV	$-\pi(0.5)^2(0.5) = -0.3927$	0.25	-1	1.5	-0.098	0.393	-0.589
	$\Sigma V = 5.286$				$\Sigma \bar{x}V = 3.048$	$\Sigma \bar{y}V = -5.047$	$\Sigma \bar{z}V = 8.555$

Thus,

$\bar{X}\Sigma V = \Sigma \bar{x} V$: $\quad \bar{X}(5.286 \text{ in}^3) = 3.048 \text{ in}^4 \qquad \bar{X} = 0.577$ in. ◄

$\bar{Y}\Sigma V = \Sigma \bar{y} V$: $\quad \bar{Y}(5.286 \text{ in}^3) = -5.047 \text{ in}^4 \qquad \bar{Y} = -0.955$ in. ◄

$\bar{Z}\Sigma V = \Sigma \bar{z} V$: $\quad \bar{Z}(5.286 \text{ in}^3) = 8.555 \text{ in}^4 \qquad \bar{Z} = 1.618$ in. ◄

SAMPLE PROBLEM 5.13

Determine the location of the centroid of the half right circular cone shown.

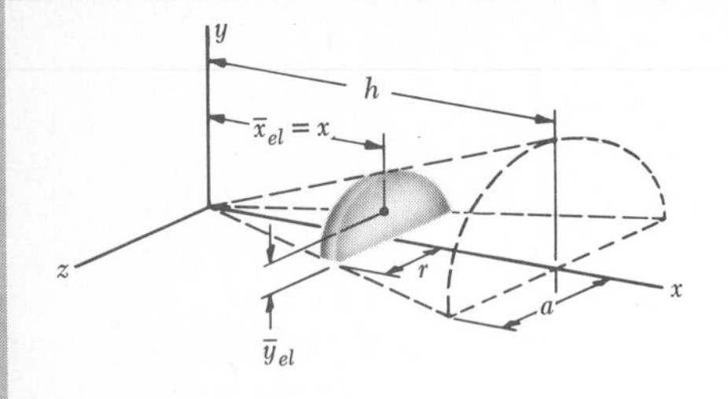

Solution. Since the xy plane is a plane of symmetry, the centroid lies in this plane and $\bar{z} = 0$. A slab of thickness dx is chosen as a differential element. The volume of this element is

$$dV = \tfrac{1}{2}\pi r^2\, dx$$

The coordinates $\bar{x}_{el}$ and $\bar{y}_{el}$ of the centroid of the element are obtained from Fig. 5.8 (semicircular area).

$$\bar{x}_{el} = x \qquad \bar{y}_{el} = \frac{4r}{3\pi}$$

We observe that r is proportional to x and write

$$\frac{r}{x} = \frac{a}{h} \qquad r = \frac{a}{h}x$$

The volume of the body is

$$V = \int dV = \int_0^h \tfrac{1}{2}\pi r^2\, dx = \int_0^h \tfrac{1}{2}\pi\left(\frac{a}{h}x\right)^2 dx = \frac{\pi a^2 h}{6}$$

The moment of the differential element with respect to the yz plane is $\bar{x}_{el}\, dV$; and the total moment of the body with respect to this plane is

$$\int \bar{x}_{el}\, dV = \int_0^h x(\tfrac{1}{2}\pi r^2)\, dx = \int_0^h x(\tfrac{1}{2}\pi)\left(\frac{a}{h}x\right)^2 dx = \frac{\pi a^2 h^2}{8}$$

Thus,

$$\bar{x}V = \int \bar{x}_{el}\, dV \qquad \bar{x}\frac{\pi a^2 h}{6} = \frac{\pi a^2 h^2}{8} \qquad \bar{x} = \tfrac{3}{4}h \blacktriangleleft$$

Likewise, the moment of the differential element with respect to the zx plane is $\bar{y}_{el}\, dV$; and the total moment is

$$\int \bar{y}_{el}\, dV = \int_0^h \frac{4r}{3\pi}(\tfrac{1}{2}\pi r^2)\, dx = \frac{2}{3}\int_0^h \left(\frac{a}{h}x\right)^3 dx = \frac{a^3 h}{6}$$

Thus,

$$\bar{y}V = \int \bar{y}_{el}\, dV \qquad \bar{y}\frac{\pi a^2 h}{6} = \frac{a^3 h}{6} \qquad \bar{y} = \frac{a}{\pi} \blacktriangleleft$$

Problems

5.101 Determine the location of the centroid of the composite body shown when (*a*) $h = 2b$, (*b*) $h = 2.5b$.

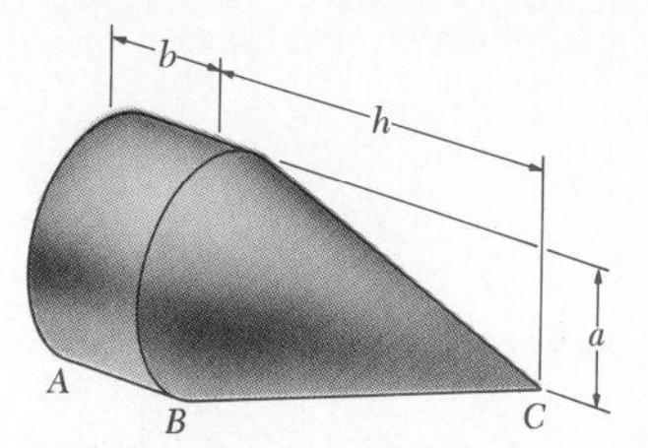

Fig. P5.101

5.102 Determine the y coordinate of the centroid of the body shown when (*a*) $b = \frac{1}{3}h$, (*b*) $b = \frac{1}{2}h$.

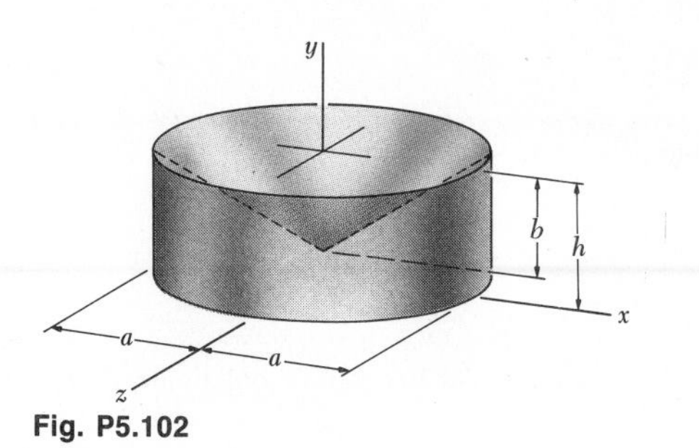

Fig. P5.102

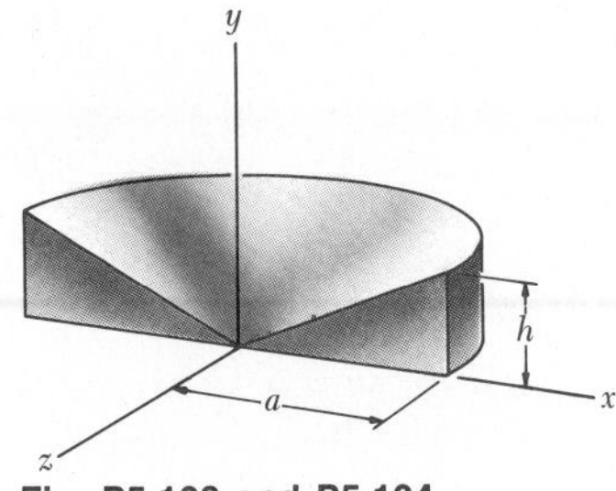

Fig. P5.103 and P5.104

5.103 Determine the y coordinate of the centroid of the body shown.

5.104 Determine the z coordinate of the centroid of the body shown. (*Hint.* Use the result of Sample Prob. 5.13.)

5.105 Determine the location of the center of gravity of the parabolic reflector shown, which is formed by machining a rectangular block so that the curved surface is a paraboloid of revolution of base radius a and height h.

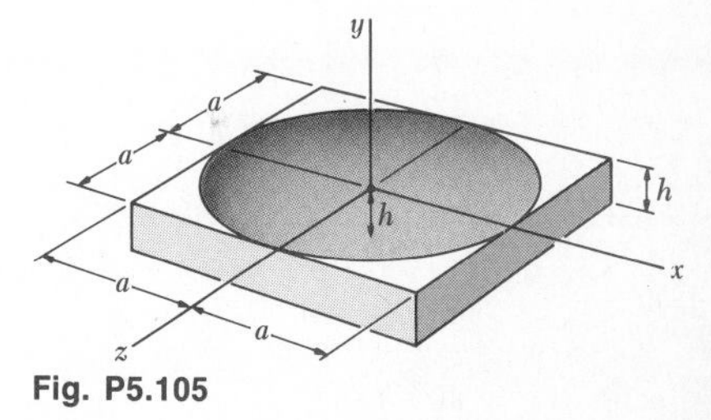

Fig. P5.105

5.106 For the machine element shown, locate the x coordinate of the center of gravity.

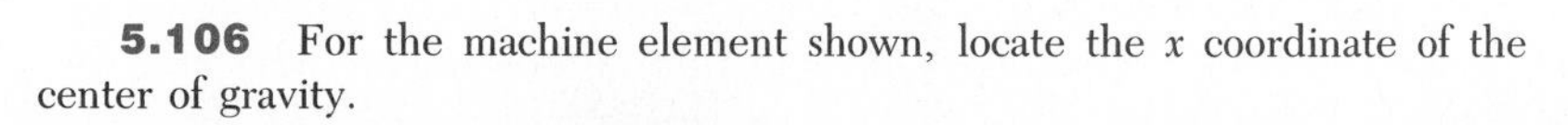

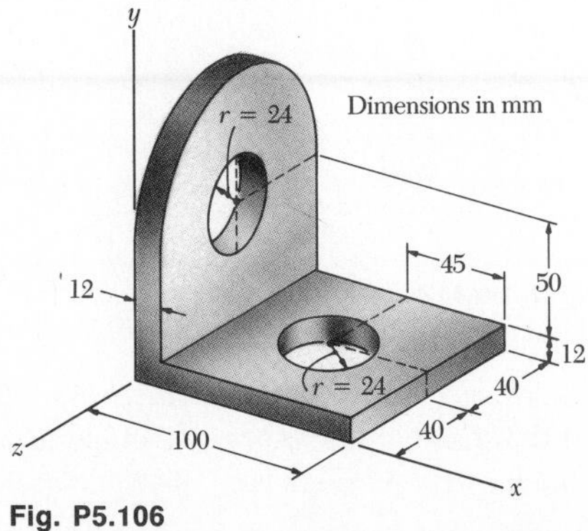

Fig. P5.106

5.107 and 5.108 For the machine element shown, locate the y coordinate of the center of gravity.

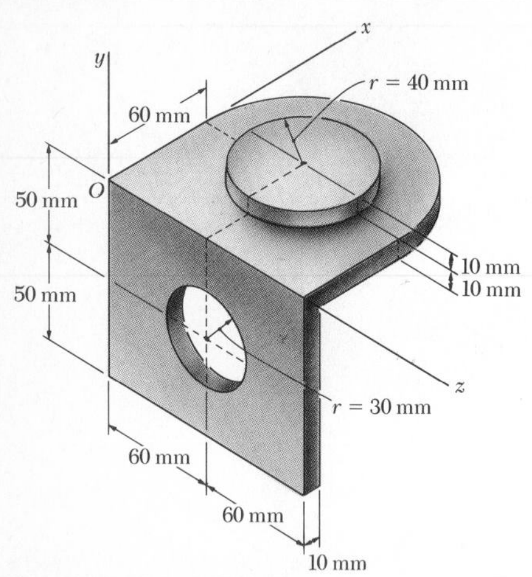

Fig. P5.107 and P5.110

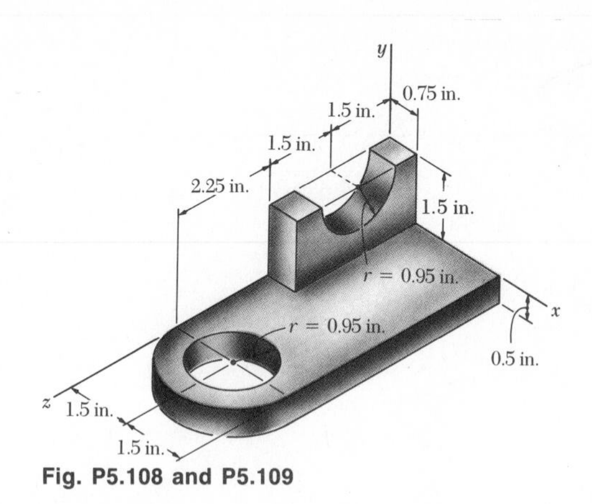

Fig. P5.108 and P5.109

5.109 For the machine element shown, locate the z coordinate of the center of gravity.

5.110 For the machine element shown, locate the x coordinate of the center of gravity.

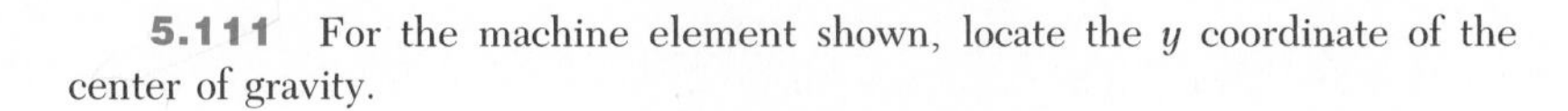

5.111 For the machine element shown, locate the y coordinate of the center of gravity.

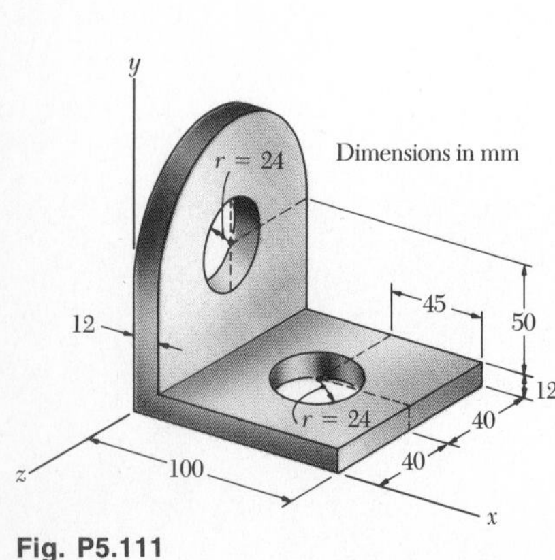

Fig. P5.111

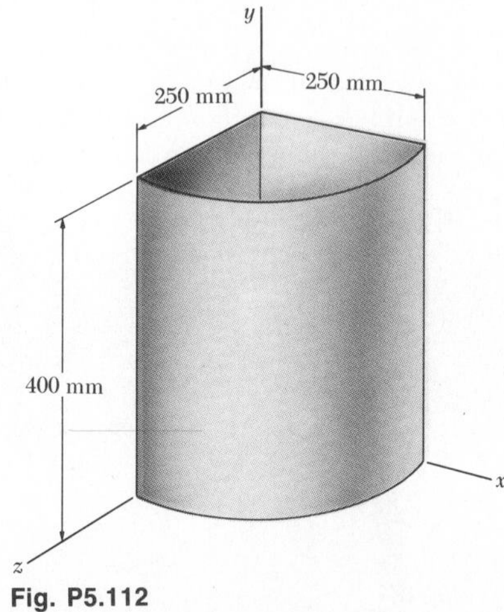

Fig. P5.112

5.112 A wastebasket, designed to fit in the corner of a room, is 400 mm high and has a base in the shape of a quarter circle of radius 250 mm. Locate the center of gravity of the wastebasket, knowing that it is made of sheet metal of uniform thickness.

5.113 and 5.114 Locate the center of gravity of the sheet-metal form shown.

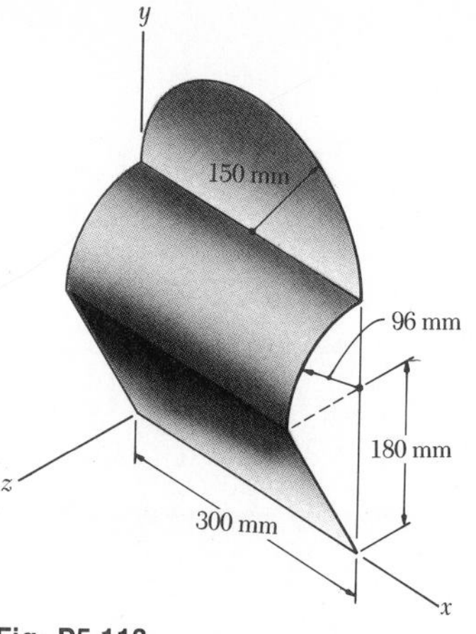

Fig. P5.113

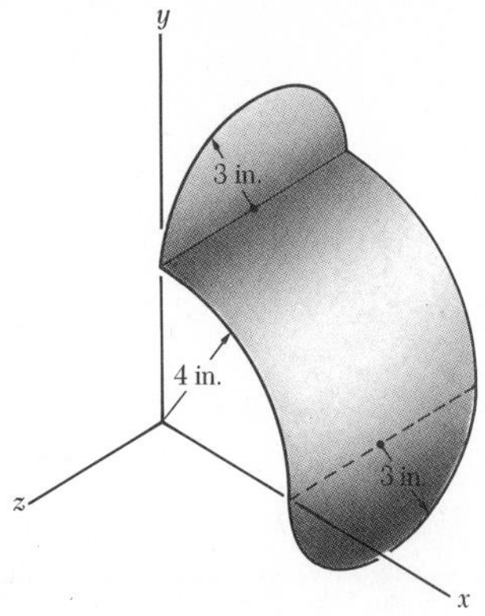

y
3 in.
4 in.
3 in.
z
x

Fig. P5.114

5.115 A corner reflector for tracking by radar has two sides in the shape of a quarter circle of radius 15 in. and one side in the shape of a triangle. Locate the center of gravity of the reflector, knowing that it is made of sheet metal of uniform thickness.

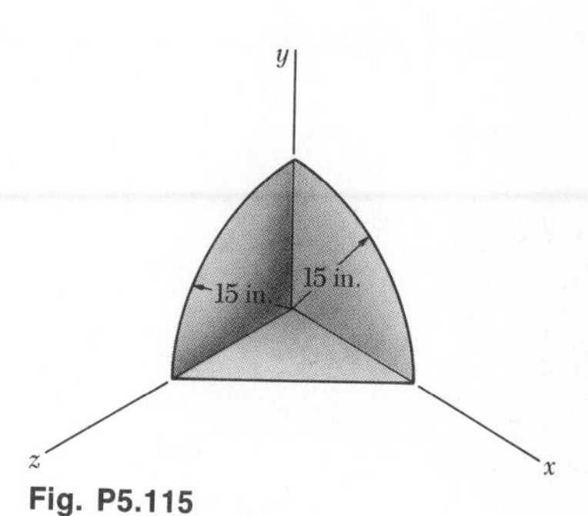

Fig. P5.115

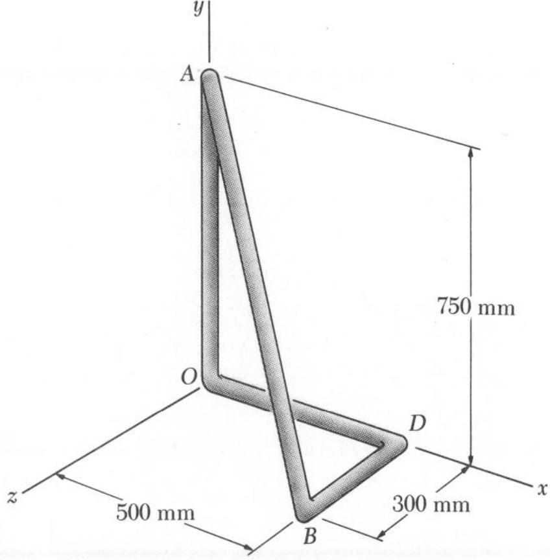

Fig. P5.116

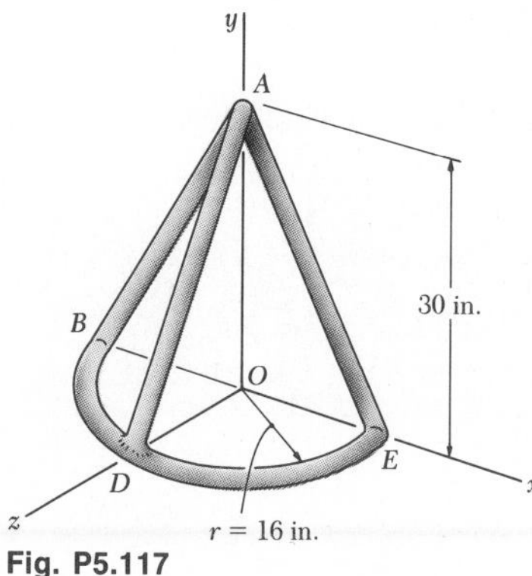

Fig. P5.117

5.116 and 5.117 Locate the center of gravity of the figure shown, knowing that it is made of thin brass rods of uniform diameter.

5.118 Three brass plates are brazed to a steel pipe to form the flagpole base shown. Knowing that the pipe has a wall thickness of 0.25 in. and that each plate is 0.2 in. thick, determine the location of the center of gravity of the base. (Specific weights: brass = 0.306 lb/in^3; steel = 0.284 lb/in^3.)

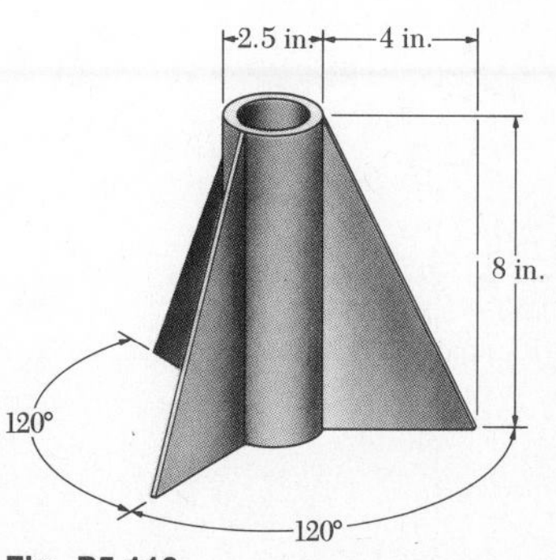

Fig. P5.118

5.119 A brass collar, of length 50 mm, is mounted on an aluminum rod of length 80 mm. Locate the center of gravity of the composite body. (Densities: brass = 8470 kg/m^3; aluminum = 2800 kg/m^3.)

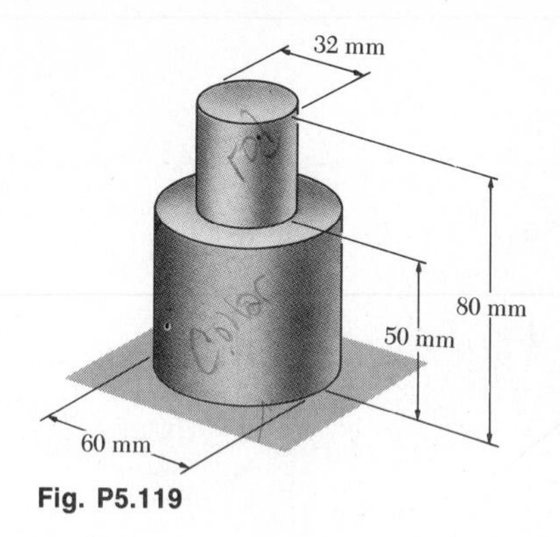

Fig. P5.119

5.120 Solve Prob. 5.119, assuming that the collar is made of aluminum and the rod is made of brass.

5.121 through 5.123 Derive by direct integration the expression given for $\bar{x}$ in Fig. 5.21 for:

5.121 A hemisphere
5.122 A semiellipsoid of revolution
5.123 A paraboloid of revolution

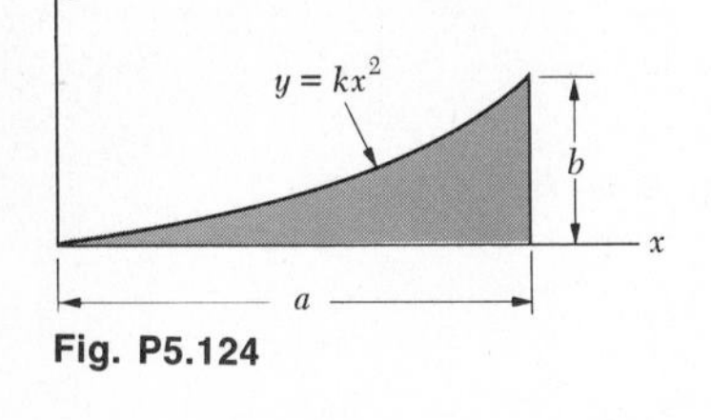

Fig. P5.124

5.124 and 5.125 Locate the centroid of the volume obtained by rotating the shaded area about the x axis.

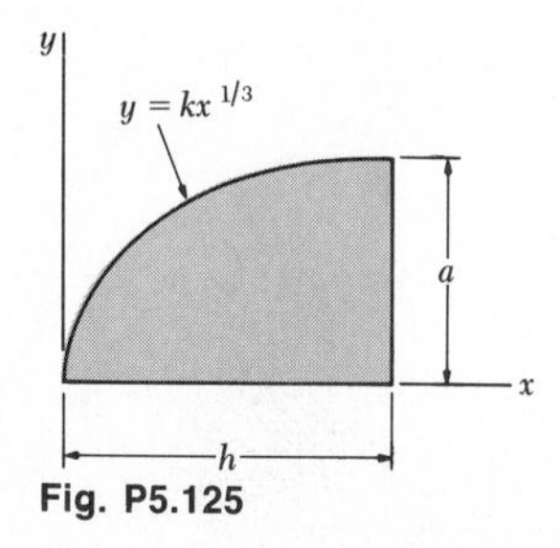

Fig. P5.125

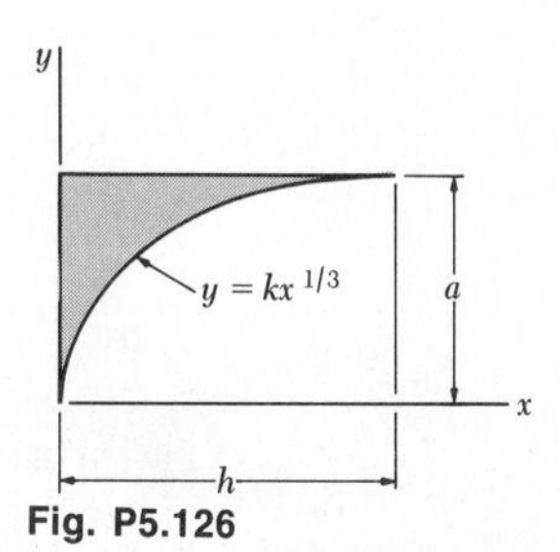

Fig. P5.126

5.126 Locate the centroid of the volume obtained by rotating the shaded area about the y axis.

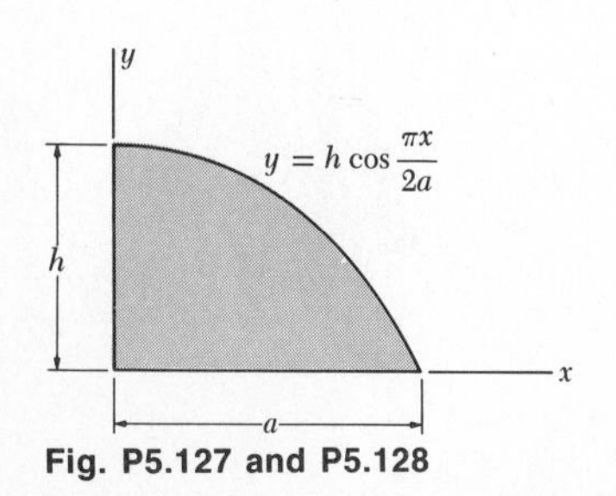

Fig. P5.127 and P5.128

***5.127** Locate the centroid of the volume generated by revolving the portion shown of the cosine curve about the x axis.

***5.128** Locate the centroid of the volume generated by revolving the portion shown of the cosine curve about the y axis. (*Hint.* Use as an element of volume a thin cylindrical shell of radius r and thickness dr.)

5.129 Locate the centroid of the volume of the irregular pyramid shown.

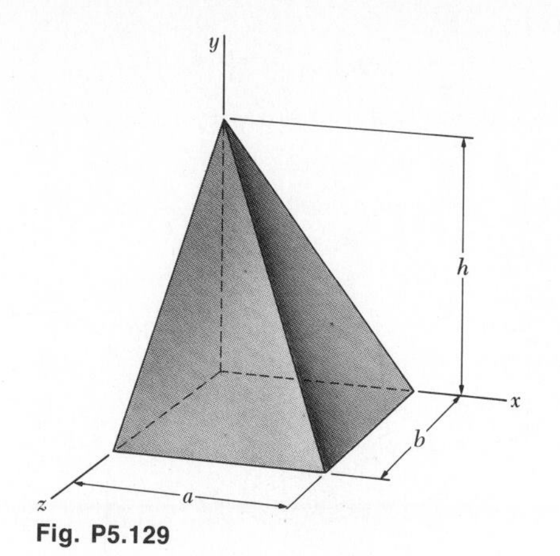

Fig. P5.129

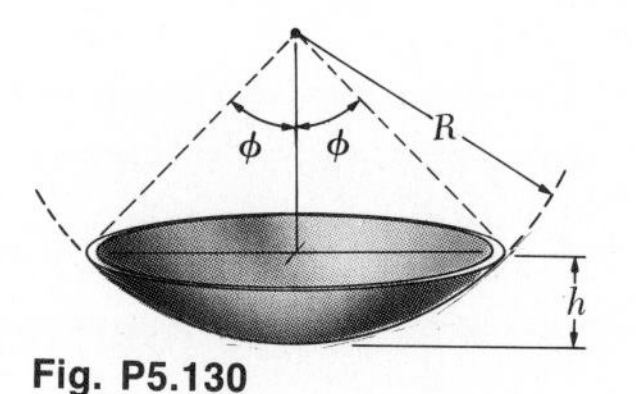

Fig. P5.130

***5.130** A thin spherical cup has a radius R and a uniform thickness t. Show by direct integration that the center of gravity of the cup is located at a distance $\frac{1}{2}h$ above the base of the cup.

5.131 Determine by direct integration the z coordinate of the centroid of the volume shown, which was cut from a rectangular prism by an oblique plane.

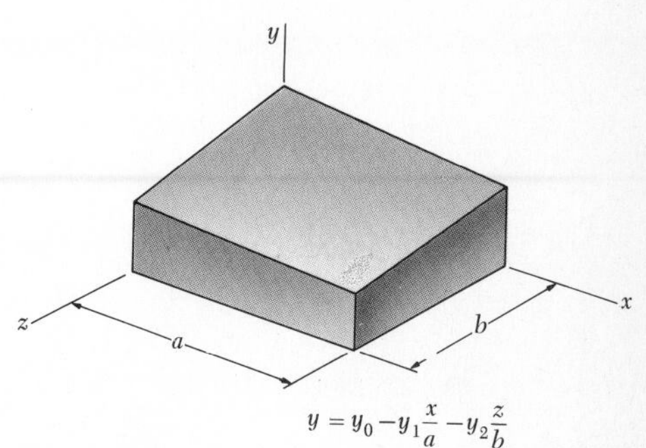

Fig. P5.131

5.132 Solve Prob. 5.131, when $a = 250$ mm, $b = 200$ mm, $y_0 = 180$ mm, $y_1 = 80$ mm, and $y_2 = 40$ mm.

5.133 Locate the centroid of the section shown, which was cut from a thin circular pipe by an oblique plane.

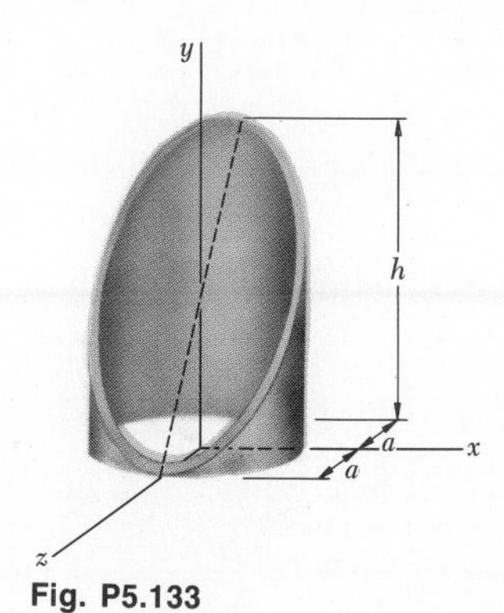

Fig. P5.133

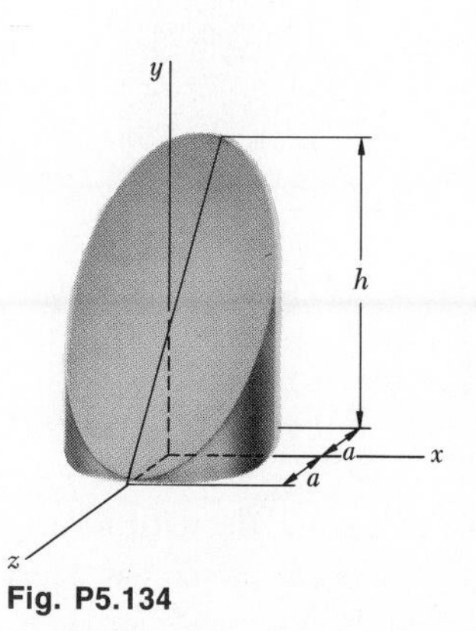

Fig. P5.134

5.134 Locate the centroid of the section shown, which was cut from a circular cylinder by an oblique plane.

Review and Summary

This chapter was devoted chiefly to the determination of the *center of gravity* of a rigid body, i.e., to the determination of the point G where a single force $\mathbf{W}$, called the *weight* of the body, may be applied to represent the effect of the earth's attraction on the body.

Center of gravity of a two-dimensional body

In the first part of the chapter, we considered *two-dimensional bodies,* such as flat plates and wires contained in the xy plane. By adding components in the vertical z direction and moments about the horizontal y and x axes [Sec. 5.2], we derived the relations

$$W = \int dW \qquad \bar{x}W = \int x\, dW \qquad \bar{y}W = \int y\, dW \tag{5.2}$$

which define the weight of the body and the coordinates $\bar{x}$ and $\bar{y}$ of its center of gravity.

Centroid of an area or line

In the case of a *homogeneous flat plate of uniform thickness* [Sec. 5.3], the center of gravity G of the plate coincides with the *centroid C of the area A* of the plate, the coordinates of which are defined by the relations

$$\bar{x}A = \int x\, dA \qquad \bar{y}A = \int y\, dA \tag{5.3}$$

Similarly, the determination of the center of gravity of a *homogeneous flat wire of uniform cross section* reduces to that of the *centroid C of the line L* representing the wire; we have

$$\bar{x}L = \int x\, dL \qquad \bar{y}L = \int y\, dL \tag{5.4}$$

First moments

The integrals in Eqs. (5.3) are referred to as the *first moments* of the area A with respect to the y and x axes and denoted by Q_y and Q_x, respectively [Sec. 5.4]. We have

$$Q_y = \bar{x}A \qquad Q_x = \bar{y}A \tag{5.6}$$

The first moments of a line may be defined in a similar way.

Properties of symmetry

The determination of the centroid C of an area or line is simplified when the area or line possesses certain *properties of symmetry.* If the area or line is symmetric with respect to an axis, its centroid C will lie on that axis; if it is symmetric with respect to two axes, C will be located at the intersection of the two axes; if it is symmetric with respect to a center O, C will coincide with O.

Center of gravity of a composite body

The *areas and centroids of various common shapes* have been tabulated in Fig. 5.8. When a flat plate may be divided into several of these shapes, the coordinates $\overline{X}$ and $\overline{Y}$ of its center of gravity G may be determined from the coordinates $\bar{x}_1, \bar{x}_2, \ldots$ and $\bar{y}_1, \bar{y}_2, \ldots$ of the centers of

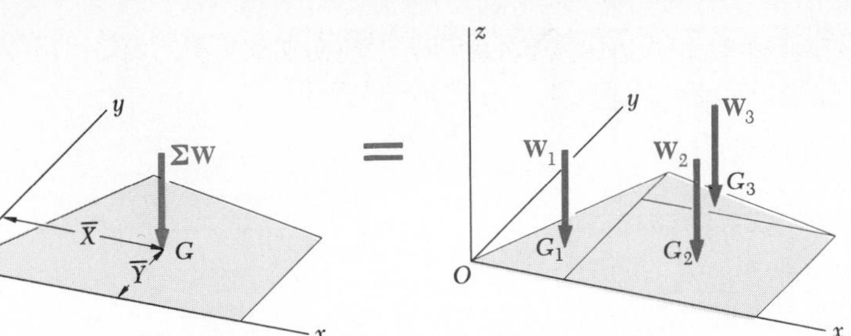
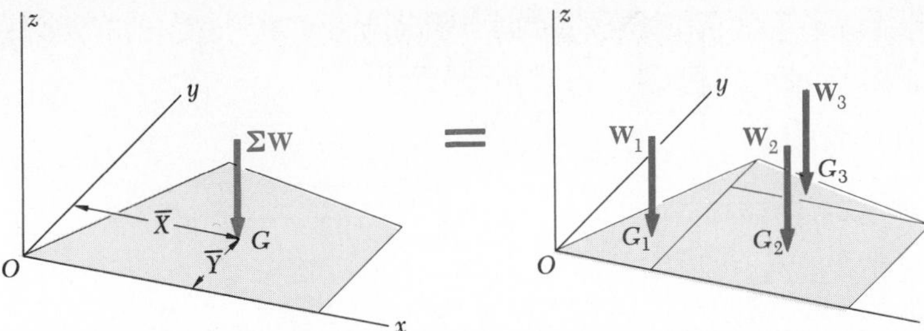

Fig. 5.24

gravity G_1, G_2, . . . of the various parts [Sec. 5.5]. Equating moments about the y and x axes, respectively (Fig. 5.24), we have

$$\overline{X}\Sigma W = \Sigma\overline{x}W \qquad \overline{Y}\Sigma W = \Sigma\overline{y}W \tag{5.7}$$

If the plate is homogeneous and of uniform thickness, the center of gravity coincides with the centroid C of the area of the plate and Eqs. (5.7) reduce to

$$Q_y = \overline{X}\Sigma A = \Sigma\overline{x}A \qquad Q_x = \overline{Y}\Sigma A = \Sigma\overline{y}A \tag{5.8}$$

These equations yield the first moments of the composite area, or they may be solved for the coordinates $\overline{X}$ and $\overline{Y}$ of its centroid [Sample Prob. 5.1]. The determination of the center of gravity of a composite wire is carried out in a similar fashion [Sample Prob. 5.2].

Determination of centroid by integration

When an area is bounded by analytical curves, the coordinates of its centroid may be determined *by integration* [Sec. 5.6]. This may be done by computing the double integrals in Eqs. (5.3), or through a *single integration,* using one of the thin rectangular or pie-shaped elements of area shown in Fig. 5.12. Denoting by $\overline{x}_{el}$ and $\overline{y}_{el}$ the coordinates of the centroid of the element dA, we have

$$Q_y = \overline{x}A = \int \overline{x}_{el}\, dA \qquad Q_x = \overline{y}A = \int \overline{y}_{el}\, dA \tag{5.9}$$

It is advantageous to use the same element of area to compute both of the first moments Q_y and Q_x; the same element may also be used to determine the area A [Sample Prob. 5.4].

Theorems of Pappus-Guldinus

The *theorems of Pappus-Guldinus* relate the determination of the area of a surface of revolution or the volume of a body of revolution to the determination of the centroid of the generating curve or area [Sec. 5.7]. The area A of the surface generated by rotating a curve of length L about a fixed axis (Fig. 5.25*a*) is

$$A = 2\pi\overline{y}L \tag{5.10}$$

where $\overline{y}$ represents the distance from the centroid C of the curve to the fixed axis. Similarly, the volume V of the body generated by rotating an area A about a fixed axis (Fig. 5.25*b*) is

$$V = 2\pi\overline{y}A \tag{5.11}$$

where $\overline{y}$ represents the distance from the centroid C of the area to the fixed axis.

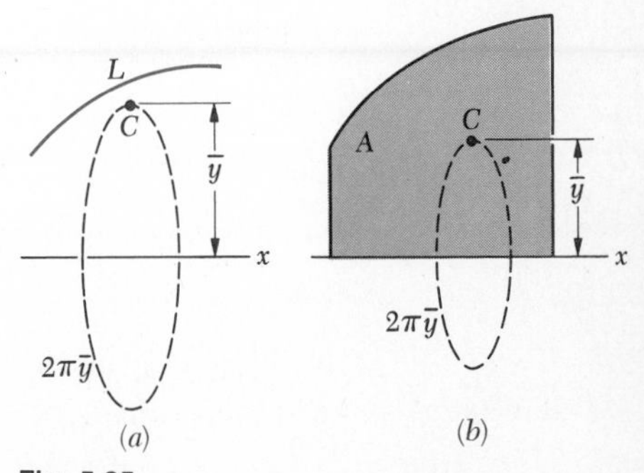

Fig. 5.25

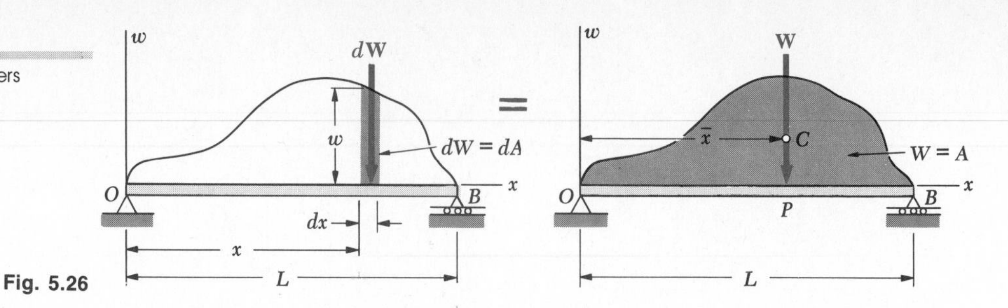

Fig. 5.26

Distributed loads

The concept of centroid of an area may be used to solve problems other than those dealing with the weight of flat plates. For example, to determine the reactions at the supports of a beam [Sec. 5.8], we may replace a *distributed load w* by a concentrated load W equal in magnitude to the area A under the load curve and passing through the centroid C of that area (Fig. 5.26). The same approach may be used to determine the resultant of the hydrostatic forces exerted on a *rectangular plate submerged in a liquid* [Sec. 5.9].

Center of gravity of a three-dimensional body

The last part of the chapter was devoted to the determination of the *center of gravity G of a three-dimensional body*. The coordinates $\bar{x}$, $\bar{y}$, $\bar{z}$ of G were defined by the relations

$$\bar{x}W = \int x\,dW \qquad \bar{y}W = \int y\,dW \qquad \bar{z}W = \int z\,dW \tag{5.16}$$

Centroid of a volume

In the case of a *homogeneous body*, the center of gravity G coincides with the *centroid C of the volume V* of the body, the coordinates of which are defined by the relations

$$\bar{x}V = \int x\,dV \qquad \bar{y}V = \int y\,dV \qquad \bar{z}V = \int z\,dV \tag{5.18}$$

If the volume possesses a *plane of symmetry*, its centroid C will lie in that plane; if it possesses two planes of symmetry, C will be located on the line of intersection of the two planes; if it possesses three planes of symmetry intersecting at one point only, C will coincide with that point [Sec. 5.10].

Center of gravity of a composite body

The *volumes and centroids of various common three-dimensional shapes* have been tabulated in Fig. 5.21. When a body may be divided into several of these shapes, the coordinates $\overline{X}$, $\overline{Y}$, $\overline{Z}$ of its center of gravity G may be determined from the corresponding coordinates of the centers of gravity of its various parts [Sec. 5.11]. We have

$$\overline{X}\Sigma W = \Sigma\bar{x}W \qquad \overline{Y}\Sigma W = \Sigma\bar{y}W \qquad \overline{Z}\Sigma W = \Sigma\bar{z}W \tag{5.19}$$

If the body is made of a homogeneous material, its center of gravity coincides with the centroid C of its volume and we write [Sample Probs. 5.11 and 5.12]

$$\overline{X}\Sigma V = \Sigma\bar{x}V \qquad \overline{Y}\Sigma V = \Sigma\bar{y}V \qquad \overline{Z}\Sigma V = \Sigma\bar{z}V \tag{5.20}$$

When a volume is bounded by analytical surfaces, the coordinates of its centroid may be determined by *integration* [Sec. 5.12]. To avoid the computation of the triple integrals in Eqs. (5.18), we may use elements of volume in the shape of thin filaments, as shown in Fig. 5.22. Denoting by $\bar{x}_{el}$, $\bar{y}_{el}$, and $\bar{z}_{el}$ the coordinates of the centroid of the element dV, we write the relations

Determination of centroid by integration

$$\bar{x}V = \int \bar{x}_{el}\, dV \qquad \bar{y}V = \int \bar{y}_{el}\, dV \qquad \bar{z}V = \int \bar{z}_{el}\, dV \qquad (5.22)$$

which involve only double integrals. If the volume possesses *two planes of symmetry*, its centroid C is located on their line of intersection. Choosing the x axis along that line and dividing the volume into thin slabs parallel to the yz plane, we determine C from the relation

$$\bar{x}V = \int \bar{x}_{el}\, dV \qquad (5.23)$$

through a *single integration* [Sample Prob. 5.13].

Review Problems

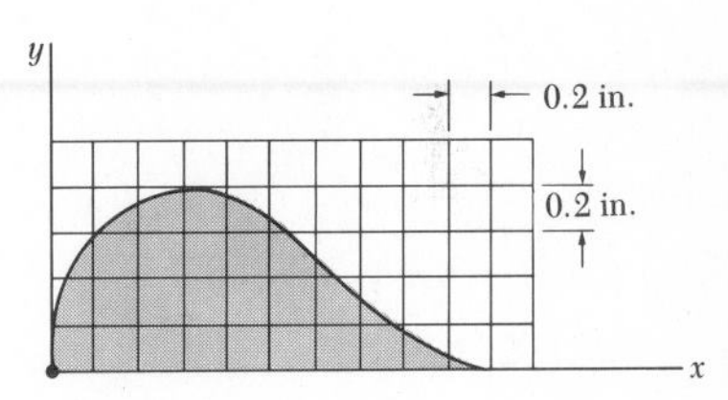

Fig. P5.135

5.135 Determine by approximate means the x and y coordinates of the centroid of the area shown.

5.136 Determine the center of gravity of the machine element shown.

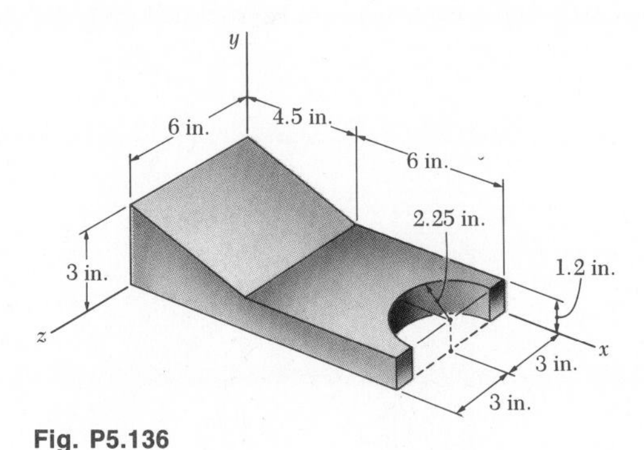

Fig. P5.136

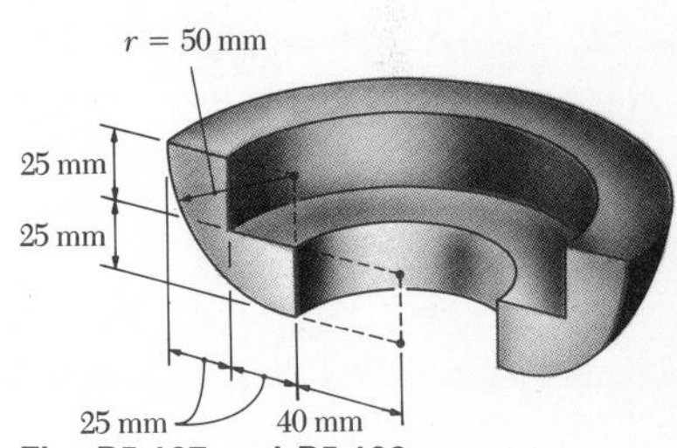

Fig. P5.137 and P5.138

5.137 Determine the volume of the body shown.

5.138 Determine the total surface area of the body shown.

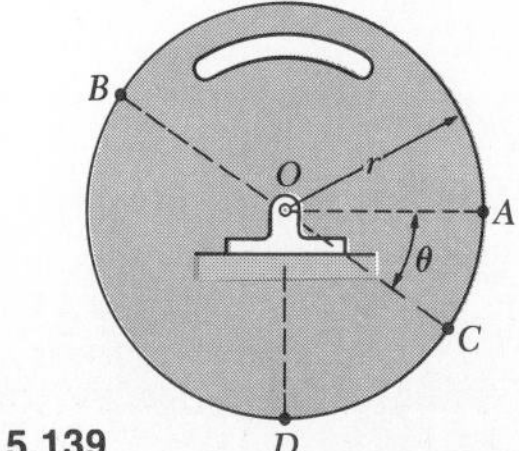

Fig. 5.139

5.139 A curved slot is cut in a uniform disk of radius r. The disk is mounted on a frictionless shaft at O and is in equilibrium in the position shown. After a weight W_1 has been attached to the rim of the disk at A, the disk takes a new position of equilibrium in which line BOC is vertical. At what point on the rim should a second weight W_2 be attached, and how large should W_2 be, if the center of gravity of the disk and the two weights is to be located at O?

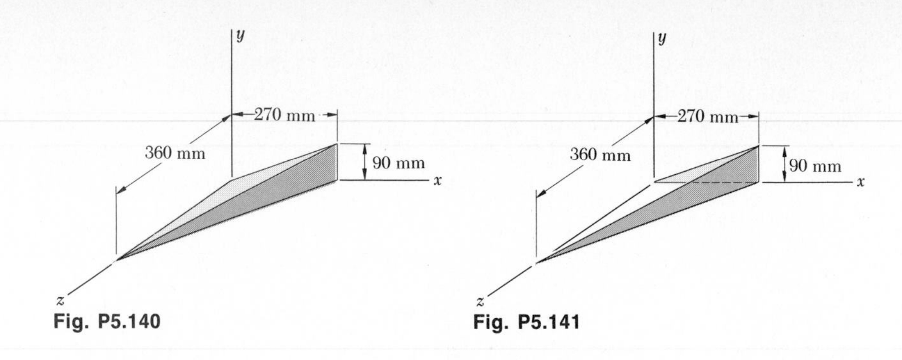

Fig. P5.140

Fig. P5.141

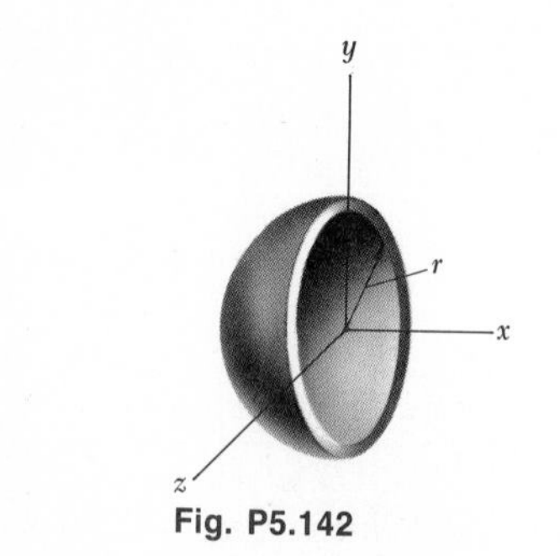

Fig. P5.142

5.140 Locate the centroid of the irregular pyramid shown. (*Hint.* See Fig. 5.21.)

5.141 Locate the center of gravity of the sheet-metal form shown.

5.142 Locate the center of gravity of a thin hemispherical shell of radius r and thickness t. (*Hint.* Consider the shell as formed by removing a hemisphere of radius r from a hemisphere of radius $r + t$; then neglect the terms containing t^2 and t^3, and keep those terms containing t.)

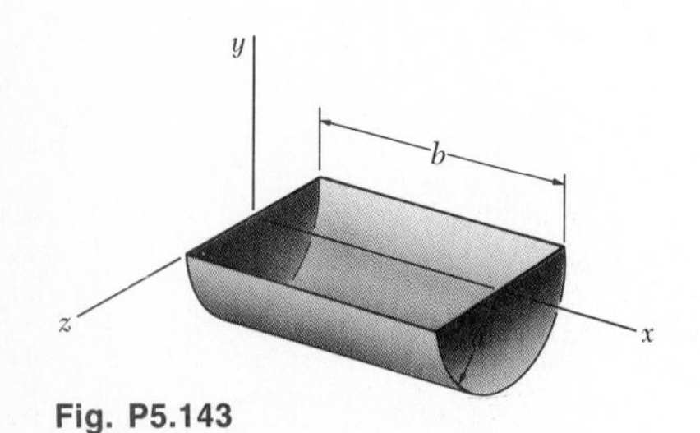

Fig. P5.143

5.143 The bin shown is made of sheet metal of uniform thickness. Determine (*a*) the center of gravity of the bin when $a = 5$ in. and $b = 15$ in., (*b*) the ratio a/b for which the y coordinate of the center of gravity is $\bar{y} = -\frac{1}{2}a$.

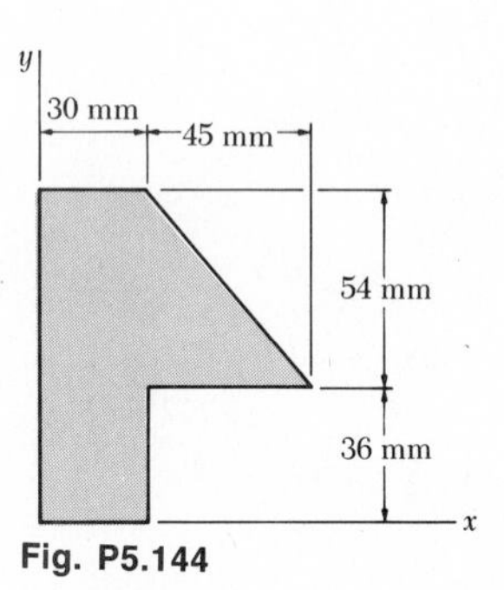

Fig. P5.144

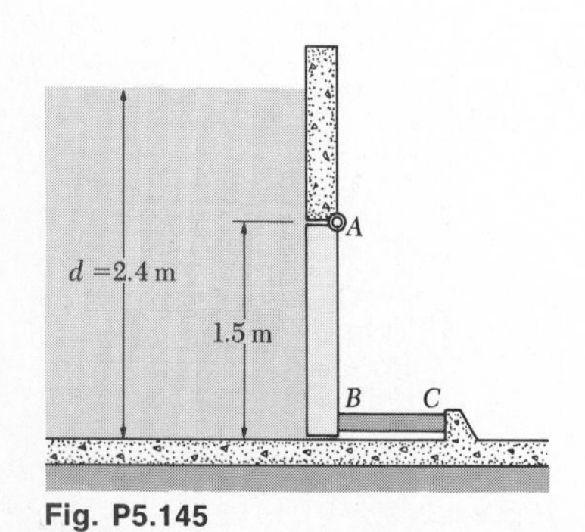

Fig. P5.145

5.144 Locate the centroid of the plane area shown.

5.145 A 1.5 × 1.5 m gate is located in a wall below water level. (*a*) Determine the magnitude and location of the resultant of the forces exerted by the water on the gate. (*b*) If the gate is hinged along its top edge A, determine the force exerted on the gate by the brace BC.

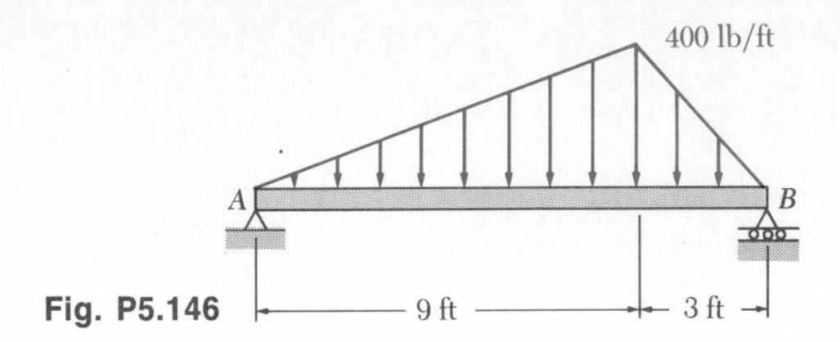

Fig. P5.146

5.146 Determine the reactions at the beam supports for the given loading condition.

The following problems are designed to be solved with a computer.

5.C1 Approximate the plane area shown by a series of rectangles of the form $bb'd'd$ and write a computer program which can be used to calculate the coordinates $\overline{X}$ and $\overline{Y}$ of the centroid of the area shown in part *a* of the figure. Use this program to calculate $\overline{X}$ and $\overline{Y}$ for the areas shown in parts *b* and *c* of the figure.

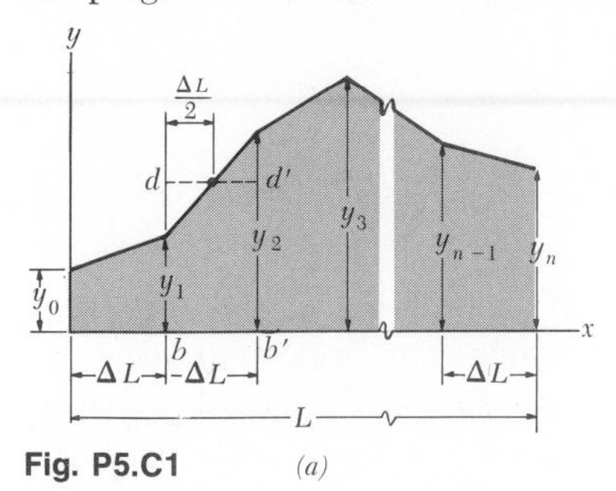

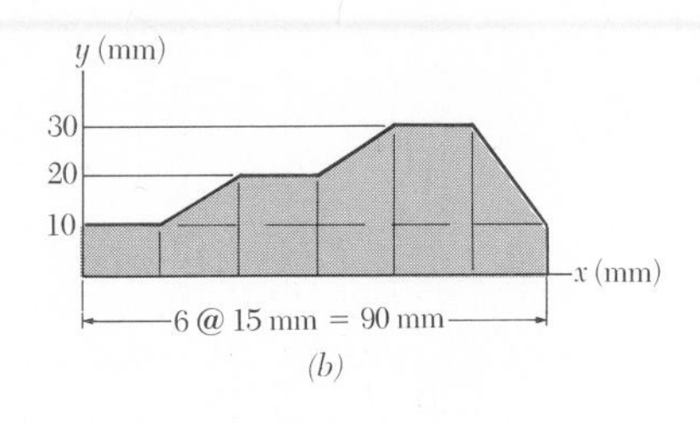

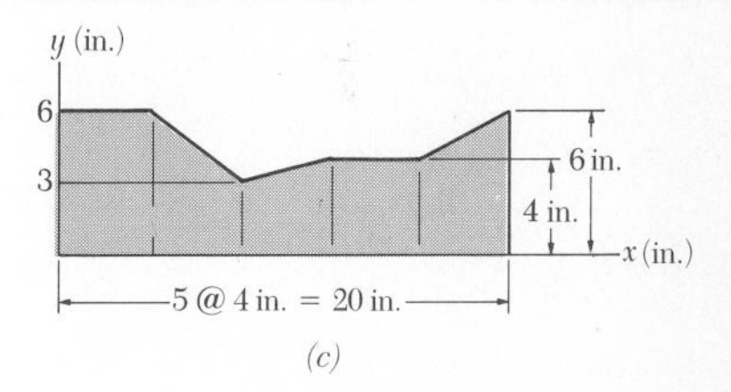

Fig. P5.C1 (*a*)

5.C2 The area of the parabolic reflector of Prob. 5.105 may be obtained by rotating the parabola shown about the y axis. Approximate the parabola by 10 straight-line segments and write a computer program which can be used to calculate the area of the parabolic reflector. Use this program to determine the area of the reflector when (*a*) $a = 10$ in., $h = 3$ in., (*b*) $a = 12$ in., $h = 8$ in., (*c*) $a = 15$ in., $h = 18$ in.

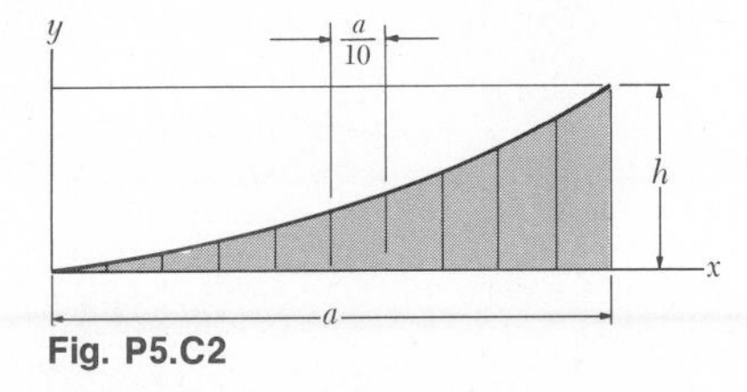

Fig. P5.C2

5.C3 Write a computer program and use it to solve Prob. 5.91 for values of d from 0 to 3 m at 0.3-m intervals.

5.C4 Sheet metal of uniform thickness is used to form the bin shown which has a base in the shape of a quarter circle and three sides. Write a computer program which can be used to calculate the coordinates of the center of gravity of the bin. Use this program to locate the center of gravity for (*a*) $a = 400$ mm, $b = c = d = 200$ mm, (*b*) $a = 400$ mm, $b = d = 600$ mm, $c = 300$ mm, (*c*) $a = 400$ mm, $b = 500$ mm, $c = 200$ mm, $d = 300$ mm.

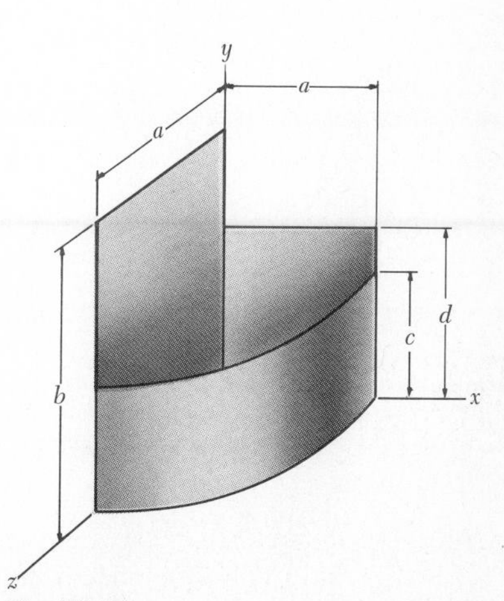

Fig. P5.C4

Analysis of Structures

CHAPTER

6

6.1. Introduction. The problems considered in the preceding chapters concerned the equilibrium of a single rigid body, and all forces involved were external to the rigid body. We shall now consider problems dealing with the equilibrium of structures made of several connected parts. These problems call for the determination not only of the external forces acting on the structure but also of the forces which hold together the various parts of the structure. From the point of view of the structure as a whole, these forces are *internal forces*.

Consider, for example, the crane shown in Fig. 6.1*a*, which carries a load W. The crane consists of three beams AD, CF, and BE connected by frictionless pins; it is supported by a pin at A and by a cable DG. The free-body diagram of the crane has been drawn in Fig. 6.1*b*. The external

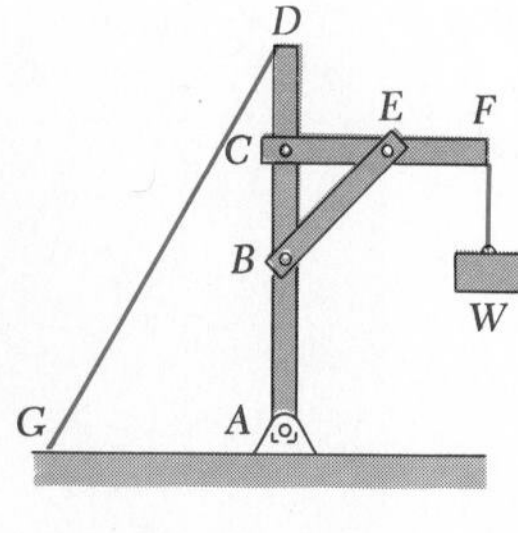

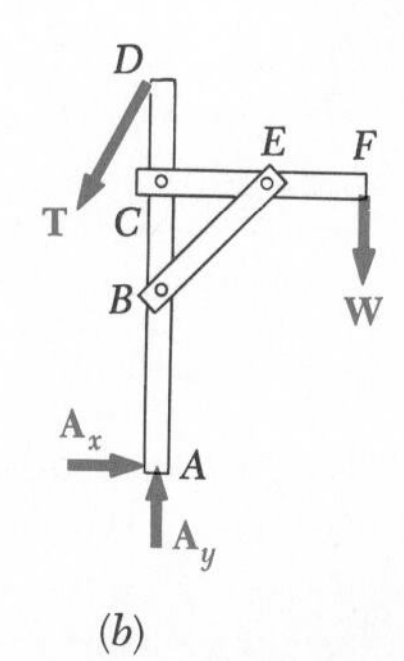

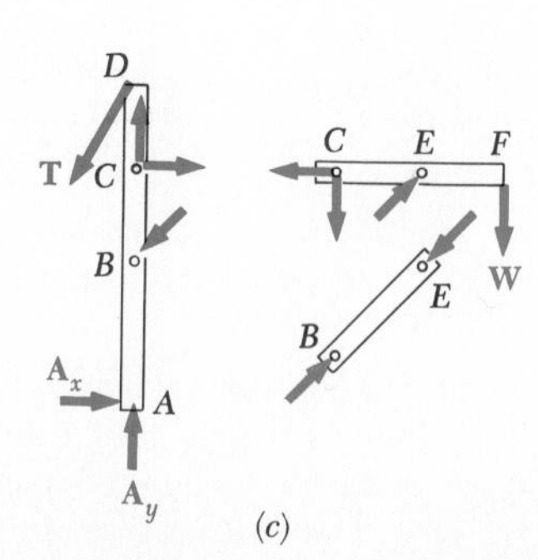

Fig. 6.1 (*a*) (*b*) (*c*)

forces are shown in the diagram and include the weight **W**, the two components $\mathbf{A}_x$ and $\mathbf{A}_y$ of the reaction at A, and the force **T** exerted by the cable at D. The internal forces holding the various parts of the crane together do not appear in the diagram. If, however, the crane is dismembered and if a free-body diagram is drawn for each of its component parts, the forces holding the three beams together must also be represented, since these forces are external forces from the point of view of each component part (Fig. 6.1*c*).

It will be noted that the force exerted at *B* by member *BE* on member *AD* has been represented as equal and opposite to the force exerted at the same point by member *AD* on member *BE*; similarly, the force exerted at *E* by *BE* on *CF* is shown equal and opposite to the force exerted by *CF* on *BE*; and the components of the force exerted at *C* by *CF* on *AD* are shown equal and opposite to the components of the force exerted by *AD* on *CF*. This is in conformity with Newton's third law, which states that *the forces of action and reaction between bodies in contact have the same magnitude, same line of action, and opposite sense.* As pointed out in Chap. 1, this law is one of the six fundamental principles of elementary mechanics and is based on experimental evidence. Its application is essential to the solution of problems involving connected bodies.

In this chapter, we shall consider three broad categories of engineering structures:

1. *Trusses*, which are designed to support loads and are usually stationary, fully constrained structures. Trusses consist exclusively of straight members connected at joints located at the ends of each member. Members of a truss, therefore, are *two-force members*, i.e., members acted upon by two equal and opposite forces directed along the member.

2. *Frames*, which are also designed to support loads and are also usually stationary, fully constrained structures. However, like the crane of Fig. 6.1, frames always contain at least one *multiforce member*, i.e., a member acted upon by three or more forces which, in general, are not directed along the member.

3. *Machines*, which are designed to transmit and modify forces and are structures containing moving parts. Machines, like frames, always contain at least one multiforce member.

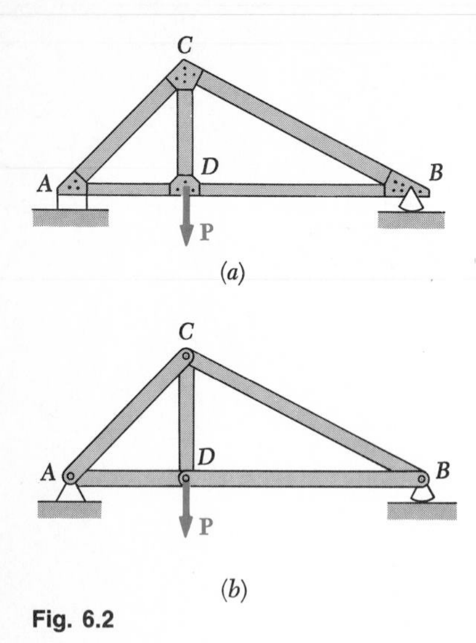

Fig. 6.2

6.2. Definition of a Truss. The truss is one of the major types of engineering structures. It provides both a practical and an economical solution to many engineering situations, especially in the design of bridges and buildings. A truss consists of straight members connected at joints; a typical truss is shown in Fig. 6.2*a*. Truss members are connected at their extremities only; thus no member is continuous through a joint. In Fig. 6.2*a*, for example, there is no member *AB*; there are instead two distinct members *AD* and *DB*. Most actual structures are made of several trusses joined together to form a space framework. Each truss is designed to carry those loads which act in its plane and thus may be treated as a two-dimensional structure.

In general, the members of a truss are slender and can support little lateral load; all loads, therefore, must be applied to the various joints, and not to the members themselves. When a concentrated load is to be applied between two joints, or when a distributed load is to be supported by the truss, as in the case of a bridge truss, a floor system must be provided which, through the use of stringers and floor beams, transmits the load to the joints (Fig. 6.3).

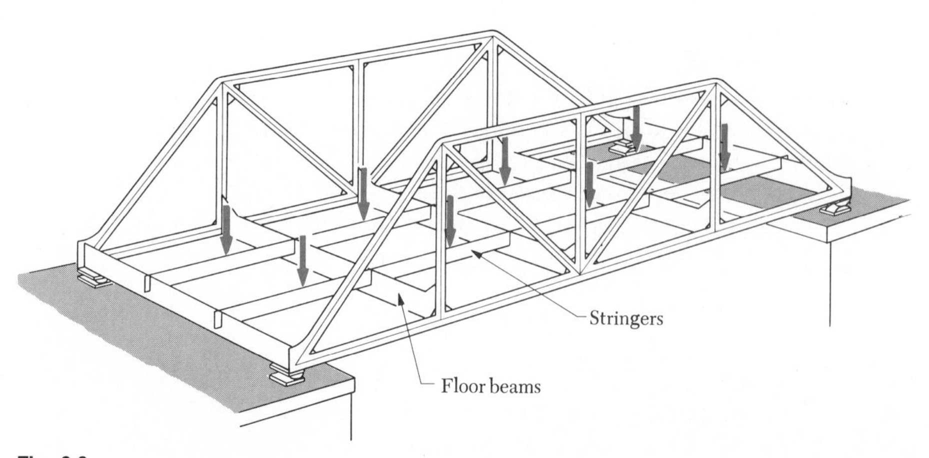

Fig. 6.3

The weights of the members of the truss are also assumed to be applied to the joints, half of the weight of each member being applied to each of the two joints the member connects. Although the members are actually joined together by means of riveted or welded connections, it is customary to assume that the members are pinned together; therefore, the forces acting at each end of a member reduce to a single force and no couple. Thus, the only forces assumed to be applied to a truss member are a single force at each end of the member. Each member may then be treated as a two-force member, and the entire truss may be considered as a group of pins and two-force members (Fig. 6.2*b*). An individual member may be acted upon as shown in either of the two sketches of Fig. 6.4. In the first sketch, the forces tend to pull the member apart, and the member is in tension, while in the second sketch, the forces tend to compress the member, and the member is in compression. Several typical trusses are shown in Fig. 6.5.

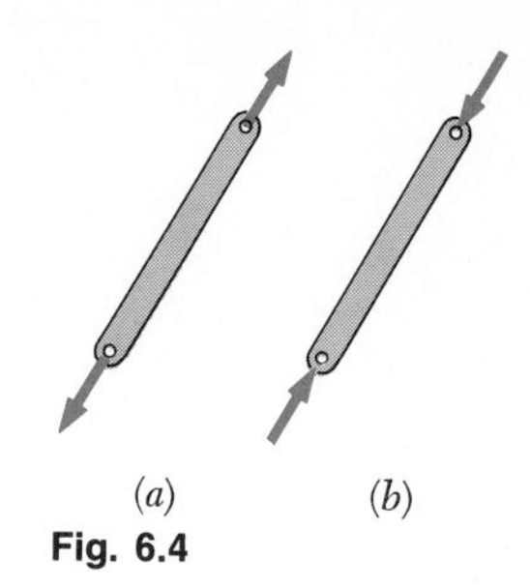

Fig. 6.4

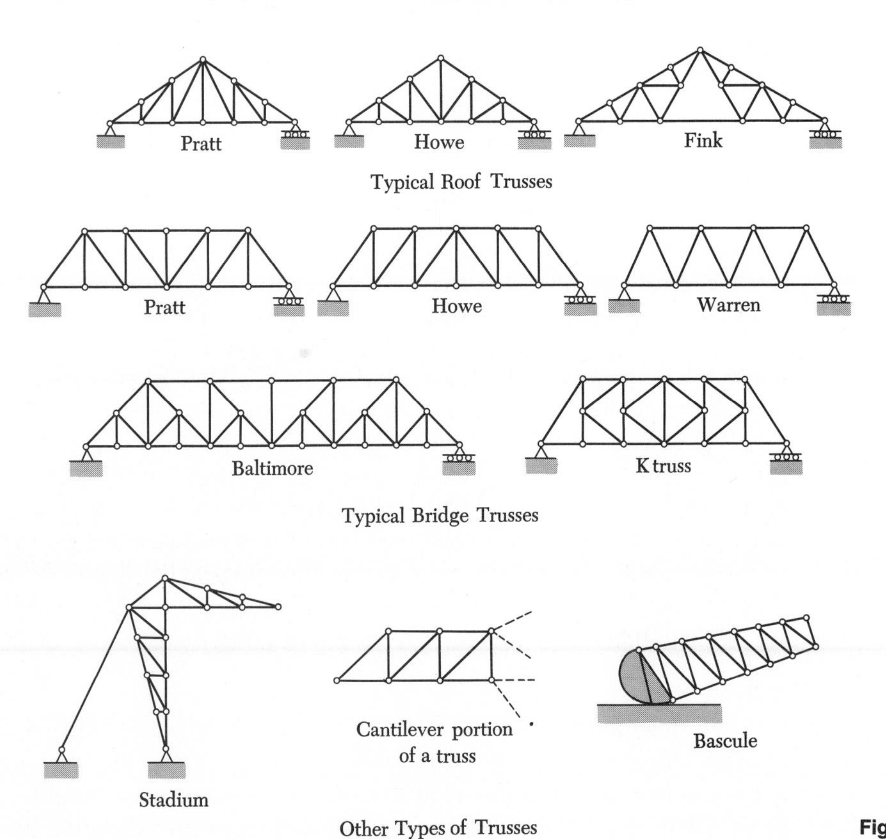

Fig. 6.5

6.3. Simple Trusses. Consider the truss of Fig. 6.6*a*, which is made of four members connected by pins at *A*, *B*, *C*, and *D*. If a load is applied at *B*, the truss will greatly deform and lose completely its original shape. On the other hand, the truss of Fig. 6.6*b*, which is made of three members connected by pins at *A*, *B*, and *C*, will deform only slightly under a load applied at *B*. The only possible deformation for this truss is one involving small changes in the length of its members. The truss of Fig. 6.6*b* is said to be a *rigid truss*, the term rigid being used here to indicate that the truss *will not collapse*.

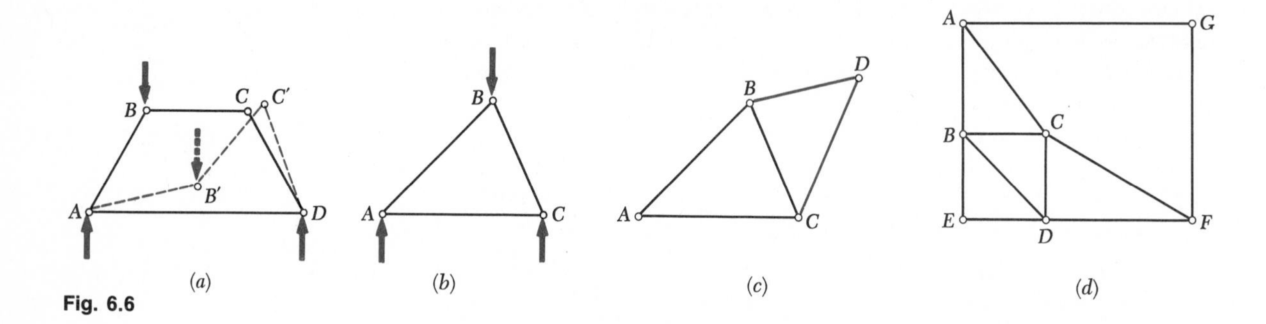

Fig. 6.6

As shown in Fig. 6.6*c*, a larger rigid truss may be obtained by adding two members *BD* and *CD* to the basic triangular truss of Fig. 6.6*b*. This procedure may be repeated as many times as desired, and the resulting truss will be rigid if each time we add two new members, we attach them to separate existing joints and connect them at a new joint.† A truss which may be constructed in this manner is called a *simple truss*.

It should be noted that a simple truss is not necessarily made only of triangles. The truss of Fig. 6.6*d*, for example, is a simple truss which was constructed from triangle *ABC* by adding successively the joints *D*, *E*, *F*, and *G*. On the other hand, rigid trusses are not always simple trusses, even when they appear to be made of triangles. The Fink and Baltimore trusses shown in Fig. 6.5, for instance, are not simple trusses, since they cannot be constructed from a single triangle in the manner described above. All the other trusses shown in Fig. 6.5 are simple trusses, as may be easily checked. (For the K truss, start with one of the central triangles.)

Returning to the basic triangular truss of Fig. 6.6*b*, we note that this truss has three members and three joints. The truss of Fig. 6.6*c* has two more members and one more joint, i.e., altogether five members and four joints. Observing that every time two new members are added, the number of joints is increased by one, we find that in a simple truss the total number of members is $m = 2n - 3$, where n is the total number of joints.

† The three joints must not be in a straight line.

6.4. Analysis of Trusses by the Method of Joints. We saw in Sec. 6.2 that a truss may be considered as a group of pins and two-force members. The truss of Fig. 6.2, whose free-body diagram is shown in Fig. 6.7*a*, may thus be dismembered, and a free-body diagram can be drawn for each pin and each member (Fig. 6.7*b*). Each member is acted upon by two forces, one at each end; these forces have the same magnitude, same line of action, and opposite sense (Sec. 4.6). Besides, Newton's third law indicates that the forces of action and reaction between a member and a pin are equal and opposite. Therefore, the forces exerted by a member on the two pins it connects must be directed along that member and be equal and opposite. The common magnitude of the forces exerted by a member on the two pins it connects is commonly referred to as the *force in the member* considered, even though this quantity is actually a scalar. Since the lines of action of all the internal forces in a truss are known, the analysis of a truss reduces to computing the forces in its various members and to determining whether each of its members is in tension or in compression.

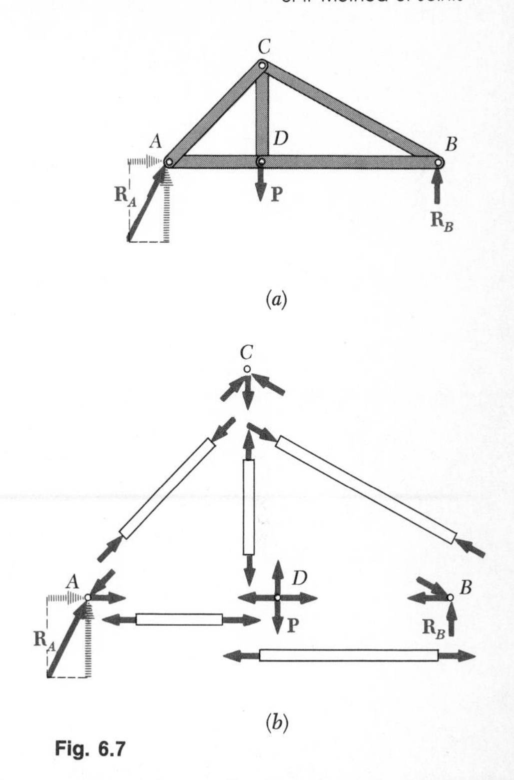

Fig. 6.7

Since the entire truss is in equilibrium, each pin must be in equilibrium. The fact that a pin is in equilibrium may be expressed by drawing its free-body diagram and writing two equilibrium equations (Sec. 2.9). If the truss contains n pins, there will be therefore $2n$ equations available, which may be solved for $2n$ unknowns. In the case of a simple truss, we have $m = 2n - 3$, that is, $2n = m + 3$, and the number of unknowns which may be determined from the free-body diagrams of the pins is thus $m + 3$. This means that the forces in all the members, as well as the two components of the reaction $\mathbf{R}_A$, and the reaction $\mathbf{R}_B$ may be found by considering the free-body diagrams of the pins.

The fact that the entire truss is a rigid body in equilibrium may be used to write three more equations involving the forces shown in the free-body diagram of Fig. 6.7*a*. Since they do not contain any new information, these equations are not independent from the equations associated with the free-body diagrams of the pins. Nevertheless, they may be used to determine immediately the components of the reactions at the supports. The arrangement of pins and members in a simple truss is such that it will then always be possible to find a joint involving only two unknown forces. These forces may be determined by the methods of Sec. 2.11 and their values transferred to the adjacent joints and treated as known quantities at these joints. This procedure may be repeated until all unknown forces have been determined.

As an example, we shall analyze the truss of Fig. 6.7 by considering successively the equilibrium of each pin, starting with a joint at which only two forces are unknown. In the truss considered, all pins are subjected to at least three unknown forces. Therefore, the reactions at the supports must first be determined by considering the entire truss as a free body and using the equations of equilibrium of a rigid body. We find in this way that $\mathbf{R}_A$ is vertical and determine the magnitudes of $\mathbf{R}_A$ and $\mathbf{R}_B$.

The number of unknown forces at joint A is thus reduced to two, and these forces may be determined by considering the equilibrium of pin A. The reaction $\mathbf{R}_A$ and the forces $\mathbf{F}_{AC}$ and $\mathbf{F}_{AD}$ exerted on pin A by members

AC and AD, respectively, must form a force triangle. First we draw $\mathbf{R}_A$ (Fig. 6.8); noting that $\mathbf{F}_{AC}$ and $\mathbf{F}_{AD}$ are directed along AC and AD, respectively, we complete the triangle and determine the magnitude and sense of $\mathbf{F}_{AC}$ and $\mathbf{F}_{AD}$. The magnitudes F_{AC} and F_{AD} represent the forces in members AC and AD. Since $\mathbf{F}_{AC}$ is directed down and to the left, that is, toward joint A, member AC pushes on pin A and is in compression. On the other hand, since $\mathbf{F}_{AD}$ is directed away from the joint, member AD pulls on pin A and is in tension.

We may now proceed to joint D, where only two forces, $\mathbf{F}_{DC}$ and $\mathbf{F}_{DB}$, are still unknown. The other forces are the load $\mathbf{P}$, which is given, and the force $\mathbf{F}_{DA}$ exerted on the pin by member AD. As indicated above, this force is equal and opposite to the force $\mathbf{F}_{AD}$ exerted by the same member on pin A. We may draw the force polygon corresponding to joint D, as shown in Fig. 6.8, and determine the forces $\mathbf{F}_{DC}$ and $\mathbf{F}_{DB}$ from that polygon. How-

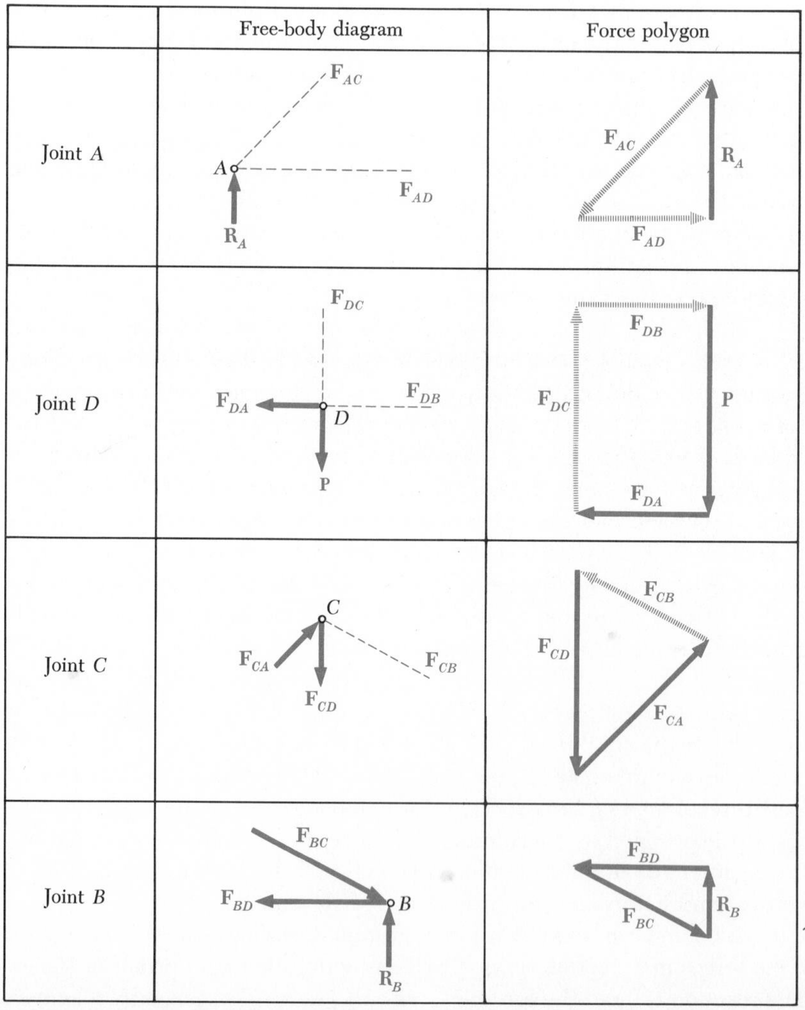

	Free-body diagram	Force polygon
Joint A	$\mathbf{F}_{AC}$, A, $\mathbf{F}_{AD}$, $\mathbf{R}_A$	$\mathbf{F}_{AC}$, $\mathbf{R}_A$, $\mathbf{F}_{AD}$
Joint D	$\mathbf{F}_{DC}$, $\mathbf{F}_{DA}$, D, $\mathbf{F}_{DB}$, $\mathbf{P}$	$\mathbf{F}_{DB}$, $\mathbf{F}_{DC}$, $\mathbf{P}$, $\mathbf{F}_{DA}$
Joint C	C, $\mathbf{F}_{CA}$, $\mathbf{F}_{CB}$, $\mathbf{F}_{CD}$	$\mathbf{F}_{CB}$, $\mathbf{F}_{CD}$, $\mathbf{F}_{CA}$
Joint B	$\mathbf{F}_{BC}$, $\mathbf{F}_{BD}$, B, $\mathbf{R}_B$	$\mathbf{F}_{BD}$, $\mathbf{F}_{BC}$, $\mathbf{R}_B$

Fig. 6.8

ever, when more than three forces are involved, it is usually more convenient to write the equations of equilibrium $\Sigma F_x = 0$, $\Sigma F_y = 0$ and solve these equations for the two unknown forces. Since both of these forces are found to be directed away from joint D, members DC and DB pull on the pin and are in tension.

Next, joint C is considered; its free-body diagram is shown in Fig. 6.8. It is noted that both $\mathbf{F}_{CD}$ and $\mathbf{F}_{CA}$ are known from the analysis of the preceding joints and that only $\mathbf{F}_{CB}$ is unknown. Since the equilibrium of each pin provides sufficient information to determine two unknowns, a check of our analysis is obtained at this joint. The force triangle is drawn, and the magnitude and sense of $\mathbf{F}_{CB}$ are determined. Since $\mathbf{F}_{CB}$ is directed toward joint C, member CB pushes on pin C and is in compression. The check is obtained by verifying that the force $\mathbf{F}_{CB}$ and member CB are parallel.

At joint B, all the forces are known. Since the corresponding pin is in equilibrium, the force triangle must close and an additional check of the analysis is obtained.

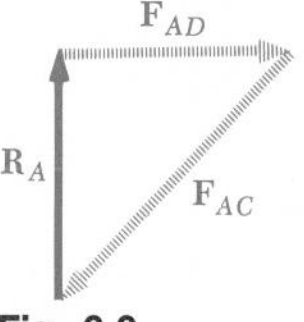

Fig. 6.9

It should be noted that the force polygons shown in Fig. 6.8 are not unique. Each of them could be replaced by an alternate configuration. For example, the force triangle corresponding to joint A could be drawn as shown in Fig. 6.9. The triangle actually shown in Fig. 6.8 was obtained by drawing the three forces $\mathbf{R}_A$, $\mathbf{F}_{AC}$, and $\mathbf{F}_{AD}$ in tip-to-tail fashion in the order in which their lines of action are encountered when moving clockwise around joint A. The other force polygons in Fig. 6.8, having been drawn in the same way, can be made to fit into a single diagram, as shown in Fig. 6.10. Such a diagram, known as *Maxwell's diagram,* greatly facilitates the *graphical analysis* of truss problems.

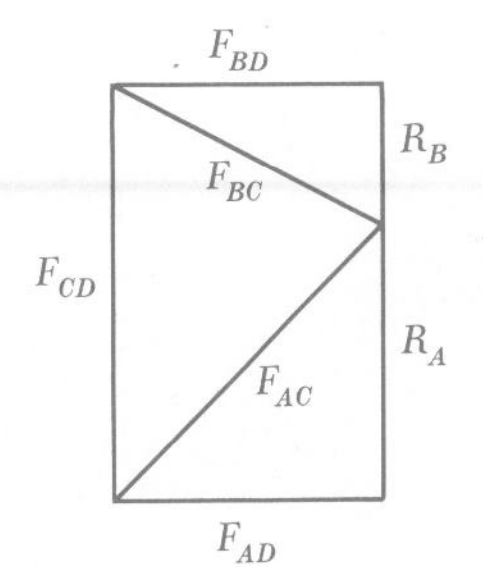

Fig. 6.10

*6.5. Joints under Special Loading Conditions.

Consider the joint shown in Fig. 6.11a, which connects four members lying in two intersecting straight lines. The free-body diagram of Fig. 6.11b shows that pin A is subjected to two pairs of directly opposite forces. The corresponding force polygon, therefore, must be a parallelogram (Fig. 6.11c), and *the forces in opposite members must be equal.*

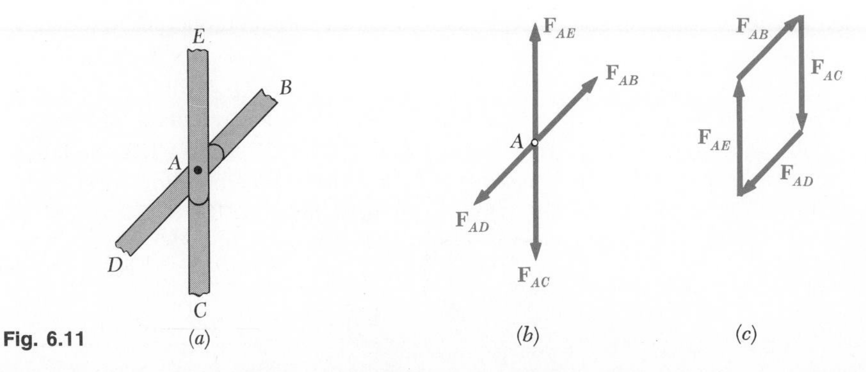

Fig. 6.11 (a) (b) (c)

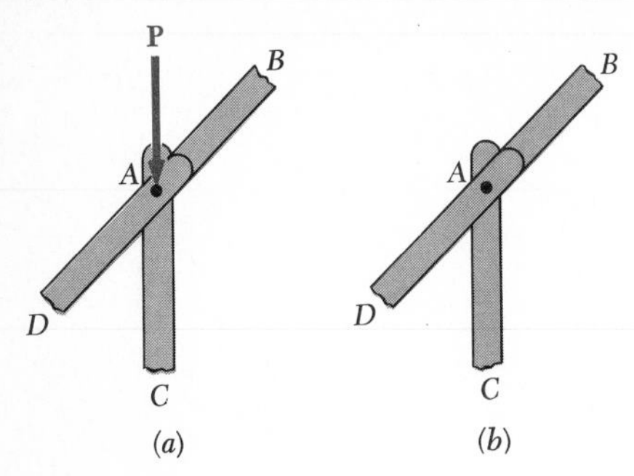

Fig. 6.12

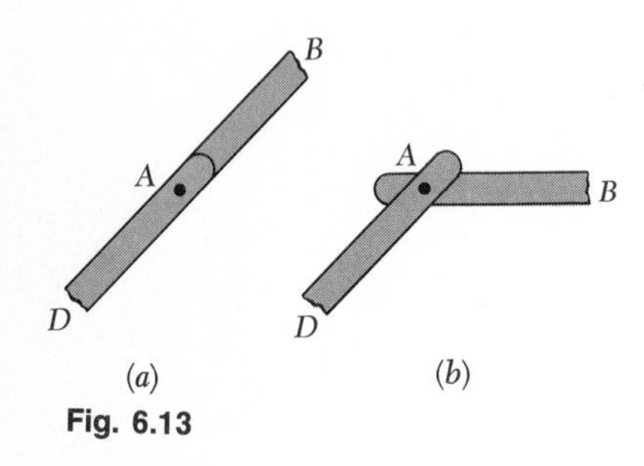

Fig. 6.13

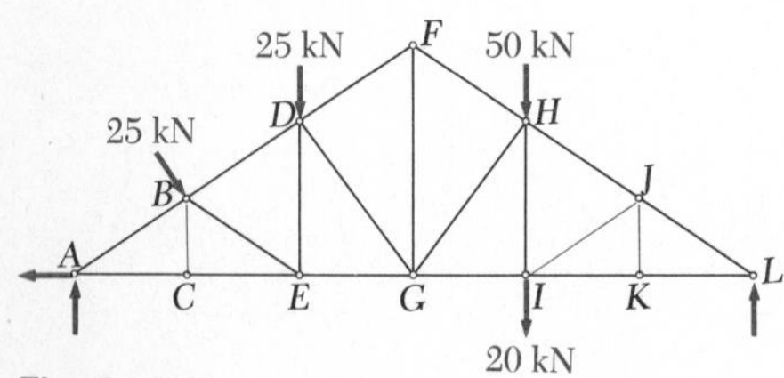

Fig. 6.14

Consider next the joint shown in Fig. 6.12*a*, which connects three members and supports a load **P**. Two of the members lie in the same line, and the load **P** acts along the third member. The free-body diagram of pin *A* and the corresponding force polygon will be as shown in Fig. 6.11*b* and *c* with $\mathbf{F}_{AE}$ replaced by the load **P**. Thus, *the forces in the two opposite members must be equal, and the force in the other member must equal P.* A particular case of special interest is shown in Fig. 6.12*b*. Since, in this case, no external load is applied to the joint, we have $P = 0$, and the force in member *AC* is zero. Member *AC* is said to be a *zero-force member.*

Consider now a joint connecting two members only. From Sec. 2.9, we know that a particle which is acted upon by two forces will be in equilibrium if the two forces have the same magnitude, same line of action, and opposite sense. In the case of the joint of Fig. 6.13*a*, which connects two members *AB* and *AD* lying in the same line, the equilibrium of pin *A* requires therefore that *the forces in the two members be equal.* In the case of the joint of Fig. 6.13*b*, the equilibrium of pin *A* is impossible unless the forces in both members are zero. Members connected as shown in Fig. 6.13*b*, therefore, must be *zero-force members.*

Spotting the joints which are under the special loading conditions listed above will expedite the analysis of a truss. Consider, for example, a Howe truss loaded as shown in Fig. 6.14. All the members represented by colored lines will be recognized as zero-force members. Joint *C* connects three members, two of which lie in the same line, and is not subjected to any external load; member *BC* is thus a zero-force member. Applying the same reasoning to joint *K*, we find that member *JK* is also a zero-force member. But joint *J* is now in the same situation as joints *C* and *K*, and member *IJ* must be a zero-force member. The examination of joints *C*, *J*, and *K* also shows that the forces in members *AC* and *CE* are equal, that the forces in members *HJ* and *JL* are equal, and that the forces in members *IK* and *KL* are equal. Furthermore, now turning our attention to joint *I*, where the 20-kN load and member *HI* are collinear, we note that the force in member *HI* is 20 kN (tension) and that the forces in members *GI* and *IK* are equal. Hence, the forces in members *GI*, *IK*, and *KL* are equal.

Students, however, should be warned against misusing the rules established in this section. For example, it would be wrong to assume that the force in member *DE* is 25 kN or that the forces in members *AB* and *BD* are equal. The conditions discussed above do not apply to joints *B* and *D*. The forces in these members and in all remaining members should be found by carrying out the analysis of joints *A*, *B*, *D*, *E*, *F*, *G*, *H*, and *L* in the usual manner. Until they have become thoroughly familiar with the conditions of application of the rules established in this section, students would be well advised to draw the free-body diagrams of all pins and to write the corresponding equilibrium equations (or draw the corresponding force polygons), whether or not the joints considered fall into the categories listed above.

A final remark concerning zero-force members: These members are not useless. While they do not carry any load under the particular loading conditions shown, the zero-force members of Fig. 6.14 will probably carry loads if the loading conditions are changed. Besides, even in the case considered, these members are needed to support the weight of the truss and to maintain the truss in the desired shape.

Skipped

***6.6. Space Trusses.** When several straight members are joined together at their extremities to form a three-dimensional configuration, the structure obtained is called a *space truss*.

We recall from Sec. 6.3 that the most elementary two-dimensional rigid truss consisted of three members joined at their extremities to form the sides of a triangle; by adding two members at a time to this basic configuration, and connecting them at a new joint, it was possible to obtain a larger rigid structure which was defined as a simple truss. Similarly, the most elementary rigid space truss consists of six members joined at their extremities to form the edges of a tetrahedron $ABCD$ (Fig. 6.15*a*). By adding three members at a time to this basic configuration, such as AE, BE, and CE, attaching them at separate existing joints, and connecting them at a new joint, we can obtain a larger rigid structure which is defined as a *simple space truss* (Fig. 6.15*b*).† Observing that the basic tetrahedron has six members and four joints and that every time three members are added, the number of joints is increased by one, we conclude that in a simple space truss the total number of members is $m = 3n - 6$, where n is the total number of joints.

If a space truss is to be completely constrained and if the reactions at its supports are to be statically determinate, the supports should consist of a combination of balls, rollers, and balls and sockets which provides six unknown reactions (see Sec. 4.8). These unknown reactions may be readily determined by solving the six equations expressing that the three-dimensional truss is in equilibrium.

Although the members of a space truss are actually joined together by means of riveted or welded connections, it is assumed that each joint consists of a ball-and-socket connection. Thus, no couple will be applied to the members of the truss, and each member may be treated as a two-force member. The conditions of equilibrium for each joint will be expressed by the three equations $\Sigma F_x = 0$, $\Sigma F_y = 0$, and $\Sigma F_z = 0$. In the case of a simple space truss containing n joints, writing the conditions of equilibrium for each joint will thus yield $3n$ equations. Since $m = 3n - 6$, these equations suffice to determine all unknown forces (forces in m members and six reactions at the supports). However, to avoid solving many simultaneous equations, care should be taken to select joints in such an order that no selected joint will involve more than three unknown forces.

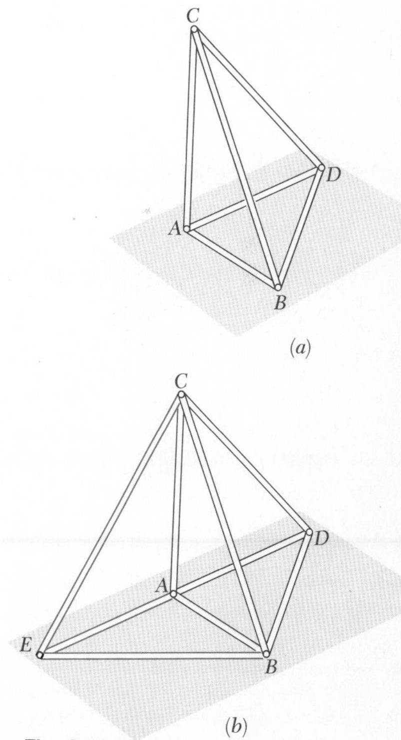

Fig. 6.15

† The four joints must not lie in a plane.

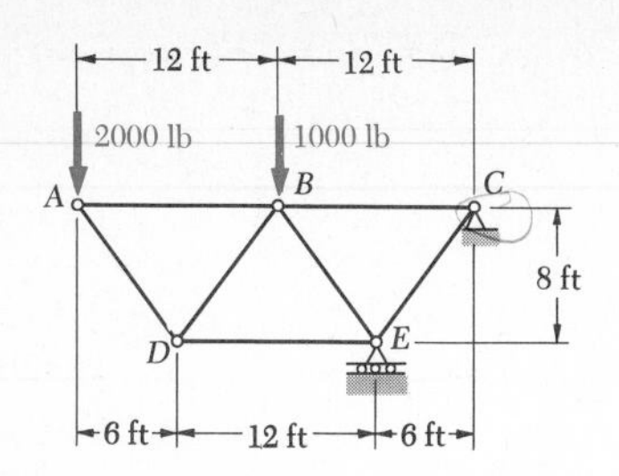

SAMPLE PROBLEM 6.1

Using the method of joints, determine the force in each member of the truss shown.

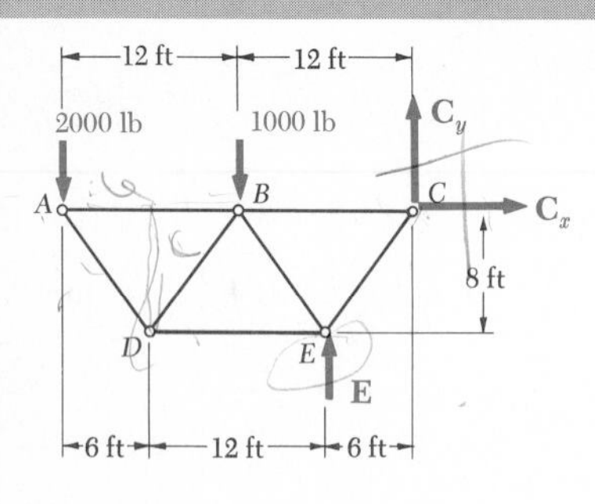

Solution. A free-body diagram of the entire truss is drawn; external forces acting on this free body consist of the applied loads and the reactions at C and E.

Equilibrium of Entire Truss.

$+\circlearrowleft \Sigma M_C = 0$: $\quad (2000 \text{ lb})(24 \text{ ft}) + (1000 \text{ lb})(12 \text{ ft}) - E(6 \text{ ft}) = 0$

$E = +10{,}000 \text{ lb}$ $\qquad\qquad$ $\mathbf{E} = 10{,}000 \text{ lb}\uparrow$

$\xrightarrow{+} \Sigma F_x = 0$: $\qquad\qquad$ $\mathbf{C}_x = 0$

$+\uparrow \Sigma F_y = 0$: $\quad -2000 \text{ lb} - 1000 \text{ lb} + 10{,}000 \text{ lb} + C_y = 0$

$C_y = -7000 \text{ lb}$ $\qquad\qquad$ $\mathbf{C}_y = 7000 \text{ lb}\downarrow$

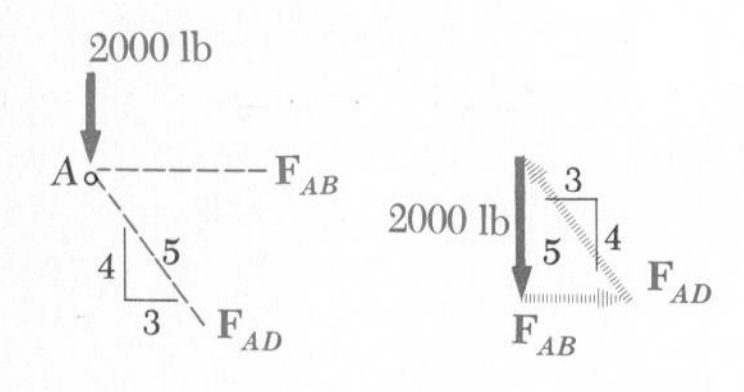

Joint A. This joint is subjected to only two unknown forces, namely, the forces exerted by members AB and AD. A force triangle is used to determine $\mathbf{F}_{AB}$ and $\mathbf{F}_{AD}$. We note that member AB pulls on the joint and thus is in tension and that member AD pushes on the joint and thus is in compression. The magnitudes of the two forces are obtained from the proportion

$$\frac{2000 \text{ lb}}{4} = \frac{F_{AB}}{3} = \frac{F_{AD}}{5}$$

$F_{AB} = 1500 \text{ lb } T$ ◄

$F_{AD} = 2500 \text{ lb } C$ ◄

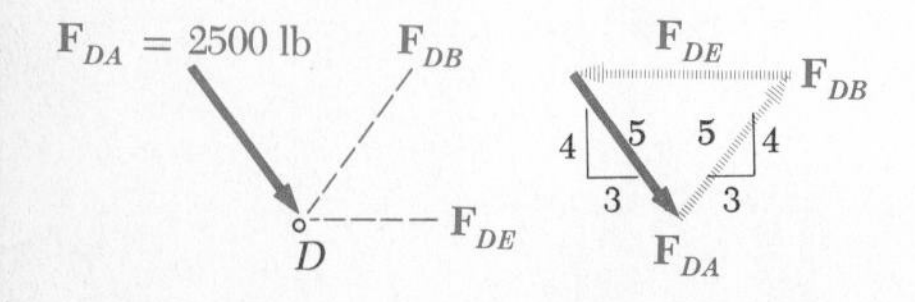

Joint D. Since the force exerted by member AD has been determined, only two unknown forces are now involved at this joint. Again, a force triangle is used to determine the unknown forces in members DB and DE.

$F_{DB} = F_{DA}$ $\qquad\qquad$ $F_{DB} = 2500 \text{ lb } T$ ◄

$F_{DE} = 2(\frac{3}{5})F_{DA}$ $\qquad\qquad$ $F_{DE} = 3000 \text{ lb } C$ ◄

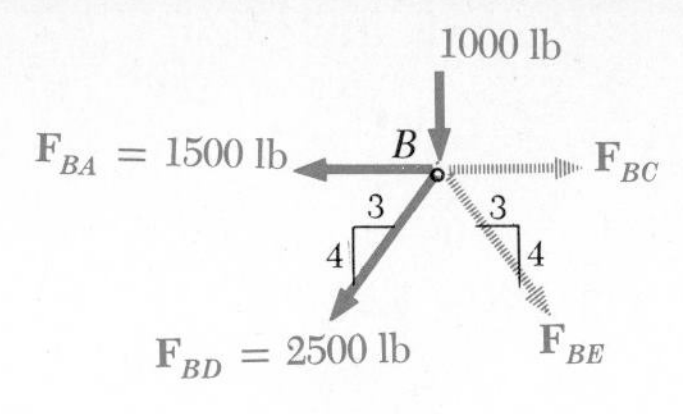

Joint B. Since more than three forces act at this joint, we determine the two unknown forces $\mathbf{F}_{BC}$ and $\mathbf{F}_{BE}$ by solving the equilibrium equations $\Sigma F_x = 0$ and $\Sigma F_y = 0$. We arbitrarily assume that both unknown forces act away from the joint, i.e., that the members are in tension. The positive value obtained for F_{BC} indicates that our assumption was correct; member BC is in tension. The negative value of F_{BE} indicates that our assumption was wrong; member BE is in compression.

$$+\uparrow\Sigma F_y = 0: \qquad -1000 - \tfrac{4}{5}(2500) - \tfrac{4}{5}F_{BE} = 0$$

$$F_{BE} = -3750 \text{ lb} \qquad F_{BE} = 3750 \text{ lb } C \blacktriangleleft$$

$$\overset{+}{\rightarrow}\Sigma F_x = 0: \qquad F_{BC} - 1500 - \tfrac{3}{5}(2500) - \tfrac{3}{5}(3750) = 0$$

$$F_{BC} = +5250 \text{ lb} \qquad F_{BC} = 5250 \text{ lb } T \blacktriangleleft$$

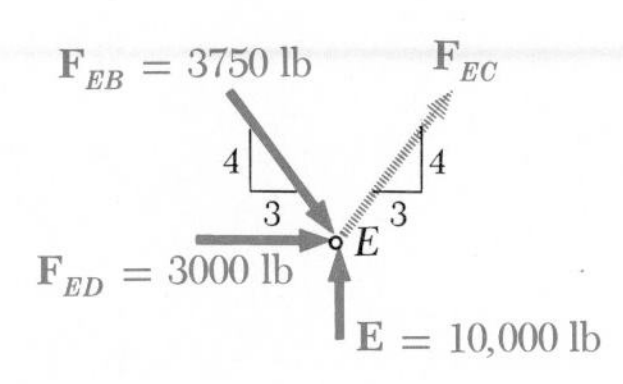

Joint E. The unknown force $\mathbf{F}_{EC}$ is assumed to act away from the joint. Summing x components, we write

$$\overset{+}{\rightarrow}\Sigma F_x = 0: \qquad \tfrac{3}{5}F_{EC} + 3000 + \tfrac{3}{5}(3750) = 0$$

$$F_{EC} = -8750 \text{ lb} \qquad F_{EC} = 8750 \text{ lb } C \blacktriangleleft$$

Summing y components, we obtain a check of our computations:

$$+\uparrow\Sigma F_y = 10{,}000 - \tfrac{4}{5}(3750) - \tfrac{4}{5}(8750)$$
$$= 10{,}000 - 3000 - 7000 = 0 \qquad \text{(checks)}$$

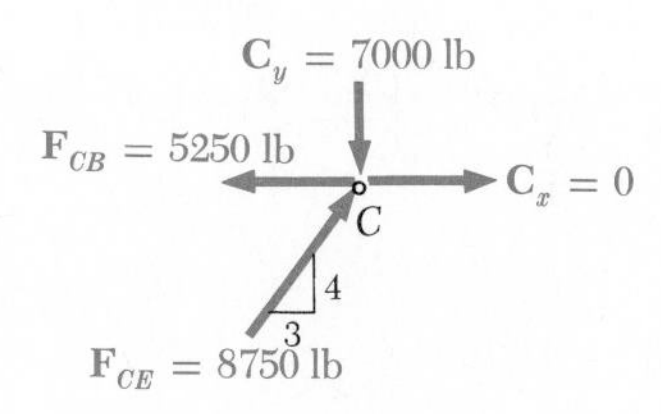

Joint C. Using the computed values of $\mathbf{F}_{CB}$ and $\mathbf{F}_{CE}$, we may determine the reactions $\mathbf{C}_x$ and $\mathbf{C}_y$ by considering the equilibrium of this joint. Since these reactions have already been determined from the equilibrium of the entire truss, we will obtain two checks of our computations. We may also merely use the computed values of all forces acting on the joint (forces in members and reactions) and check that the joint is in equilibrium:

$$\overset{+}{\rightarrow}\Sigma F_x = -5250 + \tfrac{3}{5}(8750) = -5250 + 5250 = 0 \qquad \text{(checks)}$$
$$+\uparrow\Sigma F_y = -7000 + \tfrac{4}{5}(8750) = -7000 + 7000 = 0 \qquad \text{(checks)}$$

Problems

6.1 through 6.12 Using the method of joints, determine the force in each member of the truss shown. State whether each member is in tension or compression.

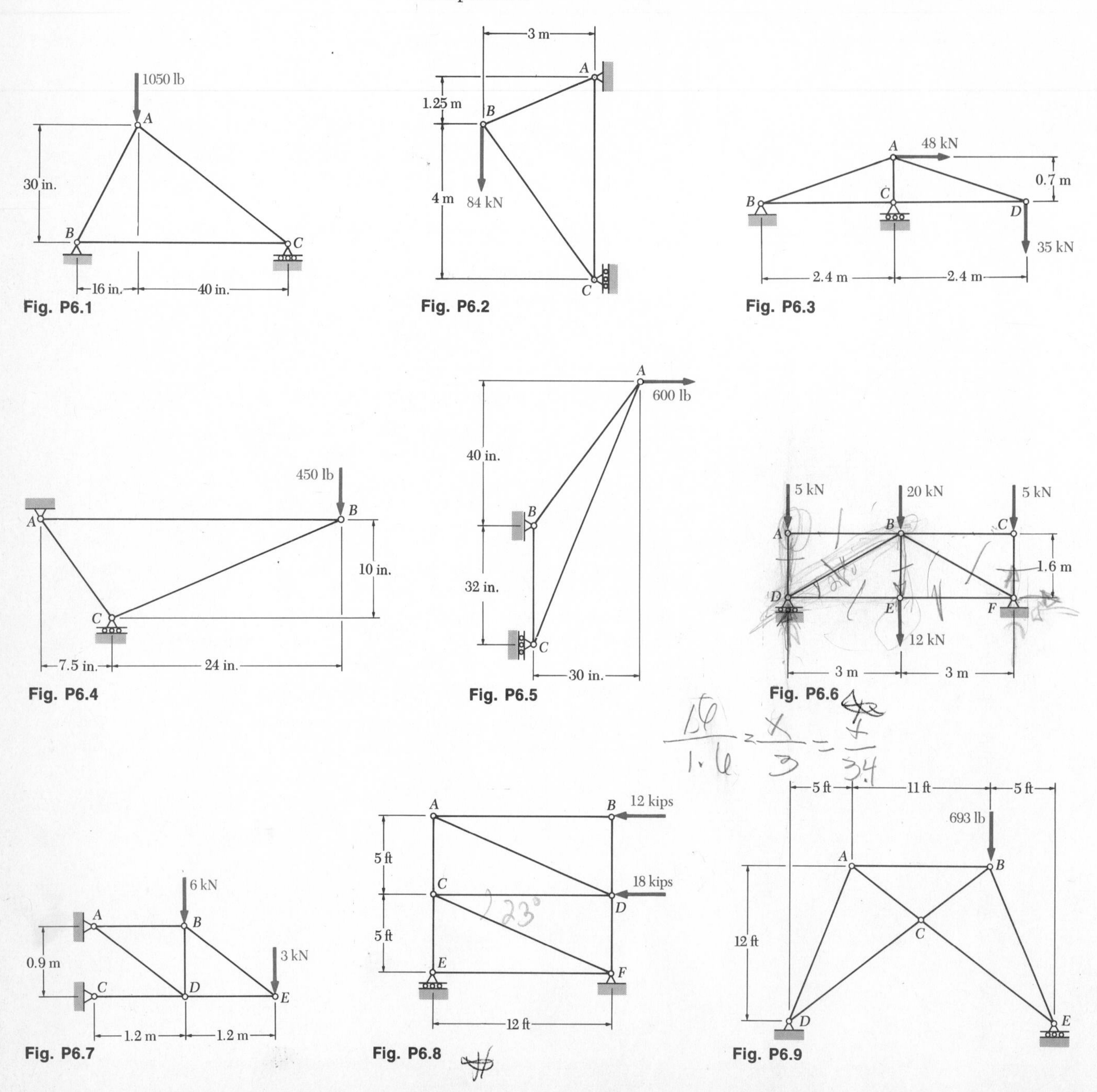

Fig. P6.1

Fig. P6.2

Fig. P6.3

Fig. P6.4

Fig. P6.5

Fig. P6.6

Fig. P6.7

Fig. P6.8

Fig. P6.9

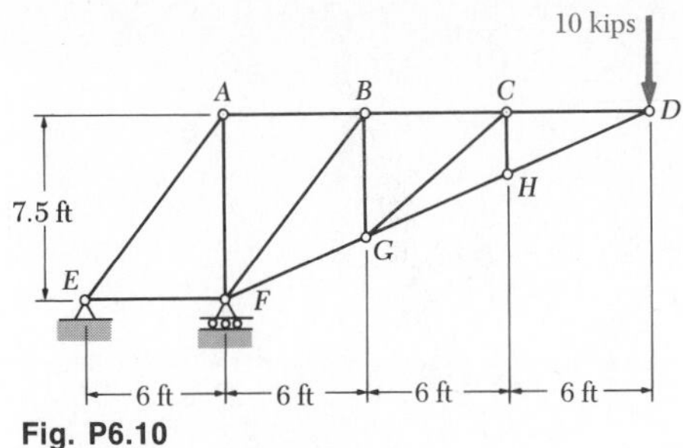

Fig. P6.10

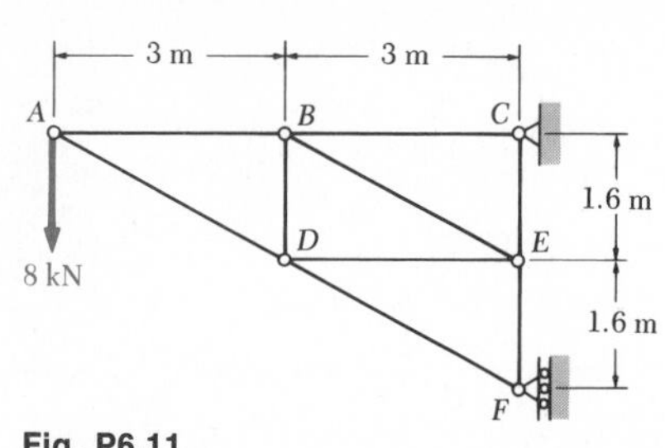

Fig. P6.11

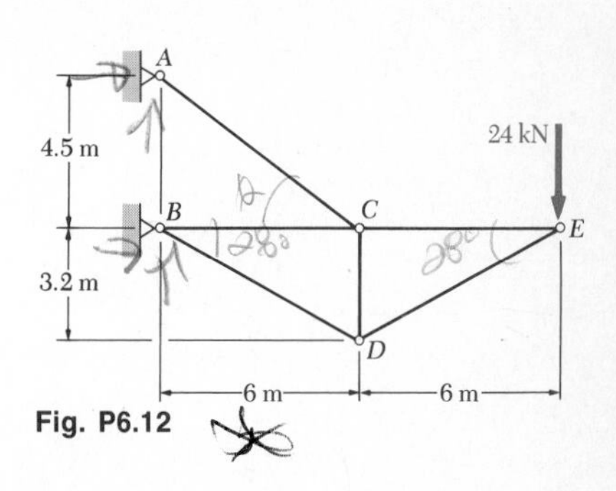

Fig. P6.12

6.13 Solve Prob. 6.10, assuming that the vertical 10-kip load is applied at joint C.

6.14 Determine whether the trusses given in Probs. 6.11, 6.15, 6.16, and 6.17 are simple trusses.

6.15 through 6.17 Determine the zero-force members in the truss shown for the given loading.

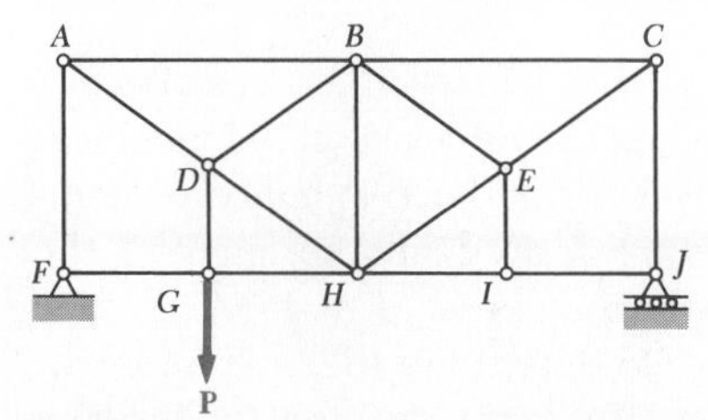

Fig. P6.15

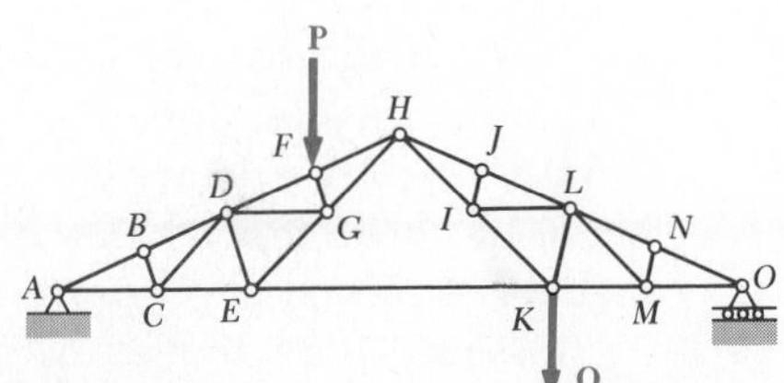

Fig. P6.16

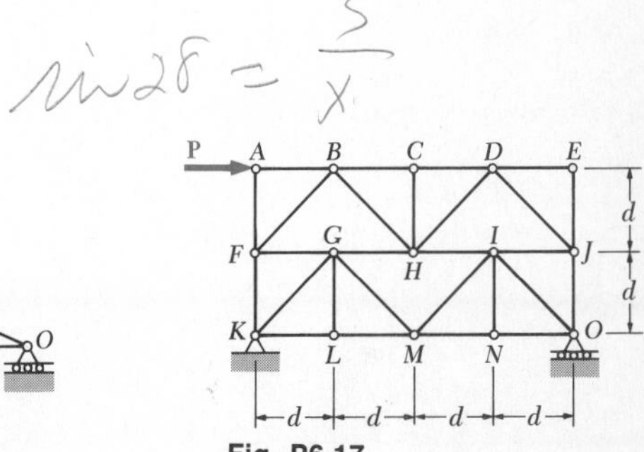

Fig. P6.17

***6.18** Six bars, each of length L, are connected to form a regular tetrahedron which rests on a frictionless horizontal surface. Determine the force in each of the six members when a vertical force **P** is applied at A.

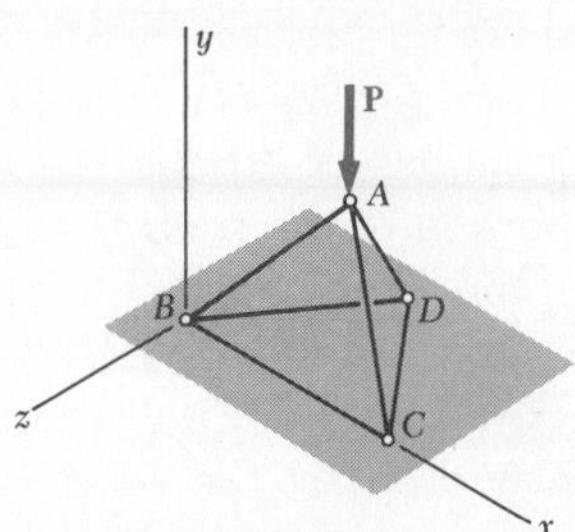

Fig. P6.18

***6.19** Twelve members, each of length L, are connected to form a regular octahedron. Determine the force in each member if two vertical loads are applied as shown.

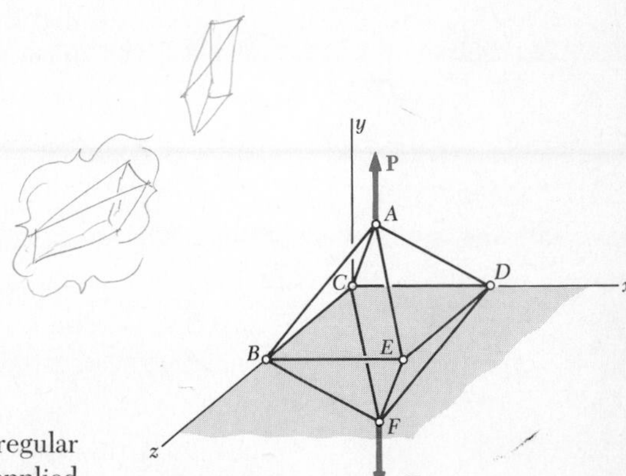

Fig. P6.19

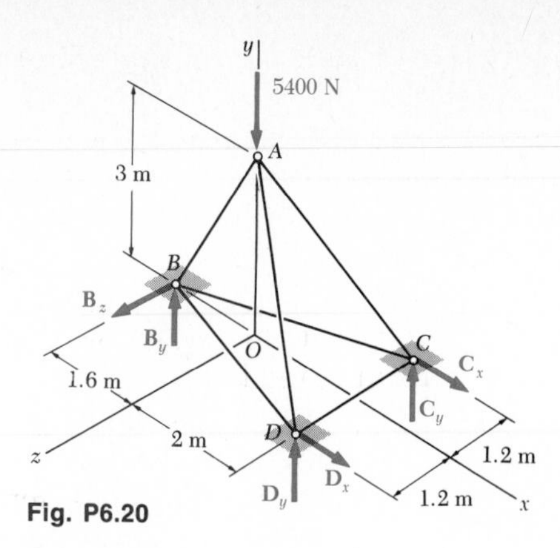

Fig. P6.20

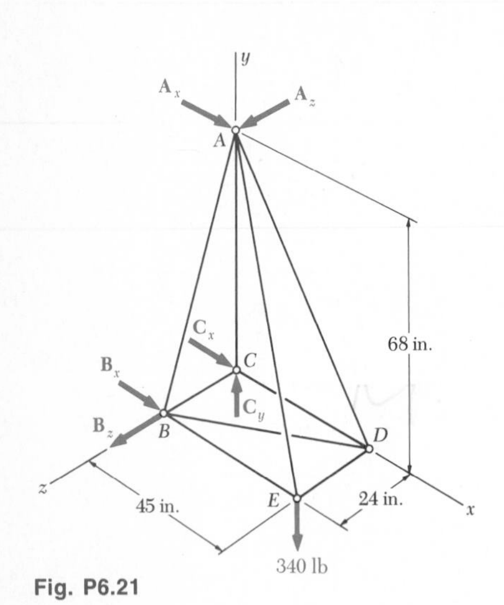

Fig. P6.21

***6.20** The three-dimensional truss is supported by the six reactions shown. If a vertical 5400-N load is applied at A, determine (*a*) the reactions, (*b*) the force in each member.

***6.21** The three-dimensional truss is supported by the six reactions shown. If a vertical 340-lb load is applied at E, determine (*a*) the reactions, (*b*) the force in each member.

6.7. Analysis of Trusses by the Method of Sections. The method of joints is most effective when the forces in all the members of a truss are to be determined. If, however, the force in only one member or the forces in a very few members are desired, another method, the method of sections, will prove more efficient.

Assume, for example, that we want to determine the force in member BD of the truss shown in Fig. 6.16*a*. To do this, we must determine the force with which member BD acts on either joint B or joint D. If we were to use the method of joints, we would choose either joint B or joint D as a free body. However, we may also choose as a free body a larger portion of the truss, composed of several joints and members, provided that the desired force is one of the external forces acting on that portion. If, in addition, the portion of the truss is chosen so that there is a total of only three unknown forces acting upon it, the desired force may be obtained by solving the equations of equilibrium for this portion of the truss. In practice, the portion of the truss to be utilized is obtained by *passing a section* through three members of the truss, one of which is the desired member, i.e., by drawing a line which divides the truss into two completely separate parts but does not intersect more than three members. Either of the two portions of the truss obtained after the intersected members have been removed may then be used as a free body.†

† In the analysis of certain trusses, sections are passed which intersect more than three members; the forces in one, or possibly two, of the intersected members may be obtained if equilibrium equations can be found, each of which involves only one unknown (see Probs. 6.38 through 6.40).

In Fig. 6.16*a*, the section *nn* has been passed through members *BD*, *BE*, and *CE*, and the portion *ABC* of the truss is chosen as the free body (Fig. 6.16*b*). The forces acting on the free body are the loads $\mathbf{P}_1$ and $\mathbf{P}_2$ at points *A* and *B* and the three unknown forces $\mathbf{F}_{BD}$, $\mathbf{F}_{BE}$, and $\mathbf{F}_{CE}$. Since it is not known whether the members removed were in tension or compression, the three forces have been arbitrarily drawn away from the free body as if the members were in tension.

The fact that the rigid body *ABC* is in equilibrium can be expressed by writing three equations which may be solved for the three unknown forces. If only the force $\mathbf{F}_{BD}$ is desired, we need write only one equation, provided that the equation does not contain the other unknowns. Thus the equation $\Sigma M_E = 0$ yields the value of the magnitude F_{BD} of the force $\mathbf{F}_{BD}$. A positive sign in the answer will indicate that our original assumption regarding the sense of $\mathbf{F}_{BD}$ was correct and that member *BD* is in tension; a negative sign will indicate that our assumption was incorrect and that *BD* is in compression.

On the other hand, if only the force $\mathbf{F}_{CE}$ is desired, an equation which does not involve $\mathbf{F}_{BD}$ or $\mathbf{F}_{BE}$ should be written; the appropriate equation is $\Sigma M_B = 0$. Again a positive sign for the magnitude F_{CE} of the desired force indicates a correct assumption, hence tension; and a negative sign indicates an incorrect assumption, hence compression.

If only the force $\mathbf{F}_{BE}$ is desired, the appropriate equation is $\Sigma F_y = 0$. Whether the member is in tension or compression is again determined from the sign of the answer.

When the force in only one member is determined, no independent check of the computation is available. However, when all the unknown forces acting on the free body are determined, the computations can be checked by writing an additional equation. For instance, if $\mathbf{F}_{BD}$, $\mathbf{F}_{BE}$, and $\mathbf{F}_{CE}$ are determined as indicated above, the computation can be checked by verifying that $\Sigma F_x = 0$.

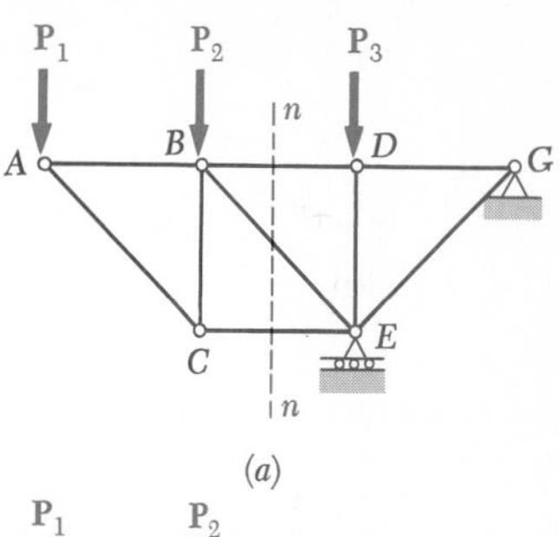

(*a*)

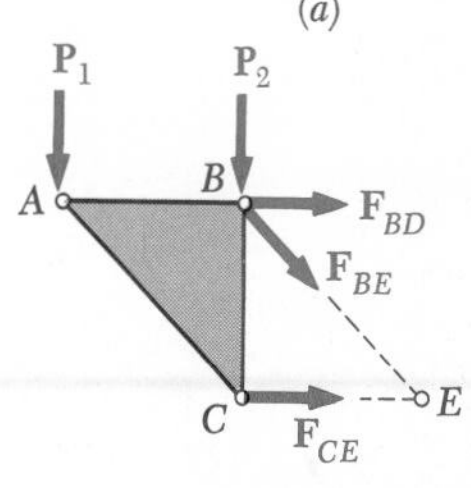

(*b*)

Fig. 6.16

*6.8. Trusses Made of Several Simple Trusses.

Consider two simple trusses *ABC* and *DEF*. If they are connected by three bars *BD*, *BE*, and *CE* as shown in Fig. 6.17*a*, they will form together a rigid truss *ABDF*. The trusses *ABC* and *DEF* can also be combined into a single rigid truss by joining joints *B* and *D* into a single joint *B* and by connecting joints *C* and *E* by a bar *CE* (Fig. 6.17*b*). The truss thus obtained is known as a Fink truss. It should be noted that the trusses of Fig. 6.17*a* and *b* are *not* simple trusses; they cannot be constructed from a triangular truss by adding successive pairs of members as prescribed in Sec. 6.3. They are rigid trusses, however, as we may check by comparing the systems of connections used to hold the simple trusses *ABC* and *DEF* together (three bars in Fig. 6.17*a*, one pin and one bar in Fig. 6.17*b*) with the systems of supports discussed in Secs. 4.4 and 4.5. Trusses made of several simple trusses rigidly connected are known as *compound trusses*.

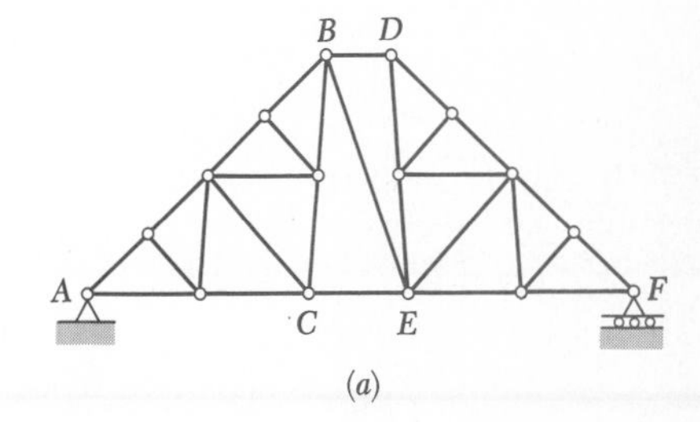

(*a*)

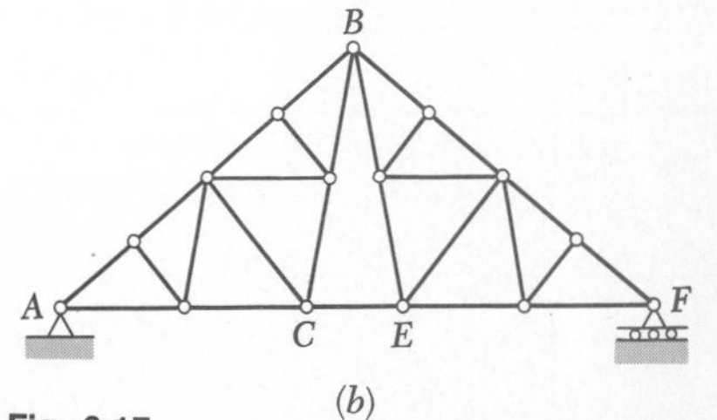

(*b*)

Fig. 6.17

It may be checked that in a compound truss the number of members m and the number of joints n are still related by the formula $m = 2n - 3$. If a compound truss is supported by a frictionless pin and a roller (involving three unknown reactions), the total number of unknowns is $m + 3$ and this number is therefore equal to the number $2n$ of equations obtained by expressing that the n pins are in equilibrium. Compound trusses supported by a pin and a roller, or by an equivalent system of supports, are *statically determinate, rigid,* and *completely constrained.* This means that all unknown reactions and forces in members can be determined by the methods of statics and that all equilibrium equations being satisfied, the truss will neither collapse nor move. The forces in the members, however, cannot all be determined by the method of joints, except by solving a large number of simultaneous equations. In the case of the compound truss of Fig. 6.17*a*, for example, it will be found more expeditious to pass a section through members *BD*, *BE*, and *CE* to determine their forces.

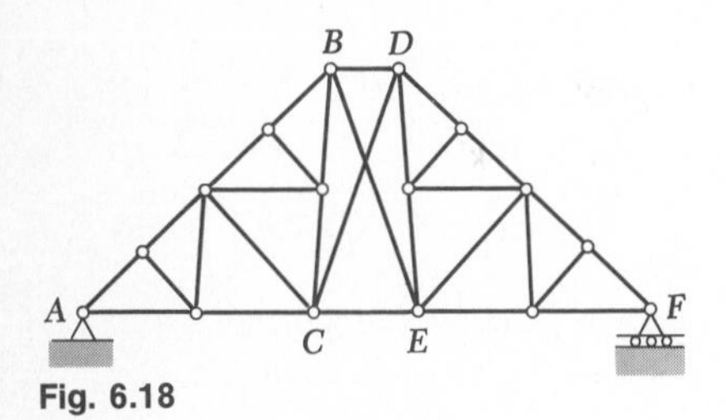

Fig. 6.18

Suppose, now, that the simple trusses *ABC* and *DEF* are connected by *four* bars *BD*, *BE*, *CD*, and *CE* (Fig. 6.18). The number of members m is now larger than $2n - 3$; the truss obtained is *overrigid*, and one of the four members *BD*, *BE*, *CD*, or *CE* is said to be *redundant.* If the truss is supported by a pin at *A* and a roller at *F*, the total number of unknowns is $m + 3$. This number is now larger than the number $2n$ of available independent equations; the truss is *statically indeterminate.*

Finally, we shall assume that the two simple trusses *ABC* and *DEF* are joined by a pin as shown in Fig. 6.19*a*. The number of members m is smaller than $2n - 3$. If the truss is supported by a pin at *A* and a roller at *F*, the total number of unknowns is $m + 3$. This number is now smaller than the number $2n$ of equilibrium equations which should be satisfied; the truss is *nonrigid* and will collapse under its own weight. However, if two pins are used to support it, the truss becomes *rigid* and will not collapse (Fig. 6.19*b*). We note that the total number of unknowns is now $m + 4$ and is equal to the number $2n$ of equations. More generally, if the reactions at the supports involve r unknowns, the condition for a compound truss to be statically determinate, rigid, and completely constrained is $m + r = 2n$. While necessary, this condition, however, is not sufficient for the equilibrium of a structure which ceases to be rigid when detached from its supports (see Sec. 6.11).

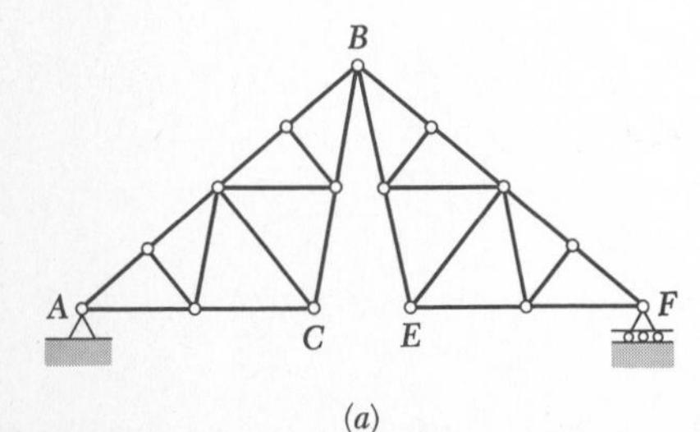

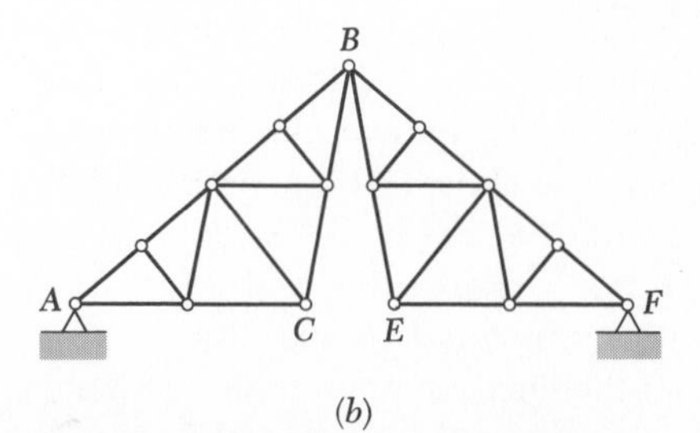

Fig. 6.19

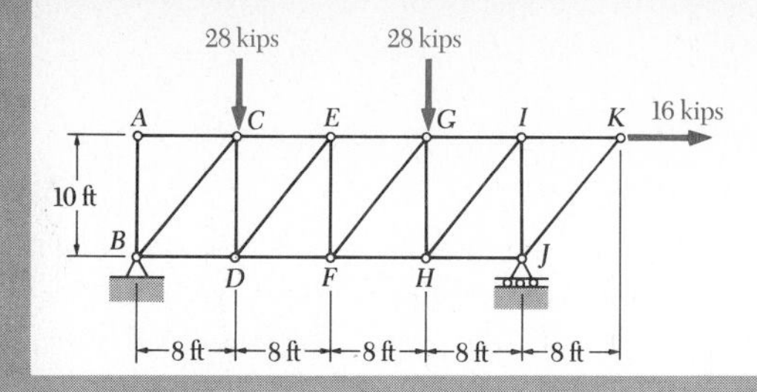

SAMPLE PROBLEM 6.2

Determine the force in members EF and GI of the truss shown.

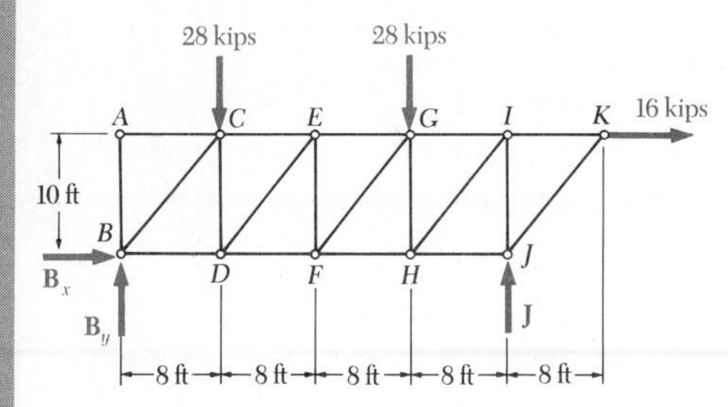

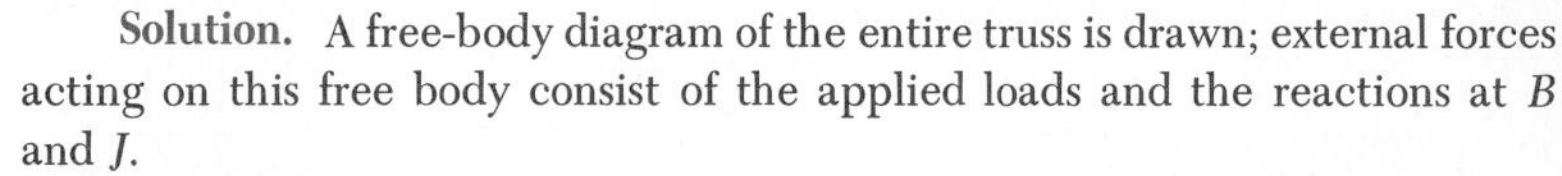

Solution. A free-body diagram of the entire truss is drawn; external forces acting on this free body consist of the applied loads and the reactions at B and J.

Equilibrium of the Entire Truss.

$$+\curvearrowleft \Sigma M_B = 0:$$
$$-(28 \text{ kips})(8 \text{ ft}) - (28 \text{ kips})(24 \text{ ft}) - (16 \text{ kips})(10 \text{ ft}) + J(32 \text{ ft}) = 0$$
$$J = +33 \text{ kips} \qquad \mathbf{J} = 33 \text{ kips}\uparrow$$

$$\overset{+}{\rightarrow}\Sigma F_x = 0: \qquad B_x + 16 \text{ kips} = 0$$
$$B_x = -16 \text{ kips} \qquad \mathbf{B}_x = 16 \text{ kips}\leftarrow$$

$$+\curvearrowleft \Sigma M_J = 0:$$
$$(28 \text{ kips})(24 \text{ ft}) + (28 \text{ kips})(8 \text{ ft}) - (16 \text{ kips})(10 \text{ ft}) - B_y(32 \text{ ft}) = 0$$
$$B_y = +23 \text{ kips} \qquad \mathbf{B}_y = 23 \text{ kips}\uparrow$$

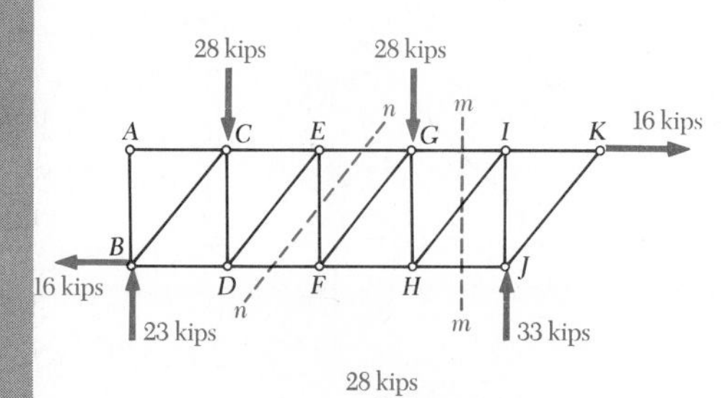

Force in Member EF. Section nn is passed through the truss so that it intersects member EF and only two additional members. After the intersected members have been removed, the left-hand portion of the truss is chosen as a free body. Three unknowns are involved; to eliminate the two horizontal forces, we write

$$+\uparrow\Sigma F_y = 0: \qquad +23 \text{ kips} - 28 \text{ kips} - F_{EF} = 0$$
$$F_{EF} = -5 \text{ kips}$$

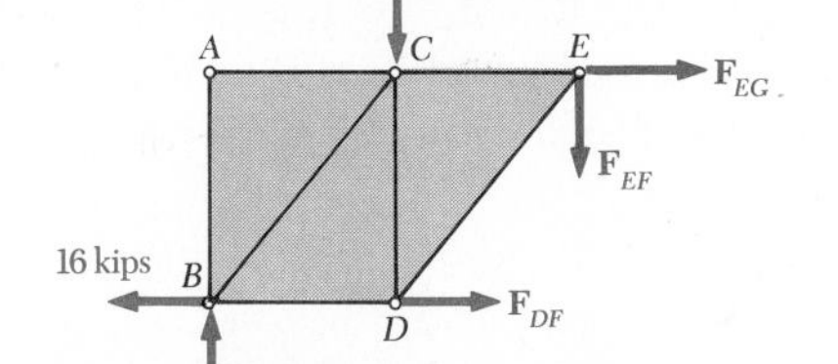

The sense of $\mathbf{F}_{EF}$ was chosen assuming member EF to be in tension; the negative sign obtained indicates that the member is in compression.

$$F_{EF} = 5 \text{ kips } C \quad ◄$$

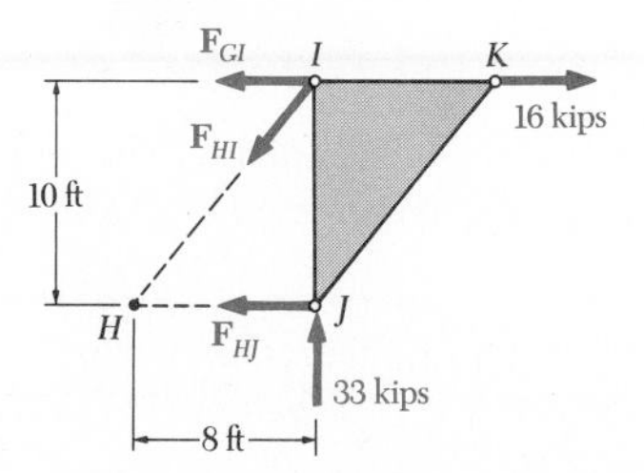

Force in Member GI. Section mm is passed through the truss so that it intersects member GI and only two additional members. After the intersected members have been removed, we choose the right-hand portion of the truss as a free body. Three unknown forces are again involved; to eliminate the two forces passing through point H, we write

$$+\curvearrowleft \Sigma M_H = 0: \qquad (33 \text{ kips})(8 \text{ ft}) - (16 \text{ kips})(10 \text{ ft}) + F_{GI}(10 \text{ ft}) = 0$$
$$F_{GI} = -10.4 \text{ kips} \qquad F_{GI} = 10.4 \text{ kips } C \quad ◄$$

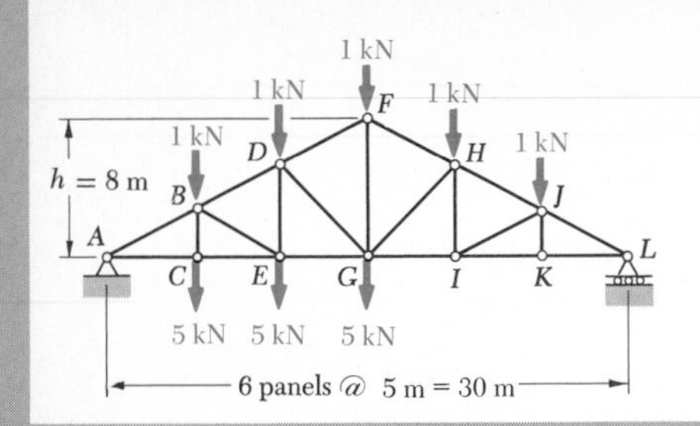

SAMPLE PROBLEM 6.3

Determine the force in members *FH*, *GH*, and *GI* of the roof truss shown.

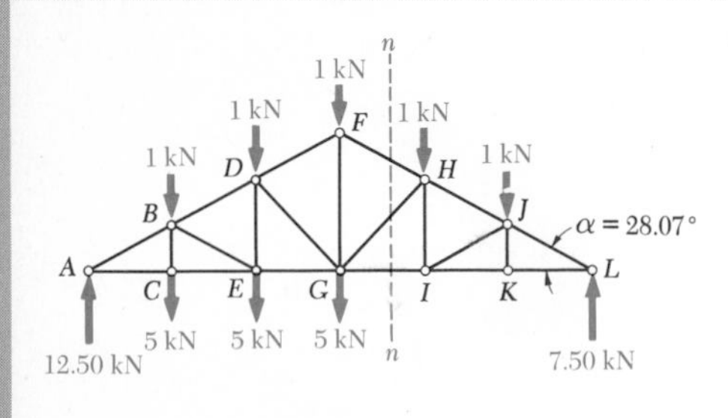

Solution. Section *nn* is passed through the truss as shown. The right-hand portion of the truss will be taken as a free body. Since the reaction at *L* acts on this free body, the value of **L** must be calculated separately, using the entire truss as a free body; the equation $\Sigma M_A = 0$ yields $\mathbf{L} = 7.50 \text{ kN}\uparrow$.

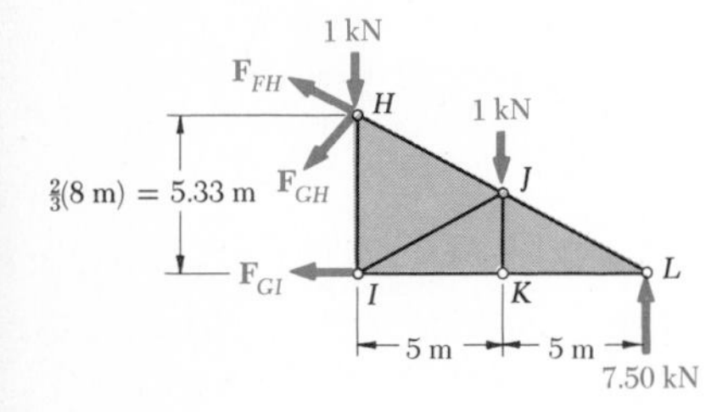

Force in Member GI. Using the portion *HLI* of the truss as a free body, the value of F_{GI} is obtained by writing

$$+\circlearrowleft \Sigma M_H = 0: \qquad (7.50 \text{ kN})(10 \text{ m}) - (1 \text{ kN})(5 \text{ m}) - F_{GI}(5.33 \text{ m}) = 0$$

$$F_{GI} = +13.13 \text{ kN} \qquad F_{GI} = 13.13 \text{ kN } T \blacktriangleleft$$

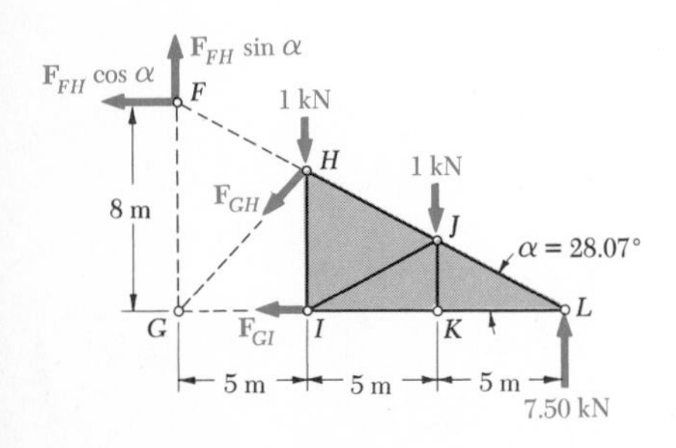

Force in Member FH. The value of F_{FH} is obtained from the equation $\Sigma M_G = 0$. We move $\mathbf{F}_{FH}$ along its line of action until it acts at point *F*, where it is resolved into its *x* and *y* components. The moment of $\mathbf{F}_{FH}$ with respect to point *G* is now equal to $(F_{FH} \cos \alpha)(8 \text{ m})$.

$$+\circlearrowleft \Sigma M_G = 0:$$

$$(7.50 \text{ kN})(15 \text{ m}) - (1 \text{ kN})(10 \text{ m}) - (1 \text{ kN})(5 \text{ m}) + (F_{FH} \cos \alpha)(8 \text{ m}) = 0$$

$$F_{FH} = -13.81 \text{ kN} \qquad F_{FH} = 13.81 \text{ kN } C \blacktriangleleft$$

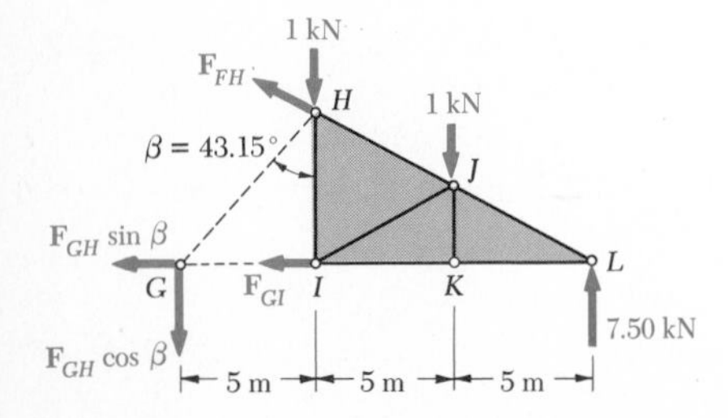

Force in Member GH. The value of F_{GH} is determined by first resolving the force $\mathbf{F}_{GH}$ into *x* and *y* components at point *G* and then solving the equation $\Sigma M_L = 0$.

$$+\circlearrowleft \Sigma M_L = 0: \qquad (1 \text{ kN})(10 \text{ m}) + (1 \text{ kN})(5 \text{ m}) + (F_{GH} \cos \beta)(15 \text{ m}) = 0$$

$$F_{GH} = -1.371 \text{ kN} \qquad F_{GH} = 1.371 \text{ kN } C \blacktriangleleft$$

Problems

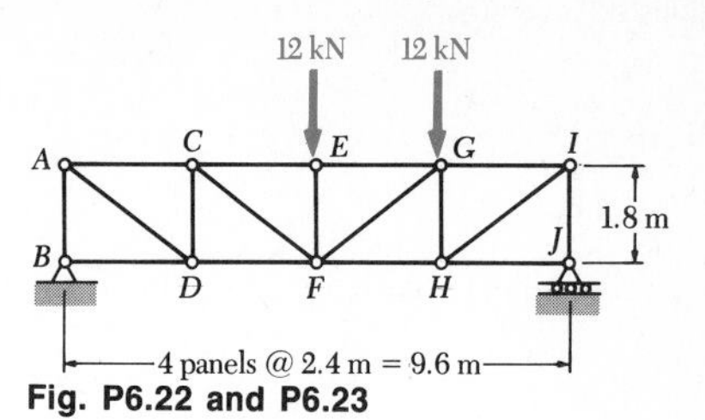

Fig. P6.22 and P6.23

6.22 Determine the force in members CD and DF of the truss shown.

6.23 Determine the force in members FG and FH of the truss shown.

6.24 Determine the force in members FG and FH of the truss shown when $P = 35$ kN.

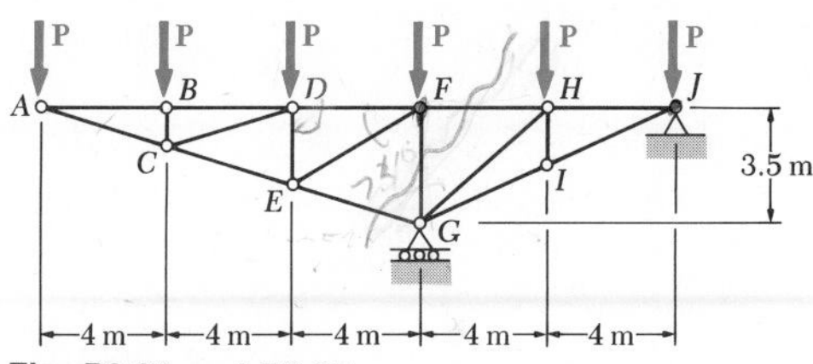

Fig. P6.24 and P6.25

6.25 Determine the force in members EF and EG of the truss shown when $P = 35$ kN.

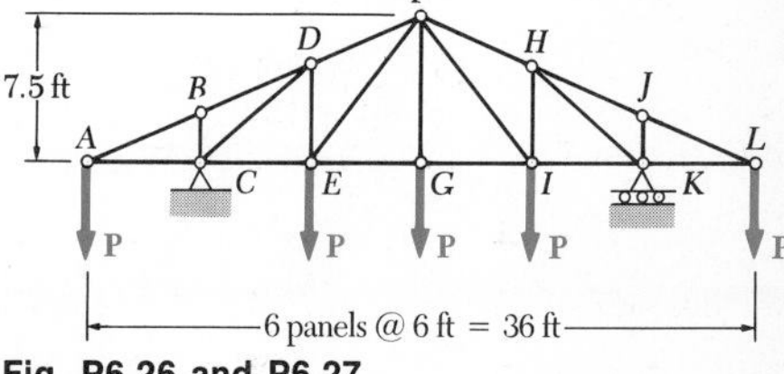

Fig. P6.26 and P6.27

6.26 Determine the force in members DE and DF of the truss shown when $P = 20$ kips.

6.27 Determine the force in members EG and EF of the truss shown when $P = 20$ kips.

6.28 Determine the force in members DF and DE of the truss shown.

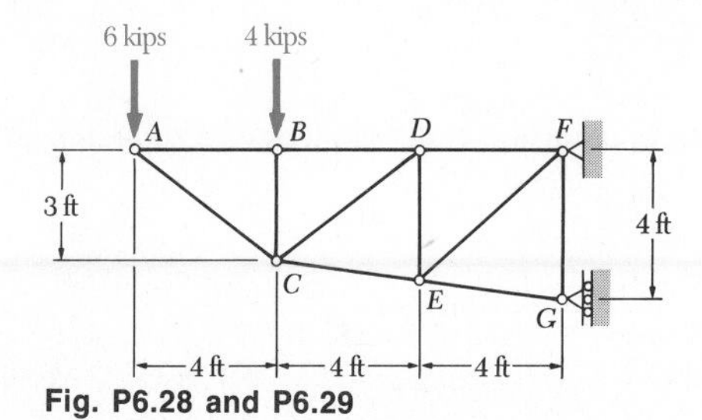

Fig. P6.28 and P6.29

6.29 Determine the force in members CD and CE of the truss shown.

6.30 Determine the force in members BD and DE of the truss shown.

6.31 Determine the force in members DG and EG of the truss shown.

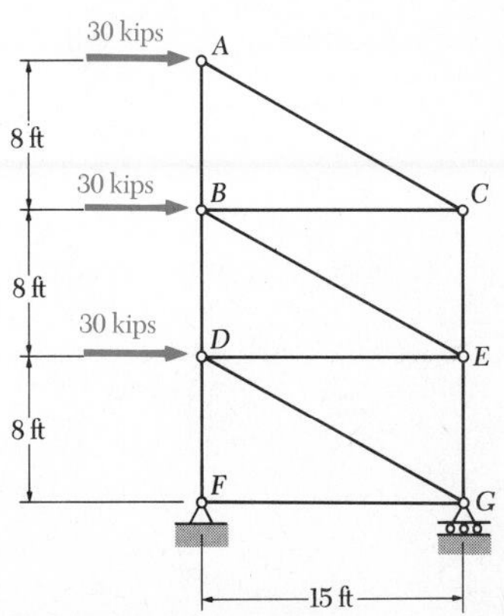

Fig. P6.30 and P6.31

6.32 Determine the force in member CD of the Fink roof truss shown when $P = 25$ kN.

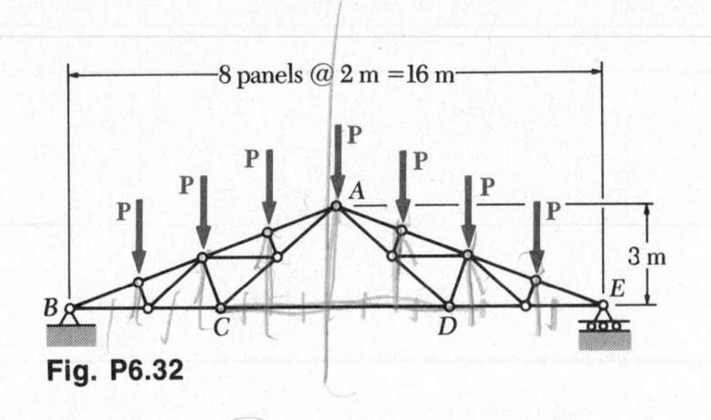

Fig. P6.32

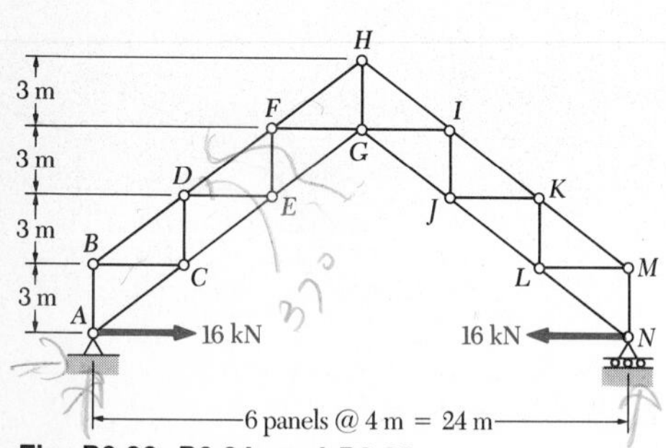

Fig. P6.33, P6.34, and P6.35

6.33 Determine the force in members CE, DE, and DF of the truss shown.

6.34 Determine the force in members DF, EF, and EG of the truss shown.

6.35 Determine the force in members GI, GJ, and HI of the truss shown.

6.36 Determine the force in members AD, CD, and CE of the truss shown.

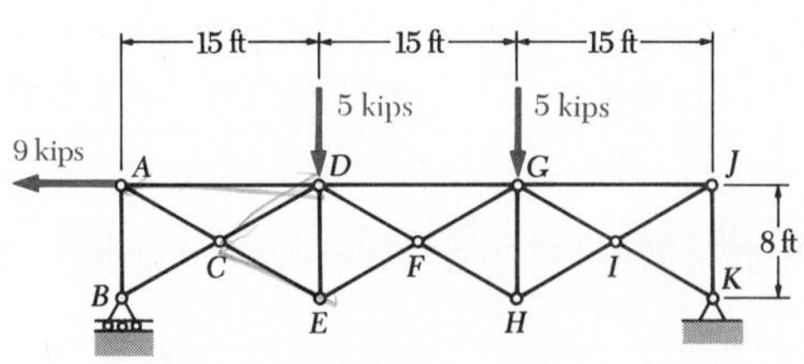

Fig. P6.36 and P6.37

6.37 Determine the force in members DG, FG, and FH of the truss shown.

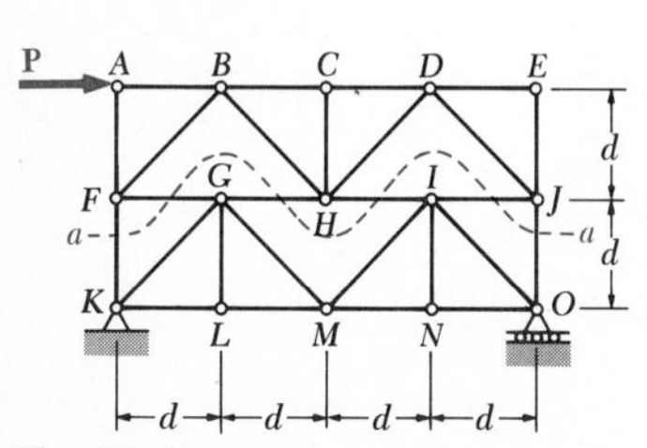

Fig. P6.38

6.38 Determine the force in members FK and JO of the truss shown. (*Hint.* Use section a-a.)

6.39 Determine the force in members DG and FH of the truss shown. (*Hint.* Use section a-a.)

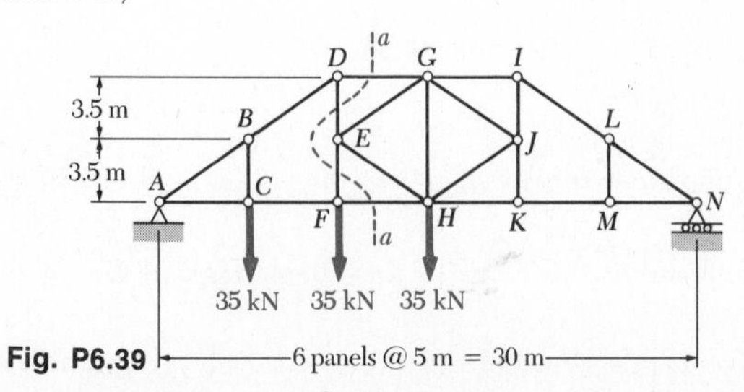

Fig. P6.39

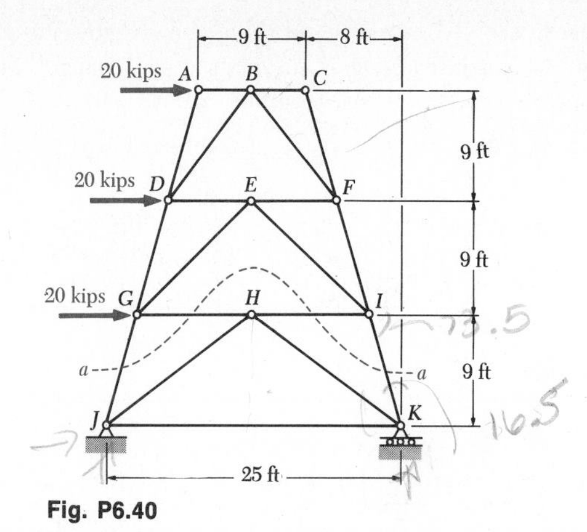

Fig. P6.40

6.40 Determine the force in member IK of the truss shown. (*Hint.* Use section a-a.)

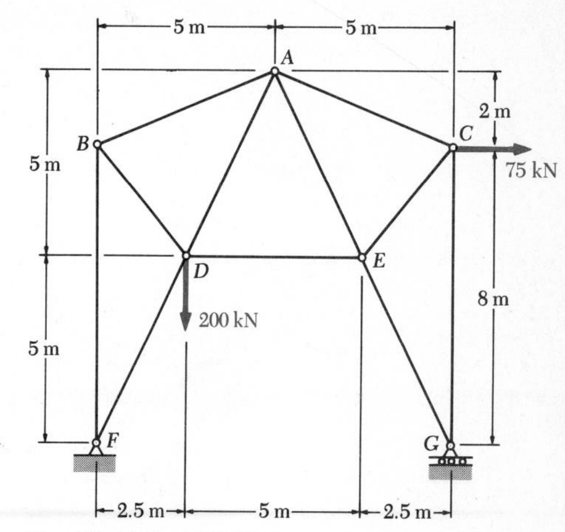

Fig. P6.41 and P6.42

6.41 Determine the force in members DE and AE of the truss shown.

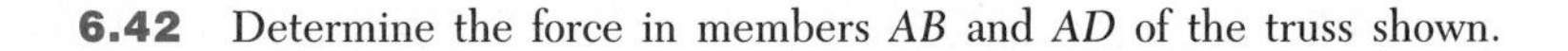

6.42 Determine the force in members AB and AD of the truss shown.

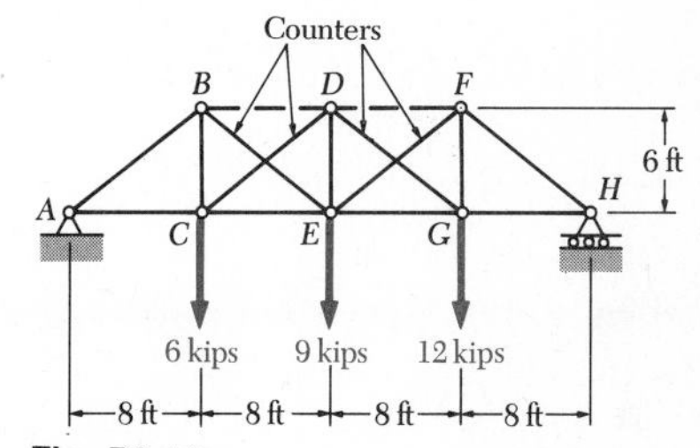

Fig. P6.43

6.43 The diagonal members in the center panel of the truss shown are very slender and can act only in tension; such members are known as *counters*. Determine the force in member DE and in the counters which are acting under the given loading.

6.44 Solve Prob. 6.43, assuming that the 6-kip load has been removed.

6.45 Solve Prob. 6.43 assuming that the 9-kip load has been removed.

6.46 and 6.47 Classify each of the given structures as completely, partially, or improperly constrained; if completely constrained, further classify as determinate or indeterminate. (All members can act both in tension and in compression.)

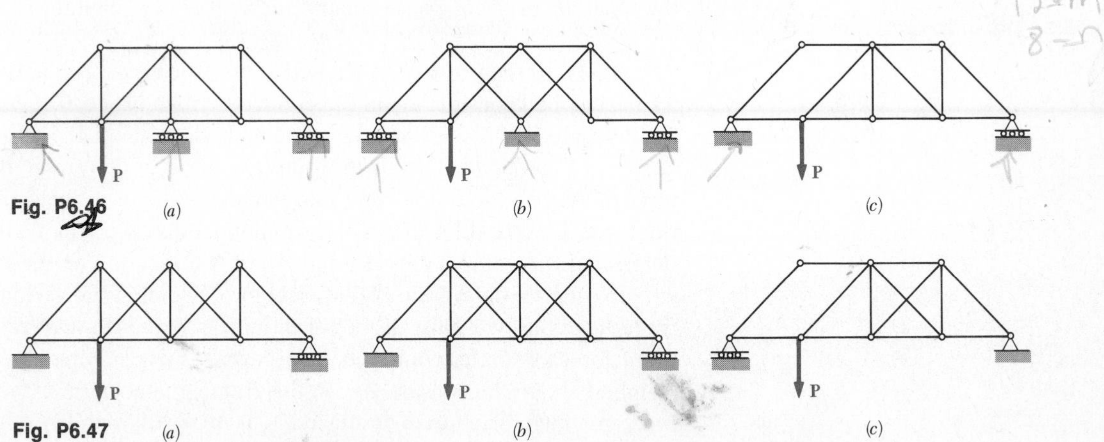

Fig. P6.46 (*a*) (*b*) (*c*)

Fig. P6.47 (*a*) (*b*) (*c*)

6.48 and 6.49 Classify each of the given structures as completely, partially, or improperly constrained; if completely constrained, further classify as determinate or indeterminate. (All members can act both in tension and in compression.)

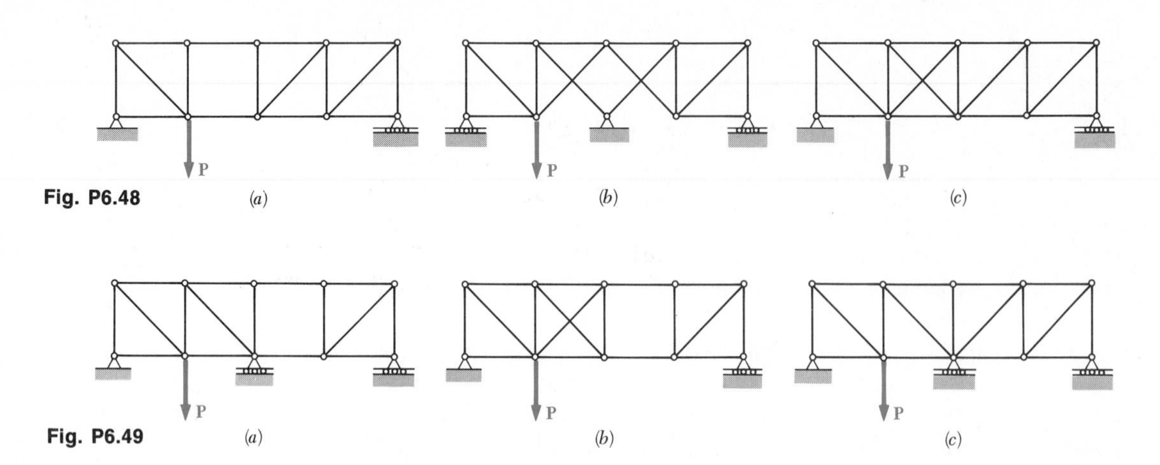

Fig. P6.48 (*a*) (*b*) (*c*)

Fig. P6.49 (*a*) (*b*) (*c*)

FRAMES AND MACHINES

6.9. Structures Containing Multiforce Members. Under Trusses, we have considered structures consisting entirely of pins and of straight two-force members. The forces acting on the two-force members were known to be directed along the members themselves. We shall now consider structures in which at least one of the members is a *multiforce* member, i.e., a member acted upon by three or more forces. These forces will generally not be directed along the members on which they act; their direction is unknown, and they should be represented therefore by two unknown components.

Frames and machines are structures containing multiforce members. *Frames* are designed to support loads and are usually stationary, fully constrained structures. *Machines* are designed to transmit and modify forces; they may or may not be stationary and will always contain moving parts.

6.10. Analysis of a Frame. As a first example of analysis of a frame, we shall consider again the crane described in Sec. 6.1, which carries a given load W (Fig. 6.20*a*). The free-body diagram of the entire frame is shown in Fig. 6.20*b*. This diagram may be used to determine the external forces acting on the frame. Summing moments about A, we first determine the force $\mathbf{T}$ exerted by the cable; summing x and y components, we then determine the components $\mathbf{A}_x$ and $\mathbf{A}_y$ of the reaction at the pin A.

In order to determine the internal forces holding the various parts of a frame together, we must dismember the frame and draw a free-body diagram for each of its component parts (Fig. 6.20*c*). First, the two-force members should be considered. In this frame, member BE is the only two-force member. The forces acting at each end of this member must have the same magnitude, same line of action, and opposite sense (Sec. 4.6). They are therefore directed along BE and will be denoted, respectively, by $\mathbf{F}_{BE}$

and $-\mathbf{F}_{BE}$. Their sense will be arbitrarily assumed as shown in Fig. 6.20*c*, and the correctness of this assumption will be checked later by the sign obtained for the common magnitude F_{BE} of the two forces.

Next, we consider the multiforce members, i.e., the members which are acted upon by three or more forces. According to Newton's third law, the force exerted at *B* by member *BE* on member *AD* must be equal and opposite to the force $\mathbf{F}_{BE}$ exerted by *AD* on *BE*. Similarly, the force exerted at *E* by member *BE* on member *CF* must be equal and opposite to the force $-\mathbf{F}_{BE}$ exerted by *CF* on *BE*. The forces that the two-force member *BE* exerts on *AD* and *CF* are therefore respectively equal to $-\mathbf{F}_{BE}$ and $\mathbf{F}_{BE}$; they have the same magnitude F_{BE} and opposite sense, and should be directed as shown in Fig. 6.20*c*.

At *C* two multiforce members are connected. Since neither the direction nor the magnitude of the forces acting at *C* is known, these forces will be represented by their *x* and *y* components. The components $\mathbf{C}_x$ and $\mathbf{C}_y$ of the force acting on member *AD* will be arbitrarily directed to the right and upward. Since, according to Newton's third law, the forces exerted by member *CF* on *AD* and by member *AD* on *CF* are equal and opposite, the components of the force acting on member *CF* *must* be directed to the left and downward; they will be denoted, respectively, by $-\mathbf{C}_x$ and $-\mathbf{C}_y$. Whether the force $\mathbf{C}_x$ is actually directed to the right and the force $-\mathbf{C}_x$ is actually directed to the left will be determined later from the sign of their common magnitude C_x, a plus sign indicating that the assumption made was correct, and a minus sign that it was wrong. The free-body diagrams of the multiforce members are completed by showing the external forces acting at *A*, *D*, and *F*.†

The internal forces may now be determined by considering the free-body diagram of either of the two multiforce members. Choosing the free-body diagram of *CF*, for example, we write the equations $\Sigma M_C = 0$, $\Sigma M_E = 0$, and $\Sigma F_x = 0$, which yield the values of the magnitudes F_{BE}, C_y, and C_x, respectively. These values may be checked by verifying that member *AD* is also in equilibrium.

It should be noted that the free-body diagrams of the pins were not shown in Fig. 6.20*c*. This was because the pins were assumed to form an integral part of one of the two members they connected. This assumption can always be used to simplify the analysis of frames and machines. When a pin connects three or more members, however, or when a pin connects a support and two or more members, or when a load is applied to a pin, a clear decision must be made in choosing the member to which the pin will be assumed to belong. (If multiforce members are involved, the pin should be attached to one of these members.) The various forces exerted on the pin should then be clearly identified. This is illustrated in Sample Prob. 6.6.

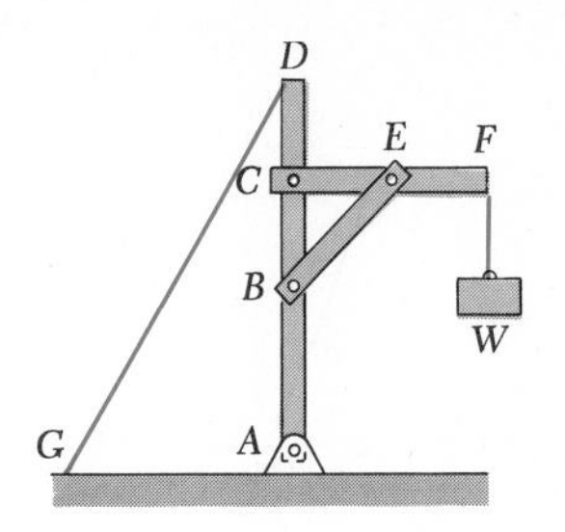

(*a*)

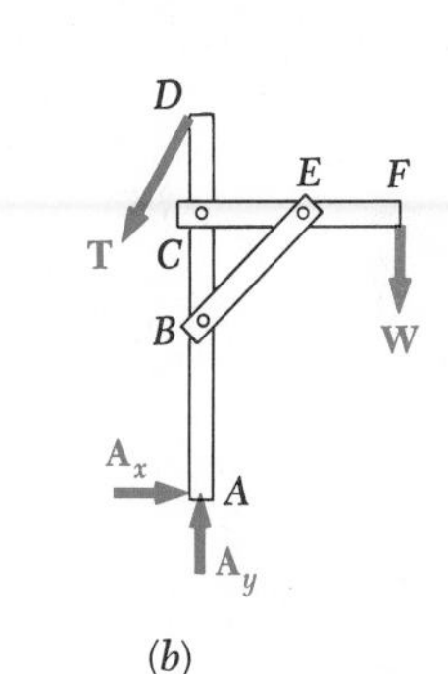

(*b*)

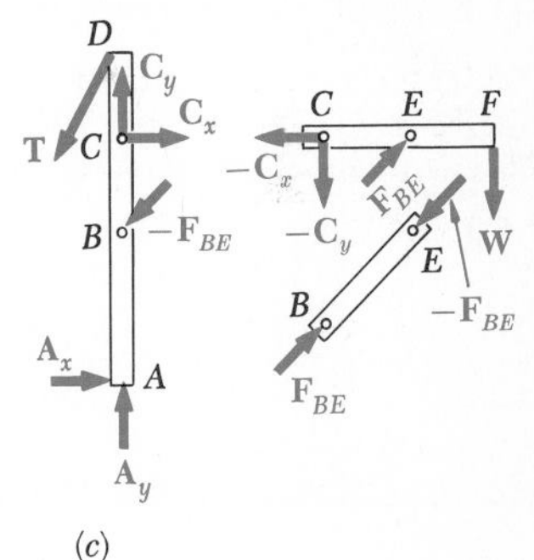

(*c*)

Fig. 6.20

† The use of a minus sign to distinguish the force exerted by one member on another from the equal and opposite force exerted by the second member on the first is not strictly necessary, since the two forces belong to different free-body diagrams and thus cannot easily be confused. In the Sample Problems, we shall represent by the same symbol equal and opposite forces which are applied to different free bodies. It should be noted that, under these conditions, the sign obtained for a given force component will not directly relate the sense of that component to the sense of the corresponding coordinate axis. Rather, a positive sign will indicate that *the sense assumed for that component in the free-body diagram* is correct, and a negative sign that it is wrong.

6.11. Frames Which Cease to Be Rigid When Detached from Their Supports. The crane analyzed in Sec. 6.10 was so constructed that it could keep the same shape without the help of its supports; it was therefore considered as a rigid body. Many frames, however, will collapse if detached from their supports; such frames cannot be considered as rigid bodies. Consider, for example, the frame shown in Fig. 6.21*a*, which consists of two members *AC* and *CB* carrying loads **P** and **Q** at their midpoints; the members are supported by pins at *A* and *B* and are connected by a pin at *C*. If detached from its supports, this frame will not maintain its shape; it should therefore be considered as made of *two distinct rigid parts AC* and *CB*.

The equations $\Sigma F_x = 0$, $\Sigma F_y = 0$, $\Sigma M = 0$ (about any given point) express the conditions for the *equilibrium of a rigid body* (Chap. 4); we should use them, therefore, in connection with the free-body diagrams of rigid bodies, namely, the free-body diagrams of members *AC* and *CB* (Fig. 6.21*b*). Since these members are multiforce members, and since pins are used at the supports and at the connection, the reactions at *A* and *B* and the

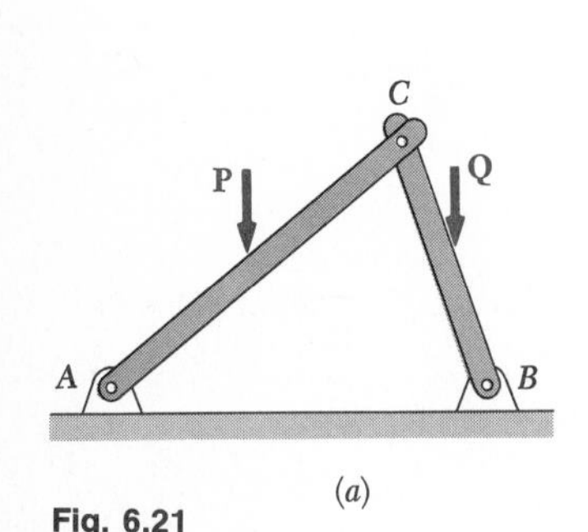

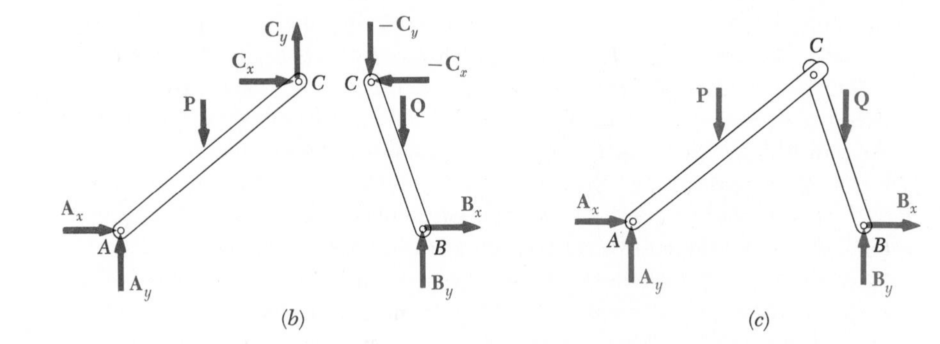

Fig. 6.21

forces at *C* will each be represented by two components. In accordance with Newton's third law, the components of the force exerted by *CB* on *AC* and the components of the force exerted by *AC* on *CB* will be represented by vectors of the same magnitude and opposite sense; thus, if the first pair of components consists of $\mathbf{C}_x$ and $\mathbf{C}_y$, the second pair will be represented by $-\mathbf{C}_x$ and $-\mathbf{C}_y$. We note that four unknown force components act on free body *AC*, while only three independent equations may be used to express that the body is in equilibrium; similarly, four unknowns, but only three equations, are associated with *CB*. However, only six different unknowns are involved in the analysis of the two members, and altogether six equations are available to express that the members are in equilibrium. Writing $\Sigma M_A = 0$ for free body *AC* and $\Sigma M_B = 0$ for *CB*, we obtain two simultaneous equations which may be solved for the common magnitude C_x of the components $\mathbf{C}_x$ and $-\mathbf{C}_x$, and for the common magnitude C_y of the components $\mathbf{C}_y$ and $-\mathbf{C}_y$. Writing, then, $\Sigma F_x = 0$ and $\Sigma F_y = 0$ for each of the two free bodies, we obtain successively the magnitudes A_x, A_y, B_x, and B_y.

We shall observe now that since the equations of equilibrium $\Sigma F_x = 0$, $\Sigma F_y = 0$, $\Sigma M = 0$ (about any given point) are satisfied by the forces acting on free body AC, and since they are also satisfied by the forces acting on free body CB, they must be satisfied when the forces acting on the two free bodies are considered simultaneously. Since the internal forces at C cancel each other, we find that the equations of equilibrium must be satisfied by the external forces shown on the free-body diagram of the frame ACB itself (Fig. 6.21*c*), although the frame is not a rigid body. These equations may be used to determine some of the components of the reactions at A and B. We shall note, however, that *the reactions cannot be completely determined from the free-body diagram of the whole frame.* It is thus necessary to dismember the frame and to consider the free-body diagrams of its component parts (Fig. 6.21*b*), even when we are interested only in finding external reactions. This may be explained by the fact that the equilibrium equations obtained for free body ACB are *necessary conditions* for the equilibrium of a nonrigid structure, *but not sufficient conditions.*

The method of solution outlined in the second paragraph of this section involved simultaneous equations. We shall now discuss a more expeditious method, which utilizes the free body ACB as well as the free bodies AC and CB. Writing $\Sigma M_A = 0$ and $\Sigma M_B = 0$ for free body ACB, we obtain B_y and A_y. Writing $\Sigma M_C = 0$, $\Sigma F_x = 0$, and $\Sigma F_y = 0$ for free body AC, we obtain successively A_x, C_x, and C_y. Finally, writing $\Sigma F_x = 0$ for ACB, we obtain B_x.

We noted above that the analysis of the frame of Fig. 6.21 involves six unknown force components and six independent equilibrium equations (the equilibrium equations for the whole frame were obtained from the original six equations and, therefore, are not independent). Moreover, we checked that all unknowns could be actually determined and that all equations could be satisfied. The frame considered is *statically determinate and rigid.*† In general, to determine whether a structure is statically determinate and rigid, we should draw a free-body diagram for each of its component parts and count the reactions and internal forces involved. We should also determine the number of independent equilibrium equations (excluding equations expressing the equilibrium of the whole structure or of groups of component parts already analyzed). If there are more unknowns than equations, the structure is *statically indeterminate.* If there are fewer unknowns than equations, the structure is *nonrigid.* If there are as many unknowns as equations, *and if all unknowns may be determined and all equations satisfied* under general loading conditions, the structure is *statically determinate and rigid;* if, however, due to an *improper arrangement* of members and supports, all unknowns cannot be determined and all equations cannot be satisfied, the structure is *statically indeterminate and nonrigid.*

† The word "rigid" is used here to indicate that the frame will maintain its shape as long as it remains attached to its supports.

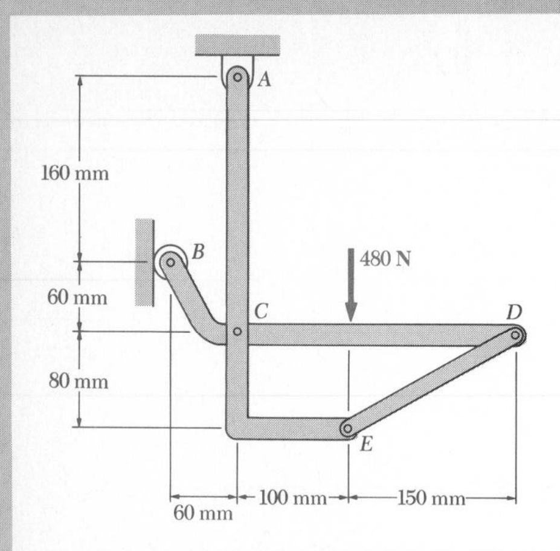

SAMPLE PROBLEM 6.4

In the frame shown, members *ACE* and *BCD* are connected by a pin at *C* and by the link *DE*. For the loading shown, determine the force in link *DE* and the components of the force exerted at *C* on member *BCD*.

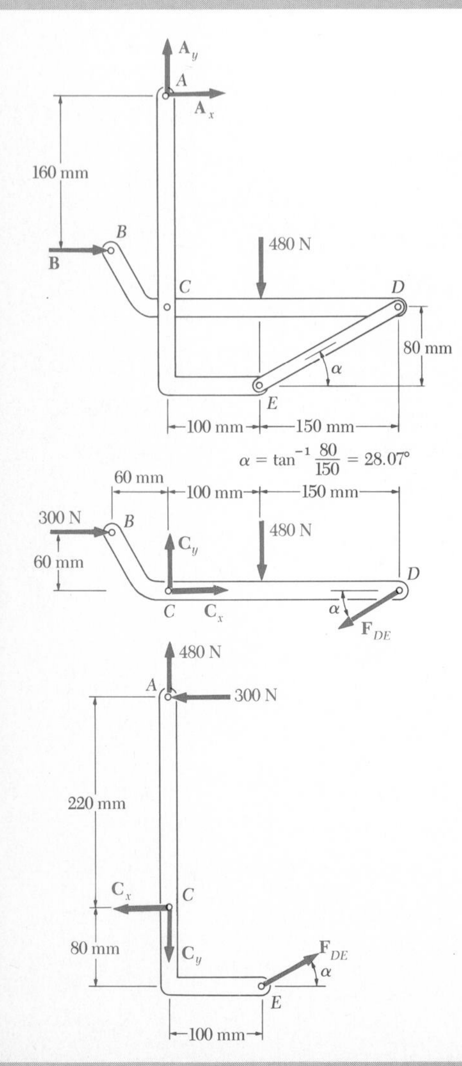

Entire Frame. Since the external reactions involve only three unknowns, we compute the reactions by considering the free-body diagram of the entire frame.

$+\uparrow\Sigma F_y = 0$: $\quad A_y - 480 \text{ N} = 0 \qquad A_y = +480 \text{ N} \qquad \mathbf{A}_y = 480 \text{ N}\uparrow$

$+\circlearrowleft\Sigma M_A = 0$: $\quad -(480 \text{ N})(100 \text{ mm}) + B(160 \text{ mm}) = 0$

$B = +300 \text{ N} \qquad \mathbf{B} = 300 \text{ N} \rightarrow$

$\overset{+}{\rightarrow}\Sigma F_x = 0$: $\quad B + A_x = 0$

$300 \text{ N} + A_x = 0 \qquad A_x = -300 \text{ N} \qquad \mathbf{A}_x = 300 \text{ N} \leftarrow$

Members. We now dismember the frame. Since only two members are connected at *C*, the components of the unknown forces acting on *ACE* and *BCD* are, respectively, equal and opposite and are assumed directed as shown. We assume that link *DE* is in tension and exerts at *D* and *E* equal and opposite forces directed as shown.

Member *BCD*. Using the free body *BCD*, we write

$+\circlearrowright\Sigma M_C = 0$: $\quad (F_{DE} \sin \alpha)(250 \text{ mm}) + (300 \text{ N})(60 \text{ mm}) + (480 \text{ N})(100 \text{ mm}) = 0$

$F_{DE} = -561 \text{ N} \qquad F_{DE} = 561 \text{ N } C$ ◄

$\overset{+}{\rightarrow}\Sigma F_x = 0$: $\quad C_x - F_{DE} \cos \alpha + 300 \text{ N} = 0$

$C_x - (-561 \text{ N}) \cos 28.07° + 300 \text{ N} = 0 \qquad C_x = -795 \text{ N}$

$+\uparrow\Sigma F_y = 0$: $\quad C_y - F_{DE} \sin \alpha - 480 \text{ N} = 0$

$C_y - (-561 \text{ N}) \sin 28.07° - 480 \text{ N} = 0 \qquad C_y = +216 \text{ N}$

From the signs obtained for C_x and C_y we conclude that the force components $\mathbf{C}_x$ and $\mathbf{C}_y$ exerted on member *BCD* are directed respectively to the left and up. We have

$$\mathbf{C}_x = 795 \text{ N} \leftarrow, \ \mathbf{C}_y = 216 \text{ N}\uparrow \text{ ◄}$$

Member *ACE* (Check). The computations are checked by considering the free body *ACE*. For example,

$$+\circlearrowleft\Sigma M_A = (F_{DE} \cos \alpha)(300 \text{ mm}) + (F_{DE} \sin \alpha)(100 \text{ mm}) - C_x(220 \text{ mm})$$
$$= (-561 \cos \alpha)(300) + (-561 \sin \alpha)(100) - (-795)(220) = 0$$

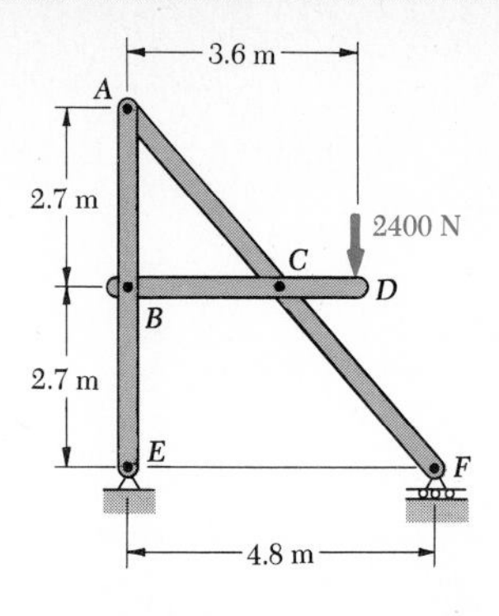

SAMPLE PROBLEM 6.5

Determine the components of the forces acting on each member of the frame shown.

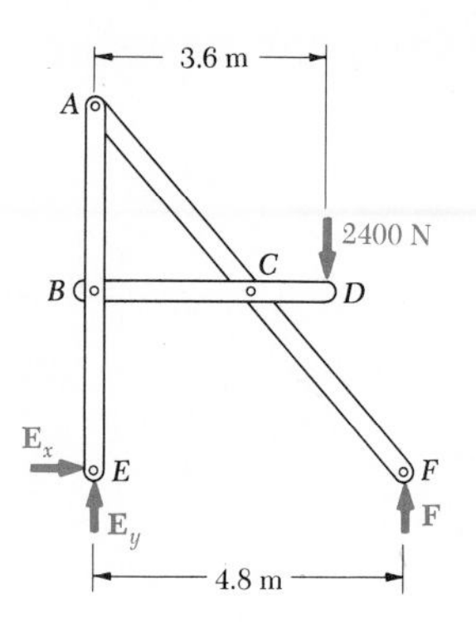

Entire Frame. Since the external reactions involve only three unknowns, we compute the reactions by considering the free-body diagram of the entire frame.

$+\circlearrowleft \Sigma M_E = 0$: $\quad -(2400\text{ N})(3.6\text{ m}) + F(4.8\text{ m}) = 0$
$F = +1800\text{ N}$ $\qquad \mathbf{F} = 1800\text{ N}\uparrow$ ◀

$+\uparrow \Sigma F_y = 0$: $\quad -2400\text{ N} + 1800\text{ N} + E_y = 0$
$E_y = +600\text{ N}$ $\qquad \mathbf{E}_y = 600\text{ N}\uparrow$ ◀

$\xrightarrow{+} \Sigma F_x = 0$: $\qquad \mathbf{E}_x = 0$ ◀

The frame is now dismembered; since only two members are connected at each joint, equal and opposite components are shown on each member at each joint.

Member *BCD*

$+\circlearrowleft \Sigma M_B = 0$: $\quad -(2400\text{ N})(3.6\text{ m}) + C_y(2.4\text{ m}) = 0 \qquad C_y = +3600\text{ N}$ ◀

$+\circlearrowleft \Sigma M_C = 0$: $\quad -(2400\text{ N})(1.2\text{ m}) + B_y(2.4\text{ m}) = 0 \qquad B_y = +1200\text{ N}$ ◀

$\xrightarrow{+} \Sigma F_x = 0$: $\quad -B_x + C_x = 0$

We note that neither B_x nor C_x can be obtained by considering only member *BCD*. The positive values obtained for B_y and C_y indicate that the force components $\mathbf{B}_y$ and $\mathbf{C}_y$ are directed as assumed.

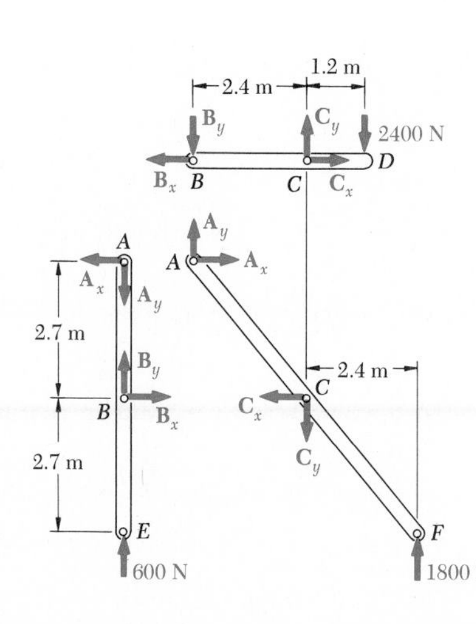

Member *ABE*

$+\circlearrowleft \Sigma M_A = 0$: $\quad B_x(2.7\text{ m}) = 0 \qquad B_x = 0$ ◀

$\xrightarrow{+} \Sigma F_x = 0$: $\quad +B_x - A_x = 0 \qquad A_x = 0$ ◀

$+\uparrow \Sigma F_y = 0$: $\quad -A_y + B_y + 600\text{ N} = 0$
$-A_y + 1200\text{ N} + 600\text{ N} = 0 \qquad A_y = +1800\text{ N}$ ◀

Member *BCD*. Returning now to member *BCD*, we write

$\xrightarrow{+} \Sigma F_x = 0$: $\quad -B_x + C_x = 0 \qquad 0 + C_x = 0 \qquad C_x = 0$ ◀

Member *ACF* (Check). All unknown components have now been found; to check the results, we verify that member *ACF* is in equilibrium.

$$+\circlearrowleft \Sigma M_C = (1800\text{ N})(2.4\text{ m}) - A_y(2.4\text{ m}) - A_x(2.7\text{ m})$$
$$= (1800\text{ N})(2.4\text{ m}) - (1800\text{ N})(2.4\text{ m}) - 0 = 0 \qquad \text{(checks)}$$

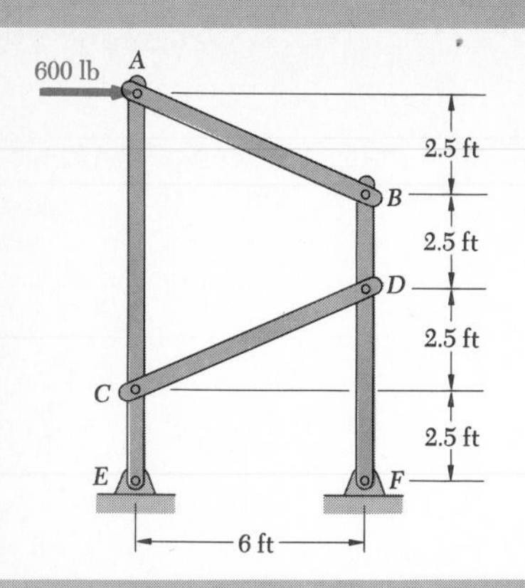

SAMPLE PROBLEM 6.6

A 600-lb horizontal force is applied to pin A of the frame shown. Determine the forces acting on the two vertical members of the frame.

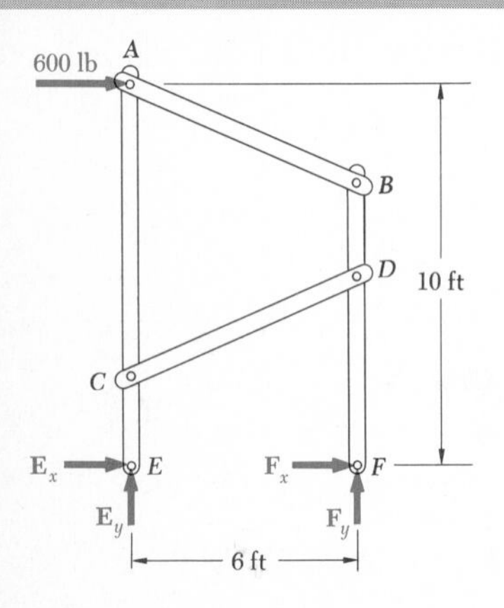

Entire Frame. The entire frame is chosen as a free body; although the reactions involve four unknowns, $\mathbf{E}_y$ and $\mathbf{F}_y$ may be determined by writing

$$+\circlearrowleft \Sigma M_E = 0: \quad -(600 \text{ lb})(10 \text{ ft}) + F_y(6 \text{ ft}) = 0$$

$$F_y = +1000 \text{ lb} \qquad \mathbf{F}_y = 1000 \text{ lb} \uparrow \blacktriangleleft$$

$$+\uparrow \Sigma F_y = 0: \quad E_y + F_y = 0$$

$$E_y = -1000 \text{ lb} \qquad \mathbf{E}_y = 1000 \text{ lb} \downarrow \blacktriangleleft$$

The equations of equilibrium of the entire frame are not sufficient to determine $\mathbf{E}_x$ and $\mathbf{F}_x$. The equilibrium of the various members must now be considered in order to proceed with the solution. In dismembering the frame we shall assume that pin A is attached to the multiforce member ACE and, thus, that the 600-lb force is applied to that member. We also note that AB and CD are two-force members.

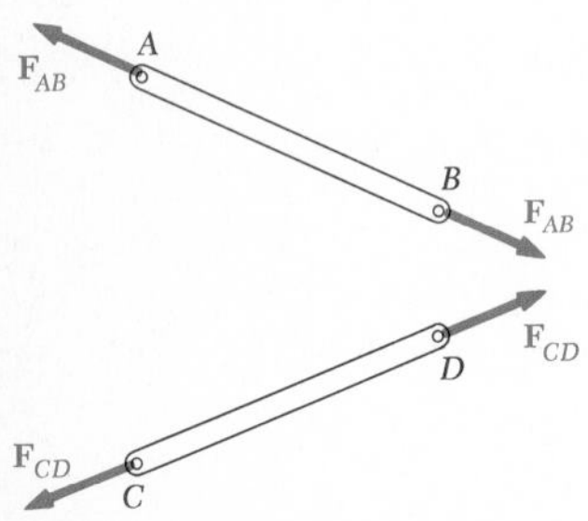

Member *ACE*

$$+\uparrow \Sigma F_y = 0: \quad -\tfrac{5}{13}F_{AB} + \tfrac{5}{13}F_{CD} - 1000 \text{ lb} = 0$$

$$+\circlearrowleft \Sigma M_E = 0: \quad -(600 \text{ lb})(10 \text{ ft}) - (\tfrac{12}{13}F_{AB})(10 \text{ ft}) - (\tfrac{12}{13}F_{CD})(2.5 \text{ ft}) = 0$$

Solving these equations simultaneously, we find

$$F_{AB} = -1040 \text{ lb} \qquad F_{CD} = +1560 \text{ lb} \blacktriangleleft$$

The signs obtained indicate that the sense assumed for F_{CD} was correct and the sense for F_{AB} incorrect. Summing now x components,

$$\overset{+}{\rightarrow} \Sigma F_x = 0: \quad 600 \text{ lb} + \tfrac{12}{13}(-1040 \text{ lb}) + \tfrac{12}{13}(+1560 \text{ lb}) + E_x = 0$$

$$E_x = -1080 \text{ lb} \qquad \mathbf{E}_x = 1080 \text{ lb} \leftarrow \blacktriangleleft$$

Entire Frame. Since $\mathbf{E}_x$ has been determined, we may return to the free-body diagram of the entire frame and write

$$\overset{+}{\rightarrow} \Sigma F_x = 0: \quad 600 \text{ lb} - 1080 \text{ lb} + F_x = 0$$

$$F_x = +480 \text{ lb} \qquad \mathbf{F}_x = 480 \text{ lb} \rightarrow \blacktriangleleft$$

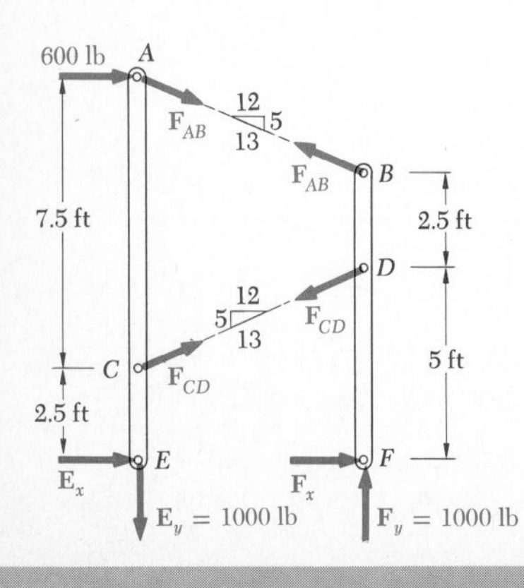

Member *BDF* (Check). We may check our computations by verifying that the equation $\Sigma M_B = 0$ is satisfied by the forces acting on member BDF.

$$\begin{aligned} +\circlearrowleft \Sigma M_B &= -(\tfrac{12}{13}F_{CD})(2.5 \text{ ft}) + (F_x)(7.5 \text{ ft}) \\ &= -\tfrac{12}{13}(1560 \text{ lb})(2.5 \text{ ft}) + (480 \text{ lb})(7.5 \text{ ft}) \\ &= -3600 \text{ lb}\cdot\text{ft} + 3600 \text{ lb}\cdot\text{ft} = 0 \qquad \text{(checks)} \end{aligned}$$

Problems

6.50 and 6.51 Determine the force in member *BD* and the components of the reaction at *C*.

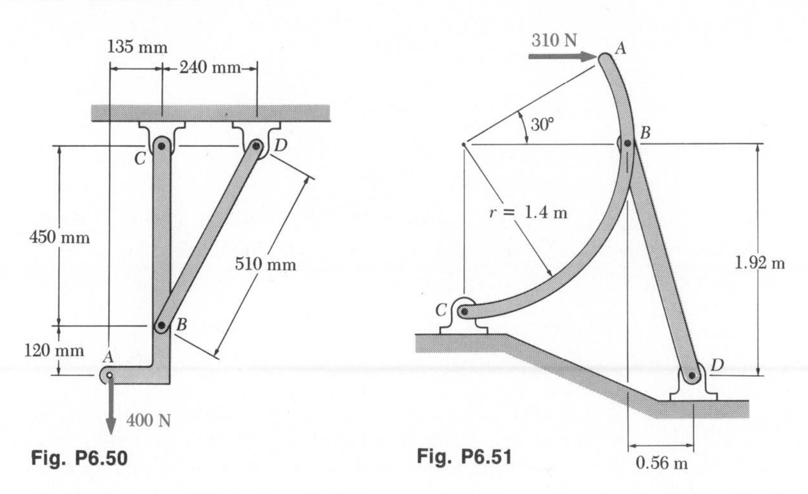

Fig. P6.50

Fig. P6.51

6.52 Determine the components of all forces acting on member *ABCD* of the assembly shown.

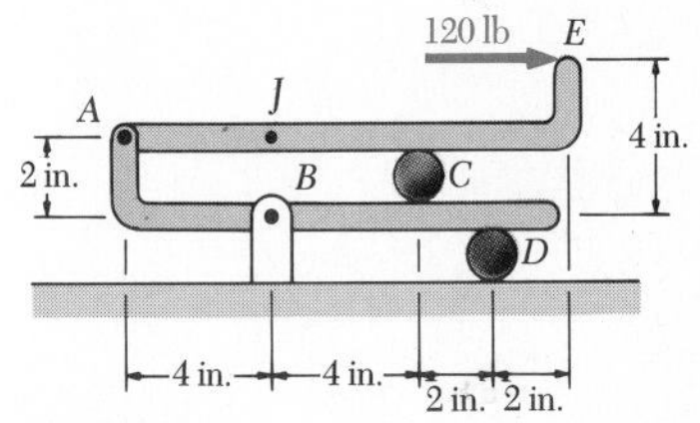

Fig. P6.52

6.53 Solve Prob. 6.52, assuming that the 120-lb force applied at *E* is directed vertically downward.

6.54 An aircraft tow bar is positioned by means of a single hydraulic cylinder *CD* which is connected to two identical arm-and-wheel units *DEF*. The entire tow bar has a mass of 200 kg and its center of gravity is located at *G*. For the position shown, determine (*a*) the force exerted by the cylinder on bracket *C*, (*b*) the force exerted on each arm by the pin at *E*.

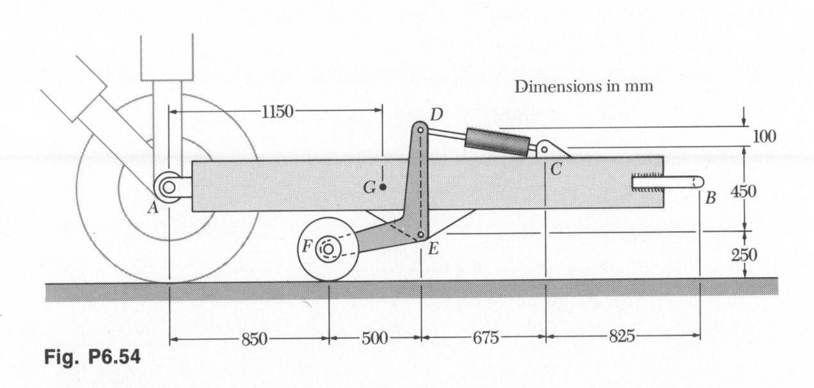

Fig. P6.54

6.55 Solve Prob. 6.54, assuming that a 70-kg mechanic is standing on the tow bar at point *B*.

6.56 Determine the components of all forces acting on member *ABCD* when $\theta = 0$.

6.57 Determine the components of all forces acting on member *ABCD* when $\theta = 90°$.

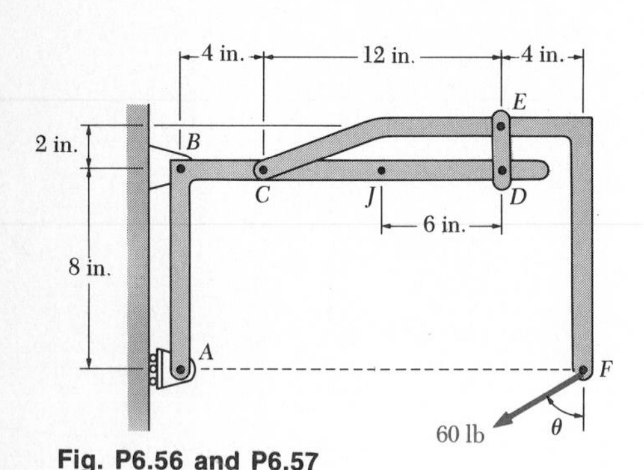

Fig. P6.56 and P6.57

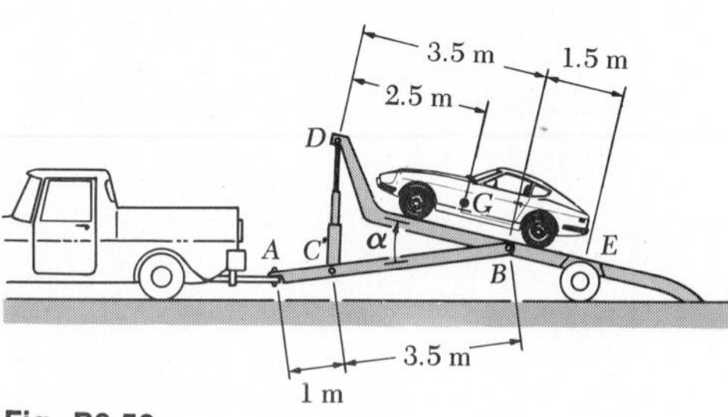

Fig. P6.58

6.58 The low-bed trailer shown is designed so that the rear end of the bed can be lowered to ground level in order to facilitate the loading of equipment or wrecked vehicles. A 1400-kg vehicle has been hauled to the position shown by a winch; the trailer is then returned to a traveling position where $\alpha = 0$ and both *AB* and *BE* are horizontal. Considering only the weight of the disabled automobile, determine the force which must be exerted by the hydraulic cylinder to maintain a position with $\alpha = 0$.

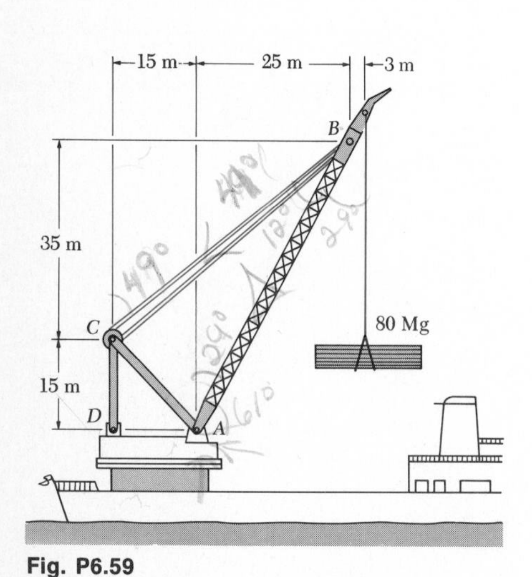

Fig. P6.59

6.59 For the marine crane shown, which is used in offshore drilling operations, determine (*a*) the force in link *CD*, (*b*) the force in the brace *AC*, (*c*) the force exerted at *A* on the boom *AB*.

6.60 Knowing that $P = 90$ lb and $Q = 60$ lb, determine the components of all forces acting on member *BCDE* of the assembly shown.

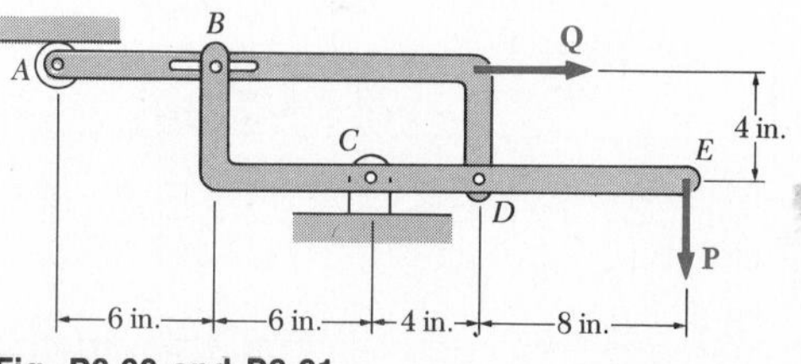

Fig. P6.60 and P6.61

6.61 Knowing that $P = 60$ lb and $Q = 90$ lb, determine the components of all forces acting on member *BCDE* of the assembly shown.

6.62 Determine the components of the reactions at D and E if the frame is loaded by a clockwise couple of magnitude 150 N · m applied (*a*) at point A, (*b*) at point B.

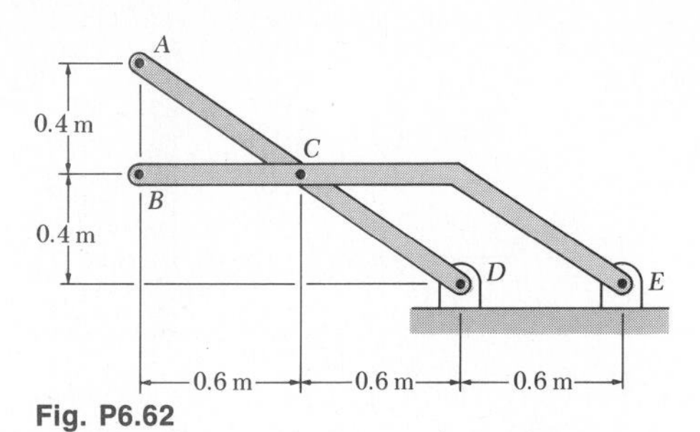

Fig. P6.62

6.63 Determine the components of the reactions at A and B, (*a*) if the 500-N load is applied as shown, (*b*) if the 500-N load is moved along its line of action and is applied at point F.

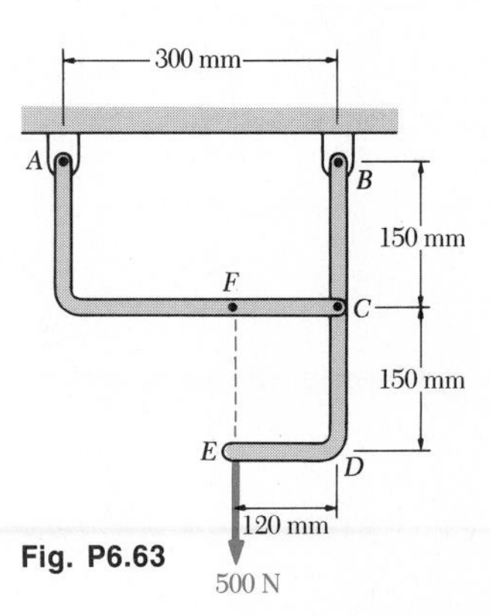

Fig. P6.63

6.64 Determine all the forces exerted on member AI if the frame is loaded by a clockwise couple of magnitude 180 lb · ft applied (*a*) at point D, (*b*) at point E.

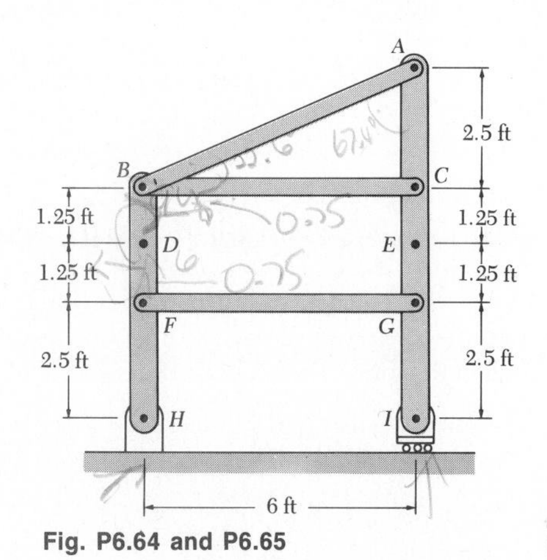

Fig. P6.64 and P6.65

6.65 Determine all the forces exerted on member AI if the frame is loaded by a 48-lb force directed horizontally to the right and applied (*a*) at point D, (*b*) at point E.

6.66 The hydraulic cylinder CF, which partially controls the position of rod DE, has been locked in the position shown. Knowing that $\theta = 60°$, determine (*a*) the force $\mathbf{P}$ for which the tension in link AB is 410 N, (*b*) the corresponding force exerted on member BCD at point C.

6.67 The hydraulic cylinder CF, which partially controls the position of rod DE, has been locked in the position shown. Knowing that $P = 400$ N and $\theta = 75°$, determine (*a*) the force in link AB, (*b*) the force exerted on member BCD at point C.

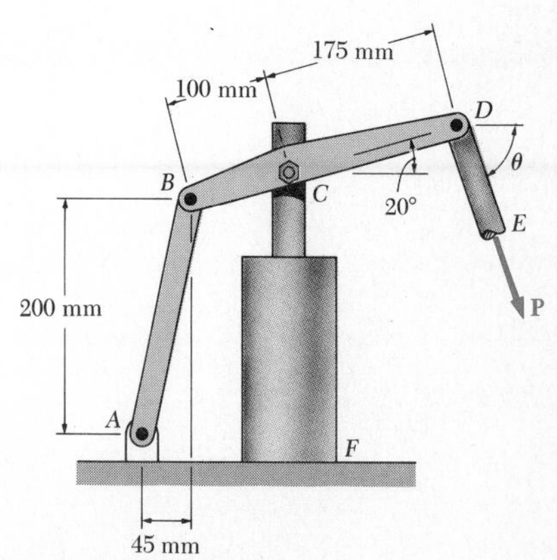

Fig. P6.66 and P6.67

6.68 Knowing that the pulley has a radius of 75 mm, determine the components of the reactions at A and B.

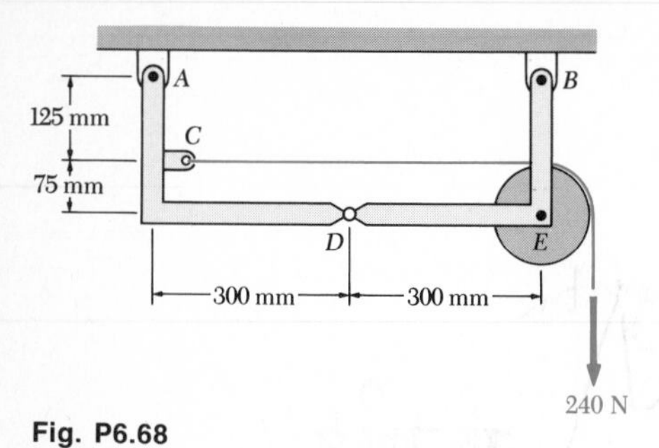

Fig. P6.68

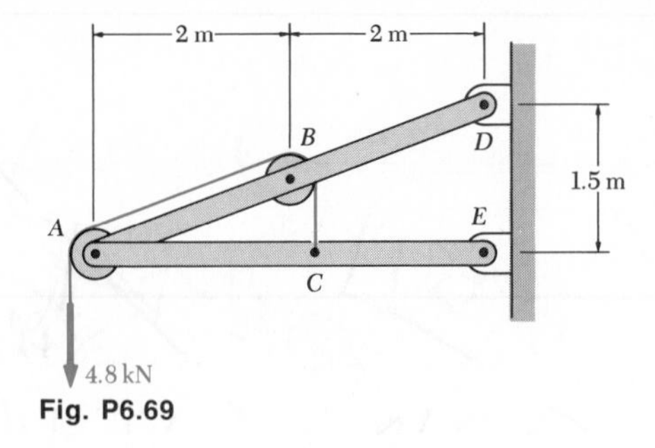

Fig. P6.69

6.69 Knowing that each pulley has a radius of 250 mm, determine the components of the reactions at D and E.

6.70 A 2-ft-diameter pipe is supported every 16 ft by a small frame; a typical frame is shown. Knowing that the combined weight of the pipe and its contents is 300 lb/ft and neglecting the effect of friction, determine the components (*a*) of the reaction at E, (*b*) of the force exerted at C on member CDE.

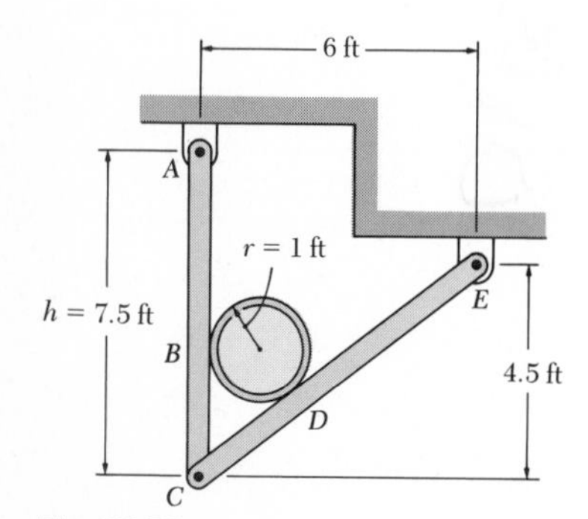

Fig. P6.70

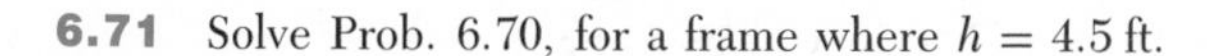

6.71 Solve Prob. 6.70, for a frame where $h = 4.5$ ft.

6.72 The cab and motor units of the front-end loader shown are connected by a vertical pin located 60 in. behind the cab wheels. The distance from C to D is 30 in. The center of gravity of the 50-kip motor unit is located at G_m, while the centers of gravity of the 18-kip cab and 16-kip load are located, respectively, at G_c and G_l. Knowing that the machine is at rest with its brakes released, determine (*a*) the reactions at each of the four wheels, (*b*) the forces exerted on the motor unit at C and D.

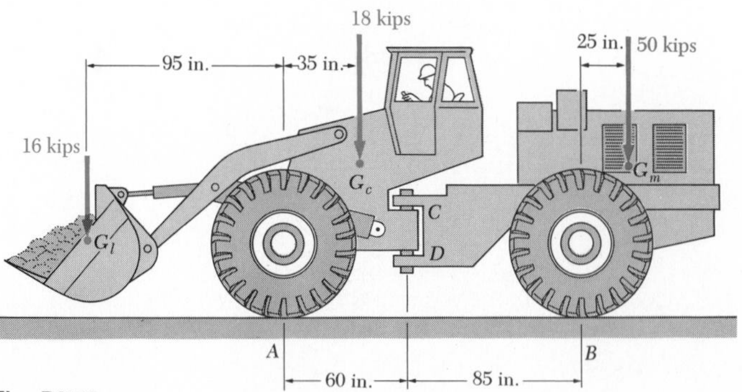

Fig. P6.72

6.73 Solve Prob. 6.72, assuming that the 16-kip load has been removed.

6.74 A 1000-kg trailer is attached to a 1250-kg automobile by a ball-and-socket trailer hitch at D. Determine (*a*) the reactions at each of the six wheels when the automobile and trailer are at rest, (*b*) the additional load on each of the automobile wheels due to the trailer.

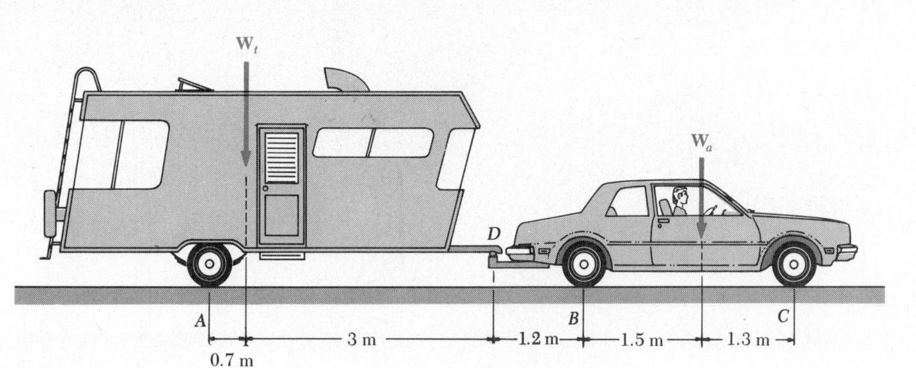

Fig. P6.74

6.75 In order to obtain a better weight distribution over the four wheels of the automobile of Prob. 6.74, a compensating hitch of the type shown is used to attach the trailer to the automobile. This hitch consists of two bar springs (only one is shown in the figure) which fit into bearings inside a support rigidly attached to the automobile. The springs are also connected by chains to the trailer frame, and specially designed hooks make it possible to place both chains under tension. (*a*) Determine the tension T required in each of the two chains if the additional load due to the trailer is to be evenly distributed over the four wheels of the automobile. (*b*) What are the corresponding reactions at each of the six wheels of the trailer-automobile combination?

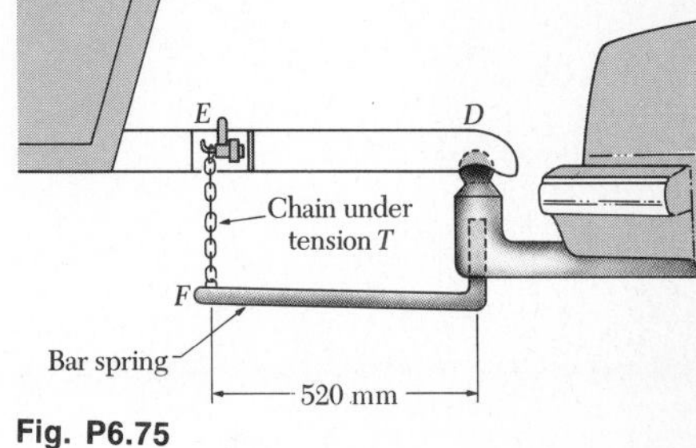

Fig. P6.75

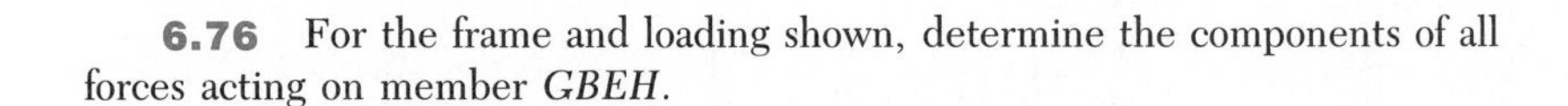

6.76 For the frame and loading shown, determine the components of all forces acting on member $GBEH$.

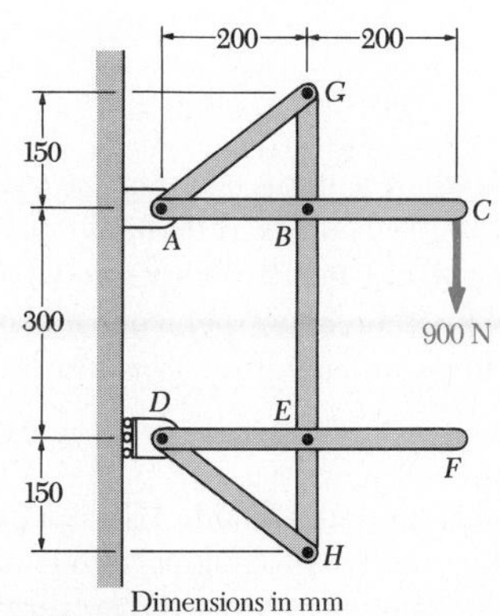

Fig. P6.76

6.77 Solve Prob. 6.76, assuming that the 900-N load is applied at F instead of C.

6.78 and 6.79 For the frame and loading shown, determine the components of all forces acting on member *ABD*.

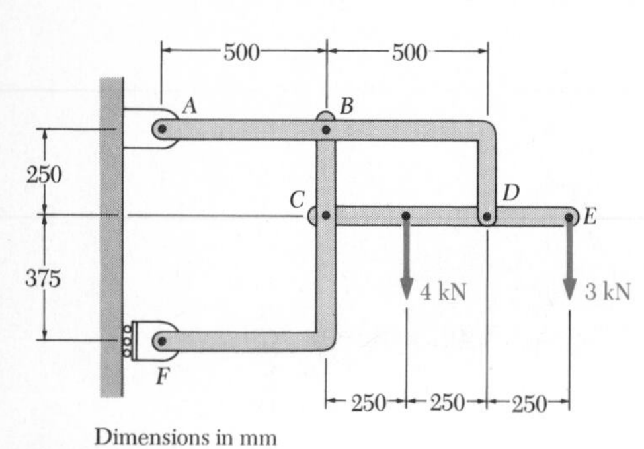

Fig. P6.78

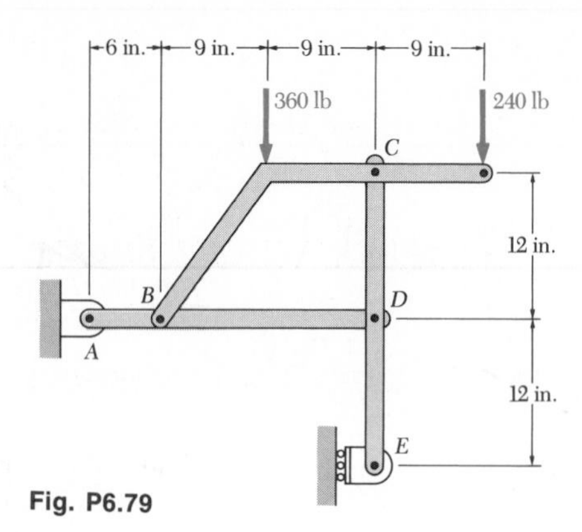

Fig. P6.79

6.80 Solve Prob. 6.79, assuming that the 360-lb load has been removed.

6.81 Solve Prob. 6.78, assuming that the 4-kN load has been removed.

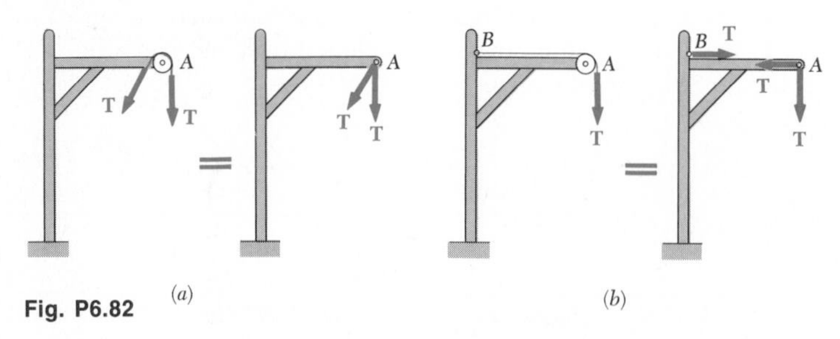

Fig. P6.82

6.82 (*a*) Show that when a frame supports a pulley at *A*, an equivalent loading of the frame and of each of its component parts may be obtained by removing the pulley and applying at *A* two forces equal and parallel to the forces of tension in the cable. (*b*) Further, show that if one end of the cable is attached to the frame at a point *B*, a force of magnitude equal to the tension should also be applied at *B*.

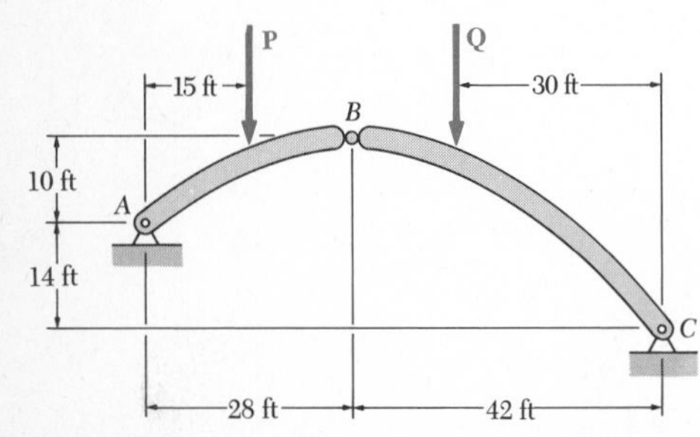

Fig. P6.83 and P6.84

6.83 The axis of the three-hinged arch *ABC* is a parabola with vertex at *B*. Knowing that $P = 109.2$ kips and $Q = 72.8$ kips, determine (*a*) the components of the reaction at *C*, (*b*) the components of the force exerted at *B* on segment *AB*.

6.84 The axis of the three-hinged arch *ABC* is a parabola with vertex at *B*. Knowing that $P = 72.8$ kips and $Q = 109.2$ kips, determine (*a*) the components of the reaction at *C*, (*b*) the components of the force exerted at *B* on segment *AB*.

6.85 Members ABC and CDE are pin-connected at C and supported by the four links AF, BG, DG, and EH. Determine the force in each link when a load **P** is applied at B as shown.

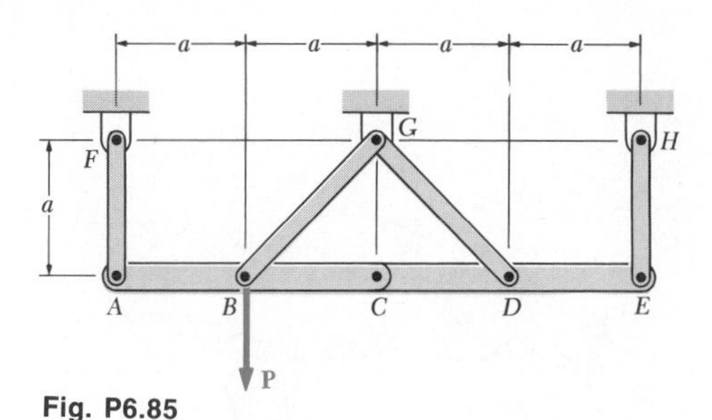

Fig. P6.85

6.86 Solve Prob. 6.85, assuming that the load **P** has been replaced by a clockwise couple of moment $\mathbf{M}_0$ applied to member ABC at B.

6.87 Members ABC and CDE are pin-connected at C and supported by the four links AF, BG, GD, and EH. For the loading shown, determine the force in each link.

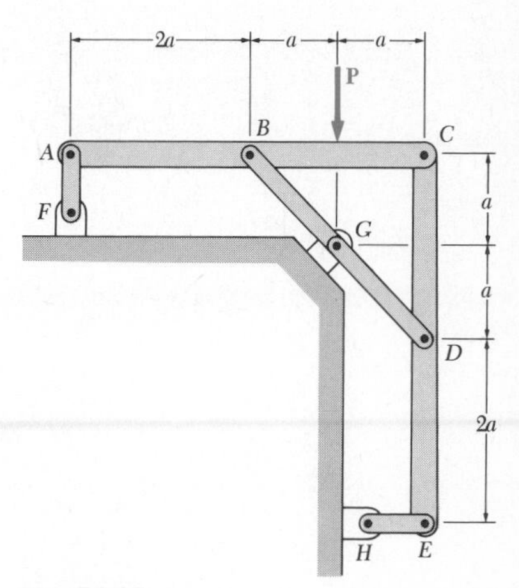

Fig. P6.87

6.88 Solve Prob. 6.87, assuming that the force **P** has been replaced by a clockwise couple of moment $\mathbf{M}_0$ applied at the same point.

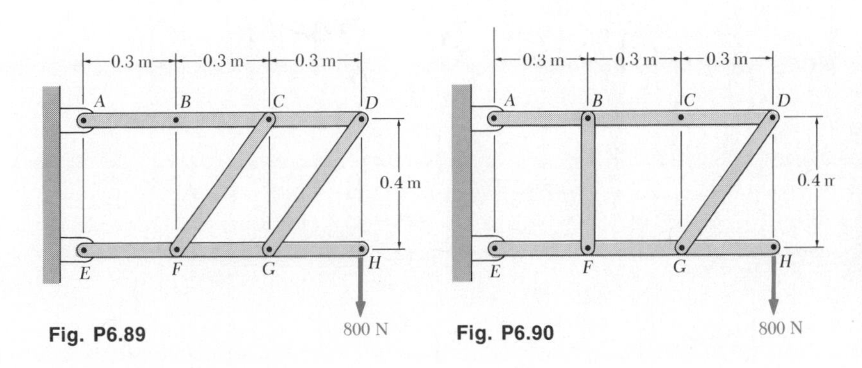

Fig. P6.89

Fig. P6.90

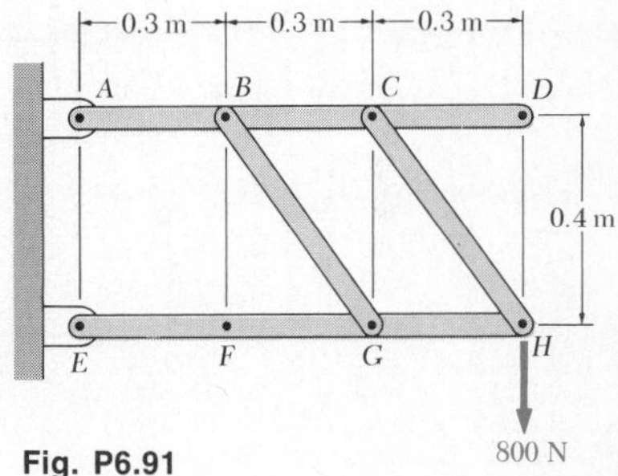

Fig. P6.91

6.89 through 6.91 The frame shown consists of members AD and EH, which are connected by two links. Determine the force in each link for the given loading.

6.92 A vertical load **P** of magnitude 750 N is applied to member ABC. Members ABC and DEF are parallel and are placed between two frictionless walls and connected by the link BE. Knowing that $a = 0.75$ m, determine the reactions at points A, C, D, and F.

6.93 The two parallel members ABC and DEF are placed between two frictionless walls and connected by a link BE. Determine the range of values of the distance a for which the load **P** can be supported.

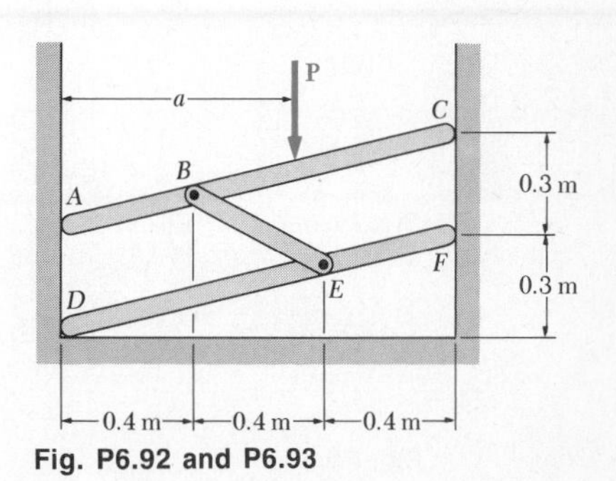

Fig. P6.92 and P6.93

6.94 Three beams are nailed together at their midpoints to form the support system shown. Assuming that only vertical forces are exerted at the connections, determine the vertical reactions at A, D, and F.

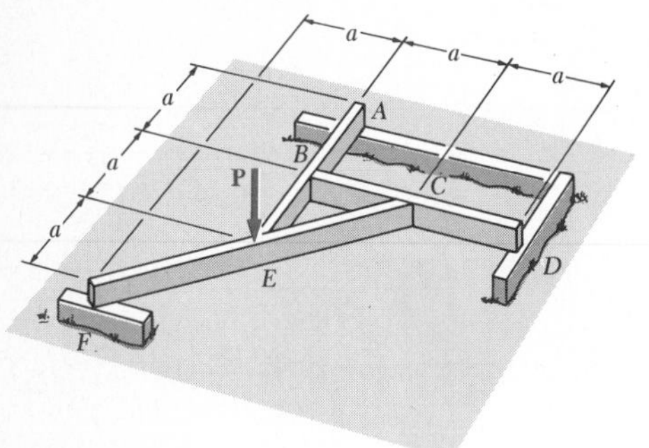

Fig. P6.94

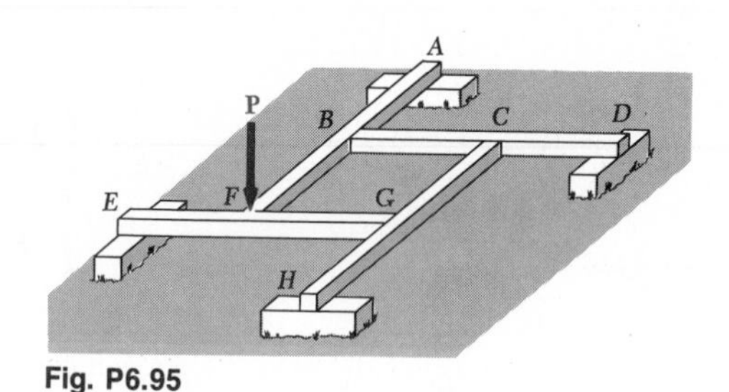

Fig. P6.95

6.95 Four beams, each of length $2a$, are nailed together at their midpoints to form the support system shown. Assuming that only vertical forces are exerted at the connections, determine the vertical reactions at A, D, E, and H.

6.96 and 6.97 Each of the frames shown consists of two L-shaped members connected by two rigid links. For each frame, determine the reactions at the supports and indicate whether the frame is rigid.

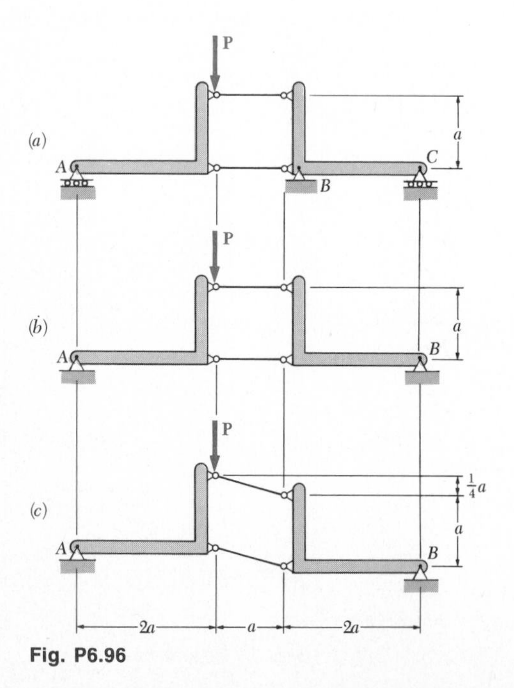

Fig. P6.96

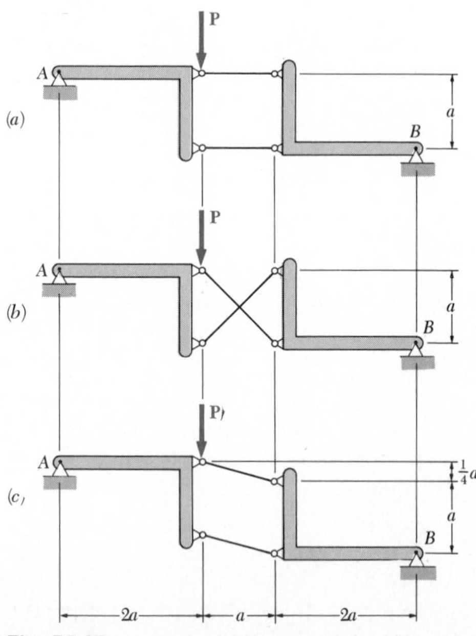

Fig. P6.97

6.12. Machines. Machines are structures designed to transmit and modify forces. Whether they are simple tools or include complicated mechanisms, their main purpose is to transform *input forces* into *output forces*. Consider, for example, a pair of cutting pliers used to cut a wire (Fig. 6.22*a*). If we apply two equal and opposite forces **P** and $-$**P** on their handles, they will exert two equal and opposite forces **Q** and $-$**Q** on the wire (Fig. 6.22*b*).

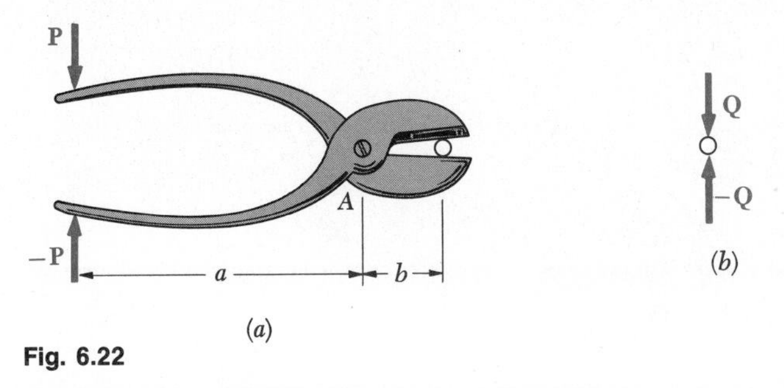

Fig. 6.22

To determine the magnitude Q of the output forces when the magnitude P of the input forces is known (or, conversely, to determine P when Q is known), we draw a free-body diagram of the pliers *alone*, showing the input forces **P** and $-$**P** and the *reactions* $-$**Q** and **Q** that the wire exerts on the pliers (Fig. 6.23). However, since a pair of pliers forms a nonrigid

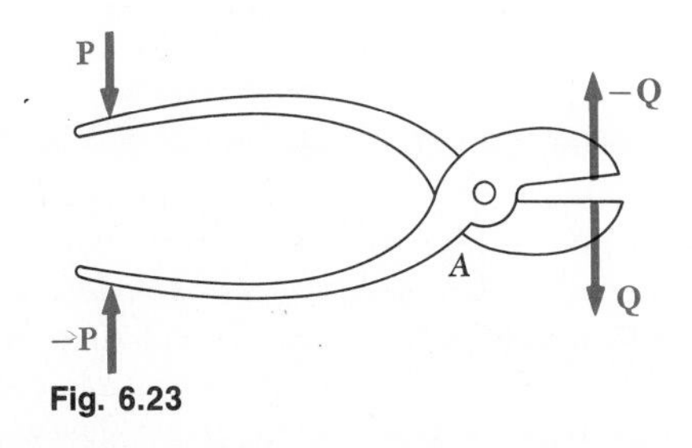

Fig. 6.23

structure, we must use one of the component parts as a free body in order to determine the unknown forces. Considering Fig. 6.24*a*, for example, and taking moments about A, we obtain the relation $Pa = Qb$, which defines the magnitude Q in terms of P or P in terms of Q. The same free-body diagram may be used to determine the components of the internal force at A; we find $A_x = 0$ and $A_y = P + Q$.

In the case of more complicated machines, it generally will be necessary to use several free-body diagrams and, possibly, to solve simultaneous equations involving various internal forces. The free bodies should be chosen to include the input forces and the reactions to the output forces, and the total number of unknown force components involved should not exceed the number of available independent equations. While it is advisable to check whether the problem is determinate before attempting to solve it, there is no point in discussing the rigidity of a machine. A machine includes moving parts and thus must be nonrigid.

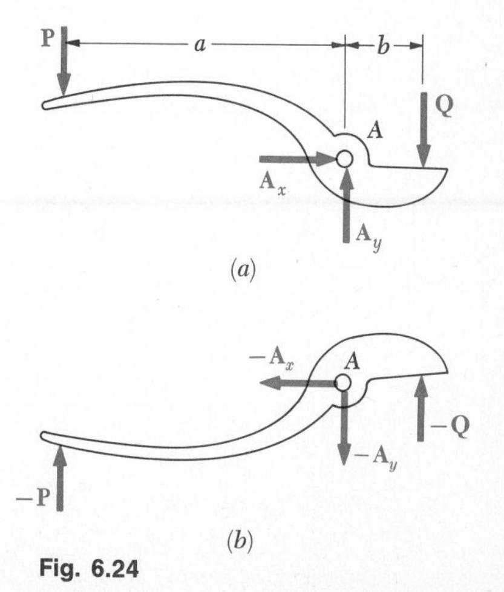

Fig. 6.24

SAMPLE PROBLEM 6.7

A hydraulic-lift table is used to raise a 1000-kg crate. It consists of a platform and two identical linkages on which hydraulic cylinders exert equal forces. (Only one linkage and one cylinder are shown.) Members *EDB* and *CG* are each of length $2a$, and member *AD* is pinned to the midpoint of *EDB*. If the crate is placed on the table, so that half of its weight is supported by the system shown, determine the force exerted by each cylinder in raising the crate for $\theta = 60°$, $a = 0.70$ m, and $L = 3.20$ m. Show that the result obtained is independent of the distance d.

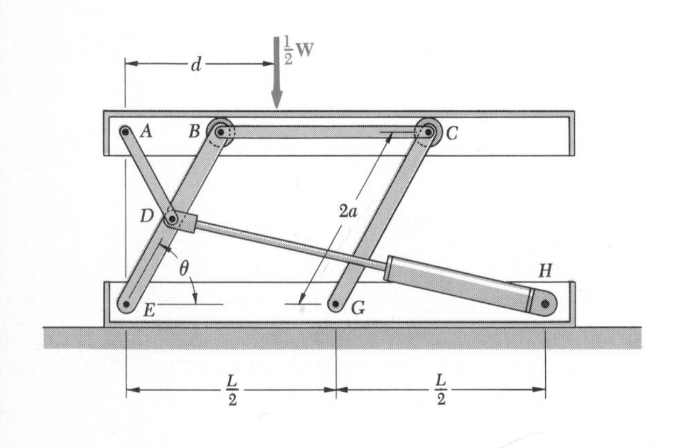

Solution. The machine considered consists of the platform and of the linkage, with an input force $\mathbf{F}_{DH}$ exerted by the cylinder and an output force equal and opposite to $\frac{1}{2}\mathbf{W}$. Since more than three unknowns would be involved, the entire mechanism is not used as a free body. The mechanism is dismembered and free-body diagrams are drawn for platform *ABC*, for roller *C*, and for member *EDB*. Since *AD*, *BC*, and *CG* are two-force members, their free-body diagrams have been omitted, and the forces they exert on the other parts of the mechanism have been drawn parallel to these members.

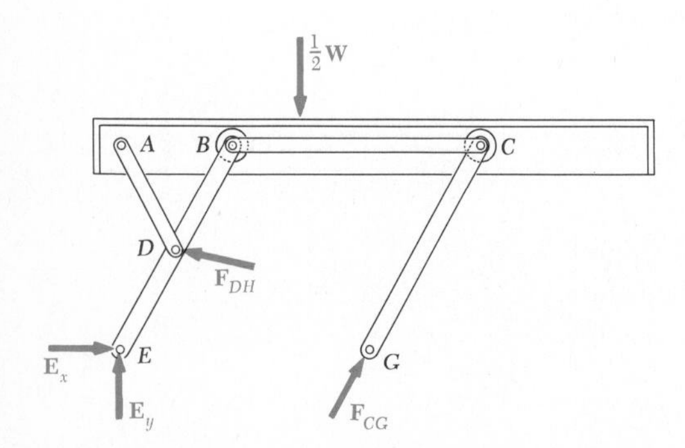

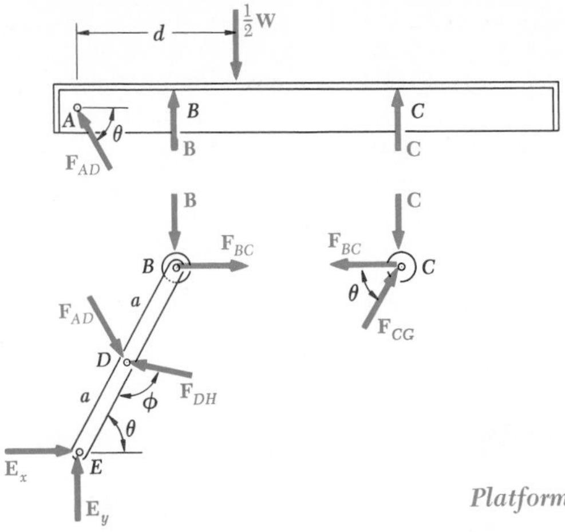

Platform ABC.

$$\xrightarrow{+}\Sigma F_x = 0: \qquad -F_{AD}\cos\theta = 0 \qquad F_{AD} = 0$$

$$+\uparrow\Sigma F_y = 0: \qquad B + C - \tfrac{1}{2}W = 0 \qquad B + C = \tfrac{1}{2}W \qquad (1)$$

Roller C. We draw a force triangle and obtain $F_{BC} = C\cot\theta$.

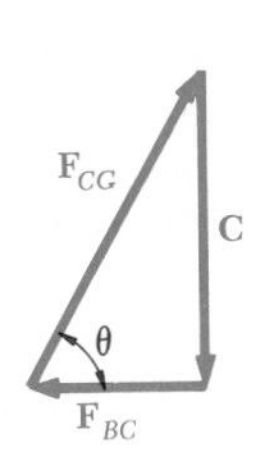

Member EDB. Recalling that $F_{AD} = 0$,

$$+\circlearrowleft\Sigma M_E = 0: \qquad F_{DH}\cos(\phi - 90°)a - B(2a\cos\theta) - F_{BC}(2a\sin\theta) = 0$$

$$F_{DH}a\sin\phi - B(2a\cos\theta) - (C\cot\theta)(2a\sin\theta) = 0$$

$$F_{DH}\sin\phi - 2(B + C)\cos\theta = 0$$

Recalling Eq. (1), we have

$$F_{DH} = W\frac{\cos\theta}{\sin\phi} \qquad (2)$$

and we observe that *the result obtained is independent of d.* ◄

Applying first the law of sines to triangle EDH, we write

$$\frac{\sin\phi}{EH} = \frac{\sin\theta}{DH} \qquad \sin\phi = \frac{EH}{DH}\sin\theta \qquad (3)$$

Using now the law of cosines, we have

$$(DH)^2 = a^2 + L^2 - 2aL\cos\theta$$

$$= (0.70)^2 + (3.20)^2 - 2(0.70)(3.20)\cos 60°$$

$$(DH)^2 = 8.49 \qquad DH = 2.91\text{ m}$$

We also note that

$$W = mg = (1000\text{ kg})(9.81\text{ m/s}^2) = 9810\text{ N} = 9.81\text{ kN}$$

Substituting for $\sin\phi$ from (3) into (2) and using the numerical data, we write

$$F_{DH} = W\frac{DH}{EH}\cot\theta = (9.81\text{ kN})\frac{2.91\text{ m}}{3.20\text{ m}}\cot 60°$$

$$F_{DH} = 5.15\text{ kN} \quad ◄$$

Problems

Fig. P6.98

6.98 The shear shown is used to cut and trim electronic-circuit-board laminates. For the position shown, determine (*a*) the vertical component of the force exerted on the shearing blade at *D*, (*b*) the reaction at *C*.

6.99 The control rod *CE* passes through a horizontal hole in the body of the toggle clamp shown. Determine (*a*) the force **Q** required to hold the clamp in equilibrium, (*b*) the corresponding force in link *BD*.

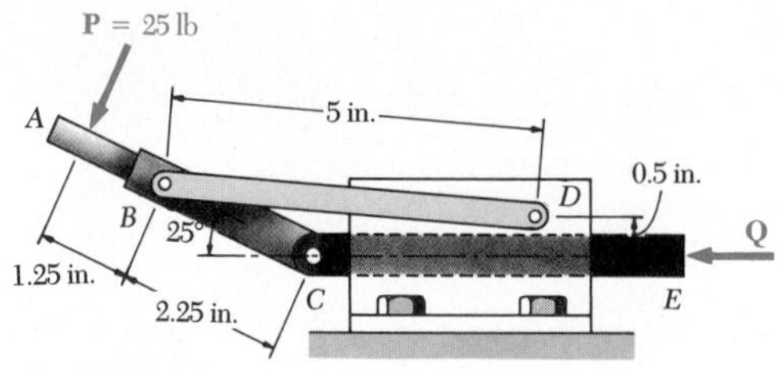

Fig. P6.99

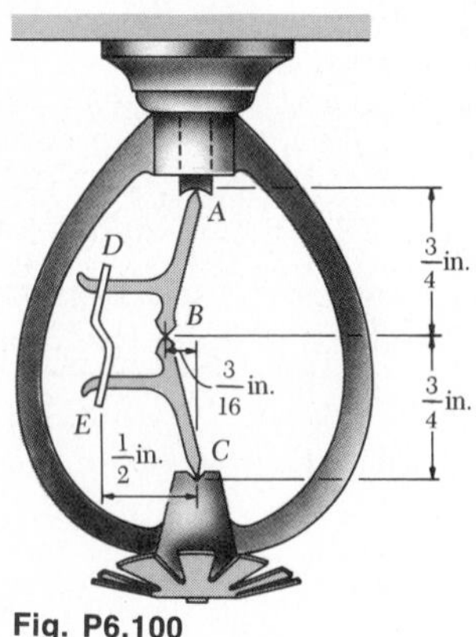

Fig. P6.100

6.100 Water pressure in the supply system exerts a downward force of 30 lb on the vertical plug at *A*. Determine the tension in the fusible link *DE* and the force exerted on member *BCE* at *B*.

6.101 Solve Prob. 6.98, assuming that the 400-N force is applied vertically downward at *A*.

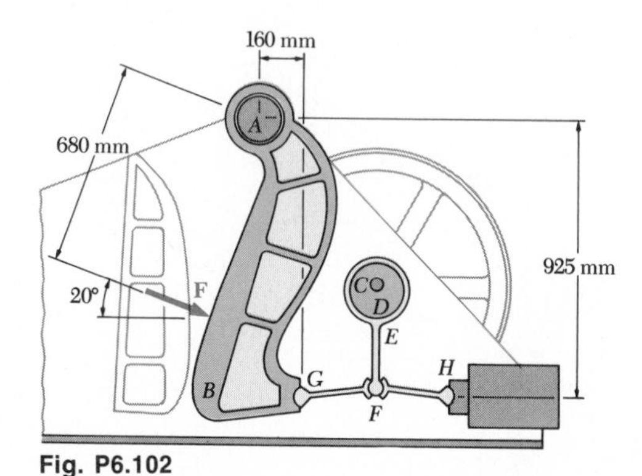

Fig. P6.102

6.102 The jaw crusher shown consists of a fixed jaw and a moving jaw *AB* which is attached to a large pin at *A*. Power is delivered to shaft *C* to which a circular disk *D* is attached eccentrically. As shaft *C* rotates, the connecting rod *EF* moves toggle *GFH*. For the position shown, the magnitude of the resultant force **F** on jaw *AB* is 360 kN and the arms *GF* and *FH* of the toggle each form an angle of 5° with the horizontal. Neglecting the effect of friction, determine the force in rod *EF*. (*Note:* At the end of each crushing cycle springs, which are not shown, return the jaw and toggle to their initial positions.)

6.103 A 9-m length of railroad track of mass 40 kg/m is lifted by the rail tongs shown. Determine the forces exerted at D and F on tong BDF.

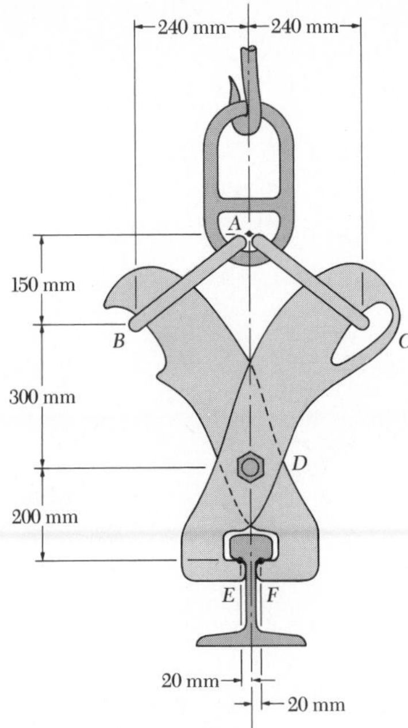

Fig. P6.103

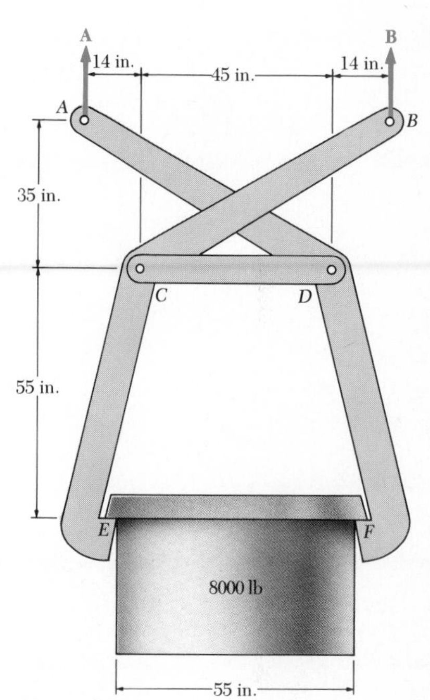

Fig. P6.104

6.104 A steel ingot weighing 8000 lb is lifted by a pair of tongs as shown. Determine the forces exerted at C and E on tong BCE.

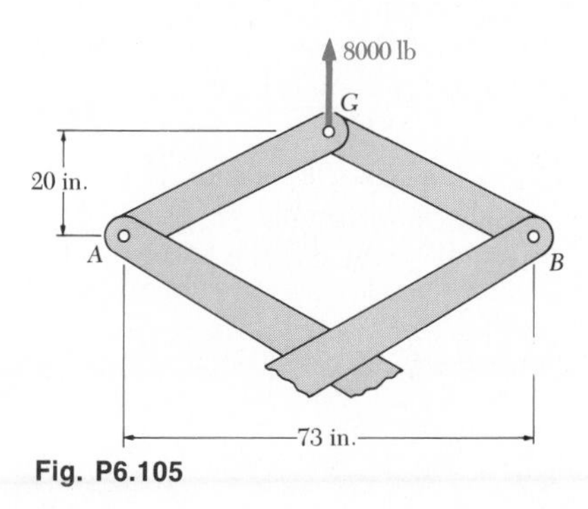

Fig. P6.105

6.105 If the toggle shown is added to the tongs of Prob. 6.104 and the load is lifted by applying a single force at G, determine the forces exerted at C and E on the tong BCE.

6.106 The action of the roll clamp is controlled by the two hydraulic cylinders shown. In order to hold firmly the paper roll, a vertical 1200-lb force is applied at the top of the roll by arm CAF. Knowing that the weight of the paper roll is 4800 lb, determine (a) the force exerted by each cylinder, (b) the force exerted at C on arm $BCEH$.

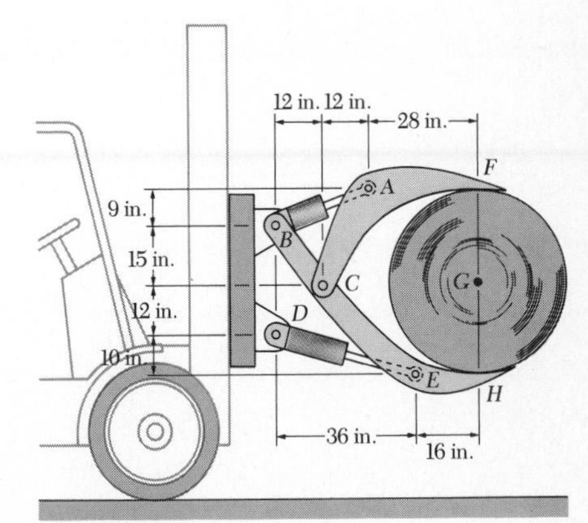

Fig. P6.106

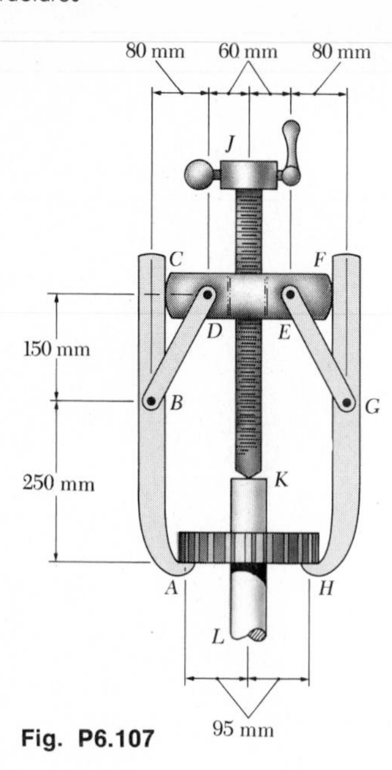

Fig. P6.107

6.107 The gear-pulling assembly shown consists of a crosshead CF, two grip arms ABC and FGH, two links BD and EG, and a threaded center rod JK. Knowing that the center rod JK must exert a 4800-N force on the vertical shaft KL in order to start the removal of the gear, determine all forces acting on grip arm ABC. Assume that the rounded ends of the crosshead are smooth and exert horizontal forces on the grip arms.

6.108 A force **P** of magnitude 2.4 kN is applied to the piston of the engine system shown. For each of the two positions shown, determine the couple **M** required to hold the system in equilibrium.

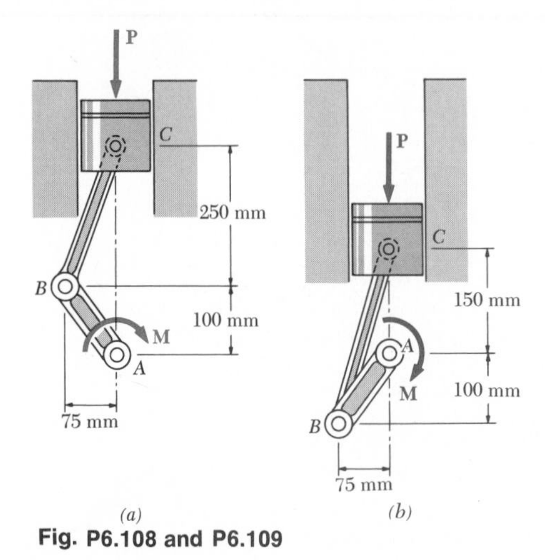

Fig. P6.108 and P6.109

6.109 A couple **M** of magnitude 315 N · m is applied to the crank of the engine system shown. For each of the two positions shown, determine the force **P** required to hold the system in equilibrium.

6.110 and 6.111 Two rods are connected by a frictionless collar B. Knowing that the magnitude of the couple $\mathbf{M}_A$ is 500 lb · in., determine (*a*) the couple $\mathbf{M}_C$ required for equilibrium, (*b*) the corresponding components of the reaction at C.

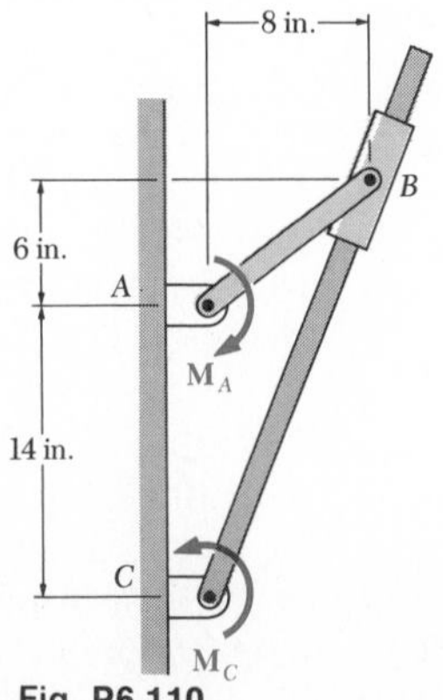

Fig. P6.110

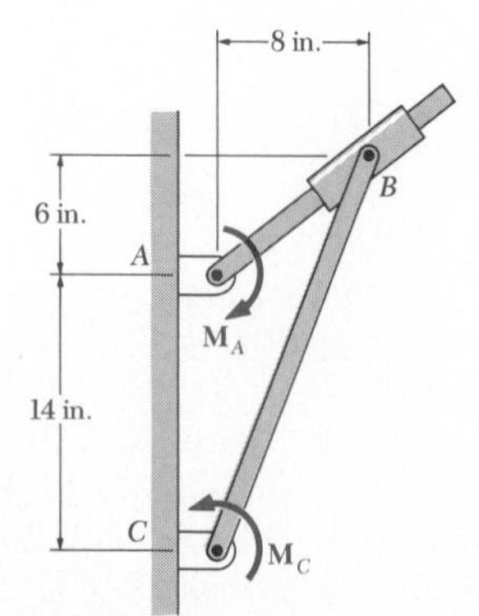

Fig. P6.111

6.112 The upper blade and lower handle of the compound-lever shears are pin-connected to the main element ABE at A and B, respectively, and to the short link CD at C and D, respectively. Determine the forces exerted on a twig when two 120-N forces are applied to the handles.

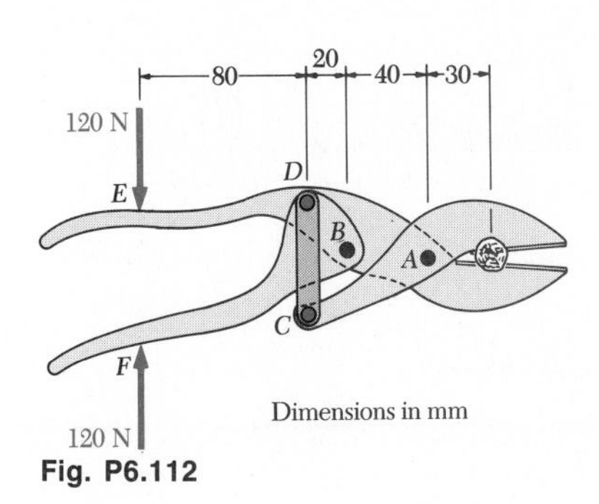

Fig. P6.112

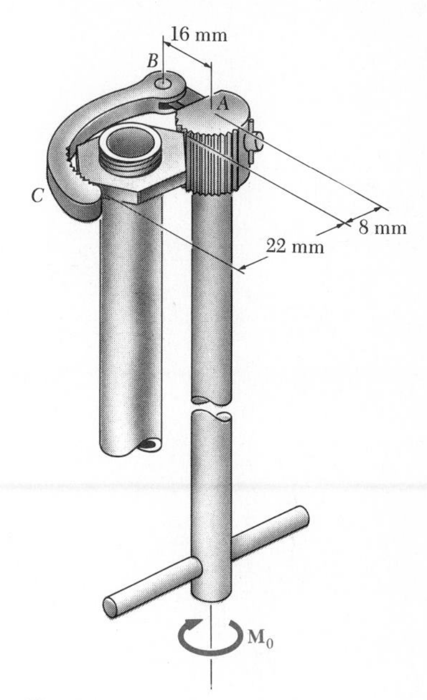

Fig. P6.113

6.113 The specialized plumbing wrench shown is used in confined areas (e.g., under a basin or sink). It consists essentially of a jaw BC pinned at B to a long rod. Knowing that the forces exerted on the nut are equivalent to a clockwise couple (when viewed from above) of magnitude 8.80 N · m, determine (*a*) the magnitude of the force exerted by pin B on jaw BC, (*b*) the couple $\mathbf{M}_0$ which is applied to the wrench.

6.114 The tool shown is used to crimp terminals onto electric wires. Knowing that a worker will apply forces of magnitude $P = 135$ N to the handles, determine the magnitude of the crimping forces which will be exerted on the terminal.

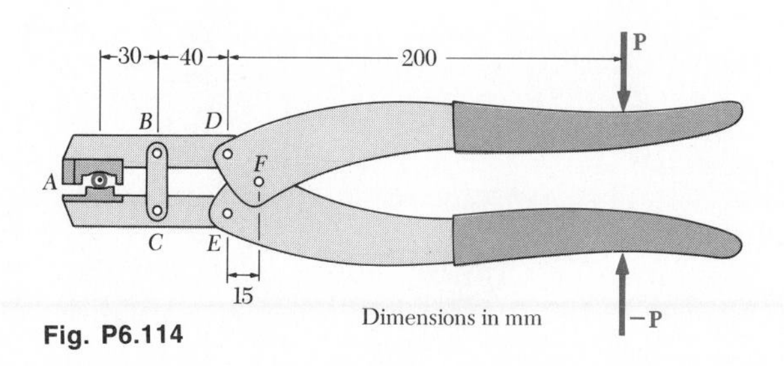

Fig. P6.114

6.115 The pliers shown are used to attach electrical connectors to flat cables which carry many separate wires. The clamping jaws remain parallel as the connectors are attached. Knowing that 380-N forces directed along line a-a are required to complete the attachment, determine the magnitude P of the forces which must be applied to the handles. Assume that pins A and D slide freely in slots cut in the jaws.

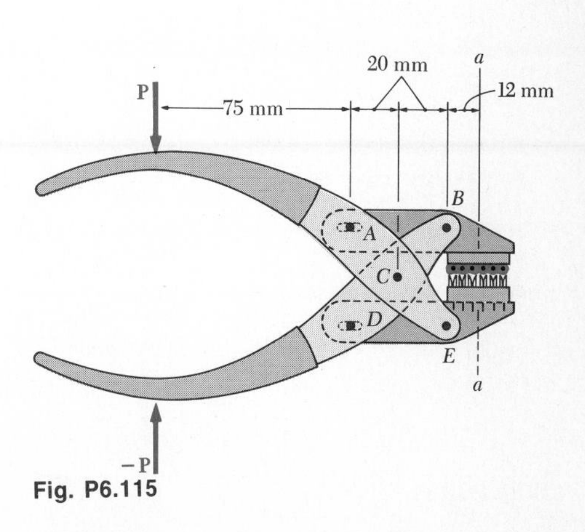

Fig. P6.115

6.116 A hand-operated hydraulic cylinder has been designed for use where space is severely limited. Determine the magnitude of the force exerted on the piston at D when two 90-lb forces are applied as shown.

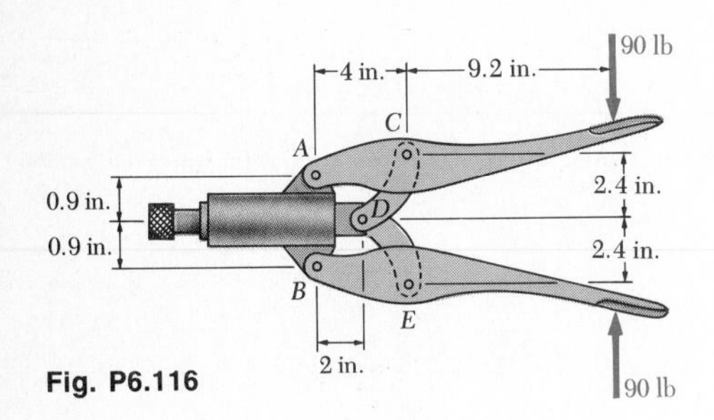

Fig. P6.116

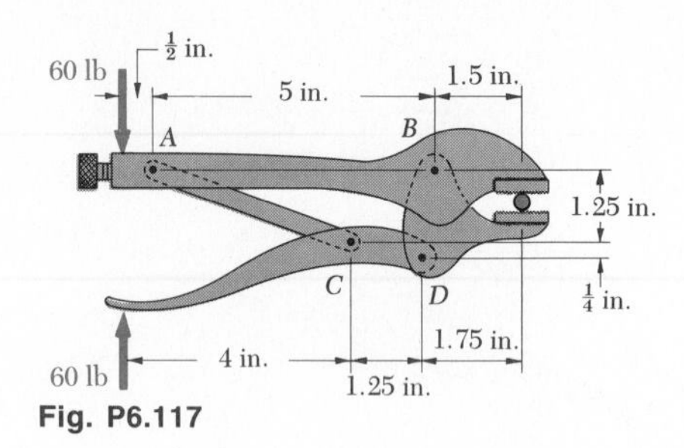

Fig. P6.117

6.117 Determine the magnitude of the gripping forces produced when two 60-lb forces are applied as shown.

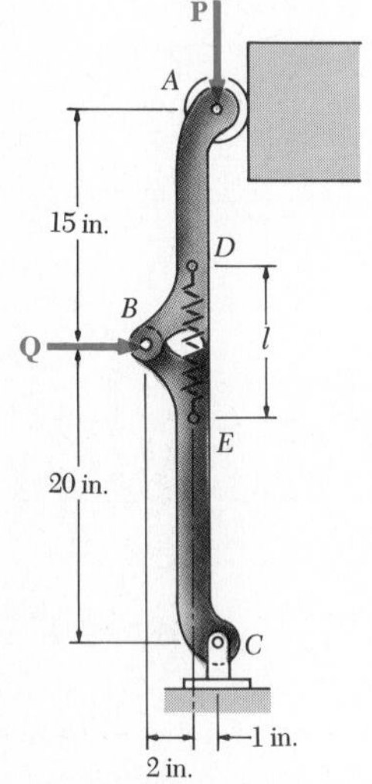

Fig. P6.118

6.118 Since the brace shown must remain in position even when the magnitude of **P** is very small, a single safety spring is attached at D and E. The spring DE has a constant of 50 lb/in. and an unstretched length of 7 in. Knowing that $l = 10$ in. and that the magnitude of **P** is 800 lb, determine the force **Q** required to release the brace.

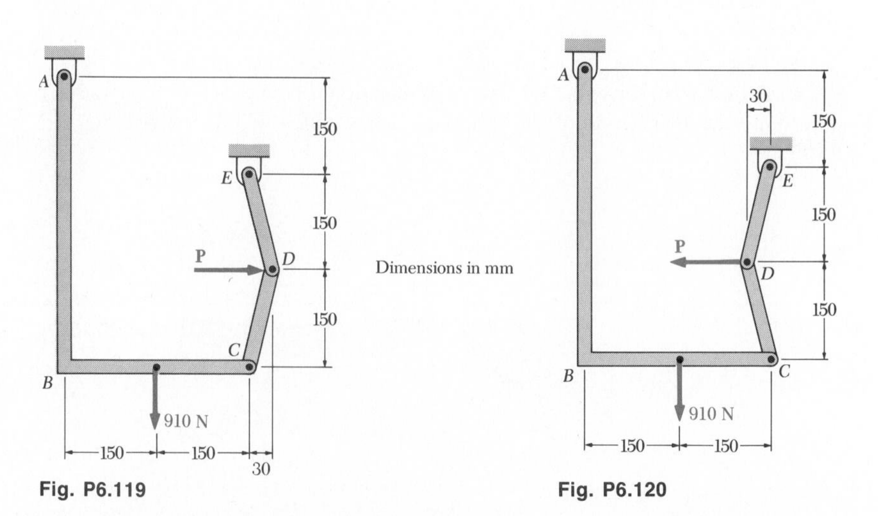

Fig. P6.119

Fig. P6.120

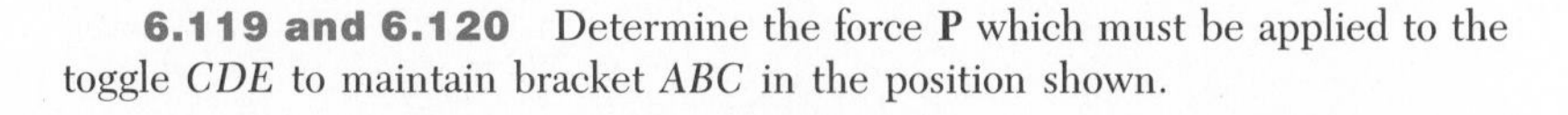
6.119 and 6.120 Determine the force **P** which must be applied to the toggle CDE to maintain bracket ABC in the position shown.

6.121 The shelf ABC is held horizontally by a self-locking brace which consists of two parts BDE and EDF hinged at E and bearing against each other at D. Knowing that the shelf is 16 in. wide and weighs 39.6 lb, determine the force **P** required to release the brace. (*Hint.* To release the brace, the forces of contact at D must be zero.)

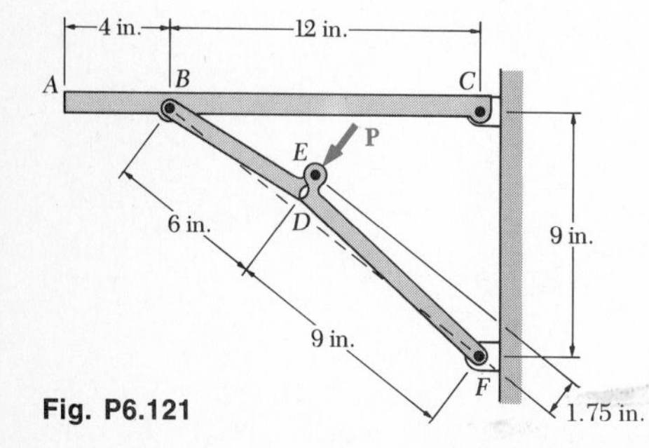

Fig. P6.121

6.122 The telescoping arm ABC is used to raise a worker to the elevation of overhead electric and telephone wires. For the extension shown, the center of gravity of the 1400-lb arm is located at point G. The worker, the bucket, and equipment attached to the bucket together weigh 450 lb and have a combined center of gravity at C. Determine the force exerted at B by the single hydraulic cylinder used when $\theta = 35°$.

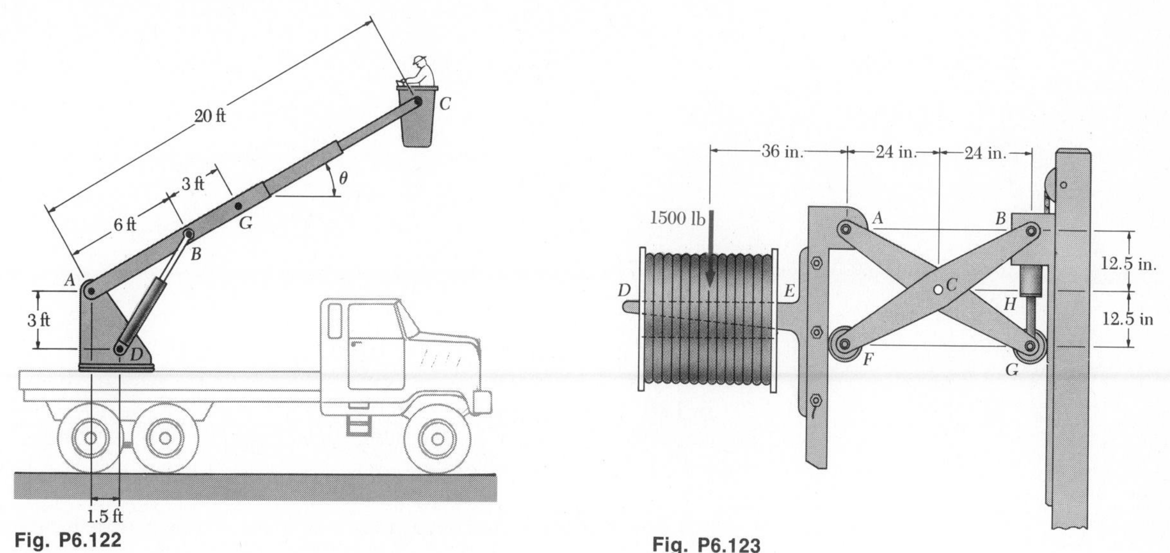

Fig. P6.122

Fig. P6.123

6.123 The position of the spindle DE of a lift truck is partially controlled by two identical linkage-and-hydraulic-cylinder systems, only one of which is shown. A 3000-lb spool of electric cable is held by the spindle in the position shown. Knowing that the load supported by the one system shown is 1500 lb, determine (*a*) the force exerted by the hydraulic cylinder on point G, (*b*) the components of the force exerted on member BCF at point C.

6.124 A 500-kg concrete slab is supported by a chain and sling attached to the bucket of the front-end loader shown. The action of the bucket is controlled by two identical mechanisms, only one of which is shown. Knowing that the mechanism shown supports one-half of the 500-kg slab, determine the force exerted (*a*) by cylinder CD, (*b*) by cylinder FH.

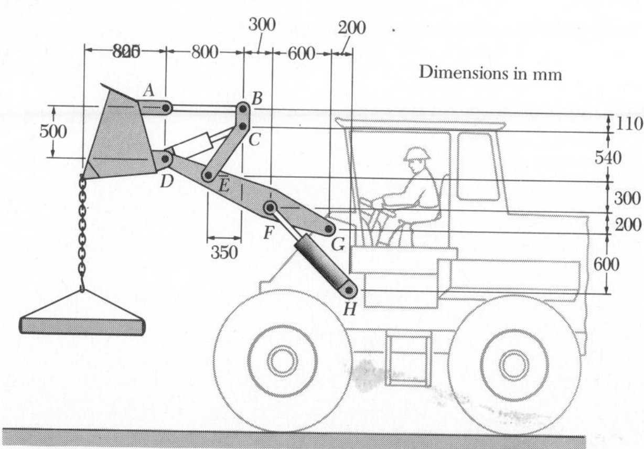

Fig. P6.124

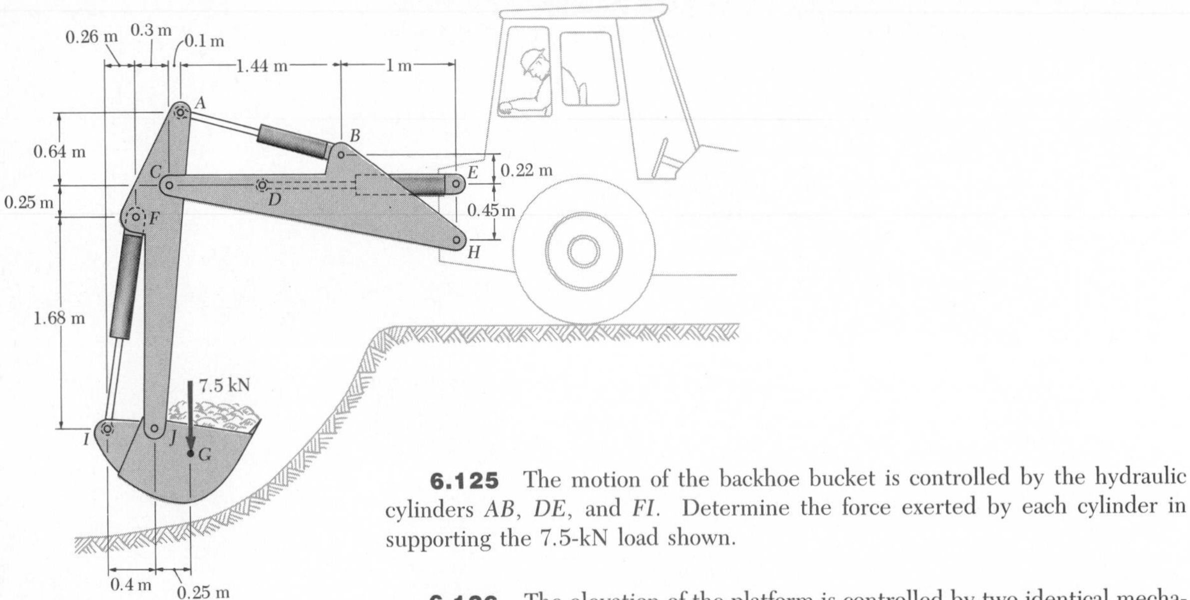

Fig. P6.125

6.125 The motion of the backhoe bucket is controlled by the hydraulic cylinders *AB*, *DE*, and *FI*. Determine the force exerted by each cylinder in supporting the 7.5-kN load shown.

6.126 The elevation of the platform is controlled by two identical mechanisms, only one of which is shown. A load of 1200 lb is applied to the mechanism shown. Knowing that the pin at *C* can transmit only a horizontal force, determine (*a*) the force in link *BE*, (*b*) the components of the force exerted by the hydraulic cylinder on pin *H*.

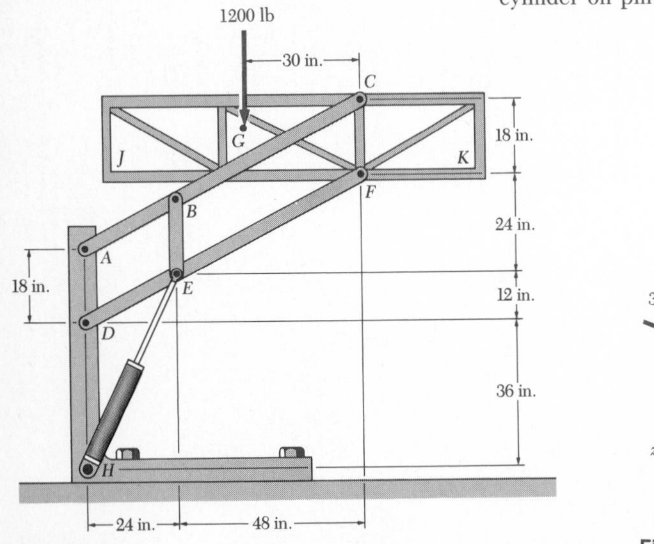

Fig. P6.126

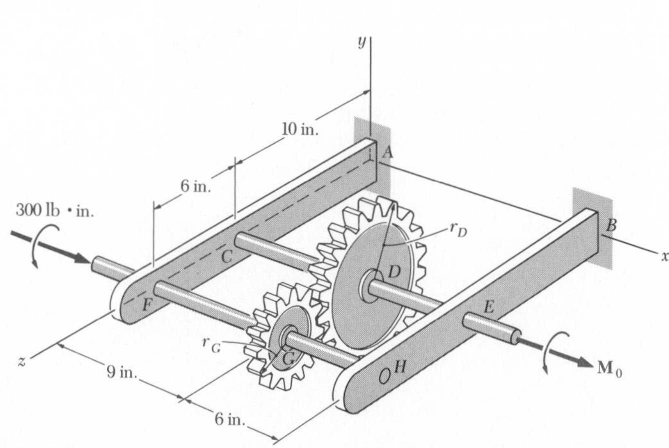

Fig. P6.127

6.127 The gears *D* and *G* are rigidly attached to shafts which are held by frictionless bearings. If $r_D = 4.5$ in. and $r_G = 1.5$ in., determine (*a*) the couple $\mathbf{M}_0$ which must be applied for equilibrium, (*b*) the reactions at *A* and *B*.

6.128 In the planetary-gear system shown, the radius of the central gear A is a, the radius of each planetary gear is b, and the radius of the outer gear E is $(a + 2b)$. In a particular gear system where $a = 50$ mm and $b = 30$ mm, a clockwise couple $\mathbf{M}_A$ is applied to gear A. If the system is to be in equilibrium, determine (a) the couple $\mathbf{M}_S$ which must be applied to the spider BCD, (b) the couple $\mathbf{M}_E$ which must be applied to the outer gear E.

6.129 In the planetary-gear system shown, the radius of the central gear A is a, the radius of each of the planetary gears is b, and the radius of the outer gear E is $(a + 2b)$. A clockwise couple of magnitude M_A is applied to the central gear A and a counterclockwise couple of magnitude $3.5M_A$ is applied to the spider BCD. If the system is to be in equilibrium, determine (a) the required ratio b/a, (b) the couple $\mathbf{M}_E$ which must be applied to the outer gear E.

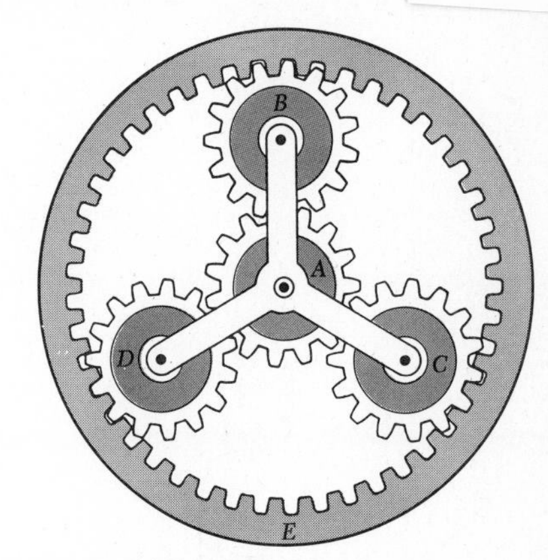

Fig. P6.128 and P6.129

***6.130** Two shafts AC and EG, which lie in the vertical yz plane, are connected by a universal joint at D. The bearings at B and E do not exert any axial force. A couple of magnitude 30 N · m (clockwise when viewed from the positive z axis) is applied to shaft AC at A. At a time when the arm of the crosspiece attached to shaft AC is vertical, determine (a) the magnitude of the couple $\mathbf{M}_G$ which must be applied to shaft EG to maintain equilibrium, (b) the reactions at B, C, and E. (*Hint.* The sum of the couples exerted on the crosspiece must be zero.)

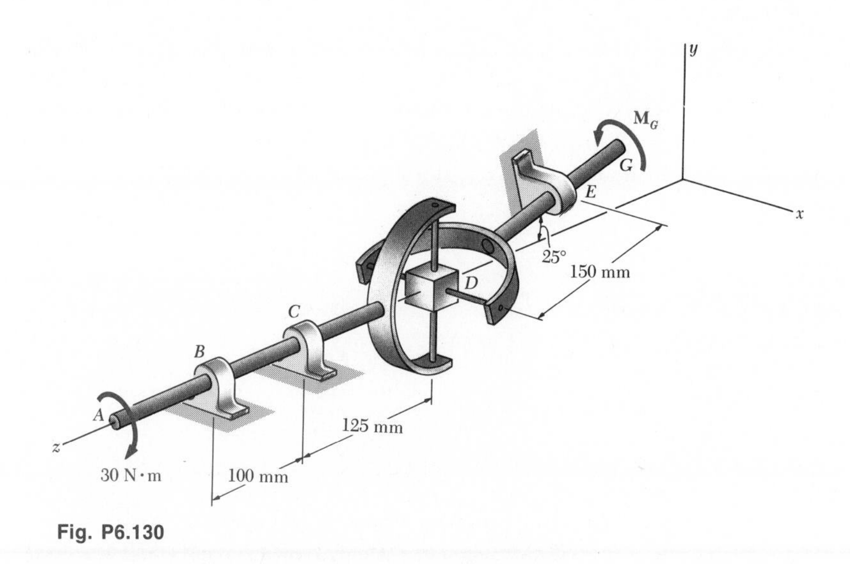

Fig. P6.130

***6.131** Solve Prob. 6.130, assuming that the arm of the crosspiece attached to shaft ABC is horizontal.

***6.132** The large mechanical tongs shown are used to grab and lift a thick 8000-kg steel slab HJ. Knowing that slipping does not occur between the tong grips and the slab at H and J, determine the components of all forces acting on member EFH. (*Hint.* Consider the symmetry of the tongs to establish relationships between the components of the force acting at E on EFH and the components of the force acting at D on CDF.)

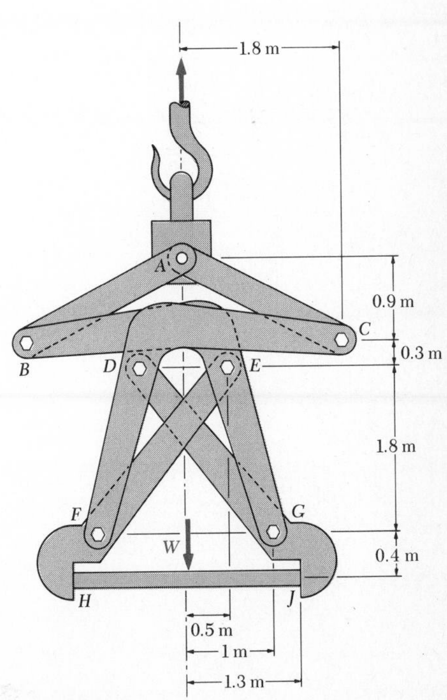

Fig. P6.132

***6.133** The mechanism shown is installed at some loading docks so that the dock edge C can be brought to the same elevation as the beds of various trucks. The weight of the platform and mechanism is roughly balanced by the force exerted by spring DE, while the final adjustment in elevation is controlled by the hydraulic cylinder GJ. Knowing that the weight of all moving parts is equivalent to the single force $\mathbf{W}$ of magnitude 1500 lb, determine the required tension in spring DE if the force exerted by the hydraulic cylinder is to be zero in the position shown.

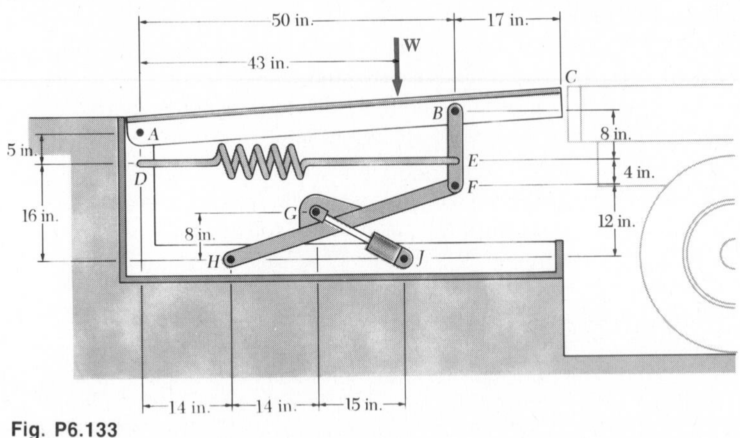

Fig. P6.133

***6.134** The weight of the platform and mechanism of Prob. 6.133 is exactly balanced by the force exerted by spring DE in the position shown and the force exerted by the hydraulic cylinder GJ is zero. A 200-lb force directed vertically downward is then applied at edge C. Determine the force which must be exerted by the hydraulic cylinder if the mechanism is to remain in the same position.

Review and Summary

In this chapter we learned to determine the *internal forces* holding together the various parts of a structure.

Analysis of trusses

The first half of the chapter was devoted to the analysis of *trusses*, i.e., to the analysis of structures consisting of *straight members connected at their extremities only*. The members being slender and unable to support lateral loads, all the loads must be applied at the joints; a truss may thus be assumed to consist of *pins and two-force members* [Sec. 6.2].

Simple trusses

A truss is said to be *rigid* if it is designed in such a way that it will not greatly deform and collapse under a small load. A triangular truss consisting of three members connected at three joints is clearly a rigid truss (Fig. 6.25*a*), and so will the truss obtained by adding two new members to the first one and connecting them at a new joint (Fig. 6.25*b*). Trusses obtained by repeating this procedure are called *simple trusses*. We may

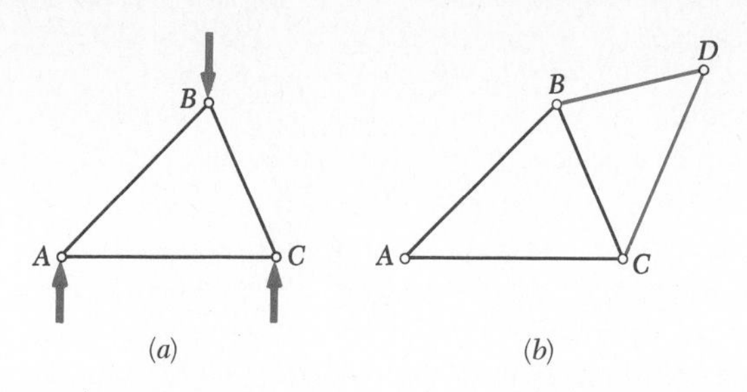

Fig. 6.25

check that in a simple truss the total number of members is $m = 2n - 3$, where n is the total number of joints [Sec. 6.3].

Method of joints

The forces in the various members of a simple truss may be determined by the *method of joints* [Sec. 6.4]. First, the reactions at the supports may be obtained by considering the entire truss as a free body. The free-body diagram of each pin is then drawn, showing the forces exerted on the pin by the members or supports it connects. Since the members are straight two-force members, the force exerted by a member on the pin is directed along that member, and only the magnitude of the force is unknown. It is always possible in the case of a simple truss to draw the free-body diagrams of the pins in such an order that only two unknown forces are included in each diagram. These forces may be obtained from the corresponding two equilibrium equations or—if only three forces are involved—from the corresponding force triangle. If the force exerted by a member on a pin is directed toward that pin, the member is in *compression;* if it is directed away from the pin, the member is in *tension* [Sample Prob. 6.1]. The analysis of a truss may sometimes be expedited by recognizing *joints under special loading conditions* [Sec. 6.5]. The method of joints may also be extended to the analysis of three-dimensional or *space trusses* [Sec. 6.6].

Method of sections

The *method of sections* is usually preferred to the method of joints when the force in only one member—or very few members—of a truss is desired [Sec. 6.7]. To determine the force in member BD of the truss of Fig. 6.26*a*, for example, we *pass a section* through members BD, BE, and CE, remove these members, and use the portion ABC of the truss as a free body (Fig. 6.26*b*). Writing $\Sigma M_E = 0$, we determine the magnitude of the force $\mathbf{F}_{BD}$, which represents the force in member BD. A positive sign indicates that the member is in *tension*, a negative sign that it is in *compression* [Sample Probs. 6.2 and 6.3].

The method of sections is particularly useful in the analysis of *compound trusses,* i.e., trusses which cannot be constructed from the basic triangular truss of Fig. 6.25*a*. These trusses may be obtained by rigidly connecting several simple trusses [Sec. 6.8]. If the component trusses have been properly connected (e.g., one pin and one link, or three nonconcurrent and nonparallel links) and if the resulting structure is properly supported (e.g., one pin and one roller), the compound truss is *statically determinate, rigid, and completely constrained.* The following

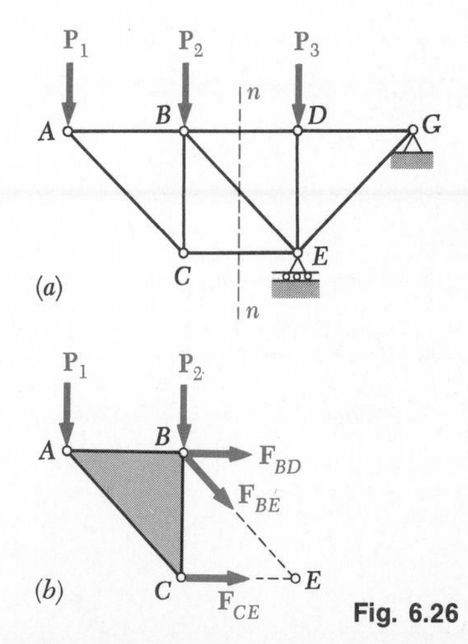

Fig. 6.26

necessary—but not sufficient—condition is then satisfied: $m + r = 2n$, where m is the number of members, r is the number of unknowns representing the reactions at the supports, and n is the number of joints.

Frames and machines

The second part of the chapter was devoted to the analysis of *frames and machines*. Frames and machines are structures which contain *multi-force members*, i.e., members acted upon by three or more forces. Frames are designed to support loads and are usually stationary, fully constrained structures. Machines are designed to transmit or modify forces and always contain moving parts [Sec. 6.9].

Analysis of a frame

To *analyze a frame*, we first consider the *entire frame as a free body* and write three equilibrium equations [Sec. 6.10]. If the frame remains rigid when detached from its supports, the reactions involve only three unknowns and may be determined from these equations [Sample Probs. 6.4 and 6.5]. On the other hand, if the frame ceases to be rigid when detached from its supports, the reactions involve more than three unknowns and cannot be completely determined [Sec. 6.11; Sample Prob. 6.6].

Multiforce members

We then *dismember the frame* and identify the various members as either two-force members or multiforce members; pins are assumed to form an integral part of one of the members they connect. We draw the free-body diagram of each of the multiforce members, noting that when two multiforce members are connected to the same two-force member, they are acted upon by that member with *equal and opposite forces of unknown magnitude but known direction*. When two multiforce members are connected by a pin, they exert on each other *equal and opposite forces of unknown direction*, which should be represented by *two unknown components*. The equilibrium equations obtained from the free-body diagrams of the multiforce members may then be solved for the various internal forces [Sample Probs. 6.4 and 6.5]. They may also be used to complete the determination of the reactions at the supports [Sample Prob. 6.6]. Actually, if the frame is *statically determinate and rigid*, the free-body diagrams of the multiforce members could provide as many equations as there are unknown forces (including the reactions) [Sec. 6.11]. However, as suggested above, it is advisable to first consider the free-body diagram of the entire frame to minimize the number of equations to be solved simultaneously.

Analysis of a machine

To *analyze a machine*, we dismember it and, following the same procedure as for a frame, draw the free-body diagram of each of the multi-force members. The corresponding equilibrium equations yield the *output forces* exerted by the machine in terms of the *input forces* applied to it, as well as the *internal forces* at the various connections [Sec. 6.12; Sample Prob. 6.7].

Review Problems

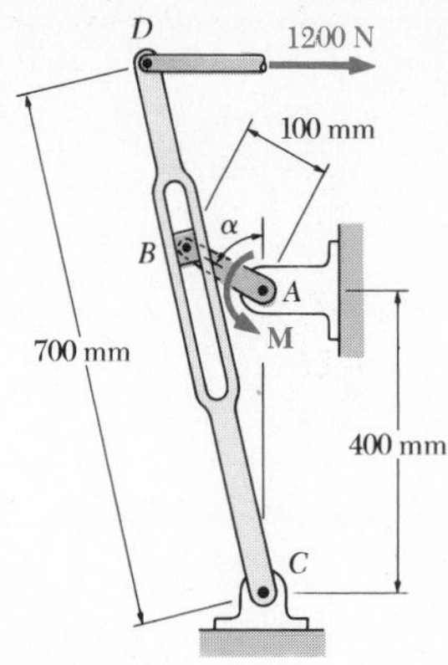

Fig. P6.135

6.135 The Whitworth mechanism shown is used to produce a quick-return motion of point D. The block at B is pinned to the crank AB and is free to slide in a slot cut in member CD. Determine the couple **M** which must be applied to the crank AB to hold the mechanism in equilibrium when (*a*) $\alpha = 0$, (*b*) $\alpha = 30°$.

6.136 Solve Prob. 6.135, when (*a*) $\alpha = 60°$, (*b*) $\alpha = 90°$.

6.137 Determine the reactions at the supports of the beams shown.

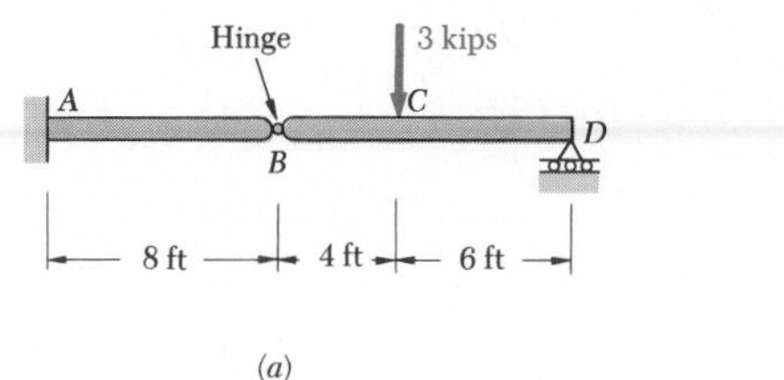

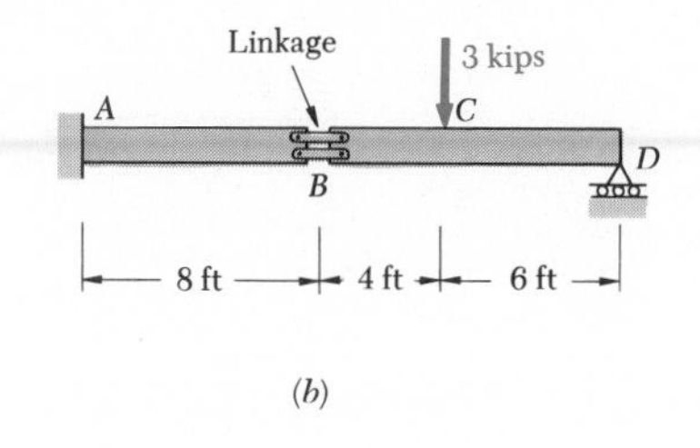

Fig. P6.137

6.138 The truck shown is used to facilitate work on overhead wires by raising a pair of buckets to the required elevation. Two tubular members AB and BC, each of length 12 ft, form the main supporting mechanism; the position of AB is controlled by means of the hydraulic cylinder DE. Two 2-ft-diameter sheaves, one on each side, are rigidly attached to member BC at B. Cables are fastened to the sheaves, pass over pulleys at H, and are fastened to a common movable block at F. The position of block F is controlled by a second hydraulic cylinder EF. Knowing that the workmen, buckets, and equipment attached to the buckets together weigh 700 lb and have a combined center of gravity at G, determine the force which must be exerted by each hydraulic cylinder to maintain the position shown. Neglect the weight of the mechanism.

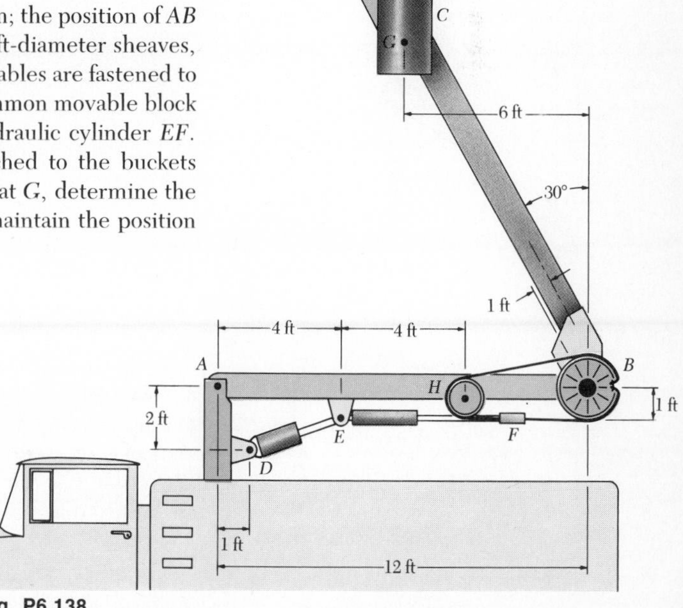

Fig. P6.138

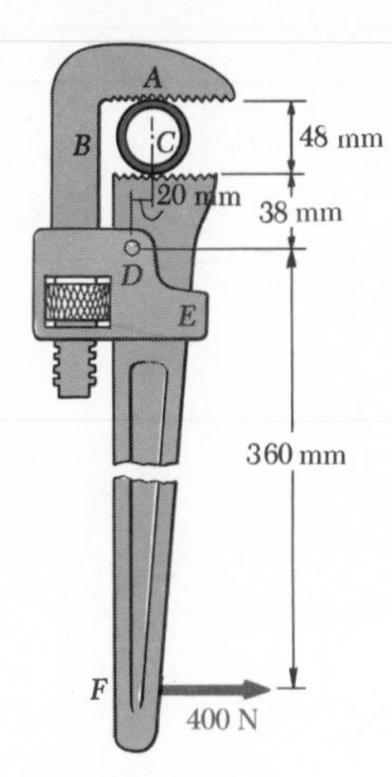

Fig. P6.139

6.139 A 48-mm-diameter pipe is gripped by the Stillson wrench shown. Portions AB and DE of the wrench are rigidly attached to each other and portion CF is connected by a pin at D. Assuming that no slipping occurs between the pipe and the wrench, determine the components of the forces exerted on the pipe at A and at C.

6.140 Knowing that the surfaces at A and D are frictionless, determine the components of all forces exerted on member ACD.

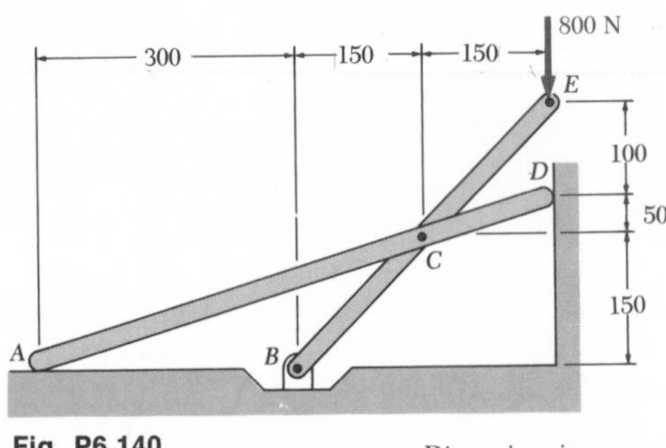

Fig. P6.140

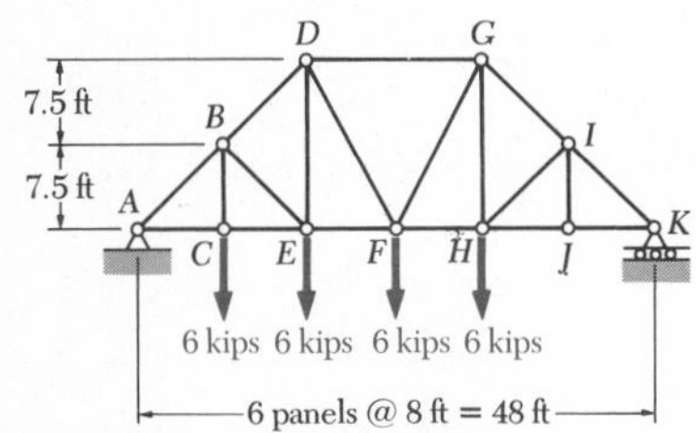

Fig. P6.141 and P6.142

6.141 Determine the force in members DE and EF of the truss shown.

6.142 Determine the force in members FG and FH of the truss shown.

6.143 Using the method of joints, determine the force in each member of the truss shown. State whether each member is in tension or compression.

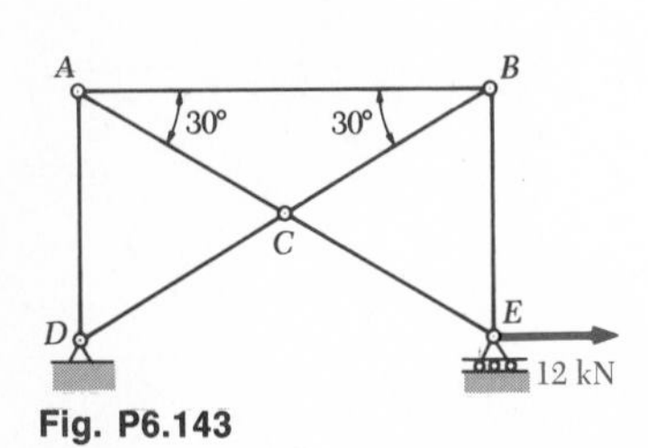

Fig. P6.143

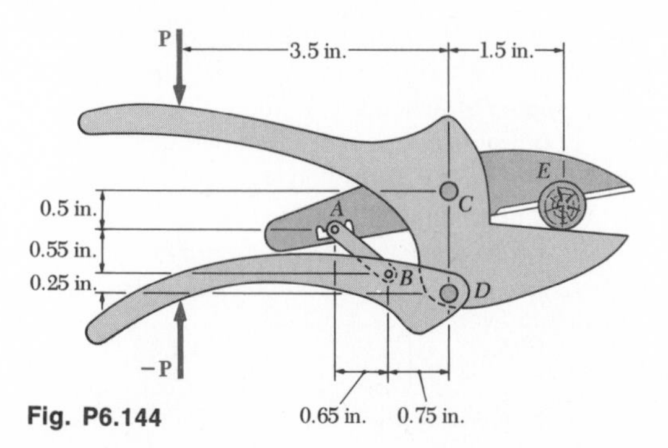

Fig. P6.144

6.144 The compound-lever pruning shears shown can be adjusted by placing pin A at various ratchet positions on blade ACE. Knowing that 292-lb vertical forces are required to complete the pruning of a twig, determine the magnitude P of the forces which must be applied to the handles when the shears are adjusted as shown.

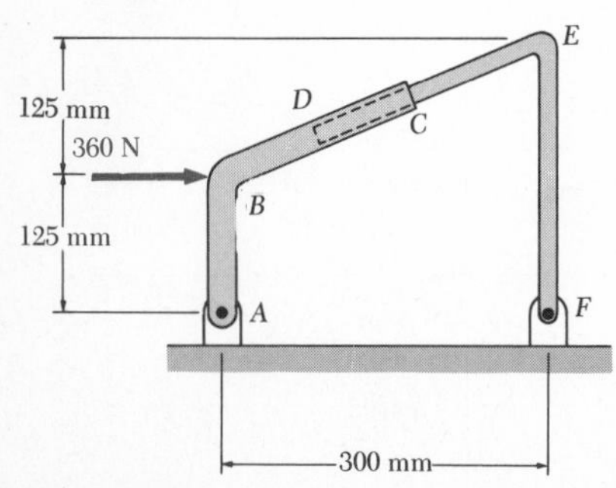

Fig. P6.145

6.145 The bent rod DEF fits into the bent pipe ABC as shown. Neglecting the effect of friction, determine the reactions at A and F due to the 360-N force applied at B.

6.146 A 400-kg block may be supported by a small frame in each of the four ways shown. The diameter of the pulley is 250 mm. For each case, determine (*a*) the force components and the couple representing the reaction at A, (*b*) the force exerted at D on the vertical member.

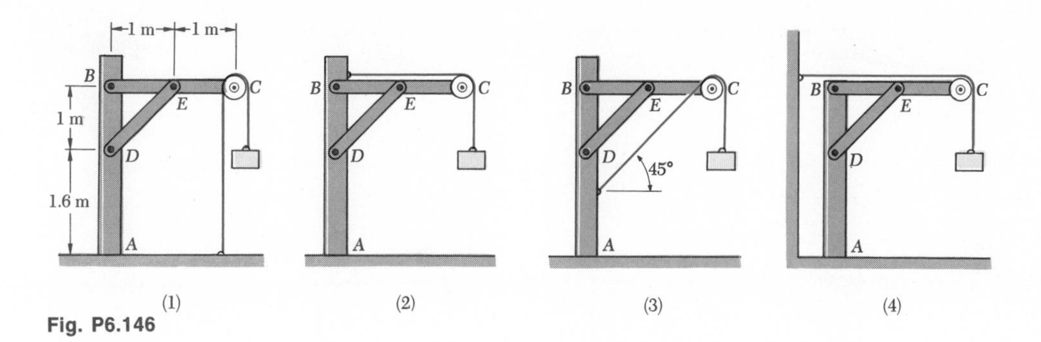

Fig. P6.146

The following problems are designed to be solved with a computer.

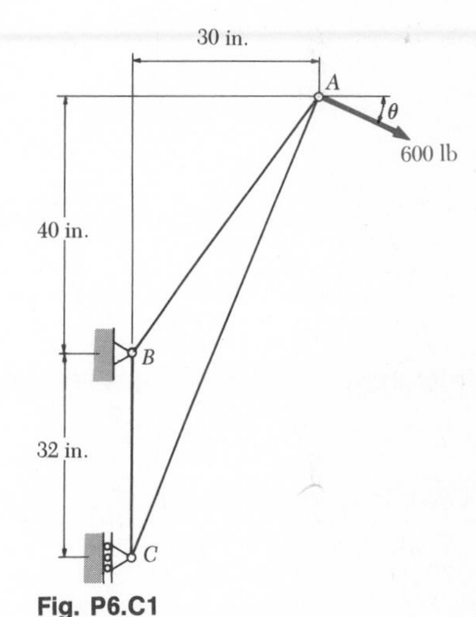

Fig. P6.C1

6.C1 For the truss and loading shown, write a computer program and calculate the force in each member for values of θ from 0 to 90° at 5° intervals.

6.C2 Several comparative designs of the truss of Sample Prob. 6.3 are to be made. Write a computer program and calculate the force in members FH, GH, and GI for values of h from 4 to 12 m at 1-m intervals.

6.C3 Various configurations of the frame shown are being investigated. (*a*) Write a computer program and calculate the force in links CF and DG for values of a from 0 to 0.3 m at 0.05-m intervals. (*b*) Explain what happens when $a = 0.15$ m.

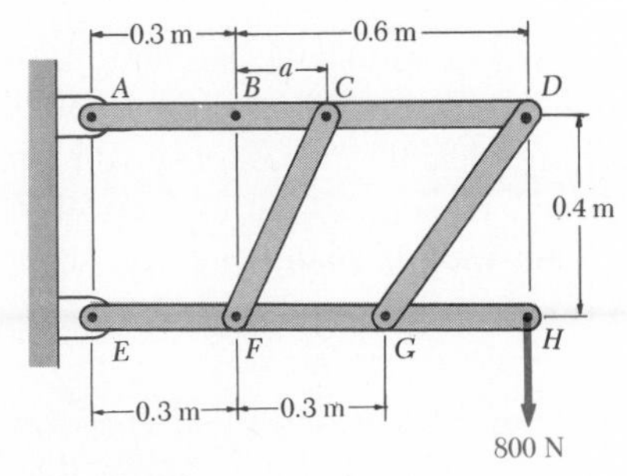

Fig. P6.C3

6.C4 Write a computer program which can be used to determine the couple **M** required to hold the system shown in equilibrium for values of θ from 0 to 180° at 15° intervals. For the case when $b = 100$ mm, $l = 350$ mm, and $P = 1.5$ kN, calculate (*a*) the magnitude M of the couple at 15° intervals, (*b*) the value of θ for which M is maximum and the corresponding value of M.

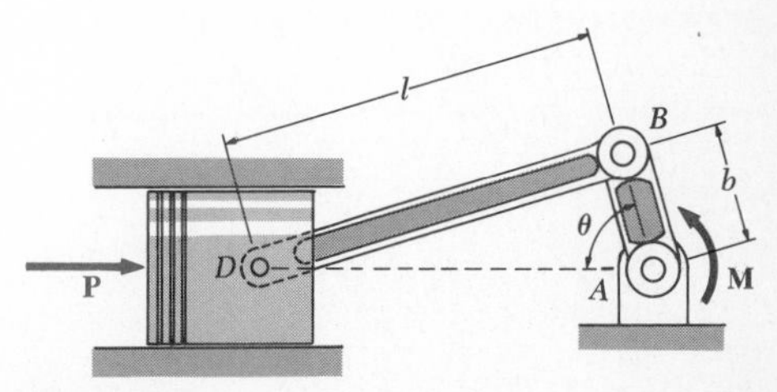

Fig. P6.C4

Forces in Beams and Cables

CHAPTER

7

* 7.1. Introduction.

In preceding chapters, two basic problems involving structures were considered: (1) the determination of the external forces acting on a structure (Chap. 4) and (2) the determination of the forces which hold together the various members forming a structure (Chap. 6). We shall now consider the problem of determining the internal forces which hold together the various parts of a given member.

We shall first analyze the internal forces in the members of a frame, such as the crane considered earlier in Chap. 6 (Secs. 6.1 and 6.10) and note that whereas the internal forces in a straight two-force member can produce only *tension* or *compression* in that member, the internal forces in any other type of member will usually produce *shear* and *bending* as well.

Most of this chapter will be devoted to the analysis of the internal forces in two important types of engineering structures, namely,

1. *Beams,* which are usually long, straight prismatic members designed to support loads applied at various points along the member.
2. *Cables,* which are flexible members capable of withstanding only tension, designed to support either concentrated or distributed loads. Cables are used in many engineering applications, such as suspension bridges and transmission lines.

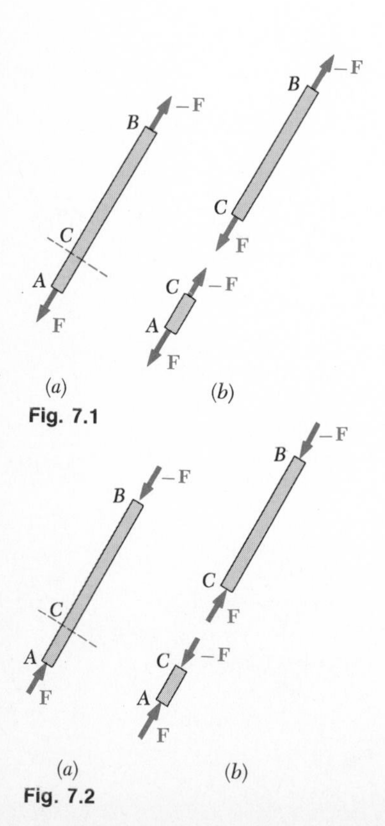

Fig. 7.1

Fig. 7.2

* 7.2. Internal Forces in Members.

We shall first consider a *straight two-force member AB* (Fig. 7.1*a*). From Sec. 4.6, we know that the forces **F** and −**F** acting at A and B, respectively, must be directed along AB in opposite sense and have the same magnitude F. Now, let us cut the member at C. To maintain the equilibrium of the free bodies AC and CB thus obtained, we must apply to AC a force −**F** equal and opposite to **F**, and to CB a force **F** equal and opposite to −**F** (Fig. 7.1*b*). These new forces are directed along AB in opposite sense and have the same magnitude F. Since the two parts AC and CB were in equilibrium before the member was cut, *internal forces* equivalent to these new forces must have

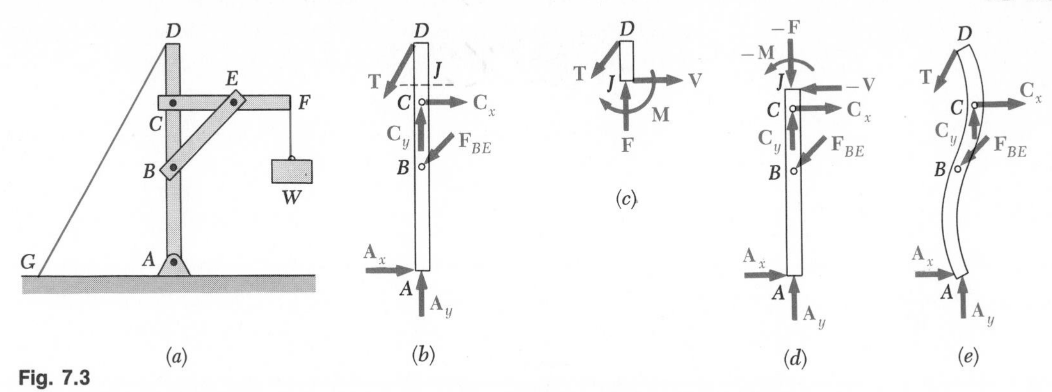

Fig. 7.3

existed in the member itself. We see that in the case of a straight two-force member, the internal forces acting on each part of the member are equivalent to an axial force. The magnitude F of this force does not depend upon the location of the section C and is referred to as the *force in member AB*. In the case considered, the member is in tension and will elongate under the action of the internal forces. In the case represented in Fig. 7.2, the member is in compression and will decrease in length under the action of the internal forces.

Next we shall consider a *multiforce member*. Take, for instance, member AD of the crane analyzed in Sec. 6.10. This crane is shown again in Fig. 7.3*a*, and the free-body diagram of member AD is drawn in Fig. 7.3*b*. We now cut member AD at J and draw a free-body diagram for each of the portions JD and AJ of the member (Fig. 7.3*c* and *d*). Considering the free body JD, we find that its equilibrium will be maintained if we apply at J a force $\mathbf{F}$ to balance the vertical component of $\mathbf{T}$, a force $\mathbf{V}$ to balance the horizontal component of $\mathbf{T}$, and a couple $\mathbf{M}$ to balance the moment of $\mathbf{T}$ about J. Again we conclude that internal forces must have existed at J before member AD was cut. The internal forces acting on the portion JD of member AD are equivalent to the force-couple system shown in Fig. 7.3*c*. According to Newton's third law, the internal forces acting on AJ must be equivalent to an equal and opposite force-couple system, as shown in Fig. 7.3*d*. It clearly appears that the action of the internal forces in member AD *is not limited to producing tension or compression* as in the case of straight two-force members; the internal forces *also produce shear and bending*. The force $\mathbf{F}$ is again in this case called an *axial force;* the force $\mathbf{V}$ is called a *shearing force;* and the moment $\mathbf{M}$ of the couple is known as the *bending moment at J*. We note that when determining internal forces in a member, we should clearly indicate on which portion of the member the forces are supposed to act. The deformation which will occur in member AD is sketched in Fig. 7.3*e*. The actual analysis of such a deformation is part of the study of mechanics of materials.

It should be noted that in a *two-force member which is not straight*, the internal forces are also equivalent to a force-couple system. This is shown in Fig. 7.4, where the two-force member ABC has been cut at D.

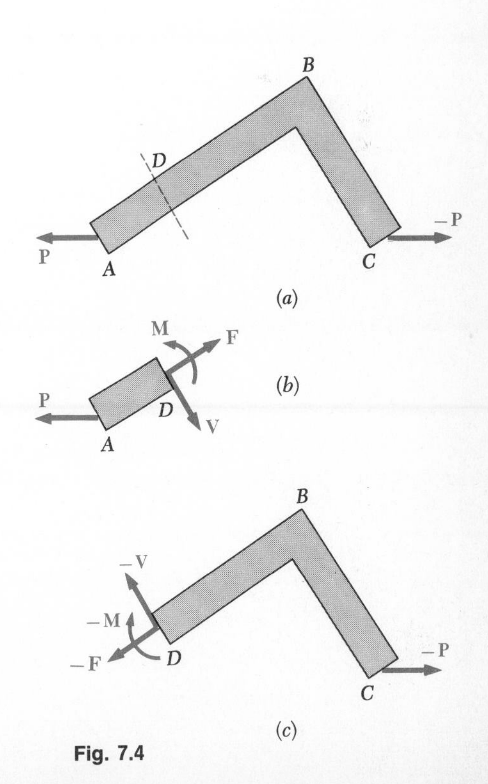

Fig. 7.4

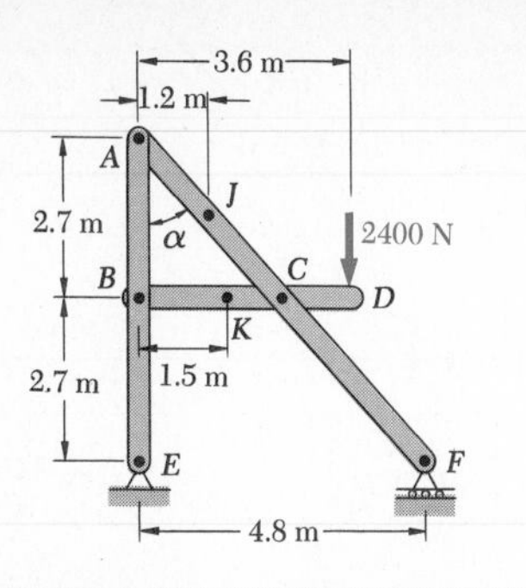

SAMPLE PROBLEM 7.1

In the frame shown, determine the internal forces (*a*) in member *ACF* at point *J* and (*b*) in member *BCD* at point *K*. This frame has been previously considered in Sample Prob. 6.5.

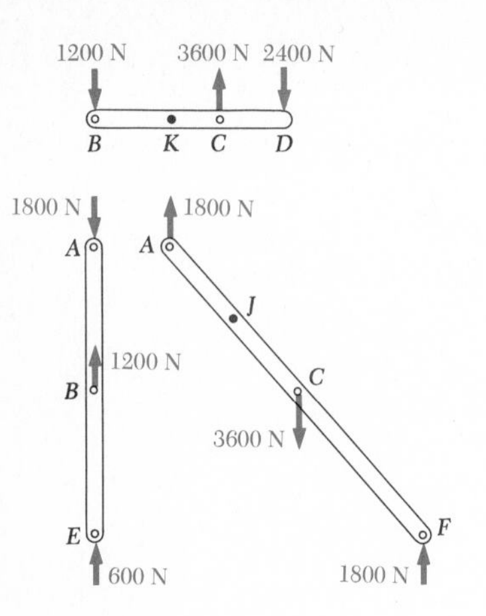

Solution. The reactions and the forces acting on each member of the frame are determined; this has been previously done in Sample Prob. 6.5, and the results are repeated here.

a. Internal Forces at J. Member *ACF* is cut at point *J*, and the two parts shown are obtained. The internal forces at *J* are represented by an equivalent force-couple system and may be determined by considering the equilibrium of either part. Considering the *free body AJ,* we write

$+\circlearrowleft \Sigma M_J = 0$: $\quad -(1800 \text{ N})(1.2 \text{ m}) + M = 0$

$M = +2160 \text{ N}\cdot\text{m} \qquad \mathbf{M} = 2160 \text{ N}\cdot\text{m} \circlearrowleft$ ◀

$+\searrow \Sigma F_x = 0$: $\quad F - (1800 \text{ N}) \cos 41.7° = 0$

$F = +1344 \text{ N} \qquad \mathbf{F} = 1344 \text{ N} \searrow$ ◀

$+\nearrow \Sigma F_y = 0$: $\quad -V + (1800 \text{ N}) \sin 41.7° = 0$

$V = +1197 \text{ N} \qquad \mathbf{V} = 1197 \text{ N} \swarrow$ ◀

The internal forces at *J* are therefore equivalent to a couple **M**, an axial force **F**, and a shearing force **V**. The internal force-couple system acting on part *JCF* is equal and opposite.

b. Internal Forces at K. We cut member *BCD* at *K* and obtain the two parts shown. Considering the *free body BK,* we write

$+\circlearrowleft \Sigma M_K = 0$: $\quad (1200 \text{ N})(1.5 \text{ m}) + M = 0$

$M = -1800 \text{ N}\cdot\text{m} \qquad \mathbf{M} = 1800 \text{ N}\cdot\text{m} \circlearrowright$ ◀

$\xrightarrow{+} \Sigma F_x = 0$: $\quad F = 0 \qquad \mathbf{F} = 0$ ◀

$+\uparrow \Sigma F_y = 0$: $\quad -1200 \text{ N} - V = 0$

$V = -1200 \text{ N} \qquad \mathbf{V} = 1200 \text{ N} \uparrow$ ◀

Problems

7.1 and 7.2 Determine the internal forces (axial force, shearing force, and bending moment) at point J of the structure indicated:

7.1 Frame and loading of Prob. 6.57

7.2 Frame and loading of Prob. 6.56.

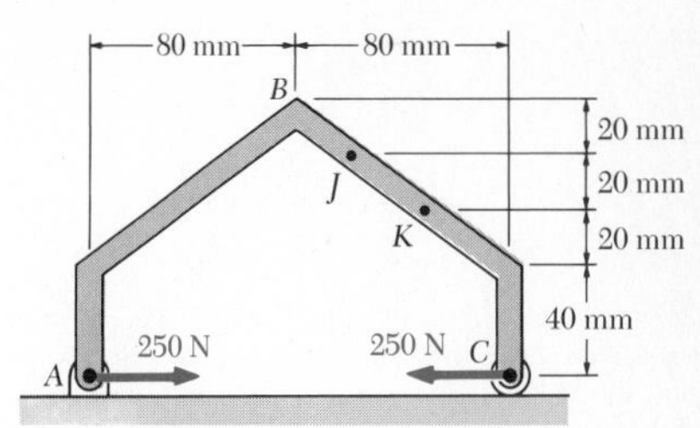

Fig. P7.3 and P7.4

7.3 and 7.4 For the frame and loading shown, determine the internal forces at the point indicated:

7.3 Point J.

7.4 Point K.

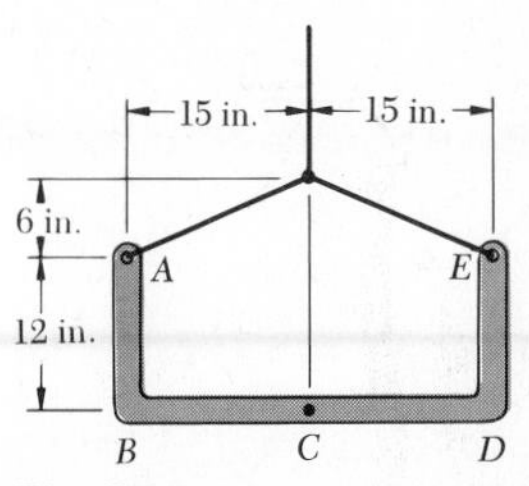

Fig. P7.5

7.5 A uniform rod $ABCDE$ has been bent and is held by a wire sling as shown. Knowing that the rod weighs 10.8 lb, determine the internal forces (a) at point C, (b) just to the right of point B.

7.6 Determine the internal forces at point J when $\alpha = 90°$.

7.7 Determine the internal forces at point J when $\alpha = 0$.

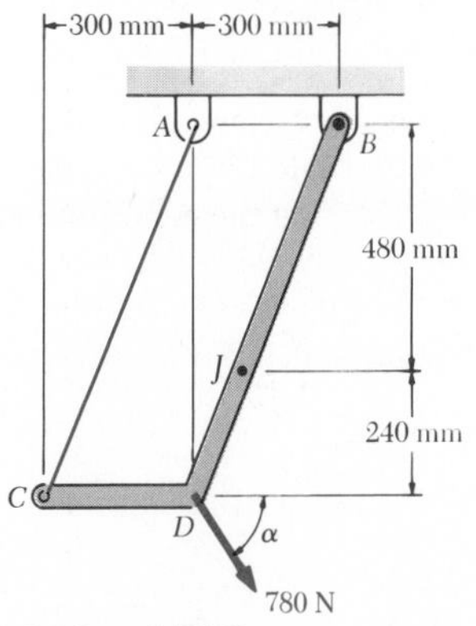

Fig. P7.6 and P7.7

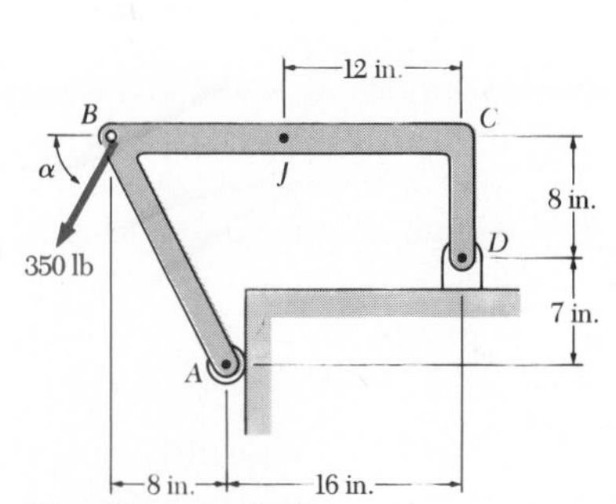

Fig. P7.8 and P7.9

7.8 Determine the internal forces at point J when $\alpha = 90°$.

7.9 Determine the internal forces at point J when $\alpha = 0$.

7.10 Knowing that the radius of each pulley is 120 mm, determine the internal forces at (a) point C, (b) point J which is 100 mm to the left of C.

7.11 Knowing that the radius of each pulley is 100 mm, determine the internal forces at (a) point C, (b) point J which is 100 mm to the left of C.

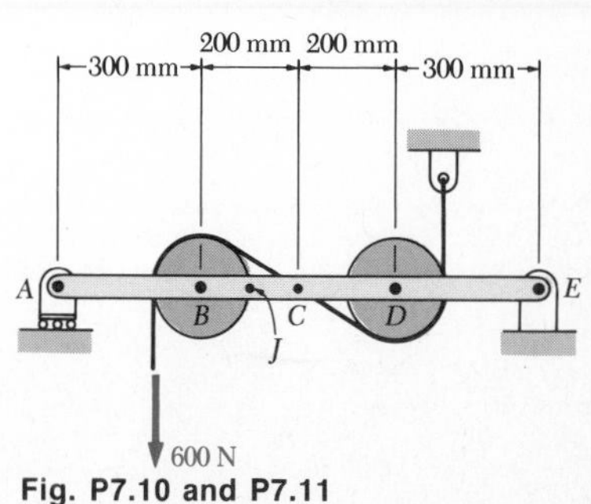

Fig. P7.10 and P7.11

7.12 A force **P** is applied to a bent rod AD which may be supported in three different ways as shown. In each case determine the internal forces just to the right of point B.

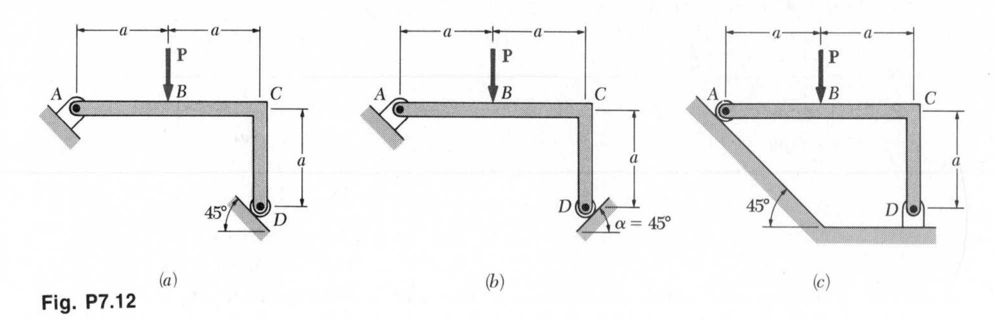

Fig. P7.12

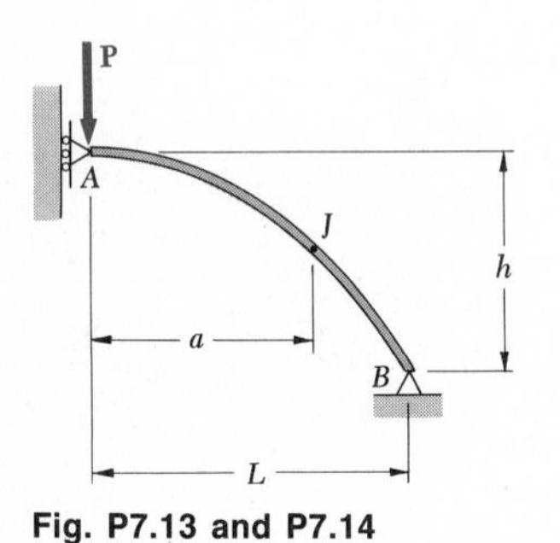

Fig. P7.13 and P7.14

7.13 The axis of the curved member AB is a parabola with vertex at A. If a vertical load **P** of magnitude 450 lb is applied at A, determine the internal forces at J when $h = 12$ in., $L = 40$ in., and $a = 24$ in.

7.14 Knowing that the axis of the curved member AB is a parabola with vertex at A, determine the magnitude and location of the maximum bending moment.

7.15 and 7.16 A half section of pipe rests on a frictionless horizontal surface as shown. If the half section of pipe has a mass of 9 kg and a diameter of 300 mm, determine the bending moment at point J when $\theta = 90°$.

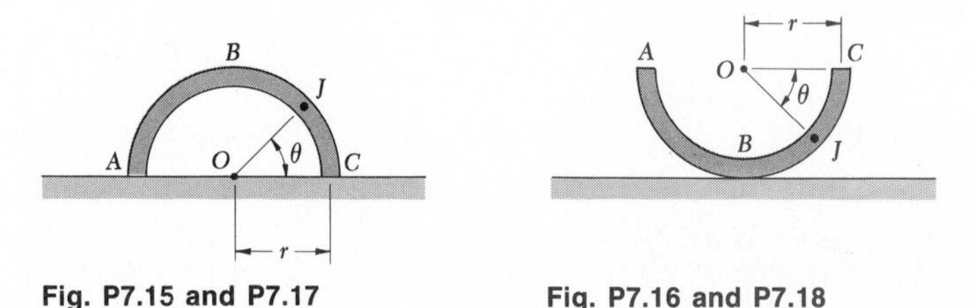

Fig. P7.15 and P7.17

Fig. P7.16 and P7.18

***7.17 and 7.18** A half section of pipe, of weight W and of unit length, rests on a frictionless horizontal surface. Determine the internal force at point J in terms of W, r, and θ.

***7.19** In Prob. 7.18, determine the magnitude and location of the maximum internal axial force.

*** 7.3. Various Types of Loading and Support.** A structural member designed to support loads applied at various points along the member is known as a *beam*. In most cases, the loads are perpendicular to the axis of the beam and will cause only shear and bending in the beam. When the loads are not at a right angle to the beam, they will also produce axial forces in the beam.

Beams are usually long, straight prismatic bars. Designing a beam consists essentially in selecting the cross section which will provide the most effective resistance to the shear and bending produced by the applied loads. The design of the beam, therefore, includes two distinct parts. In the first part, the shearing forces and bending moments produced by the loads are determined. The second part is concerned with the selection of the cross section best suited to resist the shearing forces and bending moments determined in the first part. This portion of the chapter, Beams, deals with the first part of the problem of beam design, namely, the determination of the shearing forces and bending moments in beams subjected to various loading conditions and supported in various ways. The second part of the problem belongs to the study of mechanics of materials.

A beam may be subjected to *concentrated loads* $\mathbf{P}_1$, $\mathbf{P}_2$, . . . , expressed in newtons, pounds, or their multiples kilonewtons and kips (Fig. 7.5*a*), to a *distributed load* w, expressed in N/m, kN/m, lb/ft, or kips/ft (Fig. 7.5*b*), or to a combination of both. When the load w per unit length has a constant value over part of the beam (as between A and B in Fig. 7.5*b*), the load is said to be *uniformly distributed* over that part of the beam. The determination of the reactions at the supports may be considerably simplified if distributed loads are replaced by equivalent concentrated loads, as explained in Sec. 5.8. This substitution, however, should not be performed, or at least should be performed with care, when internal forces are being computed (see Sample Prob. 7.3).

Beams are classified according to the way in which they are supported. Several types of beams frequently used are shown in Fig. 7.6. The distance L between supports is called the *span*. It should be noted that the

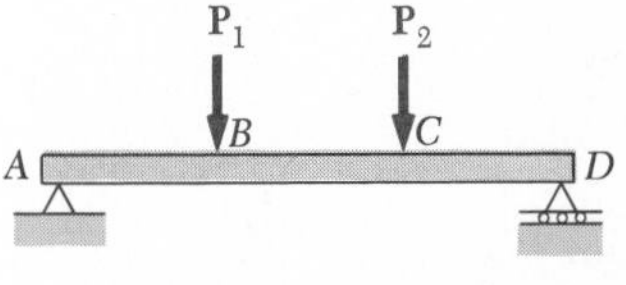

(*a*) Concentrated loads

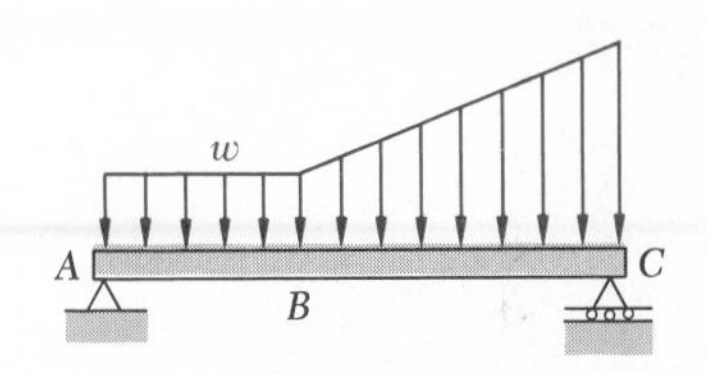

(*b*) Distributed load

Fig. 7.5 Types of loadings.

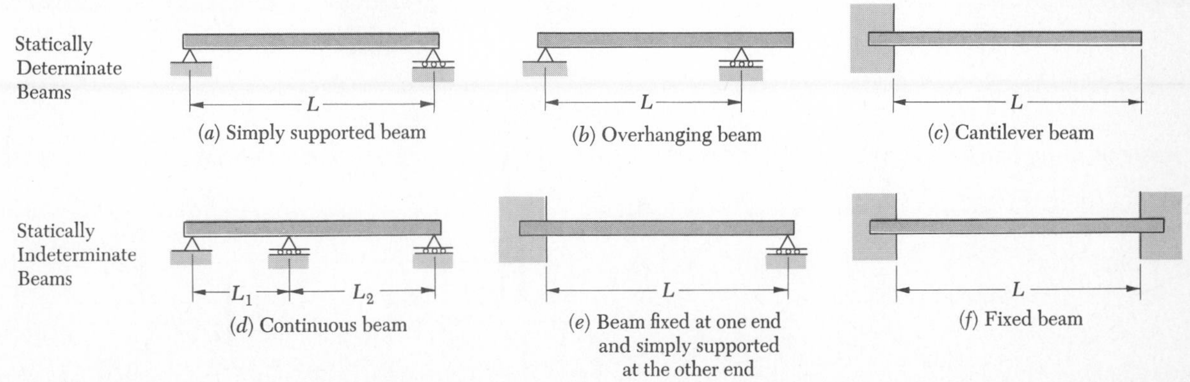

Fig. 7.6 Types of beams.

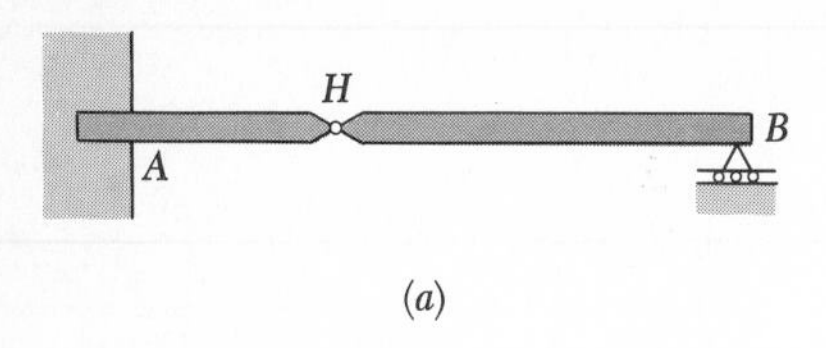

(a)

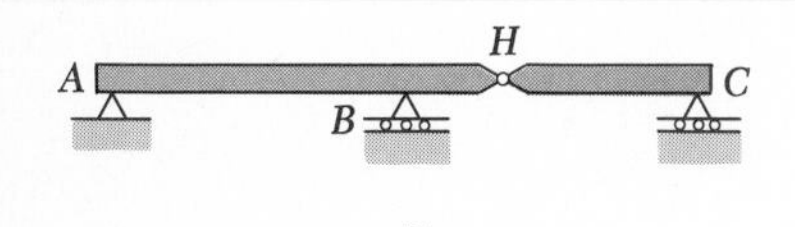

(b)

Fig. 7.7 Combined beams.

reactions will be determinate if the supports involve only three unknowns. The reactions will be statically indeterminate if more unknowns are involved; the methods of statics are not sufficient then to determine the reactions, and the properties of the beam with regard to its resistance to bending must be taken into consideration. Beams supported by two rollers are not shown here; such beams are only partially constrained and will move under certain loading conditions.

Sometimes two or more beams are connected by hinges to form a single continuous structure. Two examples of beams hinged at a point H are shown in Fig. 7.7. It will be noted that the reactions at the supports involve four unknowns and cannot be determined from the free-body diagram of the two-beam system. They can be determined, however, by considering the free-body diagram of each beam separately; six unknowns are involved (including two force components at the hinge), and six equations are available.

*** 7.4. Shear and Bending Moment in a Beam.** Consider a beam AB subjected to various concentrated and distributed loads (Fig. 7.8*a*). We propose to determine the shearing force and bending moment at any point of the beam. In the example considered here, the beam is simply supported, but the method used could be applied to any type of statically determinate beam.

First we determine the reactions at A and B by choosing the entire beam as a free body (Fig. 7.8*b*); writing $\Sigma M_A = 0$ and $\Sigma M_B = 0$, we obtain, respectively, $\mathbf{R}_B$ and $\mathbf{R}_A$.

To determine the internal forces at C, we cut the beam at C and draw the free-body diagrams of the portions AC and CB of the beam (Fig. 7.8*c*).

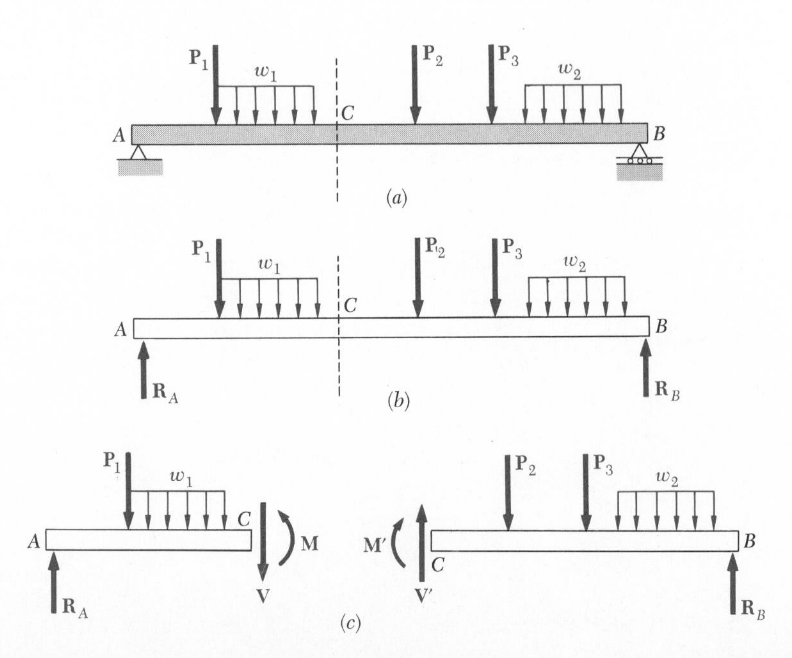

Fig. 7.8

Using the free-body diagram of AC, we may determine the shearing force **V** at C by equating to zero the sum of the vertical components of all forces acting on AC. Similarly, the bending moment **M** at C may be found by equating to zero the sum of the moments about C of all forces and couples acting on AC. We could have used just as well, however, the free-body diagram of CB† and determined the shearing force **V**′ and the bending moment **M**′ by equating to zero the sum of the vertical components and the sum of the moments about C of all forces and couples acting on CB. While this possible choice of alternate free bodies may facilitate the computation of the numerical values of the shearing force and bending moment, it makes it necessary to indicate on which portion of the beam the internal forces considered are acting. If the shearing force and bending moment, however, are to be computed at every point of the beam and efficiently recorded, we should not have to specify every time which portion of the beam is used as a free body. We shall adopt, therefore, the following conventions:

To determine the shearing force in a beam, *we shall always assume* that the internal forces **V** and **V**′ are directed as shown in Fig. 7.8*c*. A positive value obtained for their common magnitude V will indicate that this assumption was correct and that the shearing forces are actually directed as shown. A negative value obtained for V will indicate that the assumption was wrong and that the shearing forces are directed in the opposite way. Thus, only the magnitude V, together with a plus or minus sign, needs to be recorded to define completely the shearing forces at a given point of the beam. The scalar V is commonly referred to as the *shear* at the given point of the beam.

Similarly, *we shall always assume* that the internal couples **M** and **M**′ are directed as shown in Fig. 7.8*c*. A positive value obtained for their magnitude M, commonly referred to as the bending moment, will indicate that this assumption was correct, and a negative value that it was wrong. Summarizing the sign conventions we have presented, we state:

The shear V and the bending moment M at a given point of a beam are said to be positive when the internal forces and couples acting on each portion of the beam are directed as shown in Fig. 7.9a.

These conventions may be more easily remembered if we note that:

1. *The shear at C is positive when the* **external** *forces (loads and reactions) acting on the beam tend to shear off the beam at C as indicated in Fig. 7.9b.*
2. *The bending moment at C is positive when the* **external** *forces acting on the beam tend to bend the beam at C as indicated in Fig. 7.9c.*

It may also help to note that the situation described in Fig. 7.9, and corresponding to positive values of the shear and of the bending moment, is

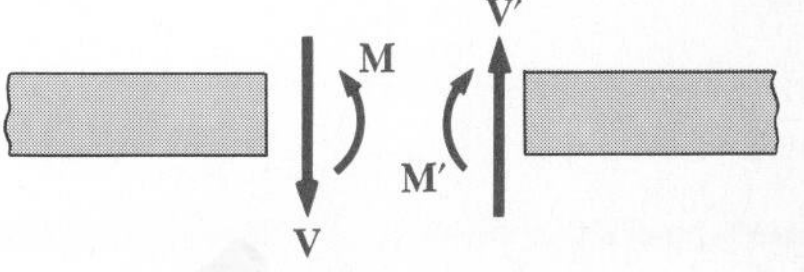

(*a*) Internal forces at section
(positive shear and positive bending moment)

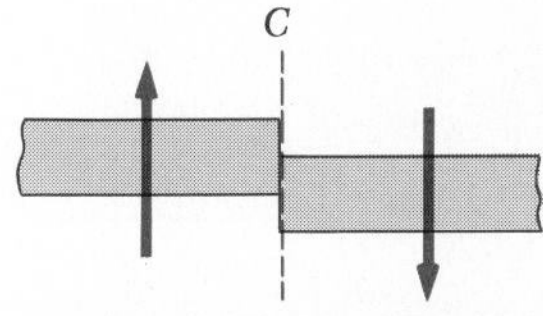

(*b*) Effect of external forces
(positive shear)

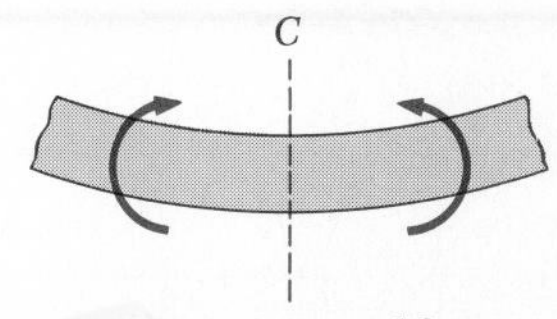

(*c*) Effect of external forces
(positive bending moment)

Fig. 7.9

†The force and couple representing the internal forces acting on CB will now be denoted by **V**′ and **M**′, rather than by −**V** and −**M** as done earlier, in order to avoid confusion when applying the sign convention which we are about to introduce.

precisely the situation which occurs in the left half of a simply supported beam carrying a single concentrated load at its midpoint. This particular example is fully discussed in the following section.

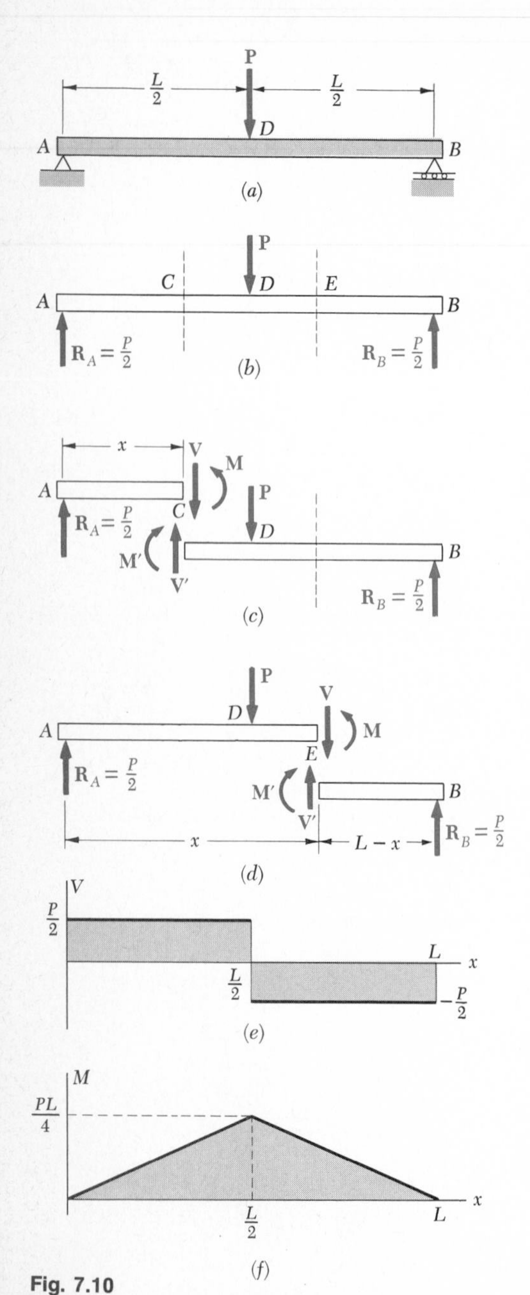

Fig. 7.10

*** 7.5. Shear and Bending-Moment Diagrams.** Now that shear and bending moment have been clearly defined in sense as well as in magnitude, we may easily record their values at any point of a beam by plotting these values against the distance x measured from one end of the beam. The graphs obtained in this way are called, respectively, the *shear diagram* and the *bending-moment diagram.* As an example, consider a simply supported beam AB of span L subjected to a single concentrated load **P** applied at its midpoint D (Fig. 7.10*a*). We first determine the reactions at the supports from the free-body diagram of the entire beam (Fig. 7.10*b*); we find that the magnitude of each reaction is equal to $P/2$.

Next we cut the beam at a point C between A and D and draw the free-body diagrams of AC and CB (Fig. 7.10*c*). *Assuming that shear and bending moment are positive,* we direct the internal forces **V** and **V′** and the internal couples **M** and **M′** as indicated in Fig. 7.9*a*. Considering the free body AC and writing that the sum of the vertical components and the sum of the moments about C of the forces acting on the free body are zero, we find $V = +P/2$ and $M = +Px/2$. Both shear and bending moment are therefore positive; this may be checked by observing that the reaction at A tends to shear off and to bend the beam at C as indicated in Fig. 7.9*b* and *c*. We may plot V and M between A and D (Fig. 7.10*e* and *f*); the shear has a constant value $V = P/2$, while the bending moment increases linearly from $M = 0$ at $x = 0$ to $M = PL/4$ at $x = L/2$.

Cutting, now, the beam at a point E between D and B and considering the free body EB (Fig. 7.10*d*), we write that the sum of the vertical components and the sum of the moments about E of the forces acting on the free body are zero. We obtain $V = -P/2$ and $M = P(L - x)/2$. The shear is therefore negative and the bending moment positive; this may be checked by observing that the reaction at B bends the beam at E as indicated in Fig. 7.9*c* but tends to shear it off in a manner opposite to that shown in Fig. 7.9*b*. We can complete, now, the shear and bending-moment diagrams of Fig. 7.10*e* and *f*; the shear has a constant value $V = -P/2$ between D and B, while the bending moment decreases linearly from $M = PL/4$ at $x = L/2$ to $M = 0$ at $x = L$.

We shall note that when a beam is subjected to concentrated loads only, the shear is of constant value between loads and the bending moment varies linearly between loads. On the other hand, when a beam is subjected to distributed loads, the shear and bending moment vary quite differently (see Sample Prob. 7.3).

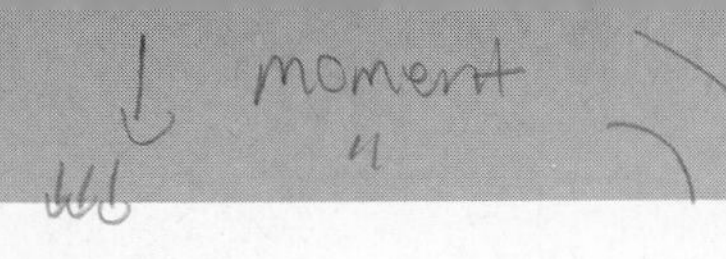

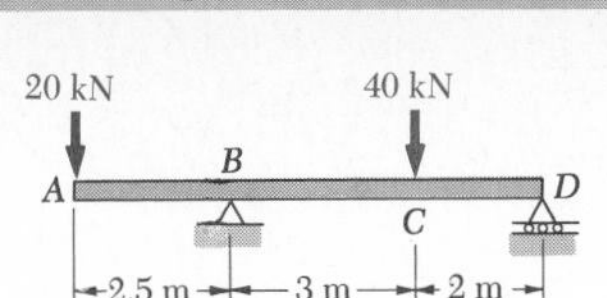

SAMPLE PROBLEM 7.2

Draw the shear and bending-moment diagram for the beam and loading shown.

Solution. The reactions are determined by considering the entire beam as a free body; they are

$$\mathbf{R}_B = 46 \text{ kN} \uparrow \qquad \mathbf{R}_D = 14 \text{ kN} \uparrow$$

We first determine the internal forces just to the right of the 20-kN load at A. Considering the stub of beam to the left of section *1* as a free body and assuming V and M to be positive (according to the standard convention), we write

$$+\uparrow \Sigma F_y = 0: \qquad -20 \text{ kN} - V_1 = 0 \qquad V_1 = -20 \text{ kN}$$

$$+\circlearrowleft \Sigma M_1 = 0: \qquad (20 \text{ kN})(0 \text{ m}) + M_1 = 0 \qquad M_1 = 0$$

We next consider as a free body the portion of beam to the left of section 2 and write

$$+\uparrow \Sigma F_y = 0: \qquad -20 \text{ kN} - V_2 = 0 \qquad V_2 = -20 \text{ kN}$$

$$+\circlearrowleft \Sigma M_2 = 0: \qquad (20 \text{ kN})(2.5 \text{ m}) + M_2 = 0 \qquad M_2 = -50 \text{ kN}\cdot\text{m}$$

The shear and bending moment at sections *3*, *4*, *5*, and *6* are determined in a similar way from the free-body diagrams shown. We obtain

$$V_3 = +26 \text{ kN} \qquad M_3 = -50 \text{ kN}\cdot\text{m}$$

$$V_4 = +26 \text{ kN} \qquad M_4 = +28 \text{ kN}\cdot\text{m}$$

$$V_5 = -14 \text{ kN} \qquad M_5 = +28 \text{ kN}\cdot\text{m}$$

$$V_6 = -14 \text{ kN} \qquad M_6 = 0$$

For several of the latter sections, the results may be more easily obtained by considering as a free body the portion of the beam to the right of the section. For example, considering the portion of the beam to the right of section *4*, we write

$$+\uparrow \Sigma F_y = 0: \qquad V_4 - 40 \text{ kN} + 14 \text{ kN} = 0 \qquad V_4 = +26 \text{ kN}$$

$$+\circlearrowleft \Sigma M_4 = 0: \qquad -M_4 + (14 \text{ kN})(2 \text{ m}) = 0 \qquad M_4 = +28 \text{ kN}\cdot\text{m}$$

We may now plot the six points shown on the shear and bending-moment diagrams. As indicated in Sec. 7.5, the shear is of constant value between concentrated loads, and the bending moment varies linearly; we obtain therefore the shear and bending-moment diagrams shown.

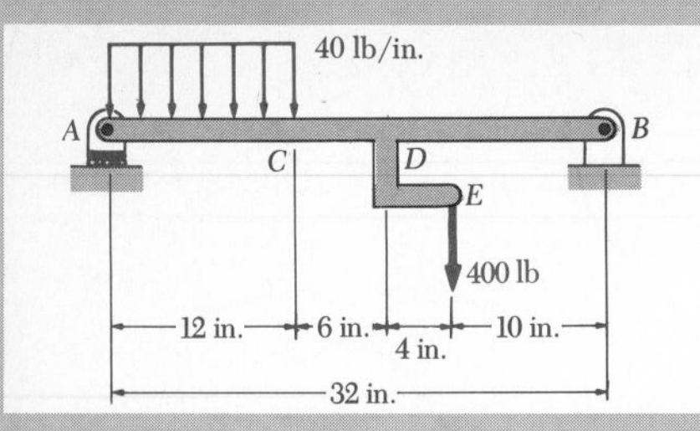

SAMPLE PROBLEM 7.3

Draw the shear and bending-moment diagrams for the beam AB. The distributed load of 40 lb/in. extends over 12 in. of the beam, from A to C, and the 400-lb load is applied at E.

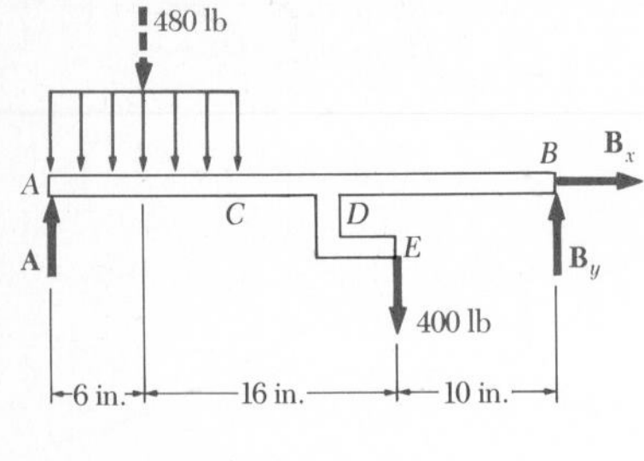

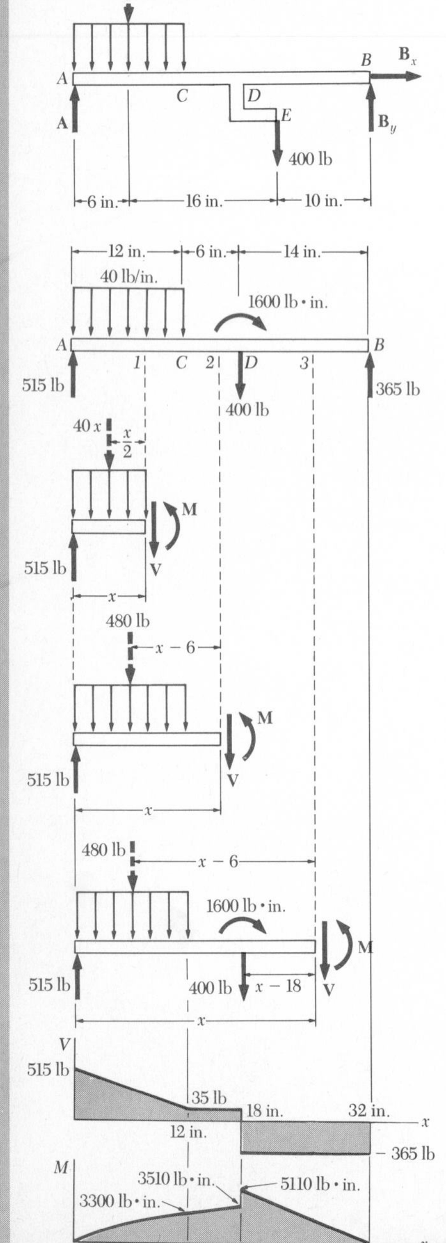

Solution. The reactions are determined by considering the entire beam as a free body.

$+\circlearrowleft \Sigma M_A = 0$: $\quad B_y(32 \text{ in.}) - (480 \text{ lb})(6 \text{ in.}) - (400 \text{ lb})(22 \text{ in.}) = 0$

$B_y = +365 \text{ lb} \qquad \mathbf{B}_y = 365 \text{ lb} \uparrow$

$+\circlearrowleft \Sigma M_B = 0$: $\quad (480 \text{ lb})(26 \text{ in.}) + (400 \text{ lb})(10 \text{ in.}) - A(32 \text{ in.}) = 0$

$A = +515 \text{ lb} \qquad \mathbf{A} = 515 \text{ lb} \uparrow$

$\xrightarrow{+} \Sigma F_x = 0$: $\quad B_x = 0 \qquad \mathbf{B}_x = 0$

The 400-lb load is now replaced by an equivalent force-couple system acting on the beam at point D.

From A to C. We determine the internal forces at a distance x from point A by considering the portion of beam to the left of section *1*. That part of the distributed load acting on the free body is replaced by its resultant, and we write

$+\uparrow \Sigma F_y = 0$: $\quad 515 - 40x - V = 0 \qquad V = 515 - 40x$

$+\circlearrowleft \Sigma M_1 = 0$: $\quad -515x + 40x(\tfrac{1}{2}x) + M = 0 \qquad M = 515x - 20x^2$

Since the free-body diagram shown may be used for all values of x smaller than 12 in., the expressions obtained for V and M are valid in the region $0 < x < 12$ in.

From C to D. Considering the portion of beam to the left of section *2* and again replacing the distributed load by its resultant, we obtain

$+\uparrow \Sigma F_y = 0$: $\quad 515 - 480 - V = 0 \qquad V = 35 \text{ lb}$

$+\circlearrowleft \Sigma M_2 = 0$: $\quad -515x + 480(x - 6) + M = 0 \qquad M = (2880 + 35x) \text{ lb·in.}$

These expressions are valid in the region 12 in. $< x <$ 18 in.

From D to B. Using the portion of beam to the left of section *3*, we obtain for the region 18 in. $< x <$ 32 in.

$+\uparrow \Sigma F_y = 0$: $\quad 515 - 480 - 400 - V = 0 \qquad V = -365 \text{ lb}$

$+\circlearrowleft \Sigma M_3 = 0$: $\quad -515x + 480(x - 6) - 1600 + 400(x - 18) + M = 0$

$M = (11{,}680 - 365x) \text{ lb·in.}$

The shear and bending-moment diagrams for the entire beam may now be plotted. We note that the couple of moment 1600 lb·in. applied at point D introduces a discontinuity into the bending-moment diagram.

Problems

7.20 through 7.25 Draw the shear and bending-moment diagrams for the beam and loading shown.

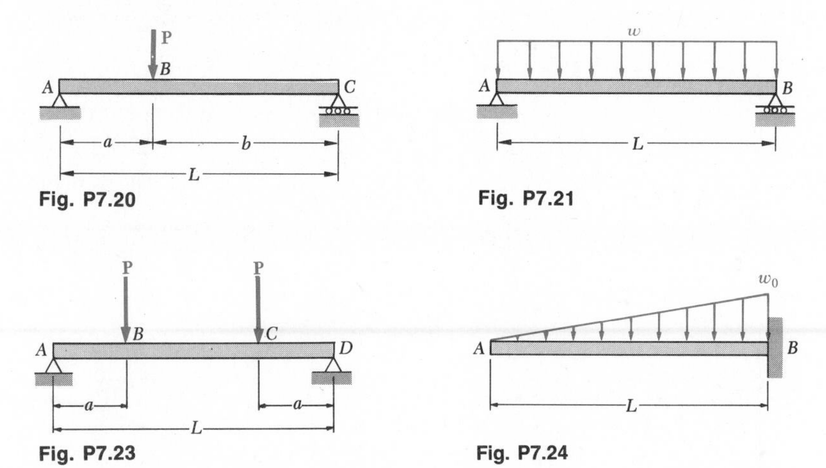

Fig. P7.20

Fig. P7.21

Fig. P7.23

Fig. P7.24

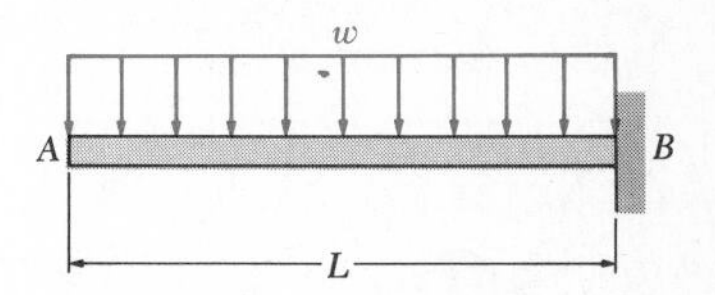

Fig. P7.22

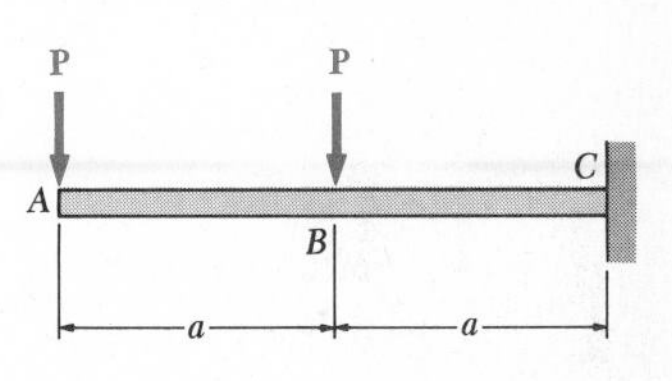

Fig. P7.25

7.26 and 7.27 Draw the shear and bending-moment diagrams for the beam AB.

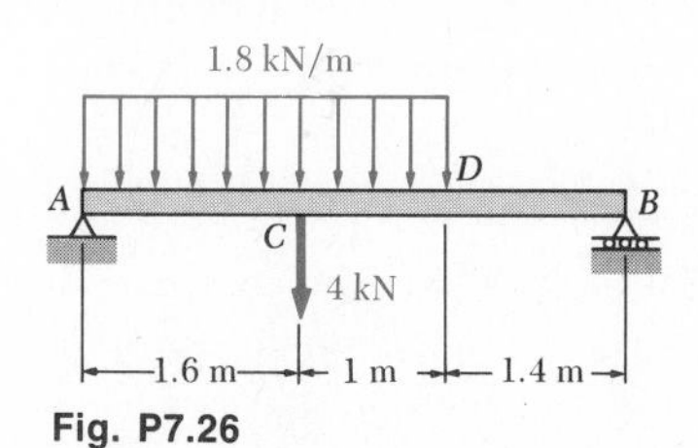

Fig. P7.26

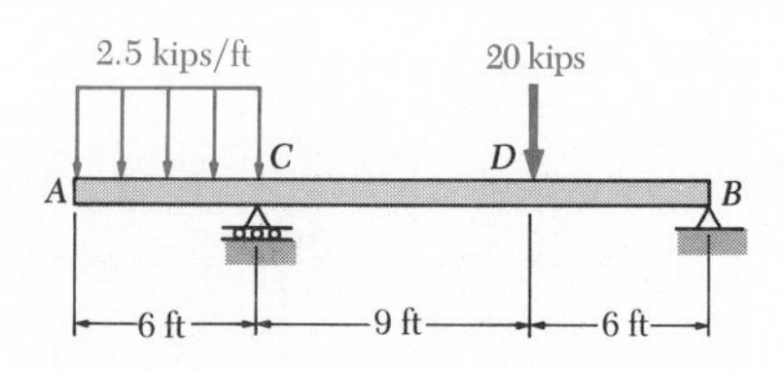

Fig. P7.27

7.28 Draw the shear and bending-moment diagrams for the beam AB if the magnitude of the distributed load is $w = 12$ lb/in.

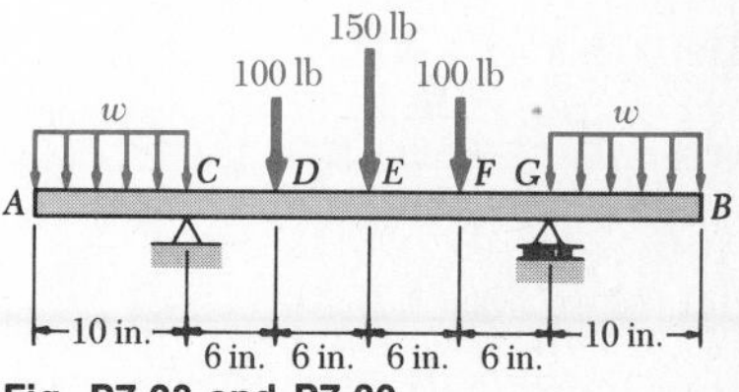

Fig. P7.28 and P7.30

7.29 Draw the shear and bending-moment diagrams for the beam AB if $a = 0.4$ m.

***7.30** Determine the distributed load w for which the maximum absolute value of the bending moment in the beam is as small as possible. (*Hint.* Draw the bending-moment diagram and then equate the absolute values of the largest positive and negative bending moments obtained.)

***7.31** Determine the distance a for which the maximum absolute value of the bending moment in the beam is as small as possible. (See hint of Prob. 7.30.)

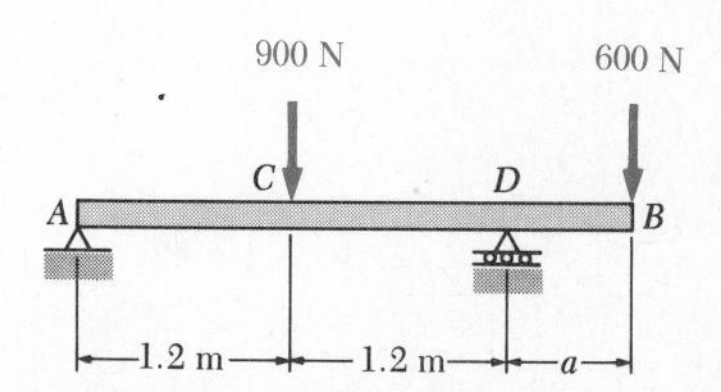

Fig. P7.29 and P7.31

7.32 through 7.34 Assuming the upward reaction of the ground to be uniformly distributed, draw the shear and bending-moment diagrams for the beam AB.

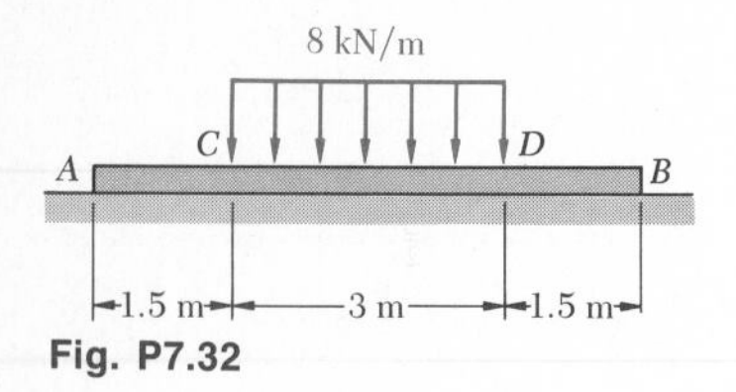

Fig. P7.32

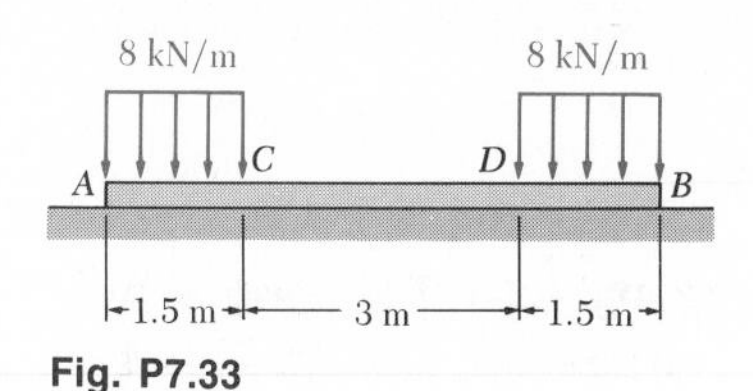

Fig. P7.33

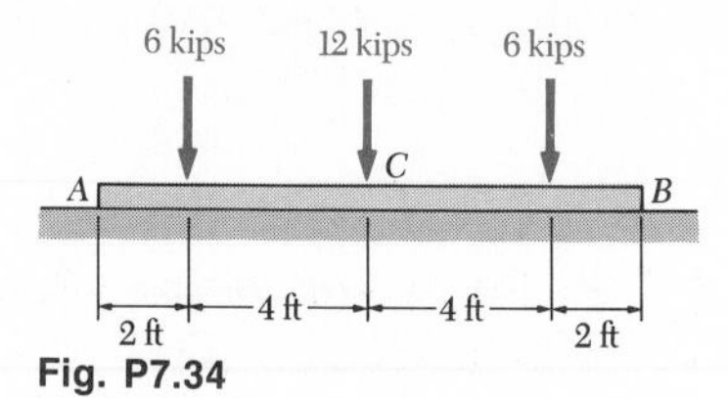

Fig. P7.34

7.35 Solve Prob. 7.34, assuming that the 12-kip load has been removed.

7.36 through 7.38 Draw the shear and bending-moment diagrams for the beam AB.

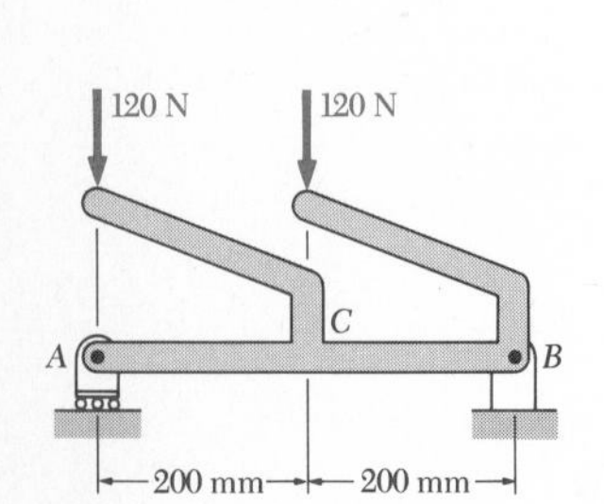

Fig. P7.36

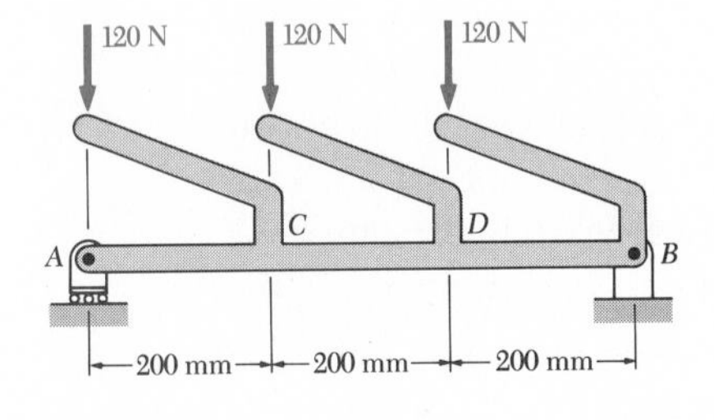

Fig. P7.37

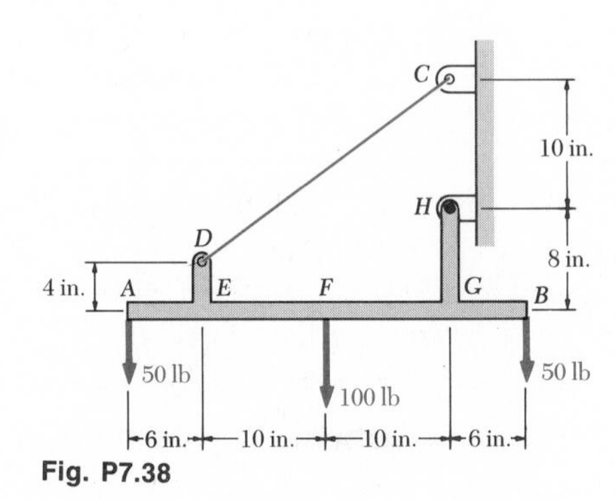

Fig. P7.38

7.39 Solve Prob. 7.38, assuming that the 100-lb load has been removed.

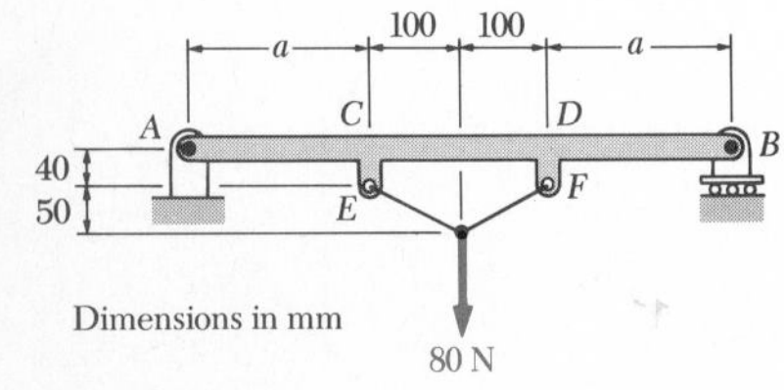

Fig. P7.40 and P7.41

7.40 Draw the shear and bending-moment diagram for beam AB if $a =$ 180 mm.

***7.41** Determine the distance a for which the maximum absolute value of the bending moment in beam AB is as small as possible. (See hint of Prob. 7.30.)

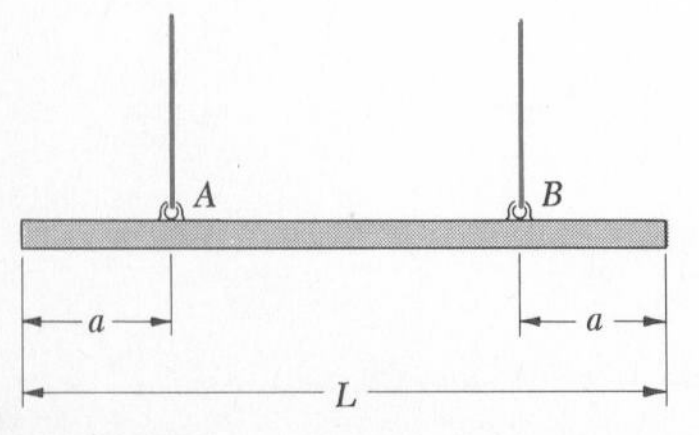

Fig. P7.42

***7.42** A uniform beam is to be picked up by crane cables attached at A and B. Determine the distance a from the ends of the beam to the points where the cables should be attached if the maximum absolute value of the bending moment in the beam is to be as small as possible. (*Hint.* Draw the bending-moment diagram in terms of a, L, and the weight w per unit length, and then equate the absolute values of the largest positive and negative bending moments obtained.)

***7.43** In order to reduce the bending moment in the cantilever beam AB, a cable and counterweight are permanently attached at end B. Determine the magnitude of the counterweight for which the maximum absolute value of the bending moment in the beam is as small as possible. (*a*) Consider only the case when the force **P** is actually applied at C. (*b*) Consider the more general case when the force **P** may either be applied at C or removed.

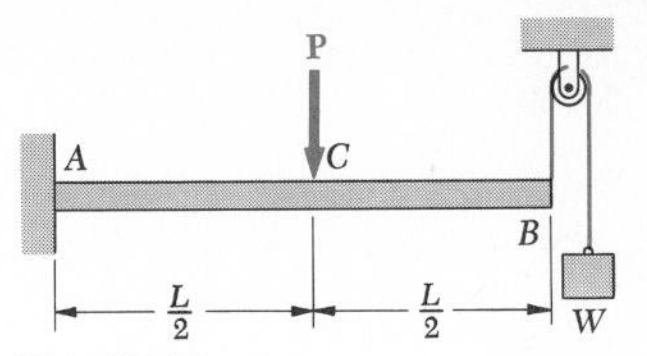

Fig. P7.43

*7.6. Relations among Load, Shear, and Bending Moment.

When a beam carries more than two or three concentrated loads, or when it carries distributed loads, the method outlined in Sec. 7.5 for plotting shear and bending moment may prove quite cumbersome. The construction of the shear diagram and, especially, of the bending-moment diagram will be greatly facilitated if certain relations existing among load, shear, and bending moment are taken into consideration.

Let us consider a simply supported beam AB carrying a distributed load w per unit length (Fig. 7.11*a*), and let C and C' be two points of the beam at a distance Δx from each other. The shear and bending moment at C will be denoted by V and M, respectively, and will be assumed positive; the shear and bending moment at C' will be denoted by $V + \Delta V$ and $M + \Delta M$.

We shall now detach the portion of beam CC' and draw its free-body diagram (Fig. 7.11*b*). The forces exerted on the free body include a load of magnitude $w\,\Delta x$ and internal forces and couples at C and C'. Since shear and bending moment have been assumed positive, the forces and couples will be directed as shown in the figure.

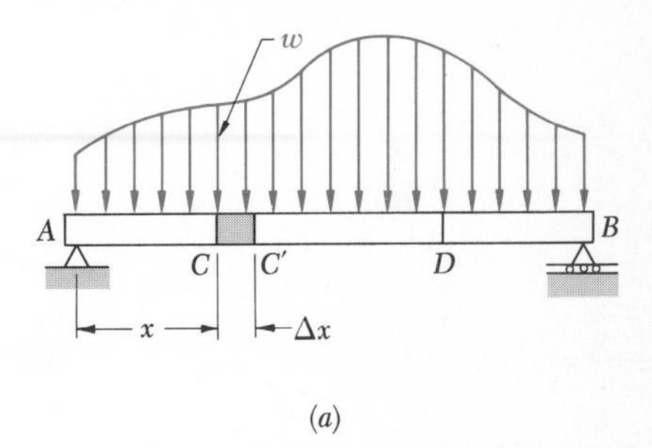

(*a*)

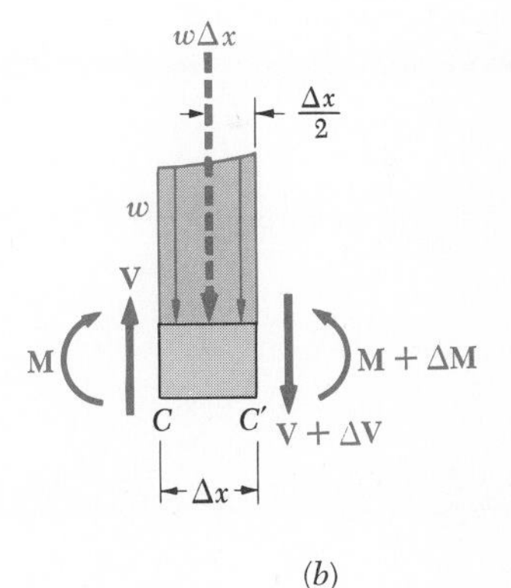

(*b*)

Fig. 7.11

Relations between Load and Shear. We write that the sum of the vertical components of the forces acting on the free body CC' is zero:

$$V - (V + \Delta V) - w\,\Delta x = 0$$
$$\Delta V = -w\,\Delta x$$

Dividing both members of the equation by Δx and then letting Δx approach zero, we obtain

$$\frac{dV}{dx} = -w \qquad (7.1)$$

Formula (7.1) indicates that for a beam loaded as shown in Fig. 7.11*a*, the slope dV/dx of the shear curve is negative; the numerical value of the slope at any point is equal to the load per unit length at that point.

Integrating (7.1) between points C and D, we obtain

$$V_D - V_C = -\int_{x_C}^{x_D} w\,dx \qquad (7.2)$$

$$V_D - V_C = -(\text{area under load curve between } C \text{ and } D) \qquad (7.2')$$

Note that this result could also have been obtained by considering the equilibrium of the portion of beam CD, since the area under the load curve represents the total load applied between C and D.

It should be observed that formula (7.1) *is not valid* at a point where a concentrated load is applied; the shear curve is discontinuous at such a point, as seen in Sec. 7.5. Similarly, formulas (7.2) and (7.2′) cease to be valid when concentrated loads are applied between C and D, since they do not take into account the sudden change in shear caused by a concentrated load. Formulas (7.2) and (7.2′), therefore, should be applied only between successive concentrated loads.

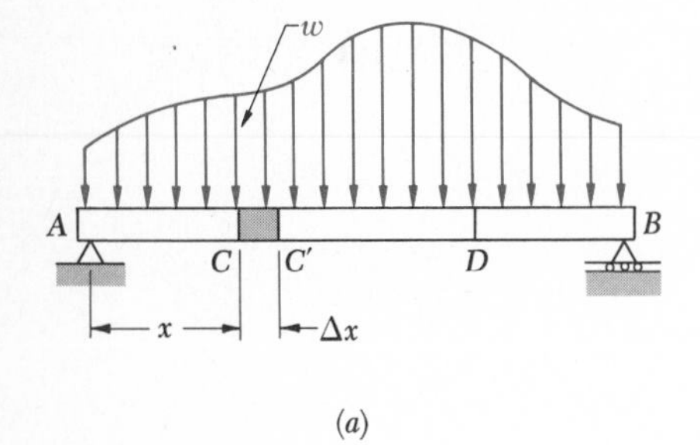

(a)

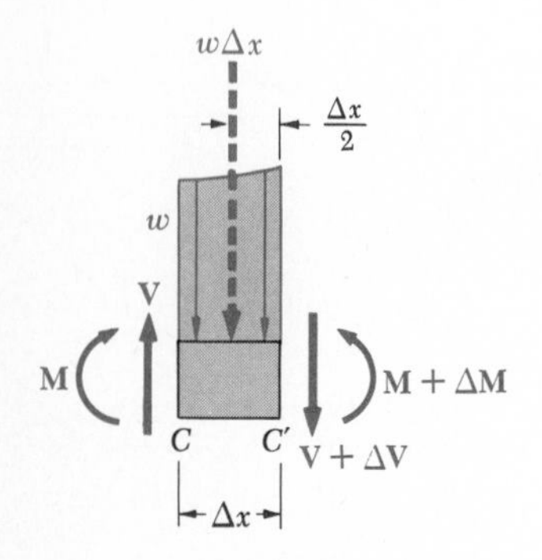

(b)

Fig. 7.11 *(repeated)*

Relations between Shear and Bending Moment. Returning to the free-body diagram of Fig. 7.11*b*, and writing now that the sum of the moments about C' is zero, we obtain

$$(M + \Delta M) - M - V\,\Delta x + w\,\Delta x \frac{\Delta x}{2} = 0$$

$$\Delta M = V\,\Delta x - \tfrac{1}{2}w(\Delta x)^2$$

Dividing both members of the equation by Δx and then letting Δx approach zero, we obtain

$$\frac{dM}{dx} = V \qquad (7.3)$$

Formula (7.3) indicates that the slope dM/dx of the bending-moment curve is equal to the value of the shear. This is true at any point where the shear has a well-defined value, i.e., at any point where no concentrated load is applied. Formula (7.3) also shows that the shear is zero at points where the bending moment is maximum. This property facilitates the determination of the points where the beam is likely to fail under bending.

Integating (7.3) between points C and D, we obtain

$$M_D - M_C = \int_{x_C}^{x_D} V\,dx \qquad (7.4)$$

$$M_D - M_C = \text{area under shear curve between } C \text{ and } D \qquad (7.4')$$

Note that the area under the shear curve should be considered positive where the shear is positive and negative where the shear is negative. Formulas (7.4) and (7.4′) are valid even when concentrated loads are applied between C and D, as long as the shear curve has been correctly drawn. The formulas cease to be valid, however, if a *couple* is applied at a point between C and D, since they do not take into account the sudden change in bending moment caused by a couple (see Sample Prob. 7.7).

Example. Let us consider a simply supported beam AB of span L carrying a uniformly distributed load w (Fig. 7.12*a*). From the free-body diagram of the entire beam we determine the magnitude of the reactions at the supports: $R_A = R_B = wL/2$ (Fig. 7.12*b*). Next, we draw the shear diagram. Close to the end A of the beam, the shear is equal to R_A, that is, to $wL/2$, as we may check by considering as a free body a very small portion of the beam. Using formula (7.2), we may then determine the shear V at any distance x from A; we write

$$V - V_A = -\int_0^x w\,dx = -wx$$

$$V = V_A - wx = \frac{wL}{2} - wx = w\left(\frac{L}{2} - x\right)$$

The shear curve is thus an oblique straight line which crosses the x axis at $x = L/2$ (Fig. 7.12*c*). Considering, now, the bending moment, we first observe that $M_A = 0$. The value M of the bending moment at any distance x from A may then be obtained from formula (7.4); we have

$$M - M_A = \int_0^x V\,dx$$

$$M = \int_0^x w\left(\frac{L}{2} - x\right) dx = \frac{w}{2}(Lx - x^2)$$

The bending-moment curve is a parabola. The maximum value of the bending moment occurs when $x = L/2$, since V (and thus dM/dx) is zero for that value of x. Substituting $x = L/2$ in the last equation, we obtain $M_{max} = wL^2/8$.

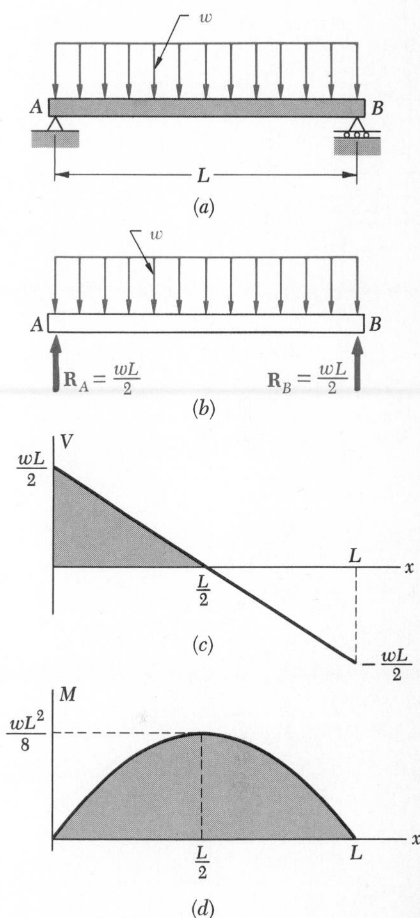

Fig. 7.12

In most engineering applications, the value of the bending moment needs to be known only at a few specific points. Once the shear diagram has been drawn, and after M has been determined at one of the ends of the beam, the value of the bending moment may then be obtained at any given point by computing the area under the shear curve and using formula (7.4′). For instance, since $M_A = 0$ for the beam of Fig. 7.12, the maximum value of the bending moment for that beam may be obtained simply by measuring the area of the shaded triangle in the shear diagram:

$$M_{max} = \frac{1}{2}\frac{L}{2}\frac{wL}{2} = \frac{wL^2}{8}$$

We note that in this example, the load curve is a horizontal straight line, the shear curve an oblique straight line, and the bending-moment curve a parabola. If the load curve had been an oblique straight line (first degree), the shear curve would have been a parabola (second degree) and the bending-moment curve a cubic (third degree). The shear and bending-moment curves will always be, respectively, one and two degrees higher than the load curve. With this in mind, we should be able to sketch the shear and bending-moment diagrams without actually determining the functions $V(x)$ and $M(x)$, once a few values of the shear and bending moment have been computed. The sketches obtained will be more accurate if we make use of the fact that at any point where the curves are continuous, the slope of the shear curve is equal to $-w$ and the slope of the bending-moment curve is equal to V.

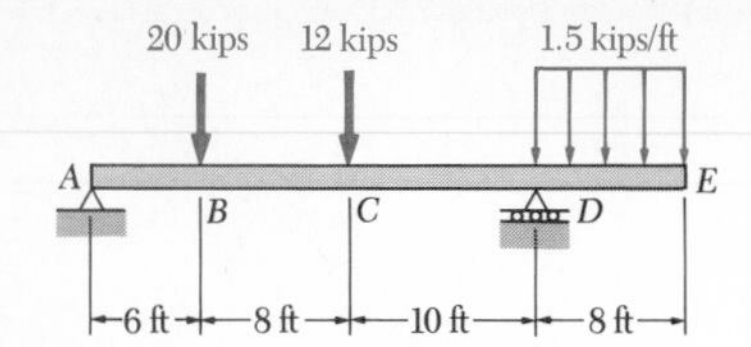

SAMPLE PROBLEM 7.4

Draw the shear and bending-moment diagrams for the beam and loading shown.

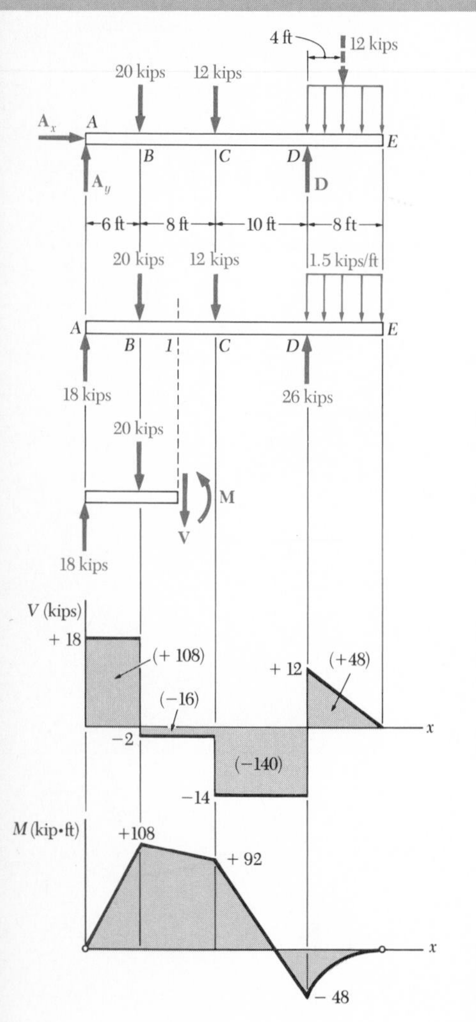

Solution. Considering the entire beam as a free body, we determine the reactions:

$+\circlearrowleft\Sigma M_A = 0$:

$$D(24 \text{ ft}) - (20 \text{ kips})(6 \text{ ft}) - (12 \text{ kips})(14 \text{ ft}) - (12 \text{ kips})(28 \text{ ft}) = 0$$

$$D = +26 \text{ kips} \qquad \mathbf{D} = 26 \text{ kips} \uparrow$$

$+\uparrow\Sigma F_y = 0$: $\quad A_y - 20 \text{ kips} - 12 \text{ kips} + 26 \text{ kips} - 12 \text{ kips} = 0$

$$A_y = +18 \text{ kips} \qquad \mathbf{A}_y = 18 \text{ kips} \uparrow$$

$\overset{+}{\rightarrow}\Sigma F_x = 0$: $\quad A_x = 0 \qquad \mathbf{A}_x = 0$

We also note that at both A and E the bending moment is zero; thus two points (indicated by dots) are obtained on the bending-moment diagram.

Shear Diagram. Since $dV/dx = -w$, we find that between concentrated loads and reactions the slope of the shear diagram is zero (i.e., the shear is constant). The shear at any point is determined by dividing the beam into two parts and considering either part as a free body. For example, using the portion of beam to the left of section *1*, we obtain the shear between B and C:

$+\uparrow\Sigma F_y = 0$: $\quad +18 \text{ kips} - 20 \text{ kips} - V = 0 \qquad V = -2 \text{ kips}$

We also find that the shear is $+12$ kips just to the right of D and zero at end E. Since the slope $dV/dx = -w$ is constant between D and E, the shear diagram between these two points is a straight line.

Bending-Moment Diagram. We recall that the area under the shear curve between two points is equal to the change in bending moment between the same two points. For convenience, the area of each portion of the shear diagram is computed and is indicated on the diagram. Since the bending moment M_A at the left end is known to be zero, we write

$$M_B - M_A = +108 \qquad M_B = +108 \text{ kip}\cdot\text{ft}$$
$$M_C - M_B = -\ 16 \qquad M_C = +\ 92 \text{ kip}\cdot\text{ft}$$
$$M_D - M_C = -140 \qquad M_D = -\ 48 \text{ kip}\cdot\text{ft}$$
$$M_E - M_D = +\ 48 \qquad M_E = 0$$

Since M_E is known to be zero, a check of the computations is obtained.

Between the concentrated loads and reactions the shear is constant, thus the slope dM/dx is constant and the bending-moment diagram is drawn by connecting the known points with straight lines. Between D and E where the shear diagram is an oblique straight line, the bending-moment diagram is a parabola.

From the V and M diagrams we note that $V_{max} = 18$ kips and $M_{max} = 108$ kip·ft.

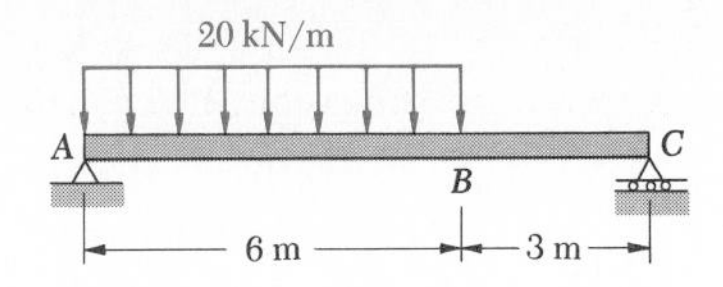

SAMPLE PROBLEM 7.5

Draw the shear and bending-moment diagrams for the beam and loading shown.

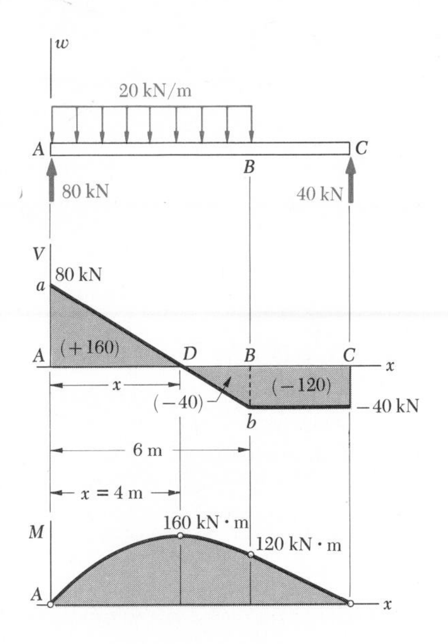

Solution. Considering the entire beam as a free body, we obtain the reactions

$$\mathbf{R}_A = 80 \text{ kN} \uparrow \qquad \mathbf{R}_C = 40 \text{ kN} \uparrow$$

Shear Diagram. The shear just to the right of A is $V_A = +80$ kN. Since the change in shear between two points is equal to *minus* the area under the load curve between the same two points, we obtain V_B by writing

$$V_B - V_A = -(20 \text{ kN/m})(6 \text{ m}) = -120 \text{ kN}$$
$$V_B = -120 + V_A = -120 + 80 = -40 \text{ kN}$$

The slope $dV/dx = -w$ being constant between A and B, the shear diagram between these two points is represented by a straight line. Between B and C, the area under the load curve is zero; therefore,

$$V_C - V_B = 0 \qquad V_C = V_B = -40 \text{ kN}$$

and the shear is constant between B and C.

Bending-Moment Diagram. We note that the bending moment at each end of the beam is zero. In order to determine the maximum bending moment, we locate the section D of the beam where $V = 0$. Considering the portion of the shear diagram between A and B, we note that the triangles DAa and DBb are similar; thus,

$$\frac{x}{80 \text{ kN}} = \frac{6 - x}{40 \text{ kN}} \qquad x = 4 \text{ m}$$

The maximum bending moment occurs at point D, where we have $dM/dx = V = 0$. The areas of the various portions of the shear diagram are computed and are given (in parentheses) on the diagram. Since the area of the shear diagram between two points is equal to the change in bending moment between the same two points, we write

$$M_D - M_A = +160 \text{ kN}\cdot\text{m} \qquad M_D = +160 \text{ kN}\cdot\text{m}$$
$$M_B - M_D = -\ 40 \text{ kN}\cdot\text{m} \qquad M_B = +120 \text{ kN}\cdot\text{m}$$
$$M_C - M_B = -120 \text{ kN}\cdot\text{m} \qquad M_C = 0$$

The bending-moment diagram consists of an arc of parabola followed by a segment of straight line; the slope of the parabola at A is equal to the value of V at that point.

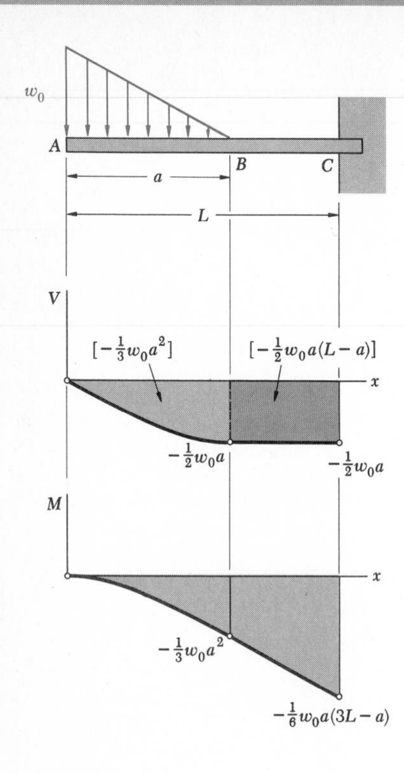

SAMPLE PROBLEM 7.6

Sketch the shear and bending-moment diagrams for the cantilever beam shown.

Solution. *Shear Diagram.* At the free end of the beam, we find $V_A = 0$. Between A and B, the area under the load curve is $\frac{1}{2}w_0a$; we find V_B by writing

$$V_B - V_A = -\tfrac{1}{2}w_0a \qquad V_B = -\tfrac{1}{2}w_0a$$

Between B and C, the beam is not loaded; thus $V_C = V_B$. At A, we have $w = w_0$ and, according to Eq. (7.1), the slope of the shear curve is $dV/dx = -w_0$, while at B the slope is $dV/dx = 0$. Between A and B, the loading decreases linearly, and the shear diagram is parabolic. Between B and C, $w = 0$, and the shear diagram is a horizontal line.

Bending-Moment Diagram. The bending moment M_A at the free end of the beam is zero. We compute the area under the shear curve and write

$$M_B - M_A = -\tfrac{1}{3}w_0a^2 \qquad M_B = -\tfrac{1}{3}w_0a^2$$
$$M_C - M_B = -\tfrac{1}{2}w_0a(L-a)$$
$$M_C = -\tfrac{1}{6}w_0a(3L-a)$$

The sketch of the bending-moment diagram is completed by recalling that $dM/dx = V$. We find that between A and B the diagram is represented by a cubic curve with zero slope at A, and between B and C by a straight line.

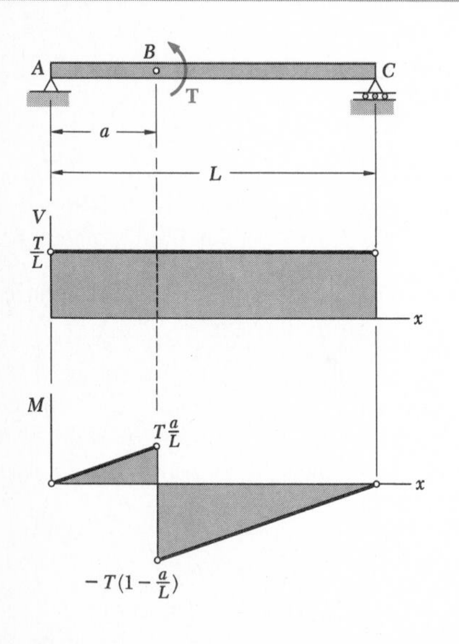

SAMPLE PROBLEM 7.7

The simple beam AC is loaded by a couple of magnitude T applied at point B. Draw the shear and bending-moment diagrams of the beam.

Solution. The entire beam is taken as a free body, and we obtain

$$\mathbf{R}_A = \frac{T}{L}\uparrow \qquad \mathbf{R}_C = \frac{T}{L}\downarrow$$

The shear at any section is constant and equal to T/L. Since a couple is applied at B, the bending-moment diagram is discontinuous at B; the bending moment decreases suddenly by an amount equal to T.

Problems

7.44 Using the methods of Sec. 7.6, solve Prob. 7.20.
7.45 Using the methods of Sec. 7.6, solve Prob. 7.21.
7.46 Using the methods of Sec. 7.6, solve Prob. 7.22.
7.47 Using the methods of Sec. 7.6, solve Prob. 7.23.
7.48 Using the methods of Sec. 7.6, solve Prob. 7.26.
7.49 Using the methods of Sec. 7.6, solve Prob. 7.27.
7.50 Using the methods of Sec. 7.6, solve Prob. 7.28.
7.51 Using the methods of Sec. 7.6, solve Prob. 7.29.
7.52 Using the methods of Sec. 7.6, solve Prob. 7.32.
7.53 Using the methods of Sec. 7.6, solve Prob. 7.33.
7.54 Using the methods of Sec. 7.6, solve Prob. 7.34.
7.55 Using the methods of Sec. 7.6, solve Prob. 7.37.

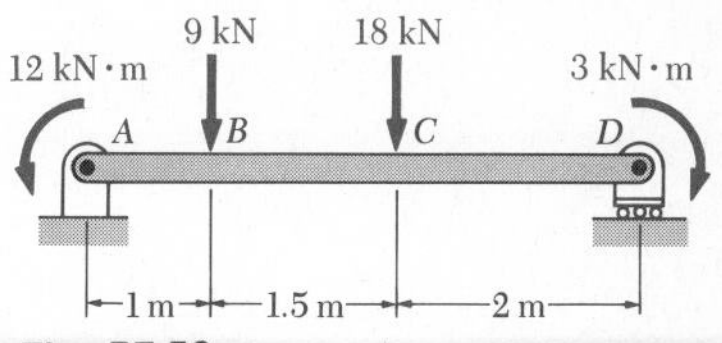

Fig. P7.56

7.56 through 7.59 Draw the shear and bending-moment diagrams for the beam and loading shown.

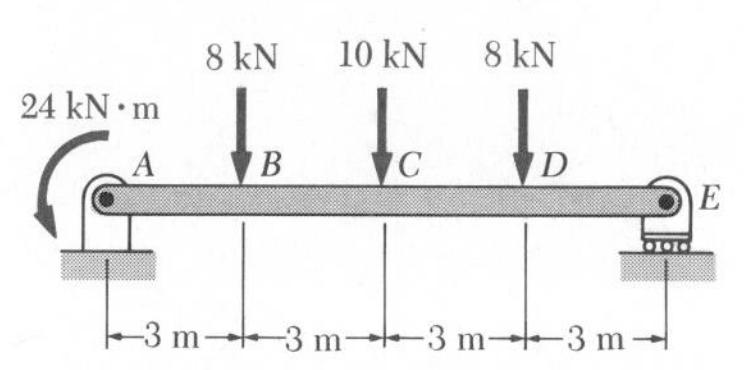

Fig. P7.57

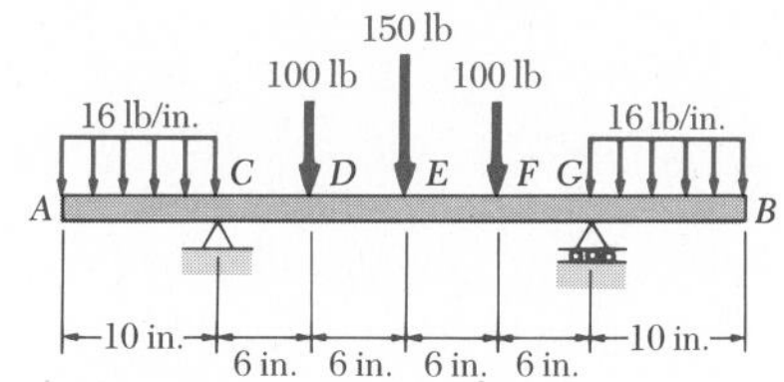

Fig. P7.58

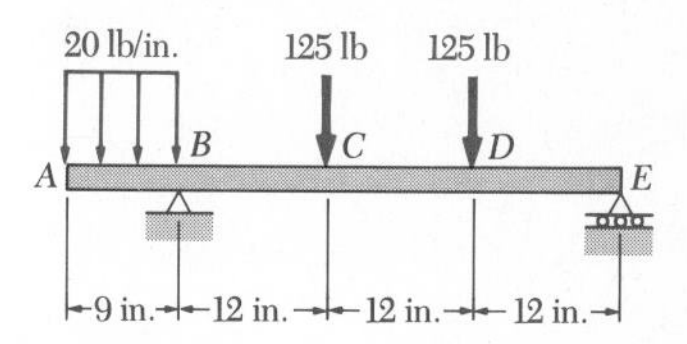

Fig. P7.59

7.60 through 7.65 Draw the shear and bending-moment diagrams for the beam and loading shown, and determine the location and magnitude of the maximum bending moment.

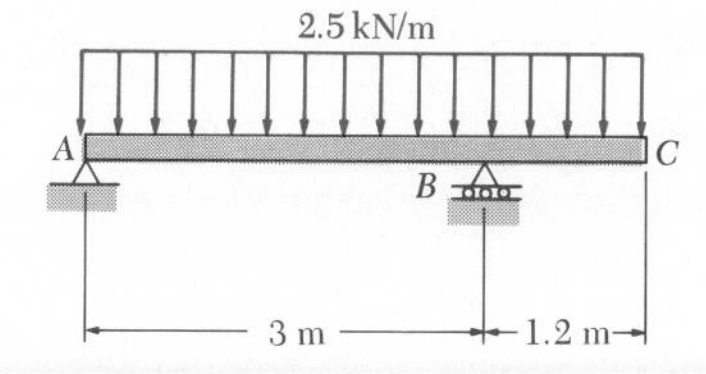

Fig. P7.60

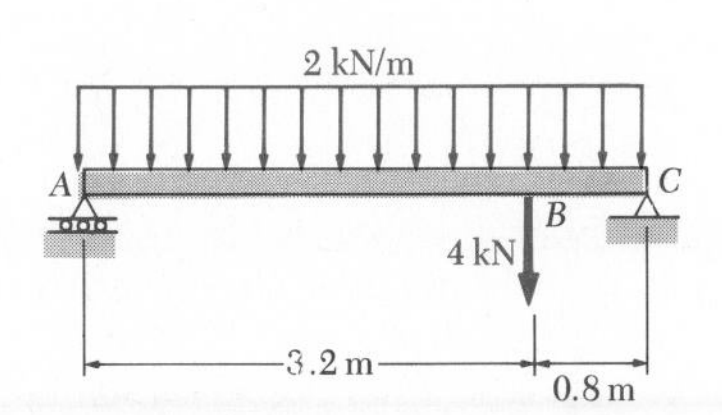

Fig. P7.61

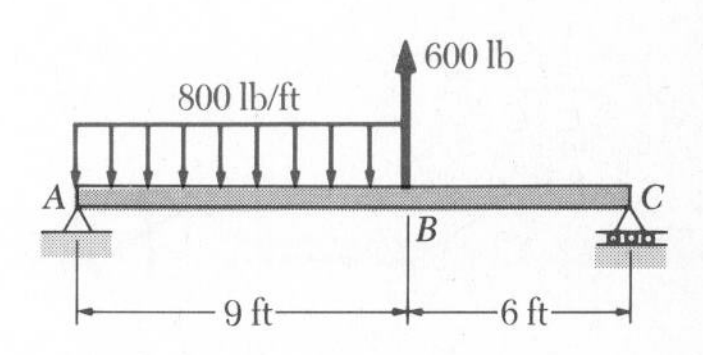

Fig. P7.62

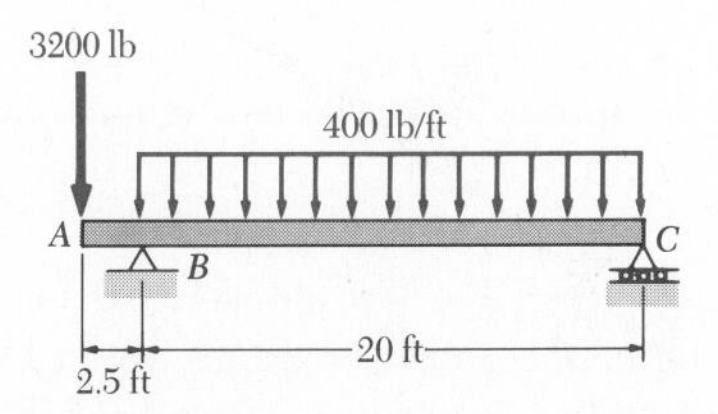

Fig. P7.63

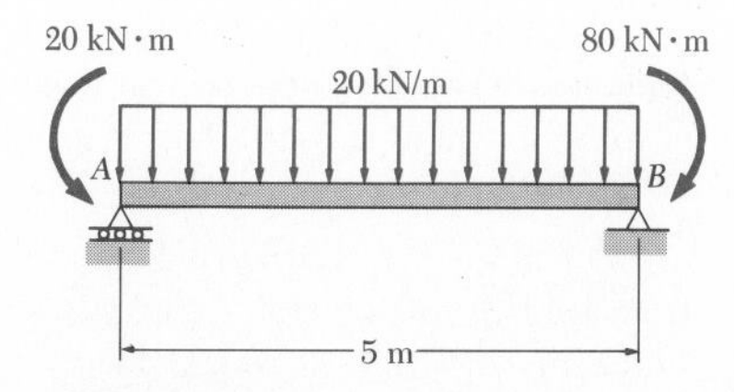

Fig. P7.64

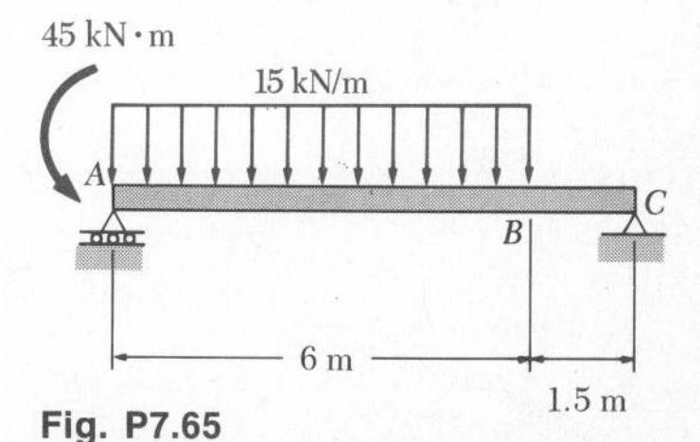

Fig. P7.65

7.66 Determine the equations of the shear and bending-moment curves for the beam and loading of Prob. 7.24. (Place the origin at point A.)

7.67 through 7.69 Determine the equations of the shear and bending-moment curves for the given beam and loading. Also determine the magnitude and location of the maximum bending moment in the beam.

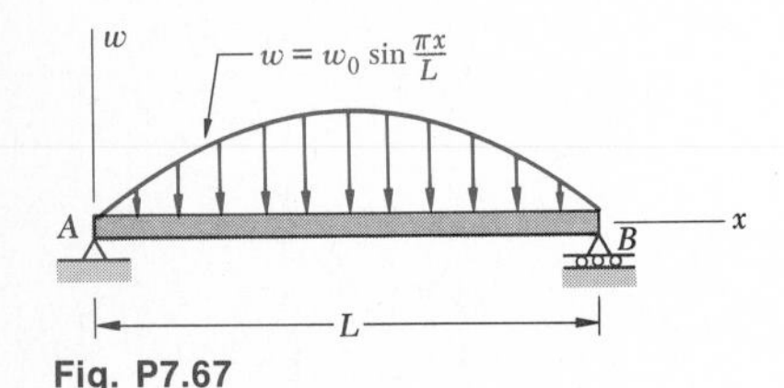

Fig. P7.67

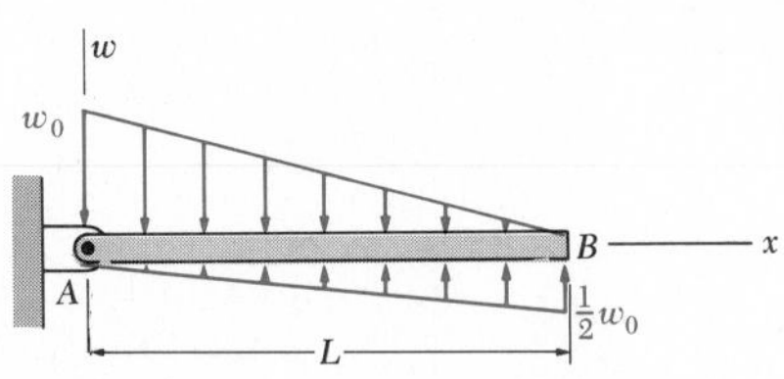

Fig. P7.68

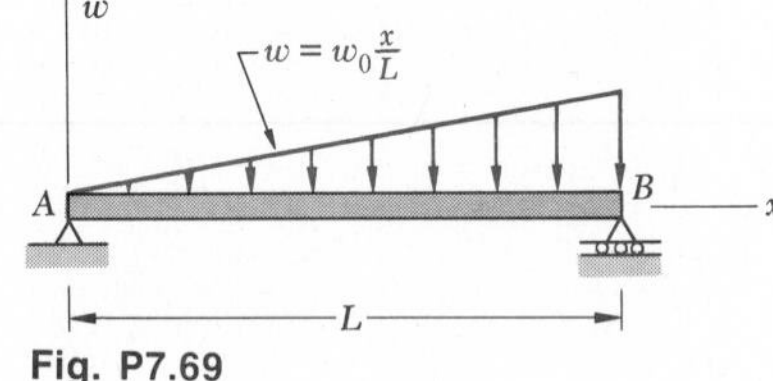

Fig. P7.69

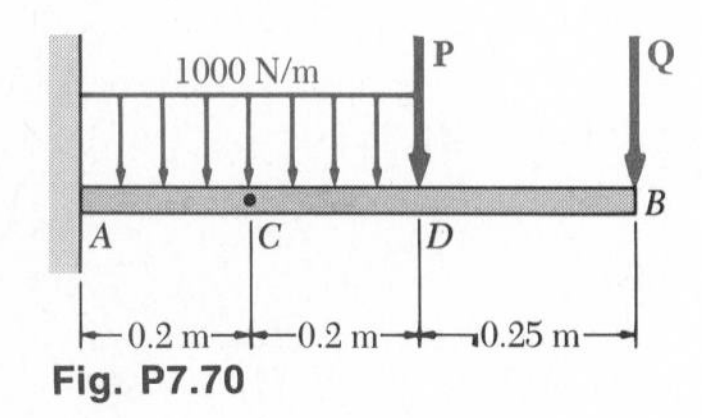

Fig. P7.70

7.70 The beam AB supports the uniformly distributed load of 1000 N/m and two concentrated loads **P** and **Q**. It has been experimentally determined that the bending moment is -395 N · m at point A and -215 N · m at point C. Determine **P** and **Q** and draw the shear and bending-moment diagrams for the beam.

***7.71** The beam AB is acted upon by the uniformly distributed load of 2 kips/ft and by two forces **P** and **Q**. It has been experimentally determined that the bending moment is +172 kip · ft at point D and +235 kip · ft at point E. Determine **P** and **Q**, and draw the shear and bending-moment diagrams for the beam.

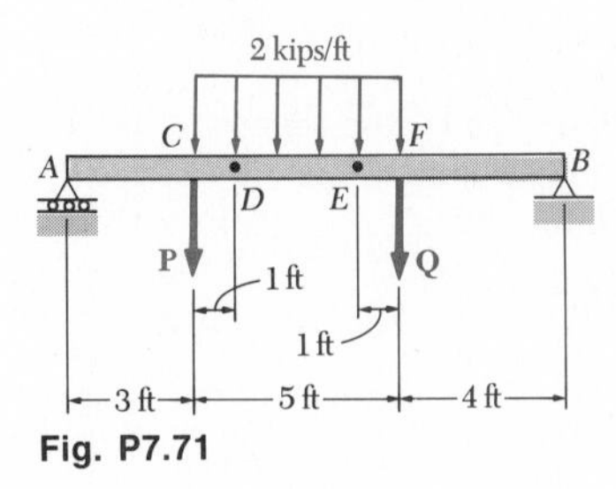

Fig. P7.71

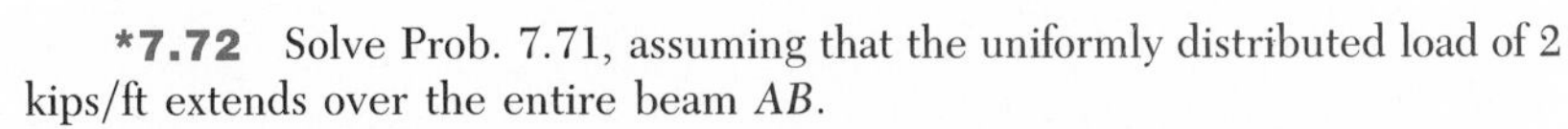
***7.72** Solve Prob. 7.71, assuming that the uniformly distributed load of 2 kips/ft extends over the entire beam AB.

***7.73** Solve Prob. 7.70, assuming that the uniformly distributed load of 1000 N/m extends over the entire beam AB.

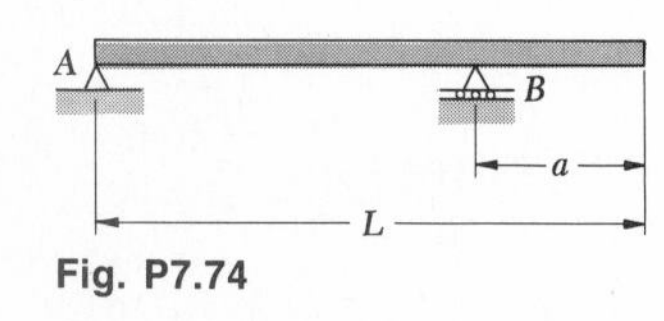

Fig. P7.74

***7.74** A uniform beam of weight w per unit length is supported as shown. Determine the distance a if the maximum absolute value of the bending moment is to be as small as possible. (See hint of Prob. 7.42.)

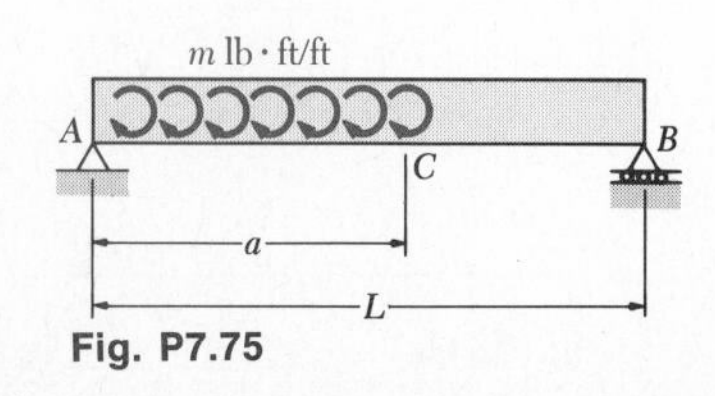

Fig. P7.75

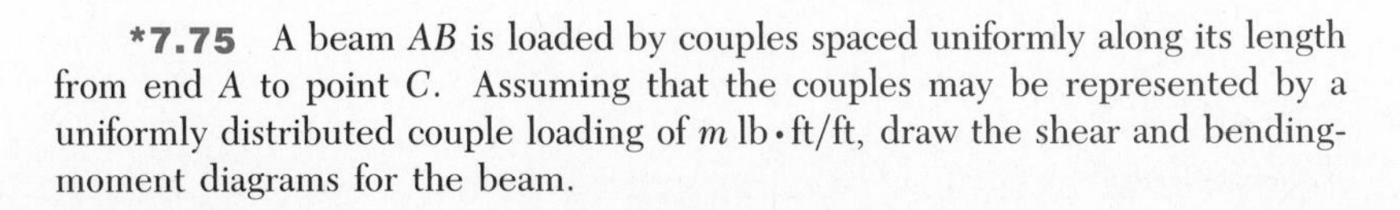
***7.75** A beam AB is loaded by couples spaced uniformly along its length from end A to point C. Assuming that the couples may be represented by a uniformly distributed couple loading of m lb · ft/ft, draw the shear and bending-moment diagrams for the beam.

***7.7. Cables with Concentrated Loads.** Cables are used in many engineering applications, such as suspension bridges, transmission lines, aerial tramways, guy wires for high towers, etc. Cables may be divided into two categories, according to their loading: (1) cables supporting concentrated loads, (2) cables supporting distributed loads. In this section, we shall examine cables of the first category.

Consider a cable attached to two fixed points A and B and supporting n given vertical concentrated loads $\mathbf{P}_1, \mathbf{P}_2, \ldots, \mathbf{P}_n$ (Fig. 7.13*a*). We assume that the cable is *flexible*, i.e., that its resistance to bending is small and may be neglected. We further assume that the *weight of the cable is negligible* compared with the loads supported by the cable. Any portion of cable between successive loads may therefore be considered as a two-force member, and the internal forces at any point in the cable reduce to a *force of tension directed along the cable.*

We assume that each of the loads lies in a given vertical line, i.e., that the horizontal distance from support A to each of the loads is known; we also assume that the horizontal and vertical distances between the supports are known. We propose to determine the shape of the cable, i.e., the vertical distance from support A to each of the points $C_1, C_2, \ldots, C_n$, and also the tension T in each portion of the cable.

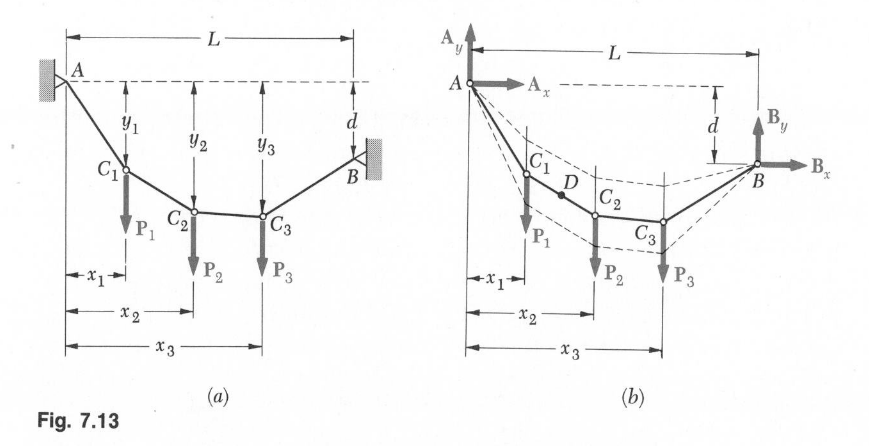

Fig. 7.13

We first draw the free-body diagram of the entire cable (Fig. 7.13*b*). Since the slope of the portions of cable attached at A and B is not known, the reactions at A and B must be represented by two components each. Thus, four unknowns are involved, and the three equations of equilibrium are not sufficient to determine the reactions at A and B.† We must therefore obtain an additional equation by considering the equilibrium of a por-

†Clearly, the cable is not a rigid body; the equilibrium equations represent, therefore, *necessary but not sufficient conditions* (see Sec. 6.11).

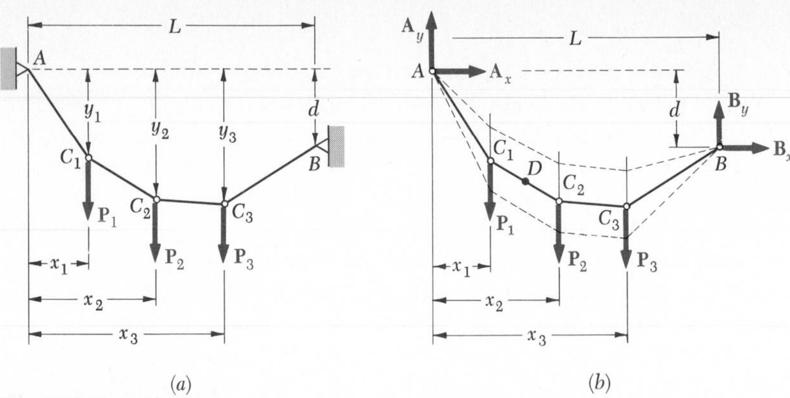

Fig. 7.13 (*repeated*)

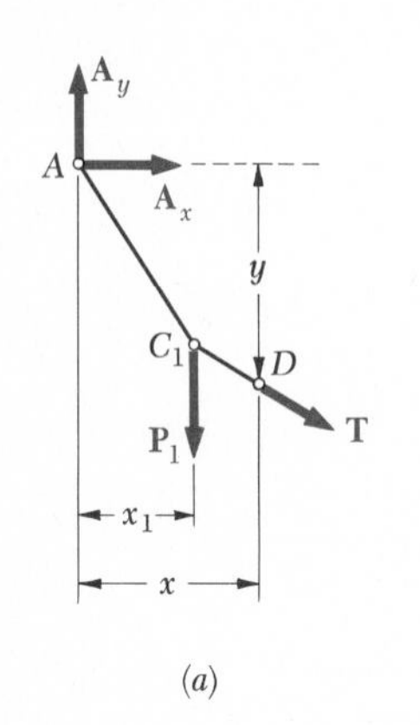

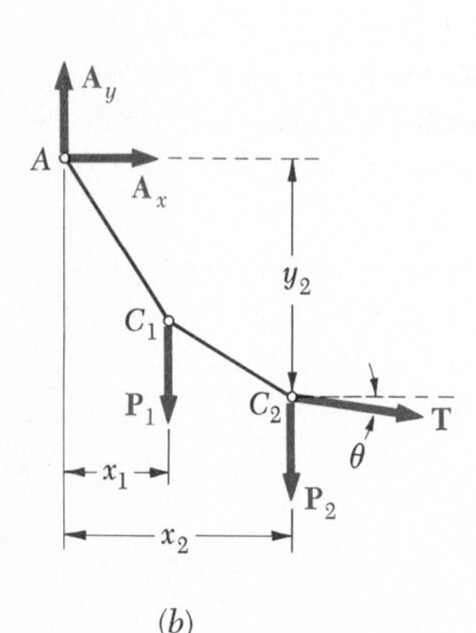

Fig. 7.14

tion of the cable. This is possible if we know the coordinates x and y of a point D of the cable. Drawing the free-body diagram of the portion of cable AD (Fig. 7.14*a*) and writing $\Sigma M_D = 0$, we obtain an additional relation between the scalar components A_x and A_y and may determine the reactions at A and B. The problem would remain indeterminate, however, if we did not know the coordinates of D, unless some other relation between A_x and A_y (or between B_x and B_y) were given. The cable might hang in any of various possible ways, as indicated by the dashed lines in Fig. 7.13*b*.

Once A_x and A_y have been determined, the vertical distance from A to any point of the cable may be easily found. Considering point C_2, for example, we draw the free-body diagram of the portion of cable AC_2 (Fig. 7.14*b*). Writing $\Sigma M_{C_2} = 0$, we obtain an equation which may be solved for y_2. Writing $\Sigma F_x = 0$ and $\Sigma F_y = 0$, we obtain the components of the force **T** representing the tension in the portion of cable to the right of C_2. We observe that $T \cos \theta = -A_x$; *the horizontal component of the tension force is the same at any point of the cable.* It follows that the tension T is maximum when $\cos \theta$ is minimum, i.e., in the portion of cable which has the largest angle of inclination θ. Clearly, this portion of cable must be adjacent to one of the two supports of the cable.

*** 7.8. Cables with Distributed Loads.** Consider a cable attached to two fixed points A and B and carrying a *distributed load* (Fig. 7.15*a*). We saw in the preceding section that for a cable supporting concentrated loads, the internal force at any point is a force of tension directed along the cable. In the case of a cable carrying a distributed load, the cable hangs in the shape of a curve, and the internal force at a point D is a force of tension **T** *directed along the tangent to the curve.* Given a certain distributed load, we propose in this section to determine the tension at any point of the cable. We shall also see in the following sections how the shape of the cable may be determined for two particular types of distributed loads.

Considering the most general case of distributed load, we draw the free-body diagram of the portion of cable extending from the lowest point

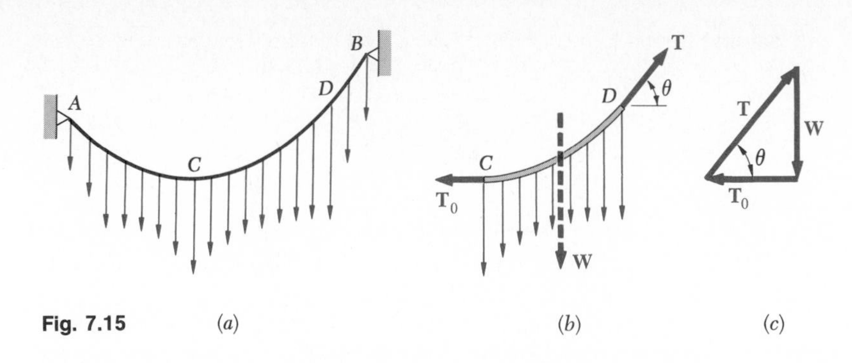

Fig. 7.15 (*a*) (*b*) (*c*)

C to a given point *D* of the cable (Fig. 7.15*b*). The forces acting on the free body are the tension force $\mathbf{T}_0$ at *C*, which is horizontal, the tension force **T** at *D*, directed along the tangent to the cable at *D*, and the resultant **W** of the distributed load supported by the portion of cable *CD*. Drawing the corresponding force triangle (Fig. 7.15*c*), we obtain the following relations:

$$T \cos \theta = T_0 \qquad T \sin \theta = W \tag{7.5}$$

$$T = \sqrt{T_0^2 + W^2} \qquad \tan \theta = \frac{W}{T_0} \tag{7.6}$$

From the relations (7.5), it appears that the horizontal component of the tension force **T** is the same at any point and that the vertical component of **T** is equal to the magnitude *W* of the load measured from the lowest point. Relations (7.6) show that the tension *T* is minimum at the lowest point and maximum at one of the two points of support.

***7.9. Parabolic Cable.** Let us assume, now, that the cable *AB* carries a load *uniformly distributed along the horizontal* (Fig. 7.16*a*). Cables of suspension bridges may be assumed loaded in this way, since the weight of the cables is small compared with the weight of the roadway. We denote by w the load per unit length (*measured horizontally*) and express it in N/m or in lb/ft. Choosing coordinate axes with origin at the lowest point *C* of the cable, we find that the magnitude *W* of the total load carried by the portion of cable extending from *C* to the point *D* of coordinates x and y is $W = wx$. The relations (7.6) defining the magnitude and direction of the tension force at *D* become

$$T = \sqrt{T_0^2 + w^2x^2} \qquad \tan \theta = \frac{wx}{T_0} \tag{7.7}$$

Moreover, the distance from *D* to the line of action of the resultant **W** is equal to half the horizontal distance from *C* to *D* (Fig. 7.16*b*). Summing moments about *D*, we write

$$+\circlearrowleft \Sigma M_D = 0: \qquad wx\frac{x}{2} - T_0 y = 0$$

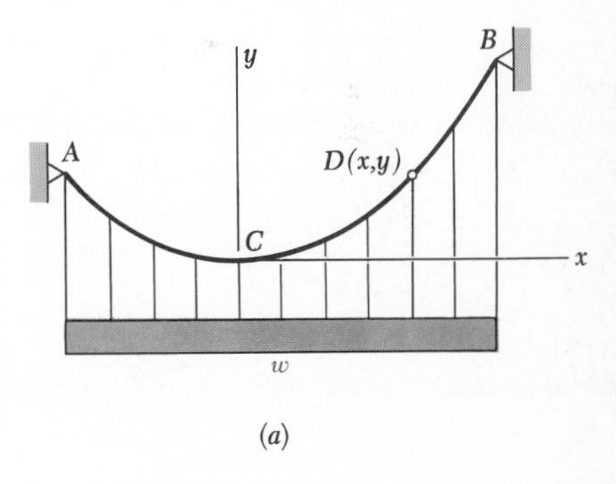

(*a*)

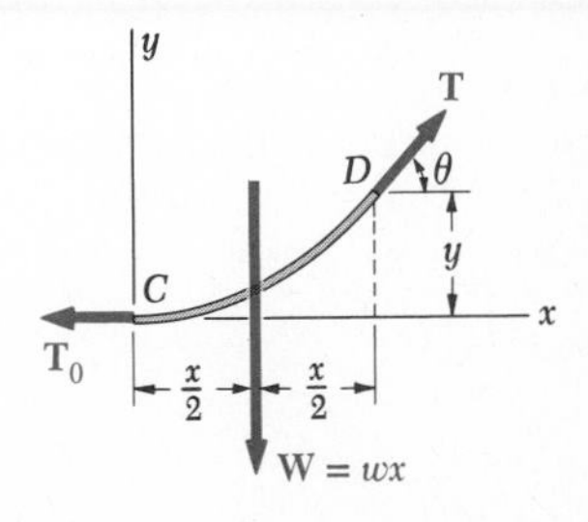

(*b*)

Fig. 7.16

and, solving for y,

$$y = \frac{wx^2}{2T_0} \tag{7.8}$$

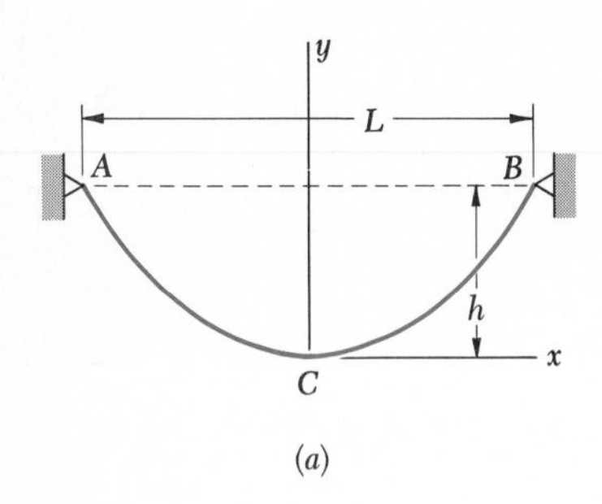

(a)

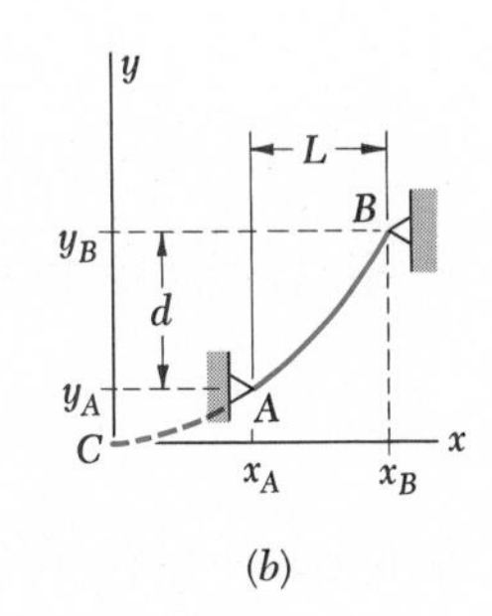

(b)

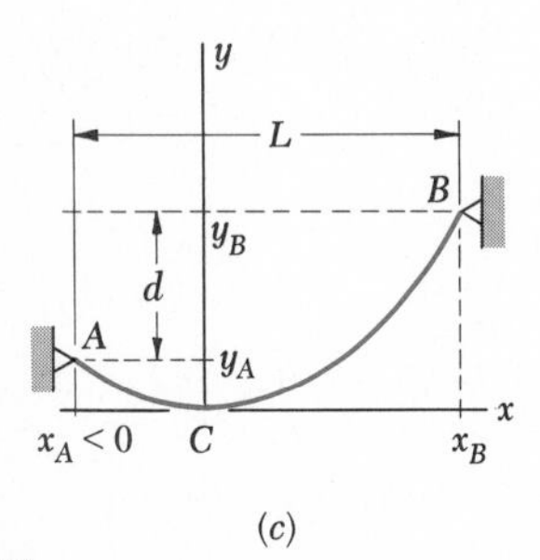

(c)

Fig. 7.17

This is the equation of a *parabola* with a vertical axis and its vertex at the origin of coordinates. The curve formed by cables loaded uniformly along the horizontal is thus a parabola.†

When the supports A and B of the cable have the same elevation, the distance L between the supports is called the *span* of the cable and the vertical distance h from the supports to the lowest point is called the *sag* of the cable (Fig. 7.17*a*). If the span and sag of a cable are known, and if the load w per unit horizontal length is given, the minimum tension T_0 may be found by substituting $x = L/2$ and $y = h$ in formula (7.8). Formulas (7.7) and (7.8) will then define the tension at any point and the shape of the cable.

When the supports have different elevations, the position of the lowest point of the cable is not known and the coordinates x_A, y_A and x_B, y_B of the supports must be determined. To this effect, we express that the coordinates of A and B satisfy Eq. (7.8) and that $x_B - x_A = L$, $y_B - y_A = d$, where L and d denote, respectively, the horizontal and vertical distances between the two supports (Fig. 7.17*b* and *c*).

The length of the cable from its lowest point C to its support B may be obtained from the formula

$$s_B = \int_0^{x_B} \sqrt{1 + \left(\frac{dy}{dx}\right)^2}\, dx \tag{7.9}$$

Differentiating (7.8), we obtain the derivative $dy/dx = wx/T_0$; substituting into (7.9) and using the binomial theorem to expand the radical in an infinite series, we have

$$s_B = \int_0^{x_B} \sqrt{1 + \frac{w^2x^2}{T_0^2}}\, dx = \int_0^{x_B} \left(1 + \frac{w^2x^2}{2T_0^2} - \frac{w^4x^4}{8T_0^4} + \cdots\right) dx$$

$$s_B = x_B\left(1 + \frac{w^2x_B^2}{6T_0^2} - \frac{w^4x_B^4}{40T_0^4} + \cdots\right)$$

and, since $wx_B^2/2T_0 = y_B$,

$$s_B = x_B\left[1 + \frac{2}{3}\left(\frac{y_B}{x_B}\right)^2 - \frac{2}{5}\left(\frac{y_B}{x_B}\right)^4 + \cdots\right] \tag{7.10}$$

The series converges for values of the ratio y_B/x_B less than 0.5; in most cases, this ratio is much smaller, and only the first two terms of the series need be computed.

† Cables hanging under their own weight are not loaded uniformly along the horizontal, and they do not form a parabola. The error introduced by assuming a parabolic shape for cables hanging under their own weight, however, is small when the cable is sufficiently taut. A complete discussion of cables hanging under their own weight is given in the next section.

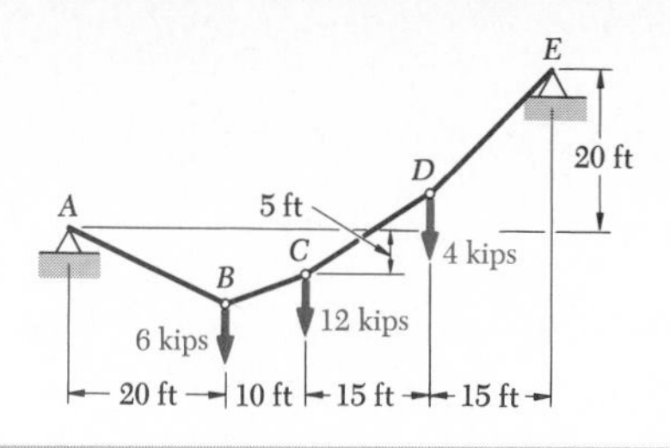

SAMPLE PROBLEM 7.8

The cable AE supports three vertical loads from the points indicated. If point C is 5 ft below the left support, determine (*a*) the elevations of points B and D, (*b*) the maximum slope and the maximum tension in the cable.

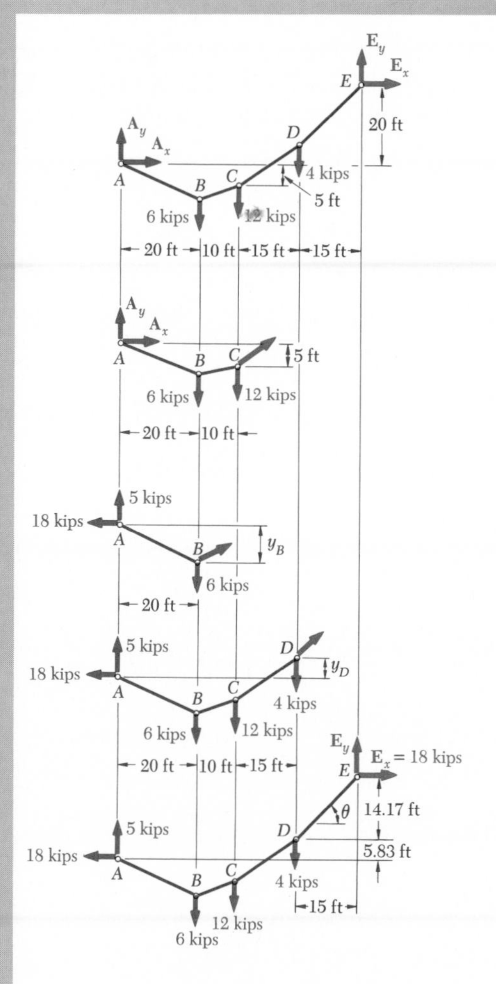

Solution. The reaction components $\mathbf{A}_x$ and $\mathbf{A}_y$ are determined as follows:

Free Body: Entire Cable

$$+\circlearrowleft \Sigma M_E = 0:$$

$$A_x(20 \text{ ft}) - A_y(60 \text{ ft}) + (6 \text{ kips})(40 \text{ ft}) + (12 \text{ kips})(30 \text{ ft}) + (4 \text{ kips})(15 \text{ ft}) = 0$$

$$20A_x - 60A_y + 660 = 0$$

Free Body: ABC

$$+\circlearrowleft \Sigma M_C = 0: \quad -A_x(5 \text{ ft}) - A_y(30 \text{ ft}) + (6 \text{ kips})(10 \text{ ft}) = 0$$

$$-5A_x - 30A_y + 60 = 0$$

Solving the two equations simultaneously, we obtain

$$A_x = -18 \text{ kips} \qquad \mathbf{A}_x = 18 \text{ kips} \leftarrow$$

$$A_y = +5 \text{ kips} \qquad \mathbf{A}_y = 5 \text{ kips} \uparrow$$

a. Elevation of Point B. Considering the portion of cable AB *as a free body*, we write

$$+\circlearrowleft \Sigma M_B = 0: \quad (18 \text{ kips})y_B - (5 \text{ kips})(20 \text{ ft}) = 0$$

$$y_B = 5.56 \text{ ft below } A \quad ◀$$

Elevation of Point D. Using the portion of cable $ABCD$ *as a free body*, we write

$$+\circlearrowleft \Sigma M_D = 0:$$

$$-(18 \text{ kips})y_D - (5 \text{ kips})(45 \text{ ft}) + (6 \text{ kips})(25 \text{ ft}) + (12 \text{ kips})(15 \text{ ft}) = 0$$

$$y_D = 5.83 \text{ ft above } A \quad ◀$$

b. Maximum Slope and Maximum Tension. We observe that the maximum slope occurs in portion DE. Since the horizontal component of the tension is constant and equal to 18 kips, we write

$$\tan\theta = \frac{14.17}{15 \text{ ft}} \qquad \theta = 43.4° \quad ◀$$

$$T_{max} = \frac{18 \text{ kips}}{\cos\theta} \qquad T_{max} = 24.8 \text{ kips} \quad ◀$$

SAMPLE PROBLEM 7.9

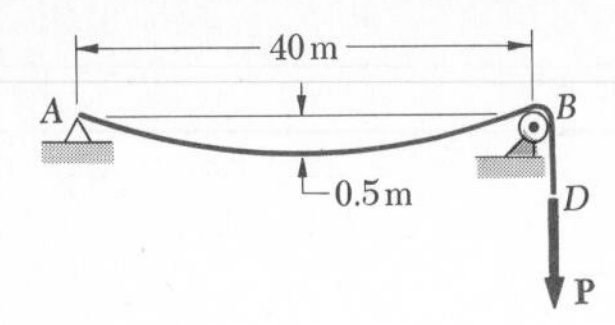

A light cable is attached to a support at A, passes over a small pulley at B, and supports a load **P**. Knowing that the sag of the cable is 0.5 m and that the mass per unit length of the cable is 0.75 kg/m, determine (*a*) the magnitude of the load **P**, (*b*) the slope of the cable at B, and (*c*) the total length of the cable from A to B. Since the ratio of the sag to the span is small, assume the cable to be parabolic. Also, neglect the weight of the portion of cable from B to D.

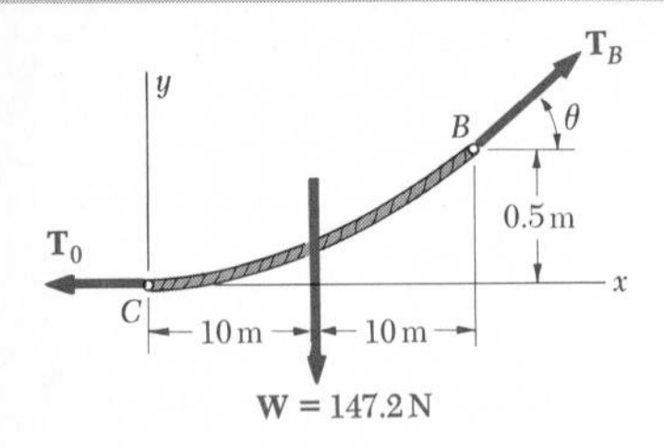

***a.* Load P.** We denote by C the lowest point of the cable and draw the free-body diagram of the portion CB of cable. Assuming the load to be uniformly distributed along the horizontal, we write

$$w = (0.75\text{ kg/m})(9.81\text{ m/s}^2) = 7.36\text{ N/m}$$

The total load for the portion CB of the cable is

$$W = wx_B = (7.36\text{ N/m})(20\text{ m}) = 147.2\text{ N}$$

and is applied halfway between C and B. Summing moments about B, we write

$$+\circlearrowleft \Sigma M_B = 0: \qquad (147.2\text{ N})(10\text{ m}) - T_0(0.5\text{ m}) = 0 \qquad T_0 = 2944\text{ N}$$

From the force triangle we obtain

$$T_B = \sqrt{T_0^2 + W^2}$$
$$= \sqrt{(2944\text{ N})^2 + (147.2\text{ N})^2} = 2948\text{ N}$$

Since the tension on each side of the pulley is the same, we find

$$P = T_B = 2948\text{ N} \quad ◄$$

***b.* Slope of Cable at *B*.** We also obtain from the force triangle

$$\tan\theta = \frac{W}{T_0} = \frac{147.2\text{ N}}{2944\text{ N}} = 0.05$$

$$\theta = 2.9° \quad ◄$$

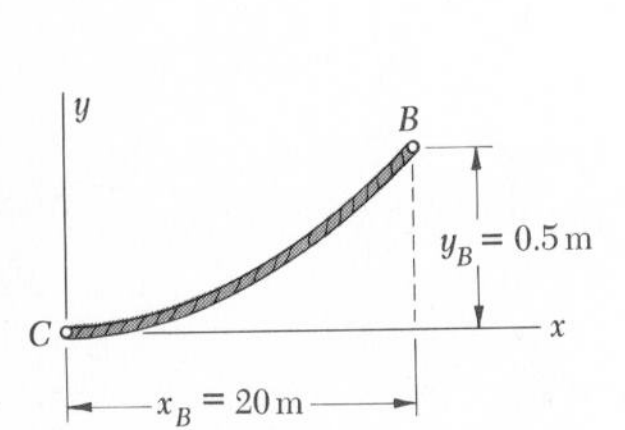

***c.* Length of Cable.** Applying Eq. (7.10) between C and B, we write

$$s_B = x_B\left[1 + \frac{2}{3}\left(\frac{y_B}{x_B}\right)^2 + \cdots\right]$$
$$= (20\text{ m})\left[1 + \frac{2}{3}\left(\frac{0.5\text{ m}}{20\text{ m}}\right)^2 + \cdots\right] = 20.00833\text{ m}$$

The total length of the cable between A and B is twice this value,

$$\text{Length} = 2s_B = 40.0167\text{ m} \quad ◄$$

Problems

7.76 Three loads are suspended as shown from the cable. Knowing that $h_C = 4$ m, determine (*a*) the components of the reaction at E, (*b*) the maximum value of the tension in the cable.

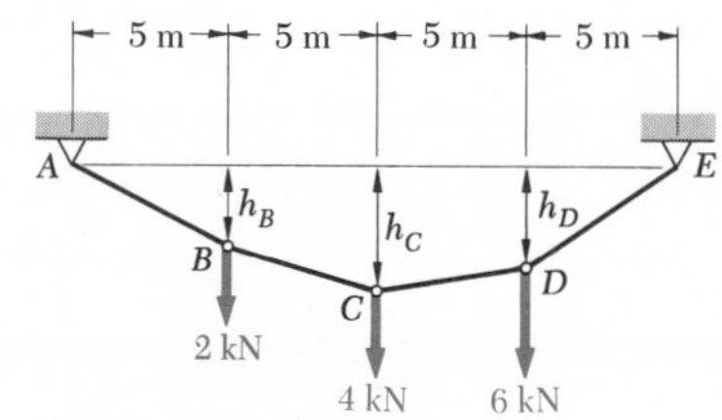

Fig. P7.76 and P7.77

7.77 Determine the sag at point C if the maximum tension in the cable is 25 kN.

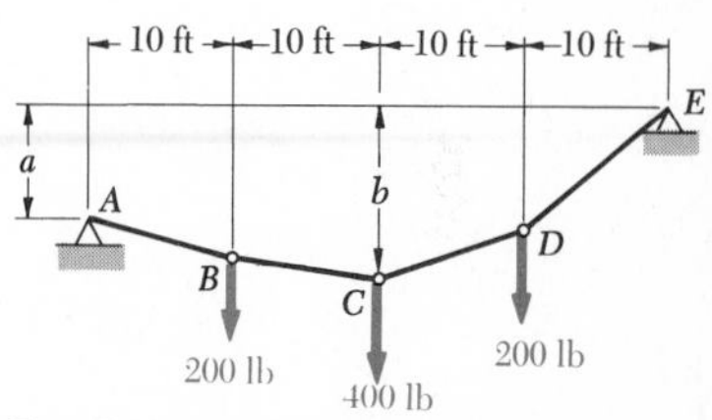

Fig. P7.78 and P7.79

7.78 If $a = 8$ ft and $b = 10$ ft, determine the components of the reaction at E for the cable and loading shown.

7.79 If $a = b = 6$ ft, determine (*a*) the components of the reaction at E, (*b*) the maximum tension in the cable.

7.80 If $h_C = 8$ m, determine (*a*) the components of the reaction at E, (*b*) the maximum tension in the cable.

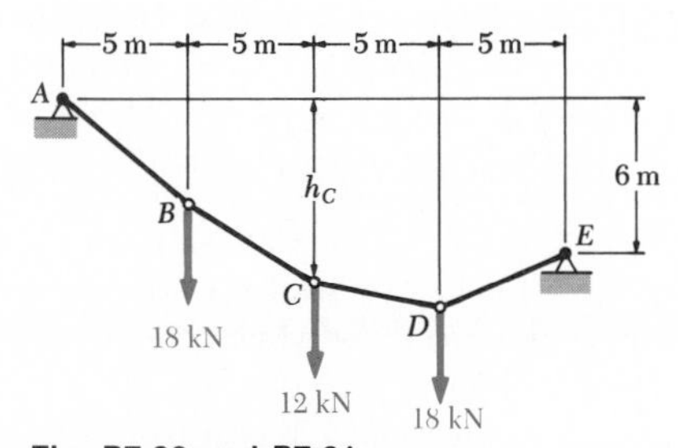

Fig. P7.80 and P7.81

7.81 Determine (*a*) the distance h_C for which the portion CD of the cable is horizontal, (*b*) the corresponding components of the reaction at E.

7.82 Cable ABC supports two loads as shown. Knowing that the distance b is to be 6 m, determine (*a*) the required magnitude of the force **P**, (*b*) the corresponding distance a.

7.83 Cable ABC supports two loads as shown. Determine the distances a and b when a horizontal force **P** of magnitude 800 N is applied at C.

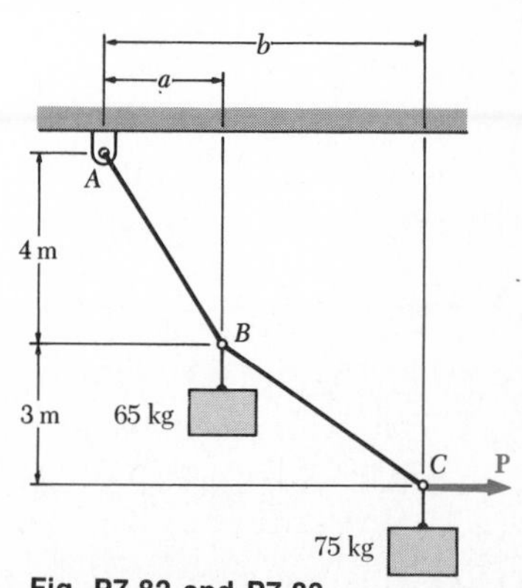

Fig. P7.82 and P7.83

7.84 Knowing that the 264-lb force applied at B and the block attached at C maintain cable $ABCD$ in the position shown, determine (a) the reaction at A, (b) the required weight W of the block, (c) the tension in each portion of the cable.

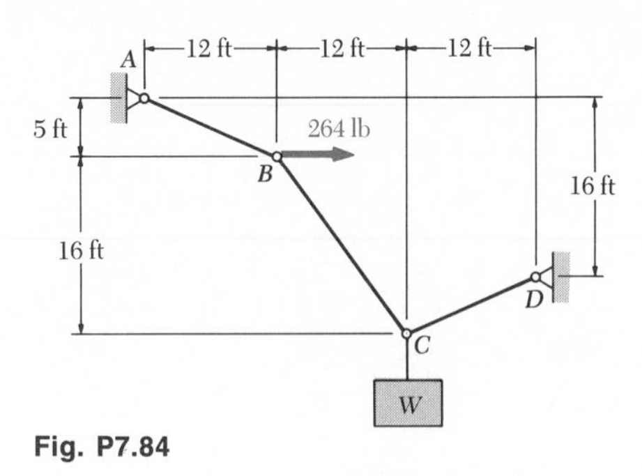

Fig. P7.84

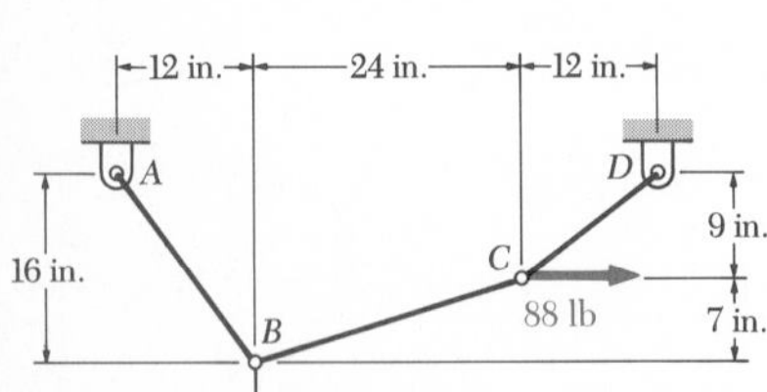
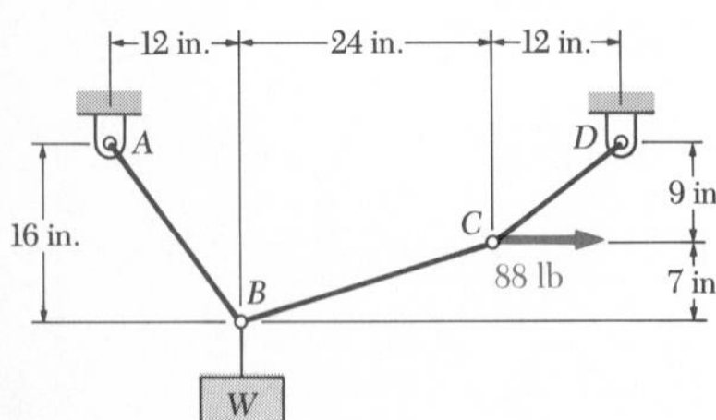

Fig. P7.85

7.85 Knowing that the 88-lb force applied at C and the block attached at B maintain the cable $ABCD$ in the position shown, determine (a) the reaction at D, (b) the required weight W of the block, (c) the tension in each portion of the cable.

7.86 An electric wire having a mass per unit length of 0.6 kg/m is strung between two insulators at the same elevation and 60 m apart. Knowing that the sag of the wire is 1.5 m, determine (a) the maximum tension in the wire, (b) the length of the wire.

7.87 The center span of the George Washington Bridge, as originally constructed, consisted of a uniform roadway suspended from four cables. The uniform load supported by each cable was $w = 9.75$ kips/ft along the horizontal. Knowing that the span L is 3500 ft and that the sag h is 316 ft, determine (a) the maximum tension in each cable, (b) the length of each cable.

7.88 The center span of the Verrazano-Narrows Bridge consists of two uniform roadways suspended from four cables. The design of the bridge included the effect of extreme temperature changes which cause the sag of the center span to vary from $h_w = 386$ ft in winter to $h_s = 394$ ft in summer. Knowing that the span is $L = 4260$ ft, determine the change in length of the cables due to extreme temperature variation.

7.89 A 76-m length of wire having a mass per unit length of 2.2 kg/m is used to span a horizontal distance of 75 m. Determine (a) the approximate sag of the wire, (b) the maximum tension in the wire. [*Hint.* Use only the first two terms of Eq. (7.10).]

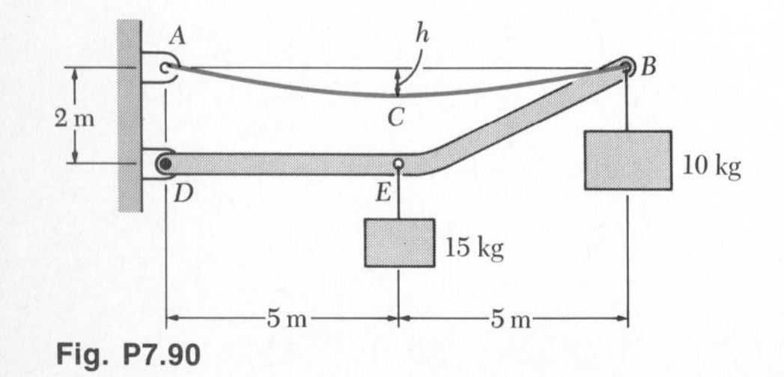

Fig. P7.90

7.90 The total mass of cable ACB is 10 kg. Assuming that the mass of the cable is distributed uniformly along the horizontal, determine (a) the sag h, (b) the slope of the cable at A.

7.91 The total mass of cable AC is 25 kg. Assuming that the mass of the cable is distributed uniformly along the horizontal, determine the sag h and the slope of the cable at A and C.

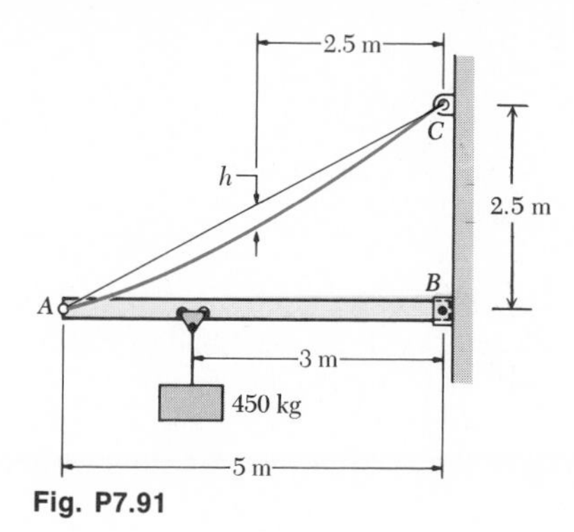

Fig. P7.91

7.92 Cable AB supports a load distributed uniformly along the horizontal as shown. The lowest point of the cable is located at a distance $a = 9$ ft below the support A. Determine the maximum and minimum values of the tension in the cable.

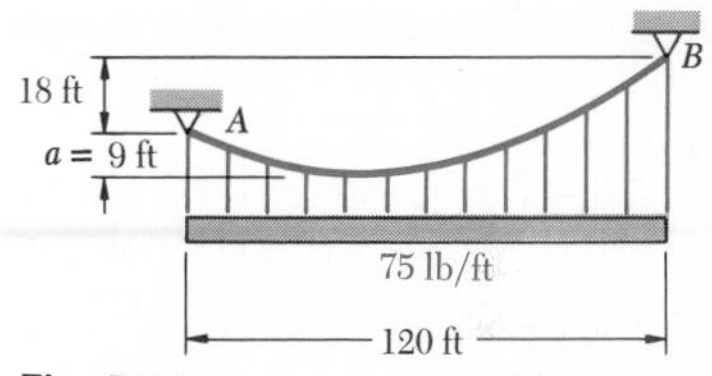

Fig. P7.92

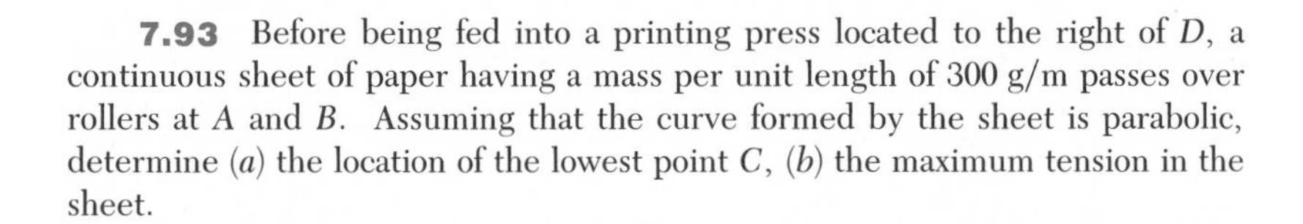

7.93 Before being fed into a printing press located to the right of D, a continuous sheet of paper having a mass per unit length of 300 g/m passes over rollers at A and B. Assuming that the curve formed by the sheet is parabolic, determine (*a*) the location of the lowest point C, (*b*) the maximum tension in the sheet.

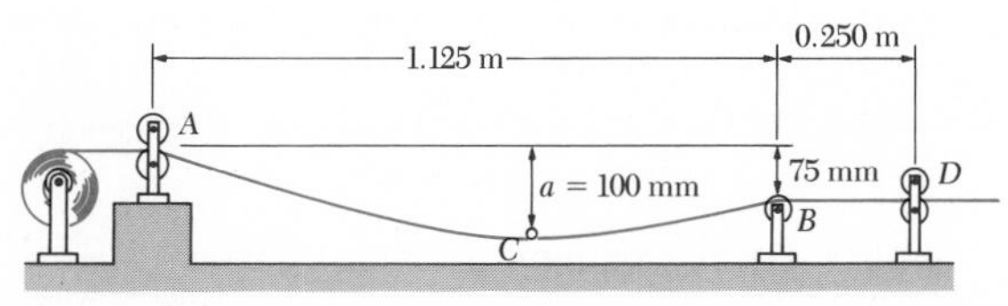

Fig. P7.93

7.94 Solve Prob. 7.93, assuming that $a = 120$ mm.

7.95 Solve Prob. 7.92, assuming that $a = 6$ ft.

***7.96** A cable AB of span L and a simple beam $A'B'$ of the same span are subjected to identical vertical loadings as shown. Show that the magnitude of the bending moment at a point C' in the beam is equal to the product of T_0h, where T_0 is the magnitude of the horizontal component of the tension force in the cable and h is the vertical distance between point C of the cable and the chord joining the points of support A and B.

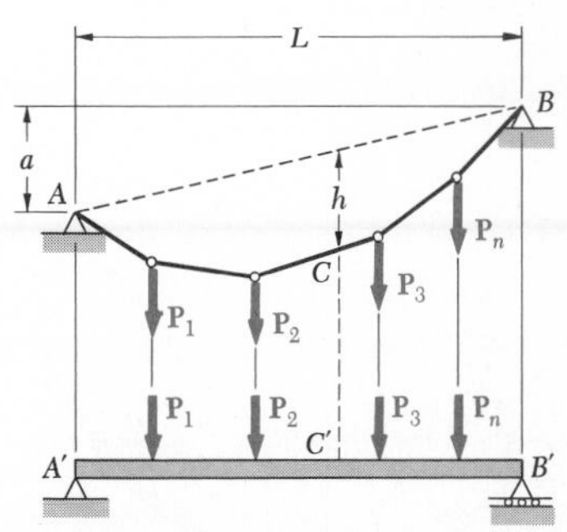

Fig. P7.96

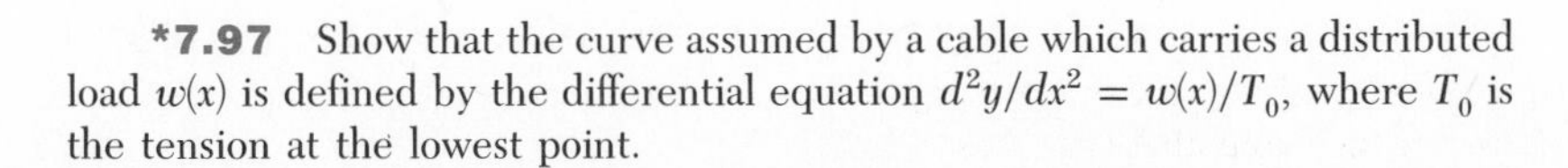

***7.97** Show that the curve assumed by a cable which carries a distributed load $w(x)$ is defined by the differential equation $d^2y/dx^2 = w(x)/T_0$, where T_0 is the tension at the lowest point.

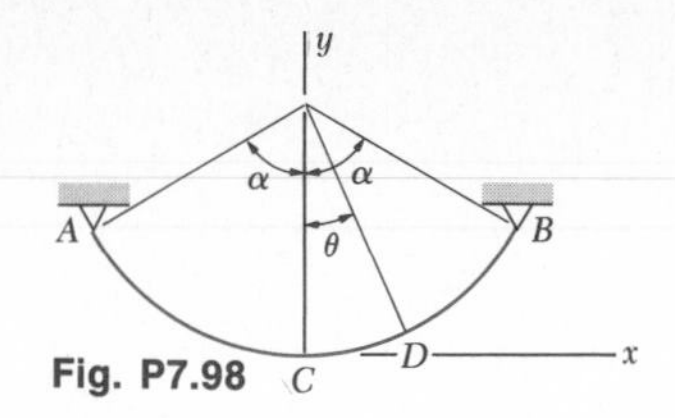

Fig. P7.98

***7.98** If the weight per unit length of the cable AB is $w_0/\cos^2\theta$, prove that the curve formed by the cable is a circular arc. (*Hint:* Use the property indicated in Prob. 7.97.)

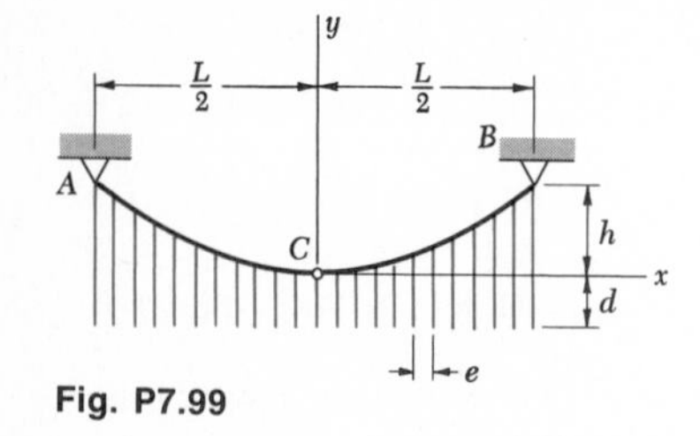

Fig. P7.99

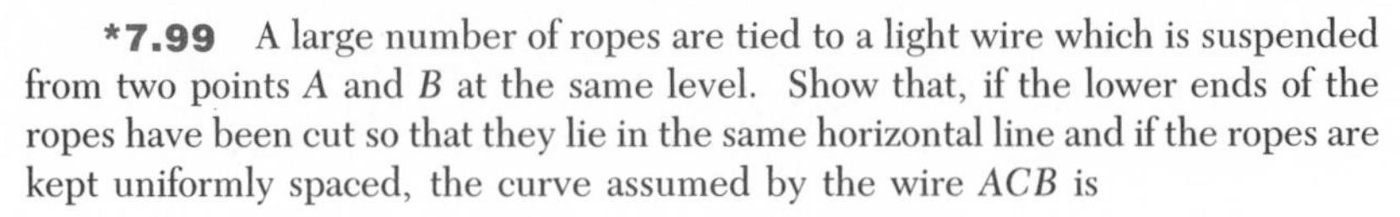

***7.99** A large number of ropes are tied to a light wire which is suspended from two points A and B at the same level. Show that, if the lower ends of the ropes have been cut so that they lie in the same horizontal line and if the ropes are kept uniformly spaced, the curve assumed by the wire ACB is

$$y + d = d\cosh\left(\frac{w_r}{T_0 e}\right)^{1/2} x$$

where w_r is the weight per unit length of the ropes, d is the length of the shortest rope, and e is the horizontal distance between adjacent ropes. (*Hint.* Use the property indicated in Prob. 7.97.)

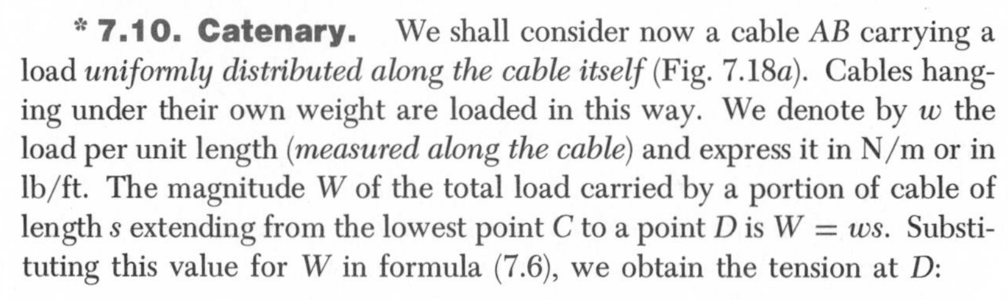

*** 7.10. Catenary.** We shall consider now a cable AB carrying a load *uniformly distributed along the cable itself* (Fig. 7.18*a*). Cables hanging under their own weight are loaded in this way. We denote by w the load per unit length (*measured along the cable*) and express it in N/m or in lb/ft. The magnitude W of the total load carried by a portion of cable of length s extending from the lowest point C to a point D is $W = ws$. Substituting this value for W in formula (7.6), we obtain the tension at D:

$$T = \sqrt{T_0^2 + w^2 s^2}$$

In order to simplify the subsequent computations, we shall introduce the constant $c = T_0/w$. We thus write

$$T_0 = wc \qquad W = ws \qquad T = w\sqrt{c^2 + s^2} \tag{7.11}$$

The free-body diagram of the portion of cable CD is shown in Fig. 7.18*b*. This diagram, however, cannot be used to obtain directly the equa-

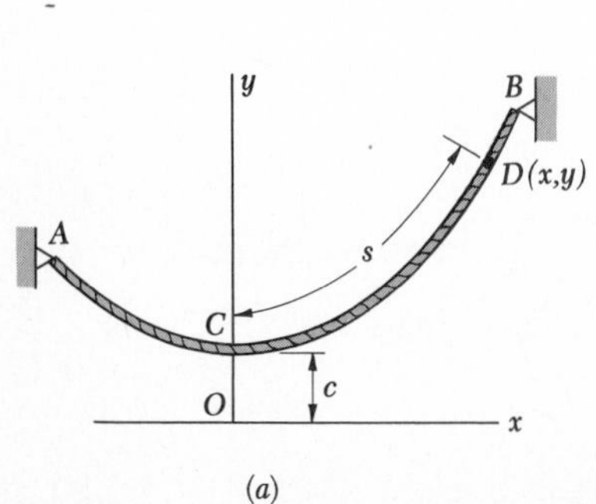

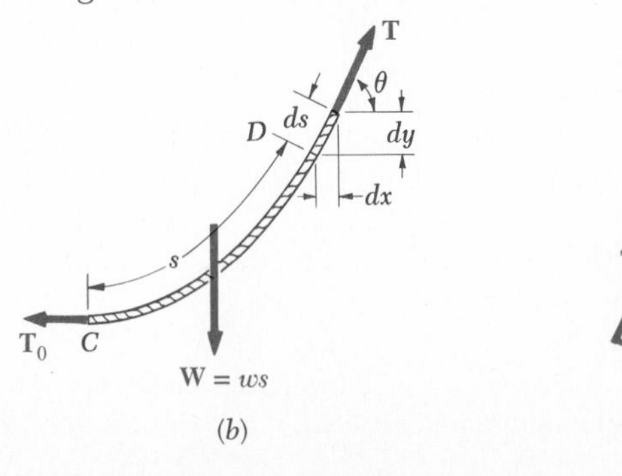

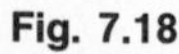
Fig. 7.18

tion of the curve assumed by the cable, since we do not know the horizontal distance from D to the line of action of the resultant $\mathbf{W}$ of the load. To obtain this equation, we shall write first that the horizontal projection of a small element of cable of length ds is $dx = ds \cos \theta$. Observing from Fig. 7.18c that $\cos \theta = T_0/T$ and using (7.11), we write

$$dx = ds \cos \theta = \frac{T_0}{T} ds = \frac{wc\, ds}{w\sqrt{c^2 + s^2}} = \frac{ds}{\sqrt{1 + s^2/c^2}}$$

Selecting the origin O of the coordinates at a distance c directly below C (Fig. 7.18a) and integrating from $C(0, c)$ to $D(x, y)$, we obtain†

$$x = \int_0^s \frac{ds}{\sqrt{1 + s^2/c^2}} = c\left[\sinh^{-1} \frac{s}{c}\right]_0^s = c \sinh^{-1} \frac{s}{c}$$

This equation, which relates the length s of the portion of cable CD and the horizontal distance x, may be written in the form

$$s = c \sinh \frac{x}{c} \tag{7.15}$$

The relation between the coordinates x and y may now be obtained by writing $dy = dx \tan \theta$. Observing from Fig. 7.18c that $\tan \theta = W/T_0$ and using (7.11) and (7.15), we write

$$dy = dx \tan \theta = \frac{W}{T_0} dx = \frac{s}{c} dx = \sinh \frac{x}{c} dx$$

Integrating from $C(0, c)$ to $D(x, y)$ and using (7.12) and (7.13), we obtain

$$y - c = \int_0^x \sinh \frac{x}{c} dx = c\left[\cosh \frac{x}{c}\right]_0^x = c\left(\cosh \frac{x}{c} - 1\right)$$

$$y - c = c \cosh \frac{x}{c} - c$$

†This integral may be found in all standard integral tables. The function

$$z = \sinh^{-1} u$$

(read "arc hyperbolic sine u") is the *inverse* of the function $u = \sinh z$ (read "hyperbolic sine z"). This function and the function $v = \cosh z$ (read "hyperbolic cosine z") are defined as follows:

$$u = \sinh z = \tfrac{1}{2}(e^z - e^{-z}) \qquad v = \cosh z = \tfrac{1}{2}(e^z + e^{-z})$$

Numerical values of the functions $\sinh z$ and $\cosh z$ are found in *tables of hyperbolic functions*. They may also be computed on most calculators either directly or from the above definitions. The student is referred to any calculus text for a complete description of the properties of these functions. In this section, we shall make use only of the following properties, which may be easily derived from the above definitions:

$$\frac{d \sinh z}{dz} = \cosh z \qquad \frac{d \cosh z}{dz} = \sinh z \tag{7.12}$$

$$\sinh 0 = 0 \qquad \cosh 0 = 1 \tag{7.13}$$

$$\cosh^2 z - \sinh^2 z = 1 \tag{7.14}$$

which reduces to

$$y = c \cosh \frac{x}{c} \tag{7.16}$$

This is the equation of a *catenary* with vertical axis. The ordinate c of the lowest point C is called the *parameter* of the catenary. Squaring both sides of Eqs. (7.15) and (7.16), subtracting, and taking (7.14) into account, we obtain the following relation between y and s:

$$y^2 - s^2 = c^2 \tag{7.17}$$

Solving (7.17) for s^2 and carrying into the last of the relations (7.11), we write these relations as follows:

$$T_0 = wc \qquad W = ws \qquad T = wy \tag{7.18}$$

The last relation indicates that the tension at any point P of the cable is proportional to the vertical distance from D to the horizontal line representing the x axis.

When the supports A and B of the cable have the same elevation, the distance L between the supports is called the *span* of the cable and the vertical distance h from the supports to the lowest point C is called the *sag* of the cable. These definitions are the same that were given in the case of parabolic cables, but it should be noted that because of our choice of coordinate axes, the sag h is now

$$h = y_A - c \tag{7.19}$$

It should also be observed that certain catenary problems involve transcendental equations which must be solved by successive approximations (see Sample Prob. 7.10). When the cable is fairly taut, however, the load may be assumed uniformly distributed *along the horizontal* and the catenary may be replaced by a parabola. The solution of the problem is thus greatly simplified, while the error introduced is small.

When the supports A and B have different elevations, the position of the lowest point of the cable is not known. The problem may be solved then in a manner similar to that indicated for parabolic cables, by expressing that the cable must pass through the supports and that $x_B - x_A = L$, $y_B - y_A = d$, where L and d denote, respectively, the horizontal and vertical distances between the two supports.

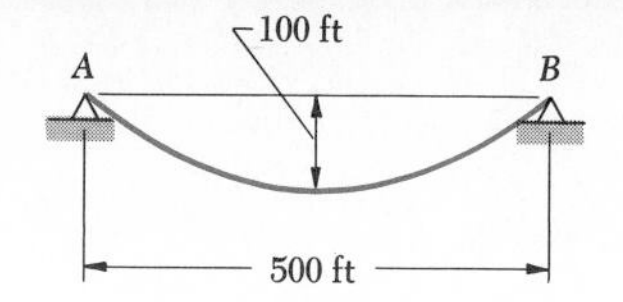

SAMPLE PROBLEM 7.10

A uniform cable weighing 3 lb/ft is suspended between two points A and B as shown. Determine (*a*) the maximum and minimum values of the tension in the cable, (*b*) the length of the cable.

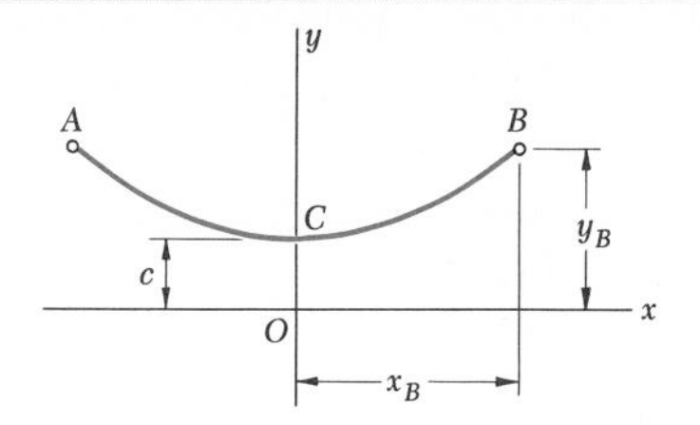

Solution. ***Equation of Cable.*** The origin of coordinates is placed at a distance c below the lowest point of the cable. The equation of the cable is given by Eq. (7.16),

$$y = c \cosh \frac{x}{c}$$

The coordinates of point B are

$$x_B = 250 \text{ ft} \qquad y_B = 100 + c$$

Substituting these coordinates into the equation of the cable, we obtain

$$100 + c = c \cosh \frac{250}{c}$$

$$\frac{100}{c} + 1 = \cosh \frac{250}{c}$$

The value of c is determined by assuming successive trial values, as shown in the following table:

c	$\frac{250}{c}$	$\frac{100}{c}$	$\frac{100}{c} + 1$	$\cosh \frac{250}{c}$
300	0.833	0.333	1.333	1.367
350	0.714	0.286	1.286	1.266
330	0.758	0.303	1.303	1.301
328	0.762	0.305	1.305	1.305

Taking $c = 328$, we have

$$y_B = 100 + c = 428 \text{ ft}$$

a. Maximum and Minimum Values of the Tension. Using Eqs. (7.18), we obtain

$$T_{\min} = T_0 = wc = (3 \text{ lb/ft})(328 \text{ ft}) \qquad T_{\min} = 984 \text{ lb} \blacktriangleleft$$
$$T_{\max} = T_B = wy_B = (3 \text{ lb/ft})(428 \text{ ft}) \qquad T_{\max} = 1284 \text{ lb} \blacktriangleleft$$

b. Length of Cable. One-half the length of the cable is found by solving Eq. (7.17):

$$y_B^2 - s_{CB}^2 = c^2 \qquad s_{CB}^2 = y_B^2 - c^2 = (428)^2 - (328)^2 \qquad s_{CB} = 275 \text{ ft}$$

The total length of the cable is therefore

$$s_{AB} = 2s_{CB} = 2(275 \text{ ft}) \qquad s_{AB} = 550 \text{ ft} \blacktriangleleft$$

Problems

7.100 A 40-m rope is strung between the roofs of two buildings, each 14 m high. The maximum tension is found to be 350 N and the lowest point of the cable is observed to be 6 m above the ground. Determine (*a*) the horizontal distance between the buildings, (*b*) the total mass of the rope.

7.101 A 50-m steel surveying tape has a mass of 1.6 kg. If the tape is stretched between two points at the same elevation and pulled until the tension at each end is 60 N, determine the horizontal distance between the ends of the tape. Neglect the elongation of the tape due to the tension.

7.102 An aerial tramway cable of length 500 ft and weighing 2.8 lb/ft is suspended between two points at the same elevation. Knowing that the sag is 125 ft, find (*a*) the horizontal distance between the supports, (*b*) the maximum tension in the cable.

7.103 A 25-ft chain of weight 30 lb is suspended between two points at the same elevation. Knowing that the sag is observed to be 10 ft, determine (*a*) the distance between the supports, (*b*) the maximum tension in the chain.

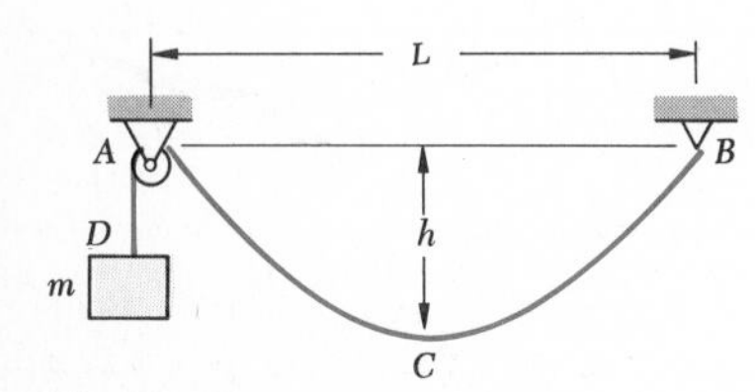

Fig. P7.104

7.104 A counterweight D of mass 40 kg is attached to a cable which passes over a small pulley at A and is attached to a support at B. Knowing that $L = 15$ m and $h = 5$ m, determine (*a*) the length of the cable from A to B, (*b*) the mass per unit length of the cable. Neglect the mass of the cable from A to D.

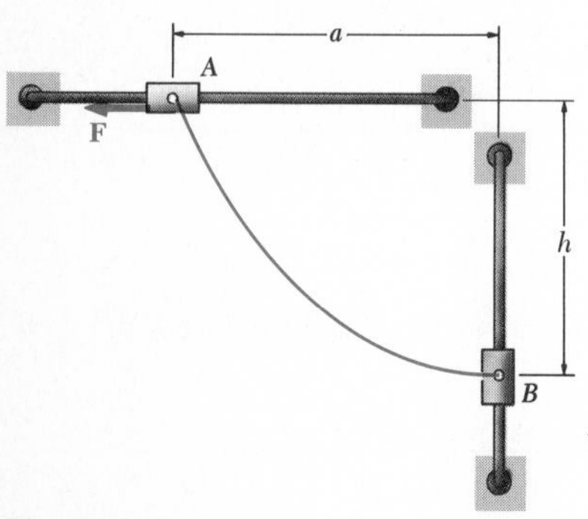

Fig. P7.106

7.105 Solve Prob. 7.104, assuming that the mass of the counterweight D is 60 kg.

7.106 The 10-ft cable AB weighs 20 lb and is attached to collars at A and B which may slide freely on the rods shown. Neglecting the weight of the collars, determine (*a*) the magnitude of the horizontal force $\mathbf{F}$ so that $h = a$, (*b*) the corresponding value of h and a, (*c*) the maximum tension in the cable.

7.107 A 300-ft wire is suspended between two points at the same elevation and 200 ft apart. Knowing that the maximum tension is 60 lb, determine the sag and the total weight of the wire.

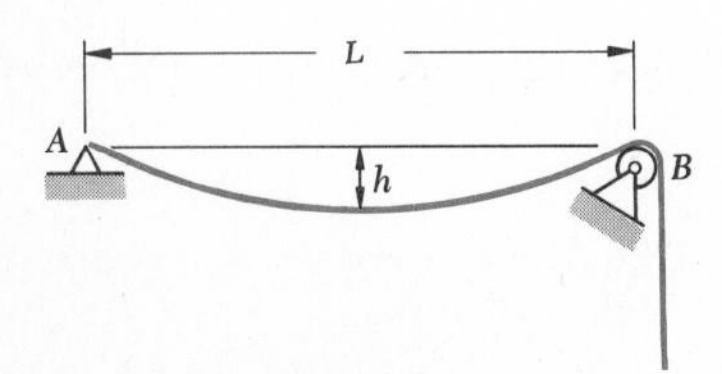

Fig. P7.108 and P7.109

7.108 A uniform cord 900 mm long passes over a frictionless pulley at B and is attached to a rigid support at A. Knowing that $L = 300$ mm, determine the smaller of the two values of h for which the cord is in equilibrium.

7.109 A uniform cord 60 in. long passes over a frictionless pulley at B and is attached to a rigid support at A. Knowing that $L = 24$ in., determine the smaller of the two values of h for which the cord is in equilibrium.

7.110 To the left of point B the long cable $ABDE$ rests on the rough horizontal surface shown. Knowing that the cable weighs 1.8 lb/ft, determine the force $\mathbf{F}$ required when $a = 9$ ft.

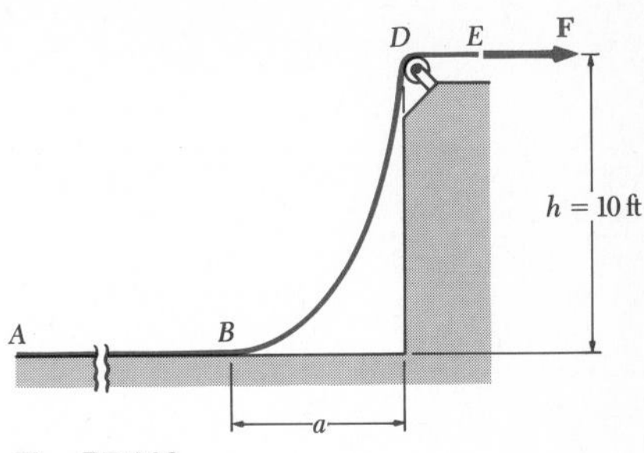

Fig. P7.110

7.111 Solve Prob. 7.110 when $a = 15$ ft.

7.112 Denoting by θ the angle formed by a uniform cable and the horizontal, show that at any point $y = c/\cos\theta$.

***7.113** (*a*) Determine the maximum allowable horizontal span for a uniform cable of weight w per unit length if the tension in the cable is not to exceed the value T_m. (*b*) Using the result of part *a*, find the maximum span of a drawn-steel wire for which $w = 0.25$ lb/ft and $T_m = 8000$ lb.

***7.114** A cable has a mass per unit length of 4 kg/m and is supported as shown. Knowing that the span L is 6 m, determine the *two* values of the sag h for which the maximum tension is 500 N.

***7.115** Determine the sag-to-span ratio for which the maximum tension in the cable is equal to the total weight of the entire cable AB.

***7.116** A cable, of weight w per unit length, is suspended between two points at the same elevation and a distance L apart. Determine the sag-to-span ratio for which the maximum tension is as small as possible. What are the corresponding values of θ_B and T_m?

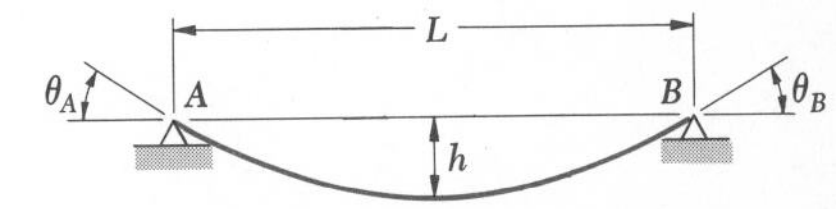

Fig. P7.114, P7.115, and P7.116

Review and Summary

In this chapter we learned to determine the internal forces which hold together the various parts of a given member in a structure.

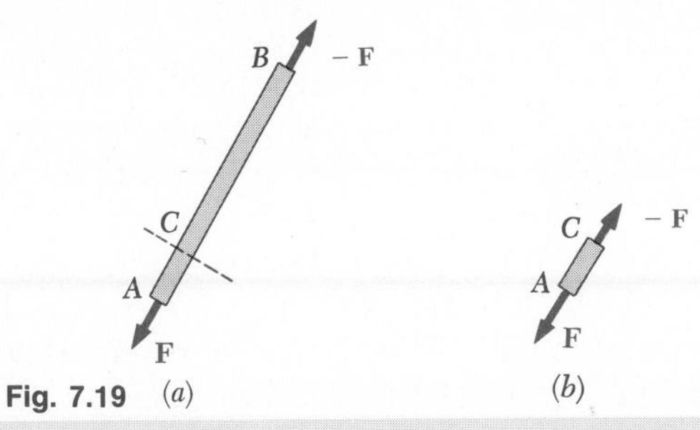

Fig. 7.19 (*a*) (*b*)

Forces in straight two-force members

Considering first a *straight two-force member AB* [Sec. 7.2], we recall that such a member is subjected at A and B to equal and opposite forces $\mathbf{F}$ and $-\mathbf{F}$ directed along AB (Fig. 7.19*a*). Cutting member AB at C and drawing the free-body diagram of portion AC, we conclude that the internal forces which existed at C in member AB are equivalent to an *axial force* $-\mathbf{F}$ equal and opposite to $\mathbf{F}$ (Fig. 7.19*b*). We note that in the case of a two-force member which is not straight, the internal forces reduce to a force-couple system and not to a single force.

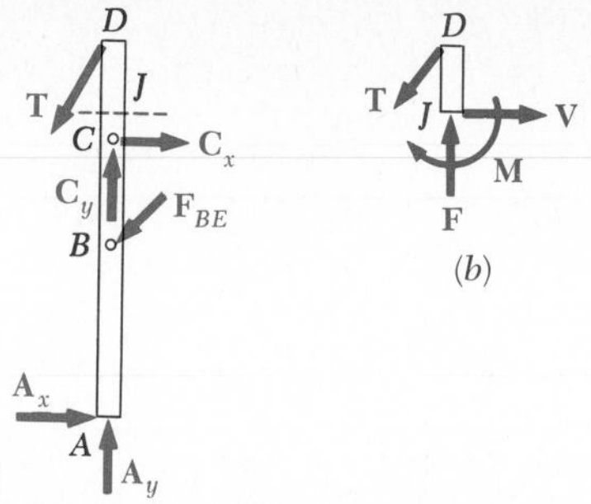

Fig. 7.20 (*a*) (*b*)

Forces in multiforce members

Considering next a *multiforce member AD* (Fig. 7.20*a*), cutting it at *J*, and drawing the free-body diagram of portion *JD*, we conclude that the internal forces at *J* are equivalent to a force-couple system consisting of the *axial force* **F**, the *shearing force* **V**, and a couple **M** (Fig. 7.20*b*). The magnitude of the shearing force measures the *shear* at point *J* and the moment of the couple is referred to as the *bending moment* at *J*. Since an equal and opposite force-couple system would have been obtained by considering the free-body diagram of portion *AJ*, it is necessary to specify which portion of member *AD* was used when recording the answers [Sample Prob. 7.1].

Forces in beams

Most of the chapter was devoted to the analysis of the internal forces in two important types of engineering structures: *beams and cables*. *Beams* are usually long, straight prismatic members designed to support loads applied at various points along the member. In general the loads are perpendicular to the axis of the beam and produce only *shear and bending* in the beam. The loads may be either *concentrated* at specific points, or *distributed* along the entire length or a portion of the beam. The beam itself may be supported in various ways; since only statically determinate beams are considered in this text, we limited our analysis to that of *simply supported beams*, *overhanging beams*, and *cantilever beams* [Sec. 7.3].

Shear and bending moment in a beam

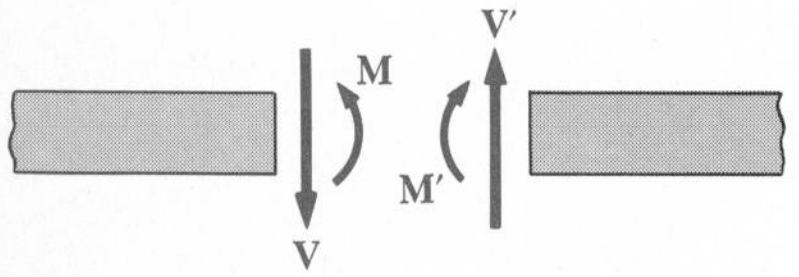

To obtain the *shear V* and *bending moment M* at a given point *C* of a beam, we first determine the reactions at the supports by considering the entire beam as a free body. We then cut the beam at *C* and use the free-body diagram of one of the two portions obtained in this fashion to determine *V* and *M*. In order to avoid any confusion regarding the sense of the shearing force **V** and couple **M** (which act in opposite directions on the two portions of the beam), the sign convention illustrated in Fig. 7.21 was adopted [Sec. 7.4]. Once the values of the shear and bending moment have been determined at a few selected points of the beam, it is usually possible to draw a *shear diagram* and a *bending-moment diagram* representing, respectively, the shear and the bending moment at any point of the beam [Sec. 7.5]. When a beam is subjected to concentrated loads only, the shear is of constant value between loads and the bending moment varies linearly between loads [Sample Prob. 7.2]. On the other hand, when a beam is subjected to distributed loads, the shear and bending moment vary quite differently [Sample Prob. 7.3].

Internal forces at section
(positive shear and positive bending moment)

Fig. 7.21

Relations among load, shear, and bending moment

The construction of the shear and bending-moment diagrams is facilitated if the following relations are taken into account. Denoting by w the distributed load per unit length (assumed positive if directed downward), we have [Sec. 7.5]:

$$\frac{dV}{dx} = -w \tag{7.1}$$

$$\frac{dM}{dx} = V \tag{7.3}$$

or, in integrated form,

$$V_D - V_C = -(\text{area under load curve between } C \text{ and } D) \tag{7.2'}$$

$$M_D - M_C = \text{area under shear curve between } C \text{ and } D \tag{7.4'}$$

Equation (7.2′) makes it possible to draw the shear diagram of a beam from the curve representing the distributed load on that beam and the value of V at one end of the beam. Similarly, Eq. (7.4′) makes it possible to draw the bending-moment diagram from the shear diagram and the value of M at one end of the beam. However, concentrated loads introduce discontinuities in the shear diagram and concentrated couples in the bending-moment diagram, none of which is accounted for in these equations [Sample Probs. 7.4 and 7.7]. Finally, we note from Eq. (7.3) that the points of the beam where the bending moment is maximum or minimum are also the points where the shear is zero [Sample Prob. 7.5].

Cables with concentrated loads

The second half of the chapter was devoted to the analysis of *flexible cables.* We first considered a cable of negligible weight supporting *concentrated loads* [Sec. 7.7]. Using the entire cable AB as a free body (Fig. 7.22), we noted that the three available equilibrium equations were not sufficient to determine the four unknowns representing the reactions at the supports A and B. However, if the coordinates of a point D of the cable are known, an additional equation may be obtained by considering the free-body diagram of the portion AD or DB of the cable. Once the reactions at the supports have been determined, the elevation of any point of the cable and the tension in any portion of the cable may be found from the appropriate free-body diagram [Sample Prob. 7.8]. It was noted that the horizontal component of the force **T** representing the tension is the same at any point of the cable.

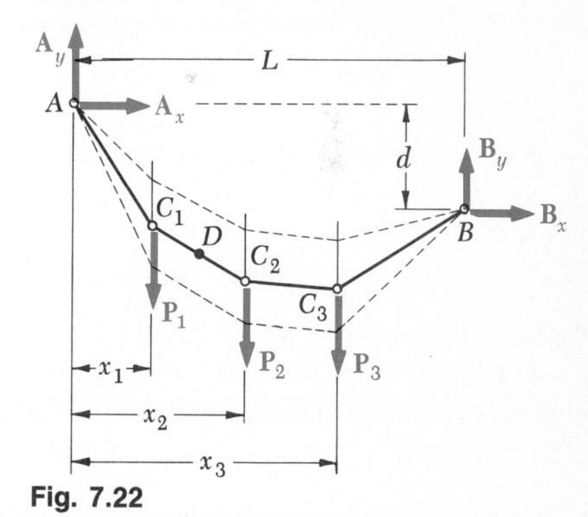

Fig. 7.22

Cables with distributed loads

We next considered cables carrying *distributed loads* [Sec. 7.8]. Using as a free body a portion of cable CD extending from the lowest point C to an arbitrary point D of the cable (Fig. 7.23), we observed that the horizontal component of the tension force **T** at D is constant and equal to the tension T_0 at C, while its vertical component is equal to the weight W of the portion of cable CD. The magnitude and direction of **T** were obtained from the force triangle:

$$T = \sqrt{T_0^2 + W^2} \qquad \tan\theta = \frac{W}{T_0} \tag{7.6}$$

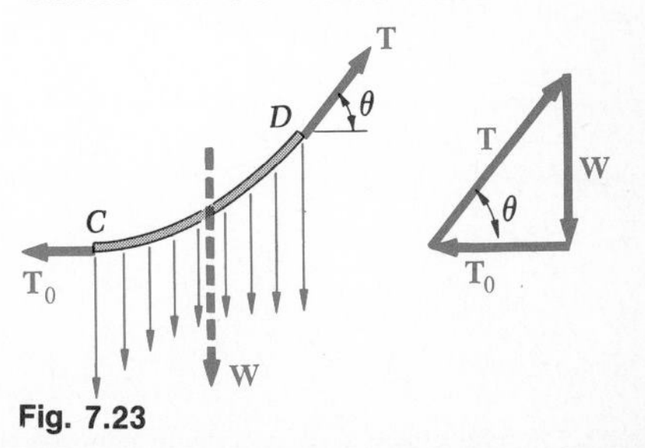

Fig. 7.23

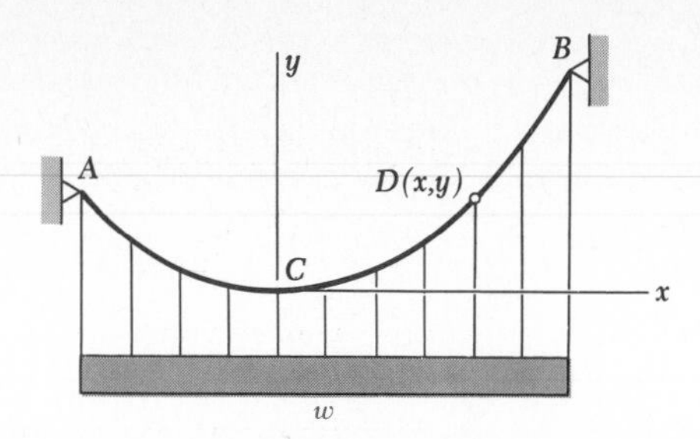

Fig. 7.24

Parabolic cable

In the case of a load *uniformly distributed along the horizontal*—as in a suspension bridge (Fig. 7.24)—the load supported by portion CD is $W = wx$, where w is the constant load per unit horizontal length [Sec. 7.9]. We also found that the curve formed by the cable is a *parabola* of equation

$$y = \frac{wx^2}{2T_0} \tag{7.8}$$

and that the length of the cable may be found by using the expansion in series given in Eq. (7.10) [Sample Prob. 7.9].

Catenary

In the case of a load *uniformly distributed along the cable itself*—e.g., a cable hanging under its own weight (Fig. 7.25)—the load supported by portion CD is $W = ws$, where s is the length measured along the cable and w is the constant load per unit length [Sec. 7.10]. Choosing the origin O of the coordinate axes at a distance $c = T_0/w$ below C, we derived the relations

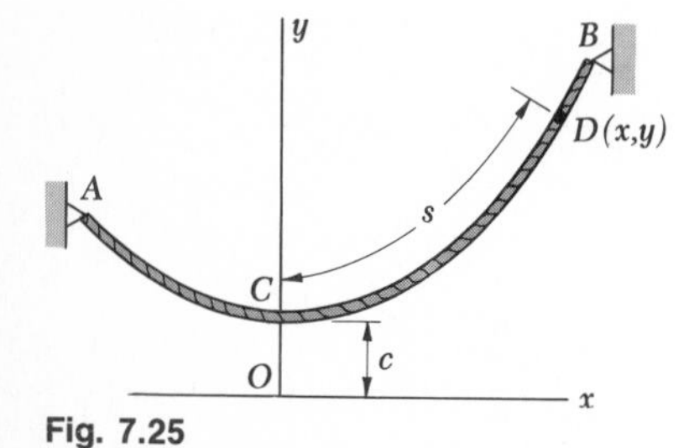

Fig. 7.25

$$s = c \sinh \frac{x}{c} \tag{7.15}$$

$$y = c \cosh \frac{x}{c} \tag{7.16}$$

$$y^2 - s^2 = c^2 \tag{7.17}$$

$$T_0 = wc \qquad W = ws \qquad T = wy \tag{7.18}$$

which may be used to solve problems involving cables hanging under their own weight [Sample Prob. 7.10]. Equation (7.16), which defines the shape of the cable, is the equation of a *catenary*.

Review Problems

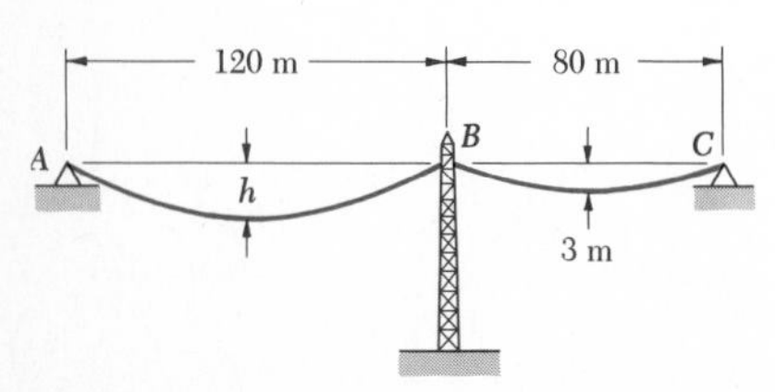

Fig. P7.117

7.117 Two cables of the same gage are attached to a transmission tower at B. Since the tower is slender, the horizontal component of the resultant of the forces exerted by the cable at B is to be zero. Assuming the cables to be parabolic, determine the required sag h of cable AB.

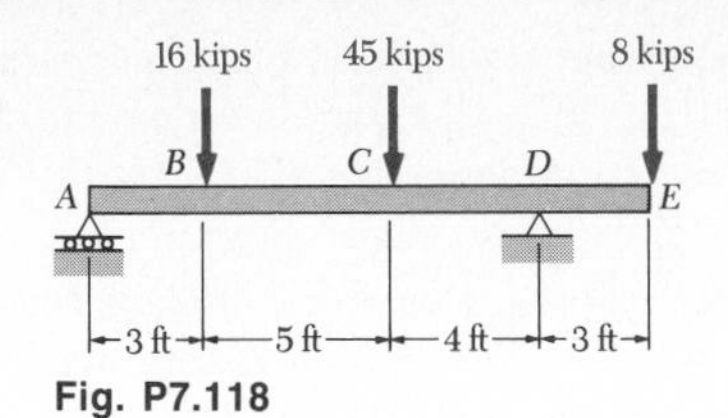

Fig. P7.118

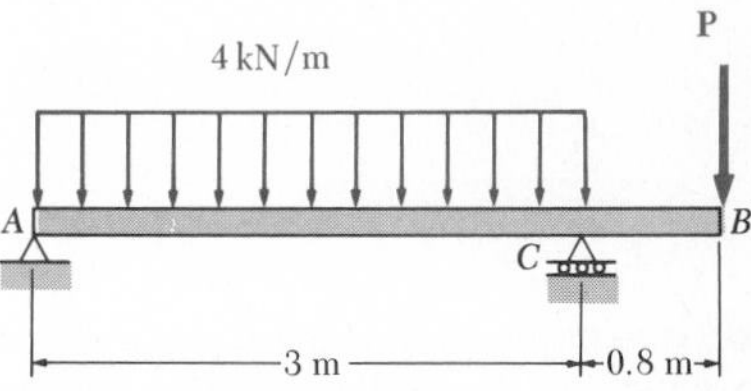

Fig. P7.119 and P7.120

7.118 Draw the shear and bending-moment diagrams for the beam and loading shown.

7.119 Draw the shear and bending-moment diagrams for the beam AB if the magnitude of the force $\mathbf{P}$ is 2.7 kN.

7.120 Determine the range of values of P for which the maximum absolute value of the bending moment in the beam is equal to or less than 3.6 kN · m.

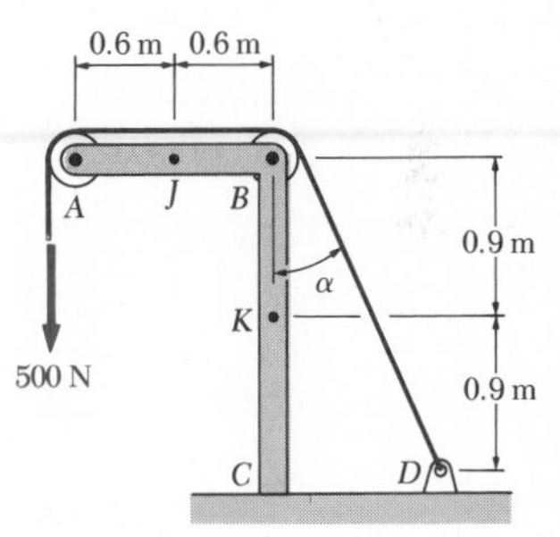

Fig. P7.122 and 7.123

7.121 A 16-mm-diameter wire rope has a mass per unit length of 0.9 kg/m and is suspended from two supports at the same elevation and 160 m apart. If the sag is 40 m, determine (*a*) the total length of the cable, (*b*) the maximum tension.

7.122 Knowing that the radius of each pulley is 150 mm and that $\alpha = 20°$, determine the internal forces at (*a*) point J, (*b*) point K.

7.123 Knowing that the radius of each pulley is 150 mm and that $\alpha = 30°$, determine the internal forces at (*a*) point J, (*b*) point K.

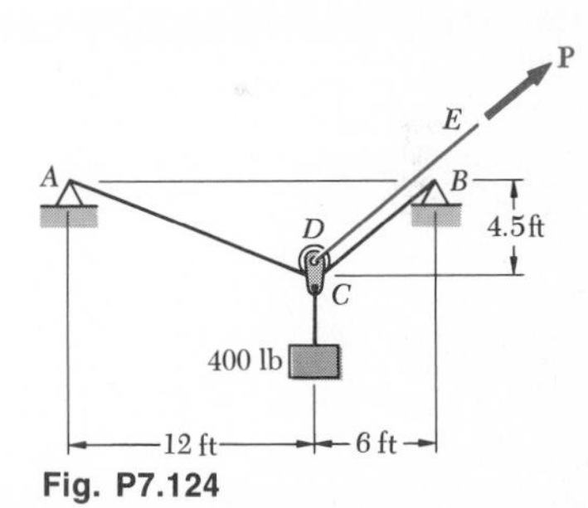

Fig. P7.124

7.124 A 400-lb load is attached to a small pulley which may roll on the cable ACB. The pulley and load are held in the position shown by a second cable DE which is parallel to the portion CB of the main cable. Determine (*a*) the reactions at A and B, (*b*) the tension in cable ACB, (*c*) the tension in cable DE. Neglect the radius of the pulleys and the weight of the cables.

7.125 A cable of length $L + \Delta$ is suspended between two points which are at the same elevation and a distance L apart. (*a*) Assuming that Δ is small compared with L and that the cable is parabolic, determine the approximate sag in terms of L and Δ. (*b*) If $L = 30$ m and $\Delta = 0.8$ m and the mass of the wire is 0.35 kg/m, use the result of part *a* to determine approximate values for the sag and the maximum tension in the wire. [*Hint.* Use only the first two terms of Eq. (7.10).]

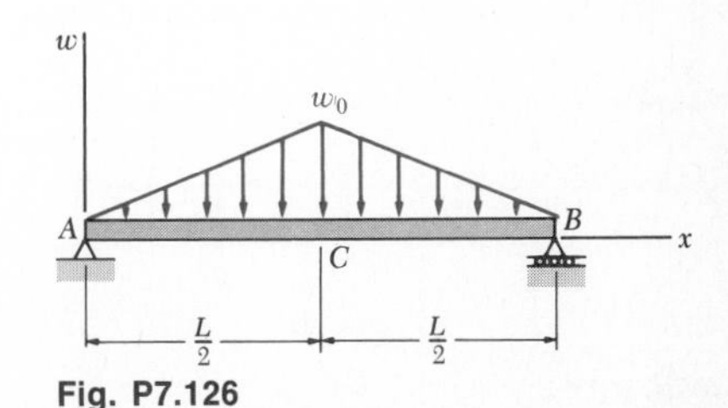

Fig. P7.126

7.126 Determine the equations of the shear and bending-moment curves for portion AC of the beam AB. Also, determine the maximum bending moment in the beam.

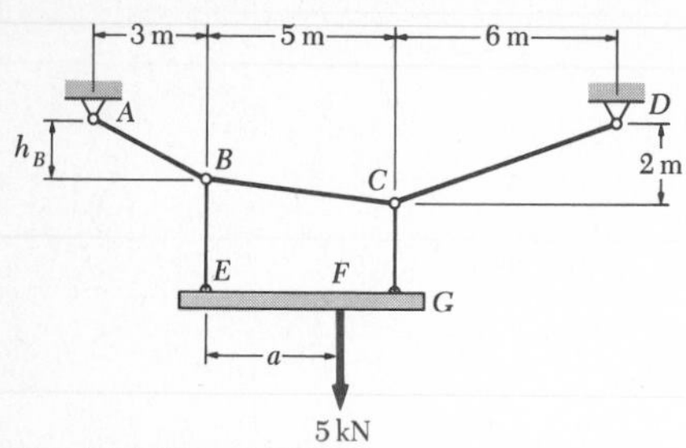

Fig. P7.127 and P7.128

7.127 A 5-kN load is applied at point F of plank EFG, which is suspended as shown from cable $ABCD$. If $h_B = 1.5$ m, determine (*a*) the distance a, (*b*) the corresponding maximum tension in cable $ABCD$.

7.128 A 5-kN load is applied at point F of plank EFG, which is suspended as shown from cable $ABCD$. If $a = 2.5$ m, determine (*a*) the sag h_B, (*b*) the corresponding maximum tension in cable $ABCD$.

The following problems are designed to be solved with a computer.

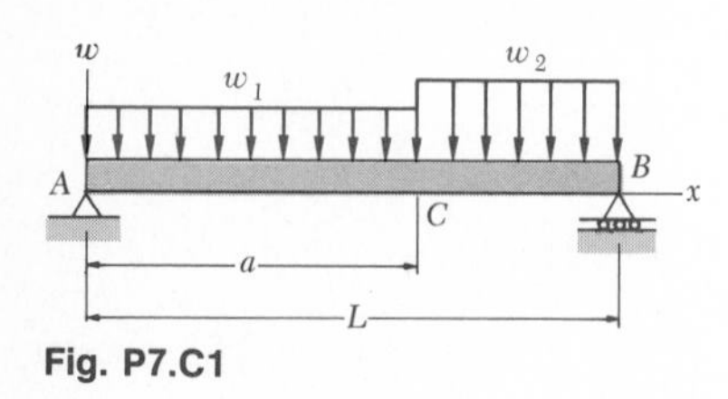

Fig. P7.C1

7.C1 Write a computer program which can be used to calculate the shearing force and the bending moment along the beam AB from $x = 0$ to $x = L$ at intervals ΔL Use this program to calculate the values of V and M for each of the following cases:

	L	ΔL	a	w_1	w_2
(*a*)	9 m	1 m	6 m	20 kN/m	0
(*b*)	9 m	1 m	6 m	0	20 kN/m
(*c*)	6 m	0.5 m	2 m	6 kN/m	15 kN/m

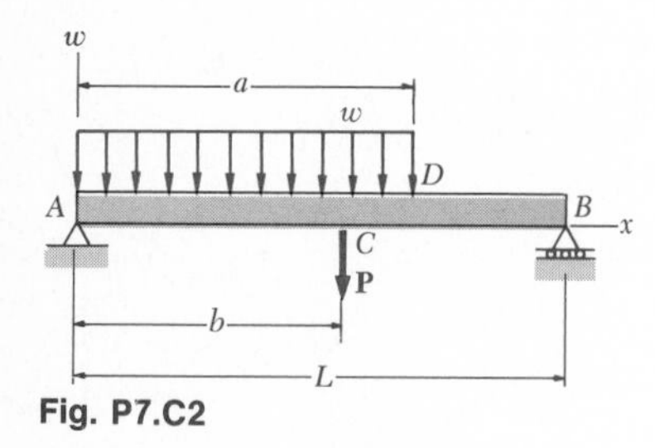

Fig. P7.C2

7.C2 Write a computer program which can be used to calculate the shearing force and the bending moment along the beam AB from $x = 0$ to $x = L$ at intervals ΔL. Use this program to calculate values of V and M for each of the following cases:

	L	ΔL	a	b	w	P	
(*a*)	9 m	1 m	6 m	—	20 kN/m	0	[Sample Prob. 7.5]
(*b*)	4 m	0.4 m	4 m	3.2 m	2 kN/m	4 kN	[Prob. 7.61]
(*c*)	4 m	0.2 m	2.6 m	1.6 m	1.8 kN/m	4 kN	[Prob. 7.26]

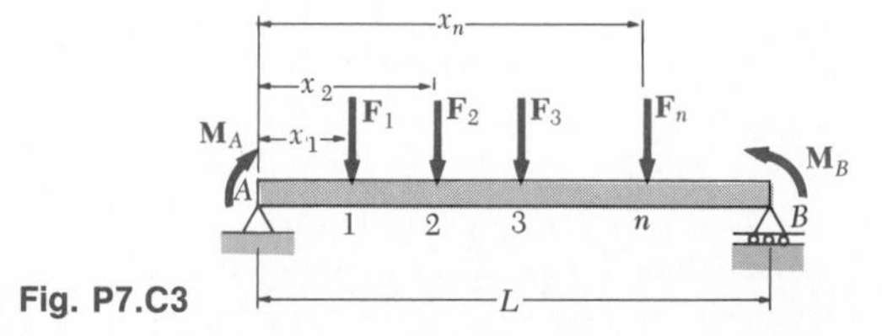

Fig. P7.C3

7.C3 For the beam and loading shown, write a computer program which can be used to calculate the bending moment in beam AB at the location of each concentrated load. Use this program to solve (*a*) Prob. 7.56, (*b*) Prob. 7.57.

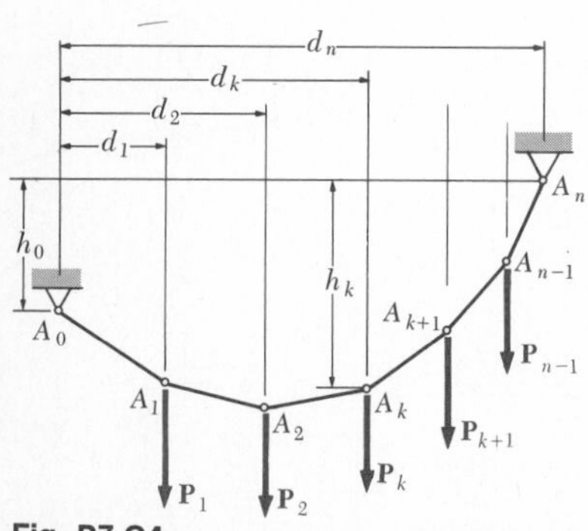

Fig. P7.C4

7.C4 Write a computer program which can be used to calculate the horizontal and vertical components of the reaction at the support A_n from the values of the loads $P_1, P_2, \ldots, P_{n-1}$, the horizontal distances $d_1, d_2, \ldots, d_n$, and the two vertical distances h_0 and h_k. Use this program to solve part *a* of Probs. 7.76, 7.78, 7.79, and 7.80.

Friction

CHAPTER

8

8.1. Introduction. In the preceding chapters, it was assumed that surfaces in contact were either *frictionless* or *rough*. If they were frictionless, the force each surface exerted on the other was normal to the surfaces and the two surfaces could move freely with respect to each other. If they were rough, it was assumed that tangential forces could develop to prevent the motion of one surface with respect to the other.

This view was a simplified one. Actually, no perfectly frictionless surface exists. When two surfaces are in contact, tangential forces, called *friction forces*, will always develop if one attempts to move one surface with respect to the other. On the other hand, these friction forces are limited in magnitude and will not prevent motion if sufficiently large forces are applied. The distinction between frictionless and rough surfaces is thus a matter of degree. This will be seen more clearly in the present chapter, which is devoted to the study of friction and of its applications to common engineering situations.

There are two types of friction: *dry friction*, sometimes called *Coulomb friction*, and *fluid friction*. Fluid friction develops between layers of fluid moving at different velocities. Fluid friction is of great importance in problems involving the flow of fluids through pipes and orifices or dealing with bodies immersed in moving fluids. It is also basic in the analysis of the motion of *lubricated mechanisms*. Such problems are considered in texts on fluid mechanics. We shall limit our present study to dry friction, i.e., to problems involving rigid bodies which are in contact along *nonlubricated* surfaces.

In the first part of this chapter, we shall analyze the equilibrium of various rigid bodies and structures, assuming dry friction at the surfaces of contact. Later we shall consider a number of specific engineering applications where dry friction plays an important role. These are wedges, square-threaded screws, journal bearings, thrust bearings, rolling resistance, and belt friction.

8.2. The Laws of Dry Friction. Coefficients of Friction.

The laws of dry friction are best understood by the following experiment. A block of weight **W** is placed on a horizontal plane surface (Fig. 8.1*a*). The forces acting on the block are its weight **W** and the reaction of the surface. Since the weight has no horizontal component, the reaction of the surface also has no horizontal component; the reaction is therefore *normal* to the surface and is represented by **N** in Fig. 8.1*a*. Suppose, now, that a horizontal force **P** is applied to the block (Fig. 8.1*b*). If **P** is small, the block will not move; some other horizontal force must therefore exist, which balances **P**. This other force is the *static-friction force* **F**, which is actually the resultant of a great number of forces acting over the entire surface of contact between the block and the plane. The nature of these forces is not known exactly, but it is generally assumed that these forces are due to the irregularities of the surfaces in contact and also, to a certain extent, to molecular attraction.

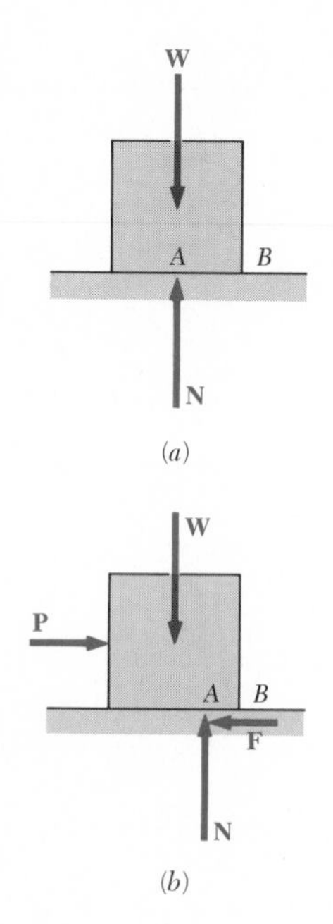

If the force **P** is increased, the friction force **F** also increases, continuing to oppose **P**, until its magnitude reaches a certain *maximum value* F_m (Fig. 8.1*c*). If **P** is further increased, the friction force cannot balance it any more and the block starts sliding.† As soon as the block has been set in motion, the magnitude of **F** drops from F_m to a lower value F_k. This is because there is less interpenetration between the irregularities of the surfaces in contact when these surfaces move with respect to each other. From then on, the block keeps sliding with increasing velocity while the friction force, denoted by $\mathbf{F}_k$ and called the *kinetic-friction force*, remains approximately constant.

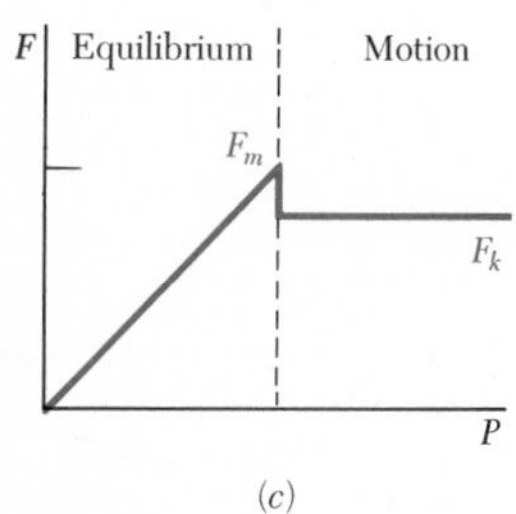

Fig. 8.1

Experimental evidence shows that the maximum value F_m of the static-friction force is proportional to the normal component N of the reaction of the surface. We have

$$F_m = \mu_s N \tag{8.1}$$

where μ_s is a constant called the *coefficient of static friction*. Similarly, the magnitude F_k of the kinetic-friction force may be put in the form

$$F_k = \mu_k N \tag{8.2}$$

where μ_k is a constant called the *coefficient of kinetic friction*. The coefficients of friction μ_s and μ_k do not depend upon the area of the surfaces in contact. Both coefficients, however, depend strongly on the *nature* of the surfaces in contact. Since they also depend upon the exact condition of the

† It should be noted that, as the magnitude F of the friction force increases from 0 to F_m, the point of application A of the resultant **N** of the normal forces of contact moves to the right, so that the couples formed, respectively, by **P** and **F** and by **W** and **N** remain balanced. If **N** reaches B before F reaches its maximum value F_m, the block will tip about B before it can start sliding (see Probs. 8.19 to 8.22).

Table 8.1 Approximate Values of Coefficient of Static Friction for Dry Surfaces

Metal on metal	0.15–0.60
Metal on wood	0.20–0.60
Metal on stone	0.30–0.70
Metal on leather	0.30–0.60
Wood on wood	0.25–0.50
Wood on leather	0.25–0.50
Stone on stone	0.40–0.70
Earth on earth	0.20–1.00
Rubber on concrete	0.60–0.90

surfaces, their value is seldom known with an accuracy greater than 5 percent. Approximate values of coefficients of static friction are given in Table 8.1 for various dry surfaces. The corresponding values of the coefficient of kinetic friction would be about 25 percent smaller. Since coefficients of friction are dimensionless quantities, the values given in Table 8.1 may be used with both SI units and U.S. customary units.

From the description given above, it appears that four different situations may occur when a rigid body is in contact with a horizontal surface:

1. The forces applied to the body do not tend to move it along the surface of contact; there is no friction force (Fig. 8.2*a*).

2. The applied forces tend to move the body along the surface of contact but are not large enough to set it in motion. The friction force **F** which has developed may be found by solving the equations of equilibrium for the body. Since there is no evidence that **F** has reached its maximum value, the equation $F_m = \mu_s N$ *cannot be used* to determine the friction force (Fig. 8.2*b*).

3. The applied forces are such that the body is just about to slide. We say that *motion is impending.* The friction force **F** has reached its maximum value F_m and, together with the normal force **N**, balances the applied forces. Both the equations of equilibrium and the equation $F_m = \mu_s N$ *may be used.* We also note that the friction force has a sense opposite to the sense of impending motion (Fig. 8.2*c*).

4. The body is sliding under the action of the applied forces, and the equations of equilibrium do not apply any more. However, **F** is now equal to $\mathbf{F}_k$ and the equation $F_k = \mu_k N$ may be used. The sense of $\mathbf{F}_k$ is opposite to the sense of motion (Fig. 8.2*d*).

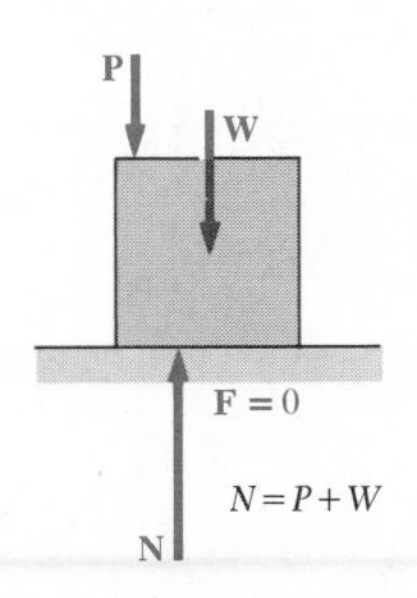

(*a*) No friction ($P_x = 0$)

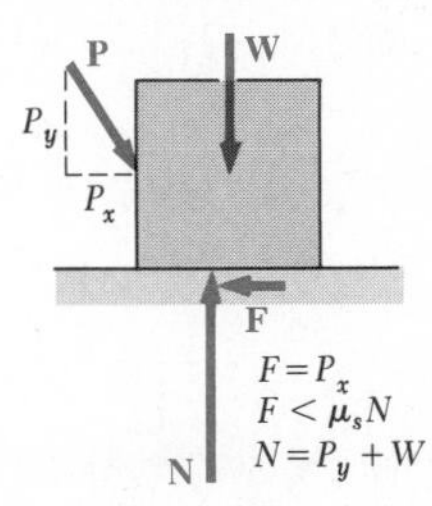

(*b*) No motion ($P_x < F_m$)

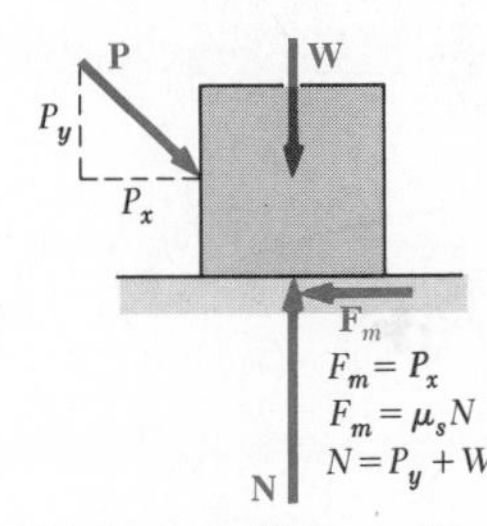

(*c*) Motion impending ⟶ ($P_x = F_m$)

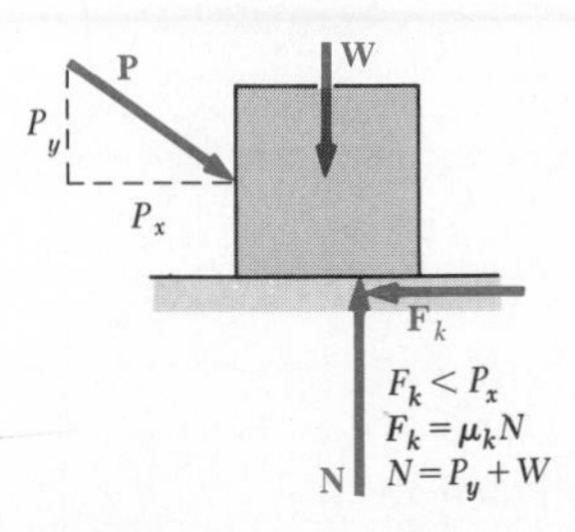

(*d*) Motion ⟶ ($P_x > F_m$)

Fig. 8.2

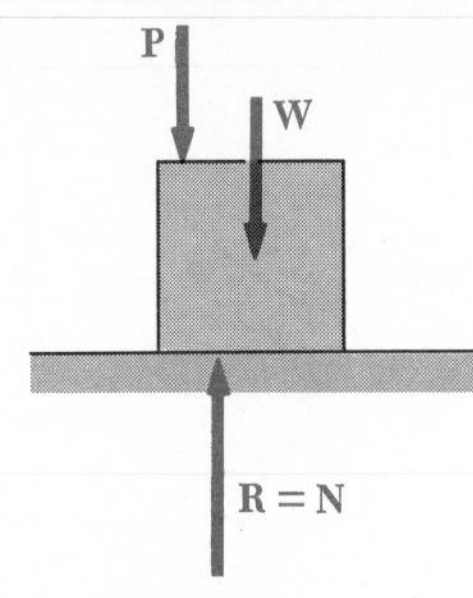

(a) No friction

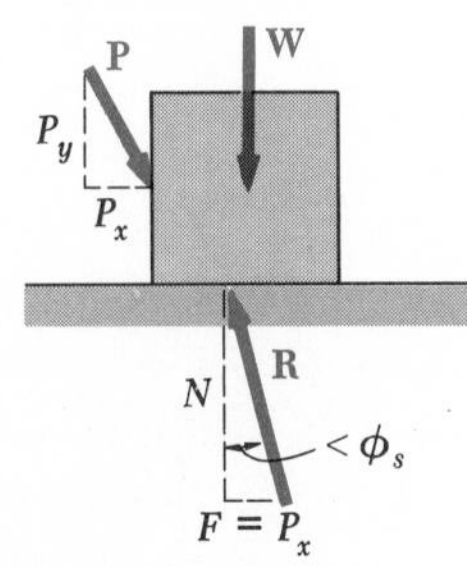

(b) No motion

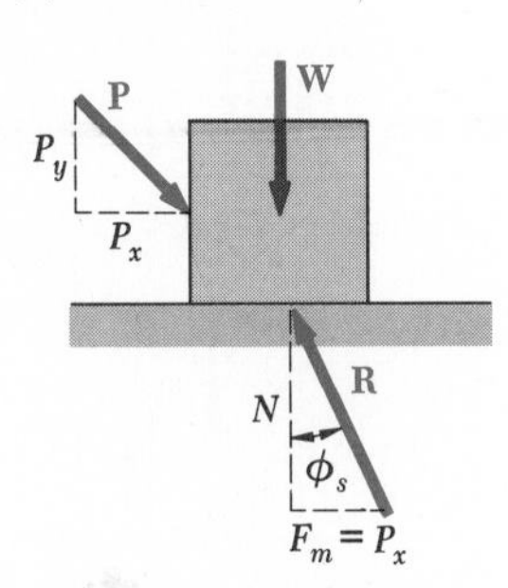

(c) Motion impending ⟶

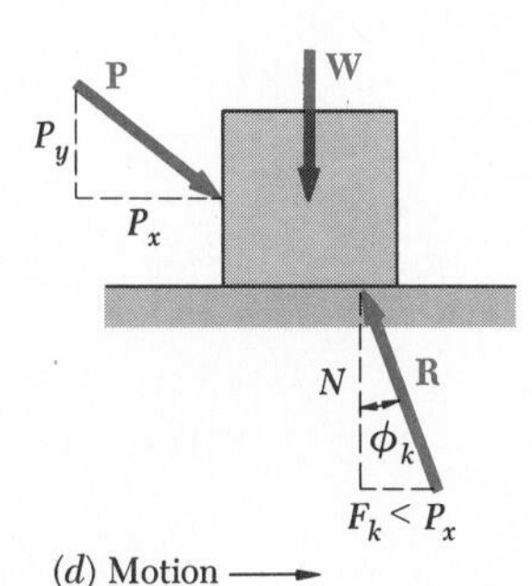

(d) Motion ⟶

Fig. 8.3

8.3. Angles of Friction. It is sometimes found convenient to replace the normal force **N** and the friction force **F** by their resultant **R**. Let us consider again a block of weight **W** resting on a horizontal plane surface. If no horizontal force is applied to the block, the resultant **R** reduces to the normal force **N** (Fig. 8.3a). However, if the applied force **P** has a horizontal component $\mathbf{P}_x$ which tends to move the block, the force **R** will have a horizontal component **F** and, thus, will form a certain angle with the vertical (Fig. 8.3b). If $\mathbf{P}_x$ is increased until motion becomes impending, the angle between **R** and the vertical grows and reaches a maximum value (Fig. 8.3c). This value is called the *angle of static friction* and is denoted by ϕ_s. From the geometry of Fig. 8.3c, we note that

$$\tan \phi_s = \frac{F_m}{N} = \frac{\mu_s N}{N}$$

$$\tan \phi_s = \mu_s \tag{8.3}$$

If motion actually takes place, the magnitude of the friction force drops to F_k; similarly, the angle between **R** and **N** drops to a lower value ϕ_k, called the *angle of kinetic friction* (Fig. 8.3d). From the geometry of Fig. 8.3d, we write

$$\tan \phi_k = \frac{F_k}{N} = \frac{\mu_k N}{N}$$

$$\tan \phi_k = \mu_k \tag{8.4}$$

Another example will show how the angle of friction may be used to advantage in the analysis of certain types of problems. Consider a block resting on a board which may be given any desired inclination; the block is subjected to no other force than its weight **W** and the reaction **R** of the board. If the board is horizontal, the force **R** exerted by the board on the block is perpendicular to the board and balances the weight **W** (Fig. 8.4a). If the board is given a small angle of inclination θ, the force **R** will deviate from the perpendicular to the board by the angle θ and will keep balancing **W** (Fig. 8.4b); it will then have a normal component **N** of magnitude $N = W \cos \theta$ and a tangential component **F** of magnitude $F = W \sin \theta$.

If we keep increasing the angle of inclination, motion will soon become impending. At that time, the angle between **R** and the normal will have reached its maximum value ϕ_s (Fig. 8.4c). The value of the angle of inclination corresponding to impending motion is called the *angle of repose*. Clearly, the angle of repose is equal to the angle of static friction ϕ_s. If the angle of inclination θ is further increased, motion starts and the angle between **R** and the normal drops to the lower value ϕ_k (Fig. 8.4d). The reaction **R** is not vertical any more, and the forces acting on the block are unbalanced.

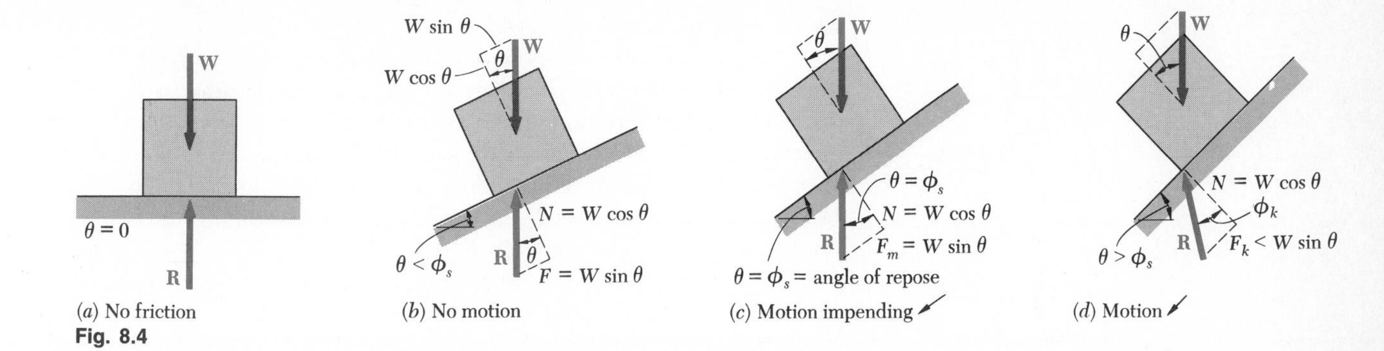

(*a*) No friction (*b*) No motion (*c*) Motion impending (*d*) Motion

Fig. 8.4

8.4. Problems Involving Dry Friction. Problems involving dry friction are found in many engineering applications. Some deal with simple situations such as the block sliding on a plane described in the preceding sections. Others involve more complicated situations as in Sample Prob. 8.3; many deal with the stability of rigid bodies in accelerated motion and will be studied in dynamics. Also, a number of common machines and mechanisms may be analyzed by applying the laws of dry friction. These include wedges, screws, journal and thrust bearings, and belt transmissions. They will be studied in the following sections.

The *methods* which should be used to solve problems involving dry friction are the same that were used in the preceding chapters. If a problem involves only a motion of translation, with no possible rotation, the body under consideration may usually be treated as a particle, and the methods of Chap. 2 may be used. If the problem involves a possible rotation, the body must be considered as a rigid body and the methods of Chap. 4 should be used. If the structure considered is made of several parts, the principle of action and reaction must be used as was done in Chap. 6.

If the body considered is acted upon by more than three forces (including the reactions at the surfaces of contact), the reaction at each surface will be represented by its components **N** and **F** and the problem will be solved from the equations of equilibrium. If only three forces act on the body under consideration, it may be found more convenient to represent each reaction by the single force **R** and to solve the problem by drawing a force triangle.

Most problems involving friction fall into one of the following *three groups:* In the *first group* of problems, all applied forces are given, and the coefficients of friction are known; we are to determine whether the body considered will remain at rest or slide. The friction force **F** *required to maintain equilibrium* is unknown (its magnitude is *not* equal to $\mu_s N$) and should be determined, together with the normal force **N**, by drawing a free-body diagram and *solving the equations of equilibrium* (Fig. 8.5*a*).

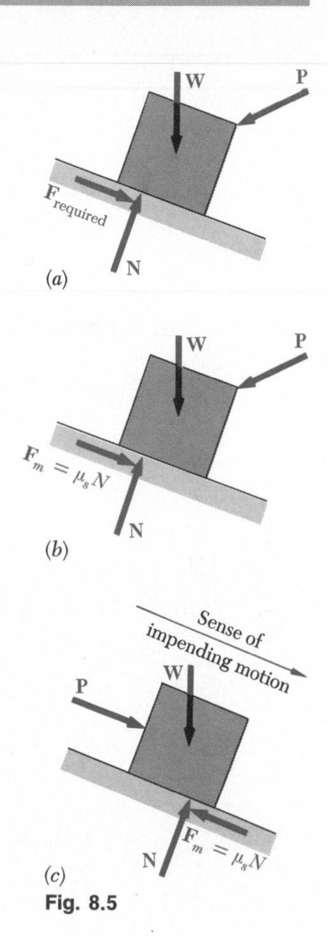

Fig. 8.5

The value found for the magnitude F of the friction force is then compared with the maximum value $F_m = \mu_s N$. If F is smaller than or equal to F_m, the body remains at rest. If the value found for F is larger than F_m, equilibrium cannot be maintained and motion takes place; the actual magnitude of the friction force is then $F_k = \mu_k N$.

In problems of the *second group,* all applied forces are given, and the motion is known to be impending; we are to determine the value of the coefficient of static friction. Here again, we determine the friction force and the normal force by drawing a free-body diagram and solving the equations of equilibrium (Fig. 8.5*b*). Since we know that the value found for F is the maximum value F_m, the coefficient of friction may be found by writing and solving the equation $F_m = \mu_s N$.

In problems of the *third group,* the coefficient of static friction is given, and it is known that the motion is impending in a given direction; we are to determine the magnitude or the direction of one of the applied forces. The friction force should be shown in the free-body diagram with a *sense opposite to that of the impending motion* and with a magnitude $F_m = \mu_s N$ (Fig. 8.5*c*). The equations of equilibrium may then be written, and the desired force may be determined.

As noted above, it may be more convenient when only three forces are involved to represent the reaction of the surface by a single force **R** and to solve the problem by drawing a force triangle. Such a solution is used in Sample Prob. 8.2.

When two bodies A and B are in contact (Fig. 8.6*a*), the forces of friction exerted, respectively, by A on B and by B on A are equal and opposite (Newton's third law). It is important in drawing the free-body diagram of one of the bodies, to include the appropriate friction force with its correct sense. The following rule should then be observed: *The sense of the friction force acting on A is opposite to that of the motion (or impending motion) of A as observed from B* (Fig. 8.6*b*).† The sense of the friction force acting on B is determined in a similar way (Fig. 8.6*c*). Note that the motion of A as observed from B is a *relative motion.* Body A may be fixed; yet it will have a relative motion with respect to B if B itself moves. Also, A may actually move down yet be observed from B to move up if B moves down faster than A.

†It is therefore *the same as that of the motion of B as observed from A.*

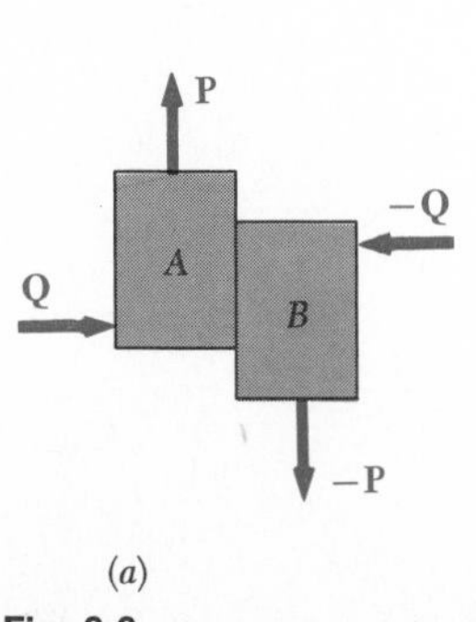

(*a*)

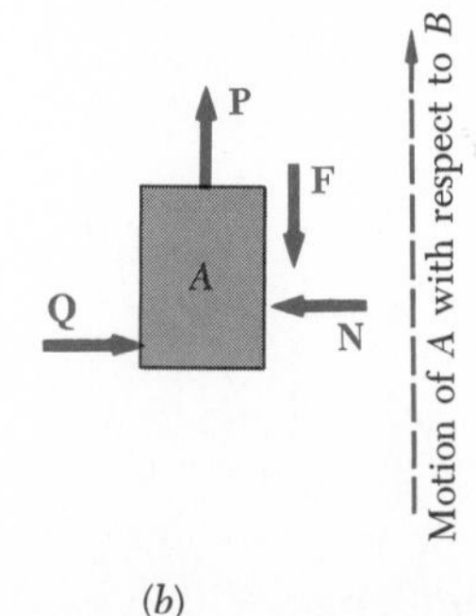

(*b*)

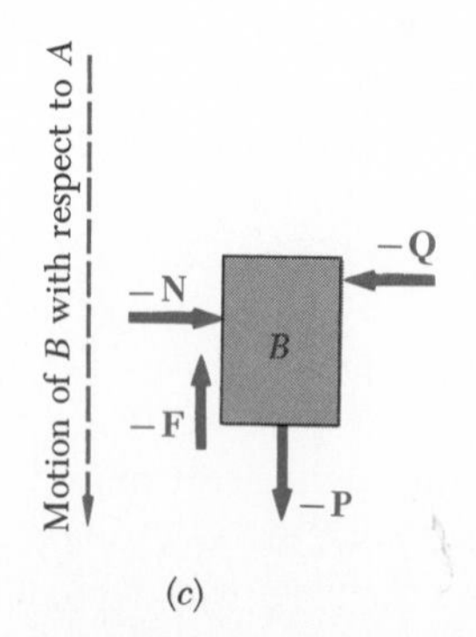

(*c*)

Fig. 8.6

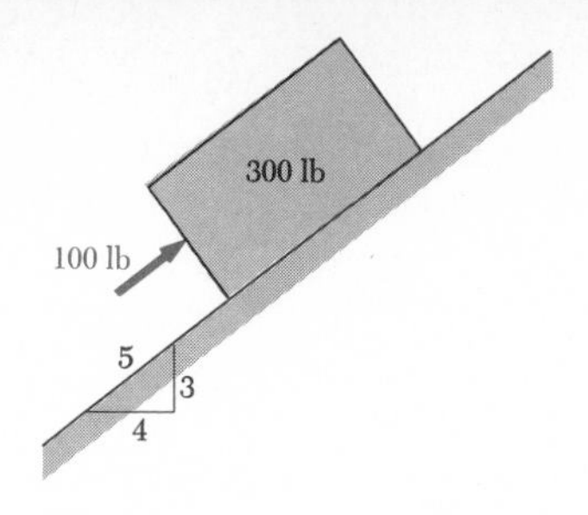

SAMPLE PROBLEM 8.1

A 100-lb force acts as shown on a 300-lb block placed on an inclined plane. The coefficients of friction between the block and the plane are $\mu_s = 0.25$ and $\mu_k = 0.20$. Determine whether the block is in equilibrium, and find the value of the friction force.

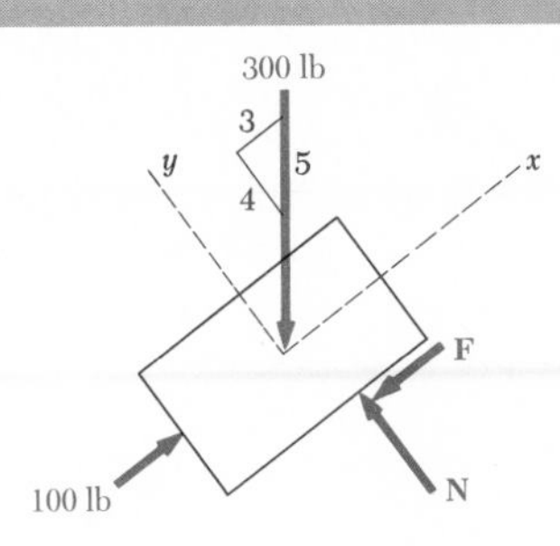

Force Required for Equilibrium. We first determine the value of the friction force *required to maintain equilibrium.* Assuming that **F** is directed down and to the left, we draw the free-body diagram of the block and write

$$+\nearrow \Sigma F_x = 0: \qquad 100 \text{ lb} - \tfrac{3}{5}(300 \text{ lb}) - F = 0$$
$$F = -80 \text{ lb} \qquad \mathbf{F} = 80 \text{ lb} \nearrow$$

$$+\nwarrow \Sigma F_y = 0: \qquad N - \tfrac{4}{5}(300 \text{ lb}) = 0$$
$$N = +240 \text{ lb} \qquad \mathbf{N} = 240 \text{ lb} \nwarrow$$

The force **F** required to maintain equilibrium is an 80-lb force directed up and to the right; the tendency of the block is thus to move down the plane.

Maximum Friction Force. The magnitude of the maximum friction force which may be developed is

$$F_m = \mu_s N \qquad F_m = 0.25(240 \text{ lb}) = 60 \text{ lb}$$

Since the value of the force required to maintain equilibrium (80 lb) is larger than the maximum value which may be obtained (60 lb), equilibrium will not be maintained and *the block will slide down the plane.*

Actual Value of Friction Force. The magnitude of the actual friction force is obtained as follows:

$$F_{\text{actual}} = F_k = \mu_k N$$
$$= 0.20(240 \text{ lb}) = 48 \text{ lb}$$

The sense of this force is opposite to the sense of motion; the force is thus directed up and to the right:

$$\mathbf{F}_{\text{actual}} = 48 \text{ lb} \nearrow \quad ◄$$

It should be noted that the forces acting on the block are not balanced; the resultant is

$$\tfrac{3}{5}(300 \text{ lb}) - 100 \text{ lb} - 48 \text{ lb} = 32 \text{ lb} \swarrow$$

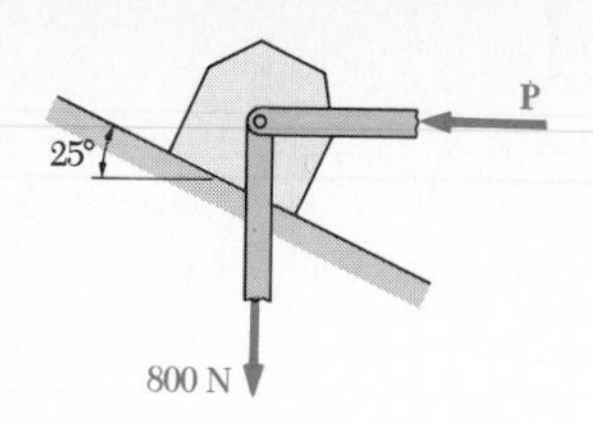

SAMPLE PROBLEM 8.2

A support block is acted upon by two forces as shown. Knowing that the coefficients of friction between the block and the incline are $\mu_s = 0.35$ and $\mu_k = 0.25$, determine the force **P** required (*a*) to start the block moving up the incline, (*b*) to keep it moving up, (*c*) to prevent it from sliding down.

Solution. For each part of the problem we draw a free-body diagram of the block and a force triangle including the 800-N vertical force, the horizontal force **P**, and the force **R** exerted on the block by the incline. The direction of **R** must be determined in each separate case. We note that since **P** is perpendicular to the 800-N force, the force triangle is a right triangle, which may easily be solved for **P**. In most other problems, however, the force triangle will be an oblique triangle and should be solved by applying the law of sines.

a. Force P *to Start Block Moving Up*

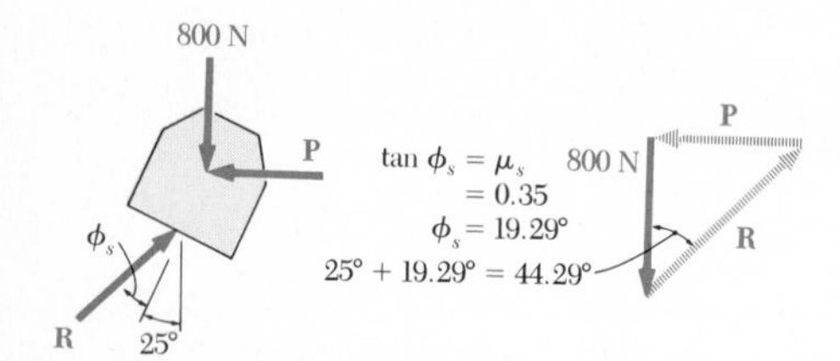

$P = (800 \text{ N}) \tan 44.29°$

$\mathbf{P} = 780 \text{ N} \leftarrow$ ◄

b. Force P *to Keep Block Moving Up*

800 N
P
ϕ_k
R
25°
$\tan \phi_k = \mu_k$
$= 0.25$
$\phi_k = 14.04°$
$25° + 14.04° = 39.04°$
800 N
P
R

$P = (800 \text{ N}) \tan 39.04°$

$\mathbf{P} = 649 \text{ N} \leftarrow$ ◄

c. Force P *to Prevent Block from Sliding Down*

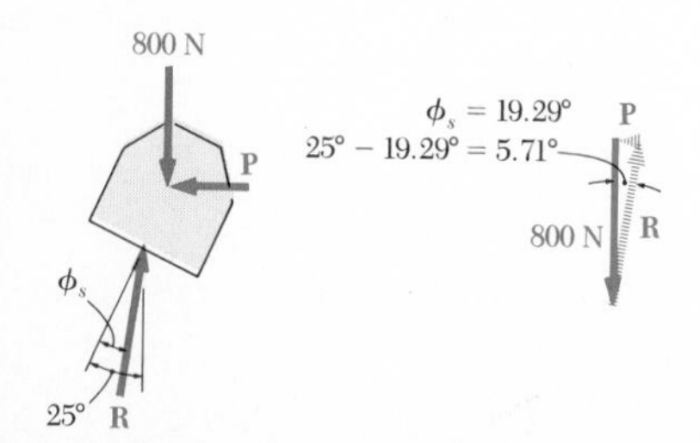

$P = (800 \text{ N}) \tan 5.71°$

$\mathbf{P} = 80.0 \text{ N} \leftarrow$ ◄

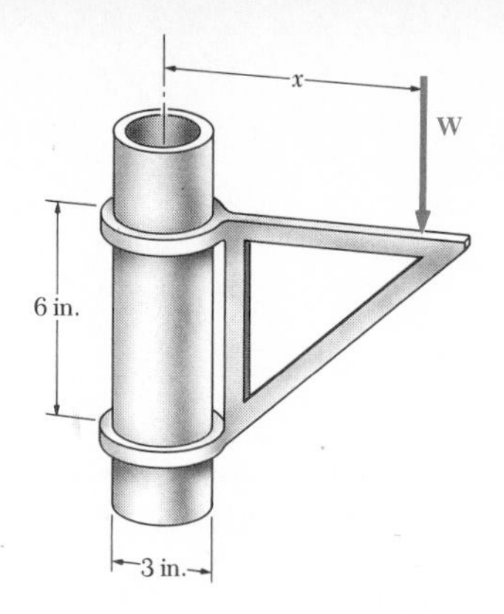

SAMPLE PROBLEM 8.3

The movable bracket shown may be placed at any height on the 3-in.-diameter pipe. If the coefficient of static friction between the pipe and bracket is 0.25, determine the minimum distance x at which the load **W** can be supported. Neglect the weight of the bracket.

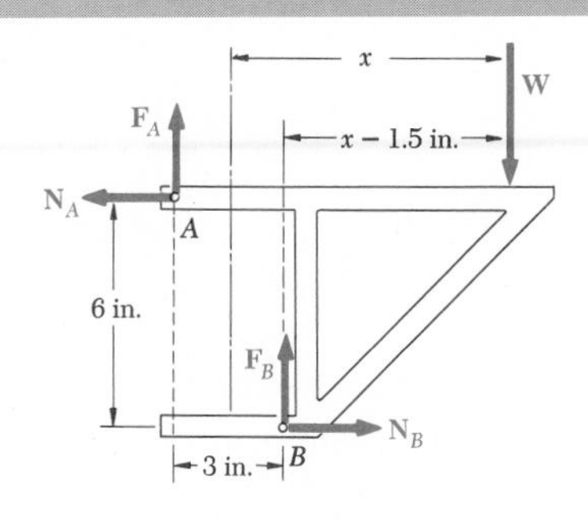

Solution. We draw the free-body diagram of the bracket. When **W** is placed at the minimum distance x from the axis of the pipe, the bracket is just about to slip, and the forces of friction at A and B have reached their maximum values:

$$F_A = \mu_s N_A = 0.25\, N_A$$
$$F_B = \mu_s N_B = 0.25\, N_B$$

Equilibrium Equations

$\overset{+}{\rightarrow}\Sigma F_x = 0$: $\quad N_B - N_A = 0$

$$N_B = N_A$$

$+\uparrow\Sigma F_y = 0$: $\quad F_A + F_B - W = 0$

$$0.25N_A + 0.25N_B = W$$

And, since N_B has been found equal to N_A,

$$0.50N_A = W$$
$$N_A = 2W$$

$+\circlearrowleft \Sigma M_B = 0$: $\quad N_A(6\text{ in.}) - F_A(3\text{ in.}) - W(x - 1.5\text{ in.}) = 0$

$$6N_A - 3(0.25N_A) - Wx + 1.5W = 0$$
$$6(2W) - 0.75(2W) - Wx + 1.5W = 0$$

Dividing through by W and solving for x,

$$x = 12\text{ in.} \blacktriangleleft$$

Problems

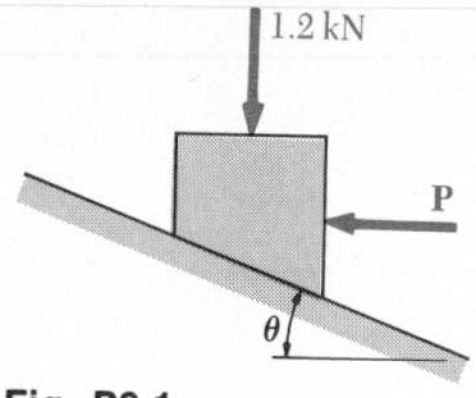

Fig. P8.1

8.1 The coefficients of friction between the block and the incline are $\mu_s = 0.35$ and $\mu_k = 0.25$. Determine whether the block is in equilibrium and find the magnitude and direction of the friction force when $\theta = 25°$ and $P = 750$ N.

8.2 Solve Prob. 8.1 when $\theta = 30°$ and $P = 150$ N.

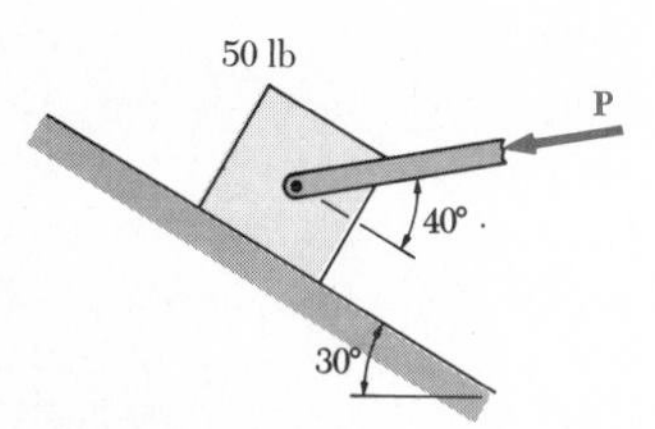

Fig. P8.3

8.3 The coefficients of friction between the 50-lb block and the incline are $\mu_s = 0.40$ and $\mu_k = 0.30$. Determine whether the block is in equilibrium and find the magnitude and direction of the friction force when $P = 120$ lb.

8.4 Solve Prob. 8.3, assuming that $P = 80$ lb.

8.5 For the block of Prob. 8.1 and $\theta = 25°$, find the smallest value of P required (*a*) to start the block up the incline, (*b*) to keep it moving up.

8.6 For the block of Prob. 8.3, find the smallest value of P required (*a*) to start the block up the incline, (*b*) to keep it moving up.

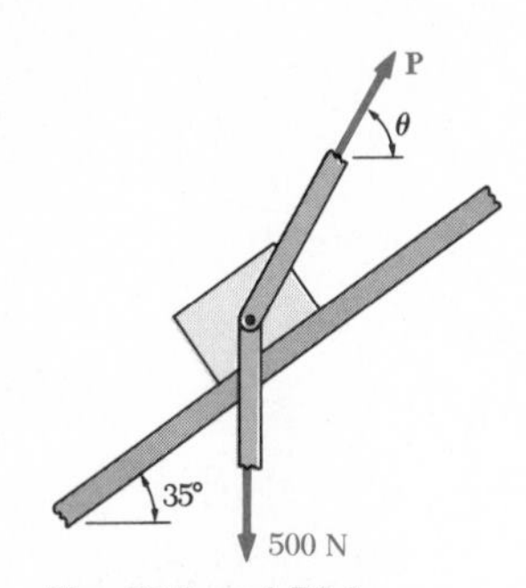

Fig. P8.7 and P8.8

8.7 The coefficients of friction between the block and the rail are $\mu_s = 0.30$ and $\mu_k = 0.25$. Knowing that $\theta = 65°$, determine the smallest value of P required (*a*) to start the block up the rail, (*b*) to keep it from moving down.

8.8 The coefficients of friction between the block and the rail are $\mu_s = 0.30$ and $\mu_k = 0.25$. Find the magnitude and direction of the smallest force **P** required (*a*) to start the block up the rail, (*b*) to keep it from moving down.

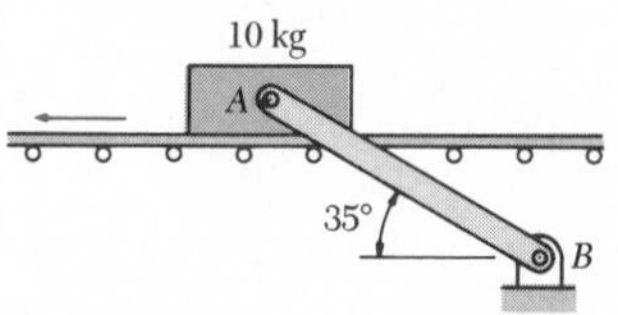

Fig. P8.9

8.9 The 10-kg block is attached to link AB and rests on a conveyor belt which is moving to the left. Knowing that the coefficients of friction between the block and the belt are $\mu_s = 0.30$ and $\mu_k = 0.25$ and neglecting the weight of the link, determine (*a*) the force in link AB, (*b*) the horizontal force **P** which should be applied to the belt to maintain its motion.

8.10 Solve Prob. 8.9, assuming that the belt is moving to the right.

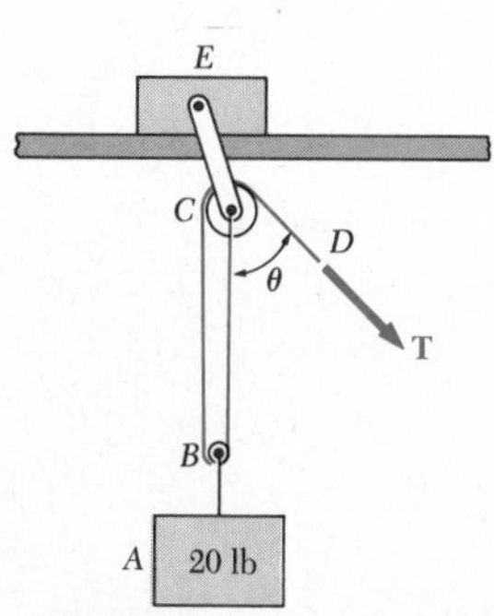

Fig. P8.11

8.11 The 20-lb block A hangs from a cable as shown. Pulley C is connected by a short link to block E, which rests on a horizontal rail. Knowing that the coefficient of static friction between block E and the rail is 0.35, and neglecting the weight of block E and the friction in the pulleys, determine the maximum allowable value of θ if the system is to remain in equilibrium.

8.12 The coefficients of friction are $\mu_s = 0.30$ and $\mu_k = 0.25$ between all surfaces of contact. Determine the smallest force **P** required to start block *D* moving if (*a*) block *C* is restrained by cable *AB* as shown, (*b*) cable *AB* is removed.

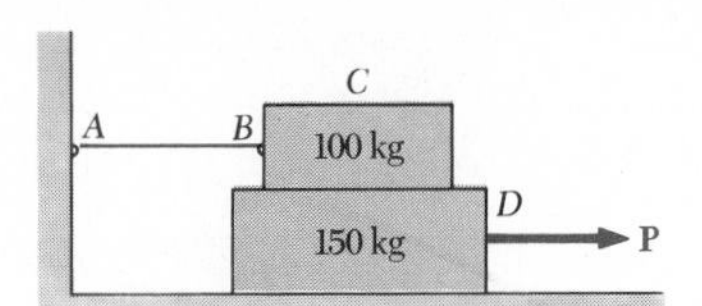

Fig. P8.12 and P8.13

8.13 The coefficients of friction are $\mu_s = 0.30$ and $\mu_k = 0.25$ between all surfaces of contact. Knowing that $P = 900$ N, determine (*a*) the resultant of the friction forces exerted on block *D* if block *C* is restrained as shown, (*b*) the friction force exerted by the ground on block *D* if cable *AB* is removed.

8.14 Three 4-kg packages *A*, *B*, and *C* are placed on a conveyor belt which is at rest. Between the belt and both packages *A* and *C* the coefficients of friction are $\mu_s = 0.30$ and $\mu_k = 0.20$; between package *B* and the belt the coefficients are $\mu_s = 0.10$ and $\mu_k = 0.08$. The packages are placed on the belt so that they are in contact with each other and at rest. Determine which, if any, of the packages will move and the friction force acting on each package.

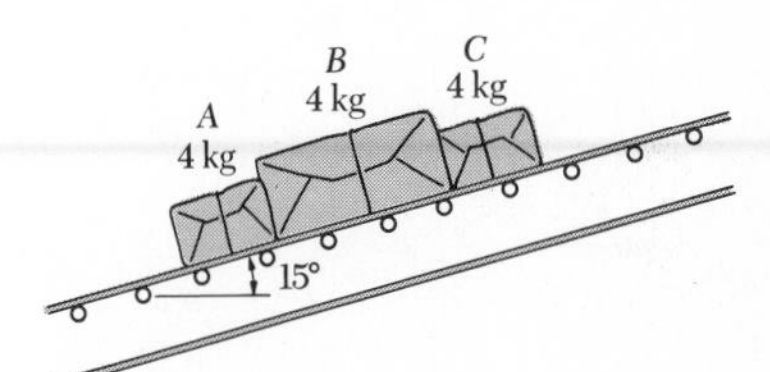

Fig. P8.14

8.15 Solve Prob. 8.14, assuming that package *B* is placed to the right of both packages *A* and *C*.

8.16 Block *A* weighs 50 lb and block *B* weighs 25 lb. Knowing that the coefficient of static friction is 0.15 between all surfaces of contact, determine the value of θ for which motion is impending.

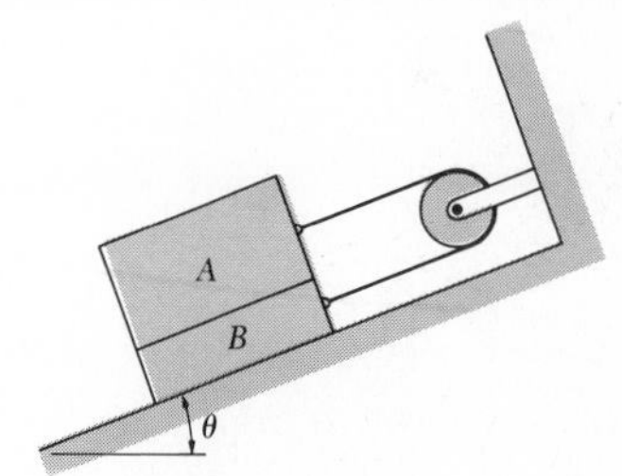

Fig. P8.16

8.17 Solve Prob. 8.16, assuming that the coefficient of static friction is 0.15 between the two blocks, and zero between block *B* and the incline.

8.18 For the system of Prob. 8.16, and assuming that the coefficients of friction are $\mu_s = 0.15$ and $\mu_k = 0.12$ between the two blocks, and zero between block *B* and the incline, determine the magnitude of the friction force between the two blocks when (*a*) $\theta = 40°$, (*b*) $\theta = 30°$.

8.19 A 60-kg cabinet is mounted on casters which can be locked to prevent their rotation. The coefficient of static friction is 0.35. If $h = 600$ mm, determine the magnitude of the force **P** required to move the cabinet to the right (*a*) if all casters are locked, (*b*) if the casters at *B* are locked and the casters at *A* are free to rotate, (*c*) if the casters at *A* are locked and the casters at *B* are free to rotate.

8.20 A 60-kg cabinet is mounted on casters which can be locked to prevent their rotation. The coefficient of static friction between the floor and each caster is 0.35. Assuming that the casters at both *A* and *B* are locked, determine (*a*) the force **P** required to move the cabinet to the right, (*b*) the largest allowable value of *h* if the cabinet is not to tip over.

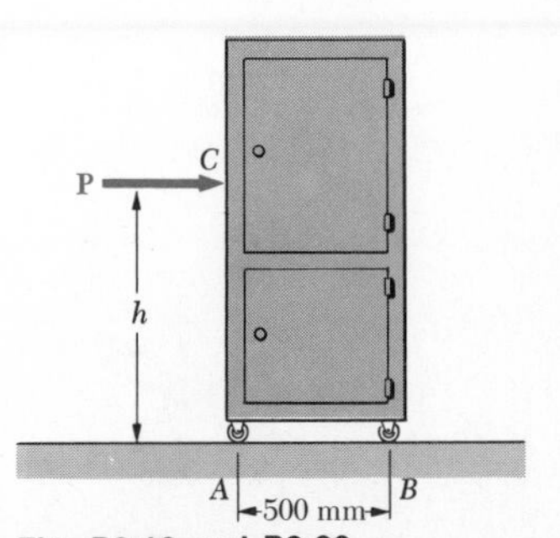

Fig. P8.19 and P8.20

8.21 A packing crate of mass 40 kg must be moved to the left along the floor without tipping. Knowing that the coefficient of static friction between the crate and the floor is 0.35, determine (*a*) the largest allowable value of α, (*b*) the corresponding magnitude of the force **P**.

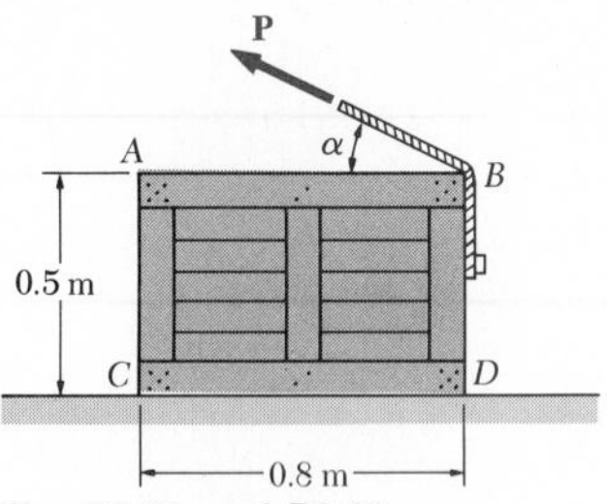

Fig. P8.21 and P8.22

8.22 A packing crate of mass 40 kg is pulled by a rope as shown. The coefficient of static friction between the crate and the floor is 0.35. If $\alpha = 40°$, determine (*a*) the magnitude of the force **P** required to move the crate, (*b*) whether the crate will slide or tip.

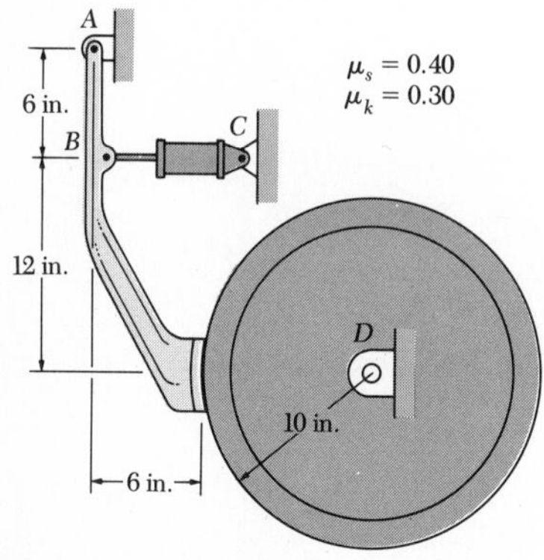

Fig. P8.23 and P8.24

8.23 A couple of magnitude 50 lb · ft is applied to the drum. Determine the smallest force which must be exerted by the hydraulic cylinder if the drum is not to rotate, when the applied couple is directed (*a*) clockwise, (*b*) counterclockwise.

8.24 The hydraulic cylinder exerts on point *B* a force of 600 lb directed to the right. Determine the moment of the friction force about the axle of the drum when the drum is rotating (*a*) clockwise, (*b*) counterclockwise.

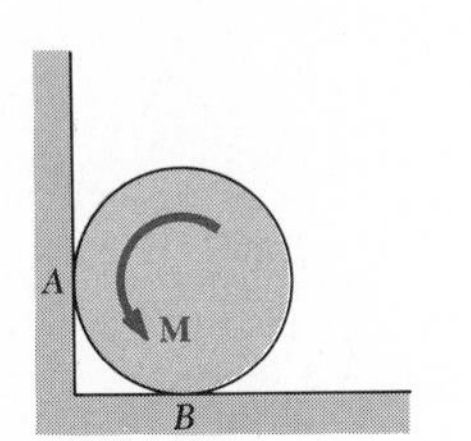

Fig. P8.25 and P8.26

8.25 Find the magnitude of the largest couple **M** which may be applied to the cylinder if it is not to spin. The cylinder has a weight *W* and a radius *r*, and the coefficient of static friction μ_s is the same at *A* and *B*.

8.26 The cylinder has a weight *W* and a radius *r*. Express in terms of *W* and *r* the magnitude of the largest couple **M** which may be applied to the cylinder if it is not to spin, assuming the coefficient of static friction to be (*a*) zero at *A* and 0.35 at *B*, (*b*) 0.28 at *A* and 0.35 at *B*.

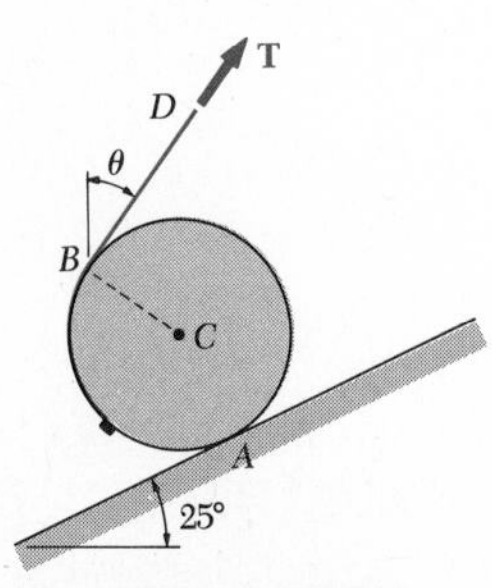

Fig. P8.27 and P8.28

***8.27** A cord is attached to and partially wound around a cylinder of weight *W* and radius *r* which rests on an incline as shown. Knowing that the coefficient of static friction between the cylinder and the incline is 0.35, find (*a*) the smallest allowable value of θ if the cylinder is to remain in equilibrium, (*b*) the corresponding value of the tension in the cord.

***8.28** A cord is attached to and partially wound around a cylinder of weight *W* and radius *r* which rests on an incline as shown. Knowing that $\theta = 30°$, find (*a*) the tension in the cord, (*b*) the smallest value of the coefficient of static friction between the cylinder and the incline for which equilibrium is maintained.

8.29 A car is stopped with its front wheels resting against a curb when its driver starts the engine and tries to drive over the curb. Knowing that the radius of the wheels is 280 mm, that the coefficient of static friction between the tires and the pavement is 0.85, and that 60 percent of the weight of the car is distributed over its front wheels and 40 percent over its rear wheels, determine the largest curb height h that the car can negotiate, assuming (*a*) front-wheel drive, (*b*) rear-wheel drive.

Fig. P8.29

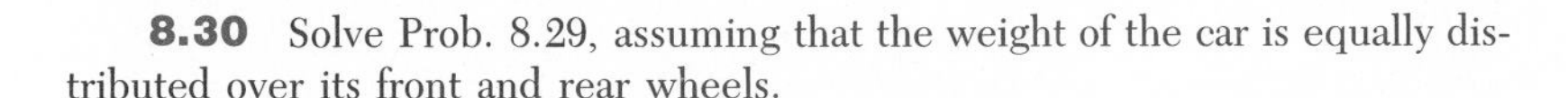

8.30 Solve Prob. 8.29, assuming that the weight of the car is equally distributed over its front and rear wheels.

8.31 The end A of a slender, uniform rod of length L and weight W bears on a horizontal surface, while its end B is supported by a cord BC. Knowing that the coefficients of friction are $\mu_s = 0.30$ and $\mu_k = 0.25$, determine (*a*) the maximum value of θ for which equilibrium is maintained, (*b*) the corresponding value of the tension in the cord.

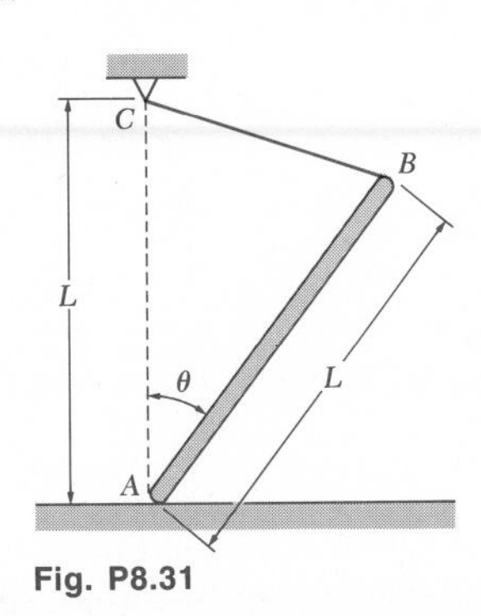

Fig. P8.31

8.32 Determine whether the rod of Prob. 8.31 is in equilibrium when $\theta = 30°$ and find the magnitude and direction of the friction force exerted on the rod.

8.33 A slender rod of length L is lodged between peg C and the vertical wall and supports a load **P** at end A. Knowing that $\theta = 35°$ and that the coefficient of static friction is 0.20 at both B and C, find the range of values of the ratio L/a for which equilibrium is maintained.

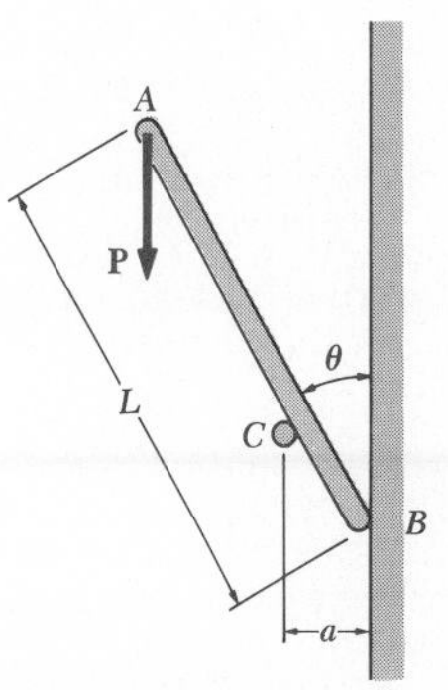

Fig. P8.33 and P8.36

8.34 Solve Prob. 8.33, assuming that the coefficient of static friction is 0.20 at B and zero at C.

8.35 Solve Prob. 8.33, assuming that the coefficient of static friction is 0.20 at C and zero at B.

8.36 A slender rod of length L is lodged between peg C and the vertical wall and supports a load **P** at end A. Knowing that $L = 12.5a$ and $\theta = 30°$ and that the coefficients of friction are $\mu_s = 0.20$ and $\mu_k = 0.15$ at C, and zero at B, determine whether the rod is in equilibrium.

8.37 Solve Prob. 8.36, assuming that $L = 6a$ and $\theta = 30°$ and that the coefficients of friction are $\mu_s = 0.20$ and $\mu_k = 0.15$ at B, and zero at C.

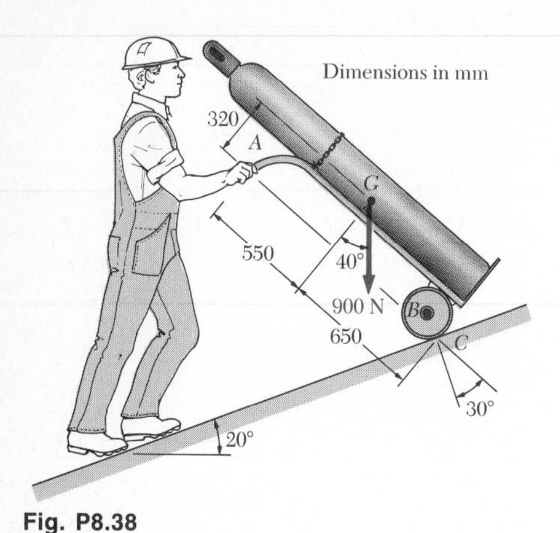

Fig. P8.38

***8.38** A worker is trying to hold in place a hand truck carrying a compressed-air cylinder. Knowing that the combined weight of the truck and cylinder is 900 N, determine the magnitude and direction of (*a*) the force the worker should exert on each handle if the wheels are free to rotate, (*b*) the smallest force he should exert on each handle if brakes have been applied to prevent the wheels from rotating. The coefficients of friction between the steel wheels and the incline are $\mu_s = 0.30$ and $\mu_k = 0.25$.

8.39 The worker of Prob. 8.38 wishes to ease the hand truck down the incline by slowly stepping backward at a constant speed. Determine the magnitude and direction of the force he should exert on each handle (*a*) if the wheels are free to rotate, (*b*) if brakes have been applied to prevent the wheels from rotating.

8.40 The shear shown is used to cut and trim electronic-circuit-board laminates. Knowing that the coefficient of kinetic friction between the blade and the vertical guide is 0.20, determine the force exerted by the edge *E* of the blade on the laminate.

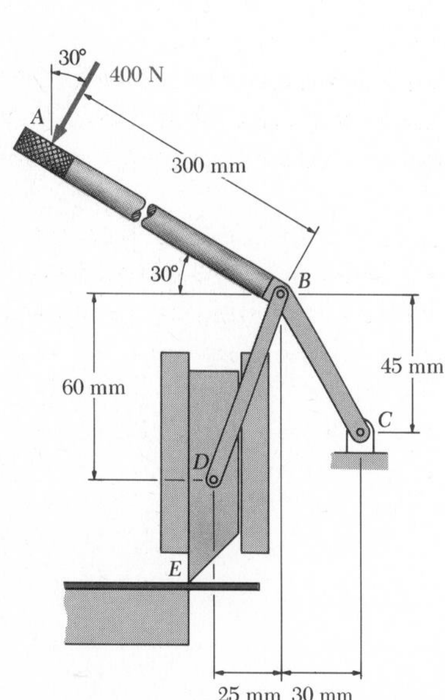

Fig. P8.40

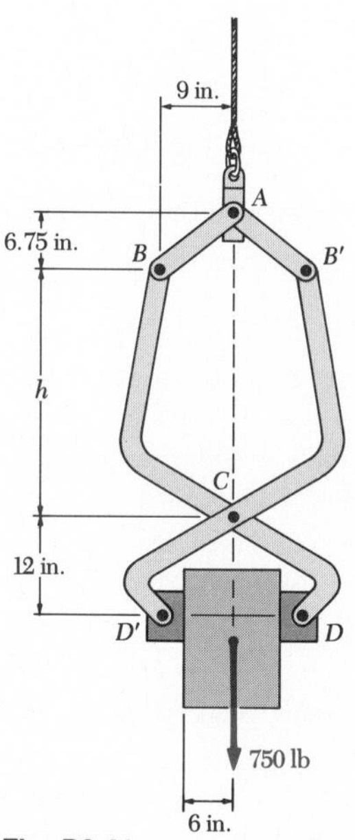

Fig. P8.41

8.41 The friction tongs shown are used to lift a 750-lb casting. Knowing that $h = 36$ in., determine the smallest allowable value of the coefficient of static friction between the casting and blocks *D* and *D'*.

8.42 For the friction tongs of Prob. 8.41, determine the smallest allowable value of h if the coefficient of static friction between the casting and blocks D and D' is 0.25.

8.43 A rod DE and a small cylinder are placed between two guides as shown. The rod is not to slip downward, however large the force **P** may be; i.e., the arrangement is to be *self-locking*. Neglecting the weight of the cylinder, determine the minimum allowable coefficients of static friction at A, B, and C.

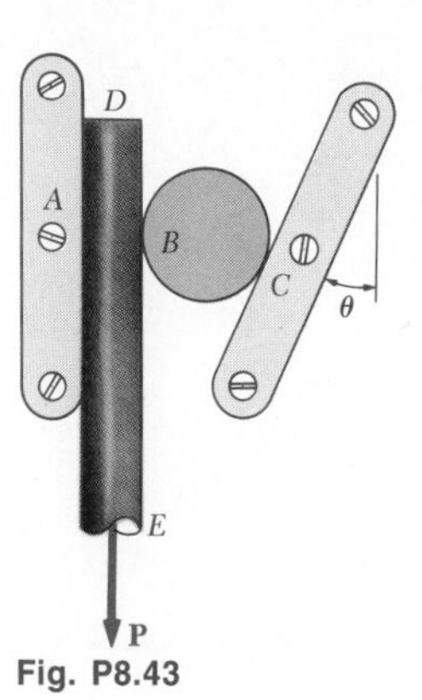

Fig. P8.43

8.44 A safety device used by workers climbing ladders fixed to high structures consists of a rail attached to the ladder and a sleeve which may slide on the flange of the rail. A chain connects the worker's belt to the end of an eccentric cam which may rotate about an axle attached to the sleeve at C. Determine the smallest allowable common value of the coefficient of static friction between the flange of the rail, the pins at A and B, and the eccentric cam if the sleeve is not to slide down when the chain is pulled vertically downward.

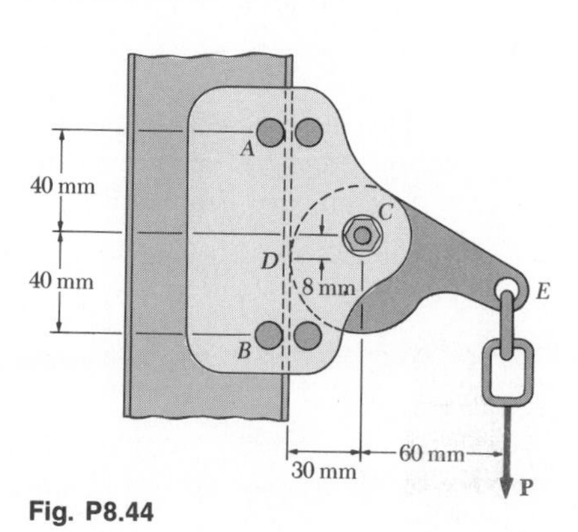

Fig. P8.44

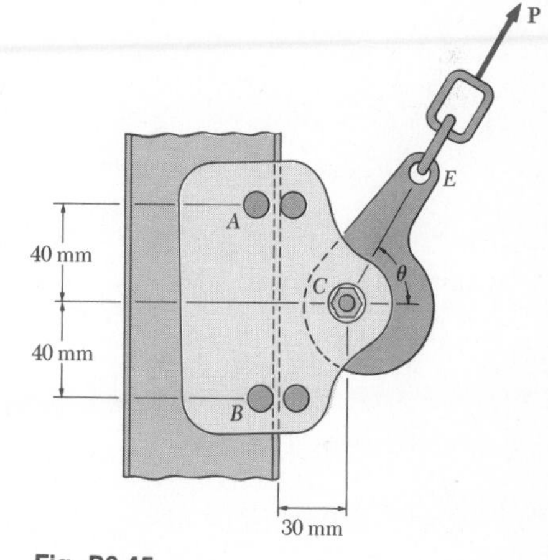

Fig. P8.45

8.45 To be of practical use, the safety sleeve described in the preceding problem must be free to slide along the rail when pulled upward. Determine the largest allowable value of the coefficient of static friction between the flange of the rail and the pins at A and B if the sleeve is to be free to slide when pulled as shown in the figure, assuming (*a*) $\theta = 60°$, (*b*) $\theta = 50°$, (*c*) $\theta = 40°$.

8.46 A light metal panel is welded to two short sleeves of 1-in. inside diameter which may slide on a horizontal rod. The coefficients of friction between the sleeves and the rod are $\mu_s = 0.40$ and $\mu_k = 0.30$. A cord attached to corner C is used to move the panel along the rod. Determine the range of values of θ for which the panel will start moving to the right.

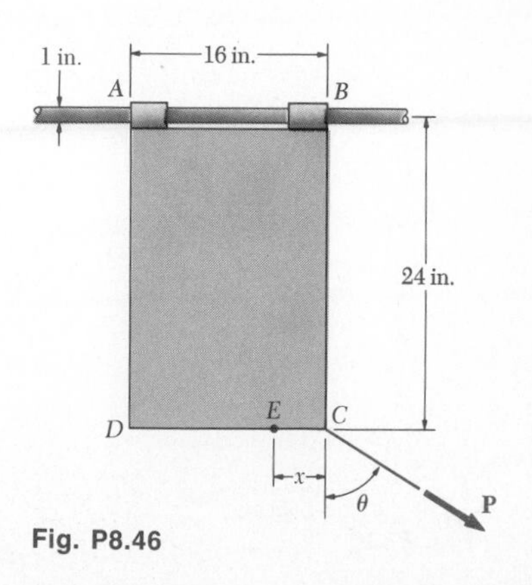

Fig. P8.46

8.47 (*a*) Solve Prob. 8.46, assuming that the cord is attached at point E at a distance $x = 4$ in. from corner C. (*b*) Determine the largest value of x for which the panel may be moved to the right.

8.48 A 10-ft beam, weighing 1200 lb, is to be moved to the left onto the platform. A horizontal force **P** is applied to the dolly, which is mounted on frictionless wheels. The coefficients of friction between all surfaces are $\mu_s = 0.30$ and $\mu_k = 0.25$ and initially $x = 2$ ft. Knowing that the top surface of the dolly is slightly higher than the platform, determine the force **P** required to start moving the beam. (*Hint.* The beam is supported at *A* and *D*.)

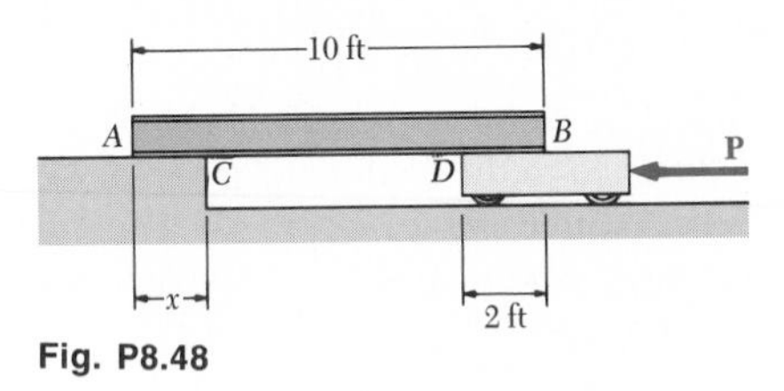

Fig. P8.48

8.49 (*a*) Show that the beam of Prob. 8.48 *cannot* be moved if the top surface of the dolly is slightly *lower* than the platform. (*b*) Show that the beam *can* be moved if two 175-lb workers stand on the beam at *B* and determine how far to the left the beam can be moved.

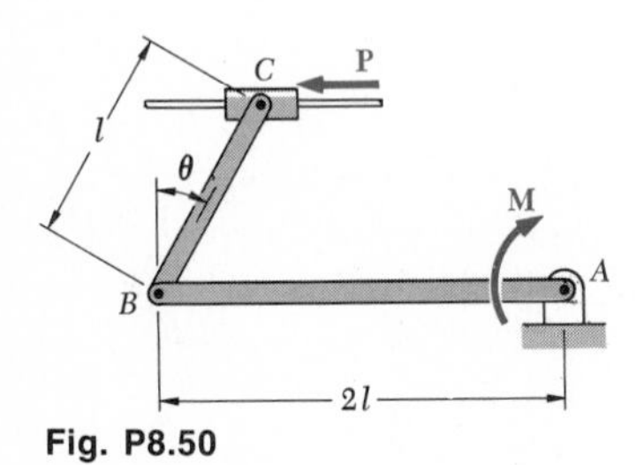

Fig. P8.50

8.50 Denoting by μ_s the coefficient of static friction between the collar *C* and the horizontal rod, determine the largest magnitude of the couple **M** for which equilibrium is maintained. Explain what happens if $\mu_s \geq \tan\theta$.

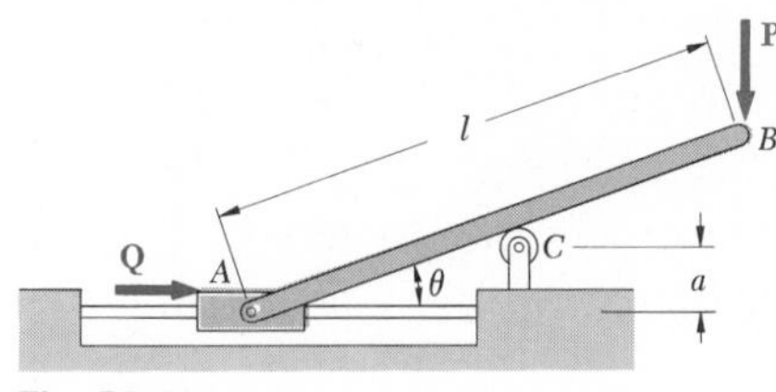

Fig. P8.51

8.51 The slender rod *AB* of length $l = 300$ mm is attached to a collar at *A* and rests on a wheel located at a vertical distance $a = 50$ mm from the horizontal rod on which the collar slides. Neglecting the friction at *C* and knowing that the coefficient of static friction between the collar and the horizontal rod is 0.25, determine the range of values of *Q* for which equilibrium is maintained when $P = 60$ N and $\theta = 20°$.

8.52 Solve Prob. 8.51, assuming that the wheel at *C* is frozen in its bearings and that the coefficient of static friction between the wheel and rod *AB* is also 0.25.

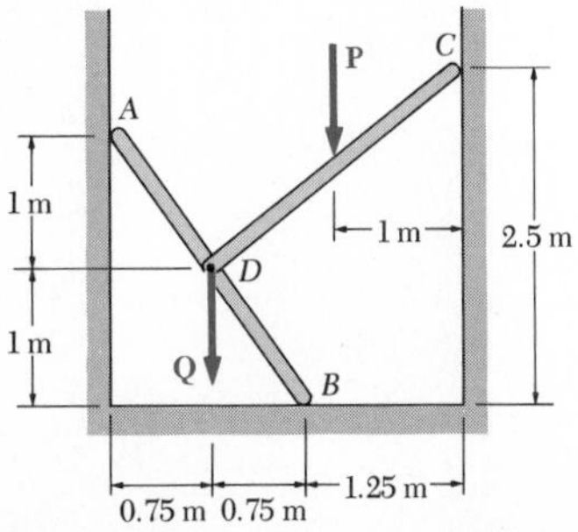

Fig. P8.53

***8.53** Two 2.5-m beams are pin-connected at *D* and support two loads **P** and **Q** as shown. Knowing that the coefficient of static friction is zero at *A* and $\frac{1}{3}$ at *B* and *C*, determine the smallest value of *P* for which equilibrium is maintained when $Q = 650$ N. (*Hint.* Note that *C* moves up when *B* moves to the right.)

***8.54** Solve Prob. 8.53, assuming that the coefficient of static friction is $\frac{1}{3}$ at all surfaces of contact.

***8.55** Two slender rods of negligible weight are pin-connected at A and attached to the 18-lb block B and the 80-lb block C as shown. The coefficient of static friction is 0.55 between all surfaces of contact. Determine the range of values of P for which equilibrium is maintained.

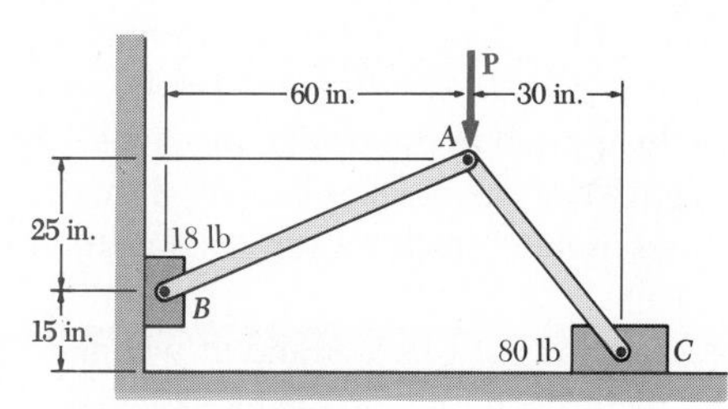

Fig. P8.55 and P8.56

***8.56** Two slender rods of negligible weight are pin-connected at A and attached to the 18-lb block B and the 80-lb block C as shown. A vertical load of magnitude $P = 300$ lb is applied at A. Determine (*a*) the minimum value of the coefficient of static friction (at both B and C) for which equilibrium is maintained, (*b*) whether sliding is impending at B or at C.

***8.57** Solve Prob. 8.56, assuming that $P = 200$ lb.

***8.58** A 3-m plank lies on two rollers A and B, located 1.5 m apart on a 10° incline. Roller A rotates slowly clockwise at a constant speed; roller B is idle. Initially, the plank is placed on the rollers with its center of gravity at A ($x = 0$) and its end C at B. Knowing that the coefficients of friction between the board and roller A are $\mu_s = 0.35$ and $\mu_k = 0.28$, (*a*) determine the value reached by x when the plank starts slipping on roller A, (*b*) describe what happens next.

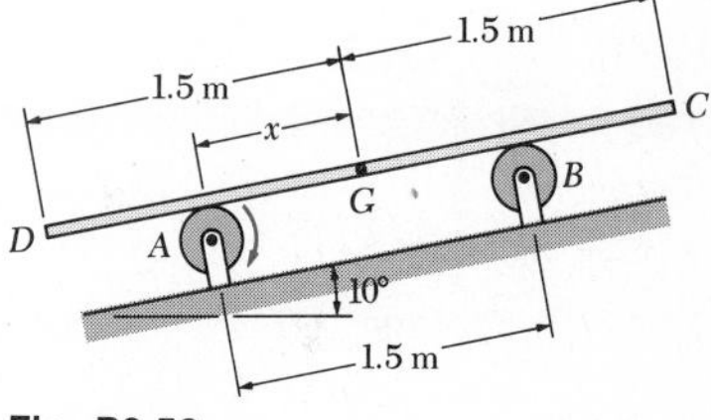

Fig. P8.58

***8.59** Solve Prob. 8.58, assuming that roller B is frozen in its bearings and that the coefficients of friction are the same at B as at A.

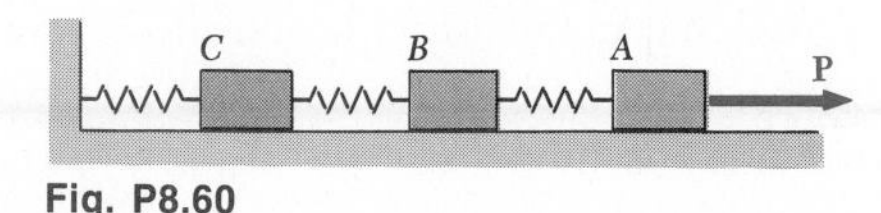

Fig. P8.60

***8.60** The mathematical model shown, consisting of blocks connected by springs, has been developed for the analysis of a certain structure. The weight of each block is $W = 5$ N, the constant of each spring is $k = 20$ N/m, and the coefficient of friction between the base and each block is 0.40. Knowing that the initial tension in each spring is zero, construct a graph showing the magnitude of the force **P** versus the position of block A as P increases from zero to 5 N and then decreases to zero.

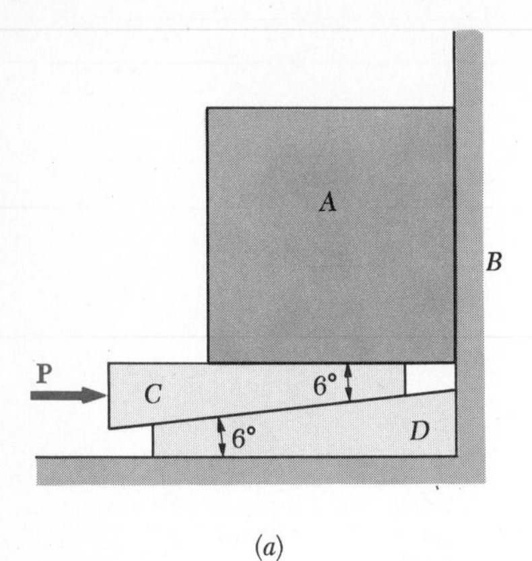

(a)

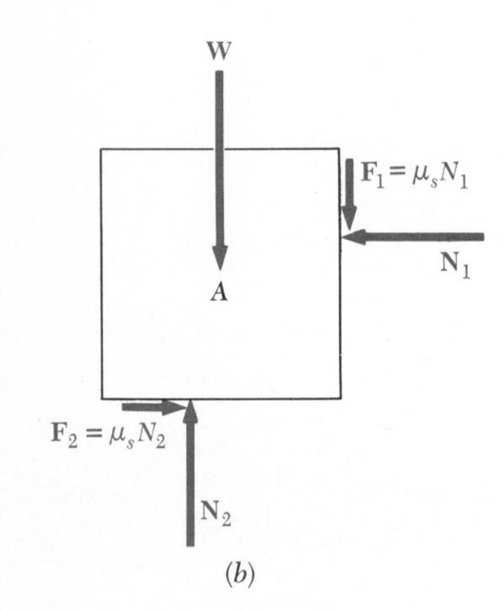

(b)

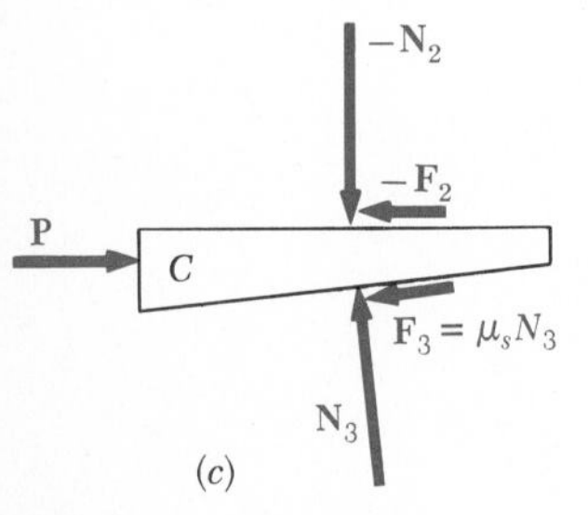

(c)

Fig. 8.7

8.5. Wedges. Wedges are simple machines used to raise large stone blocks and other heavy loads. These loads may be raised by applying to the wedge a force usually considerably smaller than the weight of the load. Besides, because of the friction existing between the surfaces in contact, a wedge, if properly shaped, will remain in place after being forced under the load. Wedges may thus be used advantageously to make small adjustments in the position of heavy pieces of machinery.

Consider the block A shown in Fig. 8.7*a*. This block rests against a vertical wall B and is to be raised slightly by forcing a wedge C between block A and a second wedge D. We want to find the minimum value of the force **P** which must be applied to the wedge C to move the block. We shall assume that the weight **W** of the block is known. It may have been given in pounds or determined in newtons from the mass of the block expressed in kilograms.

The free-body diagrams of block A and of wedge C have been drawn in Fig. 8.7*b* and *c*. The forces acting on the block include its weight and the normal and friction forces at the surfaces of contact with wall B and wedge C. The magnitudes of the friction forces $\mathbf{F}_1$ and $\mathbf{F}_2$ are equal, respectively, to $\mu_s N_1$ and $\mu_s N_2$ since the motion of the block must be started. It is important to show the friction forces with their correct sense. Since the block will move upward, the force $\mathbf{F}_1$ exerted by the wall on the block must be directed downward. On the other hand, since the wedge C moves to the right, the relative motion of A with respect to C is to the left and the force $\mathbf{F}_2$ exerted by C on A must be directed to the right.

Considering now the free body C in Fig. 8.7*c*, we note that the forces acting on C include the applied force **P** and the normal and friction forces at the surfaces of contact with A and D. The weight of the wedge is small compared with the other forces involved and may be neglected. The forces acting on C are equal and opposite to the forces $\mathbf{N}_2$ and $\mathbf{F}_2$ acting on A and are denoted, respectively, by $-\mathbf{N}_2$ and $-\mathbf{F}_2$; the friction force $-\mathbf{F}_2$ must therefore be directed to the left. We check that the force $\mathbf{F}_3$ is also directed to the left.

The total number of unknowns involved in the two free-body diagrams may be reduced to four if the friction forces are expressed in terms of the normal forces. Expressing that block A and wedge C are in equilibrium will provide four equations which may be solved to obtain the magnitude of **P**. It should be noted that in the example considered here, it will be more convenient to replace each pair of normal and friction forces by their resultant. Each free body is then subjected to only three forces, and the problem may be solved by drawing the corresponding force triangles (see Sample Prob. 8.4).

8.6. Square-Threaded Screws. Square-threaded screws are frequently used in jacks, presses, and other mechanisms. Their analysis is similar to that of a block sliding along an inclined plane.

Consider the jack shown in Fig. 8.8. The screw carries a load **W** and is supported by the base of the jack. Contact between screw and base takes place along a portion of their threads. By applying a force **P** on the handle, the screw may be made to turn and to raise the load **W**.

The thread of the base has been unwrapped and shown as a straight line in Fig. 8.9a. The correct slope was obtained by plotting horizontally the product $2\pi r$, where r is the mean radius of the thread, and vertically the *lead* L of the screw, i.e., the distance through which the screw advances in one turn. The angle θ this line forms with the horizontal is the *lead angle.* Since the force of friction between two surfaces in contact does not depend upon the area of contact, the two threads may be assumed to be in contact over a much smaller area than they actually are and the screw may be represented by the block shown in Fig. 8.9a. It should be noted, however, that in this analysis of the jack, the friction between cap and screw is neglected.

The free-body diagram of the block should include the load **W**, the reaction **R** of the base thread, and a horizontal force **Q** having the same effect as the force **P** exerted on the handle. The force **Q** should have the same moment as **P** about the axis of the screw and its magnitude should thus be $Q = Pa/r$. The force **Q**, and thus the force **P** required to raise the load **W**, may be obtained from the free-body diagram shown in Fig. 8.9a. The friction angle is taken equal to ϕ_s since the load will presumably be raised through a succession of short strokes. In mechanisms providing for the continuous rotation of a screw, it may be desirable to distinguish between the force required to start motion (using ϕ_s) and that required to maintain motion (using ϕ_k).

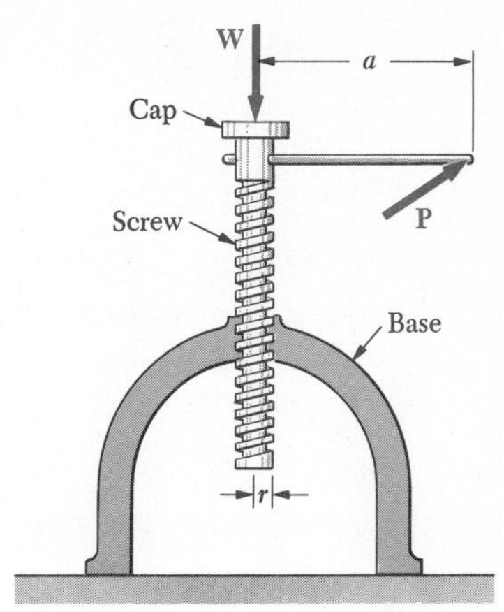

Fig. 8.8

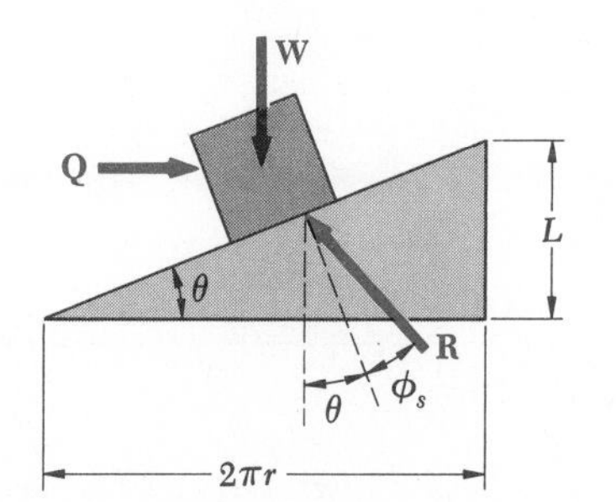

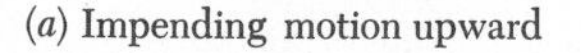

(*a*) Impending motion upward

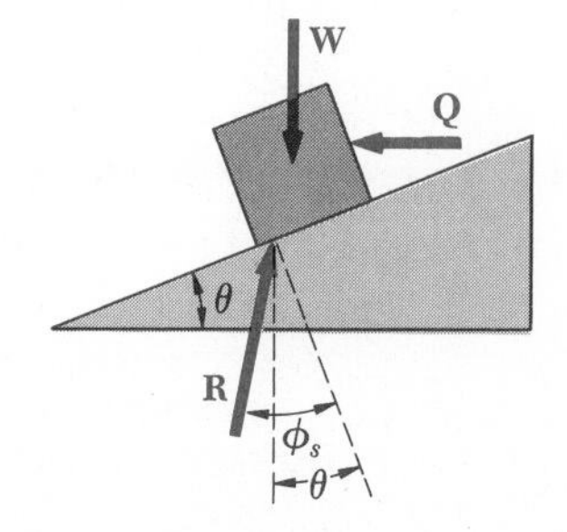

(*b*) Impending motion downward with $\phi_s > \theta$

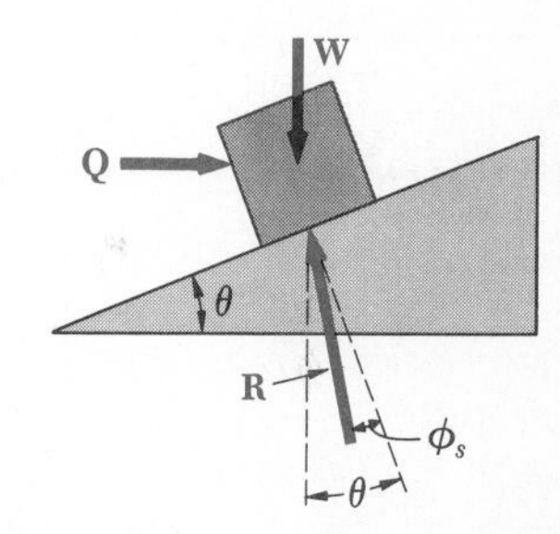

(*c*) Impending motion downward with $\phi_s < \theta$

Fig. 8.9 Block-and-incline analysis of a screw.

If the friction angle ϕ_s is larger than the lead angle θ, the screw is said to be *self-locking;* it will remain in place under the load. To lower the load, we must then apply the force shown in Fig. 8.9b. If ϕ_s is smaller than θ, the screw will unwind under the load; it is then necessary to apply the force shown in Fig. 8.9c to maintain equilibrium.

The lead of a screw should not be confused with its *pitch.* The lead was defined as the distance through which the screw advances in one turn; the pitch is the distance measured between two consecutive threads. While lead and pitch are equal in the case of *single-threaded* screws, they are different in the case of *multiple-threaded* screws, i.e., screws having several independent threads. It is easily verified that for double-threaded screws, the lead is twice as large as the pitch; for triple-threaded screws, it is three times as large as the pitch; etc.

SAMPLE PROBLEM 8.4

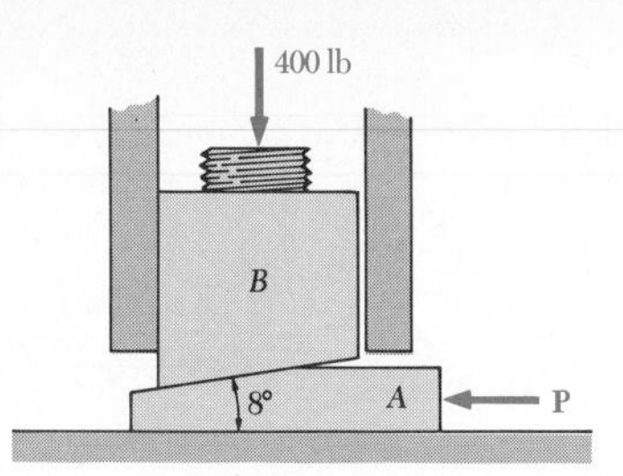

The position of the machine block B is adjusted by moving the wedge A. Knowing that the coefficient of static friction is 0.35 between all surfaces of contact, determine the force **P** required (*a*) to raise block B, (*b*) to lower block B.

Solution. For each part, the free-body diagrams of block B and wedge A are drawn, together with the corresponding force triangles, and the law of sines is used to find the desired forces. We note that since $\mu_s = 0.35$, the angle of friction is

$$\phi_s = \tan^{-1} 0.35 = 19.3°$$

a. Force P to Raise Block

Free Body: Block B

$$\frac{R_1}{\sin 109.3°} = \frac{400 \text{ lb}}{\sin 43.4°}$$

$$R_1 = 549 \text{ lb}$$

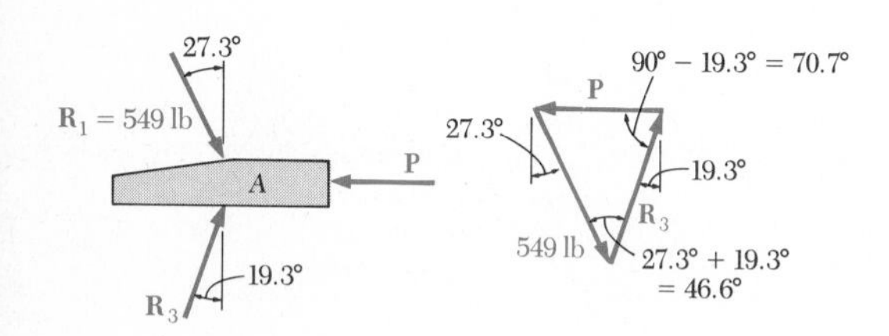

Free Body: Wedge A

$$\frac{P}{\sin 46.6°} = \frac{549 \text{ lb}}{\sin 70.7°}$$

$$P = 423 \text{ lb}$$

$\mathbf{P} = 423 \text{ lb} \leftarrow$ ◄

b. Force P to Lower Block

Free Body: Block B

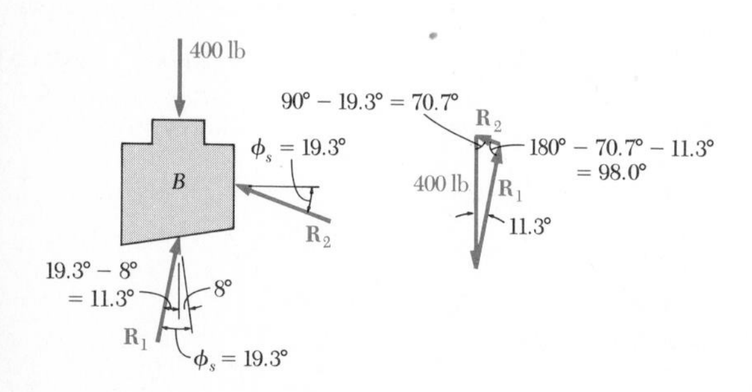

$$\frac{R_1}{\sin 70.7°} = \frac{400 \text{ lb}}{\sin 98.0°}$$

$$R_1 = 381 \text{ lb}$$

Free Body: Wedge A

$$\frac{P}{\sin 30.6°} = \frac{381 \text{ lb}}{\sin 70.7°}$$

$$P = 206 \text{ lb}$$

$\mathbf{P} = 206 \text{ lb} \rightarrow$ ◄

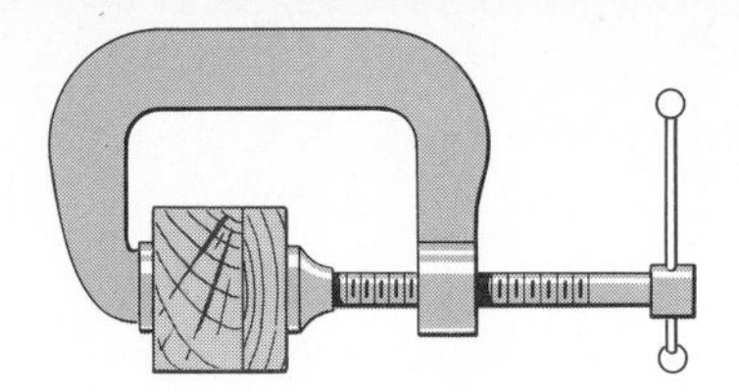

SAMPLE PROBLEM 8.5

A clamp is used to hold two pieces of wood together as shown. The clamp has a double square thread of mean diameter equal to 10 mm and with a pitch of 2 mm. The coefficient of friction between threads is $\mu_s = 0.30$. If a maximum torque of 40 N·m is applied in tightening the clamp, determine (*a*) the force exerted on the pieces of wood, (*b*) the torque required to loosen the clamp.

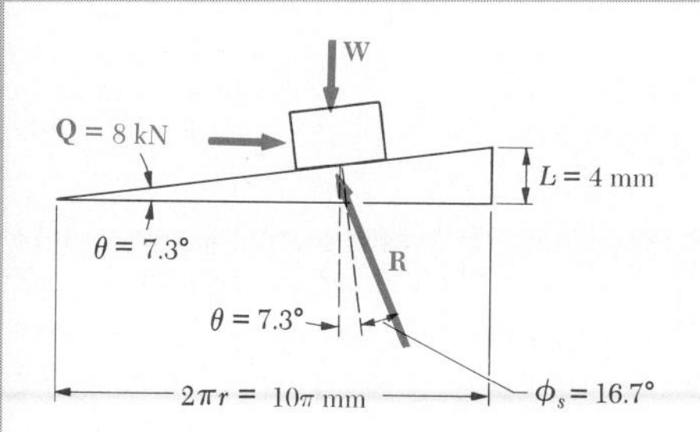

***a.* Force Exerted by Clamp.** The mean radius of the screw is $r = 5$ mm. Since the screw is double-threaded, the lead L is equal to twice the pitch: $L = 2(2 \text{ mm}) = 4$ mm. The lead angle θ and the friction angle ϕ_s are obtained by writing

$$\tan\theta = \frac{L}{2\pi r} = \frac{4 \text{ mm}}{10\pi \text{ mm}} = 0.1273 \qquad \theta = 7.3°$$

$$\tan\phi_s = \mu_s = 0.30 \qquad \phi_s = 16.7°$$

The force **Q** which should be applied to the block representing the screw is obtained by expressing that its moment Qr about the axis of the screw is equal to the applied torque.

$$Q(5 \text{ mm}) = 40 \text{ N·m}$$

$$Q = \frac{40 \text{ N·m}}{5 \text{ mm}} = \frac{40 \text{ N·m}}{5 \times 10^{-3} \text{ m}} = 8000 \text{ N} = 8 \text{ kN}$$

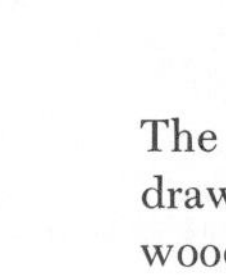

The free-body diagram and the corresponding force triangle may now be drawn for the block; the magnitude of the force **W** exerted on the pieces of wood is obtained by solving the triangle.

$$W = \frac{Q}{\tan(\theta + \phi_s)} = \frac{8 \text{ kN}}{\tan 24.0°}$$

$$W = 17.97 \text{ kN} \blacktriangleleft$$

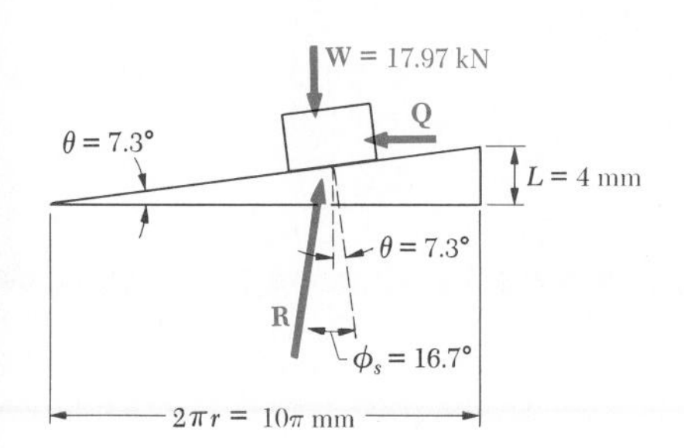

***b.* Torque Required to Loosen Clamp.** The force **Q** required to loosen the clamp and the corresponding torque are obtained from the free-body diagram and force triangle shown.

$$Q = W\tan(\phi_s - \theta) = (17.97 \text{ kN}) \tan 9.4°$$
$$= 2.975 \text{ kN}$$

$$\text{Torque} = Qr = (2.975 \text{ kN})(5 \text{ mm})$$
$$= (2.975 \times 10^3 \text{ N})(5 \times 10^{-3} \text{ m}) = 14.87 \text{ N·m}$$

$$\text{Torque} = 14.87 \text{ N·m} \blacktriangleleft$$

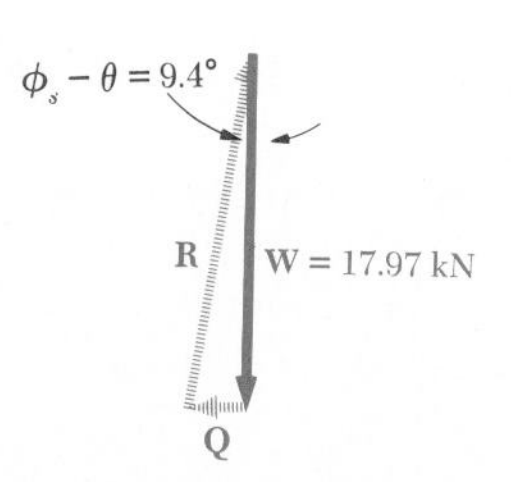

Problems

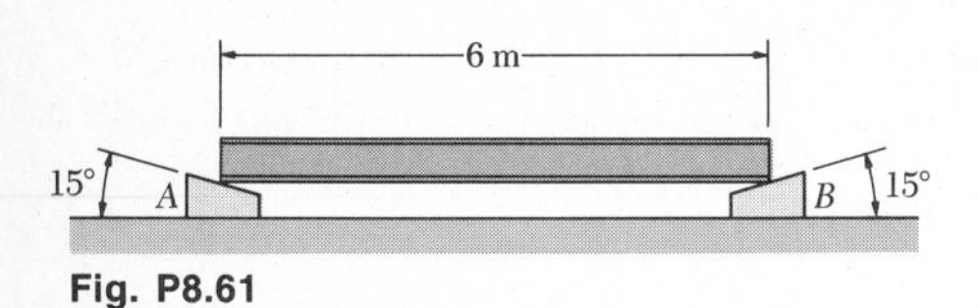

Fig. P8.61

8.61 A 6-m long wide-flange beam weighing 6650 N is supported by a 15° wedge at each end. Wedge *A* is fairly smooth ($\mu_s = 0.20$ at all surfaces of contact) and may be forced under the beam to adjust its position. Wedge *B* is rough ($\mu_s = 0.50$) and the right end of the beam may be considered as fixed. Determine the horizontal force **P** which should be applied to wedge *A* to raise the left end of the beam.

8.62 In Prob. 8.61, determine the horizontal force **P** which should be applied to wedge *A* to lower the left end of the beam.

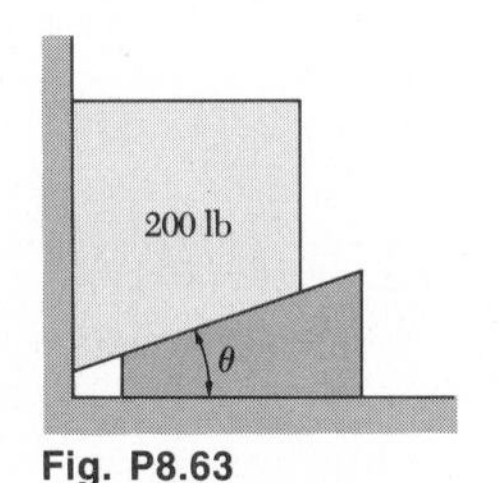

Fig. P8.63

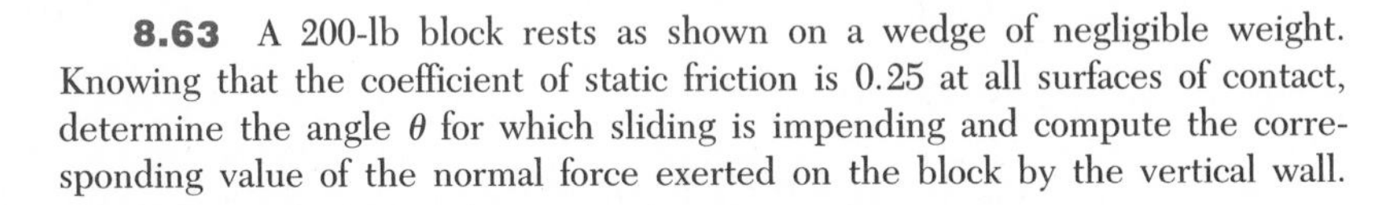

8.63 A 200-lb block rests as shown on a wedge of negligible weight. Knowing that the coefficient of static friction is 0.25 at all surfaces of contact, determine the angle θ for which sliding is impending and compute the corresponding value of the normal force exerted on the block by the vertical wall.

8.64 Solve Prob. 8.63, assuming (*a*) that the coefficient of static friction is zero between the block and the vertical wall and 0.25 at the other two surfaces of contact, (*b*) that the coefficient of static friction is zero between the block and the wedge and 0.25 at the other two surfaces of contact.

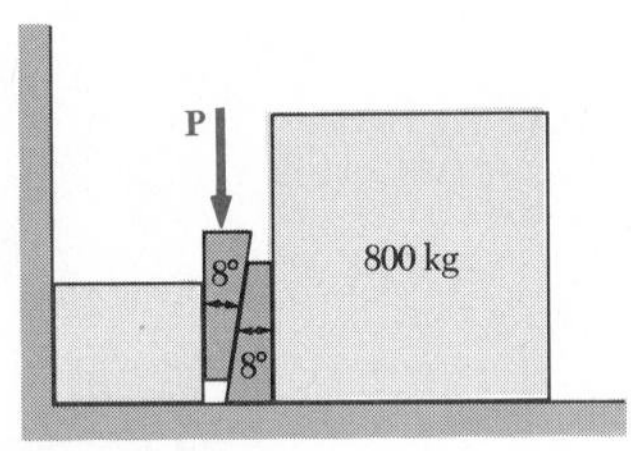

Fig. P8.65

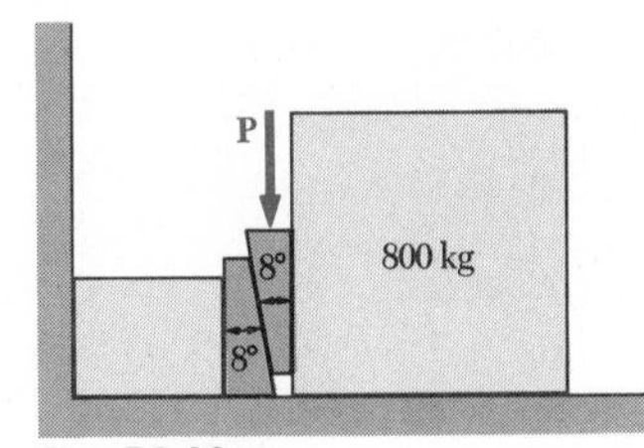

Fig. P8.66

8.65 and 8.66 Two 8° wedges of negligible weight are used to move and position the 800-kg block. Knowing that the coefficient of static friction is 0.30 at all surfaces of contact, determine the smallest force **P** which should be applied as shown to one of the wedges.

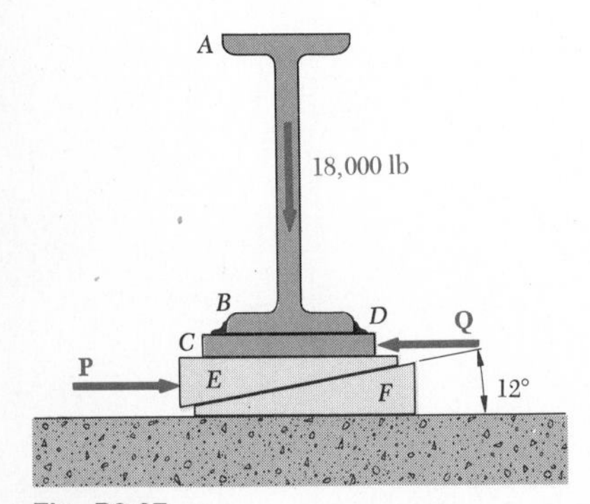

Fig. P8.67

8.67 The elevation of the end of the steel floor beam is adjusted by means of the steel wedges *E* and *F*. The steel base plate *CD* has been welded to the lower flange of the beam; the end reaction of the beam is known to be 18,000 lb. The coefficient of static friction is 0.30 between the two steel surfaces and 0.60 between the steel and concrete. If horizontal motion of the base plate is prevented by the force **Q**, determine (*a*) the force **P** required to raise the beam, (*b*) the corresponding force **Q**.

8.68 Solve Prob. 8.67, assuming that the end of the beam is to be lowered.

8.69 High-strength bolts are used in the construction of many modern steel structures. For a 1-in. nominal-diameter bolt the required minimum bolt tension is 47,250 lb. Assuming the coefficient of static friction to be 0.35, determine the required torque which must be applied to the bolt and nut. The mean diameter of the thread is 0.94 in., and the lead is 0.125 in. Neglect friction between the nut and washer, and assume the bolt to be square-threaded.

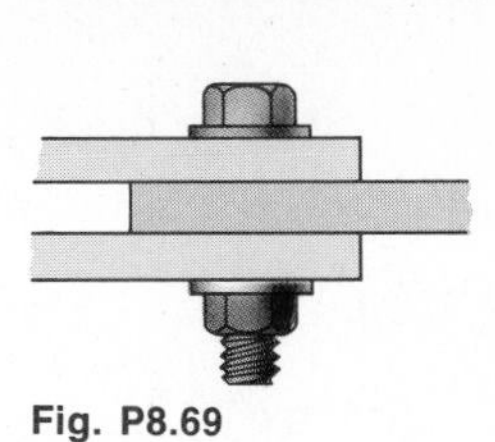
Fig. P8.69

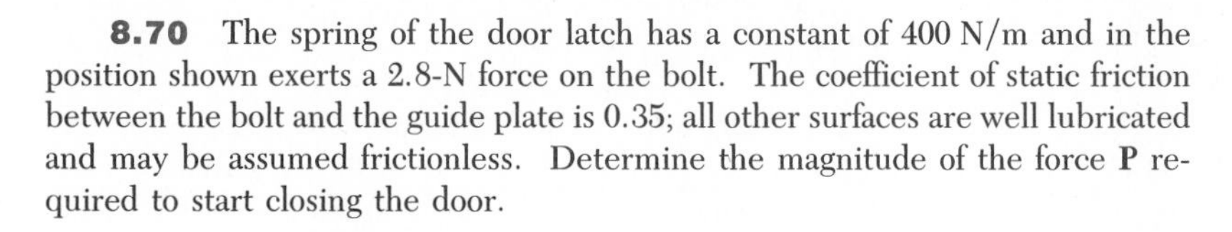

8.70 The spring of the door latch has a constant of 400 N/m and in the position shown exerts a 2.8-N force on the bolt. The coefficient of static friction between the bolt and the guide plate is 0.35; all other surfaces are well lubricated and may be assumed frictionless. Determine the magnitude of the force **P** required to start closing the door.

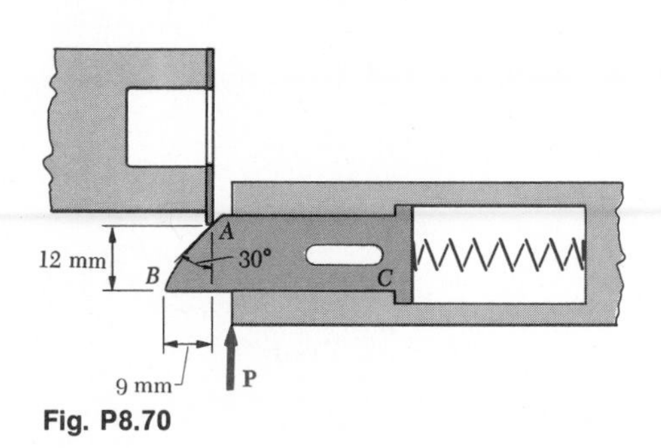

Fig. P8.70

8.71 In Prob. 8.70, determine the angle which the face of the bolt should form with the line *BC* if the force **P** required to close the door is to be the same for both the position shown and the position when *B* is almost at the guide plate.

8.72 A 12° wedge is used to split a log. The coefficient of friction between the wedge and the wood is 0.40. Knowing that a force **P** of magnitude 3.2 kN was required to insert the wedge, determine the magnitude of the forces exerted on the log by the wedge after it has been inserted.

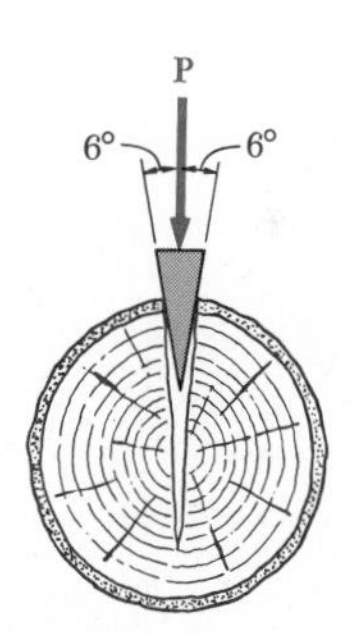

Fig. P8.72

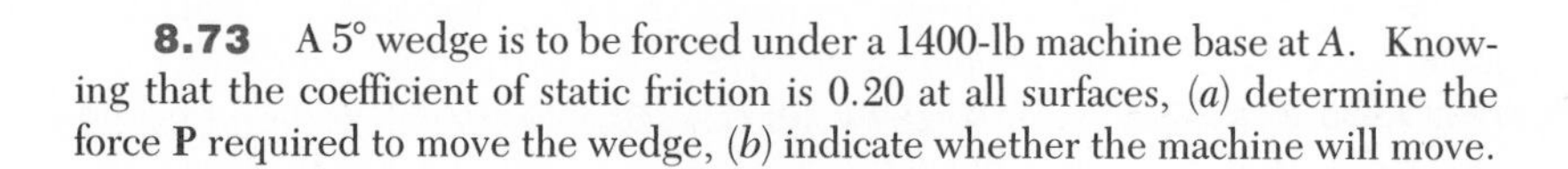

8.73 A 5° wedge is to be forced under a 1400-lb machine base at *A*. Knowing that the coefficient of static friction is 0.20 at all surfaces, (*a*) determine the force **P** required to move the wedge, (*b*) indicate whether the machine will move.

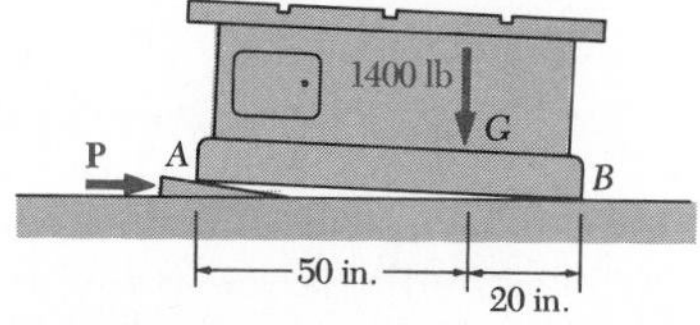

Fig. P8.73

8.74 Solve Prob. 8.73, assuming that the wedge is to be forced under the machine base at *B* instead of *A*.

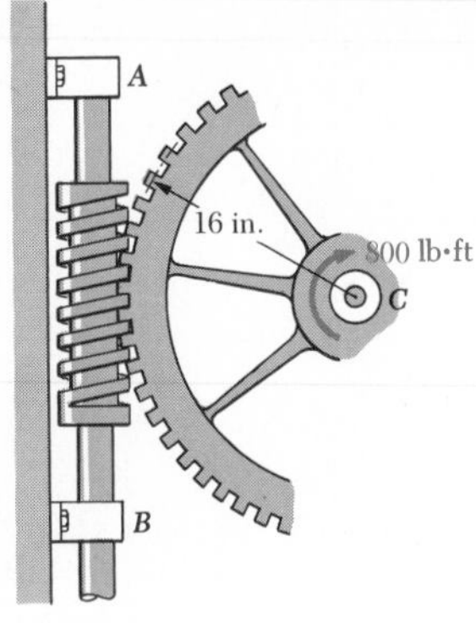

Fig. P8.75

8.75 The square-threaded worm gear shown has a mean radius of 2 in. and a pitch of $\frac{1}{2}$ in. The large gear is subjected to a constant clockwise torque of 800 lb · ft. Knowing that the coefficient of static friction between gear teeth is 0.12, determine the torque which must be applied to shaft *AB* in order to rotate the large gear counterclockwise. Neglect friction in the bearings at *A*, *B*, and *C*.

8.76 In Prob. 8.75, determine the torque which must be applied to shaft *AB* in order to rotate the large gear clockwise.

8.77 The main features of a screw-luffing crane are shown. Distances *AD* and *CD* are each 9 m. The position of the 6-Mg boom *CDE* is controlled by the screw *ABC*, which is double-threaded at each end (left-handed thread at *A*, right-handed thread at *C*). Each thread has a pitch of 16 mm and a mean diameter of 200 mm. If the coefficient of static friction is 0.08, determine the magnitude of the couple which must be applied to the screw (*a*) to raise the boom, (*b*) to lower the boom.

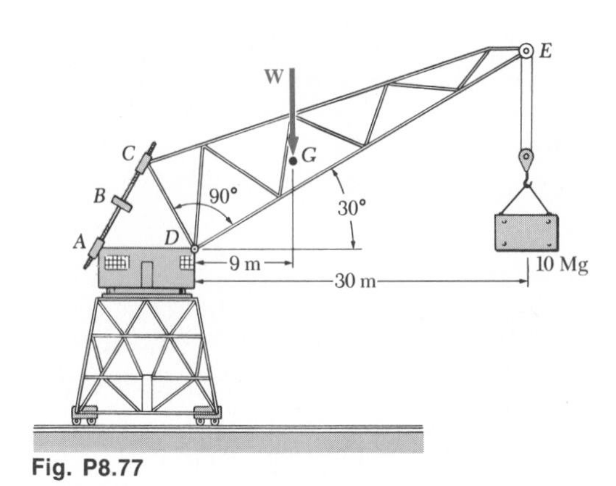

Fig. P8.77

8.78 Derive the following formulas relating the load **W** and the force **P** exerted on the handle of the jack discussed in Sec. 8.6: (*a*) $P = (Wr/a) \tan (\theta + \phi_s)$, to raise the load; (*b*) $P = (Wr/a) \tan (\phi_s - \theta)$, to lower the load if the screw is self-locking; (*c*) $P = (Wr/a) \tan (\theta - \phi_s)$, to hold the load if the screw is not self-locking.

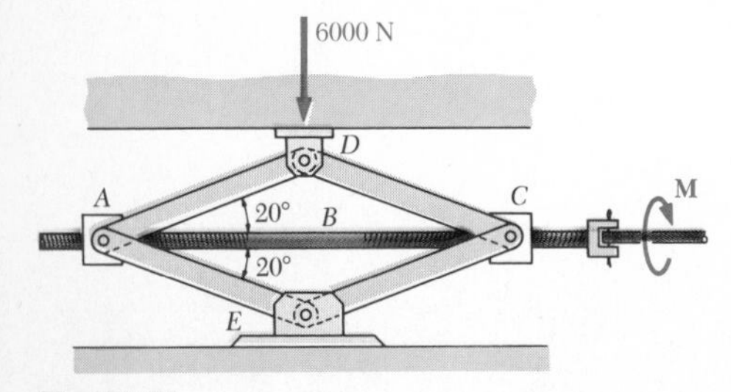

Fig. P8.79

8.79 The position of the automobile jack shown is controlled by a screw *ABC* which is single-threaded at each end (right-handed thread at *A*, left-handed thread at *C*). Each thread has a pitch of 2.5 mm and a mean diameter of 9 mm. If the coefficient of static friction is 0.15, determine the magnitude and sense of the couple **M** which must be applied to the screw to raise the automobile.

8.80 For the jack of Prob. 8.79, determine the magnitude and sense of the couple **M** which must be applied to lower the automobile.

8.81 The vise shown consists of two members connected by two double-threaded screws of mean radius 0.25 in. and pitch 0.08 in. The lower member is threaded at A and B ($\mu_s = 0.35$), but the upper member is not threaded. It is desired to apply two equal and opposite forces of 120 lb on the blocks held between the jaws. (*a*) What screw should be adjusted first? (*b*) What is the maximum torque applied in tightening the second screw?

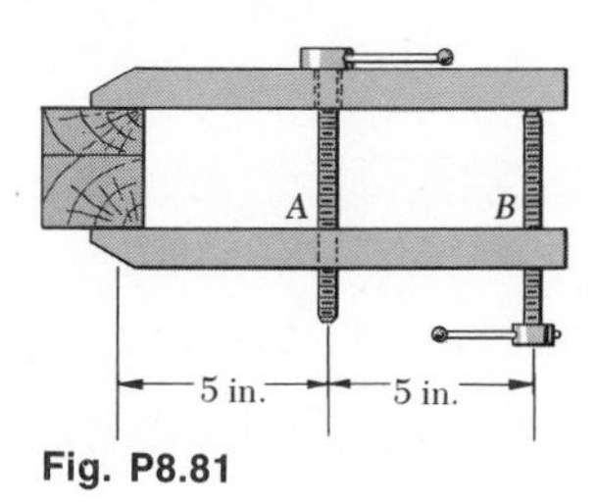

Fig. P8.81

8.82 Solve part *b* of Prob. 8.81, assuming that the wrong screw is adjusted first.

8.83 In the machinist's vise shown, the movable jaw D is rigidly attached to the tongue AB which fits loosely into the fixed body of the vise. The screw is single-threaded into the fixed base and has a mean diameter of 18 mm and a pitch of 6 mm. The coefficient of static friction is 0.25 between the threads and also between the tongue and the body. Neglecting bearing friction between the screw and the movable head, determine the torque which must be applied to the handle in order to produce a clamping force of 5 kN.

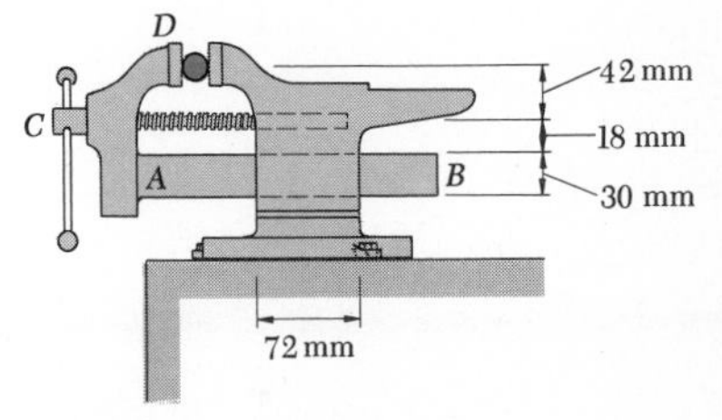

Fig. P8.83

8.84 In Prob. 8.83, a clamping force of 5 kN was obtained by tightening the vise. Determine the torque which must be applied to the screw to loosen the vise.

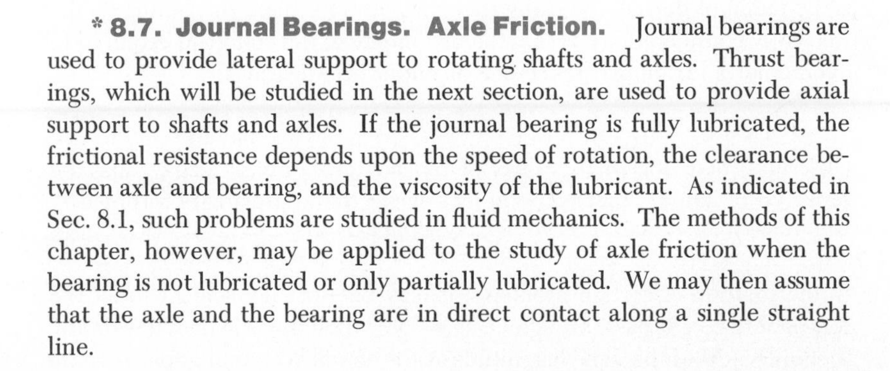

* 8.7. Journal Bearings. Axle Friction.

Journal bearings are used to provide lateral support to rotating shafts and axles. Thrust bearings, which will be studied in the next section, are used to provide axial support to shafts and axles. If the journal bearing is fully lubricated, the frictional resistance depends upon the speed of rotation, the clearance between axle and bearing, and the viscosity of the lubricant. As indicated in Sec. 8.1, such problems are studied in fluid mechanics. The methods of this chapter, however, may be applied to the study of axle friction when the bearing is not lubricated or only partially lubricated. We may then assume that the axle and the bearing are in direct contact along a single straight line.

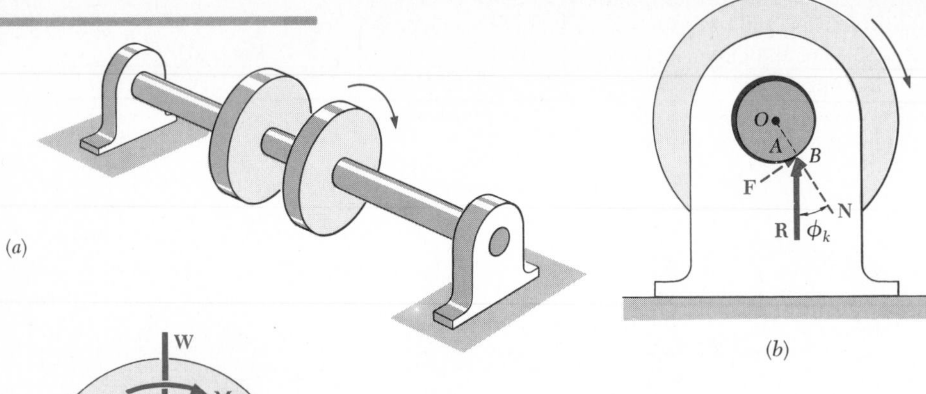

(a)

(b)

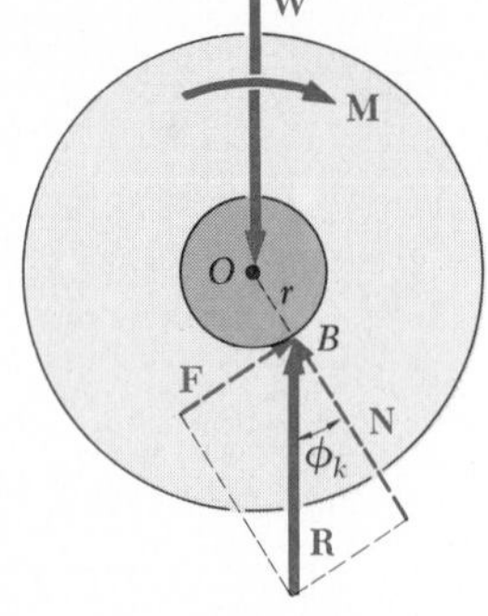

(c)

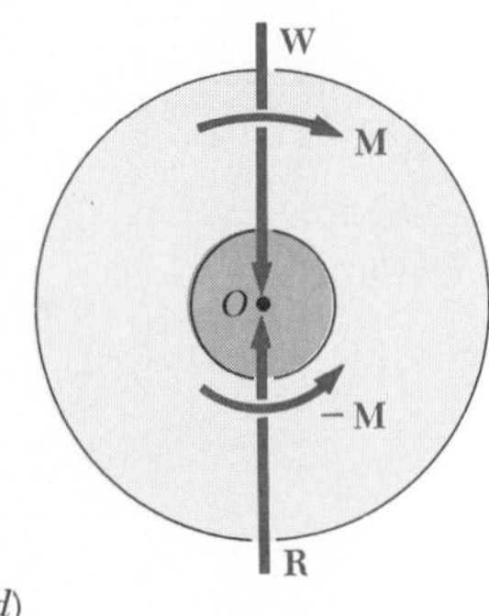

(d)

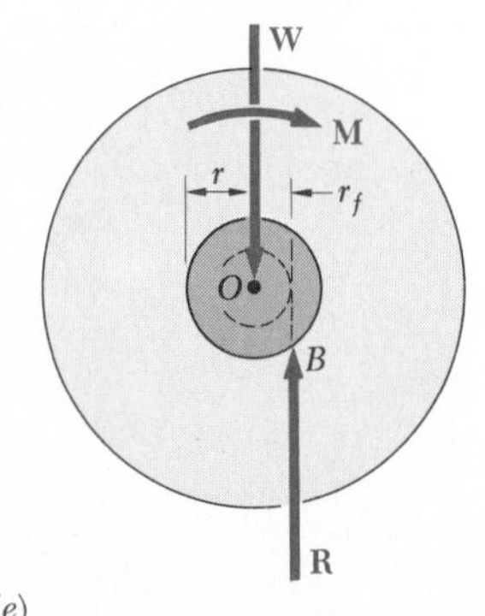

(e)
Fig. 8.10

Consider two wheels, each of weight **W**, rigidly mounted on an axle supported symmetrically by two journal bearings (Fig. 8.10*a*). If the wheels rotate, we find that to keep them rotating at constant speed, it is necessary to apply to each of them a couple **M**. A free-body diagram has been drawn in Fig. 8.10*c*, which represents one of the wheels and the corresponding half axle in projection on a plane perpendicular to the axle. The forces acting on the free body include the weight **W** of the wheel, the couple **M** required to maintain its motion, and a force **R** representing the reaction of the bearing. This force is vertical, equal, and opposite to **W** but does not pass through the center O of the axle; **R** is located to the right of O at a distance such that its moment about O balances the moment **M** of the couple. Contact between the axle and bearing, therefore, does not take place at the lowest point A when the axle rotates. It takes place at point B (Fig. 8.10*b*) or, rather, along a straight line intersecting the plane of the figure at B. Physically, this is explained by the fact that when the wheels are set in motion, the axle "climbs" in the bearings until slippage occurs. After sliding back slightly, the axle settles more or less in the position shown. This position is such that the angle between the reaction **R** and the normal to the surface of the bearing is equal to the angle of kinetic friction ϕ_k. The distance from O to the line of action of **R** is thus $r \sin \phi_k$, where r is the radius of the axle. Writing that $\Sigma M_O = 0$ for the forces acting on the free body considered, we obtain the magnitude of the couple **M** required to overcome the frictional resistance of one of the bearings:

$$M = Rr \sin \phi_k \tag{8.5}$$

Observing that, for small values of the angle of friction, $\sin \phi_k$ may be replaced by $\tan \phi_k$, that is, by μ_k, we write the approximate formula

$$M \approx Rr\mu_k \tag{8.6}$$

In the solution of certain problems, it may be more convenient to let the line of action of **R** pass through O, as it does when the axle does not rotate. A couple $-$**M** of the same magnitude as the couple **M** but of opposite sense must then be added to the reaction **R** (Fig. 8.10*d*). This couple represents the frictional resistance of the bearing.

In case a graphical solution is preferred, the line of action of **R** may be readily drawn (Fig. 8.10*e*) if we note that it must be tangent to a circle centered at O and of radius

$$r_f = r \sin \phi_k \approx r\mu_k \tag{8.7}$$

This circle is called the *circle of friction* of the axle and bearing and is independent of the loading conditions of the axle.

***8.8. Thrust Bearings. Disk Friction.** Thrust bearings are used to provide axial support to rotating shafts and axles. They are of two types: (1) *end bearings* and (2) *collar bearings* (Fig. 8.11). In the case of collar bearings, friction forces develop between the two ring-shaped areas which are in contact. In the case of end bearings, friction takes place over full circular areas, or over ring-shaped areas when the end of the shaft is hollow. Friction between circular areas, called *disk friction*, also occurs in other mechanisms, such as *disk clutches*.

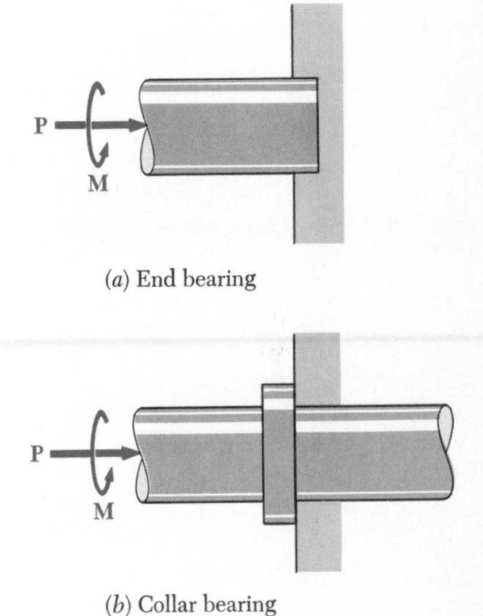

(*a*) End bearing

(*b*) Collar bearing

Fig. 8.11 Thrust bearings.

To obtain a formula which is valid in the most general case of disk friction, we shall consider a rotating hollow shaft. A couple **M** keeps the shaft rotating at constant speed while a force **P** maintains it in contact with a fixed bearing (Fig. 8.12). Contact between the shaft and the bearing takes place over a ring-shaped area of inner radius R_1 and outer radius R_2. Assuming that the pressure between the two surfaces in contact is uniform, we find that the magnitude of the normal force $\Delta\mathbf{N}$ exerted on an element of area ΔA is $\Delta N = P\,\Delta A/A$, where $A = \pi(R_2^2 - R_1^2)$, and that the magnitude of the friction force $\Delta\mathbf{F}$ acting on ΔA is $\Delta F = \mu_k\,\Delta N$. Denoting by r the distance from the axis of the shaft to the element of area ΔA, we express as follows the moment ΔM of $\Delta\mathbf{F}$ about the axis of the shaft:

$$\Delta M = r\,\Delta F = \frac{r\mu_k P\,\Delta A}{\pi(R_2^2 - R_1^2)}$$

The equilibrium of the shaft requires that the moment **M** of the couple applied to the shaft be equal in magnitude to the sum of the moments of the friction forces $\Delta\mathbf{F}$. Replacing ΔA by the infinitesimal element $dA = r\,d\theta\,dr$ used with polar coordinates, and integrating over the area of contact, we thus obtain the following expression for the magnitude of the couple **M**

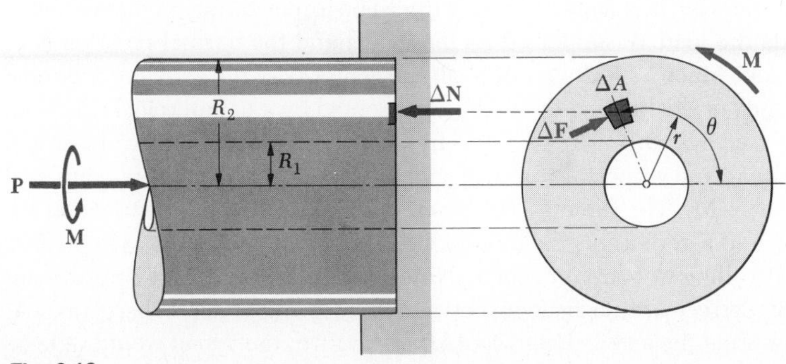

Fig. 8.12

required to overcome the frictional resistance of the bearing:

$$M = \frac{\mu_k P}{\pi(R_2^2 - R_1^2)} \int_0^{2\pi} \int_{R_1}^{R_2} r^2\, dr\, d\theta$$

$$= \frac{\mu_k P}{\pi(R_2^2 - R_1^2)} \int_0^{2\pi} \tfrac{1}{3}(R_2^3 - R_1^3)\, d\theta$$

$$M = \tfrac{2}{3}\mu_k P \frac{R_2^3 - R_1^3}{R_2^2 - R_1^2} \tag{8.8}$$

When contact takes place over a full circle of radius R, formula (8.8) reduces to

$$M = \tfrac{2}{3}\mu_k PR \tag{8.9}$$

The value of M is then the same as would be obtained if contact between shaft and bearing took place at a single point located at a distance $2R/3$ from the axis of the shaft.

The largest torque which may be transmitted by a disk clutch without causing slippage is given by a formula similar to (8.9), where μ_k has been replaced by the coefficient of static friction μ_s.

*** 8.9. Wheel Friction. Rolling Resistance.** The wheel is one of the most important inventions of our civilization. Its use makes it possible to move heavy loads with relatively little effort. Because the point of the wheel in contact with the ground at any given instant has no relative motion with respect to the ground, the wheel eliminates the large friction forces which would arise if the load were in direct contact with the ground. In practice, however, the wheel is not perfect, and some resistance to its motion exists. This resistance has two distinct causes. It is due (1) to a combined effect of axle friction and friction at the rim and (2) to the fact that the wheel and the ground deform, with the result that contact between wheel and ground takes place, not at a single point, but over a certain area.

To understand better the first cause of resistance to the motion of a wheel, we shall consider a railroad car supported by eight wheels mounted on axles and bearings. The car is assumed to be moving to the right at constant speed along a straight horizontal track. The free-body diagram of one of the wheels is shown in Fig. 8.13*a*. The forces acting on the free body include the load **W** supported by the wheel and the normal reaction **N** of the track. Since **W** is drawn through the center O of the axle, the frictional resistance of the bearing should be represented by a counterclockwise couple **M** (see Sec. 8.7). To keep the free body in equilibrium, we must add two equal and opposite forces **P** and **F**, forming a clockwise couple of moment $-$**M**. The force **F** is the friction force exerted by the track on the wheel, and **P** represents the force which should be applied to the wheel to keep it rolling at constant speed. Note that the forces **P** and **F** would not exist if there were no friction between wheel and track. The couple **M** representing the axle friction would then be zero; the wheel would slide on the track without turning in its bearing.

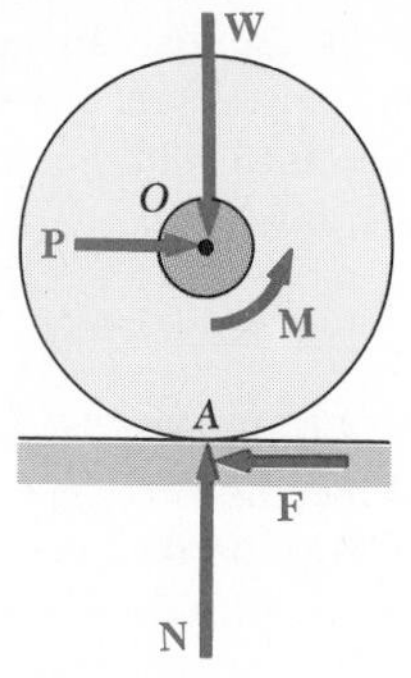

(*a*) Effect of axle friction

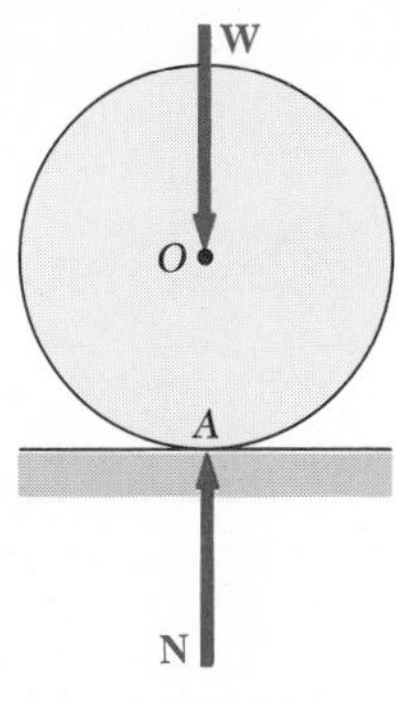

(*b*) Free wheel

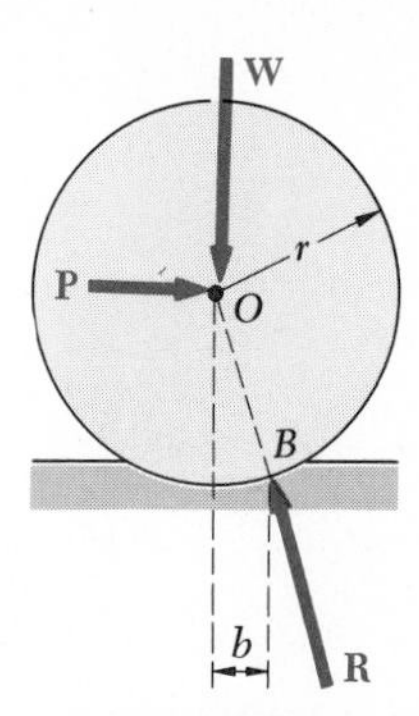

(*c*) Rolling resistance

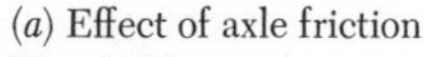

Fig. 8.13

The couple **M** and the forces **P** and **F** also reduce to zero when there is no axle friction. For example, a wheel which is not held in bearings and rolls freely and at constant speed on horizontal ground (Fig. 8.13*b*) will be subjected to only two forces: its own weight **W** and the normal reaction **N** of the ground. No friction force will act on the wheel, regardless of the value of the coefficient of friction between wheel and ground. A wheel rolling freely on horizontal ground should thus keep rolling indefinitely.

Experience, however, indicates that the wheel will slow down and eventually come to rest. This is due to the second type of resistance mentioned at the beginning of this section, known as the *rolling resistance.* Under the load **W**, both the wheel and the ground deform slightly, causing the contact between wheel and ground to take place over a certain area. Experimental evidence shows that the resultant of the forces exerted by the ground on the wheel over this area is a force **R** applied at a point B, which is not located directly under the center O of the wheel, but slightly in front of it (Fig. 8.13*c*). To balance the moment of **W** about B and to keep the wheel rolling at constant speed, it is necessary to apply a horizontal force **P** at the center of the wheel. Writing $\Sigma M_B = 0$, we obtain

$$Pr = Wb \tag{8.10}$$

where r = radius of wheel
b = horizontal distance between O and B

The distance b is commonly called the *coefficient of rolling resistance.* It should be noted that b is not a dimensionless coefficient since it represents a length; b is usually expressed in inches or in millimeters. The value of b depends upon several parameters in a manner which has not yet been clearly established. Values of the coefficient of rolling resistance vary from about 0.01 in. or 0.25 mm for a steel wheel on a steel rail to 5.0 in. or 125 mm for the same wheel on soft ground.

SAMPLE PROBLEM 8.6

A pulley of diameter 4 in. can rotate about a fixed shaft of diameter 2 in. The coefficients of static and kinetic friction between the pulley and shaft are both assumed equal to 0.20. Determine (*a*) the smallest vertical force **P** required to raise a 500-lb load, (*b*) the smallest vertical force **P** required to hold the load, (*c*) the smallest horizontal force **P** required to raise the same load.

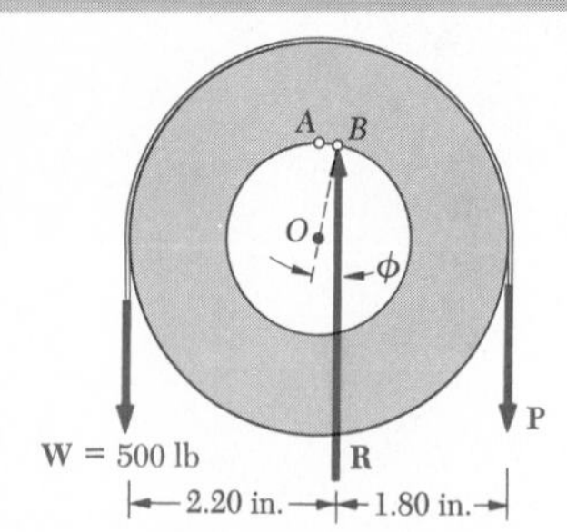

***a.* Vertical Force P Required to Raise the Load.** When the forces in both parts of the rope are equal, contact between the pulley and shaft takes place at *A*. When **P** is increased, the pulley rolls around the shaft slightly and contact takes place at *B*. The free-body diagram of the pulley when motion is impending is drawn. The perpendicular distance from the center *O* of the pulley to the line of action of **R** is

$$r_f = r \sin \phi \approx r\mu \qquad r_f \approx (1 \text{ in.})0.20 = 0.20 \text{ in.}$$

Summing moments about *B*, we write

$$+\circlearrowleft \Sigma M_B = 0: \quad (2.20 \text{ in.})(500 \text{ lb}) - (1.80 \text{ in.})P = 0$$
$$P = 611 \text{ lb} \qquad \mathbf{P} = 611 \text{ lb}\downarrow \blacktriangleleft$$

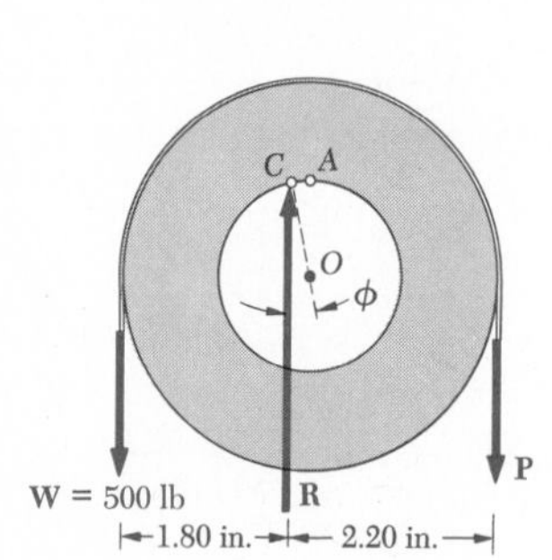

***b.* Vertical Force P to Hold the Load.** As the force **P** is decreased, the pulley rolls around the shaft and contact takes place at *C*. Considering the pulley as a free body and summing moments about *C*, we write

$$+\circlearrowleft \Sigma M_C = 0: \quad (1.80 \text{ in.})(500 \text{ lb}) - (2.20 \text{ in.})P = 0$$
$$P = 409 \text{ lb} \qquad \mathbf{P} = 409 \text{ lb}\downarrow \blacktriangleleft$$

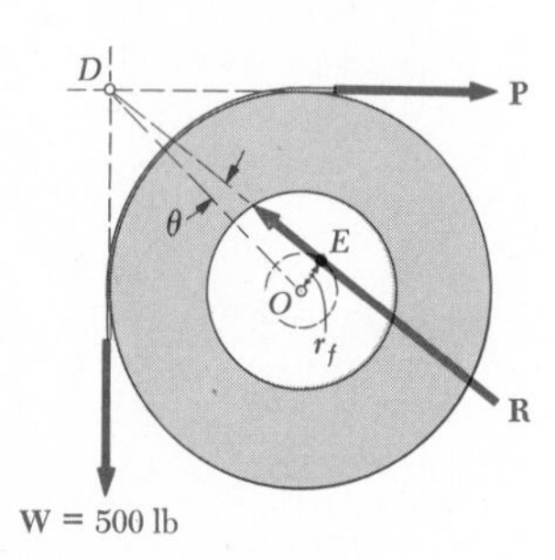

***c.* Horizontal Force P to Raise the Load.** Since the three forces **W**, **P**, and **R** are not parallel, they must be concurrent. The direction of **R** is thus determined from the fact that its line of action must pass through the point of intersection *D* of **W** and **P**, and must be tangent to the circle of friction. Recalling that the radius of the circle of friction is $r_f = 0.20$ in., we write

$$\sin \theta = \frac{OE}{OD} = \frac{0.20 \text{ in.}}{(2 \text{ in.})\sqrt{2}} = 0.0707 \qquad \theta = 4.1°$$

From the force triangle, we obtain

$$P = W \cot (45° - \theta) = (500 \text{ lb}) \cot 40.9°$$
$$= 577 \text{ lb} \qquad \mathbf{P} = 577 \text{ lb}\rightarrow \blacktriangleleft$$

P
45° − θ
R
W = 500 lb

Problems

8.85 A windlass of 10-in. diameter is used to raise or lower a 160-lb load. It is supported by two poorly lubricated bearings ($\mu_s = 0.50$) of 3-in. diameter. For each of the two positions shown, determine the magnitude of the force **P** required to start raising the load.

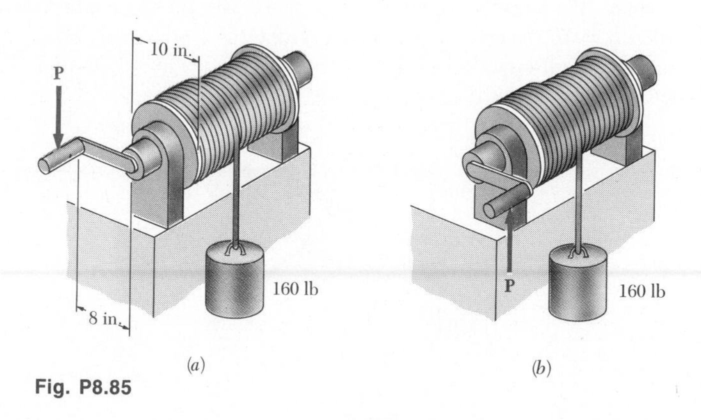

Fig. P8.85

8.86 For the windlass of Prob. 8.85, and for each of the two positions shown, determine the magnitude of the smallest force **P** which will hold the load in place.

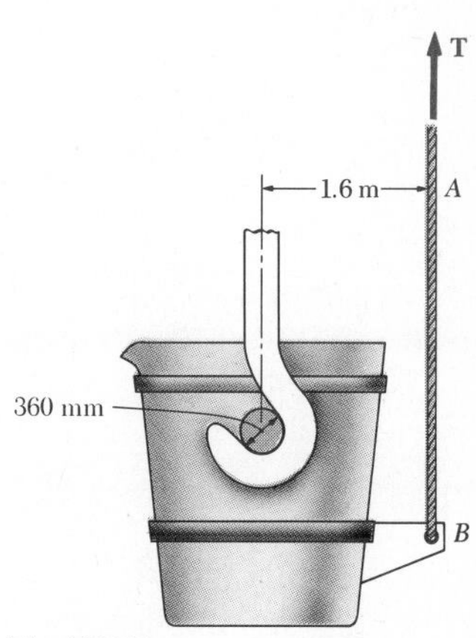

Fig. P8.87

8.87 A hot-metal ladle and its contents have a mass of 50 Mg. Knowing that the coefficient of static friction between the hooks and the pinion is 0.30, determine the tension in cable AB required to start tipping the ladle.

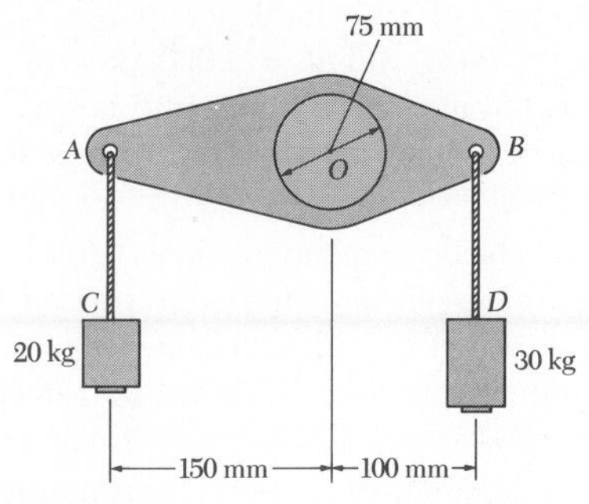

Fig. P8.88

8.88 A lever of negligible weight is loosely fitted onto a 75-mm-diameter fixed shaft. It is observed that the lever will just start rotating if a 3-kg mass is added at C. Determine the coefficient of static friction between the shaft and the lever.

8.89 An 800-N load is to be raised by the block and tackle shown. Each of the 50-mm-diameter pulleys rotates on a 10-mm-diameter axle. Knowing that $\mu = 0.25$, determine the tension in each portion of the rope as the load is slowly raised.

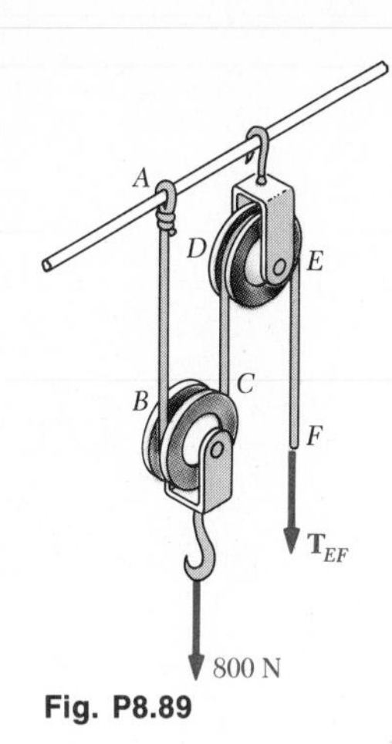

Fig. P8.89

8.90 In Prob. 8.89, determine the tension in each portion of the rope as the 800-N load is slowly lowered.

8.91 A bushing of 3-in. outside diameter fits loosely on a horizontal 2-in.-diameter shaft. If a force **P** forming an angle of 30° with the horizontal and having a magnitude $P = 160$ lb is required to start raising the 100-lb load, determine the coefficient of static friction between the shaft and the bushing. Assume that the rope does not slip on the bushing.

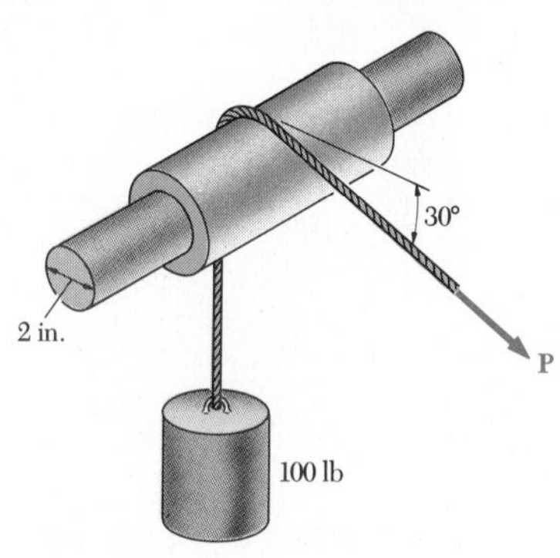

Fig. P8.91

8.92 Knowing that the coefficient of static friction between the shaft and the bushing of Prob. 8.91 is 0.35, determine the magnitude of the smallest force **P** required (*a*) to start raising the 100-lb load, (*b*) to hold the 100-lb load in place.

8.93 A loaded railroad car has a mass of 25 Mg and is supported by eight wheels of 800-mm diameter with 125-mm-diameter axles. Knowing that the coefficients of friction are $\mu_s = 0.025$ and $\mu_k = 0.020$, determine the horizontal force required (*a*) to get the car in motion, (*b*) to keep it moving with a constant velocity. Neglect the rolling resistance between the wheels and the track.

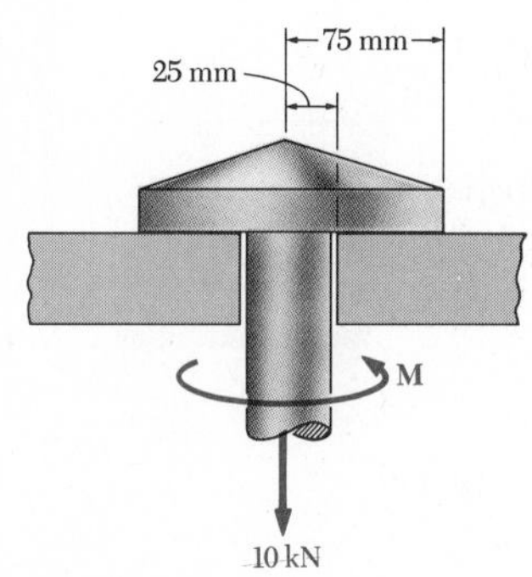

Fig. P8.95

8.94 A scooter is to be designed to roll down a 3 percent slope at a constant speed. Assuming that the coefficient of kinetic friction between the 25-mm-diameter axles and the bearings is 0.12, determine the required diameter of the wheels. Neglect the rolling resistance between the wheels and the ground.

8.95 A couple of magnitude $M = 150$ N · m is required to start the vertical shaft rotating. Determine the coefficient of static friction.

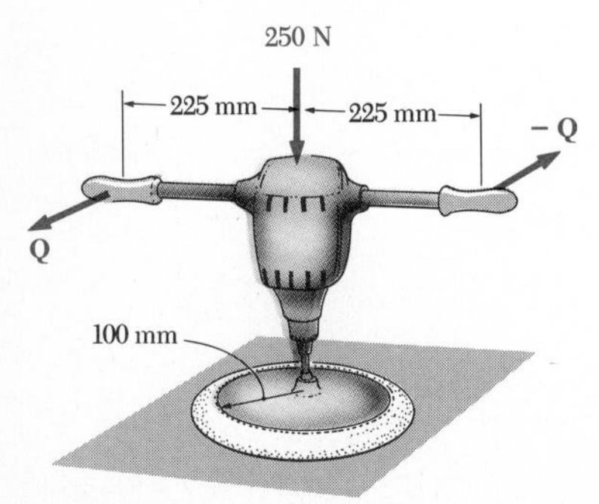

Fig. P8.96

8.96 A vertical force of 250 N is applied to a 6-kg electric polisher as it is operated on a horizontal surface. Knowing that the coefficient of kinetic friction is 0.30, determine the magnitude Q of the forces required to prevent rotation of the handles. Assume the force between the disk and the surface to be uniformly distributed.

8.97 Four springs, each of constant 40 lb/in., are used to press plate AB against the head on the vertical rod. In the position shown each spring has been compressed 2 in. The plate AB may move vertically, but it is constrained against rotation by bolts which pass through holes cut in the plate. Knowing that $\mathbf{P} = 0$ and that the coefficient of friction is 0.40, determine the magnitude of the couple **M** required to rotate the rod.

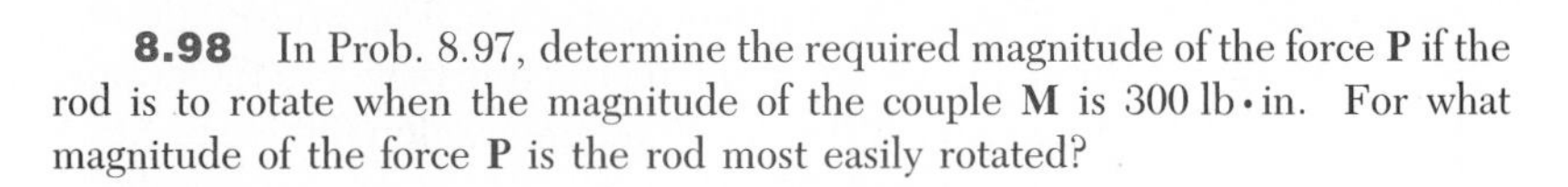

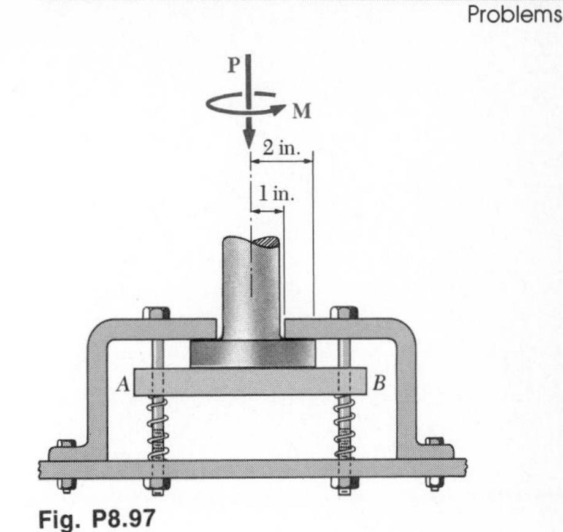

Fig. P8.97

8.98 In Prob. 8.97, determine the required magnitude of the force **P** if the rod is to rotate when the magnitude of the couple **M** is 300 lb·in. For what magnitude of the force **P** is the rod most easily rotated?

***8.99** As the surfaces of shaft and bearing wear out, the frictional resistance of a thrust bearing decreases. It is generally assumed that the wear is directly proportional to the distance traveled by any given point of the shaft, and thus to the distance r from the point to the axis of the shaft. Assuming, then, that the normal force per unit area is inversely proportional to r, show that the magnitude M of the couple required to overcome the frictional resistance of a worn-out end bearing (with contact over the full circular area) is equal to 75 percent of the value given by formula (8.9) for a new bearing.

***8.100** Assuming that bearings wear out as indicated in Prob. 8.99, show that the magnitude M of the couple required to overcome the frictional resistance of a worn-out collar bearing is

$$M = \tfrac{1}{2}\mu_k P(R_1 + R_2)$$

where P = magnitude of the total axial force
R_1, R_2 = inner and outer radii of collar

***8.101** Assuming that the pressure between the surfaces of contact is uniform, show that the magnitude M of the couple required to overcome frictional resistance for the conical pivot shown is

$$M = \frac{2}{3}\frac{\mu_k P}{\sin\theta}\frac{R_2^3 - R_1^3}{R_2^2 - R_1^2}$$

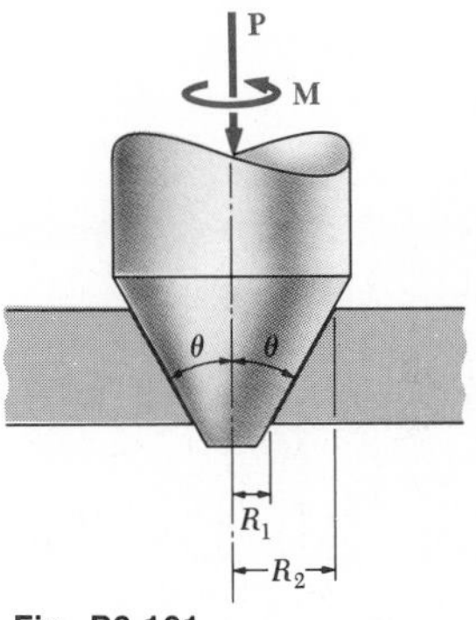

Fig. P8.101

8.102 Solve Prob. 8.96, assuming that the normal force per unit area between the disk and the surface varies linearly from a maximum at the center to zero at the circumference of the disk.

8.103 A circular disk of 8-in. diameter rolls at a constant velocity down a 3 percent incline. Determine the coefficient of rolling resistance.

8.104 Determine the horizontal force required to move a 3000-lb automobile along a horizontal road at a constant velocity. Neglect all forms of friction except rolling resistance, and assume the coefficient of rolling resistance to be 0.06 in. The diameter of each tire is 23 in.

8.105 Solve Prob. 8.93, including the effect of a coefficient of rolling resistance of 0.5 mm.

8.106 Solve Prob. 8.94, including the effect of a coefficient of rolling resistance of 1.75 mm.

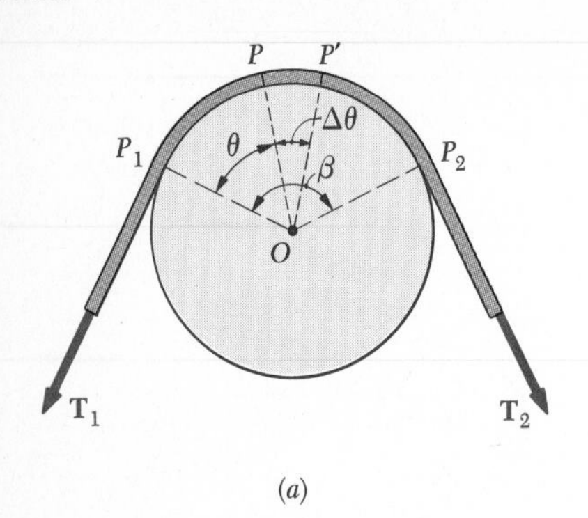

(a)

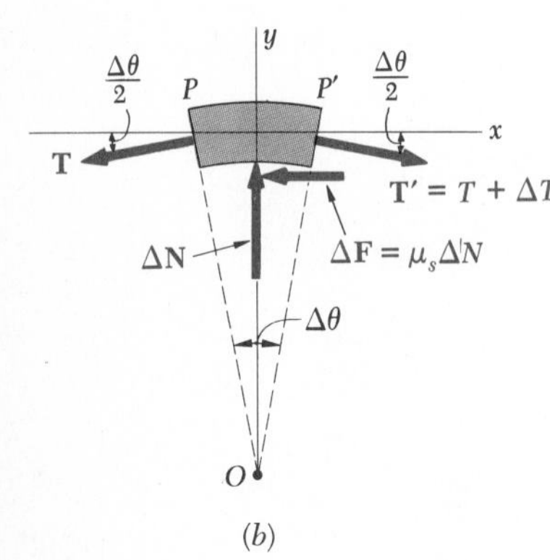

(b)

Fig. 8.14

8.10. Belt Friction.

Consider a flat belt passing over a fixed cylindrical drum (Fig. 8.14*a*). We propose to determine the relation existing between the values T_1 and T_2 of the tension in the two parts of the belt when the belt is just about to slide toward the right.

Let us detach from the belt a small element PP' subtending an angle $\Delta\theta$. Denoting by T the tension at P and by $T + \Delta T$ the tension at P', we draw the free-body diagram of the element of the belt (Fig. 8.14*b*). Besides the two forces of tension, the forces acting on the free body are the normal component $\Delta\mathbf{N}$ of the reaction of the drum and the friction force $\Delta\mathbf{F}$. Since motion is assumed to be impending, we have $\Delta F = \mu_s \Delta N$. It should be noted that if $\Delta\theta$ is made to approach zero, the magnitudes ΔN, ΔF, and the *difference* ΔT between the tension at P and the tension at P' will also approach zero; the value T of the tension at P, however, will remain unchanged. This observation helps in understanding our choice of notations.

Choosing the coordinate axes shown in Fig. 8.14*b*, we write the equations of equilibrium for the element PP':

$$\Sigma F_x = 0: \qquad (T + \Delta T)\cos\frac{\Delta\theta}{2} - T\cos\frac{\Delta\theta}{2} - \mu_s \Delta N = 0 \tag{8.11}$$

$$\Sigma F_y = 0: \qquad \Delta N - (T + \Delta T)\sin\frac{\Delta\theta}{2} - T\sin\frac{\Delta\theta}{2} = 0 \tag{8.12}$$

Solving Eq. (8.12) for ΔN and substituting into (8.11), we obtain after reductions

$$\Delta T\cos\frac{\Delta\theta}{2} - \mu_s(2T + \Delta T)\sin\frac{\Delta\theta}{2} = 0$$

We shall now divide both terms by $\Delta\theta$; as far as the first term is concerned, this will be simply done by dividing ΔT by $\Delta\theta$. The division of the second term is carried out by dividing the terms in the parentheses by 2 and the sine by $\Delta\theta/2$. We write

$$\frac{\Delta T}{\Delta\theta}\cos\frac{\Delta\theta}{2} - \mu_s\left(T + \frac{\Delta T}{2}\right)\frac{\sin(\Delta\theta/2)}{\Delta\theta/2} = 0$$

If we now let $\Delta\theta$ approach 0, the cosine approaches 1 and $\Delta T/2$ approaches zero as noted above. On the other hand, the quotient of $\sin(\Delta\theta/2)$ over $\Delta\theta/2$ approaches 1, according to a lemma derived in all calculus textbooks. Since the limit of $\Delta T/\Delta\theta$ is by definition equal to the derivative $dT/d\theta$, we write

$$\frac{dT}{d\theta} - \mu_s T = 0 \qquad \frac{dT}{T} = \mu_s\, d\theta$$

We shall now integrate both members of the last equation obtained from P_1 to P_2 (Fig. 8.14*a*). At P_1, we have $\theta = 0$ and $T = T_1$; at P_2, we have $\theta = \beta$ and $T = T_2$. Integrating between these limits, we write

$$\int_{T_1}^{T_2}\frac{dT}{T} = \int_0^{\beta}\mu_s\, d\theta$$

$$\ln T_2 - \ln T_1 = \mu_s\beta$$

or, noting that the left-hand member is equal to the natural logarithm of the quotient of T_2 and T_1,

$$\ln \frac{T_2}{T_1} = \mu_s \beta \qquad (8.13)$$

This relation may also be written in the form

$$\frac{T_2}{T_1} = e^{\mu_s \beta} \qquad (8.14)$$

The formulas we have derived apply equally well to problems involving flat belts passing over fixed cylindrical drums and to problems involving ropes wrapped around a post or capstan. They may also be used to solve problems involving band brakes. In such problems, it is the drum which is about to rotate, while the band remains fixed. The formulas may also be applied to problems involving belt drives. In these problems, both the pulley and the belt rotate; our concern is then to find whether the belt will slip, i.e., whether it will move *with respect* to the pulley.

Formulas (8.13) and (8.14) should be used only if the belt, rope, or brake is *about to slip*. Formula (8.14) will be used if T_1 or T_2 is desired; formula (8.13) will be preferred if either μ_s or the angle of contact β is desired. We should note that T_2 is always larger than T_1; T_2 therefore represents the tension in that part of the belt or rope which *pulls*, while T_1 is the tension in the part which *resists*. We should also observe that the angle of contact β must be expressed in *radians*. The angle β may be larger than 2π; for example, if a rope is wrapped n times around a post, β is equal to $2\pi n$.

If the belt, rope, or brake is actually slipping, formulas similar to (8.13) and (8.14), but involving the coefficient of kinetic friction μ_k, should be used. If the belt, rope, or brake does not slip and is not about to slip, none of these formulas may be used.

The belts used in belt drives are often V-shaped. Such a belt, called a *V belt*, is shown in Fig. 8.15*a*. It is seen that contact between belt and pulley takes place along the sides of the groove. The relation existing between the values T_1 and T_2 of the tension in the two parts of the belt when the belt is just about to slip may again be obtained by drawing the free-body diagram of an element of belt (Fig. 8.15*b* and *c*). Equations similar to (8.11) and (8.12) are derived, but the magnitude of the total friction force acting on the element is now $2\,\Delta F$, and the sum of the y components of the normal forces is $2\,\Delta N \sin(\alpha/2)$. Proceeding as above, we obtain

$$\frac{T_2}{T_1} = e^{\mu_s \beta / \sin(\alpha/2)} \qquad (8.15)$$

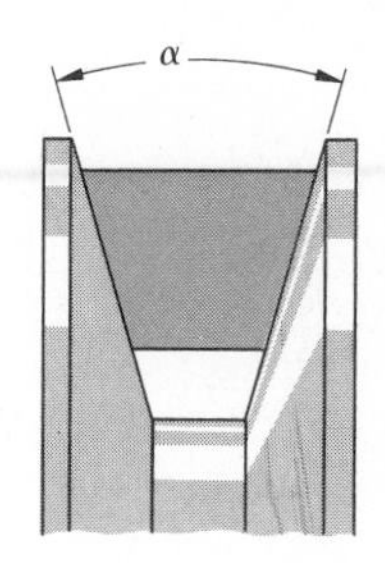

(*a*)

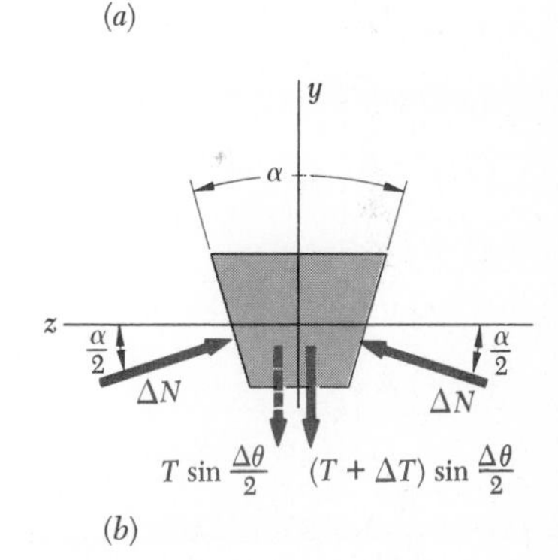

(*b*)

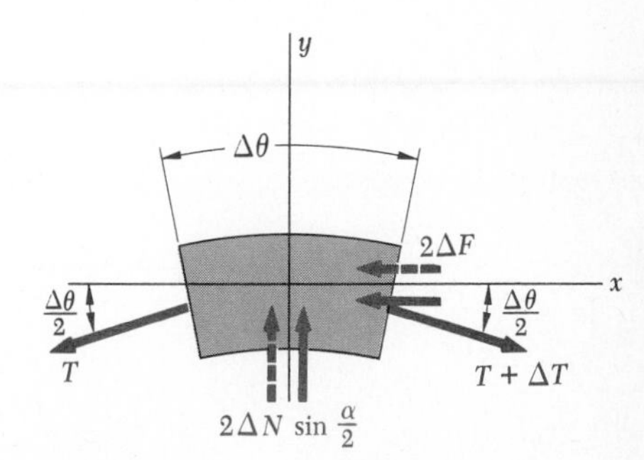

(*c*)

Fig. 8.15

SAMPLE PROBLEM 8.7

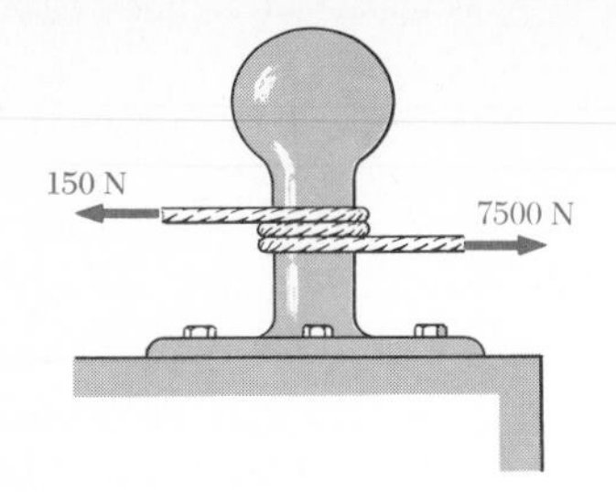

A hawser thrown from a ship to a pier is wrapped two full turns around a capstan. The tension in the hawser is 7500 N; by exerting a force of 150 N on its free end, a dock worker can just keep the hawser from slipping. (*a*) Determine the coefficient of friction between the hawser and the capstan. (*b*) Determine the tension in the hawser that could be resisted by the 150-N force if the hawser were wrapped three full turns around the capstan.

***a.* Coefficient of Friction.** Since slipping of the hawser is impending, we use Eq. (8.13):

$$\ln \frac{T_2}{T_1} = \mu_s \beta$$

Since the hawser is wrapped two full turns around the capstan, we have

$$\beta = 2(2\pi \text{ rad}) = 12.6 \text{ rad}$$
$$T_1 = 150 \text{ N} \qquad T_2 = 7500 \text{ N}$$

Therefore,

$$\mu_s \beta = \ln \frac{T_2}{T_1}$$

$$\mu_s (12.6 \text{ rad}) = \ln \frac{7500 \text{ N}}{150 \text{ N}} = \ln 50 = 3.91$$

$$\mu_s = 0.31 \blacktriangleleft$$

***b.* Hawser Wrapped Three Turns around Capstan.** Using the value of μ_s obtained in part *a*, we have now

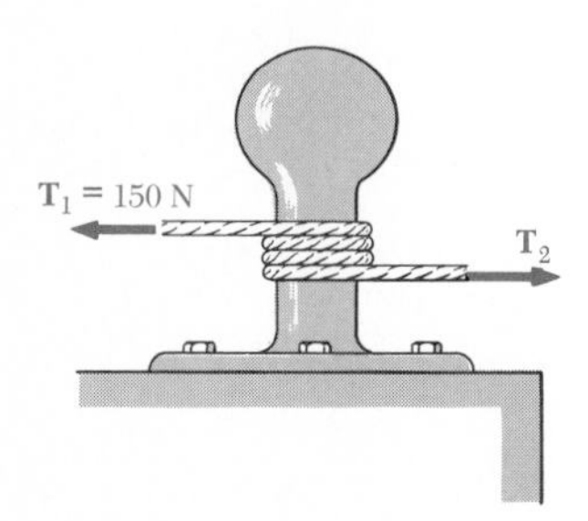

$$\beta = 3(2\pi \text{ rad}) = 18.9 \text{ rad}$$
$$T_1 = 150 \text{ N} \qquad \mu_s = 0.31$$

Substituting these values into Eq. (8.14), we obtain

$$\frac{T_2}{T_1} = e^{\mu_s \beta}$$

$$\frac{T_2}{150 \text{ N}} = e^{(0.31)(18.9)} = e^{5.86} = 350$$

$$T_2 = 52\,500 \text{ N} \qquad T_2 = 52.5 \text{ kN} \blacktriangleleft$$

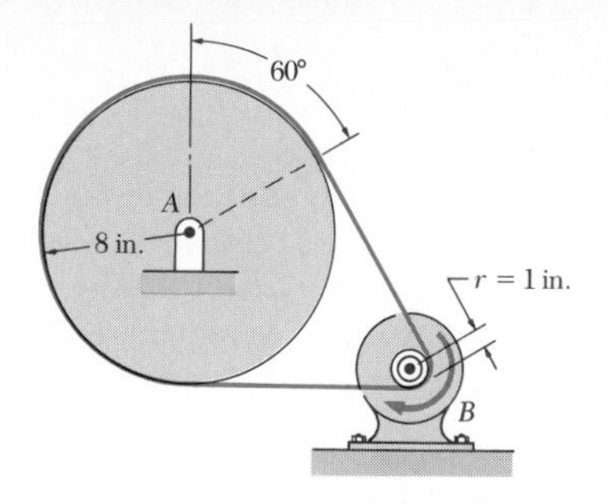

SAMPLE PROBLEM 8.8

A flat belt connects pulley A, which drives a machine tool, to pulley B, which is attached to the shaft of an electric motor. The coefficients of friction are $\mu_s = 0.25$ and $\mu_k = 0.20$ between both pulleys and the belt. Knowing that the maximum allowable tension in the belt is 600 lb, determine the largest torque which can be exerted by the belt on pulley A.

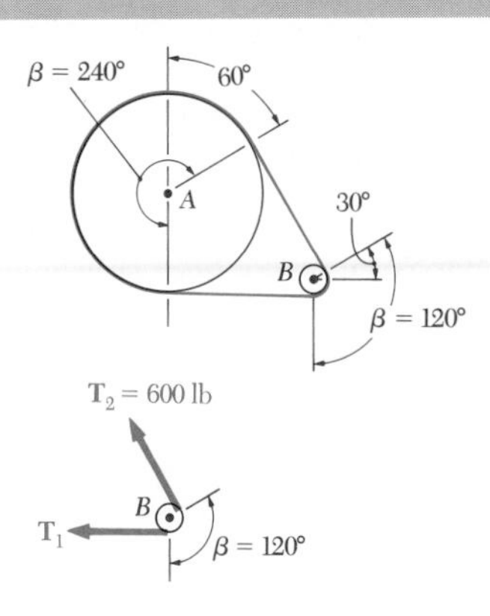

Solution. Since the resistance to slippage depends upon the angle of contact β between pulley and belt, as well as upon the coefficient of static friction μ_s, and since μ_s is the same for both pulleys, slippage will occur first on pulley B, for which β is smaller.

Pulley B. Using Eq. (8.14) with $T_2 = 600$ lb, $\mu_s = 0.25$, and $\beta = 120° = 2\pi/3$ rad, we write

$$\frac{T_2}{T_1} = e^{\mu_s \beta} \qquad \frac{600 \text{ lb}}{T_1} = e^{0.25(2\pi/3)} = 1.688$$

$$T_1 = \frac{600 \text{ lb}}{1.688} = 355.4 \text{ lb}$$

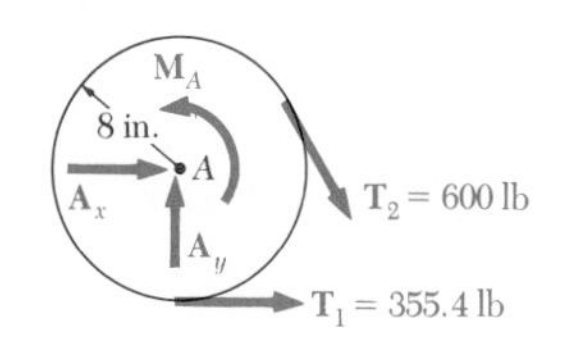

Pulley A. We draw the free-body diagram of pulley A. The couple $\mathbf{M}_A$ is applied to the pulley by the machine tool to which it is attached and is equal and opposite to the torque exerted by the belt. We write

$$+\circlearrowleft \Sigma M_A = 0: \qquad M_A - (600 \text{ lb})(8 \text{ in.}) + (355.4 \text{ lb})(8 \text{ in.}) = 0$$

$$M_A = 1957 \text{ lb} \cdot \text{in.} \qquad M_A = 163.1 \text{ lb} \cdot \text{ft} \blacktriangleleft$$

Note. We may check that the belt does not slip on pulley A by computing the value of μ_s required to prevent slipping at A and verifying that it is smaller than the actual value of μ_s. From Eq. (8.13) we have

$$\mu_s \beta = \ln \frac{T_2}{T_1} = \ln \frac{600 \text{ lb}}{355.4 \text{ lb}} = 0.524$$

and, since $\beta = 240° = 4\pi/3$ rad,

$$\frac{4\pi}{3}\mu_s = 0.524 \qquad \mu_s = 0.125 < 0.25$$

Problems

8.107 A hawser is wrapped two full turns around a capstan head. By exerting a force of 750 N on the free end of the hawser, a seaman can resist a force of 45 kN on the other end of the hawser. Determine (*a*) the coefficient of static friction, (*b*) the number of times the hawser should be wrapped around the capstan if a 300-kN force is to be resisted by the same 750-N force.

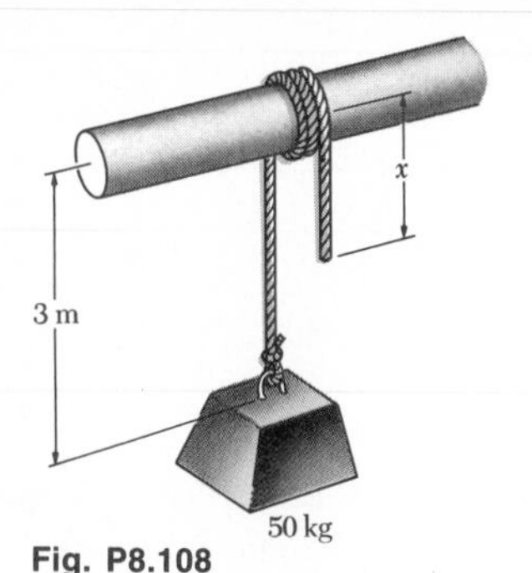

Fig. P8.108

8.108 A rope having a mass per unit length of 0.6 kg/m is wound $2\frac{1}{2}$ times around a horizontal rod. What length x of rope should be left hanging if a 50-kg load is to be supported? The coefficient of static friction between the rope and the rod is 0.30.

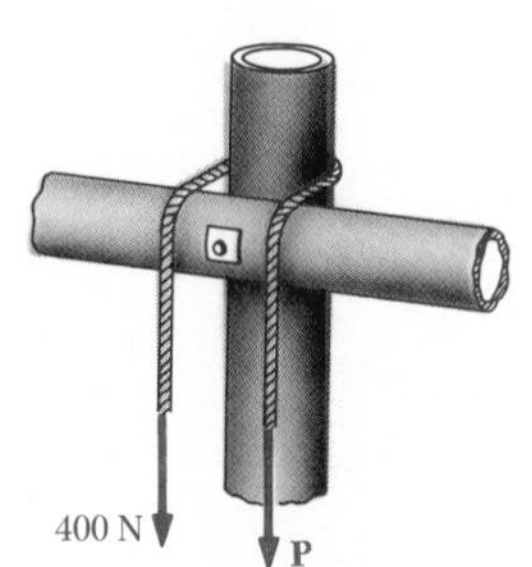

Fig. P8.109 and P8.110

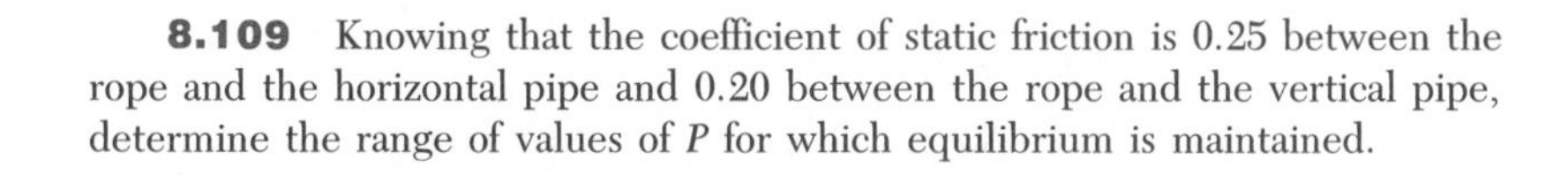
8.109 Knowing that the coefficient of static friction is 0.25 between the rope and the horizontal pipe and 0.20 between the rope and the vertical pipe, determine the range of values of P for which equilibrium is maintained.

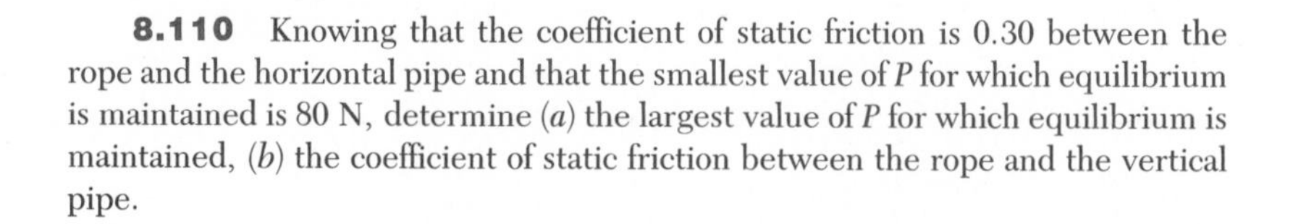
8.110 Knowing that the coefficient of static friction is 0.30 between the rope and the horizontal pipe and that the smallest value of P for which equilibrium is maintained is 80 N, determine (*a*) the largest value of P for which equilibrium is maintained, (*b*) the coefficient of static friction between the rope and the vertical pipe.

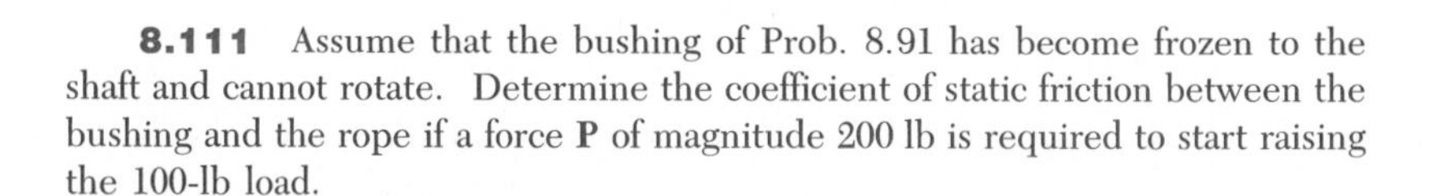
8.111 Assume that the bushing of Prob. 8.91 has become frozen to the shaft and cannot rotate. Determine the coefficient of static friction between the bushing and the rope if a force **P** of magnitude 200 lb is required to start raising the 100-lb load.

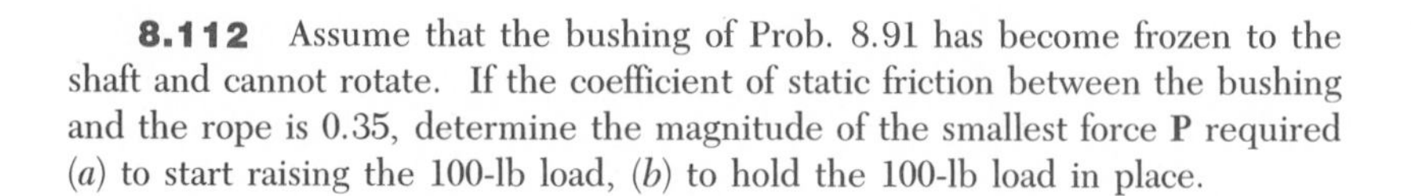
8.112 Assume that the bushing of Prob. 8.91 has become frozen to the shaft and cannot rotate. If the coefficient of static friction between the bushing and the rope is 0.35, determine the magnitude of the smallest force **P** required (*a*) to start raising the 100-lb load, (*b*) to hold the 100-lb load in place.

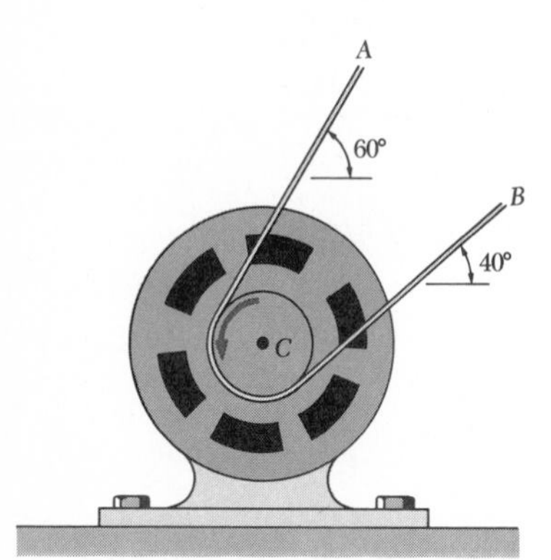

Fig. P8.113

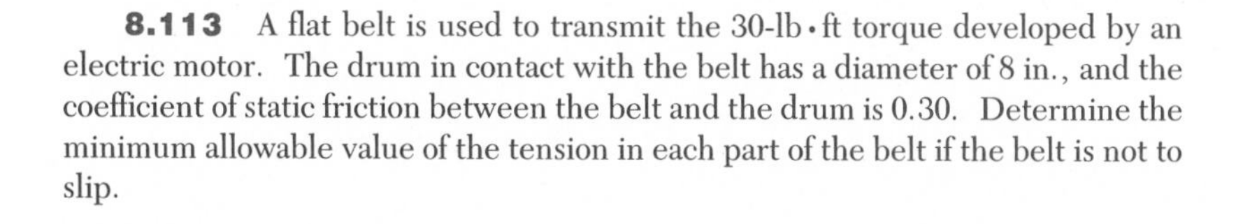
8.113 A flat belt is used to transmit the 30-lb · ft torque developed by an electric motor. The drum in contact with the belt has a diameter of 8 in., and the coefficient of static friction between the belt and the drum is 0.30. Determine the minimum allowable value of the tension in each part of the belt if the belt is not to slip.

8.114 A flat belt is used to transmit a torque from pulley A to pulley B. The radius of each pulley is 50 mm and the coefficient of static friction is 0.30. Determine the largest torque which can be transmitted if the allowable belt tension is 3 kN.

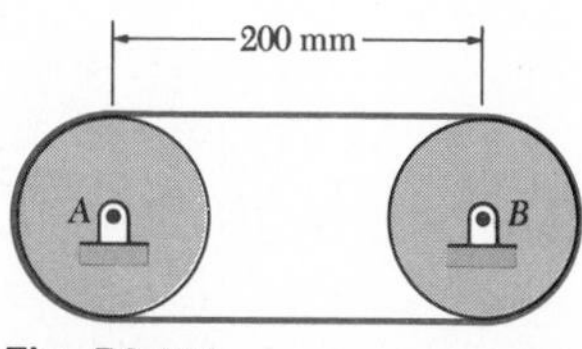

Fig. P8.114

8.115 Solve Prob. 8.114, assuming that the belt is looped around the pulleys in a figure 8.

8.116 The setup shown is used to measure the output of a small turbine. When the flywheel is at rest, the reading of each spring scale is 70 N. If a 12.60-N · m couple must be applied to the flywheel to keep it rotating clockwise at a constant speed, determine (*a*) the reading of each scale at that time, (*b*) the coefficient of kinetic friction. Assume that the length of the belt does not change.

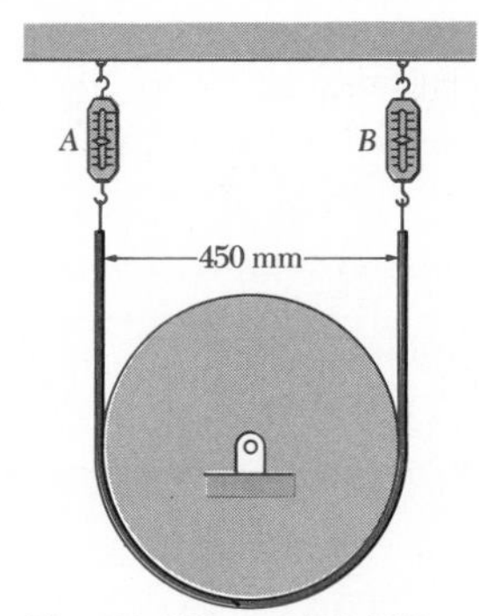

Fig. P8.116 and P8.117

8.117 The setup shown is used to measure the output of a small turbine. The coefficient of kinetic friction is 0.20 and the reading of each spring scale is 80 N when the flywheel is at rest. Determine (*a*) the reading of each scale when the flywheel is rotating clockwise at a constant speed, (*b*) the couple which must be applied to the flywheel. Assume that the length of the belt does not change.

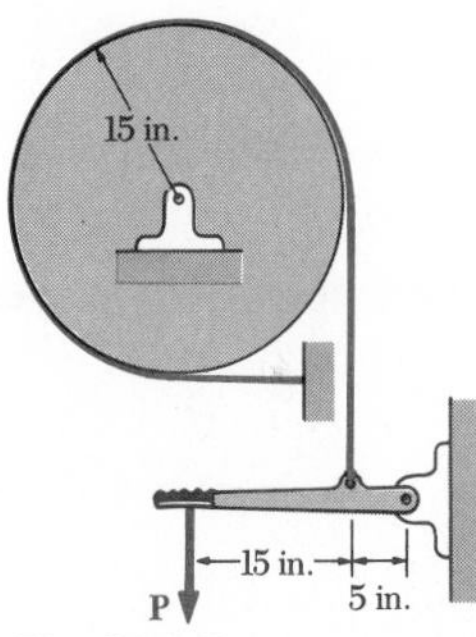

Fig. P8.118

8.118 A band brake is used to control the speed of a flywheel as shown. The coefficients of friction are $\mu_s = 0.30$ and $\mu_k = 0.25$. What couple should be applied to the flywheel to keep it rotating counterclockwise at a constant speed when $P = 10$ lb?

8.119 Solve Prob. 8.118, assuming that the flywheel rotates clockwise.

8.120 A brake drum of radius $r = 150$ mm is rotating counterclockwise when a force **P** of magnitude 60 N is applied at A. Knowing that the coefficient of kinetic friction is 0.40, determine the moment about O of the friction forces applied to the drum when $a = 250$ mm and $b = 300$ mm.

8.121 Knowing that $r = 150$ mm and $a = 250$ mm, determine the maximum value of the coefficient of kinetic friction for which the brake is not self-locking when the drum rotates counterclockwise.

8.122 Knowing that the coefficient of kinetic friction is 0.40, determine the minimum value of the ratio a/r for which the brake is not self-locking. Assume that $a > r$ and that the drum rotates counterclockwise.

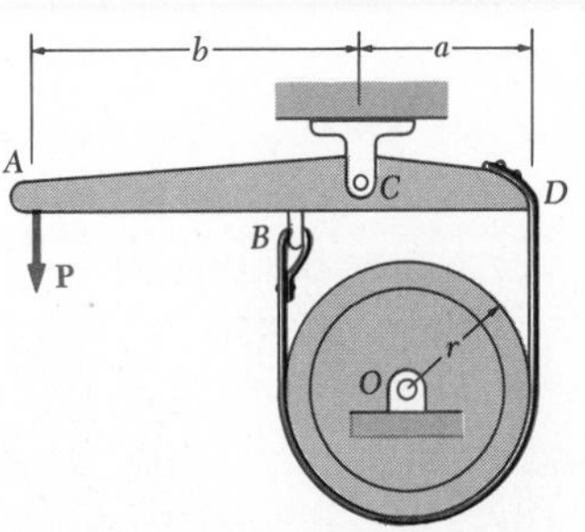

Fig. P8.120, P8.121, and P8.122

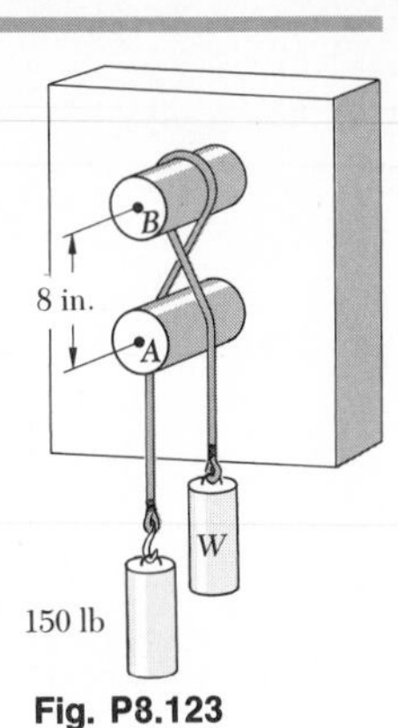

Fig. P8.123

8.123 A cord is placed over two cylinders, each of 4-in. diameter. Knowing that the coefficients of friction are $\mu_s = 0.30$ and $\mu_k = 0.25$, determine the largest weight W which can be raised when cylinder B is rotated slowly and cylinder A is kept fixed.

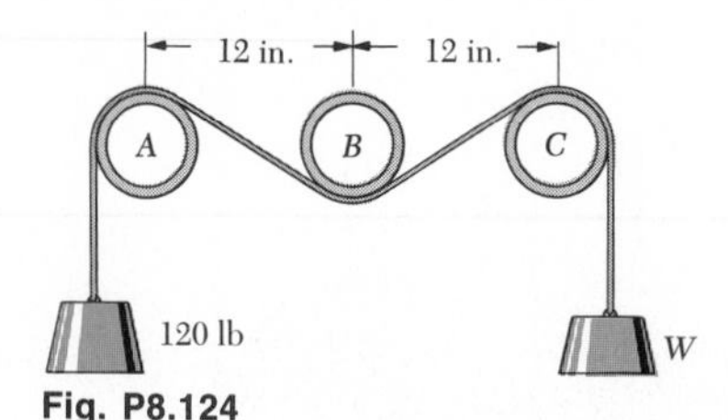

Fig. P8.124

8.124 A cable is placed around three pipes, each of 6-in. outside diameter, located in the same horizontal plane. Two of the pipes are fixed and do not rotate; the third pipe is rotated slowly. Knowing that the coefficients of friction are $\mu_s = 0.25$ and $\mu_k = 0.20$ for each pipe, determine the largest weight W which can be raised (*a*) if only pipe A is rotated, (*b*) if only pipe B is rotated, (*c*) if only pipe C is rotated.

8.125 The strap wrench shown is used to grip the pipe firmly and at the same time not mar the external surface of the pipe. Knowing that $a = 360$ mm and $r = 40$ mm, determine the minimum coefficient of static friction for which the wrench will be self-locking.

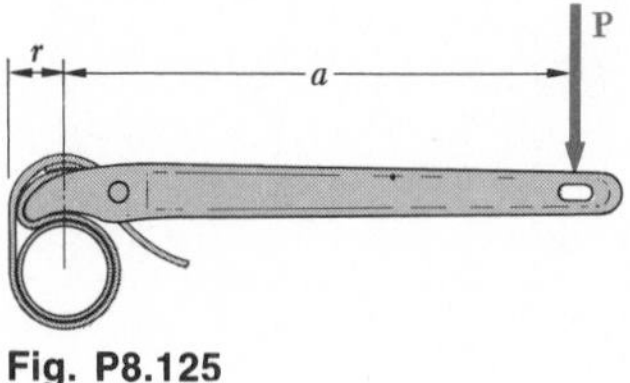

Fig. P8.125

8.126 For the strap wrench of Prob. 8.125, determine the minimum value of the ratio r/a for which the strap will be self-locking if the coefficient of static friction between the strap and the pipe is 0.25.

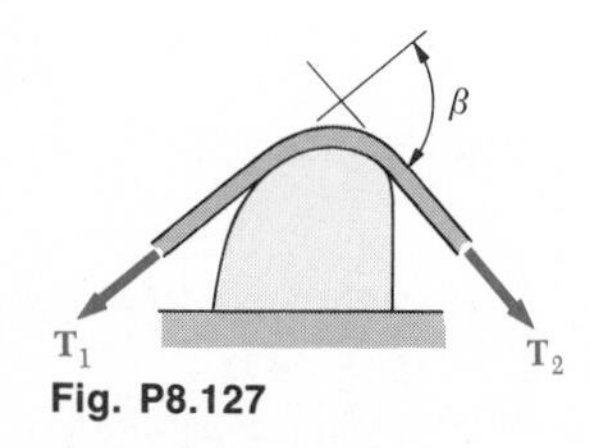

Fig. P8.127

8.127 Prove that Eqs. (8.13) and (8.14) are valid for any shape of surface provided that the coefficient of friction is the same at all points of contact.

8.128 Complete the derivation of Eq. (8.15), which relates the tension in both parts of a V belt.

8.129 Solve Prob. 8.113, assuming that the flat belt and pulley are replaced by a V belt and V pulley with $\alpha = 36°$. (The angle α is as shown in Fig. 8.15*a*.)

Review and Summary

This chapter was devoted to the study of *dry friction*, i.e., to problems involving rigid bodies which are in contact along *nonlubricated surfaces.*

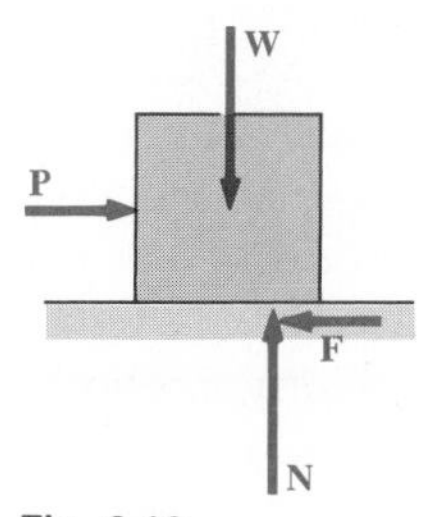

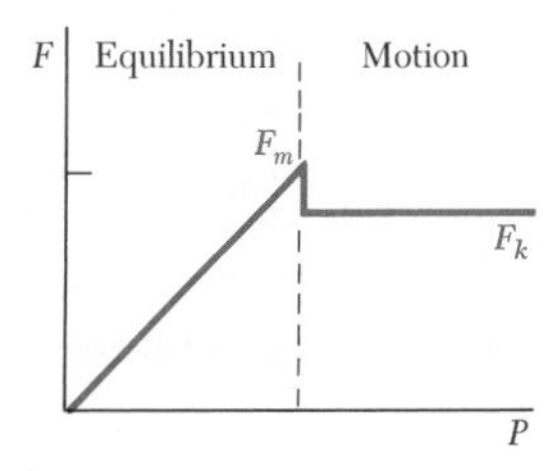

Fig. 8.16

Static and kinetic friction

Considering a block resting on a horizontal surface [Sec. 8.2] and applying to it a horizontal force **P**, we note that the block at first does not move. This shows that a *friction force* **F** must have developed to balance **P** (Fig. 8.16). As the magnitude of **P** is increased, the magnitude of **F** also increases until it reaches a maximum value F_m. If **P** is further increased, the block starts sliding and the magnitude of **F** drops from F_m to a lower value F_k. Experimental evidence shows that F_m and F_k are proportional to the normal component N of the reaction of the surface. We have

$$F_m = \mu_s N \qquad F_k = \mu_k N \qquad (8.1,\ 8.2)$$

where μ_s and μ_k are called, respectively, the *coefficient of static friction* and the *coefficient of kinetic friction.* These coefficients depend upon the nature and the condition of the surfaces in contact. Approximate values of the coefficients of static friction were given in Table 8.1.

Angles of friction

It is sometimes convenient to replace the normal force **N** and the friction force **F** by their resultant **R** (Fig. 8.17). As the friction force increases and reaches its maximum value $F_m = \mu_s N$, the angle ϕ that **R** forms with the normal to the surface increases and reaches a maximum value ϕ_s, called the *angle of static friction.* If motion actually takes place, the magnitude of **F** drops to F_k; similarly the angle ϕ drops to a lower value ϕ_k, called the *angle of kinetic friction.* As shown in Sec. 8.3, we have

$$\tan \phi_s = \mu_s \qquad \tan \phi_k = \mu_k \qquad (8.3,\ 8.4)$$

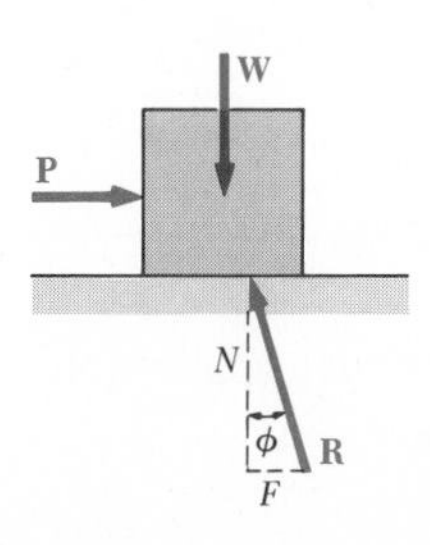

Fig. 8.17

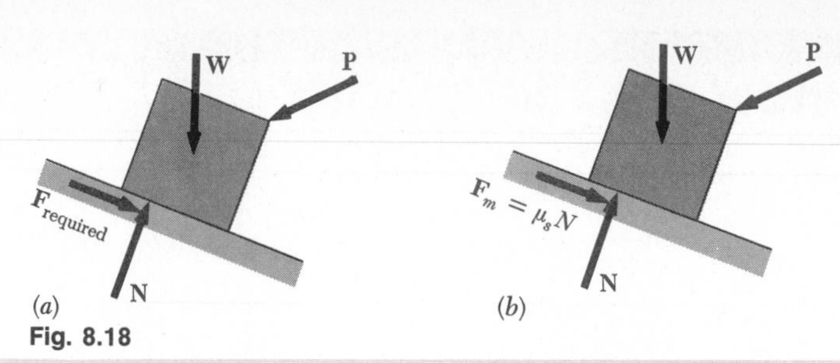

Fig. 8.18

Problems involving friction

When solving equilibrium problems involving friction, we should keep in mind that the magnitude F of the friction force is equal to $F_m = \mu_s N$ *only if the body is about to slide* [Sec. 8.4]. *If motion is not impending*, F and N should be considered as *independent unknowns* to be determined from the equilibrium equations (Fig. 8.18*a*). We should also check that the value of F required to maintain equilibrium is not larger than F_m; if it were, the body would move and the magnitude of the friction force would be $F_k = \mu_k N$ [Sample Prob. 8.1]. On the other hand, *if motion is known to be impending*, F has reached its maximum value $F_m = \mu_s N$ (Fig. 8.18*b*) and this expression may be substituted for F in the equilibrium equations [Sample Prob. 8.3]. When only three forces are involved in a free-body diagram, including the reaction **R** of the surface in contact with the body, it is usually more convenient to solve the problem by drawing a force triangle [Sample Prob. 8.2].

When a problem involves the analysis of the forces exerted on each other by *two bodies A and B*, it is important to show the friction forces with their correct sense. The correct sense for the friction force exerted by B on A, for instance, is opposite to that of the *relative motion* (or impending motion) of A with respect to B [Fig. 8.6 of Sec. 8.4].

Wedges and screws

In the second part of the chapter we considered a number of specific engineering applications where dry friction plays an important role. In the case of *wedges*, which are simple machines used to raise heavy loads [Sec. 8.5], two or more free-body diagrams were drawn and care was taken to show each friction force with its correct sense [Sample Prob. 8.4]. The analysis of *square-threaded screws*, which are frequently used in jacks, presses, and other mechanisms, was reduced to the analysis of a block sliding on an incline by unwrapping the thread of the screw and showing it as a straight line [Sec. 8.6]. This is done again in Fig. 8.19, where r denotes the *mean radius* of the thread, L the *lead* of the screw, i.e., the distance through which the screw advances in one turn, W the load, and where Qr is equal to the torque exerted on the screw. It was noted that in the case of multiple-threaded screws the lead L of the screw is *not* equal to its pitch, which is the distance measured between two consecutive threads.

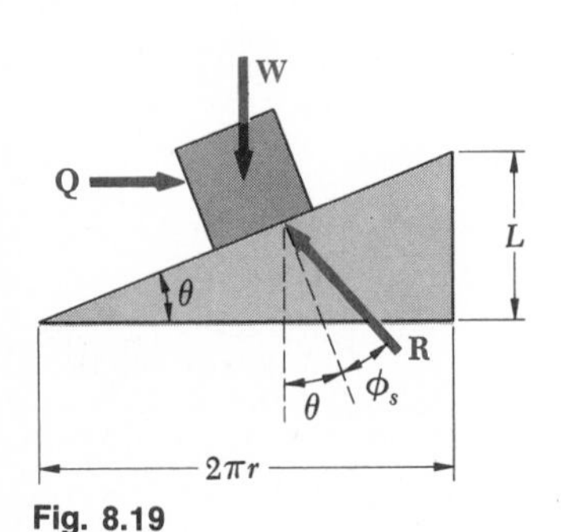

Fig. 8.19

Other applications

Other engineering applications considered in this chapter were *journal bearings* and *axle friction* [Sec. 8.7], *thrust bearings* and *disk friction* [Sec. 8.8], *wheel friction* and *rolling resistance* [Sec. 8.9], and *belt friction* [Sec. 8.10].

In solving a problem involving a *flat belt* passing over a fixed cylinder, it is important to first determine the direction in which the belt slips or is about to slip. If the drum is rotating, the motion or impending motion of the belt should be determined *relative* to the rotating drum. For instance, if the belt of Fig. 8.20 is about to slip to the right relative to the drum, the friction forces exerted by the drum on the belt will be directed to the left and the tension will be larger in the right-hand portion of the belt than in the left-hand portion. Denoting by T_2 the larger tension, by T_1 the smaller tension, by μ_s the coefficient of static friction, and by β the angle (in radians) subtended by the belt, we derived in Sec. 8.10 the formulas

$$\ln \frac{T_2}{T_1} = \mu_s \beta \qquad (8.13)$$

$$\frac{T_2}{T_1} = e^{\mu_s \beta} \qquad (8.14)$$

which were used in solving Sample Probs. 8.7 and 8.8. If the belt actually slips on the drum, the coefficient of static friction μ_s should be replaced by the coefficient of kinetic friction μ_k in both of these formulas.

Belt friction

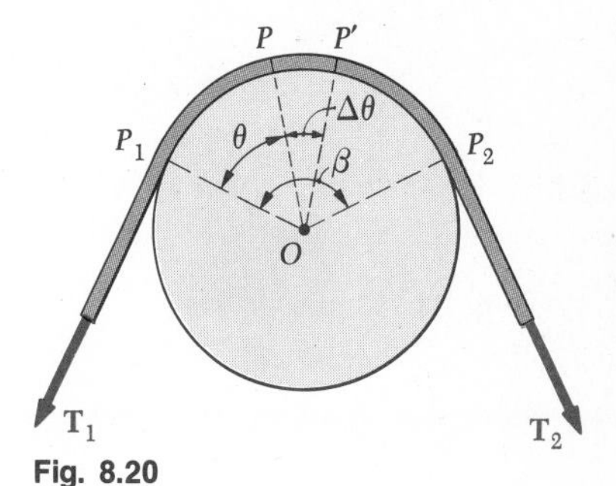

Fig. 8.20

Review Problems

8.130 Two large cylinders, each of radius $r = 500$ mm, rotate in opposite directions and form the main elements of a crusher for stone aggregate. The distance d is set equal to the maximum desired size of the crushed aggregate. If $d = 20$ mm and $\mu_s = 0.30$, determine the size s of the largest stones which will be pulled through the crusher by friction alone.

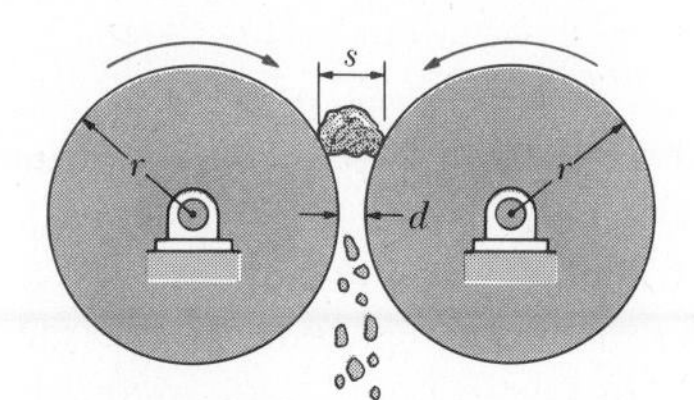

Fig. P8.130

8.131 In the vise shown, the screw is single-threaded in the upper member; it passes through the lower member and is held by a frictionless washer. The pitch of the screw is 3 mm, its mean radius is 12 mm, and the coefficient of static friction is 0.15. Determine the magnitude P of the forces exerted by the jaws when a 75-N · m couple is applied to the screw.

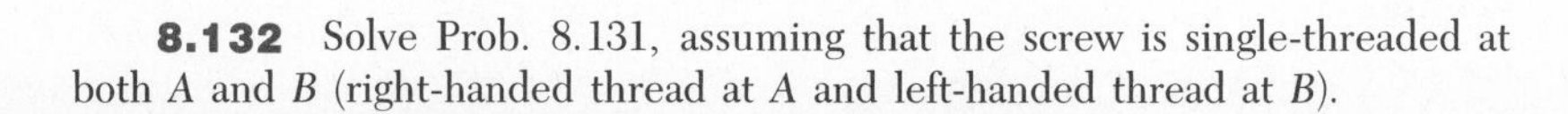
8.132 Solve Prob. 8.131, assuming that the screw is single-threaded at both A and B (right-handed thread at A and left-handed thread at B).

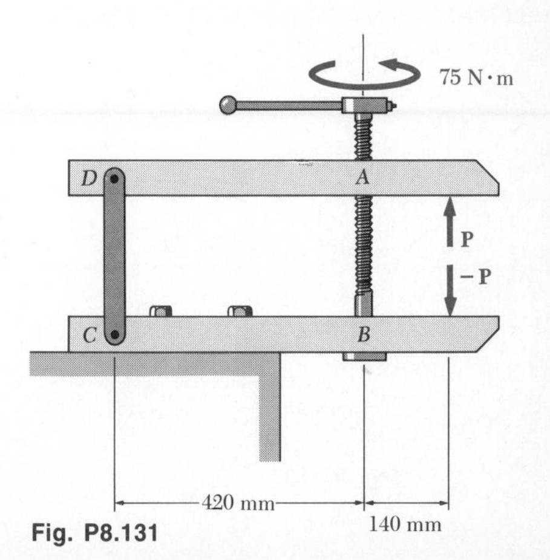

Fig. P8.131

8.133 The uniform rod AB lies in a vertical plane with its ends resting against the surfaces AC and BC. Determine the range of values of θ corresponding to equilibrium in terms of α and the angle of static friction ϕ_s between the rod and the two surfaces. Consider the case when (*a*) $\alpha > 2\phi_s$, (*b*) $\alpha \leq 2\phi_s$.

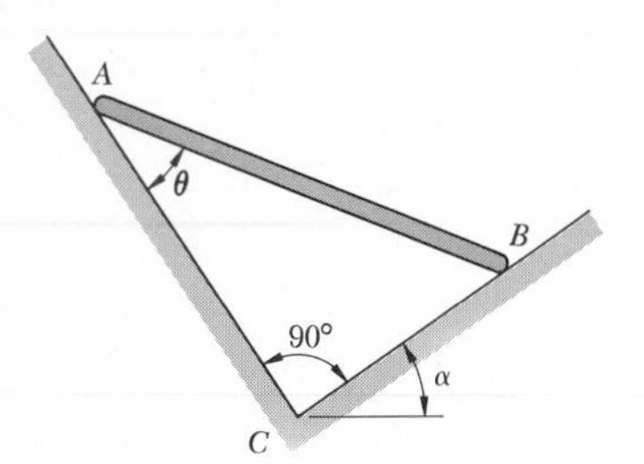

Fig. P8.133 and P8.134

8.134 The uniform rod AB lies in a vertical plane with its ends resting against the surfaces AC and BC. Knowing that the coefficient of static friction is 0.25 between the rod and each of the surfaces, determine the range of values of θ corresponding to equilibrium when (*a*) $\alpha = 45°$, (*b*) $\alpha = 25°$.

8.135 A uniform bar of length 600 mm is supported by a horizontal shaft of diameter 50 mm. The shaft fits loosely in a circular hole in the bar and the coefficient of static friction is 0.40. If $x = 100$ mm and the shaft is slowly rotated, determine the angle at which the bar will slip.

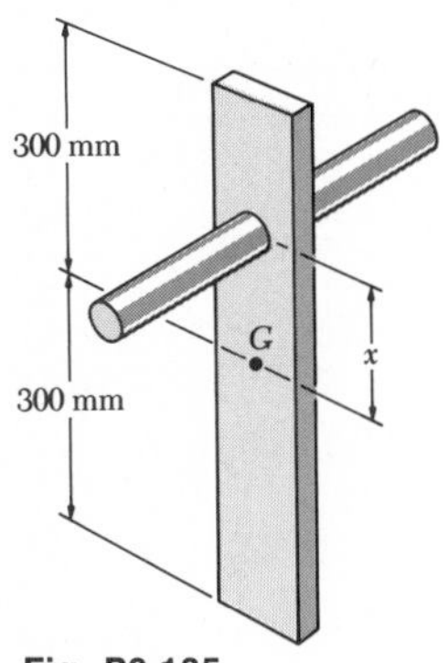

Fig. P8.135

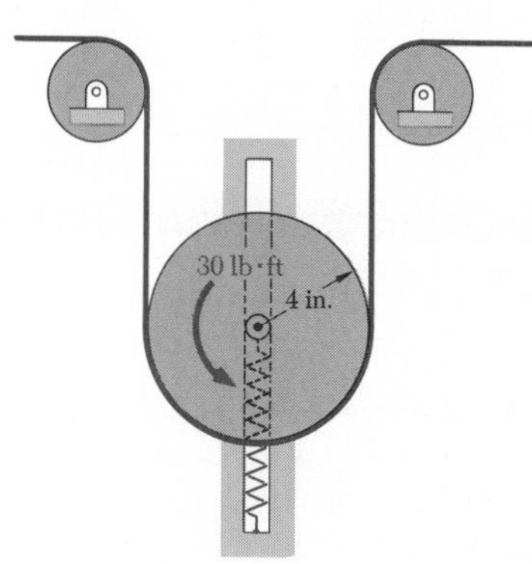

Fig. P8.136

8.136 A flat belt passes over two idler pulleys and under a rotating drum of diameter 8 in. The axle of the drum is free to move vertically in a slot, and a spring keeps the drum in contact with the belt. What is the minimum force which should be exerted by the spring if slippage is not to occur when a 30-lb·ft couple is applied to the drum? The coefficient of static friction between belt and drum is 0.30.

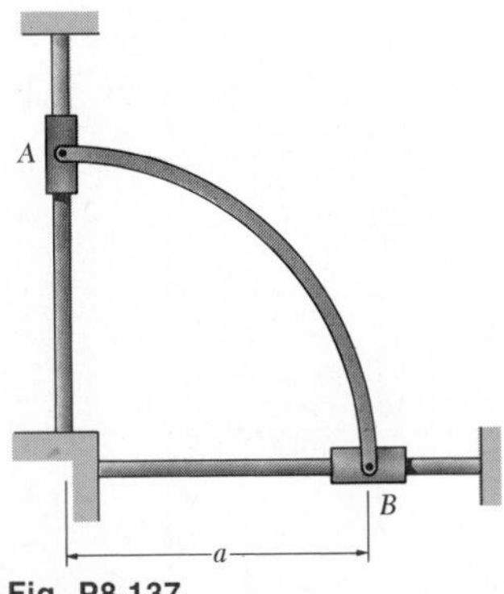

Fig. P8.137

8.137 The uniform rod AB of radius a and weight W is attached to collars at A and B which may slide on the rods shown. Denoting by μ_s the coefficient of static friction between each collar and the rod upon which it may slide, determine the smallest value of μ_s for which the rod will remain in equilibrium in the position shown.

8.138 A half section of pipe, weighing 250 lb, is to be moved to the right along the floor without tipping. Knowing that the coefficient of friction between the pipe and the floor is 0.35, determine (*a*) the largest allowable value of α, (*b*) the corresponding tension T.

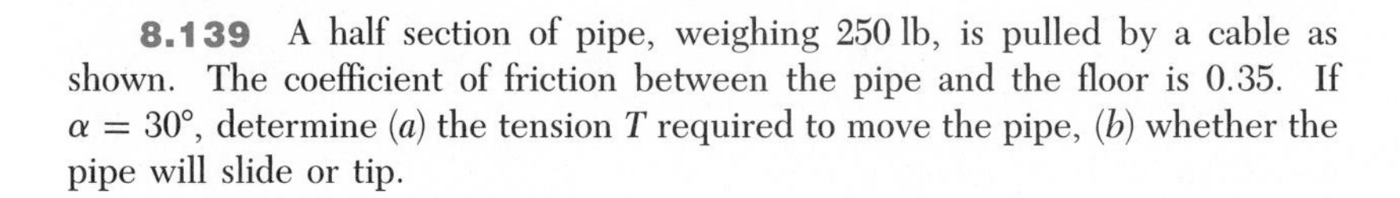

8.139 A half section of pipe, weighing 250 lb, is pulled by a cable as shown. The coefficient of friction between the pipe and the floor is 0.35. If $\alpha = 30°$, determine (*a*) the tension T required to move the pipe, (*b*) whether the pipe will slide or tip.

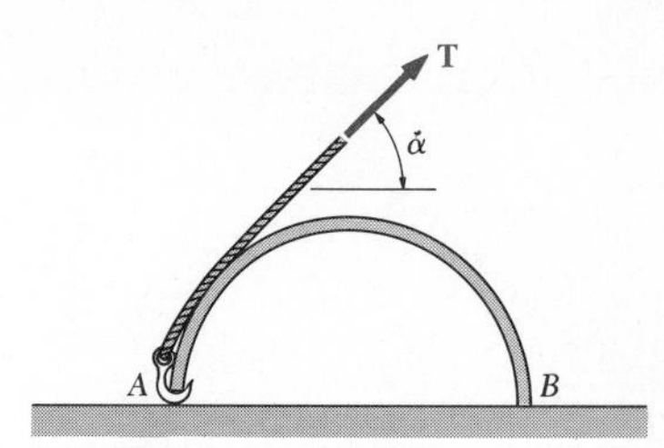

Fig. P8.138 and P8.139

8.140 A slender rod of length L is lodged between peg C and the horizontal ceiling and supports a load **P** at end B. Knowing that $\theta = 30°$ and that the coefficient of static friction is 0.40 at both A and C, find the range of values of the ratio L/a for which equilibrium is maintained.

8.141 A slender rod of length L is lodged between peg C and the horizontal ceiling and supports a load **P** at end B. Knowing that $L = 8a$, $\theta = 15°$, and that the coefficients of friction are $\mu_s = 0.60$ and $\mu_k = 0.50$ at A, and zero at C, determine whether the rod is in equilibrium and find the value of the friction force at A.

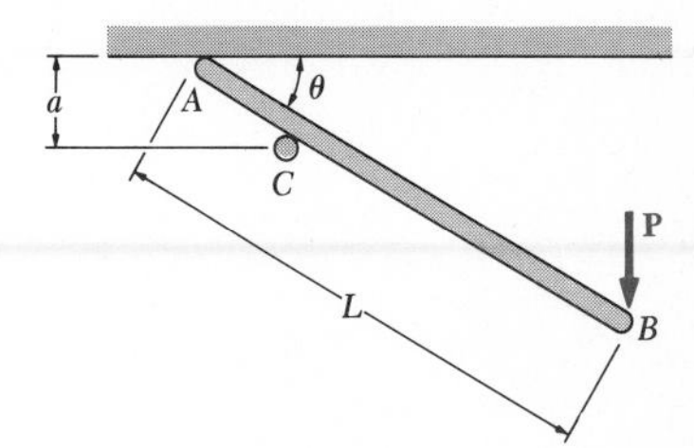

Fig. P8.140 and P8.141

The following problems are designed to be solved with a computer.

8.C1 The 40-kg block A hangs from a cable as shown. Pulley C is connected by a short link to block E, which rests on a horizontal rail. Denoting by μ_s the coefficient of static friction between block E and the rail, and neglecting the weight of block E and the friction in the pulleys, write a computer program and use it to calculate the value of θ for which motion impends for values of μ_s from 0 to 0.55 at 0.05 intervals.

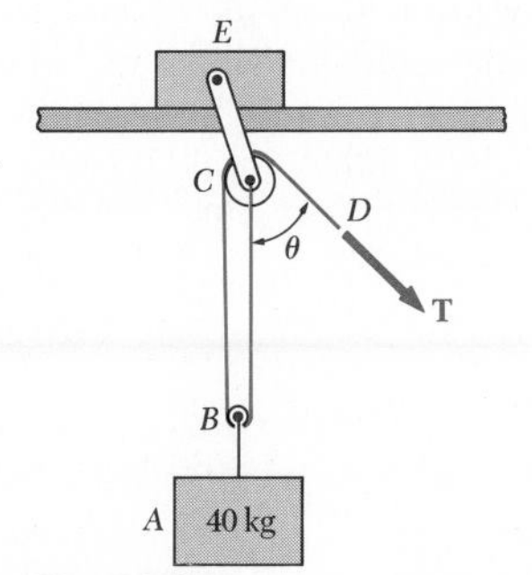

Fig. P8.C1

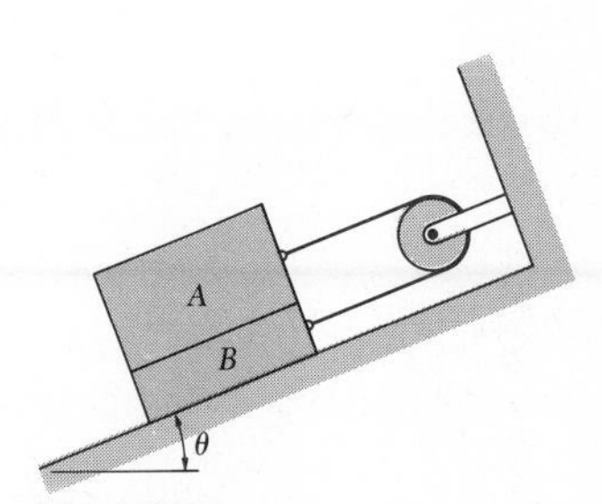

Fig. P8.C2

8.C2 The coefficient of static friction between all surfaces of contact is 0.15. Knowing that block B weighs 25 lb, write a computer program and use it to calculate the value of θ for which motion is impending for weights of block A from 0 to 60 lb at 5-lb intervals.

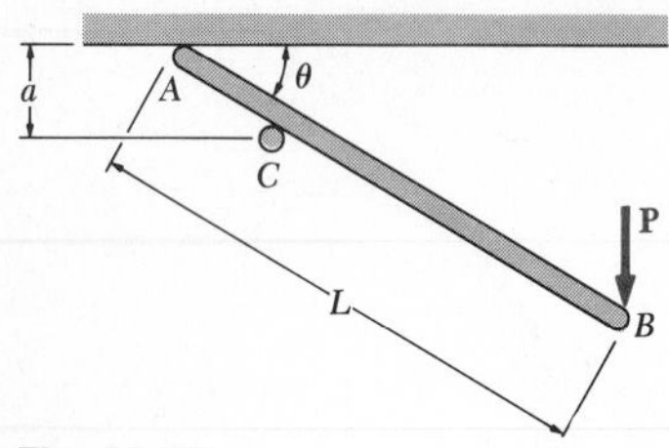

Fig. P8.C3

8.C3 A slender rod of length L is lodged between peg C and the horizontal ceiling and supports a load **P** at end B. (*a*) Knowing that the coefficient of static friction μ_s is the same at A and C, show that the following relation exists between the angle of static friction $\phi_s = \tan^{-1} \mu_s$ and the angle θ for which sliding is impending to the right:

$$\cot 2\phi_s = \cot \theta - \frac{a}{2L \sin^2 \theta \cos \theta}$$

(*b*) Use this relation to write a computer program and calculate μ_s for values of θ from 15 to 60° at 5° intervals, assuming (1) $L = 4a$, (2) $L = 8a$. (*c*) Determine to three significant figures the value of θ for which sliding impends to the right assuming $\mu_s = 0.300$ and (1) $L = 4a$, (2) $L = 8a$.

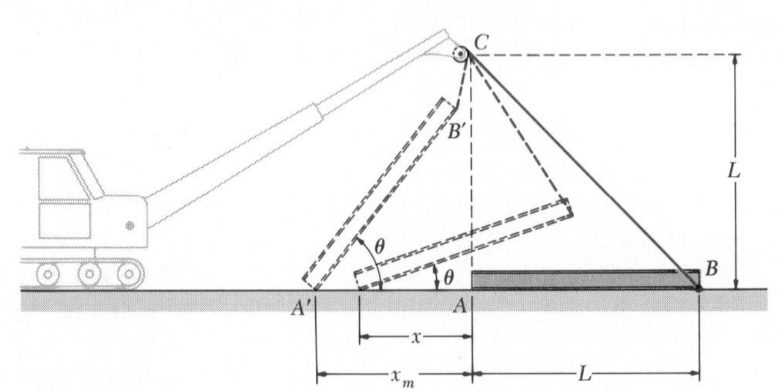

Fig. P8.C4

8.C4 The end B of a uniform beam of length L is being pulled by a stationary crane. Initially the beam lies on the ground with A directly under pulley C. As the cable is slowly pulled in, the beam first slides to the left with $\theta = 0$ until it has moved through a distance x_0. In a second phase, end B is raised, while end A keeps sliding to the left until x reaches its maximum value x_m and θ the corresponding value θ_1. The beam then rotates about A' while θ keeps increasing. As θ reaches the value θ_2, end A starts sliding to the right and keeps sliding in an irregular manner until B reaches C. Knowing that the coefficients of friction between the beam and the ground are $\mu_s = 0.50$ and $\mu_k = 0.40$, (*a*) write a program to compute x for any value of θ while the beam is sliding to the left and use this program to determine x_0, x_m, and θ_1, (*b*) modify the program to compute for any θ the value of x for which sliding would be impending to the right and use this new program to determine the value θ_2 of θ corresponding to $x = x_m$.

Distributed Forces: Moments of Inertia

CHAPTER

9

9.1. Introduction. In Chap. 5, we analyzed various systems of forces distributed over an area or volume. The three main types of forces considered were (1) weights of homogeneous plates of uniform thickness (Secs. 5.3 through 5.6), (2) distributed loads on beams and hydrostatic forces (Secs. 5.8 and 5.9), and (3) weights of homogeneous three-dimensional bodies (Secs. 5.10 and 5.11). In the case of homogeneous plates, the magnitude ΔW of the weight of an element of plate was proportional to the area ΔA of the element. In the case of distributed loads on beams, the magnitude ΔW of each elementary weight was represented by an element of area $\Delta A = \Delta W$ under the load curve; in the case of hydrostatic forces on submerged rectangular surfaces, a similar procedure was followed. In the case of homogeneous three-dimensional bodies, the magnitude ΔW of the weight of an element of the body was proportional to the volume ΔV of the element. Thus, in all cases considered in Chap. 5, the distributed forces were proportional to the elementary areas or volumes associated with them. The resultant of these forces, therefore, could be obtained by summing the corresponding areas or volumes and the moment of the resultant about any given axis could be determined by computing the first moments of the areas or volumes about that axis.

In the first part of this chapter, we shall consider distributed forces $\Delta\mathbf{F}$ whose magnitudes depend not only upon the element of area ΔA on which they act but also upon the distance from ΔA to some given axis. More precisely, the magnitude of the force per unit area $\Delta F/\Delta A$ will vary linearly with the distance to the axis. As we shall see in the next section, forces of this type are found in the study of the bending of beams and in problems involving submerged nonrectangular surfaces. Assuming that the elementary forces involved are distributed over an area A and vary linearly with the distance y to the x axis, we shall find that while the magnitude of their resultant $\mathbf{R}$ depends upon the first moment $Q_x = \int y\,dA$ of the area A, the location of the point where $\mathbf{R}$ should be applied depends upon the *second* ment, or *moment of inertia*, $I_x = \int y^2\, dA$ of the same area with respect the x axis. We shall learn to compute the moments of inertia of various eas with respect to given x and y axes. We shall also introduce the *polar* oment of inertia $J_O = \int r^2\, dA$ of an area, where r is the distance from the element of area dA to a point O. To facilitate our computations, we shall establish a relation between the moment of inertia I_x of an area A with respect to a given x axis and its moment of inertia $I_{x'}$ with respect to a parallel centroidal x' axis (parallel-axis theorem). We shall also study the transformation of the moments of inertia of a given area under a rotation of axes (Secs. 9.9 and 9.10).

In the second part of the chapter, we shall determine the moments of inertia of various *masses* with respect to a given axis. As we shall see in Sec. 9.11, the moment of inertia of a given mass about an axis AA' is defined as the integral $I = \int r^2\, dm$, where r is the distance from the axis AA' to the element of mass dm. Moments of inertia of masses are encountered in dynamics in problems involving the rotation of a rigid body about an axis. To facilitate the computation of mass moments of inertia, we shall introduce the parallel-axis theorem (Sec. 9.12) and analyze the transformation of moments of inertia of masses under a rotation of axes (Secs. 9.16 and 9.17).

MOMENTS OF INERTIA OF AREAS

9.2. Second Moment, or Moment of Inertia, of an Area. In the first part of this chapter, we shall consider distributed forces $\Delta\mathbf{F}$ of magnitude ΔF proportional to the element of area ΔA on which they act and varying linearly with the distance from ΔA to some given axis.

Consider, for example, a beam of uniform cross section, subjected to two equal and opposite couples applied at each end of the beam. Such a beam is said to be in *pure bending*, and it is shown in mechanics of materials that the internal forces in any section of the beam are distributed forces whose magnitudes $\Delta F = ky\, \Delta A$ vary linearly with the distance y from an axis passing through the centroid of the section. This axis, represented by the x axis in Fig. 9.1, is known as the *neutral axis* of the section. The forces on one side of the neutral axis are forces of compression and on the other side, forces of tension, while on the neutral axis itself the forces are zero.

The magnitude of the resultant $\mathbf{R}$ of the elementary forces $\Delta\mathbf{F}$ over the entire section is

$$R = \int ky\, dA = k \int y\, dA$$

The last integral obtained is recognized as the *first moment* Q_x of the section about the x axis; it is equal to $\bar{y}A$ and to zero, since the centroid of the section is located on the x axis. The system of the forces $\Delta\mathbf{F}$ thus reduces to a couple. The magnitude M of this couple (bending moment) must be equal to the sum of the moments $\Delta M_x = y\, \Delta F = ky^2\, \Delta A$ of the elementary forces. Integrating over the entire section, we obtain

$$M = \int ky^2\, dA = k \int y^2\, dA$$

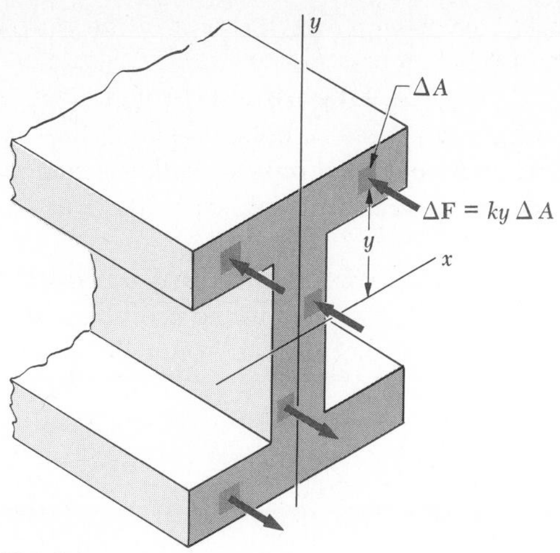

Fig. 9.1

The last integral is known as the *second moment,* or *moment of inertia,*† of the beam section with respect to the x axis and is denoted by I_x. It is obtained by multiplying each element of area dA by the *square of its distance* from the x axis and integrating over the beam section. Since each product $y^2\,dA$ is positive, whether y is itself positive or negative (or zero if y is zero), the integral I_x will always be different from zero and positive.

Another example of second moment, or moment of inertia, of an area is provided by the following problem of hydrostatics: A vertical circular gate used to close the outlet of a large reservoir is submerged under water as shown in Fig. 9.2. What is the resultant of the forces exerted by the water on the gate, and what is the moment of the resultant about the line of intersection of the plane of the gate with the water surface (x axis)?

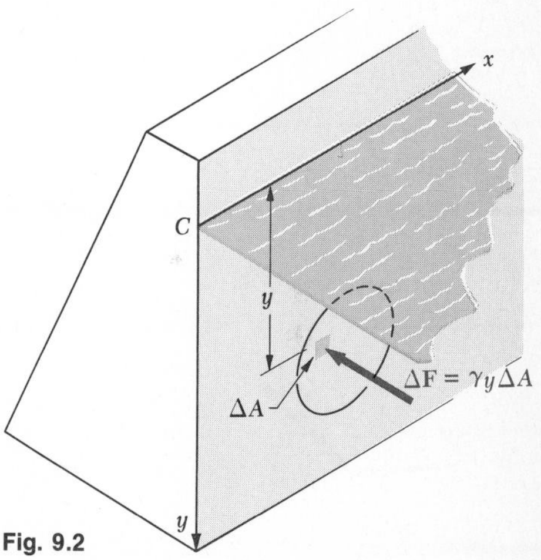

If the gate were rectangular, the resultant of the forces of pressure could be determined from the pressure curve, as was done in Sec. 5.9. Since the gate is circular, however, a more general method must be used. Denoting by y the depth of an element of area ΔA and by γ the specific weight of water, the pressure at the element is $p = \gamma y$, and the magnitude of the elementary force exerted on ΔA is $\Delta F = p\,\Delta A = \gamma y\,\Delta A$. The magnitude of the resultant of the elementary forces is thus

$$R = \int \gamma y\,dA = \gamma \int y\,dA$$

and may be obtained by computing the first moment $Q_x = \int y\,dA$ of the area of the gate with respect to the x axis. The moment M_x of the resultant must be equal to the sum of the moments $\Delta M_x = y\,\Delta F = \gamma y^2\,\Delta A$ of the

Fig. 9.2

† The term second moment is more proper than the term moment of inertia, since, logically, the latter should be used only to denote integrals of mass (see Sec. 9.11). In common engineering practice, however, moment of inertia is used in connection with areas as well as masses.

elementary forces. Integrating over the area of the gate, we have

$$M_x = \int \gamma y^2\, dA = \gamma \int y^2\, dA$$

Here again, the integral obtained represents the second moment, or moment of inertia, I_x of the area with respect to the x axis.

9.3. Determination of the Moment of Inertia of an Area by Integration. We have defined in the preceding section the second moment, or moment of inertia, of an area A with respect to the x axis.

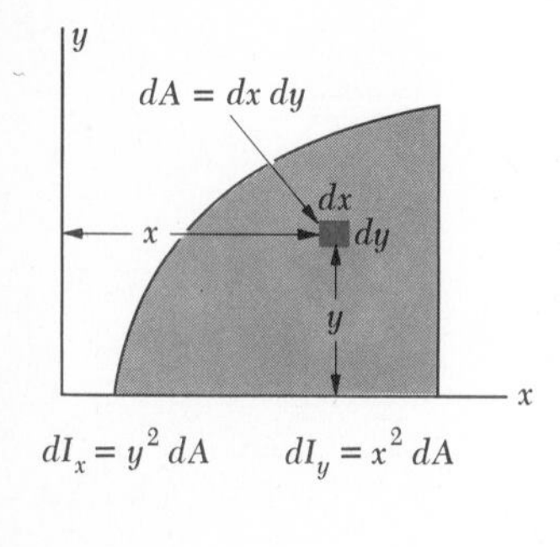

(a)

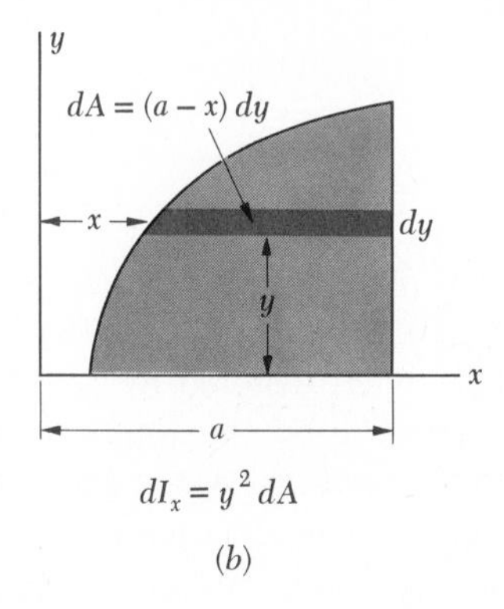

(b)

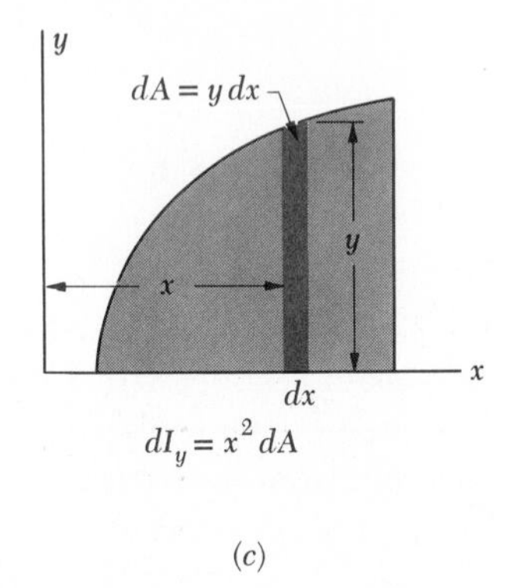

(c)

Fig. 9.3

Defining in a similar way the moment of inertia I_y of the area A with respect to the y axis, we write (Fig. 9.3a)

$$I_x = \int y^2\, dA \qquad I_y = \int x^2\, dA \tag{9.1}$$

These integrals, known as the *rectangular moments of inertia* of the area A, may be more easily computed if we choose for dA a thin strip parallel to one of the axes of coordinates. To compute I_x, the strip is chosen parallel to the x axis, so that all the points forming the strip are at the same distance y from the x axis (Fig. 9.3b); the moment of inertia dI_x of the strip is then obtained by multiplying the area dA of the strip by y^2. To compute I_y, the strip is chosen parallel to the y axis so that all the points forming the strip are at the same distance x from the y axis (Fig. 9.3c); the moment of inertia dI_y of the strip is $x^2\, dA$.

Moment of Inertia of a Rectangular Area. As an example, we shall determine the moment of inertia of a rectangle with respect to its base (Fig. 9.4). Dividing the rectangle into strips parallel to the x axis, we obtain

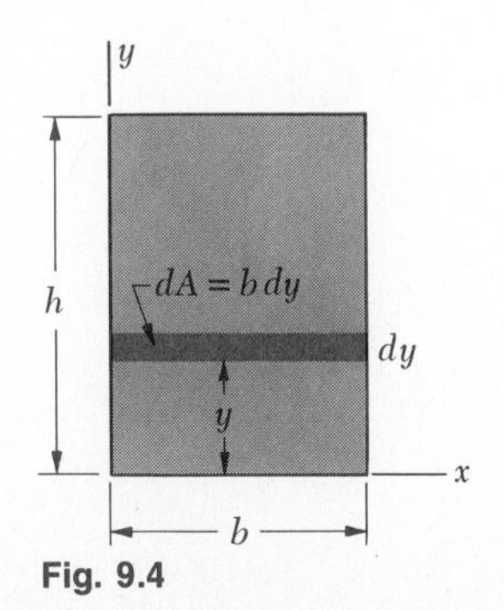

Fig. 9.4

$$dA = b\, dy \qquad dI_x = y^2 b\, dy \qquad I_x = \int_0^h by^2\, dy = \tfrac{1}{3}bh^3 \tag{9.2}$$

Computing I_x and I_y from the Same Elementary Strips. The formula just derived may be used to determine the moment of inertia dI_x with respect to the x axis of a rectangular strip parallel to the y axis such as the one shown in Fig. 9.3*c*. Making $b = dx$ and $h = y$ in formula (9.2), we write

$$dI_x = \tfrac{1}{3}y^3\,dx$$

On the other hand, we have

$$dI_y = x^2\,dA = x^2y\,dx$$

The same element may thus be used to compute the moments of inertia I_x and I_y of a given area (Fig. 9.5).

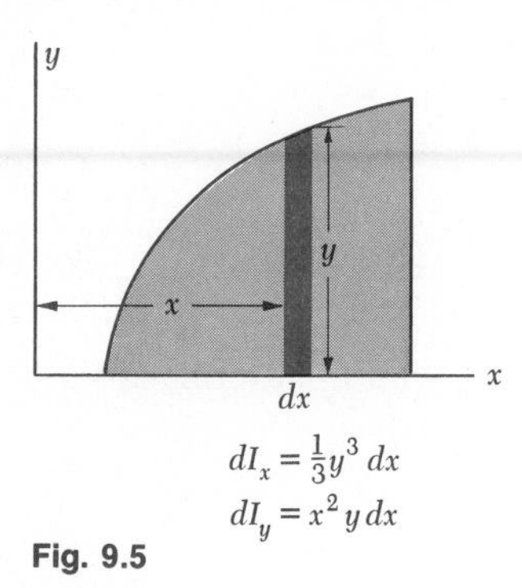

Fig. 9.5

9.4. Polar Moment of Inertia. An integral of great importance in problems concerning the torsion of cylindrical shafts and in problems dealing with the rotation of slabs is

$$J_O = \int r^2\,dA \tag{9.3}$$

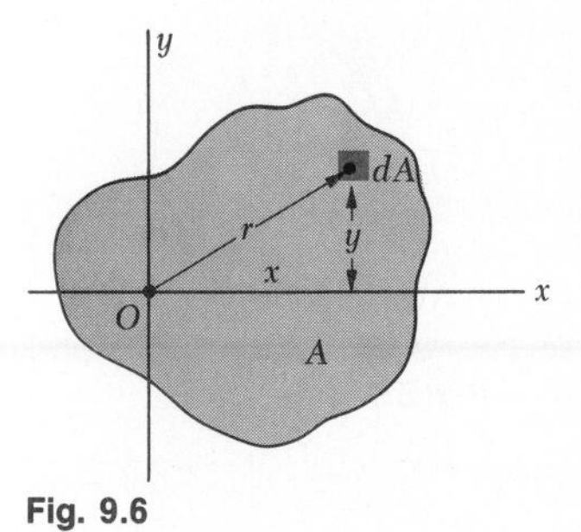

Fig. 9.6

where r is the distance from the element of area dA to the pole O (Fig. 9.6). This integral is the *polar moment of inertia* of the area A with respect to O.

The polar moment of inertia of a given area may be computed from the rectangular moments of inertia I_x and I_y of the area if these integrals are already known. Indeed, noting that $r^2 = x^2 + y^2$, we write

$$J_O = \int r^2\,dA = \int (x^2 + y^2)\,dA = \int y^2\,dA + \int x^2\,dA$$

that is,

$$J_O = I_x + I_y \tag{9.4}$$

9.5. Radius of Gyration of an Area. Consider an area A which has a moment of inertia I_x with respect to the x axis (Fig. 9.7*a*). Let us imagine that we concentrate this area into a thin strip parallel to the x axis (Fig. 9.7*b*). If the area A, thus concentrated, is to have the same moment of inertia with respect to the x axis, the strip should be placed at a distance k_x from the x axis, defined by the relation

$$I_x = k_x^2 A$$

Solving for k_x, we write

$$k_x = \sqrt{\frac{I_x}{A}} \tag{9.5}$$

The distance k_x is referred to as the *radius of gyration* of the area with respect to the x axis. We may define in a similar way the radii of gyration k_y and k_O (Fig. 9.7*c* and *d*); we write

$$I_y = k_y^2 A \qquad k_y = \sqrt{\frac{I_y}{A}} \tag{9.6}$$

$$J_O = k_O^2 A \qquad k_O = \sqrt{\frac{J_O}{A}} \tag{9.7}$$

Substituting for J_O, I_x, and I_y in terms of the radii of gyration in the relation (9.4), we observe that

$$k_O^2 = k_x^2 + k_y^2 \tag{9.8}$$

Example. As an example, let us compute the radius of gyration k_x of the rectangle shown in Fig. 9.8. Using formulas (9.5) and (9.2), we write

$$k_x^2 = \frac{I_x}{A} = \frac{\frac{1}{3}bh^3}{bh} = \frac{h^2}{3} \qquad k_x = \frac{h}{\sqrt{3}}$$

The radius of gyration k_x of the rectangle is shown in Fig. 9.8. It should not be confused with the ordinate $\bar{y} = h/2$ of the centroid of the area. While k_x depends upon the *second moment*, or moment of inertia, of the area, the ordinate $\bar{y}$ is related to the *first moment* of the area.

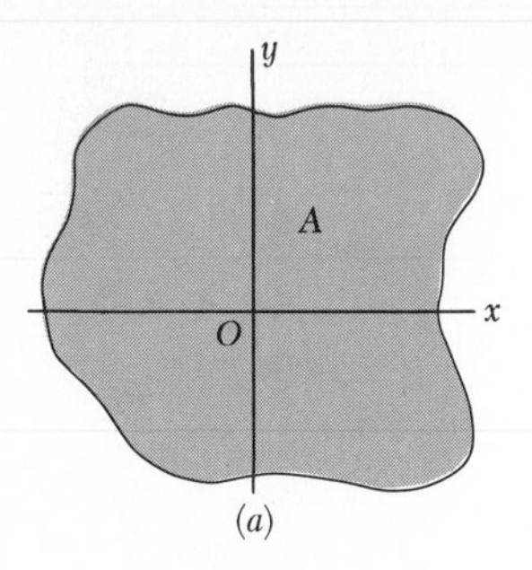

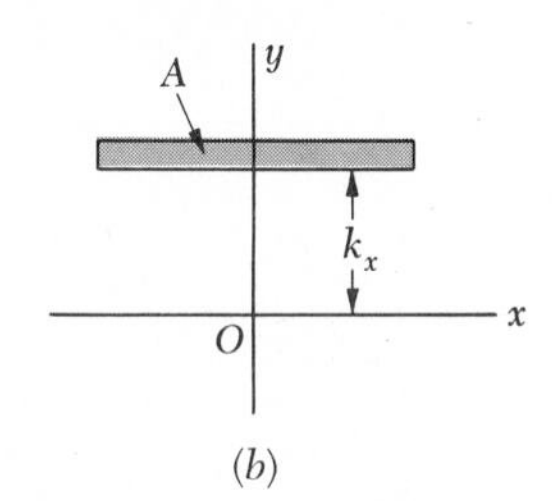

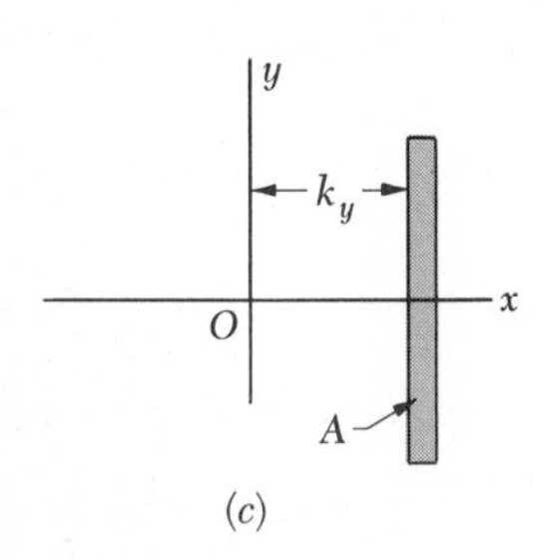

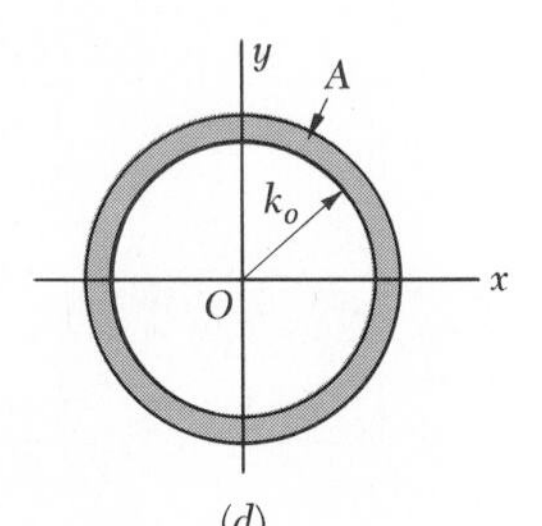

Fig. 9.7

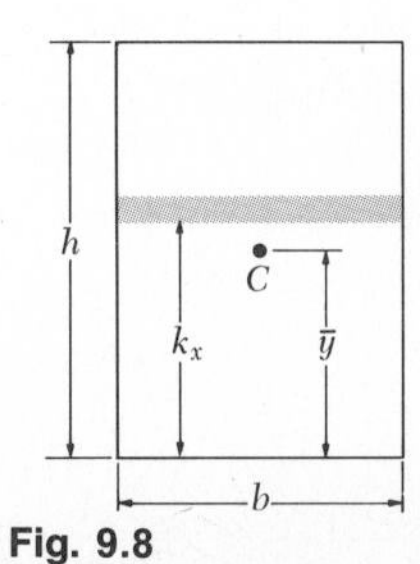

Fig. 9.8

SAMPLE PROBLEM 9.1

Determine the moment of inertia of a triangle with respect to its base.

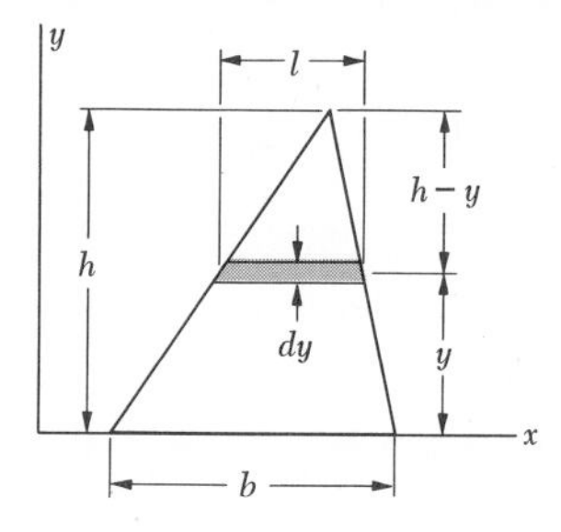

Solution. A triangle of base b and height h is drawn; the x axis is chosen to coincide with the base. A differential strip parallel to the x axis is chosen. Since all portions of the strip are at the same distance from the x axis, we write

$$dI_x = y^2\,dA \qquad dA = l\,dy$$

From the similar triangles, we have

$$\frac{l}{b} = \frac{h-y}{h} \qquad l = b\frac{h-y}{h} \qquad dA = b\frac{h-y}{h}\,dy$$

Integrating dI_x from $y = 0$ to $y = h$, we obtain

$$I_x = \int y^2\,dA = \int_0^h y^2 b\frac{h-y}{h}\,dy = \frac{b}{h}\int_0^h (hy^2 - y^3)\,dy$$

$$= \frac{b}{h}\left[h\frac{y^3}{3} - \frac{y^4}{4}\right]_0^h \qquad I_x = \frac{bh^3}{12} \blacktriangleleft$$

SAMPLE PROBLEM 9.2

(*a*) Determine the centroidal polar moment of inertia of a circular area by direct integration. (*b*) Using the result of part *a*, determine the moment of inertia of a circular area with respect to a diameter.

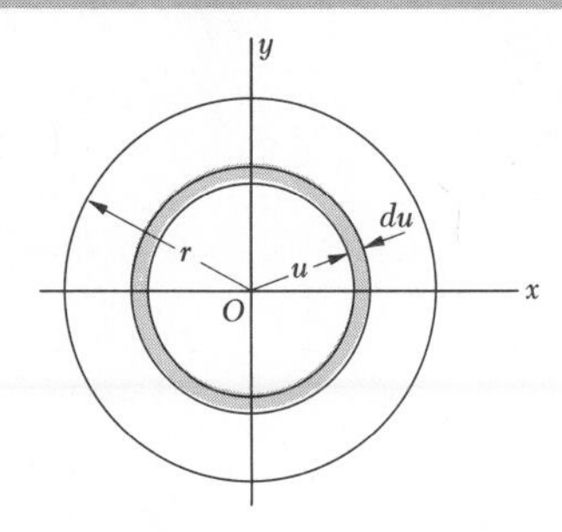

***a.* Polar Moment of Inertia.** An annular differential element of area is chosen. Since all portions of the differential area are at the same distance from the origin, we write

$$dJ_O = u^2\,dA \qquad dA = 2\pi u\,du$$

$$J_O = \int dJ_O = \int_0^r u^2(2\pi u\,du) = 2\pi\int_0^r u^3\,du$$

$$J_O = \frac{\pi}{2}r^4 \blacktriangleleft$$

***b.* Moment of Inertia with Respect to a Diameter.** Because of the symmetry of the circular area we have $I_x = I_y$. We then write

$$J_O = I_x + I_y = 2I_x \qquad \frac{\pi}{2}r^4 = 2I_x \qquad I_{\text{diameter}} = I_x = \frac{\pi}{4}r^4 \blacktriangleleft$$

SAMPLE PROBLEM 9.3

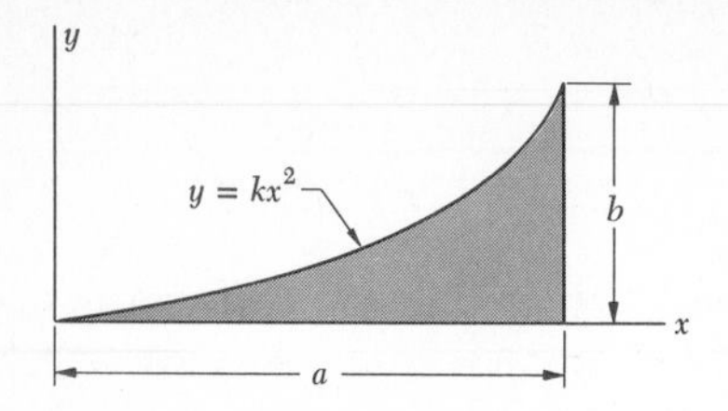

(a) Determine the moment of inertia of the shaded area shown with respect to each of the coordinate axes. This area has also been considered in Sample Prob. 5.4. (b) Using the results of part a, determine the radius of gyration of the shaded area with respect to each of the coordinate axes.

Solution. Referring to Sample Prob. 5.4, we obtain the following expressions for the equation of the curve and the total area:

$$y = \frac{b}{a^2}x^2 \qquad A = \tfrac{1}{3}ab$$

Moment of Inertia I_x. A vertical differential element of area is chosen. Since all portions of this element are *not* at the same distance from the x axis, we must treat the element as a thin rectangle. The moment of inertia of the element with respect to the x axis is

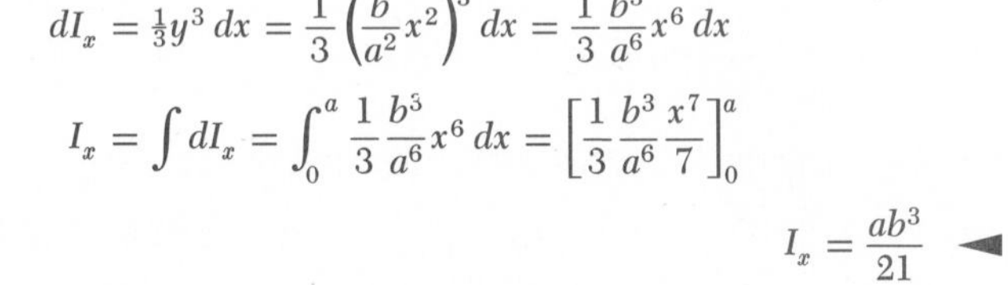

$$dI_x = \tfrac{1}{3}y^3\,dx = \frac{1}{3}\left(\frac{b}{a^2}x^2\right)^3 dx = \frac{1}{3}\frac{b^3}{a^6}x^6\,dx$$

$$I_x = \int dI_x = \int_0^a \frac{1}{3}\frac{b^3}{a^6}x^6\,dx = \left[\frac{1}{3}\frac{b^3}{a^6}\frac{x^7}{7}\right]_0^a$$

$$I_x = \frac{ab^3}{21} \quad ◄$$

Moment of Inertia I_y. The same vertical differential element of area is used. Since all portions of the element are at the same distance from the y axis, we write

$$dI_y = x^2\,dA = x^2(y\,dx) = x^2\left(\frac{b}{a^2}x^2\right)dx = \frac{b}{a^2}x^4\,dx$$

$$I_y = \int dI_y = \int_0^a \frac{b}{a^2}x^4\,dx = \left[\frac{b}{a^2}\frac{x^5}{5}\right]_0^a$$

$$I_y = \frac{a^3b}{5} \quad ◄$$

Radii of Gyration k_x and k_y

$$k_x^2 = \frac{I_x}{A} = \frac{ab^3/21}{ab/3} = \frac{b^2}{7} \qquad k_x = \sqrt{\tfrac{1}{7}}\,b \quad ◄$$

$$k_y^2 = \frac{I_y}{A} = \frac{a^3b/5}{ab/3} = \tfrac{3}{5}a^2 \qquad k_y = \sqrt{\tfrac{3}{5}}\,a \quad ◄$$

Problems

9.1 Determine the moment of inertia of the triangle of Sample Prob. 9.1 with respect to a line which passes through the upper vertex and is parallel to the x axis.

9.2 through 9.5 Determine by direct integration the moment of inertia of the shaded area with respect to the y axis.

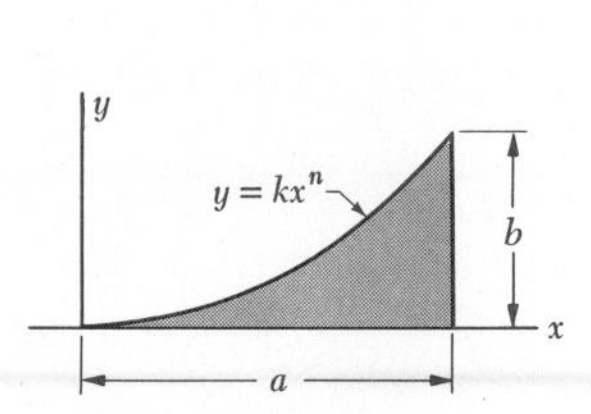

Fig. P9.2 and P9.6

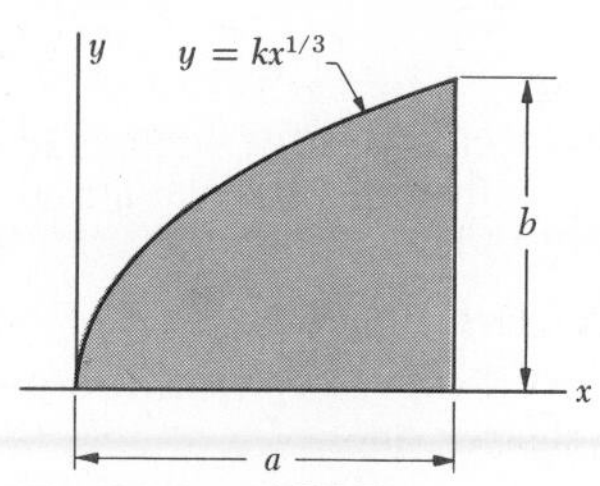

Fig. P9.3 and P9.7

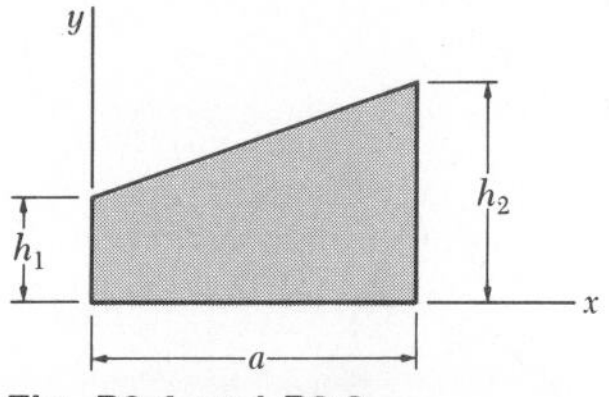

Fig. P9.4 and P9.8

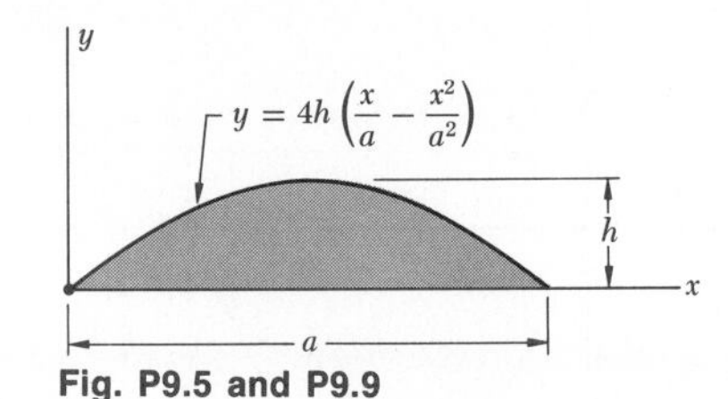

Fig. P9.5 and P9.9

9.6 through 9.9 Determine by direct integration the moment of inertia of the shaded area with respect to the x axis.

9.10 and 9.11 Determine the moment of inertia and radius of gyration of the shaded area shown with respect to the x axis.

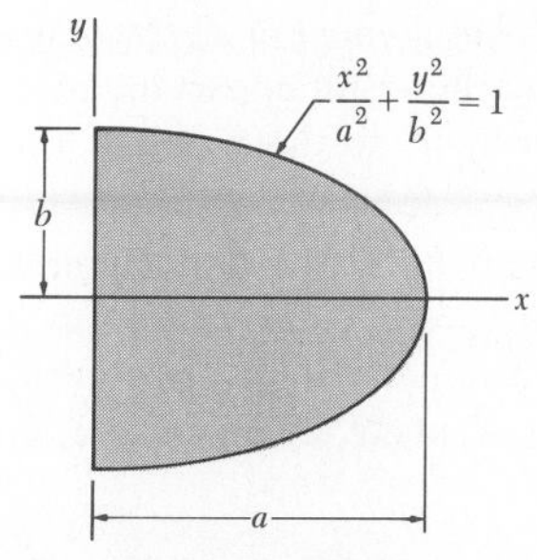

Fig. P9.10 and P9.12

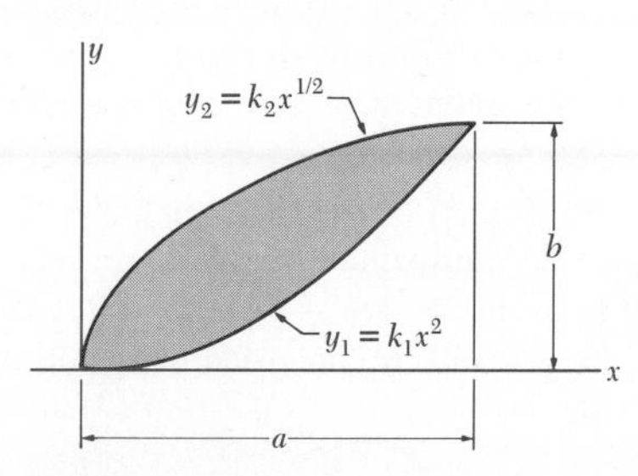

Fig. P9.11 and P9.13

9.12 and 9.13 Determine the moment of inertia and radius of gyration of the shaded area shown with respect to the y axis.

9.14 Determine the moment of inertia and radius of gyration of the shaded area shown with respect to the x axis.

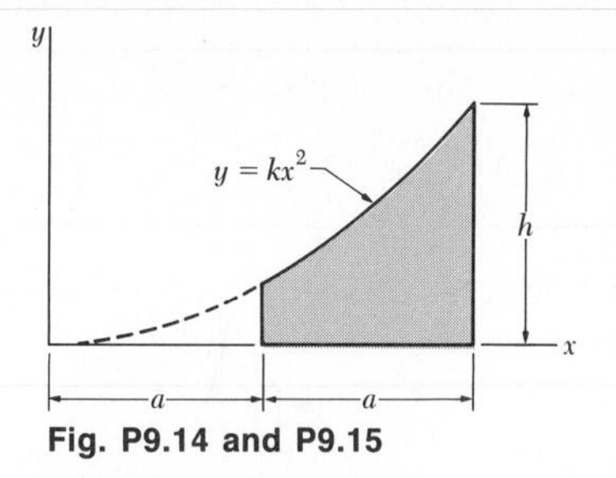

Fig. P9.14 and P9.15

9.15 Determine the moment of inertia and radius of gyration of the shaded area shown with respect to the y axis.

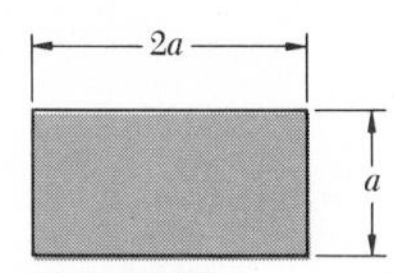

Fig. P9.16

9.16 Determine the polar moment of inertia and the polar radius of gyration of the rectangle shown with respect to the midpoint of one of its (*a*) longer sides, (*b*) shorter sides.

9.17 Determine the polar moment of inertia and the polar radius of gyration of the trapezoid shown with respect to point P.

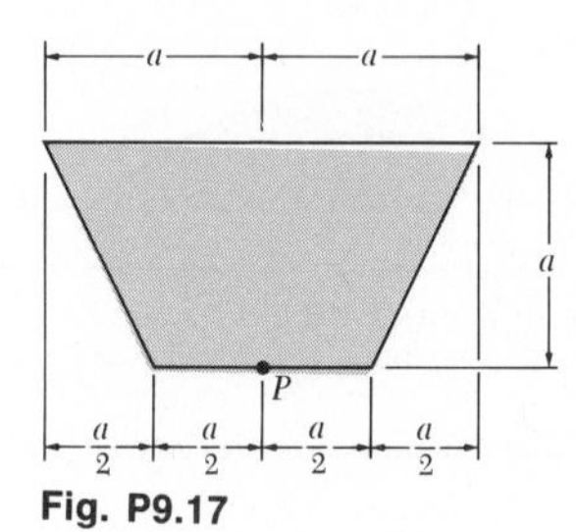

Fig. P9.17

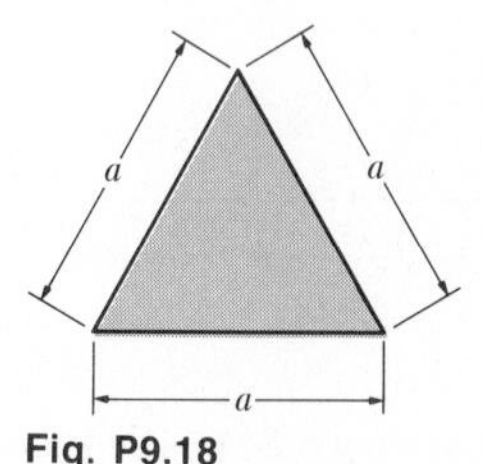

Fig. P9.18

9.18 Determine the polar moment of inertia and the polar radius of gyration of an equilateral triangle of side a with respect to one of its vertices.

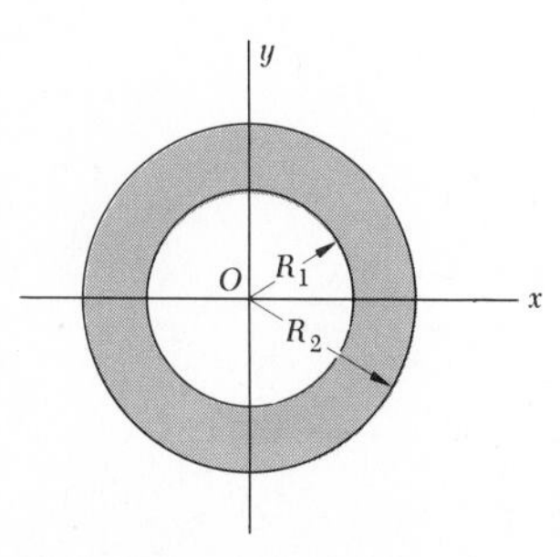

Fig. P9.19 and P9.20

9.19 (*a*) Determine by direct integration the polar moment of inertia of the annular area shown with respect to point O. (*b*) Using the result of part *a*, determine the moment of inertia of the given area with respect to the x axis.

9.20 (*a*) Show that the polar radius of gyration k_O of the annular area shown is approximately equal to the mean radius $R_m = (R_1 + R_2)/2$ for small values of the thickness $t = R_2 - R_1$. (*b*) Determine the percentage error introduced by using R_m in place of k_O for values of t/R_m respectively equal to 1, $\frac{1}{2}$, and $\frac{1}{10}$.

***9.21** Prove that the centroidal polar moment of inertia of a given area A cannot be smaller than $A^2/2\pi$. (*Hint.* Compare the moment of inertia of the given area with the moment of inertia of a circle of the same area and same centroid.)

9.6. Parallel-Axis Theorem. Consider the moment of inertia I of an area A with respect to an axis AA' (Fig. 9.9). Denoting by y the distance from an element of area dA to AA', we write

$$I = \int y^2\, dA$$

Let us now draw an axis BB' parallel to AA' through the centroid C of the area; this axis is called a *centroidal axis*. Denoting by y' the distance from

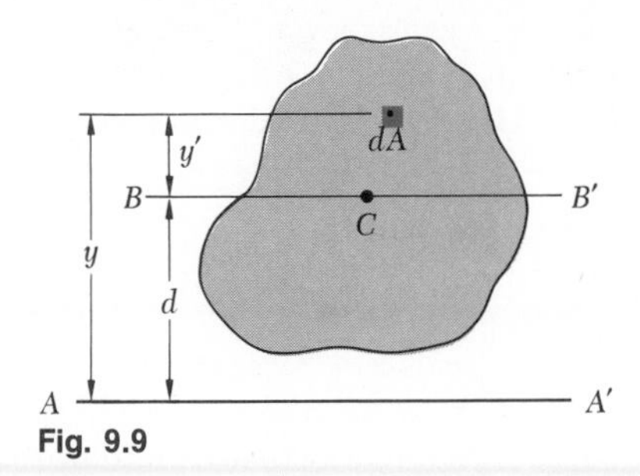

Fig. 9.9

the element dA to BB', we write $y = y' + d$, where d is the distance between the axes AA' and BB'. Substituting for y in the integral representing I, we write

$$\begin{aligned} I &= \int y^2\, dA = \int (y' + d)^2\, dA \\ &= \int y'^2\, dA + 2d \int y'\, dA + d^2 \int dA \end{aligned}$$

The first integral represents the moment of inertia $\bar{I}$ of the area with respect to the centroidal axis BB'. The second integral represents the first moment of the area with respect to BB'; since the centroid C of the area is located on that axis, the second integral must be zero. Finally, we observe that the last integral is equal to the total area A. We write therefore

$$I = \bar{I} + Ad^2 \tag{9.9}$$

This formula expresses that the moment of inertia I of an area with respect to any given axis AA' is equal to the moment of inertia $\bar{I}$ of the area with respect to a centroidal axis BB' parallel to AA' *plus* the product Ad^2 of the area A and of the square of the distance d between the two axes. This theorem is known as the *parallel-axis theorem*. Substituting k^2A for I and $\bar{k}^2A$ for $\bar{I}$, the theorem may also be expressed in the following way:

$$k^2 = \bar{k}^2 + d^2 \tag{9.10}$$

A similar theorem may be used to relate the polar moment of inertia J_O of an area about a point O and the polar moment of inertia $\bar{J}_C$ of the same area about its centroid C. Denoting by d the distance between O and C, we write

$$J_O = \bar{J}_C + Ad^2 \qquad \text{or} \qquad k_O^2 = \bar{k}_C^2 + d^2 \tag{9.11}$$

Example 1. As an application of the parallel-axis theorem, we shall determine the moment of inertia I_T of a circular area with respect to a line tangent to the circle (Fig. 9.10). We found in Sample Prob. 9.2 that the moment of inertia of a circular area about a centroidal axis is $\bar{I} = \frac{1}{4}\pi r^4$. We may write, therefore,

$$I_T = \bar{I} + Ad^2 = \tfrac{1}{4}\pi r^4 + \pi r^2 r^2 = \tfrac{5}{4}\pi r^4$$

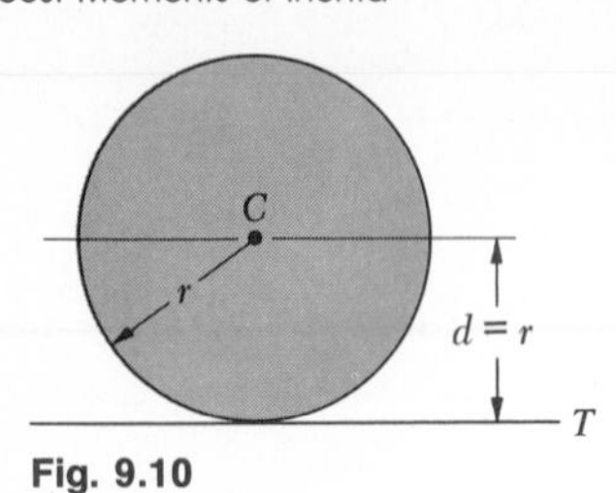

Fig. 9.10

Example 2. The parallel-axis theorem may also be used to determine the centroidal moment of inertia of an area when the moment of inertia of this area with respect to some parallel axis is known. Consider, for instance, a triangular area (Fig. 9.11). We found in Sample Prob. 9.1 that the moment of inertia of a triangle with respect to its base AA' is equal to $\frac{1}{12}bh^3$. Using the parallel-axis theorem, we write

$$I_{AA'} = \bar{I}_{BB'} + Ad^2$$
$$\bar{I}_{BB'} = I_{AA'} - Ad^2 = \tfrac{1}{12}bh^3 - \tfrac{1}{2}bh(\tfrac{1}{3}h)^2 = \tfrac{1}{36}bh^3$$

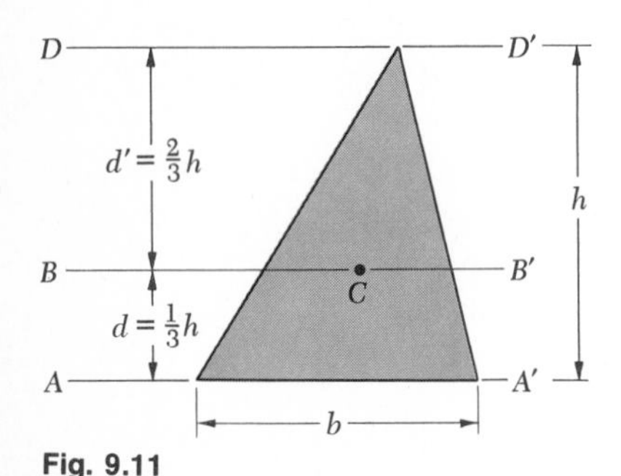

Fig. 9.11

It should be observed that the product Ad^2 was *subtracted* from the given moment of inertia in order to obtain the centroidal moment of inertia of the triangle. While this product is *added* in transferring *from* a centroidal axis to a parallel axis, it should be *subtracted* in transferring *to* a centroidal axis. In other words, the moment of inertia of an area is always smaller with respect to a centroidal axis than with respect to any other parallel axis.

Returning to Fig. 9.11, we observe that the moment of inertia of the triangle with respect to a line DD' drawn through a vertex may be obtained by writing

$$I_{DD'} = \bar{I}_{BB'} + Ad'^2 = \tfrac{1}{36}bh^3 + \tfrac{1}{2}bh(\tfrac{2}{3}h)^2 = \tfrac{1}{4}bh^3$$

Note that $I_{DD'}$ *could not* have been obtained directly from $I_{AA'}$. The parallel-axis theorem can be applied only if one of the two parallel axes passes through the centroid of the area.

9.7. Moments of Inertia of Composite Areas. Consider a composite area A made of several component areas A_1, A_2, etc. Since the integral representing the moment of inertia of A may be subdivided into integrals computed over A_1, A_2, etc., the moment of inertia of A with respect to a given axis will be obtained by adding the moments of inertia of the areas A_1, A_2, etc., with respect to the same axis. The moment of inertia of an area made of several of the common shapes shown in Fig. 9.12 may thus be obtained from the formulas given in that figure. Before adding the moments of inertia of the component areas, however, the parallel-axis theorem should be used to transfer each moment of inertia to the desired axis. This is shown in Sample Probs. 9.4 and 9.5.

The properties of the cross sections of various structural shapes are given in Fig. 9.13. As noted in Sec. 9.2, the moment of inertia of a beam section about its neutral axis is closely related to the value of the internal forces. The determination of moments of inertia is thus a prerequisite to the analysis and design of structural members.

It should be noted that the radius of gyration of a composite area is *not* equal to the sum of the radii of gyration of the component areas. In order to determine the radius of gyration of a composite area, it is necessary first to compute the moment of inertia of the area.

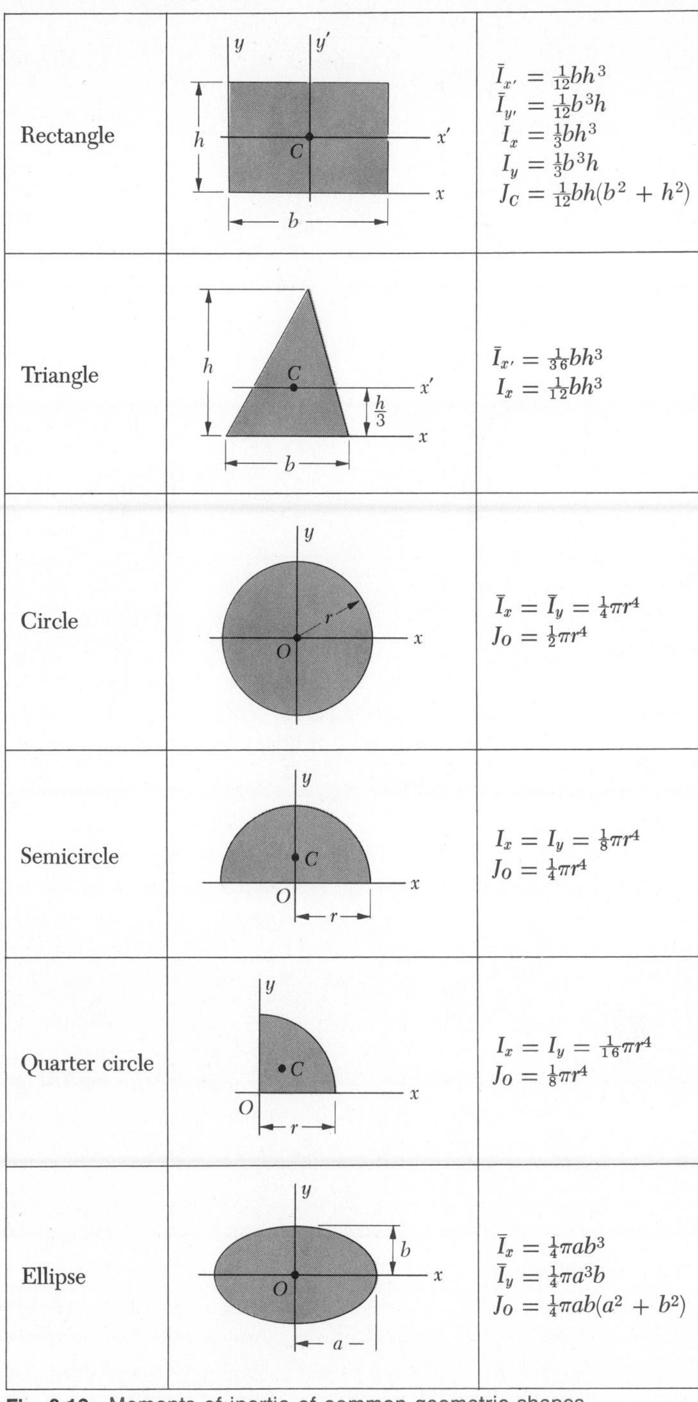

Rectangle	$\bar{I}_{x'} = \frac{1}{12}bh^3$ $\bar{I}_{y'} = \frac{1}{12}b^3h$ $I_x = \frac{1}{3}bh^3$ $I_y = \frac{1}{3}b^3h$ $J_C = \frac{1}{12}bh(b^2 + h^2)$
Triangle	$\bar{I}_{x'} = \frac{1}{36}bh^3$ $I_x = \frac{1}{12}bh^3$
Circle	$\bar{I}_x = \bar{I}_y = \frac{1}{4}\pi r^4$ $J_O = \frac{1}{2}\pi r^4$
Semicircle	$I_x = I_y = \frac{1}{8}\pi r^4$ $J_O = \frac{1}{4}\pi r^4$
Quarter circle	$I_x = I_y = \frac{1}{16}\pi r^4$ $J_O = \frac{1}{8}\pi r^4$
Ellipse	$\bar{I}_x = \frac{1}{4}\pi ab^3$ $\bar{I}_y = \frac{1}{4}\pi a^3b$ $J_O = \frac{1}{4}\pi ab(a^2 + b^2)$

Fig. 9.12 Moments of inertia of common geometric shapes.

Fig. 9.13A Properties of Rolled-Steel Shapes (U.S. Customary Units).*

	Designation	Area in^2	Depth in.	Width in.	Axis *X-X*			Axis *Y-Y*		
					$\bar{I}_x$, in^4	$\bar{k}_x$, in.	$\bar{y}$, in.	$\bar{I}_y$, in^4	$\bar{k}_y$, in.	$\bar{x}$, in.
W Shapes (Wide-Flange Shapes) 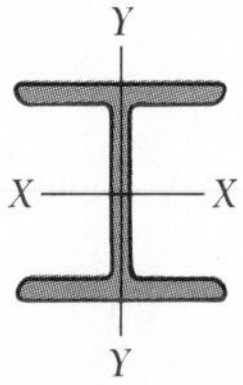	W18 × 76†	22.3	18.21	11.035	1330	7.73		152	2.61	
	W16 × 57	16.8	16.43	7.120	758	6.72		43.1	1.60	
	W14 × 38	11.2	14.10	6.770	385	5.88		26.7	1.55	
	W8 × 31	9.13	8.00	7.995	110	3.47		37.1	2.02	
S Shapes (American Standard Shapes) 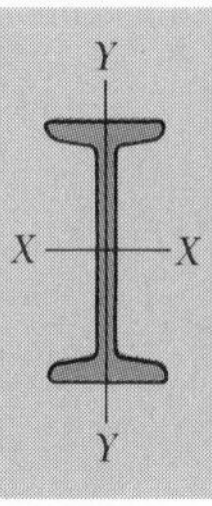	S18 × 55.7†	16.1	18.00	6.001	804	7.07		20.8	1.14	
	S12 × 31.8	9.35	12.00	5.000	218	4.83		9.36	1.00	
	S10 × 25.4	7.46	10.00	4.661	124	4.07		6.79	0.954	
	S6 × 12.5	3.67	6.00	3.332	22.1	2.45		1.82	0.705	
C Shapes (American Standard Channels) 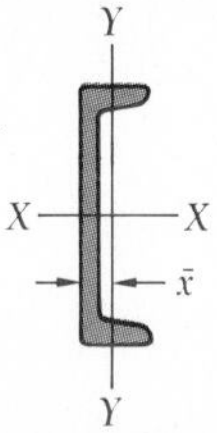	C12 × 20.7†	6.09	12.00	2.942	129	4.61		3.88	0.799	0.698
	C10 × 15.3	4.49	10.00	2.600	67.4	3.87		2.28	0.713	0.634
	C8 × 11.5	3.38	8.00	2.260	32.6	3.11		1.32	0.625	0.571
	C6 × 8.2	2.40	6.00	1.920	13.1	2.34		0.692	0.537	0.512
Angles 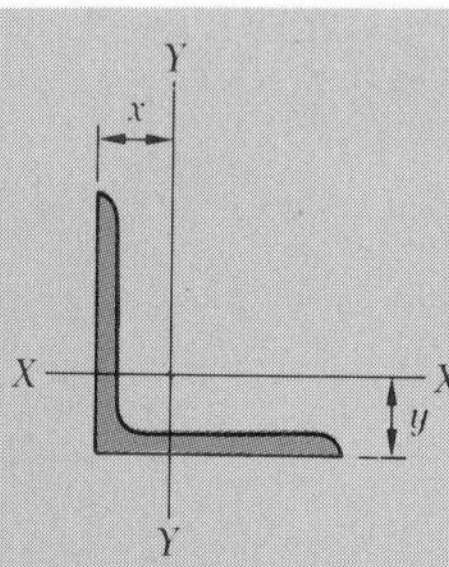	L6 × 6 × 1‡	11.00			35.5	1.80	1.86	35.5	1.80	1.86
	L4 × 4 × $\frac{1}{2}$	3.75			5.56	1.22	1.18	5.56	1.22	1.18
	L3 × 3 × $\frac{1}{4}$	1.44			1.24	0.930	0.842	1.24	0.930	0.842
	L6 × 4 × $\frac{1}{2}$	4.75			17.4	1.91	1.99	6.27	1.15	0.987
	L5 × 3 × $\frac{1}{2}$	3.75			9.45	1.59	1.75	2.58	0.829	0.750
	L3 × 2 × $\frac{1}{4}$	1.19			1.09	0.957	0.993	0.392	0.574	0.493

*Courtesy of the American Institute of Steel Construction, Chicago, Illinois

† Nominal depth in inches and weight in pounds per foot.

‡ Depth, width, and thickness in inches.

Fig. 9.13B Properties of Rolled-Steel Shapes (SI Units).

					Axis *X-X*			Axis *Y-Y*		
	Designation	Area mm^2	Depth mm	Width mm	$\bar{I}_x$ 10^6 mm^4	$\bar{k}_x$ mm	$\bar{y}$ mm	$\bar{I}_y$ 10^6 mm^4	$\bar{k}_y$ mm	$\bar{x}$ mm
W Shapes (Wide-Flange Shapes)	W460 × 113†	14400	463	280	554	196.3		63.3	66.3	
	W410 × 85	10800	417	181	316	170.7		17.94	40.6	
	W360 × 57	7230	358	172	160.2	149.4		11.11	39.4	
	W200 × 46.1	5890	203	203	45.8	88.1		15.44	51.3	
S Shapes (American Standard Shapes)	S460 × 81.4†	10390	457	152	335	179.6		8.66	29.0	
	S310 × 47.3	6032	305	127	90.7	122.7		3.90	25.4	
	S250 × 37.8	4806	254	118	51.6	103.4		2.83	24.2	
	S150 × 18.6	2362	152	84	9.2	62.2		0.758	17.91	
C Shapes (American Standard Channels)	C310 × 30.8†	3929	305	74	53.7	117.1		1.615	20.29	17.73
	C250 × 22.8	2897	254	65	28.1	98.3		0.949	18.11	16.10
	C200 × 17.1	2181	203	57	13.57	79.0		0.549	15.88	14.50
	C150 × 12.2	1548	152	48	5.45	59.4		0.288	13.64	13.00
Angles	L152 × 152 × 25.4‡	7100			14.78	45.6	47.2	14.78	45.6	47.2
	L102 × 102 × 12.7	2420			2.31	30.9	30.0	2.31	30.9	30.0
	L76 × 76 × 6.4	929			0.516	23.6	21.4	0.516	23.6	21.4
	L152 × 102 × 12.7	3060			7.24	48.6	50.5	2.61	29.2	25.1
	L127 × 76 × 12.7	2420			3.93	40.3	44.5	1.074	21.1	19.05
	L76 × 51 × 6.4	768			0.454	24.3	25.2	0.163	14.58	12.52

†Nominal depth in millimeters and mass in kilograms per meter.

‡Depth, width, and thickness in millimeters.

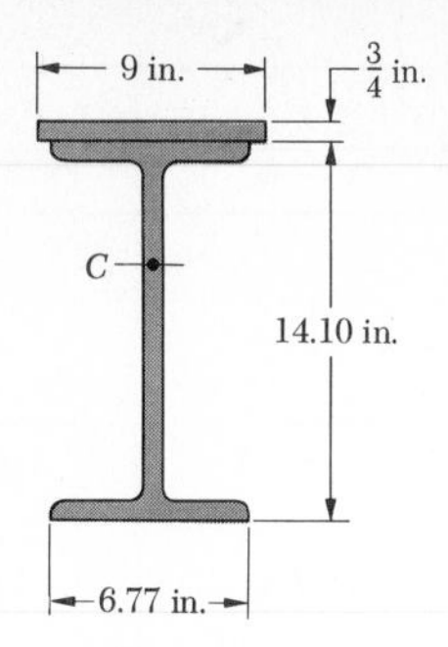

SAMPLE PROBLEM 9.4

The strength of a W14 × 38 rolled-steel beam is increased by attaching a $9 \times \frac{3}{4}$ in. plate to its upper flange as shown. Determine the moment of inertia and the radius of gyration of the composite section with respect to an axis through its centroid C and parallel to the plate.

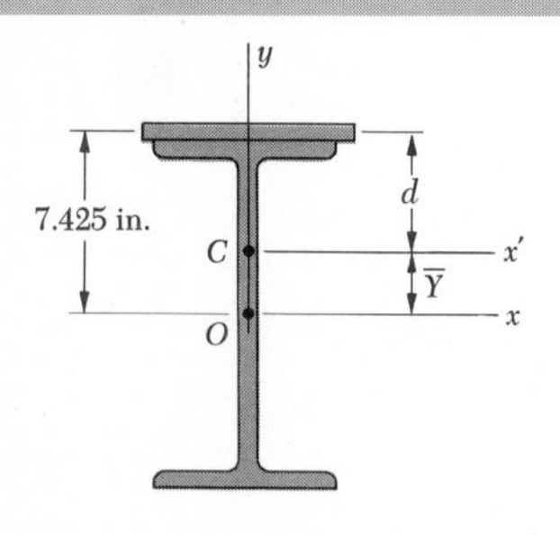

Solution. The origin of coordinates O is placed at the centroid of the wide-flange shape, and the distance $\overline{Y}$ to the centroid of the composite section is computed by the methods of Chap. 5. The area of the wide-flange shape is found by referring to Fig. 9.13*A*. The area and y coordinate of the centroid of the plate are

$$A = (9 \text{ in.})(0.75 \text{ in.}) = 6.75 \text{ in}^2$$
$$\overline{y} = \tfrac{1}{2}(14.10 \text{ in.}) + \tfrac{1}{2}(0.75 \text{ in.}) = 7.425 \text{ in.}$$

Section	Area, in^2	$\overline{y}$, in.	$\overline{y}A$, in^3
Plate	6.75	7.425	50.12
Wide-flange shape	11.20	0	0
	17.95	. . .	50.12

$$\overline{Y}\Sigma A = \Sigma\overline{y}A \qquad \overline{Y}(17.95) = 50.12 \qquad \overline{Y} = 2.792 \text{ in.}$$

Moment of Inertia. The parallel-axis theorem is used to determine the moments of inertia of the wide-flange shape and of the plate with respect to the x' axis. This axis is a centroidal axis for the composite section, but *not* for either of the elements considered separately. The value of $\overline{I}_x$ for the wide-flange shape is obtained from Fig. 9.13*A*.

For the wide-flange shape,

$$I_{x'} = \overline{I}_x + A\overline{Y}^2 = 385 + (11.20)(2.792)^2 = 472.3 \text{ in}^4$$

For the plate,

$$I_{x'} = \overline{I}_x + Ad^2 = (\tfrac{1}{12})(9)(\tfrac{3}{4})^3 + (6.75)(7.425 - 2.792)^2 = 145.2 \text{ in}^4$$

For the composite area,

$$I_{x'} = 472.3 + 145.2 = 617.5 \text{ in}^4$$

$$I_{x'} = 618 \text{ in}^4 \quad ◄$$

Radius of Gyration

$$k_{x'}^2 = \frac{I_{x'}}{A} = \frac{617.5 \text{ in}^4}{17.95 \text{ in}^2}$$

$$k_{x'} = 5.87 \text{ in.} \quad ◄$$

SAMPLE PROBLEM 9.5

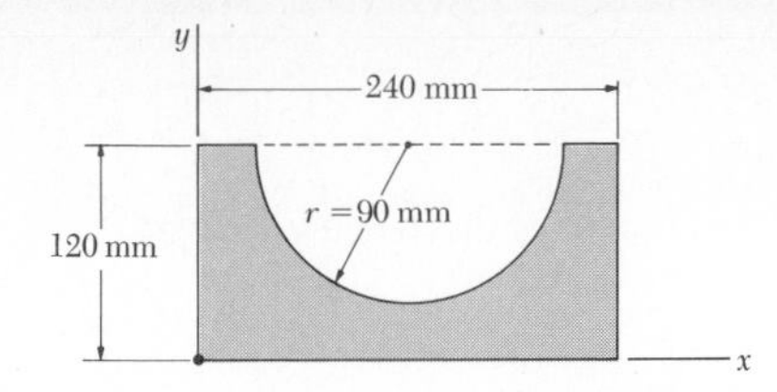

Determine the moment of inertia of the shaded area with respect to the x axis.

Solution. The given area may be obtained by subtracting a half circle from a rectangle. The moments of inertia of the rectangle and of the half circle will be computed separately.

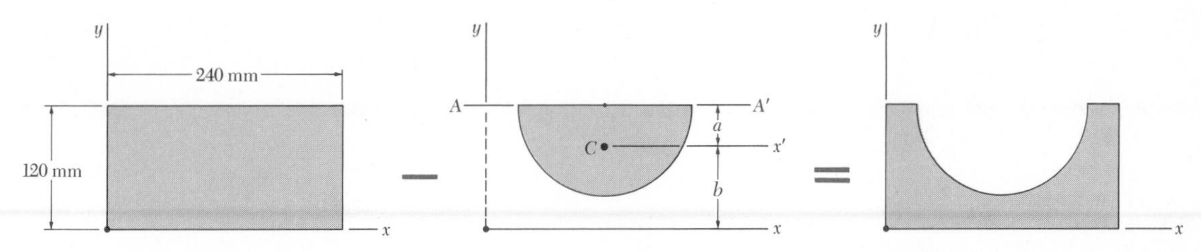

I_x for Rectangle. Referring to Fig. 9.12, we obtain

$$I_x = \tfrac{1}{3}bh^3 = \tfrac{1}{3}(240 \text{ mm})(120 \text{ mm})^3 = 138.2 \times 10^6 \text{ mm}^4$$

I_x for the Half Circle. Referring to Fig. 5.8, we locate the centroid C of the half circle with respect to diameter AA'.

$$a = \frac{4r}{3\pi} = \frac{(4)(90 \text{ mm})}{3\pi} = 38.2 \text{ mm}$$

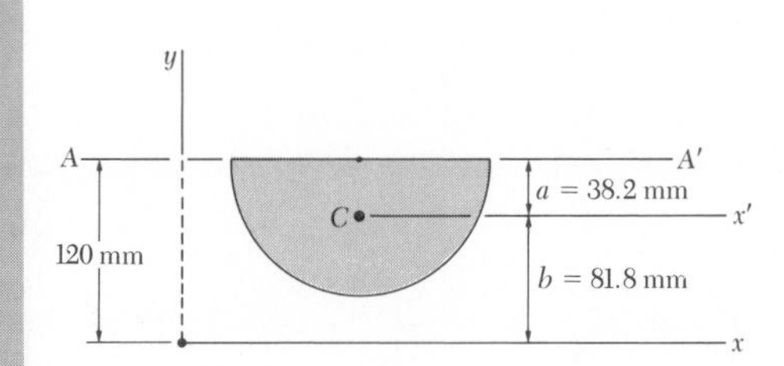

The distance b from the centroid C to the x axis is

$$b = 120 \text{ mm} - a = 120 \text{ mm} - 38.2 \text{ mm} = 81.8 \text{ mm}$$

Referring now to Fig. 9.12, we compute the moment of inertia of the half circle with respect to diameter AA'; we also compute the area of the half circle.

$$I_{AA'} = \tfrac{1}{8}\pi r^4 = \tfrac{1}{8}\pi(90 \text{ mm})^4 = 25.76 \times 10^6 \text{ mm}^4$$
$$A = \tfrac{1}{2}\pi r^2 = \tfrac{1}{2}\pi(90 \text{ mm})^2 = 12.72 \times 10^3 \text{ mm}^2$$

Using the parallel-axis theorem, we obtain the value of $\bar{I}_{x'}$:

$$I_{AA'} = \bar{I}_{x'} + Aa^2$$
$$25.76 \times 10^6 \text{ mm}^4 = \bar{I}_{x'} + (12.72 \times 10^3 \text{ mm}^2)(38.2 \text{ mm})^2$$
$$\bar{I}_{x'} = 7.20 \times 10^6 \text{ mm}^4$$

Again using the parallel-axis theorem, we obtain the value of I_x:

$$I_x = \bar{I}_{x'} + Ab^2 = 7.20 \times 10^6 \text{ mm}^4 + (12.72 \times 10^3 \text{ mm}^2)(81.8 \text{ mm})^2$$
$$= 92.3 \times 10^6 \text{ mm}^4$$

I_x for Given Area. Subtracting the moment of inertia of the half circle from that of the rectangle, we obtain

$$I_x = 138.2 \times 10^6 \text{ mm}^4 - 92.3 \times 10^6 \text{ mm}^4$$
$$I_x = 45.9 \times 10^6 \text{ mm}^4 \quad ◄$$

Problems

9.22 and 9.24 Determine the moment of inertia and the radius of gyration of the shaded area with respect to the x axis.

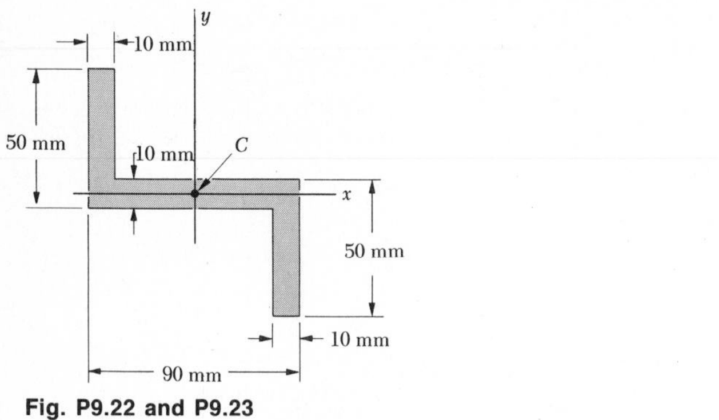

Fig. P9.22 and P9.23

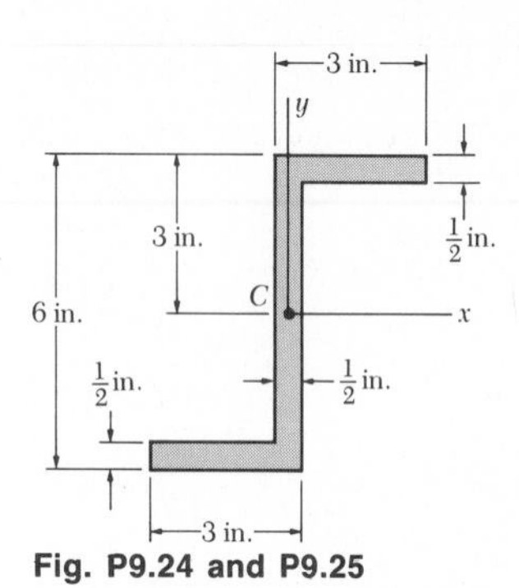

Fig. P9.24 and P9.25

9.23 and 9.25 Determine the moment of inertia and the radius of gyration of the shaded area with respect to the y axis.

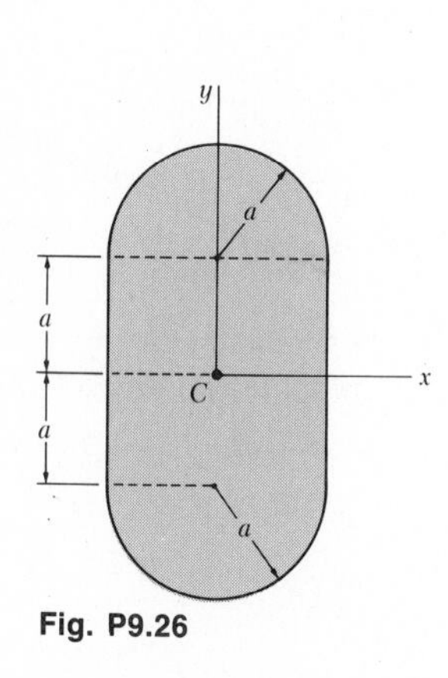

Fig. P9.26

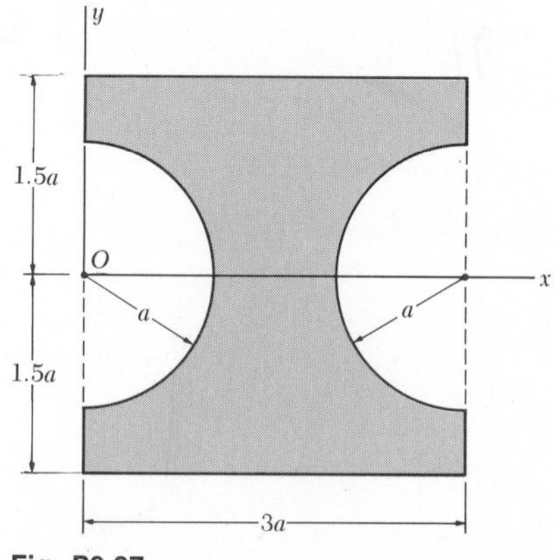

Fig. P9.27

9.26 and 9.27 Determine the moments of inertia of the shaded area shown with respect to the x and y axes when $a = 20$ mm.

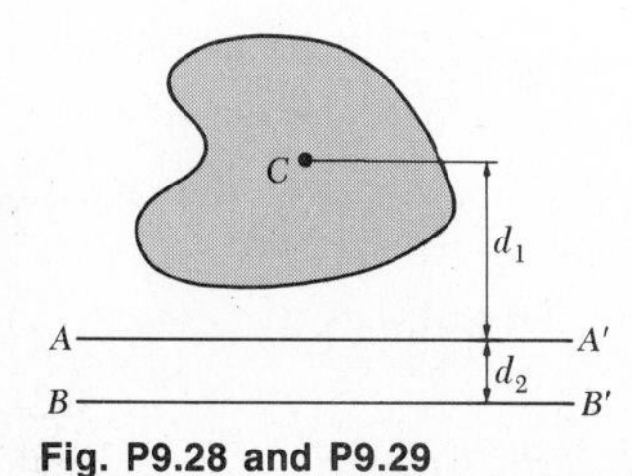

Fig. P9.28 and P9.29

9.28 Determine the shaded area and its moment of inertia with respect to a centroidal axis parallel to AA', knowing that its moments of inertia with respect to AA' and BB' are respectively 2.2×10^6 mm^4 and 4×10^6 mm^4, and that $d_1 = 25$ mm and $d_2 = 10$ mm.

9.29 Knowing that the shaded area is equal to 6000 mm^2 and that its moment of inertia with respect to AA' is 18×10^6 mm^4, determine its moment of inertia with respect to BB', for $d_1 = 50$ mm and $d_2 = 10$ mm.

9.30 The shaded area is equal to 5000 mm^2. Determine its centroidal moments of inertia $\bar{I}_x$ and $\bar{I}_y$, knowing that $\bar{I}_y = 2\bar{I}_x$ and that the polar moment of inertia of the area about point A is $J_A = 22.5 \times 10^6$ mm^4.

9.31 The polar moments of inertia of the shaded area with respect to points A, B, and D are, respectively, $J_A = 28.8 \times 10^6$ mm^4, $J_B = 67.2 \times 10^6$ mm^4, and $J_D = 45.6 \times 10^6$ mm^4. Determine the shaded area, its centroidal moment of inertia $\bar{J}_C$, and the distance d from C to D.

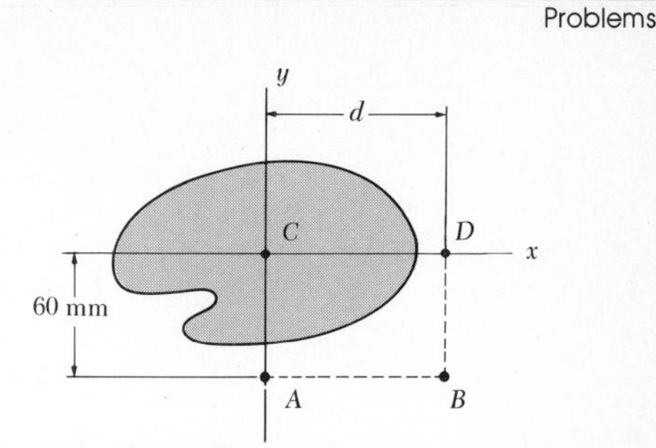

Fig. P9.30 and P9.31

9.32 and 9.33 Determine the moments of inertia $\bar{I}_x$ and $\bar{I}_y$ of the area shown with respect to centroidal axes respectively parallel and perpendicular to the side AB.

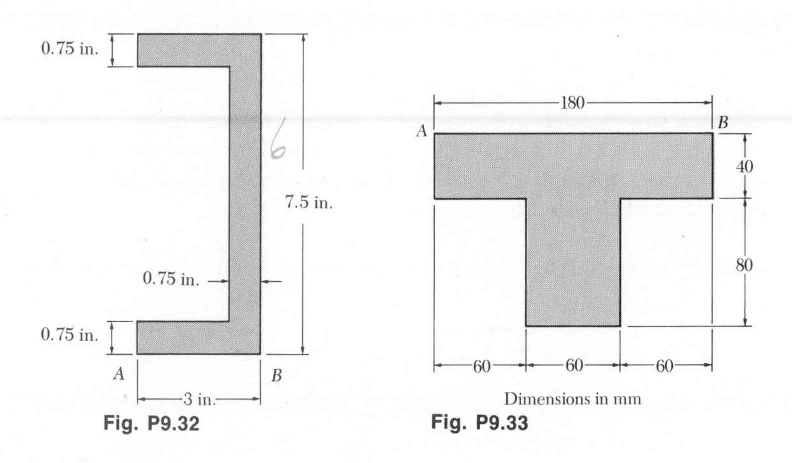

Fig. P9.32

Fig. P9.33

9.34 and 9.35 Determine the polar moment of inertia of the area shown with respect to (a) point O, (b) the centroid of the area.

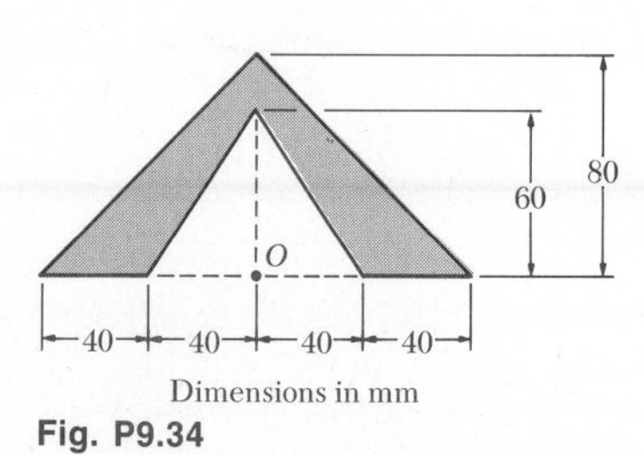

Fig. P9.34

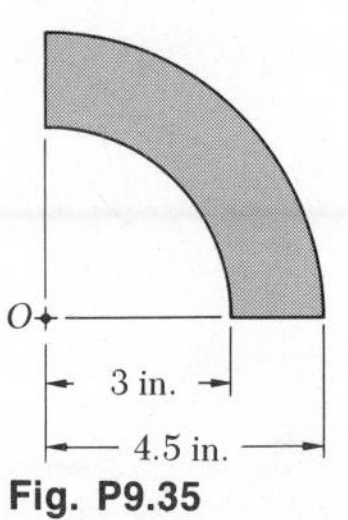

Fig. P9.35

9.36 Two W8 × 31 rolled sections may be welded at A and B in either of the two ways shown. For each arrangement, determine the moment of inertia of the section with respect to the horizontal centroidal axis.

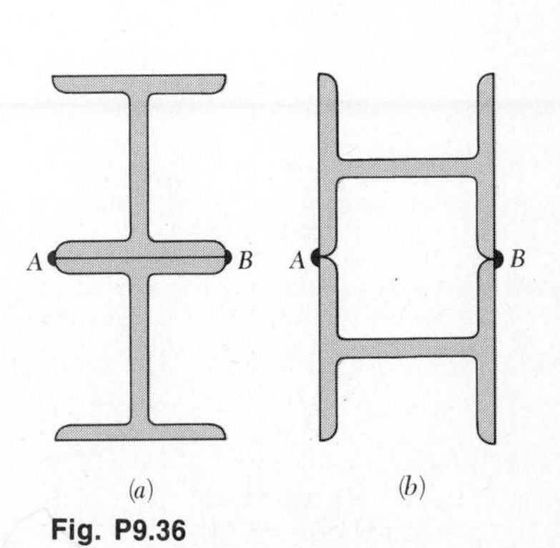

Fig. P9.36

9.37 Two $6 \times 4 \times \frac{1}{2}$ in. angles are welded together to form the section shown. Determine the moments of inertia and the radii of gyration of the section with respect to the centroidal axes shown.

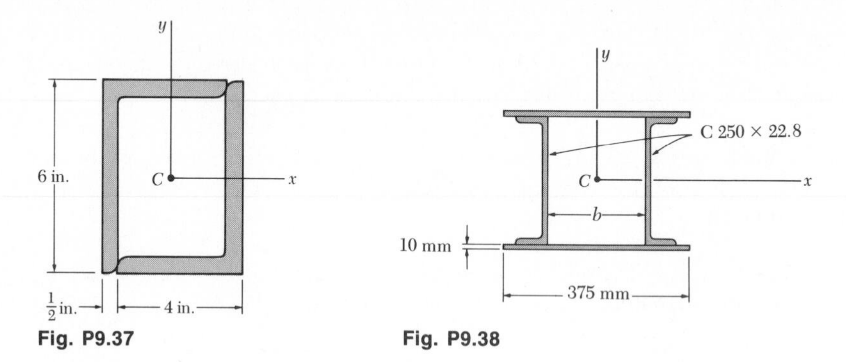

Fig. P9.37

Fig. P9.38

9.38 Two channels and two plates are used to form the column section shown. For $b = 200$ mm, determine the moments of inertia and the radii of gyration of the combined section with respect to the centroidal axes.

9.39 In Prob. 9.38, determine the distance b for which the centroidal moments of inertia $\bar{I}_x$ and $\bar{I}_y$ of the column section are equal.

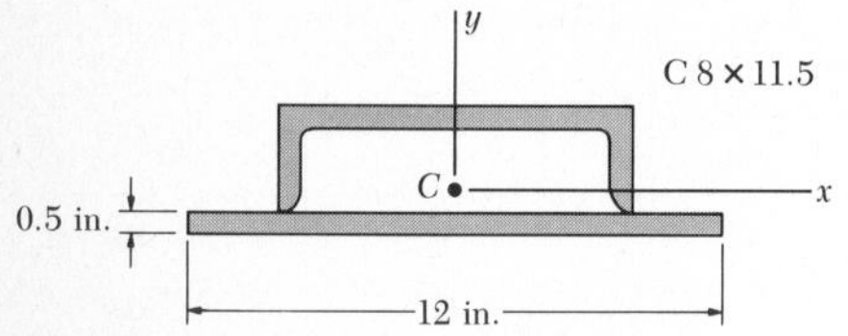

y
C 8 × 11.5
C
x
0.5 in.
12 in.

Fig. P9.40

9.40 A channel and a plate are welded together as shown to form a section which is symmetrical with respect to the y axis. Determine the moments of inertia of the section with respect to its centroidal x and y axes.

9.41 Two $102 \times 102 \times 12.7$ mm angles are welded to a 12-mm steel plate as shown. For $b = 250$ mm, determine the moments of inertia of the combined section with respect to centroidal axes respectively parallel and perpendicular to the plate.

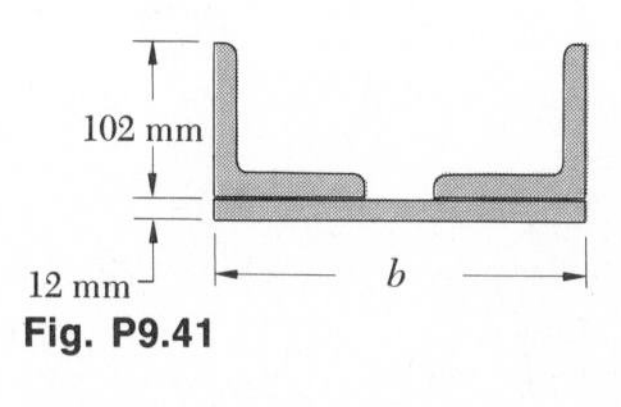

Fig. P9.41

9.42 Solve Prob. 9.41, assuming that $b = 300$ mm.

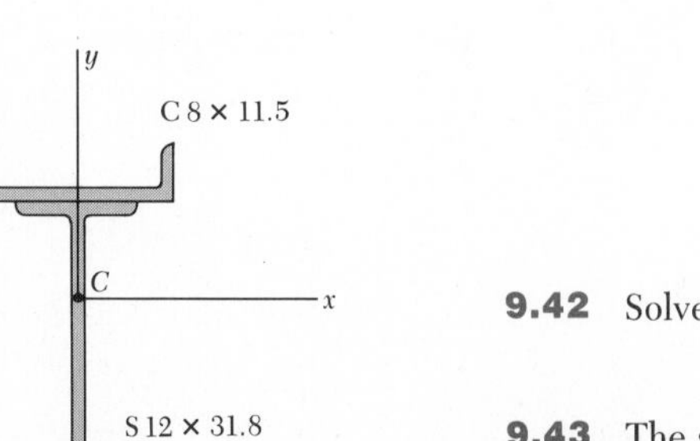

Fig. P9.43

9.43 The strength of the rolled S section shown is increased by welding a channel to its upper flange. Determine the moments of inertia of the combined section with respect to its centroidal x and y axes.

9.44 through 9.46 The panel shown forms the end of a trough which is filled with water to the line AA'. Referring to Sec. 9.2, determine the depth of the point of application of the resultant of the hydrostatic forces acting on the panel (center of pressure).

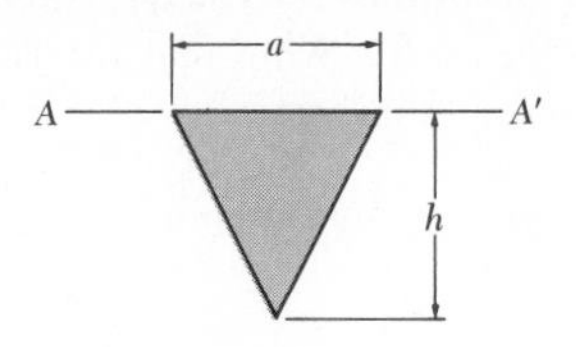

Fig. P9.44

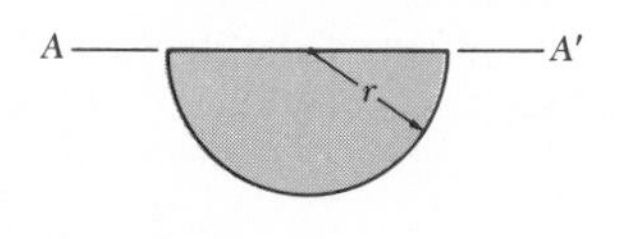

Fig. P9.45

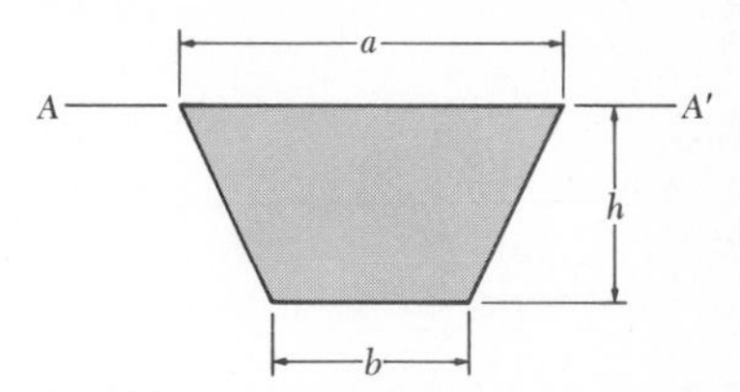

Fig. P9.46

9.47 The center of a vertical circular gate, 1.2 m in diameter, is located 1.8 m below the water surface. The gate is held by three bolts equally spaced as shown. Determine the force on each bolt.

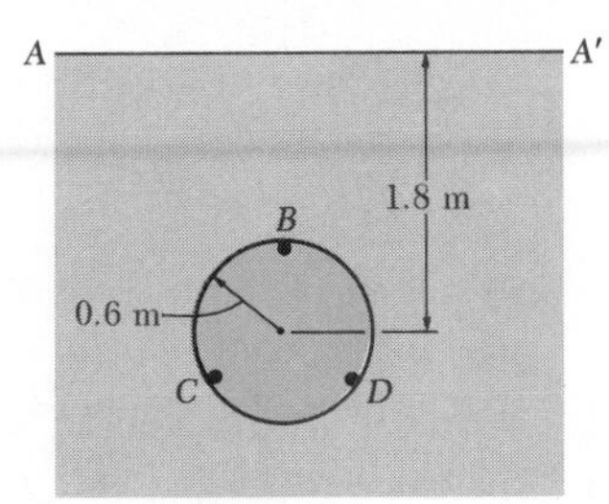

Fig. P9.47

9.48 Solve Prob. 9.47, assuming that the center of the gate is located 1.5 m below the water surface.

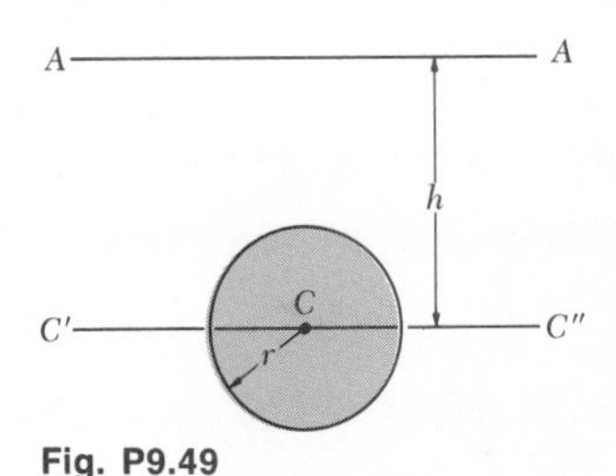

Fig. P9.49

9.49 A vertical circular gate of radius r is hinged about its diameter $C'C''$. Denoting the specific weight of water by γ, determine (*a*) the reaction at each of the two hinges, (*b*) the moment of the couple required to keep the gate closed.

***9.50** Determine the x coordinate of the centroid of the volume shown; this volume was obtained by intersecting a circular cylinder by an oblique plane. (*Hint.* The height of the volume is proportional to the x coordinate; consider an analogy between this height and the water pressure on a submerged surface.)

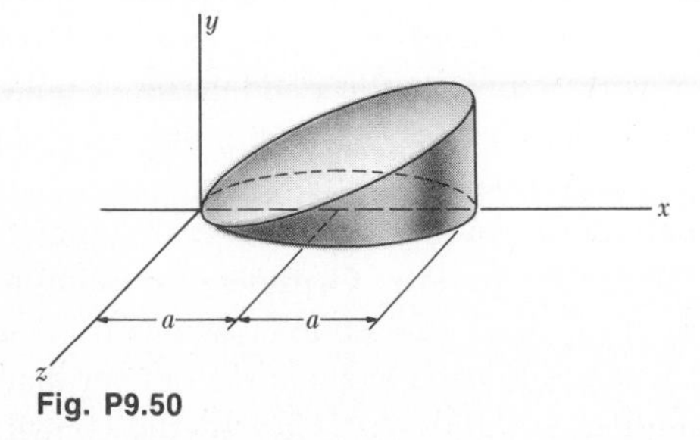

Fig. P9.50

***9.51** Determine the x coordinate of the centroid of the volume shown. (See hint of Prob. 9.50.)

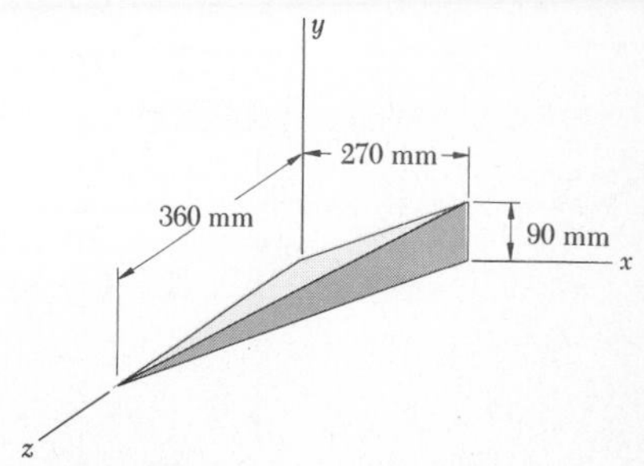

Fig. P9.51

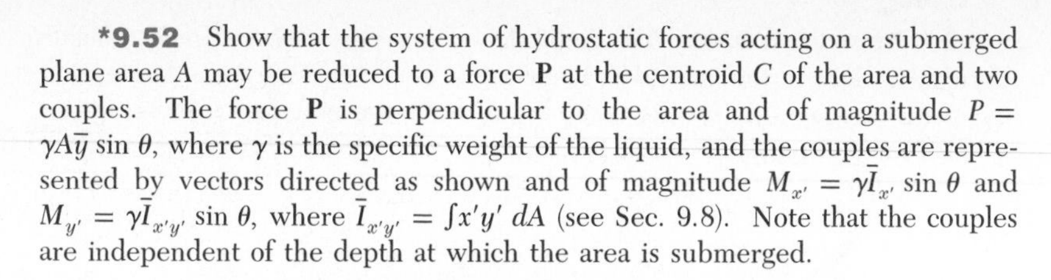

***9.52** Show that the system of hydrostatic forces acting on a submerged plane area A may be reduced to a force **P** at the centroid C of the area and two couples. The force **P** is perpendicular to the area and of magnitude $P = \gamma A\bar{y} \sin\theta$, where γ is the specific weight of the liquid, and the couples are represented by vectors directed as shown and of magnitude $M_{x'} = \gamma \bar{I}_{x'} \sin\theta$ and $M_{y'} = \gamma \bar{I}_{x'y'} \sin\theta$, where $\bar{I}_{x'y'} = \int x'y'\, dA$ (see Sec. 9.8). Note that the couples are independent of the depth at which the area is submerged.

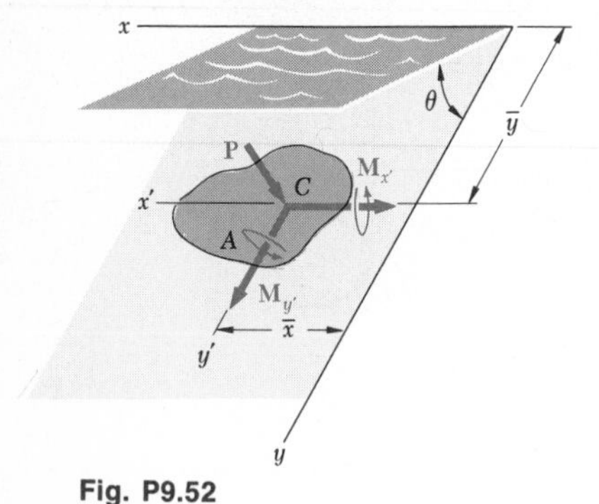

Fig. P9.52

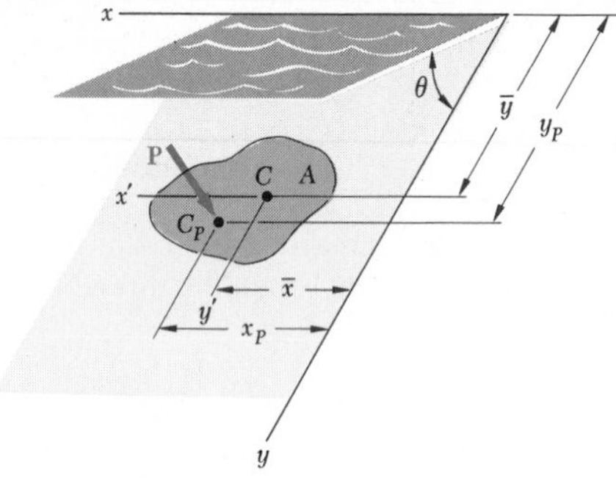

Fig. P9.53

***9.53** Show that the resultant of the hydrostatic forces acting on a submerged plane area A is a force **P** perpendicular to the area and of magnitude $P = \gamma A\bar{y} \sin\theta = \bar{p}A$, where γ is the specific weight of the liquid and $\bar{p}$ the pressure at the centroid C of the area. Show that **P** is applied at a point C_P, called the center of pressure, of coordinates $x_P = I_{xy}/A\bar{y}$ and $y_P = I_x/A\bar{y}$, where $I_{xy} = \int xy\, dA$ (see Sec. 9.8). Show also that the difference of ordinates $y_P - \bar{y}$ is equal to $\bar{k}^2_{x'}/\bar{y}$ and thus depends upon the depth at which the area is submerged.

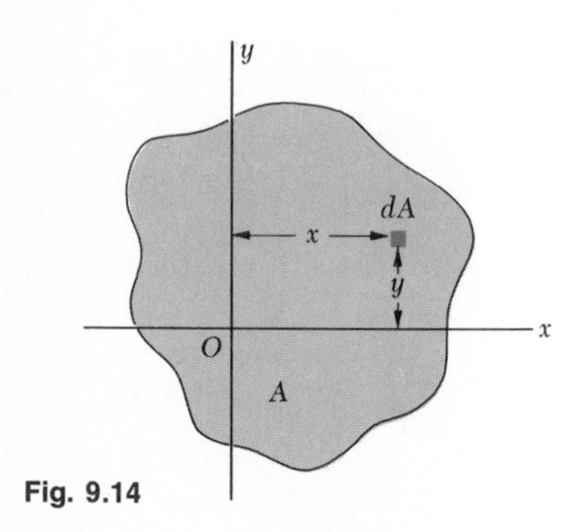

Fig. 9.14

Fig. 9.15

*9.8. Product of Inertia.

The integral

$$I_{xy} = \int xy\, dA \tag{9.12}$$

obtained by multiplying each element dA of an area A by its coordinates x and y and integrating over the area (Fig. 9.14), is known as the *product of inertia* of the area A with respect to the x and y axes. Unlike the moments of inertia I_x and I_y, the product of inertia I_{xy} may be either positive or negative.

When one or both of the x and y axes are axes of symmetry for the area A, the product of inertia I_{xy} is zero. Consider, for example, the channel section shown in Fig. 9.15. Since this section is symmetrical with respect to the x axis, we can associate to each element dA of coordinates x and y an element dA' of coordinates x and $-y$. Clearly, the contributions of any pair of elements chosen in this way cancel out, and the integral (9.12) reduces to zero.

A parallel-axis theorem similar to the one established in Sec. 9.6 for moments of inertia may be derived for products of inertia. Consider an area A and a system of rectangular coordinates x and y (Fig. 9.16). Through the centroid C of the area, of coordinates $\bar{x}$ and $\bar{y}$, we draw two *centroidal axes* x' and y' parallel, respectively, to the x and y axes. Denoting by x and y the coordinates of an element of area dA with respect to the original axes, and by x' and y' the coordinates of the same element with respect to the centroidal axes, we write $x = x' + \bar{x}$ and $y = y' + \bar{y}$. Substituting into (9.12), we obtain the following expression for the product of inertia I_{xy}:

$$\begin{aligned} I_{xy} &= \int xy\,dA = \int (x' + \bar{x})(y' + \bar{y})\,dA \\ &= \int x'y'\,dA + \bar{y}\int x'\,dA + \bar{x}\int y'\,dA + \bar{x}\bar{y}\int dA \end{aligned}$$

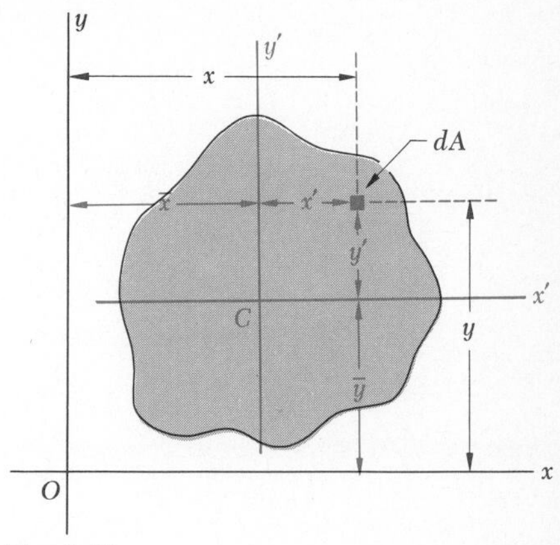

Fig. 9.16

The first integral represents the product of inertia $\bar{I}_{x'y'}$ of the area A with respect to the centroidal axes x' and y'. The next two integrals represent first moments of the area with respect to the centroidal axes; they reduce to zero, since the centroid C is located on these axes. Finally, we observe that the last integral is equal to the total area A. We write therefore

$$I_{xy} = \bar{I}_{x'y'} + \bar{x}\bar{y}A \tag{9.13}$$

* 9.9. Principal Axes and Principal Moments of Inertia.

Consider the area A and the coordinate axes x and y (Fig. 9.17). We assume that the moments and product of inertia

$$I_x = \int y^2\,dA \qquad I_y = \int x^2\,dA \qquad I_{xy} = \int xy\,dA \tag{9.14}$$

of the area A are known, and we propose to determine the moments and product of inertia $I_{x'}$, $I_{y'}$, and $I_{x'y'}$ of A with respect to new axes x' and y' obtained by rotating the original axes about the origin through an angle θ.

We first note the following relations between the coordinates x', y' and x, y of an element of area dA:

$$x' = x\cos\theta + y\sin\theta \qquad y' = y\cos\theta - x\sin\theta$$

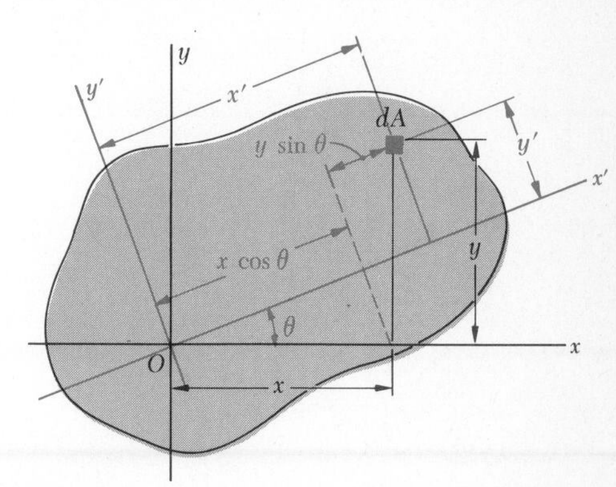

Fig. 9.17

Substituting for y' in the expression for $I_{x'}$, we write

$$\begin{aligned} I_{x'} &= \int (y')^2\,dA = \int (y\cos\theta - x\sin\theta)^2\,dA \\ &= \cos^2\theta\int y^2\,dA - 2\sin\theta\cos\theta\int xy\,dA + \sin^2\theta\int x^2\,dA \end{aligned}$$

Taking the relations (9.14) into account, we write

$$I_{x'} = I_x\cos^2\theta - 2I_{xy}\sin\theta\cos\theta + I_y\sin^2\theta \tag{9.15}$$

Similarly, we obtain for $I_{y'}$ and $I_{x'y'}$ the expressions

$$I_{y'} = I_x\sin^2\theta + 2I_{xy}\sin\theta\cos\theta + I_y\cos^2\theta \tag{9.16}$$

$$I_{x'y'} = (I_x - I_y)\sin\theta\cos\theta + I_{xy}(\cos^2\theta - \sin^2\theta) \tag{9.17}$$

Recalling the trigonometric relations

$$\sin 2\theta = 2 \sin\theta \cos\theta \qquad \cos 2\theta = \cos^2\theta - \sin^2\theta$$

and

$$\cos^2\theta = \frac{1 + \cos 2\theta}{2} \qquad \sin^2\theta = \frac{1 - \cos 2\theta}{2}$$

we may write (9.15), (9.16), and (9.17) as follows:

$$I_{x'} = \frac{I_x + I_y}{2} + \frac{I_x - I_y}{2}\cos 2\theta - I_{xy}\sin 2\theta \tag{9.18}$$

$$I_{y'} = \frac{I_x + I_y}{2} - \frac{I_x - I_y}{2}\cos 2\theta + I_{xy}\sin 2\theta \tag{9.19}$$

$$I_{x'y'} = \frac{I_x - I_y}{2}\sin 2\theta + I_{xy}\cos 2\theta \tag{9.20}$$

We observe, by adding (9.18) and (9.19) member by member, that

$$I_{x'} + I_{y'} = I_x + I_y \tag{9.21}$$

This result could have been anticipated, since both members of (9.21) are equal to the polar moment of inertia J_O.

Equations (9.18) and (9.20) are the parametric equations of a circle. This means that if we choose a set of rectangular axes and plot a point M of abscissa $I_{x'}$ and ordinate $I_{x'y'}$ for any given value of the parameter θ, all the points thus obtained will lie on a circle. To establish this property we shall eliminate θ from Eqs. (9.18) and (9.20); this is done by transposing $(I_x + I_y)/2$ in Eq. (9.18), squaring both members of Eqs. (9.18) and (9.20), and adding. We write

$$\left(I_{x'} - \frac{I_x + I_y}{2}\right)^2 + I_{x'y'}^2 = \left(\frac{I_x - I_y}{2}\right)^2 + I_{xy}^2 \tag{9.22}$$

Setting

$$I_{\text{ave}} = \frac{I_x + I_y}{2} \qquad \text{and} \qquad R = \sqrt{\left(\frac{I_x - I_y}{2}\right)^2 + I_{xy}^2} \tag{9.23}$$

we write the identity (9.22) in the form

$$(I_{x'} - I_{\text{ave}})^2 + I_{x'y'}^2 = R^2 \tag{9.24}$$

which is the equation of a circle of radius R centered at the point C of abscissa I_{ave} and ordinate 0 (Fig. 9.18*a*). It may be observed that Eqs. (9.19) and (9.20) are the parametric equations of the same circle. Furthermore, due to the symmetry of the circle about the horizontal axis, the same result would have been obtained if instead of plotting M, we had plotted a point N of abscissa $I_{y'}$ and ordinate $-I_{x'y'}$ (Fig. 9.18*b*). This property will be used in Sec. 9.10.

The two points A and B where the circle obtained intersects the axis of abscissas are of special interest: Point A corresponds to the maximum value of the moment of inertia $I_{x'}$, while point B corresponds to its minimum

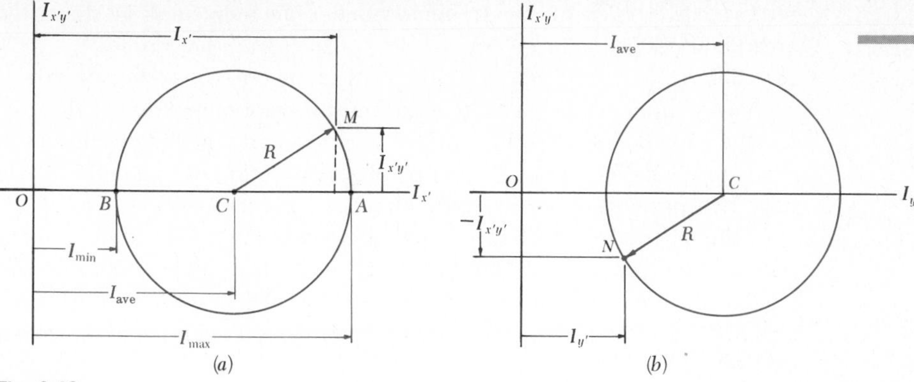

Fig. 9.18

value. Besides, both points correspond to a zero value of the product of inertia $I_{x'y'}$. Thus, the values θ_m of the parameter θ which correspond to the points A and B may be obtained by setting $I_{x'y'} = 0$ in Eq. (9.20). We obtain†

$$\tan 2\theta_m = -\frac{2I_{xy}}{I_x - I_y} \tag{9.25}$$

This equation defines two values $2\theta_m$ which are 180° apart and thus two values θ_m which are 90° apart. One of them corresponds to point A in Fig. 9.18*a* and to an axis through O in Fig. 9.17 with respect to which the moment of inertia of the given area is maximum; the other value corresponds to point B and an axis through O with respect to which the moment of inertia of the area is minimum. The two axes thus defined, which are perpendicular to each other, are called the *principal axes of the area about O*, and the corresponding values I_{max} and I_{min} of the moment of inertia are called the *principal moments of inertia of the area about O*. Since the two values θ_m defined by Eq. (9.25) were obtained by setting $I_{x'y'} = 0$ in Eq. (9.20), it is clear that the product of inertia of the given area with respect to its principal axes is zero.

We observe from Fig. 9.18*a* that

$$I_{max} = I_{ave} + R \qquad \text{and} \qquad I_{min} = I_{ave} - R \tag{9.26}$$

Substituting for I_{ave} and R from formulas (9.23), we write

$$I_{max,min} = \frac{I_x + I_y}{2} \pm \sqrt{\left(\frac{I_x - I_y}{2}\right)^2 + I_{xy}^2} \tag{9.27}$$

Unless it is possible to tell by inspection which of the two principal axes corresponds to I_{max} and which corresponds to I_{min}, it is necessary to substitute one of the values of θ_m into Eq. (9.18) in order to determine which of

†This relation may also be obtained by differentiating $I_{x'}$ in Eq. (9.18) and setting $dI_{x'}/d\theta = 0$.

the two corresponds to the maximum value of the moment of inertia of the area about O.

Referring to Sec. 9.8, we note that if an area possesses an axis of symmetry through a point O, this axis must be a principal axis of the area about O. On the other hand, a principal axis does not need to be an axis of symmetry; whether or not an area possesses properties of symmetry, it will have two principal axes of inertia about any point O.

The properties established here hold for any point O located inside or outside the given area. If the point O is chosen to coincide with the centroid of the area, any axis through O is a centroidal axis; the two principal axes of the area about its centroid are referred to as the *principal centroidal axes of the area.*

***9.10. Mohr's Circle for Moments and Products of Inertia.** The circle used in the preceding section to illustrate the relations existing between the moments and products of inertia of a given area with respect to axes passing through a fixed point O was first introduced by the German engineer Otto Mohr (1835–1918) and is known as *Mohr's circle.* We shall see that if the moments and product of inertia of an area A are known with respect to two rectangular x and y axes through a point O, Mohr's circle may be used to determine graphically (*a*) the principal axes and principal moments of inertia of the area about O, or (*b*) the moments and product of inertia of the area with respect to any other pair of rectangular axes x' and y' through O.

Consider a given area A and two rectangular coordinate axes x and y (Fig. 9.19*a*). We shall assume that the moments of inertia I_x and I_y and the product of inertia I_{xy} are known, and we shall represent them on a diagram by plotting a point X of coordinates I_x and I_{xy} and a point Y of coordinates I_y and $-I_{xy}$ (Fig. 9.19*b*). If I_{xy} is positive, as assumed in Fig. 9.19*a*, point X is located above the horizontal axis and point Y below, as shown in Fig. 9.19*b*. If I_{xy} is negative, X is located below the horizontal axis and Y above. Joining X and Y by a straight line, we define the point C of intersection of line XY with the horizontal axis and draw the circle of center C and diameter XY. Noting that the abscissa of C and the radius of the circle are respectively equal to the quantities I_{ave} and R defined by the formulas (9.23), we conclude that the circle obtained is Mohr's circle for the given area about point O. Thus the abscissas of the points A and B where the circle intersects the horizontal axes represent respectively the principal moments of inertia I_{max} and I_{min} of the area.

We also note that, since $\tan(XCA) = 2I_{xy}/(I_x - I_y)$, the angle XCA is equal in magnitude to one of the angles $2\theta_m$ which satisfy Eq. (9.25); thus the angle θ_m which defines in Fig. 9.19*a* the principal axis Oa corresponding to point A in Fig. 9.19*b* may be obtained by dividing in half the angle XCA measured on Mohr's circle. We further observe that if $I_x > I_y$ and $I_{xy} > 0$, as in the case considered here, the rotation which brings CX into CA is clockwise. But, in that case, the angle θ_m obtained from Eq. (9.25) and defining the principal axis Oa in Fig. 9.19*a* is negative; thus the rota-

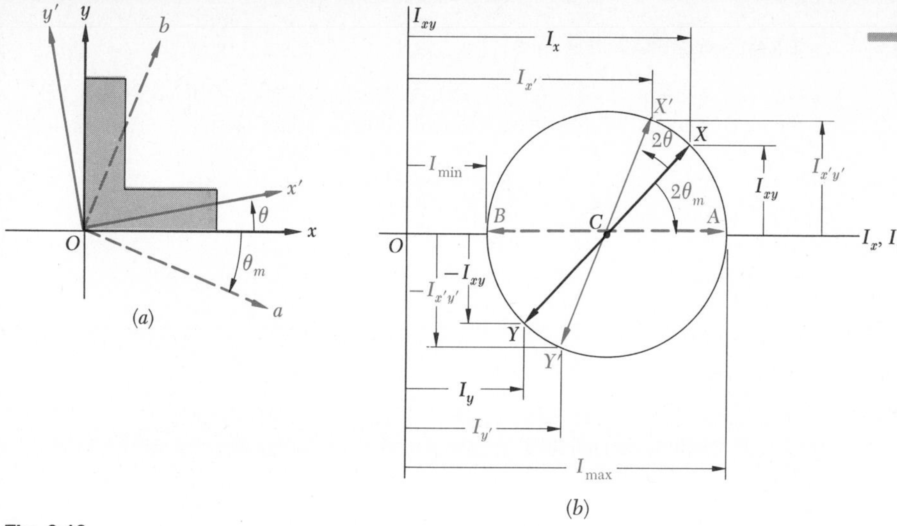

Fig. 9.19

tion bringing Ox into Oa is also clockwise. We conclude that the senses of rotation in both parts of Fig. 9.19 are the same; if a clockwise rotation through $2\theta_m$ is required to bring CX into CA on Mohr's circle, a clockwise rotation through θ_m will bring Ox into the corresponding principal axis Oa in Fig. 9.19*a*.

Since Mohr's circle is uniquely defined, the same circle may be obtained by considering the moments and product of inertia of the area A with respect to the rectangular axes x' and y' (Fig. 9.19*a*). The point X' of coordinates $I_{x'}$ and $I_{x'y'}$ and the point Y' of coordinates $I_{y'}$ and $-I_{x'y'}$ are therefore located on Mohr's circle, and the angle $X'CA$ in Fig. 9.19*b* must be equal to twice the angle $x'Oa$ in Fig. 9.19*a*. Since, as noted above, the angle XCA is twice the angle xOa, it follows that the angle XCX' in Fig. 9.19*b* is twice the angle xOx' in Fig. 9.19*a*. The diameter $X'Y'$ defining the moments and product of inertia $I_{x'}$, $I_{y'}$, and $I_{x'y'}$ of the given area with respect to rectangular axes x' and y' forming an angle θ with the x and y axes may be obtained by rotating through an angle 2θ the diameter XY corresponding to the moments and product of inertia I_x, I_y, and I_{xy}. We note that the rotation which brings the diameter XY into the diameter $X'Y'$ in Fig. 9.19*b* has the same sense as the rotation which brings the xy axes into the $x'y'$ axes in Fig. 9.19*a*.

It should be noted that the use of Mohr's circle is not limited to graphical solutions, i.e., to solutions based on the careful drawing and measuring of the various parameters involved. By merely sketching Mohr's circle and using trigonometry, one may easily derive the various relations required for a numerical solution of a given problem. Actual computations may then be carried out on a calculator (see Sample Prob. 9.8).

SAMPLE PROBLEM 9.6

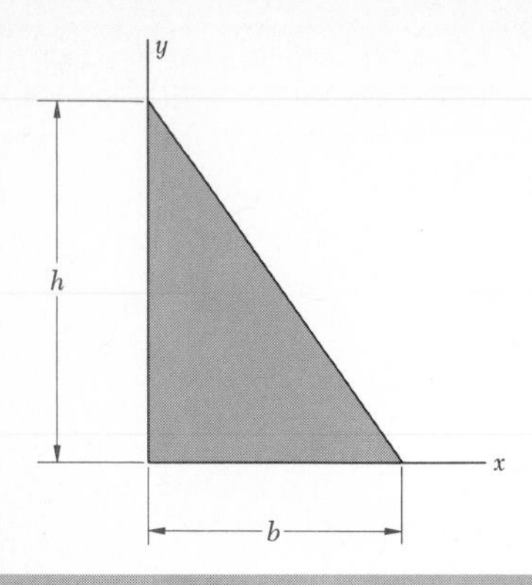

Determine the product of inertia of the right triangle shown (*a*) with respect to the x and y axes and (*b*) with respect to centroidal axes parallel to the x and y axes.

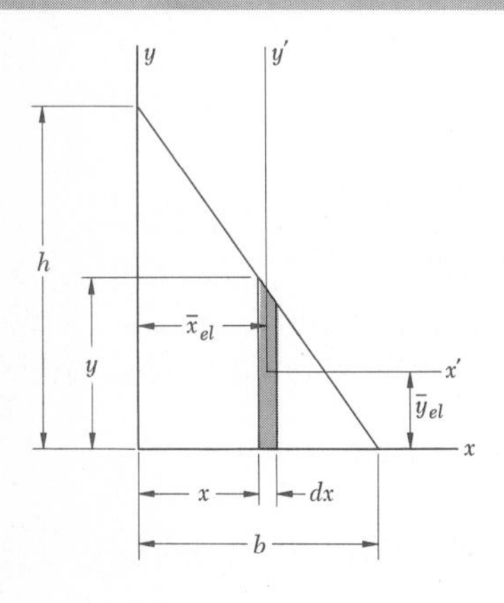

a. Product of Inertia I_{xy}**.** A vertical rectangular strip is chosen as the differential element of area. Using the parallel-axis theorem, we write

$$dI_{xy} = dI_{x'y'} + \bar{x}_{el}\bar{y}_{el}\, dA$$

Since the element is symmetrical with respect to the x' and y' axes, we note that $dI_{x'y'} = 0$. From the geometry of the triangle, we obtain

$$y = h\left(1 - \frac{x}{b}\right) \qquad dA = y\,dx = h\left(1 - \frac{x}{b}\right)dx$$

$$\bar{x}_{el} = x \qquad \bar{y}_{el} = \tfrac{1}{2}y = \tfrac{1}{2}h\left(1 - \frac{x}{b}\right)$$

Integrating dI_{xy} from $x = 0$ to $x = b$, we obtain

$$I_{xy} = \int dI_{xy} = \int \bar{x}_{el}\bar{y}_{el}\, dA = \int_0^b x(\tfrac{1}{2})h^2\left(1 - \frac{x}{b}\right)^2 dx$$

$$= h^2 \int_0^b \left(\frac{x}{2} - \frac{x^2}{b} + \frac{x^3}{2b^2}\right) dx = h^2\left[\frac{x^2}{4} - \frac{x^3}{3b} + \frac{x^4}{8b^2}\right]_0^b$$

$$I_{xy} = \tfrac{1}{24}b^2h^2 \quad ◀$$

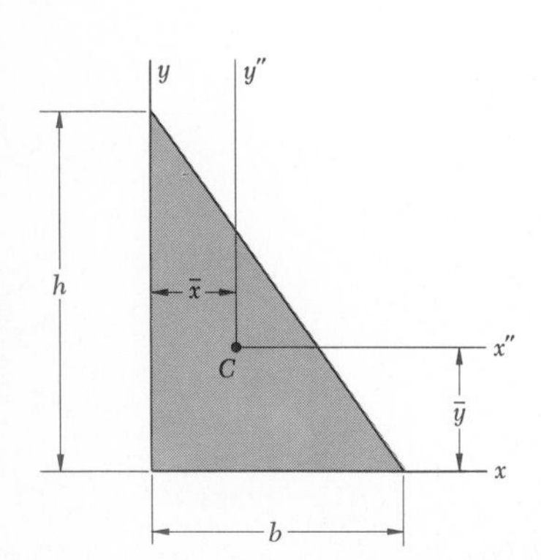

b. Product of Inertia $\bar{I}_{x''y''}$**.** The coordinates of the centroid of the triangle are

$$\bar{x} = \tfrac{1}{3}b \qquad \bar{y} = \tfrac{1}{3}h$$

Using the expression for I_{xy} obtained in part *a*, we apply the parallel-axis theorem and write

$$I_{xy} = \bar{I}_{x''y''} + \bar{x}\bar{y}A$$

$$\tfrac{1}{24}b^2h^2 = \bar{I}_{x''y''} + (\tfrac{1}{3}b)(\tfrac{1}{3}h)(\tfrac{1}{2}bh)$$

$$\bar{I}_{x''y''} = \tfrac{1}{24}b^2h^2 - \tfrac{1}{18}b^2h^2$$

$$\bar{I}_{x''y''} = -\tfrac{1}{72}b^2h^2 \quad ◀$$

SAMPLE PROBLEM 9.7

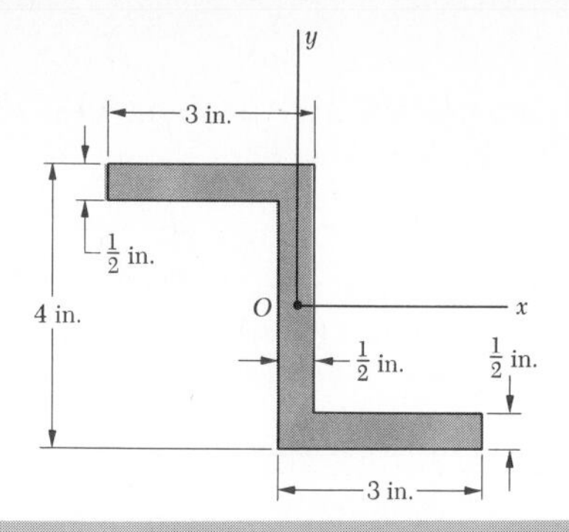

For the section shown, the moments of inertia with respect to the x and y axes have been computed and are known to be

$$I_x = 10.38 \text{ in}^4 \qquad I_y = 6.97 \text{ in}^4$$

Determine (*a*) the principal axes of the section about O, (*b*) the values of the principal moments of inertia of the section about O.

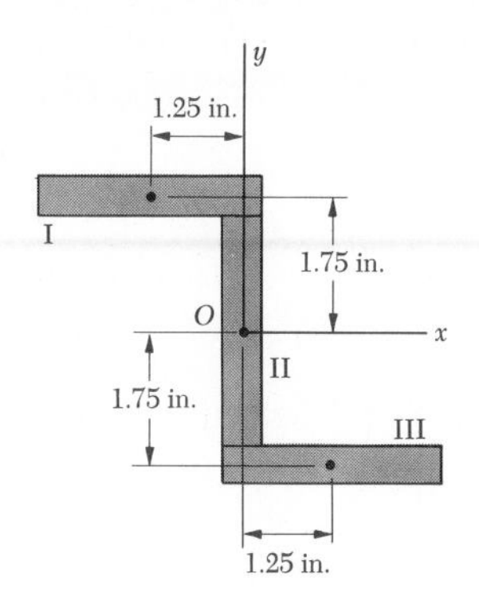

Solution. We first compute the product of inertia with respect to the x and y axes. The area is divided into three rectangles as shown. We note that the product of inertia $\bar{I}_{x'y'}$ with respect to centroidal axes parallel to the x and y axes is zero for each rectangle. Using the parallel-axis theorem $I_{xy} = \bar{I}_{x'y'} + \bar{x}\bar{y}A$, we thus find that for each rectangle, I_{xy} reduces to $\bar{x}\bar{y}A$.

Rectangle	Area, in^2	$\bar{x}$, in.	$\bar{y}$, in.	$\bar{x}\bar{y}A$, in^4
I	1.5	−1.25	+1.75	−3.28
II	1.5	0	0	0
III	1.5	+1.25	−1.75	−3.28
				−6.56

$$I_{xy} = \Sigma\bar{x}\bar{y}A = -6.56 \text{ in}^4$$

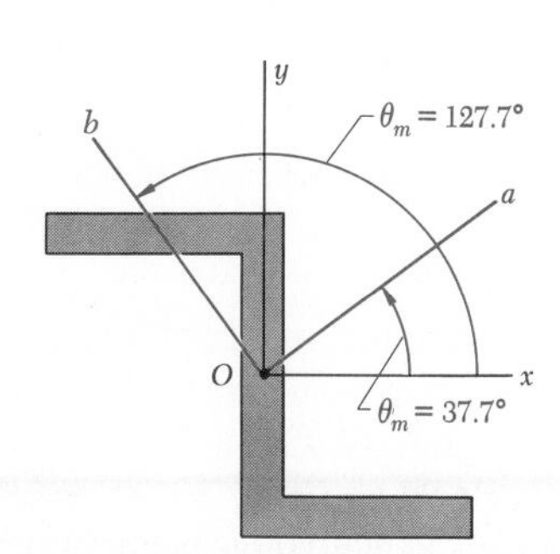

a. Principal Axes. Since the magnitudes of I_x, I_y, and I_{xy} are known, Eq. (9.25) is used to determine the values of θ_m:

$$\tan 2\theta_m = -\frac{2I_{xy}}{I_x - I_y} = -\frac{2(-6.56)}{10.38 - 6.97} = +3.85$$

$$2\theta_m = 75.4° \text{ and } 255.4°$$

$$\theta_m = 37.7° \quad \text{and} \quad \theta_m = 127.7° \blacktriangleleft$$

b. Principal Moments of Inertia. Using Eq. (9.27), we write

$$I_{\max,\min} = \frac{I_x + I_y}{2} \pm \sqrt{\left(\frac{I_x - I_y}{2}\right)^2 + I_{xy}^2}$$

$$= \frac{10.38 + 6.97}{2} \pm \sqrt{\left(\frac{10.38 - 6.97}{2}\right)^2 + (-6.56)^2}$$

$$I_{\max} = 15.45 \text{ in}^4 \qquad I_{\min} = 1.897 \text{ in}^4 \blacktriangleleft$$

Noting that the area of the section is farther away from the a axis than from the b axis, we conclude that $I_a = I_{\max} = 15.45$ in^4 and $I_b = I_{\min} = 1.897$ in^4. This conclusion may be verified by substituting $\theta = 37.7°$ into Eqs. (9.18) and (9.19).

SAMPLE PROBLEM 9.8

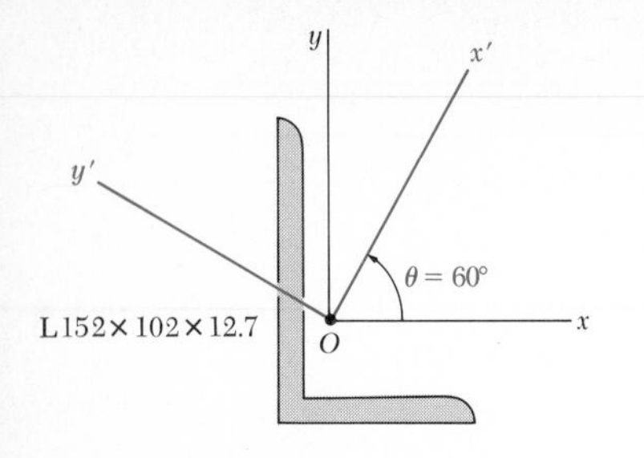

For the section shown, the moments and product of inertia with respect to the x and y axes have been computed and are known to be

$$I_x = 7.24 \times 10^6 \text{ mm}^4 \qquad I_y = 2.61 \times 10^6 \text{ mm}^4 \qquad I_{xy} = -2.54 \times 10^6 \text{mm}^4$$

Using Mohr's circle, determine (*a*) the principal axes of the section about O, (*b*) the values of the principal moments of inertia of the section about O, (*c*) the moments and product of inertia of the section with respect to the x' and y' axes forming an angle of 60° with the x and y axes.

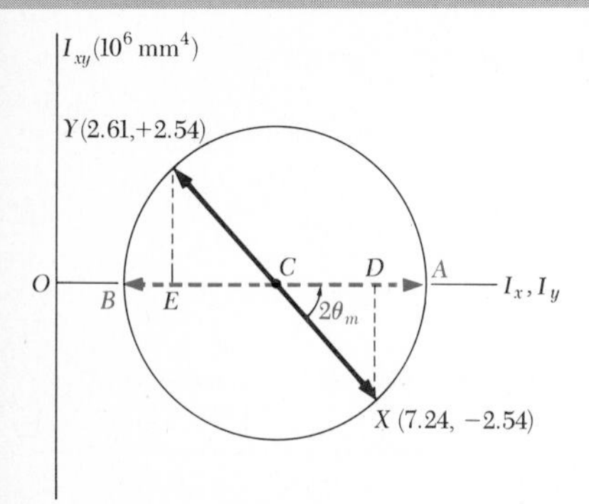

Solution. We first plot point X of coordinates $I_x = 7.24$, $I_{xy} = -2.54$, and point Y of coordinates $I_y = 2.61$, $-I_{xy} = +2.54$. Joining X and Y by a straight line, we define the center C of Mohr's circle. The abscissa of C, which represents I_{ave}, and the radius R of the circle may be measured directly or calculated as follows:

$$I_{\text{ave}} = OC = \tfrac{1}{2}(I_x + I_y) = \tfrac{1}{2}(7.24 \times 10^6 + 2.61 \times 10^6) = 4.925 \times 10^6 \text{ mm}^4$$
$$CD = \tfrac{1}{2}(I_x - I_y) = \tfrac{1}{2}(7.24 \times 10^6 - 2.61 \times 10^6) = 2.315 \times 10^6 \text{ mm}^4$$
$$R = \sqrt{(CD)^2 + (DX)^2} = \sqrt{(2.315 \times 10^6)^2 + (2.54 \times 10^6)^2} = 3.437 \times 10^6 \text{ mm}^4$$

a. Principal Axes. The principal axes of the section correspond to points A and B on Mohr's circle and the angle through which we should rotate CX to bring it into CA defines $2\theta_m$. We have

$$\tan 2\theta_m = \frac{DX}{CD} = \frac{2.54}{2.315} = 1.097 \qquad 2\theta_m = 47.6°\circlearrowleft \qquad \theta_m = 23.8°\circlearrowleft \quad ◀$$

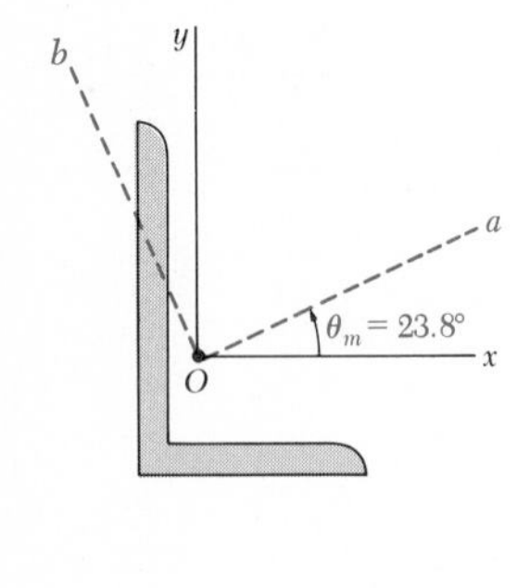

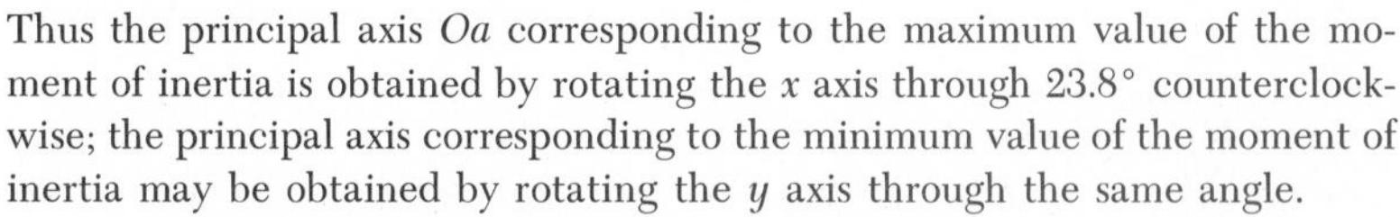

Thus the principal axis Oa corresponding to the maximum value of the moment of inertia is obtained by rotating the x axis through 23.8° counterclockwise; the principal axis corresponding to the minimum value of the moment of inertia may be obtained by rotating the y axis through the same angle.

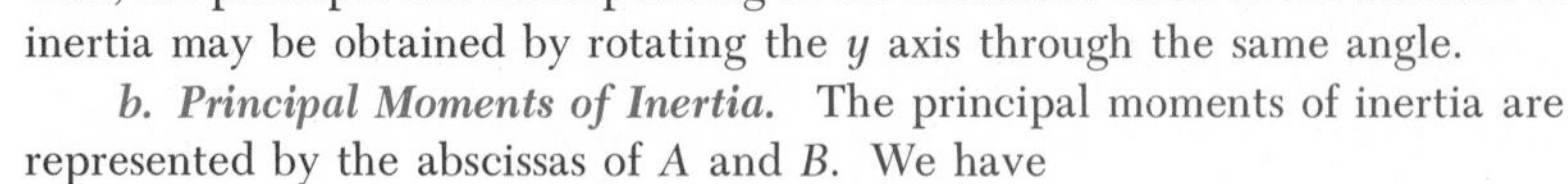

b. Principal Moments of Inertia. The principal moments of inertia are represented by the abscissas of A and B. We have

$$I_{\text{max}} = OA = OC + CA = I_{\text{ave}} + R = (4.925 + 3.437)10^6 \text{ mm}^4$$
$$I_{\text{max}} = 8.36 \times 10^6 \text{ mm}^4 \quad ◀$$
$$I_{\text{min}} = OB = OC - BC = I_{\text{ave}} - R = (4.925 - 3.437)10^6 \text{ mm}^4$$
$$I_{\text{min}} = 1.49 \times 10^6 \text{ mm}^4 \quad ◀$$

c. Moments and Product of Inertia with Respect to x′y′ Axes. The points X' and Y' on Mohr's circle which correspond to the x' and y' axes are obtained by rotating CX and CY through an angle $2\theta = 2(60°) = 120°$ counterclockwise. The coordinates of X' and Y' yield the desired moments and product of inertia. Noting that the angle that CX' forms with the horizontal axis is $\phi = 120° - 47.6° = 72.4°$, we write

$$I_{x'} = OF = OC + CF = 4.925 \times 10^6 \text{ mm}^4 + (3.437 \times 10^6 \text{ mm}^4) \cos 72.4°$$
$$I_{x'} = 5.96 \times 10^6 \text{ mm}^4 \quad ◀$$
$$I_{y'} = OG = OC - GC = 4.925 \times 10^6 \text{ mm}^4 - (3.437 \times 10^6 \text{ mm}^4) \cos 72.4°$$
$$I_{y'} = 3.89 \times 10^6 \text{ mm}^4 \quad ◀$$
$$I_{x'y'} = FX' = (3.437 \times 10^6 \text{ mm}^4) \sin 72.4°$$
$$I_{x'y'} = 3.28 \times 10^6 \text{ mm}^4 \quad ◀$$

Problems

9.54 through 9.57 Determine by direct integration the product of inertia of the given area with respect to the x and y axes.

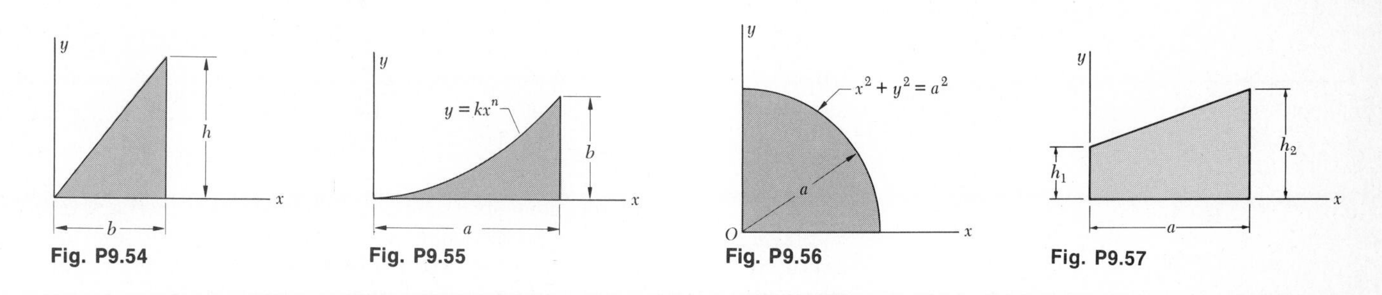

Fig. P9.54

Fig. P9.55

Fig. P9.56

Fig. P9.57

9.58 through 9.63 Using the parallel-axis theorem, determine the product of inertia of the area shown with respect to the centroidal x and y axes.

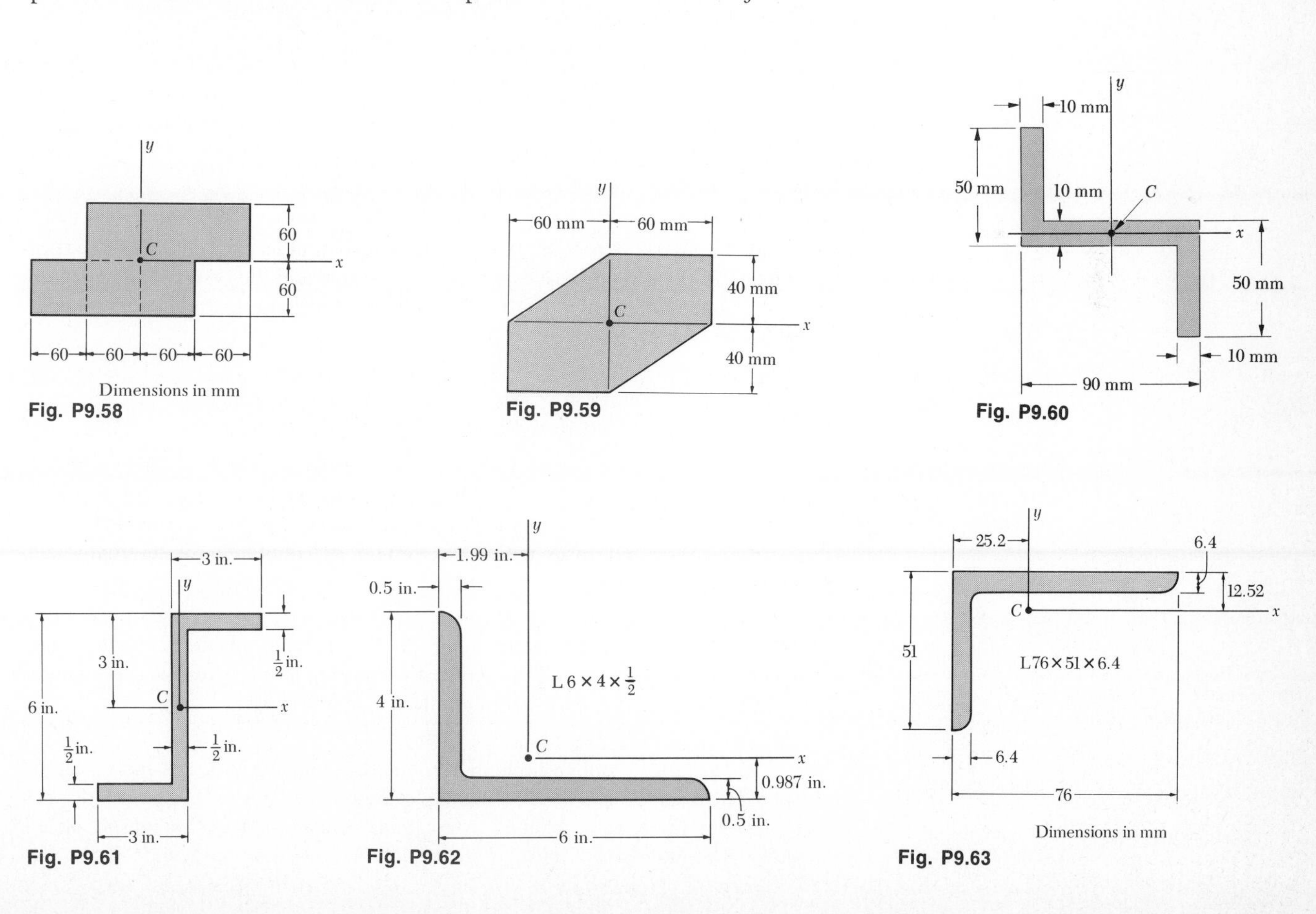

Fig. P9.58

Fig. P9.59

Fig. P9.60

Fig. P9.61

Fig. P9.62

Fig. P9.63

9.64 Determine the moments of inertia and the product of inertia of the quarter circle of Prob. 9.56 with respect to new axes obtained by rotating the x and y axes about O (*a*) through 30° counterclockwise, (*b*) through 45° counterclockwise.

9.65 Determine the moments of inertia and the product of inertia of the area of Prob. 9.61 with respect to new centroidal axes obtained by rotating the x and y axes through 30° counterclockwise.

9.66 Determine the moments of inertia and the product of inertia of the L $6 \times 4 \times \frac{1}{2}$ in. angle cross section of Prob. 9.62 with respect to new centroidal axes obtained by rotating the x and y axes through 30° clockwise. (The moments of inertia $\bar{I}_x$ and $\bar{I}_y$ of the section are given in Fig. 9.13*A*.)

9.67 Determine the moments of inertia and the product of inertia of the L $76 \times 51 \times 6.4$ mm angle cross section of Prob. 9.63 with respect to new centroidal axes obtained by rotating the x and y axes through 30° clockwise. (The moments of inertia $\bar{I}_x$ and $\bar{I}_y$ of the section are given in Fig. 9.13*B*.)

9.68 through 9.71 For the area indicated, determine the orientation of the principal axes through the centroid and the corresponding values of the moment of inertia.

9.68 Area of Prob. 9.58.
9.69 Area of Prob. 9.59.
9.70 Area of Prob. 9.60.
9.71 Area of Prob. 9.61.

9.72 and 9.73 For the angle cross section indicated, determine the orientation of the principal axes through the centroid and the corresponding values of the moment of inertia. (The moments of inertia $\bar{I}_x$ and $\bar{I}_y$ of the cross sections are given in Fig. 9.13.)

9.72 The L $6 \times 4 \times \frac{1}{2}$ in. angle of Prob. 9.62.
9.73 The L $76 \times 51 \times 6.4$ mm angle of Prob. 9.63.

9.74 Using Mohr's circle, determine the moments of inertia and the product of inertia of the quarter circle of Prob. 9.56 with respect to new axes obtained by rotating the x and y axes about O (*a*) through 30° counterclockwise, (*b*) through 45° counterclockwise.

9.75 Using Mohr's circle, determine the moments of inertia and the product of inertia of the area of Prob. 9.61 with respect to new centroidal axes obtained by rotating the x and y axes through 30° counterclockwise.

9.76 Using Mohr's circle, determine the moments of inertia and the product of inertia of the L $6 \times 4 \times \frac{1}{2}$ in. angle cross section of Prob. 9.62 with respect to new centroidal axes obtained by rotating the x and y axes through 30° clockwise. (The moments of inertia $\bar{I}_x$ and $\bar{I}_y$ of the section are given in Fig. 9.13*A*.)

9.77 Using Mohr's circle, determine the moments of inertia and the product of inertia of the L 76 × 51 × 6.4 mm angle cross section of Prob. 9.63 with respect to new centroidal axes obtained by rotating the x and y axes through 30° clockwise. (The moments of inertia $\bar{I}_x$ and $\bar{I}_y$ of the section are given in Fig. 9.13*B*.)

9.78 through 9.81 Using Mohr's circle, determine for the area indicated the orientation of the principal axes through the centroid and the corresponding values of the moment of inertia.

9.78 Area of Prob. 9.58.
9.79 Area of Prob. 9.59.
9.80 Area of Prob. 9.60.
9.81 Area of Prob. 9.61.

9.82 and 9.83 Using Mohr's circle, determine for the angle cross section indicated the orientation of the principal axes through the centroid and the corresponding values of the moment of inertia. (The moments of inertia $\bar{I}_x$ and $\bar{I}_y$ of the cross sections are given in Fig. 9.13.)

9.82 The L 6 × 4 × $\frac{1}{2}$ in. angle of Prob. 9.62.
9.83 The L 76 × 51 × 6.4 mm angle of Prob. 9.63.

9.84 Using Mohr's circle, show that for any regular polygon (such as a pentagon), (*a*) the moment of inertia with respect to every axis through the centroid is the same, (*b*) the product of inertia with respect to any pair of rectangular axes through the centroid is zero.

9.85 The moments and product of inertia of an L 127 × 76 × 12.7 mm angle cross section with respect to two rectangular axes x and y through C are, respectively, $\bar{I}_x = 3.93 \times 10^6 \text{ mm}^4$, $\bar{I}_y = 1.074 \times 10^6 \text{ mm}^4$, and $\bar{I}_{xy} < 0$, with the minimum value of the moment of inertia of the area with respect to any axis through C being $\bar{I}_{\text{min}} = 0.655 \times 10^6 \text{ mm}^4$. Using Mohr's circle, determine (*a*) the product of inertia $\bar{I}_{xy}$ of the area, (*b*) the orientation of the principal axes, (*c*) the value of $\bar{I}_{\text{max}}$.

***9.86** Prove that the expression $I_{x'}I_{y'} - I^2_{x'y'}$, where $I_{x'}$, $I_{y'}$, and $I_{x'y'}$ represent, respectively, the moments and product of inertia of a given area with respect to two rectangular axes x' and y' through a given point O, is independent of the orientation of the x' and y' axes. Considering the particular case when the x' and y' axes correspond to the maximum value of $I_{x'y'}$, show that the given expression represents the square of the tangent drawn from the origin of the coordinates to Mohr's circle.

***9.87** Using the invariance property established in the preceding problem, express the product of inertia I_{xy} of an area A with respect to two rectangular axes through O in terms of the moments of inertia I_x and I_y of A and of the principal moments of inertia I_{min} and I_{max} of A about O. Apply the formula obtained to calculate the product of inertia $\bar{I}_{xy}$ of the L 5 × 3 × $\frac{1}{2}$ in. angle cross section shown in Fig. 9.13*A*, knowing that its minimum moment of inertia is 1.577 in^4.

MOMENTS OF INERTIA OF MASSES

9.11. Moment of Inertia of a Mass. Consider a small mass Δm mounted on a rod of negligible mass which may rotate freely about an axis AA' (Fig. 9.20*a*). If a couple is applied to the system, the rod and mass, assumed initially at rest, will start rotating about AA'. The details of this motion will be studied later in dynamics. At present, we wish only to indicate that the time required for the system to reach a given speed of rotation is proportional to the mass Δm and to the square of the distance r. The product $r^2 \Delta m$ provides, therefore, a measure of the *inertia* of the system, i.e., of the resistance the system offers when we try to set it in motion. For this reason, the product $r^2 \Delta m$ is called the *moment of inertia* of the mass Δm with respect to the axis AA'.

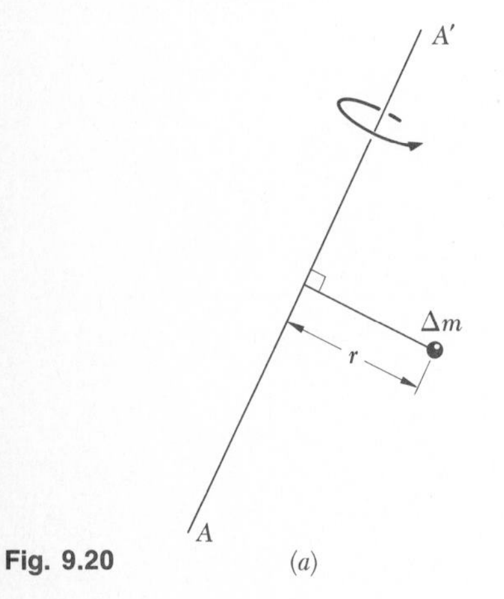

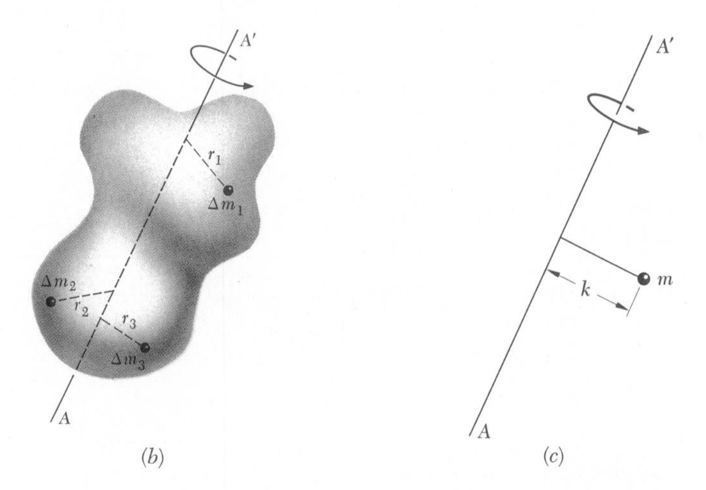

Fig. 9.20 (*a*) (*b*) (*c*)

Consider now a body of mass m which is to be rotated about an axis AA' (Fig. 9.20*b*). Dividing the body into elements of mass Δm_1, Δm_2, etc., we find that the resistance offered by the body is measured by the sum $r_1^2 \Delta m_1 + r_2^2 \Delta m_2 + \cdots$. This sum defines, therefore, the moment of inertia of the body with respect to the axis AA'. Increasing the number of elements, we find that the moment of inertia is equal, at the limit, to the integral

$$I = \int r^2 \, dm \tag{9.28}$$

The *radius of gyration* k of the body with respect to the axis AA' is defined by the relation

$$I = k^2 m \qquad \text{or} \qquad k = \sqrt{\frac{I}{m}} \tag{9.29}$$

The radius of gyration k represents, therefore, the distance at which the entire mass of the body should be concentrated if its moment of inertia with respect to AA' is to remain unchanged (Fig. 9.20*c*). Whether it is kept in its original shape (Fig. 9.20*b*) or whether it is concentrated as shown in Fig. 9.20*c*, the mass m will react in the same way to a rotation, or *gyration*, about AA'.

If SI units are used, the radius of gyration k is expressed in meters and the mass m in kilograms. The moment of inertia of a mass, therefore, will be expressed in $\text{kg}\cdot\text{m}^2$. If U.S. customary units are used, the radius of gyration is expressed in feet and the mass in slugs, i.e., in $\text{lb}\cdot\text{s}^2/\text{ft}$. The moment of inertia of a mass, then, will be expressed in $\text{lb}\cdot\text{ft}\cdot\text{s}^2$.†

The moment of inertia of a body with respect to a coordinate axis may easily be expressed in terms of the coordinates x, y, z of the element of mass dm (Fig. 9.21). Noting, for example, that the square of the distance r from

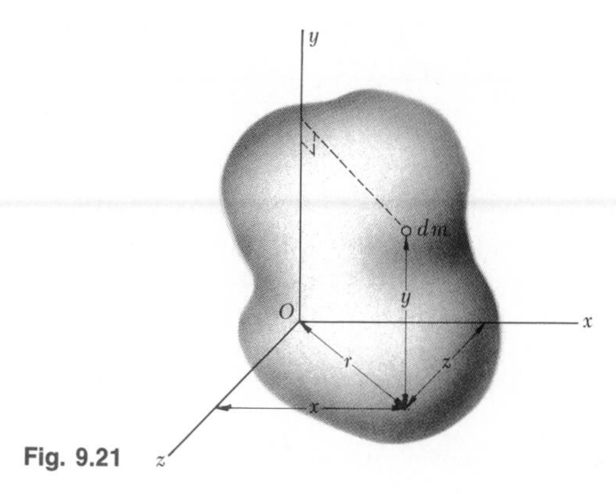

Fig. 9.21

the element dm to the y axis is $z^2 + x^2$, we express the moment of inertia of the body with respect to the y axis as

$$I_y = \int r^2\, dm = \int (z^2 + x^2)\, dm$$

Similar expressions may be obtained for the moments of inertia with respect to the x and z axes. We write

$$\begin{aligned} I_x &= \int (y^2 + z^2)\, dm \\ I_y &= \int (z^2 + x^2)\, dm \\ I_z &= \int (x^2 + y^2)\, dm \end{aligned} \qquad (9.30)$$

†It should be kept in mind when converting the moment of inertia of a mass from U.S. customary units to SI units that the base unit pound used in the derived unit $\text{lb}\cdot\text{ft}\cdot\text{s}^2$ is a unit of force (*not* of mass) and should therefore be converted into newtons. We have

$$1\ \text{lb}\cdot\text{ft}\cdot\text{s}^2 = (4.45\ \text{N})(0.3048\ \text{m})(1\ \text{s})^2 = 1.356\ \text{N}\cdot\text{m}\cdot\text{s}^2$$

or, since $\text{N} = \text{kg}\cdot\text{m/s}^2$,

$$1\ \text{lb}\cdot\text{ft}\cdot\text{s}^2 = 1.356\ \text{kg}\cdot\text{m}^2$$

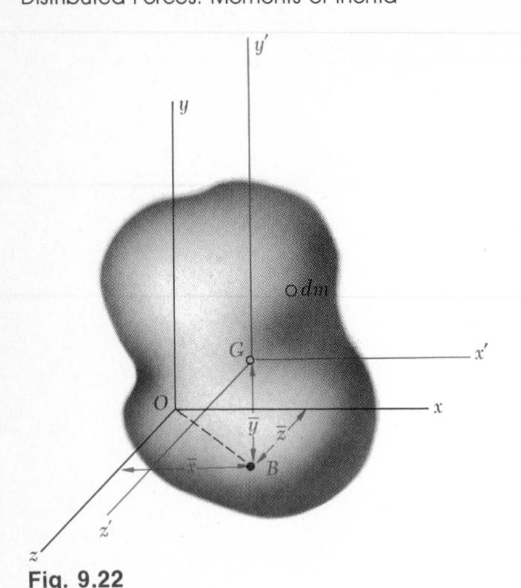

Fig. 9.22

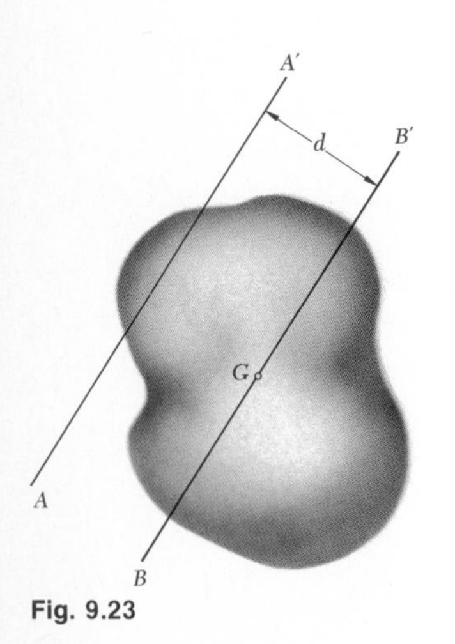

Fig. 9.23

9.12. Parallel-Axis Theorem. Consider a body of mass m. Let $Oxyz$ be a system of rectangular coordinates with origin at an arbitrary point O, and $Gx'y'z'$ a system of parallel *centroidal axes*, i.e., a system with origin at the center of gravity G of the body† and with axes x', y', z', respectively, parallel to x, y, z (Fig. 9.22). Denoting by $\bar{x}$, $\bar{y}$, $\bar{z}$ the coordinates of G with respect to $Oxyz$, we write the following relations between the coordinates x, y, z of the element dm with respect to $Oxyz$ and its coordinates x', y', z' with respect to the centroidal axes $Gx'y'z'$:

$$x = x' + \bar{x} \qquad y = y' + \bar{y} \qquad z = z' + \bar{z} \tag{9.31}$$

Referring to Eqs. (9.30), we may express the moment of inertia of the body with respect to the x axis as follows:

$$\begin{aligned} I_x &= \int (y^2 + z^2)\, dm = \int [(y' + \bar{y})^2 + (z' + \bar{z})^2]\, dm \\ &= \int (y'^2 + z'^2)\, dm + 2\bar{y} \int y'\, dm + 2\bar{z} \int z'\, dm + (\bar{y}^2 + \bar{z}^2) \int dm \end{aligned}$$

The first integral in the expression obtained represents the moment of inertia $\bar{I}_{x'}$ of the body with respect to the centroidal axis x'; the second and third integrals represent the first moment of the body with respect to the $z'x'$ and $x'y'$ planes, respectively, and, since both planes contain G, the two integrals are zero; the last integral is equal to the total mass m of the body. We write, therefore,

$$I_x = \bar{I}_{x'} + m(\bar{y}^2 + \bar{z}^2) \tag{9.32}$$

and, similarly,

$$I_y = \bar{I}_{y'} + m(\bar{z}^2 + \bar{x}^2) \qquad I_z = \bar{I}_{z'} + m(\bar{x}^2 + \bar{y}^2) \tag{9.32'}$$

We easily verify from Fig. 9.22 that the sum $\bar{z}^2 + \bar{x}^2$ represents the square of the distance OB between the y and y' axis. Similarly, $\bar{y}^2 + \bar{z}^2$ and $\bar{x}^2 + \bar{y}^2$ represent the squares of the distances between the x and x' axes, and the z and z' axes, respectively. Denoting by d the distance between an arbitrary axis AA' and a parallel centroidal axis BB' (Fig. 9.23), we may, therefore, write the following general relation between the moment of inertia I of the body with respect to AA' and its moment of inertia $\bar{I}$ with respect to BB':

$$I = \bar{I} + md^2 \tag{9.33}$$

Expressing the moments of inertia in terms of the corresponding radii of gyration, we may also write

$$k^2 = \bar{k}^2 + d^2 \tag{9.34}$$

where k and $\bar{k}$ represent the radii of gyration about AA' and BB', respectively.

†Note that the term centroidal is used to define an axis passing through the center of gravity G of the body, whether or not G coincides with the centroid of the volume of the body.

9.13. Moments of Inertia of Thin Plates. Consider a thin plate of uniform thickness t, made of a homogeneous material of density ρ (density = mass per unit volume). The mass moment of inertia of the plate with respect to an axis AA' *contained in the plane* of the plate (Fig. 9.24a) is

$$I_{AA',\text{mass}} = \int r^2\, dm$$

Since $dm = \rho t\, dA$, we write

$$I_{AA',\text{mass}} = \rho t \int r^2\, dA$$

But r represents the distance of the element of area dA to the axis AA'; the

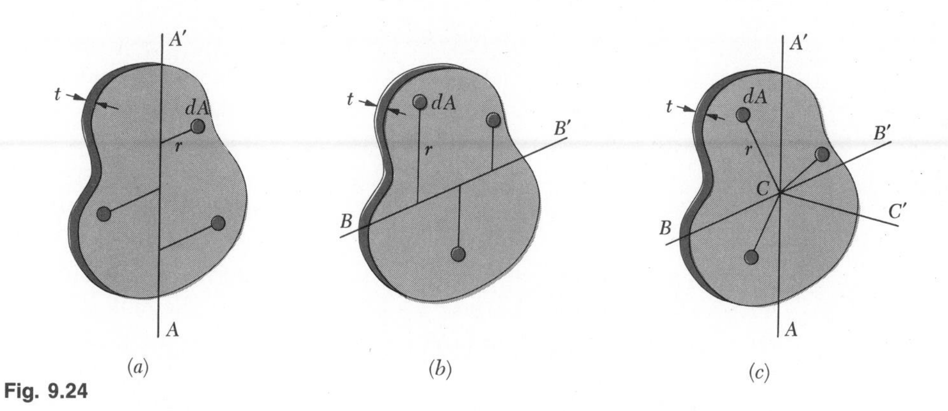

Fig. 9.24

integral is therefore equal to the moment of inertia of the area of the plate with respect to AA'. We have

$$I_{AA',\text{mass}} = \rho t I_{AA',\text{area}} \tag{9.35}$$

Similarly, we have with respect to an axis BB' perpendicular to AA' (Fig. 9.24b)

$$I_{BB',\text{mass}} = \rho t I_{BB',\text{area}} \tag{9.36}$$

Considering now the axis CC' *perpendicular* to the plate through the point of intersection C of AA' and BB' (Fig. 9.24c), we write

$$I_{CC',\text{mass}} = \rho t J_{C,\text{area}} \tag{9.37}$$

where J_C is the *polar* moment of inertia of the area of the plate with respect to point C.

Recalling the relation $J_C = I_{AA'} + I_{BB'}$ existing between polar and rectangular moments of inertia of an area, we write the following relation between the mass moments of inertia of a thin plate:

$$I_{CC'} = I_{AA'} + I_{BB'} \tag{9.38}$$

Rectangular Plate. In the case of a reactangular plate of sides a and b (Fig. 9.25), we obtain the following mass moments of inertia with respect to axes through the center of gravity of the plate:

$$I_{AA',\text{mass}} = \rho t I_{AA',\text{area}} = \rho t(\tfrac{1}{12}a^3 b)$$
$$I_{BB',\text{mass}} = \rho t I_{BB',\text{area}} = \rho t(\tfrac{1}{12}ab^3)$$

Observing that the product ρabt is equal to the mass m of the plate, we write the mass moments of inertia of a thin rectangular plate as follows:

$$I_{AA'} = \tfrac{1}{12}ma^2 \qquad I_{BB'} = \tfrac{1}{12}mb^2 \tag{9.39}$$
$$I_{CC'} = I_{AA'} + I_{BB'} = \tfrac{1}{12}m(a^2 + b^2) \tag{9.40}$$

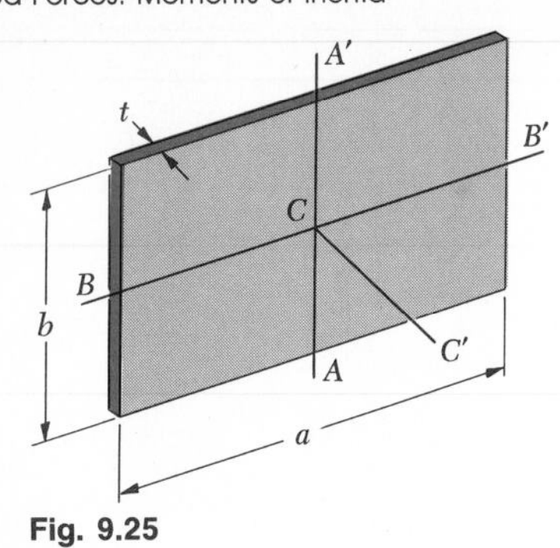

Fig. 9.25

Circular Plate. In the case of a circular plate, or disk, of radius r (Fig. 9.26), we write

$$I_{AA',\text{mass}} = \rho t I_{AA',\text{area}} = \rho t(\tfrac{1}{4}\pi r^4)$$

Observing that the product $\rho\pi r^2 t$ is equal to the mass m of the plate and that $I_{AA'} = I_{BB'}$, we write the mass moments of inertia of a circular plate as follows:

$$I_{AA'} = I_{BB'} = \tfrac{1}{4}mr^2 \tag{9.41}$$
$$I_{CC'} = I_{AA'} + I_{BB'} = \tfrac{1}{2}mr^2 \tag{9.42}$$

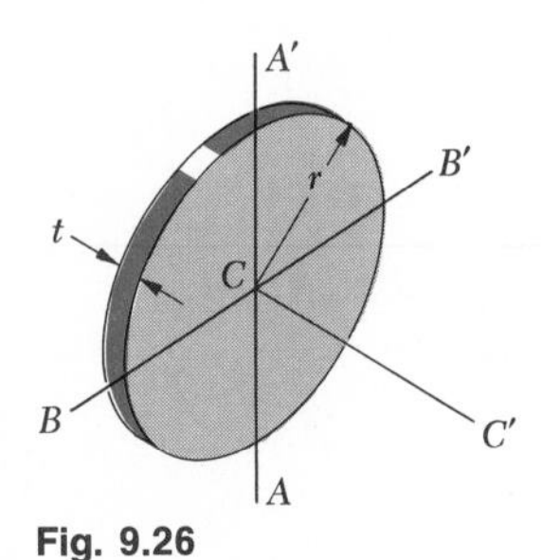

Fig. 9.26

9.14. Determination of the Moment of Inertia of a Three-Dimensional Body by Integration. The moment of inertia of a three-dimensional body is obtained by computing the integral $I = \int r^2\,dm$. If the body is made of a homogeneous material of density ρ, we have $dm = \rho\,dV$ and write $I = \rho\int r^2\,dV$. This integral depends only upon the shape of the body. Thus, in order to compute the moment of inertia of a three-dimensional body, it will generally be necessary to perform a triple, or at least a double, integration.

However, if the body possesses two planes of symmetry, it is usually possible to determine the body's moment of inertia through a single integration by choosing as an element of mass dm the mass of a thin slab perpendicular to the planes of symmetry. In the case of bodies of revolution, for example, the element of mass should be a thin disk (Fig. 9.27). Using formula (9.42), the moment of inertia of the disk with respect to the axis of revolution may be readily expressed as indicated in Fig. 9.27. Its moment of inertia with respect to each of the other two axes of coordinates will be obtained by using formula (9.41) and the parallel-axis theorem. Integration of the expressions obtained will yield the desired moment of inertia of the body of revolution.

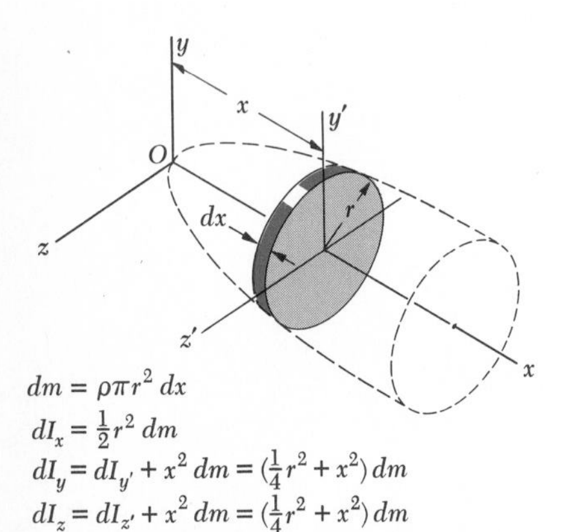

$dm = \rho\pi r^2\,dx$
$dI_x = \frac{1}{2}r^2\,dm$
$dI_y = dI_{y'} + x^2\,dm = (\frac{1}{4}r^2 + x^2)\,dm$
$dI_z = dI_{z'} + x^2\,dm = (\frac{1}{4}r^2 + x^2)\,dm$

Fig. 9.27 Determination of the moment of inertia of a body of revolution.

9.15. Moments of Inertia of Composite Bodies. The moments of inertia of a few common shapes are shown in Fig. 9.28. The moment of inertia with respect to a given axis of a body made of several of these simple shapes may be obtained by computing the moments of inertia of its component parts about the desired axis and adding them together. We should note, as we have already noted in the case of areas, that the radius of gyration of a composite body *cannot* be obtained by adding the radii of gyration of its component parts.

Slender rod	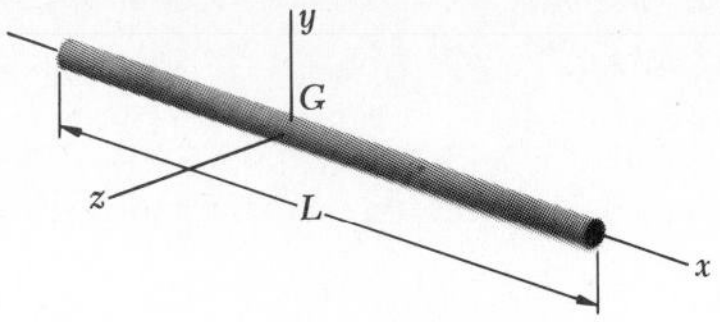	$I_y = I_z = \frac{1}{12}mL^2$
Thin rectangular plate	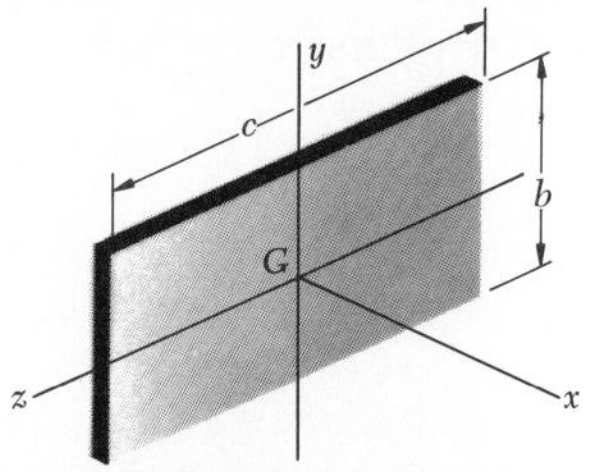	$I_x = \frac{1}{12}m(b^2 + c^2)$ $I_y = \frac{1}{12}mc^2$ $I_z = \frac{1}{12}mb^2$
Rectangular prism	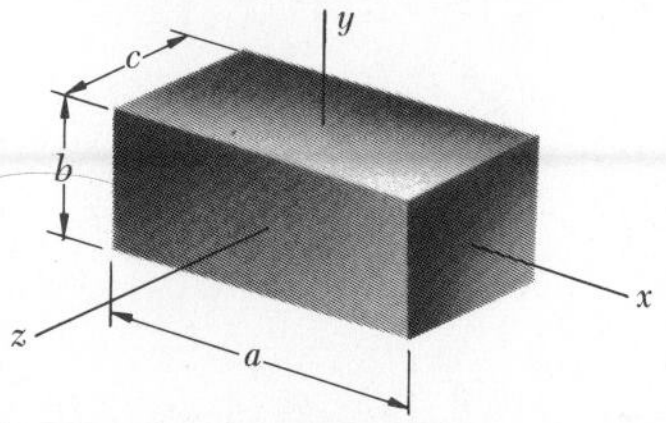	$I_x = \frac{1}{12}m(b^2 + c^2)$ $I_y = \frac{1}{12}m(c^2 + a^2)$ $I_z = \frac{1}{12}m(a^2 + b^2)$
Thin disk	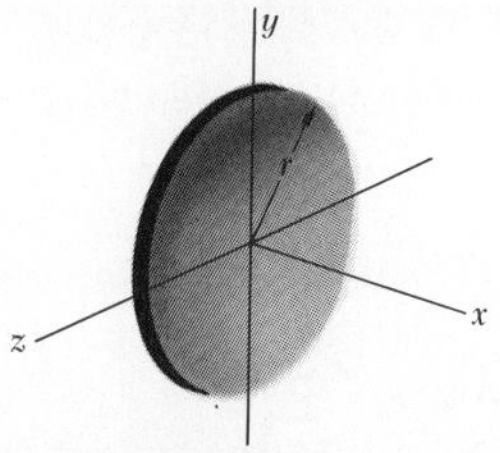	$I_x = \frac{1}{2}mr^2$ $I_y = I_z = \frac{1}{4}mr^2$
Circular cylinder	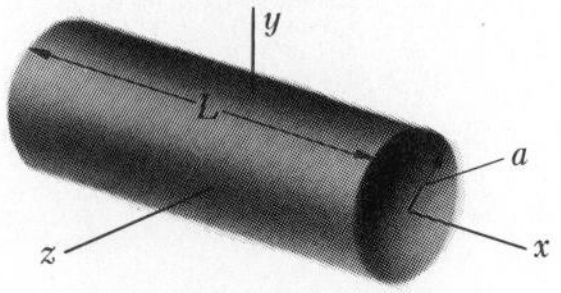	$I_x = \frac{1}{2}ma^2$ $I_y = I_z = \frac{1}{12}m(3a^2 + L^2)$
Circular cone	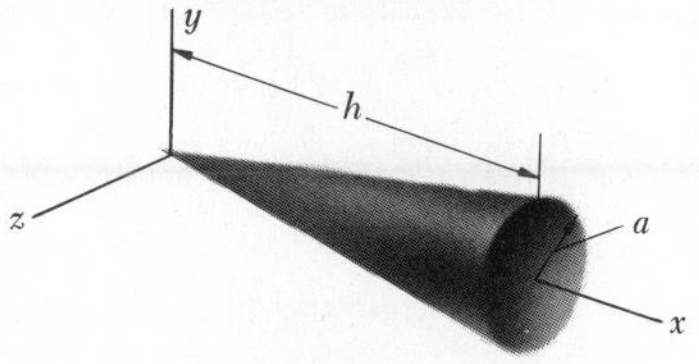	$I_x = \frac{3}{10}ma^2$ $I_y = I_z = \frac{3}{5}m(\frac{1}{4}a^2 + h^2)$
Sphere	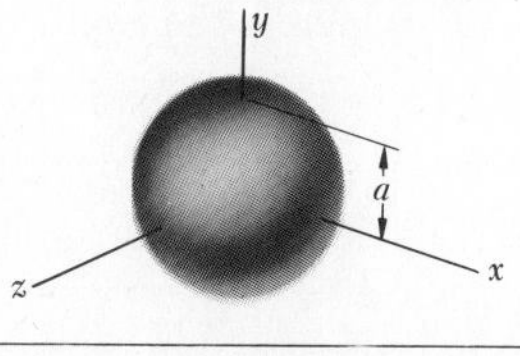	$I_x = I_y = I_z = \frac{2}{5}ma^2$

Fig. 9.28 Mass moments of inertia of common geometric shapes.

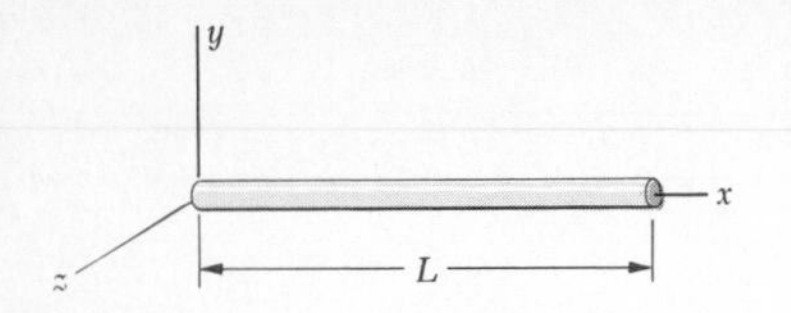

SAMPLE PROBLEM 9.9

Determine the mass moment of inertia of a slender rod of length L and mass m with respect to an axis perpendicular to the rod and passing through one end of the rod.

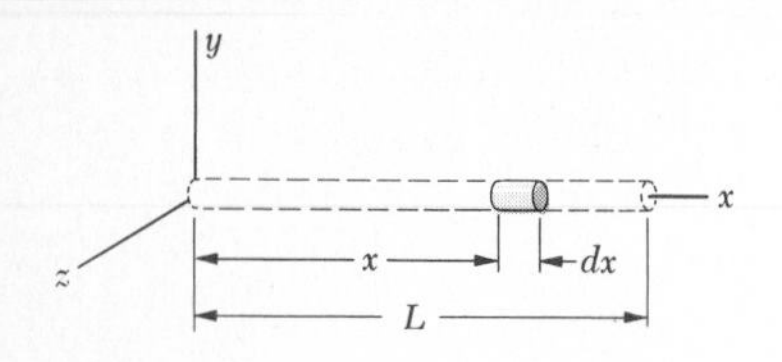

Solution. Choosing the differential element of mass shown, we write

$$dm = \frac{m}{L}\,dx$$

$$I_y = \int x^2\,dm = \int_0^L x^2 \frac{m}{L}\,dx = \left[\frac{m}{L}\frac{x^3}{3}\right]_0^L \qquad I_y = \tfrac{1}{3}mL^2 \blacktriangleleft$$

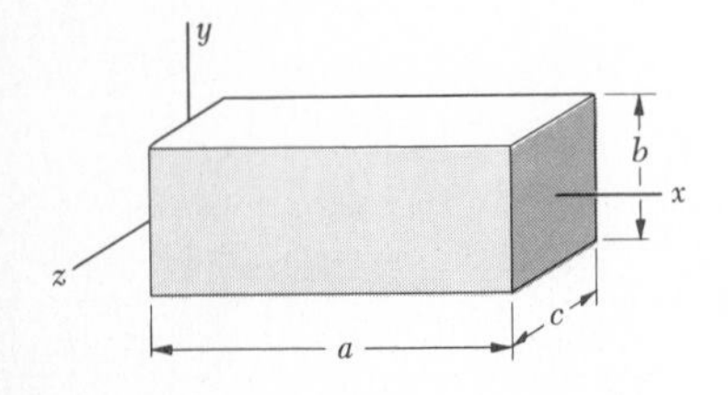

SAMPLE PROBLEM 9.10

Determine the mass moment of inertia of the homogeneous rectangular prism shown with respect to the z axis.

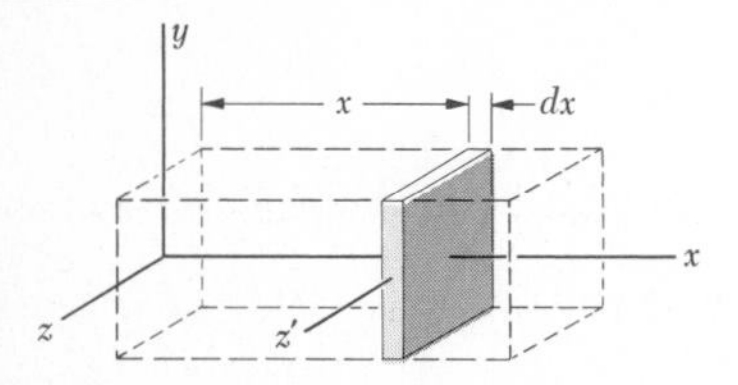

Solution. We choose as a differential element of mass the thin slab shown for which

$$dm = \rho bc\,dx$$

Referring to Sec. 9.13, we find that the moment of inertia of the element with respect to the z' axis is

$$dI_{z'} = \tfrac{1}{12}b^2\,dm$$

Applying the parallel-axis theorem, we obtain the mass moment of inertia of the slab with respect to the z axis.

$$dI_z = dI_{z'} + x^2\,dm = \tfrac{1}{12}b^2\,dm + x^2\,dm = (\tfrac{1}{12}b^2 + x^2)\,\rho bc\,dx$$

Integrating from $x = 0$ to $x = a$, we obtain

$$I_z = \int dI_z = \int_0^a (\tfrac{1}{12}b^2 + x^2)\,\rho bc\,dx = \rho abc\,(\tfrac{1}{12}b^2 + \tfrac{1}{3}a^2)$$

Since the total mass of the prism is $m = \rho abc$, we may write

$$I_z = m(\tfrac{1}{12}b^2 + \tfrac{1}{3}a^2) \qquad I_z = \tfrac{1}{12}m(4a^2 + b^2) \blacktriangleleft$$

We note that if the prism is slender, b is small compared to a and the expression for I_z reduces to $\frac{1}{3}ma^2$, which is the result obtained in Sample Prob. 9.9 when $L = a$.

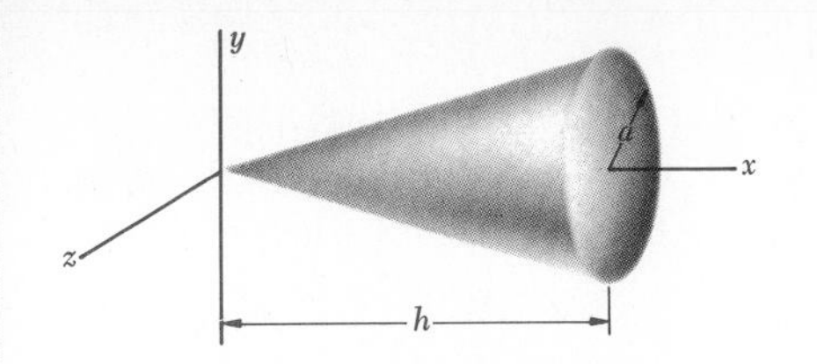

SAMPLE PROBLEM 9.11

Determine the mass moment of inertia of a right circular cone with respect to (*a*) its longitudinal axis, (*b*) an axis through the apex of the cone and perpendicular to its longitudinal axis, (*c*) an axis through the centroid of the cone and perpendicular to its longitudinal axis.

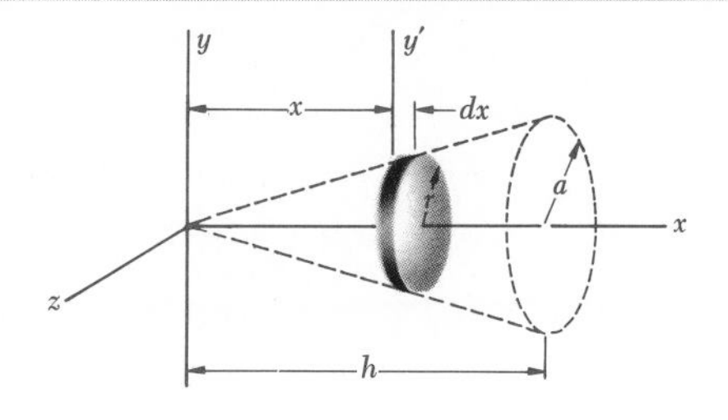

Solution. We choose the differential element of mass shown.

$$r = a\frac{x}{h} \qquad dm = \rho\pi r^2\,dx = \rho\pi\frac{a^2}{h^2}x^2\,dx$$

a. Moment of Inertia I_x. Using the expression derived in Sec. 9.13 for a thin disk, we compute the mass moment of inertia of the differential element with respect to the x axis.

$$dI_x = \tfrac{1}{2}r^2\,dm = \tfrac{1}{2}\left(a\frac{x}{h}\right)^2\left(\rho\pi\frac{a^2}{h^2}x^2\,dx\right) = \tfrac{1}{2}\rho\pi\frac{a^4}{h^4}x^4\,dx$$

Integrating from $x = 0$ to $x = h$, we obtain

$$I_x = \int dI_x = \int_0^h \tfrac{1}{2}\rho\pi\frac{a^4}{h^4}x^4\,dx = \tfrac{1}{2}\rho\pi\frac{a^4}{h^4}\frac{h^5}{5} = \tfrac{1}{10}\rho\pi a^4 h$$

Since the total mass of the cone is $m = \tfrac{1}{3}\rho\pi a^2 h$, we may write

$$I_x = \tfrac{1}{10}\rho\pi a^4 h = \tfrac{3}{10}a^2(\tfrac{1}{3}\rho\pi a^2 h) = \tfrac{3}{10}ma^2 \qquad I_x = \tfrac{3}{10}ma^2 \blacktriangleleft$$

b. Moment of Inertia I_y. The same differential element will be used. Applying the parallel-axis theorem and using the expression derived in Sec. 9.13 for a thin disk, we write

$$dI_y = dI_{y'} + x^2\,dm = \tfrac{1}{4}r^2\,dm + x^2\,dm = (\tfrac{1}{4}r^2 + x^2)\,dm$$

Substituting the expressions for r and dm, we obtain

$$dI_y = \left(\frac{1}{4}\frac{a^2}{h^2}x^2 + x^2\right)\left(\rho\pi\frac{a^2}{h^2}x^2\,dx\right) = \rho\pi\frac{a^2}{h^2}\left(\frac{a^2}{4h^2} + 1\right)x^4\,dx$$

$$I_y = \int dI_y = \int_0^h \rho\pi\frac{a^2}{h^2}\left(\frac{a^2}{4h^2} + 1\right)x^4\,dx = \rho\pi\frac{a^2}{h^2}\left(\frac{a^2}{4h^2} + 1\right)\frac{h^5}{5}$$

Introducing the total mass of the cone m, we rewrite I_y as follows:

$$I_y = \tfrac{3}{5}(\tfrac{1}{4}a^2 + h^2)\tfrac{1}{3}\rho\pi a^2 h \qquad I_y = \tfrac{3}{5}m(\tfrac{1}{4}a^2 + h^2) \blacktriangleleft$$

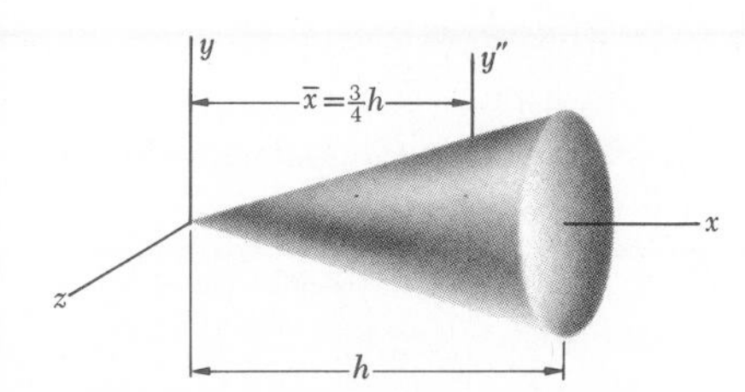

c. Moment of Inertia $\bar{I}_{y''}$. We apply the parallel-axis theorem and write

$$I_y = \bar{I}_{y''} + m\bar{x}^2$$

Solving for $\bar{I}_{y''}$ and recalling that $\bar{x} = \tfrac{3}{4}h$, we have

$$\bar{I}_{y''} = I_y - m\bar{x}^2 = \tfrac{3}{5}m(\tfrac{1}{4}a^2 + h^2) - m(\tfrac{3}{4}h)^2$$

$$\bar{I}_{y''} = \tfrac{3}{20}m(a^2 + \tfrac{1}{4}h^2) \blacktriangleleft$$

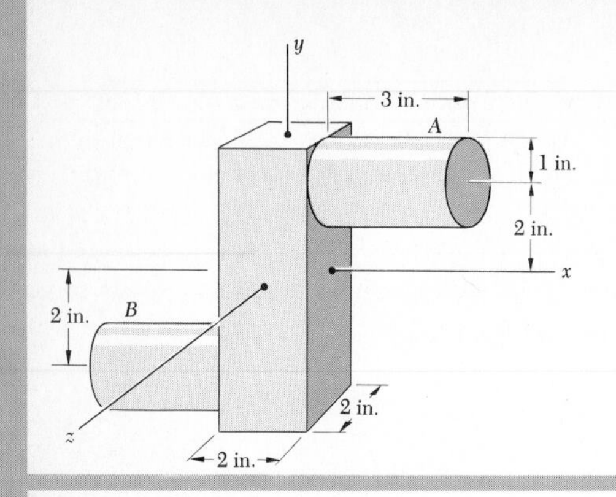

SAMPLE PROBLEM 9.12

A steel forging consists of a rectangular prism $6 \times 2 \times 2$ in. and of two cylinders of diameter 2 in. and length 3 in., as shown. Determine the mass moments of inertia with respect to the coordinate axes. (Specific weight of steel = 490 lb/ft^3.)

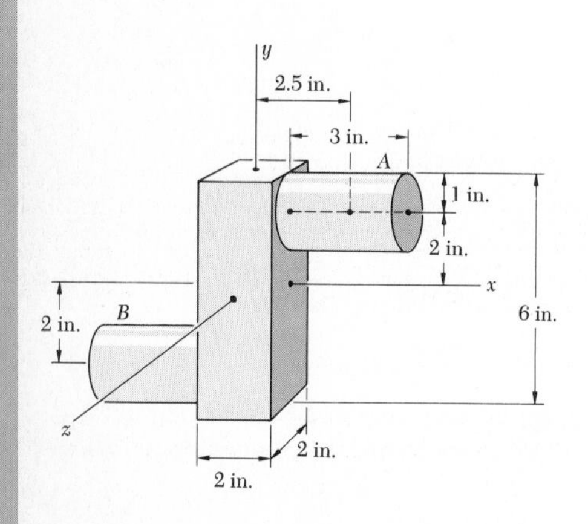

Computation of Masses
Prism

$$V = (2 \text{ in.})(2 \text{ in.})(6 \text{ in.}) = 24 \text{ in}^3$$

$$W = \frac{(24 \text{ in}^3)(490 \text{ lb/ft}^3)}{1728 \text{ in}^3/\text{ft}^3} = 6.81 \text{ lb}$$

$$m = \frac{6.81 \text{ lb}}{32.2 \text{ ft/s}^2} = 0.211 \text{ lb}\cdot\text{s}^2/\text{ft}$$

Each Cylinder

$$V = \pi(1 \text{ in.})^2(3 \text{ in.}) = 9.42 \text{ in}^3$$

$$W = \frac{(9.42 \text{ in}^3)(490 \text{ lb/ft}^3)}{1728 \text{ in}^3/\text{ft}^3} = 2.67 \text{ lb}$$

$$m = \frac{2.67 \text{ lb}}{32.2 \text{ ft/s}^2} = 0.0829 \text{ lb}\cdot\text{s}^2/\text{ft}$$

Mass Moments of Inertia. The mass moments of inertia of each component are computed from Fig. 9.28, using the parallel-axis theorem when necessary. Note that all lengths should be expressed in feet.

Prism

$$I_x = I_z = \tfrac{1}{12}(0.211 \text{ lb}\cdot\text{s}^2/\text{ft})[(\tfrac{6}{12} \text{ ft})^2 + (\tfrac{2}{12} \text{ ft})^2] = 4.88 \times 10^{-3} \text{ lb}\cdot\text{ft}\cdot\text{s}^2$$
$$I_y = \tfrac{1}{12}(0.211 \text{ lb}\cdot\text{s}^2/\text{ft})[(\tfrac{2}{12} \text{ ft})^2 + (\tfrac{2}{12} \text{ ft})^2] = 0.977 \times 10^{-3} \text{ lb}\cdot\text{ft}\cdot\text{s}^2$$

Each Cylinder

$$I_x = \tfrac{1}{2}ma^2 + m\bar{y}^2 = \tfrac{1}{2}(0.0829 \text{ lb}\cdot\text{s}^2/\text{ft})(\tfrac{1}{12} \text{ ft})^2 + (0.0829 \text{ lb}\cdot\text{s}^2/\text{ft})(\tfrac{2}{12} \text{ ft})^2 = 2.59 \times 10^{-3} \text{ lb}\cdot\text{ft}\cdot\text{s}^2$$
$$I_y = \tfrac{1}{12}m(3a^2 + L^2) + m\bar{x}^2 = \tfrac{1}{12}(0.0829 \text{ lb}\cdot\text{s}^2/\text{ft})[3(\tfrac{1}{12} \text{ ft})^2 + (\tfrac{3}{12} \text{ ft})^2] + (0.0829 \text{ lb}\cdot\text{s}^2/\text{ft})(\tfrac{2.5}{12} \text{ ft})^2 = 4.17 \times 10^{-3} \text{ lb}\cdot\text{ft}\cdot\text{s}^2$$
$$I_z = \tfrac{1}{12}m(3a^2 + L^2) + m(\bar{x}^2 + \bar{y}^2) = \tfrac{1}{12}(0.0829)[3(\tfrac{1}{12})^2 + (\tfrac{3}{12})^2] + (0.0829)[(\tfrac{2.5}{12})^2 + (\tfrac{2}{12})^2] = 6.48 \times 10^{-3} \text{ lb}\cdot\text{ft}\cdot\text{s}^2$$

Entire Body. Adding the values obtained,

$$I_x = 4.88 \times 10^{-3} + 2(2.59 \times 10^{-3}) \qquad I_x = 10.06 \times 10^{-3} \text{ lb}\cdot\text{ft}\cdot\text{s}^2 \blacktriangleleft$$
$$I_y = 0.977 \times 10^{-3} + 2(4.17 \times 10^{-3}) \qquad I_y = 9.32 \times 10^{-3} \text{ lb}\cdot\text{ft}\cdot\text{s}^2 \blacktriangleleft$$
$$I_z = 4.88 \times 10^{-3} + 2(6.48 \times 10^{-3}) \qquad I_z = 17.84 \times 10^{-3} \text{ lb}\cdot\text{ft}\cdot\text{s}^2 \blacktriangleleft$$

SAMPLE PROBLEM 9.13

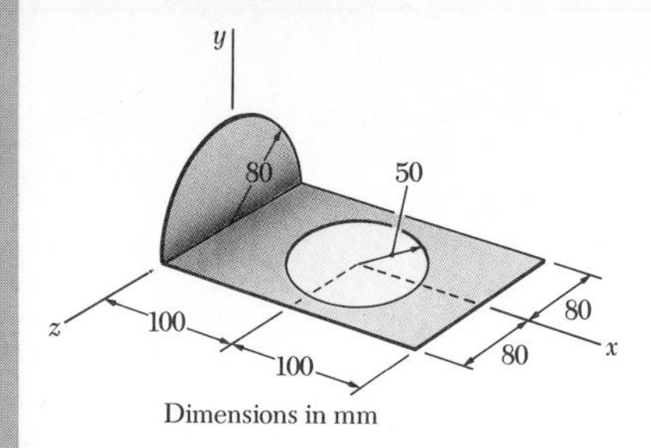

A thin steel plate 4 mm thick is cut and bent to form the machine part shown. Knowing that the density of steel is 7850 kg/m³, determine the mass moment of inertia of the machine part with respect to the coordinate axes.

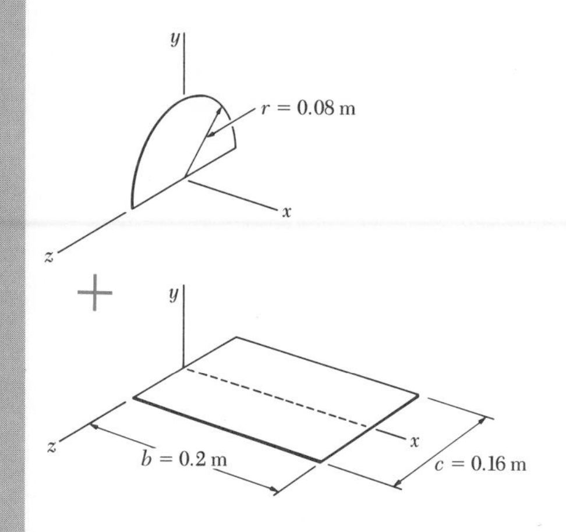

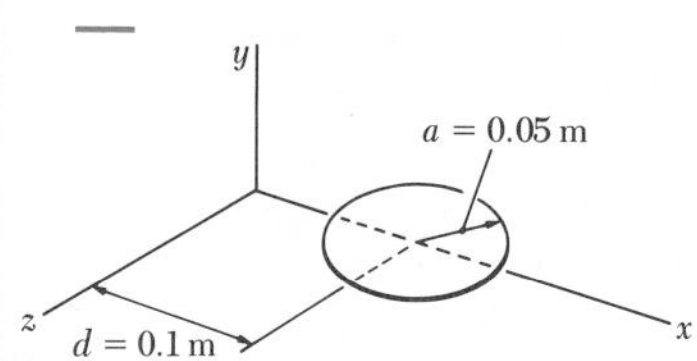

Solution. We observe that the machine part consists of a semicircular plate, plus a rectangular plate, minus a circular plate.

Computation of Masses. *Semicircular Plate*

$$V_1 = \tfrac{1}{2}\pi r^2 t = \tfrac{1}{2}\pi(0.08 \text{ m})^2(0.004 \text{ m}) = 40.21 \times 10^{-6} \text{ m}^3$$
$$m_1 = \rho V_1 = (7.85 \times 10^3 \text{ kg/m}^3)(40.21 \times 10^{-6} \text{ m}^3) = 0.3156 \text{ kg}$$

Rectangular Plate

$$V_2 = (0.200 \text{ m})(0.160 \text{ m})(0.004 \text{ m}) = 128 \times 10^{-6} \text{ m}^3$$
$$m_2 = \rho V_2 = (7.85 \times 10^3 \text{ kg/m}^3)(128 \times 10^{-6} \text{ m}^3) = 1.005 \text{ kg}$$

Circular Plate

$$V_3 = \pi a^2 t = \pi(0.050 \text{ m})^2(0.004 \text{ m}) = 31.42 \times 10^{-6} \text{ m}^3$$
$$m_3 = \rho V_3 = (7.85 \times 10^3 \text{ kg/m}^3)(31.42 \times 10^{-6} \text{ m}^3) = 0.2466 \text{ kg}$$

Mass Moments of Inertia. Using the method presented in Sec. 9.13, we compute the mass moment of inertia of each component.

Semicircular Plate. From Fig. 9.28, we observe that for a circular plate of mass m and radius r

$$I_x = \tfrac{1}{2}mr^2 \qquad I_y = I_z = \tfrac{1}{4}mr^2$$

Because of symmetry, we note that for a semicircular plate

$$I_x = \tfrac{1}{2}(\tfrac{1}{2}mr^2) \qquad I_y = I_z = \tfrac{1}{2}(\tfrac{1}{4}mr^2)$$

Since the mass of the semicircular plate is $m_1 = \tfrac{1}{2}m$, we have

$$I_x = \tfrac{1}{2}m_1 r^2 = \tfrac{1}{2}(0.3156 \text{ kg})(0.08 \text{ m})^2 = 1.010 \times 10^{-3} \text{ kg}\cdot\text{m}^2$$
$$I_y = I_z = \tfrac{1}{4}(\tfrac{1}{2}mr^2) = \tfrac{1}{4}m_1 r^2 = \tfrac{1}{4}(0.3156 \text{ kg})(0.08 \text{ m})^2 = 0.505 \times 10^{-3} \text{ kg}\cdot\text{m}^2$$

Rectangular Plate

$$I_x = \tfrac{1}{12}m_2 c^2 = \tfrac{1}{12}(1.005 \text{ kg})(0.16 \text{ m})^2 = 2.144 \times 10^{-3} \text{ kg}\cdot\text{m}^2$$
$$I_z = \tfrac{1}{3}m_2 b^2 = \tfrac{1}{3}(1.005 \text{ kg})(0.2 \text{ m})^2 = 13.400 \times 10^{-3} \text{ kg}\cdot\text{m}^2$$
$$I_y = I_x + I_z = (2.144 + 13.400)(10^{-3}) = 15.544 \times 10^{-3} \text{ kg}\cdot\text{m}^2$$

Circular Plate

$$I_x = \tfrac{1}{4}m_3 a^2 = \tfrac{1}{4}(0.2466 \text{ kg})(0.05 \text{ m})^2 = 0.154 \times 10^{-3} \text{ kg}\cdot\text{m}^2$$
$$I_y = \tfrac{1}{2}m_3 a^2 + m_3 d^2$$
$$= \tfrac{1}{2}(0.2466 \text{ kg})(0.05 \text{ m})^2 + (0.2466 \text{ kg})(0.1 \text{ m})^2 = 2.774 \times 10^{-3} \text{ kg}\cdot\text{m}^2$$
$$I_z = \tfrac{1}{4}m_3 a^2 + m_3 d^2 = \tfrac{1}{4}(0.2466 \text{ kg})(0.05 \text{ m})^2 + (0.2466 \text{ kg})(0.1 \text{ m})^2$$
$$= 2.620 \times 10^{-3} \text{ kg}\cdot\text{m}^2$$

Entire Machine Part

$$I_x = (1.010 + 2.144 - 0.154)(10^{-3}) \text{ kg}\cdot\text{m}^2 \qquad I_x = 3.00 \times 10^{-3} \text{ kg}\cdot\text{m}^2 \blacktriangleleft$$
$$I_y = (0.505 + 15.544 - 2.774)(10^{-3}) \text{ kg}\cdot\text{m}^2 \qquad I_y = 13.28 \times 10^{-3} \text{ kg}\cdot\text{m}^2 \blacktriangleleft$$
$$I_z = (0.505 + 13.400 - 2.620)(10^{-3}) \text{ kg}\cdot\text{m}^2 \qquad I_z = 11.29 \times 10^{-3} \text{ kg}\cdot\text{m}^2 \blacktriangleleft$$

Problems

9.88 A thin semicircular plate has a radius a and a mass m. Determine the mass moment of inertia of the plate with respect to (a) the centroidal axis BB', (b) the centroidal axis CC'.

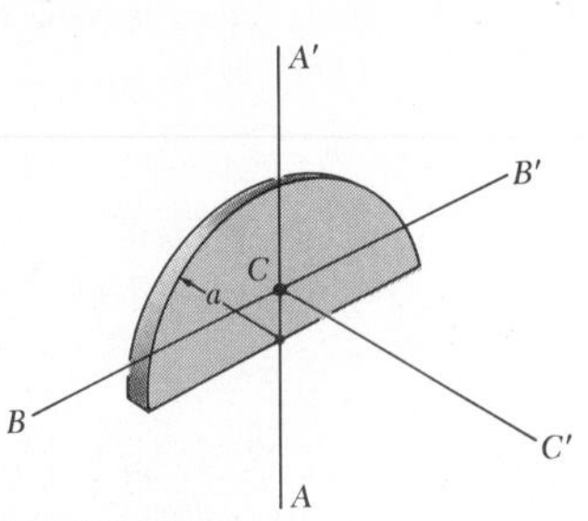

Fig. P9.88

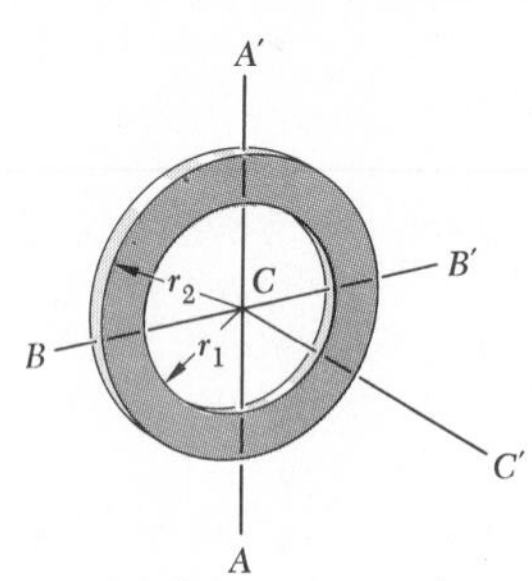

Fig. P9.89

9.89 Determine the mass moment of inertia of a ring of mass m, cut from a thin uniform plate, with respect to (a) the diameter AA' of the ring, (b) the axis CC' perpendicular to the plane of the ring.

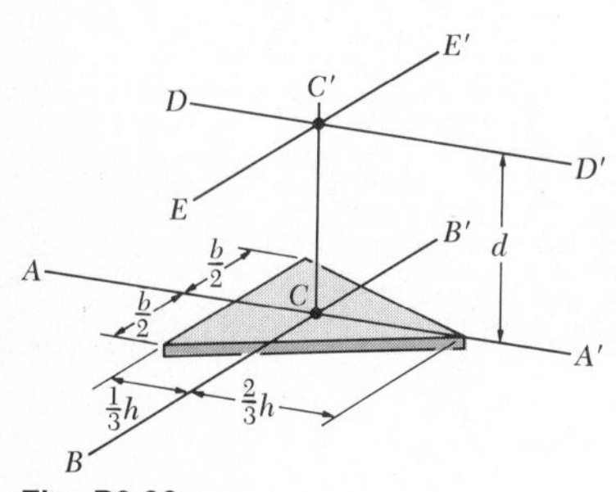

Fig. P9.90

9.90 A thin plate of mass m is cut in the shape of an isosceles triangle of base b and height h. Determine the mass moment of inertia of the plate with respect to (a) the centroidal axes AA' and BB' in the plane of the plate, (b) the centroidal axis CC' perpendicular to the plate.

9.91 Determine the mass moments of inertia of the plate of Prob. 9.90 with respect to axes DD' and EE' parallel to the centroidal axes AA' and BB' and located at a distance d from the plane of the plate.

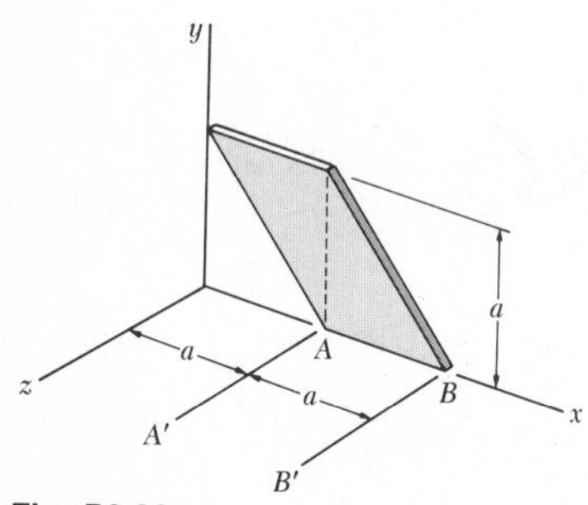

Fig. P9.92

9.92 A thin plate of mass m is cut in the shape of a parallelogram as shown. Determine the mass moment of inertia of the plate with respect to (a) the x axis, (b) the axis BB' perpendicular to the plate.

9.93 Determine the mass moment of inertia of the plate of Prob. 9.92 with respect to (a) the y axis, (b) the axis AA' perpendicular to the plate.

9.94 The area shown is revolved about the x axis to form a homogeneous solid of revolution of mass m. Express the mass moment of inertia of the solid with respect to the x axis in terms of m, a, and n.

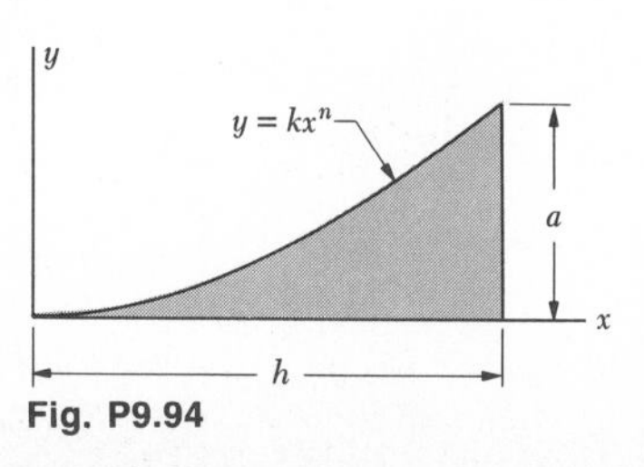

Fig. P9.94

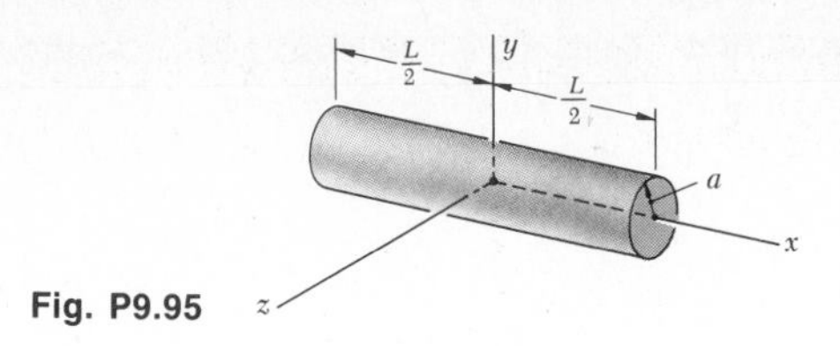

Fig. P9.95

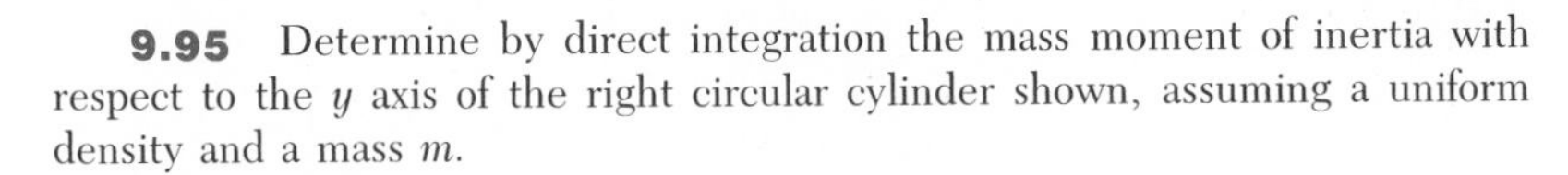

9.95 Determine by direct integration the mass moment of inertia with respect to the y axis of the right circular cylinder shown, assuming a uniform density and a mass m.

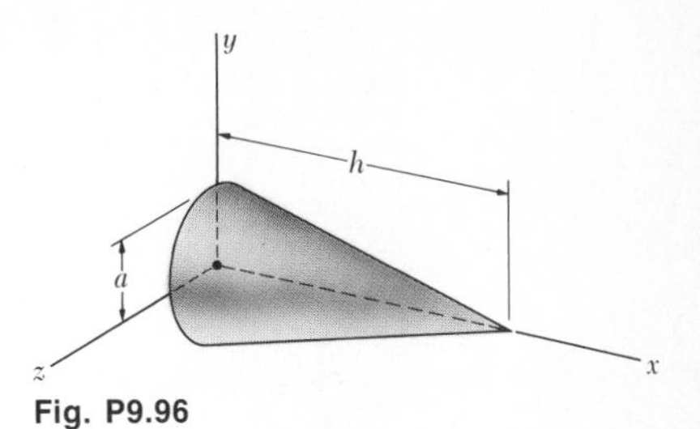

Fig. P9.96

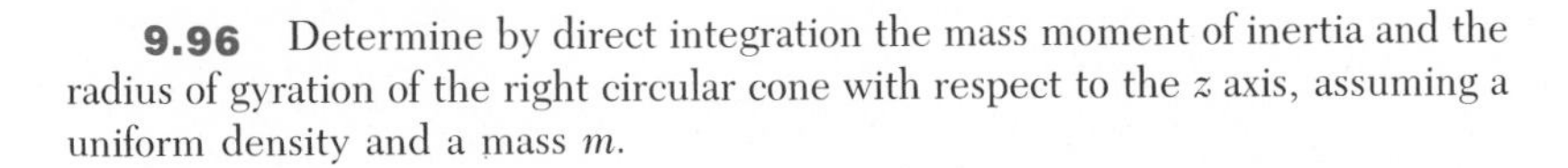

9.96 Determine by direct integration the mass moment of inertia and the radius of gyration of the right circular cone with respect to the z axis, assuming a uniform density and a mass m.

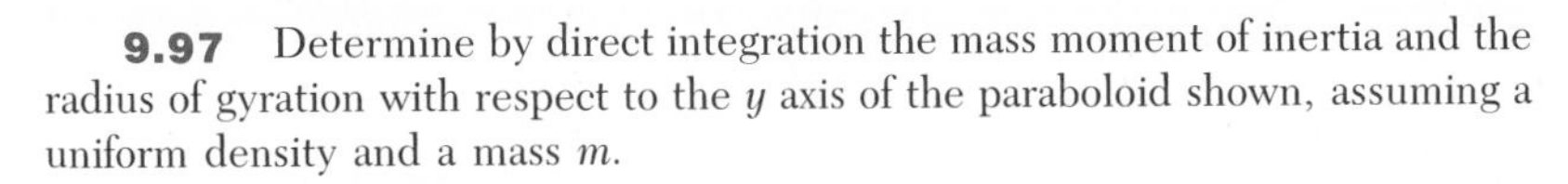

9.97 Determine by direct integration the mass moment of inertia and the radius of gyration with respect to the y axis of the paraboloid shown, assuming a uniform density and a mass m.

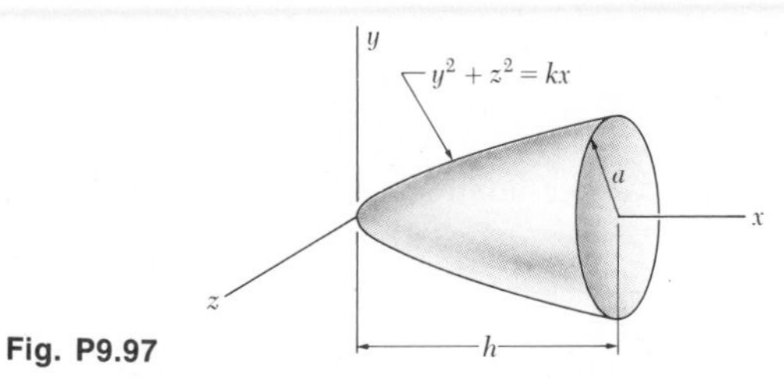

Fig. P9.97

9.98 Determine by direct integration the mass moment of inertia with respect to the x axis of the pyramid shown, assuming a uniform density and a mass m.

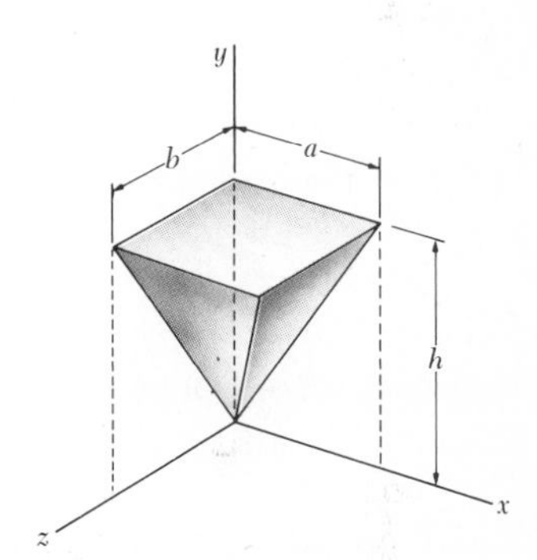

Fig. P9.98 and P9.99

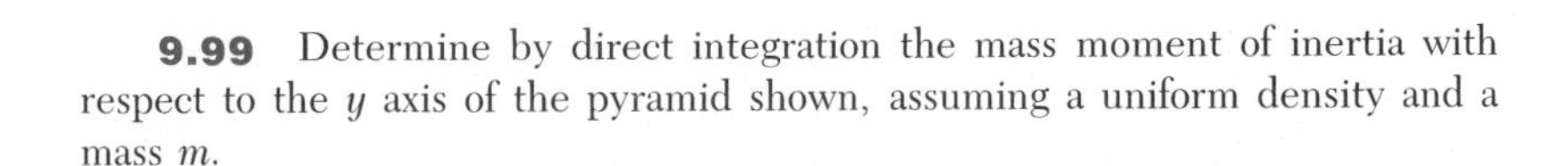

9.99 Determine by direct integration the mass moment of inertia with respect to the y axis of the pyramid shown, assuming a uniform density and a mass m.

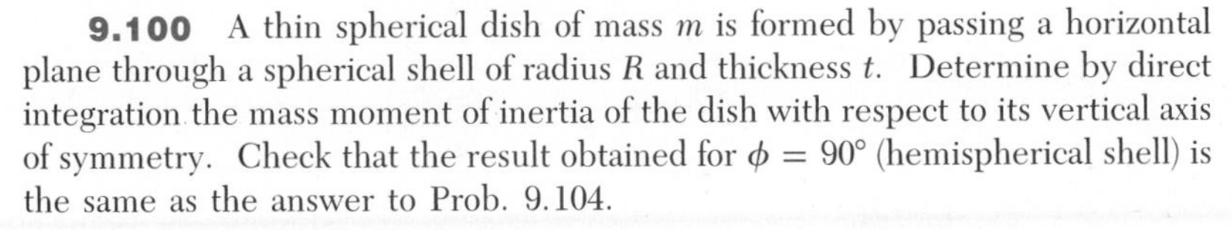

9.100 A thin spherical dish of mass m is formed by passing a horizontal plane through a spherical shell of radius R and thickness t. Determine by direct integration the mass moment of inertia of the dish with respect to its vertical axis of symmetry. Check that the result obtained for $\phi = 90°$ (hemispherical shell) is the same as the answer to Prob. 9.104.

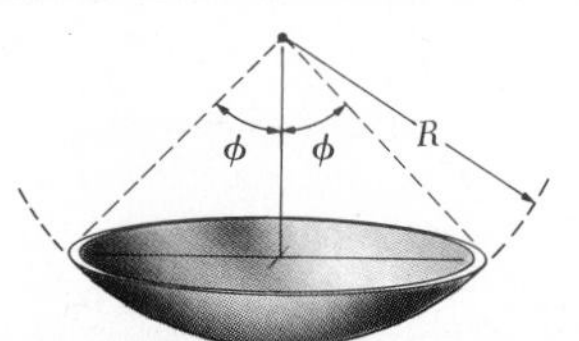

Fig. P9.100

9.101 A thin steel wire is bent into the shape of a circular arc as shown. Denoting by m' the mass per unit length of the wire, determine the mass moment of inertia of the wire with respect to each of the coordinate axes.

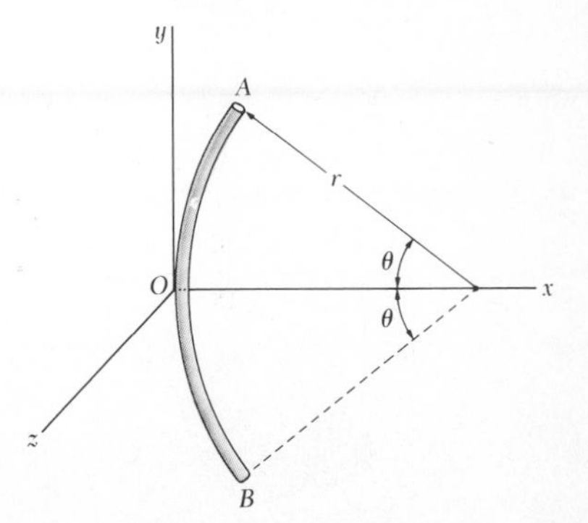

Fig. P9.101

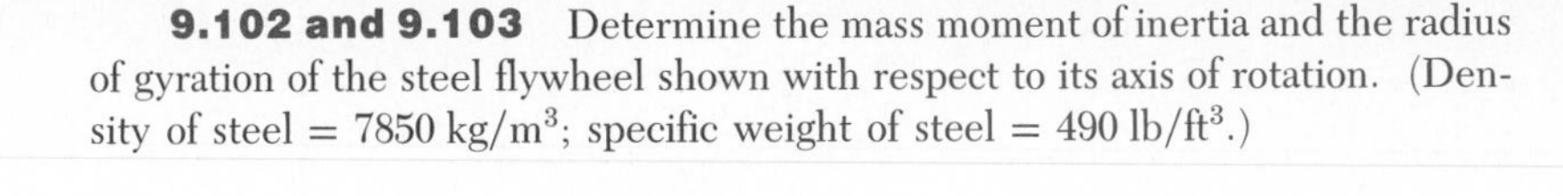

9.102 and 9.103 Determine the mass moment of inertia and the radius of gyration of the steel flywheel shown with respect to its axis of rotation. (Density of steel = 7850 kg/m^3; specific weight of steel = 490 lb/ft^3.)

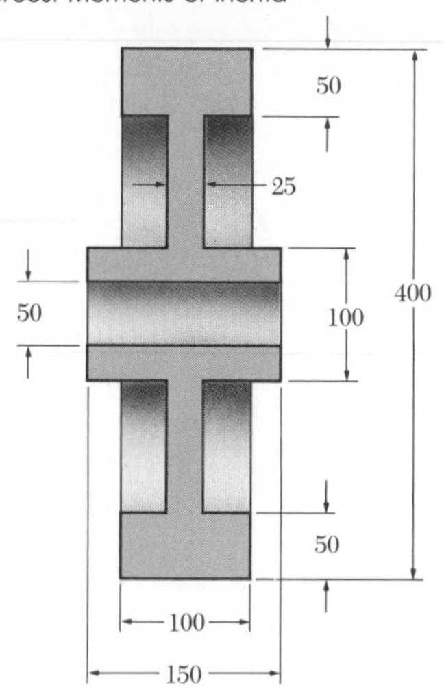

Fig. P9.102 Dimensions in mm

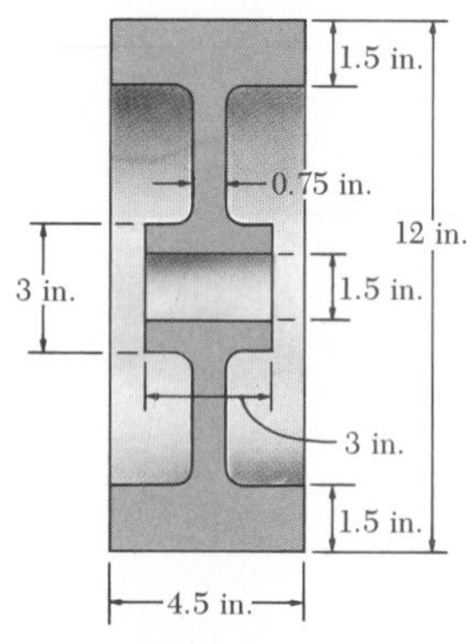

Fig. P9.103

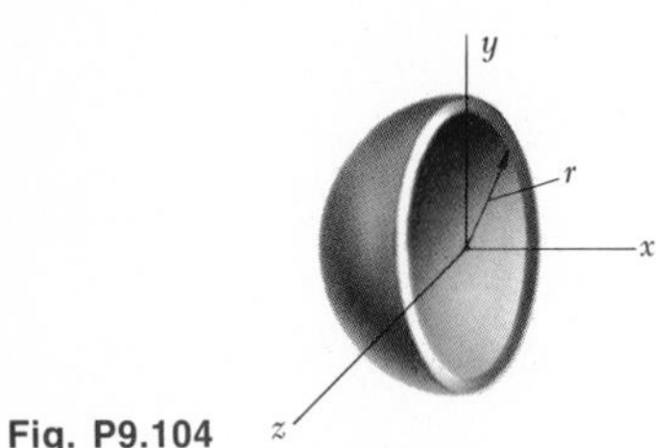

Fig. P9.104

9.104 Knowing that the thin hemispherical shell shown is of mass m and thickness t, determine the mass moment of inertia and the radius of gyration of the shell with respect to the x axis. (*Hint.* Consider the shell as formed by removing a hemisphere of radius r from a hemisphere of radius $r + t$; then neglect the terms containing t^2 and t^3 and keep those terms containing t.)

9.105 The machine part shown is formed by machining a conical surface into a circular cylinder. For $b = \frac{1}{2}h$, determine the mass moment of inertia and the radius of gyration of the machine part with respect to the y axis.

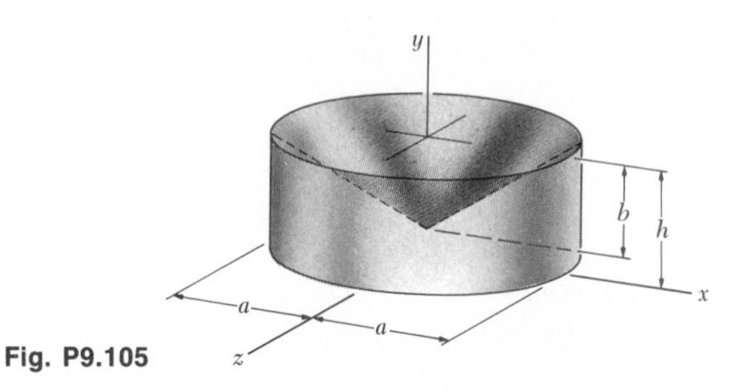

Fig. P9.105

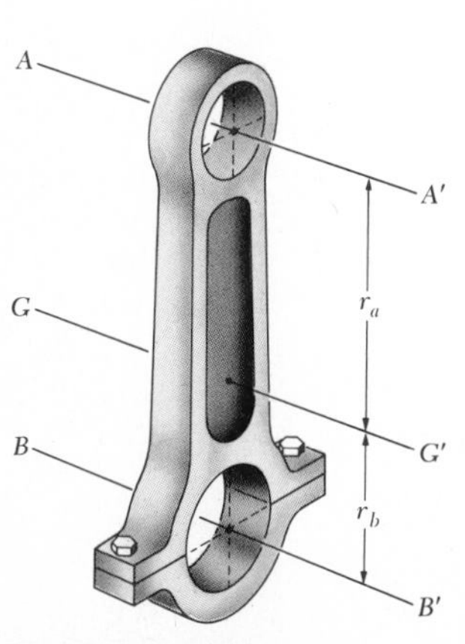

Fig. P9.106 and P9.107

9.106 For the 2-kg connecting rod shown, it has been experimentally determined that the mass moments of inertia of the rod with respect to the centerline axes of the bearings AA' and BB' are, respectively, $I_{AA'} = 78 \text{ g} \cdot \text{m}^2$ and $I_{BB'} = 41 \text{ g} \cdot \text{m}^2$. Knowing that $r_a + r_b = 290$ mm, determine (*a*) the location of the centroidal axis GG', (*b*) the radius of gyration with respect to axis GG'.

9.107 Knowing that for the 3.25-kg connecting rod shown the mass moment of inertia with respect to axis AA' is $205 \text{ g} \cdot \text{m}^2$, determine the mass moment of inertia with respect to axis BB', for $r_a = 225$ mm and $r_b = 140$ mm.

9.108 For the homogeneous ring shown, which is of density ρ, determine (*a*) the mass moment of inertia with respect to the axis BB', (*b*) the value of a_1 for which, given a_2 and h, $I_{BB'}$ is maximum, (*c*) the corresponding value of $I_{BB'}$.

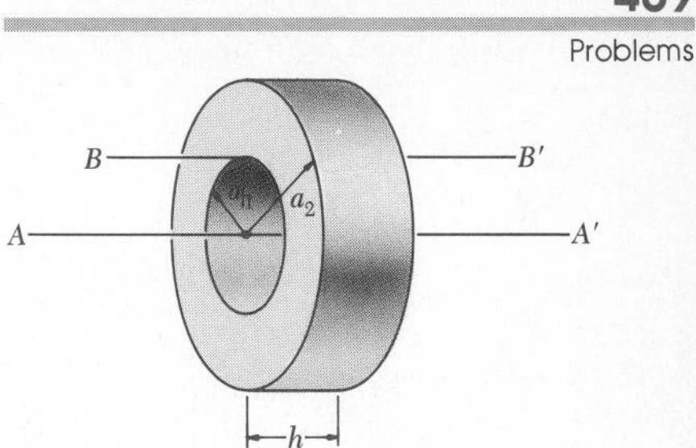

Fig. P9.108 and P9.109

9.109 Determine the mass moment of inertia and the radius of gyration of the steel ring shown with respect to the axis BB' with $a_1 = 3$ in., $a_2 = 5$ in., and $h = 4$ in. (Specific weight of steel $= 490$ lb/ft^3.)

9.110 and 9.111 A section of sheet steel 2 mm thick is cut and bent into the machine component shown. Knowing that the density of steel is 7850 kg/m^3, determine the mass moment of inertia of the component with respect to (*a*) the x axis, (*b*) the y axis, (*c*) the z axis.

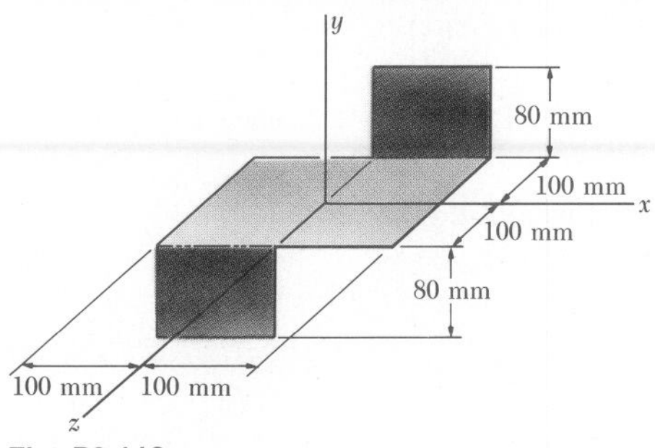

Fig. P9.110

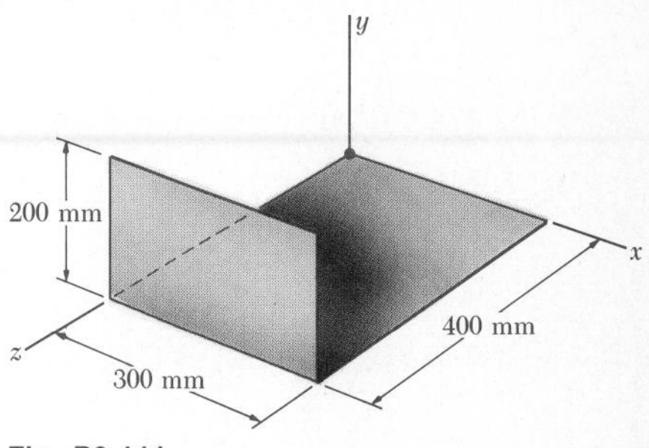

Fig. P9.111

9.112 A corner reflector for tracking by radar has two sides in the shape of a quarter circle of radius 15 in. and one side in the shape of a triangle. Each part of the reflector is formed from aluminum plate of uniform 0.05-in. thickness. Knowing that the specific weight of the aluminum used is 170 lb/ft^3, determine the mass moment of inertia of the reflector with respect to each of the coordinate axes.

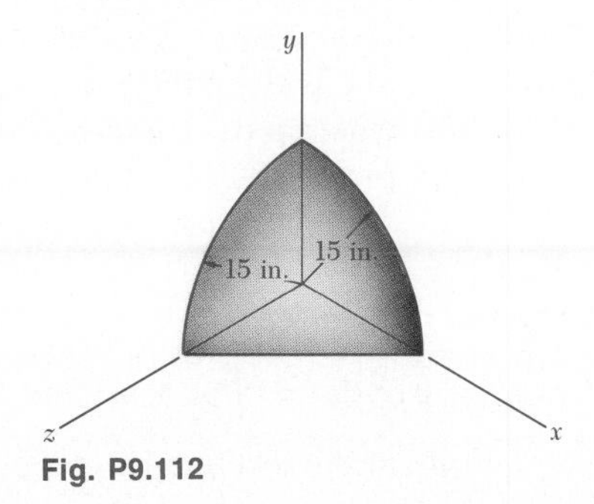

Fig. P9.112

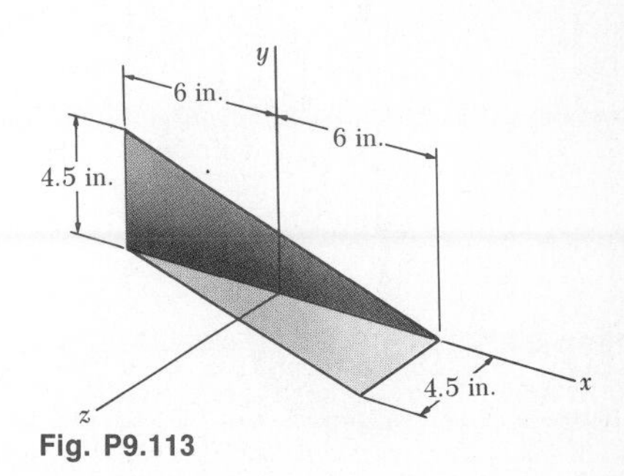

Fig. P9.113

9.113 A section of sheet steel 0.03 in. thick is cut and bent into the sheet-metal machine component shown. Knowing that the specific weight of steel is 490 lb/ft^3, determine the mass moment of inertia of the component with respect to each of the coordinate axes.

9.114 and 9.115 Determine the mass moment of inertia of the steel machine element shown with respect to the x axis. (Specific weight of steel = 490 lb/ft^3; density of steel = 7850 kg/m^3.)

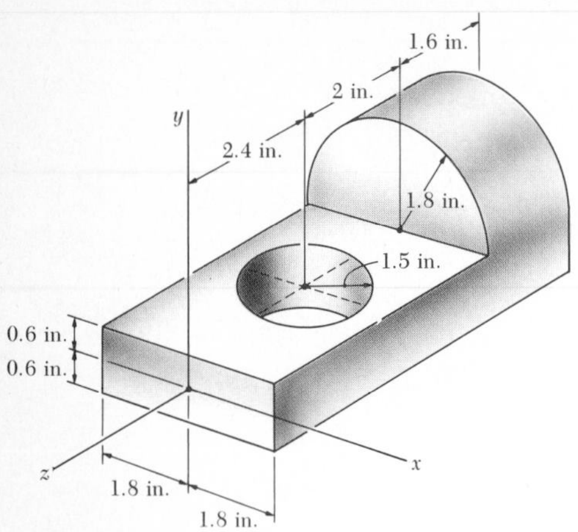

Fig. P9.115 and P9.116

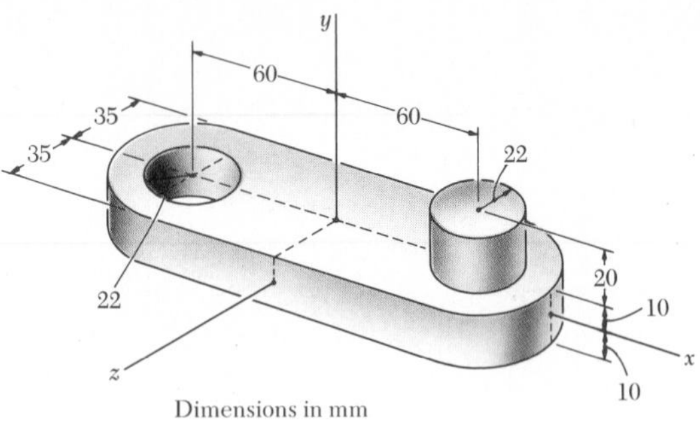

Fig. P9.114 and P9.117

9.116 and 9.117 Determine the mass moment of inertia of the steel machine element shown with respect to the y axis. (Specific weight of steel = 490 lb/ft^3; density of steel = 7850 kg/m^3.)

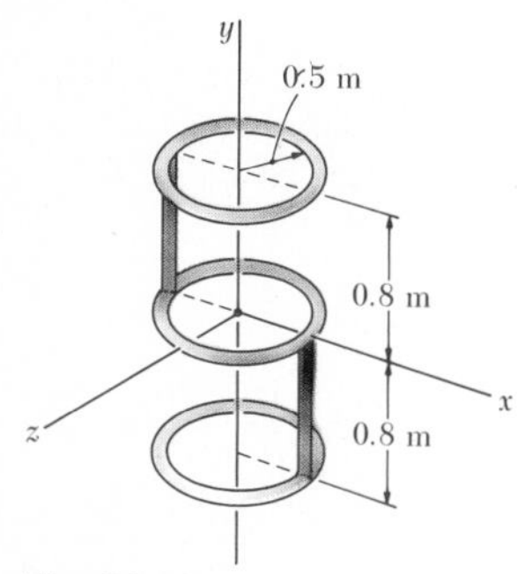

Fig. P9.118

9.118 A homogeneous wire with a mass per unit length of 1.8 kg/m is used to form the figure shown. Determine the mass moment of inertia of the wire figure with respect to (*a*) the x axis, (*b*) the y axis.

9.119 In Prob. 9.118, determine the mass moment of inertia of the wire figure with respect to the z axis.

***9.16. Moment of Inertia of a Body with Respect to an Arbitrary Axis through *O*. Mass Products of Inertia.** We shall see in this section how the moment of inertia of a body may be determined with respect to an arbitrary axis OL through the origin (Fig. 9.29) if we have computed beforehand its moments of inertia with respect to the three coordinate axes, as well as certain other quantities to be defined below.

The moment of inertia of the body with respect to OL is represented by the integral $I_{OL} = \int p^2\,dm$, where p denotes the perpendicular distance from the element of mass dm to the axis OL. But, denoting by $\boldsymbol{\lambda}$ the unit vector along OL and by $\mathbf{r}$ the position vector of the element dm, we observe that the perpendicular distance p is equal to the magnitude $r \sin\theta$ of the vector product $\boldsymbol{\lambda} \times \mathbf{r}$. We therefore write

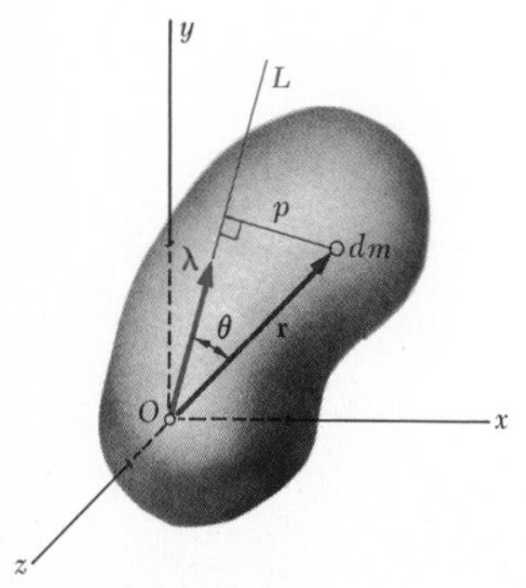

Fig. 9.29

$$I_{OL} = \int p^2\,dm = \int |\boldsymbol{\lambda} \times \mathbf{r}|^2\,dm \tag{9.43}$$

Expressing the square in terms of the rectangular components of the vector product, we have

$$I_{OL} = \int [(\lambda_x y - \lambda_y x)^2 + (\lambda_y z - \lambda_z y)^2 + (\lambda_z x - \lambda_x z)^2]\,dm$$

where the components λ_x, λ_y, λ_z of the unit vector $\boldsymbol{\lambda}$ represent the direction cosines of the axis OL, and the components x, y, z of $\mathbf{r}$ represent the coordinates of the element of mass dm. Expanding the squares in the expression obtained and rearranging the terms, we write

$$I_{OL} = \lambda_x^2 \int (y^2 + z^2)\, dm + \lambda_y^2 \int (z^2 + x^2)\, dm + \lambda_z^2 \int (x^2 + y^2)\, dm$$
$$-2\lambda_x\lambda_y \int xy\, dm - 2\lambda_y\lambda_z \int yz\, dm - 2\lambda_z\lambda_x \int zx\, dm \qquad (9.44)$$

Referring to Eqs. (9.30), we note that the first three integrals in (9.44) represent, respectively, the moments of inertia I_x, I_y, and I_z of the body with respect to the coordinate axes. The last three integrals in (9.44), which involve products of coordinates, are called the *products of inertia* of the body with respect to the x and y axes, the y and z axes, and the z and x axes, respectively. We write

$$I_{xy} = \int xy\, dm \qquad I_{yz} = \int yz\, dm \qquad I_{zx} = \int zx\, dm \qquad (9.45)$$

Substituting for the various integrals from (9.30) and (9.45) into (9.44), we have

$$I_{OL} = I_x\lambda_x^2 + I_y\lambda_y^2 + I_z\lambda_z^2 - 2I_{xy}\lambda_x\lambda_y - 2I_{yz}\lambda_y\lambda_z - 2I_{zx}\lambda_z\lambda_x \qquad (9.46)$$

We note that the definition of the products of inertia of a mass given in Eqs. (9.45) is an extension of the definition of the product of inertia of an area (Sec. 9.8). Mass products of inertia reduce to zero under the same conditions of symmetry as products of inertia of areas, and the parallel-axis theorem for mass products of inertia is expressed by relations similar to the formula derived for the product of inertia of an area. Substituting for x, y, z from Eqs. (9.31) into Eqs. (9.45), we verify that

$$\begin{aligned} I_{xy} &= \bar{I}_{x'y'} + m\bar{x}\bar{y} \\ I_{yz} &= \bar{I}_{y'z'} + m\bar{y}\bar{z} \\ I_{zx} &= \bar{I}_{z'x'} + m\bar{z}\bar{x} \end{aligned} \qquad (9.47)$$

where $\bar{x}$, $\bar{y}$, $\bar{z}$ are the coordinates of the center of gravity G of the body, and $\bar{I}_{x'y'}$, $\bar{I}_{y'z'}$, $\bar{I}_{z'x'}$ denote the products of inertia with respect to the centroidal axes x', y', z' (Fig. 9.22).

* 9.17. Ellipsoid of Inertia. Principal Axes of Inertia.

Let us assume that the moment of inertia of the body considered in the preceding section has been determined with respect to a large number of axes OL through the fixed point O, and that a point Q has been plotted on each axis OL at a distance $OQ = 1/\sqrt{I_{OL}}$ from O. The locus of the points Q thus obtained forms a surface (Fig. 9.30). The equation of that surface may be obtained by substituting $1/(OQ)^2$ for I_{OL} in (9.46) and multiplying both sides of the equation by $(OQ)^2$. Observing that

$$(OQ)\lambda_x = x \qquad (OQ)\lambda_y = y \qquad (OQ)\lambda_z = z$$

where x, y, z denote the rectangular coordinates of a point Q of the surface, we write

$$I_x x^2 + I_y y^2 + I_z z^2 - 2I_{xy}xy - 2I_{yz}yz - 2I_{zx}zx = 1 \qquad (9.48)$$

The equation obtained is that of a *quadric*. Since the moment of inertia I_{OL} is different from zero for every axis OL, no point Q may be at an infinite

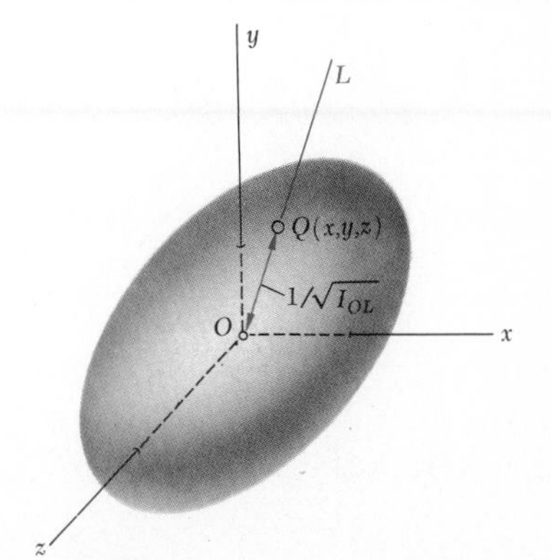

Fig. 9.30

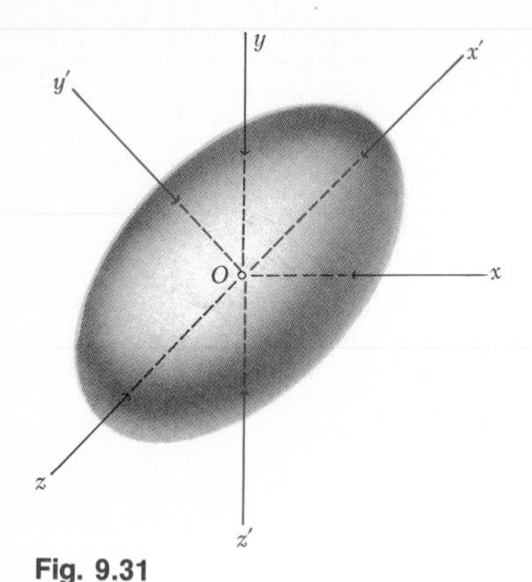

Fig. 9.31

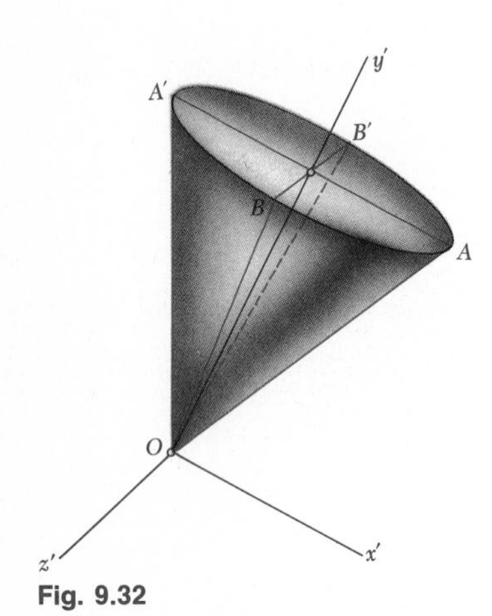

Fig. 9.32

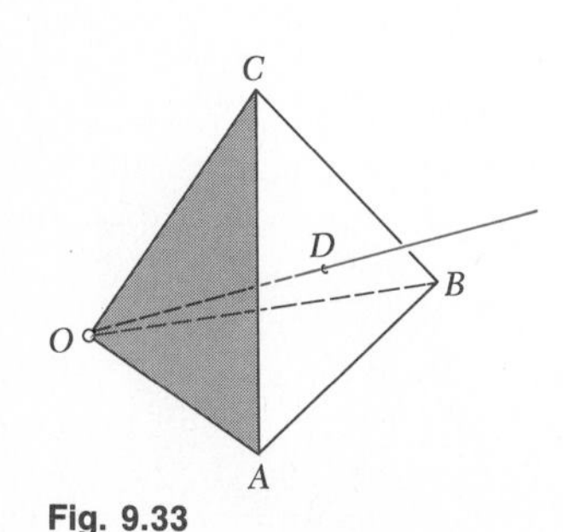

Fig. 9.33

distance from O. Thus, the quadric obtained is an *ellipsoid*. This ellipsoid, which defines the moment of inertia of the body with respect to any axis through O, is known as the *ellipsoid of inertia* of the body at O.

We observe that if the axes in Fig. 9.30 are rotated, the coefficients of the equation defining the ellipsoid change, since these are equal to the moments and products of inertia of the body with respect to the rotated coordinate axes. However, the *ellipsoid itself remains unaffected*, since its shape depends only upon the distribution of mass in the body considered. Suppose now that we choose as coordinate axes the principal axes x', y', z' of the ellipsoid of inertia (Fig. 9.31). The equation of the ellipsoid with respect to these coordinate axes will be of the form

$$I_{x'}x'^2 + I_{y'}y'^2 + I_{z'}z'^2 = 1 \tag{9.49}$$

which does not contain any product of coordinates. Thus, the products of inertia of the body with respect to the x', y', z' axes are zero. The x', y', z' axes are known as the *principal axes of inertia* of the body at O, and the coefficients $I_{x'}$, $I_{y'}$, $I_{z'}$ as the *principal moments of inertia* of the body at O. Note that, given a body of arbitrary shape and a point O, it is always possible to find axes which are the principal axes of inertia of the body at O, that is, axes with respect to which the products of inertia of the body are zero. Indeed, no matter how odd or irregular the shape of the body may be, the moments of inertia of the body with respect to axes through O will define an ellipsoid, and this ellipsoid will have principal axes which, by definition, are the principal axes of the body at O.

If the principal axes of inertia x', y', z' are used as coordinate axes, the expression obtained in Eq. (9.46) for the moment of inertia of a body with respect to an arbitrary axis through O reduces to

$$I_{OL} = I_{x'}\lambda_{x'}^2 + I_{y'}\lambda_{y'}^2 + I_{z'}\lambda_{z'}^2 \tag{9.50}$$

While the determination of the principal axes of inertia of a body of arbitrary shape is somewhat involved and would require solving a cubic equation, there are many cases when these axes may be spotted immediately. Consider, for instance, the homogeneous cone of elliptical base shown in Fig. 9.32; this cone possesses two mutually perpendicular planes of symmetry OAA' and OBB'. We check from the definition (9.45) that if the $x'y'$ and $y'z'$ planes are chosen to coincide with the two planes of symmetry, all the products of inertia are zero. The x', y', and z' axes thus selected are therefore the principal axes of inertia of the cone at O. In the case of the homogeneous regular tetrahedron $OABC$ shown in Fig. 9.33, the line joining the corner O to the center D of the opposite face is a principal axis of inertia at O and any line through O perpendicular to OD is also a principal axis of inertia at O. This property may be recognized if we observe that a rotation through 120° about OD leaves the shape and the mass distribution of the tetrahedron unchanged. It follows that the ellipsoid of inertia at O also remains unchanged under this rotation. The ellipsoid, therefore, is of revolution about OD, and the line OD, as well as any perpendicular line through O, must be a principal axis of the ellipsoid.

SAMPLE PROBLEM 9.14

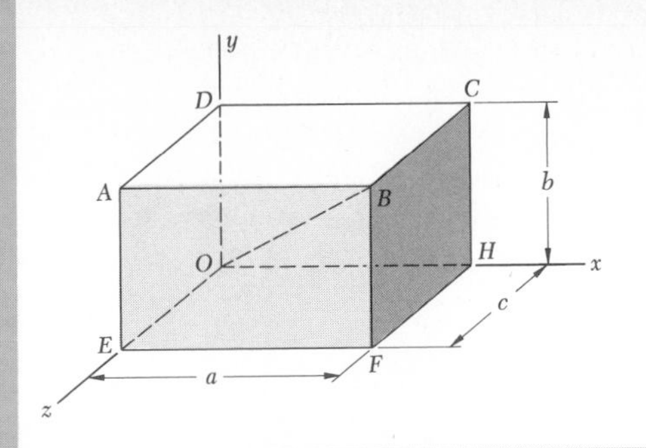

Consider a rectangular prism of mass m and sides a, b, c. Determine (*a*) the mass moments and products of inertia of the prism with respect to the coordinate axes shown, (*b*) its moment of inertia with respect to the diagonal OB.

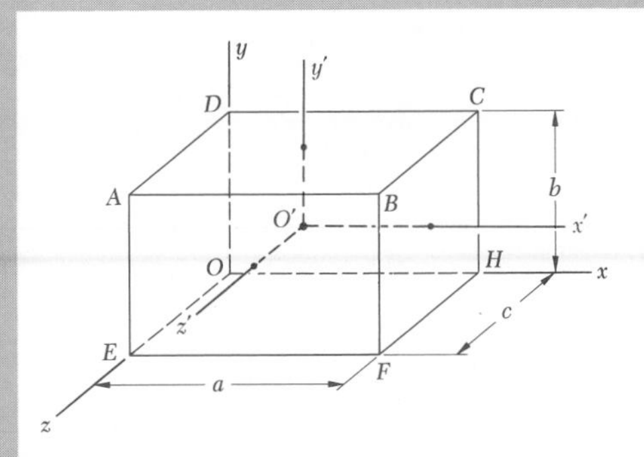

a.* Moments and Products of Inertia with Respect to the Coordinate Axes.** ***Moments of Inertia. Introducing the centroidal axes x', y', z', with respect to which the moments of inertia are given in Fig. 9.28, we apply the parallel-axis theorem:

$$I_x = \bar{I}_{x'} + m(\bar{y}^2 + \bar{z}^2) = \tfrac{1}{12}m(b^2 + c^2) + m(\tfrac{1}{4}b^2 + \tfrac{1}{4}c^2)$$

$$I_x = \tfrac{1}{3}m(b^2 + c^2) \quad ◀$$

Similarly, $\quad I_y = \tfrac{1}{3}m(c^2 + a^2) \qquad I_z = \tfrac{1}{3}m(a^2 + b^2) \quad ◀$

Products of Inertia. Because of symmetry, the products of inertia with respect to the centroidal axes x', y', z' are zero and these axes are principal axes of inertia. Using the parallel-axis theorem, we have

$$I_{xy} = \bar{I}_{x'y'} + m\bar{x}\bar{y} = 0 + m(\tfrac{1}{2}a)(\tfrac{1}{2}b) \qquad I_{xy} = \tfrac{1}{4}mab \quad ◀$$

Similarly, $\quad I_{yz} = \tfrac{1}{4}mbc \qquad I_{zx} = \tfrac{1}{4}mca \quad ◀$

***b.* Moment of Inertia with Respect to *OB*.** We recall Eq. (9.46):

$$I_{OB} = I_x\lambda_x^2 + I_y\lambda_y^2 + I_z\lambda_z^2 - 2I_{xy}\lambda_x\lambda_y - 2I_{yz}\lambda_y\lambda_z - 2I_{zx}\lambda_z\lambda_x$$

where the direction cosines of OB are

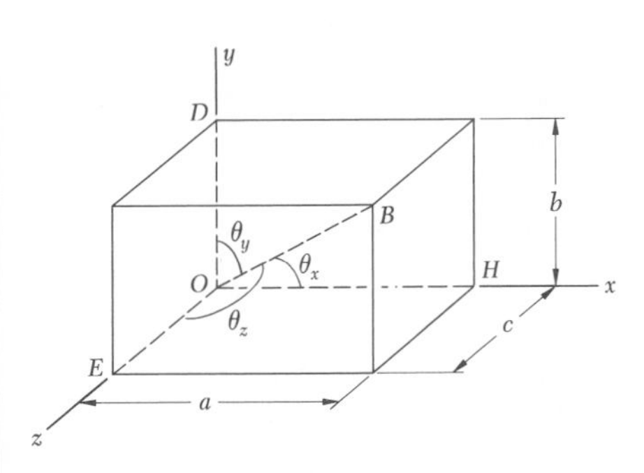

$$\lambda_x = \cos\theta_x = \frac{OH}{OB} = \frac{a}{(a^2 + b^2 + c^2)^{1/2}}$$

$$\lambda_y = \frac{b}{(a^2 + b^2 + c^2)^{1/2}} \qquad \lambda_z = \frac{c}{(a^2 + b^2 + c^2)^{1/2}}$$

Substituting the values obtained for the moments and products of inertia and for the direction cosines:

$$I_{OB} = \frac{1}{a^2 + b^2 + c^2}\left[\tfrac{1}{3}m(b^2 + c^2)a^2 + \tfrac{1}{3}m(c^2 + a^2)b^2 + \tfrac{1}{3}m(a^2 + b^2)c^2 - \tfrac{1}{2}ma^2b^2 - \tfrac{1}{2}mb^2c^2 - \tfrac{1}{2}mc^2a^2\right]$$

$$I_{OB} = \frac{m}{6}\,\frac{a^2b^2 + b^2c^2 + c^2a^2}{a^2 + b^2 + c^2} \quad ◀$$

Alternate Solution. The moment of inertia I_{OB} may be obtained directly from the principal moments of inertia $\bar{I}_{x'}$, $\bar{I}_{y'}$, $\bar{I}_{z'}$, since the line OB passes through the centroid O'. Since the x', y', z' axes are principal axes of inertia, we use Eq. (9.50) and write

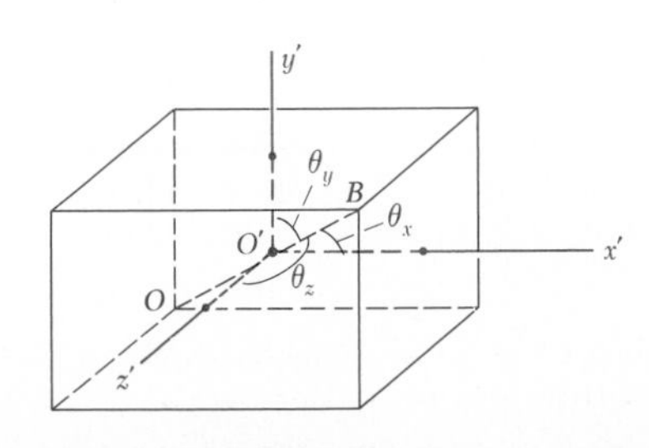

$$I_{OB} = \bar{I}_{x'}\lambda_x^2 + \bar{I}_{y'}\lambda_y^2 + \bar{I}_{z'}\lambda_z^2$$

$$= \frac{1}{a^2 + b^2 + c^2}\left[\frac{m}{12}(b^2 + c^2)a^2 + \frac{m}{12}(c^2 + a^2)b^2 + \frac{m}{12}(a^2 + b^2)c^2\right]$$

$$I_{OB} = \frac{m}{6}\,\frac{a^2b^2 + b^2c^2 + c^2a^2}{a^2 + b^2 + c^2} \quad ◀$$

Problems

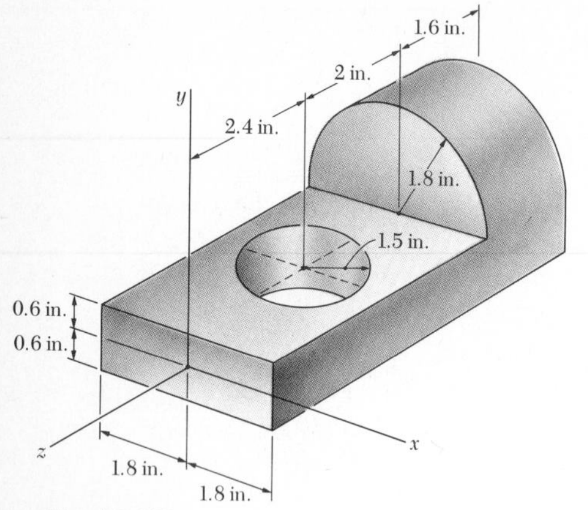

Fig. P9.120

9.120 and 9.121 Determine the mass products of inertia I_{xy}, I_{yz}, and I_{zx} of the steel machine element shown. (Specific weight of steel = 490 lb/ft^3; density of steel = 7850 kg/m^3.)

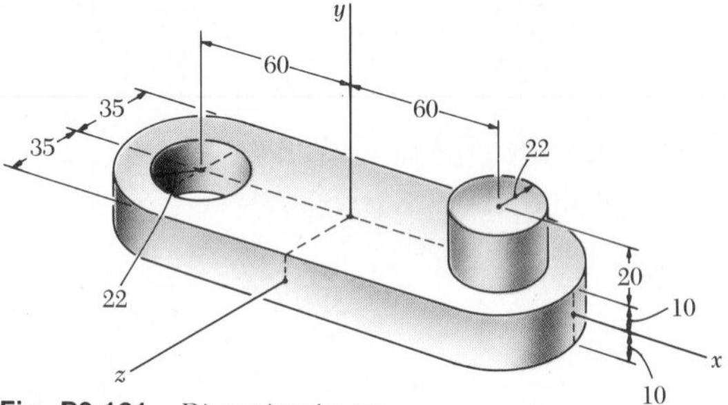

Fig. P9.121 Dimensions in mm

9.122 A section of sheet steel, 2 mm thick, is cut and bent into the machine component shown. Knowing that the density of steel is 7850 kg/m^3, determine the mass products of inertia I_{xy}, I_{yz}, and I_{zx} of the component.

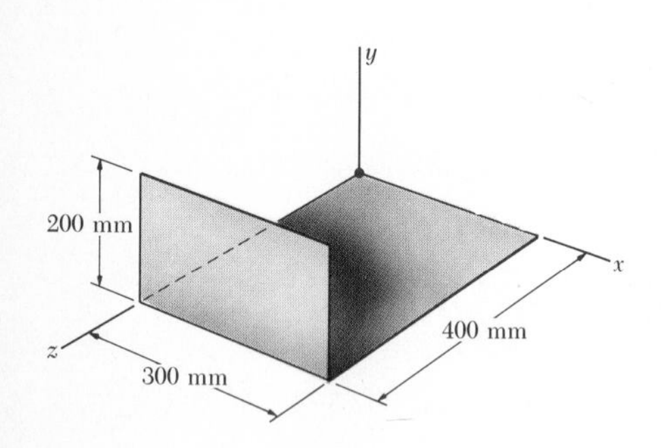

Fig. P9.122

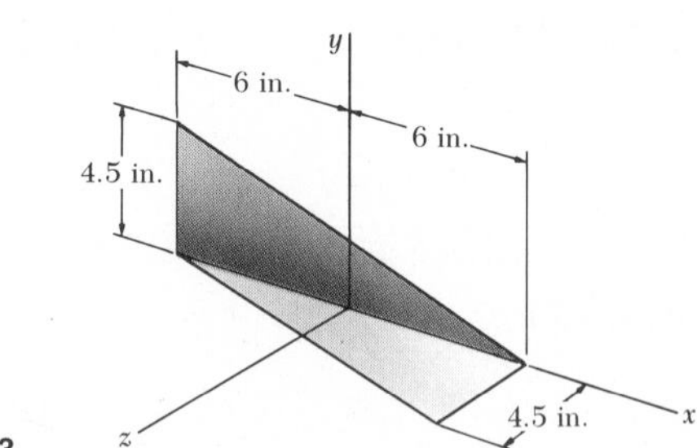

Fig. P9.123

9.123 A section of sheet steel, 0.03 in. thick, is cut and bent into the sheet-metal machine component shown. Knowing that the specific weight of steel is 490 lb/ft^3, determine the mass products of inertia I_{xy}, I_{yz}, and I_{zx} of the component. (*Hint.* See results of Sample Prob. 9.6.)

9.124 For the corner reflector of Prob. 9.112, determine the mass products of inertia I_{xy}, I_{yz}, and I_{zx}. (*Hint.* See results of Sample Prob. 9.6 and Prob. 9.56.)

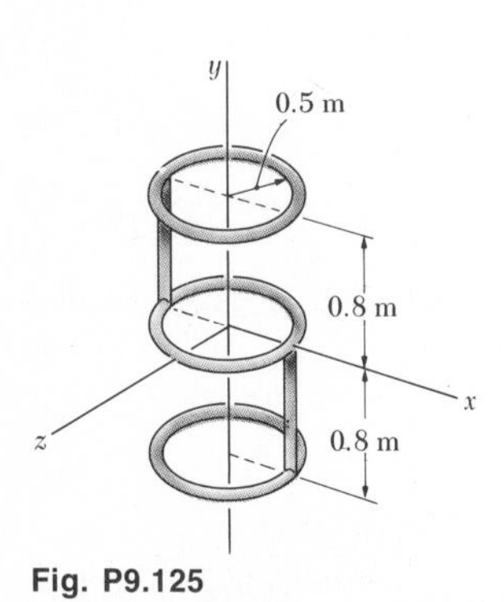

Fig. P9.125

9.125 A homogeneous wire with a mass per unit length of 1.8 kg/m is used to form the figure shown. Determine the mass products of inertia I_{xy}, I_{yz}, and I_{zx} of the wire figure.

9.126 Complete the derivation of Eqs. (9.47), which express the parallel-axis theorem for mass products of inertia.

9.127 For the homogeneous tetrahedron of mass m which is shown, (*a*) determine by direct integration the mass product of inertia I_{zx}, (*b*) deduce I_{yz} and I_{xy} from the result obtained in part *a*.

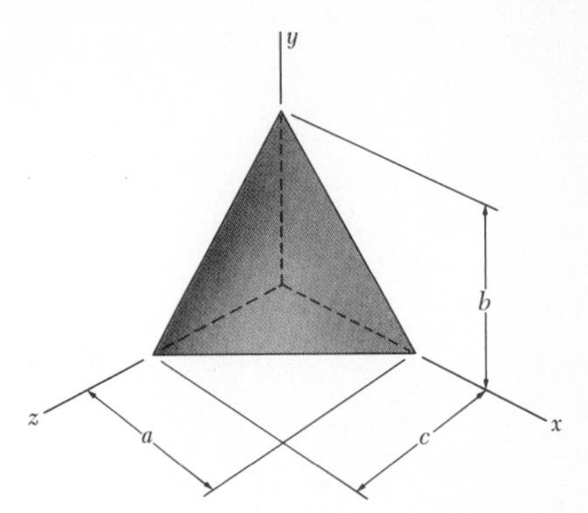

Fig. P9.127

9.128 Determine the mass moment of inertia of the right circular cone of Sample Prob. 9.11 with respect to a generator of the cone.

9.129 Determine the mass moment of inertia of the rectangular prism of Sample Prob. 9.14 with respect to the diagonal OF of its base.

9.130 Determine the mass moment of inertia of the corner reflector of Probs. 9.112 and 9.124 with respect to the axis through the origin which forms equal angles with the x, y, and z axes.

9.131 Determine the mass moment of inertia of the machine part of Sample Prob. 9.13 with respect to an axis through the origin characterized by the unit vector $\boldsymbol{\lambda} = \frac{2}{3}\mathbf{i} + \frac{1}{3}\mathbf{j} + \frac{2}{3}\mathbf{k}$.

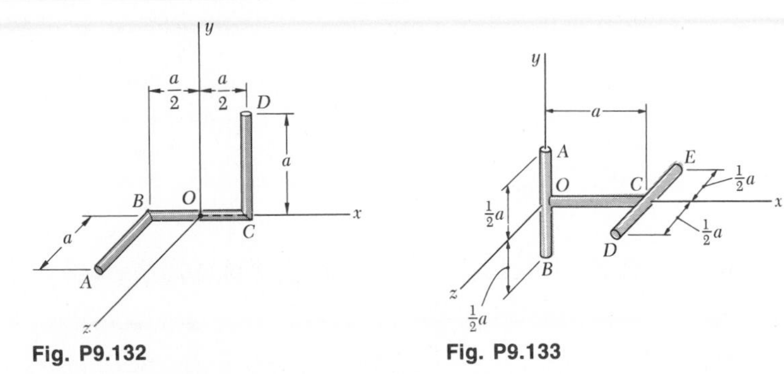

Fig. P9.132

Fig. P9.133

9.132 and 9.133 Three uniform rods, each of mass m, are welded together as shown. Determine (*a*) the mass moments of inertia and the mass products of inertia with respect to the coordinate axes, (*b*) the mass moment of inertia with respect to a line joining the origin O and point D.

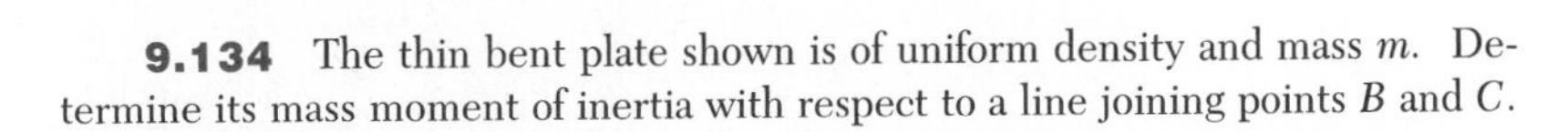

9.134 The thin bent plate shown is of uniform density and mass m. Determine its mass moment of inertia with respect to a line joining points B and C.

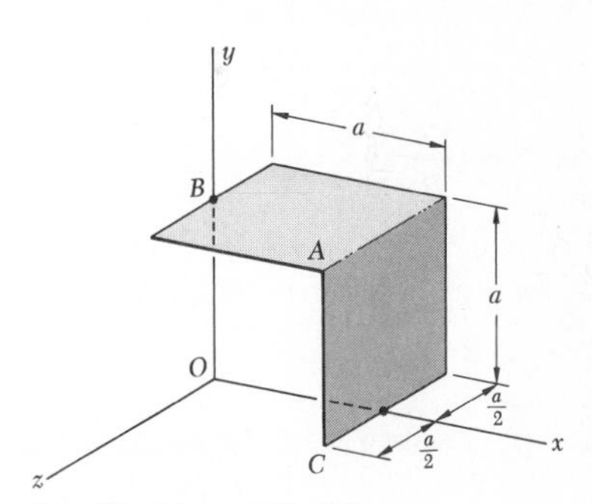

Fig. P9.134 and P9.135

9.135 The thin bent plate shown is of uniform density and mass m. Determine its mass moment of inertia with respect to a line joining the origin O and point A.

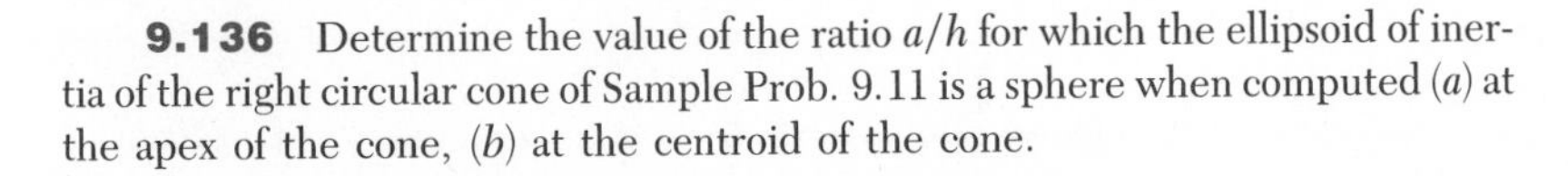

9.136 Determine the value of the ratio a/h for which the ellipsoid of inertia of the right circular cone of Sample Prob. 9.11 is a sphere when computed (*a*) at the apex of the cone, (*b*) at the centroid of the cone.

9.137 Consider a homogeneous circular cylinder of radius a and length L. Determine the value of the ratio a/L for which the ellipsoid of inertia of the cylinder is a sphere when computed (*a*) at the centroid of the cylinder, (*b*) at the center of one of its bases.

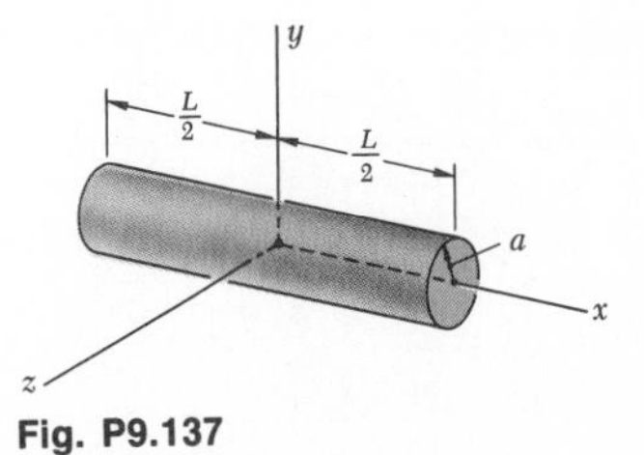

Fig. P9.137

9.138 Given an arbitrary solid and three rectangular axes x, y, and z, prove that the mass moment of inertia of the solid with respect to any one of the three axes cannot be larger than the sum of the moments of inertia of the solid with respect to the other two axes; i.e., prove that the inequality $I_x \leq I_y + I_z$ is satisfied, as well as two similar inequalities. Further prove that, if the solid is homogeneous and of revolution, and if x is the axis of revolution and y a transverse axis, then $I_y \geq \frac{1}{2}I_x$.

9.139 Consider a cube of mass m and side a. (*a*) Show that the ellipsoid of inertia at the center of the cube is a sphere, and use this property to determine the mass moment of inertia of the cube with respect to one of its diagonals. (*b*) Show that the ellipsoid of inertia at one of the corners of the cube is an ellipsoid of revolution, and determine the principal moments of inertia of the cube at that point.

9.140 Given a homogeneous solid of mass m and of arbitrary shape, and three rectangular axes x, y, and z of origin O, prove that the sum $I_x + I_y + I_z$ of the mass moments of inertia of the solid cannot be smaller than the similar sum computed for a sphere of the same mass and same material centered at O. Further prove, using the result of Prob. 9.138, that if the solid is of revolution and if x is the axis of revolution, then its moment of inertia I_y about a transverse axis y cannot be smaller than $\frac{3}{10}ma^2$, where a is the radius of the sphere of the same mass and same material.

Review and Summary

In the first half of this chapter we were concerned with the determination of the resultant **R** of forces $\Delta\mathbf{F}$ distributed over a plane area A when their magnitude is proportional to the element of area ΔA on which they act and, at the same time, varies linearly with the distance y from ΔA to a given x axis; we thus had $\Delta F = ky\,\Delta A$. We found that the magnitude of the resultant **R** was proportional to the first moment $Q_x = \int y\,dA$ of the area A, while the moment of **R** about the x axis was proportional to the *second moment,* or *moment of inertia,* $I_x = \int y^2\,dA$ of A with respect to the same axis [Sec. 9.2].

Rectangular moments of inertia

The *rectangular moments of inertia I_x and I_y of an area* [Sec. 9.3] were obtained by computing the integrals

$$I_x = \int y^2\,dA \qquad I_y = \int x^2\,dA \tag{9.1}$$

These computations may be reduced to single integrations by choosing for dA a thin strip parallel to one of the axes of coordinates. We also recall that it is possible to compute I_x and I_y from the same elementary strip (Fig. 9.34) by making use of the formula for the moment of inertia of a rectangular area [Sample Prob. 9.3].

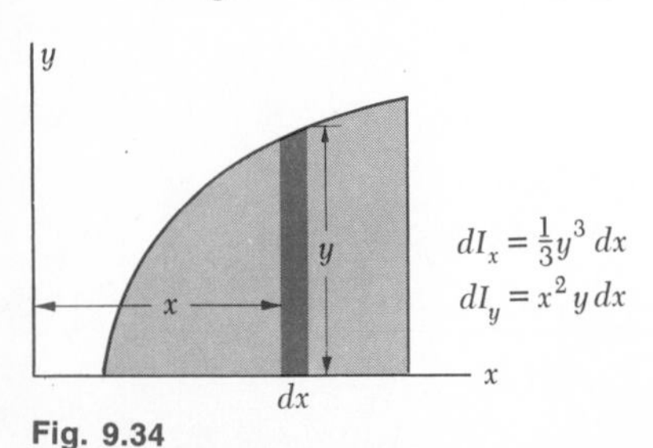

Fig. 9.34

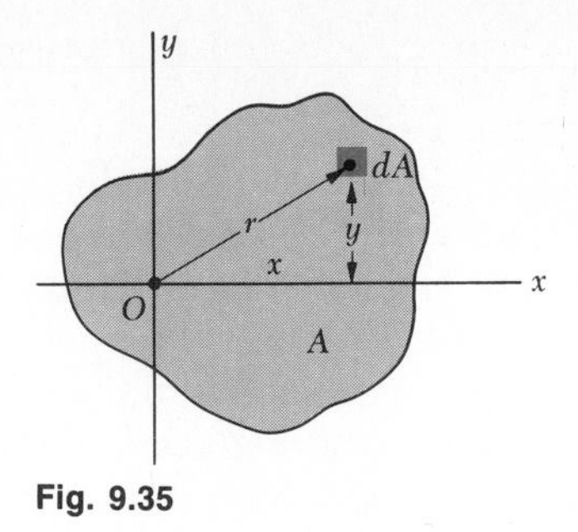

Fig. 9.35

Polar moment of inertia

The *polar moment of inertia of an area* A with respect to the pole O [Sec. 9.4] was defined as

$$J_O = \int r^2\, dA \tag{9.3}$$

where r is the distance from O to the element of area dA (Fig. 9.35). Observing that $r^2 = x^2 + y^2$, we established the relation

$$J_O = I_x + I_y \tag{9.4}$$

Radius of gyration

The *radius of gyration of an area* A with respect to the x axis [Sec. 9.5] was defined as the distance k_x such that $I_x = k_x^2 A$. With similar definitions for the radii of gyration of A with respect to the y axis and with respect to O, we had

$$k_x = \sqrt{\frac{I_x}{A}} \qquad k_y = \sqrt{\frac{I_y}{A}} \qquad k_O = \sqrt{\frac{J_O}{A}} \qquad (9.5\text{–}9.7)$$

Parallel-axis theorem

The *parallel-axis theorem* was presented in Sec. 9.6. It states that the moment of inertia I of an area with respect to any given axis AA' (Fig. 9.36) is equal to the moment of inertia $\bar{I}$ of the area with respect to a centroidal axis BB' parallel to AA' *plus* the product of the area A and the square of the distance d between the two axes:

$$I = \bar{I} + Ad^2 \tag{9.9}$$

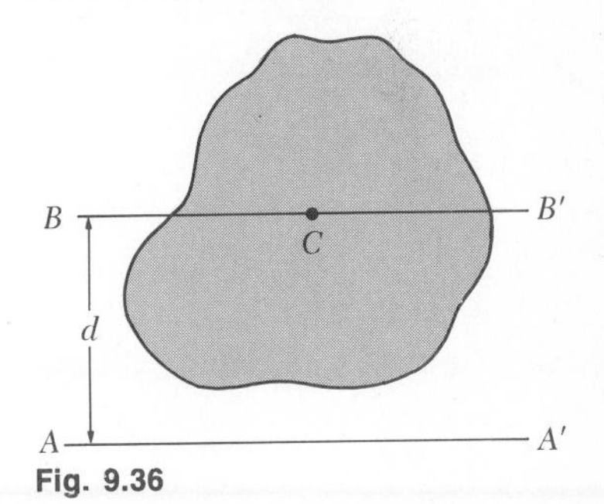

Fig. 9.36

This formula may also be used to determine the moment of inertia $\bar{I}$ of an area with respect to a centroidal axis BB' when its moment of inertia I with respect to a parallel axis AA' is known. In this case, however, the product Ad^2 should be *subtracted* from the known moment of inertia I.

A similar relation holds between the polar moment of inertia J_O of an area about a point O and the polar moment of inertia $\bar{J}_C$ of the same area about its centroid C. Denoting by d the distance between O and C, we have

$$J_O = \bar{J}_C + Ad^2 \tag{9.11}$$

Composite areas

The parallel-axis theorem may be used very effectively in the computation of the *moment of inertia of a composite area* with respect to a given axis [Sec. 9.7]. Considering each component area separately, we first compute the moment of inertia of each area with respect to a centroidal axis, using whenever possible the data provided in Figs. 9.12 and 9.13; the parallel-axis theorem is then applied to determine the moment of inertia of each component area with respect to the desired axis and the various values obtained are added [Sample Probs. 9.4 and 9.5].

Product of inertia

Sections 9.8 to 9.10 were devoted to the transformation of the moments of inertia of an area *under a rotation of the coordinate axes.* First we defined the *product of inertia of an area A* as

$$I_{xy} = \int xy\, dA \qquad (9.12)$$

and showed that $I_{xy} = 0$ if the area A is symmetrical with respect to either or both of the coordinate axes. We also derived the *parallel-axis theorem for products of inertia.* We had

$$I_{xy} = \bar{I}_{x'y'} + \bar{x}\bar{y}A \qquad (9.13)$$

where $\bar{I}_{x'y'}$ is the product of inertia of the area with respect to centroidal axes $x'y'$ parallel to the axes xy, and $\bar{x}$ and $\bar{y}$ are the coordinates of the centroid of the area [Sec. 9.8].

Rotation of axes

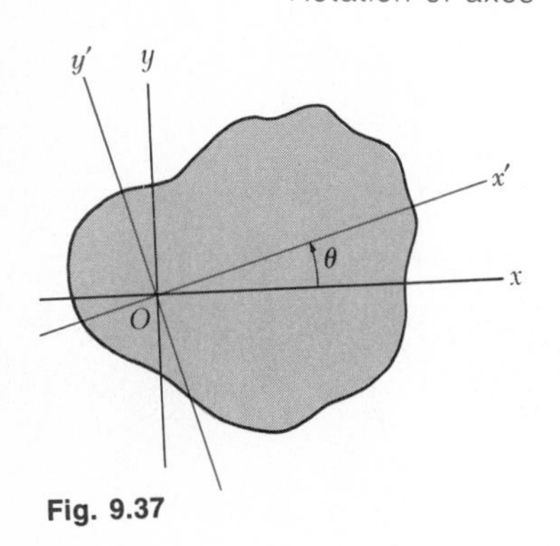

Fig. 9.37

In Sec. 9.9 we computed the moments and product of inertia $I_{x'}$, $I_{y'}$, and $I_{x'y'}$ of an area with respect to axes $x'y'$ obtained by rotating the original coordinate axes xy through an angle θ counterclockwise (Fig. 9.37), and expressed them in terms of the moments and product of inertia I_x, I_y, and I_{xy} computed with respect to the original xy axes. We had

$$I_{x'} = \frac{I_x + I_y}{2} + \frac{I_x - I_y}{2}\cos 2\theta - I_{xy}\sin 2\theta \qquad (9.18)$$

$$I_{y'} = \frac{I_x + I_y}{2} - \frac{I_x - I_y}{2}\cos 2\theta + I_{xy}\sin 2\theta \qquad (9.19)$$

$$I_{x'y'} = \frac{I_x - I_y}{2}\sin 2\theta + I_{xy}\cos 2\theta \qquad (9.20)$$

Principal axes

The *principal axes of the area about O* were defined as the two axes perpendicular to each other, with respect to which the moments of inertia of the area are, respectively, maximum and minimum. The corresponding values of θ were denoted by θ_m and obtained from the formula

$$\tan 2\theta_m = -\frac{2I_{xy}}{I_x - I_y} \qquad (9.25)$$

Principal moments of inertia

The corresponding maximum and minimum values of I are called the *principal moments of inertia* of the area about O; we had

$$I_{\max,\min} = \frac{I_x + I_y}{2} \pm \sqrt{\left(\frac{I_x - I_y}{2}\right)^2 + I_{xy}^2} \qquad (9.27)$$

We also noted that the corresponding value of the product of inertia is zero.

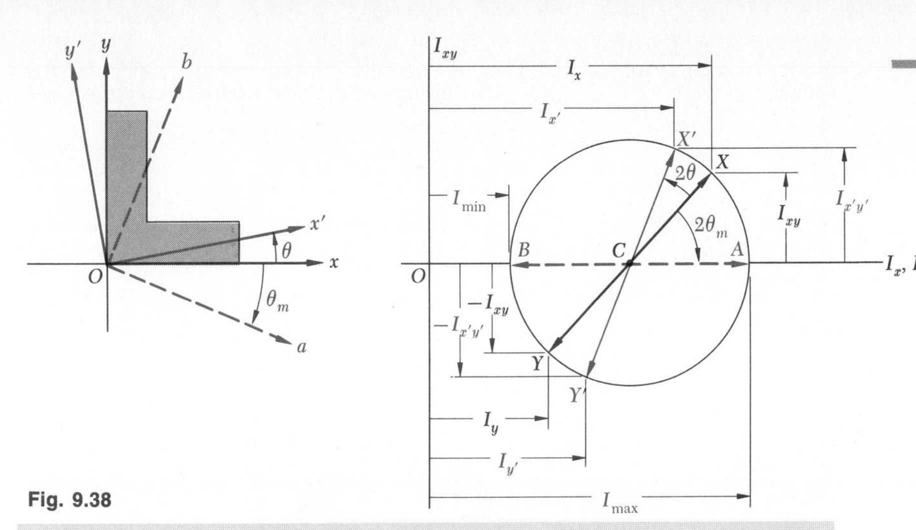

Fig. 9.38

Mohr's circle

The transformation of the moments and product of inertia of an area under a rotation of axes may be represented graphically by drawing *Mohr's circle* [Sec. 9.10]. Given the moments and product of inertia I_x, I_y, and I_{xy} of the area with respect to the coordinate axes xy, we plot points X (I_x, I_{xy}) and Y $(I_y, -I_{xy})$ and draw the line joining these two points (Fig. 9.38). This line is a diameter of Mohr's circle and thus defines this circle. As the coordinate axes are rotated through θ, the diameter rotates through *twice that angle* and the coordinates of X' and Y' yield the new values $I_{x'}$, $I_{y'}$, and $I_{x'y'}$ of the moments and product of inertia of the area. Also, the angle θ_m and the coordinates of points A and B define the principal axes a and b and the principal moments of inertia of the area [Sample Prob. 9.8].

Moments of inertia of masses

The second half of the chapter was devoted to the determination of *moments of inertia of masses,* which are encountered in dynamics in problems involving the rotation of a rigid body about an axis. The mass moment of inertia of a body with respect to an axis AA' (Fig. 9.39) was defined as

$$I = \int r^2\, dm \tag{9.28}$$

where r is the distance from AA' to the element of mass [Sec. 9.11], and its *radius of gyration* as

$$k = \sqrt{\frac{I}{m}} \tag{9.29}$$

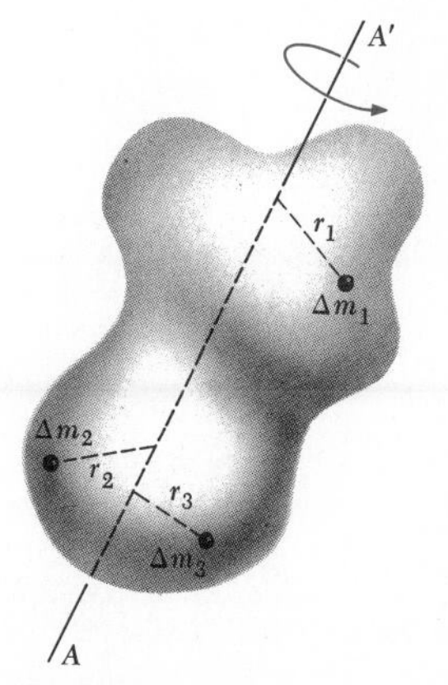

Fig. 9.39

The moments of inertia of a body with respect to the coordinates axes were expressed as

$$\begin{aligned} I_x &= \int (y^2 + z^2)\, dm \\ I_y &= \int (z^2 + x^2)\, dm \\ I_z &= \int (x^2 + y^2)\, dm \end{aligned} \tag{9.30}$$

Parallel-axis theorem

We saw that the *parallel-axis theorem* also applies to mass moments of inertia [Sec. 9.12]. Thus, the moment of inertia I of a body with respect to an arbitrary axis AA' (Fig. 9.40), may be expressed as

$$I = \bar{I} + md^2 \tag{9.33}$$

where $\bar{I}$ is the moment of inertia of the body with respect to a centroidal axis BB' parallel to the axis AA', m the mass of the body, and d the distance between the two axes.

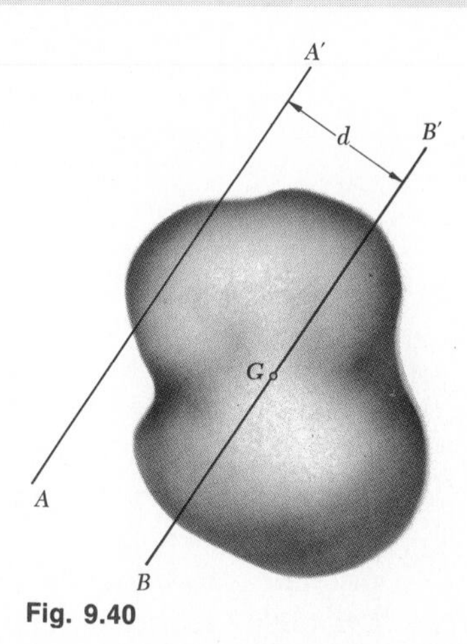

Fig. 9.40

Moments of inertia of thin plates

The moments of inertia of *thin plates* may be readily obtained from the moments of inertia of their areas [Sec. 9.13]. We found that for a *rectangular plate* the moments of inertia with respect to the axes shown (Fig. 9.41) are

$$I_{AA'} = \tfrac{1}{12}ma^2 \qquad I_{BB'} = \tfrac{1}{12}mb^2 \tag{9.39}$$
$$I_{CC'} = I_{AA'} + I_{BB'} = \tfrac{1}{12}m(a^2 + b^2) \tag{9.40}$$

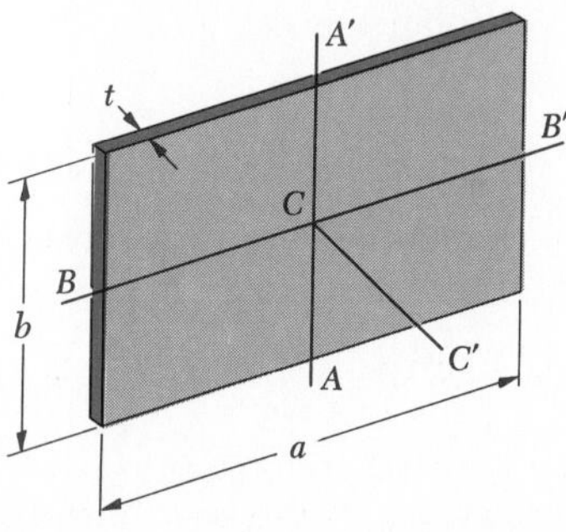

Fig. 9.41

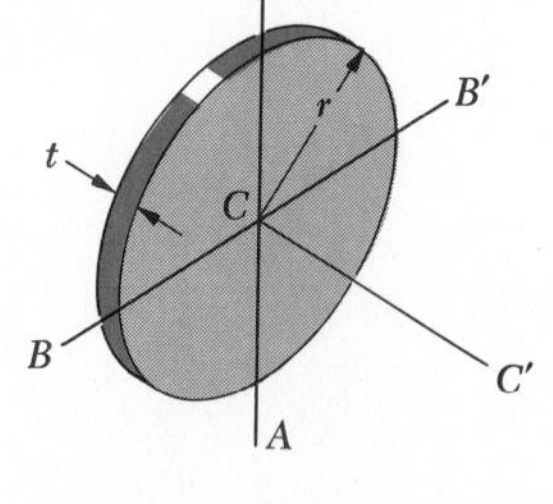

Fig. 9.42

while for a *circular plate* (Fig. 9.42) they are

$$I_{AA'} = I_{BB'} = \tfrac{1}{4}mr^2 \tag{9.41}$$
$$I_{CC'} = I_{AA'} + I_{BB'} = \tfrac{1}{2}mr^2 \tag{9.42}$$

When a body possesses *two planes of symmetry*, it is usually possible to determine its moment of inertia with respect to a given axis through a *single integration* by considering an element of mass dm in the shape of a thin plate [Sample Probs. 9.10 and 9.11]. On the other hand, when a body is made of *several common geometric shapes*, its moment of inertia with respect to a given axis may be obtained from the values given in Fig. 9.28 and the application of the parallel-axis theorem [Sample Probs. 9.12 and 9.13].

Composite bodies

Moment of inertia with respect to an arbitrary axis

In the last portion of the chapter, we learned to determine the moment of inertia of a body *with respect to an arbitrary axis OL* drawn through the origin O [Sec. 9.16]. Denoting by λ_x, λ_y, λ_z the components of the unit vector $\boldsymbol{\lambda}$ along OL (Fig. 9.43) and introducing the *products of inertia*

$$I_{xy} = \int xy\, dm \qquad I_{yz} = \int yz\, dm \qquad I_{zx} = \int zx\, dm \qquad (9.45)$$

we found that the moment of inertia of the body with respect to OL could be expressed as

$$I_{OL} = I_x\lambda_x^2 + I_y\lambda_y^2 + I_z\lambda_z^2 - 2I_{xy}\lambda_x\lambda_y - 2I_{yz}\lambda_y\lambda_z - 2I_{zx}\lambda_z\lambda_x \qquad (9.46)$$

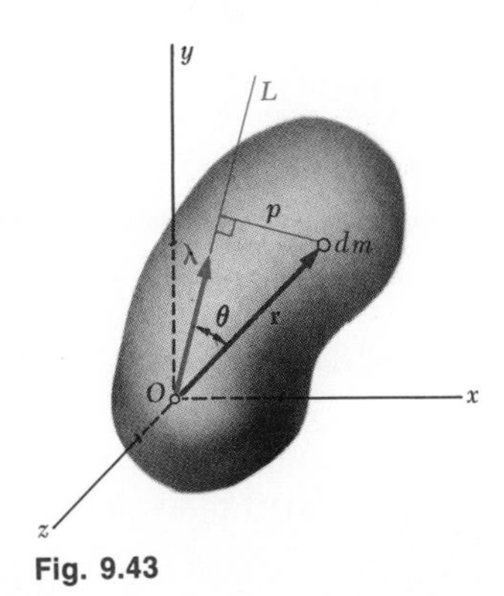

Fig. 9.43

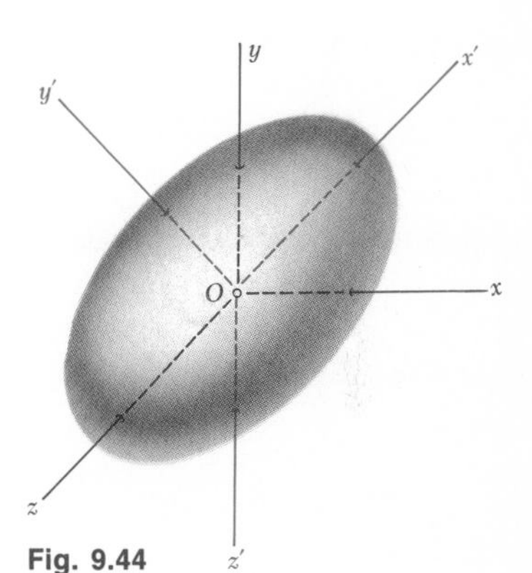

Fig. 9.44

Ellipsoid of inertia

By plotting a point Q along each axis OL at a distance $OQ = 1/\sqrt{I_{OL}}$ from O [Sec. 9.17], we obtained the surface of an ellipsoid, known as the *ellipsoid of inertia* of the body at point O. The principal axes x', y', z' of this ellipsoid (Fig. 9.44), are the *principal axes of inertia* of the body; that is, the products of inertia $I_{x'y'}$, $I_{y'z'}$, $I_{z'x'}$ of the body with respect to these axes are all zero. There are many situations where the principal axes of inertia of a body may be spotted from properties of symmetry of the body. Choosing these axes as coordinate axes, we may then express I_{OL} as

$$I_{OL} = I_{x'}\lambda_{x'}^2 + I_{y'}\lambda_{y'}^2 + I_{z'}\lambda_{z'}^2 \qquad (9.50)$$

where $I_{x'}$, $I_{y'}$, $I_{z'}$ are the *principal moments of inertia* of the body at O.

Review Problems

9.141 A thin wire has been bent to form a half ring of radius r. Denoting the mass of the wire by m, determine the mass moment of inertia of the wire with respect to (*a*) the axis AA', (*b*) the axis BB'.

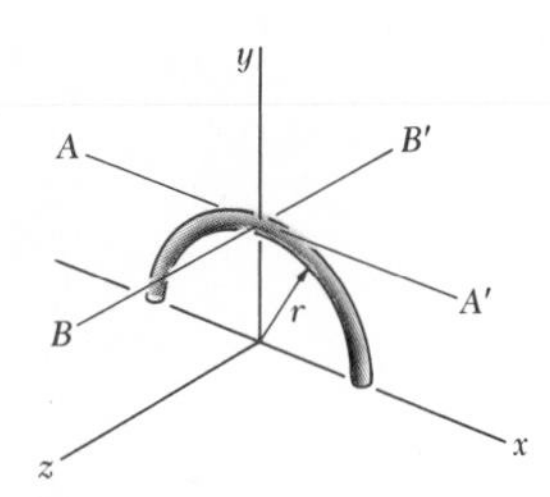

Fig. P9.141

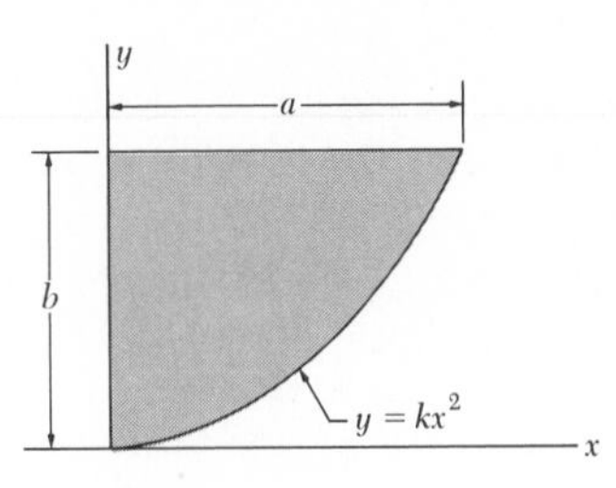

Fig. P9.142

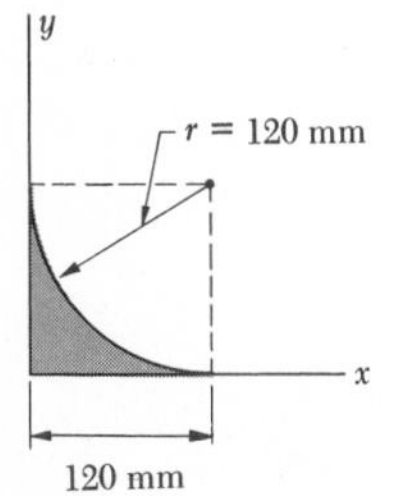

Fig. P9.143 and P9.144

9.142 Determine the moment of inertia of the shaded area with respect to (*a*) the x axis, (*b*) the y axis.

9.143 Determine the moment of inertia of the shaded area with respect to the y axis.

9.144 Determine the product of inertia of the shaded area with respect to the x and y axes. (*Hint.* Use answer to Prob. 9.56.)

9.145 Determine the polar moment of inertia and the polar radius of gyration of the shaded area shown with respect to its centroid C.

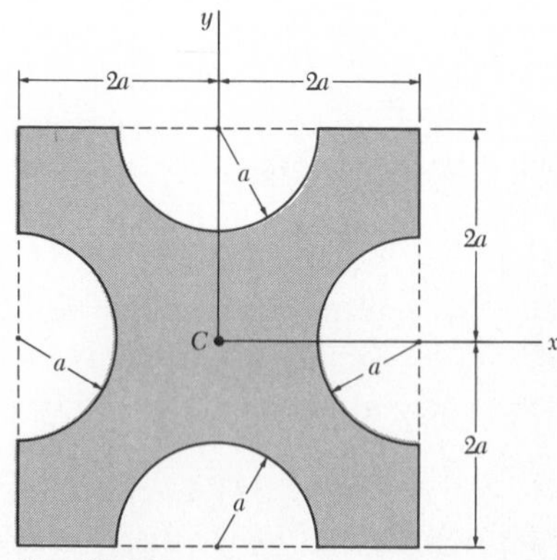

Fig. P9.145

9.146 Determine the orientation of the principal axes through the centroid C and the corresponding values of the moment of inertia for the angle cross section shown. (The moments of inertia $\bar{I}_x$ and $\bar{I}_y$ of the cross section are given in Fig. 9.13.)

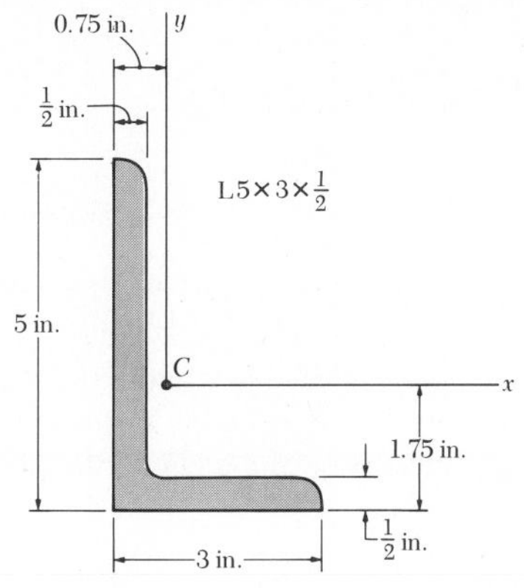

Fig. P9.146

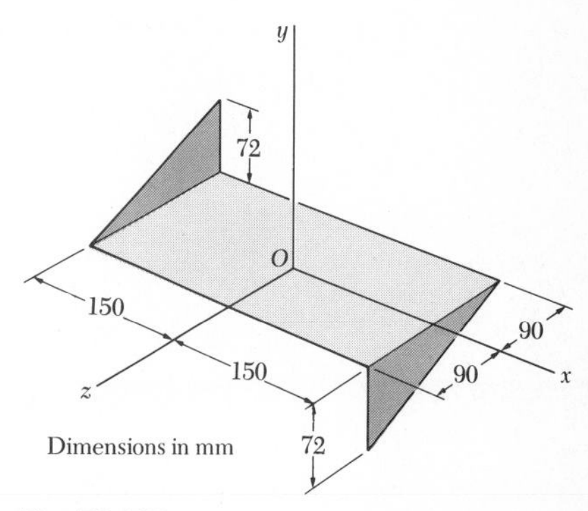

Fig. P9.147

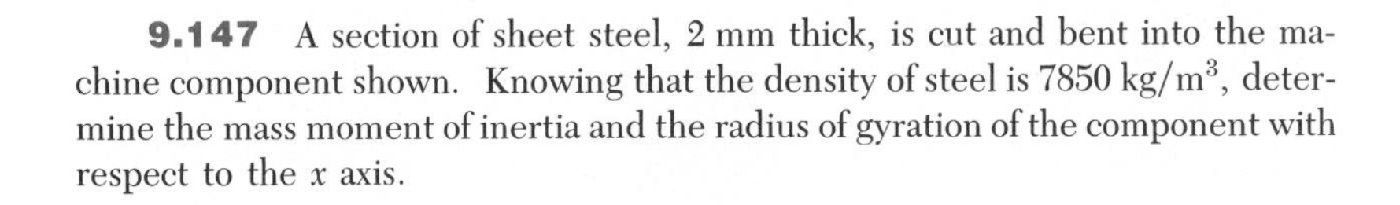
9.147 A section of sheet steel, 2 mm thick, is cut and bent into the machine component shown. Knowing that the density of steel is 7850 kg/m^3, determine the mass moment of inertia and the radius of gyration of the component with respect to the x axis.

9.148 For the machine component of Prob. 9.147, determine the mass moment of inertia and the radius of gyration with respect to the y axis.

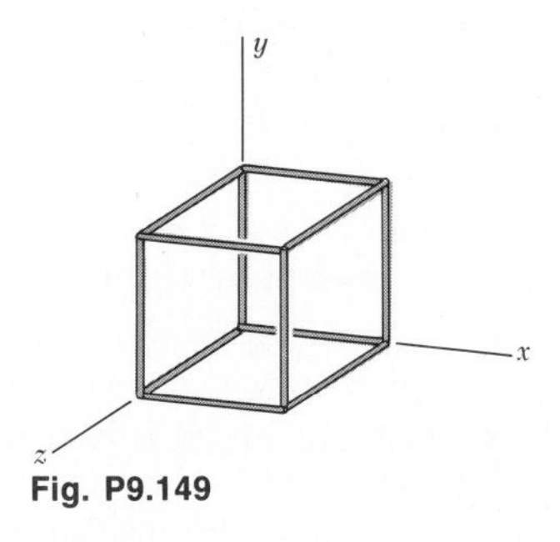

Fig. P9.149

9.149 Twelve uniform slender rods, each of length l, are welded together to form the cubical figure shown. Denoting by m the total mass of the twelve rods, determine the mass moment of inertia and the radius of gyration of the figure with respect to the x axis.

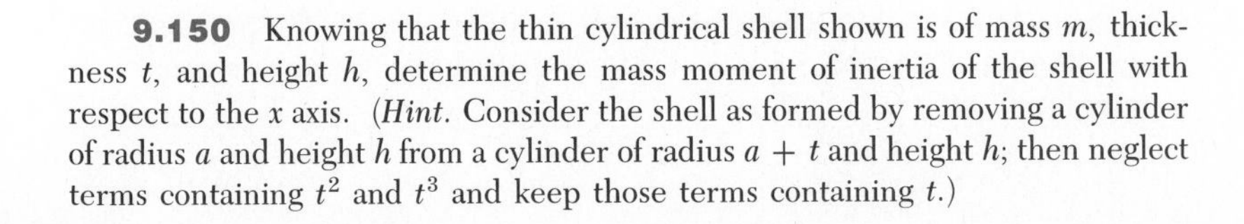
9.150 Knowing that the thin cylindrical shell shown is of mass m, thickness t, and height h, determine the mass moment of inertia of the shell with respect to the x axis. (*Hint.* Consider the shell as formed by removing a cylinder of radius a and height h from a cylinder of radius $a + t$ and height h; then neglect terms containing t^2 and t^3 and keep those terms containing t.)

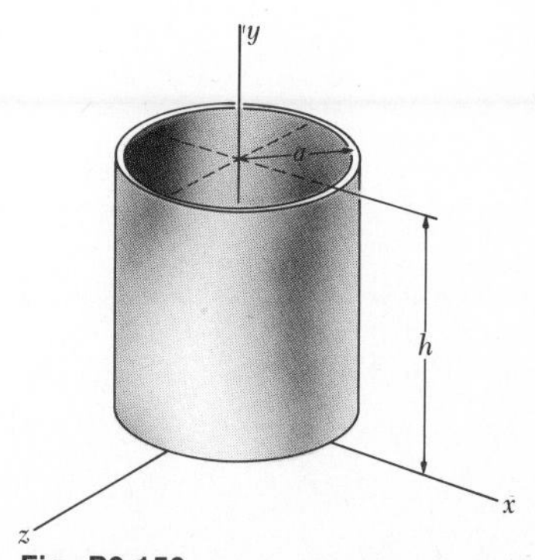

Fig. P9.150

9.151 Determine by approximate means the moment of inertia of the shaded area with respect to (*a*) the x axis, (*b*) the y axis.

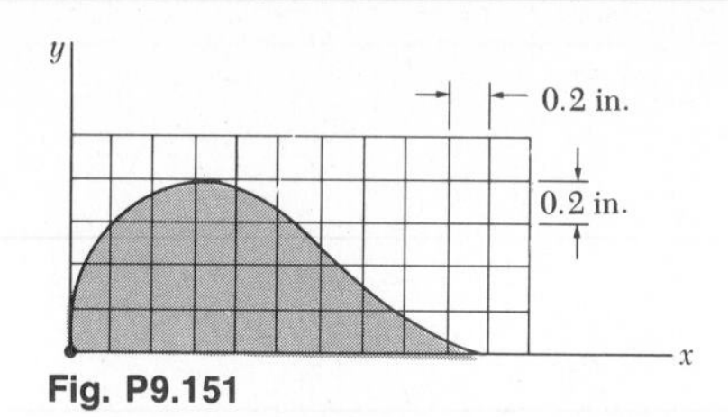

Fig. P9.151

9.152 A thin plate of mass m is cut in the shape of an equilateral triangle of side a. Determine the mass moment of inertia of the plate with respect to (*a*) the centroidal axes AA' and BB' in the plane of the plate, (*b*) the centroidal axis CC' perpendicular to the plate.

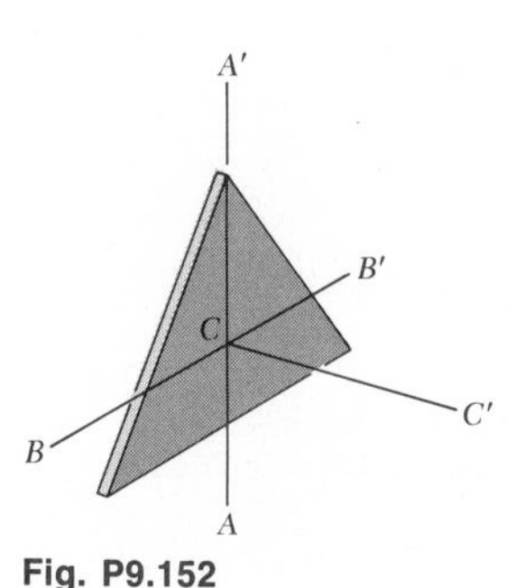

Fig. P9.152

The following problems are designed to be solved with a computer.

9.C1 Write a computer program which, for an area with known moments and product of inertia I_x, I_y, and I_{xy}, can be used to calculate the moments of inertia $I_{x'}$, $I_{y'}$, and $I_{x'y'}$ of the area with respect to axes x' and y' obtained by rotating the original axes counterclockwise through an angle θ. Use this program to compute $I_{x'}$, $I_{y'}$, and $I_{x'y'}$ for the section of Sample Prob. 9.8 for values of θ from 0 to 90° at 5° intervals.

9.C2 Write a computer program which, for an area with known moments and product of inertia I_x, I_y, and I_{xy}, can be used to calculate the orientation of the principal axes of the area and the corresponding values of the principal moments of inertia. Use this program to solve (*a*) Prob. 9.73, (*b*) parts *a* and *b* of Sample Prob. 9.8.

9.C3 Approximate the plane area shown by a series of rectangles of the form $bb'd'd$ and write a computer program which can be used to calculate the moment of inertia and radius of gyration of the area shown in part *a* of the figure with respect to each of the coordinate axes. Use this program to compute I_x, k_x, I_y, and k_y for the areas shown in parts *b* and *c* of the figure.

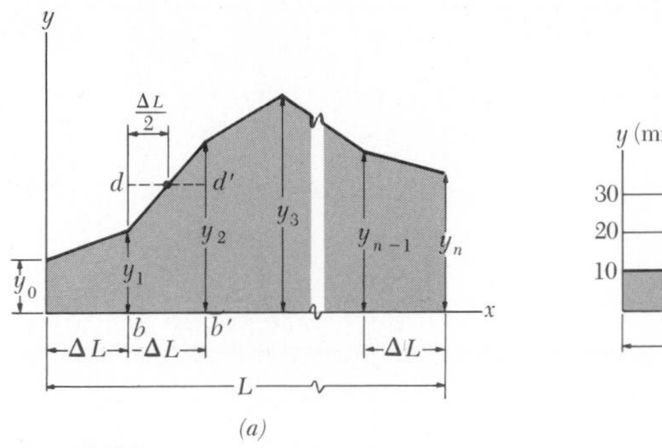

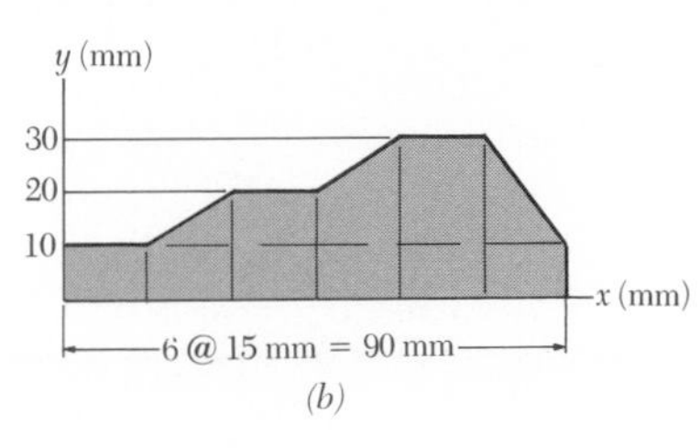

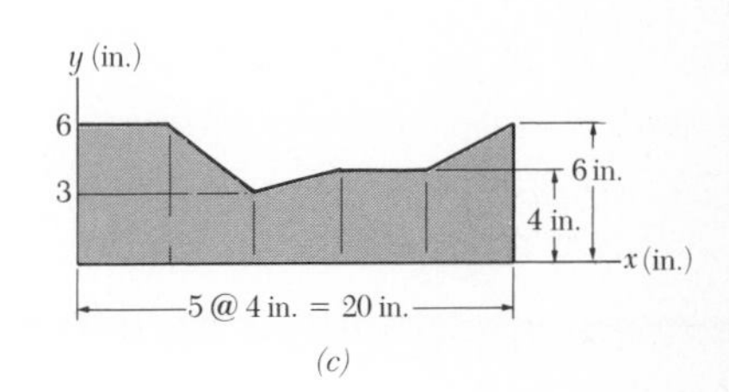

Fig. P9.C3

9.C4 The web of the steel flywheel shown consists of a solid plate of thickness t_2. Write a computer program which can be used to calculate the mass moment of inertia and the radius of gyration of the flywheel with respect to the axis of rotation. Use this program to solve (*a*) Prob. 9.102, (*b*) Prob. 9.103.

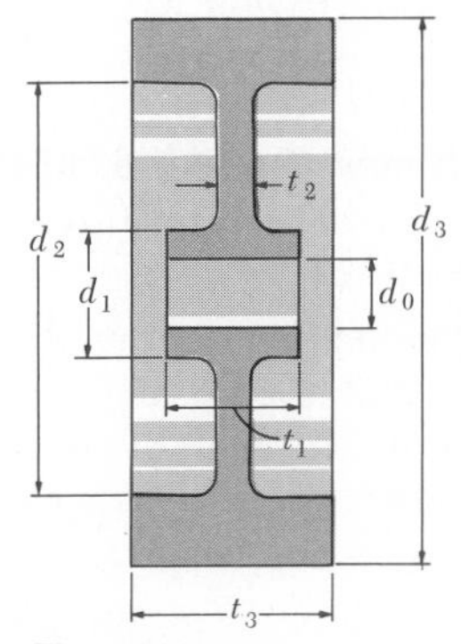

Fig. P9.C4

Method of Virtual Work

CHAPTER

10

*** 10.1. Introduction.** In the preceding chapters, problems involving the equilibrium of rigid bodies were solved by expressing that the external forces acting on the bodies were balanced. The equations of equilibrium $\Sigma F_x = 0$, $\Sigma F_y = 0$, $\Sigma M_A = 0$ were written and solved for the desired unknowns. We shall now consider a different method, which will prove more effective for solving certain types of equilibrium problems. This method is based on the *principle of virtual work* and was first formally used by the Swiss mathematician Jean Bernoulli in the eighteenth century.

As we shall see in Sec. 10.3, the principle of virtual work states that if a particle or rigid body, or, more generally, if a system of connected rigid bodies is in equilibrium under various external forces, and if the system is given an arbitrary displacement from that position of equilibrium, the total work done by the external forces during the displacement is zero. This principle may be applied particularly effectively to the solution of problems involving the equilibrium of machines or mechanisms consisting of several connected members.

In the second part of the chapter, the method of virtual work will be applied in an alternate form based on the concept of *potential energy*. As we shall see in Sec. 10.8, if a particle, rigid body, or system of rigid bodies is in equilibrium, then the derivative of its potential energy with respect to a variable defining its position must be zero.

In this chapter, we shall also learn to evaluate the mechanical efficiency of a machine (Sec. 10.5), and to determine whether a given position of equilibrium is stable, unstable, or neutral (Sec. 10.9).

***10.2. Work of a Force.** We shall first define the terms *displacement* and *work* as they are used in mechanics. Consider a particle which moves from a point A to a neighboring point A' (Fig. 10.1). If $\mathbf{r}$ denotes the position vector corresponding to point A, the small vector joining A and A' may be denoted by the differential $d\mathbf{r}$; the vector $d\mathbf{r}$ is called the *displacement* of the particle. Now let us assume that a force $\mathbf{F}$ is acting on the particle. The *work of the force* $\mathbf{F}$ *corresponding to the displacement* $d\mathbf{r}$ is defined as the quantity

$$dU = \mathbf{F} \cdot d\mathbf{r} \tag{10.1}$$

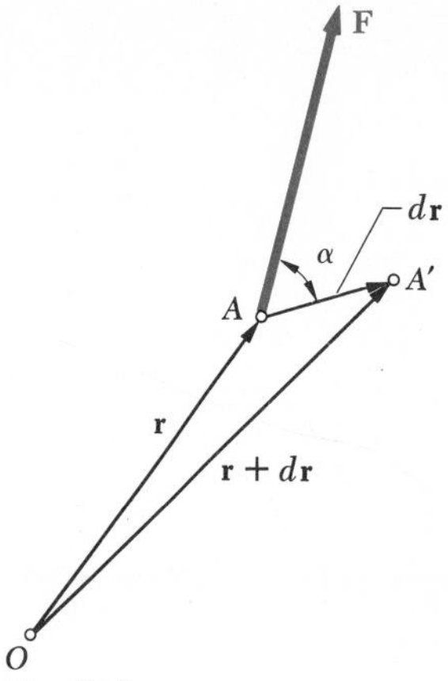

Fig. 10.1

obtained by forming the scalar product of the force $\mathbf{F}$ and of the displacement $d\mathbf{r}$. Denoting respectively by F and ds the magnitudes of the force and of the displacement, and by α the angle formed by $\mathbf{F}$ and $d\mathbf{r}$, and recalling the definition of the scalar product of two vectors (Sec. 3.9), we write

$$dU = F\,ds \cos\alpha \tag{10.1'}$$

Being a *scalar quantity*, work has a magnitude and a sign, but no direction. We also note that work should be expressed in units obtained by multiplying units of length by units of force. Thus, if U.S. customary units are used, work should be expressed in ft·lb or in·lb. If SI units are used, work should be expressed in N·m. The unit of work N·m is called a *joule* (J).†

It follows from (10.1′) that the work dU is positive if the angle α is acute and negative if α is obtuse. Three particular cases are of special interest. If the force $\mathbf{F}$ has the same direction as $d\mathbf{r}$, the work dU reduces to $F\,ds$. If $\mathbf{F}$ has a direction opposite to that of $d\mathbf{r}$, the work is $dU = -F\,ds$. Finally, if $\mathbf{F}$ is perpendicular to $d\mathbf{r}$, the work dU is zero.

The work dU of a force $\mathbf{F}$ during a displacement $d\mathbf{r}$ may also be considered as the product of F and of the component $ds \cos\alpha$ of the displacement $d\mathbf{r}$ along $\mathbf{F}$ (Fig. 10.2*a*). This view is particularly useful in the computation of the work done by the weight $\mathbf{W}$ of a body (Fig. 10.2*b*). The work of $\mathbf{W}$ is equal to the product of W and of the vertical displacement dy of the center of gravity G of the body. If the displacement is downward, the work is positive; if it is upward, the work is negative.

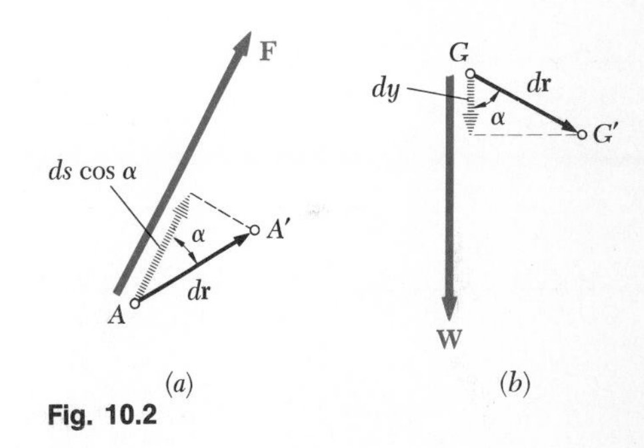

Fig. 10.2

A number of forces frequently encountered in statics *do no work*. They are forces applied to fixed points ($ds = 0$) or acting in a direction perpendicular to the displacement ($\cos\alpha = 0$). Among the forces which do no work are the following: the reaction at a frictionless pin when the body

†The joule is the SI unit of *energy*, whether in mechanical form (work, potential energy, kinetic energy) or in chemical, electrical, or thermal form. We should note that even though N·m = J, the moment of a force must be expressed in N·m, and not in joules, since the moment of a force is not a form of energy.

supported rotates about the pin, the reaction at a frictionless surface when the body in contact moves along the surface, the reaction at a roller moving along its track, the weight of a body when its center of gravity moves horizontally, the friction force acting on a wheel rolling without slipping (since at any instant the point of contact does not move). Examples of forces which *do work* are the weight of a body (except in the case considered above), the friction force acting on a body sliding on a rough surface, and most forces applied on a moving body.

In certain cases, the sum of the work done by several forces is zero. Consider, for example, two rigid bodies AC and BC connected at C by a *frictionless pin* (Fig. 10.3*a*). Among the forces acting on AC is the force $\mathbf{F}$

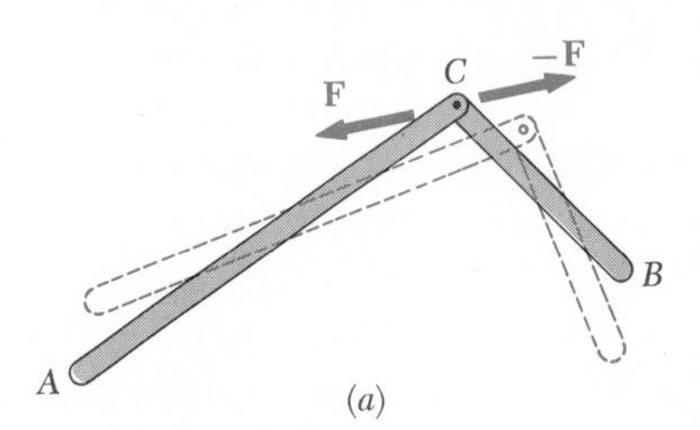

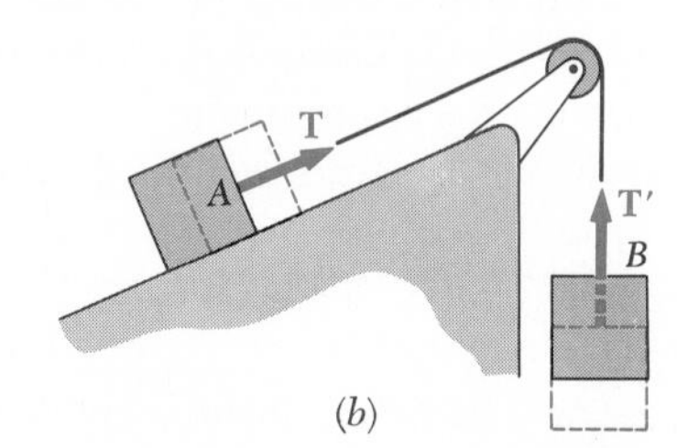

Fig. 10.3

exerted at C by BC. In general, the work of this force will not be zero, but it will be equal in magnitude and opposite in sign to the work of the force $-\mathbf{F}$ exerted by AC on BC, since these forces are equal and opposite and are applied to the same particle. Thus, when the total work done by all the forces acting on AB and BC is considered, the work of the two internal forces at C cancels out. A similar result is obtained if we consider a system consisting of two blocks connected by an *inextensible cord AB* (Fig. 10.3*b*). The work of the tension force $\mathbf{T}$ at A is equal in magnitude to the work of the tension force $\mathbf{T}'$ at B, since these forces have the same magnitude and the points A and B move through the same distance; but in one case the work is positive, and in the other it is negative. Thus, the work of the internal forces again cancels out.

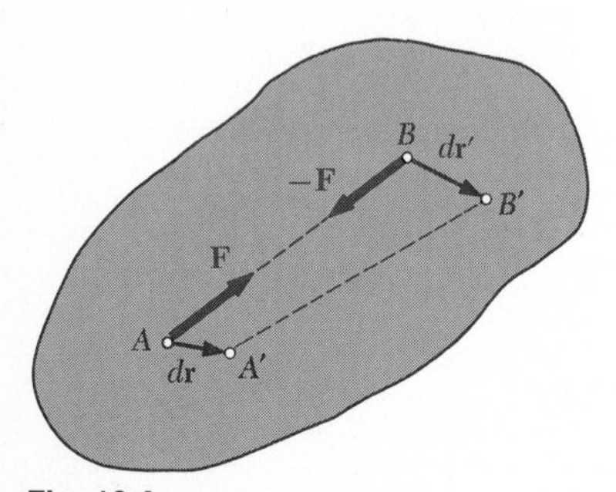

Fig. 10.4

It may be shown that the total work of the internal forces holding together the particles of a rigid body is zero. Consider two particles A and B of a rigid body and the two equal and opposite forces $\mathbf{F}$ and $-\mathbf{F}$ they exert on each other (Fig. 10.4). While, in general, small displacements $d\mathbf{r}$ and $d\mathbf{r}'$ of the two particles are different, the components of these displacements along AB must be equal; otherwise, the particles would not remain at the same distance from each other, and the body would not be rigid. Therefore, the work of $\mathbf{F}$ is equal in magnitude and opposite in sign to the work of $-\mathbf{F}$, and their sum is zero.

In computing the work of the external forces acting on a rigid body, it is often convenient to determine the work of a couple without considering separately the work of each of the two forces forming the couple. Consider the two forces $\mathbf{F}$ and $-\mathbf{F}$ forming a couple of moment $\mathbf{M}$ and acting on a rigid body (Fig. 10.5). Any small displacement of the rigid body bringing A

and B, respectively, into A' and B'' may be divided into two parts, one in which points A and B undergo equal displacements $d\mathbf{r}_1$, the other in which A' remains fixed while B' moves into B'' through a displacement $d\mathbf{r}_2$ of magnitude $ds_2 = r\,d\theta$. In the first part of the motion, the work of $\mathbf{F}$ is equal in magnitude and opposite in sign to the work of $-\mathbf{F}$, and their sum is zero. In the second part of the motion, only force $\mathbf{F}$ works, and its work is $dU = F\,ds_2 = Fr\,d\theta$. But the product Fr is equal to the magnitude M of the moment of the couple. Thus, the work of a couple of moment $\mathbf{M}$ acting on a rigid body is

$$dU = M\,d\theta \tag{10.2}$$

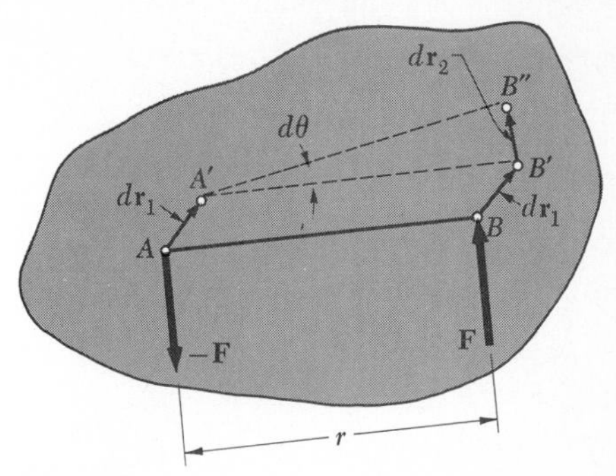

Fig. 10.5

where $d\theta$ is the small angle expressed in radians through which the body rotates. We again note that work should be expressed in units obtained by multiplying units of force by units of length.

*10.3. Principle of Virtual Work.

Consider a particle acted upon by several forces $\mathbf{F}_1, \mathbf{F}_2, \ldots, \mathbf{F}_n$ (Fig. 10.6). We shall assume that the particle undergoes a small displacement from A to A'. This displacement is possible, but it will not necessarily take place. The forces may be balanced and the particle at rest, or the particle may move under the action of the given forces in a direction different from that of AA'. The displacement considered is therefore an imaginary displacement; it is called a *virtual displacement* and is denoted by $\delta\mathbf{r}$. The symbol $\delta\mathbf{r}$ represents a differential of the first order; it is used to distinguish the virtual displacement from the displacement $d\mathbf{r}$ which would take place under actual motion. As we shall see, virtual displacements may be used to determine whether the conditions of equilibrium of a particle are satisfied.

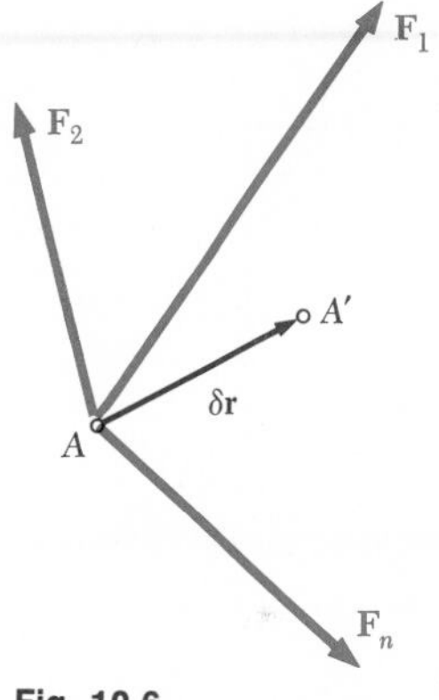

Fig. 10.6

The work of each of the forces $\mathbf{F}_1, \mathbf{F}_2, \ldots, \mathbf{F}_n$ during the virtual displacement $\delta\mathbf{r}$ is called *virtual work*. The virtual work of all the forces acting on the particle of Fig. 10.6 is

$$\begin{aligned}\delta U &= \mathbf{F}_1 \cdot \delta\mathbf{r} + \mathbf{F}_2 \cdot \delta\mathbf{r} + \cdots + \mathbf{F}_n \cdot \delta\mathbf{r} \\ &= (\mathbf{F}_1 + \mathbf{F}_2 + \cdots + \mathbf{F}_n) \cdot \delta\mathbf{r}\end{aligned}$$

or

$$\delta U = \mathbf{R} \cdot \delta\mathbf{r} \tag{10.3}$$

where $\mathbf{R}$ is the resultant of the given forces. Thus, the total virtual work of the forces $\mathbf{F}_1, \mathbf{F}_2, \ldots, \mathbf{F}_n$ is equal to the virtual work of their resultant $\mathbf{R}$.

The principle of virtual work for a particle states that, *if a particle is in equilibrium, the total virtual work of the forces acting on the particle is zero for any virtual displacement of the particle.* This condition is necessary: if the particle is in equilibrium, the resultant $\mathbf{R}$ of the forces is zero, and it follows from (10.3) that the total virtual work δU is zero. The condition is also sufficient: if the total virtual work δU is zero for any virtual displacement, the scalar product $\mathbf{R} \cdot \delta\mathbf{r}$ is zero for any $\delta\mathbf{r}$, and the resultant $\mathbf{R}$ must be zero.

In the case of a rigid body, the principle of virtual work states that *if a rigid body is in equilibrium, the total virtual work of the external forces acting on the rigid body is zero for any virtual displacement of the body.* The condition is necessary: if the body is in equilibrium, all the particles forming the body are in equilibrium and the total virtual work of the forces acting on all the particles must be zero; but we have seen in the preceding section that the total work of the internal forces is zero; the total work of the external forces must therefore also be zero. The condition may also be proved to be sufficient.

The principle of virtual work may be extended to the case of a *system of connected rigid bodies.* If the system remains connected during the virtual displacement, *only the work of the forces external to the system need be considered,* since the total work of the internal forces at the various connections is zero.

*10.4. Applications of the Principle of Virtual Work.

The principle of virtual work is particularly effective when applied to the solution of problems involving machines or mechanisms consisting of several connected rigid bodies. Consider, for instance, the toggle vise ACB of Fig. 10.7*a*, used to compress a wooden block. We wish to determine the

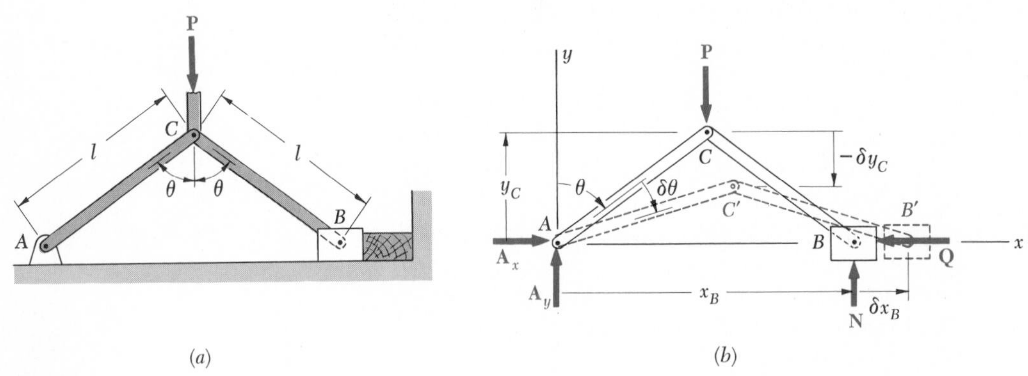

Fig. 10.7

force exerted by the vise on the block when a given force **P** is applied at C, assuming that there is no friction. Denoting by **Q** the reaction of the block on the vise, we draw the free-body diagram of the vise and consider the virtual displacement obtained by giving to the angle θ a positive increment $\delta\theta$ (Fig. 10.7*b*). Choosing a system of coordinate axes with origin at A, we note that x_B increases while y_C decreases. This is indicated in the figure by means of the positive increment δx_B and the negative increment $-\delta y_C$. The reactions $\mathbf{A}_x$, $\mathbf{A}_y$, and **N** will do no work during the virtual displacement considered, and we need only compute the work of **P** and **Q**. Since **Q** and δx_B have opposite senses, the virtual work of **Q** is $\delta U_Q = -Q\,\delta x_B$. Since **P** and the increment shown $(-\delta y_C)$ have the same sense, the virtual work of **P** is $\delta U_P = +P(-\delta y_C) = -P\,\delta y_C$. The minus signs obtained could have been predicted by simply noting that the forces **Q** and **P** are directed opposite to the positive x and y axes, respectively. Expressing the coordinates x_B and y_C in terms of the angle θ and differentiating, we obtain

$$\begin{aligned} x_B &= 2l \sin\theta \qquad & y_C &= l \cos\theta \\ \delta x_B &= 2l \cos\theta\, \delta\theta \qquad & \delta y_C &= -l \sin\theta\, \delta\theta \end{aligned} \tag{10.4}$$

The total virtual work of the forces **Q** and **P** is thus

$$\begin{aligned} \delta U = \delta U_Q + \delta U_P &= -Q\,\delta x_B - P\,\delta y_C \\ &= -2Ql \cos\theta\,\delta\theta + Pl \sin\theta\,\delta\theta \end{aligned}$$

Making $\delta U = 0$, we obtain

$$2Ql \cos\theta\,\delta\theta = Pl \sin\theta\,\delta\theta \tag{10.5}$$

$$Q = \tfrac{1}{2}P \tan\theta \tag{10.6}$$

The superiority of the method of virtual work over the conventional equilibrium equations in the problem considered here is clear: by using the method of virtual work, we were able to eliminate all unknown reactions, while the equation $\Sigma M_A = 0$ would have eliminated only two of the unknown reactions. We may take advantage of this characteristic of the method of virtual work to solve many problems involving machines and mechanisms. *If the virtual displacement considered is consistent with the constraints imposed by the supports and connections, all reactions and internal forces are eliminated and only the work of the loads, applied forces, and friction forces need be considered.*

We shall observe that the method of virtual work may also be used to solve problems involving completely constrained structures, although the virtual displacements considered will never actually take place. Consider, for example, the frame ACB shown in Fig. 10.8a. If point A is kept fixed, while B is given a horizontal virtual displacement (Fig. 10.8b), we need consider only the work of **P** and $\mathbf{B}_x$. We may thus determine the reaction

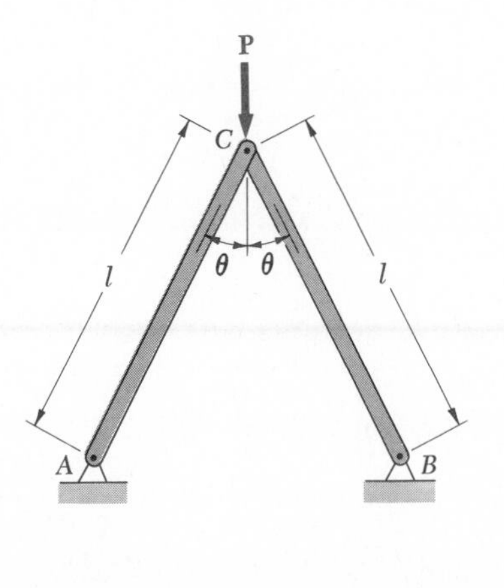

(a)

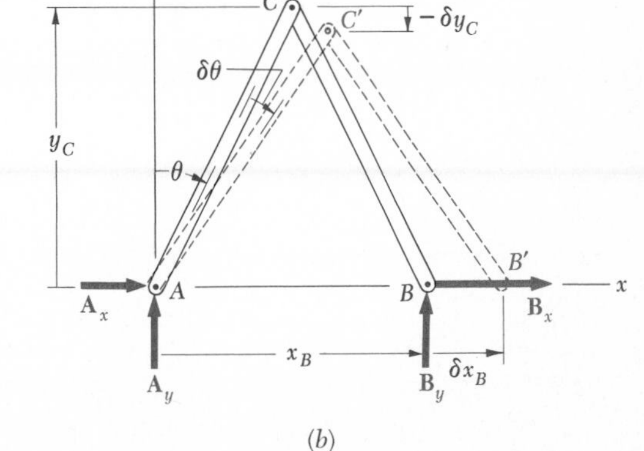

(b)

Fig. 10.8

component $\mathbf{B}_x$ in the same way as the force $\mathbf{Q}$ of the preceding example (Fig. 10.7*b*); we have

$$B_x = -\tfrac{1}{2}P \tan \theta$$

Keeping B fixed and giving to A a horizontal virtual displacement, we may similarly determine the reaction component $\mathbf{A}_x$. The components $\mathbf{A}_y$ and $\mathbf{B}_y$ may be determined by rotating the frame ACB as a rigid body about B and A, respectively.

The method of virtual work may also be used to determine the configuration of a system in equilibrium under given forces. For example, the value of the angle θ for which the linkage of Fig. 10.7 is in equilibrium under two given forces $\mathbf{P}$ and $\mathbf{Q}$ may be obtained by solving Eq. (10.6) for $\tan \theta$.

It should be noted, however, that the attractiveness of the method of virtual work depends to a large extent upon the existence of simple geometric relations between the various virtual displacements involved in the solution of a given problem. When no such simple relations exist, it is usually advisable to revert to the conventional method of Chap. 6.

*** 10.5. Real Machines. Mechanical Efficiency.** In analyzing the toggle vise in the preceding section, we assumed that no friction forces were involved. Thus, the virtual work consisted only of the work of the applied force $\mathbf{P}$ and of the reaction $\mathbf{Q}$. But the work of the reaction $\mathbf{Q}$ is equal in magnitude and opposite in sign to the work of the force exerted by the vise on the block. Equation (10.5), therefore, expresses that the *output work* $2Ql \cos \theta \, \delta\theta$ is equal to the *input work* $Pl \sin \theta \, \delta\theta$. A machine in which input and output work are equal is said to be an "ideal" machine. In a "real" machine, friction forces will always do some work, and the output work will be smaller than the input work.

Consider, for example, the toggle vise of Fig. 10.7*a*, and assume now that a friction force $\mathbf{F}$ develops between the sliding block B and the horizontal plane (Fig. 10.9). Using the conventional methods of statics and

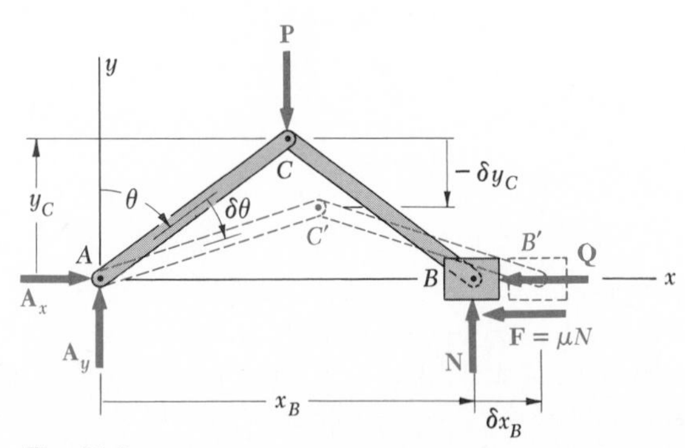

Fig. 10.9

summing moments about A, we find $N = P/2$. Denoting by μ the coefficient of friction between block B and the horizontal plane, we have $F = \mu N = \mu P/2$. Recalling formulas (10.4), we find that the total virtual work of the forces **Q**, **P**, and **F** during the virtual displacement shown in Fig. 10.9 is

$$\begin{aligned}\delta U &= -Q\,\delta x_B - P\,\delta y_C - F\,\delta x_B \\ &= -2Ql\cos\theta\,\delta\theta + Pl\sin\theta\,\delta\theta - \mu Pl\cos\theta\,\delta\theta\end{aligned}$$

Making $\delta U = 0$, we obtain

$$2Ql\cos\theta\,\delta\theta = Pl\sin\theta\,\delta\theta - \mu Pl\cos\theta\,\delta\theta \tag{10.7}$$

which expresses that the output work is equal to the input work minus the work of the friction force. Solving for Q, we have

$$Q = \tfrac{1}{2}P(\tan\theta - \mu) \tag{10.8}$$

We note that $Q = 0$ when $\tan\theta = \mu$, that is, when θ is equal to the angle of friction ϕ, and that $Q < 0$ when $\theta < \phi$. The toggle vise may thus be used only for values θ larger than the angle of friction.

The *mechanical efficiency* of a machine is defined as the ratio

$$\eta = \frac{\text{output work}}{\text{input work}} \tag{10.9}$$

Clearly, the mechanical efficiency of an ideal machine is $\eta = 1$, since input and output work are then equal, while the mechanical efficiency of a real machine will always be less than 1.

In the case of the toggle vise we have just analyzed, we write

$$\eta = \frac{\text{output work}}{\text{input work}} = \frac{2Ql\cos\theta\,\delta\theta}{Pl\sin\theta\,\delta\theta}$$

Substituting from (10.8) for Q, we obtain

$$\eta = \frac{P(\tan\theta - \mu)l\cos\theta\,\delta\theta}{Pl\sin\theta\,\delta\theta} = 1 - \mu\cot\theta \tag{10.10}$$

We check that in the absence of friction forces, we would have $\mu = 0$ and $\eta = 1$. In the general case, when μ is different from zero, the efficiency η becomes zero for $\mu\cot\theta = 1$, that is, for $\tan\theta = \mu$, or $\theta = \tan^{-1}\mu = \phi$. We check again that the toggle vise may be used only for values of θ larger than the angle of friction ϕ.

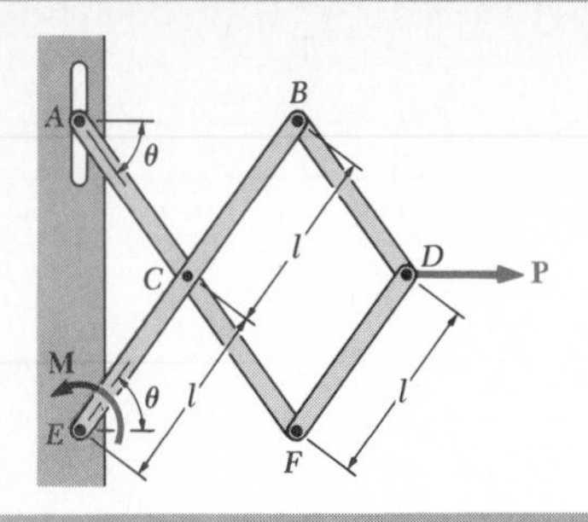

SAMPLE PROBLEM 10.1

Using the method of virtual work, determine the magnitude of the couple **M** required to maintain the equilibrium of the mechanism shown.

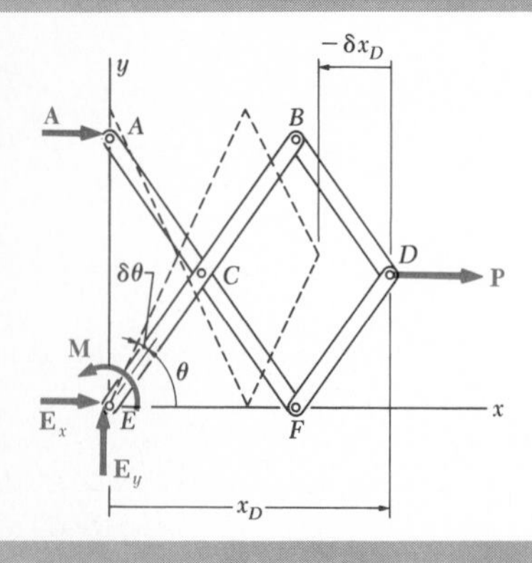

Solution. Choosing a coordinate system with origin at E, we write

$$x_D = 3l\cos\theta \qquad \delta x_D = -3l\sin\theta\,\delta\theta$$

Principle of Virtual Work. Since the reactions **A**, $\mathbf{E}_x$, and $\mathbf{E}_y$ will do no work during the virtual displacement, the total virtual work done by **M** and **P** must be zero. Noting that **P** acts in the positive x direction and **M** acts in the positive θ direction, we write

$\delta U = 0$:

$$+M\,\delta\theta + P\,\delta x_D = 0$$
$$+M\,\delta\theta + P(-3l\sin\theta\,\delta\theta) = 0$$

$$M = 3Pl\sin\theta \quad ◄$$

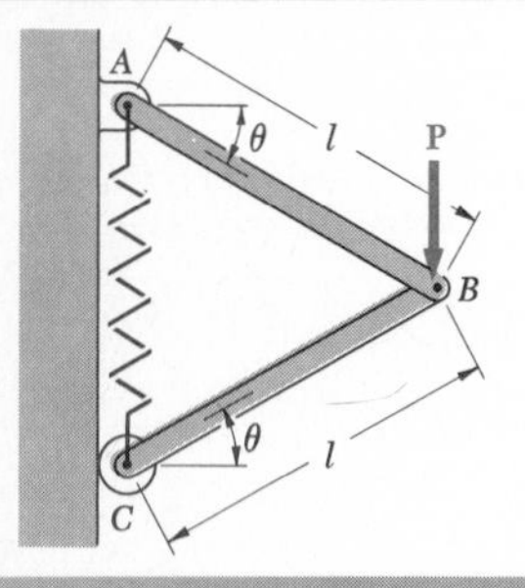

SAMPLE PROBLEM 10.2

Determine the expressions for θ and for the tension in the spring which correspond to the equilibrium position of the mechanism. The unstretched length of the spring is h, and the constant of the spring is k. Neglect the weight of the mechanism.

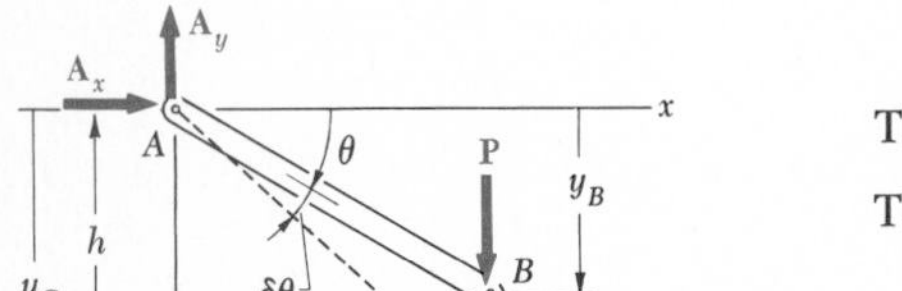

Solution. With the coordinate system shown

$$y_B = l\sin\theta \qquad y_C = 2l\sin\theta$$
$$\delta y_B = l\cos\theta\,\delta\theta \qquad \delta y_C = 2l\cos\theta\,\delta\theta$$

The elongation of the spring is $\quad s = y_C - h = 2l\sin\theta - h$

The magnitude of the force exerted at C by the spring is

$$F = ks = k(2l\sin\theta - h) \tag{1}$$

Principle of Virtual Work. Since the reactions $\mathbf{A}_x$, $\mathbf{A}_y$, and **C** do no work, the total virtual work done by **P** and **F** must be zero.

$\delta U = 0$:

$$P\,\delta y_B - F\,\delta y_C = 0$$
$$P(l\cos\theta\,\delta\theta) - k(2l\sin\theta - h)(2l\cos\theta\,\delta\theta) = 0$$

$$\sin\theta = \frac{P + 2kh}{4kl} \quad ◄$$

Substituting this expression into (1), we obtain $\qquad F = \tfrac{1}{2}P \quad ◄$

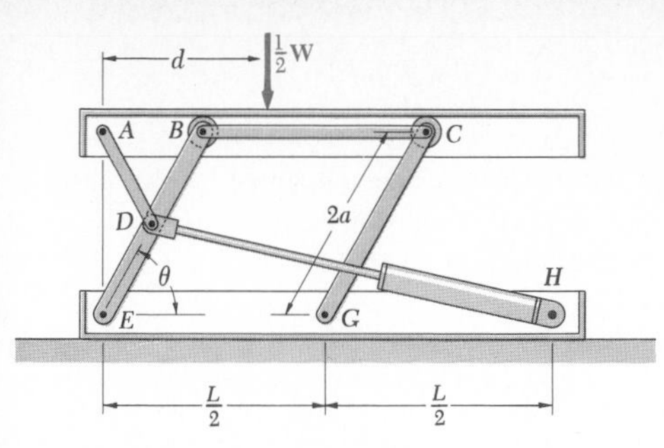

SAMPLE PROBLEM 10.3

A hydraulic-lift table is used to raise a 1000-kg crate. It consists of a platform and of two identical linkages on which hydraulic cylinders exert equal forces. (Only one linkage and one cylinder are shown.) Members *EDB* and *CG* are each of length $2a$, and member *AD* is pinned to the midpoint of *EDB*. If the crate is placed on the table, so that half of its weight is supported by the system shown, determine the force exerted by each cylinder in raising the crate for $\theta = 60°$, $a = 0.70$ m, and $L = 3.20$ m. This mechanism has been previously considered in Sample Prob. 6.7.

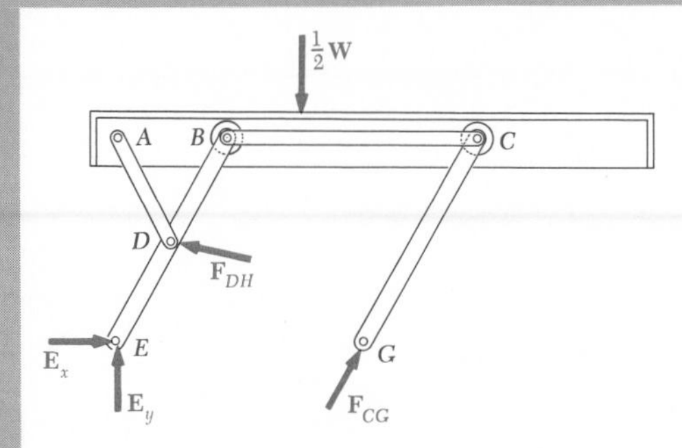

Solution. The machine considered consists of the platform and of the linkage, with an input force $\mathbf{F}_{DH}$ exerted by the cylinder and an output force equal and opposite to $\frac{1}{2}\mathbf{W}$.

Principle of Virtual Work. We first observe that the reactions at E and G do no work. Denoting by y the elevation of the platform above the base, and by s the length DH of the cylinder-and-piston assembly, we write

$$\delta U = 0: \qquad -\tfrac{1}{2}W\,\delta y + F_{DH}\,\delta s = 0 \qquad (1)$$

The vertical displacement δy of the platform is expressed in terms of the angular displacement $\delta\theta$ of EDB as follows:

$$y = (EB)\sin\theta = 2a\sin\theta$$
$$\delta y = 2a\cos\theta\,\delta\theta$$

To express δs similarly in terms of $\delta\theta$, we first note that by the law of cosines,

$$s^2 = a^2 + L^2 - 2aL\cos\theta$$

Differentiating,

$$2s\,\delta s = -2aL(-\sin\theta)\,\delta\theta$$
$$\delta s = \frac{aL\sin\theta}{s}\,\delta\theta$$

Substituting for δy and δs into (1), we write

$$(-\tfrac{1}{2}W)2a\cos\theta\,\delta\theta + F_{DH}\frac{aL\sin\theta}{s}\,\delta\theta = 0$$

$$F_{DH} = W\frac{s}{L}\cot\theta$$

With the given numerical data, we have

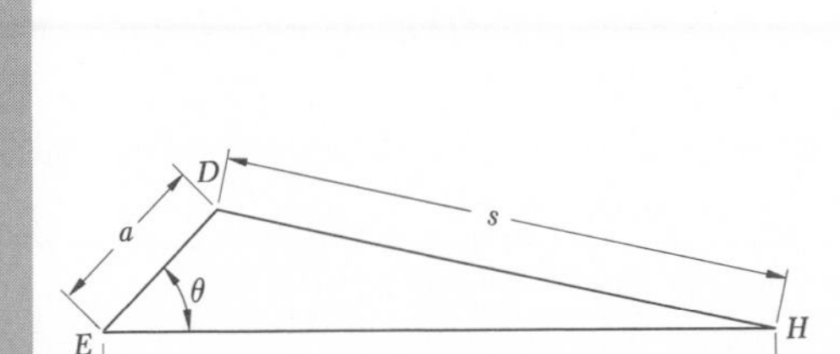

$$W = mg = (1000\text{ kg})(9.81\text{ m/s}^2) = 9810\text{ N} = 9.81\text{ kN}$$
$$s^2 = a^2 + L^2 - 2aL\cos\theta$$
$$= (0.70)^2 + (3.20)^2 - 2(0.70)(3.20)\cos 60° = 8.49$$
$$s = 2.91\text{ m}$$

$$F_{DH} = W\frac{s}{L}\cot\theta = (9.81\text{ kN})\frac{2.91\text{ m}}{3.20\text{ m}}\cot 60°$$

$$F_{DH} = 5.15\text{ kN} \quad ◄$$

Problems

10.1 Knowing that the maximum friction force exerted by the bottle on the cork is 360 N, determine (*a*) the force **P** which must be applied to the corkscrew to open the bottle, (*b*) the maximum force exerted by the base of the corkscrew on the top of the bottle.

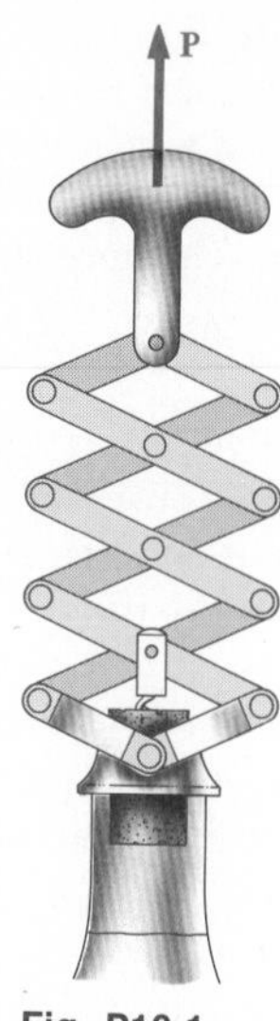

Fig. P10.1

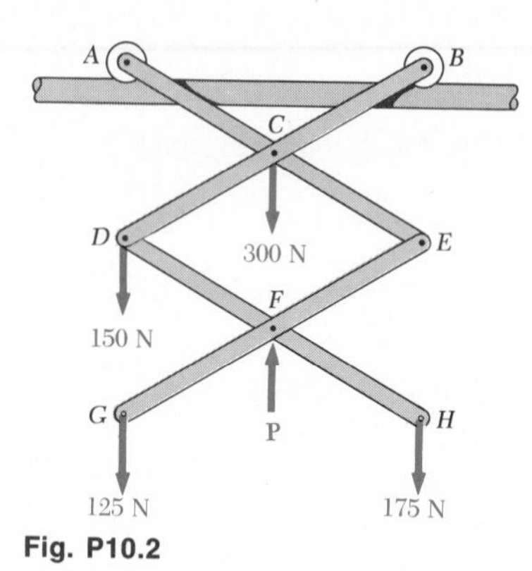

Fig. P10.2

10.2 Determine the force **P** required to maintain the equilibrium of the linkage.

10.3 Determine the horizontal force **P** which must be applied at *A* to maintain the equilibrium of the linkage.

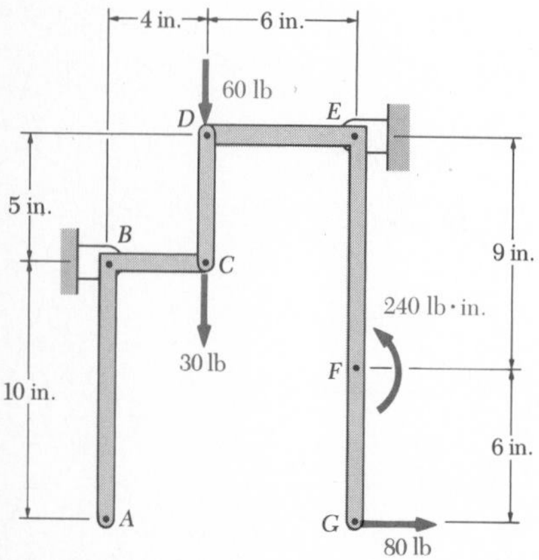

Fig. P10.3 and P10.6

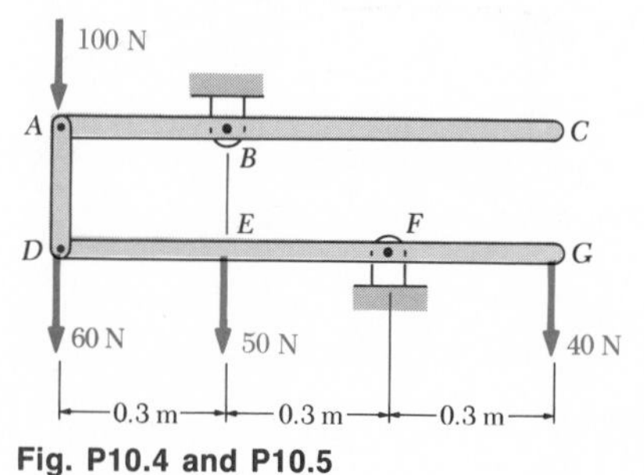

Fig. P10.4 and P10.5

10.4 Determine the vertical force **P** which must be applied at *C* to maintain the equilibrium of the linkage.

10.5 and 10.6 Determine the couple **M** which must be applied to member *ABC* to maintain the equilibrium of the linkage.

10.7 The mechanism shown is acted upon by the force **P**; derive an expression for the magnitude of the force **Q** required to maintain equilibrium.

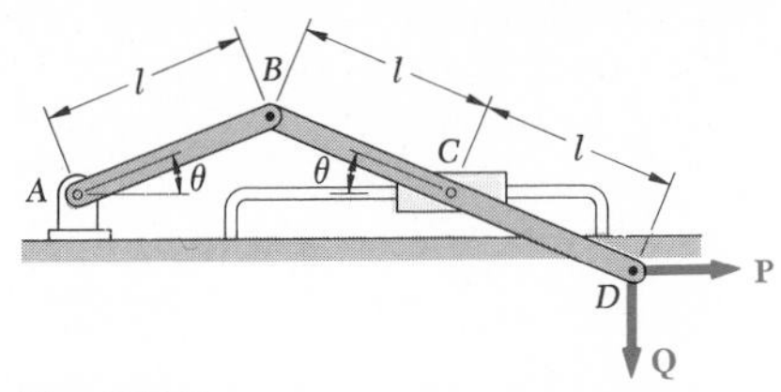

Fig. P10.7

10.8 Knowing that the line of action of the force **Q** passes through point D, derive an expression for the magnitude of **Q** required to maintain equilibrium.

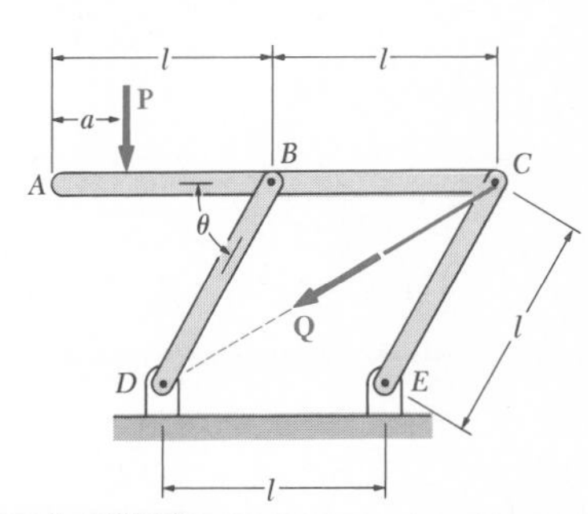

Fig. P10.8

10.9 An overhead garage door of weight W consists of a uniform rectangular panel AC supported by a cable AE attached at the middle of the upper edge of the door and by two sets of frictionless rollers A and B which may slide in horizontal and vertical channels. Express the tension T in cable AE in terms of W, a, b, and θ.

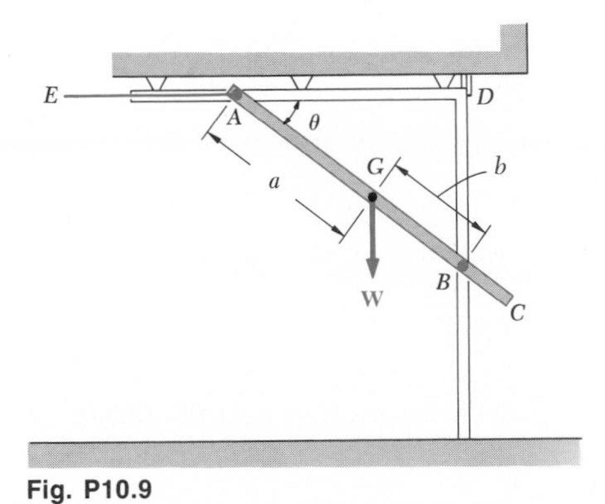

Fig. P10.9

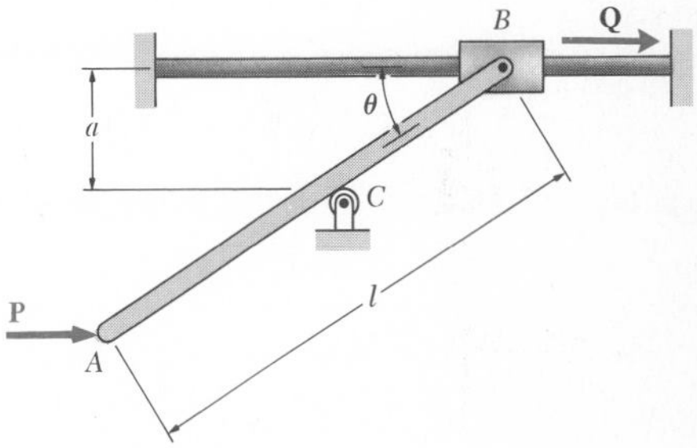

Fig. P10.10

10.10 The slender rod AB is attached to a collar B and rests on a wheel at C. Neglecting the effect of friction, derive an expression for the magnitude of the force **Q** required to maintain the equilibrium of the rod.

10.11 Solve Prob. 10.10, assuming that the force **P** applied at A is vertical and directed downward.

10.12 A uniform rod AB of length l and weight W is suspended from two cords AC and BC of equal length. Derive an expression for the magnitude of the couple **M** required to maintain the equilibrium of the rod in the position shown.

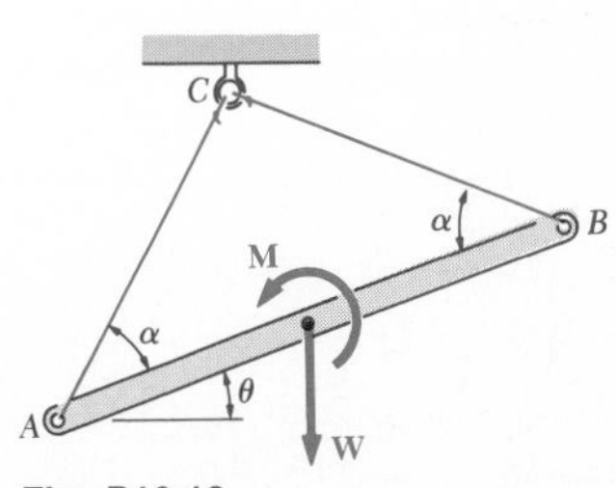

Fig. P10.12

10.13 Derive an expression for the magnitude of the couple **M** required to maintain the equilibrium of the linkage in the position shown.

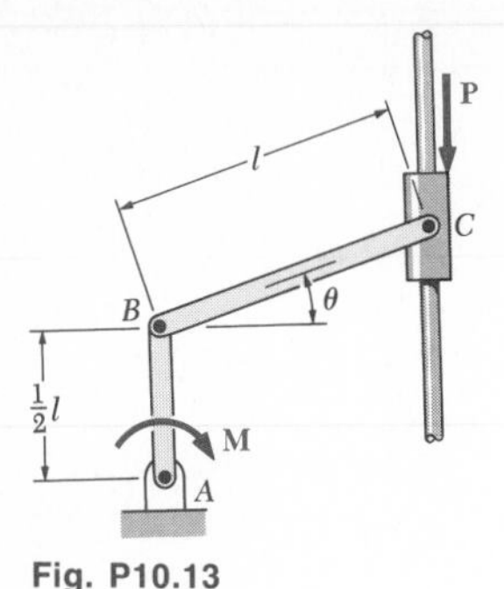

Fig. P10.13

10.14 An 800-lb force **P** is applied to the piston of the engine system shown. Knowing that $AB = 3$ in. and $BC = 8$ in., determine the couple **M** required to maintain the equilibrium of the system when (*a*) $\theta = 30°$, (*b*) $\theta = 150°$.

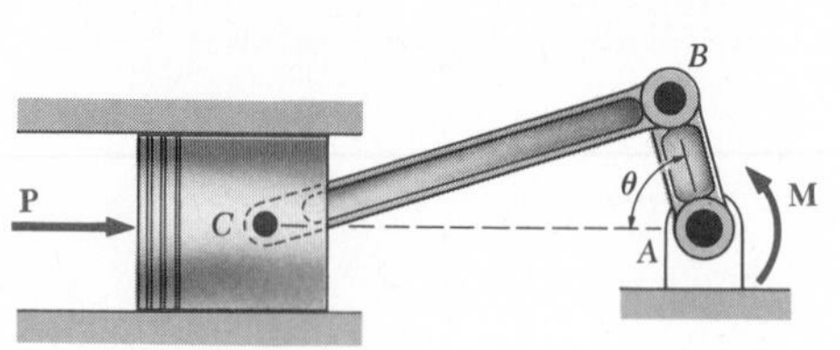

Fig. P10.14 and P10.15

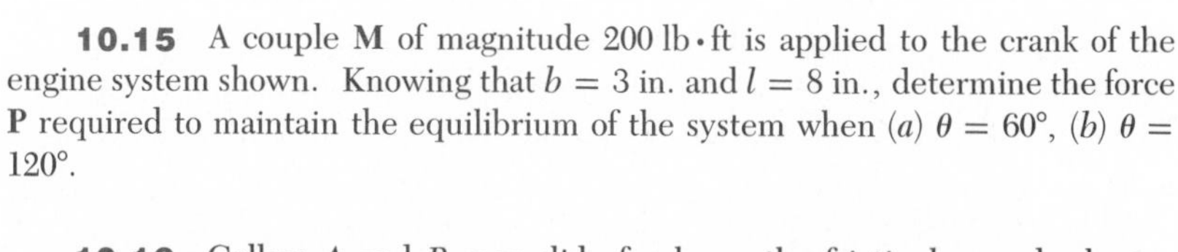
10.15 A couple **M** of magnitude 200 lb·ft is applied to the crank of the engine system shown. Knowing that $b = 3$ in. and $l = 8$ in., determine the force **P** required to maintain the equilibrium of the system when (*a*) $\theta = 60°$, (*b*) $\theta = 120°$.

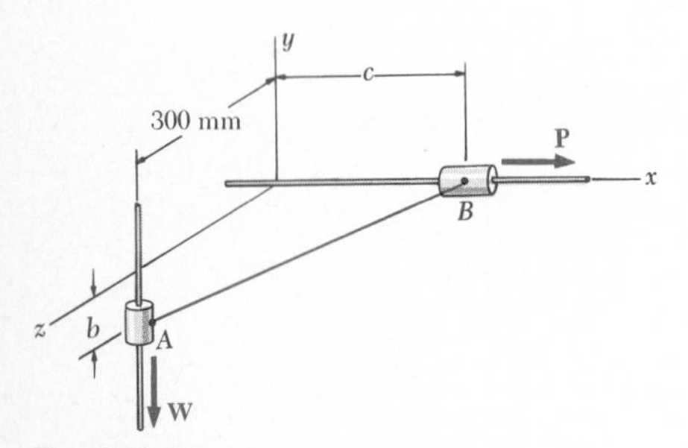

Fig. P10.16

10.16 Collars *A* and *B* may slide freely on the frictionless rods shown. Knowing that the length of the wire *AB* is 550 mm and that the weight of collar *A* is $W = 63$ N, determine the magnitude of the force **P** required to maintain the equilibrium of the system when (*a*) $c = 100$ mm, (*b*) $c = 300$ mm.

10.17 Solve Prob. 10.16 when (*a*) $c = 350$ mm, (*b*) $c = 450$ mm.

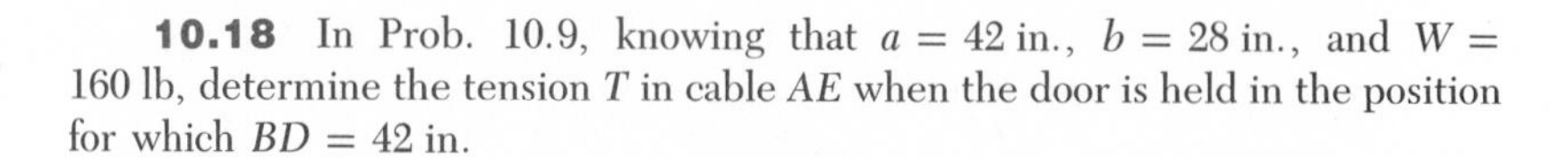
10.18 In Prob. 10.9, knowing that $a = 42$ in., $b = 28$ in., and $W = 160$ lb, determine the tension *T* in cable *AE* when the door is held in the position for which $BD = 42$ in.

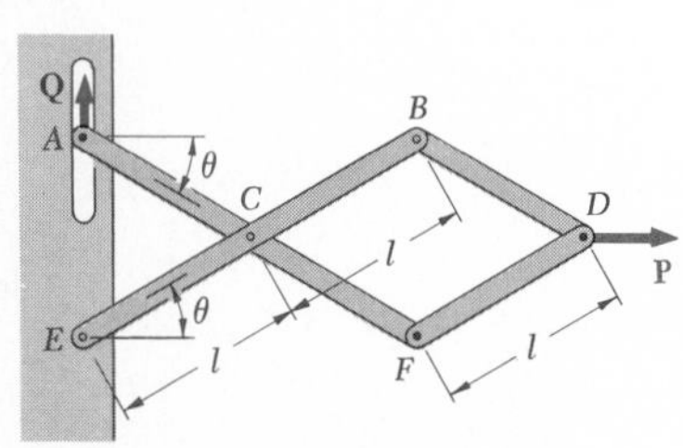

Fig. 10.20

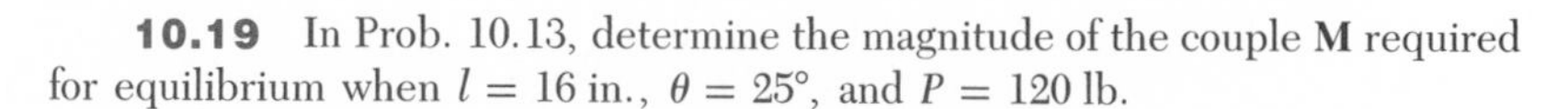
10.19 In Prob. 10.13, determine the magnitude of the couple **M** required for equilibrium when $l = 16$ in., $\theta = 25°$, and $P = 120$ lb.

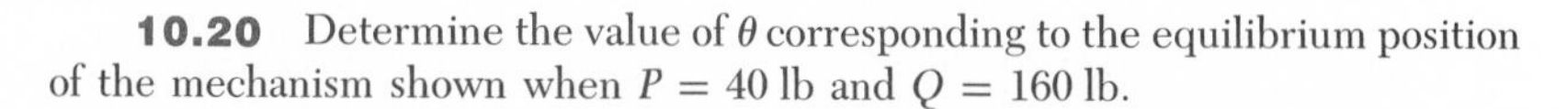
10.20 Determine the value of θ corresponding to the equilibrium position of the mechanism shown when $P = 40$ lb and $Q = 160$ lb.

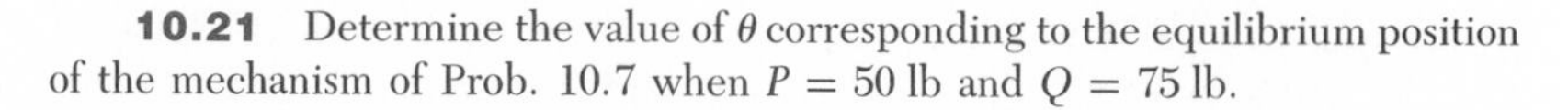
10.21 Determine the value of θ corresponding to the equilibrium position of the mechanism of Prob. 10.7 when $P = 50$ lb and $Q = 75$ lb.

10.22 Determine the value of θ corresponding to the equilibrium position of the mechanism of Prob. 10.10 when $P = 250$ N, $Q = 500$ N, $l = 400$ mm, and $a = 80$ mm.

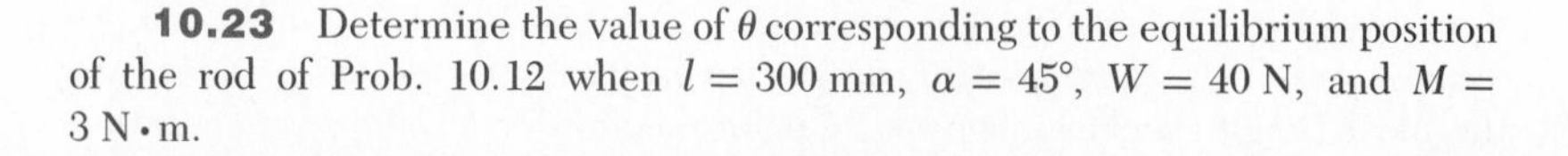
10.23 Determine the value of θ corresponding to the equilibrium position of the rod of Prob. 10.12 when $l = 300$ mm, $\alpha = 45°$, $W = 40$ N, and $M = 3$ N·m.

10.24 A vertical load **W** is applied to the linkage at B. The constant of the spring is k, and the spring is unstretched when AB and BC are horizontal. Neglecting the weight of the linkage, derive an equation in θ, W, l, and k, which must be satisfied when the linkage is in equilibrium.

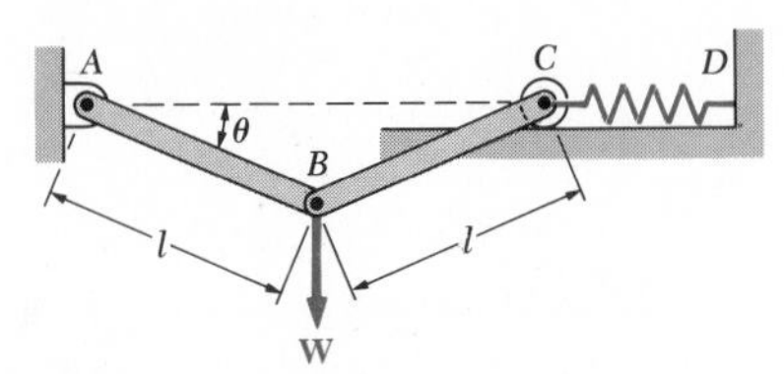

Fig. P10.24 and P10.25

10.25 A load **W** of magnitude 900 N is applied to the linkage at B. Neglecting the weight of the linkage and knowing that $l = 225$ mm, determine the value of θ corresponding to equilibrium. The constant of the spring is $k = 2$ kN/m, and the spring is unstretched when AB and BC are horizontal. (*Hint.* Solve the equation obtained for θ by trial and error.)

10.26 Two bars AD and DG are connected by a pin at D and by a spring AG. Knowing that the spring is 300 mm long when unstretched and that the constant of the spring is 5 kN/m, determine the value of x corresponding to equilibrium when a 900-N load is applied at E as shown.

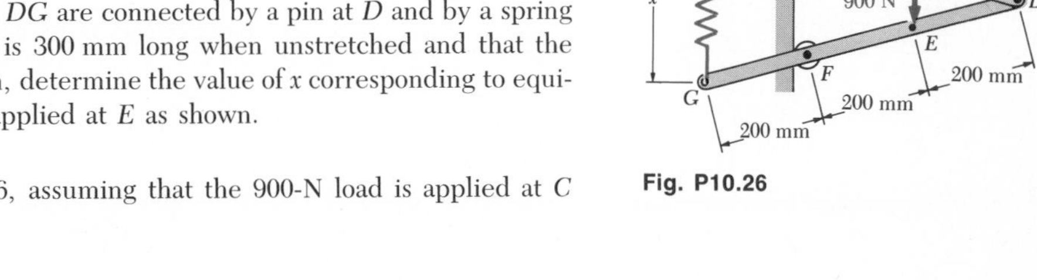

Fig. P10.26

10.27 Solve Prob. 10.26, assuming that the 900-N load is applied at C instead of E.

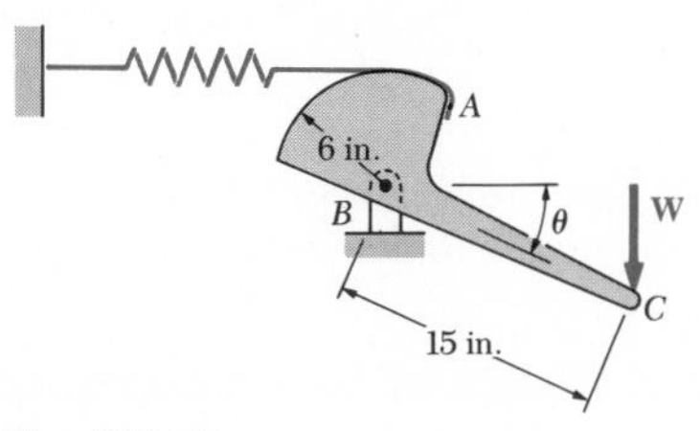

Fig. P10.28

10.28 A load **W** of magnitude 72 lb is applied to the mechanism at C. Neglecting the weight of the mechanism, determine the value of θ corresponding to equilibrium. The constant of the spring is $k = 20$ lb/in., and the spring is unstretched when $\theta = 0$.

10.29 A force **P** of magnitude 60 lb is applied to end E of cable CDE, which passes under pulley D and is attached to the mechanism at C. Neglecting the weight of the mechanism and the radius of the pulley, determine the value of θ corresponding to equilibrium. The constant of the spring is $k = 20$ lb/in., and the spring is unstretched when $\theta = 90°$.

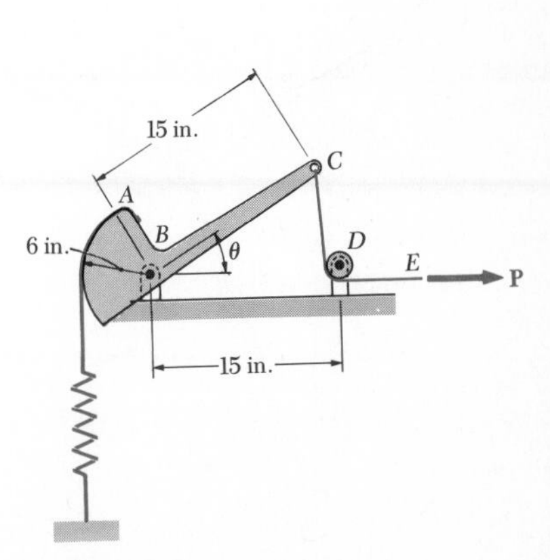

Fig. P10.29

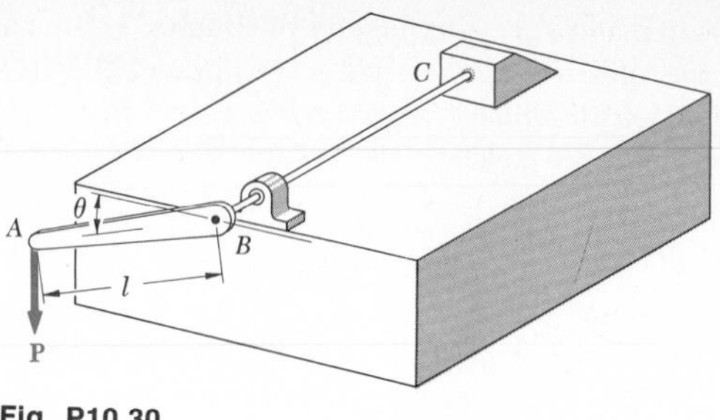

Fig. P10.30

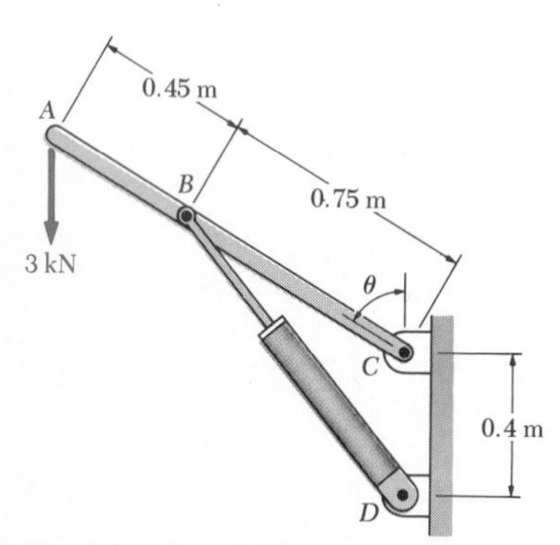

Fig. P10.32 and P10.33

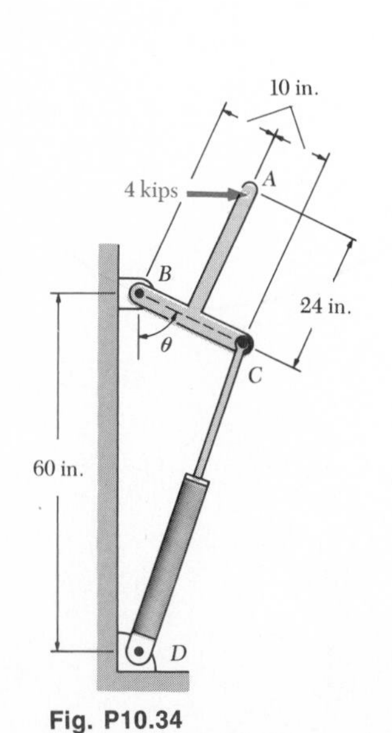

Fig. P10.34

10.30 The lever AB is attached to the horizontal shaft BC which passes through a bearing and is welded to a fixed support at C. The torsional spring constant of the shaft BC is K; that is, a couple of magnitude K is required to rotate end B through 1 radian. Knowing that the shaft is untwisted when AB is horizontal, determine the value of θ corresponding to the position of equilibrium, if $P = 125$ N, $l = 200$ mm, and $K = 15$ N · m/rad.

10.31 Solve Prob. 10.30, if $P = 600$ N, $l = 200$ mm, and $K = 15$ N · m/rad. Obtain an answer in each of the following quadrants: $0 < \theta < 90°$, $270° < \theta < 360°$, and $360° < \theta < 450°$. (It is assumed that K remains constant for the angles of twist considered.)

10.32 The position of boom ABC is controlled by the hydraulic cylinder BD. For the loading shown, determine the force exerted by the hydraulic cylinder on pin B when $\theta = 65°$.

10.33 The position of boom ABC is controlled by the hydraulic cylinder BD. For the loading shown, (*a*) express the force exerted by the hydraulic cylinder on pin B as a function of the length BD, (*b*) determine the smallest possible value of the angle θ if the maximum force that the cylinder can exert on pin B is 12.5 kN.

10.34 The position of member ABC is controlled by the hydraulic cylinder CD. For the loading shown, determine the force exerted by the hydraulic cylinder on pin C when $\theta = 60°$.

10.35 Solve Prob. 10.34 when $\theta = 120°$.

10.36 A block of weight W is pulled up a plane forming an angle α with the horizontal by a force **P** directed along the plane. If μ is the coefficient of friction between the block and the plane, derive an expression for the mechanical efficiency of the system. Show that the mechanical efficiency cannot exceed $\frac{1}{2}$ if the block is to remain in place when the force **P** is removed.

10.37 Derive an expression for the mechanical efficiency of the jack discussed in Sec. 8.6. Show that if the jack is to be self-locking, the mechanical efficiency cannot exceed $\frac{1}{2}$.

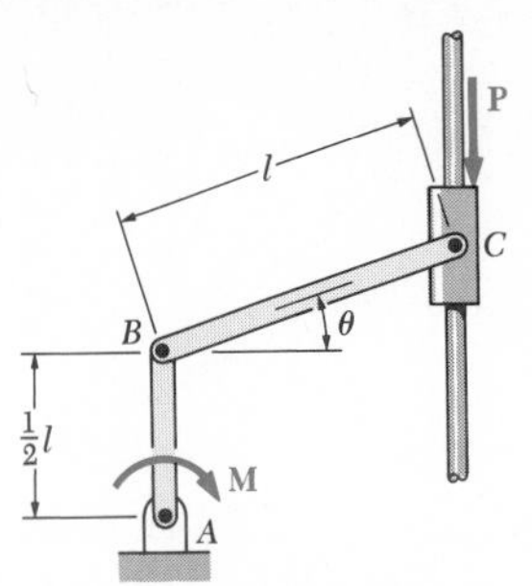

Fig. P10.38 and P10.39

10.38 Denoting by μ_s the coefficient of static friction between collar C and the vertical rod, derive an expression for the magnitude of the largest couple **M** for which equilibrium is maintained in the position shown. Explain what happens if $\mu_s \geq \tan\theta$.

10.39 Knowing that the coefficient of static friction between collar C and the vertical rod is 0.30, determine the magnitude of the largest and smallest couple **M** for which equilibrium is maintained in the position shown, when $\theta = 35°$, $l = 500$ mm, and $P = 400$ N.

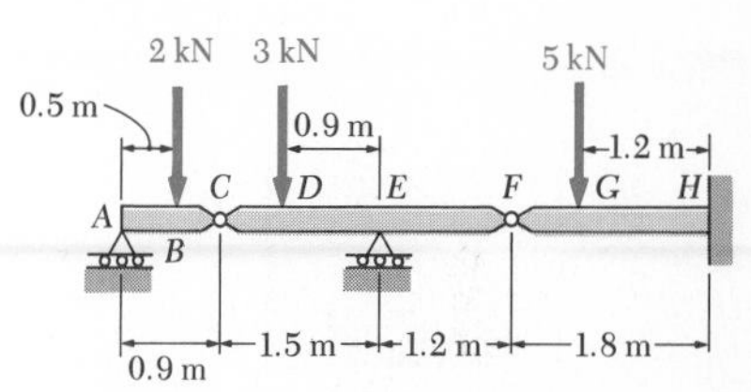

Fig. P10.40 and P10.41

10.40 Using the method of virtual work, determine the reaction at E.

10.41 Using the method of virtual work, determine separately the force and couple representing the reaction at H.

10.42 Referring to Prob. 6.122 and using the value found for F_{BD}, determine the change in length of the cylinder-and-piston assembly BD required to raise bucket C by 3 in.

10.43 Referring to Prob. 6.124 and using the value found for F_{CD}, determine the change in length of each of the two cylinder-and-piston assemblies CD required to raise the 500-kg slab by 50 mm if the other two cylinders are not activated.

10.44 Referring to Prob. 6.124 and using the value found for F_{FH}, determine the change in length of each of the two cylinder-and-piston assemblies FH required to raise the 500-kg slab by 50 mm if the other two cylinders are not activated.

10.45 Referring to Prob. 6.126 and using the answer obtained for the force exerted by the cylinder, determine the change in length of the cylinder-and-piston assembly required to raise the platform by 2 in.

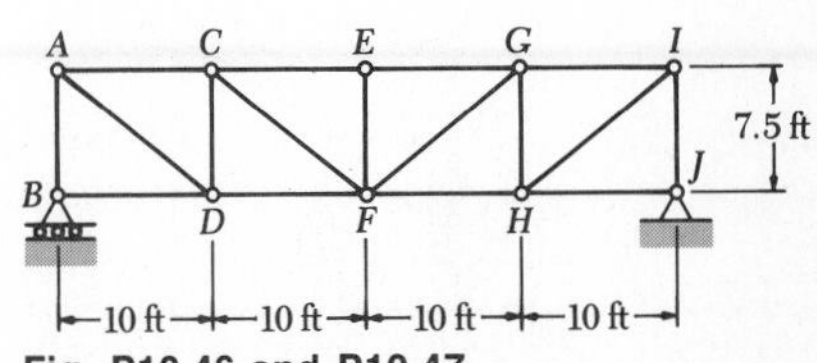

Fig. P10.46 and P10.47

10.46 Determine the vertical movement of joint C if member FG is lengthened by 1.5 in. (*Hint.* Apply a vertical load at joint C, and using the methods of Chap. 6, compute the force exerted by member FG on joints F and G. Then apply the method of virtual work for a virtual displacement making member FG longer. This method should be used only for small changes in the length of members.)

10.47 Determine the horizontal movement of joint C if member FG is lengthened by 1.5 in. (Follow hint of Prob. 10.46, applying a *horizontal* force at joint C.)

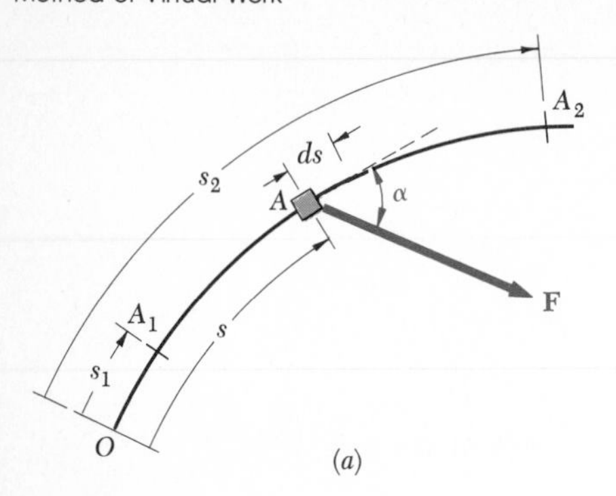

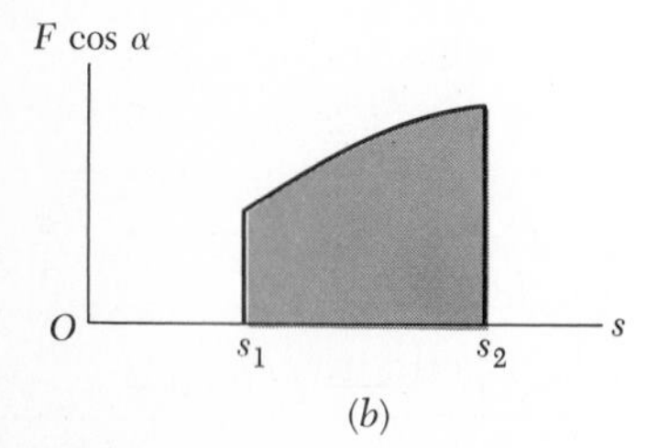

Fig. 10.10

*10.6. Work of a Force during a Finite Displacement.

Consider a force **F** acting on a particle. The work of **F** corresponding to an infinitesimal displacement $d\mathbf{r}$ of the particle was defined in Sec. 10.2 as

$$dU = \mathbf{F} \cdot d\mathbf{r} \tag{10.1}$$

The work of **F** corresponding to a finite displacement of the particle from A_1 to A_2 (Fig. 10.10*a*) is denoted by $U_{1\to2}$ and is obtained by integrating (10.1) along the curve described by the particle:

$$U_{1\to2} = \int_{A_1}^{A_2} \mathbf{F} \cdot d\mathbf{r} \tag{10.11}$$

Using the alternate expression

$$dU = F\,ds \cos\alpha \tag{10.1'}$$

given in Sec. 10.2 for the elementary work dU, we may also express the work $U_{1\to2}$ as

$$U_{1\to2} = \int_{s_1}^{s_2} (F\cos\alpha)\,ds \tag{10.11'}$$

where the variable of integration s measures the distance along the path traveled by the particle. The work $U_{1\to2}$ is represented by the area under the curve obtained by plotting $F \cos\alpha$ against s (Fig. 10.10*b*). In the case of a force **F** of constant magnitude acting in the direction of motion, formula (10.11′) yields $U_{1\to2} = F(s_2 - s_1)$.

Recalling from Sec. 10.2 that the work of a couple of moment **M** during an infinitesimal rotation $d\theta$ of a rigid body is

$$dU = M\,d\theta \tag{10.2}$$

we express as follows the work of the couple during a finite rotation of the body:

$$U_{1\to2} = \int_{\theta_1}^{\theta_2} M\,d\theta \tag{10.12}$$

In the case of a constant couple, formula (10.12) yields

$$U_{1\to2} = M(\theta_2 - \theta_1)$$

Work of a Weight. It was stated in Sec. 10.2 that the work of the weight **W** of a body during an infinitesimal displacement of the body is equal to the product of W and of the vertical displacement of the center of gravity of the body. With the y axis pointing upward, the work of **W** during a finite displacement of the body (Fig. 10.11) is obtained by writing

$$dU = -W\,dy$$

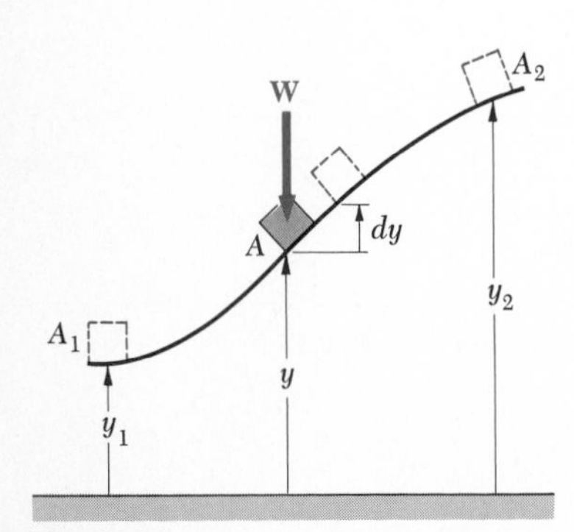

Fig. 10.11

Integrating from A_1 to A_2, we have

$$U_{1\to2} = -\int_{y_1}^{y_2} W\,dy = Wy_1 - Wy_2 \tag{10.13}$$

or

$$U_{1\to2} = -W(y_2 - y_1) = -W\,\Delta y \tag{10.13'}$$

where Δy is the vertical displacement from A_1 to A_2. The work of the weight **W** is thus equal to *the product of W and of the vertical displacement of the center of gravity of the body*. The work is *positive* when $\Delta y < 0$, that is, *when the body moves down*.

Work of the Force Exerted by a Spring. Consider a body A attached to a fixed point B by a spring; it is assumed that the spring is undeformed when the body is at A_0 (Fig. 10.12*a*). Experimental evidence shows that the magnitude of the force **F** exerted by the spring on a body A is proportional to the deflection x of the spring measured from the position A_0. We have

$$F = kx \tag{10.14}$$

where k is the *spring constant*, expressed in N/m if SI units are used, and in lb/ft or lb/in. if U.S. customary units are used. The work of the force **F** exerted by the spring during a finite displacement of the body from $A_1(x = x_1)$ to $A_2(x = x_2)$ is obtained by writing

$$dU = -F\,dx = -kx\,dx$$

$$U_{1\to2} = -\int_{x_1}^{x_2} kx\,dx = \tfrac{1}{2}kx_1^2 - \tfrac{1}{2}kx_2^2 \tag{10.15}$$

Care should be taken to express k and x in consistent units. For example, if U.S. customary units are used, k should be expressed in lb/ft and x in feet, or k in lb/in. and x in inches; in the first case, the work is obtained in ft·lb; in the second case, in in·lb. We note that the work of the force **F** exerted by the spring on the body is *positive* when $x_2 < x_1$, that is, *when the spring is returning to its undeformed position.*

Since Eq. (10.14) is the equation of a straight line of slope k passing through the origin, the work $U_{1\to2}$ of **F** during the displacement from A_1 to A_2 may be obtained by evaluating the area of the trapezoid shown in Fig. 10.12*b*. This is done by computing the values F_1 and F_2 and multiplying the base Δx of the trapezoid by its mean height $\frac{1}{2}(F_1 + F_2)$. Since the work of the force **F** exerted by the spring is positive for a negative value of Δx, we write

$$U_{1\to2} = -\tfrac{1}{2}(F_1 + F_2)\,\Delta x \tag{10.16}$$

Formula (10.16) is usually more convenient to use than (10.15) and affords fewer chances of confusing the units involved.

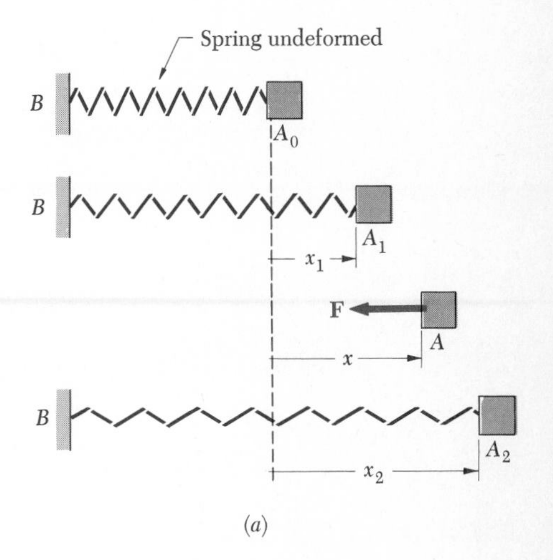

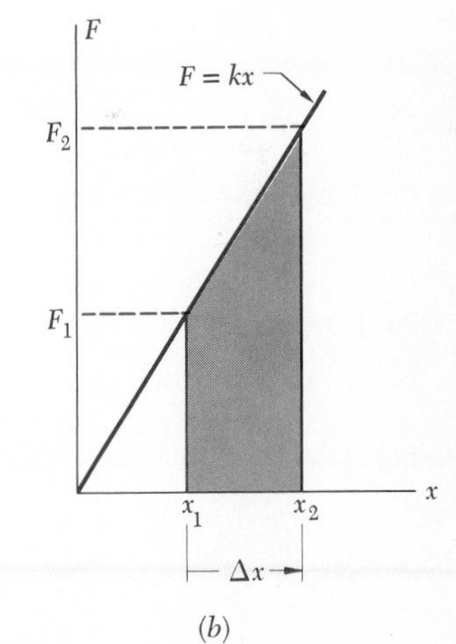

Fig. 10.12

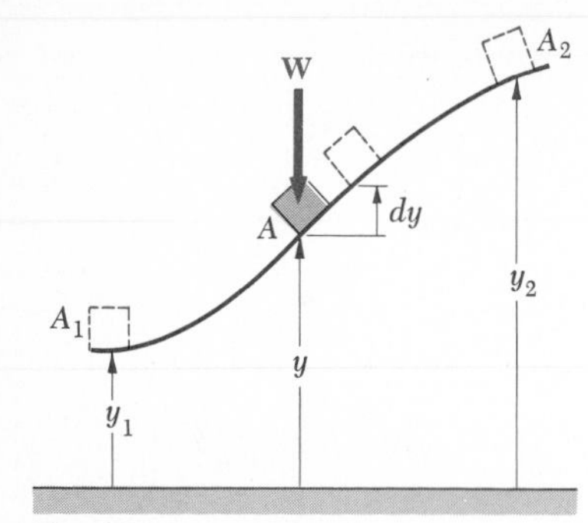

Fig. 10.11 (*repeated*)

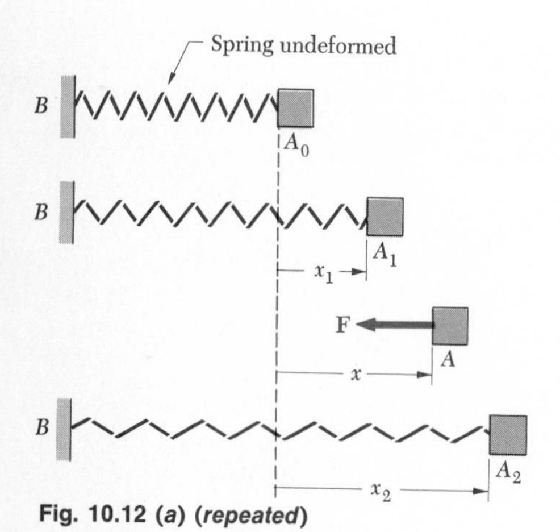

Fig. 10.12 (*a*) (*repeated*)

***10.7. Potential Energy.** Considering again the body of Fig. 10.11, we note from (10.13) that the work of the weight **W** during a finite displacement is obtained by subtracting the value of the function Wy corresponding to the second position of the body from its value corresponding to the first position. The work of **W** is thus independent of the actual path followed; it depends only upon the initial and final values of the function Wy. This function is called the *potential energy* of the body with respect to the *force of gravity* **W** and is denoted by V_g. We write

$$U_{1\rightarrow 2} = (V_g)_1 - (V_g)_2 \qquad \text{with } V_g = Wy \tag{10.17}$$

We note that if $(V_g)_2 > (V_g)_1$, that is, *if the potential energy increases* during the displacement (as in the case considered here), *the work $U_{1\rightarrow 2}$ is negative.* If, on the other hand, the work of **W** is positive, the potential energy decreases. Therefore, the potential energy V_g of the body provides a measure of *the work which may be done* by its weight **W**. Since only the *change* in potential energy, and not the actual value of V_g, is involved in formula (10.17), an arbitrary constant may be added to the expression obtained for V_g. In other words, the level from which the elevation y is measured may be chosen arbitrarily. Note that potential energy is expressed in the same units as work, i.e., in joules (J) if SI units are used,† and in ft·lb or in·lb if U.S. customary units are used.

Considering now the body of Fig. 10.12*a*, we note from formula (10.15) that the work of the elastic force **F** is obtained by subtracting the value of the function $\frac{1}{2}kx^2$ corresponding to the second position of the body from its value corresponding to the first position. This function is denoted by V_e and is called the *potential energy* of the body with respect to the *elastic force* **F**. We write

$$U_{1\rightarrow 2} = (V_e)_1 - (V_e)_2 \qquad \text{with } V_e = \tfrac{1}{2}kx^2 \tag{10.18}$$

and observe that during the displacement considered, the work of the force **F** exerted by the spring on the body is negative and the potential energy V_e increases. We should note that the expression obtained for V_e is valid only if the deflection of the spring is measured from its undeformed position.

The concept of potential energy may be used when forces other than gravity forces and elastic forces are involved. It remains valid as long as the elementary work dU of the force considered is an *exact differential.* It is then possible to find a function V, called potential energy, such that

$$dU = -dV \tag{10.19}$$

Integrating (10.19) over a finite displacement, we obtain the general formula

$$U_{1\rightarrow 2} = V_1 - V_2 \tag{10.20}$$

†See footnote, page 427.

which expresses that *the work of the force is independent of the path followed and is equal to minus the change in potential energy*. A force which satisfies Eq. (10.20) is said to be a *conservative force*.†

*** 10.8. Potential Energy and Equilibrium.** The application of the principle of virtual work is considerably simplified when the potential energy of a system is known. In the case of a virtual displacement, formula (10.19) becomes $\delta U = -\delta V$. Besides, if the position of the system is defined by a single independent variable θ, we may write $\delta V = (dV/d\theta)\,\delta\theta$. Since $\delta\theta$ must be different from zero, the condition $\delta U = 0$ for the equilibrium of the system becomes

$$\frac{dV}{d\theta} = 0 \tag{10.21}$$

In terms of potential energy, the principle of virtual work states therefore that *if a system is in equilibrium, the derivative of its total potential energy is zero.* If the position of the system depends upon several independent variables (the system is then said to possess *several degrees of freedom*), the partial derivatives of V with respect to each of the independent variables must be zero.

Consider, for example, a structure made of two members AC and CB and carrying a load W at C. The structure is supported by a pin at A and a roller at B, and a spring BD connects B to a fixed point D (Fig. 10.13*a*). The constant of the spring is k, and it is assumed that the natural length of the spring is equal to AD and thus that the spring is undeformed when B coincides with A. Neglecting the friction forces and the weight of the members, we find that the only forces which work during a displacement of the structure are the weight **W** and the force **F** exerted by the spring at point B (Fig. 10.13*b*). The total potential energy of the system will thus be obtained by adding the potential energy V_g corresponding to the gravity force **W** and the potential energy V_e corresponding to the elastic force **F**.

Choosing a coordinate system with origin at A and noting that the deflection of the spring, measured from its undeformed position, is $AB = x_B$, we write

$$V_e = \tfrac{1}{2}kx_B^2 \qquad V_g = Wy_C$$

Expressing the coordinates x_B and y_C in terms of the angle θ, we have

$$x_B = 2l\sin\theta \qquad y_C = l\cos\theta$$
$$V_e = \tfrac{1}{2}k(2l\sin\theta)^2 \qquad V_g = W(l\cos\theta)$$
$$V = V_e + V_g = 2kl^2\sin^2\theta + Wl\cos\theta \tag{10.22}$$

The positions of equilibrium of the system are obtained by equating to zero the derivative of the potential energy V. We write

$$\frac{dV}{d\theta} = 4kl^2\sin\theta\cos\theta - Wl\sin\theta = 0$$

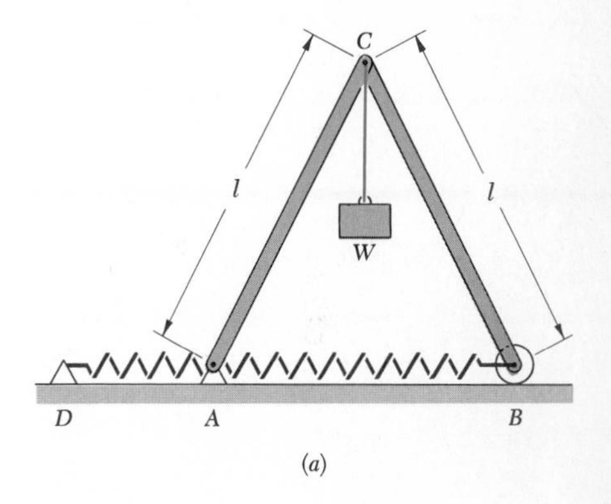

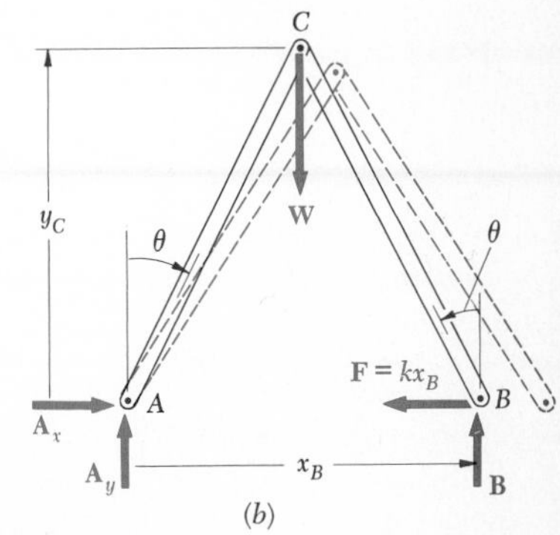

Fig. 10.13

†A detailed discussion of conservative forces is given in Sec. 13.7 of *Dynamics*.

or, factoring $l \sin \theta$,

$$\frac{dV}{d\theta} = l \sin \theta(4kl \cos \theta - W) = 0$$

There are therefore two positions of equilibrium, corresponding respectively to the values $\theta = 0$ and $\theta = \cos^{-1}(W/4kl)$.†

* 10.9. Stability of Equilibrium.

Consider the three uniform rods of length $2a$ and weight $\mathbf{W}$ shown in Fig. 10.14. While each rod is in equilibrium, there is an important difference between the three cases considered. Suppose that each rod is slightly disturbed from its position of equilibrium and then released: rod a will move back toward its original position, rod b will keep moving away from its original position, and rod c will remain in its new position. In case a, the equilibrium of the rod is said to be *stable;* in case b, the equilibrium is said to be *unstable;* and, in case c, to be *neutral.*

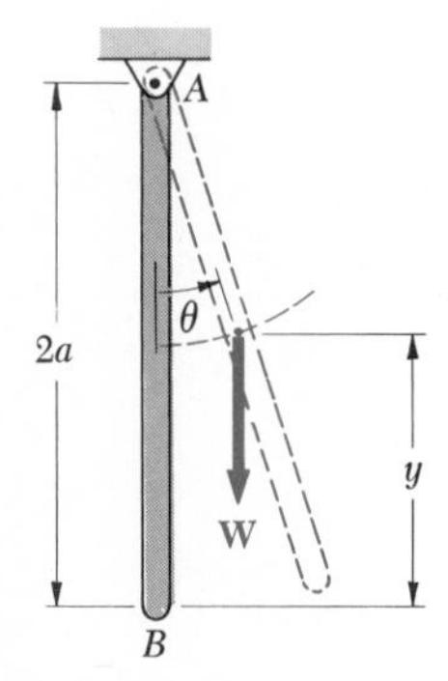

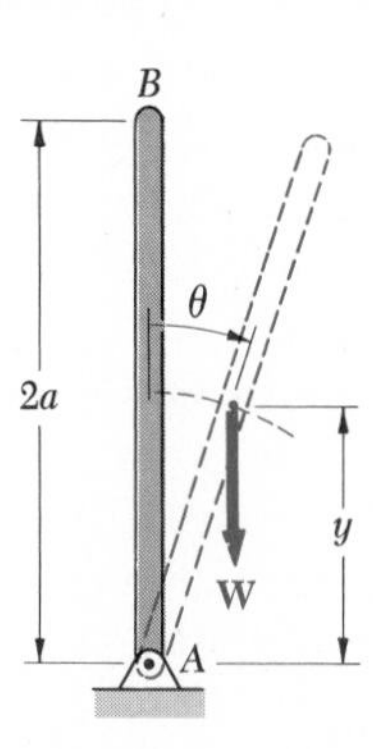

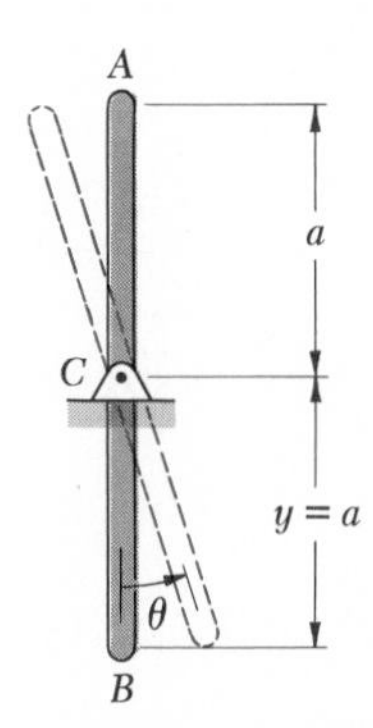

Fig. 10.14 (*a*) Stable equilibrium (*b*) Unstable equilibrium (*c*) Neutral equilibrium

Recalling from Sec. 10.7 that the potential energy V_g with respect to gravity is equal to Wy, where y is the elevation of the point of application of $\mathbf{W}$ measured from an arbitrary level, we observe that the potential energy of rod a is minimum in the position of equilibrium considered, that the potential energy of rod b is maximum, and that the potential energy of rod c is constant. Equilibrium is thus *stable, unstable,* or *neutral* according to whether the potential energy is *minimum, maximum,* or *constant* (Fig. 10.15).

That the result obtained is quite general may be seen as follows: We first observe that a force always tends to do positive work and thus to decrease the potential energy of the system on which it is applied. Therefore, when a system is disturbed from its position of equilibrium, the forces acting on the system will tend to bring it back to its original position if V is minimum (Fig. 10.15*a*) and to move it farther away if V is maximum (Fig. 10.15*b*). If V is constant (Fig. 10.15*c*), the forces will not tend to move the system either way.

†The second position does not exist if $W > 4kl$ (see Prob. 10.80 for a further discussion of the equilibrium of this system).

Recalling from calculus that a function is minimum or maximum according to whether its second derivative is positive or negative, we may summarize as follows the conditions for the equilibrium of a system with one degree of freedom (i.e., a system the position of which is defined by a single independent variable θ):

$$\frac{dV}{d\theta} = 0 \qquad \frac{d^2V}{d\theta^2} > 0\text{: stable equilibrium}$$
$$\frac{dV}{d\theta} = 0 \qquad \frac{d^2V}{d\theta^2} < 0\text{: unstable equilibrium} \tag{10.23}$$

If both the first and the second derivatives of V are zero, it is necessary to examine derivatives of a higher order to determine whether the equilibrium is stable, unstable, or neutral. The equilibrium will be neutral if all

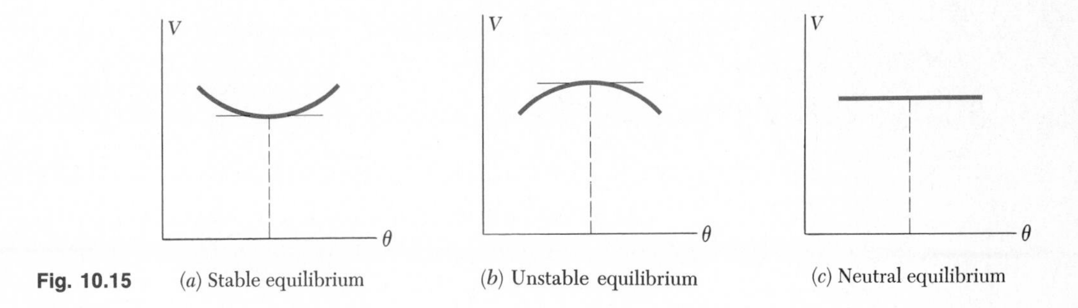

Fig. 10.15 (*a*) Stable equilibrium (*b*) Unstable equilibrium (*c*) Neutral equilibrium

derivatives are zero, since the potential energy V is then a constant. The equilibrium will be stable if the first derivative found to be different from zero is of even order, and if that derivative is positive. In all other cases the equilibrium will be unstable.

If the system considered possesses *several degrees of freedom*, the potential energy V depends upon several variables and it is thus necessary to apply the theory of functions of several variables to determine whether V is minimum. It may be verified that a system with two degrees of freedom will be stable, and the corresponding potential energy $V(\theta_1, \theta_2)$ will be minimum, if the following relations are satisfied simultaneously:

$$\frac{\partial V}{\partial \theta_1} = \frac{\partial V}{\partial \theta_2} = 0$$
$$\left(\frac{\partial^2 V}{\partial \theta_1\, \partial \theta_2}\right)^2 - \frac{\partial^2 V}{\partial \theta_1^2}\frac{\partial^2 V}{\partial \theta_2^2} < 0 \tag{10.24}$$
$$\frac{\partial^2 V}{\partial \theta_1^2} > 0 \quad \text{or} \quad \frac{\partial^2 V}{\partial \theta_2^2} > 0$$

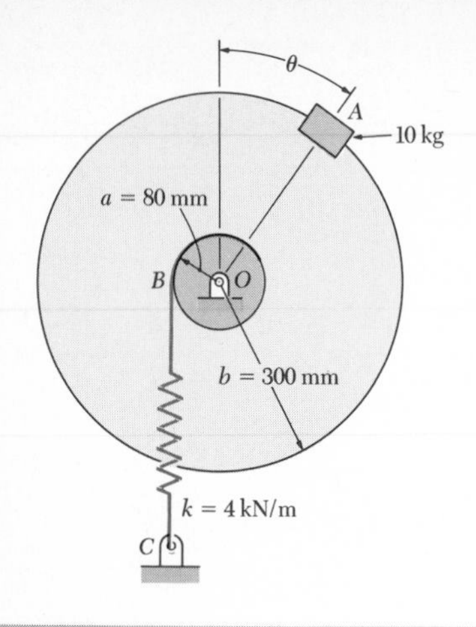

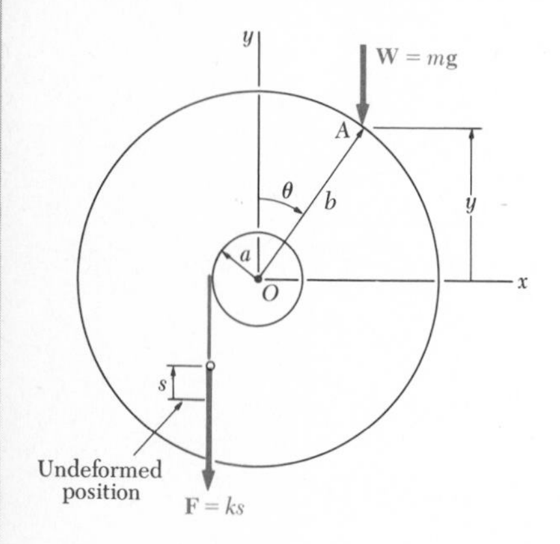

SAMPLE PROBLEM 10.4

A 10-kg block is attached to the rim of a 300-mm-radius disk as shown. Knowing that spring BC is unstretched when $\theta = 0$, determine the position or positions of equilibrium, and state in each case whether the equilibrium is stable, unstable, or neutral.

Potential Energy. Denoting by s the deflection of the spring from its undeformed position and placing the origin of coordinates at O, we obtain

$$V_e = \tfrac{1}{2}ks^2 \qquad V_g = Wy = mgy$$

Measuring θ in radians, we have

$$s = a\theta \qquad y = b\cos\theta$$

Substituting for s and y in the expressions for V_e and V_g, we write

$$V_e = \tfrac{1}{2}ka^2\theta^2 \qquad V_g = mgb\cos\theta$$
$$V = V_e + V_g = \tfrac{1}{2}ka^2\theta^2 + mgb\cos\theta$$

Positions of Equilibrium. Setting $dV/d\theta = 0$, we write

$$\frac{dV}{d\theta} = ka^2\theta - mgb\sin\theta = 0$$

$$\sin\theta = \frac{ka^2}{mgb}\theta$$

Substituting $a = 0.08$ m, $b = 0.3$ m, $k = 4$ kN/m, and $m = 10$ kg, we obtain

$$\sin\theta = \frac{(4\text{ kN/m})(0.08\text{ m})^2}{(10\text{ kg})(9.81\text{ m/s}^2)(0.3\text{ m})}\theta$$
$$\sin\theta = 0.8699\,\theta$$

where θ is expressed in radians. Solving by trial for θ, we find

$$\theta = 0 \qquad \text{and} \qquad \theta = 0.902\text{ rad}$$
$$\theta = 0 \qquad \text{and} \qquad \theta = 51.7° \quad ◀$$

Stability of Equilibrium. The second derivative of the potential energy V with respect to θ is

$$\frac{d^2V}{d\theta^2} = ka^2 - mgb\cos\theta$$
$$= (4\text{ kN/m})(0.08\text{ m})^2 - (10\text{ kg})(9.81\text{ m/s}^2)(0.3\text{ m})\cos\theta$$
$$= 25.6 - 29.43\cos\theta$$

For $\theta = 0$: $$\frac{d^2V}{d\theta^2} = 25.6 - 29.43\cos 0° = -3.83 < 0$$

The equilibrium is unstable for $\theta = 0°$ ◀

For $\theta = 51.7°$: $$\frac{d^2V}{d\theta^2} = 25.6 - 29.43\cos 51.7° = +7.36 > 0$$

The equilibrium is stable for $\theta = 51.7°$ ◀

Problems

10.48 Using the method of Sec. 10.8, solve Prob. 10.28.

10.49 Using the method of Sec. 10.8, solve Prob. 10.29.

10.50 Using the method of Sec. 10.8, solve Prob. 10.26.

10.51 Using the method of Sec. 10.8, solve Prob. 10.25.

10.52 Show that the equilibrium is neutral in Prob. 10.2.

10.53 Show that the equilibrium is neutral in Prob. 10.4.

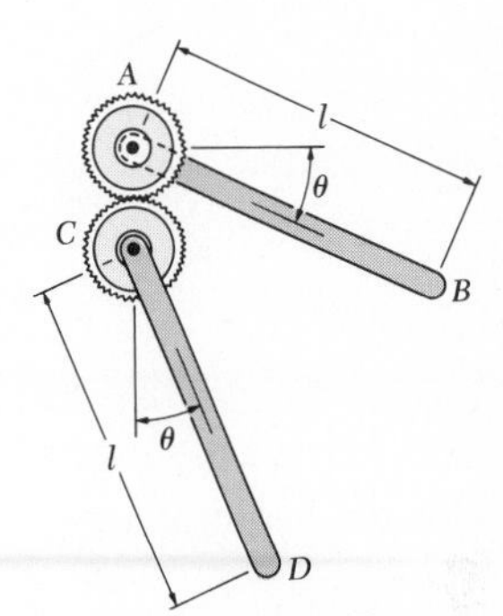

Fig. P10.54 and P10.55

10.54 Two uniform rods, each of mass m, are attached to gears of equal radii as shown. Determine the positions of equilibrium of the system and state in each case whether the equilibrium is stable, unstable, or neutral.

10.55 Two uniform rods, AB and CD, are attached to gears of equal radii as shown. Knowing that $m_{AB} = 300$ g and $m_{CD} = 500$ g, determine the positions of equilibrium of the system and state in each case whether the equilibrium is stable, unstable, or neutral.

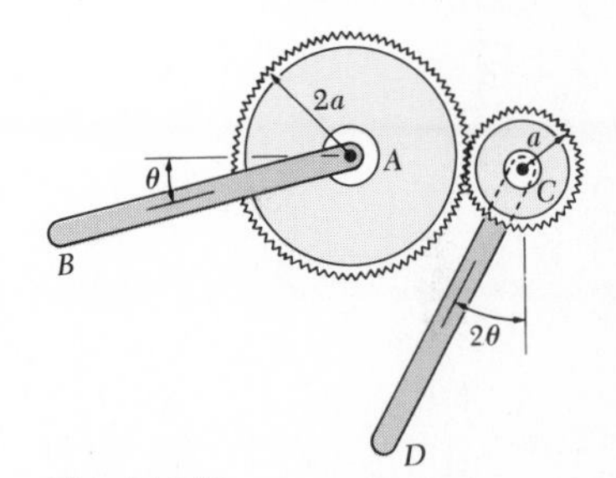

Fig. P10.56

10.56 Two uniform rods, AB and CD, of the same length l, are attached to gears as shown. Knowing that rod AB weighs 3 lb and that rod CD weighs 2 lb, determine the positions of equilibrium of the system and state in each case whether the equilibrium is stable, unstable, or neutral.

10.57 Solve Prob. 10.56, assuming that rods AB and CD weigh 3 lb each.

10.58 Using the method of Sec. 10.8, solve Prob. 10.30. Determine whether the equilibrium is stable, unstable, or neutral. (*Hint.* The potential energy corresponding to the couple exerted by a torsion spring is $\frac{1}{2}K\theta^2$, where K is the torsional spring constant and θ is the angle of twist.)

10.59 In Prob. 10.31, determine whether each of the positions of equilibrium is stable, unstable, or neutral. (See hint of Prob. 10.58.)

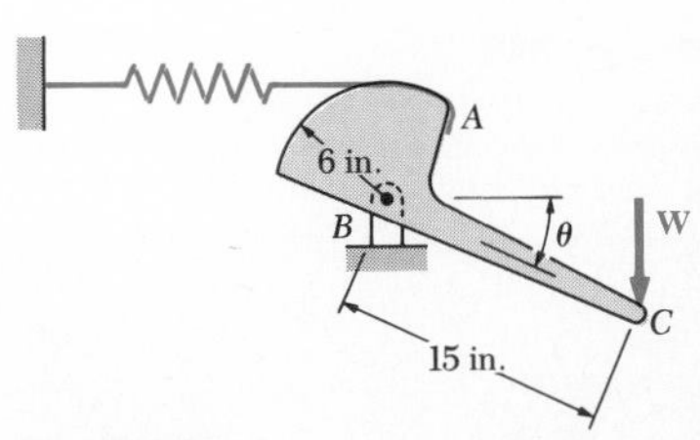

Fig. P10.60

10.60 A load **W** of magnitude 144 lb is applied to the mechanism at C. Neglecting the weight of the mechanism, determine the value of θ corresponding to equilibrium and check that the equilibrium is stable. The constant of the spring is $k = 20$ lb/in., and the spring is unstretched when $\theta = 0$.

10.61 Solve Prob. 10.60, assuming that the spring is unstretched when $\theta = 30°$.

10.62 A collar B, of weight W, may move freely along the vertical rod shown. The constant of the spring is k and the spring is unstretched when $y = 0$. (a) Derive an equation in y, W, a, and k which must be satisfied when the collar is in equilibrium. (b) Determine the value of y corresponding to equilibrium when $W = 80$ N, $a = 300$ mm, and $k = 500$ N/m, and check that the equilibrium is stable.

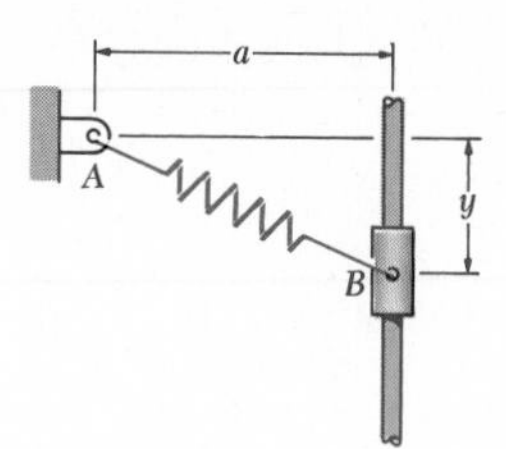

Fig. P10.62

Fig. P10.63

10.63 A slender rod AB, of weight W, is attached to blocks A and B which may move freely in the guides shown. The constant of the spring is k and the spring is unstretched when AB is horizontal. Neglecting the weight of the blocks, derive an equation in θ, W, l, and k which must be satisfied when the rod is in equilibrium.

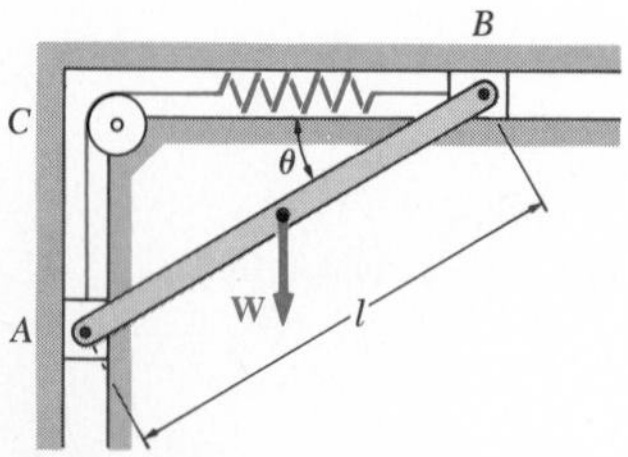

10.64 In Prob. 10.63, determine three values of θ corresponding to equilibrium when $W = 45$ N, $k = 300$ N/m, and $l = 0.5$ m. State in each case whether the equilibrium is stable, unstable, or neutral.

10.65 The constant of spring AB is k and the spring is unstretched when $\theta = 0$. (a) Neglecting the weight of the rigid arm BCD, derive an equation in θ, k, a, l, and W which must be satisfied when the arm is in equilibrium. (b) Determine three values of θ corresponding to equilibrium when $k = 75$ lb/in., $a = 10$ in., $l = 15$ in., and $W = 100$ lb. State in each case whether the equilibrium is stable, unstable, or neutral.

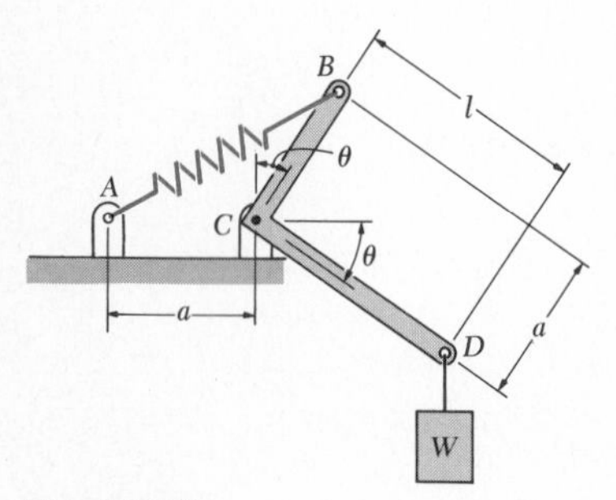

Fig. P10.65

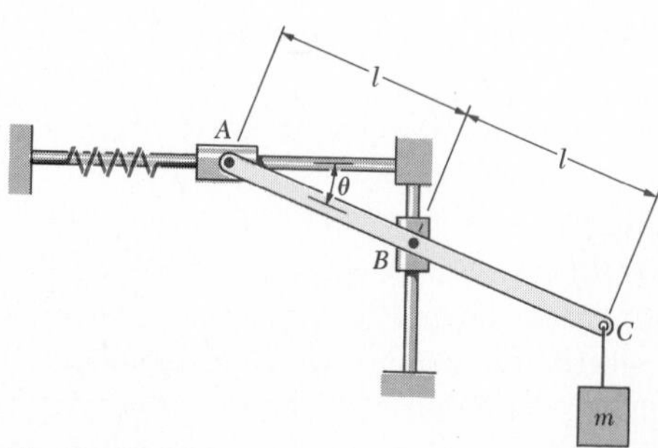

Fig. P10.66

10.66 Bar ABC is attached to collars A and B which may move freely on the rods shown. The constant of the spring is k and the spring is unstretched when $\theta = 0$. (a) Neglecting the weight of bar ABC, derive an equation in θ, m, k, and l which must be satisfied when bar ABC is in equilibrium. (b) Determine the value of θ corresponding to equilibrium when $m = 5$ kg, $k = 800$ N/m, and $l = 250$ mm, and check that the equilibrium is stable.

10.67 Solve Prob. 10.66, assuming that the spring is unstretched when $\theta = 30°$.

10.68 A vertical load **P** is applied at end B of rod BC. The constant of the spring is k and the spring is unstretched when $\theta = 60°$. (*a*) Neglecting the weight of the rod, express the angle θ corresponding to equilibrium in terms of P, k, and l. (*b*) Determine the values of θ corresponding to equilibrium when $P = 50$ lb, $k = 25$ lb/in., and $l = 8$ in. State in each case whether the equilibrium is stable, unstable, or neutral.

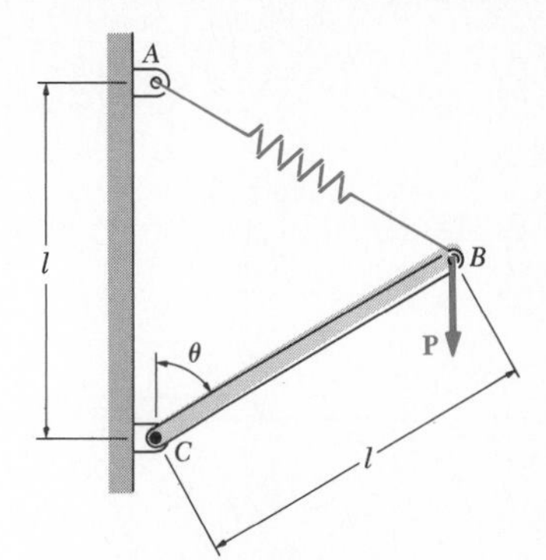

Fig. P10.68 and P10.69

10.69 A vertical load **P** of magnitude 120 lb is applied at end B of rod BC. The constant of the spring is $k = 25$ lb/in., and the spring is unstretched when $\theta = 60°$. Neglecting the weight of the rod and knowing that $l = 8$ in., determine the value of θ corresponding to equilibrium. State whether the equilibrium is stable, unstable, or neutral.

10.70 A vertical load **P** is applied at end B of rod BC. The constant of the spring is k and the spring is unstretched when $\theta = 0$. (*a*) Neglecting the weight of the rod, express the angle θ corresponding to the equilibrium position in terms of P, k, and l. (*b*) Determine the value of θ corresponding to equilibrium when $P = 400$ N, $k = 800$ N/m, and $l = 250$ mm.

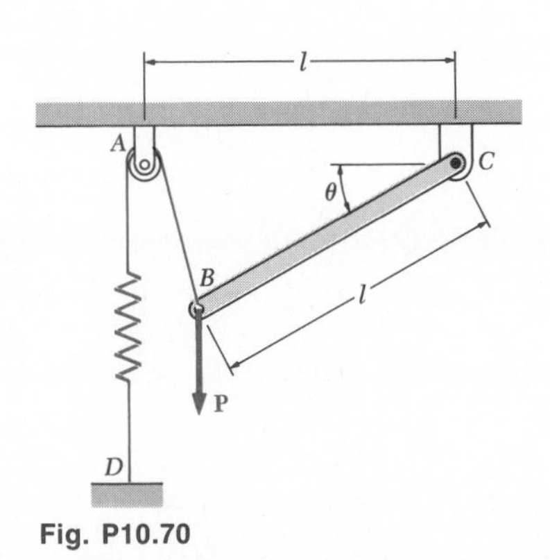

Fig. P10.70

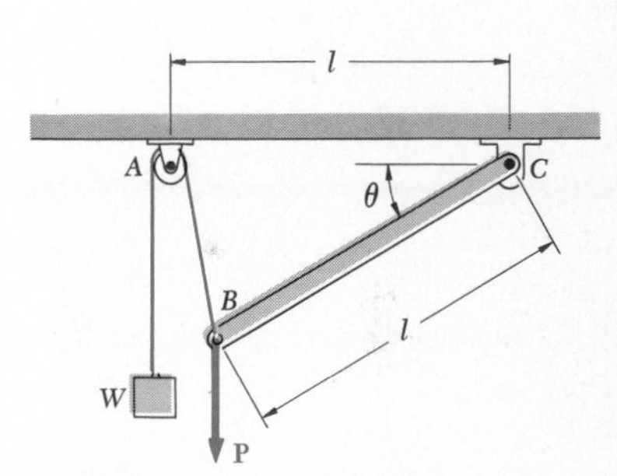

Fig. P10.71

10.71 A vertical load **P** is applied at end B of rod BC. (*a*) Neglecting the weight of the rod, express the angle θ corresponding to the equilibrium position in terms of P, l, and the counterweight W. (*b*) Determine the value of θ corresponding to equilibrium when $P = 400$ N, $W = 200$ N, and $l = 250$ mm.

10.72 The rod AB is attached to a hinge at A and to two springs, each of constant k. If $h = 600$ mm, $d = 300$ mm, and $m = 75$ kg, determine the range of values of k for which the equilibrium of rod AB is stable in the position shown. Each spring can act in either tension or compression.

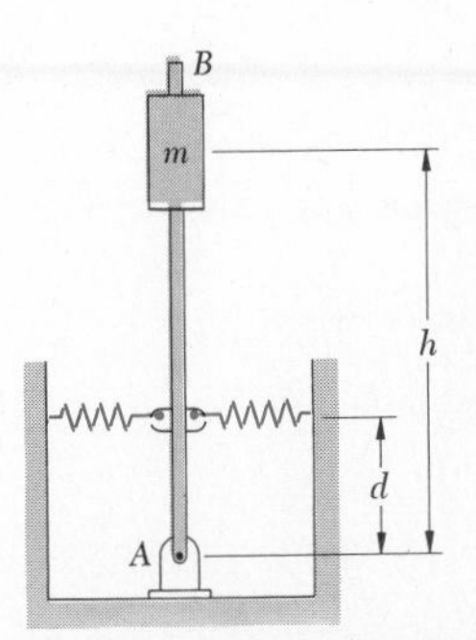

Fig. P10.72 and P10.73

10.73 If $m = 60$ kg, $h = 900$ mm, and the constant of each spring is $k = 2.5$ kN/m, determine the range of values of the distance d for which the equilibrium of rod AB is stable in the position shown. Each spring can act in either tension or compression.

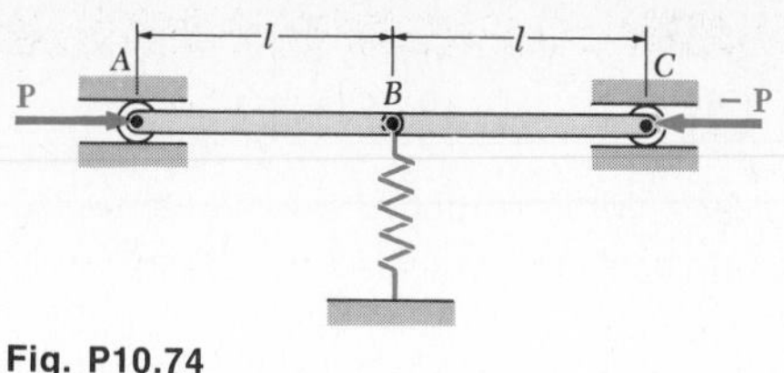

Fig. P10.74

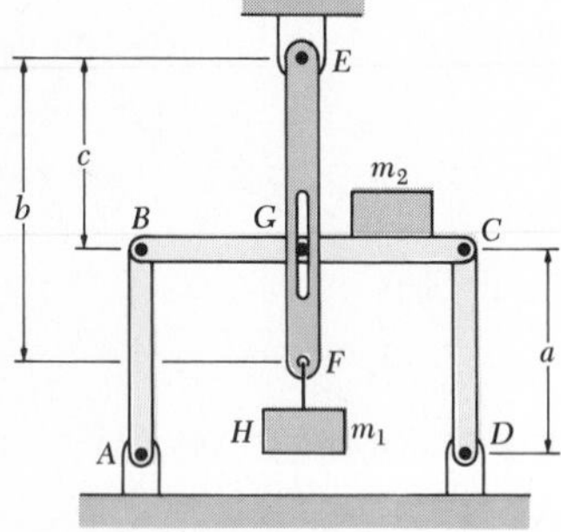

Fig. P10.75

10.74 Two bars AB and BC of negligible weight are attached to a single spring of constant k which is undeformed when the bars are horizontal. Determine the range of values of the magnitude P of the two equal and opposite forces **P** and $-$**P** for which the equilibrium of the system is stable in the position shown.

10.75 Bars BC and EF are connected at G by a pin which is attached to BC and may slide freely in a slot cut in EF. Determine the smallest mass m_1 for which the equilibrium of the mechanism is stable in the position shown.

***10.76 and 10.77** The bars shown, each of length l and of negligible weight, are attached to springs each of constant k. The springs are undeformed and the system is in equilibrium when $\theta_1 = \theta_2 = 0$. Determine the range of values of P for which the equilibrium position is stable.

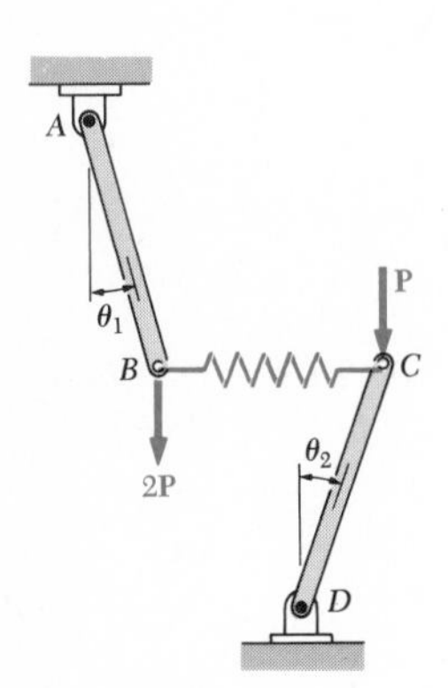

Fig. P10.76

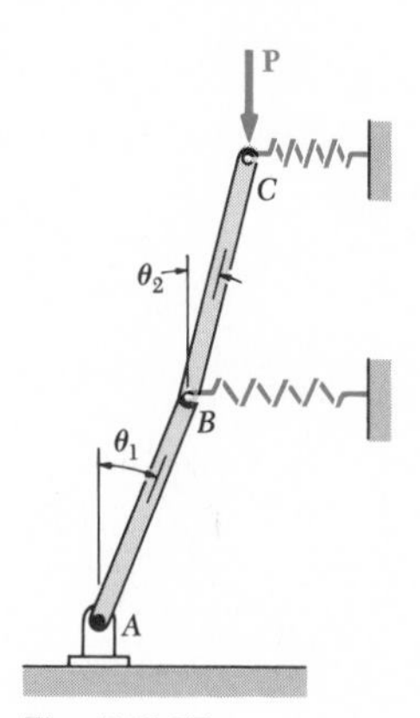

Fig. P10.77

***10.78** Solve Prob. 10.77, knowing that $l = 30$ in. and $k = 50$ lb/in.

***10.79** Solve Prob. 10.76, assuming that the vertical force applied at B is increased to 5**P**.

10.80 In Sec. 10.8, two positions of equilibrium were obtained for the system shown in Fig. 10.13, namely, $\theta = 0$ and $\theta = \cos^{-1}(W/4kl)$. Show that (*a*) if $W < 4kl$, the equilibrium is stable in the first position ($\theta = 0$) and unstable in the second, (*b*) if $W = 4kl$, the two positions coincide and the equilibrium is unstable, (*c*) if $W > 4kl$, the equilibrium is unstable in the first position ($\theta = 0$) and the second position does not exist. (*Note.* It is assumed that the system must deform as shown and that the system cannot rotate as a single rigid body about A when A and B coincide.)

Review and Summary

Work of a force

The first part of this chapter was devoted to the *principle of virtual work* and to its direct application to the solution of equilibrium problems. We first defined the *work of a force* $\mathbf{F}$ *corresponding to the small displacement* $d\mathbf{r}$ [Sec. 10.2] as the quantity

$$dU = \mathbf{F} \cdot d\mathbf{r} \tag{10.1}$$

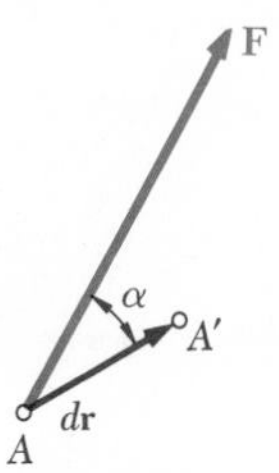

Fig. 10.16

obtained by forming the scalar product of the force $\mathbf{F}$ and of the displacement $d\mathbf{r}$ (Fig. 10.16). Denoting respectively by F and ds the magnitudes of the force and of the displacement, and by α the angle formed by $\mathbf{F}$ and $d\mathbf{r}$, we wrote

$$dU = F\, ds \cos \alpha \tag{10.1'}$$

The work dU is positive if $\alpha < 90°$, zero if $\alpha = 90°$, and negative if $\alpha > 90°$. We also found that the *work of a couple of moment* $\mathbf{M}$ acting on a rigid body is

$$dU = M\, d\theta \tag{10.2}$$

where $d\theta$ is the small angle expressed in radians through which the body rotates.

Virtual displacement

Considering a particle located at A and acted upon by several forces $\mathbf{F}_1, \mathbf{F}_2, \ldots, \mathbf{F}_n$ [Sec. 10.3], we imagined that the particle moved to a new position A' (Fig. 10.17). Since this displacement did not actually take place, it was referred to as a *virtual displacement* and denoted by $\delta\mathbf{r}$, while the corresponding work of the forces was called *virtual work* and denoted by δU. We had

Principle of virtual work

$$\delta U = \mathbf{F}_1 \cdot \delta\mathbf{r} + \mathbf{F}_2 \cdot \delta\mathbf{r} + \cdots + \mathbf{F}_n \cdot \delta\mathbf{r}$$

The *principle of virtual work* states that, *if a particle is in equilibrium, the total virtual work* δU *of the forces acting on the particle is zero for any virtual displacement of the particle.*

The principle of virtual work may be extended to the case of rigid bodies and systems of rigid bodies. Since it involves *only forces which do work*, its application provides a useful alternative to the use of the equilibrium equations in the solution of many engineering problems. It is particularly effective in the case of machines and mechanisms consisting of connected rigid bodies, since the work of the reactions at the supports is zero and the work of the internal forces at the pin connections cancels out [Sec. 10.4; Sample Probs. 10.1, 10.2, and 10.3].

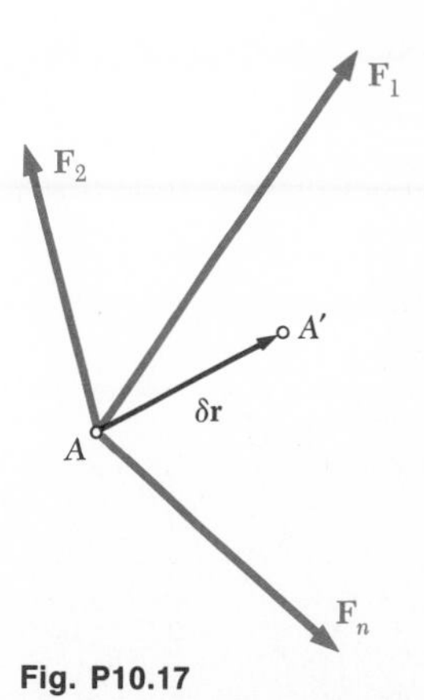

Fig. P10.17

Mechanical efficiency

In the case of *real machines*, however [Sec. 10.5], the work of the friction forces should be taken into account, with the result that the *output work will be less than the input work.* Defining the *mechanical efficiency* of a machine as the ratio

$$\eta = \frac{\text{output work}}{\text{input work}} \tag{10.9}$$

we also noted that for an ideal machine (no friction) $\eta = 1$, while for a real machine $\eta < 1$.

Work of a force over a finite displacement

In the second part of the chapter we considered the *work of forces corresponding to finite displacements* of their points of application. The work $U_{1\to2}$ of the force **F** corresponding to a displacement of the particle A from A_1 to A_2 (Fig. 10.18) was obtained by integrating the right-hand member of Eq. (10.1) or (10.1′) along the curve described by the particle [Sec. 10.6]:

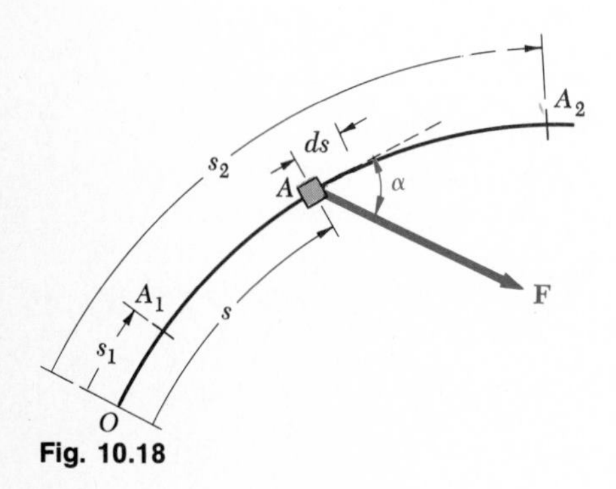

Fig. 10.18

$$U_{1\to2} = \int_{A_1}^{A_2} \mathbf{F} \cdot d\mathbf{r} \tag{10.11}$$

or

$$U_{1\to2} = \int_{s_1}^{s_2} (F \cos \alpha)\, ds \tag{10.11′}$$

Similarly, the work of a couple of moment **M** corresponding to a finite rotation from θ_1 to θ_2 of a rigid body was expressed as

$$U_{1\to2} = \int_{\theta_1}^{\theta_2} M\, d\theta \tag{10.12}$$

Work of a weight

The *work of the weight* **W** *of a body* as its center of gravity moves from the elevation y_1 to y_2 (Fig. 10.19) may be obtained by making $F = W$ and $\alpha = 180°$ in Eq. (10.11′):

$$U_{1\to2} = -\int_{y_1}^{y_2} W\, dy = Wy_1 - Wy_2 \tag{10.13}$$

The work of **W** is therefore positive *when the elevation y decreases.*

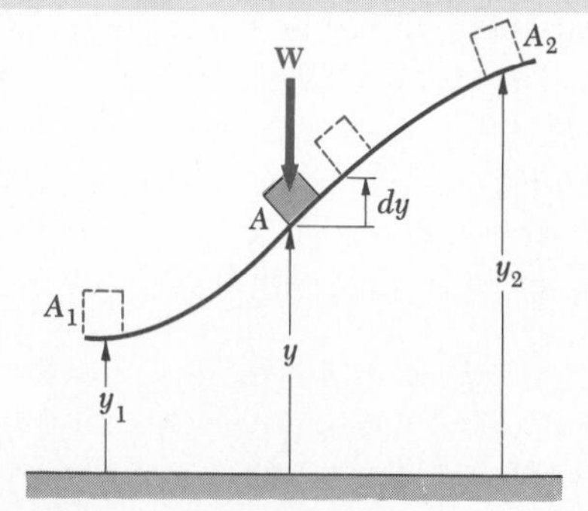

Fig. 10.19

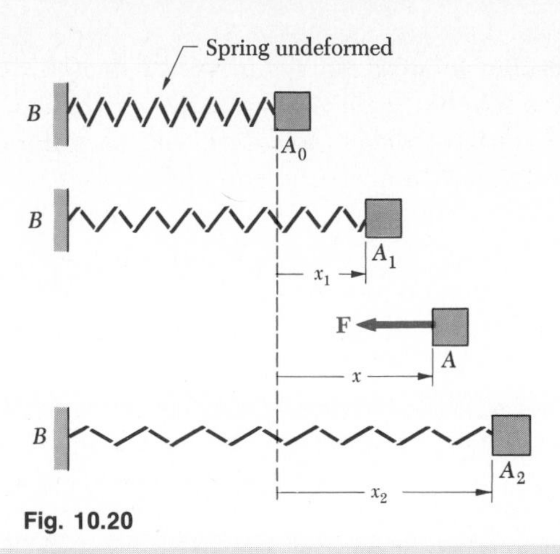

Fig. 10.20

Work of the force exerted by a spring

The *work of the force* **F** *exerted by a spring* on a body A as the spring is stretched from x_1 to x_2 (Fig. 10.20) may be obtained by making $F = kx$, where k is the constant of the spring, and $\alpha = 180°$ in Eq. (10.11′):

$$U_{1\to2} = -\int_{x_1}^{x_2} kx\,dx = \tfrac{1}{2}kx_1^2 - \tfrac{1}{2}kx_2^2 \qquad (10.15)$$

The work of **F** is therefore positive *when the spring is returning to its undeformed position.*

Potential energy

When the work of a force **F** is independent of the path actually followed between A_1 and A_2, the force is said to be a *conservative force* and its work may be expressed as

$$U_{1\to2} = V_1 - V_2 \qquad (10.20)$$

where V is the *potential energy* associated with **F**, and V_1 and V_2 represent the values of V at A_1 and A_2, respectively [Sec. 10.7]. The potential energies associated, respectively, with the *force of gravity* **W** and the *elastic force* **F** exerted by a spring were found to be

$$V_g = Wy \qquad \text{and} \qquad V_e = \tfrac{1}{2}kx^2 \qquad (10.17,\ 10.18)$$

Alternative expression for the principle of virtual work

When the position of a mechanical system depends upon a single independent variable θ, the potential energy of the system is a function $V(\theta)$ of that variable and it follows from Eq. (10.20) that $\delta U = -\delta V = -(dV/d\theta)\,\delta\theta$. The condition $\delta U = 0$ required by the principle of virtual work for the equilibrium of the system may thus be replaced by the condition

$$\frac{dV}{d\theta} = 0 \qquad (10.21)$$

The use of Eq. (10.21) may be preferred to the direct application of the principle of virtual work when all the forces involved are conservative [Sec. 10.8; Sample Prob. 10.4].

Stability of equilibrium

This approach presents another advantage, since it is possible to determine from the sign of the second derivative of V whether the equilibrium of the system is *stable, unstable,* or *neutral* [Sec. 10.9]. If $d^2V/d\theta^2 > 0$, V is *minimum* and the equilibrium is *stable;* if $d^2V/d\theta^2 < 0$, V is *maximum* and the equilibrium is *unstable;* if $d^2V/d\theta^2 = 0$, it is necessary to examine derivatives of a higher order.

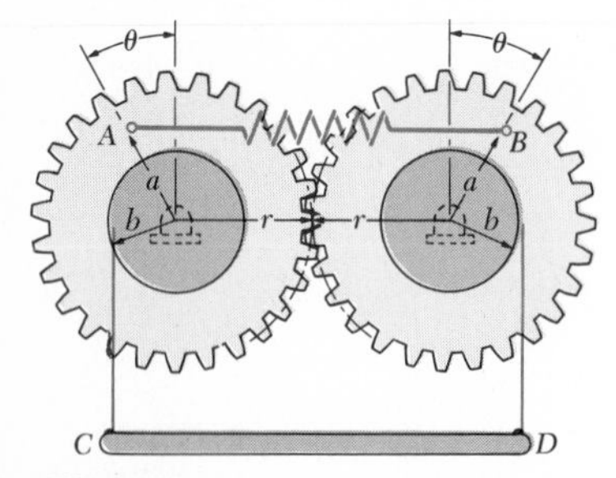

Fig. P10.81

Review Problems

10.81 A spring AB of constant k is attached to two identical gears as shown. A uniform bar CD of mass m is supported by cords wrapped around drums of radius b which are attached to the gears. If the spring is undeformed when $\theta = 0$, obtain an equation defining the angle θ corresponding to the equilibrium position.

10.82 For the mechanism of Prob. 10.81 the following numerical values are given: $k = 5$ kN/m, $a = 80$ mm, $b = 60$ mm, $r = 120$ mm, and $m = 60$ kg. Determine the values of θ corresponding to equilibrium positions and state in each case whether the equilibrium is stable, unstable, or neutral.

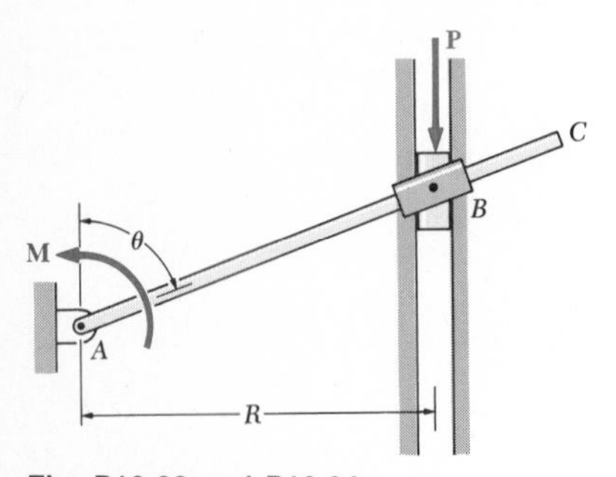

Fig. P10.83 and P10.84

10.83 Collar B may slide along rod AC and is attached by a pin to a block which may slide in the vertical slot. Derive an expression for the magnitude of the couple **M** required to maintain equilibrium.

10.84 Determine the value of θ corresponding to the equilibrium position of rod AC when $R = 12$ in., $P = 60$ lb, and $M = 80$ lb·ft.

10.85 Determine the mass m which balances the 6-kg block.

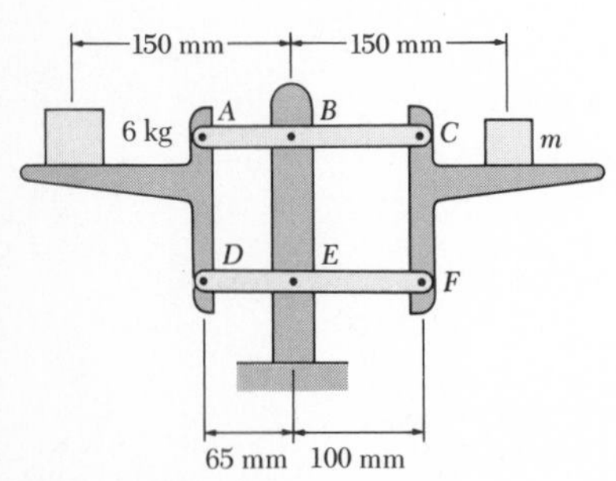

Fig. P10.85

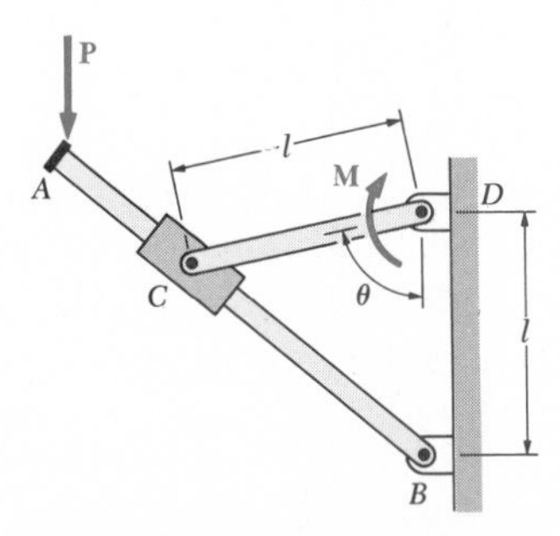

Fig. P10.86

10.86 Knowing that rod AB is of length $2l$, derive an expression for the magnitude of the couple **M** required to maintain equilibrium.

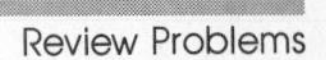

10.87 The mechanism shown is acted upon by the force **P**; derive an expression for the magnitude of the force **Q** required to maintain equilibrium.

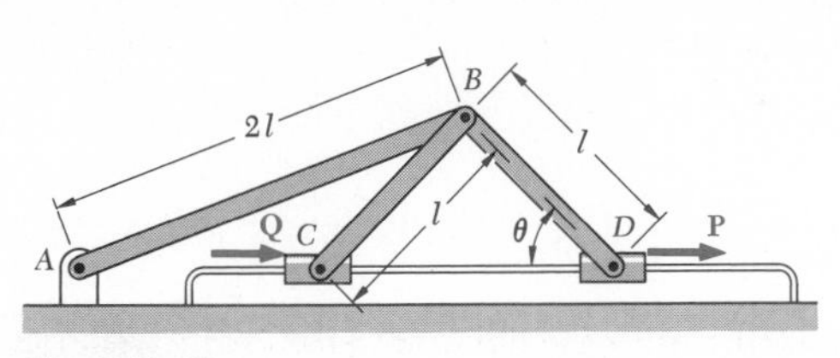

Fig. P10.87

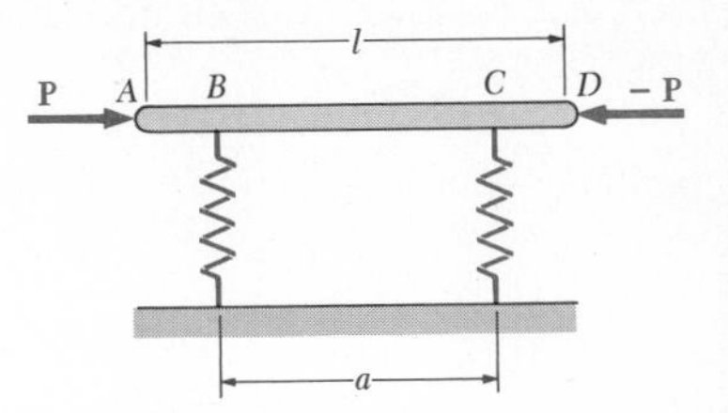

Fig. P10.88

10.88 The horizontal bar AD is attached to two springs of constant k and is in equilibrium in the position shown. Determine the range of values of the magnitude P of the two equal and opposite *horizontal* forces **P** and −**P** for which the equilibrium position is stable (*a*) if $AB = CD$, (*b*) if $AB = 2(CD)$.

10.89 (*a*) Derive an equation defining the angle θ corresponding to the equilibrium position. (*b*) Determine the angle θ corresponding to equilibrium if $P = 2W$.

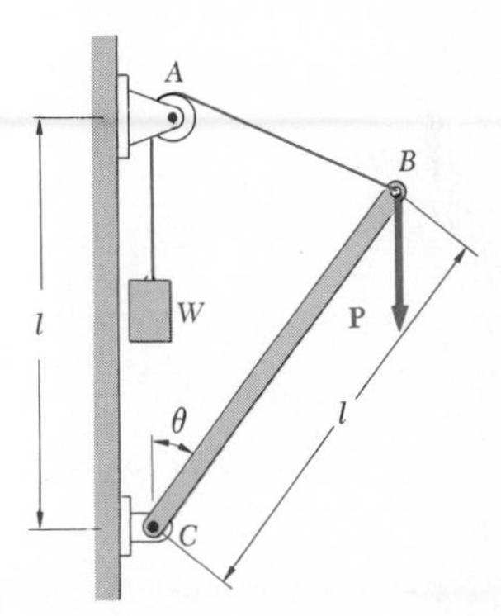

Fig. P10.89

10.90 Two bars ABC and CD are attached to a single spring of constant k which is unstretched when the bars are vertical. Determine the range of values of P for which the equilibrium of the system is stable in the position shown.

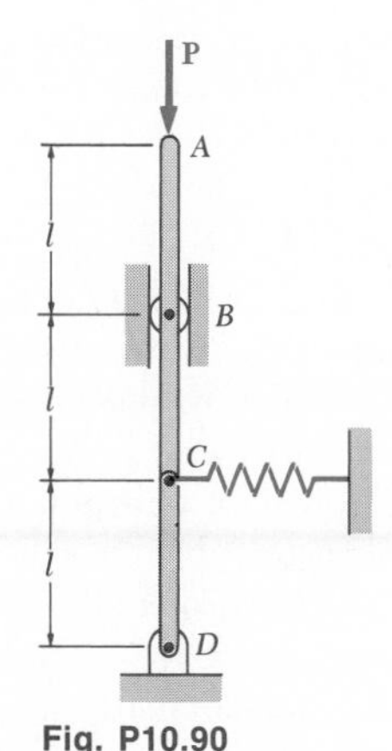

Fig. P10.90

10.91 Rod AD is acted upon by a vertical force **P** at end A and by two equal and opposite horizontal forces of magnitude Q at points B and C. Derive an expression for the magnitude Q of the horizontal forces required for equilibrium.

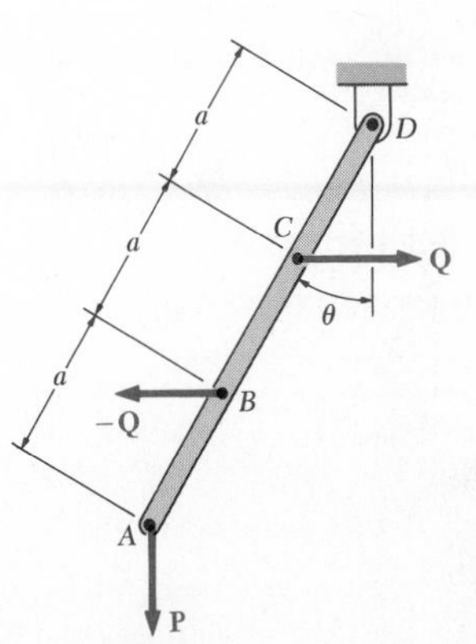

Fig. P10.91

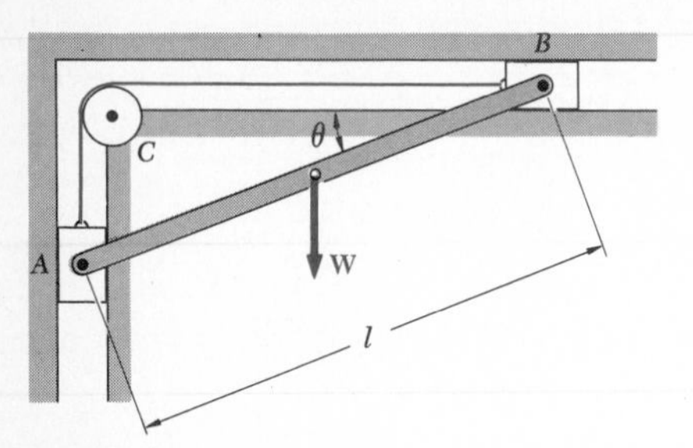

Fig. P10.92

10.92 A slender rod AB, of weight W, is attached to blocks A and B which move freely in the guides shown. The blocks are connected by an elastic cord which passes over a pulley at C. (*a*) Express the tension in the cord in terms of W and θ. (*b*) Determine the value of θ for which the tension in the cord is equal to $2W$.

The following problems are designed to be solved with a computer.

10.C1 Using the principle of virtual work, write a computer program and use it to calculate the force in member CD for values of θ from 5 to 120° at 5° intervals.

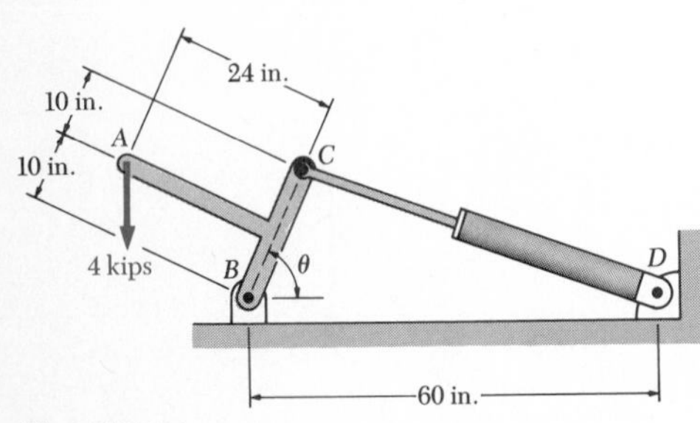

Fig. P10.C1

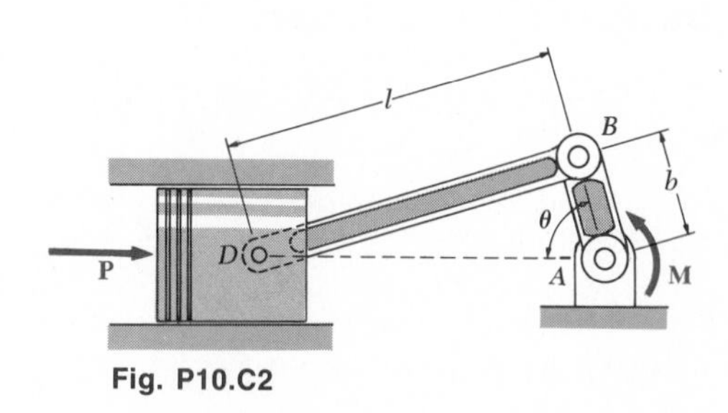

Fig. P10.C2

10.C2 A couple **M** must be applied to crank AB to maintain the equilibrium of the engine system shown when a force **P** is applied to the piston. Knowing that $b = 3$ in. and $l = 8$ in., write a computer program and use it to calculate the ratio M/P for values of θ from 0 to 180° at 10° intervals.

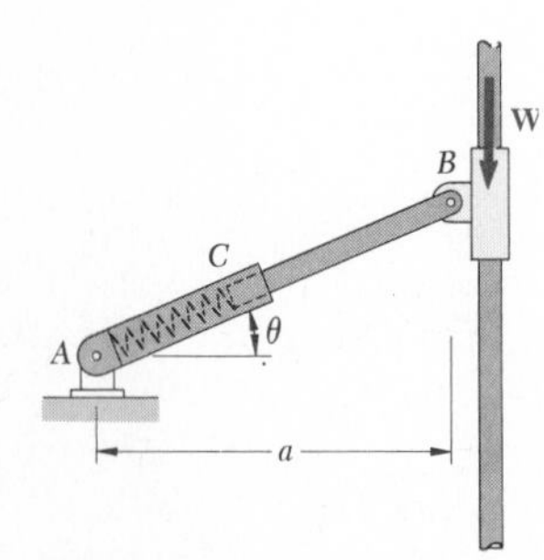

Fig. P10.C3

10.C3 The internal spring AC is of constant k and is undeformed when $\theta = 45°$. (*a*) Write a computer program which can be used to calculate the potential energy V of this system and its derivative $dV/d\theta$. (*b*) For $W = 100$ N, $a = 0.5$ m, $k = 3$ kN/m, calculate and plot the potential energy versus θ for values of θ from 0 to 50° at 5° intervals. (*c*) By examining the V-θ curve, determine the approximate values of θ corresponding to equilibrium and state in each case whether the equilibrium is stable, unstable, or neutral. (*d*) Determine, to three significant figures, the values of θ corresponding to equilibrium.

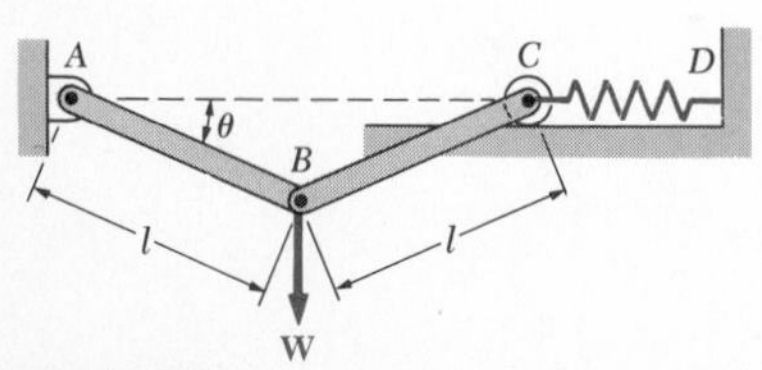

Fig. P10.C4

10.C4 A vertical load **W** is applied to the linkage at B. The constant of the spring is k, and the spring is unstretched when AB and BC are horizontal. (*a*) Neglecting the weight of the linkage, write a computer program which can be used to calculate the load W for which AB forms a given angle θ with the horizontal. (*b*) Use this program to compute W for values of θ from 0 to 80° at 5° intervals, assuming $l = 225$ mm and $k = 2$ kN/m. (*c*) Determine, to three significant figures, the value of θ for which the linkage is in equilibrium when $W = 900$ N, $l = 225$ mm, and $k = 2$ kN/m.

APPENDIX

How to Use the Interactive Tutorial Software

With Software Designed for an IBM PC™ or Compatible Computer:

A.1. Introduction. A diskette containing interactive software has been placed in the pocket inside the rear cover of this text. This software includes 12 "Tutorials," each consisting of a problem in statics for which students are more likely to require additional help. All the tutorials were designed by the authors of this text, programmed by EnginComp Software, Inc., and should be easy to use and understand.

A.2. Equipment. This software is designed for an IBM PC or compatible computer with at least 256K of memory and DOS 2.0 and higher. Your IBM PC compatible must also have a graphics display card equivalent to the IBM color/graphics adapter.

A.3. Getting Started. In order to start the Supplementary Tutorial program, your system must be activated and the DOS version you are using must be loaded into the computer. Insert your tutorial disk into the A drive and type "A:". The computer should then prompt you with "A)". Simply type "TUTORIAL" following this prompt to enter the program.

The program will display a title page, and at the bottom of the screen it will type "PRESS ANY KEY TO CONTINUE". Throughout the tutorials, all inputs required and key strokes will be specified at the bottom of the screen in a similar fashion.

After striking any key, the program will display the main menu, from which all the tutorials can be accessed. The cursor can be moved by using the [↓] and [↑] arrow keys or by typing the number corresponding to your choice. Selections are made by striking the [F10] key when your choice is highlighted.

A.4. Inside the Tutorials. Once you have made your choice of tutorial on the main menu and pressed [F10], the program will show you the tutorial title page on the bottom half of the screen, and a "Help" screen on the top half. This "Help" screen lists all the special function keys and an explanation of their action. It can be displayed at any time during the program by simply striking [F1].

In a similar fashion, you can exit the tutorial at any time by striking the [Esc] key. The program will then ask you to strike [F10] if you want to exit to the main menu, or any other key to continue. This prevents accidental termination of a tutorial.

The title page of each tutorial is followed by the statement of the problem, and you will be asked to select part of the data involved. This is done by entering each item and striking the [F10] key after all the data has been entered.

In several instances, and after giving a wrong answer, you will be invited to strike the [F2] key to obtain an explanatory comment before trying again or proceeding with the solution.

You may review your progress at any time by striking the [Home] key and then [F10]. This will return you to the beginning of the tutorial. Striking [F10] repeatedly will then enable you to review rapidly your solution up to the point that you had reached.

In the course of a tutorial you will sometimes encounter calculations requiring the use of a hand calculator. You will then be asked whether you wish to perform these calculations yourself, thus getting additional practical experience, or whether you wish your "Tutor"—i.e., the computer—to do the calculations for you, saving you time and effort.

At the end of each tutorial, you will be asked if you wish to try again the tutorial with some different data. Entering a "Y" (for yes) will take you to the beginning of the tutorial, while entering an "N" (for no) will take you back to the main menu.

Sometimes it is convenient to have a hard copy of tutorial screens. In order to do this, you must have a printer that can handle text and graphics information. Also, you must run the DOS Graphics program before starting the tutorials. In order to get a hard copy, simply strike the [Shift] [Prt Sc] keys.

A.5. Keeping Score of Your Correct Answers. The computer will keep score of your correct answers in the upper right-hand corner of the screen. The use of the [F9] key to supply the correct answer will count against your score, while the use of the [F10] key to review your progress will have no effect on your score at all. Your score will not be affected either, if you choose to have your "Tutor" perform the calculations for you when given that choice. However, if you choose to do the calculations yourself, your score will reflect the accuracy of your computations.

A.6. Exiting the Tutorials. To exit the tutorials, simply get back to the main menu, select the option "Exit Statics Tutorial," using the arrow keys or striking the [Esc] key, and then press [F10]. Always exit the tutorials before turning off your computer.

A.7. Introduction. A diskette containing interactive software has been placed in the pocket inside the rear cover of this text. This software includes 12 "Tutorials," each consisting of a problem in statics for which students are more likely to require additional help. All the tutorials were designed by the authors of this text, programmed by EnginComp Software, Inc., and should be easy to use and understand.

A.8. Equipment. This software is supplied on a double-sided 800K disk and is designed for a Macintosh computer with at least 512K of RAM, i.e., a Mac 512E, Mac Plus, Mac SE, or Mac II.

A.9. Getting Started. In order to start the Supplementary Tutorial program, your system must be activated with a Startup disk, either a floppy or hard disk with a System Folder on it. (A system version of 4.1 or higher is recommended.) Note that the Statics Tutorials disk contains a System Folder and can be used as a Startup Disk.

After starting up and inserting your tutorial disk in the computer, you will notice a disk icon in the upper-right-hand corner of the screen labeled "Vector Mechanics." Double clicking (moving the pointer over the disk icon and clicking the mouse button twice) will open the disk's directory window. Double clicking the icon labeled "Statics Tutorials" will launch the program by displaying the title page.

Clicking "Tutorials" in the lower-right-hand corner of the title page will bring a message at the bottom of the screen asking you to choose a lesson from the tutorials menu. Clicking on the heading "Tutorials" will cause the full menu to drop down. Moving the pointer down until the desired tutorial has been reached and releasing the mouse button will display the statement of the problem.

A.10. Inside the Tutorials. After selecting a tutorial from the Tutorials menu and clicking "Continue" at the statement of the problem, you will see an introductory figure and be asked to select part of the data involved. As you proceed through the tutorial you will be prompted to input more data, answer questions, or supply the missing elements of formulas and equations.

If you answer incorrectly, the bottom prompt will notify you when the correct answer and/or an explanatory comment are available in the Help menu. Note that all selections in this menu have shortcut command-key equivalents; "Answer," for instance, may be chosen by holding down the Cloverleaf (command) key and striking the [A] key.

You may review your progress at any time by selecting "Restart" from the Help menu. This will return you to the beginning of the tutorial. Striking [Enter] repeatedly will then enable you to rapidly review your solution up to the point that you have reached.

In the course of a tutorial you will sometimes encounter calculations requiring the use of a hand calculator. You will then be asked whether you wish to perform these calculations yourself, thus getting additional practical experience, or whether you wish your "Tutor"—i.e., the computer—to do the calculations for you, saving you time and effort.

At the end of each tutorial, you will be asked if you wish to try the tutorial again with some different data. Entering [Y] (for yes) will take you to the beginning of the tutorial, while entering [N] (for no) will take you back to the title screen.

Sometimes it is convenient to have a hard copy of tutorial screens. In order to do this, you must have an Imagewriter (I or II) printer connected to your Macintosh, turned on, and selected. Then choose "Print Screen" from the File menu.

A.11. Keeping Score of Your Correct Answers. The computer will keep score of your correct answers in the upper-right-hand corner of the screen. The use of "Answer" to supply the correct answer will count against your score, while the use of "Restart" to review your progress will have no effect on your score at all. Your score will not be affected either, if you choose to have your "Tutor" perform the calculations for you when given that choice. However, if you choose to do the calculations yourself, your score will reflect the accuracy of your computations.

A.12. Exiting the Tutorials. To exit the tutorials, choose "Quit" from the File menu. You will be able to select whether you wish to return to the title screen to begin another tutorial or completely quit the tutorial program. You may quit from either the title screen or from within a tutorial. Always exit the tutorial program before turning off your computer.

Index

Answers to Even-Numbered Problems†

CHAPTER 2

2.2 1391 N ∠47.8°.
2.4 8.03 kips ∠3.8°.
2.6 76.1°.
2.8 (*a*) 3660 N. (*b*) 3730 N.
2.10 (*a*) 5.85 kN. (*b*) 8.96 kN.
2.12 $T_{AC} = 666$ lb; $\alpha = 34.3°$.
2.16 (*a*) 90°. (*b*) 1545 N.
2.18 (350 N) +317 N, +147.9 N; (800 N) +274 N, +752 N; (600 N) −300 N, +520 N.
2.20 (145 lb) +100 lb, +105 lb; (200 lb) +192 lb, −56 lb.
2.22 (*a*) 2190 N. (*b*) 2060 N.
2.24 −180 lb, +525 lb.
2.28 253 N ∠86.6°.
2.30 296 lb ∠9.5°.
2.32 (*a*) 177.9 lb. (*b*) 410 lb.
2.34 (*a*) 26.8°. (*b*) 539 N.
2.36 $T_{AC} = 586$ N; $T_{BC} = 2190$ N.
2.38 $T_{AC} = 2860$ lb; $T_{BC} = 1460$ lb.
2.40 $T_{AC} = 5.22$ kN; $T_{BC} = 3.45$ kN.
2.42 $T_B = 16.73$ kips; $T_D = 14.00$ kips.
2.44 287 N $\leq P \leq$ 1600 N.
2.46 (*a*) 784 N. (*b*) 71.0°.
2.48 50 in.
2.50 (*b*) 817 N. (*d*) 613 N.
2.52 166.0 lb ∠53.4° or 407 lb ∠53.4°.
2.54 $27.4° < \alpha < 222.6°$.
2.56 (*a*) −130.1 N, +816 N, +357 N. (*b*) 98.3°, 25.0°, 66.6°.
2.58 (*a*) +628 N, −400 N, +293 N. (*b*) 38.3°, 120.0°, 68.5°.
2.60 (*a*) +16.38 lb. +34.6 lb, −11.47 lb. (*b*) 65.8°, 30.0°, 106.7°.
2.62 $F = 950$ lb; $\theta_x = 43.4°$, $\theta_y = 71.6°$, $\theta_z = 127.6°$.
2.64 (*a*) $F_y = +654$ N, $F_z = +1186$ N; $F = 1549$ N. (*b*) 119.0°.
2.66 +250 N, +500 N, −500 N.
2.68 −270 lb, +180 lb, −276 lb.
2.70 $R = 1171$ N; $\theta_x = 89.5°$, $\theta_y = 36.2°$, $\theta_z = 126.2°$.
2.72 (*a*) $R = 82.2$ lb; $\theta_x = 87.1°$, $\theta_y = 161.5°$, $\theta_z = 71.7°$. (*b*) $x = 1.269$ in.; $z = 7.95$ in.
2.74 548 N.
2.76 3375 lb.
2.78 9.71 kN.
2.80 $T_{AB} = T_{AC} = 13.50$ kN; $T_{AD} = 20.0$ kN.
2.82 (*a*) $-(25.2\text{ kN})\mathbf{i}$. (*b*) $T_{AB} = 2.25$ kN; $T_{AC} = 16.65$ kN.
2.84 $T_{AD} = 14.42$ lb; $T_{BD} = T_{CD} = 13.27$ lb.
2.86 $T_{AB} = 81.3$ lb; $T_{AC} = 22.2$ lb.
2.88 (*a*) 135 N. (*b*) $T_{AB} = 156$ N; $T_{AC} = 261$ N.
2.90 1372 N.
2.92 (*a*) 305 lb. (*b*) $T_{AD} = 40.9$ lb; $T_{BAC} = 117.0$ lb.
2.94 (*a*) 550 N. (*b*) 100 N.
2.96 (*a*) $W = \frac{1}{2}P\sqrt{1 + (d/h)^2}$. (*b*) 75.6 mm.
2.98 168.3 lb ∠13.5°.
2.100 1210 N.
2.102 $T_{AC} = 134.6$ N; $T_{BC} = 110.4$ N.
2.104 41.9°.
2.106 61.0°; +507 N, +919 N, +582 N.
2.C2 $\beta = 10°$: $F = 258$ lb, $\alpha = 69.7°$, or $F = 727$ lb, $\alpha = -69.7°$.
2.C3 $a = 0.8$ *m*: $h = 2.25$ m, $P = 189.1$ N, $T = 544$ N; $a = 2$ *m*: $h = 2.95$ m, $P = 634$ N, $T = 729$ N.
2.C4 (*b*) $R = 205$ lb; $\theta_x = 59.8°$, $\theta_y = 66.8°$, $\theta_z = 39.7°$.

CHAPTER 3

3.2 115.7 lb·in. ↻.
3.4 $P = 400$ N; $\alpha = 22.6°$.
3.6 (*a*) 41.7 N·m ↺. (*b*) 147.4 N ∠45°.
3.8 140.0 N·m ↺.
3.10 270 lb·ft ↻.
3.14 $\mathbf{M}_O = \dfrac{P(x_1y_2 - x_2y_1)\mathbf{k}}{[(x_2 - x_1)^2 + (y_2 - y_1)^2]^{1/2}}$.

†Partial answers are provided or referenced for *all* problems designed to be solved with a computer.

3.16 (*a*) $2\mathbf{i} - 7\mathbf{j} + 5\mathbf{k}$. (*b*) $30\mathbf{i} + 12\mathbf{k}$. (*c*) 0.
3.18 $(1200\ \text{N}\cdot\text{m})\mathbf{i} - (1500\ \text{N}\cdot\text{m})\mathbf{j} - (900\ \text{N}\cdot\text{m})\mathbf{k}$.
3.20 $(2400\ \text{lb}\cdot\text{ft})\mathbf{j} + (1440\ \text{lb}\cdot\text{ft})\mathbf{k}$.
3.22 1.491 m.
3.24 4.86 ft.
3.30 55.4°.
3.32 (*a*) 70.5°. (*b*) 180.0 N.
3.34 33.3°.
3.36 +2.
3.38 $M_x = -1283$ N·m; $M_y = 770$ N·m; $M_z = 1824$ N·m.
3.40 6.23 ft.
3.42 +23.0 N·m.
3.44 −176.6 lb·ft.
3.46 −222 N·m.
3.48 $+Pa/\sqrt{2}$.
3.50 13.06 in.
3.52 199.4 mm.
3.54 (*a*) 7.33 N·m ↺. (*b*) 91.6 mm.
3.56 1170 lb·in. ↺. (*b*) Either *A* and *D* ∠53.1° or *B* and *C* ∠53.1°. (*c*) 70.9 lb.
3.58 $M = 13.00$ lb·ft; $\theta_x = 67.4°$, $\theta_y = 90°$, $\theta_z = 22.6°$.
3.60 $M = 2.72$ N·m; $\theta_x = 134.9°$, $\theta_y = 58.0°$, $\theta_z = 61.9°$.
3.62 (*a*) $\mathbf{F} = 60$ lb ↓; $\mathbf{M}_B = 450$ lb·in. ↺. (*b*) $\mathbf{B} = 100$ lb ←; $\mathbf{C} = 100$ lb →.
3.64 (*a*) $\mathbf{F} = 300$ N ∠30°; $\mathbf{M} = 75$ N·m ↺. (*b*) $\mathbf{B} = 800$ N ∠30°; $\mathbf{C} = 500$ N ∠30°.
3.68 (*a*) 60 lb ∠50°; 3.24 in. from *A*. (*b*) 60 lb ∠50°; 3.87 in. below *A*.
3.70 $\mathbf{F} = -(250\ \text{kN})\mathbf{j}$; $\mathbf{M} = (15\ \text{kN}\cdot\text{m})\mathbf{i} + (7.5\ \text{kN}\cdot\text{m})\mathbf{k}$.
3.72 $\mathbf{F} = -(2.4\ \text{kips})\mathbf{j} + (1\ \text{kip})\mathbf{k}$; $\mathbf{M} = (15\ \text{kip}\cdot\text{ft})\mathbf{i} - (10\ \text{kip}\cdot\text{ft})\mathbf{j} - (24\ \text{kip}\cdot\text{ft})\mathbf{k}$.
3.74 (*a*) $\mathbf{F} = -(75\ \text{lb})\mathbf{j}$. (*b*) $x = -3.20$ in., $z = 0.640$ in.
3.76 Force-couple system at *E*.
3.78 Loading (*f*).
3.80 (*a*) 2.00 m from front axle. (*b*) 50 kN; 2.80 m from front axle.
3.82 $d = 0.75 + 0.6x$.
3.84 (*a*) 1308 N ∠66.6°. (*b*) 412 mm to the right of *A* and 250 mm to the right of *C*.
3.86 (*a*) 60.2 lb·in. ↺. (*b*) 200 lb·in. ↺. (*c*) 20.0 lb·in. ↻.
3.88 (*a*) $\mathbf{R} = P$ ∠θ; $\mathbf{M}_D^R = Pa \sin 2\theta \cos \theta$ ↻. (*b*) 35.3°.
3.90 $\mathbf{R} = -(300\ \text{N})\mathbf{i} - (240\ \text{N})\mathbf{j} + (25\ \text{N})\mathbf{k}$; $\mathbf{M}_A^R = -(3\ \text{N}\cdot\text{m})\mathbf{i} + (13.5\ \text{N}\cdot\text{m})\mathbf{j} + (9\ \text{N}\cdot\text{m})\mathbf{k}$.
3.92 $\mathbf{R} = (20\ \text{lb})\mathbf{i} - (34.6\ \text{lb})\mathbf{j}$; $\mathbf{M}_C^R = (520\ \text{lb}\cdot\text{in.})\mathbf{i} - (500\ \text{lb}\cdot\text{in.})\mathbf{k}$. (*a*) Neither. (*b*) Tighten.
3.94 500 kN; 2.56 m from *AD* and 2.00 m from *DC*.
3.96 $F_B = 20$ kips; $F_F = 15$ kips.
3.98 (*a*) $R = P\sqrt{3}$; $\theta_x = \theta_y = \theta_z = 54.7°$. (*b*) $-a$. (*c*) Diagonal *OC*.
3.100 (*a*) $\mathbf{R} = (2Pa/\sqrt{a^2 + c^2})\mathbf{i}$; $\mathbf{M}_O^R = (P/\sqrt{a^2 + c^2})(bc\mathbf{i} + ca\mathbf{j} - ab\mathbf{k})$. (*b*) $p = bc/2a$; axis of wrench is parallel to *x* axis at $y = \frac{1}{2}b$, $z = \frac{1}{2}c$.
3.102 (*a*) $R = 70.7$ N; $\theta_x = 90°$, $\theta_y = \theta_z = 135°$. (*b*) Line defined by equations $x = -2$ mm, $y = z$.
3.104 (*a*) $R = 56.6$ N; $\theta_x = 45°$, $\theta_y = 90°$, $\theta_z = 45°$. (*b*) −25 mm. (*c*) $y = +125$ mm, $z = +137.5$ mm.
3.106 (*a*) $R = 40.0$ lb; $\theta_x = 60°$, $\theta_y = 150°$, $\theta_z = 90°$. (*b*) 6.50 in. (*c*) Halfway between *A* and *B*.
3.108 (*a*) $\mathbf{A} = (a + b)(R/b)\mathbf{j} + (M/b)\mathbf{k}$; $\mathbf{B} = -(aR/b)\mathbf{j} - (M/b)\mathbf{k}$. (*b*) $\mathbf{A} = (525\ \text{N})\mathbf{j} + (400\ \text{N})\mathbf{k}$; $\mathbf{B} = -(375\ \text{N})\mathbf{j} - (400\ \text{N})\mathbf{k}$.
3.112 $M_x = 78.9$ kN·m, $M_y = 13.15$ kN·m, $M_z = -9.86$ kN·m.
3.114 (*a*) $-(1200\ \text{lb}\cdot\text{in.})\mathbf{i} + (4800\ \text{lb}\cdot\text{in.})\mathbf{j} + (7200\ \text{lb}\cdot\text{in.})\mathbf{k}$. (*b*) 3090 lb·in.
3.116 875 lb·in. ↻.
3.118 (*a*) $\mathbf{F} = 500$ N ∠60°; $\mathbf{M} = 86.2$ N·m ↻. (*b*) $\mathbf{A} = 689$ N ↑; $\mathbf{B} = 1150$ N ∠77.4°.
3.120 325 kN; $x = -0.923$ m, $z = -0.615$ m.
3.122 (*a*) $\mathbf{R} = -(250\ \text{N})\mathbf{k}$; $\mathbf{M}_A^R = (30.8\ \text{N}\cdot\text{m})\mathbf{j} - (12.00\ \text{N}\cdot\text{m})\mathbf{k}$. (*b*) $p = +48.0$ mm; axis parallel to *z* axis at $x = 123.2$ mm, $y = 0$.
3.C1 (*b*) 1.50 ft; 3.15 ft; 5.55 ft.
3.C2 *β = 30°*: $\mathbf{R} = 165.2$ N ∠21.3°, $\mathbf{M}_D = 50.9$ N·m ↻; *β = 150°*: $\mathbf{R} = 80.7$ N ∠48.1°, $\mathbf{M}_D = 22.6$ N·m ↺.
3.C3 *z = 5 in.*: (*a*) $M_x = 1800$ lb·in., $M_y = 2400$ lb·in., $M_z = -3600$ lb·in., $M_O = 4690$ lb·in. (*b*) 2120 lb·in. *z = 10 in.*: (*a*) $M_x = 2400$ lb·in., $M_y = 4800$ lb·in., $M_z = -3600$ lb·in., $M_O = 6460$ lb·in. (*b*) 3090 lb·in.
3.C4 (*c*) 70 kips; $x = +2.50$ ft, $z = -0.619$ ft.

CHAPTER 4

4.2 8.40 lb ↑.
4.4 (*a*) 34.0 kN ↑. (*b*) 4.96 kN ↑.
4.6 (*a*) 2196 lb ↑. (*b*) 804 lb ↑.
4.8 6 kips ≤ *P* ≤ 27 kips.
4.10 1.25 kN ≤ *Q* ≤ 10.25 kN.
4.12 31.0 kN.
4.14 (*a*) 150.0 lb. (*b*) 313 lb ∠69.8°.
4.16 (*a*) 400 N. (*b*) 458 N ∠49.1°.

4.18 (*a*) **A** = **B** = 37.5 lb ↑.
(*b*) **A** = 97.6 lb ∠50.2°; **B** = 62.5 lb ←.
(*c*) **A** = 49.8 lb ∠71.2°; **B** = 32.2 lb ∠60°.

4.20 (*a*) 225 N ∠45°. (*b*) 486 N ∠70.9°.

4.22 (*a*) 17.32 lb. (*b*) **B** = 69.3 lb ∠60°;
C = 34.6 lb ∠60°.

4.24 (*a*) 600 N. (*b*) **A** = 4 kN ←; **B** = 4 kN →.

4.26 (*a*) 240 N ↓. (*b*) 1768 N →.

4.28 11.06 in.

4.30 (*a*) **E** = 8.80 kips ↑; $\mathbf{M}_E$ = 36.0 kip · ft ↺.
(*b*) **E** = 4.80 kips ↑; $\mathbf{M}_E$ = 51.0 kip · ft ↺.

4.32 58.3 kN $\leq T \leq$ 141.7 kN.

4.34 $\theta = \sin^{-1}(2M \cot \alpha / Wl)$.

4.36 (*a*) $\theta = 2\cos^{-1}\left[\frac{1}{4}\left(\frac{W}{P} + \sqrt{\frac{W^2}{P^2} + 8}\right)\right]$. (*b*) 65.1°.

4.38 19.4°.

4.40 (*a*) $2 \sin \theta - \tan \theta = W/ka$. (*b*) 704 N/m.

4.42 (1) Completely constrained; determinate;
A = **C** = 196.2 N ↑. (2) Completely constrained; determinate; **B** = 0, **C** = **D** = 196.2 N ↑.
(3) Completely constrained; indeterminate;
$\mathbf{A}_x$ = 294 N →, $\mathbf{D}_x$ = 294 N ←. (4) Partially constrained; determinate; equil.;
C = **D** = 196.2 N ↑. (5) Completely constrained; determinate; **B** = 294 N →, **D** = 491 N ∠53.1°.
(6) Partially constrained; no equil. (7) Improperly constrained; indeterminate; no equil.
(8) Improperly constrained; indeterminate; no equil. (9) Completely constrained; indeterminate;
B = 196.2 N ↑, $\mathbf{D}_y$ = 196.2 N ↑.

4.44 **B** = 501 N ∠56.3°; **C** = 324 N ∠31.0°.

4.48 **A** = 185.3 N ∠62.4°; T = 92.8 N.

4.52 α = 73.9°; T_A = 4160 lb; T_B = 2310 lb.

4.54 **A** = 170 N ∠56.1°; **C** = 300 N ∠28.1°.

4.56 (*a*) $T = 4P - 3Q$. (*b*) 30 lb.

4.58 $\theta = \sin^{-1}(a/L)^{1/3}$.

4.60 (*a*) 19.90 lb ∠39.3°. (*b*) 99.6 lb ∠72.0°.

4.62 23.2°.

4.64 **A** = (120.0 N)**j** + (133.3 N)**k**;
D = (60.0 N)**j** + (166.7 N)**k**.

4.66 (*a*) 96.0 lb. (*b*) **A** = (2.40 lb)**j**; **B** = (213.6 lb)**j**.

4.68 (*a*) 1039 N. (*b*) **C** = (346 N)**i** + (1200 N)**j**;
D = −(1386 N)**i** − (480 N)**j**.

4.70 T_A = 30 lb; T_B = 10 lb; T_C = 40 lb.

4.72 10 kg; x = 500 mm, z = 375 mm.

4.74 T_{BE} = 6.62 kN; T_{BF} = 25.1 kN;
A = −(6.34 kN)**i** + (20.3 kN)**j** + (2.96 kN)**k**.

4.76 (*a*) T_{BC} = 1950 lb; T_{BD} = 1400 lb.
(*b*) **A** = (3000 lb)**i**.

4.78 (*a*) 38.7°. (*b*) 51.3°.

4.80 (*a*) 216 lb. (*b*) **A** = −(111.0 lb)**j** − (156.5 lb)**k**;
B = (37.5 lb)**j**.

4.82 (*a*) 2.40 kN. (*b*) −0.600 kN.

4.84 (*a*) 2330 N. (*b*) **A** = −(141.5 N)**j** + (1415 N)**k**;
B = −(736 N)**i** + (1368 N)**j** + (424 N)**k**.

4.86 (*a*) T_{CE} = 2110 N; T_{DE} = 272 N.
(*b*) **A** = (1415 N)**k**; **B** = −(490 N)**i** + (1226 N)**j** + (424 N)**k**.

4.88 **A** = −(112.5 lb)**i** + (150.0 lb)**j** − (75.0 lb)**k**;
$\mathbf{M}_A$ = (600 lb · ft)**i** + (225 lb · ft)**j**; F_{CE} = 201.9 lb.

4.90 (*a*) 1292 N.
(*b*) **A** = (429 N)**i** + (1226 N)**j** + (1839 N)**k**;
$\mathbf{M}_A$ = − (1104 N · m)**j** + (1809 N · m)**k**.

4.92 T_{BD} = 780 lb; T_{BE} = 650 lb; T_{CF} = 650 lb;
A = (1920 lb)**i** − (300 lb)**k**.

4.94 x = −1.144 m; z = +1.925 m.

4.96 **B** = −(1.5 lb)**i** + (15 lb)**k**; **C** = (2 lb)**j** − (15 lb)**k**;
D = (1.5 lb)**i** − (2 lb)**j**.

4.98 T_{BD} = 0; T_{BE} = 3.96 kN; T_{CD} = 3.00 kN.

4.100 60.9 lb.

4.102 450 N.

4.104 (375 N)**i**.

4.106 (*a*) 1.530 kN. (*b*) **A** = 2.42 kN ∠21.8°.

4.108 (*a*) 3.00 lb. (*b*) **A** = −(2.94 lb)**i** + (6.90 lb)**j**;
B = (7.50 lb)**j**.

4.110 (*a*) 96.2 lb ∠4.45°. (*b*) 234 lb ∠35.1°.

4.112 (*a*) 59.2°: (*b*) $T_{AB} = 0.596\, W$; $T_{CD} = 1.164\, W$.

4.114 (*a*) **A** = 0.745 P ∠63.4°; **D** = 0.471 P ∠45°.
(*b*) **A** = P →; **D** = 1.414 P ∠45°.
(*c*) **A** = 0.471 P ∠45°; **D** = 0.745 P ∠63.4°.
(*d*) **A** = 0.707 P ∠45°; **C** = 0.707 P ∠45°;
D = P ↑.

4.116 (*a*) α = 60°. (*b*) a = 84 mm. (*c*) a = 40 mm.
(*d*) α = 63.4°. (*e*) 63.4°.

4.C1 (*a*) θ = 25°: α = 40.0° ∠; θ = 75°: α = 60.0° ∠.
(*b*) θ = 25°: α = 22.0° ∠; θ = 75°: α = 67.4° ∠.

4.C2 θ = 25°: P = 10.77 lb; θ = 65°: P = 19.96 lb.

4.C3 ϕ = 12°: T_{BD} = 11.63 kN, T_{BE} = 20.8 kN;
ϕ = 24°: T_{BD} = 6.16 kN, T_{BE} = 24.2 kN.

4.C4 y = 400 mm: T = 650 N;
y = 800 mm: T = 523 N.

CHAPTER 5

5.2 $\bar{X}$ = 3.27 in., $\bar{Y}$ = 4.18 in.

5.4 $\bar{X}$ = 175.6 mm, $\bar{Y}$ = 94.4 mm.

5.6 $\bar{X}$ = 46.2 mm, $\bar{Y}$ = 47.8 mm.

5.8 $\bar{X}$ = 0, $\bar{Y}$ = 3.17 in.

5.10 $\bar{X}$ = −62.4 mm, $\bar{Y}$ = 0.

5.12 $\bar{X}$ = 120 mm, $\bar{Y}$ = 60 mm.

5.14 $\bar{X}$ = 10.11 in., $\bar{Y}$ = 3.87 in.

5.16 $\bar{X} = \frac{4r}{3} \frac{\sin^3 \alpha}{2\alpha - \sin 2\alpha}$.

5.18 $\bar{X} = -(4/3\pi)[(r_2^2 + r_1 r_2 + r_1^2)/(r_2 + r_1)]$, $\bar{Y} = 0$.

5.20 0.508.
5.22 648 N.
5.24 0.2352 in^3 and -0.2352 in^3.
5.26 $\overline{X} = 3.38$ in., $\overline{Y} = 4.07$ in.
5.28 $\overline{X} = 172.5$ mm, $\overline{Y} = 97.5$ mm.
5.30 (*a*) 5.09 lb. (*b*) 9.48 lb ⦨57.5°.
5.32 249 mm.
5.34 (*a*) $0.513b$. (*b*) 0.691 b.
5.36 50 mm.
5.38 $0.7667a$; 2.23%.
5.40 $\bar{x} = 2b/3$, $\bar{y} = h/3$.
5.42 $\bar{x} = 8a/15$, $\bar{y} = 8b/21$.
5.50 $\bar{x} = 2L/5$.
5.52 $\bar{x} = 0.549\,R$, $\bar{y} = 0$.
5.54 $\bar{x} = \bar{y} = 1.027$ in.
5.56 (*a*) $\bar{x} = 4.59 \times 10^6 \text{ mm}^3$. (*b*) $3.68 \times 10^6 \text{ mm}^3$.
5.58 (*a*) $\pi a h^2/5$. (*b*) $\pi a^2 h/6$.
5.60 314 in^2.
5.62 (*a*) 0.226 m^3. (*b*) 131.2 kg.
5.64 $255 \times 10^3 \text{ mm}^3$; $37.5 \times 10^3 \text{ mm}^2$.
5.66 3.96 in^3; 1.211 lb.
5.68 4.76 in.
5.70 (*a*) $120 \times 10^3 \text{ mm}^3$. (*b*) $12.2 \times 10^3 \text{ mm}^2$.
5.72 (*a*) $\pi R h^2$. (*b*) $2\pi R h^2/3$. (*c*) $8\pi R h^2/15$. (*d*) $\pi R h^2/3$. (*e*) $\pi R h^2/5$.
5.74 **R** = 4400 N ↓, 1.727 m to the right of *A*; **A** = 2500 N ↑; **B** = 1900 N ↑.
5.76 **A** = 4.60 kN ↑; $\mathbf{M}_A = 1.520$ kN·m ↺.
5.78 **A** = 2860 lb ↑; **B** = 740 lb ↑.
5.80 **A** = 675 N ↑; **B** = 1350 N ↓.
5.82 (*a*) 200 lb/ft. (*b*) 4800 lb ↑.
5.84 (*a*) 0.420 m. (*b*) 42.0 kN/m.
5.86 (*a*) **H** = 1766 kN →, **V** = 910 kN ↑; 2.32 m to the right of *A*. (*b*) **R** = 236 kN ⦪41.6°.
5.88 5.88 ft.
5.90 4 in.
5.92 **A** = 118.1 kN ⦪68.5°; **B** = 113.9 kN ↑.
5.94 5.18 ft.
5.96 4.47 ft.
5.98 10.20 ft.
5.100 462 mm.
5.102 (*a*) 0.448 *h*. (*b*) 0.425 *h*.
5.104 $-3a/2\pi$.
5.106 30.2 mm.
5.108 -0.0656 in.
5.110 40.3 mm.
5.112 $\overline{X} = \overline{Z} = 105.2$ mm, $\overline{Y} = 175.8$ mm.
5.114 $\overline{X} = \overline{Y} = 2.31$ in., $\overline{Z} = -3$ in.
5.116 $\overline{X} = 205$ mm, $\overline{Y} = 255$ mm, $\overline{Z} = 75$ mm.
5.118 3.44 in. above lower edge of base.
5.120 34.9 mm above the base.
5.124 $\bar{x} = 5h/6$, $\bar{y} = \bar{z} = 0$.
5.126 $\bar{x} = \bar{z} = 0$, $\bar{y} = 7a/8$.
5.128 $\bar{x} = \bar{z} = 0$, $\bar{y} = (\pi + 2)h/16$.
5.132 94.4 mm.
5.134 $\bar{x} = 0$, $\bar{y} = 5h/16$, $z = -a/4$.
5.136 $\overline{X} = 3.79$ in., $\overline{Y} = 0.923$ in., $\overline{Z} = 3$ in.
5.138 $38.9 \times 10^3 \text{ mm}^2$.
5.140 $\bar{x} = 135$ mm; $\bar{y} = 22.5$ mm; $\bar{z} = 90$ mm.
5.142 $\bar{x} = -r/2$, $\bar{y} = \bar{z} = 0$.
5.144 $\overline{X} = 24.3$ mm, $\overline{Y} = 47.8$ mm.
5.146 **A** = 1000 lb ↑; **B** = 1400 lb ↑.
5.C1 (*b*) $\overline{X} = 51.25$ mm, $\overline{Y} = 11.04$ mm. (*c*) $\overline{X} = 9.57$ in., $\overline{Y} = 2.38$ in.
5.C2 (*a*) 341 in^2. (*b*) 615 in^2. (*c*) 1354 in^2.
5.C3 *d = 0.6 m:* **A** = 11.54 kN ⦪72.2°, **B** = 10.99 kN ↑; *d = 2.1 m:* **A** = 118.1 kN ⦪68.5°, **B** = 113.9 kN ↑.
5.C4 (*a*) $\overline{X} = \overline{Z} = 168.6$ mm, $\overline{Y} = 69.4$ mm. (*b*) $\overline{X} = \overline{Z} = 147.7$ mm, $\overline{Y} = 217$ mm. (*c*) $\overline{X} = 135.4$ mm, $\overline{Y} = 141.0$ mm; $\overline{Z} = 163.4$ mm.

CHAPTER 6

6.2 $F_{AB} = 52$ kN *T*; $F_{AC} = 64$ kN *T*; $F_{BC} = 80$ kN *C*.
6.4 $F_{AB} = 1080$ lb *T*; $F_{AC} = 1800$ lb *C*; $F_{BC} = 1170$ lb *C*.
6.6 $F_{AB} = F_{BC} = 0$; $F_{AD} = F_{CF} = 5$ kN *C*; $F_{BD} = F_{BF} = 34$ kN *C*; $F_{BE} = 12$ kN *T*; $F_{DE} = F_{EF} = 30$ kN *T*.
6.8 $F_{AB} = 12$ kips *C*; $F_{AC} = 5$ kips *C*; $F_{AD} = 13$ kips *T*; $F_{BD} = 0$; $F_{CD} = 30$ kips *C*; $F_{CE} = 17.5$ kips *C*; $F_{CF} = 32.5$ kips *T*; $F_{DF} = 5$ kips *T*; $F_{EF} = 0$.
6.10 $F_{AB} = F_{BC} = F_{CD} = 24$ kips *T*; $F_{AE} = 38.4$ kips *T*; $F_{AF} = 30$ kips *C*; $F_{BF} = F_{BG} = F_{CG} = F_{CH} = 0$; $F_{DH} = F_{GH} = F_{FG} = 26$ kips *C*; $F_{EF} = 24$ kips *C*.
6.12 $F_{AC} = 80$ kN *T*; $F_{BC} = 19$ kN *C*; $F_{BD} = F_{DE} = 51$ kN *C*; $F_{CE} = 45$ kN *T*; $F_{CD} = 48$ kN *T*.
6.14 Trusses of Probs. 6.11 and 6.15 are simple trusses; others are not simple trusses.
6.16 *BC*, *CD*, *IJ*, *IL*, *KL*, *LM*, *MN*.
6.18 $F_{AB} = F_{AC} = F_{AD} = P/\sqrt{6}$ *comp.*; $F_{BC} = F_{CD} = F_{DB} = P/3\sqrt{6}$ *ten.*
6.20 (*a*) **B** = (3000 N)**j**; **C** = **D** = (1200 N)**j**. (*b*) $F_{AB} = 3400$ N *C*; $F_{AC} = F_{AD} = 1520$ N *C*; $F_{BC} = F_{BD} = 843$ N *T*; $F_{CD} = 213$ N *T*.
6.22 $F_{CD} = 9$ kN *C*; $F_{DF} = 12$ kN *T*.
6.24 $F_{FG} = 70$ kN *C*; $F_{FH} = 240$ kN *T*.
6.26 $F_{DE} = 25$ kips *T*; $F_{DF} = 13$ kips *C*.

6.28 $F_{DE} = 7.71$ kips C; $F_{DF} = 18.29$ kips T.
6.30 $F_{BD} = 48$ kips T; $F_{DE} = 60$ kips T.
6.32 133.3 kN T.
6.34 $F_{DF} = 40$ kN T; $F_{EF} = 12$ kN T; $F_{EG} = 60$ kN C.
6.36 $F_{AD} = 3.38$ kips C; $F_{CD} = 0$; $F_{CE} = 14.03$ kips T.
6.38 $F_{FK} = \frac{1}{4}P$ *ten.*; $F_{JO} = \frac{1}{4}P$ *comp.*
6.40 28.6 kips C.
6.42 $F_{AB} = 161.6$ kN C; $F_{AD} = 190.1$ kN T.
6.44 $F_{DE} = 0$; $F_{BE} = 12.5$ kips T; $F_{EF} = 2.5$ kips T.
6.46 (*a*) Completely constrained, indeterminate.
(*b*) Completely constrained, determinate.
(*c*) Partially constrained.
6.48 (*a*) Partially constrained.
(*b*) Completely constrained, determinate.
(*c*) Completely constrained, indeterminate.
6.50 $F_{BD} = 255$ N C; $\mathbf{C}_x = 120.0$ N →, $\mathbf{C}_y = 625$ N ↑.
6.52 $\mathbf{A}_x = 120$ lb →, $\mathbf{A}_y = 30$ lb ↑; $\mathbf{B}_x = 120$ lb ←,
$\mathbf{B}_y = 80$ lb ↓; $\mathbf{C} = 30$ lb ↓; $\mathbf{D} = 80$ lb ↑.
6.54 (*a*) 2.44 kN ∠8.4°. (*b*) 1.930 kN ∠51.3°.
6.56 $\mathbf{A} = 150$ lb →; $\mathbf{B}_x = 150$ lb ←, $\mathbf{B}_y = 60$ lb ↑;
$\mathbf{C} = 20$ lb ↑; $\mathbf{D} = 80$ lb ↓.
6.58 7.36 kN C.
6.60 $\mathbf{B} = 152$ lb ↓; $\mathbf{C}_x = 60$ lb ←, $\mathbf{C}_y = 200$ lb ↑;
$\mathbf{D}_x = 60$ lb →, $\mathbf{D}_y = 42$ lb ↑.
6.62 (*a*) $\mathbf{D}_x = 750$ N →, $\mathbf{D}_y = 250$ N ↓;
$\mathbf{E}_x = 750$ N ←, $\mathbf{E}_y = 250$ N ↑.
(*b*) $\mathbf{D}_x = 375$ N →, $\mathbf{D}_y = 250$ N ↓;
$\mathbf{E}_x = 375$ N ←, $\mathbf{E}_y = 250$ N ↑.
6.64 (*a*) $\mathbf{A} = 78$ lb ∠22.6°; $\mathbf{C} = 144$ lb →;
$\mathbf{G} = 72$ lb ←; $\mathbf{I} = 30$ lb ↑. (*b*) $\mathbf{A} = 78$ lb ∠22.6°;
$\mathbf{C} = 72$ lb →; $\mathbf{G} = 0$; $\mathbf{I} = 30$ lb ↑.
6.66 (*a*) 200 N ∠60°. (*b*) 574 N ∠89.0°.
6.68 $\mathbf{A}_x = 45$ N ←, $\mathbf{A}_y = 30$ N ↓; $\mathbf{B}_x = 45$ N →,
$\mathbf{B}_y = 270$ N ↑.
6.70 (*a*) $\mathbf{E}_x = 960$ lb ←, $\mathbf{E}_y = 1280$ lb ↑.
(*b*) $\mathbf{C}_x = 2640$ lb ←, $\mathbf{C}_y = 3520$ lb ↑.
6.72 (*a*) $\mathbf{A} = 15.76$ kips ↑; $\mathbf{B} = 26.2$ kips ↑.
(*b*) $\mathbf{C} = 34.6$ kips ←; $\mathbf{D}_x = 34.6$ kips →,
$\mathbf{D}_y = 2.48$ kips ↓.
6.74 (*a*) $\mathbf{A} = 3980$ N ↑; $\mathbf{B} = 4170$ N ↑; $\mathbf{C} = 2890$ N ↑.
(*b*) $\Delta B = +1326$ N; $\Delta C = -398$ N.
6.76 $\mathbf{B}_x = 3600$ N ←, $\mathbf{B}_y = 1800$ N ↓; $\mathbf{E}_x = 1200$ N →,
$\mathbf{E}_y = 0$; $\mathbf{G}_x = 2400$ N →, $\mathbf{G}_y = 1800$ N ↑; $\mathbf{H} = 0$.
6.78 $\mathbf{A}_x = 10.80$ kN ←, $\mathbf{A}_y = 7.00$ kN ↑;
$\mathbf{B}_x = 16.20$ kN ←, $\mathbf{B}_y = 0.50$ kN ↓;
$\mathbf{D}_x = 27.0$ kN →, $\mathbf{D}_y = 6.50$ kN ↓.
6.80 $\mathbf{A}_x = 660$ lb ←, $\mathbf{A}_y = 240$ lb ↑; $\mathbf{B}_x = 660$ lb ←,
$\mathbf{B}_y = 320$ lb ↓; $\mathbf{D}_x = 1320$ lb →, $\mathbf{D}_y = 80$ lb ↑.
6.84 (*a*) $\mathbf{C}_x = 126.0$ kips ←, $\mathbf{C}_y = 103.2$ kips ↑.
(*b*) $\mathbf{B}_x = 126.0$ kips ←, $\mathbf{B}_y = 6.00$ kips ↓.
6.86 $F_{AF} = 3M_0/4a$ *comp.*; $F_{BG} = M_0/\sqrt{2}a$ *ten.*;
$F_{DG} = M_0/\sqrt{2}a$ *ten.*; $F_{EH} = M_0/4a$ *comp.*
6.88 $F_{AF} = M_0/6a$ *ten.*; $F_{BG} = \sqrt{2}M_0/6a$ *ten.*;
$F_{DG} = \sqrt{2}M_0/3a$ *ten.*; $F_{EH} = M_0/6a$ *comp.*
6.90 $F_{BF} = 7.20$ kN T; $F_{DG} = 3.00$ kN C.
6.92 $\mathbf{A} = 1875$ N →; $\mathbf{C} = 375$ N ←;
$\mathbf{D}_x = 1500$ N →, $\mathbf{D}_y = 750$ N ↑; $\mathbf{F} = 3000$ N ←.
6.94 $\mathbf{A} = P/7$ ↑; $\mathbf{D} = 2P/7$ ↑; $\mathbf{F} = 4P/7$ ↑.
6.96 (*a*) $\mathbf{A} = P$ ↑; $\mathbf{B} = P$ ↓; $\mathbf{C} = P$ ↑; frame is rigid.
(*b*) Frame is not rigid.
(*c*) $\mathbf{A}_x = 2P$ →, $\mathbf{A}_y = P/2$ ↑;
$\mathbf{B}_x = 2P$ ←, $\mathbf{B}_y = P/2$ ↑; frame is rigid.
6.98 (*a*) 2860 N ↓. (*b*) $\mathbf{C} = 2700$ N ∠68.5°.
6.100 $T_{DE} = 18.00$ lb; $\mathbf{B} = 48.0$ lb ↓.
6.102 45.6 kN C.
6.104 $\mathbf{C} = 4.65$ kips →; $\mathbf{E} = 6.14$ kips ∠40.7°.
6.106 (*a*) $F_{AB} = 2630$ lb C; $F_{DE} = 9590$ lb C.
(*b*) $\mathbf{C} = 3250$ lb ∠40.8°.
6.108 (*a*) 252 N·m ↻. (*b*) 108.0 N·m ↻.
6.110 (*a*) $\mathbf{M}_C = 1261$ lb·in. ↺.
(*b*) $\mathbf{C}_x = 54.3$ lb ←, $\mathbf{C}_y = 21.7$ lb ↑.
6.112 1200 N.
6.114 2220 N.
6.116 720 lb.
6.118 $\mathbf{Q} = 315$ lb →.
6.120 $\mathbf{P} = 260$ N ←.
6.122 6480 lb ∠62.1°.
6.124 (*a*) 7.68 kN C. (*b*) 21.7 kN C.
6.126 (*a*) 3000 lb T.
(*b*) $\mathbf{H}_x = 2400$ lb ←, $\mathbf{H}_y = 4800$ lb ↓.
6.128 (*a*) $3.2M_A$ ↺. (*b*) $2.2M_A$ ↻.
6.130 (*a*) 27.2 N·m. (*b*) $\mathbf{B} = -(105.7\text{ N})\mathbf{i}$;
$\mathbf{C} = (190.2\text{ N})\mathbf{i}$; $\mathbf{E} = -(84.5\text{ N})\mathbf{i}$.
6.132 $\mathbf{E}_x = 108.4$ kN →, $\mathbf{E}_y = 159.1$ kN ↑;
$\mathbf{F}_x = 29.9$ kN →, $\mathbf{F}_y = 119.9$ kN ↓;
$\mathbf{H}_x = 138.3$ kN ←, $\mathbf{H}_y = 39.2$ kN ↓.
6.134 707 lb C.
6.136 (*a*) 117.8 N·m ↺. (*b*) 47.9 N·m ↺.
6.138 $F_{DE} = 1897$ lb C; $F_{EF} = 4200$ lb T.
6.140 $\mathbf{A} = 160.0$ N ↑; $\mathbf{C}_x = 1440$ N →, $\mathbf{C}_y = 160.0$ N ↓;
$\mathbf{D} = 1440$ N ←.
6.142 $F_{FG} = 4.53$ kips T; $F_{FH} = 10.67$ kips T.
6.144 28.6 lb.
6.146 (1) (*a*) $\mathbf{A}_x = 0$, $\mathbf{A}_y = 7.85$ kN ↑,
$\mathbf{M}_A = 15.70$ kN·m ↺. (*b*) $\mathbf{D} = 22.2$ kN ∠45°.
(2) (*a*) $\mathbf{A}_x = 0$, $\mathbf{A}_y = 3.92$ kN ↑,
$\mathbf{M}_A = 8.34$ kN·m ↺. (*b*) $\mathbf{D} = 11.10$ kN ∠45°.
(3) (*a*) $\mathbf{A}_x = 0$, $\mathbf{A}_y = 3.92$ kN ↑,
$\mathbf{M}_A = 8.34$ kN·m ↺. (*b*) $\mathbf{D} = 18.95$ kN ∠45°.
(4) (*a*) $\mathbf{A}_x = 3.92$ kN →, $\mathbf{A}_y = 3.92$ kN ↑,
$\mathbf{M}_A = 2.35$ kN·m ↻. (*b*) $\mathbf{D} = 11.10$ kN ∠45°.
6.C1 $\theta = 15°$: $F_{AB} = 2420$ lb T, $F_{AC} = 2260$ lb C,
$F_{BC} = 2090$ lb T; $\theta = 65°$: $F_{AB} = 1801$ lb T,
$F_{AC} = 2150$ lb C, $F_{BC} = 1984$ lb T.

6.C2 *h = 6 m:* F_{FH} = 17.50 kN *C*; F_{GH} = 1.601 kN *C*; F_{GI} = 17.50 kN *T*.
6.C3 (*a*) *a = 0.1 m:* F_{CF} = 22.3 kN *T*, F_{DG} = 12.00 kN *C*; *a = 0.2 m:* F_{CF} = 24.2 kN *C*, F_{DG} = 15.00 kN *T*. (*b*) Improper constraints.
6.C4 (*a*) *θ = 60°:* **M** = 149.1 N·m ↺; *θ = 135°:* **M** = 84.2 N·m ↺. (*b*) 75.1°, $\mathbf{M}_{max}$ = 156.0 N·m ↺.

CHAPTER 7

7.2 (On *JD*) **F** = 0; **V** = 80 lb ↑; **M** = 480 lb·in. ↺.
7.4 (On *KC*) **F** = 200 N ↘; **V** = 150 N ↗; **M** = 15 N·m ↺.
7.6 (On *JB*) **F** = 125 N ↗; **V** = 300 N ↘; **M** = 156 N·m ↻.
7.8 (On *JCD*) **F** = 1200 lb ←; **V** = 350 lb ↓; **M** = 13.8 kip·in. ↻.
7.10 (*a*) (On *CA*) **F** = 480 N ←; **V** = 576 N ↑; **M** = 0. (*b*) (On *JA*) **F** = 480 N ←; **V** = 576 N ↑; **M** = 57.6 N·m ↺.
7.12 (*a*) (On *BA*) **F** = 0.333 *P* →; **V** = 0.333 *P* ↑; **M** = 0.667 *Pa* ↺. (*b*) (On *BA*) **F** = *P* ←; **V** = *P* ↑; **M** = 0. (*c*) (On *BA*) **F** = 0.333 *P* ←; **V** = 0.667 *P* ↑; **M** = 0.333 *Pa* ↺.
7.14 $M_{max} = \frac{1}{4}PL$ at $a = \frac{1}{2}L$.
7.16 (On *JC*) **M** = 4.22 N·m ↺.
7.18 (On *JC*) **F** = $(W\theta/\pi)\cos\theta$ ↗; **V** = $(W\theta/\pi)\sin\theta$ ↖; **M** = $(Wr/\pi)(\sin\theta - \theta\cos\theta)$ ↺.
7.20 $M_B = +Pab/L$.
7.22 $M_B = -\frac{1}{2}wL^2$.
7.24 $M_B = -\frac{1}{6}w_0L^2$.
7.26 M_C = +6.59 kN·m.
7.28 M_E = +900 lb·in.
7.30 15 lb/in.
7.32 *M* = +9 kN·m at center.
7.34 M_C = +12 kip·ft.
7.36 *Just to the right of C: V* = +60 N; *M* = +12 N·m.
7.38 *Just to the right of F: V* = −10 lb; *M* = +1400 lb·in.
7.40 *Just to the right of C: V* = 0; *M* = +4 N·m.
7.42 0.207 *L*.
7.56 M_C = +17 kN·m.
7.58 M_E = +700 lb·in.
7.60 *M* = +1.985 kN·m, 1.260 m from *A*.
7.62 *M* = +14.40 kip·ft, 6 ft from *A*.
7.64 *M* = +16.10 kN·m, 1.900 m from *A*; M_B = −80.0 kN·m.
7.66 $V = -\frac{1}{2}w_0x^2/L$; $M = -\frac{1}{6}w_0x^3/L$.
7.68 $V = \frac{1}{4}w_0L[1 - 4(x/L) + 3(x/L)^2]$; $M = \frac{1}{4}w_0L^2[(x/L) - 2(x/L)^2 + (x/L)^3]$. $M_{max} = w_0L^2/27$, at $x = L/3$.
7.70 **P** = 300 N ↓, **Q** = 300 N ↓; V_C = +800 N, M_D = −75 N·m.
7.72 **P** = 20 kips ↓, **Q** = 75 kips ↓; V_E = +18 kips, M_F = +252 kip·ft.
7.74 0.293 *L*.
7.76 (*a*) $\mathbf{E}_x$ = 10 kN →, $\mathbf{E}_y$ = 7 kN ↑. (*b*) 12.21 kN.
7.78 $\mathbf{E}_x$ = 1000 lb →, $\mathbf{E}_y$ = 600 lb ↑.
7.80 (*a*) $\mathbf{E}_x$ = 30 kN →, $\mathbf{E}_y$ = 15 kN ↑. (*b*) 44.6 kN.
7.82 (*a*) 858 N. (*b*) 2.5 m.
7.84 (*a*) 416 lb ◺22.6°. (*b*) 210 lb. (*c*) T_{AB} = 416 lb; T_{BC} = 200 lb; T_{CD} = 130 lb.
7.86 (*a*) 1775 N. (*b*) 60.1 m.
7.88 3.749 ft.
7.90 (*a*) 111.1 mm. (*b*) 2.54° ◹
7.92 T_{max} = 9.86 kips; T_{min} = 8.04 kips.
7.94 (*a*) 698 mm from *A*. (*b*) 6.31 N.
7.100 (*a*) 35.6 m. (*b*) 49.2 kg.
7.102 (*a*) 412 ft. (*b*) 875 lb.
7.104 (*a*) 18.78 m. (*b*) 3.53 kg/m.
7.106 (*a*) 8.27 lb. (*b*) 6.69 ft. (*c*) 21.6 lb.
7.108 19.6 mm.
7.110 27.3 lb.
7.114 11.845 m; 0.365 m.
7.116 h/L = 0.338; θ_B = 56.5°; T_m = 0.755 *wL*.
7.118 M_C = +120 kip·ft.
7.120 2.38 kN ≤ *P* ≤ 4.5 kN.
7.122 (*a*) (On *JA*) **F** = 500 N ←; **V** = 500 N ↑; **M** = 300 N·m ↻. (*b*) (On *KBA*) **F** = 970 N ↑; **V** = 171.0 N ←; **M** = 446 N·m ↺.
7.124 (*a*) **A** = 380 lb ◺20.6°; **B** = 380 lb ∠36.9°. (*b*) 380 lb. (*c*) 64.5 lb.
7.126 $V = w_0L[\frac{1}{4} - (x/L)^2]$; $M = w_0L^2[\frac{1}{4}(x/L) - \frac{1}{3}(x/L)^3]$; $M_{max} = w_0L^2/12$.
7.128 (*a*) 1.546 m. (*b*) 6.63 kN.
7.C1 (*a*) V_C = −40 kN; M_C = +120 kN·m. (*b*) V_C = +10 kN; M_C = +60 kN·m. (*c*) V_C = +18 kN; M_C = +48 kN·m.
7.C2 (*b*) *x = 2.8 m: V* = −0.8 kN; *M* = +5.6 kN·m.
7.C3 (*b*) M_C = 42.0 kN·m.
7.C4 *Prob. 7.79:* $\mathbf{E}_x$ = 2000 lb →; $\mathbf{E}_y$ = 700 lb ↑.

CHAPTER 8

8.2 Block moves down; **F** = 279 N ◺30°.
8.4 Block in equilibrium; **F** = 36.3 lb ◹30°.
8.6 (*a*) 83.2 lb. (*b*) 66.3 lb.
8.8 (*a*) **P** = 392 N ∠51.7°. (*b*) *P* = 157.0 N ∠18.3°.
8.10 (*a*) 25.5 N *C*. (*b*) **P** = 20.9 N →.
8.12 (*a*) **P** = 1030 N →. (*b*) **P** = 736 N →.
8.14 *A* and *B* move; $\mathbf{F}_A$ = 7.58 N ↗; $\mathbf{F}_B$ = 3.03 N ↗; $\mathbf{F}_C$ = 10.16 N ↗.
8.16 46.4°.
8.18 (*a*) 4.60 lb. (*b*) 6.25 lb.

8.20 (*a*) $\mathbf{P} = 206$ N $\rightarrow$. (*b*) 714 mm.
8.22 (*a*) 138.6 N. (*b*) Crate will slide.
8.24 (*a*) 55.6 lb·ft. (*b*) 45.5 lb·ft.
8.26 (*a*) $0.350\ Wr$. (*b*) $0.408\ Wr$.
8.28 (*a*) $0.232W$. (*b*) 0.300.
8.30 (*a*) and (*b*) 66.7 mm.
8.32 Rod in equilibrium; $\mathbf{F} = W/4 \rightarrow$.
8.34 $4.12 \leq L/a \leq 7.42$.
8.36 Rod is not in equilibrium ($F_{req} = 0.6495P$ and $F_m = 0.6250P$).
8.38 (*a*) 161.0 N $\measuredangle$2.88°. (*b*) 109.6 N $\measuredangle$50°.
8.40 2620 N $\downarrow$.
8.42 24.75 in.
8.44 0.0533.
8.46 $21.8° \leq \theta \leq 62.9°$.
8.48 135.0 lb.
8.50 $2Pl/(\tan\theta - \mu_s)$.
8.52 $-2.26\text{ N} \leq Q \leq 81.4\text{ N}$.
8.54 $P = 0$.
8.56 (*a*) 0.556. (*b*) Sliding impending at C.
8.58 (*a*) 0.744 m. (*b*) Plank oscillates about $x = 0.555$ m.
8.60 Max. displ. of A: $x = 0.2$ m (for $P = 5$ N); final displ.: $x = 0.15$ m (for $P = 0$).
8.62 $\mathbf{P} = 451$ N $\leftarrow$.
8.64 (*a*) 28.1°; 50.0 lb. (*b*) 14.0°; 47.1 lb.
8.66 1966 N $\downarrow$.
8.68 (*a*) $\mathbf{P} = 6880$ lb $\leftarrow$. (*b*) $\mathbf{Q} = 5400$ lb $\rightarrow$.
8.70 3.25 N.
8.72 3.03 kN.
8.74 (*a*) $\mathbf{P} = 280$ lb $\leftarrow$. (*b*) Machine will move with wedge.
8.76 7.98 lb·ft.
8.80 9.02 N·m counterclockwise.
8.82 28.1 lb·in.
8.84 3.92 N·m.
8.86 (*a*) 79.9 lb. (*b*) 94.5 lb.
8.88 0.226.
8.90 $T_{AB} = 419$ N; $T_{CD} = 381$ N; $T_{EF} = 345$ N.
8.92 (*a*) 147.5 lb. (*b*) 67.8 lb.
8.94 100 mm.
8.96 13.73 N.
8.98 (*a*) 112.2 lb. (*b*) 320 lb.
8.102 10.30 N.
8.104 15.65 lb.
8.106 217 mm.
8.108 0.776 m.
8.110 (*a*) 2000 N. (*b*) 0.212.
8.112 (*a*) 208 lb. (*b*) 48.0 lb.
8.114 91.6 N·m.
8.116 (*a*) $T_A = 42.0$ N; $T_B = 98.0$ N. (*b*) 0.270.
8.118 34.6 lb·ft $\circlearrowleft$.
8.120 91.3 N·m $\circlearrowright$.
8.122 1.557
8.124 (*a*) 108.1 lb. (*b*) 67.5 lb. (*c*) 108.1 lb.
8.126 0.231.
8.130 62.2 mm.
8.132 12.27 kN.
8.134 (*a*) $16.9° \leq \theta \leq 73.1°$. (*b*) $0 \leq \theta \leq 53.1°$.
8.136 205 lb $\downarrow$.
8.138 (*a*) 69.5°. (*b*) 129.1 lb.
8.140 $L/a \geq 3.39$.
8.C1 $\mu_s = 0.15$: $\theta = 25.8°$; $\mu_s = 0.50$: $\theta = 90.0°$.
8.C2 $W_A = 10$ lb: $\theta = 28.8°$; $W_A = 40$ lb: $\theta = 55.4°$.
8.C3 (*b*) (1) $\theta = 20°$: $\mu_s = 0.285$; (2) $\theta = 20°$: $\mu_s = 0.219$. (*c*) (1) $\theta = 22.0°$; (2) $\theta = 28.7°$.
8.C4 (*a*) $x_0 = 0.600L$; $x_m = 0.604L$; $\theta_1 = 5.06°$. (*b*) $\theta_2 = 55.4°$.

CHAPTER 9

9.2 $a^3b/(n+3)$.
9.4 $a^3(h_1 + 3h_2)/12$.
9.6 $(ab^3/3)/(3n+1)$
9.8 $a(h_1^2 + h_2^2)(h_1 + h_2)/12$.
9.10 $\pi ab^3/8$; $\frac{1}{2}b$.
9.12 $\pi a^3b/8$; $\frac{1}{2}a$.
9.14 $0.0945ah^3$; $0.402h$.
9.16 (*a*) $4a^4/3$; $a\sqrt{\frac{2}{3}}$. (*b*) $17a^4/6$; $a\sqrt{\frac{17}{12}}$.
9.18 $0.1804a^4$; $0.645a$.
9.20 (*b*) −10.56%; −2.99%; −0.125%.
9.22 614×10^3 mm^4; 19.01 mm.
9.24 28.0 in^4; 2.25 in.
9.26 $\bar{I}_x = 1.268 \times 10^6$ mm^4; $\bar{I}_y = 339 \times 10^3$ mm^4.
9.28 3000 mm^2; 325×10^3 mm^4.
9.30 $\bar{I}_x = 1.500 \times 10^6$ mm^4; $\bar{I}_y = 3.00 \times 10^6$ mm^4.
9.32 $\bar{I}_x = 65.0$ in^4; $\bar{I}_y = 6.43$ in^4.
9.34 (*a*) 11.57×10^6 mm^4. (*b*) 7.81×10^6 mm^4.
9.36 (*a*) 512 in^4. (*b*) 366 in^4.
9.38 $\bar{I}_x = 186.9 \times 10^6$ mm^4, $\bar{I}_y = 167.9 \times 10^6$ mm^4; $\bar{k}_x = 118.6$ mm, $\bar{k}_y = 112.4$ mm.
9.40 $\bar{I}_x = 9.57$ in^4; $\bar{I}_y = 104.6$ in^4.
9.42 $\bar{I}_x = 7.34 \times 10^6$ mm^4; $\bar{I}_y = 101.3 \times 10^6$ mm^4.
9.44 $\frac{1}{2}h$.
9.46 $(a + 3b)h/(2a + 4b)$.
9.48 $B = 4.44$ kN; $C = D = 6.10$ kN.
9.50 $5a/4$.
9.54 $b^2h^2/8$.
9.56 $a^4/8$.
9.58 $+19.4 \times 10^6$ mm^4.
9.60 -800×10^3 mm^4.
9.62 -6.08 in^4.
9.64 (*a*) $(\pi - \sqrt{3})a^4/16$; $(\pi + \sqrt{3})a^4/16$; $+a^4/16$. (*b*) $(\pi - 2)a^4/16$; $(\pi + 2)a^4/16$; 0.
9.66 3.79 in^4; 19.88 in^4; 1.779 in^4.
9.68 +18.4°; 84.2×10^6 mm^4, 19.46×10^6 mm^4.

9.70 $-25.7°$; 2.28×10^6 mm^4, 0.230×10^6 mm^4.
9.72 $-23.8°$; 20.1 in^4, 3.59 in^4.
9.74 (a) $(\pi - \sqrt{3})a^4/16$; $(\pi + \sqrt{3})a^4/16$; $+ a^4/16$. (b) $(\pi - 2)a^4/16$; $(\pi + 2)a^4/16$; 0.
9.76 3.79 in^4; 19.88 in^4; 1.779 in^4.
9.78 $+18.4°$; 84.2×10^6 mm^4, 19.46×10^6 mm^4.
9.80 $-25.7°$; 2.28×10^6 mm^4, 0.230×10^6 mm^4.
9.82 $-23.8°$; 20.1 in^4, 3.59 in^4.
9.88 (a) $0.0699\ ma^2$. (b) $0.320\ ma^2$.
9.90 (a) $I_{AA'} = mb^2/24$; $I_{BB'} = mh^2/18$. (b) $m(3b^2 + 4h^2)/72$.
9.92 (a) $ma^2/3$. (b) $3ma^2/2$.
9.94 $\frac{1}{2}ma^2(2n + 1)/(4n + 1)$.
9.96 $m(3a^2 + 2h^2)/20$; $\sqrt{(3a^2 + 2h^2)/20}$.
9.98 $m(b^2 + 3h^2)/5$.
9.100 $\frac{1}{3}mR^2(1 - \cos\phi)(2 + \cos\phi)$
9.102 1.514 kg·m^2; 155.7 mm.
9.104 $2mr^2/3$; $0.816r$.
9.106 (a) $r_a = 176.9$ mm. (b) 87.8 mm.
9.108 (a) $\frac{1}{2}\rho\pi h(a_2^4 + 2a_2^2a_1^2 - 3a_1^4)$. (b) $a_2/\sqrt{3}$. (c) $\frac{2}{3}\rho\pi h a_2^4$.
9.110 (a) 5.14×10^{-3} kg·m^2. (b) 7.54×10^{-3} kg·m^2. (c) 3.47×10^{-3} kg·m^2.
9.112 $I_x = I_z = 36.1 \times 10^{-3}$ lb·ft·s^2; $I_y = 30.0 \times 10^{-3}$ lb·ft·s^2.
9.114 0.883 g·m^2.
9.116 31.1×10^{-3} lb·ft·s^2.
9.118 (a) 9.97 kg·m^2. (b) 4.96 kg·m^2.
9.120 $I_{xy} = I_{zx} = 0$; $I_{yz} = -3.53 \times 10^{-3}$ lb·ft·s^2.
9.122 $I_{xy} = 14.13$ g·m^2; $I_{yz} = 37.7$ g·m^2; $I_{zx} = 113.0$ g·m^2.
9.124 $I_{xy} = I_{yz} = 6.71 \times 10^{-3}$ lb·ft·s^2; $I_{zx} = 2.24 \times 10^{-3}$ lb·ft·s^2.
9.128 $(3ma^2/20)(a^2 + 6h^2)/(a^2 + h^2)$.
9.130 23.6×10^{-3} lb·ft·s^2.
9.132 (a) $I_{xy} = \frac{1}{4}ma^2$; $I_{yz} = 0$; $I_{zx} = -\frac{1}{4}ma^2$. (b) $2ma^2/3$.
9.134 $17ma^2/54$.
9.136 (a) 2. (b) $\frac{1}{2}$.
9.142 (a) $2ab^3/7$. (b) $2a^3b/15$.
9.144 1.300×10^6 mm^4.
9.146 $+19.6°$; 10.45 in^4, 1.577 in^4.
9.148 13.78×10^{-3} kg·m^2; 114.5 mm.
9.150 $m(3a^2 + 2h^2)/6$.
9.152 (a) $I_{AA'} = ma^2/24$; $I_{BB'} = ma^2/24$. (c) $ma^2/12$.
9.C2 (a) $+23.8°$, 524×10^3 mm^4, 92.5×10^3 mm^4.
9.C3 (b) $I_x = 315 \times 10^3$ mm^4, $k_x = 13.23$ mm; $I_y = 5.70 \times 10^6$ mm^4, $k_y = 56.3$ mm. (c) $I_x = 719$ in^4, $k_x = 2.79$ in.; $I_y = 11{,}880$ in^4, $k_y = 11.36$ in.
9.C4 (b) 0.4135 lb·ft·s^2; 4.91 in.

CHAPTER 10

10.2 $\mathbf{P} = 600$ N ↑.
10.4 $\mathbf{P} = 82.5$ N ↓.
10.6 $\mathbf{M} = 1320$ lb·in. ↰.
10.8 $Q = (P\cos\theta)/\sin(\theta/2)$.
10.10 $Q = P[(l/a)\sin^3\theta - 1]$.
10.12 $M = \frac{1}{2}Wl\tan\alpha\sin\theta$.
10.14 (a) 1597 lb·in. ↰. (b) 803 lb·in. ↰.
10.16 (a) 14.00 N. (b) 54.0 N.
10.18 128.0 lb.
10.20 69.4°.
10.22 57.5°.
10.24 $(1 - \cos\theta)\tan\theta = W/4kl$.
10.26 390 mm.
10.28 52.4°.
10.30 54.9°.
10.32 11.86 kN.
10.34 3.21 kips ↗
10.36 $\eta = 1/(1 + \mu\cot\alpha)$.
10.38 $M_{max} = Pl/[2(\tan\theta - \mu)]$.
10.40 7.75 kN ↑.
10.42 0.500 in.
10.44 5.65 mm.
10.46 0.625 in. ↓.
10.54 45°, stable; $-135°$, unstable.
10.56 22.0° and 158.0°, stable; 90° and 270°, unstable.
10.58 54.9°, stable
10.60 67.0°, stable.
10.62 (a) $y(1 - a/\sqrt{a^2 + y^2}) = W/k$. (b) 400 mm, stable.
10.64 12.5° and 90°, stable; 30.8°, unstable.
10.66 (a) $(1 - \cos\theta)\tan\theta = 2mg/kl$. (b) 52.0°, stable.
10.68 (a) $\theta = 2\sin^{-1}[kl/2(kl - P)]$ and $\theta = 180°$. (b) 83.6°, stable; 180°, unstable.
10.70 (a) $\theta = \tan^{-1}(P/kl)$. (b) 63.4°, stable; 243.4°, unstable.
10.72 $k > 2.45$ kN/m.
10.74 $P < kl/2$.
10.76 $P < kl/2$.
10.78 $P < 573$ lb.
10.82 16.7°, stable; 73.3°, unstable.
10.84 30.0°.
10.86 $M = Pl\cos(\theta/2)$.
10.88 (a) and (b) $P < ka^2/2l$.
10.90 $P < kl/3$.
10.92 (a) $T = W/2(1 - \tan\theta)$. (b) 36.9°.
10.C1 $\theta = 40°$: $F_{CD} = 1.872$ kips T; $\theta = 80°$: $F_{CD} = 4.44$ kips T.
10.C2 $\theta = 40°$: 2.50 in.; $\theta = 80°$: 3.16 in.; $\theta = 120°$: 2.08 in.
10.C3 (b) $\theta = 20°$: $V = 64.1$ J; $\theta = 40°$: $V = 46.4$ J. (c) and (d) 9.59°, unstable; 40.3°, stable.
10.C4 (b) $\theta = 40°$: $W = 353$ N; $\theta = 60°$: $W = 1559$ N. (c) 52.2°.

SI Prefixes

Multiplication Factor	Prefix†	Symbol
$1\,000\,000\,000\,000 = 10^{12}$	tera	T
$1\,000\,000\,000 = 10^{9}$	giga	G
$1\,000\,000 = 10^{6}$	mega	M
$1\,000 = 10^{3}$	kilo	k
$100 = 10^{2}$	hecto‡	h
$10 = 10^{1}$	deka‡	da
$0.1 = 10^{-1}$	deci‡	d
$0.01 = 10^{-2}$	centi‡	c
$0.001 = 10^{-3}$	milli	m
$0.000\,001 = 10^{-6}$	micro	μ
$0.000\,000\,001 = 10^{-9}$	nano	n
$0.000\,000\,000\,001 = 10^{-12}$	pico	p
$0.000\,000\,000\,000\,001 = 10^{-15}$	femto	f
$0.000\,000\,000\,000\,000\,001 = 10^{-18}$	atto	a

† The first syllable of every prefix is accented so that the prefix will retain its identity. Thus, the preferred pronunciation of kilometer places the accent on the first syllable, not the second.

‡ The use of these prefixes should be avoided, except for the measurement of areas and volumes and for the nontechnical use of centimeter, as for body and clothing measurements.

Principal SI Units Used in Mechanics

Quantity	Unit	Symbol	Formula
Acceleration	Meter per second squared	. . .	m/s^2
Angle	Radian	rad	†
Angular acceleration	Radian per second squared	. . .	rad/s^2
Angular velocity	Radian per second	. . .	rad/s
Area	Square meter	. . .	m^2
Density	Kilogram per cubic meter	. . .	kg/m^3
Energy	Joule	J	N·m
Force	Newton	N	$kg \cdot m/s^2$
Frequency	Hertz	Hz	s^{-1}
Impulse	Newton-second	. . .	kg·m/s
Length	Meter	m	‡
Mass	Kilogram	kg	‡
Moment of a force	Newton-meter	. . .	N·m
Power	Watt	W	J/s
Pressure	Pascal	Pa	N/m^2
Stress	Pascal	Pa	N/m^2
Time	Second	s	‡
Velocity	Meter per second	. . .	m/s
Volume, solids	Cubic meter	. . .	m^3
Liquids	Liter	L	$10^{-3}\ m^3$
Work	Joule	J	N·m

† Supplementary unit (1 revolution = 2π rad = 360°).

‡ Base unit.

U.S. Customary Units and Their SI Equivalents

Quantity	U.S. Customary Unit	SI Equivalent
Acceleration	ft/s^2	$0.3048\ m/s^2$
	$in./s^2$	$0.0254\ m/s^2$
Area	ft^2	$0.0929\ m^2$
	in^2	$645.2\ mm^2$
Energy	ft·lb	1.356 J
Force	kip	4.448 kN
	lb	4.448 N
	oz	0.2780 N
Impulse	lb·s	4.448 N·s
Length	ft	0.3048 m
	in.	25.40 mm
	mi	1.609 km
Mass	oz mass	28.35 g
	lb mass	0.4536 kg
	slug	14.59 kg
	ton	907.2 kg
Moment of a force	lb·ft	1.356 N·m
	lb·in.	0.1130 N·m
Moment of inertia		
Of an area	in^4	$0.4162 \times 10^6\ mm^4$
Of a mass	$lb \cdot ft \cdot s^2$	$1.356\ kg \cdot m^2$
Momentum	lb·s	4.448 kg·m/s
Power	ft·lb/s	1.356 W
	hp	745.7 W
Pressure or stress	lb/ft^2	47.88 Pa
	lb/in^2 (psi)	6.895 kPa
Velocity	ft/s	0.3048 m/s
	in./s	0.0254 m/s
	mi/h (mph)	0.4470 m/s
	mi/h (mph)	1.609 km/h
Volume	ft^3	$0.02832\ m^3$
	in^3	$16.39\ cm^3$
Liquids	gal	3.785 L
	qt	0.9464 L
Work	ft·lb	1.356 J

Fundamentals of Business Law: Summarized Ca

Seventh Edition

ROGER LEROY MILLER
Institute for University Studies
Arlington, Texas

GAYLORD A. JENTZ
Herbert D. Kelleher Emeritus Professor in Business Law
MSIS Department, University of Texas at Austin

SOUTH-WESTERN
CENGAGE Learning

Australia • Brazil • Japan • Korea • Mexico • Singapore • Spain • United Kingdom • United States

Fundamentals of Business Law: Summarized Cases

Seventh Edition

ROGER LEROY MILLER GAYLORD A. JENTZ

ice President and Editorial Director:
ack Calhoun

ublisher, Business Law and Accounting:
ob Dewey

cquisition Editor:
teve Silverstein

enior Developmental Editor:
an Lamar

xecutive Marketing Manager:
isa L. Lysne

Production Manager:
Bill Stryker

Technology Project Editor:
Pam Wallace

Manufacturing Coordinator:
Charlene Taylor

Compositor:
Parkwood Composition, New Richmond, WI

Printer:
West Group

Art Director:
Linda Helcher

Internal Designer:
Bill Stryker

Cover Designer:
Larry Hanes/Cincinnati

Web Coordinator:
Brian Courter

Cover Image:
© Corbis

rinted in the United States of America
4 5 6 13 12 11 10 09 08

SBN-13: 978-0-324-38172-6 (Text Only)
SBN-10: 0-324-38172-7 (Text Only)
SBN-13: 978-0-324-38168-9 (Package)
SBN-10: 0-324-38168-9 (Package)

For more information, contact:
South-Western Cengage Learning
5191 Natorp Boulevard
Mason, Ohio, 45040
USA

Or you can visit our Internet site at
academic.cengage.com/blaw

ibrary of Congress Control Number:
006930562

NTERNATIONAL LOCATIONS

ASIA (including India)
Cengage Learning
5 Shenton Way
#01-01 UIC Building
Singapore 068808

AUSTRALIA/NEW ZEALAND
Cengage Learning Australia
102 Dodds Street
Southbank, Victoria 3006
Australia

LATIN AMERICA
Cengage Learning
Seneca, 53
Colonia Polanco
11560 Mexico
D.F. Mexico

CANADA
Nelson Education, Ltd.
1120 Birchmount Road
Toronto, Ontario
Canada M1K 5G4

UK/EUROPE/MIDDLE
EAST/AFRICA
Cengage Learning
High Holborn House
50-51 Bedford Road
London WC1R 4LR
United Kingdom

SPAIN (includes Portugal)
Paraninfo Cengage Learning
Calle Magallanes, 25
28015 Madrid, Spain

CONTENTS

UNIT SIX

Debtor-Creditor Relationships 387

UNIT SEVEN

Employment Relations 441

UNIT EIGHT

Business Organizations 489

APPENDICES

PREFACE

Preface to the Instructor

Now, more than ever before, a fundamental knowledge of the tenets of business law is crucial for anyone contemplating a career in business. Consequently, we have written *Fundamentals of Business Law: Summarized Cases,* Seventh Edition, with this goal in mind: to present a clear and comprehensive treatment of what every student should know about commercial law. While some of this may change, the fundamentals rarely do—and that's what students reading this text will acquire.

What's New in the Seventh Edition

Instructors have come to rely on the up-to-date coverage, accuracy, and applicability of *Fundamentals of Business Law.* To make sure that our text engages your students' interests, solidifies their understanding of the legal concepts presented, and provides the best teaching tools available, we now offer the following items either in the text or in conjunction with the text.

NEW BANKRUPTCY REFORM ACT OF 2005

The chapter on bankruptcy (Chapter 21) has been significantly revamped due to the passage of the 2005 bankruptcy reform legislation. Bankruptcy reform has been a topic of major debate for many years. Now that the reform act has been enacted, we have overhauled the content of this chapter to reflect the changes.

AMENDMENTS TO ARTICLES 2 AND 2A OF THE UCC

To ensure that our text offers the most up-to-date coverage possible, we have rewritten parts of the chapters covering sales and lease contracts (Chapters 14 through 17) to incorporate the 2003 amendments to Articles 2 and 2A of the Uniform Commercial Code (UCC). These amendments were promulgated largely to accommodate electronic commerce. At the time this book went to press, no state had as yet adopted these amendments. Thus, instead of basing the text of these chapters on the amended version of Articles 2 and 2A, we refer in footnotes to any amendments that significantly change the UCC provisions currently in effect in most states. We include excerpts from these amendments in Appendix E.

BUSINESS LAW DIGITAL VIDEO LIBRARY

For this Seventh Edition of *Fundamentals of Business Law: Summarized Cases,* we have added special new *Video Questions* at the end of selected chapters. Each of these questions directs students to the text's Web site (at academic.cengage.com/blaw/wbl) to view a video relevant to a topic covered in the chapter. This instruction is followed by a series of questions based on the video. The questions are again repeated on the Web site, when the student accesses the video. An access code for the videos can be packaged with each new copy of this textbook for no additional charge. If Business Law Digital Video Library access did not come packaged with the textbook, students can purchase it online at http://digitalvideolibrary.westbuslaw.com.

These videos can be used for homework assignments, discussion starters, or classroom demonstrations and are useful for generating student interest. Some of the videos are clips from actual movies, such as *The Money Pit* and *Jaws.* By watching a video and answering the questions, students will gain an understanding of how the legal concepts they have studied in the chapter apply to the real-life situation portrayed in the video. **Suggested answers for all of the video questions are given in both the *Instructor's Manual* and the *Answers Manual* that accompany this text.**

EMPHASIS ON INTERNET LAW

This Seventh Edition of *Fundamentals of Business Law: Summarized Cases* is truly up to date and reflects current law to the fullest extent possible. Throughout the text, we have included sections discussing the most recent developments in the law as it is being applied to transactions and commerce conducted via the Internet. For example, in Chapter 5, which focuses on intellectual property, we point out how traditional laws—and some newly enacted laws—are being applied to online issues relating to *copyrights, trademarks, patents,* and *trade secrets.* Other chapters include sections on *privacy rights* in the online world, *jurisdictional issues* as they arise in cyberspace, *cyber torts* and *cyber crimes, online securities offerings,* and a number of other topics relating to the online legal environment. We have also devoted an entire chapter (Chapter 13) solely to the topic of *electronic contracts,* or e-contracts.

EMPHASIS ON ETHICS AND CORPORATE GOVERNANCE

Fundamentals of Business Law: Summarized Cases emphasizes the ethical dimensions of business decision making in a number of ways. First, we devote an entire chapter (Chapter 3) to ethics and business decision making. The chapter includes illustrations of how some real-world companies, such as Enron Corporation, have learned the hard way that unethical behavior can have serious consequences.

Additionally, selected cases presented in the text conclude with an *Ethical Consideration* that encourages students to probe the ethical ramifications of the court's decision. In one chapter of each unit, we also provide *A Question of Ethics,* which is located at the end of the *Questions and Case Problems* section. Finally, the questions concluding each unit-ending *Extended Case Study* generally include a question exploring the ethical dimensions of the case.

Today's companies are placing a greater emphasis on developing internal rules of corporate governance to prevent corporate misconduct, encourage accountability, and balance potentially conflicting interests. We have therefore added coverage of this topic as appropriate. For example, the chapter on ethics (Chapter 3) includes a subsection discussing corporate governance principles. The chapter covering securities law (Chapter 27) now includes an entire section on corporate governance.

APPENDIX ON THE SARBANES-OXLEY ACT OF 2002

In a number of places in this text, we refer to the Sarbanes-Oxley Act of 2002 and the corporate scandals that led to the passage of that legislation. For example, Chapter 3 contains a section examining the requirements of the Sarbanes-Oxley Act relating to confidential reporting systems. In Chapter 27, we discuss this act in the context of securities law and present an exhibit (Exhibit 27–4) containing some of the key provisions of the act relating to corporate accountability with respect to securities transactions.

Because the act is a topic of significant concern in today's business climate, for the Seventh Edition we have added **excerpts and explanatory comments on the Sarbanes-Oxley Act of 2002 as Appendix** F. Students and instructors alike will find it useful to have the provisions of the act immediately available for reference.

The Web Connection

In addition to incorporating cyberlaw throughout the basic text of the book, this Seventh Edition of *Fundamentals of Business Law: Summarized Cases* offers several other components focusing on technology.

FUNDAMENTALS OF BUSINESS LAW ON THE WEB

This text is accompanied by a streamlined Web site that allows users to easily locate the resources they seek. When you visit our Web site at academic.cengage.com/blaw/wbl, you will find a broad array of teaching/learning resources, including the following:

- ***Relevant Web sites*** for all of the *Landmark and Classic Cases* that are presented in this text.
- ***Sample answers*** to the *Case Problem with Sample Answer* problems that are included in the *Questions and Case Problems* section at the end of every chapter. This problem-answer set is designed to help your students learn how to answer case problems by acquainting them with model answers to selected case problems.
- ***Videos*** referenced in the *Video Questions* (discussed previously) that have been included in selected chapters.
- ***Internet exercises*** for every chapter in the text (at least two per chapter). These exercises help familiarize students with online legal resources while introducing them to additional information on topics covered in the chapters.
- ***Interactive quizzes*** for every chapter in the text. At the end of each chapter, a *Before the Test* section directs students to the text's Web site, where they will find a number of interactive questions relating to the topics covered in the chapter.
- ***Court case updates*** that present summaries of new cases from various legal publications, all specifically keyed to chapters in this text.
- ***Links to other important legal resources*** available for free on the Web.

ONLINE LEGAL RESEARCH GUIDE

With every new book, your students will receive a free copy of the *Online Legal Research Guide.* Written by text co-author Roger LeRoy Miller, this is the most complete brief guide to using the Internet that exists today. It even includes an appendix on how to evaluate information obtained from the Internet.

Case Presentation

In each chapter, we present cases that illustrate the principles of law discussed in the text. The cases are numbered sequentially for easy referencing in class discussions, homework assignments, and examinations. In selecting the cases to be included in this edition, our goal has been to choose cases that reflect the most current law or that represent a significant precedent in case law.

Each case is presented in the following special format:

- *Case title and full case citation.*
- *Facts.*
- *Issue.*
- *Decision.*
- *Reason.*

In addition, each case normally concludes with a *For Critical Analysis* section, which consists of a question that requires the student to think critically about a particular issue raised by the case. For selected cases, however, we have included a section titled *Why Is This Case Important?* This section clearly sets forth the importance of the court's decision for businesspersons today. Whenever possible, we also include a URL, or Internet address, just below the case citation, that can be used to access the case online (a footnote to the URL explains how to find the specific case at that Web site).

We give special emphasis to *Landmark and Classic Cases* by setting them off with a special heading and logo. These cases also include a section titled *Impact of This Case on Today's Law* that explains the significance of that particular decision for the evolution of the law in that area. Finally, we present a section titled *Relevant Web Sites* at the conclusion of each *Landmark and Classic Case* that directs students to additional online resources.

Other Special Features and Pedagogy

In addition to the components of the *Fundamentals of Business Law: Summarized Cases* teaching/learning package already described, the text offers a number of other special features and pedagogy.

Adapting the Law to the Online Environment

As well as the Web elements described earlier, we also include a special feature in the text that focuses on the online world. Most chapters in this text contain this feature, which is titled *Adapting the Law to the Online Environment.* In the feature, we discuss actual cases in which the courts have had to decide how to apply traditional laws or newly enacted statutes to specific situations that have arisen in the online environment. A concluding *For Critical Analysis* section asks the student to think critically about some aspect of the issue discussed in the feature. **Suggested answers to these questions are included in both the *Instructor's Manual* and the *Answers Manual* that accompany this text.**

Management Perspective

In selected chapters, a feature titled *Management Perspective* is included. In this feature, your students can read about issues facing managers in today's business world and learn how the courts have dealt with those issues. Each feature concludes with a section summarizing the implications of the courts' decisions for managers.

Pedagogical Devices within Each Chapter

- ***Learning Objectives*** (a series of brief questions at the beginning of each chapter designed to provide a framework for the student as he or she reads through the chapter).
- **Highlighted and numbered examples illustrating legal principles** (to clarify legal concepts).
- **URLs for cases** (as described earlier).
- **Exhibits and forms** (including many exhibits illustrating basic principles of contract law).

Chapter-Ending Pedagogy

- ***Terms and Concepts*** (with appropriate page references).
- ***Chapter Summary*** (in graphic format with page references).
- ***For Review*** (the questions set forth in the chapter-opening *Learning Objectives* section are presented again to aid the student in reviewing the chapter; answers to the even-numbered questions for each chapter are provided in Appendix A).
- ***Questions and Case Problems*** (including hypotheticals and case problems; many of the case problems are based on cases from the 2000s).
- ***Question with Sample Answer*** (a hypothetical question for which a sample answer is provided in Appendix B).
- ***Case Problem with Sample Answer*** (as discussed earlier, each chapter contains one of these case problems; the answer for the problem is provided on the text's Web site at academic.cengage.com/blaw/wbl).
- ***Video Question*** (in selected chapters).
- ***Accessing the Internet*** (including Internet exercises and interactive quizzes for each chapter).

Unit-Ending Pedagogy

At the end of each unit of *Fundamentals of Business Law: Summarized Cases* is a two-page feature titled *Extended Case Study.* This feature provides an in-depth examination of a court case relating to a topic covered in the unit. Each end-of-unit feature opens with an introductory section, which discusses the background and significance of the case being presented. Then we present excerpts from the court's

majority opinion as well as from a dissenting opinion in the case. The feature concludes with a series of questions, under the heading *Questions for Analysis,* that prompt the student to think critically about the legal, ethical, economic, international, or general business implications of the case. **Suggested answers to these questions are included in both the *Instructor's Manual* and the *Answers Manual.***

APPENDICES

To help students learn how to find and analyze case law, we have included a special appendix at the end of Chapter 1. There your students will find information, including an exhibit, on how to read case citations, how to locate cases in case reporters, and how to interpret the meaning of the different components of URLs. *The appendix to Chapter 1 also includes an annotated sample court case to help your students learn how to read and understand the cases presented in this text.*

At the end of the book, as mentioned earlier in this preface, we also include appendices providing answers to the even-numbered *For Review* questions (Appendix A) and sample answers for the end-of-chapter *Questions with Sample Answer* (Appendix B). In Appendix C, we present a helpful appendix that instructs students on how to brief cases and analyze case problems. Other appendices include the following:

- The Constitution of the United States (Appendix D).
- Excerpts from the Uniform Commercial Code (Appendix E).
- Excerpts from the Sarbanes-Oxley Act of 2002, as well as explanatory comments (Appendix F), as noted previously.

A Flexible Teaching/Learning Package

We realize that different people have different teaching philosophies and learning goals. We believe that the Seventh Edition of *Fundamentals of Business Law: Summarized Cases* and its supplements offer business law instructors a flexible teaching/learning package. For example, although we have attempted to make the materials flow from chapter to chapter, most of the chapters are self-contained. In other words, you can use the chapters in any order you wish. Additionally, the supplements accompanying this text allow instructors to choose those supplements that will most effectively complement classroom instruction.

Furthermore, each chapter of the *Instructor's Manual* contains teaching suggestions, possible discussion questions, and additional information on key statutes or other legal sources that you may wish to use in your classroom. These and other supplementary materials (including printed and multimedia supplements) all contribute to the goal of making *Fundamentals of Business Law: Summarized Cases* a flexible teaching/learning package.

Supplemental Teaching Materials

Individually and in conjunction with a number of our colleagues, we have developed supplementary teaching materials for *Fundamentals of Business Law: Summarized Cases* that we believe are the best available today. Each component of the supplements package is listed below.

PRINTED SUPPLEMENTS

- ***Online Legal Research Guide*** (free with every new copy of the text).
- ***Instructor's Manual*** (includes additional cases on point with selected case summaries, answers to the questions concluding the *Adapting the Law to the Online Environment* and *Extended Case Study* features, and answers for the *Video Questions* at the end of selected chapters; also available on the *Instructor's Resource CD-ROM,* or IRCD.
- ***Study Guide.***
- A comprehensive ***Test Bank*** (also available on the IRCD).
- ***Answers Manual*** (includes answers to the *Questions and Case Problems,* answers to the feature-ending questions, and answers for the *Video Questions* that conclude selected chapters; also available on the IRCD).

SOFTWARE, VIDEO, AND MULTIMEDIA SUPPLEMENTS

- ***Instructor's Resource CD-ROM* (IRCD)**—Includes the following supplements: *Instructor's Manual, Answers Manual,* Case-Problem Cases, Case Printouts, ExamView, PowerPoint slides, and the *Test Bank.*
- **ExamView Testing Software.**
- **PowerPoint slides.**
- **Transparency acetates.**
- **Westlaw®**—Ten free hours for qualified adopters.
- **Business Law Digital Video Library**—Provides access to over sixty-five videos, including the *Drama of the Law* videos and video clips from actual Hollywood movies. Access to Business Law Digital Video Library is available in an optional package with a new text at no additional cost. If Business Law Digital Video Library access did not come packaged with the textbook, students can purchase it online at **http://digitalvideolibrary.westbuslaw.com**.
- **VHS videotapes**—Qualified adopters using this text have access to the entire library of Business Law videos in VHS format, a vast selection covering most business law issues. For more information about the videos, visit **http://wdvl.westbuslaw.com**.

Acknowledgments for Previous Editions

Kenneth Anderson
Mott Community College

Janie Blankenship
Del Mar College

Daniel R. Cahoy
The Pennsylvania State University

Len Callahan
Embry-Riddle Aeronautical University

Anniken Davenport
Harrisburg Area Community College

Philip E. De Marco
Mission College

Carol Docan
California State University, Northridge

James T. Foster
Florence Darlington Technical College

Frank Giesber
Texas Lutheran University

Thomas F. Goldman
Bucks County Community College

Edward M. Kissling
Ocean County College

Percy L. Lambert
Borough of Manhattan Community College

Daniel A. Levin
University of Colorado, Boulder

Jane A. Malloy
Delaware County Community College

John F. Mastriani
El Paso Community College

Russell A. Meade
Gardner-Webb University

Michael W. Pearson
Arizona State University

Steven M. Platau
The University of Tampa

Lee Ruck
George Mason University

Gayle L. Terry
Mary Washington College

Sheila Vagle
Northwest Technical College

Alan L. Weldy
Goshen College

John O. Wheeler
University of Virginia

Paula York
Northern Maine Technical College

Acknowledgments for the Seventh Edition

Leonard Axelrod
Metropolitan State University

Denise A. Bartles
Missouri Western State College

Mikelle Calhoun
Valparaiso University

William V. Cheek
Embry-Riddle Aeronautical University

Felipe Chia
Harrisburg Area Community College

Jacqueline S. Groover
Piedmont College

Percy L. Lambert
Borough of Manhattan CC

Russell A. Walden
College of the Canyons

Stephen J. Willis
Vance-Granville Community College

We are greatly indebted to the many individuals at South-Western Legal Studies in Business who worked on this project. We especially wish to thank Steve Silverstein, Rob Dewey, and Jan Lamar, our longtime editors, for their encouragement and advice. Our production editor and designer, Bill Stryker, ensured that we came out with an error-free edition. We will always be in his debt. Ann Borman and Anne Sheroff also helped with the production of this book, and we thank them for their contributions.

Additionally, we thank William Eric Hollowell, co-author of the *Instructor's Manual, Study Guide,* and *Test Bank*, for his contributions to this edition. We also must thank Lavina Leed Miller, who provided expert research, editing, and proofing services for the project, as well as Vickie Reierson, who provided valuable editorial assistance to the authors. The copyediting and proofreading skills of Suzie Franklin DeFazio and Pat Lewis will not go unnoticed. Roxanna Lee and Suzanne Jasin also made many special efforts on this project. We are indebted to the staff at Parkwood Composition, our compositor. Their ability to generate the pages for this text quickly and accurately made it possible for us to meet our ambitious printing schedule.

Lastly, we know we are not perfect. If you or your students find something you don't like or want us to change, write to us and let us know via e-mail, using the "Talk to Us" feature on this text's Web site. That is how we can make *Fundamentals of Business Law: Summarized Cases* an even better book in the future.

Roger LeRoy Miller

Gaylord A. Jentz

Dedication

To Isabelle Tosi,

I applaud your smiling disposition,
your academic accomplishments,
and your bright future teaching law.

R.L.M.

To my wife, JoAnn;
my children, Kathy,
Gary, Lori, and Rory;
and my grandchildren,
Erin, Megan, Eric,
Emily, Michelle, Javier,
Carmen, and Steve.

G.A.J.

UNIT 1

The Legal Environment of Business

Unit Contents

CHAPTER 1

The Legal and Constitutional Environment of Business

LEARNING OBJECTIVES

After reading this chapter, you should be able to answer the following questions:

1. What is the Uniform Commercial Code?
2. What is the common law tradition?
3. What is a precedent? When might a court depart from precedent?
4. What are some important differences between civil law and criminal law?
5. How does the U.S. Constitution affect business activities in the United States?

The law is of interest to all persons, not just to lawyers. Those entering the world of business will find themselves subject to numerous laws and government regulations. A basic knowledge of these laws and regulations is beneficial—if not essential—to anyone contemplating a successful career in today's business world.

In this introductory chapter, after a brief look at the nature of law, we describe the relationship between business activities and the legal environment, the basic sources of American law, the common law tradition, and some important classifications of law. We conclude the chapter with a discussion of the U.S. Constitution as it affects business.

The Nature of Law

There have been, and will continue to be, different definitions of law. Although the definitions vary in their particulars, they all are based on the general observation that, at a minimum, **law** consists of *enforceable rules governing relationships among individuals and between individuals and their society.* These "enforceable rules" may consist of unwritten principles of behavior established by a nomadic tribe. They may be set forth in an ancient or a contemporary law code. They may consist of written laws and court decisions created by modern legislative and judicial bodies, as in the United States. Regardless of how such rules are created, they all have one thing in common: they establish rights, duties, and privileges that are consistent with the values and beliefs of their society or its ruling group. In the study of law, often referred to as **jurisprudence**, this broad statement about the nature of law is the point of departure for all legal scholars and philosophers.

Business Activities and the Legal Environment

As those entering the world of business will learn, laws and government regulations affect virtually all business activities—from hiring and firing decisions to workplace safety, the manufacturing and marketing of products, and business financing, for example. To make good business decisions, a basic knowledge of the laws and regulations governing these activities is beneficial, if not essential. In today's world, though, a knowledge of "black-letter" law is not enough. Businesspersons are also pressured to make ethical decisions. Thus, the study of business law necessarily involves an ethical dimension.

MANY DIFFERENT LAWS MAY AFFECT A SINGLE BUSINESS TRANSACTION

As you will note, each chapter in this text covers a specific area of the law and shows how the legal rules in that area affect business activities. Although compartmentalizing the law in this fashion facilitates learning, it does not indicate the extent to which many different laws may apply to just one transaction. It is important for businesspersons to be aware of this fact and to understand enough about the law to know when to hire an expert for advice.

■ **EXAMPLE 1.1** Suppose that you are the president of NetSys, Inc., a company that creates and maintains computer network systems for its clients, including business firms. NetSys also markets software for customers who need an internal computer network but cannot afford an individually designed intranet. One day, Janet Foxx, an operations officer for Southwest Distribution Corporation (SDC), contacts you by e-mail about a possible contract involving SDC's computer network. In deciding whether to enter into a contract with SDC, you need to consider, among other things, the legal requirements for an enforceable contract. Are the requirements different for a contract for services and a contract for products? What are your options if SDC **breaches** (breaks, or fails to perform) the contract? The answers to these questions are part of contract law and sales law.

Other questions might concern payment under the contract. How can you guarantee that NetSys will be paid? For example, if SDC pays with a check that is returned for insufficient funds, what are your options? Answers to these questions can be found in the laws that relate to negotiable instruments (such as checks) and creditors' rights. Additionally, a dispute may occur over the rights to NetSys's software, or there may be a question of liability if the software is defective. Questions may arise as to whether you and Janet had the authority to make the deal in the first place. Resolutions of these questions may be found in areas of the law that relate to intellectual property, e-commerce, torts, product liability, agency, or business organizations. ■ Finally, if any dispute cannot be resolved amicably, then the laws and the rules concerning courts and court procedures spell out the steps of a lawsuit. Exhibit 1–1 illustrates the various areas of the law that may influence business decision making.

ETHICS AND BUSINESS DECISION MAKING

Merely knowing the areas of law that may affect a business decision is not sufficient in today's business world. Businesspersons must also take ethics into account. As you will learn in Chapter 3, *ethics* is generally defined as the study of what constitutes right or wrong behavior. Today, business decision makers need to consider not just

EXHIBIT 1–1 AREAS OF THE LAW THAT MAY AFFECT BUSINESS DECISION MAKING

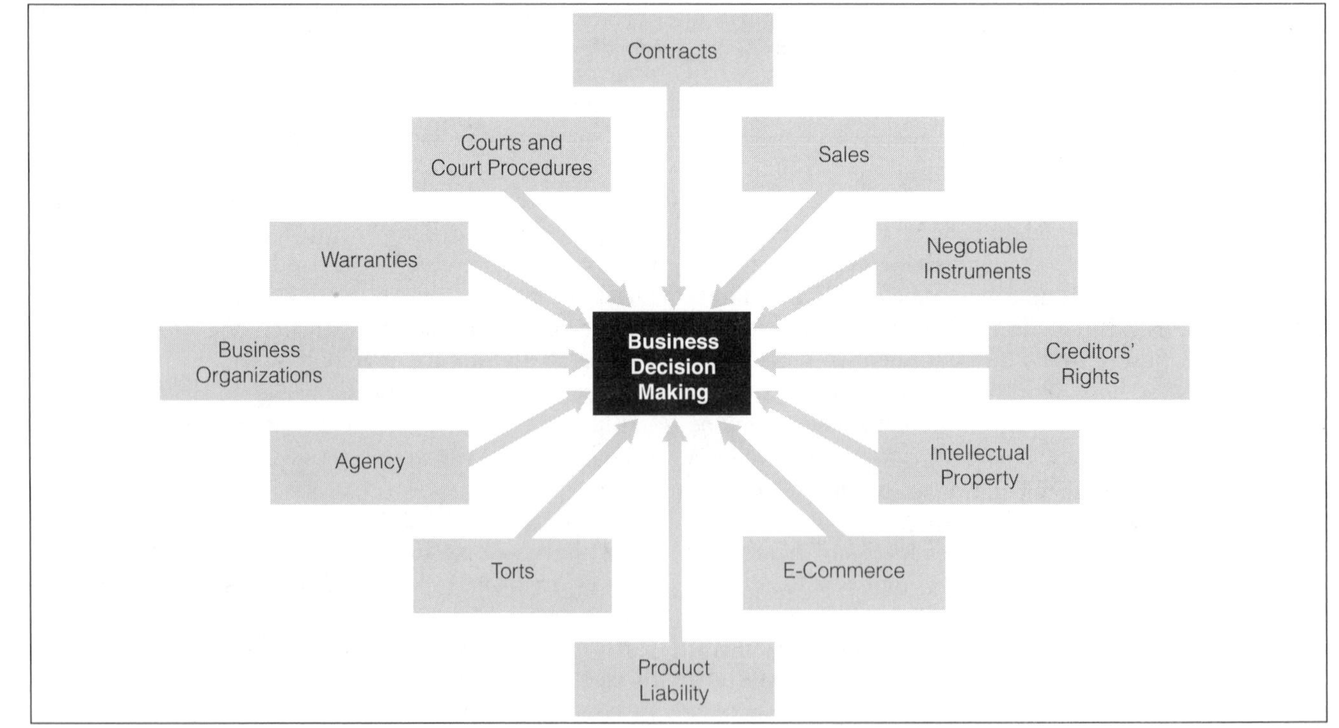

whether a decision is legal, but also whether it is ethical. Chapter 3 offers a detailed look at the importance of ethical considerations in business decision making.

Sources of American Law

There are numerous sources of American law. **Primary sources of law**, or sources that establish the law, include the following:

1. The U.S. Constitution and the constitutions of the various states.
2. Statutes, or laws, passed by Congress and by state legislatures.
3. Regulations created by administrative agencies, such as the federal Food and Drug Administration.
4. Case law (court decisions).

We describe each of these important primary sources of law in the following pages. (See the appendix following this chapter for a discussion of how to find statutes, regulations, and case law.)

Secondary sources of law are books and articles that summarize and clarify the primary sources of law. Legal encyclopedias, compilations (such as *Restatements of the Law*, which summarize court decisions on a particular topic), official comments to statutes, treatises, articles in law reviews published by law schools, and articles in other legal journals are examples of secondary sources of law. Courts often refer to secondary sources of law for guidance in interpreting and applying the primary sources of law discussed here.

Constitutional Law

The federal government and the states have separate written constitutions that set forth the general organization, powers, and limits of their respective governments. **Constitutional law** is the law as expressed in these constitutions.

The U.S. Constitution is the supreme law of the land. As such, it is the basis of all law in the United States. A law in violation of the Constitution, if challenged, will be declared unconstitutional and will not be enforced no matter what its source. Because of its paramount importance in the American legal system, we discuss the U.S. Constitution at length later in this chapter and present the complete text of the Constitution in Appendix D.

The Tenth Amendment to the U.S. Constitution reserves to the states all powers not granted to the federal government. Each state in the union has its own constitution. Unless it conflicts with the U.S. Constitution or a federal law, a state constitution is supreme within the state's borders.

Statutory Law

Statutes enacted by legislative bodies at any level of government make up another source of law, which is generally referred to as **statutory law.**

Federal Statutes Federal statutes are laws that are enacted by the U.S. Congress. Any federal statute that violates the U.S. Constitution will be held unconstitutional if it is challenged.

Examples of federal statutes that affect business operations include laws regulating the purchase and sale of securities (corporate stocks and bonds—discussed in Chapter 25) and statutes prohibiting employment discrimination (discussed in Chapter 23). Whenever a particular statute is mentioned in this text, we usually provide a footnote showing its **citation** (a reference to a publication in which a legal authority—such as a statute or a court decision—or other source can be found). In the appendix following this chapter, we explain how you can use these citations to find statutory law.

State and Local Statutes and Ordinances State statutes are laws enacted by state legislatures. Any state law that is found to conflict with the U.S. Constitution, with federal laws enacted by Congress, or with the state's constitution will be declared invalid if challenged. Statutory law also includes the **ordinances** passed by cities and counties, none of which can violate the U.S. Constitution, the relevant state constitution, or federal or state laws.

State statutes include state criminal statutes (discussed in Chapter 6), state corporation statutes (discussed in Chapters 25 through 27), state laws governing wills and trusts (discussed in Chapter 30), and state versions of the Uniform Commercial Code (to be discussed shortly). Local ordinances include zoning ordinances and local laws regulating housing construction and such matters as the overall appearance of a community.

A federal statute, of course, applies to all states. A state statute, in contrast, applies only within the state's borders. State laws thus vary from state to state.

Uniform Laws The differences among state laws were particularly notable in the 1800s, when conflicting state statutes frequently created difficulties for the rapidly developing trade and commerce among the states. To deal with these problems, a group of legal scholars and lawyers formed the National Conference of Commissioners on Uniform State Laws (NCCUSL) in 1892 to draft uniform

(model) statutes for adoption by the states. The NCCUSL still exists today and continues to issue uniform statutes.

Each state has the option of adopting or rejecting a uniform law. *Only if a state legislature adopts a uniform law does that law become part of the statutory law of that state.* Note that a state legislature may adopt all or part of a uniform law as it is written, or the legislature may rewrite the law however the legislature wishes. Hence, even when a uniform law is said to have been adopted in many states, those states' laws may not be entirely "uniform."

The earliest uniform law, the Uniform Negotiable Instruments Law, was completed by 1896 and was adopted in every state by the early 1920s (although not all states used exactly the same wording). Over the following decades, other acts were drawn up in a similar manner. In all, more than two hundred uniform acts have been issued by the NCCUSL since its inception. The most ambitious uniform act of all, however, was the Uniform Commercial Code.

The Uniform Commercial Code (UCC) The Uniform Commercial Code (UCC), which was created through the joint efforts of the NCCUSL and the American Law Institute,[1] was first issued in 1952. The UCC has been adopted in all fifty states,[2] the District of Columbia, and the Virgin Islands. The UCC facilitates commerce among the states by providing a uniform, yet flexible, set of rules governing commercial transactions. The UCC assures businesspersons that their contracts, if validly entered into, normally will be enforced.

Because of its importance in the area of commercial law, we cite the UCC frequently in this text. We also present excerpts from the version of the UCC that is in effect in most states today in Appendix E.

ADMINISTRATIVE LAW

Another important source of American law is known as **administrative law,** which consists of the rules and decisions of administrative agencies. An **administrative agency** is a federal, state, or local government agency established to perform a specific function. Administrative law and procedures constitute a dominant element in the regulatory environment of business. Rules issued by various administrative agencies now affect virtually every aspect of a business's operation, including the firm's capital structure and financing, its hiring and firing procedures, its relations with employees and unions, and the way it manufactures and markets its products.

1. This institute was formed in the 1920s and consists of practicing attorneys, legal scholars, and judges.

2. Louisiana has adopted only Articles 1, 3, 4, 5, 7, 8, and 9.

Types of Agencies At the national level, numerous **executive agencies** exist within the cabinet departments of the executive branch. For example, the Food and Drug Administration is within the Department of Health and Human Services. Executive agencies are subject to the authority of the president, who has the power to appoint and remove officers of federal agencies. There are also major **independent regulatory agencies** at the federal level, including the Federal Trade Commission, the Securities and Exchange Commission, and the Federal Communications Commission. The president's power is less pronounced in regard to independent agencies, whose officers serve for fixed terms and cannot be removed without just cause.

There are administrative agencies at the state and local levels as well. Commonly, a state agency (such as a state pollution-control agency) is created as a parallel to a federal agency (such as the Environmental Protection Agency). Just as federal statutes take precedence over conflicting state statutes, so do federal agency regulations take precedence over conflicting state regulations.

Agency Creation Because Congress cannot possibly oversee the actual implementation of all the laws it enacts, it must delegate such tasks to others, particularly when the issues relate to highly technical areas, such as air and water pollution. Congress creates an administrative agency by enacting **enabling legislation,** which specifies the name, composition, purpose, and powers of the agency being created.

■ **EXAMPLE 1.2** The Federal Trade Commission (FTC) was created in 1914 by the Federal Trade Commission Act.[3] This act prohibits unfair and deceptive trade practices. It also describes the procedures the agency must follow to charge persons or organizations with violations of the act, and it provides for judicial review (review by the courts) of agency orders. Other portions of the act grant the agency powers to "make rules and regulations for the purpose of carrying out the Act," to conduct investigations of business practices, to obtain reports from interstate corporations concerning their business practices, to investigate possible violations of the act, to publish findings of its investigations, and to recommend new legislation. The act also empowers the FTC to hold trial-like hearings and to **adjudicate** (resolve judicially) certain kinds of disputes that involve FTC regulations. ■

Note that the FTC's grant of power incorporates functions associated with the legislative branch of government (rulemaking), the executive branch (investigation and enforcement), and the judicial branch (adjudication). Taken together, these functions constitute **administrative process,**

3. 15 U.S.C. Sections 45–58.

which is the administration of law by administrative agencies.

CASE LAW AND COMMON LAW DOCTRINES

The rules of law announced in court decisions constitute another basic source of American law. These rules of law include interpretations of constitutional provisions, of statutes enacted by legislatures, and of regulations created by administrative agencies. Today, this body of judge-made law is referred to as **case law**, or the *common law*. The common law—the doctrines and principles embodied in case law—governs all areas not covered by statutory law or administrative law. Because of the importance of the common law in our legal system, we look at the origins and characteristics of the common law tradition in some detail in the pages that follow.

The Common Law Tradition

Because of our colonial heritage, much of American law is based on the English legal system. A knowledge of this tradition is crucial to understanding our legal system today because judges in the United States still apply common law principles when deciding cases.

EARLY ENGLISH COURTS

After the Normans conquered England in 1066, William the Conqueror and his successors began the process of unifying the country under their rule. One of the means they used to do this was the establishment of the king's courts, or *curiae regis*. Before the Norman Conquest, disputes had been settled according to the local legal customs and traditions in various regions of the country. The king's courts sought to establish a uniform set of rules for the country as a whole. What evolved in these courts was the beginning of the **common law**—a body of general rules that applied throughout the entire English realm. Eventually, the common law tradition became part of the heritage of all nations that were once British colonies, including the United States.

Courts developed the common law rules from the principles underlying judges' decisions in actual legal controversies. Judges attempted to be consistent, and whenever possible, they based their decisions on the principles suggested by earlier cases. They sought to decide similar cases in a similar way and considered new cases with care, because they knew that their decisions would make new law. Each interpretation became part of the law on the subject and served as a legal **precedent**—that is, a decision that furnished an example or authority for deciding subsequent cases involving similar legal principles or facts.

In the early years of the common law, there was no single place or publication where court opinions, or written decisions, could be found. Beginning in the late thirteenth and early fourteenth centuries, however, each year portions of significant decisions of that year were gathered together and recorded in *Year Books*. The *Year Books* were useful references for lawyers and judges. In the sixteenth century, the *Year Books* were discontinued, and other reports of cases became available. (See the appendix following this chapter for a discussion of how cases are reported, or published, in the United States today.)

STARE DECISIS

The practice of deciding new cases with reference to former decisions, or precedents, eventually became a cornerstone of the English and American judicial systems. The practice forms a doctrine called ***stare decisis***[4] ("to stand on decided cases").

The Importance of Precedents in Judicial Decision Making The doctrine of *stare decisis* means that once a court has set forth a principle of law as being applicable to a certain set of facts, that court and courts of lower rank must adhere to that principle and apply it in future cases involving similar fact patterns.

■ EXAMPLE 1.3 Suppose that the lower state courts in California have reached conflicting conclusions on whether drivers are liable for accidents they cause while merging into freeway traffic, even though the drivers looked and did not see any oncoming vehicles and even though witnesses (passengers in their cars) testified to that effect. To settle the law on this issue, the California Supreme Court decides to review a case involving this fact pattern. The court rules that in such a situation, the driver who is merging into traffic is liable for any accidents caused by the driver's failure to yield to freeway traffic—even if the driver looked carefully and did not see an approaching vehicle. The California Supreme Court's decision on the matter will influence the outcome of all future cases on this issue brought before the California state courts. Similarly, a decision on a given issue by the United States Supreme Court (the nation's highest court) is binding on all other courts. ■

Controlling precedents in a *jurisdiction* (an area in which a court or courts have the power to apply the law—see Chapter 2) are referred to as *binding authorities*. A **binding authority** is any source of law that a court must

4. Pronounced *ster*-ay dih-*si*-ses.

follow when deciding a case. Binding authorities include constitutions, statutes, and regulations that govern the issue being decided, as well as court decisions that are controlling precedents within the jurisdiction.

***Stare Decisis* and Legal Stability** The doctrine of *stare decisis* helps the courts to be more efficient, because if other courts have carefully reasoned through a similar case, their legal reasoning and opinions can serve as guides. *Stare decisis* also makes the law more stable and predictable. If the law on a given subject is well settled, someone bringing a case to court can usually rely on the court to make a decision based on what the law has been.

Departures from Precedent Although courts are obligated to follow precedents, sometimes a court will depart from the rule of precedent if it decides that a given precedent should no longer be followed. If a court decides that a precedent is simply incorrect or that technological or social changes have rendered the precedent inapplicable, the court might rule contrary to the precedent. Cases that overturn precedent often receive a great deal of publicity.

■ EXAMPLE 1.4 In *Brown v. Board of Education of Topeka,*[5] the United States Supreme Court expressly overturned precedent when it concluded that separate educational facilities for whites and blacks, which had been upheld as constitutional in numerous previous cases,[6] were inherently unequal. The Supreme Court's departure from precedent in *Brown* received a tremendous amount of publicity as people began to realize the ramifications of this change in the law. ■

When There Is No Precedent At times, courts hear cases for which there are no precedents within their jurisdictions on which to base their decisions. When hearing such cases, called "cases of first impression," courts often look to precedents set in other jurisdictions for guidance. Precedents from other jurisdictions, because they are not binding on the court, are referred to as **persuasive authorities.** A court may also consider a number of factors, including legal principles and policies underlying previous court decisions or existing statutes, fairness, social values and customs, public policy, and data and concepts drawn from the social sciences.

Equitable Remedies and Courts of Equity

A **remedy** is the means given to a party to enforce a right or to compensate for the violation of a right. ■ EXAMPLE 1.5 Suppose that Shem is injured because of Rowan's wrongdoing. A court may order Rowan to compensate Shem for the harm by paying Shem a certain amount of money. ■

In the early king's courts of England, the kinds of remedies that could be granted were severely restricted. If one person wronged another, the king's courts could award as compensation either money or property, including land. These courts became known as *courts of law,* and the remedies were called *remedies at law.* Even though this system introduced uniformity in the settling of disputes, when plaintiffs wanted a remedy other than economic compensation, the courts of law could do nothing, so "no remedy, no right."

Remedies in Equity *Equity* refers to a branch of the law, founded in justice and fair dealing, that seeks to supply a fair and adequate remedy when no remedy is available at law. In medieval England, when individuals could not obtain an adequate remedy in a court of law, they petitioned the king for relief. Most of these petitions were decided by an adviser to the king called the *chancellor.* The chancellor was said to be the "keeper of the king's conscience." When the chancellor thought that the claim was a fair one, new and unique remedies were granted. In this way, a new body of rules and remedies came into being, and eventually formal *chancery courts,* or *courts of equity,* were established. The remedies granted by these courts were called *remedies in equity.* Thus, two distinct court systems were created, each having a different set of judges and a different set of remedies.

Plaintiffs (those bringing lawsuits) had to specify whether they were bringing an "action at law" or an "action in equity," and they chose their courts accordingly. ■ EXAMPLE 1.6 A plaintiff might ask a court of equity to order a **defendant** (a person against whom a lawsuit is brought) to perform within the terms of a contract. A court of law could not issue such an order, because its remedies were limited to payment of money or property as compensation for damages. A court of equity, however, could issue a decree for *specific performance*—an order to perform what was promised. A court of equity could also issue an *injunction,* directing a party to do or refrain from doing a particular act. In certain cases, a court of equity could allow for the *rescission,* or cancellation, of the contract so that the parties would be returned to the positions that they held prior to the contract's formation. ■ Equitable remedies will be discussed in greater detail in Chapter 12.

The Merging of Law and Equity Today, in most states, the courts of law and equity have merged, and thus the distinction between the two courts has largely disappeared. A

5. 347 U.S. 483, 74 S.Ct. 686, 98 L.Ed. 873 (1954). See the appendix following this chapter for an explanation of how to read legal citations.
6. See *Plessy v. Ferguson,* 163 U.S. 537, 16 S.Ct. 1138, 41 L.Ed. 256 (1896).

plaintiff may now request both legal and equitable remedies in the same action, and the trial court judge may grant either form—or both forms—of relief. The merging of law and equity, however, does not diminish the importance of distinguishing legal remedies from equitable remedies. To request the proper remedy, a businessperson (or her or his attorney) must know what remedies are available for the specific kinds of harms suffered. Today, as a rule, courts will grant an equitable remedy only when the remedy at law (money damages) is inadequate.

Classifications of Law

The huge body of the law may be broken down according to several classification systems. For example, one classification system divides law into **substantive law** (all laws that define, describe, regulate, and create legal rights and obligations) and **procedural law** (all laws that establish the methods of enforcing the rights established by substantive law). Other classification systems divide law into federal law and state law or private law (dealing with relationships between persons) and public law (addressing the relationship between persons and their governments).

Frequently, people use the term **cyberlaw** to refer to the emerging body of law that governs transactions conducted via the Internet. Cyberlaw is not really a classification of law, nor is it a new *type* of law. Rather, it is an informal term used to describe traditional legal principles that have been modified and adapted to fit situations that are unique to the online world. Of course, in some areas new statutes have been enacted, at both the federal and state levels, to cover specific types of problems stemming from online communications. Throughout this book, you will read how the law in a given area is evolving to govern specific legal issues that arise in the online context. We look at one important issue in this chapter's *Adapting the Law to the Online Environment* feature.

Civil Law and Criminal Law

Civil law spells out the rights and duties that exist between persons and between persons and their governments, and the relief available when a person's rights are violated. Typically, in a civil case, a private party sues another private party (although the government can also sue a party for a civil law violation) to make that other party comply with a duty or pay for the damage caused by the failure to comply with a duty. ■ **EXAMPLE 1.7** If a seller fails to perform a contract with a buyer, the buyer may bring a lawsuit against the seller. The purpose of the lawsuit will be either to compel the seller to perform as promised or, more commonly, to obtain money damages for the seller's failure to perform. ■

Much of the law that we discuss in this text is civil law. Contract law, for example, which we will discuss in Chapters 7 through 13, is civil law. The whole body of tort law (see Chapter 4) is civil law. Note that *civil law* is not the same as a *civil law system*. As you will read shortly, in the subsection discussing national law, a civil law system is a legal system based on a written code of laws.

Criminal law has to do with wrongs committed against society for which society demands redress. Criminal acts are proscribed (prohibited) by local, state, or federal government statutes. Thus, criminal defendants are prosecuted by public officials, such as a district attorney (D.A.), on behalf of the state, and not by their victims or other private parties. Whereas in a civil case the object is to obtain remedies (such as money damages) to compensate the injured party, the object in a criminal case is to punish the wrongdoer and to deter others from similar actions. Penalties for violations of criminal statutes consist of fines and/or imprisonment—and, in some cases, death. We will discuss the differences between civil and criminal law in greater detail in Chapter 6.

National and International Law

Although the focus of this book is U.S. business law, increasingly businesspersons in this country engage in transactions that extend beyond our national borders. In these situations, the laws of other nations or the laws governing relationships among nations may come into play. For this reason, those who pursue a career in business today should have an understanding of the global legal environment.

National Law The law of a particular nation, such as the United States or Sweden, is **national law.** National law, of course, varies from country to country because each country's law reflects the interests, customs, activities, and values that are unique to that nation's culture. Even though the laws and legal systems of various countries differ substantially, broad similarities do exist.

Basically, there are two dominant legal systems in the world today. One is the common law system of England and the United States, which we have already discussed. The other system is based on Roman civil law, or "code law." The term *civil law,* as used here, refers not to civil as opposed to criminal law but to codified law—an ordered grouping of legal principles enacted into law by a legislature or governing body. In a **civil law system,** the primary source of law is a statutory code, and case precedents are

ADAPTING THE LAW TO THE ONLINE ENVIRONMENT

International Jurisdiction and the Internet

As you will learn in Chapter 2, *jurisdiction* is an important legal concept that relates to the authority of a court to hear and decide a case. Within the United States, there is a federal court system, which has jurisdiction over specific types of cases. There are also fifty state court systems, each having jurisdiction over certain types of cases. In today's interconnected world, the issue of jurisdiction has become critical. Specifically, businesses using the Internet can reach individuals in any part of the world. Does that mean that every court everywhere has jurisdiction over, say, an Internet-based company in Chicago? Clearly, this question has significant implications for any business that owns or operates a Web site.

THE MINIMUM-CONTACTS REQUIREMENT

Domestically, jurisdiction over individuals and businesses is based on the requirement of minimum contacts as outlined in *International Shoe Co. v. State of Washington.*[a] Essentially, this requirement means that a business must have a minimum level of contacts with residents of a particular state for that state's courts to exercise jurisdiction over the firm. For example, suppose that a Wisconsin driver, while on vacation in California, crashes into a California resident's car. If the crash resulted from the Wisconsin driver's negligence, a *tort,* or civil wrong (see Chapter 4), will have been committed. This contact will be sufficient to allow a California court to exercise jurisdiction over a lawsuit brought by the California resident. In the international arena, other countries' courts are applying the requirement of minimum contacts as developed in the U.S. courts.

SPECIAL INTERNATIONAL JURISDICTIONAL CHALLENGES

Suppose that a firm located in New Jersey posts a statement on its Web site that is defamatory. (As you will read in Chapter 4, *defamation* is a tort that occurs when someone publishes or publicly makes a false statement that harms another's good name, reputation, or character.) Further suppose that the person who has been defamed, or injured by the statement, lives in Australia. Can the Australian citizen sue the New Jersey firm in an Australian court? Perhaps, but at a minimum, jurisdictional authority would depend on whether the tort occurred in Australia. Thus, the question essentially is as follows: Did the tort occur when the defamatory statement was posted on the Web site (in New Jersey) or when it was downloaded and viewed (in Australia)?

To date, only one nation's highest court—that of Australia—has ruled on this issue. The case involved Joseph Gutnick, a resident of Melbourne, Australia, who sued Dow Jones & Company, the U.S. publisher of the *Wall Street Journal,* for defamation. Dow Jones argued that the Australian court could not exercise jurisdiction over its U.S. servers, which were located in New Jersey. The case ultimately reached Australia's highest court, which rejected Dow Jones's argument. The court concluded that the tort had occurred in Australia, where the article had been downloaded and viewed, and thus the Australian court had jurisdiction over the dispute.[b]

FOR CRITICAL ANALYSIS

If the opinion of the Australian high court becomes widely accepted, how might this have a "chilling effect" on Internet communications and on the dissemination of ideas?

a. 326 U.S. 310, 66 S.Ct. 154, 90 L.Ed. 95 (1945).

b. *Dow Jones & Co., Inc. v. Gutnick,* HCA 56 (December 10, 2002).

not judicially binding, as they normally are in a common law system. Although judges in a civil law system commonly refer to previous decisions as sources of legal guidance, they are not bound by precedent; in other words, the doctrine of *stare decisis* does not apply.

Generally, countries that were once colonies of Great Britain retained their English common law heritage after they achieved independence. Similarly, the civil law system, which is followed in most continental European countries, was retained in the Latin American, African, and Asian countries that were once colonies of those nations. Japan and South Africa also have civil law systems, and ingredients of the civil law system are found in the Islamic courts of predominantly Muslim countries. In the United States, the state of Louisiana, because of its historical ties to France, has in part a civil law system. The legal systems of Puerto Rico, Québec, and Scotland are similarly characterized as having elements of the civil law system.

International Law In contrast to national law, international law applies to more than one nation. **International law** can be defined as a body of written and unwritten laws observed by independent nations and governing the acts of individuals as well as governments. International law is an intermingling of rules and constraints derived from a variety of sources, including the laws of individual nations, the customs that have evolved among nations in their relations with one another, treaties, and the resolutions of international organizations. In essence, international law is the

result of centuries-old attempts to reconcile the traditional need of each nation to be the final authority over its own affairs with the desire of nations to benefit economically from trade and harmonious relations with one another.

The key difference between national law and international law is that national law can be enforced by government authorities. If a nation violates an international law, however, the most that other countries or international organizations can do (if persuasive tactics fail) is to resort to coercive actions against the violating nation. (See Chapter 32 for a more detailed treatment of international law.)

The Constitution as It Affects Business

Each of the sources of law discussed earlier helps to frame the legal environment of business. Because laws that govern business have their origin in the lawmaking authority granted by the U.S. Constitution, we examine that document more closely here. We focus on two areas of the Constitution of particular concern to business—the commerce clause and the Bill of Rights. As mentioned earlier, the Constitution provides the legal basis for both state and federal (national) powers. It is the supreme law in this country, and any law that conflicts with the Constitution, if challenged in court, will be declared invalid by the court.

THE COMMERCE CLAUSE

To prevent states from establishing laws and regulations that would interfere with trade and commerce among the states, the Constitution expressly delegated to the national government the power to regulate interstate commerce. Article I, Section 8, of the U.S. Constitution expressly permits Congress "[t]o regulate Commerce with foreign Nations, and among the several States, and with the Indian Tribes." This clause, referred to as the **commerce clause,** has had a greater impact on business than any other provision in the Constitution.

For some time, the commerce power was interpreted as being limited to *interstate* commerce (commerce among the states) and not applicable to *intrastate* commerce (commerce within a state). In 1824, however, in *Gibbons v. Ogden,*[7] the United States Supreme Court held that commerce within a state could also be regulated by the national government as long as the commerce *substantially affected* commerce involving more than one state.

The Commerce Clause and the Expansion of National Powers In *Gibbons v. Ogden,* the commerce clause was expanded to regulate activities that "substantially affect interstate commerce." As the nation grew and faced new kinds of problems, the commerce clause became a vehicle for the additional expansion of the national government's regulatory powers. Even activities that seemed purely local came under the regulatory reach of the national government if those activities were deemed to substantially affect interstate commerce. ■ **EXAMPLE 1.8** In 1942, in *Wickard v. Filburn,*[8] the Supreme Court held that wheat production by an individual farmer intended wholly for consumption on his own farm was subject to federal regulation. The Court reasoned that the home consumption of wheat reduced the demand for wheat and thus could have a substantial effect on interstate commerce. ■

The following landmark case involved a challenge to the scope of the national government's constitutional authority to regulate local activities.

7. 22 U.S. (9 Wheat.) 1, 6 L.Ed. 23 (1824).
8. 317 U.S. 111, 63 S.Ct. 82, 87 L.Ed. 122 (1942).

LANDMARK AND CLASSIC CASES

CASE 1.1 Heart of Atlanta Motel v. United States

Supreme Court of the United States, 1964.
379 U.S. 241,
85 S.Ct. 348,
13 L.Ed.2d 258.
http://straylight.law.cornell.edu/supct/cases/name.htm[a]

FACTS Congress passed the Civil Rights Act of 1964 to prohibit racial discrimination in "establishments affecting interstate commerce." These facilities included "places of public accommodation." The owner of the Heart of Atlanta Motel, in violation of the Civil Rights Act of 1964, refused to rent rooms to African Americans. The motel owner brought an action in a federal district court to have the Civil Rights Act declared unconstitutional, alleging that Congress had exceeded its constitutional authority to regulate commerce by enacting the act. The owner argued that his motel was not engaged in interstate commerce but was "of a purely local character." The motel, however, was accessible to state and interstate highways. The owner advertised nationally, maintained billboards throughout the state, and accepted convention trade from

a. This is the "Historic Supreme Court Decisions—by Party Name" page within the "Caselists" collection of the Legal Information Institute available at its site on the Web. Click on the "H" link or scroll down the list of cases to the entry for the *Heart of Atlanta* case, then click on the case name.

CASE 1.1—CONTINUED

outside the state (75 percent of the guests were residents of other states). The court ruled that the act did not violate the Constitution and enjoined (prohibited) the owner from discriminating on the basis of race. The owner appealed. The case ultimately went to the United States Supreme Court.

ISSUE Did Congress exceed its constitutional power to regulate interstate commerce by enacting the Civil Rights Act of 1964?

DECISION No. The United States Supreme Court upheld the constitutionality of the act.

REASON The Court noted that the act was passed to correct "the deprivation of personal dignity" accompanying the denial of equal access to "public establishments." Testimony before Congress leading to the passage of the act indicated that African Americans in particular experienced substantial discrimination in attempting to secure lodging while traveling. This discrimination impeded interstate travel and thus impeded interstate commerce. As for the owner's argument that his motel was "of a purely local character," the Court said that even if this was true, the motel affected interstate commerce. According to the Court, "if it is interstate commerce that feels the pinch, it does not matter how local the operation that applies the squeeze." Therefore, under the commerce clause, "the power of Congress to promote interstate commerce also includes the power to regulate the local incidents thereof, including local activities."

IMPACT OF THIS CASE ON TODAY'S LAW If the Supreme Court had invalidated the Civil Rights Act of 1964, the legal landscape of the United States would be much different today. The act prohibited discrimination based on race, color, national origin, religion, or gender in all "public accommodations" as well as discrimination in employment based on these criteria. Although state laws now prohibit many of these forms of discrimination as well, the protections available vary from state to state—and it is not certain when (and if) such laws would have been passed had the 1964 federal Civil Rights Act been deemed unconstitutional.

RELEVANT WEB SITES *To locate information on the Web concerning the* Heart of Atlanta Motel *case, go to this text's Web site at* **academic.cengage.com/blaw/wbl**, *select "Chapter 1," and click on "URLs for Landmark Cases."*

The Commerce Power Today Today, at least theoretically, the power over commerce authorizes the national government to regulate every commercial enterprise in the United States. Federal (national) legislation governs virtually every major activity conducted by businesses—from hiring and firing decisions to workplace safety, competitive practices, and financing.

In the last decade, however, the Supreme Court has begun to curb somewhat the national government's regulatory authority under the commerce clause. In 1995, the Court held—for the first time in sixty years—that Congress had exceeded its regulatory authority under the commerce clause. The Court struck down an act that banned the possession of guns within one thousand feet of any school because the act attempted to regulate an area that had "nothing to do with commerce."[9] Subsequently, the Court invalidated key portions of two other federal acts on the ground that they exceeded Congress's commerce clause authority.[10]

The current trend of not allowing the federal government to regulate noncommercial activities that take place wholly within a state's borders has led to some controversial decisions in the lower courts. In 2003, for example, the U.S. Court of Appeals for the Ninth Circuit decided a case involving marijuana use on commerce clause grounds. Eleven states, including California, have adopted "medical marijuana" laws that legalize marijuana for medical purposes. Marijuana possession, however, is illegal under the federal Controlled Substances Act (CSA).[11]

The case arose after the federal government seized the marijuana that two seriously ill California women were using on the advice of their physicians. The women argued that it was unconstitutional for the federal act to prohibit them from using marijuana for medical purposes that were legal within the state. The federal appellate court agreed. In 2005, however, the United States Supreme Court held that Congress has the authority to prohibit the *intra*state possession and noncommercial cultivation of marijuana as part of a larger regulatory scheme (the CSA).[12]

9. The United States Supreme Court held the Gun-Free School Zones Act of 1990 to be unconstitutional in *United States v. Lopez,* 514 U.S. 549, 115 S.Ct. 1624, 131 L.Ed.2d 626 (1995).

10. See, for example, *Printz v. United States,* 521 U.S. 898, 117 S.Ct. 2365, 138 L.Ed.2d 914 (1997), involving the Brady Handgun Violence Prevention Act of 1993; and *United States v. Morrison,* 529 U.S. 598, 120 S.Ct. 1740, 146 L.Ed.2d 658 (2000), concerning the federal Violence Against Women Act of 1994.

11. 21 U.S.C. Sections 801 *et seq.*

12. *Gonzales v. Raich,* __ U.S. __, 125 S.Ct. 2195, 162 L.Ed.2d 1 (2005).

The Regulatory Powers of the States As part of their inherent sovereignty, state governments have the authority to regulate affairs within their borders. This authority stems in part from the Tenth Amendment to the Constitution, which reserves to the states all powers not delegated to the national government. State regulatory powers are often referred to as **police powers.** The term encompasses not only the enforcement of criminal law but also the right of state governments to regulate private activities to protect or promote the public order, health, safety, morals, and general welfare. Fire and building codes, antidiscrimination laws, parking regulations, zoning restrictions, licensing requirements, and thousands of other state statutes covering virtually every aspect of daily life have been enacted pursuant to a state's police powers. Local governments, including cities, also exercise police powers.[13] Generally, state laws enacted pursuant to a state's police powers carry a strong presumption of validity.

The "Dormant" Commerce Clause The United States Supreme Court has interpreted the commerce clause to mean that the national government has the *exclusive* authority to regulate commerce that substantially affects trade and commerce among the states. This express grant of authority to the national government, which is often referred to as the "positive" aspect of the commerce clause, implies a negative aspect—that the states do *not* have the authority to regulate interstate commerce. This negative aspect of the commerce clause is often referred to as the "dormant" (implied) commerce clause.

The dormant commerce clause comes into play when state regulations affect interstate commerce. In this situation, the courts normally weigh the state's interest in regulating a certain matter against the burden that the state's regulation places on interstate commerce. Because courts balance the interests involved, it can be extremely difficult to predict the outcome in a particular case.

At one time, many states regulated the sale of alcoholic beverages, including wine, through a "three-tier" system. This system required separate licenses for producers, wholesalers, and retailers, subject to a complex set of overlapping regulations that effectively banned wineries from selling directly to consumers who lived outside of their state. In-state wineries, by contrast, could obtain a license for direct sales to consumers. Did these laws violate the dormant commerce clause? That was the question in the following case.

13. Local governments derive their authority to regulate their communities from the state, because they are creatures of the state. In other words, they cannot come into existence unless authorized by the state to do so.

CASE 1.2 Granholm v. Heald

Supreme Court of the United States, 2005.
__ U.S. __,
125 S.Ct. 1885,
161 L.Ed.2d 796.
http://www.findlaw.com/casecode/supreme.html[a]

FACTS In 2005, consumer spending on direct wine shipments accounted for more than 3 percent of all wine sales. Because it was not economical for every wholesaler to carry every winery's products, many small wineries relied on direct shipping to reach consumers. Domaine Alfred, a small winery in California, received requests for its wine from Michigan consumers but could not fill the orders because of that state's direct-shipment ban. The Swedenburg Estate Vineyard, a small winery in Virginia, was unable to fill orders from New York because of that state's laws. Domaine and others filed a suit in a federal district court against Michigan and others, contending that the state's laws violated the commerce clause. The court upheld the laws, but on appeal, the U.S. Court of Appeals for the Sixth Circuit reversed this ruling. Swedenburg and others filed a suit in a different federal district court against New York and others, arguing that New York's laws violated the commerce clause. The court issued a judgment in the plaintiffs' favor, but on appeal, the U.S. Court of Appeals for the Second Circuit reversed this judgment. Both cases were appealed to the United States Supreme Court.

a. In the "Browsing" section, click on "2005 Decisions." In the result, click on the name of the case to access the opinion. FindLaw is a part of West Group, the foremost provider of e-information and solutions to the U.S. legal market.

ISSUE Do state restrictions on out-of-state wineries' direct shipments to consumers violate the dormant commerce clause?

DECISION Yes. The United States Supreme Court concluded that "New York, like Michigan, discriminates against interstate commerce through its direct-shipping laws." The Court affirmed the judgment of the U.S. Court of Appeals for the Sixth Circuit, which invalidated the Michigan laws, and reversed the judgment of the U.S. Court of Appeals for the Second Circuit, which upheld the New York laws.

REASON The Supreme Court held that "state laws violate the Commerce Clause if they mandate differential treatment

CASE 1.2—CONTINUED

of in-state and out-of-state economic interests that benefits the former and burdens the latter." The laws of Michigan and New York at issue in this case contradicted this principle. The Michigan statute banned out-of-state shipments altogether. New York's regulations required an out-of-state winery to set up a distribution system in the state, subject to a licensing scheme. These laws "deprive citizens of their right to have access to the markets of other States on equal terms. * * * Allowing States to discriminate against out-of-state wine invites a multiplication of preferential trade areas destructive of the very purpose of the Commerce Clause." The Court stated that it viewed "with particular suspicion state statutes requiring business operations to be performed in the home State that could more efficiently be performed elsewhere. New York's in-state presence requirement runs contrary to our admonition that States cannot require an out-of-state firm to become a resident in order to compete on equal terms."

FOR CRITICAL ANALYSIS—Economic Consideration *Would the result have been different if the states had only required the out-of-state wineries to obtain a special license that was readily available?*

THE BILL OF RIGHTS

The importance of a written declaration of the rights of individuals eventually caused the first Congress of the United States to submit twelve amendments to the Constitution to the states for approval. The first ten of these amendments, commonly known as the **Bill of Rights,** were adopted in 1791 and embody a series of protections for the individual against various types of interference by the federal government.[14] Some constitutional protections apply to business entities as well. For example, corporations exist as separate legal entities, or legal persons, and enjoy many of the same rights and privileges as natural persons do. Summarized here are the protections guaranteed by these ten amendments (see the Constitution in Appendix D for the complete text of each amendment).

1. The First Amendment guarantees the freedoms of religion, speech, and the press and the rights to assemble peaceably and to petition the government.
2. The Second Amendment guarantees the right to keep and bear arms.
3. The Third Amendment prohibits, in peacetime, the lodging of soldiers in any house without the owner's consent.
4. The Fourth Amendment prohibits unreasonable searches and seizures of persons or property.
5. The Fifth Amendment guarantees the rights to formal accusation (see Chapter 6) by grand jury, to due process of law, and to fair payment when private property is taken for public use. Additionally, the Fifth Amendment prohibits compulsory *self-incrimination* and double jeopardy (trial for the same crime twice).
6. The Sixth Amendment guarantees the accused in a criminal case the right to a speedy and public trial by an impartial jury and with counsel. The accused has the right to cross-examine witnesses against him or her and to solicit testimony from witnesses in his or her favor.
7. The Seventh Amendment guarantees the right to a trial by jury in a civil (noncriminal) case involving at least twenty dollars.[15]
8. The Eighth Amendment prohibits excessive bail and fines, as well as cruel and unusual punishment.
9. The Ninth Amendment establishes that the people have rights in addition to those specified in the Constitution.
10. The Tenth Amendment establishes that those powers neither delegated to the federal government nor denied to the states are reserved for the states.

As originally intended, the Bill of Rights limited only the powers of the national government. Over time, however, the Supreme Court "incorporated" most of these rights into the protections against state actions afforded by the Fourteenth Amendment to the Constitution. That amendment, passed in 1868 after the Civil War, provides in part that "[n]o State shall . . . deprive any person of life, liberty, or property, without due process of law." Starting in 1925, the Supreme Court began to define various rights and liberties guaranteed in the national Constitution as constituting "due process of law," which was required of state governments under the Fourteenth Amendment. Today, most of the rights and liberties set forth in the Bill of Rights apply to state governments as well as the national government.

14. One of these proposed amendments was ratified 203 years later (in 1992) and became the Twenty-seventh Amendment to the U.S. Constitution. See Appendix D.

15. Twenty dollars was forty days' pay for the average person when the Bill of Rights was written.

We will look closely at several of the amendments in the preceding list in Chapter 6, in the context of criminal law and procedures. Here we examine two important guarantees of the First Amendment—freedom of speech and freedom of religion. These and other First Amendment freedoms (of the press, assembly, and petition) have all been applied to the states through the due process clause of the Fourteenth Amendment, which we also examine here. As you read through the following pages, keep in mind that none of these (or other) constitutional freedoms confers an absolute right. Ultimately, it is the United States Supreme Court, as the final interpreter of the Constitution, that gives meaning to these rights and determines their boundaries.

The First Amendment—Freedom of Speech Freedom of speech is the most prized freedom that Americans have. Indeed, it forms the basis for our democratic form of government, which could not exist if people were not allowed to express their political opinions freely and criticize government actions or policies. Because of its importance, the courts traditionally have protected this right to the fullest extent possible.

Speech often includes not only what we say, but also what we do to express our political, social, and religious views. The courts generally protect **symbolic speech**—gestures, movements, articles of clothing, and other forms of nonverbal expressive conduct. ■ EXAMPLE 1.9 In 1989, the Supreme Court held that the burning of the American flag to protest government policies is a constitutionally protected form of expression.[16] Similarly, participating in a hunger strike or wearing a black armband would be protected as symbolic speech. ■

—Corporate Political Speech. Political speech by corporations also falls within the protection of the First Amendment. ■ EXAMPLE 1.10 In *First National Bank of Boston v. Bellotti,*[17] national banking associations and business corporations sought United States Supreme Court review of a Massachusetts statute that prohibited corporations from making political contributions or expenditures that individuals were permitted to make. The Court ruled that the Massachusetts law was unconstitutional because it violated the right of corporations to freedom of speech. ■ Similarly, the Court has held that a law prohibiting a corporation from placing inserts in its billing to express its views on controversial issues violates the First Amendment.[18] Although a more conservative Supreme Court subsequently reversed this trend somewhat,[19] corporate political speech continues to be given significant protection under the First Amendment.

—Commercial Speech. The courts also give substantial protection to "commercial" speech, which consists of communications—primarily advertising and marketing—made by business firms that involve only their commercial interests. The protection given to commercial speech under the First Amendment is not as extensive as that afforded to noncommercial speech, however. A state may restrict certain kinds of advertising, for example, in the interest of protecting consumers from being misled by the advertising practices. States also have a legitimate interest in the beautification of roadsides, and this interest allows states to place restraints on billboard advertising. ■ EXAMPLE 1.11 Café Erotica, a nude dancing establishment, sued the state after being denied a permit to erect a billboard along an interstate highway in Florida. The state appellate court decided that because the law directly advanced a substantial government interest in highway beautification and safety, it was not an unconstitutional restraint on commercial speech.[20] ■

Generally, a restriction on commercial speech will be considered valid as long as it meets the following three criteria: (1) it must seek to implement a substantial government interest, (2) it must directly advance that interest, and (3) it must go no further than necessary to accomplish its objective. ■ EXAMPLE 1.12 The South Carolina Supreme Court held that a state statute banning ads for video gambling violated the First Amendment because the statute did not directly advance a substantial government interest. Although the court acknowledged that the state had a substantial interest in minimizing gambling, there was no evidence that a reduction in video gambling ads would result in a reduction in gambling.[21] ■

—Unprotected Speech. The United States Supreme Court has made it clear that certain types of speech will not be given any protection under the First Amendment. Speech that harms the good reputation of another, or

16. See *Texas v. Johnson,* 491 U.S. 397, 109 S.Ct. 2533, 105 L.Ed.2d 342 (1989).
17. 435 U.S. 765, 98 S.Ct. 1407, 55 L.Ed.2d 707 (1978).
18. *Consolidated Edison Co. v. Public Service Commission,* 447 U.S. 530, 100 S.Ct. 2326, 65 L.Ed.2d 319 (1980).
19. See *Austin v. Michigan Chamber of Commerce,* 494 U.S. 652, 110 S.Ct. 1391, 108 L.Ed.2d 652 (1990), in which the Court upheld a state law prohibiting corporations from using general corporate funds for independent expenditures in state political campaigns.
20. *Café Erotica v. Florida Department of Transportation,* 830 So.2d 181 (Fla.App. 1 Dist. 2002); review denied, *Café Erotica/We Dare to Bare v. Florida Department of Transportation,* 845 So.2d 888 (2003).
21. *Evans v. State,* 344 S.C. 60, 543 S.E.2d 547 (2001).

defamatory speech (see Chapter 4), will not be protected. Speech that violates criminal laws (such as threatening speech) is not constitutionally protected. Other unprotected speech includes "fighting words," or words that are likely to incite others to respond violently.

The Supreme Court has also held that obscene speech is not protected by the First Amendment. The Court has grappled from time to time with the problem of trying to establish an objective definition of obscene speech. In a 1973 case, *Miller v. California,*[22] the Supreme Court created a test for legal obscenity, which involved a set of requirements that must be met for material to be legally obscene. Under this test, material is obscene if (1) the average person finds that it violates contemporary community standards; (2) the work taken as a whole appeals to a prurient (arousing or obsessive) interest in sex; (3) the work shows patently offensive sexual conduct; and (4) the work lacks serious redeeming literary, artistic, political, or scientific merit.

Because community standards vary widely, the *Miller* test has had inconsistent applications, and obscenity remains a constitutionally unsettled issue. Numerous state and federal statutes make it a crime to disseminate obscene materials, however, and the Supreme Court has often upheld such laws, including laws prohibiting the sale and possession of child pornography.[23]

—Online Obscenity. Congress first attempted to protect minors from pornographic materials on the Internet by passing the Communications Decency Act (CDA) of 1996. The CDA declared it a crime to make available to minors online any "obscene or indecent" message that "depicts or describes, in terms patently offensive as measured by contemporary community standards, sexual or excretory activities or organs."[24] The act was challenged as an unconstitutional restraint on speech, and ultimately the United States Supreme Court ruled that portions of the act were unconstitutional. The Court held that the terms *indecent* and *patently offensive* covered large amounts of nonpornographic material with serious educational or other value.[25]

The Child Online Protection Act (COPA) of 1998[26] banned material "harmful to minors" distributed without some kind of age-verification system to separate adult and minor users. Although the COPA was more narrowly tailored than its predecessor, the CDA, it still used "contemporary community standards" to define which material was obscene and harmful to minors. When the act was challenged as violating free speech, the federal court hearing the case granted an injunction suspending the COPA, and the appellate court upheld the injunction.[27] In 2002, however, the Supreme Court reversed that decision and remanded the case, concluding that the reference to contemporary community standards alone was not sufficient to render the entire act unconstitutional.[28] On remand, the appellate court again held that the COPA violated free speech guarantees.[29] In June 2004, the Supreme Court affirmed the appellate court's decision and remanded the case for trial. The Court concluded that the claim that the COPA violated the right to free speech would likely prevail and hence left the injunction in place.[30]

In 2000, Congress enacted the Children's Internet Protection Act (CIPA),[31] which requires public schools and libraries to install *filtering software* to prevent children from accessing adult content from the Internet. **Filtering software** is designed to prevent persons from viewing certain Web sites by responding to a site's Internet address or its **meta tags,** or key words. The CIPA was also challenged on constitutional grounds, but in 2003 the Supreme Court held that the act did not violate the First Amendment. The Court concluded that because libraries can disable the filters for any patrons who ask, the system was reasonably flexible and did not burden free speech to an unconstitutional extent.[32]

The First Amendment—Freedom of Religion The First Amendment states that the government may neither establish any religion nor prohibit the free exercise of religious practices. The first part of this constitutional provision is referred to as the **establishment clause,** and the second part is known as the **free exercise clause.** Government action, both federal and state, must be consistent with this constitutional mandate.

—The Establishment Clause. The establishment clause prohibits the government from establishing a state-sponsored religion, as well as from passing laws that

22. 413 U.S. 15, 93 S.Ct. 2607, 37 L.Ed.2d 419 (1973).

23. For example, see *Osborne v. Ohio,* 495 U.S. 103, 110 S.Ct. 1691, 109 L.Ed.2d 98 (1990).

24. 47 U.S.C. Section 223(a)(1)(B)(ii).

25. *Reno v. American Civil Liberties Union,* 521 U.S. 844, 117 S.Ct. 2329, 138 L.Ed.2d 874 (1997).

26. 47 U.S.C. Section 231.

27. *American Civil Liberties Union v. Reno,* 31 F.Supp.2d 473 (E.D.Pa. 1999); aff'd, 217 F.3d 162 (3d Cir. 2002). The term *aff'd* in a citation is an abbreviation for *affirmed;* it indicates that the appellate court upheld the ruling of the lower court.

28. *Ashcroft v. American Civil Liberties Union,* 535 U.S. 564, 122 S.Ct. 1700, 152 L.Ed.2d 771 (2002).

29. *American Civil Liberties Union v. Ashcroft,* 322 F.3d 240 (3d Cir. 2003).

30. *American Civil Liberties Union v. Ashcroft,* 542 U.S. 656, 124 S.Ct. 2783, 159 L.Ed.2d 690 (2004).

31. 17 U.S.C. Sections 1701–1741.

32. *United States v. American Library Association,* 539 U.S. 194, 123 S.Ct. 2297, 156 L.Ed.2d 221 (2003).

promote (aid or endorse) religion or that show a preference for one religion over another. The establishment clause does not require a complete separation of church and state, though. On the contrary, it requires the government to accommodate religions.[33]

The establishment clause covers all conflicts about such matters as the legality of state and local government support for a particular religion, government aid to religious organizations and schools, the government's allowing or requiring school prayers, and the teaching of evolution versus fundamentalist theories of creation. The Supreme Court has held that for a government law or policy to be constitutional, it must be secular in aim, must not have the primary effect of advancing or inhibiting religions, and must not create "an excessive government entanglement with religion."[34] Generally, federal or state regulation that does not promote religion or place a significant burden on religion is constitutional even if it has some impact on religion.

—The Free Exercise Clause. The free exercise clause guarantees that a person can hold any religious belief that she or he wants; or a person can have no religious belief. When religious *practices* work against public policy and the public welfare, however, the government can act. ■ EXAMPLE 1.13 Regardless of a child's or parent's religious beliefs, the government can require certain types of vaccinations. Similarly, although children of Jehovah's Witnesses are not required to say the Pledge of Allegiance at school, their parents cannot prevent them from accepting medical treatment (such as blood transfusions) if in fact their lives are in danger. Additionally, public school students can be required to study from textbooks chosen by school authorities. ■

For business firms, an important issue involves the accommodation that businesses must make for the religious beliefs of their employees. ■ EXAMPLE 1.14 Suppose that an employee's religion prohibits him or her from working on a certain day of the week or at a certain type of job. The employer must make a reasonable attempt to accommodate these religious requirements. ■ Employers must reasonably accommodate an employee's religious beliefs even if the beliefs are not based on the tenets or dogma of a particular church, sect, or denomination. The only requirement is that the belief be religious in nature and sincerely held by the employee. (We will look further at this issue in Chapter 23, in the context of employment discrimination.)

Due Process Both the Fifth and the Fourteenth Amendments provide that no person shall be deprived of "life, liberty, or property, without due process of law." The **due process clause** of each of these constitutional amendments has two aspects—procedural and substantive.

—Procedural Due Process. Procedural due process requires that any government decision to take life, liberty, or property must be made fairly. For example, fair procedures must be used in determining whether a person will be subjected to punishment or have some burden imposed on him or her. Fair procedure has been interpreted as requiring that the person have at least an opportunity to object to a proposed action before a fair, neutral decision maker (which need not be a judge). Thus, if a driver's license is construed as a property interest, some sort of opportunity to object to its suspension or termination by the state must be provided.

—Substantive Due Process. Substantive due process focuses on the content, or substance, of legislation. If a law or other governmental action limits a *fundamental right,* it will be held to violate substantive due process unless it promotes a compelling or an overriding state interest. Fundamental rights include interstate travel, privacy, voting, and all First Amendment rights. A compelling state interest could be the public's safety. ■ EXAMPLE 1.15 Laws setting speed limits may be upheld even though they affect interstate travel, if they are shown to reduce highway fatalities. The courts uphold these laws because the state has a compelling interest in protecting the lives of its citizens. ■

In situations not involving fundamental rights, a law or action does not violate substantive due process if it rationally relates to any legitimate governmental end. It is almost impossible for a law or action to fail the *rationality test.* Under this test, virtually any business regulation will be upheld as reasonable—the United States Supreme Court has sustained insurance regulations, price and wage controls, banking limitations, and restrictions of unfair competition and trade practices against substantive due process challenges.

■ EXAMPLE 1.16 If a state legislature enacted a law imposing a fifteen-year term of imprisonment without a trial on all businesspersons who appeared in their own television commercials, the law would be unconstitutional on both substantive and procedural grounds. Substantive review would invalidate the legislation because it abridges (constrains) freedom of speech. Procedurally, the law is unfair because it imposes the penalty without giving the accused a chance to defend her or his actions. ■ The lack of procedural due process will cause a court to invalidate any statute or prior court decision. Similarly, the courts will overrule any state or federal law that violates the Constitution by denying substantive due process.

33. *Zorach v. Clauson,* 343 U.S. 306, 72 S.Ct. 679, 96 L.Ed. 954 (1952).
34. *Lemon v. Kurtzman,* 403 U.S. 602, 91 S.Ct. 2105, 29 L.Ed.2d 745 (1971).

TERMS AND CONCEPTS

CHAPTER SUMMARY The Legal and Constitutional Environment of Business

The Nature of Law (See page 2.)	Law can be defined as a body of rules of conduct with legal force and effect, prescribed by the controlling authority (the government) of a society.
Sources of American Law (See pages 4–6.)	1. *Constitutional law*—The law as expressed in the U.S. Constitution and the various state constitutions. The U.S. Constitution is the supreme law of the land. State constitutions are supreme within state borders to the extent that they do not violate the U.S. Constitution or a federal law. 2. *Statutory law*—Laws or ordinances created by federal, state, and local legislatures and governing bodies. None of these laws can violate the U.S. Constitution or the relevant state constitutions. Uniform laws, when adopted by a state legislature, become statutory law in that state. 3. *Administrative law*—The rules, orders, and decisions of federal or state government administrative agencies. 4. *Case law and common law doctrines*—Judge-made law, including interpretations of constitutional provisions, of statutes enacted by legislatures, and of regulations created by administrative agencies. The common law—the doctrines and principles embodied in case law—governs all areas not covered by statutory law (or agency regulations issued to implement various statutes).
The Common Law Tradition (See pages 6–8.)	1. *Common law*—Law that originated in medieval England with the creation of the king's courts, or *curiae regis,* and the development of a body of rules that were common to (or applied throughout) the land. 2. *Stare decisis*—A doctrine under which judges "stand on decided cases"—or follow the rule of precedent—in deciding cases. *Stare decisis* is the cornerstone of the common law tradition. 3. *Remedies*— a. Remedies at law—Money or something else of value. b. Remedies in equity—Remedies that are granted when the remedies at law are unavailable or inadequate. Equitable remedies include specific performance, an injunction, and contract rescission (cancellation).

(Continued)

CHAPTER SUMMARY The Legal and Constitutional Environment of Business—Continued

Classifications of Law (See pages 8–10.)	The law may be broken down according to several classification systems, such as substantive or procedural law, federal or state law, and private or public law. Two broad classifications are civil and criminal law, and national and international law.
The Constitution as It Affects Business (See pages 10–16.)	1. *Commerce clause*—Expressly permits Congress to regulate commerce. That power authorizes the national government, at least theoretically, to regulate every commercial enterprise in the United States. Under their police powers, state governments may regulate private activities to protect or promote the public order, health, safety, morals, and general welfare. 2. *Bill of Rights*—The first ten amendments to the U.S. Constitution. They embody a series of protections for individuals—and in some cases, business entities—against various types of interference by the federal government. One of the freedoms guaranteed by the Bill of Rights that affects businesses is the freedom of speech guaranteed by the First Amendment. Also important are the protections of the Fifth and the Fourteenth Amendments, which provide that no person shall be deprived of "life, liberty, or property, without due process of law."

FOR REVIEW

Answers for the even-numbered questions in this For Review *section can be found in Appendix A at the end of this text.*

1. What is the Uniform Commercial Code?
2. What is the common law tradition?
3. What is a precedent? When might a court depart from precedent?
4. What are some important differences between civil law and criminal law?
5. How does the U.S. Constitution affect business activities in the United States?

QUESTIONS AND CASE PROBLEMS

1–1. Legal Systems. What are the key differences between a common law system and a civil law system? Why do some countries have common law systems and others have civil law systems?

QUESTION WITH SAMPLE ANSWER

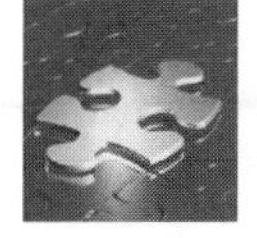

1–2. This chapter discussed a number of sources of American law. Which source of law takes priority in each of the following situations, and why?

(a) A federal statute conflicts with the U.S. Constitution.
(b) A federal statute conflicts with a state constitution.
(c) A state statute conflicts with the common law of that state.
(d) A state constitutional amendment conflicts with the U.S. Constitution.
(e) A federal administrative regulation conflicts with a state constitution.

For a sample answer to this question, go to Appendix B at the end of this text.

1–3. *Stare Decisis.* In the text of this chapter, we stated that the doctrine of *stare decisis* "became a cornerstone of the English and American judicial systems." What does *stare decisis* mean, and why has this doctrine been so fundamental to the development of our legal tradition?

1–4. Binding versus Persuasive Authority. A county court in Illinois is deciding a case involving an issue that has never been addressed before in that state's courts. The Iowa Supreme Court, however, recently decided a case involving a very similar fact pattern. Is the Illinois court obligated to follow the Iowa Supreme Court's decision on the issue? If the United States Supreme Court had decided a similar case, would that decision be binding on the Illinois court? Explain.

1–5. Remedies. Arthur Rabe is suing Xavier Sanchez for breaching a contract in which Sanchez promised to sell Rabe a Van Gogh painting for $150,000.

(a) In this lawsuit, who is the plaintiff and who is the defendant?
(b) If Rabe wants Sanchez to perform the contract as promised, what remedy should Rabe seek?
(c) Suppose that Rabe wants to cancel the contract because Sanchez fraudulently misrepresented the painting as an original Van Gogh when in fact it is a copy. In this situation, what remedy should Rabe seek?

(d) Will the remedy Rabe seeks in either situation be a remedy at law or a remedy in equity?

(e) Suppose that the court finds in Rabe's favor and grants one of these remedies. Sanchez then appeals the decision to a higher court. Read the subsection entitled "Appellants and Appellees" in the appendix following this chapter. On appeal, which party in the Rabe-Sanchez case will be the appellant (or petitioner), and which party will be the appellee (or respondent)?

1–6. Commercial Speech. A mayoral election is about to be held in a large U.S. city. One of the candidates is Luis Delgado, and his campaign supporters wish to post campaign signs on lampposts and utility posts throughout the city. A city ordinance, however, prohibits the posting of any signs on public property. Delgado's supporters contend that the city ordinance is unconstitutional because it violates their rights to free speech. What factors might a court consider in determining the constitutionality of this ordinance?

1–7. Commerce Clause. Suppose that Georgia enacts a law requiring the use of contoured rear-fender mudguards on trucks and trailers operating within its state lines. The statute further makes it illegal for trucks and trailers to use straight mudguards. In thirty-five other states, straight mudguards are legal. Moreover, in the neighboring state of Florida, straight mudguards are explicitly required by law. There is some evidence suggesting that contoured mudguards might be a little safer than straight mudguards. Discuss whether this Georgia statute would violate the commerce clause of the U.S. Constitution.

1–8. Freedom of Religion. A business has a backlog of orders, and to meet its deadlines, management decides to run the firm seven days a week, eight hours a day. One of the employees, Marjorie Tollens, refuses to work on Saturday on religious grounds. Her refusal to work means that the firm may not meet its production deadlines and may therefore suffer a loss of future business. The firm fires Tollens and replaces her with an employee who is willing to work seven days a week. Tollens claims that her employer, in terminating her employment, violated her constitutional right to the free exercise of her religion. Do you agree? Why or why not?

CASE PROBLEM WITH SAMPLE ANSWER

1–9. The city of Tacoma, Washington, enacted an ordinance that prohibited the playing of car sound systems at a volume that would be "audible" at a distance greater than fifty feet. Dwight Holland was arrested and convicted for violating the ordinance. The conviction was later dismissed, but Holland filed a civil suit in a Washington state court against the city. He claimed in part that the ordinance violated his freedom of speech under the First Amendment. On what basis might the court conclude that this ordinance is constitutional? (Hint: In playing a sound system, was Holland actually expressing himself?) [*Holland v. City of Tacoma,* 90 Wash.App. 533, 954 P.2d 290 (1998)]

After you have answered this problem, compare your answer with the sample answer given on the Web site that accompanies this text. Go to academic.cengage.com/blaw/wbl, select "Chapter 1," and click on "Case Problem with Sample Answer."

1–10. Due Process. The Board of County Commissioners of Yellowstone County, Montana, in 1994, created Zoning District 17 in a rural area of the county and a Planning and Zoning Commission for the district. The commission adopted zoning regulations, which provided, among other things, that "dwelling units" could be built only through "on-site construction." Later, county officials could not identify any health or safety concerns that the on-site construction provision addressed, and there was no indication that homes built off-site would affect property values or any other general welfare interest of the community. In December 1999, Francis and Anita Yurczyk bought two forty-acre tracts in District 17. The Yurczyks also bought a modular home and moved it onto the property the following spring. Within days, the county advised the Yurczyks that the home violated the on-site construction regulation and would have to be removed. The Yurczyks filed a suit in a Montana state court against the county, alleging in part that the regulation violated the Yurczyks' due process rights. Should the court rule in the plaintiffs' favor? Explain. [*Yurczyk v. Yellowstone County,* 2004 MT 3, 319 Mont. 169, 83 P.3d 266 (2004)]

The Legal and Constitutional Environment of Business

Today, business law professors and students can go online to access information on almost every topic covered in this text. A good point of departure for online legal research is this text's Web site at

academic.cengage.com/blaw/wbl

There you will find numerous materials relevant to this text and to business law generally, including links to various legal resources on the Web. Additionally, every chapter in this text ends with an *Accessing the Internet* feature that contains selected Web addresses.

You can access many of the sources of law discussed in this chapter at the FindLaw Web site, which is probably the most comprehensive source of free legal information on the Internet. Go to

http://www.findlaw.com

The Legal Information Institute (LII) at Cornell Law School, which offers extensive information about U.S. law, is also a good starting point for legal research. The URL for this site is

http://www.law.cornell.edu

The Library of Congress offers extensive links to state and federal government resources at

http://www.loc.gov

For an online version of the Constitution that provides hypertext links to amendments and other changes, go to

http://www.law.cornell.edu/constitution/constitution.table.html

For discussions of current issues involving the rights and liberties contained in the Bill of Rights, go to the Web site of the American Civil Liberties Union at

http://www.aclu.org

Summaries and the full texts of constitutional law decisions by the United States Supreme Court are included at the Oyez site, which also features audio clips of arguments before the Court. Go to

http://www.oyez.org/oyez/frontpage

ONLINE LEGAL RESEARCH

Go to the *Fundamentals of Business Law* home page at academic.cengage.com/blaw/wbl, select "Chapter 1," and click on "Internet Exercises." There you will find the following Internet research exercises that you can perform to learn more about topics covered in this chapter.

Activity 1–1: LEGAL PERSPECTIVE—Internet Sources of Law

Activity 1–2: MANAGEMENT PERSPECTIVE—Online Assistance from Government Agencies

Activity 1–3: SOCIAL PERSPECTIVE—The Case of the Speluncean Explorers

BEFORE THE TEST

Go to the *Fundamentals of Business Law* home page at academic.cengage.com/blaw/wbl, select "Chapter 1," and click on "Interactive Quizzes." You will find a number of interactive questions relating to this chapter.

Appendix to Chapter 1: Finding and Analyzing the Law

The statutes, agency regulations, and case law referred to in this text establish the rights and duties of businesspersons engaged in various types of activities. The cases presented in the following chapters provide you with concise, real-life illustrations of how the courts interpret and apply these laws. Because of the importance of knowing how to find statutory, administrative, and case law, this appendix offers a brief introduction to how these laws are published and to the legal "shorthand" employed in referencing these legal sources.

Finding Statutory and Administrative Law

When Congress passes laws, they are collected in a publication titled *United States Statutes at Large*. When state legislatures pass laws, they are collected in similar state publications. Most frequently, however, laws are referred to in their codified form—that is, the form in which they appear in the federal and state codes. In these codes, laws are compiled by subject.

UNITED STATES CODE

The *United States Code* (U.S.C.) arranges all existing federal laws of a public and permanent nature by subject. Each of the fifty subjects into which the U.S.C. arranges the laws is given a title and a title number. For example, laws relating to commerce and trade are collected in "Title 15, Commerce and Trade." Titles are subdivided by sections. A citation to the U.S.C. includes title and section numbers. Thus, a reference to "15 U.S.C. Section 1" means that the statute can be found in Section 1 of Title 15. ("Section" may also be designated by the symbol §, and "Sections" by §§.)

Sometimes a citation includes the abbreviation *et seq.*—as in "15 U.S.C. Sections 1 *et seq.*" The term is an abbreviated form of *et sequitur,* which in Latin means "and the following"; when used in a citation, it refers to sections that concern the same subject as the numbered section and follow it in sequence.

Commercial publications of these laws and regulations are available and are widely used. For example, West Group publishes the *United States Code Annotated* (U.S.C.A.). The U.S.C.A. contains the complete text of laws included in the U.S.C., as well as notes of court decisions that interpret and apply specific sections of the statutes, plus the text of presidential proclamations and executive orders. The U.S.C.A. also includes research aids, such as cross-references to related statutes, historical notes, and library references. A citation to the U.S.C.A. is similar to a citation to the U.S.C.: "15 U.S.C.A. Section 1."

STATE CODES

State codes follow the U.S.C. pattern of arranging law by subject. The state codes may be called codes, revisions, compilations, consolidations, general statutes, or statutes, depending on the preferences of the states. In some codes, subjects are designated by number. In others, they are designated by name. For example, "13 Pennsylvania Consolidated Statutes Section 1101" means that the statute can be found in Title 13, Section 1101, of the Pennsylvania code. "California Commercial Code Section 1101" means the statute can be found under the subject heading "Commercial Code" of the California code in Section 1101. Abbreviations may be used. For example, "13 Pennsylvania Consolidated Statutes Section 1101" may be abbreviated as "13 Pa. C.S. § 1101," and "California Commercial Code Section 1101" may be abbreviated "Cal. Com. Code § 1101."

ADMINISTRATIVE RULES

Rules and regulations adopted by federal administrative agencies are compiled in the *Code of Federal Regulations* (C.F.R.). Like the U.S.C., the C.F.R. is divided into fifty titles. Rules within each title are assigned section numbers. A full citation to the C.F.R. includes title and section numbers. For example, a reference to "17 C.F.R. Section 230.504" means that the rule can be found in Section 230.504 of Title 17.

Finding Case Law

Before discussing the case reporting system, we need to look briefly at the court system (which will be discussed in detail in Chapter 2). There are two types of courts in the United States, federal courts and state courts. Both the federal and state court systems consist of several levels, or tiers, of courts. *Trial courts,* in which evidence is presented and testimony given, are on the bottom tier (which also includes lower courts handling specialized issues). Decisions from a trial court can be appealed to a higher court, which commonly would be an intermediate *court of appeals,* or an *appellate court.* Decisions from these intermediate courts of appeals may be appealed to an even higher court, such as a state supreme court or the United States Supreme Court.

State Court Decisions

Most state trial court decisions are not published. Except in New York and a few other states that publish selected opinions of their trial courts, decisions from state trial courts are merely filed in the office of the clerk of the court, where the decisions are available for public inspection. (Sometimes they can be found online as well.) Written decisions of the appellate, or reviewing, courts, however, are published and distributed. As you will note, most of the state court cases presented in this book are from state appellate courts. The reported appellate decisions are published in volumes called *reports* or *reporters,* which are numbered consecutively. State appellate court decisions are found in the state reporters of that particular state.

Additionally, state court opinions appear in regional units of the *National Reporter System,* published by West Group. Most lawyers and libraries have the West reporters because they report cases more quickly and are distributed more widely than the state-published reports. In fact, many states have eliminated their own reporters in favor of West's National Reporter System. The National Reporter System divides the states into the following geographic areas: *Atlantic* (A. or A.2d), *North Eastern* (N.E. or N.E.2d), *North Western* (N.W. or N.W.2d), *Pacific* (P., P.2d, or P.3d) *South Eastern* (S.E. or S.E.2d), *South Western* (S.W., S.W.2d, or S.W.3d), and *Southern* (So. or So.2d). (The *2d* and *3d* in the abbreviations refer to *Second Series* and *Third Series,* respectively.) The states included in each of these regional divisions are indicated in Exhibit 1A–1, which illustrates West's National Reporter System.

After appellate decisions have been published, they are normally referred to (cited) by the name of the case; the volume, name, and page number of the state's official reporter (if different from West's National Reporter System); the volume, name, and page number of the *National Reporter;* and the volume, name, and page number of any other selected reporter. This information is included in the *citation.* (Citing a reporter by volume number, name, and page number, in that order, is common to all citations.) When more than one reporter is cited for the same case, each reference is called a *parallel citation.* For example, consider the following case: *Yale Diagnostic Radiology v. Estate of Harun Fountain,* 267 Conn. 351, 838 A.2d 179 (2004). We see that the opinion in this case may be found in Volume 267 of the official *Connecticut Reports,* on page 351. The parallel citation is to Volume 838 of the *Atlantic Reporter, Second Series,* page 179. In presenting appellate opinions in this text, in addition to the reporter, we give the name of the court hearing the case and the year of the court's decision.

A few states—including those with intermediate appellate courts, such as California, Illinois, and New York—have more than one reporter for opinions issued by their courts. Sample citations from these courts, as well as others, are listed and explained in Exhibit 1A–2 beginning on page 24.

Federal Court Decisions

Federal district court decisions are published unofficially in West's *Federal Supplement* (F. Supp. or F.Supp.2d), and opinions from the circuit courts of appeals (federal reviewing courts) are reported unofficially in West's *Federal Reporter* (F., F.2d, or F.3d). Cases concerning federal bankruptcy law are published unofficially in West's *Bankruptcy Reporter* (Bankr.). The official edition of United States Supreme Court decisions is the *United States Reports* (U.S.), which is published by the federal government. Unofficial editions of Supreme Court cases include West's *Supreme Court Reporter* (S.Ct.) and the *Lawyers' Edition of the Supreme Court Reports* (L.Ed. or L.Ed.2d). Sample citations for federal court decisions are also listed and explained in Exhibit 1A–2.

Unpublished Opinions and Old Cases

Many court opinions that are not yet published or that are not intended for publication can be accessed through Westlaw® (abbreviated in citations as "WL"), an online legal database maintained by West Group. When no citation to a published reporter is available for cases cited in this text, we give the WL citation (see Exhibit 1A–2 for an example).

On a few occasions, this text cites opinions from old, classic cases dating to the nineteenth century or earlier; some of these are from the English courts. The citations to

EXHIBIT 1A–1 NATIONAL REPORTER SYSTEM—REGIONAL/FEDERAL

Regional Reporters	Coverage Beginning	Coverage
Atlantic Reporter (A. or A.2d)	1885	Connecticut, Delaware, Maine, Maryland, New Hampshire, New Jersey, Pennsylvania, Rhode Island, Vermont, and District of Columbia.
North Eastern Reporter (N.E. or N.E.2d)	1885	Illinois, Indiana, Massachusetts, New York, and Ohio.
North Western Reporter (N.W. or N.W.2d)	1879	Iowa, Michigan, Minnesota, Nebraska, North Dakota, South Dakota, and Wisconsin.
Pacific Reporter (P., P.2d, or P.3d)	1883	Alaska, Arizona, California, Colorado, Hawaii, Idaho, Kansas, Montana, Nevada, New Mexico, Oklahoma, Oregon, Utah, Washington, and Wyoming.
South Eastern Reporter (S.E. or S.E.2d)	1887	Georgia, North Carolina, South Carolina, Virginia, and West Virginia.
South Western Reporter (S.W., S.W.2d, or S.W.3d)	1886	Arkansas, Kentucky, Missouri, Tennessee, and Texas.
Southern Reporter (So. or So.2d)	1887	Alabama, Florida, Louisiana, and Mississippi.
Federal Reporters		
Federal Reporter (F., F.2d, or F.3d)	1880	U.S. Circuit Court from 1880 to 1912; U.S. Commerce Court from 1911 to 1913; U.S. District Courts from 1880 to 1932; U.S. Court of Claims (now called U.S. Court of Federal Claims) from 1929 to 1932 and since 1960; U.S. Courts of Appeals since 1891; U.S. Court of Customs and Patent Appeals since 1929; U.S. Emergency Court of Appeals since 1943.
Federal Supplement (F.Supp. or F.Supp.2d)	1932	U.S. Court of Claims from 1932 to 1960; U.S. District Courts since 1932; and U.S. Customs Court since 1956.
Federal Rules Decisions (F.R.D.)	1939	U.S. District Courts involving the Federal Rules of Civil Procedure since 1939 and Federal Rules of Criminal Procedure since 1946.
Supreme Court Reporter (S.Ct.)	1882	U.S. Supreme Court since the October term of 1882.
Bankruptcy Reporter (Bankr.)	1980	Bankruptcy decisions of U.S. Bankruptcy Courts, U.S. District Courts, U.S. Courts of Appeals, and U.S. Supreme Court.
Military Justice Reporter (M.J.)	1978	U.S. Court of Military Appeals and Courts of Military Review for the Army, Navy, Air Force, and Coast Guard.

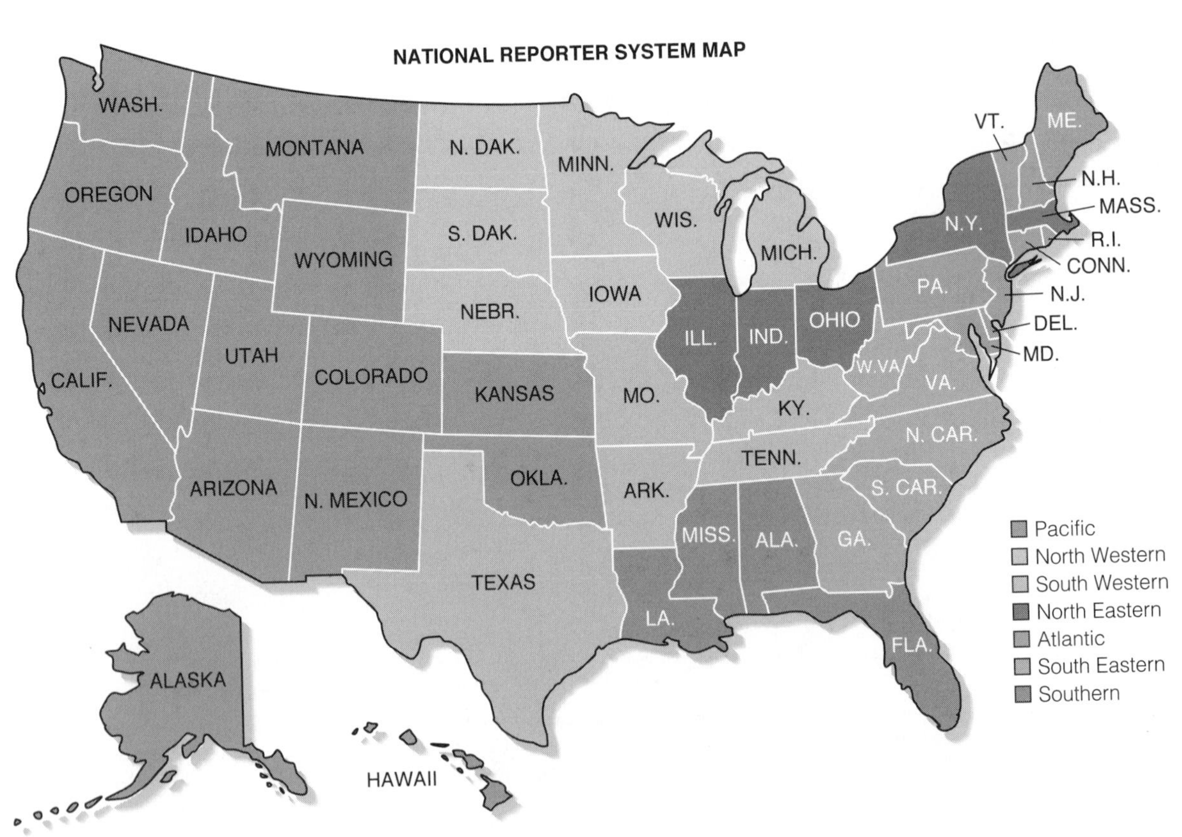

EXHIBIT 1A–2 HOW TO READ CITATIONS

State Courts

269 Neb. 82, 690 N.W.2d 778 (2005)[a]

N.W. is the abbreviation for West's publication of state court decisions rendered in the *North Western Reporter* of the National Reporter System. *2d* indicates that this case was included in the *Second Series* of that reporter. The number 690 refers to the volume number of the reporter; the number 778 refers to the first page in that volume on which this case can be found.

Neb. is an abbreviation for *Nebraska Reports,* Nebraska's official reports of the decisions of its highest court, the Nebraska Supreme Court.

125 Cal.App.4th 949, 23 Cal.Rptr.3d 233 (2005)

Cal.Rptr. is the abbreviation for West's unofficial reports—titled *California Reporter*—of the decisions of the California Supreme Court and California appellate courts.

1 N.Y.3d 280, 803 N.E.2d 757, 771 N.Y.S.2d 484 (2003)

N.Y.S. is the abbreviation for West's unofficial reports—titled *New York Supplement*—of the decisions of New York courts.

N.Y. is the abbreviation for *New York Reports,* New York's official reports of the decisions of its court of appeals. The New York Court of Appeals is the state's highest court, analogous (similar) to other states' supreme courts. (In New York, a supreme court is a trial court.)

267 Ga.App. 832, 600 S.E.2d 800 (2004)

Ga.App. is the abbreviation for *Georgia Appeals Reports,* Georgia's official reports of the decisions of its court of appeals.

Federal Courts

___ U.S. ___, 125 S.Ct. 847, ___ L.Ed.2d ___ (2005)

L.Ed. is an abbreviation for *Lawyers' Edition of the Supreme Court Reports,* an unofficial edition of decisions of the United States Supreme Court.

S.Ct. is the abbreviation for West's unofficial reports—titled *Supreme Court Reporter*—of decisions of the United States Supreme Court.

U.S. is the abbreviation for *United States Reports,* the official edition of the decisions of the United States Supreme Court.

a. The case names have been deleted from these citations to emphasize the publications. It should be kept in mind, however, that the name of a case is as important as the specific numbers of the volumes in which it is found. If a citation is incorrect, the correct citation may be found in a publication's index of case names. The date of a case is also important because, in addition to providing a check on error in citations, the value of a recent case as an authority is likely to be greater than that of an earlier case.

EXHIBIT 1A–2 HOW TO READ CITATIONS—CONTINUED

Federal Courts (continued)

394 F.3d 520 (7th Cir. 2005)

7th Cir. is an abbreviation denoting that this case was decided in the United States Court of Appeals for the Seventh Circuit.

340 F.Supp.2d 1051 (D.S.D. 2004)

D.S.D. is an abbreviation indicating that the United States District Court for the District of South Dakota decided this case.

English Courts

9 Exch. 341, 156 Eng.Rep. 145 (1854)

Eng.Rep. is an abbreviation for *English Reports, Full Reprint,* a series of reports containing selected decisions made in English courts between 1378 and 1865.

Exch. is an abbreviation for *English Exchequer Reports,* which included the original reports of cases decided in England's Court of Exchequer.

Statutory and Other Citations

18 U.S.C. Section 1961(1)(A)

U.S.C. denotes *United States Code,* the codification of *United States Statutes at Large.* The number 18 refers to the statute's U.S.C. title number and 1961 to its section number within that title. The number 1 refers to a subsection within the section and the letter A to a subdivision within the subsection.

UCC 2–206(1)(b)

UCC is an abbreviation for *Uniform Commercial Code.* The first number 2 is a reference to an article of the UCC and 206 refers to a section within that article. The number 1 refers to a subsection within the section and the letter b refers to a subdivision within the subsection.

Restatement (Second) of Contracts, Section 162

Restatement (Second) of Contracts refers to the second edition of the American Law Institute's *Restatement of the Law of Contracts.* The number 162 refers to a specific section.

17 C.F.R. Section 230.505

C.F.R. is an abbreviation for *Code of Federal Regulations,* a compilation of federal administrative regulations. The number 17 designates the regulation's title number, and 230.505 designates a specific section within that title.

(Continued)

EXHIBIT 1A–2 HOW TO READ CITATIONS—CONTINUED

Westlaw® Citations[b]

2005 WL 27554

WL is an abbreviation for "Westlaw®." The number 2005 is the year of the document that can be found with this citation in the Westlaw® database. The number 27554 is a number assigned to a specific document. A higher number indicates that a document was added to the Westlaw® database later in the year.

Uniform Resource Locators (URLs)[c]

http://www.westlaw.com[d]

The suffix *com* is the top-level domain (TLD) for this Web site. The TLD *com* is an abbreviation for "commercial," which normally means that a for-profit entity hosts (maintains or supports) this Web site.

westlaw is the host name—the part of the domain name selected by the organization that registered the name. In this case, West Group registered the name. This Internet site is the Westlaw database on the Web.

www is an abbreviation for "World Wide Web." The Web is a system of Internet servers that support documents formatted in *HTML* (hypertext markup language). HTML supports links to text, graphics, and audio and video files.

http://www.uscourts.gov

This is "The Federal Judiciary Home Page." The host is the Administrative Office of the U.S. Courts. The TLD *gov* is an abbreviation for "government." This Web site includes information and links from, and about, the federal courts.

http://www.law.cornell.edu/index.html

This part of a URL points to a Web page or file at a specific location within the host's domain. This page is a menu with links to documents within the domain and to other Internet resources.

This is the host name for a Web site that contains the Internet publications of the Legal Information Institute (LII), which is a part of Cornell Law School. The LII site includes a variety of legal materials and links to other legal resources on the Internet. The TLD *edu* is an abbreviation for "educational institution" (a school or a university).

http://www.ipl.org/div/news

This part of the Web site points to a static *news* page at this Web site, which provides links to online newspapers from around the world.

div is an abbreviation for division, which is the way that the Internet Public Library tags the contents on its Web site as relating to a specific topic.

ipl is an abbreviation for "Internet Public Library," which is an online service that provides reference resources and links to other information services on the Web. The IPL is supported chiefly by the School of Information at the University of Michigan. The TLD *org* is an abbreviation for "organization" (normally nonprofit).

b. Many court decisions that are not yet published or that are not intended for publication can be accessed through Westlaw®, an online legal database.
c. URLs are frequently changed as sites are redesigned and may not be working for other reasons, such as when a Web site has been deleted. If you are unable to find sites in this text with the specified URLs, go to the text's Web site at **http://fbl.westbuslaw.com**, where you may find an updated URL for the site or a URL for a similar site.
d. The basic form for a URL is "service://hostname/path." The Internet service for all of the URLs in this text is *http* (hypertext transfer protocol). Most Web browsers will add this prefix automatically when a user enters a host name or a hostname/path.

these cases may not conform to the descriptions given on page 22 because the reporters in which they were published have since been replaced.

Reading and Understanding Case Law

All of the court cases in this text present summaries of the courts' opinions. For those wishing to review court cases for future research projects or to gain additional legal information, the following sections will provide useful insights into how to read and understand case law.

Case Titles and Terminology

The title of a case, such as *Adams v. Jones,* indicates the names of the parties to the lawsuit. The *v.* in the case title stands for *versus,* which means "against." In the trial court, Adams was the plaintiff—the person who filed the suit. Jones was the defendant. If the case is appealed, however, the appellate court will sometimes place the name of the party appealing the decision first, so the case may be called *Jones v. Adams.* Because some reviewing courts retain the trial court order of names, it is often impossible to distinguish the plaintiff from the defendant in the title of a reported appellate court decision. You must carefully read the facts of each case to identify the parties.

The following terms and phrases are frequently encountered in court opinions and legal publications. Because it is important to understand what these terms and phrases mean, we define and discuss them here.

Plaintiffs and Defendants As mentioned in Chapter 1, the plaintiff in a lawsuit is the party that initiates the action. The defendant is the party against which a lawsuit is brought. Lawsuits frequently involve more than one plaintiff and/or defendant.

Appellants and Appellees The *appellant* is the party that appeals a case to another court or jurisdiction from the court or jurisdiction in which the case was originally brought. Sometimes an appellant is referred to as the *petitioner.* The *appellee* is the party against which the appeal is taken. The appellee may also be referred to as the *respondent.*

Judges and Justices The terms *judge* and *justice* are usually synonymous and represent two designations given to judges in various courts. All members of the United States Supreme Court, for example, are referred to as justices. And justice is the formal title usually given to judges of appellate courts, although this is not always the case. In New York, a justice is a judge of the trial court (which is called the Supreme Court), and a member of the Court of Appeals (the state's highest court) is called a judge. The term *justice* is commonly abbreviated to J., and *justices* to JJ. A Supreme Court case might refer to Justice Kennedy as Kennedy, J., or to Chief Justice Roberts as Roberts, C.J.

Decisions and Opinions Most decisions reached by reviewing, or appellate, courts are explained in written *opinions.* The opinion contains the court's reasons for its decision, the rules of law that apply, and the judgment. When all judges or justices unanimously agree on an opinion, the opinion is written for the entire court and can be deemed a *unanimous opinion.* When there is not unanimous agreement, a *majority opinion* is written, outlining the views of the majority of the judges or justices deciding the case.

Often, a judge or justice who feels strongly about making or emphasizing a point that was not made or emphasized in the unanimous or majority opinion will write a *concurring opinion.* That means the judge or justice agrees (concurs) with the judgment given in the unanimous or majority opinion but for different reasons. When there is not a unanimous opinion, a *dissenting opinion* is usually written by a judge or justice who does not agree with the majority. (See the *Extended Case Study* following Chapter 3 for an example of a dissenting opinion.) The dissenting opinion is important because it may form the basis of the arguments used years later in overruling the precedential majority opinion. Occasionally, a court issues a *per curiam* (Latin for "of the court") opinion, which does not indicate which judge or justice authored the opinion.

A Sample Court Case

Knowing how to read and analyze a court opinion is an essential step in undertaking accurate legal research. A further step involves "briefing" the case. Legal researchers routinely brief cases by summarizing and reducing the texts of the opinions to their essential elements. (For instructions on how to brief a case, go to Appendix C at the end of this text.) The cases contained within the chapters of this text have already been analyzed and partially briefed by the authors, and the essential aspects of each case are presented in a convenient format consisting of four basic sections: *Facts, Issue, Decision,* and *Reason.* Each case is followed by either a brief *For Critical Analysis* section, which presents a question regarding some issue raised by the case, or a *Why Is This Case Important?* section, which explains the significance of the case.

To illustrate the elements in a court opinion, we present a sample court case in Exhibit 1A–3 beginning on the next page. The case was decided by the U.S. Court of Appeals for the Sixth Circuit in 2005.

EXHIBIT 1A–3 A SAMPLE COURT CASE

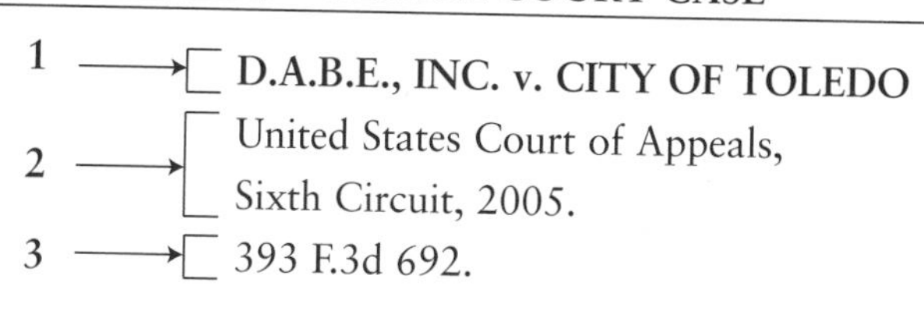

1 → **D.A.B.E., INC. v. CITY OF TOLEDO**

2 → United States Court of Appeals,
Sixth Circuit, 2005.

3 → 393 F.3d 692.

4 → **FACTS** The city of Toledo, Ohio, has regulated smoking in public places since 1987. In 2003, Toledo's city council enacted a new Clean Indoor Air Ordinance. The ordinance restricts the ability to smoke in public places—stores, theaters, courtrooms, libraries, museums, health-care facilities, restaurants, and bars. In enclosed public places, smoking is generally prohibited except in a "separate smoking lounge" that is designated for this purpose. D.A.B.E., Inc., a group composed of the owners of bars, restaurants, and bowling alleys, filed a suit in a federal district court, claiming that the ordinance constituted a taking of their property in violation of the Fifth Amendment to the U.S. Constitution. The plaintiffs also argued that the ordinance was preempted by a state statute that regulated smoking "in places of public assembly," excluding restaurants, bowling alleys, and bars. The court ruled in favor of the city. The plaintiffs appealed to the U.S. Court of Appeals for the Sixth Circuit.

5 → **ISSUE** Does the ordinance deny the plaintiffs "economically viable use of their property," as required to prove a taking? Does the state indoor smoking statute preempt the ordinance?

6 → **DECISION** No, to both questions. The U.S. Court of Appeals for the Sixth Circuit affirmed the lower court's ruling. The ordinance did not prevent the beneficial use of the plaintiffs' property, because it did not categorically prohibit smoking, but only regulated it. The state indoor smoking statute did not cover the excluded businesses, and the legislature did not indicate an intent to bar a city from restricting smoking in those places.

7 → **REASON** The Fifth Amendment provides that private property shall not "be taken for public use, without just compensation." A taking occurs when an ordinance denies an owner economically viable use of his or her property. In this case, the plaintiffs alleged that they lost customers because of the ordinance. The court reasoned that the ordinance's only effect on the plaintiffs' businesses is to restrict the areas in which customers can smoke and the conditions under which smoking is permitted. This might "require some financial investment, but an ordinance does not effect a taking merely because compliance with it requires the expenditure of money." Besides, the owners could elect to make other uses of their property. As for the preemption issue, a state statute takes precedence over a local ordinance when they conflict. In this case, a statute prohibits smoking in certain locations, but "it does not contain the slightest hint that the legislature intended to create a positive right to smoke in all public places where it did not expressly forbid smoking. Nothing in the [statute] is inconsistent with a local jurisdiction's decision to impose greater limits on public smoking."

8 → **FOR CRITICAL ANALYSIS—Economic Consideration** *How might the result have been different if the ordinance had prevented the plaintiffs' beneficial use of their property?*

EXHIBIT 1A–3 A SAMPLE COURT CASE—CONTINUED

REVIEW OF THE SAMPLE COURT CASE

1. The name of the case is *D.A.B.E., Inc. v. City of Toledo*. D.A.B.E. is the plaintiff, or petitioner; Toledo is the defendant, or respondent.
2. The court deciding this case is the U.S. Court of Appeals for the Sixth Circuit.
3. The citation is to volume 393 of the *Federal Reporter, Third Series*, page 692.
4. The *Facts* section identifies the parties to the suit, describes the events leading up to it, the allegations made by the plaintiff in the initial suit, and (because this case is an appellate court decision) the lower court's ruling and the party appealing the ruling. The appellant's contention on appeal is also sometimes included here.
5. The *Issue* section presents the central issue (or issues) to be decided by the court. In this case, the U.S. Court of Appeals for the Sixth Circuit considers whether the ordinance constitutes a taking in violation of the Fifth Amendment and whether a state statute preempts the ordinance.
6. The *Decision* section, as the term indicates, contains the court's decision on the issue or issues before the court. The decision reflects the opinion of the majority of the judges or justices hearing the case. Decisions by appellate courts are frequently phrased in reference to the lower court's decision—that is, the appellate court may "affirm" the lower court's ruling or "reverse" it. Here, the court determined that the ordinance did not effect a taking because it did not prevent the plaintiffs' beneficial use of their property. The statute did not preempt the ordinance because it did not indicate that the legislature intended to prohibit a city from restricting smoking in places excluded from the statute. The court affirmed the lower court's ruling.
7. The *Reason* section indicates the relevant laws and judicial principles that were applied in forming the particular conclusion arrived at in the case at bar ("before the court"). In this case, the relevant law included the Fifth Amendment, the state indoor smoking statute, and the principles derived from the courts' interpretations and applications of those laws. This section also explains the court's application of the law to the facts in this case.
8. The *For Critical Analysis—Economic Consideration* section raises a question to be considered in relation to the case just presented. Here the question involves an "economic" consideration. In other cases presented in this text, the "consideration" may involve a cultural, environmental, ethical, international, political, social, or technological consideration.

CHAPTER 2

Traditional and Online Dispute Resolution

LEARNING OBJECTIVES

After reading this chapter, you should be able to answer the following questions:

1. What is judicial review? How and when was the power of judicial review established?
2. Before a court can hear a case, it must have jurisdiction. Over what must it have jurisdiction? How are the courts applying traditional jurisdictional concepts to cases involving Internet transactions?
3. What is the difference between a trial court and an appellate court?
4. In a lawsuit, what are the pleadings? What is discovery, and how does electronic discovery differ from traditional discovery? What is electronic filing?
5. How are online forums being used to resolve disputes?

Ultimately, we are all affected by what the courts say and do. This is particularly true in the business world—nearly every businessperson will face either a potential or an actual lawsuit at some time or another. For this reason, anyone contemplating a career in business will benefit from an understanding of American court systems, including the mechanics of lawsuits.

In this chapter, after examining the judiciary's overall role in the American governmental scheme, we discuss some basic requirements that must be met before a party may bring a lawsuit before a particular court. We then look at the court systems of the United States in some detail and, to clarify judicial procedures, follow a hypothetical case through a state court system. Even though there are fifty-two court systems—one for each of the fifty states, one for the District of Columbia, plus a federal system—similarities abound. Keep in mind that the federal courts are not superior to the state courts; they are simply an independent system of courts, which derives its authority from Article III, Sections 1 and 2, of the U.S. Constitution. The chapter concludes with an overview of some alternative methods of settling disputes, including methods for settling disputes in online forums.

Note that technological developments are affecting court procedures just as they are having an impact on all other areas of the law. In this chapter, we will also indicate how court doctrines and procedures, as well as alternative methods of dispute settlement, are being adapted to the needs of a cyber age.

The Judiciary's Role in American Government

As you learned in Chapter 1, the entirety of American law includes the federal and state constitutions, statutes passed by legislative bodies, administrative law, and the case decisions and legal principles that form the common law. These laws would be meaningless, however, without the courts to interpret and apply them. This is the essential role of the judiciary—the courts—in the American governmental system: to interpret and apply the law.

As the branch of government entrusted with interpreting the laws, the judiciary can decide, among other things, whether the laws or actions of the other two branches are constitutional. The process for making such a determination is known as **judicial review.** The power of judicial review enables the judicial branch to act as a check on the other two branches of government, in line with the checks-and-balances system established by the U.S. Constitution.

The power of judicial review was not mentioned in the Constitution, but the concept was not new at the time the nation was founded. The doctrine of judicial review was not legally established, however, until 1803, when the United States Supreme Court rendered its decision in *Marbury v. Madison.*[1] In that case, the Supreme Court stated, "It is emphatically the province and duty of the Judicial Department to say what the law is. . . . If two laws conflict with each other, the courts must decide on the operation of each. . . . So if the law be in opposition to the Constitution . . . [t]he Court must determine which of these conflicting rules governs the case. This is the very essence of judicial duty." Since the *Marbury v. Madison* decision, the power of judicial review has remained unchallenged. Today, this power is exercised by both federal and state courts.

Basic Judicial Requirements

Before a court can hear a lawsuit, certain requirements must first be met. These requirements relate to jurisdiction, venue, and standing to sue. We examine each of these important concepts here.

JURISDICTION

In Latin, *juris* means "law," and *diction* means "to speak." Thus, "the power to speak the law" is the literal meaning of the term **jurisdiction.** Before any court can hear a case, it must have jurisdiction over the person against whom the suit is brought or over the property involved in the suit. The court must also have jurisdiction over the subject matter.

Jurisdiction over Persons Generally, a court can exercise personal jurisdiction (*in personam* jurisdiction) over any person or business that resides in a certain geographic area. A state trial court, for example, normally has jurisdictional authority over residents (including businesses) in a particular area of the state, such as a county or district. A state's highest court (often called the state supreme court, but there are exceptions—as will be discussed shortly) has jurisdiction over all residents of that state.

In addition, under the authority of a state **long arm statute,** a court can exercise personal jurisdiction over certain out-of-state defendants based on activities that took place within the state. Before exercising long arm jurisdiction over a nonresident, however, the court must be convinced that the defendant had sufficient contacts, or *minimum contacts,* with the state to justify the jurisdiction.[2] Generally, this means that the defendant must have enough of a connection to the state for the judge to conclude that it is fair for the state to exercise power over the defendant. ■ EXAMPLE 2.1 If an out-of-state defendant caused an automobile accident or sold defective goods within the state, a court will usually find that minimum contacts exist to exercise jurisdiction over that defendant. Similarly, a state may exercise personal jurisdiction over a nonresident defendant who is sued for breaching a contract that was formed within the state. ■

In regard to corporations,[3] the minimum-contacts requirement is usually met if the corporation does business within the state. ■ EXAMPLE 2.2 Suppose that a Maine corporation has a branch office or a manufacturing plant in Georgia. Does this Maine corporation have sufficient minimum contacts with the state of Georgia to allow a Georgia court to exercise jurisdiction over it? Yes, it does. If the Maine corporation advertises and sells its products in Georgia, those activities will also likely suffice to meet the minimum-contacts requirement, even if the corporate headquarters are located in a different state. ■

The following case involved a suit filed by a California resident against a Nevada hotel and others for failing to provide notice of an energy surcharge imposed on the guests. The hotel advertised heavily in California and obtained a significant percentage of its business from California residents, but it had no bank accounts or employees in California. Based on these activities, could a California state court exercise personal jurisdiction over the hotel?

1. 5 U.S. (1 Cranch) 137, 2 L.Ed. 60 (1803).

2. The minimum-contacts standard was established in *International Shoe Co. v. State of Washington,* 326 U.S. 310, 66 S.Ct. 154, 90 L.Ed. 95 (1945).

3. In the eyes of the law, corporations are "legal persons"—entities that can sue and be sued. See Chapter 25.

CASE 2.1 Snowney v. Harrah's Entertainment, Inc.

California Supreme Court, 2005.
35 Cal.4th 1054,
112 P.3d 28,
29 Cal.Rptr.3d 33.

FACTS Harrah's Entertainment, Inc., owns and operates hotels in Nevada. In 2001, Frank Snowney, a California resident, reserved a room in one of Harrah's hotels by phone from his home. The reservations clerk told him that the cost was $50 per night plus the room tax. When Snowney checked out, however, the bill included a $3 energy surcharge. Snowney filed a suit in a California state court against Harrah's and other hotel owners, alleging in part fraud for levying the surcharge without notice. The defendants filed a motion to dismiss for lack of personal jurisdiction, claiming that they did not do business in California. Snowney argued in part that the hotels advertised extensively to California residents through billboards, newspapers, and radio and television stations in the state, and through an interactive Web site that accepted reservations from state residents. The court granted the motion to dismiss, but on Snowney's appeal, a state intermediate appellate court reversed this ruling. The hotels appealed to the California Supreme Court.

ISSUE Could a California state court exercise personal jurisdiction over out-of-state hotels, based on the hotels' in-state advertising and significant business with California residents?

DECISION Yes. The California Supreme Court affirmed the judgment of the state intermediate appellate court, holding that the Nevada defendants were subject to personal jurisdiction in California state courts.

REASON The state supreme court identified three requirements for a court's exercise of jurisdiction over a nonresident defendant: the defendant must avail itself of "forum benefits," the case must relate to the defendant's contacts with the state, and the assertion of jurisdiction must be fair. The court reasoned that through a Web site, the hotels had "availed" themselves "of forum benefits. * * * By touting the proximity of their hotels to California and providing driving directions from California to their hotels, defendants' site specifically targeted residents of California. Defendants also concede that many of their patrons come from California and that some of these patrons undoubtedly made reservations using their Web site." The hotels advertised extensively in the state through direct mail, billboards, newspapers, and radio and television stations located in the state. "Given the intensity of defendants' activities in California, we * * * have little difficulty in finding a substantial connection between the two." Finally, "defendants do not contend the exercise of jurisdiction would be unfair or unreasonable, and we see no reason to conclude otherwise."

FOR CRITICAL ANALYSIS—Economic Consideration *If the Nevada hotel had not extensively advertised for California residents' business, would the outcome in this case have been the same?*

Jurisdiction over Property A court can also exercise jurisdiction over property that is located within its boundaries. This kind of jurisdiction is known as *in rem* jurisdiction, or "jurisdiction over the thing." ■ EXAMPLE 2.3 Suppose that a dispute arises over the ownership of a boat in dry dock in Fort Lauderdale, Florida. The boat is owned by an Ohio resident, over whom a Florida court normally cannot exercise personal jurisdiction. The other party to the dispute is a resident of Nebraska. In this situation, a lawsuit concerning the boat could be brought in a Florida state court on the basis of the court's *in rem* jurisdiction. ■

Jurisdiction over Subject Matter Jurisdiction over subject matter is a limitation on the types of cases a court can hear. In both the federal and state court systems, there are courts of *general* (unlimited) *jurisdiction* and courts of *limited jurisdiction.* An example of a court of general jurisdiction is a state trial court or a federal district court. An example of a state court of limited jurisdiction is a probate court. **Probate courts** are state courts that handle only matters relating to the transfer of a person's assets and obligations after that person's death, including matters relating to the custody and guardianship of children. An example of a federal court of limited subject-matter jurisdiction is a bankruptcy court. **Bankruptcy courts** handle only bankruptcy proceedings, which are governed by federal bankruptcy law (discussed in Chapter 21). In contrast, a court of general jurisdiction can decide a broad array of cases.

A court's jurisdiction over subject matter is usually defined in the statute or constitution creating the court. In both the federal and state court systems, a court's subject-matter jurisdiction can be limited not only by the subject of the lawsuit but also by the amount in controversy, by whether a case is a felony (a more serious type of crime)

or a misdemeanor (a less serious type of crime), or by whether the proceeding is a trial or an appeal.

Original and Appellate Jurisdiction The distinction between courts of original jurisdiction and courts of appellate jurisdiction normally lies in whether the case is being heard for the first time. Courts having original jurisdiction are courts of the first instance, or trial courts—that is, courts in which lawsuits begin, trials take place, and evidence is presented. In the federal court system, the *district courts* are trial courts. In the various state court systems, the trial courts are known by differing names, as will be discussed shortly.

The key point here is that, normally, any court having original jurisdiction is known as a trial court. Courts having appellate jurisdiction act as reviewing courts, or appellate courts. In general, cases can be brought before appellate courts only on appeal from an order or a judgment of a trial court or other lower court.

Jurisdiction of the Federal Courts Because the federal government is a government of limited powers, the jurisdiction of the federal courts is limited. Article III of the U.S. Constitution establishes the boundaries of federal judicial power. Section 2 of Article III states that "[t]he judicial Power shall extend to all Cases, in Law and Equity, arising under this Constitution, the Laws of the United States, and Treaties made, or which shall be made, under their Authority."

Whenever a plaintiff's cause of action is based, at least in part, on the U.S. Constitution, a treaty, or a federal law, then a **federal question** arises, and the case comes under the judicial power of the federal courts. Any lawsuit involving a federal question can originate in either a state court or a federal court. People who claim that their rights under the U.S. Constitution have been violated can begin their suits in a federal court.

Federal district courts can also exercise original jurisdiction over cases involving **diversity of citizenship.** Such cases may arise between (1) citizens of different states, (2) a foreign country and citizens of a state or of different states, or (3) citizens of a state and citizens or subjects of a foreign country. The amount in controversy must be more than $75,000 before a federal court can take jurisdiction in such cases. For purposes of diversity jurisdiction, a corporation is a citizen of both the state in which it is incorporated and the state in which its principal place of business is located. A case involving diversity of citizenship can be filed in the appropriate federal district court, or, if the case starts in a state court, it can sometimes be transferred to a federal court. A large percentage of the cases filed in federal courts each year are based on diversity of citizenship.

Note that in a case based on a federal question, a federal court will apply federal law. In a case based on diversity of citizenship, however, a federal court will apply the relevant state law (which is often the law of the state in which the court sits).

Exclusive versus Concurrent Jurisdiction When both federal and state courts have the power to hear a case, as is true in suits involving diversity of citizenship, **concurrent jurisdiction** exists. When cases can be tried only in federal courts or only in state courts, **exclusive jurisdiction** exists. Federal courts have exclusive jurisdiction in cases involving federal crimes, bankruptcy, patents, and copyrights; in suits against the United States; and in some areas of admiralty law (law governing transportation on the seas and ocean waters). States also have exclusive jurisdiction over certain subject matters—for example, divorce and adoption. The concepts of exclusive and concurrent jurisdiction are illustrated in Exhibit 2–1 on the next page.

When concurrent jurisdiction exists, a party has a choice of whether to bring a suit in, for example, a federal or a state court. The party's lawyer will consider several factors in counseling the party as to which choice is preferable. The lawyer may prefer to litigate the case in a state court because he or she is more familiar with the state court's procedures, or perhaps the attorney believes that the state court's judge or jury would be more sympathetic to the client and the case. Alternatively, the lawyer may advise the client to sue in federal court. Perhaps the state court's **docket** (the court's schedule listing the cases to be heard) is crowded, and the case could be brought to trial sooner in a federal court. Perhaps some feature of federal practice or procedure could offer an advantage in the client's case. Other important considerations include the law in the particular jurisdiction, how that law has been applied in the jurisdiction's courts, and what the results in similar cases have been in that jurisdiction.

JURISDICTION IN CYBERSPACE

The Internet's capacity to bypass political and geographic boundaries undercuts the traditional basic limitations on a court's authority to exercise jurisdiction. These limits include a party's contacts with a court's geographic jurisdiction. As already discussed, for a court to compel a defendant to come before it, there must be at least minimum contacts—the presence of a salesperson within the state, for example. Are there sufficient minimum contacts

EXHIBIT 2–1 EXCLUSIVE AND CONCURRENT JURISDICTION

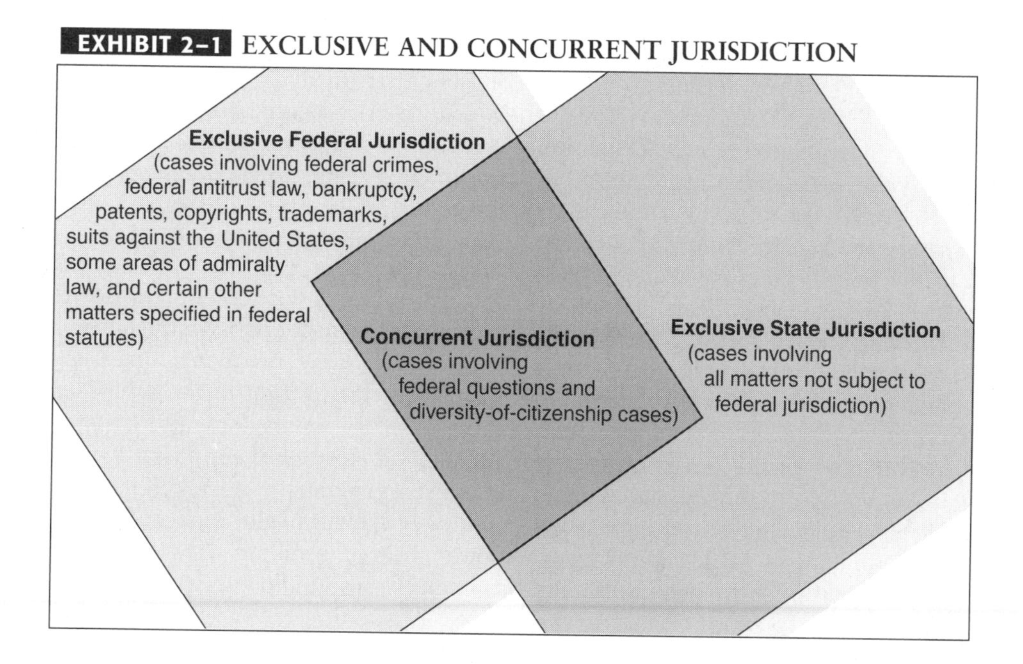
Exclusive Federal Jurisdiction
(cases involving federal crimes, federal antitrust law, bankruptcy, patents, copyrights, trademarks, suits against the United States, some areas of admiralty law, and certain other matters specified in federal statutes)
Concurrent Jurisdiction
(cases involving federal questions and diversity-of-citizenship cases)
Exclusive State Jurisdiction
(cases involving all matters not subject to federal jurisdiction)

if the only connection to a jurisdiction is an ad on the Web originating from a remote location?

The "Sliding-Scale" Standard Gradually, the courts are developing a standard—called a "sliding-scale" standard—for determining when the exercise of jurisdiction over an out-of-state defendant is proper. In developing this standard, the courts have identified three types of Internet business contacts: (1) substantial business conducted over the Internet (with contracts and sales, for example); (2) some interactivity through a Web site; and (3) passive advertising. Jurisdiction is proper for the first category, improper for the third, and may or may not be appropriate for the second.[4] An Internet communication is typically considered passive if people have to voluntarily access it to read the message and active if it is sent to specific individuals.

In certain situations, even a single contact can satisfy the minimum-contacts requirement. ■ EXAMPLE 2.4 A Texas resident, Davis, sent an unsolicited e-mail message to numerous Mississippi residents advertising a pornographic Web site. Davis falsified the "from" header in the e-mail so that it appeared that Internet Doorway had sent the mail. Internet Doorway filed a lawsuit against Davis in Mississippi claiming that its reputation and goodwill in the community had been harmed. The U.S. district court in Mississippi held that Davis's single e-mail to Mississippi residents satisfied the minimum-contacts requirement for jurisdiction. The court concluded that Davis, by sending the e-mail solicitation, should reasonably have expected that she could be "haled into court in a distant jurisdiction to answer for the ramifications."[5] ■

International Jurisdictional Issues As noted in Chapter 1, because the Internet is international in scope, international jurisdictional issues understandably have come to the fore. What seems to be emerging in the world's courts is a standard that echoes the "minimum-contacts" requirement applied by the U.S. courts. Most courts are indicating that minimum contacts—doing business within the jurisdiction, for example—are enough to compel a defendant to appear and that a physical presence is not necessary. The effect of this standard is that a business firm must comply with the laws in any jurisdiction in which it targets customers for its products.

VENUE

Jurisdiction has to do with whether a court has authority to hear a case involving specific persons, property, or subject matter. **Venue**[6] is concerned with the most appropriate location for a trial. Two state courts (or two federal courts) may have the authority to exercise jurisdiction over a case, but it may be more appropriate or convenient to hear the case in one court than in the other.

4. For a leading case on this issue, see *Zippo Manufacturing Co. v. Zippo Dot Com, Inc.,* 952 F.Supp. 1119 (W.D.Pa. 1997).

5. *Internet Doorway, Inc. v. Parks,* 138 F.Supp.2d 773 (S.D.Miss. 2001).

6. Pronounced *ven*-yoo.

Basically, the concept of venue reflects the policy that a court trying a suit should be in the geographic neighborhood (usually the county) where the incident leading to the lawsuit occurred or where the parties involved in the lawsuit reside. Venue in a civil case typically is where the defendant resides, whereas venue in a criminal case is normally where the crime occurred. Pretrial publicity or other factors, though, may require a change of venue to another community, especially in criminal cases when the defendant's right to a fair and impartial jury has been impaired. ■ EXAMPLE 2.5 A change of venue from Oklahoma City to Denver, Colorado, was ordered for the trials of Timothy McVeigh and Terry Nichols, who had been indicted in connection with the 1995 bombing of the Alfred P. Murrah Federal Building in Oklahoma City. (At trial, both McVeigh and Nichols were convicted. McVeigh received the death penalty and was put to death by lethal injection in early 2001. Nichols was sentenced to life imprisonment.) ■

Standing to Sue

Before a person can bring a lawsuit before a court, the party must have **standing to sue,** or a sufficient "stake" in a matter to justify seeking relief through the court system. In other words, to have standing, a party must have a legally protected and tangible interest at stake in the litigation. The party bringing the lawsuit must have suffered a harm, or have been threatened by a harm, as a result of the action about which she or he has complained. At times, a person will have standing to sue on behalf of another person. ■ EXAMPLE 2.6 Suppose that a child suffered serious injuries as a result of a defectively manufactured toy. Because the child is a minor, a lawsuit could be brought on his or her behalf by another person, such as the child's parent or legal guardian. ■

Standing to sue also requires that the controversy at issue be a **justiciable**[7] **controversy**—a controversy that is real and substantial, as opposed to hypothetical or academic. ■ EXAMPLE 2.7 In the above example, the child's parent could not sue the toy manufacturer merely on the ground that the toy was defective. The issue would become justiciable only if the child had actually been injured due to the defect in the toy as marketed. In other words, the parent normally could not ask the court to determine, for example, what damages might be obtained if the child had been injured, because this would be merely a hypothetical question. ■

7. Pronounced jus-*tish*-uh-bul.

The State and Federal Court Systems

As mentioned earlier in this chapter, each state has its own court system. Additionally, there is a system of federal courts. Although state court systems differ, Exhibit 2–2 on the following page illustrates the basic organizational structure characteristic of the court systems in many states. The exhibit also shows how the federal court system is organized. We turn now to an examination of these court systems, beginning with the state courts.

State Court Systems

Typically, a state court system will include several levels, or tiers, of courts. As indicated in Exhibit 2–2, state courts may include (1) trial courts of limited jurisdiction, (2) trial courts of general jurisdiction, (3) appellate courts, and (4) the state's highest court (often called the state supreme court). Generally, any person who is a party to a lawsuit has the opportunity to plead the case before a trial court and then, if he or she loses, before at least one level of appellate court. Finally, if a federal statute or federal constitutional issue is involved in the decision of the state supreme court, that decision may be further appealed to the United States Supreme Court.

Judges in the state court system are usually elected by the voters for a specified term. State judicial elections or appointments vary significantly, however, from state to state. For example, in Iowa the governor appoints judges, and then the general population decides whether to confirm their appointments in the next general election. The states usually specify the number of years that a judge will serve. In contrast, as you will read shortly, judges in the federal court system are appointed by the president of the United States and, if they are confirmed by the Senate, hold office for life—unless they engage in blatantly illegal conduct.

Trial Courts Trial courts are exactly what their name implies—courts in which trials are held and testimony taken. State trial courts have either general or limited jurisdiction. Trial courts that have general jurisdiction as to subject matter may be called county, district, superior, or circuit courts.[8] The jurisdiction of these courts is often determined by the size of the county in which the court sits. State trial courts of general jurisdiction have jurisdiction over a wide variety of subjects, including both civil disputes and criminal prosecutions.

Some courts of limited jurisdiction are referred to as *special inferior trial courts* or *minor judiciary courts.*

8. The name in Ohio is court of common pleas; the name in New York is supreme court.

EXHIBIT 2–2 THE STATE AND FEDERAL COURT SYSTEMS

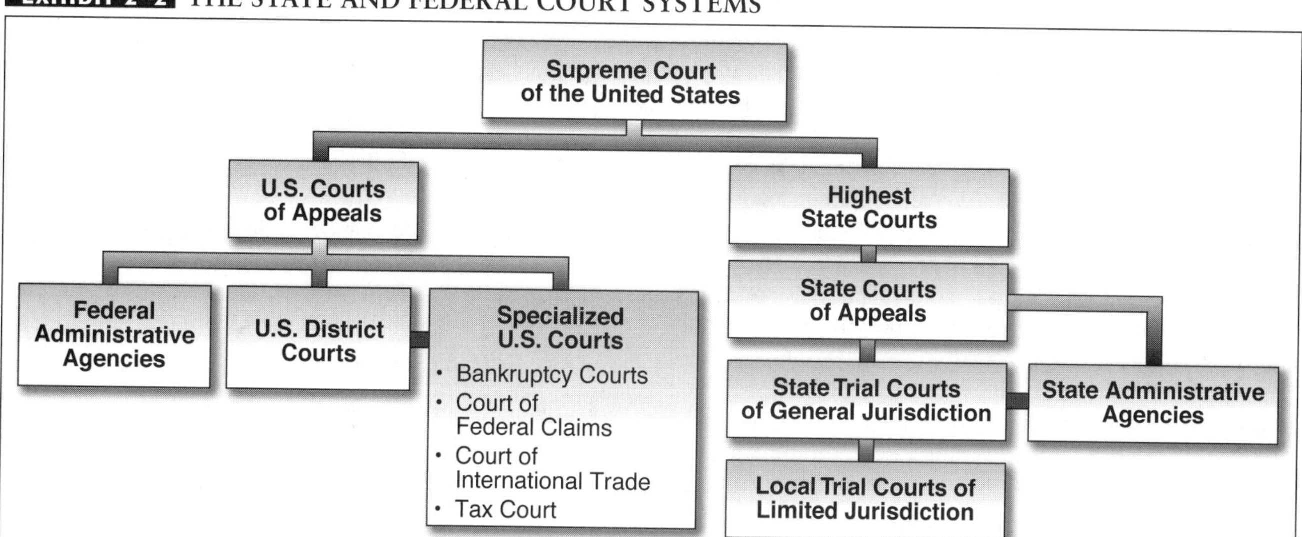

Small claims courts are inferior trial courts that hear only civil cases involving claims of less than a certain amount, such as $5,000 (the amount varies from state to state). Suits brought in small claims courts are generally conducted informally, and lawyers are not required. In a minority of states, lawyers are not even allowed to represent people in small claims courts for most purposes. Another example of an inferior trial court is a local municipal court that hears mainly traffic cases. Decisions of small claims courts and municipal courts may sometimes be appealed to a state trial court of general jurisdiction.

Other courts of limited jurisdiction as to subject matter include domestic relations courts, which handle primarily divorce actions and child-custody disputes, and probate courts, as mentioned earlier.

Appellate, or Reviewing, Courts Every state has at least one court of appeals (appellate court, or reviewing court), which may be an intermediate appellate court or the state's highest court. About three-fourths of the states have intermediate appellate courts. Generally, courts of appeals do not conduct new trials, in which evidence is submitted to the court and witnesses are examined. Rather, an appellate court panel of three or more judges reviews the record of the case on appeal, which includes a transcript of the trial proceedings, and determines whether the trial court committed an error.

Usually, appellate courts do not look at questions of *fact* (such as whether a party did, in fact, commit a certain action, such as burning a flag) but at questions of *law* (such as whether the act of flag-burning is a form of speech protected by the First Amendment to the Constitution). Only a judge, not a jury, can rule on questions of law. Appellate courts normally defer to a trial court's findings on questions of fact because the trial court judge and jury were in a better position to evaluate testimony—by directly observing witnesses' gestures, demeanor, and nonverbal behavior during the trial. At the appellate level, the judges review the written transcript of the trial, which does not include these nonverbal elements.

An appellate court will challenge a trial court's finding of fact only when the finding is clearly erroneous (that is, when it is contrary to the evidence presented at trial) or when there is no evidence to support the finding. ■ EXAMPLE 2.8 Suppose that a jury concluded that a manufacturer's product harmed the plaintiff but no evidence was submitted to the court to support that conclusion. In that situation, the appellate court would hold that the trial court's decision was erroneous. (The options exercised by appellate courts will be further discussed later in this chapter.) ■

Highest State Courts The highest appellate court in a state is usually called the supreme court but may be called by some other name. For example, in both New York and Maryland, the highest state court is called the court of appeals. The decisions of each state's highest court are final on all questions of state law. Only when issues of federal law are involved can a decision made by a state's highest court be overruled by the United States Supreme Court.

THE FEDERAL COURT SYSTEM

The federal court system is basically a three-tiered model consisting of (1) U.S. district courts (trial courts of general jurisdiction) and various courts of limited jurisdiction, (2) U.S. courts of appeals (intermediate courts of appeals), and (3) the United States Supreme Court.

Unlike state court judges, who are usually elected, federal court judges—including the justices of the Supreme Court—are appointed by the president of the United States and confirmed by the U.S. Senate. All federal judges receive lifetime appointments (because under Article III they "hold their offices during Good Behavior").

U.S. District Courts At the federal level, the equivalent of a state trial court of general jurisdiction is the district court. There is at least one federal district court in every state. The number of judicial districts can vary over time, primarily owing to population changes and corresponding caseloads. Currently, there are ninety-four federal judicial districts.

U.S. district courts have original jurisdiction in federal matters. Federal cases typically originate in district courts. There are other courts with original, but special (or limited), jurisdiction, such as the federal bankruptcy courts and others shown in Exhibit 2–2.

U.S. Courts of Appeals In the federal court system, there are thirteen U.S. courts of appeals—also referred to as U.S. circuit courts of appeals. The federal courts of appeals for twelve of the circuits, including the U.S. Court of Appeals for the District of Columbia Circuit, hear appeals from the federal district courts located within their respective judicial circuits. The Court of Appeals for the Thirteenth Circuit, called the Federal Circuit, has national appellate jurisdiction over certain types of cases, such as cases involving patent law and cases in which the U.S. government is a defendant.

The decisions of the circuit courts of appeals are final in most cases, but appeal to the United States Supreme Court is possible. Exhibit 2–3 shows the geographic boundaries of the U.S. circuit courts of appeals and the boundaries of the U.S. district courts within each circuit.

EXHIBIT 2–3 BOUNDARIES OF THE U.S. COURTS OF APPEALS AND U.S. DISTRICT COURTS

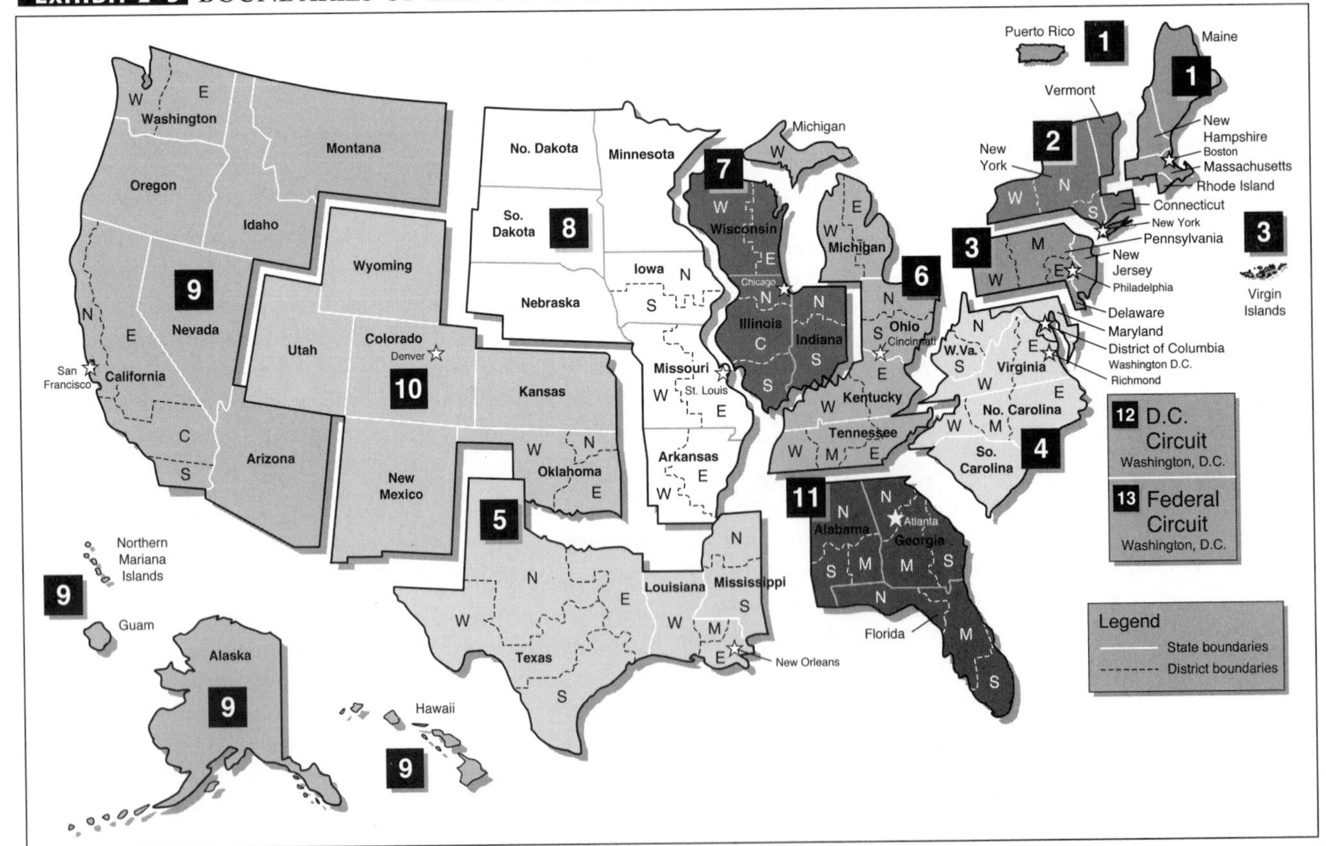

Source: Administrative Office of the United States Courts.

The United States Supreme Court The highest level of the three-tiered model of the federal court system is the United States Supreme Court. According to the language of Article III of the U.S. Constitution, there is only one national Supreme Court. All other courts in the federal system are considered "inferior." Congress is empowered to create other inferior courts as it deems necessary. The inferior courts that Congress has created include the second tier in Exhibit 2–2 on page 36—the U.S. courts of appeals—as well as the district courts and any other courts of limited, or specialized, jurisdiction.

The United States Supreme Court consists of nine justices. Although the Supreme Court has original, or trial, jurisdiction in rare instances (set forth in Article III, Section 2), most of its work is as an appeals court. The Supreme Court can review any case decided by any of the federal courts of appeals, and it also has appellate authority over some cases decided in the state courts.

—Appeals to the Supreme Court. To bring a case before the Supreme Court, a party requests the Court to issue a writ of *certiorari*. A **writ of *certiorari***[9] is an order issued by the Supreme Court to a lower court requiring the latter to send it the record of the case for review. The Court will not issue a writ unless at least four of the nine justices approve of it. This is called the **rule of four.** Whether the Court will issue a writ of *certiorari* is entirely within its discretion. The Court is not required to issue one, and most petitions for writs are denied. (Thousands of cases are filed with the Supreme Court each year; yet it hears, on average, fewer than one hundred of these cases.[10]) A denial is not a decision on the merits of a case, nor does it indicate agreement with the lower court's opinion. Furthermore, a denial of the writ has no value as a precedent.

—Petitions Granted by the Court. Typically, the Court grants petitions when cases raise important constitutional questions or when the lower courts are issuing conflicting decisions on a signficant issue. Similarly, if federal appellate courts are rendering inconsistent opinions on an important issue, the Supreme Court may review the case and hand down a decision to define the law on the matter. The justices, however, never explain their reasons for hearing certain cases and not others, so it is difficult to predict which type of case the Court might select.

9. Pronounced sur-shee-uh-*rah*-ree.

10. From the mid-1950s through the early 1990s, the Supreme Court reviewed more cases per year than it has in the last few years. In the Court's 1982–1983 term, for example, the Court issued opinions in 151 cases. In contrast, in its 2005–2006 term, the Court issued opinions in only 87 cases.

Following a State Court Case

To illustrate the procedures that would be followed in a civil lawsuit brought in a state court, we present a hypothetical case and follow it through the state court system. The case involves an automobile accident in which Kevin Anderson, driving a Mercedes, struck Lisa Marconi, driving a Ford Taurus. The accident occurred at the intersection of Wilshire Boulevard and Rodeo Drive in Beverly Hills, California. Marconi suffered personal injuries, incurring medical and hospital expenses as well as lost wages for four months. Anderson and Marconi are unable to agree on a settlement, and Marconi sues Anderson. Marconi is the plaintiff, and Anderson is the defendant. Both are represented by lawyers.

During each phase of the **litigation** (the process of working a lawsuit through the court system), Marconi and Anderson will be required to observe strict procedural requirements. A large body of law—procedural law—establishes the rules and standards for determining disputes in courts. Procedural rules are very complex, and they vary from court to court and from state to state. There is a set of federal rules of procedure as well as various sets of rules for state courts. Additionally, the applicable procedures will depend on whether the case is a civil or criminal proceeding. Generally, the Marconi-Anderson civil lawsuit will involve the procedures discussed in the following subsections. Keep in mind that attempts to settle the case may be ongoing throughout the trial.

The Pleadings

The *complaint* and *answer* (and the *counterclaim* and *reply*)—all of which are discussed below—taken together are called the **pleadings.** The pleadings inform each party of the other's claims and specify the issues (disputed questions) involved in the case. Because the rules of procedure vary depending on the jurisdiction of the court, the style and form of the pleadings may look quite different in different states.

The Plaintiff's Complaint Marconi's suit against Anderson commences when her lawyer files a **complaint** with the appropriate court. The complaint contains a statement alleging (asserting to the court, in a pleading) the facts necessary for the court to take jurisdiction, a brief summary of the facts necessary to show that the plaintiff is entitled to a remedy, and a statement of the remedy the plaintiff is seeking. Exhibit 2–4 illustrates how the complaint might read in the Marconi-Anderson case. Complaints may be lengthy or brief, depending on the complexity of the case and the rules of the jurisdiction.

EXHIBIT 2–4 EXAMPLE OF A TYPICAL COMPLAINT

IN THE LOS ANGELES SUPERIOR COURT
COUNTY OF LOS ANGELES, STATE OF CALIFORNIA

Lisa Marconi Plaintiff, v. Kevin Anderson Defendant.	CIVIL NO. 8–1026 COMPLAINT

Comes now the plaintiff and for her cause of action against the defendant alleges and states as follows:

1. The jurisdiction of this court is based on Section 86 of the California Civil Code.
2. This action is between plaintiff, a California resident living at 1434 Palm Drive, Anaheim, California, and defendant, a California resident living at 6950 Garrison Avenue, Los Angeles, California.
3. On September 10, 2007, plaintiff, Lisa Marconi, was exercising good driving habits and reasonable care in driving her car through the intersection of Rodeo Drive and Wilshire Boulevard when defendant, Kevin Anderson, negligently drove his vehicle through a red light at the intersection and collided with plaintiff's vehicle. Defendant was negligent in the operation of the vehicle as to:
   a. Speed,
   b. Lookout,
   c. Management and control.
4. As a result of the collision, plaintiff suffered severe physical injury that prevented her from working and property damage to her car. The costs she incurred included $10,000 in medical bills, $9,000 in lost wages, and $5,000 for automobile repairs.

WHEREFORE, plaintiff demands judgment against the defendant for the sum of $24,000 plus interest at the maximum legal rate and the costs of this action.

By *Roger Harrington*

Roger Harrington
Attorney for the Plaintiff
800 Orange Avenue
Anaheim, CA 91426

After the complaint has been filed, the sheriff, a deputy of the county, or another *process server* (one who delivers a complaint and summons) serves a **summons** and a copy of the complaint on defendant Anderson. The summons notifies Anderson that he must file an answer to the complaint with both the court and the plaintiff's attorney within a specified time period (usually twenty to thirty days). The summons also informs Anderson that failure to answer may result in a **default judgment** for the plaintiff, meaning the plaintiff will be awarded the damages alleged in her complaint.

The Defendant's Answer The defendant's **answer** either admits the statements or allegations set forth in the complaint or denies them and outlines any defenses that the defendant may have. If Anderson admits to all of

Marconi's allegations in his answer, the court will enter a judgment for Marconi. If Anderson denies any of Marconi's allegations, the litigation will go forward.

Anderson can deny Marconi's allegations and set forth his own claim that Marconi was in fact negligent and therefore owes him for the damage to his Mercedes. This is appropriately called a **counterclaim.** If Anderson files a counterclaim, Marconi will have to answer it with a pleading, normally called a **reply,** which has the same characteristics as an answer.

Anderson can also admit the truth of Marconi's complaint but raise new facts that may result in dismissal of the action. This is called *raising an affirmative defense.* For example, Anderson could assert the expiration of the time period under the relevant statute of limitations (a state or federal statute that sets the maximum time period during which a certain action can be brought or rights enforced) as an affirmative defense.

Motion to Dismiss A **motion to dismiss** requests the court to dismiss the case for stated reasons. A defendant often makes a motion to dismiss before filing an answer to the plaintiff's complaint. Grounds for dismissal of a case include improper delivery of the complaint and summons, improper venue, and the plaintiff's failure to state a claim for which a court could grant relief (a remedy). For example, if Marconi had suffered no injuries or losses as a result of Anderson's negligence, Anderson could move to have the case dismissed because Marconi had not stated a claim for which relief could be granted.

If the judge grants the motion to dismiss, the plaintiff generally is given time to file an amended complaint. If the judge denies the motion, the suit will go forward, and the defendant must then file an answer. Note that if Marconi wishes to discontinue the suit because, for example, an out-of-court settlement has been reached, she can likewise move for dismissal. The court can also dismiss the case on its own motion.

Pretrial Motions

Either party may attempt to get the case dismissed before trial through the use of various pretrial motions. We have already mentioned the motion to dismiss. Two other important pretrial motions are the *motion for judgment on the pleadings* and the *motion for summary judgment.*

At the close of the pleadings, either party may make a **motion for judgment on the pleadings,** or on the merits of the case. The judge will grant the motion only when there is no dispute over the facts of the case and the sole issue to be resolved is a question of law. In deciding on the motion, the judge may consider only the evidence contained in the pleadings.

In contrast, in a **motion for summary judgment** the court may consider evidence outside the pleadings, such as sworn statements (affidavits) by parties or witnesses or other documents relating to the case. A motion for summary judgment can be made by either party. As with the motion for judgment on the pleadings, a motion for summary judgment will be granted only if there are no genuine questions of fact and the sole question is a question of law.

Discovery

Before a trial begins, each party can use a number of procedural devices to obtain information and gather evidence about the case from the other party or from third parties. The process of obtaining such information is known as **discovery.** Discovery includes gaining access to witnesses, documents, records, and other types of evidence.

The Federal Rules of Civil Procedure and similar rules in the states set forth the guidelines for discovery activity. The rules governing discovery are designed to make sure that a witness or a party is not unduly harassed, that privileged material (communications that need not be presented in court) is safeguarded, and that only matters relevant to the case at hand are discoverable.

Discovery prevents surprises at trial by giving the parties access to evidence that might otherwise be hidden. This allows both parties to learn as much as they can about what to expect at a trial before they reach the courtroom. It also serves to narrow the issues so that trial time is spent on the main questions in the case.

Depositions and Interrogatories Discovery can involve the use of depositions or interrogatories, or both. A **deposition** is sworn testimony by a party to the lawsuit or any witness. The person being deposed (the deponent) answers questions asked by the attorneys, and the questions and answers are recorded by an authorized court official and sworn to and signed by the deponent. (Occasionally, written depositions are taken when witnesses are unable to appear in person.) The answers given to depositions will, of course, help the attorneys prepare their cases. They can also be used in court to impeach (challenge the credibility of) a party or a witness who changes testimony at the trial. In addition, the answers given in a deposition can be used as testimony if the witness is not available at trial.

Interrogatories are written questions for which written answers are prepared and then signed under oath. The main difference between interrogatories and written de-

positions is that interrogatories are directed to a party to the lawsuit (the plaintiff or the defendant), not to a witness, and the party can prepare answers with the aid of an attorney. The scope of interrogatories is broader because parties are obligated to answer questions, even if that means disclosing information from their records and files.

Other Information A party can serve a written request on the other party for an admission of the truth of matters relating to the trial. Any matter admitted under such a request is conclusively established for the trial. For example, Marconi can ask Anderson to admit that he was driving at a speed of forty-five miles an hour. A request for admission saves time at trial because the parties will not have to spend time proving facts on which they already agree.

A party can also gain access to documents and other items not in her or his possession in order to inspect and examine them. Likewise, a party can gain "entry upon land" to inspect the premises. Anderson's attorney, for example, normally can gain permission to inspect and duplicate Marconi's car repair bills.

When the physical or mental condition of one party is in question, the opposing party can ask the court to order a physical or mental examination. If the court is willing to make the order, which it will do only if the need for the information outweighs the right to privacy of the person to be examined, the opposing party can obtain the results of the examination.

Electronic Discovery Any relevant material, including information stored electronically, can be the object of a discovery request. Electronic evidence, or **e-evidence,** consists of all types of computer-generated or electronically recorded information, including e-mail, voice mail, spreadsheets, word-processing documents, and other data. E-evidence is important because it can reveal significant facts that are not discoverable by other means. For example, whenever a person is working on a computer, information is being recorded on the hard drive disk without ever being saved by the user. This information includes the file's location, path, creator, date created, date last accessed, concealed notes, earlier versions, passwords, and formatting. It reveals information about how, when, and by whom a document was created, accessed, modified, and transmitted. This information can be obtained from the file only in its electronic format—not from printed-out versions.

The federal rules and most state rules (as well as court decisions) now specifically allow parties to obtain discovery of electronic "data compilations" (or e-evidence). Although traditional means, such as interrogatories and depositions, may still be used to find out about the e-evidence, the parties must usually hire an expert to retrieve the evidence in its electronic format. Using special software, the expert can reconstruct e-mail exchanges to establish who knew what and when they knew it. The expert can even recover files from a computer that the user thought had been deleted. Reviewing back-up copies of documents and e-mail can provide useful—and often quite damaging—information about how a particular matter progressed over several weeks or months.

Although electronic discovery has significant advantages over paper discovery, it is also time consuming and expensive. Who should pay the costs associated with electronic discovery? This chapter's *Adapting the Law to the Online Environment* feature on the next page discusses how the law is evolving to address this issue.

PRETRIAL CONFERENCE

Either party or the court can request a pretrial conference, or hearing. Usually, the hearing consists of an informal discussion between the judge and the opposing attorneys after discovery has taken place. The purpose of the hearing is to explore the possibility of a settlement without trial and, if this is not possible, to identify the matters that are in dispute and to plan the course of the trial.

JURY SELECTION

A trial can be held with or without a jury. The Seventh Amendment to the U.S. Constitution guarantees the right to a jury trial for cases in federal courts when the amount in controversy exceeds $20. Most states have similar guarantees in their own constitutions (although the threshold dollar amount is higher than $20). The right to a trial by jury does not have to be exercised, and many cases are tried without a jury. In most states and in federal courts, one of the parties must request a jury, or the right is presumed to be waived.

Before a jury trial commences, a jury must be selected. The jury selection process is known as ***voir dire.***[11] During *voir dire* in most jurisdictions, attorneys for the plaintiff and the defendant ask prospective jurors oral questions to determine whether a potential jury member is biased or has any connection with a party to the action or with a prospective witness. In some jurisdictions, the judge may do all or part of the questioning based on written questions submitted by counsel for the parties.

During *voir dire,* a party may challenge a certain number of prospective jurors *peremptorily*—that is, ask that

11. Pronounced vwahr *deehr.*

ADAPTING THE LAW TO THE ONLINE ENVIRONMENT

Who Bears the Costs of Electronic Discovery?

Traditionally, the party responding to a discovery request must pay the expenses involved in obtaining the requested materials. If compliance would be too burdensome or too costly, however, the judge can either limit the scope of the request or shift some or all of the costs to the requesting party. How do these traditional rules governing discovery apply to requests for electronic evidence?

WHY COURTS MIGHT SHIFT THE COSTS OF ELECTRONIC DISCOVERY

Electronic discovery has dramatically increased the costs associated with complying with discovery requests. It is no longer simply a matter of photocopying paper documents. Now the responding party may need to hire computer forensics experts to make "image" copies of desktop, laptop, and server hard drives, as well as removable storage media (including CD-ROMs, DVDs, and Zip drives), back-up tapes, voice mail, cell phones, and any other device that digitally stores data.

In cases involving multiple parties or large corporations with many offices and employees, the electronic discovery process can easily run into hundreds of thousands—if not millions—of dollars. For example, in one case concert promoters alleged that thirty separate defendant companies had engaged in discriminatory practices. The federal district court hearing the case found that the complete restoration of the back-up tapes of just one of those defendants would cost $9.75 million. Acquiring 200,000 e-mail messages from another defendant would cost between $43,000 and $84,000, with an additional $247,000 to have an attorney review the retrieved documents. Restoring the 523 back-up tapes of a third defendant would cost $395,000, plus another $120,000 for the attorney to review them. The judge hearing the case decided that both plaintiffs and defendants would share in these discovery costs.[a]

WHAT FACTORS DO COURTS CONSIDER IN DECIDING WHETHER TO SHIFT COSTS?

Increasingly, the courts are shifting part of the costs of obtaining electronic discovery to the party requesting it (which is usually the plaintiff). At what point, however, should this cost-shifting occur? In *Zubulake v. UBS Warburg LLC*,[b] the court identified a three-step analysis for deciding disputes over discovery costs. First, if the data are kept in an accessible format, the usual rules of discovery apply: the responding party should pay the costs of producing responsive data. A court should consider cost-shifting *only* when electronic data are in a relatively inaccessible form, such as in back-up tapes or deleted files. Second, the court should determine what data may be found on the inaccessible media. Requiring the responding party to restore and produce responsive documents from a small sample of the requested medium is a sensible approach in most cases. Third, the court should consider a series of other factors, including, for example, the availability of the information from other sources, the total cost of production compared to the amount in controversy, and each party's ability to pay these costs.

FOR CRITICAL ANALYSIS

The court in the Zubulake *case noted that "as large companies increasingly move to entirely paper-free environments, the frequent use of cost-shifting will have the effect of crippling discovery," especially in cases in which private parties are suing large corporations. Why might cost-shifting thwart discovery? Who would benefit if courts considered cost-shifting in every case involving electronic discovery?*

a. *Rowe Entertainment, Inc. v. William Morris Agency*, 2002 WL 975713 (S.D.N.Y. 2002).
b. 2003 WL 21087884 (S.D.N.Y. 2003).

an individual not be sworn in as a juror without providing any reason. Alternatively, a party may challenge a prospective juror *for cause*—that is, provide a reason why an individual should not be sworn in as a juror. If the judge grants the challenge, the individual is asked to step down. A prospective juror may not be excluded from the jury by the use of discriminatory challenges, however, such as those based on racial criteria[12] or gender.[13]

12. *Batson v. Kentucky*, 476 U.S. 79, 106 S.Ct. 1712, 90 L.Ed.2d 69 (1986).
13. *J.E.B. v. Alabama ex rel. T.B.*, 511 U.S. 127, 114 S.Ct. 1419, 128 L.Ed.2d 89 (1994). (*Ex rel.* is Latin for *ex relatione.* The phrase refers to an action brought on behalf of the state, by the attorney general, at the instigation of an individual who has a private interest in the matter.)

AT THE TRIAL

At the beginning of the trial, the attorneys present their opening arguments, setting forth the facts that they expect to provide during the trial. Then the plaintiff's case is presented. In our hypothetical case, Marconi's lawyer would introduce evidence (relevant documents, exhibits, and the testimony of witnesses) to support Marconi's position. The defendant has the opportunity to challenge any evidence introduced and to cross-examine any of the plaintiff's witnesses.

At the end of the plaintiff's case, the defendant's attorney has the opportunity to ask the judge to direct a ver-

dict for the defendant on the ground that the plaintiff has presented no evidence that would justify the granting of the plaintiff's remedy. This is called a **motion for a directed verdict** (known in federal courts as a *motion for judgment as a matter of law*). If the motion is not granted (it seldom is), the defendant's attorney then presents the evidence and witnesses for the defendant's case. At the conclusion of the defendant's case, the defendant's attorney has another opportunity to make a motion for a directed verdict. The plaintiff's attorney can challenge any evidence introduced and cross-examine the defendant's witnesses.

After the defense concludes its presentation, the attorneys present their closing arguments, each urging a verdict in favor of her or his client. The judge instructs the jury in the law that applies to the case (these instructions are often called *charges*), and the jury retires to the jury room to deliberate a verdict. In the Marconi-Anderson case, the jury will not only decide for the plaintiff or for the defendant but, if it finds for the plaintiff, will also decide on the amount of the **award** (the money to be paid to her).

Posttrial Motions

After the jury has rendered its verdict, either party may make a posttrial motion. If Marconi wins, and Anderson's attorney has previously moved for a directed verdict, Anderson's attorney may make a **motion for judgment *n.o.v.*** (from the Latin *non obstante veredicto,* which means "notwithstanding the verdict"—called a *motion for judgment as a matter of law* in the federal courts) in Anderson's favor on the ground that the jury's verdict in favor of Marconi was unreasonable and erroneous. If the judge decides that the jury's verdict was reasonable in light of the evidence presented at trial, the motion will be denied. If the judge agrees with Anderson's attorney, then he or she will set the jury's verdict aside and enter a judgment in favor of Anderson.

Alternatively, Anderson could make a **motion for a new trial,** requesting the judge to set aside the adverse verdict and to hold a new trial. The motion will be granted if the judge is convinced, after looking at all the evidence, that the jury was in error but does not feel it is appropriate to grant judgment for the other side. A new trial may also be granted on the ground of newly discovered evidence, misconduct by the participants or the jury during the trial, or error by the judge.

The Appeal

Assume here that any posttrial motion is denied and that Anderson appeals the case. (If Marconi wins but receives a smaller money award than she sought, she can appeal also.) A notice of appeal must be filed with the clerk of the trial court within a prescribed time. Anderson now becomes the appellant, or petitioner, and Marconi becomes the appellee, or respondent.

Filing the Appeal Anderson's attorney files with the appellate court the record on appeal, which includes the pleadings, the trial transcript, the judge's rulings on motions made by the parties, and other trial-related documents. Anderson's attorney will also provide a condensation of the record, known as an *abstract,* which is filed with the reviewing court along with the brief. The **brief** is a formal legal document outlining the facts and issues of the case, the judge's rulings or jury's findings that should be reversed or modified, the applicable law, and arguments on Anderson's behalf (citing applicable statutes and relevant cases as precedents).

Marconi's attorney will file an answering brief. Anderson's attorney can file a reply to Marconi's brief, although it is not required. The reviewing court then considers the case.

Appellate Review As mentioned earlier, a court of appeals does not hear evidence. Rather, it reviews the record for errors of law. Its decision concerning a case is based on the record on appeal, the abstracts, and the attorneys' briefs. The attorneys can present oral arguments, after which the case is taken under advisement. In general, as mentioned earlier, appellate courts do not reverse findings of fact unless the findings are unsupported or contradicted by the evidence.

If the reviewing court believes that an error was committed during the trial or that the jury was improperly instructed, the judgment will be *reversed.* Sometimes the case will be *remanded* (sent back to the court that originally heard the case) for a new trial. Even when a case is remanded to a trial court for further proceedings, however, the appellate court normally spells out how the relevant law should be interpreted and applied to the case.

■ **EXAMPLE 2.9** A case may be remanded for several reasons. For instance, if the appellate court decides that a judge improperly granted summary judgment, the case will be remanded for a trial. If the appellate court decides that the trial judge erroneously applied the law, the case will be remanded for a new trial, with instructions to the trial court to apply the law as clarified by the appellate court. If the appellate court decides that the trial jury's award of damages was too high, the case will be remanded with instructions to reduce the damages award. ■ In most cases, the judgment of the lower court is

affirmed, resulting in the enforcement of the court's judgment or decree.

Appeal to a Higher Appellate Court If the reviewing court is an intermediate appellate court, the losing party normally may appeal to the state supreme court (the highest state court). Such a petition corresponds to a petition for a writ of *certiorari* from the United States Supreme Court. If the petition is granted (in some states, a petition is automatically granted), new briefs must be filed before the state supreme court, and the attorneys may be allowed or requested to present oral arguments. Like the intermediate appellate court, the supreme court may reverse or affirm the appellate court's decision or remand the case. At this point, unless a federal question is at issue, the case has reached its end.

ENFORCING THE JUDGMENT

The uncertainties of the litigation process are compounded by the lack of guarantees that any judgment will be enforceable. Even if a plaintiff wins an award of damages in court, the defendant may not have sufficient assets or insurance to cover that amount. Usually, one of the factors considered before a lawsuit is initiated is whether the defendant has sufficient assets to cover the amount of damages sought, should the plaintiff win the case.

The Courts Adapt to the Online World

We have already mentioned that the courts have attempted to adapt traditional jurisdictional concepts to the online world. Not surprisingly, the Internet has also brought about changes in court procedures and practices, including new methods for filing pleadings and other documents and issuing decisions and opinions. Several courts are experimenting with electronic delivery, such as via the Internet or CD-ROM. Some jurisdictions are exploring the possibility of cyber courts, in which legal proceedings could be conducted totally online.

ELECTRONIC FILING

The federal court system first experimented with an electronic filing system in January 1996, in an asbestos case heard by the U.S. District Court for the Northern District of Ohio. Today, a number of federal courts permit attorneys to file documents electronically in certain types of cases. At last count, more than 130,000 documents in approximately 10,000 cases had been filed electronically in federal courts. The Administrative Office of the U.S. Courts is considering permitting electronic filing in all U.S. district courts nationwide.

State and local courts are also setting up electronic filing systems. Since the late 1990s, the court system in Pima County, Arizona, has been accepting pleadings via e-mail. The supreme court of the state of Washington accepts online filings of litigation documents. In addition, electronic filing projects are being developed in other states, including California, Idaho, Kansas, Maryland, Michigan, Texas, Utah, and Virginia. The state of Colorado implemented the first statewide court e-filing system and now allows e-filing in over sixty courts. Generally, when electronic filing is made available, it is optional. In early 2001, however, a trial court judge in the District of Columbia launched a pilot project that *required* attorneys to use electronic filing for all documents relating to certain types of civil cases.

The expenses associated with an appeal can be considerable, and e-filing can add substantially to the cost. In some cases, appellants who successfully appeal a judgment are entitled to be awarded their costs, including an amount for printing the copies of the record on appeal and the briefs. In the following case, the appellants spent $16,112 for the paper copies and an additional $16,065 to prepare and submit briefs and other documents in an electronic format. Should the appellants be reimbursed for these expenses?

CASE 2.2 Phansalkar v. Andersen, Weinroth & Co.

United States Court of Appeals,
Second Circuit, 2004.
356 F.3d 188.

FACTS Andersen, Weinroth & Company (AW) is a small firm that finds and creates investment opportunities for itself, its partners, and other investors. AW's income includes returns on its investments, fees paid by its investors, and the compensation and other benefits earned by its employees for their service on boards of directors of the companies with which AW does business. Some AW employees receive stock and "investment opportunities" rather than salaries. Rohit Phansalkar worked for AW from February 1998 until June 2000, when he became the chairman and chief executive officer of Osicom Technologies, Inc. After Phansalkar left, AW refused to pay him the returns on certain "investment opportunities" that he had been given while at AW. Phansalkar filed a suit in a federal district court against AW, alleging in part breach of contract. The court awarded Phansalkar more than

CASE 2.2-CONTINUED

$4.4 million. AW appealed to the U.S. Court of Appeals for the Second Circuit, which reversed this judgment on the ground that Phansalkar had acted with disloyalty during his employment with AW by failing to disclose compensation and benefits that he received for serving on various boards. AW then asked for an award of the amount that it spent to create copies of the briefs and other documents involved in the appeal in hyperlinked CD-ROM format.

ISSUE Could AW recover the costs associated with preparing and submitting electronic copies of the appeal documents?

DECISION No. The U.S. Court of Appeals for the Second Circuit ruled that AW could not recover these costs.

REASON The court recognized that it had been one of the first courts to encourage the submission of electronic briefs. "The submission of an electronic version of a paper brief very likely entails small incremental costs. CD-ROM submissions that hyperlink briefs to relevant sections of the appellate record are more versatile, more useful, and considerably more expensive," however. The court could find "no local rule or holding * * * that allocates CD-ROM costs," and "[n]o guidance" in the Federal Rules of Civil Procedure, which authorize an award of costs incurred to produce "necessary" copies of briefs and other costs of an appeal. Factors that are important in determining if a cost is authorized by the rules include "whether the party seeking disallowance has clearly consented to the expense; whether a court has previously approved the expense; and whether the alternative arrangement costs less than the expense specifically authorized." None of these factors was present in this case. In fact, because AW "incurred costs both to produce hard copies of their appellate materials *and* to produce hyperlinked CD-ROM copies, * * * the CD-ROM costs in this case were duplicative." The court also noted that "there is no written stipulation or understanding between the parties concerning the allocation of the incremental costs of this useful technology."

FOR CRITICAL ANALYSIS—Economic Consideration *How might the result in this case have been different if the court had required, rather than merely encouraged, the submission of electronic copies of the appeal documents?*

COURTS ONLINE

Most courts today have sites on the Web. Of course, each court decides what to make available at its site. Some courts display only the names of court personnel and office phone numbers. Others add court rules and forms. Many include judicial decisions, although generally the sites do not feature archives of old decisions. Instead, decisions are usually available online only for a limited time. For example, California keeps opinions online for only sixty days. In addition, in some states, such as California and Florida, court clerks offer docket information and other searchable databases online.

Appellate court decisions are often posted online immediately after they are rendered. Recent decisions of the U.S. courts of appeals, for example, are available online at their Web sites. The United States Supreme Court also has an official Web site and publishes its opinions there immediately after they are announced to the public. (These Web sites are listed at the end of this chapter in the *Accessing the Internet* feature.) In fact, even decisions that are designated as unpublished opinions by the appellate courts are often published online.

CYBER COURTS AND PROCEEDINGS

Someday, litigants may be able to use cyber courts, in which judicial proceedings take place only on the Internet. The parties to a case could meet online to make their arguments and present their evidence. This might be done with e-mail submissions, through video cameras, in designated "chat" rooms, at closed sites, or through the use of other Internet facilities. These courtrooms could be efficient and economical. We might also see the use of virtual lawyers, judges, and juries—and possibly the replacement of court personnel with computer software. Already the state of Michigan has passed legislation creating cyber courts that will hear cases involving technology issues and high-tech businesses. Many lawyers predict that other states will follow suit.

The courts may also use the Internet in other ways. In a groundbreaking decision in early 2001, for example, a Florida county court granted "virtual" visitation rights in a couple's divorce proceeding. Although the court granted custody of the couple's ten-year-old daughter to the father, the court also ordered each parent to buy a computer and a videoconferencing system so that the mother could "visit" with her child via the Internet at any time.[14]

Alternative Dispute Resolution

Litigation is expensive. It is also time consuming. Because of the backlog of cases pending in many courts, several years may pass before a case is actually tried. For these

14. For a discussion of this case, see Shelley Emling, "After the Divorce, Internet Visits?" *Austin American-Statesman,* January 30, 2001, pp. A1 and A10.

and other reasons, more and more businesspersons are turning to **alternative dispute resolution** (ADR) as a means of settling their disputes.

Methods of ADR range from neighbors sitting down over a cup of coffee in an attempt to work out their differences to huge multinational corporations agreeing to resolve a dispute through a formal hearing before a panel of experts. The great advantage of ADR is its flexibility. Normally, the parties themselves can control how the dispute will be settled, what procedures will be used, and whether the decision reached (either by themselves or by a neutral third party) will be legally binding or nonbinding.

Today, approximately 95 percent of cases are settled before trial through some form of ADR. Indeed, most states either require or encourage parties to undertake ADR prior to trial. Several federal courts have instituted ADR programs as well. In the following pages, we examine various forms of ADR. Keep in mind, though, that new methods of ADR—and new combinations of existing methods—are constantly being devised and employed. In addition, ADR services are now being offered via the Internet. After looking at traditional forms of ADR, we examine some of the ways in which disputes are being resolved in various online forums.

NEGOTIATION

One of the simplest forms of ADR is **negotiation,** a process in which the parties attempt to settle their dispute informally, with or without attorneys to represent them. Attorneys frequently advise their clients to negotiate a settlement voluntarily before they proceed to trial.

Negotiation traditionally involves just the parties themselves and (typically) their attorneys. The attorneys, though, are advocates—they are obligated to put their clients' interests first. Often parties find it helpful to have the opinion and guidance of a neutral (unbiased) third party when deciding whether or how to negotiate a settlement of their dispute. The methods of ADR discussed next all involve neutral third parties.

MEDIATION

In the **mediation** process, the parties themselves attempt to negotiate an agreement, but with the assistance of a neutral third party, a *mediator.* In mediation, the mediator talks with the parties separately as well as jointly. The mediator emphasizes points of agreement, helps the parties evaluate their positions, and proposes solutions. The mediator, however, does not make a decision on the matter being disputed. The mediator, who need not be a lawyer, usually charges a fee for his or her services (which can be split between the parties). States that require parties to undergo ADR before trial often offer mediation as one of the ADR options or (as in Florida) the only option.

Mediation is not adversarial in nature, as lawsuits are. In litigation, the parties "do battle" with each other in the courtroom, while the judge is the neutral party. Because of its nonadversarial nature, the mediation process tends to reduce the antagonism between the disputants and to allow them to resume their former relationship. For this reason, mediation is often the preferred form of ADR for disputes involving business partners, employers and employees, or other parties involved in long-term relationships. ■ EXAMPLE 2.10 Suppose that two business partners have a dispute over how the profits of their firm should be distributed. If the dispute is litigated, the parties will be adversaries, and their respective attorneys will emphasize how the parties' positions differ, not what they have in common. In contrast, when a dispute is mediated, the mediator emphasizes the common ground shared by the parties and helps them work toward agreement. ■

Today, characteristics of mediation are being combined with those of arbitration (to be discussed next). In *binding mediation,* for example, the parties agree that if they cannot resolve the dispute, the mediator can make a legally binding decision on the issue. In *mediation-arbitration,* or "med-arb," the parties agree to first attempt to settle their dispute through mediation. If no settlement is reached, the dispute will be arbitrated.

ARBITRATION

A more formal method of ADR is **arbitration,** in which an arbitrator (a neutral third party or a panel of experts) hears a dispute and renders a decision. The key difference between arbitration and the forms of ADR just discussed is that in arbitration, the third party hearing the dispute makes the decision for the parties. Usually, the parties in arbitration agree that the third party's decision will be *legally binding,* although the parties can also agree to *nonbinding* arbitration. Additionally, arbitration that is mandated by the courts often is not binding on the parties. If the parties do not agree with the arbitrator's decision, they can go forward with the lawsuit.

In some respects, formal arbitration resembles a trial, although usually the procedural rules are much less restrictive than those governing litigation. In the typical hearing format, the parties present opening arguments to the arbitrator and state what remedies should or should not be granted. Evidence is then presented, and witnesses

may be called and examined by both sides. The arbitrator then renders a decision, which is called an *award*.

An arbitrator's award is usually the final word on the matter. Although the parties may appeal an arbitrator's decision, a court's review of the decision will be much more restricted in scope than an appellate court's review of a trial court's decision. The general view is that because the parties were free to frame the issues and set the powers of the arbitrator at the outset, they cannot complain about the results. The award will be set aside only if the arbitrator's conduct or "bad faith" substantially prejudiced the rights of one of the parties, if the award violates an established public policy, or if the arbitrator exceeded her or his powers (arbitrated issues that the parties did not agree to submit to arbitration).

Arbitration Clauses and Statutes Virtually any commercial matter can be submitted to arbitration. Frequently, parties include an **arbitration clause** in a contract (a written agreement—see Chapter 7); the clause provides that any dispute that arises under the contract will be resolved through arbitration rather than through the court system. Parties can also agree to arbitrate a dispute after a dispute arises.

Most states have statutes (often based in part on the Uniform Arbitration Act of 1955) under which arbitration clauses will be enforced, and some state statutes compel arbitration of certain types of disputes, such as those involving public employees. At the federal level, the Federal Arbitration Act (FAA), enacted in 1925, enforces arbitration clauses in contracts involving maritime activity and interstate commerce (though its applicability to employment contracts has been controversial, as discussed later in this chapter). Because of the breadth of the commerce clause (see Chapter 1), arbitration agreements involving transactions only slightly connected to the flow of interstate commerce may fall under the FAA.

The Issue of Arbitrability When a dispute arises as to whether the parties have agreed in an arbitration clause to submit a particular matter to arbitration, one party may file a suit to compel arbitration. The court before which the suit is brought will decide *not* the basic controversy but rather the issue of arbitrability—that is, whether the matter is one that must be resolved through arbitration. If the court finds that the subject matter in controversy is covered by the agreement to arbitrate, then a party may be compelled to arbitrate the dispute. Even when a claim involves a violation of a statute passed to protect a certain class of people, such as employees, a court may determine that the parties must nonetheless abide by their agreement to arbitrate the dispute. Usually, a court will allow the claim to be arbitrated if the court, in interpreting the statute, can find no legislative intent to the contrary.

No party, however, will be ordered to submit a particular dispute to arbitration unless the court is convinced that the party consented to do so.[15] Additionally, the courts will not compel arbitration if it is clear that the prescribed arbitration rules and procedures are inherently unfair to one of the parties. (See this chapter's *Management Perspective* feature on page 49 for a further discussion of this issue.)

Mandatory Arbitration in the Employment Context A significant question in the last several years has concerned mandatory arbitration clauses in employment contracts. Many claim that employees' rights are not sufficiently protected when they are forced, as a condition of hiring, to agree to arbitrate all disputes and thus waive their rights under statutes specifically designed to protect employees. The United States Supreme Court, however, has generally held that mandatory arbitration clauses in employment contracts are enforceable.

■ **EXAMPLE 2.11** In a landmark 1991 decision, *Gilmer v. Interstate/Johnson Lane Corp.*,[16] the Supreme Court held that a claim brought under a federal statute prohibiting age discrimination (see Chapter 23) could be subject to arbitration. The Court concluded that the employee had waived his right to sue when he agreed, as part of a required registration application to be a securities representative with the New York Stock Exchange, to arbitrate "any dispute, claim, or controversy" relating to his employment. ■ The *Gilmer* decision was controversial and generated much discussion during the 1990s. By the early 2000s, some lower courts began to question whether Congress intended the Federal Arbitration Act—which expressly excludes the employment contracts of "seamen, railroad employees, or any other class of workers engaged in foreign or interstate commerce"—to apply to any employment contracts. In the following case, the United States Supreme Court addressed this issue.

15. See, for example, *Wright v. Universal Maritime Service Corp.*, 525 U.S. 70, 119 S.Ct. 391, 142 L.Ed.2d 361 (1998).

16. 500 U.S. 20, 111 S.Ct. 1647, 114 L.Ed.2d 26 (1991).

CASE 2.3 Circuit City Stores, Inc. v. Adams

Supreme Court of the United States, 2001.
532 U.S. 105,
121 S.Ct. 1302,
149 L.Ed.2d 234.
http://straylight.law.cornell.edu/supct/index.html[a]

FACTS Saint Clair Adams applied for a job at Circuit City Stores, Inc. Adams signed an employment application that contained a clause requiring the arbitration of any employment-related disputes, including claims under federal and state law. Adams was hired as a sales counselor in a Circuit City store in Santa Rosa, California. Two years later, Adams filed a suit in a California state court against Circuit City, alleging employment discrimination in violation of state law. Circuit City immediately filed a suit against Adams in a federal district court, asking the court to compel arbitration of Adams's claim. Adams argued that Section 1 of the Federal Arbitration Act (FAA), which excludes from coverage "contracts of employment of seamen, railroad employees, or any other class of workers engaged in foreign or interstate commerce," excluded all employment contracts. The court entered an order in favor of Circuit City. Adams appealed to the U.S. Court of Appeals for the Ninth Circuit, which held that the arbitration agreement between Adams and Circuit City was contained in a "contract of employment" and thus was not subject to the FAA. The court interpreted the language in Section 1 to exclude all employment contracts. Circuit City appealed to the United States Supreme Court.

a. In the "Search" box, type in "Circuit City Stores" and then click on "Submit." In the result, click on the name of the case to access the opinion. The Legal Information Institute of Cornell Law School in Ithaca, New York, maintains this Web site.

ISSUE Are all employment contracts excluded from coverage under the FAA?

DECISION No. The United States Supreme Court reversed the judgment of the lower court and remanded the case, holding that the FAA applies to most employment contracts (excluding only those involving interstate transportation workers).

REASON The Supreme Court found that the federal appellate court's decision to exclude all employment contracts was inconsistent with the Court's previous decisions. The Court also reasoned that interpreting "the residual phrase to exclude all employment contracts fails to give independent effect to the statute's enumeration of the specific categories of workers which precedes it; there would be no need for Congress to use the phrases 'seamen' and 'railroad employees' if those same classes of workers were subsumed within the meaning of the 'engaged in * * * commerce' residual clause. * * * [W]here general words follow specific words in a statutory enumeration, the general words are construed to embrace only objects similar in nature to those objects enumerated by the preceding specific words. Under this rule of construction the residual clause should be read to give effect to the terms 'seamen' and 'railroad employees,' and should itself be controlled and defined by reference to the enumerated categories of workers which are recited just before it."

FOR CRITICAL ANALYSIS—Cultural Consideration *What does the decision in this case indicate about the courts' interpretation of the phrases used in statutes?*

Other Types of ADR

The three forms of ADR just discussed are the oldest and traditionally the most commonly used. In recent years, a variety of new types of ADR have emerged, some of which were mentioned earlier in the discussion of mediation. Other ADR forms that are used today are sometimes referred to as "assisted negotiation" because they involve a third party in what is essentially a negotiation process. For example, in **early neutral case evaluation,** the parties select a neutral third party (generally an expert in the subject matter of the dispute) to evaluate their respective positions. The parties explain their positions to the case evaluator in any manner they choose. The case evaluator then assesses the strengths and weaknesses of the parties' positions, and this evaluation forms the basis for negotiating a settlement.

Another form of assisted negotiation that is often used by business parties is the **mini-trial,** in which each party's attorney briefly argues the party's case before representatives of each firm who have the authority to settle the dispute. Typically, a neutral third party (usually an expert in the area being disputed) acts as an adviser. If the parties fail to reach an agreement, the adviser renders an opinion as to how a court would likely decide the issue. The proceeding assists the parties in determining whether they should negotiate a settlement or take the dispute to court.

Today's courts are also experimenting with a variety of ADR alternatives to speed up (and reduce the cost of) justice. Numerous federal courts now hold **summary jury trials (SJTs),** in which the parties present their arguments and evidence and the jury renders a verdict. The jury's verdict is not binding, but it does act as a guide to both sides

MANAGEMENT PERSPECTIVE

Arbitration Clauses in Employment Contracts

MANAGEMENT FACES A LEGAL ISSUE

Arbitration is normally simpler, speedier, and less costly than litigation. For that reason, business owners and managers today often include arbitration clauses in their contracts, including employment contracts. What happens, though, if a job candidate whom you wish to hire (or an existing employee whose contract is being renewed) objects to one or more of the provisions in an arbitration clause? If you insist that signing the agreement to arbitrate future disputes is a mandatory condition of employment, will such a clause be enforceable?

WHAT THE COURTS SAY

As you will read elsewhere in this chapter, the United States Supreme Court has consistently taken the position that because the Federal Arbitration Act (FAA) favors the arbitration of disputes, arbitration clauses in employment contracts should generally be enforced. Nonetheless, some courts have held that arbitration clauses in employment contracts should not be enforced if they are too one sided and unfair to the employee.

In one case, for example, the U.S. Court of Appeals for the Ninth Circuit refused to enforce an arbitration clause on the ground that the agreement was *unconscionable*—so one sided and unfair as to be unenforceable under "ordinary principles of state contract law." The agreement was a standard-form contract drafted by the employer (the party with superior bargaining power), and the employee had to sign it without any modification as a prerequisite to employment. Moreover, only the employees were required to arbitrate their disputes, while the employer remained free to litigate any claims it had against its employees in court. Among other things, the contract also severely limited the relief that was available to employees. For these reasons, the court held the entire arbitration agreement unenforceable.[a] Other courts have cited similar reasons for deciding not to enforce one-sided arbitration clauses.[b]

IMPLICATIONS FOR MANAGERS

Although the United States Supreme Court has made it clear that arbitration clauses in employment contracts are enforceable under the FAA, business owners and managers would be wise to exercise caution when drafting such clauses. It is especially important to make sure that the terms of the agreement are not so one sided that a court could declare the entire agreement unconscionable.

a. *Circuit City Stores, Inc. v. Adams,* 279 F.3d 889 (9th Cir. 2002). (This was the Ninth Circuit's decision, on remand, after the United States Supreme Court reviewed the case—see Case 2.3.)

b. See, for example, *Hooters of America, Inc. v. Phillips,* 173 F.3d 933 (4th Cir. 1999); and *Hardwick v. Sherwin Williams Co.,* 2002 WL 31992364 (Ohio App. 8 Dist. 2003).

in reaching an agreement during the mandatory negotiations that immediately follow the trial. Other alternatives being employed by the courts include summary procedures for commercial litigation and the appointment of special masters to assist judges in deciding complex issues.

Providers of ADR Services

ADR services are provided by both government agencies and private organizations. A major provider of ADR services is the American Arbitration Association (AAA), which was founded in 1926 and now handles over 200,000 claims a year in its numerous offices around the country. Most of the largest U.S. law firms are members of this nonprofit association.

Cases brought before the AAA are heard by an expert or a panel of experts in the area relating to the dispute and are usually settled quickly. Generally, about half of the panel members are lawyers. To cover its costs, the AAA charges a fee, paid by the party filing the claim. In addition, each party to the dispute pays a specified amount for each hearing day, as well as a special additional fee for cases involving personal injuries or property loss.

Hundreds of for-profit firms around the country also provide various forms of dispute-resolution services. Typically, these firms hire retired judges to conduct arbitration hearings or otherwise assist parties in settling their disputes. The judges follow procedures similar to those of the federal courts and use similar rules. Usually, each party to the dispute pays a filing fee and a designated fee for a hearing session or conference.

Online Dispute Resolution

An increasing number of companies and organizations offer dispute-resolution services using the Internet. The settlement of disputes in these online forums is known as **online dispute resolution (ODR)**. To date, the disputes resolved in these forums have most commonly involved disagreements over the rights to domain names (Web site addresses—see Chapter 5) and disagreements over the

quality of goods sold via the Internet, including goods sold through Internet auction sites.

ODR may be best for resolving small- to medium-sized business liability claims, which may not be worth the expense of litigation or traditional ADR methods. Rules being developed in online forums, however, may ultimately become a code of conduct for all of those who do business in cyberspace. Most online forums do not automatically apply the law of any specific jurisdiction. Instead, results are often based on general, universal legal principles. As with offline methods of dispute resolution, any party may appeal to a court at any time.

Negotiation and Mediation Services

The online negotiation of a dispute is generally simpler and more practical than litigation. Typically, one party files a complaint, and the other party is notified by e-mail. Password-protected access is possible twenty-four hours a day, seven days a week. Fees are generally low (often 2 to 4 percent, or less, of the disputed amount).

CyberSettle.com, OnlineResolution.com, and other Web-based firms offer online forums for negotiating monetary settlements. The parties to a dispute may agree to submit offers; if the offers fall within a previously agreed-on range, they will end the dispute, and the parties will split the difference. Special software keeps secret any offers that are not within the range. If there is no agreed-on range, typically an offer includes a deadline within which the other party must respond before the offer expires. The parties can drop the negotiations at any time.

Mediation providers have also tried resolving disputes online. SquareTrade, for example, has provided mediation services for the online auction site eBay and also resolves disputes among other parties. SquareTrade uses Web-based software that walks participants through a five-step e-resolution process. Negotiation between the parties occurs on a secure page within SquareTrade's Web site. The parties may consult a mediator. The entire process takes as little as ten to fourteen days, and at present no fee is charged unless the parties use a mediator.

Arbitration Programs

A number of organizations, including the American Arbitration Association, offer online arbitration programs. The Internet Corporation for Assigned Names and Numbers (ICANN), a nonprofit corporation that the federal government set up to oversee the distribution of domain names, has issued special rules for the resolution of domain name disputes.[17] ICANN has also authorized several organizations to arbitrate domain name disputes in accordance with ICANN's rules.

Resolution Forum, Inc. (RFI), a nonprofit organization associated with the Center for Legal Responsibility at South Texas College of Law, offers arbitration services through its CAN-WIN conferencing system. Using standard browser software and an RFI password, the parties to a dispute access an online conference room. When multiple parties are involved, private communications and breakout sessions are possible via private messaging facilities. RFI also offers mediation services.

The Virtual Magistrate Project (VMAG) is affiliated with the American Arbitration Association, Chicago-Kent College of Law, Cyberspace Law Institute, National Center for Automated Information Research, and other organizations. VMAG offers arbitration for disputes involving users of online systems; victims of wrongful messages, postings, and files; and system operators subject to complaints or similar demands. VMAG also arbitrates intellectual property, personal property, real property, and tort disputes related to online contracts. VMAG attempts to resolve a dispute within seventy-two hours. The proceedings occur in a password-protected online newsgroup setting, and private e-mail among the participants is possible. A VMAG arbitrator's decision is issued in a written opinion. A party may appeal the outcome to a court.

17. ICANN's Rules for Uniform Domain Name Dispute Resolution Policy are online at **http://www.icann.org/dndr/udrp/uniform-rules.htm**. Domain names will be discussed in more detail in Chapter 5, in the context of trademark law.

TERMS AND CONCEPTS

alternative dispute resolution (ADR) 46
answer 39
arbitration 46
arbitration clause 47
award 43
bankruptcy court 32
brief 43
complaint 38
concurrent jurisdiction 33
counterclaim 40
default judgment 39
deposition 40
discovery 40
diversity of citizenship 33
docket 33
early neutral case evaluation 48
e-evidence 41
exclusive jurisdiction 33
federal question 33
interrogatories 40

CHAPTER SUMMARY Traditional and Online Dispute Resolution

The Judiciary's Role in American Government (See pages 30–31.)	The role of the judiciary—the courts—in the American governmental system is to interpret and apply the law. Through the process of judicial review—determining the constitutionality of laws—the judicial branch acts as a check on the executive and legislative branches of government.
Basic Judicial Requirements (See pages 31–35.)	1. *Jurisdiction*—Before a court can hear a case, it must have jurisdiction over the person against whom the suit is brought or the property involved in the suit, as well as jurisdiction over the subject matter. a. Limited versus general jurisdiction—Limited jurisdiction exists when a court is limited to a specific subject matter, such as probate or divorce. General jurisdiction exists when a court can hear any kind of case. b. Original versus appellate jurisdiction—Original jurisdiction exists with courts that have authority to hear a case for the first time (trial courts). Appellate jurisdiction exists with courts of appeals, or reviewing courts; generally, appellate courts do not have original jurisdiction. c. Federal jurisdiction—Arises (1) when a federal question is involved (when the plaintiff's cause of action is based, at least in part, on the U.S. Constitution, a treaty, or a federal law) or (2) when a case involves diversity of citizenship (citizens of different states, for example) and the amount in controversy exceeds $75,000. d. Concurrent versus exclusive jurisdiction—Concurrent jurisdiction exists when two different courts have authority to hear the same case. Exclusive jurisdiction exists when only state courts or only federal courts have authority to hear a case. 2. *Jurisdiction in cyberspace*—Because the Internet does not have physical boundaries, traditional jurisdictional concepts have been difficult to apply in cases involving activities conducted via the Web. Gradually, the courts are developing standards to use in determining when jurisdiction over a Web owner or operator in another state is proper. 3. *Venue*—Venue has to do with the most appropriate location for a trial, which is usually the geographic area where the event leading to the dispute took place or where the parties reside. 4. *Standing to sue*—A requirement that a party must have a legally protected and tangible interest at stake sufficient to justify seeking relief through the court system. The controversy at issue must also be a justiciable controversy—one that is real and substantial, as opposed to hypothetical or academic.
The State and Federal Court Systems (See pages 35–38.)	1. *Trial courts*—Courts of original jurisdiction, in which legal actions are initiated. a. State—Courts of general jurisdiction can hear any case; courts of limited jurisdiction include divorce courts, probate courts, traffic courts, small claims courts, and so on. b. Federal—The federal district court is the equivalent of the state trial court. Federal courts of limited jurisdiction include the U.S. Tax Court, the U.S. Bankruptcy Court, and the U.S. Court of Federal Claims.

(Continued)

CHAPTER SUMMARY Traditional and Online Dispute Resolution—Continued

The State and Federal Court Systems—Continued	2. *Intermediate appellate courts*—Courts of appeals, or reviewing courts; generally without original jurisdiction. Many states have an intermediate appellate court; in the federal court system, the U.S. circuit courts of appeals are the intermediate appellate courts. 3. *Supreme (highest) courts*—Each state has a supreme court, although it may be called by some other name; appeal from the state supreme court to the United States Supreme Court is possible only if a federal question is involved. The United States Supreme Court is the highest court in the federal court system and the final arbiter of the Constitution and federal law.
Following a State Court Case (See pages 38–44.)	Rules of procedure prescribe the way in which disputes are handled in the courts. Rules differ from court to court, and separate sets of rules exist for federal and state courts, as well as for criminal and civil cases. A sample civil court case in a state court would involve the following procedures: 1. *The pleadings*— a. Complaint—Filed by the plaintiff with the court to initiate the lawsuit; served with a summons on the defendant. b. Answer—Admits or denies allegations made by the plaintiff; may assert a counterclaim or an affirmative defense. c. Motion to dismiss—A request to the court to dismiss the case for stated reasons, such as the plaintiff's failure to state a claim for which relief can be granted. 2. *Pretrial motions (in addition to the motion to dismiss)*— a. Motion for judgment on the pleadings—May be made by either party; will be granted if the parties agree on the facts and the only question is how the law applies to the facts. The judge bases the decision solely on the pleadings. b. Motion for summary judgment—May be made by either party; will be granted only if the parties agree on the facts. The judge applies the law in rendering a judgment. The judge can consider evidence outside the pleadings when evaluating the motion. 3. *Discovery*—The process of gathering evidence concerning the case. Discovery involves depositions (sworn testimony by parties to the lawsuit or any witnesses), interrogatories (written questions and answers to these questions made by parties to the action with the aid of their attorneys), and various requests (for admissions, documents, and medical examinations, for example). Discovery may also involve electronically recorded information, such as e-mail, voice mail, word-processing documents, and other data compilations. Although electronic discovery has significant advantages over paper discovery, it is also more time consuming and expensive and often requires the parties to hire experts. 4. *Pretrial conference*—Either party or the court can request a pretrial conference to identify the matters in dispute after discovery has taken place and to plan the course of the trial. 5. *Trial*—Following jury selection (*voir dire*), the trial begins with opening statements from both parties' attorneys. The following events then occur: a. The plaintiff's introduction of evidence (including the testimony of witnesses) supporting the plaintiff's position. The defendant's attorney can challenge evidence and cross-examine witnesses. b. The defendant's introduction of evidence (including the testimony of witnesses) supporting the defendant's position. The plaintiff's attorney can challenge evidence and cross-examine witnesses. c. Closing arguments by the attorneys in favor of their respective clients, the judge's instructions to the jury, and the jury's verdict. 6. *Posttrial motions*— a. Motion for judgment *n.o.v.* ("notwithstanding the verdict")—Will be granted if the judge is convinced that the jury was in error. b. Motion for a new trial—Will be granted if the judge is convinced that the jury was in error; can also be granted on the grounds of newly discovered evidence, misconduct by the participants during the trial, or error by the judge.

CHAPTER SUMMARY Traditional and O

Following a State Court Case—Continued	7. *Appeal*—Either party can appeal the trial ... reviewing the record on appeal, the abstra... hearing and renders its opinion.
The Courts Adapt to the Online World (See pages 44–45.)	A number of state and federal courts now allow ... courts via the Internet or other electronic means. ... of electronic filing systems in all federal district cou... information about the court and its procedures, and ... online. In the future, we may see "cyber courts," in w...
Alternative Dispute Resolution (See pages 45–49.)	1. *Negotiation*—The parties come together, with or wit... reach a settlement without the involvement of a thir... 2. *Mediation*—The parties themselves reach an agreeme... party, called a mediator, who proposes solutions. At the parties' req... ...ake a legally binding decision. 3. *Arbitration*—A more formal method of ADR in which the p...ies submit their dispute to a neutral third party, the arbitrator, who renders a decision. The decision may or may not be legally binding, depending on the circumstances. 4. *Other types of ADR*—These include early neutral case evaluation, mini-trials, and summary jury trials (SJTs); generally, these are forms of "assisted negotiation." 5. *Providers of ADR services*—The leading nonprofit provider of ADR services is the American Arbitration Association. Hundreds of for-profit firms also provide ADR services.
Online Dispute Resolution (See pages 49–50.)	A number of organizations and firms are now offering negotiation, mediation, and arbitration services through online forums. To date, these forums have been a practical alternative for the resolution of domain name disputes and e-commerce disputes in which the amount in controversy is relatively small.

FOR REVIEW

Answers for the even-numbered questions in this For Review *section can be found in Appendix A at the end of this text.*

1. What is judicial review? How and when was the power of judicial review established?
2. Before a court can hear a case, it must have jurisdiction. Over what must it have jurisdiction? How are the courts applying traditional jurisdictional concepts to cases involving Internet transactions?
3. What is the difference between a trial court and an appellate court?
4. In a lawsuit, what are the pleadings? What is discovery, and how does electronic discovery differ from traditional discovery? What is electronic filing?
5. How are online forums being used to resolve disputes?

QUESTIONS AND CASE PROBLEMS

2–1. Arbitration. In an arbitration proceeding, the arbitrator need not be a judge or even a lawyer. How, then, can the arbitrator's decision have the force of law and be binding on the parties involved?

QUESTION WITH SAMPLE ANSWER

2–2. Marya Callais, a citizen of Florida, was walking near a busy street in Tallahassee one day when a large crate fell off a passing truck and hit her, resulting in several injuries. She incurred a great deal of pain and suffering plus numerous medical expenses, and she could not work for six months. She wishes to sue the trucking firm for $300,000 in damages. The firm's headquarters are in Georgia, although the company does business in Florida. In what court may Callais bring suit—a Florida state court, a Georgia state court, or a federal court? What factors might influence her decision?

For a sample answer to this question, go to Appendix B at the end of this text.

2–3. Standing. Blue Cross and Blue Shield insurance companies (the Blues) provide 68 million Americans with health-care

dollars for care attrib- ...co use. In an attempt to ...e Blues filed a suit in a federal ...o companies and others, alleging ...gs. The Blues claimed that beginning in ... conspired to addict millions of Americans, ...rs of Blue Cross plans, to cigarettes and other ...ducts. The conspiracy involved misrepresentation ...e safety of nicotine and its addictive properties, market- ...efforts targeting children, and agreements not to produce or market safer cigarettes. As a result of the defendants' efforts, many tobacco users developed lung, throat, and other cancers, as well as heart disease, stroke, emphysema, and other illnesses. The defendants asked the court to dismiss the case on the ground that the plaintiffs did not have standing to sue. Do the Blues have standing in this case? Why or why not? [*Blue Cross and Blue Shield of New Jersey, Inc. v. Philip Morris, Inc.*, 36 F.Supp.2d 560 (E.D.N.Y. 1999)]

2–4. Jurisdiction. George Noonan, a Boston police detective and a devoted nonsmoker, has spent most of his career educating Bostonians about the health risks of tobacco use. In 1992, an ad for Winston cigarettes featuring Noonan's image appeared in several French magazines. Some of the magazines were on sale at newsstands in Boston. Noonan filed a suit in a federal district court against The Winston Co., Lintas:Paris (the French ad agency that created the ads), and others. Lintas:Paris and the other French defendants claimed that they did not know the magazines would be sold in Boston and filed a motion to dismiss the suit for lack of personal jurisdiction. Does the court have jurisdiction? Why or why not? [*Noonan v. The Winston Co.*, 135 F.3d 85 (1st Cir. 1998)]

CASE PROBLEM WITH SAMPLE ANSWER

2–5. Ms. Thompson filed a suit in a federal district court against her employer, Altheimer & Gray, seeking damages for alleged racial discrimination in violation of federal law. During *voir dire*, the judge asked the prospective jurors whether "there is something about this kind of lawsuit for money damages that would start any of you leaning for or against a particular party?" Ms. Leiter, one of the prospective jurors, raised her hand and explained that she had "been an owner of a couple of businesses and am currently an owner of a business, and I feel that as an employer and owner of a business that will definitely sway my judgment in this case." She explained, "I am constantly faced with people that want various benefits or different positions in the company or better contacts or, you know, a myriad of issues that employers face on a regular basis, and I have to decide whether or not that person should get them." Asked by Thompson's lawyer whether "you believe that people file lawsuits just because they don't get something they want," Leiter answered, "I believe there are some people that do." In answer to another question, she said, "I think I bring a lot of background to this case, and I can't say that it's not going to cloud my judgment. I can try to be as fair as I can, as I do every day." Thompson filed a motion to strike Leiter for cause. Should the judge grant the motion? Explain. [*Thompson v. Altheimer & Gray*, 248 F.3d 621 (7th Cir. 2001)]

After you have answered this problem, compare your answer with the sample answer given on the Web site that accompanies this text. Go to **academic.cengage.com/blaw/wbl**, select "Chapter 2," and click on "Case Problem with Sample Answer."

2–6. Arbitration. Alexander Little worked for Auto Stiegler, Inc., an automobile dealership in Los Angeles County, California, eventually becoming the service manager. While employed, Little signed an arbitration agreement that required all employment-related disputes to be submitted to arbitration. The agreement also provided that any award over $50,000 could be appealed to a second arbitrator. Little was later demoted and terminated. Alleging that these actions were in retaliation for investigating and reporting warranty fraud and thus were in violation of public policy, Little filed a suit in a California state court against Auto Stiegler. The defendant filed a motion with the court to compel arbitration. Little responded that the arbitration agreement should not be enforced in part because the appeal provision was unfairly one sided. Is this provision enforceable? Should the court grant Auto Stiegler's motion? Why or why not? [*Little v. Auto Stiegler, Inc.*, 29 Cal.4th 1064, 63 P.3d 979, 130 Cal.Rptr.2d 892 (2003)]

2–7. Standing to Sue. Lamar Advertising of Penn, LLC, an outdoor advertising business, wanted to erect billboards of varying sizes in a multiphase operation throughout the town of Orchard Park, New York. An Orchard Park ordinance restricted the signs to certain sizes in certain areas, to advertising products and services available for sale only on the premises, and to other limits. Lamar asked Orchard Park for permission to build signs in some areas larger than the ordinance allowed in those locations (but not as large as allowed in other areas). When the town refused, Lamar filed a suit in a federal district court, claiming that the ordinance violated the First Amendment. Did Lamar have standing to challenge the ordinance? If the court could sever the provisions of the ordinance restricting a sign's content from the provisions limiting a sign's size, would your answer be the same? Explain. [*Lamar Advertising of Penn, LLC v. Town of Orchard Park, New York*, 356 F.3d 365 (2d Cir. 2004)]

VIDEO QUESTION

2–8. Go to this text's Web site at academic.cengage.com/blaw/wbl and select "Chapter 2." Click on "Video Questions" and view the video titled *Jurisdiction in Cyberspace*. Then answer the following questions.

1. What standard would a court apply to determine whether it has jurisdiction over the out-of-state computer firm in the video?
2. What factors is a court likely to consider in assessing whether sufficient contacts existed when the only connection to the jurisdiction is through a Web site?
3. How do you think the court would resolve the issue in this situation?

Traditional and Online Dispute Resolution

For updated links to resources available on the Web, as well as a variety of other materials, visit this text's Web site at

academic.cengage.com/blaw/wbl

For the decisions of the United States Supreme Court, as well as information about the Supreme Court, go to

http://supremecourtus.gov

The Web site for the federal courts offers information on the federal court system and links to all federal courts at

http://www.uscourts.gov

The National Center for State Courts (NCSC) offers links to the Web pages of all state courts. Go to

http://www.ncsconline.org

For information on alternative dispute resolution, go to the American Arbitration Association's Web site at

http://www.adr.org

To learn more about online dispute resolution, go to the following Web sites:

http://onlineresolution.com
http://cybersettle.com
http://squaretrade.com

For an example of a typical set of state rules governing attorney conduct, you can access Idaho's Rules of Professional Conduct Governing Lawyers at

http://www2.state.id.us/isb/rules/irpc.htm

Picking the "right" jury can be an important part of litigation strategy, and several firms now specialize in jury consulting services. To learn more about this topic, go to the Jury Research Institute Web site at

http://www.jri-inc.com

ONLINE LEGAL RESEARCH

Go to the *Fundamentals of Business Law* home page at academic.cengage.com/blaw/wbl, select "Chapter 2," and click on "Internet Exercises." There you will find the following Internet research exercises that you can perform to learn more about topics covered in this chapter.

Activity 2–1: LEGAL PERSPECTIVE—The Judiciary's Role in American Government

Activity 2–2: MANAGEMENT PERSPECTIVE—Alternative Dispute Resolution

Activity 2–3: SOCIAL PERSPECTIVE—Resolve a Dispute Online

BEFORE THE TEST

Go to the *Fundamentals of Business Law* home page at academic.cengage.com/blaw/wbl, select "Chapter 2," and click on "Interactive Quizzes." You will find a number of interactive questions relating to this chapter.

CHAPTER 3

Ethics and Business Decision Making

LEARNING OBJECTIVES

After reading this chapter, you should be able to answer the following questions:

1. What is ethics? What is business ethics? Why is business ethics important?
2. How can business leaders encourage their companies to act ethically?
3. What are corporate compliance programs?
4. How do duty-based ethical standards differ from outcome-based ethical standards?
5. What types of ethical issues might arise in the context of international business transactions?

During the early part of the 2000s, the American public was shocked as one business ethics scandal after another became headline news. Each scandal involved serious consequences. Certainly, those responsible for grossly inflating the reported profits at WorldCom, Inc., ended up not only destroying shareholder value in a great company but also facing prison terms. Those officers and directors at Enron Corporation who utilized a system of complicated off-the-books transactions to inflate current earnings saw their company go bankrupt—one of the largest bankruptcies in U.S. history. They harmed not only their employees and shareholders but also the communities in which they worked—and themselves (some of them have since received prison sentences, and others have been indicted). Some of the officers and directors of Tyco International who used corporate funds to pay for lavish personal lifestyles are also facing prison sentences. The shareholders of that company suffered dearly, too.

In response to the public's outrage over these scandals, Congress passed the Sarbanes-Oxley Act of 2002, which will be discussed in detail in Chapters 27 and 31. This act imposed requirements on corporations that are designed to deter similar unethical and illegal business behavior in the future. Indeed, the ethics scandals of the early 2000s taught businesses throughout the country that corporate governance is not to be taken lightly. Nevertheless, new allegations of unethical business conduct continue to surface.

Business ethics, the focus of this chapter, is not just theory. It is practical, useful, and essential. Although a good understanding of business law and the legal environment is critical, it is not enough. Understanding how one should act in her or his business dealings is equally—if not more—important in today's business arena.

Business Ethics

Before we look at business ethics, we need to discuss what is meant by ethics generally. **Ethics** can be defined as the study of what constitutes right or wrong behavior. It is the branch of philosophy that focuses on morality and the way in which moral principles are derived or the way in which a given set of moral principles applies to one's conduct in daily life. Ethics has to do with questions relating to the fairness, justness, rightness, or wrongness of an action. What is fair? What is just? What is the right thing to do in this situation? These are essentially ethical questions.

WHAT IS BUSINESS ETHICS?

Business ethics focuses on what constitutes right or wrong behavior in the business world and on how moral and ethical principles are applied by businesspersons to situations that arise in their daily activities in the workplace. Note that business ethics is not a separate *kind* of ethics. The ethical standards that guide our behavior as, say, mothers, fathers, or students apply equally well to our activities as businesspersons. Business decision makers, though, must often address more complex ethical issues and conflicts in the workplace than they face in their personal lives.

WHY IS BUSINESS ETHICS IMPORTANT?

Why is business ethics important? The answer to this question is clear from this chapter's introduction. A keen and in-depth understanding of business ethics is vital to the long-run viability of a corporation. A thorough knowledge of business ethics is also essential to the well-being of the individual officers and directors of the corporation, as well as to the welfare of the firm's employees and various "stakeholders" in the entity's well-being. Certainly, corporate decisions and activities can significantly affect not only those who own, operate, or work for the company but also such groups as suppliers, the community, and society as a whole.

Note that questions concerning ethical and responsible behavior are not confined to the corporate context. Business ethics applies to *all* businesses, regardless of their organizational forms. In a business partnership, for example, partners owe a *fiduciary duty* (a duty of trust and loyalty) to each other and to their firm. This duty can sometimes conflict with what a partner sees as his or her own best interest. Partners who act solely in their own interests may violate their duties to the other partners and the firm, however. By violating this duty, they may end up paying steep penalties—as the following case illustrates.

CASE 3.1 Time Warner Entertainment Co. v. Six Flags Over Georgia, L.L.C.

Georgia Court of Appeals, 2002.
254 Ga.App. 598,
563 S.E.2d 178.
http://www.ganet.org/appeals/opinions/index.cgi[a]

FACTS The Six Flags Over Georgia theme park in Atlanta, Georgia, was developed in 1967 as a limited partnership known as Six Flags Over Georgia, L.L.C. (Flags). The sole limited partner was Six Flags Fund, Limited (Fund). The general partner was Six Flags Over Georgia, Inc. (SFOG). In 1991, Time Warner Entertainment Company (TWE) became the majority shareholder of SFOG. The next year, TWE secretly bought 13.7 acres of land next to the park, limiting the park's expansion opportunities. Over the next couple of years, using confidential business information from the park, TWE began plans to develop a competing park. Meanwhile, TWE installed no major new attractions at the park, deferred basic maintenance, withheld financial information from Fund (the limited partner), and began signing future employment contracts with SFOG officers. TWE also charged Flags for unrelated expenses, including more than $4 million for lunches in New York City and luxury automobiles for TWE officers. Flags and Fund filed a suit in a Georgia state court against TWE and SFOG, alleging, among other things, breach of fiduciary duty. A jury awarded the plaintiffs $197,296,000 in compensatory damages and $257,000,000 in punitive damages. TWE appealed to a state intermediate appellate court, alleging in part that the amount of the punitive damages was excessive.

ISSUE Was the amount of punitive damages excessive?

DECISION No. The state intermediate appellate court affirmed the judgment of the lower court.

REASON The appellate court held that the award of punitive damages was not excessive, considering the amount of compensatory damages ("the ratio of compensatory to punitive damages is 1 to 1.3"), the defendants' financial status "with collective assets measured in billions of dollars," and their "reprehensible" (blameworthy) conduct toward the plaintiffs. The appellate court stated, "In examining the degree of reprehensibility of a defendant's conduct, [there are] a number of aggravating factors [to consider], including whether the harm was more than purely economic in nature, and whether the defendant's behavior evinced [showed] indifference to or reckless disregard for the health and safety of others. Here, although the harm to Flags and Fund was primarily economic, it was caused by conduct we find especially reprehensible. Appellants' intentional breach of its fiduciary duty revealed a callous indifference to the financial well-being of its limited

(Continued)

a. At the search screen, click on "Search Court of Appeals Opinions." On the appellate court's screen, set the search year to "all." You can go to a case most directly if you use as key words the full name of a party—for example, "Time Warner Entertainment Co." This Web site is sponsored by the state of Georgia.

CASE 3.1—CONTINUED

partners and their individual investors." The court concluded that "[a]ppellants' conduct was, in short, the kind of behavior we find deserving of reproof, rebuke, or censure; blameworthy—the very definition of reprehensible. * * * Trickery and deceit are reprehensible wrongs, especially when done intentionally through affirmative acts of misconduct."

FOR CRITICAL ANALYSIS—Ethical Consideration *If TWE had proceeded with its plans to build a competing park but had not otherwise acted "reprehensibly" toward Flags and Fund, how might the decision in this case have been different?*

Setting the Right Ethical Tone

Many unethical business decisions are made simply because they *can* be made. In other words, the decision makers not only have the opportunity to make such decisions but also are not too concerned about being seriously sanctioned for their unethical actions. Perhaps one of the most difficult challenges for business leaders today is to create the right "ethical tone" in their workplaces so as to deter unethical conduct.

The Importance of Ethical Leadership

Talking about ethical business decision making means nothing if management does not set standards. Moreover, managers must apply those standards to themselves and to the employees in the company.

Attitude of Top Management One of the most important factors in creating and maintaining an ethical workplace is the attitude of top management. Managers who are not totally committed to maintaining an ethical workplace will rarely succeed in creating one. Surveys of business executives indicate that management's behavior, more than anything else, sets the ethical tone of a firm. In other words, employees take their cue from management. If a firm's managers adhere to obvious ethical norms in their business dealings, employees will likely follow their example. In contrast, if managers act unethically, employees will see no reason not to do so themselves. ■ **EXAMPLE 3.1** Suppose that Kevin observes his manager cheating on her expense account. Kevin quickly understands that such behavior is acceptable. Later, when Kevin is promoted to a managerial position, he "pads" his expense account as well—knowing that he is unlikely to face sanctions for doing so. ■

Looking the Other Way A manager who looks the other way when she or he knows about an employee's unethical behavior also sets an example—one indicating that ethical transgressions will be accepted. Managers must show that they will not tolerate unethical business behavior. Although this may seem harsh, managers have found that discharging even one employee for ethical reasons has a tremendous impact as a deterrent to unethical behavior in the workplace. The following case illustrates what can happen when managers look the other way.

CASE 3.2 In re the Exxon Valdez

United States District Court,
District of Alaska, 2004.
296 F.Supp.2d 1071.

FACTS Exxon Shipping Company owned the *Exxon Valdez,* an oil supertanker as long as three football fields with the capacity to hold 53 million gallons of crude oil. The captain of the *Valdez* was Joseph Hazelwood, an alcoholic, who had sought treatment in 1985 but had relapsed before the next spring. Exxon knew that Hazelwood had relapsed and that he drank while onboard ship, but nevertheless allowed him to command the *Valdez.* On March 24, 1989, the *Valdez* ran aground on Bligh Reef in Prince William Sound, Alaska. About 11 million gallons of crude oil leaked from the ship and spread around the sound. Commercial fisheries closed for the rest of the year. Subsistence fishing and shore-based businesses dependent on the fishing industry were disrupted. Exxon spent $2.1 billion to clean up the spilled oil and paid $303 million to those whose livelihoods were disrupted. Thousands of claims were consolidated into a single case tried in a federal district court. The jury awarded, in part, $5 billion in punitive damages against Exxon. Exxon appealed to the U.S. Court of Appeals for the Ninth Circuit, which remanded the case for reconsideration of this award, according to the reprehensibility of the defendant's conduct and other factors. The court also instructed the lower court to reduce the amount if the award was upheld.

ISSUE Did the $5 billion punitive damages award against Exxon constitute grossly excessive or arbitrary punishment?

CASE 3.2–CONTINUED

DECISION No. The court determined that Exxon's conduct was "intentionally malicious" and "highly reprehensible." Concluding that "[p]unitive damages should reflect the enormity of the defendant's offense," the court upheld the award but reduced the amount to $4.5 billion "as the means of resolving the conflict between its conclusion and the directions of the court of appeals."

REASON The court pointed out that "Exxon's conduct did not simply cause economic harm to the plaintiffs. Exxon's decision to leave Captain Hazelwood in command of the *Exxon Valdez* demonstrated reckless disregard for a broad range of legitimate Alaska concerns: the livelihood, health, and safety of the residents of Prince William Sound, the crew of the *Exxon Valdez,* and others. Exxon's conduct targeted some financially vulnerable individuals, namely subsistence fishermen [those who make their living by fishing]. Plaintiffs' harm was not the result of an isolated incident but was the result of Exxon's repeated decisions, over a period of approximately three years, to allow Captain Hazelwood to remain in command despite Exxon's knowledge that he was drinking and driving again." The court compared Exxon's conduct to other cases in which punitive damages were awarded and concluded that "Exxon's conduct was many degrees of magnitude more egregious [conspicuously wrongful]. For approximately three years, Exxon management, with knowledge that Captain Hazelwood had fallen off the wagon, willfully permitted him to operate a fully loaded, crude oil tanker in and out of Prince William Sound—a body of water which Exxon knew to be highly valuable for its fisheries resources. Exxon's argument that its conduct in permitting a relapsed alcoholic to operate an oil tanker should be characterized as less reprehensible * * * suggests that Exxon, even today, has not come to grips with the opprobrium [disgrace, contempt] which society rightly attaches to drunk driving."

WHY IS THIS CASE IMPORTANT? *This case is an excellent illustration of the consequences that a business may face when it ignores a serious risk that has been created by its action (or inaction). By allowing Captain Hazelwood, a relapsed alcoholic, to remain in charge of the* Exxon Valdez, *Exxon created serious risks to the environment and the residents of Prince William Sound. These risks led to severe harms. The consequences for Exxon—$4.5 billion in punitive damages—were also severe.*

Setting Realistic Goals Helps Managers can reduce the probability that employees will act unethically by setting realistic production or sales goals. ■ EXAMPLE 3.2 Suppose that a sales quota can be met only through high-pressure and unethical sales tactics. Employees trying to act "in the best interests of the firm" may think that management is implicitly asking them to behave unethically. ■

Periodic Evaluation Some companies require their managers to meet individually with employees and to grade them on their ethical (or unethical) behavior. ■ EXAMPLE 3.3 One company asks its employees to fill out ethical checklists each month and return them to their supervisors. This practice serves two purposes: First, it demonstrates to employees that ethics matters. Second, employees have an opportunity to reflect on how well they have measured up in terms of ethical performance. ■

Creating Ethical Codes of Conduct

One of the most effective ways of setting a tone of ethical behavior within an organization is to create an ethical code of conduct. A well-written code of ethics explicitly states a company's ethical priorities.

A Necessity—Clear Communication to Employees For an ethical code to be effective, its provisions must be clearly communicated to employees. Most large companies have implemented ethics training programs, in which managers discuss with employees on a face-to-face basis the firm's policies and the importance of ethical conduct. Some firms hold periodic ethics seminars during which employees can openly discuss any ethical problems that they may be experiencing and learn how the firm's ethical policies apply to those specific problems.

Johnson & Johnson—An Example of Web-Based Ethics Training Creating a code of conduct and implementing it are two different activities. In many companies, codes of conduct are simply documents that have very little relevance to day-to-day operations. When Johnson & Johnson wanted to "do better" than other companies with respect to ethical business decision making, it created a Center for Legal and Credo Awareness. (Its code of ethical conduct is called its *credo.*)

The center created a Web-based set of instructions designed to enhance the corporation's efforts to train employees in the importance of its code of conduct. Given that Johnson & Johnson has more than 110,000 employees in fifty-seven countries around the world, reinforcing its code of conduct and its values has not been easy, but Web-based training has helped. The company established a Web-based legal and compliance center, which uses a set

of interactive modules to train employees in areas of law and ethics.

Corporate Compliance Programs

In large corporations, ethical codes of conduct are usually just one part of a comprehensive corporate compliance program. Other components of such a program, some of which were already mentioned, include a corporation's ethics committee, ethical training programs, and internal audits to monitor compliance with applicable laws and the company's standards of ethical conduct.

The Sarbanes-Oxley Act and Web-Based Reporting Systems The Sarbanes-Oxley Act of 2002[1] requires companies to set up confidential systems so that employees and others may "raise red flags" about suspected illegal or unethical auditing and accounting practices. The act required publicly traded companies to have such systems in place by April 2003.

Some companies have implemented online reporting systems. In one Web-based reporting system, employees can click on an icon on their computers that anonymously links them with Ethicspoint, an organization based in Vancouver, Washington. Through Ethicspoint, employees may report suspicious accounting practices, sexual harassment, and other possibly unethical behavior. Ethicspoint, in turn, alerts management personnel or the audit committee at the designated company to the possible problem. Those who have used the system say that it is less inhibiting than calling a company's toll-free number.

Corporate Governance Principles Implementation of the Sarbanes-Oxley Act has prompted many companies to create new rules of *corporate governance*. As you will read in Chapter 27, corporate governance refers to the internal principles establishing the rights and responsibilities of a corporation's management, board of directors, shareholders, and *stakeholders* (those affected by corporate decisions, including employees, customers, suppliers, and creditors, for example). Corporate governance principles usually go beyond what is required to comply with existing laws. The goal is to set up a system of fair procedures and accurate disclosures that keeps all parties well informed and accountable to each other and provides a mechanism for the corporation to resolve any problems that arise. Ultimately, good corporate governance should attract investors and stimulate growth while discouraging unethical behavior and fraud.

Compliance Programs Must Be Integrated To be effective, a compliance program must be integrated throughout the firm. For large corporations, such integration is essential. Ethical policies and programs need to be coordinated and monitored by a committee that is separate from the various corporate departments. Otherwise, unethical behavior in one department can easily escape the attention of those in control of the corporation or the corporate officials responsible for implementing and monitoring the company's compliance program.

Conflicts and Trade-Offs

Management constantly faces ethical trade-offs, some of which may lead to legal problems. As mentioned earlier, firms have implied ethical (and legal) duties to a number of groups, including shareholders and employees.

When a company decides to reduce costs by downsizing and restructuring, the decision may benefit shareholders, but it will harm those employees who are laid off or fired. When downsizing occurs, which employees should be laid off first? Cost-cutting considerations might dictate firing the most senior employees, who generally have higher salaries, and retaining less senior employees, whose salaries are much lower. A company does not necessarily act illegally when it does so. Yet the decision to be made by management clearly involves an important ethical question: Which group's interests—those of the shareholders or those of employees who have been loyal to the firm for a long period of time—should take priority in this situation?

■ **EXAMPLE 3.4** In one case, an employer facing a dwindling market and decreasing sales decided to reduce its costs by eliminating some of its obligations to its employees. It did this by establishing a subsidiary corporation and then transferring some of its employees, and the administration of their retirement benefits, to that entity. The company expected the subsidiary to fail, and when it did, some employees and retirees who were left with no retirement benefits sued the company. The plaintiffs claimed that the company had breached a fiduciary duty under a federal law governing employer-provided pensions. Ultimately, the United States Supreme Court agreed with the plaintiffs, stating, among other things, that "[l]ying is inconsistent with the duty of loyalty owed by all fiduciaries."[2] ■

1. 15 U.S.C. Sections 7201 *et seq.*

2. *Varity Corp. v. Howe,* 516 U.S. 489, 116 S.Ct. 1065, 134 L.Ed.2d 130 (1996).

Companies That Defy the Rules

One of the best ways to learn the ethical responsibilities inherent in operating a business is to look at the mistakes made by other companies. In the following subsections, we describe some of the ethical failures of companies that have raised public awareness of corporate misconduct and highlighted the need for ethical leadership in business.

Enron's Growth and Demise in a Nutshell

The Enron Corporation was one of the first companies to benefit from the deregulated electricity market. By 1998, Enron was the largest energy trader in the market. When competition in energy trading increased, Enron diversified into water, power plants, and eventually high-speed Internet and fiber optics (the value of which soon became negligible). Because Enron's managers received bonuses based on whether they met earnings goals, they had an incentive to inflate the anticipated earnings on energy contracts, which they did. Enron included these anticipated earnings in its current earnings profits reports, which vastly overstated the company's actual profit. Then, to artificially maintain and even increase its reported earnings, Enron created a complex network of subsidiaries that enabled it to move losses to its subsidiaries and hide its debts.

The overall effect of these actions was to increase Enron's apparent net worth. These "off-the-books" transactions were also frequently carried out in the Cayman Islands to avoid paying federal income taxes. In addition, Enron's chief executive officer (CEO) engaged in a pattern of self-dealing by doing business with companies owned by his son and daughter. Enron's management was informed about these incidents of misconduct on numerous occasions, yet the company concealed the financial improprieties for several years—until Enron was bankrupt.

The Enron Legacy

Deceptive accounting practices were at the heart of the Enron debacle, which led to one of the largest bankruptcies in the history of U.S. business. For years to come, the Enron scandal will remain a symbol of the cost of unethical behavior to management, employees, suppliers, shareholders, the community, society, and indeed the world. Enron's shareholders lost $62 billion of value in a very short period of time in the early 2000s as a result of management's deceptive accounting practices, conflicts of interest, and deviation from accepted ethical standards of business.

Merck & Company—A Brief History of Vioxx

In 1999, Merck & Company, Inc., the maker of Vioxx, received approval from the U.S. Food and Drug Administration (FDA) to market Vioxx for the treatment of acute pain in adults. The FDA gave Vioxx a six-month priority review because it was thought that Vioxx caused fewer gastrointestinal side effects, such as bleeding, than other painkillers (including ibuprofen and aspirin). Merck spent millions of dollars persuading physicians and consumers to use Vioxx for pain, especially arthritis pain, instead of less expensive alternatives, which could cause stomach bleeding. Many people who used Vioxx found that it provided more effective short-term relief for pain, particularly from athletic injuries, than any other painkiller on the market at the time. At its peak, Vioxx had more than 20 million users.

Shortly after the drug's debut, however, troubling signs began to appear. In March 2000, Merck reported the results of a study of eight thousand people who had used Vioxx. The study compared the gastrointestinal effects of Vioxx to naproxen, another popular painkiller. Although the study ultimately found that patients taking Vioxx had less stomach bleeding than those taking naproxen, the study also indicated that patients taking Vioxx for eight months or longer had up to four times as many heart attacks and strokes as patients using naproxen. These results occurred even though the study had excluded patients with risk factors for heart disease.

Independent studies of the drug—conducted in 2001, 2002, and 2004—all suggested correlations between Vioxx and increased risk of heart attack. Finally, Merck's own study revealed that Vioxx increased cardiovascular risks after eighteen months of daily use. Shortly after that, in September 2004, Merck voluntarily removed Vioxx from the worldwide market in the largest drug recall in history.

Merck's Awareness of the Risks of Vioxx

As mentioned, the initial 2000 study on Vioxx and naproxen showed that patients taking Vioxx for an extended period had up to four times as many heart attacks and strokes as those who took naproxen. Because the drug was often prescribed on a long-term basis for arthritis patients, this was a significant finding. Merck attributed the result to naproxen's strong protective effect on the heart. Merck never tested this theory, however, and scientists outside the company who found this explanation unlikely began to conduct independent studies of the drug.

In 2001, a cardiologist proposed to Merck a study of Vioxx in patients with severe chest pain, but Merck

declined. When a 2002 study found that patients who took high doses of Vioxx had significantly more heart attacks and strokes than similar patients, Merck stated that it still had confidence in the drug's safety. Merck maintained this stance even after receiving a warning letter from the FDA in 2001 reprimanding the company for minimizing the drug's potentially serious cardiovascular effects. The FDA required Merck to send letters to physicians across the country to correct false or misleading impressions and information.

MERCK'S CHOICE

In May 2000, Merck's top research and marketing executives met to consider ways to defend Vioxx against allegations that it posed cardiovascular risks. One suggestion was to develop a study that would directly test whether Vioxx posed these risks. That idea was rejected. Merck's marketing executives were apparently afraid that conducting a study would send the wrong signal about the company's faith in Vioxx. The company's position over the following years stayed the same: Vioxx was safe unless proved otherwise. At the time this book went to press, the company continued to maintain that stance, at least with respect to the thousands of lawsuits filed blaming Vioxx as the cause of patients' injuries or deaths. In August 2005, Merck lost its first Vioxx lawsuit, in which the jury awarded $253 million (reduced to $25 million due to Texas's cap on punitive damages) to Carol Ernst, widow of Robert Ernst.

THE DEBATE CONTINUES

The debate over the safety of Vioxx and whether Merck's conduct was ethical poses an interesting question—at what point does a corporation have an ethical duty to act when presented with evidence that its product may be harmful? Various studies estimate that as many as 139,000 people who used Vioxx suffered injury or death. This figure may seem large, but it accounts for less than 1 percent of Vioxx's total users. Some would argue that even one death is too many and that Merck should be responsible for compensating all those who were injured. Others would counter that there are risks involved in the use of any drug and that Merck did nothing wrong by waiting for conclusive evidence of harm before recalling Vioxx. It is likely that the outcome of the many lawsuits that have already been filed—which could ultimately cost Merck billions of dollars in damages—and the litigation still to come will decide whether Merck's conduct was ethical. What is clear is that Merck's shareholders have lost billions of dollars in value since the company recalled Vioxx and suspicions arose about Merck's conduct.

Shortly after Vioxx was recalled in 2004, questions were posed about the safety of other drugs in the same class, called COX-2 inhibitors, which were made by different companies and competed with Vioxx. Concerns over drug safety, unethical practices by pharmaceutical companies, and mass consumer advertising of new medications have prompted many to criticize the FDA and recommend an overhaul of its drug-approval system. Even if the FDA eventually adopts revised procedures, however, questions remain over what exactly a corporation must do to fulfill its ethical duties with regard to notifying the public about the potential risks of using a product.

Business Ethics and the Law

Today, legal compliance is regarded by many as a **moral minimum**—the minimum acceptable standard for ethical business behavior. Had Enron Corporation strictly complied with existing laws and generally accepted accounting practices, very likely the "Enron scandal," which came to light in the early 2000s, would never have happened. Simply obeying the law does not fulfill all business ethics obligations, however. In the interests of preserving personal freedom, as well as for practical reasons, the law does not—and cannot—codify all ethical requirements. No law says, for example, that it is illegal to lie to one's family, but it may be unethical to do so.

It might seem that determining the legality of a given action should be simple. Either something is legal or it is not. In fact, one of the major challenges businesspersons face is that the legality of a particular action is not always clear. In part, this is because business is regulated by so many laws that it is possible to violate one of them without realizing it. The law also contains numerous "gray areas," making it difficult to predict with certainty how a court will apply a given law to a particular action. This is especially true when technological developments have raised new types of questions. For example, if a business's trademark—the distinctive name or symbol that the business uses to identify its products—is used in a domain name followed by "fraud" or "sucks," is this an infringement of the owner's trademark? For a discussion of this question, see this chapter's *Management Perspective* feature on page 64.

The following case illustrates that legal liability and good corporate ethics—represented in this case by corporate governance practices—may not be the same.

CASE 3.3 In re Walt Disney Co. Derivative Litigation

Court of Chancery of Delaware, 2005.
__ A.2d __.

FACTS Frank Wells, the president and chief operating officer of The Walt Disney Company (Disney), died in a helicopter crash in April 1994. Michael Eisner, Disney's chief executive officer (CEO), assumed the presidency. Three months later, Eisner was diagnosed with heart disease and underwent quadruple bypass surgery. Disney was forced to consider who might succeed Eisner. Michael Ovitz, one of the founders of Creative Artist Agency, a theatrical talent agency, and regarded as one of the most powerful figures in Hollywood, was Eisner's choice. Eisner did not call a board meeting to discuss the possibility of hiring Ovitz but negotiated an employment agreement himself. Eisner then contacted each of the directors individually to tell them of the deal. The board met and voted unanimously to elect Ovitz as president in September 1995. Ovitz had no real success at Disney, however, and by late 1996 its officers became critical of his performance. Eisner terminated Ovitz in December, without a board meeting to discuss or vote on the issue. Under the terms of his agreement with Eisner, Ovitz received severance pay and benefits that were valued at more than $100 million. Disney shareholders filed a suit in a Delaware state court against Eisner and others, alleging, among other things, a breach of a duty to act in good faith with respect to Disney.

ISSUE Did Eisner fulfill his obligations to act in good faith and with "honesty of purpose" with respect to Disney and its shareholders?

DECISION Yes. The court cautioned that much of the conduct of the defendants, including Eisner, "fell significantly short of the best practices of ideal corporate governance" but concluded that Eisner had not acted in bad faith.

REASON The court explained, "As a general rule, a CEO has no obligation to continuously inform the board of his actions as CEO, or to receive prior authorization for those actions. Nevertheless, a reasonably prudent CEO * * * would not have acted in as unilateral a manner as did Eisner when essentially committing the corporation to hire a second-in-command, appoint that person to the board, and provide him with one of the largest and richest employment contracts ever enjoyed by a non-CEO." The court added, "Eisner's actions in connection with Ovitz's hiring should not serve as a model for fellow executives and fiduciaries to follow. His lapses were many. He failed to keep the board as informed as he should have. He stretched the outer boundaries of his authority as CEO by acting without specific board direction or involvement." The court stated, "To my mind, these actions fall far short of what shareholders expect and demand from those entrusted with a fiduciary position. Eisner's failure to better involve the board in the process of Ovitz's hiring, usurping that role for himself, although not in violation of law, does not comport with how fiduciaries of * * * corporations are expected to act."

FOR CRITICAL ANALYSIS—Ethical Consideration *Should corporate directors be liable for a failure to comply with an ideal of "best," rather than "competent" or "standard," practices?*

Laws Regulating Business

Today's business firms are subject to extensive government regulation. As mentioned in Chapter 1, virtually every action a firm undertakes—from the initial act of going into business, to hiring and firing personnel, to selling products in the marketplace—is subject to statutory law and to numerous rules and regulations issued by administrative agencies. Furthermore, these rules and regulations are changed or supplemented frequently.

Determining whether a planned action is legal thus requires that decision makers keep abreast of the law. Normally, large business firms have attorneys on their staffs to assist them in making key decisions. Small firms must also seek legal advice before making important business decisions because the consequences of just one violation of a regulatory rule may be costly.

Ignorance of the law will not excuse a business owner or manager from liability for violating a statute or regulation. ■ **EXAMPLE 3.5** In one case, the court imposed criminal fines, as well as imprisonment, on a company's supervisory employee for violating a federal environmental act. This punishment was imposed even though the employee was completely unaware of what was required under the provisions of that act.[3] ■

"Gray Areas" in the Law

In many situations, business firms can predict with a fair amount of certainty whether a given action would be legal. For instance, firing an employee solely because of that person's race or gender would clearly violate federal

3. *United States v. Hanousek,* 176 F.3d 1116 (9th Cir. 1999).

MANAGEMENT PERSPECTIVE

"Sucks" Sites—Can They Be Shut Down?

MANAGEMENT FACES A LEGAL ISSUE

In today's online environment, a recurring challenge for businesses is how to deal with *cybergripers*—those who complain in cyberspace about corporate products, services, or activities. For trademark owners, the issue becomes particularly thorny when cybergriping sites add "sucks," "fraud," "scam," "ripoff," or some other disparaging term as a suffix to the domain name of a particular company. These sites, sometimes collectively referred to as "sucks" sites, are established solely for the purpose of criticizing the products or services sold by the companies that own the marks. In some cases, they have been used maliciously to harm the reputation of a competitor. Can businesses do anything to ward off these cyber attacks on their reputations and goodwill?

WHAT THE COURTS SAY

A number of companies have sued the owners of "sucks" sites for trademark infringement in the hope that a court or an arbitrating panel will order the owner of that site to cease using the domain name. To date, however, companies have had little success pursuing this alternative. In one case, Bear Stearns Companies, Inc., sued a cybergriper, Nye Lavalle, alleging that Lavalle infringed its trademark by creating Web sites that included "Bear Stearns" in the domain names. Two of these sites were called "BearStearnsFrauds.com" and "BearStearnsCriminals.com."

As will be discussed in detail in Chapter 5, one of the tests for trademark infringement is whether consumers would be confused by the use of a similar or identical trademark. Would consumers mistakenly believe that Lavalle's sites were operated by Bear Stearns? In the court's eyes, no. The court concluded that Lavalle's "Frauds.com" and "Criminals.com" sites were "unmistakenly critical" of the target company and that no Internet user would conclude that Bear Stearns sponsored the sites.[a]

For cybergripers, the message seems to be clear: the more outrageous or obnoxious the suffix added to a target company's trademark, the less likely it is that the use will constitute trademark infringement. This point is underscored in decisions reached by other courts as well. In *Taubman Co. v. Webfeats,*[b] for example, the court stressed that Internet users were unlikely to be confused by "sucks" sites using the Taubman Company name. Because the allegedly infringing domain names all ended with "sucks.com," the court concluded that they were unlikely to mislead Web site visitors into believing that the trademark owner was the source or sponsor of the complaint. The court also noted in its opinion that, generally, the more vicious an attack site's domain name, the less likely that a cybergriper will be found liable for trademark infringement.

IMPLICATIONS FOR MANAGERS

Business owners are unlikely to be successful in suits against the owners of "sucks" sites for trademark infringement. (Nor will they succeed in suing individuals who publish negative opinions about them in printed publications for the tort of trade libel—to be discussed in Chapter 4.) Many businesses have concluded that the best defense against cybergripers is to make it more difficult for them by buying up insulting Internet domain names before the cybergripers can do so. In fact, this has become standard practice for many large corporations. Another strategy would be to have someone monitor critical opinion on Web sites for ideas on how the business might improve its products and practices.

a. *Bear Stearns Companies, Inc. v. Lavalle,* 2002 WL 31757771 (N.D.Tex. 2002).
b. 319 F.3d 770 (6th Cir. 2003).

laws prohibiting employment discrimination. In some situations, though, the legality of a particular action may be less clear.

■ **EXAMPLE 3.6** Suppose that a firm decides to launch a new advertising campaign. How far can the firm go in making claims for its products or services? Federal and state laws prohibit firms from engaging in "deceptive advertising." At the federal level, the test for deceptive advertising normally used by the Federal Trade Commission is whether an advertising claim would deceive a "reasonable consumer."[4] At what point, though, would a reasonable consumer be deceived by a particular ad? ■

4. The Federal Trade Commission regulates deceptive trade practices, including misleading advertising.

In short, business decision makers need to proceed with caution and evaluate an action and its consequences from an ethical perspective. Generally, if a company can demonstrate that it acted in good faith and responsibly in the circumstances, it has a better chance of successfully defending its action in court or before an administrative law judge.

TECHNOLOGICAL DEVELOPMENTS AND LEGAL UNCERTAINTIES

Uncertainties concerning how particular laws may apply to specific factual situations have been compounded in the cyber age. As noted in earlier chapters, the widespread use of the Internet has given rise to situations never before faced by the courts.

In one case, for example, a court had to decide whether an online forum for employees was part of the employer's workplace. The case involved an airline pilot who claimed that defamatory, gender-based messages posted by her co-workers in the online forum created a hostile working environment.[5] As will be discussed in Chapter 23, federal law prohibits harassment in the workplace, including "hostile-environment harassment," which occurs when an employee is subjected to sexual conduct or comments that he or she perceives as offensive. Generally, employers are expected to take immediate and appropriate corrective action in response to employees' complaints of sexual harassment or abuse. Otherwise, they may be held liable for the harassing actions of an employee's co-workers or supervisors.

Approaches to Ethical Reasoning

Each individual, when faced with a particular ethical dilemma, engages in **ethical reasoning**—that is, a reasoning process in which the individual examines the situation at hand in light of her or his moral convictions or ethical standards. Businesspersons do likewise when making decisions with ethical implications.

How do business decision makers decide whether a given action is the "right" one for their firms? What ethical standards should they apply? Broadly speaking, ethical reasoning relating to business traditionally has been characterized by two fundamental approaches. One approach defines ethical behavior in terms of duty, which also implies certain rights. The other approach determines what is ethical in terms of the consequences, or outcome, of any given action. We examine each of these approaches here.

Duty-Based Ethics

Duty-based ethical standards are often derived from revealed truths, such as religious precepts. They can also be derived through philosophical reasoning.

Religious Ethical Standards In the Judeo-Christian tradition, which is the dominant religious tradition in the United States, the Ten Commandments of the Old Testament establish fundamental rules for moral action. Other religions have their own sources of revealed truth. Religious rules generally are absolute with respect to the behavior of their adherents. ■ EXAMPLE 3.7 The commandment "Thou shalt not steal" is an absolute mandate for a person, such as a Jew or a Christian, who believes that the Ten Commandments reflect revealed truth. Even a benevolent motive for stealing (such as Robin Hood's) cannot justify the act because the act itself is inherently immoral and thus wrong. ■

Ethical standards based on religious teachings also involve an element of *compassion.* ■ EXAMPLE 3.8 Even though it might be profitable for a firm to lay off a less productive employee, decision makers would give substantial weight to the potential suffering that the employee's family might experience if the employee found it difficult to find employment elsewhere. ■ Compassionate treatment of others is also mandated—to a certain extent, at least—by the Golden Rule of the ancients ("Do unto others as you would have them do unto you"), which has been adopted by most religions.

Kantian Ethics Duty-based ethical standards may also be derived solely from philosophical reasoning. The German philosopher Immanuel Kant (1724–1804), for example, identified some general guiding principles for moral behavior based on what he believed to be the fundamental nature of human beings. Kant held that it is rational to assume that human beings are qualitatively different from other physical objects occupying space. Persons are endowed with moral integrity and the capacity to reason and conduct their affairs rationally. Therefore, their thoughts and actions should be respected. When human beings are treated merely as a means to an end, they are being viewed as the equivalent of objects and are being denied their basic humanity.

A central postulate in Kantian ethics is that individuals should evaluate their actions in light of the consequences that would follow if *everyone* in society acted in the same way. This **categorical imperative** can be applied to any action. ■ EXAMPLE 3.9 Suppose that you are deciding whether to cheat on an examination. If you have adopted Kant's categorical imperative, you will decide not to cheat because if everyone cheated, the examination would be meaningless. ■

The Principle of Rights The principle that human beings have certain fundamental rights (to life, freedom, and the pursuit of happiness, for example) is deeply embedded in Western culture. The *natural law* tradition embraces the concept that certain actions (such as killing another person) are morally wrong because they are contrary to nature (the natural desire to continue living). Those who adhere to this **principle of rights,** or "rights theory," believe that a key factor in determining whether a business decision is ethical is how that decision affects the rights of others. These others include the firm's owners, its employees, the consumers of

5. *Blakey v. Continental Airlines, Inc.,* 164 N.J. 38, 751 A.2d 538 (2000).

its products or services, its suppliers, the community in which it does business, and society as a whole.

A potential dilemma for those who support rights theory, however, is that they may disagree on which rights are most important. When considering all those affected by a business decision, for example, how much weight should be given to employees relative to shareholders, customers relative to the community, or employees relative to society as a whole?

In general, rights theorists believe that whichever right is stronger in a particular circumstance takes precedence. ■ EXAMPLE 3.10 Suppose that a firm can shut down a plant to avoid dumping pollutants into a river, which would negatively affect the health of thousands of people. Alternatively, the firm could keep the plant open (and pollute the river) but save the jobs of the twelve workers in the plant. In this situation, a rights theorist can easily choose which group to favor. (Not all choices are so clear-cut, however.) ■

OUTCOME-BASED ETHICS: UTILITARIANISM

"Thou shalt act so as to generate the greatest good for the greatest number." This is a paraphrase of the major premise of the utilitarian approach to ethics. **Utilitarianism** is a philosophical theory developed by Jeremy Bentham (1748–1832) and then advanced, with some modifications, by John Stuart Mill (1806–1873)—both British philosophers. In contrast to duty-based ethics, utilitarianism is outcome oriented. It focuses on the consequences of an action, not on the nature of the action itself or on any set of preestablished moral values or religious beliefs.

Under a utilitarian model of ethics, an action is morally correct, or "right," when, among the people it affects, it produces the greatest amount of good for the greatest number. When an action affects the majority adversely, it is morally wrong. Applying the utilitarian theory thus requires (1) a determination of which individuals will be affected by the action in question; (2) a **cost-benefit analysis,** which involves an assessment of the negative and positive effects of alternative actions on these individuals; and (3) a choice among alternative actions that will produce maximum societal utility (the greatest positive net benefits for the greatest number of individuals).

The utilitarian approach to decision making commonly is employed by businesses, as well as by individuals. Weighing the consequences of a decision in terms of its costs and benefits for everyone affected by it is a useful analytical tool in the decision-making process. At the same time, utilitarianism is often criticized because its objective, calculated approach to problems tends to reduce the welfare of human beings to plus and minus signs on a cost-benefit worksheet and to "justify" human costs that many find totally unacceptable.

Business Ethics on a Global Level

Given the various cultures and religions throughout the world, it is not surprising that conflicts in ethics frequently arise between foreign and U.S. businesspersons. ■ EXAMPLE 3.11 In certain countries, the consumption of alcohol and specific foods is forbidden for religious reasons. Under such circumstances, it would be thoughtless and imprudent for a U.S. businessperson to invite a local business contact out for a drink. ■

The role played by women in other countries may also present some difficult ethical problems for firms doing business internationally. Equal employment opportunity is a fundamental public policy in the United States, and Title VII of the Civil Rights Act of 1964 prohibits discrimination against women in the employment context (see Chapter 23). Some other countries, however, offer little protection for women against gender discrimination in the workplace, including sexual harassment.

Laws governing workers in other countries have created some especially difficult ethical problems for U.S. sellers of goods manufactured in foreign countries. For example, what if a foreign company exploits its workers—by hiring women and children at below-minimum-wage rates or by requiring its employees to work long hours in a workplace full of health hazards? Given today's global communications network, few companies can assume that their actions in other nations will go unnoticed by "corporate watch" groups that discover and publicize unethical corporate behavior. As a result, U.S. businesses today usually take steps to avoid such adverse publicity—either by refusing to deal with certain suppliers or by arranging to monitor their suppliers' workplaces to make sure that the employees are not being mistreated.

THE FOREIGN CORRUPT PRACTICES ACT

Another ethical problem in international business dealings has to do with the legitimacy of certain side payments to government officials. In the United States, the majority of contracts are formed within the private sector. In many foreign countries, however, decisions on most major construction and manufacturing contracts are made by government officials because of extensive government regulation and control over trade and industry. Side payments to government officials in exchange for favorable business contracts are not unusual in such countries, nor are they considered

to be unethical. In the past, U.S. corporations doing business in these countries largely followed the dictum, "When in Rome, do as the Romans do."

In the 1970s, however, the U.S. press, and government officials as well, uncovered a number of business scandals involving large side payments by U.S. corporations to foreign representatives for the purpose of securing advantageous international trade contracts. In response to this unethical behavior, in 1977 Congress passed the Foreign Corrupt Practices Act (FCPA), which prohibits U.S. businesspersons from bribing foreign officials to secure advantageous contracts.

Prohibition against the Bribery of Foreign Officials The first part of the FCPA applies to all U.S. companies and their directors, officers, shareholders, employees, and agents. This part prohibits the bribery of most officials of foreign governments if the purpose of the payment is to get the officials to act in their official capacity to provide business opportunities.

The FCPA does not prohibit payment of substantial sums to minor officials whose duties are ministerial. These payments are often referred to as "grease," or facilitating payments. They are meant to accelerate the performance of administrative services that might otherwise be carried out at a slow pace. Thus, for example, if a firm makes a payment to a minor official to speed up an import licensing process, the firm has not violated the FCPA.

Generally, the act, as amended, permits payments to foreign officials if such payments are lawful within the foreign country. The act also does not prohibit payments to private foreign companies or other third parties unless the U.S. firm knows that the payments will be passed on to a foreign government in violation of the FCPA.

Accounting Requirements In the past, bribes were often concealed in corporate financial records. Thus, the second part of the FCPA is directed toward accountants. All companies must keep detailed records that "accurately and fairly" reflect the company's financial activities. In addition, all companies must have an accounting system that provides "reasonable assurance" that all transactions entered into by the company are accounted for and legal. These requirements assist in detecting illegal bribes. The FCPA further prohibits any person from making false statements to accountants or false entries in any record or account.

Penalties for Violations In 1988, the FCPA was amended to provide that business firms that violate the act may be fined up to $2 million. Individual officers or directors who violate the FCPA may be fined up to $100,000 (the fine cannot be paid by the company) and may be imprisoned for up to five years.

Other Nations Denounce Bribery

For twenty years, the FCPA was the only law of its kind in the world, despite attempts by U.S. political leaders to convince other nations to pass similar legislation. That situation is now changing. In 1997, the Organization for Economic Cooperation and Development created a convention (treaty) that made the bribery of foreign public officials a serious crime. By 2006, at least thirty-five countries had adopted the convention, which obligates them to enact legislation within their nations in accordance with the treaty. In addition, other international institutions, including the European Union, the Organization of American States, and the United Nations, have either passed or are in the process of negotiating rules against bribery in business transactions.

TERMS AND CONCEPTS

business ethics 57
categorical imperative 65
cost-benefit analysis 66
ethical reasoning 65
ethics 56
moral minimum 62
principle of rights 65
utilitarianism 66

CHAPTER SUMMARY  Ethics and Business Decision Making

Business Ethics (See pages 56–58.)	*Ethics* can be defined as the study of what constitutes right or wrong behavior. *Business ethics* focuses on how moral and ethical principles are applied in the business context.

(Continued)

CHAPTER SUMMARY Ethics and Business Decision Making—Continued

Setting the Right Ethical Tone (See pages 58–60.)	1. *Role of management*—Management's commitment and behavior are essential to creating an ethical workplace. Most large firms have ethical codes or policies and corporate compliance programs to help employees determine whether certain actions are ethical. 2. *Ethical trade-offs*—Management constantly faces ethical trade-offs because firms have ethical and legal duties to a number of groups, including shareholders and employees.
Business Ethics and the Law (See pages 62–65.)	1. *The moral minimum*—Lawful behavior is a moral minimum. The law has its limits, though, and some actions may be legal but not ethical. 2. *Legal uncertainties*—It may be difficult to predict with certainty whether particular actions are legal given the numerous and frequent changes in the laws regulating business and the "gray areas" in the law. 3. *Technological developments and legal uncertainties*—Technological developments can also lead to legal uncertainties until it is clear how the law will be applied to the questions raised by these developments.
Approaches to Ethical Reasoning (See pages 65–66.)	1. *Duty-based ethics*—Ethics based on religious beliefs; philosophical reasoning, such as that of Immanuel Kant; and the basic rights of human beings (the principle of rights). 2. *Outcome-based ethics (utilitarianism)*—Ethics based on philosophical reasoning, such as that of John Stuart Mill, and focusing on the consequences of actions rather than the actions themselves.
Business Ethics on a Global Level (See pages 66–67.)	Businesses must take account of the many cultural, religious, and legal differences among nations. Notable differences relate to the role of women in society, employment laws governing workplace conditions, and the practice of giving side payments to foreign officials to secure favorable contracts.

FOR REVIEW

Answers for the even-numbered questions in this For Review *section can be found in Appendix A at the end of this text.*

1. What is ethics? What is business ethics? Why is business ethics important?
2. How can business leaders encourage their companies to act ethically?
3. What are corporate compliance programs?
4. How do duty-based ethical standards differ from outcome-based ethical standards?
5. What types of ethical issues might arise in the context of international business transactions?

QUESTIONS AND CASE PROBLEMS

3–1. Business Ethics. Some business ethicists maintain that whereas personal ethics has to do with right or wrong behavior, business ethics is concerned with appropriate behavior. In other words, ethical behavior in business has less to do with moral principles than with what society deems to be appropriate behavior in the business context. Do you agree with this distinction? Do personal and business ethics ever overlap? Should personal ethics play any role in business ethical decision making?

3–2. Business Ethics and Public Opinion. Assume that you are a high-level manager for a shoe manufacturer. You know that your firm could increase its profit margin by producing shoes in Indonesia, where you could hire women for $100 a month to assemble them. You also know, however, that human rights advocates recently accused a competing shoe manufacturer of engaging in exploitative labor practices because the manufacturer sold shoes made by Indonesian women for similarly low wages. You personally do not believe that paying $100 a month to Indonesian women is unethical because you know that in their country, $100 a month is a better-than-average wage rate. Assuming that the decision is yours to make, should you have

the shoes manufactured in Indonesia and make higher profits for your company? Should you instead avoid the risk of negative publicity and the consequences of that publicity for the firm's reputation and subsequent profits? Are there other alternatives? Discuss fully.

QUESTION WITH SAMPLE ANSWER

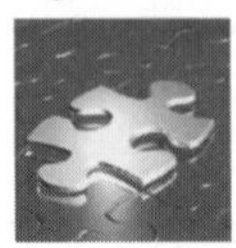

3–3. Human rights groups, environmental activists, and other interest groups concerned with unethical business practices have often conducted publicity campaigns against various corporations that those groups feel have engaged in unethical practices. Do you believe that a small group of well-organized activists should dictate how a major corporation conducts its affairs? Discuss fully.

For a sample answer to this question, go to Appendix B at the end of this text.

3–4. **Ethical Decision Making.** Shokun Steel Co. owns many steel plants. One of its plants is much older than the others. Equipment at that plant is outdated and inefficient, and the costs of production there are now twice as high as at any of Shokun's other plants. The company cannot raise the price of steel because of competition, both domestic and international. The plant employs over a thousand workers and is located in Twin Firs, Pennsylvania, which has a population of about 45,000. Shokun is contemplating whether to close the plant. What factors should the firm consider in making its decision? Will the firm violate any ethical duties if it closes the plant? Analyze these questions from the two basic perspectives on ethical reasoning discussed in this chapter.

3–5. **Ethical Conduct.** Richard and Suzanne Weinstein owned Elm City Cheese Co. Elm City sold its products to three major customers that used the cheese as a "filler" to blend into their cheeses. In 1982, Mark Federico, a certified public accountant, became Elm City's accountant and the Weinsteins' personal accountant. The Weinsteins had known Federico since he was seven years old, and even before he became their accountant, he knew the details of Elm City's business. Federico's duties went beyond typical accounting work, and when the Weinsteins were absent, Federico was put in charge of operations. In 1992, Federico was made a vice president of the company, and a year later he was placed in charge of day-to-day operations. He also continued to serve as Elm City's accountant. The relationship between Federico and the Weinsteins deteriorated, and in 1995, he resigned as Elm City's employee and as its accountant. Less than two years later, Federico opened Lomar Foods, Inc., to make the same products as Elm City by the same process and to sell the products to the same customers. Federico located Lomar closer to Elm City's suppliers. Elm City filed a suit in a Connecticut state court against Federico and Lomar, alleging, among other things, misappropriation of trade secrets. Elm City argued that it was entitled to punitive damages because Federico's conduct was "willful and malicious." Federico responded in part that he did not act willfully and maliciously because he did not know that Elm City's business details were trade secrets. Were Federico's actions "willful and malicious"? Were they ethical? Explain. [*Elm City Cheese Co. v. Federico*, 251 Conn. 59, 752 A.2d 1037 (1999)]

CASE PROBLEM WITH SAMPLE ANSWER

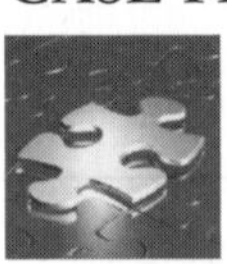

3–6. Richard Fraser was an "exclusive career insurance agent" under a contract with Nationwide Mutual Insurance Co. Fraser leased computer hardware and software from Nationwide for his business. During a dispute between Nationwide and the Nationwide Insurance Independent Contractors Association, an organization representing Fraser and other exclusive career agents, Fraser prepared a letter to Nationwide's competitors asking whether they were interested in acquiring the represented agents' policyholders. Nationwide obtained a copy of the letter and searched its electronic file server for e-mail indicating that the letter had been sent. It found a stored e-mail that Fraser had sent to a co-worker indicating that the letter had been sent to at least one competitor. The e-mail was retrieved from the co-worker's file of already received and discarded messages stored on the server. When Nationwide canceled its contract with Fraser, he filed a suit in a federal district court against the firm, alleging, among other things, violations of various federal laws that prohibit the interception of electronic communications during transmission. In whose favor should the court rule, and why? Did Nationwide act ethically in retrieving the e-mail? Explain. [*Fraser v. Nationwide Mutual Insurance Co.*, 352 F.3d 107 (3d Cir. 2004)]

After you have answered this problem, compare your answer with the sample answer given on the Web site that accompanies this text. Go to academic.cengage.com/blaw/wbl, select "Chapter 3," and click on "Case Problem with Sample Answer."

3–7. **Ethical Conduct.** Unable to pay more than $1.2 billion in debt, Big Rivers Electric Corp. filed a petition to declare bankruptcy in a federal bankruptcy court in September 1996. Big Rivers' creditors included Bank of New York (BONY), Chase Manhattan Bank, Mapco Equities, and others. The court appointed J. Baxter Schilling to work as a "disinterested" (neutral) party with Big Rivers and the creditors to resolve their disputes; the court set an hourly fee as Schilling's compensation. Schilling told Chase, BONY, and Mapco that he wanted them to pay him an additional percentage fee based on the "success" he attained in finding "new value" to pay Big Rivers' debts. He said that without such a deal, he would not perform his mediation duties. Chase agreed; the others disputed the deal, but no one told the court. In October 1998, Schilling asked the court for nearly $4.5 million in compensation, including the hourly fees, which totaled about $531,000, and the percentage fees. Big Rivers and others asked the court to deny Schilling any fees on the basis that he had improperly negotiated "secret side agreements." How did Schilling violate his duties as a "disinterested" party? Should he be denied compensation? Why or why not? [*In re Big Rivers Electric Corp.*, 355 F.3d 415 (6th Cir. 2004)]

3–8. **Ethical Conduct.** Eden Electrical, Ltd., owned twenty-five appliance stores throughout Israel, at least some of which sold

refrigerators made by Amana Co. Eden bought the appliances from Amana's Israeli distributor, Pan El A/Yesh Shem, which approached Eden about taking over the distributorship. Eden representatives met with Amana executives. The executives made assurances about Amana's good faith, its hope of having a long-term business relationship with Eden, and its willingness to have Eden become its exclusive distributor in Israel. Eden signed a distributorship agreement and paid Amana $2.4 million. Amana failed to deliver this amount in inventory to Eden, continued selling refrigerators to other entities for the Israeli market, and represented to others that it was still looking for a long-term distributor. Less than three months after signing the agreement with Eden, Amana terminated it, without explanation. Eden filed a suit in a federal district court against Amana, alleging fraud. The court awarded Eden $12.1 million in damages. Is this amount warranted? Why or why not? How does this case illustrate why business ethics is important? [*Eden Electrical, Ltd. v. Amana Co.,* 370 F.3d 824 (8th Cir. 2004)]

A QUESTION OF ETHICS

3–9. Hazen Paper Co. manufactured paper and paperboard for use in such products as cosmetic wrap, lottery tickets, and pressure-sensitive items. Walter Biggins, a chemist hired by Hazen in 1977, developed a water-based paper coating that was both environmentally safe and of superior quality. By the mid-1980s, the company's sales had increased dramatically as a result of its extensive use of "Biggins Acrylic." Because of this, Biggins thought he deserved a substantial raise in salary, and from 1984 to 1986, Biggins's persistent requests for a raise became a bone of contention between him and his employers. Biggins ran a business on the side, which involved cleaning up hazardous wastes for various companies. Hazen told Biggins that unless he signed a "confidentiality agreement" promising to restrict his outside activities during the time he was employed by Hazen and for a limited time afterward, he would be fired. Biggins said he would sign the agreement only if Hazen raised his salary to $100,000. Hazen refused to do so, fired Biggins, and hired a younger man to replace him. At the time of his discharge in 1986, Biggins was sixty-two years old, had worked for the company nearly ten years, and was just a few weeks away from being entitled to pension rights worth about $93,000. In view of these circumstances, evaluate and answer the following questions. [*Hazen Paper Co. v. Biggins,* 507 U.S. 604, 113 S.Ct. 1701, 123 L.Ed.2d 338 (1993)]

1. Did the company owe an ethical duty to Biggins to increase his salary, given that its sales increased dramatically as a result of Biggins's efforts and ingenuity in developing the coating? If you were one of the company's executives, would you have raised Biggins's salary? Why or why not?
2. Generally, what public policies come into conflict in cases involving employers who, for reasons of cost and efficiency of operations, fire older, higher-paid workers and replace them with younger, lower-paid workers? If you were an employer facing the need to cut back on personnel to save costs, what would you do, and on what ethical premises would you justify your decision?

VIDEO QUESTION

3–10. Go to this text's Web site at academic.cengage.com/blaw/wbl and select "Chapter 3." Click on "Video Questions" and view the video titled *Ethics: Business Ethics an Oxymoron?* Then answer the following questions.

1. According to the instructor in the video, what is the primary reason why businesses act ethically?
2. Which of the two approaches to ethical reasoning that were discussed in the chapter seems to have had more influence on the instructor in the discussion of how business activities are related to societies? Explain your answer.
3. The instructor asserts that "[i]n the end, it is the unethical behavior that becomes costly, and conversely ethical behavior creates its own competitive advantage." Do you agree with this statement? Why or why not?

Ethics and Business Decision Making

For updated links to resources available on the Web, as well as a variety of other materials, visit this text's Web site at

academic.cengage.com/blaw/wbl

You can find articles on shareholders and corporate accountability at the Corporate Governance Web site. Go to

http://ww.corpgov.net

Global Exchange offers information on global business activities, including some of the ethical issues stemming from those activities, at

http://www.globalexchange.org

ONLINE LEGAL RESEARCH

Go to the *Fundamentals of Business Law* home page at academic.cengage.com/blaw/wbl, select "Chapter 3," and click on "Internet Exercises." There you will find the following Internet research exercises that you can perform to learn more about topics covered in this chapter.

Activity 3–1: LEGAL PERSPECTIVE—Ethics in Business

Activity 3–2: MANAGEMENT PERSPECTIVE—Environmental Self-Audits

BEFORE THE TEST

Go to the *Fundamentals of Business Law* home page at academic.cengage.com/blaw/wbl, select "Chapter 3," and click on "Interactive Quizzes." You will find a number of interactive questions relating to this chapter.

UNIT 1 • EXTENDED CASE STUDY

Vo v. City of Garden Grove

As explained in Chapter 1, the First Amendment protects the freedom of speech. This protection is not unlimited, however. Expression—oral, written, or symbolized by conduct—is subject to reasonable restrictions on time, place, and manner. A restriction is valid if it can be justified without reference to the content of the regulated speech. It also must be narrowly tailored (go no further than necessary) to serve a significant governmental interest and leave open ample alternative channels for communication of information. The requirement of narrow tailoring is satisfied as long as the regulation promotes a substantial governmental interest that would be achieved less effectively without the regulation. In this extended case study, we examine *Vo v. City of Garden Grove,*[1] a decision in which these principles were applied.

1. 115 Cal.App.4th 425, 9 Cal.Rptr.3d 257 (4 Dist. 2004).

Case Background

On December 30, 2001, Phong Ly, a minor, was stabbed to death with a screwdriver while standing outside PC Café, a cybercafé in the city of Garden Grove, California. A *cybercafé* is an establishment that provides Internet access to customers for a fee. The number of cybercafés in Garden Grove rose from three to twenty-two between 2000 and 2002. Over the same period, gang-related activities, including violent crime, increased near some of the cybercafés. Within a few days of Ly's murder, on the basis of a report from Joseph Polisar (the city's chief of police) who detailed the accelerating criminal activity, the city enacted a temporary ordinance to halt the establishment of new cybercafés.

In July 2002, on the strength of an updated report from Polisar, the city enacted a permanent ordinance that, as amended the following December, required cybercafé owners to provide security guards and install video surveillance systems, among other things. Minors were prohibited from visiting the cybercafés during school hours unless accompanied by a parent or guardian. Thanh Thuy Vo and four other owners filed a suit in a California state court against the city, alleging in part that the ordinance violated their right to freedom of speech. The court agreed with the owners and enjoined the city from enforcing the ordinance. The city appealed to a state intermediate appellate court.

Majority Opinion

IKOLA, J. [Judge]

* * * *

From the information provided [by Chief Polisar], the city concluded that excluding minors from Cybercafes during school hours would advance its significant public interest in their protection and safety. That conclusion is reasonable. Although parents presumably believe their minor children are in school while it is in session, they are not in a position to assert direct supervision and control during school hours. As noted by the chief of police, if Cybercafes allow minor children on the premises during school hours, the potential that gang members will recruit minors is increased, as well as the potential that minors will become witnesses or victims of gang violence. *Thus, the regulation promotes a substantial government interest that would be achieved less effectively absent the regulation.* This is all that is required to meet the narrow tailoring requirement. [Emphasis added.]

Further, the means chosen to advance the city's interest are not substantially broader than necessary. The city perceived that danger to minors existed in the risky environment of the Cybercafes. The daytime curfew is limited to Cybercafes with their risky environment, and to those times when the students are not under the presumed direct control and supervision of their parents. * * *

Finally, plaintiffs presented no evidence to establish the lack of open ample alternative channels for communication. It is common knowledge that alternative channels for communication over the Internet are abundant. Many have Internet access at home. Schools (where the minors should be in any event) commonly provide Internet access, as do public libraries. And, of course, the Cybercafes themselves are open to minors, even without parental supervision, for seven hours each day.

* * * *

* * * [T]he staffing requirements make no reference to the content of any communication and are thus content neutral. And, as with the daytime curfew, ample alternative channels for communication are available. The remaining question is whether the staffing requirements are narrowly tailored to advance the city's substantial interest in public safety.

* * * *

Given the well-demonstrated criminal activity observed at Cybercafes, and their tendency to attract gang members, the * * * staffing requirements are * * * narrowly tailored to serve a significant governmental interest. Ample alternative means of communication remain open, and the requirements are not substantially broader than necessary. * * *

* * * *

* * * [W]e are not persuaded the video surveillance system affects First Amendment activity any more than does the presence of an adult employee and/or security guard. The ordinance does not require video surveillance of e-mail or images from the Internet appearing on the customer's computer screens. The ordinance requires only that the system be capable of showing "the activity and physical features of persons or areas within the premises." This is no more than can be observed by employees, security guards, or indeed, other customers. * * * For the reasons discussed * * * in connection with the employee and security guard requirements, the video surveillance requirement is a content-neutral manner restriction, narrowly tailored to advance the city's legitimate interest in public safety and deterrence of gang violence. * * *

* * * *

* * * The court's order * * * enjoining [preventing] the enforcement of * * * [the cybercafé ordinance is] reversed.

DISSENTING OPINION

SILLS, P.J. [Presiding Judge], * * * Dissenting.

I respectfully dissent to the most important part of the majority opinion, in which it holds that Garden Grove may *require* video surveillance in *every* cybercafe in the city, regardless of whether that cybercafe has experienced any gang-related violence, or, indeed, even any problems of the most minor nature.

* * * [O]nly 3 of 22 cybercafes have experienced "gang-related" violence, only 2 more have experienced serious crime of any kind (one of the two was a drug deal), * * * [and] the city's own evidence concerning cybercafes in other cities showed *no* gang-related crimes at cybercafes outside of Garden Grove. That leaves 17 cybercafes which have experienced no serious problems, a fact which should be enough to require this court to affirm the trial court's injunction * * * .

It is the video surveillance issue, though, that is the most problematic[.] * * *

* * * *

* * * There are any number of substantial means by which the city's interest in protecting against gang violence could be realized without video surveillance. Police patrols could be increased. Owners could be supplied with a list of gang members who could be refused service. Security guards could be posted at those cybercafes which have already experienced gang-related violence. Yet the majority steadfastly refuse to confront such possibilities, all in the name of deference to " * * * facts" found by the city council. Whatever that is, it isn't trying to minimize any burden on * * * speech to what is reasonably necessary.

* * * *

* * * This is the way Constitutional rights are lost. Not in the thunder of a tyrant's edict, but in the soft judicial whispers of deference.

QUESTIONS FOR ANALYSIS

1. **Law.** What is the significant, or substantial, governmental interest that is supported by the city ordinance in the *Vo* case?
2. **Law.** How exactly does this ordinance implicate First Amendment activities involving "speech"?
3. **Ethics.** Would an ordinance that allowed city officials to deny a business license to a cybercafé whose "operation does not comport with good morals" be constitutional under the First Amendment?
4. **Technological Dimensions.** Why might a cybercafé attract more criminal activity and gang-related violence than other business establishments?
5. **Implications for the Business Manager.** Before installing a video surveillance system, should a business consider any other legal issues besides those relating to freedom of speech?

UNI

Torts and Crimes

2

Unit Contents

CHAPTER 4
Torts and Cyber Torts

LEARNING OBJECTIVES

After reading this chapter, you should be able to answer the following questions:

1. What is a tort?
2. What is the purpose of tort law? What are two basic categories of torts?
3. What are the four elements of negligence?
4. What is meant by strict liability? In what circumstances is strict liability applied?
5. What is a cyber tort, and how are tort theories being applied in cyberspace?

Torts are wrongful actions.[1] Through tort law, society compensates those who have suffered injuries as a result of the wrongful conduct of others. Although some torts, such as assault and trespass, originated in the English common law, the field of tort law continues to expand. As new ways to commit wrongs are discovered, such as the use of the Internet to commit wrongful acts, the courts are extending tort law to cover these wrongs.

As you will see in later chapters of this book, many of the lawsuits brought by or against business firms are based on the tort theories discussed in this chapter. Some of the torts examined here can occur in any context, including the business environment. Others traditionally have been referred to as **business torts,** which are defined as wrongful interferences with the business rights of others. Included in business torts are such vague concepts as *unfair competition* and *wrongfully interfering with the business relations of others.*

1. The word *tort* is French for "wrong."

The Basis of Tort Law

Two notions serve as the basis of all torts: wrongs and compensation. Tort law is designed to compensate those who have suffered a loss or injury due to another person's wrongful act. In a tort action, one person or group brings a personal suit against another person or group to obtain compensation (money **damages**) or other relief for the harm suffered.

THE PURPOSE OF TORT LAW

Generally, the purpose of tort law is to provide remedies for the invasion of various *protected interests.* Society recognizes an interest in personal physical safety, and tort law provides remedies for acts that cause physical injury or that interfere with physical security and freedom of movement. Society recognizes an interest in protecting real and personal property, and tort law provides remedies for acts that cause destruction or damage to property. Society also recognizes an interest in protecting certain intangible interests (such as personal privacy, family relations, reputation, and dignity), and tort law provides remedies for invasion of these protected interests.

CLASSIFICATIONS OF TORTS

There are two broad classifications of torts: *intentional torts* and *unintentional torts* (torts involving negligence). The classification of a particular tort depends largely on how the tort occurs (intentionally or negligently) and the surrounding circumstances. In the following pages, you will read about these two classifications of torts.

Torts committed via the Internet are sometimes referred to as *cyber torts*. We look at how the courts have applied traditional tort law to wrongful actions in the online environment in the concluding pages of this chapter.

Intentional Torts against Persons

An **intentional tort,** as the term implies, requires *intent*. The **tortfeasor** (the one committing the tort) must intend to commit an act, the consequences of which interfere with the personal or business interests of another in a way not permitted by law. An evil or harmful motive is not required—in fact, the actor may even have a beneficial motive for committing what turns out to be a tortious act. In tort law, intent means only that the actor intended the consequences of his or her act or knew with substantial certainty that specific consequences would result from the act. The law generally assumes that individuals intend the *normal* consequences of their actions. Thus, forcefully pushing another—even if done in jest and without any evil motive—is an intentional tort (if injury results), because the object of a strong push can ordinarily be expected to fall down.

This section discusses intentional torts against persons, which include assault and battery, false imprisonment, intentional infliction of emotional distress, defamation, invasion of the right to privacy, appropriation, misrepresentation (fraud), and wrongful interference.

ASSAULT AND BATTERY

Any intentional, unexcused act that creates in another person a reasonable apprehension or fear of immediate harmful or offensive contact is an **assault.** Apprehension is not the same as fear. If a contact is such that a reasonable person would want to avoid it, and if there is a reasonable basis for believing that the contact will occur, then the plaintiff suffers apprehension whether or not he or she is afraid. The interest protected by tort law concerning assault is the freedom from having to expect harmful or offensive contact. The occurrence of apprehension is enough to justify compensation.

The *completion* of the act that caused the apprehension, if it results in harm to the plaintiff, is a **battery,** which is defined as an unexcused and harmful or offensive physical contact *intentionally* performed. For example, suppose that Ivan threatens Jean with a gun, then shoots her. The pointing of the gun at Jean is an assault; the firing of the gun (if the bullet hits Jean) is a battery. The interest protected by tort law concerning battery is the right to personal security and safety. The contact can be harmful, or it can be merely offensive (such as an unwelcome kiss). Physical injury need not occur. The contact can involve any part of the body or anything attached to it, such as a hat or other item of clothing, a purse, or a chair or an automobile in which one is sitting. Whether the contact is offensive or not is determined by the *reasonable person standard.*[2] The contact can be made by the defendant or by some force the defendant sets in motion—for example, a rock thrown, food poisoned, or a stick swung.

Compensation If the plaintiff shows that there was contact, and the jury agrees that the contact was offensive, the plaintiff has a right to compensation. There is no need to show that the defendant acted out of malice; the person could have just been joking or playing around. The underlying motive does not matter, only the intent to bring about the harmful or offensive contact to the plaintiff. In fact, proving a motive is never necessary (but is sometimes relevant). A plaintiff may be compensated for the emotional harm or loss of reputation resulting from a battery, as well as for physical harm.

Defenses to Assault and Battery A number of legally recognized **defenses** (reasons why plaintiffs should not obtain what they are seeking) can be raised by a defendant who is sued for assault or battery, or both:

1. *Consent.* When a person consents to the act that damages her or him, there is generally no liability (legal responsibility) for the damage done.
2. *Self-defense.* An individual who is defending his or her life or physical well-being can claim self-defense. In situations of both real and apparent danger, a person may use whatever force is reasonably necessary to prevent harmful contact.
3. *Defense of others.* An individual can act in a reasonable manner to protect others who are in real or apparent danger.

2. The *reasonable person standard* is an objective test of how a reasonable person would have acted under the same circumstances. See "The Duty of Care and Its Breach" later in this chapter.

4 *Defense of property.* Reasonable force may be used in attempting to remove intruders from one's home, although force that is likely to cause death or great bodily injury can never be used just to protect property.

FALSE IMPRISONMENT

False imprisonment is the intentional confinement or restraint of another person's activities without justification. False imprisonment interferes with the freedom to move without restraint. The confinement can be accomplished through the use of physical barriers, physical restraint, or threats of physical force. Moral pressure or threats of future harm do not constitute false imprisonment. It is essential that the person being restrained not comply with the restraint willingly.

Businesspersons are often confronted with suits for false imprisonment after they have attempted to confine a suspected shoplifter for questioning. Under the "privilege to detain" granted to merchants in some states, a merchant can use the defense of *probable cause* to justify delaying a suspected shoplifter. Probable cause exists when there is sufficient evidence to support the belief that a person is guilty. Although laws governing false imprisonment vary from state to state, generally they require that any detention be conducted in a *reasonable* manner and for only a *reasonable* length of time.

INTENTIONAL INFLICTION OF EMOTIONAL DISTRESS

The tort of *intentional infliction of emotional distress* can be defined as an intentional act that amounts to extreme and outrageous conduct resulting in severe emotional distress to another. ■ EXAMPLE 4.1 A prankster telephones an individual and says that the individual's spouse has just been in a horrible accident. As a result, the individual suffers intense mental pain or anxiety. The caller's behavior is deemed to be extreme and outrageous conduct that exceeds the bounds of decency accepted by society and is therefore **actionable** (capable of serving as the ground for a lawsuit). ■

The tort of intentional infliction of emotional distress poses several problems for the courts. One problem is the difficulty of proving the existence of emotional suffering. For this reason, courts in some jurisdictions require that the emotional distress be evidenced by some physical symptom or illness or some emotional disturbance that can be documented by a psychiatric consultant or other medical professional.

Another problem is that emotional distress claims must be subject to some limitation, or they could flood the courts with lawsuits. A society in which individuals are rewarded if they are unable to endure the normal emotional stresses of day-to-day living is obviously undesirable. Therefore, the law usually holds that indignity or annoyance alone is not enough to support a lawsuit based on the intentional infliction of emotional distress. Repeated annoyances (such as those experienced by a person who is being stalked), however, coupled with threats, are enough. In the business context, the repeated use of extreme methods to collect a delinquent account may be actionable.

DEFAMATION

Defamation of character involves wrongfully hurting a person's good reputation. The law has imposed a general duty on all persons to refrain from making false, defamatory statements about others. Breaching this duty in writing involves the tort of **libel.** Breaching this duty orally involves the tort of **slander.** The tort of defamation also arises when a false statement is made about a person's product, business, or title to property. We deal with these torts later in the chapter.

The Publication Requirement The basis of the tort of defamation is the publication of a statement or statements that hold an individual up to contempt, ridicule, or hatred. *Publication* here means that the defamatory statements are communicated to persons other than the defamed party. ■ EXAMPLE 4.2 If Thompson writes Andrews a private letter accusing him of embezzling funds, the action does not constitute libel. If Peters calls Gordon dishonest, two faced, and incompetent when no one else is around, the action does not constitute slander. In neither instance was the message communicated to a third party. ■

The courts have generally held that even dictating a letter to an administrative assistant constitutes publication, although the publication may be privileged (privileged communications will be discussed shortly). Moreover, if a third party overhears defamatory statements by chance, the courts usually hold that this also constitutes publication. Defamatory statements made via the Internet are also actionable, as you will read later in this chapter. Note further that any individual who republishes or repeats defamatory statements is liable even if that person reveals the source of such statements.

Damages for Defamation Generally, in a case alleging slander, the plaintiff must prove "special damages" to establish the defendant's liability. The plaintiff must show that the slanderous statement caused her or him to suffer actual economic or monetary losses. This requirement is

imposed in cases involving slander because slanderous statements have a temporary quality. In contrast, a libelous (written) statement has the quality of permanence, can be circulated widely, and usually results from some degree of deliberation on the part of the author. For that reason, a plaintiff can recover "general damages" for libel and need not prove any special damages.

Exceptions to the burden of proving special damages in cases alleging slander are made for certain types of slanderous statements. If a false statement constitutes "slander *per se,*" no proof of special damages is required for it to be actionable. The following four types of utterances are considered to be slander *per se:*

1. A statement that another has a loathsome communicable disease.
2. A statement that another has committed improprieties while engaging in a profession or trade.
3. A statement that another has committed or has been imprisoned for a serious crime.
4. A statement that a woman is unchaste or has engaged in serious sexual misconduct.

Defenses against Defamation Truth is normally an absolute defense against a defamation charge. In other words, if the defendant in a defamation suit can prove that his or her allegedly defamatory statements were true, the defendant will not be liable.

—Privileged Communications. Another defense that is sometimes raised is that the statements were **privileged** communications, and thus the defendant is immune from liability. Privileged communications are of two types: absolute and qualified. Only in judicial proceedings and certain government proceedings is *absolute* privilege granted. For example, statements made in the courtroom by attorneys and judges during a trial are absolutely privileged. So are statements made by government officials during legislative debate, even if the officials make such statements maliciously—that is, knowing them to be untrue. An absolute privilege is granted in these situations because government personnel deal with matters that are so much in the public interest that the parties involved should be able to speak out fully and freely without restriction.

A *qualified,* or *conditional,* privilege applies when a statement is related to a matter of public interest or when the statement is necessary to protect a person's private interest and is made to another person with an interest in the same subject matter. ■ EXAMPLE 4.3 Jorge applies for membership at the local country club. After the country club's board rejects his application, Jorge sues the club's office manager for making allegedly defamatory statements to the board concerning a conversation she had with Jorge. Assuming that the office manager had simply relayed what she considered was her duty to convey to the club's board, her statements would likely be protected by a qualified privilege. ■

—Public Figures. In general, false and defamatory statements that are made about *public figures* (public officials who exercise substantial governmental power and any persons in the public limelight) and that are published in the press are privileged if they are made without **actual malice.**[3] To be made with actual malice, a statement must be made *with either knowledge of its falsity or a reckless disregard of the truth.* Statements made about public figures, especially when they are made via a public medium, are usually related to matters of general interest; they are made about people who substantially affect all of us. Furthermore, public figures generally have some access to a public medium for answering disparaging (belittling, discrediting) falsehoods about themselves; private individuals do not. For these reasons, public figures have a greater burden of proof in defamation cases (they must prove actual malice) than do private individuals.

Invasion of the Right to Privacy

A person has a right to solitude and freedom from prying public eyes—in other words, to privacy. Although the U.S. Constitution does not contain an explicit guarantee of a right to privacy, the United States Supreme Court has held that a fundamental right to privacy is implied by various amendments to the Constitution. Some state constitutions explicitly provide for privacy rights.[4] Tort law also safeguards these rights through the tort of *invasion of privacy.* Four acts qualify as an invasion of privacy:

1. *The use of a person's name, picture, or other likeness for commercial purposes without permission.* This tort, which is usually referred to as the tort of appropriation, will be examined shortly.
2. *Intrusion into an individual's affairs or seclusion.* For example, invading someone's home or illegally searching someone's briefcase is an invasion of privacy. The tort has been held to extend to eavesdropping by wiretap, the unauthorized scanning of a bank account, compulsory blood testing, and window peeping.

3. *New York Times Co. v. Sullivan,* 376 U.S. 254, 84 S.Ct. 710, 11 L.Ed.2d 686 (1964).

4. Note also that a number of federal and state statutes protect individual rights in specific areas.

3 *Publication of information that places a person in a false light.* This could be a story attributing to the person ideas not held or actions not taken by the person. (Publishing such a story could involve the tort of defamation as well.)
4 *Public disclosure of private facts about an individual that an ordinary person would find objectionable.* A newspaper account of a private citizen's sex life or financial affairs could be an actionable invasion of privacy.

APPROPRIATION

The use by one person of another person's name, likeness, or other identifying characteristic, without permission and for the benefit of the user, constitutes the tort of **appropriation.** Under the law, an individual's right to privacy normally includes the right to the exclusive use of her or his identity.

■ **EXAMPLE 4.4** Vanna White, the hostess of the popular television game show *Wheel of Fortune,* brought a case against Samsung Electronics America, Inc. Without White's permission, Samsung included in an advertisement for its videocassette recorders (VCRs) a depiction of a robot dressed in a wig, gown, and jewelry, posed in a scene that resembled the *Wheel of Fortune* set, in a stance for which White is famous. The court held in White's favor, holding that the tort of appropriation does not require the use of a celebrity's name or likeness. The court stated that Samsung's robot ad left "little doubt" as to the identity of the celebrity whom the ad was meant to depict.[5] ■

Cases of wrongful appropriation, or misappropriation, may also involve the rights of those who invest time and funds in the creation of a special system, such as a method of broadcasting sports events. Commercial misappropriation may also occur when a person takes and uses the property of another for the sole purpose of capitalizing unfairly on the goodwill or reputation of the property owner.

MISREPRESENTATION (FRAUD)

A misrepresentation leads another to believe in a condition that is different from the condition that actually exists. This is often accomplished through a false or incorrect statement. Misrepresentations may be innocently made by someone who is unaware of the existing facts, but the tort of **fraudulent misrepresentation,** or fraud, involves intentional deceit for personal gain. The tort includes several elements:

1 The misrepresentation of facts or conditions with knowledge that they are false or with reckless disregard for the truth.
2 An intent to induce another to rely on the misrepresentation.
3 Justifiable reliance by the deceived party.
4 Damages suffered as a result of the reliance.
5 A causal connection between the misrepresentation and the injury suffered.

For fraud to occur, more than mere **puffery,** or *seller's talk,* must be involved. Fraud exists only when a person represents as a fact something he or she knows is untrue. For example, it is fraud to claim that a roof does not leak when one knows it does. Facts are objectively ascertainable, whereas seller's talk is not. "I am the best accountant in town" is seller's talk. The speaker is not trying to represent something as fact because the term *best* is a subjective, not an objective, term.[6]

Normally, the tort of misrepresentation or fraud occurs only when there is reliance on a *statement of fact.* Sometimes, however, reliance on a *statement of opinion* may involve the tort of misrepresentation if the individual making the statement of opinion has a superior knowledge of the subject matter. For example, when a lawyer makes a statement of opinion about the law in a state in which the lawyer is licensed to practice, a court would construe reliance on such a statement to be equivalent to reliance on a statement of fact. We examine fraudulent misrepresentation in further detail in Chapter 10, in the context of contract law.

WRONGFUL INTERFERENCE

Business torts involving wrongful interference are generally divided into two categories: wrongful interference with a contractual relationship and wrongful interference with a business relationship.

Wrongful Interference with a Contractual Relationship The body of tort law relating to *intentional interference with a contractual relationship* has expanded greatly in recent years. A landmark case involved an opera singer, Joanna Wagner, who was under contract to sing for a man named Lumley for a specified period of years. A man named Gye, who knew of this contract, nonetheless "enticed" Wagner to refuse to carry out the agreement, and Wagner began to sing for Gye. Gye's action consti-

5. *White v. Samsung Electronics America, Inc.,* 971 F.2d 1395 (9th Cir. 1992).

6. In contracts for the sale of goods, Article 2 of the Uniform Commercial Code distinguishes, for warranty purposes, between statements of opinion (*puffery*) and statements of fact. See Chapter 17 for a further discussion of this issue.

tuted a tort because it wrongfully interfered with the contractual relationship between Wagner and Lumley.[7] (Of course, Wagner's refusal to carry out the agreement also entitled Lumley to sue Wagner for breach of contract.)

Three elements are necessary for wrongful interference with a contractual relationship to occur:

1. A valid, enforceable contract must exist between two parties.
2. A third party must know that this contract exists.
3. The third party must *intentionally* cause either of the two parties to breach the contract.

7. *Lumley v. Gye,* 118 Eng.Rep. 749 (1853).

The contract may be between a firm and its employees or a firm and its customers. Sometimes a competitor of a firm draws away one of the firm's key employees. If the original employer can show that the competitor induced the breach—that is, that the former employee would not otherwise have broken the contract—damages can be recovered from the competitor.

The following case illustrates the elements of the tort of wrongful interference with a contractual relationship in the context of a contract between an independent sales representative and his agent (agency relationships will be discussed in Chapter 22). The case was complicated by the existence of a second contract between the sales representative and the third party.

CASE 4.1 Mathis v. Liu

United States Court of Appeals,
Eighth Circuit, 2002.
276 F.3d 1027.

FACTS Ching and Alex Liu own Pacific Cornetta, Inc. In 1997, Pacific Cornetta entered into a contract with Lawrence Mathis, under which Mathis agreed to solicit orders for Pacific Cornetta's products from Kmart Corporation for a commission of 5 percent on net sales. Under the terms, either party could terminate the contract at any time. The next year, Mathis entered into a one-year contract with John Evans, under which Evans agreed to serve as Mathis's agent to solicit orders from Kmart for the product lines that Mathis represented, including Pacific Cornetta, for a commission of 1 percent on net sales. Under the terms of this contract, either party could terminate it only on written notice of six months. A few months later, Pacific Cornetta persuaded Evans to break his contract with Mathis and enter into a contract with Pacific Cornetta to be its sales representative to Kmart. Evans terminated his contract with Mathis without notice. Two days later, Pacific Cornetta terminated its contract with Mathis. Mathis filed a suit in a federal district court against the Lius and Pacific Cornetta, alleging in part wrongful interference with a contractual relationship. The court issued a judgment that included a ruling in Mathis's favor on this claim, but Mathis appealed the amount of damages to the U.S. Court of Appeals for the Eighth Circuit.

ISSUE Had the defendants wrongfully interfered with Mathis's contract with Evans?

DECISION Yes. The appellate court affirmed the lower court's judgment, holding that the defendants had wrongfully interfered with Mathis's contractual relationship with Evans.

REASON According to the court, the factors used to determine whether a defendant's interference is improper include the following: the nature of the actor's conduct, the actor's motive, the interests of the other with which the actor's conduct interferes, the interests sought to be advanced by the actor, the social interests in protecting the freedom of action of the actor and the contractual interests of the other, the proximity or remoteness of the actor's conduct to the interference, and the relations between the parties. Because Evans's contract with Mathis required Evans to give Mathis six months' notice prior to terminating his employment, the court noted that this was not "a simple at-will arrangement" [a common law doctrine under which an employment contract can be terminated at any time by either party for any, or no, reason—see Chapter 23]. In these circumstances, the court reasoned, "the jury was entitled to conclude that Pacific Cornetta's blandishments [enticements] were improper, especially since inducing a breach of contract absent compelling justification is, in and of itself, improper." Evans's sales of Pacific Cornetta products after Pacific Cornetta terminated its contract with Mathis did not furnish a basis for damages on this claim, however, because the contract with Mathis was terminable at will.

FOR CRITICAL ANALYSIS—Ethical Consideration
Does the ruling in this case mean that Mathis is entirely without recourse? Could he sue Evans for anything?

Wrongful Interference with a Business Relationship Businesspersons devise countless schemes to attract customers, but they are forbidden by the courts to interfere unreasonably with another's business in their attempts to gain a share of the market. There is a difference between competitive methods and **predatory behavior**—actions undertaken with the intention of unlawfully driving competitors completely out of the market.

The distinction usually depends on whether a business is attempting to attract customers in general or to solicit only those customers who have shown an interest in a similar product or service of a specific competitor. If a shopping center contains two shoe stores, an employee of Store A cannot be positioned at the entrance of Store B for the purpose of diverting customers to Store A. This type of activity constitutes the tort of wrongful interference with a business relationship, which is commonly considered to be an unfair trade practice. If this type of activity were permitted, Store A would reap the benefits of Store B's advertising.

Defenses to Wrongful Interference A person will not be liable for the tort of wrongful interference with a contractual or business relationship if it can be shown that the interference was justified, or permissible. Bona fide competitive behavior is a permissible interference even if it results in the breaking of a contract. ■ EXAMPLE 4.5 If Antonio's Meats advertises so effectively that it induces Beverly's Restaurant Chain to break its contract with Otis Meat Company, Otis Meat Company will be unable to recover against Antonio's Meats on a wrongful interference theory. After all, the public policy that favors free competition in advertising outweighs any possible instability that such competitive activity might cause in contractual relations. ■

Intentional Torts against Property

Intentional torts against property include trespass to land, trespass to personal property, conversion, and disparagement of property. These torts are wrongful actions that interfere with individuals' legally recognized rights with regard to their land or personal property. The law distinguishes real property from personal property (see Chapters 28 and 29). *Real property* is land and things "permanently" attached to the land. *Personal property* consists of all other items, which are basically movable. Thus, a house and lot are real property, whereas the furniture inside a house is personal property. Money and stocks and bonds are also personal property.

Trespass to Land

A **trespass to land** occurs whenever a person, without permission, enters onto, above, or below the surface of land that is owned by another; causes anything to enter onto the land; remains on the land; or permits anything to remain on it. Actual harm to the land is not an essential element of this tort because the tort is designed to protect the right of an owner to exclusive possession of his or her property. Common types of trespass to land include walking or driving on someone else's land, shooting a gun over the land, throwing rocks at a building that belongs to someone else, building a dam across a river and thereby causing water to back up on someone else's land, and constructing a building so that part of it is on an adjoining landowner's property.

Trespass Criteria, Rights, and Duties Before a person can be a trespasser, the owner of the real property (or other person in actual and exclusive possession of the property) must establish that person as a trespasser. For example, "posted" trespass signs expressly establish as a trespasser a person who ignores these signs and enters onto the property. A guest in your home is not a trespasser—unless she or he has been asked to leave but refuses. Any person who enters onto your property to commit an illegal act (such as a thief entering a lumberyard at night to steal lumber) is established impliedly as a trespasser, without posted signs.

At common law, a trespasser is liable for damages caused to the property and generally cannot hold the owner liable for injuries sustained on the premises. This common law rule is being abandoned in many jurisdictions in favor of a "reasonable duty of care" rule that varies depending on the status of the parties; for example, a landowner may have a duty to post a notice that the property is patrolled by guard dogs. Furthermore, under the "attractive nuisance" doctrine, children do not assume the risks of the premises if they are attracted to the property by some object, such as a swimming pool, an abandoned building, or a sandpile. Trespassers normally can be removed from the premises through the use of reasonable force without the owner's being liable for assault and battery.

Defenses against Trespass to Land Trespass to land involves wrongful interference with another person's real property rights. One defense against a trespass claim is to show that the trespass was warranted, as when a trespasser enters to assist someone in danger. Another defense exists when the trespasser can show that he or she had a license to come onto the land. A *licensee* is one who is

invited (or allowed to enter) onto the property of another for the licensee's benefit. A person who enters another's property to read an electric meter, for example, is a licensee. When you purchase a ticket to attend a movie or sporting event, you are licensed to go onto the property of another to view that movie or event. Note that licenses to enter on another's property are *revocable* by the property owner. If a property owner asks a meter reader to leave and the meter reader refuses to do so, the meter reader at that point becomes a trespasser.

TRESPASS TO PERSONAL PROPERTY

Whenever an individual unlawfully harms the personal property of another or otherwise interferes with the personal property owner's right to exclusive possession and enjoyment of that property, **trespass to personal property**—also called *trespass to personalty*[8]—occurs. If a student takes a classmate's business law book as a practical joke and hides it so that the owner is unable to find it for several days prior to the final examination, the student has engaged in a trespass to personal property. (The student has also committed the tort of *conversion*—to be discussed shortly.)

If it can be shown that the trespass to personal property was warranted, then a complete defense exists. Most states, for example, allow automobile repair shops to hold a customer's car (under what is called an *artisan's lien,* discussed in Chapter 21) when the customer refuses to pay for repairs already completed. Trespass to personal property was one of the allegations in the following case. (For a discussion of whether *spamming* constitutes trespass to personal property, see the examination of cyber torts later in this chapter.)

8. Pronounced *per*-sun-ul-tee.

CASE 4.2 Register.com, Inc. v. Verio, Inc.

United States Court of Appeals,
Second Circuit, 2004.
356 F.3d 393.

FACTS The Internet Corporation for Assigned Names and Numbers (ICANN) administers the Internet domain name system. (Domain names will be discussed in detail in Chapter 5.) ICANN appoints registrars to issue the names to persons preparing to establish Web sites on the Internet. An applicant for a name must provide certain information, including an e-mail address. An agreement between ICANN and its registrars refers to this information as "WHOIS information" and requires the registrars to provide public access to it through the Internet "for any lawful purposes" except "the transmission of mass unsolicited, commercial advertising or solicitations via e-mail (spam)." A party who wishes to obtain this information must also agree not to use it for this purpose. Register.com, Inc., is an ICANN registrar. Verio, Inc., sells Web site design and other services. Verio devised an automated software program (robot, or bot) to submit daily queries for Register's WHOIS data. Verio would then send ads by e-mail and other methods to the identified parties. Despite Register's request, Verio refused to stop. Register filed a suit in a federal district court against Verio, alleging in part trespass to personal property. The court ordered Verio to, among other things, stop accessing Register's computers by bot. Verio appealed to the U.S. Court of Appeals for the Second Circuit.

ISSUE Did Verio's use of its search bot constitute trespass to personal property?

DECISION Yes. The U.S. Court of Appeals for the Second Circuit affirmed the lower court's order. Trespass to personal property occurs when one party impairs the condition, quality, or value of another's computer.

REASON The court explained that trespass to personal property is committed when one intentionally uses or interferes with a chattel [an item of personal property] in the possession of another in such a way as to impair its condition, quality, or value. In this case, Verio used search robots, which are software programs that perform multiple automated successive queries. Verio's use of these robots consumed a significant portion of the capacity of Register's computer systems. While Verio's robots alone might not incapacitate Register's systems, the court found that "if Verio were permitted to continue to access Register's computers through such robots, it was highly probable that other Internet service providers would devise similar programs to access Register's data, and that the system would be overtaxed and would crash." The court dismissed Verio's argument that it could not be liable for trespass because Register had never instructed it not to use its robot programs. Register had requested Verio to stop and

(Continued)

CASE 4.2—CONTINUED

this, in the court's view, was sufficient to advise Verio that its use of robots was not authorized and could cause harm to Register's systems.

WHY IS THIS CASE IMPORTANT? *This case provides an interesting example of how courts today apply traditional law—the law of trespass to personal property—in nontraditional contexts. The court found that Verio's use of automated search bots impaired a significant portion of the capacity of Register's computer system. Although the property involved was intangible (system capacity), the court reasoned that Verio's actions were legally the same as if they had taken or interfered with someone else's physical property.*

CONVERSION

Whenever personal property is wrongfully taken from its rightful owner or possessor and placed in the service of another, the act of *conversion* occurs. **Conversion** is defined as any act depriving an owner of personal property without that owner's permission and without just cause. When conversion occurs, the lesser offense of trespass to personal property usually occurs as well. If the initial taking of the property was unlawful, there is trespass; retention of that property is conversion. If the initial taking of the property was permitted by the owner (or for some other reason is not a trespass), failure to return it may still be conversion. Conversion is the civil side of crimes related to theft. A store clerk who steals merchandise from the store commits a crime and engages in the tort of conversion at the same time.

Even if a person mistakenly believed that she or he was entitled to the goods, the tort of conversion may occur. In other words, good intentions are not a defense against conversion; in fact, conversion can be an entirely innocent act. Someone who buys stolen goods, for example, is guilty of conversion even if he or she did not know that the goods were stolen. If the true owner brings a tort action against the buyer, the buyer must either return the property to the owner or pay the owner the full value of the property, despite having already paid the thief.

A successful defense against the charge of conversion is that the purported owner does not in fact own the property or does not have a right to possess it that is superior to the right of the holder. Necessity is another possible defense against conversion. ■ EXAMPLE 4.6 If Abrams takes Mendoza's cat, Abrams is guilty of conversion. If Mendoza sues Abrams, Abrams must return the cat or pay damages. If, however, the cat has rabies and Abrams took the cat to protect the public, Abrams has a valid defense—necessity (and perhaps even self-defense, if he can prove that he was in danger because of the cat). ■

DISPARAGEMENT OF PROPERTY

Disparagement of property occurs when economically injurious falsehoods are made about another's product or property, not about another's reputation. Disparagement of property is a general term for torts that can be more specifically referred to as *slander of quality* or *slander of title.*

Slander of Quality Publication of false information about another's product, alleging that it is not what its seller claims, constitutes the tort of **slander of quality,** or **trade libel.** The plaintiff must prove that actual damages proximately resulted from the slander of quality. In other words, the plaintiff must show not only that a third person refrained from dealing with the plaintiff because of the improper publication but also that there were associated damages. The economic calculation of such damages—they are, after all, conjectural—is often extremely difficult.

An improper publication may be both a slander of quality and a defamation. For example, a statement that disparages the quality of a product may also, by implication, disparage the character of the person who would sell such a product.

Slander of Title When a publication denies or casts doubt on another's legal ownership of any property, and this results in financial loss to that property's owner, the tort of **slander of title** may exist. Usually, this is an intentional tort in which someone knowingly publishes an untrue statement about property with the intent of discouraging a third person from dealing with the person slandered. For example, it would be difficult for a car dealer to attract customers after competitors published a notice that the dealer's stock consisted of stolen autos.

Unintentional Torts (Negligence)

The tort of **negligence** occurs when someone suffers injury because of another's failure to live up to a required *duty of care.* In contrast to intentional torts, in torts involving negligence, the tortfeasor neither wishes to bring about the consequences of the act nor believes that they will occur. The actor's conduct merely creates a *risk* of such consequences. If no risk is created, there is no negligence.

Many of the actions discussed in the section on intentional torts constitute negligence if the element of intent is missing. ■ EXAMPLE 4.7 If Juarez intentionally shoves Natsuyo, who falls and breaks an arm as a result, Juarez will have committed the intentional tort of assault and battery. If Juarez carelessly bumps into Natsuyo, however, and she falls and breaks an arm as a result, Juarez's action will constitute negligence. In either situation, Juarez has committed a tort. ■

In examining a question of negligence, one should ask the following four questions (each of these elements of negligence will be discussed below):

1. Did the defendant owe a duty of care to the plaintiff?
2. Did the defendant breach that duty?
3. Did the plaintiff suffer a legally recognizable injury as a result of the defendant's breach of the duty of care?
4. Did the defendant's breach cause the plaintiff's injury?

The Duty of Care and Its Breach

The concept of a **duty of care** arises from the notion that if we are to live in society with other people, some actions can be tolerated and some cannot; some actions are right and some are wrong; and some actions are reasonable and some are not. The basic principle underlying the duty of care is that people are free to act as they please so long as their actions do not infringe on the interests of others.

When someone fails to comply with the duty to exercise reasonable care, a potentially tortious act may have been committed. Failure to live up to a standard of care may be an act (setting fire to a building) or an omission (neglecting to put out a campfire). It may be a careless act or a carefully performed but nevertheless dangerous act that results in injury. Courts consider the nature of the act (whether it is outrageous or commonplace), the manner in which the act is performed (cautiously versus heedlessly), and the nature of the injury (whether it is serious or slight) in determining whether the duty of care has been breached.

The Reasonable Person Standard Tort law measures duty by the **reasonable person standard.** In determining whether a duty of care has been breached, the courts ask how a reasonable person would have acted in the same circumstances. The reasonable person standard is said to be (though in an absolute sense it cannot be) objective. It is not necessarily how a particular person would act. It is society's judgment on how people *should* act. If the so-called reasonable person existed, he or she would be careful, conscientious, even tempered, and honest. The courts frequently use this hypothetical reasonable person in decisions relating to other areas of law as well.

That individuals are required to exercise a reasonable standard of care in their activities is a pervasive concept in business law, and many of the issues discussed in subsequent chapters of this text have to do with this duty. What constitutes reasonable care varies, of course, with the circumstances.

The Duty of Landowners Landowners are expected to exercise reasonable care to protect persons coming onto their property from harm. As mentioned earlier, in some jurisdictions, landowners are held to owe a duty to protect even trespassers against certain risks. Landowners who rent or lease premises to tenants (see Chapter 29) are expected to exercise reasonable care to ensure that the tenants and their guests are not harmed in common areas, such as stairways, entryways, laundry rooms, and the like.

Retailers and other firms that explicitly or implicitly invite persons to come onto their premises are usually charged with a duty to exercise reasonable care to protect those persons, who are considered **business invitees.** For example, if you entered a supermarket, slipped on a wet floor, and sustained injuries as a result, the owner of the supermarket would be liable for damages if when you slipped there was no sign warning that the floor was wet. A court would hold that the business owner was negligent because the owner failed to exercise a reasonable degree of care in protecting the store's customers against foreseeable risks about which the owner knew or *should have known.* That a patron might slip on the wet floor and be injured as a result was a foreseeable risk, and the owner should have taken care to avoid this risk or to warn the customer of it. The landowner also has a duty to discover and remove any hidden dangers that might injure a customer or other invitee.

Some risks, of course, are so obvious that the owner need not warn of them. For instance, a business owner does not have to warn customers to open a door before attempting to walk through it. Other risks, however, even

though they may seem obvious to a business owner, may not be so in the eyes of another, such as a child. For example, a hardware store owner may not think it is necessary to warn customers not to climb a stepladder leaning against the back wall of the store. It is possible, though, that a child could climb up and tip the ladder over and be hurt as a result and that the store could be held liable.

The Duty of Professionals If an individual has knowledge, skill, or intelligence superior to that of an ordinary person, the individual's conduct must be consistent with that status. Professionals—including accountants, architects, dentists, engineers, lawyers, physicians, and others—are required to have a standard minimum level of special knowledge and ability. Therefore, when professionals are involved, their training and expertise are taken into account in determining what constitutes reasonable care. In other words, an accountant cannot defend against a lawsuit for negligence by stating, "But I was not familiar with that principle of accounting."

If a professional violates her or his duty of care toward a client, the professional may be sued for **malpractice.** For example, a patient might sue a physician for *medical malpractice.* A client might sue an attorney for *legal malpractice.*

The Injury Requirement and Damages

For a tort to have been committed, the plaintiff must have suffered a *legally recognizable* injury. To recover damages (receive compensation), the plaintiff must have suffered some loss, harm, wrong, or invasion of a protected interest. Essentially, the purpose of tort law is to compensate for legally recognized injuries resulting from wrongful acts. If no harm or injury results from a given negligent action, there is nothing to compensate—and no tort exists. ■ EXAMPLE 4.8 If you carelessly bump into a passerby, who stumbles and falls as a result, you may be liable in tort if the passerby is injured in the fall. If the person is unharmed, however, there normally could be no suit for damages, because no injury was suffered. ■

As already mentioned, the purpose of tort law is not to punish people for tortious acts but to compensate the injured parties for harm suffered by awarding damages. **Compensatory damages** are intended to reimburse a plaintiff for actual losses—to make the plaintiff whole. Occasionally, **punitive damages** are also awarded to punish the wrongdoer and deter others from similar wrongdoing. Punitive damages are rarely awarded in lawsuits for ordinary negligence and usually are given only in cases involving intentional torts. They may be awarded, however, in suits that involve a higher degree of negligence, called *gross negligence,* in which a person acts in reckless disregard of the consequences.

In *State Farm Mutual Automobile Insurance Co. v. Campbell,*[9] the United States Supreme Court held that to the extent an award of punitive damages is grossly excessive, it furthers no legitimate purpose and violates due process requirements (discussed in Chapter 1). Although this case dealt with intentional torts (fraud and intentional infliction of emotional distress), the Court's holding applies equally to punitive damages awards in gross negligence cases.

Causation

Another element necessary to a tort is *causation.* If a person fails in a duty of care and someone suffers injury, the wrongful activity must have caused the harm for a tort to have been committed. In deciding whether there is causation, the court must address two questions:

1. *Is there causation in fact?* Did the injury occur because of the defendant's act, or would it have occurred anyway? If an injury would not have occurred without the defendant's act, then there is causation in fact. **Causation in fact** can usually be determined by the use of the *but for* test: "but for" the wrongful act, the injury would not have occurred. Theoretically, causation in fact is limitless. One could claim, for example, that "but for" the creation of the world, a particular injury would not have occurred. Thus, as a practical matter, the law has to establish limits, and it does so through the concept of proximate cause.
2. *Was the act the proximate cause of the injury?* **Proximate cause,** or legal cause, exists when the connection between an act and an injury is strong enough to justify imposing liability. ■ EXAMPLE 4.9 Ackerman carelessly leaves a campfire burning. The fire not only burns down the forest but also sets off an explosion in a nearby chemical plant that spills chemicals into a river, killing all the fish for a hundred miles downstream and ruining the economy of a tourist resort. Should Ackerman be liable to the resort owners? To the tourists whose vacations were ruined? These are questions of proximate cause that a court must decide. ■

The courts use *foreseeability* as the test for proximate cause. If the victim of the harm or the consequences of the harm done are unforeseeable, there is no proximate cause.

9. 538 U.S. 408, 123 S.Ct. 1513, 155 L.Ed.2d 585 (2003).

It is difficult to predict when a court will say that something is foreseeable and when it will decide that something is not. How far a court stretches foreseeability is determined in part by the extent to which the court is willing to stretch the defendant's duty of care.

DEFENSES TO NEGLIGENCE

Defendants often defend against negligence claims by asserting that the plaintiffs failed to prove the existence of one or more of the required elements for negligence. Additionally, there are three basic *affirmative* defenses in negligence cases (defenses that defendants can use to avoid liability even if the facts are as the plaintiffs state): (1) assumption of risk, (2) superseding cause, and (3) contributory and comparative negligence.

Assumption of Risk A plaintiff who voluntarily enters into a risky situation, knowing the risk involved, will not be allowed to recover. This is the defense of **assumption of risk.** The requirements of this defense are (1) knowledge of the risk and (2) voluntary assumption of the risk.

The risk can be assumed by express agreement, or the assumption of risk can be implied by the plaintiff's knowledge of the risk and subsequent conduct. For example, a race-car driver entering a race knows that there is a risk of being killed or injured in a crash. Of course, the plaintiff does not assume a risk different from or greater than the risk normally carried by the activity. In our example, the driver would not assume the risk that the banking in the curves of the racetrack will give way during the race because of a construction defect.

Risks are not deemed to be assumed in situations involving emergencies. Neither are they assumed when a statute protects a class of people from harm and a member of the class is injured by the harm. For instance, employees are protected by statute from harmful working conditions and therefore do not assume the risks associated with the workplace. An employee who is injured will generally be compensated regardless of fault under state workers' compensation statutes (discussed in Chapter 23).

The question in the following case was whether a baseball thrown into the stands during a baseball game is an inherent risk of attending the game.

CASE 4.3 Loughran v. The Phillies

Superior Court of Pennsylvania, 2005.
__ A.2d __,
2005 PA Super 396.

FACTS On July 5, 2003, Jeremy Loughran attended a baseball game between the Philadelphia Phillies and the Florida Marlins. At the end of the top half of the seventh inning, Philadelphia center fielder Marlon Byrd, after catching a ball for the last out, threw the ball into the stands. The ball hit Loughran, who suffered injuries—bleeding around his left eye, a concussion, facial contusions, and abrasions—requiring immediate treatment at the Veterans Stadium Infirmary and St. Mary's Medical Center. Later, Loughran received treatment for a variety of symptoms, including headaches, vomiting, hallucinations, loss of balance, pain, and eye spasms. Loughran filed a suit in a Pennsylvania state court against Byrd and the Phillies, alleging negligence. The court issued a summary judgment in favor of the defendants. Loughran appealed to the state intermediate appellate court.

ISSUE Is an injury caused by a ball thrown into a stadium during a baseball game an inherent risk of attending the game?

DECISION Yes. The state intermediate appellate court affirmed the judgment of the lower court, holding that a spectator at a major league baseball game is not owed a duty by either the team or an individual player to protect against a ball thrown into the stands.

REASON The court explained that the state's "no duty" rule barred a plaintiff's recovery for an injury that results from a common, frequent, and expected risk inherent in an activity. A spectator at a baseball game, for example, assumes the risk of being hit by batted balls, wildly thrown balls, foul balls, and, in some instances, bats. Loughran argued that because the third out had been made, the inning was over, and therefore Byrd's throw was unexpected and not part of the game. The court, however, considered as "customary" those activities that "although not specifically sanctioned by baseball authorities, have become as integral a part of attending a game as hot dogs, cracker jack, and seventh inning stretches." For instance, "both outfielders and infielders routinely toss caught balls to fans at the end of an inning." During the game at the center of this case, "there were at least twenty (20) occasions of a ball entering the stands. At least two of those balls were thrown to fans near appellant [Loughran] by players. Appellant admits to having attended numerous baseball games in the past and to having witnessed balls tossed into the stands on previous occasions. * * * [E]ven a first-time

(Continued)

CASE 4.3—CONTINUED

spectator at a baseball game is imputed with the common or neighborhood knowledge of the risks of the game."

FOR CRITICAL ANALYSIS—Social Consideration
Based on the decision in this case, could a baseball fan be said to assume the risk of an injury from a resin bag, baseball glove, spiked shoe, catcher's mask, or some other object that a player might intentionally throw into the stands?

Superseding Cause An unforeseeable intervening event may break the connection between a wrongful act and an injury to another. If so, the event acts as a *superseding cause*—that is, it relieves a defendant of liability for injuries caused by the intervening event. ■ EXAMPLE 4.10 Suppose that Derrick keeps a can of gasoline in the trunk of his car. The presence of the gasoline creates a foreseeable risk and is thus a negligent act. If Derrick's car skids and crashes into a tree, causing the gasoline can to explode, Derrick would be liable for injuries sustained by passing pedestrians because of his negligence. If the explosion had been caused by lightning striking the car, however, the lightning would supersede Derrick's original negligence as a cause of the damage, because the lightning was not foreseeable. ■

Contributory and Comparative Negligence All individuals are expected to exercise a reasonable degree of care in looking out for themselves. In the past, under the common law doctrine of **contributory negligence,** a plaintiff who was also negligent (failed to exercise a reasonable degree of care) could not recover anything from the defendant. Under this rule, no matter how insignificant the plaintiff's negligence was relative to the defendant's negligence, the plaintiff would be precluded from recovering any damages. Today, only a few jurisdictions still hold to this doctrine. In the majority of states, the doctrine of contributory negligence has been replaced by a **comparative negligence** standard.

Under the comparative negligence standard, both the plaintiff's and the defendant's negligence are computed, and the liability for damages is distributed accordingly. Some jurisdictions have adopted a "pure" form of comparative negligence that allows the plaintiff to recover, even if the extent of his or her fault is greater than that of the defendant. For example, if the plaintiff was 80 percent at fault and the defendant 20 percent at fault, the plaintiff may recover 20 percent of his or her damages. Many states' comparative negligence statutes, however, contain a "50 percent" rule that precludes the plaintiff from any recovery if she or he was more than 50 percent at fault.

SPECIAL NEGLIGENCE DOCTRINES AND STATUTES

There are a number of special doctrines and statutes relating to negligence. We examine a few of them here.

Res Ipsa Loquitur Generally, in lawsuits involving negligence, the plaintiff has the burden of proving that the defendant was negligent. In certain situations, however, when negligence is very difficult or impossible to prove, the courts may infer that negligence has occurred; then the burden of proof rests on the defendant—to prove he or she was *not* negligent. The inference of the defendant's negligence is known as the doctrine of ***res ipsa loquitur,***[10] which translates as "the facts speak for themselves."

This doctrine is applied when the event creating the damage or injury is one that ordinarily would occur only as a result of negligence. ■ EXAMPLE 4.11 If a person undergoes knee surgery and following the surgery has a severed nerve in the knee area, that person can sue the surgeon under a theory of *res ipsa loquitur.* In this instance, the injury would never have occurred in the absence of the surgeon's negligence. ■ For the doctrine of *res ipsa loquitur* to apply, the event must have been within the defendant's power to control, and it must not have been due to any voluntary action or contribution on the part of the plaintiff.

Negligence *Per Se* Certain conduct, whether it consists of an action or a failure to act, may be treated as **negligence *per se*** (*per se* means "in or of itself"). Negligence *per se* may occur if an individual violates a statute or an ordinance providing for a criminal penalty and that violation causes another to be injured. The injured person must prove (1) that the statute clearly sets out what standard of conduct is expected, when and where it is expected, and of whom it is expected; (2) that he or she is in the class intended to be protected by the statute; and (3) that the statute was designed to prevent the type of injury that he or she suffered. The standard of conduct required by the statute is the duty that the defendant owes to the plaintiff, and a violation of the statute is the breach of that duty.

10. Pronounced *rehz ihp*-suh *low*-kwuh-tuhr.

■ EXAMPLE 4.12 A statute may require a landowner to maintain a building in safe condition and may also subject the owner to a criminal penalty, such as a fine, if the building is not kept safe. The statute is meant to protect those who are rightfully in the building. Thus, if the owner, without a sufficient excuse, violates the statute and a tenant is thereby injured, a majority of courts will hold that the owner's unexcused violation of the statute conclusively establishes a breach of a duty of care—that is, that the owner's violation is negligence *per se*. ■

"Danger Invites Rescue" Doctrine Typically, if an individual takes a defensive action, such as swerving to avoid an oncoming car, the original wrongdoer will not be relieved of liability even if the injury actually resulted from the attempt to escape harm. The same is true under the "danger invites rescue" doctrine. ■ EXAMPLE 4.13 If Lemming commits an act that endangers Salter, and Yokem sustains an injury trying to protect Salter, then Lemming will be liable for Yokem's injury. Lemming will also be liable for any injuries Salter may sustain. ■ Rescuers can injure themselves, or the person rescued, or even a stranger, but the original wrongdoer will still be liable.

Special Negligence Statutes A number of states have enacted statutes prescribing duties and responsibilities in certain circumstances. For example, most states now have what are called **Good Samaritan statutes.** Under these statutes, someone who is aided voluntarily by others cannot turn around and sue the "Good Samaritans" for negligence. These laws were passed largely to protect physicians and medical personnel who voluntarily render services in emergency situations to those in need, such as individuals hurt in car accidents.

Many states have also passed **dram shop acts,** under which a tavern owner or bartender may be held liable for injuries caused by a person who became intoxicated while drinking at the bar or who was already intoxicated when served by the bartender. In some states, statutes impose liability on *social hosts* (persons hosting parties) for injuries caused by guests who became intoxicated at the hosts' homes. Under these statutes, it is unnecessary to prove that the tavern owner, bartender, or social host was negligent.

Strict Liability

Another category of torts is called **strict liability,** or *liability without fault.* Intentional torts and torts of negligence involve acts that depart from a reasonable standard of care and cause injuries. Under the doctrine of strict liability, liability for injury is imposed for reasons other than fault. Strict liability for damages proximately caused by an abnormally dangerous or exceptional activity is one application of this doctrine. Courts apply the doctrine of strict liability in such cases because of the extreme risk of the activity. Even if blasting with dynamite is performed with all reasonable care, there is still a risk of injury. Balancing that risk against the potential for harm, it seems reasonable to ask the person engaged in the activity to pay for injuries caused by that activity. Although there is no fault, there is still responsibility because of the dangerous nature of the undertaking.

There are other applications of the strict liability principle. Persons who keep dangerous animals, for example, are strictly liable for any harm inflicted by the animals. A significant application of strict liability is in the area of *product liability*—liability of manufacturers and sellers for harmful or defective products. Liability here is a matter of social policy and is based on two factors: (1) the manufacturer or seller can better bear the cost of injury because it can spread the cost throughout society by increasing prices of goods and services, and (2) the manufacturer or seller is making a profit from its activities and therefore should bear the cost of injury as an operating expense. We will discuss product liability in greater detail in Chapter 17.

Cyber Torts

It should come as no surprise that torts can also be committed in the online environment. Torts committed via the Internet are often called **cyber torts.** Over the last ten years, the courts have had to decide how to apply traditional tort law to torts committed in cyberspace. Consider, for example, issues of proof. How can it be proved that an online defamatory remark was "published" (which requires that a third party see or hear it)? How can the identity of the person who made the remark be discovered? Can an Internet service provider (ISP), such as Yahoo, be forced to reveal the source of an anonymous comment made by one of its subscribers? We explore some of these questions in this section, as well as some of the legal issues that have arisen with respect to bulk e-mail advertising.

DEFAMATION ONLINE

Recall from the discussion of defamation earlier in this chapter that one who repeats or otherwise republishes a defamatory statement can be subject to liability as if he or she had originally published it. Thus, publishers generally can be held liable for defamatory contents in the books and periodicals that they publish. Now consider online forums.

These forums allow anyone—customers, employees, or crackpots—to complain about a firm's personnel, policies, practices, or products. Whatever the truth of the complaint is, it might have an impact on the business of the firm. One of the early questions in the online legal arena was whether the providers of such forums could be held liable, as publishers, for defamatory statements made in those forums.

Liability of Internet Service Providers Prior to the passage of the Communications Decency Act (CDA) of 1996, the courts grappled on several occasions with the question of whether ISPs should be regarded as publishers and thus be held liable for defamatory messages made by users of their services. The CDA resolved the issue by stating that "[n]o provider or user of an interactive computer service shall be treated as the publisher or speaker of any information provided by another information content provider."[11] In a number of key cases, the ISP provisions of the CDA have been invoked to shield ISPs from liability for defamatory postings on their bulletin boards.

■ EXAMPLE 4.14 In a leading case on this issue, decided the year after the CDA was enacted, America Online, Inc. (AOL, now part of Time Warner, Inc.), was not held liable even though it did not promptly remove defamatory messages of which it had been made aware. In upholding a U.S. district court's ruling in AOL's favor, a federal appellate court stated that the CDA "plainly immunizes computer service providers like AOL from liability for information that originates with third parties." The court explained that the purpose of the statute is "to maintain the robust nature of Internet communication and, accordingly, to keep government interference in the medium to a minimum."[12] ■

In subsequent cases, the courts have reached similar conclusions.[13] The courts have also extended the immunity to liability provided by the CDA to auction houses, such as eBay.[14]

Piercing the Veil of Anonymity A threshold barrier to anyone who seeks to bring an action for online defamation is discovering the identity of the person who posted the defamatory message online. ISPs can disclose personal information about their customers only when ordered to do so by a court. Consequently, businesses and individuals are increasingly resorting to lawsuits against "John Does." Then, using the authority of the courts, they can obtain from the ISPs the identities of the persons responsible for the messages.

■ EXAMPLE 4.15 In one case, Eric Hvide, a former chief executive of a company called Hvide Marine, sued a number of "John Does" who had posted allegedly defamatory statements about his company on various online message boards. Hvide, who eventually lost his job, sued the John Does for libel in a Florida court. The court ruled that Yahoo and AOL had to reveal the identities of the defendant Does.[15] ■

In other cases, however, the courts have refused to order an ISP to disclose the identity of subscribers.[16] Generally, in these cases the courts must decide which right should take priority: the right to free (anonymous) speech under the First Amendment or the right not to be defamed.

SPAM

Bulk, unsolicited e-mail ("junk" e-mail) sent to all of the users on a particular e-mailing list is often called **spam.**[17] Typically, spam consists of a product ad sent to all of the users on an e-mailing list or all of the members of a newsgroup.

Spam wastes user time and network bandwidth (the amount of data that can be transmitted within a certain time). It also imposes a burden on an ISP's equipment as well as on e-mail recipients' computer systems. Does this mean that spamming may constitute a form of trespass to personal property? For a discussion of this question, see this chapter's *Adapting the Law to the Online Environment* feature.

In the last few years, the number of e-mail ads traveling through cyberspace has escalated significantly, and today spam accounts for an estimated 60 percent of all e-mail. Indeed, many contend that unwanted e-mail is threatening the viability of e-mail as a communications tool and an instrument of commerce. Because of the problems associated with spam, a majority of the states now have laws regulating spam. In 2003, the U.S. Congress also enacted a law to regulate the use of spam.

11. 47 U.S.C. Section 230.
12. *Zeran v. America Online, Inc.,* 129 F.3d 327 (4th Cir. 1997); *cert.* denied, 524 U.S. 934, 118 S.Ct. 2341, 141 L.Ed.2d 712 (1998).
13. See, for example, *Noah v. AOL Time Warner, Inc.,* 261 F.Supp.2d 532 (E.D.Va. 2003).
14. *Stoner v. eBay, Inc.,* 2000 WL 1705637 (Cal.Super.Ct. 2000).
15. *Does v. Hvide,* 770 So.2d 1237 (Fla.App.3d 2000).
16. See, for example, *Graham v. Oppenheimer,* 2000 WL 33381418 (E.D.Va. 2000).
17. The term *spam* is said to come from a Monty Python song with the lyrics, "Spam spam spam spam, spam spam spam spam, lovely spam, wonderful spam." Like these lyrics, spam online is often considered to be a repetition of worthless text.

ADAPTING THE LAW TO THE ONLINE ENVIRONMENT

Can Spamming Constitute a Tort?

Spam-related issues are, of course, unique to the cyber age. Can traditional tort law apply to such issues? For example, when spam overloads a business's computer system or impairs its computer equipment, can the recipient of the spam bring an action under tort law? Since the latter half of the 1990s, a number of plaintiffs have done so, with some success, by claiming that spamming constitutes a trespass to personal property, or *chattels*—a term used under the common law to denote all forms of personal property. The term *trespass to chattels* is usually defined as any unauthorized interference with the use of the personal property of another.

SPAMMING AS A FORM OF TRESPASS

In a leading case on this issue, decided in 1997, Cyber Promotions, Inc., sent bulk e-mail to subscribers of CompuServe, Inc., an ISP. CompuServe subscribers complained to the service about the ads, and many canceled their subscriptions. Handling the ads also placed a tremendous burden on CompuServe's equipment.

CompuServe told Cyber Promotions to stop using CompuServe's equipment to process and store the ads—in effect, to stop sending the ads to CompuServe subscribers. Ignoring the demand, Cyber Promotions stepped up the volume of its ads. After CompuServe attempted unsuccessfully to block the flow with screening software, it filed a suit against Cyber Promotions in a federal district court, seeking an injunction on the ground that the ads constituted trespass to personal property. The court agreed and ordered Cyber Promotions to stop sending its ads to e-mail addresses maintained by CompuServe.[a]

NEW DIRECTIONS IN CALIFORNIA

A 2003 decision by the California Supreme Court may make it more difficult for plaintiffs in that state's courts to combat spammers using tort law. The case was brought by Intel Corporation against one of its former employees, Ken Hamidi. After he was fired, Hamidi sent a series of six e-mail messages to 35,000 Intel employees over a twenty-one-month period. In the messages, Hamidi criticized the company's labor practices and urged employees to leave the company.

Intel sought a court order to stop the e-mail campaign, arguing that Hamidi's actions constituted a trespass to chattels because the e-mail significantly interfered with productivity, thus causing economic damage. When the case reached the California Supreme Court, the court held that under California law, the tort of trespass to chattels requires some evidence of injury to the plaintiff's personal property. Because Hamidi's e-mail had neither damaged Intel's computer system nor impaired its functioning, Hamidi's actions did not amount to a trespass to chattels.

The court distinguished Intel's case from cases in which spamming actually interfered with a computer system or threatened to do so. The court did not reject the idea that trespass theory could apply to cyberspace. Rather, the court simply held that to succeed in a lawsuit for trespass to chattels, a plaintiff must demonstrate that some concrete harm resulted from the unwanted e-mail. Although Hamidi's e-mail messages may have diverted employees' time and attention, thereby interfering with their productivity, the e-mails did not overburden Intel's computer systems—and thus did not amount to a concrete harm in the eyes of the court.[b]

FOR CRITICAL ANALYSIS

Three dissenting justices in the California case maintained that intruding on private property, regardless of injury, should be sufficient to demonstrate trespass to chattels. Do you agree? Why or why not?

a. *CompuServe, Inc. v. Cyber Promotions, Inc.*, 962 F.Supp. 1015 (S.D. Ohio 1997).

b. *Intel Corp. v. Hamidi*, 30 Cal.4th 1342, 71 P.3d 296, 1 Cal.Rptr.3d 32 (2003).

State Regulation of Spam Because of the problems associated with spam, thirty-six states have enacted laws that prohibit or regulate its use. A few states, such as Washington, prohibit unsolicited e-mail that is promoting goods, services, or real estate for sale or lease. In some other states, including Minnesota, an unsolicited e-mail ad must state in its subject line that it is an ad ("ADV:"). Many state laws regulating spam require the senders of e-mail ads to instruct the recipients on how they can "opt out" of further e-mail ads from the same sources. For instance, in some states an unsolicited e-mail ad must include a toll-free phone number or return e-mail address through which the recipient can contact the sender to request that no more ads be e-mailed.

The most stringent state law is California's antispam law, which went into effect on January 1, 2004. That law follows the "opt-in" model favored by consumer groups and antispam advocates. In other words, the law prohibits any person or business from sending e-mail ads to or from any e-mail address in California unless the recipient has

expressly agreed to receive e-mails from the sender. An exemption is made for e-mail sent to consumers with whom the advertiser has a "preexisting or current business relationship."

The Federal CAN-SPAM Act In 2003, Congress enacted the Controlling the Assault of Non-Solicited Pornography and Marketing (CAN-SPAM) Act, which took effect on January 1, 2004. The legislation applies to any "commercial electronic mail messages" that are sent to promote a commercial product or service. Significantly, the statute preempts state antispam laws except for those provisions in state laws that prohibit false and deceptive e-mailing practices.

Generally, the act permits the use of unsolicited commercial e-mail but prohibits certain types of spamming activities, including the use of a false return address and the use of false, misleading, or deceptive information when sending e-mail. The statute also prohibits the use of "dictionary attacks"—sending messages to randomly generated e-mail addresses—and the "harvesting" of e-mail addresses from Web sites through the use of specialized software. Additionally, the law requires senders of commercial e-mail to do the following:

1. Include a return address on the e-mail.
2. Include a clear notification that the message is an ad and provide a valid physical postal address.
3. Provide a mechanism that allows recipients to "opt out" of further e-mail ads from the same source.
4. Take action on a recipient's "opt-out" request within ten days.
5. Label any sexually oriented materials as such.

TERMS AND CONCEPTS

actionable 78
actual malice 79
appropriation 80
assault 77
assumption of risk 87
battery 77
business invitee 85
business tort 76
causation in fact 86
comparative negligence 88
compensatory damages 86
contributory negligence 88
conversion 84
cyber tort 89
damages 76
defamation 78
defense 77
disparagement of property 84
dram shop act 89
duty of care 85
fraudulent misrepresentation 80
Good Samaritan statute 89
intentional tort 77
libel 78
malpractice 86
negligence 85
negligence *per se* 88
predatory behavior 82
privilege 79
proximate cause 86
puffery 80
punitive damages 86
reasonable person standard 85
res ipsa loquitur 88
slander 78
slander of quality 84
slander of title 84
spam 90
strict liability 89
tort 76
tortfeasor 77
trade libel 84
trespass to land 82
trespass to personal property 83

CHAPTER SUMMARY Torts and Cyber Torts

Intentional Torts against Persons (See pages 77–82.)	1. *Assault and battery*—An assault is an unexcused and intentional act that causes another person to be apprehensive of immediate harm. A battery is an assault that results in physical contact. 2. *False imprisonment*—The intentional confinement or restraint of another person's movement without justification. 3. *Intentional infliction of emotional distress*—An intentional act that amounts to extreme and outrageous conduct resulting in severe emotional distress to another.

CHAPTER SUMMARY Torts and Cyber Torts—Continued

Intentional Torts against Persons—Continued	4. *Defamation (libel or slander)*—A false statement of fact, not made under privilege, that is communicated to a third person and that causes damage to a person's reputation. For public figures, the plaintiff must also prove actual malice. 5. *Invasion of the right to privacy*—The use of a person's name or likeness for commercial purposes without permission, wrongful intrusion into a person's private activities, publication of information that places a person in a false light, or disclosure of private facts that an ordinary person would find objectionable. 6. *Appropriation*—The use of another person's name, likeness, or other identifying characteristic, without permission and for the benefit of the user. 7. *Misrepresentation (fraud)*—A false representation made by one party, through misstatement of facts or through conduct, with the intention of deceiving another and on which the other reasonably relies to his or her detriment. 8. *Wrongful interference*—The knowing, intentional interference by a third party with an enforceable contractual relationship or an established business relationship between other parties for the purpose of advancing the economic interests of the third party.
Intentional Torts against Property (See pages 82–84.)	1. *Trespass to land*—The invasion of another's real property without consent or privilege. Specific rights and duties apply once a person is expressly or impliedly established as a trespasser. 2. *Trespass to personal property*—Unlawfully damaging or interfering with the owner's right to use, possess, or enjoy her or his personal property. 3. *Conversion*—Wrongfully taking personal property from its rightful owner or possessor and placing it in the service of another. 4. *Disparagement of property*—Any economically injurious falsehood that is made about another's product or property; an inclusive term for the torts of *slander of quality* and *slander of title.*
Unintentional Torts (Negligence) (See pages 85–89.)	1. *Negligence*—The careless performance of a legally required duty or the failure to perform a legally required act. Elements that must be proved are that a legal duty of care exists, that the defendant breached that duty, and that the breach caused damage or injury to another. 2. *Defenses to negligence*—The basic affirmative defenses in negligence cases are (a) assumption of risk, (b) superseding cause, and (c) contributory or comparative negligence. 3. *Special negligence doctrines and statutes*— a. *Res ipsa loquitur*—A doctrine under which a plaintiff need not prove negligence on the part of the defendant because "the facts speak for themselves." b. Negligence *per se*—A type of negligence that may occur if a person violates a statute or an ordinance providing for a criminal penalty and the violation causes another to be injured. c. Special negligence statutes—State statutes that prescribe duties and responsibilities in certain circumstances, the violation of which will impose civil liability. Dram shop acts and Good Samaritan statutes are examples of special negligence statutes.
Strict Liability (See page 89.)	Under the doctrine of strict liability, a person may be held liable, regardless of the degree of care exercised, for damages or injuries caused by her or his product or activity. Strict liability includes liability for harms caused by abnormally dangerous activities, by dangerous animals, and by defective products (product liability).
Cyber Torts (See pages 89–92.)	General tort principles are being extended to cover cyber torts, or torts that occur in cyberspace, such as online defamation and spamming (which may constitute trespass to personal property). Federal and state statutes may also apply to certain forms of cyber torts. For example, under the federal Communications Decency Act of 1996, Internet service providers (ISPs) are not liable for defamatory messages posted by their subscribers. A majority of the states and the federal government now regulate unwanted e-mail ads (spam).

FOR REVIEW

Answers for the even-numbered questions in this For Review *section can be found in Appendix A at the end of this text.*

1. What is a tort?
2. What is the purpose of tort law? What are two basic categories of torts?
3. What are the four elements of negligence?
4. What is meant by strict liability? In what circumstances is strict liability applied?
5. What is a cyber tort, and how are tort theories being applied in cyberspace?

QUESTIONS AND CASE PROBLEMS

4–1. Defenses to Negligence. Corinna was riding her bike on a city street. While she was riding, she frequently looked back to verify that the books that she had fastened to the rear part of her bike were still attached. On one occasion while she was looking behind her, she failed to notice a car that was entering an intersection just as she was crossing it. The car hit her, causing her to sustain numerous injuries. Three eyewitnesses stated that the driver of the car had failed to stop at the stop sign before entering the intersection. Corinna sued the driver of the car for negligence. What defenses might the defendant driver raise in this lawsuit? Discuss fully.

4–2. Liability to Business Invitees. Kim went to Ling's Market to pick up a few items for dinner. It was a rainy, windy day, and the wind had blown water through the door of Ling's Market each time the door opened. As Kim entered through the door, she slipped and fell in the approximately one-half inch of rainwater that had accumulated on the floor. The manager knew of the weather conditions but had not posted any sign to warn customers of the water hazard. Kim injured her back as a result of the fall and sued Ling's for damages. Can Ling's be held liable for negligence in this situation? Discuss.

QUESTION WITH SAMPLE ANSWER

4–3. In which of the following situations will the acting party be liable for the tort of negligence? Explain fully.

(a) Mary goes to the golf course on Sunday morning, eager to try out a new set of golf clubs she has just purchased. As she tees off on the first hole, the head of her club flies off and injures a nearby golfer.

(b) Mary's doctor gives her some pain medication and tells her not to drive after she takes it, as the medication induces drowsiness. In spite of the doctor's warning, Mary decides to drive to the store while on the medication. Owing to her lack of alertness, she fails to stop at a traffic light and crashes into another vehicle, injuring a passenger.

For a sample answer to this question, go to Appendix B at the end of this text.

4–4. Causation. Ruth carelessly parks her car on a steep hill, leaving the car in neutral and failing to engage the parking brake. The car rolls down the hill, knocking down an electric line. The sparks from the broken line ignite a grass fire. The fire spreads until it reaches a barn one mile away. The barn houses dynamite, and the burning barn explodes, causing part of the roof to fall on and injure Jim, a passing motorist. Can Jim recover from Ruth? Why or why not?

4–5. Duty of Care. As pedestrians exited at the close of an arts and crafts show, Jason Davis, an employee of the show's producer, stood near the exit. Suddenly and without warning, Davis turned around and collided with Yvonne Esposito, an eighty-year-old woman. Esposito was knocked to the ground, fracturing her hip. After hip-replacement surgery, she was left with a permanent physical impairment. Esposito filed a suit in a federal district court against Davis and others, alleging negligence. What are the factors that indicate whether Davis owed Esposito a duty of care? What do those factors indicate in these circumstances? [*Esposito v. Davis,* 47 F.3d 164 (5th Cir. 1995)]

CASE PROBLEM WITH SAMPLE ANSWER

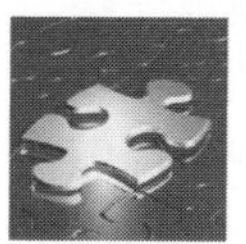

4–6. During the spring and summer of 1999, Edward and Geneva Irvine received numerous "hang-up" phone calls, including three calls in the middle of the night. With the help of their local phone company, the Irvines learned that many of the calls were from the telemarketing department of the *Akron Beacon Journal* in Akron, Ohio. The *Beacon*'s sales force was equipped with an automatic dialing machine. During business hours, the dialer was used to maximize productivity by calling multiple phone numbers at once and connecting a call to a sales representative only after it was answered. After business hours, the dialer was used to dial a list of disconnected numbers to determine whether they had been reconnected. If the dialer detected a ring, it recorded the information and dropped the call. If the automated dialing system crashed, which it did frequently, it redialed the entire list. The Irvines filed a suit in an Ohio state court against the *Beacon* and others, alleging in part invasion of privacy. In whose favor should the court rule, and why? [*Irvine v. Akron Beacon Journal,* 147 Ohio App.3d 428, 770 N.E.2d 1105 (9 Dist. 2002)]

After you have answered this problem, compare your answer with the sample answer given on the Web site that accompanies this text. Go to **academic.cengage.com/blaw/wbl**, select "Chapter 4," and click on "Case Problem with Sample Answer."

4–7. Defamation. Lydia Hagberg went to her bank, California Federal Bank, FSB, to cash a check made out to her by Smith Barney (SB), an investment services firm. Nolene Showalter, a bank employee, suspected that the check was counterfeit. Showalter phoned SB and was told that the check was not valid. As she phoned the police, Gary Wood, a bank security officer, contacted SB again and was told that its earlier statement was "erroneous" and that the check was valid. Meanwhile, a police officer arrived, drew Hagberg away from the teller's window, spread her legs, patted her down, and handcuffed her. The officer searched her purse, asked her whether she had any weapons or stolen property and whether she was driving a stolen vehicle, and arrested her. Hagberg filed a suit in a California state court against the bank and others, alleging, among other things, slander. Should the absolute privilege for communications made in judicial or other official proceedings apply to statements made when a citizen contacts the police to report suspected criminal activity? Why or why not? [*Hagberg v. California Federal Bank, FSB,* 32 Cal.4th 350, 81 P.3d 244, 7 Cal.Rptr.3d 803 (2004)]

VIDEO QUESTION

4–8. Go to this text's Web site at **academic.cengage.com/blaw/wbl** and select "Chapter 4." Click on "Video Questions" and view the video titled *Jaws.* Then answer the following questions.

1. In the video, the mayor (Murray Hamilton) and a few other men try to persuade Chief Brody (Roy Scheider) not to close the town's beaches. If Brody keeps the beaches open and a swimmer is injured or killed because he failed to warn swimmers about the potential shark danger, has Brody committed a tort? If so, what kind of tort (intentional tort against persons, intentional tort against property, negligence)? Explain your answer.
2. Can Chief Brody be held liable for any injuries or deaths to swimmers under the doctrine of strict liability? Why or why not?
3. If Chief Brody goes against the mayor's instructions and warns swimmers to stay off the beach, and the town suffers economic damages as a result, has he committed the tort of disparagement of property? Why or why not?

Torts and Cyber Torts

For updated links to resources available on the Web, as well as a variety of other materials, visit this text's Web site at

academic.cengage.com/blaw/wbl

You can find cases and articles on torts, including business torts, at the Internet Law Library's Web site at

http://www.lawguru.com/ilawlib/110.htm

For information on the *Restatements of the Law,* including the *Restatement (Second) of Torts* and the *Restatement (Third) of Torts: Products Liability,* go to the Web site of the American Law Institute at

http://www.ali.org

ONLINE LEGAL RESEARCH

Go to the *Fundamentals of Business Law* home page at **academic.cengage.com/blaw/wbl**, select "Chapter 4," and click on "Internet Exercises." There you will find the following Internet research exercises that you can perform to learn more about topics covered in this chapter.

Activity 4–1: LEGAL PERSPECTIVE—Negligence and the *Titanic*

Activity 4–2: MANAGEMENT PERSPECTIVE—Legal and Illegal Uses of Spam

BEFORE THE TEST

Go to the *Fundamentals of Business Law* home page at **academic.cengage.com/blaw/wbl**, select "Chapter 4," and click on "Interactive Quizzes." You will find a number of interactive questions relating to this chapter.

CHAPTER 5
Intellectual Property

LEARNING OBJECTIVES

After reading this chapter, you should be able to answer the following questions:

1. What is intellectual property?
2. Why are trademarks and patents protected by the law?
3. What laws protect authors' rights in the works they generate?
4. What are trade secrets, and what laws offer protection for this form of intellectual property?
5. What steps have been taken to protect intellectual property rights in today's digital age?

Of significant concern to businesspersons today is the need to protect their rights in intellectual property. **Intellectual property** is any property resulting from intellectual, creative processes—the products of an individual's mind. Although it is an abstract term for an abstract concept, intellectual property is nonetheless wholly familiar to virtually everyone. The information contained in books and computer files is intellectual property. The software you use, the movies you see, and the music you listen to are all forms of intellectual property. In fact, in today's information age, it should come as no surprise that the value of the world's intellectual property may now exceed the value of physical property, such as machines and houses.

The need to protect creative works was voiced by the framers of the U.S. Constitution more than two hundred years ago: Article I, Section 8, of the Constitution authorized Congress "[t]o promote the Progress of Science and useful Arts, by securing for limited Times to Authors and Inventors the exclusive Right to their respective Writings and Discoveries." Laws protecting patents, trademarks, and copyrights are explicitly designed to protect and reward inventive and artistic creativity. Exhibit 5–1 offers a comprehensive summary of these forms of intellectual property, as well as intellectual property that consists of *trade secrets.*

An understanding of intellectual property law is important because intellectual property has taken on increasing significance, not only in the United States but globally as well. Today, ownership rights in intangible intellectual property are more important to the prosperity of many U.S. companies than are their tangible assets. As you will read in this chapter, protecting these assets in today's online world has proved particularly challenging. This is because the Internet's capability is profoundly different from anything we have had in the past.

Trademarks and Related Property

A **trademark** is a distinctive word, name, mark, motto, device, or emblem that a manufacturer stamps, prints, or otherwise affixes to the goods it produces so that the goods

EXHIBIT 5–1 FORMS OF INTELLECTUAL PROPERTY

	DEFINITION	HOW ACQUIRED	DURATION	REMEDY FOR INFRINGEMENT
Patent	A grant from the government that gives an inventor exclusive rights to an invention.	By filing a patent application with the U.S. Patent and Trademark Office and receiving its approval.	Twenty years from the date of the application; for design patents, fourteen years.	Money damages, including royalties and lost profits, *plus* attorneys' fees. Damages may be tripled for intentional infringements.
Copyright	The right of an author or originator of a literary or artistic work, or other production that falls within a specified category, to have the exclusive use of that work for a given period of time.	Automatic (once the work or creation is put in tangible form). Only the *expression* of an idea (and not the idea itself) can be protected by copyright.	For authors: the life of the author plus 70 years. For publishers: 95 years after the date of publication or 120 years after creation.	Actual damages plus profits received by the party who infringed *or* statutory damages under the Copyright Act, *plus* costs and attorneys' fees in either situation.
Trademark (Service Mark and Trade Dress)	Any distinctive word, name, symbol, or device (image or appearance), or combination thereof, that an entity uses to distinguish its goods or services from those of others. The owner has the exclusive right to use that mark or trade dress.	1. At common law, ownership created by the use of the mark. 2. Registration with the appropriate federal or state office gives notice and is permitted if the mark is currently in use or will be within the next six months.	Unlimited, as long as it is in use. To continue notice by registration, owner must renew by filing between the fifth and sixth years, and thereafter, every ten years.	1. Injunction prohibiting the future use of the mark. 2. Actual damages plus profits received by the party who infringed (can be increased under the Lanham Act). 3. Destruction of articles that infringed. 4. *Plus* costs and attorneys' fees.
Trade Secret	Any information that a business possesses and that gives the business an advantage over competitors (including formulas, lists, patterns, plans, processes, and programs).	Through the originality and development of the information and processes that constitute the business secret and are unknown to others.	Unlimited, so long as not revealed to others. Once revealed to others, they are no longer trade secrets.	Money damages for misappropriation (the Uniform Trade Secrets Act also permits punitive damages if willful), *plus* costs and attorneys' fees.

and their origins may be identified in the marketplace. At common law, the person who used a symbol or mark to identify a business or product was protected in the use of that trademark. Clearly, by using another's trademark, a business could lead consumers to believe that its goods were made by the other business. The law seeks to avoid this kind of confusion. In the following classic case concerning Coca-Cola, the defendants argued that the Coca-Cola trademark was entitled to no protection under the law because the term did not accurately represent the product.

LANDMARK AND CLASSIC CASES

CASE 5.1 The Coca-Cola Co. v. Koke Co. of America

Supreme Court of the United States, 1920.
254 U.S. 143,
41 S.Ct. 113,
65 L.Ed. 189.
http://www.findlaw.com/casecode/supreme.html[a]

FACTS The Coca-Cola Company brought an action in a federal district court to prevent other beverage companies from using the words *Koke* and *Dope* for the defendants' products. The defendants contended that the Coca-Cola trademark was a fraudulent representation and that Coca-Cola was therefore not entitled to any help from the courts. By use of the Coca-Cola name, the defendants alleged, the Coca-Cola Company represented that the beverage contained cocaine (from coca leaves). The district court granted the injunction, but the federal appellate court reversed the lower court's decision. The Coca-Cola Company appealed to the United States Supreme Court.

ISSUE Did the marketing of products called Koke and Dope by the Koke Company of America and other firms constitute an unauthorized use of Coca-Cola's trademark?

DECISION Yes for Koke, but no for Dope. The Supreme Court enjoined [prevented] the competing beverage companies from calling their products Koke but did not prevent them from calling their products Dope.

a. This is the "U.S. Supreme Court Opinions" page within the Web site of the "FindLaw Internet Legal Resources" database. This page provides several options for accessing an opinion. Because you know the citation for this case, you can go to the "Citation Search" box, type in the appropriate volume and page numbers for the *United States Reports* ("254" and "143," respectively, for the *Coca-Cola case*), and click on "Get It."

REASON The Court noted that, to be sure, prior to 1900 the Coca-Cola beverage had contained a small amount of cocaine, but this ingredient had been deleted from the formula by 1906 at the latest, and the Coca-Cola Company had advertised to the public that no cocaine was present in its drink. Coca-Cola was a widely popular drink "to be had at almost any soda fountain." Because of the public's widespread familiarity with Coca-Cola, the retention of the name of the beverage (referring to coca leaves and kola nuts) was not misleading: "Coca-Cola probably means to most persons the plaintiff's familiar product to be had everywhere rather than a compound of particular substances." The name "Coke" was found to be so common a term for the trademarked product Coca-Cola that the defendants' use of the similar-sounding "Koke" as a name for their beverages was disallowed. The Court could find no reason to restrain the defendants from using the name "Dope," however.

IMPACT OF THIS CASE ON TODAY'S LAW
In this classic case, the United States Supreme Court made it clear that trademarks and trade names (and nicknames for those marks and names, such as the nickname "Coke" for "Coca-Cola") that are in everyday use receive protection under the common law. This holding is historically significant because the federal statute later passed to protect trademark rights (the Lanham Act of 1946, to be discussed shortly) in many ways represented a codification of common law principles governing trademarks.

RELEVANT WEB SITES *To locate information on the Web concerning the* Coca-Cola Co. *decision, go to this text's Web site at* **academic.cengage.com/blaw/wbl**, *select "Chapter 5," and click on "URLs for Landmark Cases."*

Statutory Protection of Trademarks

Statutory protection of trademarks and related property is provided at the federal level by the Lanham Act of 1946.[1] The Lanham Act was enacted in part to protect manufacturers from losing business to rival companies that used confusingly similar trademarks. The Lanham Act incorporates the common law of trademarks and provides remedies for owners of trademarks who wish to enforce their claims in federal court. Many states also have trademark statutes.

In 1995, Congress amended the Lanham Act by passing the Federal Trademark Dilution Act,[2] which extended the protection available to trademark owners by creating a federal cause of action for trademark **dilution.** Until the passage of this amendment, federal trademark law only prohibited the unauthorized use of the same mark on competing—or on noncompeting but "related"—goods or services when such use would likely confuse consumers as to the origin of those goods and services. Trademark dilution laws protect "distinctive" or "famous" trademarks (such as Jergens, McDonald's, RCA, and Macintosh) from certain unauthorized uses of the marks *regardless* of a showing of competition or a likelihood of confusion.

In 2003, the United States Supreme Court held that to establish dilution under the federal act, some evidence must establish that the allegedly infringing user's mark actually reduces the value of the famous mark or lessens its

1. 15 U.S.C. Sections 1051–1128.
2. 15 U.S.C. Section 1125.

capacity to identify goods and services.[3] A famous mark may be diluted not only by the use of an *identical* mark but also by the use of a *similar* mark.[4] More than half of the states have also enacted trademark dilution laws.

3. *Moseley v. V Secret Catalogue, Inc.,* 537 U.S. 418, 123 S.Ct. 1115, 155 L.Ed.2d 1 (2003).
4. See, for example, *Ringling Bros.–Barnum & Bailey, Combined Shows, Inc. v. Utah Division of Travel Development,* 935 F.Supp. 763 (E.D.Va. 1996).

In the following case, the owner of a famous trademark claimed that a small business owner's use of a similar mark diluted the famous mark in violation of the Lanham Act.

CASE 5.2 Starbucks Corp. v. Lundberg

United States District Court,
District of Oregon, 2005.
__ F.Supp.2d __.

FACTS Starbucks Corporation is the largest and best-known purveyor of specialty coffees and coffee products in North America. Starbucks does business in nearly 10,000 retail locations in the United States and thirty-four foreign countries and territories. Starbucks also supplies premium, fresh-roasted coffee to restaurants, airlines, sports and entertainment venues, movie theaters, hotels, and cruise ship lines throughout the world. Since 1971, Starbucks has done business under the trade name "Starbucks," displaying its mark on signs outside and within each of its stores and on its products. Starbucks' success is predicated on the consistently high quality of its coffees and the other products and services it provides. Starbucks has a reputation for excellence and is recognized for its knowledgeable staff and service. Since 2001, Samantha Lundberg has operated a coffee shop under the name "Sambuck's Coffeehouse" in Astoria, Oregon. When Lundberg (whose maiden name was Buck) chose the "Sambuck's" name, she knew that Starbucks was one of the best-known trademarks in the United States and was closely associated with coffee and stores that serve coffee. Starbucks asked Lundberg to stop using the name "Sambuck's," but she persisted. Starbucks filed a suit in a federal district court against Lundberg, alleging in part a violation of the Lanham Act.

ISSUE Did Lundberg's use of "Sambuck's" dilute the Starbucks trademark?

DECISION Yes. The court issued an injunction in favor of Starbucks, ordering Lundberg to stop using the name "Sambuck's," and any other variation on Starbucks' trademark, as the name of her coffee shop and its products.

REASON The Lanham Act prohibits the use of "any reproduction, counterfeit, copy, or colorable [reasonable] imitation of a registered mark" in a manner that is "likely to cause confusion, or to cause mistake, or to deceive." The court concluded that Lundberg's use of the "Sambuck's" mark created such confusion. The court identified several determining factors: "(1) the similarity of the marks; (2) the relatedness of the two companies' services; (3) the marketing channels used; (4) the strength of plaintiff's marks; (5) defendant's intent in selecting its marks; * * * (7) the likelihood of expansion into other markets; and (8) the degree of care likely to be exercised by the purchasers." Among other things, the court emphasized that consumers perceive "Sambuck's" and "Starbucks" to be similar marks. Over 70 percent of the persons who were shown the "Sambuck's" name in a survey indicated that it brought Starbucks to mind because of the "high degree" of similarity between the two terms. As for the "marketing channels," both parties sold their products through "stand-alone" retail stores. The court also pointed out that there was "no restriction on Lundberg's ability to open additional Sambuck's Coffeehouse locations" or "to license the mark to various competitors of Starbucks."

FOR CRITICAL ANALYSIS—Economic Consideration
Why might Starbucks have believed that its mark was threatened by a sole proprietor's use of a similar mark in a small town in a remote part of the United States?

TRADEMARK REGISTRATION

Trademarks may be registered with the state or with the federal government. To register for protection under federal trademark law, a person must file an application with the U.S. Patent and Trademark Office in Washington, D.C. Under current law, a mark can be registered (1) if it is currently in commerce or (2) if the applicant intends to put the mark into commerce within six months.

In special circumstances, the six-month period can be extended by thirty months, giving the applicant a total of three years from the date of notice of trademark approval

to make use of the mark and file the required use statement. Registration is postponed until the mark is actually used. Nonetheless, during this waiting period, any applicant can legally protect his or her trademark against a third party who previously has neither used the mark nor filed an application for it. Registration is renewable between the fifth and sixth years after the initial registration and every ten years thereafter (every twenty years for trademarks registered before 1990).

TRADEMARK INFRINGEMENT

Registration of a trademark with the U.S. Patent and Trademark Office gives notice on a nationwide basis that the trademark belongs exclusively to the registrant. The registrant is also allowed to use the symbol ® to indicate that the mark has been registered. Whenever that trademark is copied to a substantial degree or used in its entirety by another, intentionally or unintentionally, the trademark has been *infringed* (used without authorization). When a trademark has been infringed, the owner of the mark has a cause of action against the infringer. A person need not have registered a trademark in order to sue for trademark infringement, but registration does furnish proof of the date of inception of the trademark's use.

Only those trademarks that are deemed sufficiently distinctive from all competing trademarks will be protected, however. The trademarks must be sufficiently distinct to enable consumers to identify the manufacturer of the goods easily and to differentiate among competing products.

Strong Marks Fanciful, arbitrary, or suggestive trademarks are generally considered to be the most distinctive (strongest) trademarks because they are normally taken from outside the context of the particular product and thus provide the best means of distinguishing one product from another.

■ **EXAMPLE 5.1** Fanciful trademarks include invented words, such as "Xerox" for one manufacturer's copiers and "Kodak" for another company's photographic products. Arbitrary trademarks include actual words that have no literal connection to the product, such as "English Leather" used as a name for an aftershave lotion (and not for leather processed in England). Suggestive trademarks are those that imply something about a product without describing the product directly. For instance, "Dairy Queen" suggests an association between its products and milk, but it does not directly describe ice cream. ■

Secondary Meaning Descriptive terms, geographic terms, and personal names are not inherently distinctive and do not receive protection under the law *until* they acquire a secondary meaning. A secondary meaning may arise when customers begin to associate a specific term or phrase, such as "London Fog," with specific trademarked items (coats with "London Fog" labels). Whether a secondary meaning becomes attached to a term or name usually depends on how extensively the product is advertised, the market for the product, the number of sales, and other factors. The United States Supreme Court has held that even a color can qualify for trademark protection.[5] Once a secondary meaning is attached to a term or name, a trademark is considered distinctive and is protected.

Generic Terms Generic terms (general, commonly used terms that refer to an entire class of products, such as *bicycle* or *computer*) receive no protection, even if they acquire secondary meanings. A particularly thorny problem arises when a trademark acquires generic use. For instance, *aspirin* and *thermos* were originally trademarked products, but today the words are used generically. Other examples are *escalator, trampoline, raisin bran, dry ice, lanolin, linoleum, nylon,* and *corn flakes.*

Note that a generic term will not be protected under trademark law even if the term has acquired a secondary meaning. ■ **EXAMPLE 5.2** In one case, America Online, Inc. (AOL), sued AT&T Corporation, claiming that AT&T's use of "You Have Mail" on its WorldNet Service infringed AOL's trademark rights in the same phrase. The court ruled, however, that because each of the three words in the phrase was a generic term, the phrase as a whole was generic. Although the phrase had become widely associated with AOL's e-mail notification service, and thus may have acquired a secondary meaning, this issue was of no significance in this case. The court stated that it would not consider whether the mark had acquired any secondary meaning because "generic marks with secondary meaning are still not entitled to protection."[6] ■

SERVICE, CERTIFICATION, AND COLLECTIVE MARKS

A **service mark** is similar to a trademark but is used to distinguish the services of one person or company from those of another. For instance, each airline has a particular mark or symbol associated with its name. Titles and character names used in radio and television are frequently registered as service marks.

Other marks protected by law include certification marks and collective marks. A *certification mark* is used by one or more persons other than the owner to certify the region, materials, mode of manufacture, quality, or accu-

5. *Qualitex Co. v. Jacobson Products Co.,* 514 U.S. 159, 115 S.Ct. 1300, 131 L.Ed.2d 248 (1995).

6. *America Online, Inc. v. AT&T Corp.,* 243 F.3d 812 (4th Cir. 2001).

racy of the owner's goods or services. When used by members of a cooperative, association, or other organization, it is referred to as a *collective mark*. ■ EXAMPLE 5.3 Certification marks include such marks as "Good Housekeeping Seal of Approval" and "UL Tested." Collective marks appear at the ends of the credits of movies to indicate the various associations and organizations that participated in making the movie. The union marks found on the tags of certain products are also collective marks. ■

Trade Names

Trademarks apply to *products*. The term **trade name** is used to indicate part or all of a business's name, whether the business is a sole proprietorship, a partnership, or a corporation. Generally, a trade name is directly related to a business and its goodwill. Trade names may be protected as trademarks if the trade name is also the name of the company's trademarked product—for example, Coca-Cola and Starbucks. Unless also used as a trademark or service mark, a trade name cannot be registered with the federal government. Trade names are protected under the common law, however. As with trademarks, words must be unusual or fancifully used if they are to be protected as trade names. The word *Safeway*, for instance, was held by the courts to be sufficiently fanciful to obtain protection as a trade name for a grocery chain.[7]

Trade Dress

The term **trade dress** refers to the image and overall appearance of a product. Basically, trade dress is subject to the same protection as trademarks. ■ EXAMPLE 5.4 The distinctive decor, menu, layout, and style of service of a particular restaurant may be regarded as the restaurant's trade dress. Similarly, if a golf course is distinguished from other golf courses by prominent features, such as a golf hole designed to look like a lighthouse, those features may be considered the golf course's trade dress. ■ In cases involving trade dress infringement, as in trademark infringement cases, a major consideration is whether consumers are likely to be confused by the allegedly infringing use. Also, features that enhance a product's function will not be protected as trade dress.

Cyber Marks

In cyberspace, trademarks are sometimes referred to as **cyber marks.** We turn now to a discussion of trademark-related issues in cyberspace and how new laws and the courts are addressing these issues. One concern relates to the rights of a trademark's owner to use the mark as part of a domain name (Internet address). Other issues have to do with *cybersquatting, meta tags,* and *trademark dilution* on the Web. The use of licensing as a way to avoid liability for infringing on another's intellectual property rights in cyberspace will also be discussed.

7. *Safeway Stores v. Suburban Foods,* 130 F.Supp. 249 (E.D.Va. 1955).

Domain Names

In the real world, one business can often use the same name as another without causing any conflict, particularly if the businesses are small, their goods or services are different, and the areas where they do business are separate. In the online world, however, there is only one area of business—cyberspace. Thus, disputes between parties over which one has the right to use a particular domain name have become common. A **domain name** is part of an Internet address, such as "westlaw.com." The top level domain (TLD) is the part of the name to the right of the period and indicates the type of entity that operates the site (for example, "com" is an abbreviation for "commercial"). The second level (the part of the name to the left of the period) is chosen by the business entity or individual registering the domain name.

Conflicts over rights to domain names emerged during the 1990s as e-commerce expanded on a worldwide scale. As e-commerce grew, the *.com* TLD came to be widely used by businesses on the Web. Competition among firms with similar names and products preceding the *.com* TLD led, understandably, to numerous disputes over domain name rights. By using the same, or a similar, domain name, parties have attempted to profit from the goodwill of a competitor, to sell pornography, to offer for sale another party's domain name, and to otherwise infringe on others' trademarks.

As noted in Chapter 2, the federal government set up the Internet Corporation for Assigned Names and Numbers (ICANN), a nonprofit corporation, to oversee the distribution of domain names. ICANN has also played a leading role in facilitating the settlement of domain name disputes worldwide. Since January 2000, ICANN has been operating an online arbitration system to resolve domain name disputes and approve dispute-resolution providers. By 2003, ICANN-approved online arbitration providers were handling more than one thousand disputes annually.

Anticybersquatting Legislation

In the late 1990s, Congress passed legislation prohibiting another practice that had given rise to numerous disputes over domain names: cybersquatting. **Cybersquatting**

occurs when a person registers a domain name that is the same as, or confusingly similar to, the trademark of another and then offers to sell the domain name back to the trademark owner. During the 1990s, cybersquatting became a contentious issue and led to much litigation. Although it was not always easy for the courts to separate cybersquatting from legitimate business activity, many cases held that cybersquatting violated trademark law.[8]

In 1999, Congress addressed this issue by passing the Anticybersquatting Consumer Protection Act (ACPA), which amended the Lanham Act—the federal law protecting trademarks, discussed earlier in this chapter. The ACPA makes it illegal for a person to "register, traffic in, or use" a domain name (1) if the name is identical or confusingly similar to the trademark of another and (2) if the one registering, trafficking in, or using the domain name has a "bad faith intent" to profit from that trademark. The act does not define what constitutes bad faith. Instead, it lists several factors that courts can consider in deciding whether bad faith exists. Some of these factors are the trademark rights of the other person, whether there is an intent to divert consumers in a way that could harm the goodwill represented by the trademark, whether there is an offer to transfer or sell the domain name to the trademark owner, and whether there is an intent to use the domain name to offer goods and services.

The ACPA applies to all domain name registrations of trademarks, even domain names registered before the passage of the act. Successful plaintiffs in suits brought under the act can collect actual damages and profits or elect to receive statutory damages of from $1,000 to $100,000.

META TAGS

Search engines compile their results by looking through a Web site's key-word field. *Meta tags,* or key words (see Chapter 1), may be inserted into this field to increase the site's inclusion in search engine results, even though the site may have nothing to do with the inserted words. Using this same technique, one site may appropriate the key words of other sites with more frequent hits, so that the appropriating site appears in the same search engine results as the more popular sites. Using another's trademark in a meta tag without the owner's permission, however, constitutes trademark infringement.

DILUTION IN THE ONLINE WORLD

As discussed earlier, trademark *dilution* occurs when a trademark is used, without authorization, in a way that diminishes the distinctive quality of the mark. Unlike trademark infringement, a dilution cause of action does not require proof that consumers are likely to be confused by a connection between the unauthorized use and the mark. For this reason, the products involved do not have to be similar. In the first case alleging dilution on the Web, a court prohibited the use of "candyland.com" as the URL for an adult site. The court held that the use of the URL would dilute the value of the "Candyland" mark owned by the maker of the "Candyland" children's game.[9]

In another case, a court issued an injunction on the ground that spamming under another's logo is trademark dilution. In that case, Hotmail, Inc., provided e-mail services and worked to dissociate itself from spam. Van$ Money Pie, Inc., and others spammed thousands of e-mail customers, using the free e-mail Hotmail as a return address. The court ordered the defendants to stop.[10]

LICENSING

One of the ways to make use of another's trademark or other form of intellectual property, while avoiding litigation, is to obtain a *license* to do so. A **license** in this context is essentially an agreement permitting the use of a trademark, copyright, patent, or trade secret for certain purposes. For instance, a *licensee* (the party obtaining the license) might be allowed to use the trademark of the *licensor* (the party issuing the license) as part of the name of its company, or as part of its domain name, without otherwise using the mark on any products or services. Like all contracts, contracts granting licenses must be carefully drafted.

Patents

A **patent** is a grant from the government that gives an inventor the exclusive right to make, use, and sell an invention for a period of twenty years from the date of filing the application for a patent. Patents for designs, as opposed to inventions, are given for a fourteen-year period. For either a regular patent or a design patent, the applicant must demonstrate to the satisfaction of the U.S. Patent and Trademark Office that the invention, discovery, process, or design is genuine, novel, useful, and not obvious in light of current technology. A patent holder gives notice to all that an article or design is patented by placing on it the word *Patent* or *Pat.* plus the patent number. In contrast to patent law in other countries, in the

8. See, for example, *Panavision International, L.P. v. Toeppen,* 141 F.3d 1316 (9th Cir. 1998).

9. *Hasbro, Inc. v. Internet Entertainment Group, Ltd.,* 1996 WL 84853 (W.D.Wash. 1996).

10. *Hotmail Corp. v. Van$ Money Pie, Inc.,* 47 U.S.P.Q.2d [*U.S. Patent Quarterly, Second Series*] 1020 (N.D.Cal. 1998).

United States patent protection is given to the first person to invent a product or process, even though someone else may have been the first to file for a patent on that product or process.

At one time, it was difficult for developers and manufacturers of software to obtain patent protection because many software products simply automate procedures that can be performed manually. In other words, the computer programs do not meet the "novel" and "not obvious" requirements previously mentioned. Also, the basis for software is often a mathematical equation or formula, which is not patentable. In 1981, however, the United States Supreme Court held that it is possible to obtain a patent for a process that incorporates a computer program—providing, of course, that the process itself is patentable.[11] Subsequently, many patents have been issued for software-related inventions.

A significant development relating to patents is the availability online of the world's patent databases. The Web site of the U.S. Patent and Trademark Office provides searchable databases covering U.S. patents granted since 1976. The Web site of the European Patent Office maintains databases covering all patent documents in sixty-five nations and the legal status of patents in twenty-two of those countries.

Patent Infringement

If a firm makes, uses, or sells another's patented design, product, or process without the patent owner's permission, it commits the tort of patent infringement. Patent infringement may exist even though the patent owner has not put the patented product in commerce. Patent infringement may also occur even though not all features or parts of an invention are copied. (With respect to a patented process, however, all steps or their equivalent must be copied for infringement to exist.)

Often, litigation for patent infringement is so costly that the patent holder will instead offer to sell to the infringer a license to use the patented design, product, or process. Indeed, in many cases the costs of detection, prosecution, and monitoring are so high that patents are valueless to their owners; the owners cannot afford to protect them.

Business Process Patents

Traditionally, patents have been granted for inventions that are "new and useful processes, machines, manufactures, or compositions of matter, or any new and useful improvements thereof." The U.S. Patent and Trademark Office routinely rejected computer systems and software applications because they were deemed not to be useful processes, machines, articles of manufacture, or compositions of matter. They were simply considered to be mathematical algorithms, abstract ideas, or "methods of doing business." In a landmark 1998 case, however, *State Street Bank & Trust Co. v. Signature Financial Group, Inc.*,[12] the U.S. Court of Appeals for the Federal Circuit ruled that only three categories of subject matter will always remain unpatentable: (1) the laws of nature, (2) natural phenomena, and (3) abstract ideas. This decision meant, among other things, that business processes were patentable.

After this decision, numerous technology firms applied for business process patents. Walker Digital applied for a business process patent for its "Dutch auction" system, which allowed consumers to make offers for airline tickets on the Internet and led to the creation of Priceline.com. Amazon.com obtained a business process patent for its "one-click" ordering system, a method of processing credit-card orders securely.

Copyrights

A **copyright** is an intangible property right granted by federal statute to the author or originator of certain literary or artistic productions. Copyrights are governed by the Copyright Act of 1976,[13] as amended. Works created after January 1, 1978, are automatically given statutory copyright protection for the life of the author plus 70 years. For copyrights owned by publishing houses, the copyright expires 95 years from the date of publication or 120 years from the date of creation, whichever is first. For works by more than one author, the copyright expires 70 years after the death of the last surviving author.

These time periods reflect the extensions of the length of copyright protection enacted by Congress in the Copyright Term Extension Act of 1998.[14] Critics challenged this act as overstepping the bounds of Congress's power and violating the constitutional requirement that copyrights endure for only a limited time. In 2003, however, the United States Supreme Court upheld the act in *Eldred v. Ashcroft*.[15] This ruling obviously favored copyright holders by preventing copyrighted works from the 1920s and 1930s from losing protection and falling into the public domain for an additional two decades.

11. *Diamond v. Diehr*, 450 U.S. 175, 101 S.Ct. 1048, 67 L.Ed.2d 155 (1981).

12. 149 F.3d 1368 (Fed. Cir. 1998).
13. 17 U.S.C. Sections 101 *et seq.*
14. 17 U.S.C.A. Section 302.
15. 537 U.S. 186, 123 S.Ct. 769, 154 L.Ed.2d 683 (2003).

Copyrights can be registered with the U.S. Copyright Office in Washington, D.C. A copyright owner no longer needs to place a © or *Copr.* or *Copyright* on the work, however, to have the work protected against infringement. Chances are that if somebody created it, somebody owns it.

What Is Protected Expression?

Works that are copyrightable include books, records, films, artworks, architectural plans, menus, music videos, product packaging, and computer software. To obtain protection under the Copyright Act, a work must be original and fall into one of the following categories: (1) literary works; (2) musical works; (3) dramatic works; (4) pantomimes and choreographic works; (5) pictorial, graphic, and sculptural works; (6) films and other audiovisual works; and (7) sound recordings. To be protected, a work must be "fixed in a durable medium" from which it can be perceived, reproduced, or communicated. Protection is automatic. Registration is not required.

Section 102 Exclusions Section 102 of the Copyright Act specifically excludes copyright protection for any "idea, procedure, process, system, method of operation, concept, principle, or discovery, regardless of the form in which it is described, explained, illustrated, or embodied." Note that it is not possible to copyright an *idea*. The underlying ideas embodied in a work may be freely used by others. What is copyrightable is the particular way in which an idea is *expressed*. Whenever an idea and an expression are inseparable—such as the *shape* of an automobile or a woman's dress—the expression cannot be copyrighted. Generally, anything that is not an original expression will not qualify for copyright protection. Facts widely known to the public are not copyrightable. Page numbers are not copyrightable because they follow a sequence known to everyone. Mathematical calculations are not copyrightable.

Compilations of Facts Compilations of facts, however, are copyrightable. Section 103 of the Copyright Act defines a *compilation* as "a work formed by the collection and assembling of preexisting materials of data that are selected, coordinated, or arranged in such a way that the resulting work as a whole constitutes an original work of authorship." The key requirement for the copyrightability of a compilation is originality. ■ **EXAMPLE 5.5** The white pages of a telephone directory do not qualify for copyright protection when the information that makes up the directory (names, addresses, and telephone numbers) is not selected, coordinated, or arranged in an original way.[16] In one case, even the Yellow Pages of a telephone directory did not qualify for copyright protection.[17] ■

Copyright Infringement

Whenever the form or expression of an idea is copied, an infringement of copyright occurs. The reproduction does not have to be exactly the same as the original, nor does it have to reproduce the original in its entirety. If a substantial part of the original is reproduced, copyright infringement has occurred.

Damages for Copyright Infringement Those who infringe copyrights may be liable for damages or criminal penalties. These range from actual damages or statutory damages, imposed at the court's discretion, to criminal proceedings for willful violations. Actual damages are based on the harm caused to the copyright holder by the infringement, while statutory damages, not to exceed $150,000, are provided for under the Copyright Act. In addition, criminal proceedings may result in fines and/or imprisonment.

The "Fair Use" Exception The "fair use" doctrine provides for an exception to liability for copyright infringement. In certain circumstances, a person or organization can reproduce copyrighted material without paying royalties (fees paid to the copyright holder for the privilege of reproducing the copyrighted material). Section 107 of the Copyright Act provides as follows:

> [T]he fair use of a copyrighted work, including such use by reproduction in copies or phonorecords or by any other means specified by [Section 106 of the Copyright Act,] for purposes such as criticism, comment, news reporting, teaching (including multiple copies for classroom use), scholarship, or research, is not an infringement of copyright. In determining whether the use made of a work in any particular case is a fair use the factors to be considered shall include—
>
> (1) the purpose and character of the use, including whether such use is of a commercial nature or is for nonprofit educational purposes;
>
> (2) the nature of the copyrighted work;
>
> (3) the amount and substantiality of the portion used in relation to the copyrighted work as a whole; and
>
> (4) the effect of the use upon the potential market for or value of the copyrighted work.

Because these guidelines are very broad, the courts determine whether a particular use is fair on a case-by-

16. *Feist Publications, Inc. v. Rural Telephone Service Co.,* 499 U.S. 340, 111 S.Ct. 1282, 113 L.Ed.2d 358 (1991).

17. *Bellsouth Advertising & Publishing Corp. v. Donnelley Information Publishing, Inc.,* 999 F.2d 1436 (11th Cir. 1993).

case basis. Thus, anyone reproducing copyrighted material may be committing a violation. In determining whether a use is fair, courts have often considered the fourth factor to be the most important.

Copyright Protection for Software

In 1980, Congress passed the Computer Software Copyright Act, which amended the Copyright Act of 1976 to include computer programs in the list of creative works protected by federal copyright law. The 1980 statute, which classifies computer programs as "literary works," defines a computer program as a "set of statements or instructions to be used directly or indirectly in a computer in order to bring about a certain result."

Because of the unique nature of computer programs, the courts have had many problems applying and interpreting the 1980 act. Generally, though, the courts have held that copyright protection extends not only to those parts of a computer program that can be read by humans, such as the high-level language of a source code, but also to the binary-language object code of a computer program, which is readable only by the computer.[18] Additionally, such elements as the overall structure, sequence, and organization of a program were deemed copyrightable.[19] The courts have disagreed as to whether the "look and feel"—the general appearance, command structure, video images, menus, windows, and other screen displays—of computer programs should also be protected by copyright. The courts have tended, however, not to extend copyright protection to look-and-feel aspects of computer programs.

Copyrights in Digital Information

Copyright law is probably the most important form of intellectual property protection on the Internet. This is because much of the material on the Internet consists of works of authorship (including multimedia presentations, software, and database information), which are the traditional focus of copyright law. Copyright law is also important because the nature of the Internet requires that data be "copied" to be transferred online. Copies are a significant part of the traditional controversies arising in this area of the law.

The Copyright Act of 1976 When Congress drafted the principal U.S. law governing copyrights, the Copyright Act of 1976, cyberspace did not exist for most of us. The threat to copyright owners was posed not by computer technology but by unauthorized *tangible* copies of works and the sale of rights to movies, television, and other media.

Some issues that were unimagined when the Copyright Act was drafted have posed thorny questions for the courts. For instance, to sell a copy of a work, permission of the copyright holder is necessary. Because of the nature of cyberspace, however, one of the early controversies involved determining at what point an intangible, electronic "copy" of a work has been made. The courts have held that loading a file or program into a computer's random access memory, or RAM, constitutes the making of a "copy" for purposes of copyright law.[20] RAM is a portion of a computer's memory into which a file, for instance, is loaded so that it can be accessed (read or written over). Thus, a copyright is infringed when a party downloads software into RAM without owning the software or otherwise having a right to download it.[21]

Other rights, including those relating to the revision of "collective works" such as magazines, were acknowledged thirty years ago but were considered to have only limited economic value. Today, technology has made some of those rights vastly more significant. How does the old law apply to these rights? That was one of the questions in the following case.

18. See *Stern Electronics, Inc. v. Kaufman*, 669 F.2d 852 (2d Cir. 1982); and *Apple Computer, Inc. v. Franklin Computer Corp.*, 714 F.2d 1240 (3d Cir. 1983).
19. *Whelan Associates, Inc. v. Jaslow Dental Laboratory, Inc.*, 797 F.2d 1222 (3d Cir. 1986).
20. *MAI Systems Corp. v. Peak Computer, Inc.*, 991 F.2d 511 (9th Cir. 1993).
21. *DSC Communications Corp. v. Pulse Communications, Inc.*, 170 F.3d 1354 (Fed. Cir. 1999).

CASE 5.3 New York Times Co. v. Tasini

Supreme Court of the United States, 2001.
533 U.S. 483,
121 S.Ct. 2381,
150 L.Ed.2d 500.
http://supct.law.cornell.edu/supct/index.html[a]

a. In the "Search" box, type in the name of the case and click on "search."

FACTS Magazines and newspapers, including the *New York Times*, buy and publish articles written by freelance writers. Besides circulating hard copies of their periodicals, these publishers sell the contents to e-publishers for inclusion in

(Continued)

CASE 5.3—CONTINUED

online and other electronic databases. Jonathan Tasini and other freelance writers filed a suit in a federal district court against the New York Times Company and other publishers, including the e-publishers, contending that the e-publication of the articles violated the Copyright Act. The publishers claimed, among other things, that the Copyright Act gave them a right to produce "revisions" of their publications. The writers argued that the Copyright Act did not cover electronic "revisions." The court granted a summary judgment in the publishers' favor, which was reversed on the writers' appeal to the U.S. Court of Appeals for the Second Circuit. The publishers appealed to the United States Supreme Court.

ISSUE To put the contents of periodicals into e-databases and onto CD-ROMs, do publishers need to obtain the permission of the writers whose contributions are included in the periodicals?

DECISION Yes. The United States Supreme Court affirmed the lower court's judgment. The Supreme Court remanded the case for a determination as to how the writers should be compensated.

REASON The Court pointed out that databases are "vast domain[s] of diverse texts," consisting of "thousands or millions of files containing individual articles from thousands of collective works." The Court found that these databases have little relationship to the articles' original publication. The databases are not "revisions," as the publishers argued, because the databases reproduce and distribute articles "clear of the context provided by the original periodical editions"—not "as part of that particular collective work" to which the author contributed, not "as part of * * * any revision," and not "as part of * * * any later collective work in the same series," as the Copyright Act provides. The Court reasoned that a database composed of such articles was no more a revision of an original work than "a 400-page novel quoting a sonnet in passing would represent a 'revision' of that poem."

FOR CRITICAL ANALYSIS—Economic Consideration
When technology creates a situation in which rights such as those in this case become more valuable, should the law be changed to redistribute the economic benefit of those rights? Why or why not?

Further Developments in Copyright Law In the last several years, Congress has enacted legislation designed specifically to protect copyright holders in a digital age. Prior to 1997, criminal penalties under copyright law could be imposed only if unauthorized copies were exchanged for financial gain. Yet much piracy of copyrighted materials was "altruistic" in nature; that is, unauthorized copies were made and distributed not for financial gain but simply for reasons of generosity—to share the copies with others.

To combat altruistic piracy and for other reasons, Congress passed the No Electronic Theft (NET) Act of 1997. This act extends criminal liability for the piracy of copyrighted materials to persons who exchange unauthorized copies of copyrighted works, such as software, even though they realize no profit from the exchange. The act also imposes penalties on those who make unauthorized electronic copies of books, magazines, movies, or music for *personal* use, thus altering the traditional "fair use" doctrine. The criminal penalties for violating the act are steep; they include fines as high as $250,000 and incarceration for up to five years.

The Digital Millennium Copyright Act of 1998 In 1996, the World Intellectual Property Organization (WIPO) enacted a treaty to upgrade global standards of copyright protection, particularly for the Internet. In 1998, Congress implemented the provisions of the WIPO treaty by updating U.S. copyright law. The new law—the Digital Millennium Copyright Act of 1998—is a landmark step in the protection of copyright owners and, because of the leading position of the United States in the creative industries, serves as a model for other nations. Among other things, the act created civil and criminal penalties for anyone who circumvents (bypasses, or gets around—through clever maneuvering, for example) encryption software or other technological antipiracy protection. Also prohibited are the manufacture, import, sale, and distribution of devices or services for circumvention.

The act provides for exceptions to fit the needs of libraries, scientists, universities, and others. In general, the law does not restrict the "fair use" of circumvention methods for educational and other noncommercial purposes. For example, circumvention is allowed to test computer security, conduct encryption research, protect personal privacy, and enable parents to monitor their children's use of the Internet. The exceptions are to be reconsidered every three years.

The 1998 act also limited the liability of Internet service providers (ISPs). Under the act, an ISP is not liable for any copyright infringement by its customer *unless* the ISP is aware of the subscriber's violation. An ISP may be held liable only if it fails to take action to shut the subscriber down after learning of the violation. A copyright holder

must act promptly, however, by pursuing a claim in court, or the subscriber has the right to be restored to online access.

MP3 and File-Sharing Technology

At one time, music fans swapped compact discs (CDs) and recorded the songs that they liked from others' CDs onto their own cassettes. This type of "file-sharing" was awkward at best. Soon after the Internet became popular, a few enterprising programmers created software to compress large data files, particularly those associated with music. The reduced file sizes make transmitting music over the Internet feasible. The most widely known compression and decompression system is MP3, which enables music fans to download songs or entire CDs onto their computers or onto a portable listening device, such as Rio or iPod. The MP3 system also made it possible for music fans to access other music fans' files by engaging in file-sharing via the Internet.

Peer-to-Peer (P2P) Networking File-sharing via the Internet is accomplished through what is called **peer-to-peer (P2P) networking.** The concept is simple. Rather than going through a central Web server, P2P involves numerous personal computers (PCs) that are connected to the Internet. Files stored on one PC can be accessed by others who are members of the same network. Sometimes this is called a **distributed network.** In other words, parts of the network are distributed all over the country or the world. File-sharing offers an unlimited number of uses for distributed networks. For instance, thousands of researchers simultaneously allow their home computers' computing power to be accessed through file-sharing software so that very large mathematical problems can be solved quickly. Additionally, persons throughout the world can work together on the same project by using file-sharing programs.

Sharing Stored Music Files When file-sharing is used to download others' stored music files, copyright issues arise. Recording artists and their labels stand to lose large amounts of royalties and revenues if relatively few CDs are purchased and then made available on distributed networks, from which everyone can get them for free. The issue of file-sharing infringement has been the subject of an ongoing debate over the last few years.

■ EXAMPLE 5.6 In the highly publicized case of *A&M Records, Inc. v. Napster, Inc.*,[22] several firms in the recording industry sued Napster, Inc., the owner of the then-popular Napster Web site. The Napster site provided registered users with free software that enabled them to transfer exact copies of the contents of MP3 files from one computer to another via the Internet. Napster also maintained centralized search indices so that users could locate specific titles or artists' recordings on the computers of other members. The firms argued that Napster should be liable for contributory and vicarious[23] copyright infringement because it assisted others in obtaining copies of copyrighted music without the copyright owners' permission. Both the federal district court and the U.S. Court of Appeals for the Ninth Circuit agreed and held Napster liable for violating copyright laws. ■

Since the 2001 *Napster* decision, the recording industry has filed and won numerous lawsuits against companies that distribute online file-sharing software. The courts have held these Napster-like companies liable based on two theories: (1) contributory infringement, which applies if the company had reason to know about a user's infringement and failed to stop it; and (2) vicarious liability, which exists if the company was able to control the users' activities and stood to benefit financially from their infringement. In the *Napster* case, the court held the company liable under both doctrines, largely because the technology that Napster had used was centralized and gave it "the ability to locate infringing material listed on its search indices, and the right to terminate users' access to the system."

New File-Sharing Technologies In the wake of the *Napster* decision, other companies developed new technologies that allow P2P network users to share stored music files, without paying a fee, more quickly and efficiently than ever. Today's file-sharing software is decentralized and does not use search indices. Thus, the companies have no ability to supervise or control which music (or other media files) their users are exchanging. Unlike the Napster system, in which the company played a role in connecting people who were downloading and uploading songs, the new systems are designed to work without the company's input.

Software such as Morpheus and KaZaA, for example, provides users with an interface that is similar to a Web browser. This technology is different from that used by Napster. Instead of the company locating songs for users on other members' computers, the software automatically annotates files with descriptive information so that the music can easily be categorized and cross-referenced (by

22. 239 F.3d 1004 (9th Cir. 2001).

23. *Vicarious (indirect) liability* exists when one person is subject to liability for another's actions. A common example occurs in the employment context, when an employer is held vicariously liable by third parties for torts committed by employees in the course of their employment.

artist and title, for instance). When a user performs a search, the software is able to locate a list of peers that have the file available for downloading. Also, to expedite the P2P transfer and ensure that the complete file is received, the software distributes the download task over the entire list of peers simultaneously. By downloading even one file, the user becomes a point of distribution for that file, which is then automatically shared with others on the network. How the courts have decided the legality of these new digital technologies is discussed in this chapter's *Management Perspective* feature.

Trade Secrets

Some business processes and information that are not or cannot be patented, copyrighted, or trademarked are nevertheless protected against appropriation by a competitor as *trade secrets*. **Trade secrets** include customer lists, plans, research and development, pricing information, marketing techniques, production methods, and generally anything that makes an individual company unique and that would have value to a competitor.

Unlike copyright and trademark protection, protection of trade secrets extends both to ideas and to their expression. (For this reason, and because a trade secret involves no registration or filing requirements, trade secret protection may be well suited for software.) Of course, the secret formula, method, or other information must be disclosed to some persons, particularly to key employees. Businesses generally attempt to protect their trade secrets by having all employees who use the process or information agree in their contracts, or in confidentiality agreements, never to divulge it.

State and Federal Law on Trade Secrets

Under Section 757 of the *Restatement of Torts,* those who disclose or use another's trade secret, without authorization, are liable to that other party if (1) they discovered the secret by improper means, or (2) their disclosure or use constitutes a breach of a duty owed to the other party. The theft of confidential business data by industrial espionage, as when a business taps into a competitor's computer, is a theft of trade secrets without any contractual violation and is actionable in itself.

Until the late 1970s, virtually all law with respect to trade secrets was common law. In an effort to reduce the unpredictability of the common law in this area, a model act, the Uniform Trade Secrets Act, was presented to the states for adoption in 1979. Parts of this act have been adopted in more than thirty states. Typically, a state that has adopted parts of the act has adopted only those parts that encompass its own existing common law. Additionally, in 1996 Congress passed the Economic Espionage Act, which made the theft of trade secrets a federal crime. We will examine the provisions and significance of this act in Chapter 6, in the context of *white-collar crimes.*

Trade Secrets in Cyberspace

The nature of new computer technology undercuts a business firm's ability to protect its confidential information, including trade secrets.[24] For instance, a dishonest employee could e-mail trade secrets in a company's computer to a competitor or a future employer. If e-mail is not an option, the employee might walk out with the information on a CD-ROM.

International Protection for Intellectual Property

For many years, the United States has been a party to various international agreements relating to intellectual property rights. For example, the Paris Convention of 1883, to which about ninety countries are signatory, allows parties in one country to file for patent and trademark protection in any of the other member countries. Other international agreements include the Berne Convention and the TRIPS agreement.

The Berne Convention

Under the Berne Convention of 1886, as amended (an international copyright agreement), if a U.S. author writes a book, every country that has signed the convention must recognize that author's copyright in the book. Also, if a citizen of a country that has not signed the convention first publishes a book in a country that has signed, all other countries that have signed the convention must recognize that author's copyright. Copyright notice is not needed to gain protection under the Berne Convention for works published after March 1, 1989.

This convention and other international agreements have given some protection to intellectual property on a worldwide level. None of them, however, has been as significant and far reaching in scope as the agreement on Trade-Related Aspects of Intellectual Property Rights, or, more simply, TRIPS.

---

24. Note that the courts have even found that customers' e-mail addresses may constitute trade secrets. See *T-N-T Motorsports, Inc. v. Hennessey Motorsports, Inc.,* 965 S.W.2d 18 (Tex.App.—Houston [1 Dist.] 1998); rehearing overruled (1998); petition dismissed (1998).

MANAGEMENT PERSPECTIVE

File-Sharing Technology and Copyright Law

MANAGEMENT FACES A LEGAL ISSUE

Clearly, any person who downloads copyrighted music without permission from the copyright holder is liable for copyright infringement. But what about the companies that provide the software that enables users to swap copyrighted music? In what circumstances will the companies be held liable? For some time, it was difficult for the courts to apply traditional doctrines of contributory and vicarious copyright liability to new file-sharing technologies.

WHAT THE COURTS SAY

In 2005, however, the United States Supreme Court addressed this issue in *Metro-Goldwyn-Mayer Studios, Inc. v. Grokster, Ltd.*[a] In that case, organizations in the music and film industry (the plaintiffs) sued several companies that distribute file-sharing software used in P2P networks, including Grokster, Ltd., and StreamCast Networks, Inc. (the defendants). The plaintiffs claimed that the companies were contributorily and vicariously liable for the infringement of their end users.

The federal district court that initially heard the case examined the technology involved and concluded that the defendants were not liable for contributory infringement because they lacked the requisite level of knowledge. It was not enough that the defendants *generally* knew that the software they provided might be used to infringe on copyrights; they also had to have *specific knowledge* of the infringement "at a time when they can use that knowledge to stop the particular infringement." Here, the companies had merely distributed free software. They had no knowledge of whether users were swapping copyrighted files and no ability to stop users from infringing activities. Further, the two defendants could not be held vicariously liable for the infringement because it was not possible for the companies to supervise or control their users' conduct. The district court's decision was affirmed on appeal, and the plaintiffs appealed to the United States Supreme Court.

The Supreme Court reversed the lower court's decision and remanded the case for further proceedings. The Court held that "one who distributes a device [software] with the object of promoting its use to infringe the copyright, as shown by clear expression or other affirmative steps taken to foster infringement, is liable for the resulting acts of infringement by third parties." The Court did not, however, specify what kind of "affirmative steps" are necessary to establish liability and left it to the lower court on remand to determine whether the defendants had actually induced their users to commit infringement. In the Court's view, however, there was ample evidence in the record that the defendants had acted with the intent to cause copyright violations by use of their software.

a. ___ U.S. ___, 125 S.Ct. 2764, 162 L.Ed.2d 781 (2005).

IMPLICATIONS FOR MANAGERS

The Supreme Court's decision in this case shifts the focus in secondary copyright infringement cases away from specific knowledge of acts of infringement to acts that induce or promote infringement. Essentially, this means that file-sharing companies that have taken affirmative steps to promote copyright infringement can be held secondarily liable for millions of infringing acts that their users commit daily. Because the Court did not define exactly what is necessary to impose liability, however, a great deal of legal uncertainty remains concerning this issue.

THE TRIPS AGREEMENT

Representatives from more than one hundred nations signed the TRIPS agreement in 1994. The agreement established, for the first time, standards for the international protection of intellectual property rights, including patents, trademarks, and copyrights for movies, computer programs, books, and music. Prior to the agreement, U.S. sellers of intellectual property in the international market faced difficulties because many other countries had no laws protecting intellectual property rights or failed to enforce existing laws. To address this problem, the TRIPS agreement provides that each member country must include in its domestic laws broad intellectual property rights and effective remedies (including civil and criminal penalties) for violations of those rights.

Generally, the TRIPS agreement provides that each member nation must not discriminate (in the administration, regulation, or adjudication of intellectual property rights) against foreign owners of such rights. In other words, a member nation cannot give its own nationals (citizens) favorable treatment without offering the same treatment to nationals of all member countries. For instance, if a U.S. software manufacturer brings a suit for the infringement of intellectual property rights under a member country's national laws, the U.S. manufacturer is entitled to receive the same treatment as a domestic manufacturer. Each member nation must also ensure that legal

procedures are available for parties who wish to bring actions for infringement of intellectual property rights. Additionally, a related document established a mechanism for settling disputes among member nations.

Particular provisions of the TRIPS agreement relate to patent, trademark, and copyright protection for intellectual property. The agreement specifically provides copyright protection for computer programs by stating that compilations of data, databases, and other materials are "intellectual creations" and that they are to be protected as copyrightable works. Other provisions relate to trade secrets and the rental of computer programs and cinematographic works.

The Madrid Protocol

In the past, one of the difficulties in protecting U.S. trademarks internationally was that it was time consuming and expensive to apply for trademark registration in foreign countries. The filing fees and procedures for trademark registration vary significantly among individual countries. The Madrid Protocol, however, which President George W. Bush signed into law in 2003, may help to resolve these problems. The Madrid Protocol is an international treaty that has been signed by sixty-one countries. Under its provisions, a U.S. company wishing to register its trademark abroad can submit a single application and designate other member countries in which it would like to register the mark. The treaty is designed to reduce the costs of obtaining international trademark protection by more than 60 percent, according to proponents.

Although the Madrid Protocol may simplify and reduce the cost of trademark registration in foreign nations, it remains to be seen whether it will provide significant benefits to trademark owners. Even with an easier registration process, the issue of whether member countries will enforce the law and protect a mark still remains.

TERMS AND CONCEPTS

copyright 103
cyber mark 101
cybersquatting 101
dilution 98
distributed network 107
domain name 101
intellectual property 96
license 102
patent 102
peer-to-peer (P2P) networking 107
service mark 100
trade dress 101
trade name 101
trade secret 108
trademark 96

CHAPTER SUMMARY Intellectual Property

Trademarks and Related Property (See pages 96–101.)	1. A *trademark* is a distinctive mark, motto, device, or emblem that a manufacturer stamps, prints, or otherwise affixes to the goods it produces so that the goods and their origin may be identified in the marketplace. 2. The major federal statutes protecting trademarks and related property are the Lanham Act of 1946 and the Federal Trademark Dilution Act of 1995. Generally, to be protected, a trademark must be sufficiently distinctive from all competing trademarks. 3. *Trademark infringement* occurs when one uses a mark that is the same as, or confusingly similar to, the protected trademark, service mark, trade name, or trade dress of another without permission when marketing goods or services.
Cyber Marks (See pages 101–102.)	A *cyber mark* is a trademark in cyberspace. Trademark infringement in cyberspace occurs when one person uses, in a domain name or in meta tags, a name that is the same as, or confusingly similar to, the protected mark of another.
Patents (See pages 102–103.)	1. A *patent* is a grant from the government that gives an inventor the exclusive right to make, use, and sell an invention for a period of twenty years from the date of filing the application for a patent. To be patentable, an invention (or a discovery, process, or design) must be genuine, novel, useful, and not obvious in light of current technology. Computer software may be patented. 2. *Patent infringement* occurs when one uses or sells another's patented design, product, or process without the patent owner's permission.

CHAPTER SUMMARY Intellectual Property—Continued

Copyrights (See pages 103–108.)	1. A *copyright* is an intangible property right granted by federal statute to the author or originator of certain literary or artistic productions. Computer software may be copyrighted. 2. *Copyright infringement* occurs whenever the form or expression of an idea is copied without the permission of the copyright holder. An exception applies if the copying is deemed a "fair use." 3. Copyrights are governed by the Copyright Act of 1976, as amended. To protect copyrights in digital information, Congress passed the No Electronic Theft Act of 1997 and the Digital Millennium Copyright Act of 1998. 4. Technology that allows users to share files via the Internet on distributed networks often raises copyright infringement issues. In 2005, the United States Supreme Court made it clear that software companies that have taken affirmative steps to induce or promote infringement may be held liable for their end users' infringement.
Trade Secrets (See page 108.)	*Trade secrets* include customer lists, plans, research and development, and pricing information, for example. Trade secrets are protected under the common law and, in some states, under statutory law against misappropriation by competitors. The Economic Espionage Act of 1996 made the theft of trade secrets a federal crime (see Chapter 6).
International Protection for Intellectual Property (See pages 108–110.)	Various international agreements provide international protection for intellectual property. A landmark agreement is the 1994 agreement on Trade-Related Aspects of Intellectual Property Rights (TRIPS), which provides for enforcement procedures in all countries signatory to the agreement.

FOR REVIEW

Answers for the even-numbered questions in this For Review *section can be found in Appendix A at the end of this text.*

1. What is intellectual property?
2. Why are trademarks and patents protected by the law?
3. What laws protect authors' rights in the works they generate?
4. What are trade secrets, and what laws offer protection for this form of intellectual property?
5. What steps have been taken to protect intellectual property rights in today's digital age?

QUESTIONS AND CASE PROBLEMS

5–1. Copyright Infringement. In which of the following situations would a court likely hold Maruta liable for copyright infringement?

(a) At the library, Maruta photocopies ten pages from a scholarly journal relating to a topic on which she is writing a term paper.

(b) Maruta makes leather handbags and sells them in her small leather shop. She advertises her handbags as "Vutton handbags," hoping that customers might mistakenly assume that they were made by Vuitton, the well-known maker of high-quality luggage and handbags.

(c) Maruta owns a small country store. She purchases one copy of several popular movie DVDs from various DVD distributors. Then, using blank DVDs, she makes copies to rent or sell to her customers.

(d) Maruta teaches Latin American history at a small university. She has a videocassette recorder and frequently tapes television programs relating to Latin America. She then takes the videos to her classroom so that her students can watch them.

QUESTION WITH SAMPLE ANSWER

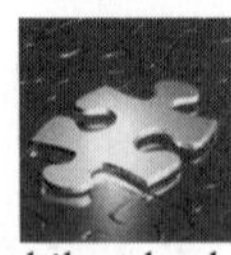

5–2. John and Andrew Doney invented a hard-bearing device for balancing rotors. Although they registered their invention with the U.S. Patent and Trademark Office, it was never used as an automobile wheel balancer. Some time later, Exetron Corp. produced an automobile wheel balancer that used a hard-bearing device with a support plate similar to that of the Doneys. Given that the Doneys had not used their device for automobile wheel balancing, does

Exetron's use of a similar hard-bearing device infringe on the Doneys' patent?

For a sample answer to this question, go to Appendix B at the end of this text.

5–3. Trademark Infringement. Elvis Presley Enterprises, Inc. (EPE), owns all of the trademarks of the Elvis Presley estate. None of these marks is registered for use in the restaurant business. Barry Capece registered "The Velvet Elvis" as a service mark for a restaurant and tavern with the U.S. Patent and Trademark Office. Capece opened a nightclub called "The Velvet Elvis" with a menu, decor, advertising, and promotional events that evoked Elvis Presley and his music. EPE filed a suit in a federal district court against Capece and others, claiming, among other things, that "The Velvet Elvis" service mark infringed on EPE's trademarks. During the trial, witnesses testified that they thought the bar was associated with Elvis Presley. Should Capece be ordered to stop using "The Velvet Elvis" mark? Why or why not? [*Elvis Presley Enterprises, Inc. v. Capece,* 141 F.3d 188 (5th Cir. 1998)]

5–4. Trademark Infringement. A&H Sportswear, Inc., a swimsuit maker, obtained a trademark for its MIRACLESUIT in 1992. The MIRACLESUIT design makes the wearer appear slimmer. The MIRACLESUIT was widely advertised and discussed in the media. The MIRACLESUIT was also sold for a brief time in the Victoria's Secret (VS) catalogue, which is published by Victoria's Secret Catalogue, Inc. In 1993, Victoria's Secret Stores, Inc., began selling a cleavage-enhancing bra, which was named THE MIRACLE BRA and for which a trademark was obtained. The next year, THE MIRACLE BRA swimwear debuted in the VS catalogue and stores. A&H filed a suit in a federal district court against VS Stores and VS Catalogue, alleging in part that THE MIRACLE BRA mark, when applied to swimwear, infringed on the MIRACLESUIT mark. A&H argued that there was a "possibility of confusion" between the marks. The VS entities contended that the appropriate standard was "likelihood of confusion" and that, in this case, there was no likelihood of confusion. In whose favor will the court rule, and why? [*A&H Sportswear, Inc. v. Victoria's Secret Stores, Inc.,* 166 F.3d 197 (3d Cir. 1999)]

CASE PROBLEM WITH SAMPLE ANSWER

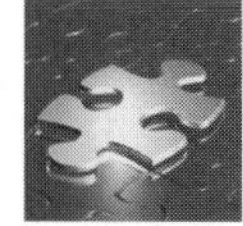

5–5. In 1999, Steve and Pierce Thumann and their father, Fred, created Spider Webs, Ltd., a partnership, to, according to Steve, "develop Internet address names." Spider Webs registered nearly two thousand Internet domain names for an average of $70 each, including the names of cities, the names of buildings, names related to a business or trade (such as air conditioning or plumbing), and the names of famous companies. It offered many of the names for sale on its Web site and through eBay.com. Spider Webs registered the domain name "ERNESTANDJULIOGALLO.COM" in Spider Webs' name. E. & J. Gallo Winery filed a suit against Spider Webs, alleging, in part, violations of the Anticybersquatting Consumer Protection Act. Gallo asked the court for, among other things, statutory damages. Gallo also sought to have the domain name at issue transferred to Gallo. During the suit, Spider Webs published anticorporate articles and opinions, and discussions of the suit, at the URL "ERNESTANDJULIOGALLO.COM." Should the court rule in Gallo's favor? Why or why not? [*E. & J. Gallo Winery v. Spider Webs, Ltd.,* 129 F.Supp.2d 1033 (S.D.Tex. 2001)]

After you have answered this problem, compare your answer with the sample answer given on the Web site that accompanies this text. Go to academic.cengage.com/blaw/wbl, select "Chapter 5," and click on "Case Problem with Sample Answer."

5–6. Trade Secrets. Four Pillars Enterprise Co. is a Taiwanese company owned by Pin Yen Yang. Avery Dennison, Inc., a U.S. corporation, is one of Four Pillars' chief competitors in the manufacture of adhesives. In 1989, Victor Lee, an Avery employee, met Yang and Yang's daughter Hwei Chen. They agreed to pay Lee $25,000 a year to serve as a consultant to Four Pillars. Over the next eight years, Lee supplied the Yangs with confidential Avery reports, including information that Four Pillars used to make a new adhesive that had been developed by Avery. The Federal Bureau of Investigation (FBI) confronted Lee, and he agreed to cooperate in an operation to catch the Yangs. When Lee next met the Yangs, he showed them documents provided by the FBI. The documents bore "confidential" stamps, and Lee said that they were Avery's confidential property. The FBI arrested the Yangs with the documents in their possession. The Yangs and Four Pillars were charged with, among other crimes, the attempted theft of trade secrets. The defendants argued in part that it was impossible for them to have committed this crime because the documents were not actually trade secrets. Should the court acquit them? Why or why not? [*United States v. Yang,* 281 F.3d 534 (6th Cir. 2002)]

5–7. Patent Infringement. As a cattle rancher in Nebraska, Gerald Gohl used handheld searchlights to find and help calving cows (cows giving birth) in harsh blizzard conditions. Gohl thought that it would be more helpful to have a portable searchlight mounted on the outside of a vehicle and remotely controlled. He and Al Gebhardt developed and patented practical applications of this idea—the Golight and the wireless, remote-controlled Radio Ray, which could rotate 360 degrees—and formed Golight, Inc., to make and market these products. In 1997, Wal-Mart Stores, Inc., began selling a portable, wireless, remote-controlled searchlight that was identical to the Radio Ray except for a stop piece that prevented the light from rotating more than 351 degrees. Golight sent Wal-Mart a letter, claiming that its device infringed Golight's patent. Wal-Mart sold its remaining inventory of the devices and stopped carrying the product. Golight filed a suit in a federal district court against Wal-Mart, alleging patent infringement. How should the court rule? Explain. [*Golight, Inc. v. Wal-Mart Stores, Inc.,* 355 F.3d 1327 (Fed. Cir. 2004)]

5–8. Trade Dress. Gateway, Inc., sells computers, computer products, computer peripherals, and computer accessories throughout the world. By 1988, Gateway had begun its first national advertising campaign using black-and-white cows and black-and-white cow spots. By 1991, black-and-white cows and spots had become Gateway's symbol. The next year, Gateway registered a black-and-white cow-spot design in association with

computers and computer peripherals as its trademark. Companion Products, Inc. (CPI), sells stuffed animals trademarked as "Stretch Pets." Stretch Pets have an animal's head and an elastic body that can wrap around the edges of computer monitors, computer cases, or televisions. CPI produces sixteen Stretch Pets, including a polar bear, a moose, several dogs, and a penguin. One of CPI's top-selling products is a black-and-white cow that CPI identifies as "Cody Cow," which was first sold in 1999. Gateway filed a suit in a federal district court against CPI, alleging trade dress infringement and related claims. What is trade dress? What is the major factor in cases involving trade dress infringement? Does that factor exist in this case? Explain. [*Gateway, Inc. v. Companion Products, Inc.*, 384 F.3d 503 (8th Cir. 2004)]

5–9. Fair Use Doctrine. Leslie Kelly is a professional photographer who has copyrighted many of his images of the American West. Some of the images can be seen on Kelly's Web site or other sites with which Kelly has a contract. Arriba Soft Corp. operates an Internet search engine that displays its results in the form of small pictures (thumbnails) rather than text. The thumbnails consist of images copied from other sites and reduced in size. By clicking on one of the thumbnails, a user can view a large version of the picture within the context of an Arriba Web page. Arriba displays the large picture by inline linking (importing the image from the other site without copying it onto Arriba's site). When Kelly discovered that his photos were displayed through Arriba's site without his permission, he filed a suit in a federal district court against Arriba, alleging copyright infringement. Arriba claimed that its use of Kelly's images was a "fair use." Considering the factors courts use to determine whether a use is fair, do Arriba's thumbnails qualify? Does Arriba's use of the larger images infringe on Kelly's copyright? Explain. [*Kelly v. Arriba Soft Corp.*, 280 F.3d 934 (9th Cir. 2002)]

VIDEO QUESTION

5–10. Go to this text's Web site at academic.cengage.com/blaw/wbl and select "Chapter 5." Click on "Video Questions" and view the video titled *The Jerk*. Then answer the following questions.

1. In the video, Navin (Steve Martin) creates a special handle for Mr. Fox's (Bill Macy's) glasses. Can Navin obtain a patent or a copyright protecting his invention? Explain your answer.
2. Suppose that after Navin legally protects his idea, Fox steals it and decides to develop it for himself, without Navin's permission. Has Fox committed infringement? If so, what kind: trademark, patent, or copyright?
3. Suppose that after Navin legally protects his idea, he realizes he doesn't have the funds to mass-produce the special handle. Navin therefore agrees to allow Fox to manufacture the product. Has Navin granted Fox a license? Explain.
4. Assume that Navin is able to manufacture his invention. What might Navin do to ensure that his product is identifiable and can be distinguished from other products on the market?

Intellectual Property

For updated links to resources available on the Web, as well as a variety of other materials, visit this text's Web site at

academic.cengage.com/blaw/wbl

An excellent overview of the laws governing various forms of intellectual property is available at FindLaw's Web site. Go to

http://profs.lp.findlaw.com

You can find answers to common questions about trademark and patent law—and links to registration forms, statutes, international patent and trademark offices, and numerous other related materials—at the Web site of the U.S. Patent and Trademark Office. Go to

http://www.uspto.gov

For information on invalid and improperly granted patents, go to

http://www.bustpatents.com

For information on copyrights, go to the U.S. Copyright Office at

http://www.loc.gov/copyright

You can find extensive information on copyright law—including United States Supreme Court decisions in this area and the texts of the Berne Convention and other international treaties on copyright issues—at the Web site of the Legal Information Institute at Cornell University's School of Law. Go to

http://www.law.cornell.edu/topics/copyright.html

ONLINE LEGAL RESEARCH

Go to the *Fundamentals of Business Law* home page at **academic.cengage.com/blaw/wbl**, select "Chapter 5," and click on "Internet Exercises." There you will find the following Internet research exercises that you can perform to learn more about topics covered in this chapter.

Activity 5–1: LEGAL PERSPECTIVE—Unwarranted Legal Threats

Activity 5–2: MANAGEMENT PERSPECTIVE—Protecting Intellectual Property across Borders

BEFORE THE TEST

Go to the *Fundamentals of Business Law* home page at **academic.cengage.com/blaw/wbl**, select "Chapter 5," and click on "Interactive Quizzes." You will find a number of interactive questions relating to this chapter.

CHAPTER 6

Criminal Law and Cyber Crimes

LEARNING OBJECTIVES

After reading this chapter, you should be able to answer the following questions:

1. What two elements must exist before a person can be held liable for a crime? Can a corporation commit crimes?
2. What are five broad categories of crimes? What is white-collar crime?
3. What defenses might be raised by criminal defendants to avoid liability for criminal acts?
4. What constitutional safeguards exist to protect persons accused of crimes? What are the basic steps in the criminal process?
5. What is cyber crime? What laws apply to crimes committed in cyberspace?

Various sanctions are used to bring about a society in which individuals engaging in business can compete and flourish. These sanctions include damages for various types of tortious conduct (as discussed in Chapter 4), damages for breach of contract (to be discussed in Chapter 12), and the equitable remedies discussed in Chapter 1. Additional sanctions are imposed under criminal law. Many statutes regulating business provide for criminal as well as civil sanctions. Therefore, criminal law joins civil law as an important element in the legal environment of business.

In this chapter, following a brief summary of the major differences between civil law and criminal law, we look at how crimes are classified and what elements must be present for criminal liability to exist. We then examine various categories of crime, the defenses that can be raised to avoid liability for criminal actions, and criminal procedural law. Criminal procedural law attempts to ensure that a criminal defendant's right to "due process of law" is enforced. This right is guaranteed by the Fourteenth Amendment to the U.S. Constitution.

Since the advent of computer networks and, more recently, the Internet, new types of crimes or new variations of traditional crimes have been committed in cyberspace. For that reason, they are often referred to as *cyber crime*. Generally, cyber crime refers more to the way particular crimes are committed than to a new category of crimes. We devote the concluding pages of this chapter to a discussion of this increasingly significant area of criminal activity.

Civil Law and Criminal Law

Remember from Chapter 1 that *civil law* spells out the duties that exist between persons or between persons and their governments, excluding the duty not to commit crimes. Contract law, for example, is part of civil law. The whole body of tort law, which deals with the infringement by one person on the legally recognized rights of another, is also an area of civil law.

Criminal law, in contrast, has to do with crime. A **crime** can be defined as a wrong against society proclaimed in a statute and punishable by society through fines and/or imprisonment—and, in some cases, death. As mentioned in Chapter 1, because crimes are *offenses against society as a whole,* they are prosecuted by a public official, such as a district attorney (D.A.), not by victims.

KEY DIFFERENCES BETWEEN CIVIL LAW AND CRIMINAL LAW

Because the state has extensive resources at its disposal when prosecuting criminal cases, there are numerous procedural safeguards to protect the rights of defendants. One of these safeguards is the higher burden of proof that applies in a criminal case. As you can see in Exhibit 6–1, which summarizes some of the key differences between civil law and criminal law, in a civil case the plaintiff usually must prove his or her case by a *preponderance of the evidence*. Under this standard, the plaintiff must convince the court that, based on the evidence presented by both parties, it is more likely than not that the plaintiff's allegation is true.

In a criminal case, however, the state must prove its case **beyond a reasonable doubt.** In addition, the verdict normally must be unanimous. For the defendant to be convicted, then, *every* juror in a criminal case must be convinced, beyond a reasonable doubt, that the defendant has committed each essential element of the offense with which she or he is charged. In contrast, in civil cases typically only three-fourths of the jurors need to agree that it is more likely than not that the defendant caused the plaintiff's harm. The higher burden of proof in criminal cases reflects a fundamental social value—the belief that it is worse to convict an innocent individual than to let a guilty person go free. We will look at other safeguards later in the chapter, in the context of criminal procedure.

CIVIL LIABILITY FOR CRIMINAL ACTS

Those who commit crimes may be subject to both civil and criminal liability. ■ EXAMPLE 6.1 Joe is walking down the street, minding his own business, when suddenly a person attacks him. In the ensuing struggle, the attacker stabs Joe several times, seriously injuring him. A police officer restrains and arrests the wrongdoer. In this situation, the attacker may be subject both to criminal prosecution by the state and to a tort lawsuit brought by Joe. ■ Exhibit 6–2 illustrates how the same act can result in both a tort action and a criminal action against the wrongdoer.

Classification of Crimes

Depending on their degree of seriousness, crimes are classified as felonies or misdemeanors. **Felonies** are serious crimes punishable by death or by imprisonment in a federal or state penitentiary for more than a year. The Model Penal Code[1] provides for four degrees of felony: (1) capital offenses, for which the maximum penalty is death; (2) first degree felonies, punishable by a maximum penalty of life imprisonment; (3) second degree felonies, punishable by a maximum of ten years' imprisonment; and (4) third degree felonies, punishable by a maximum of five years' imprisonment.

Under federal law and in most states, any crime that is not a felony is considered a *misdemeanor.* **Misdemeanors** are crimes punishable by a fine or by confinement for up to a year. If incarcerated (imprisoned), the guilty party goes to a local jail instead of a prison. Disorderly conduct and trespass are common misdemeanors. Some states have different classes of misdemeanors. For example, in Illinois misdemeanors are either Class A (confinement for up to a year), Class B (not more than six months), or Class C (not more than thirty days). Whether a crime is a felony or a misdemeanor can also determine whether the case is tried in a magistrate's court (for example, by a justice of the peace) or in a general trial court.

1. The American Law Institute issued the Official Draft of the Model Penal Code in 1962. The Model Penal Code is not a uniform code. Uniformity of criminal law among the states is not as important as uniformity in other areas of the law. Types of crimes vary with local circumstances, and it is appropriate that punishments vary accordingly. The Model Penal Code contains four parts: (1) general provisions, (2) definitions of special crimes, (3) provisions concerning treatment and corrections, and (4) provisions on the organization of corrections.

EXHIBIT 6–1 KEY DIFFERENCES BETWEEN CIVIL LAW AND CRIMINAL LAW

ISSUE	CIVIL LAW	CRIMINAL LAW
Party who brings suit	Person who suffered harm	The state
Burden of proof	Preponderance of the evidence	Beyond a reasonable doubt
Verdict	Three-fourths majority (typically)	Unanimous
Remedy	Damages to compensate for the harm, or a decree to achieve an equitable result	Punishment (fine, imprisonment, or death)

EXHIBIT 6–2 TORT LAWSUIT AND CRIMINAL PROSECUTION FOR THE SAME ACT

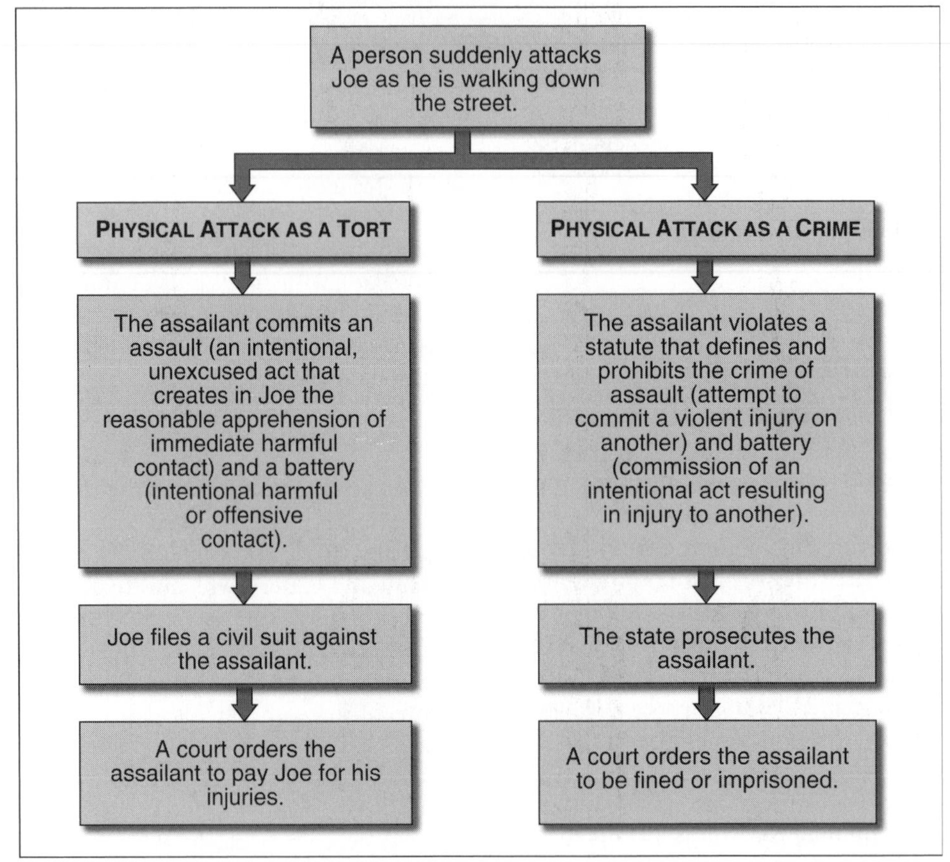

A person suddenly attacks Joe as he is walking down the street.
PHYSICAL ATTACK AS A TORT
PHYSICAL ATTACK AS A CRIME
The assailant commits an assault (an intentional, unexcused act that creates in Joe the reasonable apprehension of immediate harmful contact) and a battery (intentional harmful or offensive contact).
The assailant violates a statute that defines and prohibits the crime of assault (attempt to commit a violent injury on another) and battery (commission of an intentional act resulting in injury to another).
Joe files a civil suit against the assailant.
The state prosecutes the assailant.
A court orders the assailant to pay Joe for his injuries.
A court orders the assailant to be fined or imprisoned.

In most jurisdictions, **petty offenses** are considered to be a subset of misdemeanors. Petty offenses are minor violations, such as jaywalking or violations of building codes. Even for petty offenses, however, a guilty party can be put in jail for a few days, fined, or both, depending on state or local law.

Criminal Liability

Two elements must exist simultaneously for a person to be convicted of a crime: (1) the performance of a prohibited act and (2) a specified state of mind or intent on the part of the actor. Every criminal statute prohibits certain behavior. Most crimes require an act of *commission*—that is, a person must *do* something in order to be accused of a crime.[2] In some instances, an act of *omission* can be a crime, but only when a person has a legal duty to perform the omitted act. Failure to file a tax return is an example of an omission that is a crime.

The *guilty act* requirement is based on one of the premises of criminal law—that a person is punished for harm done to society. Thinking about killing someone or about stealing a car may be wrong, but the thoughts do no harm until they are translated into action. Of course, a person can be punished for attempting murder or robbery, but normally only if he or she took substantial steps toward the criminal objective.

A *wrongful mental state*[3] is generally required to establish criminal liability. What constitutes such a mental state varies according to the wrongful action. For murder, the act is the taking of a life, and the mental state is the intent to take life. For theft, the guilty act is the taking of another person's property, and the mental state involves both the knowledge that the property belongs to another and the intent to deprive the owner of it.

Criminal liability typically arises for actions that violate state criminal statutes. Federal criminal jurisdiction is normally limited to crimes that occur outside the jurisdiction

2. Called the *actus reus* (pronounced *ak*-tuhs *ray*-uhs), or "guilty act."

3. Called the *mens rea* (pronounced mehns *ray*-uh), or "evil intent."

of any state, crimes involving interstate commerce or communications, crimes that interfere with the operation of the federal government or its agents, and crimes directed at citizens or property located outside the United States. Federal jurisdiction also exists if a federal law or a federal government agency (such as the U.S. Department of Justice or the federal Environmental Protection Agency) defines a certain type of action as a crime. Today, businesspersons are subject to criminal penalties under numerous federal laws and regulations. We will examine many of these laws in later chapters of this text.

Corporate Criminal Liability

At one time, it was thought that a corporation could not incur criminal liability because, although a corporation is a legal person, it can act only through its agents (corporate directors, officers, and employees). Therefore, the corporate entity itself could not "intend" to commit a crime. Under modern criminal law, however, a corporation may be held liable for crimes. Obviously, corporations cannot be imprisoned, but they can be fined or denied certain legal privileges (such as a license). Today, corporations are normally liable for the crimes committed by their agents and employees within the course and scope of their employment.

Corporate directors and officers are personally liable for the crimes they commit, regardless of whether the crimes were committed for their personal benefit or on the corporation's behalf. Additionally, corporate directors and officers may be held liable for the actions of employees under their supervision. Under what has become known as the "responsible corporate officer" doctrine, a court may impose criminal liability on a corporate officer regardless of whether she or he participated in, directed, or even knew about a given criminal violation.

■ **EXAMPLE 6.2** In *United States v. Park,*[4] the chief executive officer of a national supermarket chain was held personally liable for sanitation violations in corporate warehouses, in which the food was exposed to contamination by rodents. The United States Supreme Court imposed personal liability on the corporate officer not because he intended the crime or even knew about it but because he was in a "responsible relationship" to the corporation and had the power to prevent the violation. ■ Since the *Park* decision, courts have applied this "responsible corporate officer" doctrine on a number of occasions to hold corporate officers liable for their employees' statutory violations.

4. 421 U.S. 658, 95 S.Ct. 1903, 44 L.Ed.2d 489 (1975).

Types of Crimes

The number of acts that are defined as criminal is nearly endless. Federal, state, and local laws provide for the classification and punishment of hundreds of thousands of different criminal acts. Traditionally, though, crimes have been grouped into five broad categories, or types: violent crime (crimes against persons), property crime, public order crime, white-collar crime, and organized crime. Cyber crime—which consists of crimes committed in cyberspace with the use of computers—is, as mentioned earlier in this chapter, less a category of crime than a new way to commit crime. We will examine cyber crime later in this chapter.

VIOLENT CRIME

Crimes against persons, because they cause others to suffer harm or death, are referred to as *violent crimes.* Murder is a violent crime. So is sexual assault, or rape. Assault and battery, which were discussed in Chapter 4 in the context of tort law, are also classified as violent crimes. **Robbery**—defined as the taking of cash, personal property, or any other article of value from a person by means of force or fear—is also a violent crime. Typically, states have more severe penalties for *aggravated robbery*—robbery with the use of a deadly weapon.

Each of these violent crimes is further classified by degree, depending on the circumstances surrounding the criminal act. These circumstances include the intent of the person committing the crime, whether a weapon was used, and (in cases other than murder) the level of pain and suffering experienced by the victim.

PROPERTY CRIME

The most common type of criminal activity is property crime—crimes in which the goal of the offender is some form of economic gain or the damaging of property. Robbery is a form of property crime, as well as a violent crime, because the offender seeks to gain the property of another. We look here at a number of other crimes that fall within the general category of property crime.

Burglary Traditionally, **burglary** was defined under the common law as breaking and entering the dwelling of another at night with the intent to commit a felony. Originally, the definition was aimed at protecting an individual's home and its occupants. Most state statutes have eliminated some of the requirements found in the common law definition. The time at which the breaking and entering occurs, for example, is usually immaterial. State

statutes frequently omit the element of breaking, and some states do not require that the building be a dwelling. Aggravated burglary, which is defined as burglary with the use of a deadly weapon, burglary of a dwelling, or both, incurs a greater penalty.

Larceny Any person who wrongfully or fraudulently takes and carries away another person's personal property is guilty of **larceny.** Larceny includes the fraudulent intent to deprive an owner permanently of property. Many business-related larcenies entail fraudulent conduct. Whereas robbery involves force or fear, larceny does not. Therefore, picking pockets is larceny. Similarly, taking company products and supplies home for personal use, if one is not authorized to do so, is larceny.

In most states, the definition of property that is subject to larceny statutes has expanded. Stealing computer programs may constitute larceny even though the "property" consists of magnetic impulses. Stealing computer time can also constitute larceny. So, too, can the theft of natural gas. Trade secrets can be subject to larceny statutes. Obtaining another's phone-card number and then using that number, without authorization, to place long-distance calls is a form of property theft. These types of larceny are covered by "theft of services" statutes in many jurisdictions.

The common law distinguishes between grand and petit larceny depending on the value of the property taken. Many states have abolished this distinction, but in those that have not, grand larceny is a felony and petit larceny, a misdemeanor.

Obtaining Goods by False Pretenses It is a criminal act to obtain goods by means of false pretenses—for example, buying groceries with a check, knowing that one has insufficient funds to cover it. Statutes dealing with such illegal activities vary widely from state to state.

Receiving Stolen Goods It is a crime to receive stolen goods. The recipient of such goods need not know the true identity of the owner or the thief. All that is necessary is that the recipient knows or should have known that the goods are stolen, which implies an intent to deprive the owner of those goods.

Arson The willful and malicious burning of a building (and in some states, personal property) owned by another is the crime of **arson.** At common law, arson traditionally applied only to burning down another person's house. The law was designed to protect human life. Today, arson statutes in most states have been extended to cover the destruction of any building, regardless of ownership, by fire or explosion.

Every state has a special statute that covers the burning of a building for the purpose of collecting insurance. ■ EXAMPLE 6.3 If Smith owns an insured apartment building that is falling apart and sets fire to it himself or pays someone else to do so, he is guilty not only of arson but also of defrauding insurers, which is an attempted larceny. ■ Of course, the insurer need not pay the claim when insurance fraud is proved.

Forgery The fraudulent making or altering of any writing in a way that changes the legal rights and liabilities of another is **forgery.** If, without authorization, Severson signs Bennett's name to the back of a check made out to Bennett, Severson is committing forgery. Forgery also includes changing trademarks, falsifying public records, counterfeiting, and altering a legal document.

PUBLIC ORDER CRIME

Historically, societies have always outlawed activities that are considered to be contrary to public values and morals. Today, the most common public order crimes include public drunkenness, prostitution, gambling, and illegal drug use. These crimes are sometimes referred to as *victimless crimes* because they normally harm only the offender. From a broader perspective, however, they are deemed detrimental to society as a whole because they might create an environment that gives rise to property and violent crimes.

WHITE-COLLAR CRIME

Crimes that typically occur only in the business context are commonly referred to as **white-collar crimes.** Although there is no official definition of *white-collar crime,* the term is popularly used to mean an illegal act or series of acts committed by an individual or business entity using some nonviolent means. Usually, this kind of crime is committed in the course of a legitimate occupation. Corporate crimes fall into this category.

Embezzlement When a person entrusted with another person's funds or property fraudulently appropriates the funds or property, **embezzlement** occurs. Typically, embezzlement involves an employee who steals funds. Banks face this problem, and so do a number of businesses in which corporate officers or accountants "jimmy" the books to cover up the fraudulent conversion of funds for their own benefit. Embezzlement is not larceny, because the wrongdoer does not physically take the

property from the possession of another, and it is not robbery, because force or fear is not used.

It does not matter whether the accused takes the funds from the victim or from a third person. If, as the financial officer of a large corporation, Saunders pockets a certain number of checks from third parties that were given to her to deposit into the corporate account, she is embezzling.

Ordinarily, an embezzler who returns what has been taken will not be prosecuted because the owner usually will not take the time to make a complaint, give depositions, and appear in court. That the accused intended eventually to return the embezzled property, however, does not constitute a sufficient defense to the crime of embezzlement.

Mail and Wire Fraud One of the most potent weapons against white-collar criminals is the Mail Fraud Act of 1990.[5] Under this act, it is a federal crime (mail fraud) to use the mails to defraud the public. Illegal use of the mails must involve (1) mailing or causing someone else to mail a writing—something written, printed, or photocopied—for the purpose of executing a scheme to defraud and (2) a contemplated or an organized scheme to defraud by false pretenses. If, for example, Johnson advertises by mail the sale of a cure for cancer that he knows to be fraudulent because it has no medical validity, he can be prosecuted for fraudulent use of the mails.

Federal law also makes it a crime to use wire (for example, the telephone), radio, or television transmissions to defraud.[6] Violators may be fined up to $1,000, imprisoned for up to five years, or both. If the violation affects a financial institution, the violator may be fined up to $1 million, imprisoned for up to thirty years, or both.

Bribery Basically, three types of bribery are considered crimes: (1) bribery of public officials, (2) commercial bribery, and (3) bribery of foreign officials. The attempt to influence a public official to act in a way that serves a private interest is a crime. As an element of this crime, intent must be present and proved. The bribe can be anything the recipient considers to be valuable. Realize that *the crime of bribery occurs when the bribe is offered.* It does not matter whether the person to whom the bribe is offered accepts the bribe or agrees to perform whatever action is desired by the person offering the bribe. *Accepting a bribe* is a separate crime.

Typically, people make commercial bribes to obtain proprietary information, cover up an inferior product, or secure new business. Industrial espionage sometimes involves commercial bribes. For example, a person in one firm may offer an employee in a competing firm some type of payoff in exchange for trade secrets or pricing schedules. So-called kickbacks, or payoffs for special favors or services, are a form of commercial bribery in some situations.

Bribing foreign officials to obtain favorable business contracts is a crime. The Foreign Corrupt Practices Act of 1977, which was discussed in Chapter 3, was passed to curb the use of bribery by American businesspersons in securing foreign contracts.

Bankruptcy Fraud Today, federal bankruptcy law (see Chapter 21) allows individuals and businesses to be relieved of oppressive debt through bankruptcy proceedings. Numerous white-collar crimes may be committed during the many phases of a bankruptcy proceeding. A creditor, for example, may file a false claim against the debtor, which is a crime. Also, a debtor may fraudulently transfer assets to favored parties before or after the petition for bankruptcy is filed. For example, a company-owned automobile may be "sold" at a bargain price to a trusted friend or relative. Closely related to the crime of fraudulent transfer of property is the crime of fraudulent concealment of property, such as hiding gold coins.

The Theft of Trade Secrets As discussed in Chapter 5, trade secrets constitute a form of intellectual property that for many businesses can be extremely valuable. The Economic Espionage Act of 1996[7] made the theft of trade secrets a federal crime. The act also made it a federal crime to buy or possess trade secrets of another person, knowing that the trade secrets were stolen or otherwise acquired without the owner's authorization.

Violations of the act can result in steep penalties. An individual who violates the act can be imprisoned for up to ten years and fined up to $500,000. If a corporation or other organization violates the act, it can be fined up to $5 million. Additionally, the law provides that any property acquired as a result of the violation and any property used in the commission of the violation are subject to criminal *forfeiture*—meaning that the government can take the property. A theft of trade secrets conducted via the Internet, for example, could result in the forfeiture of every computer, printer, or other device used to commit or facilitate the violation.

Insider Trading An individual who obtains "inside information" about the plans of a publicly listed corporation can often make stock-trading profits by using the

5. 18 U.S.C. Sections 1341–1342.
6. 18 U.S.C. Section 1343.
7. 18 U.S.C. Sections 1831–1839.

information to guide decisions relating to the purchase or sale of corporate securities. *Insider trading* is a violation of securities law and will be considered more fully in Chapter 27. Generally, the rule is that a person who possesses inside information and has a duty not to disclose it to outsiders may not profit from the purchase or sale of securities based on that information until the information is available to the public.

ORGANIZED CRIME

As mentioned, white-collar crime takes place within the confines of the legitimate business world. *Organized crime,* in contrast, operates *illegitimately* by, among other things, providing illegal goods and services. For organized crime, the traditional preferred markets are gambling, prostitution, illegal narcotics, and loan sharking (lending money at higher-than-legal-maximum interest rates), along with more recent ventures into counterfeiting and credit-card scams.

Money Laundering The profits from illegal activities amount to billions of dollars a year, particularly the profits from illegal drug transactions and, to a lesser extent, from racketeering, prostitution, and gambling. Under federal law, banks, savings and loan associations, and other financial institutions are required to report currency transactions involving more than $10,000. Consequently, those who engage in illegal activities face difficulties in depositing their cash profits from illegal transactions.

As an alternative to simply storing cash from illegal transactions in a safe-deposit box, wrongdoers and racketeers have invented ways to launder "dirty" money to make it "clean." This **money laundering** is done through legitimate businesses.

■ EXAMPLE 6.4 Matt, a successful drug dealer, becomes a partner with a restaurateur. Little by little, the restaurant shows an increasing profit. As a partner in the restaurant, Matt is able to report the "profits" of the restaurant as legitimate income on which he pays federal and state income taxes. He can then spend those after-tax funds without worrying that his lifestyle may exceed the level possible with his reported income. ■

The Federal Bureau of Investigation estimates that organized crime has invested tens of billions of dollars in as many as a hundred thousand business establishments in the United States for the purpose of money laundering. Globally, it is estimated that more than $700 billion in illegal money moves through the world banking system every year.

The Racketeer Influenced and Corrupt Organizations Act In 1970, in an effort to curb the apparently increasing entry of organized crime into the legitimate business world, Congress passed the Racketeer Influenced and Corrupt Organizations Act (RICO).[8] The act, which was enacted as part of the Organized Crime Control Act, makes it a federal crime to (1) use income obtained from racketeering activity to purchase any interest in an enterprise, (2) acquire or maintain an interest in an enterprise through racketeering activity, (3) conduct or participate in the affairs of an enterprise through racketeering activity, or (4) conspire to do any of the preceding activities.

Racketeering activity is not a new type of substantive crime created by RICO; rather, RICO incorporates by reference twenty-six separate types of federal crimes and nine types of state felonies[9] and declares that if a person commits two of these offenses, he or she is guilty of "racketeering activity." Additionally, RICO is more often used today to attack white-collar crimes than organized crime.

In the event of a violation, the statute permits the government to seek civil penalties, including the divestiture of a defendant's interest in a business (called *forfeiture*) or the dissolution of the business. Perhaps the most controversial aspect of RICO is that, in some cases, private individuals are allowed to recover three times their actual losses (treble damages), plus attorneys' fees, for business injuries caused by a violation of the statute. Under criminal provisions of RICO, any individual found guilty of a violation is subject to a fine of up to $25,000 per violation, imprisonment for up to twenty years, or both. Additionally, the statute provides that those who violate RICO may be required to forfeit (give up) any assets, in the form of property or cash, that were acquired as a result of the illegal activity or that were "involved in" or an "instrumentality of" the activity. (See Exhibit 6–3 on the next page for an overview of the types of crimes.)

Defenses to Criminal Liability

Among the most important defenses to criminal liability are infancy, intoxication, insanity, mistake, consent, duress, justifiable use of force, entrapment, and the statute of limitations. Many of these defenses involve assertions that the intent requirement for criminal liability is lacking. Also, in some cases, defendants are given immunity and thus relieved, at least in part, of criminal liability for crimes they committed. We look at each of these defenses here.

Note that procedural violations, such as obtaining evidence without a valid search warrant, may also operate as defenses. As you will read later in this chapter, evidence obtained in violation of a defendant's constitutional rights

8. 18 U.S.C. Sections 1961–1968.
9. See 18 U.S.C. Section 1961(1)(A).

EXHIBIT 6–3 TYPES OF CRIMES

CRIME CATEGORY	DEFINITIONS AND EXAMPLES
Violent Crime	**1.** *Definition*—Crimes that cause others to suffer harm or death. **2.** *Examples*—Murder, assault and battery, sexual assault (rape), and robbery.
Property Crime	**1.** Definition—Crimes in which the goal of the offender is some form of economic gain or the damaging of property; the most common form of crime. **2.** *Examples*—Burglary, larceny, arson, receiving stolen goods, forgery, and obtaining goods by false pretenses.
Public Order Crime	**1.** *Definition*—Crimes contrary to public values and morals. **2.** *Examples*—Public drunkenness, prostitution, gambling, and illegal drug use.
White-Collar Crime	**1.** *Definition*—An illegal act or series of acts committed by an individual or business entity using some nonviolent means to obtain a personal or business advantage; usually committed in the course of a legitimate occupation. **2.** *Examples*—Embezzlement, mail and wire fraud, bribery, bankruptcy fraud, theft of trade secrets, and insider trading.
Organized Crime	**1.** *Definition*—A form of crime conducted by groups operating illegitimately to provide the public with illegal goods and services (such as gambling or illegal narcotics). **2.** *Money laundering*—The establishment of legitimate enterprises through which "dirty" money (obtained through criminal activities, such as organized crime) can be "laundered" (made to appear to be legitimate income). **3.** *RICO*—The Racketeer Influenced and Corrupt Organizations Act (RICO) of 1970 makes it a federal crime to (a) use income obtained from racketeering activity to purchase any interest in an enterprise, (b) acquire or maintain an interest in an enterprise through racketeering activity, (c) conduct or participate in the affairs of an enterprise through racketeering activity, or (d) conspire to do any of the preceding activities. RICO provides for both civil and criminal liability.

normally may not be admitted in court. If the evidence is suppressed, then there may be no basis for prosecuting the defendant.

INFANCY

The term *infant,* as used in the law, refers to any person who has not yet reached the age of majority (see Chapter 9). In all states, certain courts handle cases involving children who are alleged to have violated the law. In some states, juvenile courts handle children's cases exclusively. In other states, however, courts that handle children's cases may also have jurisdiction over additional matters.

Originally, juvenile court hearings were informal, and lawyers were rarely present. Since 1967, however, when the United States Supreme Court ruled that a child charged with delinquency must be allowed to consult with an attorney before being committed to a state institution,[10] juvenile court hearings have become more formal. In some states, a child may be treated as an adult and tried in a regular court if she or he is above a certain age (usually fourteen) and is charged with a felony, such as rape or murder.

10. *In re Gault,* 387 U.S. 1, 87 S.Ct. 1428, 18 L.Ed.2d 527 (1967).

INTOXICATION

The law recognizes two types of intoxication, whether from drugs or from alcohol: *involuntary* and *voluntary.* Involuntary intoxication occurs when a person either is physically forced to ingest or inject an intoxicating substance or is unaware that a substance contains drugs or alcohol. Involuntary intoxication is a defense to a crime if its effect was to make a person incapable of obeying the law or incapable of understanding that the act committed was wrong. Voluntary intoxication is rarely a defense, but it may be effective in cases in which the defendant was *extremely* intoxicated when committing the wrong.

INSANITY

Someone suffering from a mental illness may be judged incapable of having the state of mind required to commit a crime. Thus, insanity may be a defense to a criminal charge. The courts have had difficulty deciding what the

test for legal insanity should be, however, and psychiatrists as well as lawyers are critical of the tests used. Almost all federal courts and some states use the relatively liberal standard set forth in the Model Penal Code:

> A person is not responsible for criminal conduct if at the time of such conduct as a result of mental disease or defect he [or she] lacks substantial capacity either to appreciate the wrongfulness of his [or her] conduct or to conform his [or her] conduct to the requirements of the law.

Some states use the *M'Naghten* test,[11] under which a criminal defendant is not responsible if, at the time of the offense, he or she did not know the nature and quality of the act or did not know that the act was wrong. Other states use the *irresistible-impulse test.* A person operating under an irresistible impulse may know an act is wrong but cannot refrain from doing it.

MISTAKE

Everyone has heard the saying, "Ignorance of the law is no excuse." Ordinarily, ignorance of the law or a mistaken idea about what the law requires is not a valid defense. In some states, however, that rule has been modified. Criminal defendants who claim that they honestly did not know that they were breaking a law may have a valid defense if (1) the law was not published or reasonably made known to the public or (2) the defendant relied on an official statement of the law that was erroneous.

A *mistake of fact,* as opposed to a *mistake of law,* operates as a defense if it negates the mental state necessary to commit a crime. ■ EXAMPLE 6.5 If Oliver Wheaton mistakenly walks off with Julie Cabrera's briefcase because he thinks it is his, there is no theft. Theft requires knowledge that the property belongs to another. (If Wheaton's act causes Cabrera to incur damages, however, Wheaton may be subject to liability for trespass to personal property or conversion, torts that were discussed in Chapter 4.) ■

CONSENT

What if a victim consents to a crime or even encourages the person intending a criminal act to commit it? Ordinarily, **consent** does not operate as a bar to criminal liability. In some rare circumstances, however, the law may allow consent to be used as a defense. In each case, the question is whether the law forbids an act that was committed against the victim's will or forbids the act without regard to the victim's wish. The law forbids murder, prostitution, and drug use regardless of whether the victim consents to it. Also, if the act causes harm to a third person who has not consented, there is no escape from criminal liability. Consent or forgiveness given after a crime has been committed is not really a defense, though it can affect the likelihood of prosecution or the severity of the sentence. Consent operates most successfully as a defense in crimes against property.

■ EXAMPLE 6.6 Barry gives Phong permission to hunt for deer on Barry's land while staying in Barry's lakeside cabin. After observing Phong carrying a gun into the cabin at night, a neighbor calls the police, and an officer subsequently arrests Phong. If charged with burglary (or aggravated burglary, because he had a weapon), Phong can assert the defense of consent. He had obtained Barry's consent to enter the premises. ■

DURESS

Duress exists when the *wrongful threat* of one person induces another person to perform an act that she or he would not otherwise perform. In such a situation, duress is said to negate the mental state necessary to commit a crime. For duress to qualify as a defense, the following requirements must be met:

1. The threat must be of serious bodily harm or death.
2. The harm threatened must be greater than the harm caused by the crime.
3. The threat must be immediate and inescapable.
4. The defendant must have been involved in the situation through no fault of his or her own.

JUSTIFIABLE USE OF FORCE

Probably the best-known defense to criminal liability is **self-defense.** Other situations, however, also justify the use of force: the defense of one's dwelling, the defense of other property, and the prevention of a crime. In all of these situations, it is important to distinguish between deadly and nondeadly force. *Deadly force* is likely to result in death or serious bodily harm. *Nondeadly force* is force that reasonably appears necessary to prevent the imminent use of criminal force.

Generally speaking, people can use the amount of nondeadly force that seems necessary to protect themselves, their dwellings, or other property or to prevent the commission of a crime. Deadly force can be used in self-defense if there is a *reasonable belief* that imminent death or grievous bodily harm will otherwise result, if the attacker is using unlawful force (an example of lawful force is that exerted by a police officer), and if the defender has not initiated or provoked the attack. Deadly force normally can be used to defend a dwelling only if the

11. A rule derived from *M'Naghten's Case,* 8 Eng.Rep. 718 (1843).

unlawful entry is violent and the person believes deadly force is necessary to prevent imminent death or great bodily harm or—in some jurisdictions—if the person believes deadly force is necessary to prevent the commission of a felony (such as arson) in the dwelling.

Entrapment

Entrapment is a defense designed to prevent police officers or other government agents from encouraging crimes in order to apprehend persons wanted for criminal acts. In the typical entrapment case, an undercover agent *suggests* that a crime be committed and somehow pressures or induces an individual to commit it. The agent then arrests the individual for the crime.

For entrapment to be considered a defense, both the suggestion and the inducement must take place. The defense is intended not to prevent law enforcement agents from setting a trap for an unwary criminal but rather to prevent them from pushing the individual into it. The crucial issue is whether the person who committed a crime was predisposed to do so or acted because the agent induced it.

Statute of Limitations

With some exceptions, such as for the crime of murder, statutes of limitations apply to crimes just as they do to civil wrongs. In other words, criminal cases must be prosecuted within a certain number of years. If a criminal action is brought after the statutory time period has expired, the accused person can raise the statute of limitations as a defense.

Immunity

At times, the state may wish to obtain information from a person accused of a crime. Accused persons are understandably reluctant to give information if it will be used to prosecute them, and they cannot be forced to do so. The privilege against self-incrimination is granted by the Fifth Amendment to the Constitution, which reads, in part, "nor shall [any person] be compelled in any criminal case to be a witness against himself." In cases in which the state wishes to obtain information from a person accused of a crime, the state can grant *immunity* from prosecution or agree to prosecute for a less serious offense in exchange for the information. Once immunity is given, the person can no longer refuse to testify on Fifth Amendment grounds, because he or she now has an absolute privilege against self-incrimination.

Often, a grant of immunity from prosecution for a serious crime is part of the **plea bargaining** between the defendant and the prosecuting attorney. The defendant may be convicted of a lesser offense, while the state uses the defendant's testimony to prosecute accomplices for serious crimes carrying heavy penalties.

Constitutional Safeguards and Criminal Procedures

Criminal law brings the power of the state, with all its resources, to bear against the individual. Criminal procedures are designed to protect the constitutional rights of individuals and to prevent the arbitrary use of power on the part of the government.

The U.S. Constitution provides specific safeguards for those accused of crimes. Most of these safeguards protect individuals against state government actions, as well as federal government actions, by virtue of the due process clause of the Fourteenth Amendment. These safeguards are set forth in the Fourth, Fifth, Sixth, and Eighth Amendments.

Fourth Amendment Protections

The Fourth Amendment protects the "right of the people to be secure in their persons, houses, papers, and effects." Before searching or seizing private property, law enforcement officers must obtain a **search warrant**—an order from a judge or other public official authorizing the search or seizure.

Search Warrants and Probable Cause To obtain a search warrant, the officers must convince a judge that they have reasonable grounds, or **probable cause,** to believe a search will reveal a specific illegality. Probable cause requires law enforcement officials to have trustworthy evidence that would convince a reasonable person that the proposed search or seizure is more likely justified than not. Furthermore, the Fourth Amendment prohibits general warrants. It requires a particular description of what is to be searched or seized. General searches through a person's belongings are impermissible. The search cannot extend beyond what is described in the warrant.

There are exceptions to the requirement of a search warrant, as when it is likely that the items sought will be removed before a warrant can be obtained. For example, if a police officer has probable cause to believe an automobile contains evidence of a crime and the vehicle is likely to be unavailable by the time a warrant is obtained, the officer can search the vehicle without a warrant.

Searches and Seizures in the Business Context To businesses and professionals, constitutional protection against unreasonable searches and seizures is important. As fed-

eral and state regulation of commercial activities increased, frequent and unannounced government inspections were conducted to ensure compliance with the regulations. Such inspections were extremely disruptive at times. In *Marshall v. Barlow's, Inc.*,[12] the United States Supreme Court held that government inspectors do not have the right to enter business premises without a warrant, although the standard of probable cause is not the same as that required in nonbusiness contexts. The existence of a general and neutral enforcement plan will justify issuance of the warrant.

Lawyers and accountants frequently possess the business records of their clients, and inspecting these documents while they are out of the hands of their true owners also requires a warrant. No warrant is required, however, for seizures of spoiled or contaminated food. Nor are warrants required for searches of businesses in such highly regulated industries as liquor, guns, and strip mining. General manufacturing is not considered to be one of these highly regulated industries, however.

Of increasing concern to many employers is how to maintain a safe and efficient workplace without jeopardizing the Fourth Amendment rights of employees "to be secure in their persons." Requiring employees to undergo random drug tests, for example, may be held to violate the Fourth Amendment.

Fifth Amendment Protections

The Fifth Amendment offers significant protections for accused persons. One is the guarantee that no one can be deprived of "life, liberty, or property without due process of law." Two other important Fifth Amendment provisions protect persons against *double jeopardy* and *self-incrimination.*

Due Process of Law Remember from Chapter 1 that *due process of law* has both procedural and substantive aspects. Procedural due process requirements underlie criminal procedures. Basically, the law must be carried out in a fair and orderly way. In criminal cases, due process means that defendants should have an opportunity to object to the charges against them before a fair, neutral decision maker, such as a judge. Defendants must also be given the opportunity to confront and cross-examine witnesses and accusers and to present their own witnesses.

Double Jeopardy The Fifth Amendment also protects persons from **double jeopardy** (being tried twice for the same criminal offense). The prohibition against double jeopardy means that once a criminal defendant is acquitted (found "not guilty") of a particular crime, the government may not reindict the person and retry him or her for the same crime.

The prohibition against double jeopardy does not preclude the crime victim from bringing a civil suit against the same person to recover damages, however. Additionally, a state's prosecution of a crime will not prevent a separate federal prosecution relating to the same activity, and vice versa. ■ **EXAMPLE 6.7** A person found "not guilty" of assault and battery in a criminal case may be sued by the victim in a civil tort case for damages. A person who is prosecuted for assault and battery in a state court may be prosecuted in a federal court for civil rights violations resulting from the same action. ■

Self-Incrimination The Fifth Amendment guarantees that no person "shall be compelled in any criminal case to be a witness against himself." Thus, in any criminal proceeding, an accused person cannot be compelled to give testimony that might subject her or him to any criminal prosecution.

The guarantee against **self-incrimination** in the Fifth Amendment extends only to natural persons. Because a corporation is a legal entity and not a natural person, the privilege against self-incrimination does not apply to it. Similarly, the business records of a partnership do not receive Fifth Amendment protection.[13] When a partnership is required to produce these records, it must do so even if the information incriminates the persons who constitute the business entity. Sole proprietors and sole practitioners (those who fully own their businesses) who have not incorporated normally cannot be compelled to produce their business records. These individuals have full protection against self-incrimination because they function in only one capacity; there is no separate business entity (see Chapter 24).

Protections under the Sixth and Eighth Amendments

The Sixth Amendment guarantees several important rights for criminal defendants: the right to a speedy trial, the right to a jury trial, the right to a public trial, the right to confront witnesses, and the right to counsel. The Eighth Amendment prohibits excessive bail and fines, as well as cruel and unusual punishment.

The Sixth Amendment right to counsel is one of the rights of which a suspect must be advised when he or she is arrested under the *Miranda* rule (discussed later in this

12. 436 U.S. 307, 98 S.Ct. 1816, 56 L.Ed.2d 305 (1978).

13. The privilege has been applied to some small family partnerships. See *United States v. Slutsky,* 352 F.Supp. 1105 (S.D.N.Y. 1972).

chapter). In many cases, a statement that a criminal suspect makes in the absence of counsel is not admissible at trial unless the suspect has knowingly and voluntarily waived this right. Is the right to counsel triggered when judicial proceedings are initiated through any preliminary step? Or is this right triggered only when a suspect is "interrogated" by the police? In the following case, the Supreme Court considered these questions.

CASE 6.1 Fellers v. United States

Supreme Court of the United States, 2004.
540 U.S. 519,
124 S.Ct. 1019,
157 L.Ed.2d 1016.
http://straylight.law.cornell.edu/supct/index.html[a]

FACTS In February 2000, an indictment was issued charging John Fellers, a resident of Lincoln, Nebraska, with conspiracy to distribute methamphetamine. Police officers Michael Garnett and Jeff Bliemeister went to Fellers's home to arrest him. They told Fellers that the purpose of their visit was to discuss his use and distribution of methamphetamine. They said that they had a warrant for his arrest and that the charges referred to his involvement with four individuals. Fellers responded that he knew the persons and had used methamphetamine with them. The officers took Fellers to jail and advised him for the first time of his right to counsel. He waived this right and repeated his earlier statements. Before Fellers's trial, the court ruled that his "jailhouse statements" could be admitted at his trial because he had waived his right to counsel before making them. After Fellers's conviction, he appealed to the U.S. Court of Appeals for the Eighth Circuit, arguing that the officers had elicited his incriminating "home statements" without advising him of his right to counsel and that his "jailhouse statements" should thus have been excluded from his trial as "fruits" of his earlier statements. The appellate court affirmed the lower court's judgment, holding that Fellers had not had a right to counsel at his home because he had not been subject to police "interrogation." Fellers appealed to the United States Supreme Court.

a. In the "Search" box, type "Fellers," select "Current decisions only," and click on "submit." In the result, scroll to the name of the case, and click on it to access the opinion. The Legal Information Institute of Cornell Law School in Ithaca, New York, maintains this Web site.

ISSUE Did Garnett and Bliemeister violate Fellers's right to counsel by deliberately eliciting information from him during their visit to his home without advising him of his right to counsel?

DECISION Yes. The United States Supreme Court reversed the lower court's decision and remanded the case for the determination of a different issue. The Supreme Court held that the Sixth Amendment bars the use at trial of a suspect's incriminating words, deliberately elicited by police after criminal proceedings have been initiated, in the absence of either counsel or a waiver of the right to counsel, regardless of whether police conduct constitutes an "interrogation."

REASON The Court explained, "The Sixth Amendment right to counsel is triggered at or after the time that judicial proceedings have been initiated * * * . [A]n accused is denied the basic protections of the Sixth Amendment when there is used against him at his trial evidence of his own incriminating words, which federal agents * * * deliberately elicited from him after he had been indicted and in the absence of his counsel." In this case, "there is no question that the officers * * * deliberately elicited information from petitioner [Fellers]. Indeed, the officers, upon arriving at petitioner's house, informed him that their purpose in coming was to discuss his involvement in the distribution of methamphetamine and his association with certain charged co-conspirators. * * * [T]he ensuing discussion took place after petitioner had been indicted, outside the presence of counsel, and in the absence of any waiver of petitioner's Sixth Amendment rights."

FOR CRITICAL ANALYSIS—Social Consideration *Should Fellers's "jailhouse statements" also have been excluded from his trial? Why or why not?*

THE EXCLUSIONARY RULE AND THE *MIRANDA* RULE

Two other procedural protections for criminal defendants are the exclusionary rule and the *Miranda* rule.

The Exclusionary Rule Under what is known as the **exclusionary rule,** all evidence obtained in violation of the constitutional rights spelled out in the Fourth, Fifth, and Sixth Amendments, as well as all evidence derived from the illegally obtained evidence, normally must be excluded from the trial. Evidence derived from illegally obtained evidence is known as the "fruit of the poisonous tree." For example, if a confession is obtained after an illegal arrest, the arrest is "the poisonous tree," and the confession, if "tainted" by the arrest, is the "fruit."

The purpose of the exclusionary rule is to deter police from conducting warrantless searches and from engaging in other misconduct. The rule is sometimes criticized

because it can lead to injustice. Many a defendant has "gotten off on a technicality" because law enforcement personnel failed to observe procedural requirements. Even though a defendant may be obviously guilty, if the evidence of that guilt was obtained improperly (without a valid search warrant, for example), it normally cannot be used against the defendant in court. In the following case, the court had to decide whether a lost computer printout should cause evidence to be excluded.

CASE 6.2 People v. McFarlan

New York Supreme Court, 2002.[a]
191 Misc.2d 531,
744 N.Y.S.2d 287.

FACTS In May 2001, Lisa Kordes saw two men picking pockets on a Lexington Avenue bus in Manhattan, in New York City. Based on Kordes's description of the men, a police department computer produced a six-photo array of possible suspects. Kordes selected a photo of Kevin McFarlan as one of the men she had seen. Five days later, on a bus, Neal Ariano, a police officer, arrested McFarlan after seeing him bump an elderly woman while placing his hand near her pocketbook. At the police station, Kordes viewed a lineup including McFarlan and identified him as the man she had seen on the Lexington Avenue bus. McFarlan was charged in a New York state court with various crimes. The printout of the computer-generated photo array that Kordes had been shown was lost, but the "People" (the state of New York) introduced into evidence a second printout to show what Kordes had seen. McFarlan argued, among other things, that the first printout was the original photo array and that because that printout had been lost, the court should presume that the photo-array procedure had been illegal. Thus, McFarlan's arrest and Kordes's identification of him in the lineup should be excluded as "fruit of the poisonous tree."

ISSUE Should the second printout of the photo array be considered "fruit of the poisonous tree" and thus inadmissible?

a. In New York, a supreme court is a trial court.

DECISION No. The New York state court held that the second printout of the photo array was not "fruit of the poisonous tree" and could be properly accepted into evidence.

REASON The court concluded that because both printouts were generated in the same format, there could be no prejudice from the fact that the defendant was selected from the first printout and that a second identical printout was later introduced into evidence. Because the printouts were identical and conveyed the full recoverable information, the court held that to decide otherwise would be absurd. "Where, for example, a witness is shown only the screen display, what must the People [public prosecutor] keep? For on such an analysis the printout of the screen could not be the 'original.'" The court pointed out that the purpose of requiring the preservation of a record was clear. According to the court, concern for the integrity of information has always led the courts to prefer the original and has led to many rules to bar or limit the use of nonoriginal material. "Here," said the court, "the original array was in electronic form in the computer memory, the testimony was unequivocal that [both] printout[s] * * * were generated in the same manner. * * * [A]s a result, defendant's argument crumbles."

FOR CRITICAL ANALYSIS—Technological Consideration *When a record is stored on a computer, which should be considered the "original" record—the version on the computer or a printout of the computer version? Why?*

The *Miranda* Rule In *Miranda v. Arizona,*[14] a case decided in 1966, the United States Supreme Court established the rule that individuals who are arrested must be informed of certain constitutional rights, including their Fifth Amendment right to remain silent and their Sixth Amendment right to counsel. If the arresting officers fail to inform a criminal suspect of these constitutional rights, any statements the suspect makes normally will not be admissible in court.

The *Miranda* decision was controversial, and Congress later attempted to overrule it. The United States Supreme Court, however, subsequently held that the *Miranda* rights enunciated by the Court in the 1966 case were constitutionally based and thus could not be overruled by a legislative act.[15]

Exceptions to the *Miranda* Rule Over time, as part of a continuing attempt to balance the rights of accused persons against the rights of society, the United States Supreme Court has carved out numerous exceptions to the *Miranda* rule. In 1984, for example, the Court recognized

14. 384 U.S. 436, 86 S.Ct. 1602, 16 L.Ed.2d 694 (1966).

15. *Dickerson v. United States,* 530 U.S. 428, 120 S.Ct. 2326, 147 L.Ed.2d 405 (2000).

a "public-safety" exception to the *Miranda* rule. The need to protect the public warranted the admissibility of statements made by the defendant (in this case, indicating where he placed the gun) as evidence at trial, even though the defendant had not been informed of his *Miranda* rights.[16]

In 1986, the Supreme Court further held that a confession need not be excluded even though the police failed to inform a suspect in custody that his attorney had tried to reach him by telephone.[17] In an important 1991 decision, the Court stated that a suspect's conviction will not be automatically overturned if the suspect was coerced into making a confession. If other, legally obtained evidence admitted at trial is strong enough to justify the conviction without the confession, then the fact that the confession was obtained illegally can, in effect, be ignored.[18]

In yet another case, in 1994, the Supreme Court ruled that a suspect must unequivocally and assertively request to exercise his or her right to counsel in order to stop police questioning. Saying, "Maybe I should talk to a lawyer" during an interrogation after being taken into custody is not enough. The Court held that police officers are not required to decipher the suspect's intentions in such situations.[19]

Criminal Process

As mentioned, a criminal prosecution differs significantly from a civil case in several respects. These differences reflect the desire to safeguard the rights of the individual against the state. Exhibit 6–4 on page 130 summarizes the major procedural steps in processing a criminal case. We discuss three phases of the criminal process—arrest, indictment or information, and trial—in more detail below.

Arrest

Before a warrant for arrest can be issued, there must be probable cause for believing that the individual in question has committed a crime. As discussed earlier, *probable cause* can be defined as a substantial likelihood that the person has committed or is about to commit a crime. Note that probable cause involves a likelihood, not just a possibility. Arrests may sometimes be made without a warrant if there is no time to get one, as when a police officer observes a crime taking place, but the action of the arresting officer is still judged by the standard of probable cause.

16. *New York v. Quarles,* 467 U.S. 649, 104 S.Ct. 2626, 81 L.Ed.2d 550 (1984).
17. *Moran v. Burbine,* 475 U.S. 412, 106 S.Ct. 1135, 89 L.Ed.2d 410 (1986).
18. *Arizona v. Fulminante,* 499 U.S. 279, 111 S.Ct. 1246, 113 L.Ed.2d 302 (1991).
19. *Davis v. United States,* 512 U.S. 452, 114 S.Ct. 2350, 129 L.Ed.2d 362 (1994).

Indictment or Information

Individuals must be formally charged with having committed specific crimes before they can be brought to trial. If issued by a grand jury, this charge is called an **indictment.**[20] A **grand jury** usually consists of more jurors than the ordinary trial jury. A grand jury does not determine the guilt or innocence of an accused party; rather, its function is to determine, after hearing the state's evidence, whether a reasonable basis (probable cause) exists for believing that a crime has been committed and whether a trial ought to be held.

Usually, grand juries are used in cases involving serious crimes, such as murder. For lesser crimes, an individual may be formally charged with a crime by what is called an **information,** or criminal complaint. An information will be issued by a government prosecutor if the prosecutor determines that there is sufficient evidence to justify bringing the individual to trial.

Trial

At a criminal trial, the accused person does not have to prove anything; the entire burden of proof is on the prosecutor (the state). As mentioned earlier, the prosecution must show that, based on all the evidence presented, the defendant's guilt is established *beyond a reasonable doubt.* If there is any reasonable doubt as to whether a criminal defendant did, in fact, commit the crime with which she or he has been charged, then the verdict must be "not guilty." Note that giving a verdict of "not guilty" is not the same as stating that the defendant is innocent; it merely means that not enough evidence was properly presented to the court to prove guilt beyond a reasonable doubt.

Courts have complex rules about what types of evidence may be presented and how the evidence may be brought out in criminal cases, especially in jury trials. These rules are designed to ensure that evidence presented at trials is relevant, reliable, and not prejudicial toward the defendant. For example, under the Sixth Amendment, persons accused of a crime have the right to confront, in open court, the witnesses against them. If a witness's testimony is presented in a document obtained in an *ex parte* examination,[21] it must be shown that the witness is unavailable to testify in the court and that the defendant had a prior opportunity to cross-examine her or him. Is a defendant's Sixth Amendment right to confront witnesses violated when, in a jury trial, the documents that are presented include the results of a state official's search of the public records at a police officer's request? That was the question in the following case.

20. Pronounced in-*dyte*-ment.
21. In this context, an *ex parte* examination is a proceeding for the benefit of the prosecution without notice to the defendant.

CASE 6.3 Michels v. Commonwealth of Virginia

Court of Appeals of Virginia, 2006.
__ S.E.2d __.
http://www.courts.state.va.us/txtcap.htm[a]

FACTS Robert Michels met Allison Formal through an online dating Web site in 2002. Michels represented himself as the retired chief executive officer of a large company that he had sold for millions of dollars. In January 2003, Michels proposed that he and Formal create a limited liability company—Formal Properties Trust, LLC—to "channel their investments in real estate."[b] Formal agreed to contribute $100,000 to the company and wrote two $50,000 checks to "Michels and Associates, LLC." Six months later, Michels told Formal that their LLC had been formed in Delaware. Later, Formal asked Michels about her investments. He responded evasively, and she demanded that an independent accountant review the LLC's records. Michels refused. Formal contacted the police. Michels was charged in a Virginia state court with obtaining money by false pretenses. The Delaware secretary of state verified, in two certified documents, that "Formal Properties Trust, L.L.C." and "Michels and Associates, L.L.C." did not exist in Delaware. After a jury trial, Michels was convicted. He appealed to a state intermediate appellate court, arguing that the certified documents were "testimonial in nature and, therefore, their admission in his criminal trial violated his Sixth Amendment rights."

ISSUE Did the admission of the Delaware secretary of state's certified documents violate Michels's rights under the Sixth Amendment?

a. Scroll to the name of the case and click on its docket number ("2008044") to access the opinion. The Judicial System of the Commonwealth of Virginia maintains this Web site.
b. A *limited liability company* is a specific form of business organization that will be discussed in detail in Chapter 24.

DECISION No. The sta affirmed Michels's conviction, certified by the Delaware state

REASON The court provided tw sion. First, the documents were not ac describe any criminal wrongdoing. They events or facts that implicated Michels in a scheme, but only verified that the Delaware searched the state's public records, with certai Second, the documents were not prepared in a resembling an *ex parte* examination. The court rec that the "principal evil at which the [Sixth Amendmen Confrontation Clause was directed was the * * * use o *ex parte* examinations as evidence against the accused." court also acknowledged that a police officer had requested the documents, but emphasized that the secretary of state was asked only to prepare a document certifying the results of "a routine search of business records." The court explained that the documents were prepared in a "non-adversarial setting in which the factors likely to cloud the perception of an official engaged in the more traditional law enforcement functions of observation and investigation of crime are simply not present."

FOR CRITICAL ANALYSIS—Social Consideration *If the Delaware secretary of state had actually met Michels and dealt with him personally with respect to the events in this case, and described these dealings in detail in the certified documents, what might have been the effect on Michels's appeal of his conviction?*

Sentencing Guidelines

Traditionally, persons who committed the same crime might receive very different sentences, depending on the judge hearing the case, the jurisdiction in which it was heard, and many other factors. Today, however, court judges typically must follow state or federal guidelines when sentencing convicted persons.

At the federal level, the Sentencing Reform Act created the U.S. Sentencing Commission, which was charged with the task of standardizing sentences for federal crimes. The commission fulfilled its task, and since 1987 its sentencing guidelines for all federal crimes have been applied by federal court judges. The guidelines establish a range of possible penalties for each federal crime. Depending on the defendant's criminal record, the seriousness of the offense, and other factors specified in the guidelines, federal judges must select a sentence from within this range.

The commission also created specific guidelines for the punishment of crimes committed by corporate employees (white-collar crimes). These guidelines established stiffer penalties for criminal violations of securities laws (see Chapter 27), antitrust laws, employment laws (see Chapter 23), mail and wire fraud, commercial bribery, and kickbacks and money laundering.[22] The guidelines allow federal judges to take into consideration a number of factors when selecting from the range of possible penalties for a specified crime. These factors include the defendant company's history of past violations, the scope of management's

22. The sentencing guidelines were amended in 2003, as required under the Sarbanes-Oxley Act of 2002, to impose stiffer penalties for corporate securities fraud—see Chapter 27.

URAL STEPS IN A CRIMINAL CASE

ARREST

Most arrests are made without a warrant. After t, who is then taken to the police station.

ING

gain, photographed, fingerprinted, and oking, charges are reviewed, and if they trate reviews the case for probable cause.

ICE

gistrate, who informs the suspect of the charges and pect requests a lawyer and cannot afford one, a lawyer is gistrate sets bail (conditions under which a suspect can obtain release position of the case).

GRAND JURY

A grand jury determines if there is probable cause to believe that the defendant committed the crime. The federal government and about half of the states require grand jury indictments for at least some felonies.

PRELIMINARY HEARING

In a court proceeding, a prosecutor presents evidence, and the judge determines if there is probable cause to hold the defendant over for trial.

INDICTMENT

An *indictment* is a written document issued by the grand jury to formally charge the suspect with a crime.

INFORMATION

An *information* is a formal criminal charge made by the prosecutor.

ARRAIGNMENT

The suspect is brought before the trial court, informed of the charges, and asked to enter a plea.

PLEA BARGAIN

A plea bargain is a prosecutor's promise to make concessions (or promise to seek concessions) in return for a suspect's guilty plea. Concessions may include a reduced charge or a lesser sentence.

GUILTY PLEA

In many jurisdictions, most cases that reach the arraignment stage do not go to trial but are resolved by a guilty plea, often as a result of a plea bargain. The judge sets the case for sentencing.

TRIAL

Trials can be either jury trials or bench trials. (In a bench trial, there is no jury, and the judge decides questions of fact as well as questions of law.) If the verdict is "guilty," the judge sets the case for sentencing. Everyone convicted of a crime has the right to an appeal.

cooperation with federal investigators, and the extent to which the firm has undertaken specific programs and procedures to prevent criminal activities by its employees.

Cyber Crimes

Some years ago, the American Bar Association defined **computer crime** as any act that is directed against computers and computer parts, that uses computers as instruments of crime, or that involves computers and constitutes abuse. Today, because much of the crime committed with the use of computers occurs in cyberspace, many computer crimes fall under the broad label of **cyber crime.**

As we mentioned earlier, most cyber crimes are not "new" crimes. Rather, they are existing crimes in which the Internet is the instrument of wrongdoing. The challenge for law enforcement is to apply traditional laws—which were designed to protect persons from physical harm or to safeguard their physical property—to crimes committed in cyberspace. Here we look at several types of activity that constitute cyber crimes against persons or property. Other cyber crimes will be discussed in later chapters of this text as they relate to particular topics.

CYBER THEFT

In cyberspace, thieves are not subject to the physical limitations of the "real" world. A thief can steal data stored in a networked computer with broadband access from anywhere on the globe. Only the speed of the connection and the thief's computer equipment limit the quantity of data that can be stolen.

Financial Crimes Computer networks also provide opportunities for employees to commit crimes that can involve serious economic losses. For example, employees of a company's accounting department can transfer funds among accounts with little effort and often with less risk than would be involved in transactions evidenced by paperwork.

Generally, the dependence of businesses on computer operations has left firms vulnerable to sabotage, fraud, embezzlement, and the theft of proprietary data, such as trade secrets or other intellectual property. As noted in Chapter 5, the piracy of intellectual property via the Internet is one of the most serious legal challenges facing lawmakers and the courts today.

Identity Theft One form of cyber theft that has increased steadily in recent years is *identity theft.* **Identity theft** occurs when the wrongdoer steals a form of identification—such as a name, date of birth, or Social Security number—and uses the information to access the victim's financial resources. This crime existed to a certain extent before the widespread use of the Internet. Thieves would "steal" calling-card numbers by watching people using public telephones, or they would rifle through garbage to find bank account or credit-card numbers. The identity thieves would then use the calling-card or credit-card numbers or would withdraw funds from the victims' accounts.

The Internet, however, has turned identity theft into perhaps the fastest-growing financial crime in the United States. From the identity thief's perspective, the Internet provides those who steal information offline with an easy medium for using items such as stolen credit-card numbers or e-mail addresses while protected by anonymity. An estimated 10 million Americans are victims of identity theft each year, and annual losses are estimated to exceed $50 billion.

CYBERSTALKING

California enacted the first stalking law in 1990, in response to the murders of six women—including Rebecca Schaeffer, a television star—by men who had harassed them. The law made it a crime to harass or follow a person while making a "credible threat" that puts that person in reasonable fear for his or her safety or the safety of the person's immediate family.[23] Since then, all other states have enacted some form of stalking laws. In about half of the states, these laws require a physical act such as following the victim.

Cyberstalkers (stalkers who commit their crimes in cyberspace), however, find their victims through Internet chat rooms, Usenet newsgroups or other bulletin boards, or e-mail. To close this "loophole" in existing stalking laws, more than three-fourths of the states now have laws specifically designed to combat cyberstalking and other forms of online harassment.

Note that cyberstalking can be even more threatening than physical stalking in some respects. While it takes a great deal of effort to physically stalk someone, it is relatively easy to harass a victim with electronic messages. Furthermore, the possibility of personal confrontation may discourage a stalker from actually following a victim. This disincentive is removed in cyberspace. Also, there is always the possibility that a cyberstalker will eventually pose a physical threat to her or his target. Finally, the Internet makes it easier to obtain information about the victim, such as where he or she lives or works. (Can Internet sellers of personal information be held responsible for crimes by stalkers who use that information to locate their victims? For a discussion of this issue, see this chapter's *Adapting the Law to the Online Environment* feature on the following page.)

23. See, for example, Cal. Penal Code Section 646.9.

ADAPTING THE LAW TO THE ONLINE ENVIRONMENT

Stalking and Internet Data Brokers

A cutting-edge issue coming before today's courts has to do with stalking and Internet data brokers—those in the business of selling personal information via the Internet. Suppose that a stalker purchases personal information (such as a home address) from a Web broker and then uses that information to commit a crime (such as murder). The stalker, of course, can be prosecuted under criminal law for the crime of murder. At issue is the broker's responsibility for the crime. Can the broker be sued under tort law (see Chapter 4) for damages?

ONE COURT WEIGHS IN

At least one court has weighed in on this issue. In *Remsburg v. Docusearch, Inc.,*[a] the Supreme Court of New Hampshire concluded that an Internet data broker had a duty to exercise reasonable care when selling personal information online. The case involved Liam Youens, a New Hampshire resident, and Docusearch, Inc., an Internet-based investigation and information service. Youens contacted Docusearch through its Web site and requested information about Amy Boyer. Youens provided his name, address, and phone number and paid Docusearch's fee by credit card. In return, Docusearch provided Boyer's home address, birth date, and Social Security number. Youens also asked for Boyer's workplace address, which Docusearch obtained for him. After Youens had obtained the address of Boyer's place of employment, he drove to the workplace, fatally shot her, and then shot and killed himself.

Helen Remsburg, Boyer's mother, sued Docusearch, claiming that the defendant had acted wrongfully. The federal district court referred the case to the New Hampshire Supreme Court for a determination of the parties' duties under the state's common law. The state supreme court held that an information broker who sells information about a third person to a client has a duty to exercise reasonable care in disclosing the information.

A FORESEEABLE RISK?

One of the key issues for the court was whether the crime committed by Youens was a foreseeable risk to Docusearch. Remember from Chapter 4 that under tort law, the test for proximate cause—and the extent of a defendant's duty of care—is the foreseeability of a risk of harm. If certain consequences of an action are not foreseeable, there is no proximate cause.

In determining whether a risk of criminal misconduct was foreseeable to Docusearch, the Supreme Court of New Hampshire found that Docusearch's information disclosure presented two foreseeable risks: stalking and identity theft. Therefore, the company had a duty to exercise reasonable care in disclosing personal information about Boyer to Youens. Because Docusearch had not exercised reasonable care (taken steps to find out if Youens's requests were for a legitimate purpose), Docusearch could be sued for damages for breaching this duty.

FOR CRITICAL ANALYSIS

What other crimes can you think of, in addition to stalking and identity theft, that might qualify as "foreseeable risks" created by the online sale of personal data by information brokers?

a. 149 N.H. 148, 816 A.2d 1001 (2003).

HACKING AND CYBERTERRORISM

Persons who use one computer to break into another are sometimes referred to as **hackers.** Hackers who break into computers without authorization often commit cyber theft. Sometimes, however, their principal aim is to prove how smart they are by gaining access to others' password-protected computers and causing random data errors or making unpaid-for telephone calls.[24] **Cyberterrorists** are hackers who, rather than trying to attract attention, strive to remain undetected so that they can exploit computers for a serious impact. Just as "real" terrorists destroyed the World Trade Center towers and a portion of the Pentagon in September 2001, cyberterrorists might explode "logic bombs" to shut down central computers. Such activities can pose a danger to national security.

Businesses may be targeted by cyberterrorists as well as hackers. The goals of a hacking operation might include a wholesale theft of data, such as a merchant's customer files, or the monitoring of a computer to discover a business firm's plans and transactions. A cyberterrorist might also want to insert false codes or data. For example, the processing control system of a food manufacturer could be changed to alter the levels of ingredients so that consumers of the food would become ill.

A cyberterrorist attack on a major financial institution such as the New York Stock Exchange or a large bank could leave securities or money markets in flux and seriously affect the daily lives of millions of citizens. Similarly, any prolonged disruption of computer, cable, satellite, or telecommunications systems due to the actions of expert hackers would have serious repercussions on business operations—and national security—on a global level. Computer viruses are another tool that

24. The total cost of crime on the Internet is estimated to be several billion dollars annually, but two-thirds of that total is said to consist of unpaid-for toll calls.

can be used by cyberterrorists to cripple communications networks.

Prosecuting Cyber Crimes

The "location" of cyber crime (cyberspace) has raised new issues in the investigation of crimes and the prosecution of offenders. A threshold issue is, of course, jurisdiction. A person who commits an act against a business in California, where the act is a cyber crime, might never have set foot in California but might instead reside in New York, or even in Canada, where the act may not be a crime. If the crime was committed via e-mail, the question arises as to whether the e-mail would constitute sufficient "minimum contacts" (see Chapter 2) for the victim's state to exercise jurisdiction over the perpetrator.

Identifying the wrongdoers can also be difficult. Cyber criminals do not leave physical traces, such as fingerprints or DNA samples, as evidence of their crimes. Even electronic "footprints" can be hard to find and follow. For example, e-mail may be sent through a remailer, an online service that guarantees that a message cannot be traced to its source.

For these reasons, laws written to protect physical property are difficult to apply in cyberspace. Nonetheless, governments at both the state and federal levels have taken significant steps toward controlling cyber crime, both by applying existing criminal statutes and by enacting new laws that specifically address wrongs committed in cyberspace.

The Computer Fraud and Abuse Act Perhaps the most significant federal statute specifically addressing cyber crime is the Counterfeit Access Device and Computer Fraud and Abuse Act of 1984 (commonly known as the Computer Fraud and Abuse Act, or CFAA). This act, as amended by the National Information Infrastructure Protection Act of 1996,[25] provides, among other things, that a person who accesses a computer online, without authority, to obtain classified, restricted, or protected data, or attempts to do so, is subject to criminal prosecution. Such data could include financial and credit records, medical records, legal files, military and national security files, and other confidential information in government or private computers. The crime has two elements: accessing a computer without authority and taking the data.

This theft is a felony if it is committed for a commercial purpose or for private financial gain, or if the value of the stolen data (or computer time) exceeds $5,000. Penalties include fines and imprisonment for up to twenty years. A victim of computer theft can also bring a civil suit against the violator to obtain damages, an injunction, and other relief.

Other Federal Statutes The federal wire fraud statute, the Economic Espionage Act of 1996, and RICO, all of which were discussed earlier in this chapter, extend to crimes committed in cyberspace as well. Other federal statutes that may apply include the Electronic Fund Transfer Act of 1978, which makes unauthorized access to an electronic fund transfer system a crime; the Anticounterfeiting Consumer Protection Act of 1996, which increased penalties for stealing copyrighted or trademarked property; and the National Stolen Property Act of 1988, which concerns the interstate transport of stolen property. The federal government has also enacted laws (many of which have been challenged on constitutional grounds) to protect minors from online pornographic materials. In later chapters of this text, you will read about other federal statutes and regulations that are designed to address wrongs committed in cyberspace in specific areas of the law.

25. 18 U.S.C. Section 1030.

TERMS AND CONCEPTS

CHAPTER SUMMARY Criminal Law and Cyber Crimes

Civil Law and Criminal Law (See pages 115–116.)	1. *Civil law*—Spells out the duties that exist between persons or between citizens and their governments, excluding the duty not to commit crimes. 2. *Criminal law*—Has to do with crimes, which are defined as wrongs against society proclaimed in statutes and punishable by society through fines and/or imprisonment—and, in some cases, death. Because crimes are *offenses against society as a whole,* they are prosecuted by a public official, not by victims. 3. *Key differences*—An important difference between civil law and criminal law is that the burden of proof is higher in criminal cases (see Exhibit 6–1 on page 116 for other differences between civil law and criminal law). 4. *Civil liability for criminal acts*—A criminal act may give rise to both criminal liability and tort liability (see Exhibit 6–2 on page 117 for an example of criminal and tort liability for the same act).
Classification of Crimes (See pages 116–117.)	1. *Felonies*—Serious crimes punishable by death or by imprisonment in a penitentiary for more than one year. 2. *Misdemeanors*—Under federal law and in most states, any crimes that are not felonies.
Criminal Liability (See pages 117–118.)	1. *Guilty act*—In general, some form of harmful act must be committed for a crime to exist. 2. *Intent*—An intent to commit a crime, or a wrongful mental state, is generally required for a crime to exist.
Corporate Criminal Liability (See page 118.)	1. *Liability of corporations*—Corporations normally are liable for the crimes committed by their agents and employees within the course and scope of their employment. Corporations cannot be imprisoned, but they can be fined or denied certain legal privileges. 2. *Liability of corporate officers and directors*—Corporate directors and officers are personally liable for the crimes they commit and may be held liable for the actions of employees under their supervision.
Types of Crimes (See pages 118–121.)	Crimes fall into five general categories: violent crime, property crime, public order crime, white-collar crime, and organized crime. See Exhibit 6–3 on page 122 for definitions and examples of these types of crimes.
Defenses to Criminal Liability (See pages 121–124.)	Defenses to criminal liability include infancy, intoxication, insanity, mistake, consent, duress, justifiable use of force, entrapment, and the statute of limitations. Also, in some cases defendants may be relieved of criminal liability, at least in part, if they are given immunity.
Constitutional Safeguards and Criminal Procedures (See pages 124–128.)	1. *Fourth Amendment*—Provides protection against unreasonable searches and seizures and requires that probable cause exist before a warrant for a search or an arrest can be issued. 2. *Fifth Amendment*—Requires due process of law, prohibits double jeopardy, and protects against self-incrimination. 3. *Sixth Amendment*—Provides guarantees of a speedy trial, a trial by jury, a public trial, the right to confront witnesses, and the right to counsel. 4. *Eighth Amendment*—Prohibits excessive bail and fines, as well as cruel and unusual punishment. 5. *Exclusionary rule*—A criminal procedural rule that prohibits the introduction at trial of all evidence obtained in violation of constitutional rights, as well as any evidence derived from the illegally obtained evidence. 6. *Miranda rule*—A rule set forth by the United States Supreme Court in *Miranda v. Arizona* that individuals who are arrested must be informed of certain constitutional rights, including their right to counsel.
Criminal Process (See pages 128–131.)	1. *Arrest, indictment, and trial*—Procedures governing arrest, indictment, and trial for a crime are designed to safeguard the rights of the individual against the state. See Exhibit 6–4 on page 130 for a summary of the procedural steps involved in prosecuting a criminal case.

CHAPTER SUMMARY Criminal Law and Cyber Crimes—Continued

Criminal Process—Continued	2. *Sentencing guidelines*—Both the federal government and the states have established sentencing laws or guidelines. The federal sentencing guidelines indicate a range of penalties for each federal crime; federal judges must abide by these guidelines when imposing sentences on those convicted of federal crimes.
Cyber Crimes (See pages 131–133.)	Cyber crime is any crime that occurs in cyberspace. Examples include cyber theft (financial crimes committed with the aid of computers, as well as identity theft), cyberstalking, hacking, and cyberterrorism. Significant federal statutes addressing cyber crimes include the Electronic Fund Transfer Act of 1978 and the Counterfeit Access Device and Computer Fraud and Abuse Act of 1984, as amended by the National Information Infrastructure Protection Act of 1996.

FOR REVIEW

Answers for the even-numbered questions in this For Review *section can be found in Appendix A at the end of this text.*

1. What two elements must exist before a person can be held liable for a crime? Can a corporation commit crimes?
2. What are five broad categories of crimes? What is white-collar crime?
3. What defenses might be raised by criminal defendants to avoid liability for criminal acts?
4. What constitutional safeguards exist to protect persons accused of crimes? What are the basic steps in the criminal process?
5. What is cyber crime? What laws apply to crimes committed in cyberspace?

QUESTIONS AND CASE PROBLEMS

6–1. Criminal versus Civil Trials. In criminal trials, the defendant must be proved guilty beyond a reasonable doubt, whereas in civil trials, the defendant need only be proved guilty by a preponderance of the evidence. Discuss why a higher burden of proof is required in criminal trials.

QUESTION WITH SAMPLE ANSWER

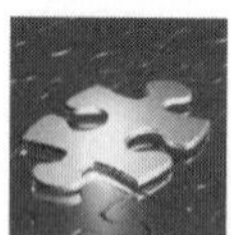

6–2. The following situations are similar (all involve the theft of Makoto's television set), yet they represent three different crimes. Identify the three crimes, noting the differences among them.

(a) While passing Makoto's house one night, Sarah sees a portable television set left unattended on Makoto's lawn. Sarah takes the television set, carries it home, and tells everyone she owns it.

(b) While passing Makoto's house one night, Sarah sees Makoto outside with a portable television set. Holding Makoto at gunpoint, Sarah forces him to give up the set. Then Sarah runs away with it.

(c) While passing Makoto's house one night, Sarah sees a portable television set in a window. Sarah breaks the front-door lock, enters, and leaves with the set.

For a sample answer to this question, go to Appendix B at the end of this text.

6–3. Types of Crimes. Which, if any, of the following crimes necessarily involve illegal activity on the part of more than one person?

(a) Bribery.
(b) Forgery.
(c) Embezzlement.
(d) Larceny.
(e) Receiving stolen property.

6–4. Double Jeopardy. Armington, while robbing a drugstore, shot and seriously injured Jennings, a drugstore clerk. Armington was subsequently convicted in a criminal trial of armed robbery and assault and battery. Jennings later brought a civil tort suit against Armington for damages. Armington contended that he could not be tried again for the same crime, as that would constitute double jeopardy, which is prohibited by the Fifth Amendment to the Constitution. Is Armington correct? Explain.

CASE PROBLEM WITH SAMPLE ANSWER

6–5. The District of Columbia Lottery Board licensed Soo Young Bae, a Washington, D.C., merchant, to operate a terminal that prints and dispenses lottery tickets for sale. Bae used the terminal to generate tickets with a face value of $525,586, for which he did not pay. The winning tickets among these had a total

redemption value of $296,153, of which Bae successfully obtained all but $72,000. Bae pleaded guilty to computer fraud, and the court sentenced him to eighteen months in prison. In sentencing a defendant for fraud, a federal court must make a reasonable estimate of the victim's loss. The court determined that the value of the loss due to the fraud was $503,650—the market value of the tickets less the commission Bae would have received from the lottery board had he sold those tickets. Bae appealed, arguing that "[a]t the instant any lottery ticket is printed," it is worth whatever value the lottery drawing later assigns to it—that is, losing tickets have no value. Bae thus calculated the loss at $296,153, the value of his winning tickets. Should the U.S. Court of Appeals for the District of Columbia Circuit affirm or reverse Bae's sentence? Why? [*United States v. Bae,* 250 F.3d 774 (C.A.D.C. 2001)]

After you have answered this problem, compare your answer with the sample answer given on the Web site that accompanies this text. Go to academic.cengage.com/blaw/wbl, select "Chapter 6," and click on "Case Problem with Sample Answer."

6–6. Larceny. In February 2001, a homeowner hired Jimmy Smith, a contractor claiming to employ a crew of thirty workers, to build a garage. The homeowner paid Smith $7,950 and agreed to make additional payments as needed to complete the project, up to $15,900. Smith promised to start the next day and finish within eight weeks. Nearly a month passed with no work, while Smith lied to the homeowner that materials were on "back order." During a second month, footings were created for the foundation, and a subcontractor poured the concrete slab, but Smith did not return the homeowner's phone calls. After eight weeks, the homeowner confronted Smith, who promised to complete the job, worked on the site that day until lunch, and never returned. Three months later, the homeowner again confronted Smith, who promised to "pay [him] off" later that day but did not do so. In March 2002, the state of Georgia filed criminal charges against Smith. While his trial was pending, he promised to pay the homeowner "next week" but again failed to make any refund. The value of the labor performed before Smith abandoned the project was between $800 and $1,000, the value of the materials was $367, and the subcontractor was paid $2,270. Did Smith commit larceny? Explain. [*Smith v. State of Georgia,* __ S.E.2d __ (Ga.App. 2004)]

A QUESTION OF ETHICS

6–7. A troublesome issue concerning the constitutional privilege against self-incrimination has to do with "jail plants"—that is, undercover police officers placed in cells with criminal suspects to gain information from the suspects. For example, in one case the police placed an undercover agent, Parisi, in a jail cell block with Lloyd Perkins, who had been imprisoned on charges unrelated to the murder that Parisi was investigating. When Parisi asked Perkins if he had ever killed anyone, Perkins made statements implicating himself in the murder. Perkins was then charged with the murder. [*Illinois v. Perkins,* 496 U.S. 292, 110 S.Ct. 2394, 110 L.Ed.2d 243 (1990)]

1. Should Perkins's statements be suppressed—that is, not be admissible as evidence at trial—because he was not "read his rights," as required by the *Miranda* decision, prior to making his self-incriminating statements? Does *Miranda* apply to Perkins's situation?
2. Do you think that it is fair for the police to resort to trickery and deception to bring those who have committed crimes to justice? Why or why not? What rights or public policies must be balanced in deciding this issue?

VIDEO QUESTION

6–8. Go to this text's Web site at academic.cengage.com/blaw/wbl and select "Chapter 6." Click on "Video Questions" and view the video titled *Casino.* Then answer the following questions.

1. In the video, a casino manager, Ace (Robert DeNiro), discusses how politicians "won their 'comp life' when they got elected." "Comps" are the free gifts that casinos give to high-stakes gamblers to keep their business. If an elected official accepts comps, is he or she committing a crime? If so, what type of crime? Explain your answers.
2. Assume that Ace committed a crime by giving politicians comps. Can the casino, Tangiers Corp., be held liable for that crime? Why or why not? How could a court punish the corporation?
3. Suppose that the Federal Bureau of Investigation wants to search the premises of Tangiers for evidence of criminal activity. If casino management refuses to consent to the search, what constitutional safeguards and criminal procedures, if any, protect Tangiers?

Criminal Law and Cyber Crimes

For updated links to resources available on the Web, as well as a variety of other materials, visit this text's Web site at

academic.cengage.com/blaw/wbl

The Bureau of Justice Statistics in the U.S. Department of Justice offers an impressive collection of statistics on crime at the following Web site:

http://www.ojp.usdoj.gov/bjs

For summaries of famous criminal cases and documents relating to these trials, go to Court TV's Web site at

http://www.courttv.com/index.html

Many criminal codes are now online. To find your state's code, go to

http://www.findlaw.com

and select "State" under the link to "Laws: Cases and Codes."

You can learn about some of the constitutional questions raised by various criminal laws and procedures by going to the Web site of the American Civil Liberties Union at

http://www.aclu.org

You can find a wealth of information on famous criminal trials at a Web site maintained by the University of Missouri–Kansas City School of Law. Go to

http://www.law.umkc.edu/faculty/projects/ftrials/ftrials.html

If you are interested in reading the Supreme Court's opinion in *Miranda v. Arizona,* go to

http://straylight.law.cornell.edu/supct/cases/name.htm

Select "M" from the menu at the top of the page, and scroll down the page that opens to the *Miranda v. Arizona* case.

To learn more about criminal procedures, access the following site and select "Anatomy of a Murder: A Trip through Our Nation's Legal Justice System":

http://library.thinkquest.org/2760/home.htm

The U.S. Sentencing Guidelines can be found online at

http://www.ussc.gov

ONLINE LEGAL RESEARCH

Go to the *Fundamentals of Business Law* home page at academic.cengage.com/blaw/wbl, select "Chapter 6," and click on "Internet Exercises." There you will find the following Internet research exercises that you can perform to learn more about topics covered in this chapter.

Activity 6–1: LEGAL PERSPECTIVE—Revisiting *Miranda*

Activity 6–2: MANAGEMENT PERSPECTIVE—Hackers

Activity 6–3: INTERNATIONAL PERSPECTIVE—Fighting Cyber Crime Worldwide

BEFORE THE TEST

Go to the *Fundamentals of Business Law* home page at academic.cengage.com/blaw/wbl, select "Chapter 6," and click on "Interactive Quizzes." You will find a number of interactive questions relating to this chapter.

UNIT 2 • EXTENDED CASE STUDY

Pinsonneault v. Merchants & Farmers Bank & Trust Co.

In Chapter 4, we discussed the common law principles of negligence. In this extended case study, we examine *Pinsonneault v. Merchants & Farmers Bank & Trust Co.,*[1] a recent decision involving an application of those principles in the context of a fatal shooting during a robbery.

1. 816 So.2d 270 (La. 2002).

CASE BACKGROUND

On October 28, 1992, Lawson Strickland and Christian Boyd escaped from the jail in Vernon Parish, Louisiana.[2] In need of money, they planned a robbery. They chose the branch of the Merchants & Farmers Bank & Trust Company in Leesville for the cover it afforded, hoping to escape through the woods behind the bank. On the night of November 3, they hid atop a hill behind a McDonald's restaurant next to the bank and waited.

At 1:30 A.M., Jesse Pinsonneault, the twenty-three-year-old assistant manager of Sambino's Pizza, left work and drove to the bank to deposit $64.06, the evening's receipts and operating cash. The bank's night deposit box was on the McDonald's side of the bank under a canopy that extended over the drive-thru lanes. As Pinsonneault's car approached, Strickland ran down the hill and hid behind the bank. When Pinsonneault got out of his car and walked up to the night deposit box, Strickland confronted him, brandishing a gun and demanding money. In the ensuing struggle, Strickland shot Pinsonneault, who died at a hospital nine hours later.

Pinsonneault's parents and brother filed a suit in a Louisiana state court against the bank and others, alleging in part that the bank negligently failed to provide adequate security to customers using the night depository. After a trial, the court issued a judgment in favor of the defendants. The Pinsonneaults appealed to a state intermediate appellate court, which reversed the trial court's judgment and awarded damages to the plaintiffs of more than $1 million. The defendants appealed to the Louisiana Supreme Court.

MAJORITY OPINION

WEIMER, Justice.

* * * *

* * * In *Posecai v. Wal-Mart Stores, Inc.*, 752 So.2d 762 (La. 1999), we held that while business owners generally have no duty to protect others from the criminal acts of third persons, "they do have a duty to implement reasonable measures to protect their patrons from criminal acts when those acts are foreseeable." Determining when a crime is foreseeable is a critical inquiry in the duty equation. This inquiry is answered employing a balancing test. * * * The foreseeability of the crime risk on the defendant's property and the gravity of the risk determine the existence and the extent of the defendant's duty. The greater the foreseeability and gravity of the harm, the greater the duty of care that will be imposed on the business. * * *

* * * *

In the instant case, there had been two armed robberies of this particular branch of Merchants Bank in the fourteen-year period prior to the attack on Jesse. The first occurred in 1984, during regular banking hours, when robbers absconded with cash from the bank. No customers were robbed. The perpetrators escaped using a helicopter they landed on the bank's front lawn. The second robbery took place in 1989. Again, this was a daytime robbery, during regular operating hours. No patrons were robbed. In this incident, the lone perpetrator escaped to the woods behind the bank through the northeast corner of the bank's property.

* * * *

Given these facts, it appears that night deposit customers at Merchants Bank faced a very low crime risk. Therefore, we concur with the trial court and find that Merchants Bank * * * did not possess the requisite foreseeability for the imposition of a duty to employ heightened security measures for the protection of patrons of its night depository.

This finding does not end our inquiry with respect to the bank's duty, however. * * * [O]ther factors, such as the location, nature, and condition of the property should also be taken into account. * * *

In this case, * * * not only did the bank recognize a duty to its patrons through the adoption of a written security plan, it took affirmative steps in furtherance of that plan, providing lighting at its night time depository, erecting fencing along vulnerable perimeters, and setting up a schedule for the installation of modern surveillance cameras at each of its branches.

Thus, we have no difficulty determining the defendant bank had a duty to implement reasonable security measures. Whether a defendant has breached a duty is a question of fact.

* * * *

[In this case] there is a reasonable factual basis for determining the lighting was adequate. There is a reasonable

2. A *parish* is a geographical and political division within the state of Louisiana that corresponds to a county in other states.

factual basis for determining the lack of security cameras was not a deterrent. The height of the shrubs was disputed as was the role of the shrubs in this incident. Fencing existed and the need for additional fencing and the deterrent effect of additional fencing was disputed. * * * [W]e find the decision of the trial court was reasonable * * * .

* * * *

We find that the Court of Appeal erred in * * * substituting its own conclusions for those of the trial court. Therefore, we reverse the Court of Appeal's judgment in favor of plaintiffs.

DISSENTING OPINION

JOHNSON, J. [Justice], dissents and assigns reasons.

I would affirm the decision of the court of appeal, assessing the bank with liability for the death of plaintiff's decedent, Jesse Pinsonneault. I agree with the court of appeal that, in accordance with * * * *Posecai v. Wal-Mart Stores, Inc.,* the bank had foreseeability of harm to its patrons sufficient to impose a duty to add adequate security measures. The bank's failure to provide such measures constitutes a breach of this duty.

* * * *

The court of appeal correctly found that although the prior crimes [at the bank] lacked similarity to the instant assault, they were nevertheless "predatory offenses which convey a risk of harm to both employees and customers of the bank." In addition, the record reveals that the bank had been informed of the hazards posed by the nature of its business through a banking newsletter, the *Advisor,* which repeatedly reported night deposit crimes and advised of the need to formulate security procedures and to install protective security measures.

* * * *

[Finally, the] record supports the court of appeal's conclusion that the bank "possessed the requisite foreseeability to impose a duty to implement, at the very least, security measures such as installation of improved lighting and functional surveillance cameras, installing fencing which would have enclosed the entire property and shield the property from the adjacent wooded area, and trimming and maintaining the shrubbery at a level which would not facilitate hiding."

QUESTIONS FOR ANALYSIS

1. **Law.** What were the principal issues before the court? What did the court rule on these issues, and why?
2. **Law.** Compare the conclusions of the majority and the dissent. What is the basis for their disagreement about the outcome of this case?
3. **Ethics.** What is an ethical basis for the majority's holding in this case? What is an ethical basis for the dissent's position?
4. **Economic Dimensions.** Is there a practical effect that the outcome in this case might have within the banking industry? If so, what are some of the possible advantages and disadvantages?
5. **Implications for the Business Owner.** What does the holding in this case indicate about the responsibility of a business to take precautions against the risk of crime?

UNIT

Contracts 3

Unit Contents

CHAPTER 7

Nature and Classification

LEARNING OBJECTIVES

After reading this chapter, you should be able to answer the following questions:

1. What is a contract? What is the objective theory of contracts?
2. What are the four basic elements necessary to the formation of a valid contract?
3. What is the difference between an implied-in-fact contract and an implied-in-law contract (quasi contract)?
4. How does a void contract differ from a voidable contract? What is an unenforceable contract?
5. Why have plain language laws been enacted? What rules guide the courts in interpreting contracts?

Keeping promises is important to a stable social order. Contract law deals with, among other things, the formation and keeping of promises. A **promise** is an assertion that something either will or will not happen in the future.

Like other types of law, contract law reflects our social values, interests, and expectations at a given point in time. It shows, for example, what kinds of promises our society thinks should be legally binding. It shows what excuses our society accepts for breaking such promises. Additionally, it shows what promises are considered to be contrary to public policy, or against the interests of society, and therefore legally void. If a promise goes against the interests of society as a whole, it will be invalid. Also, if it was made by a child or a mentally incompetent person, or on the basis of false information, a question will arise as to whether the promise should be enforced. Resolving such questions is the essence of contract law.

In the business world, questions and disputes concerning contracts arise daily. Although aspects of contract law vary from state to state, much of it is based on the common law. In 1932, the American Law Institute compiled the *Restatement of the Law of Contracts*. This work is a nonstatutory, authoritative exposition of the present law on the subject of contracts and is now in its second edition (a third edition is in the process of being drafted). Throughout the following chapters on contracts, we will refer to the second edition of the *Restatement of the Law of Contracts* as simply the *Restatement (Second) of Contracts*.

The Uniform Commercial Code (UCC), which governs contracts and other transactions relating to the sale and lease of goods, occasionally departs from common law contract rules. Generally, the different treatment of contracts falling under the UCC stems from the general policy of encouraging commerce. The ways in which the UCC changes common law contract rules will be discussed extensively in later chapters. In this unit covering the common law of contracts (Chapters 7 through 13), we only indicate briefly or in footnotes those common law rules that have been altered significantly by the UCC for sales and lease contracts.[1]

1. Throughout the chapters on the common law of contracts, we cite in footnotes the section numbers of the Uniform Commercial Code (UCC) version that is now in effect in most states. As you will read in later chapters covering sales contracts, the UCC was amended in 2003, and thus some of the section numbers may have been changed by those amendments.

The Function of Contracts

No aspect of modern life is entirely free of contractual relationships. You acquire rights and obligations, for example, when you borrow funds, when you buy or lease a house, when you procure insurance, when you form a business, and when you purchase goods or services. Contract law is designed to provide stability and predictability for both buyers and sellers in the marketplace.

Contract law assures the parties to private agreements that the promises they make will be enforceable. Clearly, many promises are kept because the parties involved feel a moral obligation to do so or because keeping a promise is in their mutual self-interest, not because the **promisor** (the person making the promise) or the **promisee** (the person to whom the promise is made) is conscious of the rules of contract law. Nevertheless, the rules of contract law are often followed in business agreements to avoid potential problems.

By supplying procedures for enforcing private agreements, contract law provides an essential condition for the existence of a market economy. Without a legal framework of reasonably assured expectations within which to plan and venture, businesspersons would be able to rely only on the good faith of others. Duty and good faith are usually sufficient, but when dramatic price changes or adverse economic conditions make it costly to comply with a promise, these elements may not be enough. Contract law is necessary to ensure compliance with a promise or to entitle the innocent party to some form of relief.

Definition of a Contract

A **contract** is an agreement that can be enforced in court. It is formed by two or more parties who agree to perform or to refrain from performing some act now or in the future. Generally, contract disputes arise when there is a promise of future performance. If the contractual promise is not fulfilled, the party who made it is subject to the sanctions of a court (see Chapter 12). That party may be required to pay money damages for failing to perform the contractual promise; in limited instances, the party may be required to perform the promised act.

In determining whether a contract has been formed, the element of intent is of prime importance. In contract law, intent is determined by what is referred to as the **objective theory of contracts,** not by the personal or subjective intent, or belief, of a party. The theory is that a party's intention to enter into a contract is judged by outward, objective facts as interpreted by a reasonable person, rather than by the party's own secret, subjective intentions. Objective facts include (1) what the party said when entering into the contract, (2) how the party acted or appeared, and (3) the circumstances surrounding the transaction. As will be discussed later in this chapter, in the subsection on express versus implied contracts, intent to form a contract may be manifested not only in words, oral or written, but also by conduct.

Requirements of a Contract

The following list describes the four requirements that must be met for a valid contract to exist. Each item will be explained more fully in the chapter indicated. Although we pair two of these requirements in some chapters, it is important to stress that each requirement is separate and independent. They are paired merely for reasons of space.

1. *Agreement.* An agreement includes an offer and an acceptance. One party must offer to enter into a legal agreement, and another party must accept the terms of the offer (Chapter 8).
2. *Consideration.* Any promises made by parties must be supported by legally sufficient and bargained-for consideration (something of value received or promised, to convince a person to make a deal) (Chapter 8).
3. *Contractual capacity.* Both parties entering into the contract must have the contractual capacity to do so; the law must recognize them as possessing characteristics that qualify them as competent parties (Chapter 9).
4. *Legality.* The contract's purpose must be to accomplish some goal that is legal and not against public policy (Chapter 9).

If any of these four elements is lacking, no contract will have been formed. Even if all of these elements exist, however, a contract may be unenforceable if the following requirements are not met. These requirements typically are raised as *defenses* to the enforceability of an otherwise valid contract.

1. *Genuineness of assent.* The apparent consent of both parties must be genuine (Chapter 10).
2. *Form.* The contract must be in whatever form the law requires; for example, some contracts must be in writing to be enforceable (Chapter 10).

Freedom of Contract and Freedom from Contract

As a general rule, the law recognizes everyone's ability to enter freely into contractual arrangements. This recognition is called *freedom of contract,* a freedom protected by the U.S. Constitution in Article I, Section 10. Because freedom of contract is a fundamental public policy of the United States, courts rarely interfere with contracts that have been voluntarily made.

Of course, as in other areas of the law, there are many exceptions to the general rule that contracts voluntarily negotiated will be enforced. For example, illegal bargains, agreements that unreasonably restrain trade, and certain unfair contracts made between one party with a great amount of bargaining power and another with little power are generally not enforced. In addition, as you will read in Chapter 9, certain contracts and clauses may not be enforceable if they are contrary to public policy, fairness, and justice. These exceptions provide *freedom from contract* for persons who may have been forced into making contracts unfavorable to themselves.

Types of Contracts

There are numerous types of contracts. They are categorized based on legal distinctions as to their formation, performance, and enforceability.

Contract Formation

As you can see in Exhibit 7–1, three classifications, or categories, of contracts are based on how and when a contract is formed. The best way to explain each type of contract is to compare one type with another, as we do in the following pages.

Bilateral versus Unilateral Contracts Every contract involves at least two parties. The **offeror** is the party making the offer. The **offeree** is the party to whom the offer is made. The offeror always promises to do or not to do something and thus is also a promisor. Whether the contract is classified as *bilateral* or *unilateral* depends on what the offeree must do to accept the offer and bind the offeror to a contract.

—Bilateral Contracts. If to accept the offer the offeree must only promise to perform, the contract is a **bilateral contract.** Hence, a bilateral contract is a "promise for a promise." An example of a bilateral contract is a contract in which one person agrees to buy another person's automobile for a specified price. No performance, such as the payment of money or delivery of goods, need take place for a bilateral contract to be formed. The contract comes into existence at the moment the promises are exchanged.

—Unilateral Contracts. If the offer is phrased so that the offeree can accept only by completing the contract performance, the contract is a **unilateral contract.** Hence, a unilateral contract is a "promise for an act." In other words, the contract is formed not at the moment when promises are exchanged but rather when the contract is *performed.* ■ EXAMPLE 7.1 Joe says to Celia, "If you drive my car from New York to Los Angeles, I'll give you $1,000." Only on Celia's completion of the act—bringing the car to Los Angeles—does she fully accept Joe's offer to pay $1,000. If she chooses not to accept the offer to drive the car to Los Angeles, there are no legal consequences. ■

Contests, lotteries, and other competitions offering prizes are also examples of offers for unilateral contracts. If a person complies with the rules of the contest—such as by submitting the right lottery number at the right place and time—a unilateral contract is formed, binding the

EXHIBIT 7–1 CLASSIFICATIONS BASED ON CONTRACT FORMATION

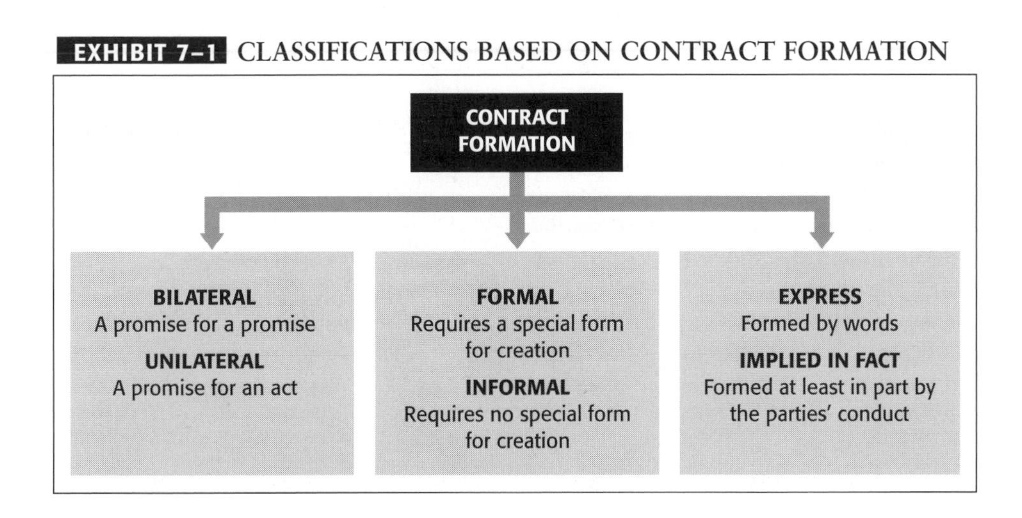
CONTRACT FORMATION
BILATERAL
A promise for a promise
UNILATERAL
A promise for an act
FORMAL
Requires a special form for creation
INFORMAL
Requires no special form for creation
EXPRESS
Formed by words
IMPLIED IN FACT
Formed at least in part by the parties' conduct

organization offering the prize to a contract to perform as promised in the offer.

Can a school's, or an employer's, letter of tentative acceptance to a prospective student, or a possible employee, qualify as a unilateral contract? That was the question in the following case.

CASE 7.1 Ardito v. City of Providence

United States District Court,
District of Rhode Island, 2003.
263 F.Supp.2d 358.

FACTS In 2001, the city of Providence, Rhode Island, decided to begin hiring police officers to fill vacancies in its police department. Because only individuals who had graduated from the Providence Police Academy were eligible, the city also decided to conduct two training sessions, the "60th and 61st Police Academies." To be admitted, an applicant had to pass a series of tests and be deemed qualified by members of the department after an interview. The applicants judged most qualified were sent a letter informing them that they had been selected to attend the academy if they successfully completed a medical checkup and a psychological examination. The letter for the applicants to the 61st Academy, dated October 15, stated that it was "a conditional offer of employment." Meanwhile, a new chief of police, Dean Esserman, decided to revise the selection process, which caused some of those who had received the letter to be rejected. Derek Ardito and thirteen other newly rejected applicants filed a suit in a federal district court against the city, seeking a halt to the 61st Academy unless they were allowed to attend. They alleged in part that the city was in breach of contract.

ISSUE Was the October 15 letter a unilateral offer that the plaintiffs had accepted by passing the required medical and psychological examinations?

DECISION Yes. The court issued an injunction to prohibit the city from conducting the 61st Police Academy unless the plaintiffs were included.

REASON The court found the October 15 letter to be "a classic example of an offer to enter into a unilateral contract. The October 15 letter expressly stated that it was a 'conditional offer of employment' and the message that it conveyed was that the recipient would be admitted into the 61st Academy if he or she successfully completed the medical and psychological examinations." The court contrasted the letter with "notices sent to applicants by the City at earlier stages of the selection process. Those notices merely informed applicants that they had completed a step in the process and remained eligible to be considered for admission into the Academy. Unlike the October 15 letter, the prior notices did not purport to extend a 'conditional offer' of admission." The court concluded that "[t]he plaintiffs accepted the City's offer of admission into the Academy by satisfying the specified conditions. Each of the plaintiffs submitted to and passed lengthy and intrusive medical and psychological examinations."

FOR CRITICAL ANALYSIS—Social Consideration *How might the city have phrased the October 15 letter to avoid its being considered a unilateral contract?*

—*Revocation of Offers for Unilateral Contracts.* A problem arises in unilateral contracts when the promisor attempts to revoke (cancel) the offer after the promisee has begun performance but before the act has been completed. ■ EXAMPLE 7.2 Suppose that Roberta offers to buy Ed's sailboat, moored in San Francisco, on delivery of the boat to Roberta's dock in Newport Beach, three hundred miles south of San Francisco. Ed rigs the boat and sets sail. Shortly before his arrival at Newport Beach, Ed receives a radio message from Roberta withdrawing her offer. Roberta's offer is an offer for a unilateral contract, and only Ed's delivery of the sailboat at her dock is an acceptance. ■

In contract law, offers are normally *revocable* (capable of being taken back, or canceled) until accepted. Under the traditional view of unilateral contracts, Roberta's revocation would terminate the offer. Because of the harsh effect on the offeree of the revocation of an offer to form a unilateral contract, the modern-day view is that once performance has been *substantially* undertaken, the offeror cannot revoke the offer. Thus, in our example, even though Ed has not yet accepted the offer by complete performance, Roberta is prohibited from revoking it. Ed can deliver the boat and bind Roberta to the contract.

Formal versus Informal Contracts Contracts that require a special form or method of creation (formation) to be enforceable are known as **formal contracts.** Formal contracts include (1) contracts under seal, (2) recognizances, (3) negotiable instruments, and (4) letters of credit.[2]

2. *Restatement (Second) of Contracts,* Section 6.

Contracts under seal are formalized writings with a special seal attached.[3] The significance of the seal has lessened, although about ten states require no consideration for formation of a valid contract when a contract is under seal. A *recognizance* is an acknowledgment in court by a person that he or she will perform some specified obligation or pay a certain sum if he or she fails to perform. One form of recognizance is the surety bond.[4] Another is the personal recognizance bond used as bail in a criminal matter. As will be discussed at length in subsequent chapters, *negotiable instruments* include checks, notes, drafts, and certificates of deposit; letters of credit are agreements to pay contingent on the purchaser's receipt of invoices and bills of lading (documents evidencing receipt of, and title to, goods shipped).

Informal contracts (also called *simple contracts*) include all other contracts. No special form is required (except for certain types of contracts that must be in writing), as the contracts are usually based on their substance rather than their form. Typically, businesspersons put their contracts in writing to ensure that there is some proof of a contract's existence should problems arise.

Express versus Implied Contracts Contracts may also be formed and categorized as express or implied by the conduct of the parties. We look here at the differences between these two types of contracts.

—Express Contracts. An **express contract** is one in which the terms of the agreement are fully and explicitly stated in words, oral or written. A signed lease for an apartment or a house is an express written contract. If a classmate accepts your offer to sell your textbooks from last semester for $75, an express oral contract has been made.

—Implied Contracts. A contract that is implied from the conduct of the parties is called an **implied-in-fact contract**, or an implied contract. This type of contract differs from an express contract in that the *conduct* of the parties, rather than their words, creates and defines at least some of the terms of the contract.

■ EXAMPLE 7.3 Suppose that you need an accountant to fill out your tax return this year. You look through the Yellow Pages and find an accounting firm located in your neighborhood. You drop by the firm's office, explain your problem to an accountant, and learn what fees will be charged. The next day you return and give the receptionist all of the necessary information and documents, such as canceled checks, W-2 forms, and so on. Then you walk out the door without saying anything expressly to the receptionist. In this situation, you have entered into an implied-in-fact contract to pay the accountant the usual and reasonable fees for the accounting services. The contract is implied by your conduct. The accountant expects to be paid for completing your tax return. By bringing in the records the accountant will need to do the work, you have implied an intent to pay for the services. ■ (For another example of how an implied-in-fact contract can arise, see this chapter's *Management Perspective* feature.)

—Requirements for an Implied-in-Fact Contract. For an implied-in-fact contract to arise, certain requirements must be met. Normally, if the following conditions exist, a court will hold that an implied contract was formed:

1. The plaintiff furnished some service or property.
2. The plaintiff expected to be paid for that service or property, and the defendant knew or should have known that payment was expected (by using the objective-theory-of-contracts test, discussed earlier in the chapter).
3. The defendant had a chance to reject the services or property and did not.

In the following case, the question before the court was whether a contract for electrical work had come into existence, given the absence of any written agreement.

3. A seal may be actual (made of wax or some other durable substance), impressed on the paper, or indicated simply by the word *seal* or the letters *L.S.* at the end of the document. *L.S.* stands for *locus sigilli,* which means "the place for the seal."

4. A *surety bond* is an obligation of a party who guarantees that he or she will pay a second party if a third party does not perform.

CASE 7.2 Homer v. Burman

Indiana Court of Appeals, 2001.
743 N.E.2d 1144.

FACTS Dave and Annette Homer owned a rental house in Marion, Indiana. When their tenant, Stephanie Clevenger, complained of malfunctioning lights, the Homers paid Burman Electric Service $2,650 to rewire the house. The parties did not sign a written contract. After the work was supposedly done, the Homers discovered, among other things, holes in the ceiling, plaster damage around electrical outlets, and exposed

CASE 7.2–CONTINUED

wires along aluminum siding on the outside of the house. When the power was reconnected, Clevenger's television, VCR, satellite receiver, and Playstation were destroyed. Over the next two months, her hair dryers, clocks, and lamps repeatedly burned out. The Homers filed a suit against Burman in an Indiana state court, alleging breach of contract. The court ruled against the Homers, who appealed to a state intermediate appellate court.

ISSUE Despite the absence of a written agreement, did the parties have a contract?

DECISION Yes. The state intermediate appellate court reversed the decision of the lower court and remanded the case to determine the amount of damages and attorneys' fees to be awarded to the Homers.

REASON The appellate court stated, "An offer, acceptance, plus consideration make up the basis for a contract. A mutual assent or a meeting of the minds on all essential elements or terms must exist in order to form a binding contract. Assent to the terms of a contract may be expressed by acts which manifest acceptance." The court concluded that the parties had a contract even though they did not put anything in writing. The court explained that "the Homers paid Burman Electric $2,650.00 to rewire their home. Burman Electric accepted the payment and began work. Therefore, because we have an offer, acceptance, consideration, and a manifestation of mutual assent, an implied-in-fact contract was in existence."

WHY IS THIS CASE IMPORTANT? *People commonly assume that to qualify as a contract, an agreement must be in writing. As this case clearly demonstrates, not all contracts need be in writing. The case also shows how an offer can be accepted by conduct, thus creating a valid contract (providing that all of the other requirements for a valid contract are met).*

MANAGEMENT PERSPECTIVE

Employment Manuals and Implied Contracts

MANAGEMENT FACES A LEGAL ISSUE

It is a common practice today for large companies or other organizations to create and distribute to their employees an employment manual, or handbook, setting forth the conditions of employment. Yet when drafting and distributing such manuals to employees, business owners and managers must consider the following question: Will statements made in an employee handbook constitute "promises" in an implied-in-fact employment contract?

WHAT THE COURTS SAY

Increasingly, courts are holding that promises made in an employment manual may create an implied-in-fact employment contract. For example, if an employment handbook states that employees will be fired only for "good cause," the employer may be held to that promise.

This is possible even if, under state law, employment is "at will." Under the employment-at-will doctrine, employers may hire and terminate employees at will, with or without cause. The at-will doctrine will not apply, however, if the terms of employment are subject to a contract between the employer and the employee. If a court holds that an implied employment contract exists, on the basis of promises made in an employment manual, the employer will be bound by the contract and liable for damages for breaching the contract.[a]

IMPLICATIONS FOR MANAGERS

To avoid being contractually bound by terms in an employment manual, you should avoid making definite statements (such as "employees will be terminated only for good cause") that would cause employees to reasonably believe that those statements are contractual promises. You should also inform employees, when initially giving them the handbook or discussing its contents with them, that it is not intended as a contract. A conspicuous written disclaimer to this effect should also be included in the employment manual. The disclaimer might read as follows: "This policy manual describes the basic personnel policies and practices of our Company. You should understand that the manual does not modify our Company's 'at will' employment doctrine or provide employees with any kind of contractual rights."

a. See, for example, *Cisco v. King*, __ S.W.3d __ (Ark.App. 2005).

Contract Performance

Contracts are also classified according to their state of performance. A contract that has been fully performed on both sides is called an **executed contract.** A contract that has not been fully performed is called an **executory contract.** If one party has fully performed but the other has not, the contract is said to be executed on the one side and executory on the other, but the contract is still classified as executory.

■ EXAMPLE 7.4 Assume that you agree to buy ten tons of coal from Western Coal Company. Further assume that Western has delivered the coal to your steel mill, where it is now being burned. At this point, the contract is an executory contract—it is executed on the part of Western and executory on your part. After you pay Western for the coal, the contract will be executed on both sides. ■

Contract Enforceability

A **valid contract** has the four elements necessary for contract formation: (1) an agreement (offer and acceptance) (2) supported by legally sufficient consideration (3) for a legal purpose and (4) made by parties who have the legal capacity to enter into the contract. As mentioned, we will discuss each of these elements in the following chapters. As you can see in Exhibit 7–2, valid contracts may be enforceable, voidable, or unenforceable. Additionally, a contract may be referred to as a *void contract.* We look next at the meaning of the terms *voidable, unenforceable,* and *void* in relation to contract enforceability.

Voidable Contracts A **voidable contract** is a *valid* contract but one that can be avoided at the option of one or both of the parties. The party having the option can elect either to avoid any duty to perform or to *ratify* (make valid) the contract. If the contract is avoided, both parties are released from it. If it is ratified, both parties must fully perform their respective legal obligations.

As you will read in Chapter 9, contracts made by minors, insane persons, and intoxicated persons may be voidable. As a general rule, for example, contracts made by minors are voidable at the option of the minor. Additionally, contracts entered into under fraudulent conditions are voidable at the option of the defrauded party. Contracts entered into under legally defined duress or undue influence are voidable (see Chapter 10).

Unenforceable Contracts An **unenforceable contract** is one that cannot be enforced because of certain legal defenses against it. It is not unenforceable because a party failed to satisfy a legal requirement of the contract; rather, it is a valid contract rendered unenforceable by some statute or law. For example, some contracts must be in writing (see Chapter 10), and if they are not, they will not be enforceable except in certain exceptional circumstances.

EXHIBIT 7–2 ENFORCEABLE, VOIDABLE, UNENFORCEABLE, AND VOID CONTRACTS

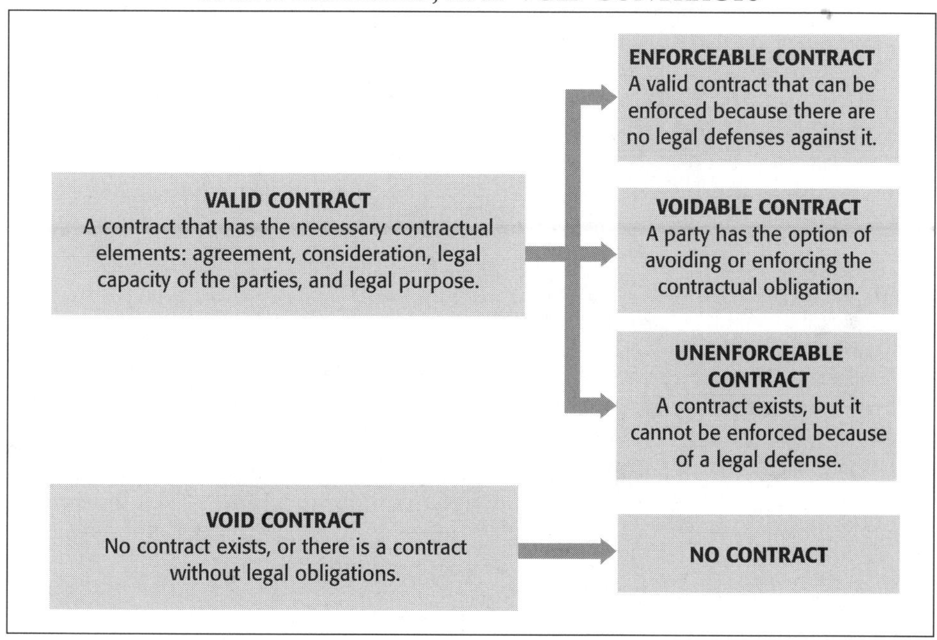

ENFORCEABLE CONTRACT
A valid contract that can be enforced because there are no legal defenses against it.
VALID CONTRACT
A contract that has the necessary contractual elements: agreement, consideration, legal capacity of the parties, and legal purpose.
VOIDABLE CONTRACT
A party has the option of avoiding or enforcing the contractual obligation.
UNENFORCEABLE CONTRACT
A contract exists, but it cannot be enforced because of a legal defense.
VOID CONTRACT
No contract exists, or there is a contract without legal obligations.
NO CONTRACT

Void Contracts A **void contract** is no contract at all. The terms *void* and *contract* are contradictory. None of the parties has any legal obligations if a contract is void. A contract can be void because, for example, one of the parties was previously determined by a court to be legally insane (and thus lacked the legal capacity to enter into a contract) or because the purpose of the contract was illegal.

Quasi Contracts

Quasi contracts, or contracts *implied in law,* are wholly different from actual contracts. Express contracts and implied-in-fact contracts are actual, or true, contracts. The word *quasi* is Latin for "as if" or "analogous to." Quasi contracts are thus not true contracts. They do not arise from any agreement, express or implied, between the parties themselves. Rather, quasi contracts are fictional contracts imposed on parties by courts "as if" the parties had entered into an actual contract. Usually, quasi contracts are imposed to avoid the *unjust enrichment* of one party at the expense of another. The doctrine of unjust enrichment is based on the theory that individuals should not be allowed to profit or enrich themselves inequitably at the expense of others.

■ EXAMPLE 7.5 Suppose that a vacationing physician is driving down the highway and encounters Emerson, who is lying unconscious on the side of the road. The physician renders medical aid that saves Emerson's life. Although the injured, unconscious Emerson did not solicit the medical aid and was not aware that the aid had been rendered, Emerson received a valuable benefit, and the requirements for a quasi contract were fulfilled. In such a situation, the law normally will impose a quasi contract, and Emerson will have to pay the physician for the reasonable value of the medical services provided. ■

Limitations on Quasi-Contractual Recovery

Although quasi contracts exist to prevent unjust enrichment, in some situations the party who obtains a benefit will not be deemed to have been unjustly enriched. Basically, the quasi-contractual principle cannot be invoked by a party who has conferred a benefit on someone else unnecessarily or as a result of misconduct or negligence.

■ EXAMPLE 7.6 You take your car to the local car wash and ask to have it run through the washer and to have the gas tank filled. While your car is being washed, you go to a nearby shopping center for two hours. In the meantime, one of the workers at the car wash mistakenly assumes that your car is the one that he is supposed to hand wax. When you come back, you are presented with a bill for a full tank of gas, a wash job, and a hand wax. Clearly, you have received a benefit, but this benefit was conferred because of a mistake by the car wash employee. You have not been *unjustly* enriched under these circumstances. People normally cannot be forced to pay for benefits "thrust" on them. ■

When a Contract Already Exists

The doctrine of quasi contract generally cannot be used when an actual contract covers the area in controversy. This is because a remedy already exists if a party is unjustly enriched as a result of a breach of contract: the nonbreaching party can sue the breaching party for breach of contract. In this instance, a court does not need to impose a quasi contract to achieve justice.

Interpretation of Contracts

To avoid disputes over contract interpretation, business managers should make sure that their intentions are clearly expressed in their contracts. Careful drafting of contracts not only helps prevent potential disputes over the meaning of certain terms but may also be crucial if the firm brings or needs to defend against a lawsuit for breach of contract.

When disputes over contract interpretation arise, the courts look to common law rules of contract interpretation for guidance. These rules include the *plain meaning rule* and various other rules that have evolved over time. In this final section, we will examine these rules. (For a summary of how the rules are applied, see Exhibit 7–3 on the next page.) First, though, we look at *plain language laws,* which regulate legal writing. Federal and state governments have enacted plain language laws in the interest of helping consumers, as well as easing the work of the courts. These laws deal with private contracts in their entirety.

Plain Language Laws

Today, a majority of the states regulate legal writing through plain language laws, particularly in consumer contracts that are primarily for personal, family, or household purposes. For example, a New York law requires residential leases and other consumer contracts to be (1) "written in a clear and coherent manner using words with common and everyday meanings" and (2) "appropriately divided and captioned by [their] various sections."[5]

5. N.Y. Gen. Oblig. Law Section 5-702.

EXHIBIT 7–3 RULES OF CONTRACT INTERPRETATION

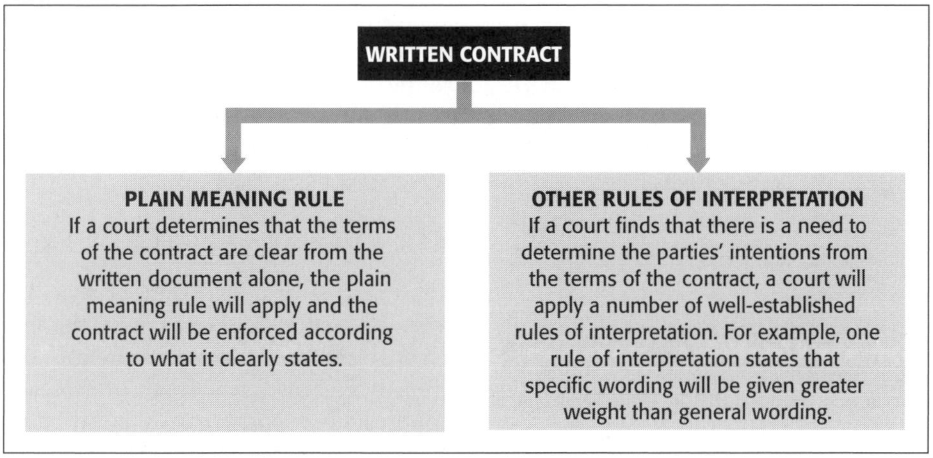

WRITTEN CONTRACT

PLAIN MEANING RULE
If a court determines that the terms of the contract are clear from the written document alone, the plain meaning rule will apply and the contract will be enforced according to what it clearly states.

OTHER RULES OF INTERPRETATION
If a court finds that there is a need to determine the parties' intentions from the terms of the contract, a court will apply a number of well-established rules of interpretation. For example, one rule of interpretation states that specific wording will be given greater weight than general wording.

If a party to a contract, such as an insurance company, violates a plain language statute, a court may refuse to enforce the contract—unless the party can show that it made a good faith effort to comply with the statute. Some state statutes even allow proposed contracts to be submitted to the state attorney general, whose approval then eliminates any liability for damages because of a supposed violation of the plain language statute.

Federal Forms and Agency Rulemaking Difficult-to-read federal government forms have also come under attack. In one case, for example, a court found that a government form giving immigrants notice of possible deportation was so difficult to read that it violated due process requirements.[6] In 1998, the federal government addressed this problem by requiring all federal agencies to use plain language in most of their forms and written communications. Since then, plain language requirements have been extended to agency rulemaking as well.

Legal Professionals Move toward Plain English In response to plain language laws, the legal profession has attempted to abandon the traditional preference for often turgid and verbose language in favor of clear, easily understandable legal writing. Judges and clients are also pressuring lawyers to make legal concepts and documents more comprehensible. At times, judges have refused to accept motions that are incoherent due to their highly technical legal language.

In some jurisdictions, judges themselves are being asked to write their opinions more clearly. For example, the chief judge of New York's highest court has encouraged judges in the New York court system to issue clearly written rulings so that other judges, lawyers, lawmakers, and everyone affected by the decisions will know what they mean.

The Plain Meaning Rule

When the terms of a written contract are clear and unequivocal, a court will enforce the contract according to its plain terms (what is clearly stated in the contract), and there is no need for the court to interpret the language of the contract. The meaning of the terms must be determined from the *face of the instrument*—from the written document alone. This is sometimes referred to as the *plain meaning rule.*

Under the plain meaning rule, if a contract's words appear to be clear and unambiguous, a court cannot consider *extrinsic evidence,* which is any evidence not contained in the document itself. Admissibility of extrinsic evidence can significantly affect how a court interprets ambiguous contractual provisions and thus can affect the outcome of litigation.

Other Rules of Interpretation

When a court finds that there is a need to determine the parties' intentions from the terms of the contract, it will interpret the language to give effect to the parties' intent as *expressed in their contract.* This is the primary purpose of the rules of interpretation—to determine the parties' intent from the language used in their agreement and to give effect to that intent. A court normally will not make or remake a contract, nor will it normally interpret the

6. *Walters v. Reno,* 145 F.3d 1032 (9th Cir. 1998).

language according to what the parties *claim* their intent was when they made the contract.[7] The courts use the following rules in interpreting contractual terms:

1. Insofar as possible, a reasonable, lawful, and effective meaning will be given to all of a contract's terms.
2. A contract will be interpreted as a whole; individual, specific clauses will be considered subordinate to the contract's general intent. All writings that are a part of the same transaction will be interpreted together.
3. Terms that were the subject of separate negotiation will be given greater consideration than standardized terms and terms that were not negotiated separately.
4. A word will be given its ordinary, commonly accepted meaning, and a technical word or term will be given its technical meaning, unless the parties clearly intended something else.
5. Specific and exact wording will be given greater consideration than general language.
6. Written or typewritten terms prevail over preprinted terms.
7. Because a contract should be drafted in clear and unambiguous language, a party that uses ambiguous expressions is held to be responsible for the ambiguities. Thus, when the language has more than one meaning, it will be interpreted *against* the party that drafted the contract.
8. Evidence of *usage of trade, course of dealing,* and *course of performance* may be admitted to clarify the meaning of an ambiguously worded contract. (We define and discuss these terms in Chapter 14.) What each of the parties does pursuant to the contract will be interpreted as consistent with what the other does and with any relevant usage of trade and course of dealing or performance. Express terms (terms expressly stated in the contract) are given the greatest weight, followed by course of performance, course of dealing, and custom and usage of trade—in that order. When considering custom and usage, a court will look at the trade customs and usage common to the particular business or industry and to the locale in which the contract was made or is to be performed.

In the following case, the ordinary meaning of a word was at the heart of a significant dispute.

7. Nevertheless, if a court finds that even after applying the rules of interpretation the terms are susceptible to more than one meaning, the court may permit extrinsic evidence to prove what the parties intended. See, for example, *Langdon v. United Restaurants, Inc.,* 105 S.W.3d 882 (Mo.Ct.App. 2003).

CASE 7.3 Citizens Communications Co. v. Trustmark Insurance

United States District Court,
District of Connecticut, 2004.
303 F.Supp.2d 197.

FACTS Citizens Communications Company is the seventh largest telephone company in the United States, with assets of nearly $7 billion. Citizens provides health insurance to its nearly five thousand employees under a self-funded health plan. Under such a plan, a company normally buys an insurance policy to cover claims that exceed a certain amount. Citizens bought a policy from Trustmark Insurance to cover claims that exceeded $100,000, subject to a $1 million maximum benefit per employee. In November 1999, Garry Lonquist, a Citizens employee, experienced complications from heart surgery that required intensive care. By January, Lonquist's medical bills exceeded $1 million. Trustmark refused to renew Citizens' policy unless, among other things, it included an amendment providing a separate $1 million deductible for Lonquist, effectively removing him from coverage. This amendment was referred to as the "Lonquist Laser." Citizens accepted this offer in a letter dated March 6, 2000. Later, Citizens filed a suit in a federal district court against Trustmark and others, seeking damages and other relief for an alleged breach of contract. Citizens claimed in part that the "Lonquist Laser" was not valid. Both parties filed motions for summary judgment.

ISSUE Was the Lonquist Laser an effective amendment of the parties' insurance contract?

DECISION Yes. The court granted Trustmark's motion for summary judgment on this issue.

REASON The court recognized that "[a]lthough the policy defines many of its terms, it does not define the term 'amendment'; nor does anything in the policy specify precisely what an amendment must look like, other than that it must be approved by Trustmark and signed by Citizens, both of which occurred here." The court further explained that "courts have consistently referred to dictionary definitions to interpret words used in insurance contracts." The court noted that according to *Webster's Unabridged Dictionary* (Random House, 2001), the ordinary meaning of the term *amendment* is " 'a change made by correction, addition, or deletion.' That is precisely what occurred here, when the parties knowingly agreed * * * to alter or change the terms" of the policy

(Continued)

CASE 7.3—CONTINUED

"by deleting or removing the Lonquist claim * * * . Accordingly, under the plain language of the contract, the Court concludes that any agreement that meets the requirements necessary to form a contract under state law and that is signed and approved by both parties satisfies the requirements of the policy and is a valid and enforceable amendment. Because the Lonquist Laser set forth in the March 6 Agreement unquestionably meets that standard, it is a valid amendment of the parties' insurance contract."

FOR CRITICAL ANALYSIS—Social Consideration
What might the parties have done to avoid this litigation?

TERMS AND CONCEPTS

CHAPTER SUMMARY Nature and Classification

The Function of Contracts (See page 143.)	Contract law establishes what kinds of promises will be legally binding and supplies procedures for enforcing legally binding promises, or agreements.
Definition of a Contract (See page 143.)	A contract is an agreement that can be enforced in court. It is formed by two or more competent parties who agree to perform or to refrain from performing some act now or in the future.
Requirements of a Contract (See page 143.)	1. *Elements of a valid contract*—Agreement, consideration, contractual capacity, and legality. 2. *Possible defenses to the enforcement of a contract*—Genuineness of assent and form.
Types of Contracts (See pages 144–149.)	1. *Bilateral*—A promise for a promise. 2. *Unilateral*—A promise for an act (acceptance is the completed—or substantial—performance of the contract by the offeree). 3. *Formal*—Requires a special form for contract formation. 4. *Informal*—Requires no special form for contract formation. 5. *Express*—Formed by words (oral, written, or a combination). 6. *Implied in fact*—Formed at least in part by the conduct of the parties. 7. *Executed*—A fully performed contract. 8. *Executory*—A contract not yet fully performed. 9. *Valid*—A contract that has the necessary contractual elements of offer and acceptance, consideration, parties with legal capacity, and having been made for a legal purpose. 10. *Voidable*—A contract in which a party has the option of avoiding or enforcing the contractual obligation. 11. *Unenforceable*—A valid contract that cannot be enforced because of a legal defense. 12. *Void*—No contract exists, or there is a contract without legal obligations.

CHAPTER SUMMARY Nature and Classification—Continued

Quasi Contracts (See page 149.)	A quasi contract, or a contract implied in law, is a contract that is imposed by law to prevent unjust enrichment.
Interpretation of Contracts (See pages 149–152.)	Increasingly, plain language laws are requiring private contracts to be written in plain language so that the terms are clear and understandable to the parties. Under the plain meaning rule, a court will enforce the contract according to its plain terms, the meaning of which must be determined from the written document alone. Other rules applied by the court when interpreting contracts include the following: 1. A reasonable, lawful, and effective meaning will be given to all contract terms. 2. A contract will be interpreted as a whole, specific clauses will be considered subordinate to the contract's general intent, and all writings that are a part of the same transaction will be interpreted together. 3. Terms that were negotiated separately will be given greater consideration than standardized terms and terms not negotiated separately. 4. Words will be given their commonly accepted meanings and technical words their technical meanings, unless the parties clearly intended otherwise. 5. Specific wording will be given greater consideration than general language. 6. Written or typewritten terms prevail over preprinted terms. 7. A party that uses ambiguous expressions is held to be responsible for the ambiguities. 8. Evidence of course of dealing, course of performance, or usage of trade is admissible to clarify an ambiguously worded contract. In these circumstances, express terms are given the greatest weight, followed by course of performance, course of dealing, and custom and usage of trade—in that order.

FOR REVIEW

Answers for the even-numbered questions in this For Review *section can be found in Appendix A at the end of this text.*

1. What is a contract? What is the objective theory of contracts?
2. What are the four basic elements necessary to the formation of a valid contract?
3. What is the difference between an implied-in-fact contract and an implied-in-law contract (quasi contract)?
4. How does a void contract differ from a voidable contract? What is an unenforceable contract?
5. Why have plain language laws been enacted? What rules guide the courts in interpreting contracts?

QUESTIONS AND CASE PROBLEMS

7–1. Express versus Implied Contracts. Suppose that a local businessperson, McDougal, is a good friend of Krunch, the owner of a local candy store. Every day on his lunch hour McDougal goes into Krunch's candy store and spends about five minutes looking at the candy. After examining Krunch's candy and talking with Krunch, McDougal usually buys one or two candy bars. One afternoon, McDougal goes into Krunch's candy shop, looks at the candy, and picks up a $1 candy bar. Seeing that Krunch is very busy, he catches Krunch's eye, waves the candy bar at Krunch without saying a word, and walks out. Is there a contract? If so, classify it within the categories presented in this chapter.

QUESTION WITH SAMPLE ANSWER

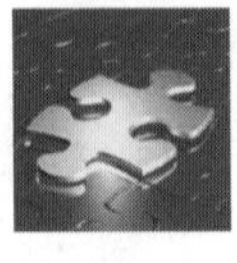

7–2. Rosalie, a wealthy widow, invited an acquaintance, Jonathan, to her home for dinner. Jonathan accepted the offer and, eager to please her, spent lavishly in preparing for the evening. His purchases

included a new blazer, new shoes, an expensive floral arrangement, and champagne. At the appointed time, Jonathan arrived at Rosalie's house only to find that she had left for the evening. Jonathan wants to sue Rosalie to recover some of his expenses. Can he? Why or why not?

For a sample answer to this question, go to Appendix B at the end of this text.

7–3. Contract Classification. High-Flying Advertising, Inc., contracted with Big Burger Restaurants to fly an advertisement above the Connecticut beaches. The advertisement offered $5,000 to any person who could swim from the Connecticut beaches to Long Island across Long Island Sound in less than a day. McElfresh saw the streamer and accepted the challenge. He started his marathon swim that same day at 10 A.M. After he had been swimming for four hours and was about halfway across the sound, McElfresh saw another plane pulling a streamer that read, "Big Burger revokes." Is there a contract between McElfresh and Big Burger? If there is a contract, what type(s) of contract is (are) formed?

7–4. Implied Contract. Thomas Rinks and Joseph Shields developed Psycho Chihuahua, a caricature of a Chihuahua dog with a "do-not-back-down" attitude. They promoted and marketed the character through their company, Wrench, L.L.C. Ed Alfaro and Rudy Pollak, representatives of Taco Bell Corp., learned of Psycho Chihuahua and met with Rinks and Shields to talk about using the character as a Taco Bell "icon." Wrench sent artwork, merchandise, and marketing ideas to Alfaro, who promoted the character within Taco Bell. Alfaro asked Wrench to propose terms for Taco Bell's use of Psycho Chihuahua. Taco Bell did not accept Wrench's terms, but Alfaro continued to promote the character within the company. Meanwhile, Taco Bell hired a new advertising agency, which proposed an advertising campaign involving a Chihuahua. When Alfaro learned of this proposal, he sent the Psycho Chihuahua materials to the agency. Taco Bell made a Chihuahua the focus of its marketing but paid nothing to Wrench. Wrench filed a suit against Taco Bell in a federal district court, claiming in part that it had an implied contract with Taco Bell, which the latter breached. Do these facts satisfy the requirements for an implied contract? Why or why not? [*Wrench L.L.C. v. Taco Bell Corp.*, 256 F.3d 446 (6th Cir. 2001)]

CASE PROBLEM WITH SAMPLE ANSWER

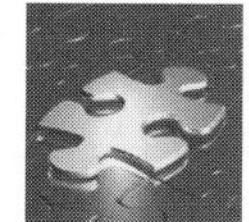

7–5. Professor Dixon was an adjunct professor at Tulsa Community College (TCC) in Tulsa, Oklahoma. Each semester, near the beginning of the term, the parties executed a written contract that always included the following provision: "It is agreed that this agreement may be cancelled by the Administration or the instructor at anytime before the first class session." In the spring semester of Dixon's seventh year, he filed a complaint with TCC alleging that one of his students, Meredith Bhuiyan, had engaged in disruptive classroom conduct. He gave her an incomplete grade and asked TCC to require her to apologize as a condition of receiving a final grade. TCC later claimed, and Dixon denied, that he was told to assign Bhuiyan a grade if he wanted to teach in the fall. Toward the end of the semester, Dixon was told which classes he would teach in the fall, but the parties did not sign a written contract. The Friday before classes began, TCC terminated him. Dixon filed a suit in an Oklahoma state court against TCC and others, alleging breach of contract. Did the parties have a contract? If so, did TCC breach it? Explain. [*Dixon v. Bhuiyan*, 10 P.3d 888 (Okla. 2000)]

After you have answered this problem, compare your answer with the sample answer given on the Web site that accompanies this text. Go to academic.cengage.com/blaw/wbl, select "Chapter 7," and click on "Case Problem with Sample Answer."

7–6. Bilateral versus Unilateral Contracts. D.L. Peoples Group (D.L.) placed an ad in a Missouri newspaper to recruit admissions representatives, who were hired to recruit Missouri residents to attend D.L.'s college in Florida. Donald Hawley responded to the ad, his interviewer recommended him for the job, and he signed, in Missouri, an "Admissions Representative Agreement," which was mailed to D.L.'s president, who signed it in his office in Florida. The agreement provided in part that Hawley would devote exclusive time and effort to the business in his assigned territory in Missouri and that D.L. would pay Hawley a commission if he successfully recruited students for the school. While attempting to make one of his first calls on his new job, Hawley was accidentally shot and killed. On the basis of his death, a claim was filed in Florida for workers' compensation. (Under Florida law, when an accident occurs outside Florida, workers' compensation benefits are payable only if the employment contract was made in Florida.) Is this admissions representative agreement a bilateral or a unilateral contract? What are the consequences of the distinction in this case? Explain. [*D.L. Peoples Group, Inc. v. Hawley*, 804 So.2d 561 (Fla.App. 1 Dist. 2002)]

7–7. Interpretation of Contracts. East Mill Associates (EMA) was developing residential "units" in East Brunswick, New Jersey, within the service area of the East Brunswick Sewerage Authority (EBSA). The sewer system required an upgrade to the Ryder's Lane Pumping Station to accommodate the new units. EMA agreed to pay "fifty-five percent (55%) of the total cost" of the upgrade. At the time, the estimated cost to EMA was $150,000 to $200,000. Impediments to the project arose, however, substantially increasing the cost. Among other things, the pumping station had to be moved to accommodate a widened road nearby. The upgrade was delayed for almost three years. When it was completed, EBSA asked EMA for $340,022.12, which represented 55 percent of the total cost. EMA did not pay. EBSA filed a suit in a New Jersey state court against EMA for breach of contract. What rule should the court apply to interpret the parties' contract? How should that rule be applied? Why? [*East Brunswick Sewerage Authority v. East Mill Associates, Inc.*, 365 N.J.Super. 120, 838 A.2d 494 (A.D. 2004)]

VIDEO QUESTION

7–8. Go to this text's Web site at academic.cengage.com/blaw/wbl and select "Chapter 7." Click on "Video Questions" and view the video titled *Bowfinger.* Then answer the following questions.

1. In the video, Renfro (Robert Downey, Jr.) says to Bowfinger (Steve Martin), "You bring me this script and Kit Ramsey and you've got yourself a 'go' picture." Assume for the purposes of this question that their agreement is a contract. Is the contract bilateral or unilateral? Is it express or implied? Is it formal or informal? Is it executed or executory? Explain your answers.
2. What criteria would a court rely on to interpret the terms of the contract?
3. Recall from the video that the contract between Bowfinger and the producer was oral. Suppose that a statute requires contracts of this type to be in writing. In that situation, would the contract be void, voidable, or unenforceable? Explain.

Nature and Classification

For updated links to resources available on the Web, as well as a variety of other materials, visit this text's Web site at

academic.cengage.com/blaw/wbl

The 'Lectric Law Library provides information on contract law, including a definition of a contract and the elements required for a contract. Go to

http://www.lectlaw.com/lay.html

and scroll down to "Contracts."

For easy-to-understand definitions of legal terms and concepts, including terms and concepts relating to contract law, go to

http://dictionary.law.com

and key in a term, such as *contract* or *consideration.*

ONLINE LEGAL RESEARCH

Go to the *Fundamentals of Business Law* home page at academic.cengage.com/blaw/wbl, select "Chapter 7," and click on "Internet Exercises." There you will find the following Internet research exercises that you can perform to learn more about topics covered in this chapter.

Activity 7–1: LEGAL PERSPECTIVE—Contracts and Contract Provisions

Activity 7–2: MANAGEMENT PERSPECTIVE—Implied Employment Contracts

Activity 7–3: INTERNATIONAL PERSPECTIVE—Contracts in Ancient Mesopotamia

BEFORE THE TEST

Go to the *Fundamentals of Business Law* home page at academic.cengage.com/blaw/wbl, select "Chapter 7," and click on "Interactive Quizzes." You will find a number of interactive questions relating to this chapter.

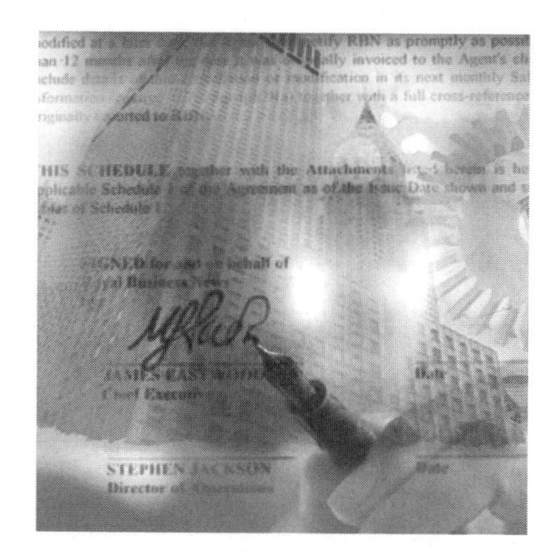

CHAPTER 8

Agreement and Consideration

LEARNING OBJECTIVES

After reading this chapter, you should be able to answer the following questions:

1. What elements are necessary for an effective offer? What are some examples of nonoffers?
2. In what circumstances will an offer be irrevocable?
3. What are the elements that are necessary for an effective acceptance?
4. What is consideration? What is required for consideration to be legally sufficient?
5. In what circumstances might a promise be enforced despite a lack of consideration?

The oft-stated saying that it is "necessity that makes laws" is certainly true in regard to contracts. In Chapter 7, we pointed out that promises and agreements, and the knowledge that some of those promises and agreements will be legally enforced, are essential to civilized society. The homes we live in, the food we eat, the clothes we wear, the cars we drive, the books we read, the videos and recordings we watch and listen to—all of these have been purchased through contractual agreements. Contract law developed over time, through the common law tradition, to meet society's need to know with certainty what kinds of promises, or contracts, will be enforced and the point at which a valid and binding contract is formed.

For a contract to be considered valid and enforceable, the requirements listed in Chapter 7 must be met. In this chapter, we look closely at two of these requirements, *agreement* and *consideration.* As you read through this chapter, keep in mind that the requirements of agreement and consideration apply to all contracts, regardless of how they are formed. Many contracts continue to be formed in the traditional way—through the exchange of paper documents. Increasingly, contracts are also being formed online—through the exchange of electronic messages or documents. Although we discuss online contracts to a limited extent in this chapter, we will look at them more closely in Chapter 13.

Agreement

An essential element for contract formation is **agreement**—the parties must agree on the terms of the contract. Ordinarily, agreement is evidenced by two events: an *offer* and an *acceptance.* One party offers a certain bargain to another party, who then accepts that bargain.

Because words often fail to convey the precise meaning intended, the law of contracts generally adheres to the *objective theory of contracts,* as discussed in Chapter 7. Under this theory, a party's words and conduct are held to mean whatever a reasonable person in the offeree's position would think they meant. The court will give words their usual meanings even if "it were proved by twenty bishops that [the] party . . . intended something else."[1]

1. Judge Learned Hand in *Hotchkiss v. National City Bank of New York,* 200 F. 287 (2d Cir. 1911); aff'd 231 U.S. 50, 34 S.Ct. 20, 58 L.Ed. 115 (1913).

REQUIREMENTS OF THE OFFER

An **offer** is a promise or commitment to perform or refrain from performing some specified act in the future. As discussed in Chapter 7, the party making an offer is called the *offeror,* and the party to whom the offer is made is called the *offeree.*

Three elements are necessary for an offer to be effective:

1. There must be a serious, objective intention by the offeror.
2. The terms of the offer must be reasonably certain, or definite, so that the parties and the court can ascertain the terms of the contract.
3. The offer must be communicated to the offeree.

Once an effective offer has been made, the offeree's acceptance of that offer creates a legally binding contract (providing the other essential elements for a valid and enforceable contract are present).

In today's e-commerce world, offers are frequently made online. Essentially, the requirements for traditional offers apply to online offers as well, as you will read in Chapter 13.

Intention The first requirement for an effective offer to exist is a serious, objective intention on the part of the offeror. Intent is not determined by the *subjective* intentions, beliefs, or assumptions of the offeror. Rather, it is determined by what a reasonable person in the offeree's position would conclude the offeror's words and actions meant. Offers made in obvious anger, jest, or undue excitement do not meet the serious-and-objective-intent test. Because these offers are not effective, an offeree's acceptance does not create an agreement.

■ EXAMPLE 8.1 You and three classmates ride to school each day in Julio's new automobile, which has a market value of $18,000. One cold morning the four of you get into the car, but Julio cannot get it started. He yells in anger, "I'll sell this car to anyone for $500!" You drop $500 in his lap. A reasonable person, taking into consideration Julio's frustration and the obvious difference in value between the car's market price and the $500 purchase price, would declare that Julio's offer was not made with serious and objective intent and that you do not have an agreement. ■

In the subsections that follow, we examine the concept of intention further as we look at the distinctions between offers and nonoffers. *Lucy v. Zehmer,* presented below, is a classic case in the area of contractual agreement. The case involved a business transaction in which boasts, brags, and dares "after a few drinks" resulted in a contract to sell certain property. The sellers claimed that the offer had been made in jest and that, in any event, the contract was voidable at their option because they were intoxicated when the offer was made and thus lacked contractual capacity (see Chapter 9). The court, however, looked to the words and actions of the parties—not their secret intentions—to determine whether a contract had been formed.

LANDMARK AND CLASSIC CASES

CASE 8.1 Lucy v. Zehmer

Supreme Court of Appeals of Virginia, 1954.
196 Va. 493,
84 S.E.2d 516.

FACTS Lucy and Zehmer had known each other for fifteen to twenty years. For some time, Lucy had been wanting to buy Zehmer's farm. Zehmer had always told Lucy that he was not interested in selling. One night, Lucy stopped in to visit with the Zehmers at a restaurant they operated. Lucy said to Zehmer, "I bet you wouldn't take $50,000 for that place." Zehmer replied, "Yes, I would, too; you wouldn't give fifty." Throughout the evening, the conversation returned to the sale of the farm. At the same time, the parties were drinking whiskey. Eventually, Zehmer wrote up an agreement, on the back of a restaurant check, for the sale of the farm, and he asked his wife to sign it—which she did. When Lucy brought an action in a Virginia state court to enforce the agreement, Zehmer argued that he had been "high as a Georgia pine" at the time and that the offer had been made in jest: "two doggoned drunks bluffing to see who could talk the biggest and say the most." Lucy claimed that he had not been intoxicated and did not think Zehmer had been, either, given the way Zehmer handled the transaction. The trial court ruled in favor of the Zehmers, and Lucy appealed.

ISSUE Can the agreement be avoided on the basis of intoxication?

DECISION No. The agreement to sell the farm was binding.

REASON The opinion of the court was that the evidence given about the nature of the conversation, the appearance and completeness of the agreement, and the signing all tended to show that a serious business transaction, not a casual jest, was intended. The court had to look into the

(Continued)

CASE 8.1—CONTINUED

objective meaning of the words and acts of the Zehmers: "An agreement or mutual assent is of course essential to a valid contract, but the law imputes to a person an intention corresponding to the reasonable meaning of his words and acts. If his words and acts, judged by a reasonable standard, manifest an intention to agree, it is immaterial what may be the real but unexpressed state of mind."

IMPACT OF THIS CASE ON TODAY'S LAW

This is a classic case in contract law because it illustrates so clearly the objective theory of contracts with respect to determining whether an offer was intended. Today, the objective theory of contracts continues to be applied by the courts, and *Lucy v. Zehmer* is routinely cited as a significant precedent in this area. Note that in cases involving contracts formed online, the issue of contractual intent rarely arises. Perhaps this is because an online offer is, by definition, "objective" in the sense that it consists of words only—the offeror's physical actions and behavior are not evidenced.

RELEVANT WEB SITES *To locate information on the Web concerning* Lucy v. Zehmer, *go to this text's Web site at* **academic.cengage.com/blaw/wbl**, *select "Chapter 8," and click on "URLs for Landmark Cases."*

—Expressions of Opinion. An expression of opinion is not an offer. It does not evidence an intention to enter into a binding agreement. ■ EXAMPLE 8.2 In *Hawkins v. McGee*,[2] Hawkins took his son to McGee, a physician, and asked McGee to operate on the son's hand. McGee said that the boy would be in the hospital three or four days and that the hand would *probably* heal a few days later. The son's hand did not heal for a month, but nonetheless the father did not win a suit for breach of contract. The court held that McGee did not make an offer to heal the son's hand in three or four days. He merely expressed an opinion as to when the hand would heal. ■

—Statements of Intention. Making a statement of an *intention* to do something in the future is not an offer. ■ EXAMPLE 8.3 If Ari says, "I *plan* to sell my stock in Novation, Inc., for $150 per share," a contract is not created if John "accepts" and tenders $150 per share for the stock. Ari has merely expressed his intention to enter into a future contract for the sale of the stock. If John accepts and tenders the $150 per share, no contract is formed because a reasonable person would conclude that Ari was only thinking about selling his stock, not promising to sell it. ■

—Preliminary Negotiations. A request or invitation to negotiate is not an offer; it only expresses a willingness to discuss the possibility of entering into a contract. Examples are statements such as "Will you sell Forest Acres?" and "I wouldn't sell my car for less than $5,000." A reasonable person in the offeree's position would not conclude that such a statement evidenced an intention to enter into a binding obligation. Likewise, when the government and private firms need to have construction work done, contractors are invited to submit bids. The *invitation* to submit bids is not an offer, and a contractor does not bind the government or private firm by submitting a bid. (The bids that the contractors submit are offers, however, and the government or private firm can bind the contractor by accepting the bid.)

—Advertisements, Catalogues, and Circulars. In general, advertisements, mail-order catalogues, price lists, and circular letters (meant for the general public) are treated as invitations to negotiate, not as offers to form a contract.[3] ■ EXAMPLE 8.4 Suppose that you put an ad in the classified section of your local newspaper offering to sell your guitar for $275. Seven people call and "accept" your "offer" before you can remove the ad from the newspaper. If the ad were truly an offer, you would be bound by seven contracts to sell your guitar. Because *initial* advertisements are treated as *invitations* to make offers rather than offers, however, you will have seven offers to choose from, and you can accept the best one without incurring any liability for the six you reject. ■ On some occasions, though, courts have construed advertisements to be offers because the ads contained definite terms that invited acceptance (such as an ad offering a reward for the return of a lost dog).[4]

Price lists are another form of invitation to negotiate or trade. A seller's price list is not an offer to sell at that price; it merely invites the buyer to offer to buy at that price. In fact, the seller usually puts "prices subject to change" on the price list. Only in rare circumstances will a price quotation be construed as an offer.[5]

—Auctions. In an auction, a seller "offers" goods for sale through an auctioneer. This is not, however, an offer to form a contract. Rather, it is an invitation asking

2. 84 N.H. 114, 146 A. 641 (1929).

3. *Restatement (Second) of Contracts,* Section 26, Comment b.

4. The classic example is *Lefkowitz v. Great Minneapolis Surplus Store, Inc.,* 251 Minn. 188, 86 N.W.2d 689 (1957).

5. See, for example, *Fairmount Glass Works v. Grunden-Martin Woodenware Co.,* 106 Ky. 659, 51 S.W. 196 (1899).

bidders to submit offers. In the context of an auction, a bidder is the offeror, and the auctioneer is the offeree. The offer is accepted when the auctioneer strikes the hammer. Before the fall of the hammer, a bidder may revoke (take back) her or his bid, or the auctioneer may reject that bid or all bids. Typically, an auctioneer will reject a bid that is below the price the seller is willing to accept.

When the auctioneer accepts a higher bid, he or she rejects all previous bids. Because rejection terminates an offer (as will be discussed later), those bids represent offers that have been terminated. Thus, if the highest bidder withdraws his or her bid before the hammer falls, none of the previous bids is reinstated. If the bid is not withdrawn or rejected, the contract is formed when the auctioneer announces, "Going once, going twice, sold!" (or something similar) and lets the hammer fall.

Traditionally, auctions have been either "with reserve" or "without reserve." In an auction with reserve, the seller (through the auctioneer) may withdraw the goods at any time before the auctioneer closes the sale by announcement or by the fall of the hammer. All auctions are assumed to be auctions with reserve unless the terms of the auction are explicitly stated to be *without reserve*. In an auction without reserve, the goods cannot be withdrawn by the seller and must be sold to the highest bidder. In auctions with reserve, the seller may reserve the right to confirm or reject the sale even after "the hammer has fallen." In this situation, the seller is obligated to notify those attending the auction that sales of goods made during the auction are not final until confirmed by the seller.[6] (For a discussion of a case involving an online auction, see this chapter's *Adapting the Law to the Online Environment* feature.)

6. These rules apply under both the common law of contracts and the Uniform Commercial Code, or UCC. See UCC 2–328.

ADAPTING THE LAW TO THE ONLINE ENVIRONMENT

Can an Online Bid Constitute an Acceptance?

Under the Uniform Commercial Code, or UCC (see Chapter 14), a bid at an auction constitutes an offer. The offer (the highest bid) is accepted when the auctioneer's hammer falls. The UCC also states that auctions are "with reserve" unless the seller specifies otherwise. As just noted, in an auction with reserve, the seller reserves the right not to sell the goods to the highest bidder. Hence, even after the hammer falls, the contract for sale remains conditioned on the seller's approval. The question of how these rules should be applied to an online auction of a domain name, in which no hammer falls, came before a California court.

THE BID (OR OFFER?)

The case involved an online auction conducted by The.TV Corporation International (DotTV) on its Web site. DotTV posted an announcement on its Web site asking for bids for rights to the "Golf.tv" domain name and stating that the name would go to the highest bidder. Je Ho Lim submitted a bid for $1,010 and authorized DotTV to charge that amount to his credit card if his bid was the highest. Later, DotTV sent Lim an e-mail message stating that he had "won the auction" and charged the bid price of $1,010 to Lim's credit card. When DotTV subsequently refused to transfer the name, Lim sued DotTV for, among other things, breach of contract. Lim argued that his bid constituted an acceptance of DotTV's offer to sell the name. DotTV contended that Lim's bid was an offer, which it had not accepted. Furthermore, even if it had accepted Lim's offer, because the auction was "with reserve," DotTV could withdraw the domain name from the auction even after acceptance. The trial court held for DotTV, and Lim appealed.

THE APPELLATE COURT'S ANALYSIS

The appellate court first looked at the UCC's provisions concerning auctions, but noted that the UCC did not apply in this case because the UCC applies only to "goods" and domain names are not goods. The court then looked at common law principles as codified in the *Restatement (Second) of Contracts.* The rules under the *Restatement* are similar to those of the UCC: a bid in an auction is an offer that is accepted when the "hammer falls," and an auction is with reserve unless otherwise specified by the seller.

The court also pointed out, however, that DotTV's charging of the bid price to Lim's credit card was inconsistent with DotTV's claim that it could withdraw the domain name from the bidding because the auction was with reserve. Furthermore, stated the court, even if it concluded that Lim's bid was an offer and not an acceptance, DotTV had accepted the offer by its e-mail to Lim stating that he had won the auction. In all, held the court, there was no evidence that a contract between DotTV and Lim had *not* been formed, and Lim had stated a valid claim against DotTV for breach of contract. The court thus reversed the lower court's decision and remanded the case for further deliberation consistent with the appellate court's opinion.[a]

FOR CRITICAL ANALYSIS

Should the UCC rules governing auctions apply to items sold on online auction sites, such as eBay? Why or why not? How can you know whether eBay's auctions are "with reserve" or "without reserve"?

a. *Lim v. The.TV Corp. International,* 99 Cal.App.4th 684, 121 Cal.Rptr.2d 333 (2d Dist. 2002).

MANAGEMENT PERSPECTIVE

The Enforcement of Preliminary Agreements

MANAGEMENT FACES A LEGAL ISSUE

Suppose that at some point during contract negotiations, the parties proclaim that they have "made a deal." Does their agreement mean that an enforceable contract has been formed even though the parties have not signed a formal contract? Or does their agreement simply mean that they have "agreed to agree" to a contract in the future? How the courts interpret such situations can, of course, have significant consequences for business owners and managers.

WHAT THE COURTS SAY

Increasingly, the courts are holding that a preliminary agreement constitutes a binding contract if all essential terms have been agreed on and no disputed issues remain to be resolved. In contrast, if the parties agree on certain major terms but leave other terms open for further negotiation, a preliminary agreement is binding only in the sense that the parties have committed themselves to negotiate the undecided terms in good faith in an effort to reach a final agreement.

Fluorogas, Ltd., learned about this distinction to its dismay when a federal district court in Texas held that a preliminary agreement that it had formed with Fluorine On Call, Ltd., was a binding contract. After executives of the two companies had enjoyed a weekend of yachting in the Florida Keys, the executives drew up a brief handwritten document stating that Fluorogas would sell to Fluorine the exclusive rights to a technology to build and sell sophisticated semiconductor equipment. When Fluorogas refused to transfer the patents and intellectual property at issue to Fluorine, Fluorine sued for breach of contract. Was there a contract? Yes, according to the court. Because the handwritten document included the essential terms of the agreement, the document constituted a contract, not an agreement to agree to form a contract at some point in the future.[a]

IMPLICATIONS FOR MANAGERS

Business owners and managers should exercise care when forming preliminary agreements. When engaging in preliminary negotiations, managers should be aware that if all material terms are agreed on, they may be bound in contract even though they have not yet drawn up a formal contract.

a. *Fluorine On Call, Ltd. v. Fluorogas Limited,* No. 01-CV-186 (W.D.Tex. 2002). This decision is not published in the *Federal Supplement.*

—Agreements to Agree. Traditionally, agreements to agree—that is, agreements to agree to the material terms of a contract at some future date—were not considered to be binding contracts. The modern view, however, is that agreements to agree may be enforceable agreements (contracts) if it is clear that the parties intend to be bound by the agreements. In other words, under the modern view the emphasis is on the parties' intent rather than on form.

■ EXAMPLE 8.5 When the Pennzoil Company discussed with the Getty Oil Company the possible purchase of Getty's stock, a memorandum of agreement was drafted to reflect the terms of the conversations. After more negotiations over the price, both companies issued press releases announcing an agreement in principle on the terms of the memorandum. The next day, Texaco, Inc., offered to buy all of Getty's stock at a higher price. The day after that, Getty's board of directors voted to accept Texaco's offer, and Texaco and Getty signed a merger agreement. When Pennzoil sued Texaco for tortious interference with its "contractual" relationship with Getty, a jury concluded that Getty and Pennzoil had intended to form a binding contract, with only the details left to be worked out, before Texaco made its offer. Texaco was held liable for wrongfully interfering with this contract.[7] ■

Suppose that two parties draw up a brief, handwritten memorandum of agreement, intending to create a more formal document later. If the arrangements are never formalized, can the memorandum constitute a binding contract? See this chapter's *Management Perspective* feature for a discussion of this question.

Definiteness The second requirement for an effective offer involves the definiteness of its terms. An offer must have reasonably definite terms so that a court can determine if a breach has occurred and give an appropriate remedy.[8]

7. *Texaco, Inc. v. Pennzoil Co.,* 729 S.W.2d 768 (Tex.App.—Houston [1st Dist.] 1987, writ ref'd n.r.e.). (Generally, a complete Texas Court of Appeals citation includes the writ-of-error history showing the Texas Supreme Court's disposition of the case. In this case, *writ ref'd n.r.e.* is an abbreviation for "writ refused, no reversible error," which means that Texas's highest court refused to grant the appellant's request to review the case, because the court did not think there was any reversible error.)

8. *Restatement (Second) of Contracts,* Section 33. The UCC has relaxed the requirements regarding the definiteness of terms in contracts for the sale of goods. See UCC 2–204(3).

An offer may invite an acceptance to be worded in such specific terms that the contract is made definite. ■ EXAMPLE 8.6 Suppose that Marcus Business Machines contacts your corporation and offers to sell "from one to ten MacCool copying machines for $1,600 each; state number desired in acceptance." Your corporation agrees to buy two copiers. Because the quantity is specified in the acceptance, the terms are definite, and the contract is enforceable. ■

In the following case, the court had to decide whether the terms of an agreement were sufficiently definite to create an enforceable contract.

CASE 8.2 Satellite Entertainment Center, Inc. v. Keaton

Superior Court of New Jersey,
Appellate Division, 2002.
347 N.J.Super. 268,
789 A.2d 662.

FACTS In 1993, John Keaton decided to open a barbecue restaurant in Jersey City, New Jersey, and entered into a six-year lease with George Williams to occupy a portion of Williams's building. After Williams died, Morris Winograd, the owner of Satellite Entertainment Center, Inc., bought the building that included Keaton's restaurant. Winograd planned to renovate the entire premises to open a new restaurant and bar. In September 1995, Winograd asked Keaton how much it would cost to buy his business. Keaton named a price of $175,000. Keaton later claimed, as corroborated by witnesses, that Winograd said he would pay that amount, that he wanted Keaton out by the end of the year, and that he wanted Keaton to manage the new enterprise. Keaton vacated the premises by December. In January, Winograd began paying Keaton a salary, but did not pay him the $175,000, despite repeated requests. In April 1997, Winograd terminated Keaton. In a subsequent claim in a New Jersey state court against Satellite and Winograd, Keaton sought the $175,000. Winograd denied agreeing to pay Keaton anything. The court ruled in Keaton's favor. Winograd appealed to a state intermediate appellate court, claiming in part that the alleged agreement should not be enforced because it did not include the essential terms of an enforceable contract.

ISSUE Did the agreement include the essential terms of an enforceable contract?

DECISION Yes. The state intermediate appellate court affirmed the judgment of the lower court on Keaton's claim for $175,000. The essential terms of the agreement for the sale of Keaton's business could be determined, and thus there was an enforceable contract between the parties.

REASON The court explained that the agreement firmly set a price of $175,000 and clearly described what Winograd was purchasing. Winograd agreed to buy all of Keaton's business, including whatever tangible assets, inventory, or "goodwill" might be involved. As the court pointed out, however, "none of those assets were particularly significant to Winograd. Thus, it is not surprising that the parties did not, for example, itemize with specificity the inventory or the furniture of Keaton's business which was to be turned over to Winograd. To Winograd, those details were unimportant. The critical point, and the real reason for Winograd's payment of $175,000, was Keaton's agreement to vacate the property by the end of 1995, which he did." Any missing incidental terms can be implied. "And that is particularly true when there has been part performance of the contract, or—as here—where one of the parties (Keaton) has fully performed his part of the bargain."

FOR CRITICAL ANALYSIS—Economic Consideration *Suppose that the assets, inventory, and "goodwill" belonging to Keaton's business had been important to Winograd. In this situation, would the court have held that an enforceable contract existed? Why or why not?*

Communication A third requirement for an effective offer is communication—the offer must be communicated to the offeree. ■ EXAMPLE 8.7 Suppose that Tolson advertises a reward for the return of her lost cat. Dirlik, not knowing of the reward, finds the cat and returns it to Tolson. Ordinarily, Dirlik cannot recover the reward because an essential element of a reward contract is that the one who claims the reward must have known it was offered. A few states would allow recovery of the reward, but not on contract principles—Dirlik would be allowed to recover on the basis that it would be unfair to deny him the reward just because he did not know about it. ■

TERMINATION OF THE OFFER

The communication of an effective offer to an offeree gives the offeree the power to transform the offer into a binding, legal obligation (a contract) by an acceptance. This power of acceptance, however, does not continue forever. It can be terminated by action of the parties or by operation of law.

Termination by Action of the Parties An offer can be terminated by the action of the parties in any of three ways: by revocation, by rejection, or by counteroffer.

—Revocation of the Offer. The offeror's act of withdrawing an offer is referred to as **revocation.** Unless an offer is irrevocable, the offeror usually can revoke the offer (even if he or she has promised to keep the offer open), as long as the revocation is communicated to the offeree before the offeree accepts. Revocation may be accomplished by an express repudiation of the offer (for example, with a statement such as "I withdraw my previous offer of October 17") or by the performance of acts that are inconsistent with the existence of the offer and that are made known to the offeree.

■ EXAMPLE 8.8 Geraldine offers to sell some land to Gary. A week passes, and Gary, who has not yet accepted the offer, learns from his friend Konstantine that Geraldine has in the meantime sold the property to Nunan. Gary's knowledge of Geraldine's sale of the land to Nunan, even though he learned of it through a third party, effectively revokes Geraldine's offer to sell the land to Gary. Geraldine's sale of the land to Nunan is inconsistent with the continued existence of the offer to Gary, and thus the offer to Gary is revoked. ■

The general rule followed by most states is that a revocation becomes effective when the offeree or the offeree's agent (a person who acts on behalf of another) actually receives it. Therefore, a letter of revocation mailed on April 1 and delivered at the offeree's residence or place of business on April 3 becomes effective on April 3.

An offer made to the general public can be revoked in the same manner in which the offer was originally communicated. ■ EXAMPLE 8.9 Suppose that a department store offers a $10,000 reward to anyone giving information leading to the apprehension of the persons who burglarized its downtown store. The offer is published in three local papers and in four papers in neighboring communities. To revoke the offer, the store must publish the revocation in all seven papers for the same number of days it published the offer. The revocation is then accessible to the general public, and the offer is revoked even if some particular offeree does not know about the revocation. ■

—Irrevocable Offers. Although most offers are revocable, some can be made irrevocable. Increasingly, courts refuse to allow an offeror to revoke an offer when the offeree has changed position because of justifiable reliance on the offer (under the doctrine of *detrimental reliance,* or *promissory estoppel,* discussed later in this chapter). In some circumstances, "firm offers" made by merchants may also be considered irrevocable. We discuss these offers in Chapter 14.

Another form of irrevocable offer is an option contract. An **option contract** is created when an offeror promises to hold an offer open for a specified period of time in return for a payment (consideration) given by the offeree. An option contract takes away the offeror's power to revoke an offer for the period of time specified in the option. If no time is specified, then a reasonable period of time is implied. ■ EXAMPLE 8.10 Suppose that you are in the business of writing movie scripts. Your agent contacts the head of development at New Line Cinema and offers to sell New Line your new movie script. New Line likes your script and agrees to pay you $25,000 for a six-month option. In this situation, you (through your agent) are the offeror, and New Line is the offeree. You cannot revoke your offer to sell New Line your script for the next six months. If after six months no contract has been formed, however, New Line loses the $25,000, and you are free to sell your movie script to another firm. ■

Option contracts are also frequently used in conjunction with the sale of real estate. ■ EXAMPLE 8.11 You might agree with a landowner to lease a home and include in the lease contract a clause stating that you will pay $15,000 for an option to purchase the home within a specified period of time. If you decide not to purchase the home after the specified period has lapsed, you lose the $15,000, and the landlord is free to sell the property to another buyer. ■

—Rejection of the Offer by the Offeree. If the offeree rejects the offer, the offer is terminated. Any subsequent attempt by the offeree to accept will be construed as a new offer, giving the original offeror (now the offeree) the power of acceptance. A rejection is ordinarily accomplished by words or by conduct evidencing an intent not to accept the offer.

As with revocation, rejection of an offer is effective only when it is actually received by the offeror or the offeror's agent. ■ EXAMPLE 8.12 Suppose that Growgood Farms mails a letter to Campbell Soup Company offering to sell carrots at ten cents a pound. (Of course, today, such offers tend to be sent electronically rather than by mail, as will be discussed in Chapter 13.) Campbell Soup Company could reject the offer by sending or faxing a letter to Growgood Farms expressly rejecting the offer, or by mailing the offer back to Growgood, evidencing an intent to reject it. ■

Merely inquiring about an offer does not constitute rejection. ■ EXAMPLE 8.13 A friend offers to buy your CD-ROM library for $300. You respond, "Is this your best offer?" or "Will you pay me $375 for it?" A reasonable person would conclude that you did not reject the

offer but merely made an inquiry for further consideration of the offer. You can still accept and bind your friend to the $300 purchase price. When the offeree merely inquires as to the firmness of the offer, there is no reason to presume that she or he intends to reject it. ■

—Counteroffer by the Offeree. A **counteroffer** is a rejection of the original offer and the simultaneous making of a new offer. ■ EXAMPLE 8.14 Suppose that Burke offers to sell his home to Lang for $170,000. Lang responds, "Your price is too high. I'll offer to purchase your house for $165,000." Lang's response is called a counteroffer because it rejects Burke's offer to sell at $170,000 and creates a new offer by Lang to purchase the home at a price of $165,000. ■

At common law, the **mirror image rule** requires that the offeree's acceptance match the offeror's offer exactly. In other words, the terms of the acceptance must "mirror" those of the offer. If the acceptance materially changes or adds to the terms of the original offer, it will be considered not an acceptance but a counteroffer—which, of course, need not be accepted. The original offeror can, however, accept the terms of the counteroffer and create a valid contract.[9]

Termination by Operation of Law The offeree's power to transform an offer into a binding, legal obligation can be terminated by operation of law if any of four conditions occur: (1) lapse of time, (2) destruction of the specific subject matter, (3) death or incompetence of the offeror or offeree, or (4) supervening illegality of the proposed contract.

—Lapse of Time. An offer terminates automatically by law when the period of time *specified in the offer* has passed. If the offer states that it will be left open until a particular date, then the offer will terminate at midnight on that day. If the offer states that it will be left open for a number of days, such as ten days, this time period normally begins to run when the offer is actually received by the offeree, not when it is formed or sent. When the offer is delayed (through the misdelivery of mail, for example), the period begins to run from the date the offeree would have received the offer, but only if the offeree knows or should know that the offer is delayed.[10]

■ EXAMPLE 8.15 Suppose that Beth offers to sell her boat to Jonah, stating that the offer will remain open until May 20. Unless Jonah accepts the offer by midnight on May 20, the offer will lapse (terminate). Now suppose that Beth writes a letter to Jonah, offering to sell him her boat if Jonah accepts the offer within twenty days of the letter's date, which is May 1. Jonah must accept within twenty days after May 1, or the offer will terminate. The same rule would apply even if Beth had used improper postage when mailing the offer, and Jonah received the letter ten days after May 1, not knowing of the improper mailing. If, however, Jonah knew about the improper mailing, the offer would lapse twenty days after the day Jonah ordinarily would have received the offer had Beth used proper postage. ■

If the offer does not specify a time for acceptance, the offer terminates at the end of a *reasonable* period of time. A reasonable period of time is determined by the subject matter of the contract, business and market conditions, and other relevant circumstances. An offer to sell farm produce, for example, will terminate sooner than an offer to sell farm equipment because farm produce is perishable and subject to greater fluctuations in market value.

—Destruction of the Subject Matter. An offer is automatically terminated if the specific subject matter of the offer is destroyed before the offer is accepted. For example, if Bekins offers to sell his prize cow to Yatsen, but the cow is struck by lightning and dies before Yatsen can accept, the offer is automatically terminated. (Note that if Yatsen accepted the offer just before lightning struck the cow, a contract would have been formed, but, because of the cow's death, a court would likely excuse Bekins's obligation to perform the contract on the basis of impossibility of performance—see Chapter 11.)

—Death or Incompetence of the Offeror or Offeree. An offeree's power of acceptance is terminated when the offeror or offeree dies or is deprived of legal capacity to enter into the proposed contract, *unless the offer is irrevocable.*[11] An offer is personal to both parties and normally cannot pass to the decedent's heirs, guardian, or estate. This rule applies whether or not one party had notice of the death or incompetence of the other party.

9. The mirror image rule has been greatly modified in regard to sales contracts. Section 2–207 [Section 2–206(3) in the 2003 amendments] of the UCC provides that a contract is formed if the offeree makes a definite expression of acceptance (such as signing the form in the appropriate location), even though the terms of the acceptance modify or add to the terms of the original offer (see Chapter 14).

10. *Restatement (Second) of Contracts,* Section 49.

11. *Restatement (Second) of Contracts,* Section 48. If the offer is irrevocable, it is not terminated when the offeror dies. Also, if the offer is such that it can be accepted by the performance of a series of acts, and those acts began before the offeror died, the offeree's power of acceptance is not terminated.

■ EXAMPLE 8.16 Kapola, who is quite ill, writes to her friend Amanda, offering to sell Amanda her grand piano for only $400. That night, Kapola dies. A few days later, Amanda, not knowing of Kapola's death, writes a letter to Kapola accepting the offer and enclosing a check for $400. Is there a contract? No. There is no contract because the offer automatically terminated on Kapola's death. ■

—Supervening Illegality of the Proposed Contract. A statute or court decision that makes an offer illegal automatically terminates the offer. ■ EXAMPLE 8.17 If Acme Finance Corporation offers to lend Jack $20,000 at 15 percent annually, and the state legislature enacts a statute prohibiting loans at interest rates greater than 12 percent before Jack can accept, the offer is automatically terminated. (If the statute is enacted after Jack accepts the offer, a valid contract is formed, but the contract may still be unenforceable—see Chapter 9.) ■

ACCEPTANCE

An **acceptance** is a voluntary act by the offeree that shows assent, or agreement, to the terms of an offer. The offeree's act may consist of words or conduct. The acceptance must be unequivocal and must be communicated to the offeror.

Who Can Accept? Generally, a third person cannot substitute for the offeree and effectively accept the offer. After all, the identity of the offeree is as much a condition of a bargaining offer as any other term contained therein. Thus, except in special circumstances, only the person to whom the offer is made or that person's agent can accept the offer and create a binding contract. For example, Lottie makes an offer to Paul. Paul is not interested, but Paul's friend José accepts the offer. No contract is formed.

Unequivocal Acceptance To exercise the power of acceptance effectively, the offeree must accept unequivocally. This is the *mirror image rule* previously discussed. If the acceptance is subject to new conditions or if the terms of the acceptance materially change the original offer, the acceptance may be deemed a counteroffer that implicitly rejects the original offer.

Certain terms, when added to an acceptance, will not qualify the acceptance sufficiently to constitute rejection of the offer. ■ EXAMPLE 8.18 Suppose that in response to a person offering to sell a painting by a well-known artist, the offeree replies, "I accept; please send a written contract." The offeree is requesting a written contract but is not making it a condition for acceptance. Therefore, the acceptance is effective without the written contract. If the offeree replies, "I accept *if* you send a written contract," however, the acceptance is expressly conditioned on the request for a writing, and the statement is not an acceptance but a counteroffer. (Notice how important each word is!)[12] ■

Silence as Acceptance Ordinarily, silence cannot constitute acceptance, even if the offeror states, "By your silence and inaction, you will be deemed to have accepted this offer." This general rule applies because an offeree should not be put under a burden of liability to act affirmatively in order to reject an offer. No consideration—that is, nothing of value—has passed to the offeree to impose such a liability.

In some instances, however, the offeree does have a duty to speak; if so, his or her silence or inaction will operate as an acceptance. Silence may be an acceptance when an offeree takes the benefit of offered services even though he or she had an opportunity to reject them and knew that they were offered with the expectation of compensation. ■ EXAMPLE 8.19 Suppose that John, a college student who earns extra income by washing store windows, taps on the window of a store and catches the attention of the store's manager. John points to the window and raises his cleaner, signaling that he will be washing the window. The manager does nothing to stop him. Here, the store manager's silence constitutes an acceptance, and an implied-in-fact contract is created. The store is bound to pay a reasonable value for John's work. ■

Silence can also operate as an acceptance when the offeree has had prior dealings with the offeror. If a merchant, for example, routinely receives shipments from a supplier and in the past has always notified the supplier when defective goods are rejected, then silence constitutes acceptance. Also, if a buyer solicits an offer specifying that certain terms and conditions are acceptable, and the seller makes the offer in response to the solicitation, the buyer has a duty to reject—that is, a duty to tell the seller that the offer is not acceptable. Failure to reject (silence) would operate as an acceptance.

Communication of Acceptance Whether the offeror must be notified of the acceptance depends on the nature of the contract. In a bilateral contract, communication of acceptance is necessary because acceptance is in the form of a promise (not performance), and the contract is formed when the promise is made (rather than when the act is performed). Communication of acceptance is not necessary, however, if the offer dispenses with the require-

12. As stated in footnote 9, in regard to sales contracts, the UCC provides that an acceptance may still be effective even if some terms are added. The new terms are simply treated as proposals for additions to the contract, unless both parties are merchants—in which case the additional terms (with some exceptions) become part of the contract [UCC 2–207(2)].

ment. Also, if the offer can be accepted by silence, no communication is necessary.[13]

In a unilateral contract, the full performance of some act is called for; therefore, acceptance is usually evident, and notification is unnecessary. Exceptions do exist, however. When the offeror requests notice of acceptance or has no adequate means of determining whether the requested act has been performed, or when the law (such as Article 2 of the UCC) requires notice of acceptance, then notice is necessary.[14]

Mode and Timeliness of Acceptance The general rule is that acceptance in a bilateral contract is timely if it is effected within the duration of the offer. Problems arise, however, when the parties involved are not dealing face to face. In such situations, the offeree may use an authorized mode of communication.

—The Mailbox Rule. Acceptance takes effect, thus completing formation of the contract, at the time the offeree sends or delivers the communication via the mode expressly or impliedly authorized by the offeror. This is the so-called **mailbox rule,** also called the "deposited acceptance rule," which the majority of courts uphold. Under this rule, if the authorized mode of communication is the mail, then an acceptance becomes valid when it is dispatched (placed in the control of the U.S. Postal Service)—*not* when it is received by the offeror.

The mailbox rule was formed to prevent the confusion that arises when an offeror sends a letter of revocation but, before it arrives, the offeree sends a letter of acceptance. Thus, whereas a revocation becomes effective only when it is *received* by the offeree, an acceptance becomes effective on *dispatch* (even if it is never received), provided that an *authorized* means of communication is used.

—Authorized Means of Communication. Authorized means of communicating an acceptance can be either expressly authorized—that is, expressly stipulated in the offer—or impliedly authorized by facts or law.[15] An acceptance sent by means not expressly or impliedly authorized is normally not effective until it is received by the offeror.

When an offeror specifies how acceptance should be made (for example, by first class mail or express delivery), *express authorization* is said to exist. Moreover, both the offeror and the offeree are bound in contract the moment that such means of acceptance are employed. Most offerors do not expressly specify the means by which the offeree is to accept. Thus, the common law recognizes the following implied authorized means of acceptance:[16]

1. The offeror's choice of a particular means in making the offer implies that the offeree is authorized to use the same *or a faster* means for acceptance.
2. When two parties are at a distance, acceptance by mail is impliedly authorized.

—Exceptions. There are three basic exceptions to the rule that a contract is formed when acceptance is sent by authorized means:

1. If the acceptance is not properly dispatched (if a letter is incorrectly addressed, for example, or lacks the proper postage), in most states it will not be effective until it is received by the offeror.
2. The offeror can stipulate in the offer that an acceptance will not be effective until it is received (usually by a specified time) by the offeror.
3. Sometimes an offeree sends a rejection first, then later changes his or her mind and sends an acceptance. Obviously, this chain of events could cause confusion and even detriment to the offeror, depending on whether the rejection or the acceptance arrives first. In such situations, the law cancels the rule of acceptance on dispatch, and the first communication received by the offeror determines whether a contract is formed. If the rejection arrives first, there is no contract.[17]

Technology and Acceptances Technology, and particularly the Internet, has all but eliminated the need for the mailbox rule because online acceptances typically are communicated instantaneously to the offeror. As you will learn in Chapter 13, while online offers are not significantly different from traditional offers contained in paper documents, online acceptances have posed some unusual problems.

13. Under the UCC, an order or other offer to buy goods that are to be promptly shipped may be treated as either a bilateral or a unilateral offer and can be accepted by a promise to ship or by actual shipment. See UCC 2–206(1)(b).

14. UCC 2–206(2).

15. *Restatement (Second) of Contracts,* Section 30, provides that an offer invites acceptance "by any medium reasonable in the circumstances," unless the offer is specific about the means of acceptance. Under Section 65, a medium is reasonable if it is one used by the offeror or one customary in similar transactions, unless the offeree knows of circumstances that would argue against the reasonableness of a particular medium (the need for speed because of rapid price changes, for example).

16. Note that UCC 2–206(1)(a) states specifically that an acceptance of an offer for the sale of goods can be made by any medium that is *reasonable* under the circumstances.

17. *Restatement (Second) of Contracts,* Section 40.

Consideration and Its Requirements

In every legal system, some promises will be enforced, and other promises will not be enforced. The simple fact that a party has made a promise, then, does not mean the promise is enforceable. Under the common law, a primary basis for the enforcement of promises is *consideration*. **Consideration** is usually defined as the value given in return for a promise. We look here at the basic elements of consideration and then at some other contract doctrines relating to consideration.

ELEMENTS OF CONSIDERATION

Often, consideration is broken down into two parts: (1) something of *legally sufficient value* must be given in exchange for the promise, and (2) there must be a *bargained-for exchange.*

Legally Sufficient Value The "something of legally sufficient value" may consist of (1) a promise to do something that one has no prior legal duty to do (to pay money on receipt of certain goods, for example), (2) the performance of an action that one is otherwise not obligated to undertake (such as providing accounting services), or (3) the refraining from an action that one has a legal right to undertake (called a **forbearance**).

Consideration in bilateral contracts normally consists of a promise in return for a promise, as explained in Chapter 7. ■ EXAMPLE 8.20 Suppose that in a contract for the sale of goods, the seller promises to ship specific goods to the buyer, and the buyer promises to pay for those goods when they are received. Each of these promises constitutes consideration for the contract. ■ In contrast, unilateral contracts involve a promise in return for a performance. ■ EXAMPLE 8.21 Suppose that Anita says to her neighbor, "If you paint my garage, I will pay you $100." Anita's neighbor paints the garage. The act of painting the garage is the consideration that creates Anita's contractual obligation to pay her neighbor $100. ■

What if, in return for a promise to pay, a person forbears to pursue harmful habits, such as the use of tobacco and alcohol? Does such forbearance create consideration for the contract? In *Hamer v. Sidway,*[18] a classic case on this issue, the court held that refraining from doing what one has a legal right to do can constitute consideration.

Bargained-for Exchange The second element of consideration is that it must provide the basis for the bargain struck between the contracting parties. The promise given by the promisor must induce the promisee to incur a legal detriment either now or in the future, and the detriment incurred must induce the promisor to make the promise. This element of bargained-for exchange distinguishes contracts from gifts.

■ EXAMPLE 8.22 Suppose that Roberto says to his son, "In consideration of the fact that you are not as wealthy as your brothers, I will pay you $5,000." This promise is not enforceable because Roberto's son has not given any return consideration for the $5,000 promised.[19] The son (the promisee) incurs no legal detriment; he does not have to promise anything or undertake (or refrain from undertaking) any action to receive the $5,000. Here, Roberto has simply stated his motive for giving his son a gift. The fact that the word *consideration* is used does not, by itself, mean that consideration has been given. ■

LEGAL SUFFICIENCY AND ADEQUACY OF CONSIDERATION

Legal sufficiency of consideration involves the requirement that consideration be something of value in the eyes of the law. Adequacy of consideration involves "how much" consideration is given. Essentially, adequacy of consideration concerns the fairness of the bargain. On the surface, fairness would appear to be an issue when the items exchanged are of unequal value. In general, however, courts do not question the adequacy of consideration if the consideration is legally sufficient. Under the doctrine of freedom of contract, parties are usually free to bargain as they wish. If people could sue merely because they had entered into an unwise contract, the courts would be overloaded with frivolous suits.

In extreme cases, however, a court of law may look to the amount or value (the adequacy) of the consideration because apparently inadequate consideration can indicate that fraud, duress, or undue influence was involved or that a gift was made (if a father "sells" a $100,000 house to his daughter for only $1,000 for example). Additionally, when the consideration is grossly inadequate, the courts may declare the contract unenforceable on the ground that it is unconscionable,[20] meaning that, generally speaking, it is so one sided under the circumstances as to be clearly unfair. (*Unconscionability* will be discussed further in Chapter 9.)

In the following case, the issue was whether consideration existed in a contract to accept lower payments for medical services than the maximum fees allowed under state regulations.

18. 124 N.Y. 538, 27 N.E. 256 (1891).

19. See *Fink v. Cox,* 18 Johns. 145, 9 Am.Dec. 191 (N.Y. 1820).

20. Pronounced un-*kon*-shun-uh-bul.

CASE 8.3 Seaview Orthopaedics v. National Healthcare Resources, Inc.

Superior Court of New Jersey,
Appellate Division, 2004.
366 N.J.Super. 501,
841 A.2d 917.
http://lawlibrary.rutgers.edu/search.shtml[a]

FACTS Consumer Health Network (CHN) is a large medical-insurance preferred provider organization (PPO) in New Jersey. CHN provides medical services providers (its clients) with a PPO network in three distinct areas: workers' compensation, group health benefits, and auto insurance. The network includes over 11,000 physicians and nearly 14,000 medical services providers (which include physicians, laboratories, and hospitals). By entering into a contract with CHN, a medical services provider gains potential access to 950,000 enrollees in exchange for accepting reimbursement at rates lower than the maximum rates permitted by New Jersey state regulations. Seaview Orthopaedics is a CHN client that renders services to auto accident victims insured by Allstate Indemnity Company. Allstate pays the CHN rates for Seaview's services through Allstate's claims administrator, National Healthcare Resources (NHR). Seaview and others filed a suit in a New Jersey state court against NHR and others to recover the difference between the CHN rates and the state's maximum rates. The court issued a summary judgment in favor of the defendants. The plaintiffs appealed to a state intermediate appellate court, claiming that the CHN contract was not enforceable, in part because it lacked consideration.

ISSUE Did the CHN contract lack consideration?

a. Click on "Search by Party Name." On that page, select "Appellate Division" as the court, and type "Seaview Orthopaedics" in the "First Name" box. Click on "Submit Form" to view a synopsis of the case, and then "click here to get this case" to read the opinion. Rutgers University School of Law in Camden, New Jersey, maintains this Web site.

DECISION No. The state intermediate appellate court affirmed the lower court's summary judgment.

REASON The court acknowledged that contracts are not enforceable in the absence of consideration—that is, both sides must get something out of the exchange. The court explained, however, that the determination of whether consideration exists does not depend on the comparative value of the things exchanged. Instead, stated the court, the consideration must merely be valuable in the sense that it is something that is bargained for in fact. The court noted, "Here, the contract provided benefits to plaintiffs in a variety of ways which either collectively or separately constituted valuable consideration for plaintiffs' promise to accept the CHN rates for reimbursement from auto accident victims (and other types of patients) and not the maximum rate permitted by the [state's] schedule. Plaintiffs, for example, obtained the benefit of marketing their businesses in a directory of providers utilized by numerous payors in the workers' compensation and health benefits markets and many thousands of potential patients. Payors make the list available to the largest PPO membership network in New Jersey and, in the health and workers' compensation settings, are generally offered substantial financial incentives when those patients use the providers on the list." The court concluded that it is the totality of the exchange of promises and benefits that must be considered, and in this case, the exchange was sufficient to create an enforceable contract.

WHY IS THIS CASE IMPORTANT? *As this case illustrates, something need not be of direct economic or financial value to be considered legally sufficient consideration. In many situations, as here, the exchange of promises and potential benefits is deemed sufficient consideration.*

Contracts That Lack Consideration

Sometimes, one or both of the parties to a contract may think that they have exchanged consideration when in fact they have not. Here we look at some situations in which the parties' promises or actions do not qualify as contractual consideration.

PREEXISTING DUTY

Under most circumstances, a promise to do what one already has a legal duty to do does not constitute legally sufficient consideration because no legal detriment is incurred.[21] The preexisting legal duty may be imposed by law or may arise out of a previous contract. A sheriff, for example, cannot collect a reward for information leading to the capture of a criminal if the sheriff already has a legal duty to capture the criminal. Likewise, if a party is already bound by contract to perform a certain duty, that duty cannot serve as consideration for a second contract.

■ **EXAMPLE 8.23** Suppose that Bauman-Bache, Inc., begins construction on a seven-story office building and after three months demands an extra $75,000 on its contract. If the extra $75,000 is not paid, it will stop working.

21. See *Foakes v. Beer*, 9 App.Cas. 605 (1884).

The owner of the land, having no one else to complete construction, agrees to pay the extra $75,000. The agreement is not enforceable because it is not supported by legally sufficient consideration; Bauman-Bache had a preexisting contractual duty to complete the building. ■

Unforeseen Difficulties The rule regarding preexisting duty is meant to prevent extortion and the so-called holdup game. What happens, though, when an honest contractor, who has contracted with a landowner to build a house, runs into extraordinary difficulties that were totally unforeseen at the time the contract was formed? In the interests of fairness and equity, the courts sometimes allow exceptions to the preexisting duty rule. In the example just mentioned, if the landowner agrees to pay extra compensation to the contractor for overcoming the unforeseen difficulties (such as having to use dynamite and special equipment to remove an unexpected rock formation to excavate for a basement), the court may refrain from applying the preexisting duty rule and enforce the agreement. When the "unforeseen difficulties" that give rise to a contract modification are the types of risks ordinarily assumed in business, however, the courts will usually assert the preexisting duty rule.[22]

Rescission and New Contract The law recognizes that two parties can mutually agree to rescind their contract, at least to the extent that it is executory (still to be carried out). **Rescission**[23] is defined as the unmaking of a contract so as to return the parties to the positions they occupied before the contract was made. When rescission and the making of a new contract take place at the same time, the courts frequently are given a choice of applying the preexisting duty rule or allowing rescission and letting the new contract stand.

Past Consideration

Promises made in return for actions or events that have already taken place are unenforceable. These promises lack consideration in that the element of bargained-for exchange is missing. In short, you can bargain for something to take place now or in the future but not for something that has already taken place. Therefore, **past consideration** is no consideration.

■ EXAMPLE 8.24 Suppose that Elsie, a real estate agent, does her friend Judy a favor by selling Judy's house and not charging any commission. Later, Judy says to Elsie, "In return for your generous act, I will pay you $3,000." This promise is made in return for past consideration and is thus unenforceable; in effect, Judy is stating her intention to give Elsie a gift. ■

Illusory Promises

If the terms of the contract express such uncertainty of performance that the promisor has not definitely promised to do anything, the promise is said to be *illusory*—without consideration and unenforceable. ■ EXAMPLE 8.25 The president of Tuscan Corporation says to his employees, "All of you have worked hard, and if profits remain high, a 10 percent bonus at the end of the year will be given—if management thinks it is warranted." This is an *illusory promise,* or no promise at all, because performance depends solely on the discretion of the president (the management). There is no bargained-for consideration. The statement declares merely that management may or may not do something in the future. ■

Option-to-cancel clauses in contracts for specified time periods sometimes present problems in regard to consideration. ■ EXAMPLE 8.26 Abe contracts to hire Chris for one year at $5,000 per month, reserving the right to cancel the contract at any time. On close examination of these words, you can see that Abe has not actually agreed to hire Chris, as Abe could cancel without liability before Chris started performance. Abe has not given up the opportunity of hiring someone else. This contract is therefore illusory. Now suppose that Abe contracts to hire Chris for a one-year period at $5,000 per month, reserving the right to cancel the contract at any time after Chris has begun performance by giving Chris thirty days' notice. Abe, by saying that he will give Chris thirty days' notice, is relinquishing the opportunity (legal right) to hire someone else instead of Chris for a thirty-day period. If Chris works for one month, at the end of which Abe gives him thirty days' notice, Chris has a valid and enforceable contractual claim for $10,000 in salary. ■

Settlement of Claims

Businesspersons or others can settle legal claims in several ways. It is important to understand the nature of the consideration given in these settlement agreements, or contracts. Claims are commonly settled through an *accord and satisfaction,* in which a debtor offers to pay a lesser amount than the creditor purports to be owed. Two other methods that are also often used to settle claims are the *release* and the *covenant not to sue.*

Accord and Satisfaction

In an **accord and satisfaction,** a debtor offers to pay, and a creditor accepts, a lesser amount than the creditor originally claimed was owed. Thus, in an accord and satisfaction, the

22. Note that under the UCC, any agreement modifying a contract within Article 2 on Sales needs no consideration to be binding. See UCC 2–209(1).

23. Pronounced reh-*sih*-zhen.

debtor attempts to terminate an existing obligation. The *accord* is the settlement agreement. In an accord, the debtor offers to give or perform something less than the parties originally agreed on, and the creditor accepts that offer in satisfaction of the claim. *Satisfaction* is the performance (usually payment), which takes place after the accord is executed. A basic rule is that there can be no satisfaction unless there is first an accord.

For accord and satisfaction to occur, the amount of the debt *must be in dispute.* If a debt is *liquidated,* accord and satisfaction cannot take place. A debt is liquidated if its amount has been ascertained, fixed, agreed on, settled, or exactly determined. An example of a liquidated debt is a loan contract in which the borrower agrees to pay a stipulated amount every month until the amount of the loan is paid. In the majority of states, acceptance of (an accord for) a lesser sum than the entire amount of a liquidated debt is not satisfaction, and the balance of the debt is still legally owed. The rationale for this rule is that the debtor has given no consideration to satisfy the obligation of paying the balance to the creditor—because the debtor has a preexisting legal obligation to pay the entire debt.

An *unliquidated debt* is the opposite of a liquidated debt. Here, reasonable persons may differ over the amount owed. It is not settled, fixed, agreed on, ascertained, or determined. In these circumstances, acceptance of payment of the lesser sum operates as a satisfaction, or discharge, of the debt. One argument to support this rule is that the parties give up a legal right to contest the amount in dispute, and thus consideration is given.

RELEASE

A **release** is a contract in which one party forfeits the right to pursue a legal claim against the other party. Releases will generally be binding if they are (1) given in good faith, (2) stated in a signed writing (required by many states), and (3) accompanied by consideration.[24] Clearly, parties are better off if they know the extent of their injuries or damages before signing releases.

■ EXAMPLE 8.27 Suppose that you are involved in an automobile accident caused by Raoul's negligence. Raoul offers to give you $1,000 if you will release him from further liability resulting from the accident. You believe that this amount will cover your damages, so you agree to and sign the release. Later you discover that the repairs to your car will cost $1,200. Can you collect the balance from Raoul? The answer is normally no; you are limited to the $1,000 in the release. Why? The reason is that a valid contract existed. You and Raoul both assented to the bargain (hence, agreement existed), and sufficient consideration was present. Your consideration for the contract was the legal detriment you suffered (by releasing Raoul from liability, you forfeited your right to sue to recover damages, should they be more than $1,000). This legal detriment was induced by Raoul's promise to give you the $1,000. Raoul's promise was, in turn, induced by your promise not to pursue your legal right to sue him for damages. ■

24. Under the UCC, a written, signed waiver or renunciation by an aggrieved party discharges any further liability for a breach, even without consideration [UCC 1–107].

COVENANT NOT TO SUE

Unlike a release, a **covenant not to sue** does not always bar further recovery. The parties simply substitute a contractual obligation for some other type of legal action based on a valid claim. Suppose (following the earlier example) that you agree with Raoul not to sue for damages in a tort action if he will pay for the damage to your car. If Raoul fails to pay, you can bring an action for breach of contract.

Promissory Estoppel

Sometimes individuals rely on promises, and such reliance may form a basis for contract rights and duties. Under the doctrine of **promissory estoppel** (also called *detrimental reliance*), a person who has reasonably relied on the promise of another can often obtain some measure of recovery. When the doctrine of promissory estoppel is applied, the promisor (the offeror) is **estopped** (barred, or impeded) from revoking the promise. For the doctrine of promissory estoppel to be applied, the following elements are required:

1. There must be a clear and definite promise.
2. The promisee must justifiably rely on the promise.
3. The reliance normally must be of a substantial and definite character.
4. Justice will be better served by the enforcement of the promise.

■ EXAMPLE 8.28 Your uncle tells you, "I'll pay you $150 a week so you won't have to work anymore." In reliance on your uncle's promise, you quit your job, but your uncle refuses to pay you. Under the doctrine of promissory estoppel, you may be able to enforce his promise.[25] Now your uncle makes a promise to give you $10,000 to buy a car. If you buy the car with your own funds and he does not pay you, you may once again be able to enforce the promise under this doctrine. ■

25. A classic example is *Ricketts v. Scothorn,* 57 Neb. 51, 77 N.W. 365 (1898).

TERMS AND CONCEPTS

acceptance 164
accord and satisfaction 168
agreement 156
consideration 166
counteroffer 163
covenant not to sue 169
estopped 169
forbearance 166
mailbox rule 165
mirror image rule 163
offer 157
option contract 162
past consideration 168
promissory estoppel 169
release 169
rescission 168
revocation 162

CHAPTER SUMMARY Agreement and Consideration

Requirements of the Offer (See pages 157–161.)	1. *Intent*—There must be a serious, objective intention by the offeror to become bound by the offer. Nonoffer situations include (a) expressions of opinion; (b) statements of intention; (c) preliminary negotiations; (d) generally, advertisements, catalogues, price lists, and circulars; (e) solicitations for bids made by an auctioneer; and (f) traditionally, agreements to agree in the future. 2. *Definiteness*—The terms of the offer must be sufficiently definite to be ascertainable by the parties or by a court. 3. *Communication*—The offer must be communicated to the offeree.
Termination of the Offer (See pages 161–164.)	1. *By action of the parties*— a. Revocation—Unless the offer is irrevocable, it can be revoked at any time before acceptance without liability. Revocation is not effective until received by the offeree or the offeree's agent. Some offers, such as a merchant's firm offer and option contracts, are irrevocable. b. Rejection—Accomplished by words or actions that demonstrate a clear intent not to accept the offer; not effective until received by the offeror or the offeror's agent. c. Counteroffer—A rejection of the original offer and the making of a new offer. 2. *By operation of law*— a. Lapse of time—The offer terminates (1) at the end of the time period specified in the offer or (2) if no time period is stated in the offer, at the end of a reasonable time period. b. Destruction of the specific subject matter of the offer—Automatically terminates the offer. c. Death or incompetence—Terminates the offer unless the offer is irrevocable. d. Illegality—Supervening illegality terminates the offer.
Acceptance (See pages 164–165.)	1. Can be made only by the offeree or the offeree's agent. 2. Must be unequivocal. Under the common law (mirror image rule), if new terms or conditions are added to the acceptance, it will be considered a counteroffer. 3. Acceptance of a unilateral offer is effective on full performance of the requested act. Generally, no communication is necessary. 4. Acceptance of a bilateral offer can be communicated by the offeree by any authorized mode of communication and is effective on dispatch. Unless the mode of communication is expressly specified by the offeror, the following methods are impliedly authorized: a. The same mode used by the offeror or a faster mode. b. Mail, when the two parties are at a distance. c. In sales contracts, by any reasonable medium.
Elements of Consideration (See page 166.)	Consideration is broken down into two parts: (1) something of *legally sufficient value* must be given in exchange for the promise, and (2) there must be a *bargained-for exchange.*

CHAPTER SUMMARY Agreement and Consideration—Continued

Legal Sufficiency and Adequacy of Consideration (See pages 166–167.)	Legal sufficiency of consideration relates to the first element of consideration—something of legal value must be given in exchange for a promise. Adequacy of consideration relates to "how much" consideration is given and whether a fair bargain was reached. Courts will inquire into the adequacy of consideration (whether the consideration is legally sufficient) only when fraud, undue influence, duress, or unconscionability may be involved.
Contracts That Lack Consideration (See pages 167–168.)	Consideration is lacking in the following situations: 1. *Preexisting duty*—Consideration is not legally sufficient if one is either by law or by contract under a preexisting duty to perform the action being offered as consideration for a new contract. 2. *Past consideration*—Actions or events that have already taken place do not constitute legally sufficient consideration. 3. *Illusory promises*—When the nature or extent of performance is too uncertain, the promise is rendered illusory (without consideration and unenforceable).
Settlement of Claims (See pages 168–169.)	1. *Accord and satisfaction*—An accord is an agreement in which a debtor offers to pay a lesser amount than the creditor purports to be owed. Satisfaction may take place when the accord is executed. 2. *Release*—An agreement in which, for consideration, a party forfeits the right to seek further recovery beyond the terms specified in the release. 3. *Covenant not to sue*—An agreement not to sue on a present, valid claim.
Promissory Estoppel (See page 169.)	The equitable doctrine of promissory estoppel applies when a promisor reasonably expects a promise to induce definite and substantial action or forbearance by the promisee, and the promisee does act in reliance on the promise. Such a promise is binding if injustice can be avoided only by enforcement of the promise. Also known as the doctrine of detrimental reliance.

FOR REVIEW

Answers for the even-numbered questions in this For Review *section can be found in Appendix A at the end of this text.*

1. What elements are necessary for an effective offer? What are some examples of nonoffers?
2. In what circumstances will an offer be irrevocable?
3. What are the elements that are necessary for an effective acceptance?
4. What is consideration? What is required for consideration to be legally sufficient?
5. In what circumstances might a promise be enforced despite a lack of consideration?

QUESTIONS AND CASE PROBLEMS

8–1. Offer. Chernek, the sole owner of a small business, has a large piece of used farm equipment for sale. He offers to sell the equipment to Bollow for $10,000. Discuss the legal effects of the following events on the offer:

(a) Chernek dies before Bollow's acceptance, and at the time she accepts, Bollow is unaware of Chernek's death.

(b) The night before Bollow accepts, a fire destroys the equipment.

(c) Bollow pays $100 for a thirty-day option to purchase the equipment. During this period, Chernek dies, and Bollow accepts the offer, knowing of Chernek's death.

(d) Bollow pays $100 for a thirty-day option to purchase the equipment. During this period, Bollow dies, and Bollow's estate accepts Chernek's offer within the stipulated time period.

QUESTION WITH SAMPLE ANSWER

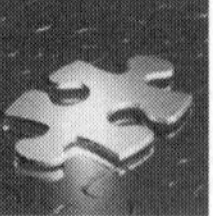

8–2. On June 1, Jason placed an ad in a local newspaper, to be run on the following Sunday, June 5, offering a reward of $100 to anyone who found his wallet. When his wallet had not been returned by June 12, he purchased another wallet and took steps to obtain

duplicates of his driver's license, credit cards, and other items that he had lost. He also placed another ad in the same newspaper revoking his offer. The second ad was the same size as the original. On June 15, Sharith, who had seen Jason's first ad in the paper, found Jason's wallet, returned it to Jason, and asked for the $100. Is Jason obligated to pay Sharith the $100? Why or why not?

For a sample answer to this question, go to Appendix B at the end of this text.

8–3. Offer and Acceptance. Carrie offered to sell a set of legal encyclopedias to Antonio for $300. Antonio said that he would think about her offer and let her know his decision the next day. Norvel, who had overheard the conversation between Carrie and Antonio, said to Carrie, "I accept your offer" and gave her $300. Carrie gave Norvel the books. The next day, Antonio, who had no idea that Carrie had already sold the books to Norvel, told Carrie that he accepted her offer. Has Carrie breached a valid contract with Antonio? Explain.

8–4. Consideration. Ben hired Lewis to drive his racing car in a race. Tuan, a friend of Lewis, promised to pay Lewis $3,000 if he won the race. Lewis won the race, but Tuan refused to pay the $3,000. Tuan contended that no legally binding contract had been formed because he had received no consideration from Lewis for his promise to pay the $3,000. Lewis sued Tuan for breach of contract, arguing that winning the race was the consideration given in exchange for Tuan's promise to pay the $3,000. What rule of law discussed in this chapter supports Tuan's claim? Explain.

CASE PROBLEM WITH SAMPLE ANSWER

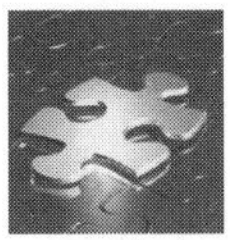

8–5. In 1995, Helikon Furniture Co. appointed Gaede as its independent sales agent for the sale of its products in parts of Texas. The parties signed a one-year contract that specified, among other things, the commissions that Gaede would receive. Over a year later, although the parties had not signed a new contract, Gaede was still representing Helikon when it was acquired by a third party. Helikon's new management allowed Gaede to continue to receive the same commissions and sent him a letter stating that it would make no changes in its sales representatives "for at least the next year." Three months later, in December 1996, the new managers sent Gaede a letter proposing new terms for a contract. Gaede continued to sell Helikon products until May 1997 when he received a letter effectively reducing the amount of his commissions. Gaede filed a suit in a Texas state court against Helikon, alleging breach of contract. Helikon argued in part that there was no contract because there was no consideration. In whose favor should the court rule, and why? [*Gaede v. SK Investments, Inc.*, 38 S.W.3d 753 (Tex.App.—Houston [14 Dist.] 2001)]

After you have answered this problem, compare your answer with the sample answer given on the Web site that accompanies this text. Go to academic.cengage.com/blaw/wbl, select "Chapter 8," and click on "Case Problem with Sample Answer."

8–6. Settlement of Claims. Shoreline Towers Condominium Owners Association in Gulf Shores, Alabama, authorized Resort Development, Inc. (RDI), to manage Shoreline's property. On Shoreline's behalf, RDI obtained a property insurance policy from Zurich American Insurance Co. In October 1995, Hurricane Opal struck Gulf Shores. RDI filed claims with Zurich regarding damage to Shoreline's property. Zurich determined that the cost of the damage was $334,901. Zurich then subtracted an applicable $40,000 deductible and sent checks totaling $294,901 to RDI. RDI disputed the amount. Zurich eventually agreed to issue a check for an additional $86,000 in return for RDI's signing a "Release of All Claims." Later, contending that the deductible had been incorrectly applied and that this was a breach of contract, among other things, Shoreline filed a suit against Zurich in a federal district court. How, if at all, should the agreement reached by RDI and Zurich affect Shoreline's claim? Explain. [*Shoreline Towers Condominium Owners Association, Inc. v. Zurich American Insurance Co.*, 196 F.Supp.2d 1210 (S.D.Ala. 2002)]

8–7. Intention. Music that is distributed on compact discs and similar media generates income in the form of "mechanical" royalties. Music that is publicly performed, such as when a song is played on the radio, included in a movie or commercial, or sampled in another song, produces "performance" royalties. Both types of royalties are divided between the songwriter and the song's publisher. Vincent Cusano is a musician and songwriter who performed under the name "Vinnie Vincent" as a guitarist with the group KISS in the early 1980s. Cusano co-wrote three songs entitled "Killer," "I Love It Loud," and "I Still Love You," which KISS recorded and released in 1982 on an album titled *Creatures of the Night.* Cusano left KISS in 1984. Eight years later, Cusano sold to Horipro Entertainment Group "one hundred (100%) percent undivided interest" of his rights in the songs "other than Songwriter's share of performance income." Later, Cusano filed a suit in a federal district court against Horipro, claiming in part that he never intended to sell the writer's share of the mechanical royalties. Horipro filed a motion for summary judgment. Should the court grant the motion? Explain. [*Cusano v. Horipro Entertainment Group*, 301 F.Supp.2d 272 (S.D.N.Y. 2004)]

VIDEO QUESTION

8–8. Go to this text's Web site at academic.cengage.com/blaw/wbl and select "Chapter 8." Click on "Video Questions" and view the video titled *Offer and Acceptance.* Then answer the following questions.

1. On the video, Vinny indicates that he can't sell his car to Oscar for four thousand dollars; then he says, "maybe five . . . ." Discuss whether Vinny has made an offer or a counteroffer.
2. Oscar then says to Vinny, "Okay, I'll take it. But you gotta let me pay you four thousand now and the other

thousand in two weeks." According to the chapter, do Oscar and Vinny have an agreement? Why or why not?

3. When Maria later says to Vinny, "I'll take it," has she accepted an offer? Why or why not?

Agreement and Consideration

For updated links to resources available on the Web, as well as a variety of other materials, visit this text's Web site at

academic.cengage.com/blaw/wbl

To learn what kinds of clauses are included in typical contracts, you can explore the collection of example contracts made available by FindLaw at

http://contracts.corporate.findlaw.com/index.html

You can find contract cases decided by the United States Supreme Court and federal appellate courts (as well as federal statutory law on contracts) at Cornell University's School of Law site:

http://www.law.cornell.edu/topics/contracts.html

To view the terms of a sample contract, go to the "forms" pages of the 'Lectric Law Library at

http://www.lectlaw.com/formb.htm

and select one of the types of contracts listed there to review.

For answers to some common questions about contract law, go to the Web site of the Legal Information Network, Inc., at

http://www.contract-law.com/TheLaw.htm

The *New Hampshire Consumer's Sourcebook* provides information on contract law from a consumer's perspective. You can access this book online at

http://doj.nh.gov/consumer/index.html

ONLINE LEGAL RESEARCH

Go to the *Fundamentals of Business Law* home page at **academic.cengage.com/blaw/wbl**, select "Chapter 8," and click on "Internet Exercises." There you will find the following Internet research exercises that you can perform to learn more about topics covered in this chapter.

Activity 8–1: LEGAL PERSPECTIVE—Contract Terms

Activity 8–2: ETHICAL PERSPECTIVE—Offers and Advertisements

Activity 8–3: MANAGEMENT PERSPECTIVE—Promissory Estoppel

BEFORE THE TEST

Go to the *Fundamentals of Business Law* home page at **academic.cengage.com/blaw/wbl**, select "Chapter 8," and click on "Interactive Quizzes." You will find a number of interactive questions relating to this chapter.

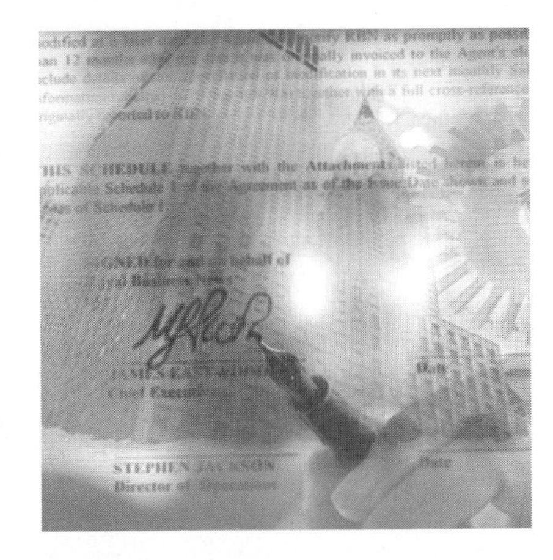

CHAPTER 9

Capacity and Legality

LEARNING OBJECTIVES

After reading this chapter, you should be able to answer the following questions:

1. What are some exceptions to the rule that a minor can disaffirm (avoid) any contract?
2. Does an intoxicated person have the capacity to enter into an enforceable contract?
3. Does the mental incompetence of one party necessarily make a contract void?
4. Under what circumstances will a covenant not to compete be enforceable? When will such covenants not be enforced?
5. What is an exculpatory clause? In what circumstances might exculpatory clauses be enforced? When will they not be enforced?

Courts generally want contracts to be enforceable, and much of the law is devoted to aiding the enforceability of contracts. Nonetheless, not all people can make legally binding contracts at all times. Contracts entered into by persons lacking the capacity to do so may be voidable. Similarly, contracts calling for the performance of an illegal act are illegal and thus void—they are not contracts at all. In this chapter, we examine contractual capacity and some aspects of illegal bargains.

Contractual Capacity

Contractual capacity is the legal ability to enter into a contractual relationship. Courts generally presume the existence of contractual capacity, but in some situations, capacity is lacking or may be questionable. A person *adjudged by a court* to be mentally incompetent, for example, cannot form a legally binding contract with another party. In other situations, a party may have the capacity to enter into a valid contract but also have the right to avoid liability under it. For example, minors—or *infants,* as they are commonly referred to in legal parlance—usually are not legally bound by contracts. In this section, we look at the effects of youth, intoxication, and mental incompetence on contractual capacity.

MINORS

Today, in virtually all states, the *age of majority* (when a person is no longer a minor) for contractual purposes is eighteen years—the so-called coming of age. (The age of majority may still be twenty-one for other purposes, however, such as the purchase and consumption of alcohol.) In addition, some states provide for the termination of minority on marriage. Subject to certain exceptions, the contracts entered into by a minor are voidable at the option of that minor.

The general rule is that a minor can enter into any contract an adult can, provided that the contract is not one prohibited by law for minors (for example, the sale of alcoholic beverages or tobacco). Although minors have the right to avoid their contracts, there are exceptions (to be discussed shortly).

Disaffirmance The legal avoidance, or setting aside, of a contractual obligation is referred to as **disaffirmance.** To disaffirm, a minor must express his or her intent, through words or conduct, not to be bound to the contract. The minor must disaffirm the entire contract, not merely a portion of it. For instance, a minor cannot decide to keep part of the goods purchased under a contract and return the remaining goods. When a minor disaffirms a contract, the minor can recover any property that he or she transferred to the adult as consideration for the contract, even if it is then in the possession of a third party.[1]

A contract can ordinarily be disaffirmed at any time during minority[2] or for a reasonable time after the minor comes of age. Though two months would probably be considered reasonable, a court may not consider it reasonable to wait a year or more after coming of age to disaffirm, depending on the circumstances. If an individual fails to disaffirm an executed contract within a reasonable time after reaching the age of majority, a court will likely hold that the contract has been ratified (*ratification* will be discussed shortly).

Note that an adult who enters into a contract with a minor cannot avoid his or her contractual duties on the ground that the minor normally can do so. Unless the minor exercises the option to disaffirm the contract, the adult party usually is bound by it.

1. The Uniform Commercial Code, in Section 2–403(1), allows an exception if the third party is a "good faith purchaser for value." See Chapter 15.

2. In some states, however, a minor who enters into a contract for the sale of land cannot disaffirm the contract until she or he reaches the age of majority.

—A Minor's Obligations on Disaffirmance. Although all states' laws permit minors to disaffirm contracts (with certain exceptions), including executed contracts, state laws differ on the extent of a minor's obligations on disaffirmance. Courts in a majority of states hold that the minor need only return the goods (or other consideration) subject to the contract, provided the goods are in the minor's possession or control. ■ EXAMPLE 9.1 Jim Garrison, a seventeen-year-old, purchases a computer from Radio Shack. While transporting the computer to his home, Garrison, through no fault of his own, is involved in a car accident. As a result of the accident, the plastic casing of the computer is broken. The next day, he returns the computer to Radio Shack and disaffirms the contract. Under the majority view, this return fulfills Garrison's duty even though the computer is now damaged. Garrison is entitled to receive a refund of the purchase price (if paid in cash) or to be relieved of any further obligations under an agreement to purchase the computer on credit. ■

A growing number of states, either by statute or by court decision, place an additional duty on the minor—the duty to restore the adult party to the position she or he held before the contract was made.

In the following case, the Tennessee Supreme Court faced the issue of whether a minor should be held responsible for damage, ordinary wear and tear, and depreciation of goods used by the minor prior to his disaffirmance of the contract. The case illustrates the trend among today's courts in regard to this issue.

CASE 9.1 Dodson v. Shrader

Supreme Court of Tennessee, 1992.
824 S.W.2d 545.

FACTS When Joseph Dodson was sixteen years old, he bought a used pickup truck for $4,900 from Shrader's Auto, which was owned by Burns and Mary Shrader. Nine months later, the truck developed mechanical problems. A mechanic informed Dodson that the problem might be a burnt valve. Without having the truck repaired, Dodson continued to drive it. One month later, the truck's engine "blew up," and the truck was rendered inoperable. Dodson disaffirmed the contract and sought to return the truck to the Shraders and obtain a full refund of the purchase price. The Shraders refused to refund the purchase price and would not accept possession of the truck. Later, while parked in Dodson's front yard, the pickup was hit by an unknown driver. Dodson filed suit against the Shraders to compel a refund of the purchase price. Although the Shraders claimed that the truck's value had been reduced to $500, the trial court granted rescission and ordered the Shraders to refund the full $4,900 purchase price to Dodson on Dodson's delivery of the truck to them. The Shraders appealed.

ISSUE Are the Shraders legally obligated to refund the full purchase price of the truck to Dodson?

DECISION No. The Supreme Court of Tennessee adopted a rule that required the seller to be compensated for the depreciated value–not the purchase price–of the pickup and remanded the case for a determination of the fairness of the contract and the fair market value of the vehicle.

(Continued)

CASE 9.1–CONTINUED

REASON The court concluded that if the minor "has not been overreached in any way, and there has been no undue influence, and the contract is a fair and reasonable one, and the minor has actually paid money on the purchase price, and taken and used the article purchased, [then] he ought not to be permitted to recover the amount actually paid, without allowing the vendor of the goods reasonable compensation for the use of, depreciation, and willful or negligent damage to the article purchased, while in his hands." The court recognized "modern conditions under which minors * * * transact a great deal of business for themselves, long before they have reached the age of legal majority." To rule otherwise, explained the court, "can only lead to the corruption of principles and encourage young people in habits of trickery and dishonesty."

FOR CRITICAL ANALYSIS—Ethical Consideration *Do you believe that the goal of protecting minors from the consequences of unwise contracts should ever outweigh the goal of encouraging minors to behave in a responsible manner?*

—Misrepresentation of Age. Suppose that a minor tells a seller she is twenty-one years old when she is really seventeen. Ordinarily, the minor can disaffirm the contract even though she has misrepresented her age. Moreover, in some jurisdictions the minor is not liable for the tort of fraudulent misrepresentation, the rationale being that such a tort judgment might indirectly force the minor to perform the contract.

In many jurisdictions, however, a minor who has misrepresented his or her age can be bound by a contract under certain circumstances. First, several states have enacted statutes for precisely this purpose. In these states, misrepresentation of age is enough to prohibit disaffirmance. Other statutes prohibit disaffirmance by a minor who has engaged in business as an adult. Second, some courts refuse to allow minors to disaffirm executed (fully performed) contracts unless they can return the consideration received. The combination of the minors' misrepresentations and their unjust enrichment has persuaded these courts to *estop* (prevent) the minors from asserting contractual incapacity.

Finally, some courts allow a misrepresenting minor to disaffirm the contract, but they hold the minor liable for damages in tort. Here, the defrauded party may sue the minor for misrepresentation or fraud. Authority is split on this point because some courts, as previously noted, have recognized that allowing a suit in tort is equivalent to indirectly enforcing the minor's contract.

—Contracts for Necessaries, Insurance, and Loans. A minor who enters into a contract for necessaries may disaffirm the contract but remains liable for the reasonable value of the goods used. **Necessaries** are basic needs, such as food, clothing, shelter, and medical services, at a level of value required to maintain the minor's standard of living or financial and social status. Thus, what will be considered a necessary for one person may be a luxury for another. Additionally, what is considered a necessary depends on whether the minor is under the care or control of his or her parents, who are required by law to provide necessaries for the minor. If a minor's parents provide the minor with shelter, for example, then a contract to lease shelter (such as an apartment) normally will not be considered a contract for necessaries.

Generally, then, to qualify as a contract for necessaries, (1) the item contracted for must be necessary to the minor's existence, (2) the value of the necessary item may be up to a level required to maintain the minor's standard of living or financial and social status, and (3) the minor must not be under the care of a parent or guardian who is required to supply this item. Unless these three criteria are met, the minor can normally disaffirm the contract *without* being liable for the reasonable value of the goods used.

Traditionally, insurance has not been viewed as a necessary, so minors can ordinarily disaffirm their insurance contracts and recover all premiums paid. Some jurisdictions, however, prohibit the disaffirmance of insurance contracts—for example, when minors contract for life insurance on their own lives. Financial loans are seldom considered to be necessaries, even if the minor spends the borrowed funds on necessaries. If, however, a lender makes a loan to a minor for the express purpose of enabling the minor to purchase necessaries, and the lender personally makes sure the money is so spent, the minor normally is obligated to repay the loan. The issue in the following case was whether a medical services provider could collect from a minor the cost of emergency services rendered to the minor when his mother did not pay.

CASE 9.2 Yale Diagnostic Radiology v. Estate of Harun Fountain

Supreme Court of Connecticut, 2004.
267 Conn. 351,
838 A.2d 179.

FACTS In March 1996, Harun Fountain was shot in the back of the head at point-blank range by a playmate. Fountain required extensive lifesaving medical services from a variety of providers, including Yale Diagnostic Radiology. Yale billed Fountain's mother, Vernetta Turner-Tucker, for the cost of its services, $17,694, but she did not pay. Instead, in January 2001, Turner-Tucker filed for bankruptcy (see Chapter 21), and all of her debts, including the amount owed to Yale, were discharged. Meanwhile, she filed a suit in a Connecticut state court against the boy who shot Fountain and obtained money for Fountain's medical care. These funds were deposited in an account—sometimes referred to as an "estate"—established on Fountain's behalf. Yale filed a suit against Fountain's estate, asking the court to order the estate to pay Yale's bill, but the court refused to grant the order. Yale appealed to a state intermediate appellate court, which reversed this ruling and ordered the payment. The estate appealed to the state supreme court.

ISSUE Can a party who provides emergency medical services to a minor collect for those services from the minor if his or her parents do not pay?

DECISION Yes. The Connecticut Supreme Court affirmed the judgment of the lower court. Yale could collect from Fountain's estate for the cost of its services under the doctrine of necessaries.

REASON The state supreme court explained that while Connecticut has long recognized the rule that a minor's contracts are voidable, a minor cannot avoid a contract for goods or services necessary for the minor's health. The court reasoned that this principle is based on the theory of quasi contract (also known as implied-in-law contracts—see Chapter 7). According to the court, two contracts arise when a necessary medical service is provided to a minor. The primary contract is between the provider and the minor's parents and is based on the parents' duty to pay for the child's necessary health expenses. A secondary contract is implied in law between the provider and the minor to prevent unjust enrichment. "Therefore," stated the court, "where necessary medical services are rendered to a minor whose parents do not pay for them, equity and justice demand that a secondary implied-in-law contract arise between the medical services provider and the minor who has received the benefits of those services."

FOR CRITICAL ANALYSIS—Social Consideration *What might have happened in future cases if the court had held that there was no implied-in-law contract between Fountain and Yale?*

Ratification In contract law, **ratification** is the act of accepting and giving legal force to an obligation that previously was not enforceable. A minor who has reached the age of majority can ratify a contract expressly or impliedly. Express ratification occurs when the minor expressly states, orally or in writing, that she or he intends to be bound by the contract. Implied ratification exists when the conduct of the minor indicates an intention to be bound by the contract. Ratification will be implied if the minor's conduct is inconsistent with disaffirmance (as when the minor enjoys the benefits of the contract) or if the minor fails to disaffirm an executed (fully performed) contract within a reasonable time after reaching the age of majority. If the contract is still executory (not yet performed or only partially performed), however, failure to disaffirm the contract will not necessarily imply ratification.

■ EXAMPLE 9.2 Lin enters a contract to sell her laptop to Arturo, a minor. If Arturo does not disaffirm the contract and, on reaching the age of majority, writes a letter to Lin stating that he still agrees to buy the laptop, he has expressly ratified the contract. If, instead, Arturo takes possession of the laptop as a minor and continues to use it well after reaching the age of majority, he has impliedly ratified the contract. ■

In determining whether implied ratification has occurred, the courts generally look at whether the minor, after reaching the age of majority, has had ample opportunity to consider the nature of the contractual obligations he or she entered into as a minor and at the extent to which the adult party to the contract has performed.

Parents' Liability As a general rule, parents are not liable for the contracts made by minor children acting on their own except contracts for necessaries, which the parents are legally required to provide. This is why businesses ordinarily require parents to cosign any contract made with a minor. The parents then become personally obligated

to perform the conditions of the contract, even if their child avoids liability.

Generally, minors are personally liable for their own torts. The parents of the minor can *also* be held liable in certain situations. In some states, parents may be liable if they failed to exercise proper parental control when they knew or should have known that this lack of control posed an unreasonable risk of harm to others. ■ **EXAMPLE 9.3** Suppose that parents allow their eleven-year-old child to drive a car on public roads. If the child drives negligently and causes someone else to be injured, the parents may be held liable for the minor's tort (negligence). ■ Other states have enacted statutes that impose liability on parents for certain kinds of tortious acts committed by their children, such as those that are willful and malicious.

Exhibit 9–1 summarizes the rules relating to contracts by minors.

Emancipation Minority status may be terminated through **emancipation,** which occurs when a child's parent or legal guardian gives up the legal right to exercise control over the child. Normally, a minor who leaves home to support himself or herself is considered emancipated. In many states, marriage terminates the status of minority. A number of jurisdictions permit minors to petition a court for emancipation. In addition, a minor may petition a court to be treated as an adult for business purposes. If the court grants the minor's request, it removes the minor's lack of contractual capacity, and the minor no longer has the right to disaffirm business contracts.

INTOXICATED PERSONS

Contractual capacity also becomes an issue when a party to a contract was intoxicated at the time the contract was made. Intoxication is a condition in which a person's normal capacity to act or think is inhibited by alcohol or some other drug. If the person was sufficiently intoxicated to lack mental capacity, the contract may be voidable even if the intoxication was purely voluntary. For the contract to be voidable, however, the person must prove that the intoxication impaired her or his reason and judgment so severely that she or he did not comprehend the legal consequences of entering into the contract. In addition, to avoid the contract in the majority of states, the person claiming intoxication must be able to return all consideration received.

If, despite intoxication, the person understood the legal consequences of the agreement, the contract is enforceable. The fact that the terms of the contract are foolish or obviously favor the other party does not make the contract voidable (unless the other party fraudulently induced the person to become intoxicated). As a practical matter, courts rarely permit contracts to be avoided on the ground of intoxication because it is difficult to determine whether a party was sufficiently intoxicated to avoid legal duties. Rather than inquire into the intoxicated person's mental state, many courts instead focus on objective indications of capacity to determine whether the contract is voidable.[3]

MENTALLY INCOMPETENT PERSONS

Contracts made by mentally incompetent persons can be void, voidable, or valid. We look here at the circumstances that determine when these classifications apply.

When the Contract Will Be Void If a court has previously determined that a person is mentally incompetent and has appointed a guardian to represent the person, any

3. See, for example, the court's decision in *Lucy v. Zehmer,* which was presented as Case 8.1 in Chapter 8.

EXHIBIT 9–1 CONTRACTS BY MINORS

General Rule	Contracts entered into by minors are voidable at the option of the minor.
Rules of Disaffirmance	A minor may disaffirm a contract at any time while still a minor and within a reasonable time after reaching the age of majority. Most states do not require restitution.
Exceptions to Basic Rules of Disaffirmance	1. *Necessaries*—Minors may disaffirm contracts for necessaries but remain liable for the reasonable value of the goods or services. 2. *Ratification*—After reaching the age of majority, a person can ratify a contract that he or she formed as a minor, becoming fully liable thereon. 3. *Fraud or misrepresentation*—In many jurisdictions, a minor who misrepresents her or his age is denied the right of disaffirmance.

contract made by that mentally incompetent person is *void*—no contract exists. Only the guardian can enter into a binding contract on behalf of the mentally incompetent person.

When the Contract Will Be Voidable If a court has not previously judged a person to be mentally incompetent but in fact the person was incompetent at the time, the contract may be *voidable*. A contract is voidable if the person did not know he or she was entering into the contract or lacked the mental capacity to comprehend its nature, purpose, and consequences. In such situations, the contract is voidable at the option of the mentally incompetent person but not the other party. The contract may then be disaffirmed or ratified (if the person regains mental competence). Like intoxicated persons, mentally incompetent persons must return any consideration and pay for the reasonable value of any necessaries they receive.

■ EXAMPLE 9.4 Milo, a mentally incompetent man who had not been previously declared incompetent by a judge, agrees to sell twenty lots in a prime residential neighborhood to Anastof. At the time he enters into the contract, Milo is confused as to which lots he is selling and how much they are worth. As a result, he contracts to sell the properties for substantially less than their value. If the court finds that Milo was unable to understand the nature and consequences of the contract, the contract is voidable. Milo can avoid the sale, provided that he returns any consideration he received. ■

When the Contract Will Be Valid A contract entered into by a mentally incompetent person (whom a court has not previously declared incompetent) may also be deemed *valid* if the person had capacity *at the time the contract was formed*. For instance, a person may be able to understand the nature and effect of entering into a certain contract yet simultaneously lack capacity to engage in other activities. In such cases, the contract ordinarily will be valid because the person is not legally mentally incompetent for contractual purposes.[4] Similarly, an otherwise mentally incompetent person may have a *lucid interval*—a temporary restoration of sufficient intelligence, judgment, and will to enter into contracts—during which he or she will be considered to have full legal capacity.

4. Modern courts no longer require a person to be completely irrational to disaffirm contracts on the basis of mental incompetence. A contract may be voidable if, by reason of a mental illness or defect, an individual was unable to act reasonably with respect to the transaction and the other party had reason to know of the condition.

Legality

To this point, we have discussed three of the requirements for a valid contract to exist—agreement, consideration, and contractual capacity. Now we examine a fourth—legality. For a contract to be valid and enforceable, it must be formed for a legal purpose. A contract to do something that is prohibited by federal or state statutory law is illegal and, as such, void from the outset and thus unenforceable. Additionally, a contract to commit a tortious act or to commit an action that is contrary to public policy is illegal and unenforceable.

CONTRACTS CONTRARY TO STATUTE

Statutes often set forth rules specifying which terms and clauses may be included in contracts and which are prohibited. We examine here several ways in which contracts may be contrary to a statute and thus illegal.

Usury Almost every state has a statute that sets the maximum rate of interest that can be charged for different types of transactions, including ordinary loans. A lender who makes a loan at an interest rate above the lawful maximum commits **usury.** The maximum rate of interest varies from state to state.

Although usury statutes place a ceiling on allowable rates of interest, exceptions are made to facilitate business transactions. For example, many states exempt corporate loans from the usury laws. In addition, almost all states have special statutes allowing much higher interest rates on small loans to help those borrowers who need funds and could not otherwise obtain loans.

The consequences for lenders who make usurious loans vary from state to state. A number of states allow the lender to recover only the principal of a loan along with interest up to the legal maximum. In effect, the lender is denied recovery of the excess interest. In other states, the lender can recover the principal amount of the loan but no interest. In a few states, a usurious loan is a void transaction, and the lender cannot recover either the principal or the interest.

Gambling All states have statutes that regulate gambling—defined as any scheme that involves the distribution of property by chance among persons who have paid valuable consideration for the opportunity (chance) to receive the property.[5] Gambling is the creation of risk for

5. See *Wishing Well Club v. City of Akron,* 112 N.E.2d 41, 66 Ohio Law Abs. 406 (1951).

the purpose of assuming it. Traditionally, the states have deemed gambling contracts illegal and thus void.

In several states, however, including Louisiana, Nevada, and New Jersey, casino gambling is legal. In other states, certain other forms of gambling are legal. California, for example, has not defined draw poker as a crime, although criminal statutes prohibit numerous other types of gambling games. A number of states allow gambling on horse races, and the majority of states have legalized state-operated lotteries, as well as lotteries (such as bingo) conducted for charitable purposes. Many states also allow gambling on Native American reservations.

Sometimes it is difficult to distinguish a gambling contract from the risk sharing inherent in almost all contracts. ■ EXAMPLE 9.5 In one case, five co-workers each received a free lottery ticket from a customer and agreed to split the winnings if one of the tickets turned out to be the winning one. At first glance, this may seem entirely legal. The court, however, noted that the oral contract in this case "was an exchange of promises to share winnings from the parties' individually owned lottery tickets upon the happening of the uncertain event" that one of the tickets would win. Consequently, concluded the court, the agreement at issue was "founded on a gambling consideration" and therefore was void.[6] ■

Sabbath (Sunday) Laws Statutes referred to as Sabbath (Sunday) laws prohibit the formation or performance of certain contracts on a Sunday. Under the common law, such contracts are legal in the absence of this statutory prohibition. According to some state and local laws, all contracts entered into on a Sunday are illegal. Laws in other states or municipalities prohibit only the sale of certain types of merchandise, such as alcoholic beverages, on a Sunday.

These laws, which date back to colonial times, are often called **blue laws.** Blue laws get their name from the blue paper on which New Haven, Connecticut, printed its new town ordinance in 1781. The ordinance prohibited all work on Sunday and required all shops to close on the "Lord's Day." A number of states and municipalities enacted laws forbidding the carrying on of "all secular labor and business on the Lord's Day." Exceptions to Sunday laws permit contracts for necessities (such as food) and works of charity. Additionally, a fully performed (executed) contract that was entered into on a Sunday normally cannot be rescinded (canceled).

Sunday laws are often not enforced, and some of these laws have been held to be unconstitutional on the ground that they are contrary to the freedom of religion. Nonetheless, as a precaution, business owners contemplating doing business in a particular locality should check to see if any Sunday statutes or ordinances will affect their business activities.

Licensing Statutes All states require that members of certain professions obtain licenses allowing them to practice. Physicians, lawyers, real estate brokers, architects, electricians, and stockbrokers are but a few of the people who must be licensed. Some licenses are obtained only after extensive schooling and examinations, which indicate to the public that a special skill has been acquired. Others require only that the particular person be of good moral character and pay a fee.

—The Purpose of Licensing Statutes. Generally, business licenses provide a means of regulating and taxing certain businesses and protecting the public against actions that could threaten the general welfare. For example, in nearly all states, a stockbroker must be licensed and must file a bond with the state to protect the public from fraudulent transactions in stock. Similarly, a plumber must be licensed and bonded to protect the public against incompetent plumbers and to protect the public health. Only persons or businesses possessing the qualifications and complying with the conditions required by statute are entitled to licenses. For instance, the owner of a tavern can be required to sell food as a condition of obtaining a license to serve liquor.

—Contracts with Unlicensed Practitioners. A contract with an unlicensed practitioner may still be enforceable, depending on the nature of the licensing statute. Some states expressly provide that the lack of a license in certain occupations bars the enforcement of work-related contracts. If the statute does not expressly state this, one must look to the underlying purpose of the licensing requirements for a particular occupation. If the purpose is to protect the public from unauthorized practitioners, a contract involving an unlicensed individual is illegal and unenforceable. If, however, the underlying purpose of the statute is to raise government revenues, a contract with an unlicensed practitioner is enforceable—although the unlicensed person is usually fined.

At issue in the following case was whether a contract entered into with an unlicensed contractor could serve as the basis for an action for wrongful interference with a contractual relationship.

6. *Dickerson v. Deno,* 770 So.2d 63 (Ala. 2000).

CASE 9.3 RCDI Construction, Inc. v. Spaceplan/Architecture, Planning & Interiors, P.A.[a]

United States District Court,
Western District of North Carolina, 2001.
148 F.Supp.2d 607.

FACTS RCDI Construction Management, Inc. (RCDI-CM), signed a contract with Dr. Anjil Patel for the construction of a hotel in North Carolina. At the time, RCDI-CM was a licensed contractor in West Virginia but not in North Carolina. The contract was assigned to RCDI Construction, Inc. (RCDI), a wholly owned subsidiary of RCDI-CM that did have a North Carolina contractor's license. RCDI began to build the hotel. When construction was nearly complete, a discharge of water damaged the hotel. On behalf of Patel, Spaceplan/Architecture, Planning & Interiors, P.A. (SAPI), assessed the condition of the building. SAPI recommended that Patel terminate the contract with RCDI, gut the building, and reconstruct the hotel. Following its termination, RCDI agreed to forfeit to Patel $421,000 owed under the construction contract and to pay Patel $6.7 million. RCDI then filed a suit in a federal district court against SAPI, alleging, among other things, wrongful interference with a contractual relationship (see Chapter 4). SAPI filed a motion for a judgment on the pleadings.[b]

a. *P.A.* is an abbreviation for "Professional Association," a form of business organization.

b. Recall from Chapter 2 that in a *motion for judgment on the pleadings,* a party asks the court to decide the issue based only on the pleadings (complaint, answer, and other documents) and no additional evidence or testimony. The motion will be granted only if no facts are in dispute and the sole issue is how the law applies to the facts.

ISSUE Is a contract with an unlicensed contractor valid for the purpose of maintaining a suit for wrongful interference with the contract?

DECISION No. The court granted SAPI's motion and dismissed RCDI's complaint.

REASON The court stated that the first element of wrongful interference with a contractual relationship is that a valid contract must exist between the plaintiff and a third person. Here, the plaintiff's claim fails because it cannot satisfy the first element. The court acknowledged that "[w]hile there is no question that a contract for the construction of a hotel was signed by RCDI-CM and Dr. Patel, it is equally uncontested that Plaintiff RCDI-CM was not a contractor licensed by the State of North Carolina at the time[.] * * * [C]ontracts entered into by unlicensed construction contractors, in violation of a statute passed for the protection of the public, are unenforceable by the contractor." The court explained that "[t]he authorities from the earliest time to the present unanimously hold that no court will lend its assistance in any way towards carrying out the terms of an illegal contract. In case any action is brought in which it is necessary to prove the illegal contract in order to maintain the action, courts will not enforce it."

FOR CRITICAL ANALYSIS—Economic Consideration *Considering the millions of dollars involved, does it seem harsh to apply the illegality doctrine to contracts such as the one in the* RCDI *case? Should it matter that the party performing the work was licensed in the state? Why or why not?*

Contracts Contrary to Public Policy

Although contracts involve private parties, some are not enforceable because of the negative impact they would have on society. These contracts are said to be *contrary to public policy.* Examples include a contract to commit an immoral act, such as selling a child, and a contract that prohibits marriage. ■ EXAMPLE 9.6 Everett offers a young man $10,000 to refrain from marrying Everett's daughter. If the young man accepts, no contract is formed (the contract is void) because it is contrary to public policy. Thus, if the man marries Everett's daughter, Everett cannot sue him for breach of contract. ■ Business contracts that may be contrary to public policy include contracts in restraint of trade and *unconscionable contracts* or *clauses* (to be discussed shortly).

Contracts in Restraint of Trade Contracts in restraint of trade (anticompetitive agreements) usually adversely affect the public policy that favors competition in the economy. Typically, such contracts also violate one or more federal or state statutes.[7] An exception is recognized when the restraint is reasonable and is part of, or supplemental to, a contract for the sale of a business or an **employment contract** (a contract stating the terms and conditions of employment). Many such exceptions involve a type of restraint called a **covenant not to compete,** or a restrictive covenant.

7. Federal statutes prohibiting anticompetitive agreements include the Sherman Antitrust Act, the Clayton Act, and the Federal Trade Commission Act.

MANAGEMENT PERSPECTIVE

Covenants Not to Compete in the Internet Context

MANAGEMENT FACES A LEGAL ISSUE

For some companies today, particularly those in high-tech industries, trade secrets are their most valuable assets. Often, to prevent departing employees from disclosing trade secrets to competing employers, business owners and managers have their key employees sign covenants not to compete. In such a covenant, the employee typically agrees not to set up a competing business or work for a competitor in a specified geographic area for a certain period of time. Generally, the time and geographic restrictions must be reasonable. A serious issue facing management today is whether time and space restrictions that have been deemed reasonable in the past serve as a guide to what might constitute reasonable restrictions in today's changing legal landscape, which includes the Internet environment.

WHAT THE COURTS SAY

There is little case law to guide management on this issue. One case involved Mark Schlack, who worked as a Web site manager for EarthWeb, Inc., in New York. Schlack signed a covenant stating that, on termination of his employment, he would not work for any competing company for one year. When he resigned and accepted an offer from a company in Massachusetts to design a Web site, EarthWeb sued to enforce the covenant not to compete. The court refused to enforce the covenant, in part because there was no evidence that Schlack had misappropriated any of EarthWeb's trade secrets or clients. The court also stated that because the Internet lacks physical borders, a covenant prohibiting an employee from working for a competitor anywhere in the world for one year is excessive in duration.[a]

In a later case, a federal district court enforced a one-year noncompete agreement against the founder of a law-related Web site business even though no geographic restriction was included in the agreement. According to the court, "Although there is no geographic limitation on the provision, this is nonetheless reasonable in light of the national, and indeed international, nature of Internet business."[b]

IMPLICATIONS FOR MANAGERS

Management in high-tech companies should avoid overreaching in terms of time and geographic restrictions in noncompete agreements. Additionally, when considering the reasonability of time and place restrictions, the courts tend to balance time restrictions against other factors, such as geographic restrictions. Because for Web-based work the geographic restriction can be worldwide in scope, the time restriction should be narrowed considerably to compensate for the extensive geographic restriction.

a. *EarthWeb, Inc. v. Schlack*, 71 F.Supp.2d 299 (S.D.N.Y. 1999).
b. *West Publishing Corp. v. Stanley*, 2004 WL 73590 (D.N.D. 2004).

—*Covenants Not to Compete and the Sale of an Ongoing Business.* Covenants not to compete are often contained in contracts concerning the sale of an ongoing business. A covenant not to compete is created when a seller agrees not to open a new store in a certain geographic area surrounding the old store. Such an agreement, when it is ancillary to a sales contract and reasonable in terms of time and geographic area, enables the seller to sell, and the purchaser to buy, the "goodwill" and "reputation" of an ongoing business. If, for example, a well-known merchant sells his or her store and opens a competing business a block away, many of the merchant's customers will likely do business at the new store. This renders less valuable the good name and reputation sold to the other merchant for a price. If a covenant not to compete is not ancillary to a sales agreement, however, it will be void because it unreasonably restrains trade and is contrary to public policy.

—*Covenants Not to Compete in Employment Contracts.* Agreements not to compete can also be included in employment contracts. People in middle-level and upper-level management positions commonly agree not to work for competitors or not to start a competing business for a specified period of time after terminating employment. Such agreements are generally legal so long as the specified period of time is not excessive in duration and the geographic restriction is reasonable. Basically, the restriction on competition must be reasonable—that is, not any greater than necessary to protect a legitimate business interest. While a time restriction of one year may be upheld by a court as reasonable, a time restriction of two, three, or five years may not be.

Determining what should be considered "reasonable" time and geographic restrictions in the online environment is a more difficult issue being addressed by the courts. The Internet environment has no physical borders, so geographic restrictions are no longer relevant. Also, a reasonable time period in the online environment may be less than what is reasonable in conventional employment contracts because the restrictions would apply worldwide. (For a further discussion of this issue, see this chapter's *Management Perspective* feature.)

—*Enforcement Problems.* The laws governing the enforceability of covenants not to compete vary significantly from state to state. In some states, including Texas, such a covenant will not be enforced unless the employee has received some benefit in return for signing the noncompete agreement. This is true even if the covenant is reasonable as to time and area. If the employee receives no benefit, the covenant will be deemed void. California prohibits the enforcement of covenants not to compete altogether.

Occasionally, depending on the jurisdiction, courts will *reform* covenants not to compete. If a covenant is found to be unreasonable in time or geographic area, the court may convert the terms into reasonable ones and then enforce the reformed covenant. This presents a problem, however, in that the judge, implicitly, has become a party to the contract. Consequently, courts usually resort to contract **reformation** only when necessary to prevent undue burdens or hardships.

Unconscionable Contracts or Clauses Ordinarily, a court does not look at the fairness or equity of a contract; for example, a court normally will not inquire into the adequacy of consideration. Persons are assumed to be reasonably intelligent, and the court does not come to their aid just because they have made unwise or foolish bargains. In certain circumstances, however, bargains are so oppressive that the courts relieve innocent parties of part or all of their duties. Such a bargain is called an **unconscionable contract** (or **unconscionable clause**). Both the Uniform Commercial Code (UCC) and the Uniform Consumer Credit Code (UCCC) embody the unconscionability concept—the former with regard to the sale of goods and the latter with regard to consumer loans and the waiver of rights.[8] A contract can be unconscionable on either procedural or substantive grounds, as discussed in the following subsections and illustrated graphically in Exhibit 9–2.

—*Procedural Unconscionability.* Procedural unconscionability may exist when a term that one of the parties does not understand becomes part of a contract. The party may not understand the term because of inconspicuous print, unintelligible language ("legalese"), lack of opportunity to read the contract, lack of opportunity to ask questions about the contract's meaning, and other factors. Procedural unconscionability sometimes relates to purported lack of voluntariness because of a disparity in bargaining power between the two parties. Contracts entered into because of one party's vastly superior bargaining power may be deemed unconscionable. These

8. See, for example, UCC Sections 2–302 and 2–719.

EXHIBIT 9–2 UNCONSCIONABILITY

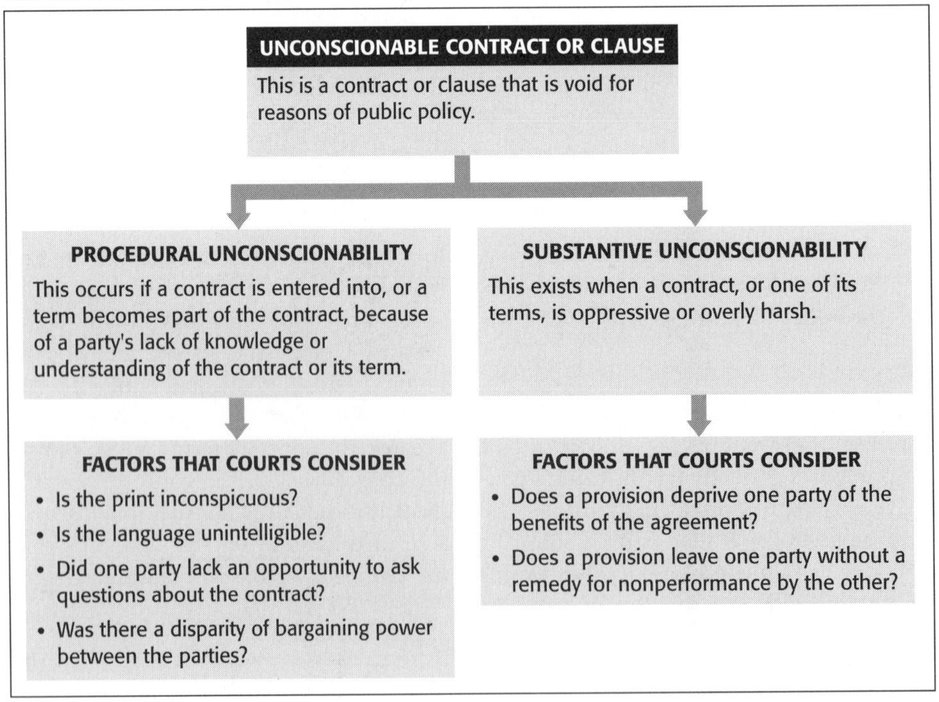

UNCONSCIONABLE CONTRACT OR CLAUSE

This is a contract or clause that is void for reasons of public policy.

PROCEDURAL UNCONSCIONABILITY

This occurs if a contract is entered into, or a term becomes part of the contract, because of a party's lack of knowledge or understanding of the contract or its term.

FACTORS THAT COURTS CONSIDER

- Is the print inconspicuous?
- Is the language unintelligible?
- Did one party lack an opportunity to ask questions about the contract?
- Was there a disparity of bargaining power between the parties?

SUBSTANTIVE UNCONSCIONABILITY

This exists when a contract, or one of its terms, is oppressive or overly harsh.

FACTORS THAT COURTS CONSIDER

- Does a provision deprive one party of the benefits of the agreement?
- Does a provision leave one party without a remedy for nonperformance by the other?

ADAPTING THE LAW TO THE ONLINE ENVIRONMENT

Is It Unconscionable for Physicians to Prescribe Medication Online?

Anyone with an e-mail address has undoubtedly received scores of messages offering to sell prescription medications, such as Viagra, online. In the past, someone who wanted a prescription for a certain medication—whether it was for allergies, weight loss, or sexual enhancement—had to visit a physician and, normally, undergo a physical examination to see if that medication was appropriate. Today, however, it is possible to enter into a contract to obtain a prescription for, and order, many medications via the Internet without ever setting foot in a physician's office. Contracting with a physician online to receive prescription drugs may be ill advised, but are such contracts unconscionable?

A VIRTUAL DIAGNOSIS

The hallmark of an unconscionable contract is that its terms are so oppressive, one sided, or unfair as to "shock the conscience" of the court. Thus, the issue is whether Internet prescription contracts are so unfair that they "shock the conscience" of the court. After all, physicians are trained to examine patients, diagnose medical conditions, and evaluate possible treatments. A physician who is prescribing medication for a person online, however, has no objective way to determine the person's health status or whether the person understands the risks involved. For example, in the online context, a physician cannot tell if a person is truthfully reporting his or her age and weight, which can significantly affect whether a medication is recommended. In addition, physicians who prescribe drugs online cannot monitor the use of these drugs and evaluate their effectiveness.

AN EMERGING ISSUE

To date, only a few courts have addressed this issue, and no court as yet has held that prescribing drugs online is unconscionable. For example, in 2003, in a case before the Kansas Supreme Court, the state attorney general claimed that it was unconscionable for an out-of-state physician to contract with residents to prescribe drugs via the Internet. The case involved three Kansas residents, including a minor, who entered into contracts on the Web to obtain prescription weight-loss drugs (Meridia and phentermine).[a]

The court held that the physician's conduct and the resulting contracts were not unconscionable because there was no evidence that the physician had deceived, oppressed, or misused superior bargaining power. The Web site used by the physician had included a great deal of general information about the specific drugs and their side effects, online questionnaires to obtain medical histories, and waivers of the physician's liability for the prescriptions. The Web site had also provided an online calculator for body weight so that a person could determine if she or he was twenty-five pounds overweight, as required to obtain a prescription for Meridia. Because the court found that the individuals had gotten exactly what they bargained for—the prescription medications they sought—the court refused to step in and declare the contracts unconscionable.

FOR CRITICAL ANALYSIS

If the physicians are not deceptive, should the courts allow all types of medications to be prescribed over the Internet? Why or why not? Explain whether the practice of prescribing medications via the Internet might reach the point at which it "shocks the conscience" of the court.

a. *State ex rel. Stovall v. DVM Enterprises, Inc.,* 275 Kan. 243, 62 P.3d 653 (2003); see also *State ex rel. Stovall v. Confimed.com, L.L.C.,* 272 Kan. 1313, 38 P.3d 707 (2002). (The Latin phrase *ex rel.* means "by or on the relation of" and is used in case names when the suit is brought by the government on the application of a private party who is interested in the matter.)

situations usually involve an **adhesion contract,** which is a contract drafted by the dominant party and then presented to the other—the adhering party—on a "take-it-or-leave-it" basis.[9]

—Substantive Unconscionability. Substantive unconscionability characterizes those contracts, or portions of contracts, that are oppressive or overly harsh. Courts generally focus on provisions that deprive one party of the benefits of the agreement or leave that party without a remedy for nonperformance by the other party. For example, suppose that a person with little income and only a fourth-grade education agrees to purchase a refrigerator for $3,000 and signs a two-year installment contract. The same type of refrigerator usually sells for $600 on the market. Despite the general rule that the courts will not inquire into the adequacy of the consideration, some courts hold that this type of contract is unconscionable because the contract terms are so oppressive as to "shock the conscience" of the court.[10] Is a contract to prescribe medications via the Internet unconscionable? For a discussion of this issue, see this chapter's *Adapting the Law to the Online Environment* feature.

9. See, for example, *Henningsen v. Bloomfield Motors, Inc.,* 32 N.J. 358, 161 A.2d 69 (1960).

10. See, for example, *Jones v. Star Credit Corp.,* 59 Misc.2d 189, 298 N.Y.S.2d 264 (1969). This case is presented in Chapter 14 as Case 14.3.

Exculpatory Clauses Closely related to the concept of unconscionability are **exculpatory clauses,** which release a party from liability in the event of monetary or physical injury, *no matter who is at fault.* Indeed, some courts refer to such clauses in terms of unconscionability. ■ EXAMPLE 9.7 Suppose, for example, that Madison Manufacturing Company hires a laborer and has him sign a contract containing the following clause:

> Said employee hereby agrees with employer, in consideration of such employment, that he will take upon himself all risks incident to his position and will in no case hold the company liable for any injury or damage he may sustain, in his person or otherwise, by accidents or injuries in the factory, or which may result from defective machinery or carelessness or misconduct of himself or any other employee in service of the employer.

This contract provision attempts to remove Madison's potential liability for injuries occurring to the employee, and it would usually be held contrary to public policy.[11] ■ Additionally, exculpatory clauses found in agreements to lease commercial property are normally held to be contrary to public policy. Such clauses are almost universally held to be illegal and unenforceable when they are included in residential property leases.

Exculpatory clauses may be enforced, however, when the parties seeking their enforcement are not involved in businesses considered important to the public interest. Businesses such as health clubs, amusement parks, horse rental concessions, golf-cart concessions, and skydiving organizations frequently use exculpatory clauses to limit their liability for patrons' injuries. Because these services are not essential, the firms offering them are sometimes considered to have no relative advantage in bargaining strength, and anyone contracting for their services is considered to do so voluntarily.

The Effect of Illegality

In general, an illegal contract is void; that is, the contract is deemed never to have existed, and the courts will not aid either party. In most illegal contracts, both parties are considered to be equally at fault—*in pari delicto.* If the contract is executory (not yet fulfilled), neither party can enforce it. If it has been executed, there can be neither contractual nor quasi-contractual recovery.

That one wrongdoer in an illegal contract is unjustly enriched at the expense of the other is of no concern to the law—except under certain circumstances (to be discussed shortly). The major justification for this hands-off attitude is that it is improper to place the machinery of justice at the disposal of a plaintiff who has broken the law by entering into an illegal bargain. Another justification is the hoped-for deterrent effect of this general rule. A plaintiff who suffers a loss because of an illegal bargain will presumably be deterred from entering into similar illegal bargains in the future.

There are exceptions to the general rule that neither party to an illegal bargain can sue for breach and neither party can recover for performance rendered. We look at these exceptions here.

Justifiable Ignorance of the Facts

When one of the parties to a contract is relatively innocent (has no reason to know that the contract is illegal), that party can often recover any benefits conferred in a partially executed contract. In this situation, the courts will not enforce the contract but will allow the parties to return to their original positions.

A court may sometimes permit an innocent party who has fully performed under a contract to enforce the contract against the guilty party. ■ EXAMPLE 9.8 A trucking company contracts with Gillespie to carry crated goods to a specific destination for a fee of $500. The trucker delivers the crates and later finds out that they contained illegal goods. Although the shipment, use, and sale of the goods are illegal under the law, the trucker, being an innocent party, can normally still legally collect the $500 from Gillespie. ■

Members of Protected Classes

When a statute protects a certain class of people, a member of that class can enforce an illegal contract even though the other party cannot. ■ EXAMPLE 9.9 Statutes prohibit certain employees (such as flight attendants) from working more than a specified number of hours per month. These employees thus constitute a class protected by statute. An employee who is required to work more than the maximum can recover for those extra hours of service. ■

Other examples of statutes designed to protect a particular class of people are **blue sky laws**—state laws that regulate and supervise investment companies for the protection of the public (see Chapter 27)—and state statutes regulating the sale of insurance. If an insurance company violates a statute when selling insurance, the purchaser can nevertheless enforce the policy and recover from the insurer.

11. For a case with similar facts, see *Little Rock & Fort Smith Railway Co. v. Eubanks,* 48 Ark. 460, 3 S.W. 808 (1887). In such a case, the exculpatory clause may also be illegal because it violates a state workers' compensation law.

WITHDRAWAL FROM AN ILLEGAL AGREEMENT

If the illegal part of a bargain has not yet been performed, the party rendering performance can withdraw from the contract and recover the performance or its value. ■ EXAMPLE 9.10 Suppose that Marta and Amil decide to wager (illegally) on the outcome of a boxing match. Each deposits money with a stakeholder, who agrees to pay the winner of the bet. At this point, each party has performed part of the agreement, but the illegal part of the agreement will not occur until the money is paid to the winner. Before such payment occurs, either party is entitled to withdraw from the agreement by giving notice to the stakeholder of his or her withdrawal. ■

SEVERABLE, OR DIVISIBLE, CONTRACTS

A contract that is *severable,* or divisible, consists of distinct parts that can be performed separately, with separate consideration provided for each part. With an *indivisible* contract, in contrast, the parties intended that complete performance by each party would be essential, even if the contract contains a number of seemingly separate provisions.

If a contract is divisible into legal and illegal portions, a court may enforce the legal portion but not the illegal one, so long as the illegal portion does not affect the essence of the bargain. This approach is consistent with the basic policy of enforcing the legal intentions of the contracting parties whenever possible. ■ EXAMPLE 9.11 If an employment contract includes an overly broad and thus illegal covenant not to compete, the court might allow the employment contract to be enforceable but reform the unreasonably broad covenant by converting its terms into reasonable ones. Alternatively, the court could declare the covenant illegal (and thus void) and enforce the remaining employment terms. ■

CONTRACTS ILLEGAL THROUGH FRAUD, DURESS, OR UNDUE INFLUENCE

Often, one party to an illegal contract is more at fault than the other. When a party has been induced to enter into an illegal bargain through fraud, duress, or undue influence on the part of the other party to the agreement, the first party will be allowed to recover for the performance or its value.

TERMS AND CONCEPTS

adhesion contract 184
blue laws 180
blue sky laws 185
contractual capacity 174
covenant not to compete 181
disaffirmance 175
emancipation 178
employment contract 181
exculpatory clause 185
necessaries 176
ratification 177
reformation 183
unconscionable contract or unconscionable clause 183
usury 179

CHAPTER SUMMARY Capacity and Legality

CONTRACTUAL CAPACITY	
Minors (See pages 174–178.)	A minor is a person who has not yet reached the age of majority. In most states, the age of majority is eighteen for contract purposes. Contracts with minors are voidable at the option of the minor. 1. *Disaffirmance*—The legal avoidance of a contractual obligation. a. Disaffirmance can take place (in most states) at any time during minority and within a reasonable time after the minor has reached the age of majority. b. If a minor disaffirms part of a contract, the entire contract must be disaffirmed. c. When disaffirming executed contracts, the minor has a duty to return received goods if they are still in the minor's control or (in some states) to pay their reasonable value. d. A minor who has committed an act of fraud (such as misrepresentation of age) will be denied the right to disaffirm by some courts. e. A minor may disaffirm a contract for necessaries but remains liable for the reasonable value of the goods.

CHAPTER SUMMARY Capacity and Legality—Continued

Minors—Continued	2. *Ratification*—The acceptance, or affirmation, of a legal obligation; may be express or implied. a. *Express ratification*—Exists when the minor, through a writing or an oral agreement, explicitly assumes the obligations imposed by the contract. b. *Implied ratification*—Exists when the conduct of the minor is inconsistent with disaffirmance or when the minor fails to disaffirm an executed contract within a reasonable time after reaching the age of majority. 3. *Parents' liability*—Generally, except for contracts for necessaries, parents are not liable for the contracts made by minor children acting on their own, nor are parents liable for minors' torts except in certain circumstances. 4. *Emancipation*—Occurs when a child's parent or legal guardian relinquishes the legal right to exercise control over the child. Normally, a minor who leaves home to support himself or herself is considered emancipated. In some jurisdictions, minors themselves are permitted to petition for emancipation for limited purposes.
Intoxicated Persons (See page 178.)	1. A contract entered into by an intoxicated person is voidable at the option of the intoxicated person if the person was sufficiently intoxicated to lack mental capacity, even if the intoxication was voluntary. 2. A contract with an intoxicated person is enforceable if, despite being intoxicated, the person understood the legal consequences of entering into the contract.
Mentally Incompetent Persons (See pages 178–179.)	1. A contract made by a person adjudged by a court to be mentally incompetent is void. 2. A contract made by a mentally incompetent person not adjudged by a court to be mentally incompetent is voidable at the option of the mentally incompetent person.
LEGALITY	
Contracts Contrary to Statute (See pages 179–180.)	1. *Usury*—Usury occurs when a lender makes a loan at an interest rate above the lawful maximum. The maximum rate of interest varies from state to state. 2. *Gambling*—Gambling contracts that contravene (go against) state statutes are deemed illegal and thus void. 3. *Sabbath (Sunday) laws*—These laws prohibit the formation or the performance of certain contracts on Sunday. Such laws vary widely from state to state, and many states do not enforce them. 4. *Licensing statutes*—Contracts entered into by persons who do not have a license, when one is required by statute, will not be enforceable *unless* the underlying purpose of the statute is to raise government revenues (and not to protect the public from unauthorized practitioners).
Contracts Contrary to Public Policy (See pages 181–185.)	1. *Contracts in restraint of trade*—Contracts to reduce or restrain free competition are illegal. Most such contracts are now prohibited by statutes. An exception is a *covenant not to compete.* It is usually enforced by the courts if the terms are ancillary to a contract (such as a contract for the sale of a business or an employment contract) and are reasonable as to time and area of restraint. Courts tend to scrutinize covenants not to compete closely. If a covenant is overbroad, a court may either reform the covenant to make the constraints more reasonable and then enforce the reformed contract or declare the covenant void and thus unenforceable. 2. *Unconscionable contracts and clauses*—When a contract or contract clause is so unfair that it is oppressive to one party, it can be deemed unconscionable; as such, it is illegal and cannot be enforced. 3. *Exculpatory clauses*—An exculpatory clause releases a party from liability in the event of monetary or physical injury, no matter who is at fault. In certain situations, exculpatory clauses may be contrary to public policy and thus unenforceable.

(Continued)

CHAPTER SUMMARY Capacity and Legality—Continued

EFFECT OF ILLEGALITY	
General Rule (See page 185.)	In general, an illegal contract is void, and the courts will not aid either party when both parties are considered to be equally at fault *(in pari delicto)*. If the contract is executory, neither party can enforce it. If the contract is executed, there can be neither contractual nor quasi-contractual recovery.
Exceptions (See pages 185–186.)	Several exceptions exist to the general rule that neither party to an illegal bargain will be able to recover. In the following situations, the court may grant recovery: 1. *Justifiable ignorance of the facts*—When one party to the contract is relatively innocent. 2. *Members of protected classes*—When one party to the contract is a member of a group of persons protected by statute, such as employees. 3. *Withdrawal from an illegal agreement*—When either party seeks to recover consideration given for an illegal contract before the illegal act is performed. 4. *Severable, or divisible, contracts*—When the court can divide the contract into illegal and legal portions and the illegal portion is not essential to the bargain. 5. *Fraud, duress, or undue influence*—When one party was induced to enter into an illegal bargain through fraud, duress, or undue influence.

FOR REVIEW

Answers for the even-numbered questions in this For Review *section can be found in Appendix A at the end of this text.*

1. What are some exceptions to the rule that a minor can disaffirm (avoid) any contract?
2. Does an intoxicated person have the capacity to enter into an enforceable contract?
3. Does the mental incompetence of one party necessarily make a contract void?
4. Under what circumstances will a covenant not to compete be enforceable? When will such covenants not be enforced?
5. What is an exculpatory clause? In what circumstances might exculpatory clauses be enforced? When will they not be enforced?

QUESTIONS AND CASE PROBLEMS

9–1. Contracts by Minors. Kalen is a seventeen-year-old minor who has just graduated from high school. He is attending a university two hundred miles from home and has contracted to rent an apartment near the university for one year at $500 per month. He is working at a convenience store to earn enough income to be self-supporting. After living in the apartment and paying monthly rent for four months, he becomes involved in a dispute with his landlord. Kalen, still a minor, moves out and returns the key to the landlord. The landlord wants to hold Kalen liable for the balance of the payments due under the lease. Discuss fully Kalen's liability in this situation.

QUESTION WITH SAMPLE ANSWER

9–2. Jermal has been the owner of a car dealership for a number of years. One day, Jermal sold one of his most expensive cars to Kessler. At the time of the sale, Jermal thought Kessler acted in a peculiar manner, but he gave the matter no further thought until four months later, when Kessler's court-appointed guardian appeared at Jermal's office, returned the car, and demanded Kessler's money back. The guardian informed Jermal that Kessler had been adjudicated mentally incompetent two months earlier by a proper court.

(a) Discuss the rights of the parties in this situation.
(b) If Kessler had been adjudicated mentally incompetent before the contract was formed, what would be the legal effect of the contract?

For a sample answer to this question, go to Appendix B at the end of this text.

9–3. Licensing Statutes. State X requires that persons who prepare and serve liquor in the form of drinks at commercial establishments be licensed by the state to do so. The only requirement for obtaining a yearly license is that the person be at least twenty-one years old. Mickey, aged thirty-five, is hired as a bar-

tender for the Southtown Restaurant. Gerald, a staunch alumnus of a nearby university, brings twenty of his friends to the restaurant to celebrate a football victory one afternoon. Gerald orders four rounds of drinks, and the bill is nearly $200. On learning that Mickey has failed to renew his bartender's license, Gerald refuses to pay, claiming that the contract is unenforceable. Discuss whether Gerald is correct.

9–4. Minority. In the 1990s, Sergei Samsonov, a Russian, was one of the top hockey players in the world. When Samsonov was seventeen years old, he signed a contract to play hockey for two seasons with the Central Sports Army Club, a Russian club known by the abbreviation CSKA. Before the start of the second season, Samsonov learned that because of a dispute between CSKA coaches, he would not be playing in Russia's premier hockey league. Samsonov hired Athletes and Artists, Inc. (A&A), an American sports agency, to make a deal with a U.S. hockey team. Samsonov signed a contract to play for the Detroit Vipers (whose corporate name was, at the time, Arena Associates, Inc.). Neither A&A nor Arena knew about the CSKA contract. CSKA filed a suit in a federal district court against Arena and others, alleging, among other things, wrongful interference with a contractual relationship. What effect will Samsonov's age have on the outcome of this suit? [*Central Sports Army Club v. Arena Associates, Inc.*, 952 F.Supp. 181 (S.D.N.Y. 1997)]

9–5. Exculpatory Clause. Norbert Eelbode applied for a job with Travelers Inn in the state of Washington. As part of the application process, Eelbode was sent to Laura Grothe, a physical therapist at Chec Medical Centers, Inc., for a preemployment physical exam. Before the exam, Eelbode signed a document that stated in part, "I hereby release Chec and the Washington Readicare Medical Group and its physicians from all liability arising from any injury to me resulting from my participation in the exam." During the exam, Grothe asked Eelbode to lift an item while bending from the waist using only his back with his knees locked. Eelbode experienced immediate sharp and burning pain in his lower back and down the back of his right leg. Eelbode filed a suit in a Washington state court against Grothe and Chec, claiming that he was injured because of an improperly administered back torso strength test. Citing the document that Eelbode signed, Grothe and Chec filed a motion for summary judgment. Should the court grant the motion? Why or why not? [*Eelbode v. Chec Medical Centers, Inc.*, 97 Wash.App. 462, 984 P.2d 436 (1999)]

CASE PROBLEM WITH SAMPLE ANSWER

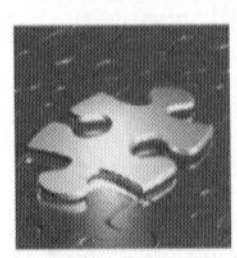

9–6. In 1993, Mutual Service Casualty Insurance Co. and its affiliates (collectively, MSI) hired Thomas Brass as an insurance agent. Three years later, Brass entered into a career agent's contract with MSI. This contract contained provisions regarding Brass's activities after termination. These provisions stated that, for a period of not less than one year, Brass could not solicit any MSI customers to "lapse, cancel, or replace" any insurance contract in force with MSI in an effort to take that business to a competitor. If he did, MSI could at any time refuse to pay the commissions that it otherwise owed him. The contract also restricted Brass from working for American National Insurance Co. for three years after termination. In 1998, Brass quit MSI and immediately went to work for American National, soliciting MSI customers. MSI filed a suit in a Wisconsin state court against Brass, claiming that he had violated the noncompete terms of his MSI contract. Should the court enforce the covenant not to compete? Why or why not? [*Mutual Service Casualty Insurance Co. v. Brass*, 242 Wis.2d 733, 625 N.W.2d 648 (2001)]

After you have answered this problem, compare your answer with the sample answer given on the Web site that accompanies this text. Go to academic.cengage.com/blaw/wbl, select "Chapter 9," and click on "Case Problem with Sample Answer."

9–7. Covenant Not to Compete. Gary Forsee was an executive officer with responsibility for the U.S. operations of BellSouth Corp., a company providing global telecommunications services. Under a covenant not to compete, Forsee agreed that for a period of eighteen months after termination from employment, he would not "provide services . . . in competition with [BellSouth] . . . to any person or entity which provides products or services identical or similar to products and services provided by [BellSouth] . . . within the territory." *Territory* was defined to include the geographic area in which Forsee provided services to BellSouth. The *services* included "management, strategic planning, business planning, administration, or other participation in or providing advice with respect to the communications services business." Forsee announced his intent to resign and accept a position as chief executive officer of Sprint Corp., a competitor of BellSouth. BellSouth filed a suit in a Georgia state court against Forsee, claiming in part that his acceptance of employment with Sprint would violate the covenant not to compete. Is the covenant legal? Should it be enforced? Why or why not? [*BellSouth Corp. v. Forsee*, 265 Ga.App. 589, 595 S.E.2d 99 (2004)]

VIDEO QUESTION

9–8. Go to this text's Web site at academic.cengage.com/blaw/wbl and select "Chapter 9." Click on "Video Questions" and view the video titled *The Money Pit*. Then answer the following questions.

1. Assume that a valid contract exists between Walter (Tom Hanks) and the plumber. Recall from the video that the plumber had at least two drinks before agreeing to take on the plumbing job. If the plumber was intoxicated, is the contract voidable? Why or why not?
2. Suppose that state law requires plumbers in Walter's state to have a plumber's license and that this plumber does not have a license. Would the contract be enforceable? Why or why not?
3. In the video, the plumber suggests that Walter has been "turned down by every other plumber in the valley." Although the plumber does not even look at the house's plumbing, he agrees to do the repairs if Walter gives him a check for $5,000 right away "before he changes his mind." If Walter later seeks to void the contract because it is contrary to public policy, what should he argue?

Capacity and Legality

For updated links to resources available on the Web, as well as a variety of other materials, visit this text's Web site at

academic.cengage.com/blaw/wbl

To find your state's statutes governing the emancipation of minors, visit the Legal Information Institute's Web site at

http://www.law.cornell.edu/topics/Table_Emancipation.htm

For a brief covering contractual capacity and mentally incompetent persons, go to

http://www.zlawyer2b.com/Contract-4-brief.html

For more information on restrictive covenants in employment contracts, you can access an article written by attorneys at Loose Brown & Associates, P.C., at

http://www.loosebrown.com/articles/art009.pdf

ONLINE LEGAL RESEARCH

Go to the *Fundamentals of Business Law* home page at academic.cengage.com/blaw/wbl, select "Chapter 9," and click on "Internet Exercises." There you will find the following Internet research exercises that you can perform to learn more about topics covered in this chapter.

Activity 9–1: MANAGEMENT PERSPECTIVE—Minors and the Law

Activity 9–2: SOCIAL PERSPECTIVE—Online Gambling

Activity 9–3: LEGAL PERSPECTIVE—Covenants Not to Compete

BEFORE THE TEST

Go to the *Fundamentals of Business Law* home page at academic.cengage.com/blaw/wbl, select "Chapter 9," and click on "Interactive Quizzes." You will find a number of interactive questions relating to this chapter.

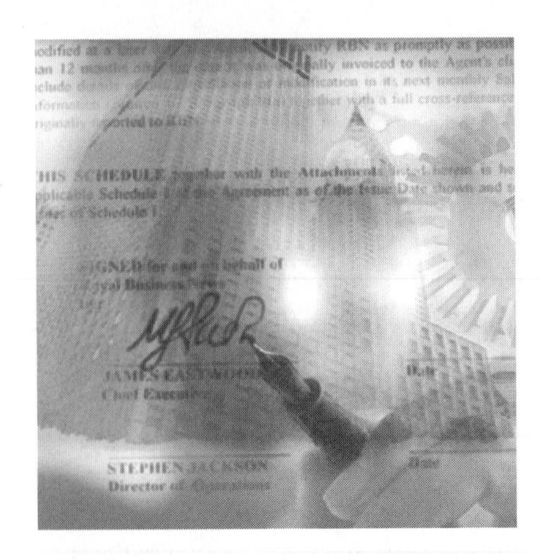

CHAPTER 10

Defenses to Contract Enforceability

LEARNING OBJECTIVES

After reading this chapter, you should be able to answer the following questions:

1. In what types of situations might genuineness of assent to a contract's terms be lacking?
2. What is the difference between a mistake of value or quality and a mistake of fact?
3. What elements must exist for fraudulent misrepresentation to occur?
4. What contracts must be in writing to be enforceable?
5. What is parol evidence? When is it admissible to clarify the terms of a written contract?

An otherwise valid contract may still be unenforceable if the parties have not genuinely assented to its terms. As mentioned in Chapter 7, lack of genuine assent is a *defense* to the enforcement of a contract. If the law were to enforce contracts not genuinely assented to by the contracting parties, injustice would result. The first part of this chapter focuses on the kinds of factors that indicate that genuineness of assent to a contract may be lacking.

A contract that is otherwise valid may also be unenforceable if it is not in the proper form. For example, if a contract is required by law to be in writing and there is no written evidence of the contract, it may not be enforceable. In the second part of this chapter, we examine the kinds of contracts that require a writing under what is called the *Statute of Frauds*. The chapter concludes with a discussion of the parol evidence rule, under which courts determine the admissibility at trial of evidence extraneous (external) to written contracts.

Genuineness of Assent

Genuineness of assent may be lacking because of mistake, fraudulent misrepresentation, undue influence, or duress. Generally, a party who demonstrates that he or she did not genuinely assent (agree) to the terms of a contract can choose either to carry out the contract or to rescind (cancel) it and thus avoid the entire transaction.

MISTAKES

We all make mistakes, so it is not surprising that mistakes are made when contracts are created. In certain circumstances, contract law allows a contract to be avoided on the basis of mistake. Realize, though, that the concept of mistake in contract law has to do with mistaken assumptions relating to contract formation. For example, the error you make when you send your monthly bank loan payment to your plumber "by mistake" is totally different from the kind of mistake that we are discussing here. In contract law, a mistake may be a defense to the enforcement of a contract if it can be proved that the parties entered into the contract, each having different assumptions relating to the subject matter of the contract.

Courts have considerable difficulty in specifying the circumstances that justify allowing a mistake to invalidate a contract. Generally, though, courts distinguish between *mistakes as to judgment of market value or conditions* and *mistakes as to fact.* Only the latter normally have legal significance.

■ EXAMPLE 10.1 Jud Wheeler contracts to buy ten acres of land because he believes that he can resell the land at a profit to Bart. Can Jud escape his contractual obligations if it later turns out that he was mistaken? Not likely. Jud's overestimation of the value of the land or of Bart's interest in purchasing it is an ordinary risk of business for which a court usually will not provide relief. Now suppose that Jud purchases a painting of a landscape from Roth's Gallery. Both Jud and Roth believe that the painting is by the artist Van Gogh. Jud later discovers that the painting is a very clever fake. Because neither Jud nor Roth was aware of this fact when they made their deal, Jud can normally rescind the contract and recover the purchase price of the painting. ■

Mistakes occur in two forms—*unilateral* and *bilateral (mutual)*. A unilateral mistake is made by only one of the contracting parties; a mutual mistake is made by both. We look next at these two types of mistakes and illustrate them graphically in Exhibit 10–1.

Unilateral Mistakes A unilateral mistake occurs when only one party is mistaken as to a *material fact*—that is, a fact important to the subject matter of the contract. Generally, a unilateral mistake does not give the mistaken party any right to relief from the contract. In other words, the contract generally is enforceable against the mistaken party. ■ EXAMPLE 10.2 Elena intends to sell her motor home for $17,500. When she learns that Chin is interested in buying a used motor home, she types a letter offering to sell her vehicle to him. When typing the letter, however, she mistakenly keys in the price of $15,700. Chin writes back, accepting Elena's offer. Even though Elena intended to sell her motor home for $17,500, she has made a unilateral mistake and is bound by contract to sell the vehicle to Chin for $15,700. ■

There are at least two exceptions to this rule.[1] First, if the *other* party to the contract knows or should have known that a mistake of fact was made, the contract may not be enforceable. In the above example, if Chin knew that Elena intended to sell her motor home for $17,500, then Elena's unilateral mistake (stating $15,700 in her offer) may render the resulting contract unenforceable. The second exception arises when a unilateral mistake of fact was due to a mathematical mistake in addition, subtraction, division, or multiplication and was made inadvertently and without gross (extreme) negligence. If a contractor's bid was significantly low because he or she made a mistake in addition when totaling the estimated costs, any contract resulting from the bid normally may be rescinded. Of course, in both situations, the mistake must still involve some *material fact*.

Bilateral (Mutual) Mistakes When both parties are mistaken about the same material fact, the contract can be rescinded by either party.[2] Note that, as with unilateral mistakes, the mistake must be about a *material fact* (one that is important and central to the contract—as was the "Van Gogh" painting in Example 10.1). If, instead, a mutual mistake concerns the future market value or quality of the object of the contract, the contract normally can be enforced by either party. This rule is based on the theory that both parties assume certain risks when they enter into a contract. Without this rule, almost any party who did not receive what she or he considered a fair bargain could argue bilateral mistake.

1. The *Restatement (Second) of Contracts,* Section 153, liberalizes the general rule to take into account the modern trend of allowing avoidance in some circumstances even though only one party has been mistaken.
2. *Restatement (Second) of Contracts,* Section 152.

EXHIBIT 10–1 MISTAKES OF FACT

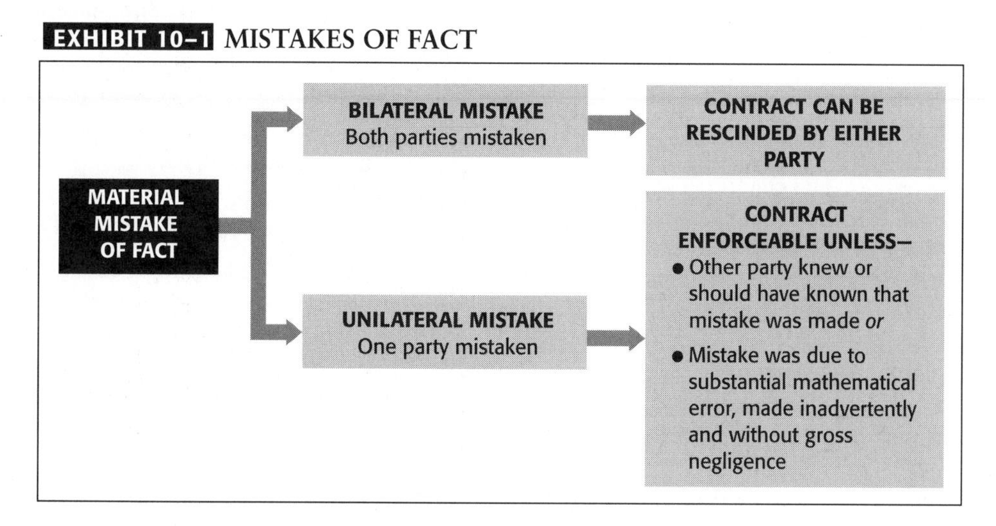

A word or term in a contract may be subject to more than one reasonable interpretation. In that situation, if the parties to the contract attach materially different meanings to the term, their mutual misunderstanding may allow the contract to be rescinded.

■ **EXAMPLE 10.3** In *Raffles v. Wichelhaus,*[3] a classic case involving a mutual mistake, Wichelhaus purchased a shipment of cotton from Raffles to arrive on a ship called the *Peerless* from Bombay, India. Wichelhaus meant a ship called the *Peerless* sailing from Bombay in October; Raffles meant another ship called the *Peerless* sailing from Bombay in December. When the goods arrived on the December *Peerless*, Raffles delivered them to Wichelhaus. By that time, however, Wichelhaus was no longer willing to accept them. The British court hearing the case stated, "There is nothing on the face of the contract to show that any particular ship called the 'Peerless' was meant; but the moment it appears that two ships called the 'Peerless' were about to sail from Bombay there is a latent ambiguity. . . . That being so, there was no consensus . . . and therefore no binding contract." ■

3. 159 Eng.Rep. 375 (1864).

In the following case, an injured worker sought to set aside a settlement agreement entered into with his employer, arguing that the agreement was based on a mutual mistake of fact—a physician's mistaken diagnosis of the worker's injury.

CASE 10.1 Roberts v. Century Contractors, Inc.

Court of Appeals of North Carolina, 2004.
162 N.C.App. 688,
592 S.E.2d 215.
http://www.aoc.state.nc.us/www/public/html/opinions.htm[a]

FACTS Bobby Roberts was an employee of Century Contractors, Inc., when a pipe struck him in a work-related accident in July 1993, causing trauma to his neck and back. Dr. James Markworth of Southeastern Orthopaedic Clinic diagnosed Roberts's injuries. After surgery and treatment, Markworth concluded that Roberts was at maximum medical improvement (MMI) and stopped treating him. Roberts agreed with Century to accept $125,000 and payment of related medical expenses, and to waive any right to make further claims in regard to his injury. In June 1998, still experiencing pain, Roberts saw Dr. Allen Friedman, who determined that Roberts was not at MMI. Markworth then admitted that his diagnosis was a mistake. Roberts filed a claim for workers' compensation (see Chapter 23), seeking compensation and medical benefits for his injury. He alleged that his agreement with Century should be set aside due to a mutual mistake of fact. The North Carolina state administrative agency authorized to rule on workers' compensation claims awarded Roberts what he sought. Century appealed to a state intermediate appellate court.

ISSUE Should the agreement between Roberts and Century be set aside on the basis of a mutual mistake of fact?

a. Click on "2004" under "Court of Appeals Opinions." In the result, scroll down to "17 February 2004" and click on the name of the case to access the opinion. The North Carolina Administrative Office of the Courts maintains this Web site.

DECISION Yes. The state intermediate appellate court affirmed the award of compensation and medical benefits to Roberts.

REASON The court explained that compromise settlement agreements, including settlement agreements in workers' compensation cases, are governed by general principles of contract law. The court stated that it is a well-settled principle of contract law that a valid contract exists only where there has been a meeting of the minds as to all essential terms of the agreement. "Therefore," said the court, "where a mistake is common to both parties and concerns a material past or presently existing fact, such that there is no meeting of the minds, a contract may be avoided." The mistake "must be as to a fact which enters into and forms the basis of the contract * * * and must be such that it animates and controls the conduct of the parties." Also, "relief from a contract due to mistake of fact will be had only where *both* parties to an agreement are mistaken." The court pointed out that Markworth's MMI diagnosis was "material to the settlement of this claim" and that both parties relied on this information in entering into settlement negotiations. Later, however, "Dr. Friedman testified, and the [state agency found] as fact, that plaintiff was not at maximum medical improvement." Thus, the court concluded that there was a mutual mistake with regard to the plaintiff's medical condition at the time of the signing of the settlement agreement.

FOR CRITICAL ANALYSIS—Social Consideration
Why did the court consider Markworth's misdiagnosis a bilateral mistake rather than a unilateral mistake?

FRAUDULENT MISREPRESENTATION

Although fraud is a tort, the presence of fraud also affects the genuineness of the innocent party's consent to a contract. When an innocent party consents to a contract with fraudulent terms, the contract usually can be avoided because he or she has not *voluntarily* consented to the terms.[4] Normally, the innocent party can either rescind (cancel) the contract and be restored to his or her original position or enforce the contract and seek damages for injuries resulting from the fraud.

Typically, fraud involves three elements:

1. A misrepresentation of a material fact must occur.
2. There must be an intent to deceive.
3. The innocent party must justifiably rely on the misrepresentation.

Additionally, to collect damages for fraud, a party must have been injured as a result of the misrepresentation.

Fraudulent misrepresentation can also occur in the online environment. Indeed, a major challenge today is how to curb Internet fraud.

Misrepresentation Must Occur The first element of proving fraud is to show that misrepresentation of a material fact has occurred. This misrepresentation can take the form of words or actions. For example, an art gallery owner's statement, "This painting is a Picasso" is a misrepresentation of fact if the painting was done by another artist.

A statement of opinion is generally not subject to a claim of fraud. For example, claims such as "This computer will never break down" and "This car will last for years and years" are statements of opinion, not fact, and contracting parties should recognize them as such and not rely on them. A fact is objective and verifiable; an opinion is usually subject to debate. Therefore, a seller is allowed to "huff and puff his [or her] wares" without being liable for fraud. In certain cases, however, particularly when a naïve purchaser relies on an expert's opinion, the innocent party may be entitled to rescission or reformation (an equitable remedy granted by a court in which the terms of a contract are altered to reflect the true intentions of the parties).

—*Misrepresentation by Conduct.* Misrepresentation can occur by conduct, as well as through express oral or written statements. For example, if a seller, by her or his actions, prevents a buyer from learning of some fact that is material to the contract, such behavior constitutes misrepresentation by conduct.[5] ■ EXAMPLE 10.4 Cummings contracts to purchase a racehorse from Garner. The horse is blind in one eye, but when Garner shows the horse, he skillfully conceals this fact by keeping the horse's head turned so that Cummings does not see the defect. The concealment constitutes fraud. ■ Another example of misrepresentation by conduct is the untruthful denial of knowledge or information concerning facts that are material to the contract when such knowledge or information is requested.

—*Misrepresentation of Law.* Misrepresentation of law does not *ordinarily* entitle a party to be relieved of a contract. ■ EXAMPLE 10.5 Debbie has a parcel of property that she is trying to sell to Barry. Debbie knows that a local ordinance prohibits building anything higher than three stories on the property. Nonetheless, she tells Barry, "You can build a condominium fifty stories high if you want to." Barry buys the land and later discovers that Debbie's statement is false. Most likely, Barry cannot avoid the contract because under the common law, people are assumed to know state and local laws. ■ Exceptions to this rule occur, however, when the misrepresenting party is in a profession known to require greater knowledge of the law than the average citizen possesses.

—*Misrepresentation by Silence.* Ordinarily, neither party to a contract has a duty to come forward and disclose facts, and a contract will not be set aside because certain pertinent information has not been volunteered. ■ EXAMPLE 10.6 Suppose that you are selling a car that has been in an accident and has been repaired. You do not need to volunteer this information to a potential buyer. If, however, the purchaser asks you if the car has had extensive bodywork and you lie, you have committed a fraudulent misrepresentation. ■

Generally, if the seller knows of a serious defect or a serious potential problem about which the buyer cannot reasonably be expected to know, the seller may have a duty to speak. In most instances, the seller must disclose only "latent" defects—that is, defects that could not readily be discovered. Thus, termites in a house may not be a latent defect because a buyer could normally discover their presence through a termite inspection. Also, when the parties are in a *fiduciary relationship* (one of trust, such as business partners, physician and patient, or attorney and client), there is a duty to disclose material facts; failure to do so may constitute fraud.

Intent to Deceive The second element of fraud is knowledge on the part of the misrepresenting party that facts have been misrepresented. This element, usually called ***scienter,***[6] or "guilty knowledge," generally signifies that

4. *Restatement (Second) of Contracts,* Sections 163 and 164.
5. *Restatement (Second) of Contracts,* Section 160.
6. Pronounced sy-*en*-ter.

there was an intent to deceive. *Scienter* clearly exists if a party knows that a fact is not as stated. *Scienter* also exists if a party makes a statement that he or she believes not to be true or makes a statement recklessly, without regard to whether it is true or false. Finally, this element is met if a party says or implies that a statement is made on some basis, such as personal knowledge or personal investigation, when it is not.

■ EXAMPLE 10.7 Suppose that Rolando, when selling a house to Cariton, tells Cariton that the plumbing pipe is of a certain quality. Rolando knows nothing about the quality of the pipe but does not believe it to be what she is representing it to be (and in fact it is not what she says it is). Rolando's statement induces Cariton to buy the house. Rolando's statement is a fraudulent misrepresentation because she does not believe that what she says is true and because she knows that she does not have any basis for making the statement. Cariton can thus avoid the contract. ■

Can an employer avoid liability for the breach of a contract that was induced by an employee's fraud during the hiring process? That was the question in the following case.

CASE 10.2 Sarvis v. Vermont State Colleges

Supreme Court of Vermont, 2001.
172 Vt. 76,
772 A.2d 494.

FACTS In 1995, Robert Sarvis was convicted of bank fraud, ordered to pay more than $12 million in restitution, and sentenced to forty-six months in prison. While incarcerated, he worked in the prison's electrical department. Two weeks after his release in 1998, he applied for an adjunct professor position at Community College of Vermont (CCV). On his résumé, he stated that during "1984–1998" he was "President and Chairman of the Board" of "CMI International, Inc., Boston, Massachusetts," where he was "[r]esponsible for all operations and financial matters." For a position as CCV's Coordinator of Academic Services, he submitted a second résumé, on which he added, "1998–present. Semi-retired. Adjunct Instructor of Business at Colby-Sawyer College and Franklin Pierce College." He stated that he was "well equipped to teach" business law and business ethics, that he had "a great interest and knowledge of business law," and that he believed he would do "an excellent job" teaching a business ethics class because this subject was "of particular concern" to him. CCV hired him as academic coordinator, teacher, and independent studies instructor. After he began work, his probation officer alerted CCV to Sarvis's criminal history, and CCV terminated his employment. Sarvis filed a suit in a Vermont state court against CCV, alleging, among other things, breach of contract. CCV filed a motion for summary judgment, in part on the ground of fraud, seeking rescission. The court granted CCV's motion, and Sarvis appealed to the Vermont Supreme Court.

ISSUE Can a partial disclosure during the preemployment process, with an intent to deceive, support the termination of an employment contract?

DECISION Yes. The Vermont Supreme Court affirmed the lower court's judgment and rescinded the contract between Sarvis and CCV.

REASON The state supreme court explained that "[t]he misrepresentation in this case occurred through plaintiff's partial disclosure of his past work history and references and his effort to limit defendant's inquiry into his past." The court emphasized that Sarvis "was not silent; he carefully drafted his résumés and supplemental materials to lead defendant to believe he had made a full disclosure about his past and his qualifications. He listed classes in business ethics and law in which he claimed he had the highest level of capability and knowledge but failed to mention his felony bank fraud conviction. Plaintiff assured defendant that making additional inquiries into his background would have revealed 'more of the same' type of information * * * . This was not true. Contact with plaintiff's probation officer or supervisor at the * * * prison would have notified defendant of plaintiff's fraud convictions, period of incarceration, and work history at the prison."

WHY IS THIS CASE IMPORTANT? *This case illustrates the long-applied legal principle that when a contract is formed as a result of fraudulent misrepresentation, the defrauded party can avoid the contract. From the facts in this case, CCV clearly would not have hired Sarvis had he fully disclosed information about his past and his qualifications.*

Reliance on the Misrepresentation The third element of fraud is *justifiable reliance* on the misrepresentation of fact. The deceived party must have a justifiable reason for relying on the misrepresentation, and the misrepresentation must be an important factor (but not necessarily the sole factor) in inducing the party to enter into the contract.

Reliance is not justified if the innocent party knows the true facts or relies on obviously extravagant statements. ■ EXAMPLE 10.8 If a used-car dealer tells you, "This old Cadillac will get over sixty miles to the gallon," you normally would not be justified in relying on this statement. Suppose, however, that Merkel, a bank director, induces O'Connell, a co-director, to sign a statement that the bank's assets will satisfy its liabilities by telling O'Connell, "We have plenty of assets to satisfy our creditors." This statement is false. If O'Connell knows the true facts or, as a bank director, should know the true facts, he is not justified in relying on Merkel's statement. If O'Connell does not know the true facts, however, *and has no way of finding them out,* he may be justified in relying on the statement. ■

Injury to the Innocent Party Most courts do not require a showing of injury when the action is to *rescind* (cancel) the contract—these courts hold that because rescission returns the parties to the positions they held before the contract was made, a showing of injury to the innocent party is unnecessary.[7]

To recover damages caused by fraud, however, proof of an injury is universally required. The measure of damages is ordinarily equal to the property's value had it been delivered as represented, less the actual price paid for the property. In actions based on fraud, courts often award *punitive,* or *exemplary, damages,* which are granted to a plaintiff over and above the compensation for the actual loss. As pointed out in Chapter 4, punitive damages are based on the public-policy consideration of punishing the defendant or setting an example to deter similar wrongdoing by others.

Undue Influence

Undue influence arises from relationships in which one party can greatly influence another party, thus overcoming that party's free will. Minors and elderly people, for example, are often under the influence of guardians. If a guardian induces a young or elderly ward (a person placed by a court under the care of a guardian) to enter into a contract that benefits the guardian, the guardian may have exerted undue influence.

Undue influence can arise from a number of confidential or fiduciary relationships, including attorney-client, physician-patient, guardian-ward, parent-child, husband-wife, and trustee-beneficiary relationships. The essential feature of undue influence is that the party being taken advantage of does not, in reality, exercise free will in entering into a contract. A contract entered into under excessive or undue influence lacks genuine assent and is therefore voidable.[8]

Duress

Assent to the terms of a contract is not genuine if one of the parties is forced into the agreement. Forcing a party to enter into a contract because of the fear created by threats is referred to as *duress.*[9] Inducing consent to a contract through blackmail or extortion also constitutes duress. Duress is both a defense to the enforcement of a contract and a ground for rescission, or cancellation, of a contract. Therefore, a party who signs a contract under duress can choose to carry out the contract or to avoid the entire transaction. (The wronged party usually has this choice in cases in which assent is not real or genuine.)

Economic need is generally not sufficient to constitute duress, even when one party exacts a very high price for an item the other party needs. If the party exacting the price also creates the need, however, economic duress may be found. ■ EXAMPLE 10.9 The Internal Revenue Service (IRS) assessed a large tax and penalty against Weller. Weller retained Eyman to contest the assessment. Two days before the deadline for filing a reply with the IRS, Eyman declined to represent Weller unless he agreed to pay a very high fee for Eyman's services. The agreement was held to be unenforceable.[10] Although Eyman had threatened only to withdraw his services, something that he was legally entitled to do, he was responsible for delaying his withdrawal until the last two days. Because Weller was forced into either signing the contract or losing his right to challenge the IRS assessment, the agreement was secured under duress. ■

The Statute of Frauds—Requirement of a Writing

Today, every state has a statute that stipulates what types of contracts must be in writing. In this text, we refer to such statutes as the **Statute of Frauds.** The primary pur-

7. For a leading case on this issue, see *Kaufman v. Jaffe,* 244 App.Div. 344, 279 N.Y.S. 392 (1935).

8. *Restatement (Second) of Contracts,* Section 177.

9. *Restatement (Second) of Contracts,* Sections 174 and 175.

10. *Thompson Crane & Trucking Co. v. Eyman,* 123 Cal.App.2d 904, 267 P.2d 1043 (1954).

pose of the statute is to ensure that, for certain types of contracts, there is reliable evidence of the contracts and their terms. These types of contracts are those deemed historically to be important or complex. Although the statutes vary slightly from state to state, the following types of contracts are normally required to be in writing or evidenced by a written memorandum:

1. Contracts involving interests in land.
2. Contracts that cannot by their terms be performed within one year from the date of formation.
3. Collateral contracts, such as promises to answer for the debt or duty of another.
4. Promises made in consideration of marriage.
5. Contracts for the sale of goods priced at $500 or more ($5,000 or more under the 2003 amendments to the Uniform Commercial Code, or UCC—see Chapter 14).

Agreements or promises that fit into one or more of these categories are said to "fall under" or "fall within" the Statute of Frauds. (Certain exceptions are made to the applicability of the Statute of Frauds in some circumstances, however, as you will read later in this section.) The actual name of the Statute of Frauds[11] is misleading because it does not apply to fraud. Rather, the statute denies enforceability to certain contracts that do not comply with its requirements.

CONTRACTS INVOLVING INTERESTS IN LAND

Land is a form of *real property,* or real estate, which includes not only land but all physical objects that are permanently attached to the soil, such as buildings, plants, trees, and the soil itself. Under the Statute of Frauds, a contract involving an interest in land must be evidenced by a writing to be enforceable.[12] If Carol, for example, contracts orally to sell Seaside Shelter to Axel but later decides not to sell, Axel cannot enforce the contract. Similarly, if Axel refuses to close the deal, Carol cannot force Axel to pay for the land by bringing a lawsuit. The Statute of Frauds is a *defense* to the enforcement of this type of oral contract.

A contract for the sale of land ordinarily involves the entire interest in the real property, including buildings, growing crops, vegetation, minerals, timber, and anything else affixed to the land. Therefore, a *fixture* (personal property so affixed or so used as to become a part of the realty—see Chapter 29) is treated as real property.

The Statute of Frauds requires written contracts not just for the sale of land but also for the transfer of other interests in land, such as mortgages and leases. We describe these other interests in Chapter 29.

THE ONE-YEAR RULE

Contracts that cannot, *by their own terms,* be performed within one year *from the day after* the contract is formed must be in writing to be enforceable. Because disputes over such contracts are unlikely to occur until some time after the contracts are made, resolution of these disputes is difficult unless the contract terms have been put in writing. The one-year period begins to run *the day after the contract is made.*

■ **EXAMPLE 10.10** Suppose that Superior University forms a contract with Kimi San, stating that San will teach three courses in history during the coming academic year (September 15 through June 15). If the contract is formed in March, it must be in writing to be enforceable—because it cannot be performed within one year. If the contract is not formed until July, however, it will not have to be in writing to be enforceable—because it can be performed within one year. ■ Exhibit 10–2 on the following page graphically illustrates the one-year rule.

Normally, the test for determining whether an oral contract is enforceable under the one-year rule of the Statute of Frauds is not whether the agreement is *likely* to be performed within one year from the day after the date of contract formation but whether performance within a year is *possible.* When performance of a contract is objectively impossible during the one-year period, the oral contract will be unenforceable.

COLLATERAL PROMISES

A **collateral promise,** or secondary promise, is one that is ancillary (subsidiary) to a principal transaction or primary contractual relationship. In other words, a collateral promise is one made by a third party to assume the debts or obligations of a primary party to a contract if that party does not perform. Any collateral promise of this nature falls under the Statute of Frauds and therefore must be in writing to be enforceable. To understand this concept, it is important to distinguish between primary and secondary promises and obligations.

Primary versus Secondary Obligations As a general rule, a contract in which a party assumes a primary obligation does not need to be in writing to be enforceable.

11. The name derives from an English act passed in 1677 titled "An Act for the Prevention of Frauds and Perjuries."

12. In some states, the contract will be enforced, however, if each party admits to the existence of the oral contract in court or admits to its existence during discovery before trial (see Chapter 2).

EXHIBIT 10–2 THE ONE-YEAR RULE

Under the Statute of Frauds, contracts that by their terms are impossible to perform within one year from the day after the date of contract formation must be in writing to be enforceable. Put another way, if it is at all possible to perform an oral contract within one year from the day after the contract is made, the contract will fall outside the Statute of Frauds and be enforceable.

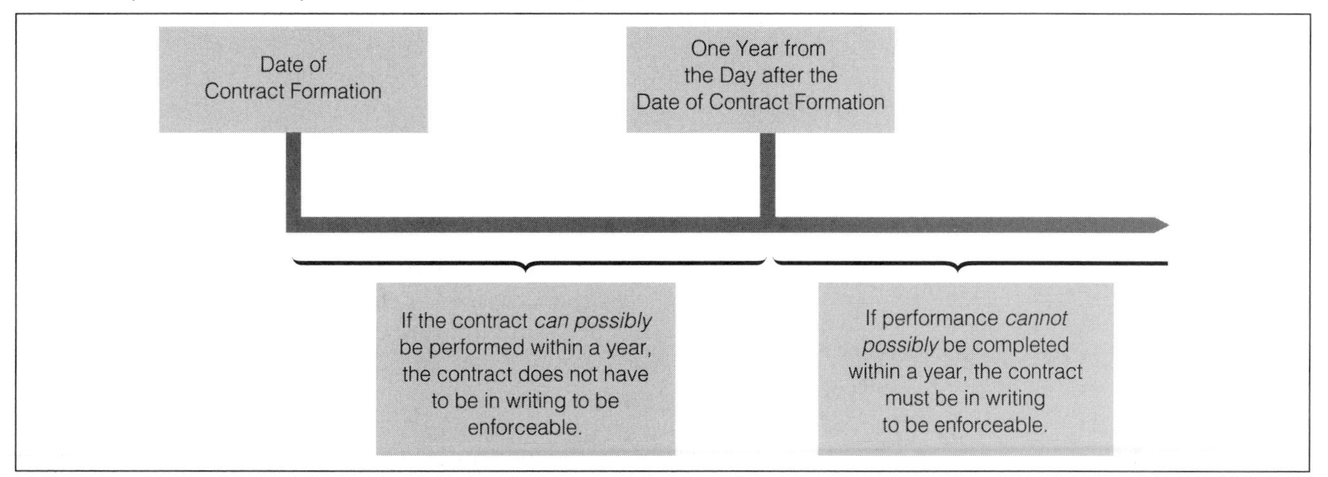

■ **EXAMPLE 10.11** Suppose that Kenneth orally contracts with Joanne's Floral Boutique to send his mother a dozen roses for Mother's Day. Kenneth promises to pay the boutique when he receives the bill for the flowers. Kenneth is a direct party to this contract and has incurred a *primary* obligation under the contract. Because he is a party to the contract and has a primary obligation to Joanne's Floral Boutique, this contract does not fall under the Statute of Frauds and does not have to be in writing to be enforceable. If Kenneth fails to pay the florist and the florist sues him for payment, Kenneth cannot raise the Statute of Frauds as a defense. He cannot claim that the contract is unenforceable because it was not in writing. ■

In contrast, a contract in which a party assumes a secondary obligation does have to be in writing to be enforceable. ■ **EXAMPLE 10.12** Suppose that Kenneth's mother borrows $10,000 from the Medford Trust Company on a promissory note payable six months later. Kenneth promises the bank officer handling the loan that he will pay the $10,000 *if his mother does not pay the loan on time.* Kenneth, in this situation, becomes what is known as a *guarantor* on the loan; that is, he is guaranteeing to the bank (the creditor) that he will pay the loan if his mother fails to do so. This kind of collateral promise, in which the guarantor states that he or she will become responsible only if the primary party does not perform, must be in writing to be enforceable. ■ We return to the concept of guaranty and the distinction between primary and secondary obligations in Chapter 21, in the context of creditors' rights.

An Exception—The "Main Purpose" Rule An oral promise to answer for the debt of another is covered by the Statute of Frauds *unless* the guarantor's purpose in accepting secondary liability is to secure a personal benefit. Under the "main purpose" rule, this type of contract need not be in writing.[13] The assumption is that a court can infer from the circumstances of a case whether a "leading objective" of the promisor was to secure a personal benefit.

■ **EXAMPLE 10.13** Carrie Oswald contracts with Machine Manufacturing Company to have some machines custom-made for her factory. To ensure that Machine Manufacturing will have the supplies it needs to make the machines, Oswald promises Allrite Materials Supply Company, Machine Manufacturing's supplier, that if Allrite continues to deliver materials to Machine Manufacturing, she will guarantee payment. This promise need not be in writing, even though the effect may be to pay the debt of another, because Oswald's main purpose is to secure a benefit for herself. ■

Another typical application of the so-called main purpose doctrine occurs when one creditor guarantees the debtor's debt to another creditor to forestall litigation. This allows the debtor to remain in business long enough to generate profits sufficient to pay *both* creditors. In this situation, the guaranty does not need to be in writing to be enforceable.

13. *Restatement (Second) of Contracts,* Section 116.

PROMISES MADE IN CONSIDERATION OF MARRIAGE

A unilateral promise to pay a sum of money or to give property in consideration of marriage must be in writing. If Mr. Baumann promises to pay Joe Villard $10,000 if Villard marries Baumann's daughter, the promise must be in writing to be enforceable. The same rule applies to **prenuptial agreements**—agreements made before marriage (also called *antenuptial agreements*) that define each partner's ownership rights in the other partner's property. For example, a prospective wife or husband may wish to limit the amount the prospective spouse can obtain if the marriage ends in divorce. Prenuptial agreements made in consideration of marriage must be in writing to be enforceable.

Generally, courts tend to give more credence to prenuptial agreements that are accompanied by consideration. ■ EXAMPLE 10.14 Maureen, who has little money, marries Kaiser, who has a net worth of $300 million. Kaiser has several children, and he wants them to receive most of his wealth on his death. Before their marriage, Maureen and Kaiser draft and sign a prenuptial agreement in which Kaiser promises to give Maureen $100,000 per year for the rest of her life if they divorce. As consideration for consenting to this amount, Kaiser offers Maureen $1 million. If Maureen consents to the agreement and accepts the $1 million, very likely a court would hold that this prenuptial agreement is valid, should it ever be contested. ■

CONTRACTS FOR THE SALE OF GOODS

The Uniform Commercial Code (UCC) contains Statute of Frauds provisions that require written evidence of a contract. Section 2–201 contains the major provision, which generally requires a writing or memorandum for the sale of goods priced at $500 or more ($5,000 or more under the 2003 amendments to the UCC—see Chapter 14). A writing that will satisfy the UCC requirement need only state the quantity term; other terms agreed on need not be stated "accurately" in the writing, as long as they adequately reflect both parties' intentions. The contract will not be enforceable, however, for any quantity greater than that set forth in the writing. In addition, the writing must be signed by the person against whom enforcement is sought. Beyond these two requirements, the writing need not designate the buyer or the seller, the terms of payment, or the price.

EXCEPTIONS TO THE STATUTE OF FRAUDS

Exceptions to the applicability of the Statute of Frauds are made in certain situations. We describe those situations here.

Partial Performance In cases involving oral contracts for the transfer of interests in land, if the purchaser has paid part of the price, taken possession, and made valuable improvements to the property, and if the parties cannot be returned to their status quo prior to the contract, a court may grant *specific performance* (performance of the contract according to its precise terms). Whether a court will enforce an oral contract for an interest in land when partial performance has taken place is usually determined by the degree of injury that would be suffered if the court chose *not* to enforce the oral contract. In some states, mere reliance on certain types of oral contracts is enough to remove them from the Statute of Frauds.

Under the UCC, an oral contract for goods priced at $500 or more ($5,000 or more under the 2003 amendments) is enforceable to the extent that a seller accepts payment or a buyer accepts delivery of the goods.[14] For example, if Ajax Corporation orders by telephone thirty crates of bleach from Cloney, Inc., and repudiates the contract after ten crates have been delivered and accepted, Cloney can enforce the contract to the extent of the ten crates Ajax accepted.

Admissions In some states, if a party against whom enforcement of an oral contract is sought admits in pleadings, testimony, or otherwise in court proceedings that a contract for sale was made, the contract will be enforceable.[15] A contract subject to the UCC will be enforceable, but only to the extent of the quantity admitted.[16] Thus, if the president of Ajax Corporation admits under oath that an oral agreement was made with Cloney, Inc., for only twenty crates of bleach, the agreement will be enforceable to that extent.

Promissory Estoppel In some states, an oral contract that would otherwise be unenforceable under the Statute of Frauds may be enforced under the doctrine of promissory estoppel, or detrimental reliance. Recall from Chapter 8 that if a promisor makes a promise on which the promisee justifiably relies to her or his detriment, a court may *estop* (prevent) the promisor from denying that a contract exists. Section 139 of the *Restatement (Second) of Contracts* provides that in these circumstances, an oral promise can be enforceable, notwithstanding the Statute of Frauds, if the reliance was foreseeable by the person making the promise and if injustice can be avoided only by enforcing the promise.

14. UCC 2–201(3)(c). See Chapter 14.
15. *Restatement (Second) of Contracts,* Section 133.
16. UCC 2–201(3)(b). See Chapter 14.

EXHIBIT 10–3 CONTRACTS SUBJECT TO THE STATUTE OF FRAUDS

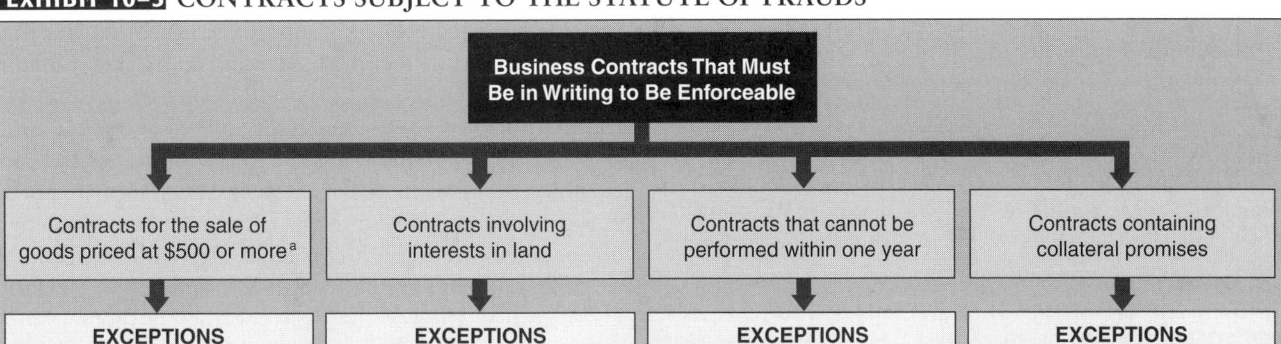

a. Under a 2003 amendment to the UCC, a contract for a sale of goods must involve goods priced at $5,000 or more to be subject to the writing requirement of the Statute of Frauds (see Chapter 14). This amendment also exempts contracts for the sale of goods from the one-year rule.
b. Some states follow Section 133 (on admissions) and Section 139 (on promissory estoppel) of the *Restatement (Second) of Contracts*.

Special Exceptions under the UCC Special exceptions to the applicability of the Statute of Frauds exist for sales contracts. Oral contracts for customized goods may be enforced in certain circumstances. Another exception has to do with oral contracts between merchants that have been confirmed in writing. We will examine these exceptions in Chapter 14. Exhibit 10–3 graphically summarizes the types of contracts that fall under the Statute of Frauds and the various exceptions that apply.

The Statute of Frauds— Sufficiency of the Writing

A written contract will satisfy the writing requirement of the Statute of Frauds. A *written memorandum* (written evidence of the oral contract) signed by the party against whom enforcement is sought will also satisfy the writing requirement.[17] The signature need not be placed at the end of the document but can be anywhere in the writing; it can even be initials rather than the full name.

A significant issue in today's business world has to do with how "signatures" can be created and verified on electronic contracts and other documents. We will examine electronic signatures in Chapter 13.

17. As mentioned earlier, under the UCC Statute of Frauds, a writing is required only for contracts for the sale of goods priced at $500 or more ($5,000 or more under the 2003 amendments to UCC Article 2). See Chapter 14.

What Constitutes a Writing?

A writing can consist of any confirmation, invoice, sales slip, check, fax, or e-mail—or such items in combination. The written contract need not consist of a single document to constitute an enforceable contract. One document may incorporate another document by expressly referring to it. Several documents may form a single contract if they are physically attached, such as by staple, paper clip, or glue. Several documents may form a single contract even if they are only placed in the same envelope.

■ **EXAMPLE 10.15** Sam orally agrees to sell some land next to a shopping mall to Terry. Sam gives Terry an unsigned memo that contains a legal description of the property, and Terry gives Sam an unsigned first draft of their contract. Sam sends Terry a signed letter that refers to the memo and to the first and final drafts of the contract. Terry sends Sam an unsigned copy of the final draft of the contract with a signed check stapled to it. Together, the documents can constitute a writing sufficient to satisfy the Statute of Frauds and bind both parties to the terms of the contract as evidenced by the writings. ■

What Must Be Contained in the Writing?

A memorandum evidencing the oral contract need only contain the essential terms of the contract. Under most provisions of the Statute of Frauds, the writing must name the parties, subject matter, consideration, and quantity.

With respect to contracts for the sale of land, some states require that the memorandum also set forth the essential terms of the contract, such as location and price, with sufficient clarity to allow the terms to be determined from the memo itself, without reference to any outside sources.[18] Under the UCC, in regard to the sale of goods, the writing need only name the quantity term and be signed by the party against whom enforcement is sought.

Because only the party against whom enforcement is sought need have signed the writing, a contract may be enforceable by one of its parties but not by the other. ■ **EXAMPLE 10.16** Rock orally agrees to buy Devlin's lake house and lot for $150,000. Devlin writes Rock a letter confirming the sale by identifying the parties and the essential terms of the sales contract—price, method of payment, and legal address—and signs the letter. Devlin has made a written memorandum of the oral land contract. Because she signed the letter, she normally can be held to the oral contract by Rock. Rock, however, because he has not signed or entered into a written contract or memorandum, can plead the Statute of Frauds as a defense, and Devlin cannot enforce the contract against him. ■

Whether a proposal to buy a pair of printing presses contained the essential terms of the contract and satisfied the Statute of Frauds was at issue in the following case.

18. *Rhodes v. Wilkins,* 83 N.M. 782, 498 P.2d 311 (1972).

CASE 10.3 Interstate Litho Corp. v. Brown

United States Court of Appeals,
First Circuit, 2001.
255 F.3d 19.

FACTS In 1995, Interstate Litho Corporation negotiated with Marc Brown, a broker in used printing equipment, and his company, Integra Technical Services, concerning Interstate's acquisition of two used printing presses. Freidel's Manufacturing, Inc., in Illinois owned one of the presses, and Graphic Engineering in Malaysia owned the other. Both were to be refurbished before they were delivered to Interstate. Interstate's president, Henry Becker, signed a proposal reflecting a $2.6 million price. Interstate advanced $75,000, of which $50,000 was wired to Freidel's. Freidel's pulled its press off the market, and Brown signed a contract to buy it. Brown never bought the press, however, because the deal with Interstate fell apart. Freidel's kept the deposit and sold the press for less than it would have received from Brown. Interstate filed a suit in a federal district court against Brown and others, seeking the return of its $75,000. Brown counterclaimed for his lost profit, asserting in part breach of contract. After a trial, the jury rejected Interstate's claim and awarded Brown $187,500 in damages on his counterclaim. Interstate appealed to the U.S. Court of Appeals for the First Circuit, arguing that, among other things, there was no enforceable contract because the proposal signed by Becker lacked the essential terms and failed to comply with the Statute of Frauds.

ISSUE Did the proposal signed by Becker contain the essential terms of the contract?

DECISION Yes. The appellate court upheld the jury's finding that the proposal contained the essential terms of the contract and constituted Interstate's agreement to buy the presses.

REASON The court explained that the writing between the parties identified in detail the two presses being purchased, identified a total price for purchasing and reworking the presses ($2.6 million, with $900,000 of this amount allocated to acquiring the two presses), and set forth a detailed delivery and payment schedule. The court emphasized that all of the terms of an agreement need not be precisely specified, and that the presence of undefined or unspecified terms will not necessarily prevent the formation of a binding contract. Here, according to the court, "the terms of the proposal—which include[d] nine pages of nitty-gritty detail on matters such as rollers and hangers, plate and blanket cylinders, and ink fountains—were sufficient for the jury to find that Interstate had agreed to purchase the two presses." The court further noted that Interstate wired $75,000 to hold the presses just a few days after signing the proposal. This action also supported the existence of a firm agreement between the parties.

FOR CRITICAL ANALYSIS—Economic Consideration *In this case, what should have formed the basis for the amount of Brown's damages?*

The Parol Evidence Rule

The **parol evidence rule** prohibits the introduction at trial of evidence of the parties' prior negotiations, prior agreements, or contemporaneous oral agreements if that evidence contradicts or varies the terms of the parties' written contract. The written contract is ordinarily assumed to be the final statement of the parties' agreement.

EXCEPTIONS TO THE PAROL EVIDENCE RULE

Because of the rigidity of the parol evidence rule, courts make several exceptions. These exceptions include the following:

1. Evidence of a *subsequent modification* of a written contract can be introduced in court. Keep in mind that the oral modifications may not be enforceable if they come under the Statute of Frauds—for example, if they increase the price of the goods for sale to $500 or more ($5,000 or more under the 2003 amendments) or increase the term for performance to more than one year. Also, oral modifications will not be enforceable if the original contract provides that any modification must be in writing.[19]
2. Oral evidence can be introduced in all cases to show that the contract was voidable or void (for example, induced by mistake, fraud, or misrepresentation). In this situation, if deception led one of the parties to agree to the terms of a written contract, oral evidence indicating fraud should not be excluded. Courts frown on bad faith and are quick to allow the introduction at trial of parol evidence when it establishes fraud.
3. When the terms of a written contract are ambiguous, evidence is admissible to show the meaning of the terms.
4. Evidence is admissible when the written contract is incomplete in that it lacks one or more of the essential terms. The courts allow evidence to "fill in the gaps" in the contract.
5. Under the UCC, evidence can be introduced to explain or supplement a written contract by showing a prior dealing, course of performance, or usage of trade.[20] We discuss these terms in further detail in Chapter 14, in the context of sales contracts. Here, it is sufficient to say that when buyers and sellers deal with each other over extended periods of time, certain customary practices develop. These practices are often overlooked in the writing of the contract, so courts allow the introduction of evidence to show how the parties have acted in the past. Usage of trade—practices and customs generally followed in a particular industry—can also shed light on the meaning of certain contract provisions, and thus evidence of trade usage may be admissible. Can usage of trade be considered in determining the meaning of the term *registration* in the context of domain name registration? For a discussion of this issue, see this chapter's *Adapting the Law to the Online Environment* feature.
6. The parol evidence rule does not apply if the existence of the entire written contract is subject to an orally agreed-on condition. Proof of the condition does not alter or modify the written terms but affects the *enforceability* of the written contract. ■ EXAMPLE 10.17 Jelek agrees to purchase Armand's car for $8,000, but only if Jelek's mechanic, Frank, inspects the car and approves of the purchase. Armand agrees to this condition, but because he is leaving town for the weekend and Jelek wants to use the car (if he buys it) before Armand returns, Jelek drafts a contract of sale, and they both sign it. Frank, the mechanic, does not approve of the purchase, and when Jelek does not buy the car, Armand sues him, alleging that he breached the contract. In this situation, Jelek's oral agreement did not alter or modify the terms of the written agreement but concerned whether the contract existed at all. Therefore, the parol evidence rule does not apply. ■
7. When an *obvious* or *gross* clerical (or typographic) error exists that clearly would not represent the agreement of the parties, parol evidence is admissible to correct the error. ■ EXAMPLE 10.18 Sempter agrees to lease 1,000 square feet of office space at the current monthly rate of $3 per square foot from Stone Enterprises. The signed written lease provides for a monthly lease payment of $300 rather than the $3,000 agreed to by the parties. Because the error is obvious, Stone Enterprises would be allowed to admit parol evidence to correct the mistake. ■

INTEGRATED CONTRACTS

The determination of whether evidence will be allowed basically depends on whether the written contract is intended to be a complete and final embodiment of the terms of the agreement. If it is so intended, it is referred to as an **integrated contract,** and extraneous evidence is excluded. If the contract is only partially integrated, evidence of consistent additional terms is admissible to supplement the written agreement.[21]

19. UCC 2–209(2), (3). See Chapter 14.
20. UCC 1–205, 2–202. See Chapter 14.
21. *Restatement (Second) of Contracts,* Section 216.

ADAPTING THE LAW TO THE ONLINE ENVIRONMENT

What Does "Registration" Mean in the Domain Name Context?

Article 2 of the Uniform Commercial Code (UCC) specifically allows evidence of trade usage to be introduced in court to explain or supplement the written terms of a contract. As mentioned earlier, this is one of the exceptions to the parol evidence rule. Article 2, however, applies to sales of *goods.* Does this mean that trade usage *cannot* be admitted to explain the meaning of terms in contracts governed by the common law—that is, contracts that do not involve sales of "goods"? Specifically, can a court consider usage of trade in determining what the term *registration* means in a contract to register a domain name?

THE QUESTION OF "EXCLUSIVE USE"

Typically, when a person or business entity registers an Internet domain name as the address for a Web site, that person or entity expects to have the exclusive right to use that name. Certainly, this was Michael Zurakov's expectation when he registered the domain name "Laborzionist.org" with Register.com, Inc., a business that provides Internet services, including the registration of domain names. Register.com established a "Coming Soon" page for Zurakov's Web site. The page, which would be accessed by anyone keying in Zurakov's domain name, contained banner ads for Register.com and other organizations, as well as a list of "Additional Services." It appeared that the ads were in some way endorsed by Zurakov and that he was the provider of the additional services.

Zurakov sued Register.com, alleging that by registering the domain name, he had obtained the exclusive right to use the name and the corresponding Web page. He claimed that Register.com's use of the page interfered with this right. Register.com asked the court to dismiss the case because, among other things, nothing in the contract stated that Zurakov would have the exclusive use of the domain name. The trial court dismissed the case after concluding that Zurakov had received "everything he bargained for" in the contract—because Register.com had indeed "registered" the domain name. Zurakov appealed.

THE MEANING OF "REGISTRATION"

In arriving at its decision, the trial court had looked at the ordinary meaning of the term *register,* which is "to make a record of." The appellate court, however, stated that "the custom and usage of 'registration' of a domain name in the Internet context is certainly more relevant than the literal definition of 'registration' found in the dictionary." According to custom and usage, the registration of a domain name conferred on the person registering the name the *exclusive* right to use that name. The court also stated that the exclusiveness of the use of a registered domain name "is already a familiar concept in the law" and cited a number of cases that illustrated this concept. In sum, concluded the appellate court, Zurakov had stated a valid claim against Register.com, and the case should go to trial.[a]

FOR CRITICAL ANALYSIS

The court also noted that if Zurakov could not have the exclusive use of the domain name, the registration contract would be "rendered illusory." What did the court mean by this statement?

a. *Zurakov v. Register.com, Inc.,* 304 A.D.2d 176, 760 N.Y.S.2d 13 (1 Dept. 2003).

TERMS AND CONCEPTS

collateral promise 197
integrated contract 202
parol evidence rule 202
prenuptial agreement 199
scienter 194
Statute of Frauds 196

CHAPTER SUMMARY  Defenses to Contract Enforceability

GENUINENESS OF ASSENT	
Mistakes (See pages 191–193.)	1. *Unilateral*—Generally, the mistaken party is bound by the contract *unless* (a) the other party knows or should have known of the mistake or (b) the mistake is an inadvertent mathematical error—such as an error in addition or subtraction—committed without gross negligence. 2. *Bilateral (mutual)*—When both parties are mistaken about the same material fact, such as identity, either party can avoid the contract. If the mistake concerns value or quality, either party can enforce the contract.

(Continued)

CHAPTER SUMMARY Defenses to Contract Enforceability—Continued

Fraudulent Misrepresentation (See pages 194–196.)	When fraud occurs, usually the innocent party can enforce or avoid the contract. The elements necessary to establish fraud are as follows: 1. A misrepresentation of a material fact must occur. 2. There must be an intent to deceive. 3. The innocent party must justifiably rely on the misrepresentation.
Undue Influence (See page 196.)	Undue influence arises from special relationships, such as fiduciary or confidential relationships, in which one party's free will has been overcome by the undue influence exerted by the other party. Usually, the contract is voidable.
Duress (See page 196.)	Duress is the tactic of forcing a party to enter into a contract under the fear of a threat—for example, the threat of violence or serious economic loss. The party forced to enter the contract can rescind the contract.
FORM	
The Statute of Frauds—Requirement of a Writing (See pages 196–200.)	*Applicability*—The following types of contracts fall under the Statute of Frauds and must be in writing to be enforceable: 1. *Contracts involving interests in land*—The statute applies to any contract for an interest in realty, such as a sale, a lease, or a mortgage. 2. *Contracts whose terms cannot be performed within one year*—The statute applies only to contracts objectively impossible to perform fully within one year from (the day after) the contract's formation. 3. *Collateral promises*—The statute applies only to express contracts made between the guarantor and the creditor whose terms make the guarantor secondarily liable. Exception: the "main purpose" rule. 4. *Promises made in consideration of marriage*—The statute applies to promises to pay money or give property in consideration of a promise to marry and to prenuptial agreements made in consideration of marriage. 5. *Contracts for the sale of goods priced at $500 or more*—Under the UCC Statute of Frauds provision in UCC 2–201. (Under a 2003 amendment to that UCC section, the threshold amount is raised from $500 to $5,000.) *Exceptions*—Partial performance, admissions, and promissory estoppel.
The Statute of Frauds—Sufficiency of the Writing (See pages 200–201.)	To constitute an enforceable contract under the Statute of Frauds, a writing must be signed by the party against whom enforcement is sought, must name the parties, must identify the subject matter, and must state with reasonable certainty the essential terms of the contract. In a sale of land, the price and a description of the property may need to be stated with sufficient clarity to allow them to be determined without reference to outside sources. Under the UCC, a contract for a sale of goods is not enforceable beyond the quantity of goods shown in the contract.
The Parol Evidence Rule (See pages 202–203.)	The parol evidence rule prohibits the introduction at trial of evidence of the parties' prior negotiations, prior agreements, or contemporaneous oral agreements that contradicts or varies the terms of the parties' written contract. The written contract is assumed to be the complete embodiment of the parties' agreement. Exceptions are made in the following circumstances: 1. To show that the contract was subsequently modified. 2. To show that the contract was voidable or void. 3. To clarify the meaning of ambiguous terms. 4. To clarify the terms of the contract when the written contract lacks one or more of its essential terms. 5. Under the UCC, to explain the meaning of contract terms in light of a prior dealing, course of performance, or usage of trade. 6. To show that the entire contract is subject to an orally agreed-on condition. 7. When an obvious clerical or typographic error was made.

FOR REVIEW

Answers for the even-numbered questions in this For Review *section can be found in Appendix A at the end of this text.*

1. In what types of situations might genuineness of assent to a contract's terms be lacking?
2. What is the difference between a mistake of value or quality and a mistake of fact?
3. What elements must exist for fraudulent misrepresentation to occur?
4. What contracts must be in writing to be enforceable?
5. What is parol evidence? When is it admissible to clarify the terms of a written contract?

QUESTIONS AND CASE PROBLEMS

10–1. Genuineness of Assent. Jerome is an elderly man who lives with his nephew, Philip. Jerome is totally dependent on Philip's support. Philip tells Jerome that unless Jerome transfers a tract of land he owns to Philip for a price 30 percent below market value, Philip will no longer support and take care of him. Jerome enters into the contract. Discuss fully whether Jerome can set aside this contract.

QUESTION WITH SAMPLE ANSWER

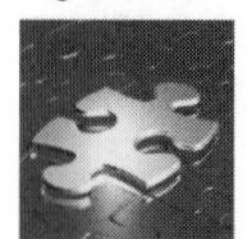

10–2. Gemma promises a local hardware store that she will pay for a lawn mower that her brother is purchasing on credit if the brother fails to pay the debt. Must this promise be in writing to be enforceable? Why or why not?

For a sample answer to this question, go to Appendix B at the end of this text.

10–3. Fraudulent Misrepresentation. Larry offered to sell Stanley his car and told Stanley that the car had been driven only 25,000 miles and had never been in an accident. Stanley hired Cohen, a mechanic, to appraise the condition of the car, and Cohen said that the car probably had at least 50,000 miles on it and probably had been in an accident. In spite of this information, Stanley still thought the car would be a good buy for the price, so he purchased it. Later, when the car developed numerous mechanical problems, Stanley sought to rescind the contract on the basis of Larry's fraudulent misrepresentation of the auto's condition. Will Stanley be able to rescind his contract? Discuss.

10–4. Genuineness of Assent. Linda Lorenzo purchased a home from Lurlene Noel in 1988 without having it inspected. The basement started leaking in 1989. In 1991, Lorenzo had the paneling removed from the basement walls and discovered that the walls were bowed inward and cracked. Lorenzo then had a civil engineer inspect the basement walls, and he found that the cracks had been caulked and painted over before the paneling was installed. He concluded that the "wall failure" had existed "for at least thirty years" and that the basement walls were "structurally unsound." Does Lorenzo have a cause of action against Noel? If so, on what ground? Discuss. [*Lorenzo v. Noel,* 206 Mich.App. 682, 522 N.W.2d 724 (1994)]

10–5. Fraudulent Misrepresentation. William Meade, Leland Stewart, Doug Vierkant, and David Girard applied for, and were offered, jobs at the El-Jay Division of Cedarapids, Inc., in Eugene, Oregon. During the interviews, each applicant asked about El-Jay's future. They were told, among other things, that El-Jay was a stable company with few downsizings and layoffs, sales were up and were expected to increase, and production was expanding. Cedarapids management had already planned to close El-Jay, however. Each applicant signed an at-will employment agreement. To take the job at El-Jay, each new employee either quit the job he was doing or passed up other employment opportunities. Each employee and his family then moved to Eugene. When El-Jay closed soon after they started their new jobs, Meade and the others filed a suit in a federal district court against Cedarapids, alleging, in part, fraudulent misrepresentation. Were the plaintiffs justified in relying on the statements made to them during their job interviews? Explain. [*Meade v. Cedarapids, Inc.,* 164 F.3d 1218 (9th Cir. 1999)]

CASE PROBLEM WITH SAMPLE ANSWER

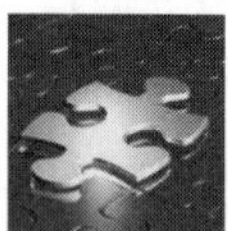

10–6. Robert Pinto, doing business as Pinto Associates, hired Richard MacDonald as an independent contractor in March 1992. The parties orally agreed on the terms of employment, including payment to MacDonald of a share of the company's income, but they did not put anything in writing. In March 1995, MacDonald quit. Pinto then told MacDonald that he was entitled to $9,602.17—25 percent of the difference between the accounts receivable and the accounts payable as of MacDonald's last day. MacDonald disagreed and demanded more than $83,500—25 percent of the revenue from all invoices, less the cost of materials and outside processing, for each of the years that he worked for Pinto. Pinto refused. MacDonald filed a suit in a Connecticut state court against Pinto, alleging breach of contract. In Pinto's response and at the trial, he testified that the parties had an oral contract under which MacDonald was entitled to 25 percent of the difference between the accounts receivable and payable as of the date of MacDonald's termination. Did the parties have an enforceable contract? How should the court rule, and why? [*MacDonald v. Pinto,* 62 Conn.App. 317, 771 A.2d 156 (2001)]

After you have answered this problem, compare your answer with the sample answer given on the Web site that accompanies this text. Go to **academic.cengage.com/blaw/wbl**, select "Chapter 10," and click on "Case Problem with Sample Answer."

10–7. The Parol Evidence Rule. Carlin Krieg owned a dairy farm in St. Joe, Indiana, appraised at $154,000 in December 1997. In August 1999, Krieg told Donald Hieber that he intended to sell the farm for $106,000. Hieber offered to buy it. Krieg also told Hieber that he wanted to retain a "right of residency" for life in the farm. In October, Krieg and Hieber executed a "Purchase Agreement" that provided Krieg "shall transfer full and complete possession" of the farm "subject to [his] right of residency." The agreement also contained an integration clause that stated "there are no conditions, representations, warranties, or agreements not stated in this instrument." In November 2000, the house was burned in a fire, rendering it unlivable. Hieber filed an insurance claim for the damage and received the proceeds, but he did not fix the house. Krieg filed a suit in an Indiana state court against Hieber, alleging breach of contract. Is there any basis on which the court can consider evidence regarding the parties' negotiations prior to their agreement for the sale of the farm? Explain. [*Krieg v. Hieber,* 802 N.E.2d 938 (Ind.App. 2004)]

VIDEO QUESTION

10–8. Go to this text's Web site at academic.cengage.com/blaw/wbl and select "Chapter 10." Click on "Video Questions" and view the video titled *Mistake.* Then answer the following questions.

1. What kind of mistake is involved in the dispute shown on the video (bilateral or unilateral, mistake of fact or mistake of value)?
2. According to the chapter, in what two situations would the supermarket be able to rescind a contract to sell peppers to Melnick at the incorrectly advertised price?
3. Does it matter if the price that was advertised was a reasonable price for the peppers? Why or why not?

Defenses to Contract Enforceability

For updated links to resources available on the Web, as well as a variety of other materials, visit this text's Web site at

academic.cengage.com/blaw/wbl

The online version of UCC Section 2–201 on the Statute of Frauds includes links to definitions of certain terms used in the section. To access this site, go to

http://www.law.cornell.edu/ucc/2/2-201.html

Professor Eric Talley of the University of Southern California provides an interesting discussion of the history and current applicability of the Statute of Frauds, both internationally and in the United States, at

http://www-bcf.usc.edu/~etalley/frauds.html

For information on the *Restatements of the Law,* including the *Restatement (Second) of Contracts,* go to the American Law Institute's Web site at

http://www.ali.org

ONLINE LEGAL RESEARCH

Go to the *Fundamentals of Business Law* home page at academic.cengage.com/blaw/wbl, select "Chapter 10," and click on "Internet Exercises." There you will find the following Internet research exercises that you can perform to learn more about topics covered in this chapter.

Activity 10–1: LEGAL PERSPECTIVE—Promissory Estoppel and the Statute of Frauds

Activity 10–2: MANAGEMENT PERSPECTIVE—Fraudulent Misrepresentation

Activity 10–3: HISTORICAL PERSPECTIVE—The English Act for the Prevention of Frauds and Perjuries

BEFORE THE TEST

Go to the *Fundamentals of Business Law* home page at academic.cengage.com/blaw/wbl, select "Chapter 10," and click on "Interactive Quizzes." You will find a number of interactive questions relating to this chapter.

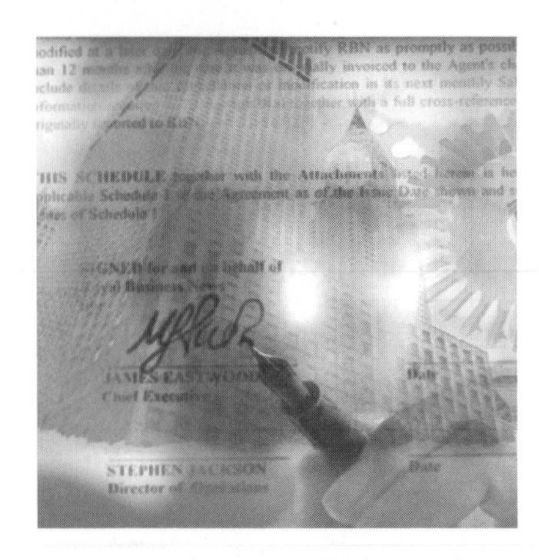

CHAPTER 11

Third Party Rights and Discharge

LEARNING OBJECTIVES

After reading this chapter, you should be able to answer the following questions:

1. What is the difference between an assignment and a delegation?
2. What rights can be assigned despite a contract clause expressly prohibiting assignment?
3. What factors indicate that a third party beneficiary is an intended beneficiary?
4. How are most contracts discharged?
5. What is a contractual condition, and how might a condition affect contractual obligations?

Because a contract is a private agreement between the parties who have entered into it, it is fitting that these parties alone should have rights and liabilities under the contract. This concept is referred to as **privity of contract,** and it establishes the basic principle that third parties have no rights in contracts to which they are not parties.

You may be convinced by now that for every rule of contract law, there is an exception. When justice cannot be served by adherence to a rule of law, exceptions to the rule must be made. In this chapter, we look at some exceptions to the rule of privity of contract. These exceptions include *assignments* and *delegations,* as well as *third party beneficiary contracts.* We also examine how contractual obligations can be discharged, or terminated. Normally, contract discharge is accomplished when both parties have performed the acts promised in the contract. In the latter part of this chapter, we look at some other ways in which contract discharge can occur.

Assignments

In a bilateral contract, normally one party has a right to require the other to perform some task, and the other has a duty to perform it. The transfer of contract *rights* to a third person is known as an **assignment.** The party assigning the rights to a third party is known as the **assignor,** and the party receiving the rights is the **assignee.** Other traditional terminology used to describe the parties in assignment relationships are the **obligee** (the person to whom a duty, or obligation, is owed) and the **obligor** (the person who is obligated to perform the duty).

When rights under a contract are assigned unconditionally, the rights of the *assignor* (the party making the assignment) are extinguished.[1] The third party (the *assignee,* or the party receiving the assignment) has a right to demand performance from the other original party to the contract (the *obligor,* the person who is obligated to perform). ■ EXAMPLE 11.1 Brent owes Alex $1,000, and Alex, the assignor, assigns to Carmen the right to receive the $1,000. Here, a valid assignment of a debt exists. Carmen, the assignee, can enforce the contract against Brent, the obligor, if Brent fails to perform. ■ Exhibit 11–1 on the next page illustrates assignment relationships.

The assignee takes only those rights that the assignor originally had. Furthermore, the assignee's rights are subject to the defenses that the obligor has against the

1. *Restatement (Second) of Contracts,* Section 317.

EXHIBIT 11–1 ASSIGNMENT RELATIONSHIPS

In the assignment relationship illustrated here, Alex assigns his *rights* under a contract that he made with Brent to a third party, Carmen. Alex thus becomes the *assignor* and Carmen the *assignee* of the contractual rights. Brent, the *obligor* (the party owing performance under the contract), now owes performance to Carmen instead of Alex. Alex's original contract rights are extinguished after the assignment.

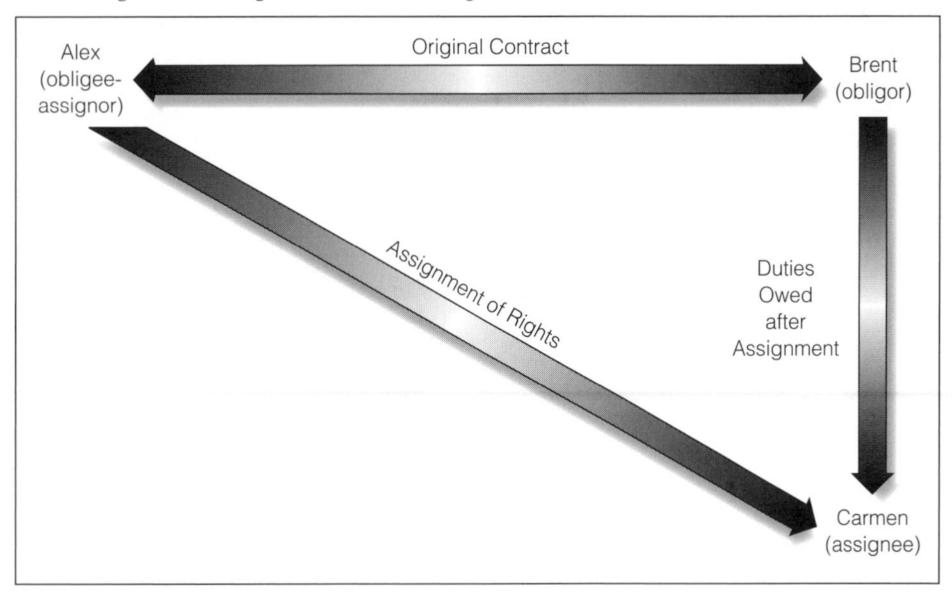

assignor. ■ **EXAMPLE 11.2** Brent owes Alex $1,000 under a contract in which Brent agreed to buy Alex's computer work station. Alex assigns his right to receive the $1,000 to Carmen. Brent, in deciding to purchase the work station, relied on Alex's fraudulent misrepresentation that the computer's hard drive had a storage capacity of 120 gigabytes. When Brent discovers that the computer can store only 20 gigabytes, he tells Alex that he is going to return the work station and cancel the contract. Even though Alex has assigned his "right" to receive the $1,000 to Carmen, Brent does not have to pay Carmen the $1,000—Brent can raise the defense of Alex's fraudulent misrepresentation to avoid payment. ■

The Importance of Assignments in the Business Context

Assignments are important because they are utilized in much business financing. Lending institutions, such as banks, frequently assign the rights to receive payments under their loan contracts to other firms, which pay for those rights. If you obtain a loan from your local bank to purchase a car, you may later receive in the mail a notice stating that your bank has transferred (assigned) its rights to receive payments on the loan to another firm and that, when the time comes to repay your loan, you must make the payments to that other firm.

Lenders that make *mortgage loans* (loans to allow prospective home buyers to purchase land or a home) often assign their rights to collect the mortgage payments to a third party, such as GMAC Mortgage Corporation. Following an assignment, the home buyer is notified that future payments must be made to the third party and not to the lender that loaned the funds. Millions of dollars change hands daily in the business world in the form of assignments of rights in contracts. If contractual rights could not be transferred (assigned), many businesses could not continue to operate.

Rights That Cannot Be Assigned

As a general rule, all rights can be assigned. Exceptions are made, however, in the following special circumstances.

When a Statute Expressly Prohibits Assignment If a statute expressly prohibits assignment, the particular right in question cannot be assigned. ■ **EXAMPLE 11.3** Marn is a new employee of CompuFuture, Inc. CompuFuture is an employer under workers' compensation statutes (see Chapter 23) in this state, and thus Marn is a covered

employee. Marn has a relatively high-risk job. In need of a loan, she borrows funds from Stark, assigning to Stark all workers' compensation benefits due her should she be injured on the job. A state statute prohibits the assignment of *future* workers' compensation benefits, and thus such rights cannot be assigned. ■

When a Contract Is Personal in Nature When a contract is for personal services, the rights under the contract normally cannot be assigned unless all that remains is a money payment.[2] ■ EXAMPLE 11.4 Brent signs a contract to be a tutor for Alex's children. Alex then attempts to assign to Carmen his right to Brent's services. Carmen cannot enforce the contract against Brent. Brent may not like Carmen's children or may not want to tutor them for some other reason. Because personal services are unique to the person rendering them, rights to receive personal services cannot be assigned. ■

When an Assignment Will Significantly Change the Risk or Duties of the Obligor A right cannot be assigned if assignment will significantly increase or alter the risks to or the duties of the obligor.[3] ■ EXAMPLE 11.5 Alex owns a hotel, and to insure it, he takes out a policy with Northwest Insurance Company. The policy insures against fire, theft, floods, and vandalism. Alex attempts to assign the insurance policy to Carmen, who also owns a hotel. The assignment is ineffective because it may substantially alter the insurance company's duty of performance and the risk that the company undertakes. An insurance company evaluates the particular risk of a certain party and tailors its policy to fit that risk. If the policy is assigned to a third party, the insurance risk is materially altered. ■

When the Contract Prohibits Assignment If a contract stipulates that the right cannot be assigned, then *ordinarily* it cannot be assigned. ■ EXAMPLE 11.6 Brent agrees to build a house for Alex. The contract between Brent and Alex states, "This contract cannot be assigned by Alex without Brent's consent. Any assignment without such consent renders this contract void, and all rights hereunder will thereupon terminate." Alex then assigns his rights to Carmen, without first obtaining Brent's consent. Carmen cannot enforce the contract against Brent. ■ This rule has several exceptions:

1. A contract cannot prevent an assignment of the right to receive money. This exception exists to encourage the free flow of money and credit in modern business settings.
2. The assignment of ownership rights in real estate often cannot be prohibited because such a prohibition is contrary to public policy in most states. Prohibitions of this kind are called restraints against **alienation** (the voluntary transfer of land ownership).
3. The assignment of negotiable instruments (see Chapter 18) cannot be prohibited.
4. In a contract for the sale of goods, the right to receive damages for breach of contract or for payment of an account owed may be assigned even though the sales contract prohibits such assignment.[4]

NOTICE OF ASSIGNMENT

Once a valid assignment of rights has been made to a third party, the third party should notify the obligor of the assignment (for example, in Exhibit 11–1 on page 208, Carmen should notify Brent). Giving notice is not legally necessary to establish the validity of the assignment because an assignment is effective immediately, whether or not notice is given. Two major problems arise, however, when notice of the assignment is *not* given to the obligor:

1. If the assignor assigns the same right to two different persons, the question arises as to which one has priority—that is, which one has the right to the performance by the obligor. Although the rule most often observed in the United States is that the first assignment in time is the first in right, some states follow the English rule, which basically gives priority to the first assignee who gives notice. ■ EXAMPLE 11.7 Brent owes Alex $5,000 on a contractual obligation. On May 1, Alex assigns this monetary claim to Carmen. Carmen gives no notice of the assignment to Brent. On June 1, for services Dorman has rendered to Alex, Alex assigns the same monetary claim (to collect $5,000 from Brent) to Dorman. Dorman immediately notifies Brent of the assignment. In the majority of states, Carmen would have priority because the assignment to her was first in time. In some states, however, Dorman would have priority because he gave first notice. ■
2. Until the obligor has notice of assignment, the obligor can discharge his or her obligation by performance to the assignor, and performance by the obligor to the assignor constitutes a discharge to the assignee. Once the obligor receives proper notice, only performance to the assignee can discharge the

2. *Restatement (Second) of Contracts,* Sections 317 and 318.
3. See Section 2–210(2) of the Uniform Commercial Code (UCC).
4. UCC 2–210(2).

obligor's obligations. ■ **EXAMPLE 11.8** In the above example, Alex assigns to Carmen his right to collect $5,000 from Brent. Carmen does not give notice to Brent. Brent subsequently pays Alex the $5,000. Although the assignment was valid, Brent's payment to Alex was a discharge of the debt, and Carmen's failure to notify Brent of the assignment caused her to lose the right to collect from Brent. If Carmen had given Brent notice of the assignment, however, Brent's payment to Alex would not have discharged the debt. ■

In the following case, the issue was whether the right to buy advertising space in certain publications at a steep discount was validly assigned from the original owner to companies that he later formed.

CASE 11.1 Gold v. Ziff Communications Co.

Appellate Court of Illinois, First District, 2001.
322 Ill.App.3d 32,
748 N.E.2d 198,
254 Ill.Dec. 752.
http://state.il.us/court/default.htm[a]

FACTS In 1982, Ziff Communications Company, a publisher of specialty magazines, bought *PC Magazine* from its founder, Anthony Gold, for more than $10 million. As part of the deal, Ziff gave Gold, or a company that he "controlled," "ad/list rights"—rights to advertise at an 80 percent discount on a limited number of pages in Ziff publications and free use of Ziff's subscriber lists. In 1983, Gold formed Software Communications, Inc. (SCI), a mail-order software business that he wholly owned, to use the ad/list rights. In 1987 and 1988, he formed two new mail-order companies, Hanson & Connors, Inc., and PC Brand, Inc. Gold told Ziff that he was allocating his ad/list rights to Hanson & Connors, which took over most of SCI's business, and to PC Brand, of which Gold owned 90 percent. Ziff's other advertisers complained about this "allocation." Ziff refused to run large ads for Hanson & Connors or to release its subscriber lists to the company. Ziff also declared PC Brand ineligible for the ad discount because it "was not controlled by Gold." Gold and his companies filed a suit in an Illinois state court against Ziff, alleging breach of contract. The court ordered Ziff to pay the plaintiffs more than $88 million in damages and interest. Ziff appealed to an intermediate state appellate court, arguing in part that Gold had not properly assigned the ad/list rights to Hanson & Connors and PC Brand.

a. On this page, click on "Appellate Court of Illinois." On the next page, in the "Appellate Court Documents" section, click on "Appellate Court opinions." In the result, in the "Appellate Court" section, click on "2001." On the next page, in the "First District" section, click on "March." Finally, scroll to the bottom of the chart and click on the case name to access the opinion. The state of Illinois maintains this Web site.

ISSUE Was there a valid assignment of rights from Gold and SCI to Hanson & Connors and PC Brand?

DECISION Yes. The state intermediate appellate court affirmed the lower court's decision on this issue. The appellate court remanded the case, however, for a new trial on the amount of the damages, reasoning that some parts of the award "were not within the reasonable contemplation of the parties."

REASON The court explained, "We agree with plaintiffs that assignments can be implied from circumstances. No particular mode or form * * * is necessary to effect a valid assignment, and any acts or words are sufficient which show an intention of transferring or appropriating the owner's interest. In the instant case, it is undisputed that Gold owned 100 [percent] of SCI. In a letter dated May 13, 1988, Gold, as president of SCI, instructed Ziff that he was allocating the ad/list rights to Hanson and PC Brand. Additionally, SCI stopped using the ad/list rights when PC Brand and Hanson were formed. * * * Gold's behavior toward his companies and his conduct toward the obligor, Ziff, implied that the ad/list rights were assigned to PC Brand and Hanson."

FOR CRITICAL ANALYSIS—Social Consideration
Would the assignments in this case have been valid if Gold had not notified Ziff?

Delegations

Just as a party can transfer rights to a third party through an assignment, a party can also transfer duties. Duties are not assigned, however; they are *delegated.* Normally, a **delegation of duties** does not relieve the party making the delegation (the **delegator**) of the obligation to perform in the event that the party to whom the duty has been delegated (the **delegatee**) fails to perform. No special form is required to create a valid delegation of duties. As long as the delegator expresses an intention to make the delegation, it is effective; the delegator need not even

use the word *delegate.* Exhibit 11–2 graphically illustrates delegation relationships.

DUTIES THAT CANNOT BE DELEGATED

As a general rule, any duty can be delegated. This rule has some exceptions, however. Delegation is prohibited in the following circumstances:

1. When performance depends on the personal skill or talents of the obligor.
2. When special trust has been placed in the obligor.
3. When performance by a third party will vary materially from that expected by the obligee (the one to whom performance is owed) under the contract.
4. When the contract expressly prohibits delegation.

The following examples will help to clarify the kinds of duties that can and cannot be delegated:

1. Brent contracts with Alex to tutor Alex in various aspects of financial underwriting and investment banking. Brent, an experienced businessperson known for his expertise in finance, delegates his duties to a third party, Carmen. This delegation is ineffective because Brent contracted to render a service that is founded on Brent's *expertise,* and the delegation changes Alex's expectancy under the contract.
2. Brent contracts with Alex to *personally* mow Alex's lawn during June, July, and August. Then Brent decides that he would rather spend the summer at the beach. Brent delegates his lawn-mowing duties to Carmen, who is in the business of mowing lawns and doing other landscaping work to earn income to pay for college. No matter how competent Carmen is, the delegation is not effective without Alex's consent. The contract was for *personal* performance.
3. Brent contracts with Alex to pick up and deliver heavy construction machinery to Alex's property. Brent delegates this duty to Carmen, who is in the business of delivering heavy machinery. This delegation is effective. The performance required is of a routine and nonpersonal nature, and the delegation does not change Alex's expectations under the contract.

EFFECT OF A DELEGATION

If a delegation of duties is enforceable, the *obligee* (the one to whom performance is owed) must accept performance from the delegatee (the one to whom the duties are delegated). ■ **EXAMPLE 11.9** In the third example in the above list, Brent delegates his duty (to pick up and deliver heavy construction machinery to Alex's property) to Carmen. In that situation, Alex (the obligee) must accept

EXHIBIT 11–2 DELEGATION RELATIONSHIPS

In the delegation relationship illustrated here, Brent delegates his *duties* under a contract that he made with Alex to a third party, Carmen. Brent thus becomes the *delegator* and Carmen the *delegatee* of the contractual duties. Carmen now owes performance of the contractual duties to Alex. Note that a delegation of duties normally does not relieve the delegator (Brent) of liability if the delegatee (Carmen) fails to perform the contractual duties.

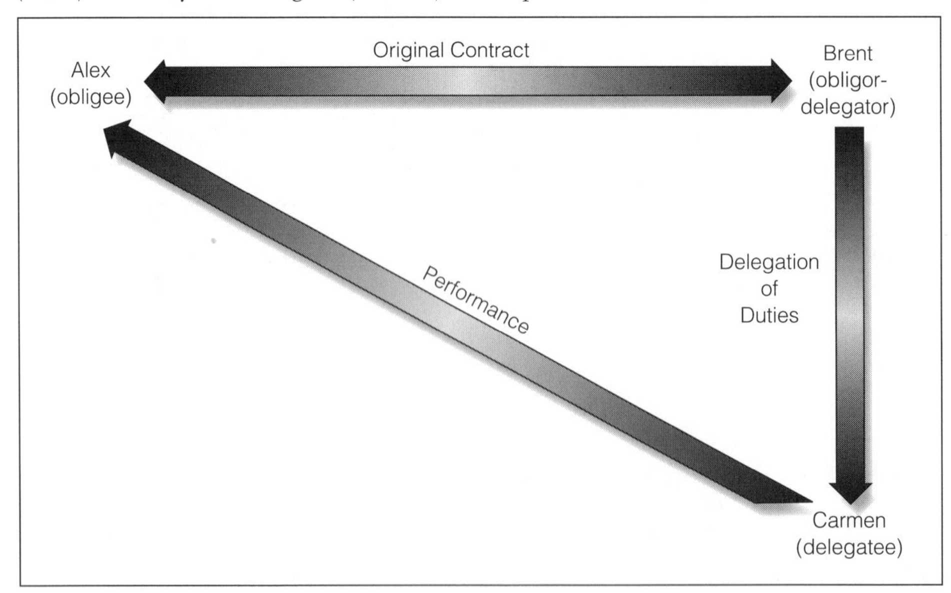

performance from Carmen (the delegatee) because the delegation was effective. The obligee can legally refuse performance from the delegatee only if the duty is one that cannot be delegated. ■

A valid delegation of duties does not relieve the delegator of obligations under the contract.[5] In the above example, if Carmen (the delegatee) fails to perform, Brent (the delegator) is still liable to Alex (the obligee). The obligee can also hold the delegatee liable if the delegatee made a promise of performance that will directly benefit the obligee. In this situation, there is an "assumption of duty" on the part of the delegatee, and breach of this duty makes the delegatee liable to the obligee. For example, if Carmen (the delegatee) promises Brent (the delegator), in a contract, to pick up and deliver the construction equipment to Alex's property but fails to do so, Alex (the obligee) can sue Brent, Carmen, or both. Although there are many exceptions, the general rule today is that the obligee can sue both the delegatee and the delegator.

Exhibit 11–3 summarizes the basic principles of the laws governing assignments and delegations.

"Assignment of All Rights"

Sometimes, a contract provides for an "assignment of all rights." The traditional view was that under this type of assignment, the assignee did not assume any duties. This view was based on the theory that the assignee's agreement to accept the benefits of the contract was not sufficient to imply a promise to assume the duties of the contract.

Modern authorities, however, take the view that the probable intent in using such general words is to create both an assignment of rights and an assumption of duties.[6] Therefore, when general words are used (for example, "I assign the contract" or "all my rights under the contract"), the contract is construed as implying both an assignment of rights and an assumption of duties.

Third Party Beneficiaries

As mentioned earlier in this chapter, to have contractual rights, a person normally must be a party to the contract. In other words, privity of contract must exist. An exception to the doctrine of privity exists when the original parties to the contract intend, at the time of contracting, that the contract performance directly benefit a third person. In this situation, the third person becomes a **third party beneficiary** of the contract. As an **intended beneficiary** of the contract, the third party has legal rights and can sue the promisor directly for breach of the contract.

Who, though, is the promisor? In bilateral contracts, both parties to the contract are promisors because they both make promises that can be enforced. In third party benefici-

5. For a classic case on this issue, see *Crane Ice Cream Co. v. Terminal Freezing & Heating Co.,* 147 Md. 588, 128 A. 280 (1925).

6. See UCC 2–210(1), (4); *Restatement (Second) of Contracts,* Section 328.

EXHIBIT 11–3 ASSIGNMENTS AND DELEGATIONS

Which rights can be assigned, and which duties can be delegated?	All rights can be assigned *unless:* **1.** A statute expressly prohibits assignment. **2.** The assignment will materially alter the obligor's risk or duties. **3.** The contract is for personal services. **4.** The contract prohibits assignment.	All duties can be delegated *unless:* **1.** Performance depends on the obligor's personal skills or talents. **2.** Special trust has been placed in the obligor. **3.** Performance by a third party will materially vary from that expected by the obligee. **4.** The contract prohibits delegation.
What if the contract prohibits assignment or delegation?	No rights can be assigned *except:* **1.** Rights to receive money. **2.** Ownership rights in real estate. **3.** Rights to negotiable instruments. **4.** Rights to sales contract payments or damages for breach of a sales contract.	No duties can be delegated.
What is the effect on the original party's rights?	On a valid assignment, effective immediately, the original party (assignor) no longer has any rights under the contract.	On a valid delegation, if the delegatee fails to perform, the original party (delegator) is liable to the obligee (who may also hold the delegatee liable).

ary contracts, courts determine the identity of the promisor by asking which party made the promise that benefits the third party—that person is the promisor. Allowing the third party to sue the promisor directly in effect circumvents the "middle person" (the promisee) and thus reduces the burden on the courts. Otherwise, the third party would sue the promisee, who would then sue the promisor.

A classic case in the area of third party beneficiary contracts is *Lawrence v. Fox*[7]—a case decided in 1859. In that case, the court set aside the traditional requirement of privity and allowed a third party to bring a suit directly against the promisor.

Types of Intended Beneficiaries

The law distinguishes between *intended* beneficiaries and *incidental* beneficiaries. Only intended beneficiaries acquire legal rights in a contract. One type of intended beneficiary is a creditor beneficiary. A *creditor beneficiary* benefits from a contract in which one party (the promisor) promises another party (the promisee) to pay a debt that the promisee owes to a third party (the creditor beneficiary). As an intended beneficiary, the creditor beneficiary can sue the promisor directly to enforce the contract.

Another type of intended beneficiary is a *donee beneficiary.* When a contract is made for the express purpose of giving a *gift* to a third party, the third party (the donee beneficiary) can sue the promisor directly to enforce the promise.[8] The most common donee beneficiary contract is a life insurance contract. ■ EXAMPLE 11.10 Akins (the promisee) pays premiums to Standard Life, a life insurance company, and Standard Life (the promisor) promises to pay a certain amount of money on Akins's death to anyone Akins designates as a beneficiary. The designated beneficiary is a donee beneficiary under the life insurance policy and can enforce the promise made by the insurance company to pay him or her on Akins's death. ■

As the law concerning third party beneficiaries evolved, numerous cases arose in which the third party beneficiary did not fit readily into either category—creditor beneficiary or donee beneficiary. Thus, the modern view, and the one adopted by the *Restatement (Second) of Contracts,* does not draw such clear lines and distinguishes only between *intended beneficiaries* (who can sue to enforce contracts made for their benefit) and *incidental beneficiaries* (who cannot sue, as will be discussed shortly).

7. 20 N.Y. 268 (1859).

8. This principle was first enunciated in *Seaver v. Ransom,* 224 N.Y. 233, 120 N.E. 639 (1918).

When the Rights of an Intended Beneficiary Vest

An intended third party beneficiary cannot enforce a contract against the original parties until the rights of the third party have *vested,* meaning that the rights have taken effect and cannot be taken away. Until these rights have vested, the original parties to the contract—the promisor and the promisee—can modify or rescind the contract without the consent of the third party. When do the rights of third parties vest? Generally, the rights vest when one of the following occurs:

1. When the third party demonstrates manifest assent to the contract, such as sending a letter or note acknowledging awareness of and consent to a contract formed for her or his benefit.
2. When the third party materially alters his or her position in detrimental reliance on the contract, such as when a donee beneficiary contracts to have a home built in reliance on the receipt of funds promised to him or her in a donee beneficiary contract.
3. When the conditions for vesting are satisfied. For example, the rights of a beneficiary under a life insurance policy vest when the insured person dies.

If the contract expressly reserves to the contracting parties the right to cancel (rescind) or modify the contract, the rights of the third party beneficiary are subject to any changes that result. In such a situation, the vesting of the third party's rights does not terminate the power of the original contracting parties to alter their legal relationships.[9]

Incidental Beneficiaries

The benefit that an **incidental beneficiary** receives from a contract between two parties is unintentional. Therefore, an incidental beneficiary cannot enforce a contract to which he or she is not a party.

■ EXAMPLE 11.11 In one case, spectators at a Mike Tyson boxing match in which Tyson was disqualified for biting his opponent's ear sued Tyson and the fight's promoters for a refund of their money on the basis of breach of contract. The spectators claimed that they had standing to sue the defendants as third party beneficiaries of the contract between Tyson and the fight's promoters. The court, however, held that the spectators did not have standing to sue because they were not in contractual privity with any of the defendants. Furthermore, any benefits they received from the contract were incidental to the contract. The court

9. Defenses raised against third party beneficiaries are given in the *Restatement (Second) of Contracts,* Section 309.

ADAPTING THE LAW TO THE ONLINE ENVIRONMENT

Government Contracts and Third Party Beneficiaries

Government entities often contract with private organizations to provide certain services to the public. Are those who benefit under such contracts intended beneficiaries? This question came before the court in a case involving a person who had registered a domain name with an organization that had contracted with the federal government to provide domain name registration services.

THE DOMAIN NAME CONFLICT

In 1994, in the early days of the Internet (as a public surfing/shopping vehicle), domain names were free for the asking. At that time, Network Solutions, Inc. (NSI), was the sole registrar of domain names. NSI had a contract with a federal government agency stating that NSI had the primary responsibility for "ensuring the quality, timeliness, and effective management" of domain name registration services.

Gary Kremen, seeing what he felt was a great opportunity, registered the name "sex.com" with NSI. Unfortunately for Kremen, Stephen Cohen also saw the potential of that domain name. Cohen, who had just gotten out of prison for impersonating a bankruptcy lawyer, knew that Kremen had already registered the name. That fact, however, in the words of the court, "was only a minor impediment for a man of Cohen's boundless resource and bounded integrity." Through forgery and deceit, Cohen succeeded in having NSI transfer the domain name to his company. When Kremen later contacted NSI, he was told that it was too late to undo the transfer. Kremen then turned to the courts for assistance.

THE LEGAL ISSUES

Kremen sued Cohen, seeking as damages the substantial profits that Cohen had made by using the name. The court held in Kremen's favor and awarded him millions of dollars in damages. Kremen could not collect the judgment, however, because Cohen had disappeared—after first transferring large sums of money to offshore accounts. Kremen then tried to hold NSI responsible for his losses by alleging, among other things, that he was an intended third party beneficiary of NSI's contract with the government. He claimed that because NSI had not "effectively managed" its duties, as it was obligated to do under the contract, his domain name had been wrongfully transferred.

Was Kremen an intended third party beneficiary of the contract? When the case ultimately reached the U.S. Court of Appeals for the Ninth Circuit, the court held that Kremen was not an intended third party beneficiary of the contract. The court noted that a third party can enforce a contract if the contract reflects an "express or implied intention of the parties to the contract to benefit the third party." The court emphasized, however, that when a contract is with a government entity, a more stringent test applies: for a third party to be an intended beneficiary, *the contract must express a clear intent to benefit the third party.* If it does not express this clear intent, then anyone who benefits under the contract is regarded as an incidental beneficiary and, as such, cannot sue to enforce the contract.[a]

FOR CRITICAL ANALYSIS

Kremen also alleged that NSI had breached an implied-in-fact contract with him, but the court dismissed this claim. Why would the court hold that no contract existed between Kremen and NSI? Was a required element for a valid contract lacking?

a. *Kremen v. Cohen,* 337 F.3d 1024 (9th Cir. 2003).

noted that the spectators got what they paid for: "the right to view whatever event transpired."[10] ■

Is a person who benefits from a contract between another party and a government entity an incidental beneficiary or an intended beneficiary? For a discussion of a case raising this question, see this chapter's *Adapting the Law to the Online Environment* feature.

INTENDED VERSUS INCIDENTAL BENEFICIARIES

In determining whether a third party beneficiary is an intended or an incidental beneficiary, the courts generally use the *reasonable person* test—that is, a beneficiary will be considered an intended beneficiary if a reasonable person in the position of the beneficiary would believe that the promisee *intended* to confer on the beneficiary the right to bring suit to enforce the contract.

In determining whether a party is an intended or an incidental beneficiary, the courts also look at a number of other factors. As you can see in Exhibit 11–4, which graphically illustrates the distinction between intended and incidental beneficiaries, the presence of one or more of the following factors strongly indicates that the third party is an intended (rather than an incidental) beneficiary to the contract:

1. Performance is rendered directly to the third party.
2. The third party has the right to control the details of performance.

10. *Castillo v. Tyson,* 268 A.D.2d 336, 701 N.Y.S.2d 423 (Sup.Ct.App.Div. 2000).

EXHIBIT 11–4 THIRD PARTY BENEFICIARIES

CONTRACT THAT BENEFITS A THIRD PARTY

INTENDED BENEFICIARY	INCIDENTAL BENEFICIARY
An intended beneficiary is a third party— • To whom performance is rendered directly • Who has the right to control the details of the performance • Who is designated a beneficiary in the contract	An incidental beneficiary is a third party— • Who benefits from a contract but whose benefit was not the reason for the contract • Who has no rights in the contract
CAN SUE TO ENFORCE THE CONTRACT	**CANNOT SUE TO ENFORCE THE CONTRACT**

3 The third party is expressly designated as a beneficiary in the contract.

Is a party who borrows funds to build a house an intended beneficiary of a contract between the lender and a party that the lender hires to monitor the progress of the home's construction? That was the question in the case presented next.

CASE 11.2 Vogan v. Hayes Appraisal Associates, Inc.

Supreme Court of Iowa, 1999.
588 N.W.2d 420.
http://www.judicial.state.ia.us/supreme/opinions/archive.asp[a]

FACTS Susan and Rollin Vogan wanted to build a home in West Des Moines, Iowa. They met with builder Gary Markley of Char Enterprises, Inc., who agreed to build the home for $169,633.59. The Vogans obtained a $170,000 construction loan from MidAmerica Savings Bank, which hired Hayes Appraisal Associates, Inc., to monitor the progress of the construction. MidAmerica was to disburse payments to Markley based on Hayes's reports. There were cost overruns on the job, and after three months, less than $2,000 of the initial loan remained. Markley said that it would take another $70,000 to finish the house. The Vogans borrowed $42,050 more, added some of their own funds, and gave the funds to the bank to continue payments to Markley based on Hayes's reports. A few weeks later, Hayes reported that the house was 90 percent complete. Seven months later, with the house still unfinished, Markley ceased working on the job. Another contractor estimated that completion would cost an additional $60,000. The Vogans filed a suit in an Iowa state court against Hayes, based in part on Hayes's contract with MidAmerica. Hayes answered in part that the Vogans were not third party beneficiaries of that contract. The court issued a judgment in favor of the Vogans. Hayes appealed to a state intermediate appellate court, which reversed the judgment. The Vogans then appealed to the Iowa Supreme Court.

a. Click on "1999," and in that list, next to "January 21, 1999," click on "Index." Scroll down the list of cases to the *Vogan* case and select "full opinion" to access the opinion. This Web site is maintained by the state of Iowa.

ISSUE Were the Vogans intended third party beneficiaries of the contract between Hayes and MidAmerica?

DECISION Yes. The Iowa Supreme Court held that the Vogans were intended third party beneficiaries of the contract between Hayes and MidAmerica, as evidenced by the statements in the contract. The state supreme court vacated the decision of the state intermediate appellate court and affirmed the judgment of the trial court.

REASON The court stated that "the primary question in a third party beneficiary case is whether the contract manifests an intent to benefit a third party. However, this intent need

(Continued)

CASE 11.2–CONTINUED

not be to benefit a third party directly." The court reasoned that in this case, "[t]he promised performance of Hayes Appraisal to MidAmerica will be of pecuniary [monetary] benefit to the Vogans, and the contract is so expressed as to give Hayes reason to know that such benefit is contemplated by MidAmerica as one of the motivating causes of making the contract." The court concluded that in these circumstances, the Vogans qualified as intended third party beneficiaries of the agreement between MidAmerica and Hayes Appraisal.

FOR CRITICAL ANALYSIS—Social Consideration *If the Vogans could not recover as third party beneficiaries of the Hayes-MidAmerica contract, would they have any legal recourse? If so, against whom?*

Contract Discharge

The most common way to **discharge,** or terminate, one's contractual duties is by the **performance** of those duties. The duty to perform under a contract may be *conditioned* on the occurrence or nonoccurrence of a certain event, or the duty may be *absolute.* As you can see in Exhibit 11–5, in addition to performance, a contract can be discharged in numerous other ways, including discharge by agreement of the parties and discharge by operation of law.

Conditions of Performance

In most contracts, promises of performance are not expressly conditioned or qualified. Instead, they are *absolute promises.* They must be performed, or the party promising the act will be in breach of contract. ■ EXAMPLE 11.12 JoAnne contracts to sell Alfonso a painting for $10,000. The parties' promises are unconditional: JoAnne's transfer of the painting to Alfonso and Alfonso's payment of $10,000 to JoAnne. The pay-

EXHIBIT 11–5 CONTRACT DISCHARGE

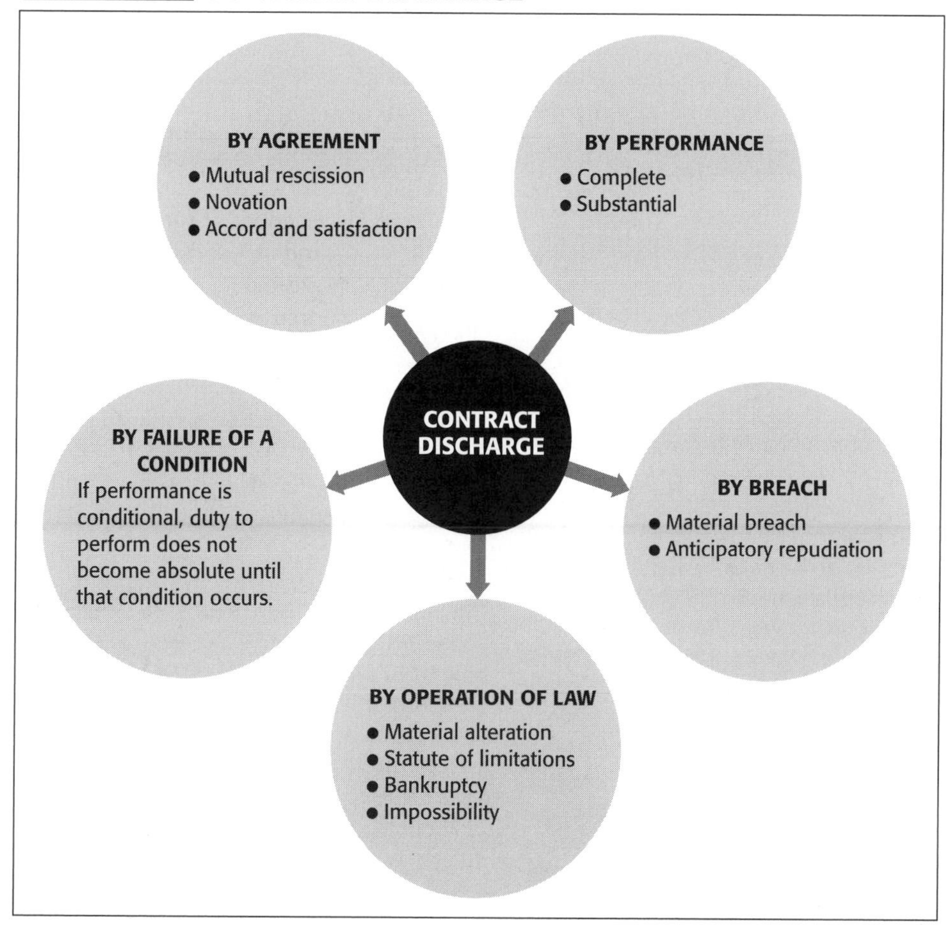

ment does not have to be made if the painting is not transferred. ■

In some situations, however, contractual promises are conditioned. A **condition** is a possible future event, the occurrence or nonoccurrence of which will trigger the performance of a legal obligation or terminate an existing obligation under a contract. If the condition is not satisfied, the obligations of the parties are discharged. ■ EXAMPLE 11.13 Suppose that Alfonso, in the previous example, offers to purchase JoAnne's painting only if an independent appraisal indicates that it is worth at least $10,000. JoAnne accepts Alfonso's offer. Their obligations (promises) are conditioned on the outcome of the appraisal. Should this condition not be satisfied (for example, if the appraiser deems the value of the painting to be only $5,000), their obligations to each other are discharged and cannot be enforced. ■

We look here at three types of conditions that can be present in any given contract: conditions precedent, conditions subsequent, and concurrent conditions.

Conditions Precedent A condition that must be fulfilled before a party's promise becomes absolute is called a **condition precedent.** The condition precedes the absolute duty to perform. ■ EXAMPLE 11.14 In the JoAnne-Alfonso example just given, Alfonso's promise is subject to the condition precedent that the appraised value of the painting must be at least $10,000. Until the condition is fulfilled, Alfonso's promise is not absolute. Insurance contracts frequently specify that certain conditions, such as passing a physical examination, must be met before the insurance company will be obligated to perform under the contract. ■

Conditions Subsequent When a condition operates to terminate a party's absolute promise to perform, it is called a **condition subsequent.** The condition follows, or is subsequent to, the absolute duty to perform. If the condition occurs, the party need not perform any further. ■ EXAMPLE 11.15 A law firm hires Julia Darby, a recent law school graduate and a newly licensed attorney. Their contract provides that the firm's obligation to continue employing Darby is discharged if she fails to maintain her license to practice law. This is a condition subsequent because a failure to maintain the license will discharge a duty that has already arisen.[11] ■

Generally, conditions precedent are common, and conditions subsequent are rare. The *Restatement (Second) of Contracts* deletes the terms *condition subsequent* and *condition precedent* and refers to both simply as "conditions."[12]

Concurrent Conditions When each party's absolute duty to perform is conditioned on the other party's absolute duty to perform, **concurrent conditions** are present. These conditions exist only when the parties expressly or impliedly are to perform their respective duties *simultaneously.* ■ EXAMPLE 11.16 If a buyer promises to pay for goods when they are delivered by the seller, each party's absolute duty to perform is conditioned on the other party's absolute duty to perform. The buyer's duty to pay for the goods does not become absolute until the seller either delivers or attempts to deliver the goods. Likewise, the seller's duty to deliver the goods does not become absolute until the buyer pays or attempts to pay for the goods. Therefore, neither can recover from the other for breach without first tendering performance. ■

DISCHARGE BY PERFORMANCE

The contract comes to an end when both parties fulfill their respective duties by performing the acts they have promised. Performance can also be accomplished by tender. **Tender** is an unconditional offer to perform by a person who is ready, willing, and able to do so. Therefore, a seller who places goods at the disposal of a buyer has tendered delivery and can demand payment according to the terms of the agreement. A buyer who offers to pay for goods has tendered payment and can demand delivery of the goods.

Once performance has been tendered, the party making the tender has done everything possible to carry out the terms of the contract. If the other party then refuses to perform, the party making the tender can consider the duty discharged and sue for **breach of contract.**

Complete versus Substantial Performance Normally, conditions expressly stated in the contract must fully occur in all aspects for *complete performance* (strict performance) of the contract to occur. Any deviation breaches the contract and discharges the other party's obligations to perform. Although in most contracts the parties fully discharge their obligations by complete performance, sometimes a party fails to fulfill all of the duties or completes the duties in a manner contrary to the terms of the contract. The issue then arises as to whether the

11. The difference between conditions precedent and conditions subsequent is relatively unimportant from a substantive point of view but very important procedurally. Usually, the plaintiff must prove conditions precedent because typically it is the plaintiff who claims that there is a duty to be performed. Similarly, the defendant must normally prove conditions subsequent because typically it is the defendant who claims that a duty no longer exists.

12. *Restatement (Second) of Contracts,* Section 224.

performance was nonetheless sufficiently substantial to discharge the contractual obligations.

To qualify as *substantial performance,* the performance must not vary greatly from the performance promised in the contract, and it must create substantially the same benefits as those promised in the contract. If performance is substantial, the other party's duty to perform remains absolute (less damages, if any, for the minor deviations).

■ EXAMPLE 11.17 A couple contracts with a construction company to build a house. The contract specifies that Brand X plasterboard be used for the walls. The builder cannot obtain Brand X plasterboard, and the buyers are on holiday in the mountains of Peru and virtually unreachable. The builder decides to install Brand Y instead, which he knows is identical in quality and durability to Brand X plasterboard. All other aspects of construction conform to the contract. Does this deviation constitute a breach of contract? Can the buyers avoid their contractual obligation to pay the builder because Brand Y plasterboard was used instead of Brand X? Very likely, a court would hold that the builder had substantially performed his end of the bargain, and therefore the couple will be obligated to pay the builder. ■

What if the plasterboard substituted for Brand X was not of the same quality as Brand X, and the value of the house was reduced by $10,000? Again, a court would likely hold that the contract was substantially performed and that the contractor should be paid the price agreed on in the contract, less that $10,000.

Performance to the Satisfaction of Another Contracts often state that completed work must personally satisfy one of the parties or a third person. The question is whether this satisfaction becomes a condition precedent, requiring actual personal satisfaction or approval for discharge, or whether the test of satisfaction is performance that would satisfy a *reasonable person* (substantial performance).

When the subject matter of the contract is personal, a contract to be performed to the satisfaction of one of the parties is conditioned, and performance must actually satisfy that party. For example, contracts for portraits, works of art, and tailoring are considered personal. Therefore, only the personal satisfaction of the party fulfills the condition—unless a court finds the party is expressing dissatisfaction only to avoid payment or otherwise is not acting in good faith.

Contracts that involve mechanical fitness, utility, or marketability need be performed only to the satisfaction of a reasonable person unless they *expressly state otherwise.* In contrast, when such contracts require performance to the satisfaction of a third party (for example, "to the satisfaction of Robert Ames, the supervising engineer"), the courts are divided. A majority of courts require the work to be satisfactory to a reasonable person, but some courts hold that the personal satisfaction of the third party designated in the contract (Robert Ames, in this example) must be met. Again, the personal judgment must be made honestly, or the condition will be excused.

Material Breach of Contract When a breach of contract is *material*[13]—that is, when performance is not deemed substantial—the nonbreaching party is excused from the performance of contractual duties and has a cause of action to sue for damages caused by the breach. If the breach is *minor* (not material), the nonbreaching party's duty to perform may sometimes be suspended until the breach is remedied, but the duty is not entirely excused. Once the minor breach is cured, the nonbreaching party must resume performance of the contractual obligations that were undertaken.

A breach entitles the nonbreaching party to sue for damages, but only a material breach discharges the nonbreaching party from the contract. The policy underlying these rules is that contracts should go forward when only minor problems occur, but contracts should be terminated if major problems arise.[14]

Did a seller's failure to repair the plumbing in an apartment building within the eight-month period specified in a contract with the buyers constitute a material breach of the parties' contract? That was the issue in the following case.

13. *Restatement (Second) of Contracts,* Section 241.

14. See UCC 2–612, which deals with installment contracts for the sale of goods.

CASE 11.3 Kim v. Park

Court of Appeals of Oregon, 2004.
192 Or.App. 365,
86 P.3d 63.
http://www.publications.ojd.state.or.us/appeals.htm[a]

a. Click on "March" under "Cases decided in 2004." In the result, scroll to the name of the case under 3/3/04 and click on it to access the opinion. The State of Oregon's Judicial Branch maintains this Web site.

FACTS Su Yong Kim sold an apartment building in Portland, Oregon, to Chon Sik Park, Bok Soon Park, Johan Cen, William Itzineag, Johnny Perea, and Patricia Maldonado. At the time, the building's plumbing violated the Portland Housing Code. The contract provided the following: "Seller shall correct the plumbing code violation . . . within eight months. . . . Buyer

shall cooperate with seller in . . . providing access to the premises to complete repairs." Kim did not make the repairs within eight months, but twelve months after the date of the contract, Kim cut holes in the walls of the apartments to expose the plumbing. Seven weeks later, early one morning, Kim sent plumbers to the building without notice to the owners, apparently in an attempt to make repairs. The owners ordered the plumbers to leave, refused to allow Kim to send others, and stopped making payments under the contract. Kim filed a suit in an Oregon state court against the buyers, seeking the amount due. The buyers asserted that Kim's failure to repair the plumbing was a material breach that excused the performance of their obligations and counterclaimed for damages for the breach. The court concluded that Kim's breach was not material and ordered relief in his favor. The buyers appealed to a state intermediate appellate court.

ISSUE Was the seller's failure to repair the plumbing according to the parties' contract a material breach excusing the buyers' performance?

DECISION Yes. The state intermediate appellate court reversed the decision of the lower court on this issue and remanded the case for a determination of the amount of damages.

REASON The appellate court reasoned that a breach is material if it "goes to the very substance of the contract and defeats the object of the parties entering into the contract." In this case, as a result of the code violations, "the City of Portland continued to assess fines against defendants and defendants lost some tenants." The court pointed out that the requirement that the seller repair the plumbing within eight months was intended to ensure that the building's plumbing would satisfy the city code within a reasonable time after the sale. Also, the buyers had purchased the building from the seller so that they could rent out the apartments in it. According to the court, "[a]lthough the repairs to the plumbing in the building would have caused some temporary inconvenience to the tenants, the failure of plaintiff to make the repairs in accordance with the contract ultimately prevented defendants from using the building as intended by the parties' agreement. In light of that evidence, we hold that as a matter of law the plaintiff's [seller's] failure to perform as promised was a material breach of the contract." The court added that because the seller's breach was material, the buyers were not obligated to continue to perform their obligation to make the payments under the contract.

WHY IS THIS CASE IMPORTANT? *This case emphasizes that when one party's failure to perform a contractual obligation causes another to suffer significant harm, the other party normally is entitled to a remedy. Recall from Chapters 7 and 8 that even when no contract exists, if a party justifiably relies to his or her detriment on the promise of another, the party may be able to obtain relief under the doctrine of quasi contract or promissory estoppel.*

Anticipatory Repudiation of a Contract Before either party to a contract has a duty to perform, one of the parties may refuse to perform her or his contractual obligations. This is called **anticipatory repudiation.**[15] When anticipatory repudiation occurs, it is treated as a material breach of contract, and the nonbreaching party is permitted to bring an action for damages immediately, even though the scheduled time for performance under the contract may still be in the future.[16] Until the nonbreaching party treats this early repudiation as a breach, however, the breaching party can retract the anticipatory repudiation by proper notice and restore the parties to their original obligations.[17]

An anticipatory repudiation is treated as a present, material breach for two reasons. First, the nonbreaching party should not be required to remain ready and willing to perform when the other party has already repudiated the contract. Second, the nonbreaching party should have the opportunity to seek a similar contract elsewhere and may have the duty to do so to minimize his or her loss.

Quite often, an anticipatory repudiation occurs when a sharp fluctuation in market prices creates a situation in which performance of the contract would be extremely unfavorable to one of the parties. ■ EXAMPLE 11.18 Shasta Manufacturing Company contracts to manufacture and sell 100,000 personal computers to New Age, Inc., a computer retailer with 500 outlet stores. Delivery is to be made two months from the date of the contract. One month later, three suppliers of computer parts raise their prices to Shasta. Because of these higher prices, Shasta stands to lose $500,000 if it sells the computers to New Age at the contract price. Shasta writes to New Age, stating that it cannot deliver the 100,000 computers at the agreed-on contract price. Even though you might

15. *Restatement (Second) of Contracts,* Section 253, and UCC 2–610.

16. The doctrine of anticipatory repudiation first arose in the landmark case of *Hochster v. De La Tour,* 2 Ellis and Blackburn Reports 678 (1853), when an English court recognized the delay and expense inherent in a rule requiring a nonbreaching party to wait until the time of performance before suing on an anticipatory repudiation.

17. See UCC 2–611.

sympathize with Shasta, its letter constitutes an anticipatory repudiation of the contract, allowing New Age the option of treating the repudiation as a material breach and proceeding immediately to pursue remedies, even though the contract delivery date is still a month away. ■

Discharge by Agreement

Any contract can be discharged by agreement of the parties. The agreement can be contained in the original contract, or the parties can form a new contract for the express purpose of discharging the original contract.

Discharge by Rescission As discussed in Chapter 8, *rescission* is the process in which the parties cancel the contract and are returned to the positions they occupied prior to the contract's formation. For *mutual rescission* to take place, the parties must make another agreement that also satisfies the legal requirements for a contract—there must be an *offer,* an *acceptance,* and *consideration.* Ordinarily, if the parties agree to rescind the original contract, their promises not to perform those acts promised in the original contract will be legal consideration for the second contract.

Mutual rescission can occur in this manner when the original contract is executory on both sides (that is, neither party has completed performance). The agreement to rescind an executory contract is generally enforceable, even if it is made orally and even if the original agreement was in writing.[18] When one party has fully performed, however, an agreement to rescind the original contract usually is not enforceable unless additional consideration or restitution is made.[19]

Discharge by Novation The process of **novation** substitutes a third party for one of the original parties. Essentially, the parties to the original contract and one or more new parties all get together and agree to the substitution. The requirements of a novation are as follows:

1. The existence of a previous, valid obligation.
2. Agreement by all of the parties to a new contract.
3. The extinguishing of the old obligation (discharge of the prior party).
4. A new, valid contract.

A novation may appear similar to an assignment or delegation. There is an important distinction, however: a novation involves a new contract, and an assignment or delegation involves the old contract.

■ EXAMPLE 11.19 Suppose that you contract with Logan Enterprises to sell it your office equipment business. Logan later decides that it should not expand at this time but knows of another party, MBI Corporation, that is interested in purchasing your business. All three of you get together and agree to a novation. As long as the new contract is supported by consideration, the novation discharges the original contract between you and Logan and replaces it with the new contract between you and MBI Corporation. Logan prefers the novation because it discharges Logan's liabilities under the contract with you. If the original contract had been an installment sales contract requiring twelve monthly payments, and Logan had merely assigned the contract (assigned its rights and delegated its duties under the contract) to MBI Corporation, Logan would have remained liable to you for the payments if MBI Corporation defaulted. ■

Discharge by Accord and Satisfaction As discussed in Chapter 8, in an *accord and satisfaction,* the parties agree to accept performance different from the performance originally promised. An *accord* is an executory contract (one that has not yet been performed) to perform some act in order to satisfy an existing contractual duty that is not yet discharged.[20] A *satisfaction* is the performance of the accord agreement. An *accord* and its *satisfaction* discharge the original contractual obligation.

Once the accord has been made, the original obligation is merely suspended until the accord agreement is fully performed. If it is not performed, the party to whom performance is owed can bring an action on the original obligation or for breach of the accord. ■ EXAMPLE 11.20 Shea obtains a judgment against Marla for $4,000. Later, both parties agree that the judgment can be satisfied by Marla's transfer of her automobile to Shea. This agreement to accept the auto in lieu of $4,000 in cash is the accord. If Marla transfers her automobile to Shea, the accord agreement is fully performed, and the $4,000 debt is discharged. If Marla refuses to transfer her car, the accord is breached. Because the original obligation is merely suspended, Shea can bring an action to enforce the judgment for $4,000 in cash or bring an action for breach of the accord. ■

Discharge by Operation of Law

Under some circumstances, contractual duties may be discharged by operation of law. These circumstances include

18. Agreements to rescind contracts involving transfers of realty, however, must be evidenced by a writing. Another exception has to do with the sale of goods under the UCC, when the sales contract requires written rescission.

19. Under UCC 2–209(1), however, no consideration is needed to modify a contract for a sale of goods. See Chapter 14. Also see UCC 1–107.

20. *Restatement (Second) of Contracts,* Section 281.

material alteration of the contract, the running of the relevant statute of limitations, bankruptcy, and impossibility of performance.

Contract Alteration To discourage parties from altering written contracts, the law allows an innocent party to be discharged when one party has materially altered a written contract without the knowledge or consent of the other party. For example, if a party alters a material term of the contract—such as the quantity term or the price term—without the knowledge or consent of the other party, the party who was unaware of the alteration can treat the contract as discharged, or terminated.

Statutes of Limitations As mentioned earlier in this text, statutes of limitations limit the period during which a party can sue on a particular cause of action. After the applicable limitations period has passed, a suit can no longer be brought. For example, the limitations period for bringing suits for breach of oral contracts is usually two to three years; for written contracts, four to five years; and for recovery of amounts awarded in judgment, ten to twenty years, depending on state law. Suits for breach of a contract for the sale of goods must be brought within four years after the cause of action has accrued. By original agreement, the parties can reduce this four-year period to a one-year period. They cannot, however, extend it beyond the four-year limitations period.

Bankruptcy A proceeding in bankruptcy attempts to allocate the debtor's assets to the creditors in a fair and equitable fashion. Once the assets have been allocated, the debtor receives a *discharge in bankruptcy* (see Chapter 21). A discharge in bankruptcy ordinarily bars enforcement of most of a debtor's contracts by the creditors.

When Performance Is Impossible After a contract has been made, performance may become impossible in an objective sense. This is known as **impossibility of performance** and may discharge the contract.[21]

—Objective Impossibility. *Objective impossibility* ("It can't be done") must be distinguished from subjective impossibility ("I'm sorry, I simply can't do it"). An example of subjective impossibility is the inability to pay funds on time because the bank is closed.[22] In effect, the nonperforming party is saying, "It is impossible for *me* to perform," rather than "It is impossible for *anyone* to perform." Accordingly, such excuses do not discharge a contract, and the nonperforming party is normally held in breach of contract. Four basic types of situations will generally qualify as grounds for the discharge of contractual obligations based on impossibility of performance:[23]

1. *When a party whose personal performance is essential to the completion of the contract dies or becomes incapacitated prior to performance.* ■ **EXAMPLE 11.21** Fred, a famous dancer, contracts with Ethereal Dancing Guild to play a leading role in its new ballet. Before the ballet can be performed, Fred becomes ill and dies. His personal performance was essential to the completion of the contract. Thus, his death discharges the contract and his estate's liability for his nonperformance. ■
2. *When the specific subject matter of the contract is destroyed.* ■ **EXAMPLE 11.22** A-1 Farm Equipment agrees to sell Gudgel the green tractor on its lot and promises to have the tractor ready for Gudgel to pick up on Saturday. On Friday night, however, a truck veers off the nearby highway and smashes into the tractor, destroying it beyond repair. Because the contract was for this specific tractor, A-1's performance is rendered impossible owing to the accident. ■
3. *When a change in the law renders performance illegal.* An example is a contract to build an apartment building, when the zoning laws are changed to prohibit the construction of residential rental property at this location. This change renders the contract impossible to perform.
4. *When performance becomes commercially impracticable.* The inclusion of this type of "impossibility" as a basis for contract discharge results from a growing trend to allow parties to discharge contracts when the originally contemplated performance turns out to be much more difficult or expensive than anticipated. In such situations, courts may excuse parties from their performance obligations under the doctrine of *commercial impracticability.* For example, in one case, a court held that a contract could be discharged because a party would have to pay ten times more than the original estimate to excavate a certain amount of gravel.[24]

—Temporary Impossibility. An occurrence or event that makes performance temporarily impossible operates to suspend performance until the impossibility ceases. Then, ordinarily, the parties must perform the contract as originally planned. If, however, the lapse of time and the

21. *Restatement (Second) of Contracts,* Section 261.
22. *Ingham Lumber Co. v. Ingersoll & Co.,* 93 Ark. 447, 125 S.W. 139 (1910).
23. *Restatement (Second) of Contracts,* Sections 262–266; and UCC 2–615.
24. *Mineral Park Land Co. v. Howard,* 172 Cal. 289, 156 P. 458 (1916).

change in circumstances surrounding the contract make it substantially more burdensome for the parties to perform the promised acts, the contract is discharged.

■ EXAMPLE 11.23 The leading case on the subject, *Autry v. Republic Productions*,[25] involved an actor who was drafted into the army in 1942. Being drafted rendered the actor's contract temporarily impossible to perform, and it was suspended until the end of the war. When the actor got out of the army, the purchasing power of the dollar had so changed that performance of the contract would have been substantially burdensome to him. Therefore, the contract was discharged. ■

25. 30 Cal.2d 144, 180 P.2d 888 (1947).

TERMS AND CONCEPTS

CHAPTER SUMMARY Third Party Rights and Discharge

THIRD PARTY RIGHTS	
Assignments (See pages 207–210.)	1. An assignment is the transfer of rights under a contract to a third party. The person assigning the rights is the *assignor*, and the party to whom the rights are assigned is the *assignee*. The assignee has a right to demand performance from the other original party to the contract. 2. Generally, all rights can be assigned, except in the following circumstances: a. When assignment is expressly prohibited by statute (for example, workers' compensation benefits). b. When a contract calls for the performance of personal services. c. When the assignment will materially increase or alter the risks or duties of the *obligor* (the party that is obligated to perform). d. When the contract itself stipulates that the rights cannot be assigned (with some exceptions). 3. The assignee should notify the obligor of the assignment. Although not legally required, notification avoids two potential problems: a. If the assignor assigns the same right to two different persons, generally the first assignment in time is the first in right, but in some states the first assignee to give notice takes priority. b. Until the obligor is notified of the assignment, the obligor can tender performance to the assignor; and if performance is accepted by the assignor, the obligor's duties under the contract are discharged without benefit to the assignee.
Delegations (See pages 210–212.)	1. A delegation is the transfer of duties under a contract to a third party (the *delegatee*), who then assumes the obligation of performing the contractual duties previously held by the one making the delegation (the *delegator*).

CHAPTER SUMMARY Third Party Rights and Discharge—Continued

Delegations—Continued	2. As a general rule, any duty can be delegated, except in the following circumstances: a. When performance depends on the personal skill or talents of the obligor. b. When special trust has been placed in the obligor. c. When performance by a third party will vary materially from that expected by the obligee (the one to whom the duty is owed) under the contract. d. When the contract expressly prohibits delegation. 3. A valid delegation of duties does not relieve the delegator of obligations under the contract. If the delegatee fails to perform, the delegator is still liable to the obligee. 4. An "assignment of all rights" or an "assignment of the contract" is often construed to mean that both the rights and the duties arising under the contract are transferred to a third party.
Third Party Beneficiaries (See pages 212–216.)	A third party beneficiary contract is one made for the purpose of benefiting a third party. 1. *Intended beneficiary*—One for whose benefit a contract is created. When the promisor (the one making the contractual promise that benefits a third party) fails to perform as promised, the third party can sue the promisor directly. Examples of third party beneficiaries are creditor and donee beneficiaries. 2. *Incidental beneficiary*—A third party who indirectly (incidentally) benefits from a contract but for whose benefit the contract was not specifically intended. Incidental beneficiaries have no rights to the benefits received and cannot sue to have the contract enforced.
CONTRACT DISCHARGE	
Conditions of Performance (See pages 216–217.)	Contract obligations may be subject to the following types of conditions: 1. *Condition precedent*—A condition that must be fulfilled before a party's promise becomes absolute. 2. *Condition subsequent*—A condition that operates to terminate a party's absolute promise to perform. 3. *Concurrent conditions*—Conditions that must be performed simultaneously. Each party's absolute duty to perform is conditioned on the other party's absolute duty to perform.
Discharge by Performance (See pages 217–220.)	A contract may be discharged by complete (strict) performance or by substantial performance. In some cases, performance must be to the satisfaction of another. Totally inadequate performance constitutes a material breach of contract. An anticipatory repudiation of a contract allows the other party to sue immediately for breach of contract.
Discharge by Agreement (See page 220.)	Parties may agree to discharge their contractual obligations in several ways: 1. *By rescission*—The parties mutually agree to rescind (cancel) the contract. 2. *By novation*—A new party is substituted for one of the primary parties to a contract. 3. *By accord and satisfaction*—The parties agree to render and accept performance different from that on which they originally agreed.
Discharge by Operation of Law (See pages 220–222.)	Parties' obligations under contracts may be discharged by operation of law owing to one of the following: 1. Contract alteration. 2. Statutes of limitations. 3. Bankruptcy. 4. Impossibility of performance.

FOR REVIEW

Answers for the even-numbered questions in this For Review *section can be found in Appendix A at the end of this text.*

1. What is the difference between an assignment and a delegation?
2. What rights can be assigned despite a contract clause expressly prohibiting assignment?
3. What factors indicate that a third party beneficiary is an intended beneficiary?
4. How are most contracts discharged?
5. What is a contractual condition, and how might a condition affect contractual obligations?

QUESTIONS AND CASE PROBLEMS

11–1. Substantial Performance. Complete performance is full performance according to the terms of a contract. Discuss the effect on the parties if there is less than full performance.

QUESTION WITH SAMPLE ANSWER

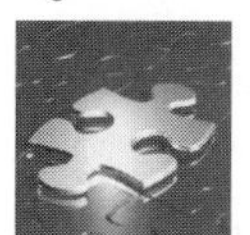

11–2. Wilken owes Rivera $2,000. Howie promises Wilken that he will pay Rivera the $2,000 in return for Wilken's promise to give Howie's children guitar lessons. Is Rivera an intended beneficiary of the Howie-Wilken contract? Explain.

For a sample answer to this question, go to Appendix B at the end of this text.

11–3. Assignments. Aron, a college student, signs a one-year lease agreement that runs from September 1 to August 31. The lease agreement specifies that the lease cannot be assigned without the landlord's consent. In late May, Aron decides not to go to summer school and assigns the balance of the lease (three months) to a close friend, Erica. The landlord objects to the assignment and denies Erica access to the apartment. Aron claims that Erica is financially sound and should be allowed the full rights and privileges of an assignee. Discuss fully whether the landlord or Aron is correct.

11–4. Impossibility of Performance. Millie contracted to sell Frank 1,000 bushels of corn to be grown on Millie's farm. Owing to a drought during the growing season, Millie's yield was much less than anticipated, and she could deliver only 250 bushels to Frank. Frank accepted the lesser amount but sued Millie for breach of contract. Can Millie defend successfully on the basis of objective impossibility of performance? Explain.

CASE PROBLEM WITH SAMPLE ANSWER

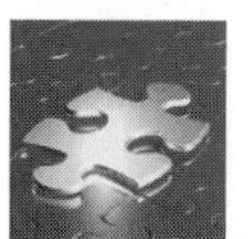

11–5. In May 1996, O'Brien-Shiepe Funeral Home, Inc., in Hempstead, New York, hired Teramo & Co. to build an addition to O'Brien's funeral home. The parties' contract did not specify a date for the completion of the work. The city of Hempstead issued a building permit for the project on June 14, and Teramo began work about two weeks later. There was some delay in construction because O'Brien asked that no work be done during funeral services, but by the end of March 1997, the work was substantially complete. The city of Hempstead issued a "Certificate of Completion" on April 15. During the construction, O'Brien made periodic payments to Teramo, but there was a balance due of $17,950, which O'Brien did not pay. To recover this amount, Teramo filed a suit in a New York state court against O'Brien. O'Brien filed a counterclaim to recover lost profits for business allegedly lost due to the time Teramo took to build the addition, and for $6,180 spent to correct problems caused by poor work. Which, if any, party is entitled to an award in this case? Explain. [*Teramo & Co. v. O'Brien-Shiepe Funeral Home, Inc.*, 283 A.D.2d 635, 725 N.Y.S.2d 87 (2001)]

After you have answered this problem, compare your answer with the sample answer given on the Web site that accompanies this text. Go to academic.cengage.com/blaw/wbl, select "Chapter 11," and click on "Case Problem with Sample Answer."

11–6. Substantial Performance. Adolf and Ida Krueger contracted with Pisani Construction, Inc., to erect a metal building as an addition to an existing structure. The two structures were to share a common wall, and the frames and panel heights of the new building were to match those of the existing structure. Shortly before completion of the project, however, it was apparent that the roofline of the new building was approximately three inches higher than that of the existing structure. Pisani modified the ridge caps of the buildings to blend the rooflines. The discrepancy had other consequences, however, including misalignment of the gutters and windows of the two buildings, which resulted in an icing problem in the winter. The Kruegers occupied the new structure, but refused to make the last payment under the contract. Pisani filed a suit in a Connecticut state court to collect. Did Pisani substantially perform its obligations? Should the Kruegers be ordered to pay? Why or why not? [*Pisani Construction, Inc. v. Krueger*, 68 Conn.App. 361, 791 A.2d 634 (2002)]

11–7. Third Party Beneficiary. Action Steel, Inc., entered into a contract with Systems Builders, Inc., a general contractor, to construct an addition to a commercial building in Indianapolis, Indiana. The contract provided that after the addition's completion, Action Steel would obtain insurance, which "shall include the interest of . . . subcontractors." The parties would then "waive all rights against . . . any of their subcontractors." Varco-Pruden

Building, a subcontractor, designed the addition, which was completed in the summer of 1995. Action Steel obtained an insurance policy from Midwestern Indemnity Co. In January 1996, a snowstorm hit the Indianapolis area, and the new addition collapsed. Midwestern paid over $1.3 million to Action Steel for the loss. Because Midwestern paid for the loss, it stood in Action Steel's place in a suit filed in an Indiana state court against Varco-Pruden and others to recover this amount. Varco-Pruden filed a motion for summary judgment, arguing in part that it was a third party beneficiary of the waiver clause in the contract between Action Steel and Systems Builders. Should the court grant this motion? Why or why not? [*Midwestern Indemnity Co. v. Systems Builders, Inc.,* 801 N.E.2d 661 (Ind.App. 2004)]

A QUESTION OF ETHICS

11–8. Bath Iron Works (BIW) offered a job to Thomas Devine, contingent on Devine's passing a drug test. The testing was conducted by NorDx, a subcontractor of Roche Biomedical Laboratories. When NorDx found that Devine's urinalysis showed the presence of opiates, a result confirmed by Roche, BIW refused to offer Devine permanent employment. Devine claimed that the ingestion of poppy seeds can lead to a positive result and that he tested positive for opiates only because of his daily consumption of poppy seed muffins. In Devine's suit against Roche, Devine argued, among other things, that he was a third party beneficiary of the contract between his employer (BIW) and NorDx (Roche). Given this factual background, consider the following questions. [*Devine v. Roche Biomedical Laboratories,* 659 A.2d 868 (Me. 1995)]

1. Is Devine an intended third party beneficiary of the BIW-NorDx contract? In deciding this issue, should the court focus on the nature of the promises made in the contract itself or on the consequences of the contract for Devine, a third party?
2. Should employees whose job security and reputation have suffered as a result of false test results be allowed to sue the drug-testing labs for the tort of negligence? In such situations, do drug-testing labs have a duty to the employees to exercise reasonable care in conducting the tests?

VIDEO QUESTION

11–9. Go to this text's Web site at academic.cengage.com/blaw/wbl and select "Chapter 11." Click on "Video Questions" and view the video titled *Third Party Beneficiaries.* Then answer the following questions.

1. Discuss whether a valid contract was formed when Oscar and Vinny bet on the outcome of a football game. Would Vinny be able to enforce the contract in court?
2. Is the Fresh Air Fund an incidental or an intended beneficiary? Why?
3. Can Maria sue to enforce Vinny's promise to donate Oscar's winnings to the Fresh Air Fund?

Third Party Rights and Discharge

For updated links to resources available on the Web, as well as a variety of other materials, visit this text's Web site at

academic.cengage.com/blaw/wbl

For a general summary of how contracts may be discharged, go to

http://www.lawyers.com/legal_topics/browse_by_topic/index.php

Click on "Business Needs." Then, under the heading "General Business," click on "Contracts" and select the appropriate links.

ONLINE LEGAL RESEARCH

Go to the *Fundamentals of Business Law* home page at academic.cengage.com/blaw/wbl, select "Chapter 11," and click on "Internet Exercises." There you will find the following Internet research exercises that you can perform to learn more about topics covered in this chapter.

Activity 11–1: LEGAL PERSPECTIVE—Anticipatory Repudiation

Activity 11–2: MANAGEMENT PERSPECTIVE—Commercial Impracticability

BEFORE THE TEST

Go to the *Fundamentals of Business Law* home page at academic.cengage.com/blaw/wbl, select "Chapter 11," and click on "Interactive Quizzes." You will find a number of interactive questions relating to this chapter.

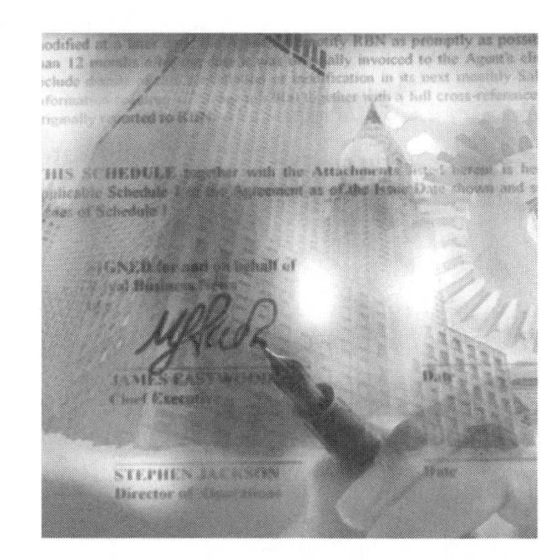

CHAPTER 12
Breach and Remedies

LEARNING OBJECTIVES

After reading this chapter, you should be able to answer the following questions:

1. What is the difference between compensatory damages and consequential damages? What are nominal damages, and when do courts award nominal damages?
2. What is the standard measure of compensatory damages when a contract is breached? How are damages computed differently in construction contracts?
3. Under what circumstances is the remedy of rescission and restitution available?
4. When do courts grant specific performance as a remedy?
5. What is the rationale underlying the doctrine of election of remedies?

Generally, a contract will not be broken as long as it is to the advantage of both parties to fulfill their contractual obligations. Normally, a person enters into a contract with another to secure an advantage. When it is no longer advantageous for a party to fulfill her or his contractual obligations, that party may breach the contract. As discussed in Chapter 11, a *breach of contract* occurs when a party fails to perform part or all of the required duties under a contract.[1] Once a party fails to perform or performs inadequately, the other party—the nonbreaching party—can choose one or more of several remedies.

The most common remedies available to a nonbreaching party under contract law include damages, rescission and restitution, specific performance, and reformation. As discussed in Chapter 1, courts distinguish between *remedies at law* and *remedies in equity.* Today, the remedy at law is normally money damages. We discuss this remedy in the first part of this chapter. Equitable remedies include rescission and restitution, specific performance, and reformation, all of which we examine later in the chapter. Usually, a court will not award an equitable remedy unless the remedy at law is inadequate. In the final pages of this chapter, we look at some special legal doctrines and concepts relating to remedies.

Damages

A breach of contract entitles the nonbreaching party to sue for money damages. As you read in Chapter 4, damages are designed to compensate a party for harm suffered as a result of another's wrongful act. In the context of contract law, damages are designed to compensate the nonbreaching party for the loss of the bargain. Often, courts say that innocent parties are to be placed in the position they would have occupied had the contract been fully performed.[2]

1. *Restatement (Second) of Contracts,* Section 235(2).

2. *Restatement (Second) of Contracts,* Section 347; and Section 1–106(1) of the Uniform Commercial Code (UCC).

TYPES OF DAMAGES

There are basically four broad categories of damages:

1. Compensatory (to cover direct losses and costs).
2. Consequential (to cover indirect and foreseeable losses).
3. Punitive (to punish and deter wrongdoing).
4. Nominal (to recognize wrongdoing when no monetary loss is shown).

Compensatory and punitive damages were discussed in Chapter 4 in the context of tort law. Here, we look at these types of damages, as well as consequential and nominal damages, in the context of contract law.

Compensatory Damages Damages compensating the nonbreaching party for the *loss of the bargain* are known as *compensatory damages.* These damages compensate the injured party only for damages actually sustained and proved to have arisen directly from the loss of the bargain caused by the breach of contract. They simply replace what was lost because of the wrong or damage.

The standard measure of compensatory damages is the difference between the value of the breaching party's promised performance under the contract and the value of her or his actual performance. This amount is reduced by any loss that the injured party has avoided, however.

■ EXAMPLE 12.1 Suppose that you contract with Marinot Industries to perform certain personal services exclusively for Marinot during August for a payment of $3,500. Marinot cancels the contract and is in breach. You are able to find another job during August but can earn only $1,000. You normally can sue Marinot for breach and recover $2,500 as compensatory damages. You may also recover from Marinot the amount that you spent to find the other job. ■ Expenses that are directly incurred because of a breach of contract—such as those incurred to obtain performance from another source—are called *incidental damages.*

The measurement of compensatory damages varies by type of contract. Certain types of contracts deserve special mention—contracts for the sale of goods, contracts for the sale of land, and construction contracts.

—Sale of Goods. In a contract for the sale of goods, the usual measure of compensatory damages is the difference between the contract price and the market price.[3] ■ EXAMPLE 12.2 MediQuick Laboratories contracts with Cal Computer Industries to purchase ten model UTS 400 network servers for $8,000 each. If Cal Computer fails to deliver the ten servers, and the current market price of the servers is $8,150, MediQuick's measure of damages is $1,500 (10 × $150) plus any incidental damages (expenses) caused by the breach. ■ If the buyer breaches and the seller has not yet produced the goods, compensatory damages normally equal the seller's lost profits on the sale rather than the difference between the contract price and the market price.

—Sale of Land. Ordinarily, because each parcel of land is unique, the remedy for a seller's breach of a contract for a sale of real estate is specific performance—that is, the buyer is awarded the parcel of property for which he or she bargained (specific performance will be discussed more fully later in this chapter). When this remedy is unavailable (because the property has been sold, for example) or when the buyer is the party in breach, the measure of damages is typically the difference between the contract price and the market price of the land. The majority of states follow this rule.

—Construction Contracts. The measure of damages in a building or construction contract varies depending on which party breaches and when the breach occurs. The owner can breach at three different stages of the construction:

1. Before performance has begun.
2. During performance.
3. After performance has been completed.

If the owner breaches *before performance has begun,* the contractor can recover only the profits that would have been made on the contract (that is, the total contract price less the cost of materials and labor). If the owner breaches *during performance,* the contractor can recover the profits plus the costs incurred in partially constructing the building. If the owner breaches *after the construction has been completed,* the contractor can recover the entire contract price plus interest.

When the contractor breaches the construction contract—either by failing to begin construction or by stopping work partway through the project—the measure of damages is the cost of completion, which includes reasonable compensation for any delay. If the contractor finishes late, the measure of damages is the loss of use. Exhibit 12–1 on page 228 summarizes the rules concerning the measurement of damages in breached construction contracts.

Consequential Damages Foreseeable damages that result from a party's breach of contract are referred to as **consequential damages,** or *special damages.* Consequential

3. That is, the difference between the contract price and the market price at the time and place at which the goods were to be delivered or tendered. [See UCC 2–708, 2–713, and 2–715(1), to be discussed in Chapter 15.]

EXHIBIT 12–1 MEASUREMENT OF DAMAGES—BREACH OF CONSTRUCTION CONTRACTS

PARTY IN BREACH	TIME OF BREACH	MEASUREMENT OF DAMAGES
Owner	Before construction has begun	Profits (contract price less cost of materials and labor)
Owner	During construction	Profits plus costs incurred up to time of breach
Owner	After construction is completed	Contract price plus interest
Contractor	Before construction has begun	Cost above contract price to complete work
Contractor	Before construction is completed	Generally, all costs incurred by owner to complete work

damages differ from compensatory damages in that they are caused by special circumstances beyond the contract itself. They flow from the consequences, or results, of a breach. When a seller fails to deliver goods, knowing that the buyer is planning to use or resell those goods immediately, consequential damages are awarded for the loss of profits from the planned resale.

■ EXAMPLE 12.3 Gilmore contracts to have a specific item shipped to her—one that she desperately needs to repair her printing press. In her contract with the shipper, Gilmore states that she must receive the item by Monday or she will not be able to print her paper and will lose $950. If the shipper is late, Gilmore normally can recover the consequential damages caused by the delay (that is, the $950 in losses). ■

To recover consequential damages, the breaching party must know (or have reason to know) that special circumstances will cause the nonbreaching party to suffer an additional loss.[4]

Punitive Damages Recall from Chapter 4 that punitive damages are designed to punish a wrongdoer and set an example to deter similar conduct in the future. Punitive damages, or *exemplary damages,* are generally not awarded in an action for breach of contract. Such damages have no legitimate place in contract law because they are, in essence, penalties, and a breach of contract is not unlawful in a criminal sense. A contract is simply a civil relationship between the parties. The law may compensate one party for the loss of the bargain—no more and no less.

In a few situations, a person's actions can cause both a breach of contract and a tort. ■ EXAMPLE 12.4 Two parties establish by contract a certain reasonable standard or duty of care. Failure to live up to that standard is a breach of the contract. The same act that breached the contract may also constitute negligence, or it may be an intentional tort if, for example, the breaching party committed fraud. In such a situation, it is possible for the nonbreaching party to recover punitive damages for the tort in addition to compensatory and consequential damages for the breach of contract. ■

Nominal Damages When no actual damage or financial loss results from a breach of contract and only a technical injury is involved, the court may award **nominal damages** to the innocent party. Nominal damages awards are often small, such as one dollar, but they do establish that the defendant acted wrongfully. Most lawsuits for nominal damages are brought as a matter of principle under the theory that a breach has occurred and some damages must be imposed regardless of actual loss.

■ EXAMPLE 12.5 Hernandez contracts to buy potatoes at fifty cents a pound from Lentz. Lentz breaches the contract and does not deliver the potatoes. Meanwhile, the price of potatoes falls. Hernandez is able to buy them in the open market at half the price he agreed to pay Lentz. Hernandez is clearly better off because of Lentz's breach. Thus, in a suit for breach of contract, Hernandez may be awarded only nominal damages for the technical injury he sustained, as no monetary loss was involved. ■

(Note that nominal damages may take on greater significance in tort cases involving defamation or trade libel. If a false statement made by a competing firm calls into question a business's integrity or good reputation, the business may find it desirable to establish the falsity of the statement in public court proceedings. If the business wins in court, the court may award nominal damages because actual damages in such cases are difficult to estimate.)

4. UCC 2–715(2). See Chapter 16.

Mitigation of Damages

In most situations, when a breach of contract occurs, the injured party is held to a duty to mitigate, or reduce, the damages that he or she suffers. Under this doctrine of **mitigation of damages,** the required action depends on the nature of the situation.

■ EXAMPLE 12.6 Some states require a landlord to use reasonable means to find a new tenant if a tenant abandons the premises and fails to pay rent. If an acceptable tenant becomes available, the landlord is required to lease the premises to this tenant to mitigate the damages recoverable from the former tenant. The former tenant is still liable for the difference between the amount of the rent under the original lease and the rent received from the new tenant. If the landlord has not used the reasonable means necessary to find a new tenant, a court will likely reduce any award by the amount of rent the landlord could have received had such reasonable means been used. ■

In the majority of states, a person whose employment has been wrongfully terminated has a duty to mitigate damages incurred because of the employer's breach of the employment contract. In other words, a wrongfully terminated employee has a duty to take a similar job if one is available. If the employee fails to do this, the damages she or he receives will be equivalent to the person's salary less the income she or he would have received in a similar job obtained by reasonable means. The employer has the burden of proving that such a job existed and that the employee could have been hired. Normally, the employee is under no duty to take a job that is not of the same type and rank.

Whether a business firm failed to mitigate its damages was at issue in the following case.

CASE 12.1 Fujitsu, Ltd. v. Federal Express Corp.

United States Court of Appeals,
Second Circuit, 2001.
247 F.3d 423.
http://www.tourolaw.edu/2ndCircuit/index.htm[a]

FACTS On May 30, 1996, Fujitsu, Ltd., shipped a container of silicon wafers (computer chips) from Narita, Japan, to Ross Technologies, Inc., in Austin, Texas, using Federal Express (FedEx) as the carrier. The next day, the container arrived in Austin and was held for clearance by the U.S. Customs Service. (FedEx cannot release imported goods until the Customs Service has approved delivery and all applicable taxes have been paid.) Meanwhile, Ross decided to reject the shipment and paid for FedEx to ship the goods back to Fujitsu.[b] The goods left Austin in good condition, but when they arrived in Japan, Fujitsu found the outer container broken and covered with an oily substance that had permeated some of the interior boxes and coated the sealed aluminum bags containing the wafers. Fujitsu reported the damage, which FedEx acknowledged. On the instructions of Fujitsu's insurance company, Fujitsu then disposed of the container and the wafers without opening any of the bags. Fujitsu filed a suit in a federal district court against FedEx, alleging in part breach of contract. The court found FedEx liable for damages in the amount of $726,640. FedEx appealed to the U.S. Court of Appeals for the Second Circuit, arguing that Fujitsu failed to mitigate its damages.

a. In the left-hand column, click on "Reported Decisions." From the menu, click on "2001." From the list, click on "April." Scroll to the name of the case and click on it to access the opinion. Touro College Jacob D. Fuchsberg Law Center in Huntington, New York, maintains this Web site.

b. Under the Warsaw Convention (an international treaty) if the shipment of goods had been a return shipment, paid for by Fujitsu, FedEx might have had a defense to liability. The court, however, found that this was a separate contract to ship the goods, paid for by Ross, and not a return of goods.

ISSUE Did Fujitsu fail to mitigate its damages by not opening the bags and attempting to salvage at least some of the wafers?

DECISION No. The U.S. Court of Appeals for the Second Circuit affirmed the judgment of the lower court.

REASON The appellate court acknowledged, "FedEx is correct that the record contains no evidence that the wafers themselves were damaged," but found that "the shipment was a total loss because the residue on the outer packaging made it impossible to access the wafers." The court explained that "the bags containing the wafers could only be opened in a specially designed and maintained 'clean room' so as to prevent dust contamination. However, because the bags themselves were coated with the oily residue, they could not be brought into a clean room for inspection, as the residue itself would contaminate the clean room." Consequently, even if the wafers had been undamaged, "Fujitsu was unable to extract them from the bags in an operable condition." The court concluded that there is "no difference between damage rendering the wafers inoperable and damage that prevents otherwise operable wafers from being used or salvaged." The court added that "efforts to salvage the wafers would have been prohibitively expensive."

WHY IS THIS CASE IMPORTANT? *Injured parties have an absolute duty to mitigate their damages. As this case illustrates, however, that duty is tempered by what actions the court considers to be reasonable under the circumstances.*

LIQUIDATED DAMAGES VERSUS PENALTIES

A **liquidated damages** provision in a contract specifies that a certain dollar amount is to be paid in the event of a future default or breach of contract. (*Liquidated* means determined, settled, or fixed.) Liquidated damages differ from penalties. A **penalty** specifies a certain amount to be paid in the event of a default or breach of contract and is designed to penalize the breaching party. Liquidated damages provisions normally are enforceable. In contrast, if a court finds that a provision calls for a penalty, the agreement as to the amount will not be enforced, and recovery will be limited to actual damages.[5]

To determine whether a particular provision is for liquidated damages or for a penalty, the court must answer two questions:

1. At the time the contract was formed, was it apparent that damages would be difficult to estimate in the event of a breach?

5. This is also the rule under the UCC. See UCC 2–718(1).

2. Was the amount set as damages a reasonable estimate of those potential damages and not excessive?[6]

If the answers to both questions are yes, the provision normally will be enforced. If either answer is no, the provision will normally not be enforced. In a construction contract, it is difficult to estimate the amount of damages that might be caused by a delay in completing construction, so liquidated damages clauses are often used.

The court in the following case addressed the liquidated damages questions just discussed in evaluating a provision in an agreement for the lease of a hotel.

6. *Restatement (Second) of Contracts,* Section 356(1).

CASE 12.2 Green Park Inn, Inc. v. Moore

North Carolina Court of Appeals, 2002.
149 N.C.App. 531,
562 S.E.2d 53.
http://www.aoc.state.nc.us/www/public/html/opinions.htm[a]

FACTS Allen and Pat McCain own Green Park Inn, Inc., which operates the Green Park Inn near Blowing Rock, North Carolina. In 1996, they leased the Inn to GMAFCO, LLC, which is owned by Gary and Gail Moore. The lease agreement provided that, in the event of a default by GMAFCO, Green Park Inn, Inc., would be entitled to $500,000 as "liquidated damages." GMAFCO defaulted on the February 2000 rent. Green Park Inn, Inc., gave GMAFCO an opportunity to cure the default, but GMAFCO made no further payments and returned possession of the property to the lessor. When Green Park Inn, Inc., sought the "liquidated damages," the Moores refused to pay. Green Park Inn, Inc., filed a suit in a North Carolina state court against the Moores, GMAFCO, and their bank to obtain the $500,000. The defendants contended in part that the lease clause requiring payment of "liquidated damages" was an unenforceable penalty provision. The court ordered the defendants to pay Green Park Inn, Inc. The defendants appealed to a state intermediate appellate court.

a. In the "Court of Appeals Opinions" section, click on "2002." On the next page, scroll to the "2 April 2002" section and click on the name of the case to access the opinion. The North Carolina Appellate Division Reporter maintains this Web site.

ISSUE Did the liquidated damages provision in the lease agreement constitute an unenforceable penalty clause?

DECISION No. The state intermediate appellate court affirmed the decision of the lower court.

REASON The appellate court quoted from the lease agreement's liquidated damages clause, which listed such items to be included in the lessor's damages as "restoration of the physical plant," "lost lease payments," and "harm to the reputation of the hotel." The clause also stated that the McCains had retired to Florida and would have to relocate back to Blowing Rock if they were "forced out of retirement to take over the operation of the hotel." The parties had agreed in the lease that $500,000 was a fair and reasonable estimate of the damages the McCains would suffer in the event of a default. The court held that these provisions satisfied the two-part test for liquidated damages: the amount of the damages would have been difficult to determine at the time the lease was signed, and the estimate of the damages was reasonable. The court noted that while some of the items listed in the clause could be determined, "others, such as the harm to the hotel's reputation or the cost to the McCains of being forced out of retirement, clearly would have been difficult to ascertain at the time the Lease Agreement was signed." The court also pointed out that after McCain and his wife came out of retirement and were again operating the hotel, he testi-

CASE 12.2–CONTINUED

fied that $500,000 was "a fair and reasonable estimate to measure the damages" and that the Moores did not attempt to show that this amount was unreasonable.

FOR CRITICAL ANALYSIS—Economic Consideration *If the lease had specified $3 million in damages, would the result in this case have been different? If so, why?*

Exhibit 12–2 summarizes the rules on the availability of the different types of damages.

Equitable Remedies

In some situations, damages are an inadequate remedy for a breach of contract. In these instances, the nonbreaching party may ask the court for an equitable remedy. Equitable remedies include rescission and restitution, specific performance, and reformation.

RESCISSION AND RESTITUTION

As discussed in Chapter 11, *rescission* is essentially an action to undo, or cancel, a contract—to return nonbreaching parties to the positions that they occupied before the transaction. When fraud, mistake, duress, or failure of consideration is present, rescission is available. The failure of one party to perform under a contract entitles the other party to rescind the contract.[7] The rescinding party must give prompt notice to the breaching party.

Restitution To rescind a contract, both parties generally must make **restitution** to each other by returning goods, property, or funds previously conveyed.[8] If the physical property or goods can be returned, they must be. If the property or goods have been consumed, restitution must be made in an equivalent dollar amount.

Essentially, restitution involves the recapture of a benefit conferred on the defendant that has unjustly enriched her or him. ■ EXAMPLE 12.7 Andrea pays $12,000 to Myles in return for his promise to design a house for her. The next day, Myles calls Andrea and tells her that he has taken a position with a large architectural firm in another state and cannot design the house. Andrea decides to hire another architect that afternoon. Andrea can require restitution of $12,000 because Myles has received an unjust benefit of $12,000. ■

Restitution Is Not Limited to Rescission Cases Restitution may be required when a contract is rescinded, but the right to restitution is not limited to rescission cases. Restitution may be sought in actions for breach of contract, tort actions, and other actions at law or in equity. Usually, restitution can be obtained when funds or property has been transferred by mistake or because of fraud.

7. The rescission discussed here refers to *unilateral* rescission, in which only one party wants to undo the contract. In *mutual* rescission, both parties agree to undo the contract. Mutual rescission discharges the contract; unilateral rescission is generally available as a remedy for breach of contract.

8. *Restatement (Second) of Contracts,* Section 370.

EXHIBIT 12–2 DAMAGES

REMEDY	AVAILABILITY	RESULT
Compensatory Damages	A party sustains and proves an injury arising directly from the loss of the bargain.	The injured party is compensated for the loss of the bargain.
Consequential Damages	Special circumstances, of which the breaching party is aware or should be aware, cause the injured party additional loss.	The injured party is given the *entire* benefit of the bargain, such as forgone profits.
Punitive Damages	Damages are normally available only when a tort is also involved.	The wrongdoer is punished, and others are deterred from committing similar acts.
Nominal Damages	There is no financial loss.	Wrongdoing is established without actual damages being suffered. The plaintiff is awarded a nominal amount (such as $1) in damages.
Liquidated Damages	A contract provides a specific amount to be paid as damages in the event that the contract is later breached.	The nonbreaching party is paid the amount stipulated in the contract for the breach, unless the amount is construed as a penalty.

An award in a case may include restitution of funds or property obtained through embezzlement, conversion, theft, copyright infringement, or misconduct by a party in a confidential or other special relationship.

Specific Performance

The equitable remedy of **specific performance** calls for the performance of the act promised in the contract. This remedy is often attractive to a nonbreaching party because it provides the exact bargain promised in the contract. It also avoids some of the problems inherent in a suit for money damages. First, the nonbreaching party need not worry about collecting the judgment.[9] Second, the nonbreaching party need not look around for another contract. Third, the actual performance may be more valuable than the money damages.

Normally, however, specific performance will not be granted unless the party's legal remedy (money damages) is inadequate.[10] For this reason, contracts for the sale of goods rarely qualify for specific performance. Money damages ordinarily are adequate in such situations because substantially identical goods can be bought or sold in the market. Only if the goods are unique will a court of equity grant specific performance. For instance, paintings, sculptures, and rare books and coins are often unique, and money damages will not enable a buyer to obtain substantially identical substitutes in the market.

Sale of Land A court will grant specific performance to a buyer in a contract for the sale of land. The legal remedy for breach of a land sales contract is inadequate because every parcel of land is considered to be unique. Money damages will not compensate a buyer adequately because the same land in the same location obviously cannot be obtained elsewhere. Only when specific performance is unavailable (for example, when the seller has sold the property to someone else) will money damages be awarded instead.

Contracts for Personal Services Personal-service contracts require one party to work personally for another party. Courts normally refuse to grant specific performance of contracts for personal services. This is because to order a party to perform personal services against his or her will amounts to a type of involuntary servitude, which is contrary to the public policy expressed in the Thirteenth Amendment to the U.S. Constitution. Moreover, the courts do not want to monitor contracts for personal services.

■ EXAMPLE 12.8 If you contract with a brain surgeon to perform brain surgery on you and the surgeon refuses to perform, the court will not compel (and you certainly would not want) the surgeon to perform under these circumstances. There is no way the court can assure meaningful performance in such a situation.[11] ■

Reformation

Reformation is an equitable remedy used when the parties have *imperfectly* expressed their agreement in writing. Reformation allows the contract to be rewritten to reflect the parties' true intentions. It applies most often when fraud or mutual mistake is present. ■ EXAMPLE 12.9 If Keshan contracts to buy a forklift from Shelley but the written contract refers to a crane, a mutual mistake has occurred. Accordingly, a court could reform the contract so that the writing conforms to the parties' original intention as to which piece of equipment is being sold. ■

Courts frequently reform contracts in two other situations. The first occurs when two parties who have made a binding oral contract agree to put the oral contract in writing but, in doing so, make an error in stating the terms. Universally, the courts allow into evidence the correct terms of the oral contract, thereby reforming the written contract. The second situation occurs when the parties have executed a written covenant not to compete (see Chapter 9). If the covenant not to compete is for a valid and legitimate purpose (such as the sale of a business) but the area or time restraints are unreasonable, some courts will reform the restraints by making them reasonable and will enforce the entire contract as reformed. Other courts, however, will throw the entire restrictive covenant out as illegal. Exhibit 12–3 summarizes the remedies, including reformation, that are available to the nonbreaching party.

Recovery Based on Quasi Contract

Recall from Chapter 7 that a *quasi contract* is not a true contract but rather a fictional contract that is imposed on the parties to prevent unjust enrichment. Hence, a quasi contract provides a basis for relief when no enforceable contract exists. The legal obligation arises because the law considers that a promise to pay for benefits received is implied by the party accepting the benefits. Generally, when one

9. Courts dispose of cases, after trials, by entering judgments. A judgment may order the losing party to pay money damages to the winning party. Collecting a judgment, however, can pose problems. For example, the judgment debtor may be insolvent (cannot pay his or her bills when they come due) or have only a small net worth, or exemption laws may prevent a creditor from seizing the debtor's assets to satisfy a debt (see Chapter 21).

10. *Restatement (Second) of Contracts,* Section 359.

11. Similarly, courts often refuse to order specific performance of construction contracts because courts are not set up to operate as construction supervisors or engineers.

EXHIBIT 12–3 REMEDIES FOR BREACH OF CONTRACT

REMEDIES AVAILABLE TO NONBREACHING PARTY			
DAMAGES • Compensatory • Consequential • Punitive • Nominal • Liquidated	**RESCISSION AND RESTITUTION**	**SPECIFIC PERFORMANCE**	**REFORMATION**

party confers a benefit on another party, justice requires that the party receiving the benefit pay a reasonable value for it.

WHEN QUASI CONTRACTS ARE USED

Quasi contract is a legal theory under which an obligation is imposed in the absence of an agreement. It allows the courts to act as if a contract exists when there is no actual contract or agreement between the parties. The courts can also use this theory when the parties have a contract, but it is unenforceable for some reason.

Quasi-contractual recovery is often granted when one party has partially performed under a contract that is unenforceable. It provides an alternative to suing for damages and allows the party to recover the reasonable value of the partial performance. ■ EXAMPLE 12.10 Ericson contracts to build two oil derricks for Petro Industries. The derricks are to be built over a period of three years, but the parties do not create a written contract. Because of the one-year rule, therefore, the Statute of Frauds will bar the enforcement of the contract (see Chapter 10).[12] After Ericson completes one derrick, Petro Industries informs him that it will not pay for the derrick. Ericson can sue Petro Industries under the theory of quasi contract. ■

THE REQUIREMENTS OF QUASI CONTRACT

To recover in quasi contract, the party seeking recovery must show the following:

1. The party conferred a benefit on the other party.
2. The party conferred the benefit with the reasonable expectation of being paid.
3. The party did not act as a volunteer in conferring the benefit.
4. The party receiving the benefit would be unjustly enriched by retaining the benefit without paying for it.

In the example just given, Ericson can sue in quasi contract because all of the conditions for quasi-contractual recovery have been fulfilled. Ericson built the oil derrick with the expectation of being paid. The derrick conferred an obvious benefit on Petro Industries, and Petro Industries would be unjustly enriched if it was allowed to keep the derrick without paying Ericson for the work. Therefore, Ericson should be able to recover the reasonable value of the oil derrick that was built (under the theory of *quantum meruit*[13]—"as much as he deserves"). The reasonable value is ordinarily equal to the fair market value.

Contract Provisions Limiting Remedies

A contract may include provisions stating that no damages can be recovered for certain types of breaches or that damages must be limited to a maximum amount. The contract may also provide that the only remedy for breach is replacement, repair, or refund of the purchase price. Provisions stating that no damages can be recovered are called *exculpatory clauses* (see Chapter 9). Provisions that affect the availability of certain remedies are called *limitation-of-liability clauses*.

Whether these contract provisions and clauses will be enforced depends on the type of breach that is excused by the provision. For example, a clause excluding liability for negligence may be enforced in some cases. When an exculpatory clause for negligence is contained in a contract

12. Under the 2003 amendments to the UCC, the one-year rule will no longer apply to sales contracts. See Chapter 14.

13. Pronounced *kwahn*-tuhm *mehr*-oo-wuht.

made between parties who have roughly equal bargaining positions, the clause usually will be enforced. The Uniform Commercial Code (UCC) specifically allows limitation-of-liability clauses to be included in contracts for the sale of goods, as will be discussed in detail in Chapter 16.[14] A provision excluding liability for fraudulent or intentional injury, however, will not be enforced. Likewise, a clause excluding liability for illegal acts or violations of the law will not be enforced.

14. UCC 2–719.

The enforceability of a limitation-of-liability clause in a home-inspection contract was at issue in the following case.

CASE 12.3 Lucier v. Williams

Superior Court of New Jersey,
Appellate Division, 2004.
366 N.J.Super. 485,
841 A.2d 907.
http://lawlibrary.rutgers.edu/search.shtml[a]

FACTS Eric Lucier and Karen Haley, first-time home buyers, contracted to buy a single-family home for $128,500 from James and Angela Williams in Berlin Township, New Jersey. The buyers asked Cambridge Associates, Limited (CAL), to perform a home inspection. CAL presented the buyers with a contract that limited CAL's liability to "$500, or 50% of fees actually paid to CAL by Client, whichever sum is smaller. Such causes include, but are not limited to, CAL's negligence, errors, omissions, . . . [or] breach of contract." Lucier reluctantly signed the contract. On CAL's behalf, Al Vasys performed the inspection and issued a report. The buyers paid CAL $385. Shortly after Lucier and Haley moved into the house, they noticed leaks, which required roof repairs estimated to cost $8,000 to $10,000. They filed a suit in a New Jersey state court against CAL and others, seeking damages for the loss. CAL filed a motion for summary judgment, claiming that under the limitation-of-liability clause, its liability, if any, was one-half the contract price, or $192.50. The court granted the motion. The plaintiffs appealed to a state intermediate appellate court.

ISSUE Was the limitation-of-liability clause enforceable?

a. In the "Search the N.J. Courts Decisions" section, type "Lucier" in the box. In the result, click on the first link to access the opinion. Rutgers Law School in Camden, New Jersey, maintains this Web site.

DECISION No. The state intermediate appellate court held that the provision was unenforceable. The court reversed the ruling of the lower court and remanded the case for further proceedings.

REASON The appellate court held that the limitation-of-liability clause was unenforceable because "(1) the contract, prepared by the home inspector, is one of adhesion" [a standard-form contract often entered into between parties with unequal bargaining power—see Chapter 9]; (2) the parties, one a consumer and the other a professional expert, have grossly unequal bargaining status; and (3) the substance of the provision eviscerates [removes an essential part of] the contract and its fundamental purpose because the potential damage level is so nominal that it has the practical effect of avoiding almost all responsibility for the professional's negligence." Also, the court held that the provision was contrary to the state's public policy of "effectuating [accomplishing] the purpose of a home inspection contract to render reliable evaluation of a home's fitness for purchase and holding professionals to certain industry standards." Imposing this contract in this setting on these consumers showed a lack of fair dealing by the professional. "If, upon the occasional dereliction [reckless disregard of duties], the home inspector's only consequence is the obligation to refund a few hundred dollars (the smaller of fifty percent of the inspection contract price or $500), there is no meaningful incentive to act diligently in the performance of [the] contracts."

FOR CRITICAL ANALYSIS—Social Consideration *What is the difference between the limitation-of-liability clause in this case and an exculpatory clause (discussed in Chapter 9)?*

Election of Remedies

In many circumstances, a nonbreaching party has several remedies available. Because the remedies may be inconsistent with one another, the common law of contracts requires the party to choose which remedy to pursue. This is called *election of remedies*. The purpose of the doctrine of election of remedies is to prevent double recovery. ■ **EXAMPLE 12.11** Suppose that Jefferson agrees to sell his land to Adams. Then Jefferson changes his mind and repudi-

ates the contract. Adams can sue for compensatory damages or for specific performance. If Adams receives damages as a result of the breach, she should not also be granted specific performance of the sales contract because that would mean she would unfairly end up with both the land and the damages. The doctrine of election of remedies requires Adams to choose the remedy she wants, and it eliminates any possibility of double recovery. ■

In contrast, remedies under the UCC are cumulative. They include all of the remedies available under the UCC for breach of a sales or lease contract.[15] We will examine the UCC provisions on limited remedies in Chapter 16, in the context of the remedies available on the breach of a contract for the sale or lease of goods.

15. See UCC 2–703 and 2–711.

TERMS AND CONCEPTS

consequential damages 227
liquidated damages 230
mitigation of damages 229
nominal damages 228
penalty 230
restitution 231
specific performance 232

CHAPTER SUMMARY Breach and Remedies

COMMON REMEDIES AVAILABLE TO NONBREACHING PARTY	
Damages (See pages 226–231.)	The legal remedy designed to compensate the nonbreaching party for the loss of the bargain. By awarding money damages, the court tries to place the parties in the positions that they would have occupied had the contract been fully performed. The nonbreaching party frequently has a duty to *mitigate* (lessen or reduce) the damages incurred as a result of the contract's breach. There are five broad categories of damages: 1. *Compensatory damages*—Damages that compensate the nonbreaching party for injuries actually sustained and proved to have arisen directly from the loss of the bargain resulting from the breach of contract. a. In breached contracts for the sale of goods, the usual measure of compensatory damages is the difference between the contract price and the market price. b. In breached contracts for the sale of land, the measure of damages is ordinarily the same as in contracts for the sale of goods. c. In breached construction contracts, the measure of damages depends on which party breaches and at what stage of construction the breach occurs. 2. *Consequential damages*—Damages resulting from special circumstances beyond the contract itself; the damages flow only from the consequences of a breach. For a party to recover consequential damages, the damages must be the foreseeable result of a breach of contract, and the breaching party must have known at the time the contract was formed that special circumstances existed that would cause the nonbreaching party to incur additional loss on breach of the contract. Also called *special damages.* 3. *Punitive damages*—Damages awarded to punish the breaching party and to deter others. Usually not awarded in an action for breach of contract unless a tort is involved. 4. *Nominal damages*—Damages small in amount (such as one dollar) that are awarded when a breach has occurred but no actual damage has been suffered. Awarded only to establish that the defendant acted wrongfully. 5. *Liquidated damages*—Damages that may be specified in a contract as the amount to be paid to the nonbreaching party in the event the contract is breached in the future. Clauses providing for liquidated damages are enforced if the damages were difficult to estimate at the time the contract was formed and if the amount stipulated is reasonable. If the amount is construed to be a penalty, the clause will not be enforced.

(Continued)

CHAPTER SUMMARY Breach and Remedies—Continued

Rescission and Restitution (See pages 231–232.)	1. *Rescission*—A remedy whereby a contract is canceled and the parties are restored to the original positions that they occupied before the transaction. Available when fraud, a mistake, duress, or failure of consideration is present. The rescinding party must give prompt notice of the rescission to the breaching party. 2. *Restitution*—When a contract is rescinded, both parties must make restitution to each other by returning the goods, property, or funds previously conveyed. Restitution prevents the unjust enrichment of the parties.
Specific Performance (See page 232.)	An equitable remedy calling for the performance of the act promised in the contract. This remedy is available only in special situations—such as those involving contracts for the sale of unique goods or land—and when monetary damages would be an inadequate remedy. Specific performance is not available as a remedy in breached contracts for personal services.
Reformation (See page 232.)	An equitable remedy allowing a contract to be "reformed," or rewritten, to reflect the parties' true intentions. Available when an agreement is imperfectly expressed in writing.
Recovery Based on Quasi Contract (See pages 232–233.)	An equitable theory imposed by the courts to obtain justice and prevent unjust enrichment in a situation in which no enforceable contract exists. The party seeking recovery must show the following: 1. A benefit was conferred on the other party. 2. The party conferring the benefit did so with the expectation of being paid. 3. The benefit was not volunteered. 4. Retaining the benefit without paying for it would result in the unjust enrichment of the party receiving the benefit.
CONTRACT DOCTRINES RELATING TO REMEDIES	
Contract Provisions Limiting Remedies (See pages 233–234.)	A contract may provide that no damages (or only a limited amount of damages) can be recovered in the event the contract is breached. Clauses excluding liability for fraudulent or intentional injury or for illegal acts cannot be enforced. Clauses excluding liability for negligence may be enforced if both parties hold roughly equal bargaining power. Under the UCC, remedies may be limited in contracts for the sale of goods.
Election of Remedies (See pages 234–235.)	A common law doctrine under which a nonbreaching party must choose one remedy from those available. This doctrine prevents double recovery. Under the UCC, remedies are cumulative for the breach of a contract for the sale of goods.

FOR REVIEW

Answers for the even-numbered questions in this For Review *section can be found in Appendix A at the end of this text.*

1. What is the difference between compensatory damages and consequential damages? What are nominal damages, and when do courts award nominal damages?
2. What is the standard measure of compensatory damages when a contract is breached? How are damages computed differently in construction contracts?
3. Under what circumstances is the remedy of rescission and restitution available?
4. When do courts grant specific performance as a remedy?
5. What is the rationale underlying the doctrine of election of remedies?

 QUESTIONS AND CASE PROBLEMS

12–1. Liquidated Damages. Carnack contracts to sell his house and lot to Willard for $100,000. The terms of the contract call for Willard to pay 10 percent of the purchase price as a deposit toward the purchase price, or as a down payment. The terms further stipulate that should the buyer breach the contract, Carnack will retain the deposit as liquidated damages. Willard pays the deposit, but because her expected financing of the $90,000 balance falls through, she breaches the contract. Two weeks later, Carnack sells the house and lot to Balkova for $105,000. Willard demands her $10,000 back, but Carnack refuses, claiming that Willard's breach and the contract terms entitle him to keep the deposit. Discuss who is correct.

QUESTION WITH SAMPLE ANSWER

12–2. In which of the following situations might a court grant specific performance as a remedy for the breach of the contract?

(a) Tarrington contracts to sell her house and lot to Rainier. Then, on finding another buyer willing to pay a higher purchase price, she refuses to deed the property to Rainier.

(b) Marita contracts to sing and dance in Horace's nightclub for one month, beginning June 1. She then refuses to perform.

(c) Juan contracts to purchase a rare coin from Edmund, who is breaking up his coin collection. At the last minute, Edmund decides to keep his coin collection intact and refuses to deliver the coin to Juan.

(d) Astro Computer Corp. has three shareholders: Coase, who owns 48 percent of the stock; De Valle, who owns 48 percent; and Cary, who owns 4 percent. Cary contracts to sell his 4 percent to De Valle but later refuses to transfer the shares to him.

For a sample answer to this question, go to Appendix B at the end of this text.

12–3. Measure of Damages. Johnson contracted to lease a house to Fox for $700 a month, beginning October 1. Fox stipulated in the contract that before he moved in, the interior of the house had to be completely repainted. On September 9, Johnson hired Keever to do the required painting for $1,000. He told Keever that the painting had to be finished by October 1 but did not explain why. On September 28, Keever quit for no reason, having completed approximately 80 percent of the work. Johnson then paid Sam $300 to finish the painting, but Sam did not finish until October 4. When Fox found that the painting had not been completed as stipulated in his contract with Johnson, he leased another home. Johnson found another tenant who would lease the property at $700 a month, beginning October 15. Johnson then sued Keever for breach of contract, claiming damages of $650. This amount included the $300 Johnson paid Sam to finish the painting and $350 for rent for the first half of October, which Johnson had lost as a result of Keever's breach. Johnson had not yet paid Keever anything for Keever's work. Can Johnson collect the $650 from Keever? Explain.

12–4. Measure of Damages. Ben owns and operates a famous candy store. He makes most of the candy sold in the store, and business is particularly heavy during the Christmas season. Ben contracts with Sweet, Inc., to purchase ten thousand pounds of sugar, to be delivered on or before November 15. Ben informs Sweet that this particular order is to be used for the Christmas season business. Because of production problems, the sugar is not tendered to Ben until December 10, at which time Ben refuses the order because it is so late. Ben has been unable to purchase the quantity of sugar needed to meet the Christmas orders and has had to turn down numerous regular customers, some of whom have indicated that they will purchase candy elsewhere in the future. The sugar that Ben has been able to purchase has cost him ten cents per pound above Sweet's price. Ben sues Sweet for breach of contract, claiming as damages the higher price paid for the sugar from other suppliers, lost profits from this year's lost Christmas sales, future lost profits from customers who have indicated that they will discontinue doing business with him, and punitive damages for failure to meet the contracted-for delivery date. Sweet claims Ben is limited to compensatory damages only. Discuss who is correct, and why.

12–5. Damages. In December 1992, Beys Specialty Contracting, Inc., contracted with New York City's Metropolitan Transportation Authority (MTA) for construction work. Beys subcontracted with Hudson Iron Works, Inc., to perform some of the work for $175,000. Under the terms of the subcontract, within seven days after the MTA approved Hudson's work and paid Beys, Beys would pay Hudson. The MTA had not yet approved any of Hudson's work when Beys submitted to the MTA invoices dated May 20 and June 21, 1993. Without proof that the MTA had paid Beys on those invoices, Hudson submitted to Beys an invoice dated September 10, claiming that the May 20 and June 21 invoices incorporated its work. Beys refused to pay, Hudson stopped working, and Beys paid another contractor $25,083 more to complete the job than if Hudson had completed its subcontract. Hudson filed a suit in a New York state court to collect on its invoice. Beys filed a counterclaim for the additional funds spent to complete Hudson's job. In whose favor should the court rule, and why? What might be the measure of damages, if any? [*Hudson Iron Works, Inc. v. Beys Specialty Contracting, Inc.*, 262 A.D.2d 360, 691 N.Y.S.2d 132 (1999)]

CASE PROBLEM WITH SAMPLE ANSWER

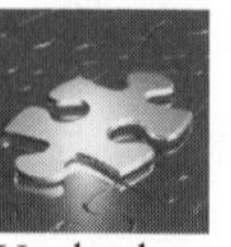

12–6. Ms. Vuylsteke, a single mother with three children, lived in Portland, Oregon. Cynthia Broan also lived in Oregon until she moved to New York City to open and operate an art gallery. Broan asked Vuylsteke to manage the gallery under a one-year contract for an annual salary of $72,000. To begin work, Vuylsteke relocated to New York. As part of the move, Vuylsteke transferred custody of

her children to her husband, who lived in London, England. In accepting the job, Vuylsteke also forfeited her husband's alimony and child-support payments, including unpaid amounts of nearly $30,000. Before Vuylsteke started work, Broan repudiated the contract. Unable to find employment for more than an annual salary of $25,000, Vuylsteke moved to London to be near her children. Vuylsteke filed a suit in an Oregon state court against Broan, seeking damages for breach of contract. Should the court hold, as Broan argued, that Vuylsteke did not take reasonable steps to mitigate her damages? Why or why not? [*Vuylsteke v. Broan*, 172 Or.App. 74, 17 P.3d 1072 (2001)]

After you have answered this problem, compare your answer with the sample answer given on the Web site that accompanies this text. Go to academic.cengage.com/blaw/wbl, select "Chapter 12," and click on "Case Problem with Sample Answer."

12–7. Liquidated Damages versus Penalties. Every homeowner in the Putnam County, Indiana, subdivision of Stardust Hills must be a member of the Stardust Hills Owners Association, Inc., and must pay annual dues of $200 for the maintenance of common areas and other community services. Under the association's rules, dues paid more than ten days late "shall bear a delinquent fee at a rate of $2.00 per day." Phyllis Gaddis failed to pay the dues on a Stardust Hills lot that she owned. Late fees began to accrue. Nearly two months later, the association filed a suit in an Indiana state court to collect the unpaid dues and the late fees. Gaddis argued in response that the delinquent fee was an unenforceable penalty. What questions should be considered in determining the status of this fee? Should the association's rule regarding assessment of the fee be enforced? Explain. [*Gaddis v. Stardust Hills Owners Association, Inc.*, 804 N.E.2d 231 (Ind.App. 2004)]

VIDEO QUESTION

12–8. Go to this text's Web site at academic.cengage.com/blaw/wbl and select "Chapter 12." Click on "Video Questions" and view the video titled *Midnight Run*. Then answer the following questions.

1. In the video, Eddie (Joe Pantoliano) and Jack (Robert DeNiro) negotiate a contract for Jack to find the Duke, a mob accountant who embezzled funds, and bring him back for trial. Assume that the contract is valid. If Jack breaches the contract by failing to bring in the Duke, what kinds of remedies, if any, can Eddie seek? Explain your answer.
2. Would the equitable remedy of specific performance be available to either Jack or Eddie in the event of a breach? Why or why not?
3. Now assume the contract between Eddie and Jack is unenforceable. Nevertheless, Jack performs his side of the bargain (brings in the Duke). Does Jack have any legal recourse in this situation? Why or why not?

Breach and Remedies

For updated links to resources available on the Web, as well as a variety of other materials, visit this text's Web site at

academic.cengage.com/blaw/wbl

The following site offers a brief summary of and several related articles on breach of contract:

http://www.legalmatch.com/law-library/article/breach-of-contract.html

ONLINE LEGAL RESEARCH

Go to the *Fundamentals of Business Law* home page at academic.cengage.com/blaw/wbl, select "Chapter 12," and click on "Internet Exercises." There you will find the following Internet research exercises that you can perform to learn more about topics covered in this chapter.

Activity 12–1: LEGAL PERSPECTIVE—Contract Damages and Contract Theory

Activity 12–2: MANAGEMENT PERSPECTIVE—The Duty to Mitigate

BEFORE THE TEST

Go to the *Fundamentals of Business Law* home page at academic.cengage.com/blaw/wbl, select "Chapter 12," and click on "Interactive Quizzes." You will find a number of interactive questions relating to this chapter.

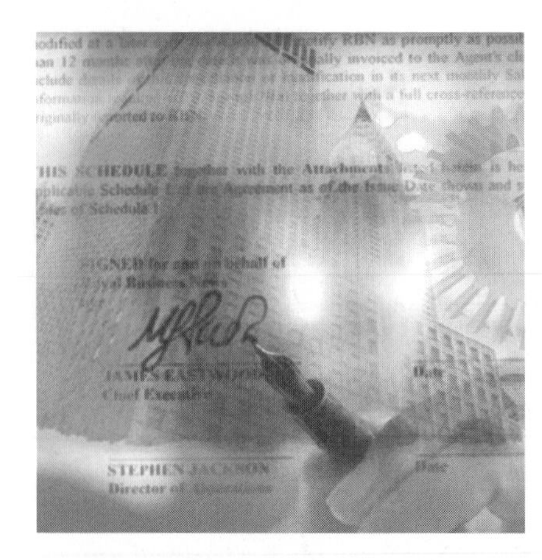

CHAPTER 13
E-Contracts

LEARNING OBJECTIVES

After reading this chapter, you should be able to answer the following questions:

1. What are some important clauses to include when making offers to form electronic contracts, or e-contracts?
2. How do shrink-wrap and click-on agreements differ from other contracts? How have traditional laws been applied to these agreements?
3. What is an electronic signature? Are electronic signatures valid?
4. What is a partnering agreement? What purpose does it serve?
5. What is the Uniform Electronic Transactions Act (UETA)? What are some of the major provisions of this act?

Electronic technology offers businesses several advantages, including speed, efficiency, and lower costs. In the 1990s, many observers argued that the development of cyberspace was revolutionary. Therefore, new legal theories, and new laws, would be needed to govern **e-contracts**, or contracts entered into electronically. To date, however, most courts have adapted traditional contract law principles and provisions of the Uniform Commercial Code, when applicable, to cases involving e-contract disputes.

In the first part of this chapter, we look at how traditional laws are being applied to contracts formed online. We then examine some new laws that have been created to apply in situations in which traditional laws governing contracts have sometimes been thought inadequate. For example, traditional laws governing signature and writing requirements are not easily adapted to contracts formed in the online environment. Thus, new laws have been created to address these issues.

Forming Contracts Online

Today, numerous contracts are being formed online. Although the medium through which these contracts are generated has changed, the age-old problems attending contract formation have not. Disputes concerning contracts formed online continue to center around contract terms and whether the parties voluntarily assented to those terms.

Note that online contracts may be formed not only for the sale of goods and services but also for *licensing*. ■ EXAMPLE 13.1 The "sale" of software generally involves a license, or a right to use the software, rather than the passage of title (ownership rights) from the seller to the buyer. ■ As you read through the following pages, keep in mind that although we typically refer to the offeror and the offeree as a *seller* and a *buyer*, in many transactions these parties would be more accurately described as a *licensor* and a *licensee*.

ONLINE OFFERS

Sellers doing business via the Internet can protect themselves against contract disputes and legal liability by creating offers that clearly spell out the terms that will govern their transactions if the offers are accepted. All important terms should be conspicuous and easily viewed by potential buyers.

An important rule to keep in mind is that the offeror controls the offer and thus the resulting contract. The seller

MANAGEMENT PERSPECTIVE

The Enforceability of Forum-Selection Clauses

MANAGEMENT FACES A LEGAL ISSUE

Because parties to contracts formed online may be located in geographically distant locations, online sellers of goods and services normally include forum-selection clauses in their contracts. These clauses can help online sellers avoid having to appear in court in many distant jurisdictions when customers are dissatisfied with their purchases. Clearly, owners and managers of Internet-based businesses benefit from such clauses, yet purchasers of online goods and services may challenge such clauses as being unfair to them. How have the courts responded to such challenges?

WHAT THE COURTS SAY

Normally, the courts will enforce clauses or contracts to which the parties have voluntarily agreed, and this principle extends to forum-selection clauses in online contracts as well. As one court stated (in a case challenging the enforceability of a forum-selection clause in Microsoft Network's online agreement), "If a forum-selection clause is clear in its purport [meaning] and has been presented to the party to be bound in a fair and forthright fashion, no . . . policies or principles have been violated."[a]

Depending on the jurisdiction, however, a court may make an exception to this rule. Consider a case brought against America Online, Inc. (AOL), by Al Mendoza and other former AOL subscribers living in California. The plaintiffs brought their case in a California state court. They sought compensatory and punitive damages, claiming that AOL had continued to debit their credit cards for monthly service fees, without authorization, for some time after they had terminated their subscriptions. AOL moved to dismiss the action on the basis of the forum-selection clause in its "Terms of Service" agreement with subscribers. That clause required all lawsuits under the agreement to be brought in Virginia, AOL's home state. A California appellate court ultimately held that the clause was unfair and unreasonable, and that public policy was best served by denying enforceability to the clause. The court also noted that Virginia law provides "significantly less" consumer protection than California law, and therefore enforcing the forum-selection clause would violate the "strong California public policy" expressed in the state's consumer protection statutes.[b]

IMPLICATIONS FOR MANAGERS

Generally, owners and managers of Web-based businesses can assume that forum-selection clauses in their online contracts will likely be enforced. Businesspersons should be aware, though, that different courts have reached varying conclusions on this issue. A court may conclude, as the California court did in the AOL case, that a particular forum-selection clause imposes an unfair burden on those who purchase goods and services from online vendors.

a. *Caspi v. Microsoft Network, L.C.C.,* 323 N.J.Super. 118, 732 A.2d 528 (1999). For another example, see *DeJohn v. The .TV Corp. International,* 245 F.Supp.2d 913 (N.D.Ill. 2003).

b. *America Online, Inc. v. Superior Court,* 90 Cal.App.4th 1, 108 Cal.Rptr.2d 699 (2001).

should therefore anticipate the terms he or she wants to include in a contract and provide for them in the offer. In some instances, a standardized contract form may suffice.

At a minimum, an online offer should include the following provisions:

1. A provision specifying the remedies available to the buyer if the goods are found to be defective or if the contract is otherwise breached. Any limitation of remedies should be clearly spelled out.
2. A clause that clearly indicates what constitutes the buyer's agreement to the terms of the offer.
3. A provision specifying how payment for the goods and of any applicable taxes must be made.
4. A statement of the seller's refund and return policies.
5. Disclaimers of liability for certain uses of the goods. For example, an online seller of business forms may add a disclaimer that the seller does not accept responsibility for the buyer's reliance on the forms rather than on an attorney's advice.
6. A statement indicating how the seller will use the information gathered about the buyer.

Dispute-Settlement Provisions In addition to the above provisions, many online offers include provisions relating to dispute settlement. For example, the offer might include an arbitration clause specifying that any dispute arising under the contract will be arbitrated in a designated forum.

Many online contracts also contain a **forum-selection clause** (indicating the forum, or location, for the resolution of any dispute arising under the contract). As discussed in Chapter 2, significant jurisdictional issues may occur when parties are at a great distance, as they often are when they form contracts via the Internet. A forum-selection clause will help to avert future jurisdictional problems and also help to ensure that the seller will not be required to appear in court in a distant state. For a discussion of forum-selection clauses in online contracts, see this chapter's *Management Perspective* feature.

Displaying the Offer The seller's Web site should include a hypertext link to a page containing the full contract so that potential buyers are made aware of the terms to which they are assenting. The contract generally must be displayed online in a readable format such as a twelve-point typeface. All provisions should be reasonably clear. ■ EXAMPLE 13.2 Suppose that Netquip sells a variety of heavy equipment, such as trucks and trailers, online at its Web site. Because Netquip's pricing schedule is very complex, the schedule must be fully provided and explained on the Web site. ■

Indicating How the Offer Can Be Accepted An online offer should also include some mechanism by which the customer may accept the offer. Typically, online sellers include boxes containing the words "I agree" or "I accept the terms of the offer" that offerees can click on to indicate acceptance. The contract resulting from such an acceptance is often called a **click-on agreement.**

ONLINE ACCEPTANCES

In many ways, click-on agreements are the Internet equivalents of *shrink-wrap agreements* (or *shrink-wrap licenses,* as they are sometimes called). A **shrink-wrap agreement** is an agreement whose terms are expressed inside a box in which the goods are packaged. (The term *shrink-wrap* refers to the plastic that covers the box.) Usually, the party who opens the box is told that she or he agrees to the terms by keeping whatever is in the box. Similarly, when the purchaser opens a software package, he or she agrees to abide by the terms of the limited license agreement.

■ EXAMPLE 13.3 John orders a new computer from a national company, which ships the computer to him. Along with the computer, the box contains an agreement setting forth the terms of the sale, including what remedies are available. The document also states that John's retention of the computer for longer than thirty days will be construed as an acceptance of the terms. ■

In most situations, a shrink-wrap agreement is not between a retailer and a buyer, but between the manufacturer of the hardware or software and the ultimate buyer-user of the product. The terms generally concern warranties, remedies, and other issues associated with the use of the product.

We look next at how the law has been applied to both shrink-wrap and click-on agreements.

Shrink-Wrap Agreements—Enforceable Contract Terms The *Restatement (Second) of Contracts*—a compilation of common law contract principles—states that parties may agree to a contract "by written or spoken words or by other action or by failure to act."[1] The Uniform Commercial Code (UCC)—the law governing sales contracts—has a similar provision. Section 2–204 of the UCC states that any contract for the sale of goods "may be made in any manner sufficient to show agreement, including conduct by both parties which recognizes the existence of such a contract." The courts have used these provisions to conclude that a binding contract can be created by conduct, including conduct accepting the terms in either a shrink-wrap agreement or a click-on agreement. Thus, a buyer's failure to object to terms contained within a shrink-wrapped software package (or an online offer) may constitute an acceptance of the terms by conduct.[2]

In many cases, the courts have enforced the terms of shrink-wrap agreements in the same way as the terms of other contracts. Some courts have reasoned that by including the terms with the product, the seller proposed a contract that the buyer could accept by using the product after having an opportunity to read the terms. Also, it seems practical from a business's point of view to enclose a full statement of the legal terms of a sale with the product rather than to read the statement over the phone, for example, when a buyer calls in an order for the product.

Shrink-Wrap Terms That May Not Be Enforced Not all of the terms included in shrink-wrap agreements have been enforced by the courts, however. One important consideration is whether the parties form their contract before or after the seller communicates the terms of the shrink-wrap agreement to the buyer. If the buyer learned of the shrink-wrap terms *after* the parties entered into a contract, a court may conclude that those terms were proposals for additional terms and were not part of the contract unless the buyer expressly agreed to them.[3] Could any other problems arise with shrink-wrap agreements? For a discussion of a case involving an issue that arose for one firm, see this chapter's *Adapting the Law to the Online Environment* feature on the next page.

Click-On Agreements As described earlier, a click-on agreement (also sometimes called a *click-on license* or *click-wrap agreement*) arises when a buyer, completing a transaction on a computer, indicates his or her assent to be bound by the terms of an offer by clicking on a box that says, for example, "I agree." The terms may be contained on a Web site through which the buyer is obtaining goods or services, or they may appear on a computer screen when software is loaded. Exhibit 13–1 on page 242 contains the

1. *Restatement (Second) of Contracts,* Section 19.
2. For a leading case on this issue, see *ProCD, Inc. v. Zeidenberg,* 86 F.3d 1447 (7th Cir. 1996).
3. See, for example, *Klocek v. Gateway, Inc.,* 104 F.Supp.2d 1332 (D.Kan. 2000).

ADAPTING THE LAW TO THE ONLINE ENVIRONMENT

Avoiding Deception in Software Sales

Sometimes, businesspersons who include shrink-wrap licenses with their products may have some terms elsewhere, such as on a disk or on a download page of the Internet. Not including all of the terms in the shrink-wrap agreement, however, can lead to problems—as one software producer learned when the state of New York brought an action against its company for fraud.

THE LAWSUIT AGAINST NETWORK ASSOCIATES, INC.

Network Associates, Inc. (NA), develops and sells software, including Gauntlet, a software firewall product, via the Internet. NA included a restrictive clause on its disks and on its Internet download page—but not in its license agreement that accompanied its products.

The restrictive clause provided that anyone installing the Gauntlet software accepted the terms and conditions of the license agreement in the box and urged users to read the license before installing the software. The clause also stated, among other things, that the customer "will not publish reviews of this product without prior consent from Network Associates." The problem was that the license agreement in the box stated that the agreement contained all of the rights and duties of the parties. How, then, did the restrictive clause apply to the sale?

When *Network World Fusion,* an online magazine, published a comparative review of firewall software products, including NA's Gauntlet, without NA's permission, NA protested. Ultimately, the state attorney general of New York brought an action against NA for fraud.

THE FRAUD ISSUE

According to the New York court hearing the case, NA's restrictive clause misled customers and was thus deceptive. First, the license agreement stated that it contained all of the terms of the agreement. Therefore, the rules and regulations listed in the restrictive clause appeared to be independent of the license contract. This could mislead purchasers of the software because they might believe that the restriction was created by some other entity, such as the federal government.

For these reasons, the court concluded that the restrictive clause was deceptive and constituted fraud. The court ordered NA to stop including the clause in its software. The court also ordered NA to reveal "the number of instances in which software was sold on disks or through the Internet containing the above-mentioned language in order for the court to determine what, if any, penalties and costs should be ordered."[a]

FOR CRITICAL ANALYSIS

What is the difference, if any, between reading a restrictive clause in a shrink-wrap agreement and accessing it through a link as part of a click-on agreement?

a. *People v. Network Associates, Inc.,* 195 Misc.2d 384, 758 N.Y.S.2d 466 (2003).

EXHIBIT 13–1 A CLICK-ON AGREEMENT

This exhibit illustrates an online offer to form a contract. To accept the offer, the user simply scrolls down the page and clicks on the "Accept" box.

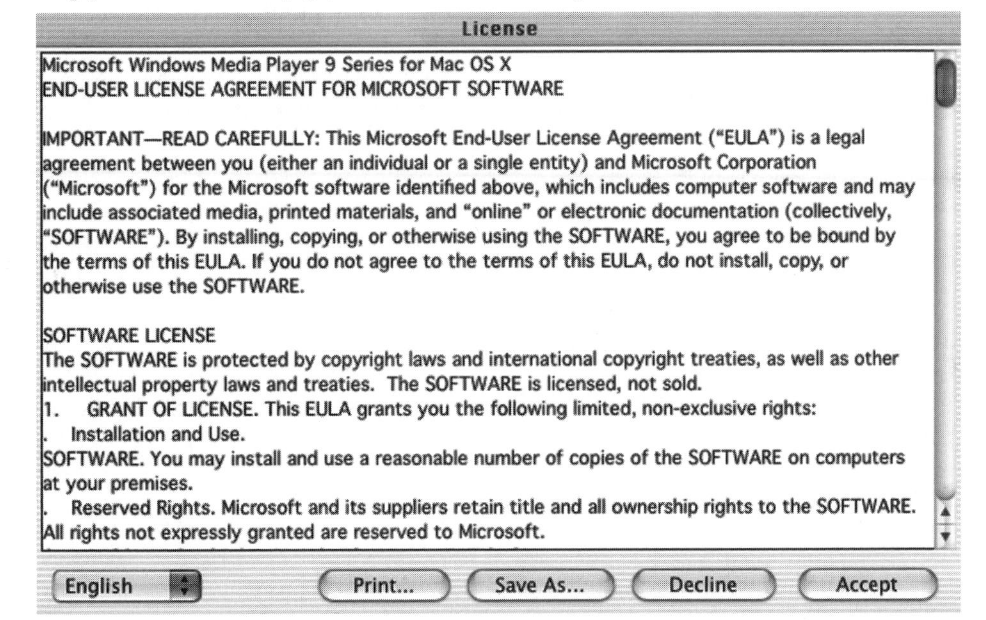

License

Microsoft Windows Media Player 9 Series for Mac OS X
END-USER LICENSE AGREEMENT FOR MICROSOFT SOFTWARE

IMPORTANT—READ CAREFULLY: This Microsoft End-User License Agreement ("EULA") is a legal agreement between you (either an individual or a single entity) and Microsoft Corporation ("Microsoft") for the Microsoft software identified above, which includes computer software and may include associated media, printed materials, and "online" or electronic documentation (collectively, "SOFTWARE"). By installing, copying, or otherwise using the SOFTWARE, you agree to be bound by the terms of this EULA. If you do not agree to the terms of this EULA, do not install, copy, or otherwise use the SOFTWARE.

SOFTWARE LICENSE
The SOFTWARE is protected by copyright laws and international copyright treaties, as well as other intellectual property laws and treaties. The SOFTWARE is licensed, not sold.
1. GRANT OF LICENSE. This EULA grants you the following limited, non-exclusive rights:
. Installation and Use.
SOFTWARE. You may install and use a reasonable number of copies of the SOFTWARE on computers at your premises.
. Reserved Rights. Microsoft and its suppliers retain title and all ownership rights to the SOFTWARE. All rights not expressly granted are reserved to Microsoft.

English | Print... | Save As... | Decline | Accept

language of a click-on agreement that accompanies a package of software made and marketed by Microsoft.

Generally, under the law governing contracts, including sales and lease contracts under the UCC, there is no requirement that all of the terms in a contract must actually have been read by all of the parties to be effective. For example, clicking on a button or box that states "I accept" to certain terms can be enough.

In the following case, the court considered the enforceability of a click-on (click-wrap) agreement under Article 2 of the UCC.

CASE 13.1 i.LAN Systems, Inc. v. Netscout Service Level Corp.

United States District Court,
District of Massachusetts, 2002.
183 F.Supp.2d 328.

FACTS i.LAN Systems, Inc., helps companies monitor their computer networks. Its technicians can solve many problems remotely over the Internet using the software tools already built into a network. These tools include software designed and sold by Netscout Service Level Corporation, formerly known as NextPoint Networks, Inc. A click-wrap provision in the NextPoint software states that the seller's liability is limited to the price paid for the software unless a different term is "specifically accepted by NextPoint in writing." In 1998, i.LAN and NextPoint signed an agreement under which i.LAN agreed to resell the software. In 1999, under a different purchase order, i.LAN bought what it thought was the unlimited right to rent, rather than sell, the software, complete with perpetual upgrades and technical support. When NextPoint disputed this interpretation, i.LAN filed a suit in a federal district court, alleging, among other things, breach of contract. i.LAN sought specific performance (a remedy in which a court orders the breaching party to perform as specified in the contract—see Chapter 12) that included unlimited upgrades and support. The defendant argued that even if the allegations were true, the click-wrap provision limited its liability to the price paid for the software. Both parties filed motions for summary judgment.

ISSUE Was the click-wrap agreement enforceable?

DECISION Yes. The court denied the plaintiff's motion for summary judgment and instead issued a summary judgment in favor of the defendant (NextPoint).

REASON The court reasoned that the plaintiff agreed to the click-wrap terms when it clicked on the "I agree" box. Those terms effectively limited the defendant's liability. In the words of the court, "pursuant to UCC 2–204, the analysis is simple: i.LAN manifested assent to the click-wrap license agreement when it clicked on the box stating 'I agree,' so the agreement is enforceable."[a] The court reasoned that "money now, terms later" forms a contract when the purchaser receives the box of software, sees the license agreement, and does not return the software. The court noted that "money now, terms later" is a practical way to form contracts, especially with purchasers of software. The court concluded that if it is "correct to enforce a shrink-wrap license agreement, where any assent is implicit, then it must also be correct to enforce a click-wrap license agreement, where the assent is explicit."

FOR CRITICAL ANALYSIS—Economic Consideration *If click-wrap agreements were not enforceable, what would be the effect on the software industry?*

a. Recall that under UCC 2–204, a contract "may be made in any manner sufficient to show agreement, including conduct by both parties which recognizes the existence of such a contract."

Browse-Wrap Terms Like the terms of a click-on agreement, **browse-wrap terms** can occur in a transaction conducted over the Internet. Unlike a click-on agreement, however, browse-wrap terms do not require an Internet user to assent to the terms before, say, downloading or using certain software. In other words, a person can install the software without clicking "I agree" to the terms of a license. Offerors of browse-wrap terms generally assert that the terms are binding without the user's active consent.

Critics contend that browse-wrap terms are not enforceable because they do not satisfy the basic elements of contract formation. Some suggest that to form a valid contract online, a user must at least be presented with the terms before indicating assent.[4] With respect to a browse-wrap term, this would require that a user navigate past it and agree to it before being able to obtain whatever is being granted to the user.

The following case involved the enforceability of a clause in an agreement that the court characterized as a browse-wrap license.

4. American Bar Association Committee on the Law of Cyberspace, "Click-Through Agreements: Strategies for Avoiding Disputes on the Validity of Assent" (document presented at the annual American Bar Association meeting in August 2001).

CASE 13.2 Specht v. Netscape Communications Corp.

United States Court of Appeals,
Second Circuit, 2002.
306 F.3d 17.

FACTS Netscape Communications Corporation's "SmartDownload" software makes it easier for users to download files from the Internet without losing progress if they pause to do some other task or their Internet connection is interrupted. Netscape offers SmartDownload free of charge on its Web site to those who indicate, by clicking the mouse in a designated box, that they wish to obtain it. John Gibson clicked in the box and downloaded the software. On the Web site's download page is a reference to a license agreement that is visible only by scrolling to the next screen. Affirmatively indicating assent to the agreement is not required to download the software. The agreement provides that any disputes arising from use of the software are to be submitted to arbitration in California. Believing that the use of SmartDownload transmits private information about its users, Gibson and others (including Specht, the nominal plaintiff) filed a suit in a federal district court in New York against Netscape, alleging violations of federal law. Netscape asked the court to order the parties to arbitration in California, as specified in the license agreement. The court denied the request, and Netscape appealed.

ISSUE Was the arbitration clause in the license agreement enforceable?

DECISION No. The U.S. Court of Appeals for the Second Circuit affirmed the judgment of the lower court.

REASON The court pointed out that "[m]utual manifestation of assent, whether by written or spoken word or by conduct, is the touchstone of contract." The court reasoned that the plaintiffs may have clicked on the SmartDownload "download" button, but "a consumer's clicking on a download button does not communicate assent to contractual terms if the offer did not make clear to the consumer that clicking on the download button would signify assent to those terms." The plaintiffs responded to an offer that did not include "an immediately visible notice" of its terms or require "unambiguous manifestation of assent" to them. The court concluded that a reference to terms that appear on a "submerged" screen, as in this case, is not sufficient notice. The design of the SmartDownload page "tended to conceal the fact that it was an express acceptance of Netscape's rules and regulations. * * * [T]here is no reason to assume that viewers will scroll down to subsequent screens simply because screens are there."

WHY IS THIS CASE IMPORTANT? *The ruling in this case is important because it marks an application of traditional contract principles to a type of dispute that can arise only in the online context. In the case, the court applied a long-standing principle of contract law: a person will not be bound to an agreement to which he or she did not consent.*

E-Signatures

In many instances, a contract cannot be enforced unless it is signed by the party against whom enforcement is sought. A significant issue in the context of e-commerce has to do with how electronic signatures, or **e-signatures,** can be created and verified on e-contracts.

In the days when many people could not write, they signed documents with an "X." Then handwritten signatures became common, followed by typed signatures, printed signatures, and, most recently, digital signatures that are transmitted electronically. Throughout the evolution of signature technology, the question of what constitutes a valid signature has arisen again and again, and with good reason—without some consensus on what constitutes a valid signature, little business or legal work could be accomplished.

E-Signature Technologies

Today, numerous technologies allow electronic documents to be signed. These include digital signatures and alternative technologies.

Digital Signatures The most prevalent e-signature technology is the *asymmetric cryptosystem,* which creates a digital signature using two different (asymmetric) cryptographic "keys." With this system, a person attaches a digital signature to a document using a private key, or code. The key has a publicly available counterpart. Anyone with the appropriate software can use the public key to verify that the digital signature was made using the private key. A **cybernotary,** or legally recognized certification authority, issues the key pair, identifies the owner of the keys, and certifies the validity of the public key. The cybernotary also serves as a repository for public keys.

Signature Dynamics With another type of signature technology, known as *signature dynamics,* a sender's signature is captured using a stylus and an electronic digitizer pad. A computer program takes the signature's measurements, the sender's identity, the time and date of the signature, and the identity of the hardware. This information is then placed in an encrypted *biometric token* attached to the document being transmitted. To verify the authenticity of the signature, the recipient of the document compares the measurements of the signature with the measurements in the token. When this type of e-signature is used, it is not necessary to have a third party verify the signatory's identity.

Other E-Signature Forms Other forms of e-signatures have been—or are now being—developed as well. ■ EXAMPLE 13.4 Some e-signatures use "smart cards." A *smart card* is a device the size of a credit card that is embedded with code and other data. Like credit and debit cards, a smart card can be inserted into computers to transfer information. Unlike those other cards, however, a smart card can be used to establish a person's identity as validly as a signature on a piece of paper. ■ In addition, technological innovations now under way will allow an e-signature to be evidenced by an image of a person's retina, fingerprint, or face that is scanned by a computer and then matched to a numeric code. The scanned image and the numeric code are registered with security companies that maintain files on an accessible server that can be used to authenticate a transaction.

STATE LAWS GOVERNING E-SIGNATURES

Most states have laws governing e-signatures. The problem is that the state e-signature laws are not uniform. Some states—California is a notable example—provide that many types of documents cannot be signed with e-signatures, while other states are more permissive. Additionally, some states recognize only digital signatures as valid, while others permit other types of e-signatures.

The National Conference of Commissioners on Uniform State Laws, in an attempt to create more uniformity among all the states, promulgated the Uniform Electronic Transactions Act (UETA) in 1999. To date, the UETA has been adopted, at least in part, by more than forty states. The UETA provides, among other things, that a signature may not be denied legal effect or enforceability solely because it is in electronic form. (Other aspects of the UETA will be discussed shortly.)

FEDERAL LAW ON E-SIGNATURES AND E-DOCUMENTS

In 2000, Congress enacted the Electronic Signatures in Global and National Commerce Act (E-SIGN Act) to provide that no contract, record, or signature may be "denied legal effect" solely because it is in an electronic form. In other words, under this law, an electronic signature is as valid as a signature on paper, and an electronic document can be as enforceable as a paper one.

For an electronic signature to be enforceable, the contracting parties must have agreed to use electronic signatures. For an electronic document to be valid, it must be in a form that can be retained and accurately reproduced.

The E-SIGN Act does not apply to all types of documents, however. Contracts and documents that are exempt include court papers, divorce decrees, evictions, foreclosures, health-insurance terminations, prenuptial agreements, and wills. Also, the only agreements governed by the UCC that fall under this law are those covered by Articles 2 and 2A and UCC 1–107 and 1–206.

Despite these limitations, the E-SIGN Act greatly expands the possibilities for contracting online. ■ EXAMPLE 13.5 From a remote location, a businessperson can now open an account with a financial institution, obtain a mortgage or other loan, buy insurance, and purchase real estate over the Internet. Payments and transfers of funds can be done entirely online. Thus, using e-contracts can avoid the time and costs associated with producing, delivering, signing, and returning paper documents. ■

At issue in the following case was whether an exchange of e-mail between the representatives of a seller and its buyers constituted a "writing."

CASE 13.3 In re Cafeteria Operators, L.P.[a]

United States Bankruptcy Court, Northern District of Texas, 2003.
299 Bankr. 411.

a. *L.P.* is an abbreviation for *limited partnership*, a type of business organization that will be discussed in Chapter 24.

FACTS Under the Perishable Agricultural Commodities Act (PACA) of 1930, a seller of fruits and vegetables has a right to the proceeds from their sale but loses this right if he or she agrees *in writing* to extend the time for payment beyond thirty

(Continued)

CASE 13.3—CONTINUED

days. Cafeteria Operators, L.P., operates cafeterias primarily in the Southwest. In 2002, Cafeteria Operators and other buyers ordered cherries from Cherrco, Inc., a wholesale food distributor, through Tim Dent of George E. Dent Sales, Inc. Cherrco bought the fruit from Peterson Farms, Inc., and others and delivered it as specified in invoices that required payment within thirty days. When the buyers failed to pay, Dent and Gene Baldwin, the buyers' representative, tried to negotiate a new payment schedule via e-mail. In January 2003, with $25,269.12 still owed, the buyers filed a petition in a federal bankruptcy court, seeking permission to pay less than this amount. Cherrco objected, asserting that it had a right to the full amount under PACA. The buyers claimed that Cherrco had waived this right in the e-mail exchanged between Dent and Baldwin. Cherrco argued in part that this e-mail was not a "writing."

ISSUE Did the e-mail exchanged between Dent and Baldwin constitute a "writing"?

DECISION Yes. The court rejected Cherrco's argument, holding that the e-mail constituted a "writing" under the E-SIGN Act. The court also held, nonetheless, that Cherrco had not waived its PACA right and awarded the seller the full amount of its claim.

REASON The court noted that an e-mail message to Dent from Sarah Peterson-Schlukebir, a Peterson Farms' representative, stated, "[B]oth Peterson Farms and Cherrco are not willing to sign" a release of the original payment terms. "If we sign the release, we are no longer covered by PACA." Dent, in turn, e-mailed the preceding message to Baldwin, who, after further negotiations, e-mailed the following, which was the last note between the parties: "[W]e do not need a formal [contract]. We will just start making the payments starting today. All we really want is the agreement from them that the payment plan is ok." The court quoted from the E-SIGN Act, which provides that "with respect to any transaction in or affecting interstate or foreign commerce * * * a contract * * * may not be denied legal effect, validity, or enforceability solely because an electronic signature or electronic record was used in its formation." In other words, in a transaction involving interstate commerce, e-mails can constitute a "writing." The court concluded, "The transactions between Cherrco and the Debtor[s] occurred in 2002, and clearly qualify for the requirement that the transaction is in or affects interstate commerce." Thus, the court concluded that the e-mails between the parties satisfied the writing requirement under PACA.

FOR CRITICAL ANALYSIS—Social Consideration
Did the e-mail between Dent and Baldwin constitute an agreement between the parties to extend the original payment terms beyond thirty days? Why or why not?

Partnering Agreements

One way that online sellers and buyers can prevent disputes over signatures in their e-contracts, as well as disputes over the terms and conditions of those contracts, is to form partnering agreements. In a **partnering agreement,** a seller and a buyer who frequently do business with each other agree in advance on the terms and conditions that will apply to all transactions subsequently conducted electronically. The partnering agreement can also establish special access and identification codes to be used by the parties when transacting business electronically.

A partnering agreement reduces the likelihood that disputes will arise under the contract because the buyer and the seller have agreed in advance to the terms and conditions that will accompany each sale. Furthermore, if a dispute does arise, a court or arbitration forum will be able to refer to the partnering agreement when determining the parties' intent with respect to subsequent contracts. Of course, even with a partnering agreement, fraud remains a possibility. If an unauthorized person uses a purchaser's designated access number and identification code, it may be some time before the problem is discovered.

The Uniform Electronic Transactions Act

As noted earlier, the Uniform Electronic Transactions Act (UETA) was promulgated in 1999. The UETA represents one of the first comprehensive efforts to create uniform laws pertaining to e-commerce.

The primary purpose of the UETA is to remove barriers to e-commerce by giving the same legal effect to electronic records and signatures as is given to paper documents and signatures. The UETA broadly defines an *e-signature* as "an electronic sound, symbol, or process attached to or logically associated with a record and executed or adopted by a person with the intent to sign the record."[5] An *e-signature* includes encrypted digital signatures, names (intended as signatures) at the ends of e-mail messages, and "clicks" on a Web page if the click includes the identification of the person. A **record** is "information that is inscribed on a tangible medium or that is stored in

5. UETA 102(8).

an electronic or other medium and is retrievable in perceivable [visual] form."[6]

THE SCOPE AND APPLICABILITY OF THE UETA

The UETA does not create new rules for electronic contracts but rather establishes that records, signatures, and contracts may not be denied enforceability solely due to their electronic form. The UETA does not apply to all writings and signatures but only to electronic records and electronic signatures *relating to a transaction.* A *transaction* is defined as an interaction between two or more people relating to business, commercial, or governmental activities.[7]

The act specifically does not apply to laws governing wills or testamentary trusts (see Chapter 30), to the UCC (other than Articles 2 and 2A), or to the Uniform Computer Information Transactions Act (discussed later in this chapter).[8] In addition, the provisions of the UETA allow the states to exclude its application to other areas of law.

As described earlier, Congress passed the E-SIGN Act in 2000, a year after the UETA was presented to the states for adoption. Thus, a significant issue is whether and to what extent the federal E-SIGN Act preempts the UETA as adopted by the states.

THE FEDERAL E-SIGN ACT AND THE UETA

The E-SIGN Act refers explicitly to the UETA and provides that if a state has enacted the uniform version of the UETA, it is not preempted by the E-SIGN Act.[9] In other words, if the state has enacted the UETA *without modification,* state law will govern. The problem is that many states have enacted nonuniform (modified) versions of the UETA, largely for the purpose of excluding other areas of state law from the UETA's terms. The E-SIGN Act specifies that those exclusions will be preempted to the extent that they are inconsistent with the E-SIGN Act's provisions.

The E-SIGN Act, however, explicitly allows the states to enact alternative requirements for the use or acceptance of electronic records or electronic signatures. Generally, however, the requirements must be consistent with the provisions of the E-SIGN Act, and the state must not give greater legal status or effect to one specific type of technology. Additionally, if a state has enacted alternative requirements *after* the E-SIGN Act was adopted, the state law must specifically refer to the E-SIGN Act. The relationship between the UETA and the E-SIGN Act is illustrated in Exhibit 13–2 on the following page.

6. UETA 102(15).
7. UETA 2(12) and 3.
8. UETA 3(b).
9. 15 U.S.C. Section 7002(2)(A)(i).

HIGHLIGHTS OF THE UETA

We look next at selected provisions of the UETA. Our discussion is, of course, based on the act's uniform provisions. Keep in mind that the states that have enacted the UETA may have adopted slightly different versions.

The Parties Must Agree to Conduct Transactions Electronically The UETA will not apply to a transaction unless each of the parties has previously agreed to conduct transactions by electronic means. The agreement need not be explicit, however, and it may be implied by the conduct of the parties and the surrounding circumstances.[10] In the comments that accompany the UETA, the drafters stated that it may be reasonable to infer that a person who gives out a business card with an e-mail address on it has consented to transact business electronically.[11] The parties' agreement may also be inferred from a letter or other writing, as well as from some verbal communications.

A person who has previously agreed to an electronic transaction can also withdraw his or her consent and refuse to conduct further business electronically. Additionally, the act expressly gives parties the power to vary the UETA's provisions by contract. In other words, *parties can opt out of some or all of the terms of the UETA.* If the parties do not opt out of the terms of the UETA, however, the UETA will govern their electronic transactions.

Attribution In the context of electronic transactions, the term *attribution* refers to the procedures that may be used to ensure that the person sending an electronic record is the same person whose e-signature accompanies the record. Under the UETA, if an electronic record or signature is the act of a particular person, the record or signature may be attributed to that person. If a person types her or his name at the bottom of an e-mail purchase order, that name would qualify as a "signature" and be attributed to the person whose name appeared. Just as in paper contracts, one may use any relevant evidence to prove that the record or signature is or is not the act of the person.[12]

Note that even if an individual's name does not appear on a record, the UETA states that the effect of the record is to be determined from the context and surrounding circumstances. In other words, a record may have legal effect even if no one has signed it. ■ **EXAMPLE 13.6** Suppose that Darby sends a fax to Corina. The fax contains a letterhead identifying Darby as the sender, but the

10. UETA 5(b).
11. UETA 5, Comment 4B.
12. UETA 9.

EXHIBIT 13–2 THE E-SIGN ACT AND THE UETA

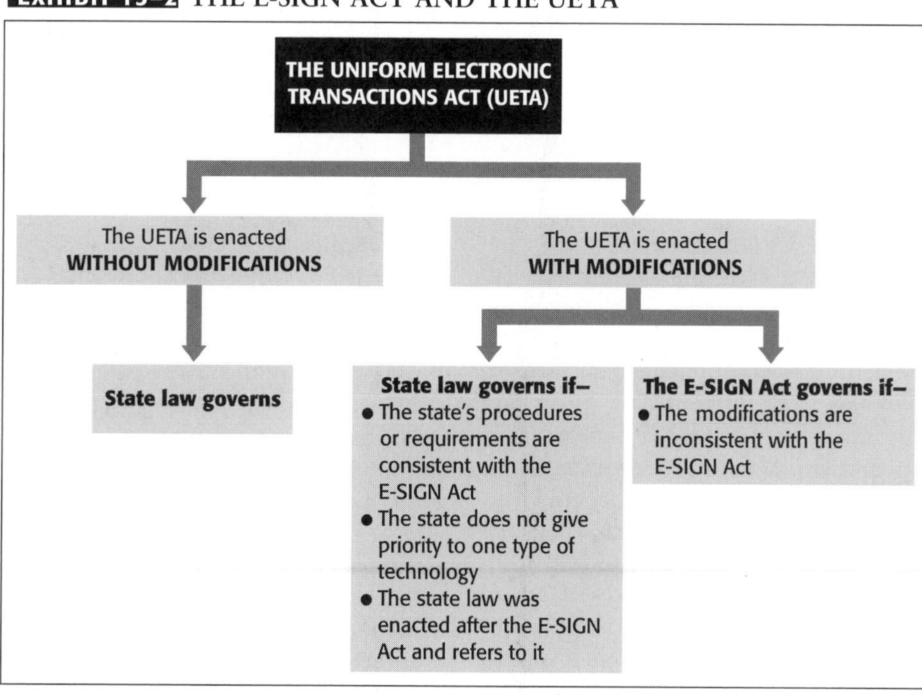

document is not signed by Darby. Depending on the circumstances, the fax may be attributed to Darby. ■

The UETA does not contain any express provisions about what constitutes fraud or whether an agent (a person who acts on behalf of another—see Chapter 22) is authorized to enter into a contract. Under the UETA, other state laws control if any issues relating to agency, authority, forgery, or contract formation arise.

Notarization If a document is required to be notarized under existing state law, the UETA provides that this requirement is satisfied by the electronic signature of a notary public or other person authorized to verify signatures. ■ EXAMPLE 13.7 Suppose that Joel intends to accept an offer to purchase real estate via e-mail. Under the UETA, the requirement is satisfied if a notary public is present to verify Joel's identity and affix an electronic signature to the e-mail acceptance. ■

The Effect of Errors The UETA encourages, but does not require, the use of security procedures (such as encryption) to verify changes to electronic documents and to correct errors. Section 10 of the UETA provides that if the parties have agreed to a security procedure and one party does not detect an error because he or she did not follow the procedure, the conforming party can legally avoid the effect of the change or error. If the parties have not agreed to use a security procedure, then other state laws (including contract law governing mistakes—see Chapter 10) will determine the effect of the error on the parties' agreement.

To avoid the effect of errors, a party must take certain steps. First, the party must promptly notify the other party of the error and of her or his intent not to be bound by the error. Second, the party must take reasonable steps to return any benefit or consideration received. Parties cannot avoid a transaction from which they have benefited. ■ EXAMPLE 13.8 Suppose that Marissa, as the result of an error, received access to valuable information. Because she cannot "give back" or make restitution for the valuable information she now knows, the transaction may be unavoidable. ■ In all other situations in which a change or error occurs in an electronic record (and the parties' agreement does not specifically address errors), the UETA states that the traditional law governing mistakes will govern.

Timing Section 15 of the UETA sets forth provisions relating to the sending and receiving of electronic records. These provisions apply unless the parties agree to differ-

ent terms. Under Section 15, an electronic record is considered *sent* when it is properly directed to the intended recipient in a form readable by the recipient's computer system. Once the electronic record leaves the control of the sender or comes under the control of the recipient, the UETA deems it to have been sent. An electronic record is considered *received* when it enters the recipient's processing system in a readable form—*even if no individual is aware of its receipt.*

Additionally, the UETA provides that, unless otherwise agreed, an electronic record is to be sent from or received at the party's principal place of business. If a party has no place of business, the provision then authorizes the place of sending or receipt to be the party's residence. If a party has multiple places of business, the record should be sent from or received at the location that has the closest relationship to the underlying transaction.

The Uniform Computer Information Transactions Act

The National Conference of Commissioners on Uniform State Laws (NCCUSL) promulgated the Uniform Computer Information Transactions Act (UCITA) in 1999. The primary purpose of the UCITA is to validate e-contracts to license or purchase software, or contracts that give access to—or allow the distribution of—**computer information.** The UCITA is controversial, and only two states (Maryland and Virginia) have adopted it, while four states (Iowa, North Carolina, Vermont, and West Virginia) have passed anti-UCITA laws. In 2003, the NCCUSL withdrew its support of the UCITA. Although the UCITA remains a legal resource, the NCCUSL will no longer seek its adoption by the states, which are thus unlikely to consider it further.

TERMS AND CONCEPTS

browse-wrap terms 243
click-on agreement 241
computer information 249
cybernotary 244
e-contract 239
e-signature 244
forum-selection clause 240
partnering agreement 246
record 246
shrink-wrap agreement 241

CHAPTER SUMMARY E-Contracts

Online Offers (See pages 239–241.)	Businesspersons who present contract offers via the Internet should keep in mind that the terms of the offer should be just as inclusive as the terms in an offer made in a written (paper) document. All possible contingencies should be anticipated and provided for in the offer. Because jurisdictional issues frequently arise with online transactions, it is particularly important to include dispute-settlement provisions in the offer, as well as a forum-selection clause. The offer should be displayed in an easily readable and clear format. An online offer should also include some mechanism, such as an "I agree" or "I accept" box, by which the customer may accept the offer.
Online Acceptances (See pages 241–244.)	1. *Shrink-wrap agreement*— a. Definition—An agreement whose terms are expressed inside a box in which the goods are packaged. The party who opens the box is informed that by keeping the goods that are in the box, he or she agrees to the terms of the shrink-wrap agreement. b. Enforceability—The courts have often enforced shrink-wrap agreements, even if the purchaser-user of the goods did not read the terms of the agreement. A court may deem a shrink-wrap agreement unenforceable, however, if the buyer learns of the shrink-wrap terms *after* the parties entered into the agreement. 2. *Click-on agreement*— a. Definition—An agreement created when a buyer, completing a transaction on a computer, is required to indicate her or his assent to be bound by the terms of an offer by clicking on a box that says, for example, "I agree." The terms of the agreement may appear on the Web site through which the buyer is obtaining goods or services, or they may appear on a computer screen when software is downloaded.

(Continued)

CHAPTER SUMMARY E-Contracts—Continued

Online Acceptances—Continued	b. Enforceability—The courts have enforced click-on agreements, holding that by clicking "I agree," the offeree has indicated acceptance by conduct. Browse-wrap terms, however (terms in a license that an Internet user does not have to read prior to downloading the product, such as software), may not be enforced on the ground that the user is not made aware that he or she is entering into a contract.
E-Signatures (See pages 244–246.)	1. *Definition*—The Uniform Electronic Transactions Act (UETA) defines the term *e-signature* as "an electronic sound, symbol, or process attached to or logically associated with a record and executed or adopted by a person with the intent to sign the record." 2. *E-signature technologies*—These include the *asymmetric cryptosystem* (which creates a digital signature using two different cryptographic "keys"); *signature dynamics* (which involves capturing a sender's signature using a stylus and an electronic digitizer pad); a *smart card* (a device the size of a credit card that is embedded with code and other data); and, probably in the near future, scanned images of retinas, fingerprints, or other physical characteristics that are linked to numeric codes. 3. *State laws governing e-signatures*—Although most states have laws governing e-signatures, these laws are not uniform. The UETA provides for the validity of e-signatures and may ultimately create more uniformity among the states in this respect. 4. *Federal law on e-signatures and e-documents*—The Electronic Signatures in Global and National Commerce Act (E-SIGN Act) of 2000 gave validity to e-signatures by providing that no contract, record, or signature may be "denied legal effect" solely because it is in an electronic form.
Partnering Agreements (See page 246.)	To reduce the likelihood that disputes will arise under their e-contracts, parties who frequently do business with each other may form a *partnering agreement.* In effect, the parties agree in advance on the terms and conditions that will apply to all transactions subsequently conducted electronically. The agreement can also establish access and identification codes to be used by the parties when transacting business electronically.
The Uniform Electronic Transactions Act (UETA) (See pages 246–249.)	1. *Definition*—A uniform act submitted to the states for adoption by the National Conference of Commissioners on Uniform State Laws (NCCUSL). 2. *Purpose*—To create rules to support the enforcement of e-contracts. Under the UETA, contracts entered into online, as well as other documents, are presumed to be valid. The UETA does not apply to transactions governed by the UCC or to wills or testamentary trusts.
The Uniform Computer Information Transactions Act (See page 249.)	1. *Definition*—A uniform act previously submitted to the states for adoption by the NCCUSL. The NCCUSL withdrew its support in 2003. 2. *Purpose*—To validate e-contracts to license or purchase software, or contracts that give access to—or allow the distribution of—computer information.

FOR REVIEW

Answers for the even-numbered questions in this For Review *section can be found in Appendix A at the end of this text.*

1. What are some important clauses to include when making offers to form electronic contracts, or e-contracts?
2. How do shrink-wrap and click-on agreements differ from other contracts? How have traditional laws been applied to these agreements?
3. What is an electronic signature? Are electronic signatures valid?
4. What is a partnering agreement? What purpose does it serve?
5. What is the Uniform Electronic Transactions Act (UETA)? What are some of the major provisions of this act?

QUESTIONS AND CASE PROBLEMS

13–1. Click-On Agreements. Paul is a financial analyst for King Investments, Inc., a brokerage firm. He uses the Internet to investigate the background and activities of companies that might be good investments for King's customers. While visiting the Web site of Business Research, Inc., Paul sees on his screen a message that reads, "Welcome to businessresearch.com. By visiting our site, you have been entered as a subscriber to our e-publication, *Companies Unlimited*. This publication will be sent to you daily at a cost of $7.50 per week. An invoice will be included with *Companies Unlimited* every four weeks. You may cancel your subscription at any time." Has Paul entered into an enforceable contract to pay for *Companies Unlimited?* Why or why not?

QUESTION WITH SAMPLE ANSWER

13–2. Anne is a reporter for the *Daily Business Journal,* a print publication consulted by investors and other businesspersons. She often uses the Internet to perform research for the articles that she writes for the publication. While visiting the Web site of Cyberspace Investments Corp., Anne reads a pop-up window that states, "Our business newsletter, *E-Commerce Weekly,* is available at a one-year subscription rate of $5 per issue. To subscribe, enter your e-mail address below and click 'SUBSCRIBE.' By subscribing, you agree to the terms of the subscriber's agreement. To read this agreement, click 'AGREEMENT.'" Anne enters her e-mail address, but does not click on "AGREEMENT" to read the terms. Has Anne entered into an enforceable contract to pay for *E-Commerce Weekly?* Explain.

For a sample answer to this question, go to Appendix B at the end of this text.

13–3. Online Acceptance. Bob, a sales representative for Central Computer Co., occasionally uses the Internet to obtain information about his customers and to look for new sales leads. While visiting the Web site of Marketing World, Inc., Bob is presented with an on-screen message that offers, "To improve your ability to make deals, read our monthly online magazine, *Sales Genius,* available at a subscription rate of $15 a month. To subscribe, fill in your name, company name, and e-mail address below, and click 'YES!' By clicking 'YES!' you agree to the terms of the subscription contract. To read this contract, click 'TERMS.'" Among those terms is a clause that allows Marketing World to charge interest for subscription bills not paid within a certain time. The terms also prohibit subscribers from copying or distributing part or all of *Sales Genius* in any form. Bob subscribes without reading the terms. Marketing World later files a suit against Bob, based on his failure to pay for his subscription. Should the court hold that Bob is obligated to pay interest on the amount? Explain.

13–4. Shrink-Wrap/Click-On Agreements. 1-A Equipment Co. signed a sales order to lease Accware 10 User NT software, which is made and marketed by ICode, Inc. Just above the signature line, the order stated: "Thank you for your order. No returns or refunds will be issued for software license and/or services. All sales are final. Please read the End User License and Service Agreement." The software was delivered in a sealed envelope inside a box. On the outside of the envelope, an "End User Agreement" provided in part, "BY OPENING THIS PACKAGING, CLICKING YOUR ACCEPTANCE OF THE AGREEMENT DURING DOWNLOAD OR INSTALLATION OF THIS PRODUCT, OR BY USING ANY PART OF THIS PRODUCT, YOU AGREE TO BE LEGALLY BOUND BY THE TERMS OF THE AGREEMENT. . . . This agreement will be governed by the laws in force in the Commonwealth of Virginia . . . and exclusive venue for any litigation shall be in Virginia." Later, dissatisfied with the software, 1-A filed a suit in a Massachusetts state court against ICode, alleging breach of contract and misrepresentation. ICode asked the court to dismiss the case on the basis of the "End User Agreement." Is the agreement enforceable? Should the court dismiss the suit? Why or why not? [*1-A Equipment Co. v. ICode, Inc.,* 43 UCC Rep.Serv.2d 807 (Mass.Dist. 2000)]

CASE PROBLEM WITH SAMPLE ANSWER

13–5. Peerless Wall & Window Coverings, Inc., is a small business in Pennsylvania. To run the cash registers in its stores, manage inventory, and link the stores electronically, in 1994 Peerless installed Point of Sale V6.5 software produced by Synchronics, Inc., a small corporation in Tennessee that develops and markets business software. Point of Sale V6.5 was written with code that used only a two-digit year field—for example, 1999 was stored as "99." This meant that all dates were interpreted as falling within the twentieth century (2001, stored as "01," would be mistaken for 1901). In other words, Point of Sale V6.5 was not "Year 2000" (Y2K) compliant. The software was licensed under a shrink-wrap agreement printed on the envelopes containing the disks. The agreement included a clause that, among other things, limited remedies to replacement within ninety days if there was a defect in the disks and stated, "The entire risk as to the quality and performance of the Software is with you." In 1995, Synchronics stopped selling and supporting Point of Sale V6.5. Two years later, Synchronics told Peerless that the software was not Y2K compliant and should be replaced. Peerless sued Synchronics in a federal district court, alleging in part breach of contract. Synchronics filed a motion for summary judgment. Who is more likely to bear the cost of replacing the software? Why? [*Peerless Wall & Window Coverings, Inc. v. Synchronics, Inc.,* 85 F.Supp.2d 519 (W.D.Pa. 2000); aff'd 234 F.3d 1265 (3d Cir. 2000)]

After you have answered this problem, compare your answer with the sample answer given on the Web site that accompanies this text. Go to academic.cengage.com/blaw/wbl, select "Chapter 13," and click on "Case Problem with Sample Answer."

13–6. Click-On Agreements. America Online, Inc. (AOL), provided e-mail service to Walter Hughes and other members under a click-on agreement titled "Terms of Service." This agreement

consisted of three parts: a "Member Agreement," "Community Guidelines," and a "Privacy Policy." The "Member Agreement" included a forum-selection clause that read, "You expressly agree that exclusive jurisdiction for any claim or dispute with AOL or relating in any way to your membership or your use of AOL resides in the courts of Virginia." When Officer Thomas McMenamon of the Methuen, Massachusetts, Police Department received threatening e-mail sent from an AOL account, he requested and obtained from AOL Hughes's name and other personal information. Hughes filed a suit in a federal district court against AOL, which filed a motion to dismiss on the basis of the forum-selection clause. Considering that the clause was a click-on provision, is it enforceable? Explain. [*Hughes v. McMenamon*, 204 F.Supp.2d 178 (D.Mass. 2002)]

13–7. Shrink-Wrap Agreements/Browse-Wrap Terms. Mary DeFontes bought a computer and a service contract from Dell Computers Corp. DeFontes was charged $950.51, of which $13.51 was identified on the invoice as "tax." This amount was paid to the state of Rhode Island. DeFontes and other Dell customers filed a suit in a Rhode Island state court against Dell, claiming that Dell was overcharging its customers by collecting a tax on service contracts and transportation costs. Dell asked the court to order DeFontes to submit the dispute to arbitration. Dell cited its "Terms and Conditions Agreement," which provides in part that by accepting delivery of Dell's products or services, a customer agrees to submit any dispute to arbitration. Customers can view this agreement through an inconspicuous link at the bottom of Dell's Web site, and Dell encloses a copy with an order when it is shipped. Dell argued that DeFontes accepted these terms by failing to return her purchase within thirty days, although the agreement did not state this. Is DeFontes bound to the "Terms and Conditions Agreement"? Should the court grant Dell's request? Why or why not? [*DeFontes v. Dell Computers Corp.*, __ A.2d __ (R.I. 2004)]

VIDEO QUESTION

13–8. Go to this text's Web site at academic.cengage.com/blaw/wbl and select "Chapter 13." Click on "Video Questions" and view the video titled *E-Contracts: Agreeing Online*. Then answer the following questions.

1. According to the instructor in the video, what is the key factor in determining whether a particular term in an online agreement is enforceable?
2. Suppose that you click on "I accept" in order to download software from the Internet. You do not read the terms of the agreement before accepting it, even though you know that such agreements often contain forum-selection and arbitration clauses. The software later causes irreparable harm to your computer system, and you want to sue. When you go to the Web site and view the agreement, however, you discover that a choice-of-law clause in the contract specified that the law of Nigeria controls. Is this term enforceable? Is it a term that should reasonably be expected in an online contract?
3. Does it matter what the term actually says if it is a type of term that one could reasonably expect to be in the contract? What arguments can be made for and against enforcing a choice-of-law clause in an online contract?

E-Contracts

For updated links to resources available on the Web, as well as a variety of other materials, visit this text's Web site at

academic.cengage.com/blaw/wbl

The Web site of the National Conference of Commissioners on Uniform State Laws (NCCUSL) includes questions and answers about the UCC, UETA, and more. Go to

http://www.nccusl.org

You can access the Uniform Commercial Code, including Article 2, at the Web site of the University of Pennsylvania Law School. Go to

http://www.law.upenn.edu/bll/ulc/ulc.htm

You can find many good resources online about the E-SIGN Act of 2000, including articles available at

http://archives.cnn.com/2000/ALLPOLITICS/stories/06/30/clinton.e.signatures.04/index.html

and

http://www.bizjournals.com/albany/stories/2000/09/11/editorial5.html

ONLINE LEGAL RESEARCH

Go to the *Fundamentals of Business Law* home page at **academic.cengage.com/blaw/wbl**, select "Chapter 13," and click on "Internet Exercises." There you will find the following Internet research exercises that you can perform to learn more about topics covered in this chapter.

Activity 13–1: LEGAL PERSPECTIVE—E-Contract Formation

Activity 13–2: MANAGEMENT PERSPECTIVE—E-Signatures

BEFORE THE TEST

Go to the *Fundamentals of Business Law* home page at **academic.cengage.com/blaw/wbl**, select "Chapter 13," and click on "Interactive Quizzes." You will find a number of interactive questions relating to this chapter.

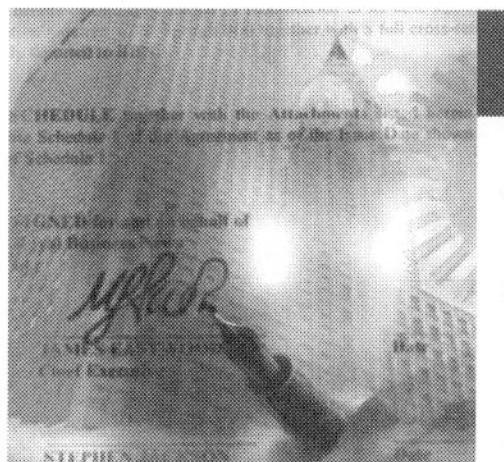

UNIT 3 • EXTENDED CASE STUDY

MPACT Construction Group, LLC v. Superior Concrete Constructors, Inc.

We set out, in Chapter 7, some of the common law rules governing the interpretation of contracts. According to these principles, when the language of a written contract is clear, a court will enforce the terms without interpreting them further. If the writing is unclear or ambiguous, however, a court will apply the rules of interpretation to give effect to the parties' intent as expressed in their contract. Among these rules is the tenet that a contract will be interpreted as a whole. Also, a term that reasonably has more than one meaning will be interpreted against the party who drafted the contract. In this extended case study, we examine *MPACT Construction Group, LLC v. Superior Concrete Constructors, Inc.*,[1] a recent decision in which these principles were applied.

1. 802 N.E.2d 901 (Ind. 2004). This opinion may be accessed online at **http://www.findlaw.com/11stategov/in/inca.html**. On that page, in the "Supreme Court" section, in the "2004" row, click on "February." On the resulting page, click on the appropriate link accompanying the name of the case to access the opinion.

CASE BACKGROUND

MPACT Construction Group, Limited Liability Company (a *limited liability company* is a special form of business organization that will be discussed in Chapter 24), entered into a contract with Flying J, Inc., to build a travel plaza in Gibson County, Indiana. The contract was an American Institute of Architects (AIA) "Standard Form Agreement Between Owner and Contractor." The form incorporated by reference the AIA "General Conditions of the Contract for Construction," which included an arbitration clause. To complete the project, MPACT entered into several contracts with subcontractors, including Superior Concrete Constructors, Inc. The subcontracts were not AIA forms but contracts that MPACT drafted.

When Flying J did not pay for all of the work, MPACT and some of the subcontractors filed mechanic's liens against Flying J. (A *mechanic's lien* is a claim on real property that ensures payment for materials and work supplied to improve the property—see Chapter 21.) Superior filed a suit based on its lien in an Indiana state court to collect the amount that it was owed. MPACT and others added claims based on their liens and on allegations of breach of contract. MPACT asked the court to order the parties to submit to arbitration under the clause referenced in the contract with Flying J. The court denied the request. MPACT appealed to a state intermediate appellate court, which affirmed the decision as to the subcontractors. MPACT appealed to the Indiana Supreme Court.

MAJORITY OPINION

SULLIVAN, Justice

* * * *

Whether MPACT and the Subcontractors agreed to arbitrate their disputes depends on whether the arbitration clause in the General Conditions of the General Contract was incorporated by reference into the subcontracts. * * *

* * * *

MPACT argues that [certain] provisions, and particularly the sentence, "The contract documents are complementary and what is required by any one shall be as binding as if required by all," [in the subcontracts' Article VI(b)] show that the General Conditions, which were incorporated into the General Contract between MPACT and Flying J, were incorporated into the subcontracts. The Subcontractors respond, and the Court of Appeals agreed, that provisions of the General Contract were incorporated for the limited purpose of governing the work to be performed. They emphasize that the sentence MPACT relies on is preceded and followed by sentences pertaining specifically to work, and that this limits the effect of that sentence.

* * * *

Courts are required to give effect to parties' contracts and to do so, courts look to the words of a contract. *In contracting, clarity of language is key.* * * * *When there is ambiguity in a contract, it is construed against its drafter.* In this instance, the AIA Standard Form of Agreement Between Contractor and Subcontractor was not used. MPACT instead drafted its own subcontracts. It was therefore MPACT's responsibility to ensure that its subcontracts * * * incorporated the arbitration clause. MPACT did not do so. [Emphasis added.]

The problem in this case seems to have resulted from poor contract drafting and inadequate contract negotiations. Each side believed at the time of contract execution that the contract provided for what it wanted—in MPACT's case, for arbitration, and in the Subcontractors' case, not for arbitra-

tion. *Regardless, it is clear that arbitration was not sufficiently discussed by the parties. This leads to one conclusion, that there was no meeting of the minds between the parties on the issue of arbitration.* Consequently, we find that there was no agreement to arbitrate between MPACT and the Subcontractors and the Subcontractors are not required to arbitrate their disputes with MPACT. [Emphasis added.]

MPACT also sought arbitration of its disputes with Flying J. The Court of Appeals found that the disputes were governed by the arbitration provision in the General Conditions of the General Contract, and held that MPACT was entitled to arbitration with Flying J. We summarily affirm the Court of Appeals on this point.

* * * *

We * * * affirm the trial court's denial of MPACT's motion as to the Subcontractors. This case is remanded to the trial court for further proceedings consistent with this opinion.

Dissenting Opinion

BOEHM, J. [Justice], dissenting.

I respectfully dissent. * * * I believe * * * that these agreements do call for arbitration of the entire multiparty dispute among the owner, the general contractor, and these several subcontractors.

The agreement between the general contractor and the owner is a standard printed form AIA construction agreement. All agree that that contract includes an enforceable arbitration clause, and an undertaking to bind subcontractors to the same terms that obligate the general. The general's agreements with the subs provide that each sub acknowledges the principal agreement and agrees to be bound by it. The principal agreement provides [among other things] that the general will impose conforming conditions on all subs. These agreements are among businesses fully familiar with this sort of arrangement, and fully cognizant that the last thing either the general or the owner wants is piecemeal litigation with different subs. The result the majority reaches produces * * * arbitration between the owner and the general and litigation with one or more subs in a separate forum. The majority concedes that the general intended to bind the subs to arbitration, but points to imprecision in the language used to accomplish that. It seems to me that the subs did understand, or should have, that arbitration was intended. They should be held to have accepted arbitration when they accepted these agreements. Accordingly, I would require arbitration of this entire dispute in one proceeding.

* * * Particularly in an industry where arbitration is widely used, ambiguity does not necessarily lead to the conclusion that no meeting of the minds occurred. Rather, I would conclude that ambiguity should be construed in favor of finding an agreement to arbitrate where that is commonplace in the industry.

Questions for Analysis

1. **Law.** What was the plaintiff's major argument? Why did the court conclude that this argument was invalid?
2. **Law.** Why did the dissent disagree with the majority's decision? What would have been the result if the dissent's opinion had prevailed in this case?
3. **Ethics.** What is an ethical basis for the majority's holding in this case? What is an ethical basis for the dissent's position?
4. **Social Dimensions.** Could the outcome in this case have a practical effect in the construction industry? If so, what are some of the possible advantages and disadvantages?
5. **Implications for the Business Manager.** How might the contracts in this case have been drafted to bind the sub-

UNIT 4

Sales and Lease Contracts

Unit Contents

CHAPTER 14

The Formation of Sales and Lease Contracts

LEARNING OBJECTIVES

After reading this chapter, you should be able to answer the following questions:

1. How do Article 2 and Article 2A of the UCC differ? What types of transactions does each article cover?
2. What is a merchant's firm offer?
3. If an offeree includes additional or different terms in an acceptance, will a contract result? If so, what happens to these terms?
4. Article 2 and Article 2A of the UCC both define several exceptions to the writing requirements of the Statute of Frauds. What are these exceptions?
5. What law governs contracts for the international sale of goods?

It is often said that the object of the law is to encourage commerce. This is particularly true with respect to the Uniform Commercial Code (UCC). The UCC facilitates commercial transactions by making the laws governing sales and lease contracts uniform, clearer, simpler, and more readily applicable to the numerous difficulties that can arise during such transactions. Recall from Chapter 1 that the UCC is one of many uniform (model) acts drafted by the National Conference of Commissioners on Uniform State Laws and submitted to the states for adoption.[1] Once a state legislature has adopted a uniform act, the act becomes statutory law in that state. Thus, when we turn to sales and lease contracts, we move away from common law principles and into the area of statutory law.

We open this chapter with a discussion of the scope of the UCC's Article 2 (on sales) and Article 2A (on leases) as a background to the focus of this chapter, which is the formation of contracts for the sale and lease of goods. Because international sales transactions are increasingly commonplace in the business world, we conclude this chapter with an examination of the United Nations Convention on Contracts for the International Sale of Goods (CISG), which governs international sales contracts.

The UCC was amended in 2003 to update its provisions to accommodate electronic commerce. As of 2005, the amendments to Articles 2 and 2A had not been adopted by any state. Throughout this chapter and the chapters that follow, however, any amendments that significantly change the UCC provisions now in effect in most states will be discussed in footnotes. Note, though, that even when the changes are not substantive, some of the section and subsection numbers may be slightly different under the amended Article 2 due to the addition of new provisions. (Excerpts from the 2003 amendments appear in Appendix E, along with the full text of Article 2.)

The Scope of the UCC

The UCC attempts to provide a consistent and integrated framework of rules to deal with all phases ordinarily arising in a commercial sales transaction from start to finish.

1. The UCC has been adopted in whole or in part by all of the states. Louisiana, however, has not adopted Articles 2 and 2A.

For example, consider the following events, all of which occur during a single sales transaction:

1. *A contract for the sale or lease of goods is formed and executed.* Article 2 and Article 2A of the UCC provide rules governing all facets of this transaction.
2. *The transaction may involve a payment—by check, electronic fund transfer, or other means.* Article 3 (on negotiable instruments), Article 4 (on bank deposits and collections), Article 4A (on fund transfers), and Article 5 (on letters of credit) cover this part of the transaction.
3. *The transaction may involve a bill of lading or a warehouse receipt that covers goods when they are shipped or stored.* Article 7 (on documents of title) deals with this subject.
4. *The transaction may involve the demand by a seller or lender for some form of security for a remaining balance owed.* Article 9 (on secured transactions) covers this part of the transaction.

Two articles of the UCC seemingly do not address the "ordinary" commercial sales transaction. Article 6, on bulk transfers, has to do with merchants who sell off the major part of their inventory. Such bulk sales are not part of the ordinary course of business. Article 8, which covers investment securities, deals with transactions involving negotiable securities (stocks and bonds)—transactions that do not involve the sale of goods. The UCC's drafters, however, considered the subject matter of Articles 6 and 8 to be *sufficiently* related to commercial transactions to warrant its inclusion in the UCC.

The Scope of Article 2—Sales

Article 2 of the UCC governs **sales contracts,** or contracts for the sale of goods. To facilitate commercial transactions, Article 2 modifies some of the common law contract requirements that were summarized in Chapter 7 and discussed in detail in Chapters 8 through 13. To the extent that it has not been modified by the UCC, however, the common law of contracts also applies to sales contracts. In general, the rule is that when a UCC provision addresses a certain issue, the UCC governs; when the UCC is silent, the common law governs.

In regard to Article 2, you should keep in mind two things. First, Article 2 deals with the sale of *goods;* it does not deal with real property (real estate), services, or intangible property such as stocks and bonds. Thus, if the subject matter of a dispute is goods, the UCC governs. If it is real estate or services, the common law applies. The relationship between general contract law and the law governing sales of goods is illustrated in Exhibit 14–1.

EXHIBIT 14–1 LAW GOVERNING CONTRACTS

This exhibit graphically illustrates the relationship between general contract law and the law governing contracts for the sale of goods. Contracts for the sale of goods are not governed exclusively by Article 2 of the Uniform Commercial Code but are also governed by general contract law whenever it is relevant and has not been modified by the UCC.

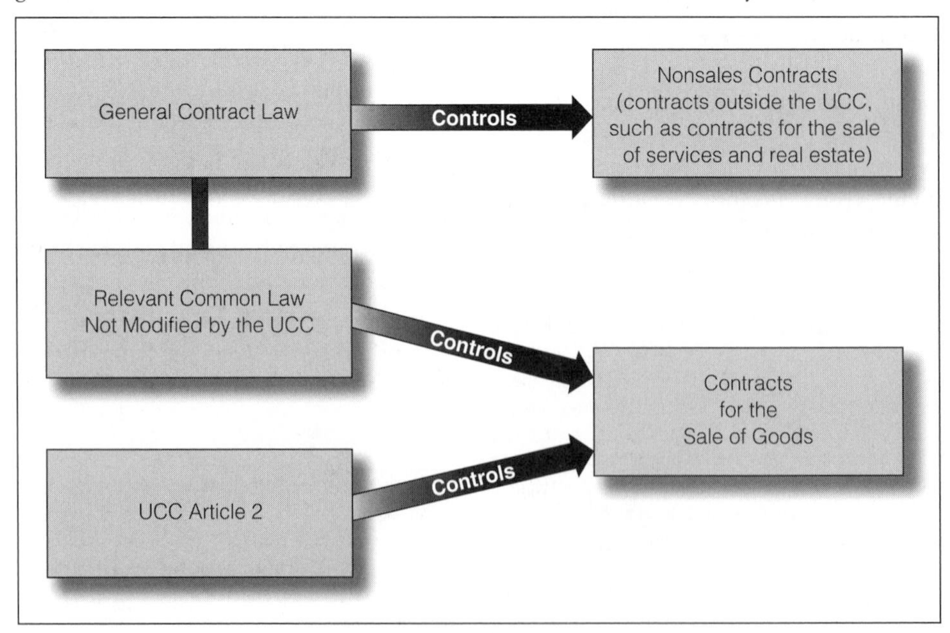

Second, in some transactions, the rules may vary quite a bit, depending on whether the buyer or the seller is a merchant. We look now at how the UCC defines three important terms: *sale, goods,* and *merchant status.*

WHAT IS A SALE?

The UCC defines a **sale** as "the passing of title from the seller to the buyer for a price" [UCC 2–106(1)]. The price may be payable in money or in other goods or services.

WHAT ARE GOODS?

To be characterized as a *good,* the item of property must be *tangible,* and it must be *movable.*[2] **Tangible property** has physical existence—it can be touched or seen. **Intangible property**—such as corporate stocks and bonds, patents and copyrights, and ordinary contract rights—has only conceptual existence and thus does not come under Article 2.[3] A movable item can be carried from place to place. Hence, real estate is excluded from Article 2.

Two areas of dispute arise in determining whether the object of a contract is goods and thus whether Article 2 is applicable. One problem has to do with *goods associated with real estate,* such as crops or timber, and the other concerns contracts involving a *combination of goods and services.*

Goods Associated with Real Estate Goods associated with real estate often fall within the scope of Article 2. Section 2–107 provides the following rules:

1. A contract for the sale of minerals or the like (including oil and gas) or a structure (such as a building) is a contract for the sale of goods if *severance,* or *separation,* is to be made by the *seller.* If the *buyer* is to sever (separate) the minerals or structure from the land, the contract is considered to be a sale of real estate governed by the principles of real property law, not the UCC.
2. A sale of growing crops[4] (such as potatoes, carrots, wheat, and the like) or timber to be cut is considered to be a contract for the sale of goods *regardless of who severs them.*
3. Other "things attached" to realty but capable of severance without material harm to the land are considered goods *regardless of who severs them.*[5] "Things attached" that are severable without harm to realty could include such items as a heater, a window air conditioner in a house, and stools in a restaurant. Thus, removal of one of these things would be considered a sale of goods. The test is whether removal will cause substantial harm to the real property to which the item is attached.

Goods and Services Combined In situations in which goods and services are combined, courts disagree. For instance, is the blood furnished to a patient during an operation a "sale of goods" or the "performance of a medical service"? Some courts say it is a good; others say it is a service. Because the UCC does not provide the answers to such questions, the courts generally try to determine which factor is predominant—the good or the service. The UCC does stipulate, however, that serving food or drink to be consumed either on or off restaurant premises is a "sale of goods," at least for the purpose of an implied warranty of merchantability (to be explained in Chapter 17) [UCC 2–314(1)]. Other special transactions, including sales of unborn animals and rare coins, are also explicitly characterized as sales of goods by the UCC.

Whether the transaction in question involves the sale of goods or services is important because the majority of courts treat services as being excluded by the UCC. If the transaction is not covered by the UCC, then UCC provisions, including those relating to implied warranties, will not apply. ■ **EXAMPLE 14.1** An Indiana company contracts to purchase customized software from Dharma Systems. The contract states that half of the purchase price is for Dharma Systems' professional services and the other half is for the goods (the software). If the court determines that the contract is predominantly for the software, rather than the services to customize the software, the court will hold that the transaction falls under Article 2.[6] ■

In the following case, the court had to determine whether Article 2 applied to a contract that involved goods and services.

2. The 2003 amendments to Article 2 change the phrasing of this section slightly. The new section states that "goods must be both *existing* and *identified* before any interest in them may pass" [Amended UCC 2–105(1)].

3. The 2003 amendments to Article 2 specifically exclude "information" that is not associated with goods [Amended UCC 2–103(1)(k)]. Nevertheless, Article 2 *may* apply to transactions involving both goods and information when a sale involves "smart goods" (for example, a toy or an automobile that contains embedded computer programs). It is up to the courts to determine whether and to what extent Article 2 should be applied to such transactions.

4. Note that the 2003 amendments moved the definition of *goods* to UCC 2–103(k), but growing crops are expressly included within the definition of *goods.* Contracts to sell timber, minerals, or structures to be removed from the land will continue to be controlled by UCC 2–107(1).

5. The UCC avoids the term *fixtures* here because of the numerous definitions of the word. A *fixture* is anything so firmly or permanently attached to land or to a building as to become a part of it. Once personal property becomes a fixture, it is governed by real estate law. See Chapter 29.

6. See *Micro Data Base Systems, Inc. v. Dharma Systems, Inc.,* 148 F.3d 649 (7th Cir. 1998).

CASE 14.1 Mécanique C.N.C., Inc. v. Durr Environmental, Inc.

United States District Court,
Southern District of Ohio, 2004.
304 F.Supp.2d 971.

FACTS Durr Environmental, Inc., contracted with GE Quartz, Inc. (GE), to design, make, install, and test a selective catalytic reduction system at GE's plant in Hebron, Ohio. The system was to control the emission of nitrogen oxides. On August 3, 2000, Mécanique C.N.C., Inc. (CNC), sent a bid to Durr offering to make and install the ductwork necessary for the system and itemizing the price. Durr responded with a signed contract. CNC also signed the contract, added three handwritten terms in an attempt to avoid certain costs, and returned it. Durr revised the contract to include only one of the terms and sent it back to CNC. Durr also sent the project's specifications, which CNC met. When problems with the ductwork developed at GE's site, Durr submitted new specifications, which CNC also met, though at substantial added expense and with some delay. Nevertheless, CNC was on schedule to complete the work when, in March 2001, Durr terminated CNC. CNC filed a suit in a federal district court against Durr and others (the defendants), alleging in part breach of contract. One of the questions was whether Article 2 applied to the CNC contract. The defendants filed a motion for summary judgment on this point.

ISSUE Was the contract between Durr and CNC predominantly a contract for a sale of goods?

DECISION Yes. The court held that the contract was predominantly for a sale of goods, with services only incidentally involved. The court granted the defendants' motion for summary judgment.

REASON The court reasoned that "virtually all commercial goods involve some type of service, whether design, assembly, installation, or manufacture," but this does not transform every contract for goods into a contract for services. That a manufacturer uses its effort and expertise to make a good does not mean that the buyer is buying the service instead of the good. The question is whether a buyer's goal is to acquire a product or procure a service. The court concluded that CNC's service, "though extensive," was incidental to Durr's purpose of acquiring the ductwork. CNC's "compensation was tied to the goods produced rather than the labor provided. There was no separate price for installation. * * * CNC played no role in the design of the product." The court added, "Even more significantly, the parties did not contemplate that CNC would be involved in any ongoing servicing or testing of the ductwork once it was installed. CNC's job was to provide a product in accordance with certain specifications. Once that product was created and installed, CNC had no further role."

WHY IS THIS CASE IMPORTANT? *This case illustrates how important it is to anticipate the factors that courts consider in determining whether Article 2 of the UCC applies. For example, even though the purchase of a product may appear to be a purchase of goods, if the contract also provides for installing and modifying the product, a court might construe the contract as predominantly for services. For the buyer, this would mean that the UCC does not apply, which may be a very important consideration in some transactions.*

WHO IS A MERCHANT?

Article 2 governs the sale of goods in general. It applies to sales transactions between all buyers and sellers. In a limited number of instances, however, the UCC presumes that certain special business standards ought to be imposed on merchants because they possess a relatively high degree of commercial expertise.[7] Such standards do not apply to the casual or inexperienced seller or buyer ("consumer"). Section 2–104 defines three ways in which merchant status can arise:

1. A merchant is a person who *deals in goods of the kind* involved in the sales contract. Thus, a retailer, a wholesaler, or a manufacturer is a merchant of those goods sold in the business. A merchant for one type of goods is not necessarily a merchant for another type. For example, a sports equipment retailer is a merchant when selling tennis equipment but not when selling a used computer.
2. A merchant is a person who, by occupation, holds himself or herself out as having knowledge and skill unique to the practices or goods involved in the transaction. Note that this broad definition may include banks or universities as merchants.

7. The provisions that apply only to merchants deal principally with the Statute of Frauds, firm offers, confirmatory memoranda, warranties, and contract modification. These special rules reflect expedient business practices commonly known to merchants in the commercial setting. They will be discussed later in this chapter.

3 A person who *employs a merchant as a broker, agent, or other intermediary* has the status of merchant in that transaction. Hence, if a "gentleman farmer" who ordinarily does not run the farm hires a broker to purchase or sell livestock, the farmer is considered a merchant in the transaction.

In summary, a person is a **merchant** when she or he, acting in a mercantile capacity, possesses or uses an expertise specifically related to the goods being sold. This basic distinction is not always clear-cut. For example, courts in most states have determined that farmers may be merchants if they sell products or livestock on a regular basis, but courts in other states have held that the drafters of the UCC did not intend to include farmers as merchants.

The Scope of Article 2A—Leases

In the past few decades, leases of personal property (goods) have become increasingly common. Article 2A of the UCC was created to fill the need for uniform guidelines in this area. Article 2A covers any transaction that creates a lease of goods, as well as subleases of goods [UCC 2A–102, 2A–103(1)(k)]. Except that it applies to leases of goods, rather than sales of goods, Article 2A is essentially a repetition of Article 2 and varies only to reflect differences between sale and lease transactions. (Note that Article 2A is not concerned with leases of real property, such as land or buildings. The laws governing these types of transactions will be examined in Chapter 29.) As previously mentioned, Article 2A was also amended in 2003, and these amendments have been recommended to the states for adoption.

Definition of a Lease

Article 2A defines a **lease agreement** as a lessor and lessee's bargain with respect to the lease of goods, as found in their language and as implied by other circumstances, including *course of dealing* and *usage of trade* or *course of performance* (to be discussed shortly) [UCC 2A–103(1)(k)]. A **lessor** is one who sells the right to the possession and use of goods under a lease [UCC 2A–103(1)(p)]. A **lessee** is one who acquires the right to the temporary possession and use of goods under a lease [UCC 2A–103(1)(o)]. In other words, the lessee is the party who is leasing the goods from the lessor. Article 2A applies to all types of leases of goods, including commercial leases and consumer leases. Special rules apply to certain types of leases, however, including consumer leases and finance leases.

Consumer Leases

A *consumer lease* involves three elements: (1) a lessor who regularly engages in the business of leasing or selling; (2) a lessee (except an organization) who leases the goods "primarily for a personal, family, or household purpose"; and (3) total lease payments that are less than a dollar amount set by state statute [UCC 2A–103(1)(e)].[8] In the interest of providing special protection for consumers, certain provisions of Article 2A apply only to consumer leases. For example, one provision states that a consumer may recover attorneys' fees if a court finds that a term in a consumer lease contract is unconscionable [UCC 2A–108(4)(a)].

Finance Leases

A *finance lease* involves a lessor, a lessee, and a supplier. The lessor buys or leases goods from a supplier and leases or subleases them to the lessee [UCC 2A–103(1)(g)].[9] Typically, in a finance lease, the lessor is simply financing the transaction. ■ EXAMPLE 14.2 Suppose that Marlin Corporation wants to lease a crane for use in its construction business. Marlin's bank agrees to purchase the equipment from Jennco, Inc., and lease the equipment to Marlin. In this situation, the bank is the lessor-financer, Marlin is the lessee, and Jennco is the supplier. ■

Article 2A, unlike ordinary contract law, makes the lessee's obligations under a commercial finance lease irrevocable and independent from the financer's obligations [UCC 2A–407]. In other words, the lessee must perform and continue to make lease payments even if the leased equipment turns out to be defective. The lessee must look almost entirely to the supplier for warranties.

The Formation of Sales and Lease Contracts

In regard to the formation of sales and lease contracts, the UCC modifies the common law in several ways. We look here at how Article 2 and Article 2A of the UCC modify common law contract rules. Remember, though, that parties to sales contracts are free to establish whatever terms they wish. The UCC comes into play only when the parties have failed to provide in their contract for a contingency that later gives rise to a dispute. The UCC makes

8. The 2003 amendments to Article 2A define a consumer lease in UCC 2A–103(1)(f). The amended section leaves it up to the states to decide whether to place a dollar limitation on the total lease payments.

9. Finance leases are defined in UCC 2A–103(1)(g) of the amended Article 2A.

this clear time and again by using such phrases as "unless the parties otherwise agree" or "absent a contrary agreement by the parties."

OFFER

In general contract law, the moment a definite offer is met by an unqualified acceptance, a binding contract is formed. In commercial sales transactions, the verbal exchanges, correspondence, and actions of the parties may not reveal exactly when a binding contractual obligation arises. The UCC states that an agreement sufficient to constitute a contract can exist even if the moment of its making is undetermined [UCC 2–204(2), 2A–204(2)].

Open Terms Remember from Chapter 8 that under the common law of contracts, an offer must be definite enough for the parties (and the courts) to ascertain its essential terms when it is accepted. In contrast, the UCC states that a sales or lease contract will not fail for indefiniteness even if one or more terms are left open as long as (1) the parties intended to make a contract and (2) there is a reasonably certain basis for the court to grant an appropriate remedy [UCC 2–204(3), 2A–204(3)].

■ EXAMPLE 14.3 Mike agrees to lease a highly specialized computer work station from CompuQuik. Mike and one of CompuQuik's sales representatives sign a lease agreement that leaves some of the details blank, to be "worked out" the following week, when the leasing manager will be back from her vacation. In the meantime, CompuQuik obtains the necessary equipment from one of its suppliers and spends several days modifying the equipment to suit Mike's needs. When the leasing manager returns, she calls Mike and tells him that his work station is ready. Mike says he is no longer interested in the work station, as he has arranged to lease the same type of equipment for a lower price from another firm. CompuQuik sues Mike to recover its costs in obtaining and modifying the equipment, and one of the issues before the court is whether the parties had an enforceable contract. The court will likely hold that they did, based on their intent and conduct, despite the "blanks" in their written agreement. ■

Relative to the common law of contracts, the UCC has radically lessened the requirement of definiteness of terms. Keep in mind, though, that the more terms left open, the less likely it is that a court will find that the parties intended to form a contract.

—Open Price Term. If the parties have not agreed on a price, the court will determine a "reasonable price at the time for delivery" [UCC 2–305(1)]. If either the buyer or the seller is to determine the price, the price is to be fixed (set) in good faith [UCC 2–305(2)]. Under the UCC, *good faith* means honesty in fact and the observance of reasonable commercial standards of fair dealing in the trade [UCC 2–103(1)(b)]. The concepts of good faith and commercial reasonableness permeate the UCC.

Sometimes the price fails to be fixed through the fault of one of the parties. In that situation, the other party can treat the contract as canceled or fix a reasonable price. ■ EXAMPLE 14.4 Perez and Merrick enter into a contract for the sale of unfinished doors and agree that Perez will determine the price. Perez refuses to specify the price. Merrick can either treat the contract as canceled or set a reasonable price [UCC 2–305(3)]. ■

—Open Payment Term. When parties do not specify payment terms, payment is due at the time and place at which the buyer is to receive the goods [UCC 2–310(a)]. The buyer can tender payment using any commercially normal or acceptable means, such as a check or credit card. If the seller demands payment in cash, however, the buyer must be given a reasonable time to obtain it [UCC 2–511(2)]. This is especially important when the contract states a definite and final time for performance.

—Open Delivery Term. When no delivery terms are specified, the buyer normally takes delivery at the seller's place of business [UCC 2–308(a)]. If the seller has no place of business, the seller's residence is used. When goods are located in some other place and both parties know it, delivery is made there. If the time for shipment or delivery is not clearly specified in the sales contract, the court will infer a "reasonable" time for performance [UCC 2–309(1)].

—Duration of an Ongoing Contract. A single contract might specify successive performances but not indicate how long the parties are required to deal with each other. In this situation, either party may terminate the ongoing contractual relationship. Principles of good faith and sound commercial practice, however, call for reasonable notification before termination to give the other party reasonable time to seek a substitute arrangement [UCC 2–309(2), (3)].

—Options and Cooperation Regarding Performance. When the contract contemplates shipment of the goods but does not specify the shipping arrangements, the *seller* has the right to make these arrangements in good faith, using commercial reasonableness in the situation [UCC 2–311].

When a sales contract omits terms relating to the assortment of goods, the *buyer* can specify the assortment.

■ EXAMPLE 14.5 Petry Drugs, Inc., agrees to purchase one thousand toothbrushes from Marconi's Dental Supply. The toothbrushes come in a variety of colors, but the contract does not specify color. Petry, the buyer, has the right to take six hundred blue toothbrushes and four hundred green ones if it wishes. Petry, however, must exercise good faith and commercial reasonableness in making its selection [UCC 2–311]. ■

—Open Quantity Term. Normally, if the parties do not specify a quantity, a court will have no basis for determining a remedy. The UCC recognizes two exceptions, however, in requirements and output contracts [UCC 2–306(1)].

In a **requirements contract,** the buyer agrees to purchase and the seller agrees to sell all or up to a stated amount of what the buyer *needs* or *requires.* ■ EXAMPLE 14.6 Umpqua Cannery forms a contract with Al Garcia. The cannery agrees to purchase from Garcia, and Garcia agrees to sell to the cannery, all of the green beans that the cannery needs or requires during the summer of 2007. ■ There is implicit consideration in a requirements contract because the buyer (the cannery, in this situation) gives up the right to buy green beans from any other seller, and this forfeited right creates a legal detriment (that is, consideration). Requirements contracts are common in the business world and are normally enforceable. If, however, the buyer promises to purchase only if the buyer *wishes* to do so, or if the buyer reserves the right to buy the goods from someone other than the seller, the promise is illusory (without consideration) and unenforceable by either party.

In an **output contract,** the seller agrees to sell and the buyer agrees to buy all or up to a stated amount of what the seller *produces.* ■ EXAMPLE 14.7 Al Garcia forms a contract with Umpqua Cannery. Garcia agrees to sell to the cannery, and the cannery agrees to purchase from Garcia, all of the beans that Garcia produces on his farm during the summer of 2007. ■ Again, because the seller essentially forfeits the right to sell goods to another buyer, there is implicit consideration in an output contract.

The UCC imposes a *good faith limitation* on requirements and output contracts. The quantity under such contracts is the amount of requirements or the amount of output that occurs during a *normal* production year. The actual quantity purchased or sold cannot be unreasonably disproportionate to normal or comparable prior requirements or output [UCC 2–306].

Merchant's Firm Offer Under regular contract principles, an offer can be revoked at any time before acceptance. The major common law exception is an *option contract* (discussed in Chapter 8), in which the offeree pays consideration for the offeror's irrevocable promise to keep the offer open for a stated period. The UCC creates a second exception for firm offers made by a merchant to sell, buy, or lease goods.

A **firm offer** arises when a merchant-offeror gives *assurances* in a *signed writing* that the offer will remain open. The merchant's firm offer is irrevocable without the necessity of consideration[10] for the stated period or, if no definite period is stated, a reasonable period (neither period to exceed three months) [UCC 2–205, 2A–205]. ■ EXAMPLE 14.8 Osaka, a used-car dealer, writes a letter to Saucedo on January 1 stating, "I have a 2004 Pontiac on the lot that I'll sell you for $12,000 any time between now and January 31." This writing creates a firm offer, and Osaka will be liable for breach if he sells the Pontiac to someone other than Saucedo before January 31. ■

It is necessary that the offer be both *written* and *signed* by the offeror.[11] When a firm offer is contained in a form contract prepared by the offeree, the offeror must also sign a separate firm offer assurance. This requirement ensures that the offeror is aware of the offer. If the firm offer is buried amid copious language in one of the pages of the offeree's form contract, the offeror may inadvertently sign the contract without realizing that it contains a firm offer, thus defeating the purpose of the rule—which is to give effect to a merchant's deliberate intent to be bound to a firm offer.

ACCEPTANCE

The following subsections examine the UCC's provisions governing acceptance. As you will see, acceptance of an offer to buy, sell, or lease goods generally may be made in any reasonable manner and by any reasonable means.[12]

Methods of Acceptance The general common law rule is that an offeror can specify, or authorize, a particular means of acceptance, making that means the only one effective for contract formation. Even an unauthorized means of communication is effective, however, as long as the acceptance is received by the specified deadline.

10. If the offeree pays consideration, then an option contract (not a merchant's firm offer) is formed.

11. "Signed" includes any symbol executed or adopted by a party with a present intention to authenticate a writing [UCC 1–201(39)]. A complete signature is not required. Therefore, initials, a thumbprint, a trade name, or any mark used in lieu of a written signature will suffice, regardless of its location on the document.

12. The UCC's rules on means of acceptance illustrate the UCC's flexibility. The rules have been adapted to new forms of communication, such as faxes and online communications.

■ **EXAMPLE 14.9** Janel offers to sell her Humvee to Arik for $48,000. The offer states, "Answer by fax within five days." If Arik sends a letter, and Janel receives it within five days, a valid contract is formed, nonetheless. ■

—*Any Reasonable Means.* When the offeror does not specify a means of acceptance, the UCC provides that acceptance can be made by any means of communication reasonable under the circumstances [UCC 2–206(1), 2A–206(1)]. This broadens the common law rules concerning authorized means of acceptance. (For a review of the requirements relating to mode and timeliness of acceptance, see Chapter 8.)

■ **EXAMPLE 14.10** Anodyne Corporation writes a letter to Bethlehem Industries offering to lease $1,000 worth of postage meters. The offer states that Anodyne will keep the offer open for only ten days from the date of the letter. Before the ten days elapse, Bethlehem sends Anodyne an acceptance by fax. Is a valid contract formed? The answer is yes, because acceptance by fax is a commercially reasonable medium of acceptance under the circumstances. Acceptance is effective on Bethlehem's transmission of the fax, which occurred before the offer expired. ■

—*Promise to Ship or Prompt Shipment.* The UCC permits a seller to accept an offer to buy goods "either by a prompt *promise* to ship or by the prompt or current shipment of conforming or nonconforming goods" [UCC 2–206(1)(b)]. *Conforming* goods are goods that accord with the contract's terms; *nonconforming* goods do not. The seller's prompt shipment of *nonconforming goods* in response to the offer constitutes both an acceptance (a contract) and a *breach* of that contract.

This rule does not apply if the seller **seasonably** (within a reasonable amount of time) notifies the buyer that the nonconforming shipment is offered only as an *accommodation,* or as a favor. The notice of accommodation must clearly indicate to the buyer that the shipment does not constitute an acceptance and that, therefore, no contract has been formed.

■ **EXAMPLE 14.11** McFarrell Pharmacy orders five cases of Johnson & Johnson 3-by-5-inch gauze pads from Halderson Medical Supply, Inc. If Halderson ships five cases of Xeroform 3-by-5-inch gauze pads instead, the shipment acts as both an acceptance of McFarrell's offer and a *breach* of the resulting contract. McFarrell may sue Halderson for any appropriate damages. If, however, Halderson notifies McFarrell that the Xeroform gauze pads are being shipped *as an accommodation*—because Halderson only has Xeroform pads in stock—the shipment will constitute a counteroffer, not an acceptance. A contract will be formed only if McFarrell accepts the Xeroform gauze pads. ■

Communication of Acceptance Under the common law, because a unilateral offer invites acceptance by a performance, the offeree need not notify the offeror of performance unless the offeror would not otherwise know about it. In other words, beginning the requested performance is an implied acceptance. The UCC is more stringent than the common law in this regard. Under the UCC, if the offeror is not notified within a reasonable time that the offeree has accepted the contract by beginning performance, then the offeror can treat the offer as having lapsed before acceptance [UCC 2–206(2), 2A–206(2)].

■ **EXAMPLE 14.12** Lee writes to Pickwick Bookstore on Monday, "Please send me a copy of *Webster's New College Dictionary* for $25.95, C.O.D.," and signs it, "Lee." Pickwick receives the request but does not ship the book for four weeks. When the book arrives, Lee rejects it, claiming that it has arrived too late to be of value. In this situation, because Lee had heard nothing from Pickwick for a month, he was justified in assuming that the store did not intend to deliver the book. Lee could consider that the offer had lapsed because of the length of time Pickwick delayed shipment. ■

Additional Terms Under the common law, if Alderman makes an offer to Beale, and Beale in turn accepts but in the acceptance makes some slight modification to the terms of the offer, there is no contract. Recall from Chapter 8 that the so-called *mirror image rule* requires that the terms of the acceptance exactly match those of the offer. The UCC, however, dispenses with the mirror image rule. The UCC generally takes the position that if the offeree's response indicates a *definite* acceptance of the offer, a contract is formed even if the acceptance includes additional or different terms from those contained in the offer [UCC 2–207(1)]. What happens to these additional terms? The answer to this question depends, in part, on whether the parties are nonmerchants or merchants.[13]

—*Rules When One Party or Both Parties Are Nonmerchants.* If one (or both) of the parties is a

13. The 2003 amendments to Article 2 do not distinguish between merchants and others in setting out rules for the effect of additional terms in sales contracts. Instead, a court is directed to determine whether (1) the terms appear in the records of both parties, (2) both parties agree to the terms even if they are not in a record, or (3) the terms are supplied or incorporated under another provision of Article 2 [Amended UCC 2–207]. Basically, the amendments give the courts more discretion to include or exclude certain additional terms.

nonmerchant, the contract is formed according to the terms of the original offer submitted by the original offeror and not according to the additional terms of the acceptance [UCC 2–207(2)]. ■ EXAMPLE 14.13 Tolsen offers in writing to sell his laptop computer, printer, and scanner to Valdez for $1,500. Valdez faxes a reply to Tolsen stating, "I accept your offer to purchase your laptop, printer, and scanner for $1,500. I would like a box of laser printer paper and two extra toner cartridges to be included in the purchase price." Valdez has given Tolsen a definite expression of acceptance (creating a contract), even though the acceptance also suggests an added term for the offer. Because Tolsen is not a merchant, the additional term is merely a proposal (suggestion), and Tolsen is not legally obligated to comply with that term. ■

—Rules When Both Parties Are Merchants. In contracts *between merchants*, the additional terms automatically become part of the contract unless (1) the original offer expressly limits acceptance to the terms of the offer, (2) the new or changed terms *materially* alter the contract, or (3) the offeror objects to the new or changed terms within a reasonable period of time [UCC 2–207(2)].

What constitutes a material alteration is frequently a question that only a court can decide. Generally, if the modification involves no unreasonable element of surprise or hardship for the offeror, the court will hold that the modification did not materially alter the contract. The issue in the following case concerned whether an additional term in the acceptance (subjecting any disputes to arbitration) constituted a material alteration of the contract terms.

CASE 14.2 Wilson Fertilizer & Grain, Inc. v. ADM Milling Co.

Court of Appeals of Indiana, 1995.
654 N.E.2d 848.

FACTS In October 1992, Wilson Fertilizer & Grain, Inc., agreed to sell grain to ADM Milling Company. ADM sent Wilson a confirmation stating that "[t]his contract is also subject to the Trade Rules of the National Grain and Feed Association [NGFA]." The NGFA rules require the arbitration of disputes and limit the time for filing a complaint to one year. Wilson did not respond to the confirmation. A dispute arose under the contract, and Wilson filed a suit against ADM in an Indiana state court. ADM moved to dismiss the action, claiming that the Trade Rules of the NGFA required the parties to arbitrate the dispute. Wilson argued that the arbitration provisions were not included in the terms of its agreement with ADM. The trial court granted ADM's motion and ordered the parties to arbitration. By that time, however, the one-year limit had expired. Wilson appealed, arguing in part that because the one-year limit imposed a hardship on Wilson, ADM's confirmation had materially altered the contract and thus had not become a part of the contract.

ISSUE Was the confirmation clause that required the arbitration of disputes a material alteration of the contract terms?

DECISION No. The Court of Appeals of Indiana affirmed the order of the lower court.

REASON The state appellate court pointed out that "the test for whether additional terms materially alter an agreement is whether their 'incorporation into the contract without express awareness by the other party would result in surprise or hardship.'" The court concluded that the arbitration provision and the requirement that claims must be filed within one year did not impose an undue hardship on Wilson. First, the court stated that the UCC "specifically permits parties to a contract for sale to reduce the time for filing claims to one year." Second, "and even more significantly," the court noted "Wilson's apparent ability to have submitted its claim for arbitration within the one-year limit." The court explained that the contract was formed in October 1992, and Wilson filed the complaint in September 1993. "If Wilson was able to file its complaint in court within one year" after ADM allegedly breached the contract, "we fail to see how a contract provision requiring Wilson to submit its claim for arbitration in the same time period imposes a hardship."

FOR CRITICAL ANALYSIS—Social Consideration *For what reasons might a company prefer to litigate a dispute rather than submit it to arbitration?*

—Conditioned on Offeror's Assent. Regardless of merchant status, the UCC provides that the offeree's expression cannot be construed as an acceptance if it contains additional or different terms that are explicitly conditioned on the offeror's assent to those terms [UCC 2–207(1)]. ■ **EXAMPLE 14.14** Philips offers to sell Hundert 650 pounds of turkey thighs at a specified price and with specified delivery terms. Hundert responds, "I accept your offer for 650 pounds of turkey thighs *on the condition that you give me ninety days to pay for them.*" Hundert's response will be construed not as an acceptance but as a counteroffer, which Philips may or may not accept. ■

—Additional Terms May Be Stricken. The UCC provides yet another option for dealing with conflicting terms in the parties' writings. Section 2–207(3) states that conduct by both parties that recognizes the existence of a contract is sufficient to establish a contract for the sale of goods even though the writings of the parties do not otherwise establish a contract. In this situation, "the terms of the particular contract will consist of those terms on which the writings of the parties agree, together with any supplementary terms incorporated under any other provisions of this Act." In a dispute over contract terms, this provision allows a court simply to strike from the contract those terms on which the parties do not agree.

■ **EXAMPLE 14.15** AAA Marketing orders goods over the phone from Brigg Sales, Inc., which ships the goods with an acknowledgment form (confirming the order) to AAA. AAA accepts and pays for the goods. The parties' writings do not establish a contract, but there is no question that a contract exists. If a dispute arises over the terms, such as the extent of any warranties, UCC 2–207(3) provides the governing rule. ■

CONSIDERATION

The common law rule that a contract requires consideration also applies to sales and lease contracts. Unlike the common law, however, the UCC does not require a contract modification to be supported by new consideration. An agreement modifying a contract for the sale or lease of goods "needs no consideration to be binding" [UCC 2–209(1), 2A–208(1)].

Modifications Must Be Made in Good Faith Of course, contract modification must be sought in good faith [UCC 1–203]. ■ **EXAMPLE 14.16** Allied, Inc., agrees to lease a new recreational vehicle (RV) to Louise for a stated monthly payment. Subsequently, a sudden shift in the market makes it difficult for Allied to lease the new RV to Louise at the contract price without suffering a loss. Allied tells Louise of the situation, and she agrees to pay an additional sum for leasing the RV. Later, Louise reconsiders and refuses to pay more than the original lease price. Under the UCC, Louise's promise to modify the contract needs no consideration to be binding. Hence, she is bound by the modified contract. ■

In this example, a shift in the market is a *good faith* reason for contract modification. What if there really was no shift in the market, however, and Allied knew that Louise needed to lease the new RV immediately but refused to deliver it unless she agreed to pay an additional sum of money? This attempt at extortion through modification without a legitimate commercial reason would be ineffective because it would violate the duty of good faith. Allied would not be permitted to enforce the higher price.

When Modification without Consideration Requires a Writing In some situations, modification of a sales or lease contract without consideration must be in writing to be enforceable. If the contract itself prohibits any changes to the contract unless they are in a signed writing, for instance, then only those changes agreed to in a signed writing are enforceable. If a consumer (nonmerchant buyer) is dealing with a merchant and the merchant supplies a form that contains a prohibition against oral modification, the consumer must sign a separate acknowledgment of such a clause [UCC 2–209(2), 2A–208(2)].

Also, under Article 2, any modification that brings a sales contract under the Statute of Frauds must usually be in writing to be enforceable. Thus, if an oral contract for the sale of goods priced at $400 is modified so that the contract goods are now priced at $600, the modification must be in writing—because sales contracts for goods priced at $500 or more must be in writing to be enforceable, as you will read shortly [UCC 2–209(3)]. If, however, the buyer accepts delivery of the goods after the modification, he or she is bound to the $600 price [UCC 2–201(3)(c)]. (Unlike Article 2, Article 2A does not say whether a modified lease needs to satisfy the Statute of Frauds.)

STATUTE OF FRAUDS

The UCC contains Statute of Frauds provisions covering sales and lease contracts. Under these provisions, sales contracts for goods priced at $500 or more and lease contracts requiring payments of $1,000 or more must be in writing to be enforceable [UCC 2–201(1), 2A–201(1)].[14]

14. Note that a 2003 amendment significantly increased the price of goods that will cause a sales contract to fall under the Statute of Frauds. Under Amended UCC 2–201(1), goods must be priced at $5,000 or more for a sales contract to be subject to the record (writing) requirement.

Sufficiency of the Writing The UCC has greatly relaxed the requirements for the sufficiency of a writing to satisfy the Statute of Frauds. A writing or a memorandum will be sufficient as long as it indicates that the parties intended to form a contract and as long as it is signed by the party (or agent of the party) against whom enforcement is sought. The contract normally will not be enforceable beyond the quantity of goods shown in the writing, however. All other terms can be proved in court by oral testimony. For leases, the writing must reasonably identify and describe the goods leased and the lease term.

Exception—Special Rules for Contracts between Merchants Once again, the UCC provides a special rule for merchants. Merchants can satisfy the requirements of a writing for the Statute of Frauds if, after the parties have agreed orally, one of the merchants sends a signed written confirmation to the other merchant within a reasonable time after the oral agreement was reached. The communication must indicate the terms of the agreement, and the merchant receiving the confirmation must have reason to know of its contents. Unless the merchant who receives the confirmation gives written notice of objection to its contents within ten days after receipt, the writing is sufficient against the receiving merchant, even though she or he has not signed anything [UCC 2–201(2)].[15] What happens if a merchant sends a written confirmation of an order that was never placed? For a discussion of this issue, see this chapter's *Adapting the Law to the Online Environment* feature.

■ EXAMPLE 14.17 Alfonso is a merchant-buyer in Cleveland. He contracts over the telephone to purchase $4,000 worth of spare aircraft parts from Goldstein, a New York City merchant-seller. Two days later, Goldstein sends written confirmation detailing the terms of the oral contract, and Alfonso subsequently receives it. If Alfonso does not notify Goldstein in writing of his objection to the contents of the confirmation within ten days of receipt, Alfonso cannot raise the Statute of Frauds as a defense against the enforcement of the oral contract. ■

Other Exceptions In addition to the special rules for merchants, the UCC defines three exceptions to the writing requirements of the Statute of Frauds. An *oral* contract for the sale of goods priced at $500 or more or the lease of goods involving total payments of $1,000 or more will be enforceable despite the absence of a writing in the circumstances discussed in the following subsections [UCC 2–201(3), 2A–201(4)]. These exceptions and other ways in which sales law differs from general contract law are summarized in Exhibit 14–2 on page 270.

—Specially Manufactured Goods. An oral contract is enforceable if (1) it is for goods that are specially manufactured for a particular buyer or specially manufactured or obtained for a particular lessee, (2) these goods are not suitable for resale or lease to others in the ordinary course of the seller's or lessor's business, and (3) the seller or lessor has substantially started to manufacture the goods or has made commitments for their manufacture or procurement. In this situation, once the seller or lessor has taken action, the buyer or lessee cannot repudiate the agreement claiming the Statute of Frauds as a defense.

■ EXAMPLE 14.18 Womach orders custom-made draperies for her new boutique. The price is $1,000, and the contract is oral. When the merchant-seller manufactures the draperies and tenders delivery to Womach, she refuses to pay for them even though the job has been completed on time. Womach claims that she is not liable because the contract was oral. Clearly, if the unique style and color of the draperies make it improbable that the seller can find another buyer, Womach is liable to the seller. Note that the seller must have made a substantial beginning in manufacturing the specialized item prior to the buyer's repudiation. (Here, the manufacture was completed.) Of course, the court must still be convinced by evidence of the terms of the oral contract. ■

—Admissions. An oral contract for the sale or lease of goods is enforceable if the party against whom enforcement of the contract is sought admits in pleadings, testimony, or other court proceedings that a contract for sale was made.[16] In this situation, the contract will be enforceable even though it was oral, but enforceability will be limited to the quantity of goods admitted.

■ EXAMPLE 14.19 Lane and Byron negotiate an agreement over the telephone. During the negotiations, Lane requests a delivery price for five hundred gallons of gasoline and a separate price for seven hundred gallons of gasoline. Byron replies that the price would be the same, $2.30 per gallon. Lane orally orders five hundred gallons. Byron honestly believes that Lane ordered seven hundred gallons and tenders that amount. Lane refuses the ship-

15. According to the comments accompanying UCC 2A–201 (Article 2A's Statute of Frauds), the "between merchants" provision was not included because the number of such transactions involving leases, as opposed to sales, was thought to be modest.

16. Any admission made under oath, including one not made in a court, satisfies UCC 2–201(3)(b) and 2A–201(4)(b) under the 2003 amendments to Articles 2 and 2A.

ADAPTING THE LAW TO THE ONLINE ENVIRONMENT

Can an Employee's E-Mail Constitute a Waiver of Contract Terms?

Under UCC 2–209, an agreement that excludes modification except by a signed writing cannot be otherwise modified. If the written-modification requirement is contained in a form supplied by one merchant to another, the other party must separately sign the form for it to be binding. This rule has an exception, though, which can be significant in the online environment. Under the UCC, *an attempt at modification that does not meet the writing requirement may operate as a waiver* [UCC 2–209(4)]. In other words, the parties can waive, or give up, the right to require that contract modifications be in a signed writing. Can an employee's e-mail communications form a waiver of a contract's written-modification requirement? This issue arose in *Cloud Corp. v. Hasbro, Inc.*[a]

THE CONTRACT TERMS AND THE PARTIES' RELATIONSHIP

Cloud Corporation contracted to supply packets of a special powder to Hasbro, Inc., for use in Hasbro's new "Wonder World Aquarium." At the time of their initial agreement, Hasbro sent a letter to Cloud containing a "terms and conditions" form, which stated that Cloud, the supplier, could not deviate from a purchase order without Hasbro's written consent. Cloud signed and returned that form to Hasbro as requested, and Hasbro began placing orders. Each time Hasbro ordered packets, Cloud sent back an "order acknowledgment" form confirming the quantity ordered.

After placing several orders, Hasbro told Cloud to change the formula in the packets. As a result, Cloud was able to produce three times as many packets using the same amount of material that it already had on hand to fill Hasbro's previous orders. Although Hasbro had not ordered any additional packets, Cloud sent Hasbro an order acknowledgment for extra packets at a lower price. Hasbro did not explicitly respond to Cloud's acknowledgment form. One of Hasbro's employees, however, referred to the additional quantities of packets at some point in her e-mail exchanges with Cloud. Several months later, after Cloud had produced the additional packets, Hasbro quit making the Wonder World Aquarium and refused to pay for the packets that it did not order. Cloud then sued Hasbro for breach of contract.

WAS THE EMPLOYEE'S E-MAIL A WAIVER?

Ultimately, a federal appellate court held that because Hasbro's employee had referred to the additional packets in at least one e-mail, Hasbro must pay for them. According to the court, the employee's e-mail alone could be sufficient to satisfy the requirement of written consent to modify the contract. Even if it did not, however, the court held that it operated as a waiver. The court stated that for the e-mail to operate as a waiver, Cloud "must show either that it reasonably relied on the other party's having waived the requirement of a writing, or that the waiver was clear and unequivocal." Here, the employee's e-mail had not clearly waived the writing requirement but there was *reasonable reliance.* According to the court, Hasbro should have advised Cloud if it did not want to be committed to buying the additional quantity rather than "leading Cloud down the primrose path."

FOR CRITICAL ANALYSIS

How might the parties to a sales contract prevent their subsequent e-mail communications from waiving the contract's explicit modification requirements? (Hint: How can the parties generally prevent contract disputes?)

a. 314 F.3d 289 (7th Cir. 2002).

ment of seven hundred gallons, and Byron sues for breach. In his pleadings and testimony, Lane admits that an oral contract was made, but only for five hundred gallons. Because Lane admits the existence of the oral contract, Lane cannot plead the Statute of Frauds as a defense. The contract is enforceable, however, only to the extent of the quantity admitted (five hundred gallons). ■

—Partial Performance. An oral contract for the sale or lease of goods is enforceable if payment has been made and accepted or goods have been received and accepted. This is the "partial performance" exception. The oral contract will be enforced at least to the extent that performance *actually* took place.

■ **EXAMPLE 14.20** Allan orally contracts to lease to Opus Enterprises a thousand chairs at $2 each to be used during a one-day concert. Before delivery, Opus sends Allan a check for $1,000, which he cashes. Later, when Allan attempts to deliver the chairs, Opus refuses delivery, claiming the Statute of Frauds as a defense, and demands the return of its $1,000. Under the UCC's partial performance rule, Allan can enforce the oral contract by tender of delivery of five hundred chairs for the $1,000 accepted. Similarly, if Opus had made no payment but had accepted the delivery of five hundred chairs from Allan, the oral contract would have been enforceable against Opus for $1,000, the lease payment due for the five hundred chairs delivered. ■

EXHIBIT 14–2 MAJOR DIFFERENCES BETWEEN CONTRACT LAW AND SALES LAW

	CONTRACT LAW	SALES LAW
Contract Terms	Contract must contain all material terms.	Open terms are acceptable, if parties intended to form a contract, but contract is not enforceable beyond quantity term.
Acceptance	Mirror image rule applies. If additional terms are added in acceptance, counteroffer is created.	Additional terms will not negate acceptance unless acceptance is made expressly conditional on assent to the additional terms.
Contract Modification	Modification requires consideration.	Modification does not require consideration.
Irrevocable Offers	Option contracts (with consideration).	Merchants' firm offers (without consideration).
Statute of Frauds Requirements	All material terms must be included in the writing.	A writing is required only for the sale of goods of $500[a] or more, but contract is not enforceable beyond quantity specified. *Exceptions:* 1. Merchants can satisfy the writing requirement by a confirmatory memorandum evidencing their agreement. 2. Contracts for specially manufactured goods are enforceable. 3. Contracts admitted to under oath by the party against whom enforcement is sought are enforceable. 4. Contracts with partial performance are enforceable.

a. Under a 2003 amendment to the UCC, a writing (record) is required only for the sale of goods priced at $5,000 or more.

PAROL EVIDENCE

If the parties to a contract set forth its terms in a confirmatory memorandum (a writing expressing offer and acceptance of the deal) or in a writing intended as their final expression, the terms of the contract cannot be contradicted by evidence of any prior agreements or contemporaneous oral agreements. The terms of the contract may, however, be explained or supplemented by *consistent additional terms or by course of dealing, usage of trade, or course of performance* [UCC 2–202, 2A–202].[17]

Ambiguous Terms If the court finds an ambiguity in a writing that is supposed to be a complete and exclusive statement of the agreement between the parties, it may accept evidence of consistent additional terms to clarify or remove the ambiguity. The court will not, however, accept evidence of contradictory terms. This is also the rule under the common law of contracts.

Course of Dealing and Usage of Trade Under the UCC, the meaning of any agreement, evidenced by the language of the parties and by their actions, must be interpreted in light of commercial practices and other surrounding circumstances. In interpreting a commercial agreement, the court will assume that the course of prior dealing between the parties and the usage of trade were taken into account when the agreement was phrased.

A **course of dealing** is a sequence of previous actions and communications between the parties to a particular transaction that establishes a common basis for their understanding [UCC 1–205(1)]. A course of dealing is restricted to the sequence of actions and communications

17. The 2003 amendments to this section substitute the word *record* for *writing* and add the stipulation that the "terms in a record may be explained by evidence of course of performance, course of dealing, or usage of trade without a preliminary determination by the court that the language used is ambiguous" [Amended UCC 2–202(2)].

between the parties that has occurred prior to the agreement in question. The UCC states, "A course of dealing between the parties and any usage of trade in the vocation or trade in which they are engaged or of which they are or should be aware give particular meaning to [the terms of the agreement] and supplement or qualify the terms of [the] agreement" [UCC 1–205(3)].

Usage of trade is any practice or method of dealing having such regularity of observance in a place, vocation, or trade as to justify an expectation that it will be observed with respect to the transaction in question [UCC 1–205(2)]. Further, the express terms of an agreement and an applicable course of dealing or usage of trade will be construed to be consistent with each other whenever reasonable. When such a construction is *unreasonable,* however, the express terms in the agreement will prevail [UCC 1–205(4)].

Course of Performance A **course of performance** is the conduct that occurs under the terms of a particular agreement. Presumably, the parties themselves know best what they meant by their words, and the course of performance actually undertaken under their agreement is the best indication of what they meant [UCC 2–208(1), 2A–207(1)].[18]

■ EXAMPLE 14.21 Janson's Lumber Company contracts with Barrymore to sell Barrymore a specified number of "two-by-fours." The lumber in fact does not measure 2 inches by 4 inches but rather 1⅞ inches by 3¾ inches. Janson's agrees to deliver the lumber in five deliveries, and Barrymore, without objection, accepts the lumber in the first three deliveries. On the fourth delivery, however, Barrymore objects that the two-by-fours do not measure 2 inches by 4 inches. The course of performance in this transaction—that is, the fact that Barrymore accepted three deliveries without objection under the agreement—is relevant in determining that here the phrase "two-by-four" actually means "1⅞ by 3¾." Janson's can also prove that two-by-fours need not be exactly 2 inches by 4 inches by applying usage of trade, course of prior dealing, or both. Janson's can, for example, show that in previous transactions, Barrymore took 1⅞-by-3¾-inch lumber without objection. In addition, Janson's can show that in the lumber trade, two-by-fours are commonly 1⅞ inches by 3¾ inches. ■

Rules of Construction The UCC provides *rules of construction* for interpreting contracts. Express terms, course of performance, course of dealing, and usage of trade are to be construed together when they do not contradict one another. When such a construction is unreasonable, however, the following order of priority controls: (1) express terms, (2) course of performance, (3) course of dealing, and (4) usage of trade [UCC 1–205(4), 2–208(2), 2A–207(2)].

UNCONSCIONABILITY

As discussed in Chapter 9, an *unconscionable contract* is one that is so unfair and one sided that it would be unreasonable to enforce it. The UCC allows the court to evaluate a contract or any clause in a contract, and if the court deems it to have been unconscionable at the time it was made, the court can (1) refuse to enforce the contract, (2) enforce the remainder of the contract without the unconscionable clause, or (3) limit the application of any unconscionable clauses to avoid an unconscionable result [UCC 2–302, 2A–108].[19] The following landmark case illustrates an early application of the UCC's unconscionability provisions.

18. UCC 2–208 and UCC 2A–207 were deleted from the UCC in the 2003 amendments. Course of performance is, however, expressly mentioned in UCC 2–202.

19. The 2003 amendments to this section changed the word *clause* to *term,* recognizing that even a single term (word) in a contract can be unconscionable and authorizing the court to strike any single term that it finds unconscionable.

LANDMARK AND CLASSIC CASES

CASE 14.3 Jones v. Star Credit Corp.

Supreme Court of New York, Nassau County, 1969.
59 Misc.2d 189,
298 N.Y.S.2d 264.

FACTS The Joneses, the plaintiffs, agreed to purchase a freezer for $900 as the result of a salesperson's visit to their home. Tax and financing charges raised the total price to $1,439.69. At trial, the freezer was found to have a maximum retail value of approximately $300. The plaintiffs, who had made payments totaling $619.88, brought a suit in a New York state court to have the purchase contract declared unconscionable under the UCC.

ISSUE Can this contract be denied enforcement on the ground of unconscionability?

(Continued)

CASE 14.3—CONTINUED

DECISION Yes. The court held that the contract was not enforceable as it stood, and the contract was reformed so that no further payments were required.

REASON The court relied on UCC 2–302(1), which states that if "the court as a matter of law finds the contract or any clause of the contract to have been unconscionable at the time it was made, the court may * * * so limit the application of any unconscionable clause as to avoid any unconscionable result." The court then examined the disparity between the $900 purchase price and the $300 retail value, as well as the fact that the credit charges alone exceeded the retail value. These excessive charges were exacted despite the seller's knowledge of the plaintiffs' limited resources. The court reformed the contract so that the plaintiffs' payments, amounting to more than $600, were regarded as payment in full.

IMPACT OF THIS CASE ON TODAY'S LAW Classical contract theory holds that a contract is a bargain in which the terms have been worked out freely between the parties. In many modern commercial transactions, this may not have happened. For example, standard-form contracts and leases are often signed by consumer-buyers who understand few of the terms used and who often do not even read them. The inclusion of Sections 2–302 and 2A–108 in the UCC gave the courts a means of policing such transactions, and the courts continue to use this means to prevent injustice.

RELEVANT WEB SITES *To locate information on the Web concerning* Jones v. Star Credit Corp., *go to this text's Web site at* **academic.cengage.com/blaw/wbl**, *select "Chapter 14," and click on "URLs for Landmark Cases."*

Contracts for the International Sale of Goods

International sales contracts between firms or individuals located in different countries are governed by the 1980 United Nations Convention on Contracts for the International Sale of Goods (CISG). The CISG governs international contracts only if the countries of the parties to the contract have ratified the CISG and if the parties have not agreed that some other law will govern their contract. As of 2006, sixty-two countries had adopted the CISG, including Canada, Mexico, the United States, some Central and South American countries, and most European nations.

Applicability of the CISG

Essentially, the CISG is to international sales contracts what Article 2 of the UCC is to domestic sales contracts. As discussed in this chapter, in domestic transactions the UCC applies when the parties to a contract for a sale of goods have failed to specify in writing some important term concerning price, delivery, or the like. Similarly, whenever the parties subject to the CISG have failed to specify in writing the precise terms of a contract for the international sale of goods, the CISG will be applied. Unlike the UCC, the CISG does not apply to consumer sales, and neither the UCC nor the CISG applies to contracts for services.

Businesspersons must take special care when drafting international sales contracts to avoid problems caused by distance, including language differences and varying national laws. The appendix at the end of this chapter, which shows an actual international sales contract used by Starbucks Coffee Company, illustrates many of the special terms and clauses that are typically contained in international contracts for the sale of goods. Annotations in the appendix explain the meaning and significance of specific clauses in the contract. (See Chapter 32 for a discussion of other laws that frame global business transactions.)

A Comparison of CISG and UCC Provisions

The provisions of the CISG, although similar for the most part to those of the UCC, differ from them in certain respects. For example, the CISG does not include any Statute of Frauds provisions. Under Article 11 of the CISG, an international sales contract does not need to be evidenced by a writing nor must it be in any particular form.

We look here at some differences between the UCC and the CISG with respect to contract formation. In the following chapters, we will continue to point out differences between the CISG and the UCC as they relate to the topics covered. These topics include risk of loss, performance, remedies, and warranties.

Offers Some differences between the UCC and the CISG concern offers. For instance, the UCC provides that a merchant's firm offer is irrevocable, even without consideration, if the merchant gives assurances in a signed writing. In contrast, under the CISG, an offer can become irrevocable without a signed writing. Article 16(2) of the CISG provides that an offer will be irrevocable if the merchant-offeror simply states orally that the offer is irrevocable or if

the offeree reasonably relies on the offer as being irrevocable. In both of these situations, the offer will be irrevocable even without a writing and without consideration.

Another difference is that, under the UCC, if the price term is left open, the court will determine "a reasonable price at the time for delivery" [UCC 2–305(1)]. Under the CISG, however, the price term must be specified, or at least provisions for its specification must be included in the agreement; otherwise, no contract normally will exist.

Acceptances Like UCC 2–207, the CISG provides that a contract can be formed even though the acceptance contains additional terms, unless the additional terms materially alter the contract. Under the CISG, however, the definition of a "material alteration" includes virtually any change in the terms. If an additional term relates to payment, quality, quantity, price, time and place of delivery, the extent of one party's liability to the other, or the settlement of disputes, the CISG considers the added term a "material alteration." In effect, then, the CISG requires that the terms of the acceptance mirror those of the offer.

Additionally, under the UCC, an acceptance is effective on dispatch. Under the CISG, however, a contract is not created until the acceptance is received by the offeror. (The offer becomes irrevocable, however, when the acceptance is sent.) Additionally, in contrast to the UCC, the CISG provides that acceptance by performance does not require that the offeror be notified of the performance.

TERMS AND CONCEPTS

CHAPTER SUMMARY The Formation of Sales and Lease Contracts*

The Scope of the UCC (See pages 258–259.)	The UCC attempts to provide a consistent, uniform, and integrated framework of rules to deal with all phases *ordinarily arising* in a commercial sales or lease transaction, including contract formation, passage of title and risk of loss, performance, remedies, payment for goods, warehoused goods, and secured transactions.
The Scope of Article 2—Sales (See pages 259–262.)	Article 2 governs contracts for the sale of goods (tangible, movable personal property). The common law of contracts also applies to sales contracts to the extent that the common law has not been modified by the UCC. If there is a conflict between a common law rule and the UCC, the UCC controls.
The Scope of Article 2A—Leases (See page 262.)	Article 2A governs contracts for the lease of goods. Except that it applies to leases, instead of sales, of goods, Article 2A is essentially a repetition of Article 2 and varies only to reflect differences between sale and lease transactions.
Offer and Acceptance (See pages 263–267.)	1. *Offer*— a. Not all terms have to be included for a contract to be formed (only the subject matter and quantity term must be specified). b. The price need not be included for a contract to be formed. c. Particulars of performance can be left open. * Although we have referenced the 2003 amendments to Articles 2 and 2A of the UCC within the text of this chapter, we do not do so in this summary—because the amendments have not as yet become law in any state.

(Continued)

CHAPTER SUMMARY The Formation of Sales and Lease Contracts—Continued

Offer and Acceptance—Continued	d. A written and signed offer by a *merchant,* covering a period of three months or less, is irrevocable without payment of consideration. 2. *Acceptance—* a. Acceptance may be made by any reasonable means of communication; it is effective when dispatched. b. The acceptance of a unilateral offer can be made by a promise to ship or by prompt shipment of conforming goods, or by prompt shipment of nonconforming goods if not accompanied by a notice of accommodation. c. Acceptance by performance requires notice within a reasonable time; otherwise, the offer can be treated as lapsed. d. A definite expression of acceptance creates a contract even if the terms of the acceptance vary from those of the offer unless the new or different terms in the acceptance are expressly conditioned on the offeror's assent to those terms.
Consideration (See page 267.)	A modification of a contract for the sale of goods does not require consideration.
Statute of Frauds (See pages 267–270.)	1. All contracts for the sale of goods priced at $500 or more must be in writing. A writing is sufficient as long as it indicates a contract between the parties and is signed by the party against whom enforcement is sought. A contract is not enforceable beyond the quantity shown in the writing. 2. When written confirmation of an oral contract *between merchants* is not objected to in writing by the receiver within ten days, the contract is enforceable. 3. Exceptions to the requirement of a writing exist in the following situations: a. When the oral contract is for specially manufactured goods not suitable for resale to others, and the seller has substantially started to manufacture the goods. b. When the defendant admits in pleadings, testimony, or other court proceedings that an oral contract for the sale of goods was made. In this situation, the contract will be enforceable to the extent of the quantity of goods admitted. c. The oral agreement will be enforceable to the extent that payment has been received and accepted by the seller or to the extent that the goods have been received and accepted by the buyer.
Parol Evidence Rule (See pages 270–271.)	1. The terms of a clearly and completely worded written contract cannot be contradicted by evidence of prior agreements or contemporaneous oral agreements. 2. Evidence is admissible to clarify the terms of a writing in the following situations: a. If the contract terms are ambiguous. b. If evidence of course of dealing, usage of trade, or course of performance is necessary to learn or to clarify the intentions of the parties to the contract.
Unconscionability (See pages 271–272.)	An unconscionable contract is one that is so unfair and one sided that it would be unreasonable to enforce it. If the court deems a contract to have been unconscionable at the time it was made, the court can (1) refuse to enforce the contract, (2) refuse to enforce the unconscionable clause of the contract, or (3) limit the application of any unconscionable clauses to avoid an unconscionable result.
Contracts for the International Sale of Goods (See pages 272–273.)	International sales contracts are governed by the United Nations Convention on Contracts for the International Sale of Goods (CISG)—if the countries of the parties to the contract have ratified the CISG (and if the parties have not agreed that some other law will govern their contract). Essentially, the CISG is to international sales contracts what Article 2 of the UCC is to domestic sales contracts. Whenever parties who are subject to the CISG have failed to specify in writing the precise terms of a contract for the international sale of goods, the CISG will be applied.

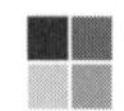

FOR REVIEW

Answers for the even-numbered questions in this For Review *section can be found in Appendix A at the end of this text.*

1. How do Article 2 and Article 2A of the UCC differ? What types of transactions does each article cover?
2. What is a merchant's firm offer?
3. If an offeree includes additional or different terms in an acceptance, will a contract result? If so, what happens to these terms?
4. Article 2 and Article 2A of the UCC both define several exceptions to the writing requirements of the Statute of Frauds. What are these exceptions?
5. What law governs contracts for the international sale of goods?

QUESTIONS AND CASE PROBLEMS

14–1. Statute of Frauds. Fresher Foods, Inc., orally agreed to purchase from Dale Vernon, a farmer, one thousand bushels of corn for $1.25 per bushel. Fresher Foods paid $125 down and agreed to pay the remainder of the purchase price on delivery, which was scheduled for one week later. When Fresher Foods tendered the balance of $1,125 on the scheduled day of delivery and requested the corn, Vernon refused to deliver it. Fresher Foods sued Vernon for damages, claiming that Vernon had breached their oral contract. Can Fresher Foods recover? If so, to what extent?

QUESTION WITH SAMPLE ANSWER

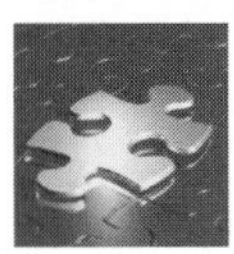

14–2. On September 1, Jennings, a used-car dealer, wrote a letter to Wheeler in which he stated, "I have a 1955 Thunderbird convertible in mint condition that I will sell you for $13,500 at any time before October 9. [Signed] Peter Jennings." By September 15, having heard nothing from Wheeler, Jennings sold the Thunderbird to another party. On September 29, Wheeler accepted Jennings's offer and tendered the $13,500. When Jennings told Wheeler he had sold the car to another party, Wheeler claimed Jennings had breached their contract. Is Jennings in breach? Explain.

For a sample answer to this question, go to Appendix B at the end of this text.

14–3. Accommodation Shipments. M. M. Salinger, Inc., a retailer of television sets, orders one hundred model Color-X sets from manufacturer Fulsom. The order specifies the price and that the television sets are to be shipped via Interamerican Freightways on or before October 30. Fulsom receives the order on October 5. On October 8, Fulsom writes Salinger a letter indicating that it has received the order and that it will ship the sets as directed, at the specified price. Salinger receives this letter on October 10. On October 28, Fulsom, in preparing the shipment, discovers it has only ninety Color-X sets in stock. Fulsom ships the ninety Color-X sets and ten television sets of a different model, stating clearly on the invoice that the ten sets are being shipped only as an accommodation. Salinger claims that Fulsom is in breach of contract. Fulsom claims that there was not an acceptance, and therefore no contract was formed. Explain who is correct, and why.

14–4. Statute of Frauds. SNK, Inc., makes video arcade games and sells them to distributors, including Entertainment Sales, Inc. (ESI). Most sales between SNK and ESI were phone orders. Over one four-month period, ESI phoned in several orders for "Samurai Showdown" games. SNK did not fill the orders. ESI filed a suit against SNK and others, alleging among other things breach of contract. There was no written contract covering the orders. ESI claimed that it had faxed purchase orders for the games to SNK but did not offer proof that the faxes had been sent or received. SNK filed a motion for summary judgment. In whose favor will the court rule, and why? [*Entertainment Sales Co. v. SNK, Inc.*, 232 Ga.App. 669, 502 S.E.2d 263 (1998)]

14–5. Goods and Services Combined. Dennis Dahlmann and Dahlmann Apartments, Ltd., entered into contracts with Sulchus Hospitality Technologies Corp. and Hospitality Management Systems, Inc. (HMS), to buy property management systems. The systems included computer hardware and software, as well as installation, training, and support services, for the Bell Tower Hotel and the Campus Inn in Ann Arbor, Michigan. The software controlled the central reservations systems at both hotels. When Dahlmann learned that the software was not Y2K compliant—that it could not be used to post reservations beyond December 31, 1999—he filed a suit against Sulchus and HMS, alleging in part breach of contract. The defendants filed a motion for summary judgment. One of the issues was whether the contracts were subject to Article 2 of the Uniform Commercial Code. Were they? Why or why not? Explain fully. [*Dahlmann v. Sulchus Hospitality Technologies Corp.*, 63 F.Supp.2d 772 (E.D.Mich. 1999)]

CASE PROBLEM WITH SAMPLE ANSWER

14–6. In 1988, International Business Machines Corp. (IBM) and American Shizuki Corp. (ASC) signed an agreement for "future purchase by IBM" of plastic film capacitors made by ASC to be used in IBM computers. The agreement stated that IBM was not

obligated to buy from ASC and that future purchase orders "shall be [ASC]'s only authorization to manufacture Items." In February 1989, IBM wrote to ASC about "the possibility of IBM purchasing 15,000,000 Plastic Capacitors per two consecutive twelve (12) month periods. . . . This quantity is a forecast only, and represents no commitment by IBM to purchase these quantities during or after this time period." ASC said that it wanted greater assurances. In a second letter, IBM reexpressed its "intent to order" from ASC 30 million capacitors over a minimum period of two years, contingent on the condition "[t]hat IBM's requirements for these capacitors continue." ASC spent about $2.6 million on equipment to make the capacitors. By 1997, the need for plastic capacitors had dissipated with the advent of new technology, and IBM told ASC that it would no longer buy them. ASC filed a suit in a federal district court against IBM, seeking $8.5 million in damages. On what basis might the court rule in favor of IBM? Explain fully. [*American Shizuki Corp. v. International Business Machines Corp.*, 251 F.3d 1206 (8th Cir. 2001)]

After you have answered this problem, compare your answer with the sample answer given on the Web site that accompanies this text. Go to academic.cengage.com/blaw/wbl, select "Chapter 14," and click on "Case Problem with Sample Answer."

14–7. Statute of Frauds. Quality Pork International is a Nebraska firm that makes and sells custom pork products. Rupari Food Services, Inc., buys and sells food products to retail operations and food brokers. In November 1999, Midwest Brokerage arranged an oral contract between Quality and Rupari, under which Quality would ship three orders to Star Food Processing, Inc., and Rupari would pay for the products. Quality shipped the goods to Star and sent invoices to Rupari. In turn, Rupari billed Star for all three orders, but paid Quality only for the first two (for $43,736.84 and $47,467.80, respectively), not for the third. Quality filed a suit in a Nebraska state court against Rupari, alleging breach of contract, to recover $44,051.98, the cost of the third order. Rupari argued that there was nothing in writing, as required by Section 2–201 of the Uniform Commercial Code (UCC), and thus there was no contract. What are the exceptions to the UCC's writing requirement? Do any of those exceptions apply here? Explain. [*Quality Pork International v. Rupari Food Services, Inc.*, 267 Neb. 474, 675 N.W.2d 642 (2004)]

VIDEO QUESTION

14–8. Go to this text's Web site at academic.cengage.com/blaw/wbl and select "Chapter 14." Click on "Video Questions" and view the video titled *Sales and Lease Contracts: Price as a Term*. Then answer the following questions.

1. Is Anna correct in assuming that a contract can exist even though the sales price for the computer equipment was not specified? Explain.
2. According to the Uniform Commercial Code (UCC), what conditions must be satisfied in order for a contract to be formed when certain terms are left open? What terms (in addition to price) can be left open?
3. Are the e-mail messages that Anna refers to sufficient proof of the contract? Would parol evidence be admissible?
4. How do the 2003 amendments to Article 2 of the UCC improve Anna's position?

The Formation of Sales and Lease Contracts

For updated links to resources available on the Web, as well as a variety of other materials, visit this text's Web site at

academic.cengage.com/blaw/wbl

For information on proposed amendments to Articles 2 and 2A of the Uniform Commercial Code (UCC), go to the Web site of the National Conference of Commissioners on Uniform State Laws (NCCUSL) at

http://www.nccusl.org

The full text of the Contracts for the International Sale of Goods (CISG) is available online at the Pace University School of Law's Institute of International Commercial Law. Go to

http://cisgw3.law.pace.edu/cisg/text/treaty.html

To read an in-depth article comparing the provisions of the CISG and the UCC, go to

http://cisgw3.law.pace.edu/cisg/thesis/Oberman.html

ONLINE LEGAL RESEARCH

Go to the *Fundamentals of Business Law* home page at academic.cengage.com/blaw/wbl, select "Chapter 14," and click on "Internet Exercises." There you will find the following Internet research exercises that you can perform to learn more about topics covered in this chapter.

Activity 14–1: LEGAL PERSPECTIVE—Is It a Contract?

Activity 14–2: MANAGEMENT PERSPECTIVE—A Checklist for Sales Contracts

BEFORE THE TEST

Go to the *Fundamentals of Business Law* home page at academic.cengage.com/blaw/wbl, select "Chapter 14," and click on "Interactive Quizzes." You will find a number of interactive questions relating to this chapter.

Appendix to Chapter 14:

An Example of a Contract for the International Sale of Coffee

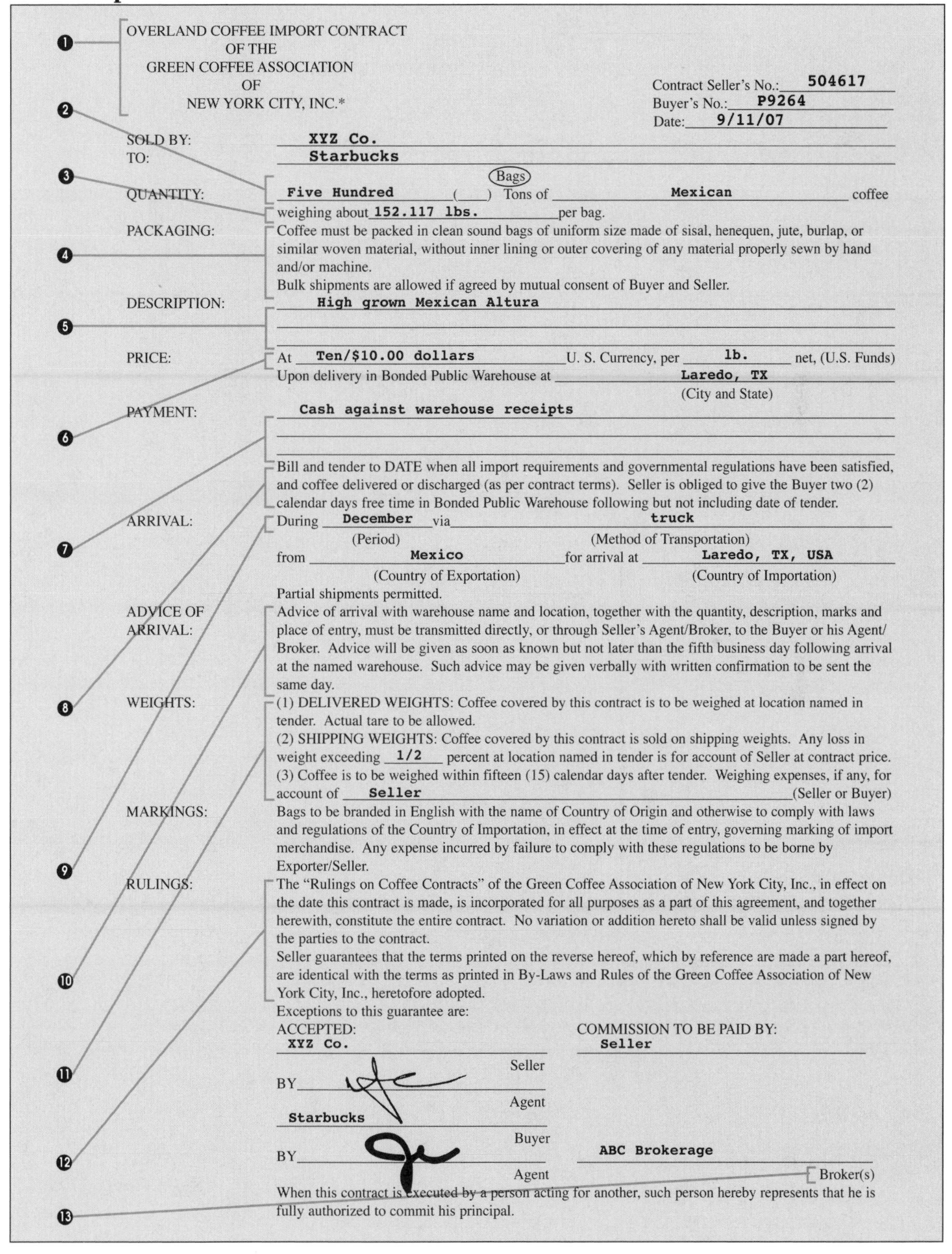

OVERLAND COFFEE IMPORT CONTRACT
OF THE
GREEN COFFEE ASSOCIATION
OF
NEW YORK CITY, INC.*

Contract Seller's No.: 504617
Buyer's No.: P9264
Date: 9/11/07

SOLD BY: XYZ Co.
TO: Starbucks

QUANTITY: Five Hundred (____) (Bags) Tons of Mexican coffee weighing about 152.117 lbs. per bag.

PACKAGING: Coffee must be packed in clean sound bags of uniform size made of sisal, henequen, jute, burlap, or similar woven material, without inner lining or outer covering of any material properly sewn by hand and/or machine.
Bulk shipments are allowed if agreed by mutual consent of Buyer and Seller.

DESCRIPTION: High grown Mexican Altura

PRICE: At Ten/$10.00 dollars U. S. Currency, per lb. net, (U.S. Funds)
Upon delivery in Bonded Public Warehouse at Laredo, TX
(City and State)

PAYMENT: Cash against warehouse receipts

Bill and tender to DATE when all import requirements and governmental regulations have been satisfied, and coffee delivered or discharged (as per contract terms). Seller is obliged to give the Buyer two (2) calendar days free time in Bonded Public Warehouse following but not including date of tender.

ARRIVAL: During December (Period) via truck (Method of Transportation)
from Mexico (Country of Exportation) for arrival at Laredo, TX, USA (Country of Importation)
Partial shipments permitted.

ADVICE OF ARRIVAL: Advice of arrival with warehouse name and location, together with the quantity, description, marks and place of entry, must be transmitted directly, or through Seller's Agent/Broker, to the Buyer or his Agent/Broker. Advice will be given as soon as known but not later than the fifth business day following arrival at the named warehouse. Such advice may be given verbally with written confirmation to be sent the same day.

WEIGHTS: (1) DELIVERED WEIGHTS: Coffee covered by this contract is to be weighed at location named in tender. Actual tare to be allowed.
(2) SHIPPING WEIGHTS: Coffee covered by this contract is sold on shipping weights. Any loss in weight exceeding 1/2 percent at location named in tender is for account of Seller at contract price.
(3) Coffee is to be weighed within fifteen (15) calendar days after tender. Weighing expenses, if any, for account of Seller (Seller or Buyer)

MARKINGS: Bags to be branded in English with the name of Country of Origin and otherwise to comply with laws and regulations of the Country of Importation, in effect at the time of entry, governing marking of import merchandise. Any expense incurred by failure to comply with these regulations to be borne by Exporter/Seller.

RULINGS: The "Rulings on Coffee Contracts" of the Green Coffee Association of New York City, Inc., in effect on the date this contract is made, is incorporated for all purposes as a part of this agreement, and together herewith, constitute the entire contract. No variation or addition hereto shall be valid unless signed by the parties to the contract.
Seller guarantees that the terms printed on the reverse hereof, which by reference are made a part hereof, are identical with the terms as printed in By-Laws and Rules of the Green Coffee Association of New York City, Inc., heretofore adopted.
Exceptions to this guarantee are:

ACCEPTED:
XYZ Co. Seller
BY ____ Agent

Starbucks Buyer
BY ____ Agent

COMMISSION TO BE PAID BY: Seller

ABC Brokerage Broker(s)

When this contract is executed by a person acting for another, such person hereby represents that he is fully authorized to commit his principal.

1. This is a contract for a sale of coffee to be *imported* internationally. If the parties have their principal places of business located in different countries, the contract may be subject to the United Nations Convention on Contracts for the International Sale of Goods (CISG). If the parties' principal places of business are located in the United States, the contract may be subject to the Uniform Commercial Code (UCC).

2. Quantity is one of the most important terms to include in a contract. Without it, a court may not be able to enforce the contract.

3. Weight per unit (bag) can be exactly stated or approximately stated. If it is not so stated, usage of trade in international contracts determines standards of weight.

4. Packaging requirements can be conditions for acceptance and payment. Bulk shipments are not permitted without the consent of the buyer.

5. A description of the coffee and the "markings" constitute express warranties. Warranties in contracts for domestic sales of goods are discussed generally in Chapter 17. International contracts rely more heavily on descriptions and models or samples.

6. Under the UCC, parties may enter into a valid contract even though the price is not set. Under the CISG, a contract must provide for an exact determination of the price.

7. The terms of payment may take one of two forms: credit or cash. Credit terms can be complicated. A cash term can be simple, and payment may be by any means acceptable in the ordinary course of business (for example, a personal check or a letter of credit). If the seller insists on actual cash, the buyer must be given a reasonable time to get it. See Chapter 16.

8. *Tender* means the seller has placed goods that conform to the contract at the buyer's disposition. What constitutes a valid tender is explained in Chapter 16. This contract requires that the coffee meet all import regulations and that it be ready for pickup by the buyer at a "Bonded Public Warehouse." (A *bonded warehouse* is a place in which goods can be stored without paying taxes until the goods are removed.)

9. The delivery date is significant because, if it is not met, the buyer may hold the seller in breach of the contract. Under this contract, the seller can be given a "period" within which to deliver the goods, instead of a specific day, which could otherwise present problems. The seller is also given some time to rectify goods that do not pass inspection (see the "Guarantee" clause on page two of the contract). For a discussion of the remedies of the buyer and seller, see Chapter 16.

10. As part of a proper tender, the seller (or its agent) must inform the buyer (or its agent) when the goods have arrived at their destination. The responsibilities of agents are set out in Chapter 22.

11. In some contracts, delivered and shipped weights can be important. During shipping, some loss can be attributed to the type of goods (spoilage of fresh produce, for example) or to the transportation itself. A seller and buyer can agree on the extent to which either of them will bear such losses.

12. Documents are often incorporated in a contract by reference, because including them word for word can make a contract difficult to read. If the document is later revised, the whole contract might have to be reworked. Documents that are typically incorporated by reference include detailed payment and delivery terms, special provisions, and sets of rules, codes, and standards.

13. In international sales transactions, and for domestic deals involving certain products, brokers are used to form the contracts. When so used, the brokers are entitled to a commission. See Chapter 22.

(Continued)

TERMS AND CONDITIONS

14

ARBITRATION: All controversies relating to, in connection with, or arising out of this contract, its modification, making or the authority or obligations of the signatories hereto, and whether involving the principals, agents, brokers, or others who actually subscribe hereto, shall be settled by arbitration in accordance with the "Rules of Arbitration" of the Green Coffee Association of New York City, Inc., as they exist at the time of the arbitration (including provisions as to payment of fees and expenses). Arbitration is the sole remedy hereunder, and it shall be held in accordance with the law of New York State, and judgment of any award may be entered in the courts of that State, or in any other court of competent jurisdiction. All notices or judicial service in reference to arbitration or enforcement shall be deemed given if transmitted as required by the aforesaid rules.

15

GUARANTEE: (a) If all or any of the coffee is refused admission into the country of importation by reason of any violation of governmental laws or acts, which violation existed at the time the coffee arrived at Bonded-Public Warehouse, seller is required, as to the amount not admitted and as soon as possible, to deliver replacement coffee in conformity to all terms and conditions of this contract, excepting only the Arrival terms, but not later than thirty (30) days after the date of the violation notice. Any payment made and expenses incurred for any coffee denied entry shall be refunded within ten (10) calendar days of denial of entry, and payment shall be made for the replacement delivery in accordance with the terms of this contract. Consequently, if Buyer removes the coffee from the Bonded Public Warehouse, Seller's responsibility as to such portion hereunder ceases.
(b) Contracts containing the overstamp "No Pass-No Sale" on the face of the contract shall be interpreted to mean: If any or all of the coffee is not admitted into the country of Importation in its original condition by reason of failure to meet requirements of the government's laws or Acts, the contract shall be deemed null and void as to that portion of the coffee which is not admitted in its original condition. Any payment made and expenses incurred for any coffee denied entry shall be refunded within ten (10) calendar days of denial of entry.

16

CONTINGENCY: This contract is not contingent upon any other contract.

CLAIMS: Coffee shall be considered accepted as to quality unless within *fifteen* (15) calendar days after delivery at Bonded Public Warehouse or within *fifteen* (15) calendar days after all Government clearances have been received, whichever is later, either:
(a) Claims are settled by the parties hereto, or,
(b) Arbitration proceedings have been filed by one of the parties in accordance with the provisions hereof.
(c) If neither (a) nor (b) has been done in the stated period or if any portion of the coffee has been removed from the Bonded Public Warehouse before representative sealed samples have been drawn by the Green Coffee Association of New York City, Inc., in accordance with its rules, Seller's responsibility for quality claims ceases for that portion so removed.
(d) Any question of quality submitted to arbitration shall be a matter of allowance only, unless otherwise provided in the contract.

17

18

DELIVERY: (a) No more than three (3) chops may be tendered for each lot of 250 bags.
(b) Each chop of coffee tendered is to be uniform in grade and appearance. All expense necessary to make coffee uniform shall be for account of seller.
(c) Notice of arrival and/or sampling order constitutes a tender, and must be given not later than the fifth business day following arrival at Bonded Public Warehouse stated on the contract.

INSURANCE: Seller is responsible for any loss or damage, or both, until Delivery and Discharge of coffee at the Bonded Public Warehouse in the Country of Importation.

All Insurance Risks, costs and responsibility are for Seller's Account until Delivery and Discharge of coffee at the Bonded Public Warehouse in the Country of Importation.

19

Buyer's insurance responsibility begins from the day of importation or from the day of tender, whichever is later.

20

FREIGHT: Seller to provide and pay for all transportation and related expenses to the Bonded Public Warehouse in the Country of Importation.

EXPORT DUTIES/TAXES: Exporter is to pay all Export taxes, duties or other fees or charges, if any, levied because of exportation.

21

IMPORT DUTIES/TAXES: Any Duty or Tax whatsoever, imposed by the government or any authority of the Country of Importation, shall be borne by the Importer/Buyer.

INSOLVENCY OR FINANCIAL FAILURE OF BUYER OR SELLER: If, at any time before the contract is fully executed, either party hereto shall meet with creditors because of inability generally to make payment of obligations when due, or shall suspend such payments, fail to meet his general trade obligations in the regular course of business, shall file a petition in bankruptcy or, for an arrangement, shall become insolvent, or commit an act of bankruptcy, then the other party may at his option, expressed in writing, declare the aforesaid to constitute a breach and default of this contract, and may, in addition to other remedies, decline to deliver further or make payment or may sell or purchase for the defaulter's account, and may collect damage for any injury or loss, or shall account for the profit, if any, occasioned by such sale or purchase.

This clause is subject to the provisions of (11 USC 365 (e) 1) if invoked.

22

23

BREACH OR DEFAULT OF CONTRACT: In the event either party hereto fails to perform, or breaches or repudiates this agreement, the other party shall subject to the specific provisions of this contract be entitled to the remedies and relief provided for by the Uniform Commercial Code of the State of New York. The computation and ascertainment of damages, or the determination of any other dispute as to relief, shall be made by the arbitrators in accordance with the Arbitration Clause herein.

Consequential damages shall not, however, be allowed.

14 Arbitration is the settling of a dispute by submitting it to a disinterested party (other than a court) that renders a decision. The procedures and costs can be provided for in an arbitration clause or incorporated through other documents. To enforce an award rendered in an arbitration, the winning party can "enter" (submit) the award in a court "of competent jurisdiction." For a general discussion of arbitration and other forms of dispute resolution (other than courts), see Chapter 2.

15 When goods are imported internationally, they must meet certain import requirements before being released to the buyer. Because of this, buyers frequently want a guaranty clause that covers the goods not admitted into the country and that either requires the seller to replace the goods within a stated time or allows the contract for those goods not admitted to be void.

16 In the "Claims" clause, the parties agree that the buyer has a certain time within which to reject the goods. The right to reject is a right by law and does not need to be stated in a contract. If the buyer does not exercise the right within the time specified in the contract, the goods will be considered accepted. See Chapter 16.

17 Many international contracts include definitions of terms so that the parties understand what they mean. Some terms are used in a particular industry in a specific way. Here, the word *chop* refers to a unit of like-grade coffee bean. The buyer has a right to inspect ("sample") the coffee. If the coffee does not conform to the contract, the seller must correct the nonconformity.

18 The "Delivery," "Insurance," and "Freight" clauses, with the "Arrival" clause on page one of the contract, indicate that this is a destination contract. The seller has the obligation to deliver the goods to the destination, not simply deliver them into the hands of a carrier. Under this contract, the destination is a "Bonded Public Warehouse" in a specific location. The seller bears the risk of loss until the goods are delivered at their destination. Typically, the seller will have bought insurance to cover the risk.

19 Delivery terms are commonly placed in all sales contracts. Such terms determine who pays freight and other costs, and, in the absence of an agreement specifying otherwise, who bears the risk of loss. International contracts may use these delivery terms or they may use INCOTERMS, which are published by the International Chamber of Commerce. For example, the INCOTERM "DDP" ("delivered duty paid") requires the seller to arrange shipment, obtain and pay for import or export permits, and get the goods through customs to a named destination.

20 Exported and imported goods are subject to duties, taxes, and other charges imposed by the governments of the countries involved. International contracts spell out who is responsible for these charges. See Chapter 32.

21 This clause protects a party if the other party should become financially unable to fulfill the obligations under the contract. Thus, if the seller cannot afford to deliver, or the buyer cannot afford to pay, for the stated reasons, the other party can consider the contract breached. This right is subject to "11 USC 365(e)(1)," which refers to a specific provision of the U.S. Bankruptcy Code dealing with executory contracts. Bankruptcy provisions are covered in Chapter 21.

22 In the "Breach or Default of Contract" clause, the parties agreed that the remedies under this contract are the remedies (except for consequential damages) provided by the UCC, as in effect in the state of New York. The amount and "ascertainment" of damages, as well as other disputes about relief, are to be determined by arbitration. Breach of contract and contractual remedies in general are explained in Chapter 16. Arbitration is discussed in Chapter 2.

23 Three clauses frequently included in international contracts are omitted here. There is no "choice-of-language" clause designating the official language to be used in interpreting the contract terms. There is no "choice-of-forum" clause designating the place in which disputes will be litigated, except for arbitration (law of New York State). Finally, there is no "*force majeure*" clause relieving the sellers or buyers from nonperformance due to events beyond their control.

CHAPTER 15
Title and Risk of Loss

LEARNING OBJECTIVES

After reading this chapter, you should be able to answer the following questions:

1. What is the significance of identifying goods to a contract?
2. If the parties to a contract do not expressly agree on when title to goods passes, what determines when title passes?
3. Risk of loss does not necessarily pass with title. If the parties to a contract do not expressly agree when risk passes and the goods are to be delivered without movement by the seller, when does risk pass?
4. Under what circumstances will the seller's title to goods being sold be void? Under what circumstances will a seller have voidable title? What is the legal effect on a good faith purchaser of the goods when the seller has a void title versus a voidable title?
5. At what point does the buyer acquire an insurable interest in goods subject to a sales contract? Can both the buyer and the seller have an insurable interest in the goods simultaneously?

A sale of goods transfers ownership rights in (title to) the goods from the seller to the buyer. Often, a sales contract is signed before the actual goods are available. For example, a sales contract for oranges might be signed in May, but the oranges may not be ready for picking and shipment until October. Any number of things can happen between the time the sales contract is signed and the time the goods are actually transferred into the buyer's possession. Fire, flood, or frost may destroy the orange groves, or the oranges may be lost or damaged in transit. The same problems may occur under a lease contract. Because of these possibilities, it is important to know the rights and liabilities of the parties between the time the contract is formed and the time the goods are actually received by the buyer or lessee.

Before the creation of the Uniform Commercial Code (UCC), *title*—the right of ownership—was the central concept in sales law, controlling all issues of rights and remedies of the parties to a sales contract. In some situations, title is still relevant under the UCC, and the UCC has special rules for determining who has title. These rules will be discussed in the sections that follow. In most situations, however, the UCC has replaced the concept of title with three other concepts: (1) identification, (2) risk of loss, and (3) insurable interest. By breaking down the transfer of ownership into these three components, the drafters of the UCC have created greater precision in the law governing sales—leaving as few points of law as possible for the courts to decide.

In lease contracts, of course, title to the goods is retained by the lessor-owner of the goods. Hence, the

UCC's provisions relating to passage of title do not apply to leased goods. Other concepts discussed in this chapter, though, including identification, risk of loss, and insurable interest, relate to lease contracts as well as to sales contracts.

Identification

Before any interest in specific goods can pass from the seller or lessor to the buyer or lessee, two conditions must prevail: (1) the goods must be in existence, and (2) they must be identified as the specific goods designated in the contract. **Identification** takes place when specific goods are designated as the subject matter of a sales or lease contract. Title and risk of loss cannot pass from seller to buyer unless the goods are identified to the contract. (As mentioned, title to leased goods remains with the lessor—or, if the owner is a third party, with that party. The lessee does not acquire title to leased goods.) Identification is significant because it gives the buyer or lessee the right to insure (or to have an insurable interest in) the goods and the right to recover from third parties who damage the goods.

In their contract, the parties can agree on when identification will take place, but identification is effective to pass title and risk of loss to the buyer only *after* the goods are considered to be in existence. If the parties do not so specify, however, the UCC provisions discussed here determine when identification takes place [UCC 2–501(1), 2A–217].

Existing Goods

If the contract calls for the sale or lease of specific and ascertained goods that are already in existence, identification takes place at the time the contract is made. For example, you contract to purchase or lease a fleet of five cars by the serial numbers listed for the cars.

Future Goods

If a sale involves unborn animals to be born within twelve months after contracting, identification takes place when the animals are conceived. If a lease involves any unborn animals, identification occurs when the animals are conceived. If a sale involves crops that are to be harvested within twelve months (or the next harvest season occurring after contracting, whichever is longer), identification takes place when the crops are planted or otherwise become growing crops. In a sale or lease of any other future goods, identification occurs when the goods are shipped, marked, or otherwise designated by the seller or lessor as the goods to which the contract refers.

Goods That Are Part of a Larger Mass

As a general rule, goods that are part of a larger mass are identified when the goods are marked, shipped, or somehow designated by the seller or lessor as the particular goods that are the subject of the contract. ■ EXAMPLE 15.1 A buyer orders 1,000 cases of beans from a 10,000-case lot. Until the seller separates the 1,000 cases of beans from the 10,000-case lot, title and risk of loss remain with the seller. ■

A common exception to this rule involves fungible goods. **Fungible goods** are goods that are alike by physical nature, by agreement, or by trade usage. Typical examples are specific grades or types of wheat, oil, and wine, usually stored in large containers. If these goods are held or intended to be held by owners as tenants in common (owners having shares undivided from the entire mass), a seller-owner can pass title and risk of loss to the buyer without an actual separation. The buyer replaces the seller as an owner in common [UCC 2–105(4)].

■ EXAMPLE 15.2 Alvarez, Braudel, and Carpenter are farmers. They deposit, respectively, 5,000 bushels, 3,000 bushels, and 2,000 bushels of grain of the same grade and quality in a bin. The three become owners in common, with Alvarez owning 50 percent of the 10,000 bushels, Braudel 30 percent, and Carpenter 20 percent. Alvarez contracts to sell her 5,000 bushels of grain to Tamur; because the goods are fungible, she can pass title and risk of loss to Tamur without physically separating the 5,000 bushels. Tamur now becomes an owner in common with Braudel and Carpenter. ■

Passage of Title

Once goods exist and are identified, the provisions of UCC 2–401 apply to the passage of title. In virtually all subsections of UCC 2–401, the words "unless otherwise explicitly agreed" appear, meaning that any explicit understanding between the buyer and the seller determines when title passes. Without an explicit agreement to the contrary, title passes to the buyer at the time and the place the seller performs by delivering the goods [UCC 2–401(2)]. The following case illustrates the significance of this event.

CASE 15.1 In re Stewart

United States Bankruptcy Court,
Western District of Arkansas, 2002.
274 Bankr. 503.

FACTS In July 1997, Gary Stewart began to buy, and occasionally sell, cattle through Barry County Livestock Auction, Inc., in Exeter, Missouri. On January 29 and February 19, 2000, Stewart bought $46,749.55 worth of cattle through Barry County, but the checks he gave in payment were returned by the bank because his account did not have sufficient funds. By March 4, Stewart had given cashier's checks to Barry County to pay for the cattle.[a] Less than forty days later, on April 11, Stewart filed for bankruptcy in a federal bankruptcy court. Some payments made to creditors within ninety days of the filing of a petition in bankruptcy (called *preferences*) can be recovered so that their amounts can be distributed more equitably among a debtor's creditors.[b] Stewart asked the court to recover his payments to Barry County. Barry County claimed that the payments were "contemporaneous exchanges for new value," which are not a preference and cannot be recovered. The question on which this issue turned was whether title to the cattle passed on the day of the sale or on the day of the payment.

ISSUE Could Barry County keep the payments it received from Stewart on the ground that they were "contemporaneous exchanges"?

DECISION No. The court held that title to the cattle passed on the dates of the sales when the cattle were physically delivered. Thus, Stewart's payment with the cashier's checks was not part of a "contemporaneous exchange"—the payment occurred after the bank returned his personal checks for insufficient funds, more than two weeks after the transfer of title.

a. A cashier's check is considered nearly the equivalent of cash.
b. Bankruptcy law will be discussed more fully in Chapter 21.

REASON The court noted that Barry County printed on each bill of sale: "ALL SALES ARE MADE, AND TITLE IS TRANSFERRED, SUBJECT TO FINAL PAYMENT * * * . IF ANY CHECK * * * TENDERED IN PAYMENT IS NOT PAID PROMPTLY ON PRESENTATION, THIS BILL OF SALE DOES NOT TRANSFER TITLE." The court stated that apparently "Barry County Livestock Auction was attempting to retain title to the cattle in the event the debtor's personal checks were not honored [paid]." The court explained, however, that "[a]ccording to the UCC, any attempt to retain title to the cattle pending the debtor's personal checks being honored by the bank is ineffective." Under UCC 2–401(2), "title passes to the buyer at the time and place at which the seller completes his performance with reference to the physical delivery of the goods." In other words, in this case, "under the UCC, title to the cattle passed upon delivery of the cattle on the day of the respective sales."

FOR CRITICAL ANALYSIS—Ethical Consideration *Often, by explicit agreement, the parties to a transaction can establish terms that differ from those provided by the UCC. In this case, why weren't the words on Barry County's bill of sale, "SUBJECT TO FINAL PAYMENT," effective in establishing the terms of the sale?*

SHIPMENT AND DESTINATION CONTRACTS

Unless otherwise agreed, delivery arrangements can determine when title passes from the seller to the buyer. In a **shipment contract,** the seller is required or authorized to ship goods by carrier, such as a trucking company. Under a shipment contract, the seller is required only to deliver conforming goods into the hands of a carrier, and title passes to the buyer at the time and place of shipment [UCC 2–401(2)(a)]. Generally, *all contracts are assumed to be shipment contracts if nothing to the contrary is stated in the contract.*

In a **destination contract,** the seller is required to deliver the goods to a particular destination (usually directly to the buyer, but sometimes to another party designated by the buyer). Title passes to the buyer when the goods are *tendered* at that destination [UCC 2–401(2)(b)]. As you will read in Chapter 16, a *tender of delivery* occurs when the seller places or holds conforming goods at the buyer's disposition (with any necessary notice), enabling the buyer to take delivery [UCC 2–503(1)].

DELIVERY WITHOUT MOVEMENT OF THE GOODS

When the sales contract does not call for the seller to ship or deliver the goods (when the buyer is to pick up the goods), the passage of title depends on whether the seller must deliver a **document of title,** such as a bill of lading or a warehouse receipt, to the buyer. A *bill of lading* is a receipt for goods that is signed by a carrier and that serves as a contract for the transportation of the goods. A *warehouse receipt* is a receipt issued by a warehouser for goods stored in a warehouse.

When a document of title is required, title passes to the buyer *when and where the document is delivered.* Thus, if the goods are stored in a warehouse, title passes to the buyer when the appropriate documents are delivered to the buyer. The goods never move. In fact, the buyer can

choose to leave the goods at the same warehouse for a period of time, and the buyer's title to those goods will be unaffected.

When no documents of title are required and delivery is made without moving the goods, title passes at the time and place the sales contract is made, if the goods have already been identified. If the goods have not been identified, title does not pass until identification occurs. ■ EXAMPLE 15.3 Juan sells lumber to Bodan. They agree that Bodan will pick up the lumber at the lumberyard. If the lumber has been identified (segregated, marked, or in any other way distinguished from all other lumber), title passes to Bodan when the contract is signed. If the lumber is still in general storage bins at the lumberyard, title does not pass to Bodan until the particular pieces of lumber to be sold under this contract are identified [UCC 2–401(3)]. ■

Sales or Leases by Nonowners

Problems occur when persons who acquire goods with *imperfect* titles attempt to sell or lease them. Sections 2–402 and 2–403 of the UCC deal with the rights of two parties who lay claim to the same goods, sold with imperfect titles. Generally, a buyer acquires at least whatever title the seller has to the goods sold.

The UCC also protects a person who leases such goods from the buyer. Of course, a lessee does not acquire whatever title the lessor has to the goods. A lessee acquires a right to possess and use the goods—that is, a *leasehold interest.* A lessee acquires whatever leasehold interest the lessor has or has the power to transfer, subject to the lease contract [UCC 2A–303, 2A–304, 2A–305].

Void Title A buyer may unknowingly purchase goods from a seller who is not the owner of the goods. If the seller is a thief, the seller's title is *void*—legally, no title exists. Thus, the buyer acquires no title, and the real owner can reclaim the goods from the buyer. If the goods were only leased, the same result would occur because the lessor has no leasehold interest to transfer.

■ EXAMPLE 15.4 If Saki steals diamonds owned by Maren, Saki has a *void title* to those diamonds. If Saki sells the diamonds to Shannon, Maren can reclaim them from Shannon even though Shannon acted in good faith and honestly was not aware that the goods were stolen. ■ (Article 2A contains similar provisions for leases.)

Voidable Title A seller has *voidable title* if the goods that she or he is selling were obtained by fraud, paid for with a check that was later dishonored, purchased from a minor, or purchased on credit when the seller was insolvent. (Under the UCC, a person is **insolvent** when that person ceases to "pay his [or her] debts in the ordinary course of business or cannot pay his [or her] debts as they become due or is insolvent within the meaning of federal bankruptcy law" [UCC 1–201(23)].)

—Good Faith Purchasers. In contrast to a seller with *void title,* a seller with *voidable title* has the power to transfer good title to a good faith purchaser for value. A **good faith purchaser** is one who buys without knowledge of circumstances that would make a person of ordinary prudence inquire about the validity of the seller's title to the goods. One who purchases *for value* gives legally sufficient consideration (value) for the goods purchased. The real, or original, owner cannot recover goods from a good faith purchaser for value [UCC 2–403(1)].[1] If the buyer of the goods is not a good faith purchaser for value, then the actual owner of the goods can reclaim them from the buyer (or from the seller, if the goods are still in the seller's possession).

The question in the following case was whether a third party buyer was a good faith purchaser in the context of a sale of timber.

1. The real owner could, of course, sue the person who initially obtained voidable title to the goods.

CASE 15.2 Memphis Hardwood Flooring Co. v. Daniel

Mississippi Supreme Court, 2000.
771 So.2d 924.
http://www.mssc.state.ms.us[a]

a. In the top row, click on the "Decisions" box. On that page, in the list, click on "Supreme Court & Court of Appeals Opinions." In the left-hand frame, in the "Year" menu, select "2000." In the "Alphabetical" menu, choose the letter "D" (for Daniel) and click on "Search." In the result, scroll to the case name and click on it to access the opinion. The state of Mississippi maintains this Web site.

FACTS Jamie Swann Daniel, a retired teacher, owned approximately 800 acres of land in Union County, Mississippi. After an ice storm, Lucky Easley and William Heppler, officers of Northern Hardwood, Inc., convinced Daniel to allow some cutting of the storm-damaged timber on the land. Daniel contracted with Northern to cut from the

(Continued)

CASE 15.2–CONTINUED

southern half of the property the "disaster hardwood timber" that was twenty inches or more in diameter.[b] More than a year later, in a second contract, Daniel agreed to the cutting of the timber down to sixteen inches, for which Easley would pay her $150,000. The same day, Easley told Robert Luther, a timber buyer for Memphis Hardwood Flooring Company, that he had an agreement to cut *all* of the timber on all of Daniel's land. Luther and Easley agreed that Easley, acting for both Northern and Memphis, would buy Daniel's timber for Northern while concealing from Daniel that Northern was actually buying for Memphis. Luther agreed to pay $410,000 to Easley. At Easley's request, Luther had Memphis's lawyer draft the agreements ("timber deeds") between Daniel and Northern and between Northern and Memphis. When Memphis proceeded to cut the timber, Daniel filed a suit in a Mississippi state court against Memphis and the others, alleging fraud. Among other things, the court canceled the agreement between Northern and Memphis.[c] Memphis appealed to the Mississippi Supreme Court, asserting that it was a good faith purchaser.

b. For the sale of this timber, Northern received $498,905, of which Daniel's share under their contract was $299,343. Daniel was paid only $134,000, however. Ultimately, she reached a settlement with Northern, Easley, and Heppler as to their liability on this contract.

c. The court also awarded Daniel more than $800,000 in damages, including reforestation costs, and issued an injunction against Memphis's further cutting. In a separate case, Northern, Easley, and Heppler pleaded guilty to charges of embezzlement and agreed to make restitution to Daniel in the amount of $250,000.

ISSUE Was Memphis a good faith purchaser?

DECISION No. The Mississippi Supreme Court affirmed the decision of the lower court.

REASON The state supreme court noted that to establish the status of a good faith purchaser "[t]he elements the innocent purchaser must prove are a valuable consideration, the presence of good faith, and the absence of notice." In this case, the record established that Memphis paid $410,000 in valuable consideration to Northern for the timber. Memphis did not adequately prove the second element, the presence of good faith, however. The court pointed out that Easley made the deal with Daniel on behalf of both Northern and Memphis and that Memphis was a knowing participant in this scheme. In the court's view, "Such actions reveal the absence of good faith. Even the circumstances surrounding the drafting of the deeds indicate the joint nature of the scheme and Memphis'[s] lack of good faith." As for the third element, "Memphis was not only on actual notice, it was a knowing and willing participant in the defrauding of Daniel. Thus, Memphis fails to establish the third element. The record supports the [lower court's] finding that Memphis was not" a good faith purchaser.

FOR CRITICAL ANALYSIS—Social Consideration *Would Memphis have been considered a participant in the fraud against Daniel if Memphis had not agreed specifically that Easley would act in its behalf? Why or why not?*

—Voidable Title and Leases. The same rules apply in circumstances involving leases. A lessor with voidable title has the power to transfer a valid leasehold interest to a good faith lessee for value. The real owner cannot recover the goods, except as permitted by the terms of the lease. The real owner can, however, receive all proceeds arising from the lease, as well as a transfer of all rights, title, and interest as lessor under the lease, including the lessor's interest in the return of the goods when the lease expires [UCC 2 A–305(1)].

The Entrustment Rule According to Section 2–403(2), entrusting goods to a merchant *who deals in goods of that kind* gives the merchant the power to transfer all rights to *a buyer in the ordinary course of business.* Entrusting includes both turning over the goods to the merchant and leaving purchased goods with the merchant for later delivery or pickup [UCC 2–403(3)]. A buyer in the ordinary course of business is a person who, in good faith and without knowledge that the sale violates the ownership rights or security interest of a third party, buys in ordinary course from a person (other than a pawnbroker) in the business of selling goods of that kind [UCC 1–201(9)]. (A *security interest* is any interest in personal property that secures payment or the performance of an obligation—see Chapter 20.)

■ **EXAMPLE 15.5** Jan leaves her watch with a jeweler to be repaired. The jeweler sells used watches. The jeweler sells Jan's watch to Kim, a customer, who does not know that the jeweler has no right to sell it. Kim, as a good faith buyer, gets good title against Jan's claim of ownership.[2] Kim, however, obtains only those rights held by the person entrusting the goods (here, Jan). Suppose that in fact Jan had stolen the watch from Greg and then left it with the jeweler to be repaired. The jeweler then sells it to Kim. In this situation, Kim gets good title against Jan, who entrusted the watch to the jeweler, but not against Greg

2. Jan, of course, can sue the jeweler for the tort of trespass to personalty or conversion (see Chapter 4) for the equivalent cash value of the watch.

(the real owner), who neither entrusted the watch to Jan nor authorized Jan to entrust it.[3] ■

Article 2A provides a similar rule for leased goods. If a lessor entrusts goods to a lessee-merchant who deals in goods of that kind, the lessee-merchant has the power to transfer all of the rights the lessor had in the goods to a buyer or sublessee in the ordinary course of business [UCC 2A–305(2)].

Risk of Loss

Under the UCC, risk of loss does not necessarily pass with title. When risk of loss passes from a seller or lessor to a buyer or lessee is generally determined by the contract between the parties. Sometimes, the contract states expressly when the risk of loss passes. At other times, it does not, and a court must interpret the performance and delivery terms of the contract to determine whether the risk has passed.

Delivery with Movement of the Goods—Carrier Cases

When the agreement does not state when risk of loss passes, the courts apply the following rules to cases involving movement of the goods (carrier cases).

Contract Terms Specific delivery terms in the contract can determine when risk of loss passes to the buyer. These terms, which are defined in Exhibit 15–1, relate generally to the determination of which party will bear the costs of delivery. *Unless otherwise agreed,* these terms also determine who has the risk of loss.

The 2003 amendments to UCC Article 2 omit these terms because they are "inconsistent with modern commercial practice." The Official Comments to the amendments, however, state that if the parties use these shipping terms without expressly agreeing on the meaning of the terms, the terms "must be interpreted in light of any appropriate usage of trade and any course of performance or course of dealing between the parties." Thus, the effect of these terms may be the same even after a state has adopted the amendments to Article 2.

Shipment Contracts In a shipment contract, if the seller or lessor is required or authorized to ship goods by carrier (but not required to deliver them to a particular final destination), risk of loss passes to the buyer or lessee when the goods are duly delivered to the carrier [UCC 2–319(1)(a), 2–509(1)(a), 2A–219(2)(a)].

■ **EXAMPLE 15.6** A seller in Texas sells five hundred cases of grapefruit to a buyer in New York, F.O.B. Houston (free on board in Houston—that is, the buyer pays the transportation charges from Houston). The contract authorizes shipment by carrier; it does not require that the seller tender the grapefruit in New York. Risk passes to the buyer when conforming goods are properly placed in the possession of the carrier. If the goods are damaged in transit, the loss is the buyer's. (Actually, buyers have recourse against carriers, subject to certain limitations, and buyers usually insure the goods from the time the goods leave the seller.) ■

3. For another example of a court's application of the entrustment rule and a discussion of the policy underlying the provisions in UCC 2–403(2), see *Madrid v. Bloomington Auto Co.,* 782 N.E.2d 386 (Ind.App. 2003).

EXHIBIT 15–1 CONTRACT TERMS—DEFINITIONS

The contract terms listed and defined in this exhibit help to determine which party will bear the costs of delivery and when risk of loss will pass from the seller to the buyer.

F.O.B. (free on board)—Indicates that the selling price of goods includes transportation costs to the specific F.O.B. place named in the contract. The seller pays the expenses and carries the risk of loss to the F.O.B. place named [UCC 2–319(1)]. If the named place is the place from which the goods are shipped (for example, the seller's city or place of business), the contract is a shipment contract. If the named place is the place to which the goods are to be shipped (for example, the buyer's city or place of business), the contract is a destination contract.

F.A.S. (free alongside)—Requires that the seller, at his or her own expense and risk, deliver the goods alongside the carrier before risk passes to the buyer [UCC 2–319(2)].

C.I.F. or C.&F. (cost, insurance, and freight or just cost and freight)—Requires, among other things, that the seller "put the goods in the possession of a carrier" before risk passes to the buyer [UCC 2–320(2)]. (These are basically pricing terms, and the contracts remain shipment contracts, not destination contracts.)

Delivery ex-ship (delivery from the carrying vessel)—Means that risk of loss does not pass to the buyer until the goods are properly unloaded from the ship or other carrier [UCC 2–322].

Destination Contracts In a destination contract, the risk of loss passes to the buyer or lessee when the goods are tendered to the buyer or lessee at the specified destination [UCC 2–319(1)(b), 2–509(1)(b), 2A–219(2)(b)]. In Example 15.6, if the contract had been F.O.B. New York, the risk of loss during transit to New York would have been the seller's.

DELIVERY WITHOUT MOVEMENT OF THE GOODS

The UCC also addresses situations in which the seller or lessor is required neither to ship nor to deliver the goods. Frequently, the buyer or lessee is to pick up the goods from the seller or lessor, or the goods are held by a *bailee*. Under the UCC, a **bailee** is a party who, by a bill of lading, warehouse receipt, or other document of title, acknowledges possession of goods and/or contracts to deliver them. A warehousing company, for example, or a trucking company that normally issues documents of title for the goods it receives is a bailee.[4]

4. Bailments will be discussed in detail in Chapter 28.

Goods Held by the Seller If the goods are held by the seller, a document of title is usually not used. If the seller is a merchant, risk of loss to goods held by the seller passes to the buyer when the buyer *actually takes physical possession of the goods* [UCC 2–509(3)]. If the seller is not a merchant, the risk of loss to goods held by the seller passes to the buyer on *tender of delivery* [UCC 2–509(3)]. (As mentioned earlier, tender of delivery occurs when the seller places conforming goods at the disposal of the buyer and gives the buyer whatever notification is reasonably necessary to enable the buyer to take possession.) With respect to leases, the risk of loss passes to the lessee on the lessee's receipt of the goods if the lessor—or supplier, in a finance lease (see Chapter 14)—is a merchant. Otherwise, the risk passes to the lessee on tender of delivery [UCC 2A–219(c)].[5]

The following case illustrates the consequences of passing the risk of loss under these principles.

5. Under the 2003 amendments to UCC 2–509(3) and 2A–219(c), the risk of loss passes to the buyer or the lessee on that party's receipt of the goods regardless of whether the seller or the lessor is a merchant.

CASE 15.3 Ganno v. Lanoga Corp.

Court of Appeals of Washington,
Division 2, 2003.
119 Wash.App. 310,
80 P.3d 180.
http://www.legalwa.org[a]

FACTS Henry Ganno went to the Lumbermen's Building Center store in Fife, Washington, where he bought a 12-foot beam weighing 100 pounds. In the lumberyard, a store employee approached Ganno, took his receipt, and used a forklift to place the beam in the open bed of Ganno's truck. The beam projected about 4 feet from the end of the truck. The employee asked Ganno if he wanted the beam flagged, Ganno said, "Yes," and the employee flagged the beam. The employee did not tie down or otherwise secure the beam, however. A sign in the lumberyard stated that it was Lumbermen's policy *not* to secure loads for customers. Ganno, who did not get out of the truck or check the load to make sure that it was secure, drove out of the lumberyard onto a public street. When he turned a corner, the beam fell off the truck. As Ganno attempted to retrieve the beam, another vehicle hit it, causing it to strike Ganno's leg and shatter his kneecap. Ganno filed a suit in a Washington state court against Lanoga Corporation, which owned the Fife Lumbermen's store, alleging negligence in failing to secure the beam. The court granted a judgment in Lanoga's favor. Ganno appealed to a state intermediate appellate court.

a. Click on the "Washington State Supreme Court and Appellate Court Decisions" link. Type "Ganno" in the search box, choose "Search case titles only," select "Washington Appellate Reports" (and *not* "Washington Reports") in the "Limit search to:" column, and click on "Search." In the result, click on the name of the case to access the opinion. Municipal Research & Services Center of Washington maintains this Web site.

ISSUE Had the risk passed to the buyer before the loss?

DECISION Yes. The state intermediate appellate court affirmed the lower court's judgment, holding that it was Ganno's duty, not Lumbermen's, to make sure his load was secure before driving onto the public streets.

REASON The appellate court noted that Lumbermen's (Lanoga) is a merchant, under UCC 2–104(1), which defines a *merchant* as "a person who deals in goods of the kind or otherwise by his occupation holds himself out as having knowledge or skill peculiar to the practices or goods involved in the transaction." Under UCC 2–509(3), "where the seller is a merchant, the risk of loss passes to the buyer on receipt of goods." Here, Ganno received the beam from Lumbermen's at its place of business. The risk of loss passed to Ganno when Lumbermen's loaded the beam onto his truck. "In the absence of a legal duty to secure a customer's load, as here,

CASE 15.3—CONTINUED

there is no liability to a customer once he is in possession of the goods." Thus, the court concluded, "Lumbermen's is not liable for the ensuing damage after Ganno took possession and left Lumbermen's property."

WHY IS THIS CASE IMPORTANT? *This case clearly illustrates how the passage of risk of loss can affect a seller's or a buyer's potential liability. The UCC's rules on passage of risk of loss vary, depending on the nature of the transaction.*

Goods Held by a Bailee When a bailee is holding goods for a person who has contracted to sell them and the goods are to be delivered without being moved, the goods are usually represented by a negotiable or nonnegotiable document of title (a bill of lading or a warehouse receipt). Risk of loss passes to the buyer when (1) the buyer receives a negotiable document of title for the goods, (2) the bailee acknowledges the buyer's right to possess the goods, or (3) the buyer receives a nonnegotiable document of title *and* has had a *reasonable time* to present the document to the bailee and demand the goods. Obviously, if the bailee refuses to honor the document, the risk of loss remains with the seller [UCC 2–503(4)(b), 2–509(2)].

In respect to leases, if goods held by a bailee are to be delivered without being moved, the risk of loss passes to the lessee on acknowledgment by the bailee of the lessee's right to possession of the goods [UCC 2A–219(2)(b)].

Conditional Sales

Buyers and sellers sometimes form sales contracts that are conditioned either on the buyer's approval of the goods or on the buyer's resale of the goods. Under such contracts, the buyer is in possession of the goods. Sometimes, however, questions arise as to whether the buyer or seller should bear the loss if, for example, the goods are damaged or stolen while in the possession of the buyer.

Sale or Return A **sale or return** (sometimes called a *sale and return*) is a type of contract by which the seller sells a quantity of goods to the buyer with the understanding that the buyer can set aside the sale by returning the goods or any portion of them. The buyer is required to pay for any goods *not* returned. When the buyer receives possession of the goods under a sale-or-return contract, the title and risk of loss pass to the buyer. Title and risk of loss remain with the buyer until the buyer returns the goods to the seller within the time period specified. If the buyer fails to return the goods within this time period, the sale is finalized. The goods are returned at the buyer's risk and expense. Goods held under a sale-or-return contract are subject to the claims of the buyer's creditors while they are in the buyer's possession (even if the buyer has not paid for the goods) [UCC 2–326, 2–327].

The UCC treats a **consignment** as a sale or return. Under a consignment, the owner of goods (the *consignor*) delivers them to another (the *consignee*) for the consignee to sell. If the consignee sells the goods, the consignee must pay the consignor for them. If the consignee does not sell the goods, they may simply be returned to the consignor. While the goods are in the possession of the consignee, the consignee holds title to them, and creditors of the consignee will prevail over the consignor in any action to repossess the goods [UCC 2–326(3)].[6]

Sale on Approval When a seller offers to sell goods to a buyer and permits the buyer to take the goods on a trial basis, a **sale on approval** is usually made. The term *sale* here is a misnomer, as only an *offer* to sell has been made, along with a bailment created by the buyer's possession. (A *bailment* is a temporary delivery of personal property into the care of another—see Chapter 28.)

Therefore, title and risk of loss (from causes beyond the buyer's control) remain with the seller until the buyer accepts (approves) the offer. Acceptance can be made expressly, by any act inconsistent with the *trial* purpose or the seller's ownership, or by the buyer's election not to return the goods within the trial period. If the buyer does not wish to accept, the buyer may notify the seller of that fact within the trial period, and the return is made at the seller's expense and risk [UCC 2–327(1)]. Goods held on approval are not subject to the claims of the buyer's creditors until acceptance.

It is often difficult to determine whether a particular transaction involves a contract for a sale on approval, a contract for a sale or return, or a contract for sale. The UCC states that (unless otherwise agreed) "if the goods are delivered primarily for use," the transaction is a sale on approval; "if the goods are delivered primarily for resale," the transaction is a sale or return [UCC 2–326(1)].

Risk of Loss When a Sales or Lease Contract Is Breached

A sales or lease contract can be breached in many ways, and the transfer of risk operates differently depending on which party breaches. Generally, the party in breach bears the risk of loss.

6. Although the 2003 amendments to UCC Article 2 retain the provisions concerning sale on approval and sale or return, the portion of this section relating to consignments is omitted. Consignments are to be covered by UCC Article 9. See, for example, UCC 9–103(d), 9–109(a)(4), and 9–319.

When the Seller or Lessor Breaches If the goods are so nonconforming that the buyer has the right to reject them, the risk of loss does not pass to the buyer until the defects are **cured**—that is, until the goods are repaired, replaced, or discounted in price by the seller—or until the buyer accepts the goods in spite of their defects (thus waiving the right to reject). ■ EXAMPLE 15.7 A buyer orders ten white refrigerators from a seller, F.O.B. the seller's plant. The seller ships amber refrigerators instead. The amber refrigerators (nonconforming goods) are damaged in transit. The risk of loss falls on the seller. Had the seller shipped white refrigerators (conforming goods) instead, the risk would have fallen on the buyer [UCC 2–510(2)]. ■

If a buyer accepts a shipment of goods and later discovers a defect, acceptance can be revoked. Revocation allows the buyer to pass the risk of loss back to the seller, at least to the extent that the buyer's insurance does not cover the loss [UCC 2–510(2)].

In regard to leases, Article 2A states a similar rule. If the lessor or supplier tenders goods that are so nonconforming that the lessee has the right to reject them, the risk of loss remains with the lessor or the supplier until cure or acceptance [UCC 2A–220(1)(a)]. If the lessee, after acceptance, revokes his or her acceptance of nonconforming goods, the revocation passes the risk of loss back to the seller or supplier, to the extent that the lessee's insurance does not cover the loss [UCC 2A–220(1)(b)].

When the Buyer or Lessee Breaches The general rule is that when a buyer or lessee breaches a contract, the risk of loss immediately shifts to the buyer or lessee. This rule has three important limitations:

1. The seller or lessor must already have identified the contract goods.
2. The buyer or lessee bears the risk for only a commercially reasonable time after the seller or lessor has learned of the breach.
3. The buyer or lessee is liable only to the extent of any deficiency in the seller's or lessor's insurance coverage [UCC 2–510(3), 2A–220(2)].

Insurable Interest

Parties to sales and lease contracts often obtain insurance coverage to protect against damage, loss, or destruction of goods. Any party purchasing insurance, however, must have a sufficient interest in the insured item to obtain a valid policy. Insurance laws—not the UCC—determine sufficiency. The UCC is helpful, however, because it contains certain rules regarding insurable interests in goods.

Insurable Interest of the Buyer or Lessee

A buyer or lessee has an **insurable interest** in identified goods. The moment the contract goods are identified by the seller or lessor, the buyer or lessee has a special property interest that allows the buyer or lessee to obtain necessary insurance coverage for those goods even before the risk of loss has passed [UCC 2–501(1), 2A–218(1)].

The rule stated in UCC 2–501(1)(c) is that buyers obtain an insurable interest in crops by identification, which occurs when the crops are planted or otherwise become growing crops, provided that the contract is for "the sale of crops to be harvested within twelve months or the next normal harvest season after contracting, whichever is longer." ■ EXAMPLE 15.8 In March, a farmer sells a cotton crop he hopes to harvest in October. When the crop is planted, the buyer acquires an insurable interest in it because those goods (the cotton crop) are identified to the sales contract between the seller and the buyer. ■

Insurable Interest of the Seller or Lessor

A seller has an insurable interest in goods if she or he retains title to the goods. Even after title passes to the buyer, a seller who has a security interest in the goods (a right to secure payment—see Chapter 20) still has an insurable interest and can insure the goods [UCC 2–501(2)]. Hence, both a buyer and a seller can have an insurable interest in identical goods at the same time. Of course, the buyer or seller must sustain an actual loss to have the right to recover from an insurance company. In regard to leases, the lessor retains an insurable interest in leased goods until the lessee exercises an option to buy and the risk of loss has passed to the lessee [UCC 2A–218(3)].

Bulk Transfers

Bulk transfers are the subject of UCC Article 6. A *bulk transfer* is defined as any transfer of a major part of the transferor's material, supplies, merchandise, or other inventory *not made in the ordinary course of the transferor's business* [UCC 6–102(1)]. Difficulties sometimes occur with bulk transfers. For example, when a business that owes debts to numerous creditors sells a substantial part of its equipment and inventories to a buyer, the business should use the proceeds to pay off the debts. What happens, though, if the seller instead spends the funds on a vacation trip, leaving the creditors without payment? Can the creditors lay any claim to the goods that were transferred in bulk to the buyer?

REQUIREMENTS OF ARTICLE 6

The purpose of Article 6 is to protect creditors in transactions involving bulk transfers. Under UCC 6–104 and 6–105, the following requirements must be met when a bulk transfer is undertaken:

1. The seller must provide the buyer with a sworn list of his or her existing creditors. The list must include those whose claims are disputed and must state names, business addresses, and amounts due.
2. The buyer and the seller must prepare a schedule of the property that is to be transferred.
3. The buyer must preserve the list of creditors and the schedule of property for six months and permit any creditor of the seller to inspect the list; alternatively, the buyer must file the list and the schedule of property in a designated public office.
4. The buyer must give notice of the proposed bulk transfer to each of the seller's creditors at least ten days before the buyer takes possession of the goods or makes payment for them, whichever happens first.

If these requirements are met, the buyer acquires title to the goods free of all claims by the seller's creditors. If the requirements are not met, goods in the possession of the buyer continue to be subject to the claims of the seller's unpaid creditors for six months [UCC 6–111].

THE MAJORITY OF STATES HAVE REPEALED ARTICLE 6

In 1988, the National Conference of Commissioners on Uniform State Laws recommended that those states that have adopted Article 6 repeal it because changes in the business and legal contexts in which bulk sales are conducted have made their regulation unnecessary. For states disinclined to do so, Article 6 has been revised to provide creditors with better protection while reducing the burden imposed on good faith purchasers. To date, at least thirty-five states have repealed Article 6, and four states have opted for the revised version. The remainder of the states have retained Article 6 in its original form.

The revised Article 6 limits its application to bulk sales by sellers whose principal business is the sale of inventory from bulk stock. It does not apply to transactions involving property valued at less than $10,000 or more than $25 million. If a seller has more than two hundred creditors, a buyer, rather than having to send individual notice to each creditor, can give notice by public filing (for example, in the office of a state's secretary of state). The notice period is increased from ten to forty-five days, and the statute of limitations is extended from six months to one year.

TERMS AND CONCEPTS

CHAPTER SUMMARY Title and Risk of Loss*

Shipment Contracts (See page 284.)	In the absence of an agreement, title and risk pass on the seller's or lessor's delivery of conforming goods to the carrier [UCC 2–319(1)(a), 2–401(2)(a), 2–509(1)(a), 2A–219(2)(a)].
Destination Contracts (See page 284.)	In the absence of an agreement, title and risk pass on the seller's or lessor's *tender* of delivery of conforming goods to the buyer or lessee at the point of destination [UCC 2–319(1)(b), 2–401(2)(b), 2–509(1)(b), 2A–219(2)(b)].

* Although we have referenced the 2003 amendments to Articles 2 and 2A of the UCC within the text of this chapter, we do not do so in this summary—because the amendments have not as yet become law in any state.

(Continued)

CHAPTER SUMMARY Title and Risk of Loss—Continued

Delivery without Movement of the Goods (See pages 284–285.)	1. In the absence of an agreement, if the goods are not represented by a document of title: a. Title passes on the formation of the contract [UCC 2–401(3)(b)]. b. Risk passes to the buyer or lessee, if the seller or lessor (or supplier, in a finance lease) is a merchant, when the buyer or lessee receives the goods or, if the seller or lessor is a nonmerchant, when the seller or lessor *tenders* delivery of the goods [UCC 2–509(3), 2A–219(c)]. 2. In the absence of an agreement, if the goods are represented by a document of title: a. If the document is negotiable and the goods are held by a bailee, title and risk pass on the buyer's *receipt* of the document [UCC 2–401(3)(a), 2–509(2)(a)]. b. If the document is nonnegotiable and the goods are held by a bailee, title passes on the buyer's receipt of the document, but risk does *not* pass until the buyer, after receipt of the document, has had a reasonable time to present the document to demand the goods [UCC 2–401(3)(a), 2–503(4)(b), 2–509(2)(c)]. 3. In the absence of an agreement, if the goods are held by a bailee and no document of title is transferred, risk passes to the buyer when the bailee acknowledges the buyer's right to the possession of the goods [UCC 2–509(2)(b)]. 4. In respect to leases, if goods held by a bailee are to be delivered without being moved, the risk of loss passes to the lessee on acknowledgment by the bailee of the lessee's right to possession of the goods [UCC 2A–219(2)(b)].
Sales or Leases by Nonowners (See pages 285–287.)	Between the owner and a good faith purchaser or between the lessee and a sublessee: 1. *Void title*—Owner prevails [UCC 2–403(1)]. 2. *Voidable title*—Buyer prevails [UCC 2–403(1)]. 3. *Entrusting to a merchant*—Buyer or sublessee prevails [UCC 2–403(2), (3); 2A–305(2)].
Sale-or-Return Contracts (See page 289.)	When the buyer receives possession of the goods, title and risk of loss pass to the buyer, but the buyer has the option of returning the goods to the seller. If the buyer returns the goods to the seller, title and risk of loss pass back to the seller [UCC 2–327(2)].
Sale-on-Approval Contracts (See page 289.)	Title and risk of loss (from causes beyond the buyer's control) remain with the seller until the buyer approves (accepts) the offer [UCC 2–327(1)].
Risk of Loss When a Sales or Lease Contract Is Breached (See pages 289–290.)	1. If the seller or lessor breaches by tendering nonconforming goods that are rejected by the buyer or lessee, the risk of loss does not pass to the buyer or lessee until the defects are cured (unless the buyer or lessee accepts the goods in spite of their defects, thus waiving the right to reject) [UCC 2–510(1), 2A–220(1)]. 2. If the buyer or lessee breaches the contract, the risk of loss immediately shifts to the buyer or lessee. Limitations to this rule are as follows [UCC 2–510(3), 2A–220(2)]: a. The seller or lessor must already have identified the contract goods. b. The buyer or lessee bears the risk for only a commercially reasonable time after the seller or lessor has learned of the breach. c. The buyer or lessee is liable only to the extent of any deficiency in the seller's or lessor's insurance coverage.
Insurable Interest (See page 290.)	1. Buyers and lessees have an insurable interest in goods the moment the goods are identified to the contract by the seller or the lessor [UCC 2–501(1), 2A–218(1)].

CHAPTER SUMMARY Title and Risk of Loss—Continued

Insurable Interest—Continued	2. Sellers have an insurable interest in goods as long as they have (1) title to the goods or (2) a security interest in the goods [UCC 2–501(2)]. Lessors have an insurable interest in leased goods until the lessee exercises an option to buy and the risk of loss has passed to the lessee [UCC 2A–218(3)].
Bulk Transfers (See pages 290–291.)	1. In a bulk transfer of assets, in those states that have not repealed Article 6 of the UCC or replaced it with the revised Article 6, the buyer acquires title to the goods free of all claims of the seller's creditors if the following requirements are met: a. The transferor (seller) provides the transferee (buyer) with a sworn list of existing creditors, stating their names, business addresses, amounts due, and any disputed claims [UCC 6–104(1)(a)]. b. The buyer and seller prepare a schedule of the property to be transferred [UCC 6–104(1)(b)]. c. The buyer preserves the list of creditors and the schedule of property for six months, allowing any creditors of the seller to inspect it, or files the list and schedule of property in a designated public office [UCC 6–104(1)(c)]. d. The buyer gives notice of the proposed bulk transfer to each creditor of the seller at least ten days before the buyer takes possession of the goods or pays for them, whichever happens first [UCC 6–105]. 2. If these requirements are not met, goods in the possession of the buyer continue to be subject to the claims of the seller's unpaid creditors for six months.

FOR REVIEW

Answers for the even-numbered questions in this For Review *section can be found in Appendix A at the end of this text.*

1. What is the significance of identifying goods to a contract?
2. If the parties to a contract do not expressly agree on when title to goods passes, what determines when title passes?
3. Risk of loss does not necessarily pass with title. If the parties to a contract do not expressly agree when risk passes and the goods are to be delivered without movement by the seller, when does risk pass?
4. Under what circumstances will the seller's title to goods being sold be void? Under what circumstances will a seller have voidable title? What is the legal effect on a good faith purchaser of the goods when the seller has a void title versus a voidable title?
5. At what point does the buyer acquire an insurable interest in goods subject to a sales contract? Can both the buyer and the seller have an insurable interest in the goods simultaneously?

QUESTIONS AND CASE PROBLEMS

15–1. Sales by Nonowners. In the following situations, two parties lay claim to the same goods sold. Explain which party would prevail in each situation.

(a) Terry steals Dom's television set and sells it to Blake, an innocent purchaser, for value. Dom learns that Blake has the set and demands its return.

(b) Karlin takes her television set for repair to Orken, a merchant who sells new and used television sets. By accident, one of Orken's employees sells the set to Grady, an innocent purchaser-customer, who takes possession. Karlin wants her set back from Grady.

QUESTION WITH SAMPLE ANSWER

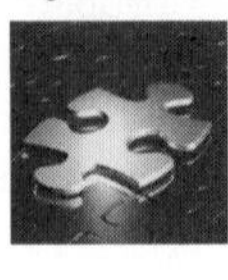

15–2. When will risk of loss pass from the seller to the buyer under each of the following contracts, assuming the parties have not expressly agreed on when risk of loss would pass?

(a) A New York seller contracts with a San Francisco buyer to ship goods to the buyer F.O.B. San Francisco.
(b) A New York seller contracts with a San Francisco buyer to ship goods to the buyer in San Francisco. There is no indication as to whether the shipment will be F.O.B. New York or F.O.B. San Francisco.
(c) A seller contracts with a buyer to sell goods located on the seller's premises. The buyer pays for the goods and arranges to pick them up the next week at the seller's place of business.
(d) A seller contracts with a buyer to sell goods located in a warehouse.

For a sample answer to this question, go to Appendix B at the end of this text.

15–3. Sales by Nonowners. Julian Makepeace, who had been declared mentally incompetent by a court, sold his diamond ring to Golding for value. Golding later sold the ring to Carmichael for value. Neither Golding nor Carmichael knew that Makepeace had been adjudged mentally incompetent by a court. Farrel, who had been appointed as Makepeace's guardian, subsequently learned that the diamond ring was in Carmichael's possession and demanded its return from Carmichael. Who has legal ownership of the ring? Why?

15–4. Sale on Approval. Chi Moy, a student, contracts to buy a television set from Ted's Electronics. Under the terms of the contract, Moy is to try out the set for thirty days, and if he likes it, he is to pay for the set at the end of the thirty-day period. If he does not want to purchase the set after thirty days, he can return the TV to Ted's Electronics with no obligation. Ten days after Moy takes the set home, it is stolen from his apartment, although he was not negligent in his care of the set in any way. Ted's Electronics claims that Moy must pay for the stolen set. Moy argues that the risk of loss falls on Ted's Electronics. Which party will prevail?

CASE PROBLEM WITH SAMPLE ANSWER

15–5. Phillip and Genevieve Carboy owned and operated Gold Hill Service Station in Fairbanks, Alaska. Gold Hill maintained underground storage tanks on its property to hold gasoline. When Gold Hill needed more fuel, Phillip placed an order with Petroleum Sales, Inc., which delivered the gasoline by filling the tanks. Gold Hill and Petroleum Sales were separately owned companies. Petroleum Sales did not oversee or operate Gold Hill and did not construct, install, or maintain the station's tanks, and Gold Hill did not tell Petroleum Sales' personnel how to fill the tanks. Parks Hiway Enterprises, LLC, owned the land next to Gold Hill. The Alaska Department of Environmental Conservation determined that benzene (a liquid, flammable hydrocarbon) had contaminated the groundwater under Parks Hiway's property and identified the gasoline in Gold Hill's tanks as the probable source. Gold Hill promptly removed the tanks, but because of the contamination, Parks Hiway stopped drawing drinking water from its well. Parks Hiway filed a suit in an Alaska state court against Petroleum Sales, among others. Should the court hold the defendant liable for the pollution? Who had title to the gasoline when it contaminated the water? Explain. [*Parks Hiway Enterprises, LLC v. CEM Leasing, Inc.*, 995 P.2d 657 (Alaska 2000)]

After you have answered this problem, compare your answer with the sample answer given on the Web site that accompanies this text. Go to **academic.cengage.com/blaw/wbl**, select "Chapter 15," and click on "Case Problem with Sample Answer."

15–6. Risk of Loss. H.S.A. II, Inc., made parts for motor vehicles. Under an agreement with Ford Motor Co., Ford provided steel to H.S.A. to make Ford parts. Ford's purchase orders for the parts contained the term "FOB Carrier Supplier's [Plant]." H.S.A. borrowed funds from GMAC Business Credit, L.L.C., under terms guaranteeing that, if funds were not otherwise available, payment would be made from H.S.A.'s inventory, raw materials, and finished goods. H.S.A. filed for bankruptcy on February 2, 2000, and ceased operations on June 20, when it had in its plant more than $1 million in finished goods for Ford. Ford sent six trucks to H.S.A. to pick up the goods. GMAC halted the removal. The parties asked the bankruptcy court to determine whose interest had priority. GMAC contended in part that Ford did not have an interest in the goods because there had not yet been a sale. Ford responded that under its purchase orders, title and risk of loss transferred on completion of the parts. In whose favor should the court rule, and why? [*In re H.S.A. II, Inc.*, 271 Bankr. 534 (E.D.Mich. 2002)]

15–7. Conditional Sales. Corvette Collection of Boston, Inc. (CCB), was a used Corvette dealership located (despite its name) in Pompano Beach, Florida. In addition to selling used Corvettes, CCB serviced Corvettes and sold Corvette parts. CCB owned some of its inventory and held the rest on consignment, although there were no signs indicating the consignments. In November 2001, CCB filed a petition for bankruptcy in a federal district court. At the time, CCB possessed six Corvettes that were consigned by Chester Finley and The Corvette Experience, Inc. (TCE). Robert Furr, on CCB's behalf, asked the court to declare that CCB held the goods under a contract for a sale or return. Finley and TCE asserted that the goods were held under a contract for a sale on approval. What difference does it make? Under what circumstances would the court rule in favor of Finley and TCE? How should the court rule under the facts as stated? Why? [*In re Corvette Collection of Boston, Inc.*, 294 Bankr. 409 (S.D.Fla. 2003)]

A QUESTION OF ETHICS

15–8. Toby and Rita Kahr accidentally included a small bag containing their sterling silver in a bag of used clothing that they donated to Goodwill Industries, Inc. The silverware, which was valued at over $3,500, had been given to them twenty-seven years earlier by Rita's father as a wedding present and had great senti-

mental value for them. The Kahrs realized what had happened shortly after Toby returned from Goodwill, but when Toby called Goodwill, he was told that the silver had immediately been sold to a customer, Karon Markland, for $15. Although Goodwill called Markland and asked her to return the silver, Markland refused to return it. The Kahrs then brought an action against Markland to regain the silver, claiming that Markland did not have good title to it. In view of these circumstances, discuss the following issues. [*Kahr v. Markland,* 187 Ill.App.3d 603, 543 N.E.2d 579, 135 Ill.Dec. 196 (1989)]

1. Did Karon Markland act wrongfully in any way by not returning the silver to Goodwill Industries when requested to do so? What would you have done in her position?
2. Goodwill argued that the entrustment rule should apply. Why would Goodwill want the rule to be applied? How might Goodwill justify its argument from an ethical point of view?

VIDEO QUESTION

15–9. Go to this text's Web site at academic.cengage.com/blaw/wbl and select "Chapter 15." Click on "Video Questions" and view the video titled *Risk of Loss.* Then answer the following questions.

1. Does Oscar have a right to refuse the shipment because the lettuce is wilted? Why or why not? What type of contract is involved in this video?
2. Does Oscar have a right to refuse the shipment because the lettuce was not organic butter crunch lettuce? Why or why not?
3. Assume that you are in Oscar's position—that is, you are buying produce for a supermarket. What different approaches might you take to avoid having to pay for a delivery of wilted produce?

Title and Risk of Loss

For updated links to resources available on the Web, as well as a variety of other materials, visit this text's Web site at

academic.cengage.com/blaw/wbl

For a brief historical review of commercial law, go to

http://www.answers.com/topic/commercial-law

For an overview of bills of lading, go to

http://www.law.cornell.edu/ucc/7/overview.html

ONLINE LEGAL RESEARCH

Go to the *Fundamentals of Business Law* home page at academic.cengage.com/blaw/wbl, select "Chapter 15," and click on "Internet Exercises." There you will find the following Internet research exercises that you can perform to learn more about topics covered in this chapter.

Activity 15–1: LEGAL PERSPECTIVE—The Entrustment Rule

Activity 15–2: MANAGEMENT PERSPECTIVE—Passage of Title

BEFORE THE TEST

Go to the *Fundamentals of Business Law* home page at academic.cengage.com/blaw/wbl, select "Chapter 15," and click on "Interactive Quizzes." You will find a number of interactive questions relating to this chapter.

CHAPTER 16

Performance and Breach of Sales and Lease Contracts

LEARNING OBJECTIVES

After reading this chapter, you should be able to answer the following questions:

1. What are the respective obligations of the parties under a contract for the sale or lease of goods?
2. What is the perfect tender rule? What are some important exceptions to this rule that apply to sales and lease contracts?
3. What options are available to the nonbreaching party when the other party to a sales or lease contract repudiates the contract prior to the time for performance?
4. What remedies are available to a seller or lessor when the buyer or lessee breaches the contract? What remedies are available to a buyer or lessee if the seller or lessor breaches the contract?
5. In contracts subject to the UCC, are parties free to limit the remedies available to the nonbreaching party on a breach of contract? If so, in what ways?

The performance that is required of the parties under a sales or lease contract consists of the duties and obligations each party has under the terms of the contract. Keep in mind that "duties and obligations" under the terms of the contract include those specified by the agreement, by custom, and by the Uniform Commercial Code (UCC). In this chapter, we examine the basic performance obligations of the parties under a sales or lease contract.

Sometimes, circumstances make it difficult for a person to carry out the promised performance, in which case the contract may be breached. When breach occurs, the aggrieved party looks for remedies—which we deal with in the second half of the chapter.

Performance Obligations

As discussed in previous chapters, the standards of *good faith* and *commercial reasonableness* are read into every contract. These standards provide a framework in which the parties can specify particulars of performance. Thus, when one party delays specifying particulars of performance for an unreasonable period of time or fails to cooperate with the other party, the innocent party is excused from any resulting delay in performance. The innocent party can proceed to perform in any reasonable manner, and the other party's failure to specify particulars or to cooperate can be treated as a breach of contract. Good faith is a question of fact for the jury. (For a discussion of the importance of good faith in the performance of contracts, see this chapter's *Management Perspective* feature.)

In the performance of a sales or lease contract, the basic obligation of the seller or lessor is to *transfer and deliver conforming goods.* The basic obligation of the buyer or lessee is to *accept and pay for conforming goods* in accordance with the contract [UCC 2–301, 2A–516(1)]. Overall performance of a sales or lease contract is controlled by the agreement between the parties. When the contract is unclear and disputes arise, the courts look to the UCC.

MANAGEMENT PERSPECTIVE
Good Faith and Fair Dealing

MANAGEMENT FACES A LEGAL ISSUE

As discussed elsewhere, all contracts governed by the Uniform Commercial Code (UCC) must meet the requirements of good faith and fair dealing. Yet do these requirements supersede the written terms of a contract? In other words, if a party adheres strictly to the express, written terms of a contract, can the party nonetheless face liability for breaching the UCC's good faith requirements?

WHAT THE COURTS SAY

Generally, the courts take the good faith provisions of the UCC very seriously. Indeed, in one case the court stated that "a party can violate the implied covenant of good faith and fair dealing without violating the express term of a contract." The issue of whether a party had breached an express provision of the contract was, in this court's view, separate and distinct from the issue of whether the implied covenant of good faith had been breached. In reaching its conclusion, the court emphasized the fact that one of the parties had superior bargaining power over the other and used this power to "prey" on the other party's lack of sophistication and financial strength.[a]

Other courts have held that good faith can be breached even when the parties have equal bargaining power. In one case, for example, the court held that although the plaintiffs were sophisticated businesspersons who had the assistance of highly competent counsel, they could still maintain an action for breach of good faith and fair dealing. The court reasoned that "the presence of bad faith is to be found in the eye of the beholder or, more to the point, in the eye of the trier of fact," indicating that it was up to a jury to determine whether the parties had performed in good faith.[b] In another case, the court held that a contract provision that gave a telephone company the right to reject potential cellular subscribers must be exercised in good faith and not selectively against the customers who were solicited by an electronics store.[c]

IMPLICATIONS FOR MANAGERS

The message for business owners and managers involved in sales contracts is clear: compliance with the literal terms of a contract is not enough—the standards of good faith and fair dealing must also be met. Although the specific standards of good faith performance are still evolving, the overriding principle is that the parties to a contract should do nothing to injure or destroy the rights of the other party to receive the fruits of the contract.

a. *Sons of Thunder, Inc. v. Borden, Inc.,* 148 N.J. 396, 690 A.2d 575 (1997).

b. *Seidenberg v. Summit Bank,* 348 N.J.Super 243, 791 A.2d 1068 (2002).

c. *Electronics Store, Inc. v. Cellco Partnership,* 127 Md.App. 385, 732 A.2d 980 (1999).

Obligations of the Seller or Lessor

The major obligation of the seller or lessor under a sales or lease contract is to tender conforming goods to the buyer or lessee.

TENDER OF DELIVERY

Tender of delivery requires that the seller or lessor have and hold *conforming goods* at the disposal of the buyer or lessee and give the buyer or lessee whatever notification is reasonably necessary to enable the buyer or lessee to take delivery [UCC 2–503(1), 2A–508(1)].

Tender must occur at a *reasonable hour* and in a *reasonable manner.* In other words, a seller cannot call the buyer at 2:00 A.M. and say, "The goods are ready. I'll give you twenty minutes to get them." Unless the parties have agreed otherwise, the goods must be tendered for delivery at a reasonable hour and kept available for a reasonable period of time to enable the buyer to take possession of them [UCC 2–503(1)(a)].

All goods called for by a contract must be tendered in a single delivery unless the parties agree otherwise or the circumstances are such that either party can rightfully request delivery in lots [UCC 2–307, 2–612, 2A–510]. Hence, an order for 1,000 shirts cannot be delivered 2 shirts at a time. If, however, the seller and the buyer contemplate that the shirts will be delivered in four orders of 250 each, as they are produced (for summer, fall, winter, and spring stock), and the price can be apportioned accordingly, it may be commercially reasonable to deliver the shirts in this way.

PLACE OF DELIVERY

The UCC provides for the place of delivery pursuant to a contract if the contract does not. Of course, the parties may agree on a particular destination, or their contract's terms or the circumstances may indicate the place of delivery.

Noncarrier Cases If the contract does not designate the place of delivery for the goods, and the buyer is expected to pick them up, the place of delivery is the *seller's place of business* or, if the seller has none, the seller's residence [UCC 2–308]. If the contract involves the sale of *identified goods,* and the parties know when they enter into the contract that these goods are located somewhere other than at the seller's place of business (such as at a warehouse), then the *location of the goods* is the place for their delivery [UCC 2–308].

■ EXAMPLE 16.1 Rogers and Aguirre live in San Francisco. In San Francisco, Rogers contracts to sell Aguirre five used trucks, which both parties know are located in a Chicago warehouse. If nothing more is specified in the contract, the place of delivery for the trucks is Chicago. ■ The seller may tender delivery either by giving the buyer a *negotiable or nonnegotiable document of title* or by obtaining the *bailee's (warehouser's) acknowledgment* that the buyer is entitled to possession.[1]

Carrier Cases In many instances, attendant circumstances or delivery terms in the contract make it apparent that the parties intend that a carrier be used to move the goods. In carrier cases, a seller can complete performance of the obligation to deliver the goods in two ways—through a shipment contract or through a destination contract.

—Shipment Contracts. Recall from Chapter 15 that a *shipment contract* requires or authorizes the seller to ship goods by a carrier. The contract does not require that the seller deliver the goods at a particular destination [UCC 2–319, 2–509].[2] Under a shipment contract, unless otherwise agreed, the seller must do the following:

1. Put the goods into the hands of the carrier.
2. Make a contract for their transportation that is reasonable according to the nature of the goods and their value. (For example, certain types of goods need refrigeration in transit.)
3. Obtain and promptly deliver or tender to the buyer any documents necessary to enable the buyer to obtain possession of the goods from the carrier.
4. Promptly notify the buyer that shipment has been made [UCC 2–504].

1. If the seller delivers a nonnegotiable document of title or merely writes instructions to the bailee to release the goods to the buyer without the bailee's *acknowledgment* of the buyer's rights, this is also a sufficient tender, unless the buyer objects [UCC 2–503(4)]. Risk of loss, however, does not pass until the buyer has a reasonable amount of time in which to present the document or to give the bailee instructions for delivery.

2. As mentioned in Chapter 15, UCC 2–319 was omitted from the 2003 amendments to Article 2.

If the seller fails to notify the buyer that shipment has been made or fails to make a proper contract for transportation, the buyer can treat the contract as breached and reject the goods, but only if a *material loss* of the goods or a significant *delay* results. Of course, the parties can agree that a lesser amount of loss or any delay will be grounds for rejection.

—Destination Contracts. Under a *destination contract,* the seller agrees to see that conforming goods will be duly tendered to the buyer at a particular destination. The goods must be tendered at a reasonable hour and held at the buyer's disposal for a reasonable length of time. The seller must also give the buyer any appropriate notice that is necessary to enable the buyer to take delivery. In addition, the seller must provide the buyer with any documents of title necessary to enable the buyer to obtain delivery from the carrier [UCC 2–503].

The Perfect Tender Rule

As previously noted, the seller or lessor has an obligation to ship or tender *conforming goods,* and the buyer or lessee is required to accept and pay for the goods according to the terms of the contract. Under the common law, the seller was obligated to deliver goods in conformity with the terms of the contract in every detail. This was called the *perfect tender* doctrine. The UCC preserves the perfect tender doctrine by stating that if goods or tender of delivery fail *in any respect* to conform to the contract, the buyer or lessee has the right to accept the goods, reject the entire shipment, or accept part and reject part [UCC 2–601, 2A–509].

■ EXAMPLE 16.2 A lessor contracts to lease fifty Vericlear monitors to be delivered at the lessee's place of business on or before October 1. On September 28, the lessor discovers that it has only thirty Vericlear monitors in inventory, but that it will have another forty Vericlear monitors within the next two weeks. The lessor tenders delivery of the thirty Vericlear monitors on October 1, with the promise that the other monitors will be delivered within two weeks. Because the lessor failed to make a perfect tender of fifty Vericlear monitors, the lessee has the right to reject the entire shipment and hold the lessor in breach. ■

Exceptions to Perfect Tender

Because of the rigidity of the perfect tender rule, several exceptions to the rule have been created, some of which are discussed here.

Agreement of the Parties Exceptions to the perfect tender rule may be established by agreement. If the parties

have agreed, for example, that defective goods or parts will not be rejected if the seller or lessor is able to repair or replace them within a reasonable period of time, the perfect tender rule does not apply.

Cure The UCC does not specifically define the term *cure,* but it refers to the right of the seller or lessor to repair, adjust, or replace defective or nonconforming goods [UCC 2–508, 2A–513]. When any tender of delivery is rejected because of nonconforming goods and the time for performance has not yet expired, the seller or lessor can notify the buyer or lessee promptly of the intention to cure and can then do so *within the contract time for performance* [UCC 2–508(1), 2A–513(1)]. Once the time for performance has expired, the seller or lessor can still, for a reasonable time, exercise the right to cure with respect to the rejected goods if he or she had, at the time of delivery, *reasonable grounds to believe that the nonconforming tender would be acceptable to the buyer or lessee* [UCC 2–508(2), 2A–513(2)].[3]

Sometimes, a seller or lessor will tender nonconforming goods with some type of price allowance. The allowance serves as the "reasonable grounds" for the seller or lessor to believe that the nonconforming tender will be acceptable to the buyer or lessee. A seller or lessor might also have other reasons for assuming that a buyer or lessee will accept a nonconforming tender. ■ EXAMPLE 16.3 Suppose that in the past the buyer, an office supply store, frequently accepted blue pens when the seller did not have black pens in stock. In this context, the seller has reasonable grounds to believe the store will again accept such a substitute. If the store rejects the substituted goods (blue pens) on a particular occasion, the seller nonetheless had reasonable grounds to believe that the blue pens would be acceptable. Therefore, the seller can cure within a reasonable time, even though the delivery of black pens will occur after the time limit for performance allowed under the contract. ■

The right to cure means that, to reject goods, the buyer or lessee must give notice to the seller or lessor of a particular defect. For example, if a lessee refuses a tender of goods as nonconforming but does not disclose the nature of the defect to the lessor, the lessee cannot later assert the defect as a defense if the defect is one that the lessor could have cured. Generally, buyers and lessees must act in good faith and state specific reasons for refusing to accept goods [UCC 2–605, 2A–514].[4]

Substitution of Carriers When an agreed-on manner of delivery (such as which carrier will be used to transport the goods) becomes impracticable or unavailable through no fault of either party, but a commercially reasonable substitute is available, the seller must use this substitute performance, which is sufficient tender to the buyer [UCC 2–614(1)]. ■ EXAMPLE 16.4 A sales contract calls for the delivery of a large generator to be shipped by Roadway Trucking Corporation on or before June 1. The contract terms clearly state the importance of the delivery date. The employees of Roadway Trucking go on strike. The seller is required to make a reasonable substitute tender, perhaps by rail if that is available. Note that the seller will normally be responsible for any additional shipping costs, unless other arrangements have been made in the sales contract. ■

Installment Contracts An **installment contract** is a single contract that requires or authorizes delivery in two or more separate lots to be accepted and paid for separately. In an installment contract, a buyer or lessee can reject an installment *only if the nonconformity substantially impairs the value* of the installment and cannot be cured [UCC 2–612(2), 2–307, 2A–510(1)].[5]

Unless the contract provides otherwise, the entire installment contract is breached only when one or more nonconforming installments *substantially* impair the value of the *whole contract.* If the buyer or lessee subsequently accepts a nonconforming installment and fails to notify the seller or lessor of cancellation, however, the contract is reinstated [UCC 2–612(3), 2A–510(2)].

A major issue to be determined is what constitutes substantial impairment of the "value of the whole contract." ■ EXAMPLE 16.5 Consider an installment contract for the sale of twenty carloads of plywood. The first carload

---

3. The 2003 amendments to UCC Articles 2 and 2A expressly exempt "consumer contracts" and "consumer leases" from these provisions [Amended UCC 2–508, 2A–508]. In other words, cure is not available as a matter of right after a justifiable revocation of acceptance under a consumer contract or lease. Also, the new provisions abandon the "reasonable grounds to believe" test, thus expanding the seller's right to cure after the time for performance has expired. Although this test has been abandoned, the requirement that the initial tender be made in good faith prevents a seller from deliberately tendering goods that the seller knows the buyer cannot use.

4. The 2003 amendments to UCC 2–605 and 2A–514 change this restriction in three ways. First, a buyer's or lessee's failure to disclose the nature of the defect affects only the right to reject or revoke acceptance, not the right to establish a breach of contract. Second, the new sections expressly require that the seller or lessor must have had a right to cure, as well as the ability to cure. Finally, these sections extend to include not only rejection but also revocation of acceptance.

5. The 2003 amendments to Articles 2 and 2A make it clear that the buyer's or lessee's right to reject an installment depends on whether the value of the installment to the buyer or lessee is substantially impaired and not on whether the seller or lessor is able to cure [Amended UCC 2–612(2), 2A–510(2)].

does not conform to the contract because 9 percent of the plywood deviates from the thickness specifications. The buyer cancels the contract, and immediately thereafter the second and third carloads of conforming plywood arrive at the buyer's place of business. If a lawsuit ensues, the court will have to grapple with the question of whether the 9 percent of nonconforming plywood *substantially* impaired the value of the whole. ■

The point to remember is that the UCC significantly alters the right of the buyer or lessee to reject the entire contract if the contract requires delivery to be made in several installments. The UCC strictly limits rejection to instances of *substantial* nonconformity.

Commercial Impracticability As mentioned in Chapter 11, occurrences that were not foreseeable by either party when a contract was made may make performance commercially impracticable. When this occurs, the rule of perfect tender no longer holds. According to UCC 2–615(a) and 2A–405(a), delay in delivery or nondelivery in whole or in part is not a breach when performance has been made impracticable "by the occurrence of a contingency the nonoccurrence of which was a basic assumption on which the contract was made." The seller or lessor must, however, notify the buyer or lessee as soon as practicable that there will be a delay or nondelivery.

—Foreseeable versus Unforeseeable Contingencies. An increase in cost resulting from inflation does not in itself excuse performance, as a seller or lessor ordinarily assumes this kind of risk when conducting business. The unforeseen contingency must be one that would have been impossible to contemplate in a given business situation. ■ EXAMPLE 16.6 A major oil company that receives its supplies from the Middle East has a contract to supply a buyer with 100,000 gallons of oil. Because of an oil embargo by the Organization of Petroleum Exporting Countries (OPEC), the seller is prevented from securing oil supplies to meet the terms of the contract. Because of the same embargo, the seller cannot secure oil from any other source. This situation comes fully under the commercial impracticability exception to the perfect tender doctrine. ■

Can unanticipated increases in a seller's costs, which make performance "impracticable," constitute a valid defense to performance on the basis of commercial impracticability? The court dealt with this question in the following case.

LANDMARK AND CLASSIC CASES

CASE 16.1 Maple Farms, Inc. v. City School District of Elmira

Supreme Court of New York, 1974.
76 Misc.2d 1080,
352 N.Y.S.2d 784.

FACTS On June 15, 1973, Maple Farms, Inc., formed an agreement with the city school district of Elmira, New York, to supply the school district with milk for the 1973–1974 school year. The agreement was in the form of a requirements contract, under which Maple Farms would sell to the school district all the milk the district required at a fixed price–which was the June market price of milk. By December 1973, the price of raw milk had increased by 23 percent over the price specified in the contract. This meant that if the terms of the contract were fulfilled, Maple Farms would lose $7,350. Because it had similar contracts with other school districts, Maple Farms stood to lose a great deal if it was held to the price stated in the contracts. When the school district would not agree to release Maple Farms from its contract, Maple Farms brought an action in a New York state court for a declaratory judgment (a determination of the parties' rights under a contract). Maple Farms contended that the substantial increase in the price of raw milk was an event not contemplated by the parties when the contract was formed and that, given the increased price, performance of the contract was commercially impracticable.

ISSUE Can Maple Farms be released from the contract on the ground of commercial impracticability?

DECISION No. The court ruled that performance in this case was not impracticable.

REASON The court reasoned that commercial impracticability arises when an event occurs that is totally unexpected and unforeseeable by the parties. The increased price of raw milk was not totally unexpected, given that in the previous year the price of milk had risen 10 percent and that the price of milk had traditionally varied. Additionally, the general inflation of prices in the United States should have been anticipated. Maple Farms had reason to know these facts and could have included a clause in its contract with the school district to protect itself from its present situation. The court also noted that the primary purpose of the contract, on the

CASE 16.1–CONTINUED

part of the school district, was to protect itself (for budgeting purposes) against price fluctuations.

IMPACT OF THIS CASE ON TODAY'S LAW This case is a classic illustration of the UCC's commercial impracticability doctrine. Under this doctrine, parties who freely enter into contracts normally will not be excused from their contractual obligations simply because changed circumstances make performance difficult or very costly. Rather, to be excused from performance, a party must show that the changed circumstances were impossible to foresee at the time the contract was formed. This principle continues to be applied today.

RELEVANT WEB SITES *To locate information on the Web concerning* Maple Farms, Inc. v. City School District of Elmira, *go to this text's Web site at* **academic.cengage.com/blaw/wbl**, *select "Chapter 16," and click on "URLs for Landmark Cases."*

—Partial Performance. Sometimes, an unforeseen event only *partially* affects the capacity of the seller or lessor to perform, and the seller or lessor is thus able to fulfill the contract *partially* but cannot tender total performance. In this event, the seller or lessor is required to allocate in a fair and reasonable manner any remaining production and deliveries among those to whom it is contractually obligated to deliver the goods, and this allocation may take into account its regular customers [UCC 2–615(b), 2A–405(b)]. The buyer or lessee must receive notice of the allocation and has the right to accept or reject the allocation [UCC 2–615(c), 2A–405(c)].

■ EXAMPLE 16.7 A Florida orange grower, Best Citrus, Inc., contracts to sell this season's crop to a number of customers, including Martin's grocery chain. Martin's contracts to purchase two thousand crates of oranges. Best Citrus has sprayed some of its orange groves with a chemical called Karmoxin. The U.S. Department of Agriculture discovers that persons who eat products sprayed with Karmoxin may develop cancer. The department issues an order prohibiting the sale of these products. Best Citrus picks all of the oranges not sprayed with Karmoxin, but the quantity does not fully meet all the contracted-for deliveries. In this situation, Best Citrus is required to allocate its production, and it notifies Martin's that it cannot deliver the full quantity agreed on in the contract and specifies the amount it will be able to deliver under the circumstances. Martin's can either accept or reject the allocation, but Best Citrus has no further contractual liability. ■

Destruction of Identified Goods The UCC provides that when an unexpected event, such as a fire, totally destroys *goods identified at the time the contract is formed* through no fault of either party and *before risk passes to the buyer or lessee,* the parties are excused from performance [UCC 2–613, 2A–221]. If the goods are only partially destroyed, however, the buyer or lessee can inspect them and either treat the contract as void or accept the goods with a reduction of the contract price.

■ EXAMPLE 16.8 Atlas Sporting Equipment agrees to lease to River Bicycles sixty bicycles of a particular model that has been discontinued. No other bicycles of that model are available. River specifies that it needs the bicycles to rent to tourists. Before Atlas can deliver the bikes, they are destroyed by a fire. In this situation, Atlas is not liable to River for failing to deliver the bikes. The goods were destroyed through no fault of either party, before the risk of loss passed to the lessee. The loss was total, so the contract is avoided. Clearly, Atlas has no obligation to tender the bicycles, and River has no obligation to pay for them. ■

Assurance and Cooperation Two other exceptions to the perfect tender doctrine apply equally to parties to sales and lease contracts: the *right of assurance* and the *duty of cooperation.*

—The Right of Assurance. The UCC provides that if one party to a contract has "reasonable grounds" to believe that the other party will not perform as contracted, he or she may *in writing* "demand adequate assurance of due performance" from the other party. Until such assurance is received, he or she may "suspend" further performance (such as payments due under the contract) without liability. What constitutes "reasonable grounds" is determined by commercial standards. If such assurances are not forthcoming within a reasonable time (not to exceed thirty days), the failure to respond may be treated as a *repudiation* of the contract [UCC 2–609, 2A–401]. The following case illustrates this principle.

CASE 16.2 Koch Materials Co. v. Shore Slurry Seal, Inc.

United States District Court,
District of New Jersey, 2002.
205 F.Supp.2d 324.
http://lawlibrary.rutgers.edu/fed/search.html[a]

FACTS Koch Materials Company is a manufacturer of road surfacing materials. In February 1998, Koch agreed to pay $5 million, payable in three installments, to Shore Slurry Seal, Inc., for an asphalt plant in New Jersey and the rights to license a specialty road surfacing substance known as Novachip. Shore also agreed that for seven years following the sale, it would buy all of its asphalt requirements from Koch, or at least 2 million gallons of asphalt per year (the Exclusive Supply Agreement). Shore promised to use at least 2.5 million square yards of Novachip annually and to pay royalties to Koch accordingly (the Sublicense Agreement). Midway through the term of the contract, Shore told Koch that it planned to sell its assets to Asphalt Paving Systems, Inc. Koch sought assurances that Asphalt Paving would continue the original deal. Shore refused to provide any more information. Koch filed a suit in a federal district court against Shore, asking in part for the right to treat Shore's failure to give assurances as a repudiation of their contract. Koch filed a motion for summary judgment on this issue.

ISSUE Did Shore in effect repudiate the contract?

DECISION Yes. The court issued a summary judgment in Koch's favor. The court concluded that Shore's failure to provide assurances to Koch constituted a repudiation of its contract, authorizing Koch to terminate the contract and seek damages.

REASON The court concluded that Koch had a commercially reasonable basis for demanding assurances. Shore planned to sell all of its assets, but retain the licensing agreement. The court stated that "any reasonable person would wonder how Shore planned to sell anything with no telephones, no computers, and no office furniture." Also, as for leasing these items, a party might ask, "Would Shore have had the financial capacity to obtain leases and hire a sales staff?" As for the requirements supply contract, the court pointed out that "[s]tart-up construction businesses * * * begin unbonded, unable to win any bid for their first year, and unable to secure sufficient bonding for large construction bids for several years.[b] Koch had no way of knowing whether Asphalt was already a going business, and, if not, whether it would be able to win sufficient subcontracting bids even to meet the minimum requirements, let alone approach the potential upside of an established enterprise like Shore."

WHY IS THIS CASE IMPORTANT? *This case clearly illustrates how the UCC comes to the aid of a party who has reasonable grounds to suspect that the other party to a contract will not perform as promised. Rather than having to "wait and see" (and possibly incur significant losses as a result), the party with such suspicions may seek adequate assurance of performance from the other party. The failure to give such assurance can be treated as an anticipatory repudiation (breach) of the contract, thus entitling the nonbreaching party to seek damages.*

a. In the "Locate a decision by citation or docket number" section, select "Civil Case," type "01-2059" in the "Enter Docket Number" box, and click on "Submit Form." From the results, click on "ca01-2059-1.html" to access the opinion. Rutgers University School of Law in Camden, New Jersey, maintains this Web site.

b. In this context, a *bond* is a guaranty to complete or to pay the cost of a construction contract if the contractor defaults.

—The Duty of Cooperation. Sometimes, the performance of one party depends on the cooperation of the other. The UCC provides that when such cooperation is not forthcoming, the other party can suspend her or his own performance without liability and hold the uncooperative party in breach or proceed to perform the contract in any reasonable manner [UCC 2–311(3)(b)].

Obligations of the Buyer or Lessee

Once the seller or lessor has adequately tendered delivery, the buyer or lessee is obligated to accept the goods and pay for them according to the terms of the contract. In the absence of any specific agreements, the buyer or lessee must make payment at the time and place the goods are received [UCC 2–310(a), 2A–516(1)].

PAYMENT

When a sale is made on credit, the buyer is obliged to pay according to the specified credit terms (for example, 60, 90, or 120 days), not when the goods are received. The credit period usually begins on the *date of shipment* [UCC 2–310(d)]. Under a lease contract, a lessee must pay the lease payment that was specified in the contract [UCC 2A–516(1)].

Payment can be made by any means agreed on by the parties—cash or any other method generally acceptable in the commercial world. If the seller demands cash when the buyer offers a check, credit card, or the like, the seller must permit the buyer reasonable time to obtain legal tender [UCC 2–511].

RIGHT OF INSPECTION

Unless otherwise agreed, or for C.O.D. (collect on delivery) transactions, the buyer or lessee has an absolute right to inspect the goods. This right allows the buyer or lessee to verify, before making payment, that the goods tendered or delivered are what were contracted for or ordered. If the goods are not what were ordered, the buyer or lessee has no duty to pay. *An opportunity for inspection is therefore a condition precedent to the right of the seller or lessor to enforce payment* [UCC 2–513(1), 2A–515(1)].

Unless otherwise agreed, inspection can take place at any reasonable place and time and in any reasonable manner. Generally, what is reasonable is determined by custom of the trade, past practices of the parties, and the like. Costs of inspecting conforming goods are borne by the buyer unless otherwise agreed [UCC 2–513(2)].

C.O.D. Shipments If a seller ships goods to a buyer C.O.D. (or under similar terms) and the buyer has not agreed to a C.O.D. shipment in the contract, the buyer can rightfully *reject* the goods. This is because a C.O.D. shipment cannot be inspected before payment, which is a denial of the buyer's right of inspection. When the buyer has agreed to a C.O.D. shipment in the contract, however, or has agreed to pay for the goods on the presentation of a bill of lading, no right of inspection exists because it was negated by the agreement [UCC 2–513(3)].[6]

Payment Due—Documents of Title Under certain contracts, payment is due on the receipt of the required documents of title even though the goods themselves may not have arrived at their destination. With C.I.F. and C.&F. contracts (see Exhibit 15–1 on page 287 in Chapter 15), payment is required on receipt of the documents unless the parties have agreed otherwise. Thus, payment may be required prior to inspection and must be made unless the buyer knows that the goods are nonconforming [UCC 2–310(b), 2–513(3)].

ACCEPTANCE

A buyer or lessee can manifest assent to, or accept, the delivered goods in the following ways, each of which constitutes *acceptance:*

1. The buyer or lessee can expressly accept the shipment by words or conduct. For example, there is an acceptance if the buyer or lessee, after having had a reasonable opportunity to inspect the goods, signifies agreement to the seller or lessor that the goods are either conforming or are acceptable despite their nonconformity [UCC 2–606(1)(a), 2A–515(1)(a)].
2. Acceptance is presumed if the buyer or lessee has had a reasonable opportunity to inspect the goods and has failed to reject them within a reasonable period of time [UCC 2–602(1), 2–606(1)(b), 2A–515(1)(b)].

Additionally, in sales contracts, the buyer will be deemed to have accepted the goods if he or she performs any act that would indicate that the seller no longer owns the goods. For example, any use or resale of the goods generally constitutes an acceptance. Limited use for the sole purpose of testing or inspecting the goods is not an acceptance, however [UCC 2–606(1)(c)].

If some of the goods delivered do not conform to the contract and the seller or lessor has failed to cure, the buyer or lessee can make a *partial* acceptance [UCC 2–601(c), 2A–509(1)]. The same is true if the nonconformity was not reasonably discoverable before acceptance. (In the latter situation, the buyer or lessee may be able to revoke the acceptance, as will be discussed later in this chapter.) A buyer or lessee cannot accept less than a single commercial unit, however. The UCC defines a *commercial unit* as a unit of goods that, by commercial usage, is viewed as a "single whole" for purposes of sale, division of which would materially impair the character of the unit, its market value, or its use [UCC 2–105(6), 2A–103(1)(c)]. A commercial unit can be a single article (such as a machine), a set of articles (such as a suite of furniture or an assortment of sizes), a quantity (such as a bale, a gross, or a carload), or any other unit treated in the trade as a single whole.

Anticipatory Repudiation

What if, before the time for contract performance, one party clearly communicates to the other the intention not to perform? As discussed in Chapter 11, such an action is a breach of the contract by anticipatory repudiation.[7]

6. References to C.O.D. and similar terms that represent commercial shorthand (such as F.O.B., C.&F., and other terms discussed in Chapter 15) have been deleted in the 2003 amendments to UCC Article 2. See, for example, Amended UCC 2–513(3)(a).

7. This doctrine was first enunciated in an English case decided in 1853, *Hochster v. De La Tour,* 2 Ellis and Blackburn Reports 678 (1853).

When anticipatory repudiation occurs, the nonbreaching party has a choice of two responses: (1) treat the repudiation as a final breach by pursuing a remedy or (2) wait to see if the repudiating party will decide to honor the contract despite the avowed intention to renege [UCC 2–610, 2A–402]. In either situation, the nonbreaching party may suspend performance.

Should the latter course be pursued, the UCC permits the breaching party (subject to some limitations) to "retract" her or his repudiation. This can be done by any method that clearly indicates an intent to perform. Once retraction is made, the rights of the repudiating party under the contract are reinstated [UCC 2–611, 2A–403].

Remedies of the Seller or Lessor

When the buyer or lessee is in breach, the seller or lessor has numerous remedies available under the UCC. Generally, the remedies available to the seller or lessor depend on the circumstances at the time of the breach, such as which party has possession of the goods, whether the goods are in transit, and whether the buyer or lessee has rejected or accepted the goods.

When the Goods Are in the Possession of the Seller or Lessor

Under the UCC, if the buyer or lessee breaches the contract before the goods have been delivered to her or him, the seller or lessor has the right to pursue the remedies discussed here.

The Right to Cancel the Contract One of the options available to a seller or lessor when the buyer or lessee breaches the contract is simply to cancel (rescind) the contract [UCC 2–703(f), 2A–523(1)(a)]. The seller or lessor must notify the buyer or lessee of the cancellation, and at that point all remaining obligations of the seller or lessor are discharged. The buyer or lessee is not discharged from all remaining obligations, however; he or she is in breach, and the seller or lessor can pursue remedies available under the UCC for breach.

The Right to Withhold Delivery In general, sellers and lessors can withhold or discontinue performance of their obligations under sales or lease contracts when the buyers or lessees are in breach. If a buyer or lessee has wrongfully rejected or revoked acceptance of contract goods (rejection and revocation of acceptance will be discussed later), failed to make proper and timely payment, or repudiated a part of the contract, the seller or lessor can withhold delivery of the goods in question [UCC 2–703(a), 2A–523(1)(c)]. If the breach results from the buyer's or the lessee's insolvency (inability to pay debts as they become due), the seller or lessor can refuse to deliver the goods unless the buyer or lessee pays in cash [UCC 2–702(1), 2A–525(1)].

The Right to Resell or Dispose of the Goods When a buyer or lessee breaches or repudiates a sales or lease contract while the seller or lessor is still in possession of the goods, the seller or lessor can resell or dispose of the goods. The seller or lessor can retain any profits made as a result of the sale and can hold the buyer or lessee liable for any loss [UCC 2–703(d), 2–706(1), 2A–523(1)(e), 2A–527(1)].[8]

When the goods contracted for are unfinished at the time of the breach, the seller or lessor can do one of two things: (1) cease manufacturing the goods and resell them for scrap or salvage value or (2) complete the manufacture and resell or dispose of them, holding the buyer or lessee liable for any deficiency. In choosing between these two alternatives, the seller or lessor must exercise reasonable commercial judgment to mitigate the loss and obtain maximum value from the unfinished goods [UCC 2–704(2), 2A–524(2)]. Any resale of the goods must be made in good faith and in a commercially reasonable manner.

In sales transactions, the seller can recover any deficiency between the resale price and the contract price, along with **incidental damages,** defined as those costs to the seller resulting from the breach [UCC 2–706(1), 2–710]. The resale can be private or public, and the goods can be sold as a unit or in parcels. The seller must give the original buyer reasonable notice of the resale, unless the goods are perishable or will rapidly decline in value [UCC 2–706(2), (3)]. A good faith purchaser in a resale takes the goods free of any of the rights of the original buyer, even if the seller fails to comply with these requirements of the UCC [UCC 2–706(5)].

In lease transactions, the lessor may lease the goods to another party and recover from the original lessee, as damages, any unpaid lease payments up to the beginning

8. Under the 2003 amendments to UCC Articles 2 and 2A, this loss includes consequential damages, except that a seller or lessor cannot recover consequential damages from a consumer under a consumer contract or lease [Amended UCC 2–706(1), 2–710, 2A–527(2), 2A–530]. Consequential damages may also be recovered, except from a consumer under a consumer contract or lease, when a seller or lessor has a right to recover the purchase price or lease payments due or to recover other damages [Amended UCC 2–708(1), 2–709(1), 2–710, 2A–528(1), 2A–529(1), 2A–530]. Subtracted from these amounts, of course, would be any expenses saved as a consequence of the buyer's or lessee's breach.

date of the lease term under the new lease. The lessor can also recover any deficiency between the lease payments due under the original lease contract and under the new lease contract, along with incidental damages [UCC 2A–527(2)].

The Right to Recover the Purchase Price or the Lease Payments Due Under the UCC, an unpaid seller or lessor can bring an action to recover the purchase price or payments due under the lease contract, plus incidental damages, if the seller or lessor is unable to resell or dispose of the goods [UCC 2–709(1), 2A–529(1)].

■ EXAMPLE 16.9 Suppose that Southern Realty contracts with Gem Point, Inc., to purchase one thousand pens with Southern Realty's name inscribed on them. Gem Point tenders delivery of the one thousand pens, but Southern Realty wrongfully refuses to accept them. In this situation, Gem Point has, as a proper remedy, an action for the purchase price. Gem Point tendered delivery of conforming goods, and Southern Realty, by failing to accept the goods, is in breach. Gem Point obviously cannot sell to anyone else the pens inscribed with the buyer's business name, so this situation falls under UCC 2–709. ■

If a seller or lessor is unable to resell or dispose of goods and sues for the contract price or lease payments due, the goods must be held for the buyer or lessee. The seller or lessor can resell or dispose of the goods at any time prior to collection (of the judgment) from the buyer or lessee but must credit the net proceeds from the sale to the buyer or lessee. This is an example of the duty to mitigate damages.

The Right to Recover Damages If a buyer or lessee repudiates a contract or wrongfully refuses to accept the goods, a seller or lessor can maintain an action to recover the damages that were sustained. Ordinarily, the amount of damages equals the difference between the contract price or lease payments and the market price or lease payments (at the time and place of tender of the goods), plus incidental damages [UCC 2–708(1), 2A–528(1)]. The time and place of tender are frequently given by such terms as F.O.B., F.A.S., and C.I.F., which determine whether there is a shipment or destination contract.[9]

In the following case, the court had to determine the proper measure of damages after a buyer breached a sales contract.

9. See Exhibit 15–1 on page 287 for a definition of these contract terms, which are omitted from the 2003 amendments to Article 2 [Amended UCC 2–319 through 2–324].

CASE 16.3 Utica Alloys, Inc. v. Alcoa, Inc.

United States District Court,
Northern District of New York, 2004.
303 F.Supp.2d 247.

FACTS Alcoa, Inc., through its business, generates scrap metal. Utica Alloys, Inc., buys and processes this type of scrap and sells it to its only user, General Electric Company (GE), which uses it in land-based power turbines. In July 2001, Utica agreed to buy all of Alcoa's scrap through August 2003. Their contract indexed the monthly price of the scrap to the monthly market price of nickel, but contemplated that the parties would review this price semiannually. In November 2001, GE reduced its production of turbines, which lowered the market value of the scrap. This change was not reflected in Alcoa's arrangement with Utica, however, because the price in their contract was based on the market value of nickel. In January 2002, the opportunity arose to review the price of the scrap, and the parties began to negotiate while they continued to ship and process the scrap. In May, when the parties were unable to agree on a price, Alcoa stated that the contract was over, retrieved the scrap processed after January, and sold it to another party. Utica filed a suit in a federal district court against Alcoa, alleging in part unjust enrichment. Alcoa counterclaimed for breach of contract. The court entered a judgment in Alcoa's favor, holding that Utica breached the agreement by failing to pay for the scrap received after January. Alcoa asked for damages based on the difference between the contract price for the *unprocessed* scrap and the price at which the *processed* scrap sold after it was retrieved.

ISSUE Was Alcoa entitled to the amount of damages that it requested?

DECISION No. The court held that the proper measure of Alcoa's damages was the difference between the contract's monthly price for the *unprocessed* scrap and the monthly fair market value of *unprocessed* scrap.

REASON The court reasoned that the amount Alcoa sought "would serve as a double penalty to Utica Alloys, Inc., for processing the scrap." The court explained that without the contract with Utica, Alcoa would have sold the unprocessed scrap at the market price for unprocessed scrap. In other

(Continued)

CASE 16.3—CONTINUED

words, whether the scrap was processed after its sale to Utica was not part of the parties' contract. The court acknowledged that processing the scrap actually decreased its value in this case because of the reduction in demand for processed scrap. But the court pointed out that when Alcoa chose to end the contract in May, it accepted the processed scrap from Utica without paying for the cost of the processing. Alcoa "cannot be permitted in one breath to denounce processing as irrelevant to the contractual relationship, while in another embrace the market change of processed scrap as the yardstick for measuring its damages under the contract."

FOR CRITICAL ANALYSIS—Economic Consideration *How, specifically, should the amount of damages in this case be determined, considering that the contract called for monthly shipments and prices?*

When the Goods Are in Transit

If the seller or lessor has delivered the goods to a carrier or a bailee but the buyer or lessee has not as yet received them, the goods are said to be in transit. If, while the goods are in transit, the seller or lessor learns that the buyer or lessee is insolvent, the seller or lessor can stop the carrier or bailee from delivering the goods, regardless of the quantity of goods shipped.

■ **EXAMPLE 16.10** Suppose that Arturo Ortega orders a truckload of lumber from Timber Products, Inc., to be shipped to Ortega six weeks later. Ortega, who owes Timber Products for a past shipment, promises to pay the debt immediately and to pay for the current shipment as soon as it is received. After the lumber has been shipped, Timber Products is notified by a bankruptcy court judge that Ortega has filed a petition in bankruptcy and listed Timber Products as one of his creditors (see Chapter 21). If the goods are still in transit, Timber Products can stop the carrier from delivering the lumber to Ortega. ■ If the buyer or lessee is in breach but is not insolvent, the seller or lessor can stop the goods in transit only if the quantity shipped is at least a carload, a truckload, a planeload, or a larger shipment [UCC 2–705(1), 2A–526(1)].[10]

Requirements for Stopping Delivery To stop delivery, the seller or lessor must *timely notify* the carrier or other bailee that the goods are to be returned or held for the seller or lessor. If the carrier has sufficient time to stop delivery, it must hold and deliver the goods according to the instructions of the seller or lessor, who is liable to the carrier for any additional costs incurred [UCC 2–705(3), 2A–526(3)].

Exceptions UCC 2–705(2) and 2A–526(2) provide that the seller or lessor loses the right to stop delivery of goods in transit when any of the following events occur:

1. The buyer or lessee obtains possession of the goods.
2. The carrier acknowledges the rights of the buyer or lessee by reshipping or storing the goods for the buyer or lessee.
3. A bailee of the goods other than a carrier acknowledges that he or she is holding the goods for the buyer or lessee.

Additionally, in sales transactions, the seller loses the right to stop delivery of goods in transit when a negotiable document of title covering the goods has been negotiated (properly transferred, giving the buyer ownership rights in the goods) to the buyer [UCC 2–705(2)].

Once the seller or lessor reclaims the goods in transit, she or he can pursue the remedies allowed to sellers and lessors when the goods are in their possession. In other words, the seller or lessor who has reclaimed goods may do the following:

1. Cancel (rescind) the contract.
2. Resell the goods and recover any deficiency.
3. Sue for any deficiency between the contract price (or lease payments due) and the market price (or market lease payments), plus incidental damages.
4. Sue to recover the purchase price or lease payments due if the goods cannot be resold, plus incidental damages.
5. Sue to recover damages.

When the Goods Are in the Possession of the Buyer or Lessee

When the buyer or lessee breaches a sales or lease contract and the goods are in the buyer's or lessee's possession, the UCC gives the seller or lessor the following limited remedies.

The Right to Recover the Purchase Price or Payments Due under the Lease Contract If the buyer or lessee has accepted the goods but refuses to pay for them, the seller or lessor can sue for the purchase price of the goods or the lease payments due, plus incidental damages [UCC 2–709(1), 2A–529(1)].

10. The 2003 amendments to UCC Articles 2 and 2A omit the restriction that prohibited the stoppage of less than "a carload, truckload, planeload, or larger shipment" because carriers can now identify a shipment as small as a single package [Amended UCC 2–705(1), 2A–526(1)].

The Right to Reclaim the Goods In regard to sales contracts, if the buyer has received goods on credit and the seller discovers that the buyer is insolvent, the seller can demand the return of the goods if the demand is made within ten days of the buyer's receipt of the goods. The seller can demand and reclaim the goods at any time if the buyer misrepresented his or her solvency in writing within three months prior to the delivery of the goods [UCC 2–702(2)].[11] The seller's right to reclaim the goods, however, is subject to the rights of a good faith purchaser or other subsequent buyer in the ordinary course of business who purchases the goods from the buyer before the seller reclaims them.

Under the UCC, a seller seeking to exercise the right to reclaim goods receives preferential treatment over the buyer's other creditors—the seller need only demand the return of the goods within ten days after the buyer has received them.[12] Because of this preferential treatment, the UCC provides that reclamation *bars* the seller from pursuing any other remedy as to these goods [UCC 2–702(3)].

In regard to lease contracts, if the lessee defaults (fails to make payments that are due, for example), the lessor may reclaim the leased goods that are in the lessee's possession [UCC 2A–525(2)].

Remedies of the Buyer or Lessee

When the seller or lessor breaches the contract, the buyer or lessee has numerous remedies available under the UCC in addition to recovery of as much of the price as has been paid. Like the remedies available to sellers and lessors, the remedies of buyers and lessees depend on the circumstances existing at the time of the breach.

When the Seller or Lessor Refuses to Deliver the Goods

If the seller or lessor refuses to deliver the goods or the buyer or lessee has rejected the goods, the remedies available to the buyer or lessee include those discussed here.

The Right to Cancel the Contract When a seller or lessor fails to make proper delivery or repudiates the contract, the buyer or lessee can cancel, or rescind, the contract. On notice of cancellation, the buyer or lessee is relieved of any further obligations under the contract but retains all rights to other remedies against the seller [UCC 2–711(1), 2A–508(1)(a)].

The Right to Recover the Goods If a buyer or lessee has made a partial or full payment for goods that remain in the possession of the seller or lessor, the buyer or lessee can recover the goods if the seller or lessor is insolvent or becomes insolvent within ten days after receiving the first payment and if the goods are identified to the contract.[13] To exercise this right, the buyer or lessee must tender to the seller or lessor any unpaid balance of the purchase price (or lease payment) [UCC 2–502, 2A–522].

The Right to Obtain Specific Performance A buyer or lessee can obtain specific performance when the goods are unique and the remedy at law is inadequate [UCC 2–716(1), 2A–521(1)]. Ordinarily, a successful suit for money damages is sufficient to place a buyer or lessee in the position he or she would have occupied if the seller or lessor had fully performed. When the contract is for the purchase of a particular work of art or a similarly unique item, however, money damages may not be sufficient. Under these circumstances, equity will require that the seller or lessor perform exactly by delivering the particular goods identified to the contract (a remedy of specific performance).

The Right of Cover In certain situations, buyers and lessees can protect themselves by obtaining **cover**—that is, by purchasing or leasing other goods to substitute for those due under the contract. This option is available when the seller or lessor repudiates the contract or fails to deliver the goods, or when a buyer or lessee has rightfully rejected goods or revoked acceptance.

In obtaining cover, the buyer or lessee must act in good faith and without unreasonable delay [UCC 2–712, 2A–518]. After purchasing or leasing substitute goods, the buyer or lessee can recover from the seller or lessor the difference between the cost of cover and the contract price (or lease payments), plus incidental and consequential damages, less the expenses (such as delivery costs) that were saved as a result of the breach [UCC 2–712, 2–715, 2A–518]. Consequential damages are any losses suffered by the buyer or lessee that the seller or lessor could have foreseen (had reason to know about) at the time of contract formation and any injury to the buyer's or lessee's person or property proximately resulting from the contract's breach [UCC 2–715(2), 2A–520(2)].

11. The 2003 amendments to UCC Article 2 omit the ten-day limitation and the three-month exception to the ten-day limitation, referring instead to "a reasonable time" [Amended UCC 2–702(2)].

12. A seller who has delivered goods to an insolvent buyer also receives preferential treatment if the buyer enters into bankruptcy proceedings (bankruptcy will be discussed in Chapter 21).

13. The 2003 amendments to UCC Articles 2 and 2A create a new right to recover goods identified to a contract when a consumer buyer or lessee makes a down payment and the seller or lessor then repudiates the contract or lease or fails to deliver the goods [Amended UCC 2–502, 2A–522].

Buyers and lessees are not required to cover, and failure to do so will not bar them from using any other remedies available under the UCC. A buyer or lessee who fails to cover, however, may *not* be able to collect consequential damages that could have been avoided by purchasing or leasing substitute goods.

The Right to Replevy Goods Buyers and lessees also have the right to replevy goods. **Replevin**[14] is an action to recover specific goods in the hands of a party who is wrongfully withholding them from the other party. Outside the UCC, the term *replevin* refers to a *prejudgment process* (a proceeding that takes place prior to a court's judgment) that permits the seizure of specific personal property in which a party claims a right or an interest. Under the UCC, the buyer or lessee can replevy goods subject to the contract if the seller or lessor has repudiated, or breached, the contract. To maintain an action to replevy goods, usually buyers and lessees must show that they are unable to cover for the goods after a reasonable effort [UCC 2–716(3), 2A–521(3)].

The Right to Recover Damages If a seller or lessor repudiates the sales contract or fails to deliver the goods, or the buyer or lessee has rightfully rejected or revoked acceptance of the goods, the buyer or lessee can sue for damages. The measure of recovery is the difference between the contract price (or lease payments) and the market price of (or lease payments that could be obtained for) the goods at the time the buyer (or lessee) *learned* of the breach.[15] The market price or market lease payments are determined at the place where the seller or lessor was supposed to deliver the goods. The buyer or lessee can also recover incidental and consequential damages, less the expenses that were saved as a result of the breach [UCC 2–713, 2A–519].

■ EXAMPLE 16.11 Schilling orders ten thousand bushels of wheat from Valdone for $5 a bushel, with delivery due on June 14 and payment due on June 20. Valdone does not deliver on June 14. On June 14, the market price of wheat is $5.50 per bushel. Schilling chooses to do without the wheat. He sues Valdone for damages for nondelivery. Schilling can recover $\$0.50 \times 10{,}000$, or $5,000, plus any expenses the breach may have caused him. The measure of damages is the market price less the contract price on the day Schilling was to have received delivery. Any expenses Schilling saved by the breach would be deducted from the damages. ■

When the Seller or Lessor Delivers Nonconforming Goods

When the seller or lessor delivers nonconforming goods, the buyer or lessee has several remedies available under the UCC.

The Right to Reject the Goods If either the goods or the tender of the goods by the seller or lessor fails to conform to the contract *in any respect,* the buyer or lessee can reject the goods. If some of the goods conform to the contract, the buyer or lessee can keep the conforming goods and reject the rest [UCC 2–601, 2A–509]. If the buyer or lessee rejects the goods, she or he may then obtain cover, cancel the contract, or sue for damages for breach of contract, just as if the seller or lessor had refused to deliver the goods (see the earlier discussion of these remedies).

—Timeliness and Reason for Rejection Required. The buyer or lessee must reject the goods within a reasonable amount of time and must notify the seller or lessor seasonably—that is, in a timely fashion or at the proper time [UCC 2–602(1), 2A–509(2)]. If the buyer or lessee fails to reject the goods within a reasonable amount of time, acceptance will be presumed.

Note that when rejecting goods, the buyer or lessee must designate defects that would have been apparent to the seller or lessor on reasonable inspection. Failure to do so precludes the buyer or lessee from using such defects to justify rejection or to establish breach when the seller or lessor could have cured the defects if they had been stated seasonably [UCC 2–605, 2A–514].[16]

—Duties of Merchant Buyers and Lessees When Goods Are Rejected. If a merchant buyer or lessee rightfully rejects goods, and the seller or lessor has no agent or business at the place of rejection, the buyer or lessee is required to follow any reasonable instructions received from the seller or lessor with respect to the goods controlled by the buyer or lessee. The buyer or lessee is entitled to reimbursement for the care and cost entailed in following the instructions [UCC 2–603, 2A–511]. The same requirements hold if the buyer or lessee rightfully

14. Pronounced ruh-*pleh*-vun.

15. The 2003 amendments to UCC Article 2 change the rule that the time for measuring damages is the time the buyer learned of the breach. Unless repudiation is involved, the buyer's damages will be based on the market price at the time for tender [Amended UCC 2–713(1)(a)].

16. The 2003 amendments to UCC 2–605 and 2A–514 change this restriction. Under the amendments, a buyer's or lessee's failure to disclose the nature of the defect affects only the right to reject or revoke acceptance, not the right to establish a breach. The new sections expressly require that the seller or lessor has had a right to cure, as well as the ability to cure.

revokes his or her acceptance of the goods at some later time [UCC 2–608(3), 2A–517(5)]. (Revocation of acceptance will be discussed shortly.)

If no instructions are forthcoming and the goods are perishable or threaten to decline in value quickly, the buyer can resell the goods in good faith, taking the appropriate reimbursement from the proceeds. In addition, the buyer is entitled to a selling commission (not to exceed 10 percent of the gross proceeds) [UCC 2–603(1), (2); 2A–511(1)]. If the goods are not perishable, the buyer or lessee may store them for the seller or lessor or reship them to the seller or lessor [UCC 2–604, 2A–512].

Buyers who rightfully reject goods that remain in their possession or control have a *security interest* in the goods (basically, a legal claim to the goods to the extent necessary to recover expenses and other costs—see Chapter 20). The security interest encompasses any payments the buyer has made for the goods, as well as any expenses incurred with regard to inspection, receipt, transportation, care, and custody of the goods [UCC 2–711(3)]. A buyer with a security interest in the goods is a "person in the position of a seller." This gives the buyer the same rights as an unpaid seller. Thus, the buyer can resell, withhold delivery of, or stop delivery of the goods. A buyer who chooses to resell must account to the seller for any amounts received in excess of the security interest [UCC 2–706(6), 2–711].

Revocation of Acceptance Acceptance of the goods precludes the buyer or lessee from exercising the right of rejection, but it does not necessarily preclude the buyer or lessee from pursuing other remedies. In certain circumstances, a buyer or lessee is permitted to *revoke* her or his acceptance of the goods. Acceptance of a lot or a commercial unit can be revoked if the nonconformity *substantially* impairs the value of the lot or unit and if one of the following factors is present:

1. If acceptance was predicated on the reasonable assumption that the nonconformity would be cured, and it has not been cured within a reasonable period of time [UCC 2–608(1)(a), 2A–517(1)(a)].[17]
2. If the buyer or lessee did not discover the nonconformity before acceptance, either because it was difficult to discover before acceptance or because assurances made by the seller or lessor that the goods were conforming kept the buyer or lessee from inspecting the goods [UCC 2–608(1)(b), 2A–517(1)(b)].

Revocation of acceptance is not effective until the seller or lessor is notified, which must occur within a reasonable time after the buyer or lessee either discovers *or should have discovered* the grounds for revocation. Additionally, revocation must occur before the goods have undergone any substantial change (such as spoilage) not caused by their own defects [UCC 2–608(2), 2A–517(4)].

The Right to Recover Damages for Accepted Goods A buyer or lessee who has accepted nonconforming goods may also keep the goods and recover damages caused by the breach. The buyer or lessee, however, must notify the seller or lessor of the breach within a reasonable time after the defect was or should have been discovered. Failure to give notice of the defects (breach) to the seller or lessor bars the buyer or lessee from pursuing any remedy [UCC 2–607(3), 2A–516(3)].[18] In addition, the parties to a sales or lease contract can insert a provision requiring that the buyer or lessee give notice of any defects in the goods within a set period.

When the goods delivered and accepted are not as warranted, the measure of damages equals the difference between the value of the goods as accepted and their value if they had been delivered as warranted, plus incidental and consequential damages if appropriate [UCC 2–714(2), 2A–519(4)]. For this and other types of breaches in which the buyer or lessee has accepted the goods, the buyer or lessee is entitled to recover for any loss "resulting in the ordinary course of events . . . as determined in any manner which is reasonable" [UCC 2–714(1), 2A–519(3)]. The UCC also permits the buyer or lessee, with proper notice to the seller or lessor, to deduct all or any part of the damages from the price or lease payments still due and payable to the seller or lessor [UCC 2–717, 2A–516(1)].

Limitation of Remedies

The parties to a sales or lease contract can vary their respective rights and obligations by contractual agreement. For example, a seller and buyer can expressly provide for remedies in addition to those provided in the UCC. They can also provide remedies in lieu of those provided in the UCC, or they can change the measure of damages. The seller can provide that the buyer's only remedy

17. Under the 2003 amendments to UCC 2–508 and 2A–513, after a justifiable revocation of acceptance cure is not available as a matter of right in a consumer contract or lease.

18. Under the 2003 amendments to UCC Articles 2 and 2A, a buyer or lessee who fails to give timely notice is barred from a remedy for the breach but *only* "to the extent that the seller is prejudiced by the failure." In other words, if the seller is not harmed by the buyer's failure to notify, the buyer may still recover. See Amended UCC 2–607(3)(a), 2A–516(3)(a).

on breach of warranty will be repair or replacement of the item, or the seller can limit the buyer's remedy to return of the goods and refund of the purchase price. In sales and lease contracts, an agreed-on remedy is in addition to those provided in the UCC unless the parties expressly agree that the remedy is exclusive of all others [UCC 2–719(1), 2A–503(1)].

WHEN AN EXCLUSIVE REMEDY FAILS IN ITS ESSENTIAL PURPOSE

If the parties state that a remedy is exclusive, then it is the sole remedy. When circumstances cause an exclusive remedy to fail in its essential purpose, however, it is no longer considered exclusive, and the buyer or lessee may pursue other remedies available under the UCC [UCC 2–719(2), 2A–503(2)]. ■ **EXAMPLE 16.12** Suppose that a sales contract limits the buyer's remedy to repair or replacement. If the goods at issue cannot be repaired and no replacements are available, the exclusive remedy has failed in its essential purpose. In this situation, the buyer normally is entitled to seek other remedies provided to a buyer by the UCC. ■

LIMITATIONS ON CONSEQUENTIAL DAMAGES

A contract can limit or exclude consequential damages, provided the limitation is not unconscionable. When the buyer or lessee is a consumer, the limitation of consequential damages for personal injuries resulting from nonconforming goods is *prima facie* (on its face) unconscionable. The limitation of consequential damages is not necessarily unconscionable, however, when the loss is commercial in nature—for example, if the loss consists of lost profits and property damage [UCC 2–719(3), 2A–503(3)].

STATUTE OF LIMITATIONS

An action for breach of contract under the UCC must be commenced *within four years after the cause of action accrues*—that is, within four years after the breach occurs. In addition to filing suit within the four-year period, a buyer or lessee who has accepted nonconforming goods usually must notify the breaching party of the breach within a reasonable time, or the aggrieved party is barred from pursuing any remedy [UCC 2–607(3)(a), 2A–516(3)]. The parties can agree in their contract to reduce this period to not less than one year, but cannot extend it beyond four years [UCC 2–725(1), 2A–506(1)]. A cause of action for breach of warranty accrues when the seller or lessor tenders delivery. This is the rule even if the aggrieved party is unaware that the cause of action has accrued [UCC 2–725(2), 2A–506(2)].[19]

19. The 2003 amendments to UCC 2–725 adjust this time limit. A cause of action may be brought within the later of four years after the right of action accrues or one year after a breach is, or should have been, discovered, but no later than five years after the time that the right has accrued. Specific rules are included for determining when rights of action accrue. In addition, the four-year limitation period cannot be reduced in a consumer contract or lease [Amended UCC 2–725, 2A–506(1)].

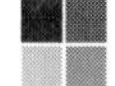

TERMS AND CONCEPTS

cover 307
incidental damages 304
installment contract 299
replevin 308

CHAPTER SUMMARY Performance and Breach of Sales and Lease Contracts*

REQUIREMENTS OF PERFORMANCE

Obligations of the Seller or Lessor (See pages 297–302.)	1. The seller or lessor must tender *conforming* goods to the buyer. Tender must take place at a *reasonable hour* and in a *reasonable manner.* Under the perfect tender doctrine, the seller or lessor must tender goods that conform exactly to the terms of the contract [UCC 2–503(1), 2A–508(1)]. 2. If the seller or lessor tenders nonconforming goods prior to the performance date and the buyer or lessee rejects them, the seller or lessor may *cure* (repair or replace the goods) within the contract time for performance [UCC 2–508(1), 2A–513(1)]. If the seller or lessor had reasonable grounds to believe the buyer or lessee would accept the tendered goods, on the buyer's or lessee's rejection the seller or lessor has a reasonable time to substitute conforming goods without liability [UCC 2–508(2), 2A–513(2)].

* Although we have referenced the 2003 amendments to Articles 2 and 2A of the UCC within the text of this chapter, we do not do so in this summary—because the amendments have not as yet become law in any state.

CHAPTER SUMMARY Performance and Breach of Sales and Lease Contracts—Continued

Obligations of the Seller or Lessor—Continued	3. If the agreed-on means of delivery becomes impracticable or unavailable, the seller must substitute an alternative means (such as a different carrier) if one is available [UCC 2–614(1)]. 4. If a seller or lessor tenders nonconforming goods in any one installment under an installment contract, the buyer or lessee may reject the installment only if its value is substantially impaired and cannot be cured. The entire installment contract is breached when one or more nonconforming installments *substantially* impair the value of the *whole* contract [UCC 2–612, 2A–510]. 5. When performance becomes commercially impracticable owing to circumstances that were not foreseeable when the contract was formed, the perfect tender rule no longer holds [UCC 2–615, 2A–405].
Obligations of the Buyer or Lessee (See pages 302–303.)	1. On tender of delivery by the seller or lessor, the buyer or lessee must pay for the goods at the time and place the buyer or lessee *receives* the goods, even if the place of shipment is the place of delivery, unless the sale is made on credit. Payment may be made by any method generally acceptable in the commercial world unless the seller demands cash [UCC 2–310, 2–511]. In lease contracts, the lessee must make lease payments in accordance with the contract [UCC 2A–516(1)]. 2. Unless otherwise agreed, the buyer or lessee has an absolute right to inspect the goods before acceptance [UCC 2–513(1), 2A–515(1)]. 3. The buyer or lessee can manifest acceptance of delivered goods expressly in words or by conduct or by failing to reject the goods after a reasonable period of time following inspection or after having had a reasonable opportunity to inspect them [UCC 2–606(1), 2A–515(1)]. A buyer will be deemed to have accepted goods if he or she performs any act inconsistent with the seller's ownership [UCC 2–606(1)(c)]. 4. Following the acceptance of delivered goods, the buyer or lessee may revoke acceptance only if the nonconformity *substantially* impairs the value of the unit or lot and if one of the following factors is present: a. Acceptance was predicated on the reasonable assumption that the nonconformity would be cured and it was not cured within a reasonable time [UCC 2–608(1)(a), 2A–517(1)(a)]. b. The buyer or lessee did not discover the nonconformity before acceptance, either because it was difficult to discover before acceptance or because the seller's or lessor's assurance that the goods were conforming kept the buyer or lessee from inspecting the goods [UCC 2–608(1)(b), 2A–517(1)(b)].
Anticipatory Repudiation (See pages 303–304.)	If, before the time for performance, either party clearly indicates to the other an intention not to perform, under UCC 2–610 and 2A–402 the aggrieved party may do the following: 1. Await performance by the repudiating party for a commercially reasonable time. 2. Resort to any remedy for breach. 3. In either situation, suspend performance.
REMEDIES FOR BREACH OF CONTRACT	
Remedies of the Seller or Lessor (See pages 304–307.)	1. *When the goods are in the possession of the seller or lessor*—The seller or lessor may do the following: a. Cancel the contract [UCC 2–703(f), 2A–523(1)(a)]. b. Withhold delivery [UCC 2–703(a), 2A–523(1)(c)]. c. Resell or dispose of the goods [UCC 2–703(d), 2–706(1), 2A–523(1)(e), 2A–527(1)]. d. Sue to recover the purchase price or lease payments due [UCC 2–703(e), 2–709(1), 2A–529(1)]. e. Sue to recover damages [UCC 2–703(e), 2–708, 2A–528].

(Continued)

CHAPTER SUMMARY Performance and Breach of Sales and Lease Contracts—Continued

Remedies of the Seller or Lessor—Continued	2. *When the goods are in transit*—The seller or lessor may stop the carrier or bailee from delivering the goods [UCC 2–705, 2A–526]. 3. *When the goods are in the possession of the buyer or lessee*—The seller or lessor may do the following: a. Sue to recover the purchase price or lease payments due [UCC 2–709(1), 2A–529(1)]. b. Reclaim the goods. A seller may reclaim goods received by an insolvent buyer if the demand is made within ten days of receipt (reclaiming goods excludes all other remedies) [UCC 2–702]; a lessor may repossess goods if the lessee is in default [UCC 2A–525(2)].
Remedies of the Buyer or Lessee (See pages 307–309.)	1. *When the seller or lessor refuses to deliver the goods*—The buyer or lessee may do the following: a. Cancel the contract [UCC 2–711(1), 2A–508(1)(a)]. b. Recover the goods if the seller or lessor becomes insolvent within ten days after receiving the first payment and the goods are identified to the contract [UCC 2–502, 2A–522]. c. Obtain specific performance (when the goods are unique and when the remedy at law is inadequate) [UCC 2–716(1), 2A–521(1)]. d. Obtain cover [UCC 2–712, 2A–518]. e. Replevy the goods (if cover is unavailable) [UCC 2–716(3), 2A–521(3)]. f. Sue to recover damages [UCC 2–713, 2A–519]. 2. *When the seller or lessor delivers or tenders delivery of nonconforming goods*—The buyer or lessee may do the following: a. Reject the goods [UCC 2–601, 2A–509]. b. Revoke acceptance (in certain circumstances) [UCC 2–608, 2A–517]. c. Accept the goods and recover damages [UCC 2–607, 2–714, 2–717, 2A–519].
Limitation of Remedies (See pages 309–310.)	Remedies may be limited in sales or lease contracts by agreement of the parties. If the contract states that a remedy is exclusive, then that is the sole remedy unless the remedy fails in its essential purpose. Sellers and lessors can also limit the rights of buyers and lessees to consequential damages unless the limitation is unconscionable [UCC 2–719, 2A–503].
Statute of Limitations (See page 310.)	The UCC has a four-year statute of limitations for actions involving breach of contract. By agreement, the parties to a sales or lease contract can reduce this period to not less than one year, but they cannot extend it beyond four years [UCC 2–725(1), 2A–506(1)].

FOR REVIEW

Answers for the even-numbered questions in this For Review *section can be found in Appendix A at the end of this text.*

1. What are the respective obligations of the parties under a contract for the sale or lease of goods?
2. What is the perfect tender rule? What are some important exceptions to this rule that apply to sales and lease contracts?
3. What options are available to the nonbreaching party when the other party to a sales or lease contract repudiates the contract prior to the time for performance?
4. What remedies are available to a seller or lessor when the buyer or lessee breaches the contract? What remedies are available to a buyer or lessee if the seller or lessor breaches the contract?
5. In contracts subject to the UCC, are parties free to limit the remedies available to the nonbreaching party on a breach of contract? If so, in what ways?

QUESTIONS AND CASE PROBLEMS

16–1. Remedies. Genix, Inc., has contracted to sell Larson five hundred washing machines of a certain model at list price. Genix is to ship the goods on or before December 1. Genix produces one thousand washing machines of this model but has not yet prepared Larson's shipment. On November 1, Larson repudiates the contract. Discuss the remedies available to Genix in this situation.

QUESTION WITH SAMPLE ANSWER

16–2. Cummings ordered two model X Super Fidelity speakers from Jamestown Wholesale Electronics, Inc. Jamestown shipped the speakers via United Parcel Service, C.O.D. (collect on delivery), although Cummings had not requested or agreed to a C.O.D. shipment of the goods. When the speakers were delivered, Cummings refused to accept them because he would not be able to inspect them before payment. Jamestown claimed that it had shipped conforming goods and that Cummings had breached their contract. Had Cummings breached the contract? Explain.

For a sample answer to this question, go to Appendix B at the end of this text.

16–3. Anticipatory Repudiation. Moore contracted in writing to sell her 1996 Ford Taurus to Hammer for $8,500. Moore agreed to deliver the car on Wednesday, and Hammer promised to pay the $8,500 on the following Friday. On Tuesday, Hammer informed Moore that he would not be buying the car after all. By Friday, Hammer had changed his mind again and tendered $8,500 to Moore. Moore, although she had not sold the car to another party, refused the tender and refused to deliver. Hammer claimed that Moore had breached their contract. Moore contended that Hammer's repudiation released her from her duty to perform under the contract. Who is correct and why?

16–4. Remedies. Rodriguez is a collector of antique cars. He contracts to purchase spare parts for a 1938 engine from Gerrard. These parts are no longer made and are scarce. To get the contract with Gerrard, Rodriguez has to pay 50 percent of the purchase price in advance. On May 1, Rodriguez sends the required payment, which is received on May 2. On May 3, Gerrard, having found another buyer willing to pay substantially more for the parts, informs Rodriguez that he will not deliver as contracted. That same day, Rodriguez learns that Gerrard is insolvent. Gerrard has the parts, and Rodriguez wants them. Discuss fully any remedies available to Rodriguez.

16–5. Limitation of Remedies. Destileria Serralles, Inc., a distributor of rum and other products, operates a rum-bottling plant in Puerto Rico. Figgie International, Inc., contracted with Serralles to provide bottle-labeling equipment capable of placing a clear label on a clear bottle of "Cristal" rum within a raised glass oval. The contract stated that Serralles's remedy, in the event of a breach of contract, was limited to repair, replacement, or refund. When the equipment was installed in the Serralles plant, problems arose immediately. Figgie attempted to repair the equipment, but when it still did not work properly several months later, Figgie refunded the purchase price and Serralles returned the equipment. Serralles asked Figgie to pay for Serralles's losses caused by the failure of the equipment and by the delay in obtaining alternative machinery. Figgie filed a suit in a federal district court, asserting that it owed nothing to Serralles because the remedy for breach was limited to repair, replacement, or refund. Serralles responded that the limitation had failed in its essential purpose. In whose favor will the court resolve this dispute? Why? [*Figgie International, Inc. v. Destileria Serralles, Inc.*, 190 F.3d 252 (4th Cir. 1999)]

CASE PROBLEM WITH SAMPLE ANSWER

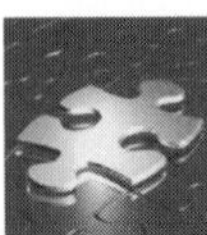

16–6. Metro-North Commuter Railroad Co. decided to install a fall-protection system for elevated walkways, roof areas, and interior catwalks in Grand Central Terminal, in New York City. The system was needed to ensure the safety of Metro-North employees when they worked at great heights on the interior and exterior of the terminal. Sinco, Inc., proposed a system called "Sayfglida," which involved a harness worn by the worker, a network of cables, and metal clips or sleeves called "Sayflinks" that connected the harness to the cables. Metro-North agreed to pay $197,325 for the installation of this system by June 26, 1999. Because the system's reliability was crucial, the contract required certain quality control processes. During a training session for Metro-North employees on June 29, the Sayflink sleeves fell apart. Within two days, Sinco manufactured and delivered two different types of replacement clips without subjecting them to the contract's quality control process, but Metro-North rejected them. Sinco suggested other possible solutions, which Metro-North did not accept. In September, Metro-North terminated its contract with Sinco and awarded the work to Surety, Inc., at a price of about $348,000. Sinco filed a suit in a federal district court, alleging breach of contract. Metro-North counterclaimed for its cost of cover. In whose favor should the court rule and why? [*Sinco, Inc. v. Metro-North Commuter Railroad Co.*, 133 F.Supp.2d 308 (S.D.N.Y. 2001)]

After you have answered this problem, compare your answer with the sample answer given on the Web site that accompanies this text. Go to **academic.cengage.com/blaw/wbl**, select "Chapter 16," and click on "Case Problem with Sample Answer."

16–7. Acceptance. In April 1996, Excalibur Oil Group, Inc., applied for credit and opened an account with Standard Distributors, Inc., to obtain snack foods and other items for Excalibur's convenience stores. For three months, Standard delivered the goods and Excalibur paid the invoices. In July, Standard

was dissolved, and its assets were distributed to J. F. Walker Co. Walker continued to deliver the goods to Excalibur, which continued to pay the invoices until November, when the firm began to experience financial difficulties. By January 1997, Excalibur owed Walker $54,241.77. Walker then dealt with Excalibur only on a collect-on-delivery basis until Excalibur's stores closed in 1998. Walker filed a suit in a Pennsylvania state court against Excalibur and its owner to recover amounts due on unpaid invoices. To successfully plead its case, Walker had to show that there was a contract between the parties. One question was whether Excalibur had manifested acceptance of the goods delivered by Walker. How does a buyer manifest acceptance? Was there an acceptance in this case? In whose favor should the court rule, and why? [*J. F. Walker Co. v. Excalibur Oil Group, Inc.*, 792 A.2d 1269 (Pa.Super. 2002)]

16–8. Right of Assurance. Advanced Polymer Sciences, Inc. (APS), based in Ohio, makes polymers and resins for use as protective coatings in industrial applications. APS also owns the technology for equipment used to make certain composite fibers. *SAVA gumarska in kemijska industria d.d.* (SAVA), based in Slovenia, makes rubber goods. In 1999, SAVA and APS contracted to form *SAVA Advanced Polymers proizvodno podjetje d.o.o.* (SAVA AP) to make and distribute APS products in Eastern Europe. Their contract provided for, among other things, the alteration of a facility to make the products using specially made equipment to be sold by APS to SAVA. Disputes arose between the parties, and in August 2000, SAVA stopped work on the new facility. APS then notified SAVA that it was stopping the manufacture of the equipment and "insist[ed] on knowing what is SAVA's intention towards this venture." In October, SAVA told APS that it was canceling their contract. In subsequent litigation, SAVA claimed that APS had repudiated the contract when it stopped making the equipment. What might APS assert in its defense? How should the court rule? Explain. [*SAVA gumarska in kemijska industria d.d. v. Advanced Polymer Sciences, Inc.*, 128 S.W.3d 304 (Tex.App.—Dallas 2004)]

Performance and Breach of Sales and Lease Contracts

For updated links to resources available on the Web, as well as a variety of other materials, visit this text's Web site at

academic.cengage.com/blaw/wbl

To view the UCC provisions discussed in this chapter, go to

http://www.law.cornell.edu/ucc/ucc.table.html

ONLINE LEGAL RESEARCH

Go to the *Fundamentals of Business Law* home page at academic.cengage.com/blaw/wbl, select "Chapter 16," and click on "Internet Exercises." There you will find the following Internet research exercises that you can perform to learn more about topics covered in this chapter.

Activity 16–1: LEGAL PERSPECTIVE—The Right to Reject Goods

Activity 16–2: MANAGEMENT PERSPECTIVE—Good Faith and Fair Dealing

BEFORE THE TEST

Go to the *Fundamentals of Business Law* home page at academic.cengage.com/blaw/wbl, select "Chapter 16," and click on "Interactive Quizzes." You will find a number of interactive questions relating to this chapter.

CHAPTER 17

Warranties and Product Liability

LEARNING OBJECTIVES

After reading this chapter, you should be able to answer the following questions:

1. What factors determine whether a seller's or lessor's statement constitutes an express warranty or mere "puffing"?
2. What implied warranties arise under the UCC?
3. Can a manufacturer be held liable to any person who suffers an injury proximately caused by the manufacturer's negligently made product?
4. What are the elements of a cause of action in strict product liability?
5. What defenses to liability can be raised in a product liability lawsuit?

Warranty is an age-old concept. In sales and lease law, a warranty is an assurance by one party of the existence of a fact on which the other party can rely. Sellers and lessors warrant to those who purchase or lease their goods that the goods are as represented or will be as promised.

The Uniform Commercial Code (UCC) has numerous rules governing product warranties as they occur in sales and lease contracts. That is the subject matter of the first part of this chapter. A natural addition to the discussion is *product liability:* Who is liable to consumers, users, and bystanders for physical harm and property damage caused by a particular good or the use thereof? Product liability encompasses the contract theory of warranty, as well as the tort theories of negligence and strict liability (discussed in Chapter 4).

Warranties

Article 2 (on sales) and Article 2A (on leases) of the UCC designate several types of warranties that can arise in a sales or lease contract, including warranties of title, express warranties, and implied warranties.

WARRANTIES OF TITLE

Title warranty arises automatically in most sales contracts. The UCC imposes three types of warranties of title.

Good Title In most instances, sellers warrant that they have good and valid title to the goods sold and that transfer of the title is rightful [UCC 2–312(1)(a)].[1] ■ **EXAMPLE 17.1** Sharon steals goods from Miguel and sells them to Carrie, who does not know that the goods are stolen. If Miguel reclaims the goods from Carrie, which he has a right to do, Carrie can then sue Sharon for breach of warranty. When Sharon sold Carrie the goods, Sharon *automatically* warranted to her that the title conveyed was valid and that its transfer was rightful. Because this was not in fact the case, Sharon breached the warranty of title imposed by UCC 2–312(1)(a) and became liable to the buyer for the appropriate damages. ■

1. Under the 2003 amendments to UCC 2–312(1)(a), good title also includes the warranty that the sale "shall not unreasonably expose the buyer to litigation because of any colorable [legitimate or reasonable] claim to or interest in the goods." Thus, the buyer is entitled not only to a good title, but also to a marketable title that is free of "colorable claims."

No Liens A second warranty of title provided by the UCC protects buyers who are unaware of any encumbrances, or **liens** (see Chapter 21)—claims, charges, or liabilities against goods at the time the contract is made [UCC 2–312(1)(b)]. This warranty protects buyers who, for example, unknowingly purchase goods that are subject to a creditor's security interest (see Chapter 20). If a creditor legally repossesses the goods from a buyer *who had no actual knowledge of the security interest,* the buyer can recover from the seller for breach of warranty.

Article 2A affords similar protection for lessees. Section 2A–211(1) provides that during the term of the lease, no claim of any third party will interfere with the lessee's enjoyment of the leasehold interest.

No Infringements A merchant-seller is also deemed to warrant that the goods delivered are free from any copyright, trademark, or patent claims of a third person[2] [UCC 2–312(3), 2A–211(2)]. If this warranty is breached and the buyer is sued by the party holding copyright, trademark, or patent rights in the goods, the buyer must notify the seller of the litigation within a reasonable time to enable the seller to decide whether to defend against the lawsuit. If the seller states in writing that she or he has decided to defend and agrees to bear all expenses, including that of an adverse judgment, then the buyer must let the seller undertake the litigation; otherwise, the buyer loses all rights against the seller if any infringement liability is established [UCC 2–607(3)(b), 2–607(5)(b)].

Article 2A provides for the same notice of litigation in situations that involve leases rather than sales [UCC 2A–516(3)(b), 2A–516(4)(b)]. There is an exception for leases to individual consumers for personal, family, or household purposes. A consumer who fails to notify the lessor within a reasonable time does not lose his or her remedy against the lessor for any liability established in the litigation [UCC 2A–516(3)(b)].

Disclaimer of Title Warranty In an ordinary sales transaction, the title warranty can be disclaimed or modified only by *specific language* in the contract [UCC 2–312(2)]. For example, sellers can assert that they are transferring only such rights, title, and interest as they have in the goods. In a lease transaction, the disclaimer must "be specific, be by a writing, and be conspicuous" [UCC 2A–214(4)].

2. Recall from Chapter 14 that a *merchant* is defined in UCC 2–104(1) as a person who deals in goods of the kind involved in the sales contract or who, by occupation, presents himself or herself as having knowledge or skill peculiar to the goods involved in the transaction.

EXPRESS WARRANTIES

A seller or lessor can create an **express warranty** by making representations concerning the quality, condition, description, or performance potential of the goods. Under UCC 2–313 and 2A–210, express warranties arise when a seller or lessor indicates any of the following:

1. That the goods conform to any affirmation (declaration that something is true) or promise of fact that the seller or lessor makes to the buyer or lessee about the goods. Such affirmations or promises are usually made during the bargaining process. Statements such as "these drill bits will penetrate stainless steel—and without dulling" are express warranties.[3]
2. That the goods conform to any description of them. For example, a label that reads "Crate contains one 150-horsepower diesel engine" or a contract that calls for the delivery of a "camel's-hair coat" creates an express warranty.
3. That the goods conform to any sample or model of the goods shown to the buyer or lessee.[4]

Basis of the Bargain To create an express warranty, a seller or lessor does not have to use formal words such as *warrant* or *guarantee* [UCC 2–313(2), 2A–210(2)]. The UCC requires that for an express warranty to be created, the affirmation, promise, description, or sample must become part of the "basis of the bargain" [UCC 2–313(1), 2A–210(1)]. Just what constitutes the basis of the bargain is difficult to say. The UCC does not define the concept, and it is a question of fact in each case whether a representation was made at such a time and in such a way that it induced the buyer or lessee to enter into the contract.

3. The 2003 amendments to UCC Article 2 introduce the term *remedial promise,* which is "a promise by the seller to repair or replace the goods or to refund all or part of the price of the goods on the happening of a specified event" [Amended UCC 2–103(1)(n), 2–313(4)]. A cause of action for the breach of a remedial promise accrues when the promise is not performed at the time performance is due [Amended UCC 2–725(2)(c)].

4. The 2003 amendments to the UCC distinguish between immediate buyers (those who enter into contracts with sellers) and remote purchasers (those who buy or lease goods from immediate buyers) and extend sellers' obligations regarding new goods to remote purchasers. For example, suppose that a manufacturer sells packaged goods to a retailer, who resells the goods to a consumer. If a reasonable person in the position of the consumer would believe that a description on the package creates an obligation, the manufacturer is liable for its breach. [See Amended UCC 2–313, 2–313A, and 2–313B.]

Statements of Opinion Statements of fact create express warranties. If the seller or lessor merely makes a statement that relates to the supposed value or worth of the goods, or makes a statement of opinion or recommendation about the goods, however, the seller or lessor is not creating an express warranty [UCC 2–313(2), 2A–210(2)].

■ EXAMPLE 17.2 A seller claims that "this is the best used car to come along in years; it has four new tires and a 250-horsepower engine just rebuilt this year." The seller has made several *affirmations of fact* that can create a warranty: the automobile has an engine; it has a 250-horsepower engine; it was rebuilt this year; there are four tires on the automobile; and the tires are new. The seller's *opinion* that the vehicle is "the best used car to come along in years," however, is known as "puffing" and creates no warranty. (*Puffing* is the expression of opinion by a seller or lessor that is not made as a representation of fact.) ■

A statement relating to the value of the goods, such as "it's worth a fortune" or "anywhere else you'd pay $10,000 for it," usually does not create a warranty. If the seller or lessor is an expert and gives an opinion as an expert to a layperson, though, then a warranty may be created.

It is not always easy to determine whether a statement constitutes an express warranty or puffing. The reasonableness of the buyer's or lessee's reliance appears to be the controlling criterion in many cases. For example, a salesperson's statements that a ladder "will never break" and will "last a lifetime" are so clearly improbable that no reasonable buyer should rely on them. Additionally, the context in which a statement is made might be relevant in determining the reasonableness of the buyer's or lessee's reliance. For example, a reasonable person is more likely to rely on a written statement made in an advertisement than on a statement made orally by a salesperson.

IMPLIED WARRANTIES

An **implied warranty** is one that *the law derives* by implication or inference because of the circumstances of a sale, rather than by the seller's express promise. In an action based on a breach of implied warranty, it is necessary to show that an implied warranty existed and that the breach of the warranty proximately caused[5] the damage sustained. We look here at some of the implied warranties that arise under the UCC.

Implied Warranty of Merchantability Every sale or lease of goods made *by a merchant* who deals in goods of the kind sold or leased automatically gives rise to an **implied warranty of merchantability** [UCC 2–314, 2A–212]. Thus, a merchant who is in the business of selling ski equipment makes an implied warranty of merchantability every time the merchant sells a pair of skis, but a neighbor selling his or her skis at a garage sale does not.

—Merchantable Goods. Goods that are deemed *merchantable* are "reasonably fit for the ordinary purposes for which such goods are used." They must be of at least average, fair, or medium-grade quality. The quality must be comparable to a level that will pass without objection in the trade or market for goods of the same description. To be merchantable, the goods must also be adequately packaged and labeled as provided by the agreement, and they must conform to the promises or affirmations of fact made on the container or label, if any.

It makes no difference whether the merchant knew or could have discovered that a product was defective (not merchantable). Of course, merchants are not absolute insurers against all accidents arising in connection with the goods. For example, a bar of soap is not unmerchantable merely because a user could slip and fall by stepping on it.

—Merchantable Food. The UCC recognizes the serving of food or drink to be consumed on or off the premises as a sale of goods subject to the implied warranty of merchantability [UCC 2–314(1)]. *Merchantable* food means food that is fit to eat. Courts generally determine whether food is fit to eat on the basis of consumer expectations. For example, the courts assume that consumers should reasonably expect on occasion to find bones in fish fillets, cherry pits in cherry pie, or a nutshell in a package of shelled nuts—because such substances are natural incidents of the food. In contrast, consumers would not reasonably expect to find an inchworm in a can of peas or a piece of glass in a soft drink—because these substances are not natural to the food product.[6] In the following classic case, the court had to determine whether a fish bone was a substance that one should reasonably expect to find in fish chowder.

5. *Proximate cause,* or legal cause, exists when the connection between an act and an injury is strong enough to justify imposing liability—see Chapter 4.

6. See, for example, *Ruvolo v. Homovich,* 149 Ohio App.3d 701, 778 N.E.2d 661 (2002).

LANDMARK AND CLASSIC CASES

CASE 17.1 Webster v. Blue Ship Tea Room, Inc.

Supreme Judicial Court of Massachusetts, 1964.
347 Mass. 421,
198 N.E.2d 309.

FACTS Blue Ship Tea Room, Inc., was located in Boston in an old building overlooking the ocean. Priscilla Webster, who had been born and raised in New England, went to the restaurant and ordered fish chowder. The chowder was milky in color. After three or four spoonfuls, she felt something lodged in her throat. As a result, she underwent two esophagoscopies; in the second esophagoscopy, a fish bone was found and removed. Webster filed suit against the restaurant in a Massachusetts state court for breach of the implied warranty of merchantability. The jury rendered a verdict for Webster, and the restaurant appealed to the state's highest court.

ISSUE Does serving fish chowder that contains a bone constitute a breach of an implied warranty of merchantability on the part of the restaurant?

DECISION No. The Supreme Judicial Court of Massachusetts held that Webster could not recover against Blue Ship Tea Room because no breach of warranty had occurred.

REASON The court, citing UCC Section 2–314, stated that "a warranty that goods shall be merchantable is implied in a contract for their sale if the seller is a merchant with respect to goods of that kind. Under this section the serving for value of food or drink to be consumed either on the premises or elsewhere is a sale. * * * Goods to be merchantable must at least be * * * fit for the ordinary purposes for which such goods are used." The question here is whether a fish bone made the chowder unfit for eating. In the judge's opinion, "the joys of life in New England include the ready availability of fresh fish chowder. We should be prepared to cope with the hazards of fish bones, the occasional presence of which in chowders is, it seems to us, to be anticipated, and which, in the light of a hallowed tradition, do not impair their fitness or merchantability."

IMPACT OF THIS CASE ON TODAY'S LAW This classic case, phrased in memorable language, was an early application of the UCC's implied warranty of merchantability to food products. The case established the rule that consumers should expect to find, on occasion, elements of food products that are natural to the product (such as fish bones in fish chowder). In such cases, the food preparers or packagers normally will not be liable. This rule continues to be applied today in cases involving similar issues.

RELEVANT WEB SITES *To locate information on the Web concerning* Webster v. Blue Ship Tea Room, Inc., *go to this text's Web site at* **academic.cengage.com/blaw/wbl**, *select "Chapter 17," and click on "URLs for Landmark Cases."*

Implied Warranty of Fitness for a Particular Purpose The **implied warranty of fitness for a particular purpose** arises when any seller or lessor (merchant or nonmerchant) knows the particular purpose for which a buyer or lessee will use the goods and knows that the buyer or lessee is relying on the skill and judgment of the seller or lessor to select suitable goods [UCC 2-315, 2A–213].

A "particular purpose" of the buyer or lessee differs from the "ordinary purpose for which goods are used" (merchantability). Goods can be merchantable but unfit for a particular purpose. ■ EXAMPLE 17.3 Suppose that you need a gallon of paint to match the color of your living room walls—a light shade somewhere between coral and peach. You take a sample to your local hardware store and request a gallon of paint of that color. Instead, you are given a gallon of bright blue paint. Here, the salesperson has not breached any warranty of implied merchantability—the bright blue paint is of high quality and suitable for interior walls—but he or she has breached an implied warranty of fitness for a particular purpose. ■

A seller or lessor does not need to have actual knowledge of the buyer's or lessee's particular purpose. It is sufficient if a seller or lessor "has reason to know" the purpose. The buyer or lessee, however, must have *relied* on the skill or judgment of the seller or lessor in selecting or furnishing suitable goods for an implied warranty to be created.

■ EXAMPLE 17.4 Bloomberg leases a computer from Future Tech, a lessor of technical business equipment. Bloomberg tells the clerk that she wants a computer that will run a complicated new engineering graphics program at a realistic speed. Future Tech leases Bloomberg an Architex One computer with a CPU speed of only 2.4 gigahertz, even though a speed of at least 3.8 gigahertz would be required to run Bloomberg's graphics program at a "realistic speed." Bloomberg, after discovering that it takes forever to run her program, wants her money back. Here, because Future Tech has breached the implied warranty of fitness for a particular purpose, Bloomberg normally will be able to recover. The clerk knew specifically that Bloomberg wanted a computer with enough speed to

run certain software. Furthermore, Bloomberg relied on the clerk to furnish a computer that would fulfill this purpose. Because Future Tech did not do so, the warranty was breached. ■

Other Implied Warranties Implied warranties can also arise (or be excluded or modified) as a result of course of dealing or usage of trade [UCC 2–314(3), 2A–212(3)]. In the absence of evidence to the contrary, when both parties to a sales or lease contract have knowledge of a well-recognized trade custom, the courts will infer that both parties intended for that trade custom to apply to their contract. For example, if an industry-wide custom is to lubricate a new car before it is delivered and a dealer fails to do so, the dealer can be held liable to a buyer for damages resulting from the breach of an implied warranty. This, of course, would also be negligence on the part of the dealer.

OVERLAPPING WARRANTIES

Sometimes, two or more warranties are made in a single transaction. An implied warranty of merchantability, an implied warranty of fitness for a particular purpose, or both can exist in addition to an express warranty. For example, when a sales contract for a new car states that "this car engine is warranted to be free from defects for 36,000 miles or thirty-six months, whichever occurs first," there is an express warranty against all defects and an implied warranty that the car will be fit for normal use.

The rule under the UCC is that express and implied warranties are construed as *cumulative* if they are consistent with one another [UCC 2–317, 2A–215]. If the warranties are inconsistent, the courts usually hold as follows:

1. *Express* warranties displace inconsistent *implied* warranties, except for implied warranties of fitness for a particular purpose.
2. Samples take precedence over inconsistent general descriptions.
3. Technical specifications displace inconsistent samples or general descriptions.

In Example 17.4 presented earlier, suppose that when Bloomberg leases the computer from Future Tech, the contract contains an express warranty concerning the speed of the CPU and the application programs that the computer is capable of running. Bloomberg does not realize that the speed expressly warranted in the contract is insufficient for her needs. Bloomberg later claims that Future Tech has breached the implied warranty of fitness for a particular purpose. Here, although the express warranty would take precedence over any implied warranty of merchantability, it normally would not take precedence over an implied warranty of fitness for a particular purpose. Bloomberg therefore has a good claim for the breach of implied warranty of fitness for a particular purpose, because she made it clear that she was leasing the computer to perform certain tasks.

THIRD PARTY BENEFICIARIES OF WARRANTIES

One of the general principles of contract law is that unless you are one of the parties to a contract, you have no rights under the contract. In other words, *privity of contract* must exist between a plaintiff and a defendant before any action based on a contract can be maintained. Two notable exceptions to the rule of privity are assignments and third party beneficiary contracts (these topics were discussed in Chapter 11). Another exception is made under warranty laws so that third parties can recover for harms suffered as a result of breached warranties.

There has been sharp disagreement among state courts as to how far warranty liability should extend, however. In view of this disagreement, the UCC offers three alternatives for liability to third parties [UCC 2–318, 2A–216].[7] All three alternatives are intended to eliminate the privity requirement with respect to certain enumerated types of injuries (personal versus property) for certain beneficiaries (for example, household members or bystanders).

WARRANTY DISCLAIMERS

Because each type of warranty is created in a special way, the manner in which warranties can be disclaimed or qualified by a seller or lessor varies depending on the type of warranty.

Express Warranties As already stated, any affirmation of fact or promise, description of the goods, or use of samples or models by a seller or lessor creates an express warranty. Obviously, then, express warranties can be excluded if the seller or lessor carefully refrains from making any promise or affirmation of fact relating to the goods, describing the goods, or using a sample or model.

The UCC does permit express warranties to be negated or limited by specific and unambiguous language, provided that this is done in a manner that protects the buyer or lessee from surprise. Therefore, a written disclaimer in language that is clear and conspicuous, and called to a buyer's or lessee's attention, could negate all oral express warranties not included in the written sales contract [UCC 2–316(1), 2A–214(1)]. This allows the seller or lessor to avoid false allegations that oral warranties were made,

7. Under the 2003 amendments to the UCC, each alternative is expanded to cover remedial promises, as well as obligations, to immediate buyers and remote purchasers [see Amended UCC 2–313, 2–313A, 2–313B, and 2–318].

and it ensures that only representations made by properly authorized individuals are included in the bargain.

Note, however, that a buyer or lessee must be made aware of any warranty disclaimers or modifications *at the time the contract is formed.* In other words, any oral or written warranties—or disclaimers—made during the bargaining process as part of a contract's formation cannot be modified at a later time by the seller or lessor.

Implied Warranties Generally speaking, unless circumstances indicate otherwise, the implied warranties of merchantability and fitness are disclaimed by the expressions "as is," "with all faults," and other similar expressions that in common understanding call the buyer's or lessee's attention to the fact that there are no implied warranties [UCC 2–316(3)(a), 2A–214(3)(a)]. The UCC also permits a seller or lessor to specifically disclaim an implied warranty of either merchantability or fitness [UCC 2–316(2), 2A–214(2)].[8]

—Disclaimer of the Implied Warranty of Merchantability. A merchantability disclaimer must specifically mention the word *merchantability.* The disclaimer does not have to be written, but if it is, the writing must be conspicuous [UCC 2–316(2), 2A–214(4)].[9] According to UCC 1–201(10),

> [a] term or clause is conspicuous when it is so written that a reasonable person against whom it is to operate ought to have noticed it. A printed heading in capitals . . . is conspicuous. Language in the body of a form is conspicuous if it is in larger or other contrasting type or color.[10]

8. The 2003 amendments to the UCC require more informative language for disclaimers of implied warranties [Amended UCC 2–316(2), 2A–214(2)].

9. Under the 2003 amendments to UCC Articles 2 and 2A, if a consumer contract or lease is set forth in a writing—physical or electronic—the implied warranty of merchantability can be disclaimed only by language also set forth in a writing [Amended UCC 2–316(3), 2A–214(3)].

10. The 2003 amendments to UCC Articles 2 and 2A expand the concept to include terms in electronic records [Amended UCC 2–103(1)(b), 2A–103(1)(d)]. These sections also add a special rule for the situation in which a sender of an e-record intends to evoke a response from an e-agent.

■ EXAMPLE 17.5 Forbes, a merchant, sells Maves a particular lawn mower selected by Forbes with the characteristics clearly requested by Maves. At the time of the sale, Forbes orally tells Maves that he does not warrant the merchantability of the mower, as it is last year's model and has been used as a demonstrator. If the mower proves to be defective and does not work, Maves can hold Forbes liable for breach of the warranty of fitness for a particular purpose but not for breach of the warranty of merchantability. Forbes's oral disclaimer mentioning the word *merchantability* is a proper disclaimer. ■

—Disclaimer of the Implied Warranty of Fitness. To disclaim an implied warranty of fitness for a particular purpose, the disclaimer *must* be in writing and be conspicuous. The word *fitness* does not have to be mentioned in the writing; it is sufficient if, for example, the disclaimer states, "THERE ARE NO WARRANTIES THAT EXTEND BEYOND THE DESCRIPTION ON THE FACE HEREOF." Thus, in Example 17.5 above, for Forbes to have disclaimed the implied warranty of fitness for a particular purpose, a conspicuous writing would have been required. Although Forbes could not be held liable for breaching the implied warranty of merchantability, he could be held liable for breaching the warranty of fitness—because he made no *written* disclaimer with respect to that warranty.

At the center of the dispute in the following case was a disclaimer of all warranties accompanying a lease transaction. In defense against a suit for amounts due under the lease, the lessee contended that the leased goods were defective when delivered and not fit for the purpose intended and that, for this reason, the consideration for the contract failed. What was the effect of the disclaimer in this situation?

CASE 17.2 International Turbine Services, Inc. v. VASP Brazilian Airlines

United States Court of Appeals,
Fifth Circuit, 2002.
278 F.3d 494.
http://www.ca5.uscourts.gov/opinions.aspx[a]

a. Under the heading "Search for opinions where:" in the box labeled "and/or Title contains text:" enter "VASP" and click on "Search." In the results, scroll to the name of the case and click on the docket number to access the opinion. The Clerk's Office of the U.S. Court of Appeals for the Fifth Circuit maintains this Web site.

FACTS On October 1, 1997, International Turbine Services, Inc. (ITS), leased an aircraft turbine engine to VASP Brazilian Airlines. The lease required ITS to deliver the engine with a Federal Aviation Administration (FAA)–approved service tag. VASP otherwise leased the engine in "'AS IS, WHERE IS' condition and with all faults." VASP bore "the risk of loss and damage to the Engine and all component parts from any and every cause whatsoever" with one exception: ITS agreed to

CASE 17.2—CONTINUED

overhaul and repair "time-controlled" and "on-condition" parts.[b] At the end of the term—which the parties extended through August 18, 1998—VASP was to return the engine in operable condition to ITS's facility in Dallas. On June 15, the pilot of a VASP plane on which the engine was mounted aborted takeoff due to strong vibrations in the engine. VASP discovered that a high-pressure turbine (HPT) blade had failed, causing severe damage to the engine. VASP disputed responsibility for the repair cost and stopped making payments under the lease. ITS filed a suit against VASP, alleging breach of contract. A federal district court awarded ITS $8,825,000 in damages, including the cost to repair the engine and the past-due lease payments. On appeal to the U.S. Court of Appeals for the Fifth Circuit, VASP argued in part that it had not bargained for a defective engine.

ISSUE Did VASP receive its bargained-for consideration when it accepted the engine?

DECISION Yes. The U.S. Court of Appeals for the Fifth Circuit affirmed the judgment of the lower court. The appellate court held that VASP could not successfully assert the defense of a failure of consideration because "[t]he Lease validly excludes all warranties with the exception of title and requires only delivery of an engine with an FAA-approved return to service tag." This exclusion meant that VASP could not legitimately complain about the condition of the leased goods.

b. A time-controlled part must be replaced or repaired after a specified number of hours or cycles. An on-condition part must be replaced or repaired whenever, on inspection, it does not comply with relevant FAA specifications.

REASON VASP "acknowledge[d] the engine arrived with the appropriate FAA tag," but argued that the tag "constitutes an implied representation that all applicable maintenance regulations and manufacturers' recommendations have been followed, including the manufacturer's recommendations regarding HPT blades." The court reiterated that "this is precisely the type of implied representation or warranty the Lease expressly excludes." The court added that "VASP representatives signed an Equipment Delivery Receipt acknowledging the engine's compliance with the terms and conditions of the Lease. VASP also was sufficiently satisfied with the engine's performance to execute a series of extensions and amendments to the original two-month lease term."

FOR CRITICAL ANALYSIS—Social Consideration *Suppose that the lease in this case had not contained a disclaimer and that the HPT blade had been defective when VASP received it. Further suppose that during a takeoff attempt by a VASP pilot with the defective HPT blade, the plane dropped suddenly and violently, causing injury to a passenger who was onboard. Can the VASP passenger who was not a party to the contract sue ITS based on the implied warranty of fitness? Why or why not?*

Buyer's or Lessee's Examination or Refusal to Inspect If a buyer or lessee actually examines the goods (or a sample or model) as fully as desired before entering into a contract, or if the buyer or lessee refuses to examine the goods on the seller's or lessor's request that he or she do so, *there is no implied warranty with respect to defects that a reasonable examination would reveal or defects that are actually found* [UCC 2–316(3)(b), 2A–214(2)(b)]. ■ **EXAMPLE 17.6** Suppose that Joplin buys an ax at Gershwin's Hardware Store. No express warranties are made. Gershwin requests that Joplin inspect the ax before buying it, but she refuses. Had she inspected the ax, she would have noticed that its handle was obviously cracked. If Joplin is later injured by the defective ax, she normally will not be able to hold Gershwin liable for breach of the warranty of merchantability because she would have spotted the defect during an inspection. ■

Warranty Disclaimers and Unconscionability The UCC sections dealing with warranty disclaimers do not refer specifically to unconscionability as a factor. Ultimately, however, the courts will test warranty disclaimers with reference to the UCC's unconscionability standards [UCC 2–302, 2A–108]. Such things as lack of bargaining position, "take-it-or-leave-it" choices, and a buyer's or lessee's failure to understand or know of a warranty disclaimer will become relevant to the issue of unconscionability.

MAGNUSON-MOSS WARRANTY ACT

The Magnuson-Moss Warranty Act of 1975[11] was designed to prevent deception in warranties by making them easier to understand. The act is mainly enforced by the Federal Trade Commission (FTC). Additionally, the U.S. attorney general or a consumer who has been injured can enforce the act if informal procedures for settling disputes prove to be ineffective. The act modifies UCC warranty rules to some extent when consumer transactions are involved. The UCC, however, remains the primary codification of warranty rules for commercial transactions.

11. 15 U.S.C. Sections 2301–2312.

Under the Magnuson-Moss Act, no seller or lessor is required to give an express written warranty for consumer goods sold. If a seller or lessor chooses to make an express written warranty, however, and the cost of the consumer goods is more than $10, the warranty must be labeled as "full" or "limited." In addition, if the cost of the goods is more than $15, by FTC regulation, the warrantor must make certain disclosures fully and conspicuously in a single document in "readily understood language." This disclosure must state the names and addresses of the warrantor(s), what specifically is warranted, procedures for enforcing the warranty, any limitations on warranty relief, and that the buyer has legal rights.

Full Warranty Although a *full warranty* may not cover every aspect of the consumer product sold, what it does cover ensures some type of consumer satisfaction in the event that the product is defective. A full warranty requires free repair or replacement of any defective part; if the product cannot be repaired within a reasonable time, the consumer has the choice of a refund or a replacement without charge. A full warranty frequently does not have a time limit on it. Any limitation on consequential damages must be *conspicuously* stated. Additionally, the warrantor need not perform warranty services if the problem with the product was caused by the consumer's unreasonable use of the product.

Limited Warranty A *limited warranty* arises when the written warranty fails to meet one of the minimum requirements of a full warranty. The fact that only a limited warranty is being given must be conspicuously stated. If the only distinction between a limited warranty and a full warranty is a time limitation, the Magnuson-Moss Warranty Act allows the warrantor to identify the warranty as a full warranty by such language as "full twelve-month warranty."

Implied Warranties Implied warranties do not arise under the Magnuson-Moss Warranty Act; they continue to be created according to UCC provisions. Implied warranties may not be disclaimed under the Magnuson-Moss Warranty Act, however. Although a warrantor can impose a time limit on the duration of an implied warranty, it must correspond to the duration of the express warranty.[12]

12. The time limit on an implied warranty occurring by virtue of the warrantor's express warranty must, of course, be reasonable, conscionable, and set forth in clear and conspicuous language on the face of the warranty.

Lemon Laws

Some purchasers of defective automobiles—called "lemons"—found that the remedies provided by the UCC were inadequate due to limitations imposed by the seller. In response to the frustrations of these buyers, all of the states have enacted *lemon laws*. Basically, lemon laws provide that if an automobile under warranty possesses a defect that significantly affects the vehicle's value or use, and the seller fails to remedy the defect within a specified number of opportunities (usually four), the buyer is entitled to a new car, replacement of defective parts, or return of all consideration paid.

In most states, lemon laws require an aggrieved new-car owner to notify the dealer or manufacturer of the problem and to give the dealer or manufacturer an opportunity to solve it. If the problem persists, the owner must then submit complaints to the arbitration program specified in the manufacturer's warranty before taking the case to court. Decisions by arbitration panels are binding on the manufacturer (that is, cannot be appealed by the manufacturer to the courts) but usually are not binding on the purchaser.

Most major automobile companies use their own arbitration panels. Some companies, however, subscribe to independent arbitration services, such as those provided by the Better Business Bureau. Although arbitration boards must meet state and/or federal standards of impartiality, industry-sponsored arbitration boards have been criticized for not being truly impartial. In response to this criticism, some states have established mandatory, government-sponsored arbitration programs for lemon-law disputes.

Product Liability

Manufacturers, sellers, and lessors of goods can be held liable to consumers, users, and bystanders for physical harm or property damage that is caused by the goods. This is called **product liability.** Product liability may be based on the warranty theories just discussed, as well as on the theories of negligence, misrepresentation, and strict liability. We look here at product liability based on negligence and misrepresentation.

NEGLIGENCE

Chapter 4 defined *negligence* as the failure to exercise the degree of care that a reasonable, prudent person would have exercised under the circumstances. If a manufacturer fails to exercise "due care" to make a product safe, a per-

MANAGEMENT PERSPECTIVE

The "Sophisticated-User" Doctrine

MANAGEMENT FACES A LEGAL ISSUE

A manufacturer's duty to warn users of the dangers associated with certain uses or misuses of its products presents an ongoing challenge for management. One issue that has come before the courts in recent years has to do with the following question: Should a manufacturer be held liable for failing to warn users of product-associated dangers when the users are just as knowledgeable about the dangers—if not more so—as the manufacturer?

For example, suppose that a foundry purchases silica sand for use in its casting operations. During the casting process, the silica sand is pulverized, creating silica dust and minute particles that, when inhaled by workers, can cause silicosis, a respiratory disease. If a foundry worker who contracts silicosis sues the sand manufacturer for failing to warn of this hazard, can the manufacturer avoid liability by claiming that the foundry was as knowledgeable as the manufacturer about the dangers of silicosis and ways to protect workers against it? How the courts answer this question obviously has significant implications for business owners and managers.

WHAT THE COURTS SAY

To date, the sophisticated-user doctrine has been adopted in a number of states. In the scenario just described, a Wisconsin court applied this doctrine for the first time and dismissed the case against the sand manufacturer.[a] In another Wisconsin case, however, a court refused to grant a manufacturer's motion for summary judgment based on the sophisticated-user defense. The case involved a manufacturer of diving boards that had sold a diving board and platform to a high school. When a student on the swim team was injured while diving off the board into a shallow part of the pool, the student sued the manufacturer for failing to warn against the dangers of locating the platform at a shallow depth. The manufacturer argued that it had no duty to warn because the high school was a sophisticated user. After all, claimed the manufacturer, the school was a member of the Wisconsin Interscholastic Athletic Association, an organization that provides high schools with information on safety practices associated with sports activities, including diving. The court ruled that the case should go to trial to determine the extent of the manufacturer's knowledge about the high school—that is, whether the manufacturer actually knew that the school was aware that the platform would be dangerous if used in less than five feet of water. If so, then the sophisticated-user doctrine would apply, and the manufacturer would not be held liable.[b]

IMPLICATIONS FOR MANAGERS

While the sophisticated-user doctrine may be a boon for manufacturers, it is not easy to predict how courts in a particular jurisdiction might apply the doctrine. Some courts have dismissed claims against manufacturers based on the doctrine (as in the case against the sand manufacturer discussed previously). Other courts, however, focus on the manufacturer's level of knowledge about the sophistication of the user (as in the case against the diving board manufacturer).

a. *Haase v. Badger Mining Corp.*, 2004 WI 97, 274 Wis.2d 143, 682 N.W.2d 389 (2004).

b. *Mohr v. St. Paul Fire & Marine Insurance Co.*, 2004 WI App. 5, 269 Wis.2d 302, 674 N.W.2d 576 (2004).

son who is injured by the product may sue the manufacturer for negligence.

Due Care Must Be Exercised The manufacturer must exercise due care in designing the product, selecting the materials, using the appropriate production process, assembling the product, and inspecting and testing any purchased products that are used in the final product sold by the manufacturer. The duty of care also extends to the placement of adequate warnings on the label informing the user of dangers of which an ordinary person might not be aware. (For a discussion of whether manufacturers have a duty to warn sophisticated and knowledgeable users of the dangers associated with the manufacturers' products, see this chapter's *Management Perspective* feature.)

Privity of Contract Not Required A product liability action based on negligence does not require privity of contract between the injured plaintiff and the negligent defendant manufacturer. Section 395 of the *Restatement (Second) of Torts* states as follows:

> A manufacturer who fails to exercise reasonable care in the manufacture of a chattel [movable good] which, unless carefully made, he [or she] should recognize as involving an unreasonable risk of causing physical harm to those who lawfully use it for a purpose for which the manufacturer should expect it to be used and to those whom he [or she] should expect to be endangered by its probable use, is subject to liability for physical harm caused to them by its lawful use in a manner and for a purpose for which it is supplied.

In other words, a manufacturer is liable for its failure to exercise due care to any person who sustains an injury proximately caused by a negligently made (defective) product, regardless of whether the injured person is in privity of contract with the negligent defendant manufacturer or lessor. Relative to the long history of the common law, this exception to the privity requirement is a fairly recent development, dating to the early part of the twentieth century.

MISREPRESENTATION

When a fraudulent misrepresentation has been made to a user or consumer, and that misrepresentation ultimately results in an injury, the basis of liability may be the tort of fraud. For example, the intentional mislabeling of packaged cosmetics and the intentional concealment of a product's defects would constitute fraudulent misrepresentation.

Strict Product Liability

Under the doctrine of strict liability (discussed in Chapter 4), people may be liable for the results of their acts regardless of their intentions or their exercise of reasonable care. Under this doctrine, liability does not depend on privity of contract. The injured party does not have to be the buyer or a third party beneficiary, as required under contract warranty theory. Indeed, the provisions of the UCC do not govern this type of liability in law because it is a tort doctrine, not a principle of the law relating to sales contracts.

STRICT PRODUCT LIABILITY AND PUBLIC POLICY

Strict product liability is imposed by law as a matter of public policy—the general principle of the law that prohibits actions that tend to be injurious to the public. With respect to strict liability, the policy rests on the threefold assumption that (1) consumers should be protected against unsafe products; (2) manufacturers and distributors should not escape liability for faulty products simply because they are not in privity of contract with the ultimate user of those products; and (3) manufacturers, sellers, and lessors of products are generally in a better position than consumers to bear the costs associated with injuries caused by their products—costs that they can ultimately pass on to all consumers in the form of higher prices.

California was the first state to impose strict product liability in tort on manufacturers. In a landmark 1963 decision, *Greenman v. Yuba Power Products, Inc.*,[13] the California Supreme Court set out the reason for applying tort law rather than contract law in cases involving consumers injured by defective products. According to the court, the "purpose of such liability is to [e]nsure that the costs of injuries resulting from defective products are borne by the manufacturers . . . rather than by the injured persons who are powerless to protect themselves."

REQUIREMENTS FOR STRICT LIABILITY

Section 402A of the *Restatement (Second) of Torts* indicates how the drafters envisioned that the doctrine of strict liability should be applied. It was issued in 1964, and during the next decade, it became a widely accepted statement of the liabilities of sellers of goods (including manufacturers, processors, assemblers, packagers, bottlers, wholesalers, distributors, retailers, and lessors). Section 402A states as follows:

> (1) One who sells any product in a defective condition unreasonably dangerous to the user or consumer or to his [or her] property is subject to liability for physical harm thereby caused to the ultimate user or consumer or to his [or her] property, if
> (a) the seller is engaged in the business of selling such a product, and
> (b) it is expected to and does reach the user or consumer without substantial change in the condition in which it is sold.
> (2) The rule stated in Subsection (1) applies although
> (a) the seller has exercised all possible care in the preparation and sale of his [or her] product, and
> (b) the user or consumer has not bought the product from or entered into any contractual relation with the seller.

The Six Requirements for Strict Liability The bases for an action in strict liability as set forth in Section 402A of the *Restatement (Second) of Torts*, and as the doctrine came to be commonly applied, can be summarized as a series of six requirements, which are listed here. Depending on the jurisdiction, if these requirements are met, a manufacturer's liability to an injured party can be virtually unlimited.

1. The product must be in a defective condition when the defendant sells it.
2. The defendant normally must be engaged in the business of selling (or otherwise distributing) that product.
3. The product must be unreasonably dangerous to the user or consumer because of its defective condition (in most states).
4. The plaintiff must incur physical harm to self or property by use or consumption of the product.
5. The defective condition must be the proximate cause of the injury or damage.
6. The goods must not have been substantially changed from the time the product was sold to the time the injury was sustained.

13. 59 Cal.2d 57, 377 P.2d 897, 27 Cal.Rptr. 697 (1963).

Unreasonably Dangerous Products Under the six requirements just listed, in any action against a manufacturer, seller, or lessor, the plaintiff does not have to show why or in what manner the product became defective. To recover damages, however, the plaintiff must show that the product was so "defective" as to be "unreasonably dangerous"; that the product caused the plaintiff's injury; and that at the time the injury was sustained, the product was essentially in the same condition as when it left the hands of the defendant manufacturer, seller, or lessor.

A court may consider a product so defective as to be an **unreasonably dangerous product** if either (1) the product is dangerous beyond the expectation of the ordinary consumer or (2) a less dangerous alternative was economically feasible for the manufacturer, but the manufacturer failed to produce it. As will be discussed next, a product may be unreasonably dangerous due to a flaw in the manufacturing process, a design defect, or an inadequate warning.

PRODUCT DEFECTS—*RESTATEMENT (THIRD) OF TORTS*

Because Section 402A of the *Restatement (Second) of Torts* did not clearly define such terms as *defective* and *unreasonably dangerous,* they were interpreted differently by different courts. In 1997, to address these concerns, the American Law Institute issued the *Restatement (Third) of Torts: Products Liability.* The *Restatement* defines the three types of product defects that have traditionally been recognized in product liability law—manufacturing defects, design defects, and inadequate warnings.

Manufacturing Defects According to Section 2(a) of the *Restatement (Third) of Torts,* a product "contains a manufacturing defect when the product departs from its intended design even though all possible care was exercised in the preparation and marketing of the product." This statement imposes liability on the manufacturer (and on the wholesaler and retailer) whether or not the manufacturer acted "reasonably." This is strict liability, or liability without fault.

Design Defects A design defect (or an inadequate warning, discussed shortly), by nature, affects all of the units of a particular product. A product "is defective in design when the foreseeable risks of harm posed by the product could have been reduced or avoided by the adoption of a reasonable alternative design by the seller or other distributor, or a predecessor in the commercial chain of distribution, and the omission of the alternative design renders the product not reasonably safe."[14]

14. *Restatement (Third) of Torts: Products Liability,* Section 2(b).

***—Problems with the* Restatement (Second) of Torts.** In the past, different states applied different tests to determine whether a product had a design defect under Section 402A of the *Restatement (Second) of Torts.* Some of the tests used were controversial, particularly one that focused on "consumer expectations" concerning a product.

***—The Test for Design Defect under the* Restatement (Third) of Torts.** The test prescribed by the *Restatement (Third) of Torts* focuses on a product's actual design and the reasonableness of that design. To succeed in a product liability suit alleging a design defect, a plaintiff has to show that there is a reasonable alternative design. In other words, a manufacturer or other defendant is liable only when the harm was reasonably preventable. According to the Official Comments accompanying the *Restatement (Third) of Torts,* factors that a court may consider on this point include

> the magnitude and probability of the foreseeable risks of harm, the instructions and warnings accompanying the product, and the nature and strength of consumer expectations regarding the product, including expectations arising from product portrayal and marketing. The relative advantages and disadvantages of the product as designed and as it alternatively could have been designed may also be considered. Thus, the likely effects of the alternative design on production costs; the effects of the alternative design on product longevity, maintenance, repair, and esthetics; and the range of consumer choice among products are factors that may be taken into account.

Can violent video games and Internet transmissions be deemed "defective products"? For a discussion of this question, see this chapter's *Adapting the Law to the Online Environment* feature on the next page.

Inadequate Warnings Product warnings and instructions alert consumers to the risks of using a product. A "reasonableness" test applies to this material. A product "is defective because of inadequate instructions or warnings when the foreseeable risks of harm posed by the product could have been reduced or avoided by the provision of reasonable instructions or warnings by the seller or other distributor, or a predecessor in the commercial chain of distribution, and the omission of the instructions or warnings renders the product not reasonably safe."[15] Generally, a seller must warn those who purchase its product of the harm that can result from the *foreseeable misuse* of the product as well.

Important factors for a court to consider under the *Restatement (Third) of Torts* include the risks of a product, the "content and comprehensibility" and "intensity

15. *Restatement (Third) of Torts: Products Liability,* Section 2(c).

ADAPTING THE LAW TO THE ONLINE ENVIRONMENT

School Shootings and Strict Liability

Over the past decade, school shootings have led to lawsuits that pose a novel question for the courts: Can the producers and distributors of violence-laden media, such as video games and Internet transmissions, be held liable for the shootings? In one case, for example, the plaintiffs were the parents of several students who were killed by their classmate, Michael Carneal, in a 1997 high school shooting in Kentucky. The plaintiffs sued Meow Media, Inc., and other companies (the defendants), alleging that the defendants should be held liable for the shootings. The plaintiffs contended that the defendants' products—including videos, video games, and Internet transmissions—"desensitized" Carneal to violence. Carneal's indifference to violence, in turn, "caused" the shootings.

THE NEGLIGENCE CLAIM

One of the plaintiffs' claims was that the defendants had breached a duty of care by distributing such violent products and were thus negligent. The court, however, did not agree with the plaintiffs that the defendants owed a duty of care to the victims. Recall from Chapter 4 that a defendant's duty of care extends only to those who are injured as a result of a *foreseeable* risk. In the court's eyes, a school shooting was not a foreseeable risk for the defendants. Thus, the court dismissed the negligence claim.

WERE THE "PRODUCTS" DEFECTIVE?

The plaintiffs also alleged that the defendants should be held liable in strict product liability because the violence contained in their products rendered those products "defective." The court never reached the issue of whether the products were defective, however, because it concluded that the violence communicated by the videos, video games, and Internet transmissions was not a "product."

Although videos and video games may be considered products for some purposes, the communications within those videos and games were not products for purposes of strict liability. The argument that an Internet transmission could constitute a product also failed. The plaintiffs had asserted that if electricity could be labeled a product, as it has been in some cases, then Internet transmissions, which can be characterized as a series of electrical impulses, should also be considered a product. The court pointed out, though, that the relevant state law defined the term *product* as something tangible—something that can be touched, felt, or otherwise perceived by the senses. The communicative element (ideas and images) of an Internet transmission was not a tangible object.

Furthermore, stated the court, even assuming that the videos, video games, and Internet transmissions were products, the plaintiffs could not succeed in a strict product liability action. For strict product liability to apply, the injuries complained of must have been caused by the products themselves. In this case, the injuries were caused not by the products but by Carneal's *reaction* to the products.[a]

FOR CRITICAL ANALYSIS

Another defense raised by the defendants in this case was that the expression in their videos, video games, and Internet transmissions was a protected form of speech under the First Amendment. Should such speech ever be restrained in the interests of protecting society against violence? Why or why not?

a. *James v. Meow Media, Inc.,* 300 F.3d 683 (6th Cir. 2002). For another case on this issue in which the court reached similar conclusions, see *Sanders v. Acclaim Entertainment, Inc.,* 188 F.Supp.2d 1264 (D.Colo. 2002).

of expression" of warnings and instructions, and the "characteristics of expected user groups."[16] For example, children will likely respond more readily to bright, bold, simple warning labels, while educated adults might need more detailed information.

There is no duty to warn about risks that are obvious or commonly known. Warnings about such risks do not add to the safety of a product and could even detract from it by making other warnings seem less significant. The obviousness of a risk and a user's decision to proceed in the face of that risk may be a defense in a product liability suit based on a warning defect. (This defense and other defenses in product liability suits will be discussed later in this chapter.)

16. *Restatement (Third) of Torts: Products Liability,* Section 2, Comment h.

MARKET-SHARE LIABILITY

Generally, in all cases involving product liability, a plaintiff must prove that the defective product that caused her or his injury was the product of a specific defendant. Since the 1980s, however, some courts have dropped this requirement when a plaintiff cannot prove which of many distributors of a harmful product supplied the particular product that caused the injuries. Instead, under a theory of market-share liability, all firms that manufactured and distributed the product during the period in question are

held liable for the plaintiff's injuries in proportion to the firms' respective shares of the market for that product during that period. ■ EXAMPLE 17.7 In one case, a plaintiff who was a hemophiliac received injections of a blood protein known as antihemophiliac factor (AHF) concentrate. The plaintiff later tested positive for HIV, the virus that causes AIDS (acquired immune deficiency syndrome). Because it was not known which manufacturer was responsible for the particular AHF received by the plaintiff, the court held that all of the manufacturers of AHF could be held liable under a theory of market-share liability.[17] ■ Many jurisdictions, however, do not apply this theory, believing that it deviates too significantly from traditional legal principles.[18]

OTHER APPLICATIONS OF STRICT LIABILITY

Although the drafters of Section 402A of the *Restatement (Second) of Torts* did not take a position on bystanders, virtually all courts extend the strict liability of manufacturers and other sellers to injured bystanders. ■ EXAMPLE 17.8 In one case, an automobile manufacturer was held liable for injuries caused by the explosion of a car's motor. A cloud of steam that resulted from the explosion caused multiple collisions because other drivers could not see well.[19] ■

The rule of strict liability is also applicable to suppliers of component parts. ■ EXAMPLE 17.9 General Motors buys brake pads from a subcontractor and puts them in Chevrolets without changing their composition. If those pads are defective, both the supplier of the brake pads and General Motors will be held strictly liable for the damages caused by the defects. ■

STATUTES OF REPOSE

As discussed earlier, *statutes of limitations* restrict the time within which an action may be brought. Many states have passed laws, called **statutes of repose,** placing outer time limits on some claims so that the defendant will not be left vulnerable to lawsuits indefinitely. These statutes may limit the time within which a plaintiff can file a product liability suit. Typically, a statute of repose begins to run at an earlier date and runs for a longer time than a statute of limitations. For example, a statute of repose may require that claims be brought within twelve years from the date of sale or manufacture of the defective product. No action can be brought if the injury occurs *after* this statutory period has lapsed. In addition, some of these legislative enactments limit the application of the doctrine of strict liability to new goods only.

17. *Smith v. Cutter Biological, Inc.,* 72 Haw. 416, 823 P.2d 717 (1991). See also *Hymowitz v. Eli Lilly and Co.,* 73 N.Y.2d 487, 539 N.E.2d 1069, 541 N.Y.S.2d 941 (1989).

18. For the Illinois Supreme Court's position on market-share liability, see *Smith v. Eli Lilly and Co.,* 137 Ill.2d 222, 560 N.E.2d 324, 148 Ill.Dec. 22 (1990).

19. For a leading case on this issue, see *Giberson v. Ford Motor Co.,* 504 S.W.2d 8 (Mo. 1974).

Defenses to Product Liability

Manufacturers, sellers, or lessors can raise several defenses to avoid liability for harms caused by their products. We look at some of these defenses in the following subsections.

ASSUMPTION OF RISK

Assumption of risk can sometimes be used as a defense in a product liability action. For example, if a buyer fails to heed a product recall by the seller, a court might conclude that the buyer assumed the risk caused by the defect that led to the recall. To establish such a defense, the defendant must show that (1) the plaintiff knew and appreciated the risk created by the product defect, and (2) the plaintiff voluntarily assumed the risk, even though it was unreasonable to do so. (See Chapter 4 for a more detailed discussion of assumption of risk.)

PRODUCT MISUSE

Similar to the defense of voluntary assumption of risk is that of misuse of the product. Here, the injured party *does not know that the product is dangerous for a particular use* (contrast this with assumption of risk), but the use is not the one for which the product was designed. The courts have severely limited this defense, however. Even if the injured party does not know about the inherent danger of using the product in a wrong way, if the misuse is *foreseeable,* the seller must take measures to guard against it.

COMPARATIVE NEGLIGENCE

Developments in the area of comparative negligence, or fault (discussed in Chapter 4), have also affected the doctrine of strict liability—the most extreme theory of product liability. Whereas previously the plaintiff's conduct was not a defense to strict liability, today many jurisdictions, when apportioning liability and damages, consider the negligent or intentional actions of both the plaintiff and the defendant. This means that even though the product is defective, the manufacturer may not be liable for all of the plaintiff's injuries caused by the product if the plaintiff misused it.

COMMONLY KNOWN DANGERS

The dangers associated with certain products (such as sharp knives and guns) are so commonly known that manufacturers need not warn users of those dangers. If a defendant succeeds in convincing the court that a plaintiff's injury resulted from a *commonly known danger,* the defendant normally will not be liable.

■ EXAMPLE 17.10 A classic case on this issue involved a plaintiff who was injured when an elastic exercise rope that she had purchased slipped off her foot and struck her in the eye, causing a detachment of the retina. The plaintiff claimed that the manufacturer should be liable because it had failed to warn users that the exerciser might slip off a foot in such a manner. The court stated that to hold the manufacturer liable in these circumstances "would go beyond the reasonable dictates of justice in fixing the liabilities of manufacturers." After all, stated the court, "[a]lmost every physical object can be inherently dangerous or potentially dangerous in a sense. . . . A manufacturer cannot manufacture a knife that will not cut or a hammer that will not mash a thumb or a stove that will not burn a finger. The law does not require [manufacturers] to warn of such common dangers."[20] ■

In the following case, the plaintiffs alleged that McDonald's, the well-known fast-food chain, should be held liable for failing to warn customers of the adverse health effects of eating its food products.

20. *Jamieson v. Woodward & Lothrop,* 247 F.2d 23, 101 D.C.App. 32 (1957).

CASE 17.3 Pelman v. McDonald's Corp.

United States District Court,
Southern District of New York, 2003.
237 F.Supp.2d 512.

FACTS McDonald's, with about 13,000 restaurants in the United States, has a 43 percent share of the U.S. fast-food market. Ashley Pelman, a New York resident, and other teenagers who often ate at McDonald's outlets became overweight and developed adverse health effects. Their parents (the plaintiffs) filed a suit in a New York state court against McDonald's and others, alleging, among other things, that the defendants failed to warn of the quantities, qualities, and levels of cholesterol, fat, salt, sugar, and other ingredients in their products, and that a diet high in fat, salt, sugar, and cholesterol could lead to obesity and health problems. The suit was transferred to a federal district court. The defendants filed a motion to dismiss the complaint.

ISSUE Were the products consumed by the plaintiffs dangerous in any way that was not open and obvious to a reasonable consumer?

DECISION No. The court dismissed the plaintiffs' complaint.

REASON The plaintiffs asserted in part that McDonald's failed to include nutritional labeling on its products. The court pointed out that "McDonald's has made its nutritional information available online and * * * upon request. Unless McDonald's has specifically promised to provide nutritional information on all its products * * * , plaintiffs do not state a claim." The court added that the plaintiffs might have alleged that McDonald's products are so extraordinarily unhealthful that they are "outside the reasonable contemplation of the consuming public." Instead, they merely alleged that the foods "contain high levels of cholesterol, fat, salt, and sugar" and that the foods are therefore unhealthful. The court pointed out that it is well known that the products offered by McDonald's "contain high levels of cholesterol, fat, salt, and sugar, and that such attributes are bad for one." The court concluded, "If a person knows or should know that eating copious orders of supersized McDonald's products is unhealthful and may result in weight gain (and its concomitant [related] problems) because of the high levels of cholesterol, fat, salt, and sugar, it is not the place of the law to protect them from their own excesses. Nobody is forced to eat at McDonald's."

WHY IS THIS CASE IMPORTANT? *The ruling in this case, which was the first of its kind, is important because, as the defendants argued and the court acknowledged, a ruling in the plaintiffs' favor "could spawn thousands of similar 'McLawsuits' against restaurants. Even if limited to that ilk of fare dubbed 'fast food,' the potential for lawsuits is great." On occasion, it is up to the courts to draw a line between individuals' responsibility to take care of themselves and society's responsibility to protect those individuals. This is one of those cases.*

OTHER DEFENSES

A defendant can also defend against product liability by showing that there is no basis for the plaintiff's claim. Suppose that a plaintiff alleges that a seller breached an implied warranty. If the seller can prove that he or she effectively disclaimed all implied warranties, the plaintiff cannot recover. Similarly, in a product liability case based on negligence, a defendant who can show that the plaintiff has not met the requirements (such as causation or the breach of a duty of care) for an action in negligence will not be liable. In regard to strict product liability, a defendant could claim that the plaintiff failed to meet one of the requirements for an action in strict liability. If, for example, the defendant establishes that the goods have been subsequently altered after they were sold, the defendant will not be held liable.

TERMS AND CONCEPTS

express warranty 316
implied warranty 317
implied warranty of fitness for a particular purpose 318
implied warranty of merchantability 317
lien 316
product liability 322
statute of repose 327
unreasonably dangerous product 325

CHAPTER SUMMARY Warranties and Product Liability*

WARRANTIES	
Warranties of Title (See pages 315–316.)	The UCC provides for the following warranties of title [UCC 2–312, 2A–211]: 1. *Good title*—A seller warrants that he or she has the right to pass good and rightful title to the goods. 2. *No liens*—A seller warrants that the goods sold are free of any encumbrances (claims, charges, or liabilities—usually called *liens*). A lessor warrants that the lessee will not be disturbed in her or his possession of the goods by the claims of a third party. 3. *No infringements*—A merchant-seller warrants that the goods are free of infringement claims (claims that a patent, trademark, or copyright has been infringed) by third parties. Lessors make similar warranties.
Express Warranties (See pages 316–317.)	1. *Under the UCC*—An express warranty arises under the UCC when a seller or lessor indicates, as part of the basis of the bargain, any of the following: a. An affirmation or promise of fact. b. A description of the goods. c. A sample shown as conforming to the contract goods [UCC 2–313, 2A–210]. 2. *Under the Magnuson-Moss Warranty Act*—Express written warranties covering consumer goods priced at more than $10, *if made,* must be labeled as one of the following: a. Full warranty—Free repair or replacement of defective parts; refund or replacement for goods if they cannot be repaired in a reasonable time. b. Limited warranty—When less than a full warranty is being offered.
Implied Warranty of Merchantability (See pages 317–318.)	When a seller or lessor is a merchant who deals in goods of the kind sold or leased, the seller or lessor warrants that the goods sold or leased are properly packaged and labeled, are of proper quality, and are reasonably fit for the ordinary purposes for which such goods are used [UCC 2–314, 2A–212].
	* Although we have referenced the 2003 amendments to Articles 2 and 2A of the UCC within the text of this chapter, we do not do so in this summary—because the amendments have not as yet become law in any state.

(Continued)

CHAPTER SUMMARY Warranties and Product Liability—Continued

Implied Warranty of Fitness for a Particular Purpose (See pages 318–319.)	Arises when the buyer's or lessee's purpose or use is expressly or impliedly known by the seller or lessor, and the buyer or lessee purchases or leases the goods in reliance on the seller's or lessor's selection [UCC 2–315, 2A–213].
Other Implied Warranties (See page 319.)	Other implied warranties can arise as a result of course of dealing or usage of trade [UCC 2–314(3), 2A–212(3)].
	PRODUCT LIABILITY
Liability Based on Negligence (See pages 322–324.)	1. The manufacturer must use due care in designing the product, selecting materials, using the appropriate production process, assembling and testing the product, and placing adequate warnings on the label or product. 2. Privity of contract is not required. A manufacturer is liable for failure to exercise due care to any person who sustains an injury proximately caused by a negligently made (defective) product.
Liability Based on Misrepresentation (See page 324.)	Fraudulent misrepresentation of a product may result in product liability based on the tort of fraud.
Strict Liability—Requirements (See pages 324–325.)	1. The defendant must sell the product in a defective condition. 2. The defendant must normally be engaged in the business of selling that product. 3. The product must be unreasonably dangerous to the user or consumer because of its defective condition (in most states). 4. The plaintiff must incur physical harm to self or property by use or consumption of the product. (Courts will also extend strict liability to include injured bystanders.) 5. The defective condition must be the proximate cause of the injury or damage. 6. The goods must not have been substantially changed from the time the product was sold to the time the injury was sustained.
Strict Liability—Product Defects (See pages 325–326.)	A product may be defective in three basic ways: 1. In its manufacture. 2. In its design. 3. In the instructions or warnings that come with it.
Market-Share Liability (See pages 326–327.)	When plaintiffs cannot prove which of many distributors of a defective product supplied the particular product that caused the plaintiffs' injuries, some courts apply market-share liability. All firms that manufactured and distributed the harmful product during the period in question are then held liable for the plaintiffs' injuries in proportion to the firms' respective shares of the market, as directed by the court.
Other Applications of Strict Liability (See page 327.)	1. Manufacturers and other sellers are liable for harms suffered by bystanders as a result of defective products. 2. Suppliers of component parts are strictly liable for defective parts that, when incorporated into a product, cause injuries to users.
Defenses to Product Liability (See pages 327–329.)	1. *Assumption of risk*—The user or consumer knew of the risk of harm and voluntarily assumed it. 2. *Product misuse*—The user or consumer misused the product in a way unforeseeable by the manufacturer. 3. *Comparative negligence*—Liability may be distributed between the plaintiff and the defendant under the doctrine of comparative negligence if the plaintiff's misuse of the product contributed to the risk of injury.

CHAPTER SUMMARY Warranties and Product Liability—Continued

Defenses to Product Liability—Continued	4. *Commonly known dangers*—If a defendant succeeds in convincing the court that a plaintiff's injury resulted from a commonly known danger, such as the danger associated with using a sharp knife, the defendant will not be liable. 5. *Other defenses*—A defendant can also defend against a product liability claim by showing that there is no basis for the plaintiff's claim (that the plaintiff has not met the requirements for an action in negligence or strict liability, for example).

FOR REVIEW

Answers for the even-numbered questions in this For Review *section can be found in Appendix A at the end of this text.*

1. What factors determine whether a seller's or lessor's statement constitutes an express warranty or mere "puffing"?
2. What implied warranties arise under the UCC?
3. Can a manufacturer be held liable to any person who suffers an injury proximately caused by the manufacturer's negligently made product?
4. What are the elements of a cause of action in strict product liability?
5. What defenses to liability can be raised in a product liability lawsuit?

QUESTIONS AND CASE PROBLEMS

17–1. Product Liability. Under what contract theory can a seller be held liable to a consumer for physical harm or property damage that is caused by the goods sold? Under what tort theories can the seller be held liable?

QUESTION WITH SAMPLE ANSWER

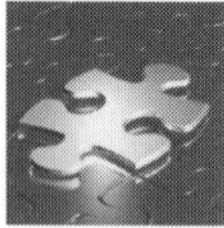

17–2. Carmen buys a television set manufactured by AKI Electronics. She is going on vacation, so she takes the set to her mother's house for her mother to use. Because the set is defective, it explodes, causing considerable damage to her mother's house. Carmen's mother sues AKI for the damage to her house. Discuss the theories under which Carmen's mother can recover from AKI.

For a sample answer to this question, go to Appendix B at the end of this text.

17–3. Warranty Disclaimers. Tandy purchased a washing machine from Marshall Appliances. The sales contract included a provision explicitly disclaiming all express or implied warranties, including the implied warranty of merchantability. The disclaimer was printed in the same size and color as the rest of the contract. The machine turned out to be a "lemon" and never functioned properly. Tandy sought a refund of the purchase price, claiming that Marshall had breached the implied warranty of merchantability. Can Tandy recover her money, notwithstanding the warranty disclaimer in the contract? Explain.

17–4. Implied Warranties. Sam, a farmer, needs to install a two-thousand-pound piece of equipment in his barn. The equipment must be lifted thirty feet into a hayloft. Sam goes to Durham Hardware and tells Durham that he needs some heavy-duty rope to be used on his farm. Durham recommends a one-inch-thick nylon rope, and Sam purchases two hundred feet of it. Sam ties the rope around the piece of equipment, puts the rope through a pulley, and with the aid of a tractor lifts the equipment off the ground. Suddenly, the rope breaks. The equipment crashes to the ground and is extensively damaged. Sam files suit against Durham for breach of the implied warranty of fitness for a particular purpose. Discuss how successful Sam will be with his suit.

CASE PROBLEM WITH SAMPLE ANSWER

17–5. In May 1995, Ms. McCathern and her daughter, together with McCathern's cousin, Ms. Sanders, and her daughter, were riding in Sanders's 1994 Toyota 4Runner. Sanders was driving, McCathern was in the front passenger seat, and the children were in the backseat. Everyone was wearing a seat belt. While the group was traveling south on Oregon State Highway 395 at

a speed of approximately 50 miles per hour, an oncoming vehicle veered into Sanders's lane of travel. When Sanders tried to steer clear, the 4Runner rolled over and landed upright on its four wheels. During the rollover, the roof over the front passenger seat collapsed, and as a result, McCathern sustained serious, permanent injuries. McCathern filed a suit in an Oregon state court against Toyota Motor Corp. and others, alleging in part that the 1994 4Runner "was dangerously defective and unreasonably dangerous in that the vehicle, as designed and sold, was unstable and prone to rollover." What is the test for product liability based on a design defect? What would McCathern have to prove to succeed under that test? [*McCathern v. Toyota Motor Corp.*, 332 Or. 59, 23 P.3d 320 (2001)]

After you have answered this problem, compare your answer with the sample answer given on the Web site that accompanies this text. Go to academic.cengage.com/blaw/wbl, select "Chapter 17," and click on "Case Problem with Sample Answer."

17–6. Implied Warranties. Shalom Malul contracted with Capital Cabinets, Inc., in August 1999 for new kitchen cabinets made by Holiday Kitchens. The price was $10,900. On Capital's recommendation, Malul hired Barry Burger to install the cabinets for $1,600. Burger finished the job in March 2000, and Malul contracted for more cabinets at a price of $2,300, which Burger installed in April. Within a couple of weeks, the doors on several of the cabinets began to "melt"—the laminate (surface covering) began to pull away from the substrate (the material underneath the surface). Capital replaced several of the doors, but the problem occurred again, affecting a total of six of thirty doors. A Holiday Kitchens representative inspected the cabinets and concluded that the melting was due to excessive heat, the result of the doors being placed too close to the stove. Malul filed a suit in a New York state court against Capital, alleging, among other things, a breach of the implied warranty of merchantability. Were these goods "merchantable"? Why or why not? [*Malul v. Capital Cabinets, Inc.*, 191 Misc.2d 399, 740 N.Y.S.2d 828 (N.Y.City Civ.Ct. 2002)]

17–7. Product Liability. In January 1999, John Clark of Clarksdale, Mississippi, bought a paintball gun. Clark practiced with the gun and knew how to screw in the carbon dioxide cartridge, pump the gun, and use its safety and trigger. He hunted and had taken a course in hunter safety education. He was aware that protective eyewear was available for purchase, but he chose not to buy it. Clark also understood that it was "common sense" not to shoot anyone in the face. Chris Rico, another Clarksdale resident, owned a paintball gun made by Brass Eagle, Inc. Rico was similarly familiar with the gun's use and its risks. Clark, Rico, and their friends played a game that involved shooting paintballs at cars whose occupants also had the guns. One night, while Clark and Rico were cruising with their guns, Rico shot at Clark's car, but hit Clark in the eye. Clark filed a suit in a Mississippi state court against Brass Eagle to recover for the injury, alleging in part that its gun was defectively designed. During the trial, Rico testified that his gun "never malfunctioned." In whose favor should the court rule? Why? [*Clark v. Brass Eagle, Inc.*, 866 So.2d 456 (Miss. 2004)]

VIDEO QUESTION

17–8. Go to this text's Web site at academic.cengage.com/blaw/wbl and select "Chapter 17." Click on "Video Questions" and view the video titled *Warranties*. Then answer the following questions.

1. Discuss whether the grocery store's label of a "Party Platter for Twenty" creates an express warranty under the Uniform Commercial Code that the platter will actually serve twenty people.
2. List and describe any implied warranties discussed in the chapter that apply to this scenario.
3. How would a court determine whether Oscar had breached any express or implied warranties concerning the quantity of food on the platter?

Warranties and Product Liability

For updated links to resources available on the Web, as well as a variety of other materials, visit this text's Web site at

academic.cengage.com/blaw/wbl

For an example of a warranty disclaimer, go to

http://www.bizguardian.com/terms.php

The Lemon Law Office Web site provides a variety of information on lemon laws. Take a look at the "Hot Lemon Tips" and "Lemon Wisdom" pages. Go to

http://www.lemonlawoffice.com

For articles explaining product liability, go to FindLaw for Business at

http://smallbusiness.findlaw.com/business-operations/insurance.html

For information on product liability litigation against tobacco companies, including defenses raised by tobacco manufacturers in trial-related documents, go to the State Tobacco Information Center's Web site at

http://stic.neu.edu/index.html

ONLINE LEGAL RESEARCH

Go to the *Fundamentals of Business Law* home page at **academic.cengage.com/blaw/wbl**, select "Chapter 17," and click on "Internet Exercises." There you will find the following Internet research exercises that you can perform to learn more about topics covered in this chapter.

Activity 17–1: LEGAL PERSPECTIVE—Remedial Promises

Activity 17–2: MANAGEMENT PERSPECTIVE—Warranties

Activity 17–3: SOCIAL PERSPECTIVE—Lemon Laws

BEFORE THE TEST

Go to the *Fundamentals of Business Law* home page at **academic.cengage.com/blaw/wbl**, select "Chapter 17," and click on "Interactive Quizzes." You will find a number of interactive questions relating to this chapter.

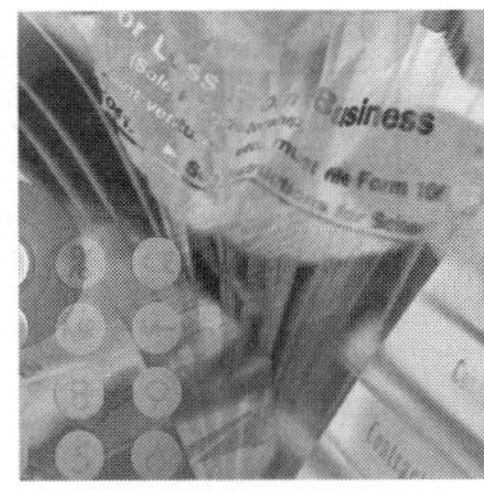

UNIT 4 • EXTENDED CASE STUDY

Mangold v. Nissan North America, Inc.

In Chapter 17, we discussed the Magnuson-Moss Warranty Act, which covers express warranties made in connection with sales of consumer goods costing more than $10. The act permits "a consumer who is damaged by the failure of a supplier, warrantor, or service contractor to comply with any obligation under this [statute], or under a written warranty, implied warranty, or service contract" to, among other things, elect repair, replacement, or refund of defective parts. Some—but not all—courts extend the protections of the act to leases. In this extended case study, we examine *Mangold v. Nissan North America, Inc.*,[1] a decision considering whether leases generally fall under the act, and specifically, whether a lessee of an automobile can maintain a cause of action for breach of warranty under the act.

1. 347 Ill.App.3d 1008, 809 N.E.2d 251, 284 Ill.Dec. 129 (2004). This opinion may be accessed online at http://www.state.il.us/court/Opinions/Search.htm. On that page, in the "Appellate Court" section, click on "2004." On the resulting page, in the "Third District" section, click on "April." On the next page, click on the name of the case to access the opinion.

CASE BACKGROUND

Nissan North America, Inc., makes Infiniti automobiles. Christopher and Erynn Mangold leased a new 2001 Infiniti I30 at Infiniti of Orland Park in Illinois. The dealer sold the vehicle to Debis Financial Services, Inc., which leased it to the Mangolds. Nissan provided a written warranty that covered the vehicle against factory defects for four years or 60,000 miles. The warranty was issued to Debis. As part of the lease, Debis assigned its rights under the warranty to the Mangolds.

Immediately after leasing the car, the Mangolds noticed excessive wind noise and brought the vehicle back to the dealership. The dealer made several attempts, which the warranty covered, to correct the problem. When these attempts proved unsuccessful, the Mangolds filed a suit in an Illinois state court against Nissan to revoke their acceptance of the car, alleging breach of warranty under the Magnuson-Moss Warranty Act.

Nissan filed a motion to dismiss, arguing that the act provides protection only in connection with a *sale* to a *consumer.* Nissan asserted that this transaction involved a *lease* and that *lessees* are not consumers. The court denied the motion. Nissan appealed to a state intermediate appellate court.

MAJORITY OPINION

Justice *LYTTON* delivered the opinion of the court.

* * * *

* * * The Act defines "consumer" as:

> [1] a buyer (other than for purposes of resale) of any consumer product, [2] any person to whom such product is transferred during the duration of an implied or written warranty (or service contract) applicable to the product, and [3] any other person who is entitled by the terms of such warranty (or service contract) * * * to enforce against the warrantor (or service contractor) the obligations of the warranty (or service contract).

* * * *

A person need only meet one of the three criteria to qualify as a consumer. Plaintiffs claim that they are consumers under both the second and third prongs of the Act. The second prong defines a consumer as any person to whom a product is transferred during a warranty's duration. The majority of courts have held that the lessees do not qualify as transferees, citing the Uniform Commercial Code's definition of transfer, which includes the "passing of title." Under this interpretation, lessees would not qualify because in a lease title does not pass to them.

Other courts have determined that the language of the Act is clear and that references to the UCC are unnecessary. Thus, when buyers transfer warranty rights to lessees, the lessees become entitled to enforce the warranty against the warrantor and therefore become "consumers."

We agree with the line of cases that finds that the Act contains sufficient language to define transfer without looking to the UCC for assistance. Since Debis transferred all its warranty rights to the plaintiffs during the duration of the warranty, plaintiffs were entitled to enforce the warranty. *Nothing in the language of the statute suggests that title must pass, and we will not add such a requirement upon review. Therefore, we find that plaintiffs are consumers under the second prong* * * * . [Emphasis added.]

* * * *

Plaintiffs also qualify as consumers under the broader third criterion in the Act * * * . To facilitate the lease, Debis assigned its warranty rights to plaintiffs. Since the warranty does not prohibit an assignment of rights, plaintiffs were entitled to enforce them to their fullest extent, including vehicle repair and replacement.

Moreover, in this case, the dealership actually serviced plaintiffs' vehicle numerous times, and each time the repairs were covered by the warranty. We find no merit in allowing plaintiffs to enforce the warranty for repairs, but denying warranty rights when a violation under the Act is asserted. *Since plaintiffs were entitled to enforce the warranty when the vehicle required repair, they qualified as consumers under the third prong.* [Emphasis added.]

* * * *

Defendants argue that even if plaintiffs qualify as consumers, the Act does not apply to them because the warranty was not part of a "sale." The Act's definitions of implied and written warranty both include the language "made in connection with [a] sale."

Defendant looks to legislative intent in arguing that Congress did not intend to cover leases * * * .

* * * [I]f legislative intent can be ascertained from the statute's plain language, that intent prevails over any other interpretive aid. Thus, we need not examine legislative intent in reviewing the Act if the plain language of the statute sufficiently represents the legislative intent.

The plain language of the statute simply requires that warranties under the Act be issued "in connection with the sale" of a consumer product. It does not require a sale to be made to the ultimate consumer with a passage of title to that party and does not forbid the subsequent assignment of the warranty. Given the plain language of the statute, "in connection with [a] sale" means the sale between the dealership and Debis Financial.

Since the vehicle was sold to the lessor with a manufacturer's written warranty, the warranty was made "in connection with" the sale, and plaintiffs, as assignees, must be allowed to enforce it. * * *

* * * [T]he case [is] remanded for further proceedings.

DISSENTING OPINION

Presiding Justice *HOLDRIDGE,* dissenting.

I respectfully dissent. * * * Here, we are asked to determine whether the Magnuson-Moss Warranty Act applies to lessees. In order to answer that question we must, of course, look to the statute itself. Does the Act clearly and unambiguously, *i.e.* expressly, apply to lessees? A review of the Act shows that it is silent as to lessees. * * *

* * * *

* * * I would then resort to standard rules of statutory construction in order to determine the intent of Congress on this issue. One such rule of statutory construction permits us to look to compare statutes that concern the same subject matter and to consider statutes on related subjects. Here, we have another consumer protection statute for comparison, the Truth in Lending Act, enacted seven years prior to Magnuson-Moss. In that statute, Congress expressly included lessees as consumers protected by that Act. One would reasonably assume that, had Congress similarly intended Magnuson-Moss to protect lessees, it would certainly have included express language as contained in the Truth in Lending Act.

I also note that when statutory language is ambiguous, it is appropriate to resort to extrinsic aids of construction such as an examination of legislative history. Here, when we consider legislative history, it is clear that Congress was warned * * * that lessees were not covered under the language as ultimately passed in the Act. Still Congress chose not to change the language of the Act in order to make it clear that lessees were covered under the Act. I would find this to be clear evidence of Congress' intent NOT to include lessees within the protections of Magnuson-Moss.

QUESTIONS FOR ANALYSIS

1. **Law.** What did the majority hold in this case? What was the reasoning supporting this decision?
2. **Law.** What was the dissent's position on the issue before the court? How did the dissent arrive at this conclusion?
3. **Ethics.** Is it fair for the court to treat *lessees* the same as it would treat *buyers* in enforcing express warranties? Why or why not?
4. **Political Dimensions.** Should there be different rules for sellers and lessors regarding warranties? If so, who should prescribe those rules?
5. **Implications for the Manufacturer.** What should a manufacturer do if it wishes to retain an express warranty in connection with a sale but avoid its application on a lease?

UNIT 5

Negotiable Instruments

Unit Contents

CHAPTER 18

Negotiability, Transferability, and Liability

LEARNING OBJECTIVES

After reading this chapter, you should be able to answer the following questions:

1. What requirements must an instrument meet to be negotiable?
2. What are the requirements for attaining the status of a holder in due course (HDC)?
3. What is the key to liability on a negotiable instrument? What is the difference between signature liability and warranty liability?
4. Certain defenses are valid against all holders, including HDCs. What are these defenses called? Name four defenses that fall within this category.
5. Certain defenses can be used to avoid payment to an ordinary holder of a negotiable instrument but are not effective against an HDC. What are these defenses called? Name four defenses that fall within this category.

Most modern commercial transactions would be inconceivable without negotiable instruments. A **negotiable instrument** is a signed writing (record) that contains an unconditional promise or order to pay an exact sum of money on demand or at a specified future time to a specific person or order, or to bearer. The checks you write to pay for groceries and other items are negotiable instruments.

A negotiable instrument can function as a substitute for cash or as an extension of credit. When a buyer writes a check to pay for goods, the check serves as a substitute for cash. When a buyer gives a seller a promissory note in which the buyer promises to pay the seller the purchase price within sixty days, the seller has essentially extended credit to the buyer for a sixty-day period. For a negotiable instrument to operate *practically* as either a substitute for cash or a credit device, or both, it is essential that the instrument be easily transferable without danger of being uncollectible. Each rule described in the following pages can be examined in light of this essential function of negotiable instruments.

It took hundreds of years for paper to become an acceptable substitute for gold or silver. In the medieval world, merchants engaging in foreign trade used negotiable instruments to finance and conduct their affairs. They did this largely because of the hazards of transporting and safekeeping gold or coins. The English king's courts of those times did not recognize the validity of negotiable instruments, however, so the merchants had to develop their own rules governing the use of these instruments. These rules were enforced by "fair" or "borough" courts. Eventually, the decisions of these courts formed a distinct set of laws that became known as the *Lex Mercatoria* (Law Merchant).

The Law Merchant was codified in England in the Bills of Exchange Act of 1882. In 1896, in the United States, the National Conference of Commissioners on Uniform State Laws (NCCUSL) drafted the Uniform Negotiable

Instruments Law. This law, which had been adopted by all of the states by 1920, was the forerunner of Article 3 of the Uniform Commercial Code (UCC).

some of the section numbers may change slightly once a state has adopted the amendments to Article 3 (subpart 9 may become subpart 12, for example).

Articles 3 and 4 of the UCC

Negotiable instruments must meet special requirements relating to form and content. These requirements, which are imposed by Article 3 of the UCC, will be discussed at length in this chapter. Article 3 also governs the process of *negotiation* (transferring an instrument from one party to another), as will be discussed. Note that UCC 3–104(b) defines *instrument* as a "negotiable instrument." For that reason, whenever the term *instrument* is used in this book, it refers to a negotiable instrument.

The 1990 Revision of Articles 3 and 4

In 1990, a revised version of Article 3 was issued for adoption by the states. Many of the changes to Article 3 simply clarified old sections, but some significantly altered the former provisions. As of this writing, all of the states except New York and South Carolina have adopted the revised article. Therefore, all references to Article 3 in this chapter and in the following chapter are to the *revised* Article 3.

Article 4 of the UCC, which governs bank deposits and collections (to be discussed in Chapter 19), was also revised in 1990. In part, these changes were necessary to reflect changes in Article 3 that affect Article 4 provisions. The revised Articles 3 and 4 are included in Appendix E.

The 2002 Amendments to Articles 3 and 4

In 2002, the NCCUSL and the American Law Institute approved a number of amendments to Articles 3 and 4 of the UCC. One of the purposes of these amendments was to update the law with respect to e-commerce. For example, the amended versions of the articles implement the policy of the Uniform Electronic Transactions Act (see Chapter 13) by removing unnecessary obstacles to electronic communications. Additionally, the word *writing* has been replaced by the term *record* throughout the articles. Other amendments relate to such topics as telephone-generated checks and the payment and discharge of negotiable instruments.

To date, only a few states have adopted these amendments. Therefore, in this chapter and the following chapter, we will indicate in footnotes whenever the amendments significantly alter existing law. Keep in mind, however, that even when the changes are not substantive,

Types of Instruments

The UCC specifies four types of negotiable instruments: *drafts, checks, promissory notes,* and *certificates of deposit* (CDs). These instruments are frequently divided into the two classifications that we will discuss in the following subsections: *orders to pay* (drafts and checks) and *promises to pay* (promissory notes and CDs).

Negotiable instruments may also be classified as either demand instruments or time instruments. A *demand instrument* is payable on demand—that is, it is payable immediately after it is issued and thereafter for a reasonable period of time. All checks are demand instruments because, by definition, they must be payable on demand. A *time instrument* is payable at a future date.

Drafts and Checks (Orders to Pay)

A **draft** is an unconditional written order that involves three parties. The party creating the draft (the **drawer**) orders another party (the **drawee**) to pay money, usually to a third party (the **payee**).

Time Drafts and Sight Drafts A *time draft* is payable at a definite future time. A *sight draft* (or demand draft) is payable on sight—that is, when it is presented to the drawee (usually a bank or financial institution) for payment. A draft can be both a time and a sight draft; such a draft is payable at a stated time after sight. A sight draft may be payable on acceptance. **Acceptance** is the drawee's written promise to pay the draft when it comes due. The usual manner of accepting an instrument is by writing the word *accepted* across the face of the instrument, followed by the date of acceptance and the signature of the drawee.

Exhibit 18–1 on the next page shows a typical time draft. For the drawee to be obligated to honor the order, the drawee must be obligated to the drawer either by agreement or through a debtor-creditor relationship. ■ **EXAMPLE 18.1** On January 16, Ourtown Real Estate Company orders $1,000 worth of office supplies from Eastman Supply Company, with payment due April 16. Also on January 16, Ourtown sends Eastman a draft drawn on its account with the First National Bank of Whiteacre as payment. In this scenario, the drawer is Ourtown, the drawee is Ourtown's bank (First National Bank of Whiteacre), and the payee is Eastman Supply Company. ■

EXHIBIT 18–1 A TYPICAL TIME DRAFT

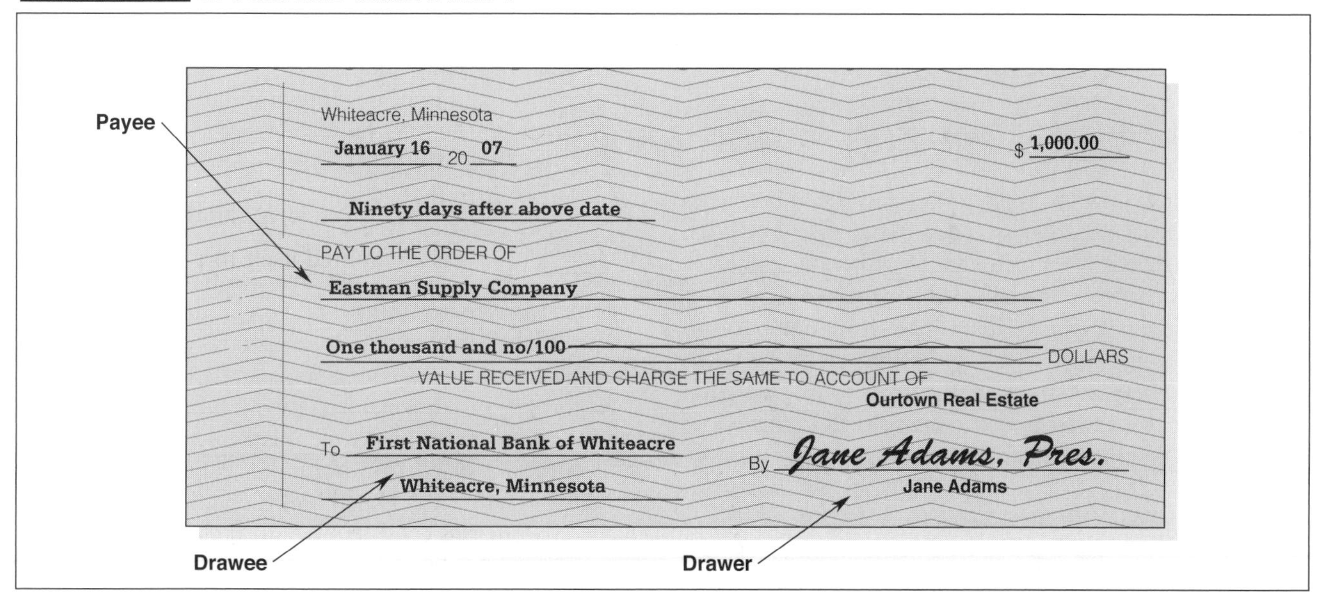
Whiteacre, Minnesota
January 16 20 07
$ 1,000.00
Ninety days after above date
PAY TO THE ORDER OF
Eastman Supply Company
One thousand and no/100 DOLLARS
VALUE RECEIVED AND CHARGE THE SAME TO ACCOUNT OF
Ourtown Real Estate
To First National Bank of Whiteacre
Whiteacre, Minnesota
By Jane Adams, Pres.
Jane Adams

Trade Acceptances A **trade acceptance** is a type of draft that is frequently used in the sale of goods. In a trade acceptance, the seller is both the drawer and the payee on this draft. Essentially, the draft orders the buyer to pay a stated amount to the seller, usually at a stated time in the future. (If the draft orders the buyer's bank to pay, it is called a *banker's acceptance*.)

■ EXAMPLE 18.2 Each year, Jackson River Fabrics sells fabric priced at $50,000 to Comfort Creations, Inc., on terms requiring payment to be made in ninety days. One year Jackson River needs cash, so it draws a *trade acceptance* (see Exhibit 18–2) that orders Comfort Creations to pay $50,000 to the order of Jackson River Fabrics ninety days hence. Jackson River presents the paper to Comfort Creations. Comfort Creations *accepts* the draft, by signing the face of the draft, and returns it to Jackson River Fabrics. The acceptance by Comfort Creations gives rise to an enforceable obligation to pay

EXHIBIT 18–2 A TYPICAL TRADE ACCEPTANCE

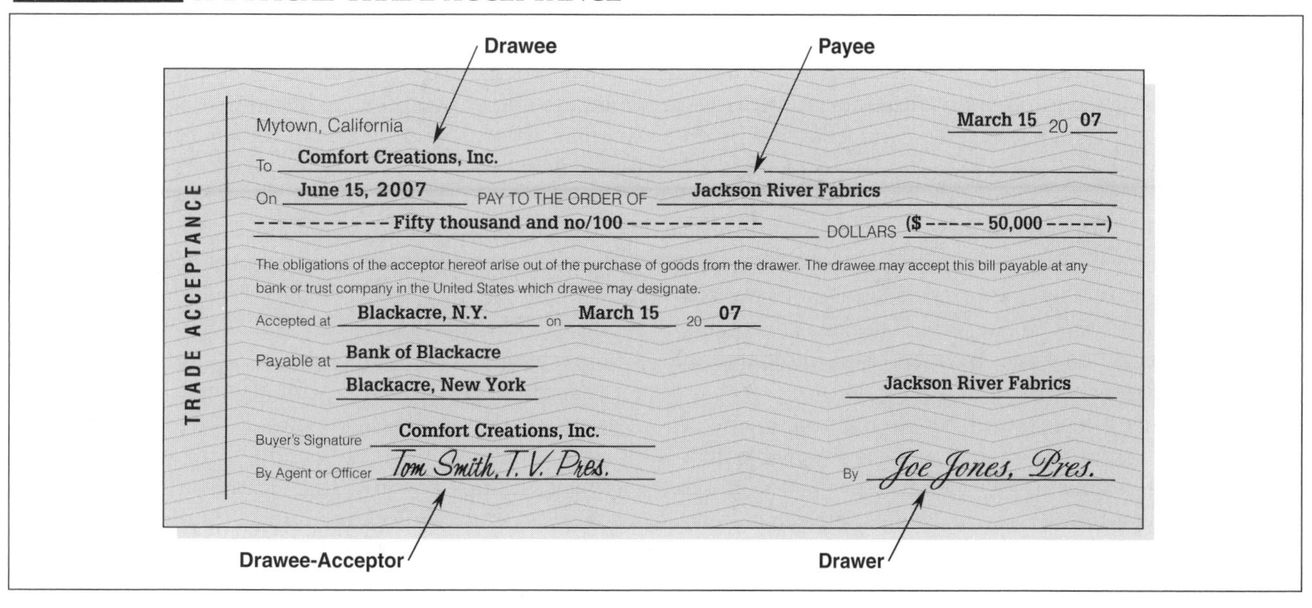
TRADE ACCEPTANCE
Mytown, California
March 15 20 07
To Comfort Creations, Inc.
On June 15, 2007 PAY TO THE ORDER OF Jackson River Fabrics
----------Fifty thousand and no/100---------- DOLLARS ($-----50,000-----)
The obligations of the acceptor hereof arise out of the purchase of goods from the drawer. The drawee may accept this bill payable at any bank or trust company in the United States which drawee may designate.
Accepted at Blackacre, N.Y. on March 15 20 07
Payable at Bank of Blackacre
Blackacre, New York
Jackson River Fabrics
Buyer's Signature Comfort Creations, Inc.
By Agent or Officer Tom Smith, T.V. Pres.
By Joe Jones, Pres.

the draft when it comes due in ninety days. Jackson River can then immediately sell the trade acceptance in the commercial money market for cash. ■

Checks The most commonly used type of draft is a **check.** The writer of the check is the drawer, the bank on which the check is drawn is the drawee, and the person to whom the check is payable is the payee. As mentioned earlier, checks are demand instruments because they are payable on demand.

Checks will be discussed more fully in Chapter 19, but it should be noted here that with certain types of checks, such as *cashier's checks*, the bank is both the drawer and the drawee. The bank customer purchases a cashier's check from the bank—that is, pays the bank the amount of the check—and indicates to whom the check should be made payable. The bank, not the customer, is the drawer of the check—as well as the drawee. The idea behind a cashier's check is that it functions the same as cash, so there is no question about whether the check will be paid—the bank has committed itself to paying the stated amount on demand. The following case illustrates what this means to the payee of a cashier's check.

CASE 18.1 Flatiron Linen, Inc. v. First American State Bank

Colorado Supreme Court, 2001.
23 P.3d 1209.

FACTS Fluffy Reed Foundation, Inc., and one of its officers, Bilgen Reed, promised to secure a $2 million loan for Flatiron Linen, Inc., for which Flatiron paid a fee. Flatiron later accused Fluffy and Reed of fraud in the deal. As a partial refund of the fee, Fluffy issued a check to Flatiron for $4,100, drawn on an account at First American State Bank. When Flatiron attempted to deposit the check, First American returned it due to insufficient funds in the account. Five months later, when the account had sufficient funds, Flatiron took the check to First American and exchanged it for a cashier's check in the amount of $4,100. In the meantime, however, Fluffy had asked First American not to pay the check. When the bank discovered its mistake, it refused to pay the cashier's check. Flatiron filed a suit in a Colorado state court against a number of parties, including First American, from which Flatiron sought to recover the amount of the cashier's check. The court granted a summary judgment in favor of First American. The state intermediate appellate court affirmed this judgment, and Flatiron appealed to the Colorado Supreme Court.

ISSUE Should a cashier's check be treated as the equivalent of cash?

DECISION Yes. The Colorado Supreme Court reversed this part of the lower court's judgment and remanded the case for proceedings consistent with this opinion. Once a cashier's check has been issued, a bank may not legitimately refuse to pay it.

REASON The state supreme court pointed out that it agreed with the majority of courts, which "hold that a cashier's check is the equivalent of cash, accepted when issued." The court explained that UCC 3–104(g) "defines a cashier's check as 'a draft with respect to which the drawer and drawee are the same bank or branches of the same bank.' Because the bank serves as both the drawer and the drawee of the cashier's check, the check becomes a promise by the bank to draw the amount of the check from its own resources and to pay the check upon demand. * * * Once the bank issues and delivers the cashier's check to the payee, the transaction is complete as far as the payee is concerned." The court also noted that "[t]he commercial world treats cashier's checks as the equivalent of cash. People accept cashier's checks as a substitute for cash because the bank, not an individual, stands behind it." To allow a bank not to pay a cashier's check "would be inconsistent with the representation it makes in issuing the check. Such a rule would undermine the public confidence in the bank and its checks and thereby deprive the cashier's check of the essential incident which makes it useful."

FOR CRITICAL ANALYSIS—Economic Consideration *What advantages might cashier's checks have over cash?*

PROMISSORY NOTES AND CERTIFICATES OF DEPOSIT (PROMISES TO PAY)

A **promissory note** is a written promise made by one person (the **maker** of the promise to pay) to another (usually a payee). A promissory note, which is often referred to simply as a *note*, can be made payable at a definite time or on demand. It can name a specific payee or merely be payable to bearer (*bearer instruments* will be discussed later in this chapter). ■ **EXAMPLE 18.3** On April 30, Laurence and

EXHIBIT 18–3 A TYPICAL PROMISSORY NOTE

Payee

☐ SECURITIES ☐ INSURANCE ☐ SAVINGS
☐ OTHER ☐ SEC. AGREEMENT
1. INV. & ACCTS. 2. CONSUMER GOODS
3. EQUIP.

$ **3,000.00** Whiteacre, Minnesota **April 30** 20 **07** Due **6/29/07**
On or before sixty days after date.
for value received, the undersigned jointly and severally promise to pay to the order of THE FIRST NATIONAL BANK OF WHITEACRE at its office in Whiteacre, Minnesota, $ **Three thousand** dollars with interest thereon from date hereof at the rate of **- 8 -** percent per annum (computed on the basis of actual days and a year of 360 days) indicated in No. **7** below.

7 INTEREST IS PAYABLE AT MATURITY
8 INTEREST IS PAID TO MATURITY
9 INTEREST IS PAYABLE ____ BEGINNING ON ____ 20__

SIGNATURE *Laurence E. Roberts* SIGNATURE *Margaret P. Roberts*
SIGNATURE ____ SIGNATURE ____

NO.
OFFICER **Clark**
BY
ACCRUAL
☐ NEW ☐ REN'L
☐ SECURED
☐ UNSECURED

Co-makers

Margaret Roberts sign a writing unconditionally promising to pay "to the order of" the First National Bank of Whiteacre $3,000 (with 8 percent interest) on or before June 29. This writing is a promissory note. ■ A typical promissory note is shown in Exhibit 18–3.

The following case illustrates how a party can collect on a promissory note—in this case, a note signed by a student to borrow funds for his education.

CASE 18.2 United States v. Durbin

United States District Court,
Southern District of Texas,
Houston Division, 1999.
64 F.Supp.2d 635.

FACTS Robert Durbin, a student, borrowed funds for his education and issued (signed) a promissory note for its repayment. The bank from which Durbin borrowed the funds lent it under a federal program to assist students at postsecondary institutions.[a] Ordinarily, repayment begins nine to twelve months after the student borrower fails to carry at least one-half of the normal full-time course load at his or her school. Under this program, the federal government guarantees that the note will be fully repaid. If the student defaults on the repayment, the lender presents the current balance—principal, interest, and costs—to the government. When the government pays the balance, it becomes the lender, and the borrower owes the government directly. After Durbin defaulted on his note and also failed to pay the government, the government filed a suit in a federal district court against Durbin to collect the amount due. The government showed that it owned the note that Durbin issued and that the note was unpaid.

a. Higher Education Act of 1965, 20 U.S.C. Section 1070.

ISSUE Did the government satisfy the requirements to collect the amount of the note from Durbin?

DECISION Yes. The court issued a judgment in favor of the government, holding Durbin liable for the unpaid balance of the note, plus interest, costs, and fees. Durbin issued the note, the government owned it, and it was unpaid.

REASON The court explained first that "[t]he practice and law of notes developed so that people could lend on the credit of a person with confidence that collecting the debt would not be complicated by side issues." In this case, "[t]he government must show three things to win: (1) the defendant is the person who issued the note; (2) the government owns the note; and (3) the note is unpaid. * * * The note may be enforced against the borrower unless he can show that he did not issue it or that he paid it. Unlike many other kinds of cases, the borrower largely has the responsibility to produce evidence. Evidence is not simply saying that something is true; evidence is specific facts of when, who, where, and how much as well as supporting records like canceled checks and tax returns. Under state law, a note is enforceable unless the borrower can show that he paid the note or that the note was

CASE 18.2–CONTINUED

forged." Here, "[t]he government has demonstrated that the defendant issued the promissory note, that the government owns the note, and that the note is in default and unpaid. The law requires that there be a judgment for the government for the principal, interest, costs, and attorney's fees."

FOR CRITICAL ANALYSIS—Social Consideration *Should students have to pay educational loans if they later decide that the education they received is not adequate to allow them to find a "good" job?*

Types of Promissory Notes Notes are used in a variety of credit transactions and often carry the name of the transaction involved. For example, a note that is secured by personal property, such as an automobile, is called a *collateral note* because the property pledged as security for the satisfaction of the debt is called *collateral* (see Chapter 20). A note payable in installments, such as for payment for a suite of furniture over a twelve-month period, is called an *installment note.*

Certificates of Deposit A **certificate of deposit** (CD) is a type of note. A CD is issued when a party deposits funds with a bank that the bank promises to repay, with interest, on a certain date [UCC 3–104(j)]. The bank is the maker of the note, and the depositor is the payee. ■ EXAMPLE 18.4 On February 15, Sara Levin deposits $5,000 with the First National Bank of Whiteacre. The bank issues a CD, in which it promises to repay the $5,000, plus 5 percent interest, on August 15. ■

Certificates of deposit in small denominations (for amounts up to $100,000) are often sold by savings and loan associations, savings banks, and commercial banks. Certificates of deposit for amounts more than $100,000 are called large or jumbo CDs. Exhibit 18–4 shows a typical small CD.

Requirements for Negotiability

For an instrument to be negotiable, it must meet the following requirements:

1. Be in writing.
2. Be signed by the maker or the drawer.
3. Be an unconditional promise or order to pay.
4. State a fixed amount of money.
5. Be payable on demand or at a definite time.
6. Be payable to order or to bearer, unless it is a check.

SIGNATURES

The UCC grants extreme latitude in regard to what constitutes a signature. UCC 1–201(39) provides that a

EXHIBIT 18–4 A TYPICAL SMALL CD

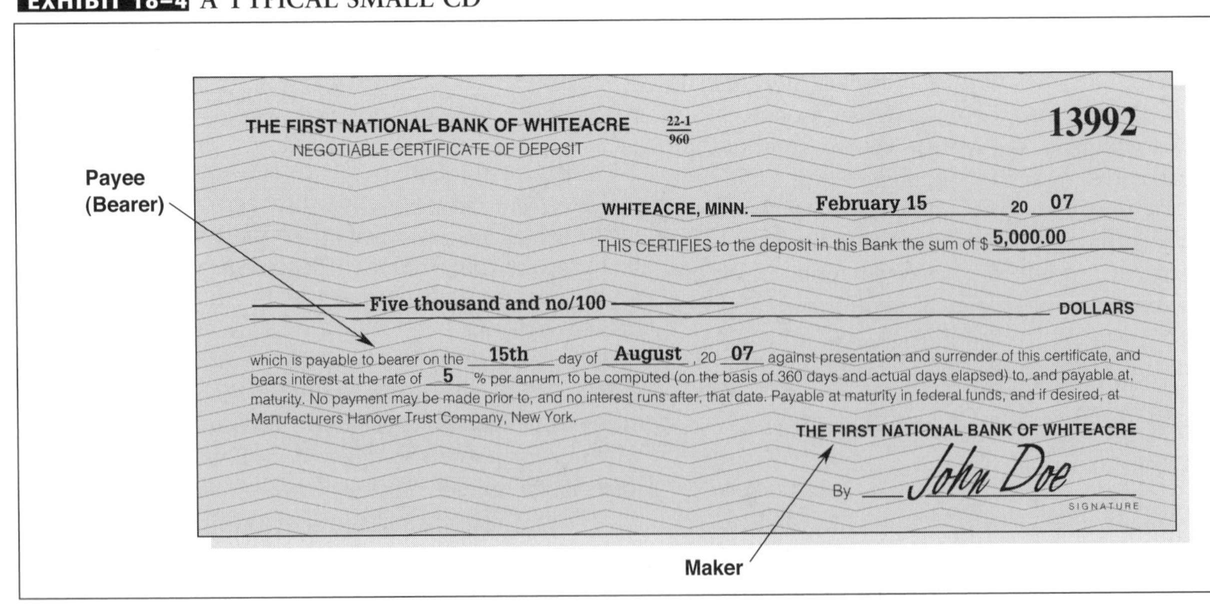

signature may include "any symbol executed or adopted by a party with a present intention to authenticate a writing." UCC 3–401(b) expands on this by stating that a "signature may be made (i) manually or by means of a device or machine, and (ii) by the use of any name, including a trade or assumed name, or by a word, mark, or symbol executed or adopted by a person with present intention to authenticate a writing." Initials, an X (if the writing is signed by a witness), or a thumbprint will normally suffice as a signature. A trade name or an assumed name is also sufficient. Signatures that are placed onto instruments by means of rubber stamps are permitted and frequently used in the business world. If necessary, parol evidence (discussed in Chapter 10) is admissible to identify the signer. When the signer is identified, the signature becomes effective.

UNCONDITIONAL PROMISE OR ORDER TO PAY

For an instrument to be negotiable, it must contain an express order or promise to pay. If a buyer executes a promissory note using the words "I promise to pay Jonas $1,000 on demand for the purchase of these goods," then this requirement for a negotiable instrument is satisfied. A mere acknowledgment of the debt, such as an I.O.U. ("I owe you"), might logically *imply* a promise, but it is not sufficient under the UCC. This is because the promise must be an affirmative (express) undertaking [UCC 3–103(a)(9)]. If such words as "to be paid on demand" or "due on demand" are added to an I.O.U., however, the need for an express promise to pay is satisfied.

A promise or order is conditional (and *not* negotiable) if it states (1) an express condition to payment, (2) that the promise or order is subject to or governed by another writing, or (3) that the rights or obligations with respect to the promise or order are stated in another writing. A mere reference to another writing, however, does not of itself make the promise or order conditional [UCC 3–106(a)]. For example, the words "As per contract" or "This debt arises from the sale of goods X and Y" do not render an instrument nonnegotiable. Similarly, a statement in the instrument that payment can be made only out of a particular fund or source will not render the instrument nonnegotiable [UCC 3–106(b)(ii)]. Finally, a simple statement in an otherwise negotiable note indicating that the note is secured by a mortgage does not destroy its negotiability [UCC 3–106(b)(i)].

A FIXED AMOUNT OF MONEY

The term *fixed amount* means an amount that is ascertainable from the face of the instrument. A demand note payable with 8 percent interest meets the requirement of a fixed amount because its amount can be determined at the time it is payable or at any time thereafter [UCC 3–104(a)]. The rate of interest may also be determined with reference to information that is not contained in the instrument if that information is readily ascertainable by reference to a formula or a source described in the instrument [UCC 3–112(b)]. For instance, an instrument that is payable at the *legal rate of interest* (a rate of interest fixed by statute) is negotiable. Mortgage notes tied to a variable rate of interest (a rate that fluctuates as a result of market conditions) are also negotiable.

PAYABLE ON DEMAND OR AT A DEFINITE TIME

Instruments that are payable on demand include those that contain the words "Payable at sight" or "Payable upon presentment." (*Presentment*—to be discussed fully later in the chapter—is simply the presenting of an instrument for payment.) The very nature of the instrument may indicate that it is payable on demand. For example, a check, by definition, is payable on demand [UCC 3–104(f)]. If no time for payment is specified and the person responsible for payment must pay on the instrument's presentment, the instrument is payable on demand [UCC 3–108(a)].

If an instrument is not payable on demand, to be negotiable it must be payable at a definite time. An instrument is payable at a definite time if it states that it is payable (1) on a specified date, (2) within a definite period of time (such as thirty days) after being presented for payment, or (3) on a date or time readily ascertainable at the time the promise or order is issued [UCC 3–108(b)]. When an instrument is payable by the maker or drawer on or before a stated date, it is clearly payable at a definite time.

PAYABLE TO ORDER OR TO BEARER

An **order instrument** is an instrument that is payable (1) "to the order of an identified person" or (2) "to an identified person or order" [UCC 3–109(b)]. An identified person is the person "to whom the instrument is initially payable" as determined by the intent of the maker or drawer [UCC 3–110(a)]. The identified person, in turn, may transfer the instrument to whomever he or she wishes. Thus, the maker or drawer is agreeing to pay either the person specified on the instrument or whomever that person might designate. In this way, the instrument retains its transferability. ■ **EXAMPLE 18.5** Suppose an instrument states, "Payable to the order of Kako Yung" or "Pay to Kako Yung or order." Clearly, the maker or drawer has indicated that a payment will be

made to Yung or to whomever Yung designates. The instrument is negotiable. ■

A **bearer instrument** is an instrument that does not designate a specific payee [UCC 3–109(a)]. The maker or drawer of a bearer instrument agrees to pay anyone who presents the instrument for payment. The term **bearer** refers to a person in possession of an instrument that is payable to bearer or indorsed in blank (with a signature only) [UCC 1–201(5), 3–109(a), 3–109(c)]. (*Indorsements* will be discussed shortly.)

Transfer of Instruments

Once issued, a negotiable instrument can be transferred by *assignment* or by *negotiation.*

Transfer by Assignment

Recall from Chapter 11 that an assignment is a transfer of rights under a contract. Under general contract principles, a transfer by assignment to an assignee gives the assignee only those rights that the assignor possessed. Any defenses that can be raised against an assignor can normally be raised against the assignee. This same principle applies when an instrument, such as a promissory note, is transferred by assignment. The transferee is then an *assignee.* Sometimes, a transfer fails to qualify as a negotiation because it fails to meet one or more of the requirements of a negotiable instrument, discussed above. When this occurs, the transfer becomes an assignment.

Transfer by Negotiation

Negotiation is the transfer of an instrument in such form that the transferee (the person to whom the instrument is transferred) becomes a holder [UCC 3–201(a)]. The UCC defines a **holder** as any person in possession of an instrument drawn, issued, or indorsed to him or her, to his or her order, to bearer, or in blank [UCC 1–201(20)]. Under UCC principles, a transfer by negotiation creates a holder who, at the very least, receives the rights of the previous possessor [UCC 3–203(b)]. Unlike an assignment, a transfer by negotiation can make it possible for a holder to receive more rights in the instrument than the prior possessor had [UCC 3–202(b), 3–305, 3–306]. A holder who receives greater rights is known as a *holder in due course,* a concept we will discuss later in this chapter.

There are two methods of negotiating an instrument so that the receiver becomes a holder: by *indorsement and delivery* and by *delivery only.* The method used depends on whether the instrument is order paper or bearer paper.

Negotiating Order Instruments If an instrument contains the name of a payee capable of indorsing it, as in "Pay to the order of Lloyd Sorenson," it is an order instrument and is negotiated by delivery with any necessary indorsements. An **indorsement**[1] is a signature placed on an instrument, such as on the back of a check, for the purpose of transferring one's ownership rights in the instrument. An *indorsement in blank* specifies no particular indorsement and can consist of a mere signature. ■ EXAMPLE 18.6 National Express Corporation issues a payroll check "to the order of Lloyd Sorenson." Sorenson takes the check to the supermarket, signs his name on the back (an indorsement), gives it to the cashier (a delivery), and receives cash. Sorenson has *negotiated* the check to the supermarket [UCC 3–201(b)]. ■ Types of indorsements and their effects are listed in Exhibit 18–5 on the following page.

Negotiating Bearer Instruments If an instrument is payable to bearer, it is negotiated by delivery—that is, by transfer into another person's possession. Indorsement is not necessary [UCC 2–301(b)]. The use of bearer instruments thus involves more risk through loss or theft than the use of order instruments.

■ EXAMPLE 18.7 Assume that Richard Kray writes a check "payable to cash" and hands it to Jessie Arnold (a delivery). Kray has issued the check (a bearer instrument) to Arnold. Arnold puts the check in her wallet, which is subsequently stolen. The thief has possession of the check. At this point, the thief has no rights to the check. If the thief "delivers" the check to an innocent third person, however, negotiation will be complete. All rights to the check will be passed absolutely to that third person, and Arnold will lose all rights to recover the proceeds of the check from that person [UCC 3–306]. Of course, Arnold could attempt to recover the money from the thief if the thief can be found. ■

Holder in Due Course (HDC)

Often, whether a holder of an instrument is entitled to obtain payment will depend on whether the holder is an ordinary holder or a *holder in due course.* An ordinary holder obtains only those rights that the transferor had in the instrument. In this respect, a holder has the same status as an assignee (see Chapter 11). Like an assignee, a holder normally is subject to the same defenses that could be asserted against the transferor.

1. Because the UCC uses the spelling *indorse* (*indorsement,* and so forth), rather than *endorse* (*endorsement,* and so forth), we adopt that spelling here and in other chapters in the text.

EXHIBIT 18–5 TYPES OF INDORSEMENTS AND THEIR CONSEQUENCES

WORDS CONSTITUTING THE INDORSEMENT	TYPE OF INDORSEMENT	INDORSER'S SIGNATURE LIABILITY[a]
"Rosemary White"	Blank	Unqualified signature liability on proper presentment and notice of dishonor.[b]
"Pay to Sam Wilson, Rosemary White"	Special	Unqualified signature liability on proper presentment and notice of dishonor.
"Without recourse, Rosemary White"	Qualified (blank for further negotiation)	No signature liability. Transfer warranty liability if breach occurs.[c]
"Pay to Sam Wilson, without recourse, Rosemary White"	Qualified (special for further negotiation)	No signature liability. Transfer warranty liability if breach occurs.
"Pay to Sam Wilson on condition he completes painting my house at 23 Elm Street by 9/1/07, Rosemary White"	Restrictive—conditional (special for further negotiation)	Signature liability only if condition is met. If condition is met, signature liability on proper presentment and notice of dishonor.
"Pay to Sam Wilson only, Rosemary White"	Restrictive—prohibitive (special for further negotiation)	Signature liability only on Sam Wilson's receiving payment. If Wilson receives payment, signature liability on proper presentment and notice of dishonor.
"For deposit, Rosemary White"	Restrictive—for deposit (blank for further negotiation)	Signature liability only on White having amount deposited in her account. If deposit is made, signature liability on proper presentment and notice of dishonor.
"Pay to Ann South in trust for John North, Rosemary White"	Restrictive—trust (special for further negotiation)	Signature liability only on payment to Ann South for John North's benefit. If restriction is met, signature liability on proper presentment and notice of dishonor.

a. *Signature liability* refers to the liability of a party who signs an instrument, as will be discussed later in this chapter. The basic questions include whether there is any liability and, if so, whether it is unqualified or restricted.
b. When an instrument is dishonored—such as when a drawer's bank refuses to cash the drawer's check on proper presentment—an indorser of the check may be liable on it if she or he is given proper *notice of dishonor.*
c. The transferor of an instrument makes certain warranties to the transferee and subsequent holders, and thus, even if the transferor's signature does not render him or her liable on the instrument, he or she may be liable for breach of a transfer warranty. Transfer warranties will be discussed later in this chapter.

In contrast, a **holder in due course** (HDC) is a holder who, by meeting certain acquisition requirements (to be discussed shortly), takes the instrument *free* of most of the defenses and claims that could be asserted against the transferor. Stated another way, an HDC can normally acquire a higher level of immunity than can an ordinary holder in regard to defenses against payment on the instrument or ownership claims to the instrument by other parties.

■ EXAMPLE 18.8 Marcia Cambry signs a $1,000 note payable to Alex Jerrod in payment for some ancient Roman coins. Jerrod negotiates the note to Alicia Larson, who promises to pay Jerrod for it in thirty days. During the next month, Larson learns that Jerrod has breached his contract with Cambry by delivering coins that were not from the Roman era, as promised, and that for this reason Cambry will not honor the $1,000 note. Whether Larson can hold Cambry liable on the note depends on whether Larson has met the requirements for HDC status. If Larson has met these requirements and thus has HDC status, Larson is entitled to payment on the note. If Larson has not met these requirements, she has the status of an ordinary holder, and Cambry's defense of breach of contract against payment to Jerrod will also be effective against Larson. ■

REQUIREMENTS FOR HDC STATUS

The basic requirements for attaining HDC status are set forth in UCC 3–302. A holder of a negotiable instrument is an HDC if she or he takes the instrument (1) for value;

(2) in good faith; and (3) without notice that it is overdue, that it has been dishonored, that any person has a defense against it or a claim to it, or that the instrument contains unauthorized signatures, contains alterations, or is so irregular or incomplete as to call into question its authenticity. We now examine each of these requirements.

Taking for Value An HDC must have given *value* for the instrument [UCC 3–302(a)(2)(i)]. A person who receives an instrument as a gift or inherits it has not met the requirement of value. In these situations, the person becomes an ordinary holder and does not possess the rights of an HDC.

—How an Instrument Is Taken for Value. Under UCC 3–303(a), a holder can take an instrument for value in one of five ways:

1. By performing the promise for which the instrument was issued or transferred.
2. By acquiring a security interest or other lien in the instrument, excluding a lien obtained by a judicial proceeding. (*Security interests* and *liens* will be discussed in Chapters 20 and 21.)
3. By taking an instrument in payment of, or as security for, a preexisting claim. ■ EXAMPLE 18.9 Zon owes Dwyer $2,000 on a past-due account. If Zon negotiates a $2,000 note signed by Gordon to Dwyer and Dwyer accepts it to discharge the overdue account balance, Dwyer has given value for the instrument. ■
4. By giving a negotiable instrument as payment. ■ EXAMPLE 18.10 Martin has issued a $500 negotiable promissory note to Paulene. The note is due six months from the date issued. Paulene needs money and does not want to wait for the maturity date to collect. She negotiates the note to her friend Kristen, who pays her $200 in cash and writes her a check—a negotiable instrument—for the balance of $300. Kristen has given full value for the note by paying $200 in cash and issuing Paulene the check for $300. ■
5. By giving an irrevocable commitment as payment.

—The Concept of Value in Negotiable Instruments Law. The concept of value in the law of negotiable instruments is not the same as the concept of *consideration* in the law of contracts. A promise to give value in the future is clearly sufficient consideration to support a contract [UCC 1–201(44)]. A promise to give value in the future, however, normally does not constitute value sufficient to make one an HDC. A holder takes an instrument for value only to the extent that the promise has been performed [UCC 3–303(a)(1)]. Therefore, if the holder plans to pay for the instrument later or plans to perform the required services at some future date, the holder has not yet given value. In that situation, the holder is not yet an HDC.

In the Larson-Cambry example presented earlier, Larson is not an HDC because she did not take the instrument (Cambry's note) for value—she has not yet paid Jerrod for the note. Thus, Cambry's defense of breach of contract is valid not only against Jerrod but also against Larson. If Larson had paid Jerrod for the note at the time of transfer (which would mean she had given value for the instrument), she would be an HDC. As an HDC, she could hold Cambry liable on the note even though Cambry has a valid defense against Jerrod on the basis of breach of contract. Exhibit 18–6 illustrates these concepts.

EXHIBIT 18–6 TAKING FOR VALUE

By exchanging defective goods for the note, Jerrod breached his contract with Cambry. Cambry could assert this defense if Jerrod presented the note to her for payment. Jerrod exchanged the note for Larson's promise to pay in thirty days, however. Because Larson did not take the note for value, she is not a holder in due course. Thus, Cambry can assert against Larson the defense of Jerrod's breach when Larson submits the note to Cambry for payment. If Larson had taken the note for value, Cambry could not assert that defense and would be liable to pay the note.

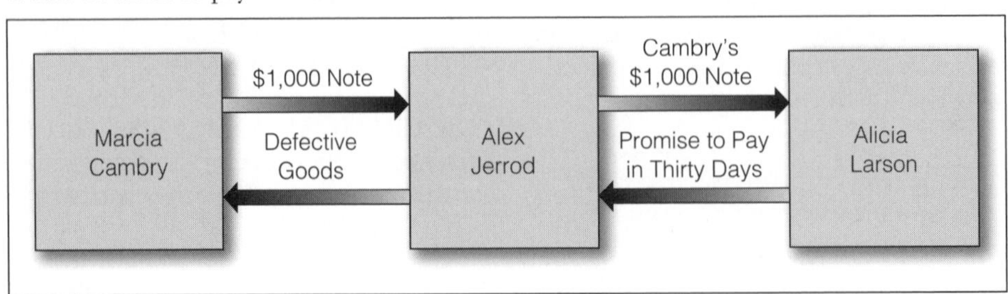

Taking in Good Faith The second requirement for HDC status is that the holder take the instrument in *good faith* [UCC 3–302(a)(2)(ii)]. This means that the holder must have acted honestly in the process of acquiring the instrument. UCC 3–103(a)(4) defines *good faith* as "honesty in fact and the observance of reasonable commercial standards of fair dealing." The good faith requirement applies only to the *holder*. It is immaterial whether the transferor acted in good faith. Thus, even a person who takes a negotiable instrument from a thief may become an HDC if the person acquires the instrument in good faith.

Because of the good faith requirement, one must ask whether the purchaser, when acquiring the instrument, honestly believed that the instrument was not defective. If a person purchases a $10,000 note for $300 from a stranger on a street corner, the issue of good faith can be raised on the grounds of both the suspicious circumstances and the grossly inadequate consideration (value).

In the following case, the court considered whether a bank observed "reasonable commercial standards of fair dealing" to fulfill the good faith requirement and become an HDC.

CASE 18.3 Mid Wisconsin Bank v. Forsgard Trading, Inc.

Wisconsin Court of Appeals, 2003.
2003 WI App. 186,
266 Wis.2d 685,
668 N.W.2d 830.
http://www.wisbar.org/WisCtApp/index.html[a]

FACTS Forsgard Trading, Inc., opened an account at Mid Wisconsin Bank in July 1999. The account agreement stated, "Any items, other than cash, accepted for deposit . . . will be given provisional credit only until collection is final." Mid Wisconsin's practice is to give immediate credit on deposits, but a bank employee may place a hold on a check if, for example, there is reasonable doubt about it. On May 7, 2001, Lakeshore Truck and Equipment Sales, Inc., wrote a check payable to Forsgard in the amount of $18,500. On May 8, Forsgard deposited the check in its account at Mid Wisconsin, which gave Forsgard immediate credit. The same day, Lakeshore issued a stop-payment order (an order to its bank not to pay the check—see Chapter 19). When Mid Wisconsin received notice on May 16 that payment had been stopped, it deducted the $18,500 from Forsgard's account. Because of transfers from the account between May 8 and May 16, the deduction resulted in a negative balance. Before this incident, Forsgard had overdrawn the account twenty-four times but, on each occasion, had deposited funds to cover the overdraft. Forsgard did not do so this time. Mid Wisconsin filed a suit in a Wisconsin state court against Forsgard, Lakeshore, and others to recover the loss. The court issued a summary judgment in Mid Wisconsin's favor. Lakeshore appealed to a state intermediate appellate court.

a. Click on "Simple Search" and then type the docket number "03-0123" in the "Keywords" box. The State Bar of Wisconsin maintains this Web site.

ISSUE Did Mid Wisconsin act in good faith?

DECISION Yes. The court affirmed the lower court's judgment on this issue, holding that Mid Wisconsin was an HDC of the check.

REASON The appellate court explained that under UCC 3–305, an HDC can recover from a drawer who places a stop-payment order on a check. Under UCC 3–302, an HDC is one who takes an instrument for value and in good faith. There was no dispute that Mid Wisconsin took the check for value. UCC 3–103(a)(4) defines good faith as "honesty in fact and the observance of reasonable commercial standards of fair dealing." Lakeshore conceded that Mid Wisconsin took the check with honesty in fact. Thus, the only question was whether Mid Wisconsin's granting Forsgard immediate credit was in line with "reasonable commercial standards of fair dealing." The court found that Mid Wisconsin's acts complied with its account agreement and that "Mid Wisconsin had no reason to suspect there would be any problem if immediate credit was extended for this check." Also, "extending immediate credit is consistent with reasonable banking standards." The court pointed out, "It would hinder commercial transactions if depository banks refused to permit the withdrawal prior to the clearance of checks. * * * [B]anking practice is to the contrary. It is clear that the Uniform Commercial Code was intended * * * to protect banks who have given credit on deposited items prior to notice of a stop payment order."

FOR CRITICAL ANALYSIS—Social Consideration
What might have been the result in this case if the account had been overdrawn when the check was deposited?

Taking without Notice The final requirement for HDC status involves *notice* [UCC 3–302]. A person will not be afforded HDC protection if he or she acquires an instrument and is *on notice* (knows or has reason to know) that it is defective in any one of the following ways [UCC 3–302(a)]:

1. It is overdue.
2. It has been dishonored.
3. There is an uncured (uncorrected) default with respect to another instrument issued as part of the same series.
4. The instrument contains an unauthorized signature or has been altered.
5. There is a defense against the instrument or a claim to the instrument.
6. The instrument is so irregular or incomplete as to call into question its authenticity.

—What Constitutes Notice? Notice of a defective instrument is given whenever the holder (1) has actual knowledge of the defect; (2) has received a notice of the defect (such as a bank's receipt of a letter listing the serial numbers of stolen bearer instruments); or (3) has reason to know that a defect exists, given all the facts and circumstances known at the time in question [UCC 1–201(25)]. The holder must also have received the notice "at a time and in a manner that gives a reasonable opportunity to act on it" [UCC 3–302(f)]. A purchaser's knowledge of certain facts, such as insolvency proceedings against the maker or drawer of the instrument, does not constitute notice that the instrument is defective [UCC 3–302(b)].

—Overdue Instruments. What constitutes notice that an instrument is overdue depends on whether it is a demand instrument (payable on demand) or a time instrument (payable at a definite time). A purchaser has notice that a *demand instrument* is overdue if she or he either takes the instrument knowing that demand has been made or takes the instrument an unreasonable length of time after its issue. For a check, a "reasonable time" is ninety days after the date of the check. For all other demand instruments, what will be considered a reasonable time depends on the circumstances [UCC 3–304(a)].

A holder of a *time instrument* who takes the instrument at any time after its expressed due date is on notice that it is overdue [UCC 3–304(b)(2)]. Nonpayment by the due date should indicate to any purchaser that the instrument may be defective. Thus, a promissory note due on May 15 must be acquired before midnight on May 15. If it is purchased on May 16, the purchaser will be an ordinary holder, not an HDC.

Sometimes an instrument reads, "Payable in thirty days." To count thirty days, you exclude the first day and count the last day. Thus, a note dated December 1 that is payable in thirty days is due by midnight on December 31. If the payment date falls on a Sunday or holiday, the instrument is payable on the next business day. If a debt is to be paid in installments or through a series of notes, the maker's default on any installment of principal or on any one note of the series will constitute notice to the purchaser that the instrument is overdue [UCC 3–304(b)(1)].

HOLDER THROUGH AN HDC

A person who does not qualify as an HDC but who derives his or her title through an HDC can acquire the rights and privileges of an HDC. This rule is sometimes called the **shelter principle.** According to UCC 3–203(b),

> Transfer of an instrument, whether or not the transfer is a negotiation, vests in the transferee any right of the transferor to enforce the instrument, including any right as a holder in due course, but the transferee cannot acquire rights of a holder in due course by a transfer, directly or indirectly, from a holder in due course if the transferee engaged in fraud or illegality affecting the instrument.

Under this rule, anyone—no matter how far removed from an HDC—who can trace his or her title ultimately back to an HDC may acquire the rights of an HDC. By extending the benefits of HDC status, the shelter principle promotes the marketability and free transferability of negotiable instruments.

There are some limitations on the shelter principle, however. Certain persons who formerly held instruments cannot improve their positions by later reacquiring the instruments from HDCs [UCC 3–203(b)]. If a holder participated in fraud or illegality affecting the instrument, or had notice of a claim or defense against an instrument, that holder is not allowed to improve her or his status by repurchasing from a later HDC.

Signature Liability

A *signature* is the key to liability on a negotiable instrument. The general rule is as follows: every party, except a qualified indorser,[2] who signs a negotiable instrument is

2. A qualified indorser—one who indorses "without recourse"—undertakes no contractual obligation to pay. A qualified indorser merely assumes warranty liability, which will be discussed later in this chapter.

either primarily or secondarily liable for payment of that instrument when it comes due. The following subsections discuss these two types of liability, as well as the conditions that must be met before liability can arise.

PRIMARY LIABILITY

A person who is primarily liable on a negotiable instrument is absolutely required to pay the instrument—unless, of course, he or she has a valid defense to payment [UCC 3–305]. Only *makers* and *acceptors* of instruments are primarily liable.

The maker of a promissory note promises to pay the note. It is the maker's promise to pay that makes the note a negotiable instrument. The words "I promise to pay" embody the maker's obligation to pay the instrument according to the terms as written at the time of the signing. If the instrument was incomplete when the maker signed it, the maker is obligated to pay it according to its stated terms or according to terms that were agreed on and later filled in to complete the instrument [UCC 3–115, 3–407(a), 3–412].

An **acceptor** is a drawee that is legally obligated to pay an instrument when it is presented later for payment. Once a drawee indicates acceptance by signing the draft, the drawee becomes an acceptor and is obligated to pay the draft when it is presented for payment [UCC 3–409(a)]. A drawee that refuses to accept a draft that *requires* the drawee's acceptance (such as a trade acceptance) has dishonored the instrument. Acceptance of a check is called *certification* (certified checks will be discussed in Chapter 19). Certification is not necessary on checks, and a bank is under no obligation to certify checks. On certification, however, the drawee bank occupies the position of an acceptor and is primarily liable on the check to any holder [UCC 3–409(d)].

SECONDARY LIABILITY

Drawers and *indorsers* are secondarily liable. On a negotiable instrument, secondary liability is similar to the liability of a guarantor in a simple contract (described in Chapter 11) in the sense that it is *contingent liability.* In other words, a drawer or an indorser will be liable only if the party that is responsible for paying the instrument refuses to do so (dishonors the instrument). In regard to drafts and checks, the drawer's secondary liability does not arise until the drawee fails to pay or to accept the instrument, whichever is required [UCC 3–412, 3–415].

Dishonor of an instrument thus triggers the liability of parties who are secondarily liable on the instrument—that is, the drawer and any *unqualified* indorsers. ■ **EXAMPLE 18.11** Nina Lee writes a check on her account at Universal Bank payable to the order of Stephen Miller. Universal Bank refuses to pay the check when Miller presents it for payment, thus dishonoring the check. In this situation, Lee will be liable to Miller on the basis of her secondary liability. ■ Drawers are secondarily liable on drafts unless they disclaim their liability by drawing the instruments "without recourse" (if the draft is a check, however, a drawer cannot disclaim liability) [UCC 3–414(e)].

Parties who are secondarily liable on a negotiable instrument promise to pay on that instrument only if the following events occur:[3]

1. The instrument is properly and timely presented.
2. The instrument is dishonored.
3. Timely notice of dishonor is given to the secondarily liable party.

Proper and Timely Presentment The formal production of a negotiable instrument for acceptance or payment is called **presentment.** The UCC requires that a holder present the instrument to the appropriate party, in a timely fashion, and give reasonable identification if demanded [UCC 3–414(f), 3–415(e), 3–501]. The party to whom the instrument must be presented depends on the type of instrument involved. A note or certificate of deposit must be presented to the maker for payment. A draft is presented to the drawee for acceptance, payment, or both. A check is presented to the drawee for payment [UCC 3–501(a), 3–502(b)].

Presentment can be made by any commercially reasonable means, including oral, written, or electronic communication [UCC 3–501(b)]. It is effective when the demand for payment or acceptance is received, although banks can treat a presentment that takes place after an established cutoff hour (such as 2 P.M.) as occurring the next business day.

One of the most crucial criteria for proper presentment is timeliness [UCC 3–414(f), 3–415(e), 3–501(b)(4)]. Failure to present an instrument on time is the most common reason that unqualified indorsers are discharged from secondary liability. The time for proper presentment for different types of instruments is shown in Exhibit 18–7.

Dishonor An instrument is dishonored when the required acceptance or payment is refused or cannot be obtained within the prescribed time. An instrument is also dishonored when required presentment is excused (as it would be, for example, if the maker had died) and the instrument is not properly accepted or paid [UCC 3–502(e), 3–504].

3. These requirements are necessary for a secondarily liable party to have signature liability on a negotiable instrument, but they are not necessary for a secondarily liable party to have warranty liability (discussed later in this chapter).

EXHIBIT 18–7 TIME FOR PROPER PRESENTMENT

TYPE OF INSTRUMENT	FOR ACCEPTANCE	FOR PAYMENT
Time	On or before due date.	On due date.
Demand	Within a reasonable time (after date or issue or after secondary party becomes liable on the instrument).	
Check	Not applicable.	Within thirty days of its date, to hold drawer secondarily liable. Within thirty days of indorsement, to hold indorser secondarily liable.

In some situations, a postponement of payment or a refusal to pay an instrument will *not* dishonor the instrument. When presentment is made after an established cut-off hour, for instance, a bank can postpone payment until the following business day without dishonoring the instrument. In addition, when the holder refuses to exhibit the instrument, to give reasonable identification, or to sign a receipt on the instrument for payment, a bank's refusal to pay does not dishonor the instrument. ■ EXAMPLE 18.12 Suppose that Deere, instead of depositing Lamar's check into his bank account, demands payment from Universal Bank in cash. The bank requests identification, which Deere refuses to provide. In this situation, the bank would be within its rights to refuse payment to Deere, and the bank's refusal to pay would not dishonor the check. ■

Proper Notice Once an instrument has been dishonored, proper notice must be given to secondary parties (drawers and indorsers) for them to be held contractually liable. Notice may be given in any reasonable manner, including an oral, written, or electronic communication, as well as notice written or stamped on the instrument itself. The bank must give any necessary notice before its midnight deadline (midnight of the next banking day after receipt). Notice by any party other than a bank must be given within thirty days following the day of dishonor or the day on which the person who is secondarily liable receives notice of dishonor [UCC 3–503].

UNAUTHORIZED SIGNATURES

People normally are not liable to pay on negotiable instruments unless their signatures appear on the instruments. If a signature is unauthorized, as occurs when a signature is forged, the general rule is that the unauthorized signature is wholly inoperative and will not bind the person whose name is forged. ■ EXAMPLE 18.13 Parra finds Dolby's checkbook lying in the street, writes out a check to himself, and forges Dolby's signature. If a bank fails to ascertain that Dolby's signature is not genuine (which banks normally have a duty to do) and cashes the check for Parra, the bank will generally be liable to Dolby for the amount. ■ (The liability of banks for paying over forged signatures will be discussed further in Chapter 19.) There are two exceptions to the general rule that an unauthorized signature will not bind the person whose name is signed:

1. An exception is made when the person whose name is signed ratifies (affirms) the signature [UCC 3–403(a)]. For example, in an agency relationship, one person—an *agent*—agrees to act for or represent another person, called the *principal* (see Chapter 22). A principal can ratify an unauthorized signature made by an agent, either expressly, by affirming the validity of the signature, or impliedly, by other conduct, such as keeping any benefits received in the transaction or failing to repudiate the signature. The parties involved need not be principal and agent. ■ EXAMPLE 18.14 Allison Malone steals several checks from her mother, Brenda Malone, makes them out to herself, and signs "Brenda Malone." Brenda, the mother, may ratify the forged signatures so that Allison will not be prosecuted for forgery. ■
2. A person whose name is forged may be precluded from denying the effectiveness of an unauthorized signature if the person's own negligence substantially contributed to the forgery [UCC 3–115, 3–401, 3–403, 3–406]. ■ EXAMPLE 18.15 Suppose that Rob, the owner of a business, leaves his signature stamp and a blank check on an office counter. An employee, using the stamp, fills in and cashes the check. Rob can be estopped (prevented), on the basis of negligence, from denying liability for payment of the check. Whatever loss occurs may be allocated, however, between certain parties on the basis of comparative negligence [UCC 3–406(b)]. For example, if Rob can demonstrate that the bank was negligent in paying the check, the bank may bear a portion of the loss. ■

An unauthorized signature operates as the signature of the unauthorized signer in favor of an HDC. A person who forges a check, for instance, can be held personally liable for payment by an HDC [UCC 3–403(a)].

SPECIAL RULES FOR UNAUTHORIZED INDORSEMENTS

Generally, when an indorsement is forged or unauthorized, the burden of loss falls on the first party to take the instrument with the forged or unauthorized indorsement. If the indorsement was made by an imposter or by a fictitious payee, however, the loss falls on the maker or drawer. We look at these two situations here.

Imposters An **imposter** is one who, by her or his personal appearance or use of the mails, Internet, telephone, or other communication, induces a maker or drawer to issue an instrument in the name of an impersonated payee. If the maker or drawer believes the imposter to be the named payee at the time of issue, the indorsement by the imposter is not treated as unauthorized when the instrument is transferred to an innocent party. This is because the maker or drawer intended the imposter to receive the instrument. In this situation, under the UCC's *imposter rule,* the imposter's indorsement will be effective—that is, not considered a forgery—insofar as the drawer or the maker is concerned [UCC 3–404(a)]. ■ EXAMPLE 18.16 Carol impersonates Donna and induces Edward to write a check payable to the order of Donna. Carol, continuing to impersonate Donna, negotiates the check to First National Bank as payment on her loan there. As the drawer of the check, Edward is liable for its amount to First National. ■

Fictitious Payees An unauthorized indorsement will also be effective when a person causes an instrument to be issued to a payee who will have *no interest* in the instrument [UCC 3–404(b), 3–405]. In this situation, the payee is referred to as a **fictitious payee.** Situations involving fictitious payees most often arise when (1) a dishonest employee deceives the employer into signing an instrument payable to a party with no right to receive payment on the instrument or (2) a dishonest employee or agent has the authority to issue an instrument on behalf of the employer. Under the UCC's *fictitious payee rule,* the payee's indorsement is not treated as a forgery, and an innocent holder or an innocent party (such as a bank that pays the instrument in good faith) can hold the employer liable on the instrument.

■ EXAMPLE 18.17 Blair Industries, Inc., gives its bookkeeper, Axel Ford, general authority to issue checks in the company name drawn on First State Bank so that Ford can pay employees' wages and other corporate bills. Ford decides to cheat Blair Industries out of $10,000 by issuing a check payable to Erica Nied, an old acquaintance. Neither Blair nor Ford intends Nied to receive any of the money, and Nied is not an employee or creditor of the company. Ford indorses the check in Nied's name, naming himself as *indorsee* (the person to whom an instrument is transferred by indorsement). He then cashes the check at a local bank, which collects payment from the drawee bank, First State Bank. First State Bank charges the Blair Industries account $10,000. Blair Industries discovers the fraud and demands that the account be recredited.

Who bears the loss? UCC 3–404(b)(2) provides the answer. Neither the local bank that first accepted the check nor First State Bank is liable. Because Ford's indorsement in the name of a payee with no interest in the instrument is "effective," there is no "forgery." Hence, the collecting bank is protected in paying on the check, and the drawee bank is protected in charging Blair's account. Thus, the employer-drawer, Blair Industries, will bear the loss. Of course, Blair Industries has recourse against Axel Ford, if Ford has not absconded with the funds. ■

Regardless of whether a dishonest employee actually signs the check or merely supplies his or her employer with names of fictitious creditors (or with true names of creditors having fictitious debts), the result is the same under the UCC. (To learn more on how to defend against liability under the fictitious payee rule, see this chapter's *Management Perspective* feature.)

Warranty Liability

In addition to the signature liability discussed in the preceding pages, transferors make certain implied warranties regarding the instruments that they are negotiating. Liability under these warranties is not subject to the conditions of proper presentment, dishonor, or notice of dishonor. These warranties arise even when a transferor does not indorse the instrument (as in the delivery of a bearer instrument) [UCC 3–416, 3–417]. Warranty liability is particularly important when a holder cannot hold a party liable on her or his signature.

Warranties fall into two categories: those that arise on the *transfer* of a negotiable instrument and those that arise on *presentment.* Both transfer and presentment warranties attempt to shift liability back to a wrongdoer or to the person who dealt face to face with the wrongdoer and thus was in the best position to prevent the wrongdoing.

TRANSFER WARRANTIES

The UCC describes five **transfer warranties** [UCC 3–416]. For transfer warranties to arise, an instrument must be transferred *for consideration.*

MANAGEMENT PERSPECTIVE

Banks Must Act in Good Faith

MANAGEMENT FACES A LEGAL ISSUE

Remember from previous chapters that the requirement of good faith underlies all transactions governed by the Uniform Commercial Code (UCC). Does this mean that a bank, to avoid liability for paying instruments with forged indorsements involving fictitious payees, must have acted in good faith when accepting the deposits? How the courts answer this question has, of course, significant implications for business owners and managers.

WHAT THE COURTS SAY

According to a number of courts, if a bank acts in bad faith in paying instruments with forged indorsements involving fictitious payees, the bank itself will end up bearing the loss. For example, in one case a Pennsylvania appellate court held that to assert the fictitious payee rule, "the bank must have acted in good faith when paying the instrument." The bank in this case had accepted 882 payroll checks generated and indorsed by Dorothy Heck, a payroll clerk employed by Pavex, Inc. The checks were made payable to various current and former Pavex employees, indorsed by Heck with the payees' names, and deposited into Heck's personal checking account at her bank.

In spite of the bank's policy that indorsements on checks must match exactly the names of the payees, the bank continued to accept Heck's deposited checks even when the indorsements did not match the payees' names. Furthermore, even though bank personnel discussed Heck's check-depositing activities on more than one occasion, they never contacted her employer to verify Heck's authority to deposit third party payroll checks. Given the bank's pattern of ignoring perceived irregularities in Heck's transactions, the trial court jury concluded that the bank had acted in bad faith and was therefore liable for approximately $170,000 of the $250,000 loss suffered by Pavex. The appellate court affirmed the trial court's decision.[a]

IMPLICATIONS FOR MANAGERS

Business owners and managers should realize that all may not always be lost when they are the victims of employees who indorse checks to fictitious payees. They should consider the possibility that the bank into which the checks were deposited may also be liable, at least in part, if the bank's employees failed to comply with its policies regarding verifying the indorsements on deposited checks.

a. *Pavex, Inc. v. York Federal Savings and Loan Association,* 716 A.2d 640 (Pa.Super.Ct. 1998).

The Five Transfer Warranties One who transfers an instrument for consideration makes the following warranties to all subsequent transferees and holders who take the instrument in good faith (with some exceptions, as will be noted shortly):

1. The transferor is entitled to enforce the instrument.
2. All signatures are authentic and authorized.
3. The instrument has not been altered.
4. The instrument is not subject to a defense or claim of any party that can be asserted against the transferor.
5. The transferor has no knowledge of any insolvency proceedings against the maker, the acceptor, or the drawer of the instrument.[4]

4. A 2002 amendment to UCC 3–416(a) adds a sixth warranty: "with respect to a remotely created consumer item, that the person on whose account the item is drawn authorized the issuance of the item in the amount for which the item is drawn." For example, a telemarketer submits an instrument to a bank for payment, claiming that the consumer on whose account the instrument purports to be drawn authorized it over the phone. Under this amendment, a bank that accepts and pays the instrument warrants to the next bank in the collection chain that the consumer authorized the item in that amount.

Parties to Whom Warranty Liability Extends The manner of transfer and the negotiation that is used determine how far and to whom a transfer warranty will run. Transfer of order paper, for consideration, by indorsement and delivery extends warranty liability to any subsequent holder who takes the instrument in good faith. The warranties of a person who transfers *without indorsement* (by the delivery of a bearer instrument), however, will extend the transferor's warranties only to the immediate transferee [UCC 3–416(a)].

■ **EXAMPLE 18.18** Wylie forges Peter's name as a maker of a promissory note. The note is made payable to Wylie. Wylie indorses the note in blank, negotiates it to Carla, and then leaves the country. Carla, without indorsement, delivers the note to Frank for consideration. Frank in turn, without indorsement, delivers the note to Ricardo for consideration. On Ricardo's presentment of the note to Peter, the forgery is discovered. Ricardo can hold Frank (the immediate transferor) liable for breach of the transfer warranty that all signatures are genuine. Ricardo cannot hold Carla liable because the transfer warranties made by Carla, who negotiated the bearer

instrument by delivery only, extend solely to Frank, the immediate transferee. ■

Note that if Wylie had added a special indorsement ("Payable to Carla") instead of a blank indorsement, the instrument would have remained an order instrument. In that situation, to negotiate the instrument to Frank, Carla would have had to indorse the instrument, and her transfer warranties would extend to all subsequent holders, including Ricardo. This example shows the importance of the distinction between a transfer by indorsement and delivery (of an order instrument) and a transfer by delivery only, without indorsement (of a bearer instrument).

PRESENTMENT WARRANTIES

Any person who presents an instrument for payment or acceptance makes the following **presentment warranties** to any other person who in good faith pays or accepts the instrument [UCC 3–417(a), (d)]:

1. The person obtaining payment or acceptance is entitled to enforce the instrument or is authorized to obtain payment or acceptance on behalf of a person who is entitled to enforce the instrument. (This is, in effect, a warranty that there are no missing or unauthorized indorsements.)
2. The instrument has not been altered.
3. The person obtaining payment or acceptance has no knowledge that the signature of the issuer of the instrument is unauthorized.[5]

These warranties are referred to as presentment warranties because they protect the person to whom the instrument is presented. The second and third warranties do not apply to makers, acceptors, and drawers. It is assumed, for example, that a drawer or a maker will recognize his or her own signature and that a maker or an acceptor will recognize whether an instrument has been materially altered.

Defenses and Discharge

Persons who would otherwise be liable on negotiable instruments may be able to avoid liability by raising certain defenses. There are two general categories of defenses—*universal defenses* and *personal defenses.*

5. As discussed in footnote 4, the 2002 amendments to Article 3 of the UCC provide additional protection for "remotely created" consumer items [see Amended UCC 3–417(a)(4)].

UNIVERSAL DEFENSES

Universal defenses (also called *real defenses*) are valid against *all* holders, including HDCs and holders who take through an HDC. Universal defenses include those described here.

Forgery Forgery of a maker's or drawer's signature cannot bind the person whose name is used unless that person *ratifies* (approves or validates) the signature or is precluded from denying it (because the forgery was made possible by the maker's or drawer's negligence, for example) [UCC 3–403(a)]. Thus, when a person forges an instrument, the person whose name is forged normally has no liability to pay any holder or any HDC the value of the forged instrument.

Fraud in the Execution If a person is deceived into signing a negotiable instrument, believing that she or he is signing something other than a negotiable instrument (such as a receipt), *fraud in the execution,* or fraud in the inception, is committed against the signer [UCC 3–305(a)(1)].

■ EXAMPLE 18.19 Gerard, a salesperson, asks a customer to sign a paper, which Gerard says is a receipt for the delivery of goods that the customer is picking up from the store. In fact, the paper is a promissory note, but the customer, Javier, is unfamiliar with the English language and does not realize this. In this situation, even if the note is negotiated to an HDC, Javier has a valid defense against payment. ■

The defense of fraud in the execution cannot be raised, however, if a reasonable inquiry would have revealed the nature and terms of the instrument.[6] Thus, the signer's age, experience, and intelligence are relevant because they frequently determine whether the signer should have known the nature of the transaction before signing.

Material Alteration An alteration is material if it changes the contract terms between any two parties *in any way.* Examples of material alterations include completing an incomplete instrument, adding words or numbers to an instrument, or making any other change to an instrument in an unauthorized manner that affects the obligation of a party to the instrument [UCC 3–407(a)]. Thus, cutting off part of the paper of a negotiable instrument, adding clauses, or making any change in the amount, the date, or the rate of interest—even if the change is only one penny, one day, or 1 percent—is material.

6. *Burchett v. Allied Concord Financial Corp.,* 74 N.M. 575, 396 P.2d 186 (1964).

It is not a material alteration, however, to correct the maker's address or to change the figures on a check so that they agree with the written amount (words outweigh figures if the written amount and the amount given in figures are different). If the alteration is not material, any holder is entitled to enforce the instrument according to its terms.

Material alteration is a *complete defense* against an ordinary holder, but only a partial defense against an HDC. An ordinary holder can recover nothing on an instrument if it has been materially altered [UCC 3–407(b)]. When the holder is an HDC, by contrast, if an original term, such as the monetary amount payable, has been *altered*, the HDC can enforce the instrument against the maker or drawer according to the original terms but not for the altered amount. If the instrument was originally incomplete and was later completed in an unauthorized manner, however, alteration no longer can be claimed as a defense against an HDC, and the HDC can enforce the instrument as completed [UCC 3–407(b)]. This is because the drawer or maker of the instrument, by issuing an incomplete instrument, will normally be held responsible for the alteration, which could have been avoided by the exercise of greater care. If the alteration is readily apparent, then obviously the holder has notice of some defect or defense and therefore cannot be an HDC [UCC 3–302(a)(1)].

Discharge in Bankruptcy Discharge in bankruptcy (see Chapter 21) is an absolute defense on any instrument, regardless of the status of the holder, because the purpose of bankruptcy is to settle finally all of the insolvent party's debts [UCC 3–305(a)(1)].

Minority Minority, or infancy, is a universal defense only to the extent that state law recognizes it as a defense to a simple contract (see Chapter 9). Because state laws on minority vary, so do determinations of whether minority is a universal defense against an HDC [UCC 3–305(a)(1)(i)].

Illegality Certain types of illegality constitute universal defenses. Other types constitute personal defenses—that is, defenses that are effective against ordinary holders but not against HDCs. If a statute provides that an illegal transaction is void, then the defense is universal—that is, absolute against both an ordinary holder and an HDC. If the law merely makes the instrument voidable, then the illegality is still a personal defense against an ordinary holder but not against an HDC [UCC 3–305(a)(1)(ii)].

Mental Incapacity If a person has been declared mentally incompetent by state court proceedings, then any instrument issued thereafter by that person is void. The instrument is void *ab initio* (from the beginning) and unenforceable by any holder or HDC [UCC 3–305(a)(1)(ii)]. Mental incapacity in these circumstances is thus a universal defense. If a person has not been judged mentally incompetent by state proceedings, mental incapacity operates as a defense against an ordinary holder but not against an HDC.

Extreme Duress When a person signs and issues a negotiable instrument under such extreme duress as an immediate threat of force or violence (for example, at gunpoint), the instrument is void and unenforceable by any holder or HDC [UCC 3–305(a)(1)(ii)]. (Ordinary duress is a defense against ordinary holders but not against HDCs.)

PERSONAL DEFENSES

Personal defenses (sometimes called *limited defenses*), such as those described here, can be used to avoid payment to an ordinary holder of a negotiable instrument, but not to an HDC or a holder with the rights of an HDC.

Breach of Contract or Breach of Warranty When there is a breach of the underlying contract for which the negotiable instrument was issued, the maker of a note can refuse to pay it, or the drawer of a check can order his or her bank to stop payment on the check. Breach of warranty can also be claimed as a defense to liability on the instrument.

■ **EXAMPLE 18.20** Rhodes agrees to purchase several sets of imported china from Livingston. The china is to be delivered in four weeks. Rhodes gives Livingston a promissory note for $2,000, which is the price of the china. The china arrives, but many of the pieces are broken, and several others are chipped or cracked. Rhodes refuses to pay the note on the basis of breach of contract and breach of warranty. (Under sales law, a seller impliedly promises that the goods are at least merchantable—see Chapter 17.) Livingston cannot enforce payment on the note because of the breach of contract and breach of warranty. If Livingston has negotiated the note to a third party, however, and the third party is an HDC, Rhodes will *not* be able to use breach of contract or warranty as a defense against liability on the note. ■

Lack or Failure of Consideration The absence of consideration (value) may be a successful personal defense in some instances [UCC 3–303(b), 3–305(a)(2)]. ■ **EXAMPLE 18.21** Tara gives Clem, as a gift, a note that states, "I promise to pay you $100,000." Clem

accepts the note. Because there is no consideration for Tara's promise, a court will not enforce the promise. ■

Fraud in the Inducement (Ordinary Fraud) A person who issues a negotiable instrument based on false statements by the other party will be able to avoid payment on that instrument, unless the holder is an HDC. ■ EXAMPLE 18.22 Jerry agrees to purchase Howard's used tractor for $24,500. Howard, knowing his statements to be false, tells Jerry that the tractor is in good working order, that it has been used for only one harvest, and that he owns the tractor free and clear of all claims. Jerry pays Howard $4,500 in cash and issues a negotiable promissory note for the balance. As it turns out, Howard still owes the original seller $10,000 on the purchase of the tractor. In addition, the tractor is three years old and has been used in three harvests. Jerry can refuse to pay the note if it is held by an ordinary holder. If Howard has negotiated the note to an HDC, however, Jerry must pay the HDC. (Of course, Jerry can then sue Howard to recover the funds paid.) ■

Illegality As mentioned, if a statute provides that an illegal transaction is void, a universal defense exists. If, however, the statute provides that an illegal transaction is voidable, the defense is personal.

Mental Incapacity As mentioned, if a maker or drawer has been declared by a court to be mentally incompetent, any instrument issued by the maker or drawer is void. Hence, mental incapacity can serve as a universal defense [UCC 3–305(a)(1)(ii)]. If a maker or drawer, in contrast, issues a negotiable instrument while mentally incompetent but before a formal court hearing has declared him or her to be so, the instrument is voidable. In this situation, mental incapacity can serve only as a personal defense.

Other Personal Defenses Other personal defenses that can be used to avoid payment to an ordinary holder of a negotiable instrument include the following:

1. Discharge by payment or cancellation [UCC 3–601(b), 3–602(a), 3–603, 3–604].
2. Unauthorized completion of an incomplete instrument [UCC 3–115, 3–302, 3–407, 4–401(d)(2)].
3. Nondelivery of the instrument [UCC 1–201(14), 3–105(b), 3–305(a)(2)].
4. Ordinary duress or undue influence rendering the contract voidable [UCC 3–305(a)(1)(ii)].

DISCHARGE FROM LIABILITY

Discharge from liability on an instrument can occur in several ways. The liability of all parties to an instrument is discharged when the party primarily liable on it pays to the holder the amount due in full [UCC 3–602, 3–603]. Payment by any other party discharges only the liability of that party and subsequent parties.

Intentional cancellation of an instrument discharges the liability of all parties [UCC 3–604]. Intentionally writing "Paid" across the face of an instrument cancels it. Intentionally tearing up an instrument cancels it. If a holder intentionally crosses out a party's signature, that party's liability and the liability of subsequent indorsers who have already indorsed the instrument are discharged. Materially altering an instrument may discharge the liability of any party affected by the alteration, as previously discussed [UCC 3–407(b)]. (An HDC may be able to enforce a materially altered instrument against its maker or drawer according to the instrument's original terms, however.)

Discharge of liability can also occur when a party's right of recourse is impaired [UCC 3–605]. A *right of recourse* is a right to seek reimbursement. Ordinarily, when a holder collects the amount of an instrument from an indorser, the indorser has a right of recourse against prior indorsers, the maker or drawer, and any accommodation parties (an *accommodation party* signs an instrument to lend her or his name as credit to another party on the instrument). If the holder has adversely affected the indorser's right to seek reimbursement from these other parties, however, the indorser is not liable on the instrument. This occurs when, for example, the holder releases or agrees not to sue a party against whom the indorser has a right of recourse.

TERMS AND CONCEPTS

acceptance 339
acceptor 350
bearer 345
bearer instrument 345
certificate of deposit (CD) 343
check 341
draft 339
drawee 339
drawer 339
fictitious payee 352
holder 345
holder in due course (HDC) 346
imposter 352
indorsement 345
maker 341

negotiable instrument 338
negotiation 345
order instrument 344
payee 339
personal defense 355
presentment 350
presentment warranty 354
promissory note 341
shelter principle 349
signature 344
trade acceptance 340
transfer warranty 352
universal defense 354

CHAPTER SUMMARY Negotiability, Transferability, and Liability

Types of Instruments (See pages 339–343.)	The UCC specifies four types of negotiable instruments: drafts, checks, promissory notes, and certificates of deposit (CDs). Two basic classification systems are used to describe negotiable instruments: 1. *Demand instruments versus time instruments*—A demand instrument is payable on demand (when the holder presents it to the maker or drawer). A time instrument is payable at a future date. 2. *Orders to pay versus promises to pay*—Checks and drafts are *orders* to pay. Promissory notes and certificates of deposit (CDs) are *promises* to pay.
Requirements for Negotiability (See pages 343–345.)	To be negotiable, an instrument must meet the following requirements: 1. Be in writing. 2. Be signed by the maker or drawer. 3. Be an unconditional promise or order to pay. 4. State a fixed amount of money. 5. Be payable on demand or at a definite time. 6. Be payable to order or bearer (unless it is a check).
Transfer of Instruments (See page 345.)	1. *Transfer by assignment*—A transfer by assignment to an assignee gives the assignee only those rights that the assignor possessed. Any defenses against payment that can be raised against an assignor can normally be raised against the assignee. 2. *Transfer by negotiation*—An order instrument is negotiated by indorsement and delivery; a bearer instrument is negotiated by delivery only.
Holder in Due Course (HDC) (See pages 345–346.)	1. *Holder*—A person in the possession of an instrument drawn, issued, or indorsed to him or her, to his or her order, to bearer, or in blank. A holder obtains only those rights that the transferor had in the instrument. 2. *Holder in due course (HDC)*—A holder who, by meeting certain acquisition requirements (summarized next), takes the instrument free of most defenses and claims to which the transferor was subject.
Requirements for HDC Status (See pages 346–349.)	To be an HDC, a holder must take the instrument: 1. *For value*—A holder can take an instrument for value in one of five ways [UCC 3–303]: a. By the complete or partial performance of the promise for which the instrument was issued or transferred. b. By acquiring a security interest or other lien in the instrument, excluding a lien obtained by a judicial proceeding. c. By taking an instrument in payment of (or as security for) a preexisting debt. d. By giving a negotiable instrument as payment. e. By giving an irrevocable commitment as payment.

(Continued)

CHAPTER SUMMARY Negotiability, Transferability, and Liability—Continued

Requirements for HDC Status—Continued	2. *In good faith*—Good faith is defined as "honesty in fact and the observance of reasonable commercial standards of fair dealing" [UCC 3–103(a)(4)]. 3. *Without notice*—To be an HDC, a holder must not be on notice that the instrument is defective in any of the following ways [UCC 3–302, 3–304]: a. The instrument is overdue. b. The instrument has been dishonored. c. There is an uncured (uncorrected) default with respect to another instrument issued as part of the same series. d. The instrument contains an unauthorized signature or has been altered. e. There is a defense against the instrument or a claim to the instrument. f. The instrument is so irregular or incomplete as to call into question its authenticity.
Holder through an HDC (See page 349.)	A holder who cannot qualify as an HDC has the *rights* of an HDC if he or she derives title through an HDC unless the holder engaged in fraud or illegality affecting the instrument [UCC 3–203(b)]. This is known as the *shelter principle.*
Signature Liability (See pages 349–352.)	Every party (except a qualified indorser) who signs a negotiable instrument is either primarily or secondarily liable for payment of the instrument when it comes due. 1. *Primary liability*—Makers and acceptors are primarily liable (an acceptor is a drawee that promises in writing to pay an instrument when it is presented for payment at a later time) [UCC 3–115, 3–407, 3–409, 3–412]. 2. *Secondary liability*—Drawers and indorsers are secondarily liable [UCC 3–412, 3–414, 3–415, 3–501, 3–502, 3–503]. Parties who are secondarily liable on an instrument promise to pay on that instrument if the following events occur: a. The instrument is properly and timely presented. b. The instrument is dishonored. c. Timely notice of dishonor is given to the secondarily liable party. 3. *Unauthorized signatures*—An unauthorized signature is wholly inoperative unless: a. The person whose name is signed ratifies (affirms) it or is precluded from denying it [UCC 3–115, 3–403, 3–406, 4–401]. b. The instrument has been negotiated to an HDC [UCC 3–403]. 4. *Special rules for unauthorized indorsements*—An unauthorized indorsement will not bind the maker or drawer except in the following circumstances: a. When an imposter induces the maker or drawer of an instrument to issue it to the imposter (imposter rule) [UCC 3–404(a)]. b. When a person signs as or on behalf of a maker or drawer, intending that the payee will have no interest in the instrument, or when an agent or employee of the maker or drawer has supplied him or her with the name of the payee, also intending the payee to have no such interest (fictitious payee rule) [UCC 3–404(b), 3–405].
Warranty Liability (See pages 352–354.)	1. *Transfer warranties*—Any person who transfers an instrument for consideration makes the following warranties to all subsequent transferees and holders who take the instrument in good faith (but when a bearer instrument is transferred by delivery only, the transferor's warranties extend only to the immediate transferee) [UCC 3–416]: a. The transferor is entitled to enforce the instrument. b. All signatures are authentic and authorized.

CHAPTER SUMMARY Negotiability, Transferability, and Liability—Continued

Warranty Liability—Continued	c. The instrument has not been altered. d. The instrument is not subject to a defense or claim of any party that can be asserted against the transferor. e. The transferor has no knowledge of any insolvency proceedings against the maker, the acceptor, or the drawer of the instrument. 2. *Presentment warranties*—Any person who presents an instrument for payment or acceptance makes the following warranties to any other person who in good faith pays or accepts the instrument [UCC 3–417(a), (d)]: a. The person obtaining payment or acceptance is entitled to enforce the instrument or is authorized to obtain payment or acceptance on behalf of a person who is entitled to enforce the instrument. (This is, in effect, a warranty that there are no missing or unauthorized indorsements.) b. The instrument has not been altered. c. The person obtaining payment or acceptance has no knowledge that the signature of the drawer of the instrument is unauthorized.
Defenses and Discharge (See pages 354–356.)	1. *Universal (real) defenses*—The following defenses are valid against all holders, including HDCs and holders with the rights of HDCs [UCC 3–305, 3–403, 3–407, 4–401]: a. Forgery. b. Fraud in the execution. c. Material alteration. d. Discharge in bankruptcy. e. Minority—if the contract is voidable under state law. f. Illegality, mental incapacity, or extreme duress—if the contract is void under state law. 2. *Personal (limited) defenses*—The following defenses are valid against ordinary holders but not against HDCs or holders with the rights of HDCs [UCC 3–105, 3–115, 3–302, 3–305, 3–407, 3–601, 3–602, 3–603, 3–604, 4–401]: a. Breach of contract or breach of warranty. b. Lack or failure of consideration (value). c. Fraud in the inducement. d. Illegality and mental incapacity—if the contract is voidable. e. Previous payment of the instrument. f. Unauthorized completion of the instrument. g. Nondelivery of the instrument. h. Ordinary duress or undue influence that renders the contract voidable.
Discharge from Liability (See page 356.)	All parties to a negotiable instrument will be discharged when the party primarily liable on it pays to a holder the amount due in full. Discharge can also occur in other circumstances (if the instrument has been canceled or materially altered, for example) [UCC 3–601 through 3–606].

FOR REVIEW

Answers for the even-numbered questions in this For Review *section can be found in Appendix A at the end of this text.*

1. What requirements must an instrument meet to be negotiable?
2. What are the requirements for attaining the status of a holder in due course (HDC)?
3. What is the key to liability on a negotiable instrument? What is the difference between signature liability and warranty liability?
4. Certain defenses are valid against all holders, including HDCs. What are these defenses called? Name four defenses that fall within this category.
5. Certain defenses can be used to avoid payment to an ordinary holder of a negotiable instrument but are not effective against an HDC. What are these defenses called? Name four defenses that fall within this category.

QUESTIONS AND CASE PROBLEMS

18–1. Parties to Negotiable Instruments. A note has two original parties. What are these parties called? A check has three original parties. What are these parties called?

QUESTION WITH SAMPLE ANSWER

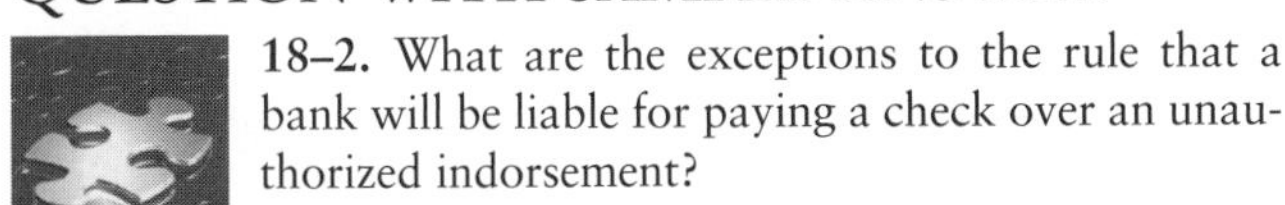

18–2. What are the exceptions to the rule that a bank will be liable for paying a check over an unauthorized indorsement?

For a sample answer to this question, go to Appendix B at the end of this text.

18–3. Defenses. Jules sold Alfred a small motorboat for $1,500; Jules maintained to Alfred that the boat was in excellent condition. Alfred gave Jules a check for $1,500, which Jules indorsed and gave to Sherry for value. When Alfred took the boat for a trial run, he discovered that the boat leaked, needed to be painted, and required a new motor. Alfred stopped payment on his check, which had not yet been cashed. Jules has disappeared. Can Sherry recover from Alfred as a holder in due course? Discuss.

18–4. Defenses. Fox purchased a used car from Emerson for $1,000. Fox paid for the car with a check, written in pencil, payable to Emerson for $1,000. Emerson, through careful erasures and alterations, changed the amount on the check to read $10,000 and negotiated the check to Sanderson. Sanderson took the check for value, in good faith, and without notice of the alteration and thus met the Uniform Commercial Code's requirements for holder in due course status. Can Fox successfully raise the universal defense of material alteration to avoid payment on the check? Explain.

18–5. Signature Liability. Marion makes a promissory note payable to the order of Perry. Perry indorses the note by writing "without recourse, Perry" and transfers the note for value to Steven. Steven, in need of cash, negotiates the note to Harriet by indorsing it with the words "Pay to Harriet, [signed] Steven." On the due date, Harriet presents the note to Marion for payment, only to learn that Marion has filed for bankruptcy and will have all debts (including the note) discharged in bankruptcy. Discuss fully whether Harriet can hold Marion, Perry, or Steven liable on the note.

18–6. Requirements for Negotiability. The following instrument was written on a sheet of paper by Jeff Nolan: "I, the undersigned, do hereby acknowledge that I owe Stephanie Craig one thousand dollars, with interest, payable out of the proceeds of the sale of my horse, Swiftfoot, next month. Payment is to be made on or before six months from date." Discuss specifically why this instrument is not negotiable.

CASE PROBLEM WITH SAMPLE ANSWER

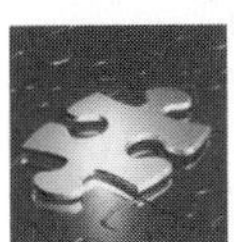

18–7. In October 1998, Somerset Valley Bank notified Alfred Hauser, president of Hauser Co., that the bank had begun to receive what appeared to be Hauser Co. payroll checks. None of the payees were Hauser Co. employees, however, and Hauser had not written the checks or authorized anyone to sign them on his behalf. Automatic Data Processing, Inc., provided payroll services for Hauser Co. and used a facsimile signature on all its payroll checks. Hauser told the bank not to cash the checks. In early 1999, Robert Triffin, who deals in negotiable instruments, bought eighteen of the checks, totaling more than $8,800, from various check-cashing agencies. The agencies stated that they had cashed the checks expecting the bank to pay them. Each check was payable to a bearer for a fixed amount, on demand, and did not state any undertaking by the person promising payment other than the payment of money. Each check bore a facsimile drawer's signature stamp identical to Hauser Co.'s authorized stamp. Each check had been returned to an agency marked "stolen check" and stamped "do not present again." When the bank refused to cash the checks, Triffin filed a suit in a New Jersey state court against Hauser Co. Were the checks negotiable instruments? Why or why not? [*Triffin v. Somerset Valley Bank,* 343 N.J.Super. 73, 777 A.2d 993 (2001)]

After you have answered this problem, compare your answer with the sample answer given on the Web site that accompanies this text. Go to **academic.cengage.com/blaw/wbl**, select "Chapter 18," and click on "Case Problem with Sample Answer."

18–8. Requirements for HDC Status. In February 2001, New York Linen Co., a party rental company, agreed to buy 550

chairs from Elite Products. On delivery of the chairs, New York Linen issued a check (dated February 27) for $13,300 to Elite. Elite's owner, Meir Shmeltzer, transferred the check to General Credit Corp., a company in the business of buying instruments from payees for cash. Meanwhile, after recounting the chairs, New York Linen discovered that delivery was not complete and stopped payment on the check. The next day, New York Linen drafted a second check, reflecting an adjusted payment of $11,275, and delivered it to Elite. A notation on the second check indicated that it was a replacement for the first check. When the first check was dishonored, General Credit filed a suit in a New York state court against New York Linen to recover the amount. New York Linen argued in part that General Credit was not a holder in due course because of the notation on the second check. In whose favor should the court rule? Why? [*General Credit Corp. v. New York Linen Co.*, __ Misc.2d __ (N.Y.City Civ.Ct. 2002)]

18–9. Defenses. On September 13, 1979, Barbara Shearer and Barbara Couvion signed a note for $22,500, with interest at 11 percent, payable in monthly installments of $232.25 to Edgar House and Paul Cook. House and Cook assigned the note to Southside Bank in Kansas City, Missouri. In 1997, the note was assigned to Midstates Resources Corp., which assigned the note to The Cadle Co. in 2000. According to the payment history that Midstates gave to Cadle, the interest rate on the note was 12 percent. A Cadle employee noticed the discrepancy and recalculated the payments at 11 percent. When Shearer and Couvion refused to make further payments on the note, Cadle filed a suit in a Missouri state court against them to collect. Couvion and Shearer responded that they had made timely payments on the note, that Cadle and the previous holders had failed to accurately apply the payments to the reduction of principal and interest, and that the note "is either paid in full and satisfied or very close to being paid in full and satisfied." Is the makers' answer sufficient to support a verdict in their favor? If so, on what ground? If not, why not? [*The Cadle Co. v. Shearer*, 69 S.W.3d 122 (Mo.App. W.D. 2002)]

18–10. Holder in Due Course. The Brown family owns several companies, including J. H. Stevedoring Co. and Penn Warehousing and Distribution, Inc. Many aspects of the companies' operations and management are intertwined. Dennis Bishop began working for J. H. and Penn in 1984. By 1997, Bishop was financial controller at J. H., where he was responsible for approving invoices for payment and reconciling the corporate checkbook. In December, Bishop began stealing from Penn and J. H. by writing checks on their accounts and usually having one of the Browns sign the checks. Over the next two years, Bishop embezzled $1,209,436. He used $370,632 of the funds to buy horses from Fasig-Tipton Co. and Fasig-Tipton Midlantic, Inc., with Penn or J. H. checks made payable to those firms. When Bishop's fraud was revealed, J. H. and Penn filed a suit in a federal district court against the Fasig-Tipton firms to recover the amounts of the checks made payable to them. In whose favor should the court rule? Why? [*J. H. Stevedoring Co. v. Fasig-Tipton Co.*, 275 F.Supp.2d 644 (E.D.Pa. 2003)]

A QUESTION OF ETHICS

18–11. Richard Caliendo, an accountant, prepared tax returns for various clients. To satisfy their tax liabilities, the clients issued checks payable to various state taxing entities and gave them to Caliendo. Between 1977 and 1979, Caliendo forged indorsements on these checks, deposited them in his own bank account, and subsequently withdrew the proceeds. In 1983, after learning of these events and after Caliendo's death, the state brought an action against Barclays Bank of New York, N.A., the successor to Caliendo's bank, to recover the amount of the checks. Barclays moved for dismissal on the ground that because the checks had never been delivered to the state, the state never acquired the status of holder and therefore never acquired any rights in the instruments. The trial court held for the state, but the appellate court reversed. The state then appealed the case to the state's highest court. That court ruled that the state could not recover the amount of the checks from the bank because, although the state was the named payee on the checks, the checks had never been delivered to the payee. [*State v. Barclays Bank of New York, N.A.*, 76 N.Y.2d 533, 563 N.E.2d 11, 561 N.Y.S.2d 697 (1990)]

1. If you were deciding this case, would you make an exception to the rule and let the state collect the funds from Barclays Bank? Why or why not? What ethical policies must be balanced in this situation?
2. Under agency law, which will be discussed in Chapter 22, delivery to the agent of a given individual or entity constitutes delivery to that person or entity. The court deemed that Caliendo was not an agent of the state but an agent of the taxpayers. Does it matter that the taxpayers may not have known this principle of agency law and might have thought that, by delivering their checks to Caliendo, they were delivering them to the state?

VIDEO QUESTION

18–12. Go to this text's Web site at **academic.cengage.com/blaw/wbl** and select "Chapter 18." Click on "Video Questions" and view the video titled *Negotiability & Transferability: Indorsing Checks*. Then answer the following questions.

1. According to the instructor in the video, what are the two reasons why banks generally require a person to indorse a check that is made out to cash (a bearer instrument), even when the check is signed in the presence of the teller?
2. Suppose that your friend makes out a check payable to cash, signs it, and hands it to you. You take the check to your bank and indorse the check with your name and the words "without recourse." What type of indorsement is this? How does this indorsement affect the bank's rights?
3. Now suppose that you go to your bank and write a check on your account payable to cash for $500. The teller gives you the cash *without* asking you to indorse the check. After you leave, the teller slips the check into

his pocket. Later, the teller delivers it (without an indorsement) to his friend Carol in payment for a gambling debt. Carol takes your check to her bank, indorses it, and deposits the money. Discuss whether Carol is a holder in due course.

Negotiability, Transferability, and Liability

For updated links to resources available on the Web, as well as a variety of other materials, visit this text's Web site at

academic.cengage.com/blaw/wbl

The National Conference of Commissioners on Uniform State Laws, in association with the University of Pennsylvania Law School, now offers an official site for final drafts of uniform and model acts. For an index of final acts, including UCC Articles 3 and 4, go to

http://www.law.upenn.edu/bll/ulc/ulc_final.htm

Cornell University's Legal Information Institute offers online access to the UCC, as well as to UCC articles as enacted by particular states and proposed revisions to articles, at

http://www.law.cornell.edu/ucc/ucc.table.html

ONLINE LEGAL RESEARCH

Go to the *Fundamentals of Business Law* home page at **academic.cengage.com/blaw/wbl**, select "Chapter 18," and click on "Internet Exercises." There you will find the following Internet research exercises that you can perform to learn more about topics covered in this chapter.

Activity 18–1: LEGAL PERSPECTIVE—Fictitious Payees

Activity 18–2: MANAGEMENT PERSPECTIVE—Holder in Due Course

BEFORE THE TEST

Go to the *Fundamentals of Business Law* home page at **academic.cengage.com/blaw/wbl**, select "Chapter 18," and click on "Interactive Quizzes." You will find a number of interactive questions relating to this chapter.

CHAPTER 19

Checks, the Banking System, and E-Money

LEARNING OBJECTIVES

After reading this chapter, you should be able to answer the following questions:

1. On what types of checks does a bank serve as both the drawer and the drawee? What type of check does a bank agree in advance to accept when the check is presented for payment?
2. When may a bank properly dishonor a customer's check without being liable to the customer? What happens if a bank wrongfully dishonors a customer's check?
3. What duties does the Uniform Commercial Code impose on a bank's customers with regard to forged and altered checks? What are the consequences of a customer's negligence in performing those duties?
4. What are the four most common types of electronic fund transfers? What is the basic purpose of the Electronic Fund Transfer Act, and how does it benefit consumers?
5. What is e-money? How is e-money stored and used? What laws apply to e-money transactions and online banking services?

Checks are the most common type of negotiable instruments that are regulated by the Uniform Commercial Code (UCC). About 50 billion personal and commercial checks are written each year in the United States. Checks are convenient to use because they serve as substitutes for cash. To be sure, most students today tend to use debit cards, rather than checks, for many of their retail transactions. Indeed, debit cards may soon account for more retail payments than checks. Nonetheless, commercial checks remain an integral part of the American economic system, and they are therefore worthy of study.

Issues relating to checks are governed by Articles 3 and 4 of the UCC. Recall from Chapter 18 that Article 3 establishes the requirements that all negotiable instruments, including checks, must meet. Article 3 also sets forth the rights and liabilities of parties to negotiable instruments. Article 4 of the UCC governs bank deposits and collections as well as bank-customer relationships. Article 4 regulates the relationships of banks with one another as they process checks for payment, and it establishes a framework for deposit and checking agreements between a bank and its customers. A check therefore may fall within the scope of Article 3 and yet be subject to the provisions of Article 4 while in the course of collection. If a conflict between Article 3 and Article 4 arises, Article 4 controls [UCC 4–102(a)].

In this chapter, we first identify the legal characteristics of checks and the legal duties and liabilities that arise when a check is issued. Then we examine the collection process—that is, the actual procedure by which the checks deposited into bank accounts move through banking channels, causing the underlying funds to be shifted from one bank account to another. Increasingly, credit cards, debit cards, and other devices and methods to transfer funds electronically are being used to pay for goods and services. In the latter part of this chapter, we look at the law governing electronic fund transfers.

Checks

A **check** is a special type of draft that is drawn on a bank, ordering the bank to pay a fixed amount of money on demand [UCC 3–104(f)]. Article 4 of the UCC defines a bank as "a person engaged in the business of banking, including a savings bank, savings and loan association, credit union, or trust company" [UCC 4–105(1)]. If any other institution (such as a brokerage firm) handles a check for payment or for collection, the check is not covered by Article 4.

Recall from the discussion of negotiable instruments in Chapter 18 that a person who writes a check is called the *drawer*. The drawer is a depositor in the bank on which the check is drawn. The person to whom the check is payable is the *payee*. The bank or financial institution on which the check is drawn is the *drawee*. If Anita Cruzak writes a check from her checking account to pay her college tuition, she is the drawer, her bank is the drawee, and her college is the payee. We now look at some special types of checks.

Cashier's Checks

Checks are usually three-party instruments, but on certain types of checks, the bank can serve as both the drawer and the drawee. For example, when a bank draws a check on itself, the check is called a **cashier's check** and is a negotiable instrument on issue (see Exhibit 19–1) [UCC 3–104(g)]. Normally, a cashier's check indicates a specific payee. In effect, with a cashier's check, the bank assumes responsibility for paying the check, thus making the check more readily acceptable as a substitute for cash.

■ **EXAMPLE 19.1** Kramer needs to pay a moving company $8,000 for moving his household goods to a new home in another state. The moving company requests payment in the form of a cashier's check. Kramer goes to a bank (he need not have an account at the bank) and purchases a cashier's check, payable to the moving company, in the amount of $8,000. Kramer has to pay the bank the $8,000 for the cashier's check, plus a small service fee. He then gives the check to the moving company. ■

Cashier's checks are sometimes used in the business community as nearly the equivalent of cash. Except in very limited circumstances, the issuing bank must honor its cashier's checks when they are presented for payment. If a bank wrongfully dishonors a cashier's check, a holder can recover from the bank all expenses incurred, interest, and consequential damages [UCC 3–411]. This same rule applies if a bank wrongfully dishonors a certified check (to be discussed shortly) or a teller's check. (A *teller's check* is a check drawn by a bank on another bank or, when drawn on a nonbank, payable at or through a bank [UCC 3–104(h)].)

Traveler's Checks

A **traveler's check** is an instrument that is payable on demand, drawn on or payable at or through a financial institution (such as a bank), and designated as a traveler's check. The institution is directly obligated to accept and pay its traveler's check according to the check's terms. The purchaser is required to sign the check at the time it is bought and again at the time it is used [UCC 3–104(i)]. Exhibit 19–2 shows an example of a traveler's check.

EXHIBIT 19–1 A CASHIER'S CHECK

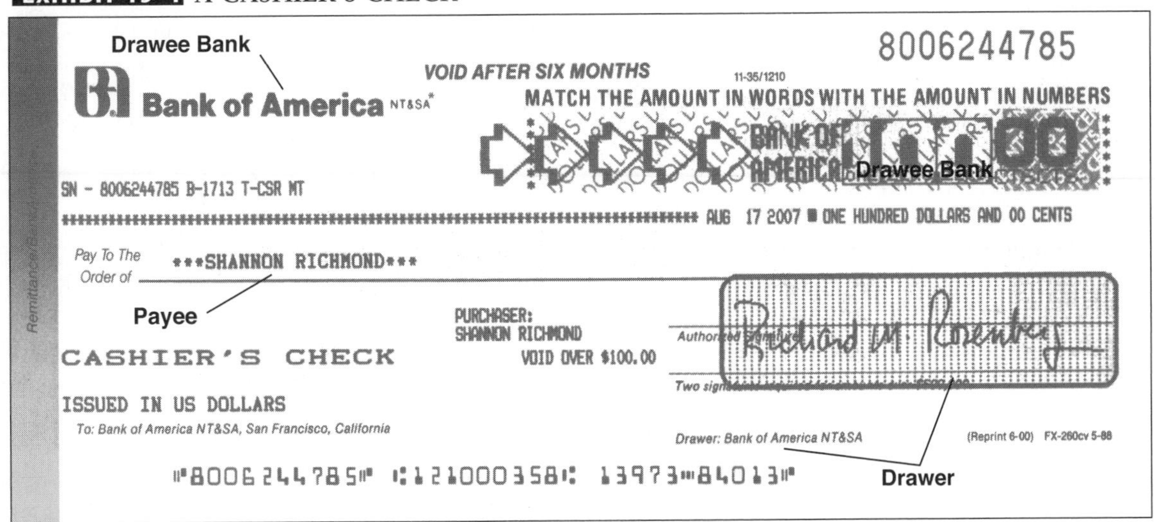

*The abbreviation *NT&SA* stands for National Trust and Savings Association. The Bank of America NT&SA is a subsidiary of Bank of America Corporation, which is engaged in financial services, insurance, investment management, and other businesses.

EXHIBIT 19–2 A TRAVELER'S CHECK

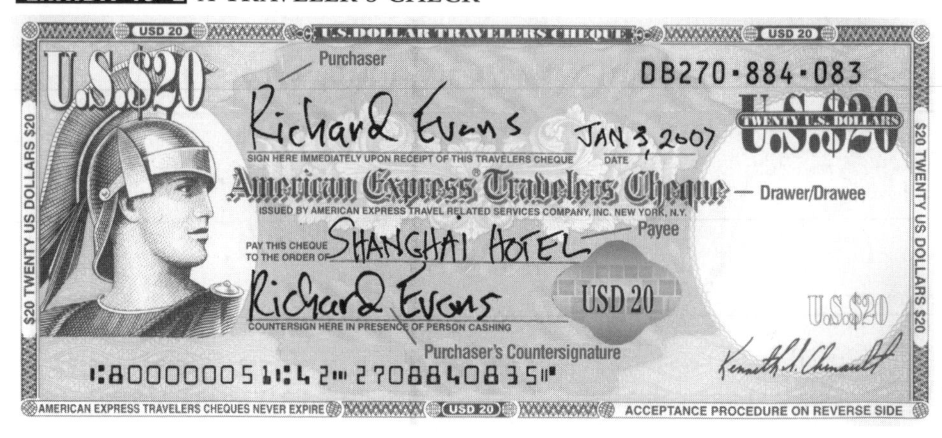

Certified Checks

A **certified check** is a check that has been *accepted* in writing by the bank on which it is drawn [UCC 3–409(d)]. When a drawee bank *certifies* (accepts) a check, it immediately charges the drawer's account with the amount of the check and transfers those funds to its own certified check account. In effect, the bank is agreeing in advance to accept that check when it is presented for payment and to make payment from those funds reserved in the certified check account. Essentially, certification prevents the bank from denying liability. It is a promise that sufficient funds are on deposit *and have been set aside* to cover the check.

A drawee bank is not obligated to certify a check, and failure to do so is not a dishonor of the check [UCC 3–409(d)]. If a bank does certify a check, however, the bank should write on the check the amount that it will pay. If the certification does not state an amount, and the amount is later increased and the instrument negotiated to a holder in due course (HDC), the obligation of the certifying bank is the amount of the instrument when it was taken by the HDC [UCC 3–413(b)].

Certification may be requested by a holder (to ensure that the check will not be dishonored for insufficient funds) or by the drawer. In either circumstance, once the check is certified, the drawer and any prior indorsers are completely discharged from liability [UCC 3–414(c), 3–415(d)].

The Bank-Customer Relationship

The bank-customer relationship begins when the customer opens a checking account and deposits funds that the bank will use to pay for checks written by the customer. Essentially, three types of relationships come into being, as discussed next.

Creditor-Debtor Relationship

A creditor-debtor relationship is created between a customer and a bank when, for example, the customer makes cash deposits into a checking account. When a customer makes a deposit, the customer becomes a creditor, and the bank a debtor, for the amount deposited.

Agency Relationship

An agency relationship also arises between the customer and the bank when the customer writes a check on his or her account. In effect, the customer is ordering the bank to pay the amount specified on the check to the holder when the holder presents the check to the bank for payment. In this situation, the bank becomes the customer's agent and is obligated to honor the customer's request. Similarly, if the customer deposits a check into her or his account, the bank, as the customer's agent, is obligated to collect payment on the check from the bank on which the check was drawn. To transfer checkbook funds among different banks, each bank acts as the agent of collection for its customer [UCC 4–201(a)]. (We will examine agency relationships in detail in Chapter 22.)

Contractual Relationship

Whenever a bank-customer relationship is established, certain contractual rights and duties arise. The specific rights and duties of the bank and its customer depend on the nature of the transaction. The respective rights and duties of banks and their customers are discussed in detail in the following pages.

Bank's Duty to Honor Checks

When a banking institution provides checking services, it agrees to honor the checks written by its customers, with the usual stipulation that the account must have sufficient funds available to pay each check [UCC 4–401(a)]. When a drawee bank *wrongfully* fails to honor a check, it is liable to its customer for damages resulting from its refusal to pay. The UCC does not attempt to specify the theory under which the customer may recover for wrongful dishonor; it merely states that the drawee (bank) is liable.

The customer's agreement with the bank includes a general obligation to keep sufficient funds on deposit to cover all checks written. The customer is liable to the payee or to the holder of a check in a civil suit if a check is dishonored for insufficient funds. If intent to defraud can be proved, the customer can also be subject to criminal prosecution for writing a bad check.

When the bank properly dishonors a check for insufficient funds, it has no liability to the customer. The bank may rightfully refuse payment on a customer's check in other circumstances as well. We look here at the rights and duties of both the bank and its customers in relation to specific situations.

Overdrafts

When the bank receives an item properly payable from its customer's checking account but the account contains insufficient funds to cover the amount of the check, the bank has two options. It can either (1) dishonor the item or (2) pay the item and charge the customer's account, thus creating an **overdraft,** providing that the customer has authorized the payment and the payment does not violate any bank-customer agreement [UCC 4–401(a)].[1] The bank can subtract the difference (plus a service charge) from the customer's next deposit or other funds belonging to the customer because the check carries with it an enforceable implied promise to reimburse the bank.

A bank can expressly agree with a customer to accept overdrafts through what is sometimes called an "overdraft protection agreement." If such an agreement is formed, any failure of the bank to honor a check because it would create an overdraft breaches this agreement and is treated as wrongful dishonor [UCC 4–402(a)].

When a check "bounces," a holder can resubmit the check, hoping that at a later date sufficient funds will be available to pay it. The holder must notify any indorsers on the check of the first dishonor, however; otherwise, they will be discharged from their signature liability.

Postdated Checks

A bank may also charge a postdated check against a customer's account, unless the customer notifies the bank, in a timely manner, not to pay the check until the stated date. The notice of postdating must be given in time to allow the bank to act on the notice before committing itself to pay on the check. The UCC states that the bank should treat a notice of postdating the same as a stop-payment order (to be discussed shortly). If the bank fails to act on the customer's notice and charges the customer's account before the date on the postdated check, the bank may be liable for any damages incurred by the customer [UCC 4–401(c)].[2]

Stale Checks

Commercial banking practice regards a check that is presented for payment more than six months from its date as a **stale check.** A bank is not obligated to pay an uncertified check presented more than six months from its date [UCC 4–404]. When receiving a stale check for payment, the bank has the option of paying or not paying the check. The bank may consult the customer before paying the check. If a bank pays a stale check in good faith without consulting the customer, however, the bank has the right to charge the customer's account for the amount of the check.

Stop-Payment Orders

A **stop-payment order** is an order by a customer to his or her bank not to pay or certify a certain check. Only a customer or a person authorized to draw on the account can order the bank not to pay the check when it is presented for payment [UCC 4–403(a)].[3] A customer has no right to stop payment on a check that has been certified or accepted by a bank, however. The customer must issue the stop-payment order within a reasonable time and in a reasonable manner to permit the bank to act on it

1. With a joint account, the bank cannot hold the nonsigning customer liable for payment of an overdraft unless that person benefited from its proceeds [UCC 4–401(b)].

2. Under the UCC, postdating does not affect the negotiability of a check. In the past, instead of treating postdated checks as checks payable on demand, some courts treated them as time drafts. Thus, regardless of whether the customer notified the bank of the postdating, a bank could not charge a customer's account for a postdated check without facing potential liability for the payment of later checks. Under the automated check-collection system now in use, however, a check is usually paid without respect to its date. Thus, today the bank can ignore the postdate on the check (treat it as a demand instrument) unless it has received notice from the customer that the check was postdated.

3. For a deceased customer, any person claiming a legitimate interest in the account may issue a stop-payment order [UCC 4–405].

EXHIBIT 19–3 A STOP-PAYMENT ORDER

Bank of America

Checking Account Stop Payment Order

To: Bank of America NT&SA
I want to stop payment on the following check(s).

ACCOUNT NUMBER: _ _ _ _ _ _ – _ _ _ _ _

SPECIFIC STOP

*ENTER DOLLAR AMOUNT: ____ *CHECK NUMBER: ____

THE CHECK WAS SIGNED BY: ____

THE CHECK IS PAYABLE TO: ____

THE REASON FOR THIS STOP PAYMENT IS: ____

STOP RANGE (Use for lost or stolen check(s) only.)

DOLLAR AMOUNT: 000

*ENTER STARTING CHECK NUMBER: ____ *END CHECK NUMBER: ____

THE REASON FOR THIS STOP PAYMENT IS: ____

BANK USE ONLY

TRANCODE:

☐ 21—ENTER STOP PAYMENT (SEE OTHER SIDE TO REMOVE)

NON READS: ____
UNPROC. STMT HIST: ____
PRIOR STMT CYCLE: ____
HOLDS ON COOLS: ____
REJECTED CHKS: ____
LARGE ITEMS: ____
FEE COLLECTED: ____
DATE ACCEPTED: ____
TIME ACCEPTED: ____

I agree that this order (1) is effective only if the above check(s) has (have) not yet been cashed or paid against my account, (2) will end six months from the date it is delivered to you unless I renew it in writing, and (3) is not valid if the check(s) was (were) accepted on the strength of my Bank of America courtesy-check guarantee card by a merchant participating in that program. I also agree (1) to notify you immediately to cancel this order if the reason for the stop payment no longer exists or (2) that closing the account on which the check(s) is (are) drawn automatically cancels this order.

IF ANOTHER BRANCH OF THIS BANK OR ANOTHER PERSON OR ENTITY BECOMES A "HOLDER IN DUE COURSE" OF THE ABOVE CHECK, I UNDERSTAND THAT PAYMENT MAY BE ENFORCED AGINST THE CHECK'S MAKER (SIGNER).

*I CERTIFY THE AMOUNT AND CHECK NUMBER(S) ABOVE ARE CORRECT.

☐ I have written a replacement check (number and date of check).

(Optional—please circle one: Mr., Ms., Mrs., Miss) CUSTOMER'S SIGNATURE **X** ____ DATE ____

[UCC 4–403(a)]. Although a stop-payment order can be given orally, usually by phone, it is binding on the bank for only fourteen calendar days unless confirmed in writing.[4] A written stop-payment order (see Exhibit 19–3) or an oral order confirmed in writing is effective for six months, at which time it must be renewed in writing [UCC 4–403(b)].

Bank's Liability for Wrongful Payment If the bank pays the check over the customer's properly instituted stop-payment order, the bank will be obligated to recredit the customer's account—but only for the amount of the actual loss suffered by the drawer because of the wrongful payment [UCC 4–403(c)]. ■ EXAMPLE 19.2 Toshio Murano orders six bamboo palms from a local nursery at $50 each and gives the nursery a check for $300. Later that day, the nursery tells Murano that it will not deliver the palms as arranged. Murano immediately calls his bank and stops payment on the check. If the bank nonetheless honors the check, the bank will be liable to Murano for the full $300. The result would be different, however, if the nursery had delivered five palms. In that situation, Murano would owe the nursery $250 for the delivered palms, and his actual losses would be only $50. Consequently, the bank would be liable to Murano for only $50. ■

4. Some states do not recognize oral stop-payment orders; they must be in writing.

Customer's Liability for Wrongful Stop-Payment Order A stop-payment order has its risks for a customer. The customer-drawer must have a *valid legal ground* for issuing such an order; otherwise, the holder can sue the drawer for payment. Moreover, defenses sufficient to refuse payment against a payee may not be valid grounds to prevent payment against a subsequent holder in due course [UCC 3–305, 3–306]. A person who wrongfully stops payment on a check not only will be liable to the payee for the amount of the check but also may be liable for consequential damages incurred by the payee as a result of the wrongful stop-payment order.

Death or Incompetence of a Customer

Neither the death nor the incompetence of a customer revokes a bank's authority to pay an item until the bank is informed of the situation and has had reasonable time to act on the notice. Thus, if, at the time a check is issued or its collection has been undertaken, a bank does not know of an adjudication of incompetence against the customer who wrote the check, the bank can pay the item without incurring liability [UCC 4–405]. Even when a bank knows of the death of its customer, for ten days after the *date of death,* it can pay or certify checks drawn on or before the date of death. An exception to this rule is made if a person claiming an interest in that account, such as an heir, orders the bank to stop payment. Without this

provision, banks would constantly be required to verify the continued life and competence of their drawers.

CHECKS BEARING FORGED DRAWERS' SIGNATURES

When a bank pays a check on which the drawer's signature is forged, generally the bank is liable. A bank may be able to recover at least some of the loss, however, from the customer (if the customer's negligence contributed to the making of the forgery), from the forger of the check (if he or she can be found), or from the holder who presented the check for payment (if the holder knew that the signature was forged).

The General Rule A forged signature on a check has no legal effect as the signature of a drawer [UCC 3–403(a)]. For this reason, banks require signature cards from each customer who opens a checking account. Signature cards allow the bank to verify whether the signatures on their customers' checks are genuine. The general rule is that the bank must recredit the customer's account when it pays a check with a forged signature. (Note that banks today normally verify signatures only on checks that exceed a certain threshold, such as $1,000, $2,500, or some higher amount. Even though a bank sometimes incurs liability costs when it has paid forged checks, the costs involved in verifying every check's signature would be much higher.)

Customer Negligence When the customer's negligence substantially contributes to the forgery, the bank normally will not be obligated to recredit the customer's account for the amount of the check [UCC 3–406].[5] The customer's liability may be reduced, however, by the amount of loss caused by negligence on the part of the bank (or other "person") paying the instrument or taking it for value if the negligence substantially contributed to the loss [UCC 3–406(b)].

■ **EXAMPLE 19.3** Gemco Corporation uses special check-writing equipment to write its payroll and business checks. Gemco discovers that one of its employees used the equipment to write himself a check for $10,000 and that the bank subsequently honored it. Gemco asks the bank to recredit $10,000 to its account for improperly paying the forged check. If the bank can show that Gemco failed to take reasonable care in controlling access to the check-writing equipment, the bank will not be required to recredit Gemco's account for the amount of the forged check. If Gemco can show that negligence on the part of the bank contributed substantially to the loss, however, then Gemco's liability may be reduced proportionately. ■

In the following case, a bank that had paid a forged check claimed that the account holder's negligence had contributed to the forgery. Specifically, the bank alleged that the account holder had failed to exercise ordinary care to prevent his wife from forging a check on the account.

5. Note that banks can shift some of the risk of forged checks to the customer by contract, such as the risk of forged signatures created by the use of facsimile or other nonmanual signature devices.

CASE 19.1 Nesper v. Bank of America

Court of Appeals
of Ohio, Sixth District,
Ottawa County, 2004.
__ Ohio App.3d __,
__ N.E.2d __.

FACTS Robert Nesper knew his wife, Patricia Nesper, had engaged in financial misconduct both before and during their marriage. The misconduct included forging Robert's name on applications for credit cards and a contract to buy a vehicle. The couple continued to live together, but Robert kept a bank account solely in his name at Bank of America. He kept the unused checks for the account hidden in their house in a room that could be locked, although the room was not kept locked all of the time. In early 2002, he became aware that Patricia had forged his name to the account's check number 275 in the amount of $2,000. Robert filed a suit in an Ohio state court against the bank, seeking the return of the $2,000 to his account. Robert argued that banks have a responsibility to refuse to honor forged checks, regardless of the marital status of the forger. The court ruled in Robert's favor. The bank appealed to a state intermediate appellate court.

ISSUE Is a bank liable for honoring a check with a forged drawer's signature even if that forged signature was made by a spouse?

DECISION Yes. The state intermediate appellate court affirmed the judgment of the lower court.

REASON The appellate court explained that a check with a forged drawer's signature is not properly payable, and if the bank pays it, the bank is generally liable. Under UCC 3–406(a),

CASE 19.1—CONTINUED

however, "[a] person whose failure to exercise ordinary care substantially contributes to * * * the making of a forged signature on an instrument is precluded from asserting * * * the forgery against a person who, in good faith, pays the instrument." Ordinary care has been found to be lacking in cases in which parties who know their signatures have been forged in the past are negligent in failing to prevent further forgeries. Also, a party may ratify a forged check by affirming it—that is, by manifesting an intent to approve the check while having knowledge of its forgery. Here, the bank failed to show that Robert did not exercise ordinary care with respect to Patricia's forgery of his signature on check number 275 or that he ratified the check. Consequently, the bank was obliged to recredit its customer's account for the amount.

FOR CRITICAL ANALYSIS—Social Consideration *Suppose that the bank had shown that Robert had failed to take reasonable care in controlling access to the blank checks for his account. Would the outcome in this case have been different? Why or why not?*

—Timely Examination of Bank Statements Required. Banks typically send or make available to their customers monthly statements detailing activity in their checking accounts. Banks are not obligated to include the canceled checks themselves with the statement sent to the customer. If the bank does not send the canceled checks (or photocopies of the canceled checks), however, it must provide the customer with information (check number, amount, and date of payment) on the statement that will allow the customer to reasonably identify the checks that the bank has paid [UCC 4–406(a), (b)]. If the bank retains the canceled checks, it must keep the checks—or legible copies of the checks—for a period of seven years [UCC 4–406(b)]. The customer may obtain a canceled check (or a copy of the check) during this period of time.

The customer has a duty to examine bank statements (and canceled checks or photocopies, if they are included with the statements) promptly and with reasonable care, and to report any alterations or forged signatures promptly [UCC 4–406(c)]. This includes forged signatures of indorsers (to be discussed later). If the customer fails to fulfill this duty and the bank suffers a loss as a result, the customer will be liable for the loss [UCC 4–406(d)]. Even if the customer can prove that she or he took reasonable care against forgeries, the UCC provides that the customer must discover the forgeries and notify the bank within a period of one year to require the bank to recredit her or his account.

—Consequences of Failing to Detect Forgeries. When the same wrongdoer has committed a series of forgeries, the UCC provides that the customer, to recover for all the forged items, must discover and report the first forged check to the bank within thirty calendar days of the receipt of the bank statement (and canceled checks or copies, if they are included) [UCC 4–406(d)(2)]. Failure to notify the bank within this period of time discharges the bank's liability for all forged checks that it pays prior to notification. (For a further discussion of this rule, see this chapter's *Management Perspective* feature on page 270.)

—When the Bank Is Also Negligent. In one situation, a bank customer can escape liability, at least in part, for failing to notify the bank of forged or altered checks promptly or within the required thirty-day period. If the customer can prove that the bank was also negligent—that is, that the bank failed to exercise ordinary care—then the bank will also be liable, and the loss will be allocated between the bank and the customer on the basis of comparative negligence [UCC 4–406(e)]. In other words, even though a customer may have been negligent, the bank may still have to recredit the customer's account for a portion of the loss if the bank failed to exercise ordinary care.

The UCC defines *ordinary care* to mean the "observance of reasonable commercial standards, prevailing in the area in which [a] person is located, with respect to the business in which that person is engaged" [UCC 3–103]. As mentioned earlier, it is customary in the banking industry to manually examine signatures only on checks over a certain amount (such as $1,000, $2,500, or some higher amount). Thus, if a bank, in accordance with prevailing banking standards, fails to examine a signature on a particular check, the bank has not necessarily breached its duty to exercise ordinary care.

Regardless of the degree of care exercised by the customer or the bank, the UCC places an absolute time limit on the liability of a bank for paying a check with a forged customer signature. A customer who fails to report a forged signature within one year from the date that the statement was made available for inspection loses the legal right to have the bank recredit his or her account [UCC 4–406(f)].

MANAGEMENT PERSPECTIVE
The Problem of Forged Checks

MANAGEMENT FACES A LEGAL ISSUE

Business owners and managers occasionally have had to deal with dishonest employees who, having access to company checks, forge checks payable to themselves. What happens when the company discovers the forgeries? Is the bank obligated to recredit the company's account?

WHAT THE COURTS SAY

As stated elsewhere, the general rule under the UCC is that a forged signature has no legal effect as the signature of the drawer. Therefore, if a drawer's bank pays a forged check, a court will order the bank to recredit the customer's account for the amount of the forged check or checks. There are exceptions to this general rule, however. An important exception arises if the bank can prove that the customer's own negligence contributed to a forgery.

In one case, for example, an employee of a company that owned a chain of coffeehouses and other businesses forged a series of checks totaling more than $330,000. The employee, who had access to blank company checks, learned how to use company software programs to print company checks on his home computer. When the bank statements arrived in the mail, the employee removed the checks that had been forged. On discovering the forgeries more than a year and a half later, the company immediately reported the problem to the bank. Could the bank be held liable in these circumstances? No, according to the court. Citing UCC 4–406(d), the court held that because the company had not reported the first forged check to the bank within thirty days, the bank's liability for payment was discharged.[a]

In another case, an employee forged a series of checks totaling $29,595 over an eleven-month period, and the employer brought an action against the bank to recover the funds. The bank argued that the customer's own negligence had contributed to the forgeries. Specifically, the customer had failed to report the first forgery to the bank within the required thirty-day period, and thus the bank was not liable. The court agreed, and, again, the customer had to bear the loss.[b]

IMPLICATIONS FOR MANAGERS

These cases illustrate that the UCC's time period of thirty days to report forgeries is not flexible. Business managers need to be aware that the consequences can be harsh if they fail to examine bank statements promptly and notify the bank of any forged items.

a. *Espresso Roma Corp. v. Bank of America, N.A.,* 100 Cal.App.4th 525, 124 Cal.Rptr.2d 549 (2002).
b. *Union Planters Bank v. Rogers,* 912 So.2d. 116 (Miss. 2005).

Checks Bearing Forged Indorsements

A bank that pays a customer's check bearing a forged indorsement must recredit the customer's account or be liable to the customer-drawer for breach of contract. ■ **EXAMPLE 19.4** Suppose that Simon issues a $500 check "to the order of Antonio." Juan steals the check, forges Antonio's indorsement, and cashes the check. When the check reaches Simon's bank, the bank pays it and debits Simon's account. The bank must recredit the $500 to Simon's account because it failed to carry out Simon's order to pay "to the order of Antonio" [UCC 4–401(a)]. Of course, Simon's bank can in turn recover—for breach of warranty (see Chapter 18)—from the bank that cashed the check when Juan presented it [UCC 4–207(a)(2)]. ■

Eventually, the loss usually falls on the first party to take the instrument bearing the forged indorsement because, as discussed in Chapter 18, a forged indorsement does not transfer title. Thus, whoever takes an instrument with a forged indorsement cannot become a holder.

The customer, in any event, has a duty to report forged indorsements promptly. Failure to report forged indorsements within a three-year period after the forged items have been made available to the customer relieves the bank of liability [UCC 4–111].

Altered Checks

The customer's instruction to the bank is to pay the exact amount on the face of the check to the holder. The bank has a duty to examine each check before making final payment. If it fails to detect an alteration, it is liable to its customer for the loss because it did not pay as the customer ordered. The loss is the difference between the original amount of the check and the amount actually paid [UCC 4–401(d)(1)]. ■ **EXAMPLE 19.5** Suppose that a check written for $11 is raised to $111. The customer's account will be charged $11 (the amount the customer ordered the bank to pay). The bank will normally be responsible for the $100. ■

Customer Negligence As in a situation involving a forged drawer's signature, a customer's negligence can shift the loss when payment is made on an altered check (unless the bank was also negligent). A common example occurs when a person carelessly writes a check and leaves large gaps around the numbers and words where additional numbers and words can be inserted (see Exhibit 19–4).

Similarly, a person who signs a check and leaves the dollar amount for someone else to fill in is barred from protesting when the bank unknowingly and in good faith pays whatever amount is shown [UCC 4–401(d)(2)]. Finally, if the bank can trace its loss on successive altered checks to the customer's failure to discover the initial alteration, then the bank can reduce its liability for reimbursing the customer's account [UCC 4–406]. The law governing the customer's duty to examine monthly statements and canceled checks, and to discover and report alterations to the bank, is the same as that applied to a forged drawer's signature.

In every situation involving a forged drawer's signature or an alteration, a bank must observe reasonable commercial standards of care in paying on a customer's checks [UCC 4–406(e)]. The customer's negligence can be used as a defense only if the bank has exercised ordinary care.

Other Parties from Whom the Bank May Recover The bank is entitled to recover the amount of loss from the transferor who, by presenting the check for payment, warrants that the check has not been materially altered. This rule has two exceptions, though. If the bank is the drawer (as it is on a cashier's check and a teller's check), it cannot recover from the presenting party if the party is an HDC acting in good faith [UCC 3–417(a)(2), 4–208(a)(2)]. The reason is that an instrument's drawer is in a better position than an HDC to know whether the instrument has been altered.

Similarly, an HDC, acting in good faith in presenting a certified check for payment, will not be held liable under warranty principles if the check was altered before the HDC acquired it [UCC 3–417(a)(2), 4–207(a)(2)]. ■ **EXAMPLE 19.6** Jordan draws a check for $500 payable to Deffen. Deffen alters the amount to $5,000. The First National Bank of Whiteacre, the drawee bank, certifies the check for $5,000. Deffen negotiates the check to Evans, an HDC. The drawee bank pays Evans $5,000. On discovering the mistake, the bank cannot recover from Evans the $4,500 paid by mistake, even though the bank was not in a superior position to detect the alteration. This is in accord with the purpose of certification, which is to obtain the definite obligation of a bank to honor a definite instrument. ■

Bank's Duty to Accept Deposits

A bank has a duty to its customer to accept the customer's deposits of cash and checks. When checks are deposited, the bank must make the funds represented by those checks available within certain time frames. A bank also has a duty to collect payment on any checks payable or indorsed to its customers and deposited by them into their accounts. Cash deposits made in U.S. currency are received into customers' accounts without being subject to further collection procedures.

EXHIBIT 19–4 A POORLY FILLED-OUT CHECK

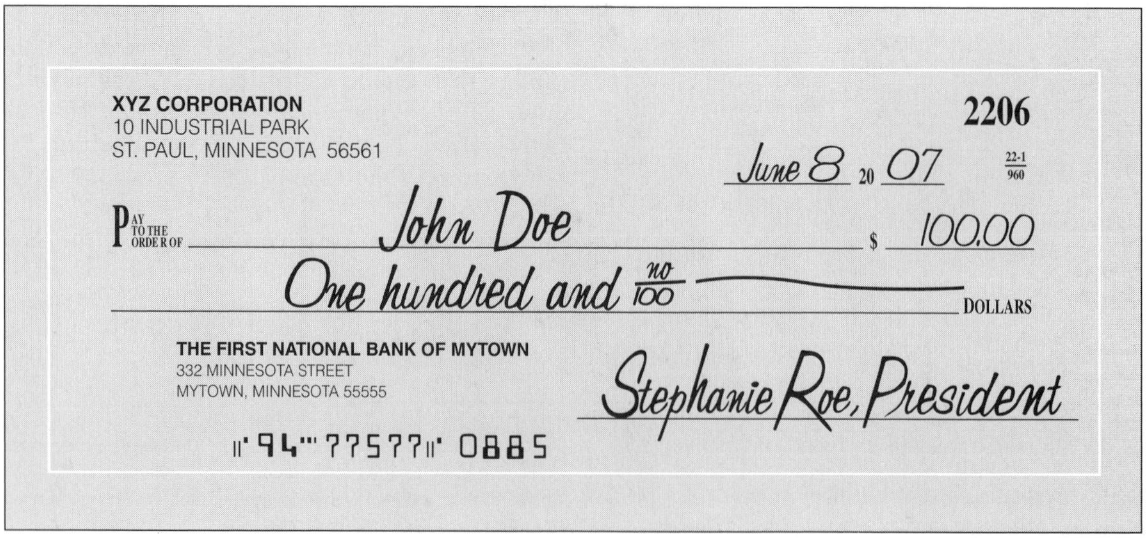

XYZ CORPORATION
10 INDUSTRIAL PARK
ST. PAUL, MINNESOTA 56561

2206

June 8 20 07 22-1/960

PAY TO THE ORDER OF John Doe $ 100.00

One hundred and no/100 DOLLARS

THE FIRST NATIONAL BANK OF MYTOWN
332 MINNESOTA STREET
MYTOWN, MINNESOTA 55555

Stephanie Roe, President

⑆94⑈77577⑆ 0885

Availability Schedule for Deposited Checks

The Expedited Funds Availability Act of 1987[6] and Regulation CC,[7] which was issued by the Federal Reserve Board of Governors (the *Federal Reserve System* will be discussed shortly) to implement the act, require that any local check deposited must be available for withdrawal by check or as cash within one business day from the date of deposit. A check is classified as a local check if the first bank to receive the check for payment and the bank on which the check is drawn are located in the same check-processing region (check-processing regions are designated by the Federal Reserve Board of Governors). For nonlocal checks, the funds must be available for withdrawal within not more than five business days.

Additional Requirements In addition to the above requirements, the Expedited Funds Availability Act requires the following:

1. That funds be available on the next business day for cash deposits and wire transfers, government checks, the first $100 of a day's check deposits, cashier's checks, certified checks, and checks for which the depositary and payor banks are branches of the same institution.
2. That the first $100 of any deposit be available for cash withdrawal on the opening of the next business day after deposit. If a local check is deposited, the next $400 is to be available for withdrawal by no later than 5:00 P.M. the next business day. If, for example, you deposit a local check for $500 on Monday, you can withdraw $100 in cash at the opening of the business day on Tuesday, and an additional $400 must be available for withdrawal by no later than 5:00 P.M. on Wednesday.

Exceptions A different availability schedule applies to deposits made at *nonproprietary* automated teller machines (ATMs). These are ATMs that are not owned or operated by the bank receiving the deposits. Basically, a five-day hold is permitted on all deposits, including cash deposits, made at nonproprietary ATMs.

Other exceptions also exist. A depository institution has eight days to make funds available in new accounts (those open less than thirty days). It has an extra four days on deposits over $5,000 (except deposits of government and cashier's checks), on accounts with repeated overdrafts, and on checks of questionable collectibility (if the institution tells the depositor it suspects fraud or insolvency).

6. 12 U.S.C. Sections 4001–4010.
7. 12 C.F.R. Sections 229.1–229.42.

The Collection Process

Usually, deposited checks involve parties that do business at different banks, but sometimes checks are written between customers of the same bank. Either situation brings into play the bank collection process as it operates within the statutory framework of Article 4 of the UCC.

The check-collection process discussed next will be modified in the future as the banking industry implements the Check Clearing in the 21st Century Act of 2003.[8] This act, which is known as "Check 21," makes an electronic copy of an original check a legal document, thereby eliminating the need to physically transport and manually process paper checks.

Designations of Banks Involved in the Collection Process The first bank to receive a check for payment is the **depositary bank.**[9] For example, when a person deposits an IRS tax-refund check into a personal checking account at the local bank, that bank is the depositary bank. The bank on which a check is drawn (the drawee bank) is called the **payor bank.** Any bank except the payor bank that handles a check during some phase of the collection process is a **collecting bank.** Any bank except the payor bank or the depositary bank to which an item is transferred in the course of this collection process is called an **intermediary bank.**

During the collection process, any bank can take on one or more of the various roles of depositary, payor, collecting, and intermediary bank. ■ **EXAMPLE 19.7** A buyer in New York writes a check on her New York bank and sends it to a seller in San Francisco. The seller deposits the check in her San Francisco bank account. The seller's bank is both a *depositary bank* and a *collecting bank*. The buyer's bank in New York is the *payor bank*. As the check travels from San Francisco to New York, any collecting bank handling the item in the collection process (other than the depositary bank and the payor bank) is also called an *intermediary bank*. Exhibit 19–5 illustrates how various banks function in the collection process in the context of this example. ■

Check Collection between Customers of the Same Bank An item that is payable by the depositary bank (also the payor bank) that receives it is called an "on-us item." If the bank does not dishonor the check by the opening of

8. 12 U.S.C. Section 5001, Pub. L. No. 108-100.
9. All definitions in this section are found in UCC 4–105. The terms *depositary* and *depository* have different meanings in the banking context. A depository bank refers to a *physical place* (a bank or other institution) in which deposits or funds are held or stored.

EXHIBIT 19–5 THE CHECK-COLLECTION PROCESS

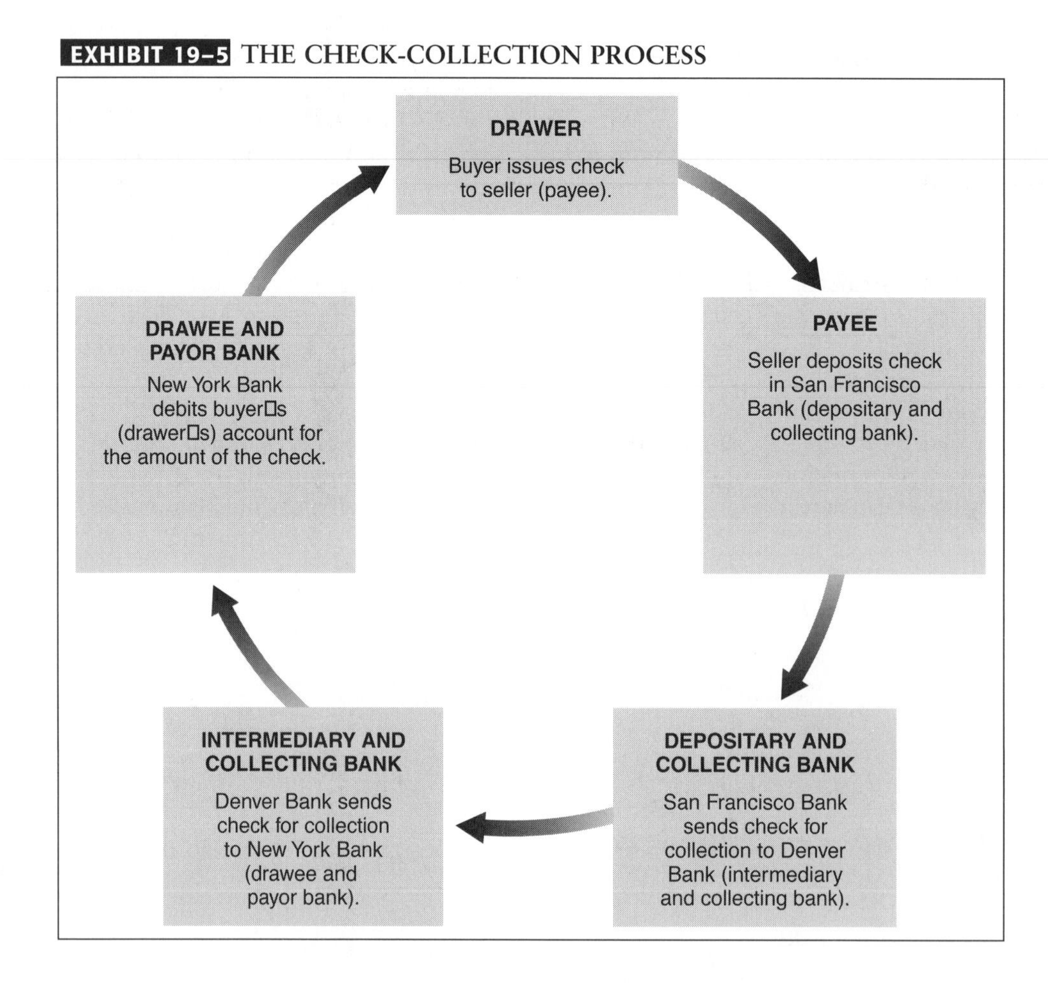
DRAWER
Buyer issues check to seller (payee).
PAYEE
Seller deposits check in San Francisco Bank (depositary and collecting bank).
DEPOSITARY AND COLLECTING BANK
San Francisco Bank sends check for collection to Denver Bank (intermediary and collecting bank).
INTERMEDIARY AND COLLECTING BANK
Denver Bank sends check for collection to New York Bank (drawee and payor bank).
DRAWEE AND PAYOR BANK
New York Bank debits buyer□s (drawer□s) account for the amount of the check.

the second banking day following its receipt, the check is considered paid [UCC 4–215(e)(2)]. ■ EXAMPLE 19.8 Williams and Merkowitz both have checking accounts at State Bank. On Monday morning, Merkowitz deposits into his own checking account a $300 check drawn by Williams. That same day, State Bank issues Merkowitz a "provisional credit" for $300. When the bank opens on Wednesday, Williams's check is considered honored, and Merkowitz's provisional credit becomes final. ■

Check Collection between Customers of Different Banks Once a depositary bank receives a check, it must arrange to present it either directly or through intermediary banks to the appropriate payor bank. Each bank in the collection chain must pass the check on before midnight of the next banking day following its receipt [UCC 4–202(b)].[10] A "banking day" is any part of a day that the bank is open to carry on substantially all of its banking functions. Thus, if a bank has only its drive-through facilities open, a check deposited on Saturday would not trigger a bank's midnight deadline until the following Monday. When the check reaches the payor bank, that bank is liable for the face amount of the check, unless the payor bank dishonors the check or returns it by midnight on the next banking day following receipt [UCC 4–302].[11]

Because of this deadline and because banks need to maintain an even work flow in the many items they handle daily, the UCC permits what is called *deferred posting*. According to UCC 4–108, "a bank may fix an afternoon hour of 2:00 P.M. or later as a cutoff hour for the handling of money and items and the making of entries on its books." Any checks received after that hour "may be

10. A bank may take a "reasonably longer time" under certain conditions, such as when the bank's computer system is down due to a power failure, but the bank must show that its action is still timely [UCC 4–202(b)].

11. Most checks are cleared by a computerized process, and communication and computer facilities may fail because of weather, equipment malfunction, or other conditions. If such conditions arise and a bank fails to meet its midnight deadline, the bank is "excused" from liability if the bank has exercised "such diligence as the circumstances require" [UCC 4–109(d)].

treated as being received at the opening of the next banking day." Thus, if a bank's "cutoff hour" is 3:00 P.M., a check received by a payor bank at 4:00 P.M. on Monday would be deferred for posting until Tuesday. In this situation, the payor bank's deadline would be midnight Wednesday.

How the Federal Reserve System Clears Checks The **Federal Reserve System** is our nation's central bank. It is a network of twelve district banks and related branches, which are located around the country and headed by the Federal Reserve Board of Governors. Most banks in the United States have Federal Reserve accounts. The Federal Reserve System has greatly simplified the check-collection process by acting as a **clearinghouse**—a system or a place where banks exchange checks and drafts drawn on each other and settle daily balances.

■ EXAMPLE 19.9 Pamela Moy of Philadelphia writes a check to Jeanne Sutton in San Francisco. When Sutton receives the check in the mail, she deposits it in her bank. Her bank then deposits the check in the Federal Reserve Bank of San Francisco, which transfers it to the Federal Reserve Bank of Philadelphia. That Federal Reserve bank then sends the check to Moy's bank, which deducts the amount of the check from Moy's account. Exhibit 19–6 illustrates this process. ■

Electronic Check Presentment In the past, most checks were processed manually—the employees of each bank in the collection chain would physically handle each check

EXHIBIT 19–6 HOW A CHECK IS CLEARED

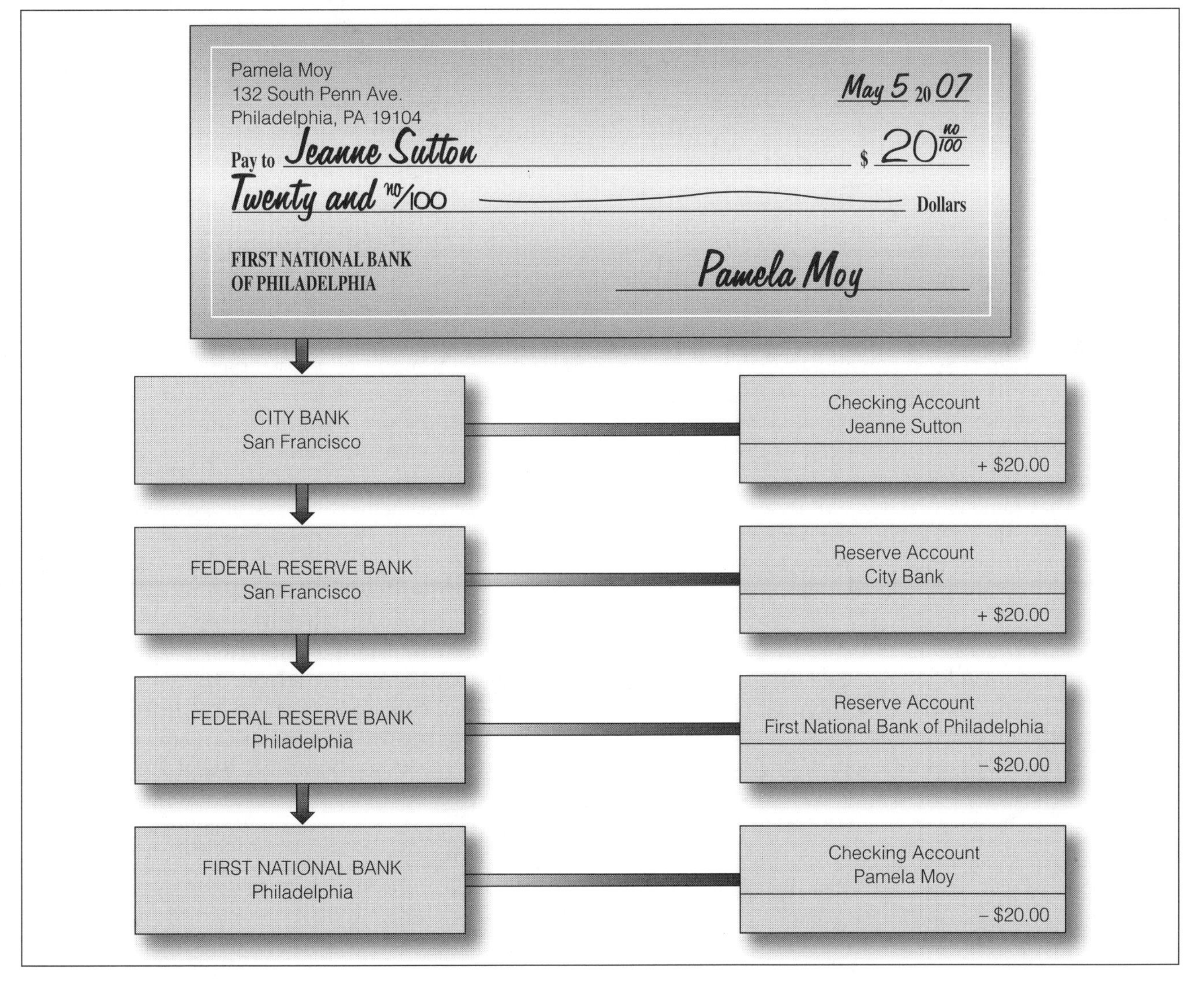

that passed through the bank for collection or payment. Today, however, most checks are processed electronically. In contrast to manual check processing, which can take days, *electronic check presentment* can be done on the day of deposit. With electronic check presentment, items may be encoded with information (such as the amount of the check) that is read and processed by other banks' computers. In some situations, a check may be retained at its place of deposit, and only its image or description is presented for payment under an electronic presentment agreement [UCC 4–110].[12]

12. This section of the UCC assumes that no bank will participate in an electronic presentment program without an agreement. See Comment 2 to UCC 4–110. For example, two banks that frequently do business with each other might enter into an agreement allowing the depositary bank to retain the physical check and send an electronic image to the other bank for presentment.

A person who encodes information on an item warrants to any subsequent bank or payor that the encoded information is correct [UCC 4–209]. This is also true for a person who retains an item while transmitting its image or information describing it as presentation for payment. This person warrants that the retention and presentment of the item comply with the electronic presentment agreement.

Regulation CC provides that a returned check must be encoded with the routing number of the depositary bank, the amount of the check, and other information and adds that this "does not affect a paying bank's responsibility to return a check within the deadlines required by the U.C.C." Under UCC 4–301(d)(2), an item is returned "when it is sent or delivered to the bank's customer or transferor or pursuant to his [or her] instructions." The question in the following case was whether an item that was not correctly encoded could be considered properly returned by its midnight deadline.

CASE 19.2 NBT Bank, N.A. v. First National Community Bank

United States Court of Appeals,
Third Circuit, 2004.
393 F.3d 404.

FACTS Human Services Consultants, Inc. (HSC), had a checking account at First National Community Bank (FNCB) in Pennsylvania. A related firm, Human Services Consultants Management, Inc., had two checking accounts at NBT Bank, N.A., in New York, one of which was under the name PA Health. On March 8, 2001, PA Health presented a check in the amount of $706,000 for deposit in its NBT account. The check was drawn on HSC's account at FNCB. NBT credited $706,000 to PA Health's account as a provisional credit and transmitted the check to the Federal Reserve Bank of Philadelphia for presentment to FNCB, which received the check on March 12. The next morning, FNCB dishonored the check because HSC's account contained insufficient funds, encoded the check for return to NBT, and sent it back to the Federal Reserve bank before 11:59 P.M. Later, on the morning of March 14, FNCB phoned NBT and also sent a letter by fax, informing it of the item's dishonor. The check's encoding was in error, however, and it was wrongly routed to another bank; as a result, NBT did not receive the physical check until March 16. Ultimately, the check was revealed to be part of a fraudulent scheme that caused NBT more than $1 million in losses. NBT filed a suit in a federal district court against FNCB, asserting that FNCB's encoding error rendered FNCB's "return" of the check ineffective. The court granted FNCB's a motion for summary judgment, and NBT appealed.

ISSUE If a dishonored check is not correctly encoded but is physically transferred back to the Federal Reserve bank before the midnight deadline, will the item be considered properly returned?

DECISION Yes. The U.S. Court of Appeals for the Third Circuit affirmed the lower court's judgment.

REASON The federal appellate court reasoned that the UCC's midnight deadline focuses on the physical return of a check, not encoding errors. The encoding requirement is found in Regulation CC, which explicitly states that the encoding requirements for routing numbers do not affect deadlines for a payor bank's return of the check. Regulation CC complements but does not necessarily replace the requirements of Article 4 of the UCC; the parties are bound by both. Under Regulation CC, the amount of damages for failure to exercise ordinary care in complying with the encoding provisions is "reduced by the amount of the loss that the [plaintiff] would have incurred even if the [defendant] bank had exercised ordinary care." Here, the parties agreed that NBT suffered no loss as a result of FNCB's encoding error. Thus, the court concluded that NBT could not recover the amount of the check from FNCB.

FOR CRITICAL ANALYSIS—Social Consideration *How might the result in this case have been different if NBT had committed the encoding error and FNCB had suffered the loss?*

Electronic Fund Transfers

The application of computer technology to banking, in the form of electronic fund transfer systems, has helped to relieve banking institutions of the burden of having to move mountains of paperwork to process fund transfers. An **electronic fund transfer (EFT)** is a transfer of funds through the use of an electronic terminal, a telephone, a computer, or magnetic tape. The law governing EFTs depends on the type of transfer involved. Consumer fund transfers are governed by the Electronic Fund Transfer Act (EFTA) of 1978.[13] Commercial fund transfers are governed by Article 4A of the UCC.

The benefits of electronic banking are obvious. Automatic payments, direct deposits, and other fund transfers are now made electronically; no physical transfers of cash, checks, or other negotiable instruments are involved.

Not surprisingly, though, electronic banking also poses difficulties on occasion. It is difficult to issue stop-payment orders with electronic banking. Also, fewer records are available to prove or disprove that a transaction took place. The possibilities for tampering with a person's private banking information have increased. Finally, customers can no longer rely on having time between the writing of a check and its deduction from an account (float time).

Types of EFT Systems

Most banks today offer EFT services to their customers. The four most common types of EFT systems used by bank customers are listed below.

1. *Automated teller machines* (ATMs)—Machines connected online to the bank's computers. Customers insert a plastic card issued by the bank and key in a personal identification number (PIN) to access their accounts and conduct banking transactions.
2. *Point-of-sale systems*—Online terminals that allow consumers to transfer funds to merchants to pay for purchases using a debit card.
3. *Direct deposits and withdrawals*—Customers can authorize the bank to allow another party—such as the government or an employer—to make direct deposits into their accounts. Similarly, a customer can direct the bank to make automatic payments to a third party at regular, recurrent intervals from the customer's funds (insurance premiums or loan payments, for example).
4. *Pay-by-Internet systems*—Some financial institutions permit their customers to access the institution's computer system via the Internet and direct a transfer of funds between accounts or pay a particular bill, such as a utility bill, for example.

Consumer Fund Transfers

The Electronic Fund Transfer Act (EFTA) provides a basic framework for the rights, liabilities, and responsibilities of users of EFT systems. Additionally, the act gave the Federal Reserve Board authority to issue rules and regulations to help implement the act's provisions. The Federal Reserve Board's implemental regulation is called **Regulation E.**

The EFTA governs financial institutions that offer electronic fund transfers involving consumer accounts. The types of accounts covered include checking accounts, savings accounts, and any other asset accounts established for personal, family, or household purposes. Note that the EFTA covers telephone transfers only if they are made in accordance with a prearranged plan under which periodic or recurring transfers are contemplated.[14] In the subsections that follow, we look closely at some of the act's provisions concerning consumers' rights and responsibilities, unauthorized transfers, and error resolution.

Consumers' Rights and Responsibilities under the EFTA

The EFTA is essentially a disclosure law benefiting consumers. The act requires financial institutions to inform consumers of their rights and responsibilities, including those listed here, with respect to EFT systems.

1. If a customer's debit card is lost or stolen and used without his or her permission, the customer may be required to pay no more than $50. The customer, however, must notify the bank of the loss or theft within two days of learning about it. Otherwise, the liability increases to $500. The customer may be liable for more than $500 if he or she does not report the unauthorized use within sixty days after it appears on the customer's statement. (If a customer voluntarily gives her or his debit card to another, who then uses it improperly, the protections just mentioned do not apply.)
2. The customer must discover any error on the monthly statement within sixty days and must notify the bank. The bank then has ten days to investigate and must report its conclusions to the customer in writing. If the bank takes longer than ten days, it must return the disputed amount to the customer's account until it finds the error. If there is no error, the customer has to give the money back to the bank.

13. 15 U.S.C. Sections 1693–1693r. The EFTA amended Title IX of the Consumer Credit Protection Act.

14. *Kashanchi v. Texas Commerce Medical Bank, N.A.*, 703 F.2d 936 (5th Cir. 1983).

3. The bank must furnish receipts for transactions made through computer terminals, but it is not obligated to do so for telephone transfers.
4. The bank must provide a monthly statement for every month in which there is an electronic transfer of funds. Otherwise, the bank must provide statements every quarter. The statement must show the amount and date of the transfer, the names of the retailers or other third parties involved, the location or identification of the terminal, and the fees. Additionally, the statement must give an address and a phone number for inquiries and error notices.
5. Any authorized prepayment for utility bills and insurance premiums can be stopped three days before the scheduled transfer.

Unauthorized Electronic Fund Transfers Unauthorized electronic fund transfers are one of the hazards of electronic banking. A paper check leaves visible evidence of a transaction, and a customer can easily detect a forgery or an alteration on a check with ordinary vigilance. Evidence of an electronic transfer, however, is often only an entry in a computer printout of the various debits and credits made to a particular account during a specified time period.

Because of the vulnerability of EFT systems to fraudulent activities, the EFTA of 1978 clearly defined what constitutes an unauthorized transfer. Under the act, a transfer is unauthorized if (1) it is initiated by a person other than the consumer who has no actual authority to initiate the transfer; (2) the consumer receives no benefit from it; and (3) the consumer did not furnish the person "with the card, code, or other means of access" to her or his account. Unauthorized access to an EFT system constitutes a federal felony, and those convicted may be sentenced to a fine of up to $10,000 and up to ten years' imprisonment.

Error Resolution Banks must strictly comply with the terms of the EFTA, or they will be held liable. For a bank's violation of the EFTA, a consumer may recover both actual damages (including attorneys' fees and costs) and punitive damages of not less than $100 and not more than $1,000. (Unlike actual damages, punitive damages are assessed to punish a defendant or to set an example for similar wrongdoers.) Failure to investigate an error in good faith makes the bank liable for treble damages (three times the amount of damages). Even when a customer has sustained no actual damage, the bank may be liable for legal costs and punitive damages if it fails to follow the proper procedures outlined by the EFTA in regard to error resolution.

COMMERCIAL TRANSFERS

Funds are also transferred electronically "by wire" between commercial parties. In fact, the dollar volume of payments by wire transfer is more than $1 trillion a day—an amount that far exceeds the dollar volume of payments made by other means. The two major wire payment systems are the Federal Reserve's wire transfer network (Fedwire) and the New York Clearing House Interbank Payments Systems (CHIPS).

Commercial wire transfers are governed by Article 4A of the UCC, which has been adopted by most states. The following example illustrates the type of fund transfer covered by Article 4A. ■ **EXAMPLE 19.10** Jellux, Inc., owes $5 million to Perot Corporation. Instead of sending Perot a check or some other instrument that would enable Perot to obtain payment, Jellux tells its bank, East Bank, to credit $5 million to Perot's account in West Bank. East Bank debits Jellux's East Bank account and wires $5 million to Perot's West Bank account. In more complex transactions, additional banks would be involved. ■

In these and similar circumstances, ordinarily a financial institution's instruction is transmitted electronically. Any means may be used, however, including first class mail. To reflect this fact, Article 4A uses the term *funds transfer* rather than wire transfer to describe the overall payment transaction. The full text of Article 4A is included in Appendix E, following Article 4 of the UCC.

E-Money and Online Banking

New forms of electronic payments (e-payments) have the potential to replace *physical* cash—coins and paper currency—with *virtual* cash in the form of electronic impulses. This is the unique promise of **digital cash,** which consists of funds stored on microchips and other computer devices. Online banking has also become a reality in today's world. In a few minutes, anybody with the proper software can access his or her account, transfer funds, write "checks," pay bills, monitor investments, and often even buy and sell stocks.

Various forms of electronic money, or **e-money,** are emerging. The simplest kind of e-money system uses *stored-value cards.* These are plastic cards embossed with magnetic strips containing magnetically encoded data. A person can use a stored-value card to purchase specific goods and services offered by the card issuer. For example, university libraries typically have copy machines that students operate by inserting a stored-value card. Each time a student makes copies, the machine deducts the per-copy fee from the card.

Another form of e-money is the smart card. **Smart cards** are plastic cards containing computer microchips that can hold more information than a magnetic strip. A smart card carries and processes security programming. This capability gives smart cards a technical advantage over stored-value cards. The microprocessors on smart cards can also authenticate the validity of transactions. Retailers can program electronic cash registers to confirm the authenticity of a smart card by examining a unique digital signature stored on its microchip. (Digital signatures were discussed in Chapter 13.)

Online Banking Services

Most online bank customers use three kinds of services. One of the most popular is bill consolidation and payment. Another is transferring funds among accounts. These online services are now offered via the Internet as well as by phone. The third is applying for loans, which many banks permit customers to do via the Internet. Customers typically have to appear in person to finalize the terms of a loan.

Two important banking activities generally are not yet available online: depositing and withdrawing funds. With smart cards, people could transfer funds on the Internet, thereby effectively transforming their personal computers into ATMs. Many observers believe that online banking is the way to introduce people to e-money and smart cards.

Since the late 1990s, several banks have operated exclusively on the Internet. These "virtual banks" have no physical branch offices. Because few people are equipped to send funds to virtual banks via smart-card technology, the virtual banks have accepted deposits through physical delivery systems, such as the U.S. Postal Service or FedEx.

Regulatory Compliance

Banks have an interest in encouraging the widespread use of online banking because of its significant potential for reducing costs and thus increasing profits. As in other areas of cyberspace, however, determining how laws apply to online banking activities can be difficult.

The Home Mortgage Disclosure Act[15] and the Community Reinvestment Act (CRA) of 1977,[16] for example, require a bank to define its market area and also to provide information to regulators about its deposits and loans. Under the CRA, banks establish market areas in communities situated next to their branch offices. The banks map these areas, using boundaries defined by counties or standard metropolitan areas, and annually review the maps. The purpose of these requirements is to prevent discrimination in lending practices.

But how does a successful "cyberbank" delineate its community? If, for instance, Bank of Internet becomes a tremendous success, does it really have a physical community? Will regulators allow a written description of a cybercommunity for Internet customers? Such regulatory issues are new, challenging, and certain to become more complicated as Internet banking widens its scope internationally.

15. 12 U.S.C. Sections 2801–2810.
16. 12 U.S.C. Sections 2901–2908.

Privacy Protection

At the present time, it is not clear which, if any, laws apply to the security of e-money payment information and e-money issuers' financial records. This is partly because it is not clear whether e-money issuers fit within the traditional definition of a financial institution.

E-Money Payment Information The Federal Reserve has decided not to impose Regulation E, which governs certain electronic fund transfers, on e-money transactions. Federal laws prohibiting unauthorized access to electronic communications might apply, however. For example, the Electronic Communications Privacy Act of 1986[17] prohibits any person from knowingly divulging to any other person the contents of an electronic communication while that communication is in transmission or in electronic storage.

E-Money Issuers' Financial Records Under the Right to Financial Privacy Act of 1978,[18] before a financial institution may give financial information about you to a federal agency, you must explicitly consent. If you do not, a federal agency wishing to obtain your financial records must obtain a warrant. A digital cash issuer may be subject to this act if that issuer is deemed to be (1) a bank by virtue of its holding customer funds or (2) any entity that issues a physical card similar to a credit or debit card.

Consumer Financial Data In 1999, Congress passed the Financial Services Modernization Act,[19] also known as the Gramm-Leach-Bliley Act, in an attempt to delineate how financial institutions can treat customer data. In general, the act and its rules[20] place restrictions and obliga-

17. 18 U.S.C. Sections 2510–2521.
18. 12 U.S.C. Sections 3401 *et seq.*
19. 12 U.S.C. Sections 24a, 248b, 1820a, 1828b, 1831v–1831y, 1848a, 2908, 4809; 15 U.S.C. Sections 80b-10a, 6701, 6711–6717, 6731–6735, 6751–6766, 6781, 6801–6809, 6821–6827, 6901–6910; and others.
20. 12 C.F.R. Part 40.

tions on financial institutions to protect consumer data and privacy. Every financial institution must provide its customers with information on its privacy policies and practices. No financial institution can disclose nonpublic personal information about a consumer to an unaffiliated third party unless the act's disclosure and opt-out requirements are met.

The Uniform Money Services Act

Over the past few years, many states have enacted various regulations that apply to money services in a rather haphazard fashion. At the same time, e-money services that operate on the Internet—which, of course, cuts across jurisdictional lines—have been asking that these regulations be made more predictable.

In 2001, the National Conference of Commissioners on Uniform State Laws recommended to state legislatures a new law that would subject all money services businesses to the same regulations that apply to other, traditional financial service businesses. This law, which is known as the Uniform Money Services Act (UMSA), has been adopted in a few states.[21]

TRADITIONAL MONEY SERVICES

Money services businesses (MSBs) have not been subject to regulation to the same extent as other financial service businesses. Unlike banks, MSBs do not accept deposits. They do, however, issue money orders, traveler's checks, and stored-value cards; exchange foreign currency; and cash checks. Immigrants often use these businesses to send money to their relatives in other countries. Because MSBs often do not have continuing relationships with their customers, their customers have sometimes evaded federal law with respect to large currency transactions or used the services to launder money (see Chapter 6). This has been particularly true with respect to financing terrorist activities.

The UMSA applies to persons engaged in money transmission, check cashing, or currency exchange. The uniform act requires an MSB involved in these activities to obtain a license from a state, to be examined by state officials, to report on its activities to the state, and to comply with certain record-keeping requirements. Each of these subjects has its provisions and exceptions. Safety and soundness measures, such as annual examinations, surety bonds (to guarantee financial soundness), and permissible investments, are also mandated by the act.

INTERNET-BASED MONEY SERVICES

Under the UMSA, Internet-based money services, and other types of e-money services, would be treated the same as other money services.[22] The drafters of the UMSA ensured that it would cover these services by referring to *monetary value* instead of simply *money* [UMSA 1–102(c)(11)].

Internet-based monetary value systems subject to the new law may include:

1. *E-money and Internet payment mechanisms*—Cash, or its substitute, that is stored as data on a chip or a personal computer so that it can be transferred over the Internet or an intranet.
2. *Internet scrip*—Monetary value that may be exchanged over the Internet but can also be redeemed for cash.
3. *Stored-value products*—Smart cards, prepaid cards, or value-added cards [UMSA 1–102(c)(21)].

21. To date, the UMSA has been adopted in Iowa, Vermont, Washington State, and the U.S. Virgin Islands.

22. The UMSA does not apply to state governments, the federal government, securities dealers, banks, businesses that incidentally transport currency and instruments in the normal course of business, payday loan businesses, and others [UMSA 1–103].

TERMS AND CONCEPTS

cashier's check 364
certified check 365
check 364
clearinghouse 374
collecting bank 372
depositary bank 372
digital cash 377
electronic fund transfer (EFT) 376
e-money 377
Federal Reserve System 374
intermediary bank 372
overdraft 366
payor bank 372
Regulation E 376
smart card 378
stale check 366
stop-payment order 366
traveler's check 364

CHAPTER SUMMARY Checks, the Banking System, and E-Money

Checks (See pages 364–365.)	1. *Cashier's check*—A check drawn by a bank on itself (the bank is both the drawer and the drawee) and purchased by a customer. In effect, the bank lends its credit to the purchaser of the check, thus making the funds available for immediate use in banking circles. 2. *Traveler's check*—An instrument on which a financial institution is both the drawer and the drawee. The purchaser must provide his or her signature as a countersignature for a traveler's check to become a negotiable instrument. 3. *Certified check*—A check for which the drawee bank certifies in writing that it will set aside funds in the drawer's account to ensure payment of the check on presentation. On certification, the drawer and all prior indorsers are completely discharged from liability on the check.
The Bank-Customer Relationship (See page 365.)	1. *Creditor-debtor relationship*—The bank and its customer have a creditor-debtor relationship (the bank is the debtor because it holds the customer's funds on deposit). 2. *Agency relationship*—Because a bank must act in accordance with the customer's orders in regard to the customer's deposited money, an agency relationship also arises—the bank is the agent for the customer, who is the principal. 3. *Contractual relationship*—The bank's relationship with its customer is also contractual; both the bank and the customer assume certain contractual duties when a customer opens a bank account.
Bank's Duty to Honor Checks (See pages 366–371.)	Generally, a bank has a duty to honor its customers' checks, provided that the customers have sufficient funds on deposit to cover the checks [UCC 4–401(a)]. The bank is liable to its customers for actual damages proved to be due to wrongful dishonor. The bank's duty to honor its customers' checks is not absolute. The following list summarizes the rights and liabilities of the bank and the customer in various situations: 1. *Overdrafts*—The bank has the right to charge a customer's account for any item properly payable, even if the charge results in an overdraft [UCC 4–401(a)]. 2. *Postdated checks*—A bank may charge a postdated check against a customer's account as a demand instrument, unless the customer notifies the bank of the postdating in time to allow the bank to act on the notice before the bank commits itself to pay on the check [UCC 4–401(c)]. 3. *Stale checks*—The bank is not obligated to pay an uncertified check presented more than six months after its date, but it may do so in good faith without liability [UCC 4–404]. 4. *Stop-payment orders*—The customer must make a stop-payment order in time for the bank to have a reasonable opportunity to act. Oral orders are binding for only fourteen days unless they are confirmed in writing. Written orders are effective for only six months unless renewed in writing. The bank is liable for wrongful payment over a timely stop-payment order, but only to the extent of the loss suffered by the drawer-customer [UCC 4–403]. 5. *Death or incompetence of a customer*—So long as the bank does not know of the death or incompetence of a customer, the bank can pay an item without liability to the customer's estate. Even with knowledge of a customer's death, a bank can honor or certify checks (in the absence of a stop-payment order) for ten days after the date of the customer's death [UCC 4–405]. 6. *Forged drawers' signatures, forged indorsements, and altered checks*—The customer has a duty to examine account statements with reasonable care on receipt and to notify the bank promptly of any forged signatures, forged or unauthorized indorsements, or alterations. On a series of unauthorized signatures or alterations by the same wrongdoer, examination and report must occur within thirty calendar days of receipt of the statement. Failure to notify the bank releases the bank from any liability unless the bank failed to exercise ordinary care. Regardless of care or lack of care, the customer is estopped (prevented) from holding the bank liable after one year for unauthorized customer signatures or alterations and after three years for unauthorized indorsements [UCC 3–403, 4–111, 4–401(a), 4–406].

CHAPTER SUMMARY Checks, the Banking System, and E-Money—Continued

Bank's Duty to Accept Deposits (See pages 371–375.)	A bank has a duty to accept deposits made by its customers into their accounts. Funds represented by checks deposited must be made available to customers according to a schedule mandated by the Expedited Funds Availability Act of 1987 and Regulation CC. A bank also has a duty to collect payment on any checks deposited by its customers. When checks deposited by customers are drawn on other banks, as they often are, the check-collection process comes into play (summarized next). 1. *Definitions of banks*—UCC 4–105 provides the following definitions of banks involved in the collection process: a. Depositary bank—The first bank to accept a check for payment. b. Payor bank—The bank on which a check is drawn. c. Collecting bank—Any bank except the payor bank that handles a check during the collection process. d. Intermediary bank—Any bank except the payor bank or the depositary bank to which an item is transferred in the course of the collection process. 2. *Check collection between customers of the same bank*—A check payable by the depositary bank that receives it is an "on-us item"; if the bank does not dishonor the check by the opening of the second banking day following its receipt, the check is considered paid [UCC 4–215(e)(2)]. 3. *Check collection between customers of different banks*—Each bank in the collection process must pass the check on to the next appropriate bank before midnight of the next banking day following its receipt [UCC 4–108, 4–202(b), 4–302]. 4. *How the Federal Reserve System clears checks*—The Federal Reserve System facilitates the check-clearing process by serving as a clearinghouse for checks. 5. *Electronic check presentment*—When checks are presented electronically, items may be encoded with information (such as the amount of the check) that is read and processed by other banks' computers. In some situations, a check may be retained at its place of deposit, and only its image or information describing it is presented for payment under a Federal Reserve agreement, clearinghouse rule, or other agreement [UCC 4–110].
Electronic Fund Transfers (See pages 376–377.)	1. *Types of EFT systems*— a. Automated teller machines (ATMs). b. Point-of-sale systems. c. Direct deposits and withdrawals. d. Pay-by-Internet systems. 2. *Consumer fund transfers*—Consumer fund transfers are governed by the Electronic Fund Transfer Act (EFTA) of 1978. The EFTA is basically a disclosure law that sets forth the rights and duties of the bank and the customer with respect to EFT systems. Banks must comply strictly with EFTA requirements. 3. *Commercial transfers*—Disputes arising as a result of unauthorized or incorrectly made fund transfers between financial institutions are not covered under the EFTA. Article 4A of the UCC, which has been adopted by almost all of the states, governs fund transfers not subject to the EFTA or other federal or state statutes.
E-Money and Online Banking (See pages 377–379.)	1. *New forms of e-payments*—These include stored-value cards and smart cards. 2. *Current online banking services*— a. Bill consolidation and payment. b. Transferring funds among accounts. c. Applying for loans.

(Continued)

CHAPTER SUMMARY Checks, the Banking System, and E-Money—Continued

E-Money and Online Banking—Continued	3. *Regulatory compliance*—Banks must define their market areas, in communities situated next to their branch offices, under the Home Mortgage Disclosure Act and the Community Reinvestment Act. It is not clear how an online bank would define its market area or its community. 4. *Privacy protection*—It is not entirely clear which, if any, laws apply to e-money and online banking. The Financial Services Modernization Act (the Gramm-Leach-Bliley Act) outlines how financial institutions can treat consumer data and privacy in general. The Right to Financial Privacy Act may also apply.
The Uniform Money Services Act (See page 379.)	The National Conference of Commissioners on Uniform State Laws has recommended the Uniform Money Services Act to state legislatures. The purpose of the act is to subject online and e-money services to the same regulations that apply to traditional financial service businesses.

FOR REVIEW

Answers for the even-numbered questions in this For Review *section can be found in Appendix A at the end of this text.*

1. On what types of checks does a bank serve as both the drawer and the drawee? What type of check does a bank agree in advance to accept when the check is presented for payment?
2. When may a bank properly dishonor a customer's check without being liable to the customer? What happens if a bank wrongfully dishonors a customer's check?
3. What duties does the Uniform Commercial Code impose on a bank's customers with regard to forged and altered checks? What are the consequences of a customer's negligence in performing those duties?
4. What are the four most common types of electronic fund transfers? What is the basic purpose of the Electronic Fund Transfer Act, and how does it benefit consumers?
5. What is e-money? How is e-money stored and used? What laws apply to e-money transactions and online banking services?

QUESTIONS AND CASE PROBLEMS

19–1. Error Resolution. Sheridan has a checking account at Gulf Bank. She frequently uses her debit card to obtain cash from the automated teller machines. She always withdraws $50 when she makes a withdrawal, but she never withdraws more than $50 in any one day. When she received the April statement on her account, she noticed that on April 13 two withdrawals for $50 each had been made from the account. Believing this to be a mistake, she went to her bank on May 10 to inform the bank of the error. A bank officer told her that the bank would investigate and inform her of the result. On May 26, the bank officer called her and said that bank personnel were having trouble locating the error but would continue to try to find it. On June 20, the bank sent her a full written report advising her that no error had been made. Sheridan, unhappy with the bank's explanation, filed a suit against the bank, alleging that it had violated the Electronic Fund Transfer Act. What was the outcome of the suit? Would it matter if the bank could show that on the day in question it had deducted $50 from Sheridan's account to cover a check that Sheridan had written to a local department store and that had cleared the bank on that day?

QUESTION WITH SAMPLE ANSWER

19–2. First Internet Bank operates exclusively on the Web with no physical branch offices. Although some of First Internet's business is transacted with smart-card technology, most of its business with its customers is conducted through the mail. First Internet offers free checking, no-fee money market accounts, mortgage refinancing, and other services. What regulation covering banks might First Internet find it difficult to comply with, and what is the difficulty?

For a sample answer to this question, go to Appendix B at the end of this text.

19–3. Forged Checks. Roy Supply, Inc., and R. M. R. Drywall, Inc., had checking accounts at Wells Fargo Bank. Both accounts required all checks to carry two signatures—that of Edward Roy and that of Twila June Moore, both of whom were executive officers of both companies. Between January 1989 and March 1991, the bank honored hundreds of checks on which Roy's signature was forged by Moore. On January 31, 1992, Roy and the

two corporations notified the bank of the forgeries and then filed a suit in a California state court against the bank, alleging negligence. Who is liable for the amounts of the forged checks? Why? [*Roy Supply, Inc. v. Wells Fargo Bank, N.A.,* 39 Cal.App.4th 1051, 46 Cal.Rptr.2d 309 (1995)]

19–4. Stale Checks. On July 15, 1986, IBP, Inc., issued to Meyer Land & Cattle Co. a check for $135,234.18 payable to both Meyer and Sylvan State Bank for the purchase of cattle. IBP wrote the check on its account at Mercantile Bank of Topeka. Someone at the Meyer firm misplaced the check. In the fall of 1995, Meyer's president, Tim Meyer, found the check behind a desk drawer. Jana Huse, Meyer's office manager, presented the check for deposit at Sylvan, which accepted it. After Mercantile received the instrument and its computers noted the absence of any stop-payment order, it paid the check with funds from IBP's checking account. IBP insisted that Mercantile recredit IBP's account. Mercantile refused. IBP filed a suit in a federal district court against Mercantile and others, claiming, among other things, that Mercantile had not acted in good faith because it had processed the check by automated means without examining it manually. Mercantile responded that its check-processing procedures adhered to its own policies, as well as reasonable commercial standards of fair dealing in the banking industry. Mercantile filed a motion for summary judgment. Should the court grant the motion? Why or why not? [*IBP, Inc. v. Mercantile Bank of Topeka,* 6 F.Supp.2d 1258 (D.Kan. 1999)]

19–5. Debit Cards. On April 20, 1999, while visiting her daughter and son-in-law Michael Dowdell, Carol Farrow asked Dowdell to fix her car. She gave him her car keys, attached to which was a small wallet containing her debit card. Dowdell repaired her car and returned the keys. Two days later, Farrow noticed that her debit card was missing and contacted Auburn Bank, which had issued the card. Farrow reviewed her automated teller machine (ATM) transaction record and noticed that a large amount of cash had been withdrawn from her checking account on April 22 and April 23. When Farrow reviewed the photos taken by the ATM cameras at the time of the withdrawals, she recognized Dowdell as the person using her debit card. Dowdell was convicted in an Alabama state court of the crime of fraudulent use of a debit card. What procedures are involved in a debit-card transaction? What problems with debit-card transactions are apparent from the facts of this case? How might these problems be prevented? [*Dowdell v. State,* 790 So.2d 359 (Ala.Crim.App. 2000)]

CASE PROBLEM WITH SAMPLE ANSWER

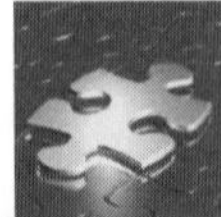

19–6. Robert Santoro was the manager of City Check Cashing, Inc., a check-cashing service in New Jersey, and Peggyann Slansky was the clerk. On July 14, Misir Koci presented Santoro with a $290,000 check signed by Melvin Green and drawn on Manufacturers Hanover Trust Co. (a bank). The check was stamped with a Manufacturers certification stamp. The date on the check had clearly been changed from August 8 to July 7. Slansky called the bank to verify the check and was told that the serial number "did not sound like one belonging to the bank." Slansky faxed the check to the bank with a query about the date, but received no reply. Slansky also called Green, who stated that the date on the check was altered before it was certified. City Check Cashing cashed and deposited the check within two hours. The drawee bank found the check to be invalid and timely returned it unpaid. City Check Cashing filed a suit in a New Jersey state court against Manufacturers and others, asserting that the bank should have responded to the fax before the midnight deadline in Section 4–302 of the Uniform Commercial Code. Did the bank violate the midnight deadline rule? Explain. [*City Check Cashing, Inc. v. Manufacturers Hanover Trust Co.,* 166 N.J. 49, 764 A.2d 411 (2001)]

After you have answered this problem, compare your answer with the sample answer given on the Web site that accompanies this text. Go to academic.cengage.com/blaw/wbl, select "Chapter 19," and click on "Case Problem with Sample Answer."

19–7. Forged Indorsement. Visiting Nurses Association of Telfair County, Inc. (VNA), maintained a checking account at Security State Bank in Valdosta, Georgia. Wanda Williamson, a VNA clerk, was responsible for making VNA bank deposits, but she was not a signatory on VNA's account. Over a four-year period, Williamson embezzled more than $250,000 from VNA by forging its indorsement on checks, cashing them at the bank, and keeping part of the proceeds. Williamson was arrested, convicted, sentenced to a prison term, and ordered to pay restitution. VNA filed a suit in a Georgia state court against the bank, alleging, among other things, negligence. The bank filed a motion for summary judgment on the ground that VNA was precluded by Section 4–406(f) of the Uniform Commercial Code from recovering on checks with forged indorsements. Should the court grant the motion? Explain. [*Security State Bank v. Visiting Nurses Association of Telfair County, Inc.,* 256 Ga.App. 374, 568 S.E.2d 491 (2002)]

19–8. Forged Signatures. Cynthia Stafford worked as an administrative professional at Gerber & Gerber, P.C. (professional corporation), a law firm, for more than two years. During that time, she stole ten checks payable to Gerber & Gerber (G&G), which she indorsed in blank by forging one of the attorney's signatures. She then indorsed the forged checks in her name and deposited them in her account at Regions Bank. Over the same period, G&G deposited in its accounts at Regions Bank thousands of checks amounting to $150 million to $200 million. Each G&G check was indorsed with a rubber stamp for deposit into the G&G account. The thefts were made possible in part because G&G kept unindorsed checks in an open file accessible to all employees and Stafford was sometimes the person assigned to stamp the checks. When the thefts were discovered, G&G filed a suit in a Georgia state court against Regions Bank to recover the stolen funds, alleging in part negligence. Regions Bank filed a motion for summary judgment. What principles apply to attribute liability between these parties? How should the court rule on the bank's motion? Explain. [*Gerber & Gerber, P.C. v. Regions Bank,* 266 Ga.App. 8, 596 S.E.2d 174 (2004)]

Checks, the Banking System, and E-Money

For updated links to resources available on the Web, as well as a variety of other materials, visit this text's Web site at

academic.cengage.com/blaw/wbl

You can obtain information about banking regulation from the Federal Deposit Insurance Corporation (FDIC) at

http://www.fdic.gov

Cornell University's Legal Information Institute provides an overview of banking, as well as a "menu of sources" of federal and state statutes and court decisions relating to banking transactions. To access this information, go to

http://www.law.cornell.edu/topics/banking.html

You can obtain extensive information about the Federal Reserve System by accessing "the Fed's" home page at

http://www.federalreserve.gov

You can find a series of articles on smart cards at the following Web site:

http://users.aol.com/pjsmart/index.htm

You can read the provisions of the Uniform Money Services Act and find out which states have adopted it by going to the Web site of the National Conference of Commissioners on Uniform State Laws at

http://www.nccusl.org

and selecting the "Money Services Act" under the "Final Acts and Legislation" link.

ONLINE LEGAL RESEARCH

Go to the *Fundamentals of Business Law* home page at academic.cengage.com/blaw/wbl, select "Chapter 19," and click on "Internet Exercises." There you will find the following Internet research exercises that you can perform to learn more about topics covered in this chapter.

Activity 19–1: MANAGEMENT PERSPECTIVE—Check Fraud

Activity 19–2: LEGAL PERSPECTIVE—Smart Cards

BEFORE THE TEST

Go to the *Fundamentals of Business Law* home page at academic.cengage.com/blaw/wbl, select "Chapter 19," and click on "Interactive Quizzes." You will find a number of interactive questions relating to this chapter.

UNIT 5 • EXTENDED CASE STUDY

Lema v. Bank of America, N.A.

In Chapter 19, we reviewed some of the laws that govern checks and the banking system, noting that the rights and duties of a bank and its customers are contractual. We discussed those rights and duties as they arise under Articles 3 and 4 of the Uniform Commercial Code (UCC). In this extended case study, we examine *Lema v. Bank of America, N.A.*,[1] a decision involving an agreement between a bank and its customer to alter the effect of the UCC as it relates to those rights and duties. The specific question before the court in this case was whether a customer may agree with a bank to forgo the protections afforded the customer under Articles 3 and 4.

1. 375 Md. 625, 826 A.2d 504 (2003).

Case Background

In 1999, Nkiambi Jean Lema, an accountant, opened a checking account with Bank of America, N.A., under the name "N. J. Lema Co." The bank's "Deposit Agreement" provided, in clause 4(e), that

> Unless prohibited by applicable law or regulation, we . . . reserve the right to charge back to your account the amount of any item deposited to your account or cashed for you which was initially paid by the payor bank and which is later returned to us due to [a] . . . claim of alteration. . . .

On November 24, Willy Amuli deposited a check for $63,000 payable to "N. J. Lema Co." into Lema's account at Bank of America. The check was drawn on an account at Bank of New York. By mid-February 2000, Lema had disbursed most of the funds to Amuli in various transactions. Meanwhile, on January 12, Bank of New York told Bank of America that the Amuli check had been altered—it was actually for $3,000. Bank of America returned $60,000 to Bank of New York and, at the end of February, debited Lema's account for the amount.

Lema filed a suit in a Maryland state court against Bank of America, seeking damages. The court awarded Lema $62,037.74. Bank of America appealed, and a state intermediate appellate court reversed the judgment. Lema appealed to the state's highest court, the Maryland Court of Appeals.

Majority Opinion

BATTAGLIA, J. [Judge]

* * * *

The official comments to UCC 1–102 and 4–301 reinforce the concept that the UCC's effect may be altered by agreement * * * . Official Comment 2 to Section 1–102 states that, *"freedom of contract is a principle of the [UCC],"* and *"the effect of its provisions may be varied by agreement."* Similarly, Official Comment 1 to Section 4–103 states that the Section "permits within wide limits variation of the effect of provisions of [Article 4] by agreement." [Emphasis added.]

The official comments also discuss the types of agreements that may vary the effect of the UCC's terms. Official Comment 2 to Section 4–103 provides that the term "agreement," as used in that Section, has the same meaning as given to it by Section 1–102(3), and specifically states that an agreement may be with respect to "all items handled for a particular customer, [for example], a general agreement between the depositary bank and the customer at the time a deposit account is opened." That, of course, is the type of agreement in issue in the present case.

* * * *

Lema contends that the Deposit Agreement did not entitle the Bank to debit Lema's account because the Agreement provides the Bank with that right only if not "prohibited by applicable law." According to Lema, [Article] 3 is the applicable law, and Section 3–401(a) * * * "provides that a person is not liable on an instrument unless the person or an authorized representative signed it." * * * [H]owever, [Article] 3 is not the *only* applicable law; [Article] 4 deals with "Bank Deposits and Collections" and also is applicable in this case.

Section 4–214 allows a collecting bank to charge back and obtain a refund from a customer's account if the bank does not receive final settlement for an item. Neither that section, nor any other section of [Article] 4 indicates that the bank's rights to charge back and reimbursement are dependent upon whether the customer indorsed a deposited item.

* * * *

* * * Thus, Lema's contention that the Bank was "prohibited by applicable law" from debiting his account because he did not sign the altered check is without merit.

* * * Lema [also] contends that the provisions of the Deposit Agreement are ambiguous as to whether they were

(Continued)

intended to modify the UCC * * * . [W]e conclude that they are not ambiguous, as asserted by Lema. The Deposit Agreement clearly states that unless prohibited by applicable law, the Bank has the right "to charge back to [Lema's] account the amount of any item deposited to [his] account or cashed for [him] which was initially paid by the payor bank and which is later returned to us due to [a] * * * claim of alteration * * * ."

The Agreement also does not terminate the Bank's rights of charge back and reimbursement, as does the UCC, once settlement becomes final and so alters the effect of Section 4–214, as permitted in the UCC. * * *

Moreover, *the Deposit Agreement is in accord with the UCC's policy of shielding transferees of items from the risk of loss created by, for instance, an unauthorized alteration.* Section 4–207(a)(3) of the UCC, for example, provides that, when a bank customer transfers an item and receives consideration, the customer warrants to the transferee that, among other things, "[t]he item has not been altered." Although Lema did not himself transfer the altered check to the Bank, the Deposit Agreement follows the general policy of this Section by providing protection for a transferee, in this case the Bank, that receives an altered item. [Emphasis added.]

* * * *

JUDGMENT * * * AFFIRMED.

Dissenting Opinion

Dissenting opinion by *HARRELL*, J. [Judge] * * * .

* * * *

* * * I disagree * * * with the conclusion of the Majority that the Deposit Agreement functions to change the legal consequences which otherwise would flow from the provisions of the [UCC]. On the contrary, the plain meaning of the unambiguous language of the [Deposit Agreement] clause in question, clause 4(e), indicates that, far from altering the [UCC] by agreement, the terms used incorporate the [UCC] in its entirety into the Agreement.

* * * *

* * * [W]e must interpret the first clause of 4(e) as reading "Unless prohibited by applicable law [which, by definition, includes the provisions of Articles 3 and 4] * * * ." Because [Article] 4 is the applicable law, and the right of charge-back granted to the Bank contained in clause 4(e) is conditioned upon such charge-back not being prohibited by the applicable law, then if a section of [Article] 4 prohibits such a charge-back, that section of [Article] 4 controls the right of the Bank to charge-back the customer's account. Section 4–214 is such a section, and states in relevant part that "rights to revoke, charge-back, and obtain a refund terminate if and when a settlement for the item received by the bank is or becomes final" * * * , as opposed to provisional.

The question comes down to whether the credit to Lema's account was provisional or final. * * * Bank of America produced no evidence as to whether its release of funds was provisional or final. * * *

* * * Bank of America is the party asserting its rights under the Deposit Agreement and, thus, Bank of America has the burden of establishing its right to do so, which includes producing evidence establishing that the release of funds to Lema's account was provisional, and that final settlement had not occurred prior to the date of the charge-back. * * * Bank of America failed to meet its burden here.

* * * [T]he judgment of the [state intermediate appellate court], in my opinion, should be reversed, and the judgment of the trial court affirmed.

Questions for Analysis

1. **Law.** How did the majority in this case answer the question stated at the beginning of this feature? What was the reasoning behind the response?
2. **Law.** Did the dissent appear to disagree with the majority in the application of the principle of freedom of contract? With what part of the majority's holding did the dissent disagree? Why?
3. **Ethics.** Is it fair for the court to hold Lema liable for the full amount of the altered check even though he did not sign or authorize the check? Why or why not?
4. **Economic Dimensions.** What is the significance of the holding in this case for a small business?
5. **Implications for the Business Owner.** Is the power to alter the provisions of the UCC unlimited? If not, what are the limits?

Debtor-Creditor Relationships

UNI 6

Unit Contents

CHAPTER 20
Secured Transactions

LEARNING OBJECTIVES

After reading this chapter, you should be able to answer the following questions:

1. What is a security interest? Who is a secured party? What is a security agreement? What is a financing statement?
2. What three requirements must be met to create an enforceable security interest?
3. What is the most common method of perfecting a security interest under Article 9?
4. If two secured parties have perfected security interests in the collateral of the debtor, which party has priority to the collateral on the debtor's default?
5. What rights does a secured creditor have on the debtor's default?

Whenever the payment of a debt is guaranteed, or *secured,* by personal property owned by the debtor or in which the debtor has a legal interest, the transaction becomes known as a **secured transaction.** The concept of the secured transaction is as basic to modern business practice as the concept of credit. Logically, sellers and lenders do not want to risk nonpayment, so they usually will not sell goods or lend funds unless the payment is somehow guaranteed. Indeed, business as we know it could not exist without laws permitting and governing secured transactions.

Article 9 of the Uniform Commercial Code (UCC) governs secured transactions as applied to personal property, fixtures by contract, accounts, instruments, commercial assignments of $1,000 or more, *chattel paper* (any writing evidencing a debt secured by personal property), agricultural liens, and what are called general intangibles (such as patents and copyrights). (Debtor-creditor transactions that are not covered under Article 9 will be discussed in Chapter 21). In 1999, the National Conference of Commissioners on Uniform State Laws promulgated a revised version of Article 9. Because the revised version, which was later amended, has been adopted by all of the states, we base this chapter's discussion of secured transactions entirely on the provisions of the revised version.

The Terminology of Secured Transactions

The UCC's terminology is now uniformly adopted in all documents used in situations involving secured transactions. A brief summary of the UCC's definitions of terms relating to secured transactions follows.

1. A **secured party** is any creditor who has a *security interest* in the *debtor's collateral.* This creditor can be a seller, a lender, a cosigner, and even a buyer of accounts or chattel paper [UCC 9–102(a)(72)].
2. A **debtor** is the "person" who *owes payment* or other performance of a secured obligation [UCC 9–102(a)(28)].
3. A **security interest** is the *interest* in the collateral (personal property, fixtures, and so forth) that *secures*

payment or performance of an obligation [UCC 1–201(37)].

4. A **security agreement** is an *agreement* that *creates* or provides for a *security interest* [UCC 9–102(a)(73)].
5. **Collateral** is the *subject* of the *security interest* [UCC 9–102(a)(12)].
6. A **financing statement**—referred to as the UCC-1 form—is the *instrument normally filed* to give *public notice* to *third parties* of the *secured party's security interest* [UCC 9–102(a)(39)].

These basic definitions form the concept under which a debtor-creditor relationship becomes a secured transaction relationship (see Exhibit 20–1).

Creating and Perfecting a Security Interest

A creditor has two main concerns if the debtor **defaults** (fails to pay the debt as promised): (1) satisfaction of the debt through the possession and (usually) sale of the collateral and (2) priority over any other creditors or buyers who may have rights in the same collateral. We look here at how these two concerns are met through the creation and perfection of a security interest.

Creating a Security Interest

To become a secured party, the creditor must obtain a security interest in the collateral of the debtor. Three requirements must be met for a creditor to have an enforceable security interest:

1. Either (a) the collateral must be in the possession of the secured party in accordance with an agreement, or (b) there must be a written or authenticated security agreement that describes the collateral subject to the security interest and is signed or authenticated by the debtor.
2. The secured party must give something of value to the debtor.
3. The debtor must have "rights" in the collateral.

Once these requirements have been met, the creditor's rights are said to *attach* to the collateral. **Attachment** gives the creditor an enforceable security interest in the collateral [UCC 9–203].[1]

Written or Authenticated Security Agreement When the collateral is *not* in the possession of the secured party, the security agreement must be either written or authenticated, and it must describe the collateral. Note here that *authentication* includes any agreement or signature inscribed on a tangible medium or stored in an electronic or other medium (called a *record*) that is retrievable [UCC 9–102(a)(7)(69)]. If the security agreement is in writing or authenticated, only the debtor's signature or authentication is required to create the security interest. The reason authentication is acceptable is to provide for electronic filing (the filing process will be discussed later).

A security agreement must contain a description of the collateral that reasonably identifies it. Generally, such words as "all the debtor's personal property" or "all the debtor's assets" would *not* constitute a sufficient description [UCC 9–108(c)]. See Exhibit 20–2 on the next page for a sample security agreement.

Secured Party Must Give Value The secured party must give to the debtor something of value. For example, a person gives value if he or she acquires a security interest

1. Note that in the context of judicial liens, discussed in Chapter 21, the term *attachment* has a different meaning. In that context, it refers to a court-ordered seizure and taking into custody of property prior to the securing of a court judgment for a past-due debt.

EXHIBIT 20–1 SECURED TRANSACTIONS—CONCEPT AND TERMINOLOGY

In a security agreement, a debtor and a creditor agree that the creditor will have a security interest in collateral in which the debtor has rights. In essence, the collateral secures the loan and ensures the creditor of payment should the debtor default.

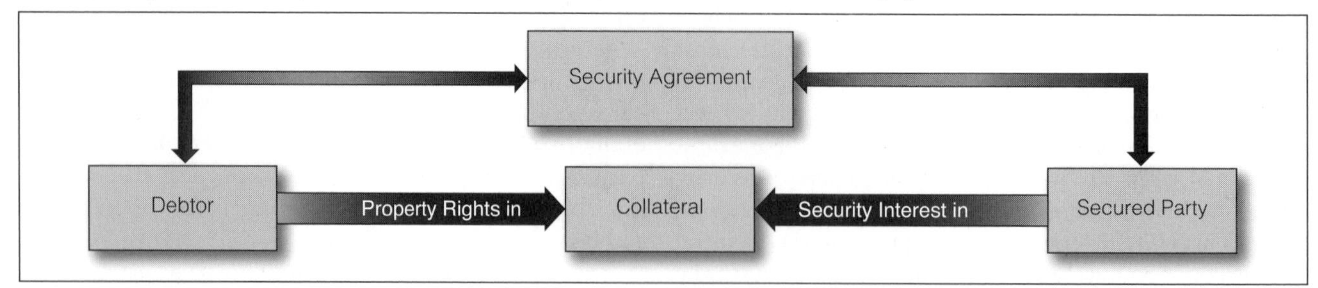

EXHIBIT 20–2 AN EXAMPLE OF A SIMPLE SECURITY AGREEMENT

____
Date

____
Name   No. and Street   City   County   State

(hereinafter called "Debtor") hereby grants to ______________________________
Name

____
No. and Street   City   County   State

(hereinafter called "Secured Party") a security interest in the following property (hereinafter called the "Collateral"): ______________________________

to secure payment and performance of obligations identified or set out as follows (hereinafter called the "Obligations"): ______________________________

Default in payment or performance of any of the Obligations or default under any agreement evidencing any of the Obligations is a default under this agreement. Upon such default, Secured Party may declare all Obligations immediately due and payable and shall have the remedies of a secured party under section(s) __________ of the Uniform Commercial Code.

Signed in (duplicate) triplicate.

______________________	______________________
Debtor	Secured Party
By______________________	By______________________

in return for a binding commitment to extend credit, as security for the satisfaction of a preexisting debt, or in return for consideration to support a simple contract [UCC 1–201(44)]. Normally, the value given by a secured party is in the form of a direct loan or a commitment to sell goods on credit.

Debtor Must Have Rights in the Collateral The debtor must have rights in the collateral—that is, the debtor must have some ownership interest or right to obtain possession of that collateral. The debtor's rights can represent either a current or a future legal interest in the collateral. For example, a retail seller-debtor can give a secured party a security interest not only in existing inventory owned by the retailer but also in *future* inventory to be acquired by the retailer.

One common misconception about having rights in the collateral is that the debtor must have title. This is not a requirement. A beneficial interest in a *trust* (trusts will be discussed in Chapter 30), when title to the trust property is held by the trustee, may be made the subject of a security interest for a loan made to the beneficiary by a secured party (a creditor).

Classifications and Definitions of Collateral

Where or how to perfect a security interest (*perfection* will be discussed shortly) sometimes depends on the classification or definition of the collateral. Collateral is generally divided into two classifications: *tangible collateral* (collateral that can be seen, felt, or touched) and *intangible collateral* (collateral that consists of or generates rights).

Exhibit 20–3 on the following two pages summarizes the various classifications of collateral and the methods of

EXHIBIT 20–3 TYPES OF COLLATERAL AND METHODS OF PERFECTION

TYPE OF COLLATERAL	DEFINITION	METHOD OF PERFECTION
TANGIBLE	All things that are movable at the time the security interest attaches (such as livestock) or that are attached to the land, including timber to be cut and growing crops.	
1. Consumer Goods [UCC 9–301, 9–303, 9–309(1), 9–310(a), 9–313(a)]	Goods used or bought primarily for personal, family, or household purposes—for example, household furniture [UCC 9–102(a)(23)].	For purchase-money security interest, attachment (that is, the creation of a security interest) is sufficient; for boats, motor vehicles, and trailers, filing or compliance with a certificate-of-title statute is required; for other consumer goods, general rules of filing or possession apply.
2. Equipment [UCC 9–301, 9–310(a), 9–313(a)]	Goods bought for or used primarily in business (and not part of inventory or farm products)—for example, a delivery truck [UCC 9–102(a)(33)].	Filing or (rarely) possession by secured party.
3. Farm Products [UCC 9–301, 9–310(a), 9–313(a)]	Crops (including aquatic goods), livestock, or supplies produced in a farming operation—for example, ginned cotton, milk, eggs, and maple syrup [UCC 9–102(a)(34)].	Filing or (rarely) possession by secured party.
4. Inventory [UCC 9–301, 9–310(a), 9–313(a)]	Goods held by a person for sale or under a contract of service or lease; raw materials held for production and work in progress [UCC 9–102(a)(48)].	Filing or (rarely) possession by secured party.
5. Accessions [UCC 9–301, 9–310(a), 9–313(a)]	Personal property that is so attached, installed, or fixed to other personal property (goods) that it becomes a part of these goods—for example, a DVD player installed in an automobile [UCC 9–102(a)(1)].	Filing or (rarely) possession by secured party (same as personal property being attached).
INTANGIBLE	Nonphysical property that exists only in connection with something else.	
1. Chattel Paper [UCC 9–301, 9–310(a), 9–312(a), 9–313(a), 9–314(a)]	A writing or writings (records) that evidence both a monetary obligation and a security interest in goods and software used in goods; for example, a security agreement or a security agreement and promissory note. *Note:* If the record or records consist of information stored in an electronic medium, the collateral is called *electronic chattel paper.* If the information is inscribed on a tangible medium, it is called *tangible chattel paper* [UCC 9–102(a)(11), (a)(31), and (a)(78)].	Filing or possession or control by secured party.
2. Instruments [UCC 9–301, 9–309(4), 9–310(a), 9–312(a) and (e), 9–313(a)]	A negotiable instrument, such as a check, note, certificate of deposit, or draft, or other writing that evidences a right to the payment of money and is not a security agreement or lease but rather a type that can ordinarily be transferred (after indorsement, if necessary) by delivery [UCC 9–102(a)(47)].	Except for temporary perfected status, filing or possession. For the sale of promissory notes, perfection can be by attachment (automatically on the creation of the security interest).

(Continued)

EXHIBIT 20-3 TYPES OF COLLATERAL AND METHODS OF PERFECTION—CONTINUED

TYPE OF COLLATERAL	DEFINITION	METHOD OF PERFECTION
3. Accounts [UCC 9–301, 9–309(2) and (5), 9–310(a)]	Any right to receive payment for the following: (a) any property, real or personal, sold, leased, licensed, assigned, or otherwise disposed of, including intellectual licensed property; (b) services rendered or to be rendered, such as contract rights; (c) insurance policies; (d) secondary obligations incurred; (e) use of a credit card; (f) winnings of a government-sponsored or government-authorized lottery or other game of chance; and (g) health-care insurance receivables, defined as an interest or claim, under an insurance policy, to payment for health-care goods or services provided [UCC 9–102(a)(2) and (a)(46)].	Filing required except for certain assignments that can be perfected by attachment (automatically on the creation of the security interest).
4. Deposit Accounts [UCC 9–104, 9–304, 9–312(b), 9–314(a)]	Any demand, time, savings, passbook, or similar account maintained with a bank [UCC 9–102(a)(29)].	Perfection by control, such as when the secured party is the bank in which the account is maintained or when the parties have agreed that the secured party can direct the disposition of funds in a particular account.
5. General Intangibles [UCC 9–301, 9–309(3), 9–310(a) and (b)(8)]	Any personal property (or debtor's obligation to pay money on such) other than that defined above [UCC 9–102(a)(42)], including software that is independent from a computer or a good [UCC 9–102(a)(44), (a)(61), and (a)(75)].	Filing only (for copyrights, with the U.S. Copyright Office), except a sale of a payment intangible by attachment (automatically on the creation of the security interest).

perfecting a security interest in collateral falling within each of those classifications.[2]

PERFECTING A SECURITY INTEREST

Perfection is the legal process by which secured parties protect themselves against the claims of third parties who may wish to have their debts satisfied out of the same collateral. Usually, perfection is accomplished by filing a financing statement with the office of the appropriate government official. In some circumstances, however, a security interest becomes perfected without the filing of a financing statement.

Perfection by Filing The most common means of perfection is by filing a *financing statement*—a document that gives public notice to third parties of the secured party's security interest—with the office of the appropriate government official. The security agreement itself can also be filed to perfect the security interest. The financing statement must provide the names of the debtor and the secured party and indicate the collateral covered by the financing statement. A uniform financing statement form, as shown in Exhibit 20–4, is now used in all states [see UCC 9–521].

Communication of the financing statement to the appropriate filing office, together with the correct filing fee, or the acceptance of the financing statement by the filing officer constitutes a filing [UCC 9–516(a)]. The word *communication* means that the filing can be accomplished electronically [UCC 9–102(a)(18)]. Once completed, filings are indexed in the name of the debtor so that they can be located by subsequent searchers. A financing statement may be filed even before a security agreement is made or a security interest attaches [UCC 9–502(d)].

—The Debtor's Name. The UCC requires that a financing statement be filed under the name of the debtor [UCC 9–502(a)(1)]. Because most states use electronic filing systems, UCC 9–503 sets out some detailed rules for determining when the debtor's name as it appears on a

2. There are additional classifications, such as agricultural liens, investment property, and commercial tort claims. For definitions of these types of collateral, see UCC 9–102(a)(5), (a)(13), and (a)(49).

EXHIBIT 20–4 THE UNIFORM FINANCING STATEMENT

UCC FINANCING STATEMENT

FOLLOW INSTRUCTIONS (front and back) CAREFULLY

A. NAME & PHONE OF CONTACT AT FILER (optional)

B. SEND ACKNOWLEDGEMENT TO: (Name and Address)

THE ABOVE SPACE IS FOR FILING OFFICE USE ONLY

1. DEBTOR'S EXACT FULL LEGAL NAME - Insert only one debtor name (1a or 1b) - do not abbreviate or combine names

1a. ORGANIZATION'S NAME			
OR 1b. INDIVIDUAL'S LAST NAME	FIRST NAME	MIDDLE NAME	SUFFIX
1c. MAILING ADDRESS	CITY	STATE / POSTAL CODE	COUNTRY
1d. TAX ID# SSN OR EIN / ADDL INFO RE ORGANIZATION DEBTOR / 1e. TYPE OF ORGANIZATION	1f. JURISDICTION OR ORGANIZATION	1g. ORGANIZATIONAL ID #, if any	☐ NONE

2. ADDITIONAL DEBTOR'S EXACT FULL LEGAL NAME - Insert only one debtor name (2a or 2b) - do not abbreviate or combine names

2a. ORGANIZATION'S NAME			
OR 2b. INDIVIDUAL'S LAST NAME	FIRST NAME	MIDDLE NAME	SUFFIX
2c. MAILING ADDRESS	CITY	STATE / POSTAL CODE	COUNTRY
2d. TAX ID# SSN OR EIN / ADDL INFO RE ORGANIZATION DEBTOR / 2e. TYPE OF ORGANIZATION	2f. JURISDICTION OR ORGANIZATION	2g. ORGANIZATIONAL ID #, if any	☐ NONE

3. SECURED PARTY'S NAME - (or NAME of TOTAL ASSIGNOR S/P) Insert only one secured party name (3a or 3b)

3a. ORGANIZATION'S NAME			
OR 3b. INDIVIDUAL'S LAST NAME	FIRST NAME	MIDDLE NAME	SUFFIX
3c. MAILING ADDRESS	CITY	STATE / POSTAL CODE	COUNTRY

4. This FINANCING STATEMENT covers the following collateral:

5. ALTERNATIVE DESIGNATION (if applicable) ☐ LESSEE/LESSOR ☐ CONSIGNEE/CONSIGNOR ☐ BAILEE/BAILOR ☐ SELLER/BUYER ☐ AG. LIEN ☐ NON-UCC FILING

6. ☐ This FINANCING STATEMENT is to be filed [for record] (or recorded) in the REAL ESTATE RECORDS. Attach Addendum (if applicable) 7. Check to REQUEST SEARCH REPORT(S) on Debtor(s) (ADDITIONAL FEE) (optional) ☐ All Debtors ☐ Debtor 1 ☐ Debtor 2

OPTIONAL FILER REFERENCE DATA

NATIONAL UCC FINANCING STATEMENT (FORM UCC1) REV. 07/29/98)

financing statement is sufficient. For corporations, which are part of "registered organizations," the debtor's name on the financing statement must be "the name of the debtor indicated on the public record of the debtor's jurisdiction of organization" [UCC 9–503(a)(1)]. Slight variations in names normally will not be considered misleading if a search of the filing office's records, using a standard computer search engine routinely used by that office, would disclose the filings [UCC 9–506(c)]. Note that if the debtor is identified by the correct name at the time of the filing of a financing statement, the secured party's interest retains its priority even if the debtor later changes his or her name.

If the debtor is a trust or a trustee with respect to property held in trust, the filed financing statement must disclose this information and must provide the trust's name as specified in its official documents [UCC 9–503(a)(3)]. For other debtors, the filed financing statement must disclose "the individual or organizational name of the debtor" [UCC 9–503(a)(4)(A)]. As used here, the word

organization includes unincorporated associations, such as clubs and some churches, as well as joint ventures and general partnerships, even when these organizations are created without obtaining any formal certificate of formation.

In general, providing only the debtor's trade name (or a fictitious name) in a financing statement is not sufficient for perfection [UCC 9–503(c)]. ■ EXAMPLE 20.1 Assume that a loan is being made to a sole proprietorship owned by Peter Jones. The trade, or fictitious, name is Pete's Plumbing. A financing statement cannot use the trade name Pete's Plumbing; rather, it must be filed under the name of the actual debtor—in this instance, Peter Jones. ■ The reason for this rule is that a sole proprietorship is not a legal entity distinct from the person who owns it. The rule also furthers an important goal of Article 9: to ensure that the debtor's name on a financing statement is one that prospective lenders can locate and recognize in future searches.

—Changes in the Debtor's Name. A problem arises when the debtor subject to a filed perfected security interest changes her or his (or its) name. What happens if a subsequent creditor extends credit to the debtor and perfects its security interest under the debtor's new name? Obviously, a search by this subsequent creditor for filed security interests under the debtor's changed name may not disclose the previously filed security interest.

The UCC's revised Article 9 attempts to prevent potential conflicts caused by changes in the debtor's name if the debtor goes into default. First, as just described, UCC 9–503 sets forth specific rules on whether a financing statement sufficiently provides the debtor's name, depending on whether the debtor is an individual, a corporation, a trust, and so on. Second, if the debtor's name is insufficient, the filing is seriously misleading *unless* a search of records by the filing officer's search engine using the debtor's correct name would disclose the security interest [UCC 9–506(b) and (c)]. Third, even if the change of name renders the financing statement misleading, the financing statement is effective as a perfection of a security interest in collateral acquired by the debtor before *or* within four months after the name change. Unless an amendment is filed within this four-month period, collateral acquired by the debtor after the four-month period is unperfected [UCC 9–507(b) and (c)].

The following case illustrates some of the complications that can arise when a debtor has used more than one name to do business and two creditors assert claims to the same collateral under those different names.

CASE 20.1 Cabool State Bank v. Radio Shack, Inc.

Missouri Court of Appeals,
Southern District, Division One, 2002.
65 S.W.3d 613.

FACTS In June 1995, Michael and Debra Boudreaux, doing business as D & J Enterprises, Inc., bought a retail electronics store operated under a franchise from Radio Shack, Inc., from Van Pamperien. To pay for the business, the Boudreauxes borrowed money from Cabool State Bank in Springfield, Missouri. The loan documents included a financing statement. On the statement's signature lines, the only capacity identified for the Boudreauxes' signatures was that of "Debtors." Elsewhere on the form, the bank listed "D & J Enterprises, Inc., Radio Shack, Dealer, Debra K. Boudreaux, Michael C. Boudreaux" as "Debtors." The statement covered, in part, the store inventory. Before the end of the year, the Boudreauxes changed the name of their business to Tri-B Enterprises, Inc. In January 1998, the store closed. The next month, Radio Shack terminated the franchise and, despite the lack of a security interest, took possession of the inventory, claiming the Boudreauxes and Tri-B owed Radio Shack $6,394.73. The bank filed a suit in a Missouri state court against Radio Shack, claiming a perfected security interest in the inventory with priority over Radio Shack's claim. The court entered a judgment for $15,529.43 in the bank's favor. Radio Shack appealed to a state intermediate appellate court.

ISSUE Did the bank's security interest take priority over Radio Shack's claim?

DECISION Yes. The state intermediate appellate court affirmed the lower court's judgment in the bank's favor.

REASON Contrary to Radio Shack's assertion that "the so-called change of name was seriously misleading," Radio Shack was not misled by the debtors' change of their business name, because the bank's financing statement was filed under the "true name" of at least one of the debtors with whom Radio Shack admitted doing business. Radio Shack also had actual knowledge of the bank loan. The appellate court noted that "(1) Boudreauxes bought the original inventory in their individual names, (2) Boudreauxes gave Bank a security interest in the original inventory and 'all inventory purchased or

CASE 20.1–CONTINUED

replaced,' (3) Bank perfected its security interest in existing and future inventory owned by Boudreauxes by filing * * * financing statements that listed Boudreauxes and D & J Enterprises, Inc., as 'Debtors,' (4) Radio Shack had actual knowledge of the loan transaction between Bank and Boudreauxes, and (5) the subject inventory was sold to both the Boudreauxes and Tri-B." The court emphasized that "[f]rom the outset, Bank listed Boudreauxes (who admittedly had an ownership interest in the inventory) as 'Debtors' on the [UCC] filings."

WHY IS THIS CASE IMPORTANT? *In today's business climate, debtors frequently change the names under which they do business. The key to determining if a security interest has been perfected in such situations is whether the financing statement that was filed adequately notifies other potential creditors that a security interest exists. As this case illustrates, if a search of the records using the debtor's correct name would disclose the interest, that is generally sufficient.*

—Description of the Collateral. The UCC requires that both the security agreement and the financing statement contain a description of the collateral in which the secured party has a security interest. The security agreement must describe the collateral because no security interest in goods can exist unless the parties agree on which goods are subject to the security interest. The financing statement must also describe the collateral because the purpose of filing the statement is to give public notice of the fact that certain goods of the debtor are subject to a security interest. Other parties who might later wish to lend funds to the debtor or buy the collateral can thus learn of the security interest by checking with the state or local office in which a financing statement for that type of collateral would be filed. For land-related security interests, a legal description of the realty is also required [UCC 9–502(b)].

Sometimes, the descriptions in the two documents vary, with the description in the security agreement being more precise than the description in the financing statement, which is allowed to be more general. ■ EXAMPLE 20.2 A security agreement for a commercial loan to a manufacturer may list all of the manufacturer's equipment subject to the loan by serial number, whereas the financing statement may simply state "all equipment owned or hereafter acquired." ■ The UCC permits broad, general descriptions in the financing statement, such as "all assets" or "all personal property." Generally, therefore, whenever the description in a financing statement accurately describes the agreement between the secured party and the debtor, the description is sufficient [UCC 9–504].

—Where to File. In most states, a financing statement must be filed centrally in the appropriate state office, such as the office of the secretary of state, in the state where the debtor is located. Filing in the county where the collateral is located is required only when the collateral consists of timber to be cut, fixtures, or collateral to be extracted—such as oil, coal, gas, and minerals [UCC 9–301(3) and (4), 9–502(b)].

The state office in which a financing statement should be filed depends on the *debtor's location,* not the location of the collateral (as was required under the unrevised Article 9) [UCC 9–301]. The debtor's location is determined as follows [UCC 9–307]:

1. For *individual debtors,* it is the state of the debtor's principal residence.
2. For an entity created by a filing (such as a corporation), it is in the state of filing. For example, if a debtor is incorporated in Maryland and has its chief executive office in New York, a secured party would file the financing statement in Maryland—the state of the debtor's organizational formation.
3. For all other entities, it is the state in which the business is located or, if the debtor has more than one office, the place from which the debtor manages its business operations and affairs.

—Consequences of an Improper Filing. Any improper filing renders the secured party unperfected and reduces the secured party's claim in bankruptcy to that of an unsecured creditor. For instance, if the debtor's name on the financing statement is inaccurate or if the collateral is not sufficiently described in the financing statement, the filing may not be effective.

Perfection without Filing In two types of situations, security interests can be perfected without filing a financing statement. The first occurs when the collateral is transferred into the possession of the secured party. The second occurs when the security interest is one of a limited number (thirteen) under the UCC that can be perfected on attachment (without a filing and without having to possess the

goods) [UCC 9–309]. The phrase *perfected on attachment* means that these security interests are automatically perfected at the time of their creation. Two of the most common security interests that are perfected on attachment are a *purchase-money security interest* in consumer goods (defined and explained below) and an assignment of a beneficial interest in a decedent's estate [UCC 9–309(1) and (13)].

—Perfection by Possession. Under the common law, one of the most usual means of obtaining financing was to **pledge** certain collateral as security for the debt and transfer the collateral into the creditor's possession. When the debt was paid, the collateral was returned to the debtor. Although the debtor usually entered into a written security agreement, an oral security agreement was also enforceable as long as the secured party possessed the collateral. Article 9 of the UCC retained the common law pledge and the principle that the security agreement need not be in writing to be enforceable if the collateral is transferred to the secured party [UCC 9–310, 9–312(b), 9–313].

For most collateral, possession by the secured party is impractical because it denies the debtor the right to use or derive income from the property to pay off the debt. ■ EXAMPLE 20.3 Suppose that a farmer takes out a loan to finance the purchase of a piece of heavy farm equipment needed to harvest crops and uses the equipment as collateral. Clearly, the purpose of the purchase would be defeated if the farmer transferred the collateral into the creditor's possession. ■ Certain items, however, such as stocks, bonds, negotiable instruments, and jewelry, are commonly transferred into the creditor's possession when they are used as collateral for loans.

—Perfection by Attachment: The Purchase-Money Security Interest in Consumer Goods. Under the UCC, thirteen types of security interests are perfected automatically at the time they are created [UCC 9–309]. The most common of these is the **purchase-money security interest (PMSI)** in consumer goods. A PMSI is created when a person buys consumer goods (items bought primarily for personal, family, or household purposes) and the seller or lender agrees to extend credit for part or all of the purchase price of the goods. The entity that extends the credit and obtains the PMSI can be either the seller (a store, for example) or a financial institution that lends the buyer the funds with which to purchase the goods [UCC 9–102(a)(2)].

A PMSI in consumer goods is perfected automatically at the time of a credit sale—that is, at the time the PMSI is created. The seller in this situation need do nothing more to perfect her or his interest. ■ EXAMPLE 20.4 Suppose that Jamie wants to purchase a new large-screen television from ABC Television, Inc. The purchase price is $2,500. Not being able to pay the entire amount in cash, Jamie signs a purchase agreement to pay $1,000 down and $100 per month until the balance plus interest is fully paid. ABC is to retain a security interest in the purchased goods until full payment has been made. Because the security interest was created as part of the purchase agreement, it is a PMSI. ■

An important exception to this rule of automatic perfection under Article 9 deals with certain types of security interests that are subject to other federal or state laws, which may require additional steps to be perfected [UCC 9–311]. For example, if a consumer purchases an automobile, trailer, or boat, the secured party will need to file a certificate of title with the appropriate state authority to perfect the PMSI.

Effective Time Duration of Perfection A financing statement is effective for five years from the date of filing [UCC 9–515)]. If a **continuation statement** is filed within six months *prior to* the expiration date, the original statement remains in effect for another five years, starting with the expiration date of the first five-year period [UCC 9–515(d) and (e)]. The effectiveness of the statement can be continued in the same manner indefinitely. Any attempt to file a continuation statement outside the six-month window will render the continuation ineffective, and the perfection will lapse at the end of the five-year period.

If a financing statement lapses, the security interest that had been perfected by the filing now becomes unperfected. A purchaser for value can acquire the collateral as if the security interest had never been perfected [UCC 9–515(c)].

The Scope of a Security Interest

In addition to covering collateral already in the debtor's possession, a security agreement can cover various other types of property, including the proceeds of the sale of collateral, after-acquired property, and future advances.

PROCEEDS

Proceeds include whatever is received when collateral is sold or disposed of in some other way [UCC 9–102(a)(64)]. A secured party's security interest in the collateral includes a security interest in the proceeds of the sale of that collateral. ■ EXAMPLE 20.5 Suppose that a bank has a perfected security interest in the inventory of a retail seller of heavy farm machinery. The retailer sells a tractor out of

this inventory to a farmer, who is by definition a *buyer in the ordinary course of business* (this term will be discussed later in the chapter). The farmer agrees, in a security agreement, to make monthly payments to the retailer for a period of twenty-four months. If the retailer should go into default on the loan from the bank, the bank is entitled to the remaining payments the farmer owes to the retailer as proceeds. ■

A security interest in proceeds perfects automatically on the *perfection* of the secured party's security interest in the original collateral and remains perfected for twenty days after receipt of the proceeds by the debtor. One way to extend the twenty-day automatic perfection period is to provide for such extended coverage in the original security agreement [UCC 9–315(c) and (d)]. This is typically done when the collateral is the type that is likely to be sold, such as a retailer's inventory—for example, of computers or DVD players. The UCC also permits a security interest in identifiable cash proceeds to remain perfected after twenty days [UCC 9–315(d)(2)].

After-Acquired Property

After-acquired property is property that the debtor acquired after the execution of the security agreement. The security agreement may provide for a security interest in after-acquired property [UCC 9–204(1)]. This is particularly useful for inventory financing arrangements because a secured party whose security interest is in existing inventory knows that the debtor will sell that inventory, thereby reducing the collateral subject to the security interest.

Generally, the debtor will purchase new inventory to replace the inventory sold. The secured party wants this newly acquired inventory to be subject to the original security interest. Thus, the after-acquired property clause continues the secured party's claim to any inventory acquired thereafter. This is not to say that the original security interest will take priority over the rights of all other creditors with regard to this after-acquired inventory, as will be discussed in the next section.

■ EXAMPLE 20.6 Amato buys factory equipment from Bronson on credit, giving as security an interest in all of her equipment—both what she is buying and what she already owns. The security interest with Bronson contains an after-acquired property clause. Six months later, Amato pays cash to another seller of factory equipment for more equipment. Six months after that, Amato goes out of business before she has paid off her debt to Bronson. Bronson has a security interest in all of Amato's equipment, even the equipment bought from the other seller. ■

Future Advances

Often, a debtor will arrange with a bank to have a *continuing line of credit* under which the debtor can borrow funds intermittently. Advances against lines of credit can be subject to a properly perfected security interest in certain collateral. The security agreement may provide that any future advances made against that line of credit are also subject to the security interest in the same collateral [UCC 9–204(c)]. Future advances do not have to be of the same type or otherwise related to the original advance to benefit from this type of cross-collateralization.[3] (*Cross-collateralization* occurs when an asset that is not the subject of a loan is used to secure that loan.)

■ EXAMPLE 20.7 Stroh is the owner of a small manufacturing plant with equipment valued at $1 million. He has an immediate need for $50,000 of working capital, so he obtains a loan from Midwestern Bank and signs a security agreement, putting up all of his equipment as security. The bank properly perfects its security interest. The security agreement provides that Stroh can borrow up to $500,000 in the future, using the same equipment as collateral for any future advances. In this situation, Midwestern Bank does not have to execute a new security agreement and perfect a security interest in the collateral each time an advance is made, up to a cumulative total of $500,000. For priority purposes, each advance is perfected as of the date of the original perfection. ■

The Floating-Lien Concept

A security agreement that provides for a security interest in proceeds, in after-acquired property, or in collateral subject to future advances by the secured party (or in all three) is often characterized as a **floating lien.** This type of security interest continues in the collateral or proceeds even if the collateral is sold, exchanged, or disposed of in some other way.

A Floating Lien in Inventory Floating liens commonly arise in the financing of inventories. A creditor is not interested in specific pieces of inventory, which are constantly changing, so the lien "floats" from one item to another, as the inventory changes.

■ EXAMPLE 20.8 Suppose that Cascade Sports, Inc., a corporation chartered in Oregon that operates as a cross-country ski dealer, has a line of credit with Portland First Bank to finance an inventory of cross-country skis. Cascade and Portland First enter into a security agreement that provides for coverage of proceeds, after-acquired

3. See official Comment 5 to UCC 9–204.

inventory, present inventory, and future advances. This security interest in inventory is perfected by filing centrally (with the office of the secretary of state in Oregon). One day, Cascade sells a new pair of the latest cross-country skis and receives a used pair in trade. That same day, Cascade purchases two new pairs of cross-country skis from a local manufacturer for cash. Later that day, to meet its payroll Cascade borrows $8,000 from Portland First Bank under the security agreement.

Portland First gets a perfected security interest in the used pair of skis under the proceeds clause, has a perfected security interest in the two new pairs of skis purchased from the local manufacturer under the after-acquired property clause, and has the new amount of funds advanced to Cascade secured on all of the above collateral by the future-advances clause. All of this is accomplished under the original perfected security interest. The various items in the inventory have changed, but Portland First still has a perfected security interest in Cascade's inventory. Hence, it has a floating lien on the inventory. ■

A Floating Lien in a Shifting Stock of Goods The concept of the floating lien can also apply to a shifting stock of goods. The lien can start with raw materials; follow them as they become finished goods and inventories; and continue as the goods are sold and are turned into accounts receivable, chattel paper, or cash.

Priorities

The importance of perfection to a secured party cannot be overemphasized, particularly when another party is claiming an interest in the same collateral covered by the perfected secured party's security interest.

THE GENERAL RULE

The general rule is that a perfected secured party's interest has priority over the interests of the following parties [UCC 9–317, 9–322]:

1. An unsecured creditor.
2. An unperfected secured party.
3. A subsequent lien creditor, such as a judgment creditor who acquires a lien on the collateral by *execution and levy*—a process discussed later in this chapter.
4. A trustee in bankruptcy (see Chapter 21)—at least, the perfected secured party has priority to the proceeds from the sale of the collateral by the trustee.
5. Most buyers who *do not* purchase the collateral in the ordinary course of a seller's business.

In addition, whether a secured party's security interest is perfected or unperfected may have serious consequences for the secured party if the debtor defaults on the debt or files for bankruptcy. For example, what if the debtor has borrowed funds from two different creditors, using the same property as collateral for both loans? If the debtor defaults on both loans, which of the two creditors has first rights to the collateral? In this situation, the creditor with a perfected security interest will prevail.

BUYERS OF THE COLLATERAL

Sometimes, the conflict is between a perfected secured party and a buyer of the collateral. The question then arises as to which party has priority to the collateral.

Types of Buyers The UCC recognizes five types of buyers whose interest in purchased goods could conflict with those of a perfected secured party on the debtor's default. These five types are as follows (see Exhibit 20–5 for details):

1. Buyers in the ordinary course of business—this type of buyer will be discussed in detail shortly.
2. Buyers *not* in the ordinary course of business of consumer goods.
3. Buyers of chattel paper [UCC 9–330].
4. Buyers of instruments, documents, or securities [UCC 9–330(d), 9–331(a)].
5. Buyers of farm products.[4]

Buyers in the Ordinary Course of Business Because buyers should not be required to find out if there is an outstanding security interest in, for example, a merchant's inventory, the UCC also provides that a person who buys "in the ordinary course of business" will take the goods free from any security interest created by the seller in the purchased collateral. This is so even if the security interest is perfected and *even if the buyer knows of its existence* [UCC 9–320(a)].[5] The UCC defines a *buyer in the ordinary course of business* as any person who in good faith, and without knowledge that the sale is in violation of the ownership rights or security interest of a third party in

4. Under the Food Security Act of 1985, buyers in the ordinary course of business include buyers of farm products from a farmer. Under this act, these buyers are protected from prior perfected security interests unless the secured parties perfected centrally by a special form called an *effective financing statement* (EFS) or the buyers received proper notice of the secured party's security interest.

5. Remember that, generally, there are three methods of perfection: by filing, by possession, and by attachment.

EXHIBIT 20–5 PRIORITY OF CLAIMS TO A DEBTOR'S COLLATERAL

PARTIES	PRIORITY
Unperfected Secured Party	An unperfected secured party prevails over unsecured creditors and creditors who have obtained judgments against the debtor but who have not begun the legal process to collect on those judgments [UCC 9–201(a)].
Purchaser of Debtor's Collateral	1. *Goods purchased in the ordinary course of the seller's business*—Buyer prevails over a secured party's security interest, even if perfected and even if the buyer knows of the security interest [UCC 9–320(a)]. 2. *Consumer goods purchased outside the ordinary course of business*—Buyer prevails over a secured party's interest, even if perfected by attachment, providing the buyer purchased as follows: a. For value. b. Without actual knowledge of the security interest. c. For use as a consumer good. d. Prior to secured party's perfection by *filing* [UCC 9–320(b)]. 3. *Buyers of chattel paper*—Buyer prevails if the buyer: a. Gave new value in making the purchase. b. Took possession in the ordinary course of the buyer's business. c. Took without knowledge of the security interest [UCC 9–330]. 4. *Buyers of instruments, documents, or securities*—Buyers who are holders in due course, holders to whom negotiable documents have been duly negotiated, or bona fide purchasers of securities have priority over a previously perfected security interest [UCC 9–330(d), 9–331(a)]. 5. *Buyers of farm products*—Buyers from a farmer take free and clear of perfected security interests unless, where permitted, a secured party centrally files an effective financing statement (EFS) or the buyer receives proper notice of the security interest before the sale.
Perfected Secured Parties to the Same Collateral	1. *The general rule*—Between two perfected secured parties in the same collateral, the general rule is that first in time of perfection is first in right to the collateral [UCC 9–322(a)(1)]. 2. *Exception for purchase-money security interest (PMSI)*—A PMSI, even if second in time of perfection, has priority providing that the following conditions are met: a. Inventory—A PMSI has priority if it is perfected and proper written or authenticated notice is given to the other security-interest holder *on* or *before* the time the debtor takes possession [UCC 9–324(b)]. b. Other collateral—A PMSI has priority, providing it is perfected within twenty days after the debtor takes possession [UCC 9–324(a)]. c. Software—Applies to a PMSI in software only if used in goods subject to a PMSI. If the goods are inventory, priority is determined the same as for inventory; if they are not, priority is determined as for goods other than inventory [UCC 9–103(c), 9–324(f)].

the goods, buys in ordinary course from a person in the business of selling goods of that kind [UCC 1–201(9)].

■ EXAMPLE 20.9 On August 1, West Bank has a perfected security interest in all of ABC Television's existing inventory and any inventory thereafter acquired. On September 1, Carla, a student at Central University, purchases one of the television sets in ABC's inventory. On December 1, ABC goes into default. Can West Bank repossess the TV set sold to Carla? The answer is no, because Carla is a buyer in the ordinary course of business (ABC is in the business of selling goods of that kind) and takes free and clear of West Bank's perfected security interest. ■

Creditors or Secured Parties

Generally, the following UCC rules apply when more than one creditor claims rights in the same collateral:

1. *Conflicting perfected security interests.* When two or more secured parties have perfected security interests in the same collateral, generally the first to perfect (file or take possession of the collateral) has priority, unless the state's statute provides otherwise [UCC 9–322(a)(1)].
2. *Conflicting unperfected security interests.* When two conflicting security interests are unperfected, the first to attach has priority [UCC 9–322(a)(3)].

3 *Conflicting perfected security interests in commingled or processed goods.* When goods to which two or more perfected security interests attach are so manufactured or commingled that they lose their identities into a product or mass, the perfected parties' security interests attach to the new product or mass "according to the ratio that the cost of goods to which each interest originally attached bears to the cost of the total product or mass" [UCC 9–336].

Under some circumstances, on the debtor's default, the perfection of a security interest will not protect a secured party against certain other third parties having claims to the collateral. We have already discussed the priority of certain types of buyers over a perfected secured party. In addition, the UCC provides that in some situations a properly perfected PMSI[6] will prevail over another security interest in after-acquired collateral, even though the other was perfected first.

An Exception to the First-in-Time Rule—The PMSI An important exception to the first-in-time rule deals with certain types of collateral, such as equipment, in which one of the secured parties has a perfected PMSI [UCC 9–324(a)]. ■ EXAMPLE 20.10 Suppose that Smith borrows funds from West Bank, signing a security agreement in which she puts up all of her present and after-acquired equipment as security. On May 1, West Bank perfects this security interest (which is not a PMSI). On July 1, Smith purchases a new piece of equipment from XYZ Company on credit, signing a security agreement.

XYZ Company, therefore, has a PMSI in the new equipment that is not a PMSI in consumer goods and thus is not automatically perfected. The delivery date for the new equipment is August 1. If Smith defaults on her payments to both West Bank and XYZ, which party—West Bank or XYZ—has priority to the new piece of equipment? Generally, West Bank would have priority because its interest perfected first in time. In this situation, however, XYZ has a PMSI, and provided that XYZ perfected its interest in the nonconsumer goods within twenty days after Smith takes possession, XYZ has priority. ■

Another Exception to the First-in-Time Rule—Security Interest in Inventory Another important exception to the first-in-time rule involves security interests in inventory [UCC 9–324(b)]. ■ EXAMPLE 20.11 On May 1, ABC borrows funds from West Bank. ABC signs a security agreement, putting up all of its present inventory and any thereafter acquired as collateral. West Bank perfects its non-PMSI on that date. On June 10, ABC buys new inventory from Martin, Inc., a manufacturer, to use for its Fourth of July sale. ABC makes a down payment for the new inventory and signs a security agreement giving Martin a PMSI in the new inventory as collateral for the remaining debt. Martin delivers the inventory to ABC on June 28. Due to a hurricane in the area, ABC's Fourth of July sale is a disaster, and most of its inventory remains unsold. In August, ABC defaults on its payments to both West Bank and Martin.

Does West Bank or Martin have priority to the new inventory delivered on June 28? If Martin has not perfected its security interest by June 28, West Bank's after-acquired collateral clause has priority because it was the first to be perfected. If, however, Martin has perfected *and* gives proper notice of its security interest to West Bank before ABC takes possession of the goods on June 28, Martin has priority. ■

The following case illustrates how the first-in-time rule and its exceptions apply to determine the priority of conflicting security interests on a debtor's default.

6. Recall that, with some exceptions (such as motor vehicles), a PMSI in consumer goods is automatically perfected—no filing is necessary.

CASE 20.2 In re Rebel Rents, Inc.

United States Bankruptcy Court,
Central District of California, 2004.
307 Bankr. 171.

FACTS Rebel Rents, Inc., was the largest independent equipment rental company in Southern California. Rebel offered a wide variety of equipment for sale or lease to construction companies, industrial firms, commercial businesses, and homeowners. In September 2000, Rebel bought certain equipment from Snorkel International, Inc., to use as inventory. Textron Financial Corporation agreed to finance the purchase, for which Rebel granted Textron a security interest in the equipment and its proceeds. On December 29, General Electric Capital Corporation (GECC) agreed to loan Rebel up to $25 million, for which Rebel granted GECC a security interest in substantially all of Rebel's assets, including its inventory and proceeds. On January 5, 2001, GECC filed a financing statement with the California secretary of state to perfect its interest. Textron filed its financing statement on January 16. Rebel filed a bankruptcy petition with a federal bankruptcy court in September 2002. Between that date and September 2003, Rebel obtained

CASE 20.2–CONTINUED

$430,661 from leasing the Snorkel equipment. Pursuant to an order of the court, Rebel surrendered the Snorkel equipment to Textron. Textron asserted that it was also owed the income from the recent leases.

ISSUE Did Textron's security interest give Textron priority to Rebel's revenue from the Snorkel equipment leases?

DECISION No. The court held that GECC's security interest had priority over Textron's interest in Rebel's income.[a]

a. Because the security interests in this case were established before the effective date of the revised version of Article 9, the court applied the previous statute, but the result would likely have been the same under the revised version.

REASON The court explained, "[C]onflicting security interests rank according to the priority in time of filing or perfection. A purchase-money lender's security interest in equipment is entitled to priority over conflicting interests in the same collateral only if the purchase-money security interest is perfected at the time the debtor receives possession of the collateral." Here, Textron's security interest attached on September 14, 2000, but was not perfected until January 16, 2001. GECC's financing statement was filed on January 5, eleven days before Textron's financing statement was filed.

FOR CRITICAL ANALYSIS—Social Consideration *What could Textron have done differently to avoid the result in this case?*

Rights and Duties of Debtors and Creditors

The security agreement itself determines most of the rights and duties of the debtor and the secured party. The UCC, however, imposes some rights and duties that are applicable in the absence of a valid security agreement that states the contrary.

Information Requests

Under UCC 9–523(a), a secured party has the option, when making the filing, of furnishing a *copy* of the financing statement being filed to the filing officer and requesting that the filing officer make a note of the file number, the date, and the hour of the original filing on the copy. The filing officer must send this copy to the person designated by the secured party or to the debtor, if the debtor makes the request. Under UCC 9–523(c) and (d), a filing officer must also give information to a person who is contemplating obtaining a security interest from a prospective debtor. The filing officer must issue a certificate that provides information on possible perfected financing statements with respect to the named debtor. The filing officer will charge a fee for the certification and for any information copies provided [UCC 9–525(d)].

Release, Assignment, and Amendment

A secured party can release all or part of any collateral described in the filing, thereby terminating its security interest in that collateral. The release is recorded by filing a uniform amendment form [UCC 9–512, 9–521(b)]. A secured party can assign all or part of the security interest to a third party (the assignee). The assignee can become the secured party of record if the assignment is filed by use of a uniform amendment form [UCC 9–514, 9–521(a)].

If the debtor and secured parties so agree, the filing can be amended—by adding new collateral if authorized by the debtor, for example—by filing a uniform amendment form that indicates by file number the initial financing statement [UCC 9–512(a)]. The amendment does not extend the time period of perfection. If, however, the amendment adds collateral, the perfection date (for priority purposes) for the new collateral begins only on the date of the filing of the amendment [UCC 9–512(b) and (c)].

Confirmation or Accounting Request by Debtor

The debtor may believe that the unpaid debt amount or the listing of the collateral subject to the security interest is inaccurate. The debtor has the right to request a confirmation of his or her view of the unpaid debt or listing of collateral. The secured party must either approve or correct this confirmation request [UCC 9–210].

The secured party must comply with the debtor's confirmation request by authenticating and sending to the debtor an accounting within fourteen days after the request is received. Otherwise, the secured party will be held liable for any loss suffered by the debtor, plus $500 [UCC 9–210, 9–625(f)].

The debtor is entitled to one request without charge every six months. For any additional requests, the secured party is entitled to the payment of a statutory fee of up to $25 per request [UCC 9–210(f)].

Termination Statement

When the debtor has fully paid the debt, if the secured party perfected the security interest by filing, the debtor is entitled to have a termination statement filed. Such a statement demonstrates to the public that the filed perfected security interest has been terminated [UCC 9–513].

Whenever consumer goods are involved, the secured party *must* file a termination statement (or, in the alternative, a release) within one month of the final payment or within twenty days of receipt of the debtor's authenticated demand, whichever is earlier [UCC 9–513(b)].

When the collateral is other than consumer goods, on an authenticated demand by the debtor, the secured party must either send a termination statement to the debtor or file such a statement within twenty days [UCC 9–513(c)]. Otherwise, the secured party is not required to file or send a termination statement. Whenever a secured party fails to file or send the termination statement as requested, the debtor can recover $500 plus any additional loss suffered [UCC 9–625(e)(4) and (f)].

Default

Article 9 defines the rights, duties, and remedies of the secured party and of the debtor on the debtor's default. Should the secured party fail to comply with her or his duties, the debtor is afforded particular rights and remedies.

The topic of default is one of great concern to secured lenders and to the lawyers who draft security agreements. What constitutes default is not always clear. In fact, Article 9 does not define the term. Consequently, parties are encouraged in practice—and by the UCC—to include in their security agreements certain standards to be applied in determining when default has actually occurred. In so doing, parties can stipulate the conditions that will constitute a default [UCC 9–601, 9–603]. Often, these critical terms are shaped by the creditor in an attempt to provide the maximum protection possible. The ultimate terms, however, may not to go beyond the limitations imposed by the good faith requirement and the unconscionability provisions of the UCC. Article 9's definition of good faith includes "honesty in fact and the observance of reasonable commercial standards of fair dealing" [UCC 9–102(a)(43)].

Although any breach of the terms of the security agreement can constitute default, default occurs most commonly when the debtor fails to meet the scheduled payments that the parties have agreed on or when the debtor becomes bankrupt.

Basic Remedies

The rights and remedies under UCC 9–601(a)(b) are *cumulative* [UCC 9–601(c)]. Therefore, if a creditor is unsuccessful in enforcing rights by one method, he or she can pursue another method. Generally, a secured party's remedies can be divided into the two basic categories discussed next.

Repossession of the Collateral—The Self-Help Remedy On the debtor's default, a secured party can take peaceful possession of the collateral without the use of judicial process [UCC 9–609(b)]. This provision is often referred to as the "self-help" provision of Article 9. The UCC does not define *peaceful possession,* however. The general rule is that the collateral has been taken peacefully if the secured party took possession without committing (1) trespass onto realty, (2) assault and/or battery, or (3) breaking and entering. On taking possession, the secured party may either retain the collateral for satisfaction of the debt [UCC 9–620] or resell the goods and apply the proceeds toward the debt [UCC 9–610].

Judicial Remedies A secured party can relinquish a security interest and use any judicial remedy available, such as proceeding to judgment on the underlying debt, followed by execution and levy. (**Execution** is the implementation of a court's decree or judgment. **Levy** is the obtaining of funds by legal process through the seizure and sale of nonsecured property, usually done after a writ of execution has been issued.) Execution and levy are rarely undertaken unless the collateral is no longer in existence or has declined so much in value that it is worth substantially less than the amount of the debt and the debtor has other assets available that may be legally seized to satisfy the debt [UCC 9–601(a)].[7]

Disposition of Collateral

Once default has occurred and the secured party has obtained possession of the collateral, the secured party may either retain the collateral in full satisfaction of the debt or sell, lease, or otherwise dispose of the collateral in any commercially reasonable manner and apply the proceeds toward satisfaction of the debt [UCC 9–602(7), 9–603, 9–610(a), 9–620]. Any sale is always subject to procedures established by state law.

Retention of Collateral by the Secured Party The UCC acknowledges that parties are sometimes better off if they

7. Some assets are exempt from creditors' claims—see Chapter 21.

do not sell the collateral. Therefore, a secured party may retain the collateral unless it consists of consumer goods subject to a PMSI and the debtor has paid 60 percent or more of the purchase price or debt—as will be discussed shortly [UCC 9–620(e)].

This general right, however, is subject to several conditions. The secured party must send notice of the proposal to the debtor if the debtor has not signed a statement renouncing or modifying her or his rights *after default* [UCC 9–620(a), 9–621]. If the collateral is consumer goods, the secured party does not need to give any other notice. In all other situations, the secured party must send notice to any other secured party from whom the secured party has received written or authenticated notice of a claim of interest in the collateral in question. The secured party must also send notice to any other junior lien claimant (one holding a lien that is subordinate to a prior lien) who has filed a statutory lien (such as a mechanic's lien—see Chapter 21) or a security interest in the collateral ten days before the debtor consented to the retention [UCC 9–621].

If, within twenty days after the notice is sent, the secured party receives an objection sent by a person entitled to receive notification, the secured party must sell or otherwise dispose of the collateral. The collateral must be disposed of in accordance with the provisions of UCC 9–602, 9–603, 9–610, and 9–613 (disposition procedures will be discussed shortly). If no such written objection is forthcoming, the secured party may retain the collateral in full or partial satisfaction of the debtor's obligation [UCC 9–620(a), 9–621].

It is clear that a creditor can retain collateral in full satisfaction of a debt. The issue in the following case was whether a creditor could retain the collateral in *partial* satisfaction of the debt.

CASE 20.3 Banks Brothers Corp. v. Donovan Floors, Inc.

Wisconsin Court of Appeals, 2000.
2000 WI App 253,
239 Wis.2d 381,
620 N.W.2d 631.
http://www.wisbar.org/WisCtApp/index.html[a]

FACTS Donovan Floors, Inc., and Breakfall, Inc., two companies controlled by James and Jo-Ann Donovan, borrowed $245,000 from Bank One, Milwaukee, N.A. The companies gave Bank One security interests in their assets, and the Donovans gave the lender a mortgage on their home. When Donovan Floors and Breakfall defaulted on the debt in 1991, Bank One filed a suit in a Wisconsin state court against the debtors. The Donovans agreed to surrender to Bank One some of the firms' assets and agreed to a foreclosure on the home. Bank One promised not to act on the agreement immediately to give the Donovans a chance to revitalize their business. In 1993, Bank One assigned the debt and its security interest to Banks Brothers Corporation. Banks, the Donovans, and the Donovans' firms (Donovan Floors and Breakfall) signed an agreement under which, among other things, Banks was given some of the firms' assets (three cars, a truck, and a van), and Breakfall was released from the debt while the others remained liable. A payment schedule was set up, but none of the payments were made. Six years later, Banks scheduled a sale of the house. The Donovans and Donovan Floors filed a suit in a Wisconsin state court against Banks, arguing that the creditor's retention of assets and release of Breakfall operated as a full satisfaction of the debt. The court denied the request, and the debtors appealed to a state intermediate appellate court.

a. In the "Court or Agency" section, select "Court of Appeals." Under the year, choose "2000" and then click on "Go." Scroll to the name of the case and click on it to access the opinion. The State Bar of Wisconsin maintains this Web site.

ISSUE Can a creditor retain collateral that secures a debt in less than full satisfaction of the debt?

DECISION Yes. The state intermediate appellate court affirmed the order of the lower court.

REASON The appellate court acknowledged that "[u]nderstandably, the Donovans and Donovan Floors would love to have their cake (the chance to save their business given to them by Banks's agreement to hold off on its right to claim the assets pledged for the debt) and eat it also (keep those assets). But that is not the way [Article 9] works. Banks had a right * * * to immediate strict foreclosure of all the pledged assets. It gave up that right in consideration for a partial payment on the debt and the concomitant [accompanying] partial satisfaction. The Donovans and Donovan Floors have no legal or moral ground to complain; they agreed to that arrangement, and did so in a statement signed after default." The court noted that although this case involved a provision in effect before the revision of Article 9, under the revised article's UCC 9–620, the result would have been the same.

FOR CRITICAL ANALYSIS—Economic Consideration *What might be the effect on a lender's willingness to loan money if the law allowed the lender to retain collateral only in full satisfaction of the debt?*

Consumer Goods When the collateral is consumer goods with a PMSI and the debtor has paid 60 percent or more of the debt or the purchase price, the secured party must sell or otherwise dispose of the repossessed collateral within ninety days [UCC 9–620(e) and (f)]. Failure to comply opens the secured party to an action for conversion or other liability under UCC 9–625(b) and (c) unless the consumer-debtor signed a written statement *after default* renouncing or modifying the right to demand the sale of the goods [UCC 9–624].

Disposition Procedures A secured party who does not choose to retain the collateral or who is required to sell it must resort to the disposition procedures prescribed under UCC 9–602(7), 9–603, 9–610(a), and 9–613. The UCC allows a great deal of flexibility with regard to disposition. UCC 9–610(a) states that after default, a secured party may sell, lease, license, or otherwise dispose of any or all of the collateral in its present condition or following any commercially reasonable preparation or processing. Although the secured party may purchase the collateral at a public sale, it may not do so at a private sale—unless the collateral is of a kind customarily sold on a recognized market or is the subject of widely distributed standard price quotations [UCC 9–610(c)].

—The Sale Must Be Accomplished in a Commercially Reasonable Manner. One of the major limitations on the disposition of the collateral is that it must be accomplished in a commercially reasonable manner. UCC 9–610(b) states as follows:

> Every aspect of a disposition of collateral, including the method, manner, time, place, and other terms, must be commercially reasonable. If commercially reasonable, a secured party may dispose of collateral by public or private proceedings, by one or more contracts, as a unit or in parcels, and at any time and place and on any terms.

—Notification Requirements. Unless the collateral is perishable or will decline rapidly in value or is a type customarily sold on a recognized market, a secured party must send to the debtor and other identified persons "a reasonable authenticated notification of disposition" [UCC 9–611(b) and (c)]. The debtor may waive the right to receive this notice, but only after default [UCC 9–624(a)].

Proceeds from the Disposition Proceeds from the disposition of collateral after default on the underlying debt are distributed in the following order:

1. Expenses incurred by the secured party in repossessing, storing, and reselling the collateral.
2. Balance of the debt owed to the secured party.
3. Junior (subordinate) lienholders who have made written or authenticated demands.
4. Unless the collateral consists of accounts, payment intangibles, promissory notes, or chattel paper, any surplus goes to the debtor [UCC 9–608(a), 9–615(a) and (e)].

Noncash Proceeds Whenever the secured party receives noncash proceeds from the disposition of collateral after default, the secured party must make a value determination and apply this value in a commercially reasonable manner [UCC 9–608(a)(3), 9–615(c)].

Deficiency Judgment Often, after proper disposition of the collateral, the secured party has not collected all that the debtor still owes. Unless otherwise agreed, the debtor is liable for any deficiency, and the creditor can obtain a **deficiency judgment** from a court to collect the deficiency. Note, however, that if the underlying transaction was, for example, a sale of accounts or of chattel paper, the debtor is entitled to any surplus or is liable for any deficiency only if the security agreement so provides [UCC 9–615(d) and (e)].

Whenever the secured party fails to conduct a disposition in a commercially reasonable manner or to give proper notice, the deficiency of the debtor is reduced to the extent that such failure affected the price received at the disposition [UCC 9–626(a)(3)].

Redemption Rights At any time before the secured party disposes of the collateral or enters into a contract for its disposition, or before the debtor's obligation has been discharged through the secured party's retention of the collateral, the debtor or any other secured party can exercise the right of *redemption* of the collateral. The debtor or other secured party can do this by tendering performance of all obligations secured by the collateral and by paying the expenses reasonably incurred by the secured party in retaking and maintaining the collateral [UCC 9–623].

TERMS AND CONCEPTS

after-acquired property 397
attachment 389
collateral 389
continuation statement 396
debtor 388
default 389

CHAPTER SUMMARY Secured Transactions

Creating a Security Interest (See pages 389–392.)	1. Unless the creditor has possession of the collateral, there must be a written or authenticated security agreement that is signed or authenticated by the debtor and describes the collateral subject to the security interest. 2. The secured party must give value to the debtor. 3. The debtor must have rights in the collateral—some ownership interest or right to obtain possession of the specified collateral.
Perfecting a Security Interest (See pages 392–396.)	1. *Perfection by filing*—The most common method of perfection is by filing a financing statement containing the names of the secured party and the debtor and indicating the collateral covered by the financing statement. a. Communication of the financing statement to the appropriate filing office, together with the correct filing fee, constitutes a filing. b. The financing statement must be filed under the name of the debtor; fictitious (trade) names normally are not accepted. c. The classification of collateral determines whether filing is necessary and where to file (see Exhibit 20–3 on pages 391 and 392). 2. *Perfection without filing*— a. By transfer of collateral—The debtor can transfer possession of the collateral to the secured party. A *pledge* is an example of this type of transfer. b. By attachment, such as the attachment of a purchase-money security interest (PMSI) in consumer goods—If the secured party has a PMSI in consumer goods (goods bought or used by the debtor for personal, family, or household purposes), the secured party's security interest is perfected automatically. In all, thirteen types of security interests can be perfected by attachment.
The Scope of a Security Interest (See pages 396–398.)	A security agreement can cover the following types of property: 1. *Collateral in the present possession or control of the debtor.* 2. *Proceeds from a sale, exchange, or disposition of secured collateral.* 3. *After-acquired property*—A security agreement may provide that property acquired after the execution of the security agreement will also be secured by the agreement. This provision often accompanies security agreements covering a debtor's inventory. 4. *Future advances*—A security agreement may provide that any future advances made against a line of credit will be subject to the initial security interest in the same collateral.
Priority of Claims to a Debtor's Collateral (See pages 398–401.)	See Exhibit 20–5 on page 399.
Rights and Duties of Debtors and Creditors (See pages 401–402.)	1. *Information request*—On request by any person, the filing officer must send a statement listing the file number, the date, and the hour of the filing of financing statements and other documents covering collateral of a particular debtor; a fee is charged.

(Continued)

CHAPTER SUMMARY Secured Transactions—Continued

<table>
<tr><td>Rights and Duties of Debtors and Creditors—Continued</td><td>
2. Release, assignment, and amendment—A secured party may (a) release part or all of the collateral described in a filed financing statement, thus ending the creditor's security interest; (b) assign part or all of the security interest to another party; and (c) amend a filed financing statement.
3. Confirmation or accounting request by debtor—If a debtor believes that the unpaid debt amount or the listing of the collateral subject to the security interest is inaccurate, the debtor has the right to request a confirmation of his or her view of the unpaid debt or listing of collateral. The secured party must authenticate and send to the debtor an accounting within fourteen days after the request is received. Only one request without charge is permitted per six-month period.
4. Termination statement—When a debt is paid, the secured party generally must send a termination statement to the debtor or file such a statement with the filing officer to whom the original financing statement was given. Failure to comply results in the secured party's liability to the debtor for $500 plus any loss suffered by the debtor.
a. If the financing statement covers consumer goods, the termination statement must be filed by the secured party within one month after the debt is paid, or if the debtor makes an authenticated demand, it must be filed within twenty days of the demand or one month after the debt is paid—whichever is earlier.
b. In all other cases, the termination statement must be filed or furnished to the debtor within twenty days after an authenticated demand is made by the debtor.
</td></tr>
<tr><td>Default
(See pages 402–404.)</td><td>
On the debtor's default, the secured party may do either of the following:
1. Take possession (peacefully or by court order) of the collateral covered by the security agreement and then pursue one of two alternatives:
a. Retain the collateral (unless the secured party has a PMSI in consumer goods and the debtor has paid 60 percent or more of the selling price or loan), in which case the secured party must
(1) Give notice to the debtor if the debtor has not signed a statement renouncing or modifying his or her rights after default. With consumer goods, no other notice is necessary.
(2) Send notice to any other secured party who has given written or authenticated notice of a claim to the same collateral or who has filed a security interest or a statutory lien ten days before the debtor consented to the retention. If an objection is received within twenty days from the debtor or any other secured party given notice, the creditor must dispose of the collateral according to the requirements of UCC 9–602, 9–603, 9–610, and 9–613. Otherwise, the creditor may retain the collateral in full or partial satisfaction of the debt.
b. Dispose of the collateral in accordance with the requirements of UCC 9–602(7), 9–603, 9–610(a), and 9–613, in which case the secured party must
(1) Dispose of (sell, lease, or license) the goods in a commercially reasonable manner.
(2) Notify the debtor and (except in sales of consumer goods) other identified persons, including those who have given notice of claims to the collateral to be sold (unless the collateral is perishable or will decline rapidly in value).
(3) Apply the proceeds in the following order:
(a) Expenses incurred by the secured party in repossessing, storing, and reselling the collateral.
(b) The balance of the debt owed to the secured party.
(c) Junior lienholders who have made written or authenticated demands.
(d) Surplus to the debtor (unless the collateral consists of accounts, payment intangibles, promissory notes, or chattel paper).
2. Relinquish the security interest and use any judicial remedy available, such as proceeding to judgment on the underlying debt, followed by execution and levy on the nonexempt assets of the debtor.
</td></tr>
</table>

FOR REVIEW

Answers for the even-numbered questions in this For Review *section can be found in Appendix A at the end of this text.*

1. What is a security interest? Who is a secured party? What is a security agreement? What is a financing statement?
2. What three requirements must be met to create an enforceable security interest?
3. What is the most common method of perfecting a security interest under Article 9?
4. If two secured parties have perfected security interests in the collateral of the debtor, which party has priority to the collateral on the debtor's default?
5. What rights does a secured creditor have on the debtor's default?

QUESTIONS AND CASE PROBLEMS

20–1. Priority Disputes. Redford is a seller of electric generators. He purchases a large quantity of generators from a manufacturer, Mallon Corp., by making a down payment and signing an agreement to pay the balance over a period of time. The agreement gives Mallon Corp. a security interest in the generators and the proceeds. Mallon Corp. properly files a financing statement on its security interest. Redford receives the generators and immediately sells one of them to Garfield on an installment contract, with payment to be made in twelve equal installments. At the time of the sale, Garfield knows of Mallon's security interest. Two months later, Redford goes into default on his payments to Mallon. Discuss Mallon's rights against purchaser Garfield in this situation.

QUESTION WITH SAMPLE ANSWER

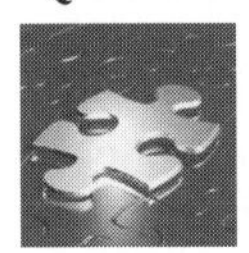

20–2. Marsh has a prize horse named Arabian Knight. Marsh is in need of working capital. She borrows $5,000 from Mendez, who takes possession of Arabian Knight as security for the loan. No written agreement is signed. Discuss whether, in the absence of a written agreement, Mendez has a security interest in Arabian Knight. If Mendez does have a security interest, is it a perfected security interest?

For a sample answer to this question, go to Appendix B at the end of this text.

20–3. The Scope of a Security Interest. Edward owned a retail sporting goods shop. A new ski resort was being created in his area, and to take advantage of the potential business, Edward decided to expand his operations. He borrowed a large amount of funds from his bank, which took a security interest in his present inventory and any after-acquired inventory as collateral for the loan. The bank properly perfected the security interest by filing a financing statement. Edward's business was profitable, so he doubled his inventory. A year later, just a few months after the ski resort had opened, an avalanche destroyed the ski slope and lodge. Edward's business consequently took a turn for the worse, and he defaulted on his debt to the bank. The bank then sought possession of his entire inventory, even though the inventory was now twice as large as it had been when the loan was made. Edward claimed that the bank had rights to only half of his inventory. Is Edward correct? Explain.

20–4. Financing Statement. In 1994, SouthTrust Bank, N.A., made a loan to Environmental Aspecs, Inc. (EAI), and its subsidiary, EAI of NC. SouthTrust perfected its security interest by filing financing statements that listed only EAI as the debtor, described only EAI's assets as collateral, and was signed only on EAI's behalf. SouthTrust believed that both companies were operating as a single business represented by EAI. In 1996, EAI of NC borrowed almost $300,000 from Advanced Analytics Laboratories, Inc. (AAL). AAL filed financing statements that listed the assets of EAI of NC as collateral but identified the debtor as EAI. The statements referred, however, to attached copies of the security agreements, which were signed by the president of EAI of NC and identified the debtor as EAI of NC. One year later, EAI and EAI of NC renegotiated their loan with SouthTrust, and the bank filed financing statements listing both companies as debtors. In 1998, EAI and EAI of NC filed for bankruptcy. One of the issues was the priority of the security interests of SouthTrust and AAL. AAL contended that its failure to identify, on its financing statements, EAI of NC as the debtor did not give SouthTrust priority. Is AAL correct? Why or why not? [*In re Environmental Aspecs, Inc.*, 235 Bankr. 378 (E.D.N.C., Raleigh Div. 1999)]

CASE PROBLEM WITH SAMPLE ANSWER

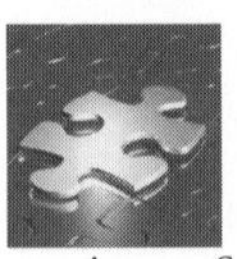

20–5. When a customer opens a credit-card account with Sears, Roebuck & Co., the customer fills out an application and sends it to Sears for review; if the application is approved, the customer receives a Sears card. The application contains a security agreement, a copy of which is also sent with the card. When a customer buys an item using the card, the customer signs a sales receipt that describes the merchandise and contains language granting Sears a purchase-money security interest (PMSI) in the

merchandise. Dayna Conry bought a variety of consumer goods from Sears on her card. When she did not make payments on her account, Sears filed a suit against her in an Illinois state court to repossess the goods. Conry filed for bankruptcy and was granted a discharge. Sears then filed a suit against her to obtain possession of the goods through its PMSI, but it could not find Conry's credit-card application to offer into evidence. Is a signed Sears sales receipt sufficient proof of its security interest? In whose favor should the court rule? Explain. [*Sears, Roebuck & Co. v. Conry,* 321 Ill.App.3d 997, 748 N.E.2d 1248, 255 Ill.Dec. 178 (3 Dist. 2001)]

After you have answered this problem, compare your answer with the sample answer given on the Web site that accompanies this text. Go to academic.cengage.com/blaw/wbl, select "Chapter 20," and click on "Case Problem with Sample Answer."

20–6. Pledge. On April 14, 1992, David and Myrna Grossman borrowed $10,000 from Brookfield Bank in Brookfield, Connecticut, and signed a note to repay the principal with interest. As collateral, the Grossmans gave the bank possession of stock certificates representing 123 shares in General Electric Co. The note was nonnegotiable and thus was not subject to Article 3 of the Uniform Commercial Code. On May 8, the bank closed its doors. The Grossmans did not make any payments on the note and refused to permit the sale of the stock to apply against the debt. The Grossmans' note and collateral were assigned to Premier Capital, Inc., which filed a suit in a Connecticut state court against them, seeking to collect the principal and interest due. The Grossmans responded in part that they were entitled to credit for the value of the stock that secured the note. By the time of the trial, the stock certificates had been lost. Should a creditor have a duty to preserve collateral that is transferred into the creditor's possession as security for a loan? How should the court rule, and why? [*Premier Capital, Inc. v. Grossman,* 68 Conn.App. 51, 789 A.2d 565 (2002)]

20–7. Priorities. PC Contractors, Inc., was an excavating business in Kansas City, Missouri. Union Bank made loans to PC, subject to a perfected security interest in its equipment and other assets, including "after-acquired property." In late 1997, PC leased heavy construction equipment from Dean Machinery Co. The lease agreements required monthly payments, which PC often made late or missed. After eighteen months, Dean demanded that PC either return the equipment or buy it. While attempting to obtain financing for the purchase, PC continued to make monthly payments. In November 2000, Dean, which had not filed a financing statement to cover the transaction, demanded full payment of the amount due. Before paying the price, PC went out of business and surrendered its assets to Union, which prepared to sell them. Dean filed a suit in a Missouri state court against Union to recover the equipment, claiming in part that the bank's security interest had not attached to the equipment because PC had not paid for it. In whose favor should the court rule, and why? [*Dean Machinery Co. v. Union Bank,* 106 S.W.3d 510 (Mo.App.W.D. 2003)]

VIDEO QUESTION

20–8. Go to this text's Web site at academic.cengage.com/blaw/wbl and select "Chapter 20." Click on "Video Questions" and view the video titled *Secured Transactions.* Then answer the following questions.

1. This chapter lists three requirements for creating a security interest. In the video, which requirement does Laura assert has not been met?
2. What, if anything, must the bank have done to perfect its interest in the editing equipment?
3. If the bank exercises its self-help remedy to repossess Onyx's editing equipment, does Laura have any chance of getting it back? Explain.
4. Assume that the bank had a perfected security interest and repossessed the editing equipment. Also assume that the purchase price (and the loan amount) for the equipment was $100,000, of which Onyx has paid $65,000. Discuss the rights and duties of the bank with regard to the collateral in this situation.

Secured Transactions

For updated links to resources available on the Web, as well as a variety of other materials, visit this text's Web site at

academic.cengage.com/blaw/wbl

To find Article 9 of the UCC as modified by a particular state on adoption, go to

http://www.law.cornell.edu/ucc/ucc.table.html

For an overview of secured transactions law and links to UCC provisions and case law on this topic, go to

http://www.law.cornell.edu/topics/secured_transactions.html

ONLINE LEGAL RESEARCH

Go to the *Fundamentals of Business Law* home page at **academic.cengage.com/blaw/wbl**, select "Chapter 20," and click on "Internet Exercises." There you will find the following Internet research exercises that you can perform to learn more about topics covered in this chapter.

Activity 20–1: LEGAL PERSPECTIVE—Repossession

Activity 20–2: MANAGEMENT PERSPECTIVE—Filing Financial Statements

BEFORE THE TEST

Go to the *Fundamentals of Business Law* home page at **academic.cengage.com/blaw/wbl**, select "Chapter 20," and click on "Interactive Quizzes." You will find a number of interactive questions relating to this chapter.

CHAPTER 21

Creditors' Rights and Bankruptcy

LEARNING OBJECTIVES

After reading this chapter, you should be able to answer the following questions:

1. What is a prejudgment attachment? What is a writ of execution? How does a creditor use these remedies?
2. What is garnishment? When might a creditor undertake a garnishment proceeding?
3. In a bankruptcy proceeding, what constitutes the debtor's estate in property? What property is exempt from the estate under federal bankruptcy law?
4. What is the difference between an exception to discharge and an objection to discharge?
5. In a Chapter 11 reorganization, what is the role of the debtor in possession?

Historically, debtors and their families were subjected to punishment, including involuntary servitude and imprisonment, for their inability to pay debts. The modern legal system, however, has moved away from a punishment philosophy in dealing with debtors. In fact, until new bankruptcy legislation was enacted in 2005, many observers said that the system had moved too far in the other direction, to the detriment of creditors.

Normally, creditors have no problem collecting the debts owed to them. When disputes arise over the amount owed, however, or when the debtor simply cannot or will not pay, what happens? What remedies are available to creditors when debtors default? We have already discussed, in Chapter 20, the remedies available to secured creditors under Article 9 of the Uniform Commercial Code (UCC). In the first part of this chapter, we focus on other laws that assist the debtor and creditor in resolving their disputes without the debtor's having to resort to bankruptcy. The second part of this chapter discusses bankruptcy as a last resort in resolving debtor-creditor problems.

Laws Assisting Creditors

Both the common law and statutory laws other than Article 9 of the UCC create various rights and remedies for creditors. We discuss here some of these rights and remedies.

LIENS

As stated in Chapter 17, a *lien* is an encumbrance on (claim against) property to satisfy a debt or protect a claim for the payment of a debt. Creditors' liens may arise under the common law or under statutory law. Statutory liens include *mechanic's liens.* Liens created at common law include *artisan's liens* and *innkeeper's liens. Judicial liens* include those that represent a creditor's efforts to collect on a debt before or after a judgment is entered by a court.

Generally, a lien creditor has priority over an unperfected secured party but not over a perfected secured party. In other words, if a person becomes a lien creditor *before* another party perfects a security interest in the same property, the lienholder has priority. If a person obtains a lien *after* another's security interest in the prop-

erty is perfected, the lienholder does not have priority. This is true for all liens except mechanic's and artisan's liens, which normally have priority over perfected security interests—unless a statute provides otherwise.

Mechanic's Liens When a person contracts for labor, services, or materials to be furnished for the purpose of making improvements on real property (land and things attached to the land, such as buildings and trees—see Chapter 29) but does not immediately pay for the improvements, the creditor can file a **mechanic's lien** on the property. This creates a special type of debtor-creditor relationship in which the real estate itself becomes security for the debt.

■ EXAMPLE 21.1 A painter agrees to paint a house for a homeowner for an agreed-on price to cover labor and materials. If the homeowner refuses to pay for the work or pays only a portion of the charges, a mechanic's lien against the property can be created. The painter is the lienholder, and the real property is encumbered (burdened) with a mechanic's lien for the amount owed. If the homeowner does not pay the lien, the property can be sold to satisfy the debt. Notice of the *foreclosure* (the process by which the creditor deprives the debtor of his or her property) and sale must be given to the debtor in advance, however. ■

Note that state law governs the procedures that must be followed to create a mechanic's lien. Generally, the lienholder must file a written notice of lien against the particular property involved. The notice of lien must be filed within a specific time period, normally measured from the last date on which materials or labor were provided (usually within 60 to 120 days). If the property owner fails to pay the debt, the lienholder is entitled to foreclose on the real estate for which the work or materials were provided and to sell it to satisfy the amount of the debt.

Artisan's Liens An **artisan's lien** is a security device created at common law through which a creditor can recover payment from a debtor for labor and materials furnished for the repair or improvement of personal property. ■ EXAMPLE 21.2 Tenetia leaves her diamond ring at the jeweler's to be repaired and to have her initials engraved on the band. In the absence of an agreement, the jeweler can keep the ring until Tenetia pays for the services. Should Tenetia fail to pay, the jeweler has a lien on Tenetia's ring for the amount of the bill and normally can sell the ring in satisfaction of the lien. ■

—A Possessory Lien. In contrast to a mechanic's lien, an artisan's lien is *possessory.* The lienholder ordinarily must have retained possession of the property and have expressly or impliedly agreed to provide the services on a cash, not a credit, basis. The lien remains in existence as long as the lienholder maintains possession, and the lien is terminated once possession is voluntarily surrendered—unless the surrender is only temporary. With a temporary surrender, there must be an agreement that the property will be returned to the lienholder. Even with such an agreement, if a third party obtains rights in that property while it is out of the possession of the lienholder, the lien is lost.

—Notice of Foreclosure and Sale Required. Modern statutes permit the holder of an artisan's lien to foreclose and sell the property subject to the lien to satisfy payment of the debt. As with the mechanic's lien, the holder of an artisan's lien is required to give notice to the owner of the property before foreclosure and sale. The sale proceeds are used to pay the debt and the costs of the legal proceedings, and the surplus, if any, is paid to the former owner.

Innkeeper's Liens An **innkeeper's lien** is another security device created at common law. An innkeeper's lien is placed on the baggage of guests for any agreed-on hotel charges that remain unpaid. If no express agreement has been made concerning the amount of those charges, then the lien will be for the reasonable value of the accommodations furnished. The innkeeper's lien is terminated either by the guest's payment of the hotel charges or by the innkeeper's surrender of the baggage to the guest, unless the surrender is temporary. Most state statutes permit the innkeeper to satisfy the debt by means of a public sale of the guest's baggage. With the use of credit cards as a means to reserve and secure motel and hotel rooms, however, innkeepers' liens are rarely, if ever, used today.

Judicial Liens When a debt is past due, a creditor can bring a legal action against the debtor to collect the debt. If the creditor is successful in the action, the court awards the creditor a judgment against the debtor (usually for the amount of the debt plus any interest and legal costs incurred in obtaining the judgment). Frequently, however, the creditor is unable to collect the awarded amount.

To ensure that a judgment in the creditor's favor will be collectible, creditors are permitted to request that certain nonexempt property of the debtor be seized to satisfy the debt. (As will be discussed later in this chapter, under state or federal statutes, certain property is exempt from attachment by creditors.) If the court orders the debtor's property to be seized prior to a judgment in the creditor's favor, the court's order is referred to as a *writ of attachment.* If the court orders the debtor's property to be seized following a judgment in the creditor's favor, the court's order is referred to as a *writ of execution.*

—Attachment. Recall from Chapter 20 that *attachment,* in the context of secured transactions, refers to the process through which a security interest in a debtor's collateral becomes enforceable. In the context of judicial liens, this word has another meaning: **attachment** is a court-ordered seizure and taking into custody of property prior to the securing of a judgment for a past-due debt. Attachment rights are created by state statutes. Attachment is a *prejudgment* remedy because it occurs either at the time of or immediately after the commencement of a lawsuit and before the entry of a final judgment. By statute, to attach before judgment, a creditor must comply with specific restrictions and requirements. The due process clause of the Fourteenth Amendment to the U.S. Constitution limits the courts' power to authorize seizure of a debtor's property without notice to the debtor or a hearing on the facts.

To use attachment as a remedy, the creditor must have an enforceable right to payment of the debt under law and must follow certain procedures. Otherwise, the creditor can be liable for damages for wrongful attachment. She or he must file with the court an *affidavit* (a written or printed statement, made under oath or sworn to) stating that the debtor is in default and indicating the statutory grounds under which attachment is sought. The creditor must also post a bond to cover at least the court costs, the value of the property attached, and the value of the loss of use of that property suffered by the debtor. When the court is satisfied that all the requirements have been met, it issues a **writ of attachment,** which directs the sheriff or other public officer to seize nonexempt property. If the creditor prevails at trial, the seized property can be sold to satisfy the judgment.

—Writ of Execution. If the debtor will not or cannot pay the judgment, the creditor is entitled to go back to the court and obtain a court order directing the sheriff to seize (levy) and sell any of the debtor's nonexempt real or personal property that is within the court's geographic jurisdiction (usually the county in which the courthouse is located). This order is called a **writ of execution.** The proceeds of the sale are used to pay off the judgment, accrued interest, and the costs of the sale. Any excess is paid to the debtor. The debtor can pay the judgment and redeem the nonexempt property at any time before the sale takes place. (Because of exemption laws and bankruptcy laws, however, many judgments are virtually uncollectible.)

GARNISHMENT

An order for **garnishment** permits a creditor to collect a debt by seizing property of the debtor that is being held by a third party. In a garnishment proceeding, the third party—the person or entity that the court is ordering to garnish an individual's property—is called the *garnishee.* Frequently, a garnishee is the debtor's employer. A creditor may seek a garnishment judgment against the debtor's employer so that part of the debtor's usual paycheck will be paid to the creditor. In some situations, however, the garnishee is a third party that holds funds belonging to the debtor (such as a bank) or has possession of, or exercises control over, funds or other types of property belonging to the debtor. Almost all types of property can be garnished, including tax refunds, pensions, and trust funds—so long as the property is not exempt from garnishment and is in the possession of a third party.

Garnishment Proceedings The legal proceeding for a garnishment action is governed by state law, and garnishment operates differently from state to state. As a result of a garnishment proceeding, as noted, the court orders a third party (such as the debtor's employer) to turn over property owned by the debtor (such as wages) to pay the debt. Garnishment can be a prejudgment remedy, requiring a hearing before a court, but is most often a postjudgment remedy. According to the laws in some states, the creditor needs to obtain only one order of garnishment, which will then apply continuously to the debtor's wages until the entire debt is paid. In other states, the judgment creditor must go back to court for a separate order of garnishment for each pay period.

Laws Limiting the Amount of Wages Subject to Garnishment Both federal and state laws limit the amount of funds that can be taken from a debtor's weekly take-home pay through garnishment proceedings.[1] Federal law provides a framework to protect debtors from suffering unduly when paying judgment debts.[2] State laws also provide dollar exemptions, and these amounts are often larger than those allowed by federal law. Under federal law, an employer cannot dismiss an employee because his or her wages are being garnished.

CREDITORS' COMPOSITION AGREEMENTS

Creditors may contract with the debtor for discharge of the debtor's liquidated debts (debts that are definite, or fixed, in amount) on payment of a sum less than that owed. This type of agreement is normally referred to as a

1. Some states (for example, Texas) do not permit garnishment of wages by private parties except under a child-support order.

2. For example, the federal Consumer Credit Protection Act of 1968, 15 U.S.C. Sections 1601–1693r, provides that a debtor can retain either 75 percent of the disposable earnings per week or a sum equivalent to thirty hours of work paid at federal minimum-wage rates, whichever is greater.

creditors' composition agreement, or simply a *composition agreement,* and is usually held to be enforceable.

MORTGAGE FORECLOSURE

Mortgage holders have the right to foreclose on mortgaged property in the event of a debtor's default. The usual method of foreclosure is by judicial sale of the property, although the statutory methods of foreclosure vary from state to state. If the proceeds of the foreclosure sale are sufficient to cover both the costs of the foreclosure and the mortgaged debt, the debtor receives any surplus. If the sale proceeds are insufficient to cover the foreclosure costs and the mortgaged debt, however, the **mortgagee** (the creditor-lender) can seek to recover the difference from the **mortgagor** (the debtor) by obtaining a deficiency judgment representing the difference between the mortgaged debt and the amount actually received from the proceeds of the foreclosure sale.

The mortgagee obtains a deficiency judgment in a separate legal action pursued subsequent to the foreclosure action. The deficiency judgment entitles the mortgagee to recover the amount of the deficiency from other property owned by the debtor.

SURETYSHIP AND GUARANTY

When a third person promises to pay a debt owed by another in the event the debtor does not pay, either a *suretyship* or a *guaranty* relationship is created. Suretyship and guaranty have a long history under the common law and provide creditors with the right to seek payment from the third party if the primary debtor defaults on her or his obligations. Exhibit 21–1 illustrates the relationship between a suretyship or guaranty party and the creditor.

Surety A contract of strict **suretyship** is a promise made by a third person to be responsible for the debtor's obligation. It is an express contract between the **surety** (the third party) and the creditor. The surety in the strictest sense is primarily liable for the debt of the principal. The creditor need not exhaust all legal remedies against the principal debtor before holding the surety responsible for payment. The creditor can demand payment from the surety from the moment the debt is due.

■ EXAMPLE 21.3 Roberto Delmar wants to borrow funds from the bank to buy a used car. Because Roberto is still in college, the bank will not lend him the funds unless his father, José Delmar, who has dealt with the

EXHIBIT 21–1 SURETYSHIP AND GUARANTY PARTIES

In a suretyship or guaranty arrangement, a third party promises to be responsible for a debtor's obligations. A third party who agrees to be responsible for the debt even if the primary debtor does not default is known as a surety; a third party who agrees to be *secondarily* responsible for the debt—that is, responsible only if the primary debtor defaults—is known as a guarantor. As noted in Chapter 10, a promise of guaranty (a collateral, or secondary, promise) normally must be in writing to be enforceable.

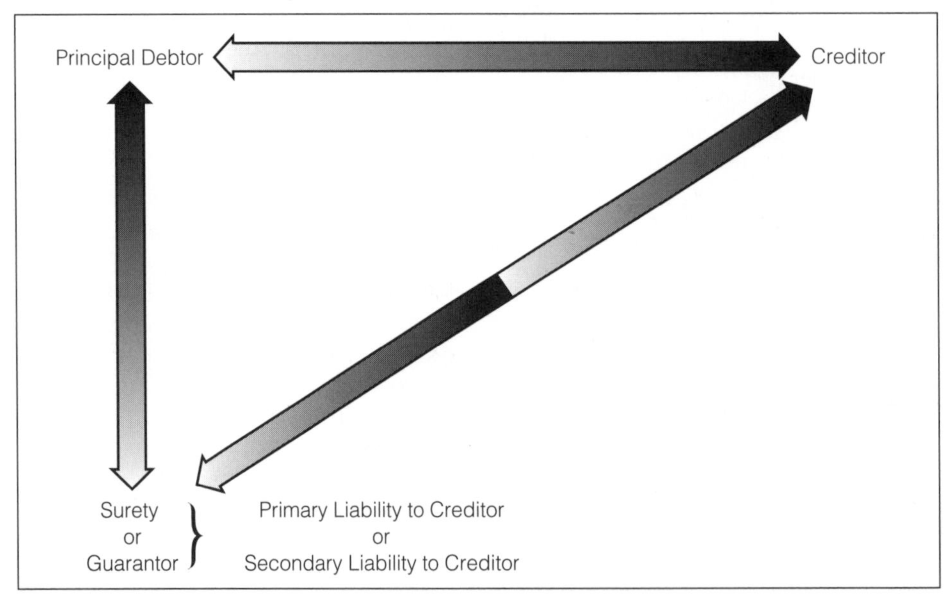

bank before, will cosign the note (add his signature to the note, thereby becoming a surety and thus jointly liable for payment of the debt). When José Delmar cosigns the note, he becomes primarily liable to the bank. On the note's due date, the bank has the option of seeking payment from either Roberto or José Delmar, or both jointly. ■

Guaranty With a suretyship arrangement, the surety is *primarily* liable for the debtor's obligation. With a guaranty arrangement, the **guarantor**—the third person making the guaranty—is *secondarily* liable. The guarantor can be required to pay the obligation *only after the principal debtor defaults*, and default usually takes place only after the creditor has made an attempt to collect from the debtor.

■ EXAMPLE 21.4 A small corporation, BX Enterprises, needs to borrow funds to meet its payroll. The bank is skeptical about the creditworthiness of BX and requires Dawson, its president, who is a wealthy businessperson and the owner of 70 percent of BX Enterprises, to sign an agreement making himself personally liable for payment if BX does not pay off the loan. As a guarantor of the loan, Dawson cannot be held liable until BX Enterprises is in default. ■

The Statute of Frauds requires that a guaranty contract between the guarantor and the creditor must be in writing to be enforceable unless the *main purpose* exception applies. (A suretyship agreement, by contrast, need not be in writing to be enforceable.) As discussed in Chapter 10, under this exception if the main purpose of the guaranty agreement is to benefit the guarantor, then the contract need not be in writing to be enforceable.

In the following case, the issue was whether a guaranty of a lease signed by an officer of a corporation was enforceable against the officer personally even though he claimed to have signed the guaranty only as a representative of the corporation.

CASE 21.1 JSV, Inc. v. Hene Meat Co.

Court of Appeals of Indiana, 2003.
794 N.E.2d 555.

FACTS On August 30, 1999, JSV, Inc., signed a lease to rent a portion of a building in Indianapolis, Indiana, from Hene Meat Company. Mark Kennedy signed the lease on behalf of JSV as one of its corporate officers. Kennedy also signed a document titled "GUARANTY," which stated that it was "an absolute and unconditional guaranty" of the lease's performance by JSV. Kennedy's printed name and signature on the document were not followed by any corporate officer designation. JSV stopped paying rent to Hene in September 2000. Hene filed a suit in an Indiana state court against JSV and Kennedy, alleging, among other things, that Kennedy was personally liable on the guaranty. He responded in part that he signed the guaranty only as an officer of JSV. The court issued a summary judgment against Kennedy for $75,041.07 in favor of Hene. Kennedy appealed to a state intermediate appellate court.

ISSUE Was Kennedy personally liable on the guaranty?

DECISION Yes. The state intermediate appellate court affirmed the judgment of the lower court, holding that the document Kennedy signed was "unambiguously a personal guaranty."

REASON The state intermediate appellate court explained that a guaranty agreement includes three parties: the obligor (the promisor or debtor), the obligee (the one to whom the debt or obligation is owed), and the guarantor. "Here, Hene as landlord under the lease was the obligee and JSV as the tenant was the obligor; the disputed issue is the identity of the guarantor." The court explained that there would have been no point in Hene's obtaining Kennedy's guaranty of the lease if he was doing so only in his official capacity as an officer of JSV. This would have been the same as JSV guaranteeing its own performance under the lease, which would have meant that JSV was both the obligor under the lease and the guarantor under the guaranty. The court felt that such a result would be "paradoxical and untenable," and concluded that the guaranty in this case is "clearly a personal one * * * because Kennedy's signature thereon is not followed by any corporate officer designation."

WHY IS THIS CASE IMPORTANT? *This case emphasizes the need to use explicit language in contracts with persons who are officers and directors of corporations. While this court ultimately concluded that the guaranty was personal even though it was not designated as such, other courts may reach different conclusions. To avoid potential problems and costly litigation, businesspersons would be wise to require persons affiliated with corporations to clearly indicate whether they are signing each document in their individual or official capacities.*

Defenses of the Surety and the Guarantor The defenses of the surety and the guarantor are basically the same. Therefore, the following discussion applies to both, although it refers only to the surety.

—Actions Releasing the Surety. Certain actions will release the surety from the obligation. For example, any binding material modification in the terms of the original contract made between the principal debtor and the creditor—including a binding agreement to extend the time for making payment—without first obtaining the consent of the surety will discharge a gratuitous surety completely and a compensated surety to the extent that the surety suffers a loss. (An example of a gratuitous surety is a father who agrees to assume responsibility for his daughter's obligation; an example of a compensated surety is a venture capitalist who will profit from a loan made to the principal debtor.)

Naturally, if the principal obligation is paid by the debtor or by another person on behalf of the debtor, the surety is discharged from the obligation. Similarly, if valid tender of payment is made, and the creditor rejects it with knowledge of the surety's existence, the surety is released from any obligation on the debt.

—Defenses of the Principal Debtor. Generally, the surety can use any defenses available to a principal debtor to avoid liability on the obligation to the creditor. Defenses available to the principal debtor that the surety *cannot* use include the principal debtor's incapacity or bankruptcy and the statute of limitations. The ability of the surety to assert any defenses the debtor may have against the creditor is the most important concept in suretyship; it means that most of the defenses available to the surety are also those of the debtor.

—Surrender or Impairment of Collateral. In addition, if a creditor surrenders the collateral to the debtor or impairs the collateral while knowing of the surety and without the surety's consent, the surety is released to the extent of any loss suffered from the creditor's actions. The primary reason for this requirement is to protect a surety who agreed to become obligated only because the debtor's collateral was in the possession of the creditor.

—Other Defenses. Obviously, a surety may also have his or her own defenses—for example, incapacity or bankruptcy. If the creditor fraudulently induced the surety to guarantee the debt of the debtor, the surety can assert fraud as a defense. In most states, the creditor has a legal duty to inform the surety, prior to the formation of the suretyship contract, of material facts known by the creditor that would substantially increase the surety's risk. Failure to so inform may constitute fraud and makes the suretyship obligation voidable.

Rights of the Surety and the Guarantor Generally, when the surety or guarantor pays the debt owed to the creditor, the surety or guarantor is entitled to certain rights. Because the rights of the surety and guarantor are basically the same, the following discussion applies to both.

—The Right of Subrogation. The surety has the legal **right of subrogation.** Simply stated, this means that any right the creditor had against the debtor now becomes the right of the surety. Included are creditor rights in bankruptcy, rights to collateral possessed by the creditor, and rights to judgments secured by the creditor. In short, the surety now stands in the shoes of the creditor and may pursue any remedies that were available to the creditor against the debtor.

—The Right of Reimbursement. The surety has a right to be reimbursed by the debtor. This **right of reimbursement** may stem either from the suretyship contract or from equity. Basically, the surety is entitled to receive from the debtor all outlays made on behalf of the suretyship arrangement. Such outlays can include expenses incurred as well as the actual amount of the debt paid to the creditor.

—The Right of Contribution. In a situation involving **co-sureties** (two or more sureties on the same obligation owed by the debtor), a surety who pays more than her or his proportionate share on a debtor's default is entitled to recover from the co-sureties the amount paid above the surety's obligation. This is the **right of contribution.** Generally, a co-surety's liability either is determined by agreement between the co-sureties or, in the absence of an agreement, can be specified in the suretyship contract itself.

■ EXAMPLE 21.5 Assume that two co-sureties are obligated under a suretyship contract to guarantee the debt of a debtor. Together, the sureties' maximum liability is $25,000. As specified in the suretyship contract, surety A's maximum liability is $15,000, and surety B's is $10,000. The debtor owes $10,000 and is in default. Surety A pays the creditor the entire $10,000. In the absence of any agreement between the two co-sureties, surety A can recover $4,000 from surety B ($10,000/$25,000 × $10,000 = $4,000). ■

Laws Assisting Debtors

The law protects debtors as well as creditors. Certain property of the debtor, for example, is exempt from creditors' actions. Probably the most familiar exemption is the **homestead exemption.** Each state permits the debtor to retain the family home, either in its entirety or up to a specified dollar amount, free from the claims of unsecured creditors or trustees in bankruptcy (a *bankruptcy trustee* is appointed by the court to hold and protect estate property, as will be discussed later in this chapter). The purpose of the homestead exemption is to ensure that the debtor will retain some form of shelter.

■ EXAMPLE 21.6 Suppose that Van Cleave owes Acosta $40,000. The debt is the subject of a lawsuit, and the court awards Acosta a judgment of $40,000 against Van Cleave. Van Cleave's home is valued at around $50,000, and the state exemption on homesteads is $25,000. There are no outstanding mortgages or other liens. To satisfy the judgment debt, Van Cleave's family home is sold at public auction for $45,000. The proceeds of the sale are distributed as follows:

1. Van Cleave is given $25,000 as his homestead exemption.
2. Acosta is paid $20,000 toward the judgment debt, leaving a $20,000 deficiency judgment that can be satisfied from any other nonexempt property (personal or real) that Van Cleave may have, if allowed by state law. ■

Various types of personal property may also be exempt from satisfaction of judgment debts. Personal property that is most often exempt includes the following:

1. Household furniture up to a specified dollar amount.
2. Clothing and certain personal possessions, such as family pictures or a Bible or other religious text.
3. A vehicle (or vehicles) for transportation (at least up to a specified dollar amount).
4. Certain classified animals, usually livestock but including pets.
5. Equipment that the debtor uses in a business or trade, such as tools or professional instruments, up to a specified dollar amount.

Bankruptcy and Reorganization

Bankruptcy law in the United States has two goals—to protect a debtor by giving him or her a fresh start, free from creditors' claims, and to ensure equitable treatment to creditors who are competing for a debtor's assets. Federal bankruptcy legislation was first enacted in 1898 and has undergone several modifications since that time.

Bankruptcy law prior to 2005 was based on the Bankruptcy Reform Act of 1978, as amended—hereinafter called the Bankruptcy Code or, more simply, the Code (not to be confused with the Uniform Commercial Code, which is also sometimes called the Code). In 2005, Congress enacted a new Bankruptcy Reform Act.[3] As you will read throughout this chapter, the 2005 act significantly overhauled certain provisions of the Bankruptcy Code—for the first time in twenty-five years. Although bankruptcy law is federal law, state laws on secured transactions, liens, judgments, and exemptions also play a role in federal bankruptcy proceedings.

The Bankruptcy Reform Act of 2005

One of the major goals of the Bankruptcy Reform Act of 2005 is to require consumers to pay as many of their debts as they possibly can instead of having those debts fully discharged in bankruptcy. Prior to the new law, only about 20 percent of personal bankruptcies were filed under Chapter 13 of the Bankruptcy Code. As you will read in a subsequent section, this chapter of the Bankruptcy Code requires the establishment of a repayment plan under which a debtor pays off as many of his or her debts as possible over a maximum period of five years. The remaining bankruptcies were filed under Chapter 7 of the Code, which permits debtors, with some exceptions, to have all of their debts discharged in bankruptcy.

The distinction has been important for all creditors. Given that most individuals who declare personal bankruptcy have few durable assets, filing for bankruptcy protection under Chapter 7 essentially means that creditors lose out. Under the new law, in contrast, whenever a debtor has an annual income in excess of the mean income in that debtor's state of residence, the debtor may be forced into a Chapter 13 repayment plan.

Another provision of the Bankruptcy Reform Act of 2005 affects the homestead exemption. Prior to the passage of the bankruptcy reform legislation, some states allowed debtors petitioning for bankruptcy to exempt all of the equity (the market value minus the outstanding mortgage owed) in their homes during bankruptcy proceedings. The 2005 act leaves these exemptions in place but puts some limits on their use. The 2005 act also includes a number of other changes. For example, one provision requires credit-card companies to make fuller disclosures about their interest rates and payment schedules. Another provision gives child-support obligations priority over other debts and allows enforcement agencies to continue efforts to collect

3. The full title of the act is the Bankruptcy Abuse Prevention and Consumer Protection Act of 2005, Pub. L. No. 109-8, 119 Stat. 23 (April 20, 2005).

child-support payments. Additionally, the act expands protections for family farmers and provides more protection for customers' personal information possessed by businesses undergoing bankruptcy.

BANKRUPTCY COURTS

Bankruptcy proceedings are held in federal bankruptcy courts, which are under the authority of the U.S. district courts, and rulings from bankruptcy courts can be appealed to the district courts. Essentially, a bankruptcy court fulfills the role of an administrative court for the federal district court concerning matters in bankruptcy. The bankruptcy court holds proceedings dealing with the procedures required to administer the estate of the debtor in bankruptcy. A bankruptcy court can conduct a jury trial if the appropriate district court has authorized it and the parties to the bankruptcy consent. Bankruptcy court judges are federally appointed for fourteen-year terms. The 2005 Bankruptcy Reform Act included a section entitled the Bankruptcy Judgeship Act of 2005, which enlarged the number of bankruptcy judges by twenty-eight (four for the Delaware District).

TYPES OF BANKRUPTCY RELIEF

Title 11 of the *United States Code* encompasses the Bankruptcy Code, which has eight chapters. Chapters 1, 3, and 5 of the Code contain general definitional provisions, as well as provisions governing case administration, creditors, the debtor, and the estate. These three chapters apply generally to all kinds of bankruptcies. The next five chapters of the Code set forth the different types of relief that debtors may seek. Chapter 7 provides for **liquidation** proceedings (the selling of all nonexempt assets and the distribution of the proceeds to the debtor's creditors). Chapter 9 governs the adjustment of a municipality's debts. Chapter 11 governs reorganizations. Chapters 12 and 13 provide for the adjustment of debts by parties with regular incomes (family farmers and family fishermen under Chapter 12 and individuals under Chapter 13).[4] A debtor (except for a municipality) need not be insolvent[5] to file for bankruptcy relief under any chapter of the Bankruptcy Code. Anyone obligated to a creditor can declare bankruptcy.

4. There are no Chapters 2, 4, 6, 8, or 10 in Title 11. Such "gaps" are not uncommon in the *United States Code*. This is because chapter numbers (or other subdivisional unit numbers) are sometimes reserved for future use when a statute is enacted. (A gap may also appear if a law has been repealed.)

5. The inability to pay debts as they become due is known as *equitable* insolvency. A *balance sheet* insolvency, which exists when a debtor's liabilities exceed assets, is not the test. Thus, it is possible for debtors to voluntarily petition for bankruptcy or to be thrown into involuntary bankruptcy even though their assets far exceed their liabilities. This may occur when a debtor's cash flow problems become severe.

SPECIAL TREATMENT OF CONSUMER-DEBTORS

To ensure that a *consumer-debtor* is fully informed of the various types of relief available, the Code requires that the clerk of the court provide certain information to all consumer-debtors prior to the commencement of a bankruptcy filing. (A **consumer-debtor** is a debtor whose debts result primarily from the purchase of goods for personal, family, or household use.) First, the clerk must give consumer-debtors written notice of the general purpose, benefits, and costs of each chapter of the Bankruptcy Code under which they might proceed. Second, under the 2005 act, the clerk must provide consumer-debtors with informational materials on the types of services available from credit counseling agencies.

In the remainder of this chapter, we deal first with liquidation proceedings under Chapter 7 of the Code. We then examine the procedures required for Chapter 11 reorganizations and Chapter 12 and 13 plans. (The latter three chapters of the Code are known as "rehabilitation" chapters.)

Chapter 7—Liquidation

Liquidation under Chapter 7 of the Bankruptcy Code is generally the most familiar type of bankruptcy proceeding and is often referred to as an *ordinary*, or *straight, bankruptcy*. Put simply, a debtor in a liquidation bankruptcy turns all assets over to a **trustee.** The trustee sells the nonexempt assets and distributes the proceeds to creditors. With certain exceptions, the remaining debts are then **discharged** (extinguished), and the debtor is relieved of the obligation to pay the debts.

Any "person"—defined as including individuals, partnerships, and corporations[6]—may be a debtor in a liquidation proceeding. Railroads, insurance companies, banks, savings and loan associations, investment companies licensed by the Small Business Administration, and credit unions cannot be debtors in a liquidation bankruptcy, however. Other chapters of the Bankruptcy Code or federal or state statutes apply to them.

A straight bankruptcy may be commenced by the filing of either a voluntary or an involuntary **petition in bankruptcy**—the document that is filed with a bankruptcy court to initiate bankruptcy proceedings. If a debtor files the petition, it is a *voluntary bankruptcy*. If one or more creditors file a petition to force the debtor into bankruptcy, it is called an *involuntary bankruptcy*. We discuss both voluntary and involuntary bankruptcy proceedings under Chapter 7 in the following subsections.

6. The definition of *corporation* includes unincorporated companies and associations. It also covers labor unions.

VOLUNTARY BANKRUPTCY

To bring a voluntary petition in bankruptcy, the debtor files official forms designated for that purpose in the bankruptcy court. The Bankruptcy Reform Act of 2005 specifies that before debtors can file a petition, they must receive credit counseling from an approved nonprofit agency within the 180-day period preceding the date of filing. The act provides detailed criteria for the **U.S. Trustee** (a government official who performs appointment and other administrative tasks that a bankruptcy judge would otherwise have to perform) to approve nonprofit budget and counseling agencies and requires that a list of approved agencies be made publicly available.[7] A debtor filing a Chapter 7 petition must include a certificate proving that he or she received an individual or group briefing from an approved counseling agency within the last 180 days (roughly six months).

The Code requires a consumer-debtor who has opted for liquidation bankruptcy proceedings to confirm the accuracy of the petition's contents. The debtor must also state in the petition, at the time of filing, that he or she understands the relief available under other chapters of the Code and has chosen to proceed under Chapter 7. If an attorney is representing the consumer-debtor, the attorney must file an affidavit stating that she or he has informed the debtor of the relief available under each chapter of the Bankruptcy Code. In addition, the 2005 act requires the attorney to reasonably attempt to verify the accuracy of the consumer-debtor's petition and schedules (described below). Failure to do so is considered perjury.

Chapter 7 Schedules The voluntary petition must contain the following schedules:

1. A list of both secured and unsecured creditors, their addresses, and the amount of debt owed to each.
2. A statement of the financial affairs of the debtor.
3. A list of all property owned by the debtor, including property that the debtor claims is exempt.
4. A list of current income and expenses.
5. A certificate from an approved credit counseling agency (as discussed previously).
6. Proof of payments received from employers within sixty days prior to the filing of the petition.
7. A statement of the amount of monthly income, itemized to show how the amount is calculated.
8. A copy of the debtor's federal income tax return (or a transcript of such a return) for the most recent year ending immediately before the filing of the petition.

As previously noted, the official forms must be completed accurately, sworn to under oath, and signed by the debtor. To conceal assets or knowingly supply false information on these schedules is a crime under the bankruptcy laws.

—Additional Information May Be Required. At the request of the court, the U.S. Trustee, or any party in interest, the debtor must file tax returns at the end of each tax year while the case is pending and provide copies to the court. This requirement also applies to Chapter 11 and 13 bankruptcies (discussed later in this chapter). Also, if requested by the U.S. Trustee or bankruptcy trustee, the debtor must provide a photo document establishing his or her identity (such as a driver's license or passport) or other personal identifying information.

—Time Period for Filing Schedules. With the exception of tax returns, failure to file the required schedules within forty-five days after the filing of the petition (unless an extension of up to forty-five days is granted) will result in an automatic dismissal of the petition. The debtor has up to seven days before the date of the first creditors' meeting to provide a copy of the most recent tax returns to the trustee.

Substantial Abuse Prior to 2005, a bankruptcy court could dismiss a Chapter 7 petition for relief (discharge of debts) if the use of Chapter 7 would constitute a "substantial abuse" of that chapter. The Bankruptcy Reform Act of 2005 established a new system of "means testing" (based on the debtor's income) to determine whether a debtor's petition is presumed to be a "substantial abuse" of Chapter 7.

—When Abuse Will Be Presumed. If the debtor's family income is greater than the median family income in the state in which the petition is filed, the trustee or any party in interest (such as a creditor) can bring a motion to dismiss the Chapter 7 petition. State median incomes vary from state to state and are calculated and reported by the U.S. Bureau of the Census.[8]

---

7. The Bankruptcy Reform Act of 2005 also required the director of the Executive Office for the U.S. Trustees to develop a curriculum for financial-management training and create materials that can be used to educate individual debtors on how to better control their finances.

8. For example, in 2004 the median family income in Kentucky was $53,319, and in West Virginia it was $49,470, according to statistics prepared by the U.S. Bureau of the Census and reported for each state annually in the *Federal Register.*

The debtor's current monthly income is calculated using the last six months' average income, less certain "allowed expenses" reflecting the basic needs of the debtor.[9] The monthly amount is then multiplied by twelve. If the resulting income exceeds the state median income by $6,000 or more,[10] abuse is presumed, and the trustee or any creditor can file a motion to dismiss the petition. A debtor can rebut (refute) the presumption of abuse "by demonstrating special circumstances that justify additional expenses or adjustments of current monthly income for which there is no reasonable alternative." (One example might be anticipated medical costs not covered by health insurance.) These additional expenses or adjustments must be itemized and their accuracy attested to under oath by the debtor.

—When Abuse Will Not Be Presumed. If the debtor's income is below the state median (or if the debtor has successfully rebutted the means-test presumption), abuse will not be presumed. In these situations, the court may still find substantial abuse, but the creditors will not have standing (see Chapter 2) to file a motion to dismiss. Basically, this leaves intact the prior law on substantial abuse, allowing the court to consider such factors as the debtor's bad faith or circumstances indicating substantial abuse.

Additional Grounds for Dismissal As noted, a debtor's voluntary petition for Chapter 7 relief may be dismissed for substantial abuse or for failing to provide the necessary documents (such as schedules and tax returns) within the specified time. In addition, a motion to dismiss a Chapter 7 filing might be granted in two other situations under the Bankruptcy Reform Act of 2005. First, if the debtor has been convicted of a violent crime or a drug-trafficking offense, the victim can file a motion to dismiss the voluntary petition.[11] Second, if the debtor fails to pay postpetition domestic-support obligations (which include child and spousal support), the court may dismiss the debtor's Chapter 7 petition.

Order for Relief If the voluntary petition for bankruptcy is found to be proper, the filing of the petition will itself constitute an *order for relief*. (An **order for relief** is a court's grant of assistance to a petitioner.) Once a consumer-debtor's voluntary petition has been filed, the clerk of the court or other appointee must give the trustee and creditors notice of the order for relief by mail not more than twenty days after entry of the order. A husband and wife may file jointly for bankruptcy under a single petition.

INVOLUNTARY BANKRUPTCY

An involuntary bankruptcy occurs when the debtor's creditors force the debtor into bankruptcy proceedings. An involuntary case cannot be commenced against a farmer[12] or a charitable institution. For an involuntary action to be filed against other debtors, the following requirements must be met: If the debtor has twelve or more creditors, three or more of these creditors having unsecured claims totaling at least $12,300 must join in the petition. If a debtor has fewer than twelve creditors, one or more creditors having a claim of $12,300 may file.

If the debtor challenges the involuntary petition, a hearing will be held, and the bankruptcy court will enter an order for relief if it finds either of the following:

1. The debtor is generally not paying debts as they become due.
2. A general receiver, assignee, or custodian took possession of, or was appointed to take charge of, substantially all of the debtor's property within 120 days before the filing of the petition.

If the court grants an order for relief, the debtor will be required to supply the same information in the bankruptcy schedules as in a voluntary bankruptcy.

An involuntary petition should not be used as an everyday debt-collection device, and the Code provides penalties for the filing of frivolous petitions against debtors. Judgment may be granted against the petitioning creditors for the costs and attorneys' fees incurred by the debtor in defending against an involuntary petition that is dismissed by the court. If the petition was filed in bad faith, damages can be awarded for injury to the debtor's reputation. Punitive damages may also be awarded.

AUTOMATIC STAY

The moment a petition, either voluntary or involuntary, is filed, an **automatic stay,** or suspension, of virtually all actions by creditors against the debtor or the debtor's property

9. Section 707 of the Bankruptcy Reform Act of 2005 describes the means test and provides a detailed listing of the expenses allowed under the act.

10. This amount ($6,000) is the equivalent of $100 per month for five years, indicating that the debtor could pay at least $100 per month under a Chapter 13 five-year repayment plan.

11. Note that the court may not dismiss a case on this ground if the debtor's bankruptcy is necessary to satisfy a claim for a domestic-support obligation.

12. The definition of *farmer* includes persons who receive more than 50 percent of their gross income from farming operations, such as tilling the soil, dairy farming, ranching, or the production or raising of crops, poultry, or livestock. Corporations and partnerships may qualify under certain conditions.

normally goes into effect. In other words, once a petition has been filed, creditors cannot contact the debtor by phone or mail or start any legal proceedings to recover debts or to repossess property. A secured creditor or other party in interest, however, may petition the bankruptcy court for relief from the automatic stay. The Code provides that if a creditor knowingly violates the automatic stay (a willful violation), any party injured, including the debtor, is entitled to recover actual damages, costs, and attorneys' fees and may be entitled to recover punitive damages as well.

Underlying the Code's automatic-stay provision for a secured creditor is a concept known as *adequate protection.* The adequate protection doctrine, among other things, protects secured creditors from losing their security as a result of the automatic stay. The bankruptcy court can provide adequate protection by requiring the debtor or trustee to make periodic cash payments or a one-time cash payment (or to provide additional collateral or replacement liens) to the extent that the stay may actually cause the value of the property to decrease. Alternatively, the court may grant other relief that protects the secured party's interest in the property, such as a guaranty by a solvent third party to cover losses suffered by the secured party as a result of the stay.

Exceptions to the Automatic Stay The 2005 Bankruptcy Reform Act provides several exceptions to the automatic stay. A new exception is created for domestic-support obligations, which include any debt owed to or recoverable by a spouse, former spouse, child of the debtor, a child's parent or guardian, or a governmental unit. In addition, proceedings against the debtor related to divorce, child custody or visitation, domestic violence, and support enforcement are not stayed. Also excepted are investigations by a securities regulatory agency, the creation or perfection of statutory liens for property taxes or special assessments on real property, eviction actions on judgments obtained prior to filing the petition, and withholding from the debtor's wages for repayment of a retirement account loan.

Limitations on the Automatic Stay Under the new Code, if a creditor or other party in interest requests relief from the stay, the stay will automatically terminate sixty days after the request, unless the court grants an extension[13] or the parties agree otherwise. Also, the automatic stay on secured debts (see Chapter 20) will terminate thirty days after the petition is filed if the debtor had filed a bankruptcy petition that was dismissed within the prior year. (This is true unless the dismissal was based on the means test and the current petition was filed under a different chapter.) Any party in interest can request the court to extend the stay by showing that the filing is in good faith.

If two or more bankruptcy petitions were dismissed during the prior year, the Code presumes bad faith, and the automatic stay does not go into effect until the court determines that the filing was made in good faith. In addition, if the petition is subsequently dismissed because the debtor failed to file the required documents within thirty days of filing, for example, the stay is terminated. Finally, the automatic stay on secured property terminates forty-five days after the creditors' meeting (to be discussed shortly) unless the debtor redeems or reaffirms certain debts (*reaffirmation* will be discussed later in this chapter). In other words, the debtor cannot keep the secured property (such as a financed automobile), even if she or he continues to make payments on it, without reinstating the rights of the secured party to collect on the debt.

Property of the Estate

On the commencement of a liquidation proceeding under Chapter 7, an *estate in property* is created. The estate consists of all the debtor's interests in property currently held, wherever located, together with community property, property transferred in a transaction voidable by the trustee, proceeds and profits from the property of the estate, and certain after-acquired property. Interests in certain property—such as gifts, inheritances, property settlements (from divorce), and life insurance death proceeds—to which the debtor becomes entitled *within 180 days after filing* may also become part of the estate. Under the 2005 act, withholdings for employee benefit plan contributions are excluded from the estate. Generally, though, the filing of a bankruptcy petition fixes a dividing line: property acquired prior to the filing of the petition becomes property of the estate, and property acquired after the filing of the petition, except as just noted, remains the debtor's.

Creditors' Meeting and Claims

Within a reasonable time after the order for relief has been granted (not less than twenty days or more than forty days), the trustee must call a meeting of the creditors listed in the schedules filed by the debtor. The bankruptcy judge does not attend this meeting.

Debtor's Presence Required The debtor is required to attend the meeting (unless excused by the court) and to submit to examination under oath by the creditors and the trustee. Failing to appear when required or making false

13. The court might grant an extension, for example, on a motion by the trustee that the property is of value to the estate.

statements under oath may result in the debtor's being denied a discharge in bankruptcy. At the meeting, the trustee ensures that the debtor is aware of the potential consequences of bankruptcy and of his or her ability to file for bankruptcy under a different chapter of the Bankruptcy Code.

Creditors' Claims To be entitled to receive a portion of the debtor's estate, each creditor normally files a *proof of claim* with the bankruptcy court clerk within ninety days of the creditors' meeting.[14] The proof of claim lists the creditor's name and address, as well as the amount that the creditor asserts is owed to the creditor by the debtor. A creditor need not file a proof of claim if the debtor's schedules list the creditor's claim as liquidated (exactly determined) and the creditor does not dispute the amount of the claim. A proof of claim is necessary if there is any dispute concerning the claim. If a creditor fails to file a proof of claim, the bankruptcy court or trustee may file the proof of claim on the creditor's behalf but is not obligated to do so.

Generally, any legal obligation of the debtor is a claim (except claims for breach of employment contracts or real estate leases for terms longer than one year). When a claim is disputed, or unliquidated, the bankruptcy court will set the value of the claim. Any creditor holding a debtor's obligation can file a claim against the debtor's estate. These claims are automatically allowed unless contested by the trustee, the debtor, or another creditor. A creditor who files a false claim commits a crime.

EXEMPTIONS

The trustee takes control over the debtor's property, but an individual debtor is entitled to exempt certain property from the bankruptcy. The Bankruptcy Code exempts the following property:[15]

1. Up to $18,450 in equity in the debtor's residence and burial plot (the homestead exemption).
2. Interest in a motor vehicle up to $2,950.
3. Interest, up to $475 for a particular item, in household goods and furnishings, wearing apparel, appliances, books, animals, crops, and musical instruments (the aggregate total of all items is limited, however, to $9,850).
4. Interest in jewelry up to $1,225.
5. Interest in any other property up to $975, plus any unused part of the $18,450 homestead exemption up to $9,250.
6. Interest in any tools of the debtor's trade up to $1,850.
7. Any unmatured life insurance contract owned by the debtor.
8. Certain interests in accrued dividends and interest under life insurance contracts owned by the debtor, not to exceed $9,850.
9. Professionally prescribed health aids.
10. The right to receive Social Security and certain welfare benefits, alimony and support, certain retirement funds and pensions, and education savings accounts held for specific periods of time.
11. The right to receive certain personal-injury and other awards up to $18,450.

Individual states have the power to pass legislation precluding debtors from using the federal exemptions within the state; a majority of the states have done this. In those states, debtors may use only state, not federal, exemptions. In the rest of the states, an individual debtor (or a husband and wife filing jointly) may choose either the exemptions provided under state law or the federal exemptions.[16]

Note also that the 2005 Bankruptcy Reform Act clarified specifically what is included in "household goods and furnishings" (referred to in number 3 in the above list). For example, the category includes one computer, one radio, one television, one videocassette recorder, educational materials or equipment primarily for use by minor dependent children, and furniture that is used exclusively by a minor dependent (or by an elderly or disabled dependent). The category does not include other items—such as works of art; electronic entertainment equipment with a fair market value of over $500; antiques and jewelry (except wedding rings) valued at more than $500; and motor vehicles, tractors, lawn mowers, watercraft, and aircraft.

THE HOMESTEAD EXEMPTION

The 2005 Bankruptcy Reform Act significantly changed the law for those debtors seeking to use state homestead

14. This ninety-day rule applies in Chapter 12 and Chapter 13 bankruptcies as well.

15. The dollar amounts stated in the Bankruptcy Code are adjusted automatically every three years on April 1 based on changes in the Consumer Price Index. The adjusted amounts are rounded to the nearest $25. The amounts stated in this chapter are in accordance with those computed on April 1, 2004.

16. State exemptions may or may not be limited with regard to value. Under state exemption laws, a debtor may enjoy an unlimited value exemption on a motor vehicle, for example, even though the federal bankruptcy scheme exempts a vehicle only up to a value of $2,950. A state's law may also define the property coming within an exemption differently than the federal law or may exclude, or except, specific items from an exemption, making it unavailable to a debtor who fits within the exception.

exemption statutes. In six states, among them Florida and Texas, homestead exemptions allow debtors petitioning for bankruptcy to shield unlimited amounts of equity in their homes from creditors. The prior Bankruptcy Code required that the debtor must have been domiciled in the state for at least six months to apply any of the state exemptions. Under the 2005 act, however, the domicile period is now two years. In other words, the debtor must have lived in the state for two years prior to filing the petition to be able to use the state homestead exemption.

In addition, if the homestead was acquired within three and a half years preceding the date of filing, the maximum equity exempted is $125,000, even if the state law would permit a higher amount. (This does not apply to equity that has been rolled over during the specified period from the sale of a previous homestead in the same state.) Also, if the debtor owes a debt arising from a violation of securities law or if the debtor committed certain criminal or tortious acts in the previous five years that indicate the filing constitutes substantial abuse, the debtor may not exempt any amount of equity.[17]

THE TRUSTEE

Promptly after the order for relief in the liquidation proceeding has been entered, an interim, or provisional, trustee is appointed by the U.S. Trustee. The interim, or provisional, trustee presides over the debtor's property until the first meeting of creditors. At this first meeting, either a permanent trustee is elected, or the interim trustee becomes the permanent trustee.

The basic duty of the trustee is to collect the debtor's available estate and reduce it to cash for distribution, preserving the interests of both the debtor and unsecured creditors. This requires that the trustee be accountable for administering the debtor's estate. To enable the trustee to accomplish this duty, the Code gives the trustee certain powers, stated in both general and specific terms. These powers must be exercised within two years of the order for relief.

New Duties under the 2005 Act The Bankruptcy Reform Act of 2005 imposes new duties on trustees (and bankruptcy administrators) with regard to means-testing all debtors who file Chapter 7 petitions. Under the new law, the U.S. Trustee or bankruptcy administrator is required to promptly review all materials filed by the debtor. Not later than ten days after the first meeting of the creditors, the trustee must file a statement as to whether the case is presumed to be an abuse under the means test. The trustee must then provide a copy of this statement concerning abuse to all creditors within five days. Not later than forty days after the first creditors' meeting, the trustee must either file a motion to dismiss the petition (or convert it to a Chapter 13 case) or file a statement setting forth the reasons why the motion would not be appropriate.

Under the 2005 act, the trustee also has new duties designed to protect domestic-support creditors (those to whom a domestic-support obligation is owed). The trustee is required to provide written notice of the bankruptcy to the claim holder (a former spouse who is owed child support, for example). The notice must also include certain information, such as the debtor's address, the name and address of the debtor's last known employer, and the address and phone number of the state child-support enforcement agency. (Note that these requirements are not limited to Chapter 7 bankruptcies, and the trustee may have additional duties in other types of bankruptcy to collect assets for distribution to the domestic-support creditor.)

The Trustee's Powers The general powers of the trustee are described by the statement that the trustee occupies a position *equivalent* in rights to that of certain other parties. ■ **EXAMPLE 21.7** The trustee has the same rights as a *lien creditor* who could have obtained a judicial lien on the debtor's property or who could have levied execution on the debtor's property. This means that a trustee has priority over an unperfected secured party to the debtor's property. ■ This right of a trustee, equivalent to that of a lien creditor, is known as the *strong-arm power.* A trustee also has power equivalent to that of a *bona fide purchaser* of real property from the debtor.

Nevertheless, in most states a creditor with a purchase-money security interest may prevail against a trustee if the creditor files within ten days (twenty days, in many states) of the debtor's receipt of the collateral, even if the bankruptcy petition is filed before the creditor perfects. ■ **EXAMPLE 21.8** Baker loaned Newbury $20,000 on January 1, taking a security interest in the machinery that Newbury purchased with the $20,000 and that was delivered on that same date. On January 27, before Baker had perfected her security interest, Newbury filed for bankruptcy. The trustee can invalidate Baker's security interest because it was unperfected when Newbury filed the bankruptcy petition. Baker can assert a claim only as

17. Specifically, the debtor may not claim the homestead exemption if the debtor has committed any criminal act, intentional tort, or willful or reckless misconduct that caused serious physical injury or death to another individual in the preceding five years. Also, if the debtor has been convicted of a felony, he or she may not be able to claim the exemption.

an unsecured creditor. But if Newbury had filed for bankruptcy on January 7, and Baker had perfected her security interest on January 8, she would have prevailed, because she would have perfected her purchase-money security interest within ten days of Newbury's receipt of the machinery. ■

The Right to Possession of the Debtor's Property The trustee has the power to require persons holding the debtor's property at the time the petition is filed to deliver the property to the trustee. A trustee usually does not take actual possession of a debtor's property. Instead, a trustee's possession is constructive. ■ EXAMPLE 21.9 Suppose that a trustee needs to obtain control of a debtor's business inventory. The trustee might change the locks on the doors to the business and hire a security guard. ■

Avoidance Powers The trustee also has specific powers of *avoidance*—that is, the trustee can set aside a sale or other transfer of the debtor's property, taking it back as a part of the debtor's estate. These powers include any voidable rights available to the debtor, preferences, certain statutory liens, and fraudulent transfers by the debtor. Each of these powers is discussed in more detail below.

The debtor shares most of the trustee's avoidance powers. Thus, if the trustee does not take action to enforce one of the rights mentioned above, the debtor in a liquidation bankruptcy can nevertheless enforce that right.[18]

Note that under the 2005 act, the trustee no longer has the power to avoid any transfer that was a bona fide payment of a domestic-support debt.

Voidable Rights A trustee steps into the shoes of the debtor. Thus, any reason that a debtor can use to obtain the return of her or his property can be used by the trustee as well. These grounds include fraud, duress, incapacity, and mutual mistake.

■ EXAMPLE 21.10 Ben sells his boat to Tara. Tara gives Ben a check, knowing that she has insufficient funds in her bank account to cover the check. Tara has committed fraud. Ben has the right to avoid that transfer and recover the boat from Tara. Once an order for relief under Chapter 7 of the Code has been entered for Ben, the trustee can exercise the same right to recover the boat from Tara, and the boat becomes a part of the debtor's estate. ■

18. Under a Chapter 11 bankruptcy (to be discussed later), for which no trustee other than the debtor generally exists, the debtor has the same avoidance powers as a trustee under Chapter 7. Under Chapters 12 and 13 (also to be discussed later), a trustee must be appointed.

Preferences A debtor is not permitted to transfer property or to make a payment that favors—or gives a **preference** to—one creditor over others. The trustee is allowed to recover payments made both voluntarily and involuntarily to one creditor in preference over another. If a *preferred creditor* (one who has received a preferential transfer from the debtor) has sold the property to an innocent third party, the trustee cannot recover the property from the innocent party. The preferred creditor, however, generally can be held accountable for the value of the property.

To have made a preferential payment that can be recovered, an *insolvent* debtor generally must have transferred property, for a *preexisting* debt, within *ninety days* prior to the filing of the petition in bankruptcy. The transfer must give the creditor more than the creditor would have received as a result of the bankruptcy proceedings. The trustee need not prove insolvency, as the Code provides that the debtor is presumed to be insolvent during this ninety-day period.

—Preferences to Insiders. Sometimes, the creditor receiving the preference is an *insider*—an individual, a partner, a partnership, a corporation, or an officer or a director of a corporation (or a relative of one of these) who has a close relationship with the debtor. In this situation, the avoidance power of the trustee is extended to transfers made within *one year* before filing; however, the *presumption* of insolvency is confined to the ninety-day period. Therefore, the trustee must prove that the debtor was insolvent at the time of a transfer that occurred prior to the ninety-day period.

—Transfers That Do Not Constitute Preferences. Not all transfers are preferences. To be a preference, the transfer must be made for something other than current consideration. Most courts generally assume that payment for services rendered within ten to fifteen days prior to the payment of the current consideration is not a preference. If a creditor receives payment in the ordinary course of business from an individual or business debtor, such as payment of last month's telephone bill, the payment cannot be recovered by the trustee in bankruptcy. To be recoverable, a preference must be a transfer for an antecedent (preexisting) debt, such as a year-old printing bill. In addition, the Code permits a consumer-debtor to transfer any property to a creditor up to a total value of $5,000, without the transfer's constituting a preference (this amount was increased from $600 to $5,000 by the 2005 act). Payment of domestic-support debts does not constitute a preference. Also, transfers that were made as part of an alternative

repayment schedule negotiated by an approved credit counseling agency are not preferences.

Liens on Debtor's Property The trustee has the power to avoid certain statutory liens against the debtor's property, such as a landlord's lien for unpaid rent. The trustee can avoid statutory liens that first became effective against the debtor when the bankruptcy petition was filed or when the debtor became insolvent. The trustee can also avoid any lien against a bona fide purchaser that was not perfected or enforceable on the date of the bankruptcy filing. Under the 2005 act, the trustee cannot avoid certain warehouser's liens (see Chapter 28), however.

Fraudulent Transfers The trustee may avoid fraudulent transfers or obligations if they were made within two years of the filing of the petition or if they were made with actual intent to hinder, delay, or defraud a creditor. Transfers made for less than a reasonably equivalent consideration are also vulnerable if by making them, the debtor became insolvent, was left engaged in business with an unreasonably small amount of capital, or intended to incur debts that he or she could not pay. When a fraudulent transfer is made outside the Code's two-year limit, creditors may seek alternative relief under state laws. Some state laws allow creditors to recover for transfers made up to three years prior to the filing of a petition.

DISTRIBUTION OF PROPERTY

The Code provides specific rules for the distribution of the debtor's property to secured and unsecured creditors. (We will examine these distributions shortly.) If any amount remains after the priority classes of creditors have been satisfied, it is turned over to the debtor. Exhibit 21–2 illustrates graphically the collection and distribution of property in most voluntary bankruptcies.

In a bankruptcy case in which the debtor has no assets,[19] creditors are notified of the debtor's petition for bankruptcy but are instructed not to file a claim. In such a case, the unsecured creditors will receive no payment, and most, if not all, of these debts will be discharged.

Distribution to Secured Creditors The rights of perfected secured creditors were discussed in Chapter 20. The Code provides that a consumer-debtor, either within thirty days of filing a liquidation petition or before the date of the first meeting of the creditors (whichever is first), must file with the clerk a statement of intention with respect to the secured collateral. The statement must indicate whether the debtor will redeem the collateral (make a single payment equal to the current value of the property), reaffirm the debt (continue making payments on the debt), or sur-

19. This type of bankruptcy is called a "no-asset" case.

EXHIBIT 21–2 COLLECTION AND DISTRIBUTION OF PROPERTY IN MOST VOLUNTARY BANKRUPTCIES

This exhibit illustrates the property that might be collected in a debtor's voluntary bankruptcy and how it might be distributed to creditors. Involuntary bankruptcies and some voluntary bankruptcies could include additional types of property and other creditors.

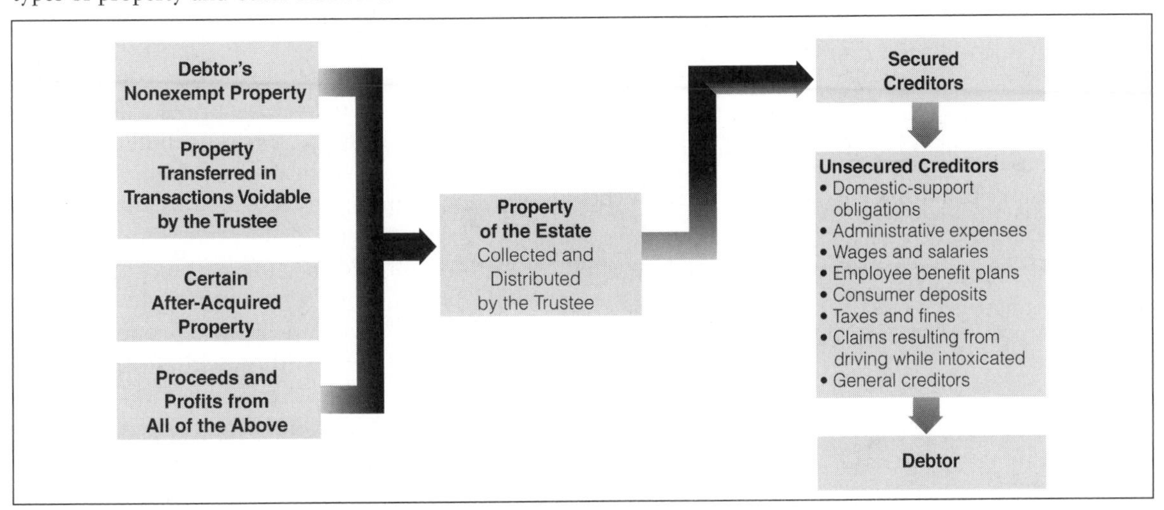

render the property to the secured party.[20] The trustee is obligated to enforce the debtor's statement within forty-five days after the meeting of the creditors. Failure of the debtor to redeem or reaffirm within forty-five days terminates the automatic stay.

If the collateral is surrendered to the perfected secured party, the secured creditor can enforce the security interest either by accepting the property in full satisfaction of the debt or by foreclosing on the collateral and using the proceeds to pay off the debt. Thus, the perfected secured party has priority over unsecured parties as to the proceeds from the disposition of the collateral. Indeed, the Code provides that if the value of the collateral exceeds the perfected secured party's claim and if the security agreement so provides, the secured party also has priority as to the proceeds in an amount that will cover reasonable fees and costs incurred because of the debtor's default. Fees include reasonable attorneys' fees. Any excess over this amount is used by the trustee to satisfy the claims of unsecured creditors. Should the collateral be insufficient to cover the secured debt owed, the secured creditor becomes an unsecured creditor for the difference.

Distribution to Unsecured Creditors Bankruptcy law establishes an order of priority for classes of debts owed to *unsecured* creditors, and they are paid in the order of their priority. Each class must be fully paid before the next class is entitled to any of the remaining proceeds. If the proceeds are insufficient to pay all the creditors in a class in full, the proceeds are distributed *proportionately* to the creditors in that class, and classes lower in priority receive nothing. The new bankruptcy law elevated domestic-support obligations to the highest priority of unsecured claims. The order of priority among classes of unsecured creditors is as follows:

1. Claims for domestic-support obligations, such as child support and alimony (subject to the priority of the administrative costs that the trustee incurred in administering assets to pay the obligations).
2. Administrative expenses including court costs, trustee fees, and attorneys' fees.
3. In an involuntary bankruptcy, expenses incurred by the debtor in the ordinary course of business from the date of the filing of the petition up to the appointment of the trustee or the court's issuance of an order for relief.
4. Unpaid wages, salaries, and commissions earned within ninety days prior to the filing of the petition, limited to $4,925 per claimant. Any claim in excess of $4,925 or earned before the ninety-day period is treated as a claim of a general creditor (listed as item 10 below).
5. Unsecured claims for contributions to be made to employee benefit plans, limited to services performed during the 180-day period prior to the filing of the bankruptcy petition and $4,925 per employee.
6. Claims by farmers and fishermen, up to $4,925, against debtor-operators of grain storage or fish storage or processing facilities.
7. Consumer deposits of up to $2,225 given to the debtor before the petition was filed in connection with the purchase, lease, or rental of property or purchase of services that were not received or provided. Any claim in excess of $2,225 is treated as a claim of a general creditor (listed as item 10 below).
8. Certain taxes and penalties due to government units, such as income and property taxes.
9. Claims for death or personal injury resulting from the operation of a motor vehicle or vessel if such operation was unlawful because the debtor was intoxicated as a result of using alcohol, a drug, or another substance. (This provision was added by the 2005 act.)
10. Claims of general creditors.

DISCHARGE

From the debtor's point of view, the primary purpose of liquidation is to obtain a fresh start through a discharge of debts.[21] As mentioned earlier, once the debtor's assets have been distributed to creditors as permitted by the Code, the debtor's remaining debts are then discharged, meaning that the debtor is not obligated to pay them. Certain debts, however, are not dischargeable in bankruptcy. Also, certain debtors may not qualify to have all debts discharged in bankruptcy. These situations are discussed below.

Exceptions to Discharge Discharge of a debt may be denied because of the nature of the claim or the conduct of the debtor. Claims that are not dischargeable in a liquidation bankruptcy include the following:

1. Claims for back taxes accruing within two years prior to bankruptcy.
2. Claims for amounts borrowed by the debtor to pay federal taxes or any nondischargeable taxes.
3. Claims against property or funds obtained by the debtor under false pretenses or by false representations.

20. Also, if applicable, the debtor must specify whether the collateral will be claimed as exempt property.

21. Discharges are granted under Chapter 7 only to individuals, not to corporations or partnerships. The latter may use Chapter 11, or they may terminate their existence under state law.

4. Claims by creditors who were not notified and did not know of the bankruptcy; these claims did not appear on the schedules the debtor was required to file.
5. Claims based on fraud or misuse of funds by the debtor while he or she was acting in a fiduciary capacity or claims involving the debtor's embezzlement or larceny.
6. Domestic-support obligations and property settlements as provided for in a separation agreement or divorce decree.
7. Claims for amounts due on a retirement account loan.
8. Claims based on willful or malicious conduct by the debtor toward another or the property of another.
9. Certain government fines and penalties, which under the 2005 act also include penalties imposed under federal election laws.
10. Certain student loans or obligations to repay funds received as an educational benefit, scholarship, or stipend—unless payment of the loans imposes an undue hardship on the debtor and the debtor's dependents.
11. Consumer debts of more than $500 for luxury goods or services owed to a single creditor incurred within ninety days of the order for relief. (Prior to the passage of the 2005 act, the amount was $1,150 and the period was sixty days.) This denial of discharge is a rebuttable presumption (that is, the denial may be challenged by the debtor), however, and any debts reasonably incurred to support the debtor or dependents are not classified as luxuries.
12. Cash advances totaling more than $750 that are extensions of open-end consumer credit obtained by the debtor within seventy days of the order for relief. (The prior law allowed $1,150 in cash advances that were obtained within sixty days.) A denial of discharge of these debts is also a rebuttable presumption.
13. Judgments or consent decrees against a debtor as a result of the debtor's operation of a motor vehicle or any vessel or aircraft while intoxicated.
14. Fees or assessments arising from a lot in a homeowners' association, as long as the debtor retained an interest in the lot.
15. Failure of the debtor to provide required or requested tax documents. (This exception to discharge also applies to Chapter 11 and Chapter 13 bankruptcies.)

In the following case, the court considered whether to order the discharge of a debtor's student loan obligations. Is it "undue hardship" if, to repay the loans, a debtor has to forgo her son's private school tuition?

CASE 21.2 In re Savage

United States Bankruptcy Appellate Panel,
First Circuit, 2004.
311 Bankr. 835.

FACTS Brenda Savage attended college in the mid-1980s—taking out five student loans—but she did not graduate. In 2003, at the age of forty-one, single, and in good health, she lived with her fifteen-year-old son in an apartment in Boston, Massachusetts. Her son attended Boston Trinity Academy, a private school. Savage worked 37.5 hours per week for Blue Cross/Blue Shield of Massachusetts. Her monthly gross wages were $3,079.79. Her employment provided health insurance, dental insurance, life insurance, a retirement savings plan, and paid vacations and personal days. She also received monthly child-support income of $180.60. After deductions, her total net monthly income was $2,030.72. Her monthly expenses included, among other things, $607 for rent, $221 for utilities, $76 for phone, $23.99 for an Internet connection, $430 for food, $75 for clothing, $12.50 for laundry and dry cleaning, $23 for medical expenses, $95.50 for transportation, $193.50 for charitable contributions, $43 for entertainment, $277.50 for her son's tuition, and $50 for his books. In February, Savage filed a petition in bankruptcy, seeking to discharge her student loan obligations to Educational Credit Management Corporation (ECMC). At the time, she owed $32,248.45. The court ordered a discharge of all but $3,120. ECMC appealed to the U.S. Bankruptcy Appellate Panel for the First Circuit.

ISSUE If a parent has to forgo private school tuition for her son in order to repay her student loan, does that constitute an "undue hardship" under bankruptcy law?

DECISION No. The U.S. Bankruptcy Appellate Panel for the First Circuit reversed the order of the bankruptcy court and remanded the case for the entry of a judgment in ECMC's favor.

REASON To obtain the discharge of a student loan in a bankruptcy proceeding, a debtor must show that "her current income is insufficient to pay" the loan and "her prospects for increasing her income in the future are too limited to afford her sufficient resources to repay the [loan] and provide herself and her dependents with a minimal (but fair) standard of living." In this case, Savage's "present income may be insufficient

CASE 21.2–CONTINUED

to pay her student loans and still maintain precisely the standard of living she now has. But * * * it would enable her to repay the loans without undue hardship. Moreover, * * * her prospects for a steady increase in income over time are promising." To prove undue hardship, a debtor must show that "her necessary and reasonable expenses leave her with too little to afford repayment." Private school tuition is not normally considered "a reasonably necessary expense." Here, Savage "has not demonstrated that the public school system cannot adequately meet her son's educational needs. * * * Given the fact that at least $322.50 (private school tuition and books) in expense can be eliminated from Ms. Savage's budget without creating undue hardship, her student loans cannot be discharged."

WHY IS THIS CASE IMPORTANT? *This case emphasizes the courts' reluctance to discharge student loan obligations in bankruptcy unless a debtor can show that repaying the loans would genuinely create an "undue hardship." Generally, a student loan will be discharged in bankruptcy only if the debtor's present and foreseeable financial circumstances indicate that it would be virtually impossible to pay the debt.*

Objections to Discharge In addition to the exceptions to discharge previously listed, a bankruptcy court may also deny the discharge of the *debtor* (as opposed to the debt). In the latter situation, the assets of the debtor are still distributed to the creditors, but the debtor remains liable for the unpaid portion of all claims. Grounds for the denial of discharge of the debtor include the following:

1. The debtor's concealment or destruction of property with the intent to hinder, delay, or defraud a creditor.
2. The debtor's fraudulent concealment or destruction of financial records.
3. The granting of a discharge to the debtor within eight years of the filing of the petition. (This period was increased from six to eight years by the 2005 act.)
4. Failure of the debtor to complete the required consumer education course (unless such a course is unavailable). (This ground for denial was provided by the 2005 act and also applies to Chapter 13 petitions.)
5. Proceedings in which the debtor could be found guilty of a felony (basically, the 2005 act states that a court may not discharge any debt until the completion of felony proceedings against the debtor).

Effect of Discharge The primary effect of a discharge is to void any judgment on a discharged debt and enjoin any action to collect a discharged debt. A discharge does not affect the liability of a co-debtor.

Revocation of Discharge On petition by the trustee or a creditor, the bankruptcy court can, within one year, revoke the discharge decree. The discharge decree will be revoked if it is discovered that the debtor acted fraudulently or dishonestly during the bankruptcy proceedings. The revocation renders the discharge void, allowing creditors not satisfied by the distribution of the debtor's estate to proceed with their claims against the debtor.

REAFFIRMATION OF DEBT

An agreement to pay a debt dischargeable in bankruptcy is called a *reaffirmation agreement*. A debtor may wish to pay a debt—such as, for example, a debt owed to a family member, physician, bank, or some other creditor—even though the debt could be discharged in bankruptcy. Also, as noted previously, under the new Code a debtor cannot retain secured property while continuing to pay without entering into a reaffirmation agreement.

To be enforceable, reaffirmation agreements must be made before the debtor is granted a discharge. The agreement must be signed and filed with the court (along with the original disclosure documents, as you will read shortly). Court approval is required unless the debtor is represented by an attorney during the negotiation of the reaffirmation and submits the proper documents and certifications. Even if the debtor is represented by an attorney, court approval may be required when it appears that the reaffirmation will result in undue hardship to the debtor. When court approval is required, a separate hearing will take place. The court will approve the reaffirmation only if it finds that the agreement will not result in undue hardship for the debtor and that the reaffirmation is consistent with the debtor's best interests.

Presumption of Undue Hardship Under the provisions of the 2005 act, if the debtor's monthly income minus the debtor's monthly expenses as shown on her or his completed and signed statement is less than the scheduled payments on the reaffirmed debt, undue hardship will be presumed. A presumption of undue hardship can be rebutted, however. The debtor can file a written statement with the court that includes an explanation identifying additional sources of funds from which to make the agreed-on payments. If the court is not satisfied with the written

explanation, it may disapprove of the reaffirmation or hold a hearing. The debtor may also rebut the presumption of undue hardship by explaining to the court in person at the hearing how she or he will be able to make future payments on the debt.

If the debtor has an attorney, the attorney must certify in writing that he or she has fully advised the debtor of the legal effect and consequences of reaffirmation. In addition, to rebut the presumption of undue hardship, the attorney must certify that, in the attorney's opinion, the debtor is able to make the payments.

New Reaffirmation Disclosures To discourage creditors from engaging in abusive reaffirmation practices, the 2005 act added new requirements for reaffirmation. The Code now provides the specific language for several pages of disclosures that must be given to debtors entering reaffirmation agreements.[22] Among other things, these disclosures explain that the debtor is not required to reaffirm any debt, but that liens on secured property, such as mortgages and cars, will remain in effect even if the debt is not reaffirmed. The reaffirmation agreement must disclose the amount of the debt reaffirmed, the rates of interest, the date payments begin, and the right to rescind. The disclosures also caution the debtor, "Only agree to reaffirm a debt if it is in your best interest. Be sure you can afford the payments you agree to make." The original disclosure documents must be signed by the debtor, certified by the debtor's attorney, and filed with the court at the same time as the reaffirmation agreement. A reaffirmation agreement that is not accompanied by the original signed disclosures will not be effective.

If the debtor is represented by an attorney and no presumption of undue hardship arises, then the reaffirmation becomes effective immediately on filing with the court. If the debtor is not represented, the reaffirmation is not effective until the court approves it. The debtor can rescind, or cancel, the agreement at any time before the court enters a discharge order, or within sixty days of the filing of the agreement, whichever is *later.*

Chapter 11—Reorganization

The type of bankruptcy proceeding most commonly used by corporate debtors is the Chapter 11 *reorganization.* In a reorganization, the creditors and the debtor formulate a plan under which the debtor pays a portion of the debts and is discharged of the remainder. The debtor is allowed to continue in business. Although this type of bankruptcy is generally a corporate reorganization, any debtors (including individuals but excluding stockbrokers and commodities brokers)[23] who are eligible for Chapter 7 relief are eligible for relief under Chapter 11.[24] In 1994, Congress established a "fast-track" Chapter 11 procedure for small-business debtors whose liabilities do not exceed $2 million and who do not own or manage real estate. This allows for bankruptcy proceedings without the appointment of committees and can save time and costs.

The same principles that govern the filing of a liquidation (Chapter 7) petition apply to reorganization (Chapter 11) proceedings. The case may be brought either voluntarily or involuntarily. The same guidelines govern the entry of the order for relief. The automatic-stay and adequate protection provisions are applicable in reorganizations as well. The 2005 Bankruptcy Reform Act's exceptions to the automatic stay also apply to Chapter 11 proceedings, as do the new provisions regarding substantial abuse and additional grounds for dismissal (or conversion) of bankruptcy petitions. Also, the 2005 act contains specific rules and limitations for individual debtors who file a Chapter 11 petition. For example, an individual debtor's postpetition acquisitions and earnings become the property of the bankruptcy estate.

MUST BE IN THE BEST INTERESTS OF THE CREDITORS

Under Section 305(a) of the Bankruptcy Code, a court, after notice and a hearing, may dismiss or suspend all proceedings in a case at any time if dismissal or suspension would better serve the interests of the creditors. Section 1112 also allows a court, after notice and a hearing, to dismiss a case under reorganization "for cause." Cause includes the absence of a reasonable likelihood of rehabilitation, the inability to effect a plan, and an unreasonable delay by the debtor that is prejudicial to (may harm the interests of) creditors.[25]

WORKOUTS

In some instances, creditors may prefer private, negotiated adjustments of creditor-debtor relations, also known as **workouts**, to bankruptcy proceedings. Often, these out-

22. Note that credit unions are exempted from these disclosure requirements.

23. In *Toibb v. Radloff,* 501 U.S. 157, 111 S.Ct. 2197, 115 L.Ed.2d 145 (1991), the United States Supreme Court ruled that a nonbusiness debtor may petition for relief under Chapter 11.

24. In addition, railroads are eligible for Chapter 11 relief.

25. See 11 U.S.C. Section 1112(b). Debtors are not prohibited from filing successive petitions, however. A debtor whose petition is dismissed, for example, can file a new Chapter 11 petition (which may be granted unless it is filed in bad faith).

of-court workouts are much more flexible and thus more conducive to a speedy settlement. Speed is critical because delay is one of the most costly elements in any bankruptcy proceeding. Another advantage of workouts is that they avoid the various administrative costs of bankruptcy proceedings.

DEBTOR IN POSSESSION

On entry of the order for relief, the debtor generally continues to operate the business as a **debtor in possession (DIP)**. The court, however, may appoint a trustee (often referred to as a *receiver*) to operate the debtor's business if gross mismanagement of the business is shown or if appointing a trustee is in the best interests of the estate.

The DIP's role is similar to that of a trustee in a liquidation. The DIP is entitled to avoid prepetition preferential payments made to creditors and prepetition fraudulent transfers of assets. The DIP has the power to decide whether to cancel or assume prepetition executory contracts (those that are not yet performed) or unexpired leases.

Under the strong-arm clause[26] of the Bankruptcy Code, a DIP can avoid any obligation or any transfer of property of the debtor that could be avoided by certain parties. These parties include (1) a creditor who extended credit to the debtor at the time of bankruptcy (petition) and who consequently obtained a lien on the debtor's property; (2) a creditor who extended credit to the debtor at the time of bankruptcy and who consequently obtained a writ of execution against the debtor that was returned unsatisfied; and (3) a bona fide purchaser of real property from the debtor if, at the time of the bankruptcy, the transfer was perfected.

COLLECTIVE BARGAINING AGREEMENTS

After the Bankruptcy Reform Act of 1978 was enacted, questions arose as to whether a reorganization debtor could reject a recently negotiated collectively bargained labor contract. In *National Labor Relations Board v. Bildisco and Bildisco*,[27] the United States Supreme Court held that a collective bargaining agreement subject to the National Labor Relations Act of 1935 (see Chapter 23) is an "executory contract" and thus is subject to *rejection* by a debtor in possession. The Court emphasized, though, that such a rejection should not be permitted unless there is a finding that the policy of Chapter 11 (successful rehabilitation of debtors) would be served by the action. Hence, when the bankruptcy court determines that rejection of a collective bargaining agreement should be permitted, it must make a reasoned finding *on the record* as to why it has determined that the rejection should be permitted.

The Code attempts to reconcile federal policies favoring collective bargaining with the need to allow a debtor company to reject executory labor contracts while trying to reorganize. The Code sets forth standards and procedures under which collective bargaining contracts can be assumed or rejected under a reorganization filing. In general, a collective bargaining contract can be rejected if the debtor has first proposed necessary contractual modifications to the union and the union has failed to adopt them without *good cause*. The company is required (1) to provide the union with the relevant information needed to evaluate this proposal and (2) to confer in *good faith* in attempting to reach a mutually satisfactory agreement on the modifications.

CREDITORS' COMMITTEES

As soon as practicable after the entry of the order for relief, a creditors' committee of unsecured creditors is appointed. If the debtor has filed a plan accepted by the creditors, however, the trustee may decide not to call a meeting of the creditors. The committee may consult with the trustee or the DIP concerning the administration of the case or the formulation of the plan. Additional creditors' committees may be appointed to represent special interest creditors. Under the 2005 act, a court may order the trustee to change the membership of a committee or to increase the number of committee members to include a small-business concern if the court deems it necessary to ensure adequate representation of the creditors.

Orders affecting the estate generally will be entered only with the consent of the committee or after a hearing in which the judge is informed of the position of the committee. As mentioned earlier, businesses with debts of less than $2 million that do not own or manage real estate can avoid creditors' committees. In these cases, orders can be entered without a committee's consent.

THE REORGANIZATION PLAN

A reorganization plan to rehabilitate the debtor is a plan to conserve and administer the debtor's assets in the hope of an eventual return to successful operation and solvency.

Filing the Plan Only the debtor may file a plan within the first 120 days after the date of the order for relief. Under the 2005 act, the 120-day period may be extended but not beyond 18 months from the date of the order for relief. If the debtor does not meet the 120-day deadline or

26. 11 U.S.C. Section 544(a).

27. 465 U.S. 513, 104 S.Ct. 1188, 79 L.Ed.2d 482 (1984).

obtain an extension, and if the debtor fails to procure the required creditor consent (discussed below) within 180 days, any party may propose a plan up to 20 months from the date of the order for relief. (In other words, the 180-day period cannot be extended beyond 20 months past the date of the order for relief.) For a small-business debtor, the time for the debtor's filing is 180 days.

The plan must be fair and equitable and must do the following:

1. Designate classes of claims and interests.
2. Specify the treatment to be afforded the classes. (The plan must provide the same treatment for all claims in a particular class.)
3. Provide an adequate means for execution. (The 2005 Bankruptcy Reform Act requires individual debtors to utilize postpetition assets as necessary to execute the plan.)
4. Provide for payment of tax claims over a five-year period.

Acceptance and Confirmation of the Plan Once the plan has been developed, it is submitted to each class of creditors for acceptance. Each class must accept the plan unless the class is not adversely affected by it. A class has accepted the plan when a majority of the creditors, representing two-thirds of the amount of the total claim, vote to approve it. Confirmation is conditioned on the debtor certifying that all postpetition domestic-support obligations have been paid in full. For small-business debtors, if the plan meets the listed requirements, the court must confirm the plan within forty-five days (unless this period is extended).

Even when all classes of creditors accept the plan, the court may refuse to confirm it if it is not "in the best interests of the creditors."[28] A former spouse or child of the debtor can block the plan if it does not provide for payment of her or his claims in cash. Under the 2005 act, if an unsecured creditor objects to the plan, specific rules apply to the value of property to be distributed under the plan. The plan can also be modified on the request of the debtor, trustee, U.S. Trustee, or holder of the unsecured claim. Tax claims must be paid over a five-year period.

Even if only one class of creditors has accepted the plan, the court may still confirm the plan under the Code's so-called **cram-down provision.** In other words, the court may confirm the plan over the objections of a class of creditors. Before the court can exercise this right of cram-down confirmation, it must be demonstrated that the plan does not discriminate unfairly against any creditors and that the plan is fair and equitable.

Chapter 11 allows debtors considerable freedom to do business. But this freedom is not without limits, as the following case illustrates.

28. The plan need not provide for full repayment to unsecured creditors. Instead, creditors receive a percentage of each dollar owed to them by the debtor.

CASE 21.3 In re Beyond.com Corp.

United States Bankruptcy Court,
Northern District of California, 2003.
289 Bankr. 138.

FACTS In 2002, Beyond.com Corporation filed a Chapter 11 petition and a reorganization plan in a federal bankruptcy court. The company also filed a disclosure statement, which set out the details underlying the plan. Among other things, the plan envisioned that the reorganized debtor would "retain all of the rights, powers, and duties of a trustee under the Bankruptcy Code." The plan appointed the debtor's former chief operating officer, John Barratt, to the position of "Liquidation Manager." In this capacity, Barratt could dispose of the debtor's property, enter into agreements on the debtor's behalf, file suits against unidentified defendants, and retain and pay advisers and other "professionals." Under most circumstances, none of these actions would be subject to court supervision or limitation. The plan also limited Barratt's personal liability for acts performed in this capacity. The debtor asked the court to approve the disclosure statement, as required before a plan is submitted to creditors.

ISSUE Did the plan comply with the notice requirements and court supervision provisions contained in Chapter 11 of the Bankruptcy Code?

DECISION No. The bankruptcy court refused to confirm Beyond.com's disclosure statement and reorganization plan.

REASON The court stated, "Conceptually, Beyond.com's plan and disclosure statement is as freewheeling with the Bankruptcy Code * * * as Enron's accountants were with the tax laws in the 1990s." The court found that "Beyond.com's proposed plan contains numerous provisions that modify the requirements of the Bankruptcy Code," including "provisions that dramatically reduce notice to creditors of matters that the drafters of the Bankruptcy Code * * * considered

CASE 21.3–CONTINUED

fundamental." These modifications to the Code, in the court's view, "are not minor, ministerial, or simply pragmatic. In effect, the plan affords the reorganized debtor the prerogative to comply selectively with the provisions of the Bankruptcy Code * * * without judicial supervision." The court concluded that these defects could not be cured and thus the plan did not comply with the notice requirements in the Code.

FOR CRITICAL ANALYSIS—Social Consideration *How much information should be revealed in a disclosure statement accompanying a reorganization plan?*

Discharge The plan is binding on confirmation; under the Bankruptcy Reform Act of 2005, however, confirmation of a plan does not discharge an individual debtor. For individual debtors, plan completion is required prior to discharge, unless the court orders otherwise. For all other debtors, the court may order discharge at any time after the plan is confirmed. The debtor is given a reorganization discharge from all claims not protected under the plan. This discharge does not apply to any claims that would be denied discharge under liquidation.

Bankruptcy Relief under Chapter 13 and Chapter 12

In addition to bankruptcy relief through liquidation and reorganization, the Code also provides for individuals' repayment plans (Chapter 13) and family-farmer and family-fishermen debt adjustments (Chapter 12). As noted previously, the 2005 Bankruptcy Reform Act includes provisions for converting Chapter 7 bankruptcies into Chapter 13 repayment plans. It is therefore likely that there will be an increase in Chapter 13 bankruptcies as a result, because those debtors who have some ability to pay their obligations will file under Chapter 13.

Individuals' Repayment Plan

Chapter 13 of the Bankruptcy Code provides for "Adjustment of Debts of an Individual with Regular Income." Individuals (not partnerships or corporations) with regular income who owe fixed unsecured debts of less than $307,675 or fixed secured debts of less than $922,975 may take advantage of bankruptcy repayment plans. Among those eligible are salaried employees; sole proprietors; and individuals who live on welfare, Social Security, fixed pensions, or investment income. Many small-business debtors have a choice of filing a plan for reorganization or for repayment. Repayment plans offer several advantages, however. One benefit is that they are less expensive and less complicated than reorganization proceedings or, for that matter, even liquidation proceedings.

Filing the Petition A repayment plan case can be initiated only by the filing of a voluntary petition by the debtor or by the conversion of a Chapter 7 petition (because of a finding of substantial abuse under the means test, for example). Certain liquidation and reorganization cases may be converted to repayment plan cases with the consent of the debtor.[29] A trustee, who will make payments under the plan, must be appointed. On the filing of a repayment plan petition, the automatic stay previously discussed takes effect. Although the stay applies to all or part of the debtor's consumer debt, it does not apply to any business debt incurred by the debtor. The automatic stay also does not apply to domestic-support obligations.

The Repayment Plan A plan of rehabilitation by repayment must provide for the following:

1. The turnover to the trustee of such future earnings or income of the debtor as is necessary for execution of the plan.
2. Full payment in deferred cash payments of all claims entitled to priority.[30]
3. Identical treatment of all claims within a particular class. (The Code permits the debtor to list codebtors, such as guarantors or sureties, as a separate class.)

—Filing the Plan. Only the debtor may file for a repayment plan. This plan may provide either for payment of all obligations in full or for payment of a lesser amount.[31] Prior to the 2005 act, the time for repayment was usually three years unless the court approved an

29. A Chapter 13 case may be converted to a Chapter 7 case either at the request of the debtor or, under certain circumstances, "for cause" by a creditor. A Chapter 13 case may be converted to a Chapter 11 case after a hearing.

30. As with a Chapter 11 reorganization plan, full repayment of all claims is not always required.

31. Under the 2005 act, a plan under Chapter 13 or Chapter 12 (to be discussed shortly) might propose to pay less than 100 percent of prepetition domestic-support obligations that had been assigned, but only if disposable income is dedicated to a five-year plan. Disposable income is also redefined to exclude the amounts reasonably necessary to pay current domestic-support obligations.

extension for up to five years. Under the new Code, the length of the payment plan (three or five years) is determined by the debtor's median family income. If the debtor's family income is greater than the state median family income under the means test (previously discussed), the proposed plan must be for five years.[32] The term may not exceed five years, however.

The Code requires the debtor to make "timely" payments from the debtor's disposable income, and the trustee must ensure that the debtor commences these payments. The plan cannot materially alter terms of repayment on a retirement loan account, however. These payment amounts must take into consideration the scheduled payments to lessors of personal property and must provide adequate protection to secured creditors of personal property. Proof of adequate insurance on personal property is required. The debtor must begin making payments under the proposed plan within thirty days after the plan has been *filed*.

If the plan has not been confirmed, the trustee is instructed to retain the payments until the plan is confirmed and then distribute them accordingly. If the plan is denied, the trustee will return the payments to the debtor less any costs. Failure of the debtor to make timely payments or to commence payments within the thirty-day period will allow the court to convert the case to a liquidation bankruptcy or to dismiss the petition.

—Confirmation of the Plan. After the plan is filed, the court holds a confirmation hearing, at which interested parties (such as creditors) may object to the plan. Under the 2005 act, the hearing must be held at least twenty days, but no more than forty-five days, after the meeting of the creditors. Confirmation of the plan is dependent on the debtor's certification that postpetition domestic-support obligations have been paid in full and that all prepetition tax returns have been filed. The court will confirm a plan with respect to each claim of a secured creditor under any of the following circumstances:

1. If the secured creditors have accepted the plan.
2. If the plan provides that secured creditors retain their liens until there is payment in full or until the debtor receives a discharge.
3. If the debtor surrenders the property securing the claims to the creditors.

In addition, for confirmation, the plan must provide that a creditor with a purchase-money security interest (PMSI—see Chapter 20) retains its lien until payment of the entire debt for a motor vehicle purchased within 910 days before filing the petition. For PMSIs on other personal property, the payment plan must cover debts incurred within a one-year period preceding the filing.

—Objection to the Plan. Unsecured creditors do not have the power to confirm a repayment plan, but they can object to it. The court can approve a plan over the objection of the trustee or any unsecured creditor only in either of the following situations:

1. When the value of the property (replacement value as of the date of filing) to be distributed under the plan is at least equal to the amount of the claims.
2. When all of the debtor's projected disposable income to be received during the plan period will be applied to making payments. Disposable income is all income received less amounts needed to pay domestic-support obligations and/or amounts needed to meet ordinary expenses to continue the operation of a business. The 2005 act also excludes from disposable income charitable contributions up to 15 percent of the debtor's gross income and the reasonable and necessary costs for health insurance for the debtor and his or her dependents.

—Modification of the Plan. Prior to completion of payments, the plan may be modified at the request of the debtor, the trustee, or an unsecured creditor. If any interested party objects to the modification, the court must hold a hearing to determine whether the modified plan will be approved.

Discharge After completion of all payments, the court grants a discharge of all debts provided for by the repayment plan. Except for allowed claims not provided for by the plan, certain long-term debts provided for by the plan, certain tax claims, payments on retirement accounts, and claims for domestic-support obligations, all other debts are dischargeable. Under prior law, a discharge of debts under a Chapter 13 repayment plan was sometimes referred to as a "superdischarge" because it allowed the discharge of fraudulently incurred debt and claims resulting from malicious or willful injury.

The 2005 Bankruptcy Reform Act, however, eliminated most of the "superdischarge" provisions, especially for debts based on fraud. Today, debts for trust fund taxes, taxes for which returns were never filed or filed late (within two years of filing), domestic-support payments, student loans, and injury or property damage from driving under the influence of alcohol or drugs are nondis-

32. See 11 U.S.C. Section 1322(d) for details on when the court will find that the Chapter 13 plan should extend to a five-year period.

chargeable. The new law also excludes fraudulent tax obligations, criminal fines and restitution, fraud by a person acting in a fiduciary capacity, and restitution for willfully and maliciously causing personal injury or death.

Even if the debtor does not complete the plan, a hardship discharge may be granted if failure to complete the plan was due to circumstances beyond the debtor's control and if the value of the property distributed under the plan was greater than what would have been paid in a liquidation. A discharge can be revoked within one year if it was obtained by fraud.

FAMILY FARMERS AND FISHERMEN

In 1986, to help relieve economic pressure on small farmers, Congress created Chapter 12 of the Bankruptcy Code. In 2005, Congress extended this protection to family fishermen,[33] modified its provisions somewhat, and made it a permanent chapter in the Bankruptcy Code (previously the statutes authorizing Chapter 12 had to be periodically renewed by Congress).

Definitions For purposes of Chapter 12, a *family farmer* is one whose gross income is at least 50 percent farm dependent and whose debts are at least 50 percent farm related. The total debt for a family farmer must not exceed $3.237 million. (Prior law required a farmer's debts to be 80 percent farm related and not to exceed $1.5 million.) A partnership or closely held corporation (at least 50 percent owned by the farm family) can also qualify as a family farmer.

A *family fisherman* is defined by the 2005 act as one whose gross income is at least 50 percent dependent on commercial fishing operations[34] and whose debts are at least 80 percent related to commercial fishing. The total debt for a family fisherman must not exceed $1.5 million. As with family farmers, a partnership or closely held corporation can also qualify.

33. Although the Code uses the terms *fishermen* and *fisherman,* Chapter 12 provisions apply equally to men and women.

34. Commercial fishing operations include catching, harvesting, or aquaculture raising fish, shrimp, lobsters, urchins, seaweed, shellfish, or other aquatic species or products.

Filing the Petition The procedure for filing a family-farmer or family-fishermen bankruptcy plan is very similar to the procedure for filing a repayment plan under Chapter 13. The debtor must file a plan not later than ninety days after the order for relief. The filing of the petition acts as an automatic stay against creditors' and co-obligors' actions against the estate.

A farmer or fisherman who has already filed a reorganization or repayment plan may convert it to a Chapter 12 plan. The debtor may also convert a Chapter 12 plan to a liquidation plan.

Content and Confirmation of the Chapter 12 Plan The content of a plan under Chapter 12 is basically the same as that of a Chapter 13 repayment plan. The plan can be modified by the debtor but, except for cause, must be confirmed or denied within forty-five days of filing.

Court confirmation of the plan is the same as for a repayment plan. In summary, the plan must provide for payment of secured debts at the value of the collateral. If the secured debt exceeds the value of the collateral, the remaining debt is unsecured. For unsecured debtors, the plan must be confirmed if either the value of the property to be distributed under the plan equals the amount of the claim or the plan provides that all of the debtor's disposable income to be received in a three-year period (or longer, by court approval) will be applied to making payments. Disposable income is all income received less amounts needed to support the farmer or fisherman and his or her family and to continue the farming or commercial fishing operation. Completion of payments under the plan discharges all debts provided for by the plan.

TERMS AND CONCEPTS

artisan's lien 411
attachment 412
automatic stay 419
consumer-debtor 417
co-surety 415
cram-down provision 430
creditors' composition agreement 413
debtor in possession (DIP) 429
discharge 417
garnishment 412
guarantor 414
homestead exemption 416
innkeeper's lien 411
liquidation 417
mechanic's lien 411
mortgagee 413
mortgagor 413
order for relief 419
petition in bankruptcy 417
preference 423
right of contribution 415
right of reimbursement 415
right of subrogation 415

surety 413
suretyship 413
trustee 417
U.S. trustee 418
workout 428
writ of attachment 412
writ of execution 412

CHAPTER SUMMARY Creditors' Rights and Bankruptcy

REMEDIES AVAILABLE TO CREDITORS	
Liens (See pages 410–412.)	1. *Mechanic's liens*—A nonpossessory, filed lien on an owner's real estate for labor, services, or materials furnished to or made on the realty. 2. *Artisan's liens*—A possessory lien on an owner's personal property for labor performed or value added. 3. *Innkeeper's liens*—A possessory lien on a hotel guest's baggage for hotel charges that remain unpaid. 4. *Judicial liens*— a. Attachment—A court-ordered seizure of property prior to a court's final determination of the creditor's rights to the property. Attachment is available only on the creditor's posting of a bond and strict compliance with the applicable state statutes. b. Writ of execution—A court order directing the sheriff to seize (levy) and sell a debtor's nonexempt real or personal property to satisfy a court's judgment in the creditor's favor.
Garnishment (See page 412.)	A collection remedy that allows the creditor to attach a debtor's funds (such as wages owed or bank accounts) and property that are held by a third person.
Creditors' Composition Agreement (See pages 412–413.)	A contract between a debtor and his or her creditors by which the debtor's debts are discharged by payment of a sum less than the amount that is actually owed.
Mortgage Foreclosure (See page 413.)	On the debtor's default, the entire mortgage debt is due and payable, allowing the creditor to foreclose on the realty by selling it to satisfy the debt.
Suretyship and Guaranty (See pages 413–415.)	Under contract, a third person agrees to be primarily or secondarily liable for the debt owed by the principal debtor. A creditor can turn to this third person for satisfaction of the debt.
LAWS ASSISTING DEBTORS	
Exemptions (See page 416.)	Numerous laws assist debtors. Additionally, state laws exempt certain types of real and personal property from levy of execution or attachment. 1. *Real property*—Each state permits a debtor to retain the family home, either in its entirety or up to a specified dollar amount, free from the claims of unsecured creditors or trustees in bankruptcy (homestead exemption). 2. *Personal property*—Personal property that is most often exempt from satisfaction of judgment debts includes the following: a. Household furniture up to a specified dollar amount. b. Clothing and certain personal possessions. c. Transportation vehicles up to a specified dollar amount. d. Certain classified animals, such as livestock and pets. e. Equipment used in a business or trade up to a specified dollar amount.

CHAPTER SUMMARY Creditors' Rights and Bankruptcy—Continued

BANKRUPTCY—A COMPARISON OF CHAPTERS 7, 11, 12, AND 13

Issue	Chapter 7	Chapter 11	Chapters 12 and 13
Purpose	Liquidation.	Reorganization.	Adjustment.
Who Can Petition	Debtor (voluntary) or creditors (involuntary).	Debtor (voluntary) or creditors (involuntary).	Debtor (voluntary) only.
Who Can Be a Debtor	Any "person" (including partnerships, corporations, and municipalities) except railroads, insurance companies, banks, savings and loan institutions, investment companies licensed by the Small Business Administration, and credit unions. Farmers and charitable institutions also cannot be involuntarily petitioned. If the court finds the petition to be a substantial abuse of the use of Chapter 7, the debtor may be required to convert to a Chapter 13 repayment plan.	Any debtor eligible for Chapter 7 relief; railroads are also eligible. Individuals have specific rules and limitations.	*Chapter 12*—Any family farmer (one whose gross income is at least 50 percent farm dependent and whose debts are at least 50 percent farm related) or family fisherman (one whose gross income is at least 50 percent dependent on commercial fishing operations and whose debts are at least 80 percent related to commercial fishing) or any partnership or closely held corporation at least 50 percent owned by a family farmer or fisherman, when total debt does not exceed a specified amount ($3.237 million for farmers and $1.5 million for fishermen). *Chapter 13*—Any individual (not partnerships or corporations) with regular income who owes fixed unsecured debts of less than $307,675 or fixed secured debts of less than $922,975.
Procedure Leading to Discharge	Nonexempt property is sold with proceeds to be distributed (in order) to priority groups. Dischargeable debts are terminated.	Plan is submitted; if it is approved and followed, debts are discharged.	Plan is submitted and must be approved if the value of the property to be distributed equals the amount of the claims or if the debtor turns over disposable income for a three-year or five-year period; if the plan is followed, debts are discharged.
Advantages	On liquidation and distribution, most debts are discharged, and the debtor has an opportunity for a fresh start.	Debtor continues in business. Creditors can either accept the plan, or it can be "crammed down" on them. The plan allows for the reorganization and liquidation of debts over the plan period.	Debtor continues in business or possession of assets. If the plan is approved, most debts are discharged after the plan period.

FOR REVIEW

Answers for the even-numbered questions in this For Review *section can be found in Appendix A at the end of this text.*

1. What is a prejudgment attachment? What is a writ of execution? How does a creditor use these remedies?
2. What is garnishment? When might a creditor undertake a garnishment proceeding?
3. In a bankruptcy proceeding, what constitutes the debtor's estate in property? What property is exempt from the estate under federal bankruptcy law?
4. What is the difference between an exception to discharge and an objection to discharge?
5. In a Chapter 11 reorganization, what is the role of the debtor in possession?

QUESTIONS AND CASE PROBLEMS

21–1. Artisan's Lien. Air Ruidoso, Ltd., operated a commuter airline and air charter service between Ruidoso, New Mexico, and airports in Albuquerque and El Paso. Executive Aviation Center, Inc., provided services for airlines at the Albuquerque International Airport. When Air Ruidoso failed to pay more than $10,000 that it owed for fuel, oil, and oxygen, Executive Aviation took possession of Air Ruidoso's plane. Executive Aviation claimed that it had a lien on the plane and filed a suit in a New Mexico state court to foreclose. Do supplies such as fuel, oil, and oxygen qualify as "materials" for the purpose of creating an artisan's lien? Why or why not?

QUESTION WITH SAMPLE ANSWER

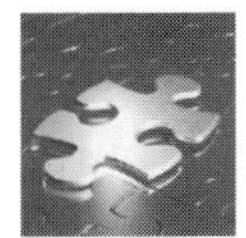

21–2. Meredith, a farmer, borrowed $5,000 from Farmer's Bank and gave the bank $4,000 in bearer bonds to hold as collateral for the loan. Meredith's neighbor, Peterson, who had known Meredith for years, signed as a surety on the note. Because of a drought, Meredith's harvest that year was only a fraction of the normal amount, and he was forced to default on his payments to Farmer's Bank. The bank did not immediately sell the bonds but instead requested $5,000 from Peterson. Peterson paid the $5,000 and then demanded that the bank give him the $4,000 in securities. Can Peterson enforce this demand? Explain.

For a sample answer to this question, go to Appendix B at the end of this text.

21–3. Liens. Sylvia takes her car to Caleb's Auto Repair Shop. A sign in the window states that all repairs must be paid for in cash unless credit is approved in advance. Sylvia and Caleb agree that Caleb will repair Sylvia's car engine and put in a new transmission. No mention is made of credit. Because Caleb is not sure how much engine repair will be necessary, he refuses to give Sylvia an estimate. He repairs the engine and puts in a new transmission. When Sylvia comes to pick up her car, she learns that the bill is $995. Sylvia is furious, refuses to pay Caleb that amount, and demands possession of her car. Caleb demands payment. Discuss the rights of the parties in this matter.

21–4. Guaranty. In 1988, Jamieson-Chippewa Investment Co. entered into a five-year commercial lease with TDM Pharmacy, Inc., for certain premises in Ellisville, Missouri, on which TDM intended to operate a small drugstore. Dennis and Tereasa McClintock ran the pharmacy business. The lease granted TDM three additional five-year options to renew. The lease was signed by TDM and by the McClintocks individually as guarantors. The lease did not state that the guaranty was continuing. In fact, there were no words of guaranty in the lease other than the single word *Guarantors* on the signature page. In 1993, Dennis McClintock, acting as the president of TDM, exercised TDM's option to renew the lease for one term. Three years later, when the pharmacy failed, TDM defaulted on the lease. Jamieson-Chippewa filed a suit in a Missouri state court against the McClintocks for the rent for the rest of the term, based on their guaranty. The McClintocks filed a motion for summary judgment, contending that they had not guaranteed any rent payments beyond the initial five-year term. How should the court rule? Why? [*Jamieson-Chippewa Investment Co. v. McClintock*, 996 S.W.2d 84 (Mo.App. E.D. 1999)]

21–5. Right of Subrogation. Levinson and Johnson, who had both signed a promissory note, did not pay the note when it was due. Instead, American Thermex, Inc., a corporation in which Johnson had a controlling interest, voluntarily paid the note. American Thermex later brought suit against Levinson, seeking reimbursement for the payment. American Thermex argued, among other things, that because it had paid the note, it had the legal right of subrogation against the note's co-maker, Levinson. Will the court agree that American Thermex has a legal right of subrogation? Why or why not? [*Levinson v. American Thermex, Inc.*, 196 Ga.App. 291, 396 S.E.2d 252 (1990)]

CASE PROBLEM WITH SAMPLE ANSWER

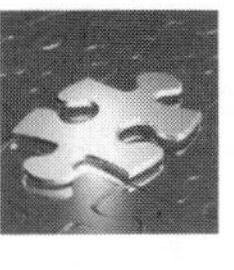

21–6. Between 1980 and 1987, Craig Hanson borrowed funds from Great Lakes Higher Education Corp. to finance his education at the University of Wisconsin. Hanson defaulted on the debt in 1989,

and Great Lakes obtained a judgment against him for $31,583.77. Three years later, Hanson filed a bankruptcy petition under Chapter 13. Great Lakes timely filed a proof of claim in the amount of $35,531.08. Hanson's repayment plan proposed to pay $135 monthly to Great Lakes over sixty months, which in total was only 19 percent of the claim, but said nothing about discharging the remaining balance. The plan was confirmed without objection. After Hanson completed the payments under the plan, without any additional proof or argument being offered, the court granted a discharge of his student loans. In 2003, Educational Credit Management Corp. (ECMC), which had taken over Great Lakes' interest in the loans, filed a motion for relief from the discharge. What is the requirement for the discharge of a student loan obligation in bankruptcy? Did Hanson meet this requirement? Should the court grant ECMC's motion? Discuss. [*In re Hanson,* 397 F.3d 482 (7th Cir. 2005)]

After you have answered this problem, compare your answer with the sample answer given on the Web site that accompanies this text. Go to academic.cengage.com/blaw/wbl, select "Chapter 21," and click on "Case Problem with Sample Answer."

21–7. Discharge in Bankruptcy. Jon Goulet attended the University of Wisconsin in Eau Claire and Regis University in Denver, Colorado, from which he earned a bachelor's degree in history in 1972. Over the next ten years, he worked as a bartender and restaurant manager. In 1984, he became a life insurance agent, and his income ranged from $20,000 to $30,000. In 1989, however, his agent's license was revoked for insurance fraud, and he was arrested for cocaine possession. From 1991 to 1995, Goulet was again at the University of Wisconsin, working toward, but failing to obtain, a master's degree in psychology. To pay for his studies, he took out student loans totaling $76,000. Goulet then returned to bartending and restaurant management and tried real estate sales. His income for the year 2000 was $1,490, and his expenses, excluding a child-support obligation, were $5,904. When the student loans came due, Goulet filed a petition for bankruptcy. On what ground might the loans be dischargeable? Should the court grant a discharge on this ground? Why or why not? [*Goulet v. Educational Credit Management Corp.,* 284 F.3d 773 (7th Cir. 2002)]

21–8. Automatic Stay. On January 22, 2001, Marlene Moffett bought a used 1998 Honda Accord from Hendrick Honda in Woodbridge, Virginia. Moffett agreed to pay $20,024.25, with interest, in sixty monthly installments, and Hendrick retained a security interest in the car. (As discussed in Chapter 20, Hendrick thus had the right to repossess the car in the event of default, subject to Moffett's right of redemption.) Hendrick assigned its rights under the sales agreement to Tidewater Finance Co., which perfected its security interest. The car was Moffett's only means of traveling the forty miles from her home to her workplace. In March and April 2002, Moffett missed two monthly payments. On April 25, Tidewater repossessed the car. On the same day, Moffett filed a Chapter 13 plan in a federal bankruptcy court. Moffett asked that the car be returned to her, in part under the Bankruptcy Code's automatic-stay provision. Tidewater asked the court to terminate the automatic stay so that it could sell the car. How can the interests of both the debtor and the creditor be fully protected in this case? What should the court rule? Explain. [*In re Moffett,* 356 F.3d 518 (4th Cir. 2004)]

A QUESTION OF ETHICS

21–9. In October 1994, Charles Edwards formed ETS Payphones, Inc., to sell and lease pay phones as investment opportunities—an investor would buy a phone from ETS, which would lease it back. ETS promised returns of 14 to 15 percent, but consistently lost money. To meet its obligations to existing investors, ETS had to continually attract new investors. Eventually, ETS defrauded thousands of investors of more than $300 million. Edwards transferred the funds from ETS to himself. In 2000, ETS filed a petition in a federal bankruptcy court to declare bankruptcy. Darryl Laddin was appointed trustee. On the debtor's behalf, Laddin filed a suit against Reliance Trust Co. and others, alleging, among other things, that the defendants helped defraud investors by "ignoring the facts" and "funneling" the investors' funds to ETS, causing it to "incur millions of dollars in additional debt." Laddin sought treble damages. [*Official Committee of Unsecured Creditors of PSA, Inc. v. Edwards,* 437 F.3d 1145 (11th Cir. 2006)]

1. The defendants argued in part that the doctrine of *in pari delicto,* which provides that a wrongdoer may not profit from his or her wrongful acts, barred Laddin's claim. Who should be considered ethically responsible for the investors' losses? Explain.
2. Laddin contended that his actions, as trustee on behalf of the debtor, should not be subject to the doctrine of *in pari delicto* because that doctrine depends on the "personal malfeasance of the individual seeking to recover." The defendants filed a motion to dismiss Laddin's complaint. Do you think that the court should rule in favor of Laddin or the defendants? Why?

VIDEO QUESTION

21–10. Go to this text's Web site at academic.cengage.com/blaw/wbl and select "Chapter 21." Click on "Video Questions" and view the video titled *The River.* Then answer the following questions.

1. In the video, a crowd (including Mel Gibson) is gathered at a farm auction in which a neighbor's (Jim Antonio's) farming goods are being sold. The people in the crowd, who are upset because they believe that the bank is selling out the farmer, begin chanting "no sale, no sale." In an effort to calm the group, the farmer tells the crowd that "they've already foreclosed" on his farm. What does he mean?
2. Assume that the auction is a result of Chapter 7 bankruptcy proceedings. Was the farmer's petition for bankruptcy voluntary or involuntary? Explain.
3. Suppose that the farmer purchased the homestead three years prior to filing a petition for bankruptcy and that the

current market value of the farm is $215,000. What is the maximum amount of equity the farmer could claim as exempt under the 2005 Bankruptcy Reform Act?

4. Compare the results of a Chapter 12 bankruptcy as opposed to a Chapter 7 bankruptcy for the farmer in the video.

Creditors' Rights and Bankruptcy

For updated links to resources available on the Web, as well as a variety of other materials, visit this text's Web site at

academic.cengage.com/blaw/wbl

The Legal Information Institute at Cornell University offers a collection of law materials concerning debtor-creditor relationships, including federal statutes and recent Supreme Court decisions on this topic, at

http://www.law.cornell.edu/wex/index.php/Debtor_and_creditor

The U.S. Department of Labor's Web site contains a page on garnishment and employees' rights in relation to garnishment proceedings at

http://www.dol.gov/dol/topic/wages/garnishments.htm

Another good resource for bankruptcy information is the American Bankruptcy Institute (ABI) at

http://www.abiworld.org

For information and news on bankruptcy reform legislation, go to the site maintained by Bankruptcy Media at

http://www.bankruptcyfinder.com/bankruptcyreformnews.html

To read a brief primer on the distribution of property in a Chapter 7 bankruptcy, go to

http://www.lawdog.com/bkrcy/lib2a8.htm

ONLINE LEGAL RESEARCH

Go to the *Fundamentals of Business Law* home page at academic.cengage.com/blaw/wbl, select "Chapter 21," and click on "Internet Exercises." There you will find the following Internet research exercises that you can perform to learn more about topics covered in this chapter.

Activity 21–1: LEGAL PERSPECTIVE—Bankruptcy

Activity 21–2: MANAGEMENT PERSPECTIVE—Bankruptcy Alternatives

Activity 21–3: MANAGEMENT PERSPECTIVE—Mechanic's Liens

BEFORE THE TEST

Go to the *Fundamentals of Business Law* home page at academic.cengage.com/blaw/wbl, select "Chapter 21," and click on "Interactive Quizzes." You will find a number of interactive questions relating to this chapter.

UNIT 6 • EXTENDED CASE STUDY

Archer v. Warner

We discussed settlements and releases in Chapter 8, novation in Chapter 11, negotiable instruments in Chapters 18 and 19, and bankruptcy law in Chapter 21. Now, in this extended case study, we examine *Archer v. Warner,*[1] a decision concerning the dischargeability of a debt in a Chapter 7 bankruptcy proceeding. The debt was owed on a promissory note in which the debtors had agreed to pay $100,000 to settle a lawsuit in which they were accused of fraud. Generally, under the Bankruptcy Code, claims for money obtained by actual fraud are not dischargeable in bankruptcy. The specific question here was whether the obligation to pay on the note could be considered a debt for money obtained by fraud.

1. 538 U.S. 314, 123 S.Ct. 1462, 155 L.Ed.2d 454 (2003). This opinion may be accessed online at **http://www.findlaw.com/casecode/supreme.html**. In the "Browsing" section, click on "2003 Decisions." On the resulting page, click on the name of the case to access the opinion.

Case Background

In 1991, Leonard and Arlene Warner bought the Warner Manufacturing Company in North Carolina for $250,000. Six months later, they sold the company to Elliott and Carol Archer for $610,000. A few months after that, the Archers filed a suit in a North Carolina state court against the Warners, alleging, among other things, fraud connected with the sale. In May 1995, the Warners agreed to pay the Archers $300,000, and the Archers agreed to "execute releases to any and all claims . . . arising out of this litigation." The Warners paid the Archers $200,000 and executed a promissory note for the remaining $100,000. The Archers executed releases "discharg[ing]" the Warners "from any and every right, claim, or demand" that the Archers "now have or might otherwise hereafter have against" them, "excepting only obligations under" the note. A few days later, the Archers dismissed their suit.

In November, the Warners failed to make the first payment on the note. The Archers filed a suit in a North Carolina state court against the Warners for payment. The Warners filed for bankruptcy under Chapter 7. The Archers asked the court to rule that the $100,000 debt was nondischargeable. The court denied the request. The Archers appealed to the U.S. Court of Appeals for the Fourth Circuit, which affirmed the lower court's decision. The Archers appealed to the United States Supreme Court.

Majority Opinion

Justice *BREYER* delivered the opinion of the Court.

The Bankruptcy Code provides that a debt shall not be dischargeable in bankruptcy "to the extent" it is "for money * * * obtained by * * * false pretenses, a false representation, or actual fraud." Can this language cover a debt embodied in a settlement agreement that settled a creditor's earlier claim "for money * * * obtained by * * * fraud"? * * *

* * * *

* * * [T]he Court of Appeals for the Fourth Circuit * * * reasoned that the settlement agreement, releases, and promissory note had worked a kind of "novation." This novation replaced (1) an original potential debt to the Archers for money obtained by fraud with (2) a new debt. The new debt was not for money obtained by fraud. It was for money promised in a settlement contract. And it was consequently dischargeable in bankruptcy.

* * * *

We agree * * * that "[t]he settlement agreement and promissory note here, coupled with the broad language of the release, completely addressed and released each and every underlying state law claim." That agreement left only one relevant debt: a debt for money promised in the settlement agreement itself. To recognize that fact, however, does not end our inquiry. We must decide whether that same debt can also amount to a debt for money obtained by fraud * * * .

* * * *

As a matter of logic, * * * the Fourth Circuit's novation theory cannot be right. * * *

* * * [T]he mere fact that a conscientious creditor has previously reduced his claim to [settlement] should not bar further inquiry into the true nature of the debt * * * .

* * * [T]he Bankruptcy Code's nondischargeability provision had originally covered only judgments sounding in fraud. Congress later changed the language so that it covered all such liabilities. This change indicated that Congress intended the fullest possible inquiry to ensure that all debts arising out of fraud are excepted from discharge, no matter what their form. Congress also intended to allow the relevant determination (whether a debt arises out of fraud) to take place in bankruptcy court, not to force it to occur earlier in state court at a time when nondischargeability concerns are not directly in issue and neither party has a full incentive to litigate them.

* * * The dischargeability provision applies to all debts that arise out of fraud. *A debt embodied in the settlement of*

(Continued)

a fraud case arises no less out of the underlying fraud than a debt embodied in a stipulation and consent decree [which occurred in a previous case, in which we held that a debt originating in fraud was not dischargeable]. Policies that favor the settlement of disputes, like those that favor repose, are neither any more nor any less at issue here than in [the previous case]. * * * [W]hat has *not* been established here * * * is that the parties meant to resolve the issue of fraud or, more narrowly, to resolve that issue for purposes of a later claim of nondischargeability in bankruptcy. In a word, we can find no significant difference between [the previous case] and the case now before us. [Emphasis added.]

* * * *

We conclude that the Archers' settlement agreement and releases may have worked a kind of novation, but that fact does not bar the Archers from showing that the settlement debt arose out of "false pretenses, a false representation, or actual fraud," and consequently is nondischargeable. We reverse the Court of Appeals' judgment to the contrary. And we remand this case for further proceedings consistent with this opinion.

Dissenting Opinion

Justice *THOMAS* * * * dissenting.

* * * Because the Court's conclusion is supported neither by the text of the Bankruptcy Code nor by any of the agreements executed by the parties, I respectfully dissent.

* * * *

* * * [I]n my view, if it is shown that a note was given and received as payment or waiver of the original debt and the parties agreed that the note was to substitute a new obligation for the old, the note fully discharges the original debt, and the nondischargeability of the original debt does not affect the dischargeability of the obligation under the note. * * *

* * * *

Based on the sweeping language of the general release, it is inaccurate for the Court to say that the parties did not "resolve the issue of fraud." To be sure, there is no legally controlling document stating that respondent [the Warners] did (or did not) commit fraud. But where it was not clear which claims were being resolved by the consent judgment, the release in this case clearly demonstrates that the parties intended to resolve conclusively not only the issue of fraud, but also any other rights, claims, or demands related to the state-court litigation, excepting only obligations under the [n]ote * * * .

* * * *

Petitioners' [the Archers] own actions in the course of this litigation support this conclusion. Throughout the proceedings below and continuing in this Court, petitioners have sought to recover only the amount of the debt set forth in the settlement agreement, which is lower than the total damages they allegedly suffered as a result of respondent's alleged fraud. * * *

* * * *

The Court today ignores the plain intent of the parties, as evidenced by a properly executed settlement agreement and general release, holding that a debt owed by respondent under a contract was "obtained by" fraud. Because I find no support for the Court's conclusion in the text of the Bankruptcy Code, or in the agreements of the parties, I respectfully dissent.

Questions for Analysis

1. **Law.** What did the majority conclude in this case? What was the reasoning leading to this conclusion?
2. **Law.** What was the dissent's contention? On what did the dissent base its position?
3. **Ethics.** What is the ethical basis for holding, as the majority does, that a claim based on a note issued under a settlement agreement, in which an underlying fraud claim was released, should not be subsequently discharged in bankruptcy?
4. **Economic Dimensions.** How might the result in this case affect a party's decision to sign a release and settlement on a fraud claim and accept a note in payment of the debt?
5. **Implications for the Business Owner.** How might the holding in this case affect an individual's decision about whether to purchase a business?

UNI 7

Employment Relations

Unit Contents

CHAPTER 22
Agency Relationships

LEARNING OBJECTIVES

After reading this chapter, you should be able to answer the following questions:

1. What is the difference between an employee and an independent contractor?
2. How do agency relationships arise?
3. What duties do agents and principals owe to each other?
4. When is a principal liable for the agent's actions with respect to third parties? When is the agent liable?
5. What are some of the ways in which an agency relationship can be terminated?

One of the most common, important, and pervasive legal relationships is that of **agency.** In an agency relationship between two parties, one of the parties, called the *agent,* agrees to represent or act for the other, called the *principal.* The principal has the right to control the agent's conduct in matters entrusted to the agent, and the agent must exercise his or her powers "for the benefit of the principal only." By using agents, a principal can conduct multiple business operations simultaneously in various locations. Thus, for example, contracts that bind the principal can be made at different places with different persons at the same time.

Agency relationships permeate the business world. Indeed, agency law is essential to the existence and operation of a corporate entity, because only through its agents can a corporation function and enter into contracts. A familiar example of an agent is a corporate officer who serves in a representative capacity for the owners of the corporation. In this capacity, the officer has the authority to bind the principal (the corporation) to a contract.

Agency Relationships

Section 1(1) of the *Restatement (Second) of Agency*[1] defines *agency* as "the fiduciary relation which results from the manifestation of consent by one person to another that the other shall act in his [or her] behalf and subject to his [or her] control, and consent by the other so to act." In other words, in a principal-agent relationship, the parties have agreed that the agent will act *on behalf and instead of* the principal in negotiating and transacting business with third parties.

The term **fiduciary** is at the heart of agency law. The term can be used both as a noun and as an adjective. When used as a noun, it refers to a person having a duty created by her or his undertaking to act primarily for another's benefit in matters connected with the undertaking. When used as an adjective, as in "fiduciary relationship," it means that the relationship involves trust and confidence.

1. The *Restatement (Second) of Agency* is an authoritative summary of the law of agency and is often referred to by judges in their decisions and opinions.

Agency relationships commonly exist between employers and employees. Agency relationships may sometimes also exist between employers and independent contractors who are hired to perform special tasks or services.

EMPLOYER-EMPLOYEE RELATIONSHIPS

Normally, all employees who deal with third parties are deemed to be agents. A salesperson in a department store, for instance, is an agent of the store's owner (the principal) and acts on the owner's behalf. Any sale of goods made by the salesperson to a customer is binding on the principal. Similarly, most representations of fact made by the salesperson with respect to the goods sold are binding on the principal.

Because employees who deal with third parties are normally deemed to be agents of their employers, agency law and employment law overlap considerably. Agency relationships, though, as will become apparent, can exist outside an employer-employee relationship and thus have a broader reach than employment laws do. Additionally, bear in mind that agency law is based on the common law. In the employment realm, many common law doctrines have been displaced by statutory law and government regulations relating to employment relationships.

Employment laws (state and federal) apply only to the employer-employee relationship. Statutes governing Social Security, withholding taxes, workers' compensation, unemployment compensation, workplace safety, employment discrimination, and the like (see Chapter 23) are applicable only if employer-employee status exists. *These laws do not apply to an independent contractor.*

EMPLOYER–INDEPENDENT CONTRACTOR RELATIONSHIPS

Independent contractors are not employees because, by definition, those who hire them have no control over the details of their physical performance. Section 2 of the *Restatement (Second) of Agency* defines an **independent contractor** as follows:

> [An independent contractor is] a person who contracts with another to do something for him [or her] but who is not controlled by the other nor subject to the other's right to control with respect to his [or her] physical conduct in the performance of the undertaking. *He [or she] may or may not be an agent.* [Emphasis added.]

Building contractors and subcontractors are independent contractors; a property owner does not control the acts of either of these professionals. Truck drivers who own their equipment and hire themselves out on a per-job basis are independent contractors, but truck drivers who drive company trucks on a regular basis are usually employees.

The relationship between a person or firm and an independent contractor may or may not involve an agency relationship. To illustrate: An owner of real estate who hires a real estate broker to negotiate a sale of the property not only has contracted with an independent contractor (the real estate broker) but also has established an agency relationship for the specific purpose of assisting in the sale of the property. Another example is an insurance agent, who is both an independent contractor and an agent of the insurance company for which she or he sells policies. (Note that an insurance *broker,* in contrast, normally is an agent of the person obtaining insurance and not of the insurance company.)

DETERMINING EMPLOYEE STATUS

The courts are frequently asked to determine whether a particular worker is an employee or an independent contractor. How a court decides this issue can have a significant effect on the rights and liabilities of the parties.

Criteria Used by the Courts In determining whether a worker has the status of an employee or an independent contractor, the courts often consider the following questions:

1. How much control can the employer exercise over the details of the work? (If an employer can exercise considerable control over the details of the work, this would indicate employee status. This is perhaps the most important factor weighed by the courts in determining employee status.)
2. Is the worker engaged in an occupation or business distinct from that of the employer? (If so, this points to independent-contractor status, not employee status.)
3. Is the work usually done under the employer's direction or by a specialist without supervision? (If the work is usually done under the employer's direction, this would indicate employee status.)
4. Does the employer supply the tools at the place of work? (If so, this would indicate employee status.)
5. For how long is the person employed? (If the person is employed for a long period of time, this would indicate employee status.)
6. What is the method of payment—by time period or at the completion of the job? (Payment by time period, such as once every two weeks or once a month, would indicate employee status.)

7 What degree of skill is required of the worker? (If little skill is required, this may indicate employee status.)

Sometimes, it is beneficial for workers to have *employee* status—to take advantage of laws protecting employees, for example. In contrast, *independent-contractor* status can sometimes be an advantage—for copyright ownership purposes, for example, as you will read shortly.

Many employers prefer to designate certain workers as independent contractors rather than as employees. After all, if a worker is an independent contractor, the worker must pay all Social Security taxes, instead of sharing them with his or her employer. Additionally, the worker will not be entitled to employer-provided benefits—such as pension plans, stock option plans, and group insurance coverage—that are often available to employees. Furthermore, as already noted, the worker will not receive the legal protections afforded to employees under such laws as those regulating safety in the workplace or those protecting employees from discrimination. Not surprisingly, some disputes between employers and workers over this issue have ended up in court.[2]

2. See, for example, *Vizcaino v. U.S. District Court for the Western District of Washington,* 173 F.3d 713 (9th Cir. 1999), a case brought by Microsoft Corporation's employees who were required to work as independent contractors instead of employees. See also *A Question of Ethics* at the end of this chapter, which focuses on this case.

Criteria Used by the IRS Often, the criteria for determining employee status are established by a statute or an administrative agency regulation. Businesspersons should be aware that the Internal Revenue Service (IRS) has established its own criteria for determining whether a worker is an independent contractor or an employee. Although the IRS once considered twenty factors in determining a worker's status, guidelines in effect since 1997 encourage IRS examiners to focus on just one of those factors—the degree of control the business exercises over the worker.

The IRS tends to closely scrutinize a firm's classification of its workers because, as mentioned, employers can avoid certain tax liabilities by hiring independent contractors instead of employees. Even when a firm classifies a worker as an independent contractor, if the IRS decides that the worker is actually an employee, then the employer will be responsible for paying any applicable Social Security, withholding, and unemployment taxes.

In contrast, when a worker is a corporate officer, the exercise of control may be the opposite of the usual situation involving an employer and an employee. In that circumstance, the question may concern the degree of control that the officer (the employee) exercises over the corporation (the employer). In the following case, the issue was whether a corporate officer was an employee of the corporation.

CASE 22.1 Nu-Look Design, Inc. v. Commissioner of Internal Revenue

United States Court of Appeals,
Third Circuit, 2004.
356 F.3d 290.

FACTS Nu-Look Design, Inc., is a home-improvement company that provides carpentry, siding installation, and general residential construction services. During 1996, 1997, and 1998, Ronald Stark was Nu-Look's president, manager, and sole shareholder. He solicited business for the company, performed the bookkeeping, handled the firm's finances, and hired and supervised its workers. Instead of paying Stark a salary or wages, Nu-Look distributed its income to him "as Mr. Stark's needs arose." Nu-Look reported on its tax returns for those years net income of $10,866.14, $14,216.37, and $7,103.60, respectively. Stark reported the same amounts as income on his tax returns. In 2001, the Internal Revenue Service (IRS) classified Stark as Nu-Look's employee and assessed federal employment taxes for 1996, 1997, and 1998. Nu-Look filed a suit in the U.S. Tax Court against the commissioner of the IRS, seeking relief from this liability. Nu-Look contended in part that Stark was not an employee because Nu-Look did not control Stark—Stark controlled Nu-Look. The court ruled against the firm, which appealed to the U.S. Court of Appeals for the Third Circuit.

ISSUE For federal tax purposes, was Stark an employee of Nu-Look?

DECISION Yes. The U.S. Court of Appeals for the Third Circuit affirmed the ruling of the lower court.

REASON The U.S. Court of Appeals for the Third Circuit focused on the nature of the services that Stark rendered and whether the income that Nu-Look distributed to Stark was payment for those services. The court pointed out that both the Federal Insurance Contributions Act and the Federal Unemployment Tax Act (see Chapter 23) impose taxes on employers based on the wages paid to individuals in their employ. *Wages,* as defined by both statutes, generally include "all remuneration for employment." *Employment* is "any service of whatever nature, performed * * * by an employee for the

CASE 22.1—CONTINUED

person employing him." Both statutes define *employee* to include "any officer of a corporation." The court acknowledged that there is an exception for an "officer of a corporation who as such does not perform any services or performs only minor services and who neither receives nor is entitled to receive, directly or indirectly, any remuneration." In this case, however, "Stark performed more than minor services and the distributions Stark received were, in fact, remuneration for his services."

WHY IS THIS CASE IMPORTANT? *Businesspersons should be aware that the mere designation of a person as either an independent contractor or a corporate officer does not mean the employer can avoid tax liability. The courts and the IRS look behind the label to ascertain the true relationship between the worker and the business entity.*

Employee Status and "Works for Hire" Under the Copyright Act of 1976, any copyrighted work created by an employee within the scope of her or his employment at the request of the employer is a "work for hire," and the employer owns the copyright to the work. When an employer hires an independent contractor—a freelance artist, writer, or computer programmer, for example—the independent contractor owns the copyright *unless* the parties agree in writing that the work is a "work for hire" and the work falls into one of nine specific categories, including audiovisual and other works.

■ EXAMPLE 22.1 Graham marketed CD-ROM discs containing compilations of software programs that are available free to the public. Graham hired James to create a file-retrieval program that allowed users to access the software on the CDs. James built into the final version of the program a notice stating that he was the author of the program and owned the copyright. Graham removed the notice. When James sold the program to another CD-ROM publisher, Graham filed a suit claiming that James's program was a "work for hire" and that Graham owned the copyright for the file-retrieval program. The court, however, decided that James—who was a skilled computer programmer who controlled the manner and method of his work—was an independent contractor and not an employee for hire. Thus, James owned the copyright for the file-retrieval program.[3] ■

How Agency Relationships Are Formed

Agency relationships normally are consensual—that is, they come about by voluntary consent and agreement between the parties. Generally, the agreement need not be in writing,[4] and consideration is not required.

A principal must have contractual capacity. A person who cannot legally enter into contracts directly should not be allowed to do so indirectly through an agent. Because an agent derives the authority to enter into contracts from the principal and because a contract made by an agent is legally viewed as a contract of the principal, it is immaterial whether the agent personally has the legal capacity to make that contract. Thus, a minor can be an agent but in some states cannot be a principal appointing an agent.[5] (When a minor is permitted to be a principal, however, any resulting contracts will be voidable by the minor principal but not by the adult third party.) In sum, any person can be an agent, regardless of whether he or she has the capacity to contract. Even a person who is legally incompetent can be appointed an agent.

An agency relationship can be created for any legal purpose. An agency relationship that is created for an illegal purpose or that is contrary to public policy is unenforceable. ■ EXAMPLE 22.2 Suppose that Sharp (as principal) contracts with Blesh (as agent) to sell illegal narcotics. This agency relationship is unenforceable because selling illegal narcotics is a felony and is contrary to public policy. ■ It is also illegal for medical doctors and other licensed professionals to employ unlicensed agents to perform professional actions.

Generally, an agency relationship can arise in four ways: by agreement of the parties, by ratification, by estoppel, and by operation of law. We look here at each of these possibilities.

AGENCY BY AGREEMENT

Most agency relationships are based on an express or implied agreement that the agent will act for the principal and that the principal agrees to have the agent so act. An agency agreement can take the form of an express written contract. ■ EXAMPLE 22.3 Renato enters into a written

3. *Graham v. James,* 144 F.3d 229 (2d Cir. 1998).

4. There are two main exceptions to the statement that agency agreements need not be in writing: (1) Whenever agency authority empowers the agent to enter into a contract that the Statute of Frauds requires to be in writing, the agent's authority from the principal must likewise be in writing (this is called the *equal dignity rule,* which will be discussed later in this chapter). (2) A power of attorney, which confers authority to an agent, must be in writing.

5. Some courts have granted exceptions to allow a minor to appoint an agent for the limited purpose of contracting for the minor's necessities of life. See *Casey v. Kastel,* 237 N.Y. 305, 142 N.E. 671 (1924).

agreement with Troy, a real estate agent, to sell Renato's house. An agency relationship exists between Renato and Troy for the sale of the house and is detailed in a document that both parties sign. ■

Many express agency agreements can be made orally. ■ EXAMPLE 22.4 Suppose that Renato asks Cary, a gardener, to contract with others for the care of his lawn on a regular basis. Cary agrees. In this situation, an agency relationship exists between Renato and Cary for the lawn care. ■

An agency agreement can also be implied by conduct. ■ EXAMPLE 22.5 A hotel expressly allows only Boris Koontz to park cars, but Boris has no employment contract there. The hotel's manager tells Boris when to work, as well as where and how to park the cars. The hotel's conduct amounts to a manifestation of its willingness to have Boris park its customers' cars, and Boris can infer from the hotel's conduct that he has authority to act as a parking valet. It can be inferred that Boris is an agent-employee for the hotel, his purpose being to provide valet parking services for hotel guests. ■

AGENCY BY RATIFICATION

On occasion, a person who is in fact not an agent (or who is an agent acting outside the scope of her or his authority) may make a contract on behalf of another (a principal). If the principal approves or affirms that contract by word or by action, an agency relationship is created by **ratification.** Ratification is a question of intent, and intent can be expressed by either words or conduct. The basic requirements for ratification are discussed later in this chapter.

AGENCY BY ESTOPPEL

When a principal causes a third person to believe that another person is his or her agent, and the third person deals with the supposed agent, the principal is "estopped (prevented) to deny" the agency relationship. In such a situation, the principal's actions create the *appearance* of an agency that does not in fact exist. The third person must prove that she or he *reasonably* believed that an agency relationship existed, though.[6] Facts and circumstances must show that an ordinary, prudent person familiar with business practice and custom would have been justified in concluding that the agent had authority.

■ EXAMPLE 22.6 Suppose that Andrew accompanies Grant, a seed sales representative, to call on a customer, Steve, the proprietor of the General Seed Store. Andrew has done independent sales work but has never signed an employment agreement with Grant. Grant boasts to Steve that he wishes he had three more assistants "just like Andrew." By making this representation, Grant creates the impression that Andrew is his agent and has authority to solicit orders. Steve has reason to believe from Grant's statements that Andrew is an agent for Grant. Steve then places seed orders with Andrew. If Grant does not correct the impression that Andrew is an agent, Grant will be bound to fill the orders just as if Andrew were really his agent. Grant's representation to Steve created the impression that Andrew was Grant's agent and had authority to solicit orders. ■

The acts or declarations of a purported *agent* in and of themselves do not create an agency by estoppel. Rather, it is the deeds or statements of the *principal* that create an agency by estoppel. ■ EXAMPLE 22.7 If Andrew walks into Steve's store and claims to be Grant's agent, when in fact he is not, and Grant has no knowledge of Andrew's representations, Grant will not be bound to any deal struck by Andrew and Steve. Andrew's acts and declarations alone do not create an agency by estoppel. ■

AGENCY BY OPERATION OF LAW

The courts may find an agency relationship in the absence of a formal agreement in other situations as well. This can occur in family relationships. For instance, suppose that one spouse purchases certain basic necessaries and charges them to the other spouse's charge account. The courts will often rule that the latter is liable for payment for the necessaries, either because of a social policy of promoting the general welfare of the spouse or because of a legal duty to supply necessaries to family members.

Agency by operation of law may also occur in emergency situations, when the agent's failure to act outside the scope of his or her authority would cause the principal substantial loss. If the agent is unable to contact the principal, the courts will often grant this emergency power. For instance, a railroad engineer may contract on behalf of her or his employer for medical care for an injured motorist hit by the train.

Duties of Agents and Principals

Once the principal-agent relationship has been created, both parties have duties that govern their conduct. As discussed previously, an agency relationship is *fiduciary*—one of trust. In a fiduciary relationship, each party owes the other the duty to act with the utmost good faith.

We now examine the various duties of agents and principals. In general, for every duty of the principal, the agent

6. These concepts also apply when a person who is in fact an agent undertakes an action that is beyond the scope of her or his authority, as will be discussed later in this chapter.

has a corresponding right, and vice versa. When one party to the agency relationship violates his or her duty to the other party, the remedies available to the nonbreaching party arise out of contract and tort law. These remedies include monetary damages, termination of the agency relationship, injunction, and required accountings.

AGENT'S DUTIES TO THE PRINCIPAL

Generally, the agent owes the principal five duties—performance, notification, loyalty, obedience, and accounting.

Performance An implied condition in every agency contract is the agent's agreement to use reasonable diligence and skill in performing the work. When an agent fails to perform her or his duties entirely, liability for breach of contract normally will result. The degree of skill or care required of an agent is usually that expected of a reasonable person under similar circumstances. Generally, this is interpreted to mean ordinary care. If an agent has represented himself or herself as possessing special skills, however, the agent is expected to exercise the degree of skill or skills claimed. Failure to do so constitutes a breach of the agent's duty.

Not all agency relationships are based on contract. In some situations, an agent acts gratuitously—that is, not in return for a monetary payment. A gratuitous agent cannot be liable for breach of contract, as there is no contract; he or she is subject only to tort liability. Once a gratuitous agent has begun to act in an agency capacity, he or she has the duty to continue to perform in that capacity in an acceptable manner and is subject to the same standards of care and duty to perform as other agents.

Notification According to a maxim in agency law, notice to the agent is notice to the principal. An agent is thus required to notify the principal of all matters that come to her or his attention concerning the subject matter of the agency. This is the *duty of notification,* or the duty to inform.

■ EXAMPLE 22.8 Suppose that Lang, an artist, is about to negotiate a contract to sell a series of paintings to Barber's Art Gallery for $15,000. Lang's agent learns that Barber is insolvent and will be unable to pay for the paintings. Lang's agent has a duty to inform Lang of this fact because it is relevant to the subject matter of the agency—the sale of Lang's paintings. ■ Generally, the law assumes that the principal knows of any information acquired by the agent that is relevant to the agency—regardless of whether the agent actually passes on this information to the principal.

Loyalty Loyalty is one of the most fundamental duties in a fiduciary relationship. Basically, the agent has the duty to act *solely for the benefit of his or her principal* and not in the interest of the agent or a third party. For example, an agent cannot represent two principals in the same transaction unless both know of the dual capacity and consent to it. The duty of loyalty also means that any information or knowledge acquired through the agency relationship is considered confidential. It would be a breach of loyalty to disclose such information either during the agency relationship or after its termination. Typical examples of confidential information are trade secrets and customer lists compiled by the principal.

In short, the agent's loyalty must be undivided. The agent's actions must be strictly for the benefit of the principal and must not result in any secret profit for the agent. ■ EXAMPLE 22.9 Suppose that Ryder contracts with Alton, a real estate agent, to sell Ryder's property. Alton knows that she can find a buyer who will pay substantially more for the property than Ryder is asking. If Alton were to secretly purchase Ryder's property, however, and then resell it at a profit to another buyer, Alton would breach her duty of loyalty as Ryder's agent. Alton has a duty to act in Ryder's best interests and can only become the purchaser in this situation with Ryder's knowledge and approval. ■

Does an agent breach the duty of loyalty if, while working for a principal, the agent solicits the principal's customers for a new competing business? That was an issue in the following case.

CASE 22.2 American Express Financial Advisors, Inc. v. Topel

United States District Court,
District of Colorado, 1999.
38 F.Supp.2d 1233.

FACTS Stephen Topel worked as a financial planner for American Express Financial Advisors, Inc. (AMEX), beginning in April 1992. More than four years later, Topel decided to resign to work for Multi-Financial Securities Corporation, an AMEX competitor. Before resigning, Topel encouraged his customers to liquidate their AMEX holdings and sent them new account forms for Multi-Financial. He ignored the request of customers

(Continued)

CASE 22.2–CONTINUED

James and Nancy Hemming to keep their investments with AMEX. In a letter on AMEX letterhead, Topel told Chris and Teresa Mammel to liquidate their AMEX holdings and invest in Multi-Financial's products. Another couple, Mr. and Ms. Rogers, changed their investments on Topel's advice. Before leaving AMEX, Topel sent a letter to all of his clients telling them that he was ending his relationship with AMEX and that their accounts would be assigned to another AMEX adviser. After Topel resigned in May 1997, he solicited the business of Theodore Benavidez, another AMEX customer. AMEX filed a suit in a federal district court against Topel, alleging, among other things, breach of fiduciary duty (duty of loyalty) and seeking damages. AMEX filed a motion for summary judgment on this issue.

ISSUE Had Topel breached his fiduciary duty of loyalty?

DECISION Yes. The court granted AMEX's motion for summary judgment in its favor with respect to this claim.

REASON The court held that Topel breached his duty of loyalty when, while working for his principal, he solicited his principal's customers for his new competing business. The court cited the principle that an agent has a duty to act solely for the benefit of the principal in all matters connected with an agency. "While an agent is entitled to make some preparations to compete with his principal after the termination of their relationship," the court acknowledged, "an agent violates his duty of loyalty if he engages in pretermination solicitation of customers for a new competing business." That Topel did not solicit Benavidez's business until after Topel left AMEX "does not negate the testimony of other customers that he solicited their business for Multi-Financial while he was still affiliated with AMEX." As for the letter that Topel sent to all of his clients before resigning, "many customers had already signed new account forms with Multi-Financial by the time this neutral letter was purportedly sent."

FOR CRITICAL ANALYSIS—Ethical Consideration *Can you think of any situations in which the duty of loyalty to one's employer could come into conflict with other duties? Explain.*

Obedience When acting on behalf of a principal, an agent has a duty to follow all lawful and clearly stated instructions of the principal. Any deviation from such instructions is a violation of this duty. During emergency situations, however, when the principal cannot be consulted, the agent may deviate from the instructions without violating this duty. Whenever instructions are not clearly stated, the agent can fulfill the duty of obedience by acting in good faith and in a manner reasonable under the circumstances.

Accounting Unless an agent and a principal agree otherwise, the agent has the duty to keep and make available to the principal an account of all property and funds received and paid out on behalf of the principal. This includes gifts from third parties in connection with the agency. For example, a gift from a customer to a salesperson for prompt deliveries made by the salesperson's firm, in the absence of a company policy to the contrary, belongs to the firm. The agent has a duty to maintain separate accounts for the principal's funds and for the agent's personal funds, and the agent should not intermingle these accounts.

PRINCIPAL'S DUTIES TO THE AGENT

The principal also owes certain duties to the agent. These duties relate to compensation, reimbursement and indemnification, cooperation, and safe working conditions.

Compensation In general, when a principal requests certain services from an agent, the agent reasonably expects payment. The principal therefore has a duty to pay the agent for services rendered. For example, when an accountant or an attorney is asked to act as an agent, an agreement to compensate the agent for such service is implied. The principal also has a duty to pay that compensation in a timely manner. Except in a gratuitous agency relationship, in which an agent does not act for payment in return, the principal must pay the agreed-on value for an agent's services. If no amount has been expressly agreed on, the principal owes the agent the customary compensation for such services.

Reimbursement and Indemnification Whenever an agent disburses funds to fulfill the request of the principal or to pay for necessary expenses in the course of a reasonable performance of his or her agency duties, the principal has the duty to reimburse the agent for these payments. Agents cannot recover for expenses incurred through their own misconduct or negligence, though.

Subject to the terms of the agency agreement, the principal has the duty to compensate, or *indemnify,* an agent for liabilities incurred because of authorized and lawful acts and transactions. For instance, if the principal fails to perform a contract formed by the agent with a third party and the third party then sues the agent, the principal is

obligated to compensate the agent for any costs incurred in defending against the lawsuit.

Additionally, the principal must indemnify (pay) the agent for the value of benefits that the agent confers on the principal. The amount of indemnification is usually specified in the agency contract. If it is not, the courts will look to the nature of the business and the type of loss to determine the amount. Note that this rule applies to acts by gratuitous agents as well. If the finder of a dog that becomes sick takes the dog to a veterinarian and pays the required fees for the veterinarian's services, the agent is entitled to be reimbursed by the owner of the dog for those fees.

Cooperation A principal has a duty to cooperate with the agent and to assist the agent in performing her or his duties. The principal must do nothing to prevent such performance.

■ EXAMPLE 22.10 Suppose that Akers (the principal) grants Johnson (the agent) an exclusive territory within which Johnson may sell Akers's products, thus creating an exclusive agency. In this situation, Akers cannot compete with Johnson within that territory—or appoint or allow another agent to so compete—because this would violate the exclusive agency. If Akers did so, he would be exposed to liability for Johnson's lost sales or profits. ■

Safe Working Conditions The common law requires the principal to provide safe working premises, equipment, and conditions for all agents and employees. The principal has a duty to inspect the working conditions and to warn agents and employees about any unsafe areas. When the agent is an employee, the employer's liability is frequently covered by state workers' compensation insurance, and federal and state statutes often require the employer to meet certain safety standards (see Chapter 23).

Agent's Authority

An agent's authority to act can be either *actual* (express or implied) or *apparent*. If an agent contracts outside the scope of his or her authority, the principal may still become liable by ratifying the contract.

ACTUAL AUTHORITY

As indicated, an agent's actual authority can be express or implied. We look here at both of these forms of actual authority.

Express Authority Authority declared in clear, direct, and definite terms is called *express authority*. Express authority can be given orally or in writing.

—The Equal Dignity Rule. In most states, the **equal dignity rule** requires that if the contract being executed is or must be in writing, then the agent's authority must also be in writing. Failure to comply with the equal dignity rule can make a contract voidable *at the option of the principal*. The law regards the contract at that point as a mere offer. If the principal decides to accept the offer, acceptance must be ratified, or affirmed, in writing.

■ EXAMPLE 22.11 Klee (the principal) orally asks Parkinson (the agent) to sell a horse ranch that Klee owns. Parkinson finds a buyer and signs a sales contract (a contract for an interest in realty must be in writing) on behalf of Klee to sell the ranch. The buyer cannot enforce the contract unless Klee subsequently ratifies Parkinson's agency status *in writing*. Once Parkinson's agency status is ratified, either party can enforce rights under the contract. ■

—Exceptions to the Equal Dignity Rule. Modern business practice allows an exception to the equal dignity rule. The equal dignity rule does not apply when an agent acts in the presence of a principal or when the agent's act of signing is merely perfunctory. Thus, if Dickens (the principal) negotiates a contract but is called out of town the day it is to be signed and orally authorizes Santini to sign the contract, the oral authorization normally is considered to be sufficient.

—Power of Attorney. Giving an agent a **power of attorney** confers express authority.[7] The power of attorney normally is a written document and is usually notarized. (A document is notarized when a **notary public**[8] signs and dates the document and imprints it with his or her seal of authority.) A power of attorney can be special (permitting the agent to do specified acts only), or it can be general (permitting the agent to transact all business for the principal). A general power of attorney grants extensive authority to an agent to act on behalf of the principal in many ways (see Exhibit 22–1 on the following page). Because of this, a general power of attorney should be used with great caution and usually only in exceptional circumstances. Ordinarily, a power of attorney terminates on the incapacity or death of the person giving the power.[9]

7. An agent who holds the power of attorney is called an *attorney-in-fact* for the principal. The holder does not have to be an attorney-at-law (and often is not).

8. A *notary public* is a public official who is authorized to attest to the authenticity of signatures.

9. A *durable* power of attorney, however, continues to be effective despite the principal's incapacity. An elderly person, for example, might grant a durable power of attorney to provide for the handling of property and investments or specific health-care needs should she or he become incompetent.

EXHIBIT 22–1 A SAMPLE POWER OF ATTORNEY

GENERAL POWER OF ATTORNEY

Know All Men by These Presents:

That I, __________ , hereinafter referred to as PRINCIPAL, in the County of __________
State of __________ , do(es) appoint __________ as my true and lawful attorney.

In principal's name, and for principal's use and benefit, said attorney is authorized hereby;

(1) To demand, sue for, collect, and receive all money, debts, accounts, legacies, bequests, interest, dividends, annuities, and demands as are now or shall hereafter become due, payable, or belonging to principal, and take all lawful means, for the recovery thereof and to compromise the same and give discharges for the same;
(2) To buy and sell land, make contracts of every kind relative to land, any interest therein or the possession thereof, and to take possession and exercise control over the use thereof;
(3) To buy, sell, mortgage, hypothecate, assign, transfer, and in any manner deal with goods, wares and merchandise, choses in action, certificates or shares of capital stock, and other property in possession or in action, and to make, do, and transact all and every kind of business of whatever nature;
(4) To execute, acknowledge, and deliver contracts of sale, escrow instructions, deeds, leases including leases for minerals and hydrocarbon substances and assignments of leases, covenants, agreements and assignments of agreements, mortgages and assignments of mortgages, conveyances in trust, to secure indebtedness or other obligations, and assign the beneficial interest thereunder, subordinations of liens or encumbrances, bills of lading, receipts, evidences of debt, releases, bonds, notes, bills, requests to reconvey deeds of trust, partial or full judgments, satisfactions of mortgages, and other debts, and other written instruments of whatever kind and nature, all upon such terms and conditions as said attorney shall approve.

GIVING AND GRANTING to said attorney full power and authority to do all and every act and thing whatsoever requisite and necessary to be done relative to any of the foregoing as fully to all intents and purposes as principal might or could do if personally present.

All that said attorney shall lawfully do or cause to be done under the authority of this power of attorney is expressly approved.

Dated: __________ /s/__________

State of California
County of __________ } SS.
On __________, before me, the undersigned, a Notary Public in and for said State, personally appeared __________
____
known to me to be the person __________ whose name __________ subscribed to the within instrument and acknowledged that __________ executed the same.

Witness my hand and official seal. (Seal) __________
Notary Public in and for said State.

Implied Authority *Implied authority* is conferred by custom, can be inferred from the position the agent occupies, or is implied by virtue of being reasonably necessary to carry out express authority. ■ EXAMPLE 22.12 Mueller is employed by Al's Supermarket to manage one of its stores. Al's has not expressly stated that Mueller has authority to contract with third persons. In this situation, though, authority to manage a business implies authority to do what is reasonably required (as is customary or can be inferred from a manager's position) to operate the business. This includes forming contracts to hire employees, to buy merchandise and equipment, and to advertise the products sold in the store. ■

APPARENT AUTHORITY

Actual authority (express or implied) arises from what the principal manifests *to the agent. Apparent authority* exists when the principal, by either words or actions, causes a *third party* reasonably to believe that an agent has authority to act, even though the agent has no express or implied authority. If the third party changes her or his position in reliance on the principal's representations, the principal may be *estopped* (prevented) from denying that the agent had authority. Note that here, in contrast to agency formation by estoppel, the issue has to do with the apparent authority of an *agent,*

not the apparent authority of a person who is in fact not an agent.

■ EXAMPLE 22.13 Suppose that a traveling salesperson, Ling (the agent), is authorized to take customers' orders. Ling does not deliver the ordered goods and is not authorized to collect payments for the goods. A customer, Byron, pays Ling for a solicited order. Ling then takes the payment to the principal's accounting department, and an accountant accepts the payment and sends Byron a receipt. This procedure is thereafter followed for other orders solicited from and paid for by Byron. Later, Ling solicits an order, and Byron pays her as before. This time, however, Ling absconds with the money. Can Byron claim that the payment to the agent was authorized and was thus, in effect, a payment to the principal?

The answer is normally yes, because the principal's *repeated* acts of accepting Byron's payment led him reasonably to believe that Ling had authority to receive payments for goods solicited. Although Ling did not have express or implied authority, the principal's conduct gave Ling *apparent* authority to collect. In this situation, the principal would be estopped from denying that Ling had authority to collect payments. ■

RATIFICATION

As already mentioned, ratification occurs when the principal affirms an agent's *unauthorized* act. When ratification occurs, the principal is bound to the agent's act, and the act is treated as if it had been authorized by the principal *from the outset.* Ratification can be either express or implied.

If the principal does not ratify the contract, the principal is not bound, and the third party's agreement with the agent is viewed as merely an unaccepted offer. Because the third party's agreement is an unaccepted offer, the third party can revoke the offer at any time, without liability, before the principal ratifies the contract.

The requirements for ratification can be summarized as follows:

1. The agent must have acted on behalf of an identified principal who subsequently ratifies the action.
2. The principal must know of all material facts involved in the transaction. If a principal ratifies a contract without knowing all of the facts, the principal can rescind (cancel) the contract.
3. The principal must affirm the agent's act in its entirety.
4. The principal must have the legal capacity to authorize the transaction at the time the agent engages in the act and at the time the principal ratifies. The third party must also have the legal capacity to engage in the transaction.
5. The principal's affirmation must occur before the third party withdraws from the transaction.
6. The principal must observe the same formalities when approving the act done by the agent as would have been required to authorize it initially.

Liability in Agency Relationships

Frequently, a question arises as to which party, the principal or the agent, should be held liable for contracts formed by the agent or for torts or crimes committed by the agent. We look here at these aspects of agency law.

LIABILITY FOR CONTRACTS

Liability for contracts formed by an agent depends on how the principal is classified and on whether the actions of the agent were authorized or unauthorized. Principals are classified as disclosed, partially disclosed, or undisclosed.[10]

A **disclosed principal** is a principal whose identity is known by the third party at the time the contract is made by the agent. A **partially disclosed principal** is a principal whose identity is not known by the third party, but the third party knows that the agent is or may be acting for a principal at the time the contract is made. ■ EXAMPLE 22.14 Sarah has contracted with a real estate agent to sell certain property. She wishes to keep her identity a secret, but the agent can make it perfectly clear to a purchaser of the real estate that the agent is acting in an agency capacity. In this situation, Sarah is a partially disclosed principal. ■ An **undisclosed principal** is a principal whose identity is totally unknown by the third party, and the third party has no knowledge that the agent is acting in an agency capacity at the time the contract is made.

Authorized Acts If an agent acts within the scope of her or his authority, normally the principal is obligated to perform the contract regardless of whether the principal was disclosed, partially disclosed, or undisclosed. Whether the agent may also be held liable under the contract, however, depends on the disclosed, partially disclosed, or undisclosed status of the principal.

—Disclosed or Partially Disclosed Principal. A disclosed or partially disclosed principal is liable to a third party for a contract made by an agent who is acting within

10. *Restatement (Second) of Agency,* Section 4.

the scope of her or his authority. If the principal is disclosed, an agent has no contractual liability for the nonperformance of the principal or the third party. If the principal is partially disclosed, in most states the agent is also treated as a party to the contract, and the third party can hold the agent liable for contractual nonperformance.[11] The following case illustrates the rules that apply to contracts signed by agents on behalf of fully disclosed principals.

11. *Restatement (Second) of Agency,* Section 321.

CASE 22.3 McBride v. Taxman Corp.

Appellate Court of Illinois,
First District, 2002.
327 Ill.App.3d 992,
765 N.E.2d 51,
262 Ill.Dec. 225.
http://state.il.us/court/default.htm[a]

FACTS Walgreens Company entered into a lease with Taxman Corporation to operate a drugstore in Kedzie Plaza, a shopping center in Chicago, Illinois, owned by Kedzie Plaza Associates; Taxman was the center's property manager. The lease required the "Landlord" to promptly remove snow and ice from the center's sidewalks. Taxman also signed, on behalf of Kedzie Associates, an agreement with Arctic Snow and Ice Control, Inc., to remove ice and snow from the sidewalks surrounding the Walgreens store. On January 27, 1996, Grace McBride, a Walgreens employee, slipped and fell on snow and ice outside the entrance to the store. McBride filed a suit in an Illinois state court against Taxman and others, alleging, among other things, that Taxman had negligently failed to remove the accumulated ice and snow.[b] Taxman filed a motion for summary judgment in its favor, which the court granted. McBride appealed to a state intermediate appellate court.

a. On this page, click on "Appellate Court of Illinois." On the next page, in the "Appellate Court Documents" section, click on "Appellate Court Opinions." In the result, in the "Appellate Court" section, click on "2002." On the next page, in the "First District" section, click on "January." Finally, scroll to the bottom of the chart and click on the case name to access the opinion. The state of Illinois maintains this Web site.

b. McBride included complaints against Walgreens and Kedzie Associates in her suit but settled these complaints before trial.

ISSUE Could Taxman be held liable for McBride's injuries?

DECISION No. The state intermediate appellate court affirmed the judgment of the lower court. The appellate court held that Taxman entered into the snow removal contracts only as the agent of the owner, whose identity was fully disclosed.

REASON The court reasoned that as the agent for a disclosed principal, Taxman had no liability for the nonperformance of the principal or the third party to the contract. The court pointed out that "the Arctic proposal and contract was signed 'Kedzie Associates by the Taxman.' The Taxman-drafted portion of the contract contained a line above the signature of Taxman's director of property management stating 'The Taxman Corporation, agent for per contracts attached.' The latter document specifically stated that the contract was not an obligation of Taxman and that all liabilities were those of the owner and not Taxman. We conclude that Taxman was the management company for the property owner and entered into the two contracts for snow and ice removal only as the owner's agent. Taxman did not assume a contractual obligation to remove snow and ice; it merely retained Arctic as a contractor on behalf of the owner."

FOR CRITICAL ANALYSIS—Economic Consideration *Suppose that the Arctic contract had not identified Kedzie as the principal. Would the court's decision in this case have been different?*

—Undisclosed Principal. When neither the fact of agency nor the identity of the principal is disclosed, the undisclosed principal is fully bound to perform just as if the principal had been fully disclosed at the time the contract was made. The agent is also liable as a party to the contract.

When a principal's identity is undisclosed and the agent is forced to pay the third party, the agent is entitled to be indemnified (compensated) by the principal. The principal had a duty to perform, even though his or her identity was undisclosed,[12] and failure to do so will make the principal ultimately liable. Once the undisclosed principal's identity is revealed, the third party generally can elect to hold either the principal or the agent liable on the contract.

12. If the agent is a gratuitous agent, and the principal accepts the benefits of the agent's contract with a third party, then the principal will be liable to the agent on the theory of quasi contract (see Chapter 7).

Conversely, the undisclosed principal can hold the third party to the contract, *unless* (1) the undisclosed principal was expressly excluded as a party in the contract, (2) the contract is a negotiable instrument signed by the agent with no indication of signing in a representative capacity, or (3) the performance of the agent is personal to the contract, allowing the third party to refuse the principal's performance.

Unauthorized Acts If an agent has no authority but nevertheless contracts with a third party, the principal cannot be held liable on the contract. It does not matter whether the principal was disclosed, partially disclosed, or undisclosed. The *agent* is liable, however. ■ **EXAMPLE 22.15** Scranton signs a contract for the purchase of a truck, purportedly acting as an agent under authority granted by Johnson. In fact, Johnson has not given Scranton any such authority. Johnson refuses to pay for the truck, claiming that Scranton had no authority to purchase it. The seller of the truck is entitled to hold Scranton liable for payment. ■

If the principal is disclosed or partially disclosed, the agent is liable to the third party as long as the third party relied on the agency status. The agent's liability here is based on the breach of an implied warranty that the agent had authority to enter the contract, not on breach of the contract itself.[13] If the third party knows at the time the contract is made that the agent is mistaken about the extent of her or his authority, though, the agent is not liable. Similarly, if the agent indicates to the third party *uncertainty* about the extent of the authority, the agent is not personally liable.

Liability for E-Agents An electronic agent, or **e-agent,** is not a person but a semiautonomous computer program that is capable of executing specific tasks. E-agents used in e-commerce include software that can search through many databases and retrieve only information that is relevant for the user.

—Authorized versus Unauthorized Acts. In the past, standard agency principles have applied only to *human* agents, who have express or implied authority to enter into specific contracts. The resolution of many disputes depends on whether the human agent acted within the scope of his or her authority. What does this concept mean when dealing with an intelligent e-agent?

Consider an example. E-agents searching the Internet may run into a wide variety of "click-on" agreements (see Chapter 13), which, by necessity, contain many different terms and conditions. If the e-agent ignores the terms and conditions of a licensing agreement outlined in the click-on setting, is the user of the agent nonetheless bound by the agreement? Conversely, many click-on agreements exempt third parties from any liability resulting from the underlying product or service. Is the user of the e-agent bound by this particular term? With respect to human agents, the courts have occasionally found that an agent could not agree to such a term without explicit authority. It remains to be seen, though, whether a court will ever conclude that an e-agent lacked the authority to enter such an agreement.

—The Uniform Electronic Transactions Act. The Uniform Electronic Transactions Act (UETA), which was discussed in detail in Chapter 13, contains several provisions relating to the principal's liability for the actions of e-agents. The majority of the states have adopted these provisions. Section 15 of the UETA states that e-agents may enter into binding agreements on behalf of their principals. Thus, if you place an order over the Internet, the company (principal) whose system took the order via an e-agent cannot claim that it did not receive your order.

The UETA also stipulates that if an e-agent does not provide an opportunity to prevent errors at the time of the transaction, the other party to the transaction can avoid the transaction. For instance, if an e-agent fails to provide an on-screen confirmation of a purchase or sale, the other party can avoid the effect of any errors.

The UETA establishes that e-agents generally have the authority to bind the principal in contract—at least in those states that have adopted the act. Presumably, this means that the principal would be bound by the terms in a contract entered into by an e-agent. Another significant issue today is whether the actions of an e-agent might create liability for other parties—a topic we treat in this chapter's *Management Perspective* feature on the next page.

LIABILITY FOR TORTS AND CRIMES

Obviously, any person, including an agent, is liable for her or his own torts and crimes. Whether a principal can also be held liable for an agent's torts and crimes depends on several factors, which we examine here. In some situations, a principal may be held liable not only for the torts of an agent but also for the torts committed by an independent contractor.

Liability for Agent's Negligence As mentioned, an agent is liable for his or her own torts. A principal may also be liable for harm an agent caused to a third party under the

13. The agent is not liable on the contract because the agent was never intended personally to be a party to the contract.

MANAGEMENT PERSPECTIVE

Online Advertising and E-Agents

MANAGEMENT FACES A LEGAL ISSUE

In the online environment, the actions of an e-agent can at times create liability (debt) for the business that hired an advertising firm. As mentioned elsewhere in this chapter, *e-agents* are computer programs that are used in e-commerce to perform certain tasks. For example, an e-agent can be used to search the Web for the best price on a particular DVD and then offer links to the appropriate Web sites. Some e-agents can locate specific products in online catalogues and actually negotiate the acquisition and delivery of the products. E-agents used for shopping on the Web are commonly referred to as "bots" or "shopping bots." The fact that numerous e-agents are out there robotically shopping for people has caused problems in the online advertising industry.

Internet advertising firms frequently charge for their services based on the number of "impressions" an ad generates. The advertising industry does not have a universal definition of the term *impression,* however. Consequently, Internet publishers use many different methods to measure ad impressions. Some companies define an impression as a single ad that appears on a Web page when the page is displayed on a (human) viewer's screen; these companies filter out the times that the ad is found by an e-agent. Other companies include visits by automated e-agents in their definition of impressions, even though no human consumer has actually viewed the ad. What happens when a dispute between parties arises over the meaning of the term *impression*?

WHAT THE COURTS SAY

To date, at least one court has addressed a dispute between parties that had differing views of what the term *impressions* means in the context of online advertising. The case involved a Web hosting company (CI Host, Inc.), which had hired an Internet advertising company (Go2Net, Inc.) to generate a certain number of "impressions" on the Internet.[a] Payment for Go2Net's services was to be based on the number of impressions. CI Host and Go2Net did not specify in their contract what they meant by "impressions," however. CI Host assumed that "impressions" referred to the number of times the ad was sent to a computer screen and viewed by a human. Go2Net counted as "impressions" all of the times that the ads were found by e-agents, such as Web crawlers or bots. When the ads did not generate sufficient sales, CI Host canceled the advertising and refused to pay Go2Net for the number of impressions it claimed to have produced. In the lawsuit that followed, a Washington state appellate court ultimately held for Go2Net because of a clause in the parties' contract stating that "all impressions billed are based on Go2Net's ad engine count." Thus, Go2Net was allowed to count as impressions the number of times that e-agents found the Internet ad.

IMPLICATIONS FOR MANAGERS

It is difficult to know what the outcome of this suit might have been if the parties had not agreed that the number of impressions would be based on Go2Net's ad engine count. The lesson for managers of companies forming contracts for online advertising, though, is very clear: unless the contract specifically defines the term impression, *the number of total impressions could include those viewed by bots, making the cost per real impression implicitly much higher. For the purchaser of Internet advertising, the true cost of an Internet ad campaign could thus be more expensive than anticipated. For the seller of Internet advertising, including impressions made by bots could lead to higher advertising revenues but also, in the long run, to more dissatisfied customers. Eventually, the seller might even find it necessary to reduce the per-impression price to advertisers to remain competitive.*

a. *Go2Net, Inc. v. CI Host, Inc.,* 115 Wash.App. 73, 60 P.3d 1245 (2003).

doctrine of ***respondeat superior,***[14] a Latin term meaning "let the master respond." This doctrine is similar to the theory of strict liability discussed in Chapter 17. The doctrine imposes **vicarious liability,** or indirect liability, on the employer—that is, liability without regard to the personal fault of the employer for torts committed by an employee in the course or scope of employment.

—Determining the Scope of Employment. The key to determining whether a principal may be liable for the torts of an agent under the doctrine of *respondeat superior* is whether the torts are committed within the scope of the agency or employment. The *Restatement (Second) of Agency,* Section 229, indicates the factors that today's courts will consider in determining whether a particular act occurred within the course and scope of employment. These factors are as follows:

1. Whether the employee's act was authorized by the employer.
2. The time, place, and purpose of the act.
3. Whether the act was one commonly performed by employees on behalf of their employers.

14. Pronounced ree-*spahn*-dee-uht soo-*peer*-ee-your.

4. The extent to which the employer's interest was advanced by the act.
5. The extent to which the private interests of the employee were involved.
6. Whether the employer furnished the means or instrumentality (for example, a truck or a machine) by which the injury was inflicted.
7. Whether the employer had reason to know that the employee would do the act in question and whether the employee had ever done it before.
8. Whether the act involved the commission of a serious crime.

A useful insight into the "scope of employment" concept may be gained from the judge's classic distinction between a "detour" and a "frolic" in the case of *Joel v. Morison* (1834).[15] In this case, the English court held that if a servant merely took a detour from his master's business, the master will be responsible. If, however, the servant was on a "frolic of his own" and not in any way "on his master's business," the master will not be liable.

—Notice of Dangerous Conditions. The employer is charged with knowledge of any dangerous conditions discovered by an employee and pertinent to the employment situation. ■ EXAMPLE 22.16 Chad, a maintenance employee in Martin's apartment building, notices a lead pipe protruding from the ground in the building's courtyard. The employee neglects either to fix the pipe or to inform the employer of the danger. John falls on the pipe and is injured. The employer is charged with knowledge of the dangerous condition regardless of whether or not Chad actually informed the employer. That knowledge is imputed to the employer by virtue of the employment relationship. ■

Liability for Agent's Intentional Torts Most intentional torts that employees commit have no relation to their employment; thus, their employers will not be held liable. Employers can face liability for an agent's intentional torts in some situations, though, including those discussed next.

—Acts Committed within the Scope of Employment. The doctrine of *respondeat superior* applies to intentional torts as well as to negligence. Thus, an employer can be held indirectly liable for the intentional torts of an employee that are committed within the course and scope of employment. For instance, an employer is liable when an employee (such as a "bouncer" at a nightclub or a security guard at a department store) commits the tort of assault and battery or false imprisonment while acting within the scope of employment.

An employer may also be liable for permitting an employee to engage in reckless actions that can injure others. ■ EXAMPLE 22.17 An employer observes an employee smoking while filling containerized trucks with highly flammable liquids. Failure to stop the employee will cause the employer to be liable for any injuries that result if a truck explodes. ■

—Acts of Misrepresentation by an Agent. A principal is exposed to tort liability whenever a third person sustains a loss due to an agent's misrepresentation. The principal's liability depends on whether the agent was actually or apparently authorized to make representations and whether such representations were made within the scope of the agency. The principal is always directly responsible for an agent's misrepresentation made within the scope of the agent's authority. When a principal has placed an agent in a position of apparent authority—making it possible for the agent to defraud a third party—the principal may also be liable for the agent's fraudulent acts.

■ EXAMPLE 22.18 Assume that Bassett is a demonstrator for Moore's products. Moore sends Bassett to a home show to demonstrate the products and to answer questions from consumers. Moore has given Bassett authority to make statements about the products. If Bassett makes only true representations, all is fine; but if she makes false claims, Moore will be liable for any injuries or damages sustained by third parties in reliance on Bassett's false representations. ■

Liability for Independent Contractor's Torts Generally, an employer is not liable for physical harm caused to a third person by the negligent act of an independent contractor in the performance of the contract. This is because the employer does not have *the right to control* the details of an independent contractor's performance. Exceptions to this rule are made in certain situations, though, such as when unusually hazardous activities are involved. Typical examples of such activities include blasting operations, the transportation of highly volatile chemicals, or the use of poisonous gases. In these situations, an employer cannot be shielded from liability merely by using an independent contractor. Strict liability is imposed on the employer-principal as a matter of law. Also, in some states, strict liability may be imposed by statute.

Liability for Agent's Crimes An agent is liable for his or her own crimes. A principal or employer is not liable for

15. 6 Car. & P. 501, 172 Eng. Reprint 1338 (1834).

an agent's crime even if the crime was committed within the scope of authority or employment—unless the principal participated by conspiracy or other action. In some jurisdictions, under specific statutes, a principal may be liable for an agent's violation, in the course and scope of employment, of regulations, such as those governing sanitation, prices, weights, and the sale of liquor.

How Agency Relationships Are Terminated

Agency law is similar to contract law in that both an agency and a contract can be terminated by an act of the parties or by operation of law. Once the relationship between the principal and the agent has ended, the agent no longer has the right to bind the principal. For an agent's apparent authority to be terminated, though, third persons may also need to be notified that the agency has been terminated.

TERMINATION BY ACT OF THE PARTIES

An agency may be terminated by act of the parties in several ways, including those discussed here.

Lapse of Time An agency agreement may specify the time period during which the agency relationship will exist. If so, the agency ends when that time period expires. For instance, if the parties agree that the agency will begin on January 1, 2007, and end on December 31, 2008, the agency is automatically terminated on December 31, 2008. If no definite time is stated, then the agency continues for a reasonable time and can be terminated at will by either party. What constitutes a "reasonable time" depends, of course, on the circumstances and the nature of the agency relationship.

Purpose Achieved An agent can be employed to accomplish a particular objective, such as the purchase of livestock for a cattle rancher. In that situation, the agency automatically ends after the cattle have been purchased. If more than one agent is employed to accomplish the same purpose, such as the sale of real estate, the first agent to complete the sale automatically terminates the agency relationship for all the others.

Occurrence of a Specific Event An agency can be created to terminate on the happening of a certain event. If Posner appoints Rubik to handle her business affairs while she is away, the agency automatically terminates when Posner returns.

Mutual Agreement Recall from the chapters on contract law that parties can cancel (rescind) a contract by mutually agreeing to terminate the contractual relationship. The same holds true in agency law regardless of whether the agency contract is in writing or whether it is for a specific duration.

Termination by One Party As a general rule, either party can terminate the agency relationship. The agent's act is called a *renunciation of authority.* The principal's act is referred to as a *revocation of authority.* Although both parties have the *power* to terminate the agency, they may not possess the *right.*

Wrongful termination can subject the canceling party to a suit for damages. ■ EXAMPLE 22.19 Rawlins has a one-year employment contract with Munro to act as an agent in return for $35,000. Munro can discharge Rawlins before the contract period expires (Munro has the *power* to breach the contract). Munro, though, will be liable to Rawlins for monetary damages because Munro has no *right* to breach the contract. ■

A special rule applies in an *agency coupled with an interest.* This type of agency is not an agency in the usual sense because it is created for the agent's benefit instead of for the principal's benefit. ■ EXAMPLE 22.20 Julie borrows $5,000 from Rob, giving Rob some of her jewelry and signing a letter giving him the power to sell the jewelry as her agent if she fails to repay the loan. After receiving the $5,000 from Rob, Julie attempts to revoke Rob's authority to sell the jewelry as her agent. Julie would not succeed in this attempt because a principal cannot revoke an agency created for the agent's benefit. ■

Notice of Termination When an agency has been terminated by act of the parties, it is the principal's duty to inform any third parties who know of the existence of the agency that it has been terminated (although notice of the termination may be given by others).

—Agent's Authority Continues until Notified. An agent's authority continues until the agent receives some notice of termination. Notice to third parties follows the general rule that an agent's *apparent authority* continues until the third party receives notice (from any source) that such authority has been terminated. If the principal knows that a third party has dealt with the agent, the principal should notify that person directly. For third parties who have heard about the agency but have not yet dealt with the agent, *constructive notice* is sufficient.[16]

16. *Constructive notice* is information or knowledge of a fact imputed by law to a person if he or she could have discovered the fact by proper diligence. Constructive notice is often accomplished by newspaper publication.

—Form of Notice. No particular form is required for notice of agency termination to be effective. The principal can notify the agent directly, or the agent can learn of the termination through some other means. ■ EXAMPLE 22.21 Manning bids on a shipment of steel, and Stone is hired as an agent to arrange transportation of the shipment. When Stone learns that Manning has lost the bid, Stone's authority to make the transportation arrangement terminates. ■

If the agent's authority is written, it must be revoked in writing, and the writing must be shown to all people who saw the original writing that established the agency relationship. Sometimes, a written authorization (such as a power of attorney) contains an expiration date. The passage of the expiration date is sufficient notice of termination for third parties.

TERMINATION BY OPERATION OF LAW

Termination of an agency by operation of law occurs in the circumstances discussed here. Note that when an agency terminates by operation of law, there is no duty to notify third persons.

Death or Insanity The general rule is that the death or mental incompetence of either the principal or the agent automatically and immediately terminates the ordinary agency relationship. Knowledge of the death is not required. ■ EXAMPLE 22.22 Suppose that Geer sends Pyron to China to purchase a rare painting. Before Pyron makes the purchase, Geer dies. Pyron's agent status is terminated at the moment of Geer's death, even though Pyron does not know that Geer has died. ■ Some states, however, have enacted statutes changing this common law rule to make knowledge of the principal's death a requirement for agency termination.

An agent's transactions that occur after the death of the principal are not binding on the principal's estate.[17] ■ EXAMPLE 22.23 Assume that Carson is hired by Perry to collect a debt from Thomas (a third party). Perry dies, but Carson, not knowing of Perry's death, still collects the funds from Thomas. Thomas's payment to Carson is no longer legally sufficient to discharge the debt to Perry because Carson's authority to collect ended on Perry's death. If Carson absconds with the funds, Thomas is still liable for the debt to Perry's estate. ■

Impossibility When the specific subject matter of an agency is destroyed or lost, the agency terminates. ■ EXAMPLE 22.24 Bullard employs Gonzalez to sell Bullard's house. Prior to any sale, the house is destroyed by fire. In this situation, Gonzalez's agency and authority to sell Bullard's house terminate. ■ Similarly, when it is impossible for the agent to perform the agency lawfully because of war or a change in the law, the agency terminates.

Changed Circumstances When an event occurs that has such an unusual effect on the subject matter of the agency that the agent can reasonably infer that the principal will not want the agency to continue, the agency terminates. ■ EXAMPLE 22.25 Roberts hires Mullen to sell a tract of land for $20,000. Subsequently, Mullen learns that there is oil under the land and that the land is worth $1 million. The agency and Mullen's authority to sell the land for $20,000 are terminated. ■

Bankruptcy If either the principal or the agent petitions for bankruptcy, the agency is *usually* terminated. In certain circumstances, as when the agent's financial status is irrelevant to the purpose of the agency, the agency relationship may continue. Insolvency (defined as the inability to pay debts when they become due or when liabilities exceed assets), as distinguished from bankruptcy, does not necessarily terminate the relationship.

War When the principal's country and the agent's country are at war with each other, the agency is terminated. In this situation, the agency is automatically suspended or terminated because there is no way to enforce the legal rights and obligations of the parties.

17. The Uniform Commercial Code (UCC) provides an exception to this rule in banking under which the bank, as the agent of the customer, can continue to exercise specific types of authority even after the customer has died or become mentally incompetent unless it has knowledge of the death or incompetence [UCC 4–405]. Even with knowledge of the customer's death, the bank has authority for ten days following the customer's death to honor checks in the absence of a stop-payment order.

TERMS AND CONCEPTS

agency 442
disclosed principal 451
e-agent 453
equal dignity rule 449
fiduciary 442
independent contractor 443
notary public 449
partially disclosed principal 451
power of attorney 449
ratification 446
respondeat superior 454
undisclosed principal 451
vicarious liability 454

CHAPTER SUMMARY Agency Relationships

Agency Relationships (See pages 442–445.)	In a *principal-agent* relationship, an agent acts on behalf of and instead of the principal in dealing with third parties. An employee who deals with third parties normally is an agent. An independent contractor is not an employee, and the employer has no control over the details of physical performance. An independent contractor may or may not be an agent.
How Agency Relationships Are Formed (See pages 445–446.)	Agency relationships may be formed by agreement, by ratification, by estoppel, and by operation of law.
Duties of Agents and Principals (See pages 446–449.)	1. *Duties of the agent*— a. Performance—The agent must use reasonable diligence and skill in performing her or his duties or use the special skills that the agent has represented to the principal that the agent possesses. b. Notification—The agent is required to notify the principal of all matters that come to his or her attention concerning the subject matter of the agency. c. Loyalty—The agent has a duty to act solely for the benefit of the principal and not in the interest of the agent or a third party. d. Obedience—The agent must follow all lawful and clearly stated instructions of the principal. e. Accounting—The agent has a duty to make available to the principal records of all property and money received and paid out on behalf of the principal. 2. *Duties of the principal*— a. Compensation—Except in a gratuitous agency relationship, the principal must pay the agreed-on value (or reasonable value) for an agent's services. b. Reimbursement and indemnification—The principal must reimburse the agent for all sums of money disbursed at the request of the principal and for all sums of money the agent disburses for necessary expenses in the course of reasonable performance of his or her agency duties. c. Cooperation—A principal must cooperate with and assist an agent in performing her or his duties. d. Safe working conditions—A principal must provide safe working conditions for the agent-employee.
Agent's Authority (See pages 449–451.)	1. *Express authority*—Can be oral or in writing. Authorization must be in writing if the agent is to execute a contract that must be in writing. 2. *Implied authority*—Authority customarily associated with the position of the agent or authority that is deemed necessary for the agent to carry out expressly authorized tasks. 3. *Apparent authority*—Exists when the principal, by word or action, causes a third party reasonably to believe that an agent has authority to act, even though the agent has no express or implied authority. 4. *Ratification*—The affirmation by the principal of an agent's unauthorized action or promise. For the ratification to be effective, the principal must be aware of all material facts.
Liability in Agency Relationships (See pages 451–456.)	1. *Liability for contracts*—If the principal's identity is disclosed or partially disclosed at the time the agent forms a contract with a third party, the principal is liable to the third party under the contract if the agent acted within the scope of his or her authority. If the principal's identity is undisclosed at the time of contract formation, the agent is personally liable to the third party, but if the agent acted within the scope of his or her authority, the principal is also bound by the contract. 2. *Liability for agent's negligence*—Under the doctrine of *respondeat superior,* the principal is liable for any harm caused to another through the agent's torts if the agent was acting within the scope of her or his employment at the time the harmful act occurred.

CHAPTER SUMMARY Agency Relationships—Continued

Liability in Agency Relationships—Continued	3. *Liability for agent's intentional torts*—Usually, employers are not liable for the intentional torts that their agents commit, unless: a. The acts are committed within the scope of employment and thus the doctrine of *respondeat superior* applies. b. The employer allowed an employee to engage in reckless acts that caused injury to another. c. The agent's misrepresentation causes a third party to sustain damage and the agent had either actual or apparent authority to act. 4. *Liability for independent contractor's torts*—A principal is not liable for harm caused by an independent contractor's negligence, unless hazardous activities are involved (in which situation the principal is strictly liable for any resulting harm) or other exceptions apply. 5. *Liability for agent's crimes*—An agent is responsible for his or her own crimes, even if the crimes were committed while the agent was acting within the scope of authority or employment. A principal will be liable for an agent's crime only if the principal participated by conspiracy or other action or (in some jurisdictions) if the agent violated certain government regulations in the course of employment.
How Agency Relationships Are Terminated (See pages 456–457.)	1. *By act of the parties*— a. Lapse of time (when the parties specified a definite time for the duration of the agency when it was established). b. Purpose achieved. c. Occurrence of a specific event. d. Mutual rescission (requires mutual consent of principal and agent). e. Termination by act of either the principal (revocation) or the agent (renunciation). (A principal cannot revoke an agency coupled with an interest.) f. Notice to third parties is required when an agency is terminated by act of the parties. Direct notice is required for those who have previously dealt with the agency; constructive notice will suffice for all other third parties. 2. *By operation of law*— a. Death or mental incompetence of either the principal or the agent (except when an agency is coupled with an interest). b. Impossibility (when the purpose of the agency cannot be achieved because of an event beyond the parties' control). c. Changed circumstances (in which it would be inequitable to require that the agency be continued). d. Bankruptcy of the principal or the agent, or war between the principal's and agent's countries. e. Notice to third parties is not required when an agency is terminated by operation of law.

FOR REVIEW

Answers for the even-numbered questions in this For Review *section can be found in Appendix A at the end of this text.*

1. What is the difference between an employee and an independent contractor?
2. How do agency relationships arise?
3. What duties do agents and principals owe to each other?
4. When is a principal liable for the agent's actions with respect to third parties? When is the agent liable?
5. What are some of the ways in which an agency relationship can be terminated?

QUESTIONS AND CASE PROBLEMS

22–1. Ratification by Principal. Springer was a political candidate running for congressional office. He was operating on a tight budget and instructed his campaign staff not to purchase any campaign materials without his explicit authorization. In spite of these instructions, one of his campaign workers ordered Dubychek Printing Co. to print some promotional materials for Springer's campaign. When the printed materials were received, Springer did not return them but instead used them during his campaign. When Springer failed to pay for the materials, Dubychek sued for recovery of the price. Springer contended that he was not liable on the sales contract because he had not authorized his agent to purchase the printing services. Dubychek argued that the campaign worker was Springer's agent and that the worker had authority to make the printing contract. Additionally, Dubychek claimed that even if the purchase was unauthorized, Springer's use of the materials constituted ratification of his agent's unauthorized purchase. Is Dubychek correct? Explain.

QUESTION WITH SAMPLE ANSWER

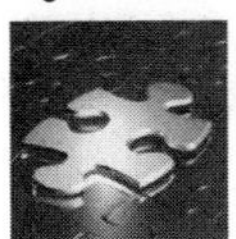

22–2. Stephen Hemmerling was a driver for the Happy Cab Co. Hemmerling paid certain fixed expenses and abided by a variety of rules relating to the use of the cab, the hours that could be worked, and the solicitation of fares, among other things. Rates were set by the state. Happy Cab did not withhold taxes from Hemmerling's pay. While driving the cab, Hemmerling was injured in an accident and filed a claim against Happy Cab in a Nebraska state court for workers' compensation benefits. Such benefits are not available to independent contractors. On what basis might the court hold that Hemmerling is an employee? Explain.

For a sample answer to this question, go to Appendix B at the end of this text.

22–3. Liability for Employee's Acts. Federated Financial Reserve Corp. leases consumer and business equipment. As part of its credit approval and debt-collection practices, Federated hires credit collectors, whom it authorizes to obtain credit reports on its customers. Janice Caylor, a Federated collector, used this authority to obtain a report on Karen Jones, who was not a Federated customer but who was the ex-wife of Caylor's roommate, Randy Lind. When Jones discovered that Lind had her address and how he had obtained it, she filed a suit in a federal district court against Federated and others. Jones claimed in part that they had violated the Fair Credit Reporting Act, the goal of which is to protect consumers from the improper use of credit reports. Under what theory might an employer be held liable for an agent-employee's violation of a statute? Does that theory apply in this case? Explain. [*Jones v. Federated Financial Reserve Corp.*, 144 F.3d 961 (6th Cir. 1998)]

22–4. Agent's Duties to Principal. Ana Barreto and Flavia Gugliuzzi asked Ruth Bennett, a real estate salesperson who worked for Smith Bell Real Estate, to list for sale their house in the Pleasant Valley area of Underhill, Vermont. Diana Carter, a California resident, visited the house as a potential buyer. Bennett worked under the supervision of David Crane, an officer of Smith Bell. Crane knew, but did not disclose to Bennett or Carter, that the house was subject to frequent and severe winds, that a window had blown in years earlier, and that other houses in the area had suffered wind damage. Crane knew of this because he lived in the Pleasant Valley area, had sold a number of nearby properties, and had been Underhill's zoning officer. Many valley residents, including Crane, had wind gauges on their homes to measure and compare wind speeds with their neighbors. Carter bought the house, and several months later, high winds blew in a number of windows and otherwise damaged the property. Carter filed a suit in a Vermont state court against Smith Bell and others, alleging fraud. She argued in part that Crane's knowledge of the winds was imputable to Smith Bell. Smith Bell responded that Crane's knowledge was obtained outside the scope of employment. What is the rule regarding how much of an agent's knowledge a principal is assumed to know? How should the court rule in this case? Why? [*Carter v. Gugliuzzi*, 168 Vt. 48, 716 A.2d 17 (1998)]

CASE PROBLEM WITH SAMPLE ANSWER

22–5. Ford Motor Credit Co. is a subsidiary of Ford Motor Co. with its own offices, officers, and directors. Ford Credit buys contracts and leases of automobiles entered into by dealers and consumers. Ford Credit also provides inventory financing for dealers' purchases of Ford and non-Ford vehicles and makes loans to Ford and non-Ford dealers. Dealers and consumers are not required to finance their purchases or leases of Ford vehicles through Ford Credit. Ford Motor is not a party to the agreements between Ford Credit and its customers and does not directly receive any payments under those agreements. Also, Ford Credit is not subject to any agreement with Ford Motor "restricting or conditioning" its ability to finance the dealers' inventories or the consumers' purchases or leases of vehicles. A number of plaintiffs filed a product liability suit in a Missouri state court against Ford Motor. Ford Motor claimed that the court did not have venue. The plaintiffs asserted that Ford Credit, which had an office in the jurisdiction, acted as Ford's "agent for the transaction of its usual and customary business" there. Is Ford Credit an agent of Ford Motor? Discuss. [*State ex rel. Ford Motor Co. v. Bacon*, 63 S.W.3d 641 (Mo. 2002)]

After you have answered this problem, compare your answer with the sample answer given on the Web site that accompanies this text. Go to **academic.cengage.com/blaw/wbl**, select "Chapter 22," and click on "Case Problem with Sample Answer."

22–6. Liability for Independent Contractor's Torts. Greif Brothers Corp., a steel drum manufacturer, owned and operated a manufacturing plant in Youngstown, Ohio. In 1987, Lowell Wilson, the

plant superintendent, hired Youngstown Security Patrol, Inc. (YSP), a security company, to guard Greif property and "deter thieves and vandals." Some YSP security guards, as Wilson knew, carried firearms. Eric Bator, a YSP security guard, was not certified as an armed guard but nevertheless took his gun, in a briefcase, to work. While working at the Greif plant on August 12, 1991, Bator fired his gun at Derrell Pusey, in the belief that Pusey was an intruder. The bullet struck and killed Pusey. Pusey's mother filed a suit in an Ohio state court against Greif and others, alleging in part that her son's death was the result of YSP's negligence, for which Greif was responsible. Greif filed a motion for a directed verdict. What is the plaintiff's best argument that Greif is responsible for YSP's actions? What is Greif's best defense? Explain. [*Pusey v. Bator,* 94 Ohio St.3d 275, 762 N.E.2d 968 (2002)]

22–7. **Principal's Duties to Agent.** Josef Boehm was an officer and the majority shareholder of Alaska Industrial Hardware, Inc. (AIH), in Anchorage, Alaska. In August 2001, Lincolnshire Management, Inc., in New York, created AIH Acquisition Corp. to buy AIH. The three firms signed a "commitment letter" to negotiate "a definitive stock purchase agreement" (SPA). In September, Harold Snow and Ronald Braley began to work, on Boehm's behalf, with Vincent Coyle, an agent for AIH Acquisition, to produce an SPA. They exchanged many drafts and dozens of e-mails. Finally, in February 2002, Braley told Coyle that Boehm would sign the SPA "early next week." That did not occur, however, and at the end of March, after more negotiations and drafts, Boehm demanded more money. AIH Acquisition agreed and, following more work by the agents, another SPA was drafted. In April, the parties met in Anchorage. Boehm still refused to sign. AIH Acquisition and others filed a suit in a federal district court against AIH. Did Boehm violate any of the duties that principals owe to their agents? If so, which duty, and how was it violated? Explain. [*AIH Acquisition Corp. v. Alaska Industrial Hardware, Inc.,* __ F.Supp.2d __ (S.D.N.Y. 2004)]

A QUESTION OF ETHICS

22–8. In 1990, the Internal Revenue Service (IRS) determined that a number of independent contractors working for Microsoft Corp. were actually employees of the company for tax purposes. The IRS arrived at this conclusion based on the significant control that Microsoft exercised over the independent contractors' work performance. As a result of the IRS's findings, Microsoft was ordered to pay back payroll taxes for hundreds of independent contractors who should have been classified as employees. Rather than contest the ruling, Microsoft required most of the workers in question, as well as a number of its other independent contractors, to become associated with employment agencies and work for Microsoft as temporary workers ("temps") or lose the opportunity to work for Microsoft. Workers who refused to register with employment agencies, as well as some who did register, sued Microsoft. The workers alleged that they were actually employees of the company and, as such, entitled to participate in Microsoft's stock option plan for employees. Microsoft countered that it need not provide such benefits because each of the workers had signed an independent-contractor agreement specifically stating that the worker was responsible for his or her own benefits. In view of these facts, consider the following questions. [*Vizcaino v. U.S. District Court for the Western District of Washington,* 173 F.3d 713 (9th Cir. 1999)]

1. Using the criteria that courts use to determine a worker's status, were the workers who sued Microsoft in this case independent contractors or employees? Explain your answer.
2. Normally, when a company hires temporary workers from an employment agency, the agency—not the employer—is responsible for paying Social Security taxes and other withholding taxes. Yet the U.S. Court of Appeals for the Ninth Circuit held that being an employee of a temporary employment agency did not preclude the employee from having the status of a common law employee of Microsoft at the same time. Is this fair to the employer? Why or why not?
3. Generally, do you believe that Microsoft was trying to "skirt the law"—and its ethical responsibilities—by requiring its employees to sign up as "temps"?
4. Each of the employees involved in this case had signed an independent-contractor agreement. In view of this fact, was the IRS's 1990 decision fair to Microsoft? Why or why not?

VIDEO QUESTION

22–9. Go to this text's Web site at **academic.cengage.com/blaw/wbl** and select "Chapter 22." Click on "Video Questions" and view the video titled *Fast Times.* Then answer the following questions.

1. Recall from the video that Brad (Judge Reinhold) is told to deliver an order of Captain Hook Fish and Chips to IBM. Is Brad an employee or an independent contractor? Why?
2. Assume that Brad is an employee and agent of Captain Hook Fish and Chips. What duties does he owe Captain Hook Fish and Chips? What duties does Captain Hook Fish and Chips, as principal, owe to Brad?
3. In the video, Brad throws part of his uniform and several bags of the food that he is supposed to deliver out of his car window while driving. If Brad is an agent-employee and his actions cause injury to a person or property, can Captain Hook Fish and Chips be held liable? Why or why not? What should Captain Hook argue to avoid liability for Brad's actions?

Agency Relationships

For updated links to resources available on the Web, as well as a variety of other materials, visit this text's Web site at

academic.cengage.com/blaw/wbl

The Legal Information Institute at Cornell University is an excellent source for information on agency law, including court cases involving agency concepts. Go to

http://www.law.cornell.edu/wex/index.php/Agency

For information on the *Restatements of the Law,* including planned revisions, go to the American Law Institute's Web site at

http://www.ali.org

ONLINE LEGAL RESEARCH

Go to the *Fundamentals of Business Law* home page at academic.cengage.com/blaw/wbl, select "Chapter 22," and click on "Internet Exercises." There you will find the following Internet research exercises that you can perform to learn more about topics covered in this chapter.

Activity 22–1: LEGAL PERSPECTIVE—Employees or Independent Contractors?

Activity 22–2: MANAGEMENT PERSPECTIVE—Liability in Agency Relationships

BEFORE THE TEST

Go to the *Fundamentals of Business Law* home page at academic.cengage.com/blaw/wbl, select "Chapter 22," and click on "Interactive Quizzes." You will find a number of interactive questions relating to this chapter.

CHAPTER 23
Employment Law

LEARNING OBJECTIVES

After reading this chapter, you should be able to answer the following questions:

1. What is the employment-at-will doctrine? When and why are exceptions to this doctrine made?
2. What federal statute governs working hours and wages? What federal statutes govern labor unions and collective bargaining?
3. What federal law was enacted to protect the health and safety of employees? What are workers' compensation laws?
4. Generally, what kind of conduct is prohibited by Title VII of the Civil Rights Act of 1964, as amended?
5. What remedies are available under Title VII of the 1964 Civil Rights Act, as amended?

Until the early 1900s, most employer-employee relationships were governed by the common law. Today, the workplace is regulated extensively by statutes and administrative agency regulations. Recall from Chapter 1 that common law doctrines apply only to areas *not* covered by statutory law. Common law doctrines have thus been displaced to a large extent by statutory law.

In the 1930s, during the Great Depression, both state and federal governments began to regulate employment relationships. Legislation during the 1930s and subsequent decades established the right of employees to form labor unions. At the heart of labor rights is the right to unionize and bargain with management for improved working conditions, salaries, and benefits. A succession of other laws during and since the 1930s provided further protection for employees. Today's employers must comply with a myriad of laws and regulations to ensure that employee rights are protected. In this chapter, we look at the most significant laws regulating employment relationships, including those that prohibit employment discrimination.

Employment at Will

Traditionally, employment relationships have generally been governed by the common law doctrine of **employment at will.** Other common law rules governing employment relationships—including rules under contract, tort, and agency law—have already been discussed at length in previous chapters of this text.

Given that many employees (those who deal with third parties) are normally deemed agents of an employer, agency concepts are especially relevant in the employment context. The distinction under agency law between employee status and independent-contractor status is also relevant to employment relationships. Generally, the laws discussed in this chapter apply only to the employer-employee relationship; they do not apply to independent contractors.

Under the employment-at-will doctrine, either party may terminate the employment relationship at any time and for any reason, unless doing so would violate the provisions of an employment contract. Nonetheless, federal and state statutes governing employment relationships

prevent the doctrine from being applied in a number of circumstances. Today, an employer is not permitted to fire an employee if to do so would violate a federal or state employment statute, such as one prohibiting employment termination for discriminatory reasons.

EXCEPTIONS TO THE EMPLOYMENT-AT-WILL DOCTRINE

Because of the harsh effects of the employment-at-will doctrine for employees, the courts have carved out various exceptions to the doctrine. These exceptions are based on contract theory, tort theory, and public policy.

Exceptions Based on Contract Theory Some courts have held that an *implied* employment contract exists between an employer and an employee. If an employee is fired outside the terms of the implied contract, he or she may succeed in an action for breach of contract even though no written employment contract exists. ■ EXAMPLE 23.1 Suppose that an employer's manual or personnel bulletin states that, as a matter of policy, workers will be dismissed only for good cause. If the employee is aware of this policy and continues to work for the employer, a court may find that there is an implied contract based on the terms stated in the manual or bulletin.[1] ■ Generally, the key consideration in determining whether an employment manual creates an implied contractual obligation is the employee's reasonable expectations.

An employer's oral promises to employees regarding discharge policy may also be considered part of an implied contract. If the employer fires a worker in a manner contrary to what was promised, a court may hold that the employer has violated the implied contract and is liable for damages. In some cases, courts have held that an implied employment contract exists even though employees agreed in writing to be employees at will.[2]

Exceptions Based on Tort Theory In a few situations, the discharge of an employee may give rise to an action for wrongful discharge under tort theories (see Chapter 4). Abusive discharge procedures may result in a suit for intentional infliction of emotional distress or defamation. In addition, some courts have permitted workers to sue their employers under the tort theory of fraud.

Exceptions Based on Public Policy The most widespread common law exception to the employment-at-will doctrine is made on the basis of public policy. Courts may apply this exception when an employer fires a worker for reasons that violate a fundamental public policy of the jurisdiction. Generally, the courts require that the public policy involved be expressed clearly in the statutory law governing the jurisdiction. ■ EXAMPLE 23.2 As you will read later in this chapter, employers with fifty or more employees are required by the Family and Medical Leave Act (FMLA) to give employees up to twelve weeks of unpaid family or medical leave per year. Mila's employer, however, has only forty employees and is thus not covered by the federal law. Nonetheless, if Mila is fired from her job because she takes three weeks of unpaid family leave to help her son through a difficult surgery, a court may deem that the employer's actions violated the public policy expressed in the FMLA. ■

Sometimes, an employer will direct employees to perform an illegal act and fire them if they refuse to do so. At other times, employers will fire or discipline employees who "blow the whistle" on the employer's wrongdoing. **Whistleblowing** occurs when an employee tells government authorities, upper-level managers, or the press that her or his employer is engaged in some unsafe or illegal activity. Whistleblowers on occasion have been protected from wrongful discharge for reasons of public policy.

Today, whistleblowers have some protection under statutory law. Most states have enacted so-called whistleblower statutes that protect a whistleblower from subsequent retaliation by the employer. On the federal level, the Whistleblower Protection Act of 1989[3] protects federal employees who blow the whistle on their employers from retaliatory actions. Whistleblower statutes may also offer an incentive to disclose information by providing the whistleblower with a monetary reward. For instance, under the federal False Claims Reform Act of 1986,[4] a whistleblower who has disclosed information relating to a fraud perpetrated against the U.S. government will receive between 15 and 25 percent of the proceeds if the government brings a suit against the wrongdoer.

WRONGFUL DISCHARGE

Whenever an employer discharges an employee in violation of an employment contract or a statute protecting employees, the employee may bring an action for **wrongful discharge.** Even if an employer's actions do not violate any provisions in an employment contract or a statute, the

1. See, for example, *Pepe v. Rival Co.,* 85 F.Supp.2d 349 (D.N.J. 1999).
2. See, for example, *Kuest v. Regent Assisted Living, Inc.,* 111 Wash.App. 36, 43 P.3d 23 (2002).
3. 5 U.S.C. Section 1201.
4. 31 U.S.C. Sections 3729–3733. This act amended the False Claims Act of 1863.

employer may still be subject to liability under a common law doctrine, such as a tort theory or agency.

Wage and Hour Laws

In the 1930s, Congress enacted several laws regulating the wages and working hours of employees. In 1931, Congress passed the Davis-Bacon Act,[5] which requires contractors and subcontractors working on government construction projects to pay "prevailing wages" to their employees. In 1936, the Walsh-Healey Act[6] was passed. This act requires that a minimum wage, as well as overtime pay at 1.5 times regular pay rates, be paid to employees of manufacturers or suppliers entering into contracts with agencies of the federal government.

In 1938, Congress passed the Fair Labor Standards Act (FLSA).[7] This act extended wage-hour requirements to cover all employers engaged in interstate commerce or in the production of goods for interstate commerce, plus selected types of other businesses. We examine here the FLSA's provisions in regard to child labor, maximum hours, and minimum wages.

Child Labor

The FLSA prohibits oppressive child labor. Children under fourteen years of age are allowed to do certain types of work, such as deliver newspapers, work for their parents, and work in the entertainment and (with some exceptions) agricultural areas. Children who are fourteen or fifteen years of age are allowed to work, but not in hazardous occupations. There are also numerous restrictions on how many hours per day and per week they can work. ■ EXAMPLE 23.3 Children under the age of sixteen cannot work during school hours, for more than three hours on a school day (or eight hours on a nonschool day), for more than eighteen hours during a school week (or forty hours during a nonschool week), or before 7 A.M. or after 7 P.M. (9 P.M. during the summer). ■ Many states require persons under sixteen years of age to obtain work permits.

Working times and hours are not restricted for persons between the ages of sixteen and eighteen, but they cannot be employed in hazardous jobs or in jobs detrimental to their health and well-being. None of these restrictions apply to persons over the age of eighteen.

Wages and Hours

The FLSA provides that a **minimum wage** of a specified amount (as of the writing of this edition, $5.15 per hour) must be paid to employees in covered industries. Congress periodically revises this minimum wage.[8] Under the FLSA, the term *wages* includes the reasonable cost of the employer in furnishing employees with board, lodging, and other facilities if they are customarily furnished by that employer.

Under the FLSA, employees who work more than forty hours per week normally must be paid 1.5 times their regular pay for all hours over forty. Note that the FLSA overtime provisions apply only after an employee has worked more than forty hours per *week*. Thus, employees who work for ten hours a day, four days per week, are not entitled to overtime pay because they do not work more than forty hours a week.

Overtime Exemptions

Certain employees are exempt from the overtime provisions of the FLSA. These exemptions include employees whose jobs are categorized as executive, administrative, or professional, as well as outside salespersons and computer employees. In the past, to fall into one of these exemptions, an employee had to earn more than a specified salary threshold and devote a certain percentage of work time to the performance of specific types of duties. Because the salary limits were low and the duties tests were complex and confusing, some employers in the last few decades have been able to avoid paying overtime wages to their employees. This prompted the U.S. Department of Labor to substantially revise the regulations pertaining to overtime for the first time in over fifty years.

New rules implemented in August 2004 expanded the number of workers eligible for overtime by nearly tripling the salary threshold.[9] Under the new provisions, workers earning less than $23,660 a year are guaranteed overtime pay for working more than forty hours per week (the previous ceiling was $8,060). The exemptions to the overtime-pay requirement do not apply to manual laborers or other workers who perform work involving repetitive operations with their hands (such as nonmanagement production-line employees, for example). The exemptions also do not apply to police, firefighters, licensed nurses, and other public-safety workers. White-collar workers who earn more than $100,000 a year, computer programmers, dental hygienists, and insurance adjusters are typically exempt—though they must also meet certain other criteria. Employers can continue to pay overtime to ineligible employees if they want to do so, but cannot waive or reduce the overtime requirements of the FLSA.

5. 40 U.S.C. Sections 276a–276a-5.
6. 41 U.S.C. Sections 35–45.
7. 29 U.S.C. Sections 201–260.
8. Note that many state and local governments also have minimum-wage laws; these laws provide for higher minimum-wage rates than required by the federal government.
9. 29 C.F.R. Section 541.

Labor Unions

In the 1930s, in addition to wage-hour laws, the government also enacted the first of several labor laws. These laws protect employees' rights to join labor unions, to bargain with management over the terms and conditions of employment, and to conduct strikes.

Federal labor laws governing union-employer relations have developed considerably since the first law was enacted in 1932. Initially, the laws were concerned with protecting the rights and interests of workers. Subsequent legislation placed some restraints on unions and granted rights to employers. We look here at four major federal statutes regulating union-employer relations.

NORRIS-LAGUARDIA ACT

In 1932, Congress protected peaceful strikes, picketing, and boycotts in the Norris-LaGuardia Act.[10] The statute restricted the power of federal courts to issue injunctions against unions engaged in peaceful strikes. In effect, this act established a national policy permitting employees to organize.

NATIONAL LABOR RELATIONS ACT

One of the foremost statutes regulating labor is the National Labor Relations Act (NLRA) of 1935.[11] This act established the rights of employees to engage in collective bargaining and to strike.

Workers Protected by the NLRA To be protected under the NLRA, an individual must be an *employee,* as that term is defined in the statute. Courts have long held that job applicants fall within the definition (otherwise, the NLRA's ban on discrimination in hiring—to be discussed shortly—would mean nothing). Additionally, the United States Supreme Court has held that individuals who are hired by a union to organize a company are to be considered employees of the company for NLRA purposes.[12]

The act specifically defined a number of employer practices as unfair to labor:

1. Interference with the efforts of employees to form, join, or assist labor organizations or with the efforts of employees to engage in concerted activities for their mutual aid or protection.
2. An employer's domination of a labor organization or contribution of financial or other support to it.
3. Discrimination in the hiring or awarding of tenure to employees based on union affiliation.
4. Discrimination against employees for filing charges under the act or giving testimony under the act.
5. Refusal to bargain collectively with the duly designated representative of the employees.

The National Labor Relations Board The NLRA also created the National Labor Relations Board (NLRB) to oversee union elections and to prevent employers from engaging in unfair and illegal union activities and unfair labor practices. The NLRB has the authority to investigate employees' charges of unfair labor practices and to file complaints against employers in response to these charges. When violations are found, the NLRB may also issue cease-and-desist orders compelling the employers to cease engaging in the unfair practices. Cease-and-desist orders can be enforced by a circuit court of appeals if necessary. Arguments over alleged unfair labor practices are first decided by the NLRB and may then be appealed to a federal court.

Under the NLRA, employers and unions have a duty to bargain in good faith. Bargaining over certain subjects is mandatory, and a party's refusal to bargain over these subjects is an unfair labor practice that can be reported to the NLRB. The question in the following case was whether an employer's use of hidden surveillance cameras was subject to mandatory bargaining.

10. 29 U.S.C. Sections 101–110, 113–115.
11. 20 U.S.C. Section 151.
12. *National Labor Relations Board v. Town & Country Electric, Inc.,* 516 U.S. 85, 116 S.Ct. 450, 133 L.Ed.2d 371 (1995).

CASE 23.1 National Steel Corp. v. NLRB

United States Court of Appeals,
Seventh Circuit, 2003.
324 F.3d 928.
http://caselaw.lp.findlaw.com/data2/cires/7th/013798p.pdf[a]

a. This is a page within the Web site maintained by FindLaw (now a part of West Group).

FACTS National Steel Corporation operates a plant in Granite City, Illinois, where it employs approximately three thousand employees, who are represented by ten different unions and covered by seven different collective bargaining agreements. National Steel uses more than a hundred video cameras in plain view to monitor areas of the plant and peri-

CASE 23.1—CONTINUED

odically installs hidden cameras to investigate suspected misconduct. In February 1999, National Steel installed a hidden camera to discover who was using a manager's office when the manager was not at work. The camera revealed a union member using the office to make long-distance phone calls. When National Steel discharged the employee, the union asked the company about other hidden cameras and indicated that it wanted to bargain over their use. National Steel refused to supply the information. The union filed a charge with the National Labor Relations Board (NLRB), which ordered National Steel to provide the information and bargain over the use of the cameras. National Steel appealed to the U.S. Court of Appeals for the Seventh Circuit.

ISSUE Can an employer be required to bargain over the use of hidden cameras in the workplace?

DECISION Yes. The U.S. Court of Appeals for the Seventh Circuit upheld the NLRB's order to National Steel to bargain over the use of hidden surveillance cameras in the workplace. The court emphasized that this order did not prohibit the use of the cameras, but only made that use a subject of collective bargaining.

REASON The court repeated the NLRB's conclusion that "the use of hidden surveillance cameras is a mandatory subject of collective bargaining because * * * the installation and use of such cameras [is] analogous to physical examinations, drug/alcohol testing requirements, and polygraph testing, all of which the Board has found to be mandatory subjects." The court noted that "hidden cameras are focused primarily on the 'working environment' that employees experience on a daily basis and are used to expose misconduct or violations of the law by employees or others." Such "changes in an employer's methods have serious implications for its employees' job security" and "the use of such devices is not entrepreneurial in character [and] is not fundamental to the basic direction of the enterprise." The court also explained that the order "only requires National Steel to negotiate with the unions over the company's installation and use of hidden surveillance cameras and * * * does not dictate how the legitimate interests of the parties are to be accommodated in the process. * * * [I]t simply directs National Steel to initiate an accommodation process, and to provide assertedly confidential information in accord with whatever accommodation the parties agree upon."

WHY IS THIS CASE IMPORTANT? *Employers should be aware that courts might require collective bargaining over any working conditions that could significantly affect the employees' daily work environment and job security.*

Labor-Management Relations Act

The Labor-Management Relations Act (LMRA) of 1947[13] was passed to proscribe certain unfair union practices, such as the *closed shop*. A **closed shop** requires union membership by its workers as a condition of employment. Although the act made the closed shop illegal, it preserved the legality of the union shop. A **union shop** does not require membership as a prerequisite for employment but can, and usually does, require that workers join the union after a specified amount of time on the job.

The LMRA also prohibited unions from refusing to bargain with employers, engaging in certain types of picketing, and *featherbedding*—causing employers to hire more employees than necessary. The act also allowed individual states to pass their own **right-to-work laws,** which make it illegal for union membership to be required for *continued* employment in any establishment. Thus, union shops are technically illegal in the twenty-three states that have right-to-work laws.

13. 29 U.S.C. Sections 141 *et seq.*

Labor-Management Reporting and Disclosure Act

In 1959, Congress enacted the Labor-Management Reporting and Disclosure Act (LMRDA).[14] The act established an employee bill of rights and reporting requirements for union activities. The act strictly regulates unions' internal business procedures, including union elections. For example, the LMRDA requires a union to hold regularly scheduled elections of officers using secret ballots. Ex-convicts are prohibited from holding union office. Moreover, union officials are accountable for union property and funds. Members have the right to attend and to participate in union meetings, to nominate officers, and to vote in most union proceedings.

The act also outlawed **hot-cargo agreements,** in which employers voluntarily agree with unions not to handle, use, or deal in goods produced by nonunion employees working for other employers. The act made all such boycotts (called **secondary boycotts**) illegal.

14. 29 U.S.C. Sections 401 *et seq.*

Worker Health and Safety

Under the common law, employees injured on the job had to rely on tort law or contract law theories in suits they brought against their employers. Additionally, workers had some recourse under the common law governing agency relationships (discussed in Chapter 22), which imposes a duty on a principal-employer to provide a safe workplace for an agent-employee. Today, numerous state and federal statutes protect employees and their families from the risk of accidental injury, death, or disease resulting from their employment. This section discusses the primary federal statute governing health and safety in the workplace, along with state workers' compensation laws.

THE OCCUPATIONAL SAFETY AND HEALTH ACT

At the federal level, the primary legislation protecting employees' health and safety is the Occupational Safety and Health Act of 1970.[15] Congress passed this act in an attempt to ensure safe and healthful working conditions for practically every employee in the country. The act requires employers to meet specific standards in addition to their general duty to keep workplaces safe.

Enforcement Agencies Three federal agencies develop and enforce the standards set by the Occupational Safety and Health Act. The Occupational Safety and Health Administration (OSHA) is part of the Department of Labor and has the authority to promulgate standards, make inspections, and enforce the act. OSHA has developed safety standards governing many workplace details, such as the structural stability of ladders and the requirements for railings. OSHA also establishes standards that protect employees against exposure to substances that may be harmful to their health.

The National Institute for Occupational Safety and Health is part of the Department of Health and Human Services. Its main duty is to conduct research on safety and health problems and to recommend standards for OSHA to adopt. Finally, the Occupational Safety and Health Review Commission is an independent agency set up to handle appeals from actions taken by OSHA administrators.

Procedures and Violations OSHA compliance officers may enter and inspect facilities of any establishment covered by the Occupational Safety and Health Act.[16] Employees may also file complaints of violations. Under the act, an employer cannot discharge an employee who files a complaint or who, in good faith, refuses to work in a high-risk area if bodily harm or death might reasonably result.

Employers with eleven or more employees are required to keep occupational injury and illness records for each employee. Each record must be made available for inspection when requested by an OSHA inspector. Whenever a work-related injury or disease occurs, employers must make reports directly to OSHA. Whenever an employee is killed in a work-related accident or when five or more employees are hospitalized as a result of one accident, the employer must notify the Department of Labor within forty-eight hours. If the company fails to do so, it will be fined. Following the accident, a complete inspection of the premises is mandatory.

Criminal penalties for willful violation of the Occupational Safety and Health Act are limited. Employers may also be prosecuted under state laws, however. In other words, the act does not preempt state and local criminal laws.[17]

STATE WORKERS' COMPENSATION LAWS

State **workers' compensation laws** establish an administrative procedure for compensating workers injured on the job. Instead of suing, an injured worker files a claim with the administrative agency or board that administers local workers' compensation claims.

Most workers' compensation statutes are similar. No state covers all employees. Typically excluded are domestic workers, agricultural workers, temporary employees, and employees of common carriers (companies that provide transportation services to the public). Typically, the statutes cover minors. Usually, the statutes allow employers to purchase insurance from a private insurer or a state fund to pay workers' compensation benefits in the event of a claim. Most states also allow employers to be self-insured—that is, employers who show an ability to pay claims do not need to buy insurance.

Requirements for Receiving Workers' Compensation In general, the right to recover benefits is predicated wholly on the existence of an employment relationship and the fact that the injury was *accidental* and *occurred on the job or in the course of employment*, regardless of fault. Intentionally inflicted self-injury, for example, would not be considered accidental and hence would not be covered. If an employee

15. 29 U.S.C. Sections 553, 651–678.

16. In the past, warrantless inspections were conducted. In 1978, however, the United States Supreme Court held that warrantless inspections violated the warrant clause of the Fourth Amendment to the Constitution. See *Marshall v. Barlow's, Inc.*, 436 U.S. 307, 98 S.Ct. 1816, 56 L.Ed.2d 305 (1978).

17. *Pedraza v. Shell Oil Co.*, 942 F.2d 48 (1st Cir. 1991); *cert.* denied, *Shell Oil Co. v. Pedraza*, 502 U.S. 1082, 112 S.Ct. 993, 117 L.Ed.2d 154 (1992).

is injured while commuting to or from work, the injury usually will not be considered to have occurred on the job or in the course of employment and hence will not be covered.

An employee must notify her or his employer promptly (usually within thirty days) of an injury. Generally, an employee must also file a workers' compensation claim with the appropriate state agency or board within a certain period (sixty days to two years) from the time the injury is first noticed, rather than from the time of the accident.

Workers' Compensation versus Litigation An employee's acceptance of workers' compensation benefits bars the employee from suing for injuries caused by the employer's negligence. By barring lawsuits for negligence, workers' compensation laws also bar employers from raising common law defenses to negligence, such as contributory negligence, assumption of risk, or injury caused by a "fellow servant" (another employee). A worker may sue an employer who *intentionally* injures the worker, however.

Income Security

Federal and state governments participate in insurance programs designed to protect employees and their families by covering the financial impact of retirement, disability, death, hospitalization, and unemployment. The key federal law on this subject is the Social Security Act of 1935.[18]

Social Security

The Social Security Act provides for old-age (retirement), survivors, and disability insurance. The act is therefore often referred to as OASDI. Both employers and employees must "contribute" under the Federal Insurance Contributions Act (FICA)[19] to help pay for benefits that will partially make up for the employees' loss of income on retirement.

The basis for the employee's and the employer's contribution is the employee's annual wage base—the maximum amount of the employee's wages that are subject to the tax. The employer withholds the employee's FICA contribution from the employee's wages and then matches this contribution. (In 2006, employers were required to withhold 6.2 percent of each employee's wages, up to a maximum wage base of $94,200, and to match this contribution.)

Retired workers are then eligible to receive monthly payments from the Social Security Administration, which administers the Social Security Act. Social Security benefits are fixed by statute but increase automatically with increases in the cost of living.

Medicare

Medicare, a federal government health-insurance program, is administered by the Social Security Administration for people sixty-five years of age and older and for some under the age of sixty-five who are disabled. It has two parts, one pertaining to hospital costs and the other to nonhospital medical costs, such as visits to physicians' offices. People who have Medicare hospital insurance can also obtain additional federal medical insurance if they pay small monthly premiums, which increase as the cost of medical care increases.

As with Social Security contributions, both the employer and the employee "contribute" to Medicare. As of this writing, both the employer and the employee pay 1.45 percent of the amount of *all* wages and salaries to finance Medicare. Unlike Social Security contributions, there is no cap on the amount of wages subject to the Medicare tax.

Private Pension Plans

Significant legislation has been enacted to regulate employee retirement plans set up by employers to supplement Social Security benefits. The major federal act covering these retirement plans is the Employee Retirement Income Security Act (ERISA) of 1974.[20] This act empowers the Labor Management Services Administration of the Department of Labor to enforce its provisions governing employers who have private pension funds for their employees. ERISA does not require an employer to establish a pension plan. When a plan exists, however, ERISA establishes standards for its management.

A key provision of ERISA concerns *vesting*. **Vesting** gives an employee a legal right to receive pension benefits at some future date when he or she stops working. Before ERISA was enacted, some employees who had worked for companies for as long as thirty years received no pension benefits when their employment terminated, because those benefits had not vested. ERISA establishes complex vesting rules. Generally, however, all employee contributions to pension plans vest immediately, and employee rights to employer contributions to a plan vest after five years of employment.

In an attempt to prevent mismanagement of pension funds, ERISA has established rules on how they must be invested. Pension managers must be cautious in their investments and refrain from investing more than 10 percent of

18. 42 U.S.C. Sections 301–1397e.
19. 26 U.S.C. Sections 3101–3125.
20. 29 U.S.C. Sections 1001 *et seq.*

the fund in securities of the employer. ERISA also contains detailed record-keeping and reporting requirements.

Unemployment Insurance

To ease the financial impact of unemployment, the United States has a system of unemployment insurance. The Federal Unemployment Tax Act (FUTA) of 1935[21] created a state-administered system that provides unemployment compensation to eligible individuals. Under this system, employers pay into a fund, and the proceeds are paid out to qualified unemployed workers. The FUTA and state laws require employers that fall under the provisions of the act to pay unemployment taxes at regular intervals.

To be eligible for unemployment compensation, a worker must be willing and able to work and be actively seeking employment. Workers who have been fired for misconduct or who have voluntarily left their jobs are not eligible for benefits. To leave a job voluntarily is to leave it without good cause. In the following case, an unemployed worker left his job. The question was whether he left the job for good cause and was therefore eligible for unemployment benefits.

21. 26 U.S.C. Sections 3301–3310.

CASE 23.2 Lewis v. Director, Employment Security Department

Court of Appeals of Arkansas,
Division I, 2004.
141 S.W.3d 896.
http://courts.state.ar.us/opinions/ca2004a.htm[a]

FACTS The jobs in the warehouse unit of Ace Hardware Corporation in Arkansas are divided between two departments: break order and full case. Break-order positions involve lifting no more than fifty pounds. The full-case department fills heavier orders (up to five thousand pounds). Employees who complete orders in less time than Ace allows earn incentive pay. Employees believe that incentive pay is lower in the full-case department. Jobs are awarded on a seniority basis: new hires start in the full-case department but bid for other positions as soon as possible. Jimmy Lewis had worked for Ace since 1984, primarily in the break-order department. Ace had a high turnover in the full-case department, however, and whenever additional workers were needed, Ace reassigned Lewis. In 1998, Lewis began to complain regularly to his superiors about the reassignments. He offered to train others to fill the position, but his managers declined. In 2003, Lewis quit and applied for unemployment benefits. Ace questioned Lewis's legal right to these benefits. An Arkansas state employment tribunal ruled in Lewis's favor, a state review board reversed the ruling, and Lewis appealed to a state intermediate appellate court.

ISSUE Was Lewis entitled to unemployment benefits?

a. Click on "January 21, 2004." On the next page, scroll to the name of the case and click on the appropriate link to access the opinion. The Arkansas Judiciary maintains this Web site.

DECISION Yes. The state intermediate appellate court reversed the review board's decision and remanded the case for an order to award benefits. The court held that Lewis left his job for good cause.

REASON The court emphasized that the only reason Ace did not train other workers to fill full-case positions was that Lewis was already trained and it was easier to reassign him. The court reasoned that Ace's staffing problems were self-created and that its reassignment policy violated its seniority policy. Lewis quit his job when he realized that Ace was never going to permanently address the underlying situation that caused his reassignment to the full-case department. Although employees regularly bid out of the full-case department, causing staffing shortages, Ace refused to train other current employees to fill those positions. After five years of complaining to management about being reassigned to a position that Lewis felt caused him to lose pay, after offering to train other employees, and after seeing management violate its seniority policy and do virtually nothing to resolve the situation, Lewis quit. The court concluded, "We agree with appellant that his circumstances would reasonably impel an average, able-bodied, qualified worker to give up his or her employment."

FOR CRITICAL ANALYSIS—Social Consideration *The court in the* Lewis *case based its decision in part on the employee's reaction to the situation on the job. What other factors should be considered in deciding whether a worker had "good cause" to quit?*

COBRA

Federal legislation also addresses the issue of health insurance for workers whose jobs have been terminated—and who are thus no longer eligible for group health-insurance plans. The Consolidated Omnibus Budget Reconciliation Act (COBRA) of 1985[22] prohibits the elimination of a worker's medical, optical, or dental insurance on the voluntary or involuntary termination of the worker's employment. The act applies to most workers who have either lost their jobs or had their hours decreased so that they are no longer eligible for coverage under the employer's health plan. Only workers fired for gross misconduct are excluded from protection.

The worker has sixty days (beginning with the date that the group coverage would stop) to decide whether to continue with the employer's group insurance plan. If the worker chooses to discontinue the coverage, the employer has no further obligation. If the worker chooses to continue coverage, though, the employer is obligated to keep the policy active for up to eighteen months. If the worker is disabled, the employer must extend coverage for up to twenty-nine months. The coverage provided must be the same as that enjoyed by the worker prior to the termination or reduction of work. If family members were originally included, for example, COBRA prohibits their exclusion. To receive continued benefits, however, the worker may be required to pay all of the premiums, as well as a 2 percent administrative charge.

Employers, with some exceptions, must comply with COBRA if they employ twenty or more workers and provide a benefit plan to those workers. An employer must inform an employee of COBRA's provisions when that worker faces termination or a reduction of hours that would affect his or her eligibility for coverage under the plan. The employer need not provide benefit coverage if the employer eliminates its group benefit plan or if the worker becomes eligible for Medicare, becomes covered under a spouse's health plan, becomes insured under a different plan (with a new employer, for example), or fails to pay the premium. An employer that does not comply with COBRA risks substantial penalties, such as a tax of up to 10 percent of the annual cost of the group plan or $500,000, whichever is less.

22. 29 U.S.C. Sections 1161–1169.

Family and Medical Leave

In 1993, Congress passed the Family and Medical Leave Act (FMLA)[23] to allow employees to take time off from work for family or medical reasons. A majority of the states also have legislation allowing for a leave from employment for family or medical reasons, and many employers maintain private family-leave plans for their workers.

Coverage and Applicability of the FMLA

The FMLA requires employers who have fifty or more employees to provide employees with up to twelve weeks of unpaid family or medical leave during any twelve-month period. Generally, an employee may take family leave to care for a newborn baby, an adopted child, or a foster child and take medical leave when the employee or the employee's spouse, child, or parent has a "serious health condition" requiring care.[24] The employer must continue the worker's health-care coverage and guarantee employment in the same position or a comparable position when the employee returns to work. An important exception to the FMLA, however, allows the employer to avoid reinstating a *key employee*—defined as an employee whose pay falls within the top 10 percent of the firm's workforce. Also, the act does not apply to part-time or newly hired employees (those who have worked for less than one year).

Employees suffering from certain chronic health conditions, such as asthma, diabetes, and pregnancy, may take FMLA leave for their own incapacities that require absences of less than three days. ■ **EXAMPLE 23.4** Estel, an employee who has asthma, suffers from periodic episodes of illness. According to regulations issued by the Department of Labor, employees with such conditions are covered by the FMLA. Thus, Estel may take a medical leave. ■

The FMLA expressly covers private and public (government) employees. Nevertheless, some states argued that public employees could not sue their state employers in federal courts to enforce their FMLA rights unless the states consented to be sued.[25] This argument came before the United States Supreme Court in the following case.

23. 29 U.S.C. Sections 2601, 2611–2619, 2651–2654.

24. The foster care must be state sanctioned before such an arrangement falls within the coverage of the FMLA.

25. Under the Eleventh Amendment to the U.S. Constitution, a state is immune from suit in a federal court unless the state agrees to be sued.

CASE 23.3 Nevada Department of Human Resources v. Hibbs

Supreme Court of the United States, 2003.
538 U.S. 721,
123 S.Ct. 1972,
155 L.Ed.2d 953.

FACTS William Hibbs worked for the Nevada Department of Human Resources. In April 1997, Hibbs asked for time off under the Family and Medical Leave Act (FMLA) to care for his sick wife, who was recovering from a car accident. The department granted Hibbs's request, allowing him to use the leave intermittently, as needed, beginning in May. Hibbs did this until August 5, after which he did not return to work. In October, the department told Hibbs that he had exhausted his FMLA leave, that no further leave would be granted, and that he must return to work by November 12. When he did not return, he was discharged. Hibbs filed a suit in a federal district court against the department. The court held that the Eleventh Amendment to the U.S. Constitution barred the suit. On Hibbs's appeal, the U.S. Court of Appeals for the Ninth Circuit reversed this holding. The department appealed to the United States Supreme Court.

ISSUE Can the FMLA, which expressly covers public employees, serve as the basis for a suit against a state employer whether or not the state consents to the suit?

DECISION Yes. The United States Supreme Court affirmed the appellate court's decision, concluding that the FMLA is "congruent and proportional" to the discrimination that Congress intended the FMLA to address.

REASON In deciding the case, the Supreme Court looked to the reason that Congress enacted the FMLA. The Court pointed out that when Congress enacted the FMLA, parental leave for fathers was rare. Even when family-leave policies did exist, men, *both in the public and private sectors,* "receive[d] notoriously discriminatory treatment in their requests for such leave." For example, the Court noted that fifteen states provided women with up to one year of extended maternity leave, while only four provided men with the same. This and other differential leave policies were not, according to the Court, attributable to any differential physical needs of men and women, but rather to the "pervasive sex-role stereotype that caring for family members is women's work. * * * By setting a minimum standard of family leave for *all* eligible employees, irrespective of gender, the FMLA attacks the formerly state-sanctioned stereotype that only women are responsible for family caregiving, thereby reducing employers' incentives to engage in discrimination by basing hiring and promotion decisions on stereotypes."

FOR CRITICAL ANALYSIS—Social Consideration *The court in this case emphasized how men frequently have received discriminatory treatment in family-leave policies. If the plaintiff had been a woman rather than a man, would that have changed the outcome of the case? Why or why not?*

VIOLATIONS OF THE FMLA

An employer who violates the FMLA may be held liable for damages to compensate an employee for unpaid wages (or salary), lost benefits, denied compensation, and actual monetary losses (such as the cost of providing for care of the family member) up to an amount equivalent to the employee's wages for twelve weeks. Supervisors may also be subject to personal liability, as employers, for violations of the act.[26] A court may require the employer to reinstate the employee in her or his job or to grant a promotion that had been denied. A successful plaintiff is entitled to court costs; attorneys' fees; and, in cases involving bad faith on the part of the employer, double damages.

Regulations issued by the Department of Labor (DOL) impose additional sanctions on employers who fail to notify employees when an absence will be counted against leave authorized under the act. Under one such rule, if an employer failed to provide notice, then the employee's absence would not count as a portion of the leave time available under the FMLA. ■ **EXAMPLE 23.5** An employee had been absent from work for thirty weeks while undergoing treatment for cancer. The employer had not designated any of the employee's time off as FMLA leave. Thus, under the DOL's regulation, the employee was entitled to an additional leave of twelve weeks. ■ In 2002, however, the United States Supreme Court invalidated this rule. The Court reasoned that it would be unjust for an employee to obtain additional protected leave as a windfall.[27]

Employment Discrimination

Out of the 1960s civil rights movement to end racial and other forms of discrimination grew a body of law protecting employees against discrimination in the workplace. This protective legislation further eroded the employment-at-will doctrine (discussed earlier in this chapter). In the

26. See, for example, *Rupnow v. TRC, Inc.,* 999 F.Supp. 1047 (N.D. Ohio 1998).

27. *Ragsdale v. Wolverine World Wide, Inc.,* 535 U.S. 81, 122 S.Ct. 1155, 152 L.Ed.2d 167 (2002).

past several decades, judicial decisions, administrative agency actions, and legislation have restricted the ability of employers, as well as unions, to discriminate against workers on the basis of race, color, religion, national origin, gender, age, or disability. A class of persons defined by one or more of these criteria is known as a **protected class.**

A number of federal statutes prohibit **employment discrimination** against members of protected classes. The most important statute is Title VII of the Civil Rights Act of 1964.[28] Title VII prohibits discrimination on the basis of race, color, religion, national origin, or gender at any stage of employment. The Age Discrimination in Employment Act of 1967[29] and the Americans with Disabilities Act of 1990[30] prohibit discrimination on the basis of age and disability, respectively.

The remainder of this chapter focuses on the kinds of discrimination prohibited by these federal statutes. Note, though, that discrimination against employees on the basis of any of these criteria may also violate state human rights statutes or other state laws or public policies prohibiting discrimination.

Title VII of the Civil Rights Act of 1964

Title VII of the Civil Rights Act of 1964 and its amendments prohibit job discrimination against employees, applicants, and union members on the basis of race, color, national origin, religion, or gender at any stage of employment. Title VII applies to employers with fifteen or more employees, labor unions with fifteen or more members, labor unions that operate hiring halls (to which members go regularly to be rationed jobs as they become available), employment agencies, and state and local governing units or agencies. A special section of the act prohibits discrimination in most federal government employment.

The Equal Employment Opportunity Commission Compliance with Title VII is monitored by the Equal Employment Opportunity Commission (EEOC). A victim of alleged discrimination, before bringing a suit against the employer, must first file a claim with the EEOC. The EEOC may investigate the dispute and attempt to obtain the parties' voluntary consent to an out-of-court settlement. If voluntary agreement cannot be reached, the EEOC may then file a suit against the employer on the employee's behalf. If the EEOC decides not to investigate the claim, the victim may bring her or his own lawsuit against the employer.

28. 42 U.S.C. Sections 2000e–2000e-17.
29. 29 U.S.C. Sections 621–634.
30. 42 U.S.C. Sections 12102–12118.

The EEOC does not investigate every claim of employment discrimination, regardless of the merits of the claim. Generally, it investigates only "priority cases," such as cases involving retaliatory discharge (firing an employee in retaliation for submitting a claim to the EEOC) and cases involving types of discrimination that are of particular concern to the EEOC.

Intentional and Unintentional Discrimination Title VII prohibits both intentional and unintentional discrimination.

—Intentional Discrimination. Intentional discrimination by an employer against an employee is known as **disparate-treatment discrimination.** Because intent may sometimes be difficult to prove, courts have established certain procedures for resolving disparate-treatment cases. Suppose that a woman applies for employment with a construction firm and is rejected. If she sues on the basis of disparate-treatment discrimination in hiring, she must show that (1) she is a member of a protected class, (2) she applied and was qualified for the job in question, (3) she was rejected by the employer, and (4) the employer continued to seek applicants for the position or filled the position with a person not in a protected class.

If the woman can meet these relatively easy requirements, she has made out a *prima facie* case of illegal discrimination. Making out a ***prima facie* case** of discrimination means that the plaintiff has met her initial burden of proof and will win in the absence of a legally acceptable employer defense. (Defenses to claims of employment discrimination will be discussed later in this chapter.) The burden then shifts to the employer-defendant, who must articulate a legal reason for not hiring the plaintiff. To prevail, the plaintiff must then show that the employer's reason is a *pretext* (not the true reason) and that discriminatory intent actually motivated the employer's decision.

—Unintentional Discrimination. Employers often use interviews and testing procedures to choose from among a large number of applicants for job openings. Minimum educational requirements are also common. These practices and procedures may have an unintended discriminatory impact on a protected class. **Disparate-impact discrimination** occurs when, as a result of educational or other job requirements or hiring procedures, an employer's workforce does not reflect the percentage of nonwhites, women, or members of other protected classes that characterizes qualified individuals in the local labor market. If a person challenging an employment practice having a discriminatory effect can show a connection between the practice and the disparity, he or she has made

out a *prima facie* case, and no evidence of discriminatory intent needs to be shown. Disparate-impact discrimination can also occur when an educational or other job requirement or hiring procedure excludes members of a protected class from an employer's workforce at a substantially higher rate than nonmembers, regardless of the racial balance in the employer's workforce.

Discrimination Based on Race, Color, and National Origin If a company's standards or policies for selecting or promoting employees have the effect of discriminating against employees or job applicants on the basis of race, color, or national origin, they are illegal—unless (except for race) they have a substantial, demonstrable relationship to realistic qualifications for the job in question. Discrimination against these protected classes in regard to employment conditions and benefits is also illegal.

■ EXAMPLE 23.6 In one case, Cynthia McCullough, an African American woman with a college degree, worked at a deli in a grocery store. More than a year later, the owner of the store promoted a white woman to the position of "deli manager." The white woman had worked in the deli for only three months, had only a sixth-grade education, and could not calculate prices or read recipes. Although the owner gave various reasons for promoting the white woman instead of McCullough, a federal appellate court held that these reasons were likely just excuses and that the real reason was discriminatory intent.[31] ■

Discrimination Based on Religion Title VII of the Civil Rights Act of 1964 also prohibits government employers, private employers, and unions from discriminating against persons because of their religion. An employer must "reasonably accommodate" the religious practices of its employees, unless to do so would cause undue hardship to the employer's business. For example, if an employee's religion prohibits him or her from working on a certain day of the week or at a certain type of job, the employer must make a reasonable attempt to accommodate these religious requirements. Employers must reasonably accommodate an employee's religious belief even if the belief is not based on the tenets or dogma of a particular church, sect, or denomination. The only requirement is that the belief be sincerely held by the employee.[32]

Discrimination Based on Gender Under Title VII, as well as other federal acts, employers are forbidden to discriminate against employees on the basis of gender. Employers are prohibited from classifying jobs as male or female and from advertising in help-wanted columns that are designated male or female unless the employer can prove that the gender of the applicant is essential to the job. Furthermore, employers cannot have separate male and female seniority lists. Generally, to succeed in a suit for gender discrimination, a plaintiff must demonstrate that gender was a determining factor in the employer's decision to hire, fire, or promote her or him. Typically, this involves looking at all of the surrounding circumstances.

The Pregnancy Discrimination Act of 1978,[33] which amended Title VII, expanded the definition of gender discrimination to include discrimination based on pregnancy. Women affected by pregnancy, childbirth, or related medical conditions must be treated—for all employment-related purposes, including the receipt of benefits under employee benefit programs—the same as other persons not so affected but similar in ability to work.

Sometimes employees claim that they were "constructively discharged" from their jobs. **Constructive discharge** occurs when the employer causes the employee's working conditions to be so intolerable that a reasonable person in the employee's position would feel compelled to quit. Title VII encompasses employer liability for constructive discharge.[34] In the case that follows, the question was whether the employer should have been aware of the employee's mistreatment and unbearable working conditions and done something about them.

31. *McCullough v. Real Foods, Inc.,* 140 F.3d 1123 (8th Cir. 1998). The federal district court had granted summary judgment for the employer in this case. The Eighth Circuit Court of Appeals reversed the district court's decision and remanded the case for trial.

32. *Frazee v. Illinois Department of Employment Security,* 489 U.S. 829, 109 S.Ct. 1514, 103 L.Ed.2d 914 (1989).

33. 42 U.S.C. Section 2000e(k).

34. *Pennsylvania State Police v. Suders,* 542 U.S. 129, 124 S.Ct. 2342, 159 L.Ed.2d 204 (2004).

CASE 23.4 Conway-Jepsen v. Small Business Administration

United States District Court,
District of Montana, 2004.
303 F.Supp.2d 1155.

FACTS In August 1992, Jo Alice Mospan took charge of the Helena, Montana, office of the U.S. Small Business Administration (SBA) as the district director. At the time, there were no other females above a certain pay level in the Helena office, and most of the senior employees and supervisors were male. Mospan was an "in-your-face" micromanager, frequently disciplining employees. In Helena, her purportedly

CASE 23.4—CONTINUED

express purpose was to harass and ultimately rid the office of certain male employees so that they could be replaced by females. In 1993, Mospan recruited Mary Conway-Jepsen from the SBA office in Santa Ana, California. Conway-Jepsen soon learned what Mospan was doing, told her that it was wrong, and refused to cooperate. Mospan retaliated. Among other things, she overloaded Conway-Jepsen with irrelevant assignments, took actions counterproductive to her projects, and set her up to make mistakes. Conway-Jepsen's physician recommended that she quit her job. She applied for a transfer, but when none was forthcoming, she resigned in August 1997. By 2000, all of the targeted males were also gone. Conway-Jepsen filed a suit in a federal district court against the SBA, claiming a violation of Title VII.

ISSUE Was Conway-Jepsen constructively discharged in violation of Title VII?

DECISION Yes. The court concluded that the hostile working conditions at the Helena SBA office were created by conduct in violation of Title VII, that Conway-Jepsen reasonably found the conditions intolerable, and that her resignation resulted from these conditions. On her behalf, the court ordered back pay, job reinstatement, compensatory damages, and attorneys' fees.

REASON The court emphasized that the retaliatory treatment spanned two and a half years. "No reasonable employee could be expected to put up with Mospan's harassment for a longer period of time." Conway-Jepsen "is an intelligent, knowledgeable, and aggressive program manager," whose "demeanor is that of a tough fighter, not a person who would easily wilt under pressure." She "was a diligent and competent SBA employee who was constructively discharged from her position by Mospan's lengthy, continuous, and pervasive pattern of retaliatory treatment." Further, Mospan ruined Conway-Jepsen's career, continuing to retaliate against her by interfering with her ability to obtain employment after her resignation. Because Mospan was the district director of the Helena SBA office, she was the SBA's agent, acting within the scope of her authority and the course of her employment. Thus, the SBA knew or should have known of the hostile working conditions but failed to remedy them.

WHY IS THIS CASE IMPORTANT? *Employers should be aware that if a manager or supervisor creates intolerable working conditions because of his or her own personal bias, the employer could be held liable under Title VII.*

Sexual Harassment Title VII also protects employees against **sexual harassment** in the workplace. Sexual harassment has often been classified as either *quid pro quo* harassment or hostile-environment harassment. *Quid pro quo* harassment[35] occurs when sexual favors are demanded in return for job opportunities, promotions, salary increases, and the like. According to the United States Supreme Court, hostile-environment harassment occurs when "the workplace is permeated with discriminatory intimidation, ridicule, and insult, that is sufficiently severe or pervasive to alter the conditions of the victim's employment and create an abusive working environment."[36]

—Harassment by Supervisors. What if an employee is harassed by a manager or supervisor of a large firm, and the firm itself (the "employer") is not aware of the harassment? Should the employer be held liable for the harassment nonetheless? In 1998, in two separate cases, the United States Supreme Court issued some significant guidelines relating to this issue. In *Faragher v. City of Boca Raton,*[37] the Court held that an employer (a city) could be held liable for a supervisor's harassment of employees even though the employer was unaware of the behavior. The Court reached this conclusion primarily because, although the city had a written policy against sexual harassment, the policy had not been distributed to city employees. Additionally, the city had not established any procedures that could be followed by employees who felt that they were victims of sexual harassment. In *Burlington Industries, Inc. v. Ellerth,*[38] the Court ruled that a company could be held liable for the harassment of an employee by one of its vice presidents even though the employee suffered no adverse job consequences.

In these two cases, the Court set forth some guidelines on workplace harassment that are helpful to employers and employees alike. On the one hand, employees benefit by the ruling that employers may be held liable for their supervisors' harassment even though the employers were unaware of the actions and even though the employees suffered no adverse job consequences. On the other hand, the Court made it clear in both decisions that employers have an affirmative defense against liability for their

35. *Quid pro quo* is a Latin phrase that is often translated to mean "something in exchange for something else."

36. *Harris v. Forklift Systems,* 510 U.S. 17, 114 S.Ct. 367, 126 L.Ed.2d 295 (1993).

37. 524 U.S. 775, 118 S.Ct. 2275, 141 L.Ed.2d 662 (1998).

38. 524 U.S. 742, 118 S.Ct. 2257, 141 L.Ed.2d 633 (1998).

supervisors' harassment of employees if the employers can show the following:

1. That they have taken "reasonable care to prevent and correct promptly any sexually harassing behavior" (by establishing effective harassment policies and complaint procedures, for example).
2. That the employees suing for harassment failed to follow these policies and procedures.

—Harassment by Co-Workers and Nonemployees. Often, employees alleging harassment complain that the actions of co-workers, not supervisors, are responsible for creating a hostile working environment. In such cases, the employee may still have a cause of action against the employer. Normally, though, the employer will be held liable only if the employer knew, or should have known, about the harassment and failed to take immediate remedial action. (For a further discussion of the importance of taking immediate and *effective* remedial action in response to complaints of sexual harassment, see this chapter's *Management Perspective* feature.)

Employers may also be liable for harassment by *nonemployees* in certain circumstances. ■ **EXAMPLE 23.7** A restaurant owner or manager knows that a certain customer repeatedly harasses a waitress and permits the harassment to continue. The restaurant owner may be liable under Title VII even though the customer is not an employee of the restaurant. The issue turns on the control that the employer exerts over a nonemployee. In one case, an owner of a Pizza Hut franchise was held liable for the harassment of a waitress by two male customers because no steps were taken to prevent the harassment.[39] ■

—Same-Gender Harassment. The courts have also had to address the issue of whether men who are harassed by other men, or women who are harassed by other women, are protected by laws that prohibit gender-based discrimination in the workplace. For example, what if the male president of a firm demands sexual favors from a male employee? Does this action qualify as sexual harassment? For some time, the courts were widely split on this issue. In 1998, in *Oncale v. Sundowner Offshore Services, Inc.*,[40] the Supreme Court resolved the issue by holding that Title VII protection extends to situations in which individuals are harassed by members of the same gender.

Remedies under Title VII Employer liability under Title VII may be extensive. If the plaintiff successfully proves that unlawful discrimination occurred, he or she may be awarded reinstatement, back pay, retroactive promotions, and damages. Compensatory damages are available only in cases of intentional discrimination. Punitive damages may be recovered against a private employer only if the employer acted with malice or reckless indifference to an individual's rights. The statute limits the total amount of compensatory and punitive damages that the plaintiff can recover from specific employers—ranging from $50,000 against employers with one hundred or fewer employees to $300,000 against employers with more than five hundred employees.

Discrimination Based on Age

Age discrimination is potentially the most widespread form of discrimination, because anyone—regardless of race, color, national origin, or gender—could be a victim at some point in life. The Age Discrimination in Employment Act (ADEA) of 1967, as amended, prohibits employment discrimination on the basis of age against individuals forty years of age or older. The act also prohibits mandatory retirement for nonmanagerial workers. For the act to apply, an employer must have twenty or more employees, and the employer's business activities must affect interstate commerce. The EEOC administers the ADEA, but the act also permits private causes of action against employers for age discrimination.

Procedures under the ADEA The burden-shifting procedure under the ADEA is similar to that under Title VII. If a plaintiff can establish that she or he (1) was a member of the protected age group, (2) was qualified for the position from which she or he was discharged, and (3) was discharged under circumstances that give rise to an inference of discrimination, the plaintiff has established a *prima facie* case of unlawful age discrimination. The burden then shifts to the employer, who must articulate a legitimate reason for the discrimination. If the plaintiff can prove that the employer's reason is only a pretext (excuse) and that the plaintiff's age was a determining factor in the employer's decision, the employer will be held liable under the ADEA.

Whether a firing of an older worker is discriminatory or simply part of a rational business decision to prune the company's ranks is not always clear. Companies often defend a decision to discharge a worker by asserting that the worker could no longer perform his or her duties or that the worker's skills were no longer needed. The employee must prove that the discharge was motivated, at

39. *Lockard v. Pizza Hut, Inc.*, 162 F.3d 1062 (10th Cir. 1998).
40. 523 U.S. 75, 118 S.Ct. 998, 140 L.Ed.2d 207 (1998).

MANAGEMENT PERSPECTIVE

Responding to Sexual Harassment in the Workplace

MANAGEMENT FACES A LEGAL ISSUE

As discussed elsewhere in this chapter, in two cases decided in 1998, the United States Supreme Court issued guidelines concerning the liability of employers for their supervisors' harassment of employees in the workplace. Generally, these guidelines allow employers to avoid liability if they take reasonable steps to prevent sexual harassment in the workplace and take prompt remedial action to correct sexually harassing behavior. Although these cases involved harassment of workers by supervisors, the principles expressed in the rulings have also been applied to harassment by co-workers. One question remains, though: When an employee complains of an incident of sexual harassment, what steps must the employer take to satisfy the requirement of "prompt remedial action"? The answer to this question has important implications for today's business owners and managers.

WHAT THE COURTS SAY

Generally, when it can be shown that a worker was sexually harassed (by either a supervisor *or* a co-worker), management must make sure that whatever action it takes will effectively prevent similar incidents of harassment in the future. Consider the outcome of two cases dealing with this precise issue. One case, decided by the Supreme Court of Hawaii, involved a plaintiff—a female worker in a restaurant—who had experienced a short, isolated instance of unwanted contact from a male co-worker (the co-worker had squeezed her buttock for approximately one second). She asked him to stop and then asked a manager, who had witnessed the incident, to ask the co-worker to not do it again. The manager did so immediately. A few weeks later, the co-worker again squeezed the plaintiff's buttock for a second or so. Management immediately suspended the co-worker and terminated him shortly thereafter. The trial court granted the employer's motion for summary judgment, concluding that the manager's response was sufficient to avoid liability. The Hawaii Supreme Court, however, reversed the trial court's decision and remanded the case for trial. The state supreme court held that reasonable minds could differ as to whether the manager's response was reasonably calculated to end the harassment. The court noted that the co-worker was given only an oral warning, which, the manager later admitted to the plaintiff, the co-worker did not take seriously. Second, the co-worker was not threatened with a future written reprimand, nor was he informed that he could be suspended or terminated if his behavior continued (even though he was, in fact, later terminated, when the harassment continued).[a]

In the other case, the plaintiff, Jamie McCurdy, worked as a radio dispatcher for the Arkansas State Police. After a police sergeant allegedly groped her breast, touched her hair, and made suggestive sexual comments to her, she immediately reported the incident to her supervisor. Shortly thereafter, supervisory officers investigated the incident and ultimately demoted the sergeant and transferred him to a different location. In this case, a federal appellate court concluded that these actions were sufficient to justify the lower court's grant of summary judgment for the employer.[b]

IMPLICATIONS FOR MANAGERS

These cases both illustrate that a key factor in a court's determination of an employer's liability for harassment in the workplace is the effectiveness of the response. Is the employer's response effective in the sense that it will help to prevent incidents of harassment? To avoid liability, managers would be wise to give more than just oral warnings to offending employees and to take definite preventive steps. These steps range from removing the offending supervisor or co-worker from a complainant's work area to terminating the offender.

a. *Arquero v. Hilton Hawaiian Village, L.L.C.,* 104 Hawaii 423, 91 P.3d 505 (2004).
b. *McCurdy v. Arkansas State Police,* 375 F.3d 762 (8th Cir. 2004).

least in part, by age bias. Proof that qualified older employees are generally discharged before younger employees or that co-workers continually made unflattering age-related comments about the discharged worker may be enough.

The plaintiff need not prove that he or she was replaced by a person outside the protected class—that is, by a person under the age of forty years.[41] Rather, the issue in all ADEA cases is whether age discrimination has, in fact, occurred, regardless of the age of the replacement worker.

State Employees Not Covered by the ADEA The United States Supreme Court has generally held that the states are immune from lawsuits brought by private individuals in federal court—unless a state consents to the suit. This immunity stems from the Supreme Court's interpretation of the Eleventh Amendment (the text of this amendment is included in Appendix D). In *Kimel v. Florida Board of*

41. *O'Connor v. Consolidated Coin Caterers Corp.,* 517 U.S. 308, 116 S.Ct. 1307, 134 L.Ed.2d 433 (1996).

Regents,[42] decided in 2000, the United States Supreme Court held that the Eleventh Amendment bars private parties (in this case, employees of two Florida state universities) from suing state employers for violations of the ADEA. According to the Court, Congress had exceeded its constitutional authority when it included in the ADEA a provision stating that "all employers," including state employers, were subject to the act.[43]

Discrimination Based on Disability

The Americans with Disabilities Act (ADA) of 1990 is designed to eliminate discriminatory employment practices that prevent otherwise qualified workers with disabilities from fully participating in the national labor force. Prior to 1990, the major federal law providing protection to those with disabilities was the Rehabilitation Act of 1973. That act covered only federal government employees and those employed under federally funded programs. The ADA extends federal protection against disability-based discrimination to all workplaces with fifteen or more workers. Basically, the ADA requires that employers "reasonably accommodate" the needs of persons with disabilities unless to do so would cause the employer to suffer an "undue hardship." The Supreme Court has generally held that lawsuits under the ADA cannot be brought against state government employers, however.[44]

Procedures under the ADA To prevail on a claim under the ADA, a plaintiff must show that he or she (1) has a disability, (2) is otherwise qualified for the employment in question, and (3) was excluded from the employment solely because of the disability. As in Title VII cases, a claim alleging a violation of the ADA may be commenced only after the plaintiff has pursued the claim through the EEOC. Plaintiffs may sue for many of the same remedies available under Title VII. The EEOC may decide to investigate and perhaps even sue the employer on behalf of the employee. If the EEOC decides not to sue, then the employee is entitled to sue.

Plaintiffs in lawsuits brought under the ADA may seek many of the same remedies available under Title VII. These include reinstatement, back pay, a limited amount of compensatory and punitive damages (for intentional discrimination), and certain other forms of relief. Repeat violators may be ordered to pay fines of up to $100,000.

What Is a Disability? The ADA defines the term *disability* as "(1) a physical or mental impairment that substantially limits one or more of the major life activities of such individuals; (2) a record of such impairment; or (3) being regarded as having such an impairment."

Health conditions that have been considered disabilities under the federal law include blindness, alcoholism, heart disease, cancer, muscular dystrophy, cerebral palsy, paraplegia, diabetes, acquired immune deficiency syndrome (AIDS), testing positive for the human immunodeficiency virus (HIV), and morbid obesity (which exists when an individual's weight is two times that of a normal person).[45] The ADA excludes from coverage certain conditions, such as kleptomania (the obsessive desire to steal).

Although the ADA's definition of disability is broad, starting in 1999 the United States Supreme Court has issued a series of decisions narrowing the definition of what constitutes a disability under the act. In 1999, for example, the Supreme Court held that severe myopia, or nearsightedness, which can be corrected with lenses, does not qualify as a disability under the ADA.[46]

In other cases decided in the early 2000s, the courts have held that plaintiffs with diabetes, bipolar disorder, epilepsy, and other such conditions do *not* fall under the ADA's protections if the conditions can be corrected. In 2002, the Supreme Court held that carpal tunnel syndrome (a repetitive-stress injury) did not constitute a disability under the ADA. The Court stated that although the employee could not perform the manual tasks associated with her job, the condition did not constitute a disability under the ADA because it did not "substantially limit" the major life activity of performing manual tasks.[47]

Reasonable Accommodation The ADA does not require that employers accommodate the needs of job applicants or employees with disabilities who are not otherwise qualified for the work. If a job applicant or an employee with a disability, with reasonable accommodation, can perform essential job functions, however, the employer must make the accommodation. Required modifications may include installing ramps for a wheelchair, establishing flexible working hours, creating or modifying job assignments, and creating or improving training materials and procedures.

Employers who do not accommodate the needs of persons with disabilities must demonstrate that the accom-

---

42. 528 U.S. 62, 120 S.Ct. 631, 145 L.Ed.2d 522 (2000).
43. State immunity under the Eleventh Amendment is not absolute, however. See *Tennessee v. Lane,* 541 U.S. 509, 124 S.Ct. 1978, 158 L.Ed.2d 820 (2004).
44. *Board of Trustees of the University of Alabama v. Garrett,* 531 U.S. 356, 121 S.Ct. 955, 148 L.Ed.2d 866 (2001).
45. *Cook v. Rhode Island Department of Mental Health,* 10 F.3d 17 (1st Cir. 1993).
46. *Sutton v. United Airlines, Inc.,* 527 U.S. 471, 119 S.Ct. 2139, 144 L.Ed.2d 450 (1999).
47. *Toyota Motor Manufacturing, Kentucky, Inc. v. Williams,* 534 U.S. 184, 122 S.Ct. 681, 151 L.Ed.2d 615 (2002).

modations will cause "undue hardship." Generally, the law offers no uniform standards for identifying what is an undue hardship other than the imposition of a "significant difficulty or expense" on the employer.

Usually, the courts decide whether an accommodation constitutes an undue hardship on a case-by-case basis. In one case, the court decided that paying for a parking space near the office for an employee with a disability was not an undue hardship.[48] In another case, the court held that accommodating the request of an employee with diabetes for indefinite leave until his disease was under control would create an undue hardship for the employer because the employer would not know when the employee was returning to work. The court stated that reasonable accommodation under the ADA means accommodation so that the employee can perform the job now or "in the immediate future" rather than at some unspecified distant time.[49]

Defenses to Employment Discrimination

The first line of defense for an employer charged with employment discrimination is, of course, to assert that the plaintiff has failed to meet his or her initial burden of proving that discrimination occurred. As noted, plaintiffs bringing cases under the ADA sometimes find it difficult to meet this initial burden because they must prove that their alleged disabilities are disabilities covered by the ADA. Furthermore, plaintiffs in ADA cases must prove that they were otherwise qualified for the job and that their disabilities were the sole reason they were not hired or were fired.

Once a plaintiff succeeds in proving that discrimination occurred, the burden shifts to the employer to justify the discriminatory practice. Often, employers attempt to justify the discrimination by claiming that it was the result of a *business necessity,* a *bona fide occupational qualification,* or a *seniority system.* In some cases, as noted earlier, an effective antiharassment policy and prompt remedial action when harassment occurs may shield employers from liability under Title VII for sexual harassment.

Business Necessity An employer may defend against a claim of disparate-impact (unintentional) discrimination by asserting that a practice that has a discriminatory effect is a **business necessity.** ■ EXAMPLE 23.8 If requiring a high school diploma is shown to have a discriminatory effect, an employer might argue that a high school education is necessary for workers to perform the job at a required level of competence. If the employer can demonstrate to the court's satisfaction that a definite connection exists between a high school education and job performance, the employer normally will succeed in this business necessity defense. ■

Bona Fide Occupational Qualification Another defense applies when discrimination against a protected class is essential to a job—that is, when a particular trait is a **bona fide occupational qualification (BFOQ).** ■ EXAMPLE 23.9 A women's clothing store might legitimately hire only female sales attendants if part of an attendant's job involves assisting clients in the store's dressing rooms. Similarly, the Federal Aviation Administration can legitimately impose age limits for airline pilots. ■

Race, however, can never be a BFOQ. Generally, courts have restricted the BFOQ defense to instances in which the employee's gender is essential to the job. The United States Supreme Court has even held that a policy that was adopted to protect the unborn children of female employees from the harmful effects of exposure to lead was an unacceptable BFOQ.[50]

Seniority Systems An employer with a history of discrimination may have no members of protected classes in upper-level positions. Even if the employer now seeks to be unbiased, it may face a lawsuit in which the plaintiff asks a court to order that minorities be promoted ahead of schedule to compensate for past discrimination. If no present intent to discriminate is shown, however, and if promotions or other job benefits are distributed according to a fair **seniority system** (in which workers with more years of service are promoted first or laid off last), the employer has a good defense against the suit. According to the Supreme Court in 2002, this defense may also apply to alleged discrimination under the ADA. If an employee with a disability requests an accommodation (such as an assignment to a particular position) that conflicts with an employer's seniority system, the accommodation will generally not be considered "reasonable" under the act.[51]

After-Acquired Evidence of Employee Misconduct In some situations, employers have attempted to avoid liability for employment discrimination on the basis of "after-acquired evidence"—that is, evidence that the employer discovers after a lawsuit is filed—of an employee's misconduct. ■ EXAMPLE 23.10 Suppose that an employer

48. *Lyons v. Legal Aid Society,* 68 F.3d 1512 (2d Cir. 1995).
49. *Myers v. Hose,* 50 F.3d 278 (4th Cir. 1995).
50. *United Auto Workers v. Johnson Controls, Inc.,* 499 U.S. 187, 111 S.Ct. 1196, 113 L.Ed.2d 158 (1991).
51. *U.S. Airways, Inc. v. Barnett,* 535 U.S. 391, 122 S.Ct. 1516, 152 L.Ed.2d 589 (2002).

fires a worker, who then sues the employer for employment discrimination. During pretrial investigation, the employer learns that the employee made material misrepresentations on his or her employment application—misrepresentations that, had the employer known about them, would have served as a ground to fire the individual. ■

According to the United States Supreme Court, after-acquired evidence of wrongdoing cannot be used to shield an employer entirely from liability for employment discrimination. It may, however, be used to limit the amount of damages for which the employer is liable.[52]

Affirmative Action

Federal statutes and regulations providing for equal opportunity in the workplace were designed to reduce or eliminate discriminatory practices with respect to hiring, retaining, and promoting employees. **Affirmative action** programs go a step further and attempt to "make up" for past patterns of discrimination by giving members of protected classes preferential treatment in hiring or promotion. During the 1960s, all federal and state government agencies, private companies that contract to do business with the federal government, and institutions that receive federal funding were required to implement affirmative action policies.

Title VII of the Civil Rights Act of 1964 neither requires nor prohibits affirmative action. Thus, most private firms have not been required to implement affirmative action policies, though many have chosen to do so.

Affirmative action programs have aroused much controversy over the last forty years, particularly when they have resulted in what is frequently called "reverse discrimination"—discrimination against "majority" individuals, such as white males. At issue is whether affirmative action programs, because of their inherently discriminatory nature, violate the equal protection clause of the Fourteenth Amendment to the Constitution.

THE *BAKKE* CASE

An early case addressing this issue, *Regents of the University of California v. Bakke*,[53] involved an affirmative action program implemented by the University of California at Davis. Allan Bakke, who had been turned down for medical school at the Davis campus, sued the university for reverse discrimination after he discovered that his academic record was better than those of some of the minority applicants who had been admitted to the program.

The United States Supreme Court held that affirmative action programs were subject to "intermediate scrutiny." To be constitutionally valid under a standard of *intermediate scrutiny*, a law or action must be substantially related to important government objectives. Applying this standard, the Court held that the university could give favorable weight to minority applicants as part of a plan to increase minority enrollment so as to achieve a more culturally diverse student body. The Court stated, however, that the use of a quota system, which explicitly reserved a certain number of places for minority applicants, violated the equal protection clause of the Fourteenth Amendment.

THE *ADARAND* CASE

Although the *Bakke* case and later court decisions alleviated the harshness of the quota system, today's courts are going even further in questioning the constitutional validity of affirmative action programs. In 1995, in its landmark decision in *Adarand Constructors, Inc. v. Peña*,[54] the United States Supreme Court held that any federal, state, or local affirmative action program that uses racial or ethnic classifications as the basis for making decisions is subject to strict scrutiny by the courts. Under a standard of *strict scrutiny*, a law or action will be upheld only if it is necessary to promote a compelling government interest.

In effect, the Court's opinion in *Adarand* means that an affirmative action program is constitutional only if it attempts to remedy past discrimination and does not make use of quotas or preferences. Furthermore, once such a program has succeeded in the goal of remedying past discrimination, it must be changed or dropped. Since then, other federal courts have followed the Supreme Court's lead by declaring affirmative action programs invalid unless they attempt to remedy specific practices of past or current discrimination.[55]

THE *HOPWOOD* CASE

In 1996, the U.S. Court of Appeals for the Fifth Circuit, in *Hopwood v. State of Texas*,[56] held that an affirmative action program at the University of Texas School of Law in Austin violated the equal protection clause. In that case, two white law school applicants sued the university when

52. *McKennon v. Nashville Banner Publishing Co.*, 513 U.S. 352, 115 S.Ct. 879, 130 L.Ed.2d 852 (1995).
53. 438 U.S. 265, 98 S.Ct. 2733, 57 L.Ed.2d 750 (1978).
54. 515 U.S. 200, 115 S.Ct. 2097, 132 L.Ed.2d 158 (1995).
55. See, for example, *Taxman v. Board of Education of the Township of Piscataway*, 91 F.3d 1547 (3d Cir. 1996); and *Schurr v. Resorts International Hotel, Inc.*, 196 F.3d 486 (3d Cir. 1999).
56. 84 F.3d 720 (5th Cir. 1996).

they were denied admission. The court decided that the affirmative action policy unlawfully discriminated in favor of minority applicants. In its opinion, the court directly challenged the *Bakke* decision by stating that the use of race even as a means of achieving diversity on college campuses "undercuts the Fourteenth Amendment." The United States Supreme Court declined to hear the case, thus letting the lower court's decision stand. Over the next years, federal courts were divided over the constitutionality of such programs.[57]

SUBSEQUENT COURT DECISIONS

In 2003, the United States Supreme Court reviewed two cases involving issues similar to that in the *Hopwood* case. Both cases involved admissions programs at the University of Michigan. In *Gratz v. Bollinger,*[58] two white applicants who were denied undergraduate admission to the university alleged reverse discrimination. The school's policy gave each applicant a score based on a number of factors, including grade point average, standardized test scores, and personal achievements. The system *automatically* awarded every "underrepresented" minority (African American, Hispanic, and Native American) applicant twenty points—one-fifth of the points needed to guarantee admission. The Court held that this policy violated the equal protection clause.

In contrast, in *Grutter v. Bollinger,*[59] the Court held that the University of Michigan Law School's admissions policy was constitutional. In that case, the Court concluded that "[u]niversities can, however, consider race or ethnicity more flexibly as a 'plus' factor in the context of individualized consideration of each and every applicant." The significant difference between the two admissions policies, in the Court's view, was that the law school's approach did not apply a mechanical formula giving "diversity bonuses" based on race or ethnicity.

57. See, for example, *Johnson v. Board of Regents of the University of Georgia,* 106 F.Supp.2d 1362 (S.D.Ga. 2000); and *Smith v. University of Washington School of Law,* 233 F.3d 1188 (9th Cir. 2000).

58. 539 U.S. 244, 123 S.Ct. 2411, 156 L.Ed.2d 257 (2003).

59. 539 U.S. 306, 123 S.Ct. 2325, 156 L.Ed.2d 304 (2003).

State Statutes

Although the focus of this chapter has been on federal legislation, most states also have statutes that prohibit employment discrimination. Generally, the same kinds of discrimination are prohibited under federal and state legislation. In addition, state statutes often provide protection for certain individuals who are not protected under federal laws. For example, a New Jersey appellate court has held that anyone over the age of eighteen is entitled to sue for age discrimination under the state law, which specifies no threshold age limit.[60]

Furthermore, state laws prohibiting discrimination may apply to firms with fewer employees than the threshold number required under federal statutes, thus offering protection to more workers. State laws may also allow for additional damages, such as damages for emotional distress, that are not available under federal statutes. Finally, some states, including California and Washington, have passed laws that end affirmative action programs in that state or modify admissions policies at state-sponsored universities.

60. *Bergen Commercial Bank v. Sisler,* 307 N.J.Super. 333, 704 A.2d 1017 (1998).

TERMS AND CONCEPTS

affirmative action 480
bona fide occupational qualification (BFOQ) 479
business necessity 479
closed shop 467
constructive discharge 474
disparate-impact discrimination 473
disparate-treatment discrimination 473
employment at will 463
employment discrimination 473
hot-cargo agreement 467
minimum wage 465
prima facie case 473
protected class 473
right-to-work law 467
secondary boycott 467
seniority system 479
sexual harassment 475
union shop 467
vesting 469
whistleblowing 464
workers' compensation laws 468
wrongful discharge 464

CHAPTER SUMMARY Employment Law

Employment at Will (See pages 463–465.)	1. *Employment-at-will doctrine*—Under this common law doctrine, either party may terminate the employment relationship at any time and for any reason ("at will"). This doctrine is still in widespread use throughout the United States, although federal and state statutes prevent the doctrine from being applied in certain circumstances. 2. *Exceptions to the employment-at-will doctrine*—To protect employees from some of the harsh results of the employment-at-will doctrine, courts have made exceptions to the doctrine on the basis of contract theory, tort theory, and public policy. 3. *Whistleblower statutes*—Most states have passed whistleblower statutes specifically to protect employees who "blow the whistle" on their employers from subsequent retaliation by those employers. The federal Whistleblower Protection Act of 1989 protects federal employees who report their employers' wrongdoing. The federal False Claims Reform Act of 1986 provides monetary rewards for whistleblowers who disclose information relating to fraud perpetrated against the U.S. government. Whistleblowers have occasionally received protection under the common law for reasons of public policy. 4. *Wrongful discharge*—Whenever an employer discharges an employee in violation of an employment contract or statutory law protecting employees, the employee may bring a suit for wrongful discharge.
Wage and Hour Laws (See page 465.)	1. *Davis-Bacon Act (1931)*—Requires contractors and subcontractors working on federal government construction projects to pay their employees "prevailing wages." 2. *Walsh-Healey Act (1936)*—Requires that employees of firms that contract with federal agencies be paid a minimum wage and overtime pay. 3. *Fair Labor Standards Act (1938)*—Extended wage-hour requirements to cover all employers whose activities affect interstate commerce plus certain other businesses. The act has specific requirements in regard to child labor, maximum hours, and minimum wages.
Labor Unions (See pages 466–467.)	1. *Norris-LaGuardia Act (1932)*—Protects peaceful strikes, picketing, and primary boycotts. 2. *National Labor Relations Act (1935)*—Established the rights of employees to engage in collective bargaining and to strike; also defined specific employer practices as unfair to labor. The National Labor Relations Board (NLRB) was created to administer and enforce the act. 3. *Labor-Management Relations Act (1947)*—Proscribes certain unfair union practices, such as the closed shop. 4. *Labor-Management Reporting and Disclosure Act (1959)*—Established an employee bill of rights and reporting requirements for union activities.
Worker Health and Safety (See pages 468–469.)	1. *Occupational Safety and Health Act (1970)*—Requires employers to meet specific safety and health standards that are established and enforced by the Occupational Safety and Health Administration (OSHA). 2. *State workers' compensation laws*—Establish an administrative procedure for compensating workers who are injured in accidents that occur on the job, regardless of fault.
Income Security (See pages 469–471.)	1. *Social Security and Medicare*—The Social Security Act of 1935 provides for old-age (retirement), survivors, and disability insurance. Both employers and employees must make contributions under the Federal Insurance Contributions Act (FICA) to help pay for benefits that will partially make up for the employees' loss of income on retirement. The Social Security Administration also administers Medicare, a health-insurance program for older or disabled persons. 2. *Private pension plans*—The federal Employee Retirement Income Security Act (ERISA) of 1974 establishes standards for the management of employer-provided pension plans.

CHAPTER SUMMARY Employment Law—Continued

Income Security—Continued	3. *Unemployment insurance*—The Federal Unemployment Tax Act of 1935 created a system that provides unemployment compensation to eligible individuals. Covered employers are taxed to help defray the costs of unemployment compensation. 4. *COBRA*—The Consolidated Omnibus Budget Reconciliation Act (COBRA) of 1985 requires employers to give employees, on termination of employment, the option of continuing their medical, optical, or dental insurance coverage for a certain period.
Family and Medical Leave (See pages 471–472.)	The Family and Medical Leave Act (FMLA) of 1993 requires employers with fifty or more employees to provide their employees (except for key employees) with up to twelve weeks of unpaid family or medical leave during any twelve-month period to care for a newborn baby, an adopted child, or a foster child; or when the employee or the employee's spouse, child, or parent has a serious health condition requiring care.
Title VII of the Civil Rights Act of 1964 (See pages 473–476.)	Title VII prohibits employment discrimination based on race, color, national origin, religion, or gender. 1. *Procedures*—Employees must file a claim with the Equal Employment Opportunity Commission (EEOC). The EEOC may sue the employer on the employee's behalf; if not, the employee may sue the employer directly. 2. *Types of discrimination*—Title VII prohibits both intentional (disparate-treatment) and unintentional (disparate-impact) discrimination. Disparate-impact discrimination occurs when an employer's practice, such as hiring only persons with a certain level of education, has the effect of discriminating against a class of persons protected by Title VII. Title VII also extends to discriminatory practices, such as various forms of sexual harassment. 3. *Remedies for discrimination under Title VII*—If a plaintiff proves that unlawful discrimination occurred, he or she may be awarded reinstatement, back pay, and retroactive promotions. Damages (both compensatory and punitive) may be awarded for intentional discrimination.
Discrimination Based on Age (See pages 476–478.)	The Age Discrimination in Employment Act (ADEA) of 1967 prohibits employment discrimination on the basis of age against individuals forty years of age or older. Procedures for bringing a case under the ADEA are similar to those for bringing a case under Title VII.
Discrimination Based on Disability (See pages 478–479.)	The Americans with Disabilities Act (ADA) of 1990 prohibits employment discrimination against persons with disabilities who are otherwise qualified to perform the essential functions of the jobs for which they apply. 1. *Procedures and remedies*—To prevail on a claim under the ADA, the plaintiff must show that she or he has a disability, is otherwise qualified for the employment in question, and was excluded from the employment solely because of the disability. Procedures under the ADA are similar to those required in Title VII cases; remedies are also similar to those under Title VII. 2. *Definition of disability*—The ADA defines the term *disability* as a physical or mental impairment that substantially limits one or more major life activities, a record of such impairment, or being regarded as having such an impairment. 3. *Reasonable accommodation*—Employers are required to reasonably accommodate the needs of persons with disabilities. Reasonable accommodations may include altering job-application procedures, modifying the physical work environment, and permitting more flexible work schedules.
Defenses to Employment Discrimination (See pages 479–480.)	If a plaintiff proves that employment discrimination occurred, employers may avoid liability by successfully asserting certain defenses. Employers may assert that the discrimination was required for reasons of business necessity, to meet a bona fide occupational qualification, or to maintain a legitimate seniority system. Evidence of prior employee misconduct acquired after the employee has been fired is not a defense to discrimination.

(Continued)

CHAPTER SUMMARY Employment Law—Continued

Affirmative Action (See pages 480–481.)	Affirmative action programs attempt to "make up" for past patterns of discrimination by giving members of protected classes preferential treatment in hiring or promotion. Increasingly, such programs are being strictly scrutinized by the courts and struck down as violating the Fourteenth Amendment.
State Statutes (See page 481.)	Generally, state laws also prohibit the kinds of discrimination prohibited by federal statutes. State laws may provide for more extensive protection and remedies than federal laws. Also, some states, such as California and Washington, have banned state-sponsored affirmative action programs.

FOR REVIEW

Answers for the even-numbered questions in this For Review *section can be found in Appendix A at the end of this text.*

1. What is the employment-at-will doctrine? When and why are exceptions to this doctrine made?
2. What federal statute governs working hours and wages? What federal statutes govern labor unions and collective bargaining?
3. What federal law was enacted to protect the health and safety of employees? What are workers' compensation laws?
4. Generally, what kind of conduct is prohibited by Title VII of the Civil Rights Act of 1964, as amended?
5. What remedies are available under Title VII of the 1964 Civil Rights Act, as amended?

QUESTIONS AND CASE PROBLEMS

QUESTION WITH SAMPLE ANSWER

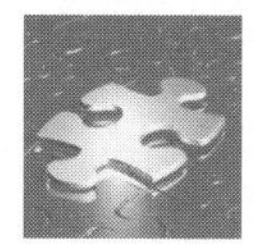

23–1. Denton and Carlo were employed at an appliance plant. Their job required them to do occasional maintenance work while standing on a wire mesh twenty feet above the plant floor. Other employees had fallen through the mesh; one was killed by the fall. When Denton and Carlo were asked by their supervisor to do work that would likely require them to walk on the mesh, they refused due to their fear of bodily harm or death. Because of their refusal to do the requested work, the two employees were fired from their jobs. Was their discharge wrongful? If so, under what federal employment law? To what federal agency or department should they turn for assistance?

For a sample answer to this question, go to Appendix B at the end of this text.

23–2. Discrimination Based on Age. Tavo Jones had worked since 1974 for Westshore Resort, where he maintained golf carts. During the first decade, he received positive job evaluations and numerous merit pay raises. He was promoted to the position of supervisor of golf-cart maintenance at three courses. Then a new employee, Ben Olery, was placed in charge of the golf courses. He demoted Jones, who was over the age of forty, to running only one of the three cart facilities, and he froze Jones's salary indefinitely. Olery also demoted five other men over the age of forty. Another cart facility was placed under the supervision of Blake Blair. Later, the cart facilities for the three courses were again consolidated, but Blair—not Jones—was put in charge. At the time, Jones was still in his forties, and Blair was in his twenties. Jones overheard Blair say that "we are going to have to do away with these . . . old and senile" men. Jones quit and sued Westshore for employment discrimination. Should he prevail? Explain.

23–3. Whistleblowing. Barbara Kraus was vice president of nursing at the New Rochelle Hospital Medical Center. She learned that a certain doctor had written on the charts of some patients that he had performed procedures for them that he had not performed. In fact, he had not obtained consent forms from the patients to perform those procedures. She reported this to the doctor's superiors, who took little action against the doctor. Some time later, Kraus was terminated. She filed a suit in a New York state court against the hospital to recover damages for wrongful termination. What is required for the court to rule in Kraus's favor? [*Kraus v. New Rochelle Hospital Medical Center*, 216 A.D.2d 360, 628 N.Y.S.2d 360 (1995)]

23–4. Wrongful Discharge. Stephen Fredrick, a pilot for Simmons Airlines Corp., criticized the safety of the aircraft that Simmons used on many of its flights and warned the airline about possible safety problems. Simmons took no action. After one of the planes crashed, Fredrick appeared on the television program *Good Morning America* to discuss his safety concerns. The same day, Fredrick refused to allow employees of Simmons to search his per-

sonal bags before a flight that he was scheduled to work. Claiming insubordination, the airline terminated Fredrick. Fredrick filed a suit in a federal district court against Simmons, claiming, among other things, that he had been discharged in retaliation for publicly expressing his concerns about the safety of the aircraft and that this discharge violated the public policy of providing for safe air travel. Simmons responded that an employee who "goes public" with his or her concerns should not be protected by the law. Will the court agree with Simmons? Explain. [*Fredrick v. Simmons Airlines Corp.*, 144 F.3d 500 (7th Cir. 1998)]

CASE PROBLEM WITH SAMPLE ANSWER

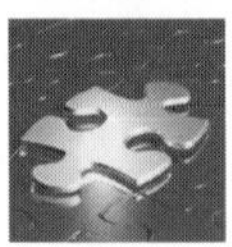

23–5. PGA Tour, Inc., sponsors professional golf tournaments. A player may enter in several ways, but the most common method is to successfully compete in a three-stage qualifying tournament known as the "Q-School." Anyone may enter the Q-School by submitting two letters of recommendation and paying $3,000 to cover greens fees and the cost of a golf cart, which is permitted during the first two stages, but is prohibited during the third stage. The rules governing the events include the "Rules of Golf," which apply at all levels of amateur and professional golf and do not prohibit the use of golf carts, and the "hard card," which applies specifically to the PGA tour and requires the players to walk the course during most of a tournament. Casey Martin is a talented golfer with a degenerative circulatory disorder that prevents him from walking golf courses. Martin entered the Q-School and asked for permission to use a cart during the third stage. PGA refused. Martin filed a suit in a federal district court against PGA, alleging a violation of the Americans with Disabilities Act. Is a golf cart in these circumstances a "reasonable accommodation" under the ADA? Why or why not? [*PGA Tour, Inc. v. Martin*, 532 U.S. 661, 121 S.Ct. 1879, 149 L.Ed.2d 904 (2001)]

After you have answered this problem, compare your answer with the sample answer given on the Web site that accompanies this text. Go to academic.cengage.com/blaw/wbl, select "Chapter 23," and click on "Case Problem with Sample Answer."

23–6. Unfair Labor Practice. The New York Department of Education's e-mail policy prohibits the use of the e-mail system for unofficial purposes, except that officials of the New York Public Employees Federation (PEF), the union representing state employees, can use the system for some limited communications, including the scheduling of union meetings and activities. In 1998, Michael Darcy, an elected PEF official, began sending mass, union-related e-mails to employees, including a summary of a union delegates' convention, a union newsletter, a criticism of proposed state legislation, and a criticism of the state governor and the Governor's Office of Employee Relations. Richard Cate, the department's chief operating officer, met with Darcy and reiterated the department's e-mail policy. When Darcy refused to stop his use of the e-mail system, Cate terminated his access to it. Darcy filed a complaint with the New York Public Employment Relations Board, alleging an unfair labor practice. Do the circumstances support Cate's action? Why or why not? [*Benson v. Cuevas*, 293 A.D.2d 927, 741 N.Y.S.2d 310 (3 Dept. 2002)]

23–7. Collective Bargaining. Verizon New York, Inc. (VNY), provides telecommunications services. Both VNY and the Communications Workers of America (CWA) are parties to collective bargaining agreements covering installation and maintenance employees. At one time, VNY supported annual blood drives. VNY, CWA, and charitable organizations jointly set dates, arranged appointments, and adjusted work schedules for the drives. For each drive, about a thousand employees, including managers, spent up to four hours traveling to a donor site, giving blood, recovering, and returning to their jobs. Employees received full pay for the time. In 2001, VNY told CWA that it would no longer allow employees to participate "on Company time," claiming that it experienced problems meeting customer requests for service during the drives. CWA filed a complaint with the National Labor Relations Board (NLRB), asking that VNY be ordered to bargain over the decision. Did VNY commit an unfair labor practice? Should the NLRB grant CWA's request? Why or why not? [*Verizon New York, Inc. v. National Labor Relations Board*, 360 F.3d 206 (D.C. Cir. 2004)]

23–8. Discrimination Based on Race. The hiring policy of Phillips Community College of the University of Arkansas (PCCUA) is to conduct an internal search for qualified applicants before advertising outside the college. Steven Jones, the university's chancellor, can determine the application and appointment process for vacant positions, however, and is the ultimate authority in hiring decisions. Howard Lockridge, an African American, was the chair of PCCUA's Technical and Industrial Department. Between 1988 and 1998, Lockridge applied for several different positions, some of which were unadvertised, some of which were unfilled for years, and some of which were filled with less qualified persons from outside the college. In 1998, when Jones advertised an opening for the position of dean of Industrial Technology and Workforce Development, Lockridge did not apply for the job. Jones hired Tracy McGraw, a white male. Lockridge filed a suit in a federal district court against the university under Title VII. The university filed a motion for summary judgment in its favor. What are the elements of a *prima facie* case of disparate-treatment discrimination? Can Lockridge pass this test, or should the court issue a judgment in the university's favor? Explain. [*Lockridge v. Board of Trustees of the University of Arkansas*, 315 F.3d 1005 (8th Cir. 2003)]

23–9. Discrimination Based on Age. The United Auto Workers (UAW) is the union that represents the employees of General Dynamics Land Systems, Inc. In 1997, a collective bargaining agreement between UAW and General Dynamics eliminated the company's obligation to provide health insurance to employees who retired after the date of the agreement, except for current workers at least fifty years old. Dennis Cline and 194 other employees, who were over forty years old but under fifty, objected to this term. They complained to the Equal Employment Opportunity Commission, claiming that the agreement violated the Age Discrimination in Employment Act (ADEA) of 1967. The ADEA forbids discriminatory preference for the "young" over the "old." Does the ADEA also prohibit favoring the old over the young? How should the court rule? Explain. [*General Dynamics Land Systems, Inc. v. Cline*, 540 U.S. 581, 124 S.Ct. 1236, 157 L.Ed.2d 1094 (2004)]

VIDEO QUESTION

23–10. Go to this text's Web site at academic.cengage.com/blaw/wbl and select "Chapter 23." Click on "Video Questions" and view the video titled *Parenthood.* Then answer the following questions.

1. In the video, Gil (Steve Martin) threatens to leave his job when he discovers that his boss is promoting another person to partner instead of him. His boss (Dennis Dugan) laughs and tells him that the threat is not realistic because if Gil leaves, he will be competing for positions with workers who are younger than he is and willing to accept lower salaries. If Gil takes his employer's advice and stays in his current position, can he sue his boss for age discrimination based on the boss's statements? Why or why not?
2. Suppose that Gil leaves his current position and applies for a job at another firm. The prospective employer refuses to hire him based on his age. What would Gil have to prove to establish a *prima facie* case of age discrimination? Explain your answer.
3. What defenses might Gil's current employer raise if Gil sues for age discrimination?

Employment Law

For updated links to resources available on the Web, as well as a variety of other materials, visit this text's Web site at

academic.cengage.com/blaw/wbl

An excellent Web site for information on employee benefits, including the full text of the Family and Medical Leave Act (FMLA), COBRA, other relevant statutes and case law, and current articles, is BenefitsLink. Go to

http://benefitslink.com/index.html

The Occupational Safety and Health Administration (OSHA) offers information related to workplace health and safety at

http://www.osha.gov

You can find the complete text of Title VII and information about the activities of the Equal Employment Opportunity Commission (EEOC) at that agency's Web site. Go to

http://www.eeoc.gov

The New York State Governor's Office of Employee Relations maintains an interactive site on sexual harassment and how to prevent sexual harassment in the workplace. Go to

http://www.goer.state.ny.us/Train/onlinelearning/SH/intro.html

The American Association for Affirmative Action provides a wealth of information about affirmative action at its Web site. Go to

http://www.affirmativeaction.org/resources/index.html

ONLINE LEGAL RESEARCH

Go to the *Fundamentals of Business Law* home page at academic.cengage.com/blaw/wbl, select "Chapter 23," and click on "Internet Exercises." There you will find the following Internet research exercises that you can perform to learn more about topics covered in this chapter.

Activity 23–1: LEGAL PERSPECTIVE—Workers' Compensation

Activity 23–2: MANAGEMENT PERSPECTIVE—Equal Employment Opportunity

Activity 23–3: SOCIAL PERSPECTIVE—Religious and National-Origin Discrimination

BEFORE THE TEST

Go to the *Fundamentals of Business Law* home page at academic.cengage.com/blaw/wbl, select "Chapter 23," and click on "Interactive Quizzes." You will find a number of interactive questions relating to this chapter.

UNIT 7 • EXTENDED CASE STUDY

Redi-Floors, Inc. v. Sonenberg Co.

In Chapter 22, we outlined the liability of principals and agents for contracts formed by agents. In this extended case study, we look at *Redi-Floors, Inc. v. Sonenberg Co.,*[1] a recent decision in which the court set out these principles.

1. 254 Ga.App. 615, 563 S.E.2d 505 (2002).

CASE BACKGROUND

Sonenberg Company managed Westchester Manor Apartments in Atlanta, Georgia. Manor Associates Limited Partnership owned the complex.[2] Manor's partners included Westchester Manor, Limited. The entry sign to the property disclosed only that Sonenberg managed it.

Sonenberg's on-site property manager contacted Redi-Floors, Inc., and requested a proposal for installing carpet in several apartments. In preparing the proposal, Redi-Floors confirmed that Sonenberg was the managing company. The property manager ordered, and Redi-Floors installed, the carpet as per the proposal. Redi-Floors sent invoices to the complex and received checks from "Westchester Manor Apartments." Believing Sonenberg owned the complex, Redi-Floors did not learn of the true owner's identity until a dispute arose concerning payment for some of the invoices.

To recover on the outstanding invoices, Redi-Floors filed a suit in a Georgia state court against Sonenberg, Manor Associates, and Westchester Manor. The court issued a directed verdict in Sonenberg's favor on the ground that Redi-Floors knew Sonenberg was only an agent. Against the other parties, Redi-Floors obtained a verdict exceeding $20,000. Unable to collect on this judgment, Redi-Floors appealed to a state intermediate appellate court.

MAJORITY OPINION

BLACKBURN, Chief Judge.

* * * *

An agent who makes a contract without identifying his principal becomes personally liable on the contract. If the agent wishes to avoid personal liability, the duty is on him to disclose his agency, and not on the party with whom he deals to discover it. The agent's disclosure of a trade name and the plaintiff's awareness of that name are not necessarily sufficient so as to protect the agent from liability. The disclosure of an agency is not complete for the purpose of relieving the agent from personal liability unless it embraces the name of the principal. [Emphasis added.]

Based on these principles, *Reed v. Burns International Security Services,* 215 Ga.App. 60, 449 S.E.2d 888 (1994), upheld a judgment in favor of a security company and against the apartment management company that contracted for security services at the apartment complex but failed to identify to the security company the name of the limited partnership owning the complex. * * * Here, at least some evidence showed that Sonenberg never disclosed the name of Manor Associates Limited Partnership to Redi-Floors. Accordingly, the trial court erred in entering a directed verdict in favor of Sonenberg.

* * * *

With respect to an undisclosed principal, the rule in Georgia is that if the buyer is in fact merely an agent and acts with the authority of an undisclosed principal, either he or such principal may be held liable at the election of the opposite party; but the contractual liability of such agent and principal is not joint, and, after an election to proceed against one, the other cannot be held. Thus, if an agent buys in his own name, without disclosing his principal, and the seller subsequently discovers that the purchase was, in fact, made for another, he may, at his choice, look for payment either to the agent or the principal * * * . On the other hand, if, at the time of the sale, the seller knows not only the person, who is nominally dealing with him, is not principal, but agent, and also knows who the principal really is, and notwithstanding all the knowledge, chooses to make the agent his debtor—dealing with him and him alone—the seller must be taken to have abandoned his recourse against the principal, and cannot afterwards, upon the failure of the agent, turn round and charge the principal, having once made his election at the time when he had the power of choosing between the one and the other. An election deliberately made, with knowledge of facts and absence of fraud, is conclusive; and the party who has once elected can claim no right to make a second choice. Thus, while it is true that a judgment against both the agent and the principal cannot stand, it is the plaintiff who is entitled to elect against which of the defendants, principal or agent, to take the judgment. * * *

* * * In the present case the trial court's erroneous granting of a directed verdict deprived the plaintiff of its right to elect which defendant it would proceed against. * * *

* * * *

This case must be remanded to the trial court with instructions to allow Redi-Floors to make an election as to which defendant it wishes to proceed against, thus restoring

2. A *limited partnership* is a form of business organization that will be discussed in Chapter 24.

(Continued)

to Redi-Floors the right to make such election. Should Redi-Floors elect to hold Manor Associates liable, nothing further is required as the existing judgment would stand against this defendant. On the other hand, should Redi-Floors seek to proceed against Sonenberg, the existing judgment is vacated and a new trial will be necessary.

DISSENTING OPINION

MILLER, Judge, * * * dissenting * * * .

* * * I dissent * * * because I believe that Redi-Floors's decision to obtain a judgment against Manor Associates Limited Partnership has rendered the error harmless. Redi-Floors had the right, as do all plaintiffs, to pursue mutually exclusive remedies prior to the verdict; but once Redi-Floors procured a judgment order which was reduced to writing against Manor Associates, that constituted an election of alternative remedies that precluded plaintiff from pursuing the excluded remedy against Sonenberg.

A corollary to the principle that a nondisclosing agent is personally liable is that the contract liability of a principal and his agent is not joint, and after election to proceed against one, the other cannot be held. As the plaintiff may not obtain judgment against both, he must make an election *prior to judgment* as to whether he wants a judgment against the agent or against the principal. If the plaintiff does not expressly announce an election, his taking a judgment against the principal constitutes an election and precludes any further action against the agent. Here, after the court entered a directed verdict in favor of Sonenberg, Redi-Floors (1) had every opportunity to reevaluate its case and to obtain court permission to dismiss its case * * * against Manor Associates instead of proceeding further or (2) could have had the court withhold entry of judgment (once the verdict against Manor Associates was obtained) until an appeal of the directed verdict was decided. In either case, Redi-Floors could have then brought this issue here as an appeal from the directed verdict *without* electing to obtain a *judgment* against Manor Associates. Redi-Floors's decision to obtain a judgment against the principal, however, was an election and precludes it from pursuing Sonenberg further.

QUESTIONS FOR ANALYSIS

1. **Law.** What did the majority conclude on the question regarding an agent's potential liability on a contract entered into on behalf of an undisclosed principal? Why?
2. **Law.** What was the dissent's major point of disagreement with the majority? On what grounds did the dissent base its disagreement on this point?
3. **Ethics.** Do agents owe to those with whom they deal on behalf of their principals any obligations besides a general duty to obey the law? If so, what is the nature of those obligations?
4. **Social Dimensions.** If the agent had not had the actual authority to enter into the contract at the center of this case, how might the outcome have been affected?
5. **Implications for Creditors.** What does the outcome in this case indicate to those who do business with others' agents on a credit basis?

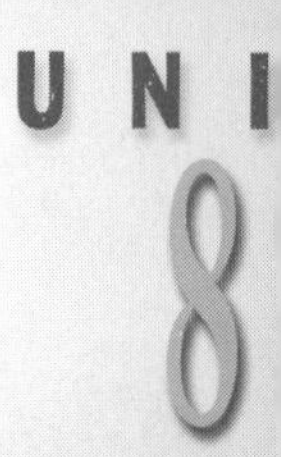

Business Organizations

Unit Contents

CHAPTER 24

Sole Proprietorships, Partnerships, and Limited Liability Companies

LEARNING OBJECTIVES

After reading this chapter, you should be able to answer the following questions:

1. Which form of business organization is the simplest? Which form arises from an agreement between two or more persons to carry on a business for profit?
2. What are the three essential elements of a partnership?
3. What is meant by joint and several liability? Why is this often considered to be a disadvantage of the partnership form of business?
4. Why do professional groups organize as a limited liability partnership? How does this form differ from a general partnership?
5. What is a limited liability company? What are some of the advantages and disadvantages of this business form?

An entrepreneur's primary motive for undertaking a business enterprise is to make profits. An *entrepreneur* is by definition one who *initiates* and *assumes the financial risks* of a new enterprise and undertakes to provide or control its management.

One of the questions faced by any entrepreneur who wishes to start a business is what form of business organization should be chosen for the business endeavor. In making this determination, a number of factors need to be considered. Four important factors are (1) ease of creation, (2) the liability of the owners, (3) tax considerations, and (4) the need for capital. In studying this unit on business organizations, keep these factors in mind as you read about the various business organizational forms available to entrepreneurs. You might also find it helpful to refer to Exhibit 26–2 in Chapter 26, on pages 555 and 556, for a comparison of the major business forms in use today with respect to formation, liability of owners, taxation, and other factors.

Traditionally, entrepreneurs have used three major forms to structure their business enterprises—the sole proprietorship, the partnership, and the corporation. In this chapter, we examine the first two of these forms. The third major traditional form—the corporation—is discussed in Chapters 25 and 26. Two relatively new forms of business enterprise—limited liability companies (LLCs) and limited liability partnerships (LLPs)—offer special advantages to businesspersons, particularly with respect to taxation and liability. We therefore look at these business forms later in this chapter.

Sole Proprietorships

The simplest form of business organization is a **sole proprietorship.** In this form, the owner is the business; thus, anyone who does business without creating a separate business organization has a sole proprietorship. Over two-thirds of all U.S. businesses are sole proprietorships.

They are usually small enterprises—about 99 percent of the sole proprietorships in the United States have revenues of less than $1 million per year. Sole proprietors can own and manage any type of business, ranging from an informal, home-office undertaking to a large restaurant or construction firm. Today, a number of online businesses that sell goods and services on a nationwide basis are organized as sole proprietorships.

ADVANTAGES OF THE SOLE PROPRIETORSHIP

A major advantage of the sole proprietorship is that the proprietor receives all of the profits (because he or she assumes all of the risk). In addition, it is often easier and less costly to start a sole proprietorship than to start any other kind of business, as few legal formalities are involved. This type of business organization also offers more flexibility than does a partnership or a corporation. The sole proprietor is free to make any decision she or he wishes concerning the business—including whom to hire, when to take a vacation, and what kind of business to pursue, for example. A sole proprietor pays only personal income taxes on the business's profits, which are reported as personal income on the proprietor's personal income tax return. Sole proprietors are also allowed to establish tax-exempt retirement accounts in the form of Keogh plans.[1]

DISADVANTAGES OF THE SOLE PROPRIETORSHIP

The major disadvantage of the sole proprietorship is that, as sole owner, the proprietor alone bears the burden of any losses or liabilities incurred by the business enterprise. In other words, the sole proprietor has unlimited liability, or legal responsibility, for all obligations incurred in doing business. This unlimited liability is a major factor to be considered in choosing a business form.

■ EXAMPLE 24.1 Sheila Fowler operates a golf shop business as a sole proprietorship. The shop is located near one of the best golf courses in the country. A professional golfer, Dean Maheesh, is seriously injured when a display of golf clubs, which one of Fowler's employees had failed to secure, falls on him. If Maheesh sues Fowler's shop (a sole proprietorship) and wins, Fowler's personal liability could easily exceed the limits of her insurance policy. In this situation, Fowler might lose not only her business, but also her house, car, and any other personal assets that can be attached to pay the judgment. ■

The sole proprietorship also has the disadvantage of lacking continuity on the death of the proprietor. When the owner dies, so does the business—it is automatically dissolved. If the business is transferred to family members or other heirs, a new proprietorship is created. Another disadvantage is that the proprietor's opportunity to raise capital is limited to personal funds and the funds of those who are willing to make loans. If the owner wishes to expand the business significantly, one way to raise more capital to finance the expansion is to join forces with another entrepreneur and establish a partnership or form a corporation.

The Law Governing Partnerships

When two or more persons agree to do business as partners, they enter into a special relationship with one another. To an extent, their relationship is similar to an agency relationship because each partner is deemed to be the agent of the other partners and of the partnership. The common law agency concepts outlined in Chapter 22 thus apply—specifically, the imputation of knowledge of, and responsibility for, acts done within the scope of the partnership relationship. In their relations with one another, partners, like agents, are bound by fiduciary ties.

The Uniform Partnership Act (UPA) governs the operation of partnerships *in the absence of express agreement* and has done much to reduce controversies in the law relating to partnerships. The UPA was originally set forth by the National Conference of Commissioners on Uniform State Laws in 1914 and has undergone several major revisions. Except for Louisiana, every state has adopted the UPA. The majority of states have adopted the most recent version of the UPA, which was established in 1994 and amended in 1997 to provide limited liability for partners in a limited liability partnership. We therefore base our discussion of the UPA on the 1997 version of the act and refer to older versions of the UPA in footnotes when appropriate.

We look next at the legal definition of a partnership and at the laws governing the formation, operation, and termination of general partnerships. Whenever relevant, we include bracketed references to specific UPA provisions. Additionally, if the 1997 amendments have significantly changed particular UPA provisions, we indicate the prior law in footnotes.

DEFINITION OF PARTNERSHIP

Conflicts commonly arise over whether a business enterprise is legally a partnership, especially in the absence of a formal, written partnership agreement. The UPA defines a **partnership** as "an association of two or more persons to carry on as co-owners a business for profit" [UPA 101(6)].

1. A *Keogh plan* is a retirement program designed for self-employed persons. The person can contribute a certain percentage of income to the plan, and interest earnings are not taxed until funds are withdrawn from the plan.

The intent to associate is a key element of a partnership, and one cannot join a partnership unless all other partners consent [UPA 401(i)].

Partnership Status In resolving disputes over whether partnership status exists, courts will usually look for the following three essential elements, which are implicit in the UPA's definition of a partnership:

1. A sharing of profits and losses.
2. A joint ownership of the business.
3. An equal right in the management of the business.

If the evidence in a particular case is insufficient to establish all three factors, the UPA provides a set of guidelines to be used. For example, the sharing of profits and losses from a business is considered *prima facie* ("on the face of it") evidence that a partnership has been created. No such inference is made, however, if the profits were received as payment of any of the following [UPA 202(c)(3)]:

1. A debt by installments or interest on a loan.
2. Wages of an employee or for the services of an independent contractor.
3. Rent to a landlord.
4. An annuity to a surviving spouse or representative of a deceased partner.
5. A sale of goodwill of a business or property.

When one of the parties disputes whether a partnership was created, the courts frequently look to other factors, such as the conduct of the parties, to determine partnership status. In the following case, the parties met and discussed buying and developing a commercial site. Using some of one party's funds, another party bought the site in the name of his business alone. The buyer refused to share ownership of the property, claiming that he had entered into only an unenforceable agreement to form a partnership and was not a partner.

CASE 24.1 Cap Care Group, Inc. v. McDonald

North Carolina Court of Appeals, 2002.
149 N.C.App. 817,
561 S.E.2d 578.
http://www.aoc.state.nc.us/www/public/html/opinions.htm[a]

FACTS Cap Care Group, Inc., PWPP Partners, and C & M Investments of High Point, Inc., buy and develop commercial real estate. Ronnel Parker is president of Cap Care and PWPP. Wayne McDonald owns and controls C & M. In November 1996, Parker and Daniel Greene, another Cap Care officer, met McDonald to discuss buying and renovating a commercial site in High Point, North Carolina. Cap Care later claimed that at the meeting the parties (Parker on behalf of Cap Care and PWPP, and McDonald on behalf of C & M) agreed that they would be partners in the purchase and development of the property, and that McDonald would make an offer to the seller on the partnership's behalf. In February 1997, McDonald signed a contract to buy the site, using PWPP funds as half of the earnest money.[b] McDonald and Greene (of Cap Care) then discussed jointly owning the property as partners. In March, McDonald bought the property in the name of C & M only. Cap Care demanded that McDonald contribute the property to the partnership, but he refused. Cap Care filed a suit in a North Carolina state court against McDonald and C & M, alleging in part breach of contract. McDonald argued that he and Parker had merely entered into an unenforceable agreement to form a partnership. The court awarded Cap Care $477,511, plus interest and fees. The defendants appealed to a state intermediate appellate court.

ISSUE Was there a valid enforceable agreement among the parties to form a partnership to purchase property?

DECISION Yes. The state intermediate appellate court affirmed the lower court's judgment.

REASON The defendants had entered into, and breached, an oral partnership contract to buy real estate as evidenced by (1) the parties' discussions, (2) the defendants' acceptance of a partner's funds, and (3) the defendants' acting according to the parties' plan until after the purchase of the property.

WHY IS THIS CASE IMPORTANT? *This case illustrates that the acceptance by conduct of an offer to create a partnership results in the formation of the partnership, whether or not the "accepting" party subjectively intended to enter into a partnership.*

a. In the "Court of Appeals Opinions" section, click on "2002." On the next page, scroll to the "16 April 2002" section and click on the name of the case to access the opinion.
b. *Earnest money* is a deposit of funds that usually accompanies an offer to buy real estate to show that the offer is earnest, or serious.

Joint Property Ownership and Partnership Status Joint ownership of property, obviously, does not in and of itself create a partnership. Therefore, if persons own real property as joint tenants or as tenants in common (forms of joint ownership, to be discussed in Chapter 28), this does not mean that they are partners in a partnership. In fact, the sharing of gross revenues and even profits from such ownership is usually not enough to create a partnership [UPA 202(c)(1) and (2)].

■ EXAMPLE 24.2 Suppose that Chiang and Burke jointly own a piece of rural property. They lease the land to a farmer, with the understanding that—in lieu of set rental payments—they will receive a share of the profits from the farming operation conducted by the farmer. This arrangement normally would not make Chiang, Burke, and the farmer partners. ■ Note that although the sharing of profits from ownership of property does not prove the existence of a partnership, sharing *both profits and losses* usually does.

Entity versus Aggregate A partnership is sometimes called a *firm* or a *company,* terms that suggest that the partnership is an entity separate and apart from its aggregate members. The law of partnership recognizes the independent entity for most purposes, but a partnership is treated as an *aggregate of the individual partners* for at least one purpose.

—Partnership as an Entity. At common law, a partnership was never treated as a separate legal entity. Thus, at common law a suit could never be brought by or against the firm in its own name; each individual partner had to sue or be sued. Today, most states provide specifically that the partnership can be treated as an entity for certain purposes. For example, a partnership usually can sue or be sued, collect judgments, and have all accounting procedures in the name of the partnership entity. In addition, the UPA clearly states, "A partnership is an entity" and "A partnership may sue and be sued in the name of the partnership" [UPA 201 and 307(a)]. As an entity, a partnership may hold the title to real or personal property in its name rather than in the names of the individual partners. Finally, federal procedural laws frequently permit the partnership to be treated as an entity in such matters as suits in federal courts and bankruptcy proceedings.

—Partnership as an Aggregate. In one circumstance, the partnership is not regarded as a separate legal entity, but is treated as an aggregate of the individual partners. For federal income tax purposes, a partnership is not a taxpaying entity. The income and losses it incurs are "passed through" the partnership framework and attributed to the partners on their individual tax returns. The partnership itself has no tax liability and is responsible only for filing an *information return* with the Internal Revenue Service. In other words, the firm itself pays no taxes. A partner's profit from the partnership (whether distributed or not) is taxed as individual income to the individual partner.

PARTNERSHIP FORMATION

As a general rule, agreements to form a partnership can be *oral, written,* or *implied by conduct.* Some partnership agreements, however, must be in writing to be legally enforceable under the Statute of Frauds (see Chapter 10 for details). For example, a partnership agreement that, by its terms, is to continue for more than one year or a partnership agreement that authorizes the partners to deal in transfers of real property must be evidenced by a sufficient writing. This section will describe some of the legal concepts related to the creation of a partnership.

The Partnership Agreement A partnership agreement, called **articles of partnership,** can include virtually any terms that the parties wish, unless they are illegal or contrary to public policy or statute [UPA 103]. The agreement usually specifies the name and location of the business, the duration of the partnership, the purpose of the business, each partner's share of the profits, how the partnership will be managed, how assets will be distributed on dissolution, and other provisions. The terms commonly included in a partnership agreement are listed in Exhibit 24–1 on the next page. (Notice that this list includes an arbitration clause, which is often included in a partnership agreement.)

Duration of the Partnership The partnership agreement can specify the duration of the partnership by stating that it will continue until a certain date or the completion of a particular project. A partnership that is specifically limited in duration is called a *partnership for a term.* If this type of partnership is dissolved (broken up) without the consent of all the partners prior to the expiration of the partnership term, the dissolution constitutes a breach of the agreement. The responsible partner can be held liable for any resulting losses. If no fixed duration is specified, the partnership is a *partnership at will.* Any partner can dissolve this type of partnership at any time without violating the agreement and without incurring liability for losses to other partners resulting from the termination.

A Corporation as Partner In a general partnership, the partners are personally liable for the debts incurred by the

EXHIBIT 24–1 COMMON TERMS INCLUDED IN A PARTNERSHIP AGREEMENT

Basic Structure	1. Name of the partnership. 2. Names of the partners. 3. Location of the business and the state law under which the partnership is organized. 4. Purpose of the partnership. 5. Duration of the partnership.
Capital Contributions	1. Amount of capital that each partner is contributing. 2. The agreed-on value of any real or personal property that is contributed instead of cash. 3. How losses and gains on contributed capital will be allocated, and whether contributions will earn interest.
Sharing of Profits and Losses	1. Percentage of the profits and losses of the business that each partner will receive. 2. When distributions of profit will be made and how net profit will be calculated.
Management and Control	1. How management responsibilities will be divided among the partners. 2. Name of the managing partner or partners, and whether other partners have voting rights.
Accounting and Partnership Records	1. Name of the bank in which the partnership will maintain its business and checking accounts. 2. Statement that an accounting of partnership records will be maintained and that any partner or her or his agent can review these records at any time. 3. The dates of the partnership's fiscal year and when the annual audit of the books will take place.
Dissolution	1. Events that will dissolve the partnership, such as the retirement, death, or incapacity of any partner. 2. How partnership property will be valued and apportioned on dissolution. 3. Whether an arbitrator will determine the value of partnership property on dissolution and whether that determination will be binding.
Miscellaneous	Whether arbitration is required for any dispute relating to the partnership agreement.

partnership. If one of the general partners is a corporation, however, what does personal liability mean? Basically, the capacity of corporations to contract is a question of corporation law. At one time, many states had restrictions on corporations becoming partners, although such restrictions have become less common over the years.

The Revised Model Business Corporation Act (discussed in Chapter 25) allows corporations generally to make contracts and incur liabilities. The UPA specifically permits a corporation to be a partner. By definition, "a partnership is an association of two or more persons," and the UPA defines *person* as including corporations [UPA 101(10)].

Partnership by Estoppel Persons who are not partners may nevertheless hold themselves out as partners and make representations that third parties rely on in dealing with them. In such a situation, a court may conclude that a *partnership by estoppel* exists. The law does not confer any partnership rights on these persons, but it may impose liability on them. This is also true when a partner represents, expressly or impliedly, that a nonpartner is a member of the firm. Whenever a third person has reasonably and detrimentally relied on the representation that a nonpartner was part of the partnership, partnership by estoppel is deemed to exist. When this occurs, the nonpartner is regarded as an agent whose acts are binding on the partnership [UPA 308].

■ **EXAMPLE 24.3** Sorento owns a small shop. Knowing that the Midland Bank will not make a loan on his credit alone, Sorento represents that Lukas, a financially secure businessperson, is a partner in Sorento's business. Lukas knows of Sorento's misrepresentation but fails to correct it. Midland Bank, relying on the strength of Lukas's reputation and credit, extends a loan to Sorento. Sorento will be liable to the bank for repaying the loan. Lukas could also be held liable to the bank in many states. Lukas has impliedly consented to the misrepresentation and normally will be estopped from denying that she is a partner of Sorento. She will be regarded as if she were in fact a partner in Sorento's business insofar as this loan is concerned. ■

RIGHTS OF PARTNERS

The rights and duties of partners are governed largely by the specific terms of their partnership agreement. In the absence of provisions to the contrary in the partnership agreement, the law imposes the rights and duties discussed here. The character and nature of the partnership business generally influence the application of these rights and duties.

The rights of partners in a partnership relate to the following areas: management, interest in the partnership, compensation, inspection of books, accounting, and property.

Management Rights In a general partnership, all partners have equal rights in managing the partnership [UPA 401(f)]. Unless the partners agree otherwise, each partner has one vote in management matters *regardless of the proportional size of his or her interest in the firm.* Often, in a large partnership, partners will agree to delegate daily management responsibilities to a management committee made up of one or more of the partners.

The majority rule controls decisions in ordinary matters connected with partnership business, unless otherwise specified in the agreement. Decisions that significantly affect the nature of the partnership or that are not apparently for carrying on the ordinary course of the partnership business, or business of the kind, however, require the *unanimous* consent of the partners [UPA 301(2), 401(i), and 401(j)]. Unanimous consent is likely to be required for a decision to undertake any of the following actions:[2]

1. To alter the essential nature of the firm's business as expressed in the partnership agreement or to alter the capital structure of the partnership.
2. To admit new partners or to enter a wholly new business.
3. To assign partnership property to a trust for the benefit of creditors.
4. To dispose of the partnership's goodwill.
5. To confess judgment against the partnership or to submit partnership claims to arbitration. (A **confession of judgment** is the act of a debtor in permitting a judgment to be entered against him or her by a creditor, for an agreed sum, without the institution of legal proceedings.)
6. To undertake any act that would make further conduct of partnership business impossible.
7. To amend the articles of the partnership agreement.

Interest in the Partnership Each partner is entitled to the proportion of business profits and losses that is designated in the partnership agreement. If the agreement does not apportion profits (indicate how the profits will be shared), the UPA provides that profits will be shared equally. If the agreement does not apportion losses, losses will be shared in the same ratio as profits [UPA 401(b)].

■ EXAMPLE 24.4 The partnership agreement for Rico and Brent provides for capital contributions of $60,000 from Rico and $40,000 from Brent, but it is silent as to how Rico and Brent will share profits or losses. In this situation, Rico and Brent will share both profits and losses equally. If their partnership agreement provided for profits to be shared in the same ratio as capital contributions, however, 60 percent of the profits would go to Rico, and 40 percent of the profits would go to Brent. If their partnership agreement was silent as to losses, losses would be shared in the same ratio as profits (60 percent/40 percent). ■

Compensation Devoting time, skill, and energy to partnership business is a partner's duty and generally is not a compensable service. Rather, as mentioned, a partner's income from the partnership takes the form of a distribution of profits according to the partner's share in the business. Partners can, of course, agree otherwise. For example, the managing partner of a law firm often receives a salary in addition to her or his share of profits for performing special administrative duties, such as managing the office or personnel.

UPA 401(h) provides that a partner is entitled to compensation for services in winding up partnership affairs (and reimbursement for expenses incurred in the process) above and apart from his or her share in the partnership profits.

Inspection of Books Partnership books and records must be kept accessible to all partners. Each partner has the right to receive (and each partner has the corresponding duty to produce) full and complete information concerning the conduct of all aspects of partnership business [UPA 403]. Each firm keeps books for recording and preserving such information. Partners contribute the information, and a bookkeeper or an accountant typically has the duty to preserve it. The books must be kept at the firm's principal business office and cannot be removed without the consent of all of the partners. Every partner, whether active or inactive, is entitled to inspect all books and records on demand and can make copies of the materials. The personal representative of a deceased partner's estate has the same right of access to partnership books and records that the decedent would have had [UPA 403].

Actions against the Partnership and among the Partners An accounting of partnership assets or profits is required to determine the value of each partner's share in the partnership. An accounting can be performed voluntarily, or it can be compelled by court order. Under UPA 405(b), a partner has the right to bring an action for an accounting during the term of the partnership, as well as on the firm's

2. The previous version of the UPA specifically listed most of these actions as requiring unanimous consent. The current version of the UPA omitted the list entirely to allow the courts more flexibility. The Official Comments explain that most of these acts, except for submitting a claim to arbitration, will likely remain outside the apparent authority of an individual partner.

dissolution and winding up.[3] A partner also has the right to bring an action against the partnership or another partner, with or without a formal accounting, in the following situations:

1. To enforce the partner's rights under the partnership agreement.
2. To enforce the partner's rights under the UPA.
3. To enforce the partner's rights and interests arising independently of the partnership relationship.

Property Rights A partner has the following three basic property rights:

1. An interest in the partnership.
2. A right in partnership property.
3. A right to participate in the management of the partnership, as previously discussed [UPA 401(f)].

—Partner's Interest in the Firm. A partner's interest in the firm is a personal asset consisting of a proportionate share of the profits earned [UPA 502] and a return of capital on the partnership's termination. A partner's interest is subject to assignment or to a judgment creditor's lien (a lien obtained through the judicial process). Judgment creditors can attach a partner's interest by petitioning the court that entered the judgment to grant the creditors a **charging order.** This order entitles the judgment creditors to the profits of the partner and to any assets available to the partner on dissolution [UPA 504]. Neither an assignment nor a court's charging order entitling a judgment creditor to receive a share of the partner's funds will cause the dissolution of the firm [UPA 503].

—Partnership Property. Property acquired by a partnership is the property of the partnership and not of the partners individually [UPA 203]. Property acquired *in the name of* the partnership or a partner is partnership property if the person's capacity as a partner or the existence of the partnership is indicated in the instrument transferring title. If the transferring instrument refers to neither of these, the property is still presumed to be partnership property when it is acquired with partnership funds [UPA 204].[4]

A partner may use or possess partnership property only on behalf of the partnership [UPA 401(g)]. A partner is not a co-owner of partnership property and has no interest in the property that can be transferred, either voluntarily or involuntarily [UPA 501]. In other words, partnership property is owned by the partnership as an entity and not by the individual partners.[5]

Duties and Liabilities of Partners

The duties and liabilities of partners are basically derived from agency law. Each partner is an agent of every other partner and acts as both a principal and an agent in any business transaction within the scope of the partnership agreement. Each partner is also a general agent of the partnership in carrying out the usual business of the firm "or business of the kind carried on by the partnership" [UPA 301(1)]. Thus, every act of a partner concerning partnership business and "business of the kind" and every contract signed in the partnership's name bind the firm. The UPA affirms general principles of agency law that pertain to the authority of a partner to bind a partnership in contract or tort.

We examine here the fiduciary duties of partners, the authority of partners, the joint and several liability that characterizes partnerships, and the limitations imposed on the liability of incoming partners for preexisting partnership debts.

Fiduciary Duties The only fiduciary duties a partner owes to the partnership and the other partners are the duty of loyalty and the duty of care, according to UPA 404(a). A partner's duty of loyalty includes accounting to the partnership for "any property, profit, or benefit" derived by the partner in the conduct of the partnership business or from a use of its property. A partner also has a duty to refrain from dealing with the firm as an adverse party or competing with it in the conduct of the partnership business [UPA 404(b)]. A partner's duty of care is limited to refraining from "grossly negligent or reckless conduct, intentional misconduct, or a knowing violation of law" [UPA 404(c)].[6]

3. Under the previous version of the UPA, a partner could bring an action for an accounting in any one of the following situations: the partnership agreement provided for an accounting, the partner was wrongfully excluded from the business or its property or books, another partner was in breach of his or her fiduciary duty, or other circumstances rendered it "just and reasonable."

4. *Partnership property* was defined in the previous version of the UPA as "all property originally brought into the partnership's stock or subsequently acquired, by purchase or otherwise, on account of the partnership" [UPA 8(1)]. This definition, unlike the current UPA, did not provide any guidance concerning when property is "acquired by" a partnership.

5. Under the previous version of the UPA, partners were *tenants in partnership.* This meant that every partner was a co-owner with all other partners of the partnership property. The current UPA does not recognize this concept.

6. The previous version of the UPA touched only briefly on the duty of loyalty and left the details of the partners' fiduciary duties to be developed under the law of agency.

These duties may not be waived or eliminated in the partnership agreement, and in fulfilling them each partner must act consistently with the obligation of good faith and fair dealing, which applies to all contracts, including partnership agreements [UPA 103(b) and 404(d)]. A partner may pursue his or her own interests, however, without automatically violating these duties [UPA 404(e)]. For example, a partner who owns a shopping mall may vote against a partnership proposal to open a competing mall.

Authority of Partners The UPA affirms general principles of agency law that pertain to the authority of a partner to bind a partnership in contract. Under the same principles, a partner may also subject a partnership to liability in tort. When a partner is apparently carrying on partnership business or business of the kind with third parties in the usual way, both the partner and the firm share liability. The partnership will not be liable, however, if the third parties know that the partner has no such authority.

—The Scope of Implied Powers. In an ordinary partnership, the partners can exercise all implied powers reasonably necessary and customary to carry on that particular business. Some customarily implied powers include the authority to make warranties on goods in the sales business, the power to convey real property in the firm's name when such conveyances are part of the ordinary course of partnership business, the power to enter into contracts consistent with the firm's regular course of business, and the power to make admissions and representations concerning partnership affairs.

The court in the following case was asked to consider whether one partner was liable for overdrafts created by another partner in one of the partnership's bank accounts "in the ordinary course of the partnership business."

CASE 24.2 Helpinstill v. Regions Bank

Texas Court of Appeals—Texarkana, 2000.
33 S.W.3d 401.

FACTS Bobby Helpinstill and Mike Brown were partners in MBO Computers. They opened a partnership bank account at Longview National Bank in Longview, Texas, each agreeing in writing that he would be individually liable for any overdrafts created on the account. Brown, who was actively managing the business, regularly wrote overdrafts on the account and covered them later with deposits. In January 1997, to pay creditors of the partnership, Brown began shuffling funds among MBO accounts at the Longview bank and two other banks in a check-kiting scheme.[a] Brown was convicted and sentenced to a term in federal prison. Regions Bank (which had acquired Longview) filed a suit in a Texas state court against Helpinstill to recover $381,011.15, the amount of the overdrafts. When the court ruled in favor of the bank, Helpinstill appealed to a state intermediate appellate court, arguing in part that he was not liable because Brown's check kiting was not within the ordinary course of partnership business.

a. *Check kiting* refers to the practice of moving funds between bank accounts for the purpose of covering account deficiencies. For example, a check drawn on an account in bank A is written to cover a deficiency in an account in bank B. Then a check drawn on an account in bank B is written to cover the deficiency in bank A—and the pattern continues.

ISSUE Is Helpinstill liable to the bank for the amount of the overdrafts?

DECISION Yes. The state intermediate appellate court affirmed the judgment of the lower court, concluding that Helpinstill's liability to the bank was established as a matter of law.

REASON According to the appellate court, "Helpinstill's liability was established by virtue of the fact that the partnership was indebted to the bank for the overdrafts that Brown created in the ordinary course of the partnership's business, and Helpinstill, as a general partner, is responsible for such partnership debts." The court reasoned that "[t]he kiting scheme, rather than creating the overdrafts, was being used to disguise them; and, while not in itself in the ordinary course of the partnership's business, the kiting scheme did not change the fact that the creation of overdrafts was in the ordinary course of the partnership's business."

FOR CRITICAL ANALYSIS—Social Consideration
What is meant by the phrase "in the ordinary course of business"?

—Authorized versus Unauthorized Actions. If a partner acts within the scope of authority, the partnership is legally bound to honor the partner's commitments to third parties. For example, a partner's authority to sell partnership products carries with it the implied authority to transfer title and to make the usual warranties. Hence, in a partnership that operates a retail tire store, any partner negotiating a contract with a customer for the sale of a set of tires can warrant that "each tire will be warranted for normal wear for 40,000 miles."

This same partner, however, does not have the authority to sell office equipment, fixtures, or the partnership office building without the consent of all of the other partners. In addition, because partnerships are formed to generate profits, a partner generally does not have the authority to make charitable contributions without the consent of the other partners. Such actions are not binding on the partnership unless they are ratified by all of the other partners.

Liability of Partners One significant disadvantage associated with a traditional partnership is that partners are *personally* liable for the debts of the partnership. Moreover, the liability is essentially unlimited, because the acts of one partner in the ordinary course of business subject the other partners to personal liability [UPA 305]. The following subsections explain the rules on a partner's liability that are followed by most states.

—Joint and Several Liability. Under UPA 306(a), partners are jointly and severally liable for partnership obligations, including contracts, torts, and breaches of trust.[7] **Joint and several[8] liability** means that a third party may sue all of the partners (jointly) or one or more of the partners separately (severally) at his or her option. This is true even if the partner did not participate in, ratify, or know about whatever it was that gave rise to the cause of action.

Generally, under UPA 307(d), however, a creditor cannot bring an action to collect a partnership debt from the partner of a nonbankrupt partnership without first attempting to collect from the partnership, or convincing a court that the attempt would be unsuccessful.

A judgment against one partner on her or his several (separate) liability does not extinguish the others' liability. (Similarly, a release of one partner does not discharge the partners' several liability.) Thus, those not sued in the first action may be sued subsequently. The first action, however, may have been conclusive on the question of liability. If, for instance, in an action against one partner, the court held that the partnership was in no way liable, the third party cannot bring an action against another partner and succeed on the issue of the partnership's liability.

If the third party is successful in a suit against a partner or partners, he or she may collect on the judgment only against the assets of those partners named as defendants. A partner who commits a tort is required to indemnify the partnership for any damages it pays.

—Liability of Incoming Partner. A newly admitted partner to an existing partnership normally has limited liability for whatever debts and obligations the partnership incurred prior to the new partner's admission. The new partner's liability can be satisfied only from partnership assets [UPA 306(b)]. This means that the new partner usually has no personal liability for these debts and obligations, but any capital contribution that he or she made to the partnership is subject to these debts.

PARTNER'S DISSOCIATION

Dissociation occurs when a partner ceases to be associated in the carrying on of the partnership business. Dissociation normally entitles the partner to have his or her interest purchased by the partnership, and terminates his or her authority to act for the partnership and to participate with the partners in running the business. Otherwise, the partnership continues to do business without the dissociating partner.[9]

Events Causing Dissociation A partner can dissociate from a partnership in a number of ways [UPA 601]. In a partnership, a partner may withdraw by giving notice of "express will to withdraw." A partnership agreement may specify an event that causes a partner's dissociation or expulsion. Also, a partner may be expelled by a unanimous vote of the other partners under certain circumstances, including the transfer of substantially all of his or her interest in the partnership.

In addition, a court or an arbitrator may expel a partner if he or she engages in wrongful conduct that affects the partnership business, breaches the partnership agreement or violates a duty owed to the partnership or the other partners, or engages in conduct that makes it "not

7. Under the previous version of the UPA, partners' liability for torts was joint and several, while their liability for contracts was joint but not several.

8. The term *several* stems from the medieval English term *severall,* which meant "separately," or "severed from" one another. As used here, *several* liability means *separate* liability, or individual liability.

9. Under the previous version of the UPA, when a partner dissociated from a partnership, the partnership was considered dissolved, its business had to be wound up, and the proceeds had to be distributed to creditors and among the partners.

reasonably practicable to carry on the business in partnership with the partner."

Among other events resulting in dissociation from a partnership, a partner dissociates by declaring bankruptcy, by assigning his or her interest in the partnership for the benefit of creditors, by incapacity through physical inability or mental incompetence, or by death.

Wrongful Dissociation A partner has the power to dissociate from a partnership at any time, but a partner's dissociation can be wrongful in a few circumstances [UPA 602]. For example, if dissociation is in breach of a partnership agreement, it is wrongful. Also, in the case of a partnership for a definite term or a particular undertaking, dissociation that occurs before the expiration of the term or the completion of the undertaking can be wrongful if the partner withdraws by express will, is expelled by a court or an arbitrator, or declares bankruptcy.

A partner who wrongfully dissociates is liable to the partnership and to the other partners for damages caused by the dissociation. This liability is in addition to any other obligation of the partner to the partnership or to the other partners. For example, a dissociating partner would be liable for any damage caused by a breach of a partnership agreement. The partner would also be liable to the partnership for costs incurred to replace the partner's expertise or obtain new financing.

Effects of Dissociation On a partner's dissociation, his or her right to participate in the management and conduct of the partnership business terminates [UPA 603]. The partner's duty of loyalty also ends. A partner's other fiduciary duties, including the duty of care, continue only with respect to events that occurred before dissociation, unless the partner participates in winding up the partnership's business (discussed later in this chapter). Thus, a partner who leaves an accounting firm, for example, may immediately compete with the firm for new clients, but must exercise care in completing ongoing client transactions and must account to the firm for any fees received from the old clients based on those transactions. After a partner's dissociation, his or her interest in the partnership must be purchased according to the rules in UPA 701.[10]

For two years after a partner dissociates from a continuing partnership, the partnership may be bound by the acts of the dissociated partner based on apparent authority [UPA 702]. In other words, the partnership may be liable to a third party with whom a dissociated partner enters into a transaction if the third party reasonably believed that the dissociated partner was still a partner. Similarly, a dissociated partner may be liable for partnership obligations entered into during a two-year period following dissociation [UPA 703].

To avoid this possible liability, a partnership should notify its creditors, and customers or clients, of a partner's dissociation and file a statement of dissociation in the appropriate state office to limit the partner's authority to ninety days after the filing [UPA 704]. A dissociated partner should also file a statement of dissociation with the state to limit his or her potential liability to ninety days.

PARTNERSHIP TERMINATION

Some changes in the relations of the partners that demonstrate unwillingness or inability to carry on partnership business dissolve the partnership, resulting in termination [UPA 801]. If any partner wishes to continue the business, she or he is free to reorganize into a new partnership with the remaining members.

The termination of a partnership has two stages, both of which must take place before termination is complete. The first stage, **dissolution,** occurs when any partner (or partners) indicates an intention to dissociate from the partnership. The second stage, **winding up,**[11] is the actual process of collecting and distributing the partnership assets.

Dissolution Dissolution of a partnership can be brought about by the acts of the partners, by the operation of law, and by judicial decree. Each of these events will be discussed here.

—Dissolution by Acts of the Partners. Dissolution of a partnership may come about through the acts of the partners in several ways. First, any partnership can be dissolved by the partners' agreement. For example, when a partnership agreement states a fixed term or a particular business objective to be accomplished, the passing of the date or the accomplishment of the objective dissolves the partnership.

Second, because a partnership at will is a voluntary association, a partner has the power to dissociate himself or herself from the partnership at any time and thus dissolve the partnership by giving notice of "express will to withdraw" [UPA 801(1)]. If, after dissolution, none of the

10. The *buyout price* is based on the amount that would have been distributed to the partner if the partnership were wound up on the date of dissociation. Offset against the price are amounts owed by the partner to the partnership, including damages for wrongful dissociation.

11. Although "winding down" would seem to describe more accurately the process of settling accounts and liquidating the assets of a partnership, "winding up" has traditionally been used in English and U.S. statutory and case law to denote this final stage of a partnership's existence.

partners, including the withdrawing partner, wants the partnership wound up, the remaining partners may continue the business [UPA 802(b)].[12]

Third, a partnership for a definite term or a particular undertaking can be dissolved if, within ninety days of a partner's dissociation caused by bankruptcy, incapacity, or death, at least half of the remaining partners decide in favor of dissolution [UPA 801(2)].

—Dissolution by Operation of Law. Any event that makes it unlawful for the partnership to continue its business will result in dissolution [UPA 801(4)]. Note, however, that if the illegality of the partnership business is a cause for dissolution, within ninety days the partners can decide to change the nature of their business and continue in the partnership.[13]

—Dissolution by Judicial Decree. For dissolution of a partnership by judicial decree to occur, an application or petition must be made in an appropriate court. The court then either denies the petition or grants a decree of dissolution.

Under UPA 801(5), a court may order dissolution when it becomes obviously impractical for the firm to continue—for example, if the business can only be operated at a loss. Additionally, a partner's impropriety involving partnership business (for example, fraud perpetrated on the other partners) or improper behavior reflecting unfavorably on the firm may provide grounds for a judicial decree of dissolution. Finally, if dissension between partners becomes so persistent and harmful as to undermine the confidence and cooperation necessary to carry on the firm's business, dissolution may also be granted.[14]

—Notice of Dissolution. The intent to dissolve or to withdraw from a firm must be communicated clearly to each partner. A partner can express this notice of intent by either actions or words. All partners will share liability for the acts of any partner who continues conducting business for the firm without knowledge that the partnership has been dissolved.

■ EXAMPLE 24.5 Alzor, Jennifer, and Carla have a partnership. Alzor tells Jennifer of her intent to withdraw. Before Carla learns of Alzor's intentions, she enters into a contract with a third party. The contract is equally binding on Alzor, Jennifer, and Carla. Unless *all* of the other partners have notice, the withdrawing partner will continue to be bound as a partner to all contracts created for the firm. ■

To avoid liability for obligations a partner incurs after dissolution of a partnership, the firm must give notice to all affected third parties [UPA 804(2)]. After dissolution, one of the partners can file a statement of dissolution with the appropriate state office, declaring that the partnership has dissolved and is winding up its business. Ninety days after this statement is filed, creditors and other nonpartners are deemed to have notice of the dissolution [UPA 805].[15]

Winding Up Once dissolution occurs and the partners have been notified, the partners cannot create new obligations on behalf of the partnership. Their only authority is to complete transactions begun but not finished at the time of dissolution and to wind up the business of the partnership [UPA 803, 804(1)]. *Winding up* includes collecting and preserving partnership assets, discharging liabilities (paying debts), and accounting to each partner for the value of her or his interest in the partnership.

Both creditors of the partnership and creditors of the individual partners can make claims on the partnership's assets. In general, partnership creditors share proportionately with the partners' individual creditors in the assets of the partners' estates, which include their interests in the partnership. A partnership's assets are distributed according to the following priorities [UPA 807]:

1. Payment of debts, including those owed to partner and nonpartner creditors.
2. Return of capital contributions and distribution of profits to partners.[16]

If the partnership's liabilities are greater than its assets, the partners bear the losses—in the absence of a contrary agreement—in the same proportion in which they shared the profits (rather than, for example, in proportion to their contributions to the partnership's capital). Partners continue in their fiduciary relationship until the winding-up process is completed.

12. Any change in a partnership caused by the withdrawal of a partner or the admission of a new partner resulted in dissolution under the previous version of the UPA. If the remaining or new partners continued the firm's business, a new partnership arose.

13. A partner's death or bankruptcy, or an event that made it illegal for the partner to carry on, also dissolved the partnership by operation of law under the previous version of the UPA.

14. Under the previous version of the UPA, a court could order dissolution based on a partner's mental incompetency.

15. Different types of notice were necessary under the previous version of the UPA, depending on the relationship of an outside party to the firm and the circumstances of dissolution. For example, partnership creditors had to receive notice personally unless dissolution resulted from an operation of law, in which case no notice was required.

16. Under the previous version of the UPA, creditors of the partnership had priority over creditors of the individual partners. Also, in distributing partnership assets, third party creditors were paid before partner creditors, and capital contributions were returned before profits.

Limited Liability Partnerships

The **limited liability partnership (LLP)** is a hybrid form of business designed mostly for professionals who normally do business as partners in a partnership. The major advantage of the LLP is that it allows a partnership to continue as a *pass-through entity* for tax purposes but limits the personal liability of the partners. A **pass-through entity** is a business entity that has no tax liability; the entity's income is passed through to the owners of the entity, who pay taxes on it.

The first state to enact an LLP statute was Texas, in 1991. Other states quickly followed suit, and by 1997, virtually all of the states had enacted LLP statutes. LLPs must be formed and operated in compliance with state statutes, which may include provisions of the UPA. The appropriate form must be filed with a central state agency, usually the secretary of state's office, and the business's name must include either "Limited Liability Partnership" or "LLP" [UPA 1001, 1002]. In addition, an LLP must file annual reports with the state to remain qualified as an LLP in that state [UPA 1003]. In most states, it is relatively easy to convert a traditional partnership into an LLP because the firm's basic organizational structure remains the same. Additionally, all of the statutory and common law rules governing partnerships still apply (apart from those modified by the LLP statute). Normally, LLP statutes are simply amendments to a state's already existing partnership law.

The LLP is especially attractive for two categories of businesses: professional services and family businesses. Professional service firms include law firms and accounting firms. *Family limited liability partnerships* (discussed later in this chapter) are basically business organizations in which all of the partners are related.

Liability in an LLP

Many professionals, such as attorneys and accountants, work together using the partnership business form. A major disadvantage of the general partnership is the unlimited personal liability of its owner-partners. Partners are also subject to joint and several (individual) liability for partnership obligations.

The LLP allows professionals to avoid personal liability for the malpractice of other partners. Although LLP statutes vary from state to state, generally each state statute limits the liability of partners in some way. For example, Delaware law protects each innocent partner from the "debts and obligations of the partnership arising from negligence, wrongful acts, or misconduct." In North Carolina, Texas, and Washington, D.C., the statutes protect innocent partners from obligations arising from "errors, omissions, negligence, incompetence, or malfeasance." The UPA more broadly exempts partners from personal liability for any partnership obligation, "whether arising in contract, tort, or otherwise" [UPA 306(c)]. Although the language of these statutes may seem to apply specifically to attorneys, virtually any group of professionals can use the LLP.

Questions remain, however, concerning the exact limits of this exemption from liability. One question is whether limits on liability apply outside the state in which the LLP was formed. Another question involves whether liability should be imposed to some extent on a negligent partner's supervising partner.

Liability outside the State of Formation Some states require that when an LLP formed in one state wants to do business in another state, the LLP must first register in the other state—for example, by filing a statement of foreign qualification [UPA 1102]. Because state LLP statutes are not uniform, a question arises in this situation. If the LLP statutes in the two states provide different liability protection, which law applies? Most states apply the law of the state in which the LLP was formed, even when the firm does business in another state, which is also the rule under UPA 1101.

Supervising Partner's Liability A partner who commits a wrongful act, such as negligence, is liable for the results of the act. Also liable is the partner who supervises the party who commits a wrongful act. This is generally true for all types of partners and partnerships, including LLPs.

When the partners are members of an LLP and more than one member is negligent, there is a question as to how liability is to be shared. Is each partner jointly and severally liable for the entire result, as a general partner would be in most states? Some states provide for proportionate liability—that is, for separate determinations of the negligence of the partners.[17] The American Institute of Certified Public Accountants supports the enactment of proportionate liability statutes.[18]

■ **EXAMPLE 24.6** Accountants Don and Jane are partners in an LLP, with Don supervising Jane. Jane negligently fails to file tax returns for their client, Centaur Tools. Centaur files a suit against Don and Jane. In a state that does not allow for proportionate liability, Don can be held liable for the entire loss. Under a proportionate

17. See, for example, Colorado Revised Statutes Annotated Section 13-21-111.5(1) and Utah Code Annotated Section 78-27-39.

18. Public Oversight Board of the SEC Practice Section, AICPA, *In the Public Interest: Issues Confronting the Accounting Profession* (New York: AICPA, March 5, 1993), Recommendation I-1.

liability statute, Don will be liable for no more than his portion of the responsibility for the missed tax deadline. (Even if Jane settles the case quickly, Don will still be liable for his portion.) ■

FAMILY LIMITED LIABILITY PARTNERSHIPS

A **family limited liability partnership (FLLP)** is a limited liability partnership in which the majority of the partners are persons related to each other, essentially as spouses, parents, grandparents, siblings, cousins, nephews, or nieces. A person acting in a fiduciary capacity for persons so related can also be a partner. All of the partners must be natural persons or persons acting in a fiduciary capacity for the benefit of natural persons.

Probably the most significant use of the FLLP form of business organization is in agriculture. Family-owned farms sometimes find this form to their benefit. The FLLP offers the same advantages as other LLPs with some additional advantages, such as, in Iowa, an exemption from real estate transfer taxes when partnership real estate is transferred among partners.[19]

Limited Partnerships

We now look at a business organizational form that limits the liability of *some* of its owners—the **limited partnership (LP).** Limited partnerships originated in medieval Europe and have been in existence in the United States since the early 1800s. In many ways, limited partnerships are like the general partnerships discussed earlier in this chapter, but they differ from general partnerships in several ways. Because of this, they are sometimes referred to as *special partnerships.*

A limited partnership consists of at least one **general partner** and one or more **limited partners.** A general partner assumes management responsibility for the partnership and so has full responsibility for the partnership and for all debts of the partnership. A limited partner contributes cash or other property and owns an interest in the firm but does not undertake any management responsibilities and is not personally liable for partnership debts beyond the amount of his or her investment. A limited partner can forfeit limited liability by taking part in the management of the business. Exhibit 24–2 compares characteristics of general and limited partnerships.[20]

19. Iowa Statutes Section 428A.2.

20. Under the UPA, a general partnership can be converted into a limited partnership and vice versa [UPA 902, 903]. The UPA also provides for the merger of a general partnership with one or more general or limited partnerships under rules that are similar to those governing corporate mergers [UPA 905].

Until 1976, the law governing limited partnerships in all states except Louisiana was the Uniform Limited Partnership Act (ULPA). Since 1976, most states and the District of Columbia have adopted the revised version of the ULPA, known as the Revised Uniform Limited Partnership Act (RULPA). Because the RULPA is the dominant law governing limited partnerships in the United States, we will refer to the RULPA in the following discussion of limited partnerships.

FORMATION OF THE LIMITED PARTNERSHIP

In contrast to the informal, private, and voluntary agreement that usually suffices for a general partnership, the formation of a limited partnership is a public and formal proceeding that must follow statutory requirements. A limited partnership must have at least one general partner and one limited partner, as mentioned previously. Additionally, the partners must sign a **certificate of limited partnership,** which requires information similar to that found in a corporate charter (see Chapter 25). The certificate must be filed with the designated state official—under the RULPA, the secretary of state. The certificate is usually open to public inspection. Does the state in which the limited partnership was formed have jurisdiction over all the partners (limited and general) in the event of a dispute involving the limited partnership? See this chapter's *Adapting the Law to the Online Environment* feature on page 504 for a discussion of a case involving this issue.

RIGHTS AND LIABILITIES OF PARTNERS

General partners, unlike limited partners, are personally liable to the partnership's creditors; thus, at least one general partner is necessary in a limited partnership so that someone has personal liability. This policy can be circumvented in states that allow a corporation to be the general partner in a partnership. Because the corporation has limited liability by virtue of corporate laws, if a corporation is the general partner, no one in the limited partnership has personal liability.

Rights of Limited Partners Subject to the limitations that will be discussed here, limited partners have essentially the same rights as general partners, including the right of access to partnership books and the right to other information regarding partnership business. On dissolution, limited partners are entitled to a return of their contributions in accordance with the partnership certificate [RULPA 201(a)(10)]. They can also assign their interests subject to the certificate [RULPA 702, 704].

EXHIBIT 24–2 A COMPARISON OF GENERAL PARTNERSHIPS AND LIMITED PARTNERSHIPS

CHARACTERISTIC	GENERAL PARTNERSHIP (UPA)	LIMITED PARTNERSHIP (RULPA)
Creation	By agreement of two or more persons to carry on a business as co-owners for profit.	By agreement of two or more persons to carry on a business as co-owners for profit. Must include one or more general partners and one or more limited partners. Filing of a certificate of limited partnership with the secretary of state is required.
Sharing of Profits and Losses	By agreement, or in the absence of agreement, profits are shared equally by the partners, and losses are shared in the same ratio as profits.	Profits are shared as required in the certificate, and losses are shared likewise, up to the amount of the limited partners' capital contributions. In the absence of a provision in the certificate, profits and losses are shared on the basis of percentages of capital contributions.
Liability	Unlimited personal liability of all partners.	Unlimited personal liability of all general partners; limited partners liable only to the extent of their capital contributions.
Capital Contribution	No minimum or mandatory amount; set by agreement.	Set by agreement.
Management	By agreement, or in the absence of agreement, all partners have an equal voice.	General partner or partners only. Limited partners have no voice or else are subject to liability as general partners (but *only* if a third party has reason to believe that the limited partner is a general partner). A limited partner may act as an agent or employee of the partnership and vote on amending the certificate or on the sale or dissolution of the partnership.
Duration	Terminated by agreement of the partners, but can continue to do business even when a partner dissociates from the partnership.	By agreement in the certificate or by retirement, death, or mental incompetence of a general partner in the absence of the right of the other general partners to continue the partnership. Death of a limited partner, unless he or she is the only remaining limited partner, does not terminate the partnership.
Distribution of Assets on Liquidation—Order of Priorities	1. Payment of debts, including those owed to partner and nonpartner creditors. 2. Return of capital contributions and distribution of profit to partners.	1. Outside creditors and partner creditors. 2. Partners and former partners entitled to distributions of partnership assets. 3. Unless otherwise agreed, return of capital contributions and distribution of profit to partners.

The RULPA provides a limited partner with the right to sue an outside party on behalf of the firm if the general partners with authority to do so have refused to file suit [RULPA 1001]. In addition, investor protection legislation, such as securities laws (discussed in Chapter 27), may give some protection to limited partners.

Liabilities of Limited Partners In contrast to the personal liability of general partners, the liability of a limited partner is limited to the capital that she or he contributes or agrees to contribute to the partnership [RULPA 502].

A limited partnership is formed by good faith compliance with the requirements for signing and filing the certificate, even if it is incomplete or defective. When a limited partner discovers a defect in the formation of the limited partnership, he or she can avoid future liability by causing an appropriate amendment or certificate to be filed or by renouncing an interest in the profits of the partnership

ADAPTING THE LAW TO THE ONLINE ENVIRONMENT

Jurisdiction Issues in Limited Partnerships

Numerous business and investment opportunities are organized as limited partnerships. Often, especially when the business involves the Internet and technology, the limited partners live in different states and have little contact with each other. In this situation, significant jurisdiction issues can arise. Which court has jurisdiction in the event of a dispute? Do the courts of the state in which a limited partnership is organized have jurisdiction over all members of the partnership, regardless of where they live?

THE *WERNER* CASE

The question of jurisdiction over limited partners came before the court in *Werner v. Miller Technology Management, L.P.*[a] A New York resident, Marc Werner, invested $250,000 as a limited partner in Interprise Technology Partners (ITP), a Delaware limited partnership. ITP was formed in 1999 to invest in information technology companies (companies engaged in creating, storing, and exchanging information on computers). Under the partnership agreement, the general partner, Miller Technology Management (MTM), was to manage the business with the advice and assistance of an advisory board that consisted of five of ITP's limited partners.

In 2002, Werner sued MTM and ITP's advisory board, claiming a pattern of self-dealing in breach of their fiduciary duties of care, disclosure, and loyalty. As it turned out, in three years of operation, ITP had invested over $45 million in companies that were affiliated with its general partner (MTM) and the limited partners on ITP's advisory board. For example, ITP paid over $17.5 million for consulting services from a company called Answerthink, Inc., which was founded and controlled by the five members of ITP's advisory board. In fact, four of the individuals on ITP's advisory board held top positions in Answerthink for which they were paid salaries of over $500,000 per year. These conflicts of interest were never disclosed to Werner or to any of the other limited partners in ITP.

"MINIMUM CONTACTS" REQUIRED

Werner contended that the defendants used their positions of control and influence over ITP to engage in transactions that benefited them personally but were detrimental to ITP. Although the self-dealing nature of these transactions seems apparent, the court first had to determine whether Delaware had jurisdiction over the defendants. The advisory board defendants claimed that they did not have "minimum contacts" (discussed in Chapter 2) with Delaware because they were not residents and did not transact any business in that state. Werner argued that because the advisory board was created to participate in the management of a Delaware limited partnership, Delaware had jurisdiction.

Ultimately, the court held that Delaware did not have jurisdiction over the limited partners on ITP's advisory board. Limited partners are not legally entitled to participate in management. Even if ITP's advisory board had done more than advise MTM, the court held that Werner had not shown that these defendants had any business contacts with Delaware, such as entering contracts or attending meetings. Thus, the court dismissed the case against the limited partners on ITP's advisory board, but it allowed the plaintiff's claim against the general partner (MTM).

FOR CRITICAL ANALYSIS

Given that the general partner and the limited partners on the advisory board engaged in the same pattern of self-dealing and nondisclosure, why did the court have jurisdiction only over the general partner? What policy considerations underlie the court's decision?

a. 831 A.2d 318 (Del. 2003).

[RULPA 304]. If the limited partner takes neither of these actions on discovering the defect, however, the partner can be held personally liable by the firm's creditors. Liability for false statements in a partnership certificate runs in favor of persons relying on the false statements and against members who know of the falsity but still sign the certificate [RULPA 207].

Limited Partners and Management Limited partners enjoy limited liability so long as they do not participate in management [RULPA 303]. A limited partner who participates in management will be just as liable as a general partner to any creditor who transacts business with the limited partnership and believes, based on a limited partner's conduct, that the limited partner is a general partner [RULPA 303]. How much actual review and advisement a limited partner can engage in before being exposed to liability is an unsettled question.[21] A limited partner who knowingly permits his or her name to be used in the name of the limited partnership is liable to creditors who extend credit

21. It is an unsettled question partly because different states have different laws. Factors to be considered under the RULPA are listed in RULPA 303(b), (c).

to the limited partnership without knowledge that the limited partner is not a general partner [RULPA 102, 303(d)].

Although limited partners cannot participate in management, this does not mean that the general partners are totally free of restrictions in running the business. The general partners in a limited partnership have fiduciary obligations to the partnership and to the limited partners, as the following case illustrates.

CASE 24.3 Smith v. Fairfax Realty, Inc.

Supreme Court of Utah, 2003.
82 P.3d 1064.
http://www.utcourts.gov/opinions[a]

FACTS In 1984, Price Development Company (PDC), a Utah corporation, bought a parcel of property in Clovis, New Mexico, from Armand and Virginia Smith. As part of the deal, the Smiths became limited partners in two limited partnerships formed to build and operate the North Plains Mall on the property. PDC became the general partner. In July 1993, PDC told the Smiths that it was forming a real estate investment trust (REIT) that might include the mall. PDC presented the Smiths with three options for their partnership interests if the mall were included. The partnership agreements required the Smiths' consent, but PDC transferred the mall into the REIT without their knowledge. Despite the Smiths' repeated requests, PDC did not disclose the transfer, led them to believe that their options were still open, and provided conflicting estimates of the value of their interests. PDC also paid high fees to itself as general partner, commingled the mall's funds with those of other PDC properties, and accrued interest to itself on its capital contributions while denying the Smiths interest on their contributions. The Smiths filed a suit in a Utah state court against PDC (now doing business as Fairfax Realty, Inc.), alleging, among other things, breach of fiduciary duty. The jury awarded the Smiths $6.6 million in damages. PDC appealed to the Utah Supreme Court.

ISSUE Did PDC breach its fiduciary duty to the Smiths?

a. In the "Supreme Court Opinions" section, click on "2003." In the result, scroll to the name of the case and click on it to access the opinion. The Utah state courts maintain this Web site.

DECISION Yes. The Utah Supreme Court affirmed the decision in the Smiths' favor, agreeing with the lower court that PDC's "actions amount to affirmative misconduct showing deliberate misrepresentation and disregard of the rights of the Smiths." The state supreme court remanded the case to the lower court for a determination of interest on the award and of reasonable costs and attorneys' fees, as required by the partnership agreements.

REASON The state supreme court analyzed PDC's conduct partly in terms of "maliciousness, reprehensibility, and wrongfulness." The court stated that "certain wrongdoings are more egregious [outrageously bad] and blameworthy than others." For that reason, "[d]eliberate false statements, acts of affirmative misconduct, and concealment of evidence of improper motive support more substantial awards, as do acts involving trickery and deceit." Also, "economic injury, especially when done intentionally through affirmative acts of misconduct, or when the target is financially vulnerable, can warrant a substantial penalty." In this case, there was "a pattern of deceit, failure to disclose, and misrepresentation." In particular, the court listed PDC's "prolonged, deliberate failure" to tell the Smiths about the mall's transfer; "the resulting unavailability of the three options that [PDC] originally gave the Smiths regarding how their interests could be handled in the REIT transaction"; and PDC's conflicting and "intentionally misleading" calculations of the value of the Smiths' interests. The court also cited PDC's other financial misconduct.

FOR CRITICAL ANALYSIS—Ethical Consideration *Why does the law impose fiduciary obligations on general partners?*

DISSOLUTION

A limited partnership is dissolved in much the same way as an ordinary partnership. The retirement, death, or mental incompetence of a general partner can dissolve the partnership, but not if the business can be continued by one or more of the other general partners in accordance with their certificate or by the consent of all of the members [RULPA 801]. The death or assignment of interest of a limited partner does not dissolve the limited partnership [RULPA 702, 704, 705]. A limited partnership can be dissolved by court decree [RULPA 802].

Bankruptcy or the withdrawal of a general partner dissolves a limited partnership. Bankruptcy of a limited partner, however, does not dissolve the partnership unless it causes the bankruptcy of the limited partnership. The

retirement of a general partner causes a dissolution unless the members consent to a continuation by the remaining general partners or unless this contingency is provided for in the certificate.

On dissolution, creditors' rights, including those of partners who are creditors, take first priority. Then partners and former partners receive unpaid distributions of partnership assets and, except as otherwise agreed, amounts representing returns on their contributions and amounts proportionate to their shares of the distributions [RULPA 804].

Limited Liability Limited Partnerships

A **limited liability limited partnership (LLLP)** is a type of limited partnership. An LLLP differs from a limited partnership in that a general partner in an LLLP has the same liability as a limited partner. In other words, the liability of all partners is limited to the amount of their investments in the firm.

A few states provide expressly for LLLPs.[22] In states that do not provide for LLLPs but do allow for limited partnerships and limited liability partnerships, a limited partnership should probably still be able to register with the state as an LLLP.

Limited Liability Companies

Thus far, we have examined sole proprietorships and various partnership forms. Here, we examine the **limited liability company (LLC)**. The LLC, like the LLP, is a hybrid that combines limited liability with the tax advantages of a partnership. Increasingly, LLCs are becoming an organizational form of choice among businesspersons, a trend encouraged by state statutes permitting their use.

22. See, for example, Colorado Revised Statutes Annotated Section 7-62-109. Other states that provide expressly for limited liability limited partnerships include Delaware, Florida, Missouri, Pennsylvania, Texas, and Virginia.

Limited liability companies are governed by state LLC statutes. These laws vary, of course, from state to state. In an attempt to create more uniformity among the states in this respect, in 1995 the National Conference of Commissioners on Uniform State Laws issued the Uniform Limited Liability Company Act (ULLCA). To date, less than one-fifth of the states have adopted the ULLCA, and thus the law governing LLCs remains far from uniform. Some provisions are common to most state statutes, however, and we base our discussion of LLCs in this section on these common elements.

The Nature of the LLC

Limited liability companies share many characteristics with corporations. Like corporations, LLCs are creatures of the state. In other words, they must be formed and operated in compliance with state law. Like shareholders in a corporation, owners of an LLC, who are called **members,** enjoy limited liability [ULLCA 303]. Also like corporations, LLCs are legal entities apart from their members. As a legal person, the LLC can sue or be sued, enter into contracts, and hold title to property [ULLCA 201]. The terminology used to describe LLCs formed in other states or nations is also similar to the terminology used in corporate law. For example, an LLC formed in one state but doing business in another state is referred to in the second state as a *foreign LLC.*

As you will read in Chapter 25, on occasion the courts will disregard the corporate entity ("pierce the corporate veil") and hold a shareholder personally liable for corporate obligations. At issue in the following case was whether this same principle should be extended to an LLC. Could the managing member of an LLC (the person designated to manage the firm—see the discussion of LLC management options later in this chapter) be held personally liable for property damage caused by the LLC?

CASE 24.4 Kaycee Land and Livestock v. Flahive

Wyoming Supreme Court, 2002.
46 P.3d 323.

FACTS Roger Flahive is the managing member of Flahive Oil & Gas, LLC. To exercise mineral rights beneath certain real property, Flahive Oil & Gas entered into a contract with Kaycee Land and Livestock in Johnson County, Wyoming, that allowed Flahive Oil & Gas to use the surface of Kaycee's land. Later, Kaycee alleged environmental contamination of its property and filed a suit in a Wyoming state court against Flahive and his LLC. On discovering that Flahive Oil & Gas had no assets at the time of the suit, Kaycee asked the court to disregard the LLC entity and hold Flahive personally liable for the contamination. Before issuing a judgment in the case, the court submitted this question to the Wyoming Supreme Court: "[I]s a claim to pierce the Limited Liability entity veil or disregard the Limited Liability Company entity in the same manner as a court would pierce a corporate veil or disregard a corporate shield, an available remedy" against an LLC? Unlike some states' statutes, Wyoming's LLC provisions do not address this issue.

CASE 24.4—CONTINUED

ISSUE May a court "pierce the veil" of an LLC and hold a member personally liable?

DECISION Yes. The Wyoming Supreme Court held that an LLC entity could be pierced. The court remanded the case for a determination as to whether piercing the veil was appropriate under the circumstances.

REASON The court pointed out that "[e]very state that has enacted LLC piercing legislation has chosen to follow corporate law standards and not develop a separate LLC standard." The overall purpose of statutes that create corporations and LLCs is to limit the liability of individual investors with a corresponding benefit to economic development. Statutes created the legal fiction of the corporation being a completely separate entity that could act independently from individual persons. "[W]hen corporations fail to follow the statutorily mandated formalities, co-mingle funds, or ignore the restrictions in their articles of incorporation regarding separate treatment of corporate property, the courts deem it appropriate to disregard the separate identity and do not permit shareholders to be sheltered from liability to third parties for damages caused by the corporations' acts." The court concluded, "If the members and officers of an LLC fail to treat it as a separate entity as contemplated by statute, they should not enjoy immunity from individual liability for the LLC's acts that cause damage to third parties."

WHY IS THIS CASE IMPORTANT? *The LLC is a "creature of statute" and thus is governed by the state's statutory law. Businesspersons should be aware, though, that courts can at times apply legal principles that are standard in corporate or partnership law to an LLC, as this case illustrates.*

LLC Formation

To form an LLC, **articles of organization** must be filed with a central state agency—usually the secretary of state's office [ULLCA 202]. Typically, the articles are required to include such information as the name of the business, its principal address, the name and address of a registered agent, the names of the members, and information on how the LLC will be managed [ULLCA 203]. The business's name must include the words *Limited Liability Company* or the initials *LLC* [ULLCA 105(a)]. In addition to filing the articles of organization, a few states require that a notice of the intention to form an LLC be published in a local newspaper.

About one-fourth of the states specifically require LLCs to have at least two members. The rest of the states usually permit one-member LLCs, although some LLC statutes are silent on this issue.

Advantages of the LLC

The LLC offers many advantages to businesspersons. A key advantage of the LLC is that the liability of members is limited to the amount of their investments. Another advantage is the flexibility of the LLC in regard to both taxation and management.

An LLC that has *two or more members* can choose to be taxed either as a partnership or as a corporation. As you will read in Chapter 25, a corporate entity must pay income taxes on its profits, and the shareholders pay personal income taxes on profits distributed as dividends. An LLC that wants to distribute profits to the members may prefer to be taxed as a partnership to avoid the "double-taxation" characteristic of the corporate entity. Unless an LLC indicates that it wishes to be taxed as a corporation, it is automatically taxed as a partnership by the Internal Revenue Service (IRS). This means that the LLC as an entity pays no taxes; rather, as in a partnership, profits are "passed through" the LLC to the members who then personally pay taxes on the profits. If an LLC's members want to reinvest the profits in the business, however, rather than distribute the profits to members, they may prefer that the LLC be taxed as a corporation. Corporate income tax rates may be lower than personal tax rates. Part of the attractiveness of the LLC is this flexibility with respect to taxation.

For federal income tax purposes, one-member LLCs are automatically taxed as sole proprietorships unless they indicate that they wish to be taxed as corporations. With respect to state taxes, most states follow the IRS rules. Still another advantage of the LLC for businesspersons is the flexibility it offers in terms of business operations and management—as will be discussed shortly.

Disadvantages of the LLC

The disadvantages of the LLC are relatively few. Although initially there was uncertainty over how LLCs would be taxed, that disadvantage no longer exists. One remaining disadvantage is that state LLC statutes are not yet uniform. Until all of the states have adopted the ULLCA, an LLC in one state will have to check the rules in the other states in which the firm does business to ensure that it retains its limited liability. Generally, though, most—if not all—states apply to a foreign LLC (an LLC formed in another state) the law of the state where the LLC was formed.

MANAGEMENT PERSPECTIVE

The Importance of Operating Agreements

MANAGEMENT FACES A LEGAL ISSUE

As mentioned elsewhere in this chapter, an operating agreement is not required to form a limited liability company (LLC). As the LLC becomes more commonly utilized as a business form, however, the value of having an operating agreement in place is becoming more evident to LLC members. Suppose, for example, that you and another individual establish a two-member LLC. The other member cashes checks payable to the LLC and pockets the funds for personal use. The LLC, of course, can sue the embezzling member to recover the funds, but if that member does not have funds available to repay the embezzled amount, does the LLC have any other recourse? Can it recover from the entity that cashed the checks? Can it recover from any of the banks involved?

WHAT THE COURTS SAY

A case involving these precise questions came before a New Jersey court in 2004. The case involved a member-managed LLC owned by two members, Clifford Kuhn, Jr., and Joseph Tumminelli. They did not form a written LLC operating agreement. Tumminelli embezzled $283,000 from the company after cashing customers' checks at Quick Cash, Inc., a local check-cashing service. Quick Cash deposited the checks in its checking account, and its bank collected on the checks from the LLC's (drawee) bank. Kuhn sued Tumminelli and the banks in a New Jersey state court to recover the embezzled funds.

Although the court ordered Tumminelli to pay Kuhn and to transfer his interest in the LLC to Kuhn, the court issued a summary judgment in favor of the banks. The court noted that under the New Jersey Limited Liability Company Act, when an LLC is managed by its members, each member has the authority to bind the company—unless the operating agreement provides otherwise. In the absence of such an agreement, held the court, Tumminelli, as a 50 percent owner of the LLC, had broad authority to bind the LLC and specific authority to indorse and cash checks payable to the LLC. In the court's eyes, this meant that Tumminelli had the actual authority to receive the checks, indorse them, and cash them at Quick Cash. Thus, neither Quick Cash nor the banks involved could be held liable for cashing and paying the checks, respectively. These institutions had no way of knowing that Tumminelli had "bad intent," and it would defy reason to hold Quick Cash and the banks liable for the wrongdoing. The court emphasized that if Kuhn had wanted to limit Tumminelli's authority, he could have easily done so in an operating agreement. Kuhn appealed the decision, but the state appellate court affirmed the trial court's ruling.[a]

IMPLICATIONS FOR MANAGERS

The implications of this case for LLC members, particularly those in member-managed LLCs, are clear. Members should spell out in an operating agreement who has authority to indorse and cash or deposit checks. By the time a dispute over this issue or other issues that arise gets to court, it may be too late to prevent losses.

a. *Kuhn v. Tumminelli*, 366 N.J.Super. 431, 841 A.2d 496 (2004).

Still another disadvantage is the lack of case law dealing with LLCs. How the courts interpret statutes provides important guidelines for businesses. Given the relative newness of the LLC as a business form in the United States, there is not, as yet, a substantial body of case law to provide this kind of guidance.

THE LLC OPERATING AGREEMENT

The members of an LLC can decide how to operate the various aspects of the business by forming an **operating agreement** [ULLCA 103(a)]. Operating agreements typically contain provisions relating to management, how profits will be divided, the transfer of membership interests, whether the LLC will be dissolved on the death or departure of a member, and other important issues.

An operating agreement need not be in writing and indeed need not even be formed for an LLC to exist. Generally, though, LLC members should protect their interests by forming a written operating agreement. As with any business arrangement, disputes may arise over any number of issues. If there is no agreement covering the topic under dispute, such as how profits will be divided, the state LLC statute will govern the outcome. For example, most LLC statutes provide that if the members have not specified how profits will be divided, they will be divided equally among the members. Generally, when an issue is not covered by an operating agreement or by an LLC statute, the courts apply the principles of partnership law. (For a discussion of a case illustrating the importance of having an operating agreement, see this chapter's *Management Perspective* feature.)

MANAGEMENT OF AN LLC

Basically, there are two options for managing an LLC. The members may decide in their operating agreement to be either a "member-managed" LLC or a "manager-managed" LLC. Most LLC statutes and the ULLCA provide that unless the articles of organization specify otherwise, an LLC is assumed to be member managed [ULLCA 203(a)(6)].

In a *member-managed* LLC, all of the members participate in management, and decisions are made by majority vote [ULLCA 404(a)]. In a *manager-managed* LLC, the members designate a group of persons to manage the firm. The management group may consist of only members, both members and nonmembers, or only nonmembers. Managers in a manager-managed LLC owe fiduciary duties to the LLC and its members, including the duty of loyalty and the duty of care [ULLCA 409(a) and (h)], just as corporate directors and officers owe fiduciary duties to the corporation and its shareholders.

OPERATING PROCEDURES

The members of an LLC can also set forth provisions governing decision-making procedures in their operating agreement. For instance, the agreement can include procedures for choosing or removing managers, an issue on which most LLC statutes are silent, although the ULLCA provides that members may choose and remove managers by majority vote [ULLCA 404(b)(3)].

The members are also free to include in the agreement provisions designating when and for what purposes formal members' meetings will be held. In contrast to state laws governing corporations, most state LLC statutes have no provisions regarding members' meetings.

Members may also specify in their agreement how voting rights will be apportioned. If they do not, LLC statutes in most states provide that voting rights are apportioned according to each member's capital contributions. Some states provide that, in the absence of an agreement to the contrary, each member has one vote.

Special Business Forms

In addition to the major business organizational forms, several special forms exist. For the most part, however, they are hybrid organizations—that is, they have characteristics similar to those of partnerships or corporations or they combine features of both.

JOINT VENTURES, SYNDICATES, AND JOINT STOCK COMPANIES

A *joint venture* is treated much like a partnership, but it differs in that it is created in contemplation of a limited activity or a single transaction. The form of a *syndicate* or an *investment group* can vary considerably. They may exist as corporations or as general or limited partnerships. In some instances, the members merely own property jointly and have no legally recognized business arrangement. The *joint stock company* is a true hybrid of a partnership and a corporation. It is similar to a corporation in that it is a shareholder organization; because of the personal liability of its members and other characteristics, however, it is usually treated like a partnership.

BUSINESS TRUSTS AND COOPERATIVES

The *business trust,* a popular form of business organization in nineteenth-century America, is somewhat similar to the corporation (see Chapter 25). Legal ownership and management of the property of the business stay with one or more of the trustees. The beneficiaries—who receive profits from the enterprise—are not personally responsible for the debts or obligations of the business trust. In some states, business trusts must pay corporate taxes. The *cooperative* is a nonprofit organization formed to provide an economic service to its members. Unincorporated cooperatives are often treated in the same way as partnerships; incorporated cooperatives, like all corporations, are subject to state corporate law.

FRANCHISES

One can also venture into business by purchasing a franchise. About 25 percent of all retail sales and an increasing part of the gross domestic product of the United States are generated by private franchises. A **franchise** is any arrangement in which the owner of a trademark, a trade name, or a copyright has licensed others to use the trademark, trade name, or copyright in selling goods or services. A **franchisee** (a purchaser of a franchise) is generally legally independent of, but economically dependent on, the integrated business system of the **franchisor** (the seller of the franchise). In other words, the franchisee can operate as an independent businessperson but still obtain the advantages of a regional or national organization. Well-known franchises include Hilton Hotels, McDonald's, and Burger King. Franchising is not so much a *form* of business organization as a way of doing business. Sole proprietorships, partnerships, and corporations can all buy and sell franchises.

TERMS AND CONCEPTS

articles of organization 507
articles of partnership 493
certificate of limited partnership 502
charging order 496
confession of judgment 495
dissociation 498
dissolution 499
family limited liability partnership (FLLP) 502
franchise 509
franchisee 509
franchisor 509
general partner 502
joint and several liability 498
limited liability company (LLC) 506
limited liability limited partnership (LLLP) 506
limited liability partnership (LLP) 501
limited partner 502
limited partnership (LP) 502
member 506
operating agreement 508
partnership 491
pass-through entity 501
sole proprietorship 490
winding up 499

CHAPTER SUMMARY Sole Proprietorships, Partnerships, and Limited Liability Companies

Sole Proprietorships (See pages 490–491.)	The simplest form of business; used by anyone who does business without creating an organization. The owner is the business. The owner pays personal income taxes on all profits and is personally liable for all business debts.
Partnerships (See pages 491–500.)	1. Created by agreement of the parties. 2. Treated as an entity except for limited purposes. 3. Partners have unlimited liability for partnership debts. 4. Each partner has an equal voice in management unless the partnership agreement provides otherwise. 5. In the absence of an agreement, partners share profits equally and share losses in the same ratio as they share profits. 6. The capital contribution of each partner is determined by agreement. 7. Each partner pays a proportionate share of income taxes on the net profits of the partnership, whether or not they are distributed; the partnership files only an information return with the Internal Revenue Service. 8. Can be terminated by agreement or can be dissolved by action of the partners (dissociation from a partnership at will), operation of law (subsequent illegality), or court decree.
Limited Liability Partnerships (LLPs) (See pages 501–502.)	1. *Formation*—A statement of qualification must be filed with the appropriate state agency, usually the secretary of state's office. Typically, an LLP is formed by professionals who work together as partners in a partnership. Under most state LLP statutes, it is relatively easy to convert a traditional partnership into an LLP. 2. *Liability of partners*—LLP statutes vary, but under the UPA, professionals generally can avoid personal liability for acts committed by other partners. Partners in an LLP continue to be liable for their own wrongful acts and for the wrongful acts of those whom they supervise. 3. *Family limited liability partnership (FLLP)*—A form of LLP in which all of the partners are family members or fiduciaries of family members; the most significant use of the FLLP is by families engaged in agricultural enterprises.

CHAPTER SUMMARY Sole Proprietorships, Partnerships, and Limited Liability Companies—Continued

Limited Partnerships (LPs) (See pages 502–506.)	1. *Formation*—A certificate of limited partnership must be filed with the secretary of state's office or other designated state official. The certificate must include information about the business, similar to the information included in a corporate charter. The partnership consists of one or more general partners and one or more limited partners. 2. *Rights and liabilities of partners*—With some exceptions, the rights of partners are the same as the rights of partners in a general partnership. General partners have unlimited liability for partnership obligations; limited partners are liable only to the extent of their contributions. 3. *Limited partners and management*—Only general partners can participate in management. Limited partners have no voice in management; if they do participate in management activities, they risk having general-partner liability. 4. *Dissolution*—Generally, a limited partnership can be dissolved in much the same way as an ordinary partnership. The death or assignment of interest of a limited partner does not dissolve the partnership; bankruptcy of a limited partner will also not dissolve the partnership unless it causes the bankruptcy of the firm. 5. *Limited liability limited partnership (LLLP)*—A special type of limited partnership in which the liability of all partners, including general partners, is limited to the amount of their investments.
Limited Liability Companies (LLCs) (See pages 506–509.)	1. *Formation*—Articles of organization must be filed with the appropriate state office—usually the office of the secretary of state—setting forth the name of the business, its principal address, the names of the owners (called members), and other relevant information. 2. *Advantages and disadvantages of the LLC*—Advantages of the LLC include limited liability, the option to be taxed as a partnership or as a corporation, and flexibility in deciding how the business will be managed and operated. Disadvantages relate mainly to the absence of uniformity in state LLC statutes and the lack of case law dealing with LLCs. 3. *Operating agreement*—When an LLC is formed, the members decide, in an operating agreement, how the business will be managed and what rules will apply to the organization. 4. *Management*—An LLC may be managed by members only, by some members and some nonmembers, or by nonmembers only.
Special Business Forms (See page 509.)	A number of special business forms exist. Typically, they are hybrid organizations having characteristics similar to partnerships or corporations, or combining features of both. Special business forms include joint ventures, syndicates or investment groups, joint stock companies, business trusts, and cooperatives. A widely used way of conducting business is the franchise.

FOR REVIEW

Answers for the even-numbered questions in this For Review *section can be found in Appendix A at the end of this text.*

1. Which form of business organization is the simplest? Which form arises from an agreement between two or more persons to carry on a business for profit?
2. What are the three essential elements of a partnership?
3. What is meant by joint and several liability? Why is this often considered to be a disadvantage of the partnership form of business?
4. Why do professional groups organize as a limited liability partnership? How does this form differ from a general partnership?
5. What is a limited liability company? What are some of the advantages and disadvantages of this business form?

QUESTIONS AND CASE PROBLEMS

QUESTION WITH SAMPLE ANSWER

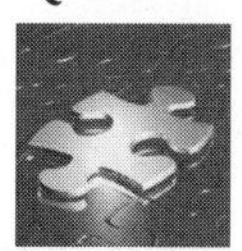

24–1. Shawna and David formed a partnership. At the time of the partnership's formation, Shawna's capital contribution was $10,000, and David's was $15,000. Later, Shawna made a $10,000 loan to the partnership when it needed working capital. The partnership agreement provided that profits were to be shared with 40 percent for Shawna and 60 percent for David. The partnership was dissolved by David's death. At the end of the dissolution and the winding up of the partnership, the partnership's assets were $50,000, and the partnership's debts were $8,000. Discuss fully how the assets should be distributed.

For a sample answer to this question, go to Appendix B at the end of this text.

24–2. Limited Liability Companies. John, Lesa, and Trevor form a limited liability company. John contributes 60 percent of the capital, and Lesa and Trevor each contribute 20 percent. Nothing is decided about how profits will be divided. John assumes that he will be entitled to 60 percent of the profits, in accordance with his contribution. Lesa and Trevor, however, assume that the profits will be divided equally. A dispute over the profits arises, and ultimately a court has to decide the issue. What law will the court apply? In most states, what will result? How could this dispute have been avoided in the first place? Discuss fully.

24–3. Liability of Limited Partners. Asher and Breem form a limited partnership with Asher as the general partner and Breem as the limited partner. Breem puts up $15,000, and Asher contributes some office equipment that he owns. A certificate of limited partnership is properly filed, and business begins. One month later, Asher becomes ill. Instead of hiring someone to manage the business, Breem takes over complete management himself. While Breem is in control, he makes a contract with Thaler involving a large amount of funds. Asher returns to work. Because of other commitments, Asher and Breem breach the Thaler contract. Thaler contends that Asher and Breem will be personally liable for damages caused by the breach if the damages cannot be satisfied out of the assets of the limited partnership. Discuss this contention.

24–4. Liability of Incoming Partner. Conklin Farm sold land to LongView Estates, a general partnership, to build condominiums. LongView gave Conklin a promissory note for $9 million as payment. A few years later, Doris Leibowitz joined LongView as a general partner. Leibowitz left the firm before the note came due, but while she was a partner, interest accrued on the balance. The condominium project failed, and LongView went out of business. Conklin filed a suit in a New Jersey state court against Leibowitz to recover some of the interest on the note. Conklin acknowledged that Leibowitz was not liable for debt incurred before she joined the firm but argued that the interest that accrued while she was a partner was "new" debt for which she was personally liable. To what extent, if any, is Leibowitz liable to Conklin? [*Conklin Farm v. Leibowitz,* 140 N.J. 417, 658 A.2d 1257 (1995)]

24–5. Indications of Partnership. Sandra Lerner was one of the original founders of Cisco Systems. When she sold her interest in Cisco, she received a substantial payment, which she invested, and she became extremely wealthy. Patricia Holmes met Lerner at Holmes's horse training facility, and they became friends. One evening in Lerner's mansion, while applying nail polish, Holmes layered a raspberry color over black to produce a new color, which Lerner liked. Later, the two created other colors with names like "Bruise," "Smog," and "Oil Slick" and titled their concept "Urban Decay." Lerner and Holmes started a firm to produce and market the polishes but never discussed the sharing of profits and losses. They agreed to build the business and then sell it. Together, they did market research, experimented with colors, worked on a logo and advertising, obtained capital from an investment firm, and hired employees. Then Lerner began working to edge Holmes out of the firm. Several months later, when Holmes was told not to attend meetings of the firm's officers, she filed a suit in a California state court against Lerner, claiming, among other things, a breach of their partnership agreement. Lerner responded in part that there was no partnership agreement because there was no agreement to divide profits. Was Lerner right? Why or why not? How should the court rule? [*Holmes v. Lerner,* 74 Cal.App.4th 442, 88 Cal.Rptr.2d 130 (1 Dist. 1999)]

CASE PROBLEM WITH SAMPLE ANSWER

24–6. In August 1998, Jea Yu contacted Cameron Eppler, president of Design88, Ltd., to discuss developing a Web site that would cater to investors and provide services to its members for a fee. Yu and Patrick Connelly invited Eppler and Ha Tran, another member of Design88, to a meeting to discuss the site. The parties agreed that Design88 would perform certain Web design, implementation, and maintenance functions for 10 percent of the profits from the site, which would be called "The Underground Trader." They signed a "Master Partnership Agreement," which was later amended to include Power Uptik Productions, LLC (PUP). The parties often referred to themselves as partners. From its offices in Virginia, Design88 designed and hosted the site, solicited members through Internet and national print campaigns, processed member applications, provided technical support, monitored access to the site, and negotiated and formed business alliances on the site's behalf. When relations among the parties soured, PUP withdrew. Design88 filed a suit against PUP and the others in a Virginia state court. Did a partnership exist among these parties? Explain. [*Design88, Ltd. v. Power Uptik Productions, LLC,* 133 F.Supp.2d 873 (W.D.Va. 2001)]

After you have answered this problem, compare your answer with the sample answer given on the Web site that accompanies

this text. Go to **academic.cengage.com/blaw/wbl**, select "Chapter 24," and click on "Case Problem with Sample Answer."

24–7. Fiduciary Duties. Charles Chaney and Lawrence Burdett were equal partners in a partnership in Georgia known as BMW Partners. Their agreement was silent as to the effect of a partner's death on the firm. The partnership's sole asset was real property, which the firm leased in 1987 to a corporation that the partners co-owned. Under the lease, the corporation was to pay the partnership $8,000 per month, but after a few years, the corporation began paying $9,000 per month. Chaney died on April 15, 1998. Burdett wanted to continue the partnership business and offered to buy Chaney's estate's interest in it. Meanwhile, claiming that the real property's fair rental value was $4,500 (not $9,000) and that the corporation had overpaid the rent by $80,000, Burdett adjusted the rental payments to recoup this amount. Bonnie Chaney, Charles's widow and his estate's legal representative, filed a suit in a Georgia state court against Burdett, alleging in part that he had breached his fiduciary duty by adjusting the amount of the rent. Did Burdett's fiduciary duty expire on Chaney's death? Explain. [*Chaney v. Burdett*, 274 Ga. 805, 560 S.E.2d 21 (2002)]

24–8. Limited Liability Companies. Michael Collins entered into a three-year employment contract with E-Magine, LLC. In business for only a brief time, E-Magine incurred significant losses. In terminating operations, which ceased before the term of the contract with Collins expired, E-Magine also terminated Collins's services. Collins signed a "final payment agreement," which purported to be a settlement of any claims that he might have against E-Magine in exchange for a payment of $24,240. Collins filed a suit in a New York state court against E-Magine, its members and managers, and others, alleging, among other things, breach of his employment contract. Collins claimed that signing the "final payment agreement" was the only means for him to obtain what he was owed for past sales commissions and asked the court to impose personal liability on the members and managers of E-Magine for breach of contract. Should the court grant this request? Why or why not? [*Collins v. E-Magine, LLC*, 291 A.D.2d 350, 739 N.Y.S.2d 15 (1 Dept. 2002)]

24–9. Partnership Status. Charlie Waugh owned and operated an auto parts junkyard in Georgia. Charlie's son, Mack, started working in the business part-time as a child and full-time when he left school at age sixteen. Mack oversaw the business's finances, depositing the profits in a bank. Charlie gave Mack a one-half interest in the business, telling him that if "something happened" to Charlie, the entire business would be his. In 1994, Charlie and his wife, Alene, transferred to Mack the land on which the junkyard was located. Two years later, however, Alene and her daughters, Gail and Jewel, falsely convinced Charlie, whose mental competence had deteriorated, that Mack had cheated him. Mack was ordered off the land. Shortly thereafter, Charlie died. Mack filed a suit in a Georgia state court against the rest of the family, asserting in part that he and Charlie had been partners and that he was entitled to Charlie's share of the business. Was the relationship between Charlie and Mack a partnership? Is Mack entitled to Charlie's "share"? Explain. [*Waugh v. Waugh*, 265 Ga.App. 799, 595 S.E.2d 647 (2004)]

Sole Proprietorships, Partnerships, and Limited Liability Companies

For updated links to resources available on the Web, as well as a variety of other materials, visit this text's Web site at

academic.cengage.com/blaw/wbl

To learn how the U.S. Small Business Administration (SBA) assists in forming, financing, and operating businesses, go to

http://www.sbaonline.sba.gov

You can find summaries, texts, and the legislative history of numerous uniform acts, including the Uniform Partnership Act (UPA), at the Web site of the National Conference of Commissioners on Uniform State Laws at

http://www.nccusl.org

For an example of a state law (that of Iowa) governing limited liability partnerships, go to

http://www.sos.state.ia.us/business/limliabpart.html

You can find information on how to form a limited liability company (LLC), including the fees charged in each state for filing LLC articles of organization, at the Web site of BIZCORP International, Inc. Go to

http://www.bizcorp.com

ONLINE LEGAL RESEARCH

Go to the *Fundamentals of Business Law* home page at academic.cengage.com/blaw/wbl, select "Chapter 24," and click on "Internet Exercises." There you will find the following Internet research exercises that you can perform to learn more about topics covered in this chapter.

Activity 24–1: LEGAL PERSPECTIVE—Liability of Dissociated Partners

Activity 24–2: LEGAL PERSPECTIVE—Limited Liability Companies

Activity 24–3: MANAGEMENT PERSPECTIVE—Limited Partnerships and Limited Liability Partnerships

BEFORE THE TEST

Go to the *Fundamentals of Business Law* home page at academic.cengage.com/blaw/wbl, select "Chapter 24," and click on "Interactive Quizzes." You will find a number of interactive questions relating to this chapter.

CHAPTER 25

Corporate Formation, Financing, and Termination

LEARNING OBJECTIVES

After reading this chapter, you should be able to answer the following questions:

1. What are the steps for bringing a corporation into existence? Who is liable for preincorporation contracts?
2. What is the difference between a *de jure* corporation and a *de facto* corporation?
3. In what circumstances might a court disregard the corporate entity ("pierce the corporate veil") and hold the shareholders personally liable?
4. What are the four steps of the merger or consolidation procedure?
5. What are two ways in which a corporation can be voluntarily dissolved? Under what circumstances might a corporation be involuntarily dissolved by state action?

A corporation is a creature of statute—an artificial being that exists only in law. Its existence generally depends on state law, although some corporations, especially public organizations, can be created under state or federal law.

Each state has its own body of corporate law, and these laws are not entirely uniform. The Model Business Corporation Act (MBCA) is a codification of modern corporation law that has been influential in the drafting and revision of state corporation statutes. Today, the majority of state statutes are guided by the revised version of the MBCA, which is often referred to as the Revised Model Business Corporation Act (RMBCA). You should keep in mind, however, that there is considerable variation among the statutes of the states that have used the MBCA or the RMBCA as a basis for their statutes, and several states do not follow either act. Consequently, individual state corporation laws should be relied on rather than the MBCA or the RMBCA.

The Nature of the Corporation

A **corporation** is a legal entity created and recognized by state law. It can consist of one or more *natural persons* (as opposed to the artificial *legal person* of the corporation) identified under a common name. A corporation can be owned by a single person, or it can have hundreds, thousands, or even millions of owners (shareholders). The corporation substitutes itself for its shareholders in conducting corporate business and in incurring liability, yet its authority to act and the liability for its actions are separate and apart from the individuals who own it.

CORPORATE PERSONNEL

In a corporation, the responsibility for the overall management of the firm is entrusted to a *board of directors,* whose members are elected by the shareholders. The board of directors hires *corporate officers* and other employees to run the daily business operations of the corporation.

When an individual purchases a share of stock in a corporation, that person becomes a shareholder and thus an owner of the corporation. Unlike the members of a partnership, the body of shareholders can change constantly without affecting the continued existence of the corporation. A shareholder can sue the corporation, and the corporation can sue a shareholder. Also, under certain circumstances, a shareholder can sue on behalf of a corporation. The rights and duties of corporate personnel will be examined in detail in Chapter 26.

The shareholder form of business organization emerged in Europe at the end of the seventeenth century. These organizations, called *joint stock companies,* frequently collapsed because their organizers absconded with the funds or proved to be incompetent. Because of this history of fraud and collapse, organizations resembling corporations were regarded with suspicion in the United States during its early years. Although several business corporations were formed after the Revolutionary War, the corporation did not come into common use for private business until the nineteenth century. Today, the corporation is one of the most important forms of business organizations in the United States.

The Constitutional Rights of Corporations

A corporation is recognized under state and federal law as a legal "person," and it enjoys many of the same rights and privileges that U.S. citizens enjoy. The Bill of Rights guarantees persons certain protections, and corporations are considered legal persons in most instances. Accordingly, a corporation as an entity has the same right of access to the courts as a natural person and can sue or be sued. It also has the right of due process before denial of life, liberty, or property, as well as freedom from unreasonable searches and seizures (see Chapter 6 for a discussion of searches and seizures in the business context) and from double jeopardy.

Under the First Amendment, corporations are entitled to freedom of speech. As we pointed out in Chapter 1, however, commercial speech (such as advertising) and political speech (such as contributions to political causes or candidates) receive significantly less protection than noncommercial speech.

Generally, a corporation is not entitled to claim the Fifth Amendment privilege against self-incrimination. Agents or officers of the corporation therefore cannot refuse to produce corporate records on the ground that it might incriminate them.[1] Additionally, the privileges and immunities clause of the Constitution (Article IV, Section 2) does not protect corporations, nor does it protect an unincorporated association.[2] This clause requires each state to treat citizens of other states equally with respect to certain rights, such as access to courts and travel rights. This constitutional clause does not apply to corporations because corporations are legal persons only, not natural citizens.

The Limited Liability of Shareholders

One of the key advantages of the corporate form is the limited liability of its owners (shareholders). Corporate shareholders normally are not personally liable for the obligations of the corporation beyond the extent of their investments. In certain limited situations, however, the "corporate veil" can be pierced and liability for the corporation's obligations extended to shareholders—a concept that will be explained later in this chapter. Additionally, to enable the firm to obtain credit, shareholders in small companies sometimes voluntarily assume personal liability, as guarantors, for corporate obligations.

Corporate Taxation

Corporate profits are taxed by various levels of government. Corporations can do one of two things with corporate profits—retain them or pass them on to shareholders in the form of **dividends.** The corporation receives no tax deduction for dividends distributed to shareholders. Dividends are again taxable (except when they represent distributions of capital) to the shareholder receiving them. This double-taxation feature of the corporation is one of its major disadvantages.

Profits that are not distributed are retained by the corporation. These **retained earnings,** if invested properly, will yield higher corporate profits in the future and thus cause the price of the company's stock to rise. Individual shareholders can then reap the benefits of these retained earnings in the capital gains they receive when they sell their shares.

The consequences of a corporation's failure to pay taxes can be severe. Indeed, the state may dissolve a corporation for this reason. Alternatively, corporate status may be suspended until the taxes are paid (see, for example, the case presented as Case 25.2 later in this chapter).

Another important corporate taxation issue is whether corporations that sell goods or services to consumers via the Internet are required to collect state sales taxes. See this chapter's *Adapting the Law to the Online Environment* feature for a discussion of this issue.

1. *Braswell v. United States,* 487 U.S. 99, 108 S.Ct. 2284, 101 L.Ed. 98 (1988). A court might allow an officer or employee to assert the Fifth Amendment privilege against self-incrimination in only a few circumstances. See, for example, *In re Three Grand Jury Subpoenas Duces Tecum Dated January 29, 1999,* 191 F.3d 173 (2nd Cir. 1999).

2. *W. C. M. Window Co. v. Bernardi,* 730 F.2d 486 (7th Cir. 1984).

ADAPTING THE LAW TO THE ONLINE ENVIRONMENT

The Internet Taxation Debate

Since the advent of the Internet, governments at the state and federal levels have debated the following question: Should state governments be able to collect sales taxes on goods sold via the Internet? Many state governments claim that sales taxes should be imposed on such transactions. They argue that their inability to tax online sales of goods to in-state customers by out-of-state corporations has caused them to suffer significant losses in sales tax revenues. Opponents of Internet taxation argue that taxing online sales will impede the growth of e-commerce. They also claim that because online sellers do not benefit from the state services that are typically paid for by tax revenues (such as fire departments and road construction), they should not be required to collect sales taxes.

THE SUPREME COURT'S APPROACH

According to a United States Supreme Court ruling in 1992, no individual state can compel an out-of-state business that lacks a substantial physical presence within that state to collect and remit state taxes.[a] If the corporation has a warehouse, office, or retail store within the state, though, the state can compel the collection of state taxes. Nevertheless, as the Court recognized in that ruling, Congress has the power to pass legislation requiring out-of-state corporations to collect and remit state sales taxes. Congress so far has chosen not to tax Internet transactions. In fact, in 1998 Congress passed the Internet Tax Freedom Act, which temporarily prohibited states from taxing sales of products conducted over the Internet. This ban expired in November 2003. Although legislation that would permanently prohibit state and local taxation of Internet sales has been proposed, it has not been enacted.

A STATE COURT'S DECISION

The issue of Internet taxation came before a Tennessee appellate court in *Prodigy Services Corp. v. Johnson.*[b] Prodigy, a Delaware corporation with its principal place of business in New York, is an Internet service provider (ISP) that offers two software programs for purchase online. A Tennessee statute imposes an obligation to collect sales taxes on anyone supplying "telecommunication services" to state residents. The Tennessee Department of Revenue determined that Prodigy's services constituted telecommunication services and assessed sales taxes. Prodigy appealed this tax assessment.

Ultimately, the state appellate court held that Prodigy did not have to charge its Tennessee customers the sales taxes. After looking closely at the wording of the statute and its legislative history, the court reasoned that the legislature had not intended the statute to apply to ISPs. The court also concluded that even if Prodigy had provided some telecommunication services, these services "were not the 'true object' of the Prodigy sale." The customer had to supply her or his own telephone services, and Prodigy had paid to use a telecommunications network to connect the customer to the main computer in New York. Thus, in the court's opinion, Prodigy was a consumer of telecommunication services rather than a provider.

FOR CRITICAL ANALYSIS

Although most states today do not require corporations that sell goods and services online to collect state sales taxes, businesspersons should be aware that the law in this area is still developing. Thus, corporations may be required to collect state taxes on Internet sales in the future.

a. See *Quill Corp. v. North Dakota,* 504 U.S. 298, 112 S.Ct. 1904, 119 L.Ed.2d 91 (1992).

b. 125 S.W.3d 413 (Tenn.Ct.App. 2003).

TORTS AND CRIMINAL ACTS

A corporation is liable for the torts committed by its agents or officers within the course and scope of their employment. This principle applies to a corporation exactly as it applies to the ordinary agency relationships discussed in Chapter 22. It follows the doctrine of *respondeat superior.*

Under modern criminal law, a corporation may be held liable for the criminal acts of its agents and employees, provided the punishment is one that can be applied to the corporation. Although corporations cannot be imprisoned, they can be fined. (Of course, corporate directors and officers can be imprisoned, and in recent years, many have faced criminal penalties for their own actions or for the actions of employees under their supervision.)

Recall from Chapter 6 that the U.S. Sentencing Commission, which was established by the Sentencing Reform Act of 1984, created standardized sentencing guidelines for federal crimes. These guidelines went into effect in 1987. The commission subsequently created specific sentencing guidelines for crimes committed by corporate employees (white-collar crimes).[3] The net effect of the guidelines has

3. Note that the Sarbanes-Oxley Act of 2002, which will be discussed in Chapter 27, stiffened the penalties for certain types of corporate crime and ordered the U.S. Sentencing Commission to revise the sentencing guidelines accordingly.

been a significant increase in criminal penalties for crimes committed by corporate personnel. Penalties depend on such factors as the seriousness of the offense, the amount of money involved, and the extent to which top company executives are implicated. Corporate lawbreakers can face fines of up to hundreds of millions of dollars, though the guidelines allow judges to impose less severe penalties in certain circumstances.

Corporate Powers

When a corporation is created, the express and implied powers necessary to achieve its purpose also come into existence. The express powers of a corporation are found in its **articles of incorporation** (a document containing information about the corporation, including its organization and functions), in the law of the state of incorporation, and in the state and federal constitutions.

Corporate **bylaws,** which are internal rules of management adopted by the corporation at its first organizational meeting, also establish the express powers of the corporation. Because state corporation statutes frequently provide default rules that apply if the company's bylaws are silent on an issue, it is important that the bylaws set forth the specific operating rules of the corporation. In addition, after the bylaws are adopted, the corporation's board of directors may pass resolutions that also grant or restrict corporate powers.

The following order of priority is used when conflicts arise among documents involving corporations:

1. The U.S. Constitution.
2. State constitutions.
3. State statutes.
4. The articles of incorporation.
5. Bylaws.
6. Resolutions of the board of directors.

Implied Powers

Certain implied powers attach when a corporation is created. Barring express constitutional, statutory, or other prohibitions, the corporation has the implied power to perform all acts reasonably appropriate and necessary to accomplish its corporate purposes. For this reason, a corporation has the implied power to borrow funds within certain limits, to lend funds, and to extend credit to those with whom it has a legal or contractual relationship.

To borrow funds, the corporation acts through its board of directors to authorize the loan. Most often, the president or chief executive officer of the corporation will execute the necessary papers on behalf of the corporation. In so doing, corporate officers have the implied power to bind the corporation in matters directly connected with the *ordinary* business affairs of the enterprise. There is a limit to what a corporate officer can do, though. A corporate officer does not have the authority to bind the corporation to an action that will greatly affect the corporate purpose or undertaking, such as the sale of substantial corporate assets.

Ultra Vires Doctrine

The term ***ultra vires*** means "beyond the powers." In corporate law, acts of a corporation that are beyond its express and implied powers are *ultra vires* acts. A majority of cases dealing with *ultra vires* acts have involved contracts made for unauthorized purposes. ■ EXAMPLE 25.1 Suarez is the chief executive officer of SOS Plumbing, Inc. He enters a contract with Carlini for the purchase of one hundred cases of brandy. It is difficult to see how this contract is reasonably related to the conduct and furtherance of the corporation's stated purpose of providing plumbing installation and services. Hence, a court would probably find such a contract to be *ultra vires.* ■

Under Section 3.04 of the RMBCA, the following remedies are available for *ultra vires* acts:

1. The shareholders may sue on behalf of the corporation to obtain an injunction (to prohibit the corporation from engaging in the *ultra vires* transactions) or to obtain damages for the harm caused by the transactions.
2. The corporation itself can sue the officers and directors who were responsible for the *ultra vires* transactions to recover damages.
3. The attorney general of the state may bring an action to obtain an injunction against the *ultra vires* transactions or to institute dissolution proceedings against the corporation for *ultra vires* acts.

Classification of Corporations

Corporations can be classified in several ways. The classification of a corporation normally depends on its location, purpose, and ownership characteristics.

Domestic, Foreign, and Alien Corporations

A corporation is referred to as a **domestic corporation** by its home state (the state in which it incorporates). A corporation formed in one state but doing business in another is referred to as a **foreign corporation** by the second state. A

corporation formed in another country (say, Mexico) but doing business in the United States is referred to in the United States as an **alien corporation.**

A corporation does not have an automatic right to do business in a state other than its state of incorporation. In some instances, it must obtain a *certificate of authority* in any state in which it plans to do business. Once the certificate has been issued, the corporation generally can exercise in that state all of the powers conferred on it by its home state. If a foreign corporation does business in a state without obtaining a certificate of authority, the state can impose substantial fines and sanctions on the corporation, and sometimes even on its officers, directors, or agents.[4]

PUBLIC AND PRIVATE CORPORATIONS

A public corporation is one formed by the government to meet some political or governmental purpose. Cities and towns that incorporate are common examples. In addition, many federal government organizations, such as the U.S. Postal Service, the Tennessee Valley Authority, and AMTRAK, are public corporations. Note that a public corporation is not the same as a *publicly held* corporation (often called a *public company*). A publicly held corporation is any corporation whose shares are publicly traded in securities markets, such as the New York Stock Exchange or the over-the-counter market.

In contrast to public corporations (*not* public companies), private corporations are created either wholly or in part for private benefit. Most corporations are private. Although they may serve a public purpose, as a public electric or gas utility does, they are owned by private persons rather than by the government.[5]

NONPROFIT CORPORATIONS

Corporations formed for purposes other than making a profit are called *nonprofit* or *not-for-profit* corporations. Private hospitals, educational institutions, charities, and religious organizations, for example, are frequently organized as nonprofit corporations. The nonprofit corporation is a convenient form of organization that allows various groups to own property and to form contracts without exposing the individual members to personal liability.

CLOSE CORPORATIONS

Many corporate enterprises in the United States fall into the category of close corporations. A **close corporation** is one whose shares are held by members of a family or by relatively few persons. Close corporations are also referred to as *closely held, family,* or *privately held* corporations. Usually, the members of the small group constituting a close corporation are personally known to each other. Because the number of shareholders is so small, there is no trading market for the shares.

In practice, a close corporation is often operated like a partnership. Some states have enacted special statutory provisions that apply to close corporations. These provisions expressly permit close corporations to depart significantly from certain formalities required by traditional corporation law.[6]

Additionally, Section 7.32 of the RMBCA, a provision added to the RMBCA in 1991 and adopted in several states, gives close corporations a substantial amount of flexibility in determining the rules by which they will operate. Under Section 7.32, if all of the shareholders of a corporation agree in writing, the corporation can operate without directors, bylaws, annual or special shareholders' or directors' meetings, stock certificates, or formal records of shareholders' or directors' decisions.[7]

Management of Close Corporations A close corporation has a single shareholder or a closely knit group of shareholders, who usually hold the positions of directors and officers. Management of a close corporation resembles that of a sole proprietorship or a partnership. As a corporation, however, the firm must meet all specific legal requirements set forth in state statutes.

To prevent a majority shareholder from dominating a close corporation, the corporation may require that more than a simple majority of the directors approve any action taken by the board. Typically, this would apply only to extraordinary actions, such as changing the amount of dividends or dismissing an employee-shareholder, and not to ordinary business decisions.

Transfer of Shares in Close Corporations By definition, a close corporation has a small number of shareholders. Thus, the transfer of one shareholder's shares to someone else can cause serious management problems. The other

4. Note that most state statutes specify certain activities, such as soliciting orders via the Internet, which are not considered doing business within the state. Thus, a foreign corporation normally does not need a certificate of authority to sell goods or services via the Internet or by mail.

5. For a leading case on the distinction between private and public corporations, see *Trustees of Dartmouth College v. Woodward,* 17 U.S. (4 Wheaton) 518, 4 L.Ed. 629 (1819).

6. For example, in some states (such as Maryland), the close corporation need not have a board of directors.

7. Shareholders cannot agree, however, to eliminate certain rights of shareholders, such as the right to inspect corporate books and records or the right to bring derivative actions (lawsuits on behalf of the corporation—see Chapter 26).

shareholders may find themselves required to share control with someone they do not know or like.

■ **EXAMPLE 25.2** Three brothers, Terry, Damon, and Henry Johnson, are the only shareholders of Johnson's Car Wash, Inc. Terry and Damon do not want Henry to sell his shares to an unknown third person. To avoid this situation, the articles of incorporation could restrict the transferability of shares to outside persons. The articles could require shareholders to offer their shares to the corporation or the other shareholders before selling them to an outside purchaser. In fact, a few states have statutes that prohibit the transfer of close corporation shares unless certain persons—including shareholders, family members, and the corporation—are first given the opportunity to purchase the shares for the same price. ■

Control of a close corporation can also be stabilized through the use of a shareholder agreement. A shareholder agreement can provide that when one of the original shareholders dies, her or his shares of stock in the corporation will be divided in such a way that the proportionate holdings of the survivors, and thus their proportionate control, will be maintained. Courts are generally reluctant to interfere with private agreements, including shareholder agreements.

The effect of a close corporation's stock transfer restriction was at the heart of the following case.

CASE 25.1 Salt Lake Tribune Publishing Co. v. AT&T Corp.

United States Court of Appeals,
Tenth Circuit, 2003.
320 F.3d 1081.
http://www.kscourts.org/ca10/wordsrch.htm[a]

FACTS In 1952, Kearns-Tribune Corporation (KTC), the owner of the *Salt Lake Tribune,* and Deseret News Publishing Company, the owner of *Deseret News* (the *Tribune*'s chief competitor in Salt Lake City, Utah), entered into a "Joint Operating Agreement" (JOA). The JOA created the Newspaper Agency Corporation (NAC) to conduct the production operations of both newspapers and restrict the transfer of NAC stock to anyone but KTC and Deseret. In 1997, the KTC shareholders, most of whom were members of the Kearns-McCarthey family, sold KTC. As part of the deal, the family formed Salt Lake Tribune Publishing Company to later repurchase the *Tribune*'s physical assets under an "Option Agreement." In 1999, AT&T Corporation acquired KTC and reorganized it as Kearns-Tribune LLC. In 2002, Salt Lake Tribune Publishing Company filed a suit in a federal district court against AT&T and others, seeking specific performance of the Option Agreement. The court ruled that a transfer of the *Tribune*'s assets would violate the JOA's stock transfer restriction. The plaintiff appealed to the U.S. Court of Appeals for the Tenth Circuit, arguing in part that the stock transfer restriction was invalid.

ISSUE Was the stock transfer restriction enforceable?

DECISION Yes. The U.S. Court of Appeals for the Tenth Circuit agreed with the lower court that the stock transfer restriction was valid. The appellate court disagreed that a transfer of the *Tribune*'s other assets would interfere with this restriction, however, and remanded the case for further proceedings.

REASON The U.S. Court of Appeals for the Tenth Circuit pointed out that under a Utah state statute, a "corporation may impose restrictions on the transfer * * * of shares of the corporation * * * for any * * * reasonable purpose." The court recognized that "[t]he obvious purpose" of the stock transfer restriction in this case was to prevent entities other than Kearns-Tribune LLC and Deseret from controlling the NAC. "The desire to limit the participation of outsiders in a close corporation like the NAC has long been recognized as a reasonable purpose for a share transfer restriction." The court added that a stock transfer restriction is "inherently more reasonable when applied to the stock of a corporation having only a few shareholders who are generally active in the business" than when imposed on the stock of a corporation having many shareholders who are not actively involved in the firm's operation. The court reasoned, however, that the restriction prevented transfer of the stock only, not of all the assets, which could be transferred without the stock. Under the circumstances, the court found that other remedies might be available to compensate Salt Lake Tribune Publishing for the stock that could not be transferred.

FOR CRITICAL ANALYSIS—Economic Consideration *The plaintiff in this case argued that the stock transfer restriction was "an unreasonable incursion [invasion] on the free flow of commerce." How do the facts of this case show that this is not a valid argument?*

a. Type "Salt Lake Tribune Publishing Co." in the box and click on "Search." In the result, scroll to the name of the case and click on it to access the opinion. Washburn University School of Law Library in Topeka, Kansas, maintains this Web site.

S Corporations

A close corporation that meets the qualifying requirements specified in Subchapter S of the Internal Revenue Code can operate as an **S corporation.** If a corporation has S corporation status, it can avoid the imposition of income taxes at the corporate level while retaining many of the advantages of a corporation, particularly limited liability.

Qualification Requirements for S Corporations Among the numerous requirements for S corporation status, the following are the most important:

1. The corporation must be a domestic corporation.
2. The corporation must not be a member of an affiliated group of corporations.
3. The shareholders of the corporation must be individuals, estates, or certain trusts. Partnerships and nonqualifying trusts cannot be shareholders. Corporations can be shareholders under certain circumstances.
4. The corporation must have one hundred or fewer shareholders.
5. The corporation must have only one class of stock, although not all shareholders need have the same voting rights.
6. No shareholder of the corporation may be a nonresident alien.

Benefits of S Corporations At times, it is beneficial for a regular corporation to elect S corporation status. Benefits include the following:

1. When the corporation has losses, the S election allows the shareholders to use the losses to offset other income.
2. When the stockholder's tax bracket is lower than the corporation's tax bracket, the S election causes the corporation's pass-through net income to be taxed in the stockholder's bracket (because it is taxed as personal income). This is particularly attractive when the corporation wants to accumulate earnings for some future business purpose.

Because of these tax benefits, many close corporations have opted for S corporation status. Today, however, the limited liability company and the limited liability partnership (see Chapter 24) offer similar advantages plus additional benefits, including more flexibility in forming and operating the business. Hence, the S corporation is losing some of its significance.

Professional Corporations

Professional persons such as physicians, lawyers, dentists, and accountants can incorporate. Professional corporations are typically identified by the letters *S.C.* (service corporation), *P.C.* (professional corporation), or *P.A.* (professional association). In general, the laws governing professional corporations are similar to those governing ordinary business corporations, but three basic areas of liability deserve special attention.

First, some courts may, for liability purposes, regard the professional corporation as a partnership in which each partner can be held liable for any malpractice liability incurred by the others within the scope of the partnership. Second, a shareholder in a professional corporation is protected from the liability imposed because of any torts (unrelated to malpractice) committed by other members. Third, many professional corporation statutes retain personal liability of professional persons for their acts and the professional acts performed under their supervision.

Corporate Formation

Up to this point, we have discussed some of the general characteristics of corporations. We now examine the process by which corporations come into existence. Generally, this process involves two steps: (1) preliminary organizational and promotional undertakings, particularly obtaining capital for the future corporation; and (2) the legal process of incorporation.

Note that one of the most common reasons for changing from a sole proprietorship or a partnership to a corporation is the need for additional capital to finance expansion. A sole proprietor can seek partners who will bring capital with them. The partnership may be able to secure more funds from potential lenders than the sole proprietor could. When a firm wants to engage in significant expansion, however, simply increasing the number of partners can result in so many partners that the firm can no longer operate effectively. Therefore, incorporation may be the best choice for an expanding business organization because a corporation can obtain more capital by issuing shares of stock.

Promotional Activities

Before a corporation becomes a reality, **promoters**—those who, for themselves or others, take the preliminary steps in organizing a corporation—frequently make contracts with investors and others on behalf of the future corporation. One of the tasks of the promoter is to issue a prospectus. A **prospectus** is a document required by federal or state securities laws (discussed in Chapter 27) that describes the financial operations of the corporation, thus allowing investors to make informed decisions. The promoter also secures the corporate charter.

In addition, a promoter may purchase or lease property with a view to selling or transferring it to the corporation when the corporation is formed. A promoter may also enter into contracts with attorneys, accountants, architects, or other professionals whose services will be needed in planning for the proposed corporation. Finally, a promoter induces people to purchase stock in the corporation.

Promoter's Liability As a general rule, a promoter is held personally liable on preincorporation contracts. Courts simply hold that promoters are not agents when a corporation has yet to come into existence. If, however, the promoter secures the contracting party's agreement to hold only the corporation (and not the promoter) liable on the contract, the promoter will not be liable in the event of any breach of contract. Basically, the personal liability of the promoter continues even after incorporation unless the third party *releases* the promoter. In most states, this rule is applied even if the promoter made the agreement in the name of, or with reference to, the proposed corporation.

Once the corporation is formed (the charter issued), the promoter remains personally liable until the corporation assumes the preincorporation contract by *novation* (discussed in Chapter 11). Novation releases the promoter and makes the corporation liable for performing the contractual obligations. In some situations, the corporation *adopts* the promoter's contract by undertaking to perform it. Most courts, however, hold that adoption in and of itself does not discharge the promoter from contractual liability. Normally, a corporation cannot *ratify* a preincorporation contract, as no principal was in existence at the time the contract was made.

Subscribers and Subscriptions Prior to the actual formation of the corporation, the promoter can contact potential investors, and they can agree to purchase capital stock in the future corporation. This agreement is often called a *subscription agreement,* and the potential investor is called a *subscriber.* Depending on state law, subscribers become shareholders as soon as the corporation is formed or as soon as the corporation accepts the agreement.

Most courts view preincorporation subscriptions as continuing offers to purchase corporate stock. On or after its formation, the corporation can choose to accept the offer to purchase stock. Many courts also treat a subscription offer as irrevocable except with the consent of all of the subscribers. A subscription is irrevocable for a period of six months unless the subscription agreement provides otherwise or unless all the subscribers agree to the revocation of the subscription [RMBCA 6.20]. In some courts and jurisdictions, the preincorporation subscriber can revoke the offer to purchase before acceptance without liability, however.

Incorporation Procedures

Exact procedures for incorporation differ among states, but the basic requirements are similar.

State Chartering The first step in the incorporation process is to select a state in which to incorporate. Because state incorporation laws differ, individuals may look for the states that offer the most advantageous tax or incorporation provisions. Delaware has historically had the least restrictive laws. Consequently, many corporations, including a number of the largest, have incorporated there. Delaware's statutes permit firms to incorporate in that state and conduct business and locate their operating headquarters elsewhere. Most other states now permit this as well. Note, though, that closely held corporations, particularly those of a professional nature, generally incorporate in the state where their principal shareholders live and work.

In the following case, the state had revoked a corporation's **corporate charter** (the document issued by a state agency or authority—usually the secretary of state—that grants a corporation legal existence and the right to function) because of the corporation's failure to pay certain taxes. The issue before the court was what effect revoking the corporate charter has on a shareholder's liability.

CASE 25.2 Bullington v. Palangio

Arkansas Supreme Court, 2001.
345 Ark. 320,
45 S.W.3d 834.
http://courts.state.ar.us/opinions/opinions.html[a]

a. Scroll down to "Search Cases by Party Name," enter "Bullington," and click on "Search." In the page showing search results, scroll down the list to 6/21/2001 and click on the case name to access the opinion. The Arkansas Judiciary maintains this Web site.

FACTS Jerry Bullington, doing business as Bullington Builders, Inc. (BBI), entered into a contract with Helen Palangio for the construction of a new house in Damascus, Arkansas. Bullington signed the contract "Jerry Bullington, [doing business as] Bullington Builders, Inc.," but did not indicate any official capacity as a corporate officer. BBI had been incorporated in 1993. Its only shareholders were Bullington, who managed the business, and his wife. About one and a half months before Palangio's house was completed, BBI's

CASE 25.2–CONTINUED

charter was revoked for failure to pay Arkansas franchise taxes,[b] and it was not reinstated. Bullington finished the house, but Palangio was not satisfied with the work or with Bullington's attempts to address her complaints. More than a year later, Palangio hired another builder to remedy the alleged defects. Palangio then filed a suit in an Arkansas state court against Bullington, alleging in part breach of contract and asserting that the corporate entity did not shield him from personal liability. The court held Bullington liable to Palangio for $19,000. Bullington appealed to the Arkansas Supreme Court.

ISSUE Is a corporate officer who assumes the firm's obligation once its charter has been revoked personally liable for the performance of the obligation?

DECISION Yes. The Arkansas Supreme Court affirmed the lower court's judgment, holding Bullington personally liable for the unsatisfactory performance of the contract with Palangio.

REASON The court stated that "to exempt any association of persons from personal liability for the debts of a proposed corporation, they must comply fully with the [law] under which the corporation is created and that partial compliance with the [law] is not sufficient." The court noted that Arkansas state law required the payment of franchise taxes and provided for the revocation of the charter of a corporation that did not pay them, which occurred in this case. Also, the court explained, the reason for holding officers and stockholders individually liable for obligations of a corporation whose charter has been revoked is that these individuals "ought not be allowed to avoid personal liability because of their nonfeasance [nonperformance of a duty or a responsibility]."

WHY IS THIS CASE IMPORTANT? *This case illustrates that once a corporation's charter has been revoked, shareholders can be held personally liable for breaching contracts entered into by the corporation.*

b. A *franchise tax* is an annual tax imposed for the privilege of doing business in a state.

Articles of Incorporation The primary document needed to begin the incorporation process is the *articles of incorporation* (see Exhibit 25–1 on the following page). The articles include basic information about the corporation and serve as a primary source of authority for its future organization and business functions. The person or persons who execute the articles are called *incorporators*. Generally, the articles of incorporation should include the elements discussed in the following subsections.

—Corporate Name. The choice of a corporate name is subject to state approval to ensure against duplication or deception. State statutes usually require that the secretary of state run a check on the proposed name in the state of incorporation. Some states require that the incorporators, at their own expense, run a check on the proposed name for the newly formed corporation. Once cleared, a name can be reserved for a short time, for a fee, pending the completion of the articles of incorporation. All corporate statutes require the corporation name to include the word *Corporation, Incorporated, Company,* or *Limited,* or abbreviations of these terms.

A corporate name cannot be the same as (or deceptively similar to) the name of an existing corporation doing business within the state. ■ EXAMPLE 25.3 Suppose that an existing corporation is named General Dynamics, Inc. The state will not allow another corporation to be called General Dynamic, Inc., because that name is deceptively similar to the first and impliedly transfers a part of the goodwill established by the first corporate user to the second corporation. ■ Note that if a future firm contemplates doing business in other states, the incorporators also need to check on existing corporate names in those states as well. Otherwise, if the firm does business under a name that is the same as or deceptively similar to an existing company's name, it may be liable for trade name infringement.

—Duration. A corporation has perpetual existence unless stated otherwise in the articles. The owners may want to prescribe a maximum duration, however, after which the corporation must formally renew its existence.

—Nature and Purpose. The articles must specify the intended business activities of the corporation, and naturally, these activities must be lawful. A general statement of corporate purpose is usually sufficient to give rise to all of the powers necessary to carry out the purpose of the organization. The articles of incorporation can state, for example, that the corporation is organized "to engage in the production and sale of agricultural products." There is a trend toward allowing corporate articles to state that the corporation is organized for "any legal business," with no mention of specifics, to avoid the need for future amendments to the corporate articles.

—Capital Structure. The articles generally set forth the capital structure of the corporation. A few state statutes require a relatively small capital investment ($1,000, for example) for ordinary business corporations but a larger capital investment for those engaged in insurance or banking. The articles must outline the number of shares of stock authorized for issuance; their valuation; the various types

EXHIBIT 25–1 ARTICLES OF INCORPORATION

ARTICLE ONE

The name of the corporation is ______________________________ .

ARTICLE TWO

The period of its duration is ________ (may be a number of years or until a certain date).

ARTICLE THREE

The purpose (or purposes) for which the corporation is organized is (are) ______________________________
______________________________ .

ARTICLE FOUR

The aggregate number of shares that the corporation shall have authority to issue is ________ of the par value of ________ dollar(s) each (or without par value).

ARTICLE FIVE

The corporation will not commence business until it has received for the issuance of its shares consideration of the value of $1,000 (can be any sum not less than $1,000).

ARTICLE SIX

The address of the corporation's registered office is ______________________________ , New Pacum, and the name of its registered agent at such address is ______________________________
______________________________ .
(Use the street or building or rural address of the registered office, not a post office box number.)

ARTICLE SEVEN

The number of initial directors is ________________ , and the names and addresses of the directors are
____
______________________________ .

ARTICLE EIGHT

The name and address of the incorporator is ______________________________
______________________________ .

(signed) ______________________________
Incorporator

Sworn to on ________________ by the above-named incorporator.
(date)

____
Notary Public ________ County, New Pacum

(Notary Seal)

or classes of stock authorized for issuance; and other relevant information concerning equity, capital, and credit.

—Internal Organization. The articles should describe the internal management structure of the corporation, although this can be included in the bylaws adopted after the corporation is formed. The articles of incorporation commence the corporation; the bylaws are formed after commencement by the board of directors. Bylaws cannot conflict with the incorporation statute or the corporation's charter [RMBCA 2.06].

Under the RMBCA, shareholders may amend or repeal the bylaws. The board of directors may also amend or repeal the bylaws unless the articles of incorporation or

provisions of the incorporation statute reserve this power to the shareholders exclusively [RMBCA 10.20]. Typical bylaw provisions describe such matters as voting requirements for shareholders, the election of the board of directors, the methods of replacing directors, and the manner and time of holding shareholders' and board meetings (these corporate activities will be discussed in Chapter 26).

—***Registered Office and Agent.*** The corporation must indicate the location and address of its registered office within the state. Usually, the registered office is also the principal office of the corporation. The corporation must also give the name and address of a specific person who has been designated as an *agent* and who can receive legal documents (such as orders to appear in court) on behalf of the corporation.

—***Incorporators.*** Each incorporator must be listed by name and must indicate an address. An incorporator is a person—often, the corporate promoter—who applies to the state on behalf of the corporation to obtain its corporate charter. The incorporator need not be a subscriber and need not have any interest at all in the corporation. Many states do not impose residency or age requirements for incorporators. States vary on the required number of incorporators; it can be as few as one or as many as three. Incorporators are required to sign the articles of incorporation when they are submitted to the state; often, this is the incorporators' only duty. In some states, they participate at the first organizational meeting of the corporation.

Certificate of Incorporation Once the articles of incorporation have been prepared, signed, and authenticated by the incorporators, they are sent to the appropriate state official, usually the secretary of state, along with the required filing fee. In many states, the secretary of state then issues a **certificate of incorporation** representing the state's authorization for the corporation to conduct business. (This may also be called the *corporate charter.*) The certificate and a copy of the articles are returned to the incorporators.

First Organizational Meeting The articles of incorporation provide for the first organizational meeting, but it is not held until after the charter has actually been granted. At this meeting, the incorporators elect the first board of directors and complete the routine business of incorporation (pass bylaws and issue stock, for example). Sometimes, the meeting is held after the election of the board, and the business transacted depends on the requirements of the state's incorporation statute, the nature of the corporation, the provisions made in the articles, and the desires of the promoters. Adoption of bylaws—the internal rules of management for the corporation—is probably the most important function of the meeting. As mentioned earlier, the shareholders, directors, and officers must abide by the bylaws in conducting corporate business.

Corporate Status

The procedures for incorporation are very specific. If they are not followed precisely, others may be able to challenge the existence of the corporation. Errors in the incorporation process can become important when, for example, a third party who is attempting to enforce a contract or bring suit for a tort injury learns of them. On the basis of improper incorporation, the plaintiff could attempt to hold the would-be shareholders personally liable. Additionally, when the corporation seeks to enforce a contract against a defaulting party, that party may be able to avoid liability on the ground of a defect in the incorporation procedure.

To prevent injustice, courts will sometimes attribute corporate status to an improperly formed corporation by holding it to be a *de jure* corporation or a *de facto* corporation. Occasionally, a corporation may be held to exist by estoppel. Additionally, in certain circumstances involving abuse of the corporate form, a court may disregard the corporate entity and hold the shareholders personally liable.

DE JURE AND *DE FACTO* CORPORATIONS

If a corporation has substantially complied with all conditions precedent to incorporation, the corporation is said to have *de jure* (rightful and lawful) existence. In most states and under the RMBCA, the certificate of incorporation is viewed as evidence that all mandatory statutory provisions have been met. This means that the corporation is properly formed, and neither the state nor a third party can attack its existence. For example, listing an incorporator's address incorrectly would technically mean that the corporation was improperly formed. The law does not regard such inconsequential procedural defects as detracting from substantial compliance, however, and the courts will uphold the *de jure* status of the corporate entity.

Sometimes, there is a defect in complying with statutory mandates—for example, the corporation charter may have expired. Under these circumstances, the corporation may have *de facto* (actual) status, meaning that the corporation in fact exists, even if not rightfully or lawfully. A corporation with *de facto* status cannot be challenged by

third parties (except for the state). The following elements are required for *de facto* status:

1. There must be a state statute under which the corporation can be validly incorporated.
2. The parties must have made a good faith attempt to comply with the statute.
3. The enterprise must already have undertaken to do business as a corporation.

CORPORATION BY ESTOPPEL

If an association that is neither an actual corporation nor a *de facto* or *de jure* corporation holds itself out as being a corporation, it normally will be estopped from denying corporate status in a lawsuit by a third party. This usually occurs when a third party contracts with an association that claims to be a corporation but does not hold a certificate of incorporation. When the third party brings suit naming the so-called corporation as the defendant, the association may not escape liability on the ground that no corporation exists. When justice requires, the courts treat an alleged corporation as if it were an actual corporation for the purpose of determining the rights and liabilities involved in a particular situation. Corporation by estoppel is thus determined by the situation. It does not extend recognition of corporate status beyond the resolution of the problem at hand.

DISREGARDING THE CORPORATE ENTITY

Occasionally, the owners use a corporate entity to perpetrate a fraud, circumvent the law, or in some other way accomplish an illegitimate objective. In these situations, the court will ignore the corporate structure by **piercing the corporate veil** and exposing the shareholders to personal liability. Generally, when the corporate privilege is abused for personal benefit or when the corporate business is treated so carelessly that the corporation and the controlling shareholder are no longer separate entities, the court will require the owner to assume personal liability to creditors for the corporation's debts.

In short, when the facts show that great injustice would result from the use of a corporation to avoid individual responsibility, a court of equity will look behind the corporate structure to the individual stockholder. The following are some of the factors that frequently cause the courts to pierce the corporate veil:

1. A party is tricked or misled into dealing with the corporation rather than the individual.
2. The corporation is set up never to make a profit or always to be insolvent, or it is too "thinly" capitalized—that is, it has insufficient capital at the time of formation to meet its prospective debts or potential liabilities.
3. Statutory corporate formalities, such as holding required corporation meetings, are not followed.
4. Personal and corporate interests are **commingled** (mixed together) to such an extent that the corporation has no separate identity.

THE COMMINGLING OF PERSONAL AND CORPORATE ASSETS

To elaborate on the fourth factor in the preceding list, consider a close corporation that is formed according to law by a single person or by a few family members. In such a situation, the separate status of the corporate entity and the sole stockholder (or family-member stockholders) must be carefully preserved. Certain practices invite trouble for the one-person or family-owned corporation: the commingling of corporate and personal funds, the failure to hold board of directors' meetings and record the minutes, or the shareholders' continuous personal use of corporate property (for example, vehicles).

In the following case, in response to a creditor's motion before a court to "pierce the corporate veil," the plaintiffs argued that "this is not a fraud, this is a family business."

CASE 25.3 Dimmitt & Owens Financial, Inc. v. Superior Sports Products, Inc.

United States District Court,
Northern District of Illinois, 2002.
196 F.Supp.2d 731.

FACTS To import sport-fishing products for wholesale distribution in North America, Donald Park incorporated Superior Sports Products, Inc. (SSP), in September 1996. Park's mother, Chong Hyok Park, was SSP's president at its incorporation, but she did not participate in its operations. Donald Park handled most of its activities out of his apartment in Schaumburg, Illinois, drawing funds from SSP when he needed to pay personal expenses. There were no employees. To engage in the same business, Superior Sports International, Inc. (SSII), was formed in February 1997, and Superior Source, Inc. (SSI), in October 1997. Chong was the

CASE 25.3—CONTINUED

president and sole shareholder of SSII, which had no employees and operated out of her home in Ontario, Canada. SSI was formed with Chong's friends, but it conducted only one sale as a wholesaler and was dissolved. None of the corporations issued stock or paid dividends, maintained corporate records, or followed other corporate formalities. All of the corporations, none of which made a profit, were financed by occasional loans by Park or his parents. This capital proved to be inadequate. In January 1998, Donald Park—representing himself as president of SSP, vice president of SSII, and manager of SSI when he was none of these—applied for, and obtained, loans from Dimmitt & Owens Financial, Inc. When the loans were not paid, Dimmitt filed a suit in a federal district court against SSP, SSII, SSI, and Donald Park, seeking in part a summary judgment to impose personal liability on Park.

ISSUE Should the court "pierce the corporate veil" and impose personal liability on Donald Park?

DECISION Yes. The court granted Dimmitt's motion and issued a summary judgment in the creditor's favor, holding that piercing the corporate veil was justified.

REASON The court noted that "Illinois courts apply a two-prong test to determine whether to pierce the corporate veil: (1) there must be such unity of interest and ownership that the separate personalities of the corporation and the individual or other corporation no longer exist; and (2) circumstances must be that such an adherence to the fiction of separate corporate existence would sanction a fraud or promote injustice." The commingling of personal and corporate funds, among other misconduct, demonstrated to the court that Donald Park would be unjustly enriched, to Dimmitt's detriment, if he were shielded from personal liability by SSP. The court reasoned that "Park exercised great personal control over the existing corporations, thereby creating the sufficient unity of interest and ownership between Park and the corporations to warrant piercing the corporate veil." Also, Park's "misrepresentations equal something akin to fraud or deception. Additionally, we find that there is a compelling public interest that individuals should not execute contracts while misrepresenting themselves and escape personal liability when monies, advanced on the basis of such contracts, disappear without a proper accounting for their disappearance."

FOR CRITICAL ANALYSIS—Social Consideration *If Park had not made the misrepresentations as to his positions with the corporations, would the "other misconduct" noted by the court have been enough to justify piercing the corporate veil in this case? Why or why not?*

Corporate Financing

Part of the process of corporate formation involves corporate financing. Corporations are financed by the issuance and sale of corporate securities. **Securities** (stocks and bonds) evidence the right to participate in earnings and the distribution of corporate property or the obligation to pay funds.

Stocks, or *equity securities,* represent the purchase of ownership in the business firm. **Bonds** (debentures), or *debt securities,* represent the borrowing of funds by firms (and governments). Of course, not all debt is in the form of debt securities. For example, some debt is in the form of accounts payable and notes payable. Accounts and notes payable are typically short-term debts. Bonds are simply a way for the corporation to split up its long-term debt so that it can market it more easily.

BONDS

Bonds are issued by business firms and by governments at all levels as evidence of the funds they are borrowing from investors. Bonds normally have a designated *maturity date*—the date when the principal, or face, amount of the bond is returned to the investor. They are sometimes referred to as *fixed-income securities* because their owners (that is, the creditors) receive fixed-dollar interest payments, usually semiannually, during the period of time prior to maturity.

Because debt financing represents a legal obligation on the part of the corporation, various features and terms of a particular bond issue are specified in a lending agreement called a **bond indenture.** A corporate trustee, often a commercial bank trust department, represents the collective well-being of all bondholders in ensuring that the corporation meets the terms of the bond issue. The bond indenture specifies the maturity date of the bond and the pattern of interest payments until maturity. The different types of corporate bonds are described in Exhibit 25–2 on the following page.

STOCKS

Issuing stocks is another way that corporations can obtain financing. The ways in which stocks differ from bonds are summarized in Exhibit 25–3 on the next page. Basically,

EXHIBIT 25–2 TYPES OF CORPORATE BONDS

Debenture Bonds	Bonds for which no specific assets of the corporation are pledged as backing. Rather, they are backed by the general credit rating of the corporation, plus any assets that can be seized if the corporation allows the debentures to go into default.
Mortgage Bonds	Bonds that pledge specific property. If the corporation defaults on the bonds, the bondholders can take the property.
Convertible Bonds	Bonds that can be exchanged for a specified number of shares of common stock under certain conditions.
Callable Bonds	Bonds that may be called in and the principal repaid at specified times or under conditions specified in the bond indenture when the bond is issued.

as mentioned, stocks represent ownership in a business firm, whereas bonds represent borrowing by the firm.

Exhibit 25–4 summarizes the types of stocks issued by corporations. We look now at the two major types of stock—*common stock* and *preferred stock*.

Common Stock The true ownership of a corporation is represented by **common stock.** Common stock provides a proportionate interest in the corporation with regard to (1) control, (2) earnings, and (3) net assets. A shareholder's interest is generally in proportion to the number of shares he or she owns out of the total number of shares issued.

Any person who purchases shares acquires voting rights—one vote per share held. Voting rights in a corporation apply to the election of the firm's board of directors and to any proposed changes in the ownership structure of the firm. For example, a holder of common stock generally has the right to vote in a decision on a proposed merger, as mergers can change the proportion of ownership. State corporation law specifies the types of actions for which shareholder approval must be obtained.

Holders of common stock are investors who assume a *residual* position in the overall financial structure of a business. In terms of receiving payment for their investments, they are last in line. They are entitled to the earnings that are left after preferred stockholders, bondholders, suppliers, employees, and other groups have been paid. Once those groups are paid, however, the owners of common stock may be entitled to *all* the remaining earnings as dividends. (The board of directors normally is not under any duty to declare the remaining earnings as dividends, however.)

Preferred Stock **Preferred stock** is stock with *preferences*. Usually, this means that holders of preferred stock have priority over holders of common stock as to dividends and as to payment on dissolution of the corporation. Holders of preferred stock may or may not have the right to vote.

EXHIBIT 25–3 HOW DO STOCKS AND BONDS DIFFER?

STOCKS	BONDS
1. Stocks represent ownership.	1. Bonds represent debt.
2. Stocks (common) do not have a fixed dividend rate.	2. Interest on bonds must always be paid, whether or not any profit is earned.
3. Stockholders can elect the board of directors, which controls the corporation.	3. Bondholders usually have no voice in, or control over, management of the corporation.
4. Stocks do not have a maturity date; the corporation usually does not repay the stockholder.	4. Bonds have a maturity date, when the corporation is to repay the bondholder the face value of the bond.
5. All corporations issue or offer to sell stocks. This is the usual definition of a corporation.	5. Corporations do not necessarily issue bonds.
6. Stockholders have a claim against the property and income of a corporation after all creditors' claims have been met.	6. Bondholders have a claim against the property and income of a corporation that must be met before the claims of stockholders.

EXHIBIT 25–4 TYPES OF STOCKS

Common Stock	Voting shares that represent ownership interest in a corporation. Common stock has the lowest priority with respect to payment of dividends and distribution of assets on the corporation's dissolution.
Preferred Stock	Shares of stock that have priority over common-stock shares as to payment of dividends and distribution of assets on dissolution. Dividend payments are usually a fixed percentage of the face value of the share.
Cumulative Preferred Stock	Required dividends not paid in a given year must be paid in a subsequent year before any common-stock dividends are paid.
Participating Preferred Stock	Stock entitling the owner to receive the preferred-stock dividend and additional dividends if the corporation has paid dividends on common stock.
Convertible Preferred Stock	Stock entitling the owners to convert their shares into a specified number of common shares either in the issuing corporation or, sometimes, in another corporation.
Redeemable, or Callable, Preferred Stock	Preferred shares issued with the express condition that the issuing corporation has the right to repurchase the shares as specified.

Preferred stock is not included among the liabilities of a business because it is equity. Like other equity securities, preferred shares have no fixed maturity date on which the firm must pay them off. Although firms occasionally buy back preferred stock, they are not legally obligated to do so. A sample cumulative convertible preferred-stock certificate is shown in Exhibit 25–5 on the following page.

Holders of preferred stock are investors who have assumed a rather cautious position in their relationship to the corporation. They have a stronger position than common shareholders with respect to dividends and claims on assets, but as a result, they will not share in the full prosperity of the firm if it grows successfully over time. This is because the value of preferred shares will not rise as rapidly as that of common shares during a period of financial success. Preferred stockholders receive fixed dividends periodically.

The return and the risk for preferred stock lie somewhere between those for bonds and those for common stock. Preferred stock is more similar to bonds than to common stock, even though preferred stock appears in the ownership section of the firm's balance sheet. As a result, preferred stock is often categorized with corporate bonds as a fixed-income security, even though the legal status is not the same.

Merger and Consolidation

Sometimes, a corporation will extend its operations through a merger or consolidation. The terms *merger* and *consolidation* often are used interchangeably, but they refer to two legally distinct proceedings. The rights and liabilities of the corporation, its shareholders, and its creditors are the same for both, however.

MERGER

A **merger** involves the legal combination of two or more corporations in such a way that only one of the corporations continues to exist. ■ EXAMPLE 25.4 Corporation A and Corporation B decide to merge. They agree that A will absorb B; so on merging, B ceases to exist as a separate entity, and A continues as the *surviving corporation.* ■ Exhibit 25–6 on page 531 illustrates this process.

After the merger, A is recognized as a single corporation, possessing all the rights, privileges, and powers of itself and B. It automatically acquires all of B's property and assets without the necessity of a formal transfer. Additionally, A becomes liable for all of B's debts and obligations. Finally, A's articles of incorporation are deemed amended to include any changes that are stated in the *articles of merger* (a document setting forth the terms and conditions of the merger that is filed with the secretary of state).

In a merger, the surviving corporation inherits the disappearing corporation's preexisting legal rights and obligations. For example, if the disappearing corporation had a right of action against a third party, the surviving corporation can bring suit after the merger to recover the disappearing corporation's damages. The corporation statutes of many states provide that a successor (surviving) corporation inherits a **chose**[8] **in action** (a right to sue for a

8. The word *chose* is French for "thing."

EXHIBIT 25–5 CUMULATIVE CONVERTIBLE PREFERRED-STOCK CERTIFICATE

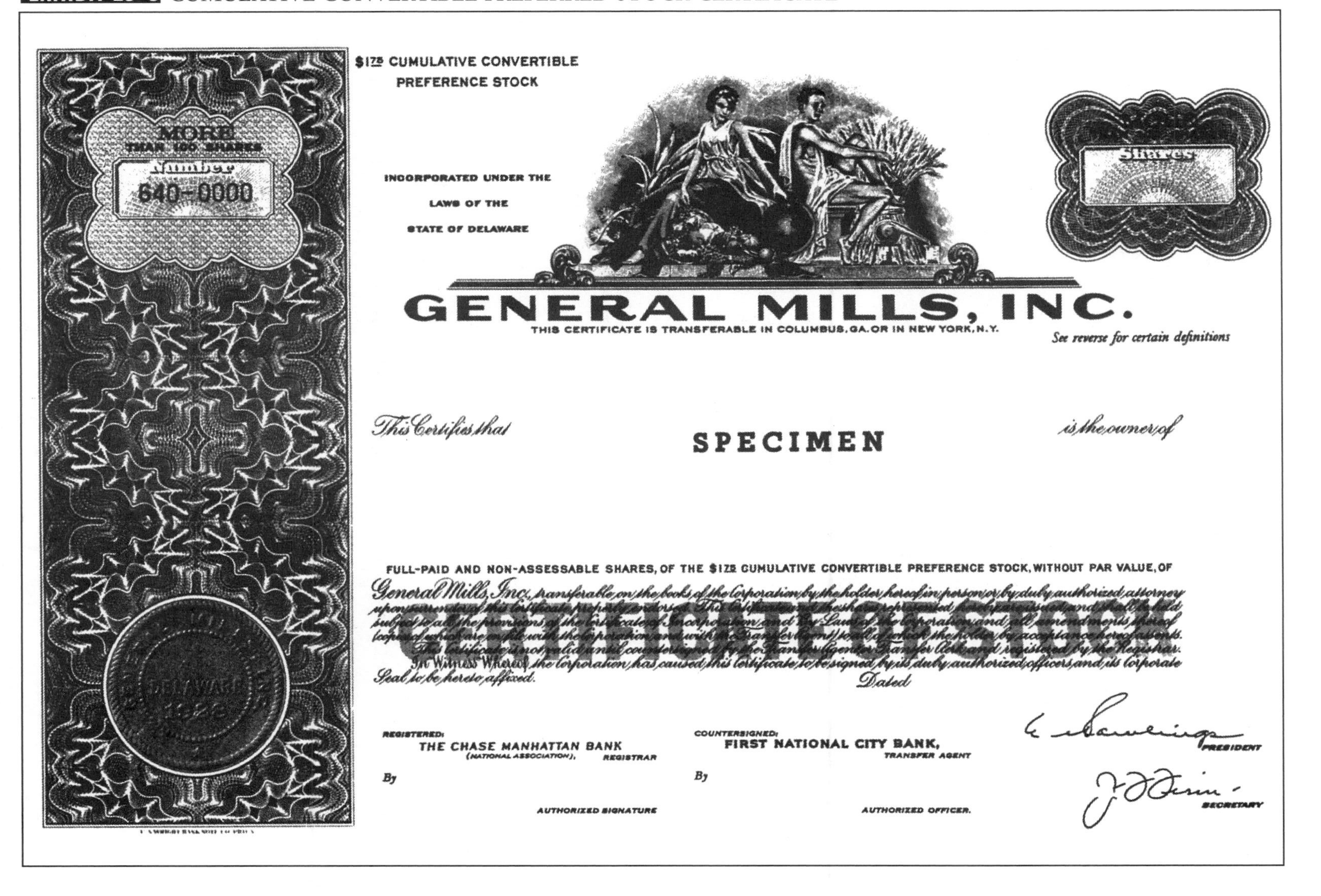

EXHIBIT 25–6 MERGER

In this illustration, Corporation A and Corporation B have decided to merge. They agree that A will absorb B; so after the merger, B no longer exists as a separate entity, and A continues as the surviving corporation.

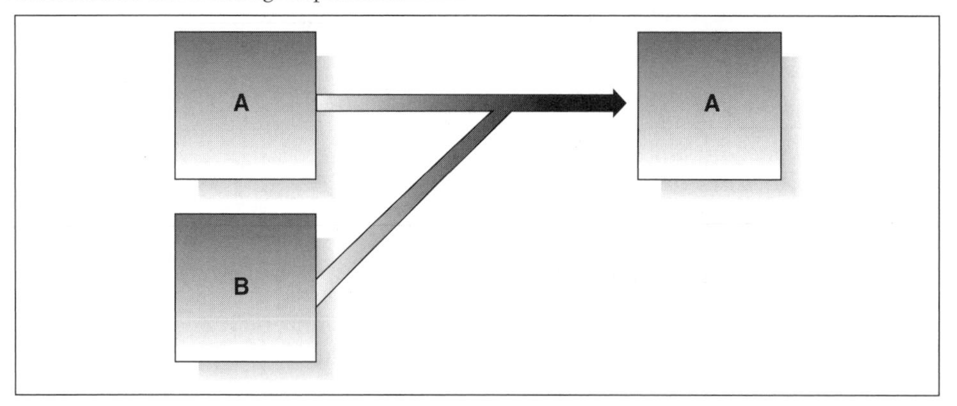

debt or sum of money) from a merging corporation as a matter of law. The common law similarly recognizes that, following a merger, a chose in action to enforce a property right will vest with the successor (surviving) corporation and no right of action will remain with the disappearing corporation.

CONSOLIDATION

In a **consolidation,** two or more corporations combine in such a way that each corporation ceases to exist and a new one emerges. ■ EXAMPLE 25.5 Corporation A and Corporation B consolidate to form an entirely new organization, Corporation C. In the process, A and B both terminate, and C comes into existence as an entirely new entity. ■ Exhibit 25–7 graphically illustrates this process.

As a result of the consolidation, C is recognized as a new corporation and a single entity; A and B cease to exist. C inherits all of the rights, privileges, and powers that A and B previously held. Title to any property and assets owned by A and B passes to C without a formal transfer. C assumes liability for all of the debts and obligations owed by A and B. The terms and conditions of the consolidation are set forth in the *articles of consolidation,* which are filed with the secretary of state. These articles *take the place of* A's and B's original corporate articles and are thereafter regarded as C's corporate articles.

EXHIBIT 25–7 CONSOLIDATION

In this illustration, Corporation A and Corporation B consolidate to form an entirely new organization, Corporation C. In the process, A and B terminate, and C comes into existence as an entirely new entity.

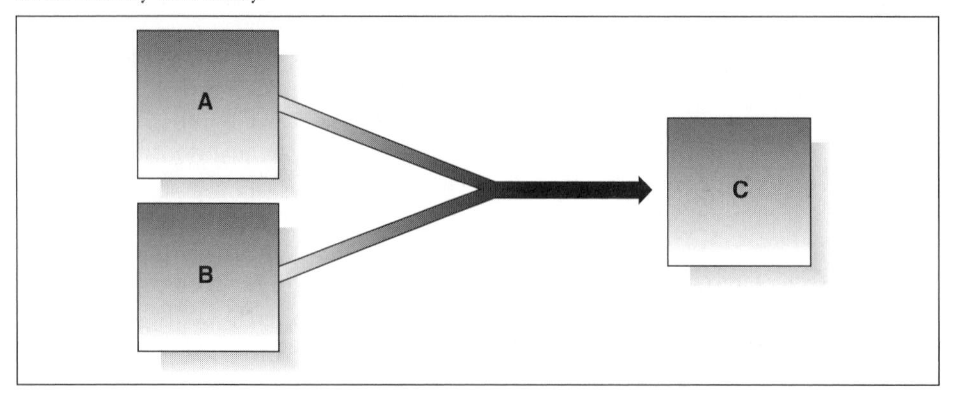

Procedure for Merger or Consolidation

All states have statutes authorizing mergers and consolidations for domestic (in-state) corporations, and most states allow the combination of domestic and foreign (out-of-state) corporations. Although the procedures vary somewhat among jurisdictions, the basic requirements for a merger or a consolidation are as follows:

1. The board of directors of each corporation involved must approve a merger or consolidation plan.
2. The shareholders of each corporation must approve the plan, by vote, at a shareholders' meeting. Most state statutes require the approval of votes representing two-thirds of the outstanding shares of the voting stock, although some states require only a simple majority, and others require a four-fifths vote. Frequently, statutes require that each class of stock approve the merger; thus, the holders of nonvoting stock must also approve. A corporation's bylaws can provide for a stricter requirement.
3. Once approved by the directors and shareholders, the plan (articles of merger or consolidation) is filed, usually with the secretary of state.
4. When state formalities are satisfied, the state issues a certificate of merger to the surviving corporation or a certificate of consolidation to the newly consolidated corporation.

Section 11.04 of the RMBCA provides for a simplified procedure for the merger of a substantially owned subsidiary corporation into its parent corporation. Under these provisions, a **short-form merger** can be accomplished *without the approval of the shareholders* of either corporation. The short-form merger can be used only when the parent corporation owns at least 90 percent of the outstanding shares of each class of stock of the subsidiary corporation. The simplified procedure requires that a plan for the merger be approved by the board of directors of the parent corporation before it is filed with the state. A copy of the merger plan must be sent to each shareholder of record of the subsidiary corporation.

Shareholder Approval

Shareholders invest in a corporate enterprise with the expectation that the board of directors will manage the enterprise and handle ordinary business matters. In contrast, actions taken on extraordinary matters must be authorized by both the board of directors and the shareholders. Often, modern statutes require that the shareholders approve certain types of extraordinary matters—such as the sale, lease, or exchange of all or substantially all corporate assets outside of the corporation's regular course of business. Other examples of matters requiring shareholder approval include amendments to the articles of incorporation, transactions concerning a merger or a consolidation, and dissolution.

Appraisal Rights

What if a shareholder disapproves of a merger or a consolidation but is outvoted by the other shareholders? The law recognizes that a dissenting shareholder should not be forced to become an unwilling shareholder in a corporation that is new or different from the one in which the shareholder originally invested. The shareholder has the right to dissent and may be entitled to be paid the fair value for the number of shares held on the date of the merger or consolidation. This right is referred to as the shareholder's **appraisal right.**

Appraisal rights are available only when a state statute specifically provides for them. Appraisal rights normally extend to regular mergers, consolidations, short-form mergers, and sales of substantially all of the corporate assets not in the ordinary course of business. Shareholders may lose their appraisal rights if they do not follow precisely the elaborate statutory procedures. Whenever they lose the right to an appraisal, dissenting shareholders must go along with the transaction despite their objections.

Purchase of Assets

When a corporation acquires all or substantially all of the assets of another corporation by direct purchase, the purchasing, or *acquiring,* corporation simply extends its ownership and control over more assets. Because no change in the legal entity occurs, the acquiring corporation is not required to obtain shareholder approval for the purchase.[9] The U.S. Department of Justice and the Federal Trade Commission, however, significantly constrain and often prohibit mergers that could result from a purchase of assets, including takeover bids.

Note that the corporation that is selling all its assets is substantially changing its business position and perhaps

9. If the acquiring corporation plans to pay for the assets with its own corporate stock and not enough authorized unissued shares are available, the shareholders must vote to approve the issuance of additional shares by amendment of the corporate articles. Additionally, acquiring corporations whose stock is traded in a national stock exchange can be required to obtain their own shareholders' approval if they plan to issue a significant number of shares, such as a number equal to 20 percent or more of the outstanding shares.

its ability to carry out its corporate purposes. For that reason, the corporation whose assets are being sold must obtain the approval of both the board of directors and the shareholders. In most states and under RMBCA 13.02, a dissenting shareholder of the selling corporation can demand appraisal rights.

Generally, a corporation that purchases the assets of another corporation is not responsible for the liabilities of the selling corporation. Exceptions to this rule are made in certain circumstances, however. In any of the following situations, the acquiring corporation will be held to have assumed *both* the assets and the liabilities of the selling corporation.

1. When the purchasing corporation impliedly or expressly assumes the seller's liabilities.
2. When the sale amounts to what is in fact a merger or consolidation.
3. When the purchaser continues the seller's business and retains the same personnel (same shareholders, directors, and officers).
4. When the sale is fraudulently executed to escape liability.

Purchase of Stock

An alternative to the purchase of another corporation's assets is the purchase of a substantial number of the voting shares of its stock. This enables the acquiring corporation to control the **target corporation** (the corporation being acquired). The acquiring corporation deals directly with the target company's shareholders in seeking to purchase the shares they hold. It does this by making a *tender offer* to all of the shareholders of the target corporation. The tender offer is publicly advertised and addressed to all shareholders of the target company. The price of the stock in the tender offer is generally higher than the market price of the target stock prior to the announcement of the tender offer. The higher price induces shareholders to tender their shares to the acquiring firm.

Termination

The termination of a corporation's existence has two phases. **Dissolution** is the legal death of the artificial "person" of the corporation. **Liquidation** is the process by which corporate assets are converted into cash and distributed among creditors and shareholders according to specific rules of preference.

DISSOLUTION

Dissolution of a corporation can be brought about in any of the following ways:

1. An act of the legislature in the state of incorporation.
2. Expiration of the time provided in the certificate of incorporation.
3. Voluntary approval of the shareholders and the board of directors.
4. Unanimous action by all shareholders.[10]
5. A court decree obtained by the attorney general of the state of incorporation for any of the following reasons: (a) failure to comply with administrative requirements (for example, failure to pay annual franchise taxes, to submit an annual report, or to have a designated registered agent), (b) procurement of a corporate charter through fraud or misrepresentation on the state, (c) abuse of corporate powers (*ultra vires* acts), (d) violation of the state criminal code after a demand to discontinue has been made by the secretary of state, (e) failure to commence business operations, or (f) abandonment of operations before starting up [RMBCA 14.20].

Sometimes a shareholder or a group of shareholders petitions a court for corporate dissolution. For example, the board of directors may be deadlocked. Courts hesitate to order involuntary dissolution in such circumstances unless there is specific statutory authorization to do so. If the shareholders cannot resolve the deadlock and if it will irreparably injure the corporation, however, the court will proceed with an involuntary dissolution. Courts can also dissolve a corporation in other circumstances, such as when the controlling shareholders or directors are committing fraudulent or oppressive acts or when management is misapplying or wasting corporate assets [RMBCA 14.30].

LIQUIDATION

When dissolution takes place by voluntary action, the members of the board of directors act as trustees of the corporate assets. As trustees, they are responsible for winding up the affairs of the corporation for the benefit of corporate creditors and shareholders. This makes the board members personally liable for any breach of their fiduciary trustee duties.

Liquidation can be accomplished without court supervision unless the members of the board do not wish to act as trustees of the corporate assets, or unless shareholders

10. This is permitted under Delaware law—see Delaware Code Section 275(c)—but not under the RMBCA.

or creditors can show cause to the court why the board should not be permitted to assume the trustee function. In either situation, the court will appoint a receiver to wind up the corporate affairs and liquidate corporate assets. A receiver is always appointed by the court if the dissolution is involuntary.

TERMS AND CONCEPTS

alien corporation 519
appraisal right 532
articles of incorporation 518
bond 527
bond indenture 527
bylaws 518
certificate of incorporation 525
chose in action 529
close corporation 519
commingle 526
common stock 528
consolidation 531
corporate charter 522
corporation 515
dissolution 533
dividend 516
domestic corporation 518
foreign corporation 518
liquidation 533
merger 529
piercing the corporate veil 526
preferred stock 528
promoter 521
prospectus 521
retained earnings 516
S corporation 521
securities 527
short-form merger 532
stock 527
target corporation 533
ultra vires 518

CHAPTER SUMMARY Corporate Formation, Financing, and Termination

The Nature of the Corporation (See pages 515–518.)	A corporation is a legal entity distinct from its owners. Formal statutory requirements, which vary somewhat from state to state, must be followed in forming a corporation. 1. *Corporate parties*—The shareholders own the corporation. They elect a board of directors to govern the corporation. The board of directors hires corporate officers and other employees to run the daily business of the firm. 2. *Corporate taxation*—The corporation pays income tax on net profits; shareholders pay income tax on the disbursed dividends that they receive from the corporation (double-taxation feature). 3. *Torts and criminal acts*—The corporation is liable for the torts committed by its agents or officers within the course and scope of their employment (under the doctrine of *respondeat superior*). In some circumstances, a corporation can be held liable (and be fined) for the criminal acts of its agents and employees. In certain situations, corporate officers may be held personally liable for corporate crimes.
Corporate Powers (See page 518.)	1. *Express powers*—The express powers of a corporation are granted by the following laws and documents (listed according to their priority): federal constitution, state constitutions, state statutes, articles of incorporation, bylaws, and resolutions of the board of directors. 2. *Implied powers*—Barring express constitutional, statutory, or other prohibitions, the corporation has the implied power to do all acts reasonably appropriate and necessary to accomplish its corporate purposes. 3. *Ultra vires doctrine*—Any act of a corporation that is beyond its express or implied powers to undertake is an *ultra vires* act. a. *Ultra vires* contracts may or may not be enforced by the courts, depending on the circumstances. b. The corporation (or shareholders on behalf of the corporation) may sue to enjoin or recover damages for *ultra vires* acts of corporate officers or directors. In addition, the state attorney general may bring an action either to obtain an injunction against the transaction or to obtain dissolution proceedings against the corporation for *ultra vires* acts.

CHAPTER SUMMARY Corporate Formation, Financing, and Termination—Continued

Classification of Corporations (See pages 518–521.)	1. *Domestic, foreign, and alien corporations*—A corporation is referred to as a *domestic corporation* within its home state (the state in which it incorporates); it is referred to as a *foreign corporation* by any state that is not its home state. An *alien corporation* is a corporation that originates in another country but does business in the United States. 2. *Public and private corporations*—A public corporation is one formed by government (for example, cities, towns, and public projects). A private corporation is one formed wholly or in part for private benefit. Most corporations are private corporations. 3. *Nonprofit corporations*—Corporations formed without a profit-making purpose (for example, charitable, educational, and religious organizations and hospitals). 4. *Close corporations*—Corporations owned by a family or a relatively small number of individuals; transfer of shares is usually restricted, and the corporation does not make a public offering of its securities. 5. *S corporations*—Small domestic corporations (must have one hundred or fewer shareholders as members) that, under Subchapter S of the Internal Revenue Code, are given special tax treatment. These corporations allow shareholders to enjoy the limited legal liability of the corporate form but avoid its double-taxation feature (corporate income is passed through to the shareholders and taxed as personal income, and the S corporation is not taxed separately). 6. *Professional corporations*—Corporations formed by professionals (for example, doctors and lawyers) to obtain the benefits of incorporation (such as tax benefits and limited liability). In most situations, the professional corporation is treated like other corporations, but sometimes the courts will disregard the corporate form and treat the shareholders as partners.
Corporate Formation (See pages 521–525.)	1. *Promotional activities*—A corporate promoter takes the preliminary steps in organizing a corporation (issues prospectus, secures charter, interests investors in the purchase of corporate stock, forms subscription agreements, makes contracts with third parties so that the corporation can immediately begin doing business on its formation, and so on). 2. *Incorporation procedures*— a. A state in which to incorporate is selected. b. The articles of incorporation are prepared and filed. The articles generally should include the corporate name, duration, nature and purpose, capital structure, internal organization, registered office and agent, and incorporators. c. The certificate of incorporation (or charter), which authorizes the corporation to conduct business, is received from the appropriate state office (usually the secretary of state). d. The first organizational meeting is held after the charter is granted. The board of directors is elected, and other business completed (bylaws passed, stock issued, and so on).
Corporate Status (See pages 525–527.)	1. *De jure or de facto corporation*—If a corporation has been improperly incorporated, courts will sometimes impute corporate status to the firm by holding that the firm is a *de jure* corporation (cannot be challenged by the state or third parties) or a *de facto* corporation (can be challenged by the state but not by third parties). 2. *Corporation by estoppel*—If a firm is neither a *de jure* nor a *de facto* corporation but represents itself to be a corporation and is sued as such by a third party, it may be held to be a corporation by estoppel.

(Continued)

CHAPTER SUMMARY Corporate Formation, Financing, and Termination—Continued

Corporate Status—Continued	3. *Disregarding the corporate entity*—To avoid injustice, courts may "pierce the corporate veil" and hold a shareholder or shareholders personally liable for a judgment against the corporation. This usually occurs only when the corporation was established to circumvent the law, when the corporate form is used for an illegitimate or fraudulent purpose, or when the controlling shareholder commingles his or her own interests with those of the corporation to such an extent that the corporation no longer has a separate identity.
Corporate Financing—Bonds (See page 527.)	Corporate bonds are securities representing *corporate debt*—money borrowed by a corporation. See Exhibit 25–2 on page 528 for a list describing the various types of corporate bonds.
Corporate Financing—Stocks (See pages 527–529.)	Stocks are equity securities issued by a corporation that represent the purchase of ownership in the business firm. 1. *Important characteristics of stockholders*— a. They need not be paid back. b. They receive dividends only when so voted by the directors. c. They are the last investors to be paid on dissolution. d. They vote for management and on major issues. 2. *Types of stock (see Exhibit 25–4 on page 529 for details)*— a. Common stock—Represents the true ownership of the firm. Holders of common stock share in the control, earning capacity, and net assets of the corporation. Common stockholders carry more risk than preferred stockholders but, if the corporation is successful, are compensated for this risk by greater returns on their investments. b. Preferred stock—Stock whose holders have a preferred status. Preferred stockholders have a stronger position than common shareholders with respect to dividends and claims on assets, but as a result, they will not share in the full prosperity of the firm if it grows successfully over time. The return and risk for preferred stock lie somewhere between those for bonds and those for common stock.
Merger and Consolidation (See pages 529–532.)	1. *Merger*—The legal combination of two or more corporations, as a result of which the surviving corporation acquires all the assets and obligations of the other corporation, which then ceases to exist. 2. *Consolidation*—The legal combination of two or more corporations, as a result of which each corporation ceases to exist and a new one emerges. The new corporation assumes all the assets and obligations of the former corporations. 3. *Procedure*—Determined by state statutes. Basic requirements are the following: a. The board of directors of each corporation involved must approve the merger or consolidation plan. b. The shareholders of each corporation must approve the merger or consolidation plan at a shareholders' meeting. c. Articles of merger or consolidation (the plan) must be filed, usually with the secretary of state. d. The state issues a certificate of merger (or consolidation) to the surviving (or newly consolidated) corporation. 4. *Short-form merger (parent-subsidiary merger)*—Possible when the parent corporation owns at least 90 percent of the outstanding shares of each class of stock of the subsidiary corporation. a. Shareholder approval is not required. b. The merger must be approved only by the board of directors of the parent corporation.

CHAPTER SUMMARY Corporate Formation, Financing, and Termination—Continued

Merger and Consolidation—Continued	c. A copy of the merger plan must be sent to each shareholder of record of the subsidiary corporation. d. The merger plan must be filed with the state. 5. *Appraisal rights*—Rights of dissenting shareholders (given by state statute) to receive the *fair value* for their shares when a merger or consolidation takes place. If the shareholder and the corporation do not agree on the fair value, a court will determine it.
Purchase of Assets (See pages 532–533.)	A purchase of assets occurs when one corporation acquires all or substantially all of the assets of another corporation. The acquiring (purchasing) corporation is not required to obtain shareholder approval; the corporation is merely increasing its assets, and no fundamental business change occurs. The acquired (purchased) corporation is required to obtain the approval of both its directors and its shareholders for the sale of its assets, because this causes a substantial change in the corporation's business position.
Purchase of Stock (See page 533.)	A purchase of stock occurs when one corporation acquires a substantial number of the voting shares of the stock of another (target) corporation.
Termination (See pages 533–534.)	The termination of a corporation involves the following two phases: 1. *Dissolution*—The legal death of the artificial "person" of the corporation. Dissolution can be brought about in any of the following ways: a. An act of the legislature in the state of incorporation. b. Expiration of the time provided in the corporate charter. c. Voluntary approval of the shareholders and the board of directors. d. Unanimous action by all shareholders. e. Court decree. 2. *Liquidation*—The process by which corporate assets are converted into cash and distributed to creditors and shareholders according to specified rules of preference. May be supervised by members of the board of directors (when dissolution is voluntary) or by a receiver appointed by the court to wind up corporate affairs.

FOR REVIEW

Answers for the even-numbered questions in this For Review *section can be found in Appendix A at the end of this text.*

1. What are the steps for bringing a corporation into existence? Who is liable for preincorporation contracts?
2. What is the difference between a *de jure* corporation and a *de facto* corporation?
3. In what circumstances might a court disregard the corporate entity ("pierce the corporate veil") and hold the shareholders personally liable?
4. What are the four steps of the merger or consolidation procedure?
5. What are two ways in which a corporation can be voluntarily dissolved? Under what circumstances might a corporation be involuntarily dissolved by state action?

QUESTIONS AND CASE PROBLEMS

25–1. Corporate Status. Three brothers inherited a small paper-supply business from their father, who had operated the business as a sole proprietorship. The brothers decided to incorporate under the name of Gomez Corp. and retained an attorney to draw up the necessary documents. The attorney drew up the papers and had the brothers sign them but neglected to send the

application for a corporate charter to the secretary of state's office. The brothers assumed that all necessary legal work had been taken care of, and they proceeded to do business as Gomez Corp. One day, a Gomez Corp. employee, while making a delivery to one of Gomez's customers, negligently ran a red light and caused a car accident. Baxter, the driver of the other vehicle, was injured as a result and sued Gomez Corp. for damages. Baxter then learned that no state charter had ever been issued to Gomez Corp., so he sued each of the brothers personally for damages. Can the brothers avoid personal liability for the tort of their employee? Explain.

QUESTION WITH SAMPLE ANSWER

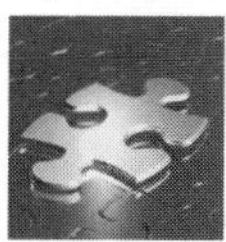

25–2. Christy, Briggs, and Dobbs are recent college graduates who want to form a corporation to manufacture and sell personal computers. Perez tells them that he will set in motion the formation of their corporation. Perez first makes a contract with Oliver for the purchase of a parcel of land for $25,000. Oliver does not know of the prospective corporate formation at the time the contract is signed. Perez then makes a contract with Kovac to build a small plant on the property being purchased. Kovac's contract is conditional on the corporation's formation. Perez secures all necessary subscription agreements and capitalization, and he files the articles of incorporation. A charter is issued.

(a) Discuss whether the newly formed corporation or Perez (or both) is liable on the contracts with Oliver and Kovac.
(b) Discuss whether the corporation, on coming into legal existence, is automatically liable to Kovac.

For a sample answer to this question, go to Appendix B at the end of this text.

25–3. Corporate Powers. Kora Nayenga and two business associates formed a corporation called Nayenga Corp. for the purpose of selling computer services. Kora, who owned 50 percent of the corporate shares, served as the corporation's president. Kora wished to obtain a personal loan from his bank for $250,000, but the bank required the note to be cosigned by a third party. Kora cosigned the note in the name of the corporation. Later, Kora defaulted on the note, and the bank sued the corporation for payment. The corporation asserted, as a defense, that Kora had exceeded his authority when he cosigned the note. Had he? Explain.

25–4. Consolidations. Determine which of the following situations describes a consolidation:

(a) Arkon Corp. purchases all of the assets of Botrek Co.
(b) Arkon Corp. and Botrek Co. combine their firms, with Arkon Corp. as the surviving corporation.
(c) Arkon Corp. and Botrek Co. agree to combine their assets, dissolve their old corporations, and form a new corporation under a new name.
(d) Arkon Corp. agrees to sell all its accounts receivable to Botrek Co.

25–5. Mergers. Tally Ho Co. was merged into Perfecto Corp., with Perfecto being the surviving corporation in the merger. Hanjo, a creditor of Tally Ho, brought suit against Perfecto Corp. for payment of the debt. The directors of Perfecto refused to pay, stating that Tally Ho no longer existed and that Perfecto had never agreed to assume any of Tally Ho's liabilities. Discuss fully whether Hanjo will be able to recover from Perfecto.

25–6. Successor Liability. In 1996, Robert McClellan, a licensed contractor doing business as McClellan Design and Construction, entered into a contract with Peppertree North Condominium Association, Inc., to do earthquake repair work on Peppertree's seventy-six-unit condominium complex in Northridge, California. McClellan completed the work, but Peppertree failed to pay. In an arbitration proceeding against Peppertree to collect the amount due, McClellan was awarded $141,000, plus 10 percent interest, attorneys' fees, and costs. McClellan filed a suit in a California state court against Peppertree to confirm the award. Meanwhile, the Peppertree board of directors filed articles of incorporation for Northridge Park Townhome Owners Association, Inc., and immediately transferred Peppertree's authority, responsibilities, and assets to the new association. Two weeks later, the court issued a judgment against Peppertree. When McClellan learned about the new association, he filed a motion asking the court to add Northridge as a debtor to the judgment. Should the court grant the motion? Why or why not? [*McClellan v. Northridge Park Townhome Owners Association, Inc.*, 89 Cal.App.4th 746, 107 Cal.Rptr.2d 702 (2 Dist. 2001)]

25–7. Corporate Dissolution. Trans-System, Inc. (TSI), is an interstate trucking business. In 1994, to provide a source of well-trained drivers, TSI formed Northwestern Career Institute, Inc., a school for persons interested in obtaining a commercial driver's license. Tim Scott, who had worked for TSI since 1987, was named chief administrative officer and director, with responsibility for day-to-day operations, which included recruiting new students, personnel matters, record keeping, and debt collecting. Scott, a Northwestern shareholder, disagreed with James Williams, the majority shareholder of both TSI and Northwestern, over four equipment leases between the two firms under which the sum of the payments exceeded the value of the equipment. Scott also disputed TSI's use, for purposes unrelated to the driving school, of $125,000 borrowed by Northwestern. Scott was terminated in 1998. He filed a suit in a Washington state court against TSI, seeking, among other things, the dissolution of Northwestern on the ground that the directors of the two firms had acted in an oppressive manner and misapplied corporate assets. Should the court grant this relief? Discuss. [*Scott v. Trans-System, Inc.*, 148 Wash.2d. 701, 64 P.3d 1 (2003)]

25–8. Torts and Criminal Acts. Greg Allen is an employee, shareholder, director, and the president of Greg Allen Construction Co. In 1996, Daniel and Sondra Estelle hired Allen's firm to renovate a home they owned in Ladoga, Indiana. To finance the cost, they obtained a line of credit from Banc One, Indiana, which required periodic inspections to disburse funds. Allen was on the job every day and supervised all of the work. He designed all of the structural changes, including a floor system for the bedroom over the living room, the floor system of

the living room, and the stairway to the second floor. He did all of the electrical, plumbing, and carpentry work and installed all of the windows. He did most of the drywall taping and finishing and most of the painting. The Estelles found much of this work to be unacceptable, and the bank's inspector agreed that it was of poor quality. When Allen failed to act on the Estelles' complaints, they filed a suit in an Indiana state court against Allen Construction and Allen personally, alleging in part that his individual work on the project was negligent. Can both Allen and his corporation be held liable for this tort? Explain. [*Greg Allen Construction Co. v. Estelle,* 798 N.E.2d 171 (Ind. 2003)]

25–9. Corporate Powers. Interbel Telephone Co-operative, Inc., is a Montana corporation organized under the Montana Rural Electric and Telephone Cooperative Act. This statute limits the purposes of such corporations to providing "adequate telephone service," but adds that this "enumeration . . . shall not be deemed to exclude like or similar objects, purposes, powers, manners, methods, or things." Mooseweb Corp. is an Internet service provider that has been owned and operated by Fred Weber since 1996. Mooseweb provides Web site hosting, modems, computer installation, technical support, and dial-up access to customers in Lincoln County, Montana. Interbel began to offer Internet service in 1999, competing with Mooseweb in Lincoln County. Weber filed a suit in a Montana state court against Interbel, alleging that its Internet service was *ultra vires.* Both parties filed motions for summary judgment. In whose favor should the court rule, and why? [*Weber v. Interbel Telephone Co-operative, Inc.,* 318 Mont. 295, 80 P.3d 88 (2003)]

CASE PROBLEM WITH SAMPLE ANSWER

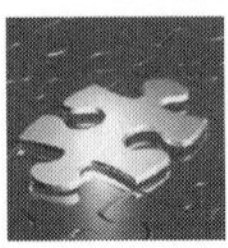

25–10. Thomas Persson and Jon Nokes founded Smart Inventions, Inc., in 1991 to market household consumer products. The success of their first product, the Smart Mop, continued with later products, which were sold through infomercials and other means. Persson and Nokes were the firm's officers and equal shareholders, with Persson responsible for product development and Nokes in charge of day-to-day operations. By 1998, they had become dissatisfied with each other's efforts. Nokes represented the firm as financially "dying," "in a grim state, . . . worse than ever," and offered to buy all of Persson's shares for $1.6 million. Persson accepted. On the day that they signed the agreement to transfer the shares, Smart Inventions began marketing a new product—the Tap Light—which was an instant success, generating millions of dollars in revenues. In negotiating with Persson, Nokes had intentionally kept the Tap Light a secret. Persson filed a suit in a California state court against Smart Inventions and others, asserting fraud and other claims. Under what principle might Smart Inventions be liable for Nokes's fraud? Is Smart Inventions liable in this case? Explain. [*Persson v. Smart Inventions, Inc.,* 125 Cal.App.4th 1141, 23 Cal.Rptr.3d 335 (2 Dist. 2005)]

After you have answered this problem, compare your answer with the sample answer given on the Web site that accompanies this text. Go to academic.cengage.com/blaw/wbl, select "Chapter 25," and click on "Case Problem with Sample Answer."

A QUESTION OF ETHICS

25–11. In a corporate merger, Diamond Shamrock retained its corporate identity, and Natomas Corp. was absorbed into Diamond's corporate hierarchy. Five inside directors (directors who are also officers of the corporation) of Natomas had "golden parachutes," which were incorporated into the merger agreement. (*Golden parachutes* are special benefits provided to a corporation's top managers in the event that the company is taken over and they are forced to leave.) The terms of the parachute agreements provided that each of the five individuals would receive a payment equal to three years' compensation in the event that they left their positions at Natomas at any time for any reason other than termination for just cause. Three of the five voluntarily left their positions after three years. Under the terms of their parachute agreements, they collected more than $10 million. A suit challenging the golden parachutes was brought by Gaillard, a Natomas shareholder. A trial court granted the defendants' motion for summary judgment; the court sustained the golden parachutes on the ground that the directors were protected by the business judgment rule (a rule under which a corporate officer or director may avoid liability to the corporation or its shareholders for poor business judgments—see Chapter 26) in effecting the agreement. The appellate court held that the business judgment rule does not apply in a review of the conduct of inside directors and remanded the case for trial. [*Gaillard v. Natomas,* 208 Cal.App.3d 1250, 256 Cal.Rptr. 702 (1989)]

1. Regardless of the legal issues, are golden parachutes ethical in a general sense? Discuss.
2. What practical considerations would lead a corporation to grant its top management such seemingly one-sided agreements?
3. In the *Gaillard* case, how would your views be affected by evidence showing that the golden parachutes had been developed and presented to the board by the very individuals who were the beneficiaries of the agreements—that is, by the five inside directors?

VIDEO QUESTION

25–12. Go to this text's Web site at academic.cengage.com/blaw/wbl and select "Chapter 25." Click on "Video Questions" and view the video titled *Corporation or LLC: Which Is Better?* Then answer the following questions.

1. Compare the liability that Anna and Caleb would be exposed to as shareholders/owners of a corporation versus as members of a limited liability company (LLC).
2. How are corporations taxed differently than LLCs?
3. Given that Anna and Caleb conduct their business (Wizard Internet) over the Internet, can you think of any drawbacks to forming an LLC?
4. If you were in the position of Anna and Caleb, would you choose to create a corporation or an LLC? Why?

Corporate Formation, Financing, and Termination

For updated links to resources available on the Web, as well as a variety of other materials, visit this text's Web site at

academic.cengage.com/blaw/wbl

Cornell University's Legal Information Institute has links to state corporation (and other) statutes at

http://www.law.cornell.edu/topics/state_statutes.html

For information on incorporation, including a list of frequently asked questions on the topic, go to

http://www.bizfilings.com

For an example of one state's (Minnesota's) statute governing corporations, go to

http://www.revisor.leg.state.mn.us/stats/302A

ONLINE LEGAL RESEARCH

Go to the *Fundamentals of Business Law* home page at academic.cengage.com/blaw/wbl, select "Chapter 25," and click on "Internet Exercises." There you will find the following Internet research exercises that you can perform to learn more about topics covered in this chapter.

Activity 25–1: LEGAL PERSPECTIVE—Corporate Law

Activity 25–2: MANAGEMENT PERSPECTIVE—Online Incorporation

Activity 25–3: LEGAL PERSPECTIVE—Mergers

BEFORE THE TEST

Go to the *Fundamentals of Business Law* home page at academic.cengage.com/blaw/wbl, select "Chapter 25," and click on "Interactive Quizzes." You will find a number of interactive questions relating to this chapter.

CHAPTER 26

Corporate Directors, Officers, and Shareholders

LEARNING OBJECTIVES

After reading this chapter, you should be able to answer the following questions:

1. What are the duties of corporate directors and officers?
2. Directors are expected to use their best judgment in managing the corporation. What must directors do to avoid liability for honest mistakes of judgment and poor business decisions?
3. What is a voting proxy? What is cumulative voting?
4. If a group of shareholders perceives that the corporation has suffered a wrong and the directors refuse to take action, can the shareholders compel the directors to act? If so, how?
5. From what sources may dividends be paid legally? In what circumstances is a dividend illegal? What happens if a dividend is illegally paid?

A corporation is not a "natural" person but a legal fiction. No one individual shareholder or director bears sole responsibility for the corporation and its actions. Rather, a corporation joins the efforts and resources of a large number of individuals for the purpose of producing greater returns than those persons could have obtained individually.

Sometimes, actions that benefit the corporation as a whole do not coincide with the separate interests of the individuals making up the corporation. In such situations, it is important to know the rights and duties of all participants in the corporate enterprise. This chapter focuses on the rights and duties of directors, officers, and shareholders and the ways in which conflicts among them are resolved.

Role of Directors

Every business corporation is governed by a board of directors. A director occupies a position of responsibility unlike that of other corporate personnel. Directors are sometimes inappropriately characterized as *agents* because they act on behalf of the corporation. No *individual* director, however, can act as an agent to bind the corporation; and as a group, directors collectively control the corporation in a way that no agent is able to control a principal. Directors are also often incorrectly characterized as *trustees* because they occupy positions of trust and control over the corporation. Unlike trustees, however, they do not own or hold title to property for the use and benefit of others.

Few legal requirements exist concerning directors' qualifications. Only a handful of states impose minimum age and residency requirements. A director is sometimes a shareholder, but this is not a necessary qualification—unless, of course, statutory provisions or corporate articles or bylaws require ownership.

ELECTION OF DIRECTORS

Subject to statutory limitations, the number of directors is set forth in the corporation's articles or bylaws. Historically, the minimum number of directors has been

three, but today many states permit fewer. Indeed, the Revised Model Business Corporation Act (RMBCA), in Section 8.01, permits corporations with fewer than fifty shareholders to eliminate the board of directors.

Normally, the incorporators appoint the initial board of directors at the time the corporation is created, or the corporation itself names the directors in the articles. The first board serves until the first annual shareholders' meeting. Subsequent directors are elected by a majority vote of the shareholders.

The term of office for a director is usually one year—from annual meeting to annual meeting. Longer and staggered terms are permissible under most state statutes. A common practice is to elect one-third of the board members each year for a three-year term. In this way, there is greater management continuity.

Removal of Directors A director can be removed *for cause* (that is, for failing to perform a required duty), either as specified in the articles or bylaws or by shareholder action. Even the board of directors itself may be given power to remove a director for cause, subject to shareholder review. In most states, a director cannot be removed without cause unless the corporation has previously authorized such an action.

Vacancies on the Board of Directors Vacancies can occur on the board of directors because of resignation or death or when a new position is created through amendment of the articles or bylaws. In these situations, either the shareholders or the board itself can fill the position, depending on state law or on the provisions of the bylaws.

More than 50 percent of the publicly traded companies in the United States are incorporated under Delaware law. Consequently, decisions of the Delaware courts on questions of corporate law have a wide impact. In the following case, a board increased the number of its members to diminish the effect that subsequently elected directors would have on the board's decisions. This may have been "legal" under the firm's bylaws, but was it valid under Delaware law?

CASE 26.1 MM Companies, Inc. v. Liquid Audio, Inc.

Delaware Supreme Court, 2003.
813 A.2d 1118.

FACTS MM Companies, Inc., a Delaware corporation with its principal place of business in New York City, owned 7 percent of the stock of Liquid Audio, Inc. In October 2001, MM sent a letter to Liquid Audio's board of directors offering to buy all of the company's stock for about $3 per share. The board rejected the offer. Liquid Audio's bylaws provide for a board of five directors divided into three classes. One class is elected each year. The next election, at which two directors would be chosen, was set for September 2002. By mid-August, it appeared that MM's nominees, Seymour Holtzman and James Mitarotonda, would win the election. The board amended the bylaws to increase the number of directors to seven and appointed Judith Frank and James Somes to fill the new positions. In September, MM's nominees were elected to the board, but their influence was diminished because there were now seven directors. MM filed a suit in a Delaware state court against Liquid Audio and others, challenging the board's actions. The court ruled in favor of the defendants. MM appealed to the Delaware Supreme Court.

ISSUE Was the board's action to amend the bylaws to increase the number of directors and fill the new positions with appointments valid?

DECISION No. The Delaware Supreme Court reversed the judgment of the lower court and remanded the case for further proceedings. The state supreme court concluded that the board acted primarily to impede the shareholders' right to vote in an impending election for successor directors.

REASON To maintain a proper balance between the stockholders' right to elect directors and the board of directors' right to manage the corporation, the court reasoned, the stockholders must have the unobstructed right to vote in an election of directors. The court determined that "[w]hen the *primary purpose* of a board of directors' [action] is to interfere with or impede the effective exercise of the shareholder franchise in a contested election for directors, the board must * * * demonstrate a compelling justification for such action." In this case, the directors' action had the primary purpose of diminishing the influence of MM's two nominees. In the court's view, this "compromised the essential role of corporate democracy in maintaining the proper allocation of power between the shareholders and the Board." Because the defendants did not demonstrate a compelling justification for their action, the court found that the amendment to Liquid Audio's bylaws that expanded the size of the board should have been invalidated.

FOR CRITICAL ANALYSIS—Political Consideration *How could MM's newly elected nominees, or any two directors, affect the decisions of a five-member board?*

BOARD OF DIRECTORS' MEETINGS

The board of directors conducts business by holding formal meetings with recorded minutes. The dates of regular meetings are usually established in the articles or bylaws or by board resolution, and no further notice is customarily required. Special meetings can be called, with notice sent to all directors. What happens if the directors do not receive proper notice? Does lack of sufficient notice of a special board meeting invalidate the resolutions adopted by the board of directors at that meeting? See this chapter's *Adapting the Law to the Online Environment* feature on the next page for a discussion of a case involving this issue.

Quorum requirements can vary among jurisdictions. (A **quorum** is the minimum number of members of a body of officials or other group that must be present in order for business to be validly transacted.) Many states leave the decision as to quorum requirements to the corporate articles or bylaws. In the absence of specific state statutes, most states provide that a quorum is a majority of the number of directors authorized in the articles or bylaws. Voting is done in person (unlike voting at shareholders' meetings, which can be done by proxy, as discussed later in this chapter).[1] The rule is one vote per director. Ordinary matters generally require a simple majority vote; certain extraordinary issues may require a greater-than-majority vote.

RIGHTS OF DIRECTORS

A director of a corporation has a number of rights, including the rights of participation, inspection, compensation, and indemnification.

Participation and Inspection A corporate director must have certain rights to function properly in that position. The main right is one of participation—meaning that the director must be notified of board of directors' meetings so as to participate in them. As pointed out earlier in this chapter, the dates of regular board meetings are usually established by the bylaws or by board resolution, and no notice of these meetings is required. If special meetings are called, however, notice is required unless waived by the director.

A director must have access to all of the corporate books and records to make decisions and to exercise the necessary supervision over corporate officers and employees. This right of inspection is virtually absolute and cannot be restricted.

1. Except in Louisiana, which allows a director to vote by proxy under certain circumstances.

Compensation and Indemnification Historically, directors have had no inherent right to compensation for their services, but have often received nominal sums as honoraria (an *honorarium* is an amount paid to someone, usually a professional, for providing a service, such as a consultation or a speech). In many corporations, directors are also chief corporate officers (president or chief executive officer, for example) and receive compensation in their managerial positions. Most directors also gain through indirect benefits, such as business contacts and prestige, and other rewards, such as stock options. There is a trend toward providing more than nominal compensation for directors, especially in large corporations where the time, work, effort, and risk involved can impose enormous burdens. Many states permit the corporate articles or bylaws to authorize compensation for directors, and in some instances the board can set its own compensation unless the articles or bylaws provide otherwise.

Corporate directors may become involved in lawsuits by virtue of their positions and their actions as directors. Most states (and RMBCA 8.51) permit a corporation to indemnify (guarantee reimbursement to) a director for legal costs, fees, and judgments involved in defending corporation-related suits. Many states specifically permit a corporation to purchase liability insurance for the directors and officers to cover indemnification. When the statutes are silent on this matter, the power to purchase such insurance is usually considered to be part of the corporation's implied power.

DIRECTORS' MANAGEMENT RESPONSIBILITIES

Directors have responsibility for all policymaking decisions necessary to the management of corporate affairs. Just as shareholders cannot act individually to bind the corporation, the directors must act as a body in carrying out routine corporate business. The general areas of responsibility of the board of directors include the following:

1. Declaration and payment of corporate dividends to shareholders.
2. Authorization for major corporate policy decisions—for example, the initiation of proceedings for the sale or lease of corporate assets outside the regular course of business, the determination of new product lines, and the oversight of major contract negotiations and major management-labor negotiations.
3. Appointment, supervision, and removal of corporate officers and other managerial employees and the determination of their compensation.
4. Financial decisions, such as the decision to issue authorized shares and bonds.

ADAPTING THE LAW TO THE ONLINE ENVIRONMENT

The Timing of Directors' Actions in an Electronic Age

As mentioned elsewhere, corporate directors can hold special board meetings to deal with extraordinary matters, provided that they give proper notice to all members of the board. If a special meeting is called without giving sufficient notice to all of the directors, are the resolutions made at that meeting invalid? If so, can the directors validate these resolutions by holding a second special board meeting with proper notice? In today's electronic age, matters of timing can be complicated and can affect a variety of other issues, including the effective date of a director's resignation and the validity of a resolution appointing a new director. These were the central issues presented by *In re Piranha, Inc.*[a]

THE ISSUE OF PROPER NOTICE

In May 2001, Edward Sample, the chairman of the board of directors of Piranha, Inc., called a special meeting for May 25 to consider restructuring Piranha's management in response to serious financial problems. The four members of the board of directors at that time were Sample, Richard Berger, Larry Greybill, and Michael Steele. Berger objected to the lack of notice of the May 25 meeting and refused to attend, claiming that the meeting was invalid. The other directors held the meeting without him and, among other things, voted to accept the resignations of two of the directors, Greybill and Steele, and appoint a new director, Mike Churchill.

After the meeting, Greybill submitted his written resignation, but Steele did not, fearing that Berger might be correct about the lack of notice rendering the meeting invalid. Nonetheless, the corporation's legal counsel filed a form with the Securities and Exchange Commission (SEC) on May 25 indicating the changes made to the board of directors. The SEC form stated that Steele *had* resigned as director and contained Steele's electronic signature.

THE SECOND MEETING

By early June, Piranha's legal counsel concluded that insufficient notice had probably rendered the May 25 meeting—and the resolutions made at that meeting—invalid. The attorneys informed the directors that, to effect the changes to the board, they would need to call another special meeting and provide sufficient notice. A second special meeting was held on June 15 with proper notice (to Berger, Sample, and Steele), and Churchill was voted in as a director. Steele, now confident that Churchill was validly appointed as a director, submitted his written resignation the following day.

Berger, however, contended that Churchill was not truly a corporate director because Steele had no right to vote at the second board meeting. At issue was whether the SEC filing of May 25 bearing Steele's electronic signature meant that Steele had effectively resigned on that date. If Churchill was not a valid director, then his subsequent vote that the corporation should file for bankruptcy would be invalid.[b]

THE COURT'S CONCLUSION

Ultimately, a federal appellate court held that Steele's electronic signature on the SEC filing did not operate as his formal resignation. Under the Uniform Electronic Transactions Act (UETA—see Chapter 13), a signature may not be denied legal effect solely because it is in an electronic form. Section 107(a) of the UETA, however, also allows a person to disavow the signature. Here, Steele claimed that he did not authorize his signature on the form submitted to the SEC. Therefore, the court found that Steele had not resigned until he submitted his written resignation following the second meeting. Thus, Churchill was a corporate director and could vote for the corporation to file for bankruptcy.

FOR CRITICAL ANALYSIS

What would the legal consequences have been if Berger had attended the first special board meeting despite the lack of sufficient notice? (Hint: See the section on the rights of directors in this chapter.)

a. 2003 WL 22922263 (5th Cir. 2003).

b. After the June 15 meeting, the directors voted that the corporation should file a petition for bankruptcy. Berger originally filed an action requesting that the court dismiss the bankruptcy petition because Churchill was not a valid director at the time of the vote. The bankruptcy court rejected this contention, and Berger appealed, resulting in the decision discussed here.

In most states, the board of directors can delegate some of its functions to an executive committee or to corporate officers. In doing so, the board is not relieved of its overall responsibility for directing the affairs of the corporation, but corporate officers and managerial personnel are empowered to make decisions relating to ordinary, daily corporate affairs within well-defined guidelines.

Role of Corporate Officers and Executives

Corporate officers and other executive employees are hired by the board of directors or, in rare instances, by the shareholders. In addition to carrying out the duties articulated in the bylaws, corporate and managerial officers

act as agents of the corporation, and the ordinary rules of agency (discussed in Chapter 22) normally apply to their employment. The qualifications required of officers and executive employees are determined at the discretion of the corporation and are included in the articles or bylaws. In most states, a person can hold more than one office and can be both an officer and a director of the corporation.

Corporate officers and other high-level managers are employees of the company, so their rights are defined by employment contracts. The board of directors, though, normally can remove corporate officers at any time with or without cause and regardless of the terms of the employment contracts—although in so doing, the corporation may be liable for breach of contract. The duties of corporate officers are the same as those of directors because both groups are involved in decision making and are in similar positions of control. Hence, officers are viewed as having the same fiduciary duties of care and loyalty in their conduct of corporate affairs as directors have, a subject to which we now turn.

Duties and Liabilities of Directors and Officers

Directors and officers are deemed *fiduciaries* of the corporation because their relationship with the corporation and its shareholders is one of trust and confidence. As fiduciaries, directors and officers owe ethical—and legal—duties to the corporation and to the shareholders as a whole. These fiduciary duties include the duty of care and the duty of loyalty.

Duty of Care

Directors and officers must exercise due care in performing their duties. The standard of *due care* has been variously described in judicial decisions and codified in many corporation codes. Generally, a director or officer is expected to act in good faith, to exercise the care that an ordinarily prudent person would exercise in similar circumstances, and to act in what he or she considers to be the best interests of the corporation [RMBCA 8.30]. Directors and officers who have not exercised the required duty of care can be held liable for the harms suffered by the corporation as a result of their negligence.

Duty to Make Informed and Reasonable Decisions Directors and officers are expected to be informed on corporate matters. To be informed, the director or officer must do what is necessary to become informed: attend presentations, ask for information from those who have it, read reports, and review other written materials such as contracts. In other words, directors and officers must carefully study a situation and its alternatives before making a decision. Depending on the nature of the business, directors and officers are often expected to act in accordance with their own knowledge and training. Nevertheless, most states (and Section 8.30 of the RMBCA) allow a director to make decisions in reliance on information furnished by competent officers or employees, professionals such as attorneys and accountants, or even an executive committee of the board without being accused of acting in bad faith or failing to exercise due care if such information turns out to be faulty.

Directors are also expected to make reasonable decisions. For example, a director should not vote to approve a *merger* (the legal combination of two or more corporations into one entity—See Chapter 25) with only a moment's consideration and on the sole basis of the price per share that is being offered.

Duty to Exercise Reasonable Supervision Directors are also expected to exercise a reasonable amount of supervision when they delegate work to corporate officers and employees. ■ **EXAMPLE 26.1** Suppose that a corporate bank director fails to attend any board of directors' meetings for five years. In addition, the director never inspects any of the corporate books or records and generally fails to supervise the efforts of the bank president and the loan committee. Meanwhile, the bank president, who is a corporate officer, makes various improper loans and permits large overdrafts. In this situation, the corporate director may be held liable to the corporation for losses resulting from the unsupervised actions of the bank president and the loan committee. ■

Dissenting Directors Directors are expected to attend board of directors' meetings, and their votes should be entered into the minutes of the meetings. Unless a dissent is entered, the director is presumed to have assented. Directors who dissent are rarely held individually liable for mismanagement of the corporation. For this reason, a director who is absent from a given meeting sometimes registers with the secretary of the board a dissent to actions taken at the meeting.

Duty of Loyalty

Loyalty can be defined as faithfulness to one's obligations and duties. In the corporate context, the duty of loyalty requires directors and officers to subordinate their personal interests to the welfare of the corporation. Among other things, this means that directors may not use corporate funds or confidential corporate information for personal advantage. They must also refrain from self-

dealing. For instance, a director should not oppose a merger that is in the corporation's best interest simply because its acceptance may cost the director her or his position. Cases dealing with fiduciary duty may involve one or more of the following:

1. Competing with the corporation.
2. Usurping (taking advantage of) a corporate opportunity.
3. Having an interest that conflicts with the interest of the corporation.
4. Engaging in insider trading (using information that is not public to make a profit trading securities, as will be discussed in Chapter 27).
5. Authorizing a corporate transaction that is detrimental to minority shareholders.

An officer or director usurps a corporate opportunity when she or he, for personal gain, takes advantage of a business opportunity that is financially within the corporation's reach and would be to the firm's practical advantage. The availability of cash to repay a corporation's debts can represent a "corporate opportunity." Does the use of that cash to repay a loan to a director constitute a "usurping" of that opportunity? That was the question in the following case.

CASE 26.2 In re Cumberland Farms, Inc.

United States Court of Appeals,
First Circuit, 2002.
284 F.3d 216.
http://www.ca1.uscourts.gov/opinions/main.php[a]

FACTS Cumberland Farms, Inc., a close corporation owned by the Haseotes family, operated a chain of gas stations. Demetrios Haseotes became the chief executive officer and chairman of the board of directors of Cumberland in 1960. In the 1970s, to provide Cumberland with a more secure gasoline supply, Haseotes acquired a refinery in Newfoundland, Canada. Because some states do not allow a company that operates a refinery to sell petroleum products retail, Haseotes chose to own the refinery through his own separate businesses, including Cumberland Crude Processing, Inc. (CCP). To operate the refinery, CCP borrowed more than $70 million from Cumberland, under an agreement that required the payment of that loan first. Haseotes also loaned money to CCP. When cash was available, Haseotes had CCP repay $5.75 million to him, without telling the Cumberland board. CCP defaulted on its debt to Cumberland, which filed a bankruptcy petition in 1992. Haseotes filed a claim for $3 million against the firm, which asserted a claim for $5.75 million against him. Cumberland argued that Haseotes breached his duty of loyalty when he had CCP pay its debt to him while ignoring its debt to Cumberland. The court disallowed Haseotes's claim. On appeal, a federal district court affirmed this ruling. Haseotes appealed to the U.S. Court of Appeals for the First Circuit.

ISSUE Did Haseotes breach his duty of loyalty to Cumberland by having CCP pay its debt to him while ignoring its debt to Cumberland?

a. In the "Opinion Number" box, type "01-1344.01A" and click on "Submit Search." In the result, click on the "Opinion" number to access the opinion. The U.S. Court of Appeals for the First Circuit maintains this Web site.

DECISION Yes. the U.S. Court of Appeals for the First Circuit affirmed the lower court's judgment. The appellate court held that Haseotes breached his duty of loyalty to Cumberland when, without informing Cumberland's board that funds had become available in CCP, he had CCP apply the funds toward its debt to himself rather than to its debt to Cumberland.

REASON The court explained that as a member of Cumberland's board of directors, Haseotes owed the corporation a fiduciary duty of loyalty and fair dealing. A director is required to act with "absolute fidelity" to the corporation and to put his or her duties to the corporation above all personal financial and business obligations. According to the court, "The fiduciary duty is especially exacting where the corporation is closely held." Thus, when a corporate director learns of an opportunity that could benefit the corporation, the director must inform the shareholders of all the material details of the opportunity. The shareholders then are able to decide whether the corporation can and should take advantage of it. In the court's view, "[I]t is inherently unfair for the director to deny the corporation that choice and instead take the opportunity for herself [or himself]." Here, "any funds that became available in CCP provided an opportunity to pay down CCP's * * * debt to Cumberland" and the loan agreement "explicitly required Haseotes to apply any available [funds] toward Cumberland's loan before paying down CCP's debt to himself." In these circumstances, "Haseotes was obligated to seek approval from Cumberland's board before acting."

WHY IS THIS CASE IMPORTANT? *As the judge in this case explained, the duty of loyalty is often considered to be more demanding for the directors and officers of a closely held corporation than a traditional one.*

Conflicts of Interest

The duty of loyalty also requires officers and directors to fully disclose to the board of directors any potential conflict of interest that might arise in any corporate transaction. The various state statutes contain different standards, but a contract will generally *not* be voidable if all of the following are true: if the contract was fair and reasonable to the corporation at the time it was made, if there was a full disclosure of the interest of the officers or directors involved in the transaction, and if the contract was approved by a majority of the disinterested directors or shareholders.

■ EXAMPLE 26.2 Southwood Corporation needs office space. Lambert Alden, one of its five directors, owns the building adjoining the corporation's main office building. He negotiates a lease with Southwood for the space, making a full disclosure to Southwood and the other four board directors. The lease arrangement is fair and reasonable, and it is unanimously approved by the corporation's board of directors. In this situation, Alden has not breached his duty of loyalty to the corporation, and the contract is thus valid. If it were otherwise, directors would be prevented from ever giving financial assistance to the corporations they serve. ■

Liability of Directors and Officers

Directors and officers are exposed to liability on many fronts. Corporate directors and officers may be held liable for the crimes and torts committed by themselves or by corporate employees under their supervision, as discussed in Chapter 6 and Chapter 22, respectively. Additionally, shareholders may perceive that the corporate directors are not acting in the best interests of the corporation and may sue the directors, in what is called a *shareholder's derivative suit,* on behalf of the corporation.[2]

The Business Judgment Rule A corporate director or officer may be able to avoid liability to the corporation or to its shareholders for poor business judgments under the **business judgment rule.** Directors and officers are expected to exercise due care and to use their best judgment in guiding corporate management, but they are not insurers of business success. Honest mistakes of judgment and poor business decisions on their part do not make them liable to the corporation for resulting damages.

The business judgment rule generally immunizes directors and officers from liability for the consequences of a decision that is within managerial authority, as long as the decision complies with management's fiduciary duties and as long as acting on the decision is within the powers of the corporation. Consequently, when there is a reasonable basis for a business decision, a court is unlikely to interfere with that decision, even if the corporation suffers as a result.

2. This type of action is discussed later in this chapter, in the context of shareholders' rights.

Requirements for the Rule to Apply To benefit from the business judgment rule, directors and officers must act in good faith, in what they consider to be the best interests of the corporation, and with the care that an ordinarily prudent person in a similar position would exercise in like circumstances. This requires an informed decision, with a rational basis, and with no conflict between the decision maker's personal interests and the interests of the corporation.

Role of Shareholders

The acquisition of a share of stock makes a person an owner and shareholder in a corporation. Shareholders thus own the corporation. Although they have no legal title to corporate property, such as buildings and equipment, they do have an equitable (ownership) interest in the firm.

As a general rule, shareholders have no responsibility for the daily management of the corporation, although they are ultimately responsible for choosing the board of directors, which does have such control. Ordinarily, corporate officers and directors owe no duty to individual shareholders unless some contract or special relationship exists between them in addition to the corporate relationship. Their duty is to act in the best interests of the corporation and its shareholder-owners as a whole. In turn, as you will read later in this chapter, controlling shareholders owe a fiduciary duty to minority shareholders. Normally, there is no legal relationship between shareholders and creditors of the corporation. Shareholders can, in fact, be creditors of the corporation and thus have the same rights of recovery against the corporation as any other creditor.

In this section, we look at the powers and voting rights of shareholders, which are generally established in the articles of incorporation and under the state's general corporation law.

Shareholders' Powers

Shareholders must approve fundamental changes affecting the corporation before the changes can be implemented. Hence, shareholders are empowered to amend the articles of incorporation (charter) and bylaws, approve a merger or the dissolution of the corporation, and approve the sale of all or substantially all of the corporation's assets. Some of these powers are subject to prior board approval.

Members of the board of directors are elected and removed by a vote of the shareholders. The first board of directors is either named in the articles of incorporation or chosen by the incorporators to serve until the first shareholders' meeting. From that time on, the selection and retention of directors are exclusively shareholder functions.

Directors usually serve their full terms; if they are unsatisfactory, they are simply not reelected. Shareholders have the inherent power, however, to remove a director from office *for cause* (breach of duty or misconduct) by a majority vote.[3] Some state statutes (and some corporate charters) even permit removal of directors without cause by the vote of a majority of the holders of outstanding shares entitled to vote.

SHAREHOLDERS' MEETINGS

Shareholders' meetings must occur at least annually. In addition, special meetings can be called to deal with urgent matters.

Notice of Meetings Each shareholder must receive written notice of the date, time, and place of a shareholders' meeting.[4] The notice must be received within a reasonable length of time prior to the date of the meeting. Notice of a special meeting must include a statement of the purpose of the meeting, and business transacted at the meeting is limited to that purpose.

Proxies Because it is usually not practical for owners of only a few shares of stock of publicly traded corporations to attend shareholders' meetings, such shareholders normally give third parties written authorization to vote their shares at the meeting. This authorization is called a **proxy** (from the Latin *procurare,* "to manage, take care of"). Proxies are often solicited by management, but any person can solicit proxies to concentrate voting power. Proxies have been used by a group of shareholders as a device for taking over a corporation (corporate takeovers were discussed in Chapter 25). Proxies are normally revocable (that is, they can be withdrawn), unless they are specifically designated as irrevocable. Under RMBCA 7.22(c), proxies last for eleven months, unless the proxy agreement provides for a longer period.

Proxy Materials and Shareholder Proposals When shareholders want to change a company policy, they can put their idea up for a shareholder vote. They can do this by submitting a shareholder proposal to the board of directors and asking the board to include the proposal in the proxy materials that are sent to all shareholders before meetings.

The Securities and Exchange Commission (SEC), which regulates the purchase and sale of securities (see Chapter 27), has special provisions relating to proxies and shareholder proposals. SEC Rule 14a-8 requires that when a company sends proxy materials to its shareholders, the company must also include whatever proposals will be considered at the meeting and provide shareholders with the opportunity to vote on the proposals by marking and returning their proxy cards. SEC Rule 14a-8 provides that all shareholders who own stock worth at least $1,000 are eligible to submit proposals for inclusion in corporate proxy materials. Only those proposals that relate to significant policy considerations rather than ordinary business operations must be included.

SHAREHOLDER VOTING

Shareholders exercise ownership control through the power of their votes. Each shareholder is entitled to one vote per share, although the power of the shareholder's vote can be enhanced by various voting techniques, as will be discussed shortly. The articles of incorporation can exclude or limit voting rights, particularly for certain classes of shares. For example, owners of preferred shares are usually denied the right to vote [RMBCA 7.21].

Quorum Requirements For shareholders to conduct business at a meeting, a quorum must be present. Generally, a quorum exists when shareholders holding more than 50 percent of the outstanding shares are in attendance. Corporate business matters are presented in the form of *resolutions,* which shareholders vote to approve or disapprove. Some state statutes have set specific voting requirements, and corporations' articles or bylaws must abide by these statutory requirements. In some states, obtaining the unanimous written consent of shareholders is a permissible alternative to holding a shareholders' meeting [RMBCA 7.25].

Once a quorum is present, voting can proceed. A majority vote of the shares represented at the meeting is usually required to pass resolutions. ■ **EXAMPLE 26.3** Assume that Novo Pictures, Inc., has 10,000 outstanding shares of voting stock. Its articles of incorporation set the quorum at 50 percent of outstanding shares and provide that a majority vote of the shares present is necessary to pass resolutions concerning ordinary matters. Therefore, for this firm,

3. A director can often demand court review of removal for cause.

4. The shareholder can waive the requirement of written notice by signing a waiver form. In some states, a shareholder who does not receive written notice, but who learns of the meeting and attends without protesting the lack of notice, is said to have waived notice by such conduct. State statutes and corporate bylaws typically set forth the time within which notice must be sent, what methods can be used, and what the notice must contain.

a quorum of shareholders representing 5,000 outstanding shares must be present at a shareholders' meeting to conduct business. If exactly 5,000 shares are represented at the meeting, a vote of at least 2,501 of those shares is needed to pass a resolution. If 6,000 shares are represented, a vote of 3,001 will be required, and so on. ■

At times, more than a simple majority vote will be required either by a state statute or by the corporate charter. Extraordinary corporate matters, such as a merger, consolidation, or dissolution of the corporation (see Chapter 25), require a higher percentage of all corporate shares entitled to vote [RMBCA 7.27].

Voting Lists The corporation prepares voting lists prior to each meeting of the shareholders. Only persons whose names appear on the corporation's shareholder records as owners are ordinarily entitled to vote.[5] The voting list contains the name and address of each shareholder as shown on the corporate records on a given cutoff date, or record date. (Under RMBCA 7.07, the record date may be as much as seventy days before the meeting.) The voting list also includes the number of voting shares held by each owner. The list is usually kept at the corporate headquarters and is available for shareholder inspection [RMBCA 7.20].

Cumulative Voting Most states permit or even require shareholders to elect directors by *cumulative voting,* a method of voting designed to allow minority shareholders representation on the board of directors.[6] With cumulative voting, the number of board members to be elected is multiplied by the number of voting shares a shareholder owns. The result equals the number of votes the shareholder has, and this total can be cast for one or more nominees for director. All nominees stand for election at the same time. When cumulative voting is not required either by statute or under the articles, the entire board can be elected by a simple majority of shares at a shareholders' meeting.

Cumulative voting can best be understood by an example. ■ EXAMPLE 26.4 Suppose that a corporation has 10,000 shares issued and outstanding. One group of shareholders (the minority shareholders) holds only 3,000 shares, and the other group of shareholders (the majority shareholders) holds the other 7,000 shares. Three members of the board are to be elected. The majority shareholders' nominees are Acevedo, Barkley, and Craycik. The minority shareholders' nominee is Drake. Can Drake be elected by the minority shareholders?

If cumulative voting is allowed, the answer is yes. The minority shareholders have 9,000 votes among them (the number of directors to be elected times the number of shares held by the minority shareholders equals 3 times 3,000, which equals 9,000 votes). All of these votes can be cast to elect Drake. The majority shareholders have 21,000 votes (3 times 7,000 equals 21,000 votes), but these votes have to be distributed among their three nominees. The principle of cumulative voting is that no matter how the majority shareholders cast their 21,000 votes, they will not be able to elect all three directors if the minority shareholders cast all of their 9,000 votes for Drake, as illustrated in Exhibit 26–1. ■

Other Voting Techniques Before a shareholders' meeting, a group of shareholders can agree in writing to vote their shares together in a specified manner. Such agreements, called *shareholder voting agreements,* are usually held to be valid and enforceable. A shareholder can also

5. When the legal owner is bankrupt, incompetent, deceased, or in some other way under a legal disability, his or her vote can be cast by a person designated by law to control and manage the owner's property.

6. See, for example, California Corporate Code Section 708. Under RMBCA 7.28, however, no cumulative voting rights exist unless the articles of incorporation so provide.

EXHIBIT 26–1 RESULTS OF CUMULATIVE VOTING

This exhibit illustrates how cumulative voting gives minority shareholders a greater chance of electing a director of their choice. By casting all of their 9,000 votes for one candidate (Drake), the minority shareholders will succeed in electing Drake to the board of directors.

BALLOT	MAJORITY SHAREHOLDERS' VOTES			MINORITY SHAREHOLDERS' VOTES	DIRECTORS ELECTED
	Acevedo	Barkley	Craycik	Drake	
1	10,000	10,000	1,000	9,000	Acevedo/Barkley/Drake
2	9,001	9,000	2,999	9,000	Acevedo/Barkley/Drake
3	6,000	7,000	8,000	9,000	Barkley/Craycik/Drake

appoint a voting agent and vote by proxy. As mentioned previously, a proxy is a written authorization to cast the shareholder's vote, and a person can solicit proxies from a number of shareholders in an attempt to concentrate voting power.

Another technique is for shareholders to enter into a **voting trust,** which is an agreement (a trust contract) under which legal title (record ownership on the corporate books) is transferred to a trustee who is responsible for voting the shares. The agreement can specify how the trustee is to vote, or it can allow the trustee to use his or her discretion. The trustee takes physical possession of the stock certificate and in return gives the shareholder a voting trust certificate. The shareholder retains all of the rights of ownership (for example, the right to receive dividend payments) except for the power to vote the shares.

Rights of Shareholders

Shareholders possess numerous rights. A significant right—the right to vote their shares—has already been discussed. We now look at some additional rights of shareholders.

Stock Certificates

A **stock certificate** is a certificate issued by a corporation that evidences ownership of a specified number of shares in the corporation. In jurisdictions that require the issuance of stock certificates, shareholders have the right to demand that the corporation issue certificates. In most states and under RMBCA 6.26, boards of directors may provide that shares of stock be uncertificated—that is, no actual, physical stock certificates will be issued. When shares are uncertificated, the corporation may be required to send each shareholder a letter or some other form of notice that contains the same information that would normally appear on the face of stock certificates.

Stock is intangible personal property, and the ownership right exists independently of the certificate itself. If a stock certificate is lost or destroyed, ownership is not destroyed with it. A new certificate can be issued to replace one that has been lost or destroyed.[7] Notice of shareholders' meetings, dividends, and operational and financial reports are all distributed according to the recorded ownership listed in the corporation's books, not on the basis of possession of the certificate.

Preemptive Rights

A **preemptive right** is a common law concept under which a shareholder is given a preference over all other purchasers to subscribe to or purchase a prorated share of a new issue of stock. In other words, a shareholder who is given preemptive rights can purchase the same percentage of the new shares being issued as she or he already holds in the company. This allows each shareholder to maintain her or his proportionate control, voting power, or financial interest in the corporation. Most statutes either (1) grant preemptive rights but allow them to be negated in the corporation's articles or (2) deny preemptive rights except to the extent that they are granted in the articles. The result is that the articles of incorporation determine the existence and scope of preemptive rights. Generally, preemptive rights apply only to additional, newly issued stock sold for cash, and the preemptive rights must be exercised within a specified time period, which is usually thirty days.

■ **EXAMPLE 26.5** Tran Corporation authorizes and issues 1,000 shares of stock. Lebow purchases 100 shares, making her the owner of 10 percent of the company's stock. Subsequently, Tran, by vote of its shareholders, authorizes the issuance of another 1,000 shares (amending the articles of incorporation). This increases its capital stock to a total of 2,000 shares. If preemptive rights have been provided, Lebow can purchase one additional share of the new stock being issued for each share she already owns—or 100 additional shares. Thus, she can own 200 of the 2,000 shares outstanding, and she will maintain her relative position as a shareholder. If preemptive rights are not allowed, her proportionate control and voting power may be diluted from that of a 10 percent shareholder to that of a 5 percent shareholder because of the issuance of the additional 1,000 shares. ■

Preemptive rights are most important in close corporations because each shareholder owns a relatively small number of shares but controls a substantial interest in the corporation. Without preemptive rights, it would be possible for a shareholder to lose his or her proportionate control over the firm.

Stock Warrants

Usually, when preemptive rights exist and a corporation is issuing additional shares, each shareholder is given **stock warrants,** which are transferable options to acquire a given number of shares from the corporation at a stated

7. The Uniform Commercial Code (UCC) provides that for a lost or destroyed certificate to be reissued, a shareholder normally must furnish an *indemnity bond.* An indemnity bond is a written promise to reimburse the holder for any actual or claimed loss caused by the issuer's or some other person's conduct. The bond protects the corporation against potential loss should the original certificate reappear at some future time in the hands of a bona fide purchaser [UCC 8–302, 8–405(2)].

price. Warrants are often publicly traded on securities exchanges. When the option to purchase is in effect for a short period of time, the stock warrants are usually referred to as *rights*.

DIVIDENDS

As mentioned in Chapter 25, a *dividend* is a distribution of corporate profits or income *ordered by the directors* and paid to the shareholders in proportion to their respective shares in the corporation. Dividends can be paid in cash, property, stock of the corporation that is paying the dividends, or stock of other corporations.[8]

State laws vary, but each state determines the general circumstances and legal requirements under which dividends are paid. State laws also control the sources of revenue to be used; only certain funds are legally available for paying dividends. Depending on state law, dividends may be paid from the following sources:

1. *Retained earnings.* All state statutes allow dividends to be paid from the undistributed net profits earned by the corporation, including capital gains from the sale of fixed assets. The undistributed net profits are called *retained earnings*.
2. *Net profits.* A few state statutes allow dividends to be issued from current net profits without regard to deficits in prior years.
3. *Surplus.* A number of statutes allow dividends to be paid out of any kind of surplus.

Illegal Dividends Sometimes, dividends are improperly paid from an unauthorized account, or their payment causes the corporation to become insolvent. Generally, in such situations, shareholders must return illegal dividends only if they knew that the dividends were illegal when the payment was received. A dividend paid while the corporation is insolvent is automatically an illegal dividend, and shareholders may be required to return the payment to the corporation or its creditors. Whenever dividends are illegal or improper, the board of directors can be held personally liable for the amount of the payment. When directors can show that a shareholder knew that a dividend was illegal when it was received, however, the directors are entitled to reimbursement from the shareholder.

Directors' Failure to Declare a Dividend When directors fail to declare a dividend, shareholders can ask a court to compel the directors to meet and to declare a dividend. To succeed, the shareholders must show that the directors have acted so unreasonably in withholding the dividend that their conduct is an abuse of their discretion.

Often, a corporation accumulates large cash reserves for a bona fide purpose, such as expansion, research, or other legitimate corporate goals. The mere fact that the firm has sufficient earnings or surplus available to pay a dividend is not enough to compel directors to distribute funds that, in the board's opinion, should not be distributed. The courts are reluctant to interfere with corporate operations and will not compel directors to declare dividends unless abuse of discretion is clearly shown.

INSPECTION RIGHTS

Shareholders in a corporation enjoy both common law and statutory inspection rights. The shareholder's right of inspection is limited, however, to the inspection and copying of corporate books and records for a *proper purpose,* provided the request is made in advance. The shareholder can inspect in person, or an attorney, accountant, or other type of assistant can do so as the shareholder's agent. The RMBCA requires the corporation to maintain an alphabetical voting list of shareholders with addresses and number of shares owned; this list must be kept open at the annual meeting for inspection by any shareholder of record [RMBCA 7.20].

The power of inspection is fraught with potential abuses, and the corporation is allowed to protect itself from them. For example, a shareholder can properly be denied access to corporate records to prevent harassment or to protect trade secrets or other confidential corporate information. Some states require that a shareholder must have held his or her shares for a minimum period of time immediately preceding the demand to inspect or must hold a minimum number of outstanding shares. The RMBCA provides, though, that every shareholder is entitled to examine specified corporate records [RMBCA 16.02]. A shareholder who is denied the right of inspection can seek a court order to compel the inspection.

TRANSFER OF SHARES

Corporate stock represents an ownership right in intangible personal property. The law generally recognizes the right to transfer stock to another person unless there are valid restrictions on its transferability. Although stock certificates are negotiable and freely transferable by indorsement and delivery, transfer of stock in closely held corporations usually is restricted. Restrictions may be found in the bylaws, stamped on the stock certificate,

8. Technically, dividends paid in stock are not dividends. They maintain each shareholder's proportionate interest in the corporation. On one occasion, a distillery declared and paid a "dividend" in bonded whiskey.

or in a shareholder agreement. The existence of any restrictions on transferability must always be noted on the face of the stock certificate, and these restrictions must be reasonable.

Sometimes, corporations or their shareholders restrict transferability by reserving the option to purchase any shares offered for resale by a shareholder. This **right of first refusal** remains with the corporation or the shareholders for only a specified period or a reasonable time. Variations on the purchase option are possible. For example, a shareholder might be required to offer the shares to other shareholders first or to the corporation first. When shares are transferred, a new entry is made in the corporate stock book to indicate the new owner. Until the corporation is notified and the entry is complete, all rights—including voting rights, the right to notice of shareholders' meetings, and the right to dividend distributions—remain with the current record owner.

Rights on Dissolution

When a corporation is dissolved and its outstanding debts and the claims of its creditors have been satisfied, the remaining assets are distributed to the shareholders in proportion to the percentage of shares owned by each shareholder. Certain classes of preferred stock can be given priority. If no class of stock has been given preferences in the distribution of assets on liquidation, then all of the stockholders share the remaining assets.

In some circumstances, shareholders may petition a court to have the corporation dissolved. Suppose, for example, that the minority shareholders know that the board of directors is mishandling corporate assets. The minority shareholders are not powerless to intervene. They can petition a court to appoint a **receiver** who will wind up corporate affairs and liquidate the business assets of the corporation. The RMBCA permits any shareholder to initiate such an action in any of the following circumstances [RMBCA 14.30]:

1. The directors are deadlocked in the management of corporate affairs. The shareholders are unable to break that deadlock, and irreparable injury to the corporation is being suffered or threatened.
2. The acts of the directors or those in control of the corporation are illegal, oppressive, or fraudulent.
3. Corporate assets are being misapplied or wasted.
4. The shareholders are deadlocked in voting power and have failed, for a specified period (usually two annual meetings), to elect successors to directors whose terms have expired or would have expired with the election of successors.

The Shareholder's Derivative Suit

When those in control of a corporation—the corporate directors—fail to sue in the corporate name to redress a wrong suffered by the corporation, shareholders are permitted to do so "derivatively" in what is known as a **shareholder's derivative suit.** Before a derivative suit can be brought, some wrong must have been done to the corporation, and the shareholders must have presented their complaint to the board of directors. Only if the directors fail to solve the problem or to take appropriate action can the derivative suit go forward.

The right of shareholders to bring a derivative action is especially important when the wrong suffered by the corporation results from the actions of corporate directors or officers. This is because the directors and officers would probably want to prevent any action against themselves.

The shareholder's derivative suit is unusual in that those suing are not pursuing rights or benefits for themselves personally but are acting as guardians of the corporate entity. Therefore, any damages recovered by the suit normally go into the corporation's treasury, not to the shareholders personally. This is true even if the company is a small, closely held corporation. ■ EXAMPLE 26.6 Zeon Corporation is owned by two shareholders, each holding 50 percent of the corporate shares. Suppose that one of the shareholders wants to sue the other for misusing corporate assets or usurping corporate opportunities. The plaintiff-shareholder will have to bring a shareholder's derivative suit (not a suit in his or her own name) because the alleged harm was suffered by Zeon, not by the plaintiff personally. Any damages awarded will go to the corporation, not to the plaintiff-shareholder. ■

Liability of Shareholders

One of the hallmarks of the corporate organization is that shareholders are not personally liable for the debts of the corporation. If the corporation fails, shareholders can lose their investments, but that is generally the limit of their liability. As discussed in Chapter 25, in certain instances of fraud, undercapitalization, or careless observance of corporate formalities, a court will *pierce the corporate veil* (disregard the corporate entity) and hold the shareholders individually liable. These situations are the exception, however, not the rule.

A shareholder can also be personally liable in certain other rare instances. One relates to illegal dividends, which were discussed previously. Two others relate to *stock subscriptions* and *watered stock,* which we discuss here.

STOCK-SUBSCRIPTION AGREEMENTS

Sometimes, stock-subscription agreements—written contracts to buy capital stock of a corporation—exist prior to incorporation. Normally, these agreements are treated as continuing offers and are irrevocable (for up to six months under RMBCA 6.20). After the corporation has been formed, it can sell shares to shareholder investors. Once the subscription agreement or stock offer is accepted, a binding contract is formed. Any refusal to pay constitutes a breach resulting in the personal liability of the shareholder.

Shares of stock can be paid for by property or by services rendered instead of cash. (Shares cannot be purchased with promissory notes, however.) The general rule is that for **par-value shares** (shares that have a specific face value, or formal cash-in value, written on them, such as one penny or one dollar), the corporation must receive a value at least equal to the par-value amount. For **no-par shares** (shares that have no face value—no specific amount printed on their face), the corporation must receive the fair market value of the shares as determined by the board or the shareholders. For either par-value or no-par shares, often the value is set based on what the corporation will receive for the shares (cash, property, or services).

WATERED STOCK

When a corporation issues shares for less than their fair market value, the shares are referred to as **watered stock.**[9] Usually, the shareholder who receives watered stock must pay the difference to the corporation (the shareholder is personally liable). In some states, the shareholder who receives watered stock may be liable to creditors of the corporation for unpaid corporate debts.

■ EXAMPLE 26.7 During the formation of a corporation, Gomez, one of the incorporators, transfers his property, Sunset Beach, to the corporation for 10,000 shares of stock. The stock has a par value of $100 per share, and thus the total price of the 10,000 shares is $1 million. After the property is transferred and the shares are issued, Sunset Beach is carried on the corporate books at a value of $1 million. On appraisal, it is discovered that the market value of the property at the time of transfer was only $500,000. The shares issued to Gomez are therefore watered stock, and he is liable to the corporation for the difference. ■

9. The phrase *watered stock* was originally used to describe cattle that were kept thirsty during a long drive and then were allowed to drink large quantities of water just prior to their sale. The increased weight of the "watered stock" allowed the seller to reap a higher profit.

Duties of Majority Shareholders

In some instances, a majority shareholder is regarded as having a fiduciary duty to the corporation and to the minority shareholders. This occurs when a single shareholder (or a few shareholders acting in concert) owns a sufficient number of shares to exercise *de facto* (actual) control over the corporation. In these situations, the majority shareholders owe a fiduciary duty to the minority shareholders.

A breach of fiduciary duty can also occur when the majority shareholders of a closely held corporation use their control to exclude the minority from certain benefits of participating in the firm. ■ EXAMPLE 26.8 Three brothers, Alfred, Carl, and Eugene, each owned a one-third interest in a corporation and had worked for the corporation for most of their adult lives. When a dispute arose concerning discrepancies in the corporation's accounting records, Carl and Eugene fired Alfred and told the company's employees that Alfred had had a nervous breakdown, which was not true. Alfred sued Carl and Eugene, alleging that they had breached their fiduciary duties. The brothers argued that because there was no reduction in the value of the corporation or the value of Alfred's shares in the company, they had not breached their fiduciary duties. The court, however, held that the brothers' conduct was unfairly prejudicial toward Alfred and supported a finding of a breach of fiduciary duty.[10] ■

Such a breach of fiduciary duties by those who control a closely held corporation may constitute *oppressive conduct.* The court in the following case was asked to review a pattern of allegedly oppressive conduct by the person in control and determine whether that conduct fell within a two-year statute of limitations.

10. *Pedro v. Pedro,* 489 N.W.2d 798 (Minn.App. 1992).

CASE 26.3 Robbins v. Sanders

Supreme Court of Alabama, 2004.
890 So.2d 998.

FACTS James and Mary Bailey owned fifty-three acres of land, subject to mortgages totaling $450,000, in Birmingham, Alabama. The Baileys rented buildings on the property and used part of the land as a landfill. In 1988, an underground fire broke out in the landfill. Pete Robbins offered to extinguish

(Continued)

CASE 26.3–CONTINUED

the fire and to pay the mortgages. The parties formed Corridor Enterprises, Inc., to which the Baileys contributed the land. Half of the stock—one thousand shares—was issued to Robbins. In 1991, the Baileys agreed to sell their thousand shares to Robbins at the rate of two shares per month. The Baileys both died in 1997. Terrill Sanders was appointed administrator of their estates. Over the next twelve months, Sanders had problems obtaining information from Robbins and uncovered discrepancies in Corridor's corporate records. On the estates' behalf, Sanders filed a suit in an Alabama state court, alleging, among other things, oppression of minority shareholders. The court assessed more than $4 million in damages against Robbins, who appealed to the Alabama Supreme Court, arguing in part that a two-year statute of limitations barred the suit.

ISSUE Were the minority shareholders entitled to recover for their claim of oppression?

DECISION Yes. The Alabama Supreme Court held that Sanders's claims on behalf of the Baileys' estates were timely. The court remanded the case, though, so that the lower court could clearly state how the damages should be apportioned among the estates and Corridor.

REASON The state supreme court rejected Robbins's statute-of-limitations argument. The court explained that the Baileys' estates became minority shareholders after the Baileys died in 1997. The court pointed out that Robbins engaged in oppressive conduct after this event. "For example, in 1998, Robbins used funds of Corridor Enterprises to purchase real estate in his own name, to invest in other businesses in his own name, and to purchase personal property for himself; he refused to provide an accounting of the corporate finances when he was requested to do so; he failed to pay the corporate property and income taxes, failed to have tax returns prepared and filed, and failed to maintain proper corporate records; he entered into a contract to sell property belonging to Corridor Enterprises without notice to or approval of the minority shareholders; and he failed to declare dividends during the entire time the estates were shareholders while he paid himself an exorbitant salary and drained the corporate funds." Sanders initiated this suit on the estates' behalf in the same year that many of these activities occurred. "Therefore, the estates' claims of * * * oppression were not time-barred to the extent those claims sought to recover damages for injuries occurring to the estates."

FOR CRITICAL ANALYSIS—Economic Consideration *What should be the basis for determining the specific amount of damages to be awarded the minority shareholders in this case?*

Major Business Forms Compared

As mentioned in Chapter 24, when deciding which form of business organization would be most appropriate, businesspersons normally take into account several factors, including ease of creation, the liability of the owners, tax considerations, and the need for capital. Each major form of business organization offers distinct advantages and disadvantages with respect to these and other factors. Exhibit 26–2 summarizes the major advantages and disadvantages of each form of business organization discussed in Chapters 24 and 25, as well as in this chapter.

EXHIBIT 26–2 MAJOR FORMS OF BUSINESS COMPARED

CHARACTERISTIC	SOLE PROPRIETORSHIP	PARTNERSHIP	CORPORATION
Method of Creation	Created at will by owner.	Created by agreement of the parties.	Charter issued by state—created by statutory authorization.
Legal Position	Not a separate entity; owner is the business.	Is a separate legal entity in most states.	Always a legal entity separate and distinct from its owners—a legal fiction for the purposes of owning property and being a party to litigation.
Liability	Unlimited liability.	Unlimited liability.	Limited liability of shareholders—shareholders are not liable for the debts of the corporation.
Duration	Determined by owner; automatically dissolved on owner's death.	Terminated by agreement of the partners, but can continue to do business even when a partner dissociates from the partnership.	Can have perpetual existence.
Transferability of Interest	Interest can be transferred, but individual's proprietorship then ends.	Although partnership interest can be assigned, assignee does not have full rights of a partner.	Shares of stock can be transferred.
Management	Completely at owner's discretion.	Each general partner has a direct and equal voice in management unless expressly agreed otherwise in the partnership agreement.	Shareholders elect directors, who set policy and appoint officers.
Taxation	Owner pays personal taxes on business income.	Each partner pays pro rata share of income taxes on net profits, whether or not they are distributed.	Double taxation—corporation pays income tax on net profits, with no deduction for dividends, and shareholders pay income tax on disbursed dividends they receive.
Organizational Fees, Annual License Fees, and Annual Reports	None.	None.	All required.
Transaction of Business in Other States	Generally no limitation.	Generally no limitation.[a]	Normally must qualify to do business and obtain certificate of authority.

a. A few states have enacted statutes requiring that foreign partnerships qualify to do business there.

(Continued)

EXHIBIT 26–2 MAJOR FORMS OF BUSINESS COMPARED—CONTINUED

CHARACTERISTIC	LIMITED PARTNERSHIP	LIMITED LIABILITY COMPANY	LIMITED LIABILITY PARTNERSHIP
Method of Creation	Created by agreement to carry on a business for a profit. At least one party must be a general partner and the other(s) limited partner(s). Certificate of limited partnership is filed. Charter must be issued by the state.	Created by an agreement of the member-owners of the company. Articles of organization are filed. Charter must be issued by the state.	Created by agreement of the partners. A statement of qualification for the limited liability partnership is filed.
Legal Position	Treated as a legal entity.	Treated as a legal entity.	Generally, treated same as a general partnership.
Liability	Unlimited liability of all general partners; limited partners are liable only to the extent of capital contributions.	Member-owners' liability is limited to the amount of capital contributions or investments.	Varies, but under the UPA, liability of a partner for acts committed by other partners is limited.
Duration	By agreement in certificate, or by termination of the last general partner (retirement, death, and the like) or last limited partner.	Unless a single-member LLC, can have perpetual existence (same as a corporation).	Remains in existence until cancellation or revocation.
Transferability of Interest	Interest can be assigned (same as general partnership), but if assignee becomes a member with consent of other partners, certificate must be amended.	Member interests are freely transferable.	Interest can be assigned same as in a general partnership.
Management	General partners have equal voice or by agreement. Limited partners may not retain limited liability if they actively participate in management.	Member-owners can fully participate in management, or management is selected by member-owners to manage on behalf of the members.	Same as a general partnership.
Taxation	Generally taxed as a partnership.	LLC is not taxed, and members are taxed personally on profits "passed through" the LLC.	Same as a general partnership.
Organizational Fees, Annual License Fees, and Annual Reports	Organizational fee required; usually not others.	Organizational fee required; others vary with states.	Fees are set by each state for filing statements of qualification, foreign qualification, and annual reports.
Transaction of Business in Other States	Generally, no limitations.	Generally, no limitation but may vary depending on state.	Must file a statement of foreign qualification before doing business in another state.

TERMS AND CONCEPTS

business judgment rule 547
no-par share 553
par-value share 553
preemptive right 550
proxy 548
quorum 543
receiver 552
right of first refusal 552
shareholder's derivative suit 552
stock certificate 550
stock warrant 550
voting trust 550
watered stock 553

CHAPTER SUMMARY Corporate Directors, Officers, and Shareholders

Role of Directors (See pages 541–544.)	1. *Directors' qualifications*—Few qualifications are required; a director can be a shareholder but is not required to be. 2. *Election of directors*—The first board of directors is usually appointed by the incorporators; thereafter, directors are elected by the shareholders. Directors usually serve a one-year term, although the term can be longer and staggered terms are permitted under most state statutes. 3. *Board of directors' meetings*—The board of directors conducts business by holding formal meetings with recorded minutes. The date of regular meetings is usually established in the corporate articles or bylaws; special meetings can be called, with notice sent to all directors. Quorum requirements vary from state to state; usually, a quorum is a majority of the corporate directors. Voting must usually be done in person, and in ordinary matters only a majority vote is required. 4. *Rights of directors*—Directors' rights include the rights of participation, inspection, compensation, and indemnification. Compensation is usually specified in the corporate articles or bylaws. 5. *Directors' management responsibilities*—Directors are responsible for declaring and paying corporate dividends to shareholders; authorizing major corporate decisions; appointing, supervising, and removing corporate officers and other managerial employees; determining employees' compensation; making financial decisions necessary to the management of corporate affairs; and issuing authorized shares and bonds. Directors may delegate some of their responsibilities to executive committees and corporate officers and executives.
Role of Corporate Officers and Executives (See pages 544–545.)	Corporate officers and other executive employees are normally hired by the board of directors. As employees, the rights of corporate officers and executives are defined by employment contracts. The duties of corporate officers are the same as those of directors.
Duties and Liabilities of Directors and Officers (See pages 545–547.)	1. *Duty of care*—Directors and officers are obligated to act in good faith, to use prudent business judgment in the conduct of corporate affairs, and to act in the corporation's best interests. If a director fails to exercise this duty of care, he or she can be answerable to the corporation and to the shareholders for breaching the duty. 2. *Duty of loyalty*—Directors and officers have a fiduciary duty to subordinate their own interests to those of the corporation in matters relating to the corporation. 3. *Conflicts of interest*—To fulfill their duty of loyalty, directors and officers must make a full disclosure of any potential conflicts between their personal interests and those of the corporation. 4. *Liability of directors and officers*—Corporate directors and officers are personally liable for their own torts and crimes; additionally, they may be held personally liable for the torts and crimes committed by corporate personnel under their direct supervision (see Chapters 6 and 22). 5. *The business judgment rule*—This rule immunizes directors and officers from liability when they acted in good faith, in the best interests of the corporation, and exercised due care. For the rule to apply, the directors and officers must have made an informed, reasonable, and loyal decision.

CHAPTER SUMMARY Corporate Directors, Officers, and Shareholders—Continued

Role of Shareholders (See pages 547–550.)	1. *Shareholders' powers*—Shareholders' powers include the approval of all fundamental changes affecting the corporation and the election of the board of directors. 2. *Shareholders' meetings*—Shareholders' meetings must occur at least annually; special meetings can be called when necessary. Notice of the date, time, and place of the meeting (and its purpose, if it is specially called) must be sent to shareholders. Shareholders may vote by proxy (authorizing someone else to vote their shares) and may submit proposals to be included in the company's proxy materials sent to shareholders before meetings. 3. *Shareholder voting*—Shareholder voting requirements and procedures are as follows: a. A minimum number of shareholders (a quorum—generally, more than 50 percent of shares held) must be present at a meeting for business to be conducted; resolutions are passed (usually) by simple majority vote. b. The corporation must prepare voting lists of shareholders of record prior to each shareholders' meeting. c. Cumulative voting may or may not be required or permitted. Cumulative voting gives minority shareholders a better chance to be represented on the board of directors. d. A shareholder voting agreement (an agreement of shareholders to vote their shares together) is usually held to be valid and enforceable. e. A shareholder may appoint a proxy (substitute) to vote her or his shares. f. A shareholder may enter into a voting trust agreement by which title (record ownership) of his or her shares is given to a trustee, and the trustee votes the shares in accordance with the trust agreement.
Rights of Shareholders (See pages 550–552.)	Shareholders have numerous rights, which may include the following: 1. The right to a stock certificate, preemptive rights, and the right to stock warrants (depending on the corporate charter). 2. The right to obtain a dividend (at the discretion of the directors). 3. Voting rights. 4. The right to inspect the corporate records. 5. The right to transfer shares (this right may be restricted in close corporations). 6. The right to a share of corporate assets when the corporation is dissolved. 7. The right to sue on behalf of the corporation (bring a shareholder's derivative suit) when the directors fail to do so.
Liability of Shareholders (See pages 552–553.)	Shareholders may be liable for the retention of illegal dividends, for breach of a stock-subscription agreement, and for the value of watered stock.
Duties of Majority Shareholders (See pages 553–554.)	In certain situations, majority shareholders may be regarded as having a fiduciary duty to minority shareholders and will be liable if that duty is breached.

FOR REVIEW

Answers for the even-numbered questions in this For Review *section can be found in Appendix A at the end of this text.*

1 What are the duties of corporate directors and officers?

2 Directors are expected to use their best judgment in managing the corporation. What must directors do to avoid liability for honest mistakes of judgment and poor business decisions?

3 What is a voting proxy? What is cumulative voting?

4 If a group of shareholders perceives that the corporation has suffered a wrong and the directors refuse to take action, can the shareholders compel the directors to act? If so, how?

5 From what sources may dividends be paid legally? In what circumstances is a dividend illegal? What happens if a dividend is illegally paid?

QUESTIONS AND CASE PROBLEMS

26–1. Voting Techniques. Algonquin Corp. has issued and has outstanding 100,000 shares of common stock. Four stockholders own 60,000 of these shares, and for the past six years they have nominated a slate of people for membership on the board, all of whom have been elected. Sergio and twenty other shareholders, owning 20,000 shares, are dissatisfied with corporate management and want a representative on the board who shares their views. Explain under what circumstances Sergio and the minority shareholders can elect their representative to the board.

QUESTION WITH SAMPLE ANSWER

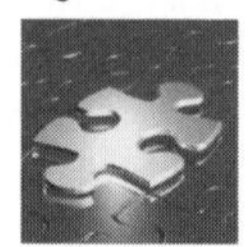

26–2. Starboard, Inc., has a board of directors consisting of three members (Ellsworth, Green, and Morino) and approximately five hundred shareholders. At a regular meeting of the board, the board selects Tyson as president of the corporation by a two-to-one vote, with Ellsworth dissenting. The minutes of the meeting do not register Ellsworth's dissenting vote. Later, during an audit, it is discovered that Tyson is a former convict and has openly embezzled $500,000 from Starboard. This loss is not covered by insurance. The corporation wants to hold directors Ellsworth, Green, and Morino liable. Ellsworth claims no liability. Discuss the personal liability of the directors to the corporation.

For a sample answer to this question, go to Appendix B at the end of this text.

26–3. Liability of Shareholders. Mallard has made a preincorporation subscription agreement to purchase 500 shares of a newly formed corporation. The shares have a par value of $100 per share. The corporation is formed, and it accepts Mallard's subscription. Mallard transfers a piece of land he owns to the corporation as payment for 250 of the shares, and the corporation issues 250 shares for it. Mallard pays for the other 250 shares with cash. One year later, with the corporation in serious financial difficulty, the board declares and pays a $5-per-share dividend. It is now learned that the land transferred by Mallard had a market value of $18,000. Discuss any liability that shareholder Mallard has to the corporation or to the creditors of the corporation.

26–4. Duty of Loyalty. Mackinac Cellular Corp. offered to sell Robert Broz a license to operate a cellular phone system in Michigan. Broz was a director of Cellular Information Systems, Inc. (CIS). CIS, as a result of bankruptcy proceedings, was in the process of selling its cellular holdings. Broz did not formally present the opportunity to the CIS board, but he told some of the firm's officers and directors, who replied that CIS was not interested. At the time, PriCellular, Inc., a firm that was interested in the Michigan license, was attempting to buy CIS. Without telling PriCellular, Broz bought the license himself. After PriCellular took over CIS, the company filed a suit in a Delaware state court against Broz, alleging that he had usurped a corporate opportunity. For what reasons might a court decide that Broz had done nothing wrong? Discuss. [*Broz v. Cellular Information Systems, Inc.*, 673 A.2d 148 (Del. 1996)]

26–5. Business Judgment Rule. Charles Pace and Maria Fuentez were shareholders of Houston Industries, Inc. (HII), and employees of Houston Lighting & Power, a subsidiary of HII, when they lost their jobs because of a company-wide reduction in its workforce. Pace, as a shareholder, three times wrote to HII, demanding that the board of directors terminate certain HII directors and officers and file a suit to recover damages for breach of fiduciary duty. Three times, the directors referred the charges to board committees and an outside law firm, which found that the facts did not support the charges. The board also received input from federal regulatory authorities about the facts behind some of the charges. The board notified Pace that it was refusing his demands. In response, Pace and Fuentez filed a shareholder's derivative suit in a Texas state court against Don Jordan and the other HII directors, contending that the board's investigation was inadequate. The defendants filed a motion for summary judgment, arguing that the suit was barred by the business judgment rule. Are the defendants right? How should the court rule? Why? [*Pace v. Jordan*, 999 S.W.2d 615 (Tex.App.—Houston [1 Dist.] 1999)]

CASE PROBLEM WITH SAMPLE ANSWER

26–6. Atlas Food Systems & Services, Inc., based in South Carolina, was a food vending service that provided refreshments to factories and other businesses. Atlas was a closely held corporation. John Kiriakides was a minority shareholder of Atlas. Alex Kiriakides was the majority shareholder. Throughout most of Atlas's history, Alex was the chairman of the board, which included John as a director. In 1995, while John was the president of the firm, the board and shareholders decided to convert Atlas to an S corporation. A few months later, however, Alex, without calling a vote, decided that the firm would not convert. In 1996, a dispute arose over Atlas's contract to buy certain property. John and others decided not to buy it. Without consulting anyone, Alex elected to go through with the sale. Within a few days, Alex refused to allow John to stay on as president. Two months later, Atlas offered to buy John's interest in the firm for almost $2 million. John refused, believing the offer was too low. John filed a suit in a South Carolina state court against Atlas and Alex, seeking, among other things, to force a buyout of John's shares. On what basis might the court grant John's request? Discuss. [*Kiriakides v. Atlas Food Systems & Services, Inc.*, 541 S.E.2d 257 (S.C. 2001)]

After you have answered this problem, compare your answer with the sample answer given on the Web site that accompanies this text. Go to **academic.cengage.com/blaw/wbl**, select "Chapter 26," and click on "Case Problem with Sample Answer."

26–7. Inspection Rights. Craig Johnson founded Distributed Solutions, Inc. (DSI), in 1991 to make software and provide consulting services, including payroll services, for small companies. Johnson was the sole officer and director and the majority shareholder. Jeffrey Hagen was a minority shareholder. In 1993, Johnson sold DSI's payroll services to himself and a few others and set up Distributed Payroll Solutions, Inc. (DPSI). In 1996, DSI had revenues of $739,034 and assets of $541,168. DSI's revenues in 1997 were $934,532. Within a year, however, all of DSI's assets were sold, and Johnson told Hagen that he was dissolving the firm because, in part, it conducted no business and had no prospects for future business. Hagen asked for corporate records to determine the value of DSI's stock, DSI's financial condition, and "whether unauthorized and oppressive acts had occurred in connection with the operation of the corporation which impacted the value of" the stock. When there was no response, Hagen filed a suit in an Illinois state court against DSI and Johnson, seeking an order to compel the inspection. The defendants filed a motion to dismiss, arguing that Hagen had failed to plead a proper purpose. Should the court grant Hagen's request? Discuss. [*Hagen v. Distributed Solutions, Inc.*, 328 Ill.App.3d 132, 764 N.E.2d 1141, 262 Ill.Dec. 24 (1 Dist. 2002)]

26–8. Duty of Loyalty. Digital Commerce, Ltd., designed software to enable its clients to sell their products or services over the Internet. Kevin Sullivan served as a Digital vice president until 2000, when he became president. Sullivan was dissatisfied that his compensation did not include stock in Digital, but he was unable to negotiate a deal that included equity (referring to shares of ownership in the company). In May, Sullivan solicited ASR Corp.'s business for Digital while he investigated employment opportunities with ASR for himself. When ASR would not include an "equity component" in a job offer, Sullivan refused to negotiate further on Digital's behalf. A few months later, Sullivan began to form his own firm to compete with Digital, conducting organizational and marketing activities on Digital's time, including soliciting ASR's business. In August, Sullivan resigned after first having all e-mail pertaining to the new firm deleted from Digital's computers. ASR signed a contract with Sullivan's new firm and paid it $400,000 for work through October 2001. Digital filed a suit in a federal district court against Sullivan, claiming that he had usurped a corporate opportunity. Did Sullivan breach his fiduciary duty to Digital? Explain. [*In re Sullivan*, 305 Bankr. 809 (W.D.Mich. 2004)]

Corporate Directors, Officers, and Shareholders

For updated links to resources available on the Web, as well as a variety of other materials, visit this text's Web site at

academic.cengage.com/blaw/wbl

One of the best sources on the Web for information on corporations, including their directors, is the EDGAR database of the Securities and Exchange Commission (SEC) at

http://www.sec.gov/edgar.shtml

You can find definitions for terms used in corporate law, as well as court decisions and articles on corporate law topics, at

http://www.law.com

For information on the SEC's rulings, including rulings on proxy materials, go to

http://www.sec.gov/rules/final.shtml

ONLINE LEGAL RESEARCH

Go to the *Fundamentals of Business Law* home page at **academic.cengage.com/blaw/wbl**, select "Chapter 26," and click on "Internet Exercises." There you will find the following Internet research exercises that you can perform to learn more about topics covered in this chapter.

Activity 26–1: LEGAL PERSPECTIVE—Liability of Directors and Officers

Activity 26–2: MANAGEMENT PERSPECTIVE—D&O Insurance

BEFORE THE TEST

Go to the *Fundamentals of Business Law* home page at **academic.cengage.com/blaw/wbl**, select "Chapter 26," and click on "Interactive Quizzes." You will find a number of interactive questions relating to this chapter.

CHAPTER 27

Investor Protection and Corporate Governance

LEARNING OBJECTIVES

After reading this chapter, you should be able to answer the following questions:

1. What is meant by the term *securities?*
2. What are the two major statutes regulating the securities industry? When was the Securities and Exchange Commission created, and what are its major purposes and functions?
3. What is insider trading? Why is it prohibited?
4. What are some of the features of state securities laws?
5. How are securities laws being applied in the online environment?

After the stock market crash of 1929, many members of Congress argued in favor of regulating securities markets. Basically, legislation for such regulation was enacted to provide investors with more information to help them make buying and selling decisions about **securities**—generally defined as any documents evidencing corporate ownership (stock) or debts (bonds)—and to prohibit deceptive, unfair, and manipulative practices. Today, the sale and transfer of securities are heavily regulated by federal and state statutes and by government agencies.

This chapter discusses the nature of federal securities regulation and its effect on the business world. We begin by looking at the federal administrative agency that regulates securities transactions. Next, we examine the major traditional laws governing securities offerings and trading. We then discuss the Sarbanes-Oxley Act of 2002,[1] which significantly affects certain types of securities transactions. In the concluding pages of this chapter, we look at how securities laws are being adapted to the online environment.

The Securities and Exchange Commission

In 1931, the Senate passed a resolution calling for an extensive investigation of securities trading. The investigation led, ultimately, to the passage by Congress of the Securities Act of 1933, which is also known as the *truth-in-securities* bill. In the following year, Congress passed

1. 15 U.S.C. Sections 7201 *et seq.*

the Securities Exchange Act. This 1934 act created the Securities and Exchange Commission (SEC).

MAJOR RESPONSIBILITIES OF THE SEC

The SEC was created as an independent regulatory agency with the function of administering the 1933 and 1934 acts. Its major responsibilities in this respect are as follows:

1. Requiring disclosure of facts concerning offerings of securities listed on national securities exchanges and of certain securities traded over the counter (OTC).
2. Regulating the trade in securities on national and regional securities exchanges and in the OTC markets.
3. Investigating securities fraud.
4. Regulating the activities of securities brokers, dealers, and investment advisers and requiring their registration.
5. Supervising the activities of mutual funds.
6. Recommending administrative sanctions, injunctive remedies, and criminal prosecution against those who violate securities laws. (The SEC can bring enforcement actions for civil violations of federal securities laws. The Fraud Section of the Criminal Division of the Department of Justice prosecutes criminal violations.)

THE SEC'S EXPANDING REGULATORY POWERS

Since its creation, the SEC's regulatory functions have gradually been increased by legislation granting it authority in different areas. For example, to further curb securities fraud, the Securities Enforcement Remedies and Penny Stock Reform Act of 1990 amended existing securities laws to allow SEC administrative law judges to hear many more types of securities violation cases; the SEC's enforcement options were also greatly expanded. Additionally, the act provides that courts can prevent persons who have engaged in securities fraud from serving as officers and directors of publicly held corporations. The Securities Acts Amendments of 1990 authorized the SEC to seek sanctions against those who violate foreign securities laws.

The National Securities Markets Improvement Act of 1996 expanded the power of the SEC to exempt persons, securities, and transactions from the requirements of the securities laws. (This part of the act is also known as the Capital Markets Efficiency Act.) The act also limited the authority of the states to regulate certain securities transactions and particular investment advisory firms. The Sarbanes-Oxley Act of 2002, which will be discussed later in this chapter, further expanded the authority of the SEC by directing the agency to issue new rules relating to corporate disclosure requirements and by creating an SEC oversight board.

Securities Act of 1933

The Securities Act of 1933[2] governs initial sales of stock by businesses. The act was designed to prohibit various forms of fraud and to stabilize the securities industry by requiring that all essential information concerning the issuance of securities be made available to the investing public. Basically, the purpose of this act is to require disclosure.

WHAT IS A SECURITY?

Section 2(1) of the Securities Act states that securities include the following:

> [A]ny note, stock, treasury stock, bond, debenture, evidence of indebtedness, certificate of interest or participation in any profit-sharing agreement, collateral-trust certificate, preorganization certificate or subscription, transferable share, investment contract, voting-trust certificate, certificate of deposit for a security, fractional undivided interest in oil, gas, or other mineral rights, or, in general, any interest or instrument commonly known as a "security," or any certificate of interest or participation in, temporary or interim certificate for, receipt for, guarantee of, or warrant or right to subscribe to or purchase, any of the foregoing.[3]

The courts have interpreted the act's definition of what constitutes a security[4] to include investment contracts. An investment contract is any transaction in which a person (1) invests (2) in a common enterprise (3) reasonably expecting profits (4) derived *primarily* or *substantially* from others' managerial or entrepreneurial efforts.[5]

For our purposes, it is probably convenient to think of securities in their most common forms—stocks and bonds issued by corporations. Bear in mind, though, that securities can take many forms and have been held to include whiskey, cosmetics, worms, beavers, boats, vacuum cleaners, muskrats, and cemetery lots, as well as investment contracts in condominiums, franchises, limited partnerships, oil or gas or other mineral rights, and farm animals accompanied by care agreements.

In the following case, the question was whether sales of pay phones and agreements to service the phones constituted sales of securities.

2. 15 U.S.C. Sections 77–77aa.
3. 15 U.S.C. Section 77b(1). Amendments in 1982 added stock options.
4. See 15 U.S.C. Section 77b(a)(1).
5. *SEC v. W. J. Howey Co.,* 328 U.S. 293, 66 S.Ct. 1100, 90 L.Ed. 1244 (1946).

CASE 27.1 SEC v. Alpha Telcom, Inc.

United States District Court,
District of Oregon, 2002.
187 F.Supp.2d 1250.

FACTS Paul Rubera started Alpha Telcom, Inc., in 1986 to sell, install, and maintain phones and business systems in Grants Pass, Oregon. In 1997, Alpha began to sell pay phones to buyers, most of whom also entered into service agreements with Alpha. Most of these buyers selected a "Level Four Service Agreement," which required Alpha to select a location for a phone, install it, obtain all licenses, maintain and clean the phone, pay the bills, and collect the revenue. Buyers were guaranteed—and were paid—a 14 percent return on the amount of their purchase. The pay-phone program was presented and promoted through American Telecommunications Company (ATC), Alpha's marketing subsidiary. From July 1998 through June 2001, Alpha's expenses for the program were $21,798,000, while revenues were $21,698,000. Despite the loss, Alpha paid investors approximately $17.9 million. To make these payments, Alpha borrowed money from ATC. Alpha filed for bankruptcy in August 2001. The Securities and Exchange Commission (SEC) filed a suit in a federal district court against Alpha and Rubera, alleging violations of the Securities Act of 1933. The defendants argued that the pay-phone program did not involve sales of securities.

ISSUE Was the pay-phone program a security?

DECISION Yes. The court concluded that the pay-phone program was a security because it involved an investment contract, which is a contract that is "(1) an investment of money; (2) in a common enterprise; (3) with the expectation of profits to be derived from the efforts of others." The court issued an injunction to prohibit further violations of the Securities Act of 1933 and ordered Rubera to disgorge profits of more than $3.7 million, plus interest.

REASON As for the individual elements of the definition of "an investment contract," the court reasoned that investors make cash investments with the expectation of receiving profits. Here, investors relied on the expertise of Alpha, as a common enterprise, to do all of the work, including making all of the business decisions with regard to the phones. Also, "ATC's only source of revenue was money from new investors. As a result, new investor money was being used to pay returns to existing investors," and "[i]nvestors would receive their 14 percent return * * * regardless of whether their particular phone actually generated that much money." Finally, "Alpha was ultimately responsible for those essential managerial efforts [that] affect the failure or success of the enterprise, and the investors retained no control over the business."

WHY IS THIS CASE IMPORTANT? *This case demonstrates that securities are not limited solely to simple stocks and bonds but can encompass a wide variety of materials. The analysis hinges on the nature of the transaction rather than the substances involved.*

REGISTRATION STATEMENT

Section 5 of the Securities Act of 1933 broadly provides that unless a security qualifies for an exemption, that security must be *registered* before it is offered to the public either through the mails or through any facility of interstate commerce, including securities exchanges. Issuing corporations must file a *registration statement* with the SEC. Investors must be provided with a prospectus that describes the security being sold, the issuing corporation, and the investment or risk attaching to the security. In principle, the registration statement and the prospectus supply sufficient information to enable unsophisticated investors to evaluate the financial risk involved.

Contents of the Registration Statement The registration statement must be written in plain English and include the following:

1. A description of the significant provisions of the security offered for sale, including the relationship between that security and the other securities of the registrant. Also, the corporation must disclose how it intends to use the proceeds of the sale.
2. A description of the corporation's properties and business.
3. A description of the management of the corporation and its security holdings; remuneration; and other benefits, including pensions and stock options. Any interests of directors or officers in any material transactions with the corporation must be disclosed.
4. A financial statement certified by an independent public accounting firm.
5. A description of pending lawsuits.

Other Requirements Before filing the registration statement and the prospectus with the SEC, the corporation is allowed to obtain an *underwriter*—a company that agrees to purchase the new issue of securities for resale to the public. There is a twenty-day waiting period (which can be accelerated by the SEC) after registration before the

securities can be sold. During this period, oral offers between interested investors and the issuing corporation concerning the purchase and sale of the proposed securities may take place, and very limited written advertising is allowed. At this time, the so-called **red herring** prospectus may be distributed. The name comes from the red legend printed across the prospectus stating that the registration has been filed but has not become effective.

After the waiting period, the registered securities can be legally bought and sold. Written advertising is allowed in the form of a **tombstone ad,** so named because historically the format resembled a tombstone. Such ads simply tell the investor where and how to obtain a prospectus. Normally, any other type of advertising is prohibited until the registration becomes effective.

■ EXAMPLE 27.1 Delphia, Inc., wants to make a public offering of its common stock. The firm files a registration statement and a prospectus with the SEC on May 1. On the same day, the company can make its first offers to sell the stock, but an actual sale can take place only *after* the SEC declares the registration to be effective. On May 10, Delphia places tombstone ads in certain trade publications to announce the offering and to inform investors where and how to get a prospectus. The firm issues a preliminary prospectus on May 15. The registration statement becomes effective—and the company makes a final prospectus available to potential buyers—on May 20. On this date, Delphia can sell the first shares in the issue. ■

Exempt Securities

A number of specific securities are exempt from the registration requirements of the Securities Act of 1933. These securities—which can also generally be resold without being registered—include the following:[6]

1. All bank securities sold prior to July 27, 1933.
2. Commercial paper (such as negotiable instruments), if the maturity date does not exceed nine months.
3. Securities of charitable organizations.
4. Securities resulting from a corporate reorganization that are issued for exchange with the issuer's existing security holders and certificates that are issued by trustees, receivers, or debtors in possession under the bankruptcy laws (bankruptcy laws were discussed in Chapter 21).
5. Securities issued exclusively for exchange with the issuer's existing security holders, provided no commission is paid (for example, stock dividends and stock splits).
6. Securities issued to finance the acquisition of railroad equipment.
7. Any insurance, endowment, or annuity contract issued by a state-regulated insurance company.
8. Government-issued securities.
9. Securities issued by banks, savings and loan associations, farmers' cooperatives, and similar institutions subject to supervision by governmental authorities.
10. In consideration of the "small amount involved,"[7] an issuer's offer of up to $5 million in securities in any twelve-month period.

For the last exemption, under Regulation A,[8] the issuer must file with the SEC a notice of the issue and an offering circular, which must also be provided to investors before the sale. This is a much simpler and less expensive process than the procedures associated with full registration. Companies are allowed to "test the waters" for potential interest before preparing the offering circular. To *test the waters* means to determine potential interest without actually selling any securities or requiring any commitment on the part of those who express interest. Small-business issuers (companies with annual revenues of less than $25 million) can also use an integrated registration and reporting system that uses simpler forms than the full registration system.

Exhibit 27–1 summarizes the securities and transactions (discussed next) that are exempt from the registration requirements under the Securities Act of 1933 and SEC regulations.

Exempt Transactions

An issuer of securities that are not exempt under one of the ten categories listed in the previous subsection can avoid the high cost and complicated procedures associated with registration by taking advantage of certain transaction exemptions. These exemptions are very broad, and thus many sales occur without registration. Because the coverage of the exemptions overlaps somewhat, an offering may qualify for more than one.

Small Offerings—Regulation D The SEC's Regulation D contains four separate exemptions from registration requirements for limited offers (offers that either involve a small amount or are made in a limited manner). Regulation D provides that any of these offerings made during any twelve-month period are exempt from the registration requirements.

6. 15 U.S.C. Section 77c.

7. 15 U.S.C. Section 77c(b).

8. 17 C.F.R. Sections 230.251–230.263.

EXHIBIT 27–1 EXEMPTIONS UNDER THE 1933 SECURITIES ACT

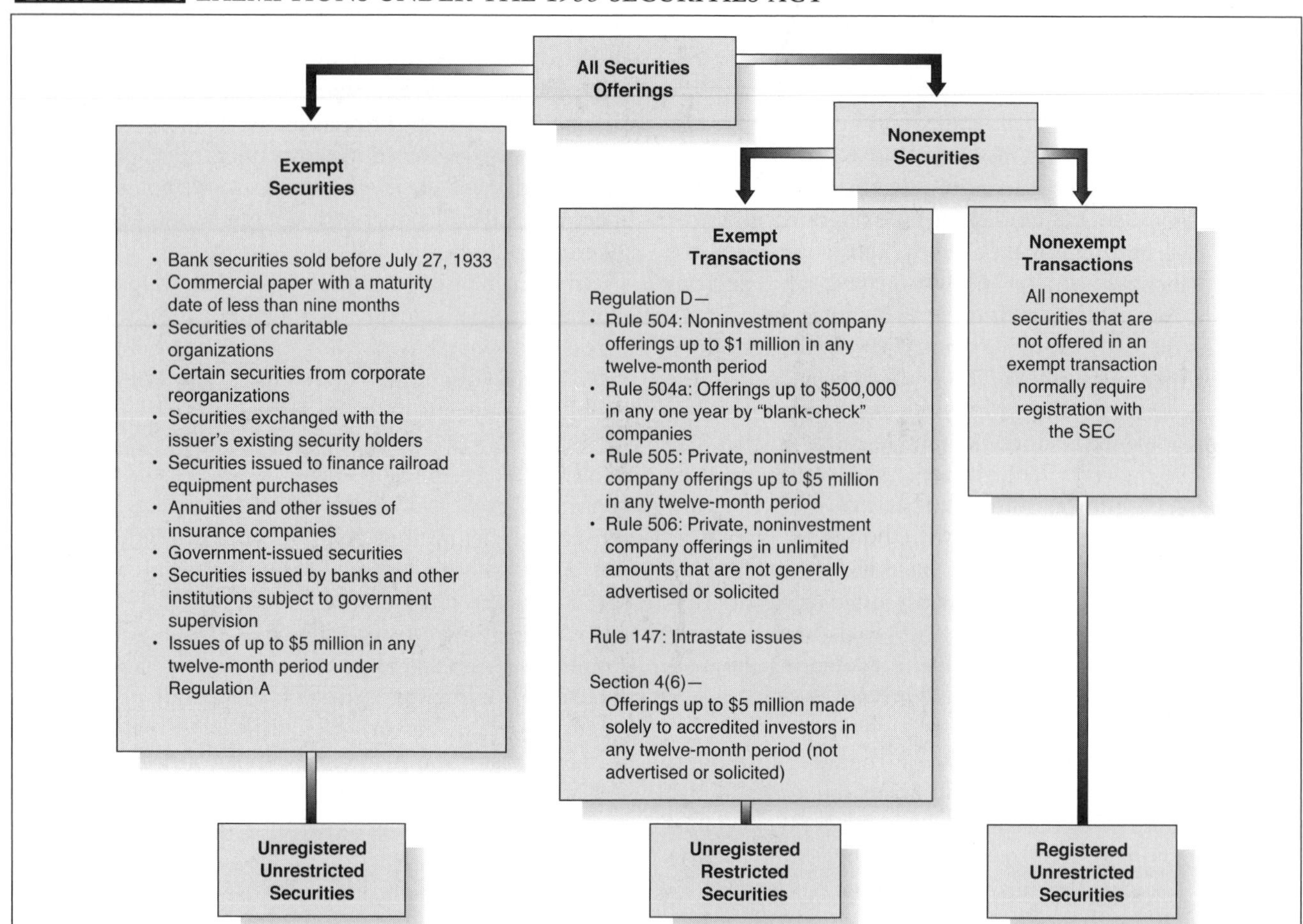

—*Rule 504.* Noninvestment company offerings up to $1 million in any twelve-month period are exempt.[9] In contrast to investment companies (discussed later in this chapter), noninvestment companies are firms that are not engaged primarily in the business of investing or trading in securities.

■ EXAMPLE 27.2 Beta, L.P., is a limited partnership that is not an investment company. Beta intends to offer $600,000 of its limited partnership interests for sale between June 1 and next May 31. The buyers will become limited partners in Beta. Because an interest in a limited partnership meets the definition of a security (discussed earlier in this chapter), its sale is subject to the registration and prospectus requirements of the Securities Act of 1933. Under Rule 504, however, the sales of Beta's interests are exempt from these requirements because Beta is a noninvestment company making an offering of less than $1 million in a twelve-month period. Therefore, Beta can sell its interests without filing a registration statement with the SEC or issuing a prospectus to any investor. ■

—*Rule 504a.* Offerings up to $500,000 in any one year by so-called *blank-check companies*—companies with no specific business plans except to locate and acquire currently unknown businesses or opportunities—are exempt if no general solicitation or advertising is used; the SEC is notified of the sales; and precaution is taken against nonexempt, unregistered resales.[10] The limits on

9. 17 C.F.R. Section 230.504. Rule 504 is the exemption used by most small businesses, but that could change under new SEC Rule 1001. This rule permits, under certain circumstances, "testing the waters" for offerings of up to $5 million *per transaction.* These offerings can be made only to "qualified purchasers" (knowledgeable, sophisticated investors), though.

10. Precautions to be taken against nonexempt, unregistered resales include asking the investor whether he or she is buying the securities for others; before the sale, disclosing to each purchaser in writing that the securities are unregistered and thus cannot be resold, except in an exempt transaction, without first being registered; and indicating on the certificates that the securities are unregistered and restricted.

advertising and unregistered resales do not apply if the offering is made solely in states that provide for registration and disclosure and the securities are sold in compliance with those provisions.[11]

—Rule 505. Private, noninvestment company offerings up to $5 million in any twelve-month period are exempt, regardless of the number of **accredited investors** (banks, insurance companies, investment companies, the issuer's executive officers and directors, and persons whose income or net worth exceeds a certain threshold), so long as there are no more than thirty-five unaccredited investors; no general solicitation or advertising is used; the SEC is notified of the sales; and precaution is taken against nonexempt, unregistered resales. If the sale involves *any* unaccredited investors, *all* investors must be given material information about the offering company, its business, and the securities before the sale. Unlike Rule 506 (discussed next), Rule 505 includes no requirement that the issuer believe each unaccredited investor "has such knowledge and experience in financial and business matters that he [or she] is capable of evaluating the merits and the risks of the prospective investment."[12]

—Rule 506. Private offerings in unlimited amounts that are not generally solicited or advertised are exempt if the SEC is notified of the sales; precaution is taken against nonexempt, unregistered resales; and the issuer believes that each unaccredited investor has sufficient knowledge or experience in financial matters to be capable of evaluating the investment's merits and risks. There may be no more than thirty-five unaccredited investors, but there are no limits on the number of accredited investors. If there are *any* unaccredited investors, the issuer must provide material information about itself, its business, and the securities before the sale to *all* purchasers.[13]

This exemption is perhaps most important to those firms that want to raise funds through the sale of securities without registering them. It is often referred to as the *private placement* exemption because it exempts "transactions not involving any public offering."[14] This provision applies to private offerings to a limited number of persons who are sufficiently sophisticated and able to assume the risk of the investment (and who thus have no need for federal registration protection). It also applies to private offerings to similarly situated institutional investors.

■ **EXAMPLE 27.3** Citco Corporation needs to raise capital to expand its operations. Citco decides to make a private $10 million offering of its common stock directly to two hundred accredited investors and a group of thirty highly sophisticated, but unaccredited, investors. Citco provides all these investors with a prospectus and material information about the firm, including its most recent financial statements. As long as Citco notifies the SEC of the sale, this offering will likely qualify as an exempt transaction under Rule 506. The offering is nonpublic and not generally advertised; there are less than thirty-five unaccredited investors, each of whom possesses sufficient knowledge and experience to evaluate the risks involved; and the issuer has provided all purchasers with the material information. Thus, Citco will not be required to comply with the registration requirements in the Securities Act of 1933. ■

Small Offerings—Section 4(6) Under Section 4(6) of the Securities Act of 1933, an offer made *solely* to accredited investors is exempt if its amount is not more than $5 million. Any number of accredited investors may participate, but no unaccredited investors may do so. No general solicitation or advertising may be used; the SEC must be notified of all sales; and precaution must be taken against nonexempt, unregistered resales. Precaution is necessary because these are *restricted* securities and may be resold only by registration or in an exempt transaction.[15] (The securities purchased and sold by most people who deal in stock are called, in contrast, *unrestricted* securities.)

Intrastate Issues—Rule 147 Also exempt are intrastate transactions involving purely local offerings.[16] This exemption applies to most offerings that are restricted to residents of the state in which the issuing company is organized and doing business. For nine months after the last sale, virtually no resales may be made to nonresidents, and precautions must be taken against this possibility. These offerings remain subject to applicable laws in the state of issue.

Resales Most securities can be resold without registration (although some resales may be subject to restrictions, as discussed above in connection with specific exemptions). The Securities Act of 1933 provides exemptions for resales by most persons other than issuers or underwriters. The average investor who sells shares of stock need not file a registration statement with the SEC. Resales of restricted securities acquired under Rule 504a, Rule 505, Rule 506, or Section 4(6), however, trigger the registration requirements unless the party selling them complies with Rule 144 or Rule 144A. These rules are sometimes referred to as "safe harbors."

11. 17 C.F.R. Section 230.504a.
12. 17 C.F.R. Section 230.505.
13. 17 C.F.R. Section 230.506.
14. 15 U.S.C. Section 77d(2).
15. 15 U.S.C. Section 77d(6).
16. 15 U.S.C. Section 77c(a)(11); 17 C.F.R. Section 230.147.

—***Rule 144.*** Rule 144 exempts restricted securities from registration on resale if there is adequate current public information about the issuer, the person selling the securities has owned them for at least one year, they are sold in certain limited amounts in unsolicited brokers' transactions, and the SEC is given notice of the resale.[17] "Adequate current public information" consists of the reports that certain companies are required to file under the Securities Exchange Act of 1934. A person who has owned the securities for at least two years is subject to none of these requirements, unless the person is an affiliate. An *affiliate* is one who controls, is controlled by, or is in common control with the issuer.

—***Rule 144A.*** Securities that at the time of issue are not of the same class as securities listed on a national securities exchange or quoted in a U.S. automated interdealer quotation system may be resold under Rule 144A.[18] They may be sold only to a qualified institutional buyer (an institution, such as an insurance company, an investment company, or a bank, that owns and invests at least $100 million in securities). The seller must take reasonable steps to ensure that the buyer knows that the seller is relying on the exemption under Rule 144A. A sample restricted stock certificate is shown in Exhibit 27–2.

17. 17 C.F.R. Section 230.144.
18. 17 C.F.R. Section 230.144A.

VIOLATIONS OF THE 1933 ACT

As mentioned, the SEC has the power to investigate and bring civil enforcement actions against companies that violate federal securities laws. It is a violation of the Securities Act of 1933 to intentionally defraud investors by misrepresenting or omitting facts in a registration statement or prospectus. Liability is also imposed on those who are negligent for not discovering the fraud. Selling securities before the effective date of the registration statement or under an exemption for which the securities do not qualify results in liability.

Criminal violations are prosecuted by the Department of Justice. Violators may be fined up to $10,000, imprisoned for up to five years, or both. The SEC is authorized to seek

EXHIBIT 27–2 A SAMPLE RESTRICTED STOCK CERTIFICATE

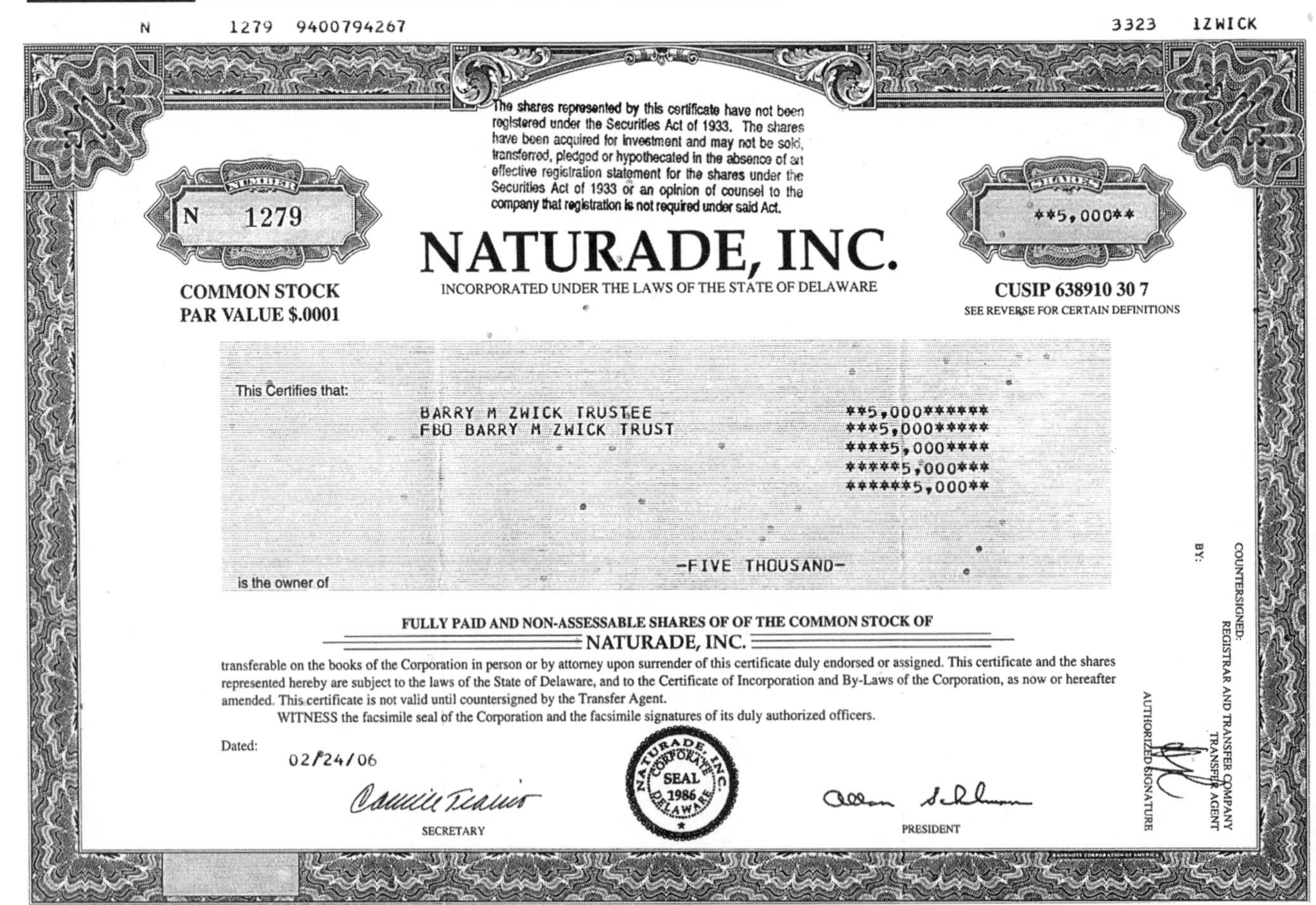

N 1279 9400794267 3323 1ZWICK

The shares represented by this certificate have not been registered under the Securities Act of 1933. The shares have been acquired for investment and may not be sold, transferred, pledged or hypothecated in the absence of an effective registration statement for the shares under the Securities Act of 1933 or an opinion of counsel to the company that registration is not required under said Act.

NUMBER N 1279

SHARES **5,000**

NATURADE, INC.

INCORPORATED UNDER THE LAWS OF THE STATE OF DELAWARE

COMMON STOCK
PAR VALUE $.0001

CUSIP 638910 30 7
SEE REVERSE FOR CERTAIN DEFINITIONS

This Certifies that:

BARRY M ZWICK TRUSTEE
FBO BARRY M ZWICK TRUST

5,000****
5,000**
****5,000****
*****5,000***
******5,000**

is the owner of -FIVE THOUSAND-

FULLY PAID AND NON-ASSESSABLE SHARES OF OF THE COMMON STOCK OF
NATURADE, INC.

transferable on the books of the Corporation in person or by attorney upon surrender of this certificate duly endorsed or assigned. This certificate and the shares represented hereby are subject to the laws of the State of Delaware, and to the Certificate of Incorporation and By-Laws of the Corporation, as now or hereafter amended. This certificate is not valid until countersigned by the Transfer Agent.

WITNESS the facsimile seal of the Corporation and the facsimile signatures of its duly authorized officers.

Dated: 02/24/06

NATURADE, INC. CORPORATE SEAL 1986 DELAWARE

SECRETARY PRESIDENT

COUNTERSIGNED:
REGISTRAR AND TRANSFER COMPANY
TRANSFER AGENT
BY:
AUTHORIZED SIGNATURE

civil sanctions against those who willfully violate the 1933 act. It can request an injunction to prevent further sales of the securities involved or ask the court to grant other relief, such as an order to a violator to refund profits. Those parties who purchase securities and suffer harm as a result of false or omitted statements may also bring suits in a federal court to recover their losses and other damages.

Securities Exchange Act of 1934

The Securities Exchange Act of 1934 provides for the regulation and registration of securities exchanges, brokers, dealers, and national securities associations such as the National Association of Securities Dealers (NASD). The SEC regulates the markets in which securities are traded by maintaining a continuous disclosure system for all corporations with securities on the securities exchanges and for those companies that have assets in excess of $10 million and five hundred or more shareholders. These corporations are referred to as Section 12 companies because they are required to register their securities under Section 12 of the 1934 act.

The act also authorizes the SEC to regulate proxy solicitations for voting (discussed in Chapter 26) and to engage in market surveillance to deter undesirable market practices such as fraud, market manipulation, and misrepresentation.

SECTION 10(b), SEC RULE 10b-5, AND INSIDER TRADING

Section 10(b) is one of the most important sections of the Securities Exchange Act of 1934. This section proscribes the use of any manipulative or deceptive device in violation of SEC rules and regulations. Among the rules that the SEC has promulgated pursuant to the 1934 act is **SEC Rule 10b-5,** which prohibits the commission of fraud in connection with the purchase or sale of any security.

One of the major goals of Section 10(b) and SEC Rule 10b-5 is to prevent so-called **insider trading.** Because of their positions, corporate directors and officers often obtain advance inside information that can affect the future market value of the corporate stock. Obviously, their positions give them a trading advantage over the general public and shareholders. The 1934 Securities Exchange Act defines inside information and extends liability to officers and directors for taking advantage of such information in their personal transactions when they know that it is unavailable to the persons with whom they are dealing.

Section 10(b) of the 1934 act and SEC Rule 10b-5 cover not only corporate officers, directors, and majority shareholders but also any persons having access to or receiving information of a nonpublic nature on which trading is based.

Disclosure under SEC Rule 10b-5 Any material omission or misrepresentation of material facts in connection with the purchase or sale of a security may violate not only the Securities Act of 1933 but also the antifraud provisions of Section 10(b) and SEC Rule 10b-5 of the 1934 act. The key to liability (which can be civil or criminal) under Section 10(b) and SEC Rule 10b-5 is whether the insider's information is *material.*

—Examples of Material Facts Calling for Disclosure. The following are some examples of material facts calling for disclosure under the rule:

1. Fraudulent trading in the company stock by a broker-dealer.
2. A dividend change (whether up or down).
3. A contract for the sale of corporate assets.
4. A new discovery, a new process, or a new product.
5. A significant change in the firm's financial condition.
6. Potential litigation against the company.

Note that any one of these facts, by itself, will not automatically be considered a material fact. Rather, it will be regarded as a material fact if it is significant enough that it will likely affect an investor's decision as to whether to purchase or sell certain securities. ■ EXAMPLE 27.4 Omega Company is the defendant in a class-action product liability suit. Omega's attorney, Paula Frasier, believes it likely that the jury will ultimately find the firm is liable, which will result in a considerable loss to the company. She advises Omega's directors, officers, and accountants of this possibility. Omega expects to make a $5 million offering of newly issued stock before the date of the projected end of the trial in the suit. Omega's potential liability in the litigation and the financial consequences to the firm are significant, material facts that should be disclosed in the financial statements and other information provided to the company's potential investors. ■

The following is one of the landmark cases interpreting SEC Rule 10b-5. The SEC sued Texas Gulf Sulphur Company for issuing a misleading press release. The release underestimated the magnitude and value of a mineral discovery. The SEC also sued several of Texas Gulf Sulphur's directors, officers, and employees under SEC Rule 10b-5 for purchasing large amounts of the corporate stock prior to the announcement of the corporation's rich ore discovery.

LANDMARK AND CLASSIC CASES

CASE 27.2 SEC v. Texas Gulf Sulphur Co.

United States Court of Appeals,
Second Circuit, 1968.
401 F.2d 833.

FACTS In 1963, Texas Gulf Sulphur Company (TGS) drilled a hole that appeared to yield a core with an exceedingly high mineral content. TGS kept secret the results of the core sample. Officers and employees of the company made substantial purchases of TGS's stock or accepted stock options after learning of the ore discovery, even though further drilling was necessary to determine whether there was enough ore to be mined commercially. On April 11, 1964, an unauthorized report of the mineral find appeared in the newspapers. On the following day, TGS issued a press release that played down the discovery. Later, TGS announced a strike of at least twenty-five million tons of ore, substantially driving up the price of TGS stock. The SEC brought suit in a federal district court against the officers and employees of TGS for violating the insider-trading prohibition of SEC Rule 10b-5. The officers and employees argued that the prohibition did not apply. They reasoned that the information on which they had traded was not material because the ore find had not been commercially proved. The court held that most of the defendants had not violated SEC Rule 10b-5, and the SEC appealed.

ISSUE Had the officers and employees of TGS violated SEC Rule 10b-5 by purchasing the stock, even though they did not know the full extent and profit potential of the mine at the time they purchased the stock?

DECISION Yes. The federal appellate court reversed the lower court's decision and remanded the case to the trial court, holding that the employees and officers had violated SEC Rule 10b-5's prohibition against insider trading.

REASON For SEC Rule 10b-5 purposes, the test of materiality is whether the information would affect the judgment of reasonable investors. Reasonable investors include speculative as well as conservative investors. "[A] major factor in determining whether the * * * discovery [of the ore] was a material fact is the importance attached to the drilling results by those who knew about it. * * * [T]he timing by those who knew of it of their stock purchases and their purchases of short-term calls [rights to buy shares at a specified price within a specified time period]—purchases in some cases by individuals who had never before purchased calls or even TGS stock—virtually compels the inference that the insiders were influenced by the drilling results. * * * We hold, therefore, that all transactions in TGS stock or calls by individuals apprised of the drilling results * * * were made in violation of Rule 10b-5."

IMPACT OF THIS CASE ON TODAY'S LAW This case affirmed the principle that the test of whether information is material, for SEC Rule 10b-5 purposes, is whether it would affect the judgment of reasonable investors. The corporate insiders' purchases of stock and stock options indicated that they were influenced by the drilling results and that the information about the drilling results was material. The courts continue to cite this case when applying SEC Rule 10b-5 to cases of alleged insider trading.

RELEVANT WEB SITES *To locate information on the Web concerning* SEC v. Texas Gulf Sulpher Co., *go to this text's Web site at* **academic.cengage.com/blaw/wbl**, *select "Chapter 27," and click on "URLs for Landmark Cases."*

—The Private Securities Litigation Reform Act of 1995. One of the unintended effects of SEC Rule 10b-5 was to deter the disclosure of some material information, such as financial forecasts. To understand why, consider an example. ■ EXAMPLE 27.5 Cross Creek Technologies announces that its projected earnings in the coming fiscal year will be $10 million. It turns out that the forecast is wrong. The earnings are in fact much lower, and the price of the company's stock is affected—negatively. The shareholders then bring a class-action suit against the company, alleging that the directors violated SEC Rule 10b-5 by disclosing misleading financial information. ■

In an attempt to rectify this problem and promote disclosure, Congress passed the Private Securities Litigation Reform Act of 1995. Among other things, the act provides a "safe harbor" for publicly held companies that make forward-looking statements, such as financial forecasts. Those who make such statements are protected against liability for securities fraud as long as the statements are accompanied by "meaningful cautionary statements identifying important factors that could cause actual results to differ materially from those in the forward-looking statement."[19]

19. 15 U.S.C. Sections 77z-2, 78u-5.

After the 1995 act was passed, a number of securities class-action suits were filed in state courts to skirt the requirements of the 1995 federal act. In response to this problem, Congress passed the Securities Litigation Uniform Standards Act of 1998.[20] The act placed stringent limits on the ability of plaintiffs to bring class-action suits in state courts against firms whose securities are traded on national stock exchanges.

Applicability of SEC Rule 10b-5 SEC Rule 10b-5 applies in virtually all cases concerning the trading of securities, whether on organized exchanges, in over-the-counter markets, or in private transactions. The rule covers, among other things, notes, bonds, agreements to form a corporation, and joint-venture agreements. Generally, it covers just about any form of security. It is immaterial whether a firm has securities registered under the 1933 act for the 1934 act to apply.

Although SEC Rule 10b-5 is applicable only when the requisites of federal jurisdiction—such as the use of the mails, of stock exchange facilities, or of any instrumentality of interstate commerce—are present, virtually no commercial transaction can be completed without such contact. In addition, the states have corporate securities laws, many of which include provisions similar to SEC Rule 10b-5.

Outsiders and SEC Rule 10b-5 The traditional insider-trading case involves true insiders—corporate officers, directors, and majority shareholders who have access to (and trade on) inside information. Increasingly, liability under Section 10(b) of the 1934 act and SEC Rule 10b-5 is being extended to include certain "outsiders"—those persons who trade on inside information acquired indirectly. Two theories have been developed under which outsiders may be held liable for insider trading: the *tipper/tippee theory* and the *misappropriation theory.*

—Tipper/Tippee Theory. Anyone who acquires inside information as a result of a corporate insider's breach of his or her fiduciary duty can be liable under SEC Rule 10b-5. This liability extends to **tippees** (those who receive "tips" from insiders) and even remote tippees (tippees of tippees).

The key to liability under this theory is that the inside information must be obtained as a result of someone's breach of a fiduciary duty to the corporation whose shares are involved in the trading. The tippee is liable under this theory only if there is a breach of a duty not to disclose inside information, the disclosure is in exchange for personal benefit, and the tippee knows (or should know) of this breach and benefits from it.[21]

—Misappropriation Theory. Liability for insider trading may also be established under the misappropriation theory. This theory holds that if an individual wrongfully obtains (misappropriates) inside information and trades on it for her or his personal gain, then the individual should be held liable because, in essence, the individual stole information rightfully belonging to another.

The misappropriation theory has been controversial because it significantly extends the reach of SEC Rule 10b-5 to outsiders who would not ordinarily be deemed fiduciaries of the corporations in whose stock they trade. The United States Supreme Court, however, has held that liability under SEC Rule 10b-5 can be based on the misappropriation theory.[22]

INSIDER REPORTING AND TRADING—SECTION 16(b)

Officers, directors, and certain large stockholders[23] of Section 12 corporations (corporations that are required to register their securities under Section 12 of the 1934 act) must file reports with the SEC concerning their ownership and trading of the corporations' securities.[24] To discourage such insiders from using nonpublic information about their companies for their personal benefit in the stock market, Section 16(b) of the 1934 act provides for the recapture by the corporation of all profits realized by an insider on any purchase and sale or sale and purchase of the corporation's stock within any six-month period.[25] It is irrelevant whether the insider actually uses inside information; *all such short-swing profits must be returned to the corporation.*

Section 16(b) applies not only to stock but also to warrants, options, and securities convertible into stock. The courts have also fashioned complex rules for determining profits. Note that the SEC exempts several transactions under Rule 16b-3.[26] For all of these reasons, corporate insiders are wise to seek specialized counsel before trading in the corporation's stock. Exhibit 27–3 compares the effects of SEC Rule 10b-5 and Section 16(b).

20. Pub. L. No. 105-353. This act amended many sections of Title 15 of the *United States Code.*

21. See, for example, *Chiarella v. United States*, 445 U.S. 222, 100 S.Ct. 1108, 63 L.Ed.2d 348 (1980); and *Dirks v. SEC*, 463 U.S. 646, 103 S.Ct. 3255, 77 L.Ed.2d 911 (1983).

22. *United States v. O'Hagan*, 521 U.S. 642, 117 S.Ct. 2199, 138 L.Ed.2d 724 (1997).

23. Those stockholders owning 10 percent of the class of equity securities registered under Section 12 of the 1934 act.

24. 15 U.S.C. Section 78*l*.

25. A person who expects the price of a particular stock to decline can realize profits by "selling short"—selling at a high price and repurchasing later at a lower price to cover the "short sale."

26. 17 C.F.R. Section 240.16b-3.

EXHIBIT 27–3 COMPARISON OF COVERAGE, APPLICATION, AND LIABILITIES UNDER SEC RULE 10b-5 AND SECTION 16(b)

AREA OF COMPARISON	SEC RULE 10b-5	SECTION 16(b)
What is the subject matter of the transaction?	Any security (does not have to be registered).	Any security (does not have to be registered).
What transactions are covered?	Purchase or sale.	Short-swing purchase and sale or short-swing sale and purchase.
Who is subject to liability?	Virtually anyone with inside information under a duty to disclose—including officers, directors, controlling stockholders, and tippees.	Officers, directors, and certain 10 percent stockholders.
Is omission or misrepresentation necessary for liability?	Yes.	No.
Are there any exempt transactions?	No.	Yes, there are a variety of exemptions.
Is direct dealing with the party necessary?	No.	No.
Who may bring an action?	A person transacting with an insider, the SEC, or a purchaser or seller damaged by a wrongful act.	A corporation or a shareholder by derivative action.

PROXY STATEMENTS

Section 14(a) of the Securities Exchange Act of 1934 regulates the solicitation of proxies from shareholders of Section 12 companies. The SEC regulates the content of proxy statements. As discussed in Chapter 26, a proxy statement is sent to shareholders when corporate officials are requesting authority to vote on behalf of the shareholders in a particular election on specified issues. Whoever solicits a proxy must fully and accurately disclose in the proxy statement all of the facts that are pertinent to the matter on which the shareholders are to vote. SEC Rule 14a-9 is similar to the antifraud provisions of SEC Rule 10b-5. Remedies for violations are extensive; they range from injunctions that prevent a vote from being taken to monetary damages.

VIOLATIONS OF THE 1934 ACT

Violations of Section 10(b) of the Securities Exchange Act of 1934 and SEC Rule 10b-5 include insider trading. This is a criminal offense, with criminal penalties. Violators of these laws may also be subject to civil liability. For any sanctions to be imposed, however, there must be *scienter*—the violator must have had an intent to defraud or knowledge of her or his misconduct (see Chapter 10). *Scienter* can be proved by showing that a defendant made false statements or wrongfully failed to disclose material facts.

Violations of Section 16(b) include the sale by insiders of stock acquired less than six months before the sale. These violations are subject to civil sanctions. Liability under Section 16(b) is strict liability. Thus, liability is imposed regardless of whether *scienter* or negligence existed.

Criminal Penalties For violations of Section 10(b) and Rule 10b-5, an individual may be fined up to $5 million, imprisoned for up to twenty years, or both.[27] A partnership or a corporation may be fined up to $25 million. Under Section 807 of the Sarbanes-Oxley Act of 2002, for a willful violation of the 1934 act the violator may, in addition to being subject to a fine, be imprisoned for up to twenty-five years.

To be found guilty of a crime under the securities laws, the jury must conclude beyond a reasonable doubt that the defendant knew he or she was acting wrongfully. The issue in the following case was whether, in light of this principle, there was enough evidence that Martha Stewart, founder of a well-known media and homemaking empire, intended to deceive investors to present the matter to a jury.

27. These numbers reflect the increased penalties imposed by the Sarbanes-Oxley Act of 2002, which will be discussed later in this chapter.

CASE 27.3 United States v. Stewart

United States District Court,
Southern District of New York, 2004.
305 F.Supp.2d 368.

FACTS Samuel Waksal, the chief executive officer of ImClone Systems, Inc., a biotechnology company, was a client of stockbroker Peter Bacanovic. Bacanovic's other clients included Martha Stewart, then the chief executive officer of Martha Stewart Living Omnimedia (MSLO). On December 27, 2001, Waksal began selling his ImClone stock. Bacanovic allegedly had Stewart informed of Waksal's sales, and she also sold her ImClone shares. The next day, ImClone announced that the Food and Drug Administration had rejected the company's application for approval of its leading product, a medication called Erbitux. The government began to investigate Stewart's ImClone trades, the media began to report on the investigation, and the value of MSLO stock began to drop. In June 2002, at a "Mid-Year Media Review" conference attended by investment professionals and investors, Stewart stated that she had previously agreed with Bacanovic to sell her ImClone stock if the price fell to $60 per share and said, "I have nothing to add on this matter today. And I'm here to talk about our terrific company." Her statements were followed by a forty-minute presentation on MSLO. Subsequently, Stewart was charged with, among other things, fraud in connection with the purchase and sale of MSLO securities in violation of the Securities Exchange Act of 1934. She filed a motion for a judgment of acquittal on this charge.

ISSUE Could a reasonable juror find beyond a reasonable doubt that Stewart lied to influence the market for the securities of her company?

DECISION No. The court concluded that in Stewart's case, "to find the essential element of criminal intent beyond a reasonable doubt, a rational juror would have to speculate."[a]

REASON The government argued that Stewart was aware that the negative publicity about her ImClone trade was having an impact on the market value of MSLO securities and that she deliberately directed her statement to investors. The government pointed out that Stewart was aware that she was speaking to analysts and investors at the conference and began by saying that she was embarking on a topic about which her audience was "probably interested." According to the government, Stewart's awareness of her audience (investors and analysts) and the timing of her statement as the price of MSLO stock was falling were sufficient to permit a jury to infer that she intended to deceive the investors when she made the statement. The court reasoned, however, that "any inference to be drawn from the makeup of the audience must also take into account the fact that Stewart was only one of several representatives of MSLO, and that MSLO was only one of several corporations making presentations at the conference. * * * There is no evidence that the negative publicity about ImClone influenced Stewart's decision to attend and take advantage of a platform from which to reach investors directly. To the contrary, her statement—a very brief portion of a much longer presentation—indicates otherwise."

FOR CRITICAL ANALYSIS—Social Consideration *How does the* scienter, *or intent, requirement in the context of criminal securities fraud differ from its counterpart in the context of civil securities fraud?*

a. Stewart was later convicted on other charges related to her sale of ImClone stock, including obstruction of justice and lying to federal officials. She was sentenced to and served five months in prison plus five months and three weeks of house arrest. In an attempt to clear her name, she appealed the decision, but a federal appeals court upheld the conviction.

Civil Sanctions The SEC can bring suit in a federal district court against anyone violating or aiding in a violation of the 1934 act or SEC rules by purchasing or selling a security while in the possession of material nonpublic information.[28] The violation must occur on or through the facilities of a national securities exchange or from or through a broker or dealer. Transactions pursuant to a public offering by an issuer of securities are excepted. The court may assess as a penalty as much as triple the profits gained or the loss avoided by the guilty party. Profit or loss is defined as "the difference between the purchase or sale price of the security and the value of that security as measured by the trading price of the security at a reasonable period of time after public dissemination of the nonpublic information."[29]

The Insider Trading and Securities Fraud Enforcement Act of 1988 enlarged the class of persons who may be subject to civil liability for insider-trading violations. This

28. The Insider Trading Sanctions Act of 1984, 15 U.S.C. Section 78u(d)(2)(A).

29. 15 U.S.C. Section 78u(d)(2)(C).

act also gave the SEC authority to award **bounty payments** (rewards given by government officials for acts beneficial to the state) to persons providing information leading to the prosecution of insider-trading violations.[30]

Private parties may also sue violators of Section 10(b) and Rule 10b-5. A private party may obtain rescission (cancellation) of a contract to buy securities or damages to the extent of the violator's illegal profits. Those found liable have a right to seek contribution from those who share responsibility for the violations, including accountants, attorneys, and corporations.[31] For violations of Section 16(b), a corporation can bring an action to recover the short-swing profits.

Corporate Governance

Corporate governance can be narrowly defined as the relationship between a corporation and its shareholders. The Organization of Economic Cooperation and Development (OECD) provides a broader definition:

> Corporate governance is the system by which business corporations are directed and controlled. The corporate governance structure specifies the distribution of rights and responsibilities among different participants in the corporation, such as the board of directors, managers, shareholders, and other stakeholders, and spells out the rules and procedures for making decisions on corporate affairs.[32]

Although this definition has no true legal value, it does set the tone for the ways in which modern corporations should be governed. In other words, effective corporate governance requires more than compliance with laws and regulations.

The Need for Good Corporate Governance

The need for effective corporate governance arises in large corporations because corporate ownership (by shareholders) is separated from corporate control (by officers and managers). In the real world, officers and managers are tempted to advance their own interests, even when such interests conflict with those of the shareholders. The recent well-publicized scandals in the corporate world illustrate the reason for concern about managerial opportunism.

Attempts at Aligning the Interests of Shareholders with Those of Officers

Some corporations have sought to align the financial interests of their officers with those of the company's shareholders by providing the officers with *stock options* for the corporation. Such options can be exercised at a set price and sold for a profit above that per-share price. When a corporation's share value increases, the options become more valuable for the officers, thereby giving them a financial stake in the share price.

Options have turned out to be an imperfect device for controlling governance, however. Executives in some companies have been tempted to "cook" the company's books in order to keep share prices higher so that they could exercise their options for a profit. Executives in other corporations experienced no losses when share prices dropped; instead, some had their options "repriced" so that they did not suffer from the share price decline and could still profit from future increases above the lowered share price. Although stock options theoretically can motivate officers to protect shareholder interests, stock option plans became a way for officers to take advantage of shareholders.

Because of numerous headline-making scandals within major corporations, there has been an outcry for more "outside" directors—the theory is that independent directors will more closely monitor the actions of corporate officers. Today, more boards have outside directors (those with no formal employment affiliation with the company). Note, though, that outside directors may not be truly independent of corporate officers; they may be friends or business associates of the leading officers. A study of board appointments found that the best way to increase one's probability of appointment was to "suck up" to the chief executive officer.[33]

Corporate Governance and Corporate Law

Good corporate governance standards are designed to address problems (such as those briefly discussed above) and to motivate officers to make decisions that promote

30. 15 U.S.C. Section 78u-1.

31. The Supreme Court has ruled that no private cause of action can be brought against those who "aid and abet" under Section 10(b) and SEC Rule 10b-5. *Central Bank of Denver, N.A. v. First Interstate Bank of Denver, N.A.,* 511 U.S. 164, 114 S.Ct. 1439, 128 L.Ed.2d 119 (1994). Only the SEC can bring actions against so-called aiders and abettors. Nevertheless, some courts have held accountants and attorneys liable as primary violators under Section 10(b), and a conflict exists in the federal circuit courts on precisely what course of conduct subjects a secondary actor to primary liability.

32. *Governance in the 21st Century: Future Studies,* OECD, 2001.

33. Jennifer Reingold, "Suck Up and Move Fast," *Fast Company,* January 2005, p. 34.

the financial interests of the company's shareholders. Generally, corporate governance entails corporate decision-making structures that monitor employees (particularly officers) to ensure that they are acting for the benefit of the shareholders. Thus, corporate governance involves, at a minimum:

1. The audited reporting of financial conditions at the corporation so that managers can be evaluated.
2. Legal protections for shareholders so that violators of the law, who attempt to take advantage of shareholders, can be punished for misbehavior and victims may recover damages for any associated losses.

The Practical Significance of Corporate Governance Effective corporate governance may have considerable practical significance. A study by researchers at Harvard University and the Wharton School of Business found that firms providing greater shareholder rights had higher profits, higher sales growth, higher firm value, and other economic advantages.[34] Better corporate governance in the form of greater accountability to investors may therefore offer the opportunity to considerably enhance institutional wealth.

Governance and Corporation Law Corporate governance is the essential purpose of corporation law in the United States. These statutes set up the legal framework for corporate governance. Under the corporate law of Delaware, where most major companies incorporate, all corporations must have in place certain structures of corporate governance. The key structure of corporate law is, of course, the board of directors. Directors make the most important decisions about the future of the corporation and monitor the actions of corporate officers. Directors are elected by shareholders to look out for their best interests.

The Board of Directors Some argue that shareholder democracy is key to improving corporate governance. If shareholders could vote on major corporate decisions, shareholders could presumably have more control over the corporation. Essential to shareholder democracy is the concept of electing the board of directors, usually at the corporation's annual meeting. Under corporate law, a corporation must have a board of directors elected by shareholders. Virtually anyone can become a director, though some organizations, such as the New York Stock Exchange, require certain standards of service for directors of their listed corporations.

Directors have the responsibility of ensuring that officers are operating wisely and in the exclusive interest of shareholders. Directors receive reports from the officers and give them managerial directions. The board in theory controls the compensation of officers (presumably tied to performance). The reality, though, is that corporate directors devote a relatively small amount of time to monitoring officers.

Ideally, shareholders would monitor the directors' supervision of officers. As one leading board monitor stated, "Boards of directors are like subatomic particles—they behave differently when they are observed." Consequently, monitoring directors, and holding them responsible for corporate failings, can induce the directors to do a better job of monitoring officers and ensuring that the company is being managed in the interest of shareholders. Although the directors can be sued for failing to effectively do their jobs, directors are rarely held personally liable.

Importance of the Audit Committee One crucial board committee is known as the *audit committee.* Members of the audit committee oversee the corporation's accounting and financial reporting processes, including both internal and outside auditors. These audit committee members must, however, have sufficient expertise and be willing to spend the time necessary to carefully examine the corporation's bookkeeping methods. Otherwise, the audit committee may be ineffective.

The audit committee also oversees the corporation's "internal controls." These are the measures taken to ensure that reported results are accurate; they are carried out largely by the company's internal auditing staff. As an example, these controls help to determine whether a corporation's debts are collectible. If the debts are not collectible, it is up to the audit committee to make sure that the corporation's financial officers cannot simply pretend that payment will eventually be made.

The Compensation Committee Another important committee of the board of directors is the *compensation committee.* This committee monitors and determines the compensation to be paid to a company's officers. In the process, it is responsible for assessing the officers' performance, and its members may try to design compensation systems that encourage better performance by the officers on behalf of the shareholders.

THE SARBANES-OXLEY ACT OF 2002

As discussed in Chapter 3, in 2002, following a series of corporate scandals, Congress passed the Sarbanes-Oxley

34. Paul A. Gompers, Joy L. Ishii, and Andrew Metrick, "Corporate Governance and Equity Prices," *Quarterly Journal of Economics,* Vol. 118 (2003), p. 107.

Act (see Appendix F for excerpts and explanatory comments). The act separately addresses certain issues relating to corporate governance. Generally, the act attempts to increase corporate accountability by imposing strict disclosure requirements and harsh penalties for violations of securities laws. Among other things, the act requires chief corporate executives to take responsibility for the accuracy of financial statements and reports that are filed with the SEC. Chief executive officers and chief financial officers must personally certify that the statements and reports are accurate and complete.

Additionally, the new rules require that certain financial and stock-transaction reports must be filed with the SEC earlier than was required under the previous rules. The act also mandates SEC oversight over a new entity, called the Public Company Accounting Oversight Board, that regulates and oversees public accounting firms. Other provisions of the act created new private civil actions and expanded the SEC's remedies in administrative and civil actions.

Because of the importance of this act for those dealing with securities transactions, we present some of its key provisions relating to corporate accountability in Exhibit 27–4. We also discuss the act and its effect on corporate governance procedures in this section. (Provisions of the act that relate to public accounting firms and accounting practices will be discussed in Chapter 31, in the context of the liability of accountants.)

EXHIBIT 27–4 SOME KEY PROVISIONS OF THE SARBANES-OXLEY ACT OF 2002 RELATING TO CORPORATE ACCOUNTABILITY

Certification Requirements—Under Section 906 of the Sarbanes-Oxley Act, the chief executive officers (CEOs) and chief financial officers (CFOs) of most major companies listed on public stock exchanges must now certify financial statements that are filed with the SEC. For virtually all filed financial reports, CEOs and CFOs have to certify that such reports "fully comply" with SEC requirements and that all of the information reported "fairly represents in all material respects, the financial conditions and results of operations of the issuer."

Under Section 302 of the act, for each quarterly and annual filing with the SEC, CEOs and CFOs of reporting companies are required to certify that a signing officer reviewed the report and that it contains no untrue statements of material fact. Also, the signing officer or officers must certify that they have established an internal control system to identify all material information, and that any deficiencies in the system were disclosed to the auditors.

Loans to Directors and Officers—Section 402 prohibits any reporting company, as well as any private company that is filing an initial public offering, from making personal loans to directors and executive officers (with a few limited exceptions, such as for certain consumer and housing loans).

Protection for Whistleblowers—Section 806 protects "whistleblowers"—those employees who report ("blow the whistle" on) securities violations by their employers—from being fired or in any way discriminated against by their employers.

Blackout Periods—Section 306 prohibits certain types of securities transactions during "blackout periods"—periods during which the issuer's ability to purchase, sell, or otherwise transfer funds in individual account plans (such as pension funds) is suspended.

Enhanced Penalties for—

- *Violations of Section 906 Certification Requirements*—A CEO or CFO who certifies a financial report or statement filed with the SEC knowing that the report or statement does not fulfill all of the requirements of Section 906 will be subject to criminal penalties of up to $1 million in fines, ten years in prison, or both. *Willful* violators of the certification requirements may be subject to $5 million in fines, twenty years in prison, or both.
- *Violations of the Securities Exchange Act of 1934*—Penalties for securities fraud under the 1934 act were also increased (as discussed earlier in this chapter). Individual violators may be fined up to $5 million, imprisoned for up to twenty years, or both. *Willful* violators may be imprisoned for up to twenty-five years in addition to being fined.
- *Destruction or Alteration of Documents*—Anyone who alters, destroys, or conceals documents or otherwise obstructs any official proceeding will be subject to fines, imprisonment for up to twenty years, or both.
- *Other Forms of White-Collar Crime*—The act stiffened the penalties for certain criminal violations, such as federal mail and wire fraud, and ordered the U.S. Sentencing Commission to revise the sentencing guidelines for white-collar crimes (see Chapter 6).

Statute of Limitations for Securities Fraud—Section 804 provides that a private right of action for securities fraud may be brought no later than two years after the discovery of the violation or five years after the violation, whichever is earlier.

More Internal Controls and Accountability The Sarbanes-Oxley Act includes some traditional securities law provisions but also introduces direct *federal* corporate governance requirements for public companies (companies whose shares are traded in the public securities markets). The law addresses many of the corporate governance procedures just discussed and creates new requirements in an attempt to make the system work more effectively. The requirements deal with independent monitoring of company officers by both the board of directors and auditors.

Sections 302 and 404 of Sarbanes-Oxley require high-level managers (the most senior officers) to establish and maintain an effective system of internal controls. Moreover, senior management must reassess the system's effectiveness on an annual basis. Some companies already had strong and effective internal control systems in place before the passage of the act, but others had to take expensive steps to bring their internal controls up to the new federal standard. These include "disclosure controls and procedures" to ensure that company financial reports are accurate and timely. Assessment must involve the documenting of financial results and accounting policies before reporting them. By 2006, hundreds of companies had reported that they had identified and corrected shortcomings in their internal control systems.

Certification and Monitoring Requirements Section 906 requires that chief executive officers (CEOs) and chief financial officers (CFOs) certify that the corporate financial statements "fairly present, in all material respects, the financial condition and results of operation of the issuer." These corporate officers are subject to both civil and criminal penalties for violation of this section.[35]

Sarbanes-Oxley also includes requirements to improve directors' monitoring of officers' activities. All members of the corporate audit committee for public companies must be outside directors. The New York Stock Exchange (NYSE) has a similar rule that also extends to the board's compensation committee. The audit committee must have a written charter that sets out its duties and provides for performance appraisal. At least one "financial expert" has to serve on the audit committee, which must hold executive meetings without company officers being present. The audit committee must establish procedures for "whistleblowers." In addition to reviewing the internal controls, the committee also monitors the actions of the outside auditor.

35. This requirement makes officers directly accountable for the accuracy of their financial reporting and precludes any "ignorance defense" if shortcomings are later discovered.

The Separation of Audit and Nonaudit Services The law includes other provisions to improve accounting accuracy. Auditors are prohibited from providing a company with substantial nonaudit services of any kind that might compromise the auditors' independence. The lead audit partner and reviewing partner must rotate off each assignment every five years. The purpose of this rotation is to prevent them from establishing unduly close relationships with the management officers they are auditing. Other rules apply to lawyers representing public companies and require them to blow the whistle on their clients when they determine that a client is engaged in illegal behavior.

Corporate Ethical Codes Sarbanes-Oxley also contains provisions for corporate ethical codes. A company regulated by the SEC must report whether it has established an ethical code governing high-level officers. The contents of that code must be publicly available. The NYSE similarly requires that each listed company adopt a code of conduct and ethics for its officers and post it on the company's Web site. This code of conduct and ethics must specifically prohibit self-dealing at the expense of shareholders. Of course, the code must also prohibit violations of the law.

The Regulation of Investment Companies

Investment companies, and mutual funds in particular, grew rapidly after World War II. **Investment companies** act on behalf of many smaller shareholders by buying a large portfolio of securities and professionally managing that portfolio. A **mutual fund** is a specific type of investment company that continually buys or sells to investors shares of ownership in a portfolio. Such companies are regulated by the Investment Company Act of 1940,[36] which provides for SEC regulation of their activities. The act was expanded by the 1970 amendments to the Investment Company Act. Further minor changes were made in the Securities Act Amendments of 1975 and in later years.

Definition of an Investment Company

For the purposes of the act, an *investment company* is any entity that (1) is engaged primarily "in the business of investing, reinvesting, or trading in securities" or (2) is engaged in such business and has more than 40 percent of its assets in investment securities. Excluded from coverage by the act are banks, insurance companies, savings and loan associations, finance companies, oil and gas drilling

36. 15 U.S.C. Sections 80a-1 to 80a-64.

firms, charitable foundations, tax-exempt pension funds, and other special types of institutions, such as closely held corporations.

Registration and Reporting Requirements

The 1940 act requires that every investment company register with the SEC by filing a notification of registration. Each year, registered investment companies must file reports with the SEC. To safeguard company assets, all securities must be held in the custody of a bank or stock exchange member, which has to follow strict procedures established by the SEC.

Restrictions on Investment Companies

The 1940 act also imposes restrictions on the activities of investment companies and persons connected with them. For example, investment companies are not allowed to purchase securities on the margin (pay only part of the total price, borrowing the rest), sell short (sell shares not yet owned), or participate in joint trading accounts. In addition, no dividends may be paid from any source other than accumulated, undistributed net income.

State Securities Laws

Today, all states have their own corporate securities laws, or "blue sky laws," that regulate the offer and sale of securities within individual state borders. (As mentioned in Chapter 9, the phrase *blue sky laws* dates to a 1917 decision by the United States Supreme Court in which the Court declared that the purpose of such laws was to prevent "speculative schemes which have no more basis than so many feet of 'blue sky.'")[37] Article 8 of the Uniform Commercial Code, which has been adopted by all of the states, also imposes various requirements relating to the purchase and sale of securities.

Requirements under State Securities Laws

Despite some differences in philosophy, all state blue sky laws have certain features. Typically, state laws have disclosure requirements and antifraud provisions, many of which are patterned after Section 10(b) of the Securities Exchange Act of 1934 and SEC Rule 10b-5. State laws also provide for the registration or qualification of securities offered or issued for sale within the state and impose disclosure requirements. Unless an exemption from registration is applicable, issuers must register or qualify their stock with the appropriate state official, often called a *corporations commissioner.* Additionally, most state securities laws regulate securities brokers and dealers.

Concurrent Regulation

State securities laws apply mainly to intrastate transactions. Since the adoption of the 1933 and 1934 federal securities acts, the state and federal governments have regulated securities concurrently. Issuers must comply with both federal and state securities laws, and exemptions from federal law are not exemptions from state laws.

The dual federal and state system has not always worked well, particularly during the early 1990s, when the securities markets underwent considerable expansion. In response, Congress passed the National Securities Markets Improvement Act of 1996, which eliminated some of the duplicate regulations and gave the SEC exclusive power to regulate most national securities activities. The National Conference of Commissioners on Uniform State Laws then substantially revised the Uniform Securities Act and recommended it to the states for adoption in 2002. Unlike the previous version of this law, the new act is designed to coordinate state and federal securities regulation and enforcement efforts. Since 2002, six states have adopted the Uniform Securities Act, and a number of other states are considering adoption.[38]

Online Securities Offerings and Disclosures

The Spring Street Brewing Company, headquartered in New York, made history when it became the first company to attempt to sell securities via the Internet. Through its online initial public offering (IPO), which ended in early 1996, Spring Street raised about $1.6 million—without having to pay any commissions to brokers or underwriters. The offering was made pursuant to Regulation A, which, as mentioned earlier in this chapter, allows small-business issuers to use a simplified registration procedure.

Such online IPOs are particularly attractive to small companies and start-up ventures that may find it difficult to raise capital from institutional investors or through underwriters. By making the offering online under Regulation A, the company can avoid both commissions and the costly

37. *Hall v. Geiger-Jones Co.,* 242 U.S. 539, 37 S.Ct. 217, 61 L.Ed. 480 (1917).

38. At the time this book went to press, the 2002 Uniform Securities Act had been adopted in Idaho, Iowa, Kansas, Missouri, Oklahoma, and South Dakota. Adoption legislation was pending in Alabama, Alaska, Nebraska, Vermont, and Virginia, as well as the U.S. Virgin Islands. You can find current information on state adoptions at **http://www.nccusl.org**.

and time-consuming filings required for a traditional IPO under federal and state law.

Regulations Governing Online Securities Offerings

One of the early questions posed by online offerings was whether the delivery of securities *information* via the Internet met the requirements of the 1933 Securities Act, which traditionally were applied to the delivery of paper documents. In an interpretive release issued in 1995, the SEC stated that "[t]he use of electronic media should be at least an equal alternative to the use of paper-based media" and that anything that can be delivered in paper form under the current securities laws might also be delivered in electronic form.[39]

Basically, there has been no change in the substantive law of disclosure; only the delivery vehicle has changed. When the Internet is used for delivery of a prospectus, the same rules apply as for the delivery of a paper prospectus. Once the three requirements listed below have been satisfied, the prospectus has been successfully delivered.

1. *Timely and adequate notice of the delivery of information is required.* Hosting a prospectus on a Web site does not constitute adequate notice, but separate e-mails or even postcards stating the URL where the prospectus can be viewed will satisfy the SEC's notice requirements.
2. *The online communication system must be easily accessible.* This is very simple to do today because virtually anyone interested in purchasing securities has access to the Web.
3. *Some evidence of delivery must be created.* This requirement is relatively easy to satisfy. Those making online offerings can require an e-mail return receipt verification of any materials sent electronically.

Potential Liability Created by Online Offering Materials

Every printed prospectus indicates that only the information given in the prospectus can be used in conjunction with making an investment decision in the securities offered. The same wording, of course, appears on Web-based offerings. What happens if an electronic prospectus contains information that conflicts with the information provided in a printed prospectus? Will such an error render the registration statement ineffective? See this chapter's *Adapting the Law to the Online Environment* feature for a discussion of a case involving this issue.

Hyperlinks to Other Web Pages Those who create such Web-based offerings may be tempted to insert hyperlinks to other Web pages. They may include links to other sites that have analyzed the future prospects of the company, the products and services sold by the company, or the offering itself. To avoid potential liability, however, online offerors (the entities making the offerings) need to exercise caution when including such hyperlinks.

■ **EXAMPLE 27.6** Suppose that a hyperlink goes to an analyst's Web page that makes optimistic statements concerning the financial outlook of the offering company. Further suppose that after the IPO, the stock price falls. By including the hyperlink on its Web site, the offering company is impliedly supporting the information presented on the linked page. If it turns out that the company knew the statements made on the analyst's Web page were false or misleading, the company may be liable for violating sections of the Securities Exchange Act of 1934.[40] ■

Regulation D Offerings Potential problems may also occur with some Regulation D offerings, if the offeror places the offering circular on its Web site for general consumption by anybody on the Internet. Because Regulation D offerings are private placements, general solicitation is restricted. If anyone can have access to the offering circular on the Web, the Regulation D exemption may be disqualified.

Online Securities Offerings by Foreign Companies

Another question raised by Internet transactions has to do with securities offerings by foreign companies. Traditionally, foreign companies have not been able to offer new shares to the U.S. public without first registering them with the SEC. Today, however, anybody in the world can offer shares of stock worldwide via the Web.

The SEC asks that foreign issuers on the Internet implement measures to warn U.S. investors. For example, a foreign company offering shares of stock on the Internet must add a disclaimer on its Web site stating that it has not gone through the registration procedure in the United States. If the SEC believes that a Web site's offering of foreign securities is targeting U.S. residents, it will pursue

39. "Use of Electronic Media for Delivery Purposes," Securities Act Release No. 33-7233 (October 6, 1995). The rules governing the use of electronic transmissions for delivery purposes were subsequently confirmed in Securities Act Release No. 33-7289 (May 9, 1996) and expanded in Securities Act Release No. 33-7856 (April 28, 2000).

40. See, for example, *In re Syntex Corp. Securities Litigation,* 95 F.3d. 922 (9th Cir. 1996), involving alleged violations of Sections 10(b) and 20(a) of the 1934 act.

ADAPTING THE LAW TO THE ONLINE ENVIRONMENT

Will Inaccurate Information in an Electronic Prospectus Invalidate the Registration?

Many companies now submit registration statements, prospectuses, and other information to the Securities and Exchange Commission (SEC) via the Internet. The SEC's Electronic Data Gathering, Analysis, and Retrieval (EDGAR) system then posts much of this information online to inform investors about the corporation, the security being sold, and the risk of investing in that security. Some corporations also send investors a printed prospectus. Theoretically, a corporation should provide the same information in electronic form as it does in a printed prospectus, but practical difficulties can arise in transmitting digital information.

THE PROBLEM WITH GRAPHICS

As anyone who is familiar with the Internet knows, the graphics, images, and audio files created by one computer are not always readable by another computer when they are exchanged online. The SEC has created Rule 304 to deal with this situation.[a] The first part of the rule states that if graphic, image, or audio material in a prospectus cannot be reproduced in an electronic form on EDGAR, the electronic prospectus must include a fair and accurate narrative description of the omitted data. The second part of Rule 304 provides that the graphic, image, and audio material contained in the version of a document delivered to investors is *deemed to be part of the electronic document* filed with the SEC.

As a result, a corporation can have two versions of a prospectus—a print version that contains graphics and an electronic version that describes the information shown in the graphics. What if the summary describing the graphics in an electronic prospectus is inaccurate but the investors received an accurate print version? That was the issue before a federal appellate court in *DeMaria v. Andersen.*[b]

THE ELECTRONIC PROSPECTUS CONTAINED INACCURACIES

In anticipation of an initial public offering (IPO), ILife.com, Inc., filed a registration statement and a prospectus with the SEC via the EDGAR database. ILife.com also distributed a printed version of the prospectus to the public. The printed prospectus contained a bar graph that provided historical financial information about the company, while the EDGAR prospectus contained a table that summarized the bar graph inaccurately (without mentioning losses). Brian DeMaria and other investors filed a suit against officers of ILife.com. The investors claimed that because of the inaccurate summary, the securities in the IPO were "unregistered" and thus were sold in violation of the Securities Act of 1933. The lower court dismissed the case, and DeMaria appealed.

THE REGISTRATION HELD VALID

Despite the inaccurate summary of the bar graph in the electronic prospectus, the appellate court had no trouble deciding that the securities sold were still registered as required by the Securities Act of 1933. Federal courts are bound to follow the SEC's interpretation of its own regulations unless the interpretations are plainly erroneous. Here, the SEC had filed a brief explaining that because the graphics in the printed prospectus are deemed to be part of the electronic registration statement, it did not matter that the narrative description was inaccurate.

FOR CRITICAL ANALYSIS

Does the part of Rule 304 that deems a printed prospectus to be part of a registration statement completely eliminate liability for any inaccuracies in the electronic materials filed? Why or why not?

a. 17 C.F.R. Section 232.304.
b. 318 F.3d 170 (2d Cir. 2003).

that company in an attempt to require it to register in the United States.[41]

ONLINE SECURITIES FRAUD

A major problem facing the SEC today is how to enforce the antifraud provisions of the securities laws in the online environment. In 1999, in the first cases involving illegal online securities offerings, the SEC filed suit against three individuals for illegally offering securities on an Internet auction site.[42] In essence, all three indicated that their companies would go public soon and attempted to sell unregistered securities via the Web auction site. All of these actions were in violation of Sections 5, 17(a)(1), and 17(a)(3) of the 1933 Securities Act. Since then, the SEC has brought a variety of Internet-related fraud cases, including cases involving investment scams and the manipulation of stock prices in Internet chat rooms.

41. International Series Release No. 1125 (March 23, 1998).

42. *In re Davis,* SEC Administrative File No. 3-10080 (October 20, 1999); *In re Haas,* SEC Administrative File No. 3-10081 (October 20, 1999); and *In re Sitaras,* SEC Administrative File No. 3-10082 (October 20, 1999).

Investment Scams An ongoing problem is how to curb online investment scams. One fraudulent investment scheme involved twenty thousand investors, who lost, in all, more than $3 million. Some cases have involved false claims about the earnings potential of home-business programs, such as the claim that one could "earn $4,000 or more each month." Others have concerned claims of "guaranteed credit repair."

Using Online Chat Rooms to Manipulate Stock Prices "Pumping and dumping" occurs when a person who has purchased a particular stock heavily promotes ("pumps up") that stock—thereby creating a great demand for it and driving up its price—and then sells ("dumps") it. The practice of pumping up a stock and then dumping it is quite old. In the online world, however, the process can occur much more quickly and efficiently.

■ EXAMPLE 27.7 The most famous case in this area involved Jonathan Lebed, a fifteen-year-old stock trader and Internet user from New Jersey. Lebed was the first minor ever charged with securities fraud by the SEC, but it is unlikely that he will be the last. The SEC charged that Lebed bought thinly traded stocks. After purchasing a stock, he would flood stock-related chat rooms, particularly at Yahoo's finance boards, with messages touting the stock's virtues. He used numerous false names so that no one would know that a single person was posting the messages. He would say that the stock was the most "undervalued stock in history" and that its price would jump by 1,000 percent "very soon." When other investors would then buy the stock, the price would go up quickly, and Lebed would sell out. The SEC forced the teenager to repay almost $300,000 in gains plus interest. He was allowed, however, to keep about $500,000 of the profits he made by trading small-company stocks that he also touted on the Internet. ■

The SEC has been bringing an increasing number of cases against those who manipulate stock prices in this way. Consider that in 1995, such fraud resulted in only six SEC cases. By 2004, the SEC had brought over two hundred actions against online perpetrators of fraudulent stock-price manipulation.

TERMS AND CONCEPTS

accredited investor 566
bounty payment 573
corporate governance 573
insider trading 568
investment company 576
mutual fund 576
red herring 564
SEC Rule 10b-5 568
security 561
tippee 570
tombstone ad 564

CHAPTER SUMMARY Investor Protection and Corporate Governance

Securities Act of 1933 (See pages 562–568.)	Prohibits fraud and stabilizes the securities industry by requiring disclosure of all essential information relating to the issuance of stocks to the investing public. 1. *Registration requirements*—Securities, unless exempt, must be registered with the SEC before being offered to the public through the mails or any facility of interstate commerce (including securities exchanges). The *registration statement* must include detailed financial information about the issuing corporation; the intended use of the proceeds of the securities being issued; and certain disclosures, such as interests of directors or officers and pending lawsuits. 2. *Prospectus*—A *prospectus* must be provided to investors, describing the security being sold, the issuing corporation, and the risk attaching to the security. 3. *Exemptions*—The SEC has exempted certain offerings from the requirements of the Securities Act of 1933. Exemptions may be determined on the basis of the size of the issue, whether the offering is private or public, and whether advertising is involved. Exemptions are summarized in Exhibit 27–1 on page 565.
Securities Exchange Act of 1934 (See pages 568–573.)	Provides for the regulation and registration of securities exchanges, brokers, dealers, and national securities associations (such as the NASD). Maintains a continuous disclosure system for all corporations with securities on the securities exchanges and for those companies that have assets in excess of $10 million and five hundred or more shareholders (Section 12 companies).

CHAPTER SUMMARY Investor Protection and Corporate Governance—Continued

Securities Exchange Act of 1934—Continued	1. *SEC Rule 10b-5 [under Section 10(b) of the 1934 act]*— a. Applies to insider trading by corporate officers, directors, majority shareholders, and any persons receiving information not available to the public who base their trading on this information. b. Liability for violations can be civil or criminal. c. May be violated by failing to disclose "material facts" that must be disclosed under this rule. d. Applies in virtually all cases concerning the trading of securities—a firm's securities do not have to be registered under the 1933 act for the 1934 act to apply. e. Liability may be based on the tipper/tippee theory or the misappropriation theory. f. Applies only when the requisites of federal jurisdiction (such as use of the mails, stock exchange facilities, or any facility of interstate commerce) are present. 2. *Insider trading [under Section 16(b) of the 1934 act]*—To prevent corporate officers and directors from taking advantage of inside information (information not available to the investing public), the 1934 act requires officers, directors, and shareholders owning 10 percent or more of the issued stock of a corporation to turn over to the corporation all short-term profits (called short-swing profits) realized from the purchase and sale or sale and purchase of corporate stock within any six-month period. 3. *Proxies [under Section 14(a) of the 1934 act]*—The SEC regulates the content of proxy statements sent to shareholders by corporate managers of Section 12 companies who are requesting authority to vote on behalf of the shareholders in a particular election on specified issues. Section 14(a) is essentially a disclosure law, with provisions similar to the antifraud provisions of SEC Rule 10b-5.
Corporate Governance (See pages 573–576.)	1. *Definition*—Corporate governance is the system by which business corporations are governed, including policies and procedures for making decisions on corporate affairs. 2. *The need for corporate governance*—Corporate governance is necessary in large corporations because corporate ownership (by the shareholders) is separated from corporate control (by officers and managers). This separation of corporate ownership and control can often result in conflicting interests. Corporate governance standards address such issues. 3. *Sarbanes-Oxley Act of 2002*—This act attempts to increase corporate accountability by imposing strict disclosure requirements and harsh penalties for violations of securities laws.
The Regulation of Investment Companies (See pages 576–577.)	The Investment Company Act of 1940 provides for SEC regulation of investment company activities. It was altered and expanded by the amendments of 1970 and 1975.
State Securities Laws (See page 577.)	All states have corporate securities laws *(blue sky laws)* that regulate the offer and sale of securities within state borders; these laws are designed to prevent "speculative schemes which have no more basis than so many feet of 'blue sky.'" States regulate securities concurrently with the federal government. The Uniform Securities Act of 2002, which is being considered by a number of states, is designed to promote coordination and reduce duplication between state and federal securities regulation.
Online Securities Offerings and Disclosures (See pages 577–579.)	In 1995, the SEC announced that anything that can be delivered in paper form under current securities laws may also be delivered in electronic form. Generally, when the Internet is used for the delivery of a prospectus, the same rules apply as for the delivery of a paper prospectus. When securities offerings are made online, the offerors should be careful that any hyperlinked materials do not mislead investors. Caution should also be used when making Regulation D offerings (private placements), because general solicitation is restricted with these offerings.
Online Securities Fraud (See pages 579–580.)	A major problem facing the SEC today is how to enforce the antifraud provisions of the securities laws in the online environment. Internet-related forms of securities fraud include investment scams and the manipulation of stock prices in online chat rooms.

FOR REVIEW

Answers for the even-numbered questions in this For Review *section can be found in Appendix A at the end of this text.*

1. What is meant by the term *securities?*
2. What are the two major statutes regulating the securities industry? When was the Securities and Exchange Commission created, and what are its major purposes and functions?
3. What is insider trading? Why is it prohibited?
4. What are some of the features of state securities laws?
5. How are securities laws being applied in the online environment?

QUESTIONS AND CASE PROBLEMS

QUESTION WITH SAMPLE ANSWER

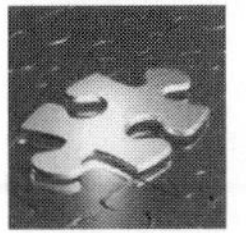

27–1. Langley Brothers, Inc., a corporation incorporated and doing business in Kansas, decides to sell no-par common stock worth $1 million to the public. The stock will be sold only within the state of Kansas. Joseph Langley, the chairman of the board, says the offering need not be registered with the Securities and Exchange Commission. His brother, Harry, disagrees. Who is right? Explain.

For a sample answer to this question, go to Appendix B at the end of this text.

27–2. Definition of a Security. The W. J. Howey Co. owned large tracts of citrus acreage in Lake County, Florida. For several years, it planted about five hundred acres annually, keeping half of the groves itself and offering the other half to the public to help finance additional development. Howey-in-the-Hills Service, Inc., was a service company engaged in cultivating and developing these groves, including harvesting and marketing the crops. Each prospective customer was offered both a land sales contract and a service contract, after being told that it was not feasible to invest in a grove unless service arrangements were made. Of the acreage sold by Howey, 85 percent was sold with a service contract with Howey-in-the-Hills Service. Howey did not register with the Securities and Exchange Commission (SEC) or meet the other administrative requirements that issuers of securities must fulfill. The SEC sued to enjoin Howey from continuing to offer the land sales and service contracts. Howey responded that no SEC violation existed because no securities had been issued. Evaluate the definition of a security given in this chapter, and then determine whether Howey or the SEC should prevail in court. [*SEC v. W. J. Howey Co.*, 328 U.S. 293, 66 S.Ct. 1100, 90 L.Ed. 1244 (1946)]

27–3. SEC Rule 10b-5. Grand Metropolitan PLC (Grand Met) planned to make a tender offer as part of an attempted takeover of the Pillsbury Co. Grand Met hired Robert Falbo, an independent contractor, to complete electrical work as part of security renovations to its offices to prevent leaks of information concerning the planned tender offer. Falbo was given a master key to access the executive offices. When an executive secretary told Falbo that a takeover was brewing, he used his key to access the offices and eavesdrop on conversations; in this way, he learned that Pillsbury was the target. Falbo bought thousands of shares of Pillsbury stock for less than $40 per share. Within two months, Grand Met made an offer for all outstanding Pillsbury stock at $60 per share and ultimately paid up to $66 per share. Falbo made a profit of more than $165,000. The Securities and Exchange Commission (SEC) filed a suit in a federal district court against Falbo and others for alleged violations of, among other things, SEC Rule 10b-5. Under what theory might Falbo be liable? Do the circumstances of this case meet all of the requirements for liability under that theory? Explain. [*SEC v. Falbo*, 14 F.Supp.2d 508 (S.D.N.Y. 1998)]

27–4. Definition of a Security. In 1997, Scott and Sabrina Levine formed Friendly Power Co. (FPC) and Friendly Power Franchise Co. (FPC-Franchise). FPC obtained a license to operate as a utility company in California. FPC granted FPC-Franchise the right to pay commissions to "operators" who converted residential customers to FPC. Each operator paid for a "franchise"—a geographic area, determined by such factors as the number of households and competition from other utilities. In exchange for 50 percent of FPC's net profits on sales to residential customers in its territory, each franchise was required to maintain a 5 percent market share of power customers in that territory. Franchises were sold to telemarketing firms, which solicited customers. The telemarketers sold interests in each franchise to between fifty and ninety-four "partners," each of whom invested funds. FPC began supplying electricity to its customers in May 1998. Less than three months later, the Securities and Exchange Commission (SEC) filed a suit in a federal district court against the Levines and others, alleging that the "franchises" were unregistered securities offered for sale to the public in violation of the Securities Act of 1933. What is the definition of a security? Should the court rule in favor of the SEC? Why or why not? [*SEC v. Friendly Power Co., LLC*, 49 F.Supp.2d 1363 (S.D.Fla. 1999)]

CASE PROBLEM WITH SAMPLE ANSWER

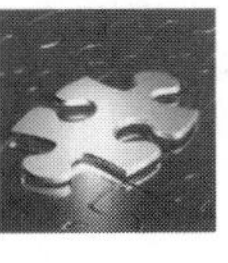

27–5. 2TheMart.com, Inc., was conceived in January 1999 to launch a Web auction site to compete with eBay, Inc. On January 19, 2TheMart announced that its Web site was in its "final

development" stages and was expected to be active by the end of July as a "preeminent" auction site. The company also said that it had "retained the services of leading Web site design and architecture consultants to design and construct" the site. Based on the announcement, investors rushed to buy 2TheMart's stock, causing a rapid increase in the price. On February 3, 2TheMart entered into an agreement with IBM to take preliminary steps to plan the site. Three weeks later, 2TheMart announced that the site was "currently in final development." On June 1, 2TheMart signed a contract with IBM to design, build, and test the site, with a target delivery date of October 8. When 2TheMart's site did not debut as announced, Mary Harrington and others who had bought the stock filed a suit in a federal district court against the firm's officers, alleging violations of the Securities Exchange Act of 1934. The defendants responded, in part, that any alleged misrepresentations were not material and asked the court to dismiss the suit. How should the court rule, and why? [*In re 2TheMart.com, Inc., Securities Litigation,* 114 F.Supp.2d 955 (C.D.Cal. 2000)]

After you have answered this problem, compare your answer with the sample answer given on the Web site that accompanies this text. Go to **academic.cengage.com/blaw/wbl**, select "Chapter 27," and click on "Case Problem with Sample Answer."

27–6. Insider Reporting and Trading. Ronald Bleakney, an officer at Natural Microsystems Corp. (NMC), a Section 12 corporation, directed NMC sales in North America, South America, and Europe. In November 1998, Bleakney sold more than 7,500 shares of NMC stock. The following March, Bleakney resigned from the firm, and the next month, he bought more than 20,000 shares of its stock. NMC provided some guidance to employees concerning the rules of insider trading, but with regard to Bleakney's transactions, the corporation said nothing about potential liability. Richard Morales, an NMC shareholder, filed a suit against NMC and Bleakney to compel recovery, under Section 16(b) of the Securities Exchange Act of 1934, of Bleakney's profits from the sale and purchase of his shares. (When Morales died, his executor, Deborah Donoghue, became the plaintiff.) Bleakney argued that he should not be liable because he relied on NMC's advice. Should the court order Bleakney to disgorge his profits? Explain. [*Donoghue v. Natural Microsystems Corp.,* 198 F.Supp.2d 487 (S.D.N.Y. 2002)]

27–7. SEC Rule 10b-5. Scott Ginsburg was chief executive officer (CEO) of Evergreen Media Corp. In 1996, Evergreen became interested in acquiring EZ Communications, Inc. Ginsburg met with EZ's CEO, Alan Box, on July 12. Evergreen and EZ executives began negotiating confidentially for the purchase of EZ at the specific price of $50 a share. Ginsburg called his brother, Mark, who spoke to their father, Jordan, about the deal. Mark and Jordan bought almost 75,000 shares of EZ stock. Evergreen's bid for EZ fell through, but in August, EZ announced its merger with another company. The price of EZ stock rose 30 percent, increasing the value of Mark and Jordan's shares by more than $1.76 million. The Securities and Exchange Commission (SEC) filed a suit in a federal district court against Ginsburg, alleging, among other things, violations of SEC Rule 10b-5 for communicating material nonpublic information to Mark and Jordan, who traded on the basis of that information. Ginsburg contended in part that the information was not material and filed a motion for a judgment as a matter of law. What is the test for materiality in this context? Does the information in this case meet the test, or should the court grant the motion? Explain. [*SEC v. Ginsburg,* 362 F.3d 1292 (11th Cir. 2004)]

VIDEO QUESTION

27–8. Go to this text's Web site at academic.cengage.com/blaw/wbl and select "Chapter 27." Click on "Video Questions" and view the video titled *Mergers and Acquisitions.* Then answer the following questions.

1. Analyze whether the purchase of Onyx Advertising is a material fact that the Quigley Co. had a duty to disclose under SEC Rule 10b-5.
2. Does it matter whether Quigley personally knew about or authorized the company spokesperson's statements? Why or why not?
3. Which case discussed in the chapter presented issues that are very similar to those raised in the video? Under the holding of that case, would Onyx Advertising be able to maintain a suit against the Quigley Co. for violation of SEC Rule 10b-5?
4. Who else might be able to bring a suit against the Quigley Co. for insider trading under SEC Rule 10b-5?

Investor Protection and Corporate Governance

For updated links to resources available on the Web, as well as a variety of other materials, visit this text's Web site at

academic.cengage.com/blaw/wbl

To access the EDGAR database of the Securities and Exchange Commission (SEC), go to

http://www.sec.gov/edgar.shtml

The Center for Corporate Law of the University of Cincinnati College of Law maintains the *Securities Lawyer's Deskbook* online. The *Deskbook* contains the basic federal securities laws and regulations and links to the principal SEC forms under those laws and regulations. Go to

http://www.law.uc.edu/CCL

For information on investor protection and securities fraud, including answers to frequently asked questions about securities fraud, go to

http://www.securitieslaw.com

ONLINE LEGAL RESEARCH

Go to the *Fundamentals of Business Law* home page at **academic.cengage.com/blaw/wbl**, select "Chapter 27," and click on "Internet Exercises." There you will find the following Internet research exercises that you can perform to learn more about topics covered in this chapter.

Activity 27–1: LEGAL PERSPECTIVE—Electronic Delivery

Activity 27–2: MANAGEMENT PERSPECTIVE—The SEC's Role

BEFORE THE TEST

Go to the *Fundamentals of Business Law* home page at **academic.cengage.com/blaw/wbl**, select "Chapter 27," and click on "Interactive Quizzes." You will find a number of interactive questions relating to this chapter.

UNIT 8 • EXTENDED CASE STUDY

Gottsacker v. Monnier

In Chapter 24, we reviewed some of the elements that distinguish limited liability companies (LLCs) from other forms of business organization. In this extended case study, we look at *Gottsacker v. Monnier,*[1] a recent decision in which the court focused on the apportionment of voting rights in an LLC and a conflict of interest among its members.

1. 281 Wis.2d 361, 697 N.W.2d 436, 2005 WI 69 (2005).

CASE BACKGROUND

In September 1998, Julie Monnier formed New Jersey LLC to buy and sell real estate. Ten days later, the company bought a warehouse at 2005 New Jersey Avenue in Sheboygan, Wisconsin, for $510,000, with the financing arranged for and guaranteed by Monnier. Paul and Gregory Gottsacker became members of New Jersey LLC in January 1999. The parties entered into a Member's Agreement, under which Monnier had "50% of the voting rights" and Paul and Gregory "collectively" had "50% of the voting rights."

Relationships among the members of New Jersey LLC subsequently became strained. On June 7, 2001, Monnier sold the warehouse to a new LLC called 2005 New Jersey LLC for the same amount as the original purchase price. The new LLC consisted of two members—Monnier and Paul. Neither one discussed the transfer with Gregory.

Monnier sent Gregory a check for $22,000, which purportedly represented his interest in the warehouse. Gregory did not cash the check but filed a suit in a Wisconsin state court against Monnier, Paul, and 2005 New Jersey LLC, alleging that they had engaged in an illegal transaction. The court ruled in Gregory's favor. The defendants appealed to a state intermediate appellate court, which affirmed the ruling. The defendants appealed to the Wisconsin Supreme Court.

MAJORITY OPINION

ANN WALSH BRADLEY, J. [Justice]

* * * *

The first issue we address is whether the [defendants] possessed the majority necessary to authorize the transfer in question. * * *

* * * *

We conclude that the Member's Agreement here is ambiguous as to the voting rights of Paul and Gregory. To begin, the term "collectively" is not defined in the document. * * *

* * * *

* * * [W]e are satisfied that the term "collectively" refers to the sum of the brothers' individual 25% interests. To conclude otherwise would require unanimous approval by the members in order to perform any act that concerns the business of the company. Here, there is no express language indicating that the parties intended such a result. Construing the Member's Agreement to allow one minority member to effectively deadlock the LLC is unreasonable absent express language.

* * * *

* * * [W]e consider next whether [the defendants] were nonetheless prohibited from voting to transfer the property because of a material conflict of interest. Here, the [trial] court found that "the conveyance [transfer] of the property by Julie Monnier and Paul Gottsacker to themselves in the guise [appearance] of a newly created LLC, unquestionably, represents a material conflict of interest." This finding is supported by the facts of the case. Not only did Monnier and Paul engage in self-dealing, but in doing so they also increased their individual interests in the new LLC which received the property. Monnier's ownership improved from 50% to 60%, while Paul's interest improved from 25% to 40%.

The question therefore becomes what, if any, impact did this conflict of interest have on Monnier and Paul's ability to vote to transfer the property? [Wisconsin Statutes Section] 183.0404 governs voting in LLCs and contemplates situations that would prevent a member from exercising that voting power. Subsection (3) of the statute explicitly states that members can be "precluded from voting." However, that subsection does not address how or when that preclusion would occur. * * *

* * * *

* * * Wis. Stat. [Section] 183.0402 * * * anticipates members having a material conflict of interest and requires them to "deal fairly" with the LLC and its other members. * * *

* * * *

Reading Wis. Stat. [Sections] 183.0404 and 183.0402 together in harmony, we determine that the [Wisconsin LLC law] does not preclude members with a material conflict of interest from voting their ownership interest with respect to a given matter. Rather, it prohibits members with a material conflict of interest from acting in a manner that constitutes a willful failure to deal fairly with the LLC or its other members. * * *

Here, the [trial] court made no express determination as to whether the petitioners willfully failed to deal fairly in spite of the conflict of interest. * * *

* * * *

Accordingly, we remand the case to the [trial] court for further findings and application of the foregoing standard.

(Continued)

* * * If it is determined by the [trial] court that this [standard] was violated, then the court will determine the appropriate remedy under the circumstances.

* * * *

In sum, we conclude that the petitioners possessed the majority necessary to authorize the transfer in question. Furthermore, we determine that the petitioners' material conflict of interest did not prohibit them from voting to make the transfer so long as they dealt fairly. However, because there was no express determination by the [trial] court as to whether the petitioners willfully failed to deal fairly with New Jersey LLC or its other member, we reverse the decision of the court of appeals and remand the case for further proceedings.

The decision of the court of appeals is reversed * * * .

Dissenting Opinion

LOUIS B. BUTLER, JR., J. [Justice] (dissenting).

At times, issues are complex and therefore are in need of complex resolutions. At times, we tend to see complexity where none exists. This, I conclude, is one of those occasions where the issue and its resolution are simple. Because there was no affirmative vote, approval, or consent to transfer the warehouse property owned by New Jersey LLC to a new limited liability company called 2005 New Jersey LLC, as required by the Member's Agreement and [the state LLC statute] no legal transfer of the property took place. As such, I would affirm the court of appeals, albeit on different grounds. * * *

* * * *

* * * It was the parties that specified [in the Member's Agreement] the two separate 50 percent interests in New Jersey LLC. It was the parties that specified the two separate voting blocks. They could have chosen to draft the agreement to take into account each person's individual interest and voting rights, but the parties chose not to do so. If the parties choose to set forth an agreement that requires the brothers to vote together as one interest, this court should not stand in their way. If the drafters of [the Wisconsin LLC law] emphasized the importance of flexibility and freedom of contract, then we ought to respect the flexibility and freedom of this agreement. In short, the trial court got it right when it concluded that Paul Gottsacker lacked the authority to act without the assent of his brother.

Gregory Gottsacker did not agree to or even know about the transfer of the warehouse engineered by petitioners in this matter. * * * I therefore agree with the trial court that the warehouse should be returned from 2005 New Jersey LLC to New Jersey LLC. Accordingly, I respectfully dissent.

Questions for Analysis

1. **Law.** What did the Wisconsin Supreme Court conclude with respect to the applicable law in this case, and what was the result?
2. **Law.** What was the dissent's position on the facts in this case? On what reasoning did the dissent base its disagreement with the majority?
3. **Economic Dimensions.** What might be an appropriate remedy in this case if the standard set out by the state supreme court is found to have been violated?
4. **Social Dimensions.** Why did the state supreme court remand the case to the trial court instead of the intermediate appellate court?
5. **Implications for Members of LLCs.** Of what should the members of an LLC—or the participants in any business organization—be aware when they have a "material conflict of interest" with other members?

Property and Its Protection

UNIT 9

Unit Contents

CHAPTER 28

Personal Property and Bailments

LEARNING OBJECTIVES

After reading this chapter, you should be able to answer the following questions:

1. What is real property? What is personal property?
2. What does it mean to own property in fee simple? What is the difference between a joint tenancy and a tenancy in common?
3. What are the three elements necessary for an effective gift? How else can property be acquired?
4. What are the three elements of a bailment?
5. What are the basic rights and duties of a bailee? What are the rights and duties of a bailor?

Property consists of the legally protected rights and interests a person has in anything with an ascertainable value that is subject to ownership. Property would have little value (and the word would have little meaning) if the law did not define the right to use it, to sell or dispose of it, and to prevent trespass on it.

Property is divided into real property and personal property. **Real property** (sometimes called *realty* or *real estate*) means the land and everything permanently attached to it. Everything else is **personal property,** or *personalty.* Attorneys sometimes refer to personal property as **chattel,** a term used under the common law to denote all forms of personal property. Personal property can be tangible or intangible. *Tangible* personal property, such as a television set or a car, has physical substance. *Intangible* personal property represents some set of rights and interests but has no real physical existence. Stocks and bonds, patents, and copyrights are examples of intangible personal property.

In the first part of this chapter, we look at the ways in which title to property is held; the methods of acquiring ownership of personal property; and issues relating to mislaid, lost, and abandoned personal property. In the second part of the chapter, we examine bailment relationships. A *bailment* is created when personal property is temporarily delivered into the care of another without a transfer of title, such as when you leave your car with a parking attendant.

Property Ownership

Property ownership[1] can be viewed as a bundle of rights, including the right to possess property and to dispose of it—by sale, gift, lease, or other means.

FEE SIMPLE

A person who holds the entire bundle of rights to property is said to be the owner in **fee simple.** The owner in fee simple is entitled to use, possess, or dispose of the property as he or she chooses during his or her lifetime, and on this owner's death, the interests in the property descend to his or her heirs. We will return to this form of property ownership in Chapter 29, in the context of ownership rights in real property.

1. The principles discussed in this section apply equally to real property ownership, which will be discussed in Chapter 29.

CONCURRENT OWNERSHIP

Persons who share ownership rights simultaneously in a particular piece of property are said to be *concurrent* owners. There are two principal types of **concurrent ownership:** *tenancy in common* and *joint tenancy.* Other types of concurrent ownership include tenancy by the entirety and community property.

Tenancy in Common The term **tenancy in common** refers to a form of co-ownership in which each of two or more persons owns an *undivided* interest in the property. The interest is undivided because each tenant has rights in the *whole* property. ■ EXAMPLE 28.1 Rosa and Chad own a rare stamp collection together as tenants in common. This does not mean that Rosa owns some particular stamps and Chad others. Rather, it means that Rosa and Chad each have rights in the *entire* collection. (If Rosa owned some of the stamps and Chad owned others, then the interest would be *divided.*) ■

On the death of a tenant in common, that tenant's interest in the property passes to her or his heirs. ■ EXAMPLE 28.2 Should Rosa die before Chad, a one-half interest in the stamp collection will become the property of Rosa's heirs. If Rosa sells her interest to Fred before she dies, Fred and Chad will be co-owners as tenants in common. If Fred dies, his interest in the personal property will pass to his heirs, and they in turn will own the property with Chad as tenants in common. ■

How should the value of the property owned by two tenants in common be apportioned when neither tenant has died but both agree that their interests should be divided? That was the question in the following case.

CASE 28.1 Clark v. Dady

Missouri Court of Appeals,
Western District, 2004.
131 S.W.3d 382.
http://www.findlaw.com/11stategov/mo/moca.html[a]

FACTS In 1998, John Dady and Mary Clark bought a mobile home in Missouri for $35,848. Clark made a $4,000 down payment. Greentree Financial financed the $31,848 balance through a promissory note signed by both parties, in whose names title was issued. They lived together until August 2001 when Dady moved out. Clark made the subsequent payments on the note and for the lot rental and utilities. Less than six months later, Clark filed a petition in a Missouri state court against Dady, asking in part that she be adjudged "the rightful owner of [the] personal property due to the care and amount of money that she has contributed to the purchase." A balance of $31,964.19, including interest, was due on the note. John Dady asked to be reimbursed for the cost of adding a $10,000 deck, a $1,500 barn, and $2,000 worth of landscaping, plus "hours of painting" and his portion of the rental value of the mobile home after he had moved out. The court awarded the mobile home to Clark and ordered Dady to pay her $3,050 for his half of the lot and loan payments that she made after he had left. Dady appealed to a state intermediate appellate court.

ISSUE Did the trial court properly determine the respective interests of the owners of the property?

a. In the "Court of Appeals" section, in the "2004" row, click on "April." On the next page, in the "04/06/2004" section, click on the docket number beside the name of the case to access the opinion.

DECISION No. The state intermediate appellate court reversed the judgment of the lower court and remanded the case for a decision as to which expenses should be credited to the parties.

REASON The court held that tenants in common own property in equal shares if the transfer that created their interests does not state otherwise, or if there is no other evidence showing the contributions of the owners toward the acquisition of the property to be unequal. Contributions toward acquisition include liability incurred in financing any part of the purchase price. Here, Clark made the down payment, but the parties jointly obligated themselves for the payment of the balance by signing a promissory note. The court explained, "As a result of the appellant's [Dady's] being jointly obligated on the note, he is considered to have contributed $15,924 to the acquisition of the mobile home ($31,848/2)." In other words, he owned "a 44.42% ($15,924/$35,848) share." The court acknowledged its responsibility to make a final determination of the parties' interests, but remanded the case because "the record is insufficient to allow us to determine what, if any, expenditures should be reimbursed to the parties."

WHY IS THIS CASE IMPORTANT? *As this case illustrates, how property is held can have a significant impact on the respective rights of the parties. In the absence of conflicting evidence, courts will presume that tenants in common own equal shares of the property, even if the parties' contributions to the property were unequal.*

Joint Tenancy In a **joint tenancy,** each of two or more persons owns an undivided interest in the property, and a deceased joint tenant's interest passes to the surviving joint tenant or tenants. The rights of a surviving joint tenant to inherit a deceased joint tenant's ownership interest—which are referred to as *survivorship rights*—distinguish the joint tenancy from the tenancy in common. A joint tenancy can be terminated by gift or by sale before a joint tenant's death; in this situation, the person who receives the property as a gift or who purchases the property becomes a tenant in common, not a joint tenant.

■ EXAMPLE 28.3 If, in the preceding example, Rosa and Chad held their stamp collection in a joint tenancy and if Rosa died before Chad, the entire collection would become the property of Chad; Rosa's heirs would receive absolutely no interest in the collection. If Rosa, while living, sold her interest to Fred, however, the sale would terminate the joint tenancy, and Fred and Chad would become owners as tenants in common. ■

Generally, it is presumed that a co-tenancy is a tenancy in common unless there is a clear intention to establish a joint tenancy. Thus, language such as "to Jerrold and Eva as joint tenants with right of survivorship, and not as tenants in common" would be necessary to create a joint tenancy.

Tenancy by the Entirety and Community Property A **tenancy by the entirety** is a less common form of ownership that can be created by a conveyance (transfer) of real property to a husband and wife. It differs from a joint tenancy only by the fact that neither spouse can make a separate lifetime transfer of his or her interest without the consent of the other spouse. In some states where statutes give the wife the right to convey her property, this form of concurrent ownership has been effectively abolished. A divorce, either spouse's death, or mutual agreement will terminate a tenancy by the entirety.

A married couple is allowed to own property as **community property** in only a limited number of states.[2] If property is held as community property, each spouse technically owns an undivided one-half interest in property acquired during the marriage. Generally, community property does not include property acquired prior to the marriage or property acquired by gift or inheritance as separate property during the marriage. After a divorce, community property is divided equally in some states and according to the discretion of the court in other states.

2. These states include Alaska, Arizona, California, Idaho, Louisiana, Nevada, New Mexico, Texas, Washington, and Wisconsin. Puerto Rico allows property to be owned as community property as well.

Acquiring Ownership of Personal Property

The most common way of acquiring personal property is by purchasing it. We have already discussed the purchase and sale of personal property (goods) in Chapters 14 through 17. Often, property is acquired by will or inheritance, a topic we will cover in Chapter 30. Here we look at additional ways in which ownership of personal property can be acquired, including acquisition by possession, production, gift, accession, and confusion.

Possession

One example of acquiring ownership by possession is the capture of wild animals. Wild animals in their natural state belong to no one, and the first person to take possession of a wild animal normally owns it. The killing of a wild animal amounts to assuming ownership of it. Merely being in hot pursuit does not give title, however. This basic rule has two exceptions. First, any wild animals captured by a trespasser are the property of the landowner, not the trespasser. Second, if wild animals are captured or killed in violation of wild-game statutes, the state, and not the capturer, obtains title to the animals.

Those who find lost or abandoned property can also acquire ownership rights through mere possession of the property, as will be discussed later in the chapter. (Ownership rights in real property can also be acquired through possession, such as *adverse possession*—see Chapter 29.)

Production

Production—the fruits of labor—is another means of acquiring ownership of personal property. For instance, writers, inventors, and manufacturers all produce personal property and thereby acquire title to it. (In some situations, though, as when a researcher is hired to invent a new product or technique, the researcher-producer may not own what is produced—see Chapter 22.)

Gifts

A **gift** is another fairly common means of acquiring and transferring ownership of real and personal property. A gift is essentially a voluntary transfer of property ownership for which no consideration is given. As discussed in Chapter 8, the presence of consideration is what distinguishes a contract from a gift. To be an effective gift, three requirements must be met—donative intent on the part of

the *donor* (the one giving the gift), delivery, and acceptance by the *donee* (the one receiving the gift).

Donative Intent When a gift is challenged in court, the court will determine whether donative intent exists by looking at the language of the donor and the surrounding circumstances. ■ EXAMPLE 28.4 A court may look at the relationship between the parties and the size of the gift in relation to the donor's other assets. A gift to a mortal enemy is viewed with suspicion. Similarly, when a gift represents a large portion of a person's assets, the court will scrutinize the transaction closely to determine the mental capacity of the donor and ascertain whether any element of fraud or duress is present. ■

Delivery The gift must be delivered to the donee. Delivery is obvious in most cases, but some objects cannot be relinquished physically. Then the question of delivery depends on the surrounding circumstances.

—Constructive Delivery. When the object itself cannot be physically delivered, a symbolic, or constructive, delivery will be sufficient. **Constructive delivery** does not confer actual possession of the object in question, only the right to take actual possession. Thus, constructive delivery is a general term used to describe an action that the law holds to be the equivalent of real delivery. ■ EXAMPLE 28.5 Suppose that you want to make a gift of various old rare coins that you have stored in a safe-deposit box at your bank. You certainly cannot deliver the box itself to the donee, and you do not want to take the coins out of the bank. In this situation, you can simply deliver the key to the box to the donee and authorize the donee's access to the box and its contents. This action constitutes a constructive delivery of the contents of the box. ■

The delivery of intangible property—such as stocks, bonds, insurance policies, and contracts, for example—must always be accomplished by symbolic, or constructive, delivery. This is because the documents represent rights and are not, in themselves, the true property.

—Delivery by Agents. Delivery may be accomplished by means of a third person who is the agent of either the donor or the donee. If the third person is the agent of the donor, the delivery is effective when the agent delivers the gift to the donee. If the third person is the agent of the donee, the gift is effectively delivered when the donor delivers the property to the donee's agent.[3] Naturally, no delivery is necessary if the gift is already in the hands of the donee. All that is necessary to complete the gift in such a situation is that the donor had the required intent and the donee accepted the gift.

—Relinquishing Control and Dominion. An effective delivery also requires giving up complete control and **dominion** (ownership rights) over the subject matter of the gift. The outcome of disputes often turns on whether control has actually been relinquished. The Internal Revenue Service scrutinizes transactions between relatives when one claims to have given income-producing property to the other. A relative who does not relinquish complete control over a piece of property will have to pay taxes on the income from that property, as opposed to the family member who received the "gift."

In the following classic case, the court focused on the requirement that a donor must relinquish complete control and dominion over property given to the donee before a gift can be effectively delivered.

3. *Bickford v. Mattocks,* 95 Me. 547, 50 A. 894 (1901).

LANDMARK AND CLASSIC CASES

CASE 28.2 In re Estate of Piper

Missouri Court of Appeals, 1984.
676 S.W.2d 897.

FACTS Gladys Piper died intestate (without a will) in 1982. At her death, she owned miscellaneous personal property worth $5,000 and had in her purse $200 in cash and two diamond rings, known as the Andy Piper rings. The contents of her purse were taken by her niece, Wanda Brown, allegedly to preserve them for the estate. Clara Kaufmann, a friend of Piper's, filed a claim against the estate for $4,800. From October 1974 until Piper's death, Kaufmann had taken Piper to the doctor, hairdresser, and grocery store; had written her checks to pay her bills; and had helped her care for her home. Kaufmann maintained that Piper had promised to pay her for these services and had given her the diamond rings as a gift. A Missouri state trial court denied her request for payment; the court found that her services had been voluntary. Kaufmann then filed a petition for delivery of personal property—the rings—which was granted by the trial court. Brown, other heirs, and the administrator of Piper's estate appealed.

(Continued)

CASE 28.2–CONTINUED

ISSUE Had Gladys Piper made an effective gift of the rings to Clara Kaufmann?

DECISION No. The state appellate court reversed the judgment of the trial court on the ground that Piper had never delivered the rings to Kaufmann.

REASON Kaufmann claimed that the rings belonged to her by reason of a "consummated gift long prior to the death of Gladys Piper." Two witnesses testified for Kaufmann at the trial that Piper had told them the rings belonged to Kaufmann but that she was going to wear them until she died. The appellate court found "no evidence of any actual delivery." The court held that the essentials of a gift are (1) a present intention to make a gift on the part of the donor, (2) a delivery of the property by the donor to the donee, and (3) an acceptance by the donee. The evidence in the case showed only an intent to make a gift. Because there was no delivery—either actual or constructive—a valid gift was not made. For Piper to have made a gift, her intention would have to have been executed by the complete and unconditional delivery of the property or the delivery of a proper written instrument evidencing the gift. As this did not occur, the court found that there had been no gift.

IMPACT OF THIS CASE ON TODAY'S LAW Although this case is relatively recent in the long span of the law governing gifts, we present it here as a classic case because it so clearly illustrates the delivery requirement when making a gift. Assuming that Piper did, indeed, intend for Kaufmann to have the rings, it was unfortunate that Kaufmann had no right to receive them after Piper's death. Yet the alternative could lead to perhaps even more unfairness. The policy behind the delivery requirement is to protect alleged donors and their heirs from fraudulent claims based solely on parol evidence. If not for this policy, an alleged donee could easily claim that a gift was made when, in fact, it was not.

RELEVANT WEB SITES *To locate information on the Web concerning* In re Estate of Piper, *go to this text's Web site at* **academic.cengage.com/blaw/wbl**, *select "Chapter 28," and click on "URLs for Landmark Cases."*

Acceptance The final requirement of a valid gift is acceptance by the donee. This rarely presents any problem, as most donees readily accept their gifts. The courts generally assume acceptance unless the circumstances indicate otherwise.

Gifts *Inter Vivos* and Gifts *Causa Mortis* A gift made during one's lifetime is termed a **gift *inter vivos*. Gifts *causa mortis*** (so-called *deathbed gifts*), in contrast, are made in contemplation of imminent death. A gift *causa mortis* does not become absolute until the donor dies from the contemplated illness, and it is automatically revoked if the donor recovers from the illness. Moreover, the donee must survive to take the gift. To be effective, a gift *causa mortis* must also meet the three requirements discussed earlier—donative intent, delivery, and acceptance by the donee.

■ EXAMPLE 28.6 Suppose that Yang is to be operated on for a cancerous tumor. Before the operation, he delivers an envelope to a close business associate. The envelope contains a letter saying, "I realize my days are numbered, and I want to give you this check for $1 million in the event of my death from this operation." The business associate cashes the check. The surgeon performs the operation and removes the tumor. Yang recovers fully. Several months later, Yang dies from a heart attack that is totally unrelated to the operation. If Yang's personal representative (the party charged with administering Yang's estate) tries to recover the $1 million, she normally will succeed. The gift *causa mortis* is automatically revoked if the donor recovers. The *specific event* that was contemplated in making the gift was death from a particular operation. Because Yang's death was not the result of this event, the gift is revoked, and the $1 million passes to Yang's estate. ■

ACCESSION

Accession means "something added." Accession occurs when someone adds value to an item of personal property by use of either labor or materials. Generally, there is no dispute about who owns the property after the accession occurs, especially when the accession is accomplished with the owner's consent. ■ EXAMPLE 28.7 A Corvette-customizing specialist comes to Hoshi's house. Hoshi has all the materials necessary to customize the car. The customizing specialist uses them to add a unique bumper to Hoshi's Corvette. Hoshi simply pays the customizer for the value of the labor, obviously retaining title to the property. ■

When accession occurs without the permission of the owner, the courts will tend to favor the owner over the improver—the one who improves the property—provided that the accession was wrongful and undertaken in bad faith. This is true even if the accession increased the value of the property substantially. In addition, many courts

will deny the improver (wrongdoer) any compensation for the value added.

If the accession is performed in good faith, however, even without the owner's consent, ownership of the improved item most often depends on whether the accession has increased the value of the property or changed its identity. The greater the increase in value, the more likely that ownership will pass to the improver. If ownership does pass, the improver must compensate the original owner for the value of the property prior to the accession. If the increase in value is not sufficient for ownership to pass to the improver, most courts will require the owner to compensate the improver for the value added.

CONFUSION

Confusion is the commingling (mixing together) of goods so that one person's personal property cannot be distinguished from another's. Confusion frequently occurs when the goods are *fungible*. *Fungible goods* are goods consisting of identical particles, such as grain or oil. For instance, if two farmers put their number 2–grade winter wheat into the same storage bin, confusion will occur and the farmers become tenants in common.

When goods are confused due to a wrongful and willful act and the wrongdoer is unable to prove what percentage of the confused goods belongs to him or her, then the innocent party ordinarily acquires title to the whole. If confusion occurs as a result of agreement, an honest mistake, or the act of some third party, the owners share ownership as tenants in common and will share any loss in proportion to their ownership interests in the property.

■ EXAMPLE 28.8 Five farmers in a small Iowa community enter into a cooperative arrangement. Each fall, the farmers harvest the same amount of number 2–grade yellow corn and store it in silos that are held by the cooperative. Each farmer thus owns one-fifth of the total corn in the silos. If, however, one farmer harvests and stores more corn than the others in the cooperative silos and wants to claim a greater ownership interest, that farmer must keep careful records. Otherwise, the courts will presume that each farmer has an equal interest in the corn. ■

Mislaid, Lost, and Abandoned Property

As already mentioned, one of the methods of acquiring ownership of property is to possess it. Simply finding something and holding on to it, however, does not necessarily give the finder any legal rights in the property. Different rules apply, depending on whether the property was mislaid, lost, or abandoned.

MISLAID PROPERTY

Property that has voluntarily been placed somewhere by the owner and then inadvertently forgotten is **mislaid property.** ■ EXAMPLE 28.9 Suppose that you go to the theater. You leave your iPod on the concession stand and then forget about it. The iPod is mislaid property, and the theater owner is entrusted with the duty of reasonable care for it. ■ When mislaid property is found, the finder does not obtain title to the goods. Instead, the owner of the place where the property was mislaid becomes the caretaker of the property because it is highly likely that the true owner will return.[4]

LOST PROPERTY

Property that is involuntarily left and forgotten is **lost property.** A finder of the property can claim title to the property against the whole world *except the true owner.*[5] If the true owner demands that the lost property be returned, the finder must return it. If a third party attempts to take possession of lost property from a finder, the third party cannot assert a better title than the finder. Under the doctrine of *relativity of title,* if two opposing parties are before the court, neither of whom can claim absolute title to the property, the one who can claim prior possession will likely have established sufficient rights to the property to win the case.

Many states have **estray statutes,** which encourage and facilitate the return of property to its true owner and then reward a finder for honesty if the property remains unclaimed. These laws provide an incentive for finders to report their discoveries by making it possible for them, after the passage of a specified period of time, to acquire legal title to the property they have found. Such statutes usually require the county clerk to advertise the property in an attempt to help the owner recover what has been lost. Some preliminary questions must always be resolved before the estray statute can be employed. The item must be lost property, not merely mislaid property. When the circumstances indicate that the property was probably lost and not mislaid or abandoned, loss is presumed as a matter of public policy, and the estray statute applies.

ABANDONED PROPERTY

Property that has been discarded by the true owner, with no intention of reclaiming title to it, is **abandoned property.** Someone who finds abandoned property acquires title to it,

4. The finder of mislaid property is an involuntary bailee (to be discussed later in this chapter).

5. See *Armory v. Delamirie,* 93 Eng.Rep. 664 (K.B. [King's Bench] 1722).

and such title is good against the whole world, *including the original owner.* The owner of lost property who eventually gives up any further attempt to find it is frequently held to have abandoned the property. If a person finds abandoned property while trespassing on the property of another, title vests in the owner of the land, not in the finder.

■ EXAMPLE 28.10 Aleka is driving with the windows down in her car. Somewhere along her route, a valuable scarf blows out the window. She retraces her route and searches for the scarf but cannot find it. She finally gives up her search and proceeds to her destination five hundred miles away. Six months later, Frye, a hitchhiker, finds the scarf. Frye has acquired title, which is good even against Aleka. By completely giving up her search, Aleka abandoned the scarf just as effectively as if she had intentionally discarded it. ■

Bailments

A **bailment** is formed by the delivery of personal property, without transfer of title, by one person, called a **bailor,** to another, called a *bailee,* usually under an agreement for a particular purpose—for example, to loan, lease, store, repair, or transport the property. On completion of the purpose, the bailee is obligated to return the bailed property in the same or better condition to the bailor or a third person or to dispose of it as directed. Bailments are usually created by agreement, but not necessarily by contract, because in many bailments not all of the elements of a contract (such as mutual assent and consideration) are present.

The law of bailments applies to many routine personal and business transactions. When individuals deal with bailments, whether they realize it or not, they are subject to the obligations and duties that arise from the bailment relationship. The number, scope, and importance of bailments created daily in the business community and in everyday life make it desirable that every person understand the elements necessary for the creation of a bailment and know what rights, duties, and liabilities flow from bailments.

Elements of a Bailment

Not all transactions involving the delivery of property from one person to another create a bailment. For such a transfer to become a bailment, three conditions must be met. We look here at each of these conditions.

Personal Property Only personal property is bailable; there can be no bailment of persons. Although a bailment of your luggage is created when it is transported by an airline, as a passenger you are not the subject of a bailment. Additionally, you cannot bail realty; thus, leasing your house to a tenant does not create a bailment. Although bailments commonly involve *tangible* items—jewelry, cattle, automobiles, and the like—*intangible* personal property, such as promissory notes and shares of corporate stock, may also be bailed.

Delivery of Possession *Delivery of possession* means the transfer of possession of the property to the bailee. Two requirements must be met for delivery of possession to occur:

1. The bailee must be given exclusive possession and control over the property.
2. The bailee must *knowingly* accept the personal property.[6] In other words, the bailee must *intend* to exercise control over it.

If either delivery of possession or knowing acceptance is lacking, there is no bailment relationship.

Two types of delivery—*physical* and *constructive*—will result in the bailee's exclusive possession of and control over the property. As discussed earlier, in the context of gifts, constructive delivery is a substitute, or symbolic, delivery. What is delivered to the bailee is not the actual property bailed (such as a car) but something so related to the property (such as the car keys) that the requirement of delivery is satisfied.

In certain unique situations, a bailment is found despite the apparent lack of the requisite elements of control and knowledge. In particular, the rental of a safe-deposit box is usually held to create a bailor-bailee relationship between the customer and the bank, despite the bank's lack of knowledge of the contents and its inability to have exclusive control of the property.[7] Another example of such a situation occurs when the bailee acquires the property accidentally or by mistake—as in finding someone else's lost or mislaid property. A bailment is created even though the bailor did not voluntarily deliver the property to the bailee. Such bailments are called *constructive* or *involuntary* bailments.

Bailment Agreement A bailment agreement, or contract, can be express or implied. Although a written agreement is not required for bailments of less than one year (that is, the Statute of Frauds does not apply—see Chapter 10), it

6. In this instance, we are dealing with *voluntary* bailments. This does not apply to *involuntary* bailments.

7. By statute or by express contract, the rental of a safe-deposit box may be regarded as a lease of space or a license (a revocable right to use the space, for a fee—see Chapter 29) instead of a bailment.

is a good idea to have one, especially when valuable property is involved.

The bailment agreement expressly or impliedly provides for the return of the bailed property to the bailor or to a third person, or it provides for disposal by the bailee. The agreement presupposes that the bailee will return the identical goods originally given by the bailor. In certain types of bailments, though, such as bailments of fungible goods, the property returned need only be equivalent property.

■ EXAMPLE 28.11 If Holman stores his grain (fungible goods) in Joe's Warehouse, a bailment is created. At the end of the storage period, however, the warehouse is not obligated to return to Holman exactly the same grain that he stored. As long as the warehouse returns goods of the same *type, grade,* and *quantity,* the warehouse—the bailee—has performed its obligation. ■

Ordinary Bailments

Bailments are either *ordinary* or *special (extraordinary).* There are three types of ordinary bailments. They are distinguished according to *which party receives a benefit from the bailment.* This factor will dictate the rights and liabilities of the parties, and the courts may use it to determine the standard of care required of the bailee in possession of the personal property. The three types of ordinary bailments are as follows:

1. *Bailment for the sole benefit of the bailor.* This is a gratuitous bailment (a bailment without consideration) for the convenience and benefit of the bailor. ■ EXAMPLE 28.12 Allen asks his friend, Sumi, to store his car in her garage while he is away. If Sumi agrees to do so, then it is a gratuitous bailment because the bailment of the car is for the sole benefit of the bailor (Allen). ■
2. *Bailment for the sole benefit of the bailee.* This type of bailment typically occurs when one person lends an item to another person (the bailee) solely for the bailee's convenience and benefit. ■ EXAMPLE 28.13 Allen asks to borrow Sumi's boat so that he can go sailing over the weekend. The bailment of the boat is for Allen's (the bailee's) sole benefit. ■
3. *Bailment for the mutual benefit of the bailee and the bailor.* This is the most common kind of bailment and involves some form of compensation for storing items or holding property while it is being serviced. It is a contractual bailment and may be referred to as a *bailment for hire* or a *commercial bailment.* ■ EXAMPLE 28.14 Allen leaves his car at a service station for an oil change. Because the service station will be paid to change Allen's oil, this is a mutual-benefit bailment. ■ Many lease arrangements in which the lease involves goods (leases were discussed in Chapters 14 through 17) also fall into this category of bailment once the lessee takes possession.

Rights of the Bailee Certain rights are implicit in the bailment agreement. Generally, the bailee has the right to take possession, to utilize the property for accomplishing the purpose of the bailment, to receive some form of compensation, and to limit her or his liability for the bailed goods. These rights of the bailee are present (with some limitations) in varying degrees in all bailment transactions.

—Right of Possession. A hallmark of the bailment agreement is that the bailee acquires the *right to control and possess the property temporarily.* The bailee's right of possession permits the bailee to recover damages from any third person for damage or loss of the property. If the property is stolen, the bailee has a legal right to regain possession of it or to obtain damages from any third person who has wrongfully interfered with the bailee's possessory rights. The bailee's right to regain possession of the property or to obtain damages is important because, as you will read shortly, a bailee is liable to the bailor for any loss or damage to bailed property resulting from the bailee's negligence.

—Right to Use Bailed Property. Depending on the type of bailment and the terms of the bailment agreement, a bailee may also have a right to use the bailed property. When no provision is made, the extent of use depends on how necessary it is for the goods to be at the bailee's disposal for the ordinary purpose of the bailment to be carried out. ■ EXAMPLE 28.15 If you borrow a friend's car to drive to the airport, you, as the bailee, would obviously be expected to use the car. In a bailment involving the long-term storage of a car, however, the bailee is not expected to use the car because the ordinary purpose of a storage bailment does not include use of the property. ■

—Right of Compensation. Except in a gratuitous bailment, a bailee has a right to be compensated as provided for in the bailment agreement, to be reimbursed for costs and services rendered in the keeping of the bailed property, or both.

To enforce the right of compensation, the bailee has a right to place a *possessory lien* (which entitles a creditor to retain possession of the debtor's goods until a debt is paid) on the specific bailed property until he or she has been fully compensated. This type of lien, sometimes

EXHIBIT 28–1 DEGREE OF CARE REQUIRED OF A BAILEE

Bailment for the Sole Benefit of the Bailor	Mutual-Benefit Bailment	Bailment for the Sole Benefit of the Bailee
DEGREE OF CARE →		
SLIGHT	REASONABLE	GREAT

referred to as an *artisan's lien* or a *bailee's lien,* was discussed in Chapter 21.

—Right to Limit Liability. In ordinary bailments, bailees have the right to limit their liability as long as the limitations are called to the attention of the bailor and are not against public policy. It is essential that the bailor be informed of the limitation in some way. Even when the bailor knows of the limitation, certain types of disclaimers of liability have been considered to be against public policy and therefore illegal. The courts carefully scrutinize *exculpatory clauses,* or clauses that limit a person's liability for her or his own wrongful acts, and in bailments they are often held to be illegal. This is particularly true in bailments for the mutual benefit of the bailor and the bailee.

Duties of the Bailee The bailee has two basic responsibilities: (1) to take appropriate care of the property and (2) to surrender the property to the bailor or dispose of it in accordance with the bailor's instructions at the end of the bailment.

—The Duty of Care. The bailee must exercise reasonable care in preserving the bailed property. What constitutes reasonable care in a bailment situation normally depends on the nature and specific circumstances of the bailment. Traditionally, the courts have determined the appropriate standard of care on the basis of the type of bailment involved. In a bailment for the sole benefit of the bailor, for example, the bailee need exercise only a slight degree of care. In a bailment for the sole benefit of the bailee, however, the bailee must exercise great care. In a mutual-benefit bailment, courts normally impose a reasonable standard of care—that is, the bailee must exercise the degree of care that a reasonable and prudent person would exercise in the same circumstances. Exhibit 28–1 illustrates these concepts. A bailee's failure to exercise appropriate care in handling the bailor's property results in tort liability.

—Duty to Return Bailed Property. At the end of the bailment, the bailee normally must hand over the original property to either the bailor or someone the bailor designates or must otherwise dispose of it as directed. This is usually a *contractual* duty arising from the bailment agreement (contract). Failure to give up possession at the time the bailment ends is a breach of contract and could result in the tort of conversion or an action based on bailee negligence. If the bailed property has been lost or is returned damaged, a court will presume that the bailee was negligent. The exception is when the obligation is excused because the goods or chattels have been destroyed, lost, or stolen through no fault of the bailee (or claimed by a third party with a superior claim).

Because the bailee has a duty to return the bailed goods to the bailor, a bailee may be liable if the goods being held or delivered are given to the wrong person. Hence, a bailee must be satisfied that a person (other than the bailor) to whom the goods are being delivered is the actual owner or has authority from the owner to take possession of the goods. Should the bailee deliver in error, then the bailee may be liable for conversion or misdelivery.

The following case involved the lease (commercial bailment) of a construction crane. The question was whether the lessee (bailee) was liable under the lease for not returning the crane.

CASE 28.3 Sunbelt Cranes Construction and Hauling, Inc. v. Gulf Coast Erectors, Inc.

United States District Court,
Middle District of Florida, 2002.
189 F.Supp.2d 1341.

FACTS In April 2000, Sunbelt Cranes Construction and Hauling, Inc., leased a construction crane to Gulf Coast Erectors, Inc., in Tampa, Florida. Clause twelve of the parties' "Operated and Maintained Equipment Lease" agreement provided as follows:

> Lessee agrees that Lessee has accepted the entire risk of loss of, loss of use of, and damage to, the Equipment. Lessee

CASE 28.3—CONTINUED

> agrees to pay Lessor for any and all loss of, loss of use of, and/or damage to, the Equipment, due to any reason or cause * * * until the Equipment has been returned to and accepted by Lessor, even though the reason or cause may be due to accident or act of God * * * .

Gulf used the crane in a residential construction project in Tampa's Ybor City neighborhood. On May 19, an accident at the construction site caused a fire that engulfed many street blocks and destroyed a great deal of property, including the crane. Sunbelt filed a suit in a federal district court against Gulf, alleging breach of contract, to recover the cost of the crane, plus interest and attorneys' fees. Sunbelt filed a motion for summary judgment in its favor.

ISSUE Was Gulf liable for the failure to return the crane?

DECISION Yes. The court granted Sunbelt's motion for summary judgment. Gulf was liable for the failure to return the crane because Gulf assumed the risk of the crane's loss under the parties' lease agreement.

REASON The court explained that a lease is a contract and that ordinary rules of contract construction apply. If the terms are not ambiguous, they must be interpreted according to the plain language of the lease. "Absent agreement or negligence, the lessor is not responsible for destruction, injury or loss of the leased property. However, the lessor may properly agree to greater responsibility than the law requires." In this case, "Sunbelt and Gulf entered into an agreement where Gulf would bear the risk of loss under any circumstance. The clause held Gulf liable for any negligence of its own, for a third party or for any acts of God. The language is clear and unambiguous on its face. Gulf chose to accept the risk of loss. The parties were equipped with the knowledge of contractual agreements and chose to model their agreement in such a manner. As such, Gulf is liable to Sunbelt according to the contract."

FOR CRITICAL ANALYSIS—Economic Consideration *What might Gulf have done to protect itself from the loss it incurred when the crane was destroyed?*

Duties of the Bailor It goes without saying that the duties of a bailor are essentially the same as the rights of a bailee. Obviously, a bailor has a duty to compensate the bailee either as agreed or as reimbursement for costs incurred by the bailee in keeping the bailed property. A bailor also has an all-encompassing duty to provide the bailee with goods or chattels that are free from known defects that could cause injury to the bailee. The bailor's duty to reveal defects to the bailee translates into two rules:

1. In a *mutual-benefit bailment,* the bailor must notify the bailee of all known defects and any hidden defects that the bailor knows of or could have discovered with reasonable diligence and proper inspection.
2. In a *bailment for the sole benefit of the bailee,* the bailor must notify the bailee of any known defects.

The bailor's duty to reveal defects is based on a negligence theory of tort law. A bailor who fails to give the appropriate notice is liable to the bailee and to any other person who might reasonably be expected to come into contact with the defective article.

A bailor can also incur *warranty liability* based on contract law (see Chapter 17) for injuries resulting from the bailment of defective articles. Property leased by a bailor must be *fit for the intended purpose of the bailment.* Warranties of fitness arise by law in sales contracts and leases, and judges have extended these warranties to situations in which the bailees are compensated for the bailment (such as when one leaves a car with a parking attendant). Article 2A of the Uniform Commercial Code (UCC) extends the implied warranties of merchantability and fitness for a particular purpose to bailments whenever the bailments include rights to use the bailed goods.[8]

Special Types of Bailments

Up to this point, our discussion of bailments has been concerned with ordinary bailments—bailments in which bailees are expected to exercise ordinary care in the handling of bailed property. Some bailment transactions warrant special consideration. These include bailments in which the bailee's duty of care is *extraordinary*—that is, the bailee's liability for loss or damage to the property is absolute—as is generally true in bailments involving common carriers and innkeepers. Warehouse companies have the same duty of care as ordinary bailees; but, like carriers, they are subject to extensive regulation under federal and state laws, including Article 7 of the UCC.

Common Carriers Transportation providers that are publicly licensed to provide transportation services to the general public are referred to as **common carriers.** They are distinguished from private carriers, which operate transportation facilities for a select clientele. A private carrier is not required to provide service to every person or company making a request. A common carrier,

8. UCC 2A–212, 2A–213.

however, must arrange carriage for all who apply, within certain limitations.[9]

The delivery of goods to a common carrier creates a bailment relationship between the shipper (bailor) and the common carrier (bailee). Unlike ordinary bailees, the common carrier is held to a standard of care based on *strict liability,* rather than reasonable care, in protecting the bailed personal property. This means that the common carrier is absolutely liable, regardless of due care, for all loss or damage to goods except damage caused by one of the following common law exceptions: (1) an act of God, (2) an act of a public enemy, (3) an order of a public authority, (4) an act of the shipper, or (5) the inherent nature of the goods.

Common carriers cannot contract away their liability for damaged goods. Subject to government regulations, however, they are permitted to limit their dollar liability to an amount stated on the shipment contract or rate filing.[10]

Warehouse Companies *Warehousing* is the business of providing storage of property for compensation.[11] Like ordinary bailees, warehouse companies are liable for loss or damage to property resulting from *negligence.* A warehouse company, however, is a professional bailee and is therefore expected to exercise a high degree of care to protect and preserve the goods. A warehouse company can limit the dollar amount of its liability, but the bailor must be given the option of paying a higher rate to the bailee to increase the amount of liability.

Unlike ordinary bailees, a warehouse company can issue *documents of title*—in particular, *warehouse receipts*—and is subject to extensive government regulation, including Article 7 of the UCC.[12] A warehouse receipt describes the bailed property and the terms of the bailment contract. It can be negotiable or nonnegotiable, depending on how it is written. It is negotiable if its terms provide that the warehouse company will deliver the goods "to the bearer" of the receipt or "to the order of" a person named on the receipt.[13] The warehouse receipt represents the goods (that is, it indicates title) and hence has value and utility in financing commercial transactions.

■ EXAMPLE 28.16 Ossip delivers 6,500 cases of canned corn to Chan, the owner of a warehouse. Chan issues a negotiable warehouse receipt payable "to bearer" and gives it to Ossip. Ossip sells and delivers the warehouse receipt to Better Foods, Inc. Better Foods is now the owner of the corn and has the right to obtain the cases by simply presenting the warehouse receipt to Chan. ■

Innkeepers At common law, innkeepers, hotel owners, and similar operators were held to the same strict liability as common carriers with respect to property brought into the rooms by guests. Today, only those who provide lodging to the public for compensation as a *regular* business are covered under this rule of strict liability. Moreover, the rule applies only to those who are guests, as opposed to lodgers. A lodger is a permanent resident of the hotel or inn, whereas a guest is a transient traveler.

In many states, innkeepers can avoid strict liability for loss of guests' valuables and funds by providing a safe in which to keep them. Each guest must be clearly notified of the availability of such a safe. Statutes often limit the liability of innkeepers with regard to articles that are not kept in the safe or are of such a nature that they normally are not kept in a safe. These statutes may limit the amount of monetary damages or even provide for no liability in the absence of innkeeper negligence.

Normally, the innkeeper (a motel owner, for example) assumes no responsibility for the safety of a guest's automobile because the guest usually retains possession and control over it. If, however, the innkeeper provides parking facilities and the guest's car is entrusted to the innkeeper or to an employee, the innkeeper will be liable under the rules that pertain to parking-lot bailments (which are ordinary bailments).

9. A common carrier is not required to take any and all property anywhere in all instances. Public regulatory agencies govern common carriers, and carriers can be restricted to geographic areas. They can also be limited to carrying certain kinds of goods or to providing only special types of transportation equipment.

10. Federal laws require common carriers to offer shippers the opportunity to obtain higher dollar limits for loss by paying a higher fee for the transport.

11. UCC 7–102(h) defines the person engaged in the storing of goods for hire as a "warehouseman."

12. A *document of title* is defined in UCC 1–201(15) as any "document which in the regular course of business or financing is treated as adequately evidencing that the person in possession of it is entitled to receive, hold, and dispose of the document and the goods it covers. To be a document of title, a document must purport to be issued by or addressed to a bailee and purport to cover goods in the bailee's possession." A *warehouse receipt* is a receipt issued by a person engaged in the business of storing goods for hire.

13. UCC 7–104.

TERMS AND CONCEPTS

abandoned property 593
accession 592
bailment 594
bailor 594
chattel 588
common carrier 597
community property 590
concurrent ownership 589
confusion 593

constructive delivery 591
dominion 591
estray statute 593
fee simple 588
gift 590
gift *causa mortis* 592
gift *inter vivos* 592
joint tenancy 590
lost property 593
mislaid property 593
personal property 588
property 588
real property 588
tenancy by the entirety 590
tenancy in common 589

CHAPTER SUMMARY Personal Property and Bailments

	PERSONAL PROPERTY
Definition of Personal Property (See page 588.)	Personal property (personalty) includes all property not classified as real property (realty). Personal property can be tangible (such as a TV set or a car) or intangible (such as stocks or bonds). Personal property may be referred to legally as *chattel*—a term used under the common law to denote all forms of personal property.
Property Ownership (See pages 588–590.)	Having the fullest ownership rights in property is called *fee simple* ownership. There are various ways of co-owning property, including *tenancy in common, joint tenancy, tenancy by the entirety,* and *community property.*
Acquiring Ownership of Personal Property (See pages 590–593.)	The most common means of acquiring ownership in personal property is by purchasing it (see Chapters 14 through 17). Another way in which personal property is often acquired is by will or inheritance (see Chapter 30). Personal property can also be acquired by possession, production, gift, accession, and confusion.
Mislaid, Lost, and Abandoned Property (See pages 593–594.)	1. *Mislaid property*—Property that is placed somewhere voluntarily by the owner and then inadvertently forgotten. A finder of mislaid property will not acquire title to the goods, and the owner of the place where the property was mislaid becomes a caretaker of the mislaid property. 2. *Lost property*—Property that is involuntarily left and forgotten. A finder of lost property can claim title to the property against the whole world *except the true owner.* 3. *Abandoned property*—Property that has been discarded by the true owner, who has no intention of claiming title to the property in the future. A finder of abandoned property can claim title to it against the whole world, *including the original owner.*
	BAILMENTS
Elements of a Bailment (See pages 594–595.)	1. *Personal property*—Bailments involve only personal property. 2. *Delivery of possession*—For an effective bailment to exist, the bailee (the one receiving the property) must be given exclusive possession and control over the property, and in a voluntary bailment, the bailee must knowingly accept the personal property. 3. *The bailment agreement*—Expressly or impliedly provides for the return of the bailed property to the bailor or a third party, or for the disposal of the bailed property by the bailee.
Ordinary Bailments (See pages 595–597.)	1. *Types of bailments*— a. Bailment for the sole benefit of the bailor—A gratuitous bailment undertaken for the sole benefit of the bailor (for example, as a favor to the bailor). b. Bailment for the sole benefit of the bailee—A gratuitous loan of an article to a person (the bailee) solely for the bailee's benefit. c. Mutual-benefit (contractual) bailment—The most common kind of bailment; involves compensation between the bailee and bailor for the service provided.

(Continued)

CHAPTER SUMMARY Personal Property and Bailments—Continued

Ordinary Bailments—Continued	2. *Rights of a bailee (duties of a bailor)*—A bailee has the right to possess the bailed goods, the right to be compensated and reimbursed for expenses, and the right to limit liability for loss or damage of the goods. 3. *Duties of a bailee (rights of a bailor)*—A bailee must exercise appropriate care over property entrusted to her or him, and bailed goods in a bailee's possession must be either returned to the bailor or disposed of according to the bailor's directions.
Special Types of Bailments (See pages 597–598.)	1. *Common carriers*—Carriers that are publicly licensed to provide transportation services to the general public. A common carrier is held to a standard of care based on *strict liability* unless the bailed property is lost or destroyed due to (a) an act of God, (b) an act of a public enemy, (c) an order of a public authority, (d) an act of the shipper, or (e) the inherent nature of the goods. 2. *Warehouse companies*—Professional bailees that differ from ordinary bailees in that they (a) can issue documents of title (warehouse receipts) and (b) are subject to state and federal statutes, including Article 7 of the UCC (as are common carriers). They must exercise a high degree of care over the bailed property and are liable for loss of or damage to property if they fail to do so. 3. *Innkeepers (hotel operators)*—Those who provide lodging to the public for compensation as a *regular* business. The common law strict liability standard to which innkeepers were once held is limited today by state statutes, which vary from state to state.

FOR REVIEW

Answers for the even-numbered questions in this For Review *section can be found in Appendix A at the end of this text.*

1. What is real property? What is personal property?
2. What does it mean to own property in fee simple? What is the difference between a joint tenancy and a tenancy in common?
3. What are the three elements necessary for an effective gift? How else can property be acquired?
4. What are the three elements of a bailment?
5. What are the basic rights and duties of a bailee? What are the rights and duties of a bailor?

QUESTIONS AND CASE PROBLEMS

28–1. Duties of the Bailee. Discuss the standard of care traditionally required of the bailee for the bailed property in each of the following situations, and determine whether the bailee breached that duty.

(a) Ricardo borrows Steve's lawn mower because his own lawn mower needs repair. Ricardo mows his front yard. To mow the backyard, he needs to move some hoses and lawn furniture. He leaves the mower in front of his house while doing so. When he returns to the front yard, he discovers that the mower has been stolen.

(b) Alicia owns a valuable speedboat. She is going on vacation and asks her neighbor, Maureen, to store the boat in one stall of Maureen's double garage. Maureen consents, and the boat is moved into the garage. Maureen needs some grocery items for dinner and drives to the store. She leaves the garage door open while she is gone, as is her custom, and the speedboat is stolen during that time.

QUESTION WITH SAMPLE ANSWER

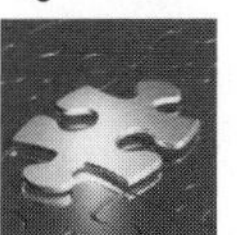

28–2. Reineken, very old and ill, wanted to make a gift to his nephew, Gerald. He had a friend obtain $2,500 in cash for him from his bank account, placed this cash in an envelope, and wrote on the envelope, "This is for my nephew, Gerald." Reineken then placed the envelope in his dresser drawer. When Reineken died a month later, his family found the envelope, and Gerald got word of the intended gift. Gerald then demanded that Reineken's daughter, the executor of Reineken's estate (the

person who was appointed by Reineken to handle his affairs after his death), turn over the gift to him. The daughter refused to do so. Discuss fully whether Gerald can successfully claim ownership rights to the $2,500.

For a sample answer to this question, go to Appendix B at the end of this text.

28–3. Bailments. Jole Liddle, a high school student in Salem School District No. 600, played varsity basketball. A letter from Monmouth College of West Long Branch, New Jersey, addressed to Liddle in care of the coach, was delivered to Liddle's school a few days after it was mailed on July 18, 1990. The letter notified Liddle that he was being recruited for a basketball scholarship. The school, which had a policy of promptly delivering any mail sent to students in care of the school, did not deliver the letter to Liddle until seven months later. Because Monmouth College had not heard from Liddle, the college ceased its efforts to recruit him. Liddle sued the school district, alleging that the coach was negligent in his duties as a bailee of the letter. The school district filed a motion to dismiss the case, arguing that the letter was not bailable property. Was the letter bailable property? Discuss fully. [*Liddle v. Salem School District No. 600,* 249 Ill.App.3d 768, 619 N.E.2d 530, 188 Ill.Dec. 905 (1993)]

28–4. Gratuitous Bailments. Raul Covarrubias, David Haro, and Javier Aguirre immigrated to the United States from Colima, Mexico, to find jobs and help their families. When they learned that Francisco Alcaraz-Garcia planned to travel to Colima, they asked him to deliver various sums, totaling more than $25,000, to their families. During customs inspections at the border, Alcaraz told officers of the U.S. Customs Service that he was not carrying more than $10,000. In fact, he carried more than $35,000. He was charged with—and convicted of—criminal currency and customs violations, and the government seized most of the cash. Covarrubias, Haro, and Aguirre filed a petition for the return of their money, arguing that Alcaraz was a gratuitous bailee and that they still had title to the money. Are they right? Explain fully. [*United States v. Alcaraz-Garcia,* 79 F.3d 769 (9th Cir. 1996)]

28–5. Gift *Inter Vivos.* Thomas Stafford owned four promissory notes. Payments on the notes were deposited into a bank account in the names of Stafford and his daughter, June Zink, "as joint tenants with right of survivorship." Stafford kept control of the notes and would not allow Zink to spend any of the proceeds. He also kept the interest on the account. On one note, Stafford indorsed "Pay to the order of Thomas J. Stafford or June S. Zink, or the survivor." The payee on each of the other notes was "Thomas J. Stafford and June S. Zink, or the survivor." When Stafford died, Zink took possession of the notes, claiming that she had been a joint tenant of the notes with her father. Stafford's son, also Thomas, filed a suit in a Virginia state court against Zink, claiming that the notes were partly his. The son argued that their father had not made a valid gift *inter vivos* of the notes to Zink. In whose favor will the court rule? Why? [*Zink v. Stafford,* 509 S.E.2d 833 (Va. 1999)]

CASE PROBLEM WITH SAMPLE ANSWER

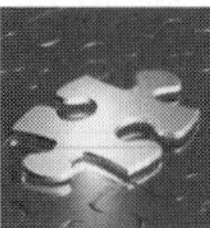

28–6. A. D. Lock owned Lock Hospitality, Inc., which in turn owned the Best Western Motel in Conway, Arkansas. Joe Terry and David Stocks were preparing the motel for renovation. As they were removing the ceiling tiles in Room 118, with Lock present in the room, they noticed a dusty cardboard box near the heating and air-supply vent where it had apparently been concealed. Terry climbed a ladder to reach the box, opened it, and handed it to Stocks. The box was filled with more than $38,000 in old currency. Lock took possession of the box and its contents. Terry and Stocks filed a suit in an Arkansas state court against Lock and his corporation to obtain the money. Should the money be characterized as lost, mislaid, or abandoned property? To whom should the court award it? Explain. [*Terry v. Lock,* 343 Ark. 452, 37 S.W.3d 202 (2001)]

After you have answered this problem, compare your answer with the sample answer given on the Web site that accompanies this text. Go to academic.cengage.com/blaw/wbl, select "Chapter 28," and click on "Case Problem with Sample Answer."

28–7. Concurrent Ownership. Vincent Slavin was a partner at Cantor Fitzgerald Securities in the World Trade Center (WTC) in New York City. In 1998, Slavin and Anna Baez became engaged and began living together. They placed both of their names on three accounts at Chase Manhattan Bank according to the bank's terms, which provided that "accounts with multiple owners are joint, payable to either owner or the survivor." Slavin arranged for the direct deposit of his salary and commissions into one of the accounts. On September 11, 2001, Slavin died when two planes piloted by terrorists crashed into the WTC towers, causing their collapse. At the time, the balance in the three accounts was $656,944.36. On September 14, Cantor Fitzgerald deposited an additional $58,264.73 into the direct-deposit account. Baez soon withdrew the entire amount from all of the accounts. Mary Jelnek, Slavin's mother, filed a suit in a New York state court against Baez to determine the ownership of the funds that had been in the accounts. In what form of ownership were the accounts held? Who is entitled to which of the funds, and why? [*In re Jelnek,* 3 Misc.3d 725, 777 N.Y.S.2d 871 (Sur. 2004)]

A QUESTION OF ETHICS

28–8. George Cook stayed at a Day's Inn motel in Nashville, Tennessee, while attending a trade show. At the trade show, Cook received orders for 225 cases of his firm's product, representing $17,336.25 in profits to the company. On the third day of his stay, Cook's room was burglarized while he was gone from the room. The burglar took Cook's order lists, as well as $174 in cash and medicine worth about $10. Cook sued the owner of the motel, Columbia Sussex Corp., alleging negligence. The motel defended by stating that it had posted a notice on the

door of Cook's room informing guests of the fact that the motel would not be liable for any valuable property not placed in the motel safe. Given these circumstances, evaluate and answer the following questions. [*Cook v. Columbia Sussex Corp.*, 807 S.W.2d 567 (Tenn. 1990)]

1. The relevant state statute governing the liability of innkeepers allowed motels to disclaim their liability by posting a notice such as the one posted by Day's Inn, but the statute also required that the notice be posted "in a conspicuous manner." The notice posted by Day's Inn on the inside of the door to Cook's room was six-by-three inches in size. In your opinion, is the notice sufficiently conspicuous? If you were the guest, would you have seen the disclaimer? Is it fair to guests to assume that they will notice such disclaimers? Discuss fully.
2. Should hotels or motels ever be allowed to disclaim liability by posting such notices? From a policy point of view, evaluate the implications of your answer.

VIDEO QUESTION

28–9. Go to this text's Web site at academic.cengage.com/blaw/wbl and select "Chapter 28." Click on "Video Questions" and view the video titled *Personal Property and Bailments*. Then answer the following questions.

1. What type of bailment is discussed in the video?
2. What were Vinny's duties with regard to the rug-cleaning machine? What standard of care should apply?
3. Did Vinny exercise the appropriate degree of care? Why or why not? How would a court decide this issue?

Personal Property and Bailments

For updated links to resources available on the Web, as well as a variety of other materials, visit this text's Web site at

academic.cengage.com/blaw/wbl

To learn whether a married person has ownership rights in a gift received by his or her spouse, go to the Web page of the Scott Law Firm at

http://www.scottlawfirm.com/property.htm

Some states and government agencies now post lists of unclaimed property online. For examples of various types of unclaimed property, go to the following Web page maintained by the state of Delaware:

http://www.state.de.us/revenue/information/Escheat.shtml

For a discussion of the origins of the term *bailment* and how bailment relationships have been defined, go to

http://www.lectlaw.com/def/b005.htm

You will find a hypertext version of Article 7 of the Uniform Commercial Code, which pertains to warehouse receipts, bills of lading, and other documents of title, at Cornell Law School's Legal Information Institute. Go to

http://www.law.cornell.edu/ucc/7/overview.html

ONLINE LEGAL RESEARCH

Go to the *Fundamentals of Business Law* home page at academic.cengage.com/blaw/wbl, select "Chapter 28," and click on "Internet Exercises." There you will find the following Internet research exercises that you can perform to learn more about topics covered in this chapter.

Activity 28–1: LEGAL PERSPECTIVE—Lost Property

Activity 28–2: MANAGEMENT PERSPECTIVE—Bailments

BEFORE THE TEST

Go to the *Fundamentals of Business Law* home page at academic.cengage.com/blaw/wbl, select "Chapter 28," and click on "Interactive Quizzes." You will find a number of interactive questions relating to this chapter.

CHAPTER 29

Real Property and Landlord-Tenant Law

LEARNING OBJECTIVES

After reading this chapter, you should be able to answer the following questions:

1. What can a person who holds property in fee simple absolute do with the property? Can a person who holds property as a life estate do the same?
2. What are the requirements for acquiring property by adverse possession?
3. What limitations may be imposed on the rights of property owners?
4. What is a leasehold estate? What types of leasehold estates, or tenancies, can be created when real property is leased?
5. What are the respective duties of the landlord and tenant concerning the use and maintenance of leased property? Is the tenant responsible for all damage that he or she causes?

From earliest times, property has provided a means for survival. Primitive peoples lived off the fruits of the land, eating the vegetation and wildlife. Later, as the wildlife was domesticated and the vegetation cultivated, property provided pasturage and farmland. In the twelfth and thirteenth centuries in Europe, the power of feudal lords was determined by the amount of land they held; the more land, the more powerful they were. After the age of feudalism passed, property continued to be an indicator of family wealth and social position. In the Western world, an individual's right to his or her property has become one of the most important rights of citizenship.

In this chapter, we first examine the nature of real property. We then look at the various ways in which real property can be owned and at how ownership rights in real property are transferred from one person to another. We conclude the chapter with a discussion of leased property and landlord-tenant relationships.

The Nature of Real Property

Real property consists of land and the buildings, plants, and trees that are on it. Real property also includes subsurface and airspace rights, as well as personal property that has become permanently attached to real property. Whereas personal property is movable, real property—also called *real estate* or *realty*—is immovable.

LAND

Land includes the soil on the surface of the earth and the natural or artificial structures that are attached to it. It further includes all the waters contained on or under the surface and much, but not necessarily all, of the airspace above it. The exterior boundaries of land extend down to the center of the earth and up to the farthest reaches of the sky (subject to certain qualifications).

AIRSPACE AND SUBSURFACE RIGHTS

The owner of real property has rights to the airspace above the land, as well as to the soil and minerals underneath it. Limitations on either airspace rights or subsurface rights normally must be indicated on the document that transfers title at the time of purchase. When no such limitations, or *encumbrances,* are noted, a purchaser normally can expect to have an unlimited right to possession of the property.

Airspace Rights Early cases involving airspace rights dealt with such matters as whether a telephone wire could be run across a person's property when the wire did not touch any of the property[1] and whether a bullet shot over a person's land constituted trespass.[2] Today, disputes concerning airspace rights may involve the right of commercial and private planes to fly over property and the right of individuals and governments to seed clouds and produce rain artificially. Flights over private land normally do not violate the property owners' rights unless the flights are low and frequent enough to cause a direct interference with the enjoyment and use of the land.[3] Leaning walls or buildings and projecting eave spouts or roofs may also violate the airspace rights of an adjoining property owner.

Subsurface Rights In many states, land ownership may be separated, in that the surface of a piece of land and the subsurface may have different owners. Subsurface rights can be extremely valuable, as these rights include the ownership of minerals, oil, and natural gas. But subsurface rights would be of little value if the owner could not use the surface to exercise those rights. Hence, a subsurface owner will have a right (called a *profit,* discussed later in this chapter) to go onto the surface of the land to, for example, discover and mine minerals.

1. *Butler v. Frontier Telephone Co.,* 186 N.Y. 486, 79 N.E. 716 (1906).
2. *Herrin v. Sutherland,* 74 Mont. 587, 241 P. 328 (1925). Shooting over a person's land constitutes trespass.
3. *United States v. Causby,* 328 U.S. 256, 66 S.Ct. 1062, 90 L.Ed. 1206 (1946).

PLANT LIFE AND VEGETATION

Plant life, both natural and cultivated, is also considered to be real property. In many instances, the natural vegetation, such as trees, adds greatly to the value of the realty. When a parcel of land is sold and the land has growing crops on it, the sale includes the crops, unless otherwise specified in the sales contract. When crops are sold by themselves, however, they are considered to be personal property or goods. Consequently, the sale of crops is a sale of goods and thus is governed by the Uniform Commercial Code (UCC) rather than by real property law.[4]

FIXTURES

Certain personal property can become so closely associated with the real property to which it is attached that the law views it as real property. Such property is known as a **fixture**—a thing *affixed* to realty, meaning that it is attached to the real property by roots; embedded in it; permanently situated on it; or permanently attached by means of cement, plaster, bolts, nails, or screws. The fixture can be physically attached to real property, be attached to another fixture, or even be without any actual physical attachment to the land (such as an extremely large and heavy statue). As long as the owner intends the property to be a fixture, it normally will be a fixture.

Fixtures are included in the sale of land if the sales contract does not provide otherwise. The sale of a house includes the land and the house and the garage on the land, as well as the cabinets, plumbing, and windows. Because these are permanently affixed to the property, they are considered to be a part of it. Certain items, such as drapes and window-unit air conditioners, are difficult to classify. Thus, a contract for the sale of a house or commercial realty should indicate which items of this sort are included in the sale.

At issue in the following case was whether an agricultural irrigation system qualified as a fixture.

4. See UCC 2–107(2).

CASE 29.1 In re Sand & Sage Farm & Ranch, Inc.

United States Bankruptcy Court,
District of Kansas, 2001.
266 Bankr. 507.

FACTS In 1988, Randolf and Sandra Ardery bought an eighty-acre tract in Edwards County, Kansas. On the land was an eight-tower center-pivot irrigation system. The system consisted of an underground well and pump connected to a pipe that ran to the pivot where the water line was attached to a further system of pipes and sprinklers suspended from the towers, extending over the land in a circular fashion. The system's engine and gearhead were bolted to a concrete slab above the pump and well and were attached to the pipe. To

CASE 29.1—CONTINUED

secure a loan to buy the land, the Arderys granted to Farmers State Bank a mortgage that covered "all buildings, improvements, and fixtures." In 1996, the Arderys, and their firm Sand & Sage Farm & Ranch, Inc., granted Ag Services of America a security interest in the farm's "equipment." Nothing in the security agreement or financing statement referred to fixtures.[a] In 2000, the Arderys and Sand & Sage filed for bankruptcy in a federal bankruptcy court and asked for permission to sell the land, with the irrigation system, to Bohn Enterprises, Limited Partnership. Ag Services claimed that it had priority to the proceeds covering the value of the irrigation system. The bank responded that it had priority because the system was a "fixture."

ISSUE Was this irrigation system a "fixture"?

DECISION Yes. The court concluded that the system was a fixture. The bank was entitled to the proceeds from its sale.

REASON The court explained that whether personal property attached to land is a fixture depends on "(i) how firmly the goods are attached or the ease of their removal (annexation); (ii) the relationship of the parties involved (intent); and (iii) how operation of the goods is related to the use of the land (adaptation). Of the three factors, intent is the controlling factor and is deduced largely from the property owner's acts and the surrounding circumstances." Here, the irrigation system was firmly attached to the realty. Disassembly and removal would be time consuming and expensive. As for intent, in each transaction all of the parties—except Ag Services, whose security agreement and financing statement did not refer to the irrigation system—"share[d] the intent that the system in question should pass with the land." Furthermore, the system "is suitably adapted to the land. There can be little dispute concerning the need for pivot irrigation in the semi-arid conditions of southwestern Kansas. * * * This alone demonstrates the relation between the operation of the goods and use of the land."

a. Security agreements and financing statements were discussed in Chapter 20.

WHY IS THIS CASE IMPORTANT? *When dealing with real property, it is crucial that the parties to a contract specifically list which fixtures they intend to be subjected to a security interest or included in a sale or transfer. In the end, it is far simpler and less expensive to itemize fixtures in a contract than to engage in litigation.*

Ownership Interests in Real Property

Ownership of property is an abstract concept that cannot exist independently of the legal system. No one can actually possess or *hold* a piece of land, the air above it, the earth below it, and all the water contained on it. The legal system therefore recognizes certain rights and duties that constitute ownership interests in real property.

Recall from Chapter 28 that property ownership is often viewed as a bundle of rights. One who possesses the entire bundle of rights is said to hold the property in *fee simple,* which is the most complete form of ownership. When only some of the rights in the bundle are transferred to another person, the effect is to limit the ownership rights of both the transferor of the rights and the recipient.

Ownership in Fee Simple

In a **fee simple absolute,** the owner has the greatest aggregation of rights, privileges, and power possible. The owner can give the property away, sell the property, or transfer the property by will to another. Also, the owner has the rights of *exclusive* possession and use of the property. The fee simple is limited to a person and his or her heirs and is assigned forever without limitation or condition.

The rights that accompany a fee simple include the right to use the land for whatever purpose the owner sees fit. Of course, other laws, including laws that prevent the owner from unreasonably interfering with another person's land and applicable zoning and environmental laws, may limit the owner's ability to use the property in certain ways. A fee simple is potentially infinite in duration and can be disposed of by *deed* (the instrument used to transfer property, which will be discussed later in this chapter) or by will. When there is no will, the fee simple passes to the owner's legal heirs.

Life Estates

A **life estate** is an estate that lasts for the life of some specified individual. A **conveyance,** or transfer of real property, "to A for his life" creates a life estate.[5] In a life estate, the life tenant's ownership rights cease to exist on the life tenant's death. The life tenant has the right to use the land, provided that he or she commits no waste (injury to the land). In other words, the life tenant cannot use the land in

5. A less common type of life estate is created by the conveyance "to A for the life of B." This is known as an estate *pur autre vie,* or an estate for the duration of the life of another.

a manner that would adversely affect its value. The life tenant is entitled to any rents generated by the land and can harvest crops from the land. If mines and oil wells are already on the land, the life tenant is entitled to any royalties and can extract minerals and oil from it, but he or she cannot exploit the land by creating new wells or mines.

The life tenant has the right to mortgage the life estate and can create liens, *easements* (discussed below), and leases, but none can extend beyond the life of the tenant. In addition, with few exceptions, the owner of a life estate has an exclusive right to possession during her or his lifetime. Along with these rights, the life tenant also has some duties—to keep the property in repair and to pay property taxes. In short, the owner of the life estate has the same rights as a fee simple owner except that the life tenant must maintain the value of the property during her or his tenancy.

NONPOSSESSORY INTERESTS

In contrast to the types of property interests just described, some interests in land do not include any rights to possess the property. These interests, known as *nonpossessory interests,* include easements, profits, and licenses.

An **easement** is the right of a person to make limited use of another person's real property without taking anything from the property. An easement, for instance, can be the right to walk or drive across another's property. In contrast, a **profit**[6] is the right to go onto land owned by another and take away some part of the land itself or some product of the land. ■ EXAMPLE 29.1 Akmed owns Sandy View. Akmed gives Carmen the right to go there to remove all the sand and gravel that she needs for her cement business. Carmen has a profit. ■

Easements and profits can be classified as either *appurtenant* or *in gross.* Because easements and profits are similar and the same rules apply to both, we discuss them together.

Easement or Profit Appurtenant With an easement or profit *appurtenant,* the owner of one piece of land has a right to go onto (or to remove things from) an *adjacent piece* of land owned by another. The easement is intended to benefit the first piece of land and runs with that land when the land is transferred. ■ EXAMPLE 29.2 Suppose Acosta has a right to drive his car across Green's land, which is adjacent to Acosta's land. This right-of-way over Green's property is an easement appurtenant to Acosta's property and can be used only by Acosta. Acosta can convey the easement when he conveys his property, though. ■

6. The term *profit,* as used here, does not refer to the "profits" made by a business firm. Rather, it means a gain or an advantage.

Easement or Profit in Gross In an easement or profit *in gross,* the right to use or take things from another's land exists even though the owner of the easement or profit does not own an adjacent tract of land. The easement is intended to benefit a particular person, not a particular piece of land, and cannot be transferred. Only one parcel of land is required for an easement or profit in gross. ■ EXAMPLE 29.3 Avery owns a parcel of land with a marble quarry. Avery conveys to Classic Stone Corporation the right to come onto her land and remove up to five hundred pounds of marble per day. Classic Stone owns a profit in gross and cannot transfer this right to another. ■ Similarly, when a utility company is granted an easement to run its power lines across another's property, it obtains an easement in gross.

Creation of an Easement or Profit Easements and profits can be created by deed or will, by contract, by implication, by necessity, or by prescription. Creation by *deed or will* simply involves the delivery of a deed or a disposition in a will by the owner of an easement or profit stating that her or his rights have been granted to the grantee (the person receiving the easement or profit). Two parties can create an easement by *contract* in which they agree that one party has the right to an easement or profit on a portion of the other party's land.

An easement or profit may arise by *implication* when the circumstances surrounding the division of a parcel of property imply its creation. ■ EXAMPLE 29.4 Barrow divides a parcel of land that has only one well for drinking water. If Barrow conveys the half without a well to Jarad, a profit by implication arises because Jarad needs drinking water. ■

An easement may also be created by *necessity.* An easement by necessity does not require a division of property for its existence. A person who rents an apartment, for example, has an easement by necessity in the private road leading up to it.

Easements and profits by *prescription* are created in essentially the same way as title to property is obtained by *adverse possession* (discussed later in this chapter). Easements, however, are nonpossessory and involve only the use of the land, whereas adverse possession occurs when one possesses someone else's land and eventually acquires title. An easement by prescription arises when one person exercises an easement, such as a right-of-way, on another person's land without the landowner's consent, and the use is apparent and continues for the length of time required by the applicable statute of limitations.

Effect of a Sale of Property When a parcel of land that is *benefited* by an easement or profit appurtenant is sold, the property carries the easement or profit along with it.

Thus, in Example 29.2, if Acosta sells his property to Thomas and includes the appurtenant right-of-way across Green's property in the deed to Thomas, Thomas will own both the property and the easement that benefits it.

When a parcel of land that has the *burden* of an easement or profit appurtenant is sold, the new owner must recognize its existence only if he or she knew or should have known of it or if it was recorded in the appropriate county office. ■ EXAMPLE 29.5 Acosta records his easement across Green's property in the appropriate county office. Green then conveys his land. The new owner of Green's property will have to allow Acosta, or any subsequent owner of Acosta's property, to continue to use the path across the land formerly owned by Green. ■

Termination of an Easement or Profit An easement or profit can be terminated or extinguished in several ways. The simplest way is to deed it back to the owner of the land that is burdened by it. Another way is to abandon it and create evidence of intent to relinquish the right to use it. Mere nonuse will not extinguish an easement or profit *unless the nonuse is accompanied by the intent to abandon.* Also, if the owner of an easement or profit becomes the owner of the property burdened by it, then it is merged into the property.

License A **license** is the revocable right of a person to come onto another person's land. It is a personal privilege that arises from the consent of the owner of the land and can be revoked by the owner. A ticket to attend a movie at a theater is an example of a license. ■ EXAMPLE 29.6 The owner of a Broadway theater issues Alena a ticket to see a play. If Alena is refused entry into the theater because she is improperly dressed, she has no right to force her way into the theater. The ticket is only a revocable license, not a conveyance of an interest in property. ■

Transfer of Ownership

Ownership of real property can pass from one person to another in several ways. Commonly, ownership interests in land are transferred by sale, and the terms of the transfer are specified in a real estate sales contract. Often, real estate brokers or agents who are licensed by the state assist the buyers and sellers during the sales transaction. Real property ownership can also be transferred by gift, by adverse possession, or by eminent domain. (Real property can be transferred by will or inheritance as well—see Chapter 30.) When ownership rights in real property are transferred, the type of interest being transferred and the conditions of the transfer normally are set forth in a *deed* executed by the person who is conveying the property.

DEEDS

Possession and title to land are passed from person to person by means of a **deed**—the instrument of conveyance of real property. A deed is a writing signed by an owner of real property that transfers title to another. Deeds must meet certain requirements. Unlike a contract, a deed does not have to be supported by legally sufficient consideration. Gifts of real property are common, and they require deeds even though there is no consideration for the gift. The necessary components of a valid deed are the following:

1. The names of the *grantor* (the giver or seller) and the *grantee* (donee or buyer).
2. Words evidencing an intent to convey the property (for example, "I hereby bargain, sell, grant, or give").
3. A legally sufficient description of the land.
4. The grantor's (and usually her or his spouse's) signature.
5. Delivery of the deed.

Warranty Deeds Different types of deeds provide different degrees of protection against defects of title. A **warranty deed** makes the greatest number of warranties and thus provides the greatest protection against defects of title. A sample warranty deed is shown in Exhibit 29–1 on the next page. In most states, special language is required to create a general warranty deed. Typically, the deed must include a written promise to protect the buyer against all claims of ownership of the property. A general warranty deed makes the grantor liable for all defects of title occurring while he or she (the grantor) and the previous titleholders held the property.

—The Usual Covenants. Warranty deeds commonly include a number of *covenants,* or promises, that the grantor makes to the grantee. A *covenant of seisin*[7] and a *covenant of the right to convey* warrant that the seller has title to, and the power to convey, the property described in the deed. A *covenant against encumbrances* is a promise that the property being sold or conveyed is not subject to any outstanding rights or interests of third parties, except as explicitly stated. Examples of common encumbrances include mortgages, liens, profits, easements, and private deed restrictions on the use of the land.

A *covenant of quiet enjoyment* guarantees that the buyer will not be disturbed in her or his possession of the land by the seller or any third persons. ■ EXAMPLE 29.7 Assume that Julio sells a two-acre lot and office building by warranty deed. Subsequently, a third person shows up

7. Pronounced see-zuhn.

EXHIBIT 29–1 A SAMPLE WARRANTY DEED

Date: May 31, 2007

Grantor: GAYLORD A. JENTZ AND WIFE, JOANN H. JENTZ

Grantor's Mailing Address (including county):
4106 North Loop Drive
Austin, Travis County, Texas

Grantee: DAVID F. FRIEND AND WIFE, JOAN E. FRIEND, AS JOINT TENANTS WITH RIGHT OF SURVIVORSHIP

Grantee's Mailing Address (including county):
5929 Fuller Drive
Austin, Travis County, Texas

Consideration:
For and in consideration of the sum of Ten and No/100 Dollars ($10.00) and other valuable consideration to the undersigned paid by the grantees herein named, the receipt of which is hereby acknowledged, and for which no lien is retained, either express or implied.

Property (including any improvements):
Lot 23, Block "A", Northwest Hills, Green Acres Addition, Phase 4, Travis County, Texas, according to the map or plat of record in volume 22, pages 331-336 of the Plat Records of Travis County, Texas.

Reservations from and Exceptions to Conveyance and Warranty:
This conveyance with its warranty is expressly made subject to the following:

Easements and restrictions of record in Volume 7863, Page 53, Volume 8430, Page 35, Volume 8133, Page 152 of the Real Property Record of Travis County, Texas; Volume 22, Pages 335-339, of the Plat Records of Travis County, Texas; and to any other restrictions and easements affecting said property which are of record in Travis County, Texas.

Grantor, for the consideration and subject to the reservations from and exceptions to conveyance and warranty, grants, sells, and conveys to Grantee the property, together with all and singular the rights and appurtenances thereto in any wise belonging, to have and hold it to Grantee, Grantee's heirs, executors, administrators, successors, or assigns forever. Grantor binds Grantor and Grantor's heirs, executors, administrators, and successors to warrant and forever defend all and singular the property to Grantee and Grantee's heirs, executors, administrators, successors, and assigns against every person whomsoever lawfully claiming or to claim the same or any part thereof, except as to the reservations from and exceptions to conveyance and warranty.

When the context requires, singular nouns and pronouns include the plural.

BY: *Gaylord A. Jentz*
Gaylord A. Jentz

BY: *JoAnn H. Jentz*
JoAnn H. Jentz

(Acknowledgment)

STATE OF TEXAS
COUNTY OF TRAVIS

This instrument was acknowledged before me on the 31st day of May, 2007
by Gaylord A. and JoAnn H. Jentz

Rosemary Potter
Notary Public, State of Texas
Notary's name (printed): *Rosemary Potter*

Notary Seal

Notary's commission expires: 1/31/2011

who has better title than Julio had and forces the buyer off the property. Here, the covenant of quiet enjoyment has been breached, and the buyer can sue Julio to recover the purchase price of the land plus any other damages incurred as a result. ■

If a contract calls for a "warranty deed" without specifying the covenants to be included in the deed, or if a deed states that the seller is providing the "usual covenants," most courts will infer that all of the covenants mentioned above are included.

—Special Warranty Deeds. In contrast to a warranty deed, a **special warranty deed,** which is frequently referred to as a *limited warranty deed,* warrants only that the grantor or seller held good title during his or her ownership of the property. In other words, the grantor is not warranting that there were no defects of title when the property was held by previous owners.

If the special warranty deed discloses all liens or other encumbrances, the seller will not be liable to the buyer if a third person subsequently interferes with the buyer's ownership. If the third person's claim arises out of, or is related to, some act of the seller, however, the seller will be liable to the buyer for damages.

Quitclaim Deeds A **quitclaim deed** offers the least amount of protection against defects in the title. Basically, a quitclaim deed conveys to the grantee whatever interest the grantor had; so, if the grantor had no interest, then the grantee receives no interest. Naturally, if the grantor had a defective title or no title at all, a conveyance by warranty deed or special warranty deed would not cure the defects. Such deeds, however, will give the buyer a cause of action to sue the seller.

A quitclaim deed can and often does serve as a release of the grantor's interest in a particular parcel of property. ■ EXAMPLE 29.8 Sandi and Jim were married for ten years and are now getting a divorce. During the marriage, Sandi purchased a parcel of waterfront property next to her grandparents' home in Louisiana. Jim helped make some improvements to the property, but he is not sure what ownership interests, if any, he has in the property because Sandi used her own funds to purchase the lot. Jim agrees to quitclaim the property to Sandi as part of the divorce settlement, releasing any interest he might have in that piece of property. ■

Recording Statutes Every jurisdiction has **recording statutes,** which allow deeds to be recorded for a fee. The grantee normally pays this fee because he or she is the one who will be protected by recording the deed. Recording a deed gives notice to the public that a certain person is now the owner of a particular parcel of real estate. Deeds are recorded in the county where the property is located. Many state statutes require that the grantor sign the deed in the presence of two witnesses before it can be recorded.

ADVERSE POSSESSION

Adverse possession is a means of obtaining title to land without delivery of a deed. Essentially, when one person possesses the property of another for a certain statutory period of time (three to thirty years, with ten years being most common), that person, called the *adverse possessor,* acquires title to the land and cannot be removed from it by the original owner. The adverse possessor may ultimately obtain a perfect title just as if there had been a conveyance by deed.

For property to be held adversely, four elements must be satisfied:

1. Possession must be actual and exclusive—that is, the possessor must take sole physical occupancy of the property.
2. The possession must be open, visible, and notorious, not secret or clandestine. The possessor must occupy the land for all the world to see.
3. Possession must be continuous and peaceable for the required period of time. This requirement means that the possessor must not be interrupted in the occupancy by the true owner or by the courts.
4. Possession must be hostile and adverse. In other words, the possessor must claim the property as against the whole world. He or she cannot be living on the property with the permission of the owner.

In the following case, the question was whether a couple had obtained title to a piece of land by *acquisitive prescription* (Louisiana's term for adverse possession).

CASE 29.2 Otwell v. Diversified Timber Services, Inc.

Court of Appeal of Louisiana,
Third Circuit, 2005.
896 So.2d 222.

FACTS In 1807, the eastern boundary of a parcel of land known as the "Charles McBride Riquet No. 39" was described

(Continued)

CASE 29.2—CONTINUED

as "the waters of Hemphill's Creek" in LaSalle Parish, Louisiana. In the 1930s, a curve in the creek was straightened, moving the bed to the west of its original path. In 1955, E. E. Jones sold 16 acres of land to Jesse Moffett under a deed that described the tract's western boundary as the "East line of the Charles McBride Riquet No. 39." In the late 1960s, Terry Brown and Margaret Otwell granted Bessie Sanders—their granddaughter and daughter, respectively—and William Sanders, Bessie's husband, title to a portion of 24 acres known as the "Terry Brown Estate." This included 3.12 acres between Hemphill's Creek and the "old slough," a natural feature that appeared to have been the creek's original bed. In 2001, Moffett sold the timber on the 3.12 acres to B & S Timber, Inc. The Sanderses filed a suit in a Louisiana state court against Moffett and others, seeking damages for "timber trespass." The court held that the plaintiffs failed to prove "just title" to the disputed land when they could not establish that the creek had flowed through the old slough in 1807, but ruled that the plaintiffs proved title through acquisitive prescription (adverse possession), and awarded damages and costs of more than $68,000. The defendants appealed to a state intermediate appellate court.

ISSUE Did the plaintiffs hold title to the disputed land by adverse possession?

DECISION Yes. The state intermediate appellate court affirmed the lower court's judgment in the plaintiffs' favor.

REASON The appellate court pointed out that "[o]wnership of immovable property may be acquired by the prescription of thirty years without the need of just title or possession in good faith. Corporeal possession sufficient to confer prescriptive title must be continuous, uninterrupted, peaceable, public, and unequivocal." The court acknowledged that "the concept of possession is neither simple nor precise," but in this case, the circumstances established that the Sanderses owned the disputed land. They assertedly acquired title from Bessie's grandfather and managed the timber. They also marked at least two hundred of the trees "with his wife's registered brand, the initials 'BO' over a half-moon," or "with the initials 'CM,' designating those trees that Mr. Sanders set aside for another individual, Chris Moss," with the intent to show "that this property was occupied by someone." The Sanderses also ran off trespassers, cut a fence to make a riding trail, shot hogs, hunted wood ducks, harvested berries, and placed the property in a hunting club that posted signs and erected deer stands. Besides, added the court, Moffett knew "as early as 1966 that Mr. and Mrs. Sanders intended to claim the property."

FOR CRITICAL ANALYSIS—Economic Consideration *Suppose that the Sanderses had done nothing involving the disputed land except to claim title. Would the result have been different? Why or why not?*

EMINENT DOMAIN

Even ownership in real property in fee simple absolute is limited by a superior ownership. Just as in medieval England the king was the ultimate landowner, so in the United States the government has an ultimate ownership right in all land. This right is called **eminent domain,** and it allows the government to take land for public use. Eminent domain gives the government the right to acquire possession of real property in the manner directed by the U.S. Constitution and the laws of the state whenever the public interest requires it. Property may be taken only for public use, not for private benefit.

The power of eminent domain is generally invoked through **condemnation** proceedings (and the power of eminent domain is sometimes referred to as the *condemnation power* of government). When a new public highway is to be built, for instance, the government must decide where to build it and how much land to condemn. After the government determines that a particular parcel of land is necessary for public use, it will first offer to buy the property. If the owner refuses the offer, the government brings a judicial (condemnation) proceeding to obtain title to the land. Then, in another proceeding, the court determines the *fair value* of the land, which is usually approximately equal to its market value.

When the government takes land owned by a private party for public use, it is referred to as a **taking,** and the government must compensate the private party. Under the so-called *takings clause* of the Fifth Amendment to the U.S. Constitution, the government may not take private property for public use without "just compensation." State constitutions contain similar provisions. Can the power of eminent domain be used to further economic development? That was the question in the following case.

CASE 29.3 Kelo v. City of New London, Connecticut

Supreme Court of the United States, 2005.
__ U.S. __ ,
125 S.Ct. 2655,
162 L.Ed.2d 439.
http://www.findlaw.com/casecode/supreme.html[a]

FACTS Decades of economic decline led a Connecticut state agency in 1990 to designate the city of New London a "distressed municipality." In 1996, the federal government closed the Naval Undersea Warfare Center, which had been located in the Fort Trumbull area of the city and had employed over 1,500 people. Within two years, the city's unemployment rate was nearly double that of the state. In 1998, Pfizer, Inc., announced that it would build a $300 million research facility on a site next to Fort Trumbull. Hoping that this would draw new business to the city, the city council approved a plan to redevelop the area that once housed the federal facility. The city bought most of the land for the project, but negotiations with some of the property owners fell through, and the city began condemnation proceedings. Susette Kelo and other affected owners filed a suit in a Connecticut state court against the city and others. The plaintiffs claimed, among other things, that the taking of their property would violate the "public use" restriction in the U.S. Constitution's Fifth Amendment. The court issued a ruling partly in favor of both sides. On appeal, the Connecticut Supreme Court held that all of the city's proposed takings were valid. The owners appealed to the United States Supreme Court.

ISSUE Can a city use its power of eminent domain to take private property for the purpose of economic development?

DECISION Yes. The United States Supreme Court affirmed the lower court's judgment. The Court held that economic development can constitute a "public use" within the meaning of the Fifth Amendment's takings clause, justifying a local government's exercise of its power of eminent domain.

REASON The Court concluded that the city's plan "unquestionably serves a public purpose," even though it would also benefit private parties. The Court explained, "Viewed as a whole, our jurisprudence has recognized that the needs of society have varied between different parts of the Nation, just as they have evolved over time in response to changed circumstances. Our earliest cases in particular embodied a strong theme of federalism, emphasizing the great respect that we owe to state legislatures and state courts in discerning local public needs. For more than a century, our public use jurisprudence has wisely eschewed [avoided] rigid formulas and intrusive scrutiny in favor of affording legislatures broad latitude in determining what public needs justify the use of the takings power." Thus, the city's "determination that the area was sufficiently distressed to justify a program of economic rejuvenation is entitled to our deference. The City has carefully formulated an economic development plan that it believes will provide appreciable benefits to the community. * * * Because that plan unquestionably serves a public purpose, the takings challenged here satisfy the public use requirement of the Fifth Amendment."

FOR CRITICAL ANALYSIS—Economic Consideration *Considering the impact of the ruling in this case, what are some arguments against the decision?*

COMMENTS The Court's ruling in the *Kelo* case generated significant controversy. Many Americans are angered by the idea that one's personal home and property can be taken by the government and given to a private party for economic gain. Responding to a public outcry, many states have introduced legislation to curb the use of eminent domain by local governments to effectuate private redevelopment projects.

a. In the "Browsing" section, click on "2005 Decisions." In the result, click on the name of the case to access the opinion.

Leasehold Estates

A **leasehold estate** is created when a real property owner (lessor, or landlord) agrees to convey the right to possess and use the property to a lessee (tenant) for a certain period of time. In every leasehold estate, the tenant has a *qualified* right to exclusive possession (qualified by the right of the landlord to enter on the premises to assure that waste is not being committed). The tenant can use the land—for example, by harvesting crops—but cannot injure it by such activities as cutting down timber for sale or extracting oil.

The respective rights and duties of the landlord and tenant that arise under a lease agreement will be discussed shortly. Here we look at the types of leasehold estates, or *tenancies*, that can be created when real property is leased.

TENANCY FOR YEARS

A **tenancy for years** is created by an express contract by which property is leased for a specified period of time, such as a day, a month, a year, or a period of years. Signing a one-year lease to occupy an apartment, for instance, creates a tenancy for years. At the end of the period specified in the lease, the lease ends (without notice), and possession of the apartment returns to the lessor. If the tenant dies during the period of the lease, the lease interest passes to the tenant's heirs as personal property. Often, leases include renewal or extension provisions.

PERIODIC TENANCY

A **periodic tenancy** is created by a lease that does not specify how long it is to last but does specify that rent is to be paid at certain intervals. This type of tenancy is automatically renewed for another rental period unless properly terminated. ■ EXAMPLE 29.9 Kayla enters a lease with Capital Properties. The lease states, "Rent is due on the tenth day of every month." This provision creates a periodic tenancy from month to month. ■ This type of tenancy can also extend from week to week or from year to year.

Under the common law, to terminate a periodic tenancy, the landlord or tenant must give at least one period's notice to the other party. If the tenancy extends from month to month, for example, one month's notice must be given prior to the last month's rent payment. State statutes may require a different period for notice of termination in a periodic tenancy, however.

TENANCY AT WILL

Suppose that a landlord rents an apartment to a tenant "for as long as both agree." In such a situation, the tenant receives a leasehold estate known as a **tenancy at will.** Under the common law, either party can terminate the tenancy without notice (that is, "at will"). This type of estate usually arises when a tenant who has been under a tenancy for years retains possession after the termination date of that tenancy with the landlord's consent. Before the tenancy has been converted into a periodic tenancy (by the periodic payment of rent), it is a tenancy at will, terminable by either party without notice. Once the tenancy is treated as a periodic tenancy, notice of termination must conform to the requirements already discussed for that type of tenancy. The death of either party or the voluntary commission of waste (harm to the premises) by the tenant will terminate a tenancy at will.

TENANCY AT SUFFERANCE

The mere possession of land without right is called a **tenancy at sufferance.** A tenancy at sufferance is not a true tenancy because it is created when a tenant *wrongfully* retains possession of property. Whenever a tenancy for years or a periodic tenancy ends and the tenant continues to retain possession of the premises without the owner's permission, a tenancy at sufferance is created.

Landlord-Tenant Relationships

In the past several decades, landlord-tenant relationships have become much more complex, as has the law governing them. Generally, the law has come to apply contract doctrines, such as those relating to implied warranties and unconscionability, to the landlord-tenant relationship. Increasingly, landlord-tenant relationships have become subject to specific state and local statutes and ordinances as well. In 1972, in an effort to create more uniformity in the law governing landlord-tenant relationships, the National Conference of Commissioners on Uniform State Laws issued the Uniform Residential Landlord and Tenant Act (URLTA). More than one-third of the states have adopted variations of the URLTA.

A landlord-tenant relationship is established by a lease contract. A lease contract may be oral or written. In most states, statutes mandate that leases be in writing for some tenancies (such as those exceeding one year). Generally, to ensure the validity of a lease agreement, it should be in writing and do the following:

1. Express an intent to establish the relationship.
2. Provide for the transfer of the property's possession to the tenant at the beginning of the term.
3. Provide for the landlord's *reversionary* (future) interest, which entitles the property owner to retake possession at the end of the term.
4. Describe the property—for example, give its street address.
5. Indicate the length of the term, the amount of the rent, and how and when it is to be paid.

ILLEGALITY

State or local law often dictates permissible lease terms. For example, a state law might prohibit gambling houses. Thus, if a landlord and tenant intend that the leased premises be used only to house an illegal betting operation, their lease is unenforceable.

A property owner cannot legally discriminate against prospective tenants on the basis of race, color, national origin, religion, gender, or disability. In addition, a tenant cannot legally promise to do something counter to laws prohibiting discrimination. A commercial tenant, for example, cannot legally promise to do business only with members of a particular race.

Rights and Duties

The rights and duties of landlords and tenants generally pertain to four broad areas of concern—the possession, use, and maintenance of leased property and, of course, rent.

Possession A landlord is obligated to give a tenant possession of the property that the tenant has agreed to lease. Many states follow the "English" rule, which requires the landlord to provide actual *physical possession* to the tenant (making sure that the previous tenant has vacated). Other states follow the "American" rule, which requires the landlord to transfer only the *legal right to possession* (thus, the new tenant is responsible for removing a previous tenant). After obtaining possession, the tenant retains the property exclusively until the lease expires, unless the lease states otherwise.

The covenant of quiet enjoyment mentioned previously also applies to leased premises. Under this covenant, the landlord promises that during the lease term, neither the landlord nor anyone having a superior title to the property will disturb the tenant's use and enjoyment of the property. This covenant forms the essence of the landlord-tenant relationship, and if it is breached, the tenant can terminate the lease and sue for damages.

If the landlord deprives the tenant of possession of the leased property or interferes with the tenant's use or enjoyment of it, an **eviction** occurs. An eviction occurs, for instance, when the landlord changes the lock and refuses to give the tenant a new key. A **constructive eviction** occurs when the landlord wrongfully performs or fails to perform any of the duties the lease requires, thereby making the tenant's further use and enjoyment of the property exceedingly difficult or impossible. Examples of constructive eviction include a landlord's failure to provide heat in the winter, light, or other essential utilities.

Use and Maintenance of the Premises If the parties do not limit by agreement the uses to which the property may be put, the tenant may make any use of it, as long as the use is legal and reasonably relates to the purpose for which the property is adapted or ordinarily used and does not injure the landlord's interest.

The tenant is responsible for any damage to the premises that he or she causes, intentionally or negligently, and may be held liable for the cost of returning the property to the physical condition it was in at the lease's inception. Unless the parties have agreed otherwise, the tenant is not responsible for ordinary wear and tear and the property's consequent depreciation in value.

In some jurisdictions, landlords of residential property are required by statute to maintain the premises in good repair. Landlords must also comply with any applicable state statutes and city ordinances regarding maintenance and repair of buildings.

Implied Warranty of Habitability The **implied warranty of habitability** requires a landlord who leases residential property to furnish premises that are in a habitable condition—that is, in a condition that is safe and suitable for people to live in. Also, the landlord must make repairs to maintain the premises in that condition for the lease's duration. Some state legislatures have enacted this warranty into law. In other jurisdictions, courts have based the warranty on the existence of a landlord's statutory duty to keep leased premises in good repair, or they have simply applied it as a matter of public policy. Generally, this warranty applies to major, or *substantial,* physical defects that the landlord knows or should know about and has had a reasonable time to repair—for example, a large hole in the roof.

Rent *Rent* is the tenant's payment to the landlord for the tenant's occupancy or use of the landlord's real property. Usually, the tenant must pay the rent even if she or he refuses to occupy the property or moves out, as long as the refusal or the move is unjustified and the lease is in force. Under the common law, if the leased premises were destroyed by fire or flood, the tenant still had to pay rent. Today, however, most state's statutes provide that if an apartment building burns down, tenants are not required to continue to pay rent.

In some situations, such as when a landlord breaches the implied warranty of habitability, a tenant may be allowed to withhold rent as a remedy. When rent withholding is authorized under a statute, the tenant must usually put the amount withheld into an *escrow account.* This account is held in the name of the depositor (the tenant) and an *escrow agent* (usually the court or a government agency), and the funds are returnable to the depositor if the third person (the landlord) fails to make the premises habitable. Generally, the tenant may withhold an amount equal to the amount by which the defect rendering the premises

unlivable reduces the property's rental value. How much that is may be determined in different ways, and a tenant who withholds more than is legally permissible is liable to the landlord for the excessive amount withheld.

Transferring Rights to Leased Property

Either the landlord or the tenant may wish to transfer her or his rights to the leased property during the term of the lease. If complete title to the leased property is transferred, the tenant becomes the tenant of the new owner. The new owner may collect subsequent rent but must abide by the terms of the existing lease agreement.

The tenant's transfer of his or her entire interest in the leased property to a third person is an *assignment of the lease*. Many leases require that the assignment have the landlord's written consent. An assignment that lacks consent can be avoided (nullified) by the landlord. State statutes may specify that the landlord may not unreasonably withhold such consent, though. Also, a landlord who knowingly accepts rent from the assignee may be held to have waived the consent requirement. When an assignment is valid, the assignee acquires all of the tenant's rights under the lease. But an assignment does not release the assigning tenant from the obligation to pay rent should the assignee default.

The tenant's transfer of all or part of the premises for a period shorter than the lease term is a **sublease.** The same restrictions that apply to an assignment of the tenant's interest in leased property apply to a sublease. If the landlord's consent is required, a sublease without such permission is ineffective. Also, a sublease does not release the tenant from her or his obligations under the lease any more than an assignment does.

TERMS AND CONCEPTS

adverse possession 609
condemnation 610
constructive eviction 613
conveyance 605
deed 607
easement 606
eminent domain 610
eviction 613
fee simple absolute 605
fixture 604
implied warranty of habitability 613
leasehold estate 611
license 607
life estate 605
periodic tenancy 612
profit 606
quitclaim deed 609
recording statutes 609
special warranty deed 609
sublease 614
taking 610
tenancy at sufferance 612
tenancy at will 612
tenancy for years 612
warranty deed 607

CHAPTER SUMMARY Real Property and Landlord-Tenant Law

The Nature of Real Property (See pages 603–605.)	Real property (also called real estate or realty) is immovable. It includes land, subsurface and airspace rights, plant life and vegetation, and fixtures.
Ownership Interests in Real Property (See pages 605–607.)	1. *Fee simple absolute*—The most complete form of ownership. 2. *Life estate*—An estate that lasts for the life of a specified individual, during which time the individual is entitled to possess, use, and benefit from the estate; ownership rights in a life estate are subject to the rights of the future-interest holder. 3. *Nonpossessory interest*—An interest that involves the right to use real property but not to possess it. Easements, profits, and licenses are nonpossessory interests.
Transfer of Ownership (See pages 607–611.)	1. *By deed*—When real property is sold or transferred as a gift, title to the property is conveyed by means of a deed. A deed must meet specific legal requirements. A *warranty deed* warrants the most extensive protection against defects of title. A *quitclaim deed* conveys to the grantee only whatever interest the grantor had in the property. A deed may be recorded in the manner prescribed by *recording statutes* in the appropriate jurisdiction to give third parties notice of the owner's interest.

CHAPTER SUMMARY Real Property and Landlord-Tenant Law—Continued

Transfer of Ownership—Continued	2. *By adverse possession*—When a person possesses the property of another for a statutory period of time (three to thirty years, with ten years being the most common), that person acquires title to the property, provided the possession is actual and exclusive, open and visible, continuous and peaceable, and hostile and adverse (without the permission of the owner). 3. *By eminent domain*—The government can take land for public use, with just compensation, when the public interest requires the taking.
Leasehold Estates (See pages 611–612.)	A leasehold estate is an interest in real property that is held for only a limited period of time, as specified in the lease agreement. Types of tenancies relating to leased property include the following: 1. *Tenancy for years*—Tenancy for a period of time stated by express contract. 2. *Periodic tenancy*—Tenancy for a period determined by the frequency of rent payments; automatically renewed unless proper notice is given. 3. *Tenancy at will*—Tenancy for as long as both parties agree; no notice of termination is required. 4. *Tenancy at sufferance*—Possession of land without legal right.
Landlord-Tenant Relationships (See pages 612–614.)	1. *Lease agreement*—The landlord-tenant relationship is created by a lease agreement. State or local laws may dictate whether the lease must be in writing and what lease terms are permissible. 2. *Rights and duties*—The rights and duties that arise under a lease agreement generally pertain to the following areas: possession, use and maintenance of the premises, and rent. 3. *Transferring rights to leased property*— a. If the landlord transfers complete title to the leased property, the tenant becomes the tenant of the new owner. The new owner may then collect the rent but must abide by the existing lease. b. Generally, in the absence of an agreement to the contrary, tenants may assign their rights (but not their duties) under a lease contract to a third person. Tenants may also sublease leased property to a third person, but the original tenant is not relieved of any obligations to the landlord under the lease. In either situation, the landlord's consent may be required, but statutes may prohibit the landlord from unreasonably withholding such consent.

FOR REVIEW

Answers for the even-numbered questions in this For Review *section can be found in Appendix A at the end of this text.*

1. What can a person who holds property in fee simple absolute do with the property? Can a person who holds property as a life estate do the same?
2. What are the requirements for acquiring property by adverse possession?
3. What limitations may be imposed on the rights of property owners?
4. What is a leasehold estate? What types of leasehold estates, or tenancies, can be created when real property is leased?
5. What are the respective duties of the landlord and tenant concerning the use and maintenance of leased property? Is the tenant responsible for all damage that he or she causes?

QUESTIONS AND CASE PROBLEMS

29–1. Property Ownership. Twenty-two years ago, Lorenz was a wanderer. At that time, he decided to settle down on an unoccupied, three-acre parcel of land that he did not own. People in the area told him that they had no idea who owned the property. Lorenz built a house on the land, got married, and raised three children while living there. He fenced in the land, installed a gate with a sign above it that read "Lorenz's Homestead," and had trespassers removed. Lorenz is now confronted by Joe Reese,

who has a deed in his name as owner of the property. Reese, claiming ownership of the land, orders Lorenz and his family off the property. Discuss who has the better "title" to the property.

QUESTION WITH SAMPLE ANSWER

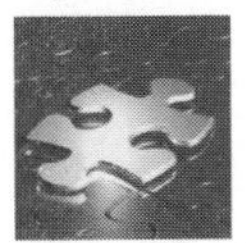

29–2. Wiley and Gemma are neighbors. Wiley's lot is extremely large, and his present and future use of it will not involve the entire area. Gemma wants to build a single-car garage and driveway along the present lot boundary. Because of ordinances requiring buildings to be set back fifteen feet from an adjoining property line, and because of the placement of her existing structures, Gemma cannot build the garage. Gemma contracts to purchase ten feet of Wiley's property along their boundary line for $3,000. Wiley is willing to sell but will give Gemma only a quitclaim deed, whereas Gemma wants a warranty deed. Discuss how the differences between these deeds would affect the rights of the parties if the title to this ten feet of land later proves to be defective.

For a sample answer to this question, go to Appendix B at the end of this text.

29–3. Taking. Richard and Jaquelyn Jackson owned property in a residential subdivision near an airport operated by the Metropolitan Knoxville Airport Authority in Blount County, Tennessee. The Airport Authority considered extending a runway near the subdivision and undertook a study that found that the noise, vibration, and pollution from aircraft using the extension would render the Jacksons' property incompatible with residential use. The airport built the extension, bringing about the predicted results, and the Jacksons filed a suit against the Airport Authority, alleging a taking of their property. The Airport Authority responded that there was no taking because there were no direct flights over the Jacksons' property. In whose favor will the court rule, and why? [*Jackson v. Metropolitan Knoxville Airport Authority,* 922 S.W.2d 860 (Tenn. 1996)]

29–4. Warranty of Habitability. Three-year-old Nkenge Lynch fell from the window of her third-floor apartment and suffered serious and permanent injuries. There were no window stops or guards on the window. The use of window stops, even if installed, is optional with the tenant. Stanley James owned the apartment building. Zsa Zsa Kinsey, Nkenge's mother, filed a suit on Nkenge's behalf in a Massachusetts state court against James, alleging in part a breach of an implied warranty of habitability. The plaintiff did not argue that the absence of stops or guards made the apartment unfit for human habitation. Instead, she asserted that their absence "endangered and materially impaired her health and safety" and, therefore, the failure to install them was a breach of warranty. Should the court rule that the absence of window stops breached a warranty of habitability? Should the court mandate that landlords provide window guards? Why or why not? [*Lynch v. James,* 44 Mass.App.Ct. 448, 692 N.E.2d 81 (1998)]

29–5. Adverse Possession. In 1972, Ted Pafundi bought a quarry in West Pawlet, Vermont, from his neighbor, Marguerite Scott. The deed vaguely described the eastern boundary of the quarry as "the westerly boundary of the lands of" the neighboring property owners. Pafundi quarried green slate from the west wall until his death in 1979, when his son Gary began to work the east wall until *his* death in 1989. Gary's daughter Connie then took over operations. All of the Pafundis used the floor of the quarry as their base of operations. In 1992, N.A.S. Holdings, Inc., bought the neighboring property. A survey revealed that virtually the entire quarry was within the boundaries of N.A.S.'s property and that twenty years earlier, Ted had actually bought only a small strip of land on the west side. When N.A.S. attempted to begin quarrying, Connie blocked its access. N.A.S. filed a suit in a Vermont state court against Connie, seeking to establish title. Connie argued that she had title to the quarry through adverse possession under a state statute with a possessory period of fifteen years. What are the elements to acquire title by adverse possession? Are they satisfied in this case? In whose favor should the court rule, and why? [*N.A.S. Holdings, Inc. v. Pafundi,* 169 Vt. 437, 736 A.2d 780 (1999)]

CASE PROBLEM WITH SAMPLE ANSWER

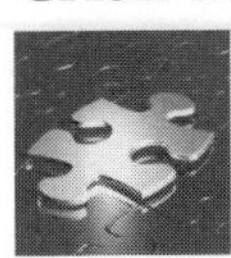

29–6. Jennifer Tribble leased an apartment from Spring Isle II, a limited partnership. The written lease agreement provided that if Tribble was forced to move because of a job transfer or because she accepted a new job, she could vacate on sixty days' notice and owe only an extra two months' rent plus no more than a $650 rerenting fee. The initial term was for one year, and the parties renewed the lease for a second one-year term. The security deposit was $900. State law allowed a landlord to withhold a security deposit for the nonpayment of rent but required timely notice stating valid reasons for the withholding or the tenant would be entitled to twice the amount of the deposit as damages. One month into the second term, Tribble notified Spring Isle in writing that she had accepted a new job and would move out within a week. She paid the extra rent required by the lease, but not the rerental fee, and vacated the apartment. Spring Isle wrote her a letter, stating that it was keeping the entire security deposit until the apartment was rerented or the lease term ended, whichever came first. Spring Isle later filed a suit in a Wisconsin state court against Tribble, claiming that she owed, among other things, the rest of the rent until the apartment had been rented again and the costs of rerenting. Tribble responded that withholding the security deposit was improper and that she was entitled to "any penalties." Does Tribble owe Spring Isle anything? Does Spring Isle owe Tribble anything? Explain. [*Spring Isle II v. Tribble,* 233 Wis.2d 274, 610 N.W.2d 229 (2000)]

After you have answered this problem, compare your answer with the sample answer given on the Web site that accompanies this text. Go to <u>academic.cengage.com/blaw/wbl</u>, select "Chapter 29," and click on "Case Problem with Sample Answer."

29–7. Commercial Lease Terms. Metropolitan Life Insurance Co. leased space in its Trail Plaza Shopping Center in Florida to Winn-Dixie Stores, Inc., to operate a supermarket. Under the lease, the landlord agreed not to permit "any [other] property located within the shopping center to be used for or occupied by any business dealing in or which shall keep in stock or sell for off-

premises consumption any staple or fancy groceries" in more than "500 square feet of sales area." In 1999, Metropolitan leased 22,000 square feet of space in Trail Plaza to 99 Cent Stuff-Trail Plaza, LLC, under a lease that prohibited it from selling "groceries" in more than 500 square feet of "sales area." Shortly after 99 Cent Stuff opened, it began selling food and other products, including soap, matches, and paper napkins. Alleging that these sales violated the parties' leases, Winn-Dixie filed a suit in a Florida state court against 99 Cent Stuff and others. The defendants argued in part that the groceries provision covered only food and the 500-square-foot restriction included only shelf space, not store aisles. How should these lease terms be interpreted? Should the court grant an injunction in Winn-Dixie's favor? Explain. [*Winn-Dixie Stores, Inc. v. 99 Cent Stuff-Trail Plaza, LLC*, 811 So.2d 719 (Fla.App. 3 Dist. 2002)]

29–8. Easements. The Wallens family owned a cabin on Lummi Island in the state of Washington. A driveway ran from the cabin across their property to South Nugent Road. In 1952, Floyd Massey bought the adjacent lot and built a cabin. To gain access to his property, he used a bulldozer to extend the driveway, without the Wallenses' permission but also without their objection. In 1975, the Wallenses sold their property to Wright Fish Co. Massey continued to use and maintain the driveway without permission or objection. In 1984, Massey sold his property to Robert Drake. Drake and his employees continued to use and maintain the driveway without permission or objection, although Drake knew it was located largely on Wright's property. In 1997, Wright sold its lot to Robert Smersh. The next year, Smersh told Drake to stop using the driveway. Drake filed a suit in a Washington state court against Smersh, claiming an easement by prescription (which is created by meeting the same requirements as adverse possession). Does Drake's use of the driveway meet all of the requirements? What should the court rule? Explain. [*Drake v. Smersh*, 122 Wash.App. 147, 89 P.3d 726 (Div. 1 2004)]

Real Property and Landlord-Tenant Law

For updated links to resources available on the Web, as well as a variety of other materials, visit this text's Web site at

academic.cengage.com/blaw/wbl

For a variety of links to real property resources, go to

http://findlaw.com/01topics/index.html

For information on condemnation procedures and rules under California law, go to

http://www.eminentdomainlaw.net/propertyguide.html

The Web site of the U.S. Department of Housing and Urban Development (HUD) has information of interest to both consumers and businesses. Go to

http://www.hud.gov

For information on Veterans Administration home loans, go to

http://www.homeloans.va.gov

Tenant Net is an advocacy Web site focusing on New York City and the state of New York. To view this site, go to

http://www.tenant.net

ONLINE LEGAL RESEARCH

Go to the *Fundamentals of Business Law* home page at academic.cengage.com/blaw/wbl, select "Chapter 29," and click on "Internet Exercises." There you will find the following Internet research exercises that you can perform to learn more about topics covered in this chapter.

Activity 29–1: LEGAL PERSPECTIVE—Eminent Domain

Activity 29–2: MANAGEMENT PERSPECTIVE—Fair Housing

Activity 29–3: SOCIAL PERSPECTIVE—The Rights of Tenants

BEFORE THE TEST

Go to the *Fundamentals of Business Law* home page at academic.cengage.com/blaw/wbl, select "Chapter 29,"and click on "Interactive Quizzes." You will find a number of interactive questions relating to this chapter.

CHAPTER 30

Insurance, Wills, and Trusts

LEARNING OBJECTIVES

After reading this chapter, you should be able to answer the following questions:

1. What is an insurable interest? When must an insurable interest exist—at the time the insurance policy is obtained, at the time the loss occurs, or both?
2. Is an insurance broker the agent of the insurance applicant or the agent of the insurer?
3. What are the basic requirements for executing a will? How may a will be revoked?
4. What is the difference between a *per stirpes* distribution and a *per capita* distribution of an estate to the grandchildren of the deceased?
5. What are the four essential elements of a trust? What is the difference between an express trust and an implied trust?

Most individuals insure both real and personal property (as well as their lives). By insuring our property, we protect ourselves against damage and loss. The first part of this chapter focuses on insurance, which is a foremost concern of all property owners. We then examine how property is transferred on the death of its owner. Certainly, the laws of succession of property are a necessary corollary to the concept of private ownership of property. Our laws require that on death, title to the property of a *decedent* (one who has recently died) must be delivered in full somewhere. In this chapter, we see that this can be done by will, through trusts, or through state laws prescribing distribution of property among heirs or next of kin.

Insurance

Insurance is a contract by which an insurance company (the insurer) promises to pay a certain amount or give something of value to another (either the insured or the beneficiary) in the event that the insured is injured, dies, or sustains damage to her or his property as a result of particular, stated contingencies. Basically, insurance is an arrangement for *transferring and allocating risk*. In many situations, **risk** can be described as a prediction concerning potential loss based on known and unknown factors. Insurance, however, involves much more than a game of chance.

Risk management normally involves the transfer of certain risks from the individual to the insurance company by a contractual agreement. The insurance contract and its provisions will be examined shortly. First, however, we look at the different types of insurance that can be obtained, insurance terminology, and the concept of insurable interest.

CLASSIFICATIONS OF INSURANCE

Insurance is classified according to the nature of the risk involved. For instance, fire insurance, casualty insurance, life insurance, and title insurance apply to different types of risk. Policies of these types protect different persons and interests. This is reasonable because the types of losses that

are expected and that are foreseeable or unforeseeable vary with the nature of the activity. Exhibit 30–1 on pages 620 and 621 presents a list of insurance classifications.

All-risk insurance is property insurance that covers damage resulting from all risks not specifically excluded. Typical exclusions are war, pollution, earthquakes, and floods. After September 11, 2001, insurance companies commonly added acts of terrorism to the list of exclusions. Separate terrorism insurance is available, though. The question in the following case was whether an agreement that required a business to carry all-risk insurance and "other reasonable insurance" included coverage against terrorism.

CASE 30.1 Omni Berkshire Corp. v. Wells Fargo Bank, N.A.

United States District Court,
Southern District of New York, 2004.
307 F.Supp.2d 534.

FACTS Omni Berkshire Corporation and its business affiliates own or operate hotels in the United States, Canada, and Mexico. In 1998, Omni borrowed $250 million from Secore Financial Corporation. The loan was secured by five of Omni's hotels, worth collectively about $349 million. The loan agreement required Omni to obtain and maintain certain insurance, including all-risk insurance, on the hotels and "such other reasonable insurance" as the lender might request. In 1998, Omni's all-risk policy provided that "damage done by terrorists . . . is insured." When Omni renewed the policy in 2002, however, it excluded coverage for acts of terrorism. Wells Fargo Bank, N.A., which collected the payments on the loan, told Omni that insurance against terrorist acts was required to comply with the agreement. Omni learned that $60 million in terrorism coverage for the five hotels for one year could be purchased for $316,000—63 percent of the cost of an all-risk policy. Omni filed a suit in a federal district court against Wells Fargo, seeking an order that it was not required to obtain insurance against terrorism.

ISSUE Is a business required to carry terrorism insurance under an agreement that requires "all-risk insurance" and "such other reasonable insurance" as the lender might request?

DECISION Yes. The court dismissed Omni's complaint and ordered it to obtain terrorism insurance. If Omni failed to do so, the court authorized Wells Fargo to obtain a policy and charge the expense to Omni.

REASON The court concluded that Omni was not required to buy insurance against terrorist acts under the "all-risk" clause, but that Wells Fargo acted reasonably in requesting the coverage under the "other reasonable insurance" clause. When the parties negotiated the loan, if they had intended to require Omni to maintain "all-risk" insurance that included coverage against terrorism, "they surely would have said so." The parties did not discuss the issue at the time, and the agreement did not define "all risk" to include terrorist acts. Nevertheless, the court concluded that Wells Fargo's concern for the risk was reasonable and noted that several recent terrorist attacks had involved hotels. For example, "[i]n the World Trade Center attacks on 9/11, one hotel was destroyed and two others were damaged." Other hotel operators and commercial property owners have bought terrorism insurance. The cost appeared to be "reasonable." Also, Omni, as well as Wells Fargo, would benefit from the coverage.

WHY IS THIS CASE IMPORTANT? *The decision in this case marked the first time that terrorism insurance was classified and required as "other reasonable insurance."*

INSURANCE TERMINOLOGY

An insurance contract is called a **policy;** the consideration paid to the insurer is called a **premium;** and the insurance company is sometimes called an **underwriter.** The parties to an insurance policy are the *insurer* (the insurance company) and the *insured* (the person covered by its provisions or the holder of the policy).

Insurance contracts usually are obtained through an *agent,* who ordinarily works for the insurance company, or through a *broker,* who is ordinarily an independent contractor. When a broker deals with an applicant for insurance, the broker is, in effect, the applicant's agent and not an agent of the insurance company. In contrast, an insurance agent is an agent of the insurance company, not of the applicant. As a general rule, the insurance company is bound by the acts of its insurance agents when they act within the agency relationship.[1] In most situations, state law determines the status of all parties writing or obtaining insurance.

1. Agency relationships were discussed in Chapter 22.

EXHIBIT 30–1 INSURANCE CLASSIFICATIONS

TYPE OF INSURANCE	COVERAGE
Accident	Covers expenses, losses, and suffering incurred by the insured because of accidents causing physical injury and any consequent disability; sometimes includes a specified payment to heirs of the insured if death results from an accident.
All-risk	Covers all losses that the insured may incur except those resulting from fraud on the part of the insured.
Automobile	May cover damage to automobiles resulting from specified hazards or occurrences (such as fire, vandalism, theft, or collision); normally provides protection against liability for personal injuries and property damage resulting from the operation of the vehicle.
Casualty	Protects against losses incurred by the insured as a result of being held liable for personal injuries or property damage sustained by others.
Credit	Pays to a creditor the balance of a debt on the disability, death, insolvency, or bankruptcy of the debtor; often offered by lending institutions.
Decreasing-term life	Provides life insurance; requires uniform payments over the life (term) of the policy, but with a decreasing face value (amount of coverage).
Employer's liability	Insures employers against liability for injuries or losses sustained by employees during the course of their employment; covers claims not covered under workers' compensation insurance.
Fidelity or guaranty	Provides indemnity against losses in trade or losses caused by the dishonesty of employees, the insolvency of debtors, or breaches of contract.
Fire	Covers losses incurred by the insured as a result of fire.
Floater	Covers movable property, as long as the property is within the territorial boundaries specified in the contract.
Group	Provides individual life, medical, or disability insurance coverage but is obtainable through a group of persons, usually employees; the policy premium is paid either entirely by the employer or partially by the employer and partially by the employee.

Insurable Interest

A person can insure anything in which she or he has an **insurable interest.** Without this insurable interest, there is no enforceable contract, and a transaction to purchase insurance coverage would have to be treated as a wager. In regard to real and personal property, an insurable interest exists when the insured derives a pecuniary benefit (a benefit consisting of or relating to money) from the preservation and continued existence of the property. Put another way, one has an insurable interest in property when one would sustain a financial loss from its destruction. In regard to life insurance, a person must have a reasonable expectation of benefit from the continued life of another in order to have an insurable interest in that person's life. The benefit may be pecuniary (as with so-called *key-person insurance,* which insures the lives of important employees, usually in small companies), or it may be founded on the relationship between the parties (by blood or affinity).

For property insurance, the insurable interest must exist at the time the loss occurs but need not exist when the policy is purchased. In contrast, for life insurance, the insurable interest must exist at the time the policy is obtained. The existence of an insurable interest is a primary concern in determining liability under an insurance policy.

The Insurance Contract

An insurance contract is governed by the general principles of contract law, although the insurance industry is heavily regulated by each state. Several aspects of the insurance contract will be treated here.

Application The filled-in application form for insurance usually is attached to the policy and made a part of the insurance contract. Thus, an insurance applicant is bound by any false statements that appear in the application (subject to certain exceptions). Because the insurance

EXHIBIT 30–1 INSURANCE CLASSIFICATIONS—CONTINUED

TYPE OF INSURANCE	COVERAGE
Health	Covers expenses incurred by the insured as a result of physical injury or illness and other expenses relating to health and life maintenance.
Homeowners'	Protects homeowners against some or all risks of loss to their residences and the residences' contents or liability arising from the use of the property.
Key-person	Protects a business in the event of the death or disability of a key employee.
Liability	Protects against liability imposed on the insured as a result of injuries to the person or property of another.
Life	Covers the death of the policyholder. On the death of the insured, an amount specified in the policy is paid by the insurer to the insured's beneficiary.
Major medical	Protects the insured against major hospital, medical, or surgical expenses.
Malpractice	Protects professionals (physicians, lawyers, and others) against malpractice claims brought against them by their patients or clients; a form of liability insurance.
Marine	Covers movable property (including ships, freight, and cargo) against certain perils or navigation risks during a specific voyage or time period.
Mortgage	Covers a mortgage loan; the insurer pays the balance of the mortgage to the creditor on the death or disability of the debtor.
No-fault auto	Covers personal injuries and (sometimes) property damage resulting from automobile accidents. The insured submits his or her claims to his or her own insurance company, regardless of who was at fault. A person may sue the party at fault or that party's insurer only when an accident results in serious medical injury and consequent high medical costs. Governed by state "no-fault" statutes.
Term life	Provides life insurance for a specified period of time (term) with no cash surrender value; usually renewable.
Title	Protects against any defects in title to real property and any losses incurred as a result of existing claims against or liens on the property at the time of purchase.

company evaluates the risk factors based on the information included in the insurance application, misstatements or misrepresentations can void a policy, especially if the insurance company can show that it would not have extended insurance if it had known the true facts.

Effective Date The effective date of an insurance contract—that is, the date on which the insurance coverage begins—is important. In some instances, the insurance applicant is not protected until a formal written policy is issued. In other situations, the applicant is protected between the time the application is received and the time the insurance company either accepts or rejects it. Four facts should be kept in mind:

1. A broker is merely the agent of an applicant. Therefore, until the broker obtains a policy, the applicant normally is not insured.
2. A person who seeks insurance from an insurance company's agent will usually be protected from the moment the application is made, provided that some form of premium has been paid. Between the time the application is received and either rejected or accepted, the applicant is covered (possibly subject to a medical examination). Usually, the agent will write a memorandum, or **binder,** indicating that a policy is pending and stating its essential terms.
3. If the parties agree that the policy will be issued and delivered at a later time, the contract is not effective until the policy is issued and delivered or sent to the applicant, depending on the agreement. Thus, any loss sustained between the time of application and the delivery of the policy is not covered.
4. The parties may agree that a life insurance policy will be binding at the time the insured pays the first premium, or the policy may be expressly contingent on the applicant's passing a physical examination. If the applicant pays the premium and passes the examination, the policy coverage is continuously in effect. If the applicant pays the premium but dies

before having the physical examination, then in order to collect, the applicant's estate must show that the applicant *would have passed* the examination had he or she not died.

Coinsurance Clauses Often, when taking out fire insurance policies, property owners insure their property for less than full value because most fires do not result in a total loss. To encourage owners to insure their property for an amount as close to full value as possible, fire insurance policies commonly include a coinsurance clause. Typically, a *coinsurance clause* provides that if the owner insures the property up to a specified percentage—usually 80 percent—of its value, she or he will recover any loss up to the face amount of the policy. If the insurance is for less than the fixed percentage, the owner is responsible for a proportionate share of the loss.

Coinsurance applies only in instances of partial loss. ■ EXAMPLE 30.1 If the owner of property valued at $100,000 takes out a policy in the amount of $40,000 and suffers a loss of $30,000, the recovery will be $15,000. The formula for calculating the recovery amount is as follows:

$$\frac{\text{amount of insurance (\$40,000)}}{\text{coinsurance percentage (80\%)} \times \text{property value (\$100,000)}} = \text{recovery percentage (50\%)}$$

$$\text{recovery percentage (50\%)} \times \text{amount of loss (\$30,000)} = \text{recovery amount (\$15,000)}$$

If the owner had taken out a policy in the amount of $80,000, then, according to the same formula, the full loss would have been recovered up to the face value of the policy. ■

Other Provisions and Clauses Some other important provisions and clauses contained in insurance contracts are listed and defined in Exhibit 30–2. The courts are aware that most people do not have the special training necessary to understand the intricate terminology used in insurance policies. Thus, the words used in an insurance contract have their ordinary meanings. They are interpreted by the courts in light of the nature of the coverage involved.

When there is an ambiguity in the policy, the provision generally is interpreted against the insurance company. Also when it is unclear whether an insurance contract actually exists because the written policy has not been delivered, the uncertainty normally is resolved against the insurance company. The court presumes that the policy is in effect unless the company can show otherwise. Similarly, an insurer must make sure that the insured is adequately notified of any change in coverage under an existing policy.

Cancellation The insured can cancel a policy at any time, and the insurer can cancel under certain circumstances. When an insurance company can cancel its insurance contract, the policy or a state statute usually requires that the insurer give advance written notice of the cancellation to the insured. The same requirement applies when only part of a policy is canceled. Any premium paid in advance may be refundable on the policy's cancellation. The insured may also be entitled to a life insurance policy's cash surrender value.

The insurer may cancel an insurance policy for various reasons, depending on the type of insurance. For example, automobile insurance can be canceled for nonpayment of

EXHIBIT 30–2 INSURANCE CONTRACT PROVISIONS AND CLAUSES

Incontestability clause	An incontestability clause provides that after a policy has been in force for a specified length of time—usually two or three years—the insurer cannot contest statements made in the application.
Appraisal clause	Insurance policies frequently provide that if the parties cannot agree on the amount of a loss covered under the policy or the value of the property lost, an appraisal, or estimate, by an impartial and qualified third party can be demanded.
Arbitration clause	Many insurance policies include clauses that call for arbitration of disputes that may arise between the insurer and the insured concerning the settlement of claims.
Antilapse clause	An antilapse clause provides that the policy will not automatically lapse if no payment is made on the date due. Ordinarily, under such a provision, the insured has a grace period of thirty or thirty-one days within which to pay an overdue premium before the policy is canceled.
Multiple insurance	Many insurance policies include a clause providing that if the insured has multiple insurance policies that cover the same property and the amount of coverage exceeds the loss, the loss will be shared proportionately by the insurance companies.

premiums or suspension of the insured's driver's license. Property insurance can be canceled for nonpayment of premiums or for other reasons, including the insured's fraud or misrepresentation, conviction for a crime that increases the hazard insured against, or gross negligence that increases the hazard insured against. Life and health policies can be canceled because of false statements made by the insured in the application, but cancellation can only take place before the effective date of an incontestability clause. An insurer cannot cancel—or refuse to renew—a policy for discriminatory reasons or other reasons that violate public policy, or because the insured has appeared as a witness in a case against the company.

Good Faith Obligations Both parties to an insurance contract are responsible for the obligation they assume under the contract (contract law was discussed in Chapters 7 through 13). In addition, both the insured and the insurer have an implied duty to act in good faith.

Good faith requires the party who is applying for insurance to reveal everything necessary for the insurer to evaluate the risk. In other words, the applicant must disclose all material facts, including all facts that an insurer would consider in determining whether to charge a higher premium or to refuse to issue a policy altogether.

Once the insurer has accepted the risk, and some event occurs that gives rise to a claim, the insurer has a duty to investigate to determine the facts. When a policy provides insurance against third party claims, the insurer is obligated to make reasonable efforts to settle such a claim. If a settlement cannot be reached, then regardless of the claim's merit, the insurer must defend any suit against the insured. Usually, a policy provides that in this situation the insured must cooperate. A policy provision may expressly require the insured to attend hearings and trials, to help in obtaining evidence and witnesses, and to assist in reaching a settlement.

The question in the following case was whether the insurer acted in bad faith in investigating and paying an insured's claim.

CASE 30.2 Columbia National Insurance Co. v. Freeman

Supreme Court of Arkansas, 2002.
347 Ark. 423,
64 S.W.3d 720.
http://courts.state.ar.us/opinions/opinions.html[a]

FACTS Gary and Peggy Freeman owned and operated Circle F Trading Company, a western wear and general store, in Arkansas. The Freemans were insured against losses to the building, its contents, continuing business expenses, and other coverage, under a policy with Columbia National Insurance Company. In October 1997, a fire damaged Circle F's building and destroyed its inventory. The Freemans filed a claim with Columbia, providing an appraisal of the lost merchandise at $107,905.13 and a list of their continuing business expenses. Columbia obtained a second appraisal of $71,231.69 and attempted to find Circle F a building to serve as a temporary office. In December, Columbia paid the Freemans $77,892.28 for inventory, supplies, and lost income. No payment was made for continuing business expenses, and no office was provided. The parties agreed on an amount of $32,725 to cover the cost of the damage to the building, but Columbia offered to pay only 80 percent of this amount. Circle F never reopened. The Freemans filed a suit in an Arkansas state court against Columbia, alleging, among other things, bad faith. A jury returned a verdict for the Freemans, awarding $170,000 in compensatory damages and $200,000 in punitive damages. Columbia filed a motion for a directed verdict, which the court denied. Columbia appealed to the Arkansas Supreme Court.

ISSUE Had Columbia acted in bad faith in investigating and paying the Freemans' claim?

DECISION Yes. The Arkansas Supreme Court affirmed the lower court's judgment. The state supreme court concluded that there was substantial evidence to support the jury's verdict that Columbia's actions constituted oppressive conduct carried out with a state of mind characterized by ill will.

REASON The court stated that "[a]n insurance company commits the tort of bad faith when it affirmatively engages in dishonest, malicious, or oppressive conduct in order to avoid a just obligation to its insured. We have defined 'bad faith' as dishonest, malicious, or oppressive conduct carried out with a state of mind characterized by hatred, ill will, or a spirit of revenge. Mere negligence or bad judgment is insufficient." In

(Continued)

a. In the "Search Cases by Party Name" section, enter "Freeman" in the "Party Name" box and select "Search by Date Range." For the date range, choose "From January 2002" and "To February 2002," and click on "Search." From the list of results, click on the name of the case to access the opinion. The Arkansas judiciary maintains this Web site.

CASE 30.2–CONTINUED

this case, the court noted there was sufficient evidence to support a finding that failure to cover the Freemans' ongoing business expenses was an act of bad faith. The court also found sufficient evidence to support the jury's findings that Columbia acted in bad faith when it failed to provide the Freemans with a temporary location for their business, when it agreed to pay "$32,725 for the cost of repairing the building but * * * tendered only eighty percent of the amount agreed on," and "when it requested that two appraisals be performed on [the Freemans'] inventory and chose to pay the [Freemans] based on the lower of the two appraisals."

FOR CRITICAL ANALYSIS—Social Consideration
Suppose that after an investigation, Columbia had simply refused to pay the Freemans' claim. Would the result in this case have been the same? Why or why not?

Defenses against Payment An insurance company can raise any of the defenses that would be valid in any ordinary action on a contract, as well as some defenses that do not apply in ordinary contract actions. If the insurance company can show that the policy was procured by fraud or misrepresentation, it may have a valid defense for not paying on a claim. (The insurance company may also have the right to disaffirm or rescind an insurance contract.) An absolute defense exists if the insurer can show that the insured lacked an insurable interest—thus rendering the policy void from the beginning. Improper actions, such as those that are against public policy or are otherwise illegal, can also give the insurance company a defense against the payment of a claim or allow it to rescind the contract.

An insurance company can be prevented from asserting some defenses that normally are available, however. ■ EXAMPLE 30.2 Suppose that a company tells an insured that information requested on a form is optional, and the insured provides it anyway. The company cannot use the information to avoid its contractual obligation under the insurance contract. Similarly, an insurance company usually cannot escape payment on the death of an insured on the ground that the person's age was stated incorrectly on the application. ■

Wills

Private ownership of property leads logically to both the protection of that property by insurance coverage while the owner is alive and the transfer of that property on the death of the owner to those designated in the owner's will. A **will** is the final declaration of how a person desires to have her or his property disposed of after death. It is a formal instrument that must follow exactly the requirements of state law to be effective. A will is referred to as a *testamentary disposition* of property, and one who dies after having made a valid will is said to have died **testate.**

A will can serve other purposes besides the distribution of property. It can appoint a guardian for minor children or incapacitated adults. It can also appoint a personal representative to settle the affairs of the deceased.

A person who dies without having created a valid will is said to have died **intestate.** In this situation, state **intestacy laws** prescribe the distribution of the property among heirs or next of kin. If no heirs or kin can be found, title to the property will be transferred to the state.

Terminology of Wills

A person who makes out a will is known as a **testator** (from the Latin *testari,* "to make a will"). The court responsible for administering any legal problems surrounding a will is called a *probate court,* as mentioned in Chapter 2. When a person dies, a personal representative administers the estate and settles finally all of the decedent's (deceased person's) affairs. An **executor** is a personal representative named in the will; an **administrator** is a personal representative appointed by the court for a decedent who dies without a will. The court will also appoint an administrator if the will does not name an executor or if the named person lacks the capacity to serve as an executor.

A gift of real estate by will is generally called a **devise,** and a gift of personal property by will is called a **bequest,** or **legacy.** The recipient of a gift by will is a **devisee** or a **legatee,** depending on whether the gift was a devise or a legacy.

Types of Gifts

Gifts by will can be specific, general, or residuary. A *specific* devise or bequest (legacy) describes particular property (such as "Eastwood Estate" or "my gold pocket watch") that can be distinguished from all the rest of the testator's property. A *general* devise or bequest (legacy) uses less restrictive terminology. For example, "I devise all my lands" is a general devise. A general bequest often specifies a sum of money instead of a particular item of property, such as a watch or an automobile. For example, "I give to my nephew, Carleton, $30,000" is a general bequest.

If the assets of an estate are insufficient to pay in full all general bequests provided for in the will, an *abatement*

takes place, meaning the legatees receive reduced benefits. ■ EXAMPLE 30.3 Yusuf's will leaves "$15,000 each to my children, Tamara and Kwame." On Yusuf's death, only $10,000 is available to honor these bequests. By abatement, each child will receive $5,000. ■ If bequests are more complicated, abatement may be more complex. The testator's intent, as expressed in the will, controls.

Sometimes, a will provides that any assets remaining after the estate's debts have been paid and specific gifts have been made—called the *residuary* (or *residuum*)—are to be distributed in a specific way, such as to the testator's spouse or descendants. Such a clause, called a *residuary clause,* is often used when the exact amount to be distributed cannot be determined until all of the other gifts and payouts have been made. If the testator has not indicated what party or parties should receive the residuary of the estate, the residuary passes according to state laws of intestacy.

Probate Procedures

Laws governing wills come into play when a will is probated. To **probate** a will means to establish its validity and to carry the administration of the estate through a court process. Probate laws vary from state to state. In 1969, however, the American Bar Association and the National Conference of Commissioners on Uniform State Laws approved the Uniform Probate Code (UPC). The UPC codifies general principles and procedures for the resolution of conflicts in settling estates and relaxes some of the requirements for a valid will contained in earlier state laws. Nearly all of the states have adopted some part of the UPC. Nevertheless, succession and inheritance laws still vary widely among the states, so one should always check the laws of the particular state involved.[2] Typically, probate procedures differ, depending on the size of the decedent's estate.

Informal Probate For smaller estates, most state statutes provide for the distribution of assets without formal probate proceedings. Faster and less expensive methods are then used. For example, property can be transferred by *affidavit* (a written statement taken before a person who has authority to affirm it), and problems or questions can be handled during an administrative hearing. In addition, some state statutes provide that title to cars, savings and checking accounts, and certain other property can be passed merely by filling out forms.

A majority of states also provide for *family settlement agreements,* which are private agreements among the beneficiaries. Once a will is admitted to probate, the family members can agree to settle among themselves the distribution of the decedent's assets. Although a family settlement agreement speeds the settlement process, a court order is still needed to protect the estate from future creditors and to clear title to the assets involved. The use of these and other types of summary procedures in estate administration can save time and expense.

Formal Probate For larger estates, formal probate proceedings normally are undertaken, and the probate court supervises every aspect of the settlement of the decedent's estate. Additionally, in some situations—such as when a guardian for minor children or for an incompetent person must be appointed and a trust has been created to protect the minor or the incompetent person—more formal probate procedures cannot be avoided. Formal probate proceedings may take several months to complete, and as a result, a sizable portion of the decedent's assets (up to perhaps 10 percent) may go toward payment of court costs and fees charged by attorneys and personal representatives.

Property Transfers outside the Probate Process In the ordinary situation, a person can employ various *will substitutes* to avoid the cost of probate. A **will substitute** is a document that attempts to dispose of an estate in the same or similar manner as a will—for example, *inter vivos* trusts (discussed later in this chapter), life insurance policies or individual retirement accounts (IRAs) with named beneficiaries, or joint-tenancy arrangements. Not all alternatives to formal probate administration are suitable for every estate, however.

Requirements for a Valid Will

A will must comply with statutory formalities designed to ensure that the testator understood his or her actions at the time the will was made. These formalities are intended to help prevent fraud. Unless they are followed, the will is declared void, and the decedent's property is distributed according to the laws of intestacy of that state. The requirements are not uniform among jurisdictions. Most states, however, uphold certain basic requirements for executing a will. We now look at these requirements.

Testamentary Capacity and Intent For a will to be valid, the testator must have testamentary capacity—that is, the testator must be of legal age and sound mind *at the time the will is made.* The legal age for executing a will varies, but in most states and under the UPC, the minimum age is eighteen years [UPC 2–501]. Thus, the will of a twenty-one-year-old decedent written when the person was sixteen is invalid if, under state law, the legal age for executing a will is eighteen.

2. For example, California law differs substantially from the UPC.

The concept of "being of sound mind" refers to the testator's ability to formulate and to comprehend a personal plan for the disposition of property. Generally, a testator must (1) intend the document to be his or her last will and testament, (2) comprehend the kind and character of the property being distributed, and (3) comprehend and remember the "natural objects of his or her bounty" (usually, family members and persons for whom the testator has affection).

A valid will is one that represents the maker's intention to transfer and distribute her or his property. When it can be shown that the decedent's plan of distribution was the result of fraud or of undue influence, the will is declared invalid. The court may sometimes infer undue influence if the testator ignored blood relatives and named as beneficiary a nonrelative who was in constant close contact with the testator and in a position to influence the making of the will.

Writing Requirements Generally, a will must be in writing. The writing itself can be informal as long as it substantially complies with the statutory requirements. In some states, a will can be handwritten in crayon or ink. It can be written on a sheet or scrap of paper, on a paper bag, or on a piece of cloth. A will that is completely in the handwriting of the testator is called a **holographic will** (sometimes referred to as an *olographic will*).

A **nuncupative will** is an oral will made before witnesses. Most states do not permit such wills. Where authorized by statute, nuncupative wills are generally valid only if made during the last illness of the testator and are therefore sometimes referred to as *deathbed wills*. Normally, only personal property can be transferred by a nuncupative will. Statutes frequently permit soldiers and sailors to make nuncupative wills when on active duty.

Signature Requirements It is a fundamental requirement that the testator's signature appear, generally at the end of the will. Each jurisdiction dictates by statute and court decision what constitutes a signature. Initials, an X or other mark, and words such as "Mom" have all been upheld as valid when it was shown that the testators *intended* them to be signatures.

Witness Requirements A will normally must be attested (sworn to) by two, and sometimes three, witnesses. The number of witnesses, their qualifications, and the manner in which the witnessing must be done are generally set out in a statute. A witness can be required to be disinterested—that is, not a beneficiary under the will. The UPC, however, provides that a will is valid even if it is attested by an interested witness [UPC 2–505]. There are no age requirements for witnesses, but they must be mentally competent.

The purpose of witnesses is to verify that the testator actually executed (signed) the will and had the requisite intent and capacity at the time. A witness does not have to read the contents of the will. Usually, the testator and all witnesses must sign in the sight or the presence of one another, but there are exceptions.[3] The UPC does not require all parties to sign in the presence of one another and deems it sufficient if the testator acknowledges her or his signature to the witnesses [UPC 2–502].

Publication Requirements A will is *published* by an oral declaration by the maker to the witnesses that the document they are about to sign is his or her "last will and testament." Publication is becoming an unnecessary formality in most states, and it is not required under the UPC.

REVOCATION OF WILLS

An executed will is revocable by the maker at any time during the maker's lifetime. The maker may revoke a will by a physical act, such as tearing up the will, or by a subsequent writing. Wills can also be revoked by operation of law. Revocation can be partial or complete, and it must follow certain strict formalities.

Revocation by a Physical Act of the Maker A testator may revoke a will by intentionally burning, tearing, canceling, obliterating, or otherwise destroying it, or by having someone else do so in the presence of the maker and at the maker's direction.[4] In some states, partial revocation by physical act of the maker is recognized. Thus, those portions of a will lined out or torn away are dropped, and the remaining parts of the will are valid. In no circumstances, however, can a provision be crossed out and an additional or substitute provision written in. Such altered portions require reexecution (re-signing) and reattestation (rewitnessing).

To revoke a will by physical act, it is necessary to follow the mandates of a state statute exactly. When a state statute prescribes the specific methods for revoking a will by physical act, only those methods can be used to revoke the will.

3. See, for example, *Slack v. Truitt,* 368 Md. 2, 791 A.2d 129 (2000).

4. The destruction cannot be inadvertent. The maker's intent to revoke must be shown. Consequently, when a will has been burned or torn accidentally, the maker normally should have a new document created to avoid any suggestion that the maker intended to revoke the will.

Revocation by a Subsequent Writing A will may also be wholly or partially revoked by a **codicil,** a written instrument separate from the will that amends or revokes provisions in the will. A codicil eliminates the necessity of redrafting an entire will merely to add to it or amend it. A codicil can also be used to revoke an entire will. The codicil must be executed with the same formalities required for a will, and it must refer expressly to the will. In effect, it updates a will because the will is "incorporated by reference" into the codicil.

A new will (second will) can be executed that may or may not revoke the first or a prior will, depending on the language used. To revoke a prior will, the second will must use language specifically revoking other wills, such as, "This will hereby revokes all prior wills." If the second will is otherwise valid and properly executed, it will revoke all prior wills. If the express *declaration of revocation* is missing, then both wills are read together. If any of the dispositions made in the second will are inconsistent with the prior will, the second will controls.

Revocation by Operation of Law Revocation by *operation of law* occurs when marriage, divorce or annulment, or the birth of a child takes place after a will has been executed. In most states, when a testator marries after executing a will that does not include the new spouse, on the testator's death the spouse can still receive the amount he or she would have taken had the testator died intestate (how an intestate's property is distributed under state laws will be discussed shortly). In effect, the will is revoked to the point of providing the spouse with an intestate share. The rest of the estate is passed under the will [UPC 2–301, 2–508]. If, however, the new spouse is otherwise provided for in the will (or by transfer of property outside the will), he or she will not be given an intestate amount.

At common law and under the UPC, divorce does not necessarily revoke the entire will. A divorce or an annulment occurring after a will has been executed will revoke those dispositions of property made under the will to the former spouse [UPC 2–508].

If a child is born after a will has been executed and if it appears that the deceased parent would have made a provision for the child, the child is entitled to receive whatever portion of the estate she or he is allowed under state laws providing for the distribution of an intestate's property. Most state laws allow a child to receive some portion of a parent's estate if no provision is made in the parent's will, unless it appears from the terms of the will that the testator intended to disinherit the child. Under the UPC, the rule is the same.

INTESTACY LAWS

As mentioned, state intestacy laws determine how property will be distributed when a person dies intestate (without a valid will). These statutes are more formally known as *statutes of descent and distribution.* Intestacy laws attempt to carry out the likely intent and wishes of the decedent. These laws assume that deceased persons would have intended that their natural heirs (spouses, children, grandchildren, or other family members) inherit their property. Therefore, intestacy statutes set out rules and priorities under which these heirs inherit the property. If no heirs exist, the state will assume ownership of the property. The rules of descent vary widely from state to state.

Surviving Spouse and Children Usually, state statutes provide that first the debts of the decedent must be satisfied out of the estate; then the remaining assets pass to the surviving spouse and to the children. A surviving spouse usually receives only a share of the estate—one-half if there is also a surviving child and one-third if there are two or more children. Only if no children or grandchildren survive the decedent will a surviving spouse succeed to the entire estate.

■ EXAMPLE 30.4 Assume that Allen dies intestate and is survived by his wife, Della, and his children, Duane and Tara. Allen's property passes according to intestacy laws. After Allen's outstanding debts are paid, Della will receive the homestead (either in fee simple or as a life estate) and ordinarily a one-third interest in all other property. The remaining real and personal property will pass to Duane and Tara in equal portions. ■ Under most state intestacy laws and under the UPC, in-laws do not share in an estate. If a child dies before his or her parents, the child's spouse will not receive an inheritance on the parents' death. For example, if Duane died before his father (Allen), Duane's spouse would not inherit Duane's share of Allen's estate.

When there is no surviving spouse or child, the order of inheritance is grandchildren, then brothers and sisters, and, in some states, parents of the decedent. These relatives are usually called *lineal descendants.* If there are no lineal descendants, then *collateral heirs*—nieces, nephews, aunts, and uncles of the decedent—make up the next group to share. If there are no survivors in any of these groups, most statutes provide for the property to be distributed among the next of kin of the collateral heirs.

Stepchildren, Adopted Children, and Illegitimate Children Under intestacy laws, stepchildren are not considered kin. Legally adopted children, however, are recognized as lawful

heirs of their adoptive parents. Statutes vary from state to state in regard to the inheritance rights of illegitimate children. Generally, an illegitimate child is treated as the child of the mother and can inherit from her and her relatives. The child is usually not regarded as the legal child of the father with the right of inheritance unless paternity was established through some legal proceeding prior to the father's death.

Distribution to Grandchildren When an intestate is survived by descendants of deceased children, a question arises as to what share these descendants (that is, grandchildren of the intestate) will receive. One method of dividing an intestate's estate is ***per stirpes.*** Under this method, within a class or group of distributees (for example, grandchildren), the children of any one descendant take the share that their deceased parent *would have been* entitled to inherit.

■ EXAMPLE 30.5 Assume that Michael, a widower, has two children, Scott and Jonathan. Scott has two children (Becky and Holly), and Jonathan has one child (Paul). Scott and Jonathan die before their father, and then Michael dies. If Michael's estate is distributed *per stirpes,* Becky and Holly each receive one-fourth of the estate (dividing Scott's one-half share). Paul receives one-half of the estate (taking Jonathan's one-half share). ■

An estate may also be distributed on a ***per capita*** basis, which means that each person in a class or group takes an equal share of the estate. If Michael's estate is distributed *per capita,* Becky, Holly, and Paul each receive a one-third share.

Trusts

A **trust** is any arrangement through which property is transferred from one person to a trustee to be administered for the transferor's or another party's benefit. It can also be defined as a right of property, real or personal, held by one party for the benefit of another. A trust can be created for any purpose that is not illegal or against public policy. Its essential elements are as follows:

1. A designated beneficiary.
2. A designated trustee.
3. A fund sufficiently identified to enable title to pass to the trustee.
4. Actual delivery by the settlor or grantor (the person creating the trust) to the trustee with the intention of passing title.

■ EXAMPLE 30.6 If James conveys his farm to the First Bank of Minnesota to be held for the benefit of his daughters, he has created a trust. James is the settlor, the First Bank of Minnesota is the trustee, and James's daughters are the beneficiaries. ■ Numerous types of trusts can be established. In this section, we look at some of the major types of trusts and their characteristics.

Express Trusts

An express trust is created or declared in explicit terms, usually in writing. Express trusts fall into two categories: *inter vivos* (living) trusts and *testamentary* trusts (trusts provided for in a last will and testament).

An ***inter vivos* trust** is executed by a grantor during her or his lifetime. The grantor (settlor) executes a *trust deed,* and legal title to the trust property passes to the named trustee. The trustee has a duty to administer the property as directed by the grantor for the benefit and in the interest of the beneficiaries. The trustee must preserve the trust property, make it productive, and, if required by the terms of the trust agreement, pay income to the beneficiaries, all in accordance with the terms of the trust. Once the *inter vivos* trust is created, the grantor has, in effect, given over the property for the benefit of the beneficiaries. Often, setting up this type of trust offers tax-related benefits.

A **testamentary trust** is created by a will to come into existence on the settlor's death. Although a testamentary trust has a trustee who maintains legal title to the trust property, the trustee's actions are subject to judicial approval. This trustee can be named in the will or be appointed by the court. Thus, a testamentary trust does not fail because a trustee has not been named in the will. The legal responsibilities of the trustee are the same as in an *inter vivos* trust. If the will setting up a testamentary trust is invalid, the trust will also be invalid. The property that was supposed to be in the trust will then pass according to intestacy laws, not according to the terms of the trust.

Implied Trusts

Sometimes, a trust will be imposed (implied) by law, even in the absence of an express trust. Implied trusts include resulting trusts and constructive trusts.

A **resulting trust** arises from the conduct of the parties. Here, the trust results, or is created, when circumstances raise an inference that the party holding legal title to the property does so for the benefit of another. ■ EXAMPLE 30.7 Suppose that Garrison wants to put one acre of land she owns on the market for sale. Because she is going out of the country for two years and would not be able to deed the property to a buyer during that period, Garrison conveys the property to her good friend Oswald. Oswald will attempt to sell the property while Garrison is gone. Because the intent of the transaction in

which Garrison conveyed the property to Oswald is neither a sale nor a gift, the property will be held in trust (a resulting trust) by Oswald for the benefit of Garrison. Therefore, on Garrison's return, Oswald will be required either to deed back the property to Garrison or, if the property has been sold, to turn over the proceeds (held in trust) to her. Here, the trust arises (results from) the *apparent intention* of the parties. ■

A **constructive trust** is an equitable trust imposed by a court in the interests of fairness and justice. In a constructive trust, the owner is declared to be a trustee for the parties who are, in equity, actually entitled to the benefits that flow from the trust. If someone wrongfully holds legal title to property—because the property was obtained through fraud or in breach of a legal duty, for example—a court may impose a constructive trust. ■ EXAMPLE 30.8 Arturo and Spring are partners in buying, developing, and selling real estate. Arturo learns through a staff member of the partnership that two hundred acres of land will soon come on the market and that the staff member will recommend that the partnership purchase the land. Arturo purchases the property secretly in his own name, thus violating his fiduciary relationship. When these facts are discovered, a court will determine that Arturo must hold the property in trust for the partnership. ■

SPECIAL TYPES OF TRUSTS

Certain trusts are created for special purposes. Three of these trusts that warrant discussion are charitable, spendthrift, and Totten trusts. A **charitable trust** is designed for the benefit of a segment of the public or of the public in general. Usually, to be deemed a charitable trust, a trust must be created for charitable, educational, religious, or scientific purposes.

A **spendthrift trust** is created to provide for the maintenance of a beneficiary by preventing him or her from being careless with the bestowed funds. Unlike other trusts, the beneficiary in a spendthrift trust is not permitted to transfer or assign his or her right to the trust's principal or future payments from the trust. Essentially, the beneficiary can draw only a certain portion of the total amount to which he or she is entitled at any one time. The majority of states allow spendthrift trust provisions that prohibit creditors from attaching such trusts.

A **Totten trust**[5] is created when one person deposits funds in her or his own name as a trustee for another. This trust is tentative in that it is revocable at will until the depositor dies or completes the gift in her or his lifetime by some unequivocal act or declaration (for example, delivery of the funds to the intended beneficiary). If the depositor dies before the beneficiary dies and if the depositor has not revoked the trust expressly or impliedly, a presumption arises that an absolute (a binding, irrevocable) trust has been created for the benefit of the beneficiary. At the death of the depositor, the beneficiary obtains property rights to the balance on hand.

5. This type of trust derives its unusual name from *In re Totten*, 179 N.Y. 112, 71 N.E. 748 (1904).

TERMS AND CONCEPTS

administrator 624
bequest 624
binder 621
charitable trust 629
codicil 627
constructive trust 629
devise 624
devisee 624
executor 624
holographic will 626
insurable interest 620
insurance 618
inter vivos trust 628
intestacy laws 624
intestate 624
legacy 624
legatee 624
nuncupative will 626
per capita 628
per stirpes 628
policy 619
premium 619
probate 625
resulting trust 628
risk 618
risk management 618
spendthrift trust 629
testamentary trust 628
testate 624
testator 624
Totten trust 629
trust 628
underwriter 619
will 624
will substitute 625

CHAPTER SUMMARY Insurance, Wills, and Trusts

INSURANCE	
Classifications (See pages 618–619.)	See Exhibit 30–1 on pages 620 and 621.
Terminology (See page 619.)	1. *Policy*—The insurance contract. 2. *Premium*—The consideration paid to the insurer for a policy. 3. *Underwriter*—The insurance company. 4. *Parties*—Include the insurer (the insurance company), the insured (the person covered by insurance), an agent (a representative of the insurance company) or a broker (ordinarily an independent contractor), and a beneficiary (a person to receive proceeds under the policy).
Insurable Interest (See page 620.)	An insurable interest exists whenever an individual or entity benefits from the preservation of the health or life of the insured or the property to be insured. For life insurance, an insurable interest must exist at the time the policy is issued. For property insurance, an insurable interest must exist at the time of the loss.
The Insurance Contract (See pages 620–624.)	1. *Laws governing*—The general principles of contract law are applied; the insurance industry is also heavily regulated by the states. 2. *Application*—An insurance applicant is bound by any false statements that appear in the application (subject to certain exceptions), which is part of the insurance contract. Misstatements or misrepresentations may be grounds for voiding the policy. 3. *Effective date*—Coverage on an insurance policy can begin when a *binder* (a written memorandum indicating that a formal policy is pending and stating its essential terms) is written; when the policy is issued; at the time of contract formation; or depending on the terms of the contract, when certain conditions are met. 4. *Provisions and clauses*—See Exhibit 30–2 on page 622. Words will be given their ordinary meanings, and any ambiguity in the policy will be interpreted against the insurance company. When the written policy has not been delivered and it is unclear whether an insurance contract actually exists, the uncertainty will be resolved against the insurance company. The court will presume that the policy is in effect unless the company can show otherwise. 5. *Defenses against payment to the insured*—Defenses include misrepresentation, fraud, or violation of warranties by the applicant.
WILLS	
Terminology (See page 624.)	1. *Intestate*—One who dies without a valid will. 2. *Testator*—A person who makes out a will. 3. *Personal representative*—A person appointed in a will or by a court to settle the affairs of a decedent. A personal representative named in the will is an *executor;* a personal representative appointed by the court for an intestate decedent is an *administrator.* 4. *Devise*—A gift of real estate by will; may be general or specific. The recipient of a devise is a *devisee.* 5. *Bequest, or legacy*—A gift of personal property by will; may be general or specific. The recipient of a bequest (legacy) is a *legatee.*
Probate Procedures (See page 625.)	To probate a will means to establish its validity and to carry the administration of the estate through a court process. Probate laws vary from state to state. Probate procedures may be informal or formal, depending on the size of the estate and other factors, such as whether a guardian for minor children must be appointed.

CHAPTER SUMMARY Insurance, Wills, and Trusts—Continued

Requirements for a Valid Will (See pages 625–626.)	1. The testator must have testamentary capacity (be of legal age and sound mind at the time the will is made). 2. A will must be in writing (except for nuncupative wills). A holographic will is completely in the handwriting of the testator. 3. A will must be signed by the testator; what constitutes a signature varies from jurisdiction to jurisdiction. 4. A nonholographic will (an attested will) must be witnessed in the manner prescribed by state statute. 5. A will may have to be *published*—that is, the testator may be required to announce to witnesses that this is his or her "last will and testament"; not required under the UPC.
Revocation of Wills (See pages 626–627.)	1. *By physical act of the maker*—Tearing up, canceling, obliterating, or deliberately destroying part or all of a will. 2. *By subsequent writing*— a. Codicil—A formal, separate document to amend or revoke an existing will. b. Second will or new will—A new, properly executed will expressly revoking the existing will. 3. *By operation of law*— a. Marriage—Generally revokes part of a will written before the marriage. b. Divorce or annulment—Revokes dispositions of property made under a will to a former spouse. c. Subsequently born child—It is *implied* that the child is entitled to receive the portion of the estate granted under intestacy distribution laws.
Intestacy Laws (See pages 627–628.)	1. Intestacy laws vary widely from state to state. Usually, the law provides that the surviving spouse and children inherit the property of the decedent (after the decedent's debts are paid). The spouse usually inherits the entire estate if there are no children, one-half of the estate if there is one child, and one-third of the estate if there are two or more children. 2. If there is no surviving spouse or child, then, in order, lineal descendants (grandchildren, brothers and sisters, and—in some states—parents of the decedent) inherit. If there are no lineal descendants, then collateral heirs (nieces, nephews, aunts, and uncles of the decedent) inherit.
	TRUSTS
Definition (See page 628.)	A trust is any arrangement through which property is transferred from one person to a trustee to be administered for another party's benefit. The essential elements of a trust are (1) a designated beneficiary, (2) a designated trustee, (3) a fund sufficiently identified to enable title to pass to the trustee, and (4) actual delivery to the trustee with the intention of passing title.
Express Trusts (See page 628.)	Express trusts are created by explicit terms, usually in writing, and fall into two categories: 1. *Inter vivos trust*—A trust executed by a grantor during her or his lifetime. 2. *Testamentary trust*—A trust created by will and coming into existence on the death of the grantor.
Implied Trusts (See pages 628–629.)	Implied trusts, which are imposed by law in the interests of fairness and justice, include the following: 1. *Resulting trust*—Arises from the conduct of the parties when an apparent intention to create a trust is present. 2. *Constructive trust*—Arises by operation of law whenever a transaction takes place in which the person who takes title to property is in equity not entitled to enjoy the beneficial interest therein.

(Continued)

CHAPTER SUMMARY Insurance, Wills, and Trusts—Continued

Special Types of Trusts (See page 629.)	1. *Charitable trust*—A trust designed for the benefit of a public group or the public in general. 2. *Spendthrift trust*—A trust created to provide for the maintenance of a beneficiary by allowing the beneficiary to receive only a certain portion of the total amount at any one time. 3. *Totten trust*—A trust created when one person deposits funds in his or her own name as a trustee for another.

FOR REVIEW

Answers for the even-numbered questions in this For Review *section can be found in Appendix A at the end of this text.*

1. What is an insurable interest? When must an insurable interest exist—at the time the insurance policy is obtained, at the time the loss occurs, or both?
2. Is an insurance broker the agent of the insurance applicant or the agent of the insurer?
3. What are the basic requirements for executing a will? How may a will be revoked?
4. What is the difference between a *per stirpes* distribution and a *per capita* distribution of an estate to the grandchildren of the deceased?
5. What are the four essential elements of a trust? What is the difference between an express trust and an implied trust?

QUESTIONS AND CASE PROBLEMS

30–1. Timing of Insurance Coverage. On October 10, Joleen Vora applied for a $50,000 life insurance policy with Magnum Life Insurance Co.; she named her husband, Jay, as the beneficiary. Joleen paid the insurance company the first year's policy premium on making the application. Two days later, before she had a chance to take the physical examination required by the insurance company and before the policy was issued, Joleen was killed in an automobile accident. Jay submitted a claim to the insurance company for the $50,000. Can Jay collect? Explain.

QUESTION WITH SAMPLE ANSWER

30–2. Benjamin is a widower who has two married children, Edward and Patricia. Patricia has two children, Perry and Paul. Edward has no children. Benjamin dies, and his typewritten will leaves all his property equally to his children, Edward and Patricia, and provides that should a child predecease him, the grandchildren are to take *per stirpes*. The will was witnessed by Patricia and by Benjamin's lawyer and was signed by Benjamin in their presence. Patricia has predeceased Benjamin. Edward claims the will is invalid.

(a) Discuss whether the will is valid.
(b) Discuss the distribution of Benjamin's estate if the will is invalid.
(c) Discuss the distribution of Benjamin's estate if the will is valid.

For a sample answer to this question, go to Appendix B at the end of this text.

30–3. Revocation. William Laneer urged his son, also William, to join the family business. The son, who was made partner, became suspicious of the handling of the business's finances. He filed a suit against the business and reported it to the Internal Revenue Service. The elder Laneer then executed a will that disinherited his son, giving him one dollar and leaving the balance of the estate equally to his four daughters, including Bellinda Barrera. Until his death more than twenty years later, Laneer harbored ill feelings toward his son. After Laneer's death, his original copy of the will could not be found. A photocopy was found in his safe-deposit box, however, and his lawyer's original copy was entered for probate in an Arkansas state court. Barrera, who wanted her brother William to share an equal portion of the inheritance, filed a petition to contest the will. Barrera claimed, among other things, that Laneer had revoked the will and that was why his original copy of the will could not be found. Was the will revoked? If so, to whom would the estate be distributed? [*Barrera v. Vanpelt*, 332 Ark. 482, 965 S.W.2d 780 (1998)]

30–4. Insurer's Defenses. In 1990, the city of Worcester, Massachusetts, adopted an ordinance that required rooming houses to be equipped with automatic sprinkler systems no later than September 25, 1995. In Worcester, James and Mark Duffy owned a forty-eight-room lodging house with two retail stores on the first floor. In 1994, the Duffys applied with General Star Indemnity Co. for an insurance policy to cover the premises. The application indicated that the premises had a sprinkler system. General issued a policy that required, among other safety features,

a sprinkler system. Within a month, the premises were inspected on behalf of General. On the inspection form forwarded to the insurer, the inspector inserted only a hyphen next to the word *sprinkler* in the list of safety systems. In July 1995, when the premises sustained over $100,000 in fire damage, General learned that there was no sprinkler system. The insurer filed a suit in a federal district court against the Duffys to rescind the policy, alleging misrepresentation in their insurance application about the presence of sprinklers. How should the court rule, and why? [*General Star Indemnity Co. v. Duffy*, 191 F.3d 55 (1st Cir. 1999)]

CASE PROBLEM WITH SAMPLE ANSWER

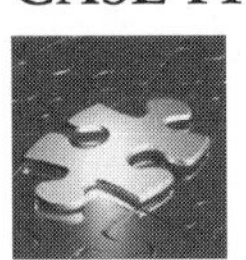

30–5. Valley Furniture & Interiors, Inc., bought an insurance policy from Transportation Insurance Co. (TIC). The policy provided coverage of $50,000 for each occurrence of property loss caused by employee dishonesty. An "occurrence" was defined as "a single act or series of related acts." Valley allowed its employees to take pay advances and to buy discounted merchandise, with the advances and the cost of the merchandise deducted from their paychecks. The payroll manager was to notify the payroll company to make the deductions. Over a period of six years, without notifying the payroll company, the payroll manager issued advances to other employees and herself and bought merchandise for herself, in amounts totaling more than $200,000. Valley filed claims with TIC for three "occurrences" of employee theft. TIC considered the acts a "series of related acts" and paid only $50,000. Valley filed a suit in a Washington state court against TIC, alleging, in part, breach of contract. What is the standard for interpreting an insurance clause? How should this court interpret "series of related acts"? Why? [*Valley Furniture & Interiors, Inc. v. Transportation Insurance Co.*, 107 Wash.App. 104, 26 P.3d 952 (Div. 1 2001)]

After you have answered this problem, compare your answer with the sample answer given on the Web site that accompanies this text. Go to academic.cengage.com/blaw/wbl, select "Chapter 30," and click on "Case Problem with Sample Answer."

30–6. Intestacy Laws. In January 1993, three and a half years after Lauren and Warren Woodward were married, they were informed that Warren had leukemia. At the time, the couple had no children, and the doctors told the Woodwards that the leukemia treatment might leave Mr. Woodward sterile. The couple arranged for Mr. Woodward's sperm to be collected and placed in a sperm bank for later use. In October 1993, Warren Woodward died. Two years later, Lauren Woodward gave birth to twin girls who had been conceived through artificial insemination using Mr. Woodward's sperm. The following year, Mrs. Woodward applied for Social Security survivor benefits for the two children. The Social Security Administration (SSA) rejected her on the ground that she had not established that the twins were the husband's children within the meaning of the Social Security Act of 1935. Mrs. Woodward then filed a paternity action in Massachusetts, and the probate court determined that Warren Woodward was the twins' father. Mrs. Woodward resubmitted her application to the SSA but was again refused survivor benefits for the twins. She then filed an action in a federal district court to determine the inheritance rights, under Massachusetts intestacy law, of children conceived from the sperm of a deceased individual and his surviving spouse. How should the court resolve this case? Should children conceived after a parent's death (by means of artificial insemination or *in vitro* fertilization) still inherit under intestate succession laws? Why or why not? [*Woodward v. Commissioner of Social Security*, 435 Mass. 536, 760 N.E.2d 257 (2002)]

30–7. Wills. In 1944, Benjamin Feinberg bought a plot in Beth Israel Cemetery in Plattsburgh, New York. A mausoleum was built on the plot to contain six crypts. In 1954, Feinberg's spouse died and was interred in one of the crypts. Feinberg, his only son, one of his two daughters, and the daughter's son, Julian Bergman, began using the mausoleum regularly as a place of prayer and meditation. When Feinberg died, he was interred in the mausoleum. His two daughters were interred in two of the remaining crypts on their deaths. Feinberg's son died in 2001 and was interred in the fifth crypt. His widow, Laurie, then changed the locks on the mausoleum and refused access to Julian, who filed a suit in a New York state court against her to obtain a key. Feinberg and all of his children died testate, but none of them made a specific bequest of their interest in the plot to anyone. Each person's will included a residuary clause, however. Who owns the plot, who has access to it, and why? [*Bergman v. Feinberg*, 6 A.D.3d 1031, 776 N.Y.S.2d 611 (3 Dept. 2004)]

VIDEO QUESTION

30–8. Go to this text's Web site at academic.cengage.com/blaw/wbl and select "Chapter 30." Click on "Video Questions" and view the video titled *Double Indemnity.* Then answer the following questions.

1. Recall from the video that Mrs. Dietrichson (Barbara Stanwyck) is attempting to take out an "accident insurance" policy (similar to life insurance) on her husband without his knowledge. Does Mrs. Dietrichson have an insurable interest in the life of her husband? Why or why not?
2. Why would Walter (Fred MacMurray), the insurance agent, refuse to sell Mrs. Dietrichson an insurance policy covering her husband's life without her husband's knowledge?
3. Suppose that Mrs. Dietrichson contacts a different insurance agent and does not tell the agent that she wants to obtain insurance on her husband without his knowledge. Instead, she asks the agent to leave an insurance application for her husband to sign. Without her husband's knowledge, Mrs. Dietrichson then fills out the application for insurance, which includes a two-year incontestability clause, and forges Mr. Dietrichson's signature. Mr. Dietrichson dies three years after the policy is issued. Will the insurance company be obligated to pay on the policy? Why or why not?

Insurance, Wills, and Trusts

For updated links to resources available on the Web, as well as a variety of other materials, visit this text's Web site at

academic.cengage.com/blaw/wbl

The Web site of the Insurance Information Institute provides a wealth of news and information on insurance-related issues, including statistical data, a glossary of insurance terms, and various PowerPoint presentations. Go to

http://www.iii.org

You can find the Uniform Probate Code, as well as links to various state probate statutes, at Cornell University's Legal Information Institute. Go to

http://www.law.cornell.edu/uniform/probate.html

Estate planning—preparing for the transfer of assets to others on one's death—involves a number of tools, including wills and trusts. These and other devices that can be used in estate planning are described by the American Academy of Estate Planning Attorneys on its Web site at

http://www.aaepa.com

To find the wills of many famous people, including Benjamin Franklin and Elvis Presley, go to

http://www.geocities.com/Athens/Acropolis/6537/wills.htm

ONLINE LEGAL RESEARCH

Go to the *Fundamentals of Business Law* home page at academic.cengage.com/blaw/wbl, select "Chapter 30," and click on "Internet Exercises." There you will find the following Internet research exercises that you can perform to learn more about topics covered in this chapter.

Activity 30–1: SOCIAL PERSPECTIVE—Disappearing Decisions

Activity 30–2: LEGAL PERSPECTIVE—Wills and Trusts

Activity 30–3: MANAGEMENT PERSPECTIVE—Risk Management in Cyberspace

BEFORE THE TEST

Go to the *Fundamentals of Business Law* home page at academic.cengage.com/blaw/wbl, select "Chapter 30," and click on "Interactive Quizzes." You will find a number of interactive questions relating to this chapter.

UNIT 9 • EXTENDED CASE STUDY

Ace Equipment Sales, Inc. v. Buccino

Real property ownership and the rights of property owners were discussed in Chapter 29. Water rights, called riparian rights, constitute a distinct property interest. The owner of land that borders a waterway, such as a river or stream, generally has the right to the reasonable and beneficial use of water on his or her land. Similar to subsurface rights (to minerals and oil, for example), riparian rights follow ownership of the land, but can be conveyed separately and often become a valuable component of real property rights. The following case explores a number of issues concerning water rights. Does an owner of the real property abutting a body of water have the right to use the water for swimming, boating, fishing, and other recreational purposes? Does it matter who owns the land under the water, or how the body of water was created? Does it make any difference whether a deed does, or does not, specifically grant the abutting owner a right to recreational uses, as opposed to industrial or other uses? What if the abutting owner is responsible for maintaining the level of the water through the operation of a dam? No Connecticut court had decided these questions prior to the case of *Ace Equipment Sales, Inc. v. Buccino*.[1] In this extended case study, we focus on the *Ace* case and examine these issues.

1. 82 Conn.App. 573, 848 A.2d 474 (2004).

Case Background

Conat Brook is connected to the Willimantic River in Willington, Connecticut. Before the 1950s, Gardiner Hall, Jr., Company, the owner of the surrounding land, built a dam and spillway to impound Conat's waters and create Hall's Pond, a twenty-acre body of water.

By the 1990s, Ace Equipment Sales, Inc., owned the land under the pond, and Ace, Willington Fish and Game Club, LLC (WFGC, LLC), and Thomas and Irma Buccino owned all of the land abutting the pond. According to their deed, the Buccinos had a right to use the pond's water to operate a mill and factory. The Buccinos were also obligated to maintain the dam.

WFGC, LLC, and Willington Fish and Game Club, Inc. (WFGC, Inc.), stocked the pond with fish and used it for recreational fishing. With the Buccinos' permission, Hall's Pond Fly Fishing Club, Inc., and others began to use the Buccinos' property to access the pond for swimming, boating, fishing, and other recreation.

Ace, WFGC, LLC, and WFGC, Inc., filed a suit in a Connecticut state court against the Buccinos and their invitees. The plaintiffs alleged that they had the exclusive right to use the pond for recreation because they owned the land under it. The court issued a summary judgment in favor of the defendants and ordered the plaintiffs to remove any obstructions to the defendants' access. The plaintiffs appealed to a state intermediate appellate court.

Majority Opinion

DUPONT, J. [Judge]

* * * *

Owners of property abutting a water surface are often called riparian owners * * * . In Connecticut, a riparian right to use non-navigable water does not create ownership of the water, but rather allows the use of the water for bathing and other recreational uses.

I

* * * *

Riparian rights exist as inherent rights incident to ownership of land contiguous to or traversed by a watercourse, and each riparian owner has an equal right with other such owners to make reasonable use of it for recreational purposes. * * * [Emphasis added.]

* * * *

If title extends only to the edge of a private watercourse but does not include the water, the titleholder has a right to use the surface in common with other riparian owners in any way that is not a trespass on the bottom of the water and can object to any obstruction of the water by another that interferes with his use.

* * * *

The [lower] court correctly held that * * * the defendants have the right to use the pond for reasonable recreational purposes in common with the owners of the bed of the pond.

II

* * * *

Both natural and artificial lakes have been used historically in Connecticut to generate power to run mills. * * * The owner of land on which a dam is constructed may use the water for any purpose, provided he does not thereby interfere with the rights of other proprietors either above or below him. * * *

We agree with the trial court that there is no distinction to be made between a natural and an artificial body of water and the riparian rights that accrue * * * . We conclude that

(Continued)

the recreational riparian rights of the Buccinos, who must maintain the dam to achieve the resultant impoundment of water, are not obliterated [destroyed] because the pond may have been artificial when created.

III

* * * *

Water rights appurtenant [attached] to abutting land pass by conveyance [transfer] of the land even though the deed does not specifically mention water rights. If the language of an easement, for example, to use water for industrial purposes, is broad enough to permit any reasonable use of a pond created by a dam, it may be used for recreational purposes *even though at the time of the creation of the easement, the parties probably contemplated that the major purpose would be to furnish water to a sawmill for power.* [Emphasis added.]

The Buccinos' deed gave them "the right * * * to take and use water from [the] pond through [a] penstock [a gate or sluice for controlling a flow of water] and the further right to take and use water from said pond that may be necessary for industrial purposes and operations in the mill and factory buildings." * * *

The deed also states that "(t)he right to draw the pond down below the natural water mark is not granted." Thus, when the grantor of the Buccinos wanted to exclude a right, it explicitly did so. There is no such exclusion in the deed for riparian rights. We interpret the entire language of the deed broadly and conclude that the deed does not impliedly preclude the defendants from using the pond for recreational as well as industrial purposes.

* * * *

The judgment is affirmed.

Dissenting Opinion

SCHALLER, J., [Judge] dissenting.

In determining that the trial court properly rendered summary judgment in favor of the defendants, Thomas Buccino and Irma Buccino, the majority concludes that the ownership of the subaqueous land was not a material fact. * * * I respectfully disagree * * * . I believe that, under the circumstances of this case, the appropriate doctrine is [a] * * * rule which has been adopted by numerous other state appellate courts * * * . I believe that [this] rule * * * produces a more sensible and fair result in situations like the present one.

* * * [Under this rule] the general public has no rights to the recreational use of a private lake, such rights being exclusive in the owner of the bed. * * * And * * * small inland lakes are susceptible of private ownership, at least to the extent that the owner or owners of the bed have the sole rights to the recreational uses of the waters.

* * * *

* * * [Under this rule] in the case of a non-navigable lake or pond where the land under the water is owned by others, no riparian rights attach to the property bordering on the water, and an attempt to exercise any such rights by invading the water is as much a trespass as if an unauthorized entry were made upon the dry land of another.

* * * *

* * * Because the construction of a man-made water body often involves the expenditure of substantial sums of money and the expense is not, as a rule, divided proportionately among the various abutting owners, the individual making the expenditure is justified in expecting that superior privileges will inure to him in return for his investment. In contrast, the abutting owners to a natural water body probably invest proportionally equal amounts for the increased value of the waterfront property. * * *

* * * *

* * * Accordingly, * * * I conclude that summary judgment was inappropriate in this case, and I would remand it for further proceedings.

Questions for Analysis

1. **Law.** How does the majority answer the questions set out in the introduction to this case?
2. **Law.** What rule does the dissent favor, and why?
3. **Social Dimensions.** Under the decision in the *Ace* case, what are the rights of the owner of the bed of a body of water abutted by public land, such as a public park or road?
4. **Economic Dimensions.** Does the majority's decision make bodies of water more available for different kinds of uses, or is the dissent's position more favorable for the use and development of bodies of water? Why?
5. **Implications for the Property Owner.** What impact might the ruling in this case have on waterfront property owners?

UNIT 10

Special Topics

Unit Contents

CHAPTER 31
Professional Liability

LEARNING OBJECTIVES

After reading this chapter, you should be able to answer the following questions:

1. Under what common law theories may professionals be liable to clients?
2. What are the rules concerning an auditor's liability to third parties?
3. How might an accountant violate federal securities laws?
4. What crimes might an accountant commit under the Internal Revenue Code?
5. What constrains professionals to keep communications with their clients confidential?

In the past decade or so, accountants, attorneys, physicians, and other professionals have found themselves increasingly subject to liability. This more extensive liability has resulted in large part from a greater public awareness of the fact that professionals are required to deliver competent services and are obligated to adhere to standards of performance commonly accepted within their professions.

Certainly, the dizzying collapse of Enron Corporation and the failure of other major companies, including WorldCom, Inc., in the early 2000s called attention to the importance of abiding by professional accounting standards. Arthur Andersen, LLP, one of the world's leading public accounting firms, ended up being indicted on criminal charges for its role in thwarting the government's investigation into Enron's accounting practices.[1] As a result, Arthur Andersen ceased to exist and roughly 85,000 employees lost their jobs. Moreover, under the Sarbanes-Oxley Act of 2002, which Congress passed in response to these events, public accounting firms throughout the nation will feel the effects for years to come. Among other things, the act imposed stricter regulation and oversight on the public accounting industry.

Considering the many potential sources of legal liability that may be imposed on them, accountants, attorneys, and other professionals should be well aware of their legal obligations. In the first part of this chapter, we look at the potential common law liability of professionals and then examine the potential liability of accountants under securities laws and the Internal Revenue Code. The chapter concludes with a brief examination of the relationships of professionals, particularly accountants and attorneys, with their clients and a discussion of limits on the liability of professionals.

Potential Common Law Liability to Clients

Under the common law, professionals may be liable to clients for breach of contract, negligence, or fraud.

1. Although Arthur Andersen, LLP, was subsequently convicted in a federal district court on the charge of obstructing justice, the United States Supreme Court reversed and remanded the case in 2005 due to erroneous jury instructions. No final decision has yet been entered on the charges. *Arthur Andersen, LLP v. United States,* ___ U.S. ___, 125 S.Ct. 2129, 161 L.Ed.2d 1008 (2005).

Liability for Breach of Contract

Accountants and other professionals face liability for any breach of contract under the common law. A professional owes a duty to his or her client to honor the terms of the contract and to perform the contract within the stated time period. If the professional fails to perform as agreed in the contract, then he or she has breached the contract, and the client has the right to recover damages from the professional. A professional may be held liable for expenses incurred by his or her client in securing another professional to provide the contracted-for services, for penalties imposed on the client for failure to meet time deadlines, and for any other reasonable and foreseeable monetary losses that arise from the professional's breach.

Liability for Negligence

Accountants and other professionals may also be held liable under the common law for negligence in the performance of their services. The elements that must be proved to establish negligence on the part of a professional are as follows:

1. A duty of care existed.
2. That duty of care was breached.
3. The plaintiff suffered an injury.
4. The injury was proximately caused by the defendant's breach of the duty of care.

All professionals are subject to standards of conduct established by codes of professional ethics, by state statutes, and by judicial decisions. They are also governed by the contracts they enter into with their clients. In their performance of contracts, professionals must exercise the established standard of care, knowledge, and judgment generally accepted by members of their professional group. We look below at the duty of care owed by two groups of professionals that frequently perform services for business firms: accountants and attorneys.

Accountant's Duty of Care Accountants play a major role in a business's financial system. Accountants have the necessary expertise and experience in establishing and maintaining accurate financial records to design, control, and audit record-keeping systems; to prepare reliable statements that reflect an individual's or a business's financial status; and to give tax advice and prepare tax returns.

—GAAP and GAAS. In the performance of their services, accountants must comply with **generally accepted accounting principles (GAAP)** and **generally accepted auditing standards (GAAS).** The Financial Accounting Standards Board (FASB, usually pronounced "faz-bee") determines what accounting conventions, rules, and procedures constitute GAAP at a given point in time. GAAS are standards concerning an auditor's professional qualities and the judgment that he or she exercises in performing an audit and report. GAAS are established by the American Institute of Certified Public Accountants. As long as an accountant conforms to generally accepted accounting principles and acts in good faith, he or she normally will not be held liable to the client for incorrect judgment.

As a general rule, an accountant is not required to discover every impropriety, **defalcation** (embezzlement), or fraud in his or her client's books. If, however, the impropriety, defalcation, or fraud has gone undiscovered because of an accountant's negligence or failure to perform an express or implied duty, the accountant will be liable for any resulting losses suffered by his or her client. Therefore, an accountant who uncovers suspicious financial transactions and fails to investigate the matter fully or to inform his or her client of the discovery can be held liable to the client for the resulting loss.

A violation of GAAP and GAAS will be considered *prima facie* evidence of negligence on the part of the accountant. Compliance with GAAP and GAAS, however, does not *necessarily* relieve an accountant from potential legal liability. An accountant may be held to a higher standard of conduct established by state statute and by judicial decisions.

—Defenses to Negligence. If an accountant is deemed guilty of negligence, the client can collect damages for losses that arose from the accountant's negligence. An accountant, however, is not without possible defenses to a cause of action for damages based on negligence. Possible defenses include the following:

1. The accountant was not negligent.
2. If the accountant was negligent, this negligence was not the proximate cause of the client's losses.
3. The client was negligent (depending on whether state law allows contributory negligence as a defense).

The following case involved a question about the second defense in the above list—whether the defendant's negligence was the proximate cause of the plaintiff's loss. The accounting firm claimed that the plaintiff's losses were due to market forces rather than a delay in the public offering of its stock.

CASE 31.1 Oregon Steel Mills, Inc. v. Coopers & Lybrand, LLP

Oregon Supreme Court, 2004.
336 Or. 329,
83 P.3d 322.

FACTS Oregon Steel Mills, Inc. (OSM), hired Coopers & Lybrand, LLP, to provide accounting services. In 1994, OSM sold some of the stock that it owned in one of its subsidiaries. On Coopers's advice, OSM reported the transaction as a $12.3 million gain on its 1994 financial statements. The next year, OSM planned a public offering of its own stock. OSM anticipated that it would file the necessary documents, which included the 1994 statements, with the Securities and Exchange Commission (SEC) in February 1996 and that the stock would be sold on May 2. The SEC concluded that the accounting treatment of the 1994 transaction was incorrect and required OSM to restate its 1994 statements. Because of the delay, the public offering did not occur on May 2, when OSM's stock was selling for $16 per share, but on June 13, when, due to unrelated factors, the price was $13.50. OSM filed a suit in an Oregon state court against Coopers, claiming that its advice regarding the 1994 transaction was incorrect, it gave the advice negligently, and this caused a delay that resulted in the stock being sold at a lower price. The court issued, in Coopers's favor, a summary judgment, which a state intermediate appellate court reversed. Coopers appealed to the Oregon Supreme Court.

ISSUE Was the accounting firm's negligence the proximate cause of the plaintiff's losses?

DECISION No. The Oregon Supreme Court reversed the lower appellate court's decision and affirmed the trial court's judgment.

REASON The state supreme court concluded that Coopers breached its duty to OSM by failing to provide competent accounting services, but that Coopers had no duty to protect OSM against fluctuations in the market price of OSM's stock. The decline in OSM's stock price in June 1996 was not reasonably foreseeable, and Coopers was not liable for damages based on that decline. The court acknowledged, "With appropriate proof, the client of a negligent accounting firm may recover damages for lost profits or lost business opportunities that result from the accounting firm's negligent acts." In this case, though, the price decline affected all steel company stocks, not just OSM's stock, and was unrelated to Coopers's misconduct. "[D]efendant's conduct caused the delay in the offering that led to an unintended adverse result. However, the intervening action of market forces on the price of plaintiff's stock was the harm-producing force, and defendant's actions did not cause the decline in the stock price so as to support liability for that decline. As a matter of law, the risk of a decline in plaintiff's stock price in June 1996 was not a reasonably foreseeable consequence of defendant's negligent acts in 1994 and early 1995."

FOR CRITICAL ANALYSIS—Economic Consideration *Based on the court's reasoning in this case, what damages might OSM recover for Coopers's negligence?*

—Unaudited Financial Statements. Sometimes accountants are hired to prepare unaudited financial statements. (A financial statement is considered *unaudited* if no procedures have been used to verify its accuracy or if insufficient procedures have been used to justify the contents.) Accountants may be subject to liability for failing, in accordance with standard accounting procedures, to delineate a balance sheet as unaudited. An accountant will also be held liable for failure to disclose to a client facts or circumstances that give reason to believe that misstatements have been made or that a fraud has been committed.

Attorney's Duty of Care The conduct of attorneys is governed by rules established by each state and by the American Bar Association's Code of Professional Responsibility and Model Rules of Professional Conduct. All attorneys owe a duty to provide competent and diligent representation. In judging an attorney's performance, the standard used will normally be that of a reasonably competent general practitioner of ordinary skill, experience, and capacity. If the attorney holds himself or herself out as having expertise in a special area of law, the standard is that of a reasonably competent specialist of ordinary skill, experience, and capacity in that area of the law. Attorneys are required to be familiar with well-settled principles of law applicable to a case and to discover law that can be found through a reasonable amount of research. The lawyer must also investigate and discover facts that could materially affect the client's legal rights.

When an attorney fails to exercise reasonable care and professional judgment, she or he breaches the duty of care. The plaintiff must then prove that the breach actually

caused him or her some injury. ■ **EXAMPLE 31.1** John Jones, an attorney, allows the statute of limitations to lapse on the claim of Karen Anderson, a client. Jones can be held liable for **malpractice** (professional negligence) because Anderson can no longer file a cause of action in this case and has lost a potential award of damages. ■

Traditionally, to establish causation, the client normally had to show that "but for" the attorney's negligence, the client would not have suffered the injury. In recent years, however, several courts have held that plaintiffs in malpractice cases need show only that the defendant's negligence was a "substantial factor" in causing the plaintiff's injury. In the following case, the Supreme Court of New Jersey addressed the issue of what standard should be applied in determining whether an attorney's malpractice was the proximate cause of the plaintiffs' injuries.

CASE 31.2 Conklin v. Hannoch Weisman

Supreme Court of New Jersey, 1996.
145 N.J. 395,
678 A.2d 1060.
http://lawlibrary.rutgers.edu/search.shtml#docket[a]

FACTS The Conklins hired the law firm of Hannoch Weisman, Professional Corporation, to represent them in a sale of one hundred acres of their farm to Longview Estates. The purchase price of the land was $12 million. Longview made a $3 million down payment and gave the Conklins a mortgage for the balance. The mortgage, however, was subordinate (second in priority) to a mortgage held by another lender: if Longview defaulted on its payments, the other lender would be paid first. When Longview defaulted, the other lender took the land, and the Conklins got nothing. They filed a suit in a New Jersey state court against Hannoch Weisman, claiming that the firm had not explained completely the risks of a subordinate mortgage. The jury was charged (instructed) to hold the firm liable only if the Conklins proved that their loss would not have occurred "but for" the firm's negligence. The jury issued a verdict in favor of the firm, but the judge decided that the jury charge had been unclear and ordered a new trial. Hannoch Weisman appealed to a state intermediate appellate court, which affirmed the order of the trial judge. The firm appealed to the Supreme Court of New Jersey.

a. On this page, enter the appropriate numbers for the citation to the *Conklin* case in the "Find by Case Citation" box, and click on "Submit Form" to access the opinion. Rutgers School of Law in Camden, New Jersey, maintains this Web site.

ISSUE Should the "substantial factor" standard be applied in determining whether an attorney's malpractice was the proximate cause of a plaintiff's injury?

DECISION Yes. The Supreme Court of New Jersey affirmed the judgment of the lower court.

REASON The state supreme court ordered a new trial "because the jury was given erroneous instructions" in the applicable law. The court pointed out that to recover for malpractice in New Jersey (and other states), a plaintiff needs to show only that the lawyer's negligence was a "substantial factor" in causing the harm. In this case and similar cases "in which there are concurrent independent causes of harm," the court stated, "[t]he question is whether the lack of adequate advice was a substantial factor in causing the Conklins' exposure to an unwanted risk of harm. * * * [T]he jury charge * * * could have confused the jury and led to an unjust result."

FOR CRITICAL ANALYSIS—Social Consideration
Should lawyers be subject to higher legal and ethical standards than other professionals? Why or why not?

PROFESSIONALS' LIABILITY FOR FRAUD

Actual fraud and constructive fraud present two different circumstances under which an accountant may be found liable. Recall from Chapter 10 that fraud, or misrepresentation, consists of the following elements:

1. A misrepresentation of a material fact has occurred.
2. There exists an intent to deceive.
3. The innocent party has justifiably relied on the misrepresentation.
4. For damages, the innocent party must have been injured.

A professional may be held liable for *actual fraud* when he or she intentionally misstates a material fact to mislead his or her client and the client justifiably relies on the misstated fact to his or her injury. A material fact is one that a reasonable person would consider important in deciding whether to act. In contrast, a professional may be held liable for *constructive fraud* whether or not he or she

acted with fraudulent intent. ■ EXAMPLE 31.2 In conducting an audit of National Computing Company (NCC), Paula, the auditor, accepts the explanations of Ron, an NCC officer, regarding certain financial irregularities, despite evidence that contradicts those explanations and indicates that the irregularities may be illegal. Paula's conduct could be characterized as an intentional failure to perform a duty in reckless disregard of the consequences of such a failure. This would constitute gross negligence and could be held to be constructive fraud. ■ Both actual and constructive fraud are potential sources of legal liability under which a client can bring an action against an accountant or other professional.

Auditors' Liability to Third Parties

Traditionally, an accountant or other professional did not owe any duty to a third person with whom he or she had no direct contractual relationship—that is, to any person not in *privity of contract* with the professional. A professional's duty was only to his or her client. Violations of statutory laws, fraud, and other intentional or reckless acts of wrongdoing were the only exceptions to this general rule.

Today, numerous third parties—including investors, shareholders, creditors, corporate managers and directors, regulatory agencies, and others—rely on professional opinions, such as those of auditors, when making decisions. In view of this extensive reliance, many courts have all but abandoned the privity requirement in regard to accountants' liability to third parties.

In this section, we focus primarily on the potential liability of auditors to third parties. Understanding an auditor's common law liability to third parties is critical, because often, when a business fails, its independent auditor (accountant) may be one of the few financially sound defendants. The majority of courts now hold that auditors can be held liable to third parties for negligence, but the standard for the imposition of this liability varies. There are generally three different views of accountants' liability to third parties, each of which we discuss below.

THE *ULTRAMARES* RULE

The traditional rule regarding an accountant's liability to third parties was enunciated by Chief Judge Benjamin Cardozo in *Ultramares Corp. v. Touche*, a case decided in 1931.[2] In *Ultramares*, Fred Stern & Company (Stern) hired the public accounting firm of Touche, Niven & Company (Touche) to review Stern's financial records and prepare a balance sheet for the year ending December 31, 1923.[3] Touche prepared the balance sheet and supplied Stern with thirty-two certified copies. According to the certified balance sheet, Stern had a net worth (assets less liabilities) of $1,070,715.26. In reality, however, Stern was insolvent—the company's records had been falsified by Stern's insiders to reflect a positive net worth. In reliance on the certified balance sheets, a lender, Ultramares Corporation, loaned substantial amounts to Stern. After Stern was declared bankrupt, Ultramares brought an action against Touche for negligence in an attempt to recover damages.

The New York Court of Appeals (that state's highest court) refused to impose liability on the accountants and concluded that they owed a duty of care only to those persons for whose "primary benefit" the statements were intended. In this case, Stern was the only person for whose primary benefit the statements were intended. The court held that in the absence of privity or a relationship "so close as to approach that of privity," a party could not recover from an accountant.

The court's requirement of privity or near privity has since been referred to as the *Ultramares* rule, or the New York rule. The rule was restated and somewhat modified in a 1985 New York case, *Credit Alliance Corp. v. Arthur Andersen & Co.*[4] In that case, the court held that if a third party has a sufficiently close relationship or nexus (link or connection) with an accountant, then the *Ultramares* privity requirement may be satisfied even if no accountant-client relationship is established. The rule enunciated in *Credit Alliance* is often referred to as the "near privity" rule. Only a minority of states have adopted this rule of accountants' liability to third parties.

Under this rule, does an accountant who is aware that a nonclient might rely on the accountant's work owe the nonclient a duty of care, when preparing reports on that party's financial status for his or her manager? Does the accountant have a duty to advise the nonclient on other financial transactions? These were the questions in the following case.

2. 255 N.Y. 170, 174 N.E. 441 (1931).

3. A *balance sheet* is often relied on by banks, creditors, stockholders, purchasers, or sellers as a basis for making decisions relating to a company's business.

4. 65 N.Y.2d 536, 483 N.E.2d 110 (1985): A "relationship sufficiently intimate to be equated with privity" is enough for a third party to sue another's accountant for negligence.

CASE 31.3 Reznor v. J. Artist Management, Inc.

United States District Court,
Southern District of New York, 2005.
365 F.Supp.2d 565.

FACTS Michael Trent Reznor met John Malm, Jr., a part-time promoter of local rock bands, in Cleveland, Ohio, in 1985. Malm became Reznor's manager and formed J. Artist Management, Inc. (JAM). Reznor became the lead singer in the band Nine Inch Nails (NIN), which performed its first show in 1988. Reznor and Malm signed a management agreement, under which JAM was to receive 20 percent of Reznor's gross compensation. Over the next few years, Reznor and Malm created other companies to sell NIN's merchandise and perform various services. In 1996, Malm hired accountant Richard Szekelyi and his firm, Navigent Group, to provide financial consulting services to JAM and the jointly owned companies. Szekelyi did not provide services to Reznor personally, but his duties included examining Reznor's financial records. Szekelyi discovered that the accounting among the parties was flawed, with Malm, for example, receiving tax benefits that should have gone to Reznor. According to Szekelyi, by 2003, Reznor owed JAM $1.56 million, and the jointly owned companies owed Reznor $5.5 million, which (as later became clear) was unlikely to be repaid. Reznor fired Malm and filed a suit in a federal district court against JAM and others, including Szekelyi and Navigent.

ISSUE Did the accountant breach his duty of care to a nonclient?

DECISION No. The court granted the motion of Szekelyi and Navigent for summary judgment and dismissed them from the case.

REASON The court recognized that an accountant can be liable to a nonclient for fraud if the accountant knows of the falsity of a client's statements, or recklessly disregards circumstances that cast doubt on the truth of the statements, and knows that the nonclient is reasonably relying on the accountant. Applying this principle, the court reasoned that "[s]ince Reznor was not Szekelyi's client, Szekelyi would be liable to Reznor for malpractice only where he was aware that the nonclient would rely on his work for a particular purpose. Szekelyi's presentations to Reznor in 2002 and 2003 meet this standard, and so Szekelyi is liable for any malpractice he may have committed in preparing and presenting these reports of Reznor's financial status. However, there is no evidence that Szekelyi breached any standard of care in preparing and presenting these reports." Furthermore, "[b]ecause Szekelyi was not Reznor's accountant or business manager, he owed Reznor in his individual capacity no such duty with respect to other transactions." Also, "there is no evidence that * * * Szekelyi knew or should have known of any impropriety" in this case involving fraud. "Indeed, there is no evidence that Szekelyi did anything but attempt to correct the only clearly flawed accounting in evidence, *i.e.*, the improper tax treatment."

WHY IS THIS CASE IMPORTANT? *The circumstances here give support to the warning that a person should take steps to become aware of the legal consequences of her or his agreements. The rock star at the center of this case might have avoided the negative impact of the result if he had sought the advice of an attorney before entering into the contract at issue.*

THE *RESTATEMENT* RULE

In the past several years, the *Ultramares* rule has been severely criticized. Auditors perform much of their work for use by persons who are not parties to the contract. Thus, it is asserted that they owe a duty to these third parties. Consequently, there has been an erosion of the *Ultramares* rule, and accountants have been exposed to potential liability to third parties.

The majority of courts have adopted the position taken by the *Restatement (Second) of Torts*, which states that accountants are subject to liability for negligence not only to their clients but also to *foreseen*, or *known*, users—or classes of users—of their reports or financial statements. Under Section 552(2) of the *Restatement (Second) of Torts*, an accountant's liability extends to those persons for whose benefit and guidance the accountant "intends to supply the information or knows that the recipient intends to supply it" and to those persons whom the accountant "intends the information to influence or knows that the recipient so intends." ■ **EXAMPLE 31.3** Steve, an accountant, prepares a financial statement for Tech Software, Inc., a client, knowing that the client will submit that statement to First National Bank to secure a loan. If Steve makes negligent misstatements or omissions in the statement, he may be held liable by the bank because he knew that the bank would rely on his work product when deciding whether to make the loan. ■

LIABILITY TO REASONABLY FORESEEABLE USERS

A small minority of courts hold accountants liable to any users whose reliance on an accountant's statements or reports was *reasonably foreseeable.* This standard has been criticized as extending liability too far. ■ **EXAMPLE 31.4** In *Raritan River Steel Co. v. Cherry, Bekaert & Holland,* the North Carolina Supreme Court stated that "in fairness accountants should not be liable in circumstances where they are unaware of the use to which their opinions will be put. Instead, their liability should be commensurate with those persons or classes of persons whom they know will rely on their work. With such knowledge the auditor can, through purchase of liability insurance, setting fees, and adopting other protective measures appropriate to the risk, prepare accordingly."[5] ■

It is the view of the majority of the courts that the *Restatement*'s approach is the more reasonable one because it allows accountants to control their exposure to liability. Liability is "fixed by the accountants' particular knowledge at the moment the audit is published," not by the foreseeability of the harm that might occur to a third party after the report is released.[6]

Even the California courts, which for several years relied on reasonable foreseeability as the standard for determining an auditor's liability to third parties, have changed their position. In a 1992 case, the California Supreme Court held that an accountant "owes no general duty of care regarding the conduct of an audit to persons other than the client." The court went on to say that if third parties rely on an auditor's opinion, "there is no liability even though the [auditor] should reasonably have foreseen such a possibility."[7]

LIABILITY OF ATTORNEYS TO THIRD PARTIES

Like accountants, attorneys may be held liable under the common law to third parties who rely on legal opinions to their detriment. Generally, an attorney is not liable to a nonclient unless there is fraud (or malicious conduct) by the attorney. The liability principles stated in Section 552 of the *Restatement (Second) of Torts,* however, may apply to attorneys just as they may apply to accountants.

5. 322 N.C. 200, 367 S.E.2d 609 (1988).

6. *Bethlehem Steel Corp. v. Ernst & Whinney,* 822 S.W.2d 592 (1991).

7. *Bily v. Arthur Young & Co.,* 3 Cal.4th 370, 834 P.2d 745, 11 Cal.Rptr.2d 51 (1992).

The Sarbanes-Oxley Act of 2002

As previously mentioned, in 2002 Congress enacted the Sarbanes-Oxley Act, which became effective on August 29, 2002. The act imposes a number of strict requirements on both domestic and foreign public accounting firms that provide auditing services to companies ("issuers") whose securities are sold to public investors. The act defines the term *issuer* as a company that has securities that are registered under Section 12 of the Securities Exchange Act of 1934; that is required to file reports under Section 15(d) of the 1934 act; or that files—or has filed—a registration statement that has not yet become effective under the Securities Act of 1933.

THE PUBLIC COMPANY ACCOUNTING OVERSIGHT BOARD

Among other things, the Sarbanes-Oxley Act calls for an increased degree of government oversight of public accounting practices. To this end, the act created the Public Company Accounting Oversight Board, which reports to the Securities and Exchange Commission. The board consists of a chair and four other members. The purpose of the board is to oversee the audit of public companies that are subject to securities laws in order to protect public investors and to ensure that public accounting firms comply with the provisions of the Sarbanes-Oxley Act.

APPLICABILITY TO PUBLIC ACCOUNTING FIRMS

Titles I and II of the act set forth the key provisions relating to the duties of the new oversight board and the requirements relating to public accounting firms—defined by the act as firms and associated persons that are "engaged in the practice of public accounting or preparing or issuing audit reports." These provisions are summarized in Exhibit 31–1. (Provisions of the act that are more directly concerned with corporate fraud and the responsibilities of corporate officers and directors were described in Chapter 27.)

REQUIREMENTS FOR MAINTAINING WORKING PAPERS

Performing an audit for a client involves an accumulation of **working papers**—the various documents used and developed during the audit. These include notes, computations, memoranda, copies, and other papers that make up the work product of an accountant's services to a client. Under the common law, which in this instance has been codified in a number of states, working papers remain the accoun-

EXHIBIT 31–1 KEY PROVISIONS OF THE SARBANES-OXLEY ACT OF 2002 RELATING TO PUBLIC ACCOUNTING FIRMS

DUTIES OF THE PUBLIC COMPANY ACCOUNTING OVERSIGHT BOARD

Title I of the Sarbanes-Oxley Act of 2002 states that the duties of the Public Company Accounting Oversight Board are as follows:

- Generally, to oversee the audit of companies ("issuers") whose securities are sold to public investors in order to protect the interests of investors and further the public interest.
- To register public accounting firms that prepare audit reports for issuers. (A nonregistered firm is prohibited from preparing, or participating in the preparation of, an audit report with respect to an issuer.)
- To establish or adopt standards relating to the preparation of audit reports for issuers.
- To enforce compliance with the Sarbanes-Oxley Act by inspecting registered public accounting firms (RPAFs) and by investigating and disciplining, by appropriate sanctions, firms that violate the act's provisions. (Sanctions range from a temporary or permanent suspension to civil penalties that can be as high as $15 million for intentional violations.)
- To perform any other duties necessary or appropriate to promote high professional standards among RPAFs and improve the quality of audit services offered by those firms.

AUDITOR INDEPENDENCE

To help ensure that auditors remain independent of the firms that they audit, Title II of the Sarbanes-Oxley Act does the following:

- Makes it unlawful for RPAFs to perform both audit and nonaudit services for the same company at the same time. Nonaudit services include the following:
  1. Bookkeeping or other services related to the accounting records or financial statements of the audit client.
  2. Financial information systems design and implementation.
  3. Appraisal or valuation services.
  4. Fairness opinions.
  5. Management functions.
  6. Broker or dealer, investment adviser, or investment banking services.
- Requires preapproval for most auditing services from the issuer's audit committee.
- Requires audit partner rotation by prohibiting RPAFs from providing audit services to an issuer if either the lead audit partner or the audit partner responsible for reviewing the audit has provided such services to the issuer in each of the prior five years.
- Requires RPAFs to make timely reports to the audit committees of the issuers. The report must indicate all critical accounting policies and practices to be used; all alternative treatments of financial information within generally accepted accounting principles that have been discussed with the issuer's management officials, the ramifications of the use of such alternative treatments, and the treatment preferred by the auditor; and other material written communications between the auditor and the issuer's management.
- Makes it unlawful for an RPAF to provide auditing services to an issuer if the issuer's chief executive officer, chief financial officer, chief accounting officer, or controller was previously employed by the auditor and participated in any capacity in the audit of the issuer during the one-year period preceding the date that the audit began.

DOCUMENT RETENTION AND DESTRUCTION

- The Sarbanes-Oxley Act provides that anyone who destroys, alters, or falsifies records with the intent to obstruct or influence a federal investigation or in relation to bankruptcy proceedings can be criminally prosecuted and sentenced to a fine, imprisoned for up to twenty years, or both.
- The act also requires accountants who audit or review publicly traded companies to retain all working papers related to the audit or review for a period of five years (now amended to seven years). Violators can be sentenced to a fine, imprisoned for up to ten years, or both.

tant's property. It is important for accountants to retain such records in the event that they need to defend against lawsuits for negligence or other actions in which their competence is challenged. But because an accountant's working papers reflect the client's financial situation, the client has a right to access them. (On a client's request, an accountant must return to the client any of the client's records or journals, and failure to do so may result in liability.)

Under Section 802(a)(1) of the Sarbanes-Oxley Act, accountants must maintain working papers relating to an

audit or review for five years (which has been amended to seven years since the passage of the act) from the end of the fiscal period in which the audit or review was concluded. A knowing violation of this requirement will subject the accountant to a fine, imprisonment for up to ten years, or both.

Potential Statutory Liability of Accountants

Both civil and criminal liability may be imposed on accountants under the Securities Act of 1933, the Securities Exchange Act of 1934, and the Private Securities Litigation Reform Act of 1995.[8]

LIABILITY UNDER THE SECURITIES ACT OF 1933

The Securities Act of 1933 requires registration statements to be filed with the Securities and Exchange Commission (SEC) prior to an offering of securities (see Chapter 27).[9] Accountants frequently prepare and certify (attest to the accuracy of) the issuer's financial statements that are included in the registration statement.

Liability under Section 11 Section 11 of the Securities Act of 1933 imposes civil liability on accountants for misstatements and omissions of material facts in registration statements. Therefore, an accountant may be found liable if she or he prepared any financial statements included in the registration statement that "contained an untrue statement of a material fact or omitted to state a material fact required to be stated therein or necessary to make the statements therein not misleading."[10]

—The Liability to Purchasers of Securities. Under Section 11, an accountant's liability for a misstatement or omission of a material fact in a registration statement extends to anyone who acquires a security covered by the registration statement. A purchaser of a security need only demonstrate that he or she has suffered a loss on the security. Proof of reliance on the materially false statement or misleading omission is not ordinarily required. Nor is there a requirement of privity between the accountant and the security purchasers.

—The Due Diligence Standard. Section 11 imposes a duty on accountants to use **due diligence** in preparing the financial statements that are included in the filed registration statements. After a purchaser has proved a loss on the security, the accountant bears the burden of showing that he or she exercised due diligence in preparing the financial statements. To avoid liability, the accountant must show that he or she had, "after reasonable investigation, reasonable grounds to believe and did believe, at the time such part of the registration statement became effective, that the statements therein were true and that there was no omission of a material fact required to be stated therein or necessary to make the statements therein not misleading."[11] Further, the failure to follow GAAP and GAAS is also proof of a lack of due diligence.

In particular, the due diligence standard places a burden on accountants to verify information furnished by a corporation's officers and directors. The burden of proving due diligence requires an accountant to demonstrate that she or he is free from negligence or fraud. Merely asking questions is not always sufficient to satisfy the requirement of due diligence. ■ EXAMPLE 31.5 In *Escott v. BarChris Construction Corp.*,[12] accountants were held liable for failing to detect danger signals in documents furnished by corporate officers that, under GAAS, required further investigation under the circumstances. ■

—Defenses to Liability. Besides proving that he or she has acted with due diligence, an accountant can raise the following defenses to Section 11 liability:

1. There were no misstatements or omissions.
2. The misstatements or omissions were not of material facts.
3. The misstatements or omissions had no causal connection to the plaintiff's loss.
4. The plaintiff purchaser invested in the securities knowing of the misstatements or omissions.

Liability under Section 12(2) Section 12(2) of the Securities Act of 1933 imposes civil liability for fraud in relation to offerings or sales of securities.[13] Liability is based on the communication to an investor, whether orally or in the written prospectus,[14] of an untrue statement or the omission of a material fact.

8. Other potential sources of civil and criminal liability that may be imposed on accountants and other professionals include provisions of the Racketeer Influenced and Corrupt Organizations Act (RICO). RICO was discussed in Chapter 6.
9. Many securities and transactions are expressly exempted from the 1933 act.
10. 15 U.S.C. Section 77k(a).
11. 15 U.S.C. Section 77k(b)(3).
12. 283 F.Supp. 643 (S.D.N.Y. 1968).
13. 15 U.S.C. Section 77*l*.
14. As discussed in Chapter 27, a *prospectus* contains financial disclosures about the corporation for the benefit of potential investors.

Penalties and Sanctions for Violations Those who purchase securities and suffer harm as a result of a false or omitted statement, or some other violation, may bring a suit in a federal court to recover their losses and other damages. The U.S. Department of Justice brings criminal actions against those who commit willful violations. The penalties include fines of up to $10,000, imprisonment for up to five years, or both. The SEC is authorized to seek an injunction against a willful violator to prevent further violations. The SEC can also ask a court to grant other relief, such as an order to a violator to refund profits derived from an illegal transaction.

LIABILITY UNDER THE SECURITIES EXCHANGE ACT OF 1934

Under Sections 18 and 10(b) of the Securities Exchange Act of 1934 and Rule 10b-5 of the SEC, an accountant may be found liable for fraud. A plaintiff has a substantially heavier burden of proof under the 1934 act than under the 1933 act. Unlike the 1933 act, the 1934 act does not require that an accountant prove due diligence to escape liability.

Liability under Section 18 Section 18 of the 1934 act imposes civil liability on an accountant who makes or causes to be made in any application, report, or document a statement that at the time and in light of the circumstances was false or misleading with respect to any material fact.[15]

Section 18 liability is narrow in that it applies only to applications, reports, documents, and registration statements filed with the SEC. This remedy is further limited in that it applies only to sellers and purchasers. Under Section 18, a seller or purchaser must prove one of the following:

1. That the false or misleading statement affected the price of the security.
2. That the purchaser or seller relied on the false or misleading statement in making the purchase or sale and was not aware of the inaccuracy of the statement.

Even if a purchaser or seller proves these two elements, an accountant can be excused from liability on proof of good faith in the preparation of the financial statement. To demonstrate good faith, an accountant must show that he or she had no knowledge that the financial statement was false and misleading. Acting in good faith requires the total absence of an intention on the part of the accountant to seek an unfair advantage over, or to defraud, another party. Proving a lack of intent to deceive, manipulate, or defraud is frequently referred to as proving a lack of *scienter* (knowledge on the part of a misrepresenting party that material facts have been misrepresented or omitted with an intent to deceive).

Absence of good faith can be demonstrated not only by proof of *scienter* but also by the accountant's reckless conduct and gross negligence. (Note that "mere" negligence in the preparation of a financial statement does not constitute liability under the 1934 act. This differs from the 1933 act, which provides that an accountant is liable for all negligent acts.) In addition to the good faith defense, accountants have available as a defense the buyer's or seller's knowledge that the financial statement was false and misleading.

Under Section 18 of the 1934 act, a court also has the discretion to assess reasonable costs, including attorneys' fees, against accountants.[16] Sellers and purchasers can maintain a cause of action "within one year after the discovery of the facts constituting the cause of action and within three years after such cause of action accrued."[17]

Liability under Section 10(b) and Rule 10b-5 The Securities Exchange Act of 1934 further subjects accountants to potential legal liability under its antifraud provisions. Section 10(b) of the 1934 act and SEC Rule 10b-5 contain the antifraud provisions. As stated in *Herman & MacLean v. Huddleston*,[18] "a private right of action under Section 10(b) of the 1934 act and Rule 10b-5 has been consistently recognized for more than 35 years."

Section 10(b) makes it unlawful for any person, including accountants, to use, in connection with the purchase or sale of any security, any manipulative or deceptive device or contrivance in contravention of SEC rules and regulations.[19] Rule 10b-5 further makes it unlawful for any person, by use of any means or instrumentality of interstate commerce, to do the following:

1. To employ any device, scheme, or artifice to defraud.
2. To make any untrue statement of a material fact or to omit to state a material fact necessary to make the statements made, in light of the circumstances, not misleading.
3. To engage in any act, practice, or course of business that operates or would operate as a fraud or deceit on any person in connection with the purchase or sale of any security.[20]

15. 15 U.S.C. Section 78r(a).
16. 15 U.S.C. Section 78r(a).
17. 15 U.S.C. Section 78r(c).
18. 459 U.S. 375, 103 S.Ct. 683, 74 L.Ed.2d 548 (1983).
19. 15 U.S.C. Section 78j(b).
20. 17 C.F.R. Section 240.10b-5.

Accountants may be held liable only to sellers or purchasers under Section 10(b) and Rule 10b-5.[21] The scope of these antifraud provisions is extremely wide. Privity is not necessary for a recovery. Under these provisions, an accountant may be found liable not only for fraudulent misstatements of material facts in written material filed with the SEC but also for any fraudulent oral statements or omissions made in connection with the purchase or sale of any security.

To recover from an accountant under the antifraud provisions of the 1934 act, a plaintiff must, in addition to establishing her or his status as a purchaser or seller, prove *scienter*,[22] a fraudulent action or deception, reliance, materiality, and causation. A plaintiff who fails to establish these elements cannot recover damages from an accountant under Section 10(b) or Rule 10b-5.

THE PRIVATE SECURITIES LITIGATION REFORM ACT OF 1995

The Private Securities Litigation Reform Act of 1995 made some changes to the potential liability of accountants and other professionals in securities fraud cases.[23] Among other things, the act imposed a new statutory obligation on accountants. An auditor must use adequate procedures in an audit to detect any illegal acts of the company being audited. If any illegality is detected, the auditor must disclose it to the company's board of directors, the audit committee, or the SEC, depending on the circumstances.[24]

In terms of liability, the 1995 act provides that in most situations, a party is liable only for that proportion of damages for which he or she is responsible.[25] An accountant who does not participate in, and is unaware of, illegal conduct may not be liable for the entire loss caused by the illegality. ■ **EXAMPLE 31.6** Nina, an accountant, helps the president and owner of Midstate Trucking Company draft financial statements that misrepresent the financial condition of Midstate, but Nina is not actually aware of the fraud. Nina might be held liable, but the amount of her liability could be proportionately less than the entire loss. ■

If an accountant knowingly aids and abets a primary violator, the SEC can seek an injunction or money damages. ■ **EXAMPLE 31.7** Smith & Jones, an accounting firm, performs an audit for ABC Sales Company that is so inadequate as to constitute gross negligence. ABC uses the materials provided by Smith & Jones as part of a scheme to defraud investors. When the scheme is uncovered, the SEC can bring an action against Smith & Jones for aiding and abetting on the ground that the firm knew or should have known of the material misrepresentations that were in its audit and on which investors were likely to rely. ■

Potential Criminal Liability

An accountant may be found criminally liable for violations of the Securities Act of 1933, the Securities Exchange Act of 1934, the Internal Revenue Code, and both state and federal criminal codes. Under both the 1933 act and the 1934 act, accountants may be subject to criminal penalties for *willful* violations—imprisonment for up to ten years and/or a fine of up to $10,000 under the 1933 act and up to $100,000 under the 1934 act. Under the Sarbanes-Oxley Act of 2002, for a securities filing that is accompanied by an accountant's false or misleading certified audit statement, the accountant may be fined up to $5 million, imprisoned for up to twenty years, or both.

The Internal Revenue Code makes aiding or assisting in the preparation of a false tax return a felony punishable by a fine of $100,000 ($500,000 in the case of a corporation) and imprisonment for up to three years.[26] Those who prepare tax returns for others also may face liability under the Internal Revenue Code. Note that one does not have to be an accountant to be subject to liability for tax-preparer penalties. The Internal Revenue Code defines a *tax preparer* as any person who prepares for compensation, or who employs one or more persons to prepare for compensation, all or a substantial portion of a tax return or a claim for a tax refund.[27]

The Internal Revenue Code also imposes on the tax preparer a penalty of $250 per return for negligent understatement of his or her client's tax liability and a penalty of $1,000 for willful understatement of tax liability or reckless or intentional disregard of rules or regulations.[28] A tax preparer may also be subject to penalties for failing to furnish the taxpayer with a copy of the return, failing to sign the return, or failing to furnish the appropriate tax identification numbers.[29]

21. See *Blue Chip Stamps v. Manor Drug Stores*, 421 U.S. 723, 95 S.Ct. 1917, 44 L.Ed.2d 539 (1975).
22. See *Ernst & Ernst v. Hochfelder*, 425 U.S. 185, 96 S.Ct. 1375, 47 L.Ed.2d 668 (1976).
23. Some parties attempted to bypass the new law by filing their suits in state, rather than federal, courts. Congress acted to block such suits by passing the Securities Litigation Uniform Standards Act of 1998.
24. 15 U.S.C. Section 78j-1.
25. 15 U.S.C. Section 78u-4(g).
26. 26 U.S.C. Section 7206(2).
27. 26 U.S.C. Section 7701(a)(36).
28. 26 U.S.C. Section 6694.
29. 26 U.S.C. Section 6695.

Additionally, the Internal Revenue Code imposes a penalty of $1,000 per document for aiding and abetting an individual's understatement of tax liability (the penalty is increased to $10,000 in corporate cases). The tax preparer's liability is limited to one penalty per taxpayer per tax year.[30]

In most states, criminal penalties may be imposed for such actions as knowingly certifying false or fraudulent reports; falsifying, altering, or destroying books of account; and obtaining property or credit through the use of false financial statements.

Confidentiality and Privilege

Professionals are constrained by the ethical tenets of their professions to keep all communications with their clients confidential.

ATTORNEY-CLIENT RELATIONSHIPS

The confidentiality of attorney-client communications is protected by law, which confers a privilege on such communications. This privilege is granted because of the need for full disclosure to the attorney of the facts of a client's case. To encourage frankness, confidential attorney-client communications relating to representation are normally held in strictest confidence and protected by law. The attorney and his or her employees may not discuss the client's case with anyone—even under court order—without the client's permission. The client holds the privilege, and only the client may waive it—by disclosing privileged information to someone outside the privilege, for example.

Note, however, that since the Sarbanes-Oxley Act was enacted in 2002, the SEC has implemented new rules requiring attorneys who become aware that a client has violated securities laws to report the violation to the SEC. Reporting a client's misconduct could be a breach of the attorney-client privilege and has caused much controversy in the legal community.

30. 26 U.S.C. Section 6701.

ACCOUNTANT-CLIENT RELATIONSHIPS

In a few states, accountant-client communications are privileged by state statute. In these states, accountant-client communications may not be revealed even in court or in court-sanctioned proceedings without the client's permission. The majority of states, however, abide by the common law, which provides that, if a court so orders, an accountant must disclose information about his or her client to the court. Physicians and other professionals may similarly be compelled to disclose in court information given to them in confidence by patients or clients.

Communications between professionals and their clients—other than those between an attorney and his or her client—are not privileged under federal law. In cases involving federal law, state-provided rights to confidentiality of accountant-client communications are not recognized. Thus, in those cases, in response to a court order, an accountant must provide the information sought.

Limiting Professionals' Liability

Accountants (and other professionals) can limit their liability to some extent by disclaiming it. Depending on the circumstances, a disclaimer that does not meet certain requirements will not be effective, and in some situations, a disclaimer may not be effective at all.

Professionals may be able to limit their liability for the misconduct of other professionals with whom they work by organizing their business as a professional corporation (P.C.) or a limited liability partnership (LLP). In some states, a professional who is a member of a P.C. is not personally liable for a co-member's misconduct unless she or he participated in it or supervised the member who acted wrongly. The innocent professional is liable only to the extent of her or his interest in the assets of the firm. This is also true for professionals who are partners in an LLP. (P.C.s were discussed in Chapter 25. LLPs were covered in Chapter 24.)

TERMS AND CONCEPTS

defalcation 639
due diligence 646
generally accepted accounting principles (GAAP) 639
generally accepted auditing standards (GAAS) 639
malpractice 641
working papers 644

CHAPTER SUMMARY Professional Liability

COMMON LAW LIABILITY	
Liability to Clients (See pages 638–642.)	1. *Breach of contract*—An accountant or other professional who fails to perform according to his or her contractual obligations can be held liable for breach of contract and resulting damages. 2. *Negligence*—An accountant or other professional, in performance of his or her duties, must use the care, knowledge, and judgment generally used by professionals in the same or similar circumstances. Failure to do so is negligence. An accountant's violation of generally accepted accounting principles or generally accepted auditing standards is *prima facie* evidence of negligence. 3. *Fraud*—Actual intent to misrepresent a material fact to a client, when the client relies on the misrepresentation, is fraud. Gross negligence in performance of duties is constructive fraud.
Auditors' Liability to Third Parties (See pages 642–644.)	An accountant may be liable for negligence to any third person the accountant knows or should have known will benefit from the accountant's work. The standard for imposing this liability varies, but generally courts follow one of the following three rules: 1. *Ultramares rule*—Liability will be imposed only if the accountant is in privity, or near privity, with the third party. 2. *Restatement rule*—Liability will be imposed only if the third party's reliance is foreseen, or known, or if the third party is among a class of foreseen, or known, users. The majority of courts follow this rule. 3. *"Reasonably foreseeable user" rule*—Liability will be imposed if the third party's use was reasonably foreseeable.
STATUTORY LIABILITY	
The Sarbanes-Oxley Act of 2002 (See pages 644–646.)	1. *Purpose*—The purpose of this act was to impose requirements on public accounting firms that provide auditing services to companies whose securities are sold to public investors. 2. *Government oversight*—Among other things, the act created the Public Company Accounting Oversight Board to create government oversight over public accounting practices. 3. *Working papers*—The act requires accountants to maintain working papers relating to an audit or review for seven years from the end of the fiscal period in which the audit or review was concluded. 4. *Other requirements*—See Exhibit 31–1 on page 645.
Securities Act of 1933, Section 11 (See page 646.)	An accountant who makes a false statement or omits a material fact in audited financial statements required for registration of securities under the law may be liable to anyone who acquires securities covered by the registration statement. The accountant's defense is basically the use of due diligence and the reasonable belief that the work was complete and correct. The burden of proof is on the accountant. Willful violations of this act may be subject to criminal penalties.
Securities Act of 1933, Section 12(2) (See pages 646–647.)	In some jurisdictions, an accountant may be liable for aiding and abetting the seller or offeror of securities when a prospectus or communication presented to an investor contained an untrue statement or omission of material fact. To be liable, the accountant must have known, or at least should have known, that an untrue statement or omission of material fact existed in the offer to sell the security.
Securities Exchange Act of 1934, Sections 10(b) and 18 (See pages 647–648.)	Accountants are held liable for false and misleading applications, reports, and documents required under the act. The burden is on the plaintiff, and the accountant has numerous defenses, including good faith and lack of knowledge that what was submitted was false. Willful violations of this act may be subject to criminal penalties.

CHAPTER SUMMARY Professional Liability—Continued

Potential Criminal Liability (See pages 648–649.)	1. Aiding or assisting in the preparation of a false tax return is a felony. Aiding and abetting an individual's understatement of tax liability is a separate crime. 2. Tax preparers who negligently or willfully understate a client's tax liability or who recklessly or intentionally disregard Internal Revenue rules or regulations are subject to criminal penalties. 3. Tax preparers who fail to provide a taxpayer with a copy of the return, fail to sign the return, or fail to furnish the appropriate tax identification numbers may also be subject to criminal penalties.
Confidentiality and Privilege (See page 649.)	The confidentiality of attorney-client communications is protected by law. The SEC, however, has implemented new rules requiring attorneys to report when a client violates securities laws. This potential breach of attorney-client privilege has caused controversy in the legal community.

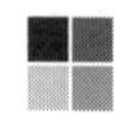

FOR REVIEW

Answers for the even-numbered questions in this For Review *section can be found in Appendix A at the end of this text.*

1. Under what common law theories may professionals be liable to clients?
2. What are the rules concerning an auditor's liability to third parties?
3. How might an accountant violate federal securities laws?
4. What crimes might an accountant commit under the Internal Revenue Code?
5. What constrains professionals to keep communications with their clients confidential?

QUESTIONS AND CASE PROBLEMS

31–1. *Ultramares* Rule. Larkin, Inc., retains Howard Perkins to manage its books and prepare its financial statements. Perkins, a certified public accountant, lives in Indiana and practices there. After twenty years, Perkins has become a bit bored with generally accepted accounting principles (GAAP) and has adopted more creative accounting methods. Now, though, Perkins has a problem, as he is being sued by Molly Tucker, one of Larkin's creditors. Tucker alleges that Perkins either knew or should have known that Larkin's financial statements would be distributed to various individuals. Furthermore, she asserts that these financial statements were negligently prepared and seriously inaccurate. What are the consequences of Perkins's failure to follow GAAP? Under the traditional *Ultramares* rule, can Tucker recover damages from Perkins? Explain.

QUESTION WITH SAMPLE ANSWER

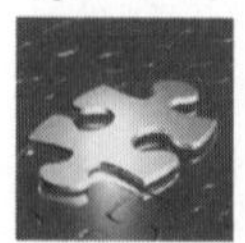

31–2. The accounting firm of Goldman, Walters, Johnson & Co. prepared financial statements for Lucy's Fashions, Inc. After reviewing the various financial statements, Happydays State Bank agreed to loan Lucy's Fashions $35,000 for expansion. When Lucy's Fashions declared bankruptcy under Chapter 11 six months later, Happydays State Bank promptly filed an action against Goldman, Walters, Johnson & Co., alleging negligent preparation of financial statements. Assuming that the court has abandoned the *Ultramares* approach, what is the result? What are the policy reasons for holding accountants liable to third parties with whom they are not in privity?

For a sample answer to this question, go to Appendix B at the end of this text.

31–3. Accountant's Liability under Rule 10b-5. In early 1995, Bennett, Inc., offered a substantial number of new common shares to the public. Harvey Helms had a long-standing interest in Bennett because his grandfather had once been president of the company. On receiving a prospectus prepared and distributed by Bennett, Helms was dismayed by the pessimism it embodied. Helms decided to delay purchasing stock in the company. Later, Helms asserted that the prospectus prepared by the accountants was overly pessimistic and contained materially misleading statements. Discuss fully how successful Helms would be in bringing a cause of action under Rule 10b-5 against the accountants of Bennett, Inc.

31–4. Negligence. Robert Potts was a partner at the accounting firm of Touche Ross and its successor, Deloitte & Touche. Potts was the concurring partner for 1988 and 1989 audits of Kahler Corp., which owned and managed the University Park Hotel.

According to the American Institute of Certified Public Accountants, a *concurring partner* assures that an audited company's financial statements conform with generally accepted accounting principles (GAAP) and generally accepted auditing standards. In 1988 and 1989, the hotel lost money, but Kahler showed a gain in 1988 by accounting for the hotel as an asset held for sale. Under GAAP, certain conditions must be met to account for property as an asset held for sale. Despite evidence that Kahler did not meet the conditions, Potts approved the audits. Potts also agreed to Kahler's valuation of the hotel, despite evidence that it was worth less than Kahler said. The Securities and Exchange Commission filed a complaint against Potts, alleging that he had acted negligently. Will the court agree? Why or why not? [*Potts v. SEC*, 151 F.3d 810 (8th Cir. 1998)]

31–5. Accountant's Liability to Third Parties. In June 1993, Sparkomatic Corp. agreed to negotiate a sale of its Kenco Engineering division to Williams Controls, Inc. At the end of July, Sparkomatic asked its accountants, Parente, Randolph, Orlando, Carey & Associates, to audit Kenco's financial statements for the previous three years and to certify interim and closing balance sheets to be included with the sale's closing documents. All of the parties knew that these documents would serve as a basis for setting the sale price. Within a few days, Williams signed an "Asset Purchase Agreement" that promised access to Parente's records with respect to Kenco. The sale closed in mid-August. In September, Williams was given the financial statements for Kenco's previous three years and the interim and closing balance sheets, all of which were certified by Parente. Williams's accountant found no errors in the closing balance sheet but did not review any of the other documents. The parties set a final purchase price. Later, however, Williams filed a suit in a federal district court against Parente, claiming negligent misrepresentation, among other things, in connection with Parente's preparation of the financial documents. Parente responded with a motion for summary judgment, asserting that the parties lacked privity. Under the *Restatement (Second) of Torts*, Section 552, how should the court rule? Explain. [*Williams Controls, Inc. v. Parente, Randolph, Orlando, Carey & Associates*, 39 F.Supp.2d 517 (M.D.Pa. 1999)]

CASE PROBLEM WITH SAMPLE ANSWER

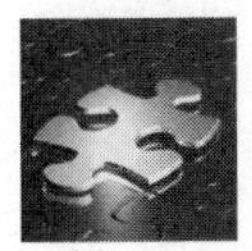

31–6. In October 1993, Marilyn Greenen, a licensed certified public accountant (CPA), began working at the Port of Vancouver, Washington (the Port), as an account manager. She was not directly engaged in public accounting at the Port, but she oversaw the preparation of financial statements and supervised employees with accounting duties. At the start of her employment, she enrolled her husband for benefits under the Port's medical plan. Her marriage was dissolved in November, but she did not notify the Port of the change. In May 1998 and April 1999, the Port confronted her about the divorce, but she did not update her insurance information. After she was terminated, she reimbursed the Port for the additional premiums it had paid for unauthorized coverage for her former spouse. The Washington State Board of Accountancy imposed sanctions on Greenen for "dishonesty and misleading representations" while, in the words of an applicable state statute, "representing oneself as a CPA." Greenen asked a Washington state court to review the case. What might be an appropriate sanction in this case? What might be Greenen's best argument against the board's action? On what reasoning might the court uphold the decision? [*Greenen v. Washington State Board of Accountancy*, 824 Wash.App. 126, 110 P.3d 224 (Div. 2 2005)]

After you have answered this problem, compare your answer with the sample answer given on the Web site that accompanies this text. Go to academic.cengage.com/blaw/wbl, select "Chapter 31," and click on "Case Problem with Sample Answer."

31–7. Accountant's Liability. In 1995, JTD Health Systems, Inc., hired Tammy Heiby as accounting coordinator. Apparently overwhelmed by the duties of the position, Heiby failed to make payroll tax payments to the Internal Revenue Service (IRS) in 1995 and 1996. Heiby tried to hide this omission by falsifying journal entries and manually writing three checks out of sequence, totaling $1.7 million and payable to a bank, from JTD's cash account (to dispose of excess funds that should have been paid in taxes). JTD hired Pricewaterhouse Coopers, LLP, to review JTD's internal accounting procedures and audit its financial statements for 1995. Coopers's inexperienced auditor was aware that the cash account had not been balanced in months and knew about the checks but never questioned them. The auditor instead mistakenly explained that the unbalanced account was due to changes in Medicaid/Medicare procedures and recommended no further investigation. In 1996, the IRS asked JTD to remit the unpaid taxes, plus interest and penalties. JTD filed a suit in an Ohio state court against Coopers, alleging common law negligence and breach of contract. Should Coopers be held liable to JTD on these grounds? Why or why not? [*JTD Health Systems, Inc. v. Pricewaterhouse Coopers, LLP*, 141 Ohio App.3d 280, 750 N.E.2d 1177 (2001)]

31–8. Accountant's Liability under the Private Securities Litigation Reform Act. Solucorp Industries, Ltd., a corporation headquartered in New York, develops and markets products for use in environmental clean-ups. Solucorp's financial statements for the six months ending December 31, 1997, recognized $1.09 million in license fees payable by Smart International, Ltd. The fees comprised about 50 percent of Solucorp's revenue for the period. At the time, however, the parties had a license agreement only "in principle," and Smart had made only one payment of $150,000. Glenn Ohlhauser, an accountant asked to audit the statements, objected to the inclusion of the fees. In February 1998, Solucorp showed Ohlhauser a license agreement backdated to September 1997 but refused to provide any financial information about Smart. Ohlhauser issued an unqualified opinion on the 1997 statements, which were included with forms filed with the Securities and Exchange Commission (SEC). The SEC sued Ohlhauser. What might be the basis in the Private Securities Litigation Reform Act for the SEC's suit? What might be Ohlhauser's defense? Discuss. [*Securities and Exchange Commission v. Solucorp Industries, Ltd.*, 197 F.Supp.2d 4 (S.D.N.Y. 2002)]

A QUESTION OF ETHICS

31–9. Crawford was a certified public accountant who prepared a financial statement for Erps Construction Co., which was seeking a loan from the First National Bank of Bluefield. Crawford knew at the time he prepared the statement that the bank would rely on the statement in making its decision on whether to extend credit to Erps. The loan was made, and Erps defaulted. The bank sued Crawford, alleging that he had been professionally negligent in preparing the financial statement, on which the bank had relied in determining whether to give the construction company a loan. Crawford defended against the suit by asserting that he could not be liable to the bank because of lack of privity. The trial court ruled that in the absence of contractual privity between the parties, the bank could not recover from the accountant. On appeal, the appellate court adopted the rule enunciated by the *Restatement (Second) of Torts* in regard to a professional's liability to third parties. [*First National Bank of Bluefield v. Crawford,* 182 W.Va. 107, 386 S.E.2d 310 (1989)]

1. What is the standard of an accountant's liability to third parties under the *Restatement (Second) of Torts?* What ethical reasoning underlies this standard?
2. Do you think that the standard of liability under the *Restatement* adequately balances the rights of accountants and the rights of third parties? Can you think of a fairer standard?
3. A few courts have adopted the principle that accountants should be liable for negligence to all persons who use and rely on their work products, provided that this use and reliance was foreseeable by the accountants at the time they prepared the documents relied on. Does such a standard of liability impose too great a burden on accountants and accounting firms? Why or why not?

VIDEO QUESTION

31–10. Go to this text's Web site at academic.cengage.com/blaw/wbl and select "Chapter 31." Click on "Video Questions" and view the video titled *Accountant's Liability.* Then answer the following questions.

1. Should Ray prepare a financial statement that values a list of assets provided by the advertising firm without verifying that the firm actually owns these assets?
2. Discuss whether Ray is in privity with the company interested in buying Laura's advertising firm.
3. Under the *Ultramares* rule, to whom does Ray owe a duty?
4. Assume that Laura did not tell Ray that she intended to give the financial statement to the potential acquirer. Would this fact change Ray's liability under the *Ultramares* rule? Explain.

Professional Liability

For updated links to resources available on the Web, as well as a variety of other materials, visit this text's Web site at

academic.cengage.com/blaw/wbl

The Web site for the Financial Accounting Standards Board can be found at

http://www.fasb.org

For information on the accounting profession, including links to the Sarbanes-Oxley Act of 2002 and articles on its impact on the accounting profession, go to the Web site of the American Institute of Certified Public Accountants at

http://www.aicpa.org

Federal tax forms and federal tax information are available from the Internal Revenue Service at its Web site, which can be found at

http://www.irs.gov

Paul R. Rice, a law professor at American University's Washington College of Law, maintains a Web site containing a large number of articles on attorney-client privilege. Go to

http://www.acprivilege.com

ONLINE LEGAL RESEARCH

Go to the *Fundamentals of Business Law* home page at **academic.cengage.com/blaw/wbl**, select "Chapter 31," and click on "Internet Exercises." There you will find the following Internet research exercises that you can perform to learn more about topics covered in this chapter.

Activity 31–1: LEGAL PERSPECTIVE—The Sarbanes-Oxley Act of 2002

Activity 31–2: MANAGEMENT PERSPECTIVE—Avoiding Legal Liability

BEFORE THE TEST

Go to the *Fundamentals of Business Law* home page at **academic.cengage.com/blaw/wbl**, select "Chapter 31," and click on "Interactive Quizzes." You will find a number of interactive questions relating to this chapter.

CHAPTER 32
International Law

LEARNING OBJECTIVES

After reading this chapter, you should be able to answer the following questions:

1. What is the principle of comity, and why do courts deciding disputes involving a foreign law or judicial decree apply this principle?
2. What is the act of state doctrine? In what circumstances is this doctrine applied?
3. Under the Foreign Sovereign Immunities Act of 1976, on what bases might a foreign state be considered subject to the jurisdiction of U.S. courts?
4. What types of provisions, or clauses, are often included in international sales contracts?
5. Do U.S. laws prohibiting employment discrimination apply in all circumstances to U.S. employees working for U.S. employers abroad?

International business transactions are not unique to the modern world. Indeed, people have always found that they can benefit from exchanging goods with others. What is new in our day is the dramatic growth in world trade and the emergence of a global business community. Because the exchange of goods, services, and ideas on a global level is now routine, students of business law should be familiar with the laws pertaining to international business transactions. In this chapter, we first examine the legal context of international business transactions. We then look at some selected areas relating to business activities in a global context, including international sales contracts, civil dispute resolution, and investment protection. We conclude the chapter with a discussion of the application of U.S. antidiscrimination laws in a transnational setting.

International Principles and Doctrines

Recall from our discussion in Chapter 1 that *international law* can be defined as a body of law—formed as a result of international customs, treaties, and organizations—that governs relations among or between nations. *National law,* in contrast, is the law of a particular nation, such as Brazil, Germany, Japan, or the United States. Here, we look at some legal principles and doctrines of international law that have evolved over time and that the courts of various nations have employed—to a greater or lesser extent—to resolve or reduce conflicts that involve a foreign element. The three important legal principles and doctrines discussed in the following subsections are based primarily on courtesy and respect and are applied in the interests of maintaining harmonious relations among nations.

THE PRINCIPLE OF COMITY

Under what is known as the principle of **comity,** one nation will defer and give effect to the laws and judicial decrees of another country, as long as those laws and judicial decrees are consistent with the law and public policy of the accommodating nation.

■ **EXAMPLE 32.1** Assume that a Swedish seller and a U.S. buyer have formed a contract, which the buyer breaches. The seller sues the buyer in a Swedish court,

which awards damages. The buyer's assets, however, are in the United States and cannot be reached unless the judgment is enforced by a U.S. court of law. In this situation, if a U.S. court determines that the procedures and laws applied in the Swedish court were consistent with U.S. national law and policy, that court will likely defer to (and enforce) the foreign court's judgment. ■

Normally, the principle of comity is extended to foreign bankruptcy proceedings. The question in the following case was whether a U.S. court is required to determine the ownership of a disputed asset before a bankruptcy case goes forward in a foreign court.

CASE 32.1 JP Morgan Chase Bank v. Altos Hornos de Mexico, S.A. de C.V.

United States Court of Appeals,
Second Circuit, 2005.
412 F.3d 418.

FACTS In April 1997, Altos Hornos de Mexico, S.A. de C.V., Mexico's largest liquid steel producer, borrowed $330 million from twenty-eight banks, including JP Morgan Chase Bank (JPMCB). JPMCB was the loan's "facility agent," distributing payments among the lenders through a "collection account" set up in the loan documents. As a term of the loan, Altos Hornos told three of its customers to pay into the collection account. In May 1999, after significant losses in the global steel market, Altos Hornos filed a petition with a Mexican court to be declared in *suspension de pagos* (suspension of payments, or SOP, which, under Mexican bankruptcy law, is similar to a Chapter 11 reorganization under U.S. bankruptcy law—see Chapter 21). The court granted the petition. In June and August, Altos Hornos's three customers paid $4.7 million into JPMCB's collection account. Altos Hornos asked the SOP court to order JPMCB to reimburse the funds for distribution in the SOP proceeding. Meanwhile, JPMCB filed a suit in a federal district court against Altos Hornos, seeking a declaration that these funds did not belong to the borrower. The court dismissed the suit on the basis of comity. JPMCB appealed to the U.S. Court of Appeals for the Second Circuit.

ISSUE Is a U.S. court required to resolve the question of the ownership of a bank account before its possible owner's bankruptcy proceeding in a Mexican court?

DECISION No. The U.S. Court of Appeals for the Second Circuit affirmed the lower court's dismissal of JPMCB's complaint. "[C]reditors may not use U.S. courts to circumvent foreign bankruptcy proceedings."

REASON The appellate court reasoned that the "orderly distribution of a debtor's property requires assembling all claims against the limited assets in a single proceeding." In that situation, deference to a foreign court under the principle of comity is appropriate so long as the foreign proceeding does not "contravene the laws or public policy of the United States." In this case, "[t]here is no dispute between the parties that the Mexican SOP proceeding is a foreign bankruptcy proceeding. There is also no question that Altos Hornos'[s] debt to [JPMCB] is properly before the Mexican SOP court; [JPMCB] appeared before the Mexican court and was acknowledged as a general unsecured creditor owed a debt of $225,355,617.25 in principal and $1,912,330.78 in interest. Accordingly, we think the salient issue (that is, the debt) has been raised before the proper Mexican court, and thus there exists a parallel foreign proceeding to which [a U.S.] court could defer. * * * Nothing in the record before us suggests that the actions taken by the Mexican bankruptcy court are not approved or allowed by American law."

FOR CRITICAL ANALYSIS—Social Consideration *Why might a Mexican court be better able to determine the ownership of the funds in the collection account in this case?*

THE ACT OF STATE DOCTRINE

The **act of state doctrine** is a judicially created doctrine that provides that the judicial branch of one country will not examine the validity of public acts committed by a recognized foreign government within its own territory. This doctrine is premised on the theory that the judicial branch should not "pass upon the validity of foreign acts when to do so would vex the harmony of our international relations with that foreign nation."[1]

The act of state doctrine can have important consequences for individuals and firms doing business with, and investing in, other countries. For example, this doctrine is frequently employed in situations involving expropriation or confiscation. **Expropriation** occurs when a government seizes a privately owned business or privately owned goods for a proper public purpose and awards just compensation. When a government seizes private property for an illegal purpose or without just compensation, the taking is referred to as a **confiscation.** The line between these two forms of taking is sometimes blurred because of differing interpretations of what is illegal and what constitutes just compensation.

1. *Libra Bank, Ltd. v. Banco Nacional de Costa Rica, S.A.*, 570 F.Supp. 870 (S.D.N.Y. 1983).

When applicable, both the act of state doctrine and the doctrine of *sovereign immunity* (to be discussed next) tend to immunize (protect) foreign governments from the jurisdiction of U.S. courts. This means that firms or individuals who own property overseas often have diminished legal protection against government actions in the countries in which they operate.

The Doctrine of Sovereign Immunity

When certain conditions are satisfied, the doctrine of **sovereign immunity** immunizes foreign nations from the jurisdiction of U.S. courts. In 1976, Congress codified this rule in the Foreign Sovereign Immunities Act (FSIA).[2] The FSIA exclusively governs the circumstances in which an action may be brought in the United States against a foreign nation, including attempts to attach a foreign nation's property.

Section 1605 of the FSIA sets forth the major exceptions to the jurisdictional immunity of a foreign state or country. A foreign state is not immune from the jurisdiction of U.S. courts when the state has "waived its immunity either explicitly or by implication" or when the action is taken "in connection with a commercial activity carried on in the United States by the foreign state" or having "a direct effect in the United States."[3]

The question frequently arises as to whether an entity falls within the category of a foreign state. The question of what is a commercial activity has also been the subject of dispute. Under Section 1603 of the FSIA, a *foreign state* includes both a political subdivision of a foreign state and an instrumentality of a foreign state. Section 1603 broadly defines a *commercial activity* as a commercial activity that is carried out by a foreign state within the United States, but it does not describe the particulars of what constitutes a commercial activity. Thus, the courts are left to decide whether a particular activity is governmental or commercial in nature.

Doing Business Internationally

A U.S. domestic firm can engage in international business transactions in a number of ways. The simplest way is to seek out foreign markets for domestically produced products or services. In other words, U.S. firms can look abroad for **export** markets for their goods and services. Alternatively, a U.S. firm can establish foreign production facilities so as to be closer to the foreign market or markets in which its products are sold. A domestic firm can also obtain revenues by licensing its technology to an existing foreign company or by selling franchises to overseas entities.

Exporting

Exporting can take two forms: direct exporting and indirect exporting. In *direct exporting,* a U.S. company signs a sales contract with a foreign purchaser that provides for the conditions of shipment and payment for the goods. (How payments are made in international transactions is discussed later in this chapter.) If sufficient business develops in a foreign country, a U.S. corporation may set up a specialized marketing organization in that foreign market by appointing a foreign agent or a foreign distributor. This is called *indirect exporting.*

When a U.S. firm desires to limit its involvement in an international market, it will typically establish an *agency relationship* with a foreign firm. In an agency relationship (discussed in Chapter 22), one person (the agent) agrees to act on behalf of another (the principal). The foreign agent is thereby empowered to enter into contracts in the agent's country on behalf of the U.S. principal.

When a substantial market exists in a foreign country, a U.S. firm may wish to appoint a distributor located in that country. The U.S. firm and the distributor enter into a **distribution agreement,** which is a contract between the seller and the distributor setting out the terms and conditions of the distributorship—for example, price, currency of payment, availability of supplies, and method of payment. The terms and conditions primarily involve contract law. Disputes concerning distribution agreements may involve jurisdictional or other issues (discussed in detail later in this chapter).

Manufacturing Abroad

An alternative to direct or indirect exporting is the establishment of foreign manufacturing facilities. Typically, U.S. firms establish manufacturing plants abroad if they believe that doing so will reduce their costs—particularly for labor, shipping, and raw materials—and enable them to compete more effectively in foreign markets. A U.S. firm can manufacture goods in other countries in several ways. They include licensing and franchising, as well as investing in a wholly owned subsidiary or a joint venture.

Licensing A U.S. firm can obtain business from abroad by licensing a foreign manufacturing company to use its copyrighted, patented, or trademarked intellectual property or trade secrets. Like any other licensing agreement (see Chapters 5 and 13), a licensing agreement with a

2. 28 U.S.C. Sections 1602–1611.

3. See, for example, *Keller v. Central Bank of Nigeria,* 277 F.3d 811 (6th Cir. 2002), in which the court held that failure to pay promised funds to a Cleveland account was an action having a direct effect in the United States.

foreign-based firm calls for a payment of royalties on some basis—such as so many cents per unit produced or a certain percentage of profits from units sold in a particular geographic territory.

Franchising Franchising is a well-known form of licensing. Recall from Chapter 24 that in a franchise arrangement the owner of a trademark, trade name, or copyright (the franchisor) licenses another (the franchisee) to use the trademark, trade name, or copyright under certain conditions or limitations in the selling of goods or services. In return, the franchisee pays a fee, which is usually based on a percentage of gross or net sales. Examples of international franchises include Holiday Inn and Hertz.

Investing in a Wholly Owned Subsidiary or a Joint Venture Another way to expand into a foreign market is to establish a wholly owned subsidiary firm in a foreign country. When a wholly owned subsidiary is established, the parent company, which remains in the United States, retains complete ownership of all the facilities in the foreign country, as well as complete authority and control over all phases of the operation. A U.S. firm can also expand into international markets through a joint venture. In a joint venture, the U.S. company owns only part of the operation; the rest is owned either by local owners in the foreign country or by another foreign entity. All of the firms involved in a joint venture share responsibilities, as well as profits and liabilities.

Commercial Contracts in an International Setting

Like all commercial contracts, an international contract should be in writing. For an example of an actual international sales contract, refer back to the appendix following Chapter 14.

CONTRACT CLAUSES

Language and legal differences among nations can create special problems for parties to international contracts when disputes arise. It is possible to avoid these problems by including in a contract special provisions designating the official language of the contract, the legal forum (court or place) in which disputes under the contract will be settled, and the substantive law that will be applied in settling any disputes. Parties to international contracts should also indicate in their contracts what acts or events will excuse the parties from performance under the contract and whether disputes under the contract will be arbitrated or litigated.

Choice of Language A deal struck between a U.S. company and a company in another country normally involves two languages. Typically, many phrases in one language are not readily translatable into another. Consequently, the complex contractual terms involved may not be understood by one party in the other party's language. To make sure that no disputes arise out of this language problem, an international sales contract should have a **choice-of-language clause** designating the official language by which the contract will be interpreted in the event of disagreement.

Choice of Forum When parties from several countries are involved, litigation may be pursued in courts in different nations. There are no universally accepted rules as to which court has jurisdiction over particular subject matter or parties to a dispute. Consequently, parties to an international transaction should always include in the contract a **forum-selection clause** indicating what court, jurisdiction, or tribunal will decide any disputes arising under the contract. It is especially important to indicate the specific court that will have jurisdiction. The forum does not necessarily have to be within the geographic boundaries of the home nation of either party. The following case involved a question about the application of a forum-selection clause.

CASE 32.2 Garware Polyester, Ltd. v. Intermax Trading Corp.

United States District Court,
Southern District of New York, 2001.
__ F.Supp.2d __.

FACTS Garware Polyester, Ltd., based in Mumbai, India, develops and makes plastics and high-tech polyester film. In 1987, Intermax Trading Corporation, based in New York, became Garware's North American sales agent. Over the next decade, the parties executed four written agreements, collectively referred to as the "Agency Agreements." Each agreement provided, "The courts at Bombay [India] alone will have juris-

CASE 32.2–CONTINUED

diction to try out suits in respect of any claim or dispute arising out of or under this agreement or in any way relating to the same." Intermax sold Garware products on a commission basis. In some transactions, Intermax arranged for customers to order products directly from Garware. In other transactions, Intermax sold Garware products through warehouse sales by which Intermax bought products from Garware, warehoused them in the United States, and resold them. When Intermax fell behind in its payments, Garware filed a suit in a U.S. district court to collect on the unpaid invoices. Garware argued that the forum-selection clause did not apply to the invoices in dispute because they involved the warehouse sales, which, Garware claimed, were not part of, and did not relate to, the Agency Agreements.

ISSUE Did the forum-selection clause require the dismissal of this suit?

DECISION Yes. The court held that each of the Agency Agreements contained a valid and enforceable forum-selection clause, which applied to this suit. The court dismissed the case for improper venue (geographic location for the trial—see Chapter 2).

REASON The court recognized that forum-selection clauses "eliminate uncertainty in international commerce and insure that the parties are not unexpectedly subjected to hostile forums and laws. Moreover, international comity dictates that American courts enforce these sorts of clauses out of respect for the integrity and respect of foreign tribunals." The court stated that a forum-selection clause applies when "claims grow out of the contractual relationship, or if the gist of those claims is a breach of that relationship." Here, "the 'gist' of Garware's claim is a breach of the Agency Agreements. The warehouse sales in question were made by Intermax for the purpose of selling Garware products in the contractually defined territory. The Agency Agreements specifically relate to Intermax's role as Garware's 'selling agent * * * in the United States of America * * * .' Thus, if these sales are not squarely within the scope of the Agency Agreements, they are, at the very least, related to the Agreements. Further, the parties' course of dealing supports the conclusion that the parties themselves believed the Agency Agreements included warehouse sales: as required by the Agreements, Garware paid commissions to Intermax on these sales."

WHY IS THIS CASE IMPORTANT? *As this case illustrates, when the underlying transaction is international, courts generally presume that forum-selection clauses are valid because such clauses eliminate uncertainty in international commerce. It takes a strong showing that the clause was unreasonable under the circumstances to overcome this presumed validity.*

Choice of Law A contractual provision designating the applicable law—such as the law of Germany or England or California—is called a **choice-of-law clause.** Every international contract typically includes a choice-of-law clause. At common law (and in European civil law systems), parties are allowed to choose the law that will govern their contractual relationship, provided that the law chosen is the law of a jurisdiction that has a substantial relationship to the parties and to the international business transaction.

Under Section 1–105 of the Uniform Commercial Code, parties may choose the law that will govern the contract as long as the choice is "reasonable." Article 6 of the United Nations Convention on Contracts for the International Sale of Goods (discussed in Chapter 14), however, imposes no limitation on the parties' choice of what law will govern the contract. The 1986 Hague Convention on the Law Applicable to Contracts for the International Sale of Goods—often referred to as the Choice-of-Law Convention—allows unlimited autonomy in the choice of law. The Hague Convention indicates that whenever a contract does not specify a choice of law, the governing law is that of the country in which the *seller's* place of business is located.

***Force Majeure* Clause** Every contract, particularly those involving international transactions, should have a ***force majeure* clause.** *Force majeure* is a French term meaning "impossible or irresistible force"—sometimes loosely identified as "an act of God." In international business contracts, *force majeure* clauses commonly stipulate that in addition to acts of God, a number of other eventualities (such as government orders or embargoes, for example) may excuse a party from liability for nonperformance.

CIVIL DISPUTE RESOLUTION

International contracts frequently include arbitration clauses. By means of such clauses, the parties agree in advance to be bound by the decision of a specified third party in the event of a dispute, as discussed in Chapter 2. The third party may be a neutral entity (such as the International Chamber of Commerce), a panel of individuals representing both parties' interests, or some other

group or organization. (For an example of an arbitration clause in an international contract, refer to the appendix following Chapter 14.) The United Nations Convention on the Recognition and Enforcement of Foreign Arbitral Awards (often referred to as the New York Convention) assists in the enforcement of arbitration clauses, as do provisions in specific treaties among nations. The New York Convention has been implemented in nearly one hundred countries, including the United States.

If a sales contract does not include an arbitration clause, litigation may occur. If the contract contains forum-selection and choice-of-law clauses, the lawsuit will be heard by a court in the specified forum and decided according to that forum's law. If no forum and choice of law have been specified, however, legal proceedings will be more complex and attended by much more uncertainty. For instance, litigation may take place in two or more countries, with each country applying its own choice-of-law rules to determine the substantive law that will be applied to the particular transactions.

Also, even if a plaintiff wins a favorable judgment in a lawsuit litigated in the plaintiff's country, there is no way to predict whether courts in the defendant's country will enforce the judgment.

Making Payment on International Transactions

Currency differences among nations and the geographic distance between parties to international sales contracts add a degree of complexity to international sales that does not exist in the domestic market. Because international contracts involve greater financial risks, special care should be taken in drafting these contracts to specify both the currency in which payment is to be made and the method of payment.

Although our national currency, the U.S. dollar, is one of the primary forms of international currency, any U.S. firm undertaking business transactions abroad must be prepared to deal with one or more other currencies. Currencies are convertible when they can be freely exchanged one for the other at some specified market rate in a **foreign exchange market.** Foreign exchange markets comprise a worldwide system for the buying and selling of foreign currencies. At any point in time, the foreign exchange rate is set by the forces of supply and demand in unrestricted foreign exchange markets. The foreign exchange rate is simply the price of a unit of one country's currency in terms of another country's currency. For example, if today's exchange rate is one hundred Japanese yen for one dollar, that means that anybody with one hundred yen can obtain one dollar, and vice versa.

Frequently, a U.S. company can rely on its domestic bank to take care of all international transfers of funds. Commercial banks often transfer funds internationally through their **correspondent banks** (banks in which they have accounts) in other countries. ■ EXAMPLE 32.2 Suppose that a customer of Citibank wishes to pay a bill in euros to a company in Paris. Citibank can draw a bank check payable in euros on its account in Crédit Lyonnais, a Paris correspondent bank, and then send the check to the French company to which its customer owes the funds. Alternatively, Citibank's customer can request a wire transfer of the funds to the French company. Citibank instructs Crédit Lyonnais by wire to pay the necessary amount in euros. ■

Regulation of Specific Business Activities

Doing business abroad can affect the economies, foreign policies, domestic policies, and other national interests of the countries involved. For this reason, nations impose laws to restrict or facilitate international business. Controls may also be imposed by international agreements. We discuss here how different types of international activities are regulated.

INVESTING

Firms that invest in foreign nations face the risk that the foreign government may take possession of the investment property. Expropriation, as already mentioned, occurs when property is taken and the owner is paid just compensation for what is taken. Expropriation does not violate generally observed principles of international law. Such principles are normally violated, however, when a government confiscates property without compensation (or without adequate compensation). Few remedies are available for confiscation of property by a foreign government. Claims are often resolved by lump-sum settlements after negotiations between the United States and the taking nation.

To counter the deterrent effect that the possibility of confiscation may have on potential investors, many countries guarantee that foreign investors will be compensated if their property is taken. A guaranty can take the form of national constitutional or statutory laws or provisions in international treaties. As further protection for foreign investments, some countries provide insurance for their citizens' investments abroad.

Export Controls

The U.S. Constitution provides in Article I, Section 9, that "No Tax or Duty shall be laid on Articles exported from any State." Thus, Congress cannot impose any export taxes. Congress can, however, use a variety of other devices to control exports. Congress may set export quotas on various items, such as grain being sold abroad. Under the Export Administration Act of 1979,[4] the flow of technologically advanced products and technical data can be restricted. In recent years, the U.S. Department of Commerce has made a controversial attempt to restrict the export of encryption software.

While restricting certain exports, the United States (and other nations) also use devices such as export incentives and subsidies to stimulate other exports and thereby aid domestic businesses. The Revenue Act of 1971,[5] for instance, gave tax benefits to firms marketing their products overseas through certain foreign sales corporations by exempting income produced by the exports. Under the Export Trading Company Act of 1982,[6] U.S. banks are encouraged to invest in export trading companies, which are formed when exporting firms join together to export a line of goods. The Export-Import Bank of the United States provides financial assistance, consisting primarily of credit guaranties given to commercial banks that in turn lend funds to U.S. exporting companies.

Import Controls

All nations have restrictions on imports, and the United States is no exception. Restrictions include strict prohibitions, quotas, and tariffs. Under the Trading with the Enemy Act of 1917,[7] for instance, no goods may be imported from nations that have been designated enemies of the United States. Other laws prohibit the importation of illegal drugs, books that urge insurrection against the United States, and agricultural products that pose dangers to domestic crops or animals.

Quotas and Tariffs Limits on the amounts of goods that can be imported are known as **quotas.** At one time, the United States had legal quotas on the number of automobiles that could be imported from Japan. Today, Japan "voluntarily" restricts the number of automobiles exported to the United States. **Tariffs** are taxes on imports. A tariff is usually a percentage of the value of the import, but it can be a flat rate per unit (for example, per barrel of oil). Tariffs raise the prices of goods, causing some consumers to purchase less expensive, domestically manufactured goods.

Dumping The United States has specific laws directed at what it sees as unfair international trade practices. **Dumping,** for example, is the sale of imported goods at "less than fair value." "Fair value" is usually determined by the price of those goods in the exporting country. Foreign firms that engage in dumping in the United States hope to undersell U.S. businesses to obtain a larger share of the U.S. market. To prevent this, an extra tariff—known as an *antidumping duty*—may be assessed on the imports.

Minimizing Trade Barriers Restrictions on imports are also known as *trade barriers.* The elimination of trade barriers is sometimes seen as essential to the world's economic well-being. Most of the world's leading trading nations are members of the World Trade Organization (WTO), which was established in 1995. To minimize trade barriers among nations, each member country of the WTO is required to grant **normal trade relations (NTR) status** (formerly known as *most-favored-nation status*) to other member countries. This means each member is obligated to treat other members at least as well as it treats the country that receives its most favorable treatment with regard to imports or exports.

Various regional trade agreements and associations also help to minimize trade barriers between nations. The European Union (EU), for example, is working to minimize or remove barriers to trade among its member countries. The EU is a single integrated trading unit made up of twenty-five European nations. Another important regional trade agreement is the North American Free Trade Agreement (NAFTA). NAFTA, which became effective on January 1, 1994, created a regional trading unit consisting of Mexico, the United States, and Canada. The primary goal of NAFTA is to eliminate tariffs among these three countries on substantially all goods over a period of fifteen to twenty years.

A more recent trade agreement is the Central America-Dominican Republic-United States Free Trade Agreement (CAFTA-DR), which was signed into law by President George W. Bush in 2005. This agreement was formed by Costa Rica, the Dominican Republic, El Salvador, Guatemala, Honduras, Nicaragua, and the United States. Once the parties agree on an effective date, CAFTA-DR will reduce trade tariffs and improve market access among all of the signatory nations.

4. 50 U.S.C. Sections 2401–2420.
5. 26 U.S.C. Sections 991–994.
6. 15 U.S.C. Sections 4001, 4003.
7. 12 U.S.C. Section 95a.

Bribing Foreign Officials

Giving cash or in-kind benefits to foreign government officials to obtain business contracts and other favors is often considered normal practice. To reduce such bribery by representatives of U.S. corporations, Congress enacted the Foreign Corrupt Practices Act in 1977.[8] This act and its implications for American businesspersons engaged in international business transactions were discussed in Chapter 3.

U.S. Antidiscrimination Laws in a Global Context

The internationalization of business raises questions about the extraterritorial application of a nation's laws—that is, the effect of the country's laws outside its boundaries. Here, we look at the extraterritorial application of U.S. laws prohibiting employment discrimination.

As explained in Chapter 23, federal laws in the United States prohibit discrimination on the basis of race, color, national origin, religion, gender, age, and disability. These laws, as they affect employment relationships, generally apply extraterritorially. Since 1984, for example, the Age Discrimination in Employment Act of 1967 has covered U.S. employees working abroad for U.S. employers. The Americans with Disabilities Act of 1990, which requires employers to accommodate the needs of workers with disabilities, also applies to U.S. nationals working abroad for U.S. firms. Title VII of the Civil Rights Act of 1964 applies to all U.S. employees working for U.S. employers abroad as well. Generally, U.S. employers must abide by U.S. antidiscrimination laws unless to do so would violate the laws of the country where their workplaces are located. This "foreign laws exception" allows employers to avoid being subjected to conflicting laws.

8. 15 U.S.C. Sections 78m–78ff.

TERMS AND CONCEPTS

act of state doctrine 656
choice-of-language clause 658
choice-of-law clause 659
comity 655
confiscation 656
correspondent bank 660
distribution agreement 657
dumping 661
export 657
expropriation 656
force majeure clause 659
foreign exchange market 660
forum-selection clause 658
normal trade relations (NTR) status 661
quota 661
sovereign immunity 657
tariff 661

CHAPTER SUMMARY International Law

International Principles and Doctrines (See pages 655–657.)	1. *The principle of comity*—Under this principle, nations give effect to the laws and judicial decrees of other nations for reasons of courtesy and international harmony. 2. *The act of state doctrine*—A doctrine under which U.S. courts avoid passing judgment on the validity of public acts committed by a recognized foreign government within its own territory. 3. *The doctrine of sovereign immunity*—When certain conditions are satisfied, foreign nations are immune from U.S. jurisdiction under the Foreign Sovereign Immunities Act of 1976. Exceptions are made (a) when a foreign state has "waived its immunity either explicitly or by implication" or (b) when the action is based on "a commercial activity carried on in the United States by the foreign state."
Doing Business Internationally (See pages 657–658.)	Ways in which U.S. domestic firms engage in international business transactions include (a) exporting, which may involve foreign agents or distributors, and (b) manufacturing abroad through licensing arrangements, franchising operations, wholly owned subsidiaries, or joint ventures.
Commercial Contracts in an International Setting (See pages 658–660.)	International business contracts often include choice-of-language, forum-selection, and choice-of-law clauses to reduce the uncertainties associated with interpreting the language of the agreement and dealing with legal differences. Most domestic and international contracts include *force majeure* clauses. They commonly stipulate that certain events, such as floods, fires, accidents, labor strikes, and government orders, may excuse a party from liability for nonperformance of the contract. Arbitration clauses are also frequently found in international contracts.

CHAPTER SUMMARY International Law—Continued

Making Payment on International Transactions (See page 660.)	1. *Currency conversion*—Because nations have different monetary systems, payment on international contracts requires currency conversion at a rate specified in a foreign exchange market. 2. *Correspondent banking*—Correspondent banks facilitate the transfer of funds from a buyer in one country to a seller in another.
Regulation of Specific Business Activities (See pages 660–662.)	National laws regulate foreign investments, exporting, and importing. The World Trade Organization attempts to minimize trade barriers among nations, as do regional trade agreements and associations, including the European Union, the North American Free Trade Agreement, and the Central America-Dominican Republic-United States Free Trade Agreement.
U.S. Antidiscrimination Laws in a Global Context (See page 662.)	The major U.S. laws prohibiting employment discrimination, including Title VII of the Civil Rights Act of 1964, the Age Discrimination in Employment Act of 1967, and the Americans with Disabilities Act of 1990, cover U.S. employees working abroad for U.S. firms—*unless* to apply the U.S. laws would violate the laws of the host country.

FOR REVIEW

Answers for the even-numbered questions in this For Review *section can be found in Appendix A at the end of this text.*

1. What is the principle of comity, and why do courts deciding disputes involving a foreign law or judicial decree apply this principle?
2. What is the act of state doctrine? In what circumstances is this doctrine applied?
3. Under the Foreign Sovereign Immunities Act of 1976, on what bases might a foreign state be considered subject to the jurisdiction of U.S. courts?
4. What types of provisions, or clauses, are often included in international sales contracts?
5. Do U.S. laws prohibiting employment discrimination apply in all circumstances to U.S. employees working for U.S. employers abroad?

QUESTIONS AND CASE PROBLEMS

QUESTION WITH SAMPLE ANSWER

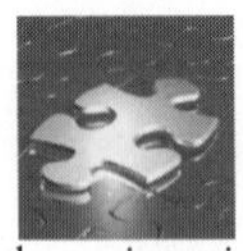

32–1. In 1995, France implemented a law making the use of the French language mandatory in certain legal documents. Documents relating to securities offerings, such as prospectuses, for example, must be written in French. So must instruction manuals and warranties for goods and services offered for sale in France. Additionally, all agreements entered into with French state or local authorities, with entities controlled by state or local authorities, and with private entities carrying out a public service (such as providing utilities) must be written in French. What kinds of problems might this law pose for U.S. businesspersons who wish to form contracts with French individuals or business firms?

For a sample answer to this question, go to Appendix B at the end of this text.

32–2. Discrimination Claims. Radio Free Europe and Radio Liberty (RFE/RL), a U.S. corporation doing business in Germany, employs more than three hundred U.S. citizens at its principal place of business in Munich, Germany. The concept of mandatory retirement is deeply embedded in German labor policy, and a contract formed in 1982 between RFE/RL and a German labor union contained a clause that required workers to be retired when they reach the age of sixty-five. When William Mahoney and other U.S. employees (the plaintiffs) reached the age of sixty-five, RFE/RL terminated their employment as required under its contract with the labor union. The plaintiffs sued RFE/RL for discriminating against them on the basis of age, in violation of the Age Discrimination in Employment Act of 1967. Will the plaintiffs succeed in their suit? Discuss fully. [*Mahoney v. RFE/RL, Inc.*, 47 F.3d 447 (D.C. Cir. 1995)]

32–3. Sovereign Immunity. Nuovo Pignone, Inc., is an Italian company that designs and manufactures turbine systems. Nuovo sold a turbine system to Cabinda Gulf Oil Co. (CABGOC). The system was manufactured, tested, and inspected in Italy; then it was sent to Louisiana for mounting on a platform by

CABGOC's contractor. Nuovo sent a representative to consult on the mounting. The platform went to a CABGOC site off the coast of West Africa. Marcus Pere, an instrument technician at the site, was killed when a turbine within the system exploded. Pere's widow filed a suit in a U.S. district court against Nuovo and others. Nuovo claimed sovereign immunity on the ground that its majority shareholder at the time of the explosion was Ente Nazionale Idrocaburi, which was created by the government of Italy to lead its oil and gas exploration and development. Is Nuovo exempt from suit under the doctrine of sovereign immunity? Is it subject to suit under the "commercial activity" exception? Why or why not? [*Pere v. Nuovo Pignone, Inc.,* 150 F.3d 477 (5th Cir. 1998)]

32–4. Dumping. In response to a petition filed on behalf of the U.S. pineapple industry, the U.S. Commerce Department initiated an investigation of canned pineapple imported from Thailand. The investigation concerned Thai producers of the canned fruit, including the Thai Pineapple Public Co. The Thai producers also turned out products, such as pineapple juice and juice concentrate, outside the scope of the investigation. These products use separate parts of the same fresh pineapple, so they share raw material costs. To determine fair value and antidumping duties, the Commerce Department had to calculate the Thai producers' cost of production and, in so doing, had to allocate a portion of the shared fruit costs to the canned fruit. These allocations were based on the producers' own financial records, which were consistent with Thai generally accepted accounting principles. The result was a determination that more than 90 percent of the canned fruit sales were below the cost of production. The producers filed a suit in the U.S. Court of International Trade against the federal government, challenging this allocation. The producers argued that their records did not reflect actual production costs, which instead should be based on the weight of fresh fruit used to make the products. Did the Commerce Department act reasonably in determining the cost of production? Why or why not? [*Thai Pineapple Public Co. v. United States,* 187 F.3d 1362 (Fed.Cir. 1999)]

CASE PROBLEM WITH SAMPLE ANSWER

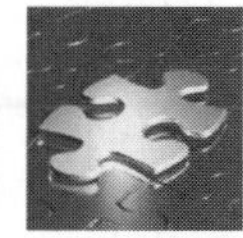

32–5. Tonoga, Ltd., doing business as Taconic Plastics, Ltd., is a manufacturer incorporated in Ireland with its principal place of business in New York. In 1997, Taconic entered into a contract with a German construction company to supply special material for a tent project designed to shelter religious pilgrims visiting holy sites in Saudi Arabia. Most of the material was made in, and shipped from, New York. The company did not pay Taconic and eventually filed for bankruptcy. Another German firm, Werner Voss Architects and Engineers, acting as an agent for the government of Saudi Arabia, guaranteed the payments due Taconic to induce it to complete the project. When it did not receive the final payment, Taconic filed a suit in a U.S. district court against the government of Saudi Arabia, claiming a breach of the guaranty and seeking to collect, in part, about $3 million. The defendant filed a motion to dismiss based, in part, on the doctrine of sovereign immunity. Under what circumstances does this doctrine apply? What are its exceptions? Should this suit be dismissed under the "commercial activity" exception? Explain. [*Tonoga, Ltd. v. Ministry of Public Works and Housing of Kingdom of Saudi Arabia,* 135 F.Supp.2d 350 (N.D.N.Y. 2001)]

After you have answered this problem, compare your answer with the sample answer given on the Web site that accompanies this text. Go to <u>academic.cengage.com/blaw/wbl</u>, select "Chapter 32," and click on "Case Problem with Sample Answer."

32–6. Imports. DaimlerChrysler Corp. makes and markets motor vehicles. DaimlerChrysler assembled the 1993 and 1994 model years of its trucks at plants in Mexico. Assembly involved sheet metal components sent from the United States. DaimlerChrysler subjected some of the parts to a complicated treatment process, which included applying coats of paint to prevent corrosion, impart color, and protect the finish. Under U.S. law, goods that are assembled abroad using U.S.-made parts can be imported tariff free. A U.S. *statute* provides that painting is "incidental" to assembly and does not affect the status of the goods. A U.S. *regulation,* however, states that "painting primarily intended to enhance the appearance of an article or to impart distinctive features or characteristics" is not incidental. The U.S. Customs Service levied a tariff on the trucks. DaimlerChrysler filed a suit in the U.S. Court of International Trade, challenging the levy. Should the court rule in DaimlerChrysler's favor? Why or why not? [*DaimlerChrysler Corp. v. United States,* 361 F.3d 1378 (Fed.Cir. 2004)]

International Law

For updated links to resources available on the Web, as well as a variety of other materials, visit this text's Web site at

academic.cengage.com/blaw/wbl

FindLaw's Web site includes an extensive array of links to international doctrines, treaties, and the laws of other nations. Go to

http://www.findlaw.com/12international

If you are interested in learning more about what is involved in exporting goods to other countries, go to the U.S. Small Business Administration's Web page for its Office of International Trade at

http://www.sba.gov/oit

For information on the legal requirements of doing business abroad, a good source is the Internet Law Library's collection of laws of other nations. Go to

http://www.lawguru.com/ilawlib/index.html

ONLINE LEGAL RESEARCH

Go to the *Fundamentals of Business Law* home page at academic.cengage.com/blaw/wbl, select "Chapter 32," and click on "Internet Exercises." There you will find the following Internet research exercises that you can perform to learn more about topics covered in this chapter.

Activity 32–1: LEGAL PERSPECTIVE—The World Trade Organization

Activity 32–2: MANAGEMENT PERSPECTIVE—Overseas Business Opportunities

BEFORE THE TEST

Go to the *Fundamentals of Business Law* home page at academic.cengage.com/blaw/wbl, select "Chapter 32,"and click on "Interactive Quizzes." You will find a number of interactive questions relating to this chapter.

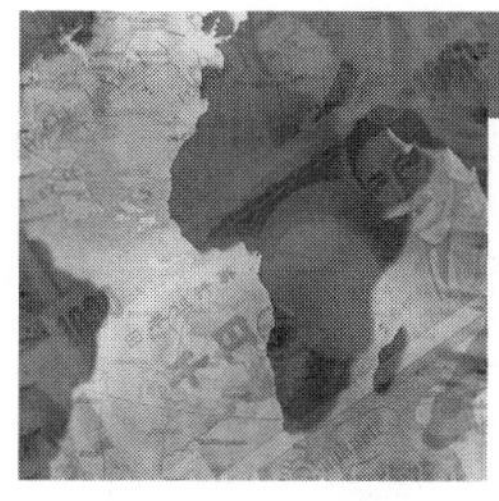

UNIT 10 • EXTENDED CASE STUDY

Dole Food Co. v. Patrickson

Chapter 32 discussed the Foreign Sovereign Immunities Act of 1976 (FSIA). Under the FSIA, foreign states, or nations, can claim certain rights in suits against them in U.S. courts, including in some circumstances immunity from the litigation. A corporate entity that is an "instrumentality" of a foreign state, as defined in the FSIA, may also avail itself of some of these rights. (Corporations and their characteristics were covered in Chapters 25 through 27.) In this extended case study, we review *Dole Food Co. v. Patrickson,*[1] a case focusing on this definition. The specific question was whether a corporate subsidiary can claim to be an instrumentality of a foreign state if the state does not own a majority of the shares of the subsidiary's stock, but does own a majority of the shares of the corporate parent.

1. 538 U.S. 468, 123 S.Ct. 1655, 155 L.Ed.2d 843 (2004). This opinion may be accessed online at **http://straylight.law.cornell.edu/supct/search/index.html**. Enter "Dole Food Company" in the "Search" box and select "All Decisions." Scroll to the name of the case and click on it to access the opinion.

CASE BACKGROUND

In 1851, Dole Food Company was founded in Hawaii. Dole is the world's largest producer and seller of fresh fruit, fresh vegetables, and fresh-cut flowers and markets a growing line of packaged foods. The firm does business in more than 90 countries, employing globally more than 33,000 full-time permanent employees and 24,000 full-time seasonal or temporary employees.

In 1997, Gerardo Patrickson and other farmworkers who worked in banana fields in Costa Rica, Ecuador, Guatemala, and Panama filed a suit in a Hawaii state court against Dole and others, seeking damages for injuries from exposure to dibromochloropropane, a chemical used as an agricultural pesticide. Dole impleaded[2] two Israeli firms—Dead Sea Bromine Company and Bromine Compounds, Ltd. (the Dead Sea companies)—that allegedly made the pesticides.

The Dead Sea companies asked a U.S. district court to hear the suit on the ground that they were instrumentalities of a foreign state as defined in the FSIA. The court denied this request, but held that it had jurisdiction on other grounds and dismissed the suit. The workers appealed to the U.S. Court of Appeals for the Ninth Circuit, which reversed the dismissal but agreed that the Dead Sea companies were not instrumentalities of a foreign state as defined in the FSIA. The Dead Sea companies appealed to the United States Supreme Court.

MAJORITY OPINION

Justice *KENNEDY* delivered the opinion of the Court.

* * * *

The State of Israel did not have direct ownership of shares in either of the Dead Sea Companies at any time pertinent to this suit. Rather, these companies were, at various times, separated from the State of Israel by one or more intermediate corporate tiers. For example, from 1984–1985, Israel wholly owned a company called Israeli Chemicals, Ltd.; which owned a majority of shares in another company called Dead Sea Works, Ltd.; which owned a majority of shares in Dead Sea Bromine Co., Ltd.; which owned a majority of shares in Bromine Compounds, Ltd.

* * * *

* * * The Dead Sea Companies urge us to ignore corporate formalities and use the colloquial [everyday] sense of that term. They ask whether, in common parlance, Israel would be said to own the Dead Sea Companies. *We reject this analysis. In issues of corporate law, structure often matters.* It is evident from the [FSIA's] text that Congress was aware of settled principles of corporate law and legislated within that context. The language of [Section] 1603(b)(2) refers to ownership of "shares," showing that *Congress intended statutory coverage to turn on formal corporate ownership.* Likewise, [Section] 1603(b)(1), another component of the definition of instrumentality, refers to a "separate legal person, corporate or otherwise." * * * [Emphasis added.]

A basic tenet [rule] of American corporate law is that the corporation and its shareholders are distinct entities. An individual shareholder, by virtue of his ownership of shares, does not own the corporation's assets and, as a result, does not own subsidiary corporations in which the corporation holds an interest. A corporate parent which owns the shares of a subsidiary does not, for that reason alone, own or have legal title to the assets of the subsidiary; and, it follows with even greater force, *the parent does not own or have legal title to the subsidiaries of the subsidiary.* The fact that the shareholder is a foreign state does not change the analysis. [Emphasis added.]

Applying these principles, it follows that Israel did not own a majority of shares in the Dead Sea Companies. The State of Israel owned a majority of shares, at various times, in companies one or more corporate tiers above the Dead Sea Companies, but at no time did Israel own a majority of

2. *Implead* is to bring a new party into a suit between others, with the allegation that the new party is liable for part of the claim to the party that impleaded it.

shares in the Dead Sea Companies. Those companies were subsidiaries of other corporations.

* * * *

The Dead Sea Companies say that the State of Israel exercised considerable control over their operations, notwithstanding Israel's indirect relationship to those companies. They appear to think that, in determining instrumentality status under the Act, control may be substituted for an ownership interest. Control and ownership, however, are distinct concepts. The terms of [Section] 1603(b)(2) are explicit and straightforward. Majority ownership by a foreign state, not control, is the benchmark of instrumentality status.

* * * *

The judgment of the Court of Appeals * * * is affirmed * * * .

DISSENTING OPINION

Justice *BREYER*, * * * dissenting * * * .

* * * Unlike the majority, I believe that the statutory phrase "other ownership interest * * * owned by a foreign state" covers a Foreign Nation's legal interest in a Corporate Subsidiary, where that interest consists of the Foreign Nation's ownership of a Corporate Parent that owns the shares of the Subsidiary.

* * * *

As far as [the FSIA] is concerned, decisions about how to incorporate, how to structure corporate entities, or whether to act through a single corporate layer or through several corporate layers are matters purely of form, not of substance. The need for federal court determination of a sovereign immunity claim is no less important where subsidiaries are involved. The need for [the FSIA's] protections is no less compelling. The risk of adverse foreign policy consequences is no less great.

That is why I doubt the majority's claim that its reading of the text of the FSIA is "[t]he better reading," leading to "[t]he better rule." The majority's rule is not better for a foreign nation, say, Mexico or Honduras, which may use a tiered corporate structure to manage and control important areas of national interest, such as natural resources, and, as a result, will find its ability to use the [U.S.] federal courts to adjudicate matters of national importance and potential sensitivity restricted. Congress is most unlikely to characterize as "better" a rule tied to legal formalities that undercuts its basic jurisdictional objectives. And working lawyers will now have to factor into complex corporate restructuring equations * * * a risk that the government might lose its previously available access to federal court.

Given these consequences, from what perspective can the Court's unnecessarily technical reading of this part of the statute produce a "better rule"? To hold, as the Court does today, that for purposes of the FSIA "other ownership interest" does not include the interest that a Foreign Nation has in a tiered Corporate Subsidiary would be not merely to depart from the primary rule that words are to be taken in their ordinary sense, but to narrow the operation of the statute to an extent that would seriously imperil the accomplishment of its purpose.

QUESTIONS FOR ANALYSIS

1. **Law.** What did the majority rule in this case, and why?
2. **Law.** Why did the dissent disagree with the majority's ruling?
3. **Social Dimensions.** Why did the majority conclude that "[m]ajority ownership by a foreign state, not control, is the benchmark of instrumentality status"?
4. **Ethical Dimensions.** Under what circumstances might a court "pierce the corporate veil" to hold a corporation's owner liable? Should the United States Supreme Court have applied these principles in this case to hold that the Dead Sea companies were instrumentalities of the state under the FSIA? Why or why not?
5. **Implications for the Investor.** How might the holding in this case affect investments in foreign "instrumentalities"?

APPENDIX A

Answers to Even-Numbered *For Review* Questions

Chapter 1

2A. ***What is the common law tradition?***
Because of our colonial heritage, much of American law is based on the English legal system. In that system, after the Norman Conquest, the king's courts sought to establish a uniform set of rules for the entire country. What evolved in these courts was the common law—a body of general legal principles that applied throughout the entire English realm. Courts developed the common law rules from the principles underlying judges' decisions in actual legal controversies.

4A. ***What are some important differences between civil law and criminal law?***
Civil law spells out the rights and duties that exist between persons and between persons and their governments, and the relief available when a person's rights are violated. In a civil case, a private party may sue another private party (the government can also sue a party for a civil law violation) to make that other party comply with a duty or pay for damages caused by a failure to comply with a duty. *Criminal law* has to do with wrongs committed against society for which society demands redress. Local, state, or federal statutes proscribe criminal acts. Public officials, such as district attorneys, not victims or other private parties, prosecute criminal defendants on behalf of the state. In a civil case, the object is to obtain remedies (such as damages) to compensate an injured party. In a criminal case, the object is to punish a wrongdoer to deter others from similar actions. Penalties for violations of criminal statutes include fines and imprisonment and, in some cases, death.

Chapter 2

2A. ***Before a court can hear a case, it must have jurisdiction. Over what must it have jurisdiction? How are the courts applying traditional jurisdictional concepts to cases involving Internet transactions?***
To hear a case, a court must have jurisdiction over the person against whom the suit is brought or over the property involved in the suit. The court must also have jurisdiction over the subject matter. Generally, courts apply a "sliding-scale" standard to determine when it is proper to exercise jurisdiction over a defendant whose only connection with the jurisdiction is the Internet.

4A. ***In a lawsuit, what are the pleadings? What is discovery, and how does electronic discover differ from traditional discovery? What is electronic filing?***
The *pleadings* include a plaintiff's complaint and a defendant's answer (and the counterclaim and reply). The pleadings inform each party of the other's claims and specify the issues involved in a case. *Discovery* is the process of obtaining information and evidence about a case from the other party or third parties. Discovery entails gaining access to witnesses, documents, records, and other types of evidence. Electronic discovery differs in its subject (e-media rather than traditional sources of information). *Electronic filing* involves the filing of court documents in electronic media, typically over the Internet.

Chapter 3

2A. ***How can business leaders encourage their companies to act ethically?***
Ethical leadership is important to create and maintain an ethical workplace. Managers can set standards and apply those standards to themselves and their firm's employees.

4A. ***How do duty-based ethical standards differ from outcome-based ethical standards?***
Duty-based ethical standards are derived from religious precepts or philosophical principles. Outcome-based ethics focuses on the consequences of an action, not on the nature of the action or on a set of preestablished moral values or religious beliefs.

CHAPTER 4

2A. ***What is the purpose of tort law? What are two basic categories of torts?***
Generally, the purpose of tort law is to provide remedies for the invasion of legally recognized and protected interests (personal safety, freedom of movement, property, and some intangibles, including privacy and reputation). The two broad categories of torts are intentional and unintentional.

4A. ***What is meant by strict liability? In what circumstances is strict liability applied?***
Strict liability is liability without fault. Strict liability for damage proximately caused by an abnormally dangerous or exceptional activity, or by the keeping of dangerous animals, is an application of this doctrine. Another significant application of strict liability is in the area of product liability.

CHAPTER 5

2A. ***Why are trademarks and patents protected by the law?***
As stated in Article I, Section 8, of the Constitution, Congress is authorized "[t]o promote the Progress of Science and useful Arts, by securing for limited Times to Authors and Inventors the exclusive Right to their respective Writings and Discoveries." Laws protecting patents and trademarks, and copyrights as well, are designed to protect and reward inventive and artistic creativity.

4A. ***What are trade secrets, and what laws offer protection for this form of intellectual property?***
Trade secrets are business processes and information that are not or cannot be patented, copyrighted, or trademarked. Trade secrets generally consist of anything that makes an individual company unique and that would have value to a competitor. The Uniform Trade Secrets Act, the Economic Espionage Act, and the common law offer trade secrets protection.

CHAPTER 6

2A. ***What are five broad categories of crime? What is white-collar crime?***
Traditionally, crimes have been grouped into the following categories: violent crime (crimes against persons), property crime, public order crime, white-collar crime, and organized crime. White-collar crime is an illegal act or series of acts committed by an individual or business entity using some nonviolent means, usually in the course of a legitimate occupation.

4A. ***What constitutional safeguards exist to protect persons accused of crimes? What are the basic steps in the criminal process?***
Under the Fourth Amendment, before searching or seizing private property, law enforcement officers must obtain a search warrant, which requires probable cause. Under the Fifth Amendment, no one can be deprived of "life, liberty, or property without due process of law." The Fifth Amendment also protects persons against double jeopardy and self-incrimination. The Sixth Amendment guarantees the right to a speedy trial, the right to a jury trial, the right to a public trial, the right to confront witnesses, and the right to counsel. All evidence obtained in violation of the Fourth, Fifth, and Sixth Amendments must be excluded from the trial, as well as all evidence derived from the illegally obtained evidence. Individuals who are arrested must be informed of certain constitutional rights, including their Fifth Amendment right to remain silent and their Sixth Amendment right to counsel. The Eighth Amendment prohibits excessive bail and fines, and cruel and unusual punishment. The basic steps in the criminal process include an arrest, the booking, the initial appearance, a preliminary hearing, a grand jury or magistrate's review, the arraignment, a plea bargain (if any), and the trial or guilty plea.

CHAPTER 7

2A. ***What are the four basic elements necessary to the formation of a valid contract?***
The basic elements for the formation of a valid contract are agreement, consideration, contractual capacity, and legality. Defenses to the enforcement of an otherwise valid contract include the lack of genuineness of assent and improper form.

4A. ***How does a void contract differ from a voidable contract? What is an unenforceable contract?***
A void contract is not a valid contract; it is no contract. A voidable contract is a valid contract, but one that can be avoided at the option of one or both of the parties. An unenforceable contract is one that cannot be enforced because of certain legal defenses against it.

CHAPTER 8

2A. ***In what circumstances will an offer be irrevocable?***
An offeror may not effectively revoke an offer if the offeree has changed position in justifiable reliance on the offer. Also, an option contract takes away the offeror's

power to revoke an offer for the period of time specified in the option (or, if unspecified, for a reasonable time).

4A. *What is consideration? What is required for consideration to be legally sufficient?*
Consideration is the value exchanged for a promise. To be legally sufficient, consideration must be "something of legal value." Something of legal value may include (1) a promise to do something that one has no prior legal duty to do, (2) the performance of an act that one is otherwise not obligated to do, or (3) the refraining from an act that one has a legal right to do.

CHAPTER 9

2A. *Does an intoxicated person have the capacity to enter into an enforceable contract?*
If a person who is sufficiently intoxicated to lack mental capacity enters into a contract, the contract is voidable at the option of that person. It must be proved that the person's reason and judgment were impaired to the extent that he or she did not comprehend the legal consequences of entering into the contract.

4A. *Under what circumstances will a covenant not to compete be enforceable? When will such covenants not be enforced?*
If a covenant not to compete is ancillary to an agreement to sell an ongoing business (enabling the seller to sell, and the purchaser to buy, the "goodwill" and "reputation" of the business), or is contained in an employment contract, and is reasonable in terms of time and geographic area, it will be enforceable. A covenant not to compete that is not ancillary to an agreement to sell an ongoing business will be void because it unreasonably restrains trade and is contrary to public policy.

CHAPTER 10

2A. *What is the difference between a mistake of value or quality and a mistake of fact?*
A mistake as to the value of a deal is an ordinary risk of business for which a court normally will not provide relief. A mistake concerning a fact important to the subject matter of a contract, however, may provide a basis for relief.

4A. *What contracts must be in writing to be enforceable?*
Contracts that are normally required to be in writing or evidenced by a written memorandum include contracts involving interests in land; contracts that cannot by their terms be performed within one year from the date of formation; collateral contracts, such as promises to answer for the debt or duty of another; promises made in consideration of marriage; and contracts for the sale of goods priced at $500 or more ($5,000 or more, under the 2003 amendments to Article 2 of the Uniform Commercial Code).

CHAPTER 11

2A. *What rights can be assigned despite a contract clause expressly prohibiting assignment?*
A contract cannot prevent an assignment of the right to receive money. The assignment of ownership rights in real estate may not be prohibited because it is contrary to public policy in most states. The assignment of negotiable instruments cannot be prohibited. In a contract for a sale of goods, the right to receive damages for breach or for payment of an account owed may be assigned even if the contract prohibits it.

4A. *How are most contracts discharged?*
The most common way to discharge, or terminate, a contract is by the performance of contractual duties.

CHAPTER 12

2A. *What is the standard measure of compensatory damages when a contract is breached? How are damages computed differently in construction contracts?*
In a contract for the sale of goods, the usual measure of compensatory damages is an amount equal to the difference between the contract price and the market price. When the buyer breaches and the seller has not yet produced the goods, compensatory damages normally equal the lost profits on the sale rather than the difference between the contract price and the market price. On the breach of a contract for a sale of land, when specific performance is not available, the measure of damages is also the difference between the contract price and the market price. The measure on the breach of a construction contract depends on who breaches, and when. Recovery for an innocent contractor is generally based on funds expended or expected profit, or both; and for a nonbreaching owner, the cost to complete the project.

4A. *When do courts grant specific performance as a remedy?*
Specific performance might be granted as a remedy when damages are an inadequate remedy and the subject matter of the contract is unique.

Chapter 13

2A. *How do shrink-wrap and click-on agreements differ from other contracts? How have traditional laws been applied to these agreements?*
A *shrink-wrap agreement* is an agreement whose terms are expressed inside a box in which the goods are packaged. A *click-wrap agreement* is formed when an offeree accepts an online offer by clicking a box on the computer screen stating "I agree" or "I accept." Generally, courts have enforced the terms of these agreements the same as the terms of other contracts, applying the traditional common law of contracts.

4A. *What is a partnering agreement? What purpose does it serve?*
A *partnering agreement* is an agreement between a seller and a buyer who often do business on the terms and conditions that apply to all of their transactions conducted electronically. Such an agreement reduces the likelihood of a dispute and provides for the resolution of any dispute that does arise.

Chapter 14

2A. *What is a merchant's firm offer?*
A firm offer by a merchant to sell, buy, or lease goods arises when a merchant-offeror gives *assurances* in a *signed writing* that an offer will remain open. This offer is irrevocable without consideration for the stated period or, if no definite period is stated, a reasonable period (neither period to exceed three months).

4A. *Article 2 and Article 2A of the UCC both define several exceptions to the writing requirements of the Statute of Frauds. What are these exceptions?*
An oral contract is enforceable if it is for (1) goods that are specially manufactured for a particular buyer or specially manufactured or obtained for a particular lessee, (2) the goods are not suitable for resale or lease to others in the ordinary course of the seller's or lessor's business, and (3) the seller or lessor has substantially started to make the goods or has committed to the manufacture or procurement of the goods. An oral contract for a sale or lease of goods is enforceable if the party against whom enforcement of the contract is sought admits in pleadings, testimony, or other court proceedings that a contract was made (but it is limited to the quantity of goods admitted). An oral contract for a sale or lease of goods is enforceable if payment has been made and accepted or goods have been received and accepted.

Chapter 15

2A. *If the parties to a contract do not expressly agree on when title to goods passes, what determines when title passes?*
In the absence of a contrary agreement, title passes to the buyer at the time and the place that the seller physically delivers the goods. Delivery terms can determine this point: under a shipment contract, title passes to the buyer at the time and place of shipment, whereas under a destination contract, title passes to the buyer when the goods are tendered at the destination.

4A. *Under what circumstances will the seller's title to goods being sold be void? Under what circumstances will a seller have voidable title? What is the legal effect on a good faith purchaser of the goods when the seller has a void title versus a voidable title?*
If a seller is a thief, the seller's title is void—legally, no title exists. A buyer would acquire no title or other interest in the goods, and the real owner can reclaim the goods. A seller has voidable title if the goods were obtained by fraud, paid for with a check that is later dishonored, purchased from a minor, or purchased on credit when the seller was insolvent. A seller with voidable title has the power to transfer good title or other interest to a good faith purchaser for value. The real, or original, owner cannot recover goods from the good faith purchaser.

Chapter 16

2A. *What is the perfect tender rule? What are some important exceptions to this rule that apply to sales and lease contracts?*
Under the perfect tender rule, the seller or lessor has an obligation to ship or tender conforming goods, and if the goods or tender of delivery fails in any respect, the buyer or lessee has the right to accept the goods, reject the entire shipment, or accept part and reject part. Exceptions to the perfect tender rule may be established by agreement. When tender is rejected because of nonconforming goods and the time for performance has not yet expired, the seller or lessor can notify the buyer or lessee promptly of the intention to cure and can then do so within the contract time for performance. Once the time for performance has expired, the seller or lessor can, for a reasonable time, exercise the right to cure if she or he had, at the time of delivery, reasonable grounds to believe that the nonconforming tender would be acceptable to the buyer or lessee. When an agreed-on manner of delivery becomes impracticable or unavailable through no fault of

either party, a seller or lessor may choose a commercially reasonable substitute. In an installment contract, a buyer or lessee can reject an installment only if the nonconformity substantially impairs the value of the installment and cannot be cured. Delay in delivery or nondelivery in whole or in part is not a breach when performance is commercially impracticable. If an unexpected event totally destroys goods identified at the time the contract is formed through no fault of either party and before risk passes to the buyer or lessee, the parties are excused from performance. If a party has reasonable grounds to believe that the other party will not perform, he or she may in writing demand assurance of performance from the other party. Until such assurance is received, he or she may suspend further performance. Finally, when required cooperation is not forthcoming, the cooperative party can suspend her or his own performance without liability.

4A. *What remedies are available to a seller or lessor when the buyer or lessee breaches the contract? What remedies are available to a buyer or lessee if the seller or lessor breaches the contract?*
Depending on the circumstances at the time of a buyer's or lessee's breach, a seller or lessor may have the right to cancel the contract, withhold delivery, resell or dispose of the goods subject to the contract, recover the purchase price (or lease payments), recover damages, stop delivery in transit, or reclaim the goods. Similarly, on a seller's or lessor's breach, a buyer or lessee may have the right to cancel the contract, recover the goods, obtain specific performance, obtain cover, replevy the goods, recover damages, reject the goods, withhold delivery, resell or dispose of the goods, stop delivery, or revoke acceptance.

Chapter 17

2A. *What implied warranties arise under the UCC?*
Implied warranties that arise under the UCC include the implied warranty of merchantability, the implied warranty of fitness for a particular purpose, and implied warranties that may arise from, or be excluded or modified by, course of dealing, course of performance, or usage of trade.

4A. *What are the elements of a cause of action in strict product liability?*
Under Section 402A of the *Restatement (Second) of Torts,* the elements of an action for strict product liability are (1) the product must be in a defective condition when the defendant sells it; (2) the defendant must normally be engaged in the business of selling (or distributing) that product; (3) the product must be unreasonably dangerous to the user or consumer because of its defective condition (in most states); (4) the plaintiff must incur physical harm to self or property by use or consumption of the product; (5) the defective condition must be the proximate cause of the injury or damage; and (6) the goods must not have been substantially changed from the time the product was sold to the time the injury was sustained.

Chapter 18

2A. *What are the requirements for attaining the status of a holder in due course (HDC)?*
A holder of a negotiable instrument becomes a holder in due course (HDC) if he or she takes the instrument (1) for value; (2) in good faith; and (3) without notice that the instrument is overdue, that it has been dishonored, that any person has a defense against it or a claim to it, or that it contains unauthorized signatures or alterations or is so irregular or incomplete as to call into question its authenticity.

4A. *Certain defenses are valid against all holders, including HDCs. What are these defenses called? Name four defenses that fall within this category.*
Universal, or real, defenses are good against the claims of all holders, including HDCs. These defenses include forgery of a maker's or drawer's signature, fraud in the execution of an instrument, material alteration of an instrument, discharge in bankruptcy, minority to the extent recognized by state law, illegality to the extent that state law makes a transaction void, mental incapacity if a person has been so adjudged in a state proceeding, and extreme duress.

Chapter 19

2A. *When may a bank properly dishonor a customer's check without being liable to the customer? What happens if a bank wrongfully dishonors a customer's check?*
A bank may dishonor a customer's check without liability to the customer when the customer's account contains insufficient funds to pay the check, providing the bank did not agree to cover overdrafts. A bank may also properly dishonor a stale check, a timely check subject to a valid stop-payment order, a check drawn after the customer's death, and forged or altered checks.

4A. *What are the four most common types of electronic fund transfers? What is the basic purpose of the Electronic Fund Transfer Act, and how does it benefit consumers?*

The four most common types of electronic fund transfer (EFT) systems used by bank customers are automated teller machines, point-of-sale systems, systems handling direct deposits and withdrawals of funds, and pay-by-Internet systems. The Electronic Fund Transfer Act (EFTA) provides a basic framework for the rights, liabilities, and responsibilities of users of these EFT systems. For consumers, the terms and conditions of EFTs must be disclosed in readily understandable language, a receipt must be provided at an e-terminal at the time of a transfer, periodic statements must describe transfers for each account to which an EFT system provides access, and some preauthorized payments can be stopped within three days before they are made.

Chapter 20

2A. *What three requirements must be met to create an enforceable security interest?*
(1) Either (a) the collateral must be in the possession of the secured party in accordance with an agreement, or (b) there must be a written or authenticated security agreement that describes the collateral subject to the security interest and is signed or authenticated by the debtor. (2) The secured party must give to the debtor something of value. (3) The debtor must have "rights" in the collateral.

4A. *If two secured parties have perfected security interests in the collateral of the debtor, which party has priority to the collateral on the debtor's default?*
When two or more secured parties have perfected security interests in the same collateral, the first to perfect has priority, unless an applicable state statute provides otherwise.

Chapter 21

2A. *What is garnishment? When might a creditor undertake a garnishment proceeding?*
Garnishment occurs when a creditor is permitted to collect a debt by seizing property of the debtor that is being held by a third party (such as a paycheck held by an employer or a checking account held by a bank). Garnishment, which is closely regulated, is used in some situations in which debts are not paid.

4A. *What is the difference between an exception to discharge and an objection to discharge?*
An *exception* to discharge is a claim that is not dischargeable in bankruptcy (such as a government claim for unpaid taxes). An *objection* to discharge is a circumstance that causes a discharge to be denied (such as concealing assets).

Chapter 22

2A. *How do agency relationships arise?*
Agency relationships are normally consensual: they arise by voluntary consent and agreement between the parties.

4A. *When is a principal liable for the agent's actions with respect to third parties? When is the agent liable?*
A disclosed or partially disclosed principal is liable to a third party for a contract made by an agent who is acting within the scope of her or his authority. If the agent exceeds the scope of authority and the principal fails to ratify the contract, the agent may be liable (and the principal may not). When neither the fact of agency nor the identity of the principal is disclosed, the agent is liable, and if the agent has acted within the scope of his or her authority, the undisclosed principal is also liable. Each party is liable for his or her own torts and crimes. A principal may also be liable for an agent's torts committed within the course or scope of employment. A principal is liable for an agent's crime if the principal participated by conspiracy or other action.

Chapter 23

2A. *What federal statute governs working hours and wages? What federal statutes govern labor unions and collective bargaining?*
The Fair Labor Standards Act is the most significant federal statute governing working hours and wages. Labor unions and collective bargaining are covered by the Norris-LaGuardia Act, the National Labor Relations Act, the Labor-Management Relations Act, and the Labor-Management Reporting and Disclosure Act.

4A. *Generally, what kind of conduct is prohibited by Title VII of the Civil Rights Act of 1964, as amended?*
Title VII of the Civil Rights Act of 1964, as amended, prohibits employment discrimination against employees, job applicants, and union members on the basis of race, color, national origin, religion, and gender at any stage of employment.

Chapter 24

2A. *What are the three essential elements of a partnership?*
The three elements are (1) a sharing of profits and losses, (2) a joint ownership of a business, and (3) an equal right in the management of the business.

4A. *Why do professional groups organize as a limited liability partnership? How does this form differ from a general partnership?*

Limited liability partnerships (LLPs) are similar to limited liability companies but are designed more for professionals who normally do business as partners in a partnership. Unlike a general partnership, the LLP allows professionals to avoid personal liability for the malpractice of other partners. Depending on state law, partners in an LLP may be exempt from liability for any partnership obligation.

CHAPTER 25

2A. ***What is the difference between a* de jure *corporation and a* de facto *corporation?***
A *de jure* corporation is one that is properly formed, and its status as a corporation is not subject to attack. A *de facto* corporation is one that exists but that has failed to comply with all of the statutory mandates and therefore could be subject to a challenge by the state (although not by a third party).

4A. ***What are the four steps of the merger or consolidation procedure?***
First, the board of directors of each corporation involved must approve the merger or consolidation plan. Second, the shareholders of each corporation must approve the plan, by vote, at a shareholders' meeting. Third, the plan (articles of merger or consolidation) is filed, usually with the secretary of state. And fourth, the state issues a certificate of merger to the surviving corporation or a certificate of consolidation to the newly consolidated corporation.

CHAPTER 26

2A. ***Directors are expected to use their best judgment in managing the corporation. What must directors do to avoid liability for honest mistakes of judgment and poor business decisions?***
Directors and officers must exercise due care in performing their duties. They are expected to be informed on corporate matters. They are expected to act in accord with their own knowledge and training. Directors are expected to exercise a reasonable amount of supervision when they delegate work to others. Directors are expected to attend board of directors' meetings. In general, directors and officers must act in good faith, in what they consider to be the best interests of the corporation, and with the care that an ordinarily prudent person in a similar position would exercise in similar circumstances. This requires an informed decision, with a rational basis, and with no conflict between the decision maker's personal interest and the interest of the corporation.

4A. ***If a group of shareholders perceives that the corporation has suffered a wrong and the directors refuse to take action, can the shareholders compel the directors to act? If so, how?***
The shareholders cannot compel the directors to act to redress a wrong suffered by the corporation, but if the directors refuse to act, the shareholders can act on behalf of the firm by filing what is known as a shareholder's derivative suit. Any damages recovered by the suit normally go into the corporation's treasury, not to the shareholders personally.

CHAPTER 27

2A. ***What are the two major statutes regulating the securities industry? When was the Securities and Exchange Commission created, and what are its major purposes and functions?***
The major statutes regulating the securities industry are the Securities Act of 1933 and the Securities Exchange Act of 1934, which created the Securities and Exchange Commission (SEC). The SEC's major functions are to (1) require the disclosure of facts concerning offerings of securities listed on national securities exchanges and of certain securities traded over the counter; (2) regulate the trade in securities on the national and regional securities exchanges and in the over-the-counter markets; (3) investigate securities fraud; (4) regulate the activities of securities brokers, dealers, and investment advisers and require their registration; (5) supervise the activities of mutual funds; and (6) recommend administrative sanctions, injunctive remedies, and criminal prosecution against those who violate securities laws.

4A. ***What are some of the features of state securities laws?***
Typically, state laws have disclosure requirements and antifraud provisions patterned after Section 10(b) of the Securities Exchange Act of 1934 and SEC Rule 10b-5. State laws provide for the registration or qualification of securities offered or issued for sale within the state with the appropriate state official. Also, most state securities laws regulate securities brokers and dealers.

CHAPTER 28

2A. ***What does it mean to own property in fee simple? What is the difference between a joint tenancy and a tenancy in common?***
An owner in *fee simple* is entitled to use, possess, or dispose of the property as he or she chooses during his or her lifetime, and on this owner's death, the interests in the property descend to his or her heirs. A *tenancy in common*

is a form of co-ownership in which each of two or more persons owns an undivided interest in the whole property. On the death of a tenant in common, that tenant's interest passes to her or his heirs. In a *joint tenancy,* each of two or more persons owns an undivided interest in the property, and a deceased joint tenant's interest passes to the surviving joint tenant or tenants. This right distinguishes the joint tenancy from the tenancy in common.

4A. ***What are the three elements of a bailment?***
A bailment is formed by the delivery of personal property, without transfer of title, by a bailor to a bailee, under an agreement, often for a particular purpose.

CHAPTER 29

2A. ***What are the requirements for acquiring property by adverse possession?***
The adverse possessor's possession must be (1) actual and exclusive; (2) open, visible, and notorious, not secret or clandestine; (3) continuous and peaceable for the statutory period of time; and (4) hostile and adverse.

4A. ***What is a leasehold estate? What types of leasehold estates, or tenancies, can be created when real property is leased?***
A *leasehold estate* is property in the possession of a tenant. Leasehold estates include tenancies for years, periodic tenancies, tenancies at will, and tenancies at sufferance.

CHAPTER 30

2A. ***Is an insurance broker the agent of the insurance applicant or the agent of the insurer?***
A broker is the agent of the insurance applicant.

4A. ***What is the difference between a* per stirpes *distribution and a* per capita *distribution of an estate to the grandchildren of the deceased?***
Per stirpes distribution dictates that grandchildren share the part of the estate that their deceased parent (and descendant of the deceased grandparent) would have been entitled to inherit. *Per capita* distribution dictates that each of these grandchildren takes an equal share of the estate.

CHAPTER 31

2A. ***What are the rules concerning an auditor's liability to third parties?***
An auditor may be liable to a third party on the ground of negligence, when the auditor knew or should have known that the third party would benefit from the auditor's work. Depending on the jurisdiction, liability may be imposed (1) only if the auditor is in privity, or near privity, with the third party; (2) only if the third party's reliance on the auditor's work was foreseen, or the third party was within a class of known or foreseeable users; or (3) if the third party's use of the auditor's work was reasonably foreseeable.

4A. ***What crimes might an accountant commit under the Internal Revenue Code?***
Crimes under the Internal Revenue Code include (1) aiding or assisting in the preparation of a false tax return; (2) aiding or abetting an individual's understatement of tax liability; (3) negligently or willfully understating a client's tax liability or recklessly or intentionally disregarding rules or regulations of the Internal Revenue Service; and (4) failing to provide a taxpayer with a copy of a tax return, failing to sign the return, or failing to furnish the appropriate tax identification numbers.

CHAPTER 32

2A. ***What is the act of state doctrine? In what circumstances is this doctrine applied?***
The *act of state doctrine* is a judicially created doctrine that provides that the judicial branch of one country will not examine the validity of public acts committed by a recognized foreign government within its own territory. This doctrine is often employed in cases involving expropriation or confiscation.

4A. ***What types of provisions, or clauses, are often included in international sales contracts?***
To avoid problems that may arise due to language and legal differences among nations, parties to an international commercial contract may include in the contract a choice-of-language clause (designating the official language by which the contract will be interpreted in the event of a disagreement); a choice-of-forum clause (indicating what court, jurisdiction, or tribunal will decide any disputes arising under the contract); a choice-of-law clause (specifying that the law of a particular nation or jurisdiction will apply to the contract); and a *force majeure* clause (indicating what events may excuse a party from liability for nonperformance).

APPENDIX B

Sample Answers for End-of-Chapter *Questions with Sample Answer*

1-2A. Question with Sample Answer

(a) The U.S. Constitution—The U.S. Constitution is the supreme law of the land. A law in violation of the Constitution, no matter what its source, will be declared unconstitutional and will not be enforced.

(b) The federal statute—Under the U.S. Constitution, when there is a conflict between federal law and state law, federal law prevails.

(c) The state statute—State statutes are enacted by state legislatures. Areas not covered by state statutory law are governed by state case law.

(d) The U.S. Constitution—State constitutions are supreme within their respective borders unless they conflict with the U.S. Constitution, which is the supreme law of the land.

(e) The federal administrative regulation—Under the U.S. Constitution, when there is a conflict between federal law and state law, federal law prevails.

2-2A. Question with Sample answer

Marya can bring suit in all three courts. The trucking firm did business in Florida, and the accident occurred there. Thus, the state of Florida would have jurisdiction over the defendant. Because the firm was headquartered in Georgia and had its principal place of business in that state, Marya could also sue in a Georgia court. Finally, because the amount in controversy exceeds $75,000, the suit could be brought in federal court on the basis of diversity of citizenship.

3-3A. Question with Sample Answer

It could be argued that businesses should not allow themselves to be dictated to by small groups of activists, just as it could be said that a government should not yield to activists who do not represent the governed. Small groups may not represent the best interests of those to whom a business owes its principal duties—consumers, employees, shareholders, and so on. It might also be asserted however, that businesses should yield to activist groups, when those groups do represent the best interests of all, or some, of those to whom a business owes a duty or the best interests of society at large. Small groups may be "cutting edge"—seeing and encouraging others to see the future, as in the case of global warming. Regardless of whether a business ultimately yields to the pressure of activists, however, the business should consider other factors before choosing which course to follow. Besides the interests of the business's stakeholders, there are the interests and the actions of those who may not have a direct stake in the business but whose interests may be parallel (competitors, for example). There are also the interests of the business itself to consider—the effect on its profitability and its continued viability of doing or not doing what certain activists urge.

4-3A. Question with Sample Answer

The correct answer is (b). The *Restatement (Second) of Torts* defines *negligence* as "conduct that falls below the standard established by law for the protection of others against unreasonable risk of harm." The standard established by law is that of a reasonable person acting with due care in the circumstances. Mary was well aware that the medication she took would make her drowsy, and her failure to observe due care (that is, refrain from driving) under the circumstances was negligent. Answer (a) is incorrect because Mary had no reason to believe the golf club was defective, and she could not have prevented the injury by the exercise of due care.

5-2A. Question with Sample Answer

Yes. A patent is not deemed useless and therefore invalid simply because it has not been used in a particular application. The Doneys' patent was valid, and Exetron Corp. infringed on the patent.

6-2A. Question with Sample Answer

(a) Sarah has wrongfully taken and carried away the personal property of another with the intent to permanently deprive the owner of such property. She has committed the crime of larceny.

(b) Sarah has unlawfully and forcibly taken the personal property of another. She has committed the crime of robbery.

(c) Sarah has broken into and entered a dwelling with the intent to commit a felony. She has committed the crime of burglary. (Most states have dispensed with the requirement that the act take place at night.)

Note the basic differences: Burglary requires breaking into and entering a building without the use of force against a person. Robbery does not involve any breaking and entering, but force is required. Larceny is the taking of personal property without force and without breaking into and entering a building. Generally, because force is used, robbery is considered the most serious of these crimes and carries the most severe penalties. Larceny involves no force or threat to human life; therefore, it carries the least severe penalty of the three. Burglary, because it involves breaking and entering, frequently where people live, carries a lesser penalty than robbery but a greater penalty than larceny.

7-2A. Question with Sample Answer

Jonathan cannot recover any damages from Rosalie, as no legally binding contract was formed. Rosalie's obligation to Jonathan was social (moral), not legal, in nature. Jonathan also cannot recover under quasi contract as Rosalie received no unjust enrichment from Jonathan's actions.

8-2A. Question with Sample Answer

No. Jason's offer was a public offer, and he revoked the offer in the same way that it had been originally communicated—by means of the newspaper advertisement. Nothing further was required for an effective revocation. Sharith did not have to know of the revocation for it to be effective. Jason was thus under no obligation to accept the wallet from Sharith or to pay Sharith the $100 reward.

9-2A. Question with Sample Answer

(a) Contracts made by a mentally incompetent person fall into two classes: those made after the person had been adjudicated mentally incompetent, and those made while the person is actually mentally incompetent but has not been so adjudicated by a court. The legal effect of the two classes is entirely different.

(b) Any contract made by a person who has been adjudicated mentally incompetent is void; it is no contract at all. Only contracts made with the court-appointed guardian are binding. If the person is actually mentally incompetent at the time the contract is made but has not been so adjudicated by the court, the contract is voidable by the mentally incompetent person or by the guardian later appointed by the court. The general rules of disaffirmance, ratification (any act of affirmation after the mentally incompetent person regains mental competence), and contracts for necessaries are basically the same as for a minor. Because Kessler was mentally incompetent, but had not been so adjudicated by the court at the time of the car sale, the guardian's disaffirmance is proper, and in most states Kessler would have to return the full purchase price. As with a minor's disaffirmance, some states would permit Jermal to deduct the reasonable value or depreciation from the purchase payment being returned. If the contract was made after Kessler had been adjudicated mentally incompetent, the contract would be void.

10-2A. Question with Sample Answer

In this situation, Gemma becomes what is known as a *guarantor* on the loan; that is, she guarantees to the hardware store that she will pay for the mower if her brother fails to do so. This kind of collateral promise, in which the guarantor states that he or she will become responsible *only* if the primary party does not perform, must be in writing to be enforceable. There is an exception, however. If the main purpose in accepting secondary liability is to secure a personal benefit—for example, if Gemma's brother bought the mower for her—the contract need not be in writing. The assumption is that a court can infer from the circumstances of the case whether the main purpose was to secure a personal benefit and thus, in effect, to answer for the guarantor's own debt.

11-2A. Question with Sample Answer

An intended beneficiary is one who can sue the promisor directly for breach of a contract made for the beneficiary's benefit. It must be clear from the contract that the parties intended the third party to benefit, the benefit in the agreement must be direct to the third party, and the liability of the promisor must arise from the language of the agreement. In this problem, Rivera is an intended beneficiary: it is clear from the contract that Howie and Wilken intended Rivera to benefit and that the benefit in the agreement is direct to Rivera (Howie is to pay Rivera directly). In this situation, Rivera is a *creditor beneficiary* (one who benefits from a contract in which one party

promises another party to pay a third party a debt owed by the promisee to the third party).

12-2A. Question with Sample Answer

Generally, the equitable remedy of specific performance will be granted only if two criteria are met: monetary damages must be inadequate as a remedy (under the circumstances), and the subject matter of the contract must be unique.

(a) In a sale of land, the buyer's contract is for a specific piece of real property. The land under contract is unique because no two pieces of real property have the same legal description. In addition, monetary damages would not compensate the buyer adequately, as the same land cannot be purchased elsewhere. Specific performance is an appropriate remedy.

(b) The basic criteria for specific performance do not apply well to personal-service contracts. If the identical service contracted for is readily available from others, the service is not unique, and monetary damages for nonperformance are adequate. If, however, the services are so personal that only the contracted party can perform them, the contract meets the test of uniqueness, but the courts will refuse to decree specific performance based on either of two theories. First, the enforcement of specific performance would require involuntary servitude (prohibited by the Thirteenth Amendment to the U.S. Constitution). Second, it is impractical to attempt to force meaningful performance by someone against his or her will. In the case of Marita and Horace, specific performance is not an appropriate remedy.

(c) A rare coin is unique, and monetary damages for breach are inadequate, as Juan cannot obtain a substantially identical substitute in the market. This is a typical case where specific performance is an appropriate remedy.

(d) The key fact for consideration here is that this is a closely held corporation. Therefore, the stock is not available in the market, and the shares become unique. The uniqueness of these shares is enhanced by the fact that if Cary sells his 4 percent of the shares to De Valle, De Valle will have a controlling voice in the corporation. Because of this, monetary damages for De Valle are totally inadequate as a remedy. Specific performance is an appropriate remedy.

13-2A. Question with Sample Answer

Anne has entered into an enforceable contract to subscribe to *E-Commerce Weekly*. In this problem, the offeror presented the offer to deliver, via e-mail, the newsletter with a statement of how to accept—by clicking on the "SUBSCRIBE" button. Consideration was in the promise to deliver the newsletter and in the price that the subscriber agreed to pay. The offeree had an opportunity to read the terms of the subscription agreement before making the contract. Whether or not she actually read those terms does not matter.

14-2A. Question with Sample Answer

Yes. Under UCC 2–205, a merchant-offeror, who in a signed writing gives assurance that an offer will remain open, creates an irrevocable offer (without payment of consideration) for the time period stated in the assurance up to a three-month period. Jennings, as a merchant, was obliged to hold the offer (which had been made in a signed writing—the letter) open until October 9. Wheeler's acceptance of the offer prior to October 9 created a valid contract, which Jennings breached when he sold the Thunderbird to a third party.

15-2A. Question with Sample Answer

(a) In a destination contract, the risk of loss passes to the buyer when the goods are tendered to the buyer at the specified destination—in this scenario, San Francisco.

(b) In a shipment contract, if the seller is required or authorized to ship goods by carrier, but the contract does not specify a locale, the risk of loss passes to the buyer when the goods are duly delivered to the carrier.

(c) If the seller is a merchant, risk of loss to goods held by the seller passes to the buyer when the buyer actually takes physical possession of the goods. If the seller is not a merchant, the risk of loss to goods held by the seller passes to the buyer on tender of delivery.

(d) When a bailee is holding goods for a person who has contracted to sell them and the goods are to be delivered without being moved, risk of loss passes to the buyer when (1) the buyer receives a negotiable document of title for the goods, (2) the bailee acknowledges the buyer's right to possess the goods, or (3) the buyer receives a nonnegotiable document of title and has had a reasonable time to present the document to the bailee and demand the goods. (If the bailee refuses to honor the document, the risk of loss remains with the seller.) If the goods are to be delivered by being moved, but the contract does not specify whether it is a destination or a shipment contract, it is presumed to be a shipment contract. If no locale is specified in the contract, risk of loss passes to the buyer when the seller delivers the goods to the carrier.

16-2A. Question with Sample Answer

No. Cummings had not breached the sales contract because the C.O.D. shipment had deprived him of his

absolute right, in the absence of agreement, to inspect the goods before accepting them. Had Cummings requested or agreed to the C.O.D. method of shipment, the result would have been different. Because he had not agreed to the C.O.D. shipment, he was fully within his rights to refuse to accept the goods because he could not inspect them prior to acceptance. In this case, it was the seller who had breached the contract by shipping the goods C.O.D. without Cummings's consent.

17-2A. QUESTION WITH SAMPLE ANSWER

Carmen's mother can bring a suit against AKI under a theory of negligence or strict liability. Under negligence theory, Carmen's mother would have to show that AKI failed to exercise due care to make the product safe and that this breach of duty was the proximate cause of the damage. If Carmen's mother brings a suit under a theory of strict liability, according to the *Restatement (Second) of Torts,* she needs to establish six basic requirements of strict product liability, which are as follows: (1) the defendant must sell the product in a defective condition; (2) the defendant must normally be engaged in the business of selling the product; (3) the product must be unreasonably dangerous to the user or consumer because of its defective condition; (4) the plaintiff must incur physical harm to self or property by use or consumption of the product; (5) the defective condition must be the proximate cause of the injury or damage; and (6) the goods must not have been substantially changed from the time the product was sold to the time the injury was sustained. Under either theory, privity of contract is not required. Some courts may not allow recovery for property damage unless personal injury also occurs.

18-2A. QUESTION WITH SAMPLE ANSWER

Generally, when there is an unauthorized indorsement, the burden of loss falls on the first party to take the instrument with the unauthorized indorsement. In two situations, however, the loss falls on the maker or drawer: (1) when an imposter induces the maker or drawer of an instrument to issue it to the imposter and (2) when a person signs as or on behalf of a maker or drawer, intending that the payee have no interest in the instrument, or when an agent or employee of the maker or drawer has supplied him or her with the name of the payee, also intending the payee to have no interest [UCC 3–404, 3–405].

19-2A. QUESTION WITH SAMPLE ANSWER

Under the Home Mortgage Disclosure Act (HMDA) and the Community Reinvestment Act of 1977, to prevent discrimination in lending practices a bank is required to define its market area. This area must be established contiguous to the bank's branch offices. It must be mapped using the existing boundaries of the counties or the standard metropolitan areas (SMSAs) in which the offices are located. A bank must delineate the community served and annually review this delineation. This issue here is how a successful Internet-only bank could delineate its community. Does an Internet bank have a physically limited market area or serve a physically distinct community? Will the Federal Reserve Board, the government agency charged with enforcing this law, allow a bank to describe its market area as a "cybercommunity"?

20-2A. QUESTION WITH SAMPLE ANSWER

Mendez has both a security interest in Arabian Knight and is a perfected secured party. He has met all the necessary criteria listed under UCC 9–203 to be a secured creditor. Mendez has given value of $5,000 and has taken possession of the collateral, Arabian Knight, owned by Marsh (who has rights in the collateral). Thus, Mendez has a security interest even though Marsh did not sign a security agreement. Once a security interest attaches, a transfer of possession of the collateral to the secured party can perfect the party's security interest without a filing [UCC 9–310(b)(6); 9–313]. Thus, a security interest was created and perfected at the time Marsh transferred Arabian Knight to Mendez as security for the loan.

21-2A. QUESTION WITH SAMPLE ANSWER

Yes. Peterson can enforce his demand. When a surety pays a debt owed to a creditor, the surety is entitled to certain rights, one of which is the legal right of subrogation. This means that the surety acquires any right against the debtor formerly held by the creditor. In other words, the surety now stands in the shoes of the creditor. Because of this right of subrogation, Peterson can enforce his demand that the bank give him the $4,000 in securities because Peterson now has acquired rights in the collateral given for the loan.

22-2A. QUESTION WITH SAMPLE ANSWER

A court might hold that Hemmerling and Happy Cab have an employment relationship primarily on the basis of control. Happy Cab clearly has the right to control the methods or means used by Hemmerling in the course of operating the taxicab by virtue of its exclusive control over the taxicab. Happy Cab exercises its control by establishing and enforcing a variety of rules relating to the

use of the cab, solicitation of fares, and so on. Other factors supporting the existence of an employment relationship include that Hemmerling is not engaged in an occupation or business distinct from that of Happy Cab and that the work is the type that can be done by employees rather than specially skilled independent contractors. Also, Happy Cab supplies the instrumentality of the trade (the cab), and given that the rates were set by the state, Hemmerling's ability to control his profit is limited. The only factor that supports an independent-contractor relationship is that Happy Cab did not withhold taxes.

23-1A. QUESTION WITH SAMPLE ANSWER

The Occupational Safety and Health Act (OSHA) requires employers to provide safe working conditions for employees. The act prohibits employers from discharging or discriminating against any employee who refuses to work when the employee believes in good faith that she or he will risk death or great bodily harm by undertaking the employment activity. Denton and Carlo had sufficient reason to believe that the maintenance job required of them by their employer involved great risk, and therefore, under OSHA, their discharge was wrongful. Denton and Carlo can turn to the Occupational Safety and Health Administration, which is part of the U.S. Department of Labor, for assistance.

24-1A. QUESTION WITH SAMPLE ANSWER

On the dissolution and winding up of the partnership, the order of liability payment of the assets is as follows:

(a) Debts owed to partnership creditors other than partners.
(b) Loans or advances made by partners.
(c) Capital contributions of partners.
(d) Profits as provided in the partnership agreement or, in the absence of an agreement, equally [UPA 40(b)].

In this situation, the partnership's creditors would be paid $8,000 first, leaving a balance of $42,000 from the $50,000. Next, Shawna would be paid $10,000 for the loan, or advance, leaving $32,000. From this amount, Shawna would receive $10,000 and David's estate $15,000 as payment for their capital contributions, leaving a balance of $7,000. The $7,000 would be split as profits, with 40 percent going to Shawna ($2,800) and 60 percent to David's estate ($4,200).

25-2A. QUESTION WITH SAMPLE ANSWER

(a) As a general rule, a promoter is personally liable for all preincorporation contracts made by the promoter. The basic theory behind such liability is that the promoter cannot be an agent for a nonexistent principal (a corporation not yet formed). It is immaterial whether the contracting party knows of the prospective existence of the corporation, and the general rule of promoter liability continues even after the corporation is formed. Three basic exceptions to promoter liability are:

(1) The promoter's contract with a third party can stipulate that the third party will look only to the new corporation, not to the promoter, for performance and liability.

(2) The third party can release the promoter from liability.

(3) After formation, the corporation can assume the contractual obligations and liability by *novation*. (If it is by *adoption*, most courts hold that the promoter is still personally liable.)

Perez is therefore personally liable on both contracts because (1) neither Oliver nor Kovac has released him from liability, (2) the corporation has not assumed contractual responsibility by novation, and (3) Perez's contract with Kovac did not limit Kovac to holding only the corporation liable. (Perez's liability was conditioned only on the corporation's formation, which did occur.)

(b) Incorporation in and of itself does not make the newly formed corporation liable for preincorporation contracts. Until the newly formed corporation assumes Perez's contracts by novation (releasing Perez from personal liability) or by adoption (undertaking to perform Perez's contracts, which makes both the corporation and Perez liable), Kovac cannot enforce Perez's contract against the corporation.

26-2A. QUESTION WITH SAMPLE ANSWER

Directors are personally answerable to the corporation and the shareholders for breach of their duty to exercise reasonable care in conducting the affairs of the corporation. *Reasonable care* is defined as the degree of care that a reasonably prudent person would use in the conduct of personal business affairs. When directors delegate the running of the corporate affairs to officers, the directors are expected to use reasonable care in selecting and supervising such officers. Failure to do so will make the directors liable for negligence or mismanagement. A director who dissents to an action by the board is not personally liable for losses resulting from that action. Unless the dissent is entered into the minutes of the board meeting, however, the director is presumed to have assented. Therefore, the first issue in the case of Starboard, Inc., is whether the board members

failed to use reasonable care in selecting Tyson as president. If so, and particularly if the board failed to provide a reasonable amount of supervision (and openly embezzled funds indicate that failure), the directors will be personally liable. This liability will include Ellsworth unless she can prove that she dissented and that she tried to reasonably supervise Tyson. Considering the facts in this case, it is questionable that Ellsworth could prove this.

27-1A. Question with Sample Answer

Langley is right. Under Section 3(a)(11) of the Securities Act of 1933, stock offerings that are restricted to residents of the state in which the issuing company is organized and doing business are exempt from registration requirements. Therefore, the Langley offering need not be registered with the SEC. The offering will, however, be subject to state securities legislation. Unless it qualifies for exemption from state registration requirements, the stock offering will have to be registered with the appropriate state official.

28-2A. Question with Sample Answer

Gerald cannot claim ownership rights to the $2,500. Three requirements must be met in order for a gift to be effective: (1) donative intent by the donor, (2) delivery of the gift to the donee, and (3) acceptance by the donee. If any of the three requirements is missing, the gift is not effective, and the property remains that of the donor. In this case, donative intent on the part of the donor, Reineken, was obviously present. The other two requirements, however, were not fulfilled. Reineken did not deliver the gift to his nephew but instead placed it in his dresser drawer. The gift thus remained in Reineken's possession. No acceptance of the gift by the nephew was possible, because the gift was never delivered. Therefore, given these circumstances, no effective gift was made, and Reineken's daughter could legally retain possession of the $2,500 as part of Reineken's estate.

29-2A. Question with Sample Answer

Gemma understandably wants a general warranty deed, as this type of deed will give her the most extensive protection against any defects of title claimed against the property transferred. The general warranty deed would have Wiley warranting the following covenants:

(a) Covenant of seisin and right to convey—a warranty that the seller has good title and power to convey.

(b) Covenant against encumbrances—a warranty by the seller that, unless stated, there are no outstanding encumbrances or liens against the property conveyed.

(c) Covenant of quiet possession—a warranty that the grantee's possession will not be disturbed by others claiming a prior legal right. Wiley, however, is conveying only ten feet along a property line that may not even be accurately surveyed. Wiley therefore does not wish to make these warranties. Consequently, he is offering a quitclaim deed, which does not convey any warranties but conveys only whatever interest, if any, the grantor owns. Although title is passed by the quitclaim deed, the quality of the title is not warranted.

Because Gemma really needs the property, it appears that she has three choices: she can accept the quitclaim deed; she can increase her offer price to obtain the general warranty deed she wants; or she can offer to have a title search made, which should satisfy both parties.

30-2A. Question with Sample Answer

(a) In most states, for a will to be valid, it must be in writing, signed by the testator, and witnessed (attested to) according to the statutes of the state. In some states, the testator is also required to publish (declare) that the document is his or her last will and testament. (Such is not required under the Uniform Probate Code.) In the case of Benjamin, the will is unquestionably written (typewritten) and signed by the testator. The only problem is with the witnesses. Some states require three witnesses, and some invalidate a will if a named beneficiary is also a witness. The Uniform Probate Code provides that a will is valid even if attested to by an interested witness. Therefore, whether the will is valid depends on the state laws dealing with witness qualifications.

(b) If the will is declared invalid, Benjamin's estate will pass in accordance with the state's intestacy laws. These statutes provide for distribution of an estate when there is no valid will. The intent of the statutes is to distribute the estate in the way that the deceased person would have wished. Generally, the estate is divided between a surviving spouse and all surviving children. Because Benjamin is a widower, if his only surviving child is Edward, the entire estate will go to Edward, and Benjamin's grandchildren, Perry and Paul, will receive nothing from the estate.

(c) If the will is valid, the estate will be divided between Benjamin's two children, Patricia and Edward. Should either or both predecease Benjamin, leaving children (Benjamin's grandchildren), the grandchildren take *per stirpes* the share that would have gone to their parent. In this case Edward, as a surviving child of Benjamin, would receive one-half of the estate, and Perry and Paul, as grandchildren, would each receive *per stirpes* one-fourth of the estate (one-half of the share that would have gone to their deceased mother, Patricia).

31-2A. Question with Sample Answer

Assuming that the circuit court has abandoned the *Ultramares* rule, it is likely that the accounting firm of Goldman, Walters, Johnson & Co. will be held liable to Happydays State Bank for negligent preparation of financial statements. This hypothetical is partially derived from *Citizens State Bank v. Timm, Schmidt & Co.* In *Citizens State Bank,* the Supreme Court of Wisconsin enunciated various policy reasons for holding accountants liable to third parties even in the absence of privity. The court suggested that this potential liability would make accountants more careful in the preparation of financial statements. Moreover, in some situations the accountants may be the only solvent defendants, and hence, unless liability is imposed on accountants, third parties who reasonably rely on financial statements may go unprotected. The court further asserted that accountants, rather than third parties, are in a better position to spread the risks. If third parties such as banks have to absorb the costs of bad loans made as a result of negligently prepared financial statements, then the cost of credit to the public in general will increase. In contrast, the court suggested accountants are in a better position to spread the risk by purchasing liability insurance.

32-1A. Question with Sample Answer

The law could pose problems for some businesspersons because certain legal terms or phrases in documents governed by U.S. law have no equivalent terms or phrases in the French legal system. Thus, businesspersons who wish their contracts to be subject to, or at least incorporate, certain U.S. legal principles that are not part of French law may have difficulty drafting those principles into the contracts. Even without differences between U.S. and French law, however, the parties may have different understandings of the contractual terms involved. Typically, many phrases in one language are not readily translatable into another.

APPENDIX C

How to Brief Cases and Analyze Case Problems

HOW TO BRIEF CASES

To fully understand the law with respect to business, you need to be able to read and understand court decisions. To make this task easier, you can use a method of case analysis that is called *briefing*. There is a fairly standard procedure that you can follow when you "brief" any court case. You must first read the case opinion carefully. When you feel you understand the case, you can prepare a brief of it.

Although the format of the brief may vary, typically it will present the essentials of the case under headings such as those listed below.

1. **Citation.** Give the full citation for the case, including the name of the case, the date it was decided, and the court that decided it.
2. **Facts.** Briefly indicate (a) the reasons for the lawsuit; (b) the identity and arguments of the plaintiff(s) and defendant(s), respectively; and (c) the lower court's decision—if appropriate.
3. **Issue.** Concisely phrase, in the form of a question, the essential issue before the court. (If more than one issue is involved, you may have two—or even more—questions here.)
4. **Decision.** Indicate here—with a "yes" or "no," if possible—the court's answer to the question (or questions) in the *Issue* section above.
5. **Reason.** Summarize as briefly as possible the reasons given by the court for its decision (or decisions) and the case or statutory law relied on by the court in arriving at its decision.

AN EXAMPLE OF A BRIEFED COURT CASE

As an example of the format used in briefing cases, we present here a briefed version of the sample court case that was presented in the appendix following Chapter 1 as Exhibit 1A–3.

D.A.B.E., INC. v. CITY OF TOLEDO
United States Court of Appeals,
Sixth Circuit, 2005.
393 F.3d 692.

FACTS The city of Toledo, Ohio, has regulated smoking in public places since 1987. In 2003, Toledo's city council enacted a new Clean Indoor Air Ordinance. The ordinance restricts the ability to smoke in public places—stores, theaters, courtrooms, libraries, museums, health-care facilities, restaurants, and bars. In enclosed public places, smoking is generally prohibited except in a "separate smoking lounge" that is designated for this purpose. D.A.B.E., Inc., a group consisting of the owners of bars, restaurants, and bowling alleys, filed a suit in a federal district court, claiming that the ordinance constituted a taking of their property in violation of the Fifth Amendment to the U.S. Constitution. The plaintiffs also argued that the ordinance was preempted (prevented from taking effect) by a state statute that regulated smoking "in places of public assembly," excluding restaurants, bowling alleys, and bars. The court ruled in favor of the city. The plaintiffs appealed to the U.S. Court of Appeals for the Sixth Circuit.

ISSUE Does the ordinance deny the plaintiffs "economically viable use of their property," as required to prove a taking? Does a state indoor smoking statute preempt the city ordinance?

DECISION No, to both questions. The U.S. Court of Appeals for the Sixth Circuit affirmed the lower court's ruling. The ordinance did not prevent the beneficial use of the plaintiffs' property, because it did not categorically prohibit smoking, but only regulated it. The state indoor smoking statute did not cover the excluded businesses, and the legislature did not indicate an intent to bar a city from restricting smoking in those places.

REASON The Fifth Amendment provides that private property shall not "be taken for public use, without just compensation." A taking occurs when an ordinance denies an owner economically viable use of his or her property. In this case, the plaintiffs alleged that they lost customers because of the ordinance. The court reasoned that the ordinance's only effect on the plaintiffs' businesses is to restrict the areas in which customers can smoke and the conditions under which smoking is permitted. This might "require some financial investment, but an ordinance does not effect a taking merely because compliance with it requires the expenditure of money." Besides, the owners could elect to make other uses of their property. As for the preemption issue, a state statute takes precedence over a local ordinance when they conflict. In this case, a statute prohibits smoking in certain locations, but "it does not contain the slightest hint that the legislature intended to create a positive right to smoke in all public places where it did not expressly forbid smoking. Nothing in the [statute] is inconsistent with a local jurisdiction's decision to impose greater limits on public smoking."

REVIEW OF SAMPLE COURT CASE

Here we provide a review of the briefed version to indicate the kind of information that is contained in each section.

CITATION The name of the case is *D.A.B.E., Inc. v. City of Toledo*. D.A.B.E. is the plaintiff; Toledo is the defendant. The U.S. Court of Appeals for the Sixth Circuit decided this case in 2005. The citation states that this case can be found in volume 393 of the *Federal Reporter, Third Series,* on page 692.

FACTS The *Facts* section identifies the plaintiff and the defendant, describes the events leading up to this suit, the allegations made by the plaintiff in the initial suit, and (because this case is an appellate court decision) the lower court's ruling and the party appealing this ruling. The appellant's contention on appeal is also sometimes included here.

ISSUE The *Issue* section presents the central issue (or issues) decided by the court. In this case, the U.S. Court of Appeals for the Sixth Circuit considers whether the ordinance constitutes a taking in violation of the Fifth Amendment and whether a state statute preempts (or supersedes) the city ordinance.

DECISION The *Decision* section includes the court's decision on the issues before it. The decision reflects the opinion of the judge or justice hearing the case. Decisions by appellate courts are frequently phrased in reference to the lower court's decision; that is, the appellate court may "affirm" the lower court's ruling or "reverse" it. Here, the court determined that the ordinance did not effect a taking because it did not prevent the plaintiffs' beneficial use of their property. The statute did not preempt the ordinance because the legislature did not indicate an intent to prohibit a city from restricting smoking in places excluded from the statute. The court affirmed the lower court's ruling.

REASON The *Reason* section includes references to the relevant laws and legal principles that the court applied in arriving at its conclusion in the case. The relevant law here included the Fifth Amendment, the state indoor smoking statute, and the principles derived from judicial interpretations and applications of those laws. This section also explains the court's application of the law to the facts in this case.

ANALYZING CASE PROBLEMS

In addition to learning how to brief cases, students of business law and the legal environment also find it helpful to know how to analyze case problems. Part of the study of business law and the legal environment usually involves analyzing case problems, such as those included in this text at the end of each chapter.

For each case problem in this book, we provide the relevant background and facts of the lawsuit and the issue before the court. When you are assigned one of these problems, your job will be to determine how the court should decide the issue, and why. In other words, you will need to engage in legal analysis and reasoning. Here we offer some suggestions on how to make this task less daunting. We begin by presenting a sample problem:

> While Janet Lawson, a famous pianist, was shopping in Quality Market, she slipped and fell on a wet floor in one of the aisles. The floor had recently been mopped by one of the store's employees, but there were no signs warning customers that the floor in that area was wet. As a result of the fall, Lawson injured her right arm and was unable to perform piano concerts for the next six months. Had she been able to perform the scheduled concerts, she would have earned approximately $60,000 over that period of time. Lawson sued Quality Market for this amount, plus another $10,000 in medical expenses. She claimed that the store's failure to warn customers of the wet floor constituted negligence and therefore the market was liable for her injuries. Will the court agree with Lawson? Discuss.

UNDERSTAND THE FACTS

This may sound obvious, but before you can analyze or apply the relevant law to a specific set of facts, you must clearly understand those facts. In other words, you should read through the case problem carefully—more than once, if necessary—to make sure you understand the identity of the plaintiff(s) and defendant(s) in the case and the progression of events that led to the lawsuit.

In the sample case just given, the identity of the parties is fairly obvious. Janet Lawson is the one bringing the suit; therefore, she is the plaintiff. Quality Market, against whom she is bringing the suit, is the defendant. Some of the case problems you may work on have multiple plaintiffs or defendants. Often, it is helpful to use abbreviations for the parties. To indicate a reference to a plaintiff, for example, the *pi* symbol—π—is often used, and a defendant is denoted by a *delta*—Δ—a triangle.

The events leading to the lawsuit are also fairly straightforward. Lawson slipped and fell on a wet floor, and she contends that Quality Market should be liable for her injuries because it was negligent in not posting a sign warning customers of the wet floor.

When you are working on case problems, realize that the facts should be accepted as they are given. For example, in our sample problem, it should be accepted that the floor was wet and that there was no sign. In other words, avoid making conjectures, such as "Maybe the floor wasn't too wet," or "Maybe an employee was getting a sign to put up," or "Maybe someone stole the sign." Questioning the facts as they are presented only adds confusion to your analysis.

LEGAL ANALYSIS AND REASONING

Once you understand the facts given in the case problem, you can begin to analyze the case. The **IRAC method** is a helpful tool to use in the legal analysis and reasoning process. IRAC is an acronym for Issue, Rule, Application, Conclusion. Applying this method to our sample problem would involve the following steps:

1. First, you need to decide what legal **issue** is involved in the case. In our sample case, the basic issue is whether Quality Market's failure to warn customers of the wet floor constituted negligence. As discussed in Chapter 4, negligence is a *tort*—a civil wrong. In a tort lawsuit, the plaintiff seeks to be compensated for another's wrongful act. A defendant will be deemed negligent if he or she breached a duty of care owed to the plaintiff and the breach of that duty caused the plaintiff to suffer harm.

2. Once you have identified the issue, the next step is to determine what **rule of law** applies to the issue. To make this determination, you will want to review carefully the text of the chapter in which the problem appears to find the relevant rule of law. Our sample case involves the tort of negligence, covered in Chapter 4. The applicable rule of law is the tort law principle that business owners owe a duty to exercise reasonable care to protect their customers ("business invitees"). Reasonable care, in this context, includes either removing—or warning customers of—*foreseeable* risks about which the owner *knew* or *should have known*. Business owners need not warn customers of "open and obvious" risks, however. If a business owner breaches this duty of care (fails to exercise the appropriate degree of care toward customers), and the breach of duty causes a customer to be injured, the business owner will be liable to the customer for the customer's injuries.

3. The next—and usually the most difficult—step in analyzing case problems is the **application** of the relevant rule of law to the specific facts of the case you are studying. In our sample problem, applying the tort law principle just discussed presents few difficulties. An employee of the store had mopped the floor in the aisle where Lawson slipped and fell, but no sign was present indicating that the floor was wet. That a customer might fall on a wet floor is clearly a foreseeable risk. Therefore, the failure to warn customers about the wet floor was a breach of the duty of care owed by the business owner to the store's customers.

4. Once you have completed step 3 in the IRAC method, you should be ready to draw your **conclusion.** In our sample case, Quality Market is liable to Lawson for her injuries, because the market's breach of its duty of care caused Lawson's injuries.

The fact patterns in the case problems presented in this text are not always as simple as those presented in our sample problem. Often, for example, there may be more than one plaintiff or defendant. A case may also involve more than one issue and more than one applicable rule of law. Furthermore, in some case problems the facts may indicate that the general rule of law should not apply. For example, suppose a store employee advised Lawson not to walk on the floor in the aisle because it was wet, but Lawson decided to walk on it anyway. This fact could alter the outcome of the case because the store could then raise the defense of assumption of risk (see Chapter 4). Nonetheless, a careful review of the chapter should always provide you with the knowledge you need to analyze the problem thoroughly and arrive at accurate conclusions.

APPENDIX D

The Constitution of the United States

Preamble

We the People of the United States, in Order to form a more perfect Union, establish Justice, insure domestic Tranquility, provide for the common defence, promote the general Welfare, and secure the Blessings of Liberty to ourselves and our Posterity, do ordain and establish this Constitution for the United States of America.

Article I

Section 1. All legislative Powers herein granted shall be vested in a Congress of the United States, which shall consist of a Senate and House of Representatives.

Section 2. The House of Representatives shall be composed of Members chosen every second Year by the People of the several States, and the Electors in each State shall have the Qualifications requisite for Electors of the most numerous Branch of the State Legislature.

No Person shall be a Representative who shall not have attained to the Age of twenty five Years, and been seven Years a Citizen of the United States, and who shall not, when elected, be an Inhabitant of that State in which he shall be chosen.

Representatives and direct Taxes shall be apportioned among the several States which may be included within this Union, according to their respective Numbers, which shall be determined by adding to the whole Number of free Persons, including those bound to Service for a Term of Years, and excluding Indians not taxed, three fifths of all other Persons. The actual Enumeration shall be made within three Years after the first Meeting of the Congress of the United States, and within every subsequent Term of ten Years, in such Manner as they shall by Law direct. The Number of Representatives shall not exceed one for every thirty Thousand, but each State shall have at Least one Representative; and until such enumeration shall be made, the State of New Hampshire shall be entitled to chuse three, Massachusetts eight, Rhode Island and Providence Plantations one, Connecticut five, New York six, New Jersey four, Pennsylvania eight, Delaware one, Maryland six, Virginia ten, North Carolina five, South Carolina five, and Georgia three.

When vacancies happen in the Representation from any State, the Executive Authority thereof shall issue Writs of Election to fill such Vacancies.

The House of Representatives shall chuse their Speaker and other Officers; and shall have the sole Power of Impeachment.

Section 3. The Senate of the United States shall be composed of two Senators from each State, chosen by the Legislature thereof, for six Years; and each Senator shall have one Vote.

Immediately after they shall be assembled in Consequence of the first Election, they shall be divided as equally as may be into three Classes. The Seats of the Senators of the first Class shall be vacated at the Expiration of the second Year, of the second Class at the Expiration of the fourth Year, and of the third Class at the Expiration of the sixth Year, so that one third may be chosen every second Year; and if Vacancies happen by Resignation, or otherwise, during the Recess of the Legislature of any State, the Executive thereof may make temporary Appointments until the next Meeting of the Legislature, which shall then fill such Vacancies.

No Person shall be a Senator who shall not have attained to the Age of thirty Years, and been nine Years a Citizen of the United States, and who shall not, when elected, be an Inhabitant of that State for which he shall be chosen.

The Vice President of the United States shall be President of the Senate, but shall have no Vote, unless they be equally divided.

The Senate shall chuse their other Officers, and also a President pro tempore, in the Absence of the Vice President, or when he shall exercise the Office of President of the United States.

The Senate shall have the sole Power to try all Impeachments. When sitting for that Purpose, they shall be on Oath or Affirmation. When the President of the United States is tried, the Chief Justice shall preside: And no Person shall be convicted without the Concurrence of two thirds of the Members present.

Judgment in Cases of Impeachment shall not extend further than to removal from Office, and disqualification to hold and enjoy any Office of honor, Trust, or Profit under the United States: but the Party convicted shall nevertheless be liable and subject to Indictment, Trial, Judgment, and Punishment, according to Law.

Section 4. The Times, Places and Manner of holding Elections for Senators and Representatives, shall be prescribed in each State by the Legislature thereof; but the Congress may at any time by Law make or alter such Regulations, except as to the Places of chusing Senators.

The Congress shall assemble at least once in every Year, and such Meeting shall be on the first Monday in December, unless they shall by Law appoint a different Day.

Section 5. Each House shall be the Judge of the Elections, Returns, and Qualifications of its own Members, and a Majority of each shall constitute a Quorum to do Business; but a smaller Number may adjourn from day to day, and may be authorized to compel the Attendance of absent Members, in such Manner, and under such Penalties as each House may provide.

Each House may determine the Rules of its Proceedings, punish its Members for disorderly Behavior, and, with the Concurrence of two thirds, expel a Member.

Each House shall keep a Journal of its Proceedings, and from time to time publish the same, excepting such Parts as may in their Judgment require Secrecy; and the Yeas and Nays of the Members of either House on any question shall, at the Desire of one fifth of those Present, be entered on the Journal.

Neither House, during the Session of Congress, shall, without the Consent of the other, adjourn for more than three days, nor to any other Place than that in which the two Houses shall be sitting.

Section 6. The Senators and Representatives shall receive a Compensation for their Services, to be ascertained by Law, and paid out of the Treasury of the United States. They shall in all Cases, except Treason, Felony and Breach of the Peace, be privileged from Arrest during their Attendance at the Session of their respective Houses, and in going to and returning from the same; and for any Speech or Debate in either House, they shall not be questioned in any other Place.

No Senator or Representative shall, during the Time for which he was elected, be appointed to any civil Office under the Authority of the United States, which shall have been created, or the Emoluments whereof shall have been increased during such time; and no Person holding any Office under the United States, shall be a Member of either House during his Continuance in Office.

Section 7. All Bills for raising Revenue shall originate in the House of Representatives; but the Senate may propose or concur with Amendments as on other Bills.

Every Bill which shall have passed the House of Representatives and the Senate, shall, before it become a Law, be presented to the President of the United States; If he approve he shall sign it, but if not he shall return it, with his Objections to the House in which it shall have originated, who shall enter the Objections at large on their Journal, and proceed to reconsider it. If after such Reconsideration two thirds of that House shall agree to pass the Bill, it shall be sent together with the Objections, to the other House, by which it shall likewise be reconsidered, and if approved by two thirds of that House, it shall become a Law. But in all such Cases the Votes of both Houses shall be determined by Yeas and Nays, and the Names of the Persons voting for and against the Bill shall be entered on the Journal of each House respectively. If any Bill shall not be returned by the President within ten Days (Sundays excepted) after it shall have been presented to him, the Same shall be a Law, in like Manner as if he had signed it, unless the Congress by their Adjournment prevent its Return in which Case it shall not be a Law.

Every Order, Resolution, or Vote, to which the Concurrence of the Senate and House of Representatives may be necessary (except on a question of Adjournment) shall be presented to the President of the United States; and before the Same shall take Effect, shall be approved by him, or being disapproved by him, shall be repassed by two thirds of the Senate and House of Representatives, according to the Rules and Limitations prescribed in the Case of a Bill.

Section 8. The Congress shall have Power To lay and collect Taxes, Duties, Imposts and Excises, to pay the Debts and provide for the common Defence and general Welfare of the United States; but all Duties, Imposts and Excises shall be uniform throughout the United States;

To borrow Money on the credit of the United States;

To regulate Commerce with foreign Nations, and among the several States, and with the Indian Tribes;

To establish an uniform Rule of Naturalization, and uniform Laws on the subject of Bankruptcies throughout the United States;

To coin Money, regulate the Value thereof, and of foreign Coin, and fix the Standard of Weights and Measures;

To provide for the Punishment of counterfeiting the Securities and current Coin of the United States;

To establish Post Offices and post Roads;

To promote the Progress of Science and useful Arts, by securing for limited Times to Authors and Inventors the exclusive Right to their respective Writings and Discoveries;

To constitute Tribunals inferior to the supreme Court;

To define and punish Piracies and Felonies committed on the high Seas, and Offenses against the Law of Nations;

To declare War, grant Letters of Marque and Reprisal, and make Rules concerning Captures on Land and Water;

To raise and support Armies, but no Appropriation of Money to that Use shall be for a longer Term than two Years;

To provide and maintain a Navy;

To make Rules for the Government and Regulation of the land and naval Forces;

To provide for calling forth the Militia to execute the Laws of the Union, suppress Insurrections and repel Invasions;

To provide for organizing, arming, and disciplining, the Militia, and for governing such Part of them as may be employed in the Service of the United States, reserving to the States respectively, the Appointment of the Officers, and the Authority of training the Militia according to the discipline prescribed by Congress;

To exercise exclusive Legislation in all Cases whatsoever, over such District (not exceeding ten Miles square) as may, by Cession of particular States, and the Acceptance of Congress, become the Seat of the Government of the United States, and to exercise like Authority over all Places purchased by the Consent of the Legislature of the State in which the Same shall be, for the Erection of Forts, Magazines, Arsenals, dock-Yards, and other needful Buildings;—And

To make all Laws which shall be necessary and proper for carrying into Execution the foregoing Powers, and all other

Powers vested by this Constitution in the Government of the United States, or in any Department or Officer thereof.

Section 9. The Migration or Importation of such Persons as any of the States now existing shall think proper to admit, shall not be prohibited by the Congress prior to the Year one thousand eight hundred and eight, but a Tax or duty may be imposed on such Importation, not exceeding ten dollars for each Person.

The privilege of the Writ of Habeas Corpus shall not be suspended, unless when in Cases of Rebellion or Invasion the public Safety may require it.

No Bill of Attainder or ex post facto Law shall be passed.

No Capitation, or other direct, Tax shall be laid, unless in Proportion to the Census or Enumeration herein before directed to be taken.

No Tax or Duty shall be laid on Articles exported from any State.

No Preference shall be given by any Regulation of Commerce or Revenue to the Ports of one State over those of another: nor shall Vessels bound to, or from, one State be obliged to enter, clear, or pay Duties in another.

No Money shall be drawn from the Treasury, but in Consequence of Appropriations made by Law; and a regular Statement and Account of the Receipts and Expenditures of all public Money shall be published from time to time.

No Title of Nobility shall be granted by the United States: And no Person holding any Office of Profit or Trust under them, shall, without the Consent of the Congress, accept of any present, Emolument, Office, or Title, of any kind whatever, from any King, Prince, or foreign State.

Section 10. No State shall enter into any Treaty, Alliance, or Confederation; grant Letters of Marque and Reprisal; coin Money; emit Bills of Credit; make any Thing but gold and silver Coin a Tender in Payment of Debts; pass any Bill of Attainder, ex post facto Law, or Law impairing the Obligation of Contracts, or grant any Title of Nobility.

No State shall, without the Consent of the Congress, lay any Imposts or Duties on Imports or Exports, except what may be absolutely necessary for executing its inspection Laws: and the net Produce of all Duties and Imposts, laid by any State on Imports or Exports, shall be for the Use of the Treasury of the United States; and all such Laws shall be subject to the Revision and Controul of the Congress.

No State shall, without the Consent of Congress, lay any Duty of Tonnage, keep Troops, or Ships of War in time of Peace, enter into any Agreement or Compact with another State, or with a foreign Power, or engage in War, unless actually invaded, or in such imminent Danger as will not admit of delay.

Article II

Section 1. The executive Power shall be vested in a President of the United States of America. He shall hold his Office during the Term of four Years, and, together with the Vice President, chosen for the same Term, be elected, as follows:

Each State shall appoint, in such Manner as the Legislature thereof may direct, a Number of Electors, equal to the whole Number of Senators and Representatives to which the State may be entitled in the Congress; but no Senator or Representative, or Person holding an Office of Trust or Profit under the United States, shall be appointed an Elector.

The Electors shall meet in their respective States, and vote by Ballot for two Persons, of whom one at least shall not be an Inhabitant of the same State with themselves. And they shall make a List of all the Persons voted for, and of the Number of Votes for each; which List they shall sign and certify, and transmit sealed to the Seat of the Government of the United States, directed to the President of the Senate. The President of the Senate shall, in the Presence of the Senate and House of Representatives, open all the Certificates, and the Votes shall then be counted. The Person having the greatest Number of Votes shall be the President, if such Number be a Majority of the whole Number of Electors appointed; and if there be more than one who have such Majority, and have an equal Number of Votes, then the House of Representatives shall immediately chuse by Ballot one of them for President; and if no Person have a Majority, then from the five highest on the List the said House shall in like Manner chuse the President. But in chusing the President, the Votes shall be taken by States, the Representation from each State having one Vote; A quorum for this Purpose shall consist of a Member or Members from two thirds of the States, and a Majority of all the States shall be necessary to a Choice. In every Case, after the Choice of the President, the Person having the greater Number of Votes of the Electors shall be the Vice President. But if there should remain two or more who have equal Votes, the Senate shall chuse from them by Ballot the Vice President.

The Congress may determine the Time of chusing the Electors, and the Day on which they shall give their Votes; which Day shall be the same throughout the United States.

No person except a natural born Citizen, or a Citizen of the United States, at the time of the Adoption of this Constitution, shall be eligible to the Office of President; neither shall any Person be eligible to that Office who shall not have attained to the Age of thirty five Years, and been fourteen Years a Resident within the United States.

In Case of the Removal of the President from Office, or of his Death, Resignation or Inability to discharge the Powers and Duties of the said Office, the same shall devolve on the Vice President, and the Congress may by Law provide for the Case of Removal, Death, Resignation or Inability, both of the President and Vice President, declaring what Officer shall then act as President, and such Officer shall act accordingly, until the Disability be removed, or a President shall be elected.

The President shall, at stated Times, receive for his Services, a Compensation, which shall neither be increased nor diminished during the Period for which he shall have been elected, and he shall not receive within that Period any other Emolument from the United States, or any of them.

Before he enter on the Execution of his Office, he shall take the following Oath or Affirmation: "I do solemnly swear (or affirm) that I will faithfully execute the Office of President of the United States, and will to the best of my Ability, preserve, protect and defend the Constitution of the United States."

Section 2. The President shall be Commander in Chief of the Army and Navy of the United States, and of the Militia of the several States, when called into the actual Service of the United States; he may require the Opinion, in writing, of the principal Officer in each of the executive Departments, upon any Subject relating to the Duties of their respective Offices, and he shall have Power to grant Reprieves and Pardons for Offenses against the United States, except in Cases of Impeachment.

He shall have Power, by and with the Advice and Consent of the Senate to make Treaties, provided two thirds of the Senators present concur; and he shall nominate, and by and with the Advice and Consent of the Senate, shall appoint Ambassadors, other public Ministers and Consuls, Judges of the supreme Court, and all other Officers of the United States, whose Appointments are not herein otherwise provided for, and which shall be established by Law; but the Congress may by Law vest the Appointment of such inferior Officers, as they think proper, in the President alone, in the Courts of Law, or in the Heads of Departments.

The President shall have Power to fill up all Vacancies that may happen during the Recess of the Senate, by granting Commissions which shall expire at the End of their next Session.

Section 3. He shall from time to time give to the Congress Information of the State of the Union, and recommend to their Consideration such Measures as he shall judge necessary and expedient; he may, on extraordinary Occasions, convene both Houses, or either of them, and in Case of Disagreement between them, with Respect to the Time of Adjournment, he may adjourn them to such Time as he shall think proper; he shall receive Ambassadors and other public Ministers; he shall take Care that the Laws be faithfully executed, and shall Commission all the Officers of the United States.

Section 4. The President, Vice President and all civil Officers of the United States, shall be removed from Office on Impeachment for, and Conviction of, Treason, Bribery, or other high Crimes and Misdemeanors.

Article III

Section 1. The judicial Power of the United States, shall be vested in one supreme Court, and in such inferior Courts as the Congress may from time to time ordain and establish. The Judges, both of the supreme and inferior Courts, shall hold their Offices during good Behaviour, and shall, at stated Times, receive for their Services a Compensation, which shall not be diminished during their Continuance in Office.

Section 2. The judicial Power shall extend to all Cases, in Law and Equity, arising under this Constitution, the Laws of the United States, and Treaties made, or which shall be made, under their Authority;—to all Cases affecting Ambassadors, other public Ministers and Consuls;—to all Cases of admiralty and maritime Jurisdiction;—to Controversies to which the United States shall be a Party;—to Controversies between two or more States;—between a State and Citizens of another State;—between Citizens of different States;—between Citizens of the same State claiming Lands under Grants of different States, and between a State, or the Citizens thereof, and foreign States, Citizens or Subjects.

In all Cases affecting Ambassadors, other public Ministers and Consuls, and those in which a State shall be a Party, the supreme Court shall have original Jurisdiction. In all the other Cases before mentioned, the supreme Court shall have appellate Jurisdiction, both as to Law and Fact, with such Exceptions, and under such Regulations as the Congress shall make.

The Trial of all Crimes, except in Cases of Impeachment, shall be by Jury; and such Trial shall be held in the State where the said Crimes shall have been committed; but when not committed within any State, the Trial shall be at such Place or Places as the Congress may by Law have directed.

Section 3. Treason against the United States, shall consist only in levying War against them, or, in adhering to their Enemies, giving them Aid and Comfort. No Person shall be convicted of Treason unless on the Testimony of two Witnesses to the same overt Act, or on Confession in open Court.

The Congress shall have Power to declare the Punishment of Treason, but no Attainder of Treason shall work Corruption of Blood, or Forfeiture except during the Life of the Person attainted.

Article IV

Section 1. Full Faith and Credit shall be given in each State to the public Acts, Records, and judicial Proceedings of every other State. And the Congress may by general Laws prescribe the Manner in which such Acts, Records and Proceedings shall be proved, and the Effect thereof.

Section 2. The Citizens of each State shall be entitled to all Privileges and Immunities of Citizens in the several States.

A Person charged in any State with Treason, Felony, or other Crime, who shall flee from Justice, and be found in another State, shall on Demand of the executive Authority of the State from which he fled, be delivered up, to be removed to the State having Jurisdiction of the Crime.

No Person held to Service or Labour in one State, under the Laws thereof, escaping into another, shall, in Consequence of any Law or Regulation therein, be discharged from such Service or Labour, but shall be delivered up on Claim of the Party to whom such Service or Labour may be due.

Section 3. New States may be admitted by the Congress into this Union; but no new State shall be formed or erected within the Jurisdiction of any other State; nor any State be formed by the Junction of two or more States, or Parts of States, without the Consent of the Legislatures of the States concerned as well as of the Congress.

The Congress shall have Power to dispose of and make all needful Rules and Regulations respecting the Territory or other Property belonging to the United States; and nothing in this Constitution shall be so construed as to Prejudice any Claims of the United States, or of any particular State.

Section 4. The United States shall guarantee to every State in this Union a Republican Form of Government, and shall pro-

tect each of them against Invasion; and on Application of the Legislature, or of the Executive (when the Legislature cannot be convened) against domestic Violence.

Article V

The Congress, whenever two thirds of both Houses shall deem it necessary, shall propose Amendments to this Constitution, or, on the Application of the Legislatures of two thirds of the several States, shall call a Convention for proposing Amendments, which, in either Case, shall be valid to all Intents and Purposes, as part of this Constitution, when ratified by the Legislatures of three fourths of the several States, or by Conventions in three fourths thereof, as the one or the other Mode of Ratification may be proposed by the Congress; Provided that no Amendment which may be made prior to the Year One thousand eight hundred and eight shall in any Manner affect the first and fourth Clauses in the Ninth Section of the first Article; and that no State, without its Consent, shall be deprived of its equal Suffrage in the Senate.

Article VI

All Debts contracted and Engagements entered into, before the Adoption of this Constitution shall be as valid against the United States under this Constitution, as under the Confederation.

This Constitution, and the Laws of the United States which shall be made in Pursuance thereof; and all Treaties made, or which shall be made, under the Authority of the United States, shall be the supreme Law of the Land; and the Judges in every State shall be bound thereby, any Thing in the Constitution or Laws of any State to the Contrary notwithstanding.

The Senators and Representatives before mentioned, and the Members of the several State Legislatures, and all executive and judicial Officers, both of the United States and of the several States, shall be bound by Oath or Affirmation, to support this Constitution; but no religious Test shall ever be required as a Qualification to any Office or public Trust under the United States.

Article VII

The Ratification of the Conventions of nine States shall be sufficient for the Establishment of this Constitution between the States so ratifying the Same.

Amendment I [1791]

Congress shall make no law respecting an establishment of religion, or prohibiting the free exercise thereof; or abridging the freedom of speech, or of the press; or the right of the people peaceably to assembly, and to petition the Government for a redress of grievances.

Amendment II [1791]

A well regulated Militia, being necessary to the security of a free State, the right of the people to keep and bear Arms, shall not be infringed.

Amendment III [1791]

No Soldier shall, in time of peace be quartered in any house, without the consent of the Owner, nor in time of war, but in a manner to be prescribed by law.

Amendment IV [1791]

The right of the people to be secure in their persons, houses, papers, and effects, against unreasonable searches and seizures, shall not be violated, and no Warrants shall issue, but upon probable cause, supported by Oath or affirmation, and particularly describing the place to be searched, and the persons or things to be seized.

Amendment V [1791]

No person shall be held to answer for a capital, or otherwise infamous crime, unless on a presentment or indictment of a Grand Jury, except in cases arising in the land or naval forces, or in the Militia, when in actual service in time of War or public danger; nor shall any person be subject for the same offence to be twice put in jeopardy of life or limb; nor shall be compelled in any criminal case to be a witness against himself, nor be deprived of life, liberty, or property, without due process of law; nor shall private property be taken for public use, without just compensation.

Amendment VI [1791]

In all criminal prosecutions, the accused shall enjoy the right to a speedy and public trial, by an impartial jury of the State and district wherein the crime shall have been committed, which district shall have been previously ascertained by law, and to be informed of the nature and cause of the accusation; to be confronted with the witnesses against him; to have compulsory process for obtaining witnesses in his favor, and to have the Assistance of Counsel for his defence.

Amendment VII [1791]

In Suits at common law, where the value in controversy shall exceed twenty dollars, the right of trial by jury shall be preserved, and no fact tried by jury, shall be otherwise re-examined in any Court of the United States, than according to the rules of the common law.

Amendment VIII [1791]

Excessive bail shall not be required, nor excessive fines imposed, nor cruel and unusual punishments inflicted.

Amendment IX [1791]

The enumeration in the Constitution, of certain rights, shall not be construed to deny or disparage others retained by the people.

Amendment X [1791]

The powers not delegated to the United States by the Constitution, nor prohibited by it to the States, are reserved to the States respectively, or to the people.

Amendment XI [1798]

The Judicial power of the United States shall not be construed to extend to any suit in law or equity, commenced or prosecuted against one of the United States by Citizens of another State, or by Citizens or Subjects of any Foreign State.

Amendment XII [1804]

The Electors shall meet in their respective states, and vote by ballot for President and Vice-President, one of whom, at least, shall not be an inhabitant of the same state with themselves; they shall name in their ballots the person voted for as President, and in distinct ballots the person voted for as Vice-President, and they shall make distinct lists of all persons voted for as President, and of all persons voted for as Vice-President, and of the number of votes for each, which lists they shall sign and certify, and transmit sealed to the seat of the government of the United States, directed to the President of the Senate;—The President of the Senate shall, in the presence of the Senate and House of Representatives, open all the certificates and the votes shall then be counted;—The person having the greatest number of votes for President, shall be the President, if such number be a majority of the whole number of Electors appointed; and if no person have such majority, then from the persons having the highest numbers not exceeding three on the list of those voted for as President, the House of Representatives shall choose immediately, by ballot, the President. But in choosing the President, the votes shall be taken by states, the representation from each state having one vote; a quorum for this purpose shall consist of a member or members from two-thirds of the states, and a majority of all states shall be necessary to a choice. And if the House of Representatives shall not choose a President whenever the right of choice shall devolve upon them, before the fourth day of March next following, then the Vice-President shall act as President, as in the case of the death or other constitutional disability of the President.—The person having the greatest number of votes as Vice-President, shall be the Vice-President, if such number be a majority of the whole number of Electors appointed, and if no person have a majority, then from the two highest numbers on the list, the Senate shall choose the Vice-President; a quorum for the purpose shall consist of two-thirds of the whole number of Senators, and a majority of the whole number shall be necessary to a choice. But no person constitutionally ineligible to the office of President shall be eligible to that of Vice-President of the United States.

Amendment XIII [1865]

Section 1. Neither slavery nor involuntary servitude, except as a punishment for crime whereof the party shall have been duly convicted, shall exist within the United States, or any place subject to their jurisdiction.

Section 2. Congress shall have power to enforce this article by appropriate legislation.

Amendment XIV [1868]

Section 1. All persons born or naturalized in the United States, and subject to the jurisdiction thereof, are citizens of the United States and of the State wherein they reside. No State shall make or enforce any law which shall abridge the privileges or immunities of citizens of the United States; nor shall any State deprive any person of life, liberty, or property, without due process of law; nor deny to any person within its jurisdiction the equal protection of the laws.

Section 2. Representatives shall be apportioned among the several States according to their respective numbers, counting the whole number of persons in each State, excluding Indians not taxed. But when the right to vote at any election for the choice of electors for President and Vice President of the United States, Representatives in Congress, the Executive and Judicial officers of a State, or the members of the Legislature thereof, is denied to any of the male inhabitants of such State, being twenty-one years of age, and citizens of the United States, or in any way abridged, except for participation in rebellion, or other crime, the basis of representation therein shall be reduced in the proportion which the number of such male citizens shall bear to the whole number of male citizens twenty-one years of age in such State.

Section 3. No person shall be a Senator or Representative in Congress, or elector of President and Vice President, or hold any office, civil or military, under the United States, or under any State, who having previously taken an oath, as a member of Congress, or as an officer of the United States, or as a member of any State legislature, or as an executive or judicial officer of any State, to support the Constitution of the United States, shall have engaged in insurrection or rebellion against the same, or given aid or comfort to the enemies thereof. But Congress may by a vote of two-thirds of each House, remove such disability.

Section 4. The validity of the public debt of the United States, authorized by law, including debts incurred for payment of pensions and bounties for services in suppressing insurrection or rebellion, shall not be questioned. But neither the United States nor any State shall assume or pay any debt or obligation incurred in aid of insurrection or rebellion against the United States, or any claim for the loss or emancipation of any slave; but all such debts, obligations and claims shall be held illegal and void.

Section 5. The Congress shall have power to enforce, by appropriate legislation, the provisions of this article.

Amendment XV [1870]

Section 1. The right of citizens of the United States to vote shall not be denied or abridged by the United States or by any State on account of race, color, or previous condition of servitude.

Section 2. The Congress shall have power to enforce this article by appropriate legislation.

Amendment XVI [1913]

The Congress shall have power to lay and collect taxes on incomes, from whatever source derived, without apportionment among the several States, and without regard to any census or enumeration.

Amendment XVII [1913]

Section 1. The Senate of the United States shall be composed of two Senators from each State, elected by the people thereof, for six years; and each Senator shall have one vote. The

electors in each State shall have the qualifications requisite for electors of the most numerous branch of the State legislatures.

Section 2. When vacancies happen in the representation of any State in the Senate, the executive authority of such State shall issue writs of election to fill such vacancies: *Provided,* That the legislature of any State may empower the executive thereof to make temporary appointments until the people fill the vacancies by election as the legislature may direct.

Section 3. This amendment shall not be so construed as to affect the election or term of any Senator chosen before it becomes valid as part of the Constitution.

Amendment XVIII [1919]

Section 1. After one year from the ratification of this article the manufacture, sale, or transportation of intoxicating liquors within, the importation thereof into, or the exportation thereof from the United States and all territory subject to the jurisdiction thereof for beverage purposes is hereby prohibited.

Section 2. The Congress and the several States shall have concurrent power to enforce this article by appropriate legislation.

Section 3. This article shall be inoperative unless it shall have been ratified as an amendment to the Constitution by the legislatures of the several States, as provided in the Constitution, within seven years from the date of the submission hereof to the States by the Congress.

Amendment XIX [1920]

Section 1. The right of citizens of the United States to vote shall not be denied or abridged by the United States or by any State on account of sex.

Section 2. Congress shall have power to enforce this article by appropriate legislation.

Amendment XX [1933]

Section 1. The terms of the President and Vice President shall end at noon on the 20th day of January, and the terms of Senators and Representatives at noon on the 3d day of January, of the years in which such terms would have ended if this article had not been ratified; and the terms of their successors shall then begin.

Section 2. The Congress shall assemble at least once in every year, and such meeting shall begin at noon on the 3d day of January, unless they shall by law appoint a different day.

Section 3. If, at the time fixed for the beginning of the term of the President, the President elect shall have died, the Vice President elect shall become President. If the President shall not have been chosen before the time fixed for the beginning of his term, or if the President elect shall have failed to qualify, then the Vice President elect shall act as President until a President shall have qualified; and the Congress may by law provide for the case wherein neither a President elect nor a Vice President elect shall have qualified, declaring who shall then act as President, or the manner in which one who is to act shall be selected, and such person shall act accordingly until a President or Vice President shall have qualified.

Section 4. The Congress may by law provide for the case of the death of any of the persons from whom the House of Representatives may choose a President whenever the right of choice shall have devolved upon them, and for the case of the death of any of the persons from whom the Senate may choose a Vice President whenever the right of choice shall have devolved upon them.

Section 5. Sections 1 and 2 shall take effect on the 15th day of October following the ratification of this article.

Section 6. This article shall be inoperative unless it shall have been ratified as an amendment to the Constitution by the legislatures of three-fourths of the several States within seven years from the date of its submission.

Amendment XXI [1933]

Section 1. The eighteenth article of amendment to the Constitution of the United States is hereby repealed.

Section 2. The transportation or importation into any State, Territory, or possession of the United States for delivery or use therein of intoxicating liquors, in violation of the laws thereof, is hereby prohibited.

Section 3. This article shall be inoperative unless it shall have been ratified as an amendment to the Constitution by conventions in the several States, as provided in the Constitution, within seven years from the date of the submission hereof to the States by the Congress.

Amendment XXII [1951]

Section 1. No person shall be elected to the office of the President more than twice, and no person who has held the office of President, or acted as President, for more than two years of a term to which some other person was elected President shall be elected to the office of President more than once. But this Article shall not apply to any person holding the office of President when this Article was proposed by the Congress, and shall not prevent any person who may be holding the office of President, or acting as President, during the term within which this Article becomes operative from holding the office of President or acting as President during the remainder of such term.

Section 2. This article shall be inoperative unless it shall have been ratified as an amendment to the Constitution by the legislatures of three-fourths of the several States within seven years from the date of its submission to the States by the Congress.

Amendment XXIII [1961]

Section 1. The District constituting the seat of Government of the United States shall appoint in such manner as the Congress may direct:

A number of electors of President and Vice President equal to the whole number of Senators and Representatives in Congress to which the District would be entitled if it were a State, but in no event more than the least populous state; they shall be in addition to those appointed by the states, but they shall be considered, for the purposes of the election of President and Vice President, to be electors appointed by a state; and they shall meet in the District and perform such duties as provided by the twelfth article of amendment.

Section 2. The Congress shall have power to enforce this article by appropriate legislation.

AMENDMENT XXIV [1964]

Section 1. The right of citizens of the United States to vote in any primary or other election for President or Vice President, for electors for President or Vice President, or for Senator or Representative in Congress, shall not be denied or abridged by the United States, or any State by reason of failure to pay any poll tax or other tax.

Section 2. The Congress shall have power to enforce this article by appropriate legislation.

AMENDMENT XXV [1967]

Section 1. In case of the removal of the President from office or of his death or resignation, the Vice President shall become President.

Section 2. Whenever there is a vacancy in the office of the Vice President, the President shall nominate a Vice President who shall take office upon confirmation by a majority vote of both Houses of Congress.

Section 3. Whenever the President transmits to the President pro tempore of the Senate and the Speaker of the House of Representatives his written declaration that he is unable to discharge the powers and duties of his office, and until he transmits to them a written declaration to the contrary, such powers and duties shall be discharged by the Vice President as Acting President.

Section 4. Whenever the Vice President and a majority of either the principal officers of the executive departments or of such other body as Congress may by law provide, transmit to the President pro tempore of the Senate and the Speaker of the House of Representatives their written declaration that the President is unable to discharge the powers and duties of his office, the Vice President shall immediately assume the powers and duties of the office as Acting President.

Thereafter, when the President transmits to the President pro tempore of the Senate and the Speaker of the House of Representatives his written declaration that no inability exists, he shall resume the powers and duties of his office unless the Vice President and a majority of either the principal officers of the executive department or of such other body as Congress may by law provide, transmit within four days to the President pro tempore of the Senate and the Speaker of the House of Representatives their written declaration that the President is unable to discharge the powers and duties of his office. Thereupon Congress shall decide the issue, assembling within forty-eight hours for that purpose if not in session. If the Congress, within twenty-one days after receipt of the latter written declaration, or, if Congress is not in session, within twenty-one days after Congress is required to assemble, determines by two-thirds vote of both Houses that the President is unable to discharge the powers and duties of his office, the Vice President shall continue to discharge the same as Acting President; otherwise, the President shall resume the powers and duties of his office.

AMENDMENT XXVI [1971]

Section 1. The right of citizens of the United States, who are eighteen years of age or older, to vote shall not be denied or abridged by the United States or by any State on account of age.

Section 2. The Congress shall have power to enforce this article by appropriate legislation.

AMENDMENT XXVII [1992]

No law, varying the compensation for the services of the Senators and Representatives, shall take effect, until an election of Representatives shall have intervened.

APPENDIX E

The Uniform Commercial Code (Excerpts)

(Adopted in fifty-two jurisdictions; all fifty States, although Louisiana has adopted only Articles 1, 3, 4, 7, 8, and 9; the District of Columbia; and the Virgin Islands.)

The Code consists of the following articles:

Art.

1. General Provisions
2. Sales

2A. Leases

3. Negotiable Instruments
4. Bank Deposits and Collections

4A. Funds Transfers

5. Letters of Credit
6. Repealer of Article 6—Bulk Transfers and [Revised] Article 6—Bulk Sales
7. Warehouse Receipts, Bills of Lading and Other Documents of Title
8. Investment Securities
9. Secured Transactions
10. Effective Date and Repealer
11. Effective Date and Transition Provisions

Article 1
GENERAL PROVISIONS

Part 1 Short Title, Construction, Application and Subject Matter of the Act

§ 1–101. Short Title.

This Act shall be known and may be cited as Uniform Commercial Code.

§ 1–102. Purposes; Rules of Construction; Variation by Agreement.

(1) This Act shall be liberally construed and applied to promote its underlying purposes and policies.

(2) Underlying purposes and policies of this Act are

(a) to simplify, clarify and modernize the law governing commercial transactions;

(b) to permit the continued expansion of commercial practices through custom, usage and agreement of the parties;

(c) to make uniform the law among the various jurisdictions.

(3) The effect of provisions of this Act may be varied by agreement, except as otherwise provided in this Act and except that the obligations of good faith, diligence, reasonableness and care prescribed by this Act may not be disclaimed by agreement but the parties may by agreement determine the standards by which the performance of such obligations is to be measured if such standards are not manifestly unreasonable.

(4) The presence in certain provisions of this Act of the words "unless otherwise agreed" or words of similar import does not imply that the effect of other provisions may not be varied by agreement under subsection (3).

(5) In this Act unless the context otherwise requires

(a) words in the singular number include the plural, and in the plural include the singular;

(b) words of the masculine gender include the feminine and the neuter, and when the sense so indicates words of the neuter gender may refer to any gender.

§ 1–103. Supplementary General Principles of Law Applicable.

Unless displaced by the particular provisions of this Act, the principles of law and equity, including the law merchant and the law relative to capacity to contract, principal and agent, estoppel, fraud, misrepresentation, duress, coercion, mistake, bankruptcy, or other validating or invalidating cause shall supplement its provisions.

§ 1–104. Construction Against Implicit Repeal.

This Act being a general act intended as a unified coverage of its subject matter, no part of it shall be deemed to be impliedly repealed by subsequent legislation if such construction can reasonably be avoided.

§ 1–105. Territorial Application of the Act; Parties' Power to Choose Applicable Law.

(1) Except as provided hereafter in this section, when a transaction bears a reasonable relation to this state and also to another state or nation the parties may agree that the law either of this state or of such other state or nation shall govern their rights and duties. Failing such agreement this Act applies to transactions bearing an appropriate relation to this state.

(2) Where one of the following provisions of this Act specifies the applicable law, that provision governs and a contrary agreement is effective only to the extent permitted by the law (including the conflict of laws rules) so specified:

Rights of creditors against sold goods. Section 2–402.

Applicability of the Article on Leases. Sections 2A–105 and 2A–106.

Applicability of the Article on Bank Deposits and Collections. Section 4–102.

Governing law in the Article on Funds Transfers. Section 4A–507.

Letters of Credit, Section 5–116.

Bulk sales subject to the Article on Bulk Sales. Section 6–103.

Applicability of the Article on Investment Securities. Section 8–106.

Law governing perfection, the effect of perfection or nonperfection, and the priority of security interests and agricultural liens. Sections 9–301 through 9–307.

As amended in 1972, 1987, 1988, 1989, 1994, 1995, and 1999.

§ 1–106. Remedies to Be Liberally Administered.

(1) The remedies provided by this Act shall be liberally administered to the end that the aggrieved party may be put in as good a position as if the other party had fully performed but neither consequential or special nor penal damages may be had except as specifically provided in this Act or by other rule of law.

(2) Any right or obligation declared by this Act is enforceable by action unless the provision declaring it specifies a different and limited effect.

§ 1–107. Waiver or Renunciation of Claim or Right After Breach.

Any claim or right arising out of an alleged breach can be discharged in whole or in part without consideration by a written waiver or renunciation signed and delivered by the aggrieved party.

§ 1–108. Severability.

If any provision or clause of this Act or application thereof to any person or circumstances is held invalid, such invalidity shall not affect other provisions or applications of the Act which can be given effect without the invalid provision or application, and to this end the provisions of this Act are declared to be severable.

§ 1–109. Section Captions.

Section captions are parts of this Act.

Part 2 General Definitions and Principles of Interpretation

§ 1–201. General Definitions.

Subject to additional definitions contained in the subsequent Articles of this Act which are applicable to specific Articles or Parts thereof, and unless the context otherwise requires, in this Act:

(1) "Action" in the sense of a judicial proceeding includes recoupment, counterclaim, set-off, suit in equity and any other proceedings in which rights are determined.

(2) "Aggrieved party" means a party entitled to resort to a remedy.

(3) "Agreement" means the bargain of the parties in fact as found in their language or by implication from other circumstances including course of dealing or usage of trade or course of performance as provided in this Act (Sections 1–205 and 2–208). Whether an agreement has legal consequences is determined by the provisions of this Act, if applicable; otherwise by the law of contracts (Section 1–103). (Compare "Contract".)

(4) "Bank" means any person engaged in the business of banking.

(5) "Bearer" means the person in possession of an instrument, document of title, or certificated security payable to bearer or indorsed in blank.

(6) "Bill of lading" means a document evidencing the receipt of goods for shipment issued by a person engaged in the business of transporting or forwarding goods, and includes an airbill. "Airbill" means a document serving for air transportation as a bill of lading does for marine or rail transportation, and includes an air consignment note or air waybill.

(7) "Branch" includes a separately incorporated foreign branch of a bank.

(8) "Burden of establishing" a fact means the burden of persuading the triers of fact that the existence of the fact is more probable than its non-existence.

(9) "Buyer in ordinary course of business" means a person that buys goods in good faith, without knowledge that the sale violates the rights of another person in the goods, and in the ordinary course from a person, other than a pawnbroker, in the business of selling goods of that kind. A person buys goods in the ordinary course if the sale to the person comports with the usual or customary practices in the kind of business in which the seller is engaged or with the seller's own usual or customary practices. A person that sells oil, gas, or other minerals at the wellhead or minehead is a person in the business of selling goods of that kind. A buyer in ordinary course of business may buy for cash, by exchange of other property, or on secured or unsecured credit, and may acquire goods or documents of title under a preexisting contract for sale. Only a buyer that takes possession of the goods or has a right to recover the goods from the seller under Article 2 may be a buyer in ordinary course of business. A person that acquires goods in a transfer in bulk or as security for or in total or partial satisfaction of a money debt is not a buyer in ordinary course of business.

(10) "Conspicuous": A term or clause is conspicuous when it is so written that a reasonable person against whom it is to operate ought to have noticed it. A printed heading in capitals (as: NON-NEGOTIABLE BILL OF LADING) is conspicuous. Language in the body of a form is "conspicuous" if it is in larger or other contrasting type or color. But in a telegram any stated term is "conspicuous". Whether a term or clause is "conspicuous" or not is for decision by the court.

(11) "Contract" means the total legal obligation which results from the parties' agreement as affected by this Act and any other applicable rules of law. (Compare "Agreement".)

(12) "Creditor" includes a general creditor, a secured creditor, a lien creditor and any representative of creditors, including an assignee for the benefit of creditors, a trustee in bankruptcy, a receiver in equity and an executor or administrator of an insolvent debtor's or assignor's estate.

(13) "Defendant" includes a person in the position of defendant in a cross-action or counterclaim.

(14) "Delivery" with respect to instruments, documents of title, chattel paper, or certificated securities means voluntary transfer of possession.

(15) "Document of title" includes bill of lading, dock warrant, dock receipt, warehouse receipt or order for the delivery of goods, and also any other document which in the regular course of business or financing is treated as adequately evidencing that the person in possession of it is entitled to receive, hold and dispose of the document and the goods it covers. To be a document of title a document must purport to be issued by or addressed to a bailee and purport to cover goods in the bailee's possession which are either identified or are fungible portions of an identified mass.

(16) "Fault" means wrongful act, omission or breach.

(17) "Fungible" with respect to goods or securities means goods or securities of which any unit is, by nature or usage of trade, the equivalent of any other like unit. Goods which are not fungible shall be deemed fungible for the purposes of this Act to the extent that under a particular agreement or document unlike units are treated as equivalents.

(18) "Genuine" means free of forgery or counterfeiting.

(19) "Good faith" means honesty in fact in the conduct or transaction concerned.

(20) "Holder" with respect to a negotiable instrument, means the person in possession if the instrument is payable to bearer or, in the cases of an instrument payable to an identified person, if the identified person is in possession. "Holder" with respect to a document of title means the person in possession if the goods are deliverable to bearer or to the order of the person in possession.

(21) To "honor" is to pay or to accept and pay, or where a credit so engages to purchase or discount a draft complying with the terms of the credit.

(22) "Insolvency proceedings" includes any assignment for the benefit of creditors or other proceedings intended to liquidate or rehabilitate the estate of the person involved.

(23) A person is "insolvent" who either has ceased to pay his debts in the ordinary course of business or cannot pay his debts as they become due or is insolvent within the meaning of the federal bankruptcy law.

(24) "Money" means a medium of exchange authorized or adopted by a domestic or foreign government and includes a monetary unit of account established by an intergovernmental organization or by agreement between two or more nations.

(25) A person has "notice" of a fact when

(a) he has actual knowledge of it; or

(b) he has received a notice or notification of it; or

(c) from all the facts and circumstances known to him at the time in question he has reason to know that it exists.

A person "knows" or has "knowledge" of a fact when he has actual knowledge of it. "Discover" or "learn" or a word or phrase of similar import refers to knowledge rather than to reason to know. The time and circumstances under which a notice

or notification may cease to be effective are not determined by this Act.

(26) A person "notifies" or "gives" a notice or notification to another by taking such steps as may be reasonably required to inform the other in ordinary course whether or not such other actually comes to know of it. A person "receives" a notice or notification when

(a) it comes to his attention; or

(b) it is duly delivered at the place of business through which the contract was made or at any other place held out by him as the place for receipt of such communications.

(27) Notice, knowledge or a notice or notification received by an organization is effective for a particular transaction from the time when it is brought to the attention of the individual conducting that transaction, and in any event from the time when it would have been brought to his attention if the organization had exercised due diligence. An organization exercises due diligence if it maintains reasonable routines for communicating significant information to the person conducting the transaction and there is reasonable compliance with the routines. Due diligence does not require an individual acting for the organization to communicate information unless such communication is part of his regular duties or unless he has reason to know of the transaction and that the transaction would be materially affected by the information.

(28) "Organization" includes a corporation, government or governmental subdivision or agency, business trust, estate, trust, partnership or association, two or more persons having a joint or common interest, or any other legal or commercial entity.

(29) "Party", as distinct from "third party", means a person who has engaged in a transaction or made an agreement within this Act.

(30) "Person" includes an individual or an organization (See Section 1–102).

(31) "Presumption" or "presumed" means that the trier of fact must find the existence of the fact presumed unless and until evidence is introduced which would support a finding of its non-existence.

(32) "Purchase" includes taking by sale, discount, negotiation, mortgage, pledge, lien, issue or re-issue, gift or any other voluntary transaction creating an interest in property.

(33) "Purchaser" means a person who takes by purchase.

(34) "Remedy" means any remedial right to which an aggrieved party is entitled with or without resort to a tribunal.

(35) "Representative" includes an agent, an officer of a corporation or association, and a trustee, executor or administrator of an estate, or any other person empowered to act for another.

(36) "Rights" includes remedies.

(37) "Security interest" means an interest in personal property or fixtures which secures payment or performance of an obligation. The term also includes any interest of a consignor and a buyer of accounts, chattel paper, a payment intangible, or a promissory note in a transaction that is subject to Article 9. The special property interest of a buyer of goods on identification of those goods to a contract for sale under Section 2–401 is not a "security interest", but a buyer may also acquire a "security interest" by complying with Article 9. Except as otherwise provided in Section 2–505, the right of a seller or lessor of goods under Article 2 or 2A to retain or acquire possession of the goods is not a "security interest", but a seller or lessor may also acquire a "security interest" by complying with Article 9. The retention or reservation of title by a seller of goods notwithstanding shipment or delivery to the buyer (Section 2–401) is limited in effect to a reservation of a "security interest".

Whether a transaction creates a lease or security interest is determined by the facts of each case; however, a transaction creates a security interest if the consideration the lessee is to pay the lessor for the right to possession and use of the goods is an obligation for the term of the lease not subject to termination by the lessee, and

(a) the original term of the lease is equal to or greater than the remaining economic life of the goods,

(b) the lessee is bound to renew the lease for the remaining economic life of the goods or is bound to become the owner of the goods,

(c) the lessee has an option to renew the lease for the remaining economic life of the goods for no additional consideration or nominal additional consideration upon compliance with the lease agreement, or

(d) the lessee has an option to become the owner of the goods for no additional consideration or nominal additional consideration upon compliance with the lease agreement.

A transaction does not create a security interest merely because it provides that

(a) the present value of the consideration the lessee is obligated to pay the lessor for the right to possession and use of the goods is substantially equal to or is greater than the fair market value of the goods at the time the lease is entered into,

(b) the lessee assumes risk of loss of the goods, or agrees to pay taxes, insurance, filing, recording, or registration fees, or service or maintenance costs with respect to the goods,

(c) the lessee has an option to renew the lease or to become the owner of the goods,

(d) the lessee has an option to renew the lease for a fixed rent that is equal to or greater than the reasonably predictable fair market rent for the use of the goods for the term of the renewal at the time the option is to be performed, or

(e) the lessee has an option to become the owner of the goods for a fixed price that is equal to or greater than the reasonably predictable fair market value of the goods at the time the option is to be performed.

For purposes of this subsection (37):

(x) Additional consideration is not nominal if (i) when the option to renew the lease is granted to the lessee the rent is

stated to be the fair market rent for the use of the goods for the term of the renewal determined at the time the option is to be performed, or (ii) when the option to become the owner of the goods is granted to the lessee the price is stated to be the fair market value of the goods determined at the time the option is to be performed. Additional consideration is nominal if it is less than the lessee's reasonably predictable cost of performing under the lease agreement if the option is not exercised;

(y) "Reasonably predictable" and "remaining economic life of the goods" are to be determined with reference to the facts and circumstances at the time the transaction is entered into; and

(z) "Present value" means the amount as of a date certain of one or more sums payable in the future, discounted to the date certain. The discount is determined by the interest rate specified by the parties if the rate is not manifestly unreasonable at the time the transaction is entered into; otherwise, the discount is determined by a commercially reasonable rate that takes into account the facts and circumstances of each case at the time the transaction was entered into.

(38) "Send" in connection with any writing or notice means to deposit in the mail or deliver for transmission by any other usual means of communication with postage or cost of transmission provided for and properly addressed and in the case of an instrument to an address specified thereon or otherwise agreed, or if there be none to any address reasonable under the circumstances. The receipt of any writing or notice within the time at which it would have arrived if properly sent has the effect of a proper sending.

(39) "Signed" includes any symbol executed or adopted by a party with present intention to authenticate a writing.

(40) "Surety" includes guarantor.

(41) "Telegram" includes a message transmitted by radio, teletype, cable, any mechanical method of transmission, or the like.

(42) "Term" means that portion of an agreement which relates to a particular matter.

(43) "Unauthorized" signature means one made without actual, implied or apparent authority and includes a forgery.

(44) "Value". Except as otherwise provided with respect to negotiable instruments and bank collections (Sections 3–303, 4–210 and 4–211) a person gives "value" for rights if he acquires them

(a) in return for a binding commitment to extend credit or for the extension of immediately available credit whether or not drawn upon and whether or not a chargeback is provided for in the event of difficulties in collection; or

(b) as security for or in total or partial satisfaction of a preexisting claim; or

(c) by accepting delivery pursuant to a preexisting contract for purchase; or

(d) generally, in return for any consideration sufficient to support a simple contract.

(45) "Warehouse receipt" means a receipt issued by a person engaged in the business of storing goods for hire.

(46) "Written" or "writing" includes printing, typewriting or any other intentional reduction to tangible form.

§ 1–202. Prima Facie Evidence by Third Party Documents.

A document in due form purporting to be a bill of lading, policy or certificate of insurance, official weigher's or inspector's certificate, consular invoice, or any other document authorized or required by the contract to be issued by a third party shall be prima facie evidence of its own authenticity and genuineness and of the facts stated in the document by the third party.

§ 1–203. Obligation of Good Faith.

Every contract or duty within this Act imposes an obligation of good faith in its performance or enforcement.

§ 1–204. Time; Reasonable Time; "Seasonably".

(1) Whenever this Act requires any action to be taken within a reasonable time, any time which is not manifestly unreasonable may be fixed by agreement.

(2) What is a reasonable time for taking any action depends on the nature, purpose and circumstances of such action.

(3) An action is taken "seasonably" when it is taken at or within the time agreed or if no time is agreed at or within a reasonable time.

§ 1–205. Course of Dealing and Usage of Trade.

(1) A course of dealing is a sequence of previous conduct between the parties to a particular transaction which is fairly to be regarded as establishing a common basis of understanding for interpreting their expressions and other conduct.

(2) A usage of trade is any practice or method of dealing having such regularity of observance in a place, vocation or trade as to justify an expectation that it will be observed with respect to the transaction in question. The existence and scope of such a usage are to be proved as facts. If it is established that such a usage is embodied in a written trade code or similar writing the interpretation of the writing is for the court.

(3) A course of dealing between parties and any usage of trade in the vocation or trade in which they are engaged or of which they are or should be aware give particular meaning to and supplement or qualify terms of an agreement.

(4) The express terms of an agreement and an applicable course of dealing or usage of trade shall be construed wherever reasonable as consistent with each other; but when such construction is unreasonable express terms control both course of dealing and usage of trade and course of dealing controls usage trade.

(5) An applicable usage of trade in the place where any part of performance is to occur shall be used in interpreting the agreement as to that part of the performance.

(6) Evidence of a relevant usage of trade offered by one party is not admissible unless and until he has given the other party such notice as the court finds sufficient to prevent unfair surprise to the latter.

§ 1–206. Statute of Frauds for Kinds of Personal Property Not Otherwise Covered.

(1) Except in the cases described in subsection (2) of this section a contract for the sale of personal property is not enforceable by way of action or defense beyond five thousand dollars in amount or value of remedy unless there is some writing which indicates that a contract for sale has been made between the parties at a defined or stated price, reasonably identifies the subject matter, and is signed by the party against whom enforcement is sought or by his authorized agent.

(2) Subsection (1) of this section does not apply to contracts for the sale of goods (Section 2–201) nor of securities (Section 8–113) nor to security agreements (Section 9–203).

As amended in 1994.

§ 1–207. Performance or Acceptance Under Reservation of Rights.

(1) A party who with explicit reservation of rights performs or promises performance or assents to performance in a manner demanded or offered by the other party does not thereby prejudice the rights reserved. Such words as "without prejudice", "under protest" or the like are sufficient.

(2) Subsection (1) does not apply to an accord and satisfaction.

As amended in 1990.

§ 1–208. Option to Accelerate at Will.

A term providing that one party or his successor in interest may accelerate payment or performance or require collateral or additional collateral "at will" or "when he deems himself insecure" or in words of similar import shall be construed to mean that he shall have power to do so only if he in good faith believes that the prospect of payment or performance is impaired. The burden of establishing lack of good faith is on the party against whom the power has been exercised.

§ 1–209. Subordinated Obligations.

An obligation may be issued as subordinated to payment of another obligation of the person obligated, or a creditor may subordinate his right to payment of an obligation by agreement with either the person obligated or another creditor of the person obligated. Such a subordination does not create a security interest as against either the common debtor or a subordinated creditor. This section shall be construed as declaring the law as it existed prior to the enactment of this section and not as modifying it. Added 1966.

Note: *This new section is proposed as an optional provision to make it clear that a subordination agreement does not create a security interest unless so intended.*

Article 2
SALES

Part 1 Short Title, General Construction and Subject Matter

§ 2–101. Short Title.

This Article shall be known and may be cited as Uniform Commercial Code—Sales.

§ 2–102. Scope; Certain Security and Other Transactions Excluded From This Article.

Unless the context otherwise requires, this Article applies to transactions in goods; it does not apply to any transaction which although in the form of an unconditional contract to sell or present sale is intended to operate only as a security transaction nor does this Article impair or repeal any statute regulating sales to consumers, farmers or other specified classes of buyers.

§ 2–103. Definitions and Index of Definitions.

(1) In this Article unless the context otherwise requires

(a) "Buyer" means a person who buys or contracts to buy goods.

(b) "Good faith" in the case of a merchant means honesty in fact and the observance of reasonable commercial standards of fair dealing in the trade.

(c) "Receipt" of goods means taking physical possession of them.

(d) "Seller" means a person who sells or contracts to sell goods.

(2) Other definitions applying to this Article or to specified Parts thereof, and the sections in which they appear are:

"Acceptance". Section 2–606.
"Banker's credit". Section 2–325.
"Between merchants". Section 2–104.
"Cancellation". Section 2–106(4).
"Commercial unit". Section 2–105.
"Confirmed credit". Section 2–325.
"Conforming to contract". Section 2–106.
"Contract for sale". Section 2–106.
"Cover". Section 2–712.
"Entrusting". Section 2–403.
"Financing agency". Section 2–104.
"Future goods". Section 2–105.
"Goods". Section 2–105.
"Identification". Section 2–501.
"Installment contract". Section 2–612.
"Letter of Credit". Section 2–325.
"Lot". Section 2–105.
"Merchant". Section 2–104.
"Overseas". Section 2–323.
"Person in position of seller". Section 2–707.

"Present sale". Section 2–106.
"Sale". Section 2–106.
"Sale on approval". Section 2–326.
"Sale or return". Section 2–326.
"Termination". Section 2–106.

(3) The following definitions in other Articles apply to this Article:

"Check". Section 3–104.
"Consignee". Section 7–102.
"Consignor". Section 7–102.
"Consumer goods". Section 9–109.
"Dishonor". Section 3–507.
"Draft". Section 3–104.

(4) In addition Article 1 contains general definitions and principles of construction and interpretation applicable throughout this Article.

As amended in 1994 and 1999.

§ 2–104. Definitions: "Merchant"; "Between Merchants"; "Financing Agency".

(1) "Merchant" means a person who deals in goods of the kind or otherwise by his occupation holds himself out as having knowledge or skill peculiar to the practices or goods involved in the transaction or to whom such knowledge or skill may be attributed by his employment of an agent or broker or other intermediary who by his occupation holds himself out as having such knowledge or skill.

(2) "Financing agency" means a bank, finance company or other person who in the ordinary course of business makes advances against goods or documents of title or who by arrangement with either the seller or the buyer intervenes in ordinary course to make or collect payment due or claimed under the contract for sale, as by purchasing or paying the seller's draft or making advances against it or by merely taking it for collection whether or not documents of title accompany the draft. "Financing agency" includes also a bank or other person who similarly intervenes between persons who are in the position of seller and buyer in respect to the goods (Section 2–707).

(3) "Between merchants" means in any transaction with respect to which both parties are chargeable with the knowledge or skill of merchants.

§ 2–105. Definitions: Transferability; "Goods"; "Future" Goods; "Lot"; "Commercial Unit".

(1) "Goods" means all things (including specially manufactured goods) which are movable at the time of identification to the contract for sale other than the money in which the price is to be paid, investment securities (Article 8) and things in action. "Goods" also includes the unborn young of animals and growing crops and other identified things attached to realty as described in the section on goods to be severed from realty (Section 2–107).

(2) Goods must be both existing and identified before any interest in them can pass. Goods which are not both existing and identified are "future" goods. A purported present sale of future goods or of any interest therein operates as a contract to sell.

(3) There may be a sale of a part interest in existing identified goods.

(4) An undivided share in an identified bulk of fungible goods is sufficiently identified to be sold although the quantity of the bulk is not determined. Any agreed proportion of such a bulk or any quantity thereof agreed upon by number, weight or other measure may to the extent of the seller's interest in the bulk be sold to the buyer who then becomes an owner in common.

(5) "Lot" means a parcel or a single article which is the subject matter of a separate sale or delivery, whether or not it is sufficient to perform the contract.

(6) "Commercial unit" means such a unit of goods as by commercial usage is a single whole for purposes of sale and division of which materially impairs its character or value on the market or in use. A commercial unit may be a single article (as a machine) or a set of articles (as a suite of furniture or an assortment of sizes) or a quantity (as a bale, gross, or carload) or any other unit treated in use or in the relevant market as a single whole.

§ 2–106. Definitions: "Contract"; "Agreement"; "Contract for Sale"; "Sale"; "Present Sale"; "Conforming" to Contract; "Termination"; "Cancellation".

(1) In this Article unless the context otherwise requires "contract" and "agreement" are limited to those relating to the present or future sale of goods. "Contract for sale" includes both a present sale of goods and a contract to sell goods at a future time. A "sale" consists in the passing of title from the seller to the buyer for a price (Section 2–401). A "present sale" means a sale which is accomplished by the making of the contract.

(2) Goods or conduct including any part of a performance are "conforming" or conform to the contract when they are in accordance with the obligations under the contract.

(3) "Termination" occurs when either party pursuant to a power created by agreement or law puts an end to the contract otherwise than for its breach. On "termination" all obligations which are still executory on both sides are discharged but any right based on prior breach or performance survives.

(4) "Cancellation" occurs when either party puts an end to the contract for breach by the other and its effect is the same as that of "termination" except that the cancelling party also retains any remedy for breach of the whole contract or any unperformed balance.

§ 2–107. Goods to Be Severed From Realty: Recording.

(1) A contract for the sale of minerals or the like (including oil and gas) or a structure or its materials to be removed from realty is a contract for the sale of goods within this Article if they are to be severed by the seller but until severance a purported present sale thereof which is not effective as a transfer of an interest in land is effective only as a contract to sell.

(2) A contract for the sale apart from the land of growing crops or other things attached to realty and capable of severance

without material harm thereto but not described in subsection (1) or of timber to be cut is a contract for the sale of goods within this Article whether the subject matter is to be severed by the buyer or by the seller even though it forms part of the realty at the time of contracting, and the parties can by identification effect a present sale before severance.

(3) The provisions of this section are subject to any third party rights provided by the law relating to realty records, and the contract for sale may be executed and recorded as a document transferring an interest in land and shall then constitute notice to third parties of the buyer's rights under the contract for sale.

As amended in 1972.

Part 2 Form, Formation and Readjustment of Contract

§ 2–201. Formal Requirements; Statute of Frauds.

(1) Except as otherwise provided in this section a contract for the sale of goods for the price of $500 or more is not enforceable by way of action or defense unless there is some writing sufficient to indicate that a contract for sale has been made between the parties and signed by the party against whom enforcement is sought or by his authorized agent or broker. A writing is not insufficient because it omits or incorrectly states a term agreed upon but the contract is not enforceable under this paragraph beyond the quantity of goods shown in such writing.

(2) Between merchants if within a reasonable time a writing in confirmation of the contract and sufficient against the sender is received and the party receiving it has reason to know its contents, its satisfies the requirements of subsection (1) against such party unless written notice of objection to its contents is given within ten days after it is received.

(3) A contract which does not satisfy the requirements of subsection (1) but which is valid in other respects is enforceable

(a) if the goods are to be specially manufactured for the buyer and are not suitable for sale to others in the ordinary course of the seller's business and the seller, before notice of repudiation is received and under circumstances which reasonably indicate that the goods are for the buyer, has made either a substantial beginning of their manufacture or commitments for their procurement; or

(b) if the party against whom enforcement is sought admits in his pleading, testimony or otherwise in court that a contract for sale was made, but the contract is not enforceable under this provision beyond the quantity of goods admitted; or

(c) with respect to goods for which payment has been made and accepted or which have been received and accepted (Sec. 2–606).

§ 2–202. Final Written Expression: Parol or Extrinsic Evidence.

Terms with respect to which the confirmatory memoranda of the parties agree or which are otherwise set forth in a writing intended by the parties as a final expression of their agreement with respect to such terms as are included therein may not be contradicted by evidence of any prior agreement or of a contemporaneous oral agreement but may be explained or supplemented

(a) by course of dealing or usage of trade (Section 1–205) or by course of performance (Section 2–208); and

(b) by evidence of consistent additional terms unless the court finds the writing to have been intended also as a complete and exclusive statement of the terms of the agreement.

§ 2–203. Seals Inoperative.

The affixing of a seal to a writing evidencing a contract for sale or an offer to buy or sell goods does not constitute the writing a sealed instrument and the law with respect to sealed instruments does not apply to such a contract or offer.

§ 2–204. Formation in General.

(1) A contract for sale of goods may be made in any manner sufficient to show agreement, including conduct by both parties which recognizes the existence of such a contract.

(2) An agreement sufficient to constitute a contract for sale may be found even though the moment of its making is undetermined.

(3) Even though one or more terms are left open a contract for sale does not fail for indefiniteness if the parties have intended to make a contract and there is a reasonably certain basis for giving an appropriate remedy.

§ 2–205. Firm Offers.

An offer by a merchant to buy or sell goods in a signed writing which by its terms gives assurance that it will be held open is not revocable, for lack of consideration, during the time stated or if no time is stated for a reasonable time, but in no event may such period of irrevocability exceed three months; but any such term of assurance on a form supplied by the offeree must be separately signed by the offeror.

§ 2–206. Offer and Acceptance in Formation of Contract.

(1) Unless other unambiguously indicated by the language or circumstances

(a) an offer to make a contract shall be construed as inviting acceptance in any manner and by any medium reasonable in the circumstances;

(b) an order or other offer to buy goods for prompt or current shipment shall be construed as inviting acceptance either by a prompt promise to ship or by the prompt or current shipment of conforming or nonconforming goods, but such a shipment of non-conforming goods does not constitute an acceptance if the seller seasonably notifies the buyer that the shipment is offered only as an accommodation to the buyer.

(2) Where the beginning of a requested performance is a reasonable mode of acceptance an offeror who is not notified of

acceptance within a reasonable time may treat the offer as having lapsed before acceptance.

§ 2–207. Additional Terms in Acceptance or Confirmation.

(1) A definite and seasonable expression of acceptance or a written confirmation which is sent within a reasonable time operates as an acceptance even though it states terms additional to or different from those offered or agreed upon, unless acceptance is expressly made conditional on assent to the additional or different terms.

(2) The additional terms are to be construed as proposals for addition to the contract. Between merchants such terms become part of the contract unless:

(a) the offer expressly limits acceptance to the terms of the offer;

(b) they materially alter it; or

(c) notification of objection to them has already been given or is given within a reasonable time after notice of them is received.

(3) Conduct by both parties which recognizes the existence of a contract is sufficient to establish a contract for sale although the writings of the parties do not otherwise establish a contract. In such case the terms of the particular contract consist of those terms on which the writings of the parties agree, together with any supplementary terms incorporated under any other provisions of this Act.

§ 2–208. Course of Performance or Practical Construction.

(1) Where the contract for sale involves repeated occasions for performance by either party with knowledge of the nature of the performance and opportunity for objection to it by the other, any course of performance accepted or acquiesced in without objection shall be relevant to determine the meaning of the agreement.

(2) The express terms of the agreement and any such course of performance, as well as any course of dealing and usage of trade, shall be construed whenever reasonable as consistent with each other; but when such construction is unreasonable, express terms shall control course of performance and course of performance shall control both course of dealing and usage of trade (Section 1–205).

(3) Subject to the provisions of the next section on modification and waiver, such course of performance shall be relevant to show a waiver or modification of any term inconsistent with such course of performance.

§ 2–209. Modification, Rescission and Waiver.

(1) An agreement modifying a contract within this Article needs no consideration to be binding.

(2) A signed agreement which excludes modification or rescission except by a signed writing cannot be otherwise modified or rescinded, but except as between merchants such a requirement on a form supplied by the merchant must be separately signed by the other party.

(3) The requirements of the statute of frauds section of this Article (Section 2–201) must be satisfied if the contract as modified is within its provisions.

(4) Although an attempt at modification or rescission does not satisfy the requirements of subsection (2) or (3) it can operate as a waiver.

(5) A party who has made a waiver affecting an executory portion of the contract may retract the waiver by reasonable notification received by the other party that strict performance will be required of any term waived, unless the retraction would be unjust in view of a material change of position in reliance on the waiver.

§ 2–210. Delegation of Performance; Assignment of Rights.

(1) A party may perform his duty through a delegate unless otherwise agreed or unless the other party has a substantial interest in having his original promisor perform or control the acts required by the contract. No delegation of performance relieves the party delegating of any duty to perform or any liability for breach.

(2) Except as otherwise provided in Section 9–406, unless otherwise agreed, all rights of either seller or buyer can be assigned except where the assignment would materially change the duty of the other party, or increase materially the burden or risk imposed on him by his contract, or impair materially his chance of obtaining return performance. A right to damages for breach of the whole contract or a right arising out of the assignor's due performance of his entire obligation can be assigned despite agreement otherwise.

(3) The creation, attachment, perfection, or enforcement of a security interest in the seller's interest under a contract is not a transfer that materially changes the duty of or increases materially the burden or risk imposed on the buyer or impairs materially the buyer's chance of obtaining return performance within the purview of subsection (2) unless, and then only to the extent that, enforcement actually results in a delegation of material performance of the seller. Even in that event, the creation, attachment, perfection, and enforcement of the security interest remain effective, but (i) the seller is liable to the buyer for damages caused by the delegation to the extent that the damages could not reasonably by prevented by the buyer, and (ii) a court having jurisdiction may grant other appropriate relief, including cancellation of the contract for sale or an injunction against enforcement of the security interest or consummation of the enforcement.

(4) Unless the circumstances indicate the contrary a prohibition of assignment of "the contract" is to be construed as barring only the delegation to the assignee of the assignor's performance.

(5) An assignment of "the contract" or of "all my rights under the contract" or an assignment in similar general terms is an assignment of rights and unless the language or the circumstances (as in an assignment for security) indicate the contrary, it is a delegation of performance of the duties of the assignor and its acceptance by the assignee constitutes a promise by him

to perform those duties. This promise is enforceable by either the assignor or the other party to the original contract.

(6) The other party may treat any assignment which delegates performance as creating reasonable grounds for insecurity and may without prejudice to his rights against the assignor demand assurances from the assignee (Section 2–609).

As amended in 1999.

Part 3 General Obligation and Construction of Contract

§ 2–301. General Obligations of Parties.

The obligation of the seller is to transfer and deliver and that of the buyer is to accept and pay in accordance with the contract.

§ 2–302. Unconscionable Contract or Clause.

(1) If the court as a matter of law finds the contract or any clause of the contract to have been unconscionable at the time it was made the court may refuse to enforce the contract, or it may enforce the remainder of the contract without the unconscionable clause, or it may so limit the application of any unconscionable clause as to avoid any unconscionable result.

(2) When it is claimed or appears to the court that the contract or any clause thereof may be unconscionable the parties shall be afforded a reasonable opportunity to present evidence as to its commercial setting, purpose and effect to aid the court in making the determination.

§ 2–303. Allocations or Division of Risks.

Where this Article allocates a risk or a burden as between the parties "unless otherwise agreed", the agreement may not only shift the allocation but may also divide the risk or burden.

§ 2–304. Price Payable in Money, Goods, Realty, or Otherwise.

(1) The price can be made payable in money or otherwise. If it is payable in whole or in part in goods each party is a seller of the goods which he is to transfer.

(2) Even though all or part of the price is payable in an interest in realty the transfer of the goods and the seller's obligations with reference to them are subject to this Article, but not the transfer of the interest in realty or the transferor's obligations in connection therewith.

§ 2–305. Open Price Term.

(1) The parties if they so intend can conclude a contract for sale even though the price is not settled. In such a case the price is a reasonable price at the time for delivery if

(a) nothing is said as to price; or

(b) the price is left to be agreed by the parties and they fail to agree; or

(c) the price is to be fixed in terms of some agreed market or other standard as set or recorded by a third person or agency and it is not so set or recorded.

(2) A price to be fixed by the seller or by the buyer means a price for him to fix in good faith.

(3) When a price left to be fixed otherwise than by agreement of the parties fails to be fixed through fault of one party the other may at his option treat the contract as cancelled or himself fix a reasonable price.

(4) Where, however, the parties intend not to be bound unless the price be fixed or agreed and it is not fixed or agreed there is no contract. In such a case the buyer must return any goods already received or if unable so to do must pay their reasonable value at the time of delivery and the seller must return any portion of the price paid on account.

§ 2–306. Output, Requirements and Exclusive Dealings.

(1) A term which measures the quantity by the output of the seller or the requirements of the buyer means such actual output or requirements as may occur in good faith, except that no quantity unreasonably disproportionate to any stated estimate or in the absence of a stated estimate to any normal or otherwise comparable prior output or requirements may be tendered or demanded.

(2) A lawful agreement by either the seller or the buyer for exclusive dealing in the kind of goods concerned imposes unless otherwise agreed an obligation by the seller to use best efforts to supply the goods and by the buyer to use best efforts to promote their sale.

§ 2–307. Delivery in Single Lot or Several Lots.

Unless otherwise agreed all goods called for by a contract for sale must be tendered in a single delivery and payment is due only on such tender but where the circumstances give either party the right to make or demand delivery in lots the price if it can be apportioned may be demanded for each lot.

§ 2–308. Absence of Specified Place for Delivery.

Unless otherwise agreed

(a) the place for delivery of goods is the seller's place of business or if he has none his residence; but

(b) in a contract for sale of identified goods which to the knowledge of the parties at the time of contracting are in some other place, that place is the place for their delivery; and

(c) documents of title may be delivered through customary banking channels.

§ 2–309. Absence of Specific Time Provisions; Notice of Termination.

(1) The time for shipment or delivery or any other action under a contract if not provided in this Article or agreed upon shall be a reasonable time.

(2) Where the contract provides for successive performances but is indefinite in duration it is valid for a reasonable time but unless otherwise agreed may be terminated at any time by either party.

(3) Termination of a contract by one party except on the happening of an agreed event requires that reasonable notification be received by the other party and an agreement dispensing with notification is invalid if its operation would be unconscionable.

§ 2–310. Open Time for Payment or Running of Credit; Authority to Ship Under Reservation.

Unless otherwise agreed

(a) payment is due at the time and place at which the buyer is to receive the goods even though the place of shipment is the place of delivery; and

(b) if the seller is authorized to send the goods he may ship them under reservation, and may tender the documents of title, but the buyer may inspect the goods after their arrival before payment is due unless such inspection is inconsistent with the terms of the contract (Section 2–513); and

(c) if delivery is authorized and made by way of documents of title otherwise than by subsection (b) then payment is due at the time and place at which the buyer is to receive the documents regardless of where the goods are to be received; and

(d) where the seller is required or authorized to ship the goods on credit the credit period runs from the time of shipment but post-dating the invoice or delaying its dispatch will correspondingly delay the starting of the credit period.

§ 2–311. Options and Cooperation Respecting Performance.

(1) An agreement for sale which is otherwise sufficiently definite (subsection (3) of Section 2–204) to be a contract is not made invalid by the fact that it leaves particulars of performance to be specified by one of the parties. Any such specification must be made in good faith and within limits set by commercial reasonableness.

(2) Unless otherwise agreed specifications relating to assortment of the goods are at the buyer's option and except as otherwise provided in subsections (1)(c) and (3) of Section 2–319 specifications or arrangements relating to shipment are at the seller's option.

(3) Where such specification would materially affect the other party's performance but is not seasonably made or where one party's cooperation is necessary to the agreed performance of the other but is not seasonably forthcoming, the other party in addition to all other remedies

(a) is excused for any resulting delay in his own performance; and

(b) may also either proceed to perform in any reasonable manner or after the time for a material part of his own performance treat the failure to specify or to cooperate as a breach by failure to deliver or accept the goods.

§ 2–312. Warranty of Title and Against Infringement; Buyer's Obligation Against Infringement.

(1) Subject to subsection (2) there is in a contract for sale a warranty by the seller that

(a) the title conveyed shall be good, and its transfer rightful; and

(b) the goods shall be delivered free from any security interest or other lien or encumbrance of which the buyer at the time of contracting has no knowledge.

(2) A warranty under subsection (1) will be excluded or modified only by specific language or by circumstances which give the buyer reason to know that the person selling does not claim title in himself or that he is purporting to sell only such right or title as he or a third person may have.

(3) Unless otherwise agreed a seller who is a merchant regularly dealing in goods of the kind warrants that the goods shall be delivered free of the rightful claim of any third person by way of infringement or the like but a buyer who furnishes specifications to the seller must hold the seller harmless against any such claim which arises out of compliance with the specifications.

§ 2–313. Express Warranties by Affirmation, Promise, Description, Sample.

(1) Express warranties by the seller are created as follows:

(a) Any affirmation of fact or promise made by the seller to the buyer which relates to the goods and becomes part of the basis of the bargain creates an express warranty that the goods shall conform to the affirmation or promise.

(b) Any description of the goods which is made part of the basis of the bargain creates an express warranty that the goods shall conform to the description.

(c) Any sample or model which is made part of the basis of the bargain creates an express warranty that the whole of the goods shall conform to the sample or model.

(2) It is not necessary to the creation of an express warranty that the seller use formal words such as "warrant" or "guarantee" or that he have a specific intention to make a warranty, but an affirmation merely of the value of the goods or a statement purporting to be merely the seller's opinion or commendation of the goods does not create a warranty.

§ 2–314. Implied Warranty: Merchantability; Usage of Trade.

(1) Unless excluded or modified (Section 2–316), a warranty that the goods shall be merchantable is implied in a contract for their sale if the seller is a merchant with respect to goods of that kind. Under this section the serving for value of food or drink to be consumed either on the premises or elsewhere is a sale.

(2) Goods to be merchantable must be at least such as

(a) pass without objection in the trade under the contract description; and

(b) in the case of fungible goods, are of fair average quality within the description; and

(c) are fit for the ordinary purposes for which such goods are used; and

(d) run, within the variations permitted by the agreement, of even kind, quality and quantity within each unit and among all units involved; and

(e) are adequately contained, packaged, and labeled as the agreement may require; and

(f) conform to the promises or affirmations of fact made on the container or label if any.

(3) Unless excluded or modified (Section 2–316) other implied warranties may arise from course of dealing or usage of trade.

§ 2–315. Implied Warranty: Fitness for Particular Purpose.

Where the seller at the time of contracting has reason to know any particular purpose for which the goods are required and that the buyer is relying on the seller's skill or judgment to select or furnish suitable goods, there is unless excluded or modified under the next section an implied warranty that the goods shall be fit for such purpose.

§ 2–316. Exclusion or Modification of Warranties.

(1) Words or conduct relevant to the creation of an express warranty and words or conduct tending to negate or limit warranty shall be construed wherever reasonable as consistent with each other; but subject to the provisions of this Article on parol or extrinsic evidence (Section 2–202) negation or limitation is inoperative to the extent that such construction is unreasonable.

(2) Subject to subsection (3), to exclude or modify the implied warranty of merchantability or any part of it the language must mention merchantability and in case of a writing must be conspicuous, and to exclude or modify any implied warranty of fitness the exclusion must be by a writing and conspicuous. Language to exclude all implied warranties of fitness is sufficient if it states, for example, that "There are no warranties which extend beyond the description on the face hereof."

(3) Notwithstanding subsection (2)

(a) unless the circumstances indicate otherwise, all implied warranties are excluded by expressions like "as is", "with all faults" or other language which in common understanding calls the buyer's attention to the exclusion of warranties and makes plain that there is no implied warranty; and

(b) when the buyer before entering into the contract has examined the goods or the sample or model as fully as he desired or has refused to examine the goods there is no implied warranty with regard to defects which an examination ought in the circumstances to have revealed to him; and

(c) an implied warranty can also be excluded or modified by course of dealing or course of performance or usage of trade.

(4) Remedies for breach of warranty can be limited in accordance with the provisions of this Article on liquidation or limitation of damages and on contractual modification of remedy (Sections 2–718 and 2–719).

§ 2–317. Cumulation and Conflict of Warranties Express or Implied.

Warranties whether express or implied shall be construed as consistent with each other and as cumulative, but if such construction is unreasonable the intention of the parties shall determine which warranty is dominant. In ascertaining that intention the following rules apply:

(a) Exact or technical specifications displace an inconsistent sample or model or general language of description.

(b) A sample from an existing bulk displaces inconsistent general language of description.

(c) Express warranties displace inconsistent implied warranties other than an implied warranty of fitness for a particular purpose.

§ 2–318. Third Party Beneficiaries of Warranties Express or Implied.

Note: If this Act is introduced in the Congress of the United States this section should be omitted. (States to select one alternative.)

Alternative A

A seller's warranty whether express or implied extends to any natural person who is in the family or household of his buyer or who is a guest in his home if it is reasonable to expect that such person may use, consume or be affected by the goods and who is injured in person by breach of the warranty. A seller may not exclude or limit the operation of this section.

Alternative B

A seller's warranty whether express or implied extends to any natural person who may reasonably be expected to use, consume or be affected by the goods and who is injured in person by breach of the warranty. A seller may not exclude or limit the operation of this section.

Alternative C

A seller's warranty whether express or implied extends to any person who may reasonably be expected to use, consume or be affected by the goods and who is injured by breach of the warranty. A seller may not exclude or limit the operation of this section with respect to injury to the person of an individual to whom the warranty extends.

As amended 1966.

§ 2–319. F.O.B. and F.A.S. Terms.

(1) Unless otherwise agreed the term F.O.B. (which means "free on board") at a named place, even though used only in connection with the stated price, is a delivery term under which

(a) when the term is F.O.B. the place of shipment, the seller must at that place ship the goods in the manner provided in this Article (Section 2–504) and bear the expense and risk of putting them into the possession of the carrier; or

(b) when the term is F.O.B. the place of destination, the seller must at his own expense and risk transport the goods to that place and there tender delivery of them in the manner provided in this Article (Section 2–503);

(c) when under either (a) or (b) the term is also F.O.B. vessel, car or other vehicle, the seller must in addition at his own expense and risk load the goods on board. If the term is F.O.B. vessel the buyer must name the vessel and in an appropriate case the seller must comply with the provisions of this Article on the form of bill of lading (Section 2–323).

(2) Unless otherwise agreed the term F.A.S. vessel (which means "free alongside") at a named port, even though used only in connection with the stated price, is a delivery term under which the seller must

(a) at his own expense and risk deliver the goods alongside the vessel in the manner usual in that port or on a dock designated and provided by the buyer; and

(b) obtain and tender a receipt for the goods in exchange for which the carrier is under a duty to issue a bill of lading.

(3) Unless otherwise agreed in any case falling within subsection (1)(a) or (c) or subsection (2) the buyer must seasonably give any needed instructions for making delivery, including when the term is F.A.S. or F.O.B. the loading berth of the vessel and in an appropriate case its name and sailing date. The seller may treat the failure of needed instructions as a failure of cooperation under this Article (Section 2–311). He may also at his option move the goods in any reasonable manner preparatory to delivery or shipment.

(4) Under the term F.O.B. vessel or F.A.S. unless otherwise agreed the buyer must make payment against tender of the required documents and the seller may not tender nor the buyer demand delivery of the goods in substitution for the documents.

§ 2–320. C.I.F. and C. & F. Terms.

(1) The term C.I.F. means that the price includes in a lump sum the cost of the goods and the insurance and freight to the named destination. The term C. & F. or C.F. means that the price so includes cost and freight to the named destination.

(2) Unless otherwise agreed and even though used only in connection with the stated price and destination, the term C.I.F. destination or its equivalent requires the seller at his own expense and risk to

(a) put the goods into the possession of a carrier at the port for shipment and obtain a negotiable bill or bills of lading covering the entire transportation to the named destination; and

(b) load the goods and obtain a receipt from the carrier (which may be contained in the bill of lading) showing that the freight has been paid or provided for; and

(c) obtain a policy or certificate of insurance, including any war risk insurance, of a kind and on terms then current at the port of shipment in the usual amount, in the currency of the contract, shown to cover the same goods covered by the bill of lading and providing for payment of loss to the order of the buyer or for the account of whom it may concern; but the seller may add to the price the amount of the premium for any such war risk insurance; and

(d) prepare an invoice of the goods and procure any other documents required to effect shipment or to comply with the contract; and

(e) forward and tender with commercial promptness all the documents in due form and with any indorsement necessary to perfect the buyer's rights.

(3) Unless otherwise agreed the term C. & F. or its equivalent has the same effect and imposes upon the seller the same obligations and risks as a C.I.F. term except the obligation as to insurance.

(4) Under the term C.I.F. or C. & F. unless otherwise agreed the buyer must make payment against tender of the required documents and the seller may not tender nor the buyer demand delivery of the goods in substitution for the documents.

§ 2–321. C.I.F. or C. & F.: "Net Landed Weights"; "Payment on Arrival"; Warranty of Condition on Arrival.

Under a contract containing a term C.I.F. or C. & F.

(1) Where the price is based on or is to be adjusted according to "net landed weights", "delivered weights", "out turn" quantity or quality or the like, unless otherwise agreed the seller must reasonably estimate the price. The payment due on tender of the documents called for by the contract is the amount so estimated, but after final adjustment of the price a settlement must be made with commercial promptness.

(2) An agreement described in subsection (1) or any warranty of quality or condition of the goods on arrival places upon the seller the risk of ordinary deterioration, shrinkage and the like in transportation but has no effect on the place or time of identification to the contract for sale or delivery or on the passing of the risk of loss.

(3) Unless otherwise agreed where the contract provides for payment on or after arrival of the goods the seller must before payment allow such preliminary inspection as is feasible; but if the goods are lost delivery of the documents and payment are due when the goods should have arrived.

§ 2–322. Delivery "Ex-Ship".

(1) Unless otherwise agreed a term for delivery of goods "ex-ship" (which means from the carrying vessel) or in equivalent language is not restricted to a particular ship and requires delivery from a ship which has reached a place at the named port of destination where goods of the kind are usually discharged.

(2) Under such a term unless otherwise agreed

(a) the seller must discharge all liens arising out of the carriage and furnish the buyer with a direction which puts the carrier under a duty to deliver the goods; and

(b) the risk of loss does not pass to the buyer until the goods leave the ship's tackle or are otherwise properly unloaded.

§ 2–323. Form of Bill of Lading Required in Overseas Shipment; "Overseas".

(1) Where the contract contemplates overseas shipment and contains a term C.I.F. or C. & F. or F.O.B. vessel, the seller unless otherwise agreed must obtain a negotiable bill of lading stating that the goods have been loaded on board or, in the case of a term C.I.F. or C. & F., received for shipment.

(2) Where in a case within subsection (1) a bill of lading has been issued in a set of parts, unless otherwise agreed if the documents are not to be sent from abroad the buyer may demand tender of the full set; otherwise only one part of the bill of lading need be tendered. Even if the agreement expressly requires a full set

(a) due tender of a single part is acceptable within the provisions of this Article on cure of improper delivery (subsection (1) of Section 2–508); and

(b) even though the full set is demanded, if the documents are sent from abroad the person tendering an incomplete set may nevertheless require payment upon furnishing an indemnity which the buyer in good faith deems adequate.

(3) A shipment by water or by air or a contract contemplating such shipment is "overseas" insofar as by usage of trade or agreement it is subject to the commercial, financing or shipping practices characteristic of international deep water commerce.

§ 2–324. "No Arrival, No Sale" Term.

Under a term "no arrival, no sale" or terms of like meaning, unless otherwise agreed,

(a) the seller must properly ship conforming goods and if they arrive by any means he must tender them on arrival but he assumes no obligation that the goods will arrive unless he has caused the non-arrival; and

(b) where without fault of the seller the goods are in part lost or have so deteriorated as no longer to conform to the contract or arrive after the contract time, the buyer may proceed as if there had been casualty to identified goods (Section 2–613).

§ 2–325. "Letter of Credit" Term; "Confirmed Credit".

(1) Failure of the buyer seasonably to furnish an agreed letter of credit is a breach of the contract for sale.

(2) The delivery to seller of a proper letter of credit suspends the buyer's obligation to pay. If the letter of credit is dishonored, the seller may on seasonable notification to the buyer require payment directly from him.

(3) Unless otherwise agreed the term "letter of credit" or "banker's credit" in a contract for sale means an irrevocable credit issued by a financing agency of good repute and, where the shipment is overseas, of good international repute. The term "confirmed credit" means that the credit must also carry the direct obligation of such an agency which does business in the seller's financial market.

§ 2–326. Sale on Approval and Sale or Return; Rights of Creditors.

(1) Unless otherwise agreed, if delivered goods may be returned by the buyer even though they conform to the contract, the transaction is

(a) a "sale on approval" if the goods are delivered primarily for use, and

(b) a "sale or return" if the goods are delivered primarily for resale.

(2) Goods held on approval are not subject to the claims of the buyer's creditors until acceptance; goods held on sale or return are subject to such claims while in the buyer's possession.

(3) Any "or return" term of a contract for sale is to be treated as a separate contract for sale within the statute of frauds section of this Article (Section 2–201) and as contradicting the sale aspect of the contract within the provisions of this Article or on parol or extrinsic evidence (Section 2–202).

As amended in 1999.

§ 2–327. Special Incidents of Sale on Approval and Sale or Return.

(1) Under a sale on approval unless otherwise agreed

(a) although the goods are identified to the contract the risk of loss and the title do not pass to the buyer until acceptance; and

(b) use of the goods consistent with the purpose of trial is not acceptance but failure seasonably to notify the seller of election to return the goods is acceptance, and if the goods conform to the contract acceptance of any part is acceptance of the whole; and

(c) after due notification of election to return, the return is at the seller's risk and expense but a merchant buyer must follow any reasonable instructions.

(2) Under a sale or return unless otherwise agreed

(a) the option to return extends to the whole or any commercial unit of the goods while in substantially their original condition, but must be exercised seasonably; and

(b) the return is at the buyer's risk and expense.

§ 2–328. Sale by Auction.

(1) In a sale by auction if goods are put up in lots each lot is the subject of a separate sale.

(2) A sale by auction is complete when the auctioneer so announces by the fall of the hammer or in other customary manner. Where a bid is made while the hammer is falling in acceptance of a prior bid the auctioneer may in his discretion reopen the bidding or declare the goods sold under the bid on which the hammer was falling.

(3) Such a sale is with reserve unless the goods are in explicit terms put up without reserve. In an auction with reserve the auctioneer may withdraw the goods at any time until he

announces completion of the sale. In an auction without reserve, after the auctioneer calls for bids on an article or lot, that article or lot cannot be withdrawn unless no bid is made within a reasonable time. In either case a bidder may retract his bid until the auctioneer's announcement of completion of the sale, but a bidder's retraction does not revive any previous bid.

(4) If the auctioneer knowingly receives a bid on the seller's behalf or the seller makes or procures such as bid, and notice has not been given that liberty for such bidding is reserved, the buyer may at his option avoid the sale or take the goods at the price of the last good faith bid prior to the completion of the sale. This subsection shall not apply to any bid at a forced sale.

Part 4 Title, Creditors and Good Faith Purchasers

§ 2–401. Passing of Title; Reservation for Security; Limited Application of This Section.

Each provision of this Article with regard to the rights, obligations and remedies of the seller, the buyer, purchasers or other third parties applies irrespective of title to the goods except where the provision refers to such title. Insofar as situations are not covered by the other provisions of this Article and matters concerning title became material the following rules apply:

(1) Title to goods cannot pass under a contract for sale prior to their identification to the contract (Section 2–501), and unless otherwise explicitly agreed the buyer acquires by their identification a special property as limited by this Act. Any retention or reservation by the seller of the title (property) in goods shipped or delivered to the buyer is limited in effect to a reservation of a security interest. Subject to these provisions and to the provisions of the Article on Secured Transactions (Article 9), title to goods passes from the seller to the buyer in any manner and on any conditions explicitly agreed on by the parties.

(2) Unless otherwise explicitly agreed title passes to the buyer at the time and place at which the seller completes his performance with reference to the physical delivery of the goods, despite any reservation of a security interest and even though a document of title is to be delivered at a different time or place; and in particular and despite any reservation of a security interest by the bill of lading

(a) if the contract requires or authorizes the seller to send the goods to the buyer but does not require him to deliver them at destination, title passes to the buyer at the time and place of shipment; but

(b) if the contract requires delivery at destination, title passes on tender there.

(3) Unless otherwise explicitly agreed where delivery is to be made without moving the goods,

(a) if the seller is to deliver a document of title, title passes at the time when and the place where he delivers such documents; or

(b) if the goods are at the time of contracting already identified and no documents are to be delivered, title passes at the time and place of contracting.

(4) A rejection or other refusal by the buyer to receive or retain the goods, whether or not justified, or a justified revocation of acceptance revests title to the goods in the seller. Such revesting occurs by operation of law and is not a "sale".

§ 2–402. Rights of Seller's Creditors Against Sold Goods.

(1) Except as provided in subsections (2) and (3), rights of unsecured creditors of the seller with respect to goods which have been identified to a contract for sale are subject to the buyer's rights to recover the goods under this Article (Sections 2–502 and 2–716).

(2) A creditor of the seller may treat a sale or an identification of goods to a contract for sale as void if as against him a retention of possession by the seller is fraudulent under any rule of law of the state where the goods are situated, except that retention of possession in good faith and current course of trade by a merchant-seller for a commercially reasonable time after a sale or identification is not fraudulent.

(3) Nothing in this Article shall be deemed to impair the rights of creditors of the seller

(a) under the provisions of the Article on Secured Transactions (Article 9); or

(b) where identification to the contract or delivery is made not in current course of trade but in satisfaction of or as security for a pre-existing claim for money, security or the like and is made under circumstances which under any rule of law of the state where the goods are situated would apart from this Article constitute the transaction a fraudulent transfer or voidable preference.

§ 2–403. Power to Transfer; Good Faith Purchase of Goods; "Entrusting".

(1) A purchaser of goods acquires all title which his transferor had or had power to transfer except that a purchaser of a limited interest acquires rights only to the extent of the interest purchased. A person with voidable title has power to transfer a good title to a good faith purchaser for value. When goods have been delivered under a transaction of purchase the purchaser has such power even though

(a) the transferor was deceived as to the identity of the purchaser, or

(b) the delivery was in exchange for a check which is later dishonored, or

(c) it was agreed that the transaction was to be a "cash sale", or

(d) the delivery was procured through fraud punishable as larcenous under the criminal law.

(2) Any entrusting of possession of goods to a merchant who deals in goods of that kind gives him power to transfer all rights of the entruster to a buyer in ordinary course of business.

(3) "Entrusting" includes any delivery and any acquiescence in retention of possession regardless of any condition expressed

between the parties to the delivery or acquiescence and regardless of whether the procurement of the entrusting or the possessor's disposition of the goods have been such as to be larcenous under the criminal law.

(4) The rights of other purchasers of goods and of lien creditors are governed by the Articles on Secured Transactions (Article 9), Bulk Transfers (Article 6) and Documents of Title (Article 7).

As amended in 1988.

Part 5 Performance

§ 2–501. Insurable Interest in Goods; Manner of Identification of Goods.

(1) The buyer obtains a special property and an insurable interest in goods by identification of existing goods as goods to which the contract refers even though the goods so identified are non-conforming and he has an option to return or reject them. Such identification can be made at any time and in any manner explicitly agreed to by the parties. In the absence of explicit agreement identification occurs

(a) when the contract is made if it is for the sale of goods already existing and identified;

(b) if the contract is for the sale of future goods other than those described in paragraph (c), when goods are shipped, marked or otherwise designated by the seller as goods to which the contract refers;

(c) when the crops are planted or otherwise become growing crops or the young are conceived if the contract is for the sale of unborn young to be born within twelve months after contracting or for the sale of crops to be harvested within twelve months or the next normal harvest season after contracting whichever is longer.

(2) The seller retains an insurable interest in goods so long as title to or any security interest in the goods remains in him and where the identification is by the seller alone he may until default or insolvency or notification to the buyer that the identification is final substitute other goods for those identified.

(3) Nothing in this section impairs any insurable interest recognized under any other statute or rule of law.

§ 2–502. Buyer's Right to Goods on Seller's Insolvency.

(1) Subject to subsections (2) and (3) and even though the goods have not been shipped a buyer who has paid a part or all of the price of goods in which he has a special property under the provisions of the immediately preceding section may on making and keeping good a tender of any unpaid portion of their price recover them from the seller if:

(a) in the case of goods bought for personal, family, or household purposes, the seller repudiates or fails to deliver as required by the contract; or

(b) in all cases, the seller becomes insolvent within ten days after receipt of the first installment on their price.

(2) The buyer's right to recover the goods under subsection (1)(a) vests upon acquisition of a special property, even if the seller had not then repudiated or failed to deliver.

(3) If the identification creating his special property has been made by the buyer he acquires the right to recover the goods only if they conform to the contract for sale.

As amended in 1999.

§ 2–503. Manner of Seller's Tender of Delivery.

(1) Tender of delivery requires that the seller put and hold conforming goods at the buyer's disposition and give the buyer any notification reasonably necessary to enable him to take delivery. The manner, time and place for tender are determined by the agreement and this Article, and in particular

(a) tender must be at a reasonable hour, and if it is of goods they must be kept available for the period reasonably necessary to enable the buyer to take possession; but

(b) unless otherwise agreed the buyer must furnish facilities reasonably suited to the receipt of the goods.

(2) Where the case is within the next section respecting shipment tender requires that the seller comply with its provisions.

(3) Where the seller is required to deliver at a particular destination tender requires that he comply with subsection (1) and also in any appropriate case tender documents as described in subsections (4) and (5) of this section.

(4) Where goods are in the possession of a bailee and are to be delivered without being moved

(a) tender requires that the seller either tender a negotiable document of title covering such goods or procure acknowledgment by the bailee of the buyer's right to possession of the goods; but

(b) tender to the buyer of a non-negotiable document of title or of a written direction to the bailee to deliver is sufficient tender unless the buyer seasonably objects, and receipt by the bailee of notification of the buyer's rights fixes those rights as against the bailee and all third persons; but risk of loss of the goods and of any failure by the bailee to honor the non-negotiable document of title or to obey the direction remains on the seller until the buyer has had a reasonable time to present the document or direction, and a refusal by the bailee to honor the document or to obey the direction defeats the tender.

(5) Where the contract requires the seller to deliver documents

(a) he must tender all such documents in correct form, except as provided in this Article with respect to bills of lading in a set (subsection (2) of Section 2–323); and

(b) tender through customary banking channels is sufficient and dishonor of a draft accompanying the documents constitutes non-acceptance or rejection.

§ 2–504. Shipment by Seller.

Where the seller is required or authorized to send the goods to the buyer and the contract does not require him to deliver them at a particular destination, then unless otherwise agreed he must

(a) put the goods in the possession of such a carrier and make such a contract for their transportation as may be reasonable having regard to the nature of the goods and other circumstances of the case; and

(b) obtain and promptly deliver or tender in due form any document necessary to enable the buyer to obtain possession of the goods or otherwise required by the agreement or by usage of trade; and

(c) promptly notify the buyer of the shipment.

Failure to notify the buyer under paragraph (c) or to make a proper contract under paragraph (a) is a ground for rejection only if material delay or loss ensues.

§ 2–505. Seller's Shipment under Reservation.

(1) Where the seller has identified goods to the contract by or before shipment:

(a) his procurement of a negotiable bill of lading to his own order or otherwise reserves in him a security interest in the goods. His procurement of the bill to the order of a financing agency or of the buyer indicates in addition only the seller's expectation of transferring that interest to the person named.

(b) a non-negotiable bill of lading to himself or his nominee reserves possession of the goods as security but except in a case of conditional delivery (subsection (2) of Section 2–507) a non-negotiable bill of lading naming the buyer as consignee reserves no security interest even though the seller retains possession of the bill of lading.

(2) When shipment by the seller with reservation of a security interest is in violation of the contract for sale it constitutes an improper contract for transportation within the preceding section but impairs neither the rights given to the buyer by shipment and identification of the goods to the contract nor the seller's powers as a holder of a negotiable document.

§ 2–506. Rights of Financing Agency.

(1) A financing agency by paying or purchasing for value a draft which relates to a shipment of goods acquires to the extent of the payment or purchase and in addition to its own rights under the draft and any document of title securing it any rights of the shipper in the goods including the right to stop delivery and the shipper's right to have the draft honored by the buyer.

(2) The right to reimbursement of a financing agency which has in good faith honored or purchased the draft under commitment to or authority from the buyer is not impaired by subsequent discovery of defects with reference to any relevant document which was apparently regular on its face.

§ 2–507. Effect of Seller's Tender; Delivery on Condition.

(1) Tender of delivery is a condition to the buyer's duty to accept the goods and, unless otherwise agreed, to his duty to pay for them. Tender entitles the seller to acceptance of the goods and to payment according to the contract.

(2) Where payment is due and demanded on the delivery to the buyer of goods or documents of title, his right as against the seller to retain or dispose of them is conditional upon his making the payment due.

§ 2–508. Cure by Seller of Improper Tender or Delivery; Replacement.

(1) Where any tender or delivery by the seller is rejected because non-conforming and the time for performance has not yet expired, the seller may seasonably notify the buyer of his intention to cure and may then within the contract time make a conforming delivery.

(2) Where the buyer rejects a non-conforming tender which the seller had reasonable grounds to believe would be acceptable with or without money allowance the seller may if he seasonably notifies the buyer have a further reasonable time to substitute a conforming tender.

§ 2–509. Risk of Loss in the Absence of Breach.

(1) Where the contract requires or authorizes the seller to ship the goods by carrier

(a) if it does not require him to deliver them at a particular destination, the risk of loss passes to the buyer when the goods are duly delivered to the carrier even though the shipment is under reservation (Section 2–505); but

(b) if it does require him to deliver them at a particular destination and the goods are there duly tendered while in the possession of the carrier, the risk of loss passes to the buyer when the goods are there duly so tendered as to enable the buyer to take delivery.

(2) Where the goods are held by a bailee to be delivered without being moved, the risk of loss passes to the buyer

(a) on his receipt of a negotiable document of title covering the goods; or

(b) on acknowledgment by the bailee of the buyer's right to possession of the goods; or

(c) after his receipt of a non-negotiable document of title or other written direction to deliver, as provided in subsection (4)(b) of Section 2–503.

(3) In any case not within subsection (1) or (2), the risk of loss passes to the buyer on his receipt of the goods if the seller is a merchant; otherwise the risk passes to the buyer on tender of delivery.

(4) The provisions of this section are subject to contrary agreement of the parties and to the provisions of this Article on sale on approval (Section 2–327) and on effect of breach on risk of loss (Section 2–510).

§ 2–510. Effect of Breach on Risk of Loss.

(1) Where a tender or delivery of goods so fails to conform to the contract as to give a right of rejection the risk of their loss remains on the seller until cure or acceptance.

(2) Where the buyer rightfully revokes acceptance he may to the extent of any deficiency in his effective insurance coverage treat the risk of loss as having rested on the seller from the beginning.

(3) Where the buyer as to conforming goods already identified to the contract for sale repudiates or is otherwise in breach before risk of their loss has passed to him, the seller may to the extent of any deficiency in his effective insurance coverage treat the risk of loss as resting on the buyer for a commercially reasonable time.

§ 2–511. Tender of Payment by Buyer; Payment by Check.

(1) Unless otherwise agreed tender of payment is a condition to the seller's duty to tender and complete any delivery.

(2) Tender of payment is sufficient when made by any means or in any manner current in the ordinary course of business unless the seller demands payment in legal tender and gives any extension of time reasonably necessary to procure it.

(3) Subject to the provisions of this Act on the effect of an instrument on an obligation (Section 3–310), payment by check is conditional and is defeated as between the parties by dishonor of the check on due presentment.

As amended in 1994.

§ 2–512. Payment by Buyer Before Inspection.

(1) Where the contract requires payment before inspection non-conformity of the goods does not excuse the buyer from so making payment unless

(a) the non-conformity appears without inspection; or

(b) despite tender of the required documents the circumstances would justify injunction against honor under this Act (Section 5–109(b)).

(2) Payment pursuant to subsection (1) does not constitute an acceptance of goods or impair the buyer's right to inspect or any of his remedies.

As amended in 1995.

§ 2–513. Buyer's Right to Inspection of Goods.

(1) Unless otherwise agreed and subject to subsection (3), where goods are tendered or delivered or identified to the contract for sale, the buyer has a right before payment or acceptance to inspect them at any reasonable place and time and in any reasonable manner. When the seller is required or authorized to send the goods to the buyer, the inspection may be after their arrival.

(2) Expenses of inspection must be borne by the buyer but may be recovered from the seller if the goods do not conform and are rejected.

(3) Unless otherwise agreed and subject to the provisions of this Article on C.I.F. contracts (subsection (3) of Section 2–321), the buyer is not entitled to inspect the goods before payment of the price when the contract provides

(a) for delivery "C.O.D." or on other like terms; or

(b) for payment against documents of title, except where such payment is due only after the goods are to become available for inspection.

(4) A place or method of inspection fixed by the parties is presumed to be exclusive but unless otherwise expressly agreed it does not postpone identification or shift the place for delivery or for passing the risk of loss. If compliance becomes impossible, inspection shall be as provided in this section unless the place or method fixed was clearly intended as an indispensable condition failure of which avoids the contract.

§ 2–514. When Documents Deliverable on Acceptance; When on Payment.

Unless otherwise agreed documents against which a draft is drawn are to be delivered to the drawee on acceptance of the draft if it is payable more than three days after presentment; otherwise, only on payment.

§ 2–515. Preserving Evidence of Goods in Dispute.

In furtherance of the adjustment of any claim or dispute

(a) either party on reasonable notification to the other and for the purpose of ascertaining the facts and preserving evidence has the right to inspect, test and sample the goods including such of them as may be in the possession or control of the other; and

(b) the parties may agree to a third party inspection or survey to determine the conformity or condition of the goods and may agree that the findings shall be binding upon them in any subsequent litigation or adjustment.

Part 6 Breach, Repudiation and Excuse

§ 2–601. Buyer's Rights on Improper Delivery.

Subject to the provisions of this Article on breach in installment contracts (Section 2–612) and unless otherwise agreed under the sections on contractual limitations of remedy (Sections 2–718 and 2–719), if the goods or the tender of delivery fail in any respect to conform to the contract, the buyer may

(a) reject the whole; or

(b) accept the whole; or

(c) accept any commercial unit or units and reject the rest.

§ 2–602. Manner and Effect of Rightful Rejection.

(1) Rejection of goods must be within a reasonable time after their delivery or tender. It is ineffective unless the buyer seasonably notifies the seller.

(2) Subject to the provisions of the two following sections on rejected goods (Sections 2–603 and 2–604),

(a) after rejection any exercise of ownership by the buyer with respect to any commercial unit is wrongful as against the seller; and

(b) if the buyer has before rejection taken physical possession of goods in which he does not have a security interest under the provisions of this Article (subsection (3) of Section 2–711), he is under a duty after rejection to hold them with reasonable care at the seller's disposition for a time sufficient to permit the seller to remove them; but

(c) the buyer has no further obligations with regard to goods rightfully rejected.

(3) The seller's rights with respect to goods wrongfully rejected are governed by the provisions of this Article on Seller's remedies in general (Section 2–703).

§ 2–603. Merchant Buyer's Duties as to Rightfully Rejected Goods.

(1) Subject to any security interest in the buyer (subsection (3) of Section 2–711), when the seller has no agent or place of business at the market of rejection a merchant buyer is under a duty after rejection of goods in his possession or control to follow any reasonable instructions received from the seller with respect to the goods and in the absence of such instructions to make reasonable efforts to sell them for the seller's account if they are perishable or threaten to decline in value speedily. Instructions are not reasonable if on demand indemnity for expenses is not forthcoming.

(2) When the buyer sells goods under subsection (1), he is entitled to reimbursement from the seller or out of the proceeds for reasonable expenses of caring for and selling them, and if the expenses include no selling commission then to such commission as is usual in the trade or if there is none to a reasonable sum not exceeding ten per cent on the gross proceeds.

(3) In complying with this section the buyer is held only to good faith and good faith conduct hereunder is neither acceptance nor conversion nor the basis of an action for damages.

§ 2–604. Buyer's Options as to Salvage of Rightfully Rejected Goods.

Subject to the provisions of the immediately preceding section on perishables if the seller gives no instructions within a reasonable time after notification of rejection the buyer may store the rejected goods for the seller's account or reship them to him or resell them for the seller's account with reimbursement as provided in the preceding section. Such action is not acceptance or conversion.

§ 2–605. Waiver of Buyer's Objections by Failure to Particularize.

(1) The buyer's failure to state in connection with rejection a particular defect which is ascertainable by reasonable inspection precludes him from relying on the unstated defect to justify rejection or to establish breach

(a) where the seller could have cured it if stated seasonably; or

(b) between merchants when the seller has after rejection made a request in writing for a full and final written statement of all defects on which the buyer proposes to rely.

(2) Payment against documents made without reservation of rights precludes recovery of the payment for defects apparent on the face of the documents.

§ 2–606. What Constitutes Acceptance of Goods.

(1) Acceptance of goods occurs when the buyer

(a) after a reasonable opportunity to inspect the goods signifies to the seller that the goods are conforming or that he will take or retain them in spite of their nonconformity; or

(b) fails to make an effective rejection (subsection (1) of Section 2–602), but such acceptance does not occur until the buyer has had a reasonable opportunity to inspect them; or

(c) does any act inconsistent with the seller's ownership; but if such act is wrongful as against the seller it is an acceptance only if ratified by him.

(2) Acceptance of a part of any commercial unit is acceptance of that entire unit.

§ 2–607. Effect of Acceptance; Notice of Breach; Burden of Establishing Breach After Acceptance; Notice of Claim or Litigation to Person Answerable Over.

(1) The buyer must pay at the contract rate for any goods accepted.

(2) Acceptance of goods by the buyer precludes rejection of the goods accepted and if made with knowledge of a non-conformity cannot be revoked because of it unless the acceptance was on the reasonable assumption that the non-conformity would be seasonably cured but acceptance does not of itself impair any other remedy provided by this Article for non-conformity.

(3) Where a tender has been accepted

(a) the buyer must within a reasonable time after he discovers or should have discovered any breach notify the seller of breach or be barred from any remedy; and

(b) if the claim is one for infringement or the like (subsection (3) of Section 2–312) and the buyer is sued as a result of such a breach he must so notify the seller within a reasonable time after he receives notice of the litigation or be barred from any remedy over for liability established by the litigation.

(4) The burden is on the buyer to establish any breach with respect to the goods accepted.

(5) Where the buyer is sued for breach of a warranty or other obligation for which his seller is answerable over

(a) he may give his seller written notice of the litigation. If the notice states that the seller may come in and defend and that if the seller does not do so he will be bound in any action against him by his buyer by any determination of fact common to the two litigations, then unless the seller after seasonable receipt of the notice does come in and defend he is so bound.

(b) if the claim is one for infringement or the like (subsection (3) of Section 2–312) the original seller may demand

in writing that his buyer turn over to him control of the litigation including settlement or else be barred from any remedy over and if he also agrees to bear all expense and to satisfy any adverse judgment, then unless the buyer after seasonable receipt of the demand does turn over control the buyer is so barred.

(6) The provisions of subsections (3), (4) and (5) apply to any obligation of a buyer to hold the seller harmless against infringement or the like (subsection (3) of Section 2–312).

§ 2–608. Revocation of Acceptance in Whole or in Part.

(1) The buyer may revoke his acceptance of a lot or commercial unit whose non-conformity substantially impairs its value to him if he has accepted it

(a) on the reasonable assumption that its nonconformity would be cured and it has not been seasonably cured; or

(b) without discovery of such non-conformity if his acceptance was reasonably induced either by the difficulty of discovery before acceptance or by the seller's assurances.

(2) Revocation of acceptance must occur within a reasonable time after the buyer discovers or should have discovered the ground for it and before any substantial change in condition of the goods which is not caused by their own defects. It is not effective until the buyer notifies the seller of it.

(3) A buyer who so revokes has the same rights and duties with regard to the goods involved as if he had rejected them.

§ 2–609. Right to Adequate Assurance of Performance.

(1) A contract for sale imposes an obligation on each party that the other's expectation of receiving due performance will not be impaired. When reasonable grounds for insecurity arise with respect to the performance of either party the other may in writing demand adequate assurance of due performance and until he receives such assurance may if commercially reasonable suspend any performance for which he has not already received the agreed return.

(2) Between merchants the reasonableness of grounds for insecurity and the adequacy of any assurance offered shall be determined according to commercial standards.

(3) Acceptance of any improper delivery or payment does not prejudice the party's right to demand adequate assurance of future performance.

(4) After receipt of a justified demand failure to provide within a reasonable time not exceeding thirty days such assurance of due performance as is adequate under the circumstances of the particular case is a repudiation of the contract.

§ 2–610. Anticipatory Repudiation.

When either party repudiates the contract with respect to a performance not yet due the loss of which will substantially impair the value of the contract to the other, the aggrieved party may

(a) for a commercially reasonable time await performance by the repudiating party; or

(b) resort to any remedy for breach (Section 2–703 or Section 2–711), even though he has notified the repudiating party that he would await the latter's performance and has urged retraction; and

(c) in either case suspend his own performance or proceed in accordance with the provisions of this Article on the seller's right to identify goods to the contract notwithstanding breach or to salvage unfinished goods (Section 2–704).

§ 2–611. Retraction of Anticipatory Repudiation.

(1) Until the repudiating party's next performance is due he can retract his repudiation unless the aggrieved party has since the repudiation cancelled or materially changed his position or otherwise indicated that he considers the repudiation final.

(2) Retraction may be by any method which clearly indicates to the aggrieved party that the repudiating party intends to perform, but must include any assurance justifiably demanded under the provisions of this Article (Section 2–609).

(3) Retraction reinstates the repudiating party's rights under the contract with due excuse and allowance to the aggrieved party for any delay occasioned by the repudiation.

§ 2–612. "Installment Contract"; Breach.

(1) An "installment contract" is one which requires or authorizes the delivery of goods in separate lots to be separately accepted, even though the contract contains a clause "each delivery is a separate contract" or its equivalent.

(2) The buyer may reject any installment which is non-conforming if the non-conformity substantially impairs the value of that installment and cannot be cured or if the non-conformity is a defect in the required documents; but if the non-conformity does not fall within subsection (3) and the seller gives adequate assurance of its cure the buyer must accept that installment.

(3) Whenever non-conformity or default with respect to one or more installments substantially impairs the value of the whole contract there is a breach of the whole. But the aggrieved party reinstates the contract if he accepts a non-conforming installment without seasonably notifying of cancellation or if he brings an action with respect only to past installments or demands performance as to future installments.

§ 2–613. Casualty to Identified Goods.

Where the contract requires for its performance goods identified when the contract is made, and the goods suffer casualty without fault of either party before the risk of loss passes to the buyer, or in a proper case under a "no arrival, no sale" term (Section 2–324) then

(a) if the loss is total the contract is avoided; and

(b) if the loss is partial or the goods have so deteriorated as no longer to conform to the contract the buyer may nevertheless demand inspection and at his option either treat the contract as voided or accept the goods with due allowance from the contract

price for the deterioration or the deficiency in quantity but without further right against the seller.

§ 2–614. Substituted Performance.

(1) Where without fault of either party the agreed berthing, loading, or unloading facilities fail or an agreed type of carrier becomes unavailable or the agreed manner of delivery otherwise becomes commercially impracticable but a commercially reasonable substitute is available, such substitute performance must be tendered and accepted.

(2) If the agreed means or manner of payment fails because of domestic or foreign governmental regulation, the seller may withhold or stop delivery unless the buyer provides a means or manner of payment which is commercially a substantial equivalent. If delivery has already been taken, payment by the means or in the manner provided by the regulation discharges the buyer's obligation unless the regulation is discriminatory, oppressive or predatory.

§ 2–615. Excuse by Failure of Presupposed Conditions.

Except so far as a seller may have assumed a greater obligation and subject to the preceding section on substituted performance:

(a) Delay in delivery or non-delivery in whole or in part by a seller who complies with paragraphs (b) and (c) is not a breach of his duty under a contract for sale if performance as agreed has been made impracticable by the occurrence of a contingency the nonoccurrence of which was a basic assumption on which the contract was made or by compliance in good faith with any applicable foreign or domestic governmental regulation or order whether or not it later proves to be invalid.

(b) Where the causes mentioned in paragraph (a) affect only a part of the seller's capacity to perform, he must allocate production and deliveries among his customers but may at his option include regular customers not then under contract as well as his own requirements for further manufacture. He may so allocate in any manner which is fair and reasonable.

(c) The seller must notify the buyer seasonably that there will be delay or non-delivery and, when allocation is required under paragraph (b), of the estimated quota thus made available for the buyer.

§ 2–616. Procedure on Notice Claiming Excuse.

(1) Where the buyer receives notification of a material or indefinite delay or an allocation justified under the preceding section he may by written notification to the seller as to any delivery concerned, and where the prospective deficiency substantially impairs the value of the whole contract under the provisions of this Article relating to breach of installment contracts (Section 2–612), then also as to the whole,

(a) terminate and thereby discharge any unexecuted portion of the contract; or

(b) modify the contract by agreeing to take his available quota in substitution.

(2) If after receipt of such notification from the seller the buyer fails so to modify the contract within a reasonable time not exceeding thirty days the contract lapses with respect to any deliveries affected.

(3) The provisions of this section may not be negated by agreement except in so far as the seller has assumed a greater obligation under the preceding section.

Part 7 Remedies

§ 2–701. Remedies for Breach of Collateral Contracts Not Impaired.

Remedies for breach of any obligation or promise collateral or ancillary to a contract for sale are not impaired by the provisions of this Article.

§ 2–702. Seller's Remedies on Discovery of Buyer's Insolvency.

(1) Where the seller discovers the buyer to be insolvent he may refuse delivery except for cash including payment for all goods theretofore delivered under the contract, and stop delivery under this Article (Section 2–705).

(2) Where the seller discovers that the buyer has received goods on credit while insolvent he may reclaim the goods upon demand made within ten days after the receipt, but if misrepresentation of solvency has been made to the particular seller in writing within three months before delivery the ten day limitation does not apply. Except as provided in this subsection the seller may not base a right to reclaim goods on the buyer's fraudulent or innocent misrepresentation of solvency or of intent to pay.

(3) The seller's right to reclaim under subsection (2) is subject to the rights of a buyer in ordinary course or other good faith purchaser under this Article (Section 2–403). Successful reclamation of goods excludes all other remedies with respect to them.

§ 2–703. Seller's Remedies in General.

Where the buyer wrongfully rejects or revokes acceptance of goods or fails to make a payment due on or before delivery or repudiates with respect to a part or the whole, then with respect to any goods directly affected and, if the breach is of the whole contract (Section 2–612), then also with respect to the whole undelivered balance, the aggrieved seller may

(a) withhold delivery of such goods;

(b) stop delivery by any bailee as hereafter provided (Section 2–705);

(c) proceed under the next section respecting goods still unidentified to the contract;

(d) resell and recover damages as hereafter provided (Section 2–706);

(e) recover damages for non-acceptance (Section 2–708) or in a proper case the price (Section 2–709);

(f) cancel.

§ 2–704. Seller's Right to Identify Goods to the Contract Notwithstanding Breach or to Salvage Unfinished Goods.

(1) An aggrieved seller under the preceding section may

(a) identify to the contract conforming goods not already identified if at the time he learned of the breach they are in his possession or control;

(b) treat as the subject of resale goods which have demonstrably been intended for the particular contract even though those goods are unfinished.

(2) Where the goods are unfinished an aggrieved seller may in the exercise of reasonable commercial judgment for the purposes of avoiding loss and of effective realization either complete the manufacture and wholly identify the goods to the contract or cease manufacture and resell for scrap or salvage value or proceed in any other reasonable manner.

§ 2–705. Seller's Stoppage of Delivery in Transit or Otherwise.

(1) The seller may stop delivery of goods in the possession of a carrier or other bailee when he discovers the buyer to be insolvent (Section 2–702) and may stop delivery of carload, truckload, planeload or larger shipments of express or freight when the buyer repudiates or fails to make a payment due before delivery or if for any other reason the seller has a right to withhold or reclaim the goods.

(2) As against such buyer the seller may stop delivery until

(a) receipt of the goods by the buyer; or

(b) acknowledgment to the buyer by any bailee of the goods except a carrier that the bailee holds the goods for the buyer; or

(c) such acknowledgment to the buyer by a carrier by reshipment or as warehouseman; or

(d) negotiation to the buyer of any negotiable document of title covering the goods.

(3) (a) To stop delivery the seller must so notify as to enable the bailee by reasonable diligence to prevent delivery of the goods.

(b) After such notification the bailee must hold and deliver the goods according to the directions of the seller but the seller is liable to the bailee for any ensuing charges or damages.

(c) If a negotiable document of title has been issued for goods the bailee is not obliged to obey a notification to stop until surrender of the document.

(d) A carrier who has issued a non-negotiable bill of lading is not obliged to obey a notification to stop received from a person other than the consignor.

§ 2–706. Seller's Resale Including Contract for Resale.

(1) Under the conditions stated in Section 2–703 on seller's remedies, the seller may resell the goods concerned or the undelivered balance thereof. Where the resale is made in good faith and in a commercially reasonable manner the seller may recover the difference between the resale price and the contract price together with any incidental damages allowed under the provisions of this Article (Section 2–710), but less expenses saved in consequence of the buyer's breach.

(2) Except as otherwise provided in subsection (3) or unless otherwise agreed resale may be at public or private sale including sale by way of one or more contracts to sell or of identification to an existing contract of the seller. Sale may be as a unit or in parcels and at any time and place and on any terms but every aspect of the sale including the method, manner, time, place and terms must be commercially reasonable. The resale must be reasonably identified as referring to the broken contract, but it is not necessary that the goods be in existence or that any or all of them have been identified to the contract before the breach.

(3) Where the resale is at private sale the seller must give the buyer reasonable notification of his intention to resell.

(4) Where the resale is at public sale

(a) only identified goods can be sold except where there is a recognized market for a public sale of futures in goods of the kind; and

(b) it must be made at a usual place or market for public sale if one is reasonably available and except in the case of goods which are perishable or threaten to decline in value speedily the seller must give the buyer reasonable notice of the time and place of the resale; and

(c) if the goods are not to be within the view of those attending the sale the notification of sale must state the place where the goods are located and provide for their reasonable inspection by prospective bidders; and

(d) the seller may buy.

(5) A purchaser who buys in good faith at a resale takes the goods free of any rights of the original buyer even though the seller fails to comply with one or more of the requirements of this section.

(6) The seller is not accountable to the buyer for any profit made on any resale. A person in the position of a seller (Section 2–707) or a buyer who has rightfully rejected or justifiably revoked acceptance must account for any excess over the amount of his security interest, as hereinafter defined (subsection (3) of Section 2–711).

§ 2–707. "Person in the Position of a Seller".

(1) A "person in the position of a seller" includes as against a principal an agent who has paid or become responsible for the price of goods on behalf of his principal or anyone who otherwise holds a security interest or other right in goods similar to that of a seller.

(2) A person in the position of a seller may as provided in this Article withhold or stop delivery (Section 2–705) and resell (Section 2–706) and recover incidental damages (Section 2–710).

§ 2–708. Seller's Damages for Non-Acceptance or Repudiation.

(1) Subject to subsection (2) and to the provisions of this Article with respect to proof of market price (Section 2–723), the measure of damages for non-acceptance or repudiation by the buyer is the difference between the market price at the time and place for tender and the unpaid contract price together with any incidental damages provided in this Article (Section 2–710), but less expenses saved in consequence of the buyer's breach.

(2) If the measure of damages provided in subsection (1) is inadequate to put the seller in as good a position as performance would have done then the measure of damages is the profit (including reasonable overhead) which the seller would have made from full performance by the buyer, together with any incidental damages provided in this Article (Section 2–710), due allowance for costs reasonably incurred and due credit for payments or proceeds of resale.

§ 2–709. Action for the Price.

(1) When the buyer fails to pay the price as it becomes due the seller may recover, together with any incidental damages under the next section, the price

(a) of goods accepted or of conforming goods lost or damaged within a commercially reasonable time after risk of their loss has passed to the buyer; and

(b) of goods identified to the contract if the seller is unable after reasonable effort to resell them at a reasonable price or the circumstances reasonably indicate that such effort will be unavailing.

(2) Where the seller sues for the price he must hold for the buyer any goods which have been identified to the contract and are still in his control except that if resale becomes possible he may resell them at any time prior to the collection of the judgment. The net proceeds of any such resale must be credited to the buyer and payment of the judgment entitles him to any goods not resold.

(3) After the buyer has wrongfully rejected or revoked acceptance of the goods or has failed to make a payment due or has repudiated (Section 2–610), a seller who is held not entitled to the price under this section shall nevertheless be awarded damages for non-acceptance under the preceding section.

§ 2–710. Seller's Incidental Damages.

Incidental damages to an aggrieved seller include any commercially reasonable charges, expenses or commissions incurred in stopping delivery, in the transportation, care and custody of goods after the buyer's breach, in connection with return or resale of the goods or otherwise resulting from the breach.

§ 2–711. Buyer's Remedies in General; Buyer's Security Interest in Rejected Goods.

(1) Where the seller fails to make delivery or repudiates or the buyer rightfully rejects or justifiably revokes acceptance then with respect to any goods involved, and with respect to the whole if the breach goes to the whole contract (Section 2–612), the buyer may cancel and whether or not he has done so may in addition to recovering so much of the price as has been paid

(a) "cover" and have damages under the next section as to all the goods affected whether or not they have been identified to the contract; or

(b) recover damages for non-delivery as provided in this Article (Section 2–713).

(2) Where the seller fails to deliver or repudiates the buyer may also

(a) if the goods have been identified recover them as provided in this Article (Section 2–502); or

(b) in a proper case obtain specific performance or replevy the goods as provided in this Article (Section 2–716).

(3) On rightful rejection or justifiable revocation of acceptance a buyer has a security interest in goods in his possession or control for any payments made on their price and any expenses reasonably incurred in their inspection, receipt, transportation, care and custody and may hold such goods and resell them in like manner as an aggrieved seller (Section 2–706).

§ 2–712. "Cover"; Buyer's Procurement of Substitute Goods.

(1) After a breach within the preceding section the buyer may "cover" by making in good faith and without unreasonable delay any reasonable purchase of or contract to purchase goods in substitution for those due from the seller.

(2) The buyer may recover from the seller as damages the difference between the cost of cover and the contract price together with any incidental or consequential damages as hereinafter defined (Section 2–715), but less expenses saved in consequence of the seller's breach.

(3) Failure of the buyer to effect cover within this section does not bar him from any other remedy.

§ 2–713. Buyer's Damages for Non-Delivery or Repudiation.

(1) Subject to the provisions of this Article with respect to proof of market price (Section 2–723), the measure of damages for non-delivery or repudiation by the seller is the difference between the market price at the time when the buyer learned of the breach and the contract price together with any incidental and consequential damages provided in this Article (Section 2–715), but less expenses saved in consequence of the seller's breach.

(2) Market price is to be determined as of the place for tender or, in cases of rejection after arrival or revocation of acceptance, as of the place of arrival.

§ 2–714. Buyer's Damages for Breach in Regard to Accepted Goods.

(1) Where the buyer has accepted goods and given notification (subsection (3) of Section 2–607) he may recover as damages for any non-conformity of tender the loss resulting in the ordinary course of events from the seller's breach as determined in any manner which is reasonable.

(2) The measure of damages for breach of warranty is the difference at the time and place of acceptance between the value of the goods accepted and the value they would have had if they had been as warranted, unless special circumstances show proximate damages of a different amount.

(3) In a proper case any incidental and consequential damages under the next section may also be recovered.

§ 2–715. Buyer's Incidental and Consequential Damages.

(1) Incidental damages resulting from the seller's breach include expenses reasonably incurred in inspection, receipt, transportation and care and custody of goods rightfully rejected, any commercially reasonable charges, expenses or commissions in connection with effecting cover and any other reasonable expense incident to the delay or other breach.

(2) Consequential damages resulting from the seller's breach include

(a) any loss resulting from general or particular requirements and needs of which the seller at the time of contracting had reason to know and which could not reasonably be prevented by cover or otherwise; and

(b) injury to person or property proximately resulting from any breach of warranty.

§ 2–716. Buyer's Right to Specific Performance or Replevin.

(1) Specific performance may be decreed where the goods are unique or in other proper circumstances.

(2) The decree for specific performance may include such terms and conditions as to payment of the price, damages, or other relief as the court may deem just.

(3) The buyer has a right of replevin for goods identified to the contract if after reasonable effort he is unable to effect cover for such goods or the circumstances reasonably indicate that such effort will be unavailing or if the goods have been shipped under reservation and satisfaction of the security interest in them has been made or tendered. In the case of goods bought for personal, family, or household purposes, the buyer's right of replevin vests upon acquisition of a special property, even if the seller had not then repudiated or failed to deliver.

As amended in 1999.

§ 2–717. Deduction of Damages From the Price.

The buyer on notifying the seller of his intention to do so may deduct all or any part of the damages resulting from any breach of the contract from any part of the price still due under the same contract.

§ 2–718. Liquidation or Limitation of Damages; Deposits.

(1) Damages for breach by either party may be liquidated in the agreement but only at an amount which is reasonable in the light of the anticipated or actual harm caused by the breach, the difficulties of proof of loss, and the inconvenience or nonfeasibility of otherwise obtaining an adequate remedy. A term fixing unreasonably large liquidated damages is void as a penalty.

(2) Where the seller justifiably withholds delivery of goods because of the buyer's breach, the buyer is entitled to restitution of any amount by which the sum of his payments exceeds

(a) the amount to which the seller is entitled by virtue of terms liquidating the seller's damages in accordance with subsection (1), or

(b) in the absence of such terms, twenty per cent of the value of the total performance for which the buyer is obligated under the contract or $500, whichever is smaller.

(3) The buyer's right to restitution under subsection (2) is subject to offset to the extent that the seller establishes

(a) a right to recover damages under the provisions of this Article other than subsection (1), and

(b) the amount or value of any benefits received by the buyer directly or indirectly by reason of the contract.

(4) Where a seller has received payment in goods their reasonable value or the proceeds of their resale shall be treated as payments for the purposes of subsection (2); but if the seller has notice of the buyer's breach before reselling goods received in part performance, his resale is subject to the conditions laid down in this Article on resale by an aggrieved seller (Section 2–706).

§ 2–719. Contractual Modification or Limitation of Remedy.

(1) Subject to the provisions of subsections (2) and (3) of this section and of the preceding section on liquidation and limitation of damages,

(a) the agreement may provide for remedies in addition to or in substitution for those provided in this Article and may limit or alter the measure of damages recoverable under this Article, as by limiting the buyer's remedies to return of the goods and repayment of the price or to repair and replacement of nonconforming goods or parts; and

(b) resort to a remedy as provided is optional unless the remedy is expressly agreed to be exclusive, in which case it is the sole remedy.

(2) Where circumstances cause an exclusive or limited remedy to fail of its essential purpose, remedy may be had as provided in this Act.

(3) Consequential damages may be limited or excluded unless the limitation or exclusion is unconscionable. Limitation of consequential damages for injury to the person in the case of consumer goods is prima facie unconscionable but limitation of damages where the loss is commercial is not.

§ 2–720. Effect of "Cancellation" or "Rescission" on Claims for Antecedent Breach.

Unless the contrary intention clearly appears, expressions of "cancellation" or "rescission" of the contract or the like shall not be construed as a renunciation or discharge of any claim in damages for an antecedent breach.

§ 2–721. Remedies for Fraud.

Remedies for material misrepresentation or fraud include all remedies available under this Article for non-fraudulent breach. Neither rescission or a claim for rescission of the contract for sale nor rejection or return of the goods shall bar or be deemed inconsistent with a claim for damages or other remedy.

§ 2–722. Who Can Sue Third Parties for Injury to Goods.

Where a third party so deals with goods which have been identified to a contract for sale as to cause actionable injury to a party to that contract

(a) a right of action against the third party is in either party to the contract for sale who has title to or a security interest or a special property or an insurable interest in the goods; and if the goods have been destroyed or converted a right of action is also in the party who either bore the risk of loss under the contract for sale or has since the injury assumed that risk as against the other;

(b) if at the time of the injury the party plaintiff did not bear the risk of loss as against the other party to the contract for sale and there is no arrangement between them for disposition of the recovery, his suit or settlement is, subject to his own interest, as a fiduciary for the other party to the contract;

(c) either party may with the consent of the other sue for the benefit of whom it may concern.

§ 2–723. Proof of Market Price: Time and Place.

(1) If an action based on anticipatory repudiation comes to trial before the time for performance with respect to some or all of the goods, any damages based on market price (Section 2–708 or Section 2–713) shall be determined according to the price of such goods prevailing at the time when the aggrieved party learned of the repudiation.

(2) If evidence of a price prevailing at the times or places described in this Article is not readily available the price prevailing within any reasonable time before or after the time described or at any other place which in commercial judgment or under usage of trade would serve as a reasonable substitute for the one described may be used, making any proper allowance for the cost of transporting the goods to or from such other place.

(3) Evidence of a relevant price prevailing at a time or place other than the one described in this Article offered by one party is not admissible unless and until he has given the other party such notice as the court finds sufficient to prevent unfair surprise.

§ 2–724. Admissibility of Market Quotations.

Whenever the prevailing price or value of any goods regularly bought and sold in any established commodity market is in issue, reports in official publications or trade journals or in newspapers or periodicals of general circulation published as the reports of such market shall be admissible in evidence. The circumstances of the preparation of such a report may be shown to affect its weight but not its admissibility.

§ 2–725. Statute of Limitations in Contracts for Sale.

(1) An action for breach of any contract for sale must be commenced within four years after the cause of action has accrued. By the original agreement the parties may reduce the period of limitation to not less than one year but may not extend it.

(2) A cause of action accrues when the breach occurs, regardless of the aggrieved party's lack of knowledge of the breach. A breach of warranty occurs when tender of delivery is made, except that where a warranty explicitly extends to future performance of the goods and discovery of the breach must await the time of such performance the cause of action accrues when the breach is or should have been discovered.

(3) Where an action commenced within the time limited by subsection (1) is so terminated as to leave available a remedy by another action for the same breach such other action may be commenced after the expiration of the time limited and within six months after the termination of the first action unless the termination resulted from voluntary discontinuance or from dismissal for failure or neglect to prosecute.

(4) This section does not alter the law on tolling of the statute of limitations nor does it apply to causes of action which have accrued before this Act becomes effective.

Article 2 Amendments (Excerpts)[1]

Part 1 Short Title, General Construction and Subject Matter

* * * *

§ 2–103. Definitions and Index of Definitions.

(1) In this article unless the context otherwise requires

* * * *

(b) "Conspicuous", with reference to a term, means so written, displayed, or presented that a reasonable person against which it is to operate ought to have noticed it. A term in an electronic record intended to evoke a response by an electronic agent is conspicuous if it is presented in a form that would enable a reasonably configured electronic agent to take it into account or react to it without review of the record by an individual. Whether a term is "conspicuous" or not is a decision for the court. Conspicuous terms include the following:

(i) for a person:

(A) a heading in capitals equal to or greater in size than the surrounding text, or in contrasting type, font, or color to the surrounding text of the same or lesser size and;

1. Additions and new wording are underlined. What follows represents only selected changes made by the 2003 amendments. Although the National Conference of Commissioners on Uniform State Laws and the American Law Institute approved the amendments in May of 2003, as of this writing, they have not as yet been adopted by any state.

(B) language in the body of a record or display in larger type than the surrounding text, or in contrasting type, font, or color to the surrounding text of the same size, or set off from surrounding text of the same size by symbols or other marks that call attention to the language; and

(ii) for a person or an electronic agent, a term that is so placed in a record or display that the person or electronic agent cannot proceed without taking action with respect to the particular term.

(c) "Consumer" means an individual who buys or contracts to buy goods that, at the time of contracting, are intended by the individual to be used primarily for personal, family, or household purposes.

(d) "Consumer contract" means a contract between a merchant seller and a consumer.

* * * *

(j) "Good faith" means honesty in fact and the observance of reasonable commercial standards of fair dealing.

(k) "Goods" means all things that are movable at the time of identification to a contract for sale. The term includes future goods, specially manufactured goods, the unborn young of animals, growing crops, and other identified things attached to realty as described in Section 2–107. The term does not include information, the money in which the price is to be paid, investment securities under Article 8, the subject matter of foreign exchange transactions, and choses in action.

* * * *

(m) "Record" means information that is inscribed on a tangible medium or that is stored in an electronic or other medium and is retrievable in perceivable form.

(n) "Remedial promise" means a promise by the seller to repair or replace the goods or to refund all or part of the price upon the happening of a specified event.

* * * *

(p) "Sign" means, with present intent to authenticate or adopt a record,

(i) to execute or adopt a tangible symbol; or

(ii) to attach to or logically associate with the record an electronic sound, symbol, or process.

* * * *

Part 2 Form, Formation, Terms and Readjustment of Contract; Electronic Contracting

§ 2–201. Formal Requirements; Statute of Frauds.

(1) A contract for the sale of goods for the price of $5,000 or more is not enforceable by way of action or defense unless there is some record sufficient to indicate that a contract for sale has been made between the parties and signed by the party against which enforcement is sought or by the party's authorized agent or broker. A record is not insufficient because it omits or incorrectly states a term agreed upon but the contract is not enforceable under this subsection beyond the quantity of goods shown in the record.

(2) Between merchants if within a reasonable time a record in confirmation of the contract and sufficient against the sender is received and the party receiving it has reason to know its contents, it satisfies the requirements of subsection (1) against the recipient unless notice of objection to its contents is given in a record within 10 days after it is received.

(3) A contract which does not satisfy the requirements of subsection (1) but which is valid in other respects is enforceable

(a) if the goods are to be specially manufactured for the buyer and are not suitable for sale to others in the ordinary course of the seller's business and the seller, before notice of repudiation is received and under circumstances which reasonably indicate that the goods are for the buyer, has made either a substantial beginning of their manufacture or commitments for their procurement; or

(b) if the party against which enforcement is sought admits in the party's pleading, or in the party's testimony or otherwise under oath that a contract for sale was made, but the contract is not enforceable under this paragraph beyond the quantity of goods admitted; or

(c) with respect to goods for which payment has been made and accepted or which have been received and accepted (Sec. 2–606).

(4) A contract that is enforceable under this section is not rendered unenforceable merely because it is not capable of being performed within one year or any other applicable period after its making.

* * * *

§ 2–207. Terms of Contract; Effect of Confirmation.

Subject to Section 2–202, if (i) conduct by both parties recognizes the existence of a contract although their records do not otherwise establish a contract, (ii) a contract is formed by an offer and acceptance, or (iii) a contract formed in any manner is confirmed by a record that contains terms additional to or different from those in the contract being confirmed, the terms of the contract, are:

(a) terms that appear in the records of both parties;

(b) terms, whether in a record or not, to which both parties agree; and

(c) terms supplied or incorporated under any provision of this Act.

* * * *

Part 3 General Obligation and Construction of Contract

* * * *

§ 2–312. Warranty of Title and Against Infringement; Buyer's Obligation Against Infringement.

(1) Subject to subsection (3) there is in a contract for sale a warranty by the seller that

(a) the title conveyed shall be, good and its transfer rightful and shall not, unreasonably expose the buyer to litigation because of any colorable claim to or interest in the goods; and

(b) the goods shall be delivered free from any security interest or other lien or encumbrance of which the buyer at the time of contracting has no knowledge.

(2) Unless otherwise agreed a seller that is a merchant regularly dealing in goods of the kind warrants that the goods shall be delivered free of the rightful claim of any third person by way of infringement or the like but a buyer that furnishes specifications to the seller must hold the seller harmless against any such claim that arises out of compliance with the specifications.

(3) A warranty under this section may be disclaimed or modified only by specific language or by circumstances that give the buyer reason to know that the seller does not claim title, that the seller is purporting to sell only the right or title as the seller or a third person may have, or that the seller is selling subject to any claims of infringement or the like.

§ 2–313. Express Warranties by Affirmation, Promise, Description, Sample; Remedial Promise.

(1) In this section, "immediate buyer" means a buyer that enters into a contract with the seller.

* * * *

(4) Any remedial promise made by the seller to the immediate buyer creates an obligation that the promise will be performed upon the happening of the specified event.

§ 2–313A. Obligation to Remote Purchaser Created by Record Packaged with or Accompanying Goods.

(1) In this section:

(a) "Immediate buyer" means a buyer that enters into a contract with the seller.

(b) "Remote purchaser" means a person that buys or leases goods from an immediate buyer or other person in the normal chain of distribution.

(2) This section applies only to new goods and goods sold or leased as new goods in a transaction of purchase in the normal chain of distribution.

(3) If in a record packaged with or accompanying the goods the seller makes an affirmation of fact or promise that relates to the goods, provides a description that relates to the goods, or makes a remedial promise, and the seller reasonably expects the record to be, and the record is, furnished to the remote purchaser, the seller has an obligation to the remote purchaser that:

(a) the goods will conform to the affirmation of fact, promise or description unless a reasonable person in the position of the remote purchaser would not believe that the affirmation of fact, promise or description created an obligation; and

(b) the seller will perform the remedial promise.

(4) It is not necessary to the creation of an obligation under this section that the seller use formal words such as "warrant" or "guarantee" or that the seller have a specific intention to undertake an obligation, but an affirmation merely of the value of the goods or a statement purporting to be merely the seller's opinion or commendation of the goods does not create an obligation.

(5) The following rules apply to the remedies for breach of an obligation created under this section:

(a) The seller may modify or limit the remedies available to the remote purchaser if the modification or limitation is furnished to the remote purchaser no later than the time of purchase or if the modification or limitation is contained in the record that contains the affirmation of fact, promise or description.

(b) Subject to a modification or limitation of remedy, a seller in breach is liable for incidental or consequential damages under Section 2–715, but not for lost profits.

(c) The remote purchaser may recover as damages for breach of a seller's obligation arising under subsection (2) the loss resulting in the ordinary course of events as determined in any reasonable manner.

(5) An obligation that is not a remedial promise is breached if the goods did not conform to the affirmation of fact, promise or description creating the obligation when the goods left the seller's control.

§ 2–313B. Obligation to Remote Purchaser Created by Communication to the Public.

(1) In this section:

(a) "Immediate buyer" means a buyer that enters into a contract with the seller.

(b) "Remote purchaser" means a person that buys or leases goods from an immediate buyer or other person in the normal chain of distribution.

(2) This section applies only to new goods and goods sold or leased as new goods in a transaction of purchase in the normal chain of distribution.

(3) If in an advertisement or a similar communication to the public a seller makes an affirmation of fact or promise that relates to the goods, provides a description that relates to the goods, or makes a remedial promise, and the remote purchaser enters into a transaction of purchase with knowledge of and with the expectation that the goods will conform to the affirmation of fact, promise, or description, or that the seller will perform the remedial promise, the seller has an obligation to the remote purchaser that:

(a) the goods will conform to the affirmation of fact, promise or description unless a reasonable person in the position of the remote purchaser would not believe that the affirmation of fact, promise or description created an obligation; and

(b) the seller will perform the remedial promise.

(4) It is not necessary to the creation of an obligation under this section that the seller use formal words such as "warrant" or "guarantee" or that the seller have a specific intention to undertake an obligation, but an affirmation merely of the value of the goods or a statement purporting to be merely the seller's opinion or commendation of the goods does not create an obligation.

(5) The following rules apply to the remedies for breach of an obligation created under this section:

(a) The seller may modify or limit the remedies available to the remote purchaser if the modification or limitation is furnished to the remote purchaser no later than the time of purchase. The modification or limitation may be furnished as part of the communication that contains the affirmation of fact, promise or description.

(b) Subject to a modification or limitation of remedy, a seller in breach is liable for incidental or consequential damages under Section 2–715, but not for lost profits.

(c) The remote purchaser may recover as damages for breach of a seller's obligation arising under subsection (2) the loss resulting in the ordinary course of events as determined in any reasonable manner.

(6) An obligation that is not a remedial promise is breached if the goods did not conform to the affirmation of fact, promise or description creating the obligation when the goods left the seller's control.

* * * *

§ 2–316. Exclusion or Modification of Warranties.

* * * *

(2) Subject to subsection (3), to exclude or modify the implied warranty of merchantability or any part of it in a consumer contract the language must be in a record, be conspicuous, and state "The seller undertakes no responsibility for the quality of the goods except as otherwise provided in this contract," and in any other contract the language must mention merchantability and in case of a record must be conspicuous. Subject to subsection (3), to exclude or modify the implied warranty of fitness the exclusion must be in a record and be conspicuous. Language to exclude all implied warranties of fitness in a consumer contract must state "The seller assumes no responsibility that the goods will be fit for any particular purpose for which you may be buying these goods, except as otherwise provided in the contract," and in any other contract the language is sufficient if it states, for example, that "There are no warranties that extend beyond the description on the face hereof." Language that satisfies the requirements of this subsection for the exclusion and modification of a warranty in a consumer contract also satisfies the requirements for any other contract.

(3) Notwithstanding subsection (2):

(a) unless the circumstances indicate otherwise, all implied warranties are excluded by expressions like "as is", "with all faults" or other language which in common understanding calls the buyer's attention to the exclusion of warranties, makes plain that there is no implied warranty, and in a consumer contract evidenced by a record is set forth conspicuously in the record; and

(b) when the buyer before entering into the contract has examined the goods or the sample or model as fully as desired or has refused to examine the goods after a demand by the seller there is no implied warranty with regard to defects which an examination ought in the circumstances to have revealed to the buyer; and

(c) an implied warranty can also be excluded or modified by course of dealing or course of performance or usage of trade.

* * * *

§ 2–318. Third Party Beneficiaries of Warranties and Obligations.

(1) In this section:

(a) "Immediate buyer" means a buyer that enters into a contract with the seller.

(b) "Remote purchaser" means a person that buys or leases goods from an immediate buyer or other person in the normal chain of distribution.

Alternative A to subsection (2)

(2) A seller's warranty to an immediate buyer, whether express or implied, a seller's remedial promise to an immediate buyer, or a seller's obligation to a remote purchaser under Section 2–313A or 2–313B extends to any natural person who is in the family or household of the immediate buyer or the remote purchaser or who is a guest in the home of either if it is reasonable to expect that the person may use, consume or be affected by the goods and who is injured in person by breach of the warranty, remedial promise or obligation. A seller may not exclude or limit the operation of this section.

Alternative B to subsection (2)

(2) A seller's warranty to an immediate buyer, whether express or implied, a seller's remedial promise to an immediate buyer, or a seller's obligation to a remote purchaser under Section 2–313A or 2–313B extends to any natural person who may reasonably be expected to use, consume or be affected by the goods and who is injured in person by breach of the warranty, remedial promise or obligation. A seller may not exclude or limit the operation of this section.

Alternative C to subsection (2)

(2) A seller's warranty to an immediate buyer, whether express or implied, a seller's remedial promise to an immediate buyer, or a seller's obligation to a remote purchaser under Section

2–313A or 2–313B extends to any person that may reasonably be expected to use, consume or be affected by the goods and that is injured by breach of the warranty, remedial promise or obligation. A seller may not exclude or limit the operation of this section with respect to injury to the person of an individual to whom the warranty, remedial promise or obligation extends.

* * * *

Part 5 Performance

* * * *

§ 2–502. Buyer's Right to Goods on Seller's Insolvency.

(1) Subject to subsections (2) and (3) and even though the goods have not been shipped a buyer that has paid a part or all of the price of goods in which the buyer has a special property under the provisions of the immediately preceding section may on making and keeping good a tender of any unpaid portion of their price recover them from the seller if:

(a) in the case of goods bought by a consumer, the seller repudiates or fails to deliver as required by the contract; or

(b) in all cases, the seller becomes insolvent within ten days after receipt of the first installment on their price.

(2) The buyer's right to recover the goods under subsection (1) vests upon acquisition of a special property, even if the seller had not then repudiated or failed to deliver.

(3) If the identification creating the special property has been made by the buyer, the buyer acquires the right to recover the goods only if they conform to the contract for sale.

* * * *

§ 2–508. Cure by Seller of Improper Tender or Delivery; Replacement.

(1) Where the buyer rejects goods or a tender of delivery under Section 2–601 or 2–612 or, except in a consumer contract, justifiably revokes acceptance under Section 2–608(1)(b) and the agreed time for performance has not expired, a seller that has performed in good faith, upon seasonable notice to the buyer and at the seller's own expense, may cure the breach of contract by making a conforming tender of delivery within the agreed time. The seller shall compensate the buyer for all of the buyer's reasonable expenses caused by the seller's breach of contract and subsequent cure.

(2) Where the buyer rejects goods or a tender of delivery under Section 2–601 or 2–612 or except in a consumer contract justifiably revokes acceptance under Section 2–608(1)(b) and the agreed time for performance has expired, a seller that has performed in good faith, upon seasonable notice to the buyer and at the seller's own expense, may cure the breach of contract, if the cure is appropriate and timely under the circumstances, by making a tender of conforming goods. The seller shall compensate the buyer for all of the buyer's reasonable expenses caused by the seller's breach of contract and subsequent cure.

§ 2–509. Risk of Loss in the Absence of Breach.

(1) Where the contract requires or authorizes the seller to ship the goods by carrier

(a) if it does not require the seller to deliver them at a particular destination, the risk of loss passes to the buyer when the goods are delivered to the carrier even though the shipment is under reservation (Section 2–505); but

(b) if it does require the seller to deliver them at a particular destination and the goods are there tendered while in the possession of the carrier, the risk of loss passes to the buyer when the goods are there so tendered as to enable the buyer to take delivery.

(2) Where the goods are held by a bailee to be delivered without being moved, the risk of loss passes to the buyer

(a) on the buyer's receipt of a negotiable document of title covering the goods; or

(b) on acknowledgment by the bailee to the buyer of the buyer's right to possession of the goods; or

(c) after the buyer's receipt of a non-negotiable document of title or other direction to deliver in a record, as provided in subsection (4)(b) of Section 2–503.

(3) In any case not within subsection (1) or (2), the risk of loss passes to the buyer on the buyer's receipt of the goods.

(4) The provisions of this section are subject to contrary agreement of the parties and to the provisions of this Article on sale on approval (Section 2–327) and on effect of breach on risk of loss (Section 2–510).

* * * *

§ 2–513. Buyer's Right to Inspection of Goods.

* * * *

(3) Unless otherwise agreed, the buyer is not entitled to inspect the goods before payment of the price when the contract provides

(a) for delivery on terms that under applicable course of performance, course of dealing, or usage of trade are interpreted to preclude inspection before payment; or

(b) for payment against documents of title, except where such payment is due only after the goods are to become available for inspection.

* * * *

Part 6 Breach, Repudiation and Excuse

* * * *

§ 2–605. Waiver of Buyer's Objections by Failure to Particularize.

(1) The buyer's failure to state in connection with rejection a particular defect or in connection with revocation of acceptance

a defect that justifies revocation precludes the buyer from relying on the unstated defect to justify rejection or revocation of acceptance if the defect is ascertainable by reasonable inspection

(a) where the seller had a right to cure the defect and could have cured it if stated seasonably; or

(b) between merchants when the seller has after rejection made a request in a record for a full and final statement in record form of all defects on which the buyer proposes to rely.

(2) A buyer's payment against documents tendered to the buyer made without reservation of rights precludes recovery of the payment for defects apparent on the face of the documents.

* * * *

§ 2–607. Effect of Acceptance; Notice of Breach; Burden of Establishing Breach After Acceptance; Notice of Claim or Litigation to Person Answerable Over.

* * * *

(3) Where a tender has been accepted

(a) the buyer must within a reasonable time after the buyer discovers or should have discovered any breach notify the seller. However, failure to give timely notice bars the buyer from a remedy only to the extent that the seller is prejudiced by the failure and

(b) if the claim is one for infringement or the like (subsection (3) of Section 2–312) and the buyer is sued as a result of such a breach the buyer must so notify the seller within a reasonable time after the buyer receives notice of the litigation or be barred from any remedy over for liability established by the litigation.

* * * *

§ 2–608. Revocation of Acceptance in Whole or in Part.

* * * *

(4) If a buyer uses the goods after a rightful rejection or justifiable revocation of acceptance, the following rules apply:

(a) Any use by the buyer which is unreasonable under the circumstances is wrongful as against the seller and is an acceptance only if ratified by the seller.

(b) Any use of the goods which is reasonable under the circumstances is not wrongful as against the seller and is not an acceptance, but in an appropriate case the buyer shall be obligated to the seller for the value of the use to the buyer.

* * * *

§ 2–612. "Installment Contract"; Breach.

* * * *

(2) The buyer may reject any installment which is non-conforming if the non-conformity substantially impairs the value of that installment to the buyer or if the non-conformity is a defect in the required documents; but if the non-conformity does not fall within subsection (3) and the seller gives adequate assurance of its cure the buyer must accept that installment.

(3) Whenever non-conformity or default with respect to one or more installments substantially impairs the value of the whole contract there is a breach of the whole. But the aggrieved party reinstates the contract if the party accepts a non-conforming installment without seasonably notifying of cancellation or if the party brings an action with respect only to past installments or demands performance as to future installments.

* * * *

Part 7 Remedies

§ 2–702. Seller's Remedies on Discovery of Buyer's Insolvency.

* * * *

(2) Where the seller discovers that the buyer has received goods on credit while insolvent the seller may reclaim the goods upon demand made within a reasonable time after the buyer's receipt of the goods. Except as provided in this subsection the seller may not base a right to reclaim goods on the buyer's fraudulent or innocent misrepresentation of solvency or of intent to pay.

* * * *

§ 2–703. Seller's Remedies in General.

(1) A breach of contract by the buyer includes the buyer's wrongful rejection or wrongful attempt to revoke acceptance of goods, wrongful failure to perform a contractual obligation, failure to make a payment when due, and repudiation.

(2) If the buyer is in breach of contract the seller, to the extent provided for by this Act or other law, may:

(a) withhold delivery of the goods:

(b) stop delivery of the goods under Section 2–705;

(c) proceed under Section 2–704 with respect to goods unidentified to the contract or unfinished;

(d) reclaim the goods under Section 2–507(2) or 2–702(2);

(e) require payment directly from the buyer under Section 2–325(c);

(f) cancel;

(g) resell and recover damages under Section 2–706;

(h) recover damages for nonacceptance or repudiation under Section 2–708(1);

(i) recover lost profits under Section 2–708(2);

(j) recover the price under Section 2–709;

(k) obtain specific performance under Section 2–716;

(l) recover liquidated damages under Section 2–718;

(m) in other cases, recover damages in any manner that is reasonable under the circumstances.

(3) If a buyer becomes insolvent, the seller may:

(a) withhold delivery under Section 2–702(1);

(b) stop delivery of the goods under Section 2–705;

(c) reclaim the goods under Section 2–702(2).

* * * *

§ 2–705. Seller's Stoppage of Delivery in Transit or Otherwise.

(1) The seller may stop delivery of goods in the possession of a carrier or other bailee when the seller discovers the buyer to be insolvent (Section 2–702) or when the buyer repudiates or fails to make a payment due before delivery or if for any other reason the seller has a right to withhold or reclaim the goods.

* * * *

§ 2–706. Seller's Resale Including Contract for Resale.

(1) In an appropriate case involving breach by the buyer, the seller may resell the goods concerned or the undelivered balance thereof. Where the resale is made in good faith and in a commercially reasonable manner the seller may recover the difference between the contract price and the resale price together with any incidental or consequential damages allowed under the provisions of this Article (Section 2–710), but less expenses saved in consequence of the buyer's breach.

* * * *

§ 2–708. Seller's Damages for Non-Acceptance or Repudiation.

(1) Subject to subsection (2) and to the provisions of this Article with respect to proof of market price (Section 2–723)

(a) the measure of damages for non-acceptance by the buyer is the difference between the contract price and the market price at the time and place for tender together with any incidental or consequential damages provided in this Article (Section 2–710), but less expenses saved in consequence of the buyer's breach; and

(b) the measure of damages for repudiation by the buyer is the difference between the contract price and the market price at the place for tender at the expiration of a commercially reasonable time after the seller learned of the repudiation, but no later than the time stated in paragraph (a), together with any incidental or consequential damages provided in this Article (Section 2–710), but less expenses saved in consequence of the buyer's breach.

(2) If the measure of damages provided in subsection (1) or in Section 2–706 is inadequate to put the seller in as good a position as performance would have done then the measure of damages is the profit (including reasonable overhead) which the seller would have made from full performance by the buyer, together with any incidental or consequential damages provided in this Article (Section 2–710).

§ 2–709. Action for the Price.

(1) When the buyer fails to pay the price as it becomes due the seller may recover, together with any incidental or consequential damages under the next section, the price

(a) of goods accepted or of conforming goods lost or damaged within a commercially reasonable time after risk of their loss has passed to the buyer; and

(b) of goods identified to the contract if the seller is unable after reasonable effort to resell them at a reasonable price or the circumstances reasonably indicate that such effort will be unavailing.

* * * *

§ 2–710. Seller's Incidental and Consequential Damages.

(1) Incidental damages to an aggrieved seller include any commercially reasonable charges, expenses or commissions incurred in stopping delivery, in the transportation, care and custody of goods after the buyer's breach, in connection with return or resale of the goods or otherwise resulting from the breach.

(2) Consequential damages resulting from the buyer's breach include any loss resulting from general or particular requirements and needs of which the buyer at the time of contracting had reason to know and which could not reasonably be prevented by resale or otherwise.

(3) In a consumer contract, a seller may not recover consequential damages from a consumer.

* * * *

§ 2–711. Buyer's Remedies in General; Buyer's Security Interest in Rejected Goods.

(1) A breach of contract by the seller includes the seller's wrongful failure to deliver or to perform a contractual obligation, making of a nonconforming tender of delivery or performance, and repudiation.

(2) If a seller is in breach of contract under subsection (1) the buyer, to the extent provided for by this Act or other law, may:

(a) in the case of rightful cancellation, rightful rejection or justifiable revocation of acceptance recover so much of the price as has been paid;

(b) deduct damages from any part of the price still due under Section 2–717;

(c) cancel;

(d) cover and have damages under Section 2–712 as to all goods affected whether or not they have been identified to the contract;

(e) recover damages for non-delivery or repudiation under Section 2–713;

(f) recover damages for breach with regard to accepted goods or breach with regard to a remedial promise under Section 2–714;

(g) recover identified goods under Section 2–502;

(h) obtain specific performance or obtain the goods by replevin or similar remedy under Section 7–716;

(i) recover liquidated damages under Section 2–718;

(j) in other cases, recover damages in any manner that is reasonable under the circumstances.

(3) On rightful rejection or justifiable revocation of acceptance a buyer has a security interest in goods in the buyer's possession or control for any payments made on their price and any expenses reasonably incurred in their inspection, receipt, transportation, care and custody and may hold such goods and resell them in like manner as an aggrieved seller (Section 2–706).

* * * *

§ 2–713. Buyer's Damages for Non-Delivery or Repudiation.

(1) Subject to the provisions of this Article with respect to proof of market price (Section 2–723), if the seller wrongfully fails to deliver or repudiates or the buyer rightfully rejects or justifiably revokes acceptance

(a) the measure of damages in the case of wrongful failure to deliver by the seller or rightful rejection or justifiable revocation of acceptance by the buyer is the difference between the market price at the time for tender under the contract and the contract price together with any incidental or consequential damages provided in this Article (Section 2–715), but less expenses saved in consequence of the seller's breach; and

(b) the measure of damages for repudiation by the seller is the difference between the market price at the expiration of a commercially reasonable time after the buyer learned of the repudiation, but no later than the time stated in paragraph (a), and the contract price together with any incidental or consequential damages provided in this Article (Section 2–715), less expenses saved in consequence of the seller's breach.

* * * *

§ 2–725. Statute of Limitations in Contracts for Sale.

(1) Except as otherwise provided in this section, an action for breach of any contract for sale must be commenced within the later of four years after the right of action has accrued under subsection (2) or (3) or one year after the breach was or should have been discovered, but no longer than five years after the right of action accrued. By the original agreement the parties may reduce the period of limitation to not less than one year but may not extend it. However, in a consumer contract, the period of limitation may not be reduced.

(2) Except as otherwise provided in subsection (3), the following rules apply:

(a) Except as otherwise provided in this subsection, a right of action for breach of a contract accrues when the breach occurs, even if the aggrieved party did not have knowledge of the breach.

(b) For breach of a contract by repudiation, a right of action accrues at the earlier of when the aggrieved party elects to treat the repudiation as a breach or when a commercially reasonable time for awaiting performance has expired.

(c) For breach of a remedial promise, a right of action accrues when the remedial promise is not performed when performance is due.

(d) In an action by a buyer against a person that is answerable over to the buyer for a claim asserted against the buyer, the buyer's right of action against the person answerable over accrues at the time the claim was originally asserted against the buyer.

(3) If a breach of a warranty arising under Section 2–312, 2–313(2), 2–314, or 2–315, or a breach of an obligation, other than a remedial promise, arising under Section 2–313A or 2–313B, is claimed the following rules apply:

(a) Except as otherwise provided in paragraph (c), a right of action for breach of a warranty arising under Section 2–313(2), 2–314 or 2–315 accrues when the seller has tendered delivery to the immediate buyer, as defined in Section 2–313, and has completed performance of any agreed installation or assembly of the goods.

(b) Except as otherwise provided in paragraph (c), a right of action for breach of an obligation other than a remedial promise arising under Section 2–313A or 2–313B accrues when the remote purchaser, as defined in Sections 2–313A and 2–313B, receives the goods.

(c) Where a warranty arising under Section 2–313(2) or an obligation, other than a remedial promise, arising under 2–313A or 2–313B explicitly extends to future performance of the goods and discovery of the breach must await the time for performance the right of action accrues when the immediate buyer as defined in Section 2–313 or the remote purchaser as defined in Sections 2–313A and 2–313B discovers or should have discovered the breach.

(d) A right of action for breach of warranty arising under Section 2–312 accrues when the aggrieved party discovers or should have discovered the breach. However, an action for breach of the warranty of non-infringement may not be commenced more than six years after tender of delivery of the goods to the aggrieved party.

* * * *

Article 2A LEASES

Part 1 General Provisions

§ 2A–101. Short Title.

This Article shall be known and may be cited as the Uniform Commercial Code—Leases.

§ 2A–102. Scope.

This Article applies to any transaction, regardless of form, that creates a lease.

§ 2A–103. Definitions and Index of Definitions.

(1) In this Article unless the context otherwise requires:

(a) "Buyer in ordinary course of business" means a person who in good faith and without knowledge that the sale to him [or her] is in violation of the ownership rights or security interest or leasehold interest of a third party in the goods buys in ordinary course from a person in the business of selling goods of that kind but does not include a pawnbroker. "Buying" may be for cash or by exchange of other property or on secured or unsecured credit and includes receiving goods or documents of title under a pre-existing contract for sale but does not include a transfer in bulk or as security for or in total or partial satisfaction of a money debt.

(b) "Cancellation" occurs when either party puts an end to the lease contract for default by the other party.

(c) "Commercial unit" means such a unit of goods as by commercial usage is a single whole for purposes of lease and division of which materially impairs its character or value on the market or in use. A commercial unit may be a single article, as a machine, or a set of articles, as a suite of furniture or a line of machinery, or a quantity, as a gross or carload, or any other unit treated in use or in the relevant market as a single whole.

(d) "Conforming" goods or performance under a lease contract means goods or performance that are in accordance with the obligations under the lease contract.

(e) "Consumer lease" means a lease that a lessor regularly engaged in the business of leasing or selling makes to a lessee who is an individual and who takes under the lease primarily for a personal, family, or household purpose [, if the total payments to be made under the lease contract, excluding payments for options to renew or buy, do not exceed $_____].

(f) "Fault" means wrongful act, omission, breach, or default.

(g) "Finance lease" means a lease with respect to which:

(i) the lessor does not select, manufacture or supply the goods;

(ii) the lessor acquires the goods or the right to possession and use of the goods in connection with the lease; and

(iii) one of the following occurs:

(A) the lessee receives a copy of the contract by which the lessor acquired the goods or the right to possession and use of the goods before signing the lease contract;

(B) the lessee's approval of the contract by which the lessor acquired the goods or the right to possession and use of the goods is a condition to effectiveness of the lease contract;

(C) the lessee, before signing the lease contract, receives an accurate and complete statement designating the promises and warranties, and any disclaimers of warranties, limitations or modifications of remedies, or liquidated damages, including those of a third party, such as the manufacturer of the goods, provided to the lessor by the person supplying the goods in connection with or as part of the contract by which the lessor acquired the goods or the right to possession and use of the goods; or

(D) if the lease is not a consumer lease, the lessor, before the lessee signs the lease contract, informs the lessee in writing (a) of the identity of the person supplying the goods to the lessor, unless the lessee has selected that person and directed the lessor to acquire the goods or the right to possession and use of the goods from that person, (b) that the lessee is entitled under this Article to any promises and warranties, including those of any third party, provided to the lessor by the person supplying the goods in connection with or as part of the contract by which the lessor acquired the goods or the right to possession and use of the goods, and (c) that the lessee may communicate with the person supplying the goods to the lessor and receive an accurate and complete statement of those promises and warranties, including any disclaimers and limitations of them or of remedies.

(h) "Goods" means all things that are movable at the time of identification to the lease contract, or are fixtures (Section 2A–309), but the term does not include money, documents, instruments, accounts, chattel paper, general intangibles, or minerals or the like, including oil and gas, before extraction. The term also includes the unborn young of animals.

(i) "Installment lease contract" means a lease contract that authorizes or requires the delivery of goods in separate lots to be separately accepted, even though the lease contract contains a clause "each delivery is a separate lease" or its equivalent.

(j) "Lease" means a transfer of the right to possession and use of goods for a term in return for consideration, but a sale, including a sale on approval or a sale or return, or retention or creation of a security interest is not a lease. Unless the context clearly indicates otherwise, the term includes a sublease.

(k) "Lease agreement" means the bargain, with respect to the lease, of the lessor and the lessee in fact as found in their language or by implication from other circumstances including course of dealing or usage of trade or course of performance as provided in this Article. Unless the context clearly indicates otherwise, the term includes a sublease agreement.

(*l*) "Lease contract" means the total legal obligation that results from the lease agreement as affected by this Article and any other applicable rules of law. Unless the context clearly indicates otherwise, the term includes a sublease contract.

(m) "Leasehold interest" means the interest of the lessor or the lessee under a lease contract.

(n) "Lessee" means a person who acquires the right to possession and use of goods under a lease. Unless the context clearly indicates otherwise, the term includes a sublessee.

(o) "Lessee in ordinary course of business" means a person who in good faith and without knowledge that the lease to him [or her] is in violation of the ownership rights or security interest or leasehold interest of a third party in the goods, leases in ordinary course from a person in the business of selling or leasing goods of that kind but does not include a pawnbroker. "Leasing" may be for cash or by exchange of other property or on secured or unsecured credit and includes receiving goods or documents of title under a pre-existing lease contract but does not include a transfer in bulk or as security for or in total or partial satisfaction of a money debt.

(p) "Lessor" means a person who transfers the right to possession and use of goods under a lease. Unless the context clearly indicates otherwise, the term includes a sublessor.

(q) "Lessor's residual interest" means the lessor's interest in the goods after expiration, termination, or cancellation of the lease contract.

(r) "Lien" means a charge against or interest in goods to secure payment of a debt or performance of an obligation, but the term does not include a security interest.

(s) "Lot" means a parcel or a single article that is the subject matter of a separate lease or delivery, whether or not it is sufficient to perform the lease contract.

(t) "Merchant lessee" means a lessee that is a merchant with respect to goods of the kind subject to the lease.

(u) "Present value" means the amount as of a date certain of one or more sums payable in the future, discounted to the date certain. The discount is determined by the interest rate specified by the parties if the rate was not manifestly unreasonable at the time the transaction was entered into; otherwise, the discount is determined by a commercially reasonable rate that takes into account the facts and circumstances of each case at the time the transaction was entered into.

(v) "Purchase" includes taking by sale, lease, mortgage, security interest, pledge, gift, or any other voluntary transaction creating an interest in goods.

(w) "Sublease" means a lease of goods the right to possession and use of which was acquired by the lessor as a lessee under an existing lease.

(x) "Supplier" means a person from whom a lessor buys or leases goods to be leased under a finance lease.

(y) "Supply contract" means a contract under which a lessor buys or leases goods to be leased.

(z) "Termination" occurs when either party pursuant to a power created by agreement or law puts an end to the lease contract otherwise than for default.

(2) Other definitions applying to this Article and the sections in which they appear are:

"Accessions". Section 2A–310(1).
"Construction mortgage". Section 2A–309(1)(d).
"Encumbrance". Section 2A–309(1)(e).
"Fixtures". Section 2A–309(1)(a).
"Fixture filing". Section 2A–309(1)(b).
"Purchase money lease". Section 2A–309(1)(c).

(3) The following definitions in other Articles apply to this Article:

"Accounts". Section 9–106.
"Between merchants". Section 2–104(3).
"Buyer". Section 2–103(1)(a).
"Chattel paper". Section 9–105(1)(b).
"Consumer goods". Section 9–109(1).
"Document". Section 9–105(1)(f).
"Entrusting". Section 2–403(3).
"General intangibles". Section 9–106.
"Good faith". Section 2–103(1)(b).
"Instrument". Section 9–105(1)(i).
"Merchant". Section 2–104(1).
"Mortgage". Section 9–105(1)(j).
"Pursuant to commitment". Section 9–105(1)(k).
"Receipt". Section 2–103(1)(c).
"Sale". Section 2–106(1).
"Sale on approval". Section 2–326.
"Sale or return". Section 2–326.
"Seller". Section 2–103(1)(d).

(4) In addition Article 1 contains general definitions and principles of construction and interpretation applicable throughout this Article.

As amended in 1990 and 1999.

§ 2A–104. Leases Subject to Other Law.

(1) A lease, although subject to this Article, is also subject to any applicable:

(a) certificate of title statute of this State: (list any certificate of title statutes covering automobiles, trailers, mobile homes, boats, farm tractors, and the like);

(b) certificate of title statute of another jurisdiction (Section 2A–105); or

(c) consumer protection statute of this State, or final consumer protection decision of a court of this State existing on the effective date of this Article.

(2) In case of conflict between this Article, other than Sections 2A–105, 2A–304(3), and 2A–305(3), and a statute or decision referred to in subsection (1), the statute or decision controls.

(3) Failure to comply with an applicable law has only the effect specified therein.

As amended in 1990.

§ 2A–105. Territorial Application of Article to Goods Covered by Certificate of Title.

Subject to the provisions of Sections 2A–304(3) and 2A–305(3), with respect to goods covered by a certificate of title issued under a statute of this State or of another jurisdiction, compliance and the effect of compliance or noncompliance with a certificate of title statute are governed by the law (including the conflict of laws rules) of the jurisdiction issuing the certificate until the earlier of (a) surrender of the certificate, or (b) four months after the goods are removed from that jurisdiction and thereafter until a new certificate of title is issued by another jurisdiction.

§ 2A–106. Limitation on Power of Parties to Consumer Lease to Choose Applicable Law and Judicial Forum.

(1) If the law chosen by the parties to a consumer lease is that of a jurisdiction other than a jurisdiction in which the lessee resides at the time the lease agreement becomes enforceable or within 30 days thereafter or in which the goods are to be used, the choice is not enforceable.

(2) If the judicial forum chosen by the parties to a consumer lease is a forum that would not otherwise have jurisdiction over the lessee, the choice is not enforceable.

§ 2A–107. Waiver or Renunciation of Claim or Right After Default.

Any claim or right arising out of an alleged default or breach of warranty may be discharged in whole or in part without consideration by a written waiver or renunciation signed and delivered by the aggrieved party.

§ 2A–108. Unconscionability.

(1) If the court as a matter of law finds a lease contract or any clause of a lease contract to have been unconscionable at the time it was made the court may refuse to enforce the lease contract, or it may enforce the remainder of the lease contract without the unconscionable clause, or it may so limit the application of any unconscionable clause as to avoid any unconscionable result.

(2) With respect to a consumer lease, if the court as a matter of law finds that a lease contract or any clause of a lease contract has been induced by unconscionable conduct or that unconscionable conduct has occurred in the collection of a claim arising from a lease contract, the court may grant appropriate relief.

(3) Before making a finding of unconscionability under subsection (1) or (2), the court, on its own motion or that of a party, shall afford the parties a reasonable opportunity to present evidence as to the setting, purpose, and effect of the lease contract or clause thereof, or of the conduct.

(4) In an action in which the lessee claims unconscionability with respect to a consumer lease:

(a) If the court finds unconscionability under subsection (1) or (2), the court shall award reasonable attorney's fees to the lessee.

(b) If the court does not find unconscionability and the lessee claiming unconscionability has brought or maintained an action he [or she] knew to be groundless, the court shall award reasonable attorney's fees to the party against whom the claim is made.

(c) In determining attorney's fees, the amount of the recovery on behalf of the claimant under subsections (1) and (2) is not controlling.

§ 2A–109. Option to Accelerate at Will.

(1) A term providing that one party or his [or her] successor in interest may accelerate payment or performance or require collateral or additional collateral "at will" or "when he [or she] deems himself [or herself] insecure" or in words of similar import must be construed to mean that he [or she] has power to do so only if he [or she] in good faith believes that the prospect of payment or performance is impaired.

(2) With respect to a consumer lease, the burden of establishing good faith under subsection (1) is on the party who exercised the power; otherwise the burden of establishing lack of good faith is on the party against whom the power has been exercised.

Part 2 Formation and Construction of Lease Contract

§ 2A–201. Statute of Frauds.

(1) A lease contract is not enforceable by way of action or defense unless:

(a) the total payments to be made under the lease contract, excluding payments for options to renew or buy, are less than $1,000; or

(b) there is a writing, signed by the party against whom enforcement is sought or by that party's authorized agent, sufficient to indicate that a lease contract has been made between the parties and to describe the goods leased and the lease term.

(2) Any description of leased goods or of the lease term is sufficient and satisfies subsection (1)(b), whether or not it is specific, if it reasonably identifies what is described.

(3) A writing is not insufficient because it omits or incorrectly states a term agreed upon, but the lease contract is not enforceable under subsection (1)(b) beyond the lease term and the quantity of goods shown in the writing.

(4) A lease contract that does not satisfy the requirements of subsection (1), but which is valid in other respects, is enforceable:

(a) if the goods are to be specially manufactured or obtained for the lessee and are not suitable for lease or sale to others in the ordinary course of the lessor's business, and the lessor, before notice of repudiation is received and under circumstances that reasonably indicate that the goods are for the lessee, has made either a substantial beginning of their manufacture or commitments for their procurement;

(b) if the party against whom enforcement is sought admits in that party's pleading, testimony or otherwise in court that a lease contract was made, but the lease contract is not enforceable under this provision beyond the quantity of goods admitted; or

(c) with respect to goods that have been received and accepted by the lessee.

(5) The lease term under a lease contract referred to in subsection (4) is:

(a) if there is a writing signed by the party against whom enforcement is sought or by that party's authorized agent specifying the lease term, the term so specified;

(b) if the party against whom enforcement is sought admits in that party's pleading, testimony, or otherwise in court a lease term, the term so admitted; or

(c) a reasonable lease term.

§ 2A–202. Final Written Expression: Parol or Extrinsic Evidence.

Terms with respect to which the confirmatory memoranda of the parties agree or which are otherwise set forth in a writing intended by the parties as a final expression of their agreement with respect to such terms as are included therein may not be contradicted by evidence of any prior agreement or of a contemporaneous oral agreement but may be explained or supplemented:

(a) by course of dealing or usage of trade or by course of performance; and

(b) by evidence of consistent additional terms unless the court finds the writing to have been intended also as a complete and exclusive statement of the terms of the agreement.

§ 2A–203. Seals Inoperative.

The affixing of a seal to a writing evidencing a lease contract or an offer to enter into a lease contract does not render the writing a sealed instrument and the law with respect to sealed instruments does not apply to the lease contract or offer.

§ 2A–204. Formation in General.

(1) A lease contract may be made in any manner sufficient to show agreement, including conduct by both parties which recognizes the existence of a lease contract.

(2) An agreement sufficient to constitute a lease contract may be found although the moment of its making is undetermined.

(3) Although one or more terms are left open, a lease contract does not fail for indefiniteness if the parties have intended to make a lease contract and there is a reasonably certain basis for giving an appropriate remedy.

§ 2A–205. Firm Offers.

An offer by a merchant to lease goods to or from another person in a signed writing that by its terms gives assurance it will be held open is not revocable, for lack of consideration, during the time stated or, if no time is stated, for a reasonable time, but in no event may the period of irrevocability exceed 3 months. Any such term of assurance on a form supplied by the offeree must be separately signed by the offeror.

§ 2A–206. Offer and Acceptance in Formation of Lease Contract.

(1) Unless otherwise unambiguously indicated by the language or circumstances, an offer to make a lease contract must be construed as inviting acceptance in any manner and by any medium reasonable in the circumstances.

(2) If the beginning of a requested performance is a reasonable mode of acceptance, an offeror who is not notified of acceptance within a reasonable time may treat the offer as having lapsed before acceptance.

§ 2A–207. Course of Performance or Practical Construction.

(1) If a lease contract involves repeated occasions for performance by either party with knowledge of the nature of the performance and opportunity for objection to it by the other, any course of performance accepted or acquiesced in without objection is relevant to determine the meaning of the lease agreement.

(2) The express terms of a lease agreement and any course of performance, as well as any course of dealing and usage of trade, must be construed whenever reasonable as consistent with each other; but if that construction is unreasonable, express terms control course of performance, course of performance controls both course of dealing and usage of trade, and course of dealing controls usage of trade.

(3) Subject to the provisions of Section 2A–208 on modification and waiver, course of performance is relevant to show a waiver or modification of any term inconsistent with the course of performance.

§ 2A–208. Modification, Rescission and Waiver.

(1) An agreement modifying a lease contract needs no consideration to be binding.

(2) A signed lease agreement that excludes modification or rescission except by a signed writing may not be otherwise modified or rescinded, but, except as between merchants, such a requirement on a form supplied by a merchant must be separately signed by the other party.

(3) Although an attempt at modification or rescission does not satisfy the requirements of subsection (2), it may operate as a waiver.

(4) A party who has made a waiver affecting an executory portion of a lease contract may retract the waiver by reasonable notification received by the other party that strict performance will be required of any term waived, unless the retraction would

be unjust in view of a material change of position in reliance on the waiver.

§ 2A–209. Lessee under Finance Lease as Beneficiary of Supply Contract.

(1) The benefit of the supplier's promises to the lessor under the supply contract and of all warranties, whether express or implied, including those of any third party provided in connection with or as part of the supply contract, extends to the lessee to the extent of the lessee's leasehold interest under a finance lease related to the supply contract, but is subject to the terms warranty and of the supply contract and all defenses or claims arising therefrom.

(2) The extension of the benefit of supplier's promises and of warranties to the lessee (Section 2A–209(1)) does not: (i) modify the rights and obligations of the parties to the supply contract, whether arising therefrom or otherwise, or (ii) impose any duty or liability under the supply contract on the lessee.

(3) Any modification or rescission of the supply contract by the supplier and the lessor is effective between the supplier and the lessee unless, before the modification or rescission, the supplier has received notice that the lessee has entered into a finance lease related to the supply contract. If the modification or rescission is effective between the supplier and the lessee, the lessor is deemed to have assumed, in addition to the obligations of the lessor to the lessee under the lease contract, promises of the supplier to the lessor and warranties that were so modified or rescinded as they existed and were available to the lessee before modification or rescission.

(4) In addition to the extension of the benefit of the supplier's promises and of warranties to the lessee under subsection (1), the lessee retains all rights that the lessee may have against the supplier which arise from an agreement between the lessee and the supplier or under other law.

As amended in 1990.

§ 2A–210. Express Warranties.

(1) Express warranties by the lessor are created as follows:

(a) Any affirmation of fact or promise made by the lessor to the lessee which relates to the goods and becomes part of the basis of the bargain creates an express warranty that the goods will conform to the affirmation or promise.

(b) Any description of the goods which is made part of the basis of the bargain creates an express warranty that the goods will conform to the description.

(c) Any sample or model that is made part of the basis of the bargain creates an express warranty that the whole of the goods will conform to the sample or model.

(2) It is not necessary to the creation of an express warranty that the lessor use formal words, such as "warrant" or "guarantee," or that the lessor have a specific intention to make a warranty, but an affirmation merely of the value of the goods or a statement purporting to be merely the lessor's opinion or commendation of the goods does not create a warranty.

§ 2A–211. Warranties Against Interference and Against Infringement; Lessee's Obligation Against Infringement.

(1) There is in a lease contract a warranty that for the lease term no person holds a claim to or interest in the goods that arose from an act or omission of the lessor, other than a claim by way of infringement or the like, which will interfere with the lessee's enjoyment of its leasehold interest.

(2) Except in a finance lease there is in a lease contract by a lessor who is a merchant regularly dealing in goods of the kind a warranty that the goods are delivered free of the rightful claim of any person by way of infringement or the like.

(3) A lessee who furnishes specifications to a lessor or a supplier shall hold the lessor and the supplier harmless against any claim by way of infringement or the like that arises out of compliance with the specifications.

§ 2A–212. Implied Warranty of Merchantability.

(1) Except in a finance lease, a warranty that the goods will be merchantable is implied in a lease contract if the lessor is a merchant with respect to goods of that kind.

(2) Goods to be merchantable must be at least such as

(a) pass without objection in the trade under the description in the lease agreement;

(b) in the case of fungible goods, are of fair average quality within the description;

(c) are fit for the ordinary purposes for which goods of that type are used;

(d) run, within the variation permitted by the lease agreement, of even kind, quality, and quantity within each unit and among all units involved;

(e) are adequately contained, packaged, and labeled as the lease agreement may require; and

(f) conform to any promises or affirmations of fact made on the container or label.

(3) Other implied warranties may arise from course of dealing or usage of trade.

§ 2A–213. Implied Warranty of Fitness for Particular Purpose.

Except in a finance of lease, if the lessor at the time the lease contract is made has reason to know of any particular purpose for which the goods are required and that the lessee is relying on the lessor's skill or judgment to select or furnish suitable goods, there is in the lease contract an implied warranty that the goods will be fit for that purpose.

§ 2A–214. Exclusion or Modification of Warranties.

(1) Words or conduct relevant to the creation of an express warranty and words or conduct tending to negate or limit a warranty must be construed wherever reasonable as consistent with each other; but, subject to the provisions of Section 2A–202 on parol or extrinsic evidence, negation or limitation is inoperative to the extent that the construction is unreasonable.

(2) Subject to subsection (3), to exclude or modify the implied warranty of merchantability or any part of it the language must mention "merchantability", be by a writing, and be conspicuous. Subject to subsection (3), to exclude or modify any implied warranty of fitness the exclusion must be by a writing and be conspicuous. Language to exclude all implied warranties of fitness is sufficient if it is in writing, is conspicuous and states, for example, "There is no warranty that the goods will be fit for a particular purpose".

(3) Notwithstanding subsection (2), but subject to subsection (4),

(a) unless the circumstances indicate otherwise, all implied warranties are excluded by expressions like "as is" or "with all faults" or by other language that in common understanding calls the lessee's attention to the exclusion of warranties and makes plain that there is no implied warranty, if in writing and conspicuous;

(b) if the lessee before entering into the lease contract has examined the goods or the sample or model as fully as desired or has refused to examine the goods, there is no implied warranty with regard to defects that an examination ought in the circumstances to have revealed; and

(c) an implied warranty may also be excluded or modified by course of dealing, course of performance, or usage of trade.

(4) To exclude or modify a warranty against interference or against infringement (Section 2A–211) or any part of it, the language must be specific, be by a writing, and be conspicuous, unless the circumstances, including course of performance, course of dealing, or usage of trade, give the lessee reason to know that the goods are being leased subject to a claim or interest of any person.

§ 2A–215. Cumulation and Conflict of Warranties Express or Implied.

Warranties, whether express or implied, must be construed as consistent with each other and as cumulative, but if that construction is unreasonable, the intention of the parties determines which warranty is dominant. In ascertaining that intention the following rules apply:

(a) Exact or technical specifications displace an inconsistent sample or model or general language of description.

(b) A sample from an existing bulk displaces inconsistent general language of description.

(c) Express warranties displace inconsistent implied warranties other than an implied warranty of fitness for a particular purpose.

§ 2A–216. Third-Party Beneficiaries of Express and Implied Warranties.

Alternative A

A warranty to or for the benefit of a lessee under this Article, whether express or implied, extends to any natural person who is in the family or household of the lessee or who is a guest in the lessee's home if it is reasonable to expect that such person may use, consume, or be affected by the goods and who is injured in person by breach of the warranty. This section does not displace principles of law and equity that extend a warranty to or for the benefit of a lessee to other persons. The operation of this section may not be excluded, modified, or limited, but an exclusion, modification, or limitation of the warranty, including any with respect to rights and remedies, effective against the lessee is also effective against any beneficiary designated under this section.

Alternative B

A warranty to or for the benefit of a lessee under this Article, whether express or implied, extends to any natural person who may reasonably be expected to use, consume, or be affected by the goods and who is injured in person by breach of the warranty. This section does not displace principles of law and equity that extend a warranty to or for the benefit of a lessee to other persons. The operation of this section may not be excluded, modified, or limited, but an exclusion, modification, or limitation of the warranty, including any with respect to rights and remedies, effective against the lessee is also effective against the beneficiary designated under this section.

Alternative C

A warranty to or for the benefit of a lessee under this Article, whether express or implied, extends to any person who may reasonably be expected to use, consume, or be affected by the goods and who is injured by breach of the warranty. The operation of this section may not be excluded, modified, or limited with respect to injury to the person of an individual to whom the warranty extends, but an exclusion, modification, or limitation of the warranty, including any with respect to rights and remedies, effective against the lessee is also effective against the beneficiary designated under this section.

§ 2A–217. Identification.

Identification of goods as goods to which a lease contract refers may be made at any time and in any manner explicitly agreed to by the parties. In the absence of explicit agreement, identification occurs:

(a) when the lease contract is made if the lease contract is for a lease of goods that are existing and identified;

(b) when the goods are shipped, marked, or otherwise designated by the lessor as goods to which the lease contract refers, if the lease contract is for a lease of goods that are not existing and identified; or

(c) when the young are conceived, if the lease contract is for a lease of unborn young of animals.

§ 2A–218. Insurance and Proceeds.

(1) A lessee obtains an insurable interest when existing goods are identified to the lease contract even though the goods identified are nonconforming and the lessee has an option to reject them.

(2) If a lessee has an insurable interest only by reason of the lessor's identification of the goods, the lessor, until default or insolvency or notification to the lessee that identification is final, may substitute other goods for those identified.

(3) Notwithstanding a lessee's insurable interest under subsections (1) and (2), the lessor retains an insurable interest until an option to buy has been exercised by the lessee and risk of loss has passed to the lessee.

(4) Nothing in this section impairs any insurable interest recognized under any other statute or rule of law.

(5) The parties by agreement may determine that one or more parties have an obligation to obtain and pay for insurance covering the goods and by agreement may determine the beneficiary of the proceeds of the insurance.

§ 2A–219. Risk of Loss.

(1) Except in the case of a finance lease, risk of loss is retained by the lessor and does not pass to the lessee. In the case of a finance lease, risk of loss passes to the lessee.

(2) Subject to the provisions of this Article on the effect of default on risk of loss (Section 2A–220), if risk of loss is to pass to the lessee and the time of passage is not stated, the following rules apply:

(a) If the lease contract requires or authorizes the goods to be shipped by carrier

(i) and it does not require delivery at a particular destination, the risk of loss passes to the lessee when the goods are duly delivered to the carrier; but

(ii) if it does require delivery at a particular destination and the goods are there duly tendered while in the possession of the carrier, the risk of loss passes to the lessee when the goods are there duly so tendered as to enable the lessee to take delivery.

(b) If the goods are held by a bailee to be delivered without being moved, the risk of loss passes to the lessee on acknowledgment by the bailee of the lessee's right to possession of the goods.

(c) In any case not within subsection (a) or (b), the risk of loss passes to the lessee on the lessee's receipt of the goods if the lessor, or, in the case of a finance lease, the supplier, is a merchant; otherwise the risk passes to the lessee on tender of delivery.

§ 2A–220. Effect of Default on Risk of Loss.

(1) Where risk of loss is to pass to the lessee and the time of passage is not stated:

(a) If a tender or delivery of goods so fails to conform to the lease contract as to give a right of rejection, the risk of their loss remains with the lessor, or, in the case of a finance lease, the supplier, until cure or acceptance.

(b) If the lessee rightfully revokes acceptance, he [or she], to the extent of any deficiency in his [or her] effective insurance coverage, may treat the risk of loss as having remained with the lessor from the beginning.

(2) Whether or not risk of loss is to pass to the lessee, if the lessee as to conforming goods already identified to a lease contract repudiates or is otherwise in default under the lease contract, the lessor, or, in the case of a finance lease, the supplier, to the extent of any deficiency in his [or her] effective insurance coverage may treat the risk of loss as resting on the lessee for a commercially reasonable time.

§ 2A–221. Casualty to Identified Goods.

If a lease contract requires goods identified when the lease contract is made, and the goods suffer casualty without fault of the lessee, the lessor or the supplier before delivery, or the goods suffer casualty before risk of loss passes to the lessee pursuant to the lease agreement or Section 2A–219, then:

(a) if the loss is total, the lease contract is avoided; and

(b) if the loss is partial or the goods have so deteriorated as to no longer conform to the lease contract, the lessee may nevertheless demand inspection and at his [or her] option either treat the lease contract as avoided or, except in a finance lease that is not a consumer lease, accept the goods with due allowance from the rent payable for the balance of the lease term for the deterioration or the deficiency in quantity but without further right against the lessor.

Part 3 Effect of Lease Contract

§ 2A–301. Enforceability of Lease Contract.

Except as otherwise provided in this Article, a lease contract is effective and enforceable according to its terms between the parties, against purchasers of the goods and against creditors of the parties.

§ 2A–302. Title to and Possession of Goods.

Except as otherwise provided in this Article, each provision of this Article applies whether the lessor or a third party has title to the goods, and whether the lessor, the lessee, or a third party has possession of the goods, notwithstanding any statute or rule of law that possession or the absence of possession is fraudulent.

§ 2A–303. Alienability of Party's Interest Under Lease Contract or of Lessor's Residual Interest in Goods; Delegation of Performance; Transfer of Rights.

(1) As used in this section, "creation of a security interest" includes the sale of a lease contract that is subject to Article 9, Secured Transactions, by reason of Section 9–109(a)(3).

(2) Except as provided in subsections (3) and Section 9–407, a provision in a lease agreement which (i) prohibits the voluntary or involuntary transfer, including a transfer by sale, sublease, creation or enforcement of a security interest, or attachment, levy, or other judicial process, of an interest of a party under the lease contract or of the lessor's residual interest in the goods, or

(ii) makes such a transfer an event of default, gives rise to the rights and remedies provided in subsection (4), but a transfer that is prohibited or is an event of default under the lease agreement is otherwise effective.

(3) A provision in a lease agreement which (i) prohibits a transfer of a right to damages for default with respect to the whole lease contract or of a right to payment arising out of the transferor's due performance of the transferor's entire obligation, or (ii) makes such a transfer an event of default, is not enforceable, and such a transfer is not a transfer that materially impairs the propsect of obtaining return performance by, materially changes the duty of, or materially increases the burden or risk imposed on, the other party to the lease contract within the purview of subsection (4).

(4) Subject to subsection (3) and Section 9–407:

(a) if a transfer is made which is made an event of default under a lease agreement, the party to the lease contract not making the transfer, unless that party waives the default or otherwise agrees, has the rights and remedies described in Section 2A–501(2);

(b) if paragraph (a) is not applicable and if a transfer is made that (i) is prohibited under a lease agreement or (ii) materially impairs the prospect of obtaining return performance by, materially changes the duty of, or materially increases the burden or risk imposed on, the other party to the lease contract, unless the party not making the transfer agrees at any time to the transfer in the lease contract or otherwise, then, except as limited by contract, (i) the transferor is liable to the party not making the transfer for damages caused by the transfer to the extent that the damages could not reasonably be prevented by the party not making the transfer and (ii) a court having jurisdiction may grant other appropriate relief, including cancellation of the lease contract or an injunction against the transfer.

(5) A transfer of "the lease" or of "all my rights under the lease", or a transfer in similar general terms, is a transfer of rights and, unless the language or the circumstances, as in a transfer for security, indicate the contrary, the transfer is a delegation of duties by the transferor to the transferee. Acceptance by the transferee constitutes a promise by the transferee to perform those duties. The promise is enforceable by either the transferor or the other party to the lease contract.

(6) Unless otherwise agreed by the lessor and the lessee, a delegation of performance does not relieve the transferor as against the other party of any duty to perform or of any liability for default.

(7) In a consumer lease, to prohibit the transfer of an interest of a party under the lease contract or to make a transfer an event of default, the language must be specific, by a writing, and conspicuous.

As amended in 1990 and 1999.

§ 2A–304. Subsequent Lease of Goods by Lessor.

(1) Subject to Section 2A–303, a subsequent lessee from a lessor of goods under an existing lease contract obtains, to the extent of the leasehold interest transferred, the leasehold interest in the goods that the lessor had or had power to transfer, and except as provided in subsection (2) and Section 2A–527(4), takes subject to the existing lease contract. A lessor with voidable title has power to transfer a good leasehold interest to a good faith subsequent lessee for value, but only to the extent set forth in the preceding sentence. If goods have been delivered under a transaction of purchase the lessor has that power even though:

(a) the lessor's transferor was deceived as to the identity of the lessor;

(b) the delivery was in exchange for a check which is later dishonored;

(c) it was agreed that the transaction was to be a "cash sale"; or

(d) the delivery was procured through fraud punishable as larcenous under the criminal law.

(2) A subsequent lessee in the ordinary course of business from a lessor who is a merchant dealing in goods of that kind to whom the goods were entrusted by the existing lessee of that lessor before the interest of the subsequent lessee became enforceable against that lessor obtains, to the extent of the leasehold interest transferred, all of that lessor's and the existing lessee's rights to the goods, and takes free of the existing lease contract.

(3) A subsequent lessee from the lessor of goods that are subject to an existing lease contract and are covered by a certificate of title issued under a statute of this State or of another jurisdiction takes no greater rights than those provided both by this section and by the certificate of title statute.

As amended in 1990.

§ 2A–305. Sale or Sublease of Goods by Lessee.

(1) Subject to the provisions of Section 2A–303, a buyer or sublessee from the lessee of goods under an existing lease contract obtains, to the extent of the interest transferred, the leasehold interest in the goods that the lessee had or had power to transfer, and except as provided in subsection (2) and Section 2A–511(4), takes subject to the existing lease contract. A lessee with a voidable leasehold interest has power to transfer a good leasehold interest to a good faith buyer for value or a good faith sublessee for value, but only to the extent set forth in the preceding sentence. When goods have been delivered under a transaction of lease the lessee has that power even though:

(a) the lessor was deceived as to the identity of the lessee;

(b) the delivery was in exchange for a check which is later dishonored; or

(c) the delivery was procured through fraud punishable as larcenous under the criminal law.

(2) A buyer in the ordinary course of business or a sublessee in the ordinary course of business from a lessee who is a merchant dealing in goods of that kind to whom the goods were entrusted by the lessor obtains, to the extent of the interest transferred, all

of the lessor's and lessee's rights to the goods, and takes free of the existing lease contract.

(3) A buyer or sublessee from the lessee of goods that are subject to an existing lease contract and are covered by a certificate of title issued under a statute of this State or of another jurisdiction takes no greater rights than those provided both by this section and by the certificate of title statute.

§ 2A–306. Priority of Certain Liens Arising by Operation of Law.

If a person in the ordinary course of his [or her] business furnishes services or materials with respect to goods subject to a lease contract, a lien upon those goods in the possession of that person given by statute or rule of law for those materials or services takes priority over any interest of the lessor or lessee under the lease contract or this Article unless the lien is created by statute and the statute provides otherwise or unless the lien is created by rule of law and the rule of law provides otherwise.

§ 2A–307. Priority of Liens Arising by Attachment or Levy on, Security Interests in, and Other Claims to Goods.

(1) Except as otherwise provided in Section 2A–306, a creditor of a lessee takes subject to the lease contract.

(2) Except as otherwise provided in subsection (3) and in Sections 2A–306 and 2A–308, a creditor of a lessor takes subject to the lease contract unless the creditor holds a lien that attached to the goods before the lease contract became enforceable.

(3) Except as otherwise provided in Sections 9–317, 9–321, and 9–323, a lessee takes a leasehold interest subject to a security interest held by a creditor of the lessor.

As amended in 1990 and 1999.

§ 2A–308. Special Rights of Creditors.

(1) A creditor of a lessor in possession of goods subject to a lease contract may treat the lease contract as void if as against the creditor retention of possession by the lessor is fraudulent under any statute or rule of law, but retention of possession in good faith and current course of trade by the lessor for a commercially reasonable time after the lease contract becomes enforceable is not fraudulent.

(2) Nothing in this Article impairs the rights of creditors of a lessor if the lease contract (a) becomes enforceable, not in current course of trade but in satisfaction of or as security for a pre-existing claim for money, security, or the like, and (b) is made under circumstances which under any statute or rule of law apart from this Article would constitute the transaction a fraudulent transfer or voidable preference.

(3) A creditor of a seller may treat a sale or an identification of goods to a contract for sale as void if as against the creditor retention of possession by the seller is fraudulent under any statute or rule of law, but retention of possession of the goods pursuant to a lease contract entered into by the seller as lessee and the buyer as lessor in connection with the sale or identification of the goods is not fraudulent if the buyer bought for value and in good faith.

§ 2A–309. Lessor's and Lessee's Rights When Goods Become Fixtures.

(1) In this section:

(a) goods are "fixtures" when they become so related to particular real estate that an interest in them arises under real estate law;

(b) a "fixture filing" is the filing, in the office where a mortgage on the real estate would be filed or recorded, of a financing statement covering goods that are or are to become fixtures and conforming to the requirements of Section 9–502(a) and (b);

(c) a lease is a "purchase money lease" unless the lessee has possession or use of the goods or the right to possession or use of the goods before the lease agreement is enforceable;

(d) a mortgage is a "construction mortgage" to the extent it secures an obligation incurred for the construction of an improvement on land including the acquisition cost of the land, if the recorded writing so indicates; and

(e) "encumbrance" includes real estate mortgages and other liens on real estate and all other rights in real estate that are not ownership interests.

(2) Under this Article a lease may be of goods that are fixtures or may continue in goods that become fixtures, but no lease exists under this Article of ordinary building materials incorporated into an improvement on land.

(3) This Article does not prevent creation of a lease of fixtures pursuant to real estate law.

(4) The perfected interest of a lessor of fixtures has priority over a conflicting interest of an encumbrancer or owner of the real estate if:

(a) the lease is a purchase money lease, the conflicting interest of the encumbrancer or owner arises before the goods become fixtures, the interest of the lessor is perfected by a fixture filing before the goods become fixtures or within ten days thereafter, and the lessee has an interest of record in the real estate or is in possession of the real estate; or

(b) the interest of the lessor is perfected by a fixture filing before the interest of the encumbrancer or owner is of record, the lessor's interest has priority over any conflicting interest of a predecessor in title of the encumbrancer or owner, and the lessee has an interest of record in the real estate or is in possession of the real estate.

(5) The interest of a lessor of fixtures, whether or not perfected, has priority over the conflicting interest of an encumbrancer or owner of the real estate if:

(a) the fixtures are readily removable factory or office machines, readily removable equipment that is not primarily used or leased for use in the operation of the real estate,

or readily removable replacements of domestic appliances that are goods subject to a consumer lease, and before the goods become fixtures the lease contract is enforceable; or

(b) the conflicting interest is a lien on the real estate obtained by legal or equitable proceedings after the lease contract is enforceable; or

(c) the encumbrancer or owner has consented in writing to the lease or has disclaimed an interest in the goods as fixtures; or

(d) the lessee has a right to remove the goods as against the encumbrancer or owner. If the lessee's right to remove terminates, the priority of the interest of the lessor continues for a reasonable time.

(6) Notwithstanding paragraph (4)(a) but otherwise subject to subsections (4) and (5), the interest of a lessor of fixtures, including the lessor's residual interest, is subordinate to the conflicting interest of an encumbrancer of the real estate under a construction mortgage recorded before the goods become fixtures if the goods become fixtures before the completion of the construction. To the extent given to refinance a construction mortgage, the conflicting interest of an encumbrancer of the real estate under a mortgage has this priority to the same extent as the encumbrancer of the real estate under the construction mortgage.

(7) In cases not within the preceding subsections, priority between the interest of a lessor of fixtures, including the lessor's residual interest, and the conflicting interest of an encumbrancer or owner of the real estate who is not the lessee is determined by the priority rules governing conflicting interests in real estate.

(8) If the interest of a lessor of fixtures, including the lessor's residual interest, has priority over all conflicting interests of all owners and encumbrancers of the real estate, the lessor or the lessee may (i) on default, expiration, termination, or cancellation of the lease agreement but subject to the agreement and this Article, or (ii) if necessary to enforce other rights and remedies of the lessor or lessee under this Article, remove the goods from the real estate, free and clear of all conflicting interests of all owners and encumbrancers of the real estate, but the lessor or lessee must reimburse any encumbrancer or owner of the real estate who is not the lessee and who has not otherwise agreed for the cost of repair of any physical injury, but not for any diminution in value of the real estate caused by the absence of the goods removed or by any necessity of replacing them. A person entitled to reimbursement may refuse permission to remove until the party seeking removal gives adequate security for the performance of this obligation.

(9) Even though the lease agreement does not create a security interest, the interest of a lessor of fixtures, including the lessor's residual interest, is perfected by filing a financing statement as a fixture filing for leased goods that are or are to become fixtures in accordance with the relevant provisions of the Article on Secured Transactions (Article 9).

As amended in 1990 and 1999.

§ 2A–310. Lessor's and Lessee's Rights When Goods Become Accessions.

(1) Goods are "accessions" when they are installed in or affixed to other goods.

(2) The interest of a lessor or a lessee under a lease contract entered into before the goods became accessions is superior to all interests in the whole except as stated in subsection (4).

(3) The interest of a lessor or a lessee under a lease contract entered into at the time or after the goods became accessions is superior to all subsequently acquired interests in the whole except as stated in subsection (4) but is subordinate to interests in the whole existing at the time the lease contract was made unless the holders of such interests in the whole have in writing consented to the lease or disclaimed an interest in the goods as part of the whole.

(4) The interest of a lessor or a lessee under a lease contract described in subsection (2) or (3) is subordinate to the interest of

(a) a buyer in the ordinary course of business or a lessee in the ordinary course of business of any interest in the whole acquired after the goods became accessions; or

(b) a creditor with a security interest in the whole perfected before the lease contract was made to the extent that the creditor makes subsequent advances without knowledge of the lease contract.

(5) When under subsections (2) or (3) and (4) a lessor or a lessee of accessions holds an interest that is superior to all interests in the whole, the lessor or the lessee may (a) on default, expiration, termination, or cancellation of the lease contract by the other party but subject to the provisions of the lease contract and this Article, or (b) if necessary to enforce his [or her] other rights and remedies under this Article, remove the goods from the whole, free and clear of all interests in the whole, but he [or she] must reimburse any holder of an interest in the whole who is not the lessee and who has not otherwise agreed for the cost of repair of any physical injury but not for any diminution in value of the whole caused by the absence of the goods removed or by any necessity for replacing them. A person entitled to reimbursement may refuse permission to remove until the party seeking removal gives adequate security for the performance of this obligation.

§ 2A–311. Priority Subject to Subordination.

Nothing in this Article prevents subordination by agreement by any person entitled to priority.

As added in 1990.

Part 4 Performance of Lease Contract: Repudiated, Substituted and Excused

§ 2A–401. Insecurity: Adequate Assurance of Performance.

(1) A lease contract imposes an obligation on each party that the other's expectation of receiving due performance will not be impaired.

(2) If reasonable grounds for insecurity arise with respect to the performance of either party, the insecure party may demand in writing adequate assurance of due performance. Until the insecure party receives that assurance, if commercially reasonable the insecure party may suspend any performance for which he [or she] has not already received the agreed return.

(3) A repudiation of the lease contract occurs if assurance of due performance adequate under the circumstances of the particular case is not provided to the insecure party within a reasonable time, not to exceed 30 days after receipt of a demand by the other party.

(4) Between merchants, the reasonableness of grounds for insecurity and the adequacy of any assurance offered must be determined according to commercial standards.

(5) Acceptance of any nonconforming delivery or payment does not prejudice the aggrieved party's right to demand adequate assurance of future performance.

§ 2A–402. Anticipatory Repudiation.

If either party repudiates a lease contract with respect to a performance not yet due under the lease contract, the loss of which performance will substantially impair the value of the lease contract to the other, the aggrieved party may:

(a) for a commercially reasonable time, await retraction of repudiation and performance by the repudiating party;

(b) make demand pursuant to Section 2A–401 and await assurance of future performance adequate under the circumstances of the particular case; or

(c) resort to any right or remedy upon default under the lease contract or this Article, even though the aggrieved party has notified the repudiating party that the aggrieved party would await the repudiating party's performance and assurance and has urged retraction. In addition, whether or not the aggrieved party is pursuing one of the foregoing remedies, the aggrieved party may suspend performance or, if the aggrieved party is the lessor, proceed in accordance with the provisions of this Article on the lessor's right to identify goods to the lease contract notwithstanding default or to salvage unfinished goods (Section 2A–524).

§ 2A–403. Retraction of Anticipatory Repudiation.

(1) Until the repudiating party's next performance is due, the repudiating party can retract the repudiation unless, since the repudiation, the aggrieved party has cancelled the lease contract or materially changed the aggrieved party's position or otherwise indicated that the aggrieved party considers the repudiation final.

(2) Retraction may be by any method that clearly indicates to the aggrieved party that the repudiating party intends to perform under the lease contract and includes any assurance demanded under Section 2A–401.

(3) Retraction reinstates a repudiating party's rights under a lease contract with due excuse and allowance to the aggrieved party for any delay occasioned by the repudiation.

§ 2A–404. Substituted Performance.

(1) If without fault of the lessee, the lessor and the supplier, the agreed berthing, loading, or unloading facilities fail or the agreed type of carrier becomes unavailable or the agreed manner of delivery otherwise becomes commercially impracticable, but a commercially reasonable substitute is available, the substitute performance must be tendered and accepted.

(2) If the agreed means or manner of payment fails because of domestic or foreign governmental regulation:

(a) the lessor may withhold or stop delivery or cause the supplier to withhold or stop delivery unless the lessee provides a means or manner of payment that is commercially a substantial equivalent; and

(b) if delivery has already been taken, payment by the means or in the manner provided by the regulation discharges the lessee's obligation unless the regulation is discriminatory, oppressive, or predatory.

§ 2A–405. Excused Performance.

Subject to Section 2A–404 on substituted performance, the following rules apply:

(a) Delay in delivery or nondelivery in whole or in part by a lessor or a supplier who complies with paragraphs (b) and (c) is not a default under the lease contract if performance as agreed has been made impracticable by the occurrence of a contingency the nonoccurrence of which was a basic assumption on which the lease contract was made or by compliance in good faith with any applicable foreign or domestic governmental regulation or order, whether or not the regulation or order later proves to be invalid.

(b) If the causes mentioned in paragraph (a) affect only part of the lessor's or the supplier's capacity to perform, he [or she] shall allocate production and deliveries among his [or her] customers but at his [or her] option may include regular customers not then under contract for sale or lease as well as his [or her] own requirements for further manufacture. He [or she] may so allocate in any manner that is fair and reasonable.

(c) The lessor seasonably shall notify the lessee and in the case of a finance lease the supplier seasonably shall notify the lessor and the lessee, if known, that there will be delay or nondelivery and, if allocation is required under paragraph (b), of the estimated quota thus made available for the lessee.

§ 2A–406. Procedure on Excused Performance.

(1) If the lessee receives notification of a material or indefinite delay or an allocation justified under Section 2A–405, the lessee may by written notification to the lessor as to any goods involved, and with respect to all of the goods if under an installment lease contract the value of the whole lease contract is substantially impaired (Section 2A–510):

(a) terminate the lease contract (Section 2A–505(2)); or

(b) except in a finance lease that is not a consumer lease, modify the lease contract by accepting the available quota in substitution, with due allowance from the rent payable

for the balance of the lease term for the deficiency but without further right against the lessor.

(2) If, after receipt of a notification from the lessor under Section 2A–405, the lessee fails so to modify the lease agreement within a reasonable time not exceeding 30 days, the lease contract lapses with respect to any deliveries affected.

§ 2A–407. Irrevocable Promises: Finance Leases.

(1) In the case of a finance lease that is not a consumer lease the lessee's promises under the lease contract become irrevocable and independent upon the lessee's acceptance of the goods.

(2) A promise that has become irrevocable and independent under subsection (1):

(a) is effective and enforceable between the parties, and by or against third parties including assignees of the parties, and

(b) is not subject to cancellation, termination, modification, repudiation, excuse, or substitution without the consent of the party to whom the promise runs.

(3) This section does not affect the validity under any other law of a covenant in any lease contract making the lessee's promises irrevocable and independent upon the lessee's acceptance of the goods.

As amended in 1990.

Part 5 Default

A. In General

§ 2A–501. Default: Procedure.

(1) Whether the lessor or the lessee is in default under a lease contract is determined by the lease agreement and this Article.

(2) If the lessor or the lessee is in default under the lease contract, the party seeking enforcement has rights and remedies as provided in this Article and, except as limited by this Article, as provided in the lease agreement.

(3) If the lessor or the lessee is in default under the lease contract, the party seeking enforcement may reduce the party's claim to judgment, or otherwise enforce the lease contract by self-help or any available judicial procedure or nonjudicial procedure, including administrative proceeding, arbitration, or the like, in accordance with this Article.

(4) Except as otherwise provided in Section 1–106(1) or this Article or the lease agreement, the rights and remedies referred to in subsections (2) and (3) are cumulative.

(5) If the lease agreement covers both real property and goods, the party seeking enforcement may proceed under this Part as to the goods, or under other applicable law as to both the real property and the goods in accordance with that party's rights and remedies in respect of the real property, in which case this Part does not apply.

As amended in 1990.

§ 2A–502. Notice After Default.

Except as otherwise provided in this Article or the lease agreement, the lessor or lessee in default under the lease contract is not entitled to notice of default or notice of enforcement from the other party to the lease agreement.

§ 2A–503. Modification or Impairment of Rights and Remedies.

(1) Except as otherwise provided in this Article, the lease agreement may include rights and remedies for default in addition to or in substitution for those provided in this Article and may limit or alter the measure of damages recoverable under this Article.

(2) Resort to a remedy provided under this Article or in the lease agreement is optional unless the remedy is expressly agreed to be exclusive. If circumstances cause an exclusive or limited remedy to fail of its essential purpose, or provision for an exclusive remedy is unconscionable, remedy may be had as provided in this Article.

(3) Consequential damages may be liquidated under Section 2A–504, or may otherwise be limited, altered, or excluded unless the limitation, alteration, or exclusion is unconscionable. Limitation, alteration, or exclusion of consequential damages for injury to the person in the case of consumer goods is prima facie unconscionable but limitation, alteration, or exclusion of damages where the loss is commercial is not prima facie unconscionable.

(4) Rights and remedies on default by the lessor or the lessee with respect to any obligation or promise collateral or ancillary to the lease contract are not impaired by this Article.

As amended in 1990.

§ 2A–504. Liquidation of Damages.

(1) Damages payable by either party for default, or any other act or omission, including indemnity for loss or diminution of anticipated tax benefits or loss or damage to lessor's residual interest, may be liquidated in the lease agreement but only at an amount or by a formula that is reasonable in light of the then anticipated harm caused by the default or other act or omission.

(2) If the lease agreement provides for liquidation of damages, and such provision does not comply with subsection (1), or such provision is an exclusive or limited remedy that circumstances cause to fail of its essential purpose, remedy may be had as provided in this Article.

(3) If the lessor justifiably withholds or stops delivery of goods because of the lessee's default or insolvency (Section 2A–525 or 2A–526), the lessee is entitled to restitution of any amount by which the sum of his [or her] payments exceeds:

(a) the amount to which the lessor is entitled by virtue of terms liquidating the lessor's damages in accordance with subsection (1); or

(b) in the absence of those terms, 20 percent of the then present value of the total rent the lessee was obligated to pay for the balance of the lease term, or, in the case of a consumer lease, the lesser of such amount or $500.

(4) A lessee's right to restitution under subsection (3) is subject to offset to the extent the lessor establishes:

(a) a right to recover damages under the provisions of this Article other than subsection (1); and

(b) the amount or value of any benefits received by the lessee directly or indirectly by reason of the lease contract.

§ 2A–505. Cancellation and Termination and Effect of Cancellation, Termination, Rescission, or Fraud on Rights and Remedies.

(1) On cancellation of the lease contract, all obligations that are still executory on both sides are discharged, but any right based on prior default or performance survives, and the cancelling party also retains any remedy for default of the whole lease contract or any unperformed balance.

(2) On termination of the lease contract, all obligations that are still executory on both sides are discharged but any right based on prior default or performance survives.

(3) Unless the contrary intention clearly appears, expressions of "cancellation," "rescission," or the like of the lease contract may not be construed as a renunciation or discharge of any claim in damages for an antecedent default.

(4) Rights and remedies for material misrepresentation or fraud include all rights and remedies available under this Article for default.

(5) Neither rescission nor a claim for rescission of the lease contract nor rejection or return of the goods may bar or be deemed inconsistent with a claim for damages or other right or remedy.

§ 2A–506. Statute of Limitations.

(1) An action for default under a lease contract, including breach of warranty or indemnity, must be commenced within 4 years after the cause of action accrued. By the original lease contract the parties may reduce the period of limitation to not less than one year.

(2) A cause of action for default accrues when the act or omission on which the default or breach of warranty is based is or should have been discovered by the aggrieved party, or when the default occurs, whichever is later. A cause of action for indemnity accrues when the act or omission on which the claim for indemnity is based is or should have been discovered by the indemnified party, whichever is later.

(3) If an action commenced within the time limited by subsection (1) is so terminated as to leave available a remedy by another action for the same default or breach of warranty or indemnity, the other action may be commenced after the expiration of the time limited and within 6 months after the termination of the first action unless the termination resulted from voluntary discontinuance or from dismissal for failure or neglect to prosecute.

(4) This section does not alter the law on tolling of the statute of limitations nor does it apply to causes of action that have accrued before this Article becomes effective.

§ 2A–507. Proof of Market Rent: Time and Place.

(1) Damages based on market rent (Section 2A–519 or 2A–528) are determined according to the rent for the use of the goods concerned for a lease term identical to the remaining lease term of the original lease agreement and prevailing at the times specified in Sections 2A–519 and 2A–528.

(2) If evidence of rent for the use of the goods concerned for a lease term identical to the remaining lease term of the original lease agreement and prevailing at the times or places described in this Article is not readily available, the rent prevailing within any reasonable time before or after the time described or at any other place or for a different lease term which in commercial judgment or under usage of trade would serve as a reasonable substitute for the one described may be used, making any proper allowance for the difference, including the cost of transporting the goods to or from the other place.

(3) Evidence of a relevant rent prevailing at a time or place or for a lease term other than the one described in this Article offered by one party is not admissible unless and until he [or she] has given the other party notice the court finds sufficient to prevent unfair surprise.

(4) If the prevailing rent or value of any goods regularly leased in any established market is in issue, reports in official publications or trade journals or in newspapers or periodicals of general circulation published as the reports of that market are admissible in evidence. The circumstances of the preparation of the report may be shown to affect its weight but not its admissibility.

As amended in 1990.

B. Default by Lessor

§ 2A–508. Lessee's Remedies.

(1) If a lessor fails to deliver the goods in conformity to the lease contract (Section 2A–509) or repudiates the lease contract (Section 2A–402), or a lessee rightfully rejects the goods (Section 2A–509) or justifiably revokes acceptance of the goods (Section 2A–517), then with respect to any goods involved, and with respect to all of the goods if under an installment lease contract the value of the whole lease contract is substantially impaired (Section 2A–510), the lessor is in default under the lease contract and the lessee may:

(a) cancel the lease contract (Section 2A–505(1));

(b) recover so much of the rent and security as has been paid and is just under the circumstances;

(c) cover and recover damages as to all goods affected whether or not they have been identified to the lease contract (Sections 2A–518 and 2A–520), or recover damages for nondelivery (Sections 2A–519 and 2A–520);

(d) exercise any other rights or pursue any other remedies provided in the lease contract.

(2) If a lessor fails to deliver the goods in conformity to the lease contract or repudiates the lease contract, the lessee may also:

(a) if the goods have been identified, recover them (Section 2A–522); or

(b) in a proper case, obtain specific performance or replevy the goods (Section 2A–521).

(3) If a lessor is otherwise in default under a lease contract, the lessee may exercise the rights and pursue the remedies provided in the lease contract, which may include a right to cancel the lease, and in Section 2A–519(3).

(4) If a lessor has breached a warranty, whether express or implied, the lessee may recover damages (Section 2A–519(4)).

(5) On rightful rejection or justifiable revocation of acceptance, a lessee has a security interest in goods in the lessee's possession or control for any rent and security that has been paid and any expenses reasonably incurred in their inspection, receipt, transportation, and care and custody and may hold those goods and dispose of them in good faith and in a commercially reasonable manner, subject to Section 2A–527(5).

(6) Subject to the provisions of Section 2A–407, a lessee, on notifying the lessor of the lessee's intention to do so, may deduct all or any part of the damages resulting from any default under the lease contract from any part of the rent still due under the same lease contract.

As amended in 1990.

§ 2A–509. Lessee's Rights on Improper Delivery; Rightful Rejection.

(1) Subject to the provisions of Section 2A–510 on default in installment lease contracts, if the goods or the tender or delivery fail in any respect to conform to the lease contract, the lessee may reject or accept the goods or accept any commercial unit or units and reject the rest of the goods.

(2) Rejection of goods is ineffective unless it is within a reasonable time after tender or delivery of the goods and the lessee seasonably notifies the lessor.

§ 2A–510. Installment Lease Contracts: Rejection and Default.

(1) Under an installment lease contract a lessee may reject any delivery that is nonconforming if the nonconformity substantially impairs the value of that delivery and cannot be cured or the nonconformity is a defect in the required documents; but if the nonconformity does not fall within subsection (2) and the lessor or the supplier gives adequate assurance of its cure, the lessee must accept that delivery.

(2) Whenever nonconformity or default with respect to one or more deliveries substantially impairs the value of the installment lease contract as a whole there is a default with respect to the whole. But, the aggrieved party reinstates the installment lease contract as a whole if the aggrieved party accepts a nonconforming delivery without seasonably notifying of cancellation or brings an action with respect only to past deliveries or demands performance as to future deliveries.

§ 2A–511. Merchant Lessee's Duties as to Rightfully Rejected Goods.

(1) Subject to any security interest of a lessee (Section 2A–508(5)), if a lessor or a supplier has no agent or place of business at the market of rejection, a merchant lessee, after rejection of goods in his [or her] possession or control, shall follow any reasonable instructions received from the lessor or the supplier with respect to the goods. In the absence of those instructions, a merchant lessee shall make reasonable efforts to sell, lease, or otherwise dispose of the goods for the lessor's account if they threaten to decline in value speedily. Instructions are not reasonable if on demand indemnity for expenses is not forthcoming.

(2) If a merchant lessee (subsection (1)) or any other lessee (Section 2A–512) disposes of goods, he [or she] is entitled to reimbursement either from the lessor or the supplier or out of the proceeds for reasonable expenses of caring for and disposing of the goods and, if the expenses include no disposition commission, to such commission as is usual in the trade, or if there is none, to a reasonable sum not exceeding 10 percent of the gross proceeds.

(3) In complying with this section or Section 2A–512, the lessee is held only to good faith. Good faith conduct hereunder is neither acceptance or conversion nor the basis of an action for damages.

(4) A purchaser who purchases in good faith from a lessee pursuant to this section or Section 2A–512 takes the goods free of any rights of the lessor and the supplier even though the lessee fails to comply with one or more of the requirements of this Article.

§ 2A–512. Lessee's Duties as to Rightfully Rejected Goods.

(1) Except as otherwise provided with respect to goods that threaten to decline in value speedily (Section 2A–511) and subject to any security interest of a lessee (Section 2A–508(5)):

(a) the lessee, after rejection of goods in the lessee's possession, shall hold them with reasonable care at the lessor's or the supplier's disposition for a reasonable time after the lessee's seasonable notification of rejection;

(b) if the lessor or the supplier gives no instructions within a reasonable time after notification of rejection, the lessee may store the rejected goods for the lessor's or the supplier's account or ship them to the lessor or the supplier or dispose of them for the lessor's or the supplier's account with reimbursement in the manner provided in Section 2A–511; but

(c) the lessee has no further obligations with regard to goods rightfully rejected.

(2) Action by the lessee pursuant to subsection (1) is not acceptance or conversion.

§ 2A–513. Cure by Lessor of Improper Tender or Delivery; Replacement.

(1) If any tender or delivery by the lessor or the supplier is rejected because nonconforming and the time for performance

has not yet expired, the lessor or the supplier may seasonably notify the lessee of the lessor's or the supplier's intention to cure and may then make a conforming delivery within the time provided in the lease contract.

(2) If the lessee rejects a nonconforming tender that the lessor or the supplier had reasonable grounds to believe would be acceptable with or without money allowance, the lessor or the supplier may have a further reasonable time to substitute a conforming tender if he [or she] seasonably notifies the lessee.

§ 2A–514. Waiver of Lessee's Objections.

(1) In rejecting goods, a lessee's failure to state a particular defect that is ascertainable by reasonable inspection precludes the lessee from relying on the defect to justify rejection or to establish default:

(a) if, stated seasonably, the lessor or the supplier could have cured it (Section 2A–513); or

(b) between merchants if the lessor or the supplier after rejection has made a request in writing for a full and final written statement of all defects on which the lessee proposes to rely.

(2) A lessee's failure to reserve rights when paying rent or other consideration against documents precludes recovery of the payment for defects apparent on the face of the documents.

§ 2A–515. Acceptance of Goods.

(1) Acceptance of goods occurs after the lessee has had a reasonable opportunity to inspect the goods and

(a) the lessee signifies or acts with respect to the goods in a manner that signifies to the lessor or the supplier that the goods are conforming or that the lessee will take or retain them in spite of their nonconformity; or

(b) the lessee fails to make an effective rejection of the goods (Section 2A–509(2)).

(2) Acceptance of a part of any commercial unit is acceptance of that entire unit.

§ 2A–516. Effect of Acceptance of Goods; Notice of Default; Burden of Establishing Default after Acceptance; Notice of Claim or Litigation to Person Answerable Over.

(1) A lessee must pay rent for any goods accepted in accordance with the lease contract, with due allowance for goods rightfully rejected or not delivered.

(2) A lessee's acceptance of goods precludes rejection of the goods accepted. In the case of a finance lease, if made with knowledge of a nonconformity, acceptance cannot be revoked because of it. In any other case, if made with knowledge of a nonconformity, acceptance cannot be revoked because of it unless the acceptance was on the reasonable assumption that the nonconformity would be seasonably cured. Acceptance does not of itself impair any other remedy provided by this Article or the lease agreement for nonconformity.

(3) If a tender has been accepted:

(a) within a reasonable time after the lessee discovers or should have discovered any default, the lessee shall notify the lessor and the supplier, if any, or be barred from any remedy against the party notified;

(b) except in the case of a consumer lease, within a reasonable time after the lessee receives notice of litigation for infringement or the like (Section 2A–211) the lessee shall notify the lessor or be barred from any remedy over for liability established by the litigation; and

(c) the burden is on the lessee to establish any default.

(4) If a lessee is sued for breach of a warranty or other obligation for which a lessor or a supplier is answerable over the following apply:

(a) The lessee may give the lessor or the supplier, or both, written notice of the litigation. If the notice states that the person notified may come in and defend and that if the person notified does not do so that person will be bound in any action against that person by the lessee by any determination of fact common to the two litigations, then unless the person notified after seasonable receipt of the notice does come in and defend that person is so bound.

(b) The lessor or the supplier may demand in writing that the lessee turn over control of the litigation including settlement if the claim is one for infringement or the like (Section 2A–211) or else be barred from any remedy over. If the demand states that the lessor or the supplier agrees to bear all expense and to satisfy any adverse judgment, then unless the lessee after seasonable receipt of the demand does turn over control the lessee is so barred.

(5) Subsections (3) and (4) apply to any obligation of a lessee to hold the lessor or the supplier harmless against infringement or the like (Section 2A–211).

As amended in 1990.

§ 2A–517. Revocation of Acceptance of Goods.

(1) A lessee may revoke acceptance of a lot or commercial unit whose nonconformity substantially impairs its value to the lessee if the lessee has accepted it:

(a) except in the case of a finance lease, on the reasonable assumption that its nonconformity would be cured and it has not been seasonably cured; or

(b) without discovery of the nonconformity if the lessee's acceptance was reasonably induced either by the lessor's assurances or, except in the case of a finance lease, by the difficulty of discovery before acceptance.

(2) Except in the case of a finance lease that is not a consumer lease, a lessee may revoke acceptance of a lot or commercial unit if the lessor defaults under the lease contract and the default substantially impairs the value of that lot or commercial unit to the lessee.

(3) If the lease agreement so provides, the lessee may revoke acceptance of a lot or commercial unit because of other defaults by the lessor.

(4) Revocation of acceptance must occur within a reasonable time after the lessee discovers or should have discovered the ground for it and before any substantial change in condition of the goods which is not caused by the nonconformity. Revocation is not effective until the lessee notifies the lessor.

(5) A lessee who so revokes has the same rights and duties with regard to the goods involved as if the lessee had rejected them.

As amended in 1990.

§ 2A–518. Cover; Substitute Goods.

(1) After a default by a lessor under the lease contract of the type described in Section 2A–508(1), or, if agreed, after other default by the lessor, the lessee may cover by making any purchase or lease of or contract to purchase or lease goods in substitution for those due from the lessor.

(2) Except as otherwise provided with respect to damages liquidated in the lease agreement (Section 2A–504) or otherwise determined pursuant to agreement of the parties (Sections 1–102(3) and 2A–503), if a lessee's cover is by lease agreement substantially similar to the original lease agreement and the new lease agreement is made in good faith and in a commercially reasonable manner, the lessee may recover from the lessor as damages (i) the present value, as of the date of the commencement of the term of the new lease agreement, of the rent under the new lease agreement applicable to that period of the new lease term which is comparable to the then remaining term of the original lease agreement minus the present value as of the same date of the total rent for the then remaining lease term of the original lease agreement, and (ii) any incidental or consequential damages, less expenses saved in consequence of the lessor's default.

(3) If a lessee's cover is by lease agreement that for any reason does not qualify for treatment under subsection (2), or is by purchase or otherwise, the lessee may recover from the lessor as if the lessee had elected not to cover and Section 2A–519 governs.

As amended in 1990.

§ 2A–519. Lessee's Damages for Non-Delivery, Repudiation, Default, and Breach of Warranty in Regard to Accepted Goods.

(1) Except as otherwise provided with respect to damages liquidated in the lease agreement (Section 2A–504) or otherwise determined pursuant to agreement of the parties (Sections 1–102(3) and 2A–503), if a lessee elects not to cover or a lessee elects to cover and the cover is by lease agreement that for any reason does not qualify for treatment under Section 2A–518(2), or is by purchase or otherwise, the measure of damages for non-delivery or repudiation by the lessor or for rejection or revocation of acceptance by the lessee is the present value, as of the date of the default, of the then market rent minus the present value as of the same date of the original rent, computed for the remaining lease term of the original lease agreement, together with incidental and consequential damages, less expenses saved in consequence of the lessor's default.

(2) Market rent is to be determined as of the place for tender or, in cases of rejection after arrival or revocation of acceptance, as of the place of arrival.

(3) Except as otherwise agreed, if the lessee has accepted goods and given notification (Section 2A–516(3)), the measure of damages for non-conforming tender or delivery or other default by a lessor is the loss resulting in the ordinary course of events from the lessor's default as determined in any manner that is reasonable together with incidental and consequential damages, less expenses saved in consequence of the lessor's default.

(4) Except as otherwise agreed, the measure of damages for breach of warranty is the present value at the time and place of acceptance of the difference between the value of the use of the goods accepted and the value if they had been as warranted for the lease term, unless special circumstances show proximate damages of a different amount, together with incidental and consequential damages, less expenses saved in consequence of the lessor's default or breach of warranty.

As amended in 1990.

§ 2A–520. Lessee's Incidental and Consequential Damages.

(1) Incidental damages resulting from a lessor's default include expenses reasonably incurred in inspection, receipt, transportation, and care and custody of goods rightfully rejected or goods the acceptance of which is justifiably revoked, any commercially reasonable charges, expenses or commissions in connection with effecting cover, and any other reasonable expense incident to the default.

(2) Consequential damages resulting from a lessor's default include:

(a) any loss resulting from general or particular requirements and needs of which the lessor at the time of contracting had reason to know and which could not reasonably be prevented by cover or otherwise; and

(b) injury to person or property proximately resulting from any breach of warranty.

§ 2A–521. Lessee's Right to Specific Performance or Replevin.

(1) Specific performance may be decreed if the goods are unique or in other proper circumstances.

(2) A decree for specific performance may include any terms and conditions as to payment of the rent, damages, or other relief that the court deems just.

(3) A lessee has a right of replevin, detinue, sequestration, claim and delivery, or the like for goods identified to the lease contract if after reasonable effort the lessee is unable to effect cover for those goods or the circumstances reasonably indicate that the effort will be unavailing.

§ 2A–522. Lessee's Right to Goods on Lessor's Insolvency.

(1) Subject to subsection (2) and even though the goods have not been shipped, a lessee who has paid a part or all of the rent and security for goods identified to a lease contract (Section 2A–217)

on making and keeping good a tender of any unpaid portion of the rent and security due under the lease contract may recover the goods identified from the lessor if the lessor becomes insolvent within 10 days after receipt of the first installment of rent and security.

(2) A lessee acquires the right to recover goods identified to a lease contract only if they conform to the lease contract.

C. Default by Lessee

§ 2A–523. Lessor's Remedies.

(1) If a lessee wrongfully rejects or revokes acceptance of goods or fails to make a payment when due or repudiates with respect to a part or the whole, then, with respect to any goods involved, and with respect to all of the goods if under an installment lease contract the value of the whole lease contract is substantially impaired (Section 2A–510), the lessee is in default under the lease contract and the lessor may:

(a) cancel the lease contract (Section 2A–505(1));

(b) proceed respecting goods not identified to the lease contract (Section 2A–524);

(c) withhold delivery of the goods and take possession of goods previously delivered (Section 2A–525);

(d) stop delivery of the goods by any bailee (Section 2A–526);

(e) dispose of the goods and recover damages (Section 2A–527), or retain the goods and recover damages (Section 2A–528), or in a proper case recover rent (Section 2A–529)

(f) exercise any other rights or pursue any other remedies provided in the lease contract.

(2) If a lessor does not fully exercise a right or obtain a remedy to which the lessor is entitled under subsection (1), the lessor may recover the loss resulting in the ordinary course of events from the lessee's default as determined in any reasonable manner, together with incidental damages, less expenses saved in consequence of the lessee's default.

(3) If a lessee is otherwise in default under a lease contract, the lessor may exercise the rights and pursue the remedies provided in the lease contract, which may include a right to cancel the lease. In addition, unless otherwise provided in the lease contract:

(a) if the default substantially impairs the value of the lease contract to the lessor, the lessor may exercise the rights and pursue the remedies provided in subsections (1) or (2); or

(b) if the default does not substantially impair the value of the lease contract to the lessor, the lessor may recover as provided in subsection (2).

As amended in 1990.

§ 2A–524. Lessor's Right to Identify Goods to Lease Contract.

(1) After default by the lessee under the lease contract of the type described in Section 2A–523(1) or 2A–523(3)(a) or, if agreed, after other default by the lessee, the lessor may:

(a) identify to the lease contract conforming goods not already identified if at the time the lessor learned of the default they were in the lessor's or the supplier's possession or control; and

(b) dispose of goods (Section 2A–527(1)) that demonstrably have been intended for the particular lease contract even though those goods are unfinished.

(2) If the goods are unfinished, in the exercise of reasonable commercial judgment for the purposes of avoiding loss and of effective realization, an aggrieved lessor or the supplier may either complete manufacture and wholly identify the goods to the lease contract or cease manufacture and lease, sell, or otherwise dispose of the goods for scrap or salvage value or proceed in any other reasonable manner.

As amended in 1990.

§ 2A–525. Lessor's Right to Possession of Goods.

(1) If a lessor discovers the lessee to be insolvent, the lessor may refuse to deliver the goods.

(2) After a default by the lessee under the lease contract of the type described in Section 2A–523(1) or 2A–523(3)(a) or, if agreed, after other default by the lessee, the lessor has the right to take possession of the goods. If the lease contract so provides, the lessor may require the lessee to assemble the goods and make them available to the lessor at a place to be designated by the lessor which is reasonably convenient to both parties. Without removal, the lessor may render unusable any goods employed in trade or business, and may dispose of goods on the lessee's premises (Section 2A–527).

(3) The lessor may proceed under subsection (2) without judicial process if that can be done without breach of the peace or the lessor may proceed by action.

As amended in 1990.

§ 2A–526. Lessor's Stoppage of Delivery in Transit or Otherwise.

(1) A lessor may stop delivery of goods in the possession of a carrier or other bailee if the lessor discovers the lessee to be insolvent and may stop delivery of carload, truckload, planeload, or larger shipments of express or freight if the lessee repudiates or fails to make a payment due before delivery, whether for rent, security or otherwise under the lease contract, or for any other reason the lessor has a right to withhold or take possession of the goods.

(2) In pursuing its remedies under subsection (1), the lessor may stop delivery until

(a) receipt of the goods by the lessee;

(b) acknowledgment to the lessee by any bailee of the goods, except a carrier, that the bailee holds the goods for the lessee; or

(c) such an acknowledgment to the lessee by a carrier via reshipment or as warehouseman.

(3) (a) To stop delivery, a lessor shall so notify as to enable the bailee by reasonable diligence to prevent delivery of the goods.

(b) After notification, the bailee shall hold and deliver the goods according to the directions of the lessor, but the lessor is liable to the bailee for any ensuing charges or damages.

(c) A carrier who has issued a nonnegotiable bill of lading is not obliged to obey a notification to stop received from a person other than the consignor.

§ 2A–527. Lessor's Rights to Dispose of Goods.

(1) After a default by a lessee under the lease contract of the type described in Section 2A–523(1) or 2A–523(3)(a) or after the lessor refuses to deliver or takes possession of goods (Section 2A–525 or 2A–526), or, if agreed, after other default by a lessee, the lessor may dispose of the goods concerned or the undelivered balance thereof by lease, sale, or otherwise.

(2) Except as otherwise provided with respect to damages liquidated in the lease agreement (Section 2A–504) or otherwise determined pursuant to agreement of the parties (Sections 1–102(3) and 2A–503), if the disposition is by lease agreement substantially similar to the original lease agreement and the new lease agreement is made in good faith and in a commercially reasonable manner, the lessor may recover from the lessee as damages (i) accrued and unpaid rent as of the date of the commencement of the term of the new lease agreement, (ii) the present value, as of the same date, of the total rent for the then remaining lease term of the original lease agreement minus the present value, as of the same date, of the rent under the new lease agreement applicable to that period of the new lease term which is comparable to the then remaining term of the original lease agreement, and (iii) any incidental damages allowed under Section 2A–530, less expenses saved in consequence of the lessee's default.

(3) If the lessor's disposition is by lease agreement that for any reason does not qualify for treatment under subsection (2), or is by sale or otherwise, the lessor may recover from the lessee as if the lessor had elected not to dispose of the goods and Section 2A–528 governs.

(4) A subsequent buyer or lessee who buys or leases from the lessor in good faith for value as a result of a disposition under this section takes the goods free of the original lease contract and any rights of the original lessee even though the lessor fails to comply with one or more of the requirements of this Article.

(5) The lessor is not accountable to the lessee for any profit made on any disposition. A lessee who has rightfully rejected or justifiably revoked acceptance shall account to the lessor for any excess over the amount of the lessee's security interest (Section 2A–508(5)).

As amended in 1990.

§ 2A–528. Lessor's Damages for Non-acceptance, Failure to Pay, Repudiation, or Other Default.

(1) Except as otherwise provided with respect to damages liquidated in the lease agreement (Section 2A–504) or otherwise determined pursuant to agreement of the parties (Section 1–102(3) and 2A–503), if a lessor elects to retain the goods or a lessor elects to dispose of the goods and the disposition is by lease agreement that for any reason does not qualify for treatment under Section 2A–527(2), or is by sale or otherwise, the lessor may recover from the lessee as damages for a default of the type described in Section 2A–523(1) or 2A–523(3)(a), or if agreed, for other default of the lessee, (i) accrued and unpaid rent as of the date of the default if the lessee has never taken possession of the goods, or, if the lessee has taken possession of the goods, as of the date the lessor repossesses the goods or an earlier date on which the lessee makes a tender of the goods to the lessor, (ii) the present value as of the date determined under clause (i) of the total rent for the then remaining lease term of the original lease agreement minus the present value as of the same date of the market rent as the place where the goods are located computed for the same lease term, and (iii) any incidental damages allowed under Section 2A–530, less expenses saved in consequence of the lessee's default.

(2) If the measure of damages provided in subsection (1) is inadequate to put a lessor in as good a position as performance would have, the measure of damages is the present value of the profit, including reasonable overhead, the lessor would have made from full performance by the lessee, together with any incidental damages allowed under Section 2A–530, due allowance for costs reasonably incurred and due credit for payments or proceeds of disposition.

As amended in 1990.

§ 2A–529. Lessor's Action for the Rent.

(1) After default by the lessee under the lease contract of the type described in Section 2A–523(1) or 2A–523(3)(a) or, if agreed, after other default by the lessee, if the lessor complies with subsection (2), the lessor may recover from the lessee as damages:

(a) for goods accepted by the lessee and not repossessed by or tendered to the lessor, and for conforming goods lost or damaged within a commercially reasonable time after risk of loss passes to the lessee (Section 2A–219), (i) accrued and unpaid rent as of the date of entry of judgment in favor of the lessor (ii) the present value as of the same date of the rent for the then remaining lease term of the lease agreement, and (iii) any incidental damages allowed under Section 2A–530, less expenses saved in consequence of the lessee's default; and

(b) for goods identified to the lease contract if the lessor is unable after reasonable effort to dispose of them at a reasonable price or the circumstances reasonably indicate that effort will be unavailing, (i) accrued and unpaid rent as of the date of entry of judgment in favor of the lessor, (ii) the present value as of the same date of the rent for the then remaining lease term of the lease agreement, and (iii) any incidental damages allowed under Section 2A–530, less expenses saved in consequence of the lessee's default.

(2) Except as provided in subsection (3), the lessor shall hold for the lessee for the remaining lease term of the lease agreement any goods that have been identified to the lease contract and are in the lessor's control.

(3) The lessor may dispose of the goods at any time before collection of the judgment for damages obtained pursuant to subsection (1). If the disposition is before the end of the remaining lease term of the lease agreement, the lessor's recovery against the lessee for damages is governed by Section 2A–527 or Section 2A–528, and the lessor will cause an appropriate credit to be provided against a judgment for damages to the extent that the amount of the judgment exceeds the recovery available pursuant to Section 2A–527 or 2A–528.

(4) Payment of the judgment for damages obtained pursuant to subsection (1) entitles the lessee to the use and possession of the goods not then disposed of for the remaining lease term of and in accordance with the lease agreement.

(5) After default by the lessee under the lease contract of the type described in Section 2A–523(1) or Section 2A–523(3)(a) or, if agreed, after other default by the lessee, a lessor who is held not entitled to rent under this section must nevertheless be awarded damages for non-acceptance under Sections 2A–527 and 2A–528.

As amended in 1990.

§ 2A–530. Lessor's Incidental Damages.

Incidental damages to an aggrieved lessor include any commercially reasonable charges, expenses, or commissions incurred in stopping delivery, in the transportation, care and custody of goods after the lessee's default, in connection with return or disposition of the goods, or otherwise resulting from the default.

§ 2A–531. Standing to Sue Third Parties for Injury to Goods.

(1) If a third party so deals with goods that have been identified to a lease contract as to cause actionable injury to a party to the lease contract (a) the lessor has a right of action against the third party, and (b) the lessee also has a right of action against the third party if the lessee:

(i) has a security interest in the goods;

(ii) has an insurable interest in the goods; or

(iii) bears the risk of loss under the lease contract or has since the injury assumed that risk as against the lessor and the goods have been converted or destroyed.

(2) If at the time of the injury the party plaintiff did not bear the risk of loss as against the other party to the lease contract and there is no arrangement between them for disposition of the recovery, his [or her] suit or settlement, subject to his [or her] own interest, is as a fiduciary for the other party to the lease contract.

(3) Either party with the consent of the other may sue for the benefit of whom it may concern.

§ 2A–532. Lessor's Rights to Residual Interest.

In addition to any other recovery permitted by this Article or other law, the lessor may recover from the lessee an amount that will fully compensate the lessor for any loss of or damage to the lessor's residual interest in the goods caused by the default of the lessee.

As added in 1990.

Revised Article 3
NEGOTIABLE INSTRUMENTS

Part 1 General Provisions and Definitions

§ 3–101. Short Title.

This Article may be cited as Uniform Commercial Code—Negotiable Instruments.

§ 3–102. Subject Matter.

(a) This Article applies to negotiable instruments. It does not apply to money, to payment orders governed by Article 4A, or to securities governed by Article 8.

(b) If there is conflict between this Article and Article 4 or 9, Articles 4 and 9 govern.

(c) Regulations of the Board of Governors of the Federal Reserve System and operating circulars of the Federal Reserve Banks supersede any inconsistent provision of this Article to the extent of the inconsistency.

§ 3–103. Definitions.

(a) In this Article:

(1) "Acceptor" means a drawee who has accepted a draft.

(2) "Drawee" means a person ordered in a draft to make payment.

(3) "Drawer" means a person who signs or is identified in a draft as a person ordering payment.

(4) "Good faith" means honesty in fact and the observance of reasonable commercial standards of fair dealing.

(5) "Maker" means a person who signs or is identified in a note as a person undertaking to pay.

(6) "Order" means a written instruction to pay money signed by the person giving the instruction. The instruction may be addressed to any person, including the person giving the instruction, or to one or more persons jointly or in the alternative but not in succession. An authorization to pay is not an order unless the person authorized to pay is also instructed to pay.

(7) "Ordinary care" in the case of a person engaged in business means observance of reasonable commercial standards, prevailing in the area in which the person is located, with respect to the business in which the person is engaged. In the case of a bank that takes an instrument for processing for collection or payment by automated means, reasonable commercial standards do not require the bank to examine the instrument if the failure to examine does not violate the bank's prescribed procedures and the bank's procedures do

not vary unreasonably from general banking usage not disapproved by this Article or Article 4.

(8) "Party" means a party to an instrument.

(9) "Promise" means a written undertaking to pay money signed by the person undertaking to pay. An acknowledgment of an obligation by the obligor is not a promise unless the obligor also undertakes to pay the obligation.

(10) "Prove" with respect to a fact means to meet the burden of establishing the fact (Section 1–201(8)).

(11) "Remitter" means a person who purchases an instrument from its issuer if the instrument is payable to an identified person other than the purchaser.

(b) [Other definitions' section references deleted.]

(c) [Other definitions' section references deleted.]

(d) In addition, Article 1 contains general definitions and principles of construction and interpretation applicable throughout this Article.

§ 3–104. Negotiable Instrument.

(a) Except as provided in subsections (c) and (d), "negotiable instrument" means an unconditional promise or order to pay a fixed amount of money, with or without interest or other charges described in the promise or order, if it:

(1) is payable to bearer or to order at the time it is issued or first comes into possession of a holder;

(2) is payable on demand or at a definite time; and

(3) does not state any other undertaking or instruction by the person promising or ordering payment to do any act in addition to the payment of money, but the promise or order may contain (i) an undertaking or power to give, maintain, or protect collateral to secure payment, (ii) an authorization or power to the holder to confess judgment or realize on or dispose of collateral, or (iii) a waiver of the benefit of any law intended for the advantage or protection of an obligor.

(b) "Instrument" means a negotiable instrument.

(c) An order that meets all of the requirements of subsection (a), except paragraph (1), and otherwise falls within the definition of "check" in subsection (f) is a negotiable instrument and a check.

(d) A promise or order other than a check is not an instrument if, at the time it is issued or first comes into possession of a holder, it contains a conspicuous statement, however expressed, to the effect that the promise or order is not negotiable or is not an instrument governed by this Article.

(e) An instrument is a "note" if it is a promise and is a "draft" if it is an order. If an instrument falls within the definition of both "note" and "draft," a person entitled to enforce the instrument may treat it as either.

(f) "Check" means (i) a draft, other than a documentary draft, payable on demand and drawn on a bank or (ii) a cashier's check or teller's check. An instrument may be a check even though it is described on its face by another term, such as "money order."

(g) "Cashier's check" means a draft with respect to which the drawer and drawee are the same bank or branches of the same bank.

(h) "Teller's check" means a draft drawn by a bank (i) on another bank, or (ii) payable at or through a bank.

(i) "Traveler's check" means an instrument that (i) is payable on demand, (ii) is drawn on or payable at or through a bank, (iii) is designated by the term "traveler's check" or by a substantially similar term, and (iv) requires, as a condition to payment, a countersignature by a person whose specimen signature appears on the instrument.

(j) "Certificate of deposit" means an instrument containing an acknowledgment by a bank that a sum of money has been received by the bank and a promise by the bank to repay the sum of money. A certificate of deposit is a note of the bank.

§ 3–105. Issue of Instrument.

(a) "Issue" means the first delivery of an instrument by the maker or drawer, whether to a holder or nonholder, for the purpose of giving rights on the instrument to any person.

(b) An unissued instrument, or an unissued incomplete instrument that is completed, is binding on the maker or drawer, but nonissuance is a defense. An instrument that is conditionally issued or is issued for a special purpose is binding on the maker or drawer, but failure of the condition or special purpose to be fulfilled is a defense.

(c) "Issuer" applies to issued and unissued instruments and means a maker or drawer of an instrument.

§ 3–106. Unconditional Promise or Order.

(a) Except as provided in this section, for the purposes of Section 3–104(a), a promise or order is unconditional unless it states (i) an express condition to payment, (ii) that the promise or order is subject to or governed by another writing, or (iii) that rights or obligations with respect to the promise or order are stated in another writing. A reference to another writing does not of itself make the promise or order conditional.

(b) A promise or order is not made conditional (i) by a reference to another writing for a statement of rights with respect to collateral, prepayment, or acceleration, or (ii) because payment is limited to resort to a particular fund or source.

(c) If a promise or order requires, as a condition to payment, a countersignature by a person whose specimen signature appears on the promise or order, the condition does not make the promise or order conditional for the purposes of Section 3–104(a). If the person whose specimen signature appears on an instrument fails to countersign the instrument, the failure to countersign is a defense to the obligation of the issuer, but the failure does not prevent a transferee of the instrument from becoming a holder of the instrument.

(d) If a promise or order at the time it is issued or first comes into possession of a holder contains a statement, required by applicable statutory or administrative law, to the effect that the rights of a holder or transferee are subject to claims or defenses that the issuer could assert against the original payee, the promise or order is not thereby made conditional for the purposes of Section 3–104(a); but if the promise or order is an instrument, there cannot be a holder in due course of the instrument.

§ 3–107. Instrument Payable in Foreign Money.

Unless the instrument otherwise provides, an instrument that states the amount payable in foreign money may be paid in the foreign money or in an equivalent amount in dollars calculated by using the current bank-offered spot rate at the place of payment for the purchase of dollars on the day on which the instrument is paid.

§ 3–108. Payable on Demand or at Definite Time.

(a) A promise or order is "payable on demand" if it (i) states that it is payable on demand or at sight, or otherwise indicates that it is payable at the will of the holder, or (ii) does not state any time of payment.

(b) A promise or order is "payable at a definite time" if it is payable on elapse of a definite period of time after sight or acceptance or at a fixed date or dates or at a time or times readily ascertainable at the time the promise or order is issued, subject to rights of (i) prepayment, (ii) acceleration, (iii) extension at the option of the holder, or (iv) extension to a further definite time at the option of the maker or acceptor or automatically upon or after a specified act or event.

(c) If an instrument, payable at a fixed date, is also payable upon demand made before the fixed date, the instrument is payable on demand until the fixed date and, if demand for payment is not made before that date, becomes payable at a definite time on the fixed date.

§ 3–109. Payable to Bearer or to Order.

(a) A promise or order is payable to bearer if it:

(1) states that it is payable to bearer or to the order of bearer or otherwise indicates that the person in possession of the promise or order is entitled to payment;

(2) does not state a payee; or

(3) states that it is payable to or to the order of cash or otherwise indicates that it is not payable to an identified person.

(b) A promise or order that is not payable to bearer is payable to order if it is payable (i) to the order of an identified person or (ii) to an identified person or order. A promise or order that is payable to order is payable to the identified person.

(c) An instrument payable to bearer may become payable to an identified person if it is specially indorsed pursuant to Section 3–205(a). An instrument payable to an identified person may become payable to bearer if it is indorsed in blank pursuant to Section 3–205(b).

§ 3–110. Identification of Person to Whom Instrument Is Payable.

(a) The person to whom an instrument is initially payable is determined by the intent of the person, whether or not authorized, signing as, or in the name or behalf of, the issuer of the instrument. The instrument is payable to the person intended by the signer even if that person is identified in the instrument by a name or other identification that is not that of the intended person. If more than one person signs in the name or behalf of the issuer of an instrument and all the signers do not intend the same person as payee, the instrument is payable to any person intended by one or more of the signers.

(b) If the signature of the issuer of an instrument is made by automated means, such as a check-writing machine, the payee of the instrument is determined by the intent of the person who supplied the name or identification of the payee, whether or not authorized to do so.

(c) A person to whom an instrument is payable may be identified in any way, including by name, identifying number, office, or account number. For the purpose of determining the holder of an instrument, the following rules apply:

(1) If an instrument is payable to an account and the account is identified only by number, the instrument is payable to the person to whom the account is payable. If an instrument is payable to an account identified by number and by the name of a person, the instrument is payable to the named person, whether or not that person is the owner of the account identified by number.

(2) If an instrument is payable to:

(i) a trust, an estate, or a person described as trustee or representative of a trust or estate, the instrument is payable to the trustee, the representative, or a successor of either, whether or not the beneficiary or estate is also named;

(ii) a person described as agent or similar representative of a named or identified person, the instrument is payable to the represented person, the representative, or a successor of the representative;

(iii) a fund or organization that is not a legal entity, the instrument is payable to a representative of the members of the fund or organization; or

(iv) an office or to a person described as holding an office, the instrument is payable to the named person, the incumbent of the office, or a successor to the incumbent.

(d) If an instrument is payable to two or more persons alternatively, it is payable to any of them and may be negotiated, discharged, or enforced by any or all of them in possession of the instrument. If an instrument is payable to two or more persons not alternatively, it is payable to all of them and may be negotiated, discharged, or enforced only by all of them. If an instrument payable to two or more persons is ambiguous as to whether it is

payable to the persons alternatively, the instrument is payable to the persons alternatively.

§ 3–111. Place of Payment.

Except as otherwise provided for items in Article 4, an instrument is payable at the place of payment stated in the instrument. If no place of payment is stated, an instrument is payable at the address of the drawee or maker stated in the instrument. If no address is stated, the place of payment is the place of business of the drawee or maker. If a drawee or maker has more than one place of business, the place of payment is any place of business of the drawee or maker chosen by the person entitled to enforce the instrument. If the drawee or maker has no place of business, the place of payment is the residence of the drawee or maker.

§ 3–112. Interest.

(a) Unless otherwise provided in the instrument, (i) an instrument is not payable with interest, and (ii) interest on an interest-bearing instrument is payable from the date of the instrument.

(b) Interest may be stated in an instrument as a fixed or variable amount of money or it may be expressed as a fixed or variable rate or rates. The amount or rate of interest may be stated or described in the instrument in any manner and may require reference to information not contained in the instrument. If an instrument provides for interest, but the amount of interest payable cannot be ascertained from the description, interest is payable at the judgment rate in effect at the place of payment of the instrument and at the time interest first accrues.

§ 3–113. Date of Instrument.

(a) An instrument may be antedated or postdated. The date stated determines the time of payment if the instrument is payable at a fixed period after date. Except as provided in Section 4–401(c), an instrument payable on demand is not payable before the date of the instrument.

(b) If an instrument is undated, its date is the date of its issue or, in the case of an unissued instrument, the date it first comes into possession of a holder.

§ 3–114. Contradictory Terms of Instrument.

If an instrument contains contradictory terms, typewritten terms prevail over printed terms, handwritten terms prevail over both, and words prevail over numbers.

§ 3–115. Incomplete Instrument.

(a) "Incomplete instrument" means a signed writing, whether or not issued by the signer, the contents of which show at the time of signing that it is incomplete but that the signer intended it to be completed by the addition of words or numbers.

(b) Subject to subsection (c), if an incomplete instrument is an instrument under Section 3–104, it may be enforced according to its terms if it is not completed, or according to its terms as augmented by completion. If an incomplete instrument is not an instrument under Section 3–104, but, after completion, the requirements of Section 3–104 are met, the instrument may be enforced according to its terms as augmented by completion.

(c) If words or numbers are added to an incomplete instrument without authority of the signer, there is an alteration of the incomplete instrument under Section 3–407.

(d) The burden of establishing that words or numbers were added to an incomplete instrument without authority of the signer is on the person asserting the lack of authority.

§ 3–116. Joint and Several Liability; Contribution.

(a) Except as otherwise provided in the instrument, two or more persons who have the same liability on an instrument as makers, drawers, acceptors, indorsers who indorse as joint payees, or anomalous indorsers are jointly and severally liable in the capacity in which they sign.

(b) Except as provided in Section 3–419(e) or by agreement of the affected parties, a party having joint and several liability who pays the instrument is entitled to receive from any party having the same joint and several liability contribution in accordance with applicable law.

(c) Discharge of one party having joint and several liability by a person entitled to enforce the instrument does not affect the right under subsection (b) of a party having the same joint and several liability to receive contribution from the party discharged.

§ 3–117. Other Agreements Affecting Instrument.

Subject to applicable law regarding exclusion of proof of contemporaneous or previous agreements, the obligation of a party to an instrument to pay the instrument may be modified, supplemented, or nullified by a separate agreement of the obligor and a person entitled to enforce the instrument, if the instrument is issued or the obligation is incurred in reliance on the agreement or as part of the same transaction giving rise to the agreement. To the extent an obligation is modified, supplemented, or nullified by an agreement under this section, the agreement is a defense to the obligation.

§ 3–118. Statute of Limitations.

(a) Except as provided in subsection (e), an action to enforce the obligation of a party to pay a note payable at a definite time must be commenced within six years after the due date or dates stated in the note or, if a due date is accelerated, within six years after the accelerated due date.

(b) Except as provided in subsection (d) or (e), if demand for payment is made to the maker of a note payable on demand, an action to enforce the obligation of a party to pay the note must be commenced within six years after the demand. If no demand for payment is made to the maker, an action to enforce the note is barred if neither principal nor interest on the note has been paid for a continuous period of 10 years.

(c) Except as provided in subsection (d), an action to enforce the obligation of a party to an unaccepted draft to pay the draft must be commenced within three years after dishonor of the

draft or 10 years after the date of the draft, whichever period expires first.

(d) An action to enforce the obligation of the acceptor of a certified check or the issuer of a teller's check, cashier's check, or traveler's check must be commenced within three years after demand for payment is made to the acceptor or issuer, as the case may be.

(e) An action to enforce the obligation of a party to a certificate of deposit to pay the instrument must be commenced within six years after demand for payment is made to the maker, but if the instrument states a due date and the maker is not required to pay before that date, the six-year period begins when a demand for payment is in effect and the due date has passed.

(f) An action to enforce the obligation of a party to pay an accepted draft, other than a certified check, must be commenced (i) within six years after the due date or dates stated in the draft or acceptance if the obligation of the acceptor is payable at a definite time, or (ii) within six years after the date of the acceptance if the obligation of the acceptor is payable on demand.

(g) Unless governed by other law regarding claims for indemnity or contribution, an action (i) for conversion of an instrument, for money had and received, or like action based on conversion, (ii) for breach of warranty, or (iii) to enforce an obligation, duty, or right arising under this Article and not governed by this section must be commenced within three years after the [cause of action] accrues.

§ 3–119. Notice of Right to Defend Action.

In an action for breach of an obligation for which a third person is answerable over pursuant to this Article or Article 4, the defendant may give the third person written notice of the litigation, and the person notified may then give similar notice to any other person who is answerable over. If the notice states (i) that the person notified may come in and defend and (ii) that failure to do so will bind the person notified in an action later brought by the person giving the notice as to any determination of fact common to the two litigations, the person notified is so bound unless after seasonable receipt of the notice the person notified does come in and defend.

Part 2 Negotiation, Transfer, and Indorsement

§ 3–201. Negotiation.

(a) "Negotiation" means a transfer of possession, whether voluntary or involuntary, of an instrument by a person other than the issuer to a person who thereby becomes its holder.

(b) Except for negotiation by a remitter, if an instrument is payable to an identified person, negotiation requires transfer of possession of the instrument and its indorsement by the holder. If an instrument is payable to bearer, it may be negotiated by transfer of possession alone.

§ 3–202. Negotiation Subject to Rescission.

(a) Negotiation is effective even if obtained (i) from an infant, a corporation exceeding its powers, or a person without capacity, (ii) by fraud, duress, or mistake, or (iii) in breach of duty or as part of an illegal transaction.

(b) To the extent permitted by other law, negotiation may be rescinded or may be subject to other remedies, but those remedies may not be asserted against a subsequent holder in due course or a person paying the instrument in good faith and without knowledge of facts that are a basis for rescission or other remedy.

§ 3–203. Transfer of Instrument; Rights Acquired by Transfer.

(a) An instrument is transferred when it is delivered by a person other than its issuer for the purpose of giving to the person receiving delivery the right to enforce the instrument.

(b) Transfer of an instrument, whether or not the transfer is a negotiation, vests in the transferee any right of the transferor to enforce the instrument, including any right as a holder in due course, but the transferee cannot acquire rights of a holder in due course by a transfer, directly or indirectly, from a holder in due course if the transferee engaged in fraud or illegality affecting the instrument.

(c) Unless otherwise agreed, if an instrument is transferred for value and the transferee does not become a holder because of lack of indorsement by the transferor, the transferee has a specifically enforceable right to the unqualified indorsement of the transferor, but negotiation of the instrument does not occur until the indorsement is made.

(d) If a transferor purports to transfer less than the entire instrument, negotiation of the instrument does not occur. The transferee obtains no rights under this Article and has only the rights of a partial assignee.

§ 3–204. Indorsement.

(a) "Indorsement" means a signature, other than that of a signer as maker, drawer, or acceptor, that alone or accompanied by other words is made on an instrument for the purpose of (i) negotiating the instrument, (ii) restricting payment of the instrument, or (iii) incurring indorser's liability on the instrument, but regardless of the intent of the signer, a signature and its accompanying words is an indorsement unless the accompanying words, terms of the instrument, place of the signature, or other circumstances unambiguously indicate that the signature was made for a purpose other than indorsement. For the purpose of determining whether a signature is made on an instrument, a paper affixed to the instrument is a part of the instrument.

(b) "Indorser" means a person who makes an indorsement.

(c) For the purpose of determining whether the transferee of an instrument is a holder, an indorsement that transfers a security interest in the instrument is effective as an unqualified indorsement of the instrument.

(d) If an instrument is payable to a holder under a name that is not the name of the holder, indorsement may be made by

the holder in the name stated in the instrument or in the holder's name or both, but signature in both names may be required by a person paying or taking the instrument for value or collection.

§ 3–205. Special Indorsement; Blank Indorsement; Anomalous Indorsement.

(a) If an indorsement is made by the holder of an instrument, whether payable to an identified person or payable to bearer, and the indorsement identifies a person to whom it makes the instrument payable, it is a "special indorsement." When specially indorsed, an instrument becomes payable to the identified person and may be negotiated only by the indorsement of that person. The principles stated in Section 3–110 apply to special indorsements.

(b) If an indorsement is made by the holder of an instrument and it is not a special indorsement, it is a "blank indorsement." When indorsed in blank, an instrument becomes payable to bearer and may be negotiated by transfer of possession alone until specially indorsed.

(c) The holder may convert a blank indorsement that consists only of a signature into a special indorsement by writing, above the signature of the indorser, words identifying the person to whom the instrument is made payable.

(d) "Anomalous indorsement" means an indorsement made by a person who is not the holder of the instrument. An anomalous indorsement does not affect the manner in which the instrument may be negotiated.

§ 3–206. Restrictive Indorsement.

(a) An indorsement limiting payment to a particular person or otherwise prohibiting further transfer or negotiation of the instrument is not effective to prevent further transfer or negotiation of the instrument.

(b) An indorsement stating a condition to the right of the indorsee to receive payment does not affect the right of the indorsee to enforce the instrument. A person paying the instrument or taking it for value or collection may disregard the condition, and the rights and liabilities of that person are not affected by whether the condition has been fulfilled.

(c) If an instrument bears an indorsement (i) described in Section 4–201(b), or (ii) in blank or to a particular bank using the words "for deposit," "for collection," or other words indicating a purpose of having the instrument collected by a bank for the indorser or for a particular account, the following rules apply:

(1) A person, other than a bank, who purchases the instrument when so indorsed converts the instrument unless the amount paid for the instrument is received by the indorser or applied consistently with the indorsement.

(2) A depositary bank that purchases the instrument or takes it for collection when so indorsed converts the instrument unless the amount paid by the bank with respect to the instrument is received by the indorser or applied consistently with the indorsement.

(3) A payor bank that is also the depositary bank or that takes the instrument for immediate payment over the counter from a person other than a collecting bank converts the instrument unless the proceeds of the instrument are received by the indorser or applied consistently with the indorsement.

(4) Except as otherwise provided in paragraph (3), a payor bank or intermediary bank may disregard the indorsement and is not liable if the proceeds of the instrument are not received by the indorser or applied consistently with the indorsement.

(d) Except for an indorsement covered by subsection (c), if an instrument bears an indorsement using words to the effect that payment is to be made to the indorsee as agent, trustee, or other fiduciary for the benefit of the indorser or another person, the following rules apply:

(1) Unless there is notice of breach of fiduciary duty as provided in Section 3–307, a person who purchases the instrument from the indorsee or takes the instrument from the indorsee for collection or payment may pay the proceeds of payment or the value given for the instrument to the indorsee without regard to whether the indorsee violates a fiduciary duty to the indorser.

(2) A subsequent transferee of the instrument or person who pays the instrument is neither given notice nor otherwise affected by the restriction in the indorsement unless the transferee or payor knows that the fiduciary dealt with the instrument or its proceeds in breach of fiduciary duty.

(e) The presence on an instrument of an indorsement to which this section applies does not prevent a purchaser of the instrument from becoming a holder in due course of the instrument unless the purchaser is a converter under subsection (c) or has notice or knowledge of breach of fiduciary duty as stated in subsection (d).

(f) In an action to enforce the obligation of a party to pay the instrument, the obligor has a defense if payment would violate an indorsement to which this section applies and the payment is not permitted by this section.

§ 3–207. Reacquisition.

Reacquisition of an instrument occurs if it is transferred to a former holder, by negotiation or otherwise. A former holder who reacquires the instrument may cancel indorsements made after the reacquirer first became a holder of the instrument. If the cancellation causes the instrument to be payable to the reacquirer or to bearer, the reacquirer may negotiate the instrument. An indorser whose indorsement is canceled is discharged, and the discharge is effective against any subsequent holder.

Part 3 Enforcement of Instruments

§ 3–301. Person Entitled to Enforce Instrument.

"Person entitled to enforce" an instrument means (i) the holder of the instrument, (ii) a nonholder in possession of the instrument

who has the rights of a holder, or (iii) a person not in possession of the instrument who is entitled to enforce the instrument pursuant to Section 3–309 or 3–418(d). A person may be a person entitled to enforce the instrument even though the person is not the owner of the instrument or is in wrongful possession of the instrument.

§ 3–302. Holder in Due Course.

(a) Subject to subsection (c) and Section 3–106(d), "holder in due course" means the holder of an instrument if:

(1) the instrument when issued or negotiated to the holder does not bear such apparent evidence of forgery or alteration or is not otherwise so irregular or incomplete as to call into question its authenticity; and

(2) the holder took the instrument (i) for value, (ii) in good faith, (iii) without notice that the instrument is overdue or has been dishonored or that there is an uncured default with respect to payment of another instrument issued as part of the same series, (iv) without notice that the instrument contains an unauthorized signature or has been altered, (v) without notice of any claim to the instrument described in Section 3–306, and (vi) without notice that any party has a defense or claim in recoupment described in Section 3–305(a).

(b) Notice of discharge of a party, other than discharge in an insolvency proceeding, is not notice of a defense under subsection (a), but discharge is effective against a person who became a holder in due course with notice of the discharge. Public filing or recording of a document does not of itself constitute notice of a defense, claim in recoupment, or claim to the instrument.

(c) Except to the extent a transferor or predecessor in interest has rights as a holder in due course, a person does not acquire rights of a holder in due course of an instrument taken (i) by legal process or by purchase in an execution, bankruptcy, or creditor's sale or similar proceeding, (ii) by purchase as part of a bulk transaction not in ordinary course of business of the transferor, or (iii) as the successor in interest to an estate or other organization.

(d) If, under Section 3–303(a)(1), the promise of performance that is the consideration for an instrument has been partially performed, the holder may assert rights as a holder in due course of the instrument only to the fraction of the amount payable under the instrument equal to the value of the partial performance divided by the value of the promised performance.

(e) If (i) the person entitled to enforce an instrument has only a security interest in the instrument and (ii) the person obliged to pay the instrument has a defense, claim in recoupment, or claim to the instrument that may be asserted against the person who granted the security interest, the person entitled to enforce the instrument may assert rights as a holder in due course only to an amount payable under the instrument which, at the time of enforcement of the instrument, does not exceed the amount of the unpaid obligation secured.

(f) To be effective, notice must be received at a time and in a manner that gives a reasonable opportunity to act on it.

(g) This section is subject to any law limiting status as a holder in due course in particular classes of transactions.

§ 3–303. Value and Consideration.

(a) An instrument is issued or transferred for value if:

(1) the instrument is issued or transferred for a promise of performance, to the extent the promise has been performed;

(2) the transferee acquires a security interest or other lien in the instrument other than a lien obtained by judicial proceeding;

(3) the instrument is issued or transferred as payment of, or as security for, an antecedent claim against any person, whether or not the claim is due;

(4) the instrument is issued or transferred in exchange for a negotiable instrument; or

(5) the instrument is issued or transferred in exchange for the incurring of an irrevocable obligation to a third party by the person taking the instrument.

(b) "Consideration" means any consideration sufficient to support a simple contract. The drawer or maker of an instrument has a defense if the instrument is issued without consideration. If an instrument is issued for a promise of performance, the issuer has a defense to the extent performance of the promise is due and the promise has not been performed. If an instrument is issued for value as stated in subsection (a), the instrument is also issued for consideration.

§ 3–304. Overdue Instrument.

(a) An instrument payable on demand becomes overdue at the earliest of the following times:

(1) on the day after the day demand for payment is duly made;

(2) if the instrument is a check, 90 days after its date; or

(3) if the instrument is not a check, when the instrument has been outstanding for a period of time after its date which is unreasonably long under the circumstances of the particular case in light of the nature of the instrument and usage of the trade.

(b) With respect to an instrument payable at a definite time the following rules apply:

(1) If the principal is payable in installments and a due date has not been accelerated, the instrument becomes overdue upon default under the instrument for nonpayment of an installment, and the instrument remains overdue until the default is cured.

(2) If the principal is not payable in installments and the due date has not been accelerated, the instrument becomes overdue on the day after the due date.

(3) If a due date with respect to principal has been accelerated, the instrument becomes overdue on the day after the accelerated due date.

(c) Unless the due date of principal has been accelerated, an instrument does not become overdue if there is default in payment of interest but no default in payment of principal.

§ 3–305. Defenses and Claims in Recoupment.

(a) Except as stated in subsection (b), the right to enforce the obligation of a party to pay an instrument is subject to the following:

(1) a defense of the obligor based on (i) infancy of the obligor to the extent it is a defense to a simple contract, (ii) duress, lack of legal capacity, or illegality of the transaction which, under other law, nullifies the obligation of the obligor, (iii) fraud that induced the obligor to sign the instrument with neither knowledge nor reasonable opportunity to learn of its character or its essential terms, or (iv) discharge of the obligor in insolvency proceedings;

(2) a defense of the obligor stated in another section of this Article or a defense of the obligor that would be available if the person entitled to enforce the instrument were enforcing a right to payment under a simple contract; and

(3) a claim in recoupment of the obligor against the original payee of the instrument if the claim arose from the transaction that gave rise to the instrument; but the claim of the obligor may be asserted against a transferee of the instrument only to reduce the amount owing on the instrument at the time the action is brought.

(b) The right of a holder in due course to enforce the obligation of a party to pay the instrument is subject to defenses of the obligor stated in subsection (a)(1), but is not subject to defenses of the obligor stated in subsection (a)(2) or claims in recoupment stated in subsection (a)(3) against a person other than the holder.

(c) Except as stated in subsection (d), in an action to enforce the obligation of a party to pay the instrument, the obligor may not assert against the person entitled to enforce the instrument a defense, claim in recoupment, or claim to the instrument (Section 3–306) of another person, but the other person's claim to the instrument may be asserted by the obligor if the other person is joined in the action and personally asserts the claim against the person entitled to enforce the instrument. An obligor is not obliged to pay the instrument if the person seeking enforcement of the instrument does not have rights of a holder in due course and the obligor proves that the instrument is a lost or stolen instrument.

(d) In an action to enforce the obligation of an accommodation party to pay an instrument, the accommodation party may assert against the person entitled to enforce the instrument any defense or claim in recoupment under subsection (a) that the accommodated party could assert against the person entitled to enforce the instrument, except the defenses of discharge in insolvency proceedings, infancy, and lack of legal capacity.

§ 3–306. Claims to an Instrument.

A person taking an instrument, other than a person having rights of a holder in due course, is subject to a claim of a property or possessory right in the instrument or its proceeds, including a claim to rescind a negotiation and to recover the instrument or its proceeds. A person having rights of a holder in due course takes free of the claim to the instrument.

§ 3–307. Notice of Breach of Fiduciary Duty.

(a) In this section:

(1) "Fiduciary" means an agent, trustee, partner, corporate officer or director, or other representative owing a fiduciary duty with respect to an instrument.

(2) "Represented person" means the principal, beneficiary, partnership, corporation, or other person to whom the duty stated in paragraph (1) is owed.

(b) If (i) an instrument is taken from a fiduciary for payment or collection or for value, (ii) the taker has knowledge of the fiduciary status of the fiduciary, and (iii) the represented person makes a claim to the instrument or its proceeds on the basis that the transaction of the fiduciary is a breach of fiduciary duty, the following rules apply:

(1) Notice of breach of fiduciary duty by the fiduciary is notice of the claim of the represented person.

(2) In the case of an instrument payable to the represented person or the fiduciary as such, the taker has notice of the breach of fiduciary duty if the instrument is (i) taken in payment of or as security for a debt known by the taker to be the personal debt of the fiduciary, (ii) taken in a transaction known by the taker to be for the personal benefit of the fiduciary, or (iii) deposited to an account other than an account of the fiduciary, as such, or an account of the represented person.

(3) If an instrument is issued by the represented person or the fiduciary as such, and made payable to the fiduciary personally, the taker does not have notice of the breach of fiduciary duty unless the taker knows of the breach of fiduciary duty.

(4) If an instrument is issued by the represented person or the fiduciary as such, to the taker as payee, the taker has notice of the breach of fiduciary duty if the instrument is (i) taken in payment of or as security for a debt known by the taker to be the personal debt of the fiduciary, (ii) taken in a transaction known by the taker to be for the personal benefit of the fiduciary, or (iii) deposited to an account other than an account of the fiduciary, as such, or an account of the represented person.

§ 3–308. Proof of Signatures and Status as Holder in Due Course.

(a) In an action with respect to an instrument, the authenticity of, and authority to make, each signature on the instrument is admitted unless specifically denied in the pleadings. If the validity of a signature is denied in the pleadings, the burden of establishing validity is on the person claiming validity, but the signature is presumed to be authentic and authorized unless the action is to enforce the liability of the purported signer and the signer is dead or incompetent at the time of

trial of the issue of validity of the signature. If an action to enforce the instrument is brought against a person as the undisclosed principal of a person who signed the instrument as a party to the instrument, the plaintiff has the burden of establishing that the defendant is liable on the instrument as a represented person under Section 3–402(a).

(b) If the validity of signatures is admitted or proved and there is compliance with subsection (a), a plaintiff producing the instrument is entitled to payment if the plaintiff proves entitlement to enforce the instrument under Section 3–301, unless the defendant proves a defense or claim in recoupment. If a defense or claim in recoupment is proved, the right to payment of the plaintiff is subject to the defense or claim, except to the extent the plaintiff proves that the plaintiff has rights of a holder in due course which are not subject to the defense or claim.

§ 3–309. Enforcement of Lost, Destroyed, or Stolen Instrument.

(a) A person not in possession of an instrument is entitled to enforce the instrument if (i) the person was in possession of the instrument and entitled to enforce it when loss of possession occurred, (ii) the loss of possession was not the result of a transfer by the person or a lawful seizure, and (iii) the person cannot reasonably obtain possession of the instrument because the instrument was destroyed, its whereabouts cannot be determined, or it is in the wrongful possession of an unknown person or a person that cannot be found or is not amenable to service of process.

(b) A person seeking enforcement of an instrument under subsection (a) must prove the terms of the instrument and the person's right to enforce the instrument. If that proof is made, Section 3–308 applies to the case as if the person seeking enforcement had produced the instrument. The court may not enter judgment in favor of the person seeking enforcement unless it finds that the person required to pay the instrument is adequately protected against loss that might occur by reason of a claim by another person to enforce the instrument. Adequate protection may be provided by any reasonable means.

§ 3–310. Effect of Instrument on Obligation for Which Taken.

(a) Unless otherwise agreed, if a certified check, cashier's check, or teller's check is taken for an obligation, the obligation is discharged to the same extent discharge would result if an amount of money equal to the amount of the instrument were taken in payment of the obligation. Discharge of the obligation does not affect any liability that the obligor may have as an indorser of the instrument.

(b) Unless otherwise agreed and except as provided in subsection (a), if a note or an uncertified check is taken for an obligation, the obligation is suspended to the same extent the obligation would be discharged if an amount of money equal to the amount of the instrument were taken, and the following rules apply:

(1) In the case of an uncertified check, suspension of the obligation continues until dishonor of the check or until it is paid or certified. Payment or certification of the check results in discharge of the obligation to the extent of the amount of the check.

(2) In the case of a note, suspension of the obligation continues until dishonor of the note or until it is paid. Payment of the note results in discharge of the obligation to the extent of the payment.

(3) Except as provided in paragraph (4), if the check or note is dishonored and the obligee of the obligation for which the instrument was taken is the person entitled to enforce the instrument, the obligee may enforce either the instrument or the obligation. In the case of an instrument of a third person which is negotiated to the obligee by the obligor, discharge of the obligor on the instrument also discharges the obligation.

(4) If the person entitled to enforce the instrument taken for an obligation is a person other than the obligee, the obligee may not enforce the obligation to the extent the obligation is suspended. If the obligee is the person entitled to enforce the instrument but no longer has possession of it because it was lost, stolen, or destroyed, the obligation may not be enforced to the extent of the amount payable on the instrument, and to that extent the obligee's rights against the obligor are limited to enforcement of the instrument.

(c) If an instrument other than one described in subsection (a) or (b) is taken for an obligation, the effect is (i) that stated in subsection (a) if the instrument is one on which a bank is liable as maker or acceptor, or (ii) that stated in subsection (b) in any other case.

§ 3–311. Accord and Satisfaction by Use of Instrument.

(a) If a person against whom a claim is asserted proves that (i) that person in good faith tendered an instrument to the claimant as full satisfaction of the claim, (ii) the amount of the claim was unliquidated or subject to a bona fide dispute, and (iii) the claimant obtained payment of the instrument, the following subsections apply.

(b) Unless subsection (c) applies, the claim is discharged if the person against whom the claim is asserted proves that the instrument or an accompanying written communication contained a conspicuous statement to the effect that the instrument was tendered as full satisfaction of the claim.

(c) Subject to subsection (d), a claim is not discharged under subsection (b) if either of the following applies:

(1) The claimant, if an organization, proves that (i) within a reasonable time before the tender, the claimant sent a conspicuous statement to the person against whom the claim is asserted that communications concerning disputed debts, including an instrument tendered as full satisfaction of a debt, are to be sent to a designated person, office, or place, and (ii) the instrument or accompanying communication was not received by that designated person, office, or place.

(2) The claimant, whether or not an organization, proves that within 90 days after payment of the instrument, the claimant tendered repayment of the amount of the instrument to the person against whom the claim is asserted. This

paragraph does not apply if the claimant is an organization that sent a statement complying with paragraph (1)(i).

(d) A claim is discharged if the person against whom the claim is asserted proves that within a reasonable time before collection of the instrument was initiated, the claimant, or an agent of the claimant having direct responsibility with respect to the disputed obligation, knew that the instrument was tendered in full satisfaction of the claim.

§ 3–312. Lost, Destroyed, or Stolen Cashier's Check, Teller's Check, or Certified Check.*

(a) In this section:

(1) "Check" means a cashier's check, teller's check, or certified check.

(2) "Claimant" means a person who claims the right to receive the amount of a cashier's check, teller's check, or certified check that was lost, destroyed, or stolen.

(3) "Declaration of loss" means a written statement, made under penalty of perjury, to the effect that (i) the declarer lost possession of a check, (ii) the declarer is the drawer or payee of the check, in the case of a certified check, or the remitter or payee of the check, in the case of a cashier's check or teller's check, (iii) the loss of possession was not the result of a transfer by the declarer or a lawful seizure, and (iv) the declarer cannot reasonably obtain possession of the check because the check was destroyed, its whereabouts cannot be determined, or it is in the wrongful possession of an unknown person or a person that cannot be found or is not amenable to service of process.

(4) "Obligated bank" means the issuer of a cashier's check or teller's check or the acceptor of a certified check.

(b) A claimant may assert a claim to the amount of a check by a communication to the obligated bank describing the check with reasonable certainty and requesting payment of the amount of the check, if (i) the claimant is the drawer or payee of a certified check or the remitter or payee of a cashier's check or teller's check, (ii) the communication contains or is accompanied by a declaration of loss of the claimant with respect to the check, (iii) the communication is received at a time and in a manner affording the bank a reasonable time to act on it before the check is paid, and (iv) the claimant provides reasonable identification if requested by the obligated bank. Delivery of a declaration of loss is a warranty of the truth of the statements made in the declaration. If a claim is asserted in compliance with this subsection, the following rules apply:

(1) The claim becomes enforceable at the later of (i) the time the claim is asserted, or (ii) the 90th day following the date of the check, in the case of a cashier's check or teller's check, or the 90th day following the date of the acceptance, in the case of a certified check.

(2) Until the claim becomes enforceable, it has no legal effect and the obligated bank may pay the check or, in the case of a teller's check, may permit the drawee to pay the check. Payment to a person entitled to enforce the check discharges all liability of the obligated bank with respect to the check.

(3) If the claim becomes enforceable before the check is presented for payment, the obligated bank is not obliged to pay the check.

(4) When the claim becomes enforceable, the obligated bank becomes obliged to pay the amount of the check to the claimant if payment of the check has not been made to a person entitled to enforce the check. Subject to Section 4–302(a)(1), payment to the claimant discharges all liability of the obligated bank with respect to the check.

(c) If the obligated bank pays the amount of a check to a claimant under subsection (b)(4) and the check is presented for payment by a person having rights of a holder in due course, the claimant is obliged to (i) refund the payment to the obligated bank if the check is paid, or (ii) pay the amount of the check to the person having rights of a holder in due course if the check is dishonored.

(d) If a claimant has the right to assert a claim under subsection (b) and is also a person entitled to enforce a cashier's check, teller's check, or certified check which is lost, destroyed, or stolen, the claimant may assert rights with respect to the check either under this section or Section 3–309.

Added in 1991.

Part 4 Liability of Parties

§ 3–401. Signature.

(a) A person is not liable on an instrument unless (i) the person signed the instrument, or (ii) the person is represented by an agent or representative who signed the instrument and the signature is binding on the represented person under Section 3–402.

(b) A signature may be made (i) manually or by means of a device or machine, and (ii) by the use of any name, including a trade or assumed name, or by a word, mark, or symbol executed or adopted by a person with present intention to authenticate a writing.

§ 3–402. Signature by Representative.

(a) If a person acting, or purporting to act, as a representative signs an instrument by signing either the name of the represented person or the name of the signer, the represented person is bound by the signature to the same extent the represented person would be bound if the signature were on a simple contract. If the represented person is bound, the signature of the representative is the "authorized signature of the represented person" and the represented person is liable on the instrument, whether or not identified in the instrument.

*[Section 3–312 was not adopted as part of the 1990 Official Text of Revised Article 3. It was officially approved and recommended for enactment in all states in August 1991 by the National Conference of Commissioners on Uniform State Laws.]

(b) If a representative signs the name of the representative to an instrument and the signature is an authorized signature of the represented person, the following rules apply:

(1) If the form of the signature shows unambiguously that the signature is made on behalf of the represented person who is identified in the instrument, the representative is not liable on the instrument.

(2) Subject to subsection (c), if (i) the form of the signature does not show unambiguously that the signature is made in a representative capacity or (ii) the represented person is not identified in the instrument, the representative is liable on the instrument to a holder in due course that took the instrument without notice that the representative was not intended to be liable on the instrument. With respect to any other person, the representative is liable on the instrument unless the representative proves that the original parties did not intend the representative to be liable on the instrument.

(c) If a representative signs the name of the representative as drawer of a check without indication of the representative status and the check is payable from an account of the represented person who is identified on the check, the signer is not liable on the check if the signature is an authorized signature of the represented person.

§ 3–403. Unauthorized Signature.

(a) Unless otherwise provided in this Article or Article 4, an unauthorized signature is ineffective except as the signature of the unauthorized signer in favor of a person who in good faith pays the instrument or takes it for value. An unauthorized signature may be ratified for all purposes of this Article.

(b) If the signature of more than one person is required to constitute the authorized signature of an organization, the signature of the organization is unauthorized if one of the required signatures is lacking.

(c) The civil or criminal liability of a person who makes an unauthorized signature is not affected by any provision of this Article which makes the unauthorized signature effective for the purposes of this Article.

§ 3–404. Impostors; Fictitious Payees.

(a) If an impostor, by use of the mails or otherwise, induces the issuer of an instrument to issue the instrument to the impostor, or to a person acting in concert with the impostor, by impersonating the payee of the instrument or a person authorized to act for the payee, an indorsement of the instrument by any person in the name of the payee is effective as the indorsement of the payee in favor of a person who, in good faith, pays the instrument or takes it for value or for collection.

(b) If (i) a person whose intent determines to whom an instrument is payable (Section 3–110(a) or (b)) does not intend the person identified as payee to have any interest in the instrument, or (ii) the person identified as payee of an instrument is a fictitious person, the following rules apply until the instrument is negotiated by special indorsement:

(1) Any person in possession of the instrument is its holder.

(2) An indorsement by any person in the name of the payee stated in the instrument is effective as the indorsement of the payee in favor of a person who, in good faith, pays the instrument or takes it for value or for collection.

(c) Under subsection (a) or (b), an indorsement is made in the name of a payee if (i) it is made in a name substantially similar to that of the payee or (ii) the instrument, whether or not indorsed, is deposited in a depositary bank to an account in a name substantially similar to that of the payee.

(d) With respect to an instrument to which subsection (a) or (b) applies, if a person paying the instrument or taking it for value or for collection fails to exercise ordinary care in paying or taking the instrument and that failure substantially contributes to loss resulting from payment of the instrument, the person bearing the loss may recover from the person failing to exercise ordinary care to the extent the failure to exercise ordinary care contributed to the loss.

§ 3–405. Employer's Responsibility for Fraudulent Indorsement by Employee.

(a) In this section:

(1) "Employee" includes an independent contractor and employee of an independent contractor retained by the employer.

(2) "Fraudulent indorsement" means (i) in the case of an instrument payable to the employer, a forged indorsement purporting to be that of the employer, or (ii) in the case of an instrument with respect to which the employer is the issuer, a forged indorsement purporting to be that of the person identified as payee.

(3) "Responsibility" with respect to instruments means authority (i) to sign or indorse instruments on behalf of the employer, (ii) to process instruments received by the employer for bookkeeping purposes, for deposit to an account, or for other disposition, (iii) to prepare or process instruments for issue in the name of the employer, (iv) to supply information determining the names or addresses of payees of instruments to be issued in the name of the employer, (v) to control the disposition of instruments to be issued in the name of the employer, or (vi) to act otherwise with respect to instruments in a responsible capacity. "Responsibility" does not include authority that merely allows an employee to have access to instruments or blank or incomplete instrument forms that are being stored or transported or are part of incoming or outgoing mail, or similar access.

(b) For the purpose of determining the rights and liabilities of a person who, in good faith, pays an instrument or takes it for value or for collection, if an employer entrusted an employee with responsibility with respect to the instrument and the employee or a person acting in concert with the employee makes a fraudulent indorsement of the instrument, the indorsement is

effective as the indorsement of the person to whom the instrument is payable if it is made in the name of that person. If the person paying the instrument or taking it for value or for collection fails to exercise ordinary care in paying or taking the instrument and that failure substantially contributes to loss resulting from the fraud, the person bearing the loss may recover from the person failing to exercise ordinary care to the extent the failure to exercise ordinary care contributed to the loss.

(c) Under subsection (b), an indorsement is made in the name of the person to whom an instrument is payable if (i) it is made in a name substantially similar to the name of that person or (ii) the instrument, whether or not indorsed, is deposited in a depositary bank to an account in a name substantially similar to the name of that person.

§ 3–406. Negligence Contributing to Forged Signature or Alteration of Instrument.

(a) A person whose failure to exercise ordinary care substantially contributes to an alteration of an instrument or to the making of a forged signature on an instrument is precluded from asserting the alteration or the forgery against a person who, in good faith, pays the instrument or takes it for value or for collection.

(b) Under subsection (a), if the person asserting the preclusion fails to exercise ordinary care in paying or taking the instrument and that failure substantially contributes to loss, the loss is allocated between the person precluded and the person asserting the preclusion according to the extent to which the failure of each to exercise ordinary care contributed to the loss.

(c) Under subsection (a), the burden of proving failure to exercise ordinary care is on the person asserting the preclusion. Under subsection (b), the burden of proving failure to exercise ordinary care is on the person precluded.

§ 3–407. Alteration.

(a) "Alteration" means (i) an unauthorized change in an instrument that purports to modify in any respect the obligation of a party, or (ii) an unauthorized addition of words or numbers or other change to an incomplete instrument relating to the obligation of a party.

(b) Except as provided in subsection (c), an alteration fraudulently made discharges a party whose obligation is affected by the alteration unless that party assents or is precluded from asserting the alteration. No other alteration discharges a party, and the instrument may be enforced according to its original terms.

(c) A payor bank or drawee paying a fraudulently altered instrument or a person taking it for value, in good faith and without notice of the alteration, may enforce rights with respect to the instrument (i) according to its original terms, or (ii) in the case of an incomplete instrument altered by unauthorized completion, according to its terms as completed.

§ 3–408. Drawee Not Liable on Unaccepted Draft.

A check or other draft does not of itself operate as an assignment of funds in the hands of the drawee available for its payment, and the drawee is not liable on the instrument until the drawee accepts it.

§ 3–409. Acceptance of Draft; Certified Check.

(a) "Acceptance" means the drawee's signed agreement to pay a draft as presented. It must be written on the draft and may consist of the drawee's signature alone. Acceptance may be made at any time and becomes effective when notification pursuant to instructions is given or the accepted draft is delivered for the purpose of giving rights on the acceptance to any person.

(b) A draft may be accepted although it has not been signed by the drawer, is otherwise incomplete, is overdue, or has been dishonored.

(c) If a draft is payable at a fixed period after sight and the acceptor fails to date the acceptance, the holder may complete the acceptance by supplying a date in good faith.

(d) "Certified check" means a check accepted by the bank on which it is drawn. Acceptance may be made as stated in subsection (a) or by a writing on the check which indicates that the check is certified. The drawee of a check has no obligation to certify the check, and refusal to certify is not dishonor of the check.

§ 3–410. Acceptance Varying Draft.

(a) If the terms of a drawee's acceptance vary from the terms of the draft as presented, the holder may refuse the acceptance and treat the draft as dishonored. In that case, the drawee may cancel the acceptance.

(b) The terms of a draft are not varied by an acceptance to pay at a particular bank or place in the United States, unless the acceptance states that the draft is to be paid only at that bank or place.

(c) If the holder assents to an acceptance varying the terms of a draft, the obligation of each drawer and indorser that does not expressly assent to the acceptance is discharged.

§ 3–411. Refusal to Pay Cashier's Checks, Teller's Checks, and Certified Checks.

(a) In this section, "obligated bank" means the acceptor of a certified check or the issuer of a cashier's check or teller's check bought from the issuer.

(b) If the obligated bank wrongfully (i) refuses to pay a cashier's check or certified check, (ii) stops payment of a teller's check, or (iii) refuses to pay a dishonored teller's check, the person asserting the right to enforce the check is entitled to compensation for expenses and loss of interest resulting from the nonpayment and may recover consequential damages if the obligated bank refuses to pay after receiving notice of particular circumstances giving rise to the damages.

(c) Expenses or consequential damages under subsection (b) are not recoverable if the refusal of the obligated bank to pay occurs because (i) the bank suspends payments, (ii) the obligated bank asserts a claim or defense of the bank that it has reasonable

grounds to believe is available against the person entitled to enforce the instrument, (iii) the obligated bank has a reasonable doubt whether the person demanding payment is the person entitled to enforce the instrument, or (iv) payment is prohibited by law.

§ 3–412. Obligation of Issuer of Note or Cashier's Check.

The issuer of a note or cashier's check or other draft drawn on the drawer is obliged to pay the instrument (i) according to its terms at the time it was issued or, if not issued, at the time it first came into possession of a holder, or (ii) if the issuer signed an incomplete instrument, according to its terms when completed, to the extent stated in Sections 3–115 and 3–407. The obligation is owed to a person entitled to enforce the instrument or to an indorser who paid the instrument under Section 3–415.

§ 3–413. Obligation of Acceptor.

(a) The acceptor of a draft is obliged to pay the draft (i) according to its terms at the time it was accepted, even though the acceptance states that the draft is payable "as originally drawn" or equivalent terms, (ii) if the acceptance varies the terms of the draft, according to the terms of the draft as varied, or (iii) if the acceptance is of a draft that is an incomplete instrument, according to its terms when completed, to the extent stated in Sections 3–115 and 3–407. The obligation is owed to a person entitled to enforce the draft or to the drawer or an indorser who paid the draft under Section 3–414 or 3–415.

(b) If the certification of a check or other acceptance of a draft states the amount certified or accepted, the obligation of the acceptor is that amount. If (i) the certification or acceptance does not state an amount, (ii) the amount of the instrument is subsequently raised, and (iii) the instrument is then negotiated to a holder in due course, the obligation of the acceptor is the amount of the instrument at the time it was taken by the holder in due course.

§ 3–414. Obligation of Drawer.

(a) This section does not apply to cashier's checks or other drafts drawn on the drawer.

(b) If an unaccepted draft is dishonored, the drawer is obliged to pay the draft (i) according to its terms at the time it was issued or, if not issued, at the time it first came into possession of a holder, or (ii) if the drawer signed an incomplete instrument, according to its terms when completed, to the extent stated in Sections 3–115 and 3–407. The obligation is owed to a person entitled to enforce the draft or to an indorser who paid the draft under Section 3–415.

(c) If a draft is accepted by a bank, the drawer is discharged, regardless of when or by whom acceptance was obtained.

(d) If a draft is accepted and the acceptor is not a bank, the obligation of the drawer to pay the draft if the draft is dishonored by the acceptor is the same as the obligation of an indorser under Section 3–415(a) and (c).

(e) If a draft states that it is drawn "without recourse" or otherwise disclaims liability of the drawer to pay the draft, the drawer is not liable under subsection (b) to pay the draft if the draft is not a check. A disclaimer of the liability stated in subsection (b) is not effective if the draft is a check.

(f) If (i) a check is not presented for payment or given to a depositary bank for collection within 30 days after its date, (ii) the drawee suspends payments after expiration of the 30-day period without paying the check, and (iii) because of the suspension of payments, the drawer is deprived of funds maintained with the drawee to cover payment of the check, the drawer to the extent deprived of funds may discharge its obligation to pay the check by assigning to the person entitled to enforce the check the rights of the drawer against the drawee with respect to the funds.

§ 3–415. Obligation of Indorser.

(a) Subject to subsections (b), (c), and (d) and to Section 3–419(d), if an instrument is dishonored, an indorser is obliged to pay the amount due on the instrument (i) according to the terms of the instrument at the time it was indorsed, or (ii) if the indorser indorsed an incomplete instrument, according to its terms when completed, to the extent stated in Sections 3–115 and 3–407. The obligation of the indorser is owed to a person entitled to enforce the instrument or to a subsequent indorser who paid the instrument under this section.

(b) If an indorsement states that it is made "without recourse" or otherwise disclaims liability of the indorser, the indorser is not liable under subsection (a) to pay the instrument.

(c) If notice of dishonor of an instrument is required by Section 3–503 and notice of dishonor complying with that section is not given to an indorser, the liability of the indorser under subsection (a) is discharged.

(d) If a draft is accepted by a bank after an indorsement is made, the liability of the indorser under subsection (a) is discharged.

(e) If an indorser of a check is liable under subsection (a) and the check is not presented for payment, or given to a depositary bank for collection, within 30 days after the day the indorsement was made, the liability of the indorser under subsection (a) is discharged.

As amended in 1993.

§ 3–416. Transfer Warranties.

(a) A person who transfers an instrument for consideration warrants to the transferee and, if the transfer is by indorsement, to any subsequent transferee that:

(1) the warrantor is a person entitled to enforce the instrument;

(2) all signatures on the instrument are authentic and authorized;

(3) the instrument has not been altered;

(4) the instrument is not subject to a defense or claim in recoupment of any party which can be asserted against the warrantor; and

(5) the warrantor has no knowledge of any insolvency proceeding commenced with respect to the maker or acceptor or, in the case of an unaccepted draft, the drawer.

(b) A person to whom the warranties under subsection (a) are made and who took the instrument in good faith may recover from the warrantor as damages for breach of warranty an amount equal to the loss suffered as a result of the breach, but not more than the amount of the instrument plus expenses and loss of interest incurred as a result of the breach.

(c) The warranties stated in subsection (a) cannot be disclaimed with respect to checks. Unless notice of a claim for breach of warranty is given to the warrantor within 30 days after the claimant has reason to know of the breach and the identity of the warrantor, the liability of the warrantor under subsection (b) is discharged to the extent of any loss caused by the delay in giving notice of the claim.

(d) A [cause of action] for breach of warranty under this section accrues when the claimant has reason to know of the breach.

§ 3–417. Presentment Warranties.

(a) If an unaccepted draft is presented to the drawee for payment or acceptance and the drawee pays or accepts the draft, (i) the person obtaining payment or acceptance, at the time of presentment, and (ii) a previous transferor of the draft, at the time of transfer, warrant to the drawee making payment or accepting the draft in good faith that:

(1) the warrantor is, or was, at the time the warrantor transferred the draft, a person entitled to enforce the draft or authorized to obtain payment or acceptance of the draft on behalf of a person entitled to enforce the draft;

(2) the draft has not been altered; and

(3) the warrantor has no knowledge that the signature of the drawer of the draft is unauthorized.

(b) A drawee making payment may recover from any warrantor damages for breach of warranty equal to the amount paid by the drawee less the amount the drawee received or is entitled to receive from the drawer because of the payment. In addition, the drawee is entitled to compensation for expenses and loss of interest resulting from the breach. The right of the drawee to recover damages under this subsection is not affected by any failure of the drawee to exercise ordinary care in making payment. If the drawee accepts the draft, breach of warranty is a defense to the obligation of the acceptor. If the acceptor makes payment with respect to the draft, the acceptor is entitled to recover from any warrantor for breach of warranty the amounts stated in this subsection.

(c) If a drawee asserts a claim for breach of warranty under subsection (a) based on an unauthorized indorsement of the draft or an alteration of the draft, the warrantor may defend by proving that the indorsement is effective under Section 3–404 or 3–405 or the drawer is precluded under Section 3–406 or 4–406 from asserting against the drawee the unauthorized indorsement or alteration.

(d) If (i) a dishonored draft is presented for payment to the drawer or an indorser or (ii) any other instrument is presented for payment to a party obliged to pay the instrument, and (iii) payment is received, the following rules apply:

(1) The person obtaining payment and a prior transferor of the instrument warrant to the person making payment in good faith that the warrantor is, or was, at the time the warrantor transferred the instrument, a person entitled to enforce the instrument or authorized to obtain payment on behalf of a person entitled to enforce the instrument.

(2) The person making payment may recover from any warrantor for breach of warranty an amount equal to the amount paid plus expenses and loss of interest resulting from the breach.

(e) The warranties stated in subsections (a) and (d) cannot be disclaimed with respect to checks. Unless notice of a claim for breach of warranty is given to the warrantor within 30 days after the claimant has reason to know of the breach and the identity of the warrantor, the liability of the warrantor under subsection (b) or (d) is discharged to the extent of any loss caused by the delay in giving notice of the claim.

(f) A [cause of action] for breach of warranty under this section accrues when the claimant has reason to know of the breach.

§ 3–418. Payment or Acceptance by Mistake.

(a) Except as provided in subsection (c), if the drawee of a draft pays or accepts the draft and the drawee acted on the mistaken belief that (i) payment of the draft had not been stopped pursuant to Section 4–403 or (ii) the signature of the drawer of the draft was authorized, the drawee may recover the amount of the draft from the person to whom or for whose benefit payment was made or, in the case of acceptance, may revoke the acceptance. Rights of the drawee under this subsection are not affected by failure of the drawee to exercise ordinary care in paying or accepting the draft.

(b) Except as provided in subsection (c), if an instrument has been paid or accepted by mistake and the case is not covered by subsection (a), the person paying or accepting may, to the extent permitted by the law governing mistake and restitution, (i) recover the payment from the person to whom or for whose benefit payment was made or (ii) in the case of acceptance, may revoke the acceptance.

(c) The remedies provided by subsection (a) or (b) may not be asserted against a person who took the instrument in good faith and for value or who in good faith changed position in reliance on the payment or acceptance. This subsection does not limit remedies provided by Section 3–417 or 4–407.

(d) Notwithstanding Section 4–215, if an instrument is paid or accepted by mistake and the payor or acceptor recovers payment or revokes acceptance under subsection (a) or (b),

the instrument is deemed not to have been paid or accepted and is treated as dishonored, and the person from whom payment is recovered has rights as a person entitled to enforce the dishonored instrument.

§ 3–419. Instruments Signed for Accommodation.

(a) If an instrument is issued for value given for the benefit of a party to the instrument ("accommodated party") and another party to the instrument ("accommodation party") signs the instrument for the purpose of incurring liability on the instrument without being a direct beneficiary of the value given for the instrument, the instrument is signed by the accommodation party "for accommodation."

(b) An accommodation party may sign the instrument as maker, drawer, acceptor, or indorser and, subject to subsection (d), is obliged to pay the instrument in the capacity in which the accommodation party signs. The obligation of an accommodation party may be enforced notwithstanding any statute of frauds and whether or not the accommodation party receives consideration for the accommodation.

(c) A person signing an instrument is presumed to be an accommodation party and there is notice that the instrument is signed for accommodation if the signature is an anomalous indorsement or is accompanied by words indicating that the signer is acting as surety or guarantor with respect to the obligation of another party to the instrument. Except as provided in Section 3–605, the obligation of an accommodation party to pay the instrument is not affected by the fact that the person enforcing the obligation had notice when the instrument was taken by that person that the accommodation party signed the instrument for accommodation.

(d) If the signature of a party to an instrument is accompanied by words indicating unambiguously that the party is guaranteeing collection rather than payment of the obligation of another party to the instrument, the signer is obliged to pay the amount due on the instrument to a person entitled to enforce the instrument only if (i) execution of judgment against the other party has been returned unsatisfied, (ii) the other party is insolvent or in an insolvency proceeding, (iii) the other party cannot be served with process, or (iv) it is otherwise apparent that payment cannot be obtained from the other party.

(e) An accommodation party who pays the instrument is entitled to reimbursement from the accommodated party and is entitled to enforce the instrument against the accommodated party. An accommodated party who pays the instrument has no right of recourse against, and is not entitled to contribution from, an accommodation party.

§ 3–420. Conversion of Instrument.

(a) The law applicable to conversion of personal property applies to instruments. An instrument is also converted if it is taken by transfer, other than a negotiation, from a person not entitled to enforce the instrument or a bank makes or obtains payment with respect to the instrument for a person not entitled to enforce the instrument or receive payment. An action for conversion of an instrument may not be brought by (i) the issuer or acceptor of the instrument or (ii) a payee or indorsee who did not receive delivery of the instrument either directly or through delivery to an agent or a co-payee.

(b) In an action under subsection (a), the measure of liability is presumed to be the amount payable on the instrument, but recovery may not exceed the amount of the plaintiff's interest in the instrument.

(c) A representative, other than a depositary bank, who has in good faith dealt with an instrument or its proceeds on behalf of one who was not the person entitled to enforce the instrument is not liable in conversion to that person beyond the amount of any proceeds that it has not paid out.

Part 5 Dishonor

§ 3–501. Presentment.

(a) "Presentment" means a demand made by or on behalf of a person entitled to enforce an instrument (i) to pay the instrument made to the drawee or a party obliged to pay the instrument or, in the case of a note or accepted draft payable at a bank, to the bank, or (ii) to accept a draft made to the drawee.

(b) The following rules are subject to Article 4, agreement of the parties, and clearing-house rules and the like:

(1) Presentment may be made at the place of payment of the instrument and must be made at the place of payment if the instrument is payable at a bank in the United States; may be made by any commercially reasonable means, including an oral, written, or electronic communication; is effective when the demand for payment or acceptance is received by the person to whom presentment is made; and is effective if made to any one of two or more makers, acceptors, drawees, or other payors.

(2) Upon demand of the person to whom presentment is made, the person making presentment must (i) exhibit the instrument, (ii) give reasonable identification and, if presentment is made on behalf of another person, reasonable evidence of authority to do so, and (. . .) sign a receipt on the instrument for any payment made or surrender the instrument if full payment is made.

(3) Without dishonoring the instrument, the party to whom presentment is made may (i) return the instrument for lack of a necessary indorsement, or (ii) refuse payment or acceptance for failure of the presentment to comply with the terms of the instrument, an agreement of the parties, or other applicable law or rule.

(4) The party to whom presentment is made may treat presentment as occurring on the next business day after the day of presentment if the party to whom presentment is made has established a cut-off hour not earlier than 2 P.M. for the receipt and processing of instruments presented for payment or acceptance and presentment is made after the cut-off hour.

§ 3–502. Dishonor.

(a) Dishonor of a note is governed by the following rules:

(1) If the note is payable on demand, the note is dishonored if presentment is duly made to the maker and the note is not paid on the day of presentment.

(2) If the note is not payable on demand and is payable at or through a bank or the terms of the note require presentment, the note is dishonored if presentment is duly made and the note is not paid on the day it becomes payable or the day of presentment, whichever is later.

(3) If the note is not payable on demand and paragraph (2) does not apply, the note is dishonored if it is not paid on the day it becomes payable.

(b) Dishonor of an unaccepted draft other than a documentary draft is governed by the following rules:

(1) If a check is duly presented for payment to the payor bank otherwise than for immediate payment over the counter, the check is dishonored if the payor bank makes timely return of the check or sends timely notice of dishonor or nonpayment under Section 4–301 or 4–302, or becomes accountable for the amount of the check under Section 4–302.

(2) If a draft is payable on demand and paragraph (1) does not apply, the draft is dishonored if presentment for payment is duly made to the drawee and the draft is not paid on the day of presentment.

(3) If a draft is payable on a date stated in the draft, the draft is dishonored if (i) presentment for payment is duly made to the drawee and payment is not made on the day the draft becomes payable or the day of presentment, whichever is later, or (ii) presentment for acceptance is duly made before the day the draft becomes payable and the draft is not accepted on the day of presentment.

(4) If a draft is payable on elapse of a period of time after sight or acceptance, the draft is dishonored if presentment for acceptance is duly made and the draft is not accepted on the day of presentment.

(c) Dishonor of an unaccepted documentary draft occurs according to the rules stated in subsection (b)(2), (3), and (4), except that payment or acceptance may be delayed without dishonor until no later than the close of the third business day of the drawee following the day on which payment or acceptance is required by those paragraphs.

(d) Dishonor of an accepted draft is governed by the following rules:

(1) If the draft is payable on demand, the draft is dishonored if presentment for payment is duly made to the acceptor and the draft is not paid on the day of presentment.

(2) If the draft is not payable on demand, the draft is dishonored if presentment for payment is duly made to the acceptor and payment is not made on the day it becomes payable or the day of presentment, whichever is later.

(e) In any case in which presentment is otherwise required for dishonor under this section and presentment is excused under Section 3–504, dishonor occurs without presentment if the instrument is not duly accepted or paid.

(f) If a draft is dishonored because timely acceptance of the draft was not made and the person entitled to demand acceptance consents to a late acceptance, from the time of acceptance the draft is treated as never having been dishonored.

§ 3–503. Notice of Dishonor.

(a) The obligation of an indorser stated in Section 3–415(a) and the obligation of a drawer stated in Section 3–414(d) may not be enforced unless (i) the indorser or drawer is given notice of dishonor of the instrument complying with this section or (ii) notice of dishonor is excused under Section 3–504(b).

(b) Notice of dishonor may be given by any person; may be given by any commercially reasonable means, including an oral, written, or electronic communication; and is sufficient if it reasonably identifies the instrument and indicates that the instrument has been dishonored or has not been paid or accepted. Return of an instrument given to a bank for collection is sufficient notice of dishonor.

(c) Subject to Section 3–504(c), with respect to an instrument taken for collection by a collecting bank, notice of dishonor must be given (i) by the bank before midnight of the next banking day following the banking day on which the bank receives notice of dishonor of the instrument, or (ii) by any other person within 30 days following the day on which the person receives notice of dishonor. With respect to any other instrument, notice of dishonor must be given within 30 days following the day on which dishonor occurs.

§ 3–504. Excused Presentment and Notice of Dishonor.

(a) Presentment for payment or acceptance of an instrument is excused if (i) the person entitled to present the instrument cannot with reasonable diligence make presentment, (ii) the maker or acceptor has repudiated an obligation to pay the instrument or is dead or in insolvency proceedings, (iii) by the terms of the instrument presentment is not necessary to enforce the obligation of indorsers or the drawer, (iv) the drawer or indorser whose obligation is being enforced has waived presentment or otherwise has no reason to expect or right to require that the instrument be paid or accepted, or (v) the drawer instructed the drawee not to pay or accept the draft or the drawee was not obligated to the drawer to pay the draft.

(b) Notice of dishonor is excused if (i) by the terms of the instrument notice of dishonor is not necessary to enforce the obligation of a party to pay the instrument, or (ii) the party whose obligation is being enforced waived notice of dishonor. A waiver of presentment is also a waiver of notice of dishonor.

(c) Delay in giving notice of dishonor is excused if the delay was caused by circumstances beyond the control of the person giving the notice and the person giving the notice exercised reasonable diligence after the cause of the delay ceased to operate.

§ 3–505. Evidence of Dishonor.

(a) The following are admissible as evidence and create a presumption of dishonor and of any notice of dishonor stated:

(1) a document regular in form as provided in subsection (b) which purports to be a protest;

(2) a purported stamp or writing of the drawee, payor bank, or presenting bank on or accompanying the instrument stating that acceptance or payment has been refused unless reasons for the refusal are stated and the reasons are not consistent with dishonor;

(3) a book or record of the drawee, payor bank, or collecting bank, kept in the usual course of business which shows dishonor, even if there is no evidence of who made the entry.

(b) A protest is a certificate of dishonor made by a United States consul or vice consul, or a notary public or other person authorized to administer oaths by the law of the place where dishonor occurs. It may be made upon information satisfactory to that person. The protest must identify the instrument and certify either that presentment has been made or, if not made, the reason why it was not made, and that the instrument has been dishonored by nonacceptance or nonpayment. The protest may also certify that notice of dishonor has been given to some or all parties.

Part 6 Discharge and Payment

§ 3–601. Discharge and Effect of Discharge.

(a) The obligation of a party to pay the instrument is discharged as stated in this Article or by an act or agreement with the party which would discharge an obligation to pay money under a simple contract.

(b) Discharge of the obligation of a party is not effective against a person acquiring rights of a holder in due course of the instrument without notice of the discharge.

§ 3–602. Payment.

(a) Subject to subsection (b), an instrument is paid to the extent payment is made (i) by or on behalf of a party obliged to pay the instrument, and (ii) to a person entitled to enforce the instrument. To the extent of the payment, the obligation of the party obliged to pay the instrument is discharged even though payment is made with knowledge of a claim to the instrument under Section 3–306 by another person.

(b) The obligation of a party to pay the instrument is not discharged under subsection (a) if:

(1) a claim to the instrument under Section 3–306 is enforceable against the party receiving payment and (i) payment is made with knowledge by the payor that payment is prohibited by injunction or similar process of a court of competent jurisdiction, or (ii) in the case of an instrument other than a cashier's check, teller's check, or certified check, the party making payment accepted, from the person having a claim to the instrument, indemnity against loss resulting from refusal to pay the person entitled to enforce the instrument; or

(2) the person making payment knows that the instrument is a stolen instrument and pays a person it knows is in wrongful possession of the instrument.

§ 3–603. Tender of Payment.

(a) If tender of payment of an obligation to pay an instrument is made to a person entitled to enforce the instrument, the effect of tender is governed by principles of law applicable to tender of payment under a simple contract.

(b) If tender of payment of an obligation to pay an instrument is made to a person entitled to enforce the instrument and the tender is refused, there is discharge, to the extent of the amount of the tender, of the obligation of an indorser or accommodation party having a right of recourse with respect to the obligation to which the tender relates.

(c) If tender of payment of an amount due on an instrument is made to a person entitled to enforce the instrument, the obligation of the obligor to pay interest after the due date on the amount tendered is discharged. If presentment is required with respect to an instrument and the obligor is able and ready to pay on the due date at every place of payment stated in the instrument, the obligor is deemed to have made tender of payment on the due date to the person entitled to enforce the instrument.

§ 3–604. Discharge by Cancellation or Renunciation.

(a) A person entitled to enforce an instrument, with or without consideration, may discharge the obligation of a party to pay the instrument (i) by an intentional voluntary act, such as surrender of the instrument to the party, destruction, mutilation, or cancellation of the instrument, cancellation or striking out of the party's signature, or the addition of words to the instrument indicating discharge, or (ii) by agreeing not to sue or otherwise renouncing rights against the party by a signed writing.

(b) Cancellation or striking out of an indorsement pursuant to subsection (a) does not affect the status and rights of a party derived from the indorsement.

§ 3–605. Discharge of Indorsers and Accommodation Parties.

(a) In this section, the term "indorser" includes a drawer having the obligation described in Section 3–414(d).

(b) Discharge, under Section 3–604, of the obligation of a party to pay an instrument does not discharge the obligation of an indorser or accommodation party having a right of recourse against the discharged party.

(c) If a person entitled to enforce an instrument agrees, with or without consideration, to an extension of the due date of the obligation of a party to pay the instrument, the extension discharges an indorser or accommodation party having a right of recourse against the party whose obligation is extended to the extent the indorser or accommodation party proves that the extension caused loss to the indorser or accommodation party with respect to the right of recourse.

(d) If a person entitled to enforce an instrument agrees, with or without consideration, to a material modification of the obligation of a party other than an extension of the due date, the modification discharges the obligation of an indorser or accommodation party having a right of recourse against the person whose obligation is modified to the extent the modification causes loss to the indorser or accommodation party with respect to the right of recourse. The loss suffered by the indorser or accommodation party as a result of the modification is equal to the amount of the right of recourse unless the person enforcing the instrument proves that no loss was caused by the modification or that the loss caused by the modification was an amount less than the amount of the right of recourse.

(e) If the obligation of a party to pay an instrument is secured by an interest in collateral and a person entitled to enforce the instrument impairs the value of the interest in collateral, the obligation of an indorser or accommodation party having a right of recourse against the obligor is discharged to the extent of the impairment. The value of an interest in collateral is impaired to the extent (i) the value of the interest is reduced to an amount less than the amount of the right of recourse of the party asserting discharge, or (ii) the reduction in value of the interest causes an increase in the amount by which the amount of the right of recourse exceeds the value of the interest. The burden of proving impairment is on the party asserting discharge.

(f) If the obligation of a party is secured by an interest in collateral not provided by an accommodation party and a person entitled to enforce the instrument impairs the value of the interest in collateral, the obligation of any party who is jointly and severally liable with respect to the secured obligation is discharged to the extent the impairment causes the party asserting discharge to pay more than that party would have been obliged to pay, taking into account rights of contribution, if impairment had not occurred. If the party asserting discharge is an accommodation party not entitled to discharge under subsection (e), the party is deemed to have a right to contribution based on joint and several liability rather than a right to reimbursement. The burden of proving impairment is on the party asserting discharge.

(g) Under subsection (e) or (f), impairing value of an interest in collateral includes (i) failure to obtain or maintain perfection or recordation of the interest in collateral, (ii) release of collateral without substitution of collateral of equal value, (iii) failure to perform a duty to preserve the value of collateral owed, under Article 9 or other law, to a debtor or surety or other person secondarily liable, or (iv) failure to comply with applicable law in disposing of collateral.

(h) An accommodation party is not discharged under subsection (c), (d), or (e) unless the person entitled to enforce the instrument knows of the accommodation or has notice under Section 3–419(c) that the instrument was signed for accommodation.

(i) A party is not discharged under this section if (i) the party asserting discharge consents to the event or conduct that is the basis of the discharge, or (ii) the instrument or a separate agreement of the party provides for waiver of discharge under this section either specifically or by general language indicating that parties waive defenses based on suretyship or impairment of collateral.

ADDENDUM TO REVISED ARTICLE 3

Notes to Legislative Counsel

1. If revised Article 3 is adopted in your state, the reference in Section 2–511 to Section 3–802 should be changed to Section 3–310.

2. If revised Article 3 is adopted in your state and the Uniform Fiduciaries Act is also in effect in your state, you may want to consider amending Uniform Fiduciaries Act § 9 to conform to Section 3–307(b)(2)(iii) and (4)(iii). See Official Comment 3 to Section 3–307.

Revised Article 4
BANK DEPOSITS AND COLLECTIONS

Part 1 General Provisions and Definitions

§ 4–101. Short Title.

This Article may be cited as Uniform Commercial Code—Bank Deposits and Collections.

As amended in 1990.

§ 4–102. Applicability.

(a) To the extent that items within this Article are also within Articles 3 and 8, they are subject to those Articles. If there is conflict, this Article governs Article 3, but Article 8 governs this Article.

(b) The liability of a bank for action or non-action with respect to an item handled by it for purposes of presentment, payment, or collection is governed by the law of the place where the bank is located. In the case of action or non-action by or at a branch or separate office of a bank, its liability is governed by the law of the place where the branch or separate office is located.

§ 4–103. Variation by Agreement; Measure of Damages; Action Constituting Ordinary Care.

(a) The effect of the provisions of this Article may be varied by agreement, but the parties to the agreement cannot disclaim a bank's responsibility for its lack of good faith or failure to exercise ordinary care or limit the measure of damages for the lack or failure. However, the parties may determine by agreement the standards by which the bank's responsibility is to be measured if those standards are not manifestly unreasonable.

(b) Federal Reserve regulations and operating circulars, clearinghouse rules, and the like have the effect of agreements under subsection (a), whether or not specifically assented to by all parties interested in items handled.

(c) Action or non-action approved by this Article or pursuant to Federal Reserve regulations or operating circulars is the exercise

of ordinary care and, in the absence of special instructions, action or non-action consistent with clearing-house rules and the like or with a general banking usage not disapproved by this Article, is prima facie the exercise of ordinary care.

(d) The specification or approval of certain procedures by this Article is not disapproval of other procedures that may be reasonable under the circumstances.

(e) The measure of damages for failure to exercise ordinary care in handling an item is the amount of the item reduced by an amount that could not have been realized by the exercise of ordinary care. If there is also bad faith it includes any other damages the party suffered as a proximate consequence.

As amended in 1990.

§ 4–104. Definitions and Index of Definitions.

(a) In this Article, unless the context otherwise requires:

(1) "Account" means any deposit or credit account with a bank, including a demand, time, savings, passbook, share draft, or like account, other than an account evidenced by a certificate of deposit;

(2) "Afternoon" means the period of a day between noon and midnight;

(3) "Banking day" means the part of a day on which a bank is open to the public for carrying on substantially all of its banking functions;

(4) "Clearing house" means an association of banks or other payors regularly clearing items;

(5) "Customer" means a person having an account with a bank or for whom a bank has agreed to collect items, including a bank that maintains an account at another bank;

(6) "Documentary draft" means a draft to be presented for acceptance or payment if specified documents, certificated securities (Section 8–102) or instructions for uncertificated securities (Section 8–102), or other certificates, statements, or the like are to be received by the drawee or other payor before acceptance or payment of the draft;

(7) "Draft" means a draft as defined in Section 3–104 or an item, other than an instrument, that is an order;

(8) "Drawee" means a person ordered in a draft to make payment;

(9) "Item" means an instrument or a promise or order to pay money handled by a bank for collection or payment. The term does not include a payment order governed by Article 4A or a credit or debit card slip;

(10) "Midnight deadline" with respect to a bank is midnight on its next banking day following the banking day on which it receives the relevant item or notice or from which the time for taking action commences to run, whichever is later;

(11) "Settle" means to pay in cash, by clearing-house settlement, in a charge or credit or by remittance, or otherwise as agreed. A settlement may be either provisional or final;

(12) "Suspends payments" with respect to a bank means that it has been closed by order of the supervisory authorities, that a public officer has been appointed to take it over, or that it ceases or refuses to make payments in the ordinary course of business.

(b) [Other definitions' section references deleted.]

(c) [Other definitions' section references deleted.]

(d) In addition, Article 1 contains general definitions and principles of construction and interpretation applicable throughout this Article.

§ 4–105. "Bank"; "Depositary Bank"; "Payor Bank"; "Intermediary Bank"; "Collecting Bank"; "Presenting Bank".

In this Article:

(1) "Bank" means a person engaged in the business of banking, including a savings bank, savings and loan association, credit union, or trust company;

(2) "Depositary bank" means the first bank to take an item even though it is also the payor bank, unless the item is presented for immediate payment over the counter;

(3) "Payor bank" means a bank that is the drawee of a draft;

(4) "Intermediary bank" means a bank to which an item is transferred in course of collection except the depositary or payor bank;

(5) "Collecting bank" means a bank handling an item for collection except the payor bank;

(6) "Presenting bank" means a bank presenting an item except a payor bank.

§ 4–106. Payable Through or Payable at Bank: Collecting Bank.

(a) If an item states that it is "payable through" a bank identified in the item, (i) the item designates the bank as a collecting bank and does not by itself authorize the bank to pay the item, and (ii) the item may be presented for payment only by or through the bank.

Alternative A

(b) If an item states that it is "payable at" a bank identified in the item, the item is equivalent to a draft drawn on the bank.

Alternative B

(b) If an item states that it is "payable at" a bank identified in the item, (i) the item designates the bank as a collecting bank and does not by itself authorize the bank to pay the item, and (ii) the item may be presented for payment only by or through the bank.

(c) If a draft names a nonbank drawee and it is unclear whether a bank named in the draft is a co-drawee or a collecting bank, the bank is a collecting bank.

As added in 1990.

§ 4–107. Separate Office of Bank.

A branch or separate office of a bank is a separate bank for the purpose of computing the time within which and determining the place at or to which action may be taken or notices or orders shall be given under this Article and under Article 3.

As amended in 1962 and 1990.

§ 4–108. Time of Receipt of Items.

(a) For the purpose of allowing time to process items, prove balances, and make the necessary entries on its books to determine its position for the day, a bank may fix an afternoon hour of 2 P.M. or later as a cutoff hour for the handling of money and items and the making of entries on its books.

(b) An item or deposit of money received on any day after a cutoff hour so fixed or after the close of the banking day may be treated as being received at the opening of the next banking day.

As amended in 1990.

§ 4–109. Delays.

(a) Unless otherwise instructed, a collecting bank in a good faith effort to secure payment of a specific item drawn on a payor other than a bank, and with or without the approval of any person involved, may waive, modify, or extend time limits imposed or permitted by this [act] for a period not exceeding two additional banking days without discharge of drawers or indorsers or liability to its transferor or a prior party.

(b) Delay by a collecting bank or payor bank beyond time limits prescribed or permitted by this [act] or by instructions is excused if (i) the delay is caused by interruption of communication or computer facilities, suspension of payments by another bank, war, emergency conditions, failure of equipment, or other circumstances beyond the control of the bank, and (ii) the bank exercises such diligence as the circumstances require.

§ 4–110. Electronic Presentment.

(a) "Agreement for electronic presentment" means an agreement, clearing-house rule, or Federal Reserve regulation or operating circular, providing that presentment of an item may be made by transmission of an image of an item or information describing the item ("presentment notice") rather than delivery of the item itself. The agreement may provide for procedures governing retention, presentment, payment, dishonor, and other matters concerning items subject to the agreement.

(b) Presentment of an item pursuant to an agreement for presentment is made when the presentment notice is received.

(c) If presentment is made by presentment notice, a reference to "item" or "check" in this Article means the presentment notice unless the context otherwise indicates.

As added in 1990.

§ 4–111. Statute of Limitations.

An action to enforce an obligation, duty, or right arising under this Article must be commenced within three years after the [cause of action] accrues.

As added in 1990.

Part 2 Collection of Items: Depositary and Collecting Banks

§ 4–201. Status of Collecting Bank as Agent and Provisional Status of Credits; Applicability of Article; Item Indorsed "Pay Any Bank".

(a) Unless a contrary intent clearly appears and before the time that a settlement given by a collecting bank for an item is or becomes final, the bank, with respect to an item, is an agent or sub-agent of the owner of the item and any settlement given for the item is provisional. This provision applies regardless of the form of indorsement or lack of indorsement and even though credit given for the item is subject to immediate withdrawal as of right or is in fact withdrawn; but the continuance of ownership of an item by its owner and any rights of the owner to proceeds of the item are subject to rights of a collecting bank, such as those resulting from outstanding advances on the item and rights of recoupment or setoff. If an item is handled by banks for purposes of presentment, payment, collection, or return, the relevant provisions of this Article apply even though action of the parties clearly establishes that a particular bank has purchased the item and is the owner of it.

(b) After an item has been indorsed with the words "pay any bank" or the like, only a bank may acquire the rights of a holder until the item has been:

(1) returned to the customer initiating collection; or

(2) specially indorsed by a bank to a person who is not a bank.

As amended in 1990.

§ 4–202. Responsibility for Collection or Return; When Action Timely.

(a) A collecting bank must exercise ordinary care in:

(1) presenting an item or sending it for presentment;

(2) sending notice of dishonor or nonpayment or returning an item other than a documentary draft to the bank's transferor after learning that the item has not been paid or accepted, as the case may be;

(3) settling for an item when the bank receives final settlement; and

(4) notifying its transferor of any loss or delay in transit within a reasonable time after discovery thereof.

(b) A collecting bank exercises ordinary care under subsection (a) by taking proper action before its midnight deadline following receipt of an item, notice, or settlement. Taking proper action within a reasonably longer time may constitute the exercise of ordinary care, but the bank has the burden of establishing timeliness.

(c) Subject to subsection (a)(1), a bank is not liable for the insolvency, neglect, misconduct, mistake, or default of another bank or

person or for loss or destruction of an item in the possession of others or in transit.

As amended in 1990.

§ 4–203. Effect of Instructions.

Subject to Article 3 concerning conversion of instruments (Section 3–420) and restrictive indorsements (Section 3–206), only a collecting bank's transferor can give instructions that affect the bank or constitute notice to it, and a collecting bank is not liable to prior parties for any action taken pursuant to the instructions or in accordance with any agreement with its transferor.

§ 4–204. Methods of Sending and Presenting; Sending Directly to Payor Bank.

(a) A collecting bank shall send items by a reasonably prompt method, taking into consideration relevant instructions, the nature of the item, the number of those items on hand, the cost of collection involved, and the method generally used by it or others to present those items.

(b) A collecting bank may send:

(1) an item directly to the payor bank;

(2) an item to a nonbank payor if authorized by its transferor; and

(3) an item other than documentary drafts to a nonbank payor, if authorized by Federal Reserve regulation or operating circular, clearing-house rule, or the like.

(c) Presentment may be made by a presenting bank at a place where the payor bank or other payor has requested that presentment be made.

As amended in 1990.

§ 4–205. Depositary Bank Holder of Unindorsed Item.

If a customer delivers an item to a depositary bank for collection:

(1) the depositary bank becomes a holder of the item at the time it receives the item for collection if the customer at the time of delivery was a holder of the item, whether or not the customer indorses the item, and, if the bank satisfies the other requirements of Section 3–302, it is a holder in due course; and

(2) the depositary bank warrants to collecting banks, the payor bank or other payor, and the drawer that the amount of the item was paid to the customer or deposited to the customer's account.

As amended in 1990.

§ 4–206. Transfer Between Banks.

Any agreed method that identifies the transferor bank is sufficient for the item's further transfer to another bank.

As amended in 1990.

§ 4–207. Transfer Warranties.

(a) A customer or collecting bank that transfers an item and receives a settlement or other consideration warrants to the transferee and to any subsequent collecting bank that:

(1) the warrantor is a person entitled to enforce the item;

(2) all signatures on the item are authentic and authorized;

(3) the item has not been altered;

(4) the item is not subject to a defense or claim in recoupment (Section 3–305(a)) of any party that can be asserted against the warrantor; and

(5) the warrantor has no knowledge of any insolvency proceeding commenced with respect to the maker or acceptor or, in the case of an unaccepted draft, the drawer.

(b) If an item is dishonored, a customer or collecting bank transferring the item and receiving settlement or other consideration is obliged to pay the amount due on the item (i) according to the terms of the item at the time it was transferred, or (ii) if the transfer was of an incomplete item, according to its terms when completed as stated in Sections 3–115 and 3–407. The obligation of a transferor is owed to the transferee and to any subsequent collecting bank that takes the item in good faith. A transferor cannot disclaim its obligation under this subsection by an indorsement stating that it is made "without recourse" or otherwise disclaiming liability.

(c) A person to whom the warranties under subsection (a) are made and who took the item in good faith may recover from the warrantor as damages for breach of warranty an amount equal to the loss suffered as a result of the breach, but not more than the amount of the item plus expenses and loss of interest incurred as a result of the breach.

(d) The warranties stated in subsection (a) cannot be disclaimed with respect to checks. Unless notice of a claim for breach of warranty is given to the warrantor within 30 days after the claimant has reason to know of the breach and the identity of the warrantor, the warrantor is discharged to the extent of any loss caused by the delay in giving notice of the claim.

(e) A cause of action for breach of warranty under this section accrues when the claimant has reason to know of the breach.

As amended in 1990.

§ 4–208. Presentment Warranties.

(a) If an unaccepted draft is presented to the drawee for payment or acceptance and the drawee pays or accepts the draft, (i) the person obtaining payment or acceptance, at the time of presentment, and (ii) a previous transferor of the draft, at the time of transfer, warrant to the drawee that pays or accepts the draft in good faith that:

(1) the warrantor is, or was, at the time the warrantor transferred the draft, a person entitled to enforce the draft or authorized to obtain payment or acceptance of the draft on behalf of a person entitled to enforce the draft;

(2) the draft has not been altered; and

(3) the warrantor has no knowledge that the signature of the purported drawer of the draft is unauthorized.

(b) A drawee making payment may recover from a warrantor damages for breach of warranty equal to the amount paid by the drawee less the amount the drawee received or is entitled to receive from the drawer because of the payment. In addition, the drawee is entitled to compensation for expenses and loss of interest resulting from the breach. The right of the drawee to recover damages under this subsection is not affected by any failure of the drawee to exercise ordinary care in making payment. If the drawee accepts the draft (i) breach of warranty is a defense to the obligation of the acceptor, and (ii) if the acceptor makes payment with respect to the draft, the acceptor is entitled to recover from a warrantor for breach of warranty the amounts stated in this subsection.

(c) If a drawee asserts a claim for breach of warranty under subsection (a) based on an unauthorized indorsement of the draft or an alteration of the draft, the warrantor may defend by proving that the indorsement is effective under Section 3–404 or 3–405 or the drawer is precluded under Section 3–406 or 4–406 from asserting against the drawee the unauthorized indorsement or alteration.

(d) If (i) a dishonored draft is presented for payment to the drawer or an indorser or (ii) any other item is presented for payment to a party obliged to pay the item, and the item is paid, the person obtaining payment and a prior transferor of the item warrant to the person making payment in good faith that the warrantor is, or was, at the time the warrantor transferred the item, a person entitled to enforce the item or authorized to obtain payment on behalf of a person entitled to enforce the item. The person making payment may recover from any warrantor for breach of warranty an amount equal to the amount paid plus expenses and loss of interest resulting from the breach.

(e) The warranties stated in subsections (a) and (d) cannot be disclaimed with respect to checks. Unless notice of a claim for breach of warranty is given to the warrantor within 30 days after the claimant has reason to know of the breach and the identity of the warrantor, the warrantor is discharged to the extent of any loss caused by the delay in giving notice of the claim.

(f) A cause of action for breach of warranty under this section accrues when the claimant has reason to know of the breach.

As amended in 1990.

§ 4–209. Encoding and Retention Warranties.

(a) A person who encodes information on or with respect to an item after issue warrants to any subsequent collecting bank and to the payor bank or other payor that the information is correctly encoded. If the customer of a depositary bank encodes, that bank also makes the warranty.

(b) A person who undertakes to retain an item pursuant to an agreement for electronic presentment warrants to any subsequent collecting bank and to the payor bank or other payor that retention and presentment of the item comply with the agreement. If a customer of a depositary bank undertakes to retain an item, that bank also makes this warranty.

(c) A person to whom warranties are made under this section and who took the item in good faith may recover from the warrantor as damages for breach of warranty an amount equal to the loss suffered as a result of the breach, plus expenses and loss of interest incurred as a result of the breach.

As added in 1990.

§ 4–210. Security Interest of Collecting Bank in Items, Accompanying Documents and Proceeds.

(a) A collecting bank has a security interest in an item and any accompanying documents or the proceeds of either:

(1) in case of an item deposited in an account, to the extent to which credit given for the item has been withdrawn or applied;

(2) in case of an item for which it has given credit available for withdrawal as of right, to the extent of the credit given, whether or not the credit is drawn upon or there is a right of charge-back; or

(3) if it makes an advance on or against the item.

(b) If credit given for several items received at one time or pursuant to a single agreement is withdrawn or applied in part, the security interest remains upon all the items, any accompanying documents or the proceeds of either. For the purpose of this section, credits first given are first withdrawn.

(c) Receipt by a collecting bank of a final settlement for an item is a realization on its security interest in the item, accompanying documents, and proceeds. So long as the bank does not receive final settlement for the item or give up possession of the item or accompanying documents for purposes other than collection, the security interest continues to that extent and is subject to Article 9, but:

(1) no security agreement is necessary to make the security interest enforceable (Section 9–203(1)(a));

(2) no filing is required to perfect the security interest; and

(3) the security interest has priority over conflicting perfected security interests in the item, accompanying documents, or proceeds.

As amended in 1990 and 1999.

§ 4–211. When Bank Gives Value for Purposes of Holder in Due Course.

For purposes of determining its status as a holder in due course, a bank has given value to the extent it has a security interest in an item, if the bank otherwise complies with the requirements of Section 3–302 on what constitutes a holder in due course.

As amended in 1990.

§ 4–212. Presentment by Notice of Item Not Payable by, Through, or at Bank; Liability of Drawer or Indorser.

(a) Unless otherwise instructed, a collecting bank may present an item not payable by, through, or at a bank by sending to the

party to accept or pay a written notice that the bank holds the item for acceptance or payment. The notice must be sent in time to be received on or before the day when presentment is due and the bank must meet any requirement of the party to accept or pay under Section 3–501 by the close of the bank's next banking day after it knows of the requirement.

(b) If presentment is made by notice and payment, acceptance, or request for compliance with a requirement under Section 3–501 is not received by the close of business on the day after maturity or, in the case of demand items, by the close of business on the third banking day after notice was sent, the presenting bank may treat the item as dishonored and charge any drawer or indorser by sending it notice of the facts.

As amended in 1990.

§ 4–213. Medium and Time of Settlement by Bank.

(a) With respect to settlement by a bank, the medium and time of settlement may be prescribed by Federal Reserve regulations or circulars, clearing-house rules, and the like, or agreement. In the absence of such prescription:

(1) the medium of settlement is cash or credit to an account in a Federal Reserve bank of or specified by the person to receive settlement; and

(2) the time of settlement is:

(i) with respect to tender of settlement by cash, a cashier's check, or teller's check, when the cash or check is sent or delivered;

(ii) with respect to tender of settlement by credit in an account in a Federal Reserve Bank, when the credit is made;

(iii) with respect to tender of settlement by a credit or debit to an account in a bank, when the credit or debit is made or, in the case of tender of settlement by authority to charge an account, when the authority is sent or delivered; or

(iv) with respect to tender of settlement by a funds transfer, when payment is made pursuant to Section 4A–406(a) to the person receiving settlement.

(b) If the tender of settlement is not by a medium authorized by subsection (a) or the time of settlement is not fixed by subsection (a), no settlement occurs until the tender of settlement is accepted by the person receiving settlement.

(c) If settlement for an item is made by cashier's check or teller's check and the person receiving settlement, before its midnight deadline:

(1) presents or forwards the check for collection, settlement is final when the check is finally paid; or

(2) fails to present or forward the check for collection, settlement is final at the midnight deadline of the person receiving settlement.

(d) If settlement for an item is made by giving authority to charge the account of the bank giving settlement in the bank receiving settlement, settlement is final when the charge is made by the bank receiving settlement if there are funds available in the account for the amount of the item.

As amended in 1990.

§ 4–214. Right of Charge-Back or Refund; Liability of Collecting Bank: Return of Item.

(a) If a collecting bank has made provisional settlement with its customer for an item and fails by reason of dishonor, suspension of payments by a bank, or otherwise to receive settlement for the item which is or becomes final, the bank may revoke the settlement given by it, charge back the amount of any credit given for the item to its customer's account, or obtain refund from its customer, whether or not it is able to return the item, if by its midnight deadline or within a longer reasonable time after it learns the facts it returns the item or sends notification of the facts. If the return or notice is delayed beyond the bank's midnight deadline or a longer reasonable time after it learns the facts, the bank may revoke the settlement, charge back the credit, or obtain refund from its customer, but it is liable for any loss resulting from the delay. These rights to revoke, charge back, and obtain refund terminate if and when a settlement for the item received by the bank is or becomes final.

(b) A collecting bank returns an item when it is sent or delivered to the bank's customer or transferor or pursuant to its instructions.

(c) A depositary bank that is also the payor may charge back the amount of an item to its customer's account or obtain refund in accordance with the section governing return of an item received by a payor bank for credit on its books (Section 4–301).

(d) The right to charge back is not affected by:

(1) previous use of a credit given for the item; or

(2) failure by any bank to exercise ordinary care with respect to the item, but a bank so failing remains liable.

(e) A failure to charge back or claim refund does not affect other rights of the bank against the customer or any other party.

(f) If credit is given in dollars as the equivalent of the value of an item payable in foreign money, the dollar amount of any charge-back or refund must be calculated on the basis of the bank-offered spot rate for the foreign money prevailing on the day when the person entitled to the charge-back or refund learns that it will not receive payment in ordinary course.

As amended in 1990.

§ 4–215. Final Payment of Item by Payor Bank; When Provisional Debits and Credits Become Final; When Certain Credits Become Available for Withdrawal.

(a) An item is finally paid by a payor bank when the bank has first done any of the following:

(1) paid the item in cash;

(2) settled for the item without having a right to revoke the settlement under statute, clearing-house rule, or agreement; or

(3) made a provisional settlement for the item and failed to revoke the settlement in the time and manner permitted by statute, clearing-house rule, or agreement.

(b) If provisional settlement for an item does not become final, the item is not finally paid.

(c) If provisional settlement for an item between the presenting and payor banks is made through a clearing house or by debits or credits in an account between them, then to the extent that provisional debits or credits for the item are entered in accounts between the presenting and payor banks or between the presenting and successive prior collecting banks seriatim, they become final upon final payment of the item by the payor bank.

(d) If a collecting bank receives a settlement for an item which is or becomes final, the bank is accountable to its customer for the amount of the item and any provisional credit given for the item in an account with its customer becomes final.

(e) Subject to (i) applicable law stating a time for availability of funds and (ii) any right of the bank to apply the credit to an obligation of the customer, credit given by a bank for an item in a customer's account becomes available for withdrawal as of right:

(1) if the bank has received a provisional settlement for the item, when the settlement becomes final and the bank has had a reasonable time to receive return of the item and the item has not been received within that time;

(2) if the bank is both the depositary bank and the payor bank, and the item is finally paid, at the opening of the bank's second banking day following receipt of the item.

(f) Subject to applicable law stating a time for availability of funds and any right of a bank to apply a deposit to an obligation of the depositor, a deposit of money becomes available for withdrawal as of right at the opening of the bank's next banking day after receipt of the deposit.

As amended in 1990.

§ 4- 216. Insolvency and Preference.

(a) If an item is in or comes into the possession of a payor or collecting bank that suspends payment and the item has not been finally paid, the item must be returned by the receiver, trustee, or agent in charge of the closed bank to the presenting bank or the closed bank's customer.

(b) If a payor bank finally pays an item and suspends payments without making a settlement for the item with its customer or the presenting bank which settlement is or becomes final, the owner of the item has a preferred claim against the payor bank.

(c) If a payor bank gives or a collecting bank gives or receives a provisional settlement for an item and thereafter suspends payments, the suspension does not prevent or interfere with the settlement's becoming final if the finality occurs automatically upon the lapse of certain time or the happening of certain events.

(d) If a collecting bank receives from subsequent parties settlement for an item, which settlement is or becomes final and the bank suspends payments without making a settlement for the item with its customer which settlement is or becomes final, the owner of the item has a preferred claim against the collecting bank.

As amended in 1990.

Part 3 Collection of Items: Payor Banks

§ 4-301. Deferred Posting; Recovery of Payment by Return of Items; Time of Dishonor; Return of Items by Payor Bank.

(a) If a payor bank settles for a demand item other than a documentary draft presented otherwise than for immediate payment over the counter before midnight of the banking day of receipt, the payor bank may revoke the settlement and recover the settlement if, before it has made final payment and before its midnight deadline, it

(1) returns the item; or

(2) sends written notice of dishonor or nonpayment if the item is unavailable for return.

(b) If a demand item is received by a payor bank for credit on its books, it may return the item or send notice of dishonor and may revoke any credit given or recover the amount thereof withdrawn by its customer, if it acts within the time limit and in the manner specified in subsection (a).

(c) Unless previous notice of dishonor has been sent, an item is dishonored at the time when for purposes of dishonor it is returned or notice sent in accordance with this section.

(d) An item is returned:

(1) as to an item presented through a clearing house, when it is delivered to the presenting or last collecting bank or to the clearing house or is sent or delivered in accordance with clearing-house rules; or

(2) in all other cases, when it is sent or delivered to the bank's customer or transferor or pursuant to instructions.

As amended in 1990.

§ 4-302. Payor Bank's Responsibility for Late Return of Item.

(a) If an item is presented to and received by a payor bank, the bank is accountable for the amount of:

(1) a demand item, other than a documentary draft, whether properly payable or not, if the bank, in any case in which it is not also the depositary bank, retains the item beyond midnight of the banking day of receipt without settling for it or, whether or not it is also the depositary bank, does not pay or return the item or send notice of dishonor until after its midnight deadline; or

(2) any other properly payable item unless, within the time allowed for acceptance or payment of that item, the bank either accepts or pays the item or returns it and accompanying documents.

(b) The liability of a payor bank to pay an item pursuant to subsection (a) is subject to defenses based on breach of a presentment warranty (Section 4–208) or proof that the person seeking enforcement of the liability presented or transferred the item for the purpose of defrauding the payor bank.

As amended in 1990.

§ 4–303. When Items Subject to Notice, Stop-Payment Order, Legal Process, or Setoff; Order in Which Items May Be Charged or Certified.

(a) Any knowledge, notice, or stop-payment order received by, legal process served upon, or setoff exercised by a payor bank comes too late to terminate, suspend, or modify the bank's right or duty to pay an item or to charge its customer's account for the item if the knowledge, notice, stop-payment order, or legal process is received or served and a reasonable time for the bank to act thereon expires or the setoff is exercised after the earliest of the following:

(1) the bank accepts or certifies the item;

(2) the bank pays the item in cash;

(3) the bank settles for the item without having a right to revoke the settlement under statute, clearing-house rule, or agreement;

(4) the bank becomes accountable for the amount of the item under Section 4–302 dealing with the payor bank's responsibility for late return of items; or

(5) with respect to checks, a cutoff hour no earlier than one hour after the opening of the next banking day after the banking day on which the bank received the check and no later than the close of that next banking day or, if no cutoff hour is fixed, the close of the next banking day after the banking day on which the bank received the check.

(b) Subject to subsection (a), items may be accepted, paid, certified, or charged to the indicated account of its customer in any order.

As amended in 1990.

Part 4 Relationship Between Payor Bank and Its Customer

§ 4–401. When Bank May Charge Customer's Account.

(a) A bank may charge against the account of a customer an item that is properly payable from the account even though the charge creates an overdraft. An item is properly payable if it is authorized by the customer and is in accordance with any agreement between the customer and bank.

(b) A customer is not liable for the amount of an overdraft if the customer neither signed the item nor benefited from the proceeds of the item.

(c) A bank may charge against the account of a customer a check that is otherwise properly payable from the account, even though payment was made before the date of the check, unless the customer has given notice to the bank of the postdating describing the check with reasonable certainty. The notice is effective for the period stated in Section 4–403(b) for stop-payment orders, and must be received at such time and in such manner as to afford the bank a reasonable opportunity to act on it before the bank takes any action with respect to the check described in Section 4–303. If a bank charges against the account of a customer a check before the date stated in the notice of postdating, the bank is liable for damages for the loss resulting from its act. The loss may include damages for dishonor of subsequent items under Section 4–402.

(d) A bank that in good faith makes payment to a holder may charge the indicated account of its customer according to:

(1) the original terms of the altered item; or

(2) the terms of the completed item, even though the bank knows the item has been completed unless the bank has notice that the completion was improper.

As amended in 1990.

§ 4–402. Bank's Liability to Customer for Wrongful Dishonor; Time of Determining Insufficiency of Account.

(a) Except as otherwise provided in this Article, a payor bank wrongfully dishonors an item if it dishonors an item that is properly payable, but a bank may dishonor an item that would create an overdraft unless it has agreed to pay the overdraft.

(b) A payor bank is liable to its customer for damages proximately caused by the wrongful dishonor of an item. Liability is limited to actual damages proved and may include damages for an arrest or prosecution of the customer or other consequential damages. Whether any consequential damages are proximately caused by the wrongful dishonor is a question of fact to be determined in each case.

(c) A payor bank's determination of the customer's account balance on which a decision to dishonor for insufficiency of available funds is based may be made at any time between the time the item is received by the payor bank and the time that the payor bank returns the item or gives notice in lieu of return, and no more than one determination need be made. If, at the election of the payor bank, a subsequent balance determination is made for the purpose of reevaluating the bank's decision to dishonor the item, the account balance at that time is determinative of whether a dishonor for insufficiency of available funds is wrongful.

As amended in 1990.

§ 4–403. Customer's Right to Stop Payment; Burden of Proof of Loss.

(a) A customer or any person authorized to draw on the account if there is more than one person may stop payment of any item drawn on the customer's account or close the account by an order to the bank describing the item or account with reasonable certainty received at a time and in a manner that affords the bank a reasonable opportunity to act on it before any action by the bank with respect to the item described in Section 4–303. If the signature of more than one person is

required to draw on an account, any of these persons may stop payment or close the account.

(b) A stop-payment order is effective for six months, but it lapses after 14 calendar days if the original order was oral and was not confirmed in writing within that period. A stop-payment order may be renewed for additional six-month periods by a writing given to the bank within a period during which the stop-payment order is effective.

(c) The burden of establishing the fact and amount of loss resulting from the payment of an item contrary to a stop-payment order or order to close an account is on the customer. The loss from payment of an item contrary to a stop-payment order may include damages for dishonor of subsequent items under Section 4–402.

As amended in 1990.

§ 4–404. Bank Not Obliged to Pay Check More Than Six Months Old.

A bank is under no obligation to a customer having a checking account to pay a check, other than a certified check, which is presented more than six months after its date, but it may charge its customer's account for a payment made thereafter in good faith.

§ 4–405. Death or Incompetence of Customer.

(a) A payor or collecting bank's authority to accept, pay, or collect an item or to account for proceeds of its collection, if otherwise effective, is not rendered ineffective by incompetence of a customer of either bank existing at the time the item is issued or its collection is undertaken if the bank does not know of an adjudication of incompetence. Neither death nor incompetence of a customer revokes the authority to accept, pay, collect, or account until the bank knows of the fact of death or of an adjudication of incompetence and has reasonable opportunity to act on it.

(b) Even with knowledge, a bank may for 10 days after the date of death pay or certify checks drawn on or before the date unless ordered to stop payment by a person claiming an interest in the account.

As amended in 1990.

§ 4–406. Customer's Duty to Discover and Report Unauthorized Signature or Alteration.

(a) A bank that sends or makes available to a customer a statement of account showing payment of items for the account shall either return or make available to the customer the items paid or provide information in the statement of account sufficient to allow the customer reasonably to identify the items paid. The statement of account provides sufficient information if the item is described by item number, amount, and date of payment.

(b) If the items are not returned to the customer, the person retaining the items shall either retain the items or, if the items are destroyed, maintain the capacity to furnish legible copies of the items until the expiration of seven years after receipt of the items. A customer may request an item from the bank that paid the item, and that bank must provide in a reasonable time either the item or, if the item has been destroyed or is not otherwise obtainable, a legible copy of the item.

(c) If a bank sends or makes available a statement of account or items pursuant to subsection (a), the customer must exercise reasonable promptness in examining the statement or the items to determine whether any payment was not authorized because of an alteration of an item or because a purported signature by or on behalf of the customer was not authorized. If, based on the statement or items provided, the customer should reasonably have discovered the unauthorized payment, the customer must promptly notify the bank of the relevant facts.

(d) If the bank proves that the customer failed, with respect to an item, to comply with the duties imposed on the customer by subsection (c), the customer is precluded from asserting against the bank:

(1) the customer's unauthorized signature or any alteration on the item, if the bank also proves that it suffered a loss by reason of the failure; and

(2) the customer's unauthorized signature or alteration by the same wrongdoer on any other item paid in good faith by the bank if the payment was made before the bank received notice from the customer of the unauthorized signature or alteration and after the customer had been afforded a reasonable period of time, not exceeding 30 days, in which to examine the item or statement of account and notify the bank.

(e) If subsection (d) applies and the customer proves that the bank failed to exercise ordinary care in paying the item and that the failure substantially contributed to loss, the loss is allocated between the customer precluded and the bank asserting the preclusion according to the extent to which the failure of the customer to comply with subsection (c) and the failure of the bank to exercise ordinary care contributed to the loss. If the customer proves that the bank did not pay the item in good faith, the preclusion under subsection (d) does not apply.

(f) Without regard to care or lack of care of either the customer or the bank, a customer who does not within one year after the statement or items are made available to the customer (subsection (a)) discover and report the customer's unauthorized signature on or any alteration on the item is precluded from asserting against the bank the unauthorized signature or alteration. If there is a preclusion under this subsection, the payor bank may not recover for breach or warranty under Section 4–208 with respect to the unauthorized signature or alteration to which the preclusion applies.

As amended in 1990.

§ 4–407. Payor Bank's Right to Subrogation on Improper Payment.

If a payor has paid an item over the order of the drawer or maker to stop payment, or after an account has been closed, or otherwise under circumstances giving a basis for objection by the drawer or maker, to prevent unjust enrichment and only to the

extent necessary to prevent loss to the bank by reason of its payment of the item, the payor bank is subrogated to the rights

(1) of any holder in due course on the item against the drawer or maker;

(2) of the payee or any other holder of the item against the drawer or maker either on the item or under the transaction out of which the item arose; and

(3) of the drawer or maker against the payee or any other holder of the item with respect to the transaction out of which the item arose.

As amended in 1990.

Part 5 Collection of Documentary Drafts

§ 4–501. Handling of Documentary Drafts; Duty to Send for Presentment and to Notify Customer of Dishonor.

A bank that takes a documentary draft for collection shall present or send the draft and accompanying documents for presentment and, upon learning that the draft has not been paid or accepted in due course, shall seasonably notify its customer of the fact even though it may have discounted or bought the draft or extended credit available for withdrawal as of right.

As amended in 1990.

§ 4–502. Presentment of "On Arrival" Drafts.

If a draft or the relevant instructions require presentment "on arrival", "when goods arrive" or the like, the collecting bank need not present until in its judgment a reasonable time for arrival of the goods has expired. Refusal to pay or accept because the goods have not arrived is not dishonor; the bank must notify its transferor of the refusal but need not present the draft again until it is instructed to do so or learns of the arrival of the goods.

§ 4–503. Responsibility of Presenting Bank for Documents and Goods; Report of Reasons for Dishonor; Referee in Case of Need.

Unless otherwise instructed and except as provided in Article 5, a bank presenting a documentary draft:

(1) must deliver the documents to the drawee on acceptance of the draft if it is payable more than three days after presentment, otherwise, only on payment; and

(2) upon dishonor, either in the case of presentment for acceptance or presentment for payment, may seek and follow instructions from any referee in case of need designated in the draft or, if the presenting bank does not choose to utilize the referee's services, it must use diligence and good faith to ascertain the reason for dishonor, must notify its transferor of the dishonor and of the results of its effort to ascertain the reasons therefor, and must request instructions.

However, the presenting bank is under no obligation with respect to goods represented by the documents except to follow any reasonable instructions seasonably received; it has a right to reimbursement for any expense incurred in following instructions and to prepayment of or indemnity for those expenses.

As amended in 1990.

§ 4–504. Privilege of Presenting Bank to Deal With Goods; Security Interest for Expenses.

(a) A presenting bank that, following the dishonor of a documentary draft, has seasonably requested instructions but does not receive them within a reasonable time may store, sell, or otherwise deal with the goods in any reasonable manner.

(b) For its reasonable expenses incurred by action under subsection (a) the presenting bank has a lien upon the goods or their proceeds, which may be foreclosed in the same manner as an unpaid seller's lien.

As amended in 1990.

Article 4A
FUNDS TRANSFERS

Part 1 Subject Matter and Definitions

§ 4A–101. Short Title.

This Article may be cited as Uniform Commercial Code—Funds Transfers.

§ 4A–102. Subject Matter.

Except as otherwise provided in Section 4A–108, this Article applies to funds transfers defined in Section 4A–104.

§ 4A–103. Payment Order–Definitions.

(a) In this Article:

(1) "Payment order" means an instruction of a sender to a receiving bank, transmitted orally, electronically, or in writing, to pay, or to cause another bank to pay, a fixed or determinable amount of money to a beneficiary if:

(i) the instruction does not state a condition to payment to the beneficiary other than time of payment,

(ii) the receiving bank is to be reimbursed by debiting an account of, or otherwise receiving payment from, the sender, and

(iii) the instruction is transmitted by the sender directly to the receiving bank or to an agent, funds-transfer system, or communication system for transmittal to the receiving bank.

(2) "Beneficiary" means the person to be paid by the beneficiary's bank.

(3) "Beneficiary's bank" means the bank identified in a payment order in which an account of the beneficiary is to be credited pursuant to the order or which otherwise is to make payment to the beneficiary if the order does not provide for payment to an account.

(4) "Receiving bank" means the bank to which the sender's instruction is addressed.

(5) "Sender" means the person giving the instruction to the receiving bank.

(b) If an instruction complying with subsection (a)(1) is to make more than one payment to a beneficiary, the instruction is a separate payment order with respect to each payment.

(c) A payment order is issued when it is sent to the receiving bank.

§ 4A–104. Funds Transfer–Definitions.

In this Article:

(a) "Funds transfer" means the series of transactions, beginning with the originator's payment order, made for the purpose of making payment to the beneficiary of the order. The term includes any payment order issued by the originator's bank or an intermediary bank intended to carry out the originator's payment order. A funds transfer is completed by acceptance by the beneficiary's bank of a payment order for the benefit of the beneficiary of the originator's payment order.

(b) "Intermediary bank" means a receiving bank other than the originator's bank or the beneficiary's bank.

(c) "Originator" means the sender of the first payment order in a funds transfer.

(d) "Originator's bank" means (i) the receiving bank to which the payment order of the originator is issued if the originator is not a bank, or (ii) the originator if the originator is a bank.

§ 4A–105. Other Definitions.

(a) In this Article:

(1) "Authorized account" means a deposit account of a customer in a bank designated by the customer as a source of payment of payment orders issued by the customer to the bank. If a customer does not so designate an account, any account of the customer is an authorized account if payment of a payment order from that account is not inconsistent with a restriction on the use of that account.

(2) "Bank" means a person engaged in the business of banking and includes a savings bank, savings and loan association, credit union, and trust company. A branch or separate office of a bank is a separate bank for purposes of this Article.

(3) "Customer" means a person, including a bank, having an account with a bank or from whom a bank has agreed to receive payment orders.

(4) "Funds-transfer business day" of a receiving bank means the part of a day during which the receiving bank is open for the receipt, processing, and transmittal of payment orders and cancellations and amendments of payment orders.

(5) "Funds-transfer system" means a wire transfer network, automated clearing house, or other communication system of a clearing house or other association of banks through which a payment order by a bank may be transmitted to the bank to which the order is addressed.

(6) "Good faith" means honesty in fact and the observance of reasonable commercial standards of fair dealing.

(7) "Prove" with respect to a fact means to meet the burden of establishing the fact (Section 1–201(8)).

(b) Other definitions applying to this Article and the sections in which they appear are:

"Acceptance"	Section 4A–209
"Beneficiary"	Section 4A–103
"Beneficiary's bank"	Section 4A–103
"Executed"	Section 4A–301
"Execution date"	Section 4A–301
"Funds transfer"	Section 4A–104
"Funds-transfer system rule"	Section 4A–501
"Intermediary bank"	Section 4A–104
"Originator"	Section 4A–104
"Originator's bank"	Section 4A–104
"Payment by beneficiary's bank to beneficiary"	Section 4A–405
"Payment by originator to beneficiary"	Section 4A–406
"Payment by sender to receiving bank"	Section 4A–403
"Payment date"	Section 4A–401
"Payment order"	Section 4A–103
"Receiving bank"	Section 4A–103
"Security procedure"	Section 4A–201
"Sender"	Section 4A–103

(c) The following definitions in Article 4 apply to this Article:

"Clearing house"	Section 4–104
"Item"	Section 4–104
"Suspends payments"	Section 4–104

(d) In addition, Article 1 contains general definitions and principles of construction and interpretation applicable throughout this Article.

§ 4A–106. Time Payment Order Is Received.

(a) The time of receipt of a payment order or communication cancelling or amending a payment order is determined by the rules applicable to receipt of a notice stated in Section 1–201(27). A receiving bank may fix a cut-off time or times on a funds-transfer business day for the receipt and processing of payment orders and communications cancelling or amending payment orders. Different cut-off times may apply to payment orders, cancellations, or amendments, or to different categories of payment orders, cancellations, or amendments. A cut-off time may apply to senders generally or different cut-off times may apply to differ-

ent senders or categories of payment orders. If a payment order or communication cancelling or amending a payment order is received after the close of a funds-transfer business day or after the appropriate cut-off time on a funds-transfer business day, the receiving bank may treat the payment order or communication as received at the opening of the next funds-transfer business day.

(b) If this Article refers to an execution date or payment date or states a day on which a receiving bank is required to take action, and the date or day does not fall on a funds-transfer business day, the next day that is a funds-transfer business day is treated as the date or day stated, unless the contrary is stated in this Article.

§ 4A–107. Federal Reserve Regulations and Operating Circulars.

Regulations of the Board of Governors of the Federal Reserve System and operating circulars of the Federal Reserve Banks supersede any inconsistent provision of this Article to the extent of the inconsistency.

§ 4A–108. Exclusion of Consumer Transactions Governed by Federal Law.

This Article does not apply to a funds transfer any part of which is governed by the Electronic Fund Transfer Act of 1978 (Title XX, Public Law 95–630, 92 Stat. 3728, 15 U.S.C. § 1693 et seq.) as amended from time to time.

Part 2 Issue and Acceptance of Payment Order

§ 4A–201. Security Procedure.

"Security procedure" means a procedure established by agreement of a customer and a receiving bank for the purpose of (i) verifying that a payment order or communication amending or cancelling a payment order is that of the customer, or (ii) detecting error in the transmission or the content of the payment order or communication. A security procedure may require the use of algorithms or other codes, identifying words or numbers, encryption, callback procedures, or similar security devices. Comparison of a signature on a payment order or communication with an authorized specimen signature of the customer is not by itself a security procedure.

§ 4A–202. Authorized and Verified Payment Orders.

(a) A payment order received by the receiving bank is the authorized order of the person identified as sender if that person authorized the order or is otherwise bound by it under the law of agency.

(b) If a bank and its customer have agreed that the authenticity of payment orders issued to the bank in the name of the customer as sender will be verified pursuant to a security procedure, a payment order received by the receiving bank is effective as the order of the customer, whether or not authorized, if (i) the security procedure is a commercially reasonable method of providing security against unauthorized payment orders, and (ii) the bank proves that it accepted the payment order in good faith and in compliance with the security procedure and any written agreement or instruction of the customer restricting acceptance of payment orders issued in the name of the customer. The bank is not required to follow an instruction that violates a written agreement with the customer or notice of which is not received at a time and in a manner affording the bank a reasonable opportunity to act on it before the payment order is accepted.

(c) Commercial reasonableness of a security procedure is a question of law to be determined by considering the wishes of the customer expressed to the bank, the circumstances of the customer known to the bank, including the size, type, and frequency of payment orders normally issued by the customer to the bank, alternative security procedures offered to the customer, and security procedures in general use by customers and receiving banks similarly situated. A security procedure is deemed to be commercially reasonable if (i) the security procedure was chosen by the customer after the bank offered, and the customer refused, a security procedure that was commercially reasonable for that customer, and (ii) the customer expressly agreed in writing to be bound by any payment order, whether or not authorized, issued in its name and accepted by the bank in compliance with the security procedure chosen by the customer.

(d) The term "sender" in this Article includes the customer in whose name a payment order is issued if the order is the authorized order of the customer under subsection (a), or it is effective as the order of the customer under subsection (b).

(e) This section applies to amendments and cancellations of payment orders to the same extent it applies to payment orders.

(f) Except as provided in this section and in Section 4A–203(a)(1), rights and obligations arising under this section or Section 4A–203 may not be varied by agreement.

§ 4A–203. Unenforceability of Certain Verified Payment Orders.

(a) If an accepted payment order is not, under Section 4A–202(a), an authorized order of a customer identified as sender, but is effective as an order of the customer pursuant to Section 4A–202(b), the following rules apply:

(1) By express written agreement, the receiving bank may limit the extent to which it is entitled to enforce or retain payment of the payment order.

(2) The receiving bank is not entitled to enforce or retain payment of the payment order if the customer proves that the order was not caused, directly or indirectly, by a person (i) entrusted at any time with duties to act for the customer with respect to payment orders or the security procedure, or (ii) who obtained access to transmitting facilities of the customer or who obtained, from a source controlled by the customer and without authority of the receiving bank, information facilitating breach of the security procedure, regardless of how the information was obtained or whether the customer was at fault. Information includes any access device, computer software, or the like.

(b) This section applies to amendments of payment orders to the same extent it applies to payment orders.

§ 4A–204. Refund of Payment and Duty of Customer to Report with Respect to Unauthorized Payment Order.

(a) If a receiving bank accepts a payment order issued in the name of its customer as sender which is (i) not authorized and not effective as the order of the customer under Section 4A–202, or (ii) not enforceable, in whole or in part, against the customer under Section 4A–203, the bank shall refund any payment of the payment order received from the customer to the extent the bank is not entitled to enforce payment and shall pay interest on the refundable amount calculated from the date the bank received payment to the date of the refund. However, the customer is not entitled to interest from the bank on the amount to be refunded if the customer fails to exercise ordinary care to determine that the order was not authorized by the customer and to notify the bank of the relevant facts within a reasonable time not exceeding 90 days after the date the customer received notification from the bank that the order was accepted or that the customer's account was debited with respect to the order. The bank is not entitled to any recovery from the customer on account of a failure by the customer to give notification as stated in this section.

(b) Reasonable time under subsection (a) may be fixed by agreement as stated in Section 1–204(1), but the obligation of a receiving bank to refund payment as stated in subsection (a) may not otherwise be varied by agreement.

§ 4A–205. Erroneous Payment Orders.

(a) If an accepted payment order was transmitted pursuant to a security procedure for the detection of error and the payment order (i) erroneously instructed payment to a beneficiary not intended by the sender, (ii) erroneously instructed payment in an amount greater than the amount intended by the sender, or (iii) was an erroneously transmitted duplicate of a payment order previously sent by the sender, the following rules apply:

(1) If the sender proves that the sender or a person acting on behalf of the sender pursuant to Section 4A–206 complied with the security procedure and that the error would have been detected if the receiving bank had also complied, the sender is not obliged to pay the order to the extent stated in paragraphs (2) and (3).

(2) If the funds transfer is completed on the basis of an erroneous payment order described in clause (i) or (iii) of subsection (a), the sender is not obliged to pay the order and the receiving bank is entitled to recover from the beneficiary any amount paid to the beneficiary to the extent allowed by the law governing mistake and restitution.

(3) If the funds transfer is completed on the basis of a payment order described in clause (ii) of subsection (a), the sender is not obliged to pay the order to the extent the amount received by the beneficiary is greater than the amount intended by the sender. In that case, the receiving bank is entitled to recover from the beneficiary the excess amount received to the extent allowed by the law governing mistake and restitution.

(b) If (i) the sender of an erroneous payment order described in subsection (a) is not obliged to pay all or part of the order, and (ii) the sender receives notification from the receiving bank that the order was accepted by the bank or that the sender's account was debited with respect to the order, the sender has a duty to exercise ordinary care, on the basis of information available to the sender, to discover the error with respect to the order and to advise the bank of the relevant facts within a reasonable time, not exceeding 90 days, after the bank's notification was received by the sender. If the bank proves that the sender failed to perform that duty, the sender is liable to the bank for the loss the bank proves it incurred as a result of the failure, but the liability of the sender may not exceed the amount of the sender's order.

(c) This section applies to amendments to payment orders to the same extent it applies to payment orders.

§ 4A–206. Transmission of Payment Order through Funds-Transfer or Other Communication System.

(a) If a payment order addressed to a receiving bank is transmitted to a funds-transfer system or other third party communication system for transmittal to the bank, the system is deemed to be an agent of the sender for the purpose of transmitting the payment order to the bank. If there is a discrepancy between the terms of the payment order transmitted to the system and the terms of the payment order transmitted by the system to the bank, the terms of the payment order of the sender are those transmitted by the system. This section does not apply to a funds-transfer system of the Federal Reserve Banks.

(b) This section applies to cancellations and amendments to payment orders to the same extent it applies to payment orders.

§ 4A–207. Misdescription of Beneficiary.

(a) Subject to subsection (b), if, in a payment order received by the beneficiary's bank, the name, bank account number, or other identification of the beneficiary refers to a nonexistent or unidentifiable person or account, no person has rights as a beneficiary of the order and acceptance of the order cannot occur.

(b) If a payment order received by the beneficiary's bank identifies the beneficiary both by name and by an identifying or bank account number and the name and number identify different persons, the following rules apply:

(1) Except as otherwise provided in subsection (c), if the beneficiary's bank does not know that the name and number refer to different persons, it may rely on the number as the proper identification of the beneficiary of the order. The beneficiary's bank need not determine whether the name and number refer to the same person.

(2) If the beneficiary's bank pays the person identified by name or knows that the name and number identify different persons, no person has rights as beneficiary except the person paid by the beneficiary's bank if that person was entitled to

receive payment from the originator of the funds transfer. If no person has rights as beneficiary, acceptance of the order cannot occur.

(c) If (i) a payment order described in subsection (b) is accepted, (ii) the originator's payment order described the beneficiary inconsistently by name and number, and (iii) the beneficiary's bank pays the person identified by number as permitted by subsection (b)(1), the following rules apply:

(1) If the originator is a bank, the originator is obliged to pay its order.

(2) If the originator is not a bank and proves that the person identified by number was not entitled to receive payment from the originator, the originator is not obliged to pay its order unless the originator's bank proves that the originator, before acceptance of the originator's order, had notice that payment of a payment order issued by the originator might be made by the beneficiary's bank on the basis of an identifying or bank account number even if it identifies a person different from the named beneficiary. Proof of notice may be made by any admissible evidence. The originator's bank satisfies the burden of proof if it proves that the originator, before the payment order was accepted, signed a writing stating the information to which the notice relates.

(d) In a case governed by subsection (b)(1), if the beneficiary's bank rightfully pays the person identified by number and that person was not entitled to receive payment from the originator, the amount paid may be recovered from that person to the extent allowed by the law governing mistake and restitution as follows:

(1) If the originator is obliged to pay its payment order as stated in subsection (c), the originator has the right to recover.

(2) If the originator is not a bank and is not obliged to pay its payment order, the originator's bank has the right to recover.

§ 4A–208. Misdescription of Intermediary Bank or Beneficiary's Bank.

(a) This subsection applies to a payment order identifying an intermediary bank or the beneficiary's bank only by an identifying number.

(1) The receiving bank may rely on the number as the proper identification of the intermediary or beneficiary's bank and need not determine whether the number identifies a bank.

(2) The sender is obliged to compensate the receiving bank for any loss and expenses incurred by the receiving bank as a result of its reliance on the number in executing or attempting to execute the order.

(b) This subsection applies to a payment order identifying an intermediary bank or the beneficiary's bank both by name and an identifying number if the name and number identify different persons.

(1) If the sender is a bank, the receiving bank may rely on the number as the proper identification of the intermediary or beneficiary's bank if the receiving bank, when it executes the sender's order, does not know that the name and number identify different persons. The receiving bank need not determine whether the name and number refer to the same person or whether the number refers to a bank. The sender is obliged to compensate the receiving bank for any loss and expenses incurred by the receiving bank as a result of its reliance on the number in executing or attempting to execute the order.

(2) If the sender is not a bank and the receiving bank proves that the sender, before the payment order was accepted, had notice that the receiving bank might rely on the number as the proper identification of the intermediary or beneficiary's bank even if it identifies a person different from the bank identified by name, the rights and obligations of the sender and the receiving bank are governed by subsection (b)(1), as though the sender were a bank. Proof of notice may be made by any admissible evidence. The receiving bank satisfies the burden of proof if it proves that the sender, before the payment order was accepted, signed a writing stating the information to which the notice relates.

(3) Regardless of whether the sender is a bank, the receiving bank may rely on the name as the proper identification of the intermediary or beneficiary's bank if the receiving bank, at the time it executes the sender's order, does not know that the name and number identify different persons. The receiving bank need not determine whether the name and number refer to the same person.

(4) If the receiving bank knows that the name and number identify different persons, reliance on either the name or the number in executing the sender's payment order is a breach of the obligation stated in Section 4A–302(a)(1).

§ 4A–209. Acceptance of Payment Order.

(a) Subject to subsection (d), a receiving bank other than the beneficiary's bank accepts a payment order when it executes the order.

(b) Subject to subsections (c) and (d), a beneficiary's bank accepts a payment order at the earliest of the following times:

(1) When the bank (i) pays the beneficiary as stated in Section 4A–405(a) or 4A–405(b), or (ii) notifies the beneficiary of receipt of the order or that the account of the beneficiary has been credited with respect to the order unless the notice indicates that the bank is rejecting the order or that funds with respect to the order may not be withdrawn or used until receipt of payment from the sender of the order;

(2) When the bank receives payment of the entire amount of the sender's order pursuant to Section 4A–403(a)(1) or 4A–403(a)(2); or

(3) The opening of the next funds-transfer business day of the bank following the payment date of the order if, at that time, the amount of the sender's order is fully covered by a withdrawable credit balance in an authorized account of the sender or the bank has otherwise received full payment from the sender, unless the order was rejected before that time or is rejected within (i) one hour after that time, or (ii) one hour after the opening of the next business day of the sender following the payment date if that time is later. If notice of rejection is received by the sender after the payment date and the authorized account of the sender does not bear interest, the bank is obliged to pay interest to the sender on the amount of the order for the number of days elapsing after the payment date to the day the sender receives notice or learns that the order was not accepted, counting that day as an elapsed day. If the withdrawable credit balance during that period falls below the amount of the order, the amount of interest payable is reduced accordingly.

(c) Acceptance of a payment order cannot occur before the order is received by the receiving bank. Acceptance does not occur under subsection (b)(2) or (b)(3) if the beneficiary of the payment order does not have an account with the receiving bank, the account has been closed, or the receiving bank is not permitted by law to receive credits for the beneficiary's account.

(d) A payment order issued to the originator's bank cannot be accepted until the payment date if the bank is the beneficiary's bank, or the execution date if the bank is not the beneficiary's bank. If the originator's bank executes the originator's payment order before the execution date or pays the beneficiary of the originator's payment order before the payment date and the payment order is subsequently cancelled pursuant to Section 4A–211(b), the bank may recover from the beneficiary any payment received to the extent allowed by the law governing mistake and restitution.

§ 4A–210. Rejection of Payment Order.

(a) A payment order is rejected by the receiving bank by a notice of rejection transmitted to the sender orally, electronically, or in writing. A notice of rejection need not use any particular words and is sufficient if it indicates that the receiving bank is rejecting the order or will not execute or pay the order. Rejection is effective when the notice is given if transmission is by a means that is reasonable in the circumstances. If notice of rejection is given by a means that is not reasonable, rejection is effective when the notice is received. If an agreement of the sender and receiving bank establishes the means to be used to reject a payment order, (i) any means complying with the agreement is reasonable and (ii) any means not complying is not reasonable unless no significant delay in receipt of the notice resulted from the use of the noncomplying means.

(b) This subsection applies if a receiving bank other than the beneficiary's bank fails to execute a payment order despite the existence on the execution date of a withdrawable credit balance in an authorized account of the sender sufficient to cover the order. If the sender does not receive notice of rejection of the order on the execution date and the authorized account of the sender does not bear interest, the bank is obliged to pay interest to the sender on the amount of the order for the number of days elapsing after the execution date to the earlier of the day the order is cancelled pursuant to Section 4A–211(d) or the day the sender receives notice or learns that the order was not executed, counting the final day of the period as an elapsed day. If the withdrawable credit balance during that period falls below the amount of the order, the amount of interest is reduced accordingly.

(c) If a receiving bank suspends payments, all unaccepted payment orders issued to it are are deemed rejected at the time the bank suspends payments.

(d) Acceptance of a payment order precludes a later rejection of the order. Rejection of a payment order precludes a later acceptance of the order.

§ 4A–211. Cancellation and Amendment of Payment Order.

(a) A communication of the sender of a payment order cancelling or amending the order may be transmitted to the receiving bank orally, electronically, or in writing. If a security procedure is in effect between the sender and the receiving bank, the communication is not effective to cancel or amend the order unless the communication is verified pursuant to the security procedure or the bank agrees to the cancellation or amendment.

(b) Subject to subsection (a), a communication by the sender cancelling or amending a payment order is effective to cancel or amend the order if notice of the communication is received at a time and in a manner affording the receiving bank a reasonable opportunity to act on the communication before the bank accepts the payment order.

(c) After a payment order has been accepted, cancellation or amendment of the order is not effective unless the receiving bank agrees or a funds-transfer system rule allows cancellation or amendment without agreement of the bank.

(1) With respect to a payment order accepted by a receiving bank other than the beneficiary's bank, cancellation or amendment is not effective unless a conforming cancellation or amendment of the payment order issued by the receiving bank is also made.

(2) With respect to a payment order accepted by the beneficiary's bank, cancellation or amendment is not effective unless the order was issued in execution of an unauthorized payment order, or because of a mistake by a sender in the funds transfer which resulted in the issuance of a payment order (i) that is a duplicate of a payment order previously issued by the sender, (ii) that orders payment to a beneficiary not entitled to receive payment from the originator, or (iii) that orders payment in an amount greater than the amount the beneficiary was entitled to receive from the originator. If the payment order is cancelled or amended, the beneficiary's bank is entitled to recover from the beneficiary

any amount paid to the beneficiary to the extent allowed by the law governing mistake and restitution.

(d) An unaccepted payment order is cancelled by operation of law at the close of the fifth funds-transfer business day of the receiving bank after the execution date or payment date of the order.

(e) A cancelled payment order cannot be accepted. If an accepted payment order is cancelled, the acceptance is nullified and no person has any right or obligation based on the acceptance. Amendment of a payment order is deemed to be cancellation of the original order at the time of amendment and issue of a new payment order in the amended form at the same time.

(f) Unless otherwise provided in an agreement of the parties or in a funds-transfer system rule, if the receiving bank, after accepting a payment order, agrees to cancellation or amendment of the order by the sender or is bound by a funds-transfer system rule allowing cancellation or amendment without the bank's agreement, the sender, whether or not cancellation or amendment is effective, is liable to the bank for any loss and expenses, including reasonable attorney's fees, incurred by the bank as a result of the cancellation or amendment or attempted cancellation or amendment.

(g) A payment order is not revoked by the death or legal incapacity of the sender unless the receiving bank knows of the death or of an adjudication of incapacity by a court of competent jurisdiction and has reasonable opportunity to act before acceptance of the order.

(h) A funds-transfer system rule is not effective to the extent it conflicts with subsection (c)(2).

§ 4A–212. Liability and Duty of Receiving Bank Regarding Unaccepted Payment Order.

If a receiving bank fails to accept a payment order that it is obliged by express agreement to accept, the bank is liable for breach of the agreement to the extent provided in the agreement or in this Article, but does not otherwise have any duty to accept a payment order or, before acceptance, to take any action, or refrain from taking action, with respect to the order except as provided in this Article or by express agreement. Liability based on acceptance arises only when acceptance occurs as stated in Section 4A–209, and liability is limited to that provided in this Article. A receiving bank is not the agent of the sender or beneficiary of the payment order it accepts, or of any other party to the funds transfer, and the bank owes no duty to any party to the funds transfer except as provided in this Article or by express agreement.

Part 3 Execution of Sender's Payment Order by Receiving Bank

§ 4A–301. Execution and Execution Date.

(a) A payment order is "executed" by the receiving bank when it issues a payment order intended to carry out the payment order received by the bank. A payment order received by the beneficiary's bank can be accepted but cannot be executed.

(b) "Execution date" of a payment order means the day on which the receiving bank may properly issue a payment order in execution of the sender's order. The execution date may be determined by instruction of the sender but cannot be earlier than the day the order is received and, unless otherwise determined, is the day the order is received. If the sender's instruction states a payment date, the execution date is the payment date or an earlier date on which execution is reasonably necessary to allow payment to the beneficiary on the payment date.

§ 4A–302. Obligations of Receiving Bank in Execution of Payment Order.

(a) Except as provided in subsections (b) through (d), if the receiving bank accepts a payment order pursuant to Section 4A–209(a), the bank has the following obligations in executing the order:

(1) The receiving bank is obliged to issue, on the execution date, a payment order complying with the sender's order and to follow the sender's instructions concerning (i) any intermediary bank or funds-transfer system to be used in carrying out the funds transfer, or (ii) the means by which payment orders are to be transmitted in the funds transfer. If the originator's bank issues a payment order to an intermediary bank, the originator's bank is obliged to instruct the intermediary bank according to the instruction of the originator. An intermediary bank in the funds transfer is similarly bound by an instruction given to it by the sender of the payment order it accepts.

(2) If the sender's instruction states that the funds transfer is to be carried out telephonically or by wire transfer or otherwise indicates that the funds transfer is to be carried out by the most expeditious means, the receiving bank is obliged to transmit its payment order by the most expeditious available means, and to instruct any intermediary bank accordingly. If a sender's instruction states a payment date, the receiving bank is obliged to transmit its payment order at a time and by means reasonably necessary to allow payment to the beneficiary on the payment date or as soon thereafter as is feasible.

(b) Unless otherwise instructed, a receiving bank executing a payment order may (i) use any funds-transfer system if use of that system is reasonable in the circumstances, and (ii) issue a payment order to the beneficiary's bank or to an intermediary bank through which a payment order conforming to the sender's order can expeditiously be issued to the beneficiary's bank if the receiving bank exercises ordinary care in the selection of the intermediary bank. A receiving bank is not required to follow an instruction of the sender designating a funds-transfer system to be used in carrying out the funds transfer if the receiving bank, in good faith, determines that it is not feasible to follow the instruction or that following the instruction would unduly delay completion of the funds transfer.

(c) Unless subsection (a)(2) applies or the receiving bank is otherwise instructed, the bank may execute a payment order by transmitting its payment order by first class mail or by any means reasonable in the circumstances. If the receiving bank is instructed to execute the sender's order by transmitting its payment order by a particular means, the receiving bank may issue its payment order by the means stated or by any means as expeditious as the means stated.

(d) Unless instructed by the sender, (i) the receiving bank may not obtain payment of its charges for services and expenses in connection with the execution of the sender's order by issuing a payment order in an amount equal to the amount of the sender's order less the amount of the charges, and (ii) may not instruct a subsequent receiving bank to obtain payment of its charges in the same manner.

§ 4A–303. Erroneous Execution of Payment Order.

(a) A receiving bank that (i) executes the payment order of the sender by issuing a payment order in an amount greater than the amount of the sender's order, or (ii) issues a payment order in execution of the sender's order and then issues a duplicate order, is entitled to payment of the amount of the sender's order under Section 4A–402(c) if that subsection is otherwise satisfied. The bank is entitled to recover from the beneficiary of the erroneous order the excess payment received to the extent allowed by the law governing mistake and restitution.

(b) A receiving bank that executes the payment order of the sender by issuing a payment order in an amount less than the amount of the sender's order is entitled to payment of the amount of the sender's order under Section 4A–402(c) if (i) that subsection is otherwise satisfied and (ii) the bank corrects its mistake by issuing an additional payment order for the benefit of the beneficiary of the sender's order. If the error is not corrected, the issuer of the erroneous order is entitled to receive or retain payment from the sender of the order it accepted only to the extent of the amount of the erroneous order. This subsection does not apply if the receiving bank executes the sender's payment order by issuing a payment order in an amount less than the amount of the sender's order for the purpose of obtaining payment of its charges for services and expenses pursuant to instruction of the sender.

(c) If a receiving bank executes the payment order of the sender by issuing a payment order to a beneficiary different from the beneficiary of the sender's order and the funds transfer is completed on the basis of that error, the sender of the payment order that was erroneously executed and all previous senders in the funds transfer are not obliged to pay the payment orders they issued. The issuer of the erroneous order is entitled to recover from the beneficiary of the order the payment received to the extent allowed by the law governing mistake and restitution.

§ 4A–304. Duty of Sender to Report Erroneously Executed Payment Order.

If the sender of a payment order that is erroneously executed as stated in Section 4A–303 receives notification from the receiving bank that the order was executed or that the sender's account was debited with respect to the order, the sender has a duty to exercise ordinary care to determine, on the basis of information available to the sender, that the order was erroneously executed and to notify the bank of the relevant facts within a reasonable time not exceeding 90 days after the notification from the bank was received by the sender. If the sender fails to perform that duty, the bank is not obliged to pay interest on any amount refundable to the sender under Section 4A–402(d) for the period before the bank learns of the execution error. The bank is not entitled to any recovery from the sender on account of a failure by the sender to perform the duty stated in this section.

§ 4A–305. Liability for Late or Improper Execution or Failure to Execute Payment Order.

(a) If a funds transfer is completed but execution of a payment order by the receiving bank in breach of Section 4A–302 results in delay in payment to the beneficiary, the bank is obliged to pay interest to either the originator or the beneficiary of the funds transfer for the period of delay caused by the improper execution. Except as provided in subsection (c), additional damages are not recoverable.

(b) If execution of a payment order by a receiving bank in breach of Section 4A–302 results in (i) noncompletion of the funds transfer, (ii) failure to use an intermediary bank designated by the originator, or (iii) issuance of a payment order that does not comply with the terms of the payment order of the originator, the bank is liable to the originator for its expenses in the funds transfer and for incidental expenses and interest losses, to the extent not covered by subsection (a), resulting from the improper execution. Except as provided in subsection (c), additional damages are not recoverable.

(c) In addition to the amounts payable under subsections (a) and (b), damages, including consequential damages, are recoverable to the extent provided in an express written agreement of the receiving bank.

(d) If a receiving bank fails to execute a payment order it was obliged by express agreement to execute, the receiving bank is liable to the sender for its expenses in the transaction and for incidental expenses and interest losses resulting from the failure to execute. Additional damages, including consequential damages, are recoverable to the extent provided in an express written agreement of the receiving bank, but are not otherwise recoverable.

(e) Reasonable attorney's fees are recoverable if demand for compensation under subsection (a) or (b) is made and refused before an action is brought on the claim. If a claim is made for breach of an agreement under subsection (d) and the agreement does not provide for damages, reasonable attorney's fees are recoverable if demand for compensation under subsection (d) is made and refused before an action is brought on the claim.

(f) Except as stated in this section, the liability of a receiving bank under subsections (a) and (b) may not be varied by agreement.

Part 4 Payment

§ 4A–401. Payment Date.

"Payment date" of a payment order means the day on which the amount of the order is payable to the beneficiary by the beneficiary's bank. The payment date may be determined by instruction of the sender but cannot be earlier than the day the order is received by the beneficiary's bank and, unless otherwise determined, is the day the order is received by the beneficiary's bank.

§ 4A–402. Obligation of Sender to Pay Receiving Bank.

(a) This section is subject to Sections 4A–205 and 4A–207.

(b) With respect to a payment order issued to the beneficiary's bank, acceptance of the order by the bank obliges the sender to pay the bank the amount of the order, but payment is not due until the payment date of the order.

(c) This subsection is subject to subsection (e) and to Section 4A–303. With respect to a payment order issued to a receiving bank other than the beneficiary's bank, acceptance of the order by the receiving bank obliges the sender to pay the bank the amount of the sender's order. Payment by the sender is not due until the execution date of the sender's order. The obligation of that sender to pay its payment order is excused if the funds transfer is not completed by acceptance by the beneficiary's bank of a payment order instructing payment to the beneficiary of that sender's payment order.

(d) If the sender of a payment order pays the order and was not obliged to pay all or part of the amount paid, the bank receiving payment is obliged to refund payment to the extent the sender was not obliged to pay. Except as provided in Sections 4A–204 and 4A–304, interest is payable on the refundable amount from the date of payment.

(e) If a funds transfer is not completed as stated in subsection (c) and an intermediary bank is obliged to refund payment as stated in subsection (d) but is unable to do so because not permitted by applicable law or because the bank suspends payments, a sender in the funds transfer that executed a payment order in compliance with an instruction, as stated in Section 4A–302(a)(1), to route the funds transfer through that intermediary bank is entitled to receive or retain payment from the sender of the payment order that it accepted. The first sender in the funds transfer that issued an instruction requiring routing through that intermediary bank is subrogated to the right of the bank that paid the intermediary bank to refund as stated in subsection (d).

(f) The right of the sender of a payment order to be excused from the obligation to pay the order as stated in subsection (c) or to receive refund under subsection (d) may not be varied by agreement.

§ 4A–403. Payment by Sender to Receiving Bank.

(a) Payment of the sender's obligation under Section 4A–402 to pay the receiving bank occurs as follows:

(1) If the sender is a bank, payment occurs when the receiving bank receives final settlement of the obligation through a Federal Reserve Bank or through a funds-transfer system.

(2) If the sender is a bank and the sender (i) credited an account of the receiving bank with the sender, or (ii) caused an account of the receiving bank in another bank to be credited, payment occurs when the credit is withdrawn or, if not withdrawn, at midnight of the day on which the credit is withdrawable and the receiving bank learns of that fact.

(3) If the receiving bank debits an account of the sender with the receiving bank, payment occurs when the debit is made to the extent the debit is covered by a withdrawable credit balance in the account.

(b) If the sender and receiving bank are members of a funds-transfer system that nets obligations multilaterally among participants, the receiving bank receives final settlement when settlement is complete in accordance with the rules of the system. The obligation of the sender to pay the amount of a payment order transmitted through the funds-transfer system may be satisfied, to the extent permitted by the rules of the system, by setting off and applying against the sender's obligation the right of the sender to receive payment from the receiving bank of the amount of any other payment order transmitted to the sender by the receiving bank through the funds-transfer system. The aggregate balance of obligations owed by each sender to each receiving bank in the funds-transfer system may be satisfied, to the extent permitted by the rules of the system, by setting off and applying against that balance the aggregate balance of obligations owed to the sender by other members of the system. The aggregate balance is determined after the right of setoff stated in the second sentence of this subsection has been exercised.

(c) If two banks transmit payment orders to each other under an agreement that settlement of the obligations of each bank to the other under Section 4A–402 will be made at the end of the day or other period, the total amount owed with respect to all orders transmitted by one bank shall be set off against the total amount owed with respect to all orders transmitted by the other bank. To the extent of the setoff, each bank has made payment to the other.

(d) In a case not covered by subsection (a), the time when payment of the sender's obligation under Section 4A–402(b) or 4A–402(c) occurs is governed by applicable principles of law that determine when an obligation is satisfied.

§ 4A–404. Obligation of Beneficiary's Bank to Pay and Give Notice to Beneficiary.

(a) Subject to Sections 4A–211(e), 4A–405(d), and 4A–405(e), if a beneficiary's bank accepts a payment order, the bank is obliged to pay the amount of the order to the beneficiary of the order. Payment is due on the payment date of the order, but if acceptance occurs on the payment date after the close of the funds-transfer business day of the bank, payment is due on the next funds-transfer business day. If the bank refuses to pay after demand by the beneficiary and receipt of notice of particular circumstances that will give rise to consequential damages as

a result of nonpayment, the beneficiary may recover damages resulting from the refusal to pay to the extent the bank had notice of the damages, unless the bank proves that it did not pay because of a reasonable doubt concerning the right of the beneficiary to payment.

(b) If a payment order accepted by the beneficiary's bank instructs payment to an account of the beneficiary, the bank is obliged to notify the beneficiary of receipt of the order before midnight of the next funds-transfer business day following the payment date. If the payment order does not instruct payment to an account of the beneficiary, the bank is required to notify the beneficiary only if notice is required by the order. Notice may be given by first class mail or any other means reasonable in the circumstances. If the bank fails to give the required notice, the bank is obliged to pay interest to the beneficiary on the amount of the payment order from the day notice should have been given until the day the beneficiary learned of receipt of the payment order by the bank. No other damages are recoverable. Reasonable attorney's fees are also recoverable if demand for interest is made and refused before an action is brought on the claim.

(c) The right of a beneficiary to receive payment and damages as stated in subsection (a) may not be varied by agreement or a funds-transfer system rule. The right of a beneficiary to be notified as stated in subsection (b) may be varied by agreement of the beneficiary or by a funds-transfer system rule if the beneficiary is notified of the rule before initiation of the funds transfer.

§ 4A–405. Payment by Beneficiary's Bank to Beneficiary.

(a) If the beneficiary's bank credits an account of the beneficiary of a payment order, payment of the bank's obligation under Section 4A–404(a) occurs when and to the extent (i) the beneficiary is notified of the right to withdraw the credit, (ii) the bank lawfully applies the credit to a debt of the beneficiary, or (iii) funds with respect to the order are otherwise made available to the beneficiary by the bank.

(b) If the beneficiary's bank does not credit an account of the beneficiary of a payment order, the time when payment of the bank's obligation under Section 4A–404(a) occurs is governed by principles of law that determine when an obligation is satisfied.

(c) Except as stated in subsections (d) and (e), if the beneficiary's bank pays the beneficiary of a payment order under a condition to payment or agreement of the beneficiary giving the bank the right to recover payment from the beneficiary if the bank does not receive payment of the order, the condition to payment or agreement is not enforceable.

(d) A funds-transfer system rule may provide that payments made to beneficiaries of funds transfers made through the system are provisional until receipt of payment by the beneficiary's bank of the payment order it accepted. A beneficiary's bank that makes a payment that is provisional under the rule is entitled to refund from the beneficiary if (i) the rule requires that both the beneficiary and the originator be given notice of the provisional nature of the payment before the funds transfer is initiated, (ii) the beneficiary, the beneficiary's bank, and the originator's bank agreed to be bound by the rule, and (iii) the beneficiary's bank did not receive payment of the payment order that it accepted. If the beneficiary is obliged to refund payment to the beneficiary's bank, acceptance of the payment order by the beneficiary's bank is nullified and no payment by the originator of the funds transfer to the beneficiary occurs under Section 4A–406.

(e) This subsection applies to a funds transfer that includes a payment order transmitted over a funds-transfer system that (i) nets obligations multilaterally among participants, and (ii) has in effect a loss-sharing agreement among participants for the purpose of providing funds necessary to complete settlement of the obligations of one or more participants that do not meet their settlement obligations. If the beneficiary's bank in the funds transfer accepts a payment order and the system fails to complete settlement pursuant to its rules with respect to any payment order in the funds transfer, (i) the acceptance by the beneficiary's bank is nullified and no person has any right or obligation based on the acceptance, (ii) the beneficiary's bank is entitled to recover payment from the beneficiary, (iii) no payment by the originator to the beneficiary occurs under Section 4A–406, and (iv) subject to Section 4A–402(e), each sender in the funds transfer is excused from its obligation to pay its payment order under Section 4A–402(c) because the funds transfer has not been completed.

§ 4A–406. Payment by Originator to Beneficiary; Discharge of Underlying Obligation.

(a) Subject to Sections 4A–211(e), 4A–405(d), and 4A–405(e), the originator of a funds transfer pays the beneficiary of the originator's payment order (i) at the time a payment order for the benefit of the beneficiary is accepted by the beneficiary's bank in the funds transfer and (ii) in an amount equal to the amount of the order accepted by the beneficiary's bank, but not more than the amount of the originator's order.

(b) If payment under subsection (a) is made to satisfy an obligation, the obligation is discharged to the same extent discharge would result from payment to the beneficiary of the same amount in money, unless (i) the payment under subsection (a) was made by a means prohibited by the contract of the beneficiary with respect to the obligation, (ii) the beneficiary, within a reasonable time after receiving notice of receipt of the order by the beneficiary's bank, notified the originator of the beneficiary's refusal of the payment, (iii) funds with respect to the order were not withdrawn by the beneficiary or applied to a debt of the beneficiary, and (iv) the beneficiary would suffer a loss that could reasonably have been avoided if payment had been made by a means complying with the contract. If payment by the originator does not result in discharge under this section, the originator is subrogated to the rights of the beneficiary to receive payment from the beneficiary's bank under Section 4A–404(a).

(c) For the purpose of determining whether discharge of an obligation occurs under subsection (b), if the beneficiary's bank accepts a payment order in an amount equal to the amount of the originator's payment order less charges of one or more receiving

banks in the funds transfer, payment to the beneficiary is deemed to be in the amount of the originator's order unless upon demand by the beneficiary the originator does not pay the beneficiary the amount of the deducted charges.

(d) Rights of the originator or of the beneficiary of a funds transfer under this section may be varied only by agreement of the originator and the beneficiary.

Part 5 Miscellaneous Provisions

§ 4A–501. Variation by Agreement and Effect of Funds-Transfer System Rule.

(a) Except as otherwise provided in this Article, the rights and obligations of a party to a funds transfer may be varied by agreement of the affected party.

(b) "Funds-transfer system rule" means a rule of an association of banks (i) governing transmission of payment orders by means of a funds-transfer system of the association or rights and obligations with respect to those orders, or (ii) to the extent the rule governs rights and obligations between banks that are parties to a funds transfer in which a Federal Reserve Bank, acting as an intermediary bank, sends a payment order to the beneficiary's bank. Except as otherwise provided in this Article, a funds-transfer system rule governing rights and obligations between participating banks using the system may be effective even if the rule conflicts with this Article and indirectly affects another party to the funds transfer who does not consent to the rule. A funds-transfer system rule may also govern rights and obligations of parties other than participating banks using the system to the extent stated in Sections 4A–404(c), 4A–405(d), and 4A–507(c).

§ 4A–502. Creditor Process Served on Receiving Bank; Setoff by Beneficiary's Bank.

(a) As used in this section, "creditor process" means levy, attachment, garnishment, notice of lien, sequestration, or similar process issued by or on behalf of a creditor or other claimant with respect to an account.

(b) This subsection applies to creditor process with respect to an authorized account of the sender of a payment order if the creditor process is served on the receiving bank. For the purpose of determining rights with respect to the creditor process, if the receiving bank accepts the payment order the balance in the authorized account is deemed to be reduced by the amount of the payment order to the extent the bank did not otherwise receive payment of the order, unless the creditor process is served at a time and in a manner affording the bank a reasonable opportunity to act on it before the bank accepts the payment order.

(c) If a beneficiary's bank has received a payment order for payment to the beneficiary's account in the bank, the following rules apply:

(1) The bank may credit the beneficiary's account. The amount credited may be set off against an obligation owed by the beneficiary to the bank or may be applied to satisfy creditor process served on the bank with respect to the account.

(2) The bank may credit the beneficiary's account and allow withdrawal of the amount credited unless creditor process with respect to the account is served at a time and in a manner affording the bank a reasonable opportunity to act to prevent withdrawal.

(3) If creditor process with respect to the beneficiary's account has been served and the bank has had a reasonable opportunity to act on it, the bank may not reject the payment order except for a reason unrelated to the service of process.

(d) Creditor process with respect to a payment by the originator to the beneficiary pursuant to a funds transfer may be served only on the beneficiary's bank with respect to the debt owed by that bank to the beneficiary. Any other bank served with the creditor process is not obliged to act with respect to the process.

§ 4A–503. Injunction or Restraining Order with Respect to Funds Transfer.

For proper cause and in compliance with applicable law, a court may restrain (i) a person from issuing a payment order to initiate a funds transfer, (ii) an originator's bank from executing the payment order of the originator, or (iii) the beneficiary's bank from releasing funds to the beneficiary or the beneficiary from withdrawing the funds. A court may not otherwise restrain a person from issuing a payment order, paying or receiving payment of a payment order, or otherwise acting with respect to a funds transfer.

§ 4A–504. Order in Which Items and Payment Orders May Be Charged to Account; Order of Withdrawals from Account.

(a) If a receiving bank has received more than one payment order of the sender or one or more payment orders and other items that are payable from the sender's account, the bank may charge the sender's account with respect to the various orders and items in any sequence.

(b) In determining whether a credit to an account has been withdrawn by the holder of the account or applied to a debt of the holder of the account, credits first made to the account are first withdrawn or applied.

§ 4A–505. Preclusion of Objection to Debit of Customer's Account.

If a receiving bank has received payment from its customer with respect to a payment order issued in the name of the customer as sender and accepted by the bank, and the customer received notification reasonably identifying the order, the customer is precluded from asserting that the bank is not entitled to retain the payment unless the customer notifies the bank of the customer's objection to the payment within one year after the notification was received by the customer.

§ 4A–506. Rate of Interest.

(a) If, under this Article, a receiving bank is obliged to pay interest with respect to a payment order issued to the bank, the amount payable may be determined (i) by agreement of the sender and receiving bank, or (ii) by a funds-transfer system

rule if the payment order is transmitted through a funds-transfer system.

(b) If the amount of interest is not determined by an agreement or rule as stated in subsection (a), the amount is calculated by multiplying the applicable Federal Funds rate by the amount on which interest is payable, and then multiplying the product by the number of days for which interest is payable. The applicable Federal Funds rate is the average of the Federal Funds rates published by the Federal Reserve Bank of New York for each of the days for which interest is payable divided by 360. The Federal Funds rate for any day on which a published rate is not available is the same as the published rate for the next preceding day for which there is a published rate. If a receiving bank that accepted a payment order is required to refund payment to the sender of the order because the funds transfer was not completed, but the failure to complete was not due to any fault by the bank, the interest payable is reduced by a percentage equal to the reserve requirement on deposits of the receiving bank.

§ 4A–507. Choice of Law.

(a) The following rules apply unless the affected parties otherwise agree or subsection (c) applies:

(1) The rights and obligations between the sender of a payment order and the receiving bank are governed by the law of the jurisdiction in which the receiving bank is located.

(2) The rights and obligations between the beneficiary's bank and the beneficiary are governed by the law of the jurisdiction in which the beneficiary's bank is located.

(3) The issue of when payment is made pursuant to a funds transfer by the originator to the beneficiary is governed by the law of the jurisdiction in which the beneficiary's bank is located.

(b) If the parties described in each paragraph of subsection (a) have made an agreement selecting the law of a particular jurisdiction to govern rights and obligations between each other, the law of that jurisdiction governs those rights and obligations, whether or not the payment order or the funds transfer bears a reasonable relation to that jurisdiction.

(c) A funds-transfer system rule may select the law of a particular jurisdiction to govern (i) rights and obligations between participating banks with respect to payment orders transmitted or processed through the system, or (ii) the rights and obligations of some or all parties to a funds transfer any part of which is carried out by means of the system. A choice of law made pursuant to clause (i) is binding on participating banks. A choice of law made pursuant to clause (ii) is binding on the originator, other sender, or a receiving bank having notice that the funds-transfer system might be used in the funds transfer and of the choice of law by the system when the originator, other sender, or receiving bank issued or accepted a payment order. The beneficiary of a funds transfer is bound by the choice of law if, when the funds transfer is initiated, the beneficiary has notice that the funds-transfer system might be used in the funds transfer and of the choice of law by the system. The law of a jurisdiction selected pursuant to this subsection may govern, whether or not that law bears a reasonable relation to the matter in issue.

(d) In the event of inconsistency between an agreement under subsection (b) and a choice-of-law rule under subsection (c), the agreement under subsection (b) prevails.

(e) If a funds transfer is made by use of more than one funds-transfer system and there is inconsistency between choice-of-law rules of the systems, the matter in issue is governed by the law of the selected jurisdiction that has the most significant relationship to the matter in issue.

Revised Article 9
SECURED TRANSACTIONS

Part 1 General Provisions

[Subpart 1. Short Title, Definitions, and General Concepts]

§ 9–101. Short Title.

This article may be cited as Uniform Commercial Code—Secured Transactions.

§ 9–102. Definitions and Index of Definitions.

(a) In this article:

(1) "Accession" means goods that are physically united with other goods in such a manner that the identity of the original goods is not lost.

(2) "Account", except as used in "account for", means a right to payment of a monetary obligation, whether or not earned by performance, (i) for property that has been or is to be sold, leased, licensed, assigned, or otherwise disposed of, (ii) for services rendered or to be rendered, (iii) for a policy of insurance issued or to be issued, (iv) for a secondary obligation incurred or to be incurred, (v) for energy provided or to be provided, (vi) for the use or hire of a vessel under a charter or other contract, (vii) arising out of the use of a credit or charge card or information contained on or for use with the card, or (viii) as winnings in a lottery or other game of chance operated or sponsored by a State, governmental unit of a State, or person licensed or authorized to operate the game by a State or governmental unit of a State. The term includes health-care insurance receivables. The term does not include (i) rights to payment evidenced by chattel paper or an instrument, (ii) commercial tort claims, (iii) deposit accounts, (iv) investment property, (v) letter-of-credit rights or letters of credit, or (vi) rights to payment for money or funds advanced or sold, other than rights arising out of the use of a credit or charge card or information contained on or for use with the card.

(3) "Account debtor" means a person obligated on an account, chattel paper, or general intangible. The term does

not include persons obligated to pay a negotiable instrument, even if the instrument constitutes part of chattel paper.

(4) "Accounting", except as used in "accounting for", means a record:

(A) authenticated by a secured party;

(B) indicating the aggregate unpaid secured obligations as of a date not more than 35 days earlier or 35 days later than the date of the record; and

(C) identifying the components of the obligations in reasonable detail.

(5) "Agricultural lien" means an interest, other than a security interest, in farm products:

(A) which secures payment or performance of an obligation for:

(i) goods or services furnished in connection with a debtor's farming operation; or

(ii) rent on real property leased by a debtor in connection with its farming operation;

(B) which is created by statute in favor of a person that:

(i) in the ordinary course of its business furnished goods or services to a debtor in connection with a debtor's farming operation; or

(ii) leased real property to a debtor in connection with the debtor's farming operation; and

(C) whose effectiveness does not depend on the person's possession of the personal property.

(6) "As-extracted collateral" means:

(A) oil, gas, or other minerals that are subject to a security interest that:

(i) is created by a debtor having an interest in the minerals before extraction; and

(ii) attaches to the minerals as extracted; or

(B) accounts arising out of the sale at the wellhead or minehead of oil, gas, or other minerals in which the debtor had an interest before extraction.

(7) "Authenticate" means:

(A) to sign; or

(B) to execute or otherwise adopt a symbol, or encrypt or similarly process a record in whole or in part, with the present intent of the authenticating person to identify the person and adopt or accept a record.

(8) "Bank" means an organization that is engaged in the business of banking. The term includes savings banks, savings and loan associations, credit unions, and trust companies.

(9) "Cash proceeds" means proceeds that are money, checks, deposit accounts, or the like.

(10) "Certificate of title" means a certificate of title with respect to which a statute provides for the security interest in question to be indicated on the certificate as a condition or result of the security interest's obtaining priority over the rights of a lien creditor with respect to the collateral.

(11) "Chattel paper" means a record or records that evidence both a monetary obligation and a security interest in specific goods, a security interest in specific goods and software used in the goods, a security interest in specific goods and license of software used in the goods, a lease of specific goods, or a lease of specific goods and license of software used in the goods. In this paragraph, "monetary obligation" means a monetary obligation secured by the goods or owed under a lease of the goods and includes a monetary obligation with respect to software used in the goods. The term does not include (i) charters or other contracts involving the use or hire of a vessel or (ii) records that evidence a right to payment arising out of the use of a credit or charge card or information contained on or for use with the card. If a transaction is evidenced by records that include an instrument or series of instruments, the group of records taken together constitutes chattel paper.

(12) "Collateral" means the property subject to a security interest or agricultural lien. The term includes:

(A) proceeds to which a security interest attaches;

(B) accounts, chattel paper, payment intangibles, and promissory notes that have been sold; and

(C) goods that are the subject of a consignment.

(13) "Commercial tort claim" means a claim arising in tort with respect to which:

(A) the claimant is an organization; or

(B) the claimant is an individual and the claim:

(i) arose in the course of the claimant's business or profession; and

(ii) does not include damages arising out of personal injury to or the death of an individual.

(14) "Commodity account" means an account maintained by a commodity intermediary in which a commodity contract is carried for a commodity customer.

(15) "Commodity contract" means a commodity futures contract, an option on a commodity futures contract, a commodity option, or another contract if the contract or option is:

(A) traded on or subject to the rules of a board of trade that has been designated as a contract market for such a contract pursuant to federal commodities laws; or

(B) traded on a foreign commodity board of trade, exchange, or market, and is carried on the books of a commodity intermediary for a commodity customer.

(16) "Commodity customer" means a person for which a commodity intermediary carries a commodity contract on its books.

(17) "Commodity intermediary" means a person that:

(A) is registered as a futures commission merchant under federal commodities law; or

(B) in the ordinary course of its business provides clearance or settlement services for a board of trade that has been designated as a contract market pursuant to federal commodities law.

(18) "Communicate" means:

(A) to send a written or other tangible record;

(B) to transmit a record by any means agreed upon by the persons sending and receiving the record; or

(C) in the case of transmission of a record to or by a filing office, to transmit a record by any means prescribed by filing-office rule.

(19) "Consignee" means a merchant to which goods are delivered in a consignment.

(20) "Consignment" means a transaction, regardless of its form, in which a person delivers goods to a merchant for the purpose of sale and:

(A) the merchant:

(i) deals in goods of that kind under a name other than the name of the person making delivery;

(ii) is not an auctioneer; and

(iii) is not generally known by its creditors to be substantially engaged in selling the goods of others;

(B) with respect to each delivery, the aggregate value of the goods is $1,000 or more at the time of delivery;

(C) the goods are not consumer goods immediately before delivery; and

(D) the transaction does not create a security interest that secures an obligation.

(21) "Consignor" means a person that delivers goods to a consignee in a consignment.

(22) "Consumer debtor" means a debtor in a consumer transaction.

(23) "Consumer goods" means goods that are used or bought for use primarily for personal, family, or household purposes.

(24) "Consumer-goods transaction" means a consumer transaction in which:

(A) an individual incurs an obligation primarily for personal, family, or household purposes; and

(B) a security interest in consumer goods secures the obligation.

(25) "Consumer obligor" means an obligor who is an individual and who incurred the obligation as part of a transaction entered into primarily for personal, family, or household purposes.

(26) "Consumer transaction" means a transaction in which (i) an individual incurs an obligation primarily for personal, family, or household purposes, (ii) a security interest secures the obligation, and (iii) the collateral is held or acquired primarily for personal, family, or household purposes. The term includes consumer-goods transactions.

(27) "Continuation statement" means an amendment of a financing statement which:

(A) identifies, by its file number, the initial financing statement to which it relates; and

(B) indicates that it is a continuation statement for, or that it is filed to continue the effectiveness of, the identified financing statement.

(28) "Debtor" means:

(A) a person having an interest, other than a security interest or other lien, in the collateral, whether or not the person is an obligor;

(B) a seller of accounts, chattel paper, payment intangibles, or promissory notes; or

(C) a consignee.

(29) "Deposit account" means a demand, time, savings, passbook, or similar account maintained with a bank. The term does not include investment property or accounts evidenced by an instrument.

(30) "Document" means a document of title or a receipt of the type described in Section 7–201(2).

(31) "Electronic chattel paper" means chattel paper evidenced by a record or records consisting of information stored in an electronic medium.

(32) "Encumbrance" means a right, other than an ownership interest, in real property. The term includes mortgages and other liens on real property.

(33) "Equipment" means goods other than inventory, farm products, or consumer goods.

(34) "Farm products" means goods, other than standing timber, with respect to which the debtor is engaged in a farming operation and which are:

(A) crops grown, growing, or to be grown, including:

(i) crops produced on trees, vines, and bushes; and

(ii) aquatic goods produced in aquacultural operations;

(B) livestock, born or unborn, including aquatic goods produced in aquacultural operations;

(C) supplies used or produced in a farming operation; or

(D) products of crops or livestock in their unmanufactured states.

(35) "Farming operation" means raising, cultivating, propagating, fattening, grazing, or any other farming, livestock, or aquacultural operation.

(36) "File number" means the number assigned to an initial financing statement pursuant to Section 9–519(a).

(37) "Filing office" means an office designated in Section 9–501 as the place to file a financing statement.

(38) "Filing-office rule" means a rule adopted pursuant to Section 9–526.

(39) "Financing statement" means a record or records composed of an initial financing statement and any filed record relating to the initial financing statement.

(40) "Fixture filing" means the filing of a financing statement covering goods that are or are to become fixtures and satisfying Section 9–502(a) and (b). The term includes the filing of a financing statement covering goods of a transmitting utility which are or are to become fixtures.

(41) "Fixtures" means goods that have become so related to particular real property that an interest in them arises under real property law.

(42) "General intangible" means any personal property, including things in action, other than accounts, chattel paper, commercial tort claims, deposit accounts, documents, goods, instruments, investment property, letter-of-credit rights, letters of credit, money, and oil, gas, or other minerals before extraction. The term includes payment intangibles and software.

(43) "Good faith" means honesty in fact and the observance of reasonable commercial standards of fair dealing.

(44) "Goods" means all things that are movable when a security interest attaches. The term includes (i) fixtures, (ii) standing timber that is to be cut and removed under a conveyance or contract for sale, (iii) the unborn young of animals, (iv) crops grown, growing, or to be grown, even if the crops are produced on trees, vines, or bushes, and (v) manufactured homes. The term also includes a computer program embedded in goods and any supporting information provided in connection with a transaction relating to the program if (i) the program is associated with the goods in such a manner that it customarily is considered part of the goods, or (ii) by becoming the owner of the goods, a person acquires a right to use the program in connection with the goods. The term does not include a computer program embedded in goods that consist solely of the medium in which the program is embedded. The term also does not include accounts, chattel paper, commercial tort claims, deposit accounts, documents, general intangibles, instruments, investment property, letter-of-credit rights, letters of credit, money, or oil, gas, or other minerals before extraction.

(45) "Governmental unit" means a subdivision, agency, department, county, parish, municipality, or other unit of the government of the United States, a State, or a foreign country. The term includes an organization having a separate corporate existence if the organization is eligible to issue debt on which interest is exempt from income taxation under the laws of the United States.

(46) "Health-care-insurance receivable" means an interest in or claim under a policy of insurance which is a right to payment of a monetary obligation for health-care goods or services provided.

(47) "Instrument" means a negotiable instrument or any other writing that evidences a right to the payment of a monetary obligation, is not itself a security agreement or lease, and is of a type that in ordinary course of business is transferred by delivery with any necessary indorsement or assignment. The term does not include (i) investment property, (ii) letters of credit, or (iii) writings that evidence a right to payment arising out of the use of a credit or charge card or information contained on or for use with the card.

(48) "Inventory" means goods, other than farm products, which:

(A) are leased by a person as lessor;

(B) are held by a person for sale or lease or to be furnished under a contract of service;

(C) are furnished by a person under a contract of service; or

(D) consist of raw materials, work in process, or materials used or consumed in a business.

(49) "Investment property" means a security, whether certificated or uncertificated, security entitlement, securities account, commodity contract, or commodity account.

(50) "Jurisdiction of organization", with respect to a registered organization, means the jurisdiction under whose law the organization is organized.

(51) "Letter-of-credit right" means a right to payment or performance under a letter of credit, whether or not the beneficiary has demanded or is at the time entitled to demand payment or performance. The term does not include the right of a beneficiary to demand payment or performance under a letter of credit.

(52) "Lien creditor" means:

(A) a creditor that has acquired a lien on the property involved by attachment, levy, or the like;

(B) an assignee for benefit of creditors from the time of assignment;

(C) a trustee in bankruptcy from the date of the filing of the petition; or

(D) a receiver in equity from the time of appointment.

(53) "Manufactured home" means a structure, transportable in one or more sections, which, in the traveling mode, is eight body feet or more in width or 40 body feet or more in length, or, when erected on site, is 320 or more square feet, and which is built on a permanent chassis and designed to be used as a dwelling with or without a permanent foundation when connected to the required utilities, and includes the plumbing, heating, air-conditioning, and electrical systems contained therein. The term includes any structure that meets all of the requirements of this paragraph except the size requirements and with respect to which the manufacturer voluntarily files a certification required by the United States Secretary of Housing and Urban Development and complies with the standards established under Title 42 of the United States Code.

(54) "Manufactured-home transaction" means a secured transaction:

(A) that creates a purchase-money security interest in a manufactured home, other than a manufactured home held as inventory; or

(B) in which a manufactured home, other than a manufactured home held as inventory, is the primary collateral.

(55) "Mortgage" means a consensual interest in real property, including fixtures, which secures payment or performance of an obligation.

(56) "New debtor" means a person that becomes bound as debtor under Section 9–203(d) by a security agreement previously entered into by another person.

(57) "New value" means (i) money, (ii) money's worth in property, services, or new credit, or (iii) release by a transferee of an interest in property previously transferred to the transferee. The term does not include an obligation substituted for another obligation.

(58) "Noncash proceeds" means proceeds other than cash proceeds.

(59) "Obligor" means a person that, with respect to an obligation secured by a security interest in or an agricultural lien on the collateral, (i) owes payment or other performance of the obligation, (ii) has provided property other than the collateral to secure payment or other performance of the obligation, or (iii) is otherwise accountable in whole or in part for payment or other performance of the obligation. The term does not include issuers or nominated persons under a letter of credit.

(60) "Original debtor", except as used in Section 9–310(c), means a person that, as debtor, entered into a security agreement to which a new debtor has become bound under Section 9–203(d).

(61) "Payment intangible" means a general intangible under which the account debtor's principal obligation is a monetary obligation.

(62) "Person related to", with respect to an individual, means:

(A) the spouse of the individual;

(B) a brother, brother-in-law, sister, or sister-in-law of the individual;

(C) an ancestor or lineal descendant of the individual or the individual's spouse; or

(D) any other relative, by blood or marriage, of the individual or the individual's spouse who shares the same home with the individual.

(63) "Person related to", with respect to an organization, means:

(A) a person directly or indirectly controlling, controlled by, or under common control with the organization;

(B) an officer or director of, or a person performing similar functions with respect to, the organization;

(C) an officer or director of, or a person performing similar functions with respect to, a person described in subparagraph (A);

(D) the spouse of an individual described in subparagraph (A), (B), or (C); or

(E) an individual who is related by blood or marriage to an individual described in subparagraph (A), (B), (C), or (D) and shares the same home with the individual.

(64) "Proceeds", except as used in Section 9–609(b), means the following property:

(A) whatever is acquired upon the sale, lease, license, exchange, or other disposition of collateral;

(B) whatever is collected on, or distributed on account of, collateral;

(C) rights arising out of collateral;

(D) to the extent of the value of collateral, claims arising out of the loss, nonconformity, or interference with the use of, defects or infringement of rights in, or damage to, the collateral; or

(E) to the extent of the value of collateral and to the extent payable to the debtor or the secured party, insurance payable by reason of the loss or nonconformity of, defects or infringement of rights in, or damage to, the collateral.

(65) "Promissory note" means an instrument that evidences a promise to pay a monetary obligation, does not evidence an order to pay, and does not contain an acknowledgment by a bank that the bank has received for deposit a sum of money or funds.

(66) "Proposal" means a record authenticated by a secured party which includes the terms on which the secured party is willing to accept collateral in full or partial satisfaction of the obligation it secures pursuant to Sections 9–620, 9–621, and 9–622.

(67) "Public-finance transaction" means a secured transaction in connection with which:

(A) debt securities are issued;

(B) all or a portion of the securities issued have an initial stated maturity of at least 20 years; and

(C) the debtor, obligor, secured party, account debtor or other person obligated on collateral, assignor or assignee of a secured obligation, or assignor or assignee of a security interest is a State or a governmental unit of a State.

(68) "Pursuant to commitment", with respect to an advance made or other value given by a secured party, means pursuant to the secured party's obligation, whether or not a subsequent event of default or other event not within the secured party's control has relieved or may relieve the secured party from its obligation.

(69) "Record", except as used in "for record", "of record", "record or legal title", and "record owner", means information that is inscribed on a tangible medium or which is stored in an electronic or other medium and is retrievable in perceivable form.

(70) "Registered organization" means an organization organized solely under the law of a single State or the

United States and as to which the State or the United States must maintain a public record showing the organization to have been organized.

(71) "Secondary obligor" means an obligor to the extent that:

(A) the obligor's obligation is secondary; or

(B) the obligor has a right of recourse with respect to an obligation secured by collateral against the debtor, another obligor, or property of either.

(72) "Secured party" means:

(A) a person in whose favor a security interest is created or provided for under a security agreement, whether or not any obligation to be secured is outstanding;

(B) a person that holds an agricultural lien;

(C) a consignor;

(D) a person to which accounts, chattel paper, payment intangibles, or promissory notes have been sold;

(E) a trustee, indenture trustee, agent, collateral agent, or other representative in whose favor a security interest or agricultural lien is created or provided for; or

(F) a person that holds a security interest arising under Section 2–401, 2–505, 2–711(3), 2A–508(5), 4–210, or 5–118.

(73) "Security agreement" means an agreement that creates or provides for a security interest.

(74) "Send", in connection with a record or notification, means:

(A) to deposit in the mail, deliver for transmission, or transmit by any other usual means of communication, with postage or cost of transmission provided for, addressed to any address reasonable under the circumstances; or

(B) to cause the record or notification to be received within the time that it would have been received if properly sent under subparagraph (A).

(75) "Software" means a computer program and any supporting information provided in connection with a transaction relating to the program. The term does not include a computer program that is included in the definition of goods.

(76) "State" means a State of the United States, the District of Columbia, Puerto Rico, the United States Virgin Islands, or any territory or insular possession subject to the jurisdiction of the United States.

(77) "Supporting obligation" means a letter-of-credit right or secondary obligation that supports the payment or performance of an account, chattel paper, a document, a general intangible, an instrument, or investment property.

(78) "Tangible chattel paper" means chattel paper evidenced by a record or records consisting of information that is inscribed on a tangible medium.

(79) "Termination statement" means an amendment of a financing statement which:

(A) identifies, by its file number, the initial financing statement to which it relates; and

(B) indicates either that it is a termination statement or that the identified financing statement is no longer effective.

(80) "Transmitting utility" means a person primarily engaged in the business of:

(A) operating a railroad, subway, street railway, or trolley bus;

(B) transmitting communications electrically, electromagnetically, or by light;

(C) transmitting goods by pipeline or sewer; or

(D) transmitting or producing and transmitting electricity, steam, gas, or water.

(b) The following definitions in other articles apply to this article:

"Applicant."	Section 5–102
"Beneficiary."	Section 5–102
"Broker."	Section 8–102
"Certificated security."	Section 8–102
"Check."	Section 3–104
"Clearing corporation."	Section 8–102
"Contract for sale."	Section 2–106
"Customer."	Section 4–104
"Entitlement holder."	Section 8–102
"Financial asset."	Section 8–102
"Holder in due course."	Section 3–302
"Issuer" (with respect to a letter of credit or letter-of-credit right).	Section 5–102
"Issuer" (with respect to a security).	Section 8–201
"Lease."	Section 2A–103
"Lease agreement."	Section 2A–103
"Lease contract."	Section 2A–103
"Leasehold interest."	Section 2A–103
"Lessee."	Section 2A–103
"Lessee in ordinary course of business."	Section 2A–103
"Lessor."	Section 2A–103
"Lessor's residual interest."	Section 2A–103
"Letter of credit."	Section 5–102
"Merchant."	Section 2–104
"Negotiable instrument."	Section 3–104
"Nominated person."	Section 5–102
"Note."	Section 3–104
"Proceeds of a letter of credit."	Section 5–114

"Prove."	Section 3–103
"Sale."	Section 2–106
"Securities account."	Section 8–501
"Securities intermediary."	Section 8–102
"Security."	Section 8–102
"Security certificate."	Section 8–102
"Security entitlement."	Section 8–102
"Uncertificated security."	Section 8–102

(c) Article 1 contains general definitions and principles of construction and interpretation applicable throughout this article.

Amended in 1999 and 2000.

§ 9–103. Purchase-Money Security Interest; Application of Payments; Burden of Establishing.

(a) In this section:

(1) "purchase-money collateral" means goods or software that secures a purchase-money obligation incurred with respect to that collateral; and

(2) "purchase-money obligation" means an obligation of an obligor incurred as all or part of the price of the collateral or for value given to enable the debtor to acquire rights in or the use of the collateral if the value is in fact so used.

(b) A security interest in goods is a purchase-money security interest:

(1) to the extent that the goods are purchase-money collateral with respect to that security interest;

(2) if the security interest is in inventory that is or was purchase-money collateral, also to the extent that the security interest secures a purchase-money obligation incurred with respect to other inventory in which the secured party holds or held a purchase-money security interest; and

(3) also to the extent that the security interest secures a purchase-money obligation incurred with respect to software in which the secured party holds or held a purchase-money security interest.

(c) A security interest in software is a purchase-money security interest to the extent that the security interest also secures a purchase-money obligation incurred with respect to goods in which the secured party holds or held a purchase-money security interest if:

(1) the debtor acquired its interest in the software in an integrated transaction in which it acquired an interest in the goods; and

(2) the debtor acquired its interest in the software for the principal purpose of using the software in the goods.

(d) The security interest of a consignor in goods that are the subject of a consignment is a purchase-money security interest in inventory.

(e) In a transaction other than a consumer-goods transaction, if the extent to which a security interest is a purchase-money security interest depends on the application of a payment to a particular obligation, the payment must be applied:

(1) in accordance with any reasonable method of application to which the parties agree;

(2) in the absence of the parties' agreement to a reasonable method, in accordance with any intention of the obligor manifested at or before the time of payment; or

(3) in the absence of an agreement to a reasonable method and a timely manifestation of the obligor's intention, in the following order:

(A) to obligations that are not secured; and

(B) if more than one obligation is secured, to obligations secured by purchase-money security interests in the order in which those obligations were incurred.

(f) In a transaction other than a consumer-goods transaction, a purchase-money security interest does not lose its status as such, even if:

(1) the purchase-money collateral also secures an obligation that is not a purchase-money obligation;

(2) collateral that is not purchase-money collateral also secures the purchase-money obligation; or

(3) the purchase-money obligation has been renewed, refinanced, consolidated, or restructured.

(g) In a transaction other than a consumer-goods transaction, a secured party claiming a purchase-money security interest has the burden of establishing the extent to which the security interest is a purchase-money security interest.

(h) The limitation of the rules in subsections (e), (f), and (g) to transactions other than consumer-goods transactions is intended to leave to the court the determination of the proper rules in consumer-goods transactions. The court may not infer from that limitation the nature of the proper rule in consumer-goods transactions and may continue to apply established approaches.

§ 9–104. Control of Deposit Account.

(a) A secured party has control of a deposit account if:

(1) the secured party is the bank with which the deposit account is maintained;

(2) the debtor, secured party, and bank have agreed in an authenticated record that the bank will comply with instructions originated by the secured party directing disposition of the funds in the deposit account without further consent by the debtor; or

(3) the secured party becomes the bank's customer with respect to the deposit account.

(b) A secured party that has satisfied subsection (a) has control, even if the debtor retains the right to direct the disposition of funds from the deposit account.

§ 9–105. Control of Electronic Chattel Paper.

A secured party has control of electronic chattel paper if the record or records comprising the chattel paper are created, stored, and assigned in such a manner that:

(1) a single authoritative copy of the record or records exists which is unique, identifiable and, except as otherwise provided in paragraphs (4), (5), and (6), unalterable;

(2) the authoritative copy identifies the secured party as the assignee of the record or records;

(3) the authoritative copy is communicated to and maintained by the secured party or its designated custodian;

(4) copies or revisions that add or change an identified assignee of the authoritative copy can be made only with the participation of the secured party;

(5) each copy of the authoritative copy and any copy of a copy is readily identifiable as a copy that is not the authoritative copy; and

(6) any revision of the authoritative copy is readily identifiable as an authorized or unauthorized revision.

§ 9–106. Control of Investment Property.

(a) A person has control of a certificated security, uncertificated security, or security entitlement as provided in Section 8–106.

(b) A secured party has control of a commodity contract if:

(1) the secured party is the commodity intermediary with which the commodity contract is carried; or

(2) the commodity customer, secured party, and commodity intermediary have agreed that the commodity intermediary will apply any value distributed on account of the commodity contract as directed by the secured party without further consent by the commodity customer.

(c) A secured party having control of all security entitlements or commodity contracts carried in a securities account or commodity account has control over the securities account or commodity account.

§ 9–107. Control of Letter-of-Credit Right.

A secured party has control of a letter-of-credit right to the extent of any right to payment or performance by the issuer or any nominated person if the issuer or nominated person has consented to an assignment of proceeds of the letter of credit under Section 5–114(c) or otherwise applicable law or practice.

§ 9–108. Sufficiency of Description.

(a) Except as otherwise provided in subsections (c), (d), and (e), a description of personal or real property is sufficient, whether or not it is specific, if it reasonably identifies what is described.

(b) Except as otherwise provided in subsection (d), a description of collateral reasonably identifies the collateral if it identifies the collateral by:

(1) specific listing;

(2) category;

(3) except as otherwise provided in subsection (e), a type of collateral defined in [the Uniform Commercial Code];

(4) quantity;

(5) computational or allocational formula or procedure; or

(6) except as otherwise provided in subsection (c), any other method, if the identity of the collateral is objectively determinable.

(c) A description of collateral as "all the debtor's assets" or "all the debtor's personal property" or using words of similar import does not reasonably identify the collateral.

(d) Except as otherwise provided in subsection (e), a description of a security entitlement, securities account, or commodity account is sufficient if it describes:

(1) the collateral by those terms or as investment property; or

(2) the underlying financial asset or commodity contract.

(e) A description only by type of collateral defined in [the Uniform Commercial Code] is an insufficient description of:

(1) a commercial tort claim; or

(2) in a consumer transaction, consumer goods, a security entitlement, a securities account, or a commodity account.

[Subpart 2. Applicability of Article]

§ 9–109. Scope.

(a) Except as otherwise provided in subsections (c) and (d), this article applies to:

(1) a transaction, regardless of its form, that creates a security interest in personal property or fixtures by contract;

(2) an agricultural lien;

(3) a sale of accounts, chattel paper, payment intangibles, or promissory notes;

(4) a consignment;

(5) a security interest arising under Section 2–401, 2–505, 2–711(3), or 2A–508(5), as provided in Section 9–110; and

(6) a security interest arising under Section 4–210 or 5–118.

(b) The application of this article to a security interest in a secured obligation is not affected by the fact that the obligation is itself secured by a transaction or interest to which this article does not apply.

(c) This article does not apply to the extent that:

(1) a statute, regulation, or treaty of the United States preempts this article;

(2) another statute of this State expressly governs the creation, perfection, priority, or enforcement of a security interest created by this State or a governmental unit of this State;

(3) a statute of another State, a foreign country, or a governmental unit of another State or a foreign country, other than a statute generally applicable to security interests, expressly governs creation, perfection, priority, or enforcement of a

security interest created by the State, country, or governmental unit; or

(4) the rights of a transferee beneficiary or nominated person under a letter of credit are independent and superior under Section 5–114.

(d) This article does not apply to:

(1) a landlord's lien, other than an agricultural lien;

(2) a lien, other than an agricultural lien, given by statute or other rule of law for services or materials, but Section 9–333 applies with respect to priority of the lien;

(3) an assignment of a claim for wages, salary, or other compensation of an employee;

(4) a sale of accounts, chattel paper, payment intangibles, or promissory notes as part of a sale of the business out of which they arose;

(5) an assignment of accounts, chattel paper, payment intangibles, or promissory notes which is for the purpose of collection only;

(6) an assignment of a right to payment under a contract to an assignee that is also obligated to perform under the contract;

(7) an assignment of a single account, payment intangible, or promissory note to an assignee in full or partial satisfaction of a preexisting indebtedness;

(8) a transfer of an interest in or an assignment of a claim under a policy of insurance, other than an assignment by or to a health-care provider of a health-care-insurance receivable and any subsequent assignment of the right to payment, but Sections 9–315 and 9–322 apply with respect to proceeds and priorities in proceeds;

(9) an assignment of a right represented by a judgment, other than a judgment taken on a right to payment that was collateral;

(10) a right of recoupment or set-off, but:

(A) Section 9–340 applies with respect to the effectiveness of rights of recoupment or set-off against deposit accounts; and

(B) Section 9–404 applies with respect to defenses or claims of an account debtor;

(11) the creation or transfer of an interest in or lien on real property, including a lease or rents thereunder, except to the extent that provision is made for:

(A) liens on real property in Sections 9–203 and 9–308;

(B) fixtures in Section 9–334;

(C) fixture filings in Sections 9–501, 9–502, 9–512, 9–516, and 9–519; and

(D) security agreements covering personal and real property in Section 9–604;

(12) an assignment of a claim arising in tort, other than a commercial tort claim, but Sections 9–315 and 9–322 apply with respect to proceeds and priorities in proceeds; or

(13) an assignment of a deposit account in a consumer transaction, but Sections 9–315 and 9–322 apply with respect to proceeds and priorities in proceeds.

§ 9–110. Security Interests Arising under Article 2 or 2A.

A security interest arising under Section 2–401, 2–505, 2–711(3), or 2A–508(5) is subject to this article. However, until the debtor obtains possession of the goods:

(1) the security interest is enforceable, even if Section 9–203(b)(3) has not been satisfied;

(2) filing is not required to perfect the security interest;

(3) the rights of the secured party after default by the debtor are governed by Article 2 or 2A; and

(4) the security interest has priority over a conflicting security interest created by the debtor.

Part 2 Effectiveness of Security Agreement; Attachment of Security Interest; Rights of Parties to Security Agreement

[Subpart 1. Effectiveness and Attachment]

§ 9–201. General Effectiveness of Security Agreement.

(a) Except as otherwise provided in [the Uniform Commercial Code], a security agreement is effective according to its terms between the parties, against purchasers of the collateral, and against creditors.

(b) A transaction subject to this article is subject to any applicable rule of law which establishes a different rule for consumers and [insert reference to (i) any other statute or regulation that regulates the rates, charges, agreements, and practices for loans, credit sales, or other extensions of credit and (ii) any consumer-protection statute or regulation].

(c) In case of conflict between this article and a rule of law, statute, or regulation described in subsection (b), the rule of law, statute, or regulation controls. Failure to comply with a statute or regulation described in subsection (b) has only the effect the statute or regulation specifies.

(d) This article does not:

(1) validate any rate, charge, agreement, or practice that violates a rule of law, statute, or regulation described in subsection (b); or

(2) extend the application of the rule of law, statute, or regulation to a transaction not otherwise subject to it.

§ 9–202. Title to Collateral Immaterial.

Except as otherwise provided with respect to consignments or sales of accounts, chattel paper, payment intangibles, or promissory notes, the provisions of this article with regard to rights and obligations apply whether title to collateral is in the secured party or the debtor.

§ 9–203. Attachment and Enforceability of Security Interest; Proceeds; Supporting Obligations; Formal Requisites.

(a) A security interest attaches to collateral when it becomes enforceable against the debtor with respect to the collateral,

unless an agreement expressly postpones the time of attachment.

(b) Except as otherwise provided in subsections (c) through (i), a security interest is enforceable against the debtor and third parties with respect to the collateral only if:

(1) value has been given;

(2) the debtor has rights in the collateral or the power to transfer rights in the collateral to a secured party; and

(3) one of the following conditions is met:

(A) the debtor has authenticated a security agreement that provides a description of the collateral and, if the security interest covers timber to be cut, a description of the land concerned;

(B) the collateral is not a certificated security and is in the possession of the secured party under Section 9–313 pursuant to the debtor's security agreement;

(C) the collateral is a certificated security in registered form and the security certificate has been delivered to the secured party under Section 8–301 pursuant to the debtor's security agreement; or

(D) the collateral is deposit accounts, electronic chattel paper, investment property, or letter-of-credit rights, and the secured party has control under Section 9–104, 9–105, 9–106, or 9–107 pursuant to the debtor's security agreement.

(c) Subsection (b) is subject to Section 4–210 on the security interest of a collecting bank, Section 5–118 on the security interest of a letter-of-credit issuer or nominated person, Section 9–110 on a security interest arising under Article 2 or 2A, and Section 9–206 on security interests in investment property.

(d) A person becomes bound as debtor by a security agreement entered into by another person if, by operation of law other than this article or by contract:

(1) the security agreement becomes effective to create a security interest in the person's property; or

(2) the person becomes generally obligated for the obligations of the other person, including the obligation secured under the security agreement, and acquires or succeeds to all or substantially all of the assets of the other person.

(e) If a new debtor becomes bound as debtor by a security agreement entered into by another person:

(1) the agreement satisfies subsection (b)(3) with respect to existing or after-acquired property of the new debtor to the extent the property is described in the agreement; and

(2) another agreement is not necessary to make a security interest in the property enforceable.

(f) The attachment of a security interest in collateral gives the secured party the rights to proceeds provided by Section 9–315 and is also attachment of a security interest in a supporting obligation for the collateral.

(g) The attachment of a security interest in a right to payment or performance secured by a security interest or other lien on personal or real property is also attachment of a security interest in the security interest, mortgage, or other lien.

(h) The attachment of a security interest in a securities account is also attachment of a security interest in the security entitlements carried in the securities account.

(i) The attachment of a security interest in a commodity account is also attachment of a security interest in the commodity contracts carried in the commodity account.

§ 9–204. After-Acquired Property; Future Advances.

(a) Except as otherwise provided in subsection (b), a security agreement may create or provide for a security interest in after-acquired collateral.

(b) A security interest does not attach under a term constituting an after-acquired property clause to:

(1) consumer goods, other than an accession when given as additional security, unless the debtor acquires rights in them within 10 days after the secured party gives value; or

(2) a commercial tort claim.

(c) A security agreement may provide that collateral secures, or that accounts, chattel paper, payment intangibles, or promissory notes are sold in connection with, future advances or other value, whether or not the advances or value are given pursuant to commitment.

§ 9–205. Use or Disposition of Collateral Permissible.

(a) A security interest is not invalid or fraudulent against creditors solely because:

(1) the debtor has the right or ability to:

(A) use, commingle, or dispose of all or part of the collateral, including returned or repossessed goods;

(B) collect, compromise, enforce, or otherwise deal with collateral;

(C) accept the return of collateral or make repossessions; or

(D) use, commingle, or dispose of proceeds; or

(2) the secured party fails to require the debtor to account for proceeds or replace collateral.

(b) This section does not relax the requirements of possession if attachment, perfection, or enforcement of a security interest depends upon possession of the collateral by the secured party.

§ 9–206. Security Interest Arising in Purchase or Delivery of Financial Asset.

(a) A security interest in favor of a securities intermediary attaches to a person's security entitlement if:

(1) the person buys a financial asset through the securities intermediary in a transaction in which the person is obligated to pay the purchase price to the securities intermediary at the time of the purchase; and

(2) the securities intermediary credits the financial asset to the buyer's securities account before the buyer pays the securities intermediary.

(b) The security interest described in subsection (a) secures the person's obligation to pay for the financial asset.

(c) A security interest in favor of a person that delivers a certificated security or other financial asset represented by a writing attaches to the security or other financial asset if:

(1) the security or other financial asset:

(A) in the ordinary course of business is transferred by delivery with any necessary indorsement or assignment; and

(B) is delivered under an agreement between persons in the business of dealing with such securities or financial assets; and

(2) the agreement calls for delivery against payment.

(d) The security interest described in subsection (c) secures the obligation to make payment for the delivery.

[Subpart 2. Rights and Duties]

§ 9–207. Rights and Duties of Secured Party Having Possession or Control of Collateral.

(a) Except as otherwise provided in subsection (d), a secured party shall use reasonable care in the custody and preservation of collateral in the secured party's possession. In the case of chattel paper or an instrument, reasonable care includes taking necessary steps to preserve rights against prior parties unless otherwise agreed.

(b) Except as otherwise provided in subsection (d), if a secured party has possession of collateral:

(1) reasonable expenses, including the cost of insurance and payment of taxes or other charges, incurred in the custody, preservation, use, or operation of the collateral are chargeable to the debtor and are secured by the collateral;

(2) the risk of accidental loss or damage is on the debtor to the extent of a deficiency in any effective insurance coverage;

(3) the secured party shall keep the collateral identifiable, but fungible collateral may be commingled; and

(4) the secured party may use or operate the collateral:

(A) for the purpose of preserving the collateral or its value;

(B) as permitted by an order of a court having competent jurisdiction; or

(C) except in the case of consumer goods, in the manner and to the extent agreed by the debtor.

(c) Except as otherwise provided in subsection (d), a secured party having possession of collateral or control of collateral under Section 9–104, 9–105, 9–106, or 9–107:

(1) may hold as additional security any proceeds, except money or funds, received from the collateral;

(2) shall apply money or funds received from the collateral to reduce the secured obligation, unless remitted to the debtor; and

(3) may create a security interest in the collateral.

(d) If the secured party is a buyer of accounts, chattel paper, payment intangibles, or promissory notes or a consignor:

(1) subsection (a) does not apply unless the secured party is entitled under an agreement:

(A) to charge back uncollected collateral; or

(B) otherwise to full or limited recourse against the debtor or a secondary obligor based on the nonpayment or other default of an account debtor or other obligor on the collateral; and

(2) subsections (b) and (c) do not apply.

§ 9–208. Additional Duties of Secured Party Having Control of Collateral.

(a) This section applies to cases in which there is no outstanding secured obligation and the secured party is not committed to make advances, incur obligations, or otherwise give value.

(b) Within 10 days after receiving an authenticated demand by the debtor:

(1) a secured party having control of a deposit account under Section 9–104(a)(2) shall send to the bank with which the deposit account is maintained an authenticated statement that releases the bank from any further obligation to comply with instructions originated by the secured party;

(2) a secured party having control of a deposit account under Section 9–104(a)(3) shall:

(A) pay the debtor the balance on deposit in the deposit account; or

(B) transfer the balance on deposit into a deposit account in the debtor's name;

(3) a secured party, other than a buyer, having control of electronic chattel paper under Section 9–105 shall:

(A) communicate the authoritative copy of the electronic chattel paper to the debtor or its designated custodian;

(B) if the debtor designates a custodian that is the designated custodian with which the authoritative copy of the electronic chattel paper is maintained for the secured party, communicate to the custodian an authenticated record releasing the designated custodian from any further obligation to comply with instructions originated by the secured party and instructing the custodian to comply with instructions originated by the debtor; and

(C) take appropriate action to enable the debtor or its designated custodian to make copies of or revisions to the authoritative copy which add or change an identified assignee of the authoritative copy without the consent of the secured party;

(4) a secured party having control of investment property under Section 8–106(d)(2) or 9–106(b) shall send to the securities intermediary or commodity intermediary with which the security entitlement or commodity contract is maintained

an authenticated record that releases the securities intermediary or commodity intermediary from any further obligation to comply with entitlement orders or directions originated by the secured party; and

(5) a secured party having control of a letter-of-credit right under Section 9–107 shall send to each person having an unfulfilled obligation to pay or deliver proceeds of the letter of credit to the secured party an authenticated release from any further obligation to pay or deliver proceeds of the letter of credit to the secured party.

§ 9–209. Duties of Secured Party If Account Debtor Has Been Notified of Assignment.

(a) Except as otherwise provided in subsection (c), this section applies if:

(1) there is no outstanding secured obligation; and

(2) the secured party is not committed to make advances, incur obligations, or otherwise give value.

(b) Within 10 days after receiving an authenticated demand by the debtor, a secured party shall send to an account debtor that has received notification of an assignment to the secured party as assignee under Section 9–406(a) an authenticated record that releases the account debtor from any further obligation to the secured party.

(c) This section does not apply to an assignment constituting the sale of an account, chattel paper, or payment intangible.

§ 9–210. Request for Accounting; Request Regarding List of Collateral or Statement of Account.

(a) In this section:

(1) "Request" means a record of a type described in paragraph (2), (3), or (4).

(2) "Request for an accounting" means a record authenticated by a debtor requesting that the recipient provide an accounting of the unpaid obligations secured by collateral and reasonably identifying the transaction or relationship that is the subject of the request.

(3) "Request regarding a list of collateral" means a record authenticated by a debtor requesting that the recipient approve or correct a list of what the debtor believes to be the collateral securing an obligation and reasonably identifying the transaction or relationship that is the subject of the request.

(4) "Request regarding a statement of account" means a record authenticated by a debtor requesting that the recipient approve or correct a statement indicating what the debtor believes to be the aggregate amount of unpaid obligations secured by collateral as of a specified date and reasonably identifying the transaction or relationship that is the subject of the request.

(b) Subject to subsections (c), (d), (e), and (f), a secured party, other than a buyer of accounts, chattel paper, payment intangibles, or promissory notes or a consignor, shall comply with a request within 14 days after receipt:

(1) in the case of a request for an accounting, by authenticating and sending to the debtor an accounting; and

(2) in the case of a request regarding a list of collateral or a request regarding a statement of account, by authenticating and sending to the debtor an approval or correction.

(c) A secured party that claims a security interest in all of a particular type of collateral owned by the debtor may comply with a request regarding a list of collateral by sending to the debtor an authenticated record including a statement to that effect within 14 days after receipt.

(d) A person that receives a request regarding a list of collateral, claims no interest in the collateral when it receives the request, and claimed an interest in the collateral at an earlier time shall comply with the request within 14 days after receipt by sending to the debtor an authenticated record:

(1) disclaiming any interest in the collateral; and

(2) if known to the recipient, providing the name and mailing address of any assignee of or successor to the recipient's interest in the collateral.

(e) A person that receives a request for an accounting or a request regarding a statement of account, claims no interest in the obligations when it receives the request, and claimed an interest in the obligations at an earlier time shall comply with the request within 14 days after receipt by sending to the debtor an authenticated record:

(1) disclaiming any interest in the obligations; and

(2) if known to the recipient, providing the name and mailing address of any assignee of or successor to the recipient's interest in the obligations.

(f) A debtor is entitled without charge to one response to a request under this section during any six-month period. The secured party may require payment of a charge not exceeding $25 for each additional response.

As amended in 1999.

Part 3 Perfection and Priority

[Subpart 1. Law Governing Perfection and Priority]

§ 9–301. Law Governing Perfection and Priority of Security Interests.

Except as otherwise provided in Sections 9–303 through 9–306, the following rules determine the law governing perfection, the effect of perfection or nonperfection, and the priority of a security interest in collateral:

(1) Except as otherwise provided in this section, while a debtor is located in a jurisdiction, the local law of that jurisdiction governs perfection, the effect of perfection or nonperfection, and the priority of a security interest in collateral.

(2) While collateral is located in a jurisdiction, the local law of that jurisdiction governs perfection, the effect of perfection or nonperfection, and the priority of a possessory security interest in that collateral.

(3) Except as otherwise provided in paragraph (4), while negotiable documents, goods, instruments, money, or tangible chattel paper is located in a jurisdiction, the local law of that jurisdiction governs:

(A) perfection of a security interest in the goods by filing a fixture filing;

(B) perfection of a security interest in timber to be cut; and

(C) the effect of perfection or nonperfection and the priority of a nonpossessory security interest in the collateral.

(4) The local law of the jurisdiction in which the wellhead or minehead is located governs perfection, the effect of perfection or nonperfection, and the priority of a security interest in as-extracted collateral.

§ 9–302. Law Governing Perfection and Priority of Agricultural Liens.

While farm products are located in a jurisdiction, the local law of that jurisdiction governs perfection, the effect of perfection or nonperfection, and the priority of an agricultural lien on the farm products.

§ 9–303. Law Governing Perfection and Priority of Security Interests in Goods Covered by a Certificate of Title.

(a) This section applies to goods covered by a certificate of title, even if there is no other relationship between the jurisdiction under whose certificate of title the goods are covered and the goods or the debtor.

(b) Goods become covered by a certificate of title when a valid application for the certificate of title and the applicable fee are delivered to the appropriate authority. Goods cease to be covered by a certificate of title at the earlier of the time the certificate of title ceases to be effective under the law of the issuing jurisdiction or the time the goods become covered subsequently by a certificate of title issued by another jurisdiction.

(c) The local law of the jurisdiction under whose certificate of title the goods are covered governs perfection, the effect of perfection or nonperfection, and the priority of a security interest in goods covered by a certificate of title from the time the goods become covered by the certificate of title until the goods cease to be covered by the certificate of title.

§ 9–304. Law Governing Perfection and Priority of Security Interests in Deposit Accounts.

(a) The local law of a bank's jurisdiction governs perfection, the effect of perfection or nonperfection, and the priority of a security interest in a deposit account maintained with that bank.

(b) The following rules determine a bank's jurisdiction for purposes of this part:

(1) If an agreement between the bank and the debtor governing the deposit account expressly provides that a particular jurisdiction is the bank's jurisdiction for purposes of this part, this article, or [the Uniform Commercial Code], that jurisdiction is the bank's jurisdiction.

(2) If paragraph (1) does not apply and an agreement between the bank and its customer governing the deposit account expressly provides that the agreement is governed by the law of a particular jurisdiction, that jurisdiction is the bank's jurisdiction.

(3) If neither paragraph (1) nor paragraph (2) applies and an agreement between the bank and its customer governing the deposit account expressly provides that the deposit account is maintained at an office in a particular jurisdiction, that jurisdiction is the bank's jurisdiction.

(4) If none of the preceding paragraphs applies, the bank's jurisdiction is the jurisdiction in which the office identified in an account statement as the office serving the customer's account is located.

(5) If none of the preceding paragraphs applies, the bank's jurisdiction is the jurisdiction in which the chief executive office of the bank is located.

§ 9–305. Law Governing Perfection and Priority of Security Interests in Investment Property.

(a) Except as otherwise provided in subsection (c), the following rules apply:

(1) While a security certificate is located in a jurisdiction, the local law of that jurisdiction governs perfection, the effect of perfection or nonperfection, and the priority of a security interest in the certificated security represented thereby.

(2) The local law of the issuer's jurisdiction as specified in Section 8–110(d) governs perfection, the effect of perfection or nonperfection, and the priority of a security interest in an uncertificated security.

(3) The local law of the securities intermediary's jurisdiction as specified in Section 8–110(e) governs perfection, the effect of perfection or nonperfection, and the priority of a security interest in a security entitlement or securities account.

(4) The local law of the commodity intermediary's jurisdiction governs perfection, the effect of perfection or nonperfection, and the priority of a security interest in a commodity contract or commodity account.

(b) The following rules determine a commodity intermediary's jurisdiction for purposes of this part:

(1) If an agreement between the commodity intermediary and commodity customer governing the commodity account expressly provides that a particular jurisdiction is the

commodity intermediary's jurisdiction for purposes of this part, this article, or [the Uniform Commercial Code], that jurisdiction is the commodity intermediary's jurisdiction.

(2) If paragraph (1) does not apply and an agreement between the commodity intermediary and commodity customer governing the commodity account expressly provides that the agreement is governed by the law of a particular jurisdiction, that jurisdiction is the commodity intermediary's jurisdiction.

(3) If neither paragraph (1) nor paragraph (2) applies and an agreement between the commodity intermediary and commodity customer governing the commodity account expressly provides that the commodity account is maintained at an office in a particular jurisdiction, that jurisdiction is the commodity intermediary's jurisdiction.

(4) If none of the preceding paragraphs applies, the commodity intermediary's jurisdiction is the jurisdiction in which the office identified in an account statement as the office serving the commodity customer's account is located.

(5) If none of the preceding paragraphs applies, the commodity intermediary's jurisdiction is the jurisdiction in which the chief executive office of the commodity intermediary is located.

(c) The local law of the jurisdiction in which the debtor is located governs:

(1) perfection of a security interest in investment property by filing;

(2) automatic perfection of a security interest in investment property created by a broker or securities intermediary; and

(3) automatic perfection of a security interest in a commodity contract or commodity account created by a commodity intermediary.

§ 9–306. Law Governing Perfection and Priority of Security Interests in Letter-of-Credit Rights.

(a) Subject to subsection (c), the local law of the issuer's jurisdiction or a nominated person's jurisdiction governs perfection, the effect of perfection or nonperfection, and the priority of a security interest in a letter-of-credit right if the issuer's jurisdiction or nominated person's jurisdiction is a State.

(b) For purposes of this part, an issuer's jurisdiction or nominated person's jurisdiction is the jurisdiction whose law governs the liability of the issuer or nominated person with respect to the letter-of-credit right as provided in Section 5–116.

(c) This section does not apply to a security interest that is perfected only under Section 9–308(d).

§ 9–307. Location of Debtor.

(a) In this section, "place of business" means a place where a debtor conducts its affairs.

(b) Except as otherwise provided in this section, the following rules determine a debtor's location:

(1) A debtor who is an individual is located at the individual's principal residence.

(2) A debtor that is an organization and has only one place of business is located at its place of business.

(3) A debtor that is an organization and has more than one place of business is located at its chief executive office.

(c) Subsection (b) applies only if a debtor's residence, place of business, or chief executive office, as applicable, is located in a jurisdiction whose law generally requires information concerning the existence of a nonpossessory security interest to be made generally available in a filing, recording, or registration system as a condition or result of the security interest's obtaining priority over the rights of a lien creditor with respect to the collateral. If subsection (b) does not apply, the debtor is located in the District of Columbia.

(d) A person that ceases to exist, have a residence, or have a place of business continues to be located in the jurisdiction specified by subsections (b) and (c).

(e) A registered organization that is organized under the law of a State is located in that State.

(f) Except as otherwise provided in subsection (i), a registered organization that is organized under the law of the United States and a branch or agency of a bank that is not organized under the law of the United States or a State are located:

(1) in the State that the law of the United States designates, if the law designates a State of location;

(2) in the State that the registered organization, branch, or agency designates, if the law of the United States authorizes the registered organization, branch, or agency to designate its State of location; or

(3) in the District of Columbia, if neither paragraph (1) nor paragraph (2) applies.

(g) A registered organization continues to be located in the jurisdiction specified by subsection (e) or (f) notwithstanding:

(1) the suspension, revocation, forfeiture, or lapse of the registered organization's status as such in its jurisdiction of organization; or

(2) the dissolution, winding up, or cancellation of the existence of the registered organization.

(h) The United States is located in the District of Columbia.

(i) A branch or agency of a bank that is not organized under the law of the United States or a State is located in the State in which the branch or agency is licensed, if all branches and agencies of the bank are licensed in only one State.

(j) A foreign air carrier under the Federal Aviation Act of 1958, as amended, is located at the designated office of the agent upon which service of process may be made on behalf of the carrier.

(k) This section applies only for purposes of this part.

[Subpart 2. Perfection]

§ 9–308. When Security Interest or Agricultural Lien Is Perfected; Continuity of Perfection.

(a) Except as otherwise provided in this section and Section 9–309, a security interest is perfected if it has attached and all of the applicable requirements for perfection in Sections 9–310 through 9–316 have been satisfied. A security interest is perfected when it attaches if the applicable requirements are satisfied before the security interest attaches.

(b) An agricultural lien is perfected if it has become effective and all of the applicable requirements for perfection in Section 9–310 have been satisfied. An agricultural lien is perfected when it becomes effective if the applicable requirements are satisfied before the agricultural lien becomes effective.

(c) A security interest or agricultural lien is perfected continuously if it is originally perfected by one method under this article and is later perfected by another method under this article, without an intermediate period when it was unperfected.

(d) Perfection of a security interest in collateral also perfects a security interest in a supporting obligation for the collateral.

(e) Perfection of a security interest in a right to payment or performance also perfects a security interest in a security interest, mortgage, or other lien on personal or real property securing the right.

(f) Perfection of a security interest in a securities account also perfects a security interest in the security entitlements carried in the securities account.

(g) Perfection of a security interest in a commodity account also perfects a security interest in the commodity contracts carried in the commodity account.

Legislative Note: Any statute conflicting with subsection (e) must be made expressly subject to that subsection.

§ 9–309. Security Interest Perfected upon Attachment.

The following security interests are perfected when they attach:

(1) a purchase-money security interest in consumer goods, except as otherwise provided in Section 9–311(b) with respect to consumer goods that are subject to a statute or treaty described in Section 9–311(a);

(2) an assignment of accounts or payment intangibles which does not by itself or in conjunction with other assignments to the same assignee transfer a significant part of the assignor's outstanding accounts or payment intangibles;

(3) a sale of a payment intangible;

(4) a sale of a promissory note;

(5) a security interest created by the assignment of a health-care-insurance receivable to the provider of the health-care goods or services;

(6) a security interest arising under Section 2–401, 2–505, 2–711(3), or 2A–508(5), until the debtor obtains possession of the collateral;

(7) a security interest of a collecting bank arising under Section 4–210;

(8) a security interest of an issuer or nominated person arising under Section 5–118;

(9) a security interest arising in the delivery of a financial asset under Section 9–206(c);

(10) a security interest in investment property created by a broker or securities intermediary;

(11) a security interest in a commodity contract or a commodity account created by a commodity intermediary;

(12) an assignment for the benefit of all creditors of the transferor and subsequent transfers by the assignee thereunder; and

(13) a security interest created by an assignment of a beneficial interest in a decedent's estate; and

(14) a sale by an individual of an account that is a right to payment of winnings in a lottery or other game of chance.

§ 9–310. When Filing Required to Perfect Security Interest or Agricultural Lien; Security Interests and Agricultural Liens to Which Filing Provisions Do Not Apply.

(a) Except as otherwise provided in subsection (b) and Section 9–312(b), a financing statement must be filed to perfect all security interests and agricultural liens.

(b) The filing of a financing statement is not necessary to perfect a security interest:

(1) that is perfected under Section 9–308(d), (e), (f), or (g);

(2) that is perfected under Section 9–309 when it attaches;

(3) in property subject to a statute, regulation, or treaty described in Section 9–311(a);

(4) in goods in possession of a bailee which is perfected under Section 9–312(d)(1) or (2);

(5) in certificated securities, documents, goods, or instruments which is perfected without filing or possession under Section 9–312(e), (f), or (g);

(6) in collateral in the secured party's possession under Section 9–313;

(7) in a certificated security which is perfected by delivery of the security certificate to the secured party under Section 9–313;

(8) in deposit accounts, electronic chattel paper, investment property, or letter-of-credit rights which is perfected by control under Section 9–314;

(9) in proceeds which is perfected under Section 9–315; or

(10) that is perfected under Section 9–316.

(c) If a secured party assigns a perfected security interest or agricultural lien, a filing under this article is not required to continue the perfected status of the security interest against creditors of and transferees from the original debtor.

§ 9–311. Perfection of Security Interests in Property Subject to Certain Statutes, Regulations, and Treaties.

(a) Except as otherwise provided in subsection (d), the filing of a financing statement is not necessary or effective to perfect a security interest in property subject to:

(1) a statute, regulation, or treaty of the United States whose requirements for a security interest's obtaining priority over the rights of a lien creditor with respect to the property preempt Section 9–310(a);

(2) [list any certificate-of-title statute covering automobiles, trailers, mobile homes, boats, farm tractors, or the like, which provides for a security interest to be indicated on the certificate as a condition or result of perfection, and any non-Uniform Commercial Code central filing statute]; or

(3) a certificate-of-title statute of another jurisdiction which provides for a security interest to be indicated on the certificate as a condition or result of the security interest's obtaining priority over the rights of a lien creditor with respect to the property.

(b) Compliance with the requirements of a statute, regulation, or treaty described in subsection (a) for obtaining priority over the rights of a lien creditor is equivalent to the filing of a financing statement under this article. Except as otherwise provided in subsection (d) and Sections 9–313 and 9–316(d) and (e) for goods covered by a certificate of title, a security interest in property subject to a statute, regulation, or treaty described in subsection (a) may be perfected only by compliance with those requirements, and a security interest so perfected remains perfected notwithstanding a change in the use or transfer of possession of the collateral.

(c) Except as otherwise provided in subsection (d) and Section 9–316(d) and (e), duration and renewal of perfection of a security interest perfected by compliance with the requirements prescribed by a statute, regulation, or treaty described in subsection (a) are governed by the statute, regulation, or treaty. In other respects, the security interest is subject to this article.

(d) During any period in which collateral subject to a statute specified in subsection (a)(2) is inventory held for sale or lease by a person or leased by that person as lessor and that person is in the business of selling goods of that kind, this section does not apply to a security interest in that collateral created by that person.

Legislative Note: This Article contemplates that perfection of a security interest in goods covered by a certificate of title occurs upon receipt by appropriate State officials of a properly tendered application for a certificate of title on which the security interest is to be indicated, without a relation back to an earlier time. States whose certificate-of-title statutes provide for perfection at a different time or contain a relation-back provision should amend the statutes accordingly.

§ 9–312. Perfection of Security Interests in Chattel Paper, Deposit Accounts, Documents, Goods Covered by Documents, Instruments, Investment Property, Letter-of-Credit Rights, and Money; Perfection by Permissive Filing; Temporary Perfection without Filing or Transfer of Possession.

(a) A security interest in chattel paper, negotiable documents, instruments, or investment property may be perfected by filing.

(b) Except as otherwise provided in Section 9–315(c) and (d) for proceeds:

(1) a security interest in a deposit account may be perfected only by control under Section 9–314;

(2) and except as otherwise provided in Section 9–308(d), a security interest in a letter-of-credit right may be perfected only by control under Section 9–314; and

(3) a security interest in money may be perfected only by the secured party's taking possession under Section 9–313.

(c) While goods are in the possession of a bailee that has issued a negotiable document covering the goods:

(1) a security interest in the goods may be perfected by perfecting a security interest in the document; and

(2) a security interest perfected in the document has priority over any security interest that becomes perfected in the goods by another method during that time.

(d) While goods are in the possession of a bailee that has issued a nonnegotiable document covering the goods, a security interest in the goods may be perfected by:

(1) issuance of a document in the name of the secured party;

(2) the bailee's receipt of notification of the secured party's interest; or

(3) filing as to the goods.

(e) A security interest in certificated securities, negotiable documents, or instruments is perfected without filing or the taking of possession for a period of 20 days from the time it attaches to the extent that it arises for new value given under an authenticated security agreement.

(f) A perfected security interest in a negotiable document or goods in possession of a bailee, other than one that has issued a negotiable document for the goods, remains perfected for 20 days without filing if the secured party makes available to the debtor the goods or documents representing the goods for the purpose of:

(1) ultimate sale or exchange; or

(2) loading, unloading, storing, shipping, transshipping, manufacturing, processing, or otherwise dealing with them in a manner preliminary to their sale or exchange.

(g) A perfected security interest in a certificated security or instrument remains perfected for 20 days without filing if the

secured party delivers the security certificate or instrument to the debtor for the purpose of:

(1) ultimate sale or exchange; or

(2) presentation, collection, enforcement, renewal, or registration of transfer.

(h) After the 20-day period specified in subsection (e), (f), or (g) expires, perfection depends upon compliance with this article.

§ 9–313. When Possession by or Delivery to Secured Party Perfects Security Interest without Filing.

(a) Except as otherwise provided in subsection (b), a secured party may perfect a security interest in negotiable documents, goods, instruments, money, or tangible chattel paper by taking possession of the collateral. A secured party may perfect a security interest in certificated securities by taking delivery of the certificated securities under Section 8–301.

(b) With respect to goods covered by a certificate of title issued by this State, a secured party may perfect a security interest in the goods by taking possession of the goods only in the circumstances described in Section 9–316(d).

(c) With respect to collateral other than certificated securities and goods covered by a document, a secured party takes possession of collateral in the possession of a person other than the debtor, the secured party, or a lessee of the collateral from the debtor in the ordinary course of the debtor's business, when:

(1) the person in possession authenticates a record acknowledging that it holds possession of the collateral for the secured party's benefit; or

(2) the person takes possession of the collateral after having authenticated a record acknowledging that it will hold possession of collateral for the secured party's benefit.

(d) If perfection of a security interest depends upon possession of the collateral by a secured party, perfection occurs no earlier than the time the secured party takes possession and continues only while the secured party retains possession.

(e) A security interest in a certificated security in registered form is perfected by delivery when delivery of the certificated security occurs under Section 8–301 and remains perfected by delivery until the debtor obtains possession of the security certificate.

(f) A person in possession of collateral is not required to acknowledge that it holds possession for a secured party's benefit.

(g) If a person acknowledges that it holds possession for the secured party's benefit:

(1) the acknowledgment is effective under subsection (c) or Section 8–301(a), even if the acknowledgment violates the rights of a debtor; and

(2) unless the person otherwise agrees or law other than this article otherwise provides, the person does not owe any duty to the secured party and is not required to confirm the acknowledgment to another person.

(h) A secured party having possession of collateral does not relinquish possession by delivering the collateral to a person other than the debtor or a lessee of the collateral from the debtor in the ordinary course of the debtor's business if the person was instructed before the delivery or is instructed contemporaneously with the delivery:

(1) to hold possession of the collateral for the secured party's benefit; or

(2) to redeliver the collateral to the secured party.

(i) A secured party does not relinquish possession, even if a delivery under subsection (h) violates the rights of a debtor. A person to which collateral is delivered under subsection (h) does not owe any duty to the secured party and is not required to confirm the delivery to another person unless the person otherwise agrees or law other than this article otherwise provides.

§ 9–314. Perfection by Control.

(a) A security interest in investment property, deposit accounts, letter-of-credit rights, or electronic chattel paper may be perfected by control of the collateral under Section 9–104, 9–105, 9–106, or 9–107.

(b) A security interest in deposit accounts, electronic chattel paper, or letter-of-credit rights is perfected by control under Section 9–104, 9–105, or 9–107 when the secured party obtains control and remains perfected by control only while the secured party retains control.

(c) A security interest in investment property is perfected by control under Section 9–106 from the time the secured party obtains control and remains perfected by control until:

(1) the secured party does not have control; and

(2) one of the following occurs:

(A) if the collateral is a certificated security, the debtor has or acquires possession of the security certificate;

(B) if the collateral is an uncertificated security, the issuer has registered or registers the debtor as the registered owner; or

(C) if the collateral is a security entitlement, the debtor is or becomes the entitlement holder.

§ 9–315. Secured Party's Rights on Disposition of Collateral and in Proceeds.

(a) Except as otherwise provided in this article and in Section 2–403(2):

(1) a security interest or agricultural lien continues in collateral notwithstanding sale, lease, license, exchange, or other disposition thereof unless the secured party authorized the disposition free of the security interest or agricultural lien; and

(2) a security interest attaches to any identifiable proceeds of collateral.

(b) Proceeds that are commingled with other property are identifiable proceeds:

(1) if the proceeds are goods, to the extent provided by Section 9–336; and

(2) if the proceeds are not goods, to the extent that the secured party identifies the proceeds by a method of tracing, including application of equitable principles, that is permitted under law other than this article with respect to commingled property of the type involved.

(c) A security interest in proceeds is a perfected security interest if the security interest in the original collateral was perfected.

(d) A perfected security interest in proceeds becomes unperfected on the 21st day after the security interest attaches to the proceeds unless:

(1) the following conditions are satisfied:

(A) a filed financing statement covers the original collateral;

(B) the proceeds are collateral in which a security interest may be perfected by filing in the office in which the financing statement has been filed; and

(C) the proceeds are not acquired with cash proceeds;

(2) the proceeds are identifiable cash proceeds; or

(3) the security interest in the proceeds is perfected other than under subsection (c) when the security interest attaches to the proceeds or within 20 days thereafter.

(e) If a filed financing statement covers the original collateral, a security interest in proceeds which remains perfected under subsection (d)(1) becomes unperfected at the later of:

(1) when the effectiveness of the filed financing statement lapses under Section 9–515 or is terminated under Section 9–513; or

(2) the 21st day after the security interest attaches to the proceeds.

§ 9–316. Continued Perfection of Security Interest Following Change in Governing Law.

(a) A security interest perfected pursuant to the law of the jurisdiction designated in Section 9–301(1) or 9–305(c) remains perfected until the earliest of:

(1) the time perfection would have ceased under the law of that jurisdiction;

(2) the expiration of four months after a change of the debtor's location to another jurisdiction; or

(3) the expiration of one year after a transfer of collateral to a person that thereby becomes a debtor and is located in another jurisdiction.

(b) If a security interest described in subsection (a) becomes perfected under the law of the other jurisdiction before the earliest time or event described in that subsection, it remains perfected thereafter. If the security interest does not become perfected under the law of the other jurisdiction before the earliest time or event, it becomes unperfected and is deemed never to have been perfected as against a purchaser of the collateral for value.

(c) A possessory security interest in collateral, other than goods covered by a certificate of title and as-extracted collateral consisting of goods, remains continuously perfected if:

(1) the collateral is located in one jurisdiction and subject to a security interest perfected under the law of that jurisdiction;

(2) thereafter the collateral is brought into another jurisdiction; and

(3) upon entry into the other jurisdiction, the security interest is perfected under the law of the other jurisdiction.

(d) Except as otherwise provided in subsection (e), a security interest in goods covered by a certificate of title which is perfected by any method under the law of another jurisdiction when the goods become covered by a certificate of title from this State remains perfected until the security interest would have become unperfected under the law of the other jurisdiction had the goods not become so covered.

(e) A security interest described in subsection (d) becomes unperfected as against a purchaser of the goods for value and is deemed never to have been perfected as against a purchaser of the goods for value if the applicable requirements for perfection under Section 9–311(b) or 9–313 are not satisfied before the earlier of:

(1) the time the security interest would have become unperfected under the law of the other jurisdiction had the goods not become covered by a certificate of title from this State; or

(2) the expiration of four months after the goods had become so covered.

(f) A security interest in deposit accounts, letter-of-credit rights, or investment property which is perfected under the law of the bank's jurisdiction, the issuer's jurisdiction, a nominated person's jurisdiction, the securities intermediary's jurisdiction, or the commodity intermediary's jurisdiction, as applicable, remains perfected until the earlier of:

(1) the time the security interest would have become unperfected under the law of that jurisdiction; or

(2) the expiration of four months after a change of the applicable jurisdiction to another jurisdiction.

(g) If a security interest described in subsection (f) becomes perfected under the law of the other jurisdiction before the earlier of the time or the end of the period described in that subsection, it remains perfected thereafter. If the security interest does not become perfected under the law of the other jurisdiction before the earlier of that time or the end of that period, it becomes unperfected and is deemed never to have been perfected as against a purchaser of the collateral for value.

[Subpart 3. Priority]

§ 9–317. Interests That Take Priority over or Take Free of Security Interest or Agricultural Lien.

(a) A security interest or agricultural lien is subordinate to the rights of:

(1) a person entitled to priority under Section 9–322; and

(2) except as otherwise provided in subsection (e), a person that becomes a lien creditor before the earlier of the time:

(A) the security interest or agricultural lien is perfected; or

(B) one of the conditions specified in Section 9–203(b)(3) is met and a financing statement covering the collateral is filed.

(b) Except as otherwise provided in subsection (e), a buyer, other than a secured party, of tangible chattel paper, documents, goods, instruments, or a security certificate takes free of a security interest or agricultural lien if the buyer gives value and receives delivery of the collateral without knowledge of the security interest or agricultural lien and before it is perfected.

(c) Except as otherwise provided in subsection (e), a lessee of goods takes free of a security interest or agricultural lien if the lessee gives value and receives delivery of the collateral without knowledge of the security interest or agricultural lien and before it is perfected.

(d) A licensee of a general intangible or a buyer, other than a secured party, of accounts, electronic chattel paper, general intangibles, or investment property other than a certificated security takes free of a security interest if the licensee or buyer gives value without knowledge of the security interest and before it is perfected.

(e) Except as otherwise provided in Sections 9–320 and 9–321, if a person files a financing statement with respect to a purchase-money security interest before or within 20 days after the debtor receives delivery of the collateral, the security interest takes priority over the rights of a buyer, lessee, or lien creditor which arise between the time the security interest attaches and the time of filing.

As amended in 2000.

§ 9–318. No Interest Retained in Right to Payment That Is Sold; Rights and Title of Seller of Account or Chattel Paper with Respect to Creditors and Purchasers.

(a) A debtor that has sold an account, chattel paper, payment intangible, or promissory note does not retain a legal or equitable interest in the collateral sold.

(b) For purposes of determining the rights of creditors of, and purchasers for value of an account or chattel paper from, a debtor that has sold an account or chattel paper, while the buyer's security interest is unperfected, the debtor is deemed to have rights and title to the account or chattel paper identical to those the debtor sold.

§ 9–319. Rights and Title of Consignee with Respect to Creditors and Purchasers.

(a) Except as otherwise provided in subsection (b), for purposes of determining the rights of creditors of, and purchasers for value of goods from, a consignee, while the goods are in the possession of the consignee, the consignee is deemed to have rights and title to the goods identical to those the consignor had or had power to transfer.

(b) For purposes of determining the rights of a creditor of a consignee, law other than this article determines the rights and title of a consignee while goods are in the consignee's possession if, under this part, a perfected security interest held by the consignor would have priority over the rights of the creditor.

§ 9–320. Buyer of Goods.

(a) Except as otherwise provided in subsection (e), a buyer in ordinary course of business, other than a person buying farm products from a person engaged in farming operations, takes free of a security interest created by the buyer's seller, even if the security interest is perfected and the buyer knows of its existence.

(b) Except as otherwise provided in subsection (e), a buyer of goods from a person who used or bought the goods for use primarily for personal, family, or household purposes takes free of a security interest, even if perfected, if the buyer buys:

(1) without knowledge of the security interest;

(2) for value;

(3) primarily for the buyer's personal, family, or household purposes; and

(4) before the filing of a financing statement covering the goods.

(c) To the extent that it affects the priority of a security interest over a buyer of goods under subsection (b), the period of effectiveness of a filing made in the jurisdiction in which the seller is located is governed by Section 9–316(a) and (b).

(d) A buyer in ordinary course of business buying oil, gas, or other minerals at the wellhead or minehead or after extraction takes free of an interest arising out of an encumbrance.

(e) Subsections (a) and (b) do not affect a security interest in goods in the possession of the secured party under Section 9–313.

§ 9–321. Licensee of General Intangible and Lessee of Goods in Ordinary Course of Business.

(a) In this section, "licensee in ordinary course of business" means a person that becomes a licensee of a general intangible in good faith, without knowledge that the license violates the rights of another person in the general intangible, and in the ordinary course from a person in the business of licensing general intangibles of that kind. A person becomes a licensee in the ordinary course if the license to the person comports with the usual or customary practices in the kind of business in which the licensor is engaged or with the licensor's own usual or customary practices.

(b) A licensee in ordinary course of business takes its rights under a nonexclusive license free of a security interest in the general intangible created by the licensor, even if the security interest is perfected and the licensee knows of its existence.

(c) A lessee in ordinary course of business takes its leasehold interest free of a security interest in the goods created by the lessor, even if the security interest is perfected and the lessee knows of its existence.

§ 9–322. Priorities among Conflicting Security Interests in and Agricultural Liens on Same Collateral.

(a) Except as otherwise provided in this section, priority among conflicting security interests and agricultural liens in the same collateral is determined according to the following rules:

(1) Conflicting perfected security interests and agricultural liens rank according to priority in time of filing or perfection. Priority dates from the earlier of the time a filing covering the collateral is first made or the security interest or agricultural lien is first perfected, if there is no period thereafter when there is neither filing nor perfection.

(2) A perfected security interest or agricultural lien has priority over a conflicting unperfected security interest or agricultural lien.

(3) The first security interest or agricultural lien to attach or become effective has priority if conflicting security interests and agricultural liens are unperfected.

(b) For the purposes of subsection (a)(1):

(1) the time of filing or perfection as to a security interest in collateral is also the time of filing or perfection as to a security interest in proceeds; and

(2) the time of filing or perfection as to a security interest in collateral supported by a supporting obligation is also the time of filing or perfection as to a security interest in the supporting obligation.

(c) Except as otherwise provided in subsection (f), a security interest in collateral which qualifies for priority over a conflicting security interest under Section 9–327, 9–328, 9–329, 9–330, or 9–331 also has priority over a conflicting security interest in:

(1) any supporting obligation for the collateral; and

(2) proceeds of the collateral if:

(A) the security interest in proceeds is perfected;

(B) the proceeds are cash proceeds or of the same type as the collateral; and

(C) in the case of proceeds that are proceeds of proceeds, all intervening proceeds are cash proceeds, proceeds of the same type as the collateral, or an account relating to the collateral.

(d) Subject to subsection (e) and except as otherwise provided in subsection (f), if a security interest in chattel paper, deposit accounts, negotiable documents, instruments, investment property, or letter-of-credit rights is perfected by a method other than filing, conflicting perfected security interests in proceeds of the collateral rank according to priority in time of filing.

(e) Subsection (d) applies only if the proceeds of the collateral are not cash proceeds, chattel paper, negotiable documents, instruments, investment property, or letter-of-credit rights.

(f) Subsections (a) through (e) are subject to:

(1) subsection (g) and the other provisions of this part;

(2) Section 4–210 with respect to a security interest of a collecting bank;

(3) Section 5–118 with respect to a security interest of an issuer or nominated person; and

(4) Section 9–110 with respect to a security interest arising under Article 2 or 2A.

(g) A perfected agricultural lien on collateral has priority over a conflicting security interest in or agricultural lien on the same collateral if the statute creating the agricultural lien so provides.

§ 9–323. Future Advances.

(a) Except as otherwise provided in subsection (c), for purposes of determining the priority of a perfected security interest under Section 9–322(a)(1), perfection of the security interest dates from the time an advance is made to the extent that the security interest secures an advance that:

(1) is made while the security interest is perfected only:

(A) under Section 9–309 when it attaches; or

(B) temporarily under Section 9–312(e), (f), or (g); and

(2) is not made pursuant to a commitment entered into before or while the security interest is perfected by a method other than under Section 9–309 or 9–312(e), (f), or (g).

(b) Except as otherwise provided in subsection (c), a security interest is subordinate to the rights of a person that becomes a lien creditor to the extent that the security interest secures an advance made more than 45 days after the person becomes a lien creditor unless the advance is made:

(1) without knowledge of the lien; or

(2) pursuant to a commitment entered into without knowledge of the lien.

(c) Subsections (a) and (b) do not apply to a security interest held by a secured party that is a buyer of accounts, chattel paper, payment intangibles, or promissory notes or a consignor.

(d) Except as otherwise provided in subsection (e), a buyer of goods other than a buyer in ordinary course of business takes free of a security interest to the extent that it secures advances made after the earlier of:

(1) the time the secured party acquires knowledge of the buyer's purchase; or

(2) 45 days after the purchase.

(e) Subsection (d) does not apply if the advance is made pursuant to a commitment entered into without knowledge of the buyer's purchase and before the expiration of the 45-day period.

(f) Except as otherwise provided in subsection (g), a lessee of goods, other than a lessee in ordinary course of business, takes the leasehold interest free of a security interest to the extent that it secures advances made after the earlier of:

(1) the time the secured party acquires knowledge of the lease; or

(2) 45 days after the lease contract becomes enforceable.

(g) Subsection (f) does not apply if the advance is made pursuant to a commitment entered into without knowledge of the lease and before the expiration of the 45-day period.

As amended in 1999.

§ 9–324. Priority of Purchase-Money Security Interests.

(a) Except as otherwise provided in subsection (g), a perfected purchase-money security interest in goods other than inventory or livestock has priority over a conflicting security interest in the same goods, and, except as otherwise provided in Section 9–327, a perfected security interest in its identifiable proceeds also has priority, if the purchase-money security interest is perfected when the debtor receives possession of the collateral or within 20 days thereafter.

(b) Subject to subsection (c) and except as otherwise provided in subsection (g), a perfected purchase-money security interest in inventory has priority over a conflicting security interest in the same inventory, has priority over a conflicting security interest in chattel paper or an instrument constituting proceeds of the inventory and in proceeds of the chattel paper, if so provided in Section 9–330, and, except as otherwise provided in Section 9–327, also has priority in identifiable cash proceeds of the inventory to the extent the identifiable cash proceeds are received on or before the delivery of the inventory to a buyer, if:

(1) the purchase-money security interest is perfected when the debtor receives possession of the inventory;

(2) the purchase-money secured party sends an authenticated notification to the holder of the conflicting security interest;

(3) the holder of the conflicting security interest receives the notification within five years before the debtor receives possession of the inventory; and

(4) the notification states that the person sending the notification has or expects to acquire a purchase-money security interest in inventory of the debtor and describes the inventory.

(c) Subsections (b)(2) through (4) apply only if the holder of the conflicting security interest had filed a financing statement covering the same types of inventory:

(1) if the purchase-money security interest is perfected by filing, before the date of the filing; or

(2) if the purchase-money security interest is temporarily perfected without filing or possession under Section 9–312(f), before the beginning of the 20-day period thereunder.

(d) Subject to subsection (e) and except as otherwise provided in subsection (g), a perfected purchase-money security interest in livestock that are farm products has priority over a conflicting security interest in the same livestock, and, except as otherwise provided in Section 9–327, a perfected security interest in their identifiable proceeds and identifiable products in their unmanufactured states also has priority, if:

(1) the purchase-money security interest is perfected when the debtor receives possession of the livestock;

(2) the purchase-money secured party sends an authenticated notification to the holder of the conflicting security interest;

(3) the holder of the conflicting security interest receives the notification within six months before the debtor receives possession of the livestock; and

(4) the notification states that the person sending the notification has or expects to acquire a purchase-money security interest in livestock of the debtor and describes the livestock.

(e) Subsections (d)(2) through (4) apply only if the holder of the conflicting security interest had filed a financing statement covering the same types of livestock:

(1) if the purchase-money security interest is perfected by filing, before the date of the filing; or

(2) if the purchase-money security interest is temporarily perfected without filing or possession under Section 9–312(f), before the beginning of the 20-day period thereunder.

(f) Except as otherwise provided in subsection (g), a perfected purchase-money security interest in software has priority over a conflicting security interest in the same collateral, and, except as otherwise provided in Section 9–327, a perfected security interest in its identifiable proceeds also has priority, to the extent that the purchase-money security interest in the goods in which the software was acquired for use has priority in the goods and proceeds of the goods under this section.

(g) If more than one security interest qualifies for priority in the same collateral under subsection (a), (b), (d), or (f):

(1) a security interest securing an obligation incurred as all or part of the price of the collateral has priority over a security interest securing an obligation incurred for value given to enable the debtor to acquire rights in or the use of collateral; and

(2) in all other cases, Section 9–322(a) applies to the qualifying security interests.

§ 9–325. Priority of Security Interests in Transferred Collateral.

(a) Except as otherwise provided in subsection (b), a security interest created by a debtor is subordinate to a security interest in the same collateral created by another person if:

(1) the debtor acquired the collateral subject to the security interest created by the other person;

(2) the security interest created by the other person was perfected when the debtor acquired the collateral; and

(3) there is no period thereafter when the security interest is unperfected.

(b) Subsection (a) subordinates a security interest only if the security interest:

(1) otherwise would have priority solely under Section 9–322(a) or 9–324; or

(2) arose solely under Section 2–711(3) or 2A–508(5).

§ 9–326. Priority of Security Interests Created by New Debtor.

(a) Subject to subsection (b), a security interest created by a new debtor which is perfected by a filed financing statement that is effective solely under Section 9–508 in collateral in which a new debtor has or acquires rights is subordinate to a security interest in the same collateral which is perfected other than by a filed financing statement that is effective solely under Section 9–508.

(b) The other provisions of this part determine the priority among conflicting security interests in the same collateral perfected by filed financing statements that are effective solely under Section 9–508. However, if the security agreements to which a new debtor became bound as debtor were not entered into by the same original debtor, the conflicting security interests rank according to priority in time of the new debtor's having become bound.

§ 9–327. Priority of Security Interests in Deposit Account.

The following rules govern priority among conflicting security interests in the same deposit account:

(1) A security interest held by a secured party having control of the deposit account under Section 9–104 has priority over a conflicting security interest held by a secured party that does not have control.

(2) Except as otherwise provided in paragraphs (3) and (4), security interests perfected by control under Section 9–314 rank according to priority in time of obtaining control.

(3) Except as otherwise provided in paragraph (4), a security interest held by the bank with which the deposit account is maintained has priority over a conflicting security interest held by another secured party.

(4) A security interest perfected by control under Section 9–104(a)(3) has priority over a security interest held by the bank with which the deposit account is maintained.

§ 9–328. Priority of Security Interests in Investment Property.

The following rules govern priority among conflicting security interests in the same investment property:

(1) A security interest held by a secured party having control of investment property under Section 9–106 has priority over a security interest held by a secured party that does not have control of the investment property.

(2) Except as otherwise provided in paragraphs (3) and (4), conflicting security interests held by secured parties each of which has control under Section 9–106 rank according to priority in time of:

(A) if the collateral is a security, obtaining control;

(B) if the collateral is a security entitlement carried in a securities account and:

(i) if the secured party obtained control under Section 8–106(d)(1), the secured party's becoming the person for which the securities account is maintained;

(ii) if the secured party obtained control under Section 8–106(d)(2), the securities intermediary's agreement to comply with the secured party's entitlement orders with respect to security entitlements carried or to be carried in the securities account; or

(iii) if the secured party obtained control through another person under Section 8–106(d)(3), the time on which priority would be based under this paragraph if the other person were the secured party; or

(C) if the collateral is a commodity contract carried with a commodity intermediary, the satisfaction of the requirement for control specified in Section 9–106(b)(2) with respect to commodity contracts carried or to be carried with the commodity intermediary.

(3) A security interest held by a securities intermediary in a security entitlement or a securities account maintained with the securities intermediary has priority over a conflicting security interest held by another secured party.

(4) A security interest held by a commodity intermediary in a commodity contract or a commodity account maintained with the commodity intermediary has priority over a conflicting security interest held by another secured party.

(5) A security interest in a certificated security in registered form which is perfected by taking delivery under Section 9–313(a) and not by control under Section 9–314 has priority over a conflicting security interest perfected by a method other than control.

(6) Conflicting security interests created by a broker, securities intermediary, or commodity intermediary which are perfected without control under Section 9–106 rank equally.

(7) In all other cases, priority among conflicting security interests in investment property is governed by Sections 9–322 and 9–323.

§ 9–329. Priority of Security Interests in Letter-of-Credit Right.

The following rules govern priority among conflicting security interests in the same letter-of-credit right:

(1) A security interest held by a secured party having control of the letter-of-credit right under Section 9–107 has priority to the extent of its control over a conflicting security interest held by a secured party that does not have control.

(2) Security interests perfected by control under Section 9–314 rank according to priority in time of obtaining control.

§ 9–330. Priority of Purchaser of Chattel Paper or Instrument.

(a) A purchaser of chattel paper has priority over a security interest in the chattel paper which is claimed merely as proceeds of inventory subject to a security interest if:

(1) in good faith and in the ordinary course of the purchaser's business, the purchaser gives new value and takes possession of the chattel paper or obtains control of the chattel paper under Section 9–105; and

(2) the chattel paper does not indicate that it has been assigned to an identified assignee other than the purchaser.

(b) A purchaser of chattel paper has priority over a security interest in the chattel paper which is claimed other than merely as proceeds of inventory subject to a security interest if the purchaser gives new value and takes possession of the chattel paper or obtains control of the chattel paper under Section 9–105 in good faith, in the ordinary course of the purchaser's business, and without knowledge that the purchase violates the rights of the secured party.

(c) Except as otherwise provided in Section 9–327, a purchaser having priority in chattel paper under subsection (a) or (b) also has priority in proceeds of the chattel paper to the extent that:

(1) Section 9–322 provides for priority in the proceeds; or

(2) the proceeds consist of the specific goods covered by the chattel paper or cash proceeds of the specific goods, even if the purchaser's security interest in the proceeds is unperfected.

(d) Except as otherwise provided in Section 9–331(a), a purchaser of an instrument has priority over a security interest in the instrument perfected by a method other than possession if the purchaser gives value and takes possession of the instrument in good faith and without knowledge that the purchase violates the rights of the secured party.

(e) For purposes of subsections (a) and (b), the holder of a purchase-money security interest in inventory gives new value for chattel paper constituting proceeds of the inventory.

(f) For purposes of subsections (b) and (d), if chattel paper or an instrument indicates that it has been assigned to an identified secured party other than the purchaser, a purchaser of the chattel paper or instrument has knowledge that the purchase violates the rights of the secured party.

§ 9–331. Priority of Rights of Purchasers of Instruments, Documents, and Securities under Other Articles; Priority of Interests in Financial Assets and Security Entitlements under Article 8.

(a) This article does not limit the rights of a holder in due course of a negotiable instrument, a holder to which a negotiable document of title has been duly negotiated, or a protected purchaser of a security. These holders or purchasers take priority over an earlier security interest, even if perfected, to the extent provided in Articles 3, 7, and 8.

(b) This article does not limit the rights of or impose liability on a person to the extent that the person is protected against the assertion of a claim under Article 8.

(c) Filing under this article does not constitute notice of a claim or defense to the holders, or purchasers, or persons described in subsections (a) and (b).

§ 9–332. Transfer of Money; Transfer of Funds from Deposit Account.

(a) A transferee of money takes the money free of a security interest unless the transferee acts in collusion with the debtor in violating the rights of the secured party.

(b) A transferee of funds from a deposit account takes the funds free of a security interest in the deposit account unless the transferee acts in collusion with the debtor in violating the rights of the secured party.

§ 9–333. Priority of Certain Liens Arising by Operation of Law.

(a) In this section, "possessory lien" means an interest, other than a security interest or an agricultural lien:

(1) which secures payment or performance of an obligation for services or materials furnished with respect to goods by a person in the ordinary course of the person's business;

(2) which is created by statute or rule of law in favor of the person; and

(3) whose effectiveness depends on the person's possession of the goods.

(b) A possessory lien on goods has priority over a security interest in the goods unless the lien is created by a statute that expressly provides otherwise.

§ 9–334. Priority of Security Interests in Fixtures and Crops.

(a) A security interest under this article may be created in goods that are fixtures or may continue in goods that become fixtures. A security interest does not exist under this article in ordinary building materials incorporated into an improvement on land.

(b) This article does not prevent creation of an encumbrance upon fixtures under real property law.

(c) In cases not governed by subsections (d) through (h), a security interest in fixtures is subordinate to a conflicting interest of an encumbrancer or owner of the related real property other than the debtor.

(d) Except as otherwise provided in subsection (h), a perfected security interest in fixtures has priority over a conflicting interest of an encumbrancer or owner of the real property if the debtor has an interest of record in or is in possession of the real property and:

(1) the security interest is a purchase-money security interest;

(2) the interest of the encumbrancer or owner arises before the goods become fixtures; and

(3) the security interest is perfected by a fixture filing before the goods become fixtures or within 20 days thereafter.

(e) A perfected security interest in fixtures has priority over a conflicting interest of an encumbrancer or owner of the real property if:

(1) the debtor has an interest of record in the real property or is in possession of the real property and the security interest:

(A) is perfected by a fixture filing before the interest of the encumbrancer or owner is of record; and

(B) has priority over any conflicting interest of a predecessor in title of the encumbrancer or owner;

(2) before the goods become fixtures, the security interest is perfected by any method permitted by this article and the fixtures are readily removable:

(A) factory or office machines;

(B) equipment that is not primarily used or leased for use in the operation of the real property; or

(C) replacements of domestic appliances that are consumer goods;

(3) the conflicting interest is a lien on the real property obtained by legal or equitable proceedings after the security interest was perfected by any method permitted by this article; or

(4) the security interest is:

(A) created in a manufactured home in a manufactured-home transaction; and

(B) perfected pursuant to a statute described in Section 9–311(a)(2).

(f) A security interest in fixtures, whether or not perfected, has priority over a conflicting interest of an encumbrancer or owner of the real property if:

(1) the encumbrancer or owner has, in an authenticated record, consented to the security interest or disclaimed an interest in the goods as fixtures; or

(2) the debtor has a right to remove the goods as against the encumbrancer or owner.

(g) The priority of the security interest under paragraph (f)(2) continues for a reasonable time if the debtor's right to remove the goods as against the encumbrancer or owner terminates.

(h) A mortgage is a construction mortgage to the extent that it secures an obligation incurred for the construction of an improvement on land, including the acquisition cost of the land, if a recorded record of the mortgage so indicates. Except as otherwise provided in subsections (e) and (f), a security interest in fixtures is subordinate to a construction mortgage if a record of the mortgage is recorded before the goods become fixtures and the goods become fixtures before the completion of the construction. A mortgage has this priority to the same extent as a construction mortgage to the extent that it is given to refinance a construction mortgage.

(i) A perfected security interest in crops growing on real property has priority over a conflicting interest of an encumbrancer or owner of the real property if the debtor has an interest of record in or is in possession of the real property.

(j) Subsection (i) prevails over any inconsistent provisions of the following statutes:

[List here any statutes containing provisions inconsistent with subsection (i).]

Legislative Note: States that amend statutes to remove provisions inconsistent with subsection (i) need not enact subsection (j).

§ 9–335. Accessions.

(a) A security interest may be created in an accession and continues in collateral that becomes an accession.

(b) If a security interest is perfected when the collateral becomes an accession, the security interest remains perfected in the collateral.

(c) Except as otherwise provided in subsection (d), the other provisions of this part determine the priority of a security interest in an accession.

(d) A security interest in an accession is subordinate to a security interest in the whole which is perfected by compliance with the requirements of a certificate-of-title statute under Section 9–311(b).

(e) After default, subject to Part 6, a secured party may remove an accession from other goods if the security interest in the accession has priority over the claims of every person having an interest in the whole.

(f) A secured party that removes an accession from other goods under subsection (e) shall promptly reimburse any holder of a security interest or other lien on, or owner of, the whole or of the other goods, other than the debtor, for the cost of repair of any physical injury to the whole or the other goods. The secured party need not reimburse the holder or owner for any diminution in value of the whole or the other goods caused by the absence of the accession removed or by any necessity for replacing it. A person entitled to reimbursement may refuse permission to remove until the secured party gives adequate assurance for the performance of the obligation to reimburse.

§ 9–336. Commingled Goods.

(a) In this section, "commingled goods" means goods that are physically united with other goods in such a manner that their identity is lost in a product or mass.

(b) A security interest does not exist in commingled goods as such. However, a security interest may attach to a product or mass that results when goods become commingled goods.

(c) If collateral becomes commingled goods, a security interest attaches to the product or mass.

(d) If a security interest in collateral is perfected before the collateral becomes commingled goods, the security interest that attaches to the product or mass under subsection (c) is perfected.

(e) Except as otherwise provided in subsection (f), the other provisions of this part determine the priority of a security interest that attaches to the product or mass under subsection (c).

(f) If more than one security interest attaches to the product or mass under subsection (c), the following rules determine priority:

(1) A security interest that is perfected under subsection (d) has priority over a security interest that is unperfected at the time the collateral becomes commingled goods.

(2) If more than one security interest is perfected under subsection (d), the security interests rank equally in proportion to the value of the collateral at the time it became commingled goods.

§ 9–337. Priority of Security Interests in Goods Covered by Certificate of Title.

If, while a security interest in goods is perfected by any method under the law of another jurisdiction, this State issues a

certificate of title that does not show that the goods are subject to the security interest or contain a statement that they may be subject to security interests not shown on the certificate:

(1) a buyer of the goods, other than a person in the business of selling goods of that kind, takes free of the security interest if the buyer gives value and receives delivery of the goods after issuance of the certificate and without knowledge of the security interest; and

(2) the security interest is subordinate to a conflicting security interest in the goods that attaches, and is perfected under Section 9–311(b), after issuance of the certificate and without the conflicting secured party's knowledge of the security interest.

§ 9–338. Priority of Security Interest or Agricultural Lien Perfected by Filed Financing Statement Providing Certain Incorrect Information.

If a security interest or agricultural lien is perfected by a filed financing statement providing information described in Section 9–516(b)(5) which is incorrect at the time the financing statement is filed:

(1) the security interest or agricultural lien is subordinate to a conflicting perfected security interest in the collateral to the extent that the holder of the conflicting security interest gives value in reasonable reliance upon the incorrect information; and

(2) a purchaser, other than a secured party, of the collateral takes free of the security interest or agricultural lien to the extent that, in reasonable reliance upon the incorrect information, the purchaser gives value and, in the case of chattel paper, documents, goods, instruments, or a security certificate, receives delivery of the collateral.

§ 9–339. Priority Subject to Subordination.

This article does not preclude subordination by agreement by a person entitled to priority.

[Subpart 4. Rights of Bank]

§ 9–340. Effectiveness of Right of Recoupment or Set-Off against Deposit Account.

(a) Except as otherwise provided in subsection (c), a bank with which a deposit account is maintained may exercise any right of recoupment or set-off against a secured party that holds a security interest in the deposit account.

(b) Except as otherwise provided in subsection (c), the application of this article to a security interest in a deposit account does not affect a right of recoupment or set-off of the secured party as to a deposit account maintained with the secured party.

(c) The exercise by a bank of a set-off against a deposit account is ineffective against a secured party that holds a security interest in the deposit account which is perfected by control under Section 9–104(a)(3), if the set-off is based on a claim against the debtor.

§ 9–341. Bank's Rights and Duties with Respect to Deposit Account.

Except as otherwise provided in Section 9–340(c), and unless the bank otherwise agrees in an authenticated record, a bank's rights and duties with respect to a deposit account maintained with the bank are not terminated, suspended, or modified by:

(1) the creation, attachment, or perfection of a security interest in the deposit account;

(2) the bank's knowledge of the security interest; or

(3) the bank's receipt of instructions from the secured party.

§ 9–342. Bank's Right to Refuse to Enter into or Disclose Existence of Control Agreement.

This article does not require a bank to enter into an agreement of the kind described in Section 9–104(a)(2), even if its customer so requests or directs. A bank that has entered into such an agreement is not required to confirm the existence of the agreement to another person unless requested to do so by its customer.

Part 4 Rights of Third Parties

§ 9–401. Alienability of Debtor's Rights.

(a) Except as otherwise provided in subsection (b) and Sections 9–406, 9–407, 9–408, and 9–409, whether a debtor's rights in collateral may be voluntarily or involuntarily transferred is governed by law other than this article.

(b) An agreement between the debtor and secured party which prohibits a transfer of the debtor's rights in collateral or makes the transfer a default does not prevent the transfer from taking effect.

§ 9–402. Secured Party Not Obligated on Contract of Debtor or in Tort.

The existence of a security interest, agricultural lien, or authority given to a debtor to dispose of or use collateral, without more, does not subject a secured party to liability in contract or tort for the debtor's acts or omissions.

§ 9–403. Agreement Not to Assert Defenses against Assignee.

(a) In this section, "value" has the meaning provided in Section 3–303(a).

(b) Except as otherwise provided in this section, an agreement between an account debtor and an assignor not to assert against an assignee any claim or defense that the account debtor may have against the assignor is enforceable by an assignee that takes an assignment:

(1) for value;

(2) in good faith;

(3) without notice of a claim of a property or possessory right to the property assigned; and

(4) without notice of a defense or claim in recoupment of the type that may be asserted against a person entitled to enforce a negotiable instrument under Section 3–305(a).

(c) Subsection (b) does not apply to defenses of a type that may be asserted against a holder in due course of a negotiable instrument under Section 3–305(b).

(d) In a consumer transaction, if a record evidences the account debtor's obligation, law other than this article requires that the record include a statement to the effect that the rights of an assignee are subject to claims or defenses that the account debtor could assert against the original obligee, and the record does not include such a statement:

(1) the record has the same effect as if the record included such a statement; and

(2) the account debtor may assert against an assignee those claims and defenses that would have been available if the record included such a statement.

(e) This section is subject to law other than this article which establishes a different rule for an account debtor who is an individual and who incurred the obligation primarily for personal, family, or household purposes.

(f) Except as otherwise provided in subsection (d), this section does not displace law other than this article which gives effect to an agreement by an account debtor not to assert a claim or defense against an assignee.

§ 9–404. Rights Acquired by Assignee; Claims and Defenses against Assignee.

(a) Unless an account debtor has made an enforceable agreement not to assert defenses or claims, and subject to subsections (b) through (e), the rights of an assignee are subject to:

(1) all terms of the agreement between the account debtor and assignor and any defense or claim in recoupment arising from the transaction that gave rise to the contract; and

(2) any other defense or claim of the account debtor against the assignor which accrues before the account debtor receives a notification of the assignment authenticated by the assignor or the assignee.

(b) Subject to subsection (c) and except as otherwise provided in subsection (d), the claim of an account debtor against an assignor may be asserted against an assignee under subsection (a) only to reduce the amount the account debtor owes.

(c) This section is subject to law other than this article which establishes a different rule for an account debtor who is an individual and who incurred the obligation primarily for personal, family, or household purposes.

(d) In a consumer transaction, if a record evidences the account debtor's obligation, law other than this article requires that the record include a statement to the effect that the account debtor's recovery against an assignee with respect to claims and defenses against the assignor may not exceed amounts paid by the account debtor under the record, and the record does not include such a statement, the extent to which a claim of an account debtor against the assignor may be asserted against an assignee is determined as if the record included such a statement.

(e) This section does not apply to an assignment of a health-care-insurance receivable.

§ 9–405. Modification of Assigned Contract.

(a) A modification of or substitution for an assigned contract is effective against an assignee if made in good faith. The assignee acquires corresponding rights under the modified or substituted contract. The assignment may provide that the modification or substitution is a breach of contract by the assignor. This subsection is subject to subsections (b) through (d).

(b) Subsection (a) applies to the extent that:

(1) the right to payment or a part thereof under an assigned contract has not been fully earned by performance; or

(2) the right to payment or a part thereof has been fully earned by performance and the account debtor has not received notification of the assignment under Section 9–406(a).

(c) This section is subject to law other than this article which establishes a different rule for an account debtor who is an individual and who incurred the obligation primarily for personal, family, or household purposes.

(d) This section does not apply to an assignment of a health-care-insurance receivable.

§ 9–406. Discharge of Account Debtor; Notification of Assignment; Identification and Proof of Assignment; Restrictions on Assignment of Accounts, Chattel Paper, Payment Intangibles, and Promissory Notes Ineffective.

(a) Subject to subsections (b) through (i), an account debtor on an account, chattel paper, or a payment intangible may discharge its obligation by paying the assignor until, but not after, the account debtor receives a notification, authenticated by the assignor or the assignee, that the amount due or to become due has been assigned and that payment is to be made to the assignee. After receipt of the notification, the account debtor may discharge its obligation by paying the assignee and may not discharge the obligation by paying the assignor.

(b) Subject to subsection (h), notification is ineffective under subsection (a):

(1) if it does not reasonably identify the rights assigned;

(2) to the extent that an agreement between an account debtor and a seller of a payment intangible limits the account debtor's duty to pay a person other than the seller and the limitation is effective under law other than this article; or

(3) at the option of an account debtor, if the notification notifies the account debtor to make less than the full amount of any installment or other periodic payment to the assignee, even if:

(A) only a portion of the account, chattel paper, or payment intangible has been assigned to that assignee;

(B) a portion has been assigned to another assignee; or

(C) the account debtor knows that the assignment to that assignee is limited.

(c) Subject to subsection (h), if requested by the account debtor, an assignee shall seasonably furnish reasonable proof that the assignment has been made. Unless the assignee complies, the account debtor may discharge its obligation by paying the assignor, even if the account debtor has received a notification under subsection (a).

(d) Except as otherwise provided in subsection (e) and Sections 2A–303 and 9–407, and subject to subsection (h), a term in an agreement between an account debtor and an assignor or in a promissory note is ineffective to the extent that it:

(1) prohibits, restricts, or requires the consent of the account debtor or person obligated on the promissory note to the assignment or transfer of, or the creation, attachment, perfection, or enforcement of a security interest in, the account, chattel paper, payment intangible, or promissory note; or

(2) provides that the assignment or transfer or the creation, attachment, perfection, or enforcement of the security interest may give rise to a default, breach, right of recoupment, claim, defense, termination, right of termination, or remedy under the account, chattel paper, payment intangible, or promissory note.

(e) Subsection (d) does not apply to the sale of a payment intangible or promissory note.

(f) Except as otherwise provided in Sections 2A–303 and 9–407 and subject to subsections (h) and (i), a rule of law, statute, or regulation that prohibits, restricts, or requires the consent of a government, governmental body or official, or account debtor to the assignment or transfer of, or creation of a security interest in, an account or chattel paper is ineffective to the extent that the rule of law, statute, or regulation:

(1) prohibits, restricts, or requires the consent of the government, governmental body or official, or account debtor to the assignment or transfer of, or the creation, attachment, perfection, or enforcement of a security interest in the account or chattel paper; or

(2) provides that the assignment or transfer or the creation, attachment, perfection, or enforcement of the security interest may give rise to a default, breach, right of recoupment, claim, defense, termination, right of termination, or remedy under the account or chattel paper.

(g) Subject to subsection (h), an account debtor may not waive or vary its option under subsection (b)(3).

(h) This section is subject to law other than this article which establishes a different rule for an account debtor who is an individual and who incurred the obligation primarily for personal, family, or household purposes.

(i) This section does not apply to an assignment of a health-care-insurance receivable.

(j) This section prevails over any inconsistent provisions of the following statutes, rules, and regulations:

[List here any statutes, rules, and regulations containing provisions inconsistent with this section.]

Legislative Note: States that amend statutes, rules, and regulations to remove provisions inconsistent with this section need not enact subsection (j).

As amended in 1999 and 2000.

§ 9–407. Restrictions on Creation or Enforcement of Security Interest in Leasehold Interest or in Lessor's Residual Interest.

(a) Except as otherwise provided in subsection (b), a term in a lease agreement is ineffective to the extent that it:

(1) prohibits, restricts, or requires the consent of a party to the lease to the assignment or transfer of, or the creation, attachment, perfection, or enforcement of a security interest in an interest of a party under the lease contract or in the lessor's residual interest in the goods; or

(2) provides that the assignment or transfer or the creation, attachment, perfection, or enforcement of the security interest may give rise to a default, breach, right of recoupment, claim, defense, termination, right of termination, or remedy under the lease.

(b) Except as otherwise provided in Section 2A–303(7), a term described in subsection (a)(2) is effective to the extent that there is:

(1) a transfer by the lessee of the lessee's right of possession or use of the goods in violation of the term; or

(2) a delegation of a material performance of either party to the lease contract in violation of the term.

(c) The creation, attachment, perfection, or enforcement of a security interest in the lessor's interest under the lease contract or the lessor's residual interest in the goods is not a transfer that materially impairs the lessee's prospect of obtaining return performance or materially changes the duty of or materially increases the burden or risk imposed on the lessee within the purview of Section 2A–303(4) unless, and then only to the extent that, enforcement actually results in a delegation of material performance of the lessor.

As amended in 1999.

§ 9–408. Restrictions on Assignment of Promissory Notes, Health-Care-Insurance Receivables, and Certain General Intangibles Ineffective.

(a) Except as otherwise provided in subsection (b), a term in a promissory note or in an agreement between an account debtor and a debtor which relates to a health-care-insurance receivable or a general intangible, including a contract, permit, license, or franchise, and which term prohibits, restricts, or requires the consent of the person obligated on the promissory note or the account debtor to, the assignment or transfer of, or creation,

attachment, or perfection of a security interest in, the promissory note, health-care-insurance receivable, or general intangible, is ineffective to the extent that the term:

(1) would impair the creation, attachment, or perfection of a security interest; or

(2) provides that the assignment or transfer or the creation, attachment, or perfection of the security interest may give rise to a default, breach, right of recoupment, claim, defense, termination, right of termination, or remedy under the promissory note, health-care-insurance receivable, or general intangible.

(b) Subsection (a) applies to a security interest in a payment intangible or promissory note only if the security interest arises out of a sale of the payment intangible or promissory note.

(c) A rule of law, statute, or regulation that prohibits, restricts, or requires the consent of a government, governmental body or official, person obligated on a promissory note, or account debtor to the assignment or transfer of, or creation of a security interest in, a promissory note, health-care-insurance receivable, or general intangible, including a contract, permit, license, or franchise between an account debtor and a debtor, is ineffective to the extent that the rule of law, statute, or regulation:

(1) would impair the creation, attachment, or perfection of a security interest; or

(2) provides that the assignment or transfer or the creation, attachment, or perfection of the security interest may give rise to a default, breach, right of recoupment, claim, defense, termination, right of termination, or remedy under the promissory note, health-care-insurance receivable, or general intangible.

(d) To the extent that a term in a promissory note or in an agreement between an account debtor and a debtor which relates to a health-care-insurance receivable or general intangible or a rule of law, statute, or regulation described in subsection (c) would be effective under law other than this article but is ineffective under subsection (a) or (c), the creation, attachment, or perfection of a security interest in the promissory note, health-care-insurance receivable, or general intangible:

(1) is not enforceable against the person obligated on the promissory note or the account debtor;

(2) does not impose a duty or obligation on the person obligated on the promissory note or the account debtor;

(3) does not require the person obligated on the promissory note or the account debtor to recognize the security interest, pay or render performance to the secured party, or accept payment or performance from the secured party;

(4) does not entitle the secured party to use or assign the debtor's rights under the promissory note, health-care-insurance receivable, or general intangible, including any related information or materials furnished to the debtor in the transaction giving rise to the promissory note, health-care-insurance receivable, or general intangible;

(5) does not entitle the secured party to use, assign, possess, or have access to any trade secrets or confidential information of the person obligated on the promissory note or the account debtor; and

(6) does not entitle the secured party to enforce the security interest in the promissory note, health-care-insurance receivable, or general intangible.

(e) This section prevails over any inconsistent provisions of the following statutes, rules, and regulations:

[List here any statutes, rules, and regulations containing provisions inconsistent with this section.]

Legislative Note: States that amend statutes, rules, and regulations to remove provisions inconsistent with this section need not enact subsection (e).

As amended in 1999.

§ 9–409. Restrictions on Assignment of Letter-of-Credit Rights Ineffective.

(a) A term in a letter of credit or a rule of law, statute, regulation, custom, or practice applicable to the letter of credit which prohibits, restricts, or requires the consent of an applicant, issuer, or nominated person to a beneficiary's assignment of or creation of a security interest in a letter-of-credit right is ineffective to the extent that the term or rule of law, statute, regulation, custom, or practice:

(1) would impair the creation, attachment, or perfection of a security interest in the letter-of-credit right; or

(2) provides that the assignment or the creation, attachment, or perfection of the security interest may give rise to a default, breach, right of recoupment, claim, defense, termination, right of termination, or remedy under the letter-of-credit right.

(b) To the extent that a term in a letter of credit is ineffective under subsection (a) but would be effective under law other than this article or a custom or practice applicable to the letter of credit, to the transfer of a right to draw or otherwise demand performance under the letter of credit, or to the assignment of a right to proceeds of the letter of credit, the creation, attachment, or perfection of a security interest in the letter-of-credit right:

(1) is not enforceable against the applicant, issuer, nominated person, or transferee beneficiary;

(2) imposes no duties or obligations on the applicant, issuer, nominated person, or transferee beneficiary; and

(3) does not require the applicant, issuer, nominated person, or transferee beneficiary to recognize the security interest, pay or render performance to the secured party, or accept payment or other performance from the secured party.

As amended in 1999.

Part 5 Filing

[Subpart 1. Filing Office; Contents and Effectiveness of Financing Statement]

§ 9–501. Filing Office.

(a) Except as otherwise provided in subsection (b), if the local law of this State governs perfection of a security interest or agricultural lien, the office in which to file a financing statement to perfect the security interest or agricultural lien is:

(1) the office designated for the filing or recording of a record of a mortgage on the related real property, if:

(A) the collateral is as-extracted collateral or timber to be cut; or

(B) the financing statement is filed as a fixture filing and the collateral is goods that are or are to become fixtures; or

(2) the office of [ ] [or any office duly authorized by [ ]], in all other cases, including a case in which the collateral is goods that are or are to become fixtures and the financing statement is not filed as a fixture filing.

(b) The office in which to file a financing statement to perfect a security interest in collateral, including fixtures, of a transmitting utility is the office of [ ]. The financing statement also constitutes a fixture filing as to the collateral indicated in the financing statement which is or is to become fixtures.

Legislative Note: The State should designate the filing office where the brackets appear. The filing office may be that of a governmental official (e.g., the Secretary of State) or a private party that maintains the State's filing system.

§ 9–502. Contents of Financing Statement; Record of Mortgage as Financing Statement; Time of Filing Financing Statement.

(a) Subject to subsection (b), a financing statement is sufficient only if it:

(1) provides the name of the debtor;

(2) provides the name of the secured party or a representative of the secured party; and

(3) indicates the collateral covered by the financing statement.

(b) Except as otherwise provided in Section 9–501(b), to be sufficient, a financing statement that covers as-extracted collateral or timber to be cut, or which is filed as a fixture filing and covers goods that are or are to become fixtures, must satisfy subsection (a) and also:

(1) indicate that it covers this type of collateral;

(2) indicate that it is to be filed [for record] in the real property records;

(3) provide a description of the real property to which the collateral is related [sufficient to give constructive notice of a mortgage under the law of this State if the description were contained in a record of the mortgage of the real property]; and

(4) if the debtor does not have an interest of record in the real property, provide the name of a record owner.

(c) A record of a mortgage is effective, from the date of recording, as a financing statement filed as a fixture filing or as a financing statement covering as-extracted collateral or timber to be cut only if:

(1) the record indicates the goods or accounts that it covers;

(2) the goods are or are to become fixtures related to the real property described in the record or the collateral is related to the real property described in the record and is as-extracted collateral or timber to be cut;

(3) the record satisfies the requirements for a financing statement in this section other than an indication that it is to be filed in the real property records; and

(4) the record is [duly] recorded.

(d) A financing statement may be filed before a security agreement is made or a security interest otherwise attaches.

Legislative Note: Language in brackets is optional. Where the State has any special recording system for real property other than the usual grantor-grantee index (as, for instance, a tract system or a title registration or Torrens system) local adaptations of subsection (b) and Section 9–519(d) and (e) may be necessary. See, e.g., Mass. Gen. Laws Chapter 106, Section 9–410.

§ 9–503. Name of Debtor and Secured Party.

(a) A financing statement sufficiently provides the name of the debtor:

(1) if the debtor is a registered organization, only if the financing statement provides the name of the debtor indicated on the public record of the debtor's jurisdiction of organization which shows the debtor to have been organized;

(2) if the debtor is a decedent's estate, only if the financing statement provides the name of the decedent and indicates that the debtor is an estate;

(3) if the debtor is a trust or a trustee acting with respect to property held in trust, only if the financing statement:

(A) provides the name specified for the trust in its organic documents or, if no name is specified, provides the name of the settlor and additional information sufficient to distinguish the debtor from other trusts having one or more of the same settlors; and

(B) indicates, in the debtor's name or otherwise, that the debtor is a trust or is a trustee acting with respect to property held in trust; and

(4) in other cases:

(A) if the debtor has a name, only if it provides the individual or organizational name of the debtor; and

(B) if the debtor does not have a name, only if it provides the names of the partners, members, associates, or other persons comprising the debtor.

(b) A financing statement that provides the name of the debtor in accordance with subsection (a) is not rendered ineffective by the absence of:

(1) a trade name or other name of the debtor; or

(2) unless required under subsection (a)(4)(B), names of partners, members, associates, or other persons comprising the debtor.

(c) A financing statement that provides only the debtor's trade name does not sufficiently provide the name of the debtor.

(d) Failure to indicate the representative capacity of a secured party or representative of a secured party does not affect the sufficiency of a financing statement.

(e) A financing statement may provide the name of more than one debtor and the name of more than one secured party.

§ 9–504. Indication of Collateral.

A financing statement sufficiently indicates the collateral that it covers if the financing statement provides:

(1) a description of the collateral pursuant to Section 9–108; or

(2) an indication that the financing statement covers all assets or all personal property.

As amended in 1999.

§ 9–505. Filing and Compliance with Other Statutes and Treaties for Consignments, Leases, Other Bailments, and Other Transactions.

(a) A consignor, lessor, or other bailor of goods, a licensor, or a buyer of a payment intangible or promissory note may file a financing statement, or may comply with a statute or treaty described in Section 9–311(a), using the terms "consignor", "consignee", "lessor", "lessee", "bailor", "bailee", "licensor", "licensee", "owner", "registered owner", "buyer", "seller", or words of similar import, instead of the terms "secured party" and "debtor".

(b) This part applies to the filing of a financing statement under subsection (a) and, as appropriate, to compliance that is equivalent to filing a financing statement under Section 9–311(b), but the filing or compliance is not of itself a factor in determining whether the collateral secures an obligation. If it is determined for another reason that the collateral secures an obligation, a security interest held by the consignor, lessor, bailor, licensor, owner, or buyer which attaches to the collateral is perfected by the filing or compliance.

§ 9–506. Effect of Errors or Omissions.

(a) A financing statement substantially satisfying the requirements of this part is effective, even if it has minor errors or omissions, unless the errors or omissions make the financing statement seriously misleading.

(b) Except as otherwise provided in subsection (c), a financing statement that fails sufficiently to provide the name of the debtor in accordance with Section 9–503(a) is seriously misleading.

(c) If a search of the records of the filing office under the debtor's correct name, using the filing office's standard search logic, if any, would disclose a financing statement that fails sufficiently to provide the name of the debtor in accordance with Section 9–503(a), the name provided does not make the financing statement seriously misleading.

(d) For purposes of Section 9–508(b), the "debtor's correct name" in subsection (c) means the correct name of the new debtor.

§ 9–507. Effect of Certain Events on Effectiveness of Financing Statement.

(a) A filed financing statement remains effective with respect to collateral that is sold, exchanged, leased, licensed, or otherwise disposed of and in which a security interest or agricultural lien continues, even if the secured party knows of or consents to the disposition.

(b) Except as otherwise provided in subsection (c) and Section 9–508, a financing statement is not rendered ineffective if, after the financing statement is filed, the information provided in the financing statement becomes seriously misleading under Section 9–506.

(c) If a debtor so changes its name that a filed financing statement becomes seriously misleading under Section 9–506:

(1) the financing statement is effective to perfect a security interest in collateral acquired by the debtor before, or within four months after, the change; and

(2) the financing statement is not effective to perfect a security interest in collateral acquired by the debtor more than four months after the change, unless an amendment to the financing statement which renders the financing statement not seriously misleading is filed within four months after the change.

§ 9–508. Effectiveness of Financing Statement If New Debtor Becomes Bound by Security Agreement.

(a) Except as otherwise provided in this section, a filed financing statement naming an original debtor is effective to perfect a security interest in collateral in which a new debtor has or acquires rights to the extent that the financing statement would have been effective had the original debtor acquired rights in the collateral.

(b) If the difference between the name of the original debtor and that of the new debtor causes a filed financing statement that is effective under subsection (a) to be seriously misleading under Section 9–506:

(1) the financing statement is effective to perfect a security interest in collateral acquired by the new debtor before, and within four months after, the new debtor becomes bound under Section 9B–203(d); and

(2) the financing statement is not effective to perfect a security interest in collateral acquired by the new debtor more than four months after the new debtor becomes bound under Section 9–203(d) unless an initial financing statement providing the name of the new debtor is filed before the expiration of that time.

(c) This section does not apply to collateral as to which a filed financing statement remains effective against the new debtor under Section 9–507(a).

§ 9–509. Persons Entitled to File a Record.

(a) A person may file an initial financing statement, amendment that adds collateral covered by a financing statement, or amendment that adds a debtor to a financing statement only if:

(1) the debtor authorizes the filing in an authenticated record or pursuant to subsection (b) or (c); or

(2) the person holds an agricultural lien that has become effective at the time of filing and the financing statement covers only collateral in which the person holds an agricultural lien.

(b) By authenticating or becoming bound as debtor by a security agreement, a debtor or new debtor authorizes the filing of an initial financing statement, and an amendment, covering:

(1) the collateral described in the security agreement; and

(2) property that becomes collateral under Section 9–315(a)(2), whether or not the security agreement expressly covers proceeds.

(c) By acquiring collateral in which a security interest or agricultural lien continues under Section 9–315(a)(1), a debtor authorizes the filing of an initial financing statement, and an amendment, covering the collateral and property that becomes collateral under Section 9–315(a)(2).

(d) A person may file an amendment other than an amendment that adds collateral covered by a financing statement or an amendment that adds a debtor to a financing statement only if:

(1) the secured party of record authorizes the filing; or

(2) the amendment is a termination statement for a financing statement as to which the secured party of record has failed to file or send a termination statement as required by Section 9–513(a) or (c), the debtor authorizes the filing, and the termination statement indicates that the debtor authorized it to be filed.

(e) If there is more than one secured party of record for a financing statement, each secured party of record may authorize the filing of an amendment under subsection (d).

As amended in 2000.

§ 9–510. Effectiveness of Filed Record.

(a) A filed record is effective only to the extent that it was filed by a person that may file it under Section 9–509.

(b) A record authorized by one secured party of record does not affect the financing statement with respect to another secured party of record.

(c) A continuation statement that is not filed within the six-month period prescribed by Section 9–515(d) is ineffective.

§ 9–511. Secured Party of Record.

(a) A secured party of record with respect to a financing statement is a person whose name is provided as the name of the secured party or a representative of the secured party in an initial financing statement that has been filed. If an initial financing statement is filed under Section 9–514(a), the assignee named in the initial financing statement is the secured party of record with respect to the financing statement.

(b) If an amendment of a financing statement which provides the name of a person as a secured party or a representative of a secured party is filed, the person named in the amendment is a secured party of record. If an amendment is filed under Section 9–514(b), the assignee named in the amendment is a secured party of record.

(c) A person remains a secured party of record until the filing of an amendment of the financing statement which deletes the person.

§ 9–512. Amendment of Financing Statement.

[Alternative A]

(a) Subject to Section 9–509, a person may add or delete collateral covered by, continue or terminate the effectiveness of, or, subject to subsection (e), otherwise amend the information provided in, a financing statement by filing an amendment that:

(1) identifies, by its file number, the initial financing statement to which the amendment relates; and

(2) if the amendment relates to an initial financing statement filed [or recorded] in a filing office described in Section 9–501(a)(1), provides the information specified in Section 9–502(b).

[Alternative B]

(a) Subject to Section 9–509, a person may add or delete collateral covered by, continue or terminate the effectiveness of, or, subject to subsection (e), otherwise amend the information provided in, a financing statement by filing an amendment that:

(1) identifies, by its file number, the initial financing statement to which the amendment relates; and

(2) if the amendment relates to an initial financing statement filed [or recorded] in a filing office described in Section 9–501(a)(1), provides the date [and time] that the initial financing statement was filed [or recorded] and the information specified in Section 9–502(b).

[End of Alternatives]

(b) Except as otherwise provided in Section 9–515, the filing of an amendment does not extend the period of effectiveness of the financing statement.

(c) A financing statement that is amended by an amendment that adds collateral is effective as to the added collateral only from the date of the filing of the amendment.

(d) A financing statement that is amended by an amendment that adds a debtor is effective as to the added debtor only from the date of the filing of the amendment.

(e) An amendment is ineffective to the extent it:

(1) purports to delete all debtors and fails to provide the name of a debtor to be covered by the financing statement; or

(2) purports to delete all secured parties of record and fails to provide the name of a new secured party of record.

Legislative Note: States whose real-estate filing offices require additional information in amendments and cannot search their records by both the name of the debtor and the file number should enact Alternative B to Sections 9–512(a), 9–518(b), 9–519(f), and 9–522(a).

§ 9–513. Termination Statement.

(a) A secured party shall cause the secured party of record for a financing statement to file a termination statement for the financing statement if the financing statement covers consumer goods and:

(1) there is no obligation secured by the collateral covered by the financing statement and no commitment to make an advance, incur an obligation, or otherwise give value; or

(2) the debtor did not authorize the filing of the initial financing statement.

(b) To comply with subsection (a), a secured party shall cause the secured party of record to file the termination statement:

(1) within one month after there is no obligation secured by the collateral covered by the financing statement and no commitment to make an advance, incur an obligation, or otherwise give value; or

(2) if earlier, within 20 days after the secured party receives an authenticated demand from a debtor.

(c) In cases not governed by subsection (a), within 20 days after a secured party receives an authenticated demand from a debtor, the secured party shall cause the secured party of record for a financing statement to send to the debtor a termination statement for the financing statement or file the termination statement in the filing office if:

(1) except in the case of a financing statement covering accounts or chattel paper that has been sold or goods that are the subject of a consignment, there is no obligation secured by the collateral covered by the financing statement and no commitment to make an advance, incur an obligation, or otherwise give value;

(2) the financing statement covers accounts or chattel paper that has been sold but as to which the account debtor or other person obligated has discharged its obligation;

(3) the financing statement covers goods that were the subject of a consignment to the debtor but are not in the debtor's possession; or

(4) the debtor did not authorize the filing of the initial financing statement.

(d) Except as otherwise provided in Section 9–510, upon the filing of a termination statement with the filing office, the financing statement to which the termination statement relates ceases to be effective. Except as otherwise provided in Section 9–510, for purposes of Sections 9–519(g), 9–522(a), and 9–523(c), the filing with the filing office of a termination statement relating to a financing statement that indicates that the debtor is a transmitting utility also causes the effectiveness of the financing statement to lapse.

As amended in 2000.

§ 9–514. Assignment of Powers of Secured Party of Record.

(a) Except as otherwise provided in subsection (c), an initial financing statement may reflect an assignment of all of the secured party's power to authorize an amendment to the financing statement by providing the name and mailing address of the assignee as the name and address of the secured party.

(b) Except as otherwise provided in subsection (c), a secured party of record may assign of record all or part of its power to authorize an amendment to a financing statement by filing in the filing office an amendment of the financing statement which:

(1) identifies, by its file number, the initial financing statement to which it relates;

(2) provides the name of the assignor; and

(3) provides the name and mailing address of the assignee.

(c) An assignment of record of a security interest in a fixture covered by a record of a mortgage which is effective as a financing statement filed as a fixture filing under Section 9–502(c) may be made only by an assignment of record of the mortgage in the manner provided by law of this State other than [the Uniform Commercial Code].

§ 9–515. Duration and Effectiveness of Financing Statement; Effect of Lapsed Financing Statement.

(a) Except as otherwise provided in subsections (b), (e), (f), and (g), a filed financing statement is effective for a period of five years after the date of filing.

(b) Except as otherwise provided in subsections (e), (f), and (g), an initial financing statement filed in connection with a public-finance transaction or manufactured-home transaction is effective for a period of 30 years after the date of filing if it indicates that it is filed in connection with a public-finance transaction or manufactured-home transaction.

(c) The effectiveness of a filed financing statement lapses on the expiration of the period of its effectiveness unless before the lapse a continuation statement is filed pursuant to subsection (d). Upon lapse, a financing statement ceases to be effective and

any security interest or agricultural lien that was perfected by the financing statement becomes unperfected, unless the security interest is perfected otherwise. If the security interest or agricultural lien becomes unperfected upon lapse, it is deemed never to have been perfected as against a purchaser of the collateral for value.

(d) A continuation statement may be filed only within six months before the expiration of the five-year period specified in subsection (a) or the 30-year period specified in subsection (b), whichever is applicable.

(e) Except as otherwise provided in Section 9–510, upon timely filing of a continuation statement, the effectiveness of the initial financing statement continues for a period of five years commencing on the day on which the financing statement would have become ineffective in the absence of the filing. Upon the expiration of the five-year period, the financing statement lapses in the same manner as provided in subsection (c), unless, before the lapse, another continuation statement is filed pursuant to subsection (d). Succeeding continuation statements may be filed in the same manner to continue the effectiveness of the initial financing statement.

(f) If a debtor is a transmitting utility and a filed financing statement so indicates, the financing statement is effective until a termination statement is filed.

(g) A record of a mortgage that is effective as a financing statement filed as a fixture filing under Section 9–502(c) remains effective as a financing statement filed as a fixture filing until the mortgage is released or satisfied of record or its effectiveness otherwise terminates as to the real property.

§ 9–516. What Constitutes Filing; Effectiveness of Filing.

(a) Except as otherwise provided in subsection (b), communication of a record to a filing office and tender of the filing fee or acceptance of the record by the filing office constitutes filing.

(b) Filing does not occur with respect to a record that a filing office refuses to accept because:

(1) the record is not communicated by a method or medium of communication authorized by the filing office;

(2) an amount equal to or greater than the applicable filing fee is not tendered;

(3) the filing office is unable to index the record because:

(A) in the case of an initial financing statement, the record does not provide a name for the debtor;

(B) in the case of an amendment or correction statement, the record:

(i) does not identify the initial financing statement as required by Section 9–512 or 9–518, as applicable; or

(ii) identifies an initial financing statement whose effectiveness has lapsed under Section 9–515;

(C) in the case of an initial financing statement that provides the name of a debtor identified as an individual or an amendment that provides a name of a debtor identified as an individual which was not previously provided in the financing statement to which the record relates, the record does not identify the debtor's last name; or

(D) in the case of a record filed [or recorded] in the filing office described in Section 9–501(a)(1), the record does not provide a sufficient description of the real property to which it relates;

(4) in the case of an initial financing statement or an amendment that adds a secured party of record, the record does not provide a name and mailing address for the secured party of record;

(5) in the case of an initial financing statement or an amendment that provides a name of a debtor which was not previously provided in the financing statement to which the amendment relates, the record does not:

(A) provide a mailing address for the debtor;

(B) indicate whether the debtor is an individual or an organization; or

(C) if the financing statement indicates that the debtor is an organization, provide:

(i) a type of organization for the debtor;

(ii) a jurisdiction of organization for the debtor; or

(iii) an organizational identification number for the debtor or indicate that the debtor has none;

(6) in the case of an assignment reflected in an initial financing statement under Section 9–514(a) or an amendment filed under Section 9–514(b), the record does not provide a name and mailing address for the assignee; or

(7) in the case of a continuation statement, the record is not filed within the six-month period prescribed by Section 9–515(d).

(c) For purposes of subsection (b):

(1) a record does not provide information if the filing office is unable to read or decipher the information; and

(2) a record that does not indicate that it is an amendment or identify an initial financing statement to which it relates, as required by Section 9–512, 9–514, or 9–518, is an initial financing statement.

(d) A record that is communicated to the filing office with tender of the filing fee, but which the filing office refuses to accept for a reason other than one set forth in subsection (b), is effective as a filed record except as against a purchaser of the collateral which gives value in reasonable reliance upon the absence of the record from the files.

§ 9–517. Effect of Indexing Errors.

The failure of the filing office to index a record correctly does not affect the effectiveness of the filed record.

§ 9–518. Claim Concerning Inaccurate or Wrongfully Filed Record.

(a) A person may file in the filing office a correction statement with respect to a record indexed there under the person's name

if the person believes that the record is inaccurate or was wrongfully filed.

[Alternative A]

(b) A correction statement must:

(1) identify the record to which it relates by the file number assigned to the initial financing statement to which the record relates;

(2) indicate that it is a correction statement; and

(3) provide the basis for the person's belief that the record is inaccurate and indicate the manner in which the person believes the record should be amended to cure any inaccuracy or provide the basis for the person's belief that the record was wrongfully filed.

[Alternative B]

(b) A correction statement must:

(1) identify the record to which it relates by:

(A) the file number assigned to the initial financing statement to which the record relates; and

(B) if the correction statement relates to a record filed [or recorded] in a filing office described in Section 9–501(a)(1), the date [and time] that the initial financing statement was filed [or recorded] and the information specified in Section 9–502(b);

(2) indicate that it is a correction statement; and

(3) provide the basis for the person's belief that the record is inaccurate and indicate the manner in which the person believes the record should be amended to cure any inaccuracy or provide the basis for the person's belief that the record was wrongfully filed.

[End of Alternatives]

(c) The filing of a correction statement does not affect the effectiveness of an initial financing statement or other filed record.

Legislative Note: States whose real-estate filing offices require additional information in amendments and cannot search their records by both the name of the debtor and the file number should enact Alternative B to Sections 9–512(a), 9–518(b), 9–519(f), and 9–522(a).

[Subpart 2. Duties and Operation of Filing Office]

§ 9–519. Numbering, Maintaining, and Indexing Records; Communicating Information Provided in Records.

(a) For each record filed in a filing office, the filing office shall:

(1) assign a unique number to the filed record;

(2) create a record that bears the number assigned to the filed record and the date and time of filing;

(3) maintain the filed record for public inspection; and

(4) index the filed record in accordance with subsections (c), (d), and (e).

(b) A file number [assigned after January 1, 2002,] must include a digit that:

(1) is mathematically derived from or related to the other digits of the file number; and

(2) aids the filing office in determining whether a number communicated as the file number includes a single-digit or transpositional error.

(c) Except as otherwise provided in subsections (d) and (e), the filing office shall:

(1) index an initial financing statement according to the name of the debtor and index all filed records relating to the initial financing statement in a manner that associates with one another an initial financing statement and all filed records relating to the initial financing statement; and

(2) index a record that provides a name of a debtor which was not previously provided in the financing statement to which the record relates also according to the name that was not previously provided.

(d) If a financing statement is filed as a fixture filing or covers as-extracted collateral or timber to be cut, [it must be filed for record and] the filing office shall index it:

(1) under the names of the debtor and of each owner of record shown on the financing statement as if they were the mortgagors under a mortgage of the real property described; and

(2) to the extent that the law of this State provides for indexing of records of mortgages under the name of the mortgagee, under the name of the secured party as if the secured party were the mortgagee thereunder, or, if indexing is by description, as if the financing statement were a record of a mortgage of the real property described.

(e) If a financing statement is filed as a fixture filing or covers as-extracted collateral or timber to be cut, the filing office shall index an assignment filed under Section 9–514(a) or an amendment filed under Section 9–514(b):

(1) under the name of the assignor as grantor; and

(2) to the extent that the law of this State provides for indexing a record of the assignment of a mortgage under the name of the assignee, under the name of the assignee.

[Alternative A]

(f) The filing office shall maintain a capability:

(1) to retrieve a record by the name of the debtor and by the file number assigned to the initial financing statement to which the record relates; and

(2) to associate and retrieve with one another an initial financing statement and each filed record relating to the initial financing statement.

[Alternative B]

(f) The filing office shall maintain a capability:

(1) to retrieve a record by the name of the debtor and:

(A) if the filing office is described in Section 9–501(a)(1), by the file number assigned to the initial financing statement to which the record relates and the date [and time] that the record was filed [or recorded]; or

(B) if the filing office is described in Section 9–501(a)(2), by the file number assigned to the initial financing statement to which the record relates; and

(2) to associate and retrieve with one another an initial financing statement and each filed record relating to the initial financing statement.

[End of Alternatives]

(g) The filing office may not remove a debtor's name from the index until one year after the effectiveness of a financing statement naming the debtor lapses under Section 9–515 with respect to all secured parties of record.

(h) The filing office shall perform the acts required by subsections (a) through (e) at the time and in the manner prescribed by filing-office rule, but not later than two business days after the filing office receives the record in question.

[(i) Subsection[s] [(b)] [and] [(h)] do[es] not apply to a filing office described in Section 9–501(a)(1).]

Legislative Notes:

1. States whose filing offices currently assign file numbers that include a verification number, commonly known as a "check digit," or can implement this requirement before the effective date of this Article should omit the bracketed language in subsection (b).

2. In States in which writings will not appear in the real property records and indices unless actually recorded the bracketed language in subsection (d) should be used.

3. States whose real-estate filing offices require additional information in amendments and cannot search their records by both the name of the debtor and the file number should enact Alternative B to Sections 9–512(a), 9–518(b), 9–519(f), and 9–522(a).

4. A State that elects not to require real-estate filing offices to comply with either or both of subsections (b) and (h) may adopt an applicable variation of subsection (i) and add "Except as otherwise provided in subsection (i)," to the appropriate subsection or subsections.

§ 9–520. Acceptance and Refusal to Accept Record.

(a) A filing office shall refuse to accept a record for filing for a reason set forth in Section 9–516(b) and may refuse to accept a record for filing only for a reason set forth in Section 9–516(b).

(b) If a filing office refuses to accept a record for filing, it shall communicate to the person that presented the record the fact of and reason for the refusal and the date and time the record would have been filed had the filing office accepted it. The communication must be made at the time and in the manner prescribed by filing-office rule but [, in the case of a filing office described in Section 9–501(a)(2),] in no event more than two business days after the filing office receives the record.

(c) A filed financing statement satisfying Section 9–502(a) and (b) is effective, even if the filing office is required to refuse to accept it for filing under subsection (a). However, Section 9–338 applies to a filed financing statement providing information described in Section 9–516(b)(5) which is incorrect at the time the financing statement is filed.

(d) If a record communicated to a filing office provides information that relates to more than one debtor, this part applies as to each debtor separately.

Legislative Note: A State that elects not to require real-property filing offices to comply with subsection (b) should include the bracketed language.

§ 9–521. Uniform Form of Written Financing Statement and Amendment.

(a) A filing office that accepts written records may not refuse to accept a written initial financing statement in the following form and format except for a reason set forth in Section 9–516(b):

[NATIONAL UCC FINANCING STATEMENT (FORM UCC1)(REV. 7/29/98)]

[NATIONAL UCC FINANCING STATEMENT ADDENDUM (FORM UCC1Ad)(REV. 07/29/98)]

(b) A filing office that accepts written records may not refuse to accept a written record in the following form and format except for a reason set forth in Section 9–516(b):

[NATIONAL UCC FINANCING STATEMENT AMENDMENT (FORM UCC3)(REV. 07/29/98)]

[NATIONAL UCC FINANCING STATEMENT AMENDMENT ADDENDUM (FORM UCC3Ad)(REV. 07/29/98)]

§ 9–522. Maintenance and Destruction of Records.

[Alternative A]

(a) The filing office shall maintain a record of the information provided in a filed financing statement for at least one year after the effectiveness of the financing statement has lapsed under Section 9–515 with respect to all secured parties of record. The record must be retrievable by using the name of the debtor and by using the file number assigned to the initial financing statement to which the record relates.

[Alternative B]

(a) The filing office shall maintain a record of the information provided in a filed financing statement for at least one year after the effectiveness of the financing statement has lapsed under Section 9–515 with respect to all secured parties of record. The record must be retrievable by using the name of the debtor and:

(1) if the record was filed [or recorded] in the filing office described in Section 9–501(a)(1), by using the file number assigned to the initial financing statement to which the record relates and the date [and time] that the record was filed [or recorded]; or

(2) if the record was filed in the filing office described in Section 9–501(a)(2), by using the file number assigned to the initial financing statement to which the record relates.

[End of Alternatives]

(b) Except to the extent that a statute governing disposition of public records provides otherwise, the filing office immediately may destroy any written record evidencing a financing statement. However, if the filing office destroys a written record, it shall maintain another record of the financing statement which complies with subsection (a).

Legislative Note: States whose real-estate filing offices require additional information in amendments and cannot search their records by both the name of the debtor and the file number should enact Alternative B to Sections 9–512(a), 9–518(b), 9–519(f), and 9–522(a).

§ 9–523. Information from Filing Office; Sale or License of Records.

(a) If a person that files a written record requests an acknowledgment of the filing, the filing office shall send to the person an image of the record showing the number assigned to the record pursuant to Section 9–519(a)(1) and the date and time of the filing of the record. However, if the person furnishes a copy of the record to the filing office, the filing office may instead:

(1) note upon the copy the number assigned to the record pursuant to Section 9–519(a)(1) and the date and time of the filing of the record; and

(2) send the copy to the person.

(b) If a person files a record other than a written record, the filing office shall communicate to the person an acknowledgment that provides:

(1) the information in the record;

(2) the number assigned to the record pursuant to Section 9–519(a)(1); and

(3) the date and time of the filing of the record.

(c) The filing office shall communicate or otherwise make available in a record the following information to any person that requests it:

(1) whether there is on file on a date and time specified by the filing office, but not a date earlier than three business days before the filing office receives the request, any financing statement that:

(A) designates a particular debtor [or, if the request so states, designates a particular debtor at the address specified in the request];

(B) has not lapsed under Section 9–515 with respect to all secured parties of record; and

(C) if the request so states, has lapsed under Section 9–515 and a record of which is maintained by the filing office under Section 9–522(a);

(2) the date and time of filing of each financing statement; and

(3) the information provided in each financing statement.

(d) In complying with its duty under subsection (c), the filing office may communicate information in any medium. However, if requested, the filing office shall communicate information by issuing [its written certificate] [a record that can be admitted into evidence in the courts of this State without extrinsic evidence of its authenticity].

(e) The filing office shall perform the acts required by subsections (a) through (d) at the time and in the manner prescribed by filing-office rule, but not later than two business days after the filing office receives the request.

(f) At least weekly, the [insert appropriate official or governmental agency] [filing office] shall offer to sell or license to the public on a nonexclusive basis, in bulk, copies of all records filed in it under this part, in every medium from time to time available to the filing office.

Legislative Notes:

1. States whose filing office does not offer the additional service of responding to search requests limited to a particular address should omit the bracketed language in subsection (c)(1)(A).

2. A State that elects not to require real-estate filing offices to comply with either or both of subsections (e) and (f) should specify in the appropriate subsection(s) only the filing office described in Section 9–501(a)(2).

§ 9–524. Delay by Filing Office.

Delay by the filing office beyond a time limit prescribed by this part is excused if:

(1) the delay is caused by interruption of communication or computer facilities, war, emergency conditions, failure of equipment, or other circumstances beyond control of the filing office; and

(2) the filing office exercises reasonable diligence under the circumstances.

§ 9–525. Fees.

(a) Except as otherwise provided in subsection (e), the fee for filing and indexing a record under this part, other than an initial financing statement of the kind described in subsection (b), is [the amount specified in subsection (c), if applicable, plus]:

(1) $[X] if the record is communicated in writing and consists of one or two pages;

(2) $[2X] if the record is communicated in writing and consists of more than two pages; and

(3) $[½X] if the record is communicated by another medium authorized by filing-office rule.

(b) Except as otherwise provided in subsection (e), the fee for filing and indexing an initial financing statement of the following kind is [the amount specified in subsection (c), if applicable, plus]:

(1) $______ if the financing statement indicates that it is filed in connection with a public-finance transaction;

(2) $______ if the financing statement indicates that it is filed in connection with a manufactured-home transaction.

[Alternative A]

(c) The number of names required to be indexed does not affect the amount of the fee in subsections (a) and (b).

[Alternative B]

(c) Except as otherwise provided in subsection (e), if a record is communicated in writing, the fee for each name more than two required to be indexed is $______.

[End of Alternatives]

(d) The fee for responding to a request for information from the filing office, including for [issuing a certificate showing] [communicating] whether there is on file any financing statement naming a particular debtor, is:

(1) $______ if the request is communicated in writing; and

(2) $______ if the request is communicated by another medium authorized by filing-office rule.

(e) This section does not require a fee with respect to a record of a mortgage which is effective as a financing statement filed as a fixture filing or as a financing statement covering as-extracted collateral or timber to be cut under Section 9–502(c). However, the recording and satisfaction fees that otherwise would be applicable to the record of the mortgage apply.

Legislative Notes:

1. To preserve uniformity, a State that places the provisions of this section together with statutes setting fees for other services should do so without modification.

2. A State should enact subsection (c), Alternative A, and omit the bracketed language in subsections (a) and (b) unless its indexing system entails a substantial additional cost when indexing additional names.

As amended in 2000.

§ 9–526. Filing-Office Rules.

(a) The [insert appropriate governmental official or agency] shall adopt and publish rules to implement this article. The filing-office rules must be[:

(1)] consistent with this article[; and

(2) adopted and published in accordance with the [insert any applicable state administrative procedure act]].

(b) To keep the filing-office rules and practices of the filing office in harmony with the rules and practices of filing offices in other jurisdictions that enact substantially this part, and to keep the technology used by the filing office compatible with the technology used by filing offices in other jurisdictions that enact substantially this part, the [insert appropriate governmental official or agency], so far as is consistent with the purposes, policies, and provisions of this article, in adopting, amending, and repealing filing-office rules, shall:

(1) consult with filing offices in other jurisdictions that enact substantially this part; and

(2) consult the most recent version of the Model Rules promulgated by the International Association of Corporate Administrators or any successor organization; and

(3) take into consideration the rules and practices of, and the technology used by, filing offices in other jurisdictions that enact substantially this part.

§ 9–527. Duty to Report.

The [insert appropriate governmental official or agency] shall report [annually on or before ______] to the [Governor and Legislature] on the operation of the filing office. The report must contain a statement of the extent to which:

(1) the filing-office rules are not in harmony with the rules of filing offices in other jurisdictions that enact substantially this part and the reasons for these variations; and

(2) the filing-office rules are not in harmony with the most recent version of the Model Rules promulgated by the International Association of Corporate Administrators, or any successor organization, and the reasons for these variations.

Part 6 Default

[Subpart 1. Default and Enforcement of Security Interest]

§ 9–601. Rights after Default; Judicial Enforcement; Consignor or Buyer of Accounts, Chattel Paper, Payment Intangibles, or Promissory Notes.

(a) After default, a secured party has the rights provided in this part and, except as otherwise provided in Section 9–602, those provided by agreement of the parties. A secured party:

(1) may reduce a claim to judgment, foreclose, or otherwise enforce the claim, security interest, or agricultural lien by any available judicial procedure; and

(2) if the collateral is documents, may proceed either as to the documents or as to the goods they cover.

(b) A secured party in possession of collateral or control of collateral under Section 9–104, 9–105, 9–106, or 9–107 has the rights and duties provided in Section 9–207.

(c) The rights under subsections (a) and (b) are cumulative and may be exercised simultaneously.

(d) Except as otherwise provided in subsection (g) and Section 9–605, after default, a debtor and an obligor have the rights provided in this part and by agreement of the parties.

(e) If a secured party has reduced its claim to judgment, the lien of any levy that may be made upon the collateral by virtue of an execution based upon the judgment relates back to the earliest of:

(1) the date of perfection of the security interest or agricultural lien in the collateral;

(2) the date of filing a financing statement covering the collateral; or

(3) any date specified in a statute under which the agricultural lien was created.

(f) A sale pursuant to an execution is a foreclosure of the security interest or agricultural lien by judicial procedure within the meaning of this section. A secured party may purchase at the sale and thereafter hold the collateral free of any other requirements of this article.

(g) Except as otherwise provided in Section 9–607(c), this part imposes no duties upon a secured party that is a consignor or is a buyer of accounts, chattel paper, payment intangibles, or promissory notes.

§ 9–602. Waiver and Variance of Rights and Duties.

Except as otherwise provided in Section 9–624, to the extent that they give rights to a debtor or obligor and impose duties on a secured party, the debtor or obligor may not waive or vary the rules stated in the following listed sections:

(1) Section 9–207(b)(4)(C), which deals with use and operation of the collateral by the secured party;

(2) Section 9–210, which deals with requests for an accounting and requests concerning a list of collateral and statement of account;

(3) Section 9–607(c), which deals with collection and enforcement of collateral;

(4) Sections 9–608(a) and 9–615(c) to the extent that they deal with application or payment of noncash proceeds of collection, enforcement, or disposition;

(5) Sections 9–608(a) and 9–615(d) to the extent that they require accounting for or payment of surplus proceeds of collateral;

(6) Section 9–609 to the extent that it imposes upon a secured party that takes possession of collateral without judicial process the duty to do so without breach of the peace;

(7) Sections 9–610(b), 9–611, 9–613, and 9–614, which deal with disposition of collateral;

(8) Section 9–615(f), which deals with calculation of a deficiency or surplus when a disposition is made to the secured party, a person related to the secured party, or a secondary obligor;

(9) Section 9–616, which deals with explanation of the calculation of a surplus or deficiency;

(10) Sections 9–620, 9–621, and 9–622, which deal with acceptance of collateral in satisfaction of obligation;

(11) Section 9–623, which deals with redemption of collateral;

(12) Section 9–624, which deals with permissible waivers; and

(13) Sections 9–625 and 9–626, which deal with the secured party's liability for failure to comply with this article.

§ 9–603. Agreement on Standards Concerning Rights and Duties.

(a) The parties may determine by agreement the standards measuring the fulfillment of the rights of a debtor or obligor and the duties of a secured party under a rule stated in Section 9–602 if the standards are not manifestly unreasonable.

(b) Subsection (a) does not apply to the duty under Section 9–609 to refrain from breaching the peace.

§ 9–604. Procedure If Security Agreement Covers Real Property or Fixtures.

(a) If a security agreement covers both personal and real property, a secured party may proceed:

(1) under this part as to the personal property without prejudicing any rights with respect to the real property; or

(2) as to both the personal property and the real property in accordance with the rights with respect to the real property, in which case the other provisions of this part do not apply.

(b) Subject to subsection (c), if a security agreement covers goods that are or become fixtures, a secured party may proceed:

(1) under this part; or

(2) in accordance with the rights with respect to real property, in which case the other provisions of this part do not apply.

(c) Subject to the other provisions of this part, if a secured party holding a security interest in fixtures has priority over all owners and encumbrancers of the real property, the secured party, after default, may remove the collateral from the real property.

(d) A secured party that removes collateral shall promptly reimburse any encumbrancer or owner of the real property, other than the debtor, for the cost of repair of any physical injury caused by the removal. The secured party need not reimburse the encumbrancer or owner for any diminution in value of the real property caused by the absence of the goods removed or by any necessity of replacing them. A person entitled to reimbursement may refuse permission to remove until the secured party gives adequate assurance for the performance of the obligation to reimburse.

§ 9–605. Unknown Debtor or Secondary Obligor.

A secured party does not owe a duty based on its status as secured party:

(1) to a person that is a debtor or obligor, unless the secured party knows:

(A) that the person is a debtor or obligor;

(B) the identity of the person; and

(C) how to communicate with the person; or

(2) to a secured party or lienholder that has filed a financing statement against a person, unless the secured party knows:

(A) that the person is a debtor; and

(B) the identity of the person.

§ 9–606. Time of Default for Agricultural Lien.

For purposes of this part, a default occurs in connection with an agricultural lien at the time the secured party becomes entitled to enforce the lien in accordance with the statute under which it was created.

§ 9–607. Collection and Enforcement by Secured Party.

(a) If so agreed, and in any event after default, a secured party:

(1) may notify an account debtor or other person obligated on collateral to make payment or otherwise render performance to or for the benefit of the secured party;

(2) may take any proceeds to which the secured party is entitled under Section 9–315;

(3) may enforce the obligations of an account debtor or other person obligated on collateral and exercise the rights of the debtor with respect to the obligation of the account debtor or other person obligated on collateral to make payment or otherwise render performance to the debtor, and with respect to any property that secures the obligations of the account debtor or other person obligated on the collateral;

(4) if it holds a security interest in a deposit account perfected by control under Section 9–104(a)(1), may apply the balance of the deposit account to the obligation secured by the deposit account; and

(5) if it holds a security interest in a deposit account perfected by control under Section 9–104(a)(2) or (3), may instruct the bank to pay the balance of the deposit account to or for the benefit of the secured party.

(b) If necessary to enable a secured party to exercise under subsection (a)(3) the right of a debtor to enforce a mortgage nonjudicially, the secured party may record in the office in which a record of the mortgage is recorded:

(1) a copy of the security agreement that creates or provides for a security interest in the obligation secured by the mortgage; and

(2) the secured party's sworn affidavit in recordable form stating that:

(A) a default has occurred; and

(B) the secured party is entitled to enforce the mortgage nonjudicially.

(c) A secured party shall proceed in a commercially reasonable manner if the secured party:

(1) undertakes to collect from or enforce an obligation of an account debtor or other person obligated on collateral; and

(2) is entitled to charge back uncollected collateral or otherwise to full or limited recourse against the debtor or a secondary obligor.

(d) A secured party may deduct from the collections made pursuant to subsection (c) reasonable expenses of collection and enforcement, including reasonable attorney's fees and legal expenses incurred by the secured party.

(e) This section does not determine whether an account debtor, bank, or other person obligated on collateral owes a duty to a secured party.

As amended in 2000.

§ 9–608. Application of Proceeds of Collection or Enforcement; Liability for Deficiency and Right to Surplus.

(a) If a security interest or agricultural lien secures payment or performance of an obligation, the following rules apply:

(1) A secured party shall apply or pay over for application the cash proceeds of collection or enforcement under Section 9–607 in the following order to:

(A) the reasonable expenses of collection and enforcement and, to the extent provided for by agreement and not prohibited by law, reasonable attorney's fees and legal expenses incurred by the secured party;

(B) the satisfaction of obligations secured by the security interest or agricultural lien under which the collection or enforcement is made; and

(C) the satisfaction of obligations secured by any subordinate security interest in or other lien on the collateral subject to the security interest or agricultural lien under which the collection or enforcement is made if the secured party receives an authenticated demand for proceeds before distribution of the proceeds is completed.

(2) If requested by a secured party, a holder of a subordinate security interest or other lien shall furnish reasonable proof of the interest or lien within a reasonable time. Unless the holder complies, the secured party need not comply with the holder's demand under paragraph (1)(C).

(3) A secured party need not apply or pay over for application noncash proceeds of collection and enforcement under Section 9–607 unless the failure to do so would be commercially unreasonable. A secured party that applies or pays over for application noncash proceeds shall do so in a commercially reasonable manner.

(4) A secured party shall account to and pay a debtor for any surplus, and the obligor is liable for any deficiency.

(b) If the underlying transaction is a sale of accounts, chattel paper, payment intangibles, or promissory notes, the debtor is not entitled to any surplus, and the obligor is not liable for any deficiency.

As amended in 2000.

§ 9–609. Secured Party's Right to Take Possession after Default.

(a) After default, a secured party:

(1) may take possession of the collateral; and

(2) without removal, may render equipment unusable and dispose of collateral on a debtor's premises under Section 9–610.

(b) A secured party may proceed under subsection (a):

(1) pursuant to judicial process; or

(2) without judicial process, if it proceeds without breach of the peace.

(c) If so agreed, and in any event after default, a secured party may require the debtor to assemble the collateral and make it available to the secured party at a place to be designated by the secured party which is reasonably convenient to both parties.

§ 9–610. Disposition of Collateral after Default.

(a) After default, a secured party may sell, lease, license, or otherwise dispose of any or all of the collateral in its present condition or following any commercially reasonable preparation or processing.

(b) Every aspect of a disposition of collateral, including the method, manner, time, place, and other terms, must be commercially reasonable. If commercially reasonable, a secured party may dispose of collateral by public or private proceedings, by one or more contracts, as a unit or in parcels, and at any time and place and on any terms.

(c) A secured party may purchase collateral:

(1) at a public disposition; or

(2) at a private disposition only if the collateral is of a kind that is customarily sold on a recognized market or the subject of widely distributed standard price quotations.

(d) A contract for sale, lease, license, or other disposition includes the warranties relating to title, possession, quiet enjoyment, and the like which by operation of law accompany a voluntary disposition of property of the kind subject to the contract.

(e) A secured party may disclaim or modify warranties under subsection (d):

(1) in a manner that would be effective to disclaim or modify the warranties in a voluntary disposition of property of the kind subject to the contract of disposition; or

(2) by communicating to the purchaser a record evidencing the contract for disposition and including an express disclaimer or modification of the warranties.

(f) A record is sufficient to disclaim warranties under subsection (e) if it indicates "There is no warranty relating to title, possession, quiet enjoyment, or the like in this disposition" or uses words of similar import.

§ 9–611. Notification before Disposition of Collateral.

(a) In this section, "notification date" means the earlier of the date on which:

(1) a secured party sends to the debtor and any secondary obligor an authenticated notification of disposition; or

(2) the debtor and any secondary obligor waive the right to notification.

(b) Except as otherwise provided in subsection (d), a secured party that disposes of collateral under Section 9–610 shall send to the persons specified in subsection (c) a reasonable authenticated notification of disposition.

(c) To comply with subsection (b), the secured party shall send an authenticated notification of disposition to:

(1) the debtor;

(2) any secondary obligor; and

(3) if the collateral is other than consumer goods:

(A) any other person from which the secured party has received, before the notification date, an authenticated notification of a claim of an interest in the collateral;

(B) any other secured party or lienholder that, 10 days before the notification date, held a security interest in or other lien on the collateral perfected by the filing of a financing statement that:

(i) identified the collateral;

(ii) was indexed under the debtor's name as of that date; and

(iii) was filed in the office in which to file a financing statement against the debtor covering the collateral as of that date; and

(C) any other secured party that, 10 days before the notification date, held a security interest in the collateral perfected by compliance with a statute, regulation, or treaty described in Section 9–311(a).

(d) Subsection (b) does not apply if the collateral is perishable or threatens to decline speedily in value or is of a type customarily sold on a recognized market.

(e) A secured party complies with the requirement for notification prescribed by subsection (c)(3)(B) if:

(1) not later than 20 days or earlier than 30 days before the notification date, the secured party requests, in a commercially reasonable manner, information concerning financing statements indexed under the debtor's name in the office indicated in subsection (c)(3)(B); and

(2) before the notification date, the secured party:

(A) did not receive a response to the request for information; or

(B) received a response to the request for information and sent an authenticated notification of disposition to each secured party or other lienholder named in that response whose financing statement covered the collateral.

§ 9–612. Timeliness of Notification before Disposition of Collateral.

(a) Except as otherwise provided in subsection (b), whether a notification is sent within a reasonable time is a question of fact.

(b) In a transaction other than a consumer transaction, a notification of disposition sent after default and 10 days or more before the earliest time of disposition set forth in the notification is sent within a reasonable time before the disposition.

§ 9–613. Contents and Form of Notification before Disposition of Collateral: General.

Except in a consumer-goods transaction, the following rules apply:

(1) The contents of a notification of disposition are sufficient if the notification:

(A) describes the debtor and the secured party;

(B) describes the collateral that is the subject of the intended disposition;

(C) states the method of intended disposition;

(D) states that the debtor is entitled to an accounting of the unpaid indebtedness and states the charge, if any, for an accounting; and

(E) states the time and place of a public disposition or the time after which any other disposition is to be made.

(2) Whether the contents of a notification that lacks any of the information specified in paragraph (1) are nevertheless sufficient is a question of fact.

(3) The contents of a notification providing substantially the information specified in paragraph (1) are sufficient, even if the notification includes:

(A) information not specified by that paragraph; or

(B) minor errors that are not seriously misleading.

(4) A particular phrasing of the notification is not required.

(5) The following form of notification and the form appearing in Section 9–614(3), when completed, each provides sufficient information:

NOTIFICATION OF DISPOSITION OF COLLATERAL

To: [*Name of debtor, obligor, or other person to which the notification is sent*]

From: [*Name, address, and telephone number of secured party*]

Name of Debtor(s): [*Include only if debtor(s) are not an addressee*]

[*For a public disposition:*]

We will sell [or lease or license, *as applicable*] the [*describe collateral*] [to the highest qualified bidder] in public as follows:

Day and Date: ______

Time: ______

Place: ______

[*For a private disposition:*]

We will sell [or lease or license, *as applicable*] the [*describe collateral*] privately sometime after [*day and date*].

You are entitled to an accounting of the unpaid indebtedness secured by the property that we intend to sell [or lease or license, *as applicable*] [for a charge of $______]. You may request an accounting by calling us at [*telephone number*].

[End of Form]

As amended in 2000.

§ 9–614. Contents and Form of Notification before Disposition of Collateral: Consumer-Goods Transaction.

In a consumer-goods transaction, the following rules apply:

(1) A notification of disposition must provide the following information:

(A) the information specified in Section 9–613(1);

(B) a description of any liability for a deficiency of the person to which the notification is sent;

(C) a telephone number from which the amount that must be paid to the secured party to redeem the collateral under Section 9–623 is available; and

(D) a telephone number or mailing address from which additional information concerning the disposition and the obligation secured is available.

(2) A particular phrasing of the notification is not required.

(3) The following form of notification, when completed, provides sufficient information:

[*Name and address of secured party*]

[*Date*]

NOTICE OF OUR PLAN TO SELL PROPERTY

[*Name and address of any obligor who is also a debtor*]

Subject: [*Identification of Transaction*]

We have your [*describe collateral*], because you broke promises in our agreement.

[*For a public disposition:*]

We will sell [*describe collateral*] at public sale. A sale could include a lease or license. The sale will be held as follows:

Date: ______

Time: ______

Place: ______

You may attend the sale and bring bidders if you want.

[*For a private disposition:*]

We will sell [*describe collateral*] at private sale sometime after [*date*]. A sale could include a lease or license.

The money that we get from the sale (after paying our costs) will reduce the amount you owe. If we get less money than you owe, you [*will or will not, as applicable*] still owe us the difference. If we get more money than you owe, you will get the extra money, unless we must pay it to someone else.

You can get the property back at any time before we sell it by paying us the full amount you owe (not just the past due payments), including our expenses. To learn the exact amount you must pay, call us at [*telephone number*].

If you want us to explain to you in writing how we have figured the amount that you owe us, you may call us at [*telephone number*] [or write us at [*secured party's address*]] and request a written explanation. [We will charge you $______ for the explanation if we sent you another written explanation of the amount you owe us within the last six months.]

If you need more information about the sale call us at [*telephone number*] [or write us at [*secured party's address*]].

We are sending this notice to the following other people who have an interest in [*describe collateral*] or who owe money under your agreement:

[*Names of all other debtors and obligors, if any*]

[End of Form]

(4) A notification in the form of paragraph (3) is sufficient, even if additional information appears at the end of the form.

(5) A notification in the form of paragraph (3) is sufficient, even if it includes errors in information not required by paragraph (1), unless the error is misleading with respect to rights arising under this article.

(6) If a notification under this section is not in the form of paragraph (3), law other than this article determines the effect of including information not required by paragraph (1).

§ 9–615. Application of Proceeds of Disposition; Liability for Deficiency and Right to Surplus.

(a) A secured party shall apply or pay over for application the cash proceeds of disposition under Section 9–610 in the following order to:

(1) the reasonable expenses of retaking, holding, preparing for disposition, processing, and disposing, and, to the extent provided for by agreement and not prohibited by law, reasonable attorney's fees and legal expenses incurred by the secured party;

(2) the satisfaction of obligations secured by the security interest or agricultural lien under which the disposition is made;

(3) the satisfaction of obligations secured by any subordinate security interest in or other subordinate lien on the collateral if:

(A) the secured party receives from the holder of the subordinate security interest or other lien an authenticated demand for proceeds before distribution of the proceeds is completed; and

(B) in a case in which a consignor has an interest in the collateral, the subordinate security interest or other lien is senior to the interest of the consignor; and

(4) a secured party that is a consignor of the collateral if the secured party receives from the consignor an authenticated demand for proceeds before distribution of the proceeds is completed.

(b) If requested by a secured party, a holder of a subordinate security interest or other lien shall furnish reasonable proof of the interest or lien within a reasonable time. Unless the holder does so, the secured party need not comply with the holder's demand under subsection (a)(3).

(c) A secured party need not apply or pay over for application noncash proceeds of disposition under Section 9–610 unless the failure to do so would be commercially unreasonable. A secured party that applies or pays over for application noncash proceeds shall do so in a commercially reasonable manner.

(d) If the security interest under which a disposition is made secures payment or performance of an obligation, after making the payments and applications required by subsection (a) and permitted by subsection (c):

(1) unless subsection (a)(4) requires the secured party to apply or pay over cash proceeds to a consignor, the secured party shall account to and pay a debtor for any surplus; and

(2) the obligor is liable for any deficiency.

(e) If the underlying transaction is a sale of accounts, chattel paper, payment intangibles, or promissory notes:

(1) the debtor is not entitled to any surplus; and

(2) the obligor is not liable for any deficiency.

(f) The surplus or deficiency following a disposition is calculated based on the amount of proceeds that would have been realized in a disposition complying with this part to a transferee other than the secured party, a person related to the secured party, or a secondary obligor if:

(1) the transferee in the disposition is the secured party, a person related to the secured party, or a secondary obligor; and

(2) the amount of proceeds of the disposition is significantly below the range of proceeds that a complying disposition to a person other than the secured party, a person related to the secured party, or a secondary obligor would have brought.

(g) A secured party that receives cash proceeds of a disposition in good faith and without knowledge that the receipt violates the rights of the holder of a security interest or other lien that is not subordinate to the security interest or agricultural lien under which the disposition is made:

(1) takes the cash proceeds free of the security interest or other lien;

(2) is not obligated to apply the proceeds of the disposition to the satisfaction of obligations secured by the security interest or other lien; and

(3) is not obligated to account to or pay the holder of the security interest or other lien for any surplus.

As amended in 2000.

§ 9–616. Explanation of Calculation of Surplus or Deficiency.

(a) In this section:

(1) "Explanation" means a writing that:

(A) states the amount of the surplus or deficiency;

(B) provides an explanation in accordance with subsection (c) of how the secured party calculated the surplus or deficiency;

(C) states, if applicable, that future debits, credits, charges, including additional credit service charges or interest, rebates, and expenses may affect the amount of the surplus or deficiency; and

(D) provides a telephone number or mailing address from which additional information concerning the transaction is available.

(2) "Request" means a record:

(A) authenticated by a debtor or consumer obligor;

(B) requesting that the recipient provide an explanation; and

(C) sent after disposition of the collateral under Section 9–610.

(b) In a consumer-goods transaction in which the debtor is entitled to a surplus or a consumer obligor is liable for a deficiency under Section 9–615, the secured party shall:

(1) send an explanation to the debtor or consumer obligor, as applicable, after the disposition and:

(A) before or when the secured party accounts to the debtor and pays any surplus or first makes written demand on the consumer obligor after the disposition for payment of the deficiency; and

(B) within 14 days after receipt of a request; or

(2) in the case of a consumer obligor who is liable for a deficiency, within 14 days after receipt of a request, send to the consumer obligor a record waiving the secured party's right to a deficiency.

(c) To comply with subsection (a)(1)(B), a writing must provide the following information in the following order:

(1) the aggregate amount of obligations secured by the security interest under which the disposition was made, and, if the amount reflects a rebate of unearned interest or credit service charge, an indication of that fact, calculated as of a specified date:

(A) if the secured party takes or receives possession of the collateral after default, not more than 35 days before the secured party takes or receives possession; or

(B) if the secured party takes or receives possession of the collateral before default or does not take possession of the collateral, not more than 35 days before the disposition;

(2) the amount of proceeds of the disposition;

(3) the aggregate amount of the obligations after deducting the amount of proceeds;

(4) the amount, in the aggregate or by type, and types of expenses, including expenses of retaking, holding, preparing for disposition, processing, and disposing of the collateral, and attorney's fees secured by the collateral which are known to the secured party and relate to the current disposition;

(5) the amount, in the aggregate or by type, and types of credits, including rebates of interest or credit service charges, to which the obligor is known to be entitled and which are not reflected in the amount in paragraph (1); and

(6) the amount of the surplus or deficiency.

(d) A particular phrasing of the explanation is not required. An explanation complying substantially with the requirements of subsection (a) is sufficient, even if it includes minor errors that are not seriously misleading.

(e) A debtor or consumer obligor is entitled without charge to one response to a request under this section during any six-month period in which the secured party did not send to the debtor or consumer obligor an explanation pursuant to subsection (b)(1). The secured party may require payment of a charge not exceeding $25 for each additional response.

§ 9–617. Rights of Transferee of Collateral.

(a) A secured party's disposition of collateral after default:

(1) transfers to a transferee for value all of the debtor's rights in the collateral;

(2) discharges the security interest under which the disposition is made; and

(3) discharges any subordinate security interest or other subordinate lien [other than liens created under [cite acts or statutes providing for liens, if any, that are not to be discharged]].

(b) A transferee that acts in good faith takes free of the rights and interests described in subsection (a), even if the secured party fails to comply with this article or the requirements of any judicial proceeding.

(c) If a transferee does not take free of the rights and interests described in subsection (a), the transferee takes the collateral subject to:

(1) the debtor's rights in the collateral;

(2) the security interest or agricultural lien under which the disposition is made; and

(3) any other security interest or other lien.

§ 9–618. Rights and Duties of Certain Secondary Obligors.

(a) A secondary obligor acquires the rights and becomes obligated to perform the duties of the secured party after the secondary obligor:

(1) receives an assignment of a secured obligation from the secured party;

(2) receives a transfer of collateral from the secured party and agrees to accept the rights and assume the duties of the secured party; or

(3) is subrogated to the rights of a secured party with respect to collateral.

(b) An assignment, transfer, or subrogation described in subsection (a):

(1) is not a disposition of collateral under Section 9–610; and

(2) relieves the secured party of further duties under this article.

§ 9–619. Transfer of Record or Legal Title.

(a) In this section, "transfer statement" means a record authenticated by a secured party stating:

(1) that the debtor has defaulted in connection with an obligation secured by specified collateral;

(2) that the secured party has exercised its post-default remedies with respect to the collateral;

(3) that, by reason of the exercise, a transferee has acquired the rights of the debtor in the collateral; and

(4) the name and mailing address of the secured party, debtor, and transferee.

(b) A transfer statement entitles the transferee to the transfer of record of all rights of the debtor in the collateral specified in the statement in any official filing, recording, registration, or certificate-of-title system covering the collateral. If a transfer statement is presented with the applicable fee and request form to the official or office responsible for maintaining the system, the official or office shall:

(1) accept the transfer statement;

(2) promptly amend its records to reflect the transfer; and

(3) if applicable, issue a new appropriate certificate of title in the name of the transferee.

(c) A transfer of the record or legal title to collateral to a secured party under subsection (b) or otherwise is not of itself a disposition of collateral under this article and does not of itself relieve the secured party of its duties under this article.

§ 9–620. Acceptance of Collateral in Full or Partial Satisfaction of Obligation; Compulsory Disposition of Collateral.

(a) Except as otherwise provided in subsection (g), a secured party may accept collateral in full or partial satisfaction of the obligation it secures only if:

(1) the debtor consents to the acceptance under subsection (c);

(2) the secured party does not receive, within the time set forth in subsection (d), a notification of objection to the proposal authenticated by:

(A) a person to which the secured party was required to send a proposal under Section 9–621; or

(B) any other person, other than the debtor, holding an interest in the collateral subordinate to the security interest that is the subject of the proposal;

(3) if the collateral is consumer goods, the collateral is not in the possession of the debtor when the debtor consents to the acceptance; and

(4) subsection (e) does not require the secured party to dispose of the collateral or the debtor waives the requirement pursuant to Section 9–624.

(b) A purported or apparent acceptance of collateral under this section is ineffective unless:

(1) the secured party consents to the acceptance in an authenticated record or sends a proposal to the debtor; and

(2) the conditions of subsection (a) are met.

(c) For purposes of this section:

(1) a debtor consents to an acceptance of collateral in partial satisfaction of the obligation it secures only if the debtor agrees to the terms of the acceptance in a record authenticated after default; and

(2) a debtor consents to an acceptance of collateral in full satisfaction of the obligation it secures only if the debtor agrees to the terms of the acceptance in a record authenticated after default or the secured party:

(A) sends to the debtor after default a proposal that is unconditional or subject only to a condition that collateral not in the possession of the secured party be preserved or maintained;

(B) in the proposal, proposes to accept collateral in full satisfaction of the obligation it secures; and

(C) does not receive a notification of objection authenticated by the debtor within 20 days after the proposal is sent.

(d) To be effective under subsection (a)(2), a notification of objection must be received by the secured party:

(1) in the case of a person to which the proposal was sent pursuant to Section 9–621, within 20 days after notification was sent to that person; and

(2) in other cases:

(A) within 20 days after the last notification was sent pursuant to Section 9–621; or

(B) if a notification was not sent, before the debtor consents to the acceptance under subsection (c).

(e) A secured party that has taken possession of collateral shall dispose of the collateral pursuant to Section 9–610 within the time specified in subsection (f) if:

(1) 60 percent of the cash price has been paid in the case of a purchase-money security interest in consumer goods; or

(2) 60 percent of the principal amount of the obligation secured has been paid in the case of a non-purchase-money security interest in consumer goods.

(f) To comply with subsection (e), the secured party shall dispose of the collateral:

(1) within 90 days after taking possession; or

(2) within any longer period to which the debtor and all secondary obligors have agreed in an agreement to that effect entered into and authenticated after default.

(g) In a consumer transaction, a secured party may not accept collateral in partial satisfaction of the obligation it secures.

§ 9–621. Notification of Proposal to Accept Collateral.

(a) A secured party that desires to accept collateral in full or partial satisfaction of the obligation it secures shall send its proposal to:

(1) any person from which the secured party has received, before the debtor consented to the acceptance, an authenticated notification of a claim of an interest in the collateral;

(2) any other secured party or lienholder that, 10 days before the debtor consented to the acceptance, held a security interest in or other lien on the collateral perfected by the filing of a financing statement that:

(A) identified the collateral;

(B) was indexed under the debtor's name as of that date; and

(C) was filed in the office or offices in which to file a financing statement against the debtor covering the collateral as of that date; and

(3) any other secured party that, 10 days before the debtor consented to the acceptance, held a security interest in the collateral perfected by compliance with a statute, regulation, or treaty described in Section 9–311(a).

(b) A secured party that desires to accept collateral in partial satisfaction of the obligation it secures shall send its proposal to any secondary obligor in addition to the persons described in subsection (a).

§ 9–622. Effect of Acceptance of Collateral.

(a) A secured party's acceptance of collateral in full or partial satisfaction of the obligation it secures:

(1) discharges the obligation to the extent consented to by the debtor;

(2) transfers to the secured party all of a debtor's rights in the collateral;

(3) discharges the security interest or agricultural lien that is the subject of the debtor's consent and any subordinate security interest or other subordinate lien; and

(4) terminates any other subordinate interest.

(b) A subordinate interest is discharged or terminated under subsection (a), even if the secured party fails to comply with this article.

§ 9–623. Right to Redeem Collateral.

(a) A debtor, any secondary obligor, or any other secured party or lienholder may redeem collateral.

(b) To redeem collateral, a person shall tender:

(1) fulfillment of all obligations secured by the collateral; and

(2) the reasonable expenses and attorney's fees described in Section 9–615(a)(1).

(c) A redemption may occur at any time before a secured party:

(1) has collected collateral under Section 9–607;

(2) has disposed of collateral or entered into a contract for its disposition under Section 9–610; or

(3) has accepted collateral in full or partial satisfaction of the obligation it secures under Section 9–622.

§ 9–624. Waiver.

(a) A debtor or secondary obligor may waive the right to notification of disposition of collateral under Section 9–611 only by an agreement to that effect entered into and authenticated after default.

(b) A debtor may waive the right to require disposition of collateral under Section 9–620(e) only by an agreement to that effect entered into and authenticated after default.

(c) Except in a consumer-goods transaction, a debtor or secondary obligor may waive the right to redeem collateral under Section 9–623 only by an agreement to that effect entered into and authenticated after default.

[Subpart 2. Noncompliance with Article]

§ 9–625. Remedies for Secured Party's Failure to Comply with Article.

(a) If it is established that a secured party is not proceeding in accordance with this article, a court may order or restrain collection, enforcement, or disposition of collateral on appropriate terms and conditions.

(b) Subject to subsections (c), (d), and (f), a person is liable for damages in the amount of any loss caused by a failure to comply with this article. Loss caused by a failure to comply may include loss resulting from the debtor's inability to obtain, or increased costs of, alternative financing.

(c) Except as otherwise provided in Section 9–628:

(1) a person that, at the time of the failure, was a debtor, was an obligor, or held a security interest in or other lien on the collateral may recover damages under subsection (b) for its loss; and

(2) if the collateral is consumer goods, a person that was a debtor or a secondary obligor at the time a secured party failed to comply with this part may recover for that failure in any event an amount not less than the credit service charge plus 10 percent of the principal amount of the obligation or the time-price differential plus 10 percent of the cash price.

(d) A debtor whose deficiency is eliminated under Section 9–626 may recover damages for the loss of any surplus. However, a debtor or secondary obligor whose deficiency is eliminated or reduced under Section 9–626 may not otherwise recover under subsection (b) for noncompliance with the provisions of this part relating to collection, enforcement, disposition, or acceptance.

(e) In addition to any damages recoverable under subsection (b), the debtor, consumer obligor, or person named as a debtor in a filed record, as applicable, may recover $500 in each case from a person that:

(1) fails to comply with Section 9–208;

(2) fails to comply with Section 9–209;

(3) files a record that the person is not entitled to file under Section 9–509(a);

(4) fails to cause the secured party of record to file or send a termination statement as required by Section 9–513(a) or (c);

(5) fails to comply with Section 9–616(b)(1) and whose failure is part of a pattern, or consistent with a practice, of noncompliance; or

(6) fails to comply with Section 9–616(b)(2).

(f) A debtor or consumer obligor may recover damages under subsection (b) and, in addition, $500 in each case from a person that, without reasonable cause, fails to comply with a request under Section 9–210. A recipient of a request under Section 9–210 which never claimed an interest in the collateral or obligations that are the subject of a request under that section has a reasonable excuse for failure to comply with the request within the meaning of this subsection.

(g) If a secured party fails to comply with a request regarding a list of collateral or a statement of account under Section 9–210, the secured party may claim a security interest only as shown in the list or statement included in the request as against a person that is reasonably misled by the failure.

As amended in 2000.

§ 9–626. Action in Which Deficiency or Surplus Is in Issue.

(a) In an action arising from a transaction, other than a consumer transaction, in which the amount of a deficiency or surplus is in issue, the following rules apply:

(1) A secured party need not prove compliance with the provisions of this part relating to collection, enforcement, disposition, or acceptance unless the debtor or a secondary obligor places the secured party's compliance in issue.

(2) If the secured party's compliance is placed in issue, the secured party has the burden of establishing that the collection, enforcement, disposition, or acceptance was conducted in accordance with this part.

(3) Except as otherwise provided in Section 9–628, if a secured party fails to prove that the collection, enforcement, disposition, or acceptance was conducted in accordance with the provisions of this part relating to collection, enforcement, disposition, or acceptance, the liability of a debtor or a secondary obligor for a deficiency is limited to an amount by which the sum of the secured obligation, expenses, and attorney's fees exceeds the greater of:

(A) the proceeds of the collection, enforcement, disposition, or acceptance; or

(B) the amount of proceeds that would have been realized had the noncomplying secured party proceeded in accordance with the provisions of this part relating to collection, enforcement, disposition, or acceptance.

(4) For purposes of paragraph (3)(B), the amount of proceeds that would have been realized is equal to the sum of the secured obligation, expenses, and attorney's fees unless the secured party proves that the amount is less than that sum.

(5) If a deficiency or surplus is calculated under Section 9–615(f), the debtor or obligor has the burden of establishing that the amount of proceeds of the disposition is significantly below the range of prices that a complying disposition to a person other than the secured party, a person related to the secured party, or a secondary obligor would have brought.

(b) The limitation of the rules in subsection (a) to transactions other than consumer transactions is intended to leave to the court the determination of the proper rules in consumer transactions. The court may not infer from that limitation the nature of the proper rule in consumer transactions and may continue to apply established approaches.

§ 9–627. Determination of Whether Conduct Was Commercially Reasonable.

(a) The fact that a greater amount could have been obtained by a collection, enforcement, disposition, or acceptance at a different time or in a different method from that selected by the secured party is not of itself sufficient to preclude the secured party from establishing that the collection, enforcement, disposition, or acceptance was made in a commercially reasonable manner.

(b) A disposition of collateral is made in a commercially reasonable manner if the disposition is made:

(1) in the usual manner on any recognized market;

(2) at the price current in any recognized market at the time of the disposition; or

(3) otherwise in conformity with reasonable commercial practices among dealers in the type of property that was the subject of the disposition.

(c) A collection, enforcement, disposition, or acceptance is commercially reasonable if it has been approved:

(1) in a judicial proceeding;

(2) by a bona fide creditors' committee;

(3) by a representative of creditors; or

(4) by an assignee for the benefit of creditors.

(d) Approval under subsection (c) need not be obtained, and lack of approval does not mean that the collection, enforcement, disposition, or acceptance is not commercially reasonable.

§ 9–628. Nonliability and Limitation on Liability of Secured Party; Liability of Secondary Obligor.

(a) Unless a secured party knows that a person is a debtor or obligor, knows the identity of the person, and knows how to communicate with the person:

(1) the secured party is not liable to the person, or to a secured party or lienholder that has filed a financing statement against the person, for failure to comply with this article; and

(2) the secured party's failure to comply with this article does not affect the liability of the person for a deficiency.

(b) A secured party is not liable because of its status as secured party:

(1) to a person that is a debtor or obligor, unless the secured party knows:

(A) that the person is a debtor or obligor;

(B) the identity of the person; and

(C) how to communicate with the person; or

(2) to a secured party or lienholder that has filed a financing statement against a person, unless the secured party knows:

(A) that the person is a debtor; and

(B) the identity of the person.

(c) A secured party is not liable to any person, and a person's liability for a deficiency is not affected, because of any act or omission arising out of the secured party's reasonable belief that a transaction is not a consumer-goods transaction or a consumer transaction or that goods are not consumer goods, if the secured party's belief is based on its reasonable reliance on:

(1) a debtor's representation concerning the purpose for which collateral was to be used, acquired, or held; or

(2) an obligor's representation concerning the purpose for which a secured obligation was incurred.

(d) A secured party is not liable to any person under Section 9–625(c)(2) for its failure to comply with Section 9–616.

(e) A secured party is not liable under Section 9–625(c)(2) more than once with respect to any one secured obligation.

Part 7 Transition

§ 9–701. Effective Date.

This [Act] takes effect on July 1, 2001.

§ 9–702. Savings Clause.

(a) Except as otherwise provided in this part, this [Act] applies to a transaction or lien within its scope, even if the transaction or lien was entered into or created before this [Act] takes effect.

(b) Except as otherwise provided in subsection (c) and Sections 9–703 through 9–709:

(1) transactions and liens that were not governed by [former Article 9], were validly entered into or created before this [Act] takes effect, and would be subject to this [Act] if they had been entered into or created after this [Act] takes effect, and the rights, duties, and interests flowing from those transactions and liens remain valid after this [Act] takes effect; and

(2) the transactions and liens may be terminated, completed, consummated, and enforced as required or permitted by this [Act] or by the law that otherwise would apply if this [Act] had not taken effect.

(c) This [Act] does not affect an action, case, or proceeding commenced before this [Act] takes effect.

As amended in 2000.

§ 9–703. Security Interest Perfected before Effective Date.

(a) A security interest that is enforceable immediately before this [Act] takes effect and would have priority over the rights of a person that becomes a lien creditor at that time is a perfected security interest under this [Act] if, when this [Act] takes effect, the applicable requirements for enforceability and perfection under this [Act] are satisfied without further action.

(b) Except as otherwise provided in Section 9–705, if, immediately before this [Act] takes effect, a security interest is enforceable and would have priority over the rights of a person that becomes a lien creditor at that time, but the applicable requirements for enforceability or perfection under this [Act] are not satisfied when this [Act] takes effect, the security interest:

(1) is a perfected security interest for one year after this [Act] takes effect;

(2) remains enforceable thereafter only if the security interest becomes enforceable under Section 9–203 before the year expires; and

(3) remains perfected thereafter only if the applicable requirements for perfection under this [Act] are satisfied before the year expires.

§ 9–704. Security Interest Unperfected before Effective Date.

A security interest that is enforceable immediately before this [Act] takes effect but which would be subordinate to the rights of a person that becomes a lien creditor at that time:

(1) remains an enforceable security interest for one year after this [Act] takes effect;

(2) remains enforceable thereafter if the security interest becomes enforceable under Section 9–203 when this [Act] takes effect or within one year thereafter; and

(3) becomes perfected:

(A) without further action, when this [Act] takes effect if the applicable requirements for perfection under this [Act] are satisfied before or at that time; or

(B) when the applicable requirements for perfection are satisfied if the requirements are satisfied after that time.

§ 9–705. Effectiveness of Action Taken before Effective Date.

(a) If action, other than the filing of a financing statement, is taken before this [Act] takes effect and the action would have resulted in priority of a security interest over the rights of a person that becomes a lien creditor had the security interest become enforceable before this [Act] takes effect, the action is effective to perfect a security interest that attaches under this [Act] within one year after this [Act] takes effect. An attached security interest becomes unperfected one year after this [Act] takes effect unless the security interest becomes a perfected security interest under this [Act] before the expiration of that period.

(b) The filing of a financing statement before this [Act] takes effect is effective to perfect a security interest to the extent the

filing would satisfy the applicable requirements for perfection under this [Act].

(c) This [Act] does not render ineffective an effective financing statement that, before this [Act] takes effect, is filed and satisfies the applicable requirements for perfection under the law of the jurisdiction governing perfection as provided in [former Section 9–103]. However, except as otherwise provided in subsections (d) and (e) and Section 9–706, the financing statement ceases to be effective at the earlier of:

(1) the time the financing statement would have ceased to be effective under the law of the jurisdiction in which it is filed; or

(2) June 30, 2006.

(d) The filing of a continuation statement after this [Act] takes effect does not continue the effectiveness of the financing statement filed before this [Act] takes effect. However, upon the timely filing of a continuation statement after this [Act] takes effect and in accordance with the law of the jurisdiction governing perfection as provided in Part 3, the effectiveness of a financing statement filed in the same office in that jurisdiction before this [Act] takes effect continues for the period provided by the law of that jurisdiction.

(e) Subsection (c)(2) applies to a financing statement that, before this [Act] takes effect, is filed against a transmitting utility and satisfies the applicable requirements for perfection under the law of the jurisdiction governing perfection as provided in [former Section 9–103] only to the extent that Part 3 provides that the law of a jurisdiction other than the jurisdiction in which the financing statement is filed governs perfection of a security interest in collateral covered by the financing statement.

(f) A financing statement that includes a financing statement filed before this [Act] takes effect and a continuation statement filed after this [Act] takes effect is effective only to the extent that it satisfies the requirements of Part 5 for an initial financing statement.

§ 9–706. When Initial Financing Statement Suffices to Continue Effectiveness of Financing Statement.

(a) The filing of an initial financing statement in the office specified in Section 9–501 continues the effectiveness of a financing statement filed before this [Act] takes effect if:

(1) the filing of an initial financing statement in that office would be effective to perfect a security interest under this [Act];

(2) the pre-effective-date financing statement was filed in an office in another State or another office in this State; and

(3) the initial financing statement satisfies subsection (c).

(b) The filing of an initial financing statement under subsection (a) continues the effectiveness of the pre-effective-date financing statement:

(1) if the initial financing statement is filed before this [Act] takes effect, for the period provided in [former Section 9–403] with respect to a financing statement; and

(2) if the initial financing statement is filed after this [Act] takes effect, for the period provided in Section 9–515 with respect to an initial financing statement.

(c) To be effective for purposes of subsection (a), an initial financing statement must:

(1) satisfy the requirements of Part 5 for an initial financing statement;

(2) identify the pre-effective-date financing statement by indicating the office in which the financing statement was filed and providing the dates of filing and file numbers, if any, of the financing statement and of the most recent continuation statement filed with respect to the financing statement; and

(3) indicate that the pre-effective-date financing statement remains effective.

§ 9–707. Amendment of Pre-Effective-Date Financing Statement.

(a) In this section, "Pre-effective-date financing statement" means a financing statement filed before this [Act] takes effect.

(b) After this [Act] takes effect, a person may add or delete collateral covered by, continue or terminate the effectiveness of, or otherwise amend the information provided in, a pre-effective-date financing statement only in accordance with the law of the jurisdiction governing perfection as provided in Part 3. However, the effectiveness of a pre-effective-date financing statement also may be terminated in accordance with the law of the jurisdiction in which the financing statement is filed.

(c) Except as otherwise provided in subsection (d), if the law of this State governs perfection of a security interest, the information in a pre-effective-date financing statement may be amended after this [Act] takes effect only if:

(1) the pre-effective-date financing statement and an amendment are filed in the office specified in Section 9–501;

(2) an amendment is filed in the office specified in Section 9–501 concurrently with, or after the filing in that office of, an initial financing statement that satisfies Section 9–706(c); or

(3) an initial financing statement that provides the information as amended and satisfies Section 9–706(c) is filed in the office specified in Section 9–501.

(d) If the law of this State governs perfection of a security interest, the effectiveness of a pre-effective-date financing statement may be continued only under Section 9–705(d) and (f) or 9–706.

(e) Whether or not the law of this State governs perfection of a security interest, the effectiveness of a pre-effective-date financing statement filed in this State may be terminated after this [Act] takes effect by filing a termination statement in the office in which the pre-effective-date financing statement is filed, unless an initial financing statement that satisfies Section 9–706(c) has been filed in the office specified by the law of the

jurisdiction governing perfection as provided in Part 3 as the office in which to file a financing statement.

As amended in 2000.

§ 9–708. Persons Entitled to File Initial Financing Statement or Continuation Statement.

A person may file an initial financing statement or a continuation statement under this part if:

(1) the secured party of record authorizes the filing; and

(2) the filing is necessary under this part:

(A) to continue the effectiveness of a financing statement filed before this [Act] takes effect; or

(B) to perfect or continue the perfection of a security interest.

As amended in 2000.

§ 9–709. Priority.

(a) This [Act] determines the priority of conflicting claims to collateral. However, if the relative priorities of the claims were established before this [Act] takes effect, [former Article 9] determines priority.

(b) For purposes of Section 9–322(a), the priority of a security interest that becomes enforceable under Section 9–203 of this [Act] dates from the time this [Act] takes effect if the security interest is perfected under this [Act] by the filing of a financing statement before this [Act] takes effect which would not have been effective to perfect the security interest under [former Article 9]. This subsection does not apply to conflicting security interests each of which is perfected by the filing of such a financing statement.

As amended in 2000.

APPENDIX F

The Sarbanes-Oxley Act of 2002 (Excerpts and Explanatory Comments)

Note: The author's explanatory comments appear in italics following the excerpt from each section.

SECTION 302

Corporate responsibility for financial reports[1]

(a) Regulations required

The Commission shall, by rule, require, for each company filing periodic reports under section 13(a) or 15(d) of the Securities Exchange Act of 1934 (15 U.S.C. 78m, 78o(d)), that the principal executive officer or officers and the principal financial officer or officers, or persons performing similar functions, certify in each annual or quarterly report filed or submitted under either such section of such Act that—

(1) the signing officer has reviewed the report;

(2) based on the officer's knowledge, the report does not contain any untrue statement of a material fact or omit to state a material fact necessary in order to make the statements made, in light of the circumstances under which such statements were made, not misleading;

(3) based on such officer's knowledge, the financial statements, and other financial information included in the report, fairly present in all material respects the financial condition and results of operations of the issuer as of, and for, the periods presented in the report;

(4) the signing officers—

(A) are responsible for establishing and maintaining internal controls;

(B) have designed such internal controls to ensure that material information relating to the issuer and its consolidated subsidiaries is made known to such officers by others within those entities, particularly during the period in which the periodic reports are being prepared;

(C) have evaluated the effectiveness of the issuer's internal controls as of a date within 90 days prior to the report; and

(D) have presented in the report their conclusions about the effectiveness of their internal controls based on their evaluation as of that date;

(5) the signing officers have disclosed to the issuer's auditors and the audit committee of the board of directors (or persons fulfilling the equivalent function)—

(A) all significant deficiencies in the design or operation of internal controls which could adversely affect the issuer's ability to record, process, summarize, and report financial data and have identified for the issuer's auditors any material weaknesses in internal controls; and

(B) any fraud, whether or not material, that involves management or other employees who have a significant role in the issuer's internal controls; and

(6) the signing officers have indicated in the report whether or not there were significant changes in internal controls or in other factors that could significantly affect internal controls subsequent to the date of their evaluation, including any corrective actions with regard to significant deficiencies and material weaknesses.

(b) Foreign reincorporations have no effect

Nothing in this section shall be interpreted or applied in any way to allow any issuer to lessen the legal force of the statement required under this section, by an issuer having reincorporated or having engaged in any other transaction that resulted in the transfer of the corporate domicile or offices of the issuer from inside the United States to outside of the United States.

(c) Deadline

The rules required by subsection (a) of this section shall be effective not later than 30 days after July 30, 2002.

1. This section of the Sarbanes-Oxley Act is codified at 15 U.S.C. Section 7241.

EXPLANATORY COMMENTS: *Section 302 requires the chief executive officer (CEO) and chief financial officer (CFO) of each public company to certify that they have reviewed the company's quarterly and annual reports to be filed with the Securities and Exchange Commission (SEC). The CEO and CFO must certify that, based on their knowledge, the reports do not contain any untrue statement of a material fact or any half-truth that would make the report misleading, and that the information contained in the reports fairly presents the company's financial condition.*

In addition, this section also requires the CEO and CFO to certify that they have created and designed an internal control system for their company and have recently evaluated that system to ensure that it is effectively providing them with relevant and accurate financial information. If the signing officers have found any significant deficiencies or weaknesses in the company's system or have discovered any evidence of fraud, they must have reported the situation, and any corrective actions they have taken, to the auditors and the audit committee.

SECTION 306

Insider trades during pension fund blackout periods[2]

(a) Prohibition of insider trading during pension fund blackout periods

(1) In general

Except to the extent otherwise provided by rule of the Commission pursuant to paragraph (3), it shall be unlawful for any director or executive officer of an issuer of any equity security (other than an exempted security), directly or indirectly, to purchase, sell, or otherwise acquire or transfer any equity security of the issuer (other than an exempted security) during any blackout period with respect to such equity security if such director or officer acquires such equity security in connection with his or her service or employment as a director or executive officer.

(2) Remedy

(A) In general

Any profit realized by a director or executive officer referred to in paragraph (1) from any purchase, sale, or other acquisition or transfer in violation of this subsection shall inure to and be recoverable by the issuer, irrespective of any intention on the part of such director or executive officer in entering into the transaction.

(B) Actions to recover profits

An action to recover profits in accordance with this subsection may be instituted at law or in equity in any court of competent jurisdiction by the issuer, or by the owner of any security of the issuer in the name and in behalf of the issuer if the issuer fails or refuses to bring such action within 60 days after the date of request, or fails diligently to prosecute the action thereafter, except that no such suit shall be brought more than 2 years after the date on which such profit was realized.

(3) Rulemaking authorized

The Commission shall, in consultation with the Secretary of Labor, issue rules to clarify the application of this subsection and to prevent evasion thereof. Such rules shall provide for the application of the requirements of paragraph (1) with respect to entities treated as a single employer with respect to an issuer under section 414(b), (c), (m), or (o) of Title 26 to the extent necessary to clarify the application of such requirements and to prevent evasion thereof. Such rules may also provide for appropriate exceptions from the requirements of this subsection, including exceptions for purchases pursuant to an automatic dividend reinvestment program or purchases or sales made pursuant to an advance election.

(4) Blackout period

For purposes of this subsection, the term "blackout period", with respect to the equity securities of any issuer—

(A) means any period of more than 3 consecutive business days during which the ability of not fewer than 50 percent of the participants or beneficiaries under all individual account plans maintained by the issuer to purchase, sell, or otherwise acquire or transfer an interest in any equity of such issuer held in such an individual account plan is temporarily suspended by the issuer or by a fiduciary of the plan; and

(B) does not include, under regulations which shall be prescribed by the Commission—

(i) a regularly scheduled period in which the participants and beneficiaries may not purchase, sell, or otherwise acquire or transfer an interest in any equity of such issuer, if such period is—

(I) incorporated into the individual account plan; and

(II) timely disclosed to employees before becoming participants under the individual account plan or as a subsequent amendment to the plan; or

(ii) any suspension described in subparagraph (A) that is imposed solely in connection with persons becoming participants or beneficiaries, or ceasing to be participants or beneficiaries, in an individual account plan by reason of a corporate merger, acquisition, divestiture, or similar transaction involving the plan or plan sponsor.

(5) Individual account plan

For purposes of this subsection, the term "individual account plan" has the meaning provided in section 1002(34) of Title 29, except that such term shall not include a one-participant retirement plan (within the meaning of section 1021(i)(8)(B) of Title 29).

(6) Notice to directors, executive officers, and the Commission

2. Codified at 15 U.S.C. Section 7244.

In any case in which a director or executive officer is subject to the requirements of this subsection in connection with a blackout period (as defined in paragraph (4)) with respect to any equity securities, the issuer of such equity securities shall timely notify such director or officer and the Securities and Exchange Commission of such blackout period.

* * * *

EXPLANATORY COMMENTS: *Corporate pension funds typically prohibit employees from trading shares of the corporation during periods when the pension fund is undergoing significant change. Prior to 2002, however, these blackout periods did not affect the corporation's executives, who frequently received shares of the corporate stock as part of their compensation. During the collapse of Enron, for example, its pension plan was scheduled to change administrators at a time when Enron's stock price was falling. Enron's employees therefore could not sell their shares while the price was dropping, but its executives could and did sell their stock, consequently avoiding some of the losses. Section 306 was Congress's solution to the basic unfairness of this situation. This section of the act required the SEC to issue rules that prohibit any director or executive officer from trading during pension fund blackout periods. (The SEC later issued these rules, entitled Regulation Blackout Trading Restriction, or Reg BTR.) Section 306 also provided shareholders with a right to file a shareholder's derivative suit against officers and directors who have profited from trading during these blackout periods (provided that the corporation has failed to bring a suit). The officer or director can be forced to return to the corporation any profits received, regardless of whether the director or officer acted with bad intent.*

SECTION 402

Periodical and other reports[3]

* * * *

(i) Accuracy of financial reports

Each financial report that contains financial statements, and that is required to be prepared in accordance with (or reconciled to) generally accepted accounting principles under this chapter and filed with the Commission shall reflect all material correcting adjustments that have been identified by a registered public accounting firm in accordance with generally accepted accounting principles and the rules and regulations of the Commission.

(j) Off-balance sheet transactions

Not later than 180 days after July 30, 2002, the Commission shall issue final rules providing that each annual and quarterly financial report required to be filed with the Commission shall disclose all material off-balance sheet transactions, arrangements, obligations (including contingent obligations), and other relationships of the issuer with unconsolidated entities or other persons, that may have a material current or future effect on financial condition, changes in financial condition, results of operations, liquidity, capital expenditures, capital resources, or significant components of revenues or expenses.

(k) Prohibition on personal loans to executives

(1) In general

It shall be unlawful for any issuer (as defined in section 7201 of this title), directly or indirectly, including through any subsidiary, to extend or maintain credit, to arrange for the extension of credit, or to renew an extension of credit, in the form of a personal loan to or for any director or executive officer (or equivalent thereof) of that issuer. An extension of credit maintained by the issuer on July 30, 2002, shall not be subject to the provisions of this subsection, provided that there is no material modification to any term of any such extension of credit or any renewal of any such extension of credit on or after July 30, 2002.

(2) Limitation

Paragraph (1) does not preclude any home improvement and manufactured home loans (as that term is defined in section 1464 of Title 12), consumer credit (as defined in section 1602 of this title), or any extension of credit under an open end credit plan (as defined in section 1602 of this title), or a charge card (as defined in section 1637(c)(4)(e) of this title), or any extension of credit by a broker or dealer registered under section 78o of this title to an employee of that broker or dealer to buy, trade, or carry securities, that is permitted under rules or regulations of the Board of Governors of the Federal Reserve System pursuant to section 78g of this title (other than an extension of credit that would be used to purchase the stock of that issuer), that is—

(A) made or provided in the ordinary course of the consumer credit business of such issuer;

(B) of a type that is generally made available by such issuer to the public; and

(C) made by such issuer on market terms, or terms that are no more favorable than those offered by the issuer to the general public for such extensions of credit.

(3) Rule of construction for certain loans

Paragraph (1) does not apply to any loan made or maintained by an insured depository institution (as defined in section 1813 of Title 12), if the loan is subject to the insider lending restrictions of section 375b of Title 12.

(l) Real time issuer disclosures

Each issuer reporting under subsection (a) of this section or section 78o(d) of this title shall disclose to the public on a rapid and current basis such additional information concerning material changes in the financial condition or operations of the issuer, in plain English, which may include trend and qualitative information and graphic presentations, as the Commission determines, by rule, is necessary or useful for the protection of investors and in the public interest.

3. This section of the Sarbanes-Oxley Act amended some of the provisions of the 1934 Securities Exchange Act and added the paragraphs reproduced here at 15 U.S.C. Section 78m.

EXPLANATORY COMMENTS: *Corporate executives during the Enron era typically received extremely large salaries, significant bonuses, and abundant stock options, even when the companies for which they worked were suffering. Executives were also routinely given personal loans from corporate funds, many of which were never paid back. The average large company during that period loaned almost $1 million a year to top executives, and some companies, including Tyco International and Adelphia Communications Corporation, loaned hundreds of millions of dollars to their executives every year. Section 402 amended the 1934 Securities Exchange Act to prohibit public companies from making personal loans to executive officers and directors. There are a few exceptions to this prohibition, such as home-improvement loans made in the ordinary course of business. Note also that while loans are forbidden, outright gifts are not. A corporation is free to give gifts to its executives, including cash, provided that these gifts are disclosed on its financial reports. The idea is that corporate directors will be deterred from making substantial gifts to their executives by the disclosure requirement—particularly if the corporation's financial condition is questionable—because making such gifts could be perceived as abusing their authority.*

SECTION 403

Directors, officers, and principal stockholders[4]

(a) Disclosures required

(1) Directors, officers, and principal stockholders required to file

Every person who is directly or indirectly the beneficial owner of more than 10 percent of any class of any equity security (other than an exempted security) which is registered pursuant to section 78l of this title, or who is a director or an officer of the issuer of such security, shall file the statements required by this subsection with the Commission (and, if such security is registered on a national securities exchange, also with the exchange).

(2) Time of filing

The statements required by this subsection shall be filed—

(A) at the time of the registration of such security on a national securities exchange or by the effective date of a registration statement filed pursuant to section 78l(g) of this title;

(B) within 10 days after he or she becomes such beneficial owner, director, or officer;

(C) if there has been a change in such ownership, or if such person shall have purchased or sold a security-based swap agreement (as defined in section 206(b) of the Gramm-Leach-Bliley Act (15 U.S.C. 78c note)) involving such equity security, before the end of the second business day following the day on which the subject transaction has been executed, or at such other time as the Commission shall establish, by rule, in any case in which the Commission determines that such 2-day period is not feasible.

(3) Contents of statements

A statement filed—

(A) under subparagraph (A) or (B) of paragraph (2) shall contain a statement of the amount of all equity securities of such issuer of which the filing person is the beneficial owner; and

(B) under subparagraph (C) of such paragraph shall indicate ownership by the filing person at the date of filing, any such changes in such ownership, and such purchases and sales of the security-based swap agreements as have occurred since the most recent such filing under such subparagraph.

(4) Electronic filing and availability

Beginning not later than 1 year after July 30, 2002—

(A) a statement filed under subparagraph (C) of paragraph (2) shall be filed electronically;

(B) the Commission shall provide each such statement on a publicly accessible Internet site not later than the end of the business day following that filing; and

(C) the issuer (if the issuer maintains a corporate website) shall provide that statement on that corporate website, not later than the end of the business day following that filing.

* * * *

EXPLANATORY COMMENTS: *This section dramatically shortens the time period provided in the Securities Exchange Act of 1934 for disclosing transactions by insiders. The prior law stated that most transactions had to be reported within ten days of the beginning of the following month, although certain transactions did not have to be reported until the following fiscal year (within the first forty-five days). Because some of the insider trading that occurred during the Enron fiasco did not have to be disclosed (and was therefore not discovered) until long after the transactions, Congress added this section to reduce the time period for making disclosures. Under Section 403, most transactions by insiders must be electronically filed with the SEC within two business days. Also, any company that maintains a Web site must post these SEC filings on its site by the end of the next business day. Congress enacted this section in the belief that if insiders are required to file reports of their transactions promptly with the SEC, companies will do more to police themselves and prevent insider trading.*

SECTION 404

Management assessment of internal controls[5]

(a) Rules required

The Commission shall prescribe rules requiring each annual report required by section 78m(a) or 78o(d) of this title to contain an internal control report, which shall—

(1) state the responsibility of management for establishing and maintaining an adequate internal control structure and procedures for financial reporting; and

4. This section of the Sarbanes-Oxley Act amended the disclosure provisions of the 1934 Securities Exchange Act, at 15 U.S.C. Section 78p.

5. Codified at 15 U.S.C. Section 7262.

(2) contain an assessment, as of the end of the most recent fiscal year of the issuer, of the effectiveness of the internal control structure and procedures of the issuer for financial reporting.

(b) Internal control evaluation and reporting

With respect to the internal control assessment required by subsection (a) of this section, each registered public accounting firm that prepares or issues the audit report for the issuer shall attest to, and report on, the assessment made by the management of the issuer. An attestation made under this subsection shall be made in accordance with standards for attestation engagements issued or adopted by the Board. Any such attestation shall not be the subject of a separate engagement.

EXPLANATORY COMMENTS: *This section was enacted to prevent corporate executives from claiming they were ignorant of significant errors in their companies' financial reports. For instance, several CEOs testified before Congress that they simply had no idea that the corporations' financial statements were off by billions of dollars. Congress therefore passed Section 404, which requires each annual report to contain a description and assessment of the company's internal control structure and financial reporting procedures. The section also requires that an audit be conducted of the internal control assessment, as well as the financial statements contained in the report. This section goes hand in hand with Section 302 (which, as discussed previously, requires various certifications attesting to the accuracy of the information in financial reports).*

Section 404 has been one of the more controversial and expensive provisions in the Sarbanes-Oxley Act because it requires companies to assess their own internal financial controls to make sure that their financial statements are reliable and accurate. A corporation might need to set up a disclosure committee and a coordinator, establish codes of conduct for accounting and financial personnel, create documentation procedures, provide training, and outline the individuals who are responsible for performing each of the procedures. Companies that were already well managed have not experienced substantial difficulty complying with this section. Other companies, however, have spent millions of dollars setting up, documenting, and evaluating their internal financial control systems. Although initially creating the internal financial control system is a onetime-only expense, the costs of maintaining and evaluating it are ongoing. Some corporations that spent considerable sums complying with Section 404 have been able to offset these costs by discovering and correcting inefficiencies or frauds within their systems. Nevertheless, it is unlikely that any corporation will find compliance with this section to be inexpensive.

SECTION 802 (A)

Destruction, alteration, or falsification of records in Federal investigations and bankruptcy[6]

Whoever knowingly alters, destroys, mutilates, conceals, covers up, falsifies, or makes a false entry in any record, document, or tangible object with the intent to impede, obstruct, or influence the investigation or proper administration of any matter within the jurisdiction of any department or agency of the United States or any case filed under title 11, or in relation to or contemplation of any such matter or case, shall be fined under this title, imprisoned not more than 20 years, or both.

Destruction of corporate audit records[7]

(a) (1) Any accountant who conducts an audit of an issuer of securities to which section 10A(a) of the Securities Exchange Act of 1934 (15 U.S.C. 78j-1(a)) applies, shall maintain all audit or review workpapers for a period of 5 years from the end of the fiscal period in which the audit or review was concluded.

(2) The Securities and Exchange Commission shall promulgate, within 180 days, after adequate notice and an opportunity for comment, such rules and regulations, as are reasonably necessary, relating to the retention of relevant records such as workpapers, documents that form the basis of an audit or review, memoranda, correspondence, communications, other documents, and records (including electronic records) which are created, sent, or received in connection with an audit or review and contain conclusions, opinions, analyses, or financial data relating to such an audit or review, which is conducted by any accountant who conducts an audit of an issuer of securities to which section 10A(a) of the Securities Exchange Act of 1934 (15 U.S.C. 78j-1(a)) applies. The Commission may, from time to time, amend or supplement the rules and regulations that it is required to promulgate under this section, after adequate notice and an opportunity for comment, in order to ensure that such rules and regulations adequately comport with the purposes of this section.

(b) Whoever knowingly and willfully violates subsection (a)(1), or any rule or regulation promulgated by the Securities and Exchange Commission under subsection (a)(2), shall be fined under this title, imprisoned not more than 10 years, or both.

(c) Nothing in this section shall be deemed to diminish or relieve any person of any other duty or obligation imposed by Federal or State law or regulation to maintain, or refrain from destroying, any document.

EXPLANATORY COMMENTS: *Section 802(a) enacted two new statutes that punish those who alter or destroy documents. The first statute is not specifically limited to securities fraud cases. It provides that anyone who alters, destroys, or falsifies records in federal investigations or bankruptcy may be criminally prosecuted and sentenced to a fine or to up to twenty years in prison, or both. The second statute requires auditors of public companies to keep all audit or review working papers for five years but expressly allows the SEC to amend or supplement these requirements as it sees fit. The SEC has, in fact, amended this section by issuing a rule that requires auditors who audit reporting companies to retain working papers for seven years from the conclusion of the review. Section 802(a) further provides that anyone who knowingly and willfully violates this statute is subject to*

6. Codified at 15 U.S.C. Section 1519.

7. Codified at 15 U.S.C. Section 1520.

criminal prosecution and can be sentenced to a fine, imprisoned for up to ten years, or both if convicted.

This portion of the Sarbanes-Oxley Act implicitly recognizes that persons who are under investigation often are tempted to respond by destroying or falsifying documents that might prove their complicity in wrongdoing. The severity of the punishment should provide a strong incentive for these individuals to resist the temptation.

SECTION 804

Time limitations on the commencement of civil actions arising under Acts of Congress[8]

(a) Except as otherwise provided by law, a civil action arising under an Act of Congress enacted after the date of the enactment of this section may not be commenced later than 4 years after the cause of action accrues.

(b) Notwithstanding subsection (a), a private right of action that involves a claim of fraud, deceit, manipulation, or contrivance in contravention of a regulatory requirement concerning the securities laws, as defined in section 3(a)(47) of the Securities Exchange Act of 1934 (15 U.S.C. 78c(a)(47)), may be brought not later than the earlier of—

(1) 2 years after the discovery of the facts constituting the violation; or

(2) 5 years after such violation.

EXPLANATORY COMMENTS: *Prior to the enactment of this section, Section 10(b) of the Securities Exchange Act of 1934 had no express statute of limitations. The courts generally required plaintiffs to have filed suit within one year from the date that they should (using due diligence) have discovered that a fraud had been committed but no later than three years after the fraud occurred. Section 804 extends this period by specifying that plaintiffs must file a lawsuit within two years after they discover (or should have discovered) a fraud but no later than five years after the fraud's occurrence. This provision has prevented the courts from dismissing numerous securities fraud lawsuits.*

SECTION 806

Civil action to protect against retaliation in fraud cases[9]

(a) Whistleblower protection for employees of publicly traded companies.—

No company with a class of securities registered under section 12 of the Securities Exchange Act of 1934 (15 U.S.C. 78l), or that is required to file reports under section 15(d) of the Securities Exchange Act of 1934 (15 U.S.C. 78o(d)), or any officer, employee, contractor, subcontractor, or agent of such company, may discharge, demote, suspend, threaten, harass, or in any other manner discriminate against an employee in the terms and conditions of employment because of any lawful act done by the employee—

(1) to provide information, cause information to be provided, or otherwise assist in an investigation regarding any conduct which the employee reasonably believes constitutes a violation of section 1341, 1343, 1344, or 1348, any rule or regulation of the Securities and Exchange Commission, or any provision of Federal law relating to fraud against shareholders, when the information or assistance is provided to or the investigation is conducted by—

(A) a Federal regulatory or law enforcement agency;

(B) any Member of Congress or any committee of Congress; or

(C) a person with supervisory authority over the employee (or such other person working for the employer who has the authority to investigate, discover, or terminate misconduct); or

(2) to file, cause to be filed, testify, participate in, or otherwise assist in a proceeding filed or about to be filed (with any knowledge of the employer) relating to an alleged violation of section 1341, 1343, 1344, or 1348, any rule or regulation of the Securities and Exchange Commission, or any provision of Federal law relating to fraud against shareholders.

(b) Enforcement action.—

(1) In general.—A person who alleges discharge or other discrimination by any person in violation of subsection (a) may seek relief under subsection (c), by—

(A) filing a complaint with the Secretary of Labor; or

(B) if the Secretary has not issued a final decision within 180 days of the filing of the complaint and there is no showing that such delay is due to the bad faith of the claimant, bringing an action at law or equity for de novo review in the appropriate district court of the United States, which shall have jurisdiction over such an action without regard to the amount in controversy.

(2) Procedure.—

(A) In general.—An action under paragraph (1)(A) shall be governed under the rules and procedures set forth in section 42121(b) of title 49, United States Code.

(B) Exception.—Notification made under section 42121(b)(1) of title 49, United States Code, shall be made to the person named in the complaint and to the employer.

(C) Burdens of proof.—An action brought under paragraph (1)(B) shall be governed by the legal burdens of proof set forth in section 42121(b) of title 49, United States Code.

(D) Statute of limitations.—An action under paragraph (1) shall be commenced not later than 90 days after the date on which the violation occurs.

(c) Remedies.—

(1) In general.—An employee prevailing in any action under subsection (b)(1) shall be entitled to all relief necessary to make the employee whole.

8. Codified at 28 U.S.C. Section 1658.

9. Codified at 18 U.S.C. Section 1514A.

(2) Compensatory damages.—Relief for any action under paragraph (1) shall include—

(A) reinstatement with the same seniority status that the employee would have had, but for the discrimination;

(B) the amount of back pay, with interest; and

(C) compensation for any special damages sustained as a result of the discrimination, including litigation costs, expert witness fees, and reasonable attorney fees.

(d) Rights retained by employee.—Nothing in this section shall be deemed to diminish the rights, privileges, or remedies of any employee under any Federal or State law, or under any collective bargaining agreement.

EXPLANATORY COMMENTS: *Section 806 is one of several provisions that were included in the Sarbanes-Oxley Act to encourage and protect whistleblowers—that is, employees who report their employer's alleged violations of securities law to the authorities. This section applies to employees, agents, and independent contractors who work for publicly traded companies or testify about such a company during an investigation. It sets up an administrative procedure at the Department of Labor for individuals who claim that their employer retaliated against them (fired or demoted them, for example) for blowing the whistle on the employer's wrongful conduct. It also allows the award of civil damages—including back pay, reinstatement, special damages, attorneys' fees, and court costs—to employees who prove that they suffered retaliation. Since this provision was enacted, whistleblowers have filed numerous complaints with the Department of Labor under this section.*

SECTION 807

Securities fraud[10]

Whoever knowingly executes, or attempts to execute, a scheme or artifice—

(1) to defraud any person in connection with any security of an issuer with a class of securities registered under section 12 of the Securities Exchange Act of 1934 (15 U.S.C. 78l) or that is required to file reports under section 15(d) of the Securities Exchange Act of 1934 (15 U.S.C. 78o(d)); or

(2) to obtain, by means of false or fraudulent pretenses, representations, or promises, any money or property in connection with the purchase or sale of any security of an issuer with a class of securities registered under section 12 of the Securities Exchange Act of 1934 (15 U.S.C. 78l) or that is required to file reports under section 15(d) of the Securities Exchange Act of 1934 (15 U.S.C. 78o(d)); shall be fined under this title, or imprisoned not more than 25 years, or both.

EXPLANATORY COMMENTS: *Section 807 adds a new provision to the federal criminal code that addresses securities fraud. Prior to 2002, federal securities law had already made it a crime—under Section 10(b) of the Securities Exchange Act of 1934 and SEC Rule 10b-5, both of which are discussed in Chapter 27—to intentionally defraud someone in connection with a purchase or sale of securities, but the offense was not listed in the federal criminal code. Also, paragraph 2 of Section 807 goes beyond what is prohibited under securities law by making it a crime to obtain by means of false or fraudulent pretenses any money or property from the purchase or sale of securities. This new provision allows violators to be punished by up to twenty-five years in prison, a fine, or both.*

SECTION 906

Failure of corporate officers to certify financial reports[11]

(a) Certification of periodic financial reports.—Each periodic report containing financial statements filed by an issuer with the Securities Exchange Commission pursuant to section 13(a) or 15(d) of the Securities Exchange Act of 1934 (15 U.S.C. 78m(a) or 78o(d)) shall be accompanied by a written statement by the chief executive officer and chief financial officer (or equivalent thereof) of the issuer.

(b) Content.—The statement required under subsection (a) shall certify that the periodic report containing the financial statements fully complies with the requirements of section 13(a) or 15(d) of the Securities Exchange Act of 1934 (15 U.S.C. 78m or 78o(d)) and that information contained in the periodic report fairly presents, in all material respects, the financial condition and results of operations of the issuer.

(c) Criminal penalties.—Whoever—

(1) certifies any statement as set forth in subsections (a) and (b) of this section knowing that the periodic report accompanying the statement does not comport with all the requirements set forth in this section shall be fined not more than $1,000,000 or imprisoned not more than 10 years, or both; or

(2) willfully certifies any statement as set forth in subsections (a) and (b) of this section knowing that the periodic report accompanying the statement does not comport with all the requirements set forth in this section shall be fined not more than $5,000,000, or imprisoned not more than 20 years, or both.

EXPLANATORY COMMENTS: *As previously discussed, under Section 302 a corporation's CEO and CFO are required to certify that they believe the quarterly and annual reports their company files with the SEC are accurate and fairly present the company's financial condition. Section 906 adds "teeth" to these requirements by authorizing criminal penalties for those officers who intentionally certify inaccurate SEC filings. Knowing violations of the requirements are punishable by a fine of up to $1 million, ten years in prison, or both. Willful violators may be fined up to $5 million, sentenced to up to twenty years in prison, or both. Although the difference between a knowing and a willful violation is not entirely clear, the section is obviously intended to remind corporate officers of the serious consequences of certifying inaccurate reports to the SEC.*

10. Codified at 18 U.S.C. Section 1348.

11. Codified at 18 U.S.C. Section 1350.

 GLOSSARY

A

ABANDONED PROPERTY ■ Property with which the owner has voluntarily parted, with no intention of recovering it.

ACCELERATION CLAUSE ■ A clause that allows a payee or other holder of a time instrument to demand payment of the entire amount due, with interest, if a certain event occurs, such as a default in the payment of an installment when due.

ACCEPTANCE ■ In contract law, a voluntary act by the offeree that shows assent, or agreement, to the terms of an offer; may consist of words or conduct. In negotiable instruments law, the drawee's signed agreement to pay a draft when presented.

ACCEPTOR ■ A drawee that is legally obligated to pay an instrument when the instrument is presented later for payment.

ACCESSION ■ Occurs when an individual adds value to personal property by either labor or materials. In some situations, a person may acquire ownership rights in another's property through accession.

ACCOMMODATION PARTY ■ A person who signs an instrument for the purpose of lending his or her name as credit to another party on the instrument.

ACCORD AND SATISFACTION ■ A common means of settling a disputed claim, in which a debtor offers to pay a lesser amount than the creditor purports to be owed. The creditor's acceptance of the offer creates an accord (agreement), and when the accord is executed, satisfaction occurs.

ACCREDITED INVESTORS ■ In the context of securities offerings, "sophisticated" investors, such as banks, insurance companies, investment companies, the issuer's executive officers and directors, and persons whose income or net worth exceeds a certain threshold.

ACT OF STATE DOCTRINE ■ A doctrine providing that the judicial branch of one country will not examine the validity of public acts committed by a recognized foreign government within its own territory.

ACTIONABLE ■ Capable of serving as the basis of a lawsuit. An actionable claim can be pursued in a lawsuit or other court action.

ACTUAL MALICE ■ A condition that exists when a person makes a statement with either knowledge of its falsity or a reckless disregard for the truth. In a defamation suit, a statement made about a public figure normally must be made with actual malice for liability to be incurred.

ADEQUATE PROTECTION DOCTRINE ■ In bankruptcy law, a doctrine that protects secured creditors from losing the value of their security as a result of the automatic stay on legal proceedings. The bankruptcy court can provide adequate protection by requiring the debtor or trustee to make cash payments to the creditor or to provide additional collateral or replacement liens in case the stay causes the property to lose value.

ADHESION CONTRACT ■ A "standard-form" contract, such as that between a large retailer and a consumer, in which the stronger party dictates the terms.

ADJUDICATE ■ To render a judicial decision. In the administrative process, the proceeding in which an administrative law judge hears and decides issues that arise when an administrative agency charges a person or a firm with violating a law or regulation enforced by the agency.

ADMINISTRATIVE AGENCY ■ A federal or state government agency established to perform a specific function. Administrative agencies are authorized by legislative acts to make and enforce rules to administer and enforce the acts.

ADMINISTRATIVE LAW ■ The body of law created by administrative agencies (in the form of rules, regulations, orders, and decisions) in order to carry out their duties and responsibilities.

ADMINISTRATIVE LAW JUDGE (ALJ) ■ One who presides over an administrative agency hearing and has the power to administer oaths, take testimony, rule on questions of evidence, and make determinations of fact.

ADMINISTRATIVE PROCESS ■ The procedure used by administrative agencies in the administration of law.

ADMINISTRATOR ■ One who is appointed by a court to handle the probate (disposition) of a person's estate if that person dies intestate (without a valid will) or if the executor named in the will cannot serve.

ADVERSE POSSESSION ■ The acquisition of title to real property by occupying it openly, without the consent of the owner, for a period of time specified by a state statute. The occupation must be actual, open, notorious, exclusive, and in opposition to all others, including the owner.

AFFIRMATIVE ACTION ■ Job-hiring policies that give special consideration to members of protected classes in an effort to overcome present effects of past discrimination.

AFTER-ACQUIRED PROPERTY ■ Property that is acquired by the debtor after the execution of a security agreement.

AGENCY ■ A relationship between two parties in which one party (the agent) agrees to represent or act for the other (the principal).

AGENT A person who agrees to represent or act for another, called the principal.

AGREEMENT ■ A meeting of two or more minds in regard to the terms of a contract; usually broken down into two events—an offer by one party to form a contract, and an acceptance of the offer by the person to whom the offer is made.

ALIEN CORPORATION ■ A designation in the United States for a corporation formed in another country but doing business in the United States.

ALIENATION ■ The process of transferring land out of one's possession (thus "alienating" the land from oneself).

ALLONGE ■ A piece of paper firmly attached to a negotiable instrument, on which transferees can make indorsements if there is no room left on the instrument itself.

ALTERNATIVE DISPUTE RESOLUTION (ADR) ■ The resolution of disputes in ways other than those involved in the traditional judicial process. Negotiation, mediation, and arbitration are forms of ADR.

ANSWER ■ Procedurally, a defendant's response to the plaintiff's complaint.

ANTICIPATORY REPUDIATION ■ An assertion or action by a party indicating that he or she will not perform an obligation that the party is contractually obligated to perform at a future time.

ANTITRUST LAWS ■ Laws protecting commerce from unlawful restraints.

APPRAISAL RIGHT ■ The right of a dissenting shareholder, who objects to an extraordinary transaction of the corporation (such as a merger or a consolidation), to have his or her shares appraised and to be paid the fair value of those shares by the corporation.

APPROPRIATION ■ In tort law, the use by one person of another person's name, likeness, or other identifying characteristic without permission and for the benefit of the user.

ARBITRATION ■ The settling of a dispute by submitting it to a disinterested third party (other than a court), who renders a decision that is (most often) legally binding.

ARBITRATION CLAUSE ■ A clause in a contract that provides that, in the event of a dispute, the parties will submit the dispute to arbitration rather than litigate the dispute in court.

ARSON ■ The intentional burning of another's dwelling. Some statutes

have expanded this to include any real property regardless of ownership and the destruction of property by other means—for example, by explosion.

ARTICLES OF INCORPORATION ■ The document filed with the appropriate governmental agency, usually the secretary of state, when a business is incorporated. State statutes usually prescribe what kind of information must be contained in the articles of incorporation.

ARTICLES OF ORGANIZATION ■ The document filed with a designated state official by which a limited liability company is formed.

ARTICLES OF PARTNERSHIP ■ A written agreement that sets forth each partner's rights and obligations with respect to the partnership.

ARTISAN'S LIEN ■ A possessory lien given to a person who has made improvements and added value to another individual's personal property as security for payment for services performed.

ASSAULT ■ Any word or action intended to make another person fearful of immediate physical harm; a reasonably believable threat.

ASSIGNEE ■ A party to whom the rights under a contract are transferred, or assigned.

ASSIGNMENT ■ The act of transferring to another all or part of one's rights arising under a contract.

ASSIGNOR ■ A party who transfers (assigns) his or her rights under a contract to another party (called the assignee).

ASSUMPTION OF RISK ■ A doctrine under which a plaintiff may not recover for injuries or damages suffered from risks he or she knows of and has voluntarily assumed.

ATTACHMENT ■ In a secured transaction, the process by which a secured creditor's interest "attaches" to the property of another (collateral) and the creditor's security interest becomes enforceable. In the context of judicial liens, a court-ordered seizure and taking into custody of property prior to the securing of a judgment for a past-due debt.

ATTEMPTED MONOPOLIZATION ■ Any actions by a firm to eliminate competition and gain monopoly power.

AUTOMATIC STAY ■ In bankruptcy proceedings, the suspension of virtually all litigation and other action by creditors against the debtor or the debtor's property. The stay is effective the moment the debtor files a petition in bankruptcy.

AWARD ■ In litigation, the amount of money awarded to a plaintiff in a civil lawsuit as damages. In the context of alternative dispute resolution, the decision rendered by an arbitrator.

B

BAILEE ■ One to whom goods are entrusted by a bailor. Under the UCC, a party who, by a bill of lading, warehouse receipt, or other document of title, acknowledges possession of goods and/or contracts to deliver them.

BAILMENT ■ A situation in which the personal property of one person (a bailor) is entrusted to another (a bailee), who is obligated to return the bailed property to the bailor or dispose of it as directed.

BAILOR ■ One who entrusts goods to a bailee.

BAIT-AND-SWITCH ADVERTISING ■ Advertising a product at a very attractive price (the "bait") and then, once the consumer is in the store, saying that the advertised product is either not available or is of poor quality; the customer is then urged to purchase ("switched" to) a more expensive item.

BANKRUPTCY COURT ■ A federal court of limited jurisdiction that handles only bankruptcy proceedings. Bankruptcy proceedings are governed by federal bankruptcy law.

BATTERY ■ The unprivileged, intentional touching of another.

BEARER ■ A person in possession of an instrument payable to bearer or indorsed in blank.

BEARER INSTRUMENT ■ Any instrument that is not payable to a specific person, including instruments payable to the bearer or to "cash."

BEQUEST ■ A gift by will of personal property (from the verb *to bequeath*).

BEYOND A REASONABLE DOUBT ■ The burden of proof used in criminal cases. If there is any reasonable doubt that a criminal defendant did not commit the crime with which she or he has been charged, then the verdict must be "not guilty."

BILATERAL CONTRACT ■ A type of contract that arises when a promise is given in exchange for a return promise.

BILL OF RIGHTS ■ The first ten amendments to the U.S. Constitution.

BINDER ■ A written, temporary insurance policy.

BINDING AUTHORITY ■ Any source of law that a court must follow when deciding a case. Binding authorities include constitutions, statutes, and regulations that govern the issue being decided, as well as court decisions that are controlling precedents within the jurisdiction.

BLANK INDORSEMENT ■ An indorsement that specifies no particular indorsee and can consist of a mere signature. An order instrument that is indorsed in blank becomes a bearer instrument.

BLUE LAWS ■ State or local laws that prohibit the performance of certain types of commercial activities on Sunday.

BLUE SKY LAWS ■ State laws that regulate the offer and sale of securities.

BONA FIDE OCCUPATIONAL QUALIFICATION (BFOQ) ■ Identifiable characteristics reasonably necessary to the normal operation of a particular business. These characteristics can include gender, national origin, and religion, but not race.

BOND ■ A certificate that evidences a corporate (or government) debt. It is a security that involves no ownership interest in the issuing entity.

BOND INDENTURE ■ A contract between the issuer of a bond and the bondholder.

BOUNTY PAYMENT ■ A reward (payment) given to a person or persons who perform a certain service, such as informing legal authorities of illegal actions.

BREACH ■ The failure to perform a legal obligation.

BREACH OF CONTRACT ■ The failure, without legal excuse, of a promisor to perform the obligations of a contract.

BRIEF ■ A formal legal document submitted by the attorney for the appellant or the appellee (in answer to the appellant's brief) to an appellate court when a case is appealed. The appellant's brief outlines the facts and issues of the case, the judge's rulings or jury's findings that should be reversed or modified, the applicable law, and the arguments on the client's behalf.

BROWSE-WRAP TERMS ■ Terms and conditions of use that are presented to an Internet user at the time certain products, such as software, are being downloaded but that need not be agreed to (by clicking "I agree," for example) before the user is able to install or use the product.

BURGLARY ■ The unlawful entry or breaking into a building with the intent to commit a felony. (Some state statutes expand this to include the intent to commit any crime.)

BUSINESS ETHICS ■ Ethics in a business context; a consensus of what constitutes right or wrong behavior in the world of business and the application of moral principles to situations that arise in a business setting.

BUSINESS INVITEE ■ A person, such as a customer or a client, who is invited onto business premises by the owner of those premises for business purposes.

BUSINESS JUDGMENT RULE ■ A rule that immunizes corporate management from liability for actions that result in corporate losses or damages if the actions are undertaken in good faith and are within both the power of the corporation and the authority of management to make.

BUSINESS NECESSITY ■ A defense to allegations of employment discrimination in which the employer demonstrates that an employment practice that discriminates against members of a protected class is related to job performance.

BUSINESS TORT ■ Wrongful interference with another's business rights.

BUSINESS TRUST ■ A form of business organization in which investors (trust beneficiaries) transfer cash or property to trustees in exchange for trust certificates that represent their investment shares. The certificate holders share in the trust's profits but have limited liability.

BUYOUT PRICE ■ The amount payable to a partner on his or her dissociation from a partnership, based on the amount distributable to that partner if the firm were wound up on that date, and offset by any damages for wrongful dissociation.

BYLAWS ■ A set of governing rules adopted by a corporation or other association.

C

CASE LAW ■ The rules of law announced in court decisions. Case law includes the aggregate of reported cases that interpret judicial precedents, statutes, regulations, and constitutional provisions.

CASHIER'S CHECK ■ A check drawn by a bank on itself.

CATEGORICAL IMPERATIVE ■ A concept developed by the philosopher Immanuel Kant as an ethical guideline for behavior. In deciding whether an action is right or wrong, or desirable or undesirable, a person should evaluate the action in terms of what would happen if everybody else in the same situation, or category, acted the same way.

CAUSATION IN FACT ■ An act or omission without which an event would not have occurred.

CEASE-AND-DESIST ORDER ■ An administrative or judicial order prohibiting a person or business firm from conducting activities that an agency or court has deemed illegal.

CERTIFICATE OF DEPOSIT (CD) ■ A note of a bank in which the bank acknowledges a receipt of money from a party and promises to repay the money, with interest, to the party on a certain date.

CERTIFICATE OF INCORPORATION ■ The primary document that evidences corporate existence (often referred to as the *corporate charter*).

CERTIFICATE OF LIMITED PARTNERSHIP ■ The basic document filed with a designated state official by which a limited partnership is formed.

CERTIFIED CHECK ■ A check that has been accepted in writing by the bank on which it is drawn. Essentially, the bank, by certifying (accepting) the check, promises to pay the check at the time the check is presented.

CHARGING ORDER ■ In partnership law, an order granted by a court to a judgment creditor that entitles the creditor to attach profits or assets of a partner on the dissolution of the partnership.

CHARITABLE TRUST ■ A trust in which the property held by the trustee must be used for a charitable purpose, such as the advancement of health, education, or religion.

CHATTEL ■ All forms of personal property.

CHECK ■ A draft drawn by a drawer ordering the drawee bank or financial institution to pay a fixed amount of money to the holder on demand.

CHECKS AND BALANCES ■ The principle under which the powers of the national government are divided among three separate branches—the executive, legislative, and judicial branches—each of which exercises a check on the actions of the others.

CHOICE-OF-LANGUAGE CLAUSE ■ A clause in a contract designating the official language by which the contract will be interpreted in the event of a future disagreement over the contract's terms.

CHOICE-OF-LAW CLAUSE ■ A clause in a contract designating the law (such as the law of a particular state or nation) that will govern the contract.

CHOSE IN ACTION ■ A right that can be enforced in court to recover a debt or to obtain damages.

CITATION ■ A reference to a publication in which a legal authority—such as a statute or a court decision—or other source can be found.

CIVIL LAW ■ The branch of law dealing with the definition and enforcement of all private or public rights, as opposed to criminal matters.

CIVIL LAW SYSTEM ■ A system of law derived from that of the Roman Empire and based on a code rather than case law; the predominant system of law in the nations of continental Europe and the nations that were once their colonies. In the United States, Louisiana, because of its historical ties to France, has in part a civil law system.

CLEARINGHOUSE ■ A system or place where banks exchange checks and drafts drawn on each other and settle daily balances.

CLICK-ON AGREEMENT ■ An agreement that arises when a buyer, engaging in a transaction on a computer, indicates his or her assent to be bound by the terms of an offer by clicking on a button that says, for example, "I agree"; sometimes referred to as a *click-on license* or a *click-wrap agreement*.

CLOSE CORPORATION ■ A corporation whose shareholders are limited to a small group of persons, often including only family members. In a close corporation, the shareholders' rights to transfer shares to others are usually restricted.

CLOSED SHOP ■ A firm that requires union membership by its workers as a condition of employment. The closed shop was made illegal by the Labor-Management Relations Act of 1947.

CODICIL ■ A written supplement or modification to a will. A codicil must be executed with the same formalities as a will.

COLLATERAL ■ Under Article 9 of the UCC, the property subject to a security interest, including accounts and chattel paper that have been sold.

COLLATERAL PROMISE ■ A secondary promise that is ancillary (subsidiary) to a principal transaction or primary contractual relationship, such as a promise made by one person to pay the debts of another if the latter fails to perform. A collateral promise normally must be in writing to be enforceable.

COLLECTING BANK ■ Any bank handling an item for collection, except the payor bank.

COLLECTIVE BARGAINING ■ The process by which labor and management negotiate the terms and conditions of employment, including working hours and workplace conditions.

COMITY ■ The principle by which one nation defers and gives effect to the laws and judicial decrees of another nation. This recognition is based primarily on respect.

COMMERCE CLAUSE ■ The provision in Article I, Section 8, of the U.S. Constitution that gives Congress the power to regulate interstate commerce.

COMMINGLE ■ To mix together; to put funds or goods together into one mass so that they are so mixed that they no longer have separate identities. In corporate law, if personal and corporate interests are commingled to the extent that the corporation has no separate identity, a court may "pierce the corporate veil" and expose the shareholders to personal liability.

COMMON CARRIER ■ An owner of a truck, railroad, airline, ship, or other vehicle that is licensed to offer transportation services to the public, generally in return for compensation or a payment.

COMMON LAW ■ That body of law developed from custom or judicial decisions in English and U.S. courts, not attributable to a legislature.

COMMON STOCK ■ Shares of ownership in a corporation that give the

owner of the stock a proportionate interest in the corporation with regard to control, earnings, and net assets. Shares of common stock are lowest in priority with respect to payment of dividends and distribution of the corporation's assets on dissolution.

COMMUNITY PROPERTY ■ A form of concurrent ownership of property in which each spouse technically owns an undivided one-half interest in property acquired during the marriage. This form of joint ownership occurs in only ten states and Puerto Rico.

COMPARATIVE NEGLIGENCE ■ A rule in tort law that reduces the plaintiff's recovery in proportion to the plaintiff's degree of fault, rather than barring recovery completely; used in the majority of states.

COMPENSATORY DAMAGES ■ A money award equivalent to the actual value of injuries or damages sustained by the aggrieved party.

COMPLAINT ■ The pleading made by a plaintiff alleging wrongdoing on the part of the defendant; the document that, when filed with a court, initiates a lawsuit.

COMPUTER CRIME ■ Any act that is directed against computers and computer parts, that uses computers as instruments of crime, or that involves computers and constitutes abuse.

COMPUTER INFORMATION ■ As defined by the Uniform Computer Information Transactions Act, "information in an electronic form obtained from or through use of a computer, or that is in digital or an equivalent form capable of being processed by a computer."

CONCURRENT CONDITIONS ■ Conditions that must occur or be performed at the same time; they are mutually dependent. No obligations arise until these conditions are simultaneously performed.

CONCURRENT JURISDICTION ■ Jurisdiction that exists when two different courts have the power to hear a case. For example, some cases can be heard in either a federal or a state court.

CONCURRENT OWNERSHIP ■ Joint ownership.

CONDEMNATION ■ The process of taking private property for public use through the government's power of eminent domain.

CONDITION ■ A qualification, provision, or clause in a contractual agreement, the occurrence or nonoccurrence of which creates, suspends, or terminates the obligations of the contracting parties.

CONDITION PRECEDENT ■ In a contractual agreement, a condition that must be met before a party's promise becomes absolute.

CONDITION SUBSEQUENT ■ A condition in a contract that, if not fulfilled, operates to terminate a party's absolute promise to perform.

CONFESSION OF JUDGMENT ■ The act or agreement of a debtor in permitting a judgment to be entered against him or her by a creditor, for an agreed sum, without the institution of legal proceedings.

CONFISCATION ■ A government's taking of a privately owned business or personal property without a proper public purpose or an award of just compensation.

CONFUSION ■ The mixing together of goods belonging to two or more owners so that the separately owned goods cannot be identified.

CONSENT ■ The voluntary agreement to a proposition or an act of another; a concurrence of wills.

CONSEQUENTIAL DAMAGES ■ Special damages that compensate for a loss that does not directly or immediately result from the breach (for example, lost profits). For the plaintiff to collect consequential damages, they must have been reasonably foreseeable at the time the breach or injury occurred.

CONSIDERATION ■ Generally, the value given in return for a promise. The consideration must result in a detriment to the promisee (something of legally sufficient value and bargained for) or a benefit to the promisor.

CONSIGNMENT ■ A transaction in which an owner of goods (the consignor) delivers the goods to another (the consignee) for the consignee to sell. The consignee pays the consignor only for the goods that are sold by the consignee.

CONSOLIDATION ■ A contractual and statutory process in which two or more corporations join to become a completely new corporation. The original corporations cease to exist, and the new corporation acquires all of their assets and liabilities.

CONSTITUTIONAL LAW ■ The body of law derived from the U.S. Constitution and the constitutions of the various states.

CONSTRUCTIVE DELIVERY ■ An act equivalent to the actual, physical delivery of property that cannot be physically delivered because of difficulty or impossibility. For example, the transfer of a key to a safe constructively delivers the contents of the safe.

CONSTRUCTIVE DISCHARGE ■ A termination of employment brought about by making the employee's working conditions so intolerable that the employee reasonably feels compelled to leave.

CONSTRUCTIVE EVICTION ■ A form of eviction that occurs when a landlord fails to perform adequately any of the undertakings (such as providing heat in the winter) required by the lease, thereby making the tenant's further use and enjoyment of the property exceedingly difficult or impossible.

CONSTRUCTIVE TRUST ■ An equitable trust that is imposed in the interests of fairness and justice when someone wrongfully holds legal title to property. A court may require the owner to hold the property in trust for the person or persons who rightfully should own the property.

CONSUMER-DEBTOR ■ An individual whose debts are primarily consumer debts (debts for purchases made primarily for personal or household use).

CONTINUATION STATEMENT ■ A statement that, if filed within six months prior to the expiration date of the original financing statement, continues the perfection of the original security interest for another five years. The perfection of a security interest can be continued in the same manner indefinitely.

CONTRACT ■ An agreement that can be enforced in court; formed by two or more competent parties who agree, for consideration, to perform or to refrain from performing some legal act now or in the future.

CONTRACTUAL CAPACITY ■ The threshold mental capacity required by law for a party who enters into a contract to be bound by that contract.

CONTRIBUTORY NEGLIGENCE ■ A rule in tort law that completely bars the plaintiff from recovering any damages if the damage suffered is partly the plaintiff's own fault; used in a minority of states.

CONVERSION ■ Wrongfully taking or retaining possession of an individual's personal property and placing it in the service of another.

CONVEYANCE ■ The transfer of a title to land from one person to another by deed; a document (such as a deed) by which an interest in land is transferred from one person to another.

"COOLING-OFF" LAWS ■ Laws that allow buyers a period of time, such as three days, in which to cancel door-to-door sales contracts.

COOPERATIVE ■ An association, which may or may not be incorporated, that is organized to provide an economic service to its members. Unincorporated cooperatives are often treated like partnerships for tax and other legal purposes. Examples of cooperatives include consumer purchasing cooperatives, credit cooperatives, and farmers' cooperatives.

COPYRIGHT ■ The exclusive right of "authors" to publish, print, or sell an intellectual production for a statutory period of time. A copyright has the same monopolistic nature as a patent or trademark, but it differs in that it applies exclusively to works of art, literature, and other works of authorship (including computer programs).

CORPORATE CHARTER ■ The document issued by a state agency or authority (usually the secretary of state) that grants a corporation legal existence and the right to function.

CORPORATION ■ A legal entity formed in compliance with statutory requirements that is distinct from its shareholder-owners.

CORRESPONDENT BANK ■ A bank in which another bank has an account (and vice versa) for the purpose of facilitating fund transfers.

COST-BENEFIT ANALYSIS ■ A decision-making technique that involves weighing the costs of a given action against the benefits of that action.

CO-SURETY ■ A joint surety; a person who assumes liability jointly with another surety for the payment of an obligation.

COUNTERADVERTISING ■ New advertising that is undertaken pursuant to a Federal Trade Commission order for the purpose of correcting earlier false claims that were made about a product.

COUNTERCLAIM ■ A claim made by a defendant in a civil lawsuit against the plaintiff. In effect, the defendant is suing the plaintiff.

COUNTEROFFER ■ An offeree's response to an offer in which the offeree rejects the original offer and at the same time makes a new offer.

COURSE OF DEALING ■ Prior conduct between the parties to a contract that establishes a common basis for their understanding.

COURSE OF PERFORMANCE ■ The conduct that occurs under the terms of a particular agreement. Such conduct indicates what the parties to an agreement intended it to mean.

COVENANT NOT TO COMPETE ■ A contractual promise of one party to refrain from conducting business similar to that of another party for a certain period of time and within a specified geographic area. Courts commonly enforce such covenants if they are reasonable in terms of time and geographic area and part of, or supplemental to, an employment contract or a contract for the sale of a business.

COVENANT NOT TO SUE ■ An agreement to substitute a contractual obligation for some other type of legal action based on a valid claim.

COVER ■ Under the UCC, a remedy that allows the buyer or lessee, on the seller's or lessor's breach, to purchase the goods, in good faith and within a reasonable time, from another seller or lessor and substitute them for the goods due under the contract. If the cost of cover exceeds the cost of the contract goods, the breaching seller or lessor will be liable to the buyer or lessee for the difference, plus incidental and consequential damages.

CRAM-DOWN PROVISION ■ A provision of the Bankruptcy Code that allows a court to confirm a debtor's Chapter 11 reorganization plan even though only one class of creditors has accepted it. To exercise the court's right under this provision, the court must demonstrate that the plan does not discriminate unfairly against any creditors and is fair and equitable.

CREDITORS' COMPOSITION AGREEMENT ■ An agreement formed between a debtor and his or her creditors in which the creditors agree to accept a lesser sum than that owed by the debtor in full satisfaction of the debt.

CRIME ■ A wrong against society proclaimed in a statute and, if committed, punishable by society through fines and/or imprisonment–and, in some cases, death.

CRIMINAL LAW ■ Law that defines and governs actions that constitute crimes. Generally, criminal law has to do with wrongful actions committed against society for which society demands redress.

CURE ■ The right of a party who tenders nonconforming performance to correct that performance within the contract period [UCC 2–508(1)].

CYBER CRIME ■ A crime that occurs online, in the virtual community of the Internet, as opposed to the physical world.

CYBER MARK ■ A trademark in cyberspace.

CYBER TORT ■ A tort committed in cyberspace.

CYBERLAW ■ An informal term used to refer to all laws governing electronic communications and transactions, particularly those conducted via the Internet.

CYBERNOTARY ■ A legally recognized authority that can certify the validity of digital signatures.

CYBERSQUATTING ■ The act of registering a domain name that is the same as, or confusingly similar to, the trademark of another and then offering to sell that domain name back to the trademark owner.

CYBERSTALKER ■ A person who commits the crime of stalking in cyberspace. Generally, stalking consists of harassing a person and putting that person in reasonable fear for his or her safety or the safety of the person's immediate family.

CYBERTERRORIST ■ A hacker whose purpose is to exploit a target computer for a serious impact, such as corrupting a program to sabotage a business.

D

DAMAGES ■ Money sought as a remedy for a breach of contract or a tortious action.

DEBTOR ■ Under Article 9 of the UCC, a debtor is any party who owes payment or performance of a secured obligation, whether or not the party actually owns or has rights in the collateral.

DEBTOR IN POSSESSION (DIP) ■ In Chapter 11 bankruptcy proceedings, a debtor who is allowed to continue in possession of the estate in property (the business) and to continue business operations.

DECEPTIVE ADVERTISING ■ Advertising that misleads consumers, either by unjustified claims concerning a product's performance or by the omission of a material fact concerning the product's composition or performance.

DEED ■ A document by which title to property (usually real property) is passed.

DEFALCATION ■ Embezzlement; the misappropriation of funds by a party, such as a corporate officer or a public official, in a fiduciary relationship with another.

DEFAMATION ■ Anything published or publicly spoken that causes injury to another's good name, reputation, or character.

DEFAULT ■ The failure to observe a promise or discharge an obligation. The term is commonly used to mean the failure to pay a debt when it is due.

DEFAULT JUDGMENT ■ A judgment entered by a court against a defendant who has failed to appear in court to answer or defend against the plaintiff's claim.

DEFENDANT ■ One against whom a lawsuit is brought; the accused person in a criminal proceeding.

DEFENSE ■ A reason offered and alleged by a defendant in an action or suit as to why the plaintiff should not recover or establish what she or he seeks.

DEFICIENCY JUDGMENT ■ A judgment against a debtor for the amount of a debt remaining unpaid after the collateral has been repossessed and sold.

DELEGATEE ■ A party to whom contractual obligations are transferred, or delegated.

DELEGATION OF DUTIES ■ The act of transferring to another all or part of one's duties arising under a contract.

DELEGATOR ■ A party who transfers (delegates) her or his obligations under a contract to another party (called the delegatee).

DEPOSITARY BANK ■ The first bank to receive a check for payment.

DEPOSITION ■ The testimony of a party to a lawsuit or a witness taken under oath before a trial.

DESTINATION CONTRACT ■ A contract for the sale of goods in which the seller is required or authorized to ship the goods by carrier and tender delivery of the goods at a particular destination. The seller assumes liability for any losses or damage to the goods until they are tendered at the destination specified in the contract.

DEVISE ■ To make a gift of real property by will.

DEVISEE ■ One designated in a will to receive a gift of real property.

DIGITAL CASH ■ Funds contained on computer software, in the form of secure programs stored on microchips and other computer devices.

DILUTION ■ With respect to trademarks, a doctrine under which distinctive or famous trademarks are protected from certain unauthorized uses of the marks regardless of a showing of competition or a likelihood of confusion. Congress created a federal cause of action for dilution in 1995 with the passage of the Federal Trademark Dilution Act.

DISAFFIRMANCE ■ The legal avoidance, or setting aside, of a contractual obligation.

DISCHARGE ■ The termination of an obligation. In bankruptcy proceedings, the extinction of the debtor's dischargeable debts. In contract law, discharge occurs when the parties have fully performed their contractual obligations or when events, conduct of the parties, or operation of law releases the parties from performance.

DISCLOSED PRINCIPAL ■ A principal whose identity is known to a third party at the time the agent makes a contract with the third party.

DISCOVERY ■ A phase in the litigation process during which the opposing parties may obtain information from each other and from third parties prior to trial.

DISPARAGEMENT OF PROPERTY ■ An economically injurious falsehood made about another's product or property. A general term for torts that are more specifically referred to as slander of quality or slander of title.

DISPARATE-IMPACT DISCRIMINATION ■ A form of employment discrimination that results from certain employer practices or procedures that, although not discriminatory on their face, have a discriminatory effect.

DISPARATE-TREATMENT DISCRIMINATION ■ A form of employment discrimination that results when an employer intentionally discriminates against employees who are members of protected classes.

DISSOCIATION ■ The severance of the relationship between a partner and a partnership when the partner ceases to be associated with the carrying on of the partnership business.

DISSOLUTION ■ The formal disbanding of a partnership or a corporation. It can take place by (1) acts of the partners or, in a corporation, of the shareholders and board of directors; (2) the subsequent illegality of the partnership business; (3) the expiration of a time period stated in a partnership agreement or a certificate of incorporation; or (4) judicial decree.

DISTRIBUTED NETWORK ■ A network that can be used by persons located (distributed) around the country or the globe to share computer files.

DISTRIBUTION AGREEMENT ■ A contract between a seller and a distributor of the seller's products setting out the terms and conditions of the distributorship.

DIVERSITY OF CITIZENSHIP ■ Under Article III, Section 2, of the Constitution, a basis for federal district court jurisdiction over a lawsuit between (1) citizens of different states, (2) a foreign country and citizens of a state or of different states, or (3) citizens of a state and citizens or subjects of a foreign country. The amount in controversy must be more than $75,000 before a federal district court can take jurisdiction in such cases.

DIVESTITURE ■ The act of selling one or more of a company's divisions or parts, such as a subsidiary or plant; often mandated by the courts in merger or monopolization cases.

DIVIDEND ■ A distribution to corporate shareholders of corporate profits or income, disbursed in proportion to the number of shares held.

DOCKET ■ The list of cases entered on a court's calendar and thus scheduled to be heard by the court.

DOCUMENT OF TITLE ■ Paper exchanged in the regular course of business that evidences the right to possession of goods (for example, a bill of lading or a warehouse receipt).

DOMAIN NAME ■ The last part of an Internet address, such as "westlaw.com." The top level (the part of the name to the right of the period) indicates the type of entity that operates the site ("com" is an abbreviation for "commercial"). The second level (the part of the name to the left of the period) is chosen by the entity.

DOMESTIC CORPORATION ■ In a given state, a corporation that does business in, and is organized under the law of, that state.

DOMINION ■ Ownership rights in property, including the right to possess and control the property.

DOUBLE JEOPARDY ■ A situation occurring when a person is tried twice for the same criminal offense; prohibited by the Fifth Amendment to the Constitution.

DRAFT ■ Any instrument drawn on a drawee that orders the drawee to pay a certain sum of money, usually to a third party (the payee), on demand or at a definite future time.

DRAM SHOP ACT ■ A state statute that imposes liability on the owners of bars and taverns, as well as those who serve alcoholic drinks to the public, for injuries resulting from accidents caused by intoxicated persons when the sellers or servers of alcoholic drinks contributed to the intoxication.

DRAWEE ■ The party that is ordered to pay a draft or check. With a check, a bank or a financial institution is always the drawee.

DRAWER ■ The party that initiates a draft (such as a check), thereby ordering the drawee to pay.

DUE DILIGENCE ■ A required standard of care that certain professionals, such as accountants, must meet to avoid liability for securities violations.

DUE PROCESS CLAUSE ■ The provisions in the Fifth and Fourteenth Amendments to the Constitution that guarantee that no person shall be deprived of life, liberty, or property without due process of law. Similar clauses are found in most state constitutions.

DUMPING ■ The selling of goods in a foreign country at a price below the price charged for the same goods in the domestic market.

DURESS ■ Unlawful pressure brought to bear on a person, causing the person to perform an act that she or he would not otherwise perform.

DUTY OF CARE ■ The duty of all persons, as established by tort law, to exercise a reasonable amount of care in their dealings with others. Failure to exercise due care, which is normally determined by the "reasonable person standard," constitutes the tort of negligence.

E-AGENT ■ A computer program that by electronic or other automated means can independently initiate an action or respond to electronic messages or data without review by an individual.

EARLY NEUTRAL CASE EVALUATION ■ A form of alternative dispute resolution in which a neutral third party evaluates the strengths and weaknesses of the disputing parties' positions. The evaluator's opinion then forms the basis for negotiating a settlement.

EASEMENT ■ A nonpossessory right to use another's property in a manner established by either express or implied agreement.

E-CONTRACT ■ A contract that is formed electronically.

E-EVIDENCE ■ A type of evidence that consists of computer-generated or electronically recorded information, including e-mail, voice mail, spreadsheets, word-processing documents, and other data.

ELECTRONIC FUND TRANSFER (EFT) ■ A transfer of funds through the use of an electronic terminal, a telephone, a computer, or magnetic tape.

EMANCIPATION ■ In regard to minors, the act of being freed from parental control; occurs when a child's parent or legal guardian relinquishes the legal right to exercise control over the child. Normally, a minor who leaves home to support himself or herself is considered emancipated.

EMBEZZLEMENT ■ The fraudulent appropriation of funds or other property by a person to whom the funds or property has been entrusted.

EMINENT DOMAIN ■ The power of a government to take land for public use from private citizens for just compensation.

E-MONEY ■ Prepaid funds recorded on a computer or a card (such as a smart card or a stored-value card).

EMPLOYMENT AT WILL ■ A common law doctrine under which either party may terminate an employment relationship at any time for any reason, unless a contract specifies otherwise.

EMPLOYMENT CONTRACT ■ A contract between an employer and an employee in which the terms and conditions of employment are stated.

EMPLOYMENT DISCRIMINATION ■ Treating employees or job applicants unequally on the basis of race, color, national origin, religion, gender, age, or disability; prohibited by federal statutes.

ENABLING LEGISLATION ■ A statute enacted by Congress that authorizes the creation of an administrative agency and specifies the name, composition, purpose, and powers of the agency being created.

ENTRAPMENT ■ In criminal law, a defense in which the defendant claims that he or she was induced by a public official—usually an undercover agent or police officer—to commit a crime that he or she would otherwise not have committed.

ENTREPRENEUR ■ One who initiates and assumes the financial risk of a new business enterprise and undertakes to provide or control its management.

ENVIRONMENTAL IMPACT STATEMENT (EIS) ■ A statement required by the National Environmental Policy Act for any major federal action that will significantly affect the quality of the environment. The statement must analyze the action's impact on the environment and explore alternative actions that might be taken.

EQUAL DIGNITY RULE ■ In most states, a rule stating that express authority given to an agent must be in writing if the contract to be made on behalf of the principal is required to be in writing.

EQUAL PROTECTION CLAUSE ■ The provision in the Fourteenth Amendment to the Constitution that guarantees that no state will "deny to any person within its jurisdiction the equal protection of the laws." This clause mandates that the state governments treat similarly situated individuals in a similar manner.

EQUITABLE PRINCIPLES AND MAXIMS ■ General propositions or principles of law that have to do with fairness (equity).

E-SIGNATURE ■ As defined by the Uniform Electronic Transactions Act, "an electronic sound, symbol, or process attached to or logically associated with a record and executed or adopted by a person with the intent to sign the record."

ESTABLISHMENT CLAUSE ■ The provision in the First Amendment to the Constitution that prohibits the government from establishing any state-sponsored religion or enacting any law that promotes religion or favors one religion over another.

ESTATE IN PROPERTY ■ In bankruptcy proceedings, all of the debtor's interests in property currently held, wherever located, together with certain jointly owned property, property transferred in transactions voidable by the trustee, proceeds and profits from the property of the estate, and certain property interests to which the debtor becomes entitled within 180 days after filing for bankruptcy.

ESTOPPED ■ Barred, impeded, or precluded.

ESTRAY STATUTE ■ A statute defining finders' rights in property when the true owners are unknown.

ETHICAL REASONING ■ A reasoning process in which an individual links his or her moral convictions or ethical standards to the particular situation at hand.

ETHICS ■ Moral principles and values applied to social behavior.

EVICTION ■ A landlord's act of depriving a tenant of possession of the leased premises.

EXCLUSIONARY RULE ■ In criminal procedure, a rule under which any evidence that is obtained in violation of the accused's constitutional rights guaranteed by the Fourth, Fifth, and Sixth Amendments, as well as any evidence derived from illegally obtained evidence, will not be admissible in court.

EXCLUSIVE DISTRIBUTORSHIP ■ A distributorship in which the seller and the distributor of the seller's products agree that the distributor has the exclusive right to distribute the seller's products in a certain geographic area.

EXCLUSIVE JURISDICTION ■ Jurisdiction that exists when a case can be heard only in a particular court or type of court.

EXCLUSIVE-DEALING CONTRACT ■ An agreement under which a seller forbids a buyer to purchase products from the seller's competitors.

EXCULPATORY CLAUSE ■ A clause that releases a contractual party from liability in the event of monetary or physical injury, no matter who is at fault.

EXECUTED CONTRACT ■ A contract that has been completely performed by both parties.

EXECUTION ■ An action to carry into effect the directions in a court decree or judgment.

EXECUTIVE AGENCY ■ An administrative agency within the executive branch of government. At the federal level, executive agencies are those within the cabinet departments.

EXECUTOR ■ A person appointed by a testator in a will to see that her or his will is administered appropriately.

EXECUTORY CONTRACT ■ A contract that has not as yet been fully performed.

EXPORT ■ To sell products to buyers located in other countries.

EXPRESS CONTRACT ■ A contract in which the terms of the agreement are stated in words, oral or written.

EXPRESS WARRANTY ■ A seller's or lessor's oral or written promise or affirmation of fact ancillary to an underlying sales or lease agreement regarding the quality, description, or performance of the goods being sold or leased.

EXPROPRIATION ■ The seizure by a government of a privately owned business or personal property for a proper public purpose and with just compensation.

EXTENSION CLAUSE ■ A clause in a time instrument that allows the instrument's date of maturity to be extended into the future.

F

FAMILY LIMITED LIABILITY PARTNERSHIP (FLLP) ■ A type of limited liability partnership owned by family members or fiduciaries of family members.

FEDERAL FORM OF GOVERNMENT ■ A system of government in which the states form a union and the sovereign power is divided between the central government and the member states.

FEDERAL QUESTION ■ A question that pertains to the U.S. Constitution, acts of Congress, or treaties. A federal question provides a basis for federal jurisdiction.

FEDERAL RESERVE SYSTEM ■ A network of twelve district banks and related branches located around the country and headed by the Federal Reserve Board of Governors. Most banks in the United States have Federal Reserve accounts.

FEE SIMPLE ■ An absolute form of property ownership entitling the property owner to use, possess, or dispose of the property as he or she chooses during his or her lifetime. On death, the interest in the property descends to the owner's heirs.

FEE SIMPLE ABSOLUTE ■ An ownership interest in land in which the owner has the greatest possible aggregation of rights, privileges, and power. Ownership in fee simple absolute is limited absolutely to a person and her or his heirs.

FELONY ■ A crime—such as arson, murder, rape, or robbery—that carries the most severe sanctions, which range from more than one year in a state or federal prison to the death penalty.

FICTITIOUS PAYEE ■ A payee on a negotiable instrument whom the

maker or drawer does not intend to have an interest in the instrument. Indorsements by fictitious payees are treated as authorized indorsements under Article 3 of the UCC.

FIDUCIARY ■ As a noun, a person having a duty created by his or her undertaking to act primarily for another's benefit in matters connected with the undertaking. As an adjective, a relationship founded on trust and confidence.

FILTERING SOFTWARE ■ A computer program that includes a pattern through which data are passed. When designed to block access to certain Web sites, the pattern blocks the retrieval of a site whose URL or key words are on a list within the program.

FINANCING STATEMENT ■ A document prepared by a secured creditor, and filed with the appropriate state or local official, to give notice to the public that the creditor has a security interest in collateral belonging to the debtor named in the statement.

FIRM OFFER ■ An offer (by a merchant) that is irrevocable without consideration for a stated period of time or, if no definite period is stated, for a reasonable time (neither period to exceed three months). A firm offer by a merchant must be in writing and must be signed by the offeror.

FIXTURE ■ A thing that was once personal property but has become attached to real property in such a way that it takes on the characteristics of real property and becomes part of that real property.

FLOATING LIEN ■ A security interest in proceeds, after-acquired property, or collateral subject to future advances by the secured party (or all three); a security interest in collateral that is retained even when the collateral changes in character, classification, or location.

FORBEARANCE ■ The act of refraining from an action that one has a legal right to undertake.

***FORCE MAJEURE* CLAUSE** ■ A provision in a contract stipulating that certain unforeseen events—such as war, political upheavals, or acts of God—will excuse a party from liability for nonperformance of contractual obligations.

FOREIGN CORPORATION ■ In a given state, a corporation that does business in the state without being incorporated therein.

FOREIGN EXCHANGE MARKET ■ A worldwide system in which foreign currencies are bought and sold.

FORGERY ■ The fraudulent making or altering of any writing in a way that changes the legal rights and liabilities of another.

FORMAL CONTRACT ■ A contract that by law requires a specific form, such as being executed under seal, to be valid.

FORUM-SELECTION CLAUSE ■ A provision in a contract designating the court, jurisdiction, or tribunal that will decide any disputes arising under the contract.

FRANCHISE ■ Any arrangement in which the owner of a trademark, trade name, or copyright licenses another to use that trademark, trade name, or copyright in the selling of goods and services.

FRANCHISEE ■ One receiving a license to use another's (the franchisor's) trademark, trade name, or copyright in the sale of goods and services.

FRANCHISOR ■ One licensing another (the franchisee) to use the owner's trademark, trade name, or copyright in the sale of goods or services.

FRAUDULENT MISREPRESENTATION ■ Any misrepresentation, either by misstatement or omission of a material fact, knowingly made with the intention of deceiving another and on which a reasonable person would and does rely to his or her detriment.

FREE EXERCISE CLAUSE ■ The provision in the First Amendment to the Constitution that prohibits the government from interfering with people's religious practices or forms of worship.

FUNGIBLE GOODS ■ Goods that are alike by physical nature, by agreement, or by trade usage. Examples of fungible goods are wheat, oil, and wine that are identical in type and quality. When owners of fungible goods hold the goods as tenants in common, title and risk can pass without actually separating the goods being sold from the mass of fungible goods.

G

GARNISHMENT ■ A legal process used by a creditor to collect a debt by seizing property of the debtor (such as wages) that is being held by a third party (such as the debtor's employer).

GENERAL PARTNER ■ In a limited partnership, a partner who assumes responsibility for the management of the partnership and liability for all partnership debts.

GENERALLY ACCEPTED ACCOUNTING PRINCIPLES (GAAP) ■ The conventions, rules, and procedures necessary to define accepted accounting practices at a particular time. The source of the principles is the Financial Accounting Standards Board (FASB).

GENERALLY ACCEPTED AUDITING STANDARDS (GAAS) ■ Standards concerning an auditor's professional qualities and the judgment exercised by him or her in the performance of an examination and report. The source of the standards is the American Institute of Certified Public Accountants.

GIFT ■ Any voluntary transfer of property made without consideration, past or present.

GIFT *CAUSA MORTIS* ■ A gift made in contemplation of death. If the donor does not die of that ailment, the gift is revoked.

GIFT *INTER VIVOS* ■ A gift made during one's lifetime and not in contemplation of imminent death, in contrast to a gift *causa mortis*.

GOOD FAITH PURCHASER ■ A purchaser who buys without notice of any circumstance that would put a person of ordinary prudence on inquiry as to whether the seller has valid title to the goods being sold.

GOOD SAMARITAN STATUTE ■ A state statute stipulating that persons who provide emergency services to, or rescue, someone in peril cannot be sued for negligence, unless they act recklessly, thereby causing further harm.

GRAND JURY ■ A group of citizens called to decide, after hearing the state's evidence, whether a reasonable basis (probable cause) exists for believing that a crime has been committed and whether a trial ought to be held.

GROUP BOYCOTT ■ The refusal by a group of competitors to deal with a particular person or firm; prohibited by the Sherman Act.

GUARANTOR ■ A person who agrees to satisfy the debt of another (the debtor) only after the principal debtor defaults. Thus, a guarantor's liability is secondary.

H

HACKER ■ A person who uses one computer to break into another. Professional computer programmers refer to such persons as "crackers."

HISTORICAL SCHOOL ■ A school of legal thought that emphasizes the evolutionary process of law and that looks to the past to discover what the principles of contemporary law should be.

HOLDER ■ Any person in possession of an instrument drawn, issued, or indorsed to him or her, to his or her order, to bearer, or in blank.

HOLDER IN DUE COURSE (HDC) ■ A holder who acquires a negotiable instrument for value; in good faith; and without notice that the instrument is overdue, that it has been dishonored, that any person has a defense against it or a claim to it, or that the instrument contains unauthorized signatures, has been altered, or is so irregular or incomplete as to call into question its authenticity.

HOLOGRAPHIC WILL ■ A will written entirely in the signer's handwriting and usually not witnessed.

HOMESTEAD EXEMPTION ■ A law permitting a debtor to retain the family home, either in its entirety or up to a specified dollar amount, free from the claims of unsecured creditors or trustees in bankruptcy.

HORIZONTAL MERGER ■ A merger between two firms that are competing in the same marketplace.

HORIZONTAL RESTRAINT ■ Any agreement that in some way restrains competition between rival firms competing in the same market.

HOT-CARGO AGREEMENT ■ An agreement in which employers voluntarily agree with unions not to handle, use, or deal in other employers' goods that were not produced by union employees; a type of secondary boycott explicitly prohibited by the Labor-Management Reporting and Disclosure Act of 1959.

I

IDENTIFICATION ■ In a sale of goods, the express designation of the goods provided for in the contract.

IDENTITY THEFT ■ The act of stealing another's identifying information—such as a name, date of birth, or Social Security number—and using that information to access the victim's financial resources.

IMPLIED-IN-FACT CONTRACT ■ A contract formed in whole or in part from the conduct of the parties (as opposed to an express contract).

IMPLIED WARRANTY ■ A warranty that arises by law because of the circumstances of a sale, rather than by the seller's express promise.

IMPLIED WARRANTY OF FITNESS FOR A PARTICULAR PURPOSE ■ A warranty that goods sold or leased are fit for a particular purpose. The warranty arises when any seller or lessor knows the particular purpose for which a buyer or lessee will use the goods and knows that the buyer or lessee is relying on the skill and judgment of the seller or lessor to select suitable goods.

IMPLIED WARRANTY OF HABITABILITY ■ An implied promise by a landlord that rented residential premises are fit for human habitation—that is, in a condition that is safe and suitable for people to live in.

IMPLIED WARRANTY OF MERCHANTABILITY ■ A warranty that goods being sold or leased are reasonably fit for the general purpose for which they are sold or leased, are properly packaged and labeled, and are of proper quality. The warranty automatically arises in every sale or lease of goods made by a merchant who deals in goods of the kind sold or leased.

IMPOSSIBILITY OF PERFORMANCE ■ A doctrine under which a party to a contract is relieved of his or her duty to perform when performance becomes objectively impossible or totally impracticable (through no fault of either party).

IMPOSTER ■ One who, by use of the mails, Internet, telephone, or personal appearance, induces a maker or drawer to issue an instrument in the name of an impersonated payee. Indorsements by imposters are treated as authorized indorsements under Article 3 of the UCC.

INCIDENTAL BENEFICIARY ■ A third party who incidentally benefits from a contract but whose benefit was not the reason the contract was formed. An incidental beneficiary has no rights in a contract and cannot sue to have the contract enforced.

INCIDENTAL DAMAGES ■ All costs resulting from a breach of contract, including all reasonable expenses incurred because of the breach.

INDEPENDENT CONTRACTOR ■ One who works for, and receives payment from, an employer but whose working conditions and methods are not controlled by the employer. An independent contractor is not an employee but may be an agent.

INDEPENDENT REGULATORY AGENCY ■ An administrative agency that is not considered part of the government's executive branch and is not subject to the authority of the president. Independent agency officials cannot be removed without cause.

INDICTMENT ■ A charge by a grand jury that a named person has committed a crime.

INDORSEE ■ The person to whom a negotiable instrument is transferred by indorsement.

INDORSEMENT ■ A signature placed on an instrument for the purpose of transferring one's ownership rights in the instrument.

INDORSER ■ A person who transfers an instrument by signing (indorsing) it and delivering it to another person.

INFORMAL CONTRACT ■ A contract that does not require a specified form or formality to be valid.

INFORMATION ■ A formal accusation or complaint (without an indictment) issued in certain types of actions (usually criminal actions involving lesser crimes) by a law officer, such as a magistrate.

INFORMATION RETURN ■ A tax return submitted by a partnership that reports only the income and losses earned by the business. The partnership as an entity does not pay taxes on the income received by the partnership. A partner's profit from the partnership (whether distributed or not) is taxed as individual income to the individual partner.

INNKEEPER'S LIEN ■ A possessory lien placed on the luggage of hotel guests for hotel charges that remain unpaid.

INSIDER TRADING ■ The purchase or sale of securities on the basis of "inside information" (information that has not been made available to the public).

INSOLVENT ■ Under the UCC, a term describing a person who ceases to pay "his [or her] debts in the ordinary course of business or cannot pay his [or her] debts as they become due or is insolvent within the meaning of federal bankruptcy law" [UCC 1–201(23)].

INSTALLMENT CONTRACT ■ Under the UCC, a contract that requires or authorizes delivery in two or more separate lots to be accepted and paid for separately.

INSURABLE INTEREST ■ An interest either in a person's life or well-being or in property that is sufficiently substantial that insuring against injury to (or the death of) the person or against damage to the property does not amount to a mere wagering (betting) contract.

INSURANCE ■ A contract in which, for a stipulated consideration, one party agrees to compensate the other for loss on a specific subject by a specified peril.

INTANGIBLE PROPERTY ■ Property that cannot be seen or touched but exists only conceptually, such as corporate stocks and bonds, patents and copyrights, and ordinary contract rights. Article 2 of the UCC does not govern intangible property.

INTEGRATED CONTRACT ■ A written contract that constitutes the final expression of the parties' agreement. If a contract is integrated, evidence extraneous to the contract that contradicts or alters the meaning of the contract in any way is inadmissible.

INTELLECTUAL PROPERTY ■ Property resulting from intellectual, creative processes.

INTENDED BENEFICIARY ■ A third party for whose benefit a contract is formed. An intended beneficiary can sue the promisor if such a contract is breached.

INTENTIONAL TORT ■ A wrongful act knowingly committed.

***INTER VIVOS* TRUST** ■ A trust created by the grantor (settlor) and effective during the grantor's lifetime; a trust not established by a will.

INTERMEDIARY BANK ■ Any bank to which an item is transferred in the course of collection, except the depositary or payor bank.

INTERNATIONAL LAW ■ The law that governs relations among nations. National laws, customs, treaties, and international conferences and organizations are generally considered to be the most important sources of international law.

INTERROGATORIES ■ A series of written questions for which written answers are prepared, usually with the assistance of the party's attorney, and then signed under oath by a party to a lawsuit.

INTESTACY LAWS ■ State statutes that specify how property will be distributed when a person dies intestate (without a valid will); also called statutes of descent and distribution.

INTESTATE ■ As a noun, one who has died without having created a valid will; as an adjective, the state of having died without a will.

INVESTMENT COMPANY ■ A company that acts on behalf of many smaller shareholders/owners by buying a large portfolio of securities and professionally managing that portfolio.

JOINT AND SEVERAL LIABILITY ■ In partnership law, a doctrine under which a plaintiff may sue, and collect a judgment from all of the partners together (jointly), or one or more of the partners separately (severally, or individually). This is true even if one of the partners sued did not participate in, ratify, or know about whatever it was that gave rise to the cause of action.

JOINT STOCK COMPANY ■ A hybrid form of business organization that combines characteristics of a corporation and a partnership. Usually, a joint stock company is regarded as a partnership for tax and other legal purposes.

JOINT TENANCY ■ The joint ownership of property by two or more co-owners in which each co-owner owns an undivided portion of the property. On the death of one of the joint tenants, his or her interest automatically passes to the surviving joint tenant(s).

JOINT VENTURE ■ A joint undertaking of a specific commercial enterprise by an association of persons. A joint venture normally is not a legal entity and is treated like a partnership for federal income tax purposes.

JUDICIAL REVIEW ■ The process by which a court decides on the constitutionality of legislative enactments and actions of the executive branch.

JURISDICTION ■ The authority of a court to hear and decide a specific action.

JURISPRUDENCE ■ The science or philosophy of law.

JUSTICIABLE CONTROVERSY ■ A controversy that is not hypothetical or academic but real and substantial; a requirement that must be satisfied before a court will hear a case.

LARCENY ■ The wrongful taking and carrying away of another person's personal property with the intent to permanently deprive the owner of the property. Some states classify larceny as either grand or petit, depending on the property's value.

LAW ■ A body of enforceable rules governing relationships among individuals and between individuals and their society.

LEASE AGREEMENT ■ In regard to the lease of goods, an agreement in which one person (the lessor) agrees to transfer the right to the possession and use of property to another person (the lessee) in exchange for rental payments.

LEASEHOLD ESTATE ■ An estate in realty held by a tenant under a lease. In every leasehold estate, the tenant has a qualified right to possess and/or use the land.

LEGACY ■ A gift of personal property under a will.

LEGAL POSITIVISM ■ A school of legal thought centered on the assumption that there is no law higher than the laws created by a national government. Laws must be obeyed, even if they are unjust, to prevent anarchy.

LEGAL REALISM ■ A school of legal thought of the 1920s and 1930s that generally advocated a less abstract and more realistic approach to the law, an approach that takes into account customary practices and the circumstances in which transactions take place. This school left a lasting imprint on American jurisprudence.

LEGATEE ■ One designated in a will to receive a gift of personal property.

LESSEE ■ A person who acquires the right to the possession and use of another's goods in exchange for rental payments.

LESSOR ■ A person who sells the right to the possession and use of goods to another in exchange for rental payments.

LETTER OF CREDIT ■ A written instrument, usually issued by a bank on behalf of a customer or other person, in which the issuer promises to honor drafts or other demands for payment by third persons in accordance with the terms of the instrument.

LEVY ■ The obtaining of money by legal process through the seizure and sale of nonsecured property, usually done after a writ of execution has been issued.

LIBEL ■ Defamation in writing or other permanent form (such as a digital recording) having the quality of permanence.

LICENSE ■ A revocable right or privilege of a person to come onto another person's land.

LIEN ■ An encumbrance on a property to satisfy a debt or protect a claim for payment of a debt.

LIFE ESTATE ■ An interest in land that exists only for the duration of the life of some person, usually the holder of the estate.

LIMITED LIABILITY COMPANY (LLC) ■ A hybrid form of business enterprise that offers the limited liability of a corporation and the tax advantages of a partnership.

LIMITED LIABILITY LIMITED PARTNERSHIP (LLLP) ■ A type of limited partnership in which the liability of all of the partners, including general partners, is limited to the amount of their investments.

LIMITED LIABILITY PARTNERSHIP (LLP) ■ A hybrid form of business organization that is used mainly by professionals who normally do business in a partnership. Like a partnership, an LLP is a pass-through entity for tax purposes, but the personal liability of the partners is limited.

LIMITED PARTNER ■ In a limited partnership, a partner who contributes capital to the partnership but has no right to participate in the management and operation of the business. The limited partner assumes no liability for partnership debts beyond the capital contributed.

LIMITED PARTNERSHIP (LP) ■ A partnership consisting of one or more general partners (who manage the business and are liable to the full extent of their personal assets for debts of the partnership) and one or more limited partners (who contribute only assets and are liable only up to the extent of their contributions).

LIQUIDATED DAMAGES ■ An amount, stipulated in a contract, that the parties to the contract believe to be a reasonable estimation of the damages that will occur in the event of a breach.

LIQUIDATION ■ In regard to corporations, the process by which corporate assets are converted into cash and distributed among creditors and shareholders according to specific rules of preference. In bankruptcy proceedings, the sale of all of the nonexempt assets of a debtor and the distribution of the proceeds to the debtor's creditors. Chapter 7 of the Bankruptcy Code provides for liquidation bankruptcy proceedings.

LITIGATION ■ The process of resolving a dispute through the court system.

LONG ARM STATUTE ■ A state statute that permits a state to obtain personal jurisdiction over nonresident defendants. A defendant must have certain "minimum contacts" with that state for the statute to apply.

LOST PROPERTY ■ Property with which the owner has involuntarily parted and then cannot find or recover.

M

MAILBOX RULE ■ A rule providing that an acceptance of an offer becomes effective on dispatch (on being placed in an official mailbox), if mail is, expressly or impliedly, an authorized means of communication of acceptance to the offeror.

MAKER ■ One who promises to pay a fixed amount of money to the holder of a promissory note or a certificate of deposit (CD).

MALPRACTICE ■ Professional misconduct or the lack of the requisite degree of skill as a professional. Negligence—the failure to exercise due care—on the part of a professional, such as a physician, is commonly referred to as malpractice.

MARKET CONCENTRATION ■ The degree to which a small number of firms control a large percentage share of a relevant market; determined by calculating the percentages held by the largest firms in that market.

MARKET POWER ■ The power of a firm to control the market price of its product. A monopoly has the greatest degree of market power.

MARKET-SHARE TEST ■ The primary measure of monopoly power. A firm's market share is the percentage of a market that the firm controls.

MECHANIC'S LIEN ■ A statutory lien on the real property of another, created to ensure payment for work performed and materials furnished in the repair or improvement of real property, such as a building.

MEDIATION ■ A method of settling disputes outside of court by using the services of a neutral third party, who acts as a communicating agent between the parties and assists them in negotiating a settlement.

MEMBER ■ The term used to designate a person who has an ownership interest in a limited liability company.

MERCHANT ■ A person who is engaged in the purchase and sale of goods. Under the UCC, a person who deals in goods of the kind involved in the sales contract or who holds herself or himself out as having skill or knowledge peculiar to the practices or use of the goods being purchased or sold. For definitions, see UCC 2–104.

MERGER ■ A contractual and statutory process in which one corporation (the surviving corporation) acquires all of the assets and liabilities of another corporation (the merged corporation). The shareholders of the merged corporation either are paid for their shares or receive shares in the surviving corporation.

META TAG ■ A key word in a document that can serve as an index reference to the document. On the Web, search engines return results based, in part, on the tags in Web documents. Words inserted into a Web site's key-word field to increase the site's inclusion in search engine results.

MINIMUM WAGE ■ The lowest wage, either by government regulation or union contract, that an employer may pay an hourly worker.

MINI-TRIAL ■ A private proceeding in which each party to a dispute argues its position before the other side and vice versa. A neutral third party may be present as an adviser and may render an opinion if the parties fail to reach an agreement.

MIRROR IMAGE RULE ■ A common law rule that requires that the terms of the offeree's acceptance adhere exactly to the terms of the offeror's offer for a valid contract to be formed.

MISDEMEANOR ■ A lesser crime than a felony, punishable by a fine or incarceration in jail for up to one year.

MISLAID PROPERTY ■ Property with which the owner has voluntarily parted and then cannot find or recover.

MITIGATION OF DAMAGES ■ A rule requiring a plaintiff to do whatever is reasonable to minimize the damages caused by the defendant.

MONEY LAUNDERING ■ Falsely reporting income that has been obtained through criminal activity as income obtained through a legitimate business enterprise—in effect, "laundering" the "dirty money."

MONOPOLIZATION ■ The possession of monopoly power in the relevant market and the willful acquisition or maintenance of that power, as distinguished from growth or development as a consequence of a superior product, business acumen, or historic accident.

MONOPOLY ■ A term generally used to describe a market in which there is a single seller or a limited number of sellers.

MONOPOLY POWER ■ The ability of a monopoly to dictate what takes place in a given market.

MORAL MINIMUM ■ The minimum degree of ethical behavior expected of a business firm, which is usually defined as compliance with the law.

MORTGAGEE ■ Under a mortgage agreement, the creditor who takes a security interest in the debtor's property.

MORTGAGOR ■ Under a mortgage agreement, the debtor who gives the creditor a security interest in the debtor's property in return for a mortgage loan.

MOTION FOR A DIRECTED VERDICT ■ In a jury trial, a motion for the judge to take the decision out of the hands of the jury and to direct a verdict for the party who filed the motion on the ground that the other party has not produced sufficient evidence to support her or his claim.

MOTION FOR A NEW TRIAL ■ A motion asserting that the trial was so fundamentally flawed (because of error, newly discovered evidence, prejudice, or other reason) that a new trial is necessary to prevent a miscarriage of justice.

MOTION FOR JUDGMENT *N.O.V.* ■ A motion requesting the court to grant judgment in favor of the party making the motion on the ground that the jury's verdict against him or her was unreasonable and erroneous.

MOTION FOR JUDGMENT ON THE PLEADINGS ■ A motion by either party to a lawsuit at the close of the pleadings requesting the court to decide the issue solely on the pleadings without proceeding to trial. The motion will be granted only if no facts are in dispute.

MOTION FOR SUMMARY JUDGMENT ■ A motion requesting the court to enter a judgment without proceeding to trial. The motion can be based on evidence outside the pleadings and will be granted only if no facts are in dispute.

MOTION TO DISMISS ■ A pleading in which a defendant asserts that the plaintiff's claim fails to state a cause of action (that is, has no basis in law) or that there are other grounds on which a suit should be dismissed.

MUTUAL FUND ■ A specific type of investment company that continually buys or sells to investors shares of ownership in a portfolio.

N

NATIONAL LAW ■ Law that pertains to a particular nation (as opposed to international law).

NATURAL LAW ■ The belief that government and the legal system should reflect universal moral and ethical principles that are inherent in human nature. The natural law school is the oldest and one of the most significant schools of legal thought.

NECESSARIES ■ Necessities required for life, such as food, shelter, clothing, and medical attention; may include whatever is believed to be necessary to maintain a person's standard of living or financial and social status.

NEGLIGENCE ■ The failure to exercise the standard of care that a reasonable person would exercise in similar circumstances.

NEGLIGENCE *PER SE* ■ An action or failure to act in violation of a statutory requirement.

NEGOTIABLE INSTRUMENT ■ A signed writing (record) that contains an unconditional promise or order to pay an exact sum of money on demand or at an exact future time to a specific person or order, or to bearer.

NEGOTIATION ■ In alternative dispute resolution, a process in which parties attempt to settle their dispute informally, with or without attorneys to represent them. In negotiable instruments law, the transfer of an

instrument in such form that the transferee (the person to whom the instrument is transferred) becomes a holder.

NOMINAL DAMAGES ■ A small monetary award (often one dollar) granted to a plaintiff when no actual damage was suffered.

NO-PAR SHARES ■ Corporate shares that have no face value; that is, no specific dollar amount is printed on their face.

NORMAL TRADE RELATIONS (NTR) STATUS ■ A status granted in an international treaty by a provision stating that the citizens of the contracting nations may enjoy the privileges accorded by either party to citizens of its NTR nations. Generally, this status is designed to establish equality of international treatment.

NOTARY PUBLIC ■ A public official authorized to attest to the authenticity of signatures.

NOVATION ■ The substitution, by agreement, of a new contract for an old one, with the rights under the old one being terminated. Typically, novation involves the substitution of a new person who is responsible for the contract and the removal of the original party's rights and duties under the contract.

NUISANCE ■ A common law doctrine under which persons may be held liable for using their property in a manner that unreasonably interferes with others' rights to use or enjoy their own property.

NUNCUPATIVE WILL ■ An oral will (often called a deathbed will) made before witnesses; usually limited to transfers of personal property.

O

OBJECTIVE THEORY OF CONTRACTS ■ A theory under which the intent to form a contract will be judged by outward, objective facts (what the party said when entering into the contract, how the party acted or appeared, and the circumstances surrounding the transaction) as interpreted by a reasonable person, rather than by the party's own secret, subjective intentions.

OBLIGEE ■ One to whom an obligation is owed.

OBLIGOR ■ One who owes an obligation to another.

OFFER ■ A promise or commitment to perform or refrain from performing some specified act in the future.

OFFEREE ■ A person to whom an offer is made.

OFFEROR ■ A person who makes an offer.

ONLINE DISPUTE RESOLUTION (ODR) ■ The resolution of disputes with the assistance of organizations that offer dispute-resolution services via the Internet.

OPERATING AGREEMENT ■ In a limited liability company, an agreement in which the members set forth the details of how the business will be managed and operated. State statutes typically give the members wide latitude in deciding for themselves the rules that will govern their organization.

OPTION CONTRACT ■ A contract under which the offeror cannot revoke his or her offer for a stipulated time period, and the offeree can accept or reject the offer during this period without fear that the offer will be made to another person. The offeree must give consideration for the option (the irrevocable offer) to be enforceable.

ORDER FOR RELIEF ■ A court's grant of assistance to a complainant. In bankruptcy proceedings, the order relieves the debtor of the immediate obligation to pay the debts listed in the bankruptcy petition.

ORDER INSTRUMENT ■ A negotiable instrument that is payable "to the order of an identified person" or "to an identified person or order."

ORDINANCE ■ A regulation enacted by a city or county legislative body that becomes part of that state's statutory law.

OUTPUT CONTRACT ■ An agreement in which a seller agrees to sell and a buyer agrees to buy all or up to a stated amount of what the seller produces.

OVERDRAFT ■ A check that is paid by the bank when the checking account on which the check is written contains insufficient funds to cover the check.

P

PARENT-SUBSIDIARY MERGER ■ A merger of companies in which one company (the parent corporation) owns most of the stock of the other company (the subsidiary corporation). A parent-subsidiary merger (short-form merger) can use a simplified procedure when the parent corporation owns at least 90 percent of the outstanding shares of each class of stock of the subsidiary corporation.

PARTIALLY DISCLOSED PRINCIPAL ■ A principal whose identity is unknown by a third party, but the third party knows that the agent is or may be acting for a principal at the time the agent and the third party form a contract.

PARTNERING AGREEMENT ■ An agreement between a seller and a buyer who frequently do business with each other concerning the terms and conditions that will apply to all subsequently formed electronic contracts.

PARTNERSHIP ■ An agreement by two or more persons to carry on, as co-owners, a business for profit.

PAR-VALUE SHARES ■ Corporate shares that have a specific face value, or formal cash-in value, written on them, such as one dollar.

PASS-THROUGH ENTITY ■ A business entity that has no tax liability; the entity's income is passed through to the owners, and the owners pay taxes on the income.

PAST CONSIDERATION ■ An act that takes place before the contract is made and that ordinarily, by itself, cannot be consideration for a later promise to pay for the act.

PATENT ■ A government grant that gives an inventor the exclusive right or privilege to make, use, or sell his or her invention for a limited time period.

PAYEE ■ A person to whom an instrument is made payable.

PAYOR BANK ■ The bank on which a check is drawn (the drawee bank).

PEER-TO-PEER (P2P) NETWORKING ■ The sharing of resources (such as files, hard drives, and processing styles) among multiple computers without necessarily requiring a central network server.

PENALTY ■ A contractual clause that states that a certain amount of money damages will be paid in the event of a future default or breach of contract. The damages are not a measure of compensation for the contract's breach but rather a punishment for a default. The agreement as to the amount will not be enforced, and recovery will be limited to actual damages.

PER CAPITA ■ A Latin term meaning "per person." In the law governing estate distribution, a method of distributing the property of an intestate's estate so that each heir in a certain class (such as grandchildren) receives an equal share.

***PER SE* VIOLATION** ■ A type of anticompetitive agreement that is considered to be so injurious to the public that there is no need to determine whether it actually injures market competition; rather, it is in itself *(per se)* a violation of the Sherman Act.

PER STIRPES ■ A Latin term meaning "by the roots." In the law governing estate distribution, a method of distributing an intestate's estate so that each heir in a certain class (such as grandchildren) takes the share to which her or his deceased ancestor (such as a mother or father) would have been entitled.

PERFECTION ■ The legal process by which secured parties protect themselves against the claims of third parties who may wish to have their debts satisfied out of the same collateral; usually accomplished by filing a financing statement with the appropriate government official.

PERFORMANCE ■ In contract law, the fulfillment of one's duties arising under a contract with another; the normal way of discharging one's contractual obligations.

PERIODIC TENANCY ■ A lease interest in land for an indefinite period

involving payment of rent at fixed intervals, such as week to week, month to month, or year to year.

PERSONAL DEFENSES ■ Defenses that can be used to avoid payment to an ordinary holder of a negotiable instrument but not a holder in due course (HDC) or a holder with the rights of an HDC.

PERSONAL PROPERTY ■ Property that is movable; any property that is not real property.

PERSUASIVE AUTHORITY ■ Any legal authority or source of law that a court may look to for guidance but on which it need not rely in making its decision. Persuasive authorities include cases from other jurisdictions and secondary sources of law.

PETITION IN BANKRUPTCY ■ The document that is filed with a bankruptcy court to initiate bankruptcy proceedings. The official forms required for a petition in bankruptcy must be completed accurately, sworn to under oath, and signed by the debtor.

PETTY OFFENSE ■ In criminal law, the least serious kind of criminal offense, such as a traffic or building-code violation.

PIERCING THE CORPORATE VEIL ■ An action in which a court disregards the corporate entity and holds the shareholders personally liable for corporate debts and obligations.

PLAINTIFF ■ One who initiates a lawsuit.

PLEA BARGAINING ■ The process by which a defendant and the prosecutor in a criminal case work out a mutually satisfactory disposition of the case, subject to court approval; usually involves the defendant's pleading guilty to a lesser offense in return for a lighter sentence.

PLEADINGS ■ Statements made by the plaintiff and the defendant in a lawsuit that detail the facts, charges, and defenses involved in the litigation. The complaint and answer are part of the pleadings.

PLEDGE ■ A common law security device (retained in Article 9 of the UCC) in which personal property is transferred into the possession of the creditor as security for the payment of a debt and retained by the creditor until the debt is paid.

POLICE POWERS ■ Powers possessed by the states as part of their inherent sovereignty. These powers may be exercised to protect or promote the public order, health, safety, morals, and general welfare.

POLICY ■ In insurance law, a contract between the insurer and the insured in which, for a stipulated consideration, the insurer agrees to compensate the insured for loss on a specific subject by a specified peril.

POSITIVE LAW ■ The body of conventional, or written, law of a particular society at a particular point in time.

POTENTIALLY RESPONSIBLE PARTY (PRP) ■ A party liable for the costs of cleaning up a hazardous waste–disposal site under the Comprehensive Environmental Response, Compensation, and Liability Act (CERCLA). Any person who generated the hazardous waste, transported it, owned or operated the waste site at the time of disposal, or currently owns or operates the site may be responsible for some or all of the clean-up costs.

POWER OF ATTORNEY ■ A written document, which is usually notarized, authorizing another to act as one's agent; can be special (permitting the agent to do specified acts only) or general (permitting the agent to transact all business for the principal).

PRECEDENT ■ A court decision that furnishes an example or authority for deciding subsequent cases involving identical or similar facts.

PREDATORY BEHAVIOR ■ Business behavior that is undertaken with the intention of unlawfully driving competitors out of the market.

PREDATORY PRICING ■ The pricing of a product below cost with the intent to drive competitors out of the market.

PREDOMINANT-FACTOR TEST ■ A test courts use to determine whether a contract is primarily for the sale of goods or for the sale of services.

PREEMPTION ■ A doctrine under which certain federal laws preempt, or take precedence over, conflicting state or local laws.

PREEMPTIVE RIGHTS ■ Rights held by shareholders that entitle them to purchase newly issued shares of a corporation's stock, equal in percentage to shares already held, before the stock is offered to any outside buyers. Preemptive rights enable shareholders to maintain their proportionate ownership and voice in the corporation.

PREFERENCE ■ In bankruptcy proceedings, property transfers or payments made by the debtor that favor (give preference to) one creditor over others. The bankruptcy trustee is allowed to recover payments made both voluntarily and involuntarily to one creditor in preference over another.

PREFERRED STOCK ■ Classes of stock that have priority over common stock as to both payment of dividends and distribution of assets on the corporation's dissolution.

PREMIUM ■ In insurance law, the price paid by the insured for insurance protection for a specified period of time.

PRENUPTIAL AGREEMENT ■ An agreement made before marriage that defines each partner's ownership rights in the other partner's property. Prenuptial agreements must be in writing to be enforceable.

PRESENTMENT ■ The act of presenting an instrument to the party liable on the instrument to collect payment. Presentment also occurs when a person presents an instrument to a drawee for a required acceptance.

PRESENTMENT WARRANTIES ■ Implied warranties, made by any person who presents an instrument for payment or acceptance, that (1) the person obtaining payment or acceptance is entitled to enforce the instrument or is authorized to obtain payment or acceptance on behalf of a person who is entitled to enforce the instrument, (2) the instrument has not been altered, and (3) the person obtaining payment or acceptance has no knowledge that the signature of the drawer of the instrument is unauthorized.

PRICE DISCRIMINATION ■ Setting prices in such a way that two competing buyers pay two different prices for an identical product or service.

PRICE-FIXING AGREEMENT ■ An agreement between competitors to fix the prices of products or services at a certain level.

***PRIMA FACIE* CASE** ■ A case in which the plaintiff has produced sufficient evidence of his or her conclusion that the case can go to a jury; a case in which the evidence compels the plaintiff's conclusion if the defendant produces no affirmative defense or evidence to disprove it.

PRIMARY SOURCE OF LAW ■ A document that establishes the law on a particular issue, such as a constitution, a statute, an administrative rule, or a court decision.

PRINCIPAL ■ In agency law, a person who agrees to have another, called the agent, act on her or his behalf.

PRINCIPLE OF RIGHTS ■ The principle that human beings have certain fundamental rights (to life, freedom, and the pursuit of happiness, for example). Those who adhere to this "rights theory" believe that a key factor in determining whether a business decision is ethical is how that decision affects the rights of various groups. These groups include the firm's owners, its employees, the consumers of its products or services, its suppliers, the community in which it does business, and society as a whole.

PRIVILEGE ■ A legal right, exemption, or immunity granted to a person or a class of persons. In the context of defamation, an absolute privilege immunizes the actor from suit, regardless of whether the actor's statements were malicious. A qualified privilege immunizes an actor from suit only when the privilege is properly exercised in the performance of a legal or moral duty.

PRIVITY OF CONTRACT ■ The relationship that exists between the promisor and the promisee of a contract.

PROBABLE CAUSE ■ Reasonable grounds for believing that a person should be arrested or searched.

PROBATE ■ The process of proving and validating a will and settling all matters pertaining to an estate.

PROBATE COURT ■ A state court of limited jurisdiction that conducts proceedings relating to the settlement of a deceased person's estate.

PROCEDURAL LAW ■ Law that establishes the methods of enforcing the rights established by substantive law.

PROCEEDS ■ Under Article 9 of the UCC, whatever is received when the collateral is sold or otherwise disposed of, such as by exchange.

PRODUCT LIABILITY ■ The legal liability of manufacturers, sellers, and lessors of goods to consumers, users, and bystanders for injuries or damages that are caused by the goods.

PROFIT ■ In real property law, the right to enter on and remove things from the property of another (for example, the right to enter onto a person's land and remove sand and gravel therefrom).

PROMISE ■ An assertion that something either will or will not happen in the future.

PROMISEE ■ A person to whom a promise is made.

PROMISOR ■ A person who makes a promise.

PROMISSORY ESTOPPEL ■ A doctrine that applies when a promisor makes a clear and definite promise on which the promisee justifiably relies; such a promise is binding if justice will be better served by the enforcement of the promise.

PROMISSORY NOTE ■ A written promise made by one person (the maker) to pay a fixed amount of money to another person (the payee or a subsequent holder) on demand or on a specified date.

PROMOTER ■ A person who takes the preliminary steps in organizing a corporation, including (usually) issuing a prospectus, procuring stock subscriptions, making contracts for necessary purchases and services, securing a corporate charter, and the like.

PROPERTY ■ Legally protected rights and interests in anything with an ascertainable value that is subject to ownership.

PROSPECTUS ■ A document required by federal or state securities laws that describes the financial operations of the corporation, thus allowing investors to make informed decisions.

PROTECTED CLASS ■ A group of persons protected by specific laws because of the group's defining characteristics. Under laws prohibiting employment discrimination, these characteristics include race, color, religion, national origin, gender, age, and disability.

PROXIMATE CAUSE ■ Legal cause; exists when the connection between an act and an injury is strong enough to justify imposing liability.

PROXY ■ In corporation law, a written agreement between a stockholder and another party under which the stockholder authorizes the other party to vote the stockholder's shares in a certain manner.

PUFFERY ■ A salesperson's often exaggerated claims concerning the quality of property offered for sale. Such claims involve opinions rather than facts and are not considered to be legally binding promises or warranties.

PUNITIVE DAMAGES ■ Money damages that may be awarded to a plaintiff to punish the defendant and deter future similar conduct.

PURCHASE-MONEY SECURITY INTEREST (PMSI) ■ A security interest that arises when a seller or lender extends credit for part or all of the purchase price of goods purchased by a buyer.

QUALIFIED INDORSEMENT ■ An indorsement on a negotiable instrument in which the indorser disclaims any contract liability on the instrument. The notation "without recourse" is commonly used to create a qualified indorsement.

QUASI CONTRACT ■ A fictional contract imposed on parties by a court in the interests of fairness and justice; usually imposed to avoid the unjust enrichment of one party at the expense of another.

QUITCLAIM DEED ■ A deed intended to pass any title, interest, or claim that the grantor may have in the property without warranting that such title is valid. A quitclaim deed offers the least amount of protection against defects in the title.

QUORUM ■ The number of members of a decision-making body that must be present before business may be transacted.

QUOTA ■ A set limit on the amount of goods that can be imported.

RATIFICATION ■ The act of accepting and giving legal force to an obligation that previously was not enforceable.

REAFFIRMATION AGREEMENT ■ An agreement between a debtor and a creditor in which the debtor voluntarily agrees to pay, or reaffirm, a debt dischargeable in bankruptcy. To be enforceable, the agreement must be made before the debtor is granted a discharge.

REAL PROPERTY ■ Land and everything attached to it, such as trees and buildings.

REASONABLE PERSON STANDARD ■ The standard of behavior expected of a hypothetical "reasonable person"; the standard against which negligence is measured and that must be observed to avoid liability for negligence.

RECEIVER ■ In a corporate dissolution, a court-appointed person who winds up corporate affairs and liquidates corporate assets.

RECORD ■ According to the Uniform Electronic Transactions Act, information that is either inscribed on a tangible medium or stored in an electronic or other media and that is retrievable. The Uniform Computer Information Transactions Act also uses the term *record* instead of *writing*.

RECORDING STATUTES ■ Statutes that allow deeds, mortgages, and other real property transactions to be recorded so as to provide notice to future purchasers or creditors of an existing claim on the property.

RED HERRING ■ A preliminary prospectus that can be distributed to potential investors after the registration statement (for a securities offering) has been filed with the Securities and Exchange Commission. The name derives from the red legend printed across the prospectus stating that the registration has been filed but has not become effective.

REFORMATION ■ A court-ordered correction of a written contract so that it reflects the true intentions of the parties.

REGULATION E ■ A set of rules issued by the Federal Reserve System's Board of Governors to protect users of elecronic fund transfer systems.

REGULATION Z ■ A set of rules promulgated by the Federal Reserve Board of Governors to implement the provisions of the Truth-in-Lending Act.

RELEASE ■ A contract in which one party forfeits the right to pursue a legal claim against the other party.

REMEDY ■ The relief given to an innocent party to enforce a right or compensate for the violation of a right.

REPLEVIN ■ An action to recover identified goods in the hands of a party who is wrongfully withholding them from the other party. Under the UCC, this remedy is usually available only if the buyer or lessee is unable to cover.

REPLY ■ Procedurally, a plaintiff's response to a defendant's answer.

REQUIREMENTS CONTRACT ■ An agreement in which a buyer agrees to purchase and the seller agrees to sell all or up to a stated amount of what the buyer needs or requires.

RES IPSA LOQUITUR ■ A doctrine under which negligence may be inferred simply because an event occurred, if it is the type of event that would never occur in the absence of negligence. Literally, the term means "the facts speak for themselves."

RESALE PRICE MAINTENANCE AGREEMENT ■ An agreement between a manufacturer and a retailer in which the manufacturer specifies what the retail prices of its products must be.

RESCISSION ■ A remedy whereby a contract is canceled and the parties are returned to the positions they occupied before the contract was

made; may be effected through the mutual consent of the parties, by the parties' conduct, or by court decree.

RESPONDEAT SUPERIOR ■ Latin for "let the master respond." A doctrine under which a principal or an employer is held liable for the wrongful acts committed by agents or employees while acting within the course and scope of their agency or employment.

RESTITUTION ■ An equitable remedy under which a person is restored to his or her original position prior to loss or injury, or placed in the position he or she would have been in had the breach not occurred.

RESTRICTIVE INDORSEMENT ■ Any indorsement on a negotiable instrument that requires the indorsee to comply with certain instructions regarding the funds involved. A restrictive indorsement does not prohibit the further negotiation of the instrument.

RESULTING TRUST ■ An implied trust arising from the conduct of the parties. A trust in which a party holds the actual legal title to another's property but only for that person's benefit.

RETAINED EARNINGS ■ The portion of a corporation's profits that has not been paid out as dividends to shareholders.

REVOCATION ■ In contract law, the withdrawal of an offer by an offeror. Unless the offer is irrevocable, it can be revoked at any time prior to acceptance without liability.

RIGHT OF CONTRIBUTION ■ The right of a co-surety who pays more than his or her proportionate share on a debtor's default to recover the excess paid from other co-sureties.

RIGHT OF FIRST REFUSAL ■ The right to purchase personal or real property—such as corporate shares or real estate—before the property is offered for sale to others.

RIGHT OF REIMBURSEMENT ■ The legal right of a person to be restored, repaid, or indemnified for costs, expenses, or losses incurred or expended on behalf of another.

RIGHT OF SUBROGATION ■ The right of a person to stand in the place of (be substituted for) another, giving the substituted party the same legal rights that the original party had.

RIGHT-TO-WORK LAW ■ A state law providing that employees may not be required to join a union as a condition of retaining employment.

RISK ■ A prediction concerning potential loss based on known and unknown factors.

RISK MANAGEMENT ■ Planning that is undertaken to protect one's interest should some event threaten to undermine its security. In the context of insurance, risk management involves transferring certain risks from the insured to the insurance company.

ROBBERY ■ The act of forcefully and unlawfully taking personal property of any value from another. Force or intimidation is usually necessary for an act of theft to be considered a robbery.

RULE OF FOUR ■ A rule of the United States Supreme Court under which the Court will not issue a writ of *certiorari* unless at least four justices approve of the decision to issue the writ.

RULE OF REASON ■ A test by which a court balances the positive effects (such as economic efficiency) of an agreement against its potentially anticompetitive effects. In antitrust litigation, many practices are analyzed under the rule of reason.

RULEMAKING ■ The process undertaken by an administrative agency when formally adopting a new regulation or amending an old one. Rulemaking involves notifying the public of a proposed rule or change and receiving and considering the public's comments.

S

S CORPORATION ■ A close business corporation that has met certain requirements set out in the Internal Revenue Code and thus qualifies for special income tax treatment. Essentially, an S corporation is taxed the same as a partnership, but its owners enjoy the privilege of limited liability.

SALE ■ The passing of title to property from the seller to the buyer for a price.

SALE ON APPROVAL ■ A type of conditional sale in which the buyer may take the goods on a trial basis. The sale becomes absolute only when the buyer approves of (or is satisfied with) the goods being sold.

SALE OR RETURN ■ A type of conditional sale in which title and possession pass from the seller to the buyer, but the buyer retains the option to return the goods during a specified period even though the goods conform to the contract.

SALES CONTRACT ■ A contract for the sale of goods under which the ownership of goods is transferred from a seller to a buyer for a price.

SCIENTER ■ Knowledge by the misrepresenting party that material facts have been falsely represented or omitted with an intent to deceive.

SEARCH WARRANT ■ An order granted by a public authority, such as a judge, that authorizes law enforcement personnel to search particular premises or property.

SEASONABLY ■ Within a specified time period or, if no period is specified, within a reasonable time.

SEC RULE 10b-5 ■ A rule of the Securities and Exchange Commission (SEC) that makes it unlawful, in connection with the purchase or sale of any security, to make any untrue statement of a material fact or to omit a material fact if such omission causes the statement to be misleading.

SECONDARY BOYCOTT ■ A union's refusal to work for, purchase from, or handle the products of a secondary employer, with whom the union has no dispute, in order to force that employer to stop doing business with the primary employer, with whom the union has a labor dispute.

SECONDARY SOURCE OF LAW ■ A publication that summarizes or interprets the law, such as a legal encyclopedia, a legal treatise, or an article in a law review.

SECURED PARTY ■ A lender, seller, or any other person in whose favor there is a security interest, including a person to whom accounts or chattel paper has been sold.

SECURED TRANSACTION ■ Any transaction in which the payment of a debt is guaranteed, or secured, by personal property owned by the debtor or in which the debtor has a legal interest.

SECURITY ■ Generally, a stock certificate, bond, note, debenture, warrant, or other document given as evidence of an ownership interest in a corporation or as a promise of repayment by a corporation.

SECURITY AGREEMENT ■ An agreement that creates or provides for a security interest between the debtor and a secured party.

SECURITY INTEREST ■ Any interest in personal property or fixtures that secures payment or performance of an obligation.

SELF-DEFENSE ■ The legally recognized privilege to protect oneself or one's property against injury by another. The privilege of self-defense protects only acts that are reasonably necessary to protect oneself, one's property, or another person.

SELF-INCRIMINATION ■ The giving of testimony that may subject the testifier to criminal prosecution. The Fifth Amendment to the Constitution protects against self-incrimination by providing that no person "shall be compelled in any criminal case to be a witness against himself."

SENIORITY SYSTEM ■ In regard to employment relationships, a system in which those who have worked longest for the company are first in line for promotions, salary increases, and other benefits; they are also the last to be laid off if the workforce must be reduced.

SERVICE MARK ■ A mark used in the sale or the advertising of services to distinguish the services of one person from those of others. Titles, character names, and other distinctive features of radio and television programs may be registered as service marks.

SEXUAL HARASSMENT ■ In the employment context, the demanding of sexual favors in return for job promotions or other benefits, or language or conduct that is so sexually offensive that it creates a hostile working environment.

SHAREHOLDER'S DERIVATIVE SUIT ■ A suit brought by a shareholder to enforce a corporate cause of action against a third person.

SHELTER PRINCIPLE ■ The principle that the holder of a negotiable instrument who cannot qualify as a holder in due course (HDC), but who derives his or her title through an HDC, acquires the rights of an HDC.

SHIPMENT CONTRACT ■ A contract for the sale of goods in which the seller is required or authorized to ship the goods by carrier. The seller assumes liability for any losses or damage to the goods until they are delivered to the carrier.

SHORT-FORM MERGER ■ A merger between a subsidiary corporation and a parent corporation that owns at least 90 percent of the outstanding shares of each class of stock issued by the subsidiary corporation. Short-form mergers can be accomplished without the approval of the shareholders of either corporation.

SHRINK-WRAP AGREEMENT ■ An agreement whose terms are expressed in a document located inside a box in which goods (usually software) are packaged; sometimes called a *shrink-wrap license*.

SIGNATURE ■ Under the UCC, "any symbol executed or adopted by a party with a present intention to authenticate a writing."

SLANDER ■ Defamation in oral form.

SLANDER OF QUALITY (TRADE LIBEL) ■ The publication of false information about another's product, alleging that it is not what its seller claims.

SLANDER OF TITLE ■ The publication of a statement that denies or casts doubt on another's legal ownership of any property, causing financial loss to that property's owner.

SMALL CLAIMS COURT ■ A special court in which parties may litigate small claims (such as those involving $5,000 or less). Attorneys are not required in small claims courts and, in some states, are not allowed to represent the parties.

SMART CARD ■ A card containing a microprocessor that permits storage of funds via security programming, can communicate with other computers, and does not require online authorization for fund transfers.

SOCIOLOGICAL SCHOOL ■ A school of legal thought that views the law as a tool for promoting justice in society.

SOLE PROPRIETORSHIP ■ The simplest form of business organization, in which the owner is the business. The owner reports business income on his or her personal income tax return and is legally responsible for all debts and obligations incurred by the business.

SOVEREIGN IMMUNITY ■ A doctrine that immunizes foreign nations from the jurisdiction of U.S. courts when certain conditions are satisfied.

SPAM ■ Bulk, unsolicited ("junk") e-mail.

SPECIAL INDORSEMENT ■ An indorsement on an instrument that indicates the specific person to whom the indorser intends to make the instrument payable; that is, it names the indorsee.

SPECIAL WARRANTY DEED ■ A deed in which the grantor only warrants that the grantor or seller held good title during his or her ownership of the property and does not warrant that there were no defects of title when the property was held by previous owners.

SPECIFIC PERFORMANCE ■ An equitable remedy requiring exactly the performance that was specified in a contract; usually granted only when money damages would be an inadequate remedy and the subject matter of the contract is unique (for example, real property).

SPENDTHRIFT TRUST ■ A trust created to protect the beneficiary from spending all the funds to which she or he is entitled. Only a certain portion of the total amount is given to the beneficiary at any one time, and most states prohibit creditors from attaching assets of the trust.

STALE CHECK ■ A check, other than a certified check, that is presented for payment more than six months after its date.

STANDING TO SUE ■ The requirement that an individual must have a sufficient stake in a controversy before he or she can bring a lawsuit. The plaintiff must demonstrate that he or she has been either injured or threatened with injury.

STARE DECISIS ■ A common law doctrine under which judges are obligated to follow the precedents established in prior decisions.

STATUTE OF FRAUDS ■ A state statute under which certain types of contracts must be in writing to be enforceable.

STATUTE OF LIMITATIONS ■ A federal or state statute setting the maximum time period during which a certain action can be brought or certain rights enforced.

STATUTE OF REPOSE ■ Basically, a statute of limitations that is not dependent on the happening of a cause of action. Statutes of repose generally begin to run at an earlier date and run for a longer period of time than statutes of limitations.

STATUTORY LAW ■ The body of law enacted by legislative bodies (as opposed to constitutional law, administrative law, or case law).

STOCK ■ An equity (ownership) interest in a corporation, measured in units of shares.

STOCK CERTIFICATE ■ A certificate issued by a corporation evidencing the ownership of a specified number of shares in the corporation.

STOCK WARRANT ■ A certificate that grants the owner the option to buy a given number of shares of stock, usually within a set time period.

STOP-PAYMENT ORDER ■ An order by a bank customer to his or her bank not to pay or certify a certain check.

STORED-VALUE CARD ■ A card bearing magnetic strips that hold magnetically encoded data, providing access to stored funds.

STRICT LIABILITY ■ Liability regardless of fault. In tort law, strict liability is imposed on a manufacturer or seller that introduces into commerce a good that is unreasonably dangerous when in a defective condition.

STRIKE ■ An action undertaken by unionized workers when collective bargaining fails; the workers leave their jobs, refuse to work, and (typically) picket the employer's workplace.

SUBLEASE ■ A lease executed by the lessee of real estate to a third person, conveying the same interest that the lessee enjoys but for a shorter term than that held by the lessee.

SUBSTANTIVE LAW ■ Law that defines, describes, regulates, and creates legal rights and obligations.

SUMMARY JURY TRIAL (SJT) ■ A method of settling disputes, used in many federal courts, in which a trial is held, but the jury's verdict is not binding. The verdict acts only as a guide to both sides in reaching an agreement during the mandatory negotiations that immediately follow the summary jury trial.

SUMMONS ■ A document informing a defendant that a legal action has been commenced against him or her and that the defendant must appear in court on a certain date to answer the plaintiff's complaint. The document is delivered by a sheriff or any other person so authorized.

SUPREMACY CLAUSE ■ The provision in Article VI of the Constitution that provides that the Constitution, laws, and treaties of the United States are "the supreme Law of the Land." Under this clause, state and local laws that directly conflict with federal law will be rendered invalid.

SURETY ■ A person, such as a cosigner on a note, who agrees to be primarily responsible for the debt of another.

SURETYSHIP ■ An express contract in which a third party to a debtor-creditor relationship (the surety) promises to be primarily responsible for the debtor's obligation.

SYMBOLIC SPEECH ■ Nonverbal expressions of beliefs. Symbolic speech,

which includes gestures, movements, and articles of clothing, is given substantial protection by the courts.

SYNDICATE ■ An investment group of persons or firms brought together for the purpose of financing a project that they would not or could not undertake independently.

T

TAKEOVER ■ The acquisition of control over a corporation through the purchase of a substantial number of the voting shares of the corporation.

TAKING ■ The taking of private property by the government for public use. The government may not take private property for public use without "just compensation."

TANGIBLE PROPERTY ■ Property that has physical existence and can be distinguished by the senses of touch, sight, and so on. A car is tangible property; a patent right is intangible property.

TARGET CORPORATION ■ The corporation to be acquired in a corporate takeover; a corporation to whose shareholders a tender offer is submitted.

TARIFF ■ A tax on imported goods.

TENANCY AT SUFFERANCE ■ A type of tenancy under which a tenant who, after rightfully being in possession of leased premises, continues (wrongfully) to occupy the property after the lease has been terminated. The tenant has no rights to possess the property and occupies it only because the person entitled to evict the tenant has not done so.

TENANCY AT WILL ■ A type of tenancy that either party can terminate without notice; usually arises when a tenant who has been under a tenancy for years retains possession, with the landlord's consent, after the tenancy for years has terminated.

TENANCY BY THE ENTIRETY ■ The joint ownership of property by a husband and wife. Neither party can transfer her or his interest in the property without the consent of the other.

TENANCY FOR YEARS ■ A type of tenancy under which property is leased for a specified period of time, such as a month, a year, or a period of years.

TENANCY IN COMMON ■ Co-ownership of property in which each party owns an undivided interest that passes to her or his heirs at death.

TENDER ■ An unconditional offer to perform an obligation by a person who is ready, willing, and able to do so.

TESTAMENTARY TRUST ■ A trust that is created by will and therefore does not take effect until the death of the testator.

TESTATE ■ Having left a will at death.

TESTATOR ■ One who makes and executes a will.

THIRD PARTY BENEFICIARY ■ One for whose benefit a promise is made in a contract but who is not a party to the contract.

TIPPEE ■ A person who receives inside information.

TOMBSTONE AD ■ An advertisement, historically in a format resembling a tombstone, of a securities offering. The ad tells potential investors where and how they may obtain a prospectus.

TORT ■ A civil wrong not arising from a breach of contract. A breach of a legal duty that proximately causes harm or injury to another.

TORTFEASOR ■ One who commits a tort.

TOTTEN TRUST ■ A trust created by the deposit of a person's own funds in his or her own name as a trustee for another. It is a tentative trust, revocable at will until the depositor dies or completes the gift in his or her lifetime by some unequivocal act or declaration.

TOXIC TORT ■ A civil wrong arising from exposure to a toxic substance, such as asbestos, radiation, or hazardous waste.

TRADE ACCEPTANCE ■ A draft that is drawn by a seller of goods ordering the buyer to pay a specified sum of money to the seller, usually at a stated time in the future. The buyer accepts the draft by signing the face of the draft, thus creating an enforceable obligation to pay the draft when it comes due. On a trade acceptance, the seller is both the drawer and the payee.

TRADE DRESS ■ The image and overall appearance of a product—for example, the distinctive decor, menu, layout, and style of service of a particular restaurant. Basically, trade dress is subject to the same protection as trademarks.

TRADE NAME ■ A term that is used to indicate part or all of a business's name and that is directly related to the business's reputation and goodwill. Trade names are protected under the common law (and under trademark law, if the name is the same as the firm's trademarked property).

TRADE SECRETS ■ Information or processes that give a business an advantage over competitors that do not know the information or processes.

TRADEMARK ■ A distinctive word, name, mark, motto, device, or emblem that a manufacturer stamps, prints, or otherwise affixes to the goods it produces so that they may be identified on the market and their origins made known. Once a trademark is established (under the common law or through registration), the owner is entitled to its exclusive use.

TRANSFER WARRANTIES ■ Implied warranties, made by any person who transfers an instrument for consideration to subsequent transferees and holders who take the instrument in good faith, that (1) the transferor is entitled to enforce the instrument; (2) all signatures are authentic and authorized; (3) the instrument has not been altered; (4) the instrument is not subject to a defense or claim of any party that can be asserted against the transferor; and (5) the transferor has no knowledge of any insolvency proceedings against the maker, the acceptor, or the drawer of the instrument.

TRAVELER'S CHECK ■ A check that is payable on demand, drawn on or payable through a financial institution (bank), and designated as a traveler's check.

TREBLE DAMAGES ■ Damages that, by statute, are three times the amount that the fact finder determines is owed.

TRESPASS TO LAND ■ The entry onto, above, or below the surface of land owned by another without the owner's permission or legal authorization.

TRESPASS TO PERSONAL PROPERTY ■ The unlawful taking or harming of another's personal property; interference with another's right to the exclusive possession of his or her personal property.

TRUST ■ An arrangement in which title to property is held by one person (a trustee) for the benefit of another (a beneficiary).

TRUST INDORSEMENT ■ An indorsement for the benefit of the indorser or a third person; also known as an agency indorsement. The indorsement results in legal title vesting in the original indorsee.

TYING ARRANGEMENT ■ An agreement between a buyer and a seller in which the buyer of a specific product or service becomes obligated to purchase additional products or services from the seller.

U

U.S. TRUSTEE ■ A government official who performs certain administrative tasks that a bankruptcy judge would otherwise have to perform.

ULTRA VIRES ■ A Latin term meaning "beyond the powers"; in corporate law, acts of a corporation that are beyond its express and implied powers to undertake.

UNCONSCIONABLE CONTRACT (OR UNCONSCIONABLE CLAUSE) ■ A contract or clause that is void on the basis of public policy because one party, as a result of disproportionate bargaining power, is forced to accept terms that are unfairly burdensome and that unfairly benefit the dominating party.

UNDERWRITER ■ In insurance law, the insurer, or the one assuming a risk in return for the payment of a premium.

UNDISCLOSED PRINCIPAL ■ A principal whose identity is unknown by a third person, and the third person has no knowledge that the agent is acting for a principal at the time the agent and the third person form a contract.

UNENFORCEABLE CONTRACT ■ A valid contract rendered unenforceable by some statute or law.

UNILATERAL CONTRACT ■ A contract that results when an offer can be accepted only by the offeree's performance.

UNION SHOP ■ A place of employment where all workers, once employed, must become union members within a specified period of time as a condition of their continued employment.

UNIVERSAL DEFENSES ■ Defenses that are valid against all holders of a negotiable instrument, including holders in due course (HDCs) and holders with the rights of HDCs.

UNREASONABLY DANGEROUS PRODUCT ■ In product liability, a product that is defective to the point of threatening a consumer's health and safety. A product will be considered unreasonably dangerous if it is dangerous beyond the expectation of the ordinary consumer or if a less dangerous alternative was economically feasible for the manufacturer, but the manufacturer failed to produce it.

USAGE OF TRADE ■ Any practice or method of dealing having such regularity of observance in a place, vocation, or trade as to justify an expectation that it will be observed with respect to the transaction in question.

USURY ■ Charging an illegal rate of interest.

UTILITARIANISM ■ An approach to ethical reasoning that evaluates behavior not on the basis of any absolute ethical or moral values but on the consequences of that behavior for those who will be affected by it. In utilitarian reasoning, a "good" decision is one that results in the greatest good for the greatest number of people affected by the decision.

V

VALID CONTRACT ■ A contract that results when the elements necessary for contract formation (agreement, consideration, legal purpose, and contractual capacity) are present.

VENUE ■ The geographic district in which an action is tried and from which the jury is selected.

VERTICAL MERGER ■ The acquisition by a company at one level in a marketing chain of a company at a higher or lower level in the chain (such as a company merging with one of its suppliers or retailers).

VERTICAL RESTRAINT ■ Any restraint on trade created by agreements between firms at different levels in the manufacturing and distribution process.

VERTICALLY INTEGRATED FIRM ■ A firm that carries out two or more functional phases (manufacture, distribution, and retailing, for example) of the chain of production.

VESTING ■ The creation of an absolute or unconditional right or power.

VICARIOUS LIABILITY ■ Legal responsibility placed on one person for the acts of another; indirect liability imposed on a supervisory party (such as an employer) for the actions of a subordinate (such as an employee) because of the relationship between the two parties.

VOID CONTRACT ■ A contract having no legal force or binding effect.

VOIDABLE CONTRACT ■ A contract that may be legally avoided (canceled, or annulled) at the option of one or both of the parties.

VOIR DIRE ■ An old French phrase meaning "to speak the truth." In legal language, the phrase refers to the process in which the attorneys question prospective jurors to learn about their backgrounds, attitudes, biases, and other characteristics that may affect their ability to serve as impartial jurors.

VOTING TRUST ■ An agreement (trust contract) under which legal title to shares of corporate stock is transferred to a trustee who is authorized by the shareholders to vote the shares on their behalf.

W

WARRANTY DEED ■ A deed in which the grantor assures (warrants to) the grantee that the grantor has title to the property conveyed in the deed, that there are no encumbrances on the property other than what the grantor has represented, and that the grantee will enjoy quiet possession of the property; a deed that provides the greatest amount of protection for the grantee.

WATERED STOCK ■ Shares of stock issued by a corporation for which the corporation receives, as payment, less than the stated value of the shares.

WETLANDS ■ Water-saturated areas of land that are designated by government agencies (such as the Army Corps of Engineers or the Environmental Protection Agency) as protected areas that support wildlife and therefore cannot be filled in or dredged by private contractors or parties without a permit.

WHISTLEBLOWING ■ An employee's disclosure to government authorities, upper-level managers, or the press that the employer is engaged in unsafe or illegal activities.

WHITE-COLLAR CRIME ■ Nonviolent crime committed by individuals or corporations to obtain a personal or business advantage.

WILL ■ An instrument directing what is to be done with the testator's property on his or her death, made by the testator and revocable during his or her lifetime. No interests in the testator's property pass until the testator dies.

WILL SUBSTITUTES ■ Various documents that attempt to dispose of an estate in the same or similar manner as a will, such as trusts or life insurance plans.

WINDING UP ■ The second of two stages in the termination of a partnership or corporation. Once the firm is dissolved, it continues to exist legally until the process of winding up all business affairs (collecting and distributing the firm's assets) is complete.

WORKERS' COMPENSATION LAWS ■ State statutes establishing an administrative procedure for compensating workers' injuries that arise out of—or in the course of—their employment, regardless of fault.

WORKING PAPERS ■ The various documents used and developed by an accountant during an audit. Working papers include notes, computations, memoranda, copies, and other papers that make up the work products of an accountant's services to a client.

WORKOUT ■ An out-of-court agreement between a debtor and creditors in which the parties work out a payment plan or schedule under which the debtor's debts can be discharged.

WRIT OF ATTACHMENT ■ A court's order, issued prior to a trial to collect a debt, directing the sheriff or other officer to seize nonexempt property of the debtor. If the creditor prevails at trial, the seized property can be sold to satisfy the judgment.

WRIT OF *CERTIORARI* ■ A writ from a higher court asking the lower court for the record of a case.

WRIT OF EXECUTION ■ A court's order, issued after a judgment has been entered against the debtor, directing the sheriff to seize (levy) and sell any of the debtor's nonexempt real or personal property. The proceeds of the sale are used to pay off the judgment, accrued interest, and costs of the sale; any surplus is paid to the debtor.

WRONGFUL DISCHARGE ■ An employer's termination of an employee's employment in violation of the law.

TABLE OF CASES

A

B

S

T

U

W

Y

Z

INDEX

A

D

E

F

G

H

I

M

N

U

V

W

Y

UNIT SEVEN
Employment Relations 441

UNIT EIGHT
Business Organizations 489

UNIT NINE
Property and Its Protection 587

UNIT TEN
Special Topics 637

APPENDICES

CONTENTS IN BRIEF